Encyclopedia
of
REPRODUCTION

Volume 2   Ep–L

Encyclopedia of REPRODUCTION

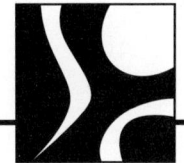

Volume 2   Ep–L

Editors-in-Chief

Ernst Knobil

The H. Wayne Hightower Professor in the Medical Sciences
and
Ashbel Smith Professor,
University of Texas Health Sciences Center
Houston, Texas

Jimmy D. Neill

Distinguished Professor
University of Alabama at Birmingham

ACADEMIC PRESS

San Diego   London   Boston   New York   Sydney   Tokyo   Toronto

Academic Press
a division of Harcourt Brace & Company
525 B Street, Suite 1900, San Diego, California 92101-4495, USA
http://www.apnet.com

Academic Press Limited
24-28 Oval Road, London NW1 7DX, UK
http://www.hbuk.co.uk/ap/

Library of Congress Catalog Card Number: 98-84463

International Standard Book Number: 0-12-227020-7 (set)
International Standard Book Number: 0-12-227021-5 (vol. 1)
International Standard Book Number: 0-12-227022-3 (vol. 2)
International Standard Book Number: 0-12-227023-1 (vol. 3)
International Standard Book Number: 0-12-227024-X (vol. 4)

PRINTED IN THE UNITED STATES OF AMERICA
98 99 00 01 02 03 MM 9 8 7 6 5 4 3 2 1

Contents

G

H

Contents of Other Volumes

V O L U M E   4

──────────── **P** ────────────

Progesterone Actions on Behavior
Anne M. Etgen

Contents by Subject Area

DOMESTIC ANIMALS

REPTILES AND AMPHIBIA

FISH, ELASMOBRANCHII, AND CYCLOSTOMES

REPRODUCTIVE BEHAVIOR

FEMALE REPRODUCTIVE SYSTEMS

MALE REPRODUCTIVE SYSTEMS

PREGNANCY

LACTATION

Preface

The publication of the *Encyclopedia of Reproduction* comes at a most opportune time. Hardly a day goes by when the news media do not report some new dimension in the treatment of infertility or, conversely, controversies associated with the control of fertility and the ethical issues raised by both. Organismal cloning is a matter of constant debate, and the pharmacological correction of erectile dysfunction has become a preoccupation of international dimensions. Procreation remains a subject of universal interest to every segment of society, from scientists to students, from science reporters to the proverbial person on the street.

The present work should serve as a convenient and comprehensive source of information encompassing all aspects of the subject of reproduction as it relates to the entire animal kingdom. It should be as useful to the expert exploring reproductive phenomena outside his or her own field as it is to students and to the educated public at large. Topics for inclusion were initially generated by forming a matrix of systems (gametes, fertilization, and early embryogenesis; reproductive behavior; female reproductive systems; male reproductive systems; pregnancy; and lactation) and of groups of animals (humans and experimental primates; domestic animals; mammals; birds; reptiles and amphibia; fish, elasmobranchii, and cyclostomes; and invertebrates).

A group of outstanding Associate Editors having expertise in one of more of these areas was then recruited. The preliminary list of entries prepared by the Editors was refined and expanded at a meeting with these Associate Editors, who then identified the appropriate authors. Manuscripts were critically reviewed by the Associate Editors and finally scrutinized by us and the editorial staff at Academic Press.

In a work of this kind, errors of omission and of commission are inevitable and we should appreciate having them called to our attention for correction in possible future editions. The 542 entries constituting the work each contain a glossary of terms, a summary introduction, cross-references to related articles, and a reading list. A standard subject index and an index of reproductive systems and zoological groupings are also provided.

Each entry was written to be self-contained, inevitably leading to some overlap of content. We do not view this as a weakness, but instead believe that it will facilitate a reader's search for information by reducing the number of entries that have to be consulted.

The completion of this project demanded the best efforts of a large number of participants. Chief among them are the 700 authors, especially those who wrote articles on short notice so that the publication deadline could be met. The stellar group of 15 Associate Editors, each of whom possesses great breadth of knowledge and who, as a group, span the spectrum of expertise from zoology to animal husbandry to obstetrics and gynecology, also rendered exceptional service.

Finally, we acknowledge the indispensable contributions of the staff at Academic Press: Jasna Markovac, Editor-in-Chief for Biomedical Science, who originally conceived of the Encyclopedia; and Chris Morris, Gail Rice, and Erika Conner, Major Reference Works editors who provided ongoing management of the project.

Ernst Knobil
Jimmy D. Neill

Guide to the Encyclopedia

ORGANIZATION

The *Encyclopedia of Reproduction* is organized to provide the maximum ease of use for its readers. All of the articles are arranged in a single alphabetical sequence by title. Articles whose titles begin with the letters A to En are in Volume 1, articles with titles from Ep through L are in Volume 2, then M through Pri in Volume 3, and Pro to Z in Volume 4.

Volume 4 also includes a complete subject index for the entire work, an alphabetical list of the contributors to the encyclopedia, and a glossary of key terms used in the articles.

Article titles generally begin with the key noun or noun phrase indicating the topic, with any descriptive terms following. For example, "Uterus, Human" is the article title rather than "Human Uterus," and "Migration, Birds" is the title rather than "Bird Migration." This is done so that the same phenomenon or feature can be studied across various groups. For example, all the articles on female reproductive systems in humans, other mammals, birds, etc., appear in one sequence in the Fe- section of the encyclopedia.

INDEX

The Subject Index in Volume 4 contains more than 20,000 entries. The subjects are listed alphabetically and indicate the volume and page number where information on this topic can be found. In addition, the Table of Contents by Subject Area also functions as an index, since it lists all the topics covered in a given area; e.g., the encyclopedia includes 90 differ-

ent articles dealing with reproduction in invertebrates.

OUTLINE

Each entry in the Encyclopedia begins with a topical outline that indicates the general content of the article. This outline serves two functions. First, it provides a brief preview of the article, so that the reader can get a sense of what is contained there without having to leaf through the pages. Second, it highlights important subtopics that will be discussed within the article. For example, the article "Fallopian Tube" includes subtopics such as "Tubal Disorders," "Tubal Sterilization," and "Assisted Reproductive Techniques Involving the Tube."

The outline is intended as an overview and thus it lists only the major headings of the article. In addition, extensive second-level and third-level headings will be found within the article.

GLOSSARY

The Glossary section contains terms that are important to an understanding of the article and that may be unfamiliar to the reader. Each term is defined in the context of the article in which it is used. Thus the same term may appear as a glossary entry in two or more articles, with the details of the definition varying slightly from one article to another. The encyclopedia has approximately 4,250 glossary entries.

In addition, Volume 4 provides a comprehensive glossary that collects all the core vocabulary of reproductive biology in one A-Z list. This section can be consulted for definitions of terms not found in the individual glossary for a given article.

DEFINING STATEMENT

The text of each article in the encyclopedia begins with an introductory paragraph that defines the topic under discussion and summarizes the content of the article. For example, the article "Energetics in Reproduction" begins with the following statement:

Energetics of reproduction is defined as the amount of energy that an animal expends to reproduce. The energetics of reproduction can include the costs of gamete manufacture, synthesis of secondary sexual characteristics and sex-attractant chemicals (pheromones), and reproductive behavior including territorial defense, nest building, courtship rituals, and parental care. . . .

CROSS-REFERENCES

Almost all articles in the Encyclopedia have cross-references to other articles. These cross-references appear at the conclusion of the article text. They indicate articles that can be consulted for further information on the same topic or for other information on a related topic. For example, the article "Osteoporosis" contains references to the articles "Estrogen Replacement Therapy" and "Menopause."

BIBLIOGRAPHY

The Bibliography section appears as the last element in an article. The reference sources listed there are the authors' recommendations of the most appropriate materials for further research on the given topic. The bibliography entries are for the benefit of the reader and thus they do not represent a complete listing of all the materials consulted by the author in preparing the article.

COMPANION WORKS

The *Encyclopedia of Reproduction* is one of a series of multivolume reference works in the life sciences published by Academic Press. Other such titles include the *Encyclopedia of Human Biology, Encyclopedia of Cancer, Encyclopedia of Toxicology, Encyclopedia of Immunology*, and *Encyclopedia of Microbiology*.

Epididymis

Trevor G. Cooper

Institute of Reproductive Medicine of the University, Münster

GLOSSARY

blood–epididymis (–epididymal tubule) barrier Anatomical and physiological attributes of the epididymal epithelium that prevent equilibration between blood and luminal fluid and maintain the unique composition of luminal fluid.

epididymal fluid (plasma) The fluid in which epididymal sperm are bathed and obtained by centrifugation of luminal contents to remove spermatozoa.

epididymal sperm Mature or immature sperm contained within the epididymal canal.

epididymis The single duct that conveys sperm from the testis to the vas deferens in which sperm mature and are stored.

extragonadal reserve The sperm stored in the epididymis, amounting to several days' worth of testicular production, available for ejaculation.

spermiophagy The phagocytic ingestion of spermatozoa.

sperm maturation The process occurring in the epididymis during which immature spermatozoa produced in the testis gain the potential to fertilize eggs.

The epididymis is a long, single, coiled tubule connecting the testicular efferent ducts to the vas deferens and is responsible for the sustenance, protection, transport, maturation, and storage of spermatozoa. It does this by maintaining a highly specialized luminal environment in which the spermatozoa are bathed. The composition of the luminal fluid is dependent on epithelial cells lining the tubule that synthesize, secrete, and transport blood-borne compounds into the lumen and remove testicular fluid so that spermatozoa and impermeant secretions are accumulated at high concentrations within the lumen. Epididymal secretions function to modify some components of testicular spermatozoa and become integral parts of the sperm's membrane so that they can initiate the events leading to fertilization. Other secretions are involved in maintaining sperm viable but quiescent prior to ejaculation.

I. INTRODUCTION

Epididymis–*epí* (Greek for "on") *dídymoi* [Greek for "twins" (testes)]–refers to the organ found on the surface of the testis of mammals, reptiles, and cartilagenous fishes that conveys spermatozoa from the efferent ducts of the testis to the ductus deferens. The term also describes the organ attached to, but not necessarily on, the testis of animals with abdominal testes (Fig. 1).

Main Latin declensions are as follows: nom. singl. epididymis, gen. singl. ductus (caput, corpus, and cauda) epididymidis; nom. pl. epididymides, gen. pl. capita (corpora and caudae) epididymidum; and adj. epididymal, caput (corpus and cauda) epididymidal. Also included are singl. ductus (vas) deferens and pl. ductuli (vasa) deferentia.

II. EVOLUTIONARY APPEARANCE, EMBRYOLOGICAL ORIGIN, AND LOCATION

An epididymis is present in vertebrates that practice internal fertilization (Elasmobranchs, Reptilia, Aves, and Mammalia) but not in Teleost fishes or

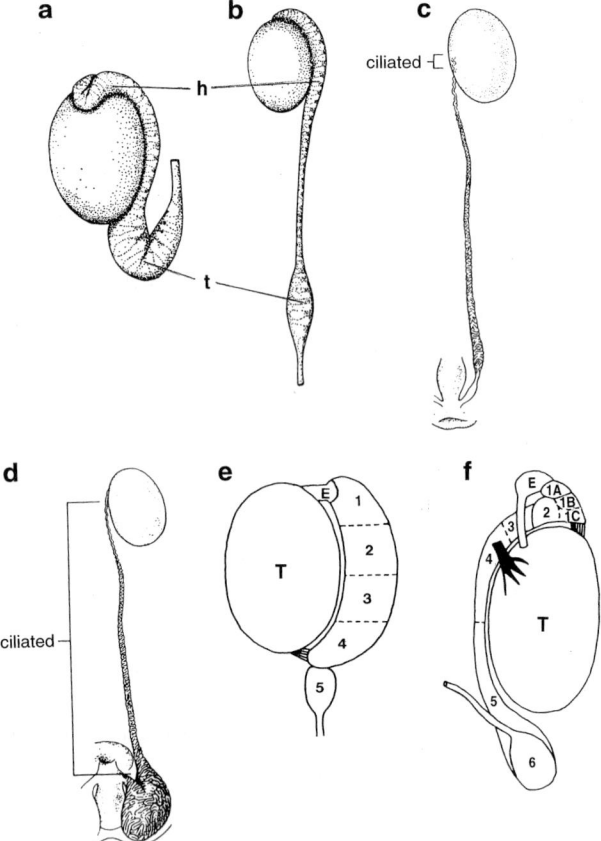

FIGURE 1 Schematic diagrams of the gross morphological form of various epididymides. (a, b) The relationship of the epididymis to the testis in a typical scrotal (a) and testicond mammal (b) is shown (h, caput; t, cauda epididymidis). (c, d) The difference in locations of ciliated cells (of the efferent ducts) in the sperm ducts of a nonpasserine (c) and passerine (d) bird are shown. (e, f) The different segmental anatomies of the epididymides of a protherian (spiny anteater) (e) and eutherian (rat) (f) are compared. (e) E, efferent ducts; 1–4, regions of the initial segment; 5, cauda epididymidis. (f) 1A, 1B, 1C, and 2, initial segments; 3, caput epididymidis; 4 and 5, corpus epididymidis; 6, cauda epididymidis [compiled from Glover (1973), Bedford (1979), and Djakiew and Jones (1982)].

Amphibia, which mainly practice external fertilization and have instead a spermatic duct. The epididymis is a convoluted structure characterized by the processes of sperm transport, maturation, and storage that occur within it. It thus differs from the spermatic duct of bony fishes and amphibians that merely transports functionally mature sperm re-

leased from the testis to the exterior. (The spermatic duct of some bony fishes displays the epididymal features of hormonally induced testicular sperm motility development and ion pumps altering luminal sodium and pH, and in some Anurans the seminal duct resembles an epididymis in being coiled to store sperm.)

The epididymis developed evolutionarily by usurping ducts of primitive urinary systems, the opisthonephros (Elasmobranchs) or mesonephros (Reptilia, Aves, and Mammalia), for the purpose of conveying germ cell products of the testis to the exterior. Embryologically, the epididymis arises from the single Mesonephric (Wolffian) duct making connections with the rete testis via the efferent ducts which are multiple and of mesonephric tubule origin.

In mammals the location of the caput epididymidis is dictated by that of the testes and can therefore be abdominal (in Testicondi), inguinal, or scrotal. In scrotal mammals it descends into the scrotum via its gubernacular connection with the tunica vaginalis, thus bringing the testis with it. Some passerine birds have cloacal protruberances beneath which lie the dilated tubule of the terminal epididymis (also called the seminal vesicles but not analogous to these organs of mammals) that contains stored spermatozoa (Fig. 1). In animals with abdominal or inguinal testicles the cauda epididymidis with its sperm store always lies closer to the exterior.

III. MACROSCOPIC APPEARANCE

In most species three gross anatomical regions of the epididymis can be discerned: a head (caput), attached to the testis by the testicular efferent ducts and incorporating the initial segments; a normally narrower body (corpus or isthmus); and a usually bulbous tail (cauda). In some species the narrow region may not be the corpus; in the sparrow it comprises ciliated efferent ducts (Fig. 1) and in the guinea pig both efferent ducts and the initial segment. Other terminology refers to initial, middle, and terminal segments of the epididymis based on function and is approximately equivalent to initial segment/caput, corpus, and cauda epididymides, where fluid absorption (initial), sperm maturation (middle), and sperm storage (terminal) occur. The ductus deferens is continuous with the cauda epididymidis and in some species contains a distal thickening (the ampulla) to aid expulsion of sperm to the urethra during seminal emission.

The caput region in some species (including man) contains within its bounding capsule the testicular efferent ducts. These differ from the single epididymal tubule in being multiple (single in rays), containing ciliated and nonciliated cells, and lacking epithelial basal cells. Since the efferent ducts can be considered part of the caput epididymidis (defined morphologically), some of its functions are discussed despite its different histology. The epididymis in sharks is connected with the Leydig gland [no connection (other than its discoverer) with testicular Leydig cells], the secretions of which may be related to sperm aggregation (see Section VII), and in protherians (platypus and spiny anteater) it is a simpler structure in which a greater portion is considered to comprise the initial segment than in metatherians (marsupials) and eutherian mammals (Fig. 1).

IV. HISTOLOGY AND CYTOLOGY

The epididymis consists of a single, highly convoluted tubule, the length of which varies from 3 m (rats) to 50 m (boars), surrounded by a connective tissue capsule with septulae that divide it into a number of lobules.

A. Peritubular Elements

Surrounding the tubule are peritubular myoid cells with more circular and longitudinal muscle distally and an interstitium containing blood, lymphatics, and nerves that supply the muscles and blood vessels. Epididymal innervation is poor proximally but dense sympathetic fibers from the inferior mesenteric ganglion control activity of the distal regions. (There is some suggestion of intraepithelial innervation in the Djungarian hamster epididymis.) Injected oil droplets are seen to undergo pendular movements that give rise to space-gaining movements proximally and retrograde mixing action distally. After ejaculation contents of the ductus deferens are returned to mix with caudal contents.

B. Epithelial Cell Types

The epithelium consists of permanent cell types continually surveyed by wandering cells of the immune system.

1. Principal Cells

The tubule is lined by a pseudostratified epithelium throughout the duct (Fig. 2) with nuclei of at least three cell types occupying different levels within it. The main cell type, the principal cell, is both secretory and absorptive. Principal cell height is tallest, microvilli length longest, and luminal diameter smallest in the initial segments of the caput epididymidis, and generally cells become progressively shorter, and the lumen larger, toward the distal region. Tight junctions between the epithelial cells largely restrict the free movement of large molecules into the lumen from blood by a paracellular route and constitute the anatomical basis of the "blood–epididymal tubule barrier." These junctions are incomplete in the efferent ducts and in the epididymis are not as tight as those in the testis, but they can be extensive in certain bats in which osmotic pressure in the lumen is five times that of blood. Tight junctions can be bypassed by transcytosis in both directions (see Section VII) and gap junctions permit communication between principal cells.

2. Basal Cells

Basal cells in the rat form a network beneath the principal cells. These cells are characteristically present along the length of the epididymis and ductus deferens of the adult animal and are absent at birth. They are more numerous and obvious in some species than others and may be round or triangular in shape. In the rat they form an interconnecting network beneath the principal cells and may extend to the lumen (Fig. 2).

3. Apical and Clear Cells and Immunocytes

Depending on species, epididymal region, and state of development, apical cells (caput) and clear cells (corpus and cauda) are present that are more absorptive in nature. Intraepithelial lymphocytes ("halo cells") and macrophages and interstitial mast cells are occasionally found, but rarely are they found in the lumen under normal conditions (Fig. 2).

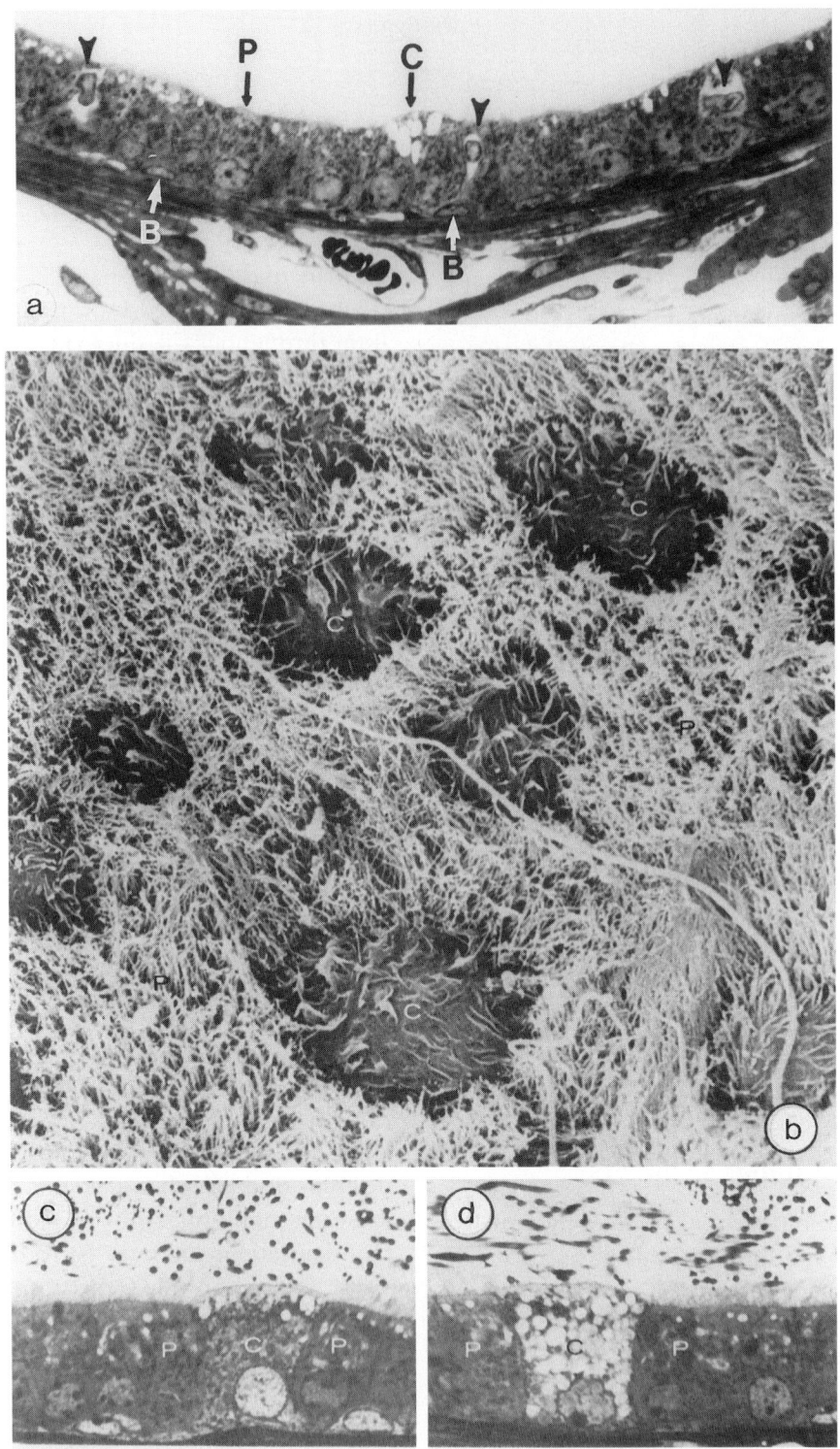

FIGURE 2 Histology of the rat epididymis. (a) Light micrograph showing principal cells (P), clear cells (C), basal cells (B), and lymphocytes (arrowheads) in the epithelium of the rat cauda epididymidis. Magnification, ×820. A subepithelial blood vessel and peritubular muscle are shown. (b) Scanning electron micrograph of the rat cauda epididymidis revealing the difference between the apical surfaces of the absorptive clear cells (C) and microvilli of principal cells. A sperm tail lies across the field of view. Magnification, ×2500. (c, d) Light micrographs of principal cells (P) on either side of clear cells (C) of differing vacuolar content and two pale-staining basal cells are visible in (c) . Magnification, ×1020 (Micrographs courtesy of C.-H. Yeung and from Hamilton *et al.* Copyright © 1977 John Wiley and Sons, Inc.).

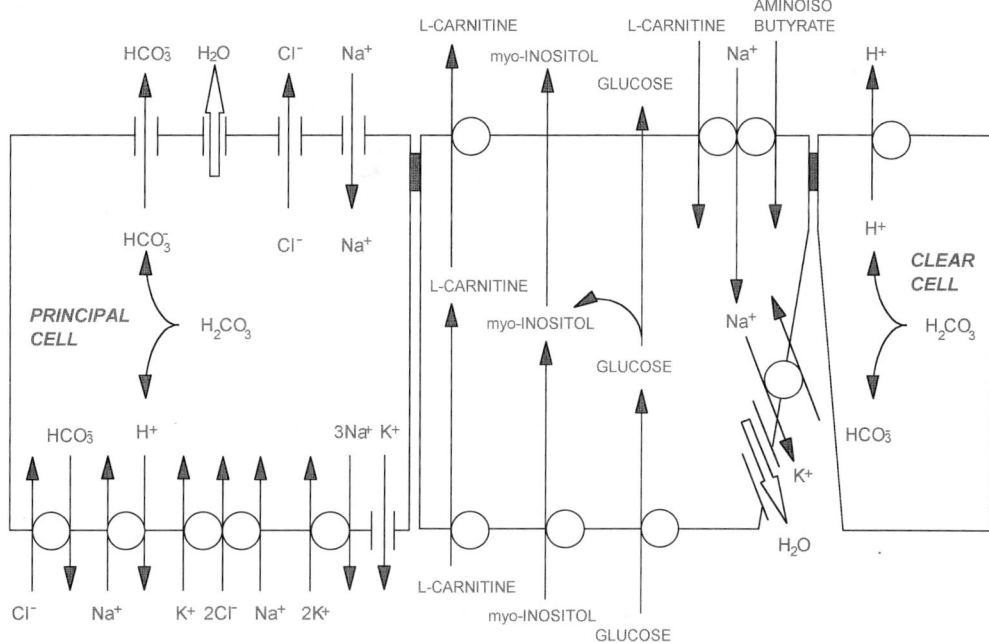

FIGURE 3 Composite scheme of ion pumps, carriers, and channels and transporters for organic solutes and the movement of water in the epididymis and efferent ducts.

V. EPITHELIAL CELL FUNCTION

A. Principal Cells

1. Antiluminal Transport of Substances

i. Resorption of Ions, Water, and Organic Solutes The majority of fluid leaving the testis is absorbed in the efferent ducts and the remainder in the proximal epididymis. As a result, spermatozoa and impermeant epididymal secretions are concentrated, luminal osmotic pressure is raised, and sperm cells are partially dehydrated (Fig. 3). Fluid resorption alone is responsible for raising the concentration of carnitine from the distal corpus epididymidis onwards. Principal cells transport Na^+ from the lumen through apical sodium channels to basolateral Na^+/K^+-ATPases, where they are exchanged for K^+. This activity is highest in the efferent ducts and in the epididymis proximal activity exceeds that distally. Water flows osmotically from the lumen through aquaporin (CHIP28) water channels located in the apical membranes of the efferent ducts and ampulla of the ductus deferens.

Both L-carnitine and amino isobutyric acid are absorbed from the proximal caput epididymal lumen by saturable, carrier-mediated, sodium-dependent mechanisms (Fig. 3). Amino acids released by the action of γ-glutamyl transpeptidase on luminal glutathione are also taken into principal cells.

ii. Endocytosis and Transcytosis Pinocytosis. Pinocytototic vesicles are present in the apical cytoplasm of principal cells, and endocytic uptake of material in the lumen can occur by fluid phase, adsorptive or carrier-mediated mechanisms (e.g., for androgen-binding protein), transferrin, oxytocin, clusterin [sulfated glycoprotein-2 (SGP-2)], or α2-macroglobulin. In this way the epididymis removes from the lumen protein components from the testis and some proteins from more proximal epididymal regions.

Phagocytosis. Uptake of large particles by principal cells is rare, but in the brush-tailed possum principal cells phagocytose the entire cytoplasmic droplet from spermatozoa and digest the spermatozoa. There are rare reports of spermiophagy by epithelial cells all along the male tract but there is no evidence for wholesale removal of sperm by this mechanism (see Section VII).

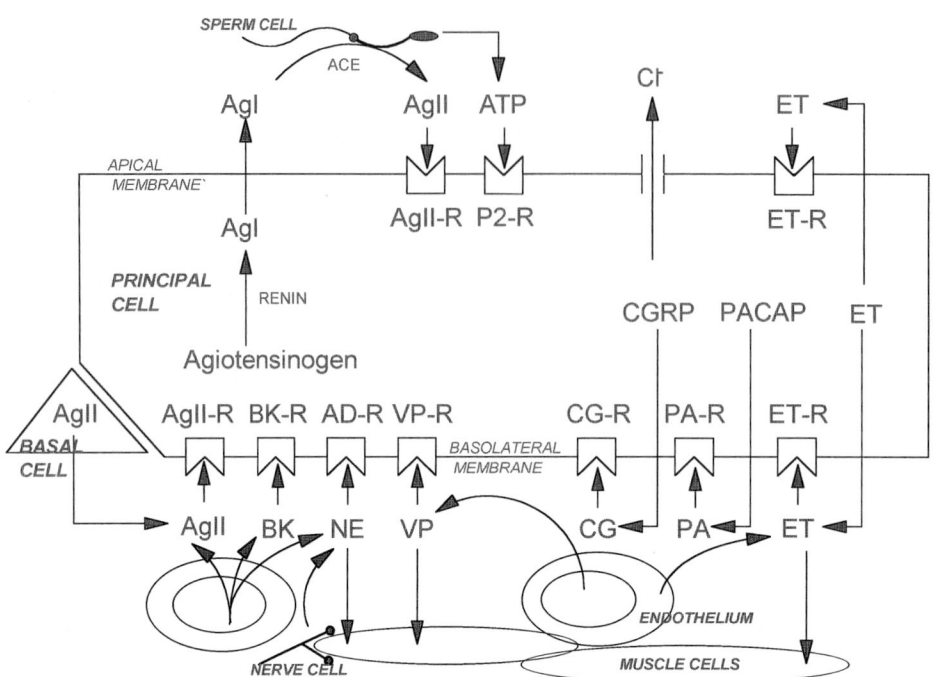

FIGURE 4 Paracrine and autocrine control of epithelial cell functions. Receptors on the apical cell membrane for angiotensin II (AGII-R), endothelin (ET-R), and purinergic transmitters (P2-R) control chloride transport and fluid secretion. Spermatozoa exert paracrine control in the luminal compartment by converting angiotensin I (AgI, secreted by principal cells that synthesize it from angiotensinogen via renin) to angiotensin II (AgII) via angiotensin-converting enzyme activity (ACE). At the basal aspect of the cells paracrine control is mediated by basal cells (via angiotensin II), possibly (intraepithelial) nerve cells via norepinephrine (NE), the endothelium (endothelin; ET), whereas the principal cells exercise autocrine control by secretion of calcitonin gene-related peptide (CGRP), pituitary adenylyl cyclase-activating peptide (PACAP), and endothelin (ET) that act on adrenergic receptors (AD-R) and receptors for angiotensin II (AgII-R), bradykinin (BK-R), vasopressin (VP-R), calcitonin gene-related peptide (CG-R), pituitary adenylyl cyclase-activating peptide (PA-R), and endothelin (ET-R) on the basolateral surface (based on Wong *et al.*, 1992).

Trancytosis. Bypass of the obstacles to paracellular movement has been demonstrated for substances (e.g., cationic ferritin) introduced into the rete testis and vas deferens but not epididymis proper. Whether natural components move from the lumen to the interstitium remains to be proven, but it has been postulated that the transfer of lipid material from the lumen to basal cells must involve passing through principal cells.

2. Proluminal Transport of Substances

i. Fluid Secretion In addition to fluid resorption, epididymal principal cells are capable of secreting fluid in response to local conditions. Water flows into the lumen secondarily to chloride and bicarbonate secretion when their intracellular concentrations ex-

ceed their electrochemical equilibriums. Basolateral Na^+-K^+-$2Cl^-$ symports, Na^+-H^+, $3Na^+$-$2K^+$, and Cl^--HCO_3^- exchangers, and K^+ channels as well as apical chloride channels (the cystic fibrosis transmembrane receptor protein) (Fig. 3) have been demonstrated in rat epididymal cells.

Since some regulators of fluid secretion (e.g., vasopressin, norepinephrine, and endothelin) are also associated with stimulation of peritubular muscle, they may form part of a local mechanism to facilitate sperm movement through the duct by making fluid less viscous and propulsive activity more efficient (Fig. 4; see Section VIII).

ii. Low-Molecular-Weight Osmolyte Transport and Transcytosis Both L-carnitine and myoinositol are transported against concentration differences from blood to lumen—the former in the distal caput and the latter in the proximal cauda—but for neither has the carrier been identified. Messenger RNA for the renal inositol transporter is absent from the epididymis but a weak signal for the renal taurine transporter is present (Fig. 3). The rate of bypass of the blood–epididymal tubule barrier by macromolecules (e.g., immunoglobulins) is low but it has been demonstrated to occur by vesicular uptake.

3. Synthetic Activity

i. Low-Molecular-Weight Compounds Epididymal principal cells synthesize glycerophosphocholine from circulating precursors and possibly from sperm. Glycerophosphocholine is synthesized and secreted into the lumen of the proximal caput with high concentration in all regions of the duct as a consequence of water resorption. Despite one report of high activity of epididymal enzymes involved in the biosynthesis of inositol, no direct demonstration of this has been forthcoming.

ii. Macromolecular Synthesis and Secretion A feature of gene expression in the epididymis is that transcription and translation are cell and region specific. Although certain cells within a tubule cross section express a gene product and adjacent cells do not (e.g., immobilin), this is unlikely to be of significance for sperm in the lumen because the thorough mixing of luminal contents would dissipate local concentration gradients in luminal fluid established between cells. However, region specificity would ensure a sequence of delivery of proteins to developing spermatozoa (Table 1; see Section VII). An additional peculiarity is that some housekeeping

TABLE 1
Proteinaceous Epididymal Secretions and Their Possible Functions

Name	Pseudonym	Regions	Species	kDa	Suggested function	Putative role
FLB1		ED, caput, corpus	Hu	94		Sperm–egg fusion
Cres		IS	R	14, 19	Proteinase inhibitor	Protection of sperm
GPX5		IS, caput	M	24	Glutathione peroxidase	Protection of sperm
GGT		IS, caput	R	71	γ-Glutamyl transpeptidase	Degrade GSSG
B/C	ESP1, RBP, MEP7	IS, caput, corpus	R, M	16, 18	Retinoic acid-binding protein	Modification of sperm
Immobilin		Caput	R, H	>400	Mucoprotein	Inhibition of *in situ* motility
SGP-2	DAG, clusterin, apo-J	Caput	R	66–70	Complement lysis inhibitor	Modification of sperm
anti-KC1		Caput	Hu			Sperm–egg fusion
PBP	MEP9	Caput	R, M	23	Phospholipid-binding protein	Modification of sperm
HE2		Caput, corpus	Hu			Sperm–egg fusion
HE1	EP1, ESP14.6	Caput, corpus, cauda	Hu		Sterol-binding protein	Modification of sperm
D/E	AEG, CRISP1	Caput, corpus, cauda	R, M	27–37	Carboxypeptidase Y-like	Sperm–egg fusion
2D6		Corpus	R	23		Sperm–egg fusion
p34h	p26h	Corpus	R, H	34, 26		Zona binding
β-NGF		Corpus	R, M	32	Nerve growth factor	Control of epithelium
HE5	CD52	Corpus, cauda	Hu	18–25		Immune defense
Fibronectin		Corpus, cauda	Hu	220	Disintegrin	Sperm–egg fusion
HE4		Corpus, cauda	Hu		Extracellular proteinase inhibitor	Decapacitation factor
ASF		Cauda	Ra	129–259	Acrosome stabilization	Decapacitation factor
His-50		Cauda	R	50		Decapacitation factor
SOD		Cauda	R		Superoxide dismutase	Protection of sperm
Catalase		Cauda	R		Catalase	Protection of sperm

Note: Abbreviations used: AEG, acidic epididymal glycoprotein; apo-J, apolipoprotein J; ASF, acrosome stabilizing factor; CRES, cysteine-rich epididymal spermatogenic protein; CRISP1, cystatin-related epididymal-specific protein; DAG, dimeric acidic glycoprotein; ED, efferent ducts; ESP1, epididymal secretory protein-1; GGT, γ-glutamyl transpeptidase; H, hamster; His-50, high tonic strength; Hu, human; IS, initial segments; MEP, mouse epididymal protein; M, mouse; NGF, nerve growth factor; PBP, phospholipid-binding protein; R, rat; Ra, rabbit; RBP, retinoic acid-binding protein; SGP-2, sulfated glycoprotein 2.

(somatic) gene products may be secreted and thus take on functional significance in the epididymis (e.g., proopiomelanocortin, proenkephalin, nerve growth factor-β, CD52, and SGP-2; Table 1).

Because of its extensive Golgi apparatus, the epididymis was the first organ from which this organelle was characterized both histochemically and biochemically. Thus, the majority of secreted proteins are glycoproteins and are secreted via merocrine secretion, the normal route of protein export. Apocrine secretion is postulated to explain the export from the vas deferens of proteins (e.g., mouse vas deferens protein) that lack signal peptides. This may also be an export route in the epididymis in which luminal vesicles are seen to arise from epithelial cells even in well-fixed epididymal material. The occasional loss of degenerate cells into the lumen (holocrine secretion) may occur.

B. Apical and Clear Cells

Both apical cells (in the caput epididymidis) and clear cells (in the corpus and cauda epididymidis) lack true apical microvilli but have luminal flaps and folds (Fig. 2) capable of engulfing particulate matter. Clear cells contain large vacuoles in the cytoplasm, the size of which depends on the luminal Na^+ concentration, and in the rat clear cells contain either glycoprotein or lipids. These cells take up foreign particulate matter introduced into the lumen by fluid-phase endocytosis and normally remove from the lumen the contents of ruptured cytoplasmic droplets from spermatozoa. They also remove some, but not all, types of protein secretions of principal cells situated more proximally. In the rat, both apical and clear cells contain carbonic anhydrase involved in luminal acidification (species lacking clear cells have this enzyme in principal cells) and receptors for vitamin D, whereas clear cells have a proton-ATPase involved in lumen acidification and receptors for aldosterone and vitamin D.

C. Basal Cells

The expression of macrophage antigens has suggested a role related to immunological protection of sperm cells. In seasonal breeders or certain experimental conditions in which degeneration of sperm or the epithelium occurs, basal cells become phagocytic. The presence in basal cells and cells in the interstitium of apolipoprotein E (involved in sterol transport) suggests the ability to recycle lipids. Despite little cytoplasm they express high levels of glutathione S-transferase and superoxide dismutase, possibly involved in detoxification.

D. Immunocytes

Under normal conditions, and depending on species, lymphocytes of the $CD4^+$ and $CD8^+$ class are present within the epididymal epithelium and MHC II-bearing cells and macrophages are present in the interstitium. Their presence within the epididymal lumen only follows rupture of the duct and an immune response to the autoantigenic spermatozoa. The epididymis may be a site of entry of leukocytes in disease.

VI. LUMINAL FLUID

The regional secretory and resorptive function of the principal cells, together with the tight junctions between them, ensures that maturing sperm experience an ever-changing microenvironment as they are propelled along the duct.

A. Fluid Composition

1. Low-Molecular-Weight Components

A loss of total inorganic cations (sodium is lost, whereas potassium is secreted into the lumen) with approximately constant amounts of inorganic anions (phosphate levels increase) is compensated for by secretion of organic solutes (Fig. 5). Exceptionally high (millimolar) concentrations of amino acids proximally (mainly glutamate from glutathione hydrolysis) that decrease distally, a high concentration of glycerophosphocholine in all regions, and increases in both L-carnitine and myoinositol along the length occur in many species (Fig. 5). The protein composition of epididymal fluid differs from that of serum and testicular fluid and shows regional variations in the nature and intensity of electrophoretic bands.

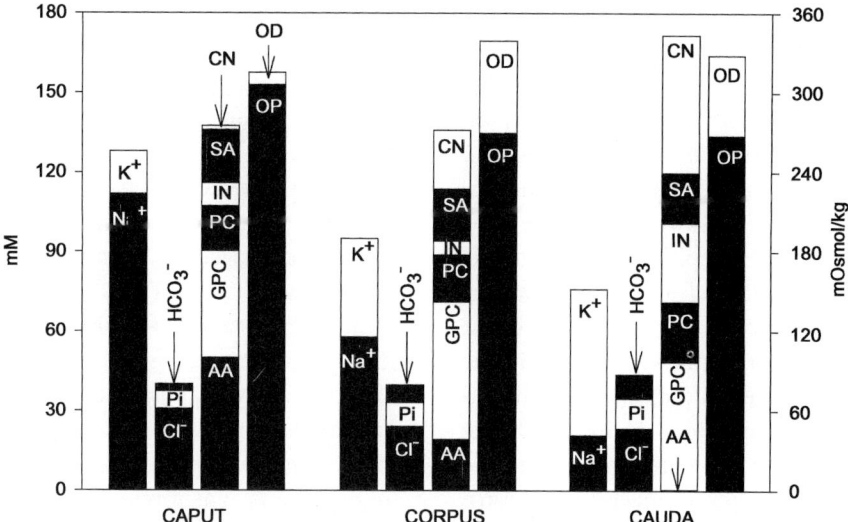

FIGURE 5 Composition (first three columns, left-hand ordinate) and osmotic pressure (fourth column, right-hand ordinate) of sperm-free luminal fluid from the three major regions in the rat epididymis (abscissa). AA, total amino acids; CN, L-carnitine; GPC, glycerophosphocholine; IN, myoinositol; OD, osmotic deficit (difference between measured osmolality and osmotic pressure); OP, osmotic pressure calculated from known constituents; PC, phosphocholine; Pi, inorganic phosphate; SA, sialic acid.

Epididymal fluid centrifuged free of spermatozoa contains no detectable reducing sugars, although circulating glucose has access to the lumen. In the rat, luminal pH is more acidic (6.5–7.1) than arterial blood (7.4). Also, respiratory gas tension of pCO_2 (52 or 53 mmHg) is higher than that in systemic blood (39 mmHg), whereas pO_2 (23 or 24 mmHg) is slightly lower than in that venous blood (38 mmHg) and much less than that in arterial (60 mmHg) blood. The luminal potential is about −20 mV.

The roles of the low-molecular-weight compounds in sperm maturation and storage have not been unequivocally defined, but many compounds are non-disturbing osmolytes found in organs with osmotic roles. In the species of bat with intraluminal osmolality of 1500 mOsmol/kg, dehydration of the sperm cells is considered to be a mechanism of quiescence (see Section VII). The lower ionic strength of luminal fluid may help maintain electrostatically held proteins on the sperm surface.

2. Macromolecules

Many potentially lytic enzymes, glycosyl transferases, and glycosidases have been detected in luminal fluid but their *in situ* activity, natural substrates,

inhibitors, and functions are not known. The presence of viral proteins in epididymal fluid of the mouse has been reported. In recent years the number of new proteins identified in the epididymal lumen has paralleled the widespread introduction of new molecular biology techniques, but there has been no parallel growth regarding insight into to their possible functions. Listed in Table 1 are some epididymal proteins that are either secreted or exposed on the luminal surface of epithelial cells that have been mooted to play some role in fertilization. The major site of expression of the genes from *in situ* hybridization or proteins by immunohistochemistry and molecular mass are given in Table 1.

B. Interaction of Spermatozoa with Epididymal Secretions

The rising carnitine content of maturing spermatozoa reflects its passive uptake into the cell. The mode of interaction of proteins with spermatozoa is likely to reflect the function of the protein (see Section VII). Some are loosely attached (easily washed off) (e.g., phospholipid-binding protein), although the remaining proteins may be functional (e.g., D/E). Alternatively, somewhat tighter binding includes both electrostatic forces (e.g., His-50) and integral membrane proteins held by glycosylphosphatidylinositol anchors (e.g., CD52). Some proteins (e.g., enzymes or phospholipid-binding proteins) may be considered only temporarily associated with the sperm surface during membrane remodeling, whereas others (e.g., decapacitation factors, ASF, and His-50) attach to sperm in the epididymis after their membranes have been rendered fragile during maturation, necessitating this protection, and they are lost upon capacitation. Other proteins whose function is expressed after leaving the epididymis remain on the sperm until zona binding and are lost with the acrosome (e.g., p34h and p26h) or remain after the acrosome reaction in order to promote fusion with the oolemma (e.g., 2D6 and D/E).

VII. FUNCTIONS OF THE EPIDIDYMIS

A. Sustenance and Protection of Spermatozoa

During their transit through the epididymis, sperm are maintained in a viable condition and their autoantigens protected from the body's immune system. Within the epididymal lumen spermatozoa have access to substrates and oxygen (see Section VI). Protection from cells of the immune system is effected by the intercellular tight junctional complexes preventing paracellular transport of intraepithelial leukocytes, immunosuppression, and other mechanisms. The potentially self-inflicted damage of reactive oxygen species that could be generated *in situ* by enzymes in the sperm's cytoplasmic droplet is thought to be prevented by protective epididymal enzymes. The transcription products of the genes for catalase and secreted forms of glutathione peroxidase and superoxide dismutase are present in epididymal cells and the latter two enzyme activities have been demonstrated in the epididymal lumen. Toxicants could be removed by glutathione S-transferase present in basal cells. Protease inhibitors entering the epididymis from the testis (e.g., α2-macroglobulin) or secreted by the epithelium [e.g., cystatin-related epididymal-specific protein (CRES)] may prevent

damage to the epithelium by lytic enzymes leaking from degenerating spermatozoa.

B. Transport of Spermatozoa

In most species the average transit time of sperm through the organ is 1 or 2 weeks, with mean rates of transit between 0.5 and 5 m per day in rats and boars, respectively. The shorter human organ allows rapid transit (2–4 days). Because the sperm entering the epididymis are poorly motile, their transport is achieved by peristaltic activity of the duct. Spontaneous muscular activity is augmented by adrenergic, cholinergic, and purinergic nerve fibers, blood-borne angiotensins, vasopressin, and oxytocin, and paracrine products of the epithelium endothelin (see Fig. 4). Androgens reduce muscle tone and prolong epididymal residence of spermatozoa so that enhanced sperm transport follows bilateral removal of the testis.

C. Sperm Maturation

Sperm maturation refers to the change in functional capacity that occurs in sperm during their sojourn in the epididymis. It was initially demonstrated in mammals by the obviously greater vigor of motility expressed *in vitro* by sperm taken from the cauda than the caput, by the nature of sperm motion (changing from circular to progressive), and by the longer retention of flagellation by mature sperm *in vitro*. It has subsequently been confirmed in the epididymides of marsupials, protherians, passerine birds, snakes (during active spermatogenesis), and sharks. Other noticeable changes included as maturational events include (i) the movement of the cytoplasmic droplet from the neck to the end of the midpiece, (ii) changes to other features such as whole cell (surface) charge as a result of changes in membrane carbohydrate residues and sperm proteins, (iii) the oxidation of intracellular disulfides, (iv) major alterations in sperm lipids structure, and (v) in some species (e.g., guinea pig) a massive remodeling of acrosomal shape.

The functional importance of maturational changes was demonstrated by the superior fertilization results stemming from insemination of sperm taken from more distal epididymal regions in mammals and birds. By inseminating sperm at different sites within the female tract (uterus or oviduct) or removing various egg investments for *in vitro* fertilization (cumulus free and zona free), it has been shown that the following functions of sperm necessary for fertilization are acquired during epididymal transit: the ability to move progressively and cross the cervix or uterotubal junction, to bind to and penetrate the zona pellucida, and to bind to and fuse with the zona-free vitellus. Other changes in maturing cells include differences in the state of chromatin condensation and methylation and the ability to form a male pronucleus and generate normal embryos when fertilization is achieved.

1. Mechanism of Sperm Maturation: Time or Specific Epididymal Secretions

Since mature sperm (from the cauda epididymidis) are physiologically older than immature spermatozoa (from the caput epididymidis), a long-standing question relating to maturational changes in sperm is whether they are due to aging alone or specific epididymal secretions. Classic experiments in which sperm were retained within the proximal epididymis by ligatures showed that sperm develop some potential for motility but not normal surface charge or fertilizing capacity. This indicates that some aspects of motility are time dependent, presumably inherent properties of the sperm cells. In some species forward progression *in vitro* can be induced even in testicular sperm in the absence of epididymal secretions. On the other hand, that fertilizing capacity only develops if immature sperm are retained more distally in the epididymis has led to the concept that this is dependent on epididymal secretions.

2. Epididymal Secretions Involved in Sperm Maturation

There is increasing evidence that epithelial cell secretions are involved directly and indirectly in the development of the sperm's ability to interact with eggs. Since caput sperm contact proximal secretions within the first day of entering the epididymis and only later contact more distal secretions, this may signify that membrane alterations are necessary for subsequent uptake of secreted proteins in certain

membrane domains. Table 1 shows that proximal secretions include those that remodel sperm membranes and are involved with sperm–egg fusion (at the equatorial region), whereas proteins involved in sperm–zona binding (over the acrosome) and decapacitation factors (also over the flagellum) are secreted more distally. A hypothetical scheme outlining this idea is given in Fig. 6. Potentially sperm protective enzymes are found throughout the lumen.

i. Direct Involvement of Proteins in Sperm Maturation **Modification of Sperm Lipids.** Some epididymal secretions related to phospho-/glycolipid binding are similar to testicular secretions that accompany testicular sperm into the posttesticular duct. For example, SGP-1 (or prosaposin, the precursor of saposins involved in the specific activation of lysosomal enzymes that degrade membrane lipids) and SGP-2 (or clusterin or apolipoprotein J, a complement lysis inhibitor active at fluid–tissue interfaces) are present in testicular fluid and on testicular spermatozoa. Both are present on sperm leaving the testis but are lost from sperm to epithelial cells of the rete testis and efferent ducts. Sperm subsequently take up an epididymal form of SGP-2 (lower molecular weight) that is secreted by principal cells and that is present in epididymal fluid in all regions. The epididymis synthesizes SGP-1, but this is located in the lysosomes in which it is involved in the disposal of ingested lipids and not secreted, although testicular SGP-1 is present in luminal fluid throughout the epididymis. A phospholipid-binding protein identified as an epididymal secretion is also present on testicular sperm and in rete testis fluid in rats and, consequently, is present in the lumen along the length. A solely epididymal secretion from the human epididymis (HE1) has sequence homology with a sterol-binding protein. It is possible that all four proteins may act in concert at sperm membrane surfaces to modify the lipid composition and thus facilitate the uptake/insertion of proteins secreted more distally (Fig. 5).

Secretion of Sperm-Coating Proteins. To date only weak evidence supports the view that the epididymal secretion proteins D/E (AEG and CRISP1), "forward motility protein" (FMP), and carnitine are

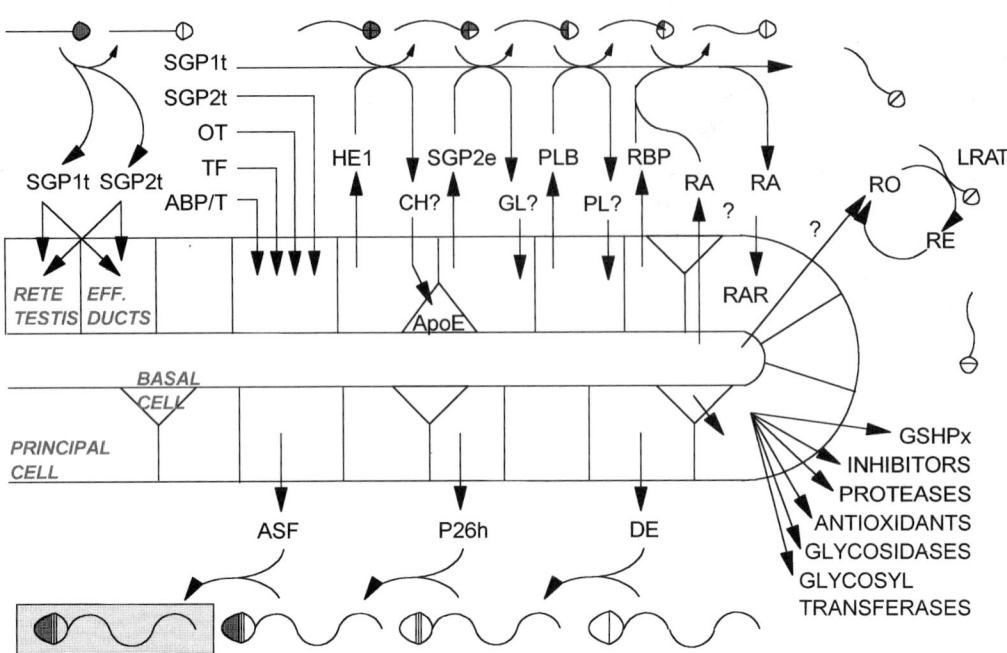

FIGURE 6 Speculative scheme of sperm maturation with sperm first being modified by proteins in testicular and epididymal fluid and later acquiring coating proteins in specific locations to facilitate sperm–egg and sperm–zona binding and to necessitate capacitation. ABP/T, androgen-binding protein/testosterone complex; ApoE, apolipoprotein E (involved in sterol transport); ASF, acrosome stabilizing factor; CH, cholesterol; DE (AEG, see Table 1); GL, glycolipid; GSHPx, glutathione peroxidase; HE1, protein with sequence homology with a sterol-binding protein; LRAT, lecithin-retinol acyltransferase; OT, oxytocin, p26h (see Table 1); PL, phospholipid; PLB, phospholipid-binding protein; RA, retinoic acid; RAR, retinoic acid receptor-α; RBP, retinoic acid-binding protein (see Table 1); RE, retinyl esters; RO, retinol; SGP-1t, testicular form of sulfated glycoprotein-1; SGP-2t/e, testicular and epididymal forms of sulfated glycoprotein-2; TF, transferrin.

implicated in the development of the potential of sperm motility. Changes in membrane lipids may well alter membrane ion pumps with consequences for motility, although proof of this is lacking. Secreted epididymal proteins that bind to sperm and are involved with zona recognition include p34h (man), p26h, EP3 (hamster), and PES (rat). Secreted epididymal proteins found on the sperm surface and postulated to be involved in the ability of acrosome-reacted sperm to bind and fuse with the vitelline membrane of zona-free eggs include FLB1, SOB2, fibronectin, HE2 (man), D/E, and possibly 2D6 (rat). All these essential attributes have to be present in the spermatozoon at the time of interaction with the egg for it to fertilize them.

ii. Indirect Involvement of Proteins in Sperm Maturation Proteins Involved in Sperm–Zona Binding.

Certain sperm proteins present on testicular sperm have been implicated in the primary binding of sperm to the zona pellucida [a sperm surface hyaluronidase (PH-20 in many species including man and 2B1 in the rat), M42, sp56, P95, and galactosyltransferase (GT) (mouse)]. During epididymal transit some of these undergo posttranslational processing involving a reduction in size (PH-20, 2B1, and M42) or a change in location on the sperm head (PH-20 and GT). Autoproteolytic activity has been considered, but for PH-20, M42, and 2B1 the changes can be mimicked by trypsin *in vitro* but the source of the enzymes involved has not been determined.

Proteins of the metalloprotease family are present in the epididymal epithelium that could be responsible for the conversion of inactive testicular forms. One, the snake venom-like epididymal apical protein-1 (EAP-1), is present apically but not on sperm;

others are secreted (e.g., carboxypeptidase Y-like D/E, plasminogen activator, and angiotensin converting enzyme). Their activity may be regulated by proteinase inhibitors from the testis (e.g., α2-macroglobulin) or secreted by the epididymis (e.g., HE4 and cystatin c) that surrounds the sperm in the lumen.

Secondary binding of sperm to the zona pellucida (after induction of the acrosome reaction) and penetration of the zona involve the sperm protease acrosin, located within the acrosome during spermatogenesis. This enzyme and acrosomal transmembrane cyritestin also undergo a reduction in size during epididymal transit. From their intraacrosomal location, the epididymis could promote autoactivation via alteration of intraacrosomal pH.

Proteins Involved in Sperm–Egg Fusion. In many species sperm–egg fusion can be blocked by peptide with the RGD (Arg-Gly-Asp) sequence and the involvement in many species of a sperm protein with both viral fusion peptide and disintegrin domains [fertilin-β (PH-30)] is indicated. This protein, present on testicular sperm, also undergoes a reduction in size and location on the sperm head upon epididymal passage that can be mimicked *in vitro* by trypsin and may be affected by the enzymes cited previously.

D. Storage of Spermatozoa

In all species, sperm are stored in a quiescent state in the tail of the epididymis before ejaculation. This extragonadal reserve permits ejaculation of larger numbers of sperm than the daily testicular production produces. Thus, the size of this extragonadal store depends on the mating strategy of the species; it is large in bulls and rams, in which multiple copulations can be performed in a day, rather meager in man, and hardly existent in one marsupial mouse that needs to limit the number of sperm per ejaculate to maximize fertile copulations.

The storage time of sperm in nonseasonal breeders normally approximates the proximal transit time but can be extended artificially. Sperm cannot be stored indefinitely in the epididymis and aging spermatozoa undergo intraluminal degeneration in the distal epididymal lumen of celibate males. Isolated reports of

spermiophagy by epithelial cells in all regions of the posttesticular tract do not provide evidence for the large-scale removal of spermatozoa from a normal, intact epididymis. Pathological conditions leading to disruption of the duct, extravasation of spermatozoa, and infiltration of leukocytes and macrophages result in extensive intra- and extratubular phagocytosis of sperm cells.

Once the caudal reserve is full, the continued passage of recently matured sperm into the cauda expels existing mature spermatozoa through the ductus deferens and the urethra into the urine (rams and man) or by spontaneous ejaculates (rodents), thus ensuring a heterogenous population of mature cells of different ages in the cauda available for ejaculation. Some seasonal animals, reptiles, and mammals store sperm in the epididymis even when the testis is regressed so that conceptions can occur following hibernation when the females ovulate. In these cases, the fertilizing sperm are several months old.

1. Mechanism of Sperm Storage

i. Enforcing Metabolic Quiescence Metabolic quiescence of sperm cells is induced by one or more of several species-dependent mechanisms. Those acting to reduce motility *in situ* include (i) the lowering of luminal sodium ion concentration (preventing proton efflux and a rise in intracellular pH that triggers motility) in all species; (ii) the high concentration of sperm (billions/ml) and (iii) secretion of a viscous mucoprotein (immobilin in rodents) both of which would restrict sperm movement; and (iv) the production of permeant acids keeping intracellular sperm pH low (bulls). It is possible that the high $p\text{CO}_2$ and dehydration of sperm (to a high degree in some bats) also contribute to sperm quiescence.

ii. Preventing Premature Sperm Activation Functional quiescence (prevention of inopportune or precocious fertilization events) within the epididymis is enforced by the (i) secretion of proteinaceous decapacitation factors that protect labile sperm membranes (e.g., acrosome stabilizing factor in rabbits); (ii) aggregation of sperm, including the pairing of sperm in New World marsupials (e.g., opossums) and rouleaux formation of New World eutherians (guinea pig sperm, flying squirrel, and naked tail armadillo); and (iii) formation of large associations of sperm aggregates within an acellular boundary of glycoproteins (spermatophore) or the less extensive agglomerations in spermazeugmata (e.g., sharks and rays). Both pairing of sperm and rouleaux formation involve apposition or stacking of acrosomal membranes and may help to protect these labile structures during epididymal transit. In the guinea pig, sperm in rouleaux are held together by a sperm protein WH-30, present on testicular spermatozoa that undergo a change during epididymal transit to facilitate aggregation. Sperm aggregation in elasmobranchs may be related to the prevention of sperm loss during aquatic copulation since it is a device employed by amphibians that deposit spermatophores under water on leaves (but, lacking an epididymis, they utilize secretions of their cloacal glands). All these decapacitating processes are mediated by epididymal epithelial cell secretions that are expected to be reversed upon ejaculation or later in the female tract.

VIII. CONTROL OF EPIDIDYMAL FUNCTION

A. Temperature Control of Epididymal Function

The scrotum acts as a cooling device for its contents, utilizing evaporative heat loss to cool venous blood in the pampiniform plexus that precools incoming arterial blood supplying the epididymis and testis. This is aided by the lack or thinning of hair or wool on scrotal skin over the protruding cauda epididymidis so that, e.g., temperatures of the rat caput, corpus, and cauda epididymidis *in situ* are 35, 34, and 32°C, respectively. The major portion of the epididymal fat pad surrounds the caput epididymidis and is situated in the inguinal region, and a smaller deposit lies between the cauda epididymidis and the testis, both well situated to act as thermal insulators. In species with abdominal testes, the extragonadal sperm reserve sperm store is found close to the exterior and in scrotal mammals it is always more exposed.

In the cryptepididymal model (in which the epididymis is reflected into the abdomen but the testis remains scrotally located so as to maintain its spermatogenic and androgenic functions), males are fertile, indicating that sperm maturation can occur at body temperature. However, when freshly matured sperm are prevented by ligatures from entering the caudal reservoir the animals are infertile, indicating that sperm cannot be stored at body temperature and thus confirming the view that the cooler scrotal environment is of functional benefit to mature sperm. Indeed, it has been convincingly argued that the epididymis was the prime mover in the evolution of testicular descent.

The poor abdominal storage of sperm reflects the responses of the epithelium to elevated temperature which include reduced absorption of water from the duct lumen and reduced gene expression, protein secretion, and enzymatic activity. Abdominal positioning of the epididymis also reduces the storage volume of the epididymis and as a result increases sperm transit rates, reduces the time needed for capacitation, and advances fertilization, probably as a result of diminished epididymal secretion of decapacitation factors.

B. Endocrine Control of Epididymal Function

1. *Androgens*

In the developing embryo the internal male genitalia, including the epididymis, are dependent on 5α-dihydrotestosterone. In the adult, the macroscopic and microscopic structure of the epididymis and the fertilizing potential of its contained spermatozoa are dependent on testicular androgens. Androgen dependence of sperm maturation and storage was first studied in hypophysectomized male mammals given pituitary hormones and androgens, and subsequent studies confirmed the androgen dependence of both sperm maturation and storage in epididymides of castrated animals and in *in vitro* tubule culture, but there is less androgen dependence of epididymal function in lower vertebrates.

Both circulating and luminal testosterone are converted by 5α-reductases (types 1 and 2) to 5α-dihydrotestosterone, the effects of which are mediated via nuclear androgen receptors. In contrast to the prostate, epididymal androgen receptors in the rat do not move from the nucleus to cytoplasm when androgen levels fall. Within the epididymis there are regional responses to androgen deprivation, e.g., after castration there is complete loss of mRNA for γ-glutamyl transpeptidase types II and III in the initial segment, protein B/C in the caput, E-cadherin in the caput, and cauda and protein D/E in all regions with complete restoration of mRNA levels with androgens alone. In contrast, castration initiates no response of γ-glutamyl transpeptidase type IV mRNA in the cauda or SGP-2 mRNA in the initial segment or caput, but an increase in SGP-2 mRNA in the corpus and cauda is depressed by androgens. Other proteins display partial response to castration (e.g., EAP-1 and 5α-reductase), indicating that both endocrine factors and exocrine factors are involved.

There is evidence in the epididymis for control exerted on transcriptional initiation, the rate of transcription, the stability of mRNA, and posttranscriptional control. There may be feedback from protein products (e.g., protein D/E is an RNA-binding protein). Further elements in the control sequence that have been identified in the epididymis include androgen-responsive elements in genes coding for androgen-dependent proteins (e.g., retinoic acid-binding protein and glutathione peroxidase in the epididymis and aldose-reductase-like protein in the vas deferens) and transcription factors with which the androgen receptor may interact [e.g., polyomavirus enhancer activator-3 (PEA-3)] upstream to the coding regions (e.g., γ-glutamyl transpeptidase and glutathione peroxidase).

Androgens also control the synthesis of nonproteinaceous secretions (e.g., glycerophosphocholine) and the transepithelial transport of L-carnitine and myoinositol. Additional effects of androgens include increasing the size of the organ and reducing the rate of sperm passage through the epididymis, thus prolonging the time sperm are in contact with secretions.

2. Other Hormones

Regulation of the epididymis by the nonreproductive hormones aldosterone (increasing water resorption) and norepinephrine (stimulating fluid secretion) has been demonstrated. Aldosterone receptors are found in epididymal clear cells and estrogen receptor types 1 and 2 are present in the efferent ducts and epididymis but, apart from possible action on water resorption, their roles and those of the demonstrated prolactin and melatonin receptors in the epididymis remain to be demonstrated. Circulating vasopressin stimulates fluid secretion by principal cells and contractile activity of peritubular myoid cells (Fig. 4).

C. Paracrine and Autocrine Control of Epididymal Function

Epididymal functions lost on castration but not restored with androgen replacement (e.g., the mRNAs of the proteins proenkephalin A, CRES, 5α-reductase type 1, γ-glutamyl transpeptidase type IV, and the transcription factor PEA-3) indicate another mode of testicular control over the epididymis involving paracrine factors. These factors are present in testicular fluid entering the epididymal lumen and include (i) growth factors that maintain epithelial cell height; (ii) androgen-binding protein, which has been implicated in stimulation of 5α-reductase type 1 activity in principal cells and thus possible modification of endocrine control of principal cell function; (iii) sperm themselves, which may promote gene expression by epithelial cells (e.g., proenkephalin A and proopiomelanocortin); and (iv) sperm products that control fluid secretion into the lumen, e.g., ATP acting on apical P2-purinergic receptors, angiotensin II (converted from angiotensin I secreted from principal cells by sperm angiotensin-converting enzyme) acting on apical angiotensin II membrane receptors, and estradiol-17β synthesized by sperm by the P450-aromatase in the cytoplasmic droplet that may regulate water resorption in the efferent ducts.

Extratubular paracrine interactions include (i) peritubular connective tissue factors that maintain epithelial cell structure in culture and (ii) basal cells that contain, and may secrete, angiotensin II to reach adjacent basolateral principal angiotensin II receptors and promote fluid secretion (Fig. 4). Autocrine control may include the production by epithelial principal cells of calcitonin gene-related peptide, pituitary adenylyl cyclase-activating polypeptide, and endothelin, all of which promote water secretion by the cells that produce them. Recent demonstrations of swelling-sensitive ion conductances in epididymal cells suggest yet another level of control of intraluminal fluid composition, with fluid secretion being promoted under conditions in which osmotic changes in epithelial cell cytoplasm are affected.

D. Environmental and Nutritional Control

In seasonal breeders the circannual activity of the epididymis largely follows testicular activity, with the epididymal epithelium regressing (moles) or undergoing conversion to a stratified epithelium (Djungarian hamster) during the inactive season in response to declining androgens. In reptiles and bats testicular decline is not accompanied by a loss of epididymal function; rather, epididymal sperm storage is maintained despite low testosterone. The epididymis contains receptors for the vitamins D (in clear cells) and retinoic acid (in principal cells). Vitamin A is involved in maintenance of the epithelium since epididymal proteins that are reduced in vitamin A-deficient animals can be increased by either vitamin A or testosterone treatment.

IX. EPIDIDYMIS AND FERTILITY

In several domestic species, some infertile animals (with the "Dag" defect) ejaculate sperm that have bent tails. The angularity of ejaculated sperm occurs at the midpiece/principal piece junction at the site of the cytoplasmic droplet, although sperm entering the epididymis from the testis are normal. Thus, the response causing flagellar bending develops during epididymal transit and could be caused by intraluminal osmotic imbalance resulting in swelling of the cell which would enforce the tail bending.

No analogous cases of the Dag defect have been reported for man; only a small percentage of infertile men suffer from epididymal anomalies. The majority of cases are azoospermic due to the absence of the entire epididymis (efferent ducts and rete remain) in cystic fibrosis or are a result of some epididymal remnants being present in the less severe, genital form of congenital bilateral absence of the vas deferens. In both diseases the cystic fibrosis gene produces an abnormal cystic fibrosis transmembrane conductance regulator (chloride channel) in the apical membrane of the developing organ, and in the absence of fluid secretion the lumen becomes occluded and the organ eventually atrophies. Isolated cases of ciliary tufts in semen reflect damage to the efferent ducts; testicular varicocele can influence epididymal function, owing to the apposition of the epididymis to blood vessels on the testis surface, and epididymitis can cause temporary reduction in epididymal secretion. Sperm storage diseases are occasionally reported.

Recently, transgenic mice models have thrown some light on the importance of epididymal function for fertility. Transgenic mice lacking a functional retinoic acid α receptor are infertile as a consequence of the replacement of distal epididymal epithelial cells with a stratified epithelium. In the absence of fluid resorption eventual blockage of the excurrent ducts leads to azoospermia. Modification of the estradiol type I receptor also leads to azoospermia due to impaired fluid resorption in the efferent ducts and subsequent swelling of testicular tubules, causing pressure damage to the spermatogeneic epithelium. Most interesting are the mice deficient in c-*ros* tyrosine kinase receptor that display infertility *in vivo* but fertility *in vitro*. These animals lacked the initial segment of the epididymis, although sperm production in the testis and transport through the epididymis appeared to be normal. The initial segment is the site of many protooncogenes (e.g., A-raf) that control the synthesis and expression of transcription factors that could be involved in epididymal development or protein secretion by the organ.

X. EPIDIDYMIS AND CONTRACEPTION

Attempts to induce infertility via an epididymal approach have failed. There have been no successful attempts at inducing infertility through promoting rapid sperm transport through the organ, with the aim of reducing the contact of sperm with epididymal secretions, or by reducing the amount of epididymal secretions to spermatozoa. Thus, where sperm transport through the epididymis is increased by estradiol, the resulting infertility could stem from hormone transferred to the female at mating. Totally abolishing epididymal glucosidase activity or reducing epididymal carnitine concentrations by half have no effect on rat fertility or sperm motility. Most successful posttesticular contraceptions are achieved in animal models through an action on epididymal spermatozoa in which the epididymis is not thought to play an active role but could be used to accumulate contraceptive drugs to the site of sperm storage.

Acknowledgment

This article is dedicated to the memory of my father, who helped in far more ways than he could have imagined.

See Also the Following Articles

INTERNAL FERTILIZATION, BIRDS AND MAMMALS; SPERMATOGENESIS, OVERVIEW; SPERMATOZOA; SPERM TRANSPORT; TESTIS, OVERVIEW

Bibliography

Bedford, J. M. (1979). Evolution of the sperm maturation and sperm storage functions of the epididymis. In *The Spermatozoon* (D. W. Fawcett and J. M. Bedford, Eds.), pp. 7–21. Urban & Schwarzenberg, Baltimore.

Brooks, D. E. (1979). Biochemical environment of maturing spermatozoa. In *The Spermatozoon* (D. W. Fawcett and J. M. Bedford, Eds.), pp. 23–34. Urban & Schwarzenberg, Baltimore.

Cooper, T. G. (1995). The epididymal influence on sperm maturation. *Reprod. Med. Rev.* **4**, 141–161.

Djakiew, D., and Jones, R. C. (1982). Stereological analysis of the epididymis of the Echidna, Tachyglossus Acilleatus and Wistar rat. *Aust. J. Zool.* **30**, 865–875.

Glover, T. D. (1973). Aspects of sperm production in some East African mammals. *J. Reprod. Fertil.* **35**, 45–53.

Hamilton, D. W. (1990). Anatomy of mammalian male accessory reproductive organs. In *Marshall's Physiology of Reproduction: Reproduction in the Male* (G. E. Lamming, Ed.), 4th ed., Vol. 2, pp. 691–746. Churchill Livingstone, Edinburgh, UK.

Hamilton, D. W., Olson, G. E., and Cooper, T. G. (1977). Regional variation in the surface morphology of the epithelium of the rat ductuli efferentes, ductus epididymidis and vas deferens. *Anat. Rec.* **188**, 13–28.

Hinton, B. T., Palladino, M. A., Rudolph, D., Lan, Z.-J., and Labus, J. C. (1996). The role of the epididymis in the protection of spermatozoa. In *Current Topics in Developmental Biology* (R. A. Pederson and G. P. Schatten, Eds.), Vol. 33, pp. 61–102. Academic Press, London.

Orgebin-Crist, M.-C., Danzo, B. J., and Davies, J. (1975). Endocrine control of the development and maintenance of sperm fertilizing ability in the epididymis. In *Handbook of Physiology, Section VII Endocrinology, Vol V., Male* (D. W. Hamilton and R. O. Greep, Eds.), pp. 319–338.

Pöllänen, P., and Cooper, T. G. (1994). Immunology of the testicular excurrent ducts. *J. Reprod. Immunol.* **26**, 167–216.

Robaire, B., and Hermo, L. (1988). Efferent ducts, epididymis, and vas deferens: Structure, function, and regulation. In *The Physiology of Reproduction* (E. Knobil and J. D. Neill, Eds.), 2nd ed., pp. 999–1080. Raven Press, New York.

Setchell, B. P., Maddocks, S., and Brooks, D. E. (1994). Anatomy, vascularity, innervation, and fluids of the male reproductive tract. In *The Physiology of Reproduction* (E. Knobil and J. D. Neill, Eds.), 2nd ed., pp. 1063–1175. Raven Press, New York.

van Tienhoven, A. (1983). Anatomy of the reproductive system. In *Reproductive Physiology of Vertebrates* (A. van Tienhoven, Ed.), pp. 96–136. Cornell Univ. Press, Ithaca, NY.

Wong, P. Y. D., Huang, S. J., Leung, A. Y. H., Fu, W. O., Chung, Y. W., Zhou, S. T., Yip, W. W. K., and Chan, W. K. L. (1992). Physiology and pathophysiology of electrolyte transport in the epididymis. In *Spermatogenesis, Fertilization, Contraception. Molecular, Cellular and Endocrine Events in Male Reproduction* (E. Nieschlag and U.-F. Habenicht, Eds.), Schering Foundation Workshop 4, pp. 319–344.

Epitheliochorial Placentation

Vibeke Dantzer

Royal Veterinary and Agricultural University, Denmark

GLOSSARY

angiogenesis The modulation and formation of new generations of capillaries from existing ones.

chorioallantois The membrane resulting from fusion between the vascular allantois and nonvascular chorion.

chorion frondosum Projections of the chorion that provide a substantial increase in surface area.

chorion laeve The smooth part of chorion.

conceptus The fertilized ovum, which through subsequent development comprise the embryo/fetus and its associated membranes including the fetal placenta.

hemotroph Nutrients transferred from maternal blood across the interhemal barrier of the placenta to the fetal circulation.

histotroph Uterine epithelial and glandular secretions and fragments of degenerated cells.

placenta The vascularized embryonic membranes and the endometrium in intimate contact.

trophoblast The outer cell layer of the blastocyst and later the outer epithelium of the chorion. It expresses paternal genetic characteristics in order to participate in establishment of the placenta.

vasculogenesis First or de novo development of blood vessels.

Epitheliochorial placentation is the establishment and development of a placenta in which the chorion is vascularized by the allantois of the conceptus and comes in intimate contact with an intact maternal uterine epithelium. In such placentae the interhemal barrier consists of four cell layers and two more or less compressed layers of connective tissue. On the fetal side, these layers consist of endothelium, mesenchyme, and trophoblast. On the maternal side these layers consist of uterine epithelium, connective tissue, and endothelium. The contact area is increased by the development of folds or villi which maximize the surface area available for the exchange of oxygen, carbon dioxide, nutrients, and metabolic products across this interhemal barrier. This increase is related to changes in shape and extension of the surface of the chorion frondosum and complementary uterine changes. This development can be diffuse, related to the whole chorionic sac, or multiplex/cotyledonary, being assembled in designated areas. Due to the noninvasiveness of the epitheliochorial placenta, there is no immediate loss of maternal tissue in the afterbirth, which consists of the fetal membranes including the fetal part of the placenta. This type of placenta is therefore designated as indeciduate. Studies of placentation are intimately related to the processes involved in the establishment of primary contact, implantation, and subsequent placental development. While this process is conserved within species, it displays marked variation among species. This process between two genetically different tissues is governed by the interaction between a wide range of signal transmitter substances such as hormones, growth factors, and cytokines.

I. INTRODUCTION

Epitheliochorial placenta can be diffuse or multiplex in shape with feto–maternal interdigitation being folded or villous in nature. It is found among a variety of species and can be divided into three types: (i) The mainly diffuse, folded type, which is charac-

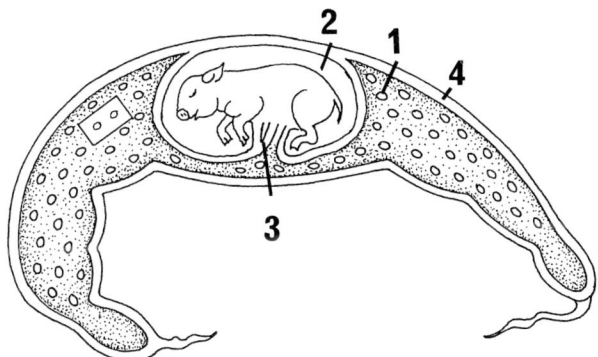

FIGURE 1 A porcine fetus in its fetal membranes showing their interrelationship and the diffuse placenta with the small rounded spots representing macroscopically visible areolae. 1, Allantois; 2, amnion; 3, yolk sac; 4, allantochorion. The marked area is shown in detail in Fig. 2.

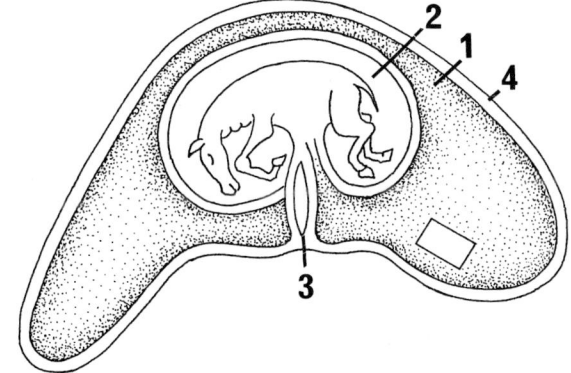

FIGURE 3 The fetus from a horse in its fetal membranes showing their interrelationship and the diffuse placenta. 1, Allantois; 2, amnion; 3, yolk sac; 4, allantochorion. The marked area is shown in detail in Fig. 4.

teristic of pigs (Figs. 1 and 2) and is also seen in the squamate type reptiles (*Chalcides chalcides*), lower primates such as Lorisidae (the bush babies), the American mole (*Scalopus aquaticus*), in some arteriodactyla such as peccaries, and hippopotamuses; (ii) the diffuse villous type in perisssodactyla such as horse (Figs. 3 and 4), donkey, and zebras, in cetacea such as dolphins, and in some arteriodactyla such as llamas, camels, and chevrotains; (iii) the multiplex or cotyledonary villous form is found in cattle, goats,

and sheep (Figs. 11, 12, and 16), deer, moose, okapis, giraffes, duikers, gnu, waterbuck, reedbuck, impala, eland, kudu, and bison.

Recognition of the morphologically different stages of development and the factors involved in the regulation thereof during the initial phases of placentation up to the establishment of the typical architecture is essential to achieve insight into the regulatory mechanisms. A correlation between morphologically well-characterized developmental stages and physiological and biochemical factors by immunohistochemistry and *in situ* hybridization can

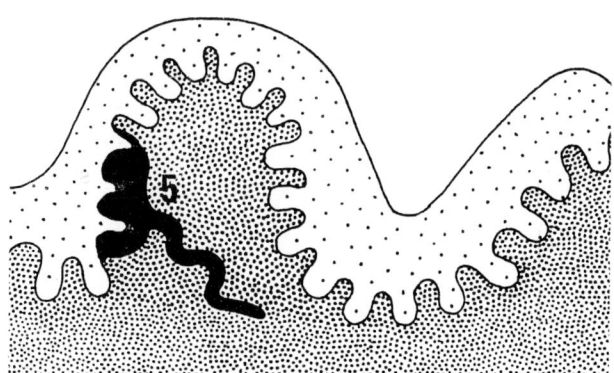

FIGURE 2 Detail from the area marked in Fig. 1 showing the folded type of pig placenta. The maternal side is dark (densely dotted) and the fetal side is bright (lightly dotted). The maternal part of the areola–uterine gland complex is black (5) (from Dellmann, 1998).

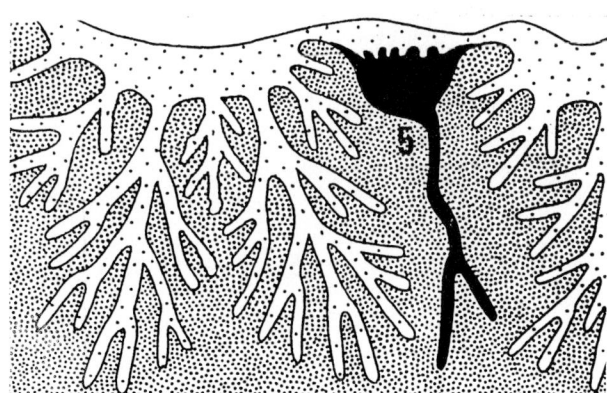

FIGURE 4 Detail from the area marked in Fig. 2. showing the villous, microcotyledonary type of the horse placenta. The maternal side is dark (densely dotted) and the fetal side is bright (lightly dotted). The maternal part of the areola–uterine gland complex is black (5) (from Dellmann, 1998).

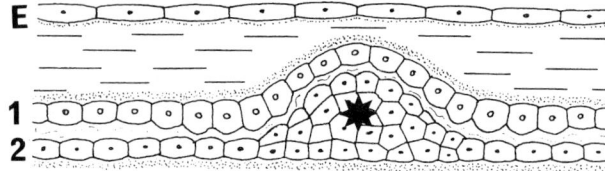

FIGURE 5 Schematic drawing of the apposition phase in the pig. The uterine epithelium (2) forms epithelial proliferations (*), giving an anchoring effect to the apposed trophoblast (1) of the chorionic membrane. E, Endoderm.

then be described to give a better understanding of the regulatory mechanisms for complementary epithelial and vascular growth and cellular differentiation during initial placentation as well as during growth and specialization of the placenta throughout gestation.

The best investigated species are pig, horse, and cattle, within each of the three groups mentioned previously, and they will be used as examples.

II. MORPHOLOGICAL EVENTS

A. Pig

The chorioallantoic placenta is diffuse, folded, and indeciduate. The gestation time is 114 days and litter size is 8–16 (Figs. 5–7). Maternal recognition of pregnancy occurs at Days 11 or 12 postcoitum (pc) and the corpora lutea persist throughout pregnancy. From Days 10 or 11 to Day 13, migration and spacing of the rapidly elongating blastocysts (1–1.5 m) take place.

1. The Yolk Sac

The yolk sac shows hemocytogenesis and angiogenesis beginning Day 14 pc. There is a fusion to the chorion and an apposition to the endometrium forming a simple, short-lasting choriovitelline placenta. Regression of the yolk sac begins about Day 20 pc, so the real functional placenta is rapidly taken

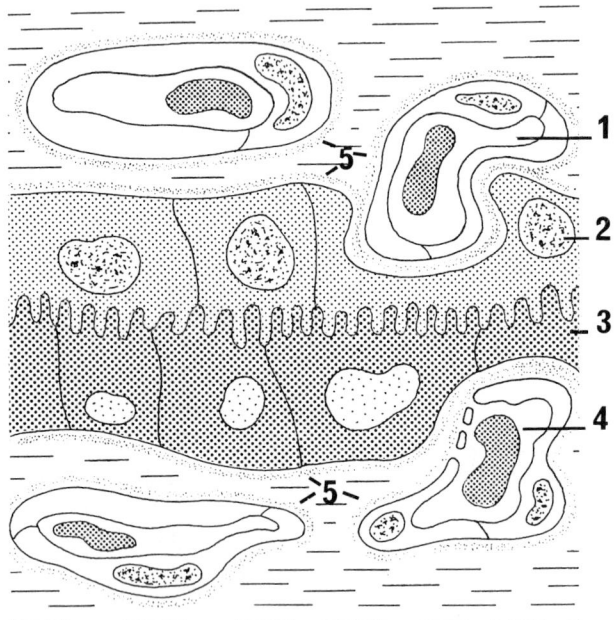

FIGURE 6 Schematic drawing of the tissue layers of the feto–maternal interhemal barrier of the epitheliochorial placenta type of swine and horse. 1, Fetal capillary; 2, trophoblast; 3, uterine epithelium; 4, maternal capillary; 5, basal laminae (modified from Leiser and Kaufmann, 1994).

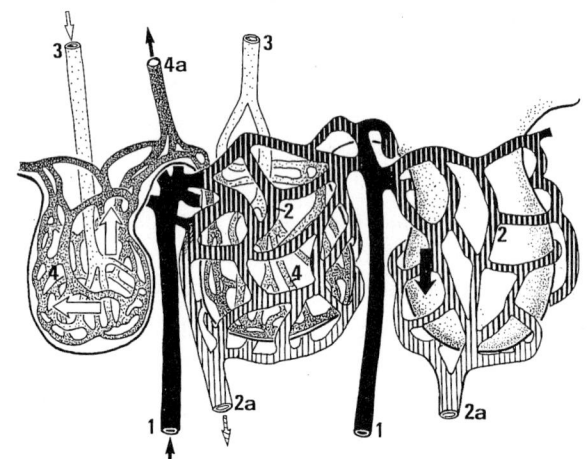

FIGURE 7 Scheme of vascularization of the placenta of the pig on Day 99 of pregnancy. Left, the fetal vasculature; right, the maternal vasculature; middle, the combined materno–fetal vasculature. The maternal arterioles (1) run to the top of maternal ridges, where they branch into the maternal capillary nets (2), which are drained by venules (2a). The fetal arterioles (3) can be followed into the fetal capillary networks (4) and into fetal venules (4a). The large arrows demonstrate a mixture of cross-current (solid vertical arrow vs white horizontal arrow) and countercurrent (solid vertical arrow vs white vertical arrow) materno–fetal blood flow interrelationship in the pig. The density of dots in the fetal venules (4a) illustrates the efficiency in diffusional exchange (e.g., oxygen) (from Dantzer *et al.*, 1988).

over by the developing chorioallantoic placenta (Fig. 1).

2. Placentation

The production of estrogens by the blastocysts is essential for the regulation of myometrial activity and for the spacing of blastocysts along the mesometrial side of the endometrium. In addition, estrogens have been implicated in the induction of other factors necessary for a whole cascade of events triggering the development required for successful placentation and embryonic development. Estrogen synthesis is highest close to the embryonic disc and it is also in this area that placentation is initiated. As gestation progresses, the developmental stages of placentation extend to the periphery of the embryonic sac as well as to the antimesometrial side.

Anchoring of the blastocysts is provided by formation of endometrial epithelial cell proliferation and complementary chorionic caps (Fig. 5). This gives a close apposition from Day 13 pc and just following this stage the rich glycocalyx of the microvilli of the uterine epithelial cells disappears along with microvilli on both uterine and trophoblast cells before new interdigitating microvilli appear between the two epithelia beginning at Day 14 pc. In the narrow interepithelial space an electron-dense material, with some glycan characteristics, constitutes part of this intimate contact. The development of characteristic microscopic foldings of different orders creating rugae separated by fossae increases the surface area for exchange (Fig. 2). This increase in surface area is further enhanced by the interdigitating microvilli mentioned previously.

It is characteristic that indentations of the subepithelial capillaries on both maternal and fetal sides increase during gestation, creating a short physical distance for exchange of 2 μm but with all cellular layers intact. However, the trophoblast of the fetal fossae remains columnar and the apposed uterine epithelium cuboidal, although their number and thus extension decreases during gestation.

In the cytoplasm, organelles are typical for secretion on the maternal side and for absorption on the fetal side. In addition, the maternal epithelium develops a very specialized lysosomal system, probably reflecting a high turnover of organelles as well as high metabolic activity. In the periods from Days 20 to 30 and after Day 70 to term, the trophoblast cells have a very well-developed smooth endoplasmic reticulum due to their differentiation to secrete high levels of estrogens, which is reflected in maternal blood plasma.

3. Vascular Interrelationship

An increase in subepithelial vacularization, angiogenesis, is seen at Day 15 pc on the uterine side and 2 or 3 days later vasculogenesis and angiogenesis occur in the fused allantochorion on the fetal side (Fig. 7).

Subsequent development of microscopic folds in the chorioallantois together with the rapidly developing vascular systems on both sides and its organization creates the characteristic architecture from Day 32 pc. Namely, the mutually folded pig placenta with maternal endometrial arterioles branch into a capillary network at the top of maternal ridges on the fetal side of the placenta, and maternal venules arise in the fossae between them. The fetal arterioles carry blood low in oxygen and nutrient concentration and high concentrations of carbon dioxide and metabolic waste products. These arterioles extend to the top of the fetal ridge, where they progressively branch into capillary networks indenting the trophoblast during pregnancy. The venules leave at the fetal side, at the sides of the fetal furrows. The combined vascular system forms a network that varies from a crosscurrent to countercurrent exchange flow system (Fig. 7). In addition the indentations ensure that the physical distance for diffusion and the gradient for different substances to be transported between the two vascular systems become optimized.

The capillary system is wider on the maternal than on the fetal side. It forms local sinusoidal dilations and bendings retarding flow to enhance exchange. These dilations may, as one of the stimulatory factors, contribute to direct growth during gestation by capillary wall tension and stress created by the blood pressure and flow.

B. Horse

The chorioallantoic placenta is diffuse, villous, and indeciduate. The length of gestation is 340 days with

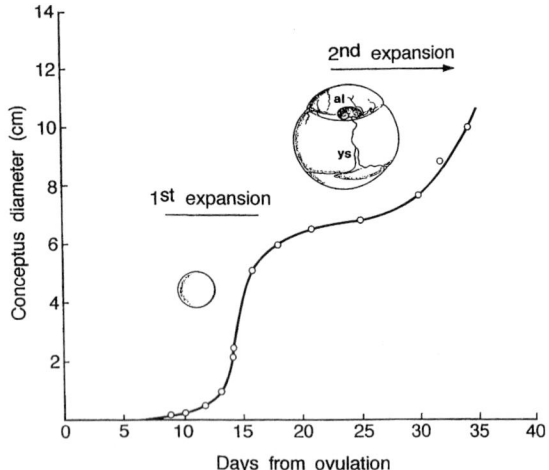

FIGURE 8 A profile of conceptus diameter in the horse up to 34 days after ovulation. The first expansion phase, associated with endoderm migration, is equivalent to the elongation phase in ruminants. The second expansion phase, which continues until full placentation is achieved, is due to growth of the allantois (al)., (ys), Yolk sac (from Stewart *et al.*, 1995).

one fetus (Figs. 3, 4, 6, 8–10). Development of the blastocyst of the mare differs from that of pig and cow in several aspects: There is no elongation of the blastocysts, there is a replacement of the zona pellucida at Days 8 or 9 pc to a tough translucent glycoprotein-rich capsule which remains until about Day 21 pc, and there is development of invasive trophoblast in a specific zone for the formation of endometrial cups (38–120 days pc). Implantation begins at Day 21, although it is superficial until Day 40 pc when more intimate contact begins to develop. The formation of typical villus/crypt interdigitations is first established about Day 110 pc.

The conceptus thus remains spherical and is held in place by uterine tone from about Day 21 to Day 50 pc (Fig. 8). The primary and the secondary corpora lutea, which are formed simultaneously with endometrial cup formation at Day 36, both undergo regression, almost at the same time as the endometrial cups, namely, between Days 100 and 200 pc. Thereafter, their hormone synthesis is taken over by the placenta, although at a lower level.

1. The Yolk Sac

The yolk sac fuses with chorion and is functional until it is replaced by the vascularized allantoic sac

at Day 40 pc (Fig. 9). Thereafter, the yolk sac persists throughout gestations without apparent importance for fetal nutrition (Fig. 3).

In the horse, most research has been focused on endometrial cup formation and to a lesser extend on placentation, vascularization, and the areolar–uterine gland complex.

2. Endometrial Cup Formation

A chorionic girdle of rapidly proliferating trophoblast cells develops about Day 36 pc at the edge between the yolk sac and the expanding allantoic sac (Fig. 9). The whole process seems to be rather closely programmed and is terminated as migration ceases at Day 40 pc. It is proposed that the chorionic girdle arises due to the unique position of the extraembryonic membranes within the conceptus and the release of mitogenic factors (such as hepatocyte growth factor-scatter factor) from the allantoic mesoderm. The epithelium becomes columnar stratified and binucleate. They invade the endometrium between the uterine glands by irregular apical projections, dominated by an abundance of microfilaments, protruding into or between adjacent uterine epithelial cells following the basal lamina until they reach a uterine gland and then penetrate into the underlying stroma, where they grow up to three times in size and often become polyploid (Fig. 10). In the endometrium, invasion of trophoblastic girdle cells induces migration of small lymphocytes into a stroma, already rich in macrophages and plasma cells, as well as numerous granular lymphocytes in the uterine epithelium. The luminal epithelium at sites of invasion is temporarily destroyed, whereas uterine gland epithelium remains intact. In the endometrial cups, the connective tissue components of the endometrium decline and are probably displaced, leaving only a few vessels, whereas connective tissue, stromal cells, vessels, and glands remain between the cups. The uterine vessels remain intact except for the large lymphatic vessels, which are invaded by girdle cells at the deeper marginal zones close to the uterine glands.

The granular lymphocytes, seen in the early stage of cup formation, probably remove the surplus of girdle cells, which have failed to migrate, and the damaged uterine epithelial cells.

The girdle cells express at least one unique tissue-

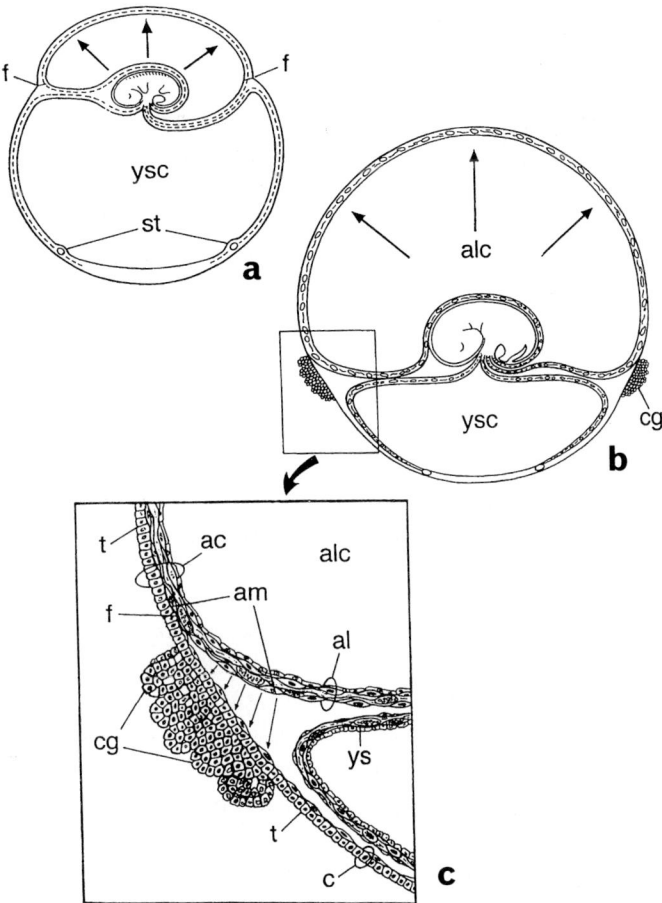

FIGURE 9 Diagrammatic representation of the equine conceptus at (a) 25 days and (b) 36 days after ovulation showing the expanding allantois (arrows), regressing yolk sac, and position of the chorionic girdle (cg). The detail in c shows how the border of the chorionic girdle on the allantoic side starts beneath the point of fusion (f) between chorion (c) and allantois(al) and how the allantoic mesoderm (am) is in close contact with it and therefore probably provides the necessary mitogenic stimulus (arrows). ysc, yolk sac cavity; alc, allantoic cavity; st, sinus terminalis; ys, yolk sac; ac, allantochorion; t, trophoblast (from Stewart *et al.*, 1995).

specific antigen and also acquire a very high concentration of paternally derived class I major histocompatibility complex antigen, which disappears progressively as the cells transform to cup cells. At the time of invasion, the trophoblast girdle cells begin to hypertrophy and differentiate to achieve the capability to synthesize equine chorion gonadotropin, a high-molecular-weight glycoprotein.

The endometrial cups, which vary from a few millimeters to 5 cm, are invaded by leukocytes and undergo degeneration between Days 80 and 120 pc, followed by rejections of their remnants into chorionic folds (Days 120–150 pc), forming allantochorionic pouches.

3. Placentation

At Days 38–40 pc the allantois has vascularized more than 90% of the chorion (Fig. 9). The vascularized chorioallantois first forms simple and later grouped and more complicated branching villi, between the glandular mouth, into the endometrium, where complementary crypts are developing (Fig. 4). Due to the separation by the numerous uterine gland openings and their shape, this surface area extension

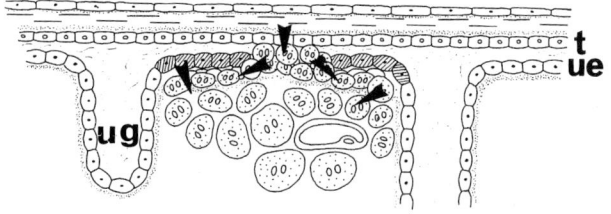

FIGURE 10 Schematic drawing showing the migration of binucleated trophoblast cells (arrowheads) between the uterine epithelium and its basal lamina, which is penetrated close to the uterine gland epithelium. The formation of an endometrial cup is then seen in the endometrium, bordered by two uterine glands. t, Trophoblast; ue, uterine epithelium; ug, uterine glands. The basal lamina is light gray.

on the fetal side is called microcotyledonary; the horse placenta in this respect is similar to the bovine uterine–placental interrelationship. In the horse, however, many thousands of microcotyledons, each 1 or 2 mm in diameter, are closely packed in the superficial endometrial layer, forming microcotyledons. The interhemal barrier has great similarities to the porcine placenta (see Section II,A). The cryptal cells of the uterine epithelium are slightly less dense than the tightly apposed trophoblast cells of the villi of the microcotyledons. The interdigitating microvilli from trophoblast and maternal cryptal epithelium is seen as a dark border by both light and electron microscopy. Both maternal and fetal epithelium, although most prominent in the trophoblast, are rich in smooth endoplasmic reticulum reflecting their high capacity for steroid synthesis of both progesterone and estrogens, which are mainly estrone and the unique equid B-ring unsaturated estrogens, named equilin and equilenin. The trophoblast also contains cisternae of rough endoplasmic reticulum situated in the basal part of the cells intermingled with large vesicles containing electron-dense bodies. Sites of local apoptosis are especially prominent in the later stages of gestation, probably reflecting reorganization and the cessation of placental growth toward term.

4. The Vascular Interrelationship

The vascular interrelationships described as a mainly countercurrent exchange system with fetal arterioles going directly to the villous tip, where it divides into the subepithelial capillary net draining back to venules arising at the villous base. The maternal arterioles run to the fetal side of cryptal septae,

branch into a meshwork, and are drained to the venous system at the base of the maternal crypts.

C. Ruminants

The bovine placenta is multiplex or cotyledonary and villous and indeciduate (Figs. 11 and 12). The gestation time is 280 days, and litter size is usually one or seldom two. The bovine placenta is a subtype of the epithelichorial placenta type because hybrid cell formations take place by fusion of migrating binucleate trophoblast cells with maternal epithelial cells. It is therefore termed a synepitheliochorial placenta.

1. The Yolk Sac

The yolk sac is very similar to that of the pig and regresses very early (see Section II,A) in gestation.

2. Placentation

The blastocyst elongates and becomes immobilized by the development of long cellular trophoblast papillae, which extend into the uterine gland openings around Day 18 pc (Fig. 13). This is followed by the migration of binucleate trophoblast cells to the uterine epithelium, where they fuse with uterine epithelial cells, which then become trinucleated. This takes place both at the glandular free endometrial prominences, the caruncles, and at the intercaruncular regions where the uterine glands are located. At

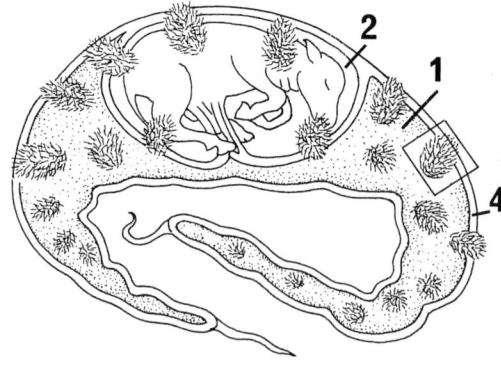

FIGURE 11 A bovine fetus in its fetal membranes, demonstrating their interrelationship and the cotyledons of the cotyledonary or multiplex placenta type. 1, Allantois; 2, amnion; yolk sac cannot be seen (see Fig. 1); 4, allantochorion. The marked area is shown in detail in Fig. 12 (modified from Dellmann, 1998).

COW

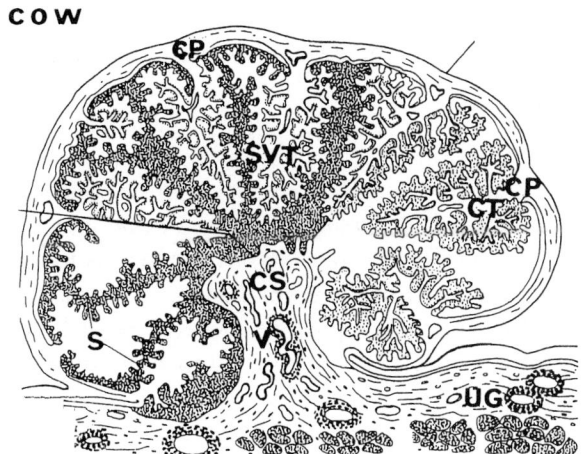

FIGURE 12 Simplified scheme of a mushroom-like shaped (convex) cow placentome. Left, the maternal side with part of the caruncle and maternal septal tissue (S); right, the fetal side with part of a cotyledon showing the villous tree of conical shape with centrally located connective tissue (CT); middle, the combined septae of the caruncular crypts and a cotyledon (SVT) with villi arising from the chorionic plate (CP) and septae from the caruncular stalk (CS). UG, uterine glands (from Leiser *et al.*, 1997). Copyright © (1997). Reprinted by permission of John Wiley & Sons, Inc.

Day 22, the blastocyst extends equally in both uterine horns and by Day 27 pc the establishment of an overall intimate epithelial contact with interdigitating microvilli has taken place. At Days 32–34 pc, simple vascularized allantochorionic villi develop apposed to the maternal caruncles, probably a maternally induced growth, and simultaneously the endo-

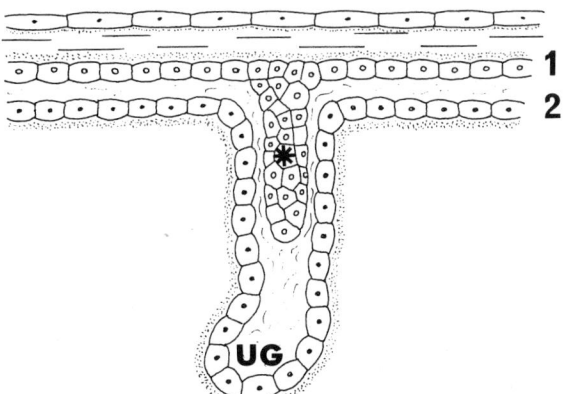

FIGURE 13 Schematic drawing of a long cellular trophoblast papillae (*) extending into an uterine gland (UG). 1, Trophoblast; 2, uterine epithelium.

metrium of the caruncle develops complementary crypts. During gestation, the villi as well as the cryptal septae become longer and deeper, respectively, as well as branched. During gestation, this leads to a successive decrease in mesenchyme in the villi and of connective tissue in the crypt septae. Their respective cores are composed mainly of vascular stem vessels and capillary networks. The interhemal barrier contains the characteristic cellular elements with organelle development suitable for epithelial transport and metabolism. The typical light, granulated, and large binucleate cells on the fetal side and on the maternal side the trinucleate hybrid cell formation will be described separately. Lipid inclusions basally in the maternal crypt cells are characteristic.

3. Binucleate Cells

Binucleate cell structure and function have been studied extensively (Fig. 14). Binucleate cells originate from columnar trophoblast, which divide into two cells. One cell remains columnar, whereas the other undergoes karyokinesis without accompanying cytokinesis to become a binucleate cell with two pairs of centrioles (Fig. 15). Binucleate cells, first seen on Day 17, constitute about 20% of the trophoblast cells with a decline during the week before term. They differentiate into voluminous cells with two spherical nuclei provided with conspicuous nucleoli. The Golgi complex and the rough endoplasmic reticulum become well developed and a regional aggregation of granules is characteristic for the mature cell. They are without microvilli and migrate, which is facilitated by their "loss" of desmosomes, across the interdigitation between maternal and fetal epithelium. Here the binucleate cells fuse with uterine epithelial cells, which then become trinucleated hybrid cells. In this way, the hormones synthesized by the binucleate cells are released close to the maternal circulation. Secretory granules accumulate basally in the trinucleated hybrid cells and are released close to the maternal capillaries. The hormones identified include placental lactogen, progesterone, and prostanoids such as prostacyclins and prostaglandins. The trinucleated cells remain in contact with neighboring epithelial cells and with the opposite trophoblast cells by interdigitating microvilli. The hybrid cells eventually degenerate.

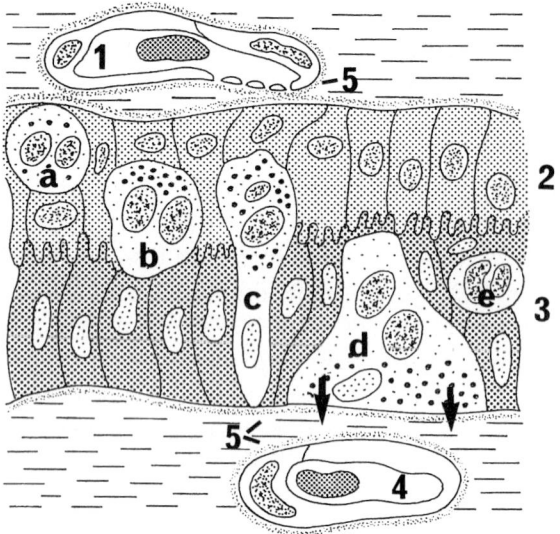

FIGURE 14 Schematic drawing of the interhemal barrier of the synepithelichorial placenta from cow. 1, Fetal capillary; 2, trophoblast; 3, uterine epithelium; 4, maternal capillary; 5, basal laminae. The cells marked a–e show the migration of binucleate trophoblast cells (a, b), fusion of binucleate cell with uterine epithelial cell (c), materno–fetal hybrid cell (d), which secretes its granular content (arrows) into maternal stroma, and degenerating hybrid cell (e) (modified from Leiser and Kaufmann, 1994).

In goat and sheep the shape of the placentome is concave compared to being convex in cattle (Fig. 16). Furthermore, the maternal epithelium becomes a syncytium as the typical binucleate trophoblast cells also fuse with the hybrid cells, thereby contributing to syncytium formation (Fig. 15).

4. Vascular Interrelationship

The architectural interrelationship between the maternal and fetal capillary systems is a villous type with cross-current to preferentially countercurrent blood flow. Thus, it is a very efficient exchange system.

III. PLACENTAL SUBUNITS

A. Areolar–Uterine Gland Complex

During pregnancy uterine gland secretions contribute to nourishment of the offspring to a variable extent. The secretions named uterine milk or histotroph consist of specialized secretory products from the glands mixed with extruded cells.

In the diffuse epithelichorial placenta the secretions are accumulated opposite the gland mouth in an areolar cavity lined by smooth uterine epithelium and a dome of allantochorion where development of villi covered by high columnar trophoblast cells gives an increased surface area for phagocytosis of histotroph. In this way, macromolecules can be transferred from the maternal uterus to the fetal circulation.

1. Pig

In the porcine placenta, the areolae develop from Day 15 pc, as intimate contact between trophoblast and endometrium becomes established. In the advanced placenta they are present throughout the fetal membranes as opaque regular spots (a few millimeters in diameter)—the regular areolae—or as translucent variable larger structures—the irregular areolae. The regular areolae receive secretions from one uterine gland, whereas the irregular ones receive secretions from several uterine glands. Their number,

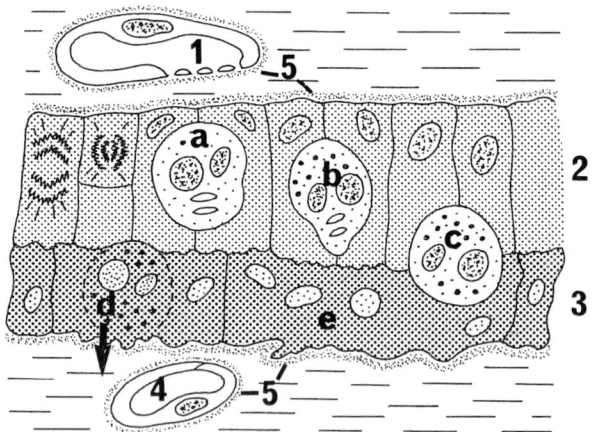

FIGURE 15 Schematic drawing of the interhemal barrier of the synepithelichorial placenta from sheep. 1, Fetal capillary; 2, trophoblast; 3, uterine epithelium and hybridized syncytium; 4, maternal capillary; 5, basal laminae. Formation and migration of binucleate cells are shown in a–e. Migration (a, b) and fusion of a binucleate cell with uterine symplasm (c). Materno–fetal hybrid cell syncytium (d, e), which secretes its granular content (arrow) into the maternal stroma (modified from Hoffmann and Wooding, 1993).

sheep

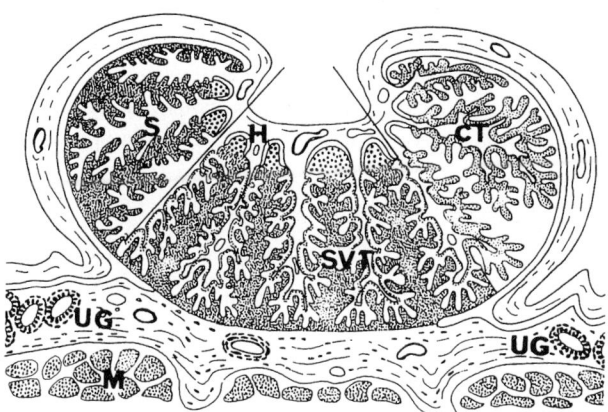

FIGURE 16 Simplified scheme of a cup-like shaped (concave) placentome from sheep. Left, the maternal side with part of the caruncle and maternal septal tissue (S); right, the fetal side with part of a cotyledon showing the villous tree of irregular cylindrical shape, with centrally located connective tissue (CT); middle, the combined septae of the caruncular crypts and a cotyledon (SVT) with villi arising from the chorionic plate and septae from the broad caruncular stalk. UG, uterine glands; M, myometrium. Note the distinct hematomata (H) at the tips of the maternal septae (from Leiser *et al.* 1997). Copyright © (1997). Reprinted by permission of John Wiley & Sons, Inc.

vascularization, and histological structure differ. It is estimated that there are about 7000 regular and 1500 irregular areolae per placenta.

2. The Horse

In the horse, the areolae are smaller and more dense than in the pig because they border the base of the microcotyledons at their origin on the fetal side. The neck of the uterine glands can therefore be followed between the microplacentomes to their openings into the areolar cavity. Because the interhemal barrier and thus the route for hemotroph is established rather late, the nutrition provided by histotroph plays a very important role in equine placentation, although this has not been thoroughly investigated.

3. Ruminants

In ruminants, the areola–uterine gland complexes are related to the chorion laeve because the caruncles are free of uterine glands. Between the placentomes, they form minute areolae. In ewes, the transfer of calcium is localized to the intercotyledonary region including the areolae.

4. Iron

In the pig and probably in the horse, the only source of iron is provided by iron bound to a glandular-secreted glycoprotein, uteroferrin. In the pig, this glycoprotein is taken up by the columnar trophoblast of the areolar villi, released into the fetal circulation, and bound in the liver (mannose binding) or stored in the allantoic fluid. In cattle a similar process also takes places but, in addition, iron is provided by the hematomata.

B. Hematomata

Hematomata are present only in the multiplex type of epitheliochorial placentation. In cattle, they are typically seen in the late stages of gestation. They are localized to the interface between uterine epithelium and trophoblast at the top of cryptal septae apposed to the base of fetal villi, where maternal leaky vessels release blood into small localized areas. The hematomata are more numerous and occur earlier in sheep and goat, in which they are localized to the invaginated central part of the placentome (Fig. 16).

IV. GLYCANS

By using a whole battery of lectins, it has been shown that among species there are similarities in the interhemal barrier but also specific differences in glycan expression among pig, horse, and cattle. The materno–fetal interface seems, therefore, to exhibits its own particular "glycotype," which could be part of a cause rather than the result of structural diversities.

V. CARBONIC ANHYDRASE ACTIVITY

Carbonic anhydrase activity shows great variability among species although it is consistent in the maternal endothelium. It is very active at the materno–fetal interface in pig but only exhibits weak activity in horse and practically no activity in cattle.

See Also the Following Articles

Endotheliochorial Placentation; Horses (Equidae); Pigs; (Suidae); Placenta: Implantation and Development; Ruminants (Ruminantia)

Bibliography

Amoroso, E. C. (1952). Placentation. In *Marshall's Physiology of Reproduction* (A. S. Parkes, Ed.), 3rd ed., Vol. II. Longmans, Green, London.

Dantzer, V., and Leiser, R. (1994). Initial Vascularisation in the Pig Placenta: I. Demonstration of Nonglandular Areas by Histology and Corrosion Casts. *Anat. Rec.* **238**, 177–190.

Dantzer, V., and Leiser, R. (1993). Microvasculature of regular and irregular areolae of the areola-gland subunit of the porcine placenta: Structural and functional aspects. *Anat. Embryol.* **188**, 257–267.

Dantzer, V., Leiser, R., Kaufmann, P., and Luckhardt, M. (1988). Comparative morphological aspects of placental vascularization. *Trophoblast Res.* **3**, 235–260.

Denker, H.-W. (1990). Trophoblast–endometrial interactions at embryo implantation: A cell biological paradox. In *Trophoblast Invasion and Endometrial Receptivity. Novel Aspects of Cell Biology of Embryo Implantation* (H.-W. Denker and J. D. Aplin, Eds.), Trophoblast Research, Vol. 4, pp. 3–29. Plenum, New York.

Enders, A. C., and Liu, I. K. M. (1991). Trophoblast–uterine interactions during equine chorionic girdle cell maturation, migration and transformation. *Am. J. Anat.* **192**, 366–381.

Enders, A. C., Lantz, K. C., Schlafke, S., and Liu, I. K. M. (1995). New and old vessels: The remodelling of the endometrial vasculature during establishment of endometrial cups. *Biol. Reprod. Monogr.* **1**, 181–190.

Hoffmann L. H., and Wooding F. B. P. (1993). Giant and binucleate trophoblast cells of mammals. *J. Exp. Zool.* **266**, 559–577.

Jones, C. J. P., Dantzer, V., Krebs, C., Leiser, R., and Stoddart, R. W. (1997). The localization of glycans in the placenta: A comparative study of epitheliochorial, endotheliochorial and haemomonochorial placentation. *Micrsc. Rec. Tech.* **38**, 100–114.

Leiser, R., Krebs, C., Ebert, B., and Dantzer, V. (1997). Placental vascular corrosion cast studies: A comparison between ruminants and humans. *Microsc. Res. Technol.* **38**, 76–87.

Leiser, R., and Dantzer, V. (1988). Structural and functional aspects of porcine placental microvasculature. *Anat. Embryol.* **177**, 409–419.

Leiser, R., and Kaufmann, P. (1994). Placental structure: In a comparative aspect. *Exp. Clin. Endocrinol.* **102**, 122–134.

Lennard, S. N., Stewart, F., Allen, W. R., and Heap, R. B. (1995). Growth factor production in pregnant equine uterus. *Biol. Reprod. Monogr.* **1**, 161–171.

Morgan, G., Wooding, F. B. P., Care, A. D., and Jones, G. V. (1997). Genetic regulation of placental function: A quantitative *in situ* hybridization study of calcium binding protein (Calbindin-D_{9k}) and calcium ATPase mRNA in sheep placenta. *Placenta* **18**, 211–218.

Mossman, H. W. (1987). *Vertebrate Fetal Membranes: Comparative Ontogeny and Morphology, Evolution, Phylogenetic Significance, Basic Functions, Research Opportunities.* Macmillan, New York.

Ramsey, E. M. (1982). *The Placenta. Human and Animal.* Praeger, New York.

Ridderstråle, Y., Persson, E., Dantzer, V., and Leiser, R. (1997). Carbonic anhydrase activity different placenta types: A comparative study of pig, horse, cow, mink, rat and human. *Micrsc. Rec. Tech.* **38**, 115–124.

Stewart, F. (1996). Roles of mesenchymal–epithelial interactions and hepatocyte growth factor-scatter factor (HGF-SF) in placental development. *Rev. Reprod.* **1**, 144.

Stewart, F., Lennard, S. N., and Allan, W. R. (1995). Mechanisms controlling formation of the equine chorionic girdle. *Biol. Reprod. Monogr.* **1**, 151–159.

Stroband, H. W. J., and Van der Lende, T. (1990). Embryonic and uterine development during early pregnancy in pigs. *J. Reprod. Fertil.* **40**(Suppl.), 261–277.

Wooding, F. B. P., and Flint, A. P. F. (1994). Placentation. In *Marshall's Physiology of Reproduction* (G. E. Lamming, Ed.), 4th ed., Vol II, Chap. 4. Chapman & Hall, London.

Equidae

see Horses

Equine Chorionic Gonadotropin

Janet F. Roser
University of California, Davis

I. Introduction
II. Source and Excretion of eCG
III. The Role of eCG
IV. Regulation and Secretion of eCG
V. Structure–Function Relationship of eCG
VI. Summary

GLOSSARY

half-life the rate of disappearance of a hormone to reach half of its concentration in the blood.

hybrids Anything of heterogeneous origin or composition.

isoelectric point: The pH at which the molecule has no net electric charge and fails to move in an electric field.

isoform Usually referring to a family of gonadotropins such as the pituitary hormone, luteinizing hormone, that are similar in amino acid makeup but have different isoelectric points due to differences in carbohydrate composition, amino acid composition, or three-dimensional structure.

major histocompatibility complex A locus on a chromosome composed of multiple genes encoding histocompatibility antigens or cell surface glycoproteins that can induce an immune response when introduced into another individual.

N-linked oligosaccharide A carbohydrate group composed of monosaccharide molecules connected to asparagine in a glycoprotein.

O-linked oligosaccharide A carbohydrate group composed of monosaccharide molecules connected to serine (threonine) in a glycoprotein.

signal transduction The biochemical pathway by which proteins send signals from the cell surface to the nucleus.

subunit A polypeptide chain linked to other polypeptide chains to form a protein.

trophectoderm A layer of ectodermic tissue developed on the outer surface of the blastodermic vesicle of many mammals.

trophoblast Trophectoderm and extraembryonic endoderm.

Equine chorionic gonadotropin (eCG) is a glycosylated placental protein hormone involved in the maintenance of early pregnancy in the mare. It is now referred to exclusively in scientific literature as eCG because of its similarities in structure and function to primate and human chorionic gonadotropin. In fact, the equids, primates, and humans are the only mammals known to express a member of this glycoprotein hormone family in their placenta.

I. INTRODUCTION

In 1930, two young investigators, the late Harold Cole and George Hart of the Department of Animal Science at the University of California at Davis, first discovered the existence of equine chorionic gonadotropin (eCG) (or what they termed at the time pregnant mare serum gonadotropin) in serum of pregnant mares. It had been only 2 years earlier that German obstetricians Aschheim and Zondek reported the presence of large quantities of a gonadotropic hormone in the serum and urine of women during their first trimester of pregnancy. This glycoprotein hormone later became known as human chorionic gonadotropin (hCG). For the next several years, endocrinologists attempted to detect gonadotropin activity in the blood of other pregnant species, including rat, cow, ewe, sow, bitch, and cat, but to no avail. In 1966, Tullner and Hertz described the pattern of CG levels in the rhesus monkey during early pregnancy. By that time it had been demonstrated that hCG activity resembled more closely the activity of an anterior pituitary fraction subsequently called luteinizing hormone (LH) and eCG resembled more

29

closely the activity of the other pituitary hormone fraction called follicle-stimulating hormone (FSH) but with some additional LH-like activity. It was shown that as little as 0.1 ml of crude serum from a 60-day pregnant mare could stimulate follicular activity and induce ovulation in an immature rat. Because eCG was available in large quantities in the blood, researchers were able to extract the material, partially purify it, and use it in a number of experiments on different animal species. These studies opened up new doors to understanding the physiology of reproductive events and the structure–function relationship of gonadotropins in general.

Cole and Hart's classic paper, announcing the discovery of eCG, presented a most remarkable and accurate description of the general pattern of secretion of eCG in the mare during pregnancy long before the advent of radioimmunoassays or enzyme immunoassays. By injecting 18- to 23-day-old prepubertal rats with the serum from 62 pregnant mares taken between mating and 222 days, the investigators were able to show that uterine–ovarian weight increased in rats injected with serum from 37- to 41-day pregnant mares. Based on their studies and others, eCG rises in serum of pregnant mares between Days 37 and 41, increases sharply to a peak between Days 60 and 80, and steadily declines to undetectable levels by Day 160 (Fig. 1A). Cole and colleagues continued to study the unique aspects of eCG for the next 30 years. They modified and refined their *in vivo* and *in vitro* techniques to learn more about the biochemistry and physiology of eCG. They discovered the source of eCG and determined its half-life in the horse. Along with the late Wadslaw Bileanski, they demonstrated the profound influence of fetal genotype on the rate of eCG production in early pregnancy. In addition, many world-class researchers have contributed further to our understanding of the role and regulation of eCG and its structure–function relationship. Among the most prominent have been Harold Papkoff and Twink Allen. Papkoff and colleagues were the first to propose that the hormone be designated eCG because of its similarities to hCG. They have led the field in describing the unique chemical, physical, and biological properties of the subunits that make up the gonadotropin. Allen and

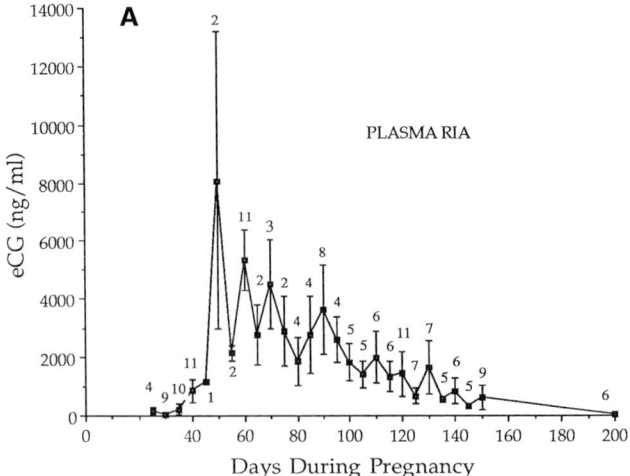

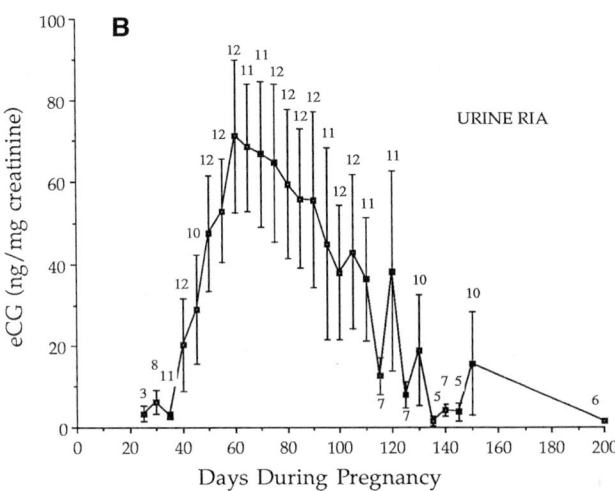

FIGURE 1 eCG concentrations in (A) plasma and (B) urine during pregnancy in the mare (*n* = 12). Numbers above the standard error bars refer to the number of mares sampled on that day (adapted with permission from Roser and Lofstedt, 1989).

associates identified the origin of the endometrial cups and further characterized the secretion and role of eCG during pregnancy in equids. This chapter reviews current knowledge of eCG: its source, excretion, role, regulation, secretion, and structure–function relationship. The review articles and updated journal articles found in the Bibliography supply the reader with specific references.

II. SOURCE AND EXCRETION OF eCG

A. The Endometrial Cups

The source of eCG was controversial and unsettled for many years. From 1934 to 1973, the source of eCG was claimed to be of fetal origin, maternal origin, and fetal origin. In 1934, Catchpole and Lyons, two of Cole's graduate students, reported that eCG was probably secreted by the trophoblast cells of the developing placenta, taken up, and accumulated selectively in the underlying endometrium. Their conclusion was based on the changes in ovarian–uterine weights of rats after treatment with extracts of fetal, placental, and endometrial tissues. Since they had not measured tissue directly from the endometrial cups, the highest amount of eCG activity appeared to be the allantochorion and not the endometrium— thus a placental source. During that same year, Cole and Goss reported that the highest concentrations of eCG were in the endometrial cups. Thus, the controversy continued. In 1954, Clegg and associates concluded that the endometrial cups were of maternal origin. Their conclusion was based on morphological studies indicating that the cups were large glycogen-filled decidual cells having no direct physical connection to the overlying epitheliochorial placenta. In 1973, Allen and associates wrote a series of classic papers that ended the controversy. They reported that the eCG-secreting endometrial cups were actually derived from specialized chorionic girdle cells that invaded the endometrium and changed shape and function. They demonstrated that these specialized equine trophoblast cells were able to secrete eCG in culture for over 180 days—100 days longer than what had been observed *in vivo*.

B. Development of the Chorionic Girdle Cells

The girdle cells were first described in 1897 by Ewart as "a complex whitish band, nearly a quarter of an inch in width, placed nearly equatorially on the embryonic sac." He further described this white band as being an important link between the embryo and the uterine surface for the purpose of providing additional nourishment to the embryo. It was not until the classic series of papers by Allen and associates that the chorionic girdle was discussed in terms of its involvement in the formation of endometrial cups and secretion of eCG. They described the white band to first appear around Day 25 of gestation as a thickening of the trophoblast that develops around the circumference of the conceptus (Fig. 2). They also described the following events. The specialized cells multiply and grow toward the endometrium over the next 10 or so days. Gland-like structures form within the folds of the girdle and secrete a mucoid material that helps to bind the girdle cells to the endometrium. Between Days 36 and 38 the girdle cells invade and destroy endometrial tissue, migrate down the openings of the endometrial glands, and pass into the stroma where they round up, enlarge, and become mature eCG-secreting endometrial cups. The structure of the equine chorionic girdle between Days 28 and 42 has been elegantly presented by Enders and Liu (Figs. 3A and 3B). The exact signal that stimulates the maturation, migration, and transformation of the chorionic girdle cells has not been determined.

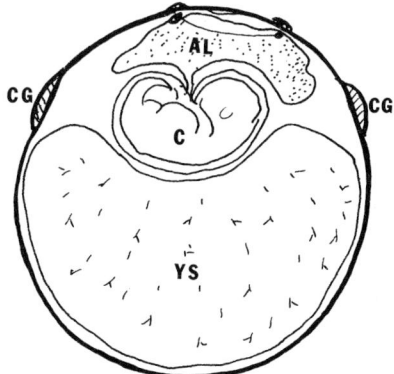

FIGURE 2 Diagrammatic representation of a horse conceptus on Day 36 after ovulation. The chorionic girdle is seen as a thickening of the trophoblast cells surrounding the conceptus. AL, allantois; CG, chorionic girdle; YS, yolk sac; C, conceptus (courtesy of Allen C. Enders).

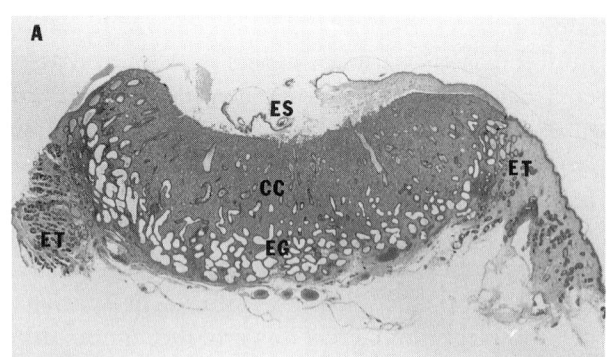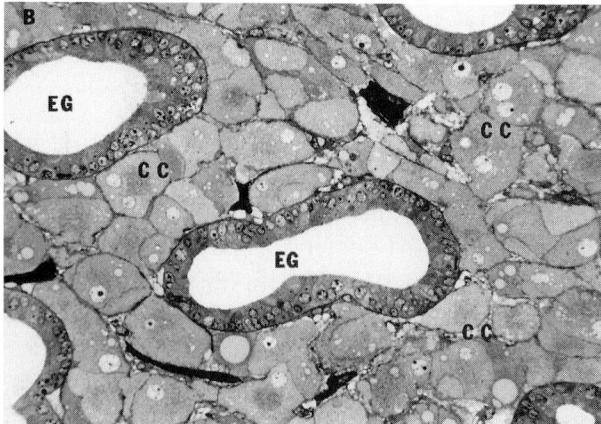

FIGURE 3 Histological sections of endometrial cups at 51 days gestation. (A) The cup is a solid mass of eCG-secreting cup cells (CC) with distended endometrial glands (EG). Exocrine secretion (ES) covers the lumenal surface. Endometrial tissues (ET) can be seen surrounding the cup (magnification, ×12). (B) Binucleated cup cells (CC) surrounding endometrial glands (EG) (magnification, ×300) (courtesy of Allen C. Enders).

C. Demise of the Endometrial Cups

Allen and associates described the demise of the endometrial cups. The large endometrial cup cells remain in tightly packed clusters for 60–80 days, secreting large quantities of eCG into the peripheral circulation via the lymphatic system. The invading cells express high concentrations of paternally derived class 1 major histocompatibility complex (MHC) antigens which stimulate a strong maternal humoral response as well as a strong cell-mediated maternal immune response. The immune reaction hastens the death and eventual disintegration of the cups by 140–160 days but does not affect the placenta or the conceptus.

D. Excretion of eCG

Although large amounts of eCG are found in the peripheral circulation (<1–8 μg/ml), very little (~1%) is excreted into the urine. Thus, eCG is generally purified from blood. In women, a much larger percentage of CG is excreted into urine, making it easier to obtain and prepare material for therapeutic, diagnostic, and research purposes. It has been demonstrated by RIA (Fig. 1B), however, that there is enough eCG in the urine of pregnant mares to follow pregnancy by using an enzyme immunoassay kit in the field. This tool has been useful in following pregnancy rates in populations of feral mares undergoing contraceptive manipulations in the wild to reduce their numbers as well as in exotic equid species undergoing intense breeding programs in captivity to perpetuate the breed.

III. THE ROLE OF eCG

A. In Equid

Classically, the role of eCG is to maintain the primary corpus luteum (CL) and to stimulate and support the development of secondary CL in order to sustain adequate progesterone levels during the first trimester of pregnancy. Daels and associates suggest that eCG also stimulates production of estrogens from the CL. Other investigators suggest that it is FSH and not eCG that stimulates ovarian follicles to mature and that the LH-like activity of eCG merely induces final maturation of the dominant follicle to ovulate or luteinize.

Since eCG does not superovulate mares and binding of a CG to equine LH receptors is only 2–7% that of LH, its role as a FSH and a luteotrophic hormone during pregnancy in the mare is curious. In other studies eCG stimulated progesterone production from slices of primary and secondary CL derived from both ovulated and unovulated equine follicles. In addition, eCG induced progesterone production by equine granulosa cells in a dose-dependent manner. Since serum levels of eCG can rise quite high (1–8 μg/ml) compared to LH levels (5–130 ng/ml), very high doses of eCG in *in vivo* or *in vitro* studies may be necessary to detect any LH/FSH effects on the nonpregnant equine ovary or the equine testis.

The necessity of eCG to maintain the primary and secondary CLs and pregnancy has been challenged by Allen and associates using interspecies pregnancies. When donkey embryos were transferred to uteri of mares, 30% were able to implant normally and the fetus developed to term in the complete absence of eCG and without development of secondary CL beyond Day 40, suggesting that eCG is not essential to maintain pregnancy. However, since abortion of donkey embryos from the horse uterus was due, at least in part, to a failure of placental formation, the validity of the model is questionable.

It had long been surmised by Antczac and associates at Cornell University that the high sialic acid content of eCG served to isolate the trophoblast cells from recognition by the maternal immune system. This does not appear to be the case since subsequent studies demonstrated that the maternal immune system of the mare does respond to paternal antigens between Days 45 and 70 of pregnancy in the presence of high concentrations of eCG.

Another role of eCG in the pregnant mare may be its ability to stimulate growth and development of the fetal gonads. It has long been known that the fetal gonads undergo tremendous hypertrophy between 100 and 240 days of gestation and that this enlargement is associated with an increase in estrogen and androgen production. One could easily assume that the high levels of eCG during this time are responsible for stimulating growth and development of the gonads. However, eCG is selectively excluded from placental absorption and eCG activity is very low in the fetal blood and fluids compared with levels in maternal blood; thus, the mechanism by which eCG stimulates the fetal gonads is unclear.

B. In Other Species

A unique characteristic of eCG is its dual LH and FSH activity in nonequids. This activity was first demonstrated using the classic Steelman–Pohley bioassay for FSH and the ovarian ascorbic acid depletion assay for LH. Equine CG binds both LH and FSH receptors in the pig and the rat. In bioassays, eCG stimulates testosterone secretion by Leydig cells, cAMP and estrogen by immature rat Sertoli cells, and mature granulosa cells and plasminogen activator by both rat Sertoli cells and granulosa cells. Equine CG was one of the first commercially available gonadotopins to be used to induce superovulation in domestic species and research animals. Usually, subcutaneous or intramuscular injections of eCG are given to the animal to stimulate additional follicular growth prior to ovulation. With current procedures, superovulation increases the yield of normal embryos about fivefold in the cow, goat, sheep, and rabbit but only slightly in pigs. It is relatively ineffective in superovulating the horse. This dual LH/FSH activity in nonequids appears to reside in the structural makeup of the eCG molecule.

IV. REGULATION AND SECRETION OF eCG

A. During Early Pregnancy

The regulatory factors and/or mechanisms for synthesis and secretion of eCG by endometrial cups are unknown. Secretory granules, like those found in gonadotrophs, have not been identified in the cup cells suggesting that eCG secretion is probably a constitutive function of the cells. Unlike pituitary gonadotropins, serum levels of eCG are not episodic or pulsatile in nature, further suggesting that eCG is secreted in more of a constant, unregulated mode. Other hormones, such as gonadotropin-releasing hormone, estrogens, and androgens, do not seem to affect the secretion pattern of eCG. Loss of the fetus does not affect the normal life span of the endometrial cups and eCG secretion appears to persist, suggesting that the pregnancy itself does not play a major role in regulating eCG secretion. Chorionic girdle cells recovered from mares prior to invasion of the endometrium at Day 36 and cultured *in vitro* will persist in a healthy state for periods in excess of their *in utero* life span secreting eCG without the addition of exogenous stimuli. It would appear as if the cells were preprogrammed to turn on, secret eCG, and turn off to complete their life cycle.

That there are other factors that may affect the secretion of eCG can only be surmised indirectly from studies of levels of eCG in response to different breeds, hybrids, sires, dams, mares with different body sizes, mares carrying different gender fetuses, and mares carrying twins. It has been reported that draft horses have lower levels of eCG than light horses . Although this observation has been attributed to the dilution effect of the larger blood volume, there are no data to confirm this hypothesis. Donkeys have higher circulating levels of eCG than horses. Genotype of the fetus influences the level of secretion of eCG which is not surprising given the fact that the endometrial cups are of fetal origin. Allen and associates have shown that the endometrial cups of a hybrid, i.e., a horse mare mated to a jack, secrete much less eCG than if the horse mare had been mated with a stallion. In contrast, the endometrial cups of a jenny crossed with a stallion secrete much more eCG than a horse × horse cross. It appears that the ratio of FSH : LH of eCG is also affected by genotype. The FSH : LH ratio is greater in a jenny carrying a hinny fetus than in a jenny carrying a donkey fetus. In addition, there is more follicular growth and luteinization on the jenny's ovaries during pregnancy than on the donkey's ovaries. The crossing of certain sires and dams appears to affect the amount of eCG that is secreted. The size of the endometrial cups in relation to the size of the mare also appears to affect circulating concentrations of eCG. Mares carrying bilateral twins have higher levels of eCG than those that carry singletons or unilateral twins, probably due to the increase in surface area of the developing cups. Although conflicting data have been reported, fetal gender appears to affect levels of eCG. In one report, mares carrying female fetuses had higher levels of eCG, whereas in another report, higher levels were associated with male fetuses. It must be kept in mind that many of these findings are confounded due to the difficulty in separating any one factor from another during pregnancy.

B. During Midgestation

Equine CG disappears from the mare's circulation around midgestation or Days 120–160 of pregnancy. Declining levels of eCG are thought to result from a maternal cell-mediated immunological response to paternal antigens in the fetal chorionic cells, resulting in leukocyte invasion around Days 45–70 of gestation followed by cytolysis of the cup cells. Other observations appear to contradict the theory that cup demise regulates eCG levels. For example, (i) peripheral levels of eCG increase despite histological evidence for an immune reaction soon after cup formation; (ii) eCG declines before invasion of the cups by immune cells; (iii) a paternal cytotoxic response may not be detected but eCG levels still decline in some mares; (iv) serum levels of eCG are not different between MHC-compatible and -incompatible pregnancies; (v) heavy leukocyte invasion observed in pregnancies established by mating a stallion and a jenny is not associated with a decline in eCG levels; and (vi) cytolysis of the endometrial cups occurs distant from leukocyte invasion.

V. STRUCTURE–FUNCTION RELATIONSHIP OF eCG

A. Structural and Physiochemical Properties

Equine CG is a member of the family of glycoproteins, including pituitary LH, FSH, and thyroid-stimulating hormone. These hormones consist of dissimilar, glycosylated α and β subunits noncovalently bound to each other. The α subunit is made up of 96 amino acids, whereas the β subunit is made up of 149 amino acids. The α subunits are identical within each species, whereas the β subunits differ and confer upon the molecule its biological specificity. In all species examined, the α subunit of these hormones has been shown to be encoded by a single gene that is highly conserved. Equine CG-α has 70–80% amino acid homology with other mammalian gonadotropin α subunits. However, it shows unique transposition of tyrosine and histidine at amino acids 87 and 93 which may explain some of the unique properties of eCG. Equine CG-β and equine LH-β are identical in their amino acid composition. Unlike human LH and human CG, the equine pituitary and placental glycoproteins both have a glycosylated C-terminal extension composed of an additional 30 residues. Not surprising, there is a single gene for eCG/LH-β in the horse, which explains why equine CG exhibits some of the same FSH-like activity that eLH displays when administered to nonequids. Homology between the equine CG-β and LH-β from other species ranges from 64% relative to human LH to 77% relative to whale LH. A portion of the cDNA for the β chain for donkey CG representing amino acids 85–146 has been cloned. The homology between the two chorionic gonadotropins within that sequence is 61%, but the C-terminal extensions are quite different.

In terms of its secondary structure, equine CG-α has 10 half-cystine residues, whereas equine CG-β contains 12 half-cystine residues. The location of the half-cystine residues is highly conserved and identical between equine and human CG as well as all the other pituitary glycoprotein hormones. In terms of carbohydrate (CHO) content, eCG is the most highly glycosylated of the mammalian pituitary and placental glycoproteins, with approximately 45% of its mass attributable to carbohydrate moieties. About 20% of the α subunit and more than 50% of the β subunit of eCG are carbohydrate moieties with a mixture of N- and O-linked oligosaccharides. The α subunit has two N-linked oligosaccharide groups at Asn56 and Asn82. The β subunit has a single N-linked and four to six O-linked carbohydrate chains. The N-linked site is at Asn13. The O-linked units are predominantly in the glycosylated tail (residues 111–149). Sialic acid makes up a major portion of the CHO chains (14%) along with hexose, hexosamine, galactose, and glucsoamine. The long half-life and potency of eCG has been attributed to its high sialic acid content. The half-life in the rat has been reported to have two half-life components, the first half-life being 12 min and the second half-life being 6 or 7 hr. Previous studies have indicated 26 hr in the rabbit, 6 days in the gelding, and 21.2 hr in the ewe.

Since the primary structures of equine CG and LH are identical, the observed variation in molecular weights between the two molecules is due to differences in CHO content. There appear to be different eCG molecules or isoforms with different N-terminal amino acid compositions or carbohydrate content in serum and from cultured trophoblast cells. The presence of such isoforms may result in a divergence in biological activity at different stages of pregnancy.

When measuring the FSH:LH bioactivity of the eCG molecule derived from serum during pregnancy, FSH activity was greater on Days 71 and 104 than on Day 39.

The molecular weight of eCG is about 60–70 kDa. It has an isoelectric point of 2.4. Initial steps in the purification process include precipitation with metaphosphoric acid or acetone, followed by further fractionation with graded concentrations of alcohol. At the end of this step, eCG is in a partially purified form with a potency of 2000–3000 IU/mg protein, good enough to superovulate laboratory and farm animals. Using additional column chromatography steps yields a final purification product with a potency of 13,500–15,000 IU/mg protein. Several noted protein biochemists such as Harold Papkoff and associates have published excellent schemes for purification of eCG.

B. Biological Activity (Subunits and Their Hybrids)

According to Papkoff (1978), when eCG is dissociated into its subunits, the α subunit and β subunit have approximately 4 and 6% of the activity of the intact molecule, respectively. Recombining the α and β subunits of equine CG resulted in a gain to 32–44% of original bioactivity. Based on work carried out by Papkoff's lab, Daryl Ward's lab, and Bousfield's lab, making hybrids of equine CG or eLH with subunits of other gonadotropins resulted in variable activity. For example, α subunits from other glycoproteins combined with equine CG-β or equine LH-β were biologically inactive. The equine α subunit combined with ovine LH-β was as potent as recombinant eCG and nearly as potent as intact equine LH. When combining the equine α subunit with porcine LH-β, the recombinant was equivalent to equine LH and nearly 50 times as potent as porcine LH. Equine LH-α alone binds to FSH receptors and inhibits FSH stimulatory activity. The intrinsic FSH activity of equine LH may be associated with its α subunit, whereas others would argue that the intrinsic FSH activity lies within the β subunit.

C. Biological Activity (in the Equid)

As indicated in the previous section on the role of eCG in the equid, eCG has very little effect on the ovary in the mare other than during pregnancy. Although eCG and eLH are identical in their amino acid makeup, eCG binding affinity is only one-tenth that of eLH for LH receptors in the horse. It has been suggested that the different carbohydrate composition of the two molecules is responsible for the lack of binding of eCG to LH receptors. Alternatively, the receptor population of the horse gonad may be heterogeneous, with some receptors recognizing only CG and others recognizing only LH depending on the mare's reproductive state. Sairam and associates have shown, using ovine LH receptors, that these binding sites can have very different affinities for the different gonadotropins.

Although eCG has intrinsic FSH-like activity in other species and binds to FSH receptors, eCG does not bind to FSH receptors in horse follicles or testis, suggesting that it is a LH-like hormone in the horse. This is not the case in the donkey, in which eCG binds equally well to FSH receptors and LH receptors. In contrast, donkey LH binds to equine testis preparations with only 10% the activity of equine LH. Zebra CG is similar to donkey CG in that it has primarily LH activity and little FSH activity in equine tissues.

D. Biological Activity
(in Other Species)

The dual activity and different potencies of eCG in other species has been demonstrated in *in vivo* studies as well as in *in vitro* studies. *In vivo* activity is strongly influenced by the half-life of the hormone which is related to its glycosylation. *In vitro* activity using radioreceptor assays or bioassays is influenced by time of incubation and carbohydrate content of the hormone. Bousfield and associates carried out several studies to determine the importance of the carbohydrate moieties on binding activity and biological activity of eCG. For example, deglycosylation of the Asn56 oligosaccharide of the α subunit of eLH, eCG, or eFSH recombined with eLH-β increased FSH receptor binding activity of eLH by two- to fourfold. In general, deglycosylation of LH-like hormones increases receptor binding activity but decreases biological activity, suggesting that the carbohydrate component is important for signal transduction and

cellular response. Work in Papkoff's lab showed that modification of histidine and tyrosine residues on eCG significantly reduced LH and FSH biological activities of the molecule.

VI. SUMMARY

Equine CG is secreted by endometrial cups, which are specialized trophoblast cells of fetal origin that invade the maternal endometrium at Day 36 of gestation. These cups secrete eCG into the peripheral circulation between Days 37 and 41 of pregnancy, with peak concentrations of eCG between Days 60 and 80 which then decline by Days 120–140 due to degeneration of the endometrial cups in response to an immune reaction. The major role of eCG in the horse is to maintain the primary CL and induce and support secondary CL for maintenance of progesterone production and pregnancy during the first half of gestation. However, there is some doubt as to whether eCG and the secondary CL are necessary for a successful pregnancy in the mare. Equine CG is a glycoprotein hormone composed of two noncovalently bound subunits and is identical in its amino acid composition to equine LH. Both eCG and eLH arise from a single gene and both have a glycosylated C-terminal extension. The α subunit of eCG consists of 96 amino acids and two N-linked glycosylation sites, whereas the β subunit consists of 149 amino acids and one N-linked glycosylation site and four to six O-linked glycosylation sites. The carbohydrate moieties may be involved in signal transduction and cellular responses. The high sialic acid content of the oligosaccharides is mainly responsible for the long half-life of eCG in the horse as well as in other species treated with exogenous eCG. Equine CG has both LH- and FSH-like activity when injected into other species or when used in *in vitro* bioassays and radioreceptor assays. It does not appear to display LH/FSH activity when injected into the horse during the estrous cycle and has very low binding activity to LH receptors in equine follicles, CL, or testes. Although very little eCG is excreted into the urine, enough is present for detection by sensitive immunoassays. Equine CG in urine has been used to follow early pregnancy in both feral horse populations and exotic endangered equids in captivity.

See Also the Following Articles

Chorionic Gonadotropin, Human; FSH (Follicle-Stimulating Hormone); Horses (Equidae); LH (Luteinizing Hormone)

Bibliography

Albrecht, B. A., and Daels, P. F. (1997). Immunolocalization of 3-beta hydroxysteroid dehydrogenase, cytochrome P450 17β-hydroxylase/17,20-lyase and cytochrome P450 aromatase in the equine corpus luteum of dioestrus and early pregnancy. *J. Reprod. Fertil.* **111**, 127–133.

Allen, W. R. (1980). Hormonal control of early pregnancy in the mare. *Vet. Clin. North Am.* **2**, 291–302.

Allen, W. R., and Stewart, F. (1993). Equine chorionic gonadotropin. In *Equine Reproduction* (A. O. McKinnon and J. L. Voss, Eds.), pp. 81–96. Lea & Febiger, Philadelphia.

Ascheim, S., and Zondek, B. (1928). Die Schwangerschaftsdiagnose aus dem Harn durch Nachweis des Hypophysenvordeslapperhormons. *Klin. Wochenschr.* **7**, 1404–1411.

Bousfield G. R., Liu, W.-K., and Ward, D. N. (1985). Hybrids from equine LH: Alpha enhances, beta diminishes activity. *Mol. Cell. Endocrinol.* **40**, 69–77.

Butnev, V. Y., Gotchall, R. R., Baker, V. L., Moore, W. T., and Bousfield, G. R. (1996). Negative influence of O-linked oligosaccarides of high molecular weight equine chorionic gonadotropin (eCG) on it LH and FSH receptor binding activities. *Endocrinology* **137**, 2530–2542.

Butnev, V. Y., Gotchall R. R., Butnev, V. Y., Baker, V. L., Moore, W. T., and Bousfield, G. R. (1998). Hormone-specific inhibitory influence of α-subunit Asn56 oligosaccharide on in vitro subunit association and follicle-stimulating hormone receptor binding of equine gonadotropins. *Biol. Reprod.* **58**, 458–469.

Cole, H. H., and Hart, G. H. (1930). The potency of blood serum of mares in progressive stages of pregnancy in effecting the sexual maturity of the immature rat. *Am. J. Physiol.* **93**, 57–68.

Enders, A. C., and Liu, I. K. M. (1991). Trophoblast–uterine interactions during equine chorionic girdle cell maturation, migration, and transformation. *Am. J. Anat.* **192**, 366–381.

Ginther, O. J. (1992). Endocrinology of pregnancy. In *Reproductive Biology of the Mare* (O. J. Ginther, Ed.), pp. 419–422. Equiservices, Cross Plains, Wisconsin.

Kirkpatrick, J. F., Lasley, J. F., Shidler, S. E., Roser, J. F., and Turner, J. W. (1993). Non-instrumented immunoassay field test for pregnancy detection in free-roaming feral horses. *J. Wildlife Management* **57**, 168–173.

Murphy, B. D., and Martinuk, S. D. (1991). Equine chorionic gonadotropin. *Endocr. Rev.* **12**, 27–44.

Papkoff, H., Bewley, T. A., and Ramachandran, J. (1978). Physiochemical and biological characterizations of pregnant mare serum gonadotropins and its subunits. *Biochem. Biophys. Acta* **532**, 185–194.

Roser, J. F., and Lofstedt, R. M. (1989). Urinary eCG patterns in the mare during pregnancy. *Theriogenology* **32**, 607–622.

Schams, D., and Papkoff, H. (1972). Chemical and immunological studies on pregnant mare serum gonadotropin. *Biochem. Biophys. Acta* **263**, 139–148.

Tullner, W. W., and Hertz, R. (1966). Chorionic gonadotropin levels in the rhesus monkey during early pregnancy. *Endocrinology* **78**, 204–207.

Erection

George J. Christ
Albert Einstein College of Medicine

GLOSSARY

arterial smooth muscle cells The smooth muscle cells that comprise the wall of penile blood vessels.

corpora cavernosa The specialized paired vascular tissues that are the primary modulators of erectile capacity.

corporal smooth muscle cells The extravascular smooth muscle cells that comprise the bulk of the parenchyma in the specialized vascular sinuses of the corpora cavernosa.

endothelial cells The cells that line the penile blood vessels and vascular sinuses of the corpora cavernosa.

erection The complex series of neurovascular events that results in arterial and corporal smooth muscle relaxation and thus increased blood flow to the penis.

flaccidity The normal state of the penis in which the smooth muscle is contracted, and blood flow and intracavernous pressure are maintained at low levels.

gap junctions Integral membrane-spanning proteins that span the extracellular space of apposed smooth muscle cells to provide partial cytoplasmic continuity between adjacent cells.

rigidity The entrapment of blood within the penis under high-pressure, low-flow conditions that are sufficient to permit coitus.

venoocclusive mechanism The closure of the emissary veins due their compression against the tunica albuginea by the relaxed corporal smooth muscle cells.

P enile erection is a complex neurovascular event that is of paramount importance not only to the propagation of the species but also to the emotional and psychological well-being of men and their sexual partners. The normal condition of the human penis is flaccidity, and this is the case for most adult men for approximately 23 hr each day. Only during rapid eye movement sleep or sexual or reflex stimulation does the penis become erect. A sustained and rigid erection is what makes the act of copulation possible. As with many other physiologic processes, erectile function is compromised by the progressive ravages of aging and disease. In light of the increased life expectancy of our aging population, it is not surprising that male sexual function and dysfunction has received much attention recently in the medical, scientific, and lay press. The goal of this article is to review the many strides that have been made in the past two decades concerning our understanding of the physiology of erection.

I. ANATOMY AND PHYSIOLOGY OF ERECTION

A. Penile Anatomy

The anatomy of the penis has been well studied and is concisely illustrated in Fig. 1a. In short, the penis consists of the paired corpora cavernosa and the corpus spongiosum. The former are enclosed by a thick nondistensible fibrous sheath known as the tunica albuginea. However, enclosure of the corpora by the tunica is incomplete and thus permits communication between the corpora via a medial septum. With respect to the latter, the corpus spongiosum surrounds the urethra and terminates as the glans penis. The blood supply to the corporal erectile bodies (i.e., the corpora cavernosa) is derived from branches of the right and left cavernous arteries, which give rise to the helicine arteries. Helicine arterioles terminate in the corporal parenchyma, which consists of specialized vascular sinuses or lacunae. The innervation of the penis is mainly from the sympathetic (thoracolumbar spinal cord, T11–L2) and

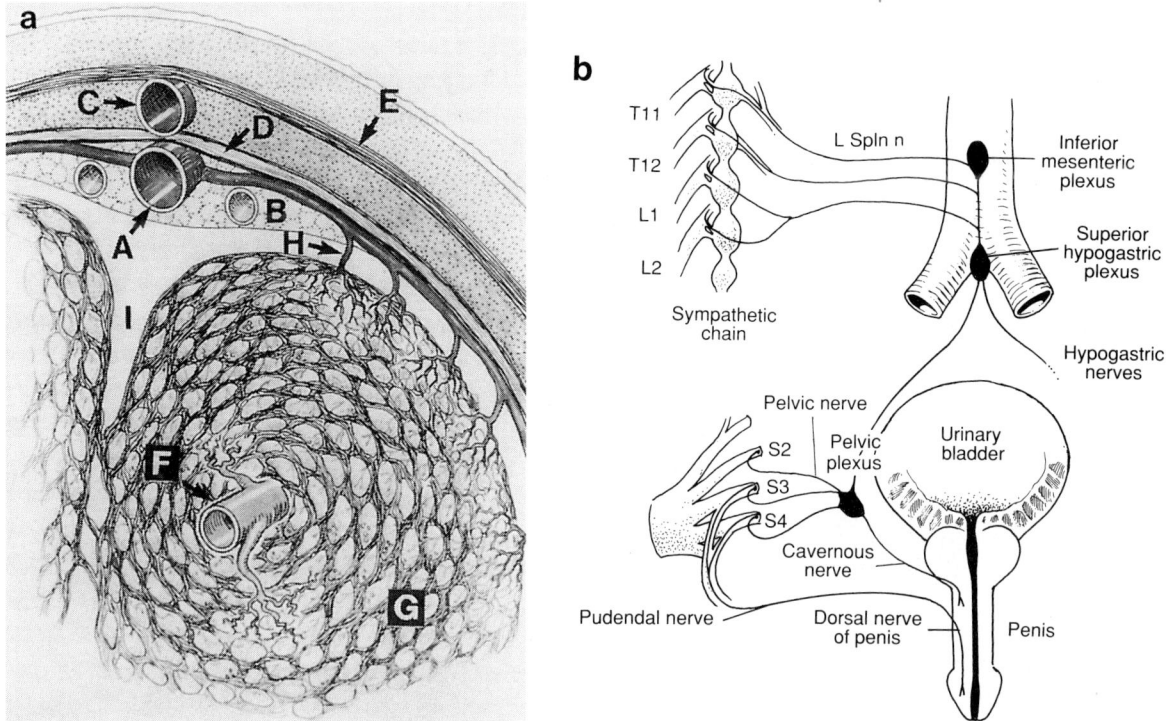

FIGURE 1 (a) Cross section of corpora cavernosa shows deep dorsal vein (A), dorsal artery (B), superficial dorsal vein (C), Buck's fascia (D), dartos fascia (E), cavernous artery (F), corporeal smooth muscle cells arranged into vascular sinuses (trabecula) (G), emissary vein (H), and medial septum of corporeal bodies (I) (adapted with permission from Lerner *et al.*, *J. Urol.* **149**, 1246–1255, Fig. 1, 1993). (b) Schematic representation of the efferent nerve supply to the penis and bladder. The upper portion of the figure shows sympathetic innervation of the pudendal artery. The lower portion of the figure shows parasympathetic nerve supply to the corpora [reproduced with permission from Melman *et al.*, In *Impotence* (A. H. Bennett, Ed.), pp. 18–30, Figs. 2 and 3, Saunders, Philadelphia, 1994].

the parasympathetic (sacral spinal cord, S2–S4) nervous systems (Fig. 1b). Somatic innervation is supplied by the pudendal nerve. The innervation density of the corporal parenchyma, that is, the ratio of nerve terminals to smooth muscle cells, has not been precisely determined. However, with respect to the sympathetic and parasympathetic effector innervation to the corporal smooth muscle, it appears certain that there is far less than one nerve terminal for each smooth muscle cell. This relative paucity of innervation has important implications to the physiology of erection and the etiology of erectile dysfunction.

B. Penile Histology

The terminal branches of the helicine arterioles end in the corporal parenchyma and thus feed into the specialized vascular sinuses that are a hallmark of the corpora cavernosa. The vascular sinuses themselves are formed by bundles of smooth muscle cells (i.e., the corporal smooth muscle cell) interspersed in a heavily collagenous extracellular matrix and lined by endothelial cells (Fig. 2A). Although areas of apposition between adjacent corporal smooth muscle cells are restricted by the architecture of the extracellular space, such contacts are frequently characterized by the presence of gap junctions (Figs. 2B and 2C). The extracellular matrix consists of an abundance of type 1 and 4 collagen, with a lesser amount of type 3 collagen present. Under normal physiological conditions these vascular sinusoids are largely continuous, forming a complex vascular network that traverses throughout the paired corpora cavernosa. The physiological integrity of these vascular

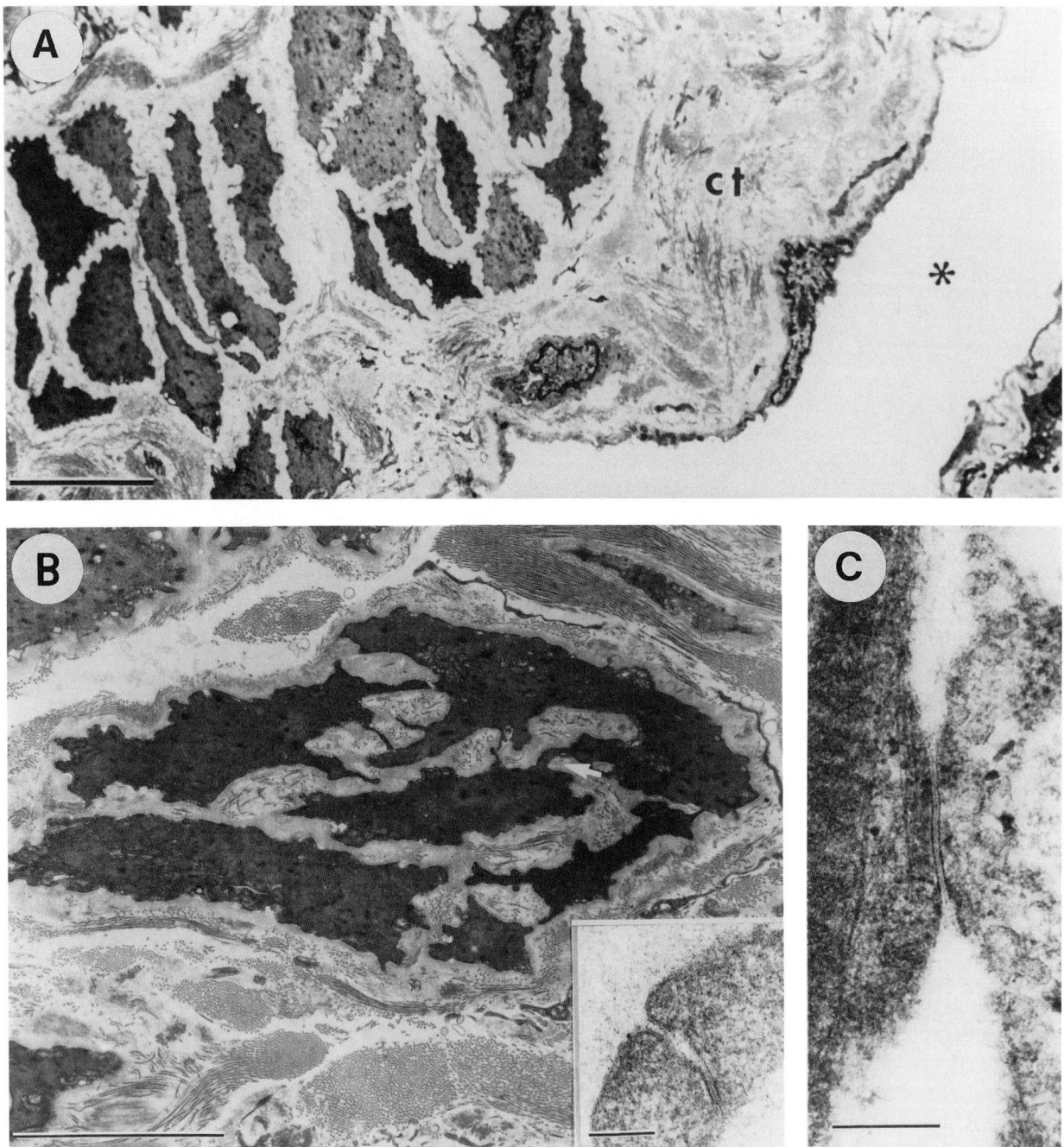

FIGURE 2 Transmission electron micrographs of thin sections of surgically removed human corpus cavernosum tissue. (A) Primarily a cross-sectional orientation depicting a bundle of smooth muscle cells interspersed in a heavily collagenous extracellular matrix. Note that the sinusoid formed by this bundle of smooth muscle cells is lined by an endothelium (*). Scale bar = 5 μM. (B) This electron micrograph of corpus cavernosum shows a region of contact between smooth muscle cells. This region is enlarged in the inset to show a gap junction. Scale bars = 5 mm in the main figure and 0.125 mm in the inset. (C) Electron micrograph of corporeal smooth muscle cells in culture shows larger gap junction between cells. Scale bar = 0.25 mm (adapted with permission from Lerner *et al., J. Urol.*, **149**, 1246–1255, Fig. 1, 1993; Campos de Carvalho *et al., J. Urol.*, **149**, 1568–1575, Fig. 1, 1993 © Williams and Wilkins).

sinuses is a primary determinant of erectile capacity. As such, the human penis is a highly complex vascular organ that is largely under the control of extravascular smooth muscle, that is, corporal smooth muscle.

C. Penile Hemodynamics: Flaccidity

The flaccid penile state is the normal condition of the human penis, and it is maintained as such in the average male for approximately 23 hr a day. Maintenance of the flaccid penile state is achieved by the tonic contraction of the corporal and arterial smooth muscle largely under the influence of the sympathetic adrenergic input (i.e., norepinephrine; Table 1). The effects of sympathetic input on arterial and trabecular smooth muscle tone are further modulated by the presence of a variety of other neural, paracrine, autocrine, and humoral factors, such as arginine vasopressin, endothelin-1, angiotensin II, prostaglandins, nitric oxide, calcitonin gene-related peptide, and neuropeptide Y. During flaccidity, the corporal parenchyma is exposed to a low basal blood flow rate of ~1.4–4.0 ml/min/100 g of tissue and a correspondingly small intracavernous pressure of 5–8 mm Hg. As shown in Figs. 3A and 3C, under these conditions corporal blood flow follows a preferential pathway that traverses from the paired cavernous arteries to the smaller helicine arterioles, and

then slowly filters through the nutrient vascular sinuses before exiting the penis through the emissary veins. Importantly, the subtunical emissary veins remain patent under these low-flow/low-pressure conditions.

In the correct hormonal milieu, and in the absence of neural or vascular disease or congenital or other structural abnormalities, the hemodynamics of penile erection are remarkably similar to that observed in any other vascular tissue; that is, smooth muscle relaxation leads to increased blood flow (Figs. 3B and 3D). Specifically, psychic stimuli that is received or generated in the brain or physical stimulation of the genital organs causes the transmission of nerve impulses from the spinal cord to the cavernous nerves (via the autonomic nervous system), thereby releasing neurotransmitters/neuromodulators (Table 1). Acetylcholine released from parasympathetic cholinergic nerve terminals is thought to be an important modulator of erectile capacity by virtue of presynaptic modulation of sympathetic input as well as direct activation of endothelial cells. However, atropine-resistant erectile responses are well described, and there is unequivocally an important contribution of nonadrenergic, noncholinergic (NANC) nerves to penile erection.

The release of nitric oxide (NO) from NANC nerves has been documented to play an important role in penile erection in diverse species. NO^- is

TABLE 1
Primary Effectors of Corporal and Arterial Smooth Muscle Tone

Neurotransmitter/ neuromodulator hormone	Source cell	Putative receptor subtype	Second messenger pathway	Smooth muscle effect
Norepinephrine	Adrenergic neurons	α_1-Adrenergic	IP_3/DAG/Ca^{2+}/PKC	Contraction
Endothelin-1	EC, SMC?	$ET_{A/B}$	IP_3/DAG/Ca^{2+}/PKC	Contraction
ACH	CHOL neurons	M	NO^-/GC/cGMP/PKG	Relaxation
NO^-	NANC neurons	GC	GC/cGMP/PKG	Relaxation
VIP	NANC neurons	VIP	AC/cAMP/PKA	Relaxation
PGE_1	EC, SMC?	EP	AC/cAMP/PKA	Relaxation

Note. Abbreviations used: EC, endothelial cells; SMC, smooth muscle cells; IP_3, inositol trisphosphate; DAG, diacylglycerol; PKC, protein kinase C; ACH, acetylcholine; CHOL, cholinergic; M, muscarinic; NO^-, nitric oxide; VIP, vasoactive intestinal polypeptide; PGE_1, prostaglandin E_1; NANC, nonadrenergic, noncholinergic; GC, guanylate cyclase; AC, adenylate cyclase; PKA, protein kinase A; PKG, protein kinase G; ET, endothelin.

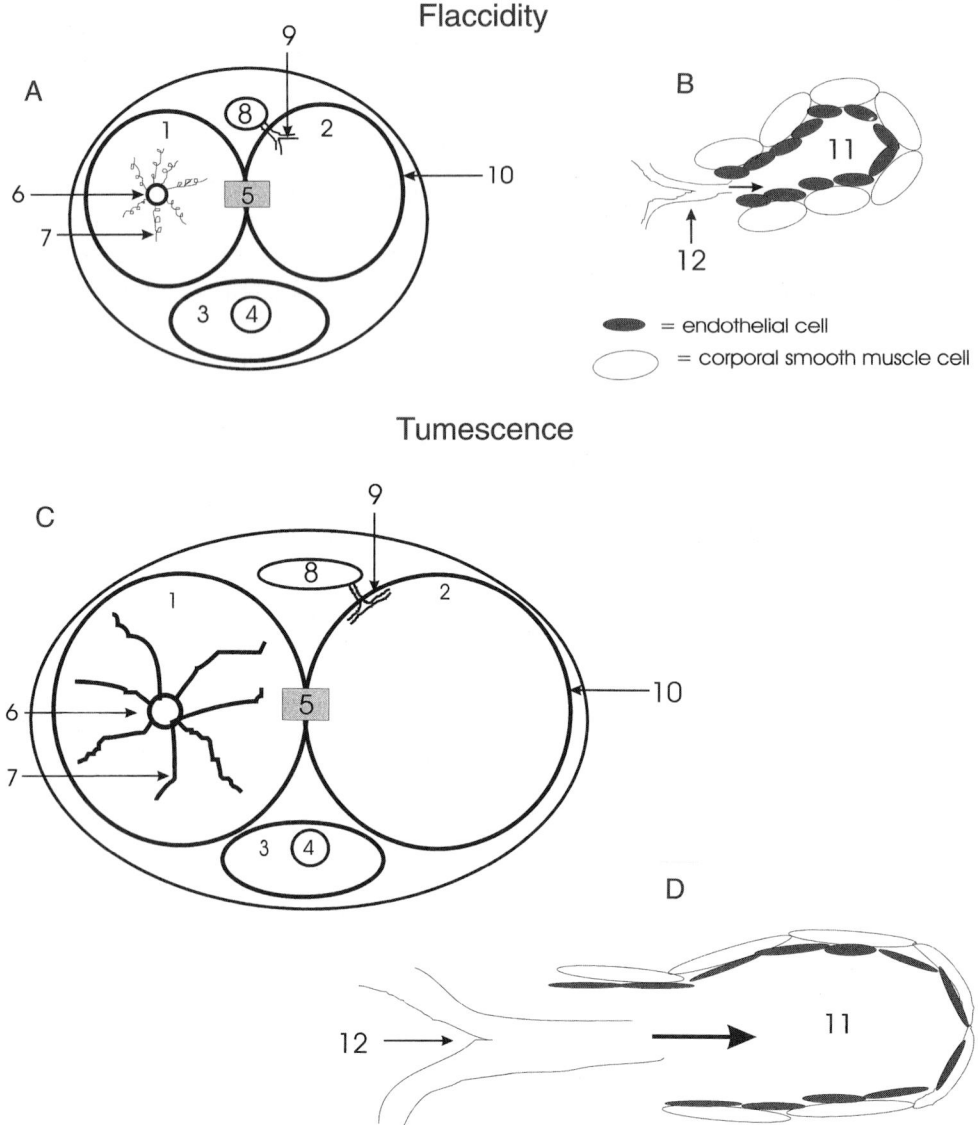

FIGURE 3 The vascular mechanism(s) modulating flaccidity and erection in the human penis. (A) The macroscopic vascular status of the penis during flaccidity. Note that the helicine arteries (7) are contracted and the emissary veins (9) are patent. (B) Microscopic vascular correlates of flaccidity. Illustrated is a representative helicine arteriole (12) feeding a representative corporal vascular sinus (11). For clarity and simplicity, only the diameter of the helicine arteriole is depicted (although the wall of the arteriole would be one smooth muscle thick at this point). In this physiological scenario both the arterial and corporal smooth muscle cells are tonically contracted. It is also important to keep in mind that these vascular sinuses exist in three dimensions and thus they represent a continuum or network of sinusoidal spaces fed by many arterioles. (C) The macroscopic vascular status of the penis during erection and rigidity. Note that in this physiological scenario the helicine arterioles (7) are fully dilated and the emissary veins (9) are compressed against the fibroelastic tunica albuginea. (D) The microscopic vascular correlates of erection and rigidity. Note that the same helicine arteriole (12) feeding the same representative corporal vascular sinus (11) as depicted in B is now fully dilated (i.e., the arterial smooth muscle cells are now fully relaxed), and the corporal smooth muscle cells are also fully relaxed. 1 and 2, the paired corpora cavernosa; 3, corpus spongiosum; 4, urethra; 5, medial septum between the paired corpora cavernosa; 6, cavernous artery; 7, helicine arterioles; 8, dorsal vein; 9, emissary veins; 10, tunica albuginea; 11, representative cavernous vascular sinus; 12, representative terminal helicine arteriole.

synthesized in NANC nerves by nitric oxide synthase. Nitric oxide synthase catalyzes the conversion of molecular oxygen and L-arginine into nitric oxide and L-citrulline. It is precisely because molecular oxygen is a substrate for the formation of NO$^-$ that the pO_2 of cavernosal blood has been postulated to be an important determinant of erectile capacity. Nonetheless, NO$^-$ released from NANC nerves diffuses into the smooth muscle cells to elicit smooth muscle relaxation (Table 1). While neurogenic NO$^-$ is certainly an important mediator of penile erection, recent experimental evidence, as well as the viability and fertility of the NOS knockout mouse, clearly indicates that the control of penile erection is physiologically redundant. Such physiologic plasticity should not be considered surprising given the importance of penile erection to the survival of the species.

Regardless of the exact neuronal pathway activated, it is the postsynaptic actions of the released neurotransmitters/neuromodulators that directly elicits relaxation of the corporal and arterial smooth muscle and thus results in substantially increased blood flow to the penis (from ≤4 ml/min/100 mg tissue to between 50 and 80 ml/min/100 mg tissue). During this time of increased blood flow, intracorporal pressure rises in parallel from the 5–8 mm Hg observed during flaccidity to a level roughly equivalent with the mean arterial blood pressure. Thus, it can be seen that it is the intricate coordination of this complicated series of neurovascular events that is required to ensure the accumulation of blood in the corpora and the resulting elevation in pressure in the corporal vascular sinuses. Ultimately, it is the combination of increased blood flow and intracorporal pressure, both of which are the direct result of relaxation of corporal and arterial smooth muscle, that combine to guarantee the compression of the subtunical emissary veins and occlusion of blood flow from the penis; the latter is referred to as the venoocclusive mechanism. Proper functioning of the venoocclusive mechanism is an absolute prerequisite to the achievement of a normal erectile response.

It is important to point out that it is during the achievement of rigidity and sustainment of erection that blood is actually stored in the sinusoids at a very low flow rate and the penis becomes fully erect.

In this sense, the corporal smooth muscle actually serves a capacitive function in which it is capable of shifting and/or storing large volumes of blood in the penis, within tens of seconds to a few minutes. Thus, it is not surprising that in the absence of severe vascular disease, relaxation of the corporal smooth muscle is both necessary and sufficient to elicit penile erection. Such observations further emphasize the critical role played by these specialized extravascular smooth muscle cells in the entire erectile process.

II. CORPORAL TISSUE FUNCTION

There are two questions of central importance to understanding the initiation, maintenance, and modulation of smooth muscle tone in the human penis: Which are the relevant neurotransmitters, neuromodulators, and hormones (i.e., the first messengers) that are responsible for altering corporal smooth muscle tone and by which receptor/effector mechanism (i.e., second messenger) is this accomplished? And, how is it that the effects of locally restricted neural signals are so rapidly spread among the arterial and corporal smooth muscle cells throughout the penis? The known sequence of events responsible for integrative erectile physiology is summarized in Fig. 4 and further detailed in the following sections.

A. Pharmacology of Corporal Smooth Muscle

With respect to the first question, corporal smooth muscle tone is modulated by a plethora of vasomodulators that are found in the corporal milieu (Table 1). In fact, both dynamic alterations in and sustained contraction of corporal smooth muscle cells result largely from the activation of specific membrane receptors by either neurotransmitters released from autonomic nerve terminals or vasoactive substances present in the adjacent plasma.

As in many other vascular tissues, an increase in intracellular calcium levels is the trigger for corporal smooth muscle contraction. While many substance undoubtedly modulate intracellular calcium levels in corporal smooth muscle cells, intracellular calcium

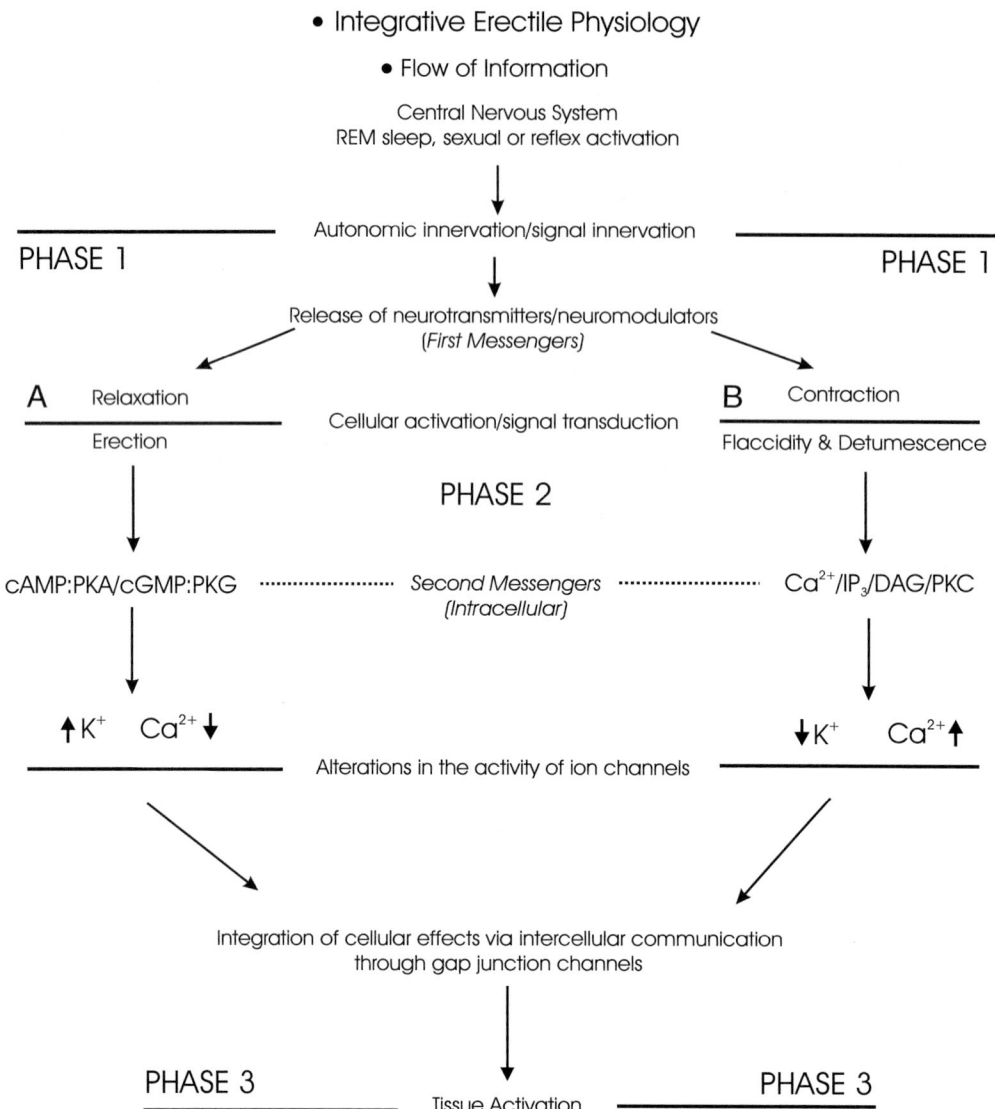

FIGURE 4 Flow chart depicting the salient features of integrative erectile physiology. As illustrated, the erectile process was arbitrarily divided into three broad phases. Phase 1 involves activation of the autonomic nervous effector pathway(s) by any one of a variety of know mechanisms, resulting in the release of neurotransmitters/neuromodulators from synaptic varicosities around blood vessels and throughout the corporal parenchyma (see Fig. 6). Phase 2 consists of diffusion of released neurotransmitter to the smooth muscle cell membrane-bound receptor sites (or in the case of NO^-, diffusion across the smooth muscle cell membrane) and activation of the corresponding second messenger pathway. For the sake of clarity and simplicity, phase 2 was subdivided into two parallel pathways: Pathway A shows the series of intracellular events that lead to arterial and corporal smooth muscle relaxation and thus erection, and pathway B shows the series of intracellular events that lead to arterial and corporal smooth muscle contraction and thus detumescence and flaccidity. As illustrated, the effects of increases in intracellular second messenger levels on the level of corporal smooth muscle tone are thought to be elicited, in large part, by activation of protein kinases and the subsequent phosphorylation of specific amino acid residues on relevant ion channel proteins, thus differentially altering the activity of K and Ca channels. It should be noted that the emphasis on the effects of second messenger molecules on K and Ca channels is not meant to exclude the importance of the phosphorylation/activation of proteins/enzymes that affect other aspects of smooth muscle regulation (e.g., myofilament calcium sensitivity or other calcium sequestration/ extrusion or release mechanisms) but merely to highlight the documented effects on known ion channel subtypes. In short, the intracellular effects of activation of both the contraction and relaxation pathways converge at the level of the nonjunctional ion channels present on corporal smooth muscle cells. Phase 3 involves the spread of the cellular activation process to adjacent, coupled smooth muscle cells via intercellular communication through gap junctions.

mobilization in response to activation of the α_1-adrenergic and $ET_{A/B}$ receptor subtypes is the best understood. Typically, activation of either receptor subtype results in a rapid (onset approximately 20–30 sec after application of drug) and transient (calcium levels return to near resting concentrations with 2 or 3 min) 3-to 10-fold increase in intracellular calcium levels in cultured human corporal smooth muscle.

These increases are dependent on both the liberation of calcium from IP_3-sensitive calcium stores in the sarcoplasmic reticulum and transmembrane calcium flux through L-type voltage-dependent calcium channels. With respect to the latter, the differential distribution of calcium ions across the cell membrane guarantees that the opening of L-type calcium channels will result in the movement of calcium ions down their electrochemical gradient from the outside of the cell to the inside (Fig. 5B). Furthermore, the sustained (tonic) contraction of corporal smooth muscle is exquisitely dependent on continuous transmembrane calcium flux through these voltage-dependent calcium channels. As such, any alteration in the entry of calcium into the cell, or in the release of calcium ions from intracellular stores (i.e., sarcoplasmic reticulum), will have a direct impact on the level of corporal smooth muscle tone. Additionally, as is the case for many smooth muscle cell types, the metabolism of phosphatidylinositolbisphosphate leads to increases in diacylglycerol (DAG) and protein kinase C (PKC), which both serve major roles in modulating the contractile status of corporal smooth muscle. In short, the increased activation of PKC by DAG is thought to result in increased open probability of the L-type Ca channel and decreased open probability for K channels.

The important contribution of transmembrane calcium flux to the contraction of corporal smooth muscle tone accounts, in large part, for the critical role played by potassium channels in modulating both smooth muscle contraction and resting membrane potential (i.e., -40 to -50 mV). More specifically, in contrast to the situation with calcium, the differential distribution of potassium ions across the cell membrane (Fig. 5B) ensures that opening of K channels leads to the outward flux of potassium ions down their electrochemical gradient. This movement of positive charge out of the cell produces cellular hyperpolarization. The resulting hyperpolarization is associated with a decreased open probability of the voltage-dependent calcium channels and thus diminished transmembrane calcium flux and corporal smooth muscle relaxation. In this regard, there are at least four K channel subtypes present in human corporal smooth muscle, with the two most predominant being the large conductance calcium-sensitive K channel (i.e., K_{Ca}) and the metabolically regulated K channel (K_{ATP}; the putative target for the class of compounds known as the K channel modulators).

Importantly, the activity of both Ca and K channels is modulated by second messenger molecules. For example, receptor-activated increases in intracellular cAMP levels due to locally released vasoactive intestinal polypeptide or prostaglandin E_1, or following increases in intracellular cGMP levels elicited by nitric oxide released from nerve terminals or endothelial cells, serve as a primary stimulus for relaxation of corporal smooth muscle. Presumably, increases in intracellular cAMP and cGMP levels lead to activation of PKA and PKG, respectively; although there is no guaranteed specificity of the subsequent kinase activation by cyclic nucleotides (e.g., there may be cross talk between these two cyclic mononucleotide pathways at the level of the protein kinase). Thus, the effects of PKA and PKG are thought to be the opposite of that of PKC. That is, their main action is thought to be decreased open probability of the L-type calcium channel and increased open probability of K channels (K_{Ca} channel subtype in particular). The mechanistic basis for these effects is presumably related to phosphorylation of specific amino acid residues on the K and Ca channels. These facts have clearly established the importance of Ca and K ion channels in modulating corporal smooth muscle tone.

B. Intercellular Communication through Gap Junctions: The Physiological Basis for the Integration of Penile Erection

As illustrated in Fig. 6, not every smooth muscle cell receives direct input from a neuronal varicosity; furthermore, the corpora cavernosa are characterized

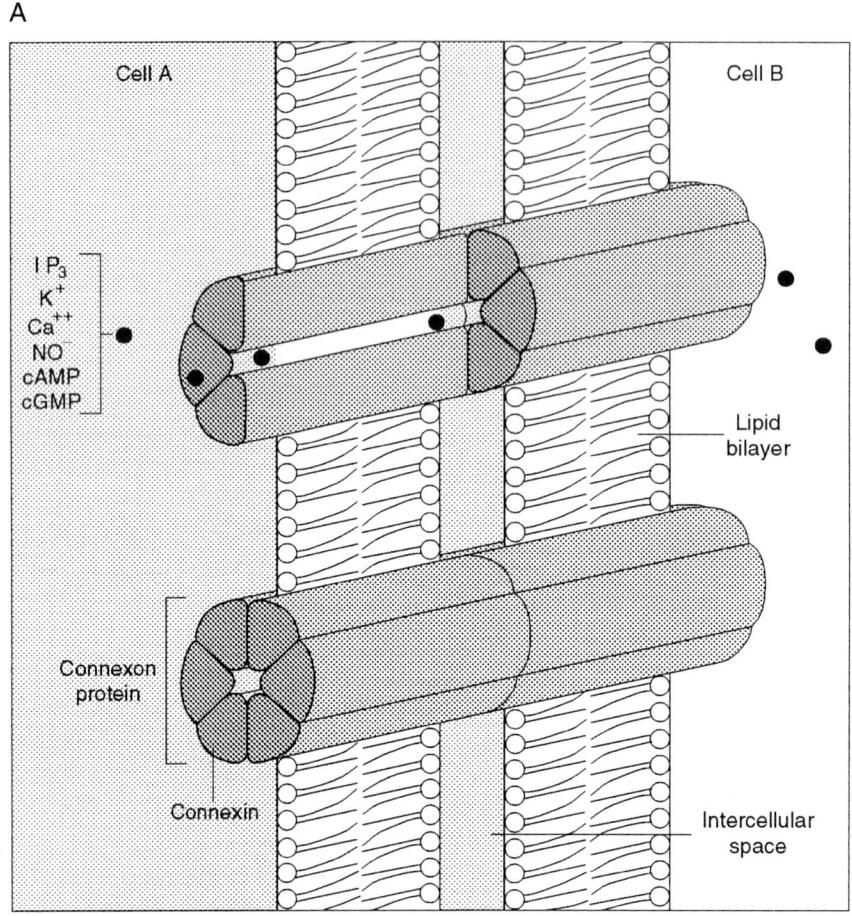

A

FIGURE 5 Intercellular communication, nonjunctional ion channels, and the control of penile erection. (a) The salient features of two representative gap junction channels. This diagram highlights the fact that many physiologically relevant second messenger molecules and ionic species are gap junction permeant. Note that the aggregation of hundreds to thousands of these individual channels is required to visualize the junctional plaques depicted in Fig. 2 (reproduced with permission from Christ *et al., Circ. Res.* 79, 634, Fig. 3, 1996). (b) The junctional (i.e., gap junctional) and nonjunctional (i.e., K and Ca channels) ion channels present on human corporal smooth muscle cells. This diagram illustrates the electrochemical basis for ion diffusion gradients present in human corporal smooth muscle as well as the role for gap junctions in participating in the coordination of corporal smooth muscle responses. The arrows depict the direction of ion flow. Note that with respect to gap junctions the arrows are clearly bidirectional. As illustrated, groups of corporal smooth muscle cells are thought to exist in a state of relative equilibrium and thus can be quite rapidly and syncytially activated.

by a highly collagenous extracellular matrix and thus a tortuous extracellular space (Fig. 2). Since neither of these conditions favor the direct activation of all smooth muscle cells via neurotransmitter diffusion through the extracellular pathway, it is relevant to ask how the rapid and syncytial corporal smooth muscle relaxation (i.e., erection) and contraction (i.e., detumescence) responses are achieved. In this

regard, there has been no clear demonstration, even under quasiphysiological conditions, that corporal smooth muscle cells possess any capacity for regenerative electrical events (i.e., propagated action potentials). Therefore, much recent attention has been paid to the role of intercellular diffusion of second messenger molecules and ions through intercellular channels formed by gap junction proteins.

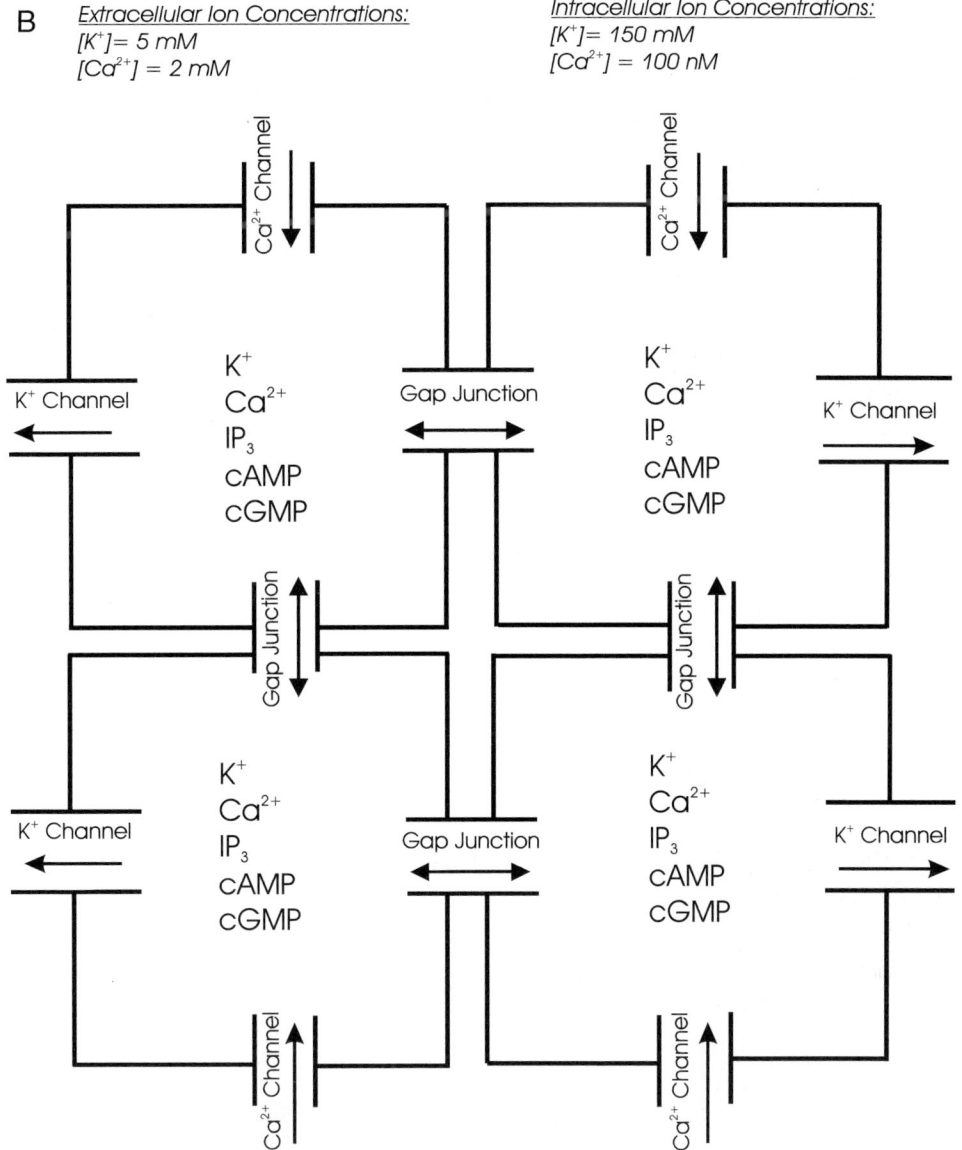

B *Extracellular Ion Concentrations:*
$[K^+]= 5 \, mM$
$[Ca^{2+}] = 2 \, mM$

Intracellular Ion Concentrations:
$[K^+]= 150 \, mM$
$[Ca^{2+}] = 100 \, nM$

FIGURE 5   (*continued*)

Briefly, gap junctions represent a diverse family of related protein, known as connexins, that are named according to their molecular weight. As illustrated in Fig. 5A, six presumably homologous membrane-spanning connexins join to form a connexon, or hemichannel, in the cell membrane. The union of these connexons across the extracellular space provides an aqueous intercellular channel and thus partial cytoplasmic continuity between adjacent cells. At areas of membrane apposition between human corporal smooth muscle cells, these integral membrane proteins are known to coalesce to form junctional plaques consisting of hundreds of these individual gap junction proteins (Figs. 2B and 2C). Of the more than 12 distinct mammalian connexins identified, connexin 43 (Cx43) is the predominant one present and functional between human corporal smooth muscle cells. As such, intercellular communication through gap junctions serves as an important anatomic substrate for the coordination of contraction and relaxation responses among corporal smooth muscle cells.

Innervation of penile blood vessel

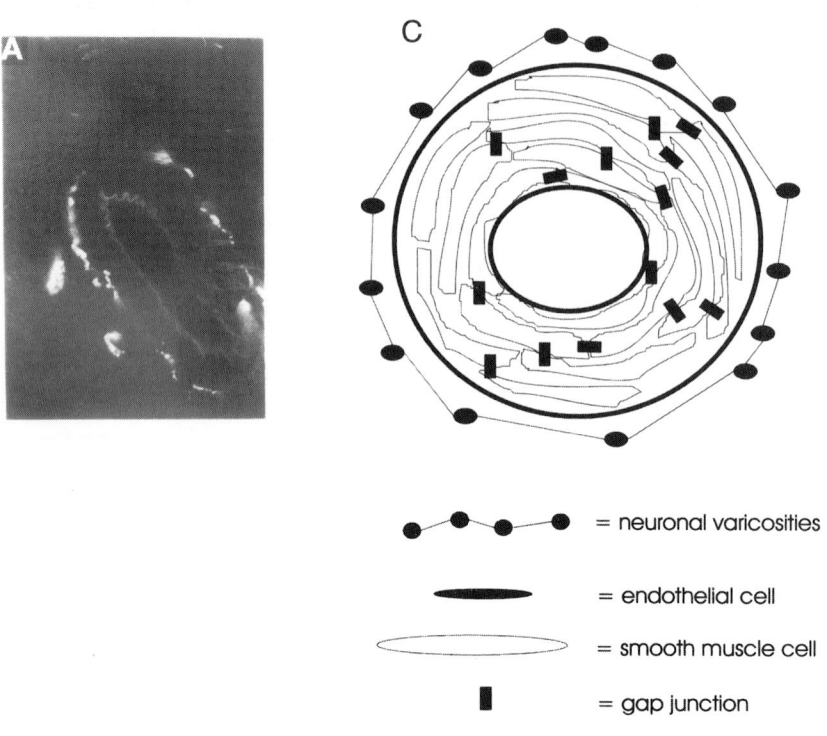

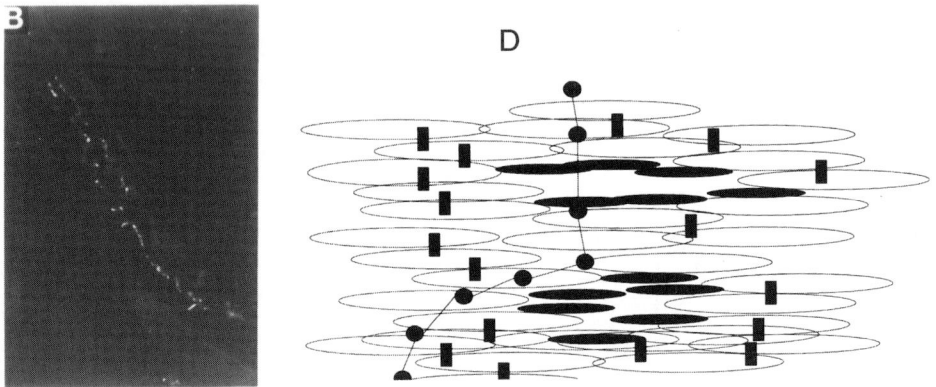

Innervation of corporal parenchyma

FIGURE 6   The effector innervation of the human penis. (A and B) Synaptophysin immunostaining of transverse sections through the corpus cavernosum of a 2-month-old F-344 rat; note that the histology of the rat penis is very similar to that of the human penis. (A) Representative example of dense synaptophysin-like immunostaining of presynaptic nerve terminals generally observed around the adventitial layer of blood vessels. Note the nonspecific fluorescence from the internal elastic lamina. (B) Representative example of less dense synaptic terminals found in the corporal parenchyma. Note that, at this magnification, the field illustrated contains dozens of smooth muscle cells. Magnification ×400 in all views. (C and D) Schematic depictions of the putative mechanistic basis for the neurovascular control of human penile erection, based on the histological information conveyed from data such as those depicted in panels A and B (reproduced with permission from Rehman, *Am. J. Physiol. Heart Circ. Physiol.* **41**, H1967, 1997).

More specifically, as illustrated in Fig. 5, the permeability of Cx43-derived gap junction channels is sufficient to permit the intercellular diffusion of the majority of physiologically relevant second messenger molecules/ions that modulate corporal smooth muscle tone. In fact, IP$_3$ and/or calcium ions have been shown to diffuse between cultured human corporal smooth muscle cells. Thus, these intercellular channels provide an important mechanism to ensure that the effects of activation of the cells most proximal to a locally restricted neuronal or hormonal signal are efficiently transmitted to more distally removed smooth muscle cells, which are not directly activated (Fig. 5). In this scenario, even relatively small numbers of nerve terminals would be sufficient to simultaneously activate groups or networks of corporal smooth muscle cells (Figs. 5 and 6), thus permitting the rapid and syncytial corporal smooth muscle responses that are an absolute prerequisite to penile erection and detumescence. Experimental, theoretical, and clinical studies all support this supposition. It is for all these reasons that intercellular solute diffusion among coupled corporal smooth muscle cells has been postulated to have a major role in the integration and coordination of penile erection and detumescence.

See Also the Following Articles

Impotence; Male Reproductive System, Overview; Penis

Bibliography

Adams, M. A., Banting, J. D., Maurice, D. H., Morales, A., and Heaton, J. P. W. (1997). Vascular control mechanisms in penile erection: Phylogeny and the inevitability of multiple and overlapping systems. *Int. J. Impot. Res.* **9,** 85–91.

Andersson, K.-E., and Wagner, G. (1995). Physiology of penile erection. *Phys. Rev.* **75,** 191–236.

Benson, G. S., McConnel, J., Lipshultz, L. J., Corriere, J. N., Jr., and Wood, G. (1980). Neuromorphology and neuropharmacology of the human penis. *J. Clin. Invest.* **65,** 506–513.

Christ, G. J. (1995). The penis as a vascular organ. *Urol. Clin. North Am.* **22,** 727–745.

Lue, T. J., and Tanagho, E. A. (1987). Physiology of erection and pharmacological management of impotence. *J. Urol.* **137,** 829–836.

Rajfer, J., Aronson, W. J., Bush, P. A., Dorey, F. J., and Ignarro, L. J. (1992). Nitric oxide as a mediator of relaxation of the corpus cavernosum in response to nonadrenergic, noncholinergic neurotransmission. *N. Engl. J. Med.* **326,** 90–94.

Saenz de Tejada, I. (1992). Mechanisms for the regulation of penile smooth muscle contractility. In *World Book of Impotence* (T. F. Lue, Ed.), Chap. 3, pp. 39–48 Smith-Gordon, London.

Erythroblastosis Fetalis

Donald J. Dudley

University of Utah School of Medicine

G. Marc Jackson

University of Pennsylvania

GLOSSARY

erythroblastosis fetalis A fetal condition in which maternal antibodies cross the placenta and destroy fetal red blood cells, leading to hydrops fetalis.

fetomaternal hemorrhage The transfer across the placenta of fetal red blood cells into the maternal circulation.

hydrops fetalis Anasarca of the fetus, often due to severe fetal anemia.

Kleihauer–Betke test A measure of the percentage of fetal blood in maternal blood, used to calculate the needed dose of Rh-immune globulin.

Liley curve A graphical representation of amniotic fluid bilirubin across gestational age, used in the management of blood group isoimmunization.

$\Delta OD450$ A measure of the amount of breakdown products of bilirubin, used in the management of erythroblastosis fetalis.

Rh-immune globulin Passive immunization of antibodies to prevent Rh isoimmunization in Rh-negative women.

Erythroblastosis fetalis occurs as the result of isoimmunization of the mother with foreign red blood cell (RBC) antigens leading to the production of specific antibody directed against protein epitopes on the surface of the RBCs. These antibodies cross the placenta, bind and destroy fetal RBCs, and start the pathophysiologic processes leading to erythroblastosis fetalis (or immune hydrops fetalis). The diagnosis and management of erythroblastosis fetalis, or blood group isoimmunization, is a triumph of modern medicine based on careful and scrupulous scientific technique. From the discovery of the blood group system to the molecular diagnosis of fetal blood type and *in utero* therapy, blood group isoimmunization has been transformed from a lethal disease afflicting up to 2 or 3% of all pregnancies to a relatively rare, but still important, condition.

Erythroblastosis fetalis was noted to be associated with fetal edema, neonatal hyperbilirubinemia, and neonatal anemia in the 1930s. Passage of maternal antibodies across the placenta with the destruction of fetal erythrocytes was the proposed pathophysiology. Although many antigens of the cell membranes of red blood cells (RBCs) have been described, only a few are clinically important causes of maternal isoimmunization leading to hemolysis of fetal RBCs. Historically, most cases of erythroblastosis fetalis were associated with antibodies directed against the Rh factor. However, with the use of Rh-immune globulin (RhIG), so-called "minor antigens" on the RBC membrane have assumed new importance as causes of blood group isoimmunization.

I. GENETICS AND BIOCHEMISTRY OF BLOOD GROUP ISOIMMUNIZATION

In 1940, Landsteiner and Wiener produced rabbit immune sera to rhesus monkey erythrocytes which coagglutinated with 85% of human RBCs. They designated this property of serum the "Rh factor." Agglutinated cells were called "Rh positive" and Rh factor is now known to be an antibody directed against a RBC surface antigen of the rhesus blood group system. The Rh gene complex encodes three different proteins and is composed of eight possible different gene combinations (CDe, cde, cDE, cDe, Cde, cdE, CDE, and CdE).

The Rh antigen complex on the RBC membrane is the final expression of a group of five possible antigens (C, D, E, c, and e). Thus, Rh positive indicates the presence of the D antigen and "Rh negative" indicates the absence of D antigen on RBCs. Most cases of Rh isoimmunization causing transfusion reactions or serious hemolytic disease of the fetus and newborn are the result of D antigen incompatibility. The genetic locus for the Rh antigen is on the short arm of chromosome 1, where there exist two distinct genes encoding products of Rh antigens. The first gene codes for the C/c and E/e antigens, and the second gene codes for the D antigen. D-negative patients lack the RhD gene on both of their chromosomes, and hence there is no gene product or protein produced. There is a relatively constant volume of Rh antigen sites available on the RBC surface, with about 100,000 sites per cell.

At least seven (and perhaps more) different D-antigen epitopes are thought to exist based on studies using human monoclonal anti-D antibodies. These epitopes serve as the site for binding with antibodies directed against the D protein (or other proteins in the case of minor antigens). These different epitopes are likely part of the same protein–lipid complex but are more or less expressed on the RBC surface depending on the depth of the polypeptide portion that is embedded in the RBC membrane lipid bilayer. It seems likely that this finding explains the immunologic variation of responses against the Rh blood group system and in fetal hemolytic disease.

II. PRINCIPLES OF BLOOD GROUP ISOIMMUNIZATION

Several circumstances must exist for blood group isoimmunization to occur. First, the fetus must have RBCs which bear the foreign antigen, and the mother must be negative for this antigen. With regard to the Rh antigen, about 15% of whites are Rh negative, while only 5–8% of American blacks and 1 or 2% of Asians and Native Americans are Rh negative. Thus, in the white population, an Rh negative woman mates with an Rh-positive man in about 85% of pairings. In men who are Rh positive, about 60% are heterozygous and 40% are homozygous for the D antigen. Therefore, since one-half of conceptions by heterozygous men will be Rh positive, the overall chance of a Rh-positive man producing a Rh-positive fetus is about 70% and a Rh-negative woman has about a 60% chance of bearing an Rh-positive fetus in pairings in which the male D status is not known. Thus, about 10% of pregnancies in the white population are Rh incompatible. However, only 20% of incompatible pregnancies cause maternal sensitization. Prior to RhIG prophylaxis, about 1% of pregnant women had anti-D antibody, and about 16% of Rh-negative women became isoimmunized after their first Rh-incompatible pregnancy. This figure has declined precipitously with the advent of RhIG prophylaxis.

Second, a sufficient number of fetal RBCs must gain access to the maternal circulation. Although transplacental passage of fetal RBCs was first proposed in the 1940s, documentation of fetal RBCs in the maternal circulation was not reported until the mid-1950s. Fetomaternal hemorrhage resulting in isoimmunization is most common at delivery, occurring in up to 50% of births. In more than half of these births, <0.1 ml of fetal blood enters the maternal circulation. However, in up to 1% of cases, the estimated volume of fetomaternal hemorrhage is more than 30 ml. Predisposing factors for significant fetomaternal hemorrhage include cesarean delivery, multiple gestation, bleeding placenta previa or abruption, and manual removal of the placenta. Fetomaternal hemorrhage leading to isoimmunization can occur after spontaneous or induced abortion, ectopic pregnancy, amniocentesis in the second and third trimester, and chorionic villus sampling.

However, most cases of excessive fetomaternal hemorrhages occur in women without these risk factors and who experience an uncomplicated vaginal delivery. Even without labor, antepartum fetomaternal hemorrhage can occur in sufficient volume to result in isoimmunization. Antepartum fetomaternal hemorrhage in this fashion causes about 1 or 2% of cases of isoimmunization before delivery, but antepartum sensitization rarely occurs before the third trimester. Thus, this possibility is the rationale for RhIG prophylaxis administered at the beginning of the third trimester in pregnancies of Rh-negative women.

The third characteristic needed for isoimmunization against a blood group antigen is that the mother have the immunologic capacity to produce antibody against the specific foreign antigen. As many as 30% of Rh-negative women are immunologic "nonresponders" who do not produce antibody even when challenged with large volumes of Rh-positive blood. Also, ABO-incompatibility appears to have a protective effect against the development of Rh sensitization.

III. PREVENTION OF Rh ISOIMMUNIZATION: THE TRIUMPH OF Rh-IMMUNE GLOBULIN

That passively administered antibody will prevent active immunization by its specific antigen was well-known to immunologists prior to the use of RhIG for the prevention of Rh disease. The most likely mechanism for this effect is known as central inhibition, whereby fetal RBCs coated with exogenously administered anti-D are removed from the circulation by the spleen and lymph nodes and suppress the primary immune response by preventing the transformation of B cells to IgG-producing plasma cells. While other mechanisms may be involved, this hypothesis has gained the most scientific support.

The general principles of preventing Rh isoimmunization with RhIG include the following: (i) 10 μg Rh-immune globulin should be given for every 1 ml of fetal blood in the maternal circulation, (ii) RhIG should be given within 72 hr of possible exposure to blood from D-positive individuals, (iii) RhIG should be used whenever there is the possibility of such exposure, (iv) RhIG should be used relatively liberally if the D status of a possible exposure is not known, and (v) RhIG should not be given to women previously shown to be sensitized to the D antigen.

RhIG should be administered to Rh-negative

women with possible D exposure after spontaneous miscarriage, induced abortion, amniocentesis, chorionic villus sampling, birth, or transfusion with blood products. The 72-hr time limit set for the administration of Rh-immune globulin is an artifact of studies using male prisoner volunteers. Since prison officials would allow investigators to visit the volunteers at 3-day intervals, the use of Rh-immune globulin at intervals of more than 3 days after a challenge with Rh-positive cells was never adequately studied. However, for best efficacy RhIG should be given before the primary immune response is established. The time for a primary immune response varies, and RhIG should be administered to appropriate mothers as soon as possible after delivery. If the neonatal Rh status is unknown by the third day after delivery, RhIG should be given to an Rh-negative mother rather than waiting until the results of neonatal screening become available.

Since 1 or 2% of susceptible women became sensitized even after appropriate postpartum RhIG prophylaxis, "prophylaxis failures" occurring from antepartum fetomaternal hemorrhage are the most likely cause. Studies have shown that antenatal sensitization can be reduced from 2 to 0.1% when 300 μg of RhIG is administered at 28 weeks' gestation.

IV. MANAGEMENT OF THE UNSENSITIZED Rh-NEGATIVE PREGNANT WOMAN

The management of the Rh-negative, unsensitized woman during pregnancy is depicted in Fig. 1. At the first prenatal visit of each pregnancy, every patient should have their ABO blood group, Rh type, and antibody status determined. These tests should be done in each subsequent pregnancy because previous maternal antibody screening is not an adequate assessment for the current pregnancy. If the patient is Rh negative and has no demonstrable antibody against D antigen, she is a candidate for RhIG prophylaxis at 28 weeks' gestation and again immediately postpartum. It is likely unnecessary to obtain a second antibody screen to ensure that the patient is not already sensitized prior to the intramuscular administration of 300 μg RhIG.

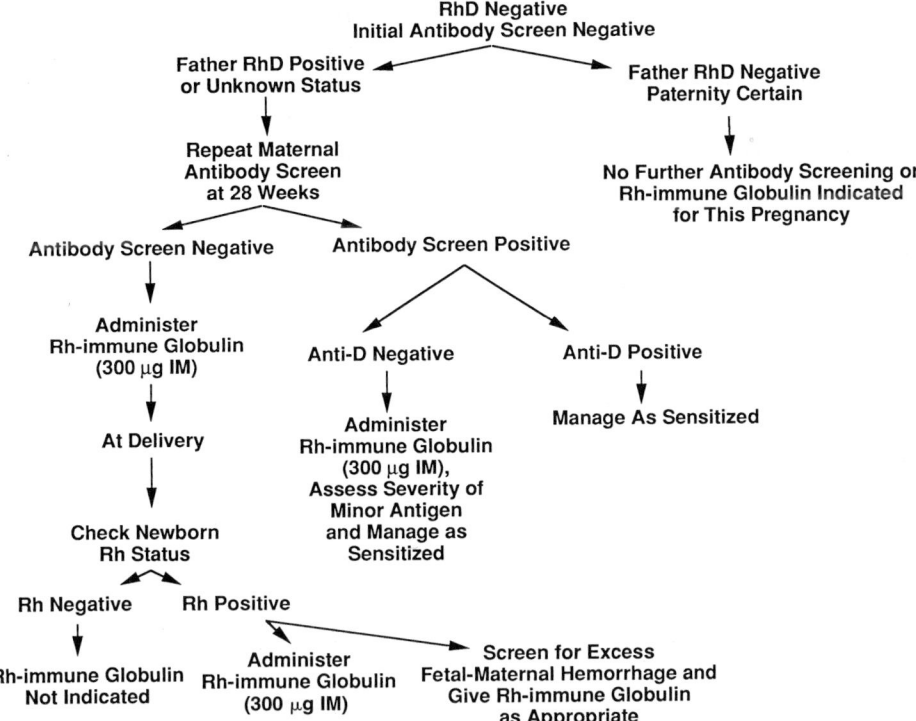

FIGURE 1   Management of unsensitized Rh-negative women. Reproduced with permission from Obstetrics: Normal and Problem Pregnancies, 3rd ed. (Gabbe, Niebyl, and Simpson, Eds.) Churchill Livingstone New York, NY.

Standard doses of RhIG (300 μg) only protect against 30 ml of fetal blood in the maternal circulation. Since about 1% of deliveries result in a fetomaternal hemorrhage of $\geqq$30 ml, Rh-negative patients with an Rh-positive newborn should be screened for a more precise quantitation of fetal blood in their circulation with the use of Kleihauer–Betke test. In this way, the appropriate amount of RhIG can be determined. If the volume of hemorrhage is estimated to be $\geqq$30 ml whole blood, the administered dose of RhIG should be based on 10 μg RhIG per milliliter of whole fetal blood in the maternal circulation. Multiple vials of RhIG may be needed to prevent Rh isoimmunization in this setting.

RhIG prophylaxis is also indicated if the patient is Rh negative, unsensitized, and experienced spontaneous or induced abortion or ectopic pregnancy. If the pregnancy loss occurs at 12 weeks gestation or less, a 50-μg dose of RhIG is adequate to cover the entire fetal blood volume, but if the gestational age is unknown or beyond 12 weeks a 300-μg dose of RhIG is required. If the gestational age is unclear or cannot be ascertained, full-dose RhIG should be given.

An Rh-negative, unsensitized patient who has antepartum bleeding or suffers an unexplained second- or third-trimester fetal death should have Kleihauer–Betke testing and a precise volume of fetal blood determined. If fetal cells are found in the maternal circulation, RhIG is indicated at a dose of 10 μg per estimated milliliter of whole fetal blood. Again, multiple vials of RhIG may be needed to prevent Rh isoimmunization in this situation.

Antenatal Rh-immune globulin is indicated at the time of chorionic villus sampling or amniocentesis in an Rh-negative, unsensitized patient. For first-trimester procedures, 50 μg of RhIG should be sufficient, but for second- or third-trimester procedures a full 300-μg dose is needed even if the procedure is not associated with obvious bleeding. When amniocentesis is performed within 72 hr of delivery (e.g., determining fetal pulmonary maturity), RhIG may be withheld and given if the infant is found to be Rh positive if delivered within 72 hr of the procedure. However, if delivery is to be delayed for more than 72 hr, Rh-immune globulin should be given. This critical step should not be overlooked in managing these pregnancies.

V. MANAGEMENT OF MILD TO MODERATE Rh ISOIMMUNIZATION

Any patient with a significant anti-D antibody titer should be considered Rh sensitized and her pregnancies managed according to contemporary guidelines so that fetal and neonatal morbidity and mortality is minimized. In general, there are two likely outcomes of fetuses from Rh-sensitized pregnancies: (i) mild to moderate cases in which it is unlikely to require intrauterine intervention and fetuses may be delivered when they achieve pulmonary maturation, and (ii) severe hemolytic disease which requires intrauterine transfusion and early delivery. Accurate assignment of gestational age using menstrual dates and early ultrasound is critical since this dictates the timing of amniocentesis, umbilical cord blood sampling, *in utero* treatment, and delivery.

The first step is to document well the obstetric history. Since fetal hemolytic disease tends to be either as severe or more severe in subsequent pregnancies, the obstetric history plays a key role in the guiding management. Prior intrauterine or neonatal deaths from hemolytic disease carries a grave progno-

sis, and if a mother has had a hydropic fetus, the chance that the next Rh-incompatible fetus will become hydropic is 80% or better. Rarely will a Rh-incompatible fetus be less severely affected than was noted in previous pregnancies.

Figure 2 details the management of the Rh-sensitized woman in her first pregnancy. The next step in managing Rh-sensitized pregnancies is to determine the fetal Rh status. The fetus might be Rh negative and therefore not need further testing. If the woman was sensitized during a pregnancy by another partner or after a mismatched blood transfusion, determination of the paternal Rh antigen status is indicated. If the father is Rh negative and paternity is certain, further assessment and intervention are unnecessary. If the father is Rh positive, zygosity testing for the D antigen can be performed by a qualified blood bank laboratory. DNA analysis can also be used to determine zygosity. If the father is homozygous for the D antigen, all children fathered by him will be Rh positive, but if he is heterozygous, there is a 50% likelihood that each pregnancy will have an Rh-negative fetus.

Recent advances in molecular genetics allow for the determination of fetal Rh status without the need for fetal red blood cells. The Rh locus on chromosome 1p34–p36 has been cloned, and polymerase chain reaction (PCR) now can detect fetal Rh status from uncultured amniocytes and chorionic villi. Typing for other minor antigens is now offered by most contemporary prenatal diagnostic genetic laboratories. Early studies have shown near 100% accuracy

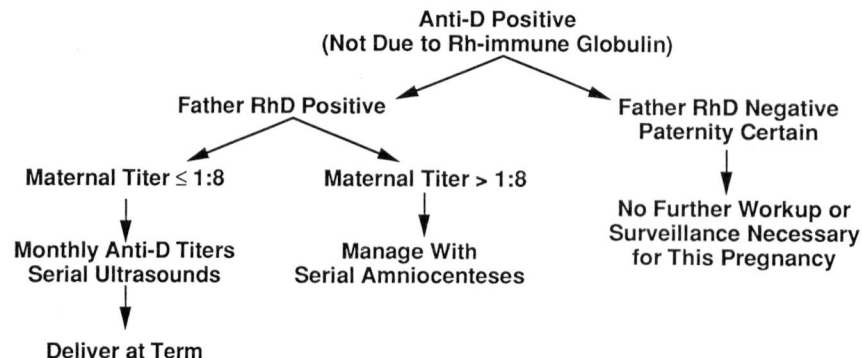

FIGURE 2 Management of Rh-sensitized women in the first pregnancy. Reproduced with permission from Obstetrics: Normal and Problem Pregnancies, 3rd ed. (Gabbe, Niebyl, and Simpson, Eds.) Churchill Livingstone New York, NY.

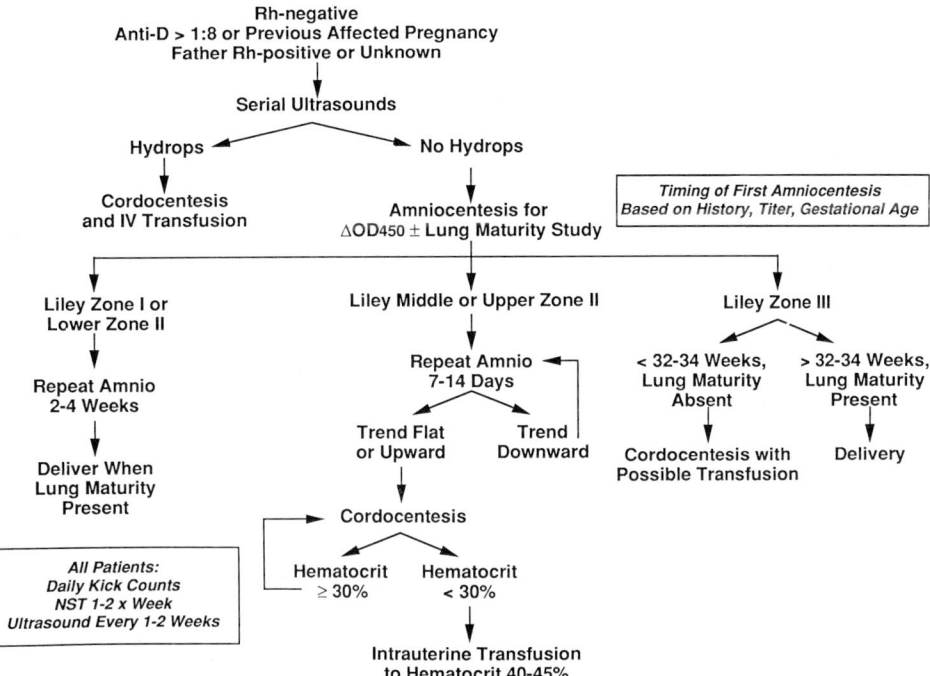

FIGURE 3 Management of Rh-sensitized women at risk for severe hydrops fetalis. Reproduced with permission from Obstetrics: Normal and Problem Pregnancies, 3rd ed. (Gabbe, Niebyl, and Simpson, Eds.) Churchill Livingstone New York, NY.

in determining fetal Rh status from cells isolated from amniotic fluid samples. While the possibility of errors in ascertaining Rh status are possible, this appears to be quite low.

The next step is to determine the amount of maternal circulating antibody against the D antigen. The level of the anti-D antibody titer determines the need for amniocentesis. Severe erythroblastosis or perinatal death rarely occurs when antibody levels remain below a certain "critical titer" which can vary from laboratory to laboratory. Since the reliability and method of antibody titration varies greatly from one laboratory to another, an anti-D titer of 1:8 or greater is usually considered an indication for an interventional management approach. If the initial anti-D titer is less than 1:8 and the patient has no history of a previously affected infant, serial antibody titer determinations every 2–4 weeks is appropriate management. Along with sonographic evaluations, this conservative management scheme should detect those fetuses at risk for severe hemolytic disease.

If after these steps the fetus is assessed as being at risk for mild to moderate hemolytic disease, amniotic fluid evaluation for evidence of RBC destruction should be performed (Fig. 3). Spectrophotometric determinations of amniotic fluid bilirubin correlate reasonably well with the severity of fetal hemolysis. Amniotic fluid bilirubin, a by-product of fetal hemolysis, reaches the amniotic fluid after excretion into fetal pulmonary and tracheal secretions and then diffusion across the fetal membranes. Bilirubin causes a shift in the spectrophotometric density with a peak at a wavelength of 450 nm. The ΔOD_{450}, or the amount of shift in optical density from linearity at 450 nm, is used to estimate the amount of fetal red cell hemolysis.

The Liley curve provides a template for the management of Rh-immunized pregnancies based on ΔOD_{450} values in the third trimester. Unaffected fetuses and those with mild anemia have ΔOD_{450} values in the lowest zone (zone I), whereas severely affected fetuses have ΔOD_{450} values in the highest zone (zone III). Fetuses with zone II values may have disease that can range from mild to severe, and degree of severity can be determined by following the trend of the amniotic fluid bilirubin in serial determinations.

Amniotic fluid bilirubin normally tends to decrease as pregnancy advances so that the boundaries of the zones slope downward as gestational age increases. Amniotic fluid ΔOD_{450} values from infants who die *in utero* from erythroblastosis usually have an upward trend such that a horizontal or rising trend is ominous and indicates the need for intervention.

In mild to moderate cases, contemporary management includes serial amniocenteses to determine the trend of ΔOD_{450} values over the latter part of gestation. A single ΔOD_{450} value is usually not sufficient since the ΔOD_{450} trend, as established by serial amniocenteses, provides more helpful data regarding fetal status. Also, amniotic fluid bilirubin levels are at best an indirect measure of fetal anemia, and reliance only on ΔOD_{450} values can lead one to believe that a mildly affected fetus may have severe hemolytic disease. Extrapolation of Liley's original graph in pregnancies before 26 weeks' gestation is an unproven practice. In recent studies of Rh-isoimmunized pregnancies less than 26 weeks' gestation, absolute values of bilirubin have been found to be predictive of pregnancy outcome. Between 16 and 20 weeks' gestation, a bilirubin value greater than or trending above 0.15 predicted severe isoimmunization, whereas ΔOD_{450} values below 0.09 indicated mild or absent disease. These data suggest that fetuses with ΔOD_{450} values below 0.15 can be managed with serial amniocenteses, whereas those fetuses with a ΔOD_{450} above 0.15 should be managed via direct fetal blood access.

Along with serial amniocenteses, a helpful adjunct in the management of mild to moderate Rh isoimmunization is provided by sonographic evaluation of the pregnancy. Findings of polyhydramnios, placental thickness $\geqq 4$ cm, pericardial effusion, dilation of the cardiac chambers, chronic enlargement of the spleen and liver, visualization of both sides of the fetal bowel wall, and dilation of the umbilical vein may be indicators of significant prehydropic fetal anemia. However, these findings are not reliable in distinguishing mild from severe hemolytic disease, and the role of ultrasound in the monitoring of fetuses with severe Rh immunization is limited to the determination of gestational age, monitoring for hydropic changes, and guidance for invasive procedures.

The goal of antepartum management in fetuses with mild to moderate disease is to determine those who will develop severe disease and thus require intrauterine transfusion. About half of susceptible infants of Rh-immunized pregnancies have mild to moderate disease and do not require intrauterine transfusion. In general, mild to moderate fetal hemolysis can be expected when the involved pregnancy is the first sensitized pregnancy or if previously delivered Rh-positive infants have been mildly to moderately affected. In these cases, sonographic evaluation of the fetus every 2–4 weeks from 18 weeks' gestation until delivery is recommended and, if the fetus shows no evidence of hydrops, amniotic fluid ΔOD_{450} determinations should commence at 24–28 weeks' gestation. The timing of serial amniocenteses and determination of the need for intrauterine transfusion is based on the trend in ΔOD_{450} values. With values falling within the low zone or the lower half of the middle zone, amniocentesis is repeated every 2–4 weeks. If ΔOD_{450} values rise into the upper quarter of the middle zone or into zone III, especially before 30 weeks' gestation, severe hemolytic disease should be suspected and intrauterine transfusion considered. In some cases, a single ΔOD_{450} value in zone III also may indicate severe anemia, and if at any time the fetus has evidence of hydrops by ultrasound a fetal hematocrit <15% is likely. The sonographic detection of hydrops fetalis in Rh-isoimmunized pregnancies constitutes a fetal emergency and mandates fetal transfusion for severe anemia in the immediate future.

VI. MANAGEMENT OF SEVERE Rh ISOIMMUNIZATION

The first successful intrauterine transfusion was reported in 1963. Early transfusions were accomplished via an intraperitoneal approach, but recently the technology for direct intravascular transfusion has been developed which has led to highly successful treatment regimens. The administration of RBCs to the affected fetus corrects fetal anemia, improving fetal oxygenation, and reduces extramedullary hematopoietic demand, subsequently resulting in a fall in portal venous pressure and improved hepatic function. Because these pathophysiologic events are thought to play a key role in the development of hydrops, appropriate replacement of blood will lead to resolution of fetal hydrops.

Direct fetal vascular access was first described in the 1980s, thus enabling accurate assessment of the degree of fetal anemia and directed fetal intravascular transfusion as treatment for hydrops. With fetal blood sampling, the presence or absence of the fetal RBC antigenic status can be determined, particularly if the zygosity of the father is unknown or uncertain. Direct fetal blood sampling is now considered by some authorities to be one of the first steps in the management of the fetus at risk for severe hemolytic disease, but the possible adverse consequences of umbilical cord blood sampling make this position controversial.

Once fetal anemia is suspected or confirmed, direct intravascular fetal transfusion may be performed using ultrasound to guide placement of a needle into the umbilical vein and is this is considered the preferred approach in the treatment of erythroblastosis fetalis. Direct access to the fetal circulation has a number of advantages over the intraperitoneal approach. First, measurement of the fetal hematocrit before transfusion can be made so that the volume of blood required to correct anemia can more precisely be calculated. Second, some fetuses will be determined to be Rh negative and, third, a posttransfusion hematocrit can be used to determine when the next transfusion should occur. Fourth, transfusion into the fetal vascular system ensures a more rapid correction of fetal anemia, a particularly important event for hydropic fetuses because they often do not adequately absorb RBCs transfused into the fetal peritoneum. However, direct intravascular access has several potential disadvantages, including the rare possibility of volume overload in the compromised fetus leading to further problems, procedure-related complications such as the inability to access the fetal vascular compartment, and the possibility of increasing the severity of maternal sensitization should fetomaternal hemorrhage occur.

Perinatal survival in pregnancies managed with intravascular transfusion is significantly improved for hydropic fetuses when compared with intraperitoneal approaches. Overall, survival rates range from 75 to 94% in some specialized centers, with a survival rate of nonhydropic fetuses exceeding 90% and a survival rate approaching 75% in fetuses with erythroblastosis fetalis. Intravascular transfusion is associated with a procedure-related complication rate of up to 15% and a mortality rate between 1 and 5%, with hydropic fetuses being at the greatest risk.

Other modalities have been used in the management of the severely affected Rh-isoimmunized pregnancy. Combining intravascular and intraperitoneal transfusions has been used in an attempt to decrease the number of transfusions needed to correct anemia, but this approach has not gained widespread acceptance. Plasmapheresis has been used to reduce the level of maternal anti-D, but with little success. Conversely, intravenous high-dose immunoglobulin (IgG) has been found to have possible positive effects in the severely affected pregnancy. One possible mechanism of action is that Fc receptor blockade in the fetal reticuloendothelial system inhibits the removal and destruction of anti-D-coated RBCs in the fetal spleen and liver. The reported success of intravenous IgG administration in ameliorating some cases of severe Rh disease indicates that this is a promising treatment in need of further testing prior to widespread acceptance.

VII. MINOR ANTIGENS

In addition to the D antigen, hundreds of other distinct antigens, known as "minor," "atypical," or "irregular" antigens, exist on the surface of RBCs. These minor antigens, prior to the introduction of RhIG, were an infrequent cause of maternal sensitization or fetal or neonatal hemolytic disease. However, with the introduction and use of RhIG prophylaxis, minor antigen sensitization has become relatively more frequent. Antibodies to minor antigens are now thought to be more common in the general population than antibodies to the D antigen. These antigens include Kell, Duffy, C, c, E, and e and can result in hydrops fetalis. Many other antigenic systems exist, some of which are considered "private" antigens in which only a few families are affected. The reader is referred to a more definitive source for a complete list of these antigens and their potential for causing severe isoimmunization.

VIII. FUTURE RESEARCH

Although the field of blood group isoimmunization is characterized by significant advances based on astute clinical observation and rigorous scientific technique, many questions remain. What is the role for intraperitoneal transfusion? Can ultrasound technology be advanced so that less invasive measures can be developed to detect the severely affected fetus? Does intravenous IgG have a role in these pregnancies? Can preimplantation genetic diagnosis be utilized to select Rh-negative embryos? Can the Rh gene be successfully deleted from sperm so that homozygous men can father Rh-negative children? Can prophylaxis be extended to minor blood group antigens (as was developed for Rh-negative women)? With further research into the prevention and treatment of erythroblastosis fetalis, there is the potential to diminish further the impact of this devastating disease.

See Also the Following Article

IMMUNOLOGY OF REPRODUCTION

Bibliography

Bennett, P. R., Warwick, R., Vaughan, J. V., *et al.* (1993). Prenatal determination of fetal RhD type by DNA amplification. *N. Engl. J. Med.* **329**, 607.

Berkowitz, R. L., Chitkara, U., Goldberg, J. D., *et al.* (1986). Intrauterine intravascular transfusions for severe red blood cell isoimmunization: Ultrasound-guided percutaneous approach. *Am. J. Obstet. Gynecol.* **155**, 574.

Bowman, J. M. (1978). The management of Rh-isoimmunization. *Obstet. Gynecol.* **52**, 1.

Bowman, J. M., and Pollock, J. M. (1978), Antenatal Rh prophylaxis: 28 weeks' gestation service program. *Can. Med. Assoc. J.* **118**, 627.

Bowman, J. M., Pollock, J. T., Manning, F. A., *et al.* (1992). Maternal Kell blood group alloimmunization. *Obstet. Gynecol.* **79**, 239.

Bowman, J. M., Pollock, L. M., Peterson, L. E., *et al.* (1994). Fetomaternal hemorrhage following funipuncture: Increase in severity of maternal red-cell alloimmunization. *Obstet. Gynecol.* **84**, 839.

Freda, V. J. (1965). The Rh problem in obstetrics and a new concept of its management using amniocenteses and spectrophotometric scanning of amniotic fluid. *Am. J. Obstet. Gynecol.* **92**, 341.

Harman, C. R., Bowman, J. M., Manning, F. A., and Menticoglou, S. M. (1990). Intrauterine transfusion—Intraperitoneal versus intravascular approach: A case-control comparison. *Am. J. Obstet. Gynecol.* **162**, 1053.

Landsteiner, K., and Wiener, A. S. (1940). An agglutinable factor in human blood recognized by immune sera for rhesus blood. *Proc. Soc. Exp. Biol. Med.* **43**, 223.

Liley, A. W. (1961). Liquor amnii analysis in the management of pregnancy complicated by rhesus sensitization. *Am. J. Obstet. Gynecol.* **82**, 1359.

Marsh, W. L., Chaganti, R. S. K., Mayer, K., *et al.* (1974). Mapping human autosomes: Evidence supporting assignment of rhesus to the short arm of chromosome number 1. *Science* **183**, 966.

Nicolaides, K. H., Fontanarosa, M., Gabbe, S. G., and Rodeck, C. H. (1988). Failure of ultrasonographic parameters to predict the severity of fetal anemia in rhesus isoimmunization. *Am. J. Obstet. Gynecol.* **158**, 920.

Queenan, J. T., Tomai, T. P., Ural, S. H., and King, J. C. (1993). Deviation in amniotic fluid optical density at a wavelength of 450 nm in Rh-immunized pregnancies from 14 to 40 weeks' gestation: A proposal for clinical management. *Am. J. Obstet. Gynecol.* **168**, 1370.

Reece, E. A., Copel, J. A., Scioscia, A. L., *et al.* (1988). Diagnostic fetal umbilical blood sampling in the management of isoimmunization. *Am. J. Obstet. Gynecol.* **159**, 1057.

Scott, J. R., Kochenour, N. K., Larkin, R. M., *et al.* (1984). Changes in the management of severely Rh-immunized patients. *Am. J. Obstet. Gynecol.* **149**, 336.

Estradiol

see Estrogens, Overview

Estriol

see Estrogens, Overview

Estrogen Action, Behavior

Lynda Uphouse and Sharmin Maswood

Texas Woman's University

GLOSSARY

analgesia Insensitivity to pain.
neonatal Referring to the period in life soon after birth.
ovariectomy The surgical removal of ovaries.
periovulatory Near to that portion of the female's reproductive cycle when ovulation occur.
postmenopausal Referring to the period in a female's life after natural cessation of menstruation.

I. INTRODUCTION

Estrogen is a steroid hormone synthesized predominantly in the ovaries (female reproductive organs) but is also produced (to a limited extent) in the adrenals and in adipose (fatty) tissue. As a hormone, secreted into the blood supply, estrogen is carried throughout the body where it acts to coordinate multiple physiological functions that are important for female reproduction. In the central nervous system (CNS), estrogen plays a dominant role in coordinating neural activity so that the behavioral and physiological events that are essential for successful reproduction occur. Although estrogen may be best known as a modulator of female mating behavior, the hormone's behavioral effects extend beyond just the mating act. In fact, one of the oldest known effects of estrogen is the hormone's enhancement of locomotor activity. The term "estrus," which refers to the periovulatory period in nonprimate mammals, is derived from the Greek word *oistros*, meaning "gadfly" or "frenzy." In today's world of "action" heroes, estrogen might be viewed as an action hormone. Estrogen increases activity, attention, and alertness (the three A's). Therefore, any behavior that involves one or more of the three A's can be influenced by estrogen. Nevertheless, mating behavior remains the best understood behavior controlled by estrogen.

II. MATING BEHAVIOR

In the early 1940s, investigators discovered that when the ovaries were removed from female rats, the rats showed an immediate cessation of mating behavior; and portions of the behavior were restored by injection with estrogen. Since these original studies, an essential role for estrogen in female mating behavior of many mammalian species has been demonstrated. Another ovarian hormone, progesterone, is important in facilitating mating behavior and is required for the entire sequence of behavioral events, but mating can occur in the absence of progesterone. It is important to note that there is a delay (16–24 hr) before estrogen increases mating behavior in rats. Thus, estrogen is said to "prime" the female for the behavior to occur. The dependency of mating behavior on female gonadal hormones explains why, in many species, female mammals only engage in mating behavior near the time of ovulation, when the female is capable of producing offspring. Humans and some other primates are an exception because, in these species, mating behavior is not synchronized with ovulation (e.g., mating behavior occurs independently of the female's ability to produce offspring). However, even in humans, libido (e.g., desire to mate) appears to be increased by estrogen.

Estrogen (or estrogen followed by progesterone) increases a variety of other behaviors that function to increase the probability that mating and offspring production will occur. For example, female rats, given a choice between spending time with a sexually active male and spending time with another female or a castrated male, will spend more time with the sexually active male. Moreover, such females will work harder performing tasks which allow access to such males. Therefore, estrogen appears to increase the female's "preference" for a mate and she increases behaviors that will move her into the proximity of the mate. Similarly, behaviors which make the female more attractive to the male are increased in sexually receptive females (and such behaviors are also increased by estrogen). In rodents, such behaviors have been referred to as "solicitation" or "proceptive" behaviors and include behaviors known as "hopping and darting." This is a sequence of behaviors in which the female approaches the male (perhaps getting his attention) and then rapidly moves away in a darting motion which is ended by the female's assumption of the lordosis posture (a posture required for successful mating to occur). Progesterone plays an essential role in such proceptive and solicitation behaviors; estrogen, while sufficient to elicit the mating posture, is not sufficient to elicit solicitation and proceptive behaviors. However, progesterone is not effective without prior estrogen treatment.

III. SENSORY AND MOTOR BEHAVIOR

In rodents, estrogen increases locomotor activity as measured by running in an activity wheel or by activity in an open field. Human females also show evidence of greater energy or activity during portions of the menstrual cycle associated with high estrogen. Estrogen has been shown to increase the female's perception (e.g., estrogen reduces the intensity of the signal that is required to produce a noticeable reaction) of various sensory stimuli, including somatosensory, visual, auditory, and olfactory signals. In humans, such intensity thresholds may also be lower at midcycle, when estrogen levels are high. When estrogen levels are high, there is also an increase in the ability to discriminate two stimuli applied to the surface of the skin (e.g., fine, two-point discrimination is enhanced). Such enhancing effects of estrogen on both sensory and motor events may be responsible for the observation that, in rodents, estrogen enhances performance on motor tasks (e.g., balance beams) that are dependent on coordinated limb movements. In the human female, there is an increase in motor coordination and manual dexterity at midcycle of the menstrual cycle.

IV. PAIN

The effect of estrogen on sensory perception may account, in part, for observations that women have lower detection, pain, and pain tolerance thresholds (although this may also depend on the nature of the stimulus). In rodents, paradigms utilized to measure such pain sensitivity have included tail-flick behavior (where a beam of light focused on the tail is used to produce a thermal stimulus) or jump behavior in a foot-shock or hot-plate procedure. The latency to produce the tail-flick response or jump response is used as an index of the female's sensitivity to pain. In general, female rats are more sensitive than male rats to shock and thermal stimuli and, during the reproductive cycle, female rats show enhanced pain sensitivity during the part of the cycle near ovulation, when estrogen levels are highest. When ovariectomized rats are injected with estrogen, acute estrogen increases pain sensitivity. However, prolonged estrogen exposure may reduce pain sensitivity. Moreover, estrogen may modulate the female's sensitivity to the production of both morphine-induced analgesia and stress-induced analgesia. When the effects of morphine on the tail-flick response or jump response are examined, female rats near the time of ovulation show a lower sensitivity to the analgesic effects of morphine, and estrogen may be responsible for this attenuation of opioid-induced analgesia. However, estrogen may simultaneously trigger a nonopioid form of analgesia. Therefore, estrogen's modulation of the behavioral response to painful stimuli is multifaceted. Moreover, the female's responsivity to painful stimuli can also be modulated by diet. Of special interest is the observation that fat consumption by female rats may attenuate reproductive cycle-dependent fluctuations in pain sensitivity. The latter observation is especially interesting since food consumption varies with hormonal state.

V. EATING BEHAVIOR

Human females testify to the impact of reproductive hormones on eating behavior. Everyone has heard tales about women's food cravings during pregnancy and many women report food cravings during the menstrual cycle. Such tales may have some physiological basis because estrogen modulates both the amount and type of food intake. Under many conditions, estrogen is reported to reduce total food intake. However, consistent with its overall activational effect, estrogen increases the preference for "energy-rich" foods, such as carbohydrates and fat, while reducing the preference for protein-rich foods. Estrogen's effect on eating behavior may also have relevance for other behavioral effects of the hormone (see Section VII).

VI. LEARNING AND MEMORY

In recent years, there has been an increasing emphasis on estrogen's effect on performance, learning, and memory. Although there is a growing body of evidence that estrogen can increase accurate performance on a variety of tasks, the hormone's positive impact is clearly task specific, may be age dependent, and probably requires prolonged hormone treatment. Nevertheless, in many cases, estrogen treatment has been associated with improved performance as well as with beneficial effects on memory. Some investigators have argued that estrogen's beneficial effect on memory may be mediated through the hormone's effect on the neurotransmitter, acetylcholine. In humans, positive effects have been seen in postmenopausal women and there is some indication that estrogen may reduce the memory impairment and/or onset of Alzheimer's disease. Interestingly, enhancing effects of estrogen on memory can be seen in male as well as in female rats. Females are most likely to develop Alzheimer's disease during the postmenopausal years when estrogen levels are low, and the incidence of Alzheimer's disease is greater in females than in males. Because the male hormone, testosterone, can be converted to estrogen in the brain, estrogen's memory-protecting effect may contribute to gender differences present after the age of menopause in the vulnerability to Alzheimer's disease.

However, estrogen may also adversely affect performance when the task includes conditions of "emotionality." This latter observation is interesting because estrogen is thought to be anxiogenic (see below). Another task feature which affects the directional effect of estrogen on performance may be the number and/or type of distracters. For example, in a spatial maze task, in which the animal must attend to the position of the goal box in relation to environmental stimuli, estrogen may reduce performance as the number of spatial cues increases. The impact of such "distracters" on performance may result from the enhancing effects of estrogen on "arousal" or "attention" rather than on memory (e.g., if estrogen-treated females attend to too many cues, any one of which would be sufficient to solve the problem, the task difficulty could be increased).

VII. MOOD DISORDERS

In recent years, ovarian hormones as modulators of behavior have attracted public attention because of media discussions of menstrual-related disorders [such as premenstrual syndrome (PMS)], of behavioral changes accompanying menopause, and of the female's vulnerability to development of mood disorders. Women are twice as likely as men to develop depression. Moreover, women are more likely than men to exhibit anxiety disorder, panic disorder, and eating disorders such as anorexia and bulimia nervosa. Animal models have provided relatively consistent evidence that estrogen is anxiogenic and that progesterone is anxiolytic. The occurrence of menstrual-related mood disorders such as PMS and the accentuation of mood dysfunction near menopause have reinforced notions that female gonadal hormones may be responsible for the human female's increased vulnerability to these mood dysfunctions. Specifically, estrogen has been linked to the female's vulnerability to anxiety disorders, whereas withdrawal from estrogen, increased progesterone, or a reduction in the relative amount of estrogen and progesterone have been linked to the development of depression. Because of the relationship between the neurotransmitters serotonin and norepinephrine and various mood disorders, estrogen may influence the female's vulnerability to mood disorders by altering these neurotransmitter systems. Many effects of estrogen on serotonin and norepinephrine have been

reported; and the effects of drugs which influence these neurotransmitters are modulated by estrogen.

However, to date, no single explanation for the difference in vulnerability to mood disorders between males and females has been clearly documented. Differences in levels of estrogen probably cannot account for the female's vulnerability to anxiety or depression. Estrogen's modulation of these behaviors must be viewed from the perspective of the state of the individual at the time of estrogen treatment (e.g., estrogen's overall behavioral effect will be superimposed on other physiological and environmental events). For example, females have more active serotonergic systems than do males, estrogen increases serotonergic functioning, and accentuated activity of serotonergic neurons may be associated with vulnerability to anxiety. However, estrogen treatment does not cause anxiety. Yet, if an environmental event that accentuates serotonin functioning occurs together with estrogen stimulation, the two events may summate and anxiety may result. While neither estrogen treatment alone nor the environmental challenge alone may be sufficient to precipitate anxiety, the combination of the two may. Therefore, while it is impossible to hold estrogen accountable for the female's enhanced vulnerability to mood disorders, it is also impossible to rule out the hormone's participation.

It is also important to note that there is a greater association between the cyclicity of gonadal hormones and the female's enhanced vulnerability to mood dysfunction than there is between absolute levels of hormones and the behaviors. In humans, affected individuals are more likely to show symptoms of PMS during ovulatory than during nonovulatory cycles. Therefore, the important factor influencing the female's vulnerability to this disorder may be the change, rather than the presence or absence, of the hormones. This cyclicity hypothesis is reinforced by observations that postmenopausal women show less evidence of psychiatric disturbance than do younger, still menstruating women. This idea may also account for observations that perimenopausal women show greater susceptibility to mood dysfunctions when estrogen levels are declining or widely fluctuating. Similarly, the eating disorder anorexia nervosa is most prevalent at puberty.

VIII. THOUGHT DISORDERS

In contrast to anxiety and depression, schizophrenia is more likely to occur in men than in women. For this disease, gender differences are thought to result from a protective action afforded by the female's higher circulating levels of estrogen. Historically, studies of the neural mechanisms of schizophrenia have focused on the neurotransmitter, dopamine. Estrogen's protective action against the development of schizophrenia has also been attributed to the hormone's modulation of dopamine in the CNS. Recently, serotonin has also been implicated in the development of schizophrenia and in the protective action of estrogen.

IX. BEHAVIORAL DEVELOPMENT

The behavioral effects of estrogen may extend far beyond the time of estrogen exposure. It is well established that the presence of estrogen during critical periods alters the development of the nervous system and contributes to gender differences that are present in adulthood. With the exception of mating behavior, few studies have focused on the importance of neonatal estrogen to gender differences in specific behaviors present in adulthood. Such an emphasis is likely to increase in the future.

X. DIVERSE BEHAVIORS OR HORMONAL INTEGRATION

To understand how (or why) a single hormone with clear reproductive importance can influence such a wide range of behaviors, it is first necessary to appreciate that estrogen acts throughout the CNS and not just in areas that are critical for control of female reproductive behavior. Moreover, estrogen can affect neural functioning in several ways. Estrogen can alter cellular functioning by binding to specific estrogen receptors inside the cell. The intracellular receptors function as cellular switches to turn on (or off) the production of estrogen-responsive proteins. Estrogen may also interact with estrogen receptors localized in the plasma membranes of cells;

estrogen may alter the functioning of other plasma membrane receptors, such as neurotransmitter receptors; or estrogen may also alter the amount of neurotransmitter released by neurons. For many neurotransmitters, their receptors consist of groups of proteins, each of which is affected by the neurotransmitter. When activated by the neurotransmitter, the different receptors can relay distinct, and even conflicting, cellular signals. However, when activated simultaneously, a totally new signal may result. Estrogen may alter the relative strength of these individual neurotransmitter receptors and thereby alter the significance of incoming signals. Because neurotransmitters link one neuron to another in an information network, estrogen may change the flow of information (increasing some signals and reducing others) through this network. The effects of estrogen on neural functioning might be equated to those of a pianist at a keyboard, where the resulting melody depends not only on the combination of piano keys selected but also on their sequence. The "melody" of neural functioning is dependent on the combination and sequence of neurotransmitter action and can be altered by estrogen.

This latter view of estrogen action is important because it provides a framework for understanding how estrogen may modulate behavior without causing the behavior to occur. From such a framework, estrogen can be seen to enhance behaviors appropriate for finding a mate while simultaneously delaying the actual act of mating until the female is physiologically prepared for mating. In a species in which finding a suitable mate is essential for successful production of offspring, an increase in activity and exploration should increase the female's opportunity for location of and discrimination among suitable mates. Similarly, accentuated sensitivity to pain may protect the female from environmental circumstances that are unfavorable to reproductive success. Finally, estrogen may shift motivational priorities so that mate acquisition is temporarily more important than vegetative functions such as the acquisition of food. In primate species and, especially in humans, behavioral effects of the action hormone can sometimes conflict with long-term goals or cultural traditions. Nevertheless, by temporarily switching behavior from a "maintenance" to an action state, estrogen accentuates a well-orchestrated pattern of behavioral change designed to increase the probability that successful mating and that survival of the species will occur.

See Also the Following Articles

MATING BEHAVIORS, MAMMALS; MENOPAUSE; PMS (PREMENSTRUAL SYNDROME)

Bibliography

Beach, F. A. (1976). Sexual attractivity, proceptivity, and receptivity in female mammals. *Horm. Behav.* 7, 105–138.

Clemens, L., and Weaver, D. (1985). The role of gonadal hormones in the activation of feminine sexual behavior. In *Handbook of Behavioral Neurobiology* (N. Adler, D. Pfaff, and R. Goy, Eds.), pp. 183–227. Plenum, New York.

Halbreich, U. (1995). Menstrually related disorders: What we do know, what we only believe that we know, and what we know that we do not know. *Crit. Rev. Neurobiol.* 9, 163–175.

Mogil, J. S., Sternberg, W. F., Kest, B., Marek, P. L., and Liebeskind, J. C. (1993). Sex differences in the antagonism of swim stress-induced analgesia: Effects of gonadectomy and estrogen replacement. *Pain* 53, 17–25.

Pfaff, D., and Modianos, D. (1985). Neural mechanisms of female reproductive behavior. In *Handbook of Behavioral Neurobiology* (N. Adler, D. Pfaff, and R. Goy, Eds.), pp. 423–493. Plenum, New York.

Smith, S. S. (1994). Female sex steroid hormones: From receptors to networks to performance—Actions on the sensorimotor system. *Prog. Neurobiol.* 44, 55–86.

Uphouse, L. (1997). Multiple serotonin receptors: Too many, not enough, or just the right number? *Neurosci. Biobehav. Rev.,* 21, 679–698.

Wade, G. N, and Schneider, J. E. (1992). Metabolic fuels and reproduction in female mammals. *Neurosci. Biobehav. Rev.* 16, 235–272.

Estrogen Action, Bone

Robert Marcus

VA Medical Center and Stanford University

GLOSSARY

osteoblast The primary bone-forming cell, derived from stem cells in marrow stroma. Osteoblasts control bone remodeling by elaborating cytokines that suppress or recruit osteoclast production and maturation.

osteoclast Multinucleated giant cells derived from macrophage/monocyte precursors. Osteoclasts respond to cytokines secreted by osteoblasts in response to hormonal or other signals.

osteoporosis A clinical disorder of global skeletal fragility that is characterized by a reduction in the amount of bone as well as disruption of its normal microscopic architecture. The consequence of osteoporosis is fracture and deformity with minimal trauma.

remodeling A continuous process of breakdown and renewal in bone. The final common pathway for adult bone loss or gain.

I. INTRODUCTION

The profound loss of bone that attends estrogen withdrawal and the skeletal protection conferred by estrogen replacement to oöphorectomized animals and postmenopausal women have long been recognized, but only recently has it been clearly established that bone itself is a target tissue for 17β-estradiol. Estrogen is required for normal pubertal acceleration in bone acquisition, for epiphyseal closure, for optimizing adequate peak bone mass at skeletal maturity, and for adult skeletal maintenance. Loss of estrogen during growth years results in deficits in peak bone mass and, during adult life, in increased bone turnover and loss, leading frequently to deficits in the quantity and quality of bone (osteoporosis) and their associated fractures. Sustained estrogen use by menopausal women conserves bone and protects against fracture. This chapter summarizes current understanding of the skeletal actions of estrogen. Although many insights in this field derive from various animal models, those with the clearest relevance to human physiology will be emphasized. This topic has been extensively reviewed by Turner *et al.* (1).

II. MOLECULAR BASIS OF ESTROGEN ACTION ON BONE

For many years, failure to document specific, high-affinity binding of 17β-estradiol in bone cells fostered the belief that estrogen acted on bone only through proximal actions on other tissues. The almost simultaneous demonstration of estradiol receptors (ERs) in rat (2) and human (3) osteoblasts revitalized the concept of bone as an estrogen target tissue. ERs are found in normal and transformed bone cell lines. The dominant species of ER in bone appears to be the recently cloned ER-β (D. McDonnell, personal communication). Postreceptor actions of estradiol are difficult to show with consistency. Estradiol directly modulates the proliferation of osteoblasts, the major bone-forming cells, and their precursors from marrow stroma. It also modulates differentiated osteoblastic functions, but with great species variability. Estradiol stimulates proliferation of fetal rat, mouse, and transformed human osteoblasts (4–6) but has either no effect or inhibits proliferation of human osteoblasts (7, 8). In bone, estrogen response represents effects on cells of varying maturity, from the most nondifferentiated precursor to various degrees of osteoblastic maturation.

Uncertainty persists regarding estradiol action on osteoclasts, the major bone-resorbing cells. Avian osteoclasts specifically bind and respond to estrogen (9), but demonstration of direct estrogen actions on mammalian osteoclasts has proven difficult to establish. The current paradigm of estrogen action on bone is that estradiol binds to osteoblast receptors, with release of cytokines or other agents that recruit or suppress recruitment of osteoclast precursors (see below).

III. ESTROGEN AND SKELETAL ACQUISITION

Before puberty, bone acquisition is slow, steady, and independent of sex steroids. With pubertal onset, increased estradiol concentrations accelerate the recruitment of chondrocytes into proliferation, leading to the familiar spurt in linear growth. Growth velocity is maximal for a few years, then tapers and stops entirely when epiphyseal growth centers ossify. Pubertal growth requires estradiol in both girls and boys, as shown by recent clinical examples. In one report (10), an extremely tall young man had profound deficits in bone mineral density (BMD) and persistent epiphyseal opening on radiographs. Androgenization and plasma testosterone concentrations were normal, but estradiol and estrone concentrations were increased. He showed no response to exogenous estrogen despite a 10-fold increase in circulating hormone. His estrogen receptor gene showed a premature stop codon mutation in exon 2. In a second report, a young man gave a similar clinical presentation but had vanishingly low concentrations of plasma estrogens and an exuberant skeletal response to administered estrogen (11). This man proved to have a mutation in exon 9 of the cytochrome P450 aromatase gene. These cases do not preclude a role for testosterone in skeletal development but rather establish the primary importance of estrogen.

IV. ESTROGEN AND ADULT BONE MAINTENANCE

Adult bone maintenance is governed by the rate and efficiency of remodeling, a coupled process in which a wave of osteoclastic bone breakdown (resorption) is followed by influx of osteoblastic cells that form new bone. Remodeling is carried out by independent "bone-remodeling units" on bone surfaces. It is initiated by signals that recruit precursor cells of monocyte lineage from the marrow to a bone surface site. Some of these signals are cytokines that are released by osteoblasts (see below). Congregated at the bone surface, precursor cells fuse into multinucleated osteoclasts that resorb a cavity out of the bone. In cortical bone, resorption creates tunnels within Haversian canals. On trabecular surfaces it erodes scalloped areas called Howship's lacunae to a depth of about 60 μm. Coupled to resorption, bone formation begins when cytokines or growth factors embedded and released from resorbed bone matrix attract preosteoblasts from marrow stroma into the resorption cavity base. These mature into osteoblasts that replace the missing bone by secreting new matrix. Duration of a complete remodeling cycle normally requires about 6 months.

If replacement of resorbed bone were completely efficient, bone would be restored to its initial state on termination of a remodeling cycle. However, remodeling is not entirely efficient, and small deficits persist with each cycle, leading consequently to age-related bone loss, a normal phenomenon that begins shortly after linear growth stops. Altered remodeling dynamics constitute the final pathway through which diverse stimuli, such as dietary insufficiency, hormones, and drugs, affect the rate of bone turnover.

Although osteoclasts have been considered the proximal cell involved in remodeling activation, it is difficult to show that receptors in mammalian osteoclasts are agents that initiate remodeling. Regulation of osteoclastic resorption *in vitro* requires the presence of osteoblasts or conditioned medium from osteblast cultures. Therefore, it appears that regulation of remodeling begins with a hormonal interaction with osteoblastic cells, whose secreted cytokines recruit osteoclasts to the remodeling site from precursor cell pools in the marrow.

Estrogen powerfully regulates production of cytokines involved with bone remodeling. These include insulin-like growth factors (IGFs) I and II, interleukins (ILs)-1, -6, and -11, and transforming growth factor-β. Estradiol inhibits IL-6 secretion by marrow stromal osteoblastic cells (13) and treatment

of mice with neutralizing IL-6 antibodies suppresses osteoclast production after oöphorectomy (14). Estradiol suppressed IL-6 production in human fetal osteoblasts transfected with estradiol receptor gene (15), and IL-6 gene knockout prevented bone loss in oöphorectomized mice (16). Thus, IL-6 is implicated as a critical molecule by which osteoblasts signal increased bone remodeling, and this suggests that this link is a major site for estrogen action in bone. Another interleukin, IL-1, is secreted by circulating monocytes (17). Production is elevated from cells obtained from postmenopausal women and is inhibited by estrogen (18). [See reviews by Oursler *et al.* (19) and Pacifici (20).]

V. MENOPAUSAL CHANGES IN SKELETAL HOMEOSTASIS

Bone mass stabilizes from age 25 years until about age 50 but decreases when circulating FSH rises and estradiol drops. With spontaneous menopause daily calcium loss increases to 60 mg from premenopausal values of 20 mg (21). After a decade, this would amount to 13% of an original whole body calcium mass of 1000 g, representing a standard deviation BMD change and constituting a two or threefold increase in fracture risk (22–24). This change reflects dominance of bone resorption over formation, due mainly to greater activation of bone remodeling. Women who take estrogen as they enter menopause maintain premenopausal calcium balance (25). Dietary compensation for menopausal changes in calcium economy would require an increase in daily calcium intake from 1000 mg to about 1500 mg. Menopausal bone loss affects the entire skeleton, but is marked in trabecular bone, reflecting its higher prevalence of surfaces.

The following is a plausible sequence for these changes: (i) Estrogen loss promotes osteoblast cytokine secretion; (ii) increased bone resorption elevates plasma ionized calcium activity, which suppresses parathyroid hormone secretion; and (iii) this decreases renal synthesis of 1,25(OH)$_2$ vitamin D, thereby blunting intestinal calcium absorption. With estrogen replacement, remodeling decreases, 1,25(OH)$_2$ vitamin D levels and intestinal absorption rise, and calcium balance is restored.

Although some decrease in bone turnover occurs several years following last menses, turnover rates remain increased into the eighth decade (26, 27). Body weight has a major influence on postmenopausal bone loss, with heavier subjects being relatively protected (28) perhaps due to their higher degree of mechanical loading or to their higher circulating concentrations of estrone.

VI. SKELETAL EFFECTS OF ESTROGEN REPLACEMENT THERAPY

A. Bone Mass

Administration of estrogen at the time of oöphorectomy or within a few years of natural menopause conserves BMD (29–33). As a means to protect bone mass in the early menopausal years, estrogen is significantly more effective than other antiresorptive agents. It has been frequently stated that little is gained by starting estrogen after the early menopausal years. Strong evidence argues against this view. Estrogen benefits bone mass well beyond early menopause (31, 34, 35), although there currently is insufficient information regarding fracture outcomes in women who start estrogen at an advanced age.

Bone changes that follow hormone replacement reflect estrogen's role as an antiresorptive hormone: BMD rises over 12–18 months and then plateaus. This pattern reflects an estrogen-induced decrease in the "remodeling space," a transient deficit in bone in which resorption has occurred but formation has not yet started or remains incomplete. The plateau indicates restoration of this remodeling transient to a new steady-state level. In a recent 3-year multicenter trial, estrogen increased lumbar spine BMD by 5% and hip BMD by 2% over 3 years (36). A prospective study (23) established that older estrogen-replaced women do lose some bone, probably because estrogen does not prevent age-related associated changes in intestinal, renal, and parathyroid function.

B. Fracture Protection

Data suggesting that estrogen protects against fragility-related fractures are sparse compared to those showing effects on BMD. An early clinical trial (37) showed less vertebral deformity in women taking conjugated estrogens than in those on placebo, but most support for an antifracture effect of estrogen is epidemiologic and suggests a reduced risk of fracture of about 50–60% (38–42). Realization of this benefit requires sustained hormone administration for 5 years or more (39). Framingham study results (41) indicate that 7 years of estrogen is required for fracture protection and suggest that even this exposure does not protect bone in women beyond 75 years, although Kanis *et al.* (40) reported a 30% reduction in hip fracture risk for estrogen users beyond 80 years of age. The importance of sustained therapy is reinforced by the report of Cauley *et al.* (42) that current users had a 50% reduction in fracture, but women who stopped estrogen had no fracture protection, even if they had previously taken it for more than 10 years.

It is generally assumed that fracture protection directly reflects preservation of BMD. While this assumption is compatible with the evidence, other factors should be considered since estrogen users show a reduced fracture risk even after data are adjusted for bone mass (42). Subjects in observational cohort studies do not randomly decide to take estrogen, and pretreatment characteristics of these women might themselves underlie a lower fracture risk. Women choosing to take estrogen may already exhibit higher than average commitment to healthful behaviors that reduce fracture risk. The possibility that long-term estrogen may reduce the risk of falls also requires investigation.

VII. CLINICAL ISSUES REGARDING THE SKELETAL ASPECTS OF ESTROGEN REPLACEMENT

Uncertainty persists regarding the optimal type, dose, and mode of administration of estrogen; about the selection of women to offer estrogen; and about the consequence of adding progestins to the treatment regimen. Space does not permit comprehensive discussion of these issues, but a few summary points can be made.

Little evidence points toward accelerated bone loss in perimenopausal women while circulating FSH values remain low, so commencing estrogen in early perimenopausal years should not be beneficial. In the United States, most estrogen-treated women during the past four decades received a single drug, conjugated equine estrogens (Premarin), at a single dose (0.625 mg/day). Fracture data thus heavily reflect this regimen, but use of other forms of estrogen in Europe and the United Kingdom show that most prescribed estrogens offer skeletal benefit. Doses sufficient to elevate plasma 17β-estradiol concentrations above 70 pg/ml suppress bone remodeling and protect bone mass at the spine. Hip protection may require higher doses. A dose of 0.625 mg of conjugated estrogens protects spine and hip BMD in at least 90% of women who adhere to treatment (36). Delivery route is not critical for skeletal response. Transdermal 17β-estradiol acts similarly to oral hormone (43). Skeletal protection by estrogen requires continued therapy. Termination of hormone results immediately in increased bone turnover and bone loss (44).

Equivalent favorable responses to oral and other forms of estrogen means physicians can choose estrogen preparations on other considerations. Beneficial changes in lipoproteins are observed primarily with oral estrogen, so women in whom such changes are important (i.e., those at high risk for cardiovascular disease with unfavorable lipoprotein profiles) may particularly benefit from oral estrogen. By contrast, some women experience headaches or other systemic side effects with one particular form of estrogen but tolerate others. Physicians can take comfort in the knowledge that, at least with respect to the skeleton, the method of delivery is not a matter of consequence.

A. The Effect of Progestins

Until recently, estrogen prescription in the United States was primarily continuous and unopposed. Multiple schedules for interposing cyclic or continuous progestins are now available for uterine protection. No interaction of any of three progestin regimens on the skeletal effects of unopposed estrogen were observed in one trial (36). These included cyclic medroxyprogesterone acetate (MPA), 10 mg/day for 12 days each month; continuous MPA, 2.5 mg/day; or micronized progesterone, 200 mg/day for 12 days

each month. Progestins other than MPA, such as norethindrone, interact with the testosterone receptor and may be anabolic on muscle and bone (45). Interest has been expressed in progesterone as sole therapy, without estrogen, but postmenopausal women treated with 20 mg/day MPA lost bone from all sites (46). No convincing evidence for an osteotropic effect of nonandrogenic progestins has been presented. The use of long-acting MPA for contraception is of great concern since this agent creates an estrogen-deprived state, in which bone loss is predictable.

B. Selective Estrogen Response Modifiers

Tamoxifen, an estrogen agonist on bone, maintains BMD in breast cancer survivors who are not candidates for replacement estrogen (47–49). Other compounds are under development which mimic 17β-estradiol on bone and on lipoproteins but do not adversely affect endometrium or breast. Raloxifene is such an agent (50). It affords skeletal protection, does not stimulate uterine growth, and antagonizes estrogen action on the breast. Clinical trials of raloxifene in postmenopausal women are currently in progress. This or similar drugs may prove tolerable and effective for many women who are unable or unwilling to take long-term estrogen.

See Also the Following Articles

Estrogen Action, Breast; Estrogen Replacement Therapy; Menopause; Progestins

Bibliography

1. Turner, R. T., Riggs, B. L., and Spelsberg, T. C. (1994). Skeletal effects of estrogen. *Endocr. Rev.* 15 275–300.
2. Komm, B. S., Terpening, C. M., Benz, D. J., Graeme, K. A., O'Malley, B. W., and Haussler, M. R. (1988). Estrogen binding receptor mRNA and biologic response in osteoblast-like osteosarcoma cells. *Science* 241, 81–84.
3. Eriksen, E. F., Colvard, D. S., Berg, N. J., Graham, M. L., Mann, K. G., Spelsberg, T. C., and Riggs, B. L. (1988). Evidence of estrogen receptors in normal human osteoblast-like cells. *Science* 241, 84–86.
4. Ernst, M., Heath, J. K., and Rodan, G. A. (1989). Estradiol effects on proliferation, messenger ribonucleic acid for collagen and insulinlike growth factor-I, and parathyroid

hormone-stimulated adenylate cyclase activity in osteoblastic cells from calvarieae and long bones. *Endocrinology* **125**, 825–833.

5. Majeska, R. J., Ryaby, J. T., and Einhorn, T. A. (1994). Direct modulation of osteoblastic activity with estrogen. *J. Bone Joint Surg.* **76**, 713–721.

6. Scheven, B. A., Damen, C. A., Hamilton, N. J., Verhaar, H. J., and Duursma, S. A. (1992). Stimulatory effects of estrogen and progesterone on proliferation and differentiation of normal human osteoblast-like cells in vitro. *Biochem. Biophys. Res. Commun.* **186**, 54–60.

7. Keeting, P. E., Scott, R. E., Colvard, D. S., Han, I. K., Spelsberg, T. C , and Rigs, B. L. (1991). Lack of a direct effect of estrogen on proliferation and differentiation of normal human osteoblast-like cells. *J. Bone Miner. Res.* **6**, 827–833.

8. Bodo, M., Venti Donti, G., Becchetti, E., Pezzetti, F., Paludetti, G., Donti, E., and Maurizi, M. (1991). Effects of steroid on human normal and otosclerotic osteoblastic cells: Influence on thymidine and leucine uptake and incorporation. *Cell. Mol. Biol.* **37**, 597–606.

9. Oursler, M. J., Osdoby, P., Patterson, J., Pyfferoen, J., Riggs, B. L., and Spelsberg, T. C. (1991). Avian osteoclasts as estrogen target cells. *Proc. Natl. Acad. Sci. USA* **88**, 6613–6617.

10. Smith, E. P., Boyd, J., Frank, G. R., Takahashi, H., Cohen, R. M., Specker, B., Williams, T. C., Lubahn, D. B., and Korach, K. S. (1994). Estrogen resistance caused by a mutation in the estrogen-receptor in a man. *N. Engl. J. Med.* **331**, 1056–1061.

11. Carani, C., Qin, K., Simoni, M., Faustini-Fustini, M., Serpente, S., Boyd, J., Korach, K. S., and Simpson, E. R. (1997). Effect of testosterone and estradiol in a man with aromatase deficiency. *N. Engl. J. Med.* **337**, 91–95.

12. Marcus, R. (1987). Normal and abnormal bone remodeling in man. *Annu. Rev. Med.* **38**, 129–141.

13. Girasole, G., Jilka, R. L., Passeri, G., Boswell, S., Boder, G., Williams, D. C., and Manolagas, S. C. (1992). Marrow-derived stromal cells and osteoblasts in vitro: A potential mechanism for the antiosteoporotic effects of estrogens. *J. Clin. Invest.* **89**, 883–891.

14. Jilka, R. L., Hangoc, G., Girasole, G., Passeri, G., Williams, D. C., Abrams, J. S., Boyce, B., Broxmeyer, H., and Manolagas, S. C. (1992). Increased osteoclast development after estrogen loss: mediation by interleukin-6. *Science* **257**, 88–91.

15. Kassem, M., Harris, S. A., Spelsberg, T. C., and Riggs, B. L. (1996). Estrogen inhibits interleukin-6 production in a human osteoblastic cell line expressing high levels of estrogen receptors. *J. Bone Miner. Res.* **11**, 193–199.

16. Balena, R., Costasitini, F., Yamamoto, M., Markatos, A., Cortese, R., Rodan, G. A., and Poli, V. (1993). Mice with IL-6 gene knockout do not lose cancellous bone after ovariectomy. *J. Bone Miner. Res.* **8**, S-130. (Abstract)

17. Pacifici, R., Rifas, L., Teitelbaum, S., Slatopolsky, E., McCraken, R., Bergfeld, M., Lee, W., Avioli, L. V., and Peck, W. A. (1987). Spontaneous release of interleukin 1 from human blood monocytes reflects bone formation in idiopathic osteoporosis. *Proc. Natl. Acad. Sci. USA* **84** 4616–4620.

18. Pacifici, R., Rifas, L., McCracken, R., Vered, I., McCurry, C., Avioli, L. V., and Peck, W. A. (1989). Ovarian steroid treatment blocks a postmenopausal increase in blood monocyte interleukin 1 release. *Proc. Natl. Acad. Sci. USA* **86**, 2398–2402.

19. Oursler, M. J., Kassem, M., Turner, R., Riggs, B. L., and Spelsberg, T. C. (1996). Regulation of bone cell function by gonadal steroids. In *Osteoporosis* (R. Marcus, D. Feldman, and J. Kelsey, Eds.), pp. 237–260. Academic Press, San Diego.

20. Pacifici, R. (1996). Postmenopausal osteoporosis: How the hormonal changes of menopause cause bone loss. In *Osteoporosis* (R. Marcus, D. Feldman, and J. Kelsey, Eds.), pp. 647–659. Academic Press, San Diego.

21. Heaney, R. P., Recker, R. R., and Saville, P. D. (1977). Calcium balance and calcium requirements in middle-aged women. *Am. J. Clin. Nutr.*, 1603–1611.

22. Hui, S. L., Slemenda, C. W., and Johnston, C. C., Jr. (1989). Baseline measurement of bone mass predicts fracture in white women. *Ann. Intern. Med.* **111**, 355–361.

23. Cummings, S. R., Black, D. M., Nevitt, M. C., Browner, W. S., Cauley, J. A., Genant, H. K., Mascioli, S. R., Scott, J. C., Seeley, D. G., Steiger, P., and Vogt, T. (for the SOF Research Group) (1990). Appendicular bone density and age predict hip fracture in women. *J. Am. Med. Assoc.* **263**, 665–668.

24. Melton, L. J., III, Atkinson, E. J., O'Fallon, W. M., Wahner, H. W., and Riggs, B. L. (1993). Long-term fracture prediction by bone mineral assessed at different skeletal sites. *J. Bone Miner. Res.* **8**, 1227–1234.

25. Heaney, R. P., Recker, R. R., and Saville, P. D. (1978). Menopausal changes in bone remodeling. *J. Lab. Clin. Med.*, 964–970.

26. Blunt, B. A., Klauber, M. R., Barrett-Connor, E. L., and Edelstein, S. L. (1994). Sex differences in bone mineral density in 1653 men and women in the sixth through tenth decades of life: The Rancho Bernardo Study. *J. Bone Miner. Res.* **9**, 1333–1338.

27. Greenspan, S., Maitland, L. A., Myers, E. R., Krasnow, M. B., and Tamiko, H. K. (1994). Femoral bone loss progresses with age: A longitudinal study in women over age 65. *J. Bone Miner. Res.* **9**, 1959–1965.

28. Harris, S., Dallal, G. E., and Dawson-Hughes, B. (1992). Influence of body weight on rates of change in bone density of the spine, hip, and radius in postmenopausal women. *Calcif. Tissue Int.* **50**, 19–23.

29. Lindsay, R., Hart, D. M., Aitken, J. M., MacDonald, E. B., and Anderson, J. B. (1976). Long-term prevention of postmenopausal osteoporosis by oestrogen. Evidence for an increased bone mass after delayed onset of oestrogen treatment. *Lancet* **1**, 1038–1040.

30. Lindsay, R., Hart, D. M., Forrest, C., and Baird, C. (1980). Prevention of spinal osteoporosis in oophorectomised women. *Lancet* **2**, 1151–1153.

31. Recker, R. R., Saville, P. D., and Heaney, R. P. (1977). Effect of estrogens and calcium carbonate on bone loss in postmenopausal women. *Ann. Int. Med.* **87**, 649–655.

32. Horsman, A., Gallagher, J. C., Simpson, M., and Nordin, B. E. C. (1977). Prospective trial of oestrogen and calcium in postmenopausal women. *Br. Med. J.* **2**, 789–792.

33. Christansen, C., Christensen, M. S., McNair, P., Hagen, C., Stocklund, K., and Transbøl, I. (1980). Prevention

of early postmenopausal bone loss: Controlled 2-years study in 315 normal females. *Eur. J. Clin. Invest.* **10**, 273–279.

34. Lindsay, R., and Tohme, J. F. (1990). Estrogen treatment of patients with established postmenopausal osteoporosis. *Obstet. Gynecol.* **76**, 290–295.

35. Quigley, M. E. T., Martin, P. L., Burnier, A. M., and Brooks, P. (1987). Estrogen therapy arrests bone loss in elderly women. *Am. J. Obstet. Gynecol.* **156**, 1516–1523.

36. The Writing Group for the PEPI Trial (1996). Effects of hormone therapy on bone mineral density. Results from the Postmenopausal Estrogen/Progestin Interventions (PEPI) Trial. *J. Am. Med. Assoc.* **276**, 1389–1396.

37. Nachtigall, L. E., Nachtigall, R. H., Nachtigall, R. D., and Beckmann, E. M. (1979). Estrogen replacement therapy I: A 10-year prospective study in the relationship to osteoporosis. *Obstet. Gynecol.* **53**, 277–281.

38. Hutchinson, T. A., Polansky, S. M., and Feinstein, A. R. (1979). Post-menopausal oestrogens protect against fractures of hip and distal radius. A case-control study. *Lancet* **2**, 705–709.

39. Weiss, N. S., Ure, C. L., Ballard, J. H., Williams, A. R., and Daling, J. R. (1980). Decreased risk of fractures of the hip and lower forearm with postmenopausal use of estrogen. *N. Engl. J. Med.* **303**, 1195–1198.

40. Kanis, J. A., Johnell, O., Gullberg, B., Allander, E., Dilsen, G., and Gennari, C. (1992). Evidence for efficacy of drugs affecting bone metabolism in preventing hip fracture. *Br. Med. J.* **305**, 1124–1128.

41. Felson, D. T., Zhang, Y., Hannon, M. T., Kiel, D. P., Wilson, P. W., and Anderson, J. J. (1993). The effect of postmenopausal estrogen therapy on bone density in elderly women. *N. Engl. J. Med.* **329**, 1141–1146.

42. Cauley, J. A., Seeley, D. G., Enstrud, K., Ettinger, B., Black, D., and Cummings, S. R. (for the Study of Osteoporotic Fractures Research Group) (1995). Estrogen replacement therapy and fractures in older women. *Ann. Int. Med.* **122**, 9–16.

43. Stevenson, J. C., Cust, M. P., Gangar, K. F., Hillard, T. C., Lees, B., and Whitehead, M. I. (1990). Effects of transdermal versus oral hormone replacement therapy on bone density in spine and proximal femur in postmenopausal women. *Lancet* **2**, 265–269.

44. Christiansen, C., Christensen, M. S., and Transbøl, I. (1981). Bone mass in postmenopausal women after withdrawal of oestrogen/gestagen replacement therapy. *Lancet* **1**, 459–461.

45. Christiansen, C., and Riis, B. J. (1990). 17β estradiol and continuous norethisterone: A unique treatment for established osteoporosis in elderly women. *J. Clin. Endocrinol. Metab.* **71**, 836–841.

46. Gallagher, J. C., Kable, W. T., and Goldgar, D. (1991). Effect of progestin therapy on cortical and trabecular bone: Comparison with estrogen. *Am. J. Med.* **90**, 171–178.

47. Gotfredsen, A., Christiansen, C., and Palshof, T. (1984). The effect of tamoxifen on bone mineral content in premenopausal women with breast cancer. *Cancer* **53**, 853–857.

48. Love, R., Mazess, R., Tormey, D., Barden, H., Newcomb, P., and Jordan, V. (1988). Bone mineral density in women with breast cancer treated with adjuvant tamoxifen for at least two years. *Breast Cancer Res. Treat.* **12**, 297–301.

49. Turken, S., Siris, E., Seldin, D., Flaster, E., Hyman, G., and Lindsay, R. (1989). Effects of tamoxifen on spinal bone density in women with breast cancer. *J. Natl. Cancer Inst.* **81**, 1086–1088.

50. Black, L. J., Sato, M., and Rowley, E. R. (1994). Raloxifene (LY139481HCL) prevents bone loss and reduces serum cholesterol without causing uterine hypertrophy in ovariectomized rats. *J. Clin. Invest.* **93**, 62–69.

Estrogen Action, Breast

Serdar E. Bulun and Evan R. Simpson

University of Texas Southwestern Medical Center

GLOSSARY

breast ducts Mainframe tubular structures of the glandular component of the breast tissue.

coactivators Adaptor proteins that bridge between nuclear receptors and general transcription factors to enhance transcription.

corepressors Adaptor proteins that bridge between nuclear receptors and general transcription factors to inhibit transcription.

end or *lateral buds* Dense bulbous ends where breast growth and differentiation take place.

hormone agonists Analogs that bind to receptors and elicit the same biological response as the natural hormone.

hormone antagonists Molecules that bind to receptors but fail to elicit the normal biological response.

lobuloalveolar units Collections of ductules and differentiated apocrine glands that secrete milk.

nuclear receptors Ligand-dependent transcription factors that regulate the expression of target genes via binding to specific response elements in genomic DNA.

signal transduction A cellular system whereby a hormonal signal gives rise by a cascade of molecular interactions to gene transcription.

I. INTRODUCTION

Estrogen action on the breast has been one of the most exciting and rewarding areas of research in medical science. The recognition of the essential role of estrogen in the development and growth of the breast glandular and stromal tissues, the widespread use of estrogen antagonists in the treatment of breast cancer (the most common female cancer), and the controversy regarding the association between postmenopausal hormone replacement and breast cancer make this subject particularly relevant to the research scientist and clinician. Although a number of natural steroids may have estrogenic activity, only estradiol-17β has been demonstrated to act as a potent estrogen both *in vivo* and *in vitro*. For example, the C_{18} steroid, estrone, is of itself only weakly estrogenic and must therefore be converted by 17β-hydroxysteroid dehydrogenase to estradiol-17β for full estrogenic action. Therefore, for practical purposes, the term estrogen in this chapter will refer to the specific steroid, estradiol-17β.

II. THE ROLE OF ESTROGEN IN BREAST DEVELOPMENT

A. Stages of Breast Development

The patterns of growth, differentiation, and the generation of various morphological structures of the breast are similar in mammals. Consequently, current knowledge on the developmental biology of human mammary glands derives primarily from experiments using rodents (Shyamala, 1997). In both species, glandular structures composed of mainly epithelial cells grow to invade a pad of adipose tissue in a controlled fashion. This occurs at various developmental stages in a discontinuous fashion. At birth, the mammary glands are rudimentary structures with simple ducts composed of epithelial cells which remain quiescent until puberty (Fig. 1). The major influence on breast growth at puberty is estrogen. In

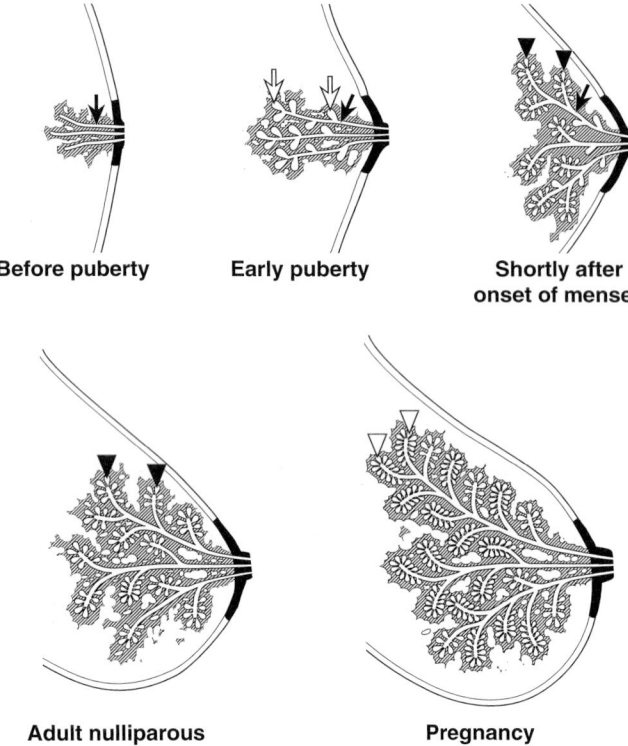

Before puberty   **Early puberty**   **Shortly after onset of menses**

Adult nulliparous woman   **Pregnancy**

FIGURE 1 Breast development. The breasts of both sexes follow a similar course of development until puberty. In girls in early puberty, the ducts grow (black arrows) and divide in a dichotomous and sympodial fashion under the influence of estrogen. Ductal growth occurs through increased mitotic activity at terminal and lateral buds (open arrows). After the first ovulatory cycle, lobular structures composed of alveolar buds appear which differentiate into lobuloalveolar units (black arrowheads). The number of these units increases with each ovulatory cycle, and in the adult woman's breast, many lobuloalveolar units are recognized. During pregnancy, the breast attains its maximal differentiation and growth. The most differentiated lobule type (open arrowheads) is observed at the end of pregnancy or during the lactational period.

most girls, the first response to the increasing levels of circulating estrogen is an increase in size and pigmentation of the areola, the formation of a mass of breast tissue just underneath the areola, and an increase in the size of the adipose stroma. Estrogen also stimulates the growth of the glandular epithelial cells to form a tree-like pattern of ducts originating from the nipple and terminating with dense bulbous structures, which are called end buds when located at the tips of ducts or lateral buds when located on the sides and spaced regularly along the ducts (Fig. 1). In the growing ducts, the site of intensive mitotic activity resides in the distal portions of the end buds that invade the fat pad with branching at regular intervals. The resulting growth is due to progressive addition of epithelial cells produced by these terminal structures. A minimal zone of fat pad, however, remains between the ducts. As such, ductal growth stops as the end bulbs approach the periphery of the fat pad or are confined by the lateral ducts. At this time, the end bulbs disappear, and the subtending ducts become mitotically quiescent. It should be pointed out again that estrogen not only stimulates glandular growth but also gives rise to the growth of the adipose stroma of the breast (Shyamala and Ferenczy, 1984). In the adult female, with each ovarian cycle the lateral buds differentiate to give rise to small alveolar buds primarily under the influence of progesterone (Fig. 1). The critical function of estrogen at this stage is believed to be the induction of progesterone receptors (PR). Although differentiation of lateral buds to lobuloalveolar structures takes place in the nonpregnant adult, from a proliferative standpoint, the glands remain quiescent until the onset of pregnancy. At the onset of pregnancy, the epithelial cell replication begins once again. This results in numerous outgrowths from the lateral and terminal buds. Branches arising from the lateral walls of the ducts and lobuloalveolar structures are indicative of overt differentiation that begins to fill the interductal spaces. During pregnancy, circulating prolactin levels as well as estrogen and progesterone levels rise to extremely high concentrations. Although the principal hormone involved in milk biosynthesis is prolactin, lactation is inhibited by the large amounts of circulating progesterone and possibly estrogen during pregnancy by interfering with prolactin action at the alveolar cell prolactin receptor level. At birth, after the removal of inhibition on prolactin action by progesterone and estrogen, the mammary epithelial cells achieve full functional differentiation characterized by secretion of milk under the influence of prolactin that is episodically secreted by suckling. Following the cessation of lactation, the glands undergo involution and become quiescent again until the onset of another pregnancy.

B. How Does Estrogen Stimulate Breast Development?

It is now well established that almost all aspects of mammary biology are under hormonal regulation. The primary effect of estrogen in rodents is to stimulate growth of the ductal portion of the gland system and the mammary adipose tissue at puberty (Shyamala, 1997). Progesterone in these animals influences growth of the lobuloalveolar components of the breast gland during adult life and pregnancy. Neither hormone, however, alone or in combination, is capable of yielding optimal breast growth and development. It should be pointed out that full differentiation of the gland requires insulin, cortisol, thyroxin, prolactin, and growth hormone. Circumstantial and experimental data recently clarified the relative roles of estrogen and progesterone in the breast growth and development at various stages. During puberty, the major direct effect of estrogen is the development and growth of breast ducts and surrounding adipose stroma. Estrogen action alone, however, is not sufficient for the development of alveolar structures. In the adult, on the other hand, it was proposed that estrogen action on the breast is necessary primarily to stimulate the expression of the PR, which is essential for lobuloalveolar differentiation at various stages. The following pieces of evidence support these conclusions: First, although their levels fluctuate, estrogen receptors (ER) are present in the mammary glands of rodents from all developmental stages, whereas PR expression varies with the developmental stages more distinctively such that the PR is present in adult nonpregnant females, but levels of PR begin to diminish during mid- to late pregnancy and are undetectable during established lactation. Moreover, in all developmental states, there is a positive correlation between the proliferative potential of the mammary epithelial cells and the presence of PR but not ER. Second, in oophorectomized mice, estrogen increases mitotic activity in the end buds to give rise to linear growth of branching ducts. On the other hand, the growth elicited in response to both estrogen and progesterone in these mice is characterized by outgrowth from terminal buds and lateral walls of the ducts that eventually results in the development of mammary glands, analogous to that seen during pregnancy, characterized by full lobuloalveolar development. Third, as a consequence of estrogen action, PRs are expressed at higher levels in the terminal and lateral buds of the growing ducts within a subpopulation of ductal epithelial cells that are selectively responsive to estrogen. These cells, as a consequence of estrogen action, develop PR and undergo proliferation and lobuloalveolar differentiation during pregnancy; that is, among the ductal epithelial cells, there is heterogeneity with regard to estrogen responsiveness, the presence of PR, and progesterone responsiveness. Finally, the absolute requirement for the PR to elicit lobuloalveolar growth is revealed from studies on transgenic mice that lack PR. In these PR knockout mice, there is a complete absence of interductal lobuloalveolar structures despite the administration of estrogen and progesterone at levels sufficient to cause full lobuloalveolar development in wild-type mice. In fact, the presence of only ductal but no lobuloalveolar development in the PR knockout mice is very similar to early pubertal breast development in humans before cyclic ovulatory periods begin (Fig. 1). These data from mice should be applied to the human breast with caution since little is known about expression of steroid receptors in the disease-free human breast. It appears that ER transcript levels in the human breast may be under developmental regulation and may be inversely related to the circulating estrogen levels (Boyd *et al.*, 1996).

III. SIGNAL TRANSDUCTION ASSOCIATED WITH ESTROGEN ACTION IN THE BREAST

A. Estrogen Receptors

Estrogen action in the breast and other tissues are mediated through the ER, a member of a large superfamily of nuclear receptors that function as ligand-activated transcription factors (Katzenellenbogen, 1996). These receptor proteins share a common structural and functional organization, with distinct areas that are responsible for ligand binding, DNA binding, and transcriptional activation. It should be pointed out that a novel ER complementary (c) DNA was cloned from the rat prostate which was named ERb subtype to distinguish it from the previously cloned ER cDNA (consequently ERα subtype) (Kuiper *et al.*, 1997). Although the rat ERβ protein is highly homologous to rat ERα protein, the role of ERβ in the human breast development and growth is not currently known. Therefore, in this section, the term ER will refer to only the classical ERα. Two highly conserved regions are observed in the ER and other nuclear receptors: one in approximately the middle of the protein (known as domain C), which is involved in interaction with DNA, and one in the carboxy-terminal region (known as domain E/F) that binds hormones and is structurally and functionally complex (Katzenellenbogen, 1996). This ligand-binding region contains many other functions, such as heat shock protein association, dimerization, nuclear localization, and hormone-dependent transactivation (AF-2). The amino-terminal region (domain A/B), on the other hand, contains a ligand-independent activation function (AF-1).

B. Estrogen-Dependent Transcriptional Activation

Estradiol-17β is transported in the circulation bound to the protein sex hormone binding globulin, with relatively high affinity. Estrogen can freely diffuse across the plasma and nuclear membranes of all cells but is sequestered only within cells that contain ERs. The ER, in the absence of its ligand, is found to be complexed with heat shock proteins within the nucleus. Binding of estrogen transforms the ER into its active form by mediating the disassociation of heat shock proteins and allowing dimerization of receptor proteins (Shibata *et al.*, 1997) The activated ER is now able to bind to estrogen response element DNA, often located in the 5′ flanking region of estrogen-responsive genes. These DNA sequences function as enhancers, conferring estrogen inducibility on the genes.

It is generally assumed that transactivators, e.g., estrogen-occupied ER homodimer complex, stimulate gene expression by facilitating the assembly of basal transcription factors into a stable preinitiation complex, thereby increasing the transcription rate and messenger (m) RNA synthesis (Fig. 2). For ex-

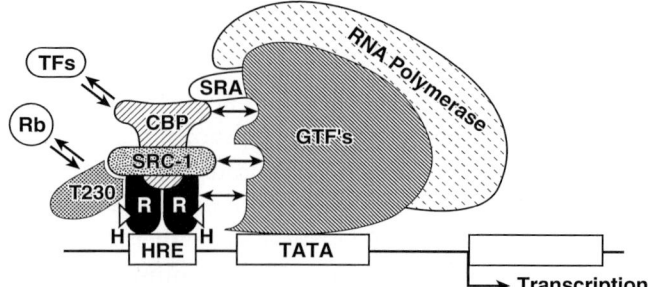

FIGURE 2 A hypothetical model for estrogen receptor-dependent regulation of transcription. This complex model involves the coactivators, SRC-1 and CBP, as the basic bridging proteins that enhance protein–protein interactions between the estrogen (H) occupied ER homodimer and the general transcription factors (GTFs). As a result, the RNA polymerase complex becomes activated to initiate transcription. This model also indicates that other transactivators (TFs) enhance transcription of the target gene in a similar way. SRA and T230 are other recently identified coactivators. Additionally, retinoblastoma protein (Rb) can interact with T230 and thus antagonize its action. Hypothetically, if these coactivators were replaced by corepressors depending on the ligand and intracellular presence of coregulators, this would inhibit transcription (copyright 1997 by The Endocrine Society).

ample, ER was shown to interact with TFIIB (Shibata *et al.*, 1997). Although ER and other nuclear receptors can directly interact with components of the general transcriptional machinery, other factors such as coactivators and corepressors are also thought to participate in this process. It is now known that coactivators are adaptor proteins that function as bridging factors between specific activators and the general transcriptional machinery and are presumably required for efficient transcription (Fig. 2). In fact, the tissue-specific action of estrogen is presumed to be regulated by the presence and absence of these coregulators in that particular tissue (Shibata *et al.*, 1997). An extensively studied coactivator is steroid receptor coactivator-1 (SRC-1). SRC-1 enhances transcriptional activities of ER as well as PR, glucocorticoid receptor, thyroid hormone receptor, and retinoic acid receptors through their cognate response elements (Fig. 2). Recently, other coactivators that are homologous to SRC-1 were isolated suggestive of the presence of a gene family of coactivators. One of these, ERAP160 (ER-associated protein 160), in fact, was found to be a product of the *SRC-1* gene as a result of an alternative splicing event. ERAP140 (ER-associated protein 140), on the other hand, is encoded by a different gene (Shibata *et al.*, 1997).

CBP [CREB (cyclic AMP response element binding protein) binding protein] is a ubiquitous, evolutionarily conserved nuclear phosphoprotein that integrates several different signal-responsive transcriptional activators to the basal transcriptional apparatus (Shibata *et al.*, 1997). Although CBP was originally identified as a coactivator of CREB, it was later shown that CBP and SRC-1 interact and synergize with each other to enhance ER-dependent gene expression (Fig. 2). In this context, CBP/coactivator complexes may be viewed as "integrators" based on their role in determining the relative transcriptional responses of a specific target gene in the face of activation of multiple signaling pathways. Although evidence for the presence and function of corepressors is accumulating, the role of corepressors in modulating estrogen action is not as well understood as in the case of coactivators.

In addition, other cell signaling pathways also modulate the transcriptional activity of the ER (Katzenellenbogen, 1996). For example, in MCF-7 human breast cancer cells, inducers of protein kinase A (PKA) and protein kinase C (PKC) markedly synergize with estradiol-17β in enhancing ER-dependent transcriptional activation. This synergistic stimulation did not appear to result from changes in ER content or in the binding affinity of the ER for ligand or DNA but rather may be a consequence of a stabilization or facilitation of interaction with target components of transcriptional machinery, possibly through phosphorylation of the ER or other proteins, e.g., coregulators. Also of interest, stimulation of PKA activates the agonistic activity of tamoxifen-like but not ICI164,384-like antiestrogens and reduces the effectiveness of tamoxifen as an estrogen antagonist.

C. Role of Coregulators in Determining Tissue-Specific Action of Estrogen Agonists and Antagonists

Although both estrogens and antiestrogens bind within the hormone binding domain of the ER, the association must differ because estrogen binding activates transcription, whereas antiestrogens fully or partially fail in this role (Katzenellenbogen, 1996). Antiestrogens are believed to act in large measure by competing for binding to the ER and altering the conformation of the ER such that the receptor fails to effectively activate gene transcription. Models of antiestrogen action at the molecular level are beginning to emerge, and recent data indicate that antiestrogens fall into at least two distinct categories: antiestrogens such as tamoxifen that are mixed or partial agonists/antagonists and compounds such as ICI164,384 that are pure antagonists. The partial agonist/antagonist compounds form complexes with the ER that bind as dimers to estrogen-response elements; there, they block ligand-dependent transcription activation (AF-2) mediated by region E of the receptor, but they are believed to have little or no effect on the ligand-independent transcription activation function (AF-1) located in region A/B of the receptor. Thus, they are generally partial or mixed agonists/antagonists, and their action must involve some subtle difference in ligand–receptor interaction, very likely associated with the basic or polar side chain that characterizes the antagonist members of this class. In the case of the more complete antagonists such as ICI164,384, ER conformation must clearly differ from that of the estradiol-17β-occupied ER because some differences in ER binding to DNA and reduction of the ER content of target cells appear to contribute to but may not fully explain the pure antagonist character of this antiestrogen.

Based on the evidence summarized previously, the following model was proposed, which hypothesizes that the function of estrogen agonists or antagonists may be determined by changes in receptor conformation and by cellular concentration of coactivators and corepressors (Fig. 3) (Shibata *et al.*, 1997). When bound by an agonist such as estradiol-17β, the receptors are activated and change conformation, resulting in the dissociation of heat shock proteins and enhancement of ligand-inducible phosphorylation. Subsequently, the receptors dimerize, bind to the estrogen response element of a target gene, and recruit nuclear coactivators that, either directly or indirectly, bridge receptors and general transcription factors. On the other hand, when bound by a complete antagonist such as ICI164,384, the change in receptor conformation is different from that of the agonist-bound ER. This causes binding of the ER to the

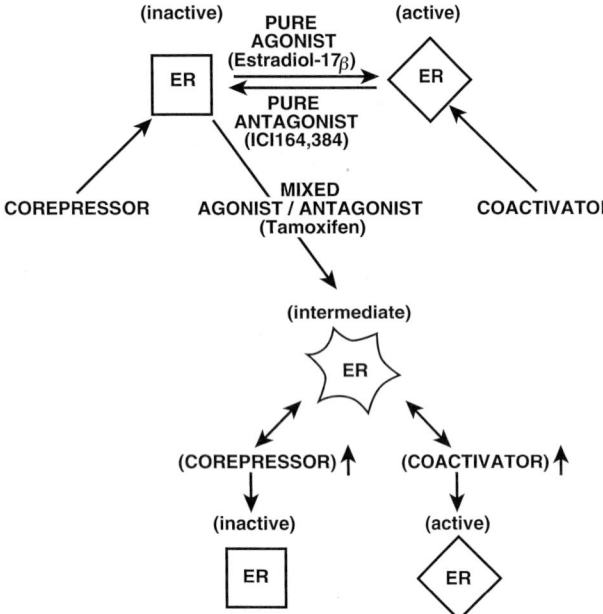

FIGURE 3 A hypothetical model that explains the mixed agonistic/antagonistic properties of tamoxifen. This figure indicates that receptor conformation is dependent on which ligand binds to the receptor (ER). The ER conformation induced by pure agonists (e.g., estradiol-17β) can make it interact with coactivators, whereas conformation induced by pure antagonists (e.g., ICI164,384) results in interaction with corepressors. For mixed agonists/antagonists (e.g., tamoxifen) the conformation varies depending on the relative ratio of coactivator and corepressor within the cell. When the intracellular corepressor is high (e.g., breast glandular epithelium), ER will be inactivated. In contrast, when the intracellular coactivator is high (e.g., endometrial glandular epithelium), the ER will shift to bind to the coactivator, resulting in an active receptor (copyright 1997 by The Endocrine Society).

estrogen response element as a nonproductive homodimer and the recruitment of corepressors or the interference with coactivator binding, resulting in the inhibition of gene transcription. Based on this model, agonists and antagonists can differentially modulate receptor conformation and recruit different cellular coactivators or corepressors to regulate gene transcription in either a positive or negative manner, analogous to an accelerator and a brake. As illustrated in Fig. 3, the function of mixed agonists/antagonists such as tamoxifen may also be explained. Tamoxifen has been used as a standard therapy for selected groups of women with breast cancer. Large randomized trials have confirmed the benefit of tamoxifen treatment in extending both the disease-free and overall survival rate of patients when treated for a period up to 5 years. Recent reports by the National Cancer Institute, however, indicated that patients receiving tamoxifen for 10 years or more experienced a shorter disease-free survival compared with women given the drug for only 5 years. Moreover, tamoxifen use was associated with development of endometrial hyperplasia. These clinical data suggest the possibility that tamoxifen is converted from an antagonist to an agonist in the breast upon long-term treatment, and that it acts as an agonist in a different tissue such as the endometrium. As discussed earlier, antiestrogens may act by interfering with coactivator binding to the liganded ER by recruiting cellular corepressors, thereby inhibiting ER function. Since the ER occupied by a mixed agonist/antagonist such as tamoxifen has an intermediate conformation (intermediate to active or inactive), it is possible that the cellular content of these cofactors determines whether tamoxifen will serve to activate or inhibit ER function (Fig. 3). Therefore, *in vivo* actions of mixed agonists/antagonists may produce unwarranted physiological and clinical complications, depending on the cell type and the intracellular concentration of coactivators and corepressors. Furthermore, the agonistic actions of mixed agonists/antagonists may be mediated through the ligand-independent AF-1 domain. Additionally, mutations of ER during long-term treatment may also cause a switch whereby cells begin to respond to tamoxifen as an agonist. Finally, the complement of genes that are expressed in a given cell type is determined at the time of cellular differentiation. It is apparent that such expressible genes exist in so-called "DNAse I-hypersensitive" regions, i.e., they reside in regions of the chromatin that are more readily digested *in vitro* by DNAse I than is the bulk of chromosomal DNA. The DNAse I-hypersensitive regions define the structural framework in a given cell for the genes available for expression. Taken together, cell type and promoter context of a gene determine whether tamoxifen acts as an agonist or antagonist to transcription in a given target tissue.

IV. ESTROGEN ACTION AND BREAST CANCER

Although there is considerable circumstantial and laboratory evidence that is suggestive of a link between ovarian steroids and breast cancer, it is still not known whether estrogen causes development of this neoplasm. Another controversial issue is whether the risk of breast cancer increases because of postmenopausal hormone replacement. Identification of the target genes of estrogen in disease-free and malignant breast tissues and characterization of the molecular determinants of cell-specific estrogen action will give rise to a better understanding of possible carcinogenic properties of estrogen. On the other hand, sufficiently large randomized trials are required to address the risk of developing clinically recognizable breast cancer associated with postmenopausal hormone replacement.

A. Does Estrogen Cause Breast Cancer?

Epidemiological studies have linked nulliparity, early menarche, and late menopause to a higher risk for developing breast cancer presumably as a result of prolonged exposure to estrogen. Moreover, treating rodents with pharmacological doses of estrogen gave rise to development of breast cancer. Despite these data, the carcinogenic effect on the breast of higher than average quantities of estrogen exposure has been challenged because no differences in urinary estrogen levels were identified between women with breast cancer and those in the normal population. Another group of investigators, however, found higher free estrogen concentrations in serum due to low steroid hormone-binding globulin levels in women with breast cancer. Additionally, progesterone agonists were shown to increase the incidence of spontaneous mammary tumors in mice and dogs, and the trend toward increased breast cancer risk with increased duration of postmenopausal use of estrogen and progestin but not estrogen alone is suggestive of a more significant role of progestins in carcinogenesis (Bergkvist *et al.*, 1989; Horwitz, 1992). Although the role of estrogen in this instance

may be the induction of PR as discussed in Section I, estrogen action appears to be essential for the end result. Finally, studies of the expression of ER and steroidogenic enzymes in the breast tissues provided more evidence in support of the role of estrogen in carcinogenesis. For example, ER levels in the upper outer quadrant of cancer-bearing breasts were much higher compared with disease-free controls (Ricketts *et al.*, 1991). Several laboratories have also shown that the adipose tissue levels of aromatase (the enzyme that catalyzes the conversion of C_{19} steroids to estrogens) are highest in the cancer-bearing quadrant compared with breast tissue distant to the tumor (Bulun *et al.*, 1993), and in the case of disease-free breast, the highest aromatase expression was found in the upper and outer regions where the highest occurrence of cancer is ordinarily seen (Bulun *et al.*, 1996). More conclusive mechanistic studies are required to verify these observations of associative nature to establish exposure to excessive estrogen as a carcinogenic factor. There is very little doubt, however, regarding the role of estrogen in promoting the growth of already established breast tumors. Tamoxifen and aromatase inhibitors have long been used with considerable success in the treatment of hormone-responsive breast cancer.

B. The Risk of Breast Cancer and Postmenopausal Estrogen Replacement

It should be pointed out again that it is inappropriate and unjustifiable to assume that known actions of estrogen and progesterone on the uterus are applicable to breast tissues (Horwitz, 1992). For example, in contrast to endometrial cancer, adding a progestin to postmenopausal administration of conjugated estrogens is not only ineffective for protection against breast cancer but also may increase breast cancer risk slightly compared with the estrogen-only regimen (Bergkvist *et al.*, 1989). Recent case–control or cohort studies found either no risk or slightly increased risk of developing breast cancer associated with hormone replacement in general (Colditz *et al.*, 1995; Newcomb *et al.*, 1995). None of the published meta-analyses, however, demonstrated a significantly increased risk in ever users compared with never users of hormone replacement. Current use, however, may be associated with a small increased risk. Among the features associated with increased risk were a positive family history, treatment for more than 15 years, use of estradiol instead of conjugated estrogens, and combination therapy including a progestin (Bergkvist *et al.*, 1989). This slightly increased risk is clearly outweighed by the benefits of estrogen replacement on overall quality of life, bone metabolism, prevention of urogenital atrophy, and cardiovascular function. Preliminary information does not suggest a detrimental effect of hormone replacement in women with a previous diagnosis of breast cancer, but these reports include few women with limited follow-up data. Because all existing data for estrogen replacement in healthy and cancer patients are derived only from observational studies, reliable information on the balance of risks and benefits must await the results of the Women's Health Initiative, a large-scale randomized clinical trial.

C. Steroid Hormone Receptors in Breast Cancer

Measurements of the ER and the estrogen-induced PR are used by most clinicians as indicators of both overall prognosis and likelihood of response to endocrine therapy. As predicted, patients with ER+/PR+ tumors (approximately two-thirds of all cases) have the highest likelihood of response; conversely, those with ER−/PR− tumors do not respond to antiestrogen treatment well. Unfortunately, most patients treated successfully with endocrine therapy eventually develop endocrine-resistant disease recurrence. The main mechanism appears to involve selection of ER− cells among the heterogeneous population of tumor cells, thereby giving rise to loss of ER expression. The emergence of mutant ER variants (e.g., forms which are constitutively active) and changes in the cellular content of the nuclear receptor coregulators are also potential explanations for this frustrating phenomenon. Finally, hormonal manipulation (antiestrogens and aromatase inhibitors) is currently being tested in large clinical trials as a chemopreventive measure for the prophylactic use against breast cancer in healthy postmenopausal women.

Acknowledgments

Preparation of this chapter was supported in part by Grant No. CA67167 from the National Cancer Institute, Grant No. AG08174 from the National Institute on Aging, and Grant No. AIBS 256 from the U.S. Army Medical Research and Materiel Command

See Also the Following Articles

Breast Cancer; Estrogen Action, Bone; Estrogen Replacement Therapy

Bibliography

Bergkvist, L., Adami, H. O., Persson, I., Hoover, R., and Schairer, C. (1989). The risk of breast cancer after estrogen and estrogen-progestin replacement. *N. Engl. J. Med.* **321,** 293–297.

Boyd, M. T., Hildebrandt, R. H., and Bartow, S. A. (1996). Expression of the estrogen receptor gene in developing and adult human breast. *Breast Cancer Res. Treat.* **37,** 243–251.

Bulun, S. E., Price, T. M., Mahendroo, M. S., Aitken, J., and Simpson, E. R. (1993). A link between breast cancer and local estrogen biosynthesis suggested by quantification of breast adipose tissue aromatase cytochrome P450 transcripts using competitive polymerase chain reaction after reverse transcription. *J. Clin. Endocrinol. Metab.* **77,** 1622–1628.

Bulun, S. E., Sharda, G., Rink, J., Sharma, S., and Simpson, E. R. (1996). Distribution of aromatase P450 transcripts and adipose fibroblasts in the human breast. *J. Clin. Endocrinol. Metab.* **81,** 1273–1277.

Colditz, G. A., Hankinson, S. E., Hunter, D. J., *et al.* (1995). The use of estrogens and progestins and the risk of breast cancer in postmenopausal women. *N. Engl. J. Med.* **332,** 1589–1593.

Horwitz, K. B. (1992). The molecular biology of RU486. Is there a role for antiprogestins in the treatment of breast cancer? *Endocrinol. Rev.* **13,** 146–163.

Katzenellenbogen, B. S. (1996). Estrogen receptors: Bioactivities and interactions with cell signaling pathways. *Biol. Reprod.* **54,** 287–293.

Kuiper, G. G. J. M., Carlsson, B., Grandien, K., *et al.* (1997). Comparison of the ligand binding specificity and transcript tissue distribution of estrogen receptors α and β. *Endocrinology* **138,** 863–870.

Newcomb, P. A., Longnecker, M. P., Storer, B. E., *et al.* (1995). Long-term hormone replacement therapy and risk of breast cancer in postmenopausal women. *Am. J. Epidemiol.* **142**, 788–795.

Ricketts, D., Turnbull, L., Ryall, G., *et al.* (1991). Estrogen and progesterone receptors in the normal female breast. *Cancer Res.* **51**, 1817–1822.

Shibata, H., Spencer, T. E., Oñate, S. A., *et al.* (1997). Role of co-activators and co-repressors in the mechanism of steroid/thyroid receptor action. In *Recent Progress in Hormone Research* (P. M. Conn, Ed.), pp. 141–165. Endocrine Society Press, Bethesda, MD.

Shyamala, G. (1997). Roles of estrogen and progesterone in normal mammary gland development. Insights from progesterone receptor null mutant mice and *in situ* localization of receptor. *Trends Endocrinol. Metab.* **8**, 34–39.

Shyamala, G., and Ferenczy, A. (1984). Mammary fat pad may be a potential site for initiation of estrogen action in normal mouse mammary glands. *Endocrinology* **115**, 1078–1081.

■

Estrogen Action on the Female Reproductive Tract

Jonathan Lindzey and Kenneth S. Korach

National Institutes of Health

I. Mechanisms of Estrogen Action
II. ERα Gene and Protein
III. Ontogeny and Anatomy of the Female Reproductive Tract
IV. Estrogen Effects on Development of the Female Reproductive Tract
V. Estrogen Regulation of Physiology of the Female Reproductive Tract
VI. Conclusion

GLOSSARY

edema Swelling of tissues related to increased water content in cells and interstitial spaces.

ERKO Estrogen receptor knockout mouse; a mouse in which the gene coding for the estrogen receptor-α has been disrupted.

hyperplasia Mitotic increase in cell numbers typically measured by increases in DNA synthesis.

stroma The supporting framework of a tissue, including connective tissue, vasculature, and nerves.

transactivation The mechanisms by which nuclear transcription factors alter rates of gene transcription.

Under the influences of ovarian and placental hormones the oviduct, uterus, cervix, and vagina serve to transport ovulated eggs, promote fertilization, provide an environment for implantation of fertilized eggs and development of the fetus, and facilitate parturition. Both the normal development and normal physiological states of the reproductive tracts of adult females are heavily influenced by the effects of steroid hormones such as estrogens and progestins. Indeed, the coordinated effects of cyclical changes in ovarian estrogens and progestins are critical for proper regulation of uterine physiology. This review discusses the molecular mechanisms by which estrogens act on target tissues and the effects of ovarian and placental estrogens on the development and function of the adult female reproductive tract.

I. MECHANISMS OF ESTROGEN ACTION

The synthesis of estradiol (E_2) is accomplished through the aromatization of androgen precursors by P450-aromatase enzyme. The predominant ovarian estrogen, estradiol, is synthesized and secreted primarily by granulosa cells of the maturing follicles and the placenta during pregnancy. These estrogens are secreted into the systemic circulation and upon reaching estrogen target tissues can exert their effects through three possible pathways of action: (i) rapid membrane receptor-mediated events, (ii) a newly described nuclear ERβ receptor, and (iii) the classical ERα receptor. The presence of membrane-mediated E_2 effects in the uterus is controversial and appears to be more important in neural tissues. ERβ mRNA is present in large amounts in some tissues but does not appear to be expressed at high levels in the uterus of mice. Thus, most of the documented uterine effects of E_2 are probably mediated by nuclear ERα receptor and, for this reason, the following discussion focuses primarily on the structure and function of ERα.

Current dogma holds that E_2 moves passively from the circulation and interstitial spaces across cell membranes and binds to and activates nuclear estrogen receptor (ER). The activated ER–ligand complex then binds as a homodimer to specific DNA sequences termed estrogen response elements (EREs) within the regulatory regions of target genes. The ERE serves to bring the transactivation domains of the activated ER into a position where these domains interact with coactivators and the transcription initiation complex on the promoters of responsive genes. In this way, activated ligand–ERα complexes can either inhibit or stimulate rates of transcription of target genes.

II. ERα GENE AND PROTEIN

Specificity of steroid hormone action lies largely in (i) hormone-specific binding by the receptor, (ii) sequence-specific DNA binding exhibited by the different types of steroid receptors, (iii) cell- and tissue-specific patterns of steroid receptor expression, and (iv) control of access of steroid receptors to genes through differential organization of chromatin and association of nuclear factors such as coactivators in the many different target cells and tissues.

ERα is a member of the ligand-dependent family of nuclear transcription factors and shares strong homology with other steroid receptors and the thyroid hormone receptor. The ERα gene contains nine exons and the 65-kDa ER protein has been divided into six domains termed A–F (Fig. 1). Many of the features that confer hormone specificity are due to the structure–function relationships found within the different domains of the ERα protein. The activities and functions of these domains are described below.

The steroid-binding domain (SBD) of the ERα (domain E) consists of a sequence of approximately 250 amino acids that forms a hydrophobic pocket capable of stereospecific, high-affinity binding of E_2 ($k_d \sim$ 0.1 nM). Binding of estrogens to the ERα initiates the sequence of events outlined previously that result in transcriptional activation or suppression of target genes. Most evidence indicates that ligand-bound ERα acts as a dimer and that the dimerization func-

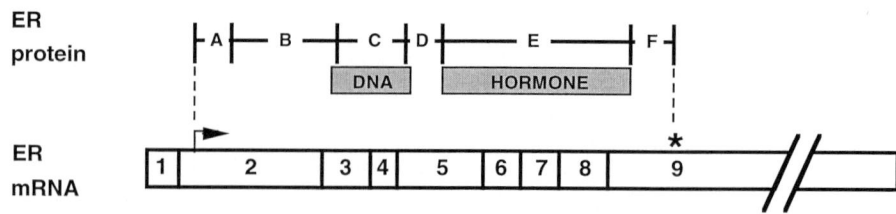

FIGURE 1 ERα protein and gene structure. Depicted are the different domains of the ERα protein and the exons that code for these domains. DNA- and steroid-binding domains are designated. Transactivation functions, AF-1 and AF-2, are localized within the B and E domains, respectively (from Lindzey and Korach, 1996).

tions reside in a small stretch of amino acids contained within the E domain. Mutations of amino acids in this hydrophobic stretch can eliminate dimerization, steroid binding, and, hence, transactivation.

The ERα is specifically targeted to the nucleus of a target cell by sequences of amino acids termed nuclear localization signals (NLSs). The NLSs of transcription factors tend to be basic and in the ER are located between amino acids 250 and 270, a region of domain D that is homologous to the NLSs of the glucocorticoid and progesterone receptors.

The DNA-binding domain (DBD) (domain C) consists of two zinc fingers that exhibit specific interactions with EREs. The first zinc finger dictates sequence-specific interactions with DNA, whereas the second appears to dictate the spacing requirements between the arms of the palindromic ERE. These fingers are critical for DNA binding but the surrounding amino acids may also influence binding. The DBD is highly conserved between different classes of steroid receptors and mutation of a few critical amino acids will alter the specificity of interactions with different steroid response elements.

The ERα; contains two regions that regulate transcriptional activity, AF-1 and AF-2, located in the A/B and E domains, respectively. The degree to which AF-1 and AF-2 act independently or additively appears to vary with the cell type and target gene. Experiments using AF-1- and AF-2-truncated ERα have revealed that growth factors and other second messenger-mediated peptides may stimulate phosphorylation of AF-1 and, consequently, activation of ER without the binding of E_2. It appears, however, that AF-2 is dependent primarily on E_2 binding. These data demonstrate that the ERα is actually a nuclear transcription factor that responds to both steroid and second messenger signaling pathways. This ability to integrate different signaling pathways provides an important link in the coordinated regulation of cellular physiology in estrogen target cells (see Section IV).

Although most E_2 effects on the female reproductive tract appear to be mediated by ERα, more detailed examinations of the newly described ERβ may demonstrate some ERβ-mediated effects on development or estrogen regulation of the female reproduc-

tive tract. All indications are that ERβ exerts its biological effects through similar molecular mechanisms to those of ERα. The ERβ contains the same modular domain structure and exhibits the highest degree of homology to ERα in the DBD and SBD. In keeping with this homology, ERβ exhibits very similar affinities and specificities of steroid binding as well as similar specificities for DNA interactions. Indeed, the two proteins are reported to form heterodimers, which raises the potential for complicated interactions in cells that coexpress the two receptor types. Future development of high-quality ERβ specific ligands and antibodies will allow us to determine what role if any ERβ plays in mediating E_2 effects on the uterus and other reproductive tract organs.

III. ONTOGENY AND ANATOMY OF THE FEMALE REPRODUCTIVE TRACT

Early in development the embryo possesses both Wolffian and Müllerian ducts which give rise to male and female accessory sex structures, respectively. At about 5 weeks gestation, the testes of a human male fetus produces Müllerian-inhibiting substance (MIS) which causes regression of the Müllerian ducts. However, in the female fetus, MIS is not produced and the Wolffian ducts regress, while the Müllerian ducts undergo morphogenesis to produce the oviduct, uterus, cervix, and parts of the vagina. The oviduct derives from cranial portions of the Müllerian ducts. Morphogenesis of the uterus involves an initial fusing of the middle portions of the Müllerian ducts that continues toward the posterior end of the Müllerian ducts. In humans, this fusion produces a single uterine compartment or lumen, whereas in mice fusion of the two ducts occurs at a lower level near the cervix and results in two distinct uterine horns. The upper vagina derives from the Müllerian ducts, whereas the lower vagina develops from components of the Müllerian duct and urogenital sinus (Fig. 2).

The oviduct consists of the isthmus (nearest the uterus), the ampulla, which is more central and somewhat dilated, and the infundibulum, which is funnel shaped with finger-like projections (fimbriae) that contact the ovary. The oviduct consists of three

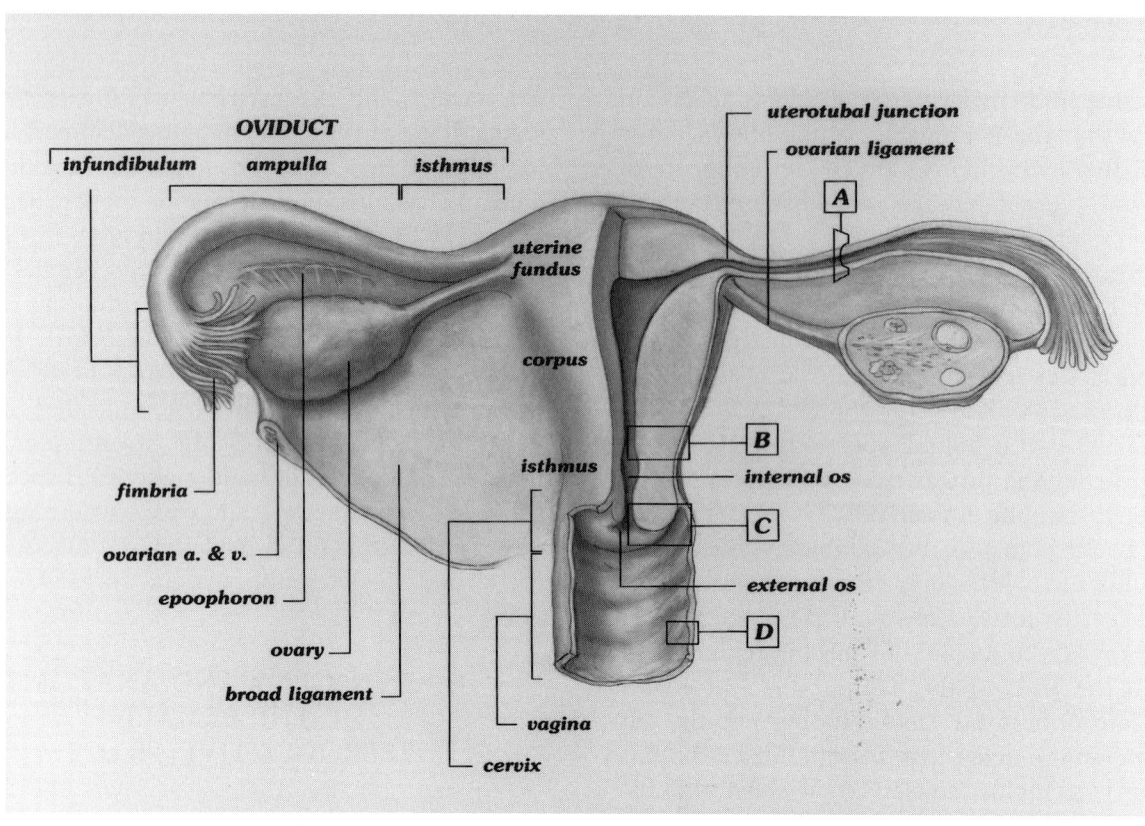

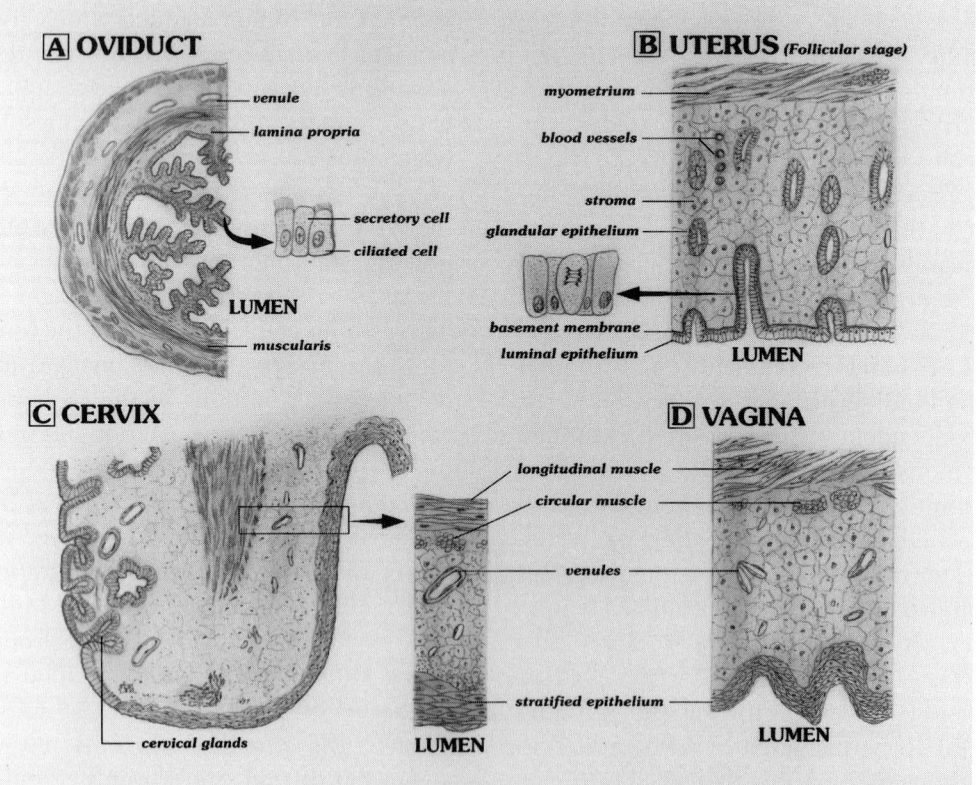

FIGURE 2 (Top) Gross morphology of the female reproductive tract. (Bottom) Histology of cross sections through the ampullary region of the oviduct (A) and longitudinal sections through the uterine lining (B), cervix (C), and vagina (D) (from Korach and Quarmby, 1985).

layers: an external serous, central muscularis, and inner mucosal layers. The mucosal layer forms folds and contains both secretory cells and ciliated cells located at the apices of the folds. Following ovulation, contractions of the muscularis layer and sweeping of the ciliated epithelial cells assist movement of the egg(s) down the oviduct toward the uterus.

The human uterus is divided into four regions running longitudinally from anterior to posterior: the fundus, corpus, isthmus, and cervix. There are two major tissue layers in the uterus: an outer muscular layer or myometrium, composed mostly of smooth muscle, and the inner layer or endometrium, which is composed of connective tissue (stromal) and an epithelial (mesenchyme) layer of cells (Fig. 2). These components of the uterus are surrounded by the peritoneum, which is a thin layer of connective tissue that surrounds the underlying myometrium and endometrium. The myometrium contains two concentric layers of smooth muscle with a thin outer layer that is continuous with the oviduct and is composed of muscle fibers oriented along the length of the uterus. The inner layer is thicker and consists of oblique spirals of fibers that form an interlocking mesh. The human endometrium contains both stromal and epithelial cells (Fig. 2) that can be divided into three layers: zona compacta, zona spongiosa, and zona basilis. The zona compacta contains both stromal and epithelial cells and forms the inner layer or lining of the uterus. The uterine lining is composed of a single layer of columnar epithelial cells which also forms extensions or invaginations into the underlying stroma. These invaginations form "endometrial glands" lined by epithelial cells. Thus, included within the zona compacta are the epithelial cells lining the lumen, the necks of the endometrial glands, and the surrounding stromal cells. Moving outward from the zona compacta, the zona spongiosa contains the main bodies of the glands and stromal cells. Together the zona compacta and spongiosa form the zona functionalis. The underlying zona basilis contains basal layers of stromal cells. While it appears that all the layers of the uterine endometrium can be regenerated from the zona basilis, the stromal and epithelial cells of the zona functionalis undergo the most dramatic morphological and physiological restructuring over the course of the ovarian cycle.

Hormonally induced changes in the myometrial and endometrial changes are involved in preparation of the uterus for implantation, gestation, and parturition.

The cervix is a cylindrical organ divided into upper or supravaginal, endocervical, and lower or ectocervical regions. It is composed of an outer fibromuscular layer containing smooth muscles and an inner epithelial layer. The muscle cells are continuous with the myometrial layer of the uterine corpus. The epithelial layer in the ectocervical region contains squamous epithelia, whereas the endocervical region is lined by columnar epithelium that is continuous with the epithelial layer of the vagina. The cervix performs or assists the following functions: synthesis and secretion of cervical mucous, transport and protection of sperm, retention of developing fetuses, and parturition and expulsion of menstrual debris.

The vagina transports sperm, promotes sperm coagulation, and functions as a canal for parturition and expulsion of menstrual debris. It is divided into upper, mid, and lower regions with the upper vagina connecting to the cervix at the fornix. The vagina is composed of an outer fascia, a central muscular layer, and an inner epithelial layer. The muscle layer merges with the cervical myometrial layer and is composed of outer and inner layers which contain muscle fibers arranged longitudinally or in a spiral configuration, respectively.

IV. ESTROGEN EFFECTS ON DEVELOPMENT OF THE FEMALE REPRODUCTIVE TRACT

The ability of E_2 to influence reproductive tract growth and physiology is due to the expression of ERα in the different tissue components of the oviduct, uterus, cervix, and vagina. Important sites of ERα expression include uterine arteries, myometrium, and stromal and epithelial cells of the oviduct and uterus. It is important to note that both the stromal and epithelial cells of the uterus express ERα both early in development and in adult animals. Therefore, both tissue types and cell types are subject to the effects of E_2 during development and adulthood. However, studies using a mouse in which the

ERα gene has been disrupted (ERKO) illustrate that, although the uteri are hypoplastic and reduced in size, the basic morphological development and differentiation of the different cell and tissue types occurs independent of the influence of the classical ERα signaling pathways. Furthermore, aromatase-deficient female mice and humans also develop smaller, hypoplastic uteri, further indicating that estradiol is not needed for the basic differentiation of these structures. Thus, differentiation of tissue and cell types in the uterus, cervix, and vagina occurs independently of ERα and E_2 effects. However, as described below, the influence of estrogen is required for full growth, differentiation, and function in the adult. In addition, it is highly likely that ERα mediates many of the carcinogenic and pathological effects that can result from exposure to abnormal levels of estrogens during critical periods of fetal development. For instance, exposure to diethylstilbestrol (DES) *in utero* can result in structural abnormalities in the uterus and vagina, vaginal adenosis, and vaginal adenocarcinomas in both mice and humans. In mice, neonatal DES exposure can also result in uterine adenocarcinomas in adult females.

V. ESTROGEN REGULATION OF PHYSIOLOGY OF THE FEMALE REPRODUCTIVE TRACT

Following menstruation, the ovary enters into the follicular phase of the ovarian cycle. In this phase, another round of ovarian follicles begin to mature and the levels of E_2 increase until a peak is reached just prior to ovulation. This E_2 peak triggers an ovulatory surge of luteinizing hormone (LH) that results in ovulation of a mature follicle. During this period of increasing E_2, the oviduct and uterus enter into a proliferative phase in which the actions of E_2 dominate oviductal and uterine physiology. Following ovulation, the ovary enters a luteal phase in which E_2 levels decrease, whereas progesterone (P) synthesized by the corpora lutea rises. During the luteal phase, the oviduct regresses and the uterus enters into a secretory phase in which P plays a dominant role. If fertilization and implantation do not occur, then levels of P drop dramatically as the corpus luteum decays and menses occurs followed by another

round of maturing follicles and uterine proliferation. It is the coordinated effects of E_2 and P over the course of the estrus cycle that prepare the oviduct, uterus, and cervix for zygote transport, fertilization, implantation, gestation, and parturition. In addition, the changing patterns of E_2 and P also alter expression of ER and PR and, hence, the abilities of different tissues to respond in an appropriate fashion.

A. Proliferative Phase

The proliferative phase of the oviduct and uterus coincides with the follicular stage of the ovarian cycle and is dominated by the effects of E_2. In the oviduct, periods of high E_2 promote an increase in epithelial mitotic activity, hypertrophy, and ciliation of the mucosal epithelium. The ciliary activity is highest near ovulation and this helps sweep the ovulated egg down the oviduct toward the uterus. As the oviduct prepares for egg transport and fertilization, the uterus undergoes proliferative changes in both the myometrium and the endometrium. These include E_2-induced synthesis of contractile proteins, creatinine phosphokinase, and ATPase in myometrial cells. In addition, estradiol also promotes glycogen buildup and cellular hypertrophy. In short, estradiol appears to prepare the physical and energetic needs of the myometrium required for the contractile processes involved in menstruation, gestation, and parturition.

As E_2 levels increase during the follicular phase, ERα expression increases first in the stromal cells and then in the epithelial cells of the endometrium. Under the influence of E_2, the endometrium undergoes extensive remodeling or proliferation which prepares the luminal epithelium and stroma for implantation. During the initial phases (1–4 hr) following E_2 exposure in mice, changes in the epithelium include hyperemia and water imbibition. In later stages of E_2 stimulation, there are increases in DNA and protein synthesis, hyperplasia, and hypertrophy. Morphologically, the epithelial cells become columnar, endometrial glands dilate, and the stromal cells exhibit some increased numbers of cells and edema. ERα-mediated effects produce substantial increases in uterine gene activity and protein synthesis which is necessary for the growth response in the uterine endometrium. Some evidence indicates that ERα expression in the epithelial cells is not readily detectable until after DNA synthesis has begun. These data suggest that the initial effects of E_2 on epithelial hyperplasia are mediated through activation of stromal ERα and subsequent influences of the stroma on the epithelium. This hypothesis is supported by experiments demonstrating that wild-type stromal tissues can mediate E_2-induced mitosis in epithelial tissues from ERKO mice. Another important effect of E_2 during uterine proliferation is the induction of progesterone receptor synthesis which allows the uterus to respond to elevated P present during the secretory phase of the uterus. During the proliferative phase, elevated E_2 also regulates uterine expression of growth factors and growth factor receptors, such as epidermal growth factor (EGF) and EGF-R, in the uterus. These elevated levels of EGF appear to be partly responsible for the uterotropic effects induced by E_2. Indeed, exogenous EGF can mimic many of the uterotropic effects of E_2 observed in mice and this "estrogenic" effect of EGF appears to be mediated through ligand-independent activation of ERα discussed earlier (see Section I). This was definitively established in experiments in which EGF failed to induce any uterotropic effects in female ERKO mice.

During the proliferative phase, the dimensions of the endocervical canal change and the cervix undergoes increased vascularization, thickening of the epithelial layers, and edema. Synthesis of cervical mucous also reaches its peak during the periovulatory period when E_2 is at its highest levels. This mucous promotes fertilization by assisting in the transport and protection of sperm from the acidic pHs of the vagina. The vaginal epithelium also undergoes thickening and cornification associated with estradiol-induced increases in cell division of the basal epithelium as well as an ER-mediated increase in keratin gene expression. As a result of the epithelial thickening, the superficial layers are further removed from a blood supply and undergo keratinization and sloughing.

B. Secretory Phase

During luteal stages of the ovarian cycle, reproductive physiology of the female reproductive tract is dominated by the effects of P. In general, the actions of E_2 tend to be antagonized by the elevated P present during this phase. In the oviduct, elevated P induces regression of the oviductal epithelium that includes atrophy and deciliation of the mucosal epithelium. It appears that this P effect may be due to downregulation of ERα and hence an inability of E_2 to continue stimulation of the mucosal epithelium.

During periods of high P, the uterus enters a secretory phase in which the uterine endometrium prepares for implantation of the blastocyst. These changes include increasing vascularization and thickening of the mucosal layer of the epithelium characterized by stromal mitosis, cellular hypertrophy, and edema. During this phase, gland cells increase in size, spiral arteries become more developed, and the synthesis and secretion of secretory products increases. Elevated P also results in suppression of both stromal and epithelial ERα expression and hence decreased responsivity to E_2. P also induces thickening of cervical mucous that promotes formation of a cervical–vaginal barrier following sperm transport. In addition, P antagonizes the actions of E_2 on vaginal epithelial proliferation and cornification.

In the absence of implantation and secretion of chorionic gonadotropins, the corpora lutea degenerates, serum P levels drop, and the endometrium degenerates (menses). Menses consists of a sloughing of the proliferated zona functionalis, in particular the zona compacta. Once the hypothalamic–pituitary–gonadal axis is freed from P suppression, pituitary gonadotropins initiate another round of folliculogenesis, estrogen levels begin to rise, and the uterus undergoes another cycle of repair and proliferation. During this round of uterine proliferation, the zona functionalis regenerates from the remnants of the zona functionalis or basal cells from the zona basilis.

C. Implantation, Gestation, and Parturition

Implantation of the hatched blastocyst requires that the endometrium be hormonally primed and receptive to the blastocyst. In most species, E_2 effects on implantation are limited to its necessary but permissive effects on uterine proliferation. In mice and rats, there is evidence that in addition to its permissive role, exposure to preovulatory and prenidatory peaks of E_2 are required for successful implantation.

However, in most species, it is exposure to elevated levels of P following ovulation that is essential for successful implantation.

If implantation occurs, continued exposure to elevated levels of P contributes to maintaining the proliferated and secretory states of the uterine endometrium. Depending on the species, gestational P may be secreted primarily by the corpus luteum, placenta, or both. During pregnancy, some of the most significant effects of P involve its role in balancing the effects of E_2 on the myometrium and maintaining a quiescent state . As mentioned earlier, E_2 increases contractility of the myometrium through altering metabolism, synthesis of contractile proteins, and sensitivity to hormonal signals. P maintains the muscle cells in less active states through antagonism of E_2 effects on myometrial contractility and thereby facilitates full-term gestations. As gestation approaches term and E_2 levels increase as P levels decrease, E_2 again begins to increase myometrial sensitivity to oxytocin and PGF-2α, presumably through induction of oxytocin and PGF-2α receptors. In addition, E_2 stimulates synthesis and formation of gap junctions between muscle cells which results in more efficient transmission of muscle action potentials and the synchronous contraction of the myometrial layer that is required for successful parturition. Thus, exposure to late gestation levels of E_2 alters both the structural characteristics and hormonal sensitivities of myometrial cells to promote parturition.

The unopposed actions of E_2 during late gestation also play a role in the cervical ripening and softening which assists in passage of the fetus. These effects may be due to E_2 reducing the amount of collagen and enhancing sensitivity to relaxin, a hormone that inhibits muscle activity. These results combine to produce a cervix that is softer and better able to accommodate the degree of stretching required for successful parturition.

V. CONCLUSION

Successful reproduction requires the coordinated actions and responses of the hypothalamus, pituitary, gonads, and reproductive tract. This is achieved by a complex system of neuroendocrine and endocrine signaling pathways that include hypothalamic-releasing factors, trophic pituitary hormones (i.e., gonadotropins), and gonadal steroids. In this article, we have provided a general overview of the molecular mechanisms that mediate the effects of estrogens on the physiological and biochemical responses of the female reproductive tract during the ovarian cycle, pregnancy, and parturition. An important point is that the proper regulation of morphological, physiological, and biochemical changes in the female reproductive tract is largely driven by the cyclical changes in circulating levels of ovarian E_2 and P. It is the coordinated effects of E_2 and P that ensure proper timing and preparation of the female reproductive tract and hence maximize the likelihood of successful pregnancy and parturition.

See Also the Following Articles

Estrogen Action, Behavior; Estrogen Action, Bone; Estrogen Action, Breast; Estrogen Secretion, Regulation of; Labor and Delivery, Human; Steroid Hormone Receptors

Bibliography

Brenner, R., and Slayden, O. (1994). Cyclic changes in the primate oviduct and endometrium. In *The Physiology of Reproduction* (E. Knobil and J. D. Neill, Eds.), pp. 541–569. Raven Press, New York.

Challis, J., and Olson, D. (1988). Parturition. In *The Physiology of Reproduction* (E. Knobil and J. D. Neill, Eds.), pp. 2177–2216. Raven Press, New York.

Curtis, S. W., Washburn, T., Sewall, C., DiAugustine, R., Lindzey, J., Couse, J. F., and Korach, K. S. (1996). Physiological coupling of growth factor and steroid receptor signaling pathways: Estrogen receptor knockout mice lack estrogen-like response to epidermal growth factor. *Proc. Natl. Acad. Sci. USA* **93**, 12626–12630.

Korach, K. S., and Quarmby, V. E. (1985). Morphological, physiological, and biochemical aspects of female reproduction. In *Reproductive Toxicology* (R. L. Dixon, Eds.), pp. 47–68. Raven Press, New York.

Korach, K., Couse, J., Curtis, S., Washburn, T., Lindzey, J., Kimbro, K. S., Eddy, E., Migliaccio, S., Snedeker, S., Lubahn, D., Schomberg, D., and Smith, E. (1996). Estrogen receptor gene disruption: Molecular characterization and experimental and clinical phenotypes. *Recent Prog. Horm. Res.* **51**, 159–188.

Lindzey, J., and Korach, K. S. (1996). Steroid hormones. In *Endocrinology: Basic and Clinical Principles* (P. M. Conn and S. Melmed, Eds.), pp. 47–62. Humana Press, Totowa, NJ.

Estrogen Effects and Receptors, Subavian Species

Marina Paolucci

Università degli Studi Di Napoli

Noemi Custodia and Ian P. Callard

Boston University

I. Structure and Biosynthesis
II. Site of Synthesis
III. Estrogen Effects
IV. Mechanism of Estrogen Action
V. Concluding Remarks

GLOSSARY

amygdala The part of the brain concerned with inner emotions (fear, anger, etc.).

elasmobranchs Cartilaginous fishes; sharks, skates, and rays.

3β-hydroxysteroid dehydrogenase/Ω-5,4 isomerase An enzyme complex that acts on certain steroid hormones to move the 5,6 double bond to the 4,5 position and to convert the 3 hydroxyl group to a ketone. It allows crossover between the Δ^5 and Δ^4 pathways of steroid hormone synthesis.

hypothalamic–pituitary unit The ventral portion of brain with anatomically attached endocrine gland (pituitary) which controls the reproductive system.

Leydig cells The principal androgen secreting cells of the testis.

lower vertebrates Reptiles, amphibians, and fish (cartilaginous and bony).

oviparous Egg laying.

teleosts The dominant bony fishes.

viviparous Live bearing.

Estrogens are classified according to their chemical structure as steroids. Steroids are synthesized from stored sterols in endocrine glands and are released into the bloodstream when signaled by pituitary tropic hormones. In the bloodstream, steroids, bound to proteins, travel to distant cellular targets which they enter by diffusion. Once in the cell, they are recognized by high-affinity molecules (receptors) within the cell. A hormone–receptor complex is formed and it translocates into the nucleus where it binds to the DNA, setting off the transcription of specific proteins. There is significant evolutionary conservation of estrogen receptor structure throughout the vertebrates.

Estrogens, although mainly secreted by the ovary, are also synthesized in the testis and other tissues, such as fat and brain. They act at several levels to determine the behavior and phenotype of the female: the gonad itself, secondary reproductive structures (e.g., uterus), and, in addition, the liver, brain, pituitary, muscle, skeleton, and vascular system. Effects on the hypothalamic–pituitary unit are important in the overall regulatory scheme for ovarian function and reproduction. Estrogens are responsible for the growth and development of secondary sexual characters and are essential for vitellogenin synthesis by the liver in females. The principles concerning the action of estrogens, and the structure of the estrogen receptor, are similar in all vertebrates.

I. STRUCTURE AND BIOSYNTHESIS

Our knowledge of steroid biosynthesis in subavian species indicates that it has much in common with birds and mammals, the biosynthetic pathways and

```
Pileup estrogen receptor sequences:

HUMAN     LQPHGQQVPYYLENEPSGYTVREAGPPAFYRPNSDNRRQGGRERLASTNDKGSMAMESAK
CHICKEN   IHHHSQQVPYYLENEQGSFGMREAAPPAFYRPSSDNRRHSIRERMSSTNEKGSLSMESTK
LIZARD    ------------------------------------------------------------
FROG      IHHHGQQVPYYLESEQGTFAVREAAPPTFYRSSSDNRRQSGRERMSSANDKGPPSMESTK

             •  •            •  •  C region         •      •          •  ∘
HUMAN     ETRYCAVCNDYASGYHYGVWSCEGCKAFFKRSIQGHNDYMCPATNQCTIDKNRRKSCQAC
CHICKEN   ETRYCAVCNDYASGYHYGVWSCEGCKAFFKRSIQGHNDYMCPATNQCTIDKNRRKSCQAC
LIZARD    ------------------------------------GHNDYMCPATNQCTIDKNRRKSCQAC
FROG      ETRYCAVCSDYASGYHYGVWSCEGCKAFFKRSIQGHNDYMCPATNQCTIDKNRRKSCQAC
                                              ****************************
                                        D region
HUMAN     RLRKCYEVGMMKGGIRKDRRGGRMLKHKRQRDDGEGRGEVGSAGDMRAANLWPSPLMIKR
CHICKEN   RLRKCYEVGMMKGGIRKDRRGGEMMKQKRQREEQDSRNGEASSTELRAPTLWTSPLVVKH
LIZARD    RLRKCYEVGMMKGGIRKDRRGGRILKHKRQREEHDNRN-AGAIVERRSPNLWPSPLMITH
FROG      RLRKCYEVGMMKGGIRKDRRGGRLLKHKRQKEEQEQKN-DVDPSEIRTASIWVNPSVKS-
          *********************..*.***... ..      . *.  .* *. .
                                        E region
HUMAN     SKKNSLALSLTADQMVSALLDAEPPILYSEYDPTRPFSEASMMGLLTNLADRELVHMINW
CHICKEN   NKKNSPALSLTAEQMVSALLEAEPPIVYSEYDPNRPFNEASMMTLLTNLADRELVHMINW
LIZARD    NKKNSPALSLTADQIVSALLEAEPPVVYSEYDPSRPFSEASMMTLLTNLADRELVHMINW
FROG      -MKLSPVLSLTAEQLISALMEAEAPIVYSEHDSTKPLSEASMMTLLTNLADRELVHMINW
          *  *    *****.*..***..** *..*** *  .*  **** ****************

HUMAN     AKRVPGFVDLTLHDQVHLLECAWLEILMIGLVWRSMEHPGKLLFAPNLLLDRNQGKCVEG
CHICKEN   AKRVPGFVDLTLHDQVHLLECAWLEILMIGLVWRSMEHPGKLLFAPNLLLDRNQGKCVEG
LIZARD    AKRVPGFVDLSLHDQVHLLECAWLEILMIGLVWRSVEHPGKLLFAPNLLLDRNQGKCVEG
FROG      AKRVPGFVDLTLHDQVHLLECAWLEILMVGLIWRSVEHPGKLSFAPNLLLDRNQGRCVEG
          **********.***************.**.***.****** *************.****
                                        E region
HUMAN     MVEIFDMLLATSSRFRMMNLQGEEFVCLKSIILLNSGVYTFLSSTLKSLEEKDHIHRVLD
CHICKEN   MVEIFDMLLATAARFRMMNLQGEEFVCLKSIILLNSGVYTFLSSTLKSLEERDYIHRVLD
LIZARD    FVEIFDMLLATSSRFRMMNVQGEEFVCLKSIILLNSGIYTFLSSTLKSLEEKDHIHRVLD
FROG      LVEIFDMLVTTATRFRMMRLRGEEFICLKSIILLNSGVYTFLSSTLESLEDTDLIHIILD
           *******..*..***** ..****.*****************.******* *** . ** .**

HUMAN     KITDTLIHLMAKAGLTLQQQHQRLAQLLLILSHIRHMSNKGMEHLYSMKCKNVVPLYDLL
CHICKEN   KITDTLIHLMAKSGLSLQQQHRRLAQLLLILSHIRHMSNKGMEHLYNMKCKNVVPLYDLL
LIZARD    KIIDTLLHLMAKSGLSLQQQHRRLAQLLLILSHFRHMSNKGM------------------
FROG      KIIDTLVHFMAKSGLSLQQQQRRLAQLLLILSHIRHMSNKGMEHLYSMKCKNVVPLYDLL
          ** ***.* ***.** ****.. ********** ******* *******

HUMAN     LEMLDAHRLHAPTSRGGASVEETDQSHLATAGSTSSHSLQKYYITG-EAEGFPATV
CHICKEN   LEMLDAHRLHAPAARSAAPMEEENRNQLTTAP-ASSHSLQSFYINSKEEESMQNTI
LIZARD    --------------------------------------------------------
FROG      LEMLDAHRIHTPKDK-TTTQEEDSRSPPTTTVNGASPCLQPYYTNT-EEVSLQSTV
```

enzymes for estrogen synthesis being highly conserved. The basic structure, common to steroid hormones, consists of three rings of six atoms of carbon (the phenanthrene ring) and one ring of five atoms of carbon (the pentane ring); this yields a complex structure called cyclopentanoperhydrophenanthrene. To this structure oxygen and carbon substitutes are added in different positions to form different steroids. All steroids are synthesized from cholesterol; this 27-carbon compound is synthesized via a biosynthetic pathway involving at least 30 steps, beginning with acetate. The conversion of cholesterol into pregnenolone is common for all classes of bioactive steroids. From pregnenolone, two metabolic pathways lead to the androgens (androstenedione and testosterone) which are then aromatized respectively to estrone or estradiol, the most potent estrogen. These two major estrogens are interconvertible. In addition to these two, there is a third quantitatively important estrogen, estriol, derived from estrone by either hepatic 16-hydroxylation of estrone or in the primate placental–fetal unit. All three major estrogens have been widely identified among lower vertebrates, but the major focus has been on estrone and estradiol. In elasmobranchs, estrone and estradiol have been found in the ovary and peripheral plasma, although estradiol predominates. In teleosts, the relative proportions of estrone and estradiol vary with the reproductive cycle. In amphibians and reptiles estrone and estradiol are the main ovarian steroids.

II. SITE OF SYNTHESIS

In vertebrates the main site of estrogen synthesis is the ovary. The ovary consists of a connective tissue matrix, the stroma, which contains blood vessels and developing eggs surrounded by follicle cells. These support the oocyte and function as a conduit for vitellogenin, the precursor of yolk. The follicle cells and the contained oocyte are enveloped by the theca. The organization of the ovarian follicle differs between mammalian and nonmammalian vertebrates. In mammals the follicle cells form several layers, termed the granulosa, and the theca can be divided into an internal and external layer. For many nonmammalian species the granulosa is not as complex as in mammals. Typically, in mature follicles from cyclostomes to amphibians, the granulosa is single layered, as in chelonians. In snakes and lizards, however, it is multilayered and complex. The theca is not always distinguishable into internal and external theca. Some amphibians and teleosts have only a single cell layer, and cyclostomes, elasmobranchs, and reptiles are more complex, with the innermost layers being considered theca interna and outermost layers considered theca externa. However, the functional contribution of granulosa and theca layers to follicular steroidogenesis is similar to that of mammals, although their cellular roles in steroidogenesis may be reversed. According to the two-cell model of steroidgenesis of mammals, androgens are produced by the theca from cholesterol and are then carried

FIGURE 1 Multiple alignment of the estrogen receptor amino acid sequence of human compared to some submammalian species. The three most important regions involved in the overall function of the receptor are the C domain, D domain, and E domain. The C domain (DNA-binding domain) is the most conserved region among species (92–100%). It is characterized by being rich in basic amino acid residues (Lys and Arg) and a large number of cysteine residues (●). There are three Cys-x2-Cys sequences and another three Cys residue that are conserved among species and play an important role in the formation of zinc fingers. The first two Cys-x2-Cys are separated by 13 amino acid residues and form a zinc finger that is similar to the *Xenopus* TFIIIA finger loop. The other Cys-x2-Cys is able to bind to other Cys residues further up to form another zinc finger. There is also a nuclear localization signal (bold) at the end of the C region. The D region (hinge region between C and E regions) is the most variable region among species. However, it plays an important role in the function of the C and E regions. It is a very hydrophilic region (rich in Arg and Lys residues). This region also displays two nuclear localization signals. The E region (the hormone-binding domain) is highly conserved but not as much as the C domain (60–95%). It has a large number of Lys residues and is a highly hydrophobic region. It has two dimerization regions and it contains two phosphorylation sites basic-basic-X-Ser (one of them underlined).

to the granulosa where they are aromatized to estrogens. A similar two-cell type model has been proposed for the production of estrogens in teleosts, amphibians, and reptiles. However, follicular steroidogenesis in the elasmobranchs (spiny dogfish and little skate) is like that of birds, with thecal elements synthesizing androgens and estrogens from granulosa-derived progesterone.

Estrogens are synthesized in tissues other than the gonad, such as the brain and fat, through the process of aromatization. Brain aromatase appears to be of particular importance in teleost fish, in which the level of activity of this enzyme is 100–1000 times that of the mammalian brain and 10 times that of the teleost ovary. Estrogens are degraded and metabolized by the liver of nonmammals, as in mammals, and are excreted in bile and urine as sulfates and glucuronides.

III. ESTROGEN EFFECTS

The biological actions of estrogens are well documented in lower vertebrates. In concert with progesterone, they regulate the ovary through actions on the hypothalamic–pituitary axis and the liver via induction of vitellogenin (yolk protein precursor) synthesis. They further stimulate growth and differentiation of the female reproductive tract and control its functions relative to the passage of the ovulated egg and maintenance of gestation in viviparous species. In addition, through actions on the central nervous system these hormones control reproductive behavior associated with courtship and mating.

A. The Gonads

The gonads perform two separate functions: production of gametes (gametogenesis) and synthesis of hormones (steroidogenesis). It is widely accepted that hormones produced by the gonad have an effect on the gonad itself, as well as via feedback actions on pituitary gonadotropin secretion. Although estrogen receptors (ERs) are present in the ovary of birds and mammals, the presence of estrogen binding sites in ovarian follicles of lower vertebrates has not yet been demonstrated. Estrogens are generally considered to be female hormones, but they are also synthesized and secreted by the testis, as demonstrated by the ability of Leydig cells to aromatize androgens to estrogens. The presence of estrogen receptors in the testis of elasmobranchs, amphibians, and reptiles indicates that the testis itself is a site of estrogen action. Although definitive functions have not yet been assigned to estrogen in the vertebrate testis, in the urodele amphibian (mudpuppy, *Necturus maculosus*), it has been suggested that estrogens are involved in the demise of Leydig cells at the end of their cycle of differentiation. In the testis of the turtle, *Chrysemys picta*, ER concentration is highest when Leydig cells are regressing and just before the start of the next spermatogenic cycle. In the lizard (*Podarcis s. sicula*) testis, the ER may be involved in the earliest stages of spermatogenesis, as proposed for dogfish, *Squalus acanthias*. The presence of ER in the testis of the reproductively inactive lizard *P. s. sicula* and frog, *Rana esculenta*, suggests a possible role for estrogens as inhibitors of testicular functions. In general, these data are consistent with mammalian studies indicating that estrogens have inhibitory effects on androgen production. Moreover, ER localization within the testis and its expression pattern throughout the reproductive cycle suggest estrogen involvement during the first stages of spermatogenesis.

B. Secondary Sexual Characters

It is well established that estrogens play an important role in eliciting the growth and development of the secondary sex characteristics. Experiments in which the gonads are removed (ovariectomy) and sex steroids are injected have shown that estrogens have an essential function in stimulating the growth of the oviducts. Typically, the oviduct has a lumen lined with an epithelial monolayer, a submucosal glandular region, circular and longitudinal smooth muscle layers, and an outer layer of peritoneum. The structural and functional differences seen in the oviducts of different species represent adaptations for different modes of reproduction. Epithelial gland structure and oviducal protein secretion vary according to function. Thus, in oviparous forms, secretion of secondary and tertiary (shell) membranes, or, in viviparous species, the development of maternal–embryonal connections are gonadal steroid dependent. In their simplest form the oviducts are ciliated tubes, with glands secreting mucus or a gelatinous covering for the eggs. Oviducts of this type are common among bony fishes and amphibians. In vertebrates with large shelled eggs (elasmobranchs, reptiles, and birds), oviducts specialize to produce egg albumen and shell components. In viviparous species the embryo is sheltered and nourished in a specialized region of the oviduct, the uterus. In general, it can be said that the effects of estrogen on the oviducts of all nonmammalian species are growth stimulatory, as in birds and mammals, and that they act to regulate both secretory and contractile activity. As in birds and mammals, estrogens act through receptors, thus confirming the oviduct as a target organ for estradiol. Binding sites for estradiol have been found in the cytosol and nuclear extract from oviducts of an elasmobranch, amphibians, and reptiles.

In these species, the putative ER fulfills the general requirements for steroid receptors binding estrogen, specifically and with high affinity and binding to DNA. As the oviduct grows during the reproductive cycle, the level of ER in the nucleus parallels the increases in plasma estradiol, suggesting a causal relationship between receptor levels and steroid ligand. This is supported by a drop in receptor level after ovariectomy in some reptiles. However, administration of estradiol, progesterone, or both hormones to ovariectomized animals does not restore receptor levels. It is likely that other factors, in addition to sex steroids, are critical for the full development of the oviduct and expression of receptors.

In summary, although effects of estrogen on the oviduct of lower vertebrates have not been investigated as thoroughly as in mammals and birds, it is reasonable to suggest that the basic actions of estrogen are similar throughout the vertebrates, and they involve interaction of the hormone with its receptor, binding to DNA, followed by cell proliferation, gland and cell differentiation, and secretion.

C. Central Nervous System

Estrogens exert a wide range of effects in the central nervous system of vertebrates, from early devel-opment through sex differentiation, maturation, and adulthood. Furthermore, the brain of all species is capable of synthesizing estrogens from androgen substrate (aromatization), thus local areas of the brain may be exposed to high local concentrations. This is particularly true for teleost species which have very high levels of aromatization and estrogen production, indicating the importance of local production and control of estrogen by the brain for a variety of central nervous functions, including neural transmission, growth, regeneration, and synaptic growth throughout vertebrates.

Based on binding of labeled estradiol and the detection of ER, the principal target regions of estrogen are the hypothalamus–preoptic area and, with the exception of teleosts, the amygdala. These regions are the primary neuroendocrine areas controlling the pituitary activity and they function to translate external stimuli (temperature, day length, food availability, and stress) into internal responses. ERs have been demonstrated in the brain of the dogfish, goldfish, lizard, and turtle. Their biochemical properties are similar to those reported for ER in other target organs (liver and oviduct). ER distribution is in agreement with a role in reproductive behavior.

D. Estrogens and Vitellogenesis

Estrogens are intimately involved in stimulating the formation of yolk in nonmammalian vertebrates. The process of yolk synthesis and deposition is referred to as vitellogenesis. During vitellogenesis, the yolk precursor or vitellogenin (VTG) is synthesized in the liver, released into the bloodstream, and taken up by growing oocytes. Yolk accumulates in oocytes as a food reservoir to be used during embryonic development. The amount of yolk within oocytes varies widely, ranging from the relatively small quantity found in most teleost fishes and amphibians to the extremely large quantity stored in elasmobranch reptile and bird oocytes. Vitellogenin is specifically induced by estrogen in subavian species. Studies carried out in the turtle and in the green frog have shown that estrogen cooperates with pituitary growth hormones in VTG stimulation. The molecular mechanism of estrogen stimulation of VTG genes has been extensively investigated in lower vertebrates, and much evidence demonstrates that estrogens stimulate VTG synthesis through an estrogen-specific receptor in hepatic cells of all nonmammalian vertebrates. Experiments using the South African clawed frog, *Xenopus laevis*, as a model have played a central role in demonstrating the molecular action of the activated ER as a transcription factor which interacts with the VTG gene. An ER has been detected in the liver of fishes, amphibians, and reptiles. The concentration of the ER varies among species, but it is in the femtomole range. Hepatic ER fluctuates according to the phase of the reproductive cycle. In the spotted seatrout *Cynoscion nebulosus*, nuclear ER levels are higher in the liver of the females actively synthesizing VTG. In the turtle ER is present in both cytosol and nuclear extract of the liver. In the snake *Nerodia sipedon*, changes in ER levels in both cytosol and nuclear extract have been reported throughout the reproductive cycle, coinciding temporally with the estradiol surge and plasma VTG levels. In the frog *R. esculenta* and the lizard *P. s. sicula*, hepatic ER, plasma estradiol, and VTG have been monitored throughout the reproductive cycle, and these studies demonstrate a positive correlation of nuclear receptor to estradiol and VTG plasma levels.

IV. MECHANISM OF ESTROGEN ACTION

About 40 years ago, Jensen detected high-affinity, low-capacity estradiol-binding molecules in the rat uterus. Since that time, much work has been directed toward understanding the mechanism of steroid hormone action and the structure of receptors. Throughout the animal kingdom, estradiol receptors, like other steroid receptors, have a highly conserved linear structure. This comprises five domains (A–E), each of which has defined functions. The A/B domain, or amino terminus of the receptor protein, is involved in the activation of transcription; the C domain (the DNA-binding domain), through its specialized "zinc fingers," is involved in interaction with the DNA of the gene; the D region is the "hinge" around which the molecule can flex; and the E domain is the ligand (steroid)-binding domain. Multiple alignment of the deduced amino acid sequences

of several subavian ERs compared with that of the human is shown in Fig. 1. The C domain (DNA binding) is most conserved among species (92–100% homology) and is characterized by basic amino acids and cysteine residues, which play an important part in organizing the zinc fingers. This region also has the nuclear localization signal. The hinge region is the most variable and the E domain is somewhat less conserved between species (60–95% homology) than the C domain. Although a second ER (ER-β) has been found in mammals, so far characterization of the ER subtype in subavian species has been made only for a teleost fish, which falls into the ER-β family.

V. CONCLUDING REMARKS

Although our knowledge of the hormonal control systems for lower vertebrate reproductive processes is still fragmentary compared to that of birds and mammals, we can say that common regulatory mechanisms are present throughout the vertebrates. Much remains to be done, particularly with regard to our understanding of the reproductive processes of endangered species and improving reproductive capacity. In addition, although the impact of environmental factors on development and fertility of all species is now being recognized, the study of the reproductive processes of a wide variety of species of all classes is required in order to gain an understanding of the dangers of foreign chemicals in the environment (xenobiotics).

Acknowledgments

This work was supported by an Italian MURST grant to MP, a fellowship from NIH (NICHD) to NC, and grants from the NSF to IPC.

See Also the Following Articles

Bibliography

Callard, I. P., and Callard, G. V. (1987). Sex steroid receptors and non-receptor binding proteins. In *Hormones and Reproduction in Fishes, Amphibians and Reptiles*, pp. 355–384. Plenum Press, New York.

Callard, I. P., and Klosterman, L. (part A); Callard, G. V. (part B) (1988). Reproductive physiology. In *Physiology of Elasmobranch Fishes*, pp. 277–312. Springer-Verlag, Berlin..

Chester-Jones, I., Ingleton, P. M., and Phillips, J. G. (1987). *Fundamentals of Comparative Vertebrate Endocrinology*. Plenum Press, New York.

Clark, J. H., and Mani, S. K. (1994). Actions of ovarian steroid hormones. In *The Physiology of Reproduction*, pp. 1011–1059. Raven Press, New York.

Fostier, A., Jalabert, B., Billard, R., Breton, B., and Zohar, Y. (1983). The gonadal steroids. In *Fish Physiology*, Vol. IX, Part A, pp. 277–346. Academic Press, New York.

Estrogen, Effects in Birds

Colin J. Saldanha and Barney A. Schlinger

University of California, Los Angeles

GLOSSARY

activational effects Transient changes in structure and function typically due to sex steroid exposure in adult animals.

aromatase The enzyme responsible for the conversion of androgens to estrogens.

estrogen receptor A ligand-activated transcription factor with high affinity for estrogens that can bind to chromatin and regulate gene expression.

estrogen response element A palindromic nucleotide sequence that when located in the upstream promoter region of a gene permits estrogenic regulation of the downstream coding region.

organizational effects Permanent changes in structure and function typically due to sex steroid exposure during discrete critical periods of development.

ovalbumin Egg-white protein synthesized and secreted in large amounts upon estrogenic stimulation of tubular gland epithelial cells in the avian oviduct.

$\mathbf{A}$ substantial portion of our knowledge of estrogen action has come from studies on avian models. This is due in part to the fact that estrogens produce some extreme responses in some avian tissues. Diurnal birds also perform dramatic visual and acoustical displays that make estrogen-dependent behaviors easy to document and measure. Additionally, since each embryo develops independently of maternal influences within the eggshell, studies investigating the effects of estrogen on development are relatively simple to perform. This article highlights several estrogenic influences on avian anatomy, physiology, and behavior. We present these estrogen effects as a "time line" from embryo to adulthood in an attempt to underscore the rich diversity of estrogenic regulation in avian biology.

I. INTRODUCTION

Few chemical messengers exert the wide range of specific effects as do members of the steroid family. Among steroids, estrogens stand out for their influence on a wide variety of tissues and biological endpoints (Fig. 1). Research on several vertebrate classes has identified species-, sex-, age-, and tissue-specific

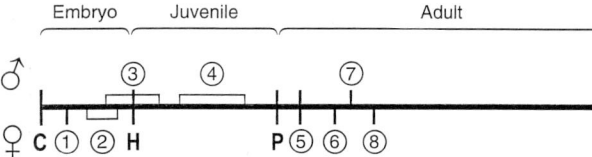

FIGURE 1. Estrogens have dramatic effects throughout the life span of birds. Depicted in the schematic (1–8) are the various documented effects of estrogens from conception (C) through hatching (H) and beyond puberty (P). As shown, estrogens are known to influence both females (below the life line) and males (above the life line). 1, Ovarian differentiation; 2, demasculinization/feminization; 3, masculinization; 4, song learning; 5, hypothalamo–pituitary–gonadal axis regulation; 6, sexual behavioral receptivity; 7, copulatory behavior and aggression; 8, yolk protein stimulation and shell deposition.

actions of estrogen. These effects span morphological, physiological, and behavioral arenas during development, adulthood, and aging. Estrogens are primarily synthesized by the gonads, in most cases in higher amounts in the female relative to the male. However, their various roles in organizing the development of the nervous system in males and in females, and in maintaining the integrity of the vascular, skeletal, and nervous systems, compel our consideration of estrogens as critical modulators of biological processes in males and females in both reproductive and nonreproductive contexts. It remains true, however, that most of what we know about estrogen synthesis, secretion, and response comes from investigations of reproductive function.

Estradiol and estrone (estrogens) are synthesized from their androgenic precursors testosterone and androstenedione respectively via the action of the enzyme aromatase (estrogen synthetase). Estrogens are primarily synthesized in ovarian tissue in female vertebrates, and via transport through the vasculature gain access to and modulate biological processes in distant target tissues. However, estrogens can also be synthesized in testes, but at smaller amounts than in ovaries. Estrogens can also be synthesized at nongonadal sites at levels sufficient to activate local estrogen-dependent functions and/or enrich the vasculature. The identification of aromatase expression in nongonadal tissues has defined important targets of

estrogen action. Thus, in some species, circulating estrogens with access to all types of central and peripheral tissues are a reflection of estrogen synthesis and secretion from multiple sites.

II. ESTROGEN RECEPTORS AND SENSITIVITY

In order to modulate physiological processes, estrogen must act via a receptor in target tissues. The presence of estrogen receptors provides information as to sites of estrogen action, at the scale of organs or cells. In addition, the presence of estrogen receptors gives an indication of the mechanism of action of estrogen on those cells or tissues. Biochemical and autoradiographic binding studies, immunohistochemistry using antibodies against estrogen receptors, and molecular studies that have demonstrated the presence of estrogen receptor mRNA have all been used to investigate the presence of estrogen receptors in peripheral and central tissues of birds. The widespread distribution of estrogen receptors in avian tissues is impressive. Estrogen receptors have been localized in the chicken primordial ovary as early as Day 5 *in ovo*. Estrogen receptors are expressed in adult tissues as diverse as epithelial cells of the oviduct; hepatocytes of the liver; endothelial cells of the vascular system; osteoclasts of the skeletal system; individual cells of smooth muscle; secondary cells of the retina, skin, and feather; and neurons and glia within the central and peripheral nervous system. The data underscore the diversity and pluripotent action of estrogen in the bird. Additionally, avian tissues can modify their sensitivity to estrogen by up- or downregulating estrogen receptor expression across ages, seasons, and individuals. For example, estrogen receptor expression in some song nuclei of the zebra finch (*Taenopygia guttata*) brain declines after the birds learn their song and approach adulthood. Additionally, species differences in estrogen sensitivity of these very brain nuclei exist as evidenced by the continued expression of estrogen receptor within song nuclei in the canary (*Serinus canarius*) throughout adulthood. Thus, the distribution of estrogen receptors can not only point to the sites of estrogen action but also suggest functional roles for estrogen via actions on its receptor.

III. MECHANISM OF ESTROGEN ACTION

The cellular responsivity of avian target tissues to estrogen has proven pivotal in our understanding of the general mechanisms whereby all steroid hormones alter cellular function and thereby exert their potent effects. It has long been known that estrogen administration increases the size and weight of the avian oviduct. Increases in circulating estrogen are typically observed early in the breeding cycle and its involvement in egg laying is described later.

The classic studies of O'Malley and colleagues (1995) on the chick oviduct established that estrogen was specifically able to bind the DNA of oviducal cells, initiate the transcription of novel RNAs, and that one of these RNAs was translated into the egg-white protein ovalbumin. Briefly, given its lipophilic nature, estrogen has potential access to all cells and all cellular compartments. Once bound to its receptor, the estrogen–estrogen receptor complex dimerizes and interacts with chromatin within the nucleus. Estrogen response elements (EREs) located within the promotor regions of specific genes are bound by the estrogen–estrogen receptor complex, thus allowing estrogen to affect gene expression. Because steroids are powerful modulators of gene expression, studies of estrogen effects on the chicken oviduct have taught biologists from a variety of disciplines about the mechanisms of steroid action and the regulation of gene expression at its most basic level.

IV. ORGANIZATIONAL EFFECTS ON EMBRYONIC BIRDS

The nature of estrogen effects on avian physiology and behavior is perhaps most vividly demonstrated by its organizational influence during the processes that mold the avian embryo into a female or male phenotype—a process involving reproductive tissues, such as the ovaries, testes, and genitalia, and nonreproductive tissues, such as the brain. The data clearly demonstrate that estrogen is a crucial signal in the sexual differentiation of birds.

A. Ovarian Development

The first known actions of estrogens in birds occur very early in development because they function as a fundamental signal initiating ovarian differentiation. In birds, as in mammals, sex is determined genetically: Female birds are heterogametic (ZW), whereas males are homogametic (ZZ). Presumably, early expression of a gene or genes on the W chromosome allows for aromatase expression in the gonadal ridge. Estrogens formed there appear to induce differentiation of the bipotential gonadal primordium to differentiate into an ovary and to prevent testicular development. This critical role of estrogen in feminine differentiation is strongly supported by the studies of Elbrecht and Smith (1992) on embryonic chickens (*Gallus domesticus*). Fertile eggs were treated with vehicle (control), an aromatase inhibitor, an aromatase inhibitor and estradiol, or just estradiol. These treatments were performed at various stages of embryonic development, including prior to Day 7 *in ovo*. The eggs were allowed to hatch and the sex of the offspring was determined using genetic or phenotypic markers. Since there is near equivalency (1:1) in the sex ratio, one would expect that vehicle-treated controls would contain an equal proportion of males and females among the offspring, and this was indeed the case based on both genotype and phenotypic markers. Eggs treated with the aromatase inhibitor also yielded a 1:1 sex ratio based on genotype, but almost 100% of the offspring were judged as male based on the presence of testes rather than an ovary. This dramatic change in the phenotypic sex ratio suggests that the inhibition of aromatase, resulting in a depletion of available estrogen, causes masculine gonadal development in the presence of a feminine genotype. This effect is specific to estrogen because coadministration of E_2 ameliorates the sex-reversing influence of aromatase inhibitors and E_2 administration causes feminine development of genetic males. Thus, estrogens are critical for female development. This pioneering work is supported by studies in the zebra finch in which genetically female embryos, treated with an aromatase inhibitor, develop large amounts of testicular tissue, whereas vehicle-treated controls develop a normal ovary. Moreover, aromatase expression in the ZW (female) gonadal ridge of chickens coincides with the process of sexual differentiation. Thus, in birds, the very first signal necessary for the transformation of the primordial gonad to an ovary, and therefore for feminine differentiation, is estrogen.

B. Secondary Sexual Characteristics

Once the ovary is formed ovarian estrogens continue to feminize and demasculinize secondary sexual characteristics in the avian female. In most birds, only the left ovary and oviduct develop and those on the right degenerate early in development. Degeneration of the right ovary and oviduct is mediated by the secretion of Müllerian-inhibiting hormone (MIH). However, preservation of left oviduct is affected by the local estrogenic signal provided by the left ovary itself. Estrogen administration to developing chickens antagonizes MIH, thus preserving the ovaries and oviduct bilaterally. As mentioned earlier, sexual differentiation also involves the degeneration of structures characteristic of the male but which are present in embryos of both sexes. Estrogen administration prevents the formation of a masculine copulatory tubercle (hemipenis) and syrinx in ducks (*Anas platyrhyncos*) and the foam gland in the japanese quail (*Coturnix japonica*). Thus, estrogens are critical for gonadal and phenotypic sexual differentiation in birds.

C. Neural Structure and Function

In addition to developmental effects on peripheral tissues, sexual differentiation includes dramatic modulation of neural structures which permit sex-specific behavioral expression in adulthood. Estrogens exert important organizational effects on the brain at the level of the diencephalon and telencephalon. In galliformes such as the Japanese quail, the hypothalamic preoptic area (POA) includes areas essential for the expression of masculine copulatory

behavior. Presumably, estrogens from the embryonic ovary naturally demasculinize and feminize the brains of developing birds. Embryonic exposure to estrogen appears to organize the neural circuits necessary for such behaviors because estrogen administration to male embryos demasculinizes the POA insofar as the expression of copulatory behavior is controlled by the POA. As adults, males exposed to estrogens *in ovo* fail to express typically masculine copulatory behaviors presumably due to estrogen acting within the developing POA. It should be noted, however, that the specific structural basis for the demasculinizing effect of estrogen on quail copulatory behavior is poorly understood.

In the brains of oscine songbirds (order Passeriformes), however, estrogens clearly organize developing neural structure in the forebrain. In these species, males possess a defined circuit of telencephalic nuclei responsible for the learning and generation of song. In songbird species in which males sing more than females, the volume of the song nuclei of males is typically larger than those of females. Several studies have shown that estrogen administration to females during a defined temporal window early in development results in masculinized (larger) song nuclei in adulthood. The cellular basis of this increase in nuclear volume is detectable in terms of an increase in neuron size, neuron number, the length of dendritic processes, and the extent to which cells accumulate androgen. Thus, estrogen masculinizes the structure of forebrain nuclei in terms of cell size, cell chemistry, and connectivity in the songbird brain. This structural effect has functional implications. Female songbird hatchlings that have been masculinized by the administration of estrogen sing as adults. Notably, this occurs in the zebra finch, a species in which normal adult females never sing, thus highlighting the dramatic effect of estrogen on the organization of neural circuits. Of prime importance in both the experiments on the quail POA and the songbird telencephalon is that the temporal cascade of estrogenic events is initiated early in development with structural and functional consequences detectable during adulthood. Unfortunately, although estrogen can masculinize the developing female zebra finch brain, the endogenous physiology underlying the presumed estrogenic masculinization of the male brain is unidentified.

V. ESTROGENIC EFFECTS ON JUVENILE BIRDS

A convincing body of literature implicates estrogen as a modulator of neural and behavioral processes in juvenile birds. Most of this literature has focused on the role of estrogen during the period of song learning that birds undergo prior to reaching adulthood. As mentioned earlier, during development, estrogens can be detected in the blood of both males and females of some songbirds species. During the juvenile stage songbirds undergo a period of song learning characterized by an initial overproduction of song syllables (plastic song) which eventually is honed into a smaller repertoire of syllables corresponding to full adult song (crystallization). Marler and associates (1987) found that estrogen titers were elevated in the plasma of sparrows during the period of song acquisition but decreased to baseline levels upon the crystallization of song in juveniles. These data suggest that estrogens are available to modulate the neural events which underlie the process of song learning in songbirds. Furthermore, it is known that juvenile zebra finches treated with tamoxifen (an estrogen receptor antagonist) demonstrate substantial impairment of adult song, suggesting that song learning may be modulated by estrogen. Although further research on the specificity of the action of estrogen on learning is required, it is exciting to consider the possibility that highly complex behavioral tasks involving learning and memory may be modulated by estrogen in the songbird.

VI. ACTIVATIONAL EFFECTS ON ADULT BIRDS

Estrogens continue to exert potent effects on adult peripheral and neural tissues in birds. Most of what we know about estrogenic modulation of adult structure and function comes from studies on reproductive physiology and behavior. Of importance is the observation that estrogenic influences are apparent in adults of both sexes and estrogen is perhaps exemplified by its role in the regulation of its own secretion via actions at the level of the hypothalamus and on the pituitary.

A. Hypothalamo–Pituitary– Gonadal Axis

In adult birds, as in other vertebrates, gonadal secretions are regulated by luteinizing hormone (LH) and follicle-stimulating hormone (FSH) release from gonadotropes within the pituitary gland. LH and FSH, in turn, are released in response to gonadotropin-releasing hormone (GnRH) release from the hypothalamus. This endocrine circuit is regulated by many external factors, including environmental and behavioral cues, but it is modulated from within via its own secretions. It is well established that circulating estrogens can feed back onto the hypothalamus and pituitary to inhibit their secretions, thereby modulating their own rate of synthesis and secretion. Thus, high levels of estrogen can inhibit GnRH and LH expression via actions directly onto hypothalamic neurons and pituitary gonadotropes, respectively, resulting in a decrease in the activity of the reproductive axis and indirectly inhibiting further increases in estrogen synthesis. Although this has been demonstrated in female chickens and quail, it is likely that the conversion of androgen to estrogen in the hypothalamus or the pituitary of the male is also involved in the feedback regulation of testicular androgen secretion. This is suggested by studies in the songbird in which estradiol treatment of adult males causes a dramatic regression of the testes in the zebra finch or inhibition of the aromatase enzyme substantially increases circulating androgen levels.

B. Female Physiology and Behavior

It is not surprising that estrogens modulate female reproductive processes in birds as they do in several vertebrate classes. Circulating estrogens are elevated early in the breeding season, suggesting an influence on sexual rather than parental aspects of avian reproduction. In most investigated species, circulating estrogens increase dramatically during courtship, copulation, and egg laying and experimental evidence strongly implicates estrogens in the mediation of many of these processes.

1. Skin and Plumage

In some bird species, steroids trigger sexually dimorphic growth of some plumage characteristics. In several species, notably the domestic chicken, the development of the female plumage is estrogen dependent. This influence is perhaps most dramatically demonstrated in the Sebright Bantam, in which males carry a dominant genetic trait resulting in henny feathering. Thus, in this species, except for typically masculine combs and wattles, no sexual dimorphism in plumage is apparent. Wilson *et al.* (1987) demonstrated that the henny plumage of male Sebrights is a direct result of high levels of aromatase expressed in dermal tissue. Dermal aromatase was high enough to elevate local estrogen levels to feminize plumage and contribute to circulating estrogen titers. These data are supported by the observation that ovarian transplants into castrated male chickens result in the development of a female-specific plumage. Estrogens have other documented effects on female skin. For example, as eggs are laid by the female, many female and some male birds develop a specialization of the skin on the ventral thorax and abdomen that is used to incubate the eggs. The skin of this brood or incubation patch is defeathered and can undergo substantial hyperplasia and can become hypervascularized and edematous. Depending on the species, estrogens, in addition to other hormones, can stimulate one or more of these characteristics of the brood patch.

2. Sexual Behavior

The sexual behavior of the adult bird has been investigated most extensively in galliformes, columbiformes, and passerine songbirds. As mentioned earlier, estrogen levels are high during the sexual phase of the breeding cycle in females of all three orders and estrogen is known to increase the incidence of female-specific sexual behaviors in chickens, ring doves, and zebra finches. In the chicken, the female typically displays a sexual crouch, thereby permitting

the male to mount and copulate. Estrogen administration increases the incidence of sexual crouching in ovariectomized chickens, suggesting that this sexual behavior may be estrogen dependent in this species. These findings are supported by studies on ovariectomized ring doves, in which low doses of estrogen increase the rate of preening, nest solicitations, and sexual crouching. In songbirds, the receptive display given by estrogen-treated females is often used to monitor the extent to which phenotypic or behavioral qualities of males are sexually stimulating to females. It is notable, however, that the neural bases for these activational effects on female sexual behavior are unknown. However, taken together the data strongly suggest that sexual behavior in the adult female bird is estrogen dependent.

3. Egg Laying

The production of an egg involves the mobilization of nutrients from the maternal body and their deposition within a protective mineral-rich covering. Both of these processes appear to be mediated by the high circulating estrogen titers early in the avian breeding cycle. Nutrients must be present in sufficient amounts to feed the developing embryo from laying to the posthatching period, a time that can be considerable especially in large birds. Most of these nutrients are synthesized in the maternal liver in the periods just prior to egg laying. The increasing estrogen levels within the maternal circulation are critical modulators of protein synthesis in the liver. Estrogens act directly on chick hepatocytes to stimulate vitellogenin synthesis. Precursors of yolk proteins, including albumin, globulin, and vitellin, are detectable in increased amounts in the circulation in chickens treated with estrogen compared to untreated controls. In addition, other necessary components of the egg, including fat-soluble vitamins such as vitamin A and water-soluble vitamins such as riboflavin and biotin, are also increased in the maternal circulation upon estrogen administration. Thus, nutrients necessary for survival of the embryo during the period *in ovo* are stimulated within the maternal body under the influence of estrogen.

As mentioned earlier, the marked effects of estrogen on the avian oviduct have long been used as a model of steroid action. The oviducal lining of the avian oviduct consists of three layers of undifferentiated epithelial cells and we have learned that the differentiation of these epithelial cells into specific subtypes with unique secretory products and the proliferation of these cells are under estrogenic control. Upon estrogen administration, the magnum portion of the chicken oviduct swells and begins to differentiate into tubular epithelium, goblet cells, and ciliated epithelium. Tubular epithelia primarily secrete the egg proteins ovalbumin and conalbumin, whereas the goblet cells secrete avidin (in response to progesterone), and the ciliated epithelia function to assist in motility of the oviducal magnum. The specificity of estrogen action is evidenced by the observation that estrogen-treated chicken oviduct demonstrates a large increase in ovalbumin expression in the tubular gland epithelia with no avidin expression. However, upon subsequent progesterone treatment, ovalbumin expression remains restricted to the tubular epithelia, and avidin expression is upregulated specifically in the goblet cells. Importantly, progesterone-induced avidin expression is itself estrogen dependent. Estrogen administration increases the concentration of progesterone receptor mRNA and protein, thus modulating further steroid-induced oviducal avidin synthesis and secretion.

In addition to the contents of the egg, the eggshell itself places energy demands on maternal physiology, particularly in terms of mineral content. High amounts of calcium must be available to the shell gland to ensure that a viable and durable egg is laid. Under the influence of high circulating estrogen, birds have evolved an elegant method of recruiting the necessary minerals, including calcium, during the period just prior to egg laying. Estrogen treatment increases the mineral content of maternal blood, including a 900% increase in calcium, a 300% increase in inorganic phosphate, and near doubling of magnesium, in preparation for shell synthesis. These increases of circulating minerals represent a direct action of estrogen on osteoclasts in the bird's skeleton. Estrogen acts directly on the bones to leach out the calcium and phosphate necessary to synthesize the eggshell. It is this leaching process which is reflected as an increase in circulating mineral titers. Thus, estrogen effects are apparent in terms of synthesizing both the protective covering of the embryo and the nutrients necessary for its survival.

C. Male Physiology and Behavior

In the quail, the medial portion of the POA (POM) is sexually dimorphic in structure, appearing two or three times larger in the male relative to the female. Several studies have shown that this locus is necessary for the complete expression of masculine copulatory behaviors in the adult male. Castration of adult male quail causes a dramatic reduction of the volume of the POM and replacement with testosterone restores its volume. The interpretation that this is an androgenic effect, however, is premature. Importantly, treatment with the nonaromatizable androgen dihydrotestosterone has minimal effects on restoring POM size and the effect of estradiol is similar to that of testosterone. These data strongly suggest that aromatization of testosterone is necessary to increase the size of the POM. Indeed, treatment of castrated, testosterone-replaced male quail with aromatase inhibitors prevents the growth of the POM to normal male levels. Thus, in the adult male quail, masculine structuring of the hypothalamus is estrogen dependent.

The estrogenic activation of the male avian diencephalon extends into the behavioral arena. Quail demonstrate a stereotyped series of sexual behaviors, including strutting, mounting, and aggressive pursuit of copulation. Castration dramatically reduces the incidence of all of these behaviors and treatment with testosterone restores the entire range of masculine sexual behaviors. As with the increase in mPOA volume, the restoration of the copulatory sequence of behaviors is estrogen dependent. Treatment of adult males with aromatase inhibitors or estrogen receptor antagonists alone dramatically lowers the incidence of these behaviors, suggesting the active hormone responsible for sexual behavior in the male is estrogen. Convincing evidence suggests that it is the aromatization of circulating testosterone in the POA itself that is capable of increasing local estrogen levels by upregulating POM aromatase directly, thereby increasing occupancy of local estrogen receptors and activating the expression of these masculine behaviors.

VII. CONCLUSION

Several aspects of avian physiology and behavior are modulated by estrogens secreted from gonadal and extragonadal sites. It is clear that these effects begin at the very early stages of embryonic development and extend into adulthood. Although little is known about the effects of estrogen on the aging bird, it would not be surprising to find that estrogen protects against age-related degeneration as it apparently does in other vertebrates including humans. The sites of estrogen synthesis, mode of secretion, and the cellular bases of its effects on various target tissues in birds continue to provide a strong system for investigating the steroidal regulation of physiology and behavior.

See Also the Following Articles

AVIAN REPRODUCTION • EGG, AVIAN • FEMALE REPRODUCTIVE SYSTEM, BIRDS • SEXUAL ATTRACTANTS • SONGBIRDS AND SINGING

Bibliography

Adkins, E. Early organizational effects of hormones. In *Neuroendocrinology of Reproduction: Physiology and Behavior* (N. T. Adler, Ed.), pp. 159–228. Plenum, New York.

Arnold, A. P. Developmental plasticity in neural circuits controlling birdsong: Sexual differentiation and the neural basis of learning. *J. Neurobiol.* 23, 1506–1528.

Balthazart, J., Tlemacani, O., and Ball, G. F. (1996). Do sex differences in the brain explain sex differences in the hormonal induction of reproductive behavior? What 25 years of research on the Japanese Quail tells us. *Horm. Behav.* 30, 627–661.

Elbrecht, A., and Smith, R. G. (1992). Aromatase enzyme activity and sex determination in chickens. *Science* 255, 467–470.

Jones, R. E. (1971). The incubation patch of birds. *Biol. Rev.* 46, 315–339.

Marler, P., Peters, S., and Wingfield, J. C. (1987). Correlations between song acquisition, song production, and plasma levels of testosterone and estradiol in sparrows. *J. Neurobiol.* 18, 531–548.

O'Malley, B. W. Thirty years of steroid hormone action: Personal recollections of an investigator. *Steroids* 60, 490–498.

Schlinger, B. A., and Arnold, A. P. (1995). Estrogen synthesis and secretion by the songbird brain. In *Neurobiological Effects of Sex Steroid Hormones* (P. E. Miceyvich and R. P. Hamner, Jr., Eds.), pp. 297–323. Cambridge Univ. Press, Cambridge, UK.

Schlinger, B. A., and Callard, G. V. (1991). Brain–steroid interactions and the control of aggressiveness in birds. In *Neuroendocrine Perspectives* (E. E. Muller and R. M. MacLeod, Eds.), pp. 1–43. Springer-Verlag, New York.

van Tienhoven, A. (1983). *The Reproductive Physiology of Vertebrates*, 2nd ed. Comstock, Ithaca, NY.

Wade, J., and Arnold, A. P. (1996). Functional testicular tissue does not masculinize development of the zebra finch song system. *Proc. Natl. Acad. Sci. USA* 93, 5264–5268.

Wilson, J. D., Leshin, M., and George, F. (1987). The Sebright bantam chicken and the genetic control of extraglandular aromatase. *Endocr. Rev.* 8, 363–376.

Wingfield, J. C., and Farner, D. S. (1993). Endocrinology of reproduction in wild species. In *Avian Biology* (D. S. Farner, J. R. King, and K. C. Parks, Eds.), Vol. IX, pp. 164–327. Academic Press, San Diego.

Estrogen Receptors

see Steroid Hormone Receptors

Estrogen Replacement Therapy

Rogerio A. Lobo

Columbia University College of Physicians and Surgeons

GLOSSARY

Alzheimer's disease (AD) A severe, progressive disorder involving major neuronal degeneration in specific regions of the brain, identified by the presence of distinctive plaques and tangles. It is characterized by loss of memory, diminished cognitive function, behavioral abnormalities, and eventual death. The main risk factor for the disease is age; early-onset AD typically occurs before the age of 50 and late-onset AD after 60.

hot flush or *hot flash* A temporary sensation of warmth, as in the face or upper body, experienced by some women during or after menopause.

hypogonadal Affected by or involving a deficiency in the secretions of the ovary or testis.

menopause The permanent termination of the menstrual cycle in the human female, occurring at a mean age of about 51 years.

osteoporosis An abnormal loss of density and weight in bone, associated with the postmenopausal stage in women and typically characterized by loss of stature, various deformities, and fractures from minimal or incidental injury.

postmenopausal Having experienced menopause; relating to or involving the period of life after menopause.

progestin A substance, either natural or artificial, that is able to induce the modifications in the uterus necessary for the implantation and development of a fertilized ovum.

Estrogen replacement therapy (ERT), or hormone replacement therapy (HRT), has been prescribed largely for relieving symptoms associated with estrogen loss. With increasing life expectancy, women now can expect to experience at least 30 years of estrogen "deficiency." In recent years we have realized that ERT affords protection against osteoporosis, cardiovascular disease, and possibly Alzheimer's disease. All causes of mortality are significantly reduced with ERT. However, the use of ERT in the United States is low (<20%). Multiple regimens involving several routes of administration are available and allow for the desirable effect of greater flexibility in prescribing practices. Risks of ERT include uterine (endometrial) cancer and possibly breast cancer. The excess in rates of endometrial cancer can be eliminated by the concomitant use of progestins (HRT). The association of ERT or HRT with breast cancer has not been proven. If this risk exists, it is likely to be in the range of a 20–30% increase. The risk is likely to be related both to dose and to duration of use. Weighing all known risks and benefits, ERT appears to have greater benefits overall and to be cost-effective. Quality of life, which is more difficult to quantify, is also improved with ERT.

I. INTRODUCTION

Estrogen replacement therapy (ERT) generally refers to the use of estrogen in postmenopausal or hypogonadal women. Some clinicians and patients have taken issue with the term "replacement" in postmenopausal women because they believe that menopause is a natural process and "replacement" denotes that women are not complete. Nevertheless, the term ERT has been widely used and for the purpose of this article will be the only term used. Also in common use, however, is hormonal replacement therapy (HRT). In general, HRT has been used as a more inclusive term, including those regimens in which a progestin is used.

ERT is prescribed for symptomatic relief of symptoms related to estrogen deficiency (hot flushes, vaginal dryness, etc.) but also affords significant benefits in terms of reducing the risk of developing diseases such as osteoporosis, cardiovascular disease, and, possibly, Alzheimer's, which typically occurs in later life.

Menopause occurs at a mean age of 51–54 years and is largely determined genetically. The age of menopause has not changed appreciably since records have been kept, whereas life expectancy has dramatically increased and now approaches 82 years. Thus, an average woman entering menopause in the 1990s can expect to have approximately 30 years of life with a relatively low level of estrogen. Even though this event is a natural one, a relative estrogen deficiency of this duration is a new phenomenon of the modern world. In previous centuries, life expectancy was not much beyond the age of menopause.

If menopause occurs prior to age 40, it is considered "premature." Here ERT may have to be considered for an even longer duration. Clearly, in younger women ERT plays a significant role in alleviating symptoms of estrogen deficiency, and symptoms such as hot flushes are more profound with acute and abrupt decreases in estrogen status, as occurs with surgical menopause (oophorectomy).

The prevalence of ERT in the United States is low, and no more than 20% of the postmenopausal population currently receive estrogen. Use is slightly higher in women who have undergone hysterectomy/oophorectomy and in those being treated for osteoporosis. The reasons more women are not receiving ERT are complex; however, continuance with ERT is a significant problem. For oral estrogen, continuance drops to 30% by 3 years after the initial prescription.

II. INDICATIONS

The major symptom of acute estrogen withdrawal or estrogen deficiency is vasomotor instability or the hot flush. Approximately 80% of estrogen-deficient women will have symptomatic hot flushes as the first sign of estrogen deficiency. This percentage decreases with time and is rarely a concern 10 years after menopause. While the etiology of this physiological event has not been determined, it is probably triggered by a catecholamine release disruption in the temperature regulatory centers of the hypothalamus in the brain in response to estrogen deficiency. Heat dissipation occurs (core temperature decreases) while peripheral vasodilation occurs and skin resistance drops from sweating. There are characteristic hormonal responses during a flush [increase in luteinizing hormone (LH) and proopiomelanocortin, peptides, etc.] but these are largely epiphenomena.

ERT significantly decreases hot flush activity in a dose-related way. A crude correlation exists between the number of hot flushes per day and serum estradiol (E2). The overlap in values and responses does not allow levels of E2 to be useful in monitoring treatment. There is also a well-known placebo effect of reduction in hot flushes (approximately 30%). The response of hot flushes to ERT is a time-related process, with the first significant reduction witnessed at 2 weeks. Markers of hypothalamic pituitary function, such as follicle-stimulating hormone (FSH) and LH, are not appropriate for monitoring the effects of ERT. Reductions in serum FSH are only slight with physiologic ERT; thus, FSH levels cannot be used to monitor ERT, unlike serum thyroid-stimulating hormone, which is an excellent monitor of thyroid replacement.

Vulvovaginal complaints (vaginal dryness, irritation, and dyspareunia) are frequently encountered in estrogen deficiency. However, these symptoms usually occur after approximately a year of estrogen deficiency and are not as acute as other symptoms. Urinary frequency, urgency, and loss of bladder control are also later signs. Some of these symptoms are related to the decline in collagen content (type I) which occurs with estrogen deficiency. All these symptoms respond well to ERT. In general, local or topical therapy is more effective than systemic ERT in that the results occur more rapidly. ERT reduces the vaginal pH, increases the percentage of superficial cells (and reduces the parabasal cells) of the vaginal mucosa, increases blood flow, and increases collagen (or prevents its further breakdown). It may decrease cholinergic bladder tone and increase sympathetic activity, thus aiding in problems of urination and stress incontinence.

Positive mood and psychological well-being are improved with ERT. Even if hot flushes and other symptoms are lacking, postmenopausal women "cope" better and have less depressive symptoms when they choose to receive ERT. Nevertheless, the role of ERT in treating depression remains unsubstantiated. Thus, psychological changes occurring in association with estrogen deficiency may be aided by ERT, but ERT should not be used to treat psychiatric disorders.

A. Prophylaxis

1. Osteoporosis

Abundant evidence exists for the influence of estrogen on bone mass. This occurs for both calcific bone and bone collagen content and is a result of bone estrogen receptor- and nonreceptor-mediated effects. Similarly, ERT has been shown to prevent osteoporosis, to stabilize bone mass and increase it temporarily in women with established osteoporosis, and to reduce fractures. Cancellous bone (spine and hip) is more sensitive to the effects of estrogen than cortical bones (long bones).

ERT exhibits a dose–response effect in preserving bone mass in postmenopausal women. Relatively low doses (minimal effective doses) have been found to have a significant effect in reducing some bone loss, whereas larger doses are needed to prevent any loss of bone or to put on new bone. Doses above a certain threshold level [approximately equivalent to 0.625 mg conjugated equine estrogen (CEE)] do not exert any additional benefit in the majority of women. However, in approximately 10–15% of women, 0.625 mg CEE will not be sufficient to maintain bone mass.

In osteoporotic women, ERT in a dose-related fashion results in an increase in bone mass (approximately 5%/year). However, this effect usually plateaus and is not normally sustained beyond 3 years.

Increases in bone mass require an adequate intake of calcium and vitamin D. There is evidence that progestins may increase bone formation as does androgen. Similarly, there appears to be a synergistic effect of ERT with bisphosphonates such as alendronate.

2. Cardiovascular Disease

It has been well established by observational studies that ERT reduces the risk of cardiovascular (CV) disease and mortality by approximately 50%. Clinical trial data will be forthcoming and are not likely to alter this fact but may give other estimates of risk reduction. Indeed, the relative risk estimates currently derived from epidemiologic studies vary widely (0.3–0.7). Recently, it has been reported that the addition of progestin (HRT) does not attenuate the benefits of ERT.

Most of the protective effects have been reported with the use of oral estrogen. However, many of the benefits listed below are related to the effect of estrogen on the coronary vasculature and are independent of the type of estrogen utilized. The major advantage of oral estrogen is the beneficial lipid/lipoprotein profile (Fig. 1). Nonoral estrogen does not lead to as beneficial a change in high-density lipoprotein (HDL) cholesterol, although triglycerides (and very low-density lipoprotein) are generally not increased as occurs with oral estrogen; total cholesterol and LDL cholesterol are lowered to a similar degree. In addition, oral and nonoral estrogen are associated with reductions in LP(a). ERT (both oral and nonoral) tends to reduce LDL oxidation. Studies have demonstrated that ERT prolongs the time it takes for LDL to be oxidized *in vivo*.

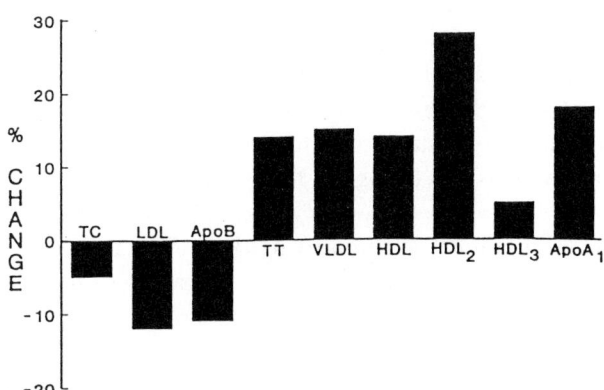

FIGURE 1 Percentage change in lipids and lipoproteins with the use of oral estrogen (equivalent to 0.625 mg for at least 3 months). TC, total cholesterol; TT, total triglycerides; VLDL, very low-density lipoprotein; LDL, low-density lipoprotein; HDL, high-density lipoprotein (adapted from Lobo, 1991, p. 925. © The Endocrine Society).

Oral estrogen favors an increase in coagulation or the tendency of blood to clot, which is still within the normal range. Similarly, there is a reduction in antithrombin III. These changes seem to be counteracted by decreases in fibrinogen and plasminogen activator inhibitor and an increase in plasminogen. In general, none of these coagulation changes are observed with nonoral estrogen.

Both oral and nonoral estrogens affect coronary artery tone and vessels in a beneficial way. The following changes have been observed: an increase in coronary blood flow, an increase in endothelial relaxing factor/nitric oxide, an increase in prostacyclin, a reduction in endothelin, an increase in circulatory angiotensin-converting enzyme levels, as well as calcium antagonistic effects. In addition, there are other vascular effects relevant to atherosclerosis such as plaque stabilization. These arterial changes may be mediated via nonclassical receptor mechanisms. Some effects are mediated through Ca^{2+} as well as other ionic channels.

ERT also improves carbohydrate tolerance and decreases insulin resistance. While both oral and nonoral estrogen have this effect, larger doses of oral ERT (>0.625 mg CEE) nullify it.

While there is little argument that ERT has a benefit effect on coronary disease and atherosclerosis, there is less consensus about stroke. Cohort studies, which have included older women (in their 70s), generally have shown a beneficial effect of ERT on stroke incidence and mortality.

In general, a threshold level of estrogen is needed to observe the full spectrum of beneficial CV changes. The approximate threshold level for these changes in serum is 50–60 pg/ml E_2.

3. Alzheimer's Risk and Cognitive Function

Potent and significant effects of estrogen on the brain have been documented in basic and animal studies. E2 enhances synaptic activity and neuronal growth, increases cerebral blood flow, decreases amyloid deposition, and has an important role in several other functional neuronal processes. ERT has been shown in at least four observational studies to reduce the risk of developing Alzheimer's. The odds ratio is in the range of 0.4, and the risk decreases the longer ERT is continued. A few less impressive

studies have also shown that ERT enhances or helps maintain deteriorating mental function in women with the diagnosis of Alzheimer's. These data are less convincing, and while estrogen deficiency is not considered to be the cause of Alzheimer's disease, ERT may play an important modulatory role.

In normal postmenopausal women, there are mixed findings on the effects of ERT on memory and cognitive function. Beneficial or no significant changes have been reported. Part of the difficulty in reaching conclusions in this area relates to the precision and reproducibility of the measures used to assess brain function.

III. REGIMENS

The aim of ERT is to provide "replacement" in a fashion which is as physiologic as possible. Follicular phase levels of E_2 during the normal menstrual cycle range between 40 and 100 pg/ml. As stated earlier, threshold levels of E_2 are in the range of 50–60 pg/ml for most women. Nevertheless, any increment of estrogen levels from baseline is expected to exert some significant effect, thus leading to the concept of a minimal effective dose.

Oral ERT results in higher levels of estrone (E_1) than E_2. This is true for oral estradiol as well as estrone products. CEE is a mixture of at least 10 conjugated estrogens derived from equine pregnant urine. Estrone sulfate is the major component, but the biological activities of equilin, 17α-dihydroequilin, and several other B ring estrogens, including Δ^{89} dihydroestrone, have been documented. Table 1 compares the standard doses of the most frequently prescribed oral estrogens and the levels of E_1 and E_2 achieved.

Synthetic estrogens, given orally, are vastly more potent. Ethinyl estradiol is used in oral contraceptives (Ocs). A dose of 5 μg is equivalent to the standard ERT doses used (0.625 mg CEE or 1 mg micronized estradiol). Standard ERT doses are five or six times less than the amount of estrogen used in OCs.

Oral estrogens have a potent hepatic ("first pass") effect which results in the loss of approximately 30% of its activity with a single passage after oral adminis-

TABLE 1
Mean Serum Estradiol (E_2) and Estrone (E_1)

Estrogen dose (mg)	Level (pg/ml)	
	E_2	E_1
CEE (0.3)[a]	18	76
CEE (0.625)	39	153
CEE (1.25)	60	220
Micronized E_2 (1)	35	190
Micronized E_2 (2)	63	300
E_1 sulfate (0.625)	34	125
E_1 sulfate (1.25)	42	220

[a] CEE contains biologically active estrogens other than E_2 and E_1.

tration. However, this results in stimulation of hepatic proteins and enzymes. Some of these changes are not particularly beneficial (an increase in procoagulation factors, e.g., V, VII, and angiotensinogen), whereas other changes are beneficial (an increase in HDL cholesterol and a decrease in fibrinogen).

Nonoral estrogen delivers estrogen in the form in which it is formulated. E_2 is administered in patches, gels, and subcutaneously. This synthetic administration is not subject to major hepatic effects as with oral therapy. Standard doses in the United States of alcohol-based or matrix patches are 0.05 or 0.1 mg. Matrix patches are preferable because there is less skin reaction and estrogen delivery is more reliable. While levels of E_2 vary widely among women, levels with transdermal therapy are more constant in individual women than with oral ERT. With the 0.05-mg patch, E_2 levels are in the 40–50 pg/ml range, and with 0.1 mg, levels are typically 70–100 pg/ml. It is not unusual, however, for some women to have levels in excess of 200 pg/ml.

As stated earlier, in women with vulvovaginal or urinary complaints, vaginal therapy is most appropriate. Creams of estradiol or CEE are available. Systemic absorption occurs but with levels which are one-fourth of that achieved after similar milligram doses administered orally. For CEE, only 0.5 g (0.3 mg) is necessary, and for micronized E_2 doses as low as 0.25 mg are sufficient. Other products (tablets and rings) are available which have been designed to limit systemic absorption. A silistic ring of E_2 is now available which delivers E_2 to the vagina for 3 months with only minimal systemic absorption.

A. Progestins

In women with a uterus, a progestin is necessary to "oppose" the proliferative effects of estrogen on the endometrium. Progestins are usually administered orally but may be used vaginally or as an intrauterine device. The dose of progestins should be kept low to prevent attenuating the beneficial effects of ERT on the CV system and brain.

There are many ways to administer progestins. The most commonly used oral progestins are medroxyprogesterone acetate (MPA) in doses of approximately 5–10 mg, norethindrone in doses of 0.3–1 mg, and micronized progesterone in doses of 100–300 mg. Equivalent doses to prevent hyperplasia when administered for at least 10 days in a women receiving ERT (equivalent to 0.625 mg CEE) are as follows: MPA, 5 mg; norethindrone, 0.7 mg; micronized progesterone, 200 mg. Larger doses of estrogen may require larger doses and more prolonged regimens of progestins. In sequential administration of progestins, the number of days (length of exposure) is more important than the dose. Thus, if a woman is receiving oral ERT continuously, a regimen of at least 10–12 days of exposure is preferable to a 7-day regimen.

When progestins are administered sequentially (10–14 days each month), withdrawal bleeding occurs in about 80% of women. Continuous administration of both estrogen and progestin (continuous–combined therapy) was developed to achieve amenorrhea. In the first 3–6 months, breakthrough bleeding and spotting is common. In some women on this regimen, amenorrhea is never completely achieved. The most common combination, marketed in the United States and approved by the FDA, is Prempro. A single tablet of Prempro contains 0.625 mg CEE and 2.5 mg MPA. Currently, the only marketed sequential regimen is Premphase, which contains 0.625 mg CEE and 5 mg MPA and is taken for 14 days each cycle. Other regimens are under review by the FDA.

Progesterone administered vaginally (in low

doses) avoids systemic effects and results in high concentrations of progesterone in the uterus. Intrauterine delivery of progestins is ideal for targeting the uterus is are not approved in the United States.

Progestins, particularly when taken orally, may lead to problems of continuance or compliance because of side effects, including mood alterations and bleeding. These have to be dealt with effectively and usually require more flexibility in prescribing habits.

IV. RISK–BENEFIT ASSESSMENT

Among the risks that have been associated with ERT are endometrial disease, breast cancer, side effects such as vaginal bleeding, somatic complaints, and idiosyncratic reactions including hypertension and thrombosis. Endometrial disease occurs with unopposed ERT in women who have a uterus. Although a woman's risk of developing endometrial cancer with unopposed estrogen use is two- to eightfold higher than that for the general population, in most patients precursor lesions, primarily endometrial hyperplasia, result from "unopposed" estrogen therapy. Thus, the risk is less for endometrial cancer than for varying degrees of hyperplasia. One recently conducted study showed that the risk of endometrial hyperplasia was 20% after 1 year of use of doses of 0.625 mg oral conjugated equine estrogens. All the hyperplasias were histologically "cystic" (the most benign type). In another recent study, the 3-year postmenopausal estrogen/progestin interventions trial, the risk was approximately 40% at the end of 3 years. No cancers were reported in either of the studies, and the addition of a progestin essentially eliminated the hyperplasia.

The risk of developing endometrial cancer is the same for a woman taking estrogen and progestin (hormone replacement therapy) as for the general population. In a woman receiving HRT, the risk of endometrial cancer is not thought to be hormonally related. In women who develop endometrial cancer with estrogen alone, the risk of death due to endometrial cancer is not high. Endometrial cancers associated with estrogen use are thought not to be as aggressive as spontaneously occurring cancers. In addition, tumors occurring in women taking estrogen are likely to be discovered and treated at an earlier stage, thus improving survival rates.

More controversial is the risk of breast cancer with ERT. Several meta-analyses have suggested either no significantly increased risk (a relative risk of approximately 1.0) or a risk as high as 1.6. It has also been suggested that there is no additional risk for women with a family history of breast cancer, although the risk for these women is already several times higher. An increased surveillance bias exists for women who see their doctors regularly. It is also possible that estrogen use causes breast cancer to occur earlier in some women, but it is not clear which women are at greatest risk. There are data to support the notion that any proposed increased risk in breast cancer with ERT is accounted for by the confounding addition of moderate alcohol consumption in those women. Nevertheless, breast cancer-related mortality does not appear to be increased, and some data suggest that it may be lower among estrogen users.

Although we can expect to see more analyses on the association between ERT and breast cancer risk, new data are not likely to be any more illuminating. Thus, we are left with the question of whether estrogen use carries any increased risk for breast cancer. Both the estrogen dose and the duration of use are implicated in risk estimates. For moderate doses of estrogen, the risk of breast cancer, if real, is probably no more than 20–30% over that for the rest of the population and pertains only to those women who are susceptible, e.g., moderate alcohol consumption and/or genetic susceptibility. Unfortunately, most women "at risk" cannot be identified before therapy is initiated.

One of the greatest concerns of women receiving estrogen is the return of menstrual bleeding. Somatic complaints such as breast tenderness and bloating may also occur with ERT but can be alleviated by alterations in dose and type of preparation. Such concerns should be discussed with the patient, and the choice of regimen should remain flexible.

Idiosyncratic reactions, including hypertension, thrombosis, and allergic manifestations, have also been observed in users of estrogen, particularly oral estrogen. Hypertension with estrogen use, the cause of which is not entirely clear, occurs in about 5% of oral estrogen users. Estrogen usually causes no change in blood pressure; it may actually reduce blood pressure, a finding that has relevance for normotensive as well as hypertensive individuals. However, an increase in both diastolic and systolic blood pressure has been noted in susceptible individuals and is rapidly reversible with discontinuation of ERT. A different form of estrogen, particularly a nonoral form, may eliminate the problem. Alterations in the route of estrogen administration and dose have resulted in improved blood pressure in such individuals.

Although for many years we have been reassured that noncontraceptive (lower) doses of ERT do not increase the risk of thrombosis, recent epidemiological studies have suggested a small but significant increase in risk with oral estrogen. This risk has not been validated, ruling out issues such as detection and reporting biases.

In women with a history of thrombosis, however, there is an increased risk with estrogen administered orally, a risk that is not easily identifiable except by reviewing the patient's history and by measuring factors such as Leiden's factor V mutation, antithrombin III, protein S, and protein C. Thus, women who have a family history of thrombosis or have had thrombotic events, particularly with oral contraceptives or any prior estrogen use, should be counseled very carefully and monitored closely. Nonoral estrogen is recommended for such patients.

A. Benefits and Alterations in Mortality

Several studies have reported that ERT results in a reduction of approximately 40% after 15 years of use in all causes of mortality. This is largely explained by a reduction in cardiovascular mortality. The nurses cohort study, which is a epidemiological study that has shown an increased risk of breast cancer, showed that the reduction in mortality may be less significant after 10 years of use (due to breast cancer risk), but this is at odds with results from other studies.

Because CV mortality is greatest among diseases potentially affected by ERT (including cancer), the reduction (approximately 50%) in CV disease accounts for most of the observed reductions in mortal-

ity. Nevertheless, CV mortality, and that attributable to cancer, has been found to be reduced in some studies. When the risk-associated mortality rates are weighed against the increased benefits (CV disease, osteoporosis, etc.), the benefits clearly outweigh the risks, particularly in women with CV risk factors and no risk factors for breast cancer.

In addition, ERT has been shown in three separate cost analyses to be a cost-effective treatment and is similar in impact to the health economics of treating hypertension. Not quantified in any of these reports is an improvement in the quality of life most women experience with ERT and the reduction in symptoms if present.

Finally, it is a woman's personal and private decision which influences whether or not she should use ERT. This decision should be made with full knowledge of the current data on risks and benefits and factoring in her individual risk profile. It is important that a women who chooses to use ERT feels good about the decision. If she does not, continuance will be a problem for her. A long duration of therapy is necessary to reap the larger benefits of this treatment.

See Also the Following Articles

Breast Cancer; Estrogen Action, Bone; Estrogens, Overview; Hypogonadism; Menopause; Osteoporosis; Progestins

Bibliography

Col, N. F., Eckman, M. H., Karas, R. H., Pauker, S. G., Goldberg, R. J., Rose, E. M., *et al.* (1997). Patient-specific decisions about hormone replacement therapy in postmenopausal women. *J. Am. Med. Assoc.* **277**, 1140–1147.

Grady, D., Rubin, S. M., Petitti, D. B., Fox, C. S., Black, D., Ettinger, B., *et al.* (1992). Hormone therapy to prevent disease and prolong life in postmenopausal women. *Ann. Intern. Med.* **117**, 1016–1037.

Grodstein, F., Stampfer, M. J., Manson, J. E., Colditz, G. A., Willett, W. C., Rosner, B., *et al.* (1996). Postmenopausal estrogen and progestin use and the risk of cardiovascular disease. *N. Engl. J. Med.* **335**, 453–461.

Lindsay, R., Bush, T. L., Grady, D., Speroff, L., and Lobo, R. A. (1996). Therapeutic controversy. Estrogen replacement in menopause. *J. Clin. Endocrinol. Metab.* **81**, 3829–3838.

Lobo, R. A. (1991). Clinical review 27. Effects of hormonal

replacement on lipids and lipoproteins in postmenopausal women. *J. Clin. Endocrinol. Metab.* **73**, 925–930.

Lobo, R. A. (1994). Treatment of the postmenopausal woman: Where we are today. In *Treatment of the Postmenopausal Woman: Basic and Clinical Aspects* (R. A. Lobo, Ed.), pp. 427–432. Raven Press, New York.

Lobo, R. A. (1995). Benefits and risks of estrogen replacement therapy. *Am. J. Obset. Gynecol.* **173**(Suppl.), 982–989.

Stafford, R. S., Saglam, D., Causino, N., and Blumenthal, D. (1997). Low rates of hormone replacement in visits to United States primary care physicians. *Am. J. Obstet. Gynecol.* **177**, 381–387.

Wiklund, I., Karlberg, J., and Mattsson, L.-A. (1993). Quality of life of postmenopausal women on a regimen of transdermal estradiol therapy: A double-blind placebo-controlled study. *Am. J. Obstet. Gynecol.* **168**, 824–830.

Estrogen Secretion, Regulation of

Koji Yoshinaga

National Institutes of Health

I. Introduction
II. Regulation of Estrogen Secretion by the Ovary
III. Estrogen Secretion during Pregnancy
IV. Estrogen Biosynthesis by Other Tissues

Chemically, the estrogens produced in the body are steroidal in nature. One of the four rings of its structure (ring A) has a hydroxyl group at carbon 3. Thus, the A-ring is an aromatic ring, and the distribution of double bonds makes the steroid phenolic. Major biologically active estrogens produced in the body are estradiol-17β, estrone, and estriol. The enzyme, aromatase complex, plays the final and essential role in production of estradiol and estrone from their precursors. Estriol is produced from estrone. The major tissues that secrete estrogens exist only temporarily in the body. Mature ovarian follicles are the major sources of estrogen found in the blood of nonpregnant women. The ovarian follicular cells proliferate and differentiate to acquire the ability to produce estrogen, and their function is activated and maintained under the influence of gonadotropins, follicle-stimulating hormone (FSH) and luteinizing hormone (LH), and other molecules. Gonadotropins that are secreted by the anterior pituitary gland stimulate granulosa cells and theca interna cells of the follicle, respectively, to produce estrogen in the follicle. FSH and LH prepare granulosa cells for steroidogenic ability and stimulate estrogen secretion from mature ovarian follicles, acting directly and indirectly through a number of intraovarian factors. In pregnant women the placenta plays a major role in estrogen secretion. Estrogen secretion by the placenta is regulated mostly by the fetal adrenal glands by providing the precursors of estrogen.

I. INTRODUCTION

Experimental animals have been used as models to understand the mechanisms of gonadotropin action in estrogen secretion in humans. Since the endocrine mechanisms in experimental animals are not always the same as those in humans, considerable attention must be paid to species differences in endocrine mechanisms. Gonadotropins, the major regulators of estrogen secretion, are regulated by the species-specific feedback mechanism of the hypothalamo–pituitary–gonadal axis. Secretion of gonadotropins is also influenced by physiological conditions such as pregnancy, lactation, and environmental factors. Therefore, these factors indirectly regulate estrogen secretion. Because of the ephemeral nature of sources of estrogen-secreting tissues in the body, regulation of estrogen secretion needs to be considered for specific physiological conditions and in different animal species. For these reasons, estrogen secretion is described following the order of the natural sequence of female reproductive process, namely, the menstrual or estrous cycle and pregnancy.

II. REGULATION OF ESTROGEN SECRETION BY THE OVARY

A. The Ovulatory Cycle

Adult females ovulate cyclically and estrogen secretion precedes each ovulation. The length of the ovulatory cycle, or the menstrual cycle in primates, varies among the species. It is on average 28 days in women, 16 or 17 days in sheep, and 4 or 5 days in rats and mice. The estrogen in circulating blood during the preovulatory phase of the cycle is mainly secreted by the follicles selected for ovulation. Rabbits, ferrets, and cats belong to a group of animals in which ovulation usually is induced by the mating stimulus. These females display nearly constant estrus with elevated estrogen level because competent follicles are present most of the time during the breeding season. This species variation in the cyclic estrogen secretion and subsequent ovulation is controlled by interactions between endocrine organs and represents the feedback mechanism of the hypothalamo–pituitary–ovarian axis.

The short estrous cycle in rats and mice is due to the incomplete cycle, i.e., the luteal phase is missing from the complete cycle. Impaired function of the hypothalamus may result in loss of the endocrine feedback mechanism and females fail to exhibit ovarian cyclicity. When the hypothalamo–pituitary axis loses its cyclic pattern of preovulatory surges of gonadotropin secretion, estrogen secretion may continue for a prolonged period of time, reflecting constantly elevated gonadotropins.

B. Follicle Development

In order for a follicle to develop to the mature stage and secrete estrogen, it must be stimulated by many factors. In humans, in the most undeveloped stage of the follicle, the primordial follicle has a size of 30–60 μm in diameter and is composed of the oocyte and surrounding layer of spindle-shaped cells which develop into granulosa cells (Fig. 1). The first step of the follicle development is the change of spindle-shaped cells to cuboidal granulosa cells, and the follicle is now called the primary follicle. Gonadotropins do not influence this early development of follicles. The oocyte of the primary follicle produces, however, a growth factor (growth differentiation factor-9) that stimulates the primary follicle to grow further. Activin has been shown to stimulate proliferation of primitive granulosa cells and make them responsive to follicle-stimulating hormone (FSH) by increasing FSH receptor. As follicles develop, they gain the ability to secrete estrogen. The estrogen-secreting mature follicle is a large vesicle that is filled with fluid (Fig. 2). The inner surface of the vesicle is lined with granulosa cells and the area of the oocyte bulges out into the cavity (antrum) of the follicle, forming the cumulus oophorus. The outer limit of the granulosa cell layer is a layer of membrane named the basement membrane and further outside of the follicle there are other cells named theca interna cells. The theca interna cell layer contains capillaries but these blood vessels do not cross the basement membrane. Therefore, there is no direct blood supply to granulosa cells. All nutrients and oxygen needed by the cells inside of the basement membrane are supplied by permeation through the basement membrane.

In women, ovulation takes place approximately 14 days after the first day of the menses. Serum concentration of estradiol starts to increase approximately 5 or 6 days before the ovulatory surge of gonadotropins (50–60 ng/ml) and reaches a peak level (200 ng/ml) at the gonadotropin surge. Ovulation takes place approximately 10 hr after the surge. The remaining tissues of the ovulated follicle form a corpus luteum. During the luteal phase the level falls to a low level of approximately 50 ng/ml but is elevated to 100 ng/ml for a week. This elevated estro-

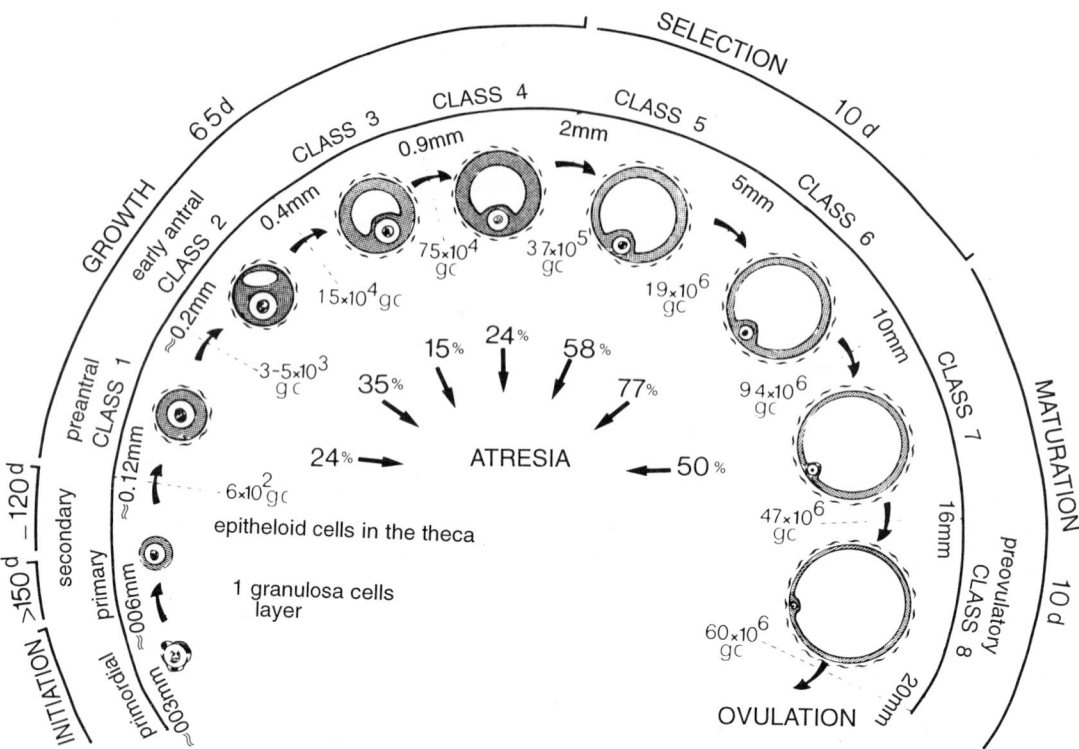

FIGURE 1 Stages of folliculogenesis in the adult human ovary and level of atresia in the eight classes of growing follicles. The classes under study are defined by the granulosa cell (gc) numbers and the corresponding estimated folliclular diameter (mm) (from A. Gougeon, Dynamics of follicular growth in the human: A model from preliminary results, *Hum. Reprod.* **1**, 81, 1986, with permission).

gen secretion during the luteal phase is not observed in the rhesus monkey. Toward the end of the luteal phase progesterone secretion by the corpus luteum declines and the endometrial lining sloughs off, resulting in menstrual discharges.

FSH acts on granulosa cells and prepares the follicle for estrogen secretion by regulating production of specific proteins. Among these are steroidogenic acute regulatory (StAR) protein and prohibitin. StAR protein plays an essential role in transporting cholesterol to the inner membrane of mitochondria where cholesterol side chain is cleaved to produce pregnenolone, which is metabolized to produce progesterone. FSH also regulates granulosa cells to proliferate by reducing production of prohibitin that has an inhibitory effect on cell proliferation. FSH, therefore, prepares developing follicles by regulating production of proteins that are responsible for further development of the follicle and for production of steroid hormones.

The rapid growth of follicles and proliferation of granulosa cells appear to be mediated by various growth factors. Epidermal growth factor (EGF) and/or other members of the EGF family, probably of theca cell origin, stimulate rapid proliferation of granulosa cells. EGF, on the other hand, inhibits estrogen secretion. Follistatin is found more in small follicles than in large follicles. Follistatin is considered to antagonize the steroidogenic effect of activin. In this way, growing follicles, although capable of estrogen secretion, do not secrete high levels of estrogen until they reach the preovulatory stage of follicular development. Gonadotropins act on granulosa cells to stimulate the production of insulin-like growth factor-I (IGF-I) as well as to suppress the release of IGF-binding protein (IGF-BP). The unbound IGF-I may act on theca cells to stimulate the production of EGF/TGF-α, which in turn diffuses back to the granulosa cells to prevent apoptosis. Insulin and IGFs appear to play important roles in the

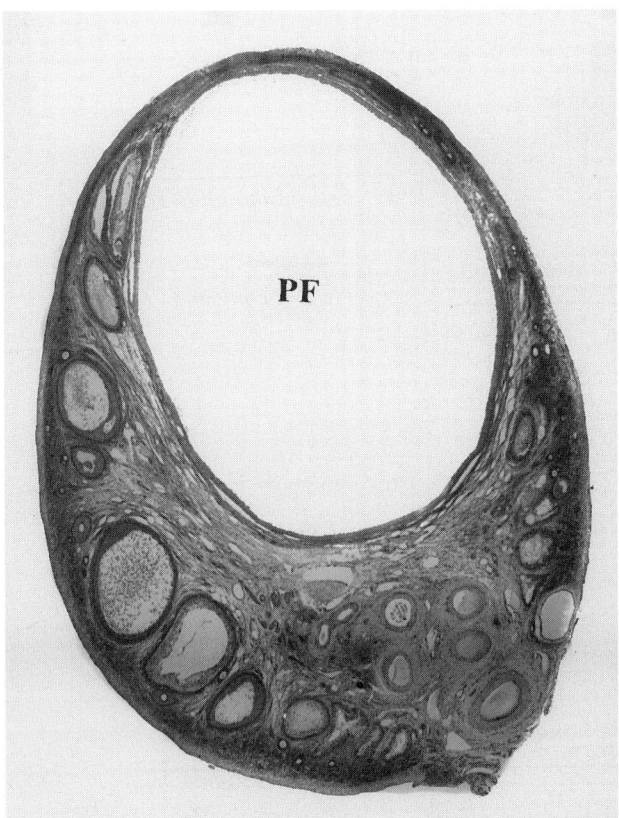

FIGURE 2 A cross section of the rhesus monkey ovary containing a large preovulatory follicle (PF; 4.6 mm in diameter) and several developing follicles. The preovulatory follicle is the major source of estrogen secretion during the follicular phase of the menstrual cycle (courtesy of Dr. Marilyn J. Koering).

follicle to amplify the effect of FSH on granulosa cells. A significant stimulatory effect of synergistic action of FSH and IGF-I on StAR protein production has been shown in porcine granulosa cells.

C. Estrogen Biosynthesis in the Ovary

Granulosa cells and theca interna cells of mature antral follicles produce steroid hormones. Granulosa cells and theca cells produce progesterone independently. These cells import the source of cholesterol in the form of low-density lipoprotein (LDL) through its receptor on the cell surface. Transport of cholesterol into the inner membrane of mitochondria is facilitated by StAR protein.

Because theca cells lack the key enzyme, the aromatase complex that converts androgen to estrogen, they do not produce estrogen. On the other hand, granulosa cells lack the key enzyme, 17α-hydroxy steroid dehydrogenase (17α-HSDH), that is an essential enzyme for androgen production (Fig. 3). Therefore, theca cells produce androgen that moves across the basement membrane to granulosa cells where androgen is converted to estrogen by aromatase. Gonadotropins, FSH and LH, stimulate, in concert, biosynthesis of estrogen. The "two-cell—two gonadotropin" theory of estrogen biosynthesis dictates that LH binds to the LH receptor of theca interna cells and stimulates 17α-HSDH enzyme to metabolize progesterone to androstenedione. The androstenedione is then transported to granulosa cells. FSH binds to the FSH receptor of granulosa cells and stimulates aromatase so that conversion of androstenedione to estrogen is accelerated (Fig. 4).

LH selectively regulates androgen synthesis in theca cells which is mediated by cyclic AMP at the level of increased 17-hydroxylase/C17-20 lyase activity. The enzyme P450c17α catalyzes two sequential reactions; the first is hydroxylation at carbon atom number 17 of Δ^4-progesterone or Δ^5-pregnenolone. This reaction can be followed by scission of the C-20,21 side chain from the steroid nucleus (17,20 lyase reaction) with the resultant formation of androstenedione or dehydroepiandrosterone (Fig. 3). These androgens serve as precursors for the production of testosterone and estrogens. The enzyme protein cytochrome P450c17α is expressed by a single gene, CPY17, and is regulated by steroidogenic factor-1. This gene is expressed in theca cells and not in granulosa cells.

The regulation of steroidogenesis in the ovary involves complex interactions of gonadotropins and intracellular signaling pathways. Gonadotropin actions are mediated by the adenylate cyclase–cyclic AMP system. In addition, LH action has been shown to be mediated by phospholipases.

Movement of cholesterol and its metabolites within and between cells is very important for estrogen biosynthesis. The cytoskeleton that maintains cellular structure and function plays an important role in the ability of steroid-producing cells to secrete

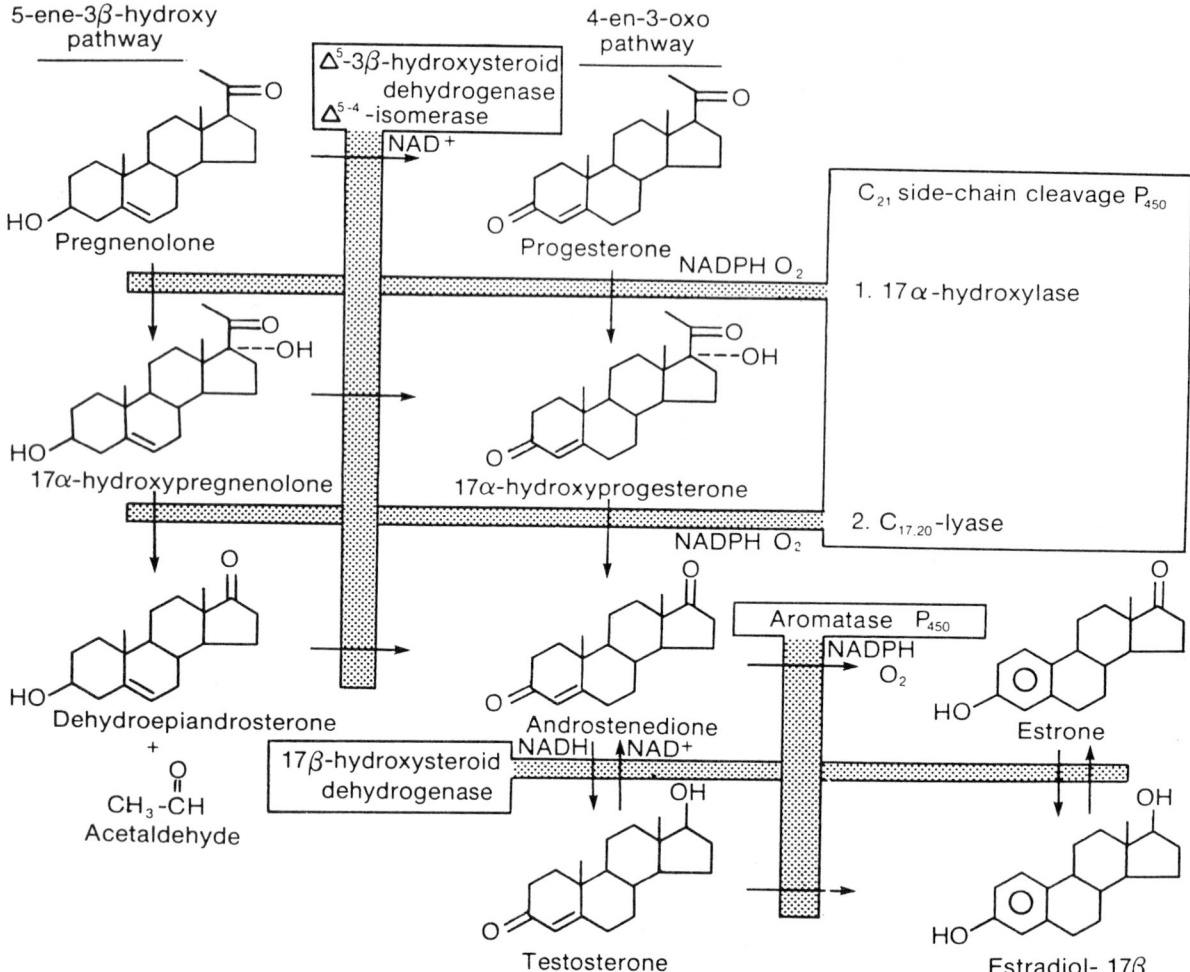

FIGURE 3   Steroidgenic pathways from pregnenolone to estrogens (from Gore-Langton and Armstrong, 1988, with permission).

steroids. This is evidenced by inhibition of secretion of estrogen and progesterone after granulosa cells are treated with drugs that prevent proper formation of the cytoskeleton.

D. Estrogen Secretion at the Preovulatory Gonadotropin Surge

During the development of preovulatory ovarian follicles, the aromatase enzyme that converts androgens to estrogens is induced in granulosa cells that line the inner surface of the preovulatory follicles. The increase in protein and mRNA of this enzyme is stimulated by increases in FSH. Estrogen secretion reaches elevated levels as the follicles become ready to ovulate. The high levels of estrogen then trigger the sudden release of LH, and estrogen secretion drops abruptly because LH terminates the production of aromatase mRNA and protein in the follicle. This abrupt change in estrogen secretion is explained as follows: Binding of FSH to its receptor in the plasma membrane of antral granulosa cells induces expression of the aromatase gene and increases aromatase production. When the LH is released during the preovulatory surge, it binds to the LH receptor of preovulatory granulosa cells. This stimulates a protein kinase (A-kinase) which, in turn, influences aromatase gene expression by activating transcription factors and coactivators that are essential for mRNA production.

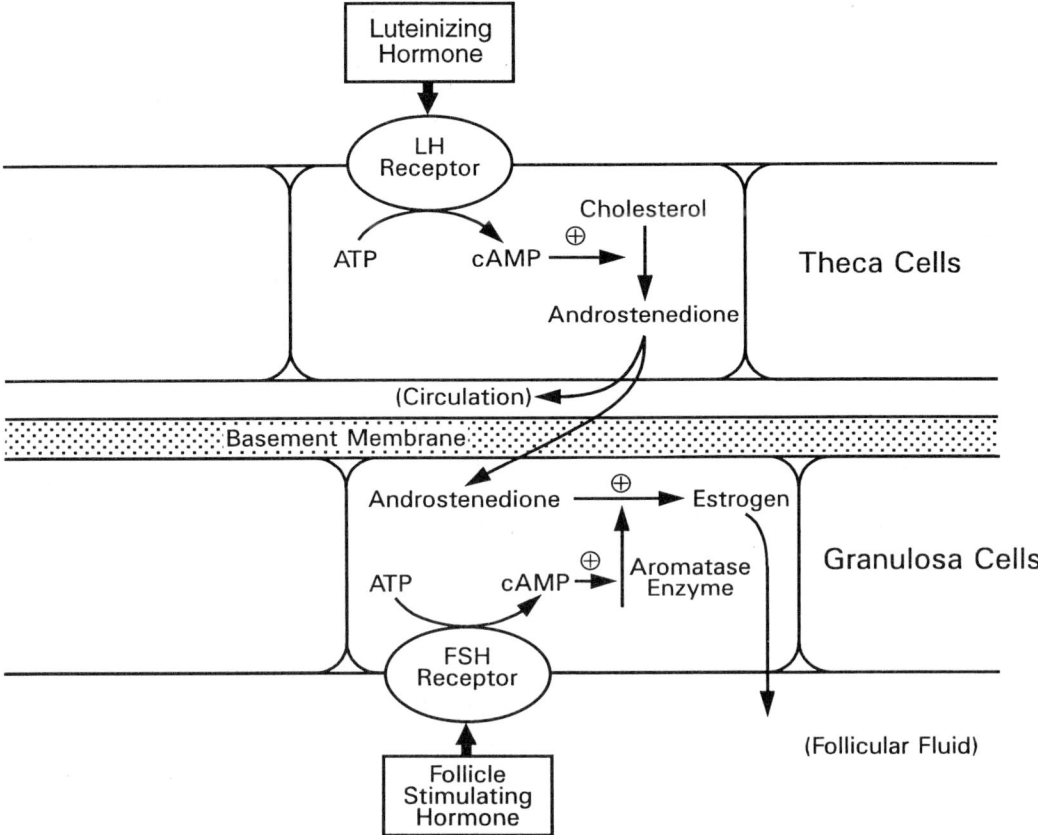

FIGURE 4 The two-cell/two gonadotropin hypothesis of follicular estrogen production. FSH, follicle-stimulating hormone; LH, luteinizing hormone; ATP, adenosin triphosphate; cAMP, cyclic adenosine monophosphate (from Adashi, 1991, with permission).

E. Luteal Phase Estrogen Secretion

After ovulation the theca and granulosa cells that remained in the follicle luteinize and form a corpus luteum. Human lutein cells are equipped with all P450 steroidogenic enzymes necessary for estrogen production. At the midluteal phase progesterone secretion reaches the peak. Estrogen also increases at the midluteal phase. Removal of the corpus luteum in women results in an immediate decline of progesterone and estrogen levels, indicating that the human corpus luteum secretes estrogen. However, the magnitude of estrogen secretion during the luteal phase in the human is much less than that of the preovulatory phase. Estrogen levels during the luteal phase in the rhesus monkey remain low and they do not peak at the midluteal phase. Detailed studies in sheep indicate that there are waves of FSH secretion and

follicles develop during the luteal phase. Estrogen levels, however, do not reflect the FSH waves but reflect LH pulses that are infrequent during the luteal phase.

During pseudopregnancy (which corresponds to the complete cycle with the luteal phase) in small rodents, such as rats and mice, prolactin secreted by the pituitary gland plays a major role in the luteotropic complex. In these small rodents, the mating stimulus applied to the vaginal–uterine cervical area results in the twice daily release of prolactin from the pituitary which stimulates newly formed corpora lutea in the ovary to increase and maintain secretion of progesterone. The estrous cycle following the mating is suppressed and ovulation is suspended, but the cyclic gonadotropin secretion is not completely inhibited. Thus, there is a subthreshold increase in gonadotropin secretion approximately 4 days after the previous ovulatory surge. This small increase of gonadotropin stimulates estrogen secretion, which serves as a preimplantation increase of estrogen (preimplantation estrogen surge) in case of pregnancy. The level of estrogen secretion, though a fraction of that seen at the preovulatory period, is sufficient to produce a proestrous vaginal smear in some rats or mice.

There is an antagonistic relationship between the gonadotropin secretory activity and that of prolactin in the pituitary. Gonadotropin secretion required for estrogen secretion is inhibited by, for example, intensive suckling stimulus of pups during lactation. Rats and mice nursing a large litter of suckling young, therefore, have high progesterone levels in circulation and very low estrogen due to reduced levels of mRNA and protein of GnRH.

The ovaries of patients with polycystic ovary syndrome (PCOS) contain many follicles which fail to ovulate. PCOS is a general term describing a heterogeneous population of patients with polycystic ovaries, amenorrhea, hirsutism, and obesity. Gonadotropin profiles indicate abnormal LH/FSH ratios with higher than normal LH values. In most cases the cystic follicles are atretic and estrogen levels are low. Some patients with PCOS have a genetic deficiency in steroidogenic enzymes including aromatase. PCOS is currently viewed as a form of functional intraovarian hyperandrogenism. Intense IGFBP-4 immunostaining is found in antral follicles from PCOS patients without hyperinsulinemia. Since IGFBP-4 has high affinity for IGF-1, its abundant presence would imply that IGF-1-mediated gonadotropin action on granulosa cells would be reduced. This results in reduced aromatase activity and a hyperandrogenic intraovarian environment.

III. ESTROGEN SECRETION DURING PREGNANCY

A. Early Pregnancy

The major pituitary hormone that maintains luteal production of progesterone in primates and many large domestic animals is LH, whereas it is prolactin in rats and mice. Thus, these small rodents represent specific and unique endocrine conditions that influence secretion of estrogen during early pregnancy. When recently mated female mice are exposed to strange males, approximately 60% of the mice terminate pregnancy and return estrus (the Bruce effect). The olfactory reproductive pheromones apparently stimulated the cyclic activity of the GnRH/gonadotropin system to abolish the recently established sustained pregnancy type activity of the prolactin/progesterone system.

In rats and mice that are pregnant during lactation, blastocyst implantation fails when lactation is intensive. Suckling stimulates prolactin secretion from the pituitary. The greater the suckling stimulus, the larger the amount of prolactin secretion. Intensive lactation suppresses pituitary gonadotropin secretion that is essential for estrogen secretion. Lactational suppression of gonadotropin secretion in rats involves increased progesterone secretion. Because prolactin is luteotropic in this species, the more suckling stimulus is applied to teats, the more prolactin is secreted and the higher is the progesterone secretion by the ovary. Reduced levels of GnRH are reported in lactating rats.

The lactation-induced suppression of gonadotropin secretion and the pheromone-induced suppression of prolactin (luteotropin) secretion in rats and mice are considered to be two extreme cases in which the delicate balance of gonadotropic activity and prolactin-secreting activity during early pregnancy is tipped to one side or the other.

B. Estrogen Biosynthesis by Blastocyst

The ability of preimplantation embryos to synthesize estrogen varies depending on species. While preimplantation embryos of mice do not synthesize estrogen, those of pigs and mares have a greater ability to synthesize estrogens compared to other species. Pig embryos gain the ability to convert progesterone into estrone and estradiol at the developmental stage at which tubular shape reaches 10–17 mm in length on Days 10.5–12 of pregnancy. Estrogen synthetic ability decreases at a later stage of development. Estrogen synthesis in preimplantation embryos is regulated by trophoblastic steroid synthetic enzyme machinery, including aromatase, as well as availability of precursors. Estradiol and estrone can be metabolized to hydroxylated catechol estrogens. Catechol estrogens are reported to be produced in uterine endometrium around the time of blastocyst implantation in mice and rabbits. Ovarian steroid hormones regulate the catechol estrogen conversion. Pig blastocysts also have the capacity to convert phenolic estrogens to their 2- and 4-hydroxylated metabolites, catechol estrogens. Catechol estrogens have been shown to be more effective than their precursors in stimulating the synthesis of prostaglandins in the uterus.

C. Estrogen Biosynthesis by the Placenta

Compartmentalization and discontinuity of the metabolic process characterize placental estrogen biosynthesis. The placenta does not have a complete set of enzymes that are needed to synthesize estrogens from acetate or cholesterol. Thus, cholesterol transported from the maternal compartment is metabolized to progesterone in the placental compartment and secreted. Dehydroepiandrosterone sulfate is transported from the maternal or fetal compartments into the placenta and dehydroepiandrosterone is aromatized to produce estrone and estradiol, which are secreted. The 16α-hydroxylated metabolite of dehydroepiandrosterone sulfate is also transported from the fetus and aromatized in the placenta to produce estriol (Fig. 5). Fetal adrenals contribute most significantly to the placental estrogen biosynthesis by providing the precursor steroids. Thus, maternal estrogen excretion drops sharply if the fetus dies, and very low estrogen excretion is observed in pregnancy with fetal anencephaly.

The ability of the placenta to secrete estrogen de-

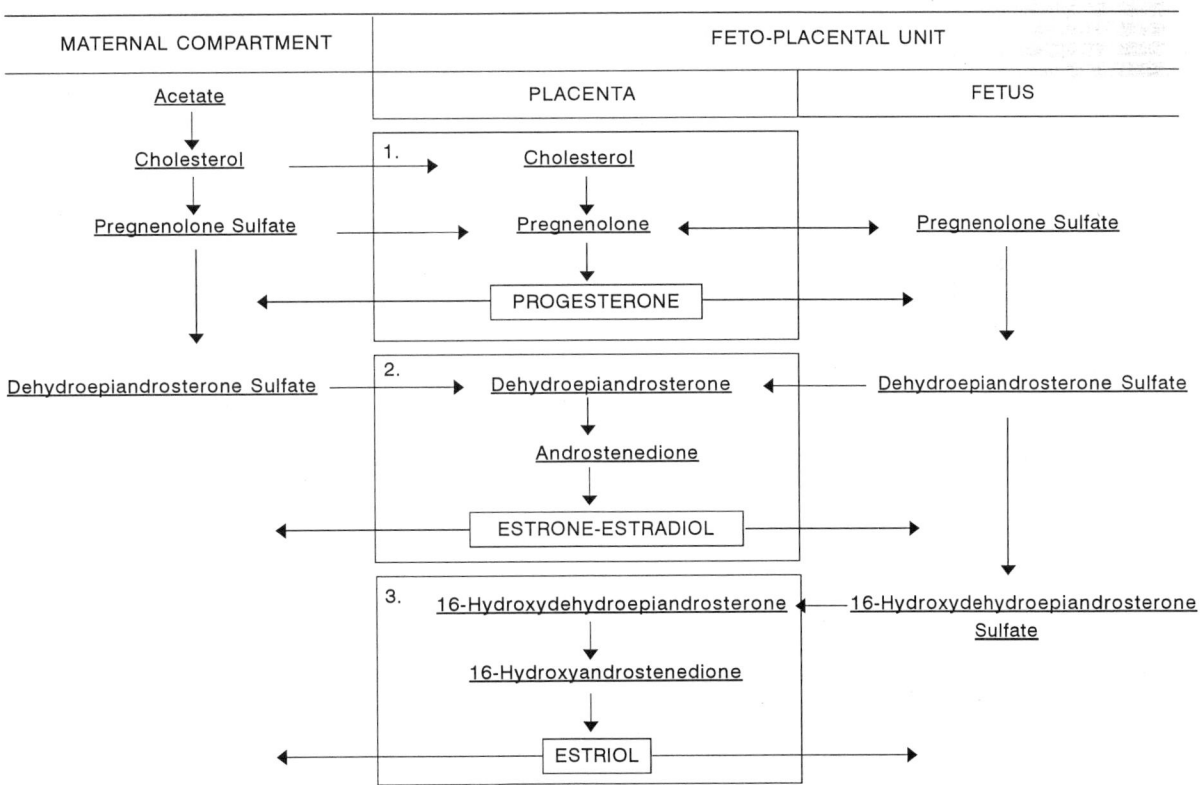

FIGURE 5 Diagrammatic representation of steroid synthesis by the placenta in the human being. Compartmentalization and discontinuity in the formation of progesterone, estradiol, and estriol from either maternal or fetal precursors are demonstrated. The pathways under No. 1, from progesterone to dehydroepiandrosterone, and under No. 2, from dehydroepiandrosterone to estrone–estradiol, have been demonstrated in a wide range of animal species [from K. J. Ryan, In *The Foeto–Placental Unit* (A. Pecile and C. Finzi, Eds.), pp. 120–131, Excerpta Medica Foundation, Amsterdam, 1969, with permission].

TABLE 1

Estrogens Formed by Mammalian Placentas from Dehydroepiandrosterone or Androstenedione *in Vitro*

Species	Estrogen formed
Human, marmoset, squirrel monkey, rhesus monkey, iris monkey, baboon, chimpanzee, organgutan	Estradiol-17β, estrone
Horse, sow	Estradiol-17β, estrone
Cow, sheep, goat	Estradiol-17β, estradiol-17α, estrone
Guinea pig, rat, dog, rabbit	None

Note. From Ryan, 1973, with permission.

pends on the presence of aromatase in the placenta. Table 1 shows those species whose placentae form estrogen from dehydroepiandrosterone or androstenedione *in vitro*. Placentae of rats, guinea pigs, rabbits, and dogs do not form estrogen. Placentae of these species lack aromatase activity. Placentae of pigs, sheep, goats, cows, horses, rhesus monkeys, and humans do have aromatase and thus secrete estrogens. The estrogen produced in the placenta acts on critical steps such as LDL receptor and mitochondrial P450scc in the biosynthetic pathway to stimulate progesterone production. After 8 or 9 weeks of pregnancy, the major source of estrogen in women's circulation is the placenta. The levels of estrogen continue to rise from the second week throughout the gestation period until term (Fig. 6).

D. Periparturition Estrogen Secretion

In humans the steroidogenic activity of the corpus luteum declines around the fourth week of pregnancy and the placenta takes over the ovarian steroidogenic role (the luteal placental shift). The luteal placental shift completes by the seventh week. Peripheral levels of estrogens steadily increase during pregnancy, reflecting the progressing placental steroidogenic activity (Fig. 6). Prior to parturition there is no consistent increase or decrease in circulating levels of estrogens

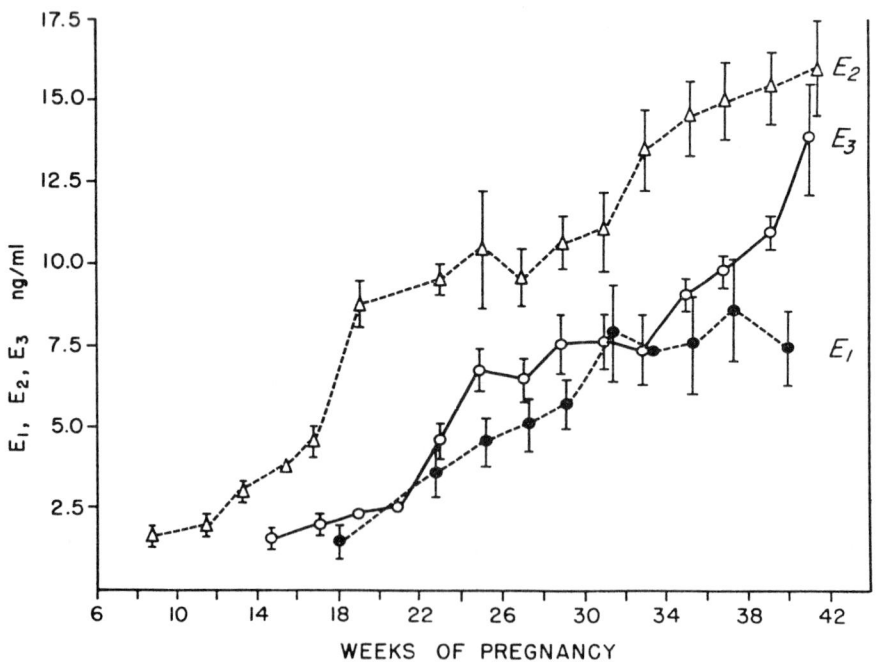

FIGURE 6 Mean plasma estrogen levels during human pregnancy [redrawn by D. Tulchinsky, In *Maternal–Fetal Endocrinology* (D. Tulchinsky and A. B. Little, Eds.), 2nd ed., Saunders, Philadelphia, 1994; originally from D. Tulchinsky, C. J. Hobel, E. Ueager, and J. R. Marshall, Plasma estrone, estradiol, progesterone and 17-hydroxyprogesterone in human pregnancy-normal pregnancy, *Am. J. Obstet. Gynecol.* **112**, 1095, 1972, with permission].

or progesterone in humans. Because the source of estrogen secretion in humans is the placenta, estrogen levels drop precipitously after parturition. Because steroidogenic activity of the placenta dominates after the luteal–placental shift, the corpus luteum remains relatively inactive. After parturition a decreased luteal function continues for 2 or 3 days and, if there is no lactation, normal ovarian function resumes in approximately 6 weeks. Term corpus luteum secretes little estrogen. In humans there is no immediate postpartum ovulation.

In contrast to humans, the placenta of small rodents such as mice and rats does not secrete estrogen because it lacks aromatase. In these animals, the ovaries secrete estrogen. Comparison of P450arom mRNA and aromatase contents in follicles and corpora lutea of preovulatory and pregnant ovaries indicates that their contents are regulated by different hormonal factors and that the capacity of follicles and corpora lutea to synthesize estradiol is not strictly related to the changes in the contents of P450arom mRNA and enzyme contained within these tissues.

Compared with low estrogen secretion levels during the first half of pregnancy, estrogen secretion levels are slightly elevated during the second half of pregnancy. This appears to reflect an increase in gonadotropin secretion coincidental with the switch of the luteotropic role from pituitary prolactin to placental luteotropin. Ovarian estrogen secretion increases 3 or 4 days prior to parturition in rats (Fig. 7). The secretion continues to rise to a peak on the day of parturition, when the level is equal to the preovulatory level on the day of proestrus of the estrous cycle. On the day of parturition, rat ovaries contain large preovulatory follicles and ovulation takes place within 24 hr after parturition. The females become pregnant again when fertilization takes place at the postpartum ovulation. Estrogen secretion is, however, suppressed by lactation, and the degree of suppression correlates with the intensity of the suckling stimulus.

Ovulation resumes in nonnursing postpartum women between 6 and 10 weeks postpartum. Breast-feeding usually delays resumption of ovulation and menses. This suppressive effect of lactation on ovarian cyclicity is related to an increase in prolactin

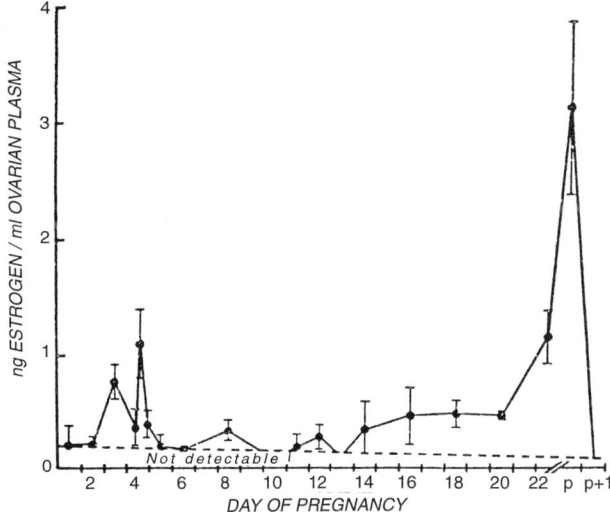

FIGURE 7 Estrogen concentration in ovarian venous plasma during pregnancy in the rat (from K. Yoshinaga, A. R. Hawkins, and J. F. Stocker, Estrogen Secretion by the Rat Ovary *in vivo* during the Estrous Cycle and Pregnancy, *Endocrinology* **85**, 103, 1969, © The Endocrine Society, with permission).

secretion that suppresses gonadotropin secretion in women.

IV. ESTROGEN BIOSYNTHESIS BY OTHER TISSUES

When a cell expresses the aromatase complex, it can synthesize estrogen provided that precursors are available and other cellular conditions are met. Endocrine organs that have the ability to produce estrogens are the testis and the adrenal gland. Although the testis is the major organ to produce androgen, the precursor of estrogen, a small quantity of estrogen is secreted by testis. The aromatase is localized in interstitial tissue. In many species of mammals, including humans, conversion of androgen to estrogen by testicular tissues has been reported to be low, as 5α-reductase converts testosterone to dihydrotestosterone. In stallions, however, the testis produces large amounts of estrogen. The concentration of conjugated estrogens in the stallion testis is about four times greater than the concentration of free estrogen.

Equine aromatase may convert estradiol to 2-hydroxy estradiol. Aromatase has been localized in the adrenal cortex. Regulatory mechanisms of estrogen secretion in these endocrine organs are not clear and abnormal conditions such as some cases of 5α-reductase deficiency provide opportunity for further investigation. Deficiency of 5α-reductase results in a potential increase in estrogen secretion because testosterone cannot be converted to dihydrotestosterone.

In human adipose tissue aromatase activity is in the stromal tissue and not in adipocytes. Growth factors are implicated to have some effect on adipose tissue aromatase activity. In view of the presence of adipose tissue surrounding mammary glands, the locally produced estrogen may be considered a factor in the development of certain forms of breast cancer.

In the brain, aromatase is localized in the medial preoptic nucleus, the bed nucleus of the stria terminalis, the medial amygdaloid nucleus, and the ventromedial nucleus. Localized estrogen production in these brain nuclei appears to play important roles in reproductive behavior and hormone secretion. Aromatase has been reported to be expressed in some neoplastic tissues, including breast cancer cells, endometriosis-derived stromal cells, endometrial carcinoma, estrogen-induced renal tumors in hamsters, and rat arterial smooth muscle cells in culture.

See Also the Following Articles

Blastocyst; Bruce Effect; DHEA (Dehydroepiandrosterone); Follicular Steroidogenesis; FSH (Follicle-Stimulating Hormone); Gonadotropin Biosynthesis; Gonadotropin Secretion, Control of; LH (Luteinizing Hormone)

Bibliography

Adashi, E. Y. (1991). The ovarian life cycle. In *Reproductive Endocrinology* (S. S. C. Yen and R. B. Jaffe, Eds.), pp. 181–237. Saunders, Philadelphia.

Falcone, T., and Little, A. B. (1994). Placental synthesis of steroid hormones. In *Maternal–Fetal Endocrinology* (D. Tulchinsky and A. B. Little, Eds.), 2nd ed., pp. 1–14. Saunders, Philadelphia.

Fitzpatrick, S. L., Carlone, D. L., Robker, R. L., and Richards, J. S. (1997). Expression of aromatase in the ovary: Down-regulation of mRNA by the ovulatory luteinizing hormone surge. *Steroids* **62**, 197–206.

Gore-Langton, R. E., and Armstrong, D. T. (1988). Follicular steroidogenesis and its control. In *The Physiology of Reproduction* (E. Knobil and J. D. Neill, Eds.), pp. 331–385. Raven Press, New York.

Gougeon, A. (1996). Regulation of ovarian follicular development in primates: Facts and hypothesis. *Endocr. Rev.* **17**, 121–155.

Hillier, S. G., Whitelaw, P. F., and Smyth, C. D. (1994). Follicular oestrogen synthesis: The "two-cell, two gonadotrophin" model revisited. *Mol. Cell. Endocrinol.* **100**, 51–54.

O'Malley, B. W., and Strott, C. A. (1991). Steroid hormones: Meatbolism and mechanism of action. In *Reproductive Endocrinology* (S. S. C. Yen and R. B. Jaffe, Eds.), pp. 156–180. Saunders, Philadelphia.

Pepe, G. J., and Albrecht, E. D. (1995). Actions of placental and fetal adrenal steroids in primate pregnancy. *Endocr. Rev.* **16**, 608–648.

Ryan, K. J. (1973). Steroid hormones in mammalian pregnancy. In *Handbook of Physiology, Section 7: Endocrinology, Vol. II. Female Reproductive System*, Part 2, pp. 285–293. American Physiological Society, Washington, DC.

Estrogen Synthesis

see Steroidogenesis, Overview

Estrogens, Overview

Carolyn L. Smith

Baylor College of Medicine

GLOSSARY

activation function One of two specific regions within the estrogen receptor that contributes to the ability of the receptor to activate target gene transcription. One activation function is located in the amino terminus of the estrogen receptor. The other overlaps the ligand-binding domain in the carboxy terminus.

cytochrome P450 A superfamily of more than 220 genes grouped into 36 gene families that encode heme-based enzymes. Three of the enzymes required for estrogen biosynthesis (cholesterol side chain cleavage enzyme, 17α-hydroxylase, and aromatase) are members of this superfamily.

estrogen receptor A 66-kDa nuclear protein that binds to estrogens with high affinity. The estrogen receptor is a ligand-regulated transcription factor.

estrogen response element A palindromic nucleotide sequence that is specifically recognized and bound by the DNA-binding domain of estrogen receptors. The consensus estrogen response element is GGTCAnnnTGACC.

sex hormone binding globulin A glycoprotein found in the blood that binds some androgens and estrogens with relatively high affinity. This protein contributes to the regulation of the amount of free estrogen present in plasma.

Estrogens are a class of steroid hormones that play an important role in female reproductive processes. They are produced primarily in the ovary and exert their biological effects via binding to and activating the estrogen receptor. Although estrogens are considered female sex steroids, they are present in low levels in men.

I. BACKGROUND

It was noted prior to the 1900s that a number of changes in the reproductive tract occur as a result of ovariectomy, and the ability of ovaries transplanted into castrated female animals to ameliorate these effects demonstrated that this gland produced a substance(s) that acted in an endocrine manner. This conclusion was supported by the observation that accessory reproductive organs, such as the vagina and uterus, underwent cyclical changes in a fashion similar to the ovaries. Because the substance produced by the ovary induced estrus, it was referred to as estrogen. Allen and Doisy were able to isolate lipoidal substances from the large ovarian follicles of swine and demonstrate that this material could induce estrous changes in the vagina of rats. Subsequently, the laboratories of Doisy and Butenandt crystallized a low-molecular-weight lipophilic steroidal substance from the urine of pregnant women which they originally termed "theelin" and later renamed estrone. Estrogens were not purified from ovarian tissue until 1935, when approximately 12 mg of 17β-estradiol was isolated from 4 tons of sow's ovaries.

It was quickly established that these purified steroids could induce growth stimulation of reproductive tract tissues, and research efforts were then focused on defining the mechanism(s) by which estrogens exerted their biological effects. An early hypothesis proposed that estrogen action is linked with its participation in the enzymatic processes of steroid metabolism. This concept was put aside after experiments with tritium-labeled estradiol, synthesized by Jensen in 1957, established that estradiol concentrates in uterine cells and induces growth without itself undergoing chemical change. The subsequent demonstration that increasing amounts of the antiestrogen nafoxidine could inhibit rat uterine uptake of estradiol as well as estrogen-stimulated cell growth corroborated that steroid binding to a bona fide receptor is required for estrogen action.

II. STRUCTURAL FEATURES

Estrogens are steroid hormones consisting of a cyclopentanoperhydrophenanthrene structure. The estrane steroid nucleus, which is common to all estrogens and distinct from other classes of steroids such as androgens and progestins, contains 18-carbon atoms arranged into four rings as shown in Fig. 1 (top). Several features of estrogens are critical for their hormonal activity, the most notable of which is the aromatic A ring with the phenolic hydroxyl group located at the C3 position. An oxygen atom

in the form of a hydroxyl or ketone group is also required at the C17 position.

The most potent, naturally occurring estrogen is 17β-estradiol, which has 3-hydroxyl and 17β-hydroxyl groups (Fig. 1, bottom). In humans, there are two other major forms of estrogen in the circulation, estrone and estriol. In contrast to 17β-estradiol, estrone has a ketone group at the C17 position instead of a hydroxyl, whereas estriol differs from 17β-estradiol by the addition of a 16α-hydroxyl group. The biological activity of estrogens is determined, in part, by their ability to bind to and activate the estrogen receptor (ER). Both estrone and estriol are less potent estrogens than 17β-estradiol, and this is reflected in the reduced affinity the ER has for these steroids.

III. SOURCES OF ESTROGENS

A. Biosynthesis

Humans produce several different estrogens whose biosynthesis occurs in a variety of tissues. The predominant estrogen and its primary site of biosynthesis varies with gender and age. In premenopausal

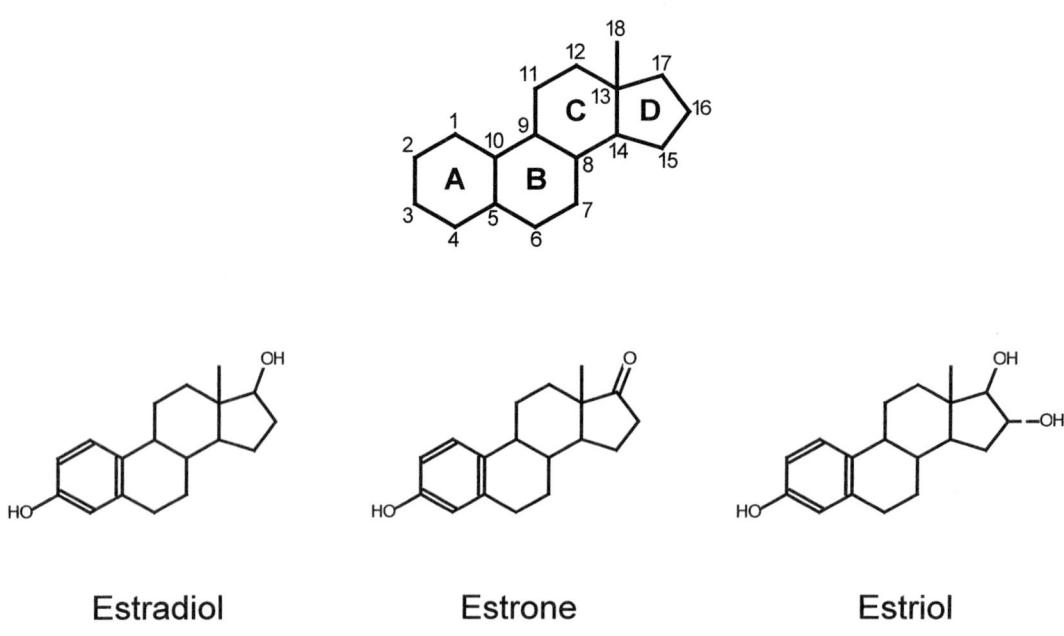

Estradiol   **Estrone**   **Estriol**

FIGURE 1   Chemical structures of the C18 estrane steroid nucleus (top) and several naturally occurring steroidal estrogens (bottom). The 18-carbon atoms in the estrane molecule are numbered.

Table 1
Plasma Estrogen Levels in Premenopausal Women

Compound	Menstrual cycle phase	Concentration in plasma	
		nmol/liter	pg/ml
17β-Estradiol	Early follicular	0.2	60
	Late follicular	1.2–2.6	330–700
	Midluteal	0.7	200
Estrone	Early follicular	0.18	50
	Late follicular	0.5–1.1	150–300
	Midluteal	0.4	110

Note. From B. R. Carr, The ovary, In *Textbook of Reproductive Medicine* (B. R. Carr and R. E. Blackwell, Eds.), p. 199, Appleton & Lange, Norwalk, CT, 1993.

women, 17β-estradiol is the most abundant estrogen, and the majority of this steroid is produced by the granulosa cells of the ovary. Relatively high levels of estrone also are present in cycling women, but little originates in the ovary. Most estrone is generated by the aromatization of androstenedione in adipose tissue. Aromatase has also been detected in skin, skeletal muscle, hair follicles, bone tissue, several sites in the brain (e.g., hypothalamus and amygdala), and the Leydig cells of the adult testes. Plasma estrogen levels vary with respect to the stage of menstrual cycle, and the highest levels are found in the late follicular stage (Table 1). During pregnancy, estriol, which is produced by the placental aromatizaton of fetal-derived androgens, is the most abundant estrogen and its measurement can be used as an index of fetal/placental functions. In postmenopausal women and in men, relatively low levels of estradiol (<20 pg/ml) are generated by the peripheral conversion of androgens to estrogens.

Estrogens, like all other steroids, are derived from cholesterol synthesized *de novo* from two-carbon units (acetyl coenzyme A) or obtained via uptake of cholesterol synthesized in the liver and transported to steroidogenic tissues by low-density lipoproteins. Estrogen biosynthesis requires three heme-based, cytochrome P450 enzyme complexes and two different dehydrogenases. The rate-limiting step in all steroid synthesis is the removal of the cholesterol side chain by the cytochrome P450 side chain cleavage enzyme

which results in the generation of the C21 steroid, pregnenolone. Subsequent processing by 17α-hydroxylase cytochrome P450, which converts the steroid from a C21 to a C19 structure, and 3β-hydroxysteroid dehydrogenase, which oxidizes the hydroxyl group at the C3 position to a ketone group, results in the synthesis of the C19 estrogen precursor, androstenedione (Fig. 2). The cytochrome P450 aromatase enzyme (P450arom) then catalyzes the unsaturation and aromatization of the A ring to produce estrone. Both androstenedione and estrone are substrates for 17β-hydroxysteroid dehydrogenase (17β-HSD), which can reversibly convert these oxosteroids to their 17β-hydroxysteroid counterparts, testosterone and 17β-estradiol, respectively. Several distinct forms of 17β-HSD have been cloned. The type I 17β-HSD is capable of both oxidation and reduction, whereas the type II enzyme has a higher oxidation activity. In ovary and placenta, the major C18 steroid produced has a 17β-hydroxyl group, suggesting that the predominant 17β-HSD isoforms present in these tissues readily reduce oxosteroids to 17β-hydroxysteroids. The relative distribution of 17β-HSD isoforms has the potential to modulate the availability of the potent 17β-hydroxy-estrogen, 17β-estradiol, to the ER in target tissues.

Most cells do not express the complete complement of steroid biosynthetic enzymes required to synthesize estrogens. Thus, the predominant C18 estrogen produced by a given tissue depends on the C19 steroid presented to the aromatase enzyme complex. For example, granulosa cells lack 17α-hydroxylase activity and are therefore dependent on an external source for the C19 steroid necessary for estradiol biosynthesis. This precursor is synthesized in the thecal cell layer of the ovarian follicle. The formation of estrone by adipose tissue utilizes androstenedione made in the adrenal cortex as its C19 precursor. Estriol is produced by the syncytiotrophoblast within the placenta from 16α-hydroxydehydroisoandrosterone sulfate produced by the combined activities of the fetal adrenal and liver.

B. Pharmacological Substances

The major therapeutic uses of estrogens are estrogen replacement therapy to postmenopausal women

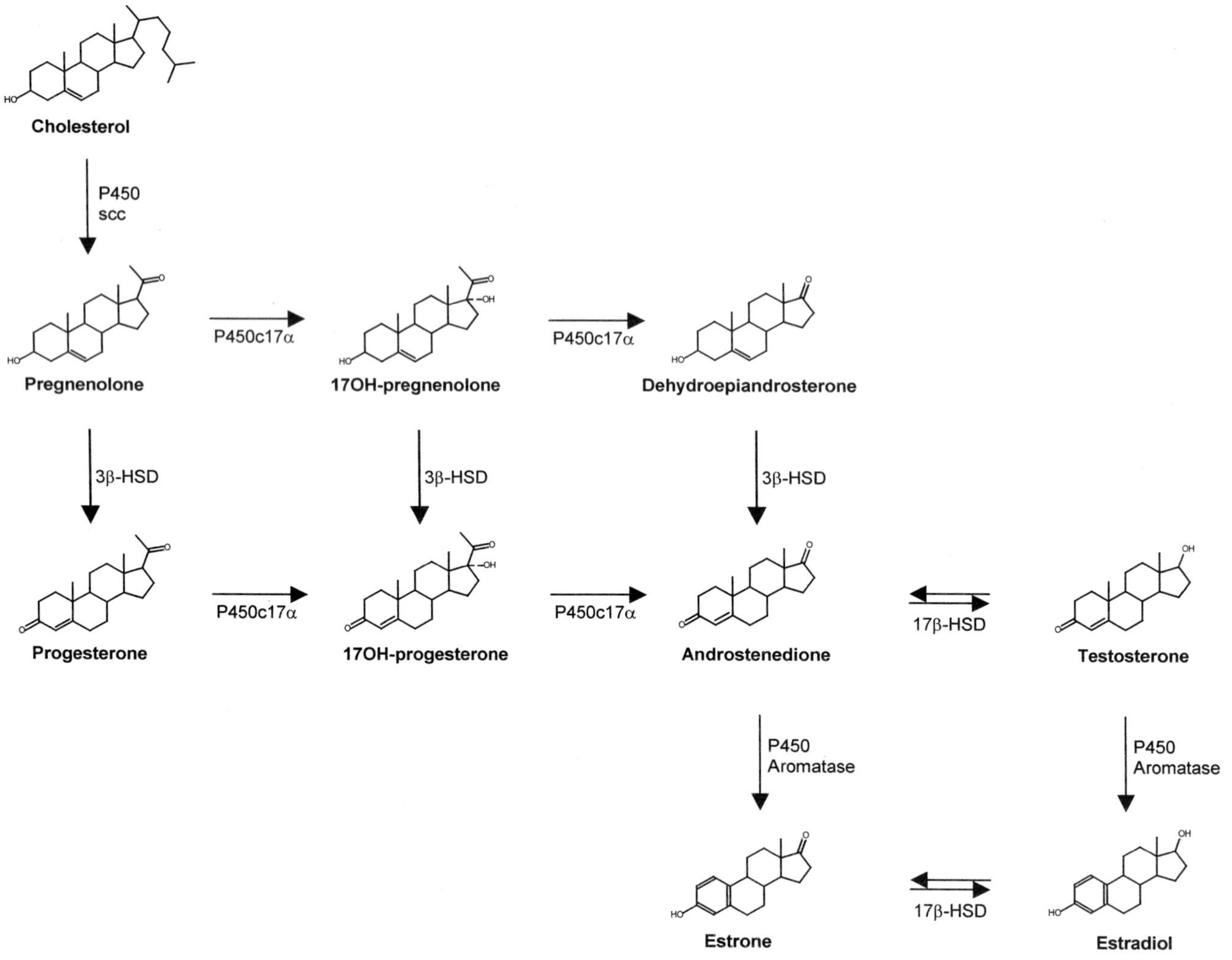

FIGURE 2 Major pathways of steroid hormone biosynthesis in the human ovary.

and, in combination with progestins, as oral contraceptives. Although 17β-estradiol is the most potent naturally occurring estrogen, it has very poor activity when administered orally. Several different types of chemical modifications can be made to estrogens to increase their oral bioactivity (Fig. 3). One class of pharmacological estrogens with prolonged activity is esterified at the C17-hydroxyl group with benzoate, valerate, or cyclopentylpropionate. Deesterification of these compounds via esterases produces biologically active estradiol *in vivo*. A second type of pharmacological estrogen is 17α-ethinyl estradiol, which has an ethinyl group (C $\equiv$ CH) linked to C17 of

the estradiol molecule. This compound is the most widely used synthetic oral estrogen and is the estrogen component found in most oral contraceptives. It is worth noting that antiestrogens also have been developed for clinical uses, such as the treatment of breast cancer. These ER antagonists may or may not be steroidal in nature.

C. Nonsteroidal Estrogens

A number of synthetic, nonsteroidal compounds (Fig. 4) are able to mimic estrogen-like effects. This group includes estrogen analogs such as diethylstil-

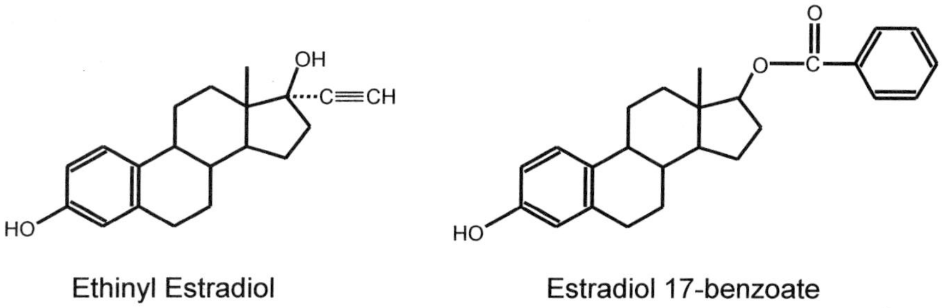

Ethinyl Estradiol   **Estradiol 17-benzoate**

FIGURE 3 Chemical structures of two synthetic steroidal estrogens.

bestrol (DES), which was generated for clinical use. DES binds to the ER with high affinity yet does not bind to the plasma transport protein, sex hormone-binding globulin (see Section IV), and is therefore a more potent estrogen than 17β-estradiol. In addition, there are a variety of chemicals found in the environment that possess weak estrogenic activity; they are referred to as "xenoestrogens." This group of compounds includes a variety of industrial products, including pesticides such as DDT and other organochlorine compounds (e.g., polychorinated biphenols) as well as the monomer used to produce polycarbonate plastics, bisphenol A. Many of these molecules have a phenolic ring structure and hydroxyl groups that probably contribute to their abil-

ity to bind to the ER with moderate affinity. Although the impact of these molecules on human health is controversial, these chemicals are able to mimic estrogen effects in laboratory tests and several have been shown to disrupt reproductive processes in wildlife, suggesting that they may impinge on estrogen signaling pathways in nature.

In addition to the man-made xenoestrogens discussed previously, there are a number of naturally occurring substances that exhibit estrogenic activity. These chemicals are produced by plants and are more specifically referred to as phytoestrogens. The most abundant phytoestrogen is genistein, which is found in high concentration in soybeans. These compounds can bind to the ER with varying degrees of effective-

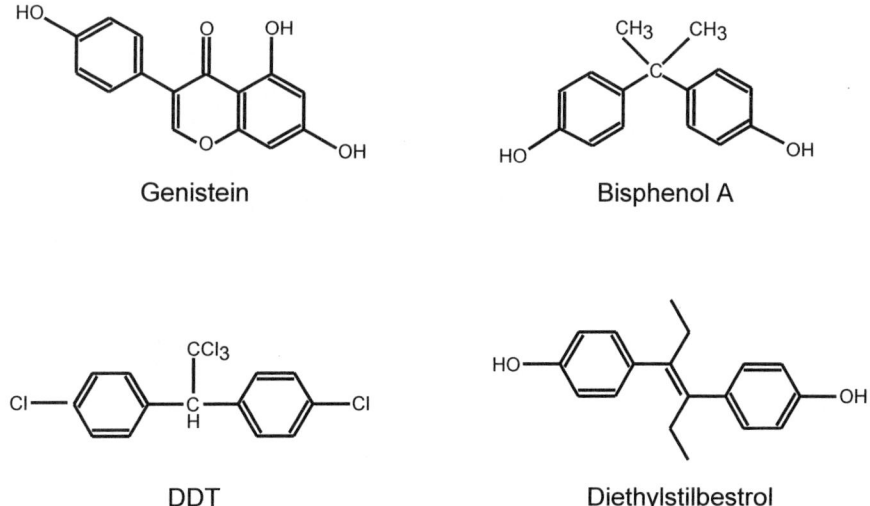

Genistein   **Bisphenol A**

DDT   **Diethylstilbestrol**

FIGURE 4 Chemical structures of selected environmental compounds with estrogenic activity. Genistein is a phytoestrogen, whereas the other three compounds are man-made.

ness. However, the mechanism by which they exert their estrogenic effects *in vivo* is a matter of ongoing investigation.

IV. TRANSPORT AND METABOLISM

After secretion into the circulation, only 2 or 3% of biosynthetic estradiol remains free in plasma. Of the remaining amount, approximately 60% is weakly bound to albumin and ~38% is bound to a 95-kDa glycoprotein, sex hormone-binding globulin (SHBG) also known as testosterone–estradiol-binding globulin. As the latter name implies, this protein also binds to testosterone and dihydrotestosterone with an affinity approximately five- and threefold greater, respectively, than estradiol. It is generally accepted that the protein-bound hormone is inactive, and that only the free fraction is able to enter target cells and exert biological effects. The metabolic clearance rate of estrogens is inversely related to their affinity for SHBG and other plasma-binding proteins. The concentration of SHBG, therefore, influences estrogen metabolism and the amount of hormone available for entry to the target cell. Plasma SHBG levels are altered in response to a number of clinical conditions, including pregnancy, diabetes, and obesity. In addition, estrogens increase SHBG levels and are likely responsible for the twofold higher levels of SHBG present in women in comparison to men.

Circulating estrogens are quickly metabolized in the liver to inactive, water-soluble compounds with the net result being that 50–80% of modified estrone and estradiol is excreted in urine and up to 20% in feces. Estrone sulfate derived from estrone and estradiol is the most abundant estrogen in blood. A second major class of catabolic products is the catechol estrogens, primarily 2-hydroxyestrogens. Catechol estrogen also can be produced at C4, but only in relatively small amounts. The majority of these compounds end up in the urine as glucuronosides. Alternatively, catechol *O*-methyl transferase metabolizes catechol estrogens to their methoxy forms, which also have a high metabolic clearance rate. Although several studies have indicated that catechol estrogens may have relatively weak biological activities in comparison to estradiol, the prevailing view

is that these compounds are primarily the catabolic products of estrogens. Estrogens also may be taken up by the liver and hydroxylated at the C16 position of the D ring. This pathway leads to the formation of 16α-hydroxylated derivatives, primarily estriol–glucuronide, which are rapidly cleared in the urine.

V. MECHANISM OF ESTROGEN ACTION

Estrogens are lipophilic substances which exert their actions by passively diffusing across the plasma membrane and binding to high-affinity receptor proteins, known as estrogen receptors. The ER is a member of the nuclear receptor superfamily, which includes receptors for other steroids (e.g., progesterone and androgens), vitamins [e.g., vitamin D and vitamin A (retinoids)], and thyroid hormone. Like other members of the superfamily, the ER is a ligand-inducible transcription factor. Upon binding to estrogen agonists, it becomes "activated" and thereby able to positively regulate the expression of target genes. The alteration in the expression of these genes and the proteins that they encode ultimately results in an estrogenic response within a given target cell and/or tissue.

Alignment of the deduced amino acid sequences of ER cDNAs cloned from a number of species, including human, rat, mouse, chicken, trout, and *Xenopus*, reveals that the ER protein is composed of six functional domains, designated A–F (Fig. 5). The ligand-binding domain (E domain) is located within the carboxyl-terminal region of the ER and this region alone (amino acids 302–595) is sufficient for

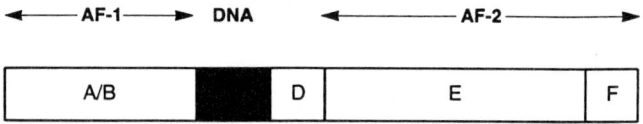

FIGURE 5   The domain structure of the estrogen receptor. The primary amino acid sequence is divided into six domains (A–F) based on sequence homology between the human and chicken ERs. Domain C (represented by the solid box) is responsible for DNA binding. The regions corresponding to the transcriptional activation function domains (AF-1 and AF-2) are shown.

high-affinity estradiol binding comparable in magnitude to that of native receptor (k_d, 0.4 nM). Experiments indicate that upon hormone binding, the ligand-binding domain undergoes a conformational change that includes folding into a more compact structure. This change in ligand-binding domain conformation varies with the nature of the ligand (agonist versus antagonist) and is thought to facilitate interactions of the ER with other proteins necessary for the activation of target gene expression.

The most highly conserved region of ER is the DNA-binding domain (region C). This relatively cysteine-rich sequence forms a structural motif proposed to form two type II zinc fingers. Each finger is composed of a peptide loop of 12 or 13 amino acids stabilized at its base by the coordinate binding of a single zinc atom by four invariant cysteine residues. It is this region in the context of the surrounding amino acids which recognizes a specific DNA sequence referred to as an estrogen response element (ERE). Estrogen response elements are *cis*-acting DNA enhancer sequences which confer steroid sensitivity regardless of orientation or position with respect to their target gene promoters. They are typically located in the 5′-regulatory regions of estrogen-responsive genes but have been found overlapping translational initiation sites or within the first coding exon of target genes. The ER binds to EREs as a homodimer, and sequences within its DNA and ligand-binding domains mediate receptor–receptor interactions. The consensus ERE sequence (GGTCAnnnTGACC) is composed of a palindromic repeat of GGTCA separated by three nucleotides. Although this sequence has been identified in some ER target genes (e.g., the *Xenopus* vitellogenin A2 gene), variations of this sequence (e.g., *TGTCAnnnTGTCC* found in the rat prolactin gene) also can serve as functional EREs.

The ER is a phosphoprotein, and under basal conditions the receptor is phosphorylated to a relatively low extent. However, upon the addition of estradiol, the overall level of ER phosphorylation is markedly increased. Five phosphorylation sites have been mapped in the human ER; four in the amino terminus (Ser104, Ser106, Ser118, and Ser167) and one within the ligand-binding domain (Tyr537). Estrogen exposure increases the relative phosphorylation of all four

serine residues, and several laboratories have identified Ser118 as the major phosphoacceptor site in ER purified from estradiol-treated cells. Estrogen treatment does not appear to enhance phosphorylation of Tyr537. Although mutation of Ser118 to an alanine residue diminishes the estradiol-induced transcriptional activity of the mutated receptor relative to wild-type receptor, the precise role ER phosphorylation plays in receptor activation is unclear.

In its role as a transcription factor, the ER possesses two activation domains. One is located in the A/B domain [activation function-1 (AF-1)] and is constitutively active. The second activation function (AF-2) is found in the carboxyl-terminal portion of the receptor (domains E and F). This latter activation domain overlaps the ligand-binding domain, and its ability to activate target gene expression is dependent on estrogen binding. Both deletion and point mutation analysis indicated that the sequences required for AF-2 function reside primarily within domain E. The F domain, while not required for AF-2 activity, does modulate the ability of estradiol to stimulate gene transcription and of antiestrogens to inhibit estrogen-activated gene expression. Depending on the cell and promoter context, ER transcriptional activity is determined primarily by AF-1, AF-2, or a synergistic combination of the activities of both domains.

Recently, it has been shown that the ER binds to a number of nuclear proteins that act as bridging factors between the ER and the general transcriptional machinery. These putative "coactivators" bind to the ER in an estrogen-dependent manner, presumably as a result of the conformational changes in the structure of the ligand-binding domain, and enhance the ability of receptor to activate gene expression. Although the precise mechanism by which coactivators potentiate the transactivation activity of the ER (and other members of the nuclear receptor family) is unknown, several coactivators possess histone acetylase activity. It is possible that core histone hyperacetylation induces alterations in chromatin structure that ultimately promote transcription initiation by the general transcriptional machinery.

The previous discussion has dealt primarily with studies of the classical ER, whose existence has been appreciated for more than 40 years. However, in

1996, Gustafsson and colleagues identified a second, distinct estrogen receptor by molecular biological approaches. This novel receptor was called ERβ, and the protein originally identified as "the estrogen receptor" is now referred to as ERα. ERβ also is a member of the nuclear receptor family and binds to 17β-estradiol with high affinity. The DNA-binding domains of the ERα and ERβ are very highly conserved (~96% identity) and the ligand-binding domains also share a significant degree of homology (~55% identity). It is notable that the A/B domain is very poorly conserved (<20% identity). In a fashion similar to the classical ERα, estradiol can activate ERβ and thereby increase the expression of ERE-containing, synthetic target genes. Current research efforts are focused towards obtaining a clearer understanding of both the similarities and differences in ERα and ERβ expression patterns, ligand-binding characteristics, and transcription factor activities.

See Also the Following Articles

ANTIESTROGENS; ESTROGEN ACTION, BEHAVIOR; HORMONAL CONTRACEPTION; SHBG (SEX HORMONE-BINDING GLOBULIN); STEROID HORMONE RECEPTORS

Bibliography

Carr, B. R. (1992). Disorders of the ovary and female reproductive tract. In *Williams Textbook of Endocrinology* (J. D. Wilson and D. W. Foster, Eds.), pp. 733–798. Saunders, Philadelphia.

Henzl, M. R. (1996). Synthetic sex steroids. In *Reproductive Endocrinology, Surgery and Technology* (E. Y. Adashi, J. A. Rock, and Z. Rosenwaks, Eds.), pp. 587–605. Lippincott-Raven, Philadelphia.

Jensen, E. V. (1997). History of the estrogen receptor concept and its relation to antiestrogens. In *Estrogens and Antiestrogens: Basic and Clinical Aspects* (R. Lindsay, D. W. Dempster, and V. C. Jordan, Eds.), pp. 3–8. Lippincott-Raven, Philadelphia.

Korach, K. S., Migliaccio, S., and Davis, V. L. (1995). Estrogens. In *Principals of Pharmacology: Basic Concepts and Clinical Applications* (P. L. Munson, R. A. Mueller, and G. R. Breese, Eds.), pp. 809–825. Chapman & Hall, New York.

Kuiper, G. G. J. M., Enmark, E., Pelto-Huikko, M., Nilsson, S., and Gustafsson, J.-A. (1996). Cloning of a novel estrogen receptor expressed in rat prostate and ovary. *Proc. Natl. Acad. Sci. USA* 93, 5925–5930.

Robinson, J. A., and Spelsberg, T. C. (1997). Mode of action at the cellular level with specific reference to bone cells. In *Estrogens and Antiestrogens: Basic and Clinical Aspects* (R. Lindsay, D. W. Dempster, and V. C. Jordan, Eds.), pp. 43–62. Lippincott-Raven, Philadelphia.

Rosner, W. (1996). Sex steroid transport: Binding proteins. In *Reproductive Endocrinology, Surgery and Technology* (E. Y. Adashi, J. A. Rock, and Z. Rosenwaks, Eds.), pp. 605–625. Lippincott-Raven, Philadelphia.

Simpson, E. R. (1996). Estrogens. In *Reproductive Endocrinology, Surgery and Technology* (E. Y. Adashi, J. A. Rock, and Z. Rosenwaks, Eds.), pp. 477–492. Lippincott-Raven, Philadelphia.

Tsai, M.-J., and O'Malley, B. W. (1994). Molecular mechanisms of action of steroid/thyroid receptor superfamily members. *Annu. Rev. Biochem.* 63, 451–486.

Estrone

see Estrogens, Overview

Estrous Cycle

Signe M. Kilen and Neena B. Schwartz

Northwestern University

GLOSSARY

atresia Degeneration of the follicle and the oocyte.

corpus luteum A structure that develops from granulosa and thecal cells following ovulation; a predominant site of progesterone secretion.

follicular stimulating hormone (FSH) A hormone secreted by the gonadotropes in the anterior pituitary gland; stimulates ovarian follicular growth and steroidogenesis.

folliculogenesis Maturation of the follicle.

gonadotropin-releasing hormone (GnRH) A decapeptide hormone secreted by hypothalamic neurons; controls the release of LH and FSH.

inhibin A glycoprotein produced primarily in the ovary; feeds back to the pituitary to selectively inhibit release of FSH.

luteinizing hormone (LH) A hormone secreted by the gonadotropes in the anterior pituitary pituitary; together with FSH stimulates ovarian follicular growth and estrogen secretion; induces ovulation and production of the corpus luteum.

oogenesis Growth and maturation of the oocyte.

prolactin A hormone secreted by the lactotropes in the anterior pituitary; controls mammary growth and lactogenesis in addition to its luteotropic activity in the rodent.

steroids Estradiol and progesterone; they are secreted by the ovarian follicle and corpus luteum, respectively; they exert negative and positive feedback activity on hypothalamus and gonadotropes; regulate growth of vagina, oviducts, and uterus; and cause sexual behavior by actions on the brain.

The estrous cycle is the sequence of reproductive processes in mammalian species and, unlike the primate menstrual cycle, is characterized overtly by mating behavior. In fact, the word "estrous," originating from the Greek word "*oistros*," means "frenzy." The behavior of a female "in heat" coincides with the event of ovulation, ensuring the opportunity for fertilization and pregnancy. At this time, she will become receptive and proceptive to the advances of the male.

I. INTRODUCTION

The estrous cycle is a cascade of hormonal and behavioral events which are progressive, highly synchronized, and repetitive. Reproductive cycles are remarkably precise in their nature but demonstrate great diversity among species. Reproductive cycles have adapted under the constraints of the various environments in which different species survive. Nowhere are ecologically correlated adaptations more prominently expressed than in the mammalian estrous cycle. This chapter will outline the elements and cyclic events common to all estrous cycles and describe some features that demonstrate aspects of diversification. We will then focus on the estrous cycle in the rat.

The underlying ovarian events of the estrous cycle are the progressive growth and maturation of follicles, ovulation of the eggs from the ovary, and the subsequent luteinization of the ruptured follicles producing corpus luteum. In the absence of pregnancy, the cycle repeats itself. The key events that occur can be described as follows: (i) The initial stages of oogenesis and folliculogenesis are independent of pituitary hormones, in contrast to the final stages of follicular growth and maturation, which are highly dependent on the two gonadotropins, luteinizing hormone (LH) and follicular stimulating hor-

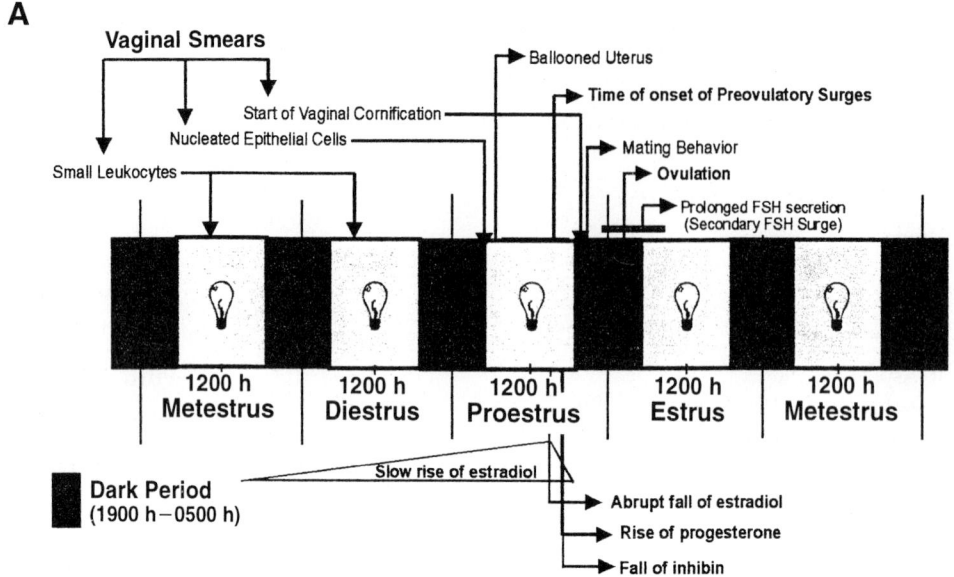

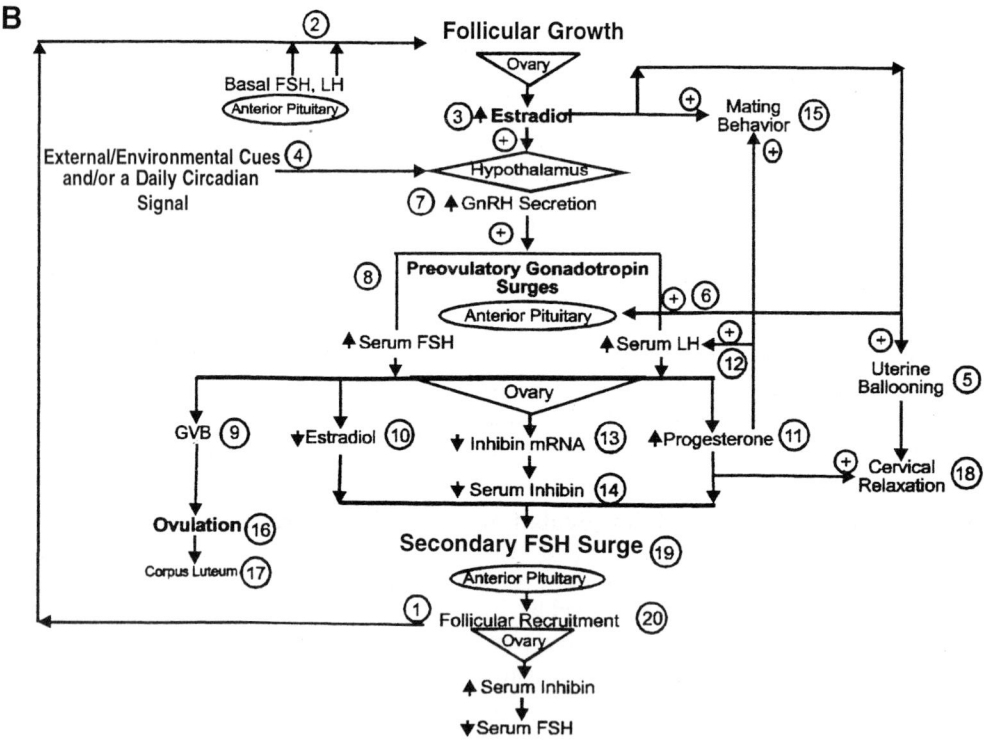

FIGURE 1 (A) Cycle stages and the timing of events in the 4-day estrous cycle of the rat. Serum estradiol starts to rise slowly on the evening of metestrus, priming the reproductive axis for the ovulation-related events that occur on the day of proestrus. On the morning of proestrus, the uterus is ballooned due to accumulation of intrauterine fluid. Serum estradiol reaches a critical "threshold" level on the day of proestrus and, at the time of the daily neural signal, triggers the preovulatory LH and FSH surges. The serum level of estradiol falls precipitously, while serum progesterone rises, inducing mating behavior later in the evening. Ovulation takes place in the early hours of estrus. Prolonged FSH secretion, resulting from a fall in serum inhibin, recruits new

mone (FSH); (ii) as the follicles mature and are prepared for ovulation, a number of the prepared follicles may become atretic; (iii) during the final stages of follicular growth, estrogen is secreted from the maturing follicle and primes the reproductive system for possible pregnancy; (iv) estrogen priming elicits gonadotropin-releasing hormone (GnRH) release from the hypothalamus; (v) preovulatory surges of LH and FSH trigger ovulation; (vi) ova appear in the oviduct; (vii) ovarian steroid secretion brings about structural cyclic changes in the uterus, preparing it for implantation in the event of fertilization; and (viii) the sequence of events is cyclic in nature, and the length of each cycle is species specific and may vary from 4 days to 1 year.

II. COMPARATIVE ASPECTS OF THE ESTROUS CYCLE

A century ago, the English zoologist Walter Heape described the progressive stages of one estrous cycle and classified the different estrous cycles observed in the mammals. The cycle itself, divided into four stages, centered around the period of estrus when mating behavior was displayed (Fig. 1A). He called the period preceding estrus "proestrus," which signified the period of follicular growth in the ovary, and he termed the period succeeding estrus "metestrus," a recovery period following ovulation, and "diestrus," a period when the ovarian secretions from the corpus luteum prepare the uterus for implantation. If fertilization does not occur, the cycle is repeated and another set of follicles is prepared for ovulation. Heape's terminology has been useful in describing the estrous cycle of smaller laboratory rodents, which

display a short and precise cycle when exposed to a strict light–dark schedule. Heape also coined the term "anestrus" to signify the time period when no mating behavior was displayed. Therefore, one full reproductive cycle consists of an anestrous and an estrous period.

Species whose estrous period is confined to a certain time of year are referred to as seasonal breeders. They have a prolonged anestrous period characterized by little or no follicular growth, small ovaries, and regressed secondary sex accessory tissues. During the breeding season, when there is follicular growth and estrogen secretion, animals may exhibit only one wave of follicular growth (seasonal monoestrus), recurring waves of follicular growth (seasonal polyestrus), or continuous follicular availability (nonseasonal polyestrus). The timing of the breeding season is highly species specific, depending significantly on dietary and climactic environmental factors. In the spring, for instance, estrus will occur in the fox, mink, and squirrel at a time when temperatures are rising, rainfall is abundant, light-to-dark ratio is increasing, and the chances for the mother and offspring to survive are good. On the other hand, sheep, displaying a long gestation period, are seasonally polyestrous and breed in the fall when temperature is falling and light-to-dark ratio is decreasing, allowing parturition to occur in the spring. Some species, like laboratory rats and mice, show cyclical activity throughout the year if not mated and are polyestrous.

A. Coitus-Induced Ovulators

Species can be separated into induced and spontaneous ovulators. Induced ovulators require coitus to elicit the preovulatory surge of the gonadotropins

---

follicles for the next cycle. (B) The secondary FSH surge recruits a cohort of follicles (1) which then matures under the influence of basal levels of FSH and LH (2). These follicles begin to secrete increasing amounts of estradiol (3). Every day, under conditions of a controlled light–dark cycle (14 hr light:10 hr dark), an excitatory signal arises in the hypothalamus (4). Estradiol sensitizes the hypothalamus (3), causes fluid retention in the uterine lumen (5), primes the anterior pituitary gland by increasing GnRH receptors (6), and triggers the release of GnRH at 1400 hr (7), causing LH and FSH preovulatory surges (8). The LH surge causes germinal vesicle breakdown in the oocytes (9), a decrease in estradiol secretion (10), and an increase in serum progesterone (11), which enhances LH release (12). Inhibin synthesis and secretion are suppressed (13, 14). Mating behavior is induced by the progesterone (15), and ovulation occurs (16), leading to formation of the corpus luteum (17). The secondary FSH surge (19) causes follicular recruitment of the next cohort of follicles (20), initiating a new cycle (1).

followed by ovulation; spontaneous ovulation is also preceded by the preovulatory gonadotropin surge but takes place in the absence of a male and coitus.

Coitus-induced ovulation, observed in species such as camel, cat, ferret, mink, squirrel, and rabbit, is triggered by a reflex initiated by vaginal stimulation, eliciting hypothalamic GnRH release and subsequently LH from the anterior pituitary. In some species, multiple acts of coitus are necessary to elicit the preovulatory gonadotropin surge, suggesting that a certain threshold of excitatory stimulation has to be reached before the reflex is activated. Induced ovulation requires the continuous presence of growing and mature follicles ready for ovulation. These species demonstrate a continuous estrus characterized by follicular growth and estrogen secretion; the estrogen, acting as an ovarian signal denoting mature follicles, induces mating behavior and heightens the cervical and hypothalamic sensitivity to coital stimulation.

The neuroendocrine reflex responsible for ovulation in the rabbit, the most extensively studied induced ovulator, is referred to as the copulaceptive reflex arc. The afferent component of the reflex arc, activated during copulation, includes nerves originating from the vaginocervical region, the perineal area, and the large cutaneous area comprising the flanks including inner and outer portion of the thighs. In addition to somatosensory stimulation, and unlike in the rat, olfactory, visual, and auditory stimulation play a contributing, but differential, role in eliciting the reflex. In fact, in the case of the mink and the vole, the mere presence of a male, without mating and intromissions, is sufficient to elicit ovulation in an estrous female.

The time interval between copulation and ovulation differs among species. Vole and rabbit ovulate within 9–12 hr after mating; ferret, cat, and mink, on the other hand, exhibit a prolonged latency period ranging from 25 to 58 hr. The increased latency period results from slow responsiveness on the part of the ovary rather than a central nervous system (CNS)–pituitary delay.

Some induced ovulators, such as rabbit, ferret, and cat, tend to breed more readily in the spring than in the other seasons. The vole is a definite seasonal breeder characterized by suppressed follicular growth during an anestrus period. As the spring approaches and the length of the photoperiod and supply of food are increased, folliculogenesis is initiated. The female starts to explore her territory and eventually encounters a male. The combination of springtime and the presence of a male induce immediate estrous behavior. She will instantly mate and subsequently ovulate. The vole exemplifies the most efficient reproductive pattern: ovulating in response to mating and adapting her breeding season to the environment.

B. Spontaneous Ovulators

Spontaneous ovulation is elicited by a preovulatory gonadotropin surge which is tightly controlled by estrogen. The list of mammals that show spontaneous ovulation is long and includes small mammals, such as rat, mouse, hamster, guinea pig, and bat, as well as larger animals, such as dog, fox, and some marsupials. In the rat, mouse, and hamster the preovulatory gonadotropin release occurs in the afternoon of proestrus, within a precise time period, with mating behavior some hours later and ovulation in the early hours of estrus morning.

Although the major events in mouse and hamster's estrous cycle are similar to those of the rat with regard to the timing of events, length of the cycle, and the mating behavior, there are some fundamental differences. Unlike in the rat, there is no evidence of postgonadectomy rise in pituitary content of LH in hamster, suggesting a major difference in the ovarian steroid feedback mechanism. Continuous light and cold induce constant estrus (constant production of follicles and atresia) in the rat but not in the hamster, in which ovulation continues for some time. After a prolonged cold period, the hamster abruptly becomes anestrous, however, due to its ability to hibernate. Sham ovariectomy of the rat, but not the hamster, interferes with cyclicity by advancing events prematurely or preventing them from occurring. Grouping of nonpregnant females in the absence of a male prevents cycling in mice. This is accounted for by a pheromone, released by the female, which inhibits follicular growth; a pheromone, released by the male, will override this inhibition. Such pheromone-induced effects have not been observed in rats.

C. Seasonal Breeder

The female sheep displays a reproductive pattern characterized by an endogenous rhythm of one or more estrous cycles in addition to an annual or seasonal rhythm (seasonal polyestrus). The ovine estrous cycle displays spontaneous ovulation, lasts 16 or 17 days, and takes place in the fall and the winter.

The estrous behavior lasts 24–48 hr, a period when the ewe displays signs of proceptivity and receptivity. The onset of estrus is coupled to the preovulatory LH surge, which is not synchronized to a critical period of the day; ovulation occurs 24–30 hr after the first sign of estrous behavior. Progesterone is the primary inhibitor and regulator of the GnRH pulse generator and is secreted solely by the corpus luteum. Progesterone is maintained at a high level during most of the cycle but falls dramatically 1 or 2 days before ovulation, thereby relaxing its inhibitory hold on the GnRH pulse regulator. The disinhibition of the GnRH pulse generator allows GnRH and LH pulse frequency to increase, inducing folliculogenesis and stimulation of ovarian secretion of estrogen. The enhanced estrogen secretion, rising as the progesterone level declines, triggers the preovulatory surge of gonadotropins followed by ovulation.

The transitory changes, necessary to induce seasonal estrus or anestrus, are accounted for by two external cues: the photoperiod and a pheromone, released by the ram. Prolonged daylight during spring activates an inhibitory neural signal inducing estrogen to suppress the GnRH pulse generator and thereby preventing ovulation, and thus inducing anestrus. Short-day photoperiod, on the other hand, relieves this estogen-induced inhibition of the pulse generator, which, when coupled to the presence of the male and his released pheromone ("ram's effect"), will increase its frequency output, bringing the ewe into estrus.

D. Corpus Luteum

Reproductive diversity is also seen in the life span and function of the corpus luteum. In most mammals the corpus luteum of nonpregnancy lasts a shorter time than that of pregnancy; this is because the placenta has developed luteotropic activity. This is not true for the dog, ferret, and kangaroo, in which the corpus luteum of the nonpregnant animal lasts as long as the corpus luteum of pregnancy. In some species of marsupials, the nonpregnant corpus luteum even outlasts the corpus luteum of pregnancy. In contrast, rat, hamster, and mouse have a nonfunctional corpus luteum if mating has not occurred; the act of coitus and intromissions trigger the release of prolactin, which induces a functional corpus luteum.

III. PRIMARY EVENTS IN THE ESTROUS CYCLE

The short and precise estrous cycle of the laboratory rat has been a useful model for reproductive studies. The laboratory rat is defined as a spontaneously ovulating, nonseasonal, polyestrous animal. The rat ovulates every 4 or 5 days throughout the year, unless interrupted by pregnancy or pseudopregnancy.

The cycle is well maintained when the animals are placed in an established light–dark schedule, for instance, 10 hr of dark followed by 14 hr of light. The sequential stages can be monitored by examining the cell types that appear in the vaginal smear (Fig. 1A). In a 4-day cycling rat, the vaginal smear predominantly shows leukocytes on metestrus and diestrus. On proestrus, the smear is characterized by mostly nucleated epithelial cells, which change to predominantly cornified, squamous epithelial cells by the morning of estrus. The 5-day cycling rat will show either an additional day of leukocytic smear (diestrus) or an additional day of cornified smear (estrus). The vaginal cytology on a single day cannot be used as the sole criterion for estrous events but should be followed over several cycles.

A. Physiological Changes in the Reproductive Organ: Ovary

The fundamental unit of the ovary is the follicle, which contains the female gamete or ovum. A dramatic maturation process, folliculogenesis, takes place as the follicle grows from a primordial follicle, consisting of a germ cell surrounded by a single layer of epithelial cells, to a large glandular hormone-producing structure, the Graafian or the tertiary follicle.

Early follicular development prior to the antrum stage is independent of gonadotropin stimulation; further developments, including steroid secretion, are highly dependent on sustained low levels of gonadotropins. At any stage of follicular development, degeneration or "atresia" of some follicles takes place. Therefore, relatively few follicles, fewer than 1% of the total numbers present at birth, are destined to ovulate. What determines which follicles will initially start to grow, which will undergo full maturation process and ovulate, and which will degenerate and become atretic is not currently known.

The physical process of ovulation occurs as the colloid osmotic pressure of the follicular fluid is increased, permitting water to enter the follicle, and as the collagenous support in the follicular wall breaks down. These events are the result of changes in enzymes induced by the LH surge. Approximately 10–14 hr after the initiation of the preovulatory gonadotropin surge, follicular rupture takes place and ova are extruded. Since sexual receptivity, or estrous behavior, takes place some hours prior to ovulation, the spermatozoa are in the oviduct by the time the ova reach the site of fertilization, and they have been capacitated.

Following ovulation, the follicles undergo dramatic internal structural changes as they develop into corpora lutea, producing progesterone. The preovulatory antral space is occupied by granulosa cells and invading thecal cells. The function of the corpus luteum is to secrete progesterone, the hormone necessary for preparing the uterine endometrium for implantation of the fertilized eggs.

If mating takes place, multiple intromissions by the male stimulate the uterine cervical area, inducing a neuroendocrine reflex, resulting in a "luteotropic" hormone (prolactin) being released by the pituitary. In the absence of prolactin in the rat, the corpora lutea do not persist and the next cycle occurs.

The major secretory products of the growing follicle and the subsequent corpus luteum are the steroid hormones, which play an important role in acting on primary and secondary target tissues of reproduction. Ovarian steroid biosynthesis is under control of the two gonadotropins, LH and FSH, which stimulate

their target tissue by interacting with specific, high-affinity binding sites or receptors on the surface of the cells. The membrane density of these receptors is important in determining the cell's responsiveness to the respective gonadotropins and therefore the fate of a given follicle at any stage of the development. Granulosa cells, the major producers of estrogen, possess FSH receptors at all stages of development and acquire LH receptors only at later stages of follicular development. In contrast, theca interna cells, the only suppliers of androgen, possess only LH receptors. Undifferentiated granulosa cells of the preantral stage respond to basal FSH stimulation by an increased activity of the aromatase enzyme complex, converting androgen to estrogen and, later, by providing the necessary enzymes for metabolizing cholesterol to progesterone. The principal androgens, testosterone and androstenedione, produced by the thecal cells under LH stimulation enter the granulosa cells by passing through the surrounding, highly vascularized tissue, and, under stimulation of FSH, are converted to estrogen. It is this joint cooperation of LH action on thecal cells, together with the action of FSH in promoting aromatase activity in the granulosa cells, that forms the "two cells, two gonadotropins" theory for controlling follicular estrogen synthesis. As the follicles mature, FSH, in synergy with the intacellularly produced estrogen in the granulosa cells, induces LH receptors on the membranes of the granulosa cells. Once the granulosa cells have acquired LH receptors, they are capable of responding to the LH surge, undergoing luteinization and forming into a corpus luteum.

In the 4-day cycling rat, estradiol is at low basal level throughout estrus and most of metestrus. In the evening of metestrus, however, the serum level starts to rise and continues to rise through diestrus, reaching a peak in the afternoon of proestrus (Fig. 1A). As the preovulatory LH/FSH surge is initiated in the afternoon of proestrus, estrogen falls dramatically and reaches a basal low level by the morning of estrus. Serum progesterone rises immediately after the initiation of the preovulatory surge and returns to basal level by the morning of estrus. This surge of progesterone, in addition to potentiating the preovulatory LH surge and preventing the recurrence of another LH/FSH surge the next day, induces mating behavior and, on the morning of estrus, relaxes the uterine cervix, allowing uterine intraluminal fluid to be released.

In addition to steroids, the follicles produce two dimeric proteins, inhibin and activin, which are structurally related by sharing common β subunits, with inhibin also containing an α subunit. The two proteins have opposing actions on the gonadotrope; inhibin suppresses and activin stimulates FSH biosynthesis and secretion. Follistatin, a structurally unrelated protein, reduces FSH biosynthesis and secretion by binding to activin, thereby neutralizing activin's stimulatory effect on FSH. Follistatin does not bioneutralize inhibin to the same extent.

B. Neuroendocrine Control Mechanisms: The Hypothalamic–Pituitary–Ovarian Axis

The estrous cycle is under the control of the CNS. Information from external signals (visual, olfactory, auditory, tactile, etc.) is received by the nervous system (Fig. 1B, 4). These signals, transmitted to groups of cells in the hypothalamus, are processed, integrated, and eventually transduced to a humoral signal, the GnRH, a decapeptide which reaches the anterior pituitary by a portal vascular system. The gonadotropes, responding to GnRH, synthesize and release LH and FSH, which induce ovarian folliculogenesis, steroidogenesis, ovulation, and formation of corpus luteum. The ovarian steroids exert feedback effects on a host of target tissues, including the brain and the pituitary gland, which contain estradiol receptors.

In addition to its classical negative feedback control system whereby estradiol maintains a suppression of GnRH, the reproductive axis in the female also features a unique positive feedback control system responsible for delivering the robust preovulatory gonadotropin surges, which induce ovulation in the morning of estrus. The rising tide of estrogen (Fig. 1B, 3), priming the reproductive axis, reaches a critical threshold level at which it will provide information of ovarian readiness for ovulation and allows for a neural signal, derived from the 24-hr neural clock (Fig. 1B, 4), to be transmitted to the hypothalamus, inducing a robust release of GnRH (Fig. 1B,

7). If the tide of estrogen has not reached its critical threshold by the early morning of proestrus, the preovulatory surge will be delayed by 24 hr, and the 4-day cycling rat will convert to a 5-day cycling rat.

1. The Hypothalamus

The hypothalamus contains clusters of neurons, or nuclei, which are symmetrically located around the cerebrospinal fluid-filled third ventricle. Axonal projections between the nuclei and other areas of the brain provide a complex, but well integrated, network of transsynaptic interactions between neural and endocrine control centers. Five structurally distinct hypothalamic nuclei are involved in the immediate control of reproduction, including signals inducing the preovulatory gonadotropin surge. The suprachiasmatic nucleus (SCN) and the medial optic nucleus participate in the timing of the LH surge; the medial preoptic area provides an important site for integration of signals controlling the reproductive cycle. The median eminence (ME), the site to which most GnRH axons are projected, forms the base of the hypothalamus; it is richly supplied with blood vessels that drain into the pituitary stalk and empty into a capillary plexus in the anterior pituitary. This vascular link between ME and the anterior pituitary is referred to as the hypophysial portal system, providing the anatomical site for neurosecretion of GnRH and other neuropeptides involved in pituitary function. In proximity to the ME is the arcuate nucleus, which plays a supportive role in releasing GnRH into the hypophysial portal system.

GnRH is released in a pulsatile pattern of rhythmic secretory bursts, whose frequency and amplitude vary according to the cycle stage. The greatest GnRH release triggers the preovulatory surge of gonadotropins on the afternoon of proestrus. The pattern of pulsatile GnRH secretion appears to be intrinsic; that is, the GnRH neuron itself is capable of releasing GnRH in a pulsatile fashion, as demonstrated in isolated, superfused hypothalamic fragments and in cell lines of immortalized GnRH-producing neurons (GT1 cells). The discovery of the episodic nature of GnRH secretion led to the concept of an oscillator, or pulse generator, residing in the CNS. GnRH neurosecretory activity is regulated by a wide variety of inputs that convey information about both the internal and the external environment. The ovarian steroid feedback activity constitutes one major input; the second set of inputs, playing the critical role of neural integration, consist of neuropeptides, neurotransmitters, and neuroactive amino acids.

GnRH neurons do not express receptors for gonadal steroids. Ovarian steroids serve as neuromodulators, indirectly affecting the GnRH neurons, by targeting a second input—the synaptic efferents of various neurotransmitters, neuropeptides, and neuroactive amino acids. Studies have shown that GnRH neurons do indeed receive synaptic input from efferents containing norepinephrine, dopamine, serotonin, γ-aminobutyric acid (GABA), galanin, corticotropin-releasing hormone, neuropeptide Y, substance P, proopiomelanocortin, and β-endorphin.

2. Anterior Pituitary

The anterior pituitary contains six types of hormone-producing cells, including the gonadotropes that secrete both LH and FSH. The gonadotropes, easily distinguished from the other cells by either immunocytochemistry or morphology, account for up to 15% of the cells in the gland. LH and FSH are heterodimers composed of two glycoprotein subunits: a common α subunit and a unique β subunit, which confers biological activity and specificity. High-frequency pulses of GnRH favor pulsatile LH release. Although lower frequency pulses elicit pulsatile FSH release, FSH synthesis and release depend much less on the GnRH signal.

LH and FSH during the rat estrous cycle are maintained at low basal levels on estrus, metestrus, diestrus, and on the morning of proestrus (Fig. 1A). This is primarily due to the effect of negative feedback by estradiol and progesterone. Removal of this inhibitory effect by ovariectomy results in higher plasma levels of LH and FSH, rising slowly over several days. Estrogen administration totally returns LH and partially returns FSH to their basal levels, suggesting that ovarian negative feedback can account for maintaining LH, but not FSH, at its basal level. The additional negative feedback component for FSH is the ovarian peptide inhibin. Inhibin selectively suppresses FSH synthesis and secretion and, during the estrous cycle, its serum level remains high from es-

trus to the evening of proestrus. To what degree the negative feedback actions of the steroids exert their effects on the pituitary and the hypothalamus, selectively or in combination, is not well defined; inhibin is known to affect the gonadotropes selectively.

At 1400–1600 hr on the afternoon of proestrus, serum levels of the two gonadotropins start to rise rapidly, generating the preovulatory surge (Fig. 1B, 8) which, in turn, induces follicular rupture and ovulation (Fig. 1B, 16) in the ovary 10–14 hr later. This dramatic surge peaks by 1800 and 1900 hr, at which time serum levels of both LH and FSH start to decline. By the morning of estrus, LH level has returned to its basal level; serum FSH, on the other hand, shows a secondary rise (Fig. 1B, 19) before declining to its basal level on the evening of estrus. This prolonged, secondary FSH surge functions to recruit a new cohort of ovarian follicles for the subsequent cycle (Fig. 1B, 1 and 20).

Increased pulsatility of LH coincides with increased pulsatility of GnRH, establishing GnRH as the primary signal for the preovulatory surge. Accompanying this enhanced GnRH release is an increased pituitary sensitivity to GnRH, provided primarily by an elevated level of GnRH receptors. The GnRH receptor, like the gonadotropin receptors, belongs to the G-protein receptor family, and its pituitary level varies across the estrous cycle, reaching a peak on the morning of proestrus. This elevated GnRH receptor level is a result of estrogen priming of the pituitary.

The classic experiments of Everett and Sawyer in 1950, which demonstrated that the LH release mechanism is under the control of a 24-hr photoperiodic circadian rhythm, established that a hypothalamic signal occurs daily at a "critical period" between 1400 and 1600 hr (Fig. 1B, 4). The timing of this daily critical period can be shifted by altering the environmental light–dark schedule; that is, the circadian rhythm can be entrained by a time cue, referred to as a "zeitgeber," thereby shifting the daily critical period for LH release.

The effects of numerous neuropeptides and neurotransmitters on pituitary secretion have been elucidated. Neuropeptide Y, present in high concentration in portal blood, exerts a positive effect on GnRH-induced LH and FSH secretion following estrogen priming; in the absence of priming, neuropeptide Y inhibits FSH secretion. Galanin, a neuropeptide found primarily in the arcuate nucleus, will stimulate LH release. Substance P, a tachykinin, may inhibit LH secretion directly at the level of pituitary but has variable effects on LH and FSH when delivered to the hypothalamus. A noradrenergic stimulatory pathway is most certainly involved in the release of the preovulatory GnRH–LH–FSH cascade; the excitatory amino acid glutamic acid will increase circulating LH, whereas GABA is a potent inhibitor of pulsatile LH secretion. Opiate pathways are also involved in the gonadotropin release; opioid peptides can inhibit ovulation and their antagonists raise LH secretion, suggesting a tonic inhibition of GnRH by opiate pathways. Recent studies have demonstrated that the tonic inhibition exerted by the opioids can be lifted through neural plasticity generated by estrogen-induced synaptic retraction of the opiate inhibitory axons in the arcuate nucleus.

Circulating steroids play a crucial role in coordinating mating behavior to ovulation (Fig. 1B, 15). The spontaneously ovulating rat exhibits mating or lordosis behavior only during late proestrus when she has been exposed for an appropriate period of time to both estrogen and progesterone.

The lordosis behavior, a somatosensory reflex, is characterized by a sequence of defined events. Similarly, the male displays a programmed pattern of motor activity necessary for successful intromissions. In fact, successful mating results from a coordinated, well-synchronized, and matched series of reflexes in the male and female rat, providing a mechanism that selectively ensures only conspecific competent members success in their reproductive efforts.

IV. NATURE OF CYCLICITY

The basis for cyclicity as expressed in the spontaneously ovulating laboratory rat is best understood by summarizing the events as they occur during one cycle period. Figure 1B delineates these events: (i) There is a daily facilitory signal from the SCN (4); (ii) there is a facilitory signal from the follicles, estrogen (3), which slowly builds up, causing both negative and positive feedback actions (6, 7, 15) on the hypothalamic–pituitary axis; (iii) when estrogen reaches a critical threshold, it triggers release of GnRH (7) at the time set by the daily neural signal; and (iv) the following events take place: the preovulatory gonadotropin surges (8); estrogen level declines (10); progesterone rises (11), facilitating the LH/FSH surge (12) and preventing the recurrence of another surge the next day; and, inhibin level falls (13, 14), inducing prolonged secretion of FSH (19), which initiates growth and maturation of a new set of follicles (1, 20). As the follicles grow, a steady rise of estrogen (3) primes the reproductive axis for generation of a new cycle if mating does not occur.

What is the mechanism by which the hypothalamic–pituitary–ovarian axis does not reach a steady level of hormones, as do the hypothalamic–pituitary–testicular or thyroid or adrenal–cortical axes? We believe that there are two critical factors: (i) The estradiol feedback, in conjunction with the "time of day" signal, is converted to a positive signal, triggering the GnRH surge; this, in turn, leads to (ii) the microorgan (follicles) that secretes the ovarian feedback hormone which disappears after ovulation, shutting off the negative feedback. Thus, the system is reset, a new crop of follicles arrives on the scene, and the cycle is repeated.

See Also the Following Articles

FSH (Follicle-Stimulating Hormone); GnRH (Gonadotropin-Releasing Hormone); LH (Luteinizing Hormone); Oogenesis, in Mammals; Prolactin, Actions of; Seasonal Reproduction, Mammals

Bibliography

Austin, C. R., and Short, R. V. (1991). *Reproduction in Mammals*, Vol. 3. Cambridge Univ. Press, Cambridge.

Bronson, F. H. (1989). *Mammalian Reproductive Biology*. Univ. of Chicago Press, Chicago.

Freeman, M. E. (1994). The neuroendocrine control of the ovarian cycle of the rat. In *The Physiology of Reproduction*

(E. Knobil and J. D. Neill, Eds.), Vol. II. Raven Press, New York.

Nequin, L. G., Alvarez, J., and Schwartz, N. B. (1979). Measurements of serum steroid and gonadotropin levels and uterine and ovarian variables throughout 4 day and 5 day estrous cycle in the rat. *Biol. Reprod.* **20**, 659–670.

Schwartz, N. B. (1973). Mechanisms controlling ovulation in small mammals. In *Handbook of Physiology*, Endocrinology,

Vol. II, Part I. American Physiological Society, Washington, DC.

Schwartz, N. B. (1995). The 1994 Stevenson Award lecture: Follicle-stimulating hormone and luteinizing hormone: A tale of two gonadotropins. *Can. J. Physiol. Pharmacol.* **73**, 675–684.

Schwartz, N. B., and Waltz, P. (1970). Role of ovulation in the regulation of the estrous cycle. *Fed. Proc.* **29**, 1907–1912.

Estrus

Janice E. Thornton
Oberlin College

Patricia D. Finn
University of Washington

I. Introduction
II. Components of Estrus
III. Patterns of Estrus
IV. Hormonal Control of Estrous Behavior
V. Neural Regulation of Estrous Behavior

Females from most vertebrate species show recurring periods of heightened sexual activity interspersed with periods of sexual inactivity. In mammalian females (except primates) these periods of sexual activity are called estrus.

GLOSSARY

hypothalamus An area located at the base of the brain that is involved in the hormonal control of sexual behavior.

lordosis A reflexive mating posture that occurs in female rodents and a variety of other mammals.

ovariectomy Removal of the ovaries.

proceptive behavior Species-specific behaviors by the female which serve to initiate or maintain sexual contact with the male.

receptive behavior Species-specific behavior by the female which is necessary for intravaginal ejaculation to occur (i.e., it is generally the species-typical mating posture).

vaginal cycle The cyclic changes in cell type that are seen in the vagina and that have been correlated to cyclic changes in the ovary.

I. INTRODUCTION

Females of most mammalian species display cyclic changes in reproductive function, including sexual behavior. These recurring periods of heightened sexual activity are referred to as heat or estrus (estrus or oestrus, noun; estrous or estrual, adjective). The term estrus is derived from the Greek word *oistros*, meaning gadfly. When gadflies buzz around cattle they drive them into a frenzy of hyperactive behavior, similar to the bellowing and hyperactive behavior of a cow in behavioral estrus. For most species, periods of estrus are separated by periods of sexual inactivity called anestrus. For example, when a female hamster is placed with a male that is vasectomized (to avoid pregnancy), she will be in estrus and allow copulations only for a few hours every 4 days. Although used with other mammals, the term estrus is generally not used in describing primates. Rather, the reproductive cycle in primates is called a menstrual cycle. Menstrual cycles are specific to primates because they are the only species that have spiral arteries in the uterine wall that periodically engorge with blood and are then sloughed at the end of the ovarian cycle to produce menses (occasionally dogs show a small amount of vaginal bleeding just before ovulation but it is a different phenomenon). Although there is some cyclicity in sexual behavior during the menstrual cycle in primates, females in many species of primates will mate throughout the cycle rather than only during a circumscribed period such as estrus.

Note that the term estrus is also used in other contexts. In describing the ovarian and vaginal cycles, the term estrus is used for the day on which ovulation occurs and a particular type of cell (i.e., cornified) predominates in the vagina. Furthermore, the ovarian cycle is often referred to as the estrous cycle. All this terminology is somewhat confusing because behavioral estrus actually occurs on the day of the ovarian and vaginal cycle called proestrus rather than on the day called estrus (see Estrous Cycle).

Research on estrus has examined the components of estrus, the patterns of estrus shown, the role of ovarian hormones in the control of estrous behaviors, and the neural elements that underlie the production of estrous behaviors.

II. COMPONENTS OF ESTRUS

Females that are in estrus emit a plethora of behavioral and nonbehavioral cues that are indicative of their estrous condition. Beach (1976) proposed that estrous behavior could be divided into two components: receptive or copulatory behaviors, which are necessary for the male to achieve intravaginal ejaculation, and courtship or proceptive behaviors, which the female uses to initiate and maintain sexual interaction. In addition, nonbehavioral cues are apparent

in estrous females and serve to, among other things, enhance the attractiveness of the female to the male.

A. Receptive/Copulatory Behavior

Receptive behavior is generally defined as the species-typical mating posture of the female. Lordosis is a common reflexive mating posture that occurs in rodents and a variety of other mammals (Fig. 1). The term lordosis is derived from a medical term that refers to the curvature of the spine as viewed from the side. The lordosis posture generally consists of a pronounced arching of the back, accompanied by deflection of the tail and immobilization. In a four-legged animal, this results in the elevation and stabilization of the hindquarters and vagina which is necessary for the male to gain intromission (i.e., insertion of the penis into the vagina). Lordosis is sometimes seen prior to male contact but usually occurs only in response to tactile stimulation of the female's flanks and anogenital region by the male during a mount. Lordosis has been quantitated in a number of ways. In some species, the frequency of lordosis and/or the amount of time the lordosis posture is held is measured (e.g., hamsters and guinea pigs).

FIGURE 1   When in estrus, females from a number of species, including rats, show the reflexive mating posture called lordosis. This behavior is generally shown in response to stimulation of the female's flanks and rump during mounting by the male. This posture, which consists of an immobile stance, an arched back, elevation of the hindquarters, and deflection of the tail, is necessary for the male to gain intromission (insertion of the penis into the vagina) (photo courtesy of Mathew Cunningham).

In a number of species, including rats, it is common to calculate the lordosis quotient (LQ) as a numerical measure of receptivity. The LQ is the ratio of the number of times lordosis is shown in response to a fixed number of mounts (usually 10) multiplied by 100. For example, a female which shows lordosis to 8 of 10 mounts would have an LQ of 80. A female showing an LQ of 20 would be considered less receptive than a female showing an LQ of 80.

B. Proceptive/Courtship Behaviors

Females take an active role in initiating sexual activity. Females in estrus are highly motivated and will expend considerable effort to seek out a male and initiate/maintain copulation. For example, female rats in estrus will repeatedly press a bar to gain access to and copulate with a male, whereas an anestrous female will not. Behaviors by the female which serve to establish or maintain sexual interaction with the male are called proceptive behaviors.

Proceptive behavior is usually indicated by affiliative behaviors that initiate sexual union and by sexually solicitous behaviors. One of the most common ways to measure affiliative behavior is to determine the amount of time the female maintains physical proximity to a male. Another behavioral measure is the number of sexual solicitations made by the female. These solicitations vary considerably between species. During estrus, a female rat will wiggle her ears (which is actually an extremely rapid vibration of the head), a female ewe will nudge a ram with her head, and a female dog will strike the male repeatedly with her forepaws. Estrous females of a number of species (e.g., rodents, certain canines, and ungulates) will show an approach–withdrawal pattern which induces the male to chase after the female.

In females of some species such as rats, proceptive behaviors also play a role in pacing sexual interaction. During copulation, male rats show multiple mounts and intromissions prior to ejaculation. Using approach–withdrawal behavior and other sexual solicitations, the female paces the timing of the intromissions so that she receives the pattern of vaginal–cervical stimulation which is optimal for induction of pregnancy.

Although estrous behavior is dependent on ovarian hormones (in most mammalian species), hormones are not "sex potions" and do not compel the female to always show sexual behavior regardless of social and environmental conditions. Ovarian hormones increase the probability that a specific stimulus, such as a male, will elicit particular behaviors, such as sexual solicitations, but they do not guarantee it. Females of some species show a great deal of selectivity in their choice of mating partners. For example, a dog in estrus will direct her proceptive/receptive behavior toward some males and may actively avoid and reject others.

C. Nonbehavioral Components

Although the term estrus usually refers to the sexual behavior shown by females, there are also many other physiological characteristics that are correlated with the occurrence of the sexual behavior. Some of these characteristics are apparent to conspecifics. For example, a female in estrus may exhibit changes in appearance or emit chemical cues that can be picked up by the olfactory system of conspecifics. These chemical cues may be from urine, vaginal secretions, or from scent glands. Male dogs can readily differentiate between balls of cotton impregnated with vaginal secretions from estrous compared to anestrous female dogs and spend a much greater amount of time sniffing the cotton with estrous secretions.

III. PATTERNS OF ESTRUS

Although some mammals show behavioral estrus at different times of the year than when they ovulate (e.g., musk shrew), in most mammals sexual behavior and ovulation occur together. However, even among these females, estrus occurs in a number of different patterns. For example, one way females have been categorized is as spontaneous vs induced ovulators. Most mammalian females are spontaneous ovulators (e.g., rats, sheep, and dogs). In spontaneous ovulators, ovarian hormones bring about behavioral estrus followed soon afterwards by "spontaneous" ovulation (usually within 12–24 hr). In contrast, females from a number of species, such as cats, rabbits, and camels, are induced ovulators. In induced ovulators, ovarian hormones cause behavioral estrus but ovulation occurs (i.e., is "induced") only if the female copulates (i.e., receives vaginal–cervical stimulation). Even among those species in which ovarian hormones cause both estrus and spontaneous ovulation, females of some species are polyestrus and show repeated cycles throughout the year (e.g., guinea pigs, mice, and rats), others are seasonally polyestrus and exhibit repeated cycles only during a particular season (e.g., sheep and cows), and still others are seasonally monestrus and show only a single cycle at a particular time of year (e.g., fox and roe deer).

Not surprisingly, for all of these species, estrus usually occurs in the female at a point in the ovulatory cycle when she is most likely to be fertile and become pregnant. Furthermore, estrous cycles are generally coordinated with environmental cues and gestational length so that offspring are produced during favorable times of the year when food and resources are more abundant.

Females of numerous species (including mink, ferrets, skunks, rabbits, pigs, mice, and rats) show postpartum estrus. This is a behavioral estrus that occurs just prior to or after giving birth. It is similar, but not identical, to the estrus seen during a normal ovarian cycle (e.g., the postpartum estrus is usually of shorter duration). In the wild, these females do not often undergo repetitive estrous cycles since for most of their reproductive lives they are either pregnant or suckling young (both of which usually inhibit estrous cycles). The postpartum estrus decreases the interval between successive litters and maximizes reproductive output because a litter that is conceived during postpartum estrus is born at approximately the same time that the previous litter is weaned.

IV. HORMONAL CONTROL OF ESTROUS BEHAVIOR

In those mammals in which ovarian activity and sexual behavior occur together, ovarian hormones generally control both the behavioral and the nonbehavioral components of estrus.

A. Historical Perspective

It has been known for centuries that removal of the ovaries (ovariectomy) can eliminate estrus. Aristotle tells of the practice of removing the ovaries of camels who were used in warfare; because camels copulate laying down, a female camel in heat could be a real liability in an armed conflict. Because estrous cycles stop in females of most species after the removal of ovaries, it was surmised early on that cycles of estrous behavior must be due to cycles of ovarian changes. In the early 1900s, Stockard and Papanicolaou found that ovarian changes caused specific changes in vaginal cytology (cell types) and developed a noninvasive technique for looking at changes in vaginal cytology (i.e., changes in the "vaginal cycle"). Use of this technique enabled researchers to correlate estrous behavior with vaginal cell types and thus with changes in the ovary. Once ovarian hormones were isolated and characterized, it became possible to measure how they change across the cycle and to administer ovarian hormones in different combinations and in various temporal patterns to determine how they affect estrous behavior. It was found that the two major hormones involved are estradiol (an estrogen) and progesterone (a progestin).

B. Role of Estrogen

During the early part of the ovarian cycle, the predominant structure on the ovary is the follicle. The follicle consists of a maturing egg and several layers of associated ovarian cells (i.e., theca and granulosa cells) that surround the egg. The maturing follicle synthesizes and releases relatively large amounts of estrogen. This estrogen is secreted in a pulsatile manner and enters the general circulation. It acts at sites throughout the body and the nervous system.

Across species, the factor which most consistently regulates female estrous behavior is the ovarian hormone estrogen. Generally, females that have had their ovarian hormones removed via ovariectomy do not show estrous behavior in response to males but, if treated with estrogen, they do. Estrogen affects all the components of estrus, including receptivity, proceptivity, and nonbehavioral components.

Estrogen has to act for a relatively long period of time before it induces estrous behavior. For example, an ovariectomized female rat that is injected with estrogen will not show lordosis until at least 16–24 hr later. Interestingly, continuous exposure during the interval from first estrogen exposure to behavioral estrus is not necessary. Sexual receptivity can be induced with two discontinuous pulses of estrogen, each of which last about an hour: one at the beginning of estrogen treatment and another during the later part of estrogen exposure.

Estrogen acts by binding to specific estrogen receptors in particular parts of the brain and spinal cord. Estrogen can alter protein expression and one of the proteins induced by estrogen action is the progesterone receptor.

C. Role of Progesterone

Like estrogen, progesterone is secreted in a pulsatile manner from cells in the ovary, travels through the bloodstream, and acts throughout the body and nervous system. In several species, the developing follicle secretes a surge of progesterone just prior to the appearance of estrous behavior and ovulation. In these species, which include rats, hamsters, mice, and guinea pigs, progesterone is necessary for the full expression of behavioral estrus. Consistent with this, progesterone administration to ovariectomized estrogen-treated females of these species produces maximal levels of sexual behavior.

The importance of progesterone in controlling estrous behavior varies with the species. In some species, such as the guinea pig, progesterone is obligatory and estrogen alone is insufficient to induce estrus. In contrast, species such as ferrets and prairie voles do not depend on progesterone for the induction of estrus. In rats, estrogen alone can produce high levels of lordosis behavior in ovariectomized females, but the combination of estrogen followed by progesterone produces a more complete display of female reproductive behavior. In particular, proceptive behaviors in the rat are strongly progesterone dependent.

Generally, progesterone is not effective in producing sexual behavior in ovariectomized females unless they have received prior estrogen treatment. Females need to be exposed to estrogen for at least 24 hr in order for progesterone to act. Once progesterone is administered it affects sexual behavior within a few hours. In many species, estrogen and progesterone are said to act synergistically because the combination of the two hormones (in sequence) is much more effective than either hormone alone. Progesterone influences estrous behavior by binding to progestin receptors in specific areas of the brain. Because estrogen increases the number of progestin receptors, this is one reason that a background of estrogen exposure is needed before progesterone can act to facilitate estrous behavior.

IV. NEURAL REGULATION OF ESTROUS BEHAVIOR

Estrogen, alone or in combination with progesterone, induces behavioral estrus in ovariectomized females of many species by acting at specific cellular receptors which are located in circumscribed areas of the brain and spinal cord. Estrogen and progesterone receptors are found in similar neuroanatomical areas in a wide range of vertebrate species, and many of these areas are important in controlling sexual behavior.

One area of the brain in which estrogen and progesterone receptors are found in high density is the hypothalamus (Fig. 2). A variety of experiments have confirmed that the hypothalamus, in particular the ventromedial nucleus of the hypothalamus (VMH), is a critical component of the neural circuitry underlying hormone-induced lordosis.

Other areas of the brain also play a role in controlling lordosis. Based on research done primarily in the female rat, a model of the neural substrate underlying lordosis has emerged (Fig. 2; see Pfaff *et al.*, 1994). Data indicate that estrogen and progesterone activate neural cells in the VMH and these estrogen/progesterone-responsive cells of the VMH project to the midbrain central gray (MCG; which may also have some estrogen-responsive cells) and to other parts of the midbrain (e.g., the midbrain reticular formation). Cells from these midbrain areas then project to the lower brain stem [e.g., to the medullary reticular formation (MeRF)]. When the flanks of a female are

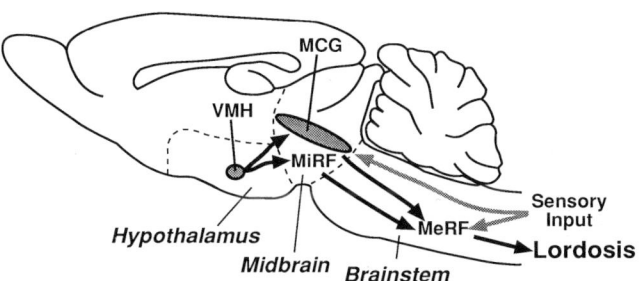

FIGURE 2 A schematic representation of the neural circuitry underlying the display of lordosis. When females are exposed to high levels of estrogen (and progesterone in some species) for the appropriate amount of time, it results in the activation of the ventromedial nucleus of the hypothalamus (VMH), which sends information to midbrain areas [e.g., the midbrain reticular formation (MiRF) and the midbrain central gray (MCG)]. The MiRF and MCG project to the brain stem (e.g., medullary reticular formation (MeRF)]. During mounting by a male, sensory receptors are activated in the female in the region of the flanks and rump, and this information is relayed through the spinal cord to the MeRF and MCG. If this neural circuitry is activated by ovarian hormones, the sensory input results in the motor output of the lordotic reflex.

stimulated by a male during copulation, cutaneous information enters the spinal cord and is sent to the lower brain stem and the MCG. If the female has been exposed to estrogen (and in some cases progesterone also) for the appropriate amount of time, the resulting stimulation of the VMH and the MCG will cause this sensory input (flank stimulation) to activate the lower brain stem (the MeRF). This area controls motor neurons involved in the lordosis reflex, and lordosis results.

Through what mechanisms do estrogen and progesterone act to stimulate the VMH and MCG cells resulting in the activation of the motor neurons that control lordosis? It is known that when estrogen and progesterone bind their respective receptors, the hormone/receptor complexes act to change DNA transcription and subsequently protein synthesis and also cause electrical changes in neural activity. There

are a number of neurotransmitter and neuropeptide systems that are affected by these changes (only some of which are described here). For example, the neurotransmitters acetylcholine (ACh), epinephrine, and norepinephrine (NE) appear to facilitate lordosis, whereas the neurotransmitter serotonin inhibits lordosis. Estrogen and progesterone act on these neurotransmitter systems in a variety of ways, such as increasing the number of receptors for a particular neurotransmitter (e.g., for ACh) and increasing the excitability of certain neurons (e.g., certain NE neurons). Two neuropeptides that facilitate lordosis behavior are gonadotropin-releasing hormone and oxytocin. Estrogen increases the number of receptors for both these neuropeptides. There is not yet a complete map of all the neural components that are changed by ovarian hormones. It remains to be determined exactly how all these (and other) neuroactive agents act and interact to produce estrus.

See Also the Following Articles

Estrogen, Effects on Behavior; Lordosis; Progesterone Actions on Behavior; Seasonal Reproduction, Mammals

Bibliography

Beach, F. A. (1976). Sexual attractivity, proceptivity, and receptivity in female mammals. *Horm. Behav.* 7, 105–138.

Carter, C. S. (1992). Neuroendocrinology of sexual behavior in the female. In *Behavioral Endocrinology* (J. B. Becker, S. M. Breedlove, and D. Crews, Eds.), pp. 71–96. MIT Press, Cambridge.

Nelson, R. J. (1995). *An Introduction to Behavioral Endocrinology*. Sinauer, Sunderland, MA.

Pfaff, D. W., and Schwartz-Giblin, S. (1994). Cellular mechanisms of female reproductive behaviors. In *The Physiology of Reproduction* (E. Knobil and J. Neill, Eds.), pp. 1487–1568. Raven Press, New York.

Pfaff, D. W., Schwartz-Giblin, S., McCarthy, M. M., and Kow, L.-M. (1994). Cellular and molecular mechanisms of female reproductive behaviors. In *The Physiology of Reproduction* (E. Knobil and J. D. Neill, Eds.), pp. 107–220. Raven Press, New York.

Eunuchoidism

Victor Y. Fujimoto and Michael R. Soules

University of Washington School of Medicine

GLOSSARY

epiphyseal plate maturation A process in which the chondrocytes of the epiphyseal plate differentiate and cease mitosis.

eunuchoidism A clinical condition in which the extremities are disproportionately greater than the torso.

hypergonadotropic hypogonadism A disease state in which eunuchoidism occurs due to primary gonadal failure.

hypogonadotropic hypogonadism A disease state in which eunuchoidism occurs due to inadequate gonadotropin secretion from the pituitary gland.

pubertal growth spurt A steroid-dependent process of accelerated bone growth during adolescence.

Eunuchoidism is a clinical condition of disproportionately excessive long bone growth that results from a state of hypogonadism occurring during adolescence. Eunuchoidism most commonly results from an isolated deficiency of gonadotropin production which was initially described soon after the development of methods to quantitate urinary gonadotropins.

I. INTRODUCTION

A eunuchoid body habitus arises as a result of a delay in exposure to sex hormones because growth of the spine is decreased more than the long bones of the extremities. Patients with a deficiency in hypothalamic gonadotropin-releasing hormone (GnRH), which classically results in deficiency of both follicle-stimulating hormone (FSH) and luteinizing hormone with growth hormone (GH) and other pituitary hormones unaffected, usually display eunuchoidal proportions. The epiphyses fail to fuse at the appropriate times and the longitudinal growth of the long bones continues, producing height:span and pubis–head:pubis–floor ratios that are significantly less than 1. Due to the delay in epiphyseal maturation and closure in eunuchoidal patients, radiographs of the hand and wrist typically reveal a delay in bone age. Skeletal maturity can be determined by comparing the patient's radiographs to the atlas of Greulich and Pyle to determine a relative bone age. The eunuchoidal habitus is typically associated with a delay in puberty and poor progression of primary and secondary sexual characteristics. Most males with eunuchoidal proportions suffer from other sequelae of hypogonadism, including infertility and adequate virilization. Although our understanding of the "process" that results in eunuchoidism remains incomplete because it occurs in both males and females, a perspective of our current understanding of the physiology and pathophysiology underlying the steroid regulation of long bone growth and ultimately epiphyseal closure will be presented.

II. PHYSIOLOGY OF THE EPIPHYSEAL GROWTH PLATE

Longitudinal bone growth is a process of chondrocyte proliferation and endochondral ossification (cartilage growth, matrix formation, and calcification) within the epiphyseal growth plates. The pro-

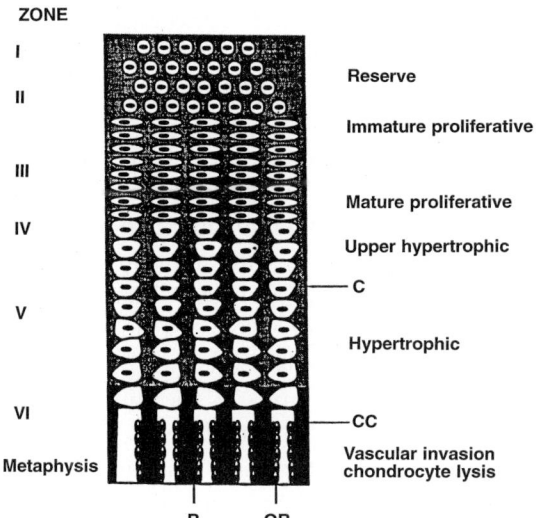

ZONE

I

II — Reserve

Immature proliferative

III

Mature proliferative

IV — Upper hypertrophic

— C

V

Hypertrophic

VI — CC

Metaphysis — Vascular invasion chondrocyte lysis

B   OB

FIGURE 1 Schematic diagram of a longitudinal section through the epiphyseal growth plate. B, bone; OB, osteoblast; CC, calcified cartilage; C, cartilage matrix (reprinted with permission from Price *et al.,* 1994, courtesy of Macmillan Press Ltd.).

cess of bone modeling, in which bone is resorbed and replaced by new bone, also contributes to the ultimate shape and length of adult bone. Although one cannot deny the important genetic component of adult limb bone length, nutrition, mechanical influences, and disease can also affect the overall growth of bone.

Structurally located between the epiphysis and the metaphysis of the long bones, the epiphyseal growth plate is composed of three separate tissue types (the cartilage component, the bony tissue of the metaphysis, and the fibrous tissue surrounding the growth plate). The cellular organization of the cartilage component in the growth plate can be divided into six different zones of maturation (Fig. 1). The reserve layer or germinal cell layer is located in zone I and surrounded by cartilage matrix, primarily composed of randomly orientated collagen type II fibrils and proteoglycans. These chondrocytes are characterized by a low rate of proliferation and low proteoglycan and collagen synthesis. Type II collagen is the most abundant of the collagens in the growth plate and found almost exclusively in cartilage. Surrounding the germinal layer in zones II and III are proliferative

chondrocytes, the cells undergoing the rapid mitosis necessary for the growth spurt. Organized in longitudinal columns, these cells secrete large quantities of type II collagen and proteoglycans. Zone IV represents the upper hypertrophic zone, which is composed of low mitotic, hypertrophic cells that actively secrete collagen and proteoglycans. In zone V, chondrocytes stop replicating and achieve their final differentiated stage as terminal chondrocytes. It is here that mineral deposition begins first within the matrix vesicles between the chondrocytes. Matrix calcification occurs in longitudinal septae with high levels of alkaline phosphatase activity. It has been demonstrated in the rat that 1 cell division in the germinal layer eventually results in 30 mature chondrocytes and a bone growth of 0.9 mm. It is estimated that 40 cell divisions in the germinal cell layer are necessary for complete growth of the rat tibia. Zone VI represents the junction of the growth plate and the metaphysis, marking the transition from cartilage to bone. Calcium phosphate formation is replaced by crystalline hydroxyapatite, whereas osteoblasts produce a new unmineralized matrix, the osteoid, rich in proteoglycans and collagen type I. Osteoclastic degradation of bony matrix and chondroclastic removal of cartilage trabeculae allow for the replacement of metaphyseal bone by lamellar bone.

While the regulation of longitudinal growth is complex, several factors have been identified to be critical to the growth process. GH secretion increases during puberty and stimulates longitudinal bone growth in a dose-dependent fashion. GH receptors have been identified in 20- and 50-day-old rabbit epiphyseal growth plate tissue, suggesting a direct hormone–receptor interaction occurring at the epiphyseal plate. GH messenger RNA and GH binding have both been demonstrated in rat epiphyseal germinal layer chondrocytes. However, recent evidence also suggests that GH indirectly stimulates bone growth via the stimulation of insulin-like growth factor-1 (IGF-1) production, which acts directly on increasing chondrocyte mitosis and matrix formation. IGF-1 mRNA and binding have both been demonstrated in the proliferative layer chondrocytes of growth plates. Since young men and women suffering from hypogonadism are tall with relatively long limbs compared to the spine, the epiphyseal growth plates are thought to be open for GH action for an extended period of time. It is this finding that initially indicated there was a role for sex steroids in the developmental regulation of epiphyseal plate growth and maturation.

III. THE PUBERTAL GROWTH SPURT

One of the key developmental elements of puberty is the growth spurt. The pubertal growth spurt is second only to the neonatal period in terms of the relative height velocity achieved. The longitudinal growth rate of bone during adolescence is approximately 1.5- to 2.0-fold greater than the prepubertal growth rate excluding infancy. Furthermore, it is unique for the dramatic maturation of the skeleton that occurs during the adolescent transition. There are gender-specific differences in the pubertal growth spurt. The pubertal growth spurt occurs approximately 2 years later in boys than in girls, providing a partial explanation for their added height (they have 2 extra years wherein they experience a slow steady longitudinal growth in their bones). At the onset of the growth spurt, boys are on average 10 cm taller than girls. The duration of the pubertal growth spurt is similar in both boys and girls (about 2 years). The peak height velocity of males averages about 10.3 cm/year, approximately 1 cm more than that of females (9 cm/year). In females, menarche occurs after the peak height velocity and marks the termination of the accelerated growth phase. In males, the peak height velocity typically follows earlier secondary sex characteristic changes such as male genital growth and pubic hair maturation. Several key endocrine changes occur during adolescence that are critical to the growth spurt: (i) Growth hormone secretion increases at puberty along with IGF-1 production as previously mentioned, and (ii) sex steroid production increases. Sex steroids indirectly stimulate IGF-1 production by increasing the secretion of GH and paracrine IGF-1 production in cartilage. The development of secondary sexual characteristics is gender specific, and the somatic growth that occurs during puberty in both males and females appears to be regulated in a similar manner.

IV. EUNUCHOIDISM: PRESENTATION AND CLINICAL ASSESSMENT

As mentioned previously, eunuchoidism is characterized by disproportionate long bone growth relative to the spine. Because hypogonadal patients have delayed epiphyseal fusion with a prolonged growth phase, their extremities "outgrow" their spine. This leads to an abnormal upper/lower segment ratio. Upper/lower segment ratio is defined as the length from the top of the pubic ramus to the top of the head divided by the distance from the top of the pubic ramus to the sole of the foot. The upper segment can be measured with the patient sitting with thighs horizontal. After subtracting the height of the stool, the sitting height represents the upper segment length. The lower segment length can be determined by measuring the distance from the floor to the top of the pubic symphysis. It is important to recognize that normal upper/lower segment ratios can vary racially, with black adults generally having lower and Asian adults having higher ratios. Although 2 standard deviation variations from the normal mean are not well documented, it is generally accepted that men and women with upper/lower segment ratios <1 are eunuchoidal in presentation (Table 1). Alternatively, height–arm span difference has been used as a measure of eunuchoidal proportions in which the arm span should exceed the height by at least 1 cm in women and 6 cm in men (Table 1). While the physical exam is critical to the definition of eunuchoidism, several key tests should be obtained to further evaluate the condition. Skeletal maturation (bone age) should be assessed by radiographs of the left hand and wrist, and serum blood testing of FSH and sex steroids can be helpful.

V. SEX STEROID REGULATION OF EPIPHYSEAL GROWTH PLATE MATURATION

There has long been a controversy revolving around the relative roles of estrogens and androgens in somatic growth regulation. Such controversy has existed due to the fact that both estrogens and andro-

TABLE 1
Average Anthropometric Measurements[a]

Age (years)	Boys					Girls					Both sexes		Span Difference span − height	
	Height		Weight (kg)	Lower segment (cm)	Ratio upper:lower	Height		Weight (kg)	Lower segment (cm)	Ratio upper:lower	Head (cm)	Chest (cm)	Male (cm)	Female (cm)
	cm	Annual increase				cm	Annual increase							
Birth	50.8		8.4	18.8	1.70	50.8		3.2	18.8	1.70	35	35	−2.5	−2.5
0.5	67.8		8.5	25.7	1.62	65.8		7.7	25.3	1.60	43.4	44	−2.5	−3.0
1	76.1	25.3	10.8	30.0	1.54	74.2	23.4	9.9	20.4	1.52	46.5	47	−2.5	−3.3
1.5	81.9		12.2	32.8	1.50	80.0		11.3	32.5	1.46	48.0	48	−2.7	−3.3
2	87.4	11.4	13.2	36.1	1.42	86.1	11.9	12.5	36.7	1.41	49.0	50	−3.0	−3.5
2.5	92.2		14.8	38.9	1.37	91.1		13.6	38.9	1.34			−3.0	−3.8
3	96.4	9.0	15.8	41.0	1.35	95.4	9.3	14.7	41.5	1.30	50.0	52	−2.7	−4.0
3.5	100.2		16.3	43.6	1.30	99.5		15.9	43.8	1.27			−2.7	−4.0
4	104.0	7.6	17.3	46.4	1.24	103.3	7.9	16.9	46.5	1.22	50.5	53	−3.0	−3.8
4.5	107.6		18.4	48.5	1.22	107.2		18.1	49.0	1.19			−3.0	−3.5
5	110.7	6.7	19.4	50.6	1.19	110.6	7.3	10.2	51.4	1.15	50.8	55	−3.3	−3.5
6	117.7	7.0	21.9	55.5	1.12	117.6	7.0	21.9	56.0	1.10	51.2	56	−2.5	−3.3
7	123.8	6.1	21.6	60.0	1.07	123.8	6.2	21.7	60.1	1.06	51.6	57	−2.5	−2.0
8	120.9	6.1	27.6	61.0	1.03	120.8	6.0	28.1	64.2	1.02	52.0	50	−1.8	−1.8
9	135.4	5.5	31.0	67.0	1.02	135.4	5.6	31.6	67.3	1.01		60	0	−1.2
10	141.0	5.0	34.8	70.8	0.99	141.0	5.6	35.4	70.5	1.00	58.0	61	0	−1.0
11	145.9	4.9	38.8	73.7	0.95	147.7	6.7	40.1	74.2	0.90			0	0
12	151.4	5.5	43.2	76.4	0.98	154.2	0.5	45.5	77.5	0.99	53.2	66	+8.0	0
13	157.5	6.1	47.9	80.0	0.97	150.5	5.3	50.1	79.7	1.00			+3.3	0
14	161.8	7.3	54.0	83.6	0.97	102.9	3.4	54.5	80.6	1.01	54.0	72[b]	+3.3	0
15	171.1	6.3	60.0	86.3	0.95	104.8	1.9	57.4	81.5	1.01			+4.3	+1.2
16	175.2	4.1	64.4	88.0	0.99	165.5	0.7	59.2	81.9	1.01	55.0	77[b]	+4.6	+1.2
17	176.6	1.4	66.9	88.0	0.99	165.5	0	60.5	81.9	1.01	55.4	82[b]	+5.8	+1.2

[a] From L. Wilkins, *The Diagnosis and Treatment of Endocrine Disorders in Childhood and Adolescence,* 1966; courtesy of Charles C Thomas, Publisher, Ltd., Springfield, Illinois.

gens are found in both pubertal males and females. Until recently, it has been difficult to separate androgens and estrogens in those clinical circumstances of abnormal sex steroid production that delay puberty and result in eunuchoidal proportions and tall stature.

A. Estrogens and Epiphyseal Growth Plate Maturation

A unique biologic opportunity to assess the role of estrogen exists in patients with complete androgen insensitivity syndrome (AIS). Complete AIS is a condition in which there is either absent or defective androgen receptors, rendering testosterone and dihydrotestosterone biologically ineffective in target tissues. Zachmann *et al.* studied pubertal bone growth in these patients and found that peak bone velocity occurred at a mean age of 12.7 years, which is more similar to normal girls (12.4 years) than normal boys (13.9 years). Similarly, mean peak height velocity approximated that of normal girls (7.4 cm/year compared to 7.3 cm/year). It can be inferred from these observations that estrogen, even without androgen action, is sufficient to support normal pubertal skeletal growth. That AIS patients achieve a height that is slightly greater than that of normal women suggests a contribution by Y-linked genes in the overall growth of these patients. Data from Turner's syndrome patients indicating that replacement estrogen accelerates the growth of these girls further suggest a primary role for estrogen in the accelerated growth of the epiphyseal plate.

The mechanisms of action of estrogen in accelerating epiphyseal fusion involve the induction of maturation in chondrocytes and osteoblasts. Estrogen has recently been shown to be the most important factor in stimulating maturation of the chondrocytes and osteoblasts. A 28-year-old man with estrogen resistance caused by a mutation in the estrogen–receptor gene has been described. The patient was tall with an arm span (213 cm) that exceeded his height (204 cm) by 9 cm. Radiography of his left wrist and hand revealed a markedly delayed bone age of 15 years despite elevated serum levels of estradiol. His serum testosterone level was 445 ng/dl, within the normal range for males. Knee films also documented open epiphyses with minimal evidence of epiphyseal maturation over a 10-year period. This unique phenotype provides evidence that estrogen is an obligatory hormone for the induction of epiphyseal maturation in both men and women. Conversely, this model supplies evidence that androgens alone are not sufficient to induce the maturation process of the epiphyseal plate.

Another demonstration of estrogen's vital importance in epiphyseal maturation has recently been made in a patient with aromatase deficiency. This patient is a female pseudohermaphrodite who had virilized genitalia at birth and increasing virilization with the absence of breast development at puberty. Due to the lack of aromatase, there was minimal conversion of testosterone to estradiol. Interestingly, despite high serum testosterone levels, serum gonadotropin levels were high, resulting in the hyperstimulated ovarian condition. In this patient there was a 4-year delay in bone maturation by age 14. Ethinyl estradiol (20 μg) administered daily to this patient reversed her ovarian cyst formation and induced breast development, menarche, and a pubertal growth spurt of 13 cm to a final height of 166.4 cm. There has also been a report of male and female siblings both diagnosed with aromatase deficiency. In the male sibling there was a clear delay in bone maturation (bone age = 14; chronological age = 24 3/12) with elevated serum testosterone and androstenedione levels and almost undetectable levels of serum estradiol. Here again, androgens in the absence of estrogen were not sufficient to cause the maturation and closure of the epiphyseal plate.

B. Androgens and Epiphyseal Growth Plate Maturation

Androgens in the pubertal growth spurt in both men and women have long been considered to be the key steroids in the pubertal growth spurt. It has been surmised that testicular androgens represent the primary stimulant in males, whereas adrenal androgens simulate the male hormone environment in females. However, as noted previously, evidence has been mounting for a significant role of estrogens in the pubertal growth process and maturation. If this is the case, what is the role of androgens in the pubertal growth spurt in both abnormal and normal conditions? There is evidence that androgens stimulate growth directly at the level of the epiphyseal chondrocyte. Androgen binding sites have been identified in human fetal chondrocytes, with DHT being the active DNA-synthesizing androgen. Androgen action may also be indirect because androgens have been shown to regulate various growth factors, including IGF-I and fibroblast growth factor, which mediate chondrocyte proliferation without further differentiation. Clinical evidence that androgens can accelerate bone growth has been provided by several studies. Oxandrolone, an androgen that is resistant to aromatization, has been shown to increase height velocity with concomitant acceleration of epiphyseal maturation in both males and females. However, androgens via conversion to estrogens increase serum GH and IGF-1, which may partially account for the increased height velocity seen in pubertal males. In the aromatase-deficient male described previously, normal growth occurred despite low-normal IGF-1 levels and absent estrogen levels, suggesting a direct role for androgens in epiphyseal plate growth. Regarding the role of androgens in the regulation of epiphyseal plate maturation, recent studies in aromatase-deficient subjects strongly refute the previous view that androgens are important in the process of normal epiphyseal closure.

VI. DISEASE STATES ASSOCIATED WITH EUNUCHOIDISM

The common thread that ties together all the clinical states associated with eunuchoidal stature is hypogonadism resulting in the absence of sex steroid production during the adolescent years. Implicit in this statement is the fact that both androgens and estrogens are typically absent in patients suffering from eunuchoidism.

A. Hypogonadotropic Hypogonadism

Among the various clinically recognized disease states, Kallmann's syndrome has become recognized as synonymous with the term "eunuch." Kallmann's syndrome was originally described by Kallmann *et al.* in 1944 in men with hypogonadism and anosmia as the cardinal features of presentation. In women, a hypogonadal state results in primary amenorrhea. Kallmann's syndrome is the most frequent cause of isolated gonadotropin deficiency. It is believed that the lack of gonadotropin secretion in Kallmann's syndrome results from an underlying absence of mature GnRH-secreting neurons in the mediobasal hypothalamus and arcuate nucleus. The origin of GnRH-secreting neurons in the olfactory placode of all mammals including the human points to a lack of neuronal migration as the cause of the disorder. Although segregation analysis in familial cases of Kallmann's syndrome has revealed several modes of inheritance, including autosomal recessive and autosomal dominant forms, recently a gene has been identified as a candidate gene for X-linked recessive Kallmann's syndrome which occurs in males. This putative gene, the *KAL* gene, is thought to encode a molecule with cell-adhesive properties that may interfere with the migration of GnRH neurons to the hypothalamus during development. Due to the lack of gonadotropin secretion in adolescence, hypogonadism results with pubertal delay. Both male and female patients suffering from Kallmann's syndrome have a reversible hypogonadism which is responsive to exogenous gonadotropin therapy or GnRH pump therapy.

Since it is recognized that Kallmann's syndrome patients suffer from anosmia, a separate but similar category of patients has been described to distinguish those patients suffering from deficient endogenous gonadotropin secretion without concomitant anosmia (idiopathic hypogonadotropic hypogonadism). These patients also fail to undergo puberty because of inadequate gonadotropin secretion which is caused by a lack of endogenous pulsatile GnRH secretion. Similar to Kallmann's syndrome, various genetic modes of inheritance have been described. Generally, men with idiopathic hypogonadotropic hypogonadism lack any historical or physical evidence of sexual maturation. They suffer from a delayed bone age with associated eunuchoidal habitus. Aside from the lack of anosmia, these patients are phenotypically similar to Kallmann's syndrome patients. Because of the lack of anosmia, it is difficult to distinguish these patients from those with severe constitutional delay of puberty. Somatic abnormalities associated with both Kallmann's syndrome and idiopathic hypogonadotropic hypogonadism include cleft palate, unilateral renal agenesis, and nerve deafness. Idiopathic hypogonadotropic hypogonadal men have been shown to have reduced cortical and trabecular bone mineral density. A role for sex steroid involvement in bone matrix development is suggested by the fact that these patients suffer from osteoporosis both before and after epiphyseal closure. The finding that testosterone enanthate supplementation increases bone density during treatment further supports this concept. It is surmised that the osteopenia of hypogonadal men may be a result of a decrease in the peak bone mass achieved during skeletal maturation with an accelerated loss of established bone. Furthermore, the persistent osteopenia into adulthood in eunuchoidal men suggests that puberty must occur during a critical period of bone development if normal peak bone mineral density is to be attained.

B. Hypergonadotropic Hypogonadism (Primary Gonadal Failure)

Klinefelter's syndrome in men represents another condition associated with eunuchoidal stature. These patients have a genetic male phenotype with genotypes that vary from 47,XXY to 49,XXXXY, with primary gonadal failure as the cause of their hypogonadal state. The testes are small and characterized by seminiferous tubule dysgenesis with limited spermatogenesis and sex steroid production during adolescence. These patients have a myriad of other problems, including major and minor congenital anomalies, delayed mental and emotional development, and adjustment disorders. Klinefelter's syndrome males present with disproportionately long legs prepubertally, which is exacerbated during puberty by the delay in epiphyseal closure that occurs due to hypogonadism. However, there may not be an associated increase in arm span postpubertally as typically seen in hypogonadotropic hypogonadal men and women.

VII. PROPOSED MODEL FOR EUNUCHOIDISM

Eunuchoidism is associated with any condition in which there is a delay in pubertal maturation and sex steroid production. Our understanding of the various roles that androgens and estrogens play in the regulation of bone growth and skeletal maturation has increased considerably with some recent unique clinical examples. The current evidence supports a hypothesis of estrogen and androgen action that is not sexually dimorphic. While the normal pubertal growth spurt appears to be mediated in males and females by both androgens and estrogens, estrogen appears to be the primary sex steroid responsible for epiphyseal maturation and closure in both sexes (Fig. 2). Estrogen can be viewed as a primary determinant of the final height of an adult, given its unique role in regulating the terminal differentiation of chondrocytes within the epiphyseal plate.

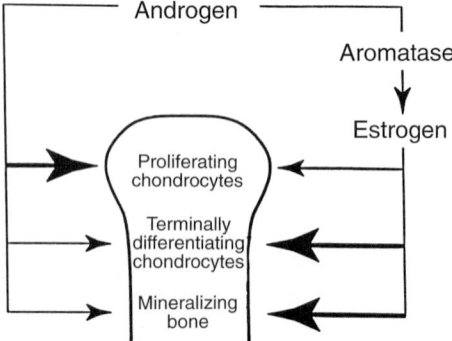

FIGURE 2 Suggested model of the relative actions of androgen and estrogen on the epiphysis (reprinted with permission from Smith and Korach, 1996, courtesy of Scandinavian University Press).

See Also the Following Articles

ANDROGENS, OVERVIEW; FSH (FOLLICLE-STIMULATING HORMONE); GnRH (GONADOTROPIN-RELEASING HORMONE); HYPOGONADISM; IGF (INSULIN-LIKE GROWTH FACTORS); KALLMANN'S SYNDROME; KLINEFELTER'S SYNDROME; LH (LUTEINIZING HORMONE)

Bibliography

Conte, F. A., Grumbach, M. M., Ito, Y., Fisher, C. R., and Simpson, E. R. (1994). A syndrome of female pseudohermaphrodism, hypergonadotropic hypogonadism, and multicystic ovaries associated with missense mutations in the gene encoding aromatase (P450arom). *J. Clin. Endocrinol. Metab.* **78**, 1287–1292.

Cutler, G. B., Jr., Cassorla, F. G., Ross, J. L., Pescovitz, O. H., Barnes, K. M., Comite, F., Feuillan, P. P., Laue, L., Foster, C. M., Kenigsberg, D., Caruso-Nicoletti, M., Garcia, H. B., Uriarte, M., Hench, K. D., Skerda, M. C., Long, L. M., and Loriaux, D. L. (1986). Pubertal growth: Physiology and pathophysiology. *Recent Prog. Horm. Res.* **42**, 443–465.

Finkelstein, J. S., Neer, R. M., Biller, B. M. K., Crawford, J. D., and Klibanski, A. (1992). Osteopenia in men with a history of delayed puberty. *N. Engl. J. Med.* **326**, 600–604.

Kallmann, F. J., Schoenfeld, W. A., and Barrera, S. E. (1944). The genetic aspects of primary eunuchoidism. *Am. J. Mental Deficiency* **48**, 203–236.

Morishima, K., Grumbach, M. M., Simpson, E. R., Fisher, C., and Qin, K. (1995). Aromatase deficiency in male and female siblings caused by a novel mutation and the physiological role of estrogens. *J. Clin. Endocrinol. Metab.* **80**, 3689–3698.

Pescovitz, O. H. (1990). The endocrinology of the pubertal growth spurt. *Acta Paediatr. Scand.* **367**(Suppl.), 119–125.

Price, J. S., Oyajobi, B. O., and Russell, R. G. G. (1994). The cell biology of bone growth. *Eur. J. Clin. Nutr.* **48**, S131–S149.

Rosenfield, R. L. (1990). Diagnosis and management of delayed puberty. *J. Clin. Endocrinol. Metab.* **70**, 559–562.

Smith, E. P., and Korach, K. S. (1996). Oestrogen receptor deficiency: Consequences for growth. *Acta Paediatr. Scand.* **417**(Suppl.), 39–43.

Smith, E. P., Boyd, J., Frank, G. R., Takahashi, H., Cohen, R. M., Specker, B., Williams, T. C., Lubahn, D. B., and Korach, K. S. (1994). Estrogen resistance caused by a mutation in the estrogen-receptor gene in a man. *N. Engl. J. Med.* **331**, 1056–1061.

Zachmann, M., Prader, A., Sobel, E. H., Crigler, J. F., Ritzen, E. M., Atares, M., and Ferrandez, M. (1986). Pubertal growth in patients with androgen insensitivity: Indirect evidence for the importance of estrogens in pubertal growth of girls. *J. Pediatr.* **108**, 694–697.

Fallopian Tube

Diaa M. El-Mowafi and Michael P. Diamond

Wayne State University

GLOSSARY

adhesions Scar fibrous tissue connecting sites at aberrant locations.

ectopic pregnancy Pregnancy occurring outside of the uterine cavity, most commonly in the Fallopian tube.

Fallopian tube A conduit allowing passage of egg from ovary to exit the abdominal cavity to enter the uterus; the site of fertilization in humans.

falloposcopy Viewing the interior of the tubal lumen with a narrow fiber scope.

tuboplasty Surgery on the tube, most commonly to restore tubal pateny or reduce scarring to promote fertility. includes procedures such as neosalpingstomy and fimbrioplasty.

The Fallopian tube is essential for unassisted procreation, serving as a conduit for the ovulated oocyte (egg) to enter the uterine cavity following fertilization. Its functions are regulated both hormonally and by the nervous system, and it is the usual site for oocyte fertilization. Tubal function can be impaired by infection, surgery, adhesions, and other pathologic processes. Function of the tube can often be restored surgically or it can be bypassed through the use of *in vitro* fertilization of oocytes followed by intrauterine embryo transfer.

I. EMBRYOLOGY

Between the fifth and sixth week after oocyte fertilization, a longitudinal groove called Müller's groove arises from the coelomic epithelium on each side lateral to the mesonephric duct (Fig. 1). The edges of this groove fuse to form a canal called the Müllerian or paramesonephric duct. The Fallopian tubes develop from the cranial parts of the paramesonephric ducts, with their cranial ends remaining open connecting the duct with the coelomic (peritoneal) cavity and the caudal end communicating with the uterine cornua. Congenital anomalies of the tube include aplasia, in which the tube fails to form; hypoplasia, in which the tube is long, narrow, and tortuous; accessory ostia, and congenital diverticulae.

II. ANATOMY

The Fallopian tubes are paired, tubular, seromuscular organs whose course runs medially from the cornua of the uterus toward the ovary laterally. The tubes are situated in the upper margins of the broad ligaments between the round and utero ovarian ligaments (Fig. 2). Each tube is about 10 cm long with variations in length from 7 to 14 cm. The abdominal ostium is situated at the base of a funnel-shaped expansion of the tube, the infundibulum, the circum-

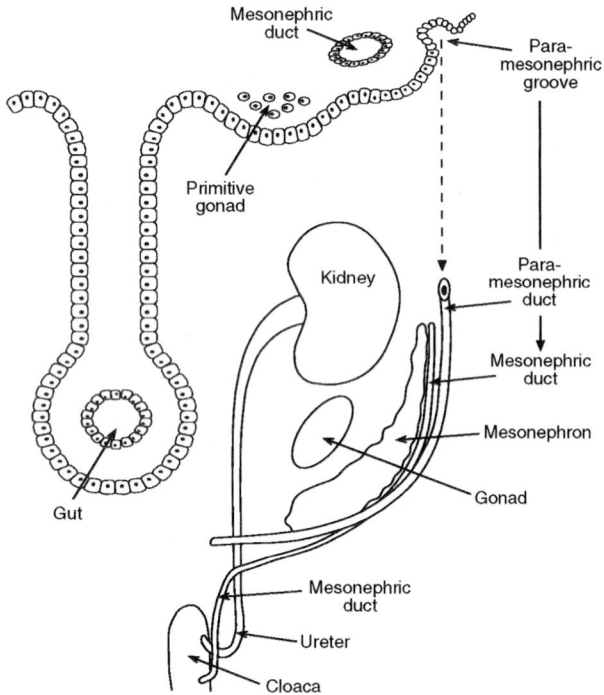

FIGURE 1 Embryology of the tubes.

ference of which is enhanced by irregular processes called fimbriae. The ovarian fimbria is longer and more deeply grooved than the others and is closely applied to the tubal pole of the ovary. Passing medially, the infundibulum opens into the thin-walled ampulla forming more than half the length of the tube and 1 or 2 cm in outer diameter; it is succeeded by the isthmus, a round and cord-like structure constituting the medial one-third of the tube and 0.5– 1 cm in outer diameter. The interstitial or cornual portion of the tube continues from the isthmus through the uterine wall to empty into the uterine cavity. This segment of the tube is about 1 cm in length and 1 mm in inner diameter.

III. HISTOLOGY

The tubal wall consists of three layers: the internal mucosa (endosalpinx), the intermediate muscular layer (myosalpinx), and the outer serosa, which is continuous with the peritoneum of the broad ligament and uterus, the upper margin of which is the mesosalpinx. The endosalpinx is thrown into longi-

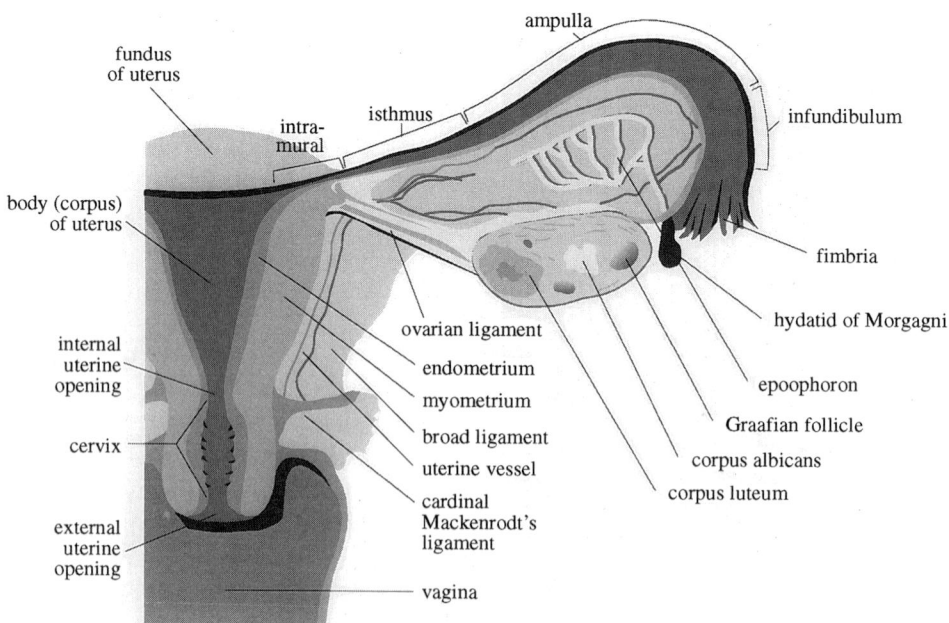

FIGURE 2 Anatomy of the tubes.

tudinal folds, called primary folds, increasing in number toward the fimbria and lined by columnar epithelium of three types: ciliated, secretory, and peg cells. In the ampullary and infundibular sections, secondary folds of the tubal mucosa also exist, markedly increasing the surface areas of these segments of the tube. The myosalpinx actually consists of an inner circular and an outer longitudinal layer to which a third layer is added in the interstitial portion of the tube.

The serosa of the tube is composed of an epithelial layer histologically indistinguishable from peritoneum elsewhere in the abdominal cavity.

IV. PHYSIOLOGY

The tubes act as ducts for sperm, oocyte, and fertilized ovum transport, in addition to being the normal site of fertilization. These functions depend mainly on three factors: tubal motility, tubal cilia, and tubal fluid.

A. Tubal Motility

Peristallic contraction of the smooth muscle fibers in the tubal wall allows the gametes (the sperm and egg) to be brought together, thus allowing fertilization and subsequent transport of the fertilized ovum from the normal site of fertilization in the ampulla to the normal site of implantation in the uterus. This movement is primarily regulated by three intrinsic systems: the estrogen–progesterone hormonal milieu, the adrenergic–nonadrenergic system, and prostaglandins.

Estrogens acting at α receptors stimulate tubal motility, whereas progesterone, which activates β receptors, inhibits tubal motility. Before ovulation, contractions are gentle, with some individual variations in rate and pattern. At ovulation, contractions become vigorous and the mesosalpinx contracts to bring the tube in more contact with the ovary while the fimbria contracts rhythmically to sweep over the ovarian surface. As progesterone level rises 4–6 days after ovulation, it inhibits tubal motility. This may lead to relaxation of the tubal musculature to allow

passage of the ovum into the uterus by the action of the tubal cilia. The effects of estrogen and progesterone on oviductal motility and morphology is mediated through these steroids' receptors. The changes in receptors levels are critical in determining the functional state of the oviduct.

Adrenergic innervations are thought to be involved in regulations of tubal motility, particularly isthmic motility changes. During menstruation and the proliferative (preovulatory) phase, the human tube is very sensitive to α-adrenergic compounds such as norepinephrine. After ovulation and during the luteal phase, the response to norepinephrine is decreased and the inhibitory effect of β-adrenergic compounds is more evident. Estrogens potentiate the activation of α receptors, whereas progesterone potentiates the activation of β receptors. Activation of the β receptors by raised progesterone level in the luteal phase leads to relaxation of the circular muscles; thus, the isthmic luminal diameter is increased and transisthmic passage of the fertilized ovum is facilitated.

Although there is a controversy regarding the role of prostaglandins in the regulation of spontaneous tubal motility, it has been found that prostaglandin $F_{2\alpha}$ ($PGF_{2\alpha}$) stimulates whereas PGE_1 and PGE_2 inhibit Fallopian tube contractions. Contrary to their differential activity on tubal motility, all three natural prostaglandins (PGF_2, PGE_1, and PGE_2) stimulate ciliary activity *in vitro*.

In summary, the initial rise in progesterone after ovulations causes β- mediated contractions of the two inner layers of the uterotubal junction, thus causing tubal locking of the ovum. After a few days, sensitivity of the muscles to adrenergic stimulation diminishes, whereas other factors, such as prostaglandins, dominate leading to relaxation of the uterotubal junction and release of the fertilized ovum into the uterine cavity.

B. Tubal Cilia

There are fewer ciliated cells in the isthmus than in the ampullary portion of the tube, whereas they are most prominent in the fimbriated infundibulum. Ciliation and deciliation is a continuous process throughout the menstrual cycle. Ciliation is maximum in the periovulatory period, particularly in the fimbria. Estrogen enhances the process of ciliation, whereas progesterone inhibits it, so significant deciliation occurs in atrophic postmenopausal tube.

Ciliary activity is responsible for the pickup of ova by the fimbrial ostium and movement through the ampulla, as well as the distribution of the tubal fluid which supports gamete maturation and fertilization and facilitates gamete and embryo transport. The close approximation between the ovary and fimbria is likely to be important for ovum pickup, although transperitoneal migration has been reported. The importance of ciliary activity is affirmed by the tubal dysfunction seen in associations with the deciliation of salpingitis. Questions are raised, however, in women suffering from Kartagener's syndrome, which is the immotile cilia syndrome, when the women are still fertile.

C. Tubal Fluid

Tubal fluid is rich in mucoproteins, electrolytes, and enzymes. This fluid is abundant in midcycle when gametes or embryos are present and may play an important role during fertilization and early cleavage. Fluid in the tubes is believed to be formed by (i) selective transudation from the blood and (ii) active secretion from the epithelial lining. The rate of fluid accumulation is 1–3 ml/24 hr and the rate of production is increased significantly around the time of ovulation.

V. TUBAL DISORDERS

A. Pelvic Inflammatory Disease

Pelvic inflammatory disease (PID) is inflammation of the upper genital tract characterized primarily by salpingitis. The disorder may coexist with endometritis or oophoritis, may spread as peritonitis, and may extend along the paracolic gutters to the liver to cause the Fitz–Hugh–Curtis syndrome.

The cervical barrier plays a crucial role in preventing the ascent of vaginal organisms to the upper genital tract. This barrier may be compromised after miscarriage, delivery, cervical surgery such as ampu-

tation, dilatation, and cauterization, or at the time of an intrauterine device insertion. The causative organism in most initial cases is *Chlamydia trachomatis; Neisseria gonorrhoeae, Mycoplasma hominis, Ureaplasma uralytica,* and *Actinomyces israelii* have also been found to be a cause in some cases. In subsequent episodes of PID, other aerobic and anaerobic organisms may be the causative agent to the pelvic infection. The infection usually starts as an asymptomatic cervicitis. If the natural barriers such as the narrow endocervical canal, the downward flow of mucus, the presence of antibacterial lysozymes, and the production of specific local IgA fail, the infection may spread to the endometrium. The monthly shedding of the endometrium is another protection against infection, aided by the mechanical effect of the uterotubal junctions and the ciliary activities of the Fallopian tubes which creates a downward flow of tubal fluid. Therefore, not all the organisms that reach the endometrial cavity necessarily spread to the Fallopian tubes. Nonetheless, the open tubal lumen allows spread of infection to the peritoneum, causing salpingooophritis and peritonitis. Consequences of PID include blockage of the tube usually either proximately at the site of insertion into the uterus or distally causing a hydrosalpinx with partial or complete distal obstruction. Less commonly, a midtubal segment of the tube may become occluded. Other sequelae may include pyosalpinx, tubal or tuboovarian abscess, and peritubal adhesions. The long-term consequences of PID include recurrent PID in almost 25% of cases after one episode of salpingitis, chronic pelvic or abdominal pain in one of every five affected patients, tuboovarian abscess in about 34% of hospitalized patients, Fits–Hugh–Curtis syndrome, deep dyspareunia in two of every five patients, and menstrual disturbances in four of every five patients. Additionally, the risk of ectopic pregnancy increases seven times that of control subjects. The risk of subsequent infertility is approximately 12% after one episode of PID, 35% after two episodes, and 75% after three or more episodes.

B. Ectopic Pregnancy

The incidence of ectopic pregnancy has significantly increased during the past two decades—more than fourfold from 4.8 to 20.9 per 1000 live births. However, the mortality rate due to ectopic decreased significantly during the same period from 53.2 to 5.3 deaths per 10,000 cases, perhaps due to more awareness by patients and health care providers. However, ectopic pregnancy remains the second leading cause of maternal mortality in the United States.

Ectopic pregnancy results from a delay in the passage of the fertilized ovum through the Fallopian tube. This delay can result from anatomical abnormalities of the tubes, such as constriction and false passage formation (e.g., diverticulum), or from tubal dysfunction as altered contractility or abnormal ciliary activity. Tubal anatomy and function can both be altered by either tubal surgery or prior PID. These are often present together in the same individual. The surgical procedures predisposed to ectopic pregnancy include salpingolysis and ovariolysis, fimbrioplasty, neosalpingostomy, and tubal anastomosis. In fact, our success of microsurgery may have allowed patency of tubes to be restored, only to predispose these patients to ectopics. Approximately 95% of ectopics occur in the tubes, in which most of them are located in its distal parts, particularly the ampulla. The other 5% of cases occur in the ovaries—a rudimentary horn of bicornuate uterus, broad ligaments, peritoneum, and cervix.

Treatment of tubal pregnancy may be expectant, medical, or surgical. Asymptomatic tubal pregnancies with low and falling β-human chorionic gonadotropin (hCG) levels 48 hr apart may be treated expectantly in reliable patients with readily available access to the hospital. Such tubal gestations frequently resolve spontaneously, but treatment has limitations and a relatively high failure rate. Medical therapy may be systemic or local. Intramuscular methotrexate at a dosage of 1 mg/kg per day on alternative days with cirtovorum factor rescue at a dosage of 0.1 mg/kg, for a total period of 8 days or a single-dose methotrexate (1 mg/kg), is the most commonly used protocol. Failure is more likely to occur if β-hCG levels exceed 6500 mIU/ml or if fetal heart motion is identified on ultrasonography. Local injection of methotrexate, $PGF_{2\alpha}$, or hypertonic glucose solution into the tubal pregnancy under laparoscopic visualization appears to be effective in selected patients. Local medical treatment is also plausible, with the drug injected transvaginally under ultrasonographic control and local anesthesia.

Surgical treatment can be achieved via laparotomy or laparoscopy. Through laparotomy, one of the following procedures may be performed:

1. Salpingostomy: Used to treat unruptured ampullary gestations. A linear incision is performed in the antimesentric border of the tube and the ectopic gestation is expressed through it. The edges of the incision can be cauterized if bleeding occurs, but this is otherwise usually not approximated by sutures. In the ampullary segment of the tube, these will usually heal by secondary intention without fistula formation or tubal obstruction. In part, this healing profile may stem from the frequent localization of ampullary ectopics not in the tubal lumen but rather in the potential space between the serosa and muscularis. A salpingotomy varies slightly from a salpingostomy in that there is suturing of the incision edges by sutures. Often, salpingostomy may be preferred because sutures may increase the incidence of postoperative adhesions.

2. Salpingectomy: Performed in a patient in whom a tube is thought to be nonsalvagable or a patient who does not desire future conception. It may be partial or complete. The removal of the ipsilateral ovary is not recommended, as was once the common practice, unless the ovary is involved with the ectopic or other ovarian pathology exists.

3. Segmental resection and anastomosis: Used most commonly for unruptured ectopic in the isthmus of the tubes since this part is narrow, and if salpingostomy is performed proximal tubal obstruction and subsequent ectopic may occur. Anastomosis is often difficult during the unscheduled times surgical treatment of ectopics occur because operating room staff may be unfamiliar with microsurgical equipment. Additionally, edema and vascularity of the tube may be increased at this time. An alternative is to perform resection, with reanastomosis at a later time. If this option is chosen, the woman must be counseled that she can develop a repeat ectopic in the blindly ending distal segment. A third option is to perform a linear salpingostomy, thereby accepting the increased risk of fistula formation or occlusion

but avoiding a repeat surgery in women in whom the tube heals without complication.

4. Fimbrial evacuation: The distally implanted tubal pregnancy is evacuated by "milking" or "suction" through the fimbrial end. The drawback of this method is that the incidence of recurrent ectopic is double that of salpingostomy and bleeding from the implantation site may continue. However, if the eccyesis is already extruding from the fimbriated end of the tube, grasping the products of conception is a reasonable alternative.

Through laparoscopy, one of the following procedures may be performed if the patient is hemodynamically stable:

1. Linear salpingostomy: After the injection of vasopressin in the mesosalpinx, an incision is made along the antimesentric border of the ectopic with the CO_2 laser, the argon laser, or by electrocoagulation. The ectopic gestation and the surrounding clot often extrude from the incision. The tissue is then grasped and removed through one of the second puncture probes. An irrigator can be placed into the incision, and irrigation under pressure may dislodge the eccyesis. Bleeding points are secured by bipolar diathermy and salpingostomy incision is left unsutured to heal by second intention.

2. Segmental resection: The bipolar electrocautery forceps is applied distal to the target segment. The tube, mesosalpinx, and blood vessels are fulgurated in successive steps toward the uterus. The specimen is then sharply resected with laparoscopic scissors.

3. Salpingectomy: The tube, together with the adjacent mesosalpinx, is grasped, electrocoagulated, and divided at the level of the uterotubal junction. The tuboovarian ligament is then grasped, electrocoagulated, and divided. The mesosalpinx is then electrocoagulated and divided until complete excision of the tube is accomplished.

C. Salpingitis Isthmica Nodosa

Salpingitis isthmica nodosa (SIN) is a disease process characterized by nodular thickenings in the intramural (interstitial) and isthmical part of the tubes. It is common around the age of 35 and more common among African Americans. The exact cause of the disease is unknown, but the inflammatory process due to tuberculosis, gonorrhea, or bilharziasis is the most accepted explanation. The condition is bilateral in about 35% of cases; the nodules vary from a few millimeters to 2.5 cm in diameter. These are firm with a smooth surface, giving a beaded appearance. Microscopically, these are gland-like spaces scattered throughout the myosalpinx with hyperplasia and hypertophy of muscle fibers. The surrounding stroma is infiltrated with plasma cells or eosinophils. This condition has the sequence of ectopic pregnancy and infertility due to occlusion at the isthmic–ampullary junction.

D. Neoplasms of the Fallopian Tubes

Although the Fallopian tubes are a common site of metastatic spread, primary tumors are rare. The benign tubal neoplasms include adenomatoid tumors, leiomyoma, teratomas, fibroma, fibroadenoma, papilloma, lipoma, hemangioma, lymphangioma, mesothelioma, and mesonephroma. The malignant tumors are either secondary or primary as adenocarcinoma, sarcoma, and choriocarcinoma. Tubal carcinoma is the rarest genital cancer, represented about 0.3% of all genital malignancy, and the age of incidence is 40–60 years. It is commonly mistaken with ovarian carcinoma but tubal carcinoma is characterized by being mainly in the tube and shows a papillary pattern; microscopically the tubal mucosa should be involved and the transition between benign and malignant tubal epithelium is demonstrated. The risk factors for tubal carcinoma are infertility, which was found in up to 70% of cases, and chronic salpingitis, which may itself be the cause of infertility. The clinical presentation which may suggest tubal cancer is the triad of pelvic pain, serosanguinous vaginal discharge escaping continuously or in gushes (hydrops tubae profluens), and pelvic mass.

VI. TUBOPLASTY

The term "tuboplasty" refers to a broad group of surgical procedures involved with the surgical correction of tubal disease, which may contribute to infertility. Most of these procedures can be performed at laparoscopy as well as laparotomy. Regardless of how the procedures are performed, the principles of microsurgery should be applied to the extent possible, including the use of magnification, fine-caliber microsurgical instruments, meticulous hemostasis, minimization of tissue handling, prevention of tissue desiccation, avoidance of introduction of foreign bodies, and use of fine suture material of low tissue reactivity. In contrast to prior inclusion of precise reapproximation of all tissue planes, we believe it may be better to apply this "tenet" in situations in which the surgeon wishes to protect or isolate underlying tissue. The following are types of tuboplasty:

1. Salpingolysis: Lysing adhesions from the tube to adjacent surfaces such as the ovary, uterus, bowel, and peritoneal surfaces.

2. Resection and anastomosis: Performed for reversal of sterilization procedures or treatment of pathologic tubal obstructions.

3. Neosalpingostomy: Creation of a new tubal ostium whether terminal, ampullary, or isthmic. The tube may be a hydrosalpinx but this is not required. In patients with a hydrosalpinx, the likelihood of conceiving following surgical correction will be related to the size of the hydrosalpinx, the thickness of the tubal wall, and, perhaps most important, the extent of damage to the tubal mucosa as assessed by loss of tubal folds and appearance of the fimbria.

4. Fimbrioplasty: Operation on the distal end of the tube, including reconstruction of existent fimbria by ostial stretching, deagglutination, lysis of perifimbrial adhesions, and incisions through the tubal wall to recover fimbriae. A fimbrioplasty is distinguished from a neosalpingostomy in that a fimbrioplasty has at least a partial ostium remaining. Because of the diversity in tubal damage which can be treated by a "fimbrioplasty" (from a filmy adhesion to the fimbria to a large hydrosalpinx with a pinpoint opening), success following treatment is extremely variable.

5. Tubouterine implantation: Suitable for cornual obstruction unamenable to anastomosis in which the blocked segment is excised and tube is reimplanted into the uterine cornu or fundus. This procedure is currently infrequently performed because of low success rates.

6. Tubal cannulation: Proximal tubal blockage as assessed by hysterosalpingogram, chromopertubation, or hysteroscopy may represent true obstruction or the presence of an occluding agent such as a mucus plug. Transcervical probing of the proximal Fallopian tube with a special fine guide wire under fluoroscopic guidance or through hysteroscopic catheterization may be effective in dislodging debris or breaking up fine adhesions which cause tubal occlusion and result in infertility. In the past, these patients would have needed laparotomy with tubal microsurgery or *in vitro* fertilization. A 58% pregnancy rate was achieved by this procedure in carefully selected patients in whom proximal tubal obstruction was identified to be the primary or sole cause of infertility.

VII. ASSISTED REPRODUCTIVE TECHNIQUES INVOLVING THE TUBE

Assisted reproductive technologies (ART) are utilized as a means of helping infertile couples to conceive. They all involve recovery (aspiration) of oocytes from the ovary (usually under transvaginal ultrasound guidance). In *in vitro* fertilization (IVF), the oocytes are incubated with spermatozoa (the male gamete) and then evaluated approximately 18 hr later to determine whether fertilization has occurred. The zygote is then cultured an additional 24–48 hr, during which time embryo cellular division occurs, followed by transcervical transfer of embryos into the uterine cavity. Alternative methods of ART are to transfer the gametes or zygotes into the Fallopian tubes using the following procedures:

1. Gamete intra-Fallopian transfer: Sperm and a limited number of oocytes (usually two to four) are loaded into a special catheter with a culture medium, and the gametes are delivered through the laparoscope or a transcervical catheter into the Fallopian tube.

2. Zygote intra-Fallopian transfer: The zygote is transferred into the ampulary part of the tube approximately 18–24 hr after fertilization between the male and female gametes *in vitro*.

VIII. TUBAL STERILIZATION

The tubes, as a target to induce female sterilization, can be approached via different routes, namely, laparotomy, minilaparotomy, laparoscopy, colpotomy, and hysteroscopy.

Laparotomy: Many procedures have been used; the following are the most common:
  Simple ligation: Ligation of the tube in continuity.
  Madlener (1919): Crushing and ligation of a loop of the tube (Fig. 3A).
  Pomeroy (1929): Ligature around a loop of the tube and excision (Fig. 3B).
  Irving (1924): Double ligature, excision of a tubal segment, and concealing of the proximal end, which is buried in the myometrium, and the distal end, which is buried in the broad ligament (Fig. 3C).
  Uchida (1975): Tubal serosa is stripped from muscular coat, tubal segment is excised, and proximal end is ligated and buried in the broad ligament (Fig. 3D).
  Kroener (1969): Fimbriectomy, ligation of distal end of tube, and mesosalpinx and excision of fimbriated end.
  Parkland: Double ligature and removal of a segment in between.
  Salpingectomy: Removal of both tubes (Fig. 3E).
  Cornual resection: Removal of the interstitial portion of the tube together with a wedge of myometrium.
Minilaparotomy: A small transverse abdominal incision 2 or 3 cm in length through which the Pomeroy, Parkland, or Kroener techniques can be utilized.
Laparoscopy
  Electrosurgery: i-monopolar and ii-bipolar; Electrical coagulation of adjacent segments of the tube with or without transection of the tube.
  Fallope ring: A Loop of tube is drawn into an applicator tube; a silastic ring is placed around the tubal loop (Fig. 3F).
  Clips (e.g., Hulka or Filshie): Plastic crushing clips are placed across the tube (Fig. 3G).
Colpotomy: Whether anterior or posterior, through which one of the previously mentioned procedures done through laparotomy can be accomplished. This option is currently infrequently performed.
Hysteroscopy: Experimental approaches have been attempted by the following:
Electrocoagulation
Chemical scarification
Intratubal plugging devices (ITD)
  Uterotubal junction devices: Placed into the uterotubal junction and held in place by lateral "wings" at the distal tip and a loop-size extension of the proximal part. The uterotubal junction is not occluded.
  Blocking uterotubal junction devices: Occlude the intramural part of the tube. However, this part of the tube is a muscular structure and expulsion or pain are frequent drawbacks. A special type of these devices is the P-block developed by Brundin. The P-block consists of a nylon skeleton with a 4-mm-long hydrogel body and two nylon anchoring wings. After insertion, the hydrogel expands and occludes the intramural part of the tube.
  Intratubal formed-in-place silicone devices (Ovabloc ITDs): Ovabloc ITDs are formed *in situ* after the intratubal administration of liquid medical-grade silicone rubber mixed with catalyst in a mixer–dispenser. Radiopaque spherical silver powder is added to the silicone so that the ITD can be identified on X ray.

The failure rate of commonly utilized methods of tubal sterilization is variable from one method to the other, but generally it is approximately 0.3–1%; this may be due to misidentification of the tubes or recanalization. The success rate of reversal of the sterilization is dependent on the

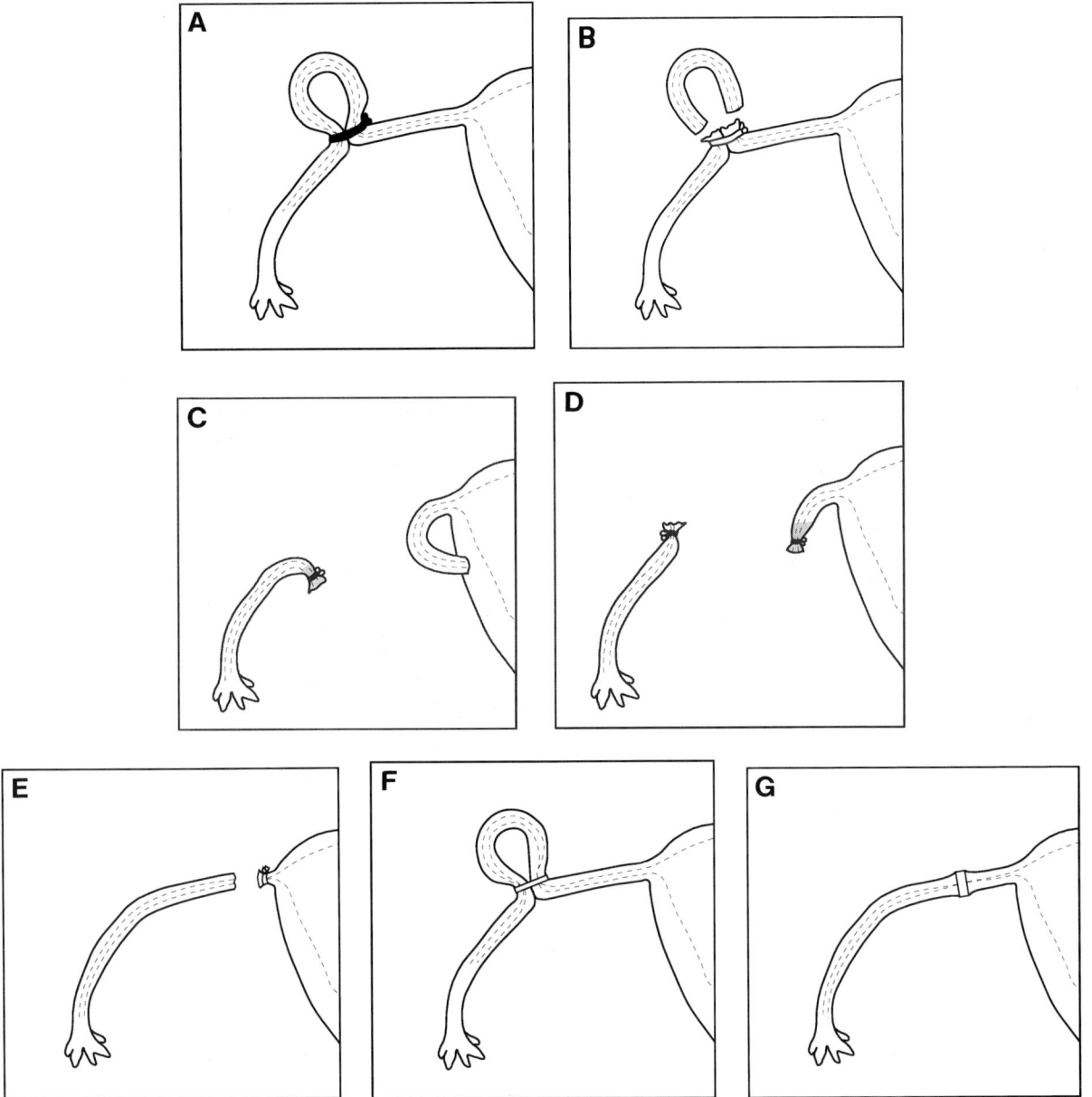

FIGURE 3 A, Madlener technique; B, Pomeroy technique; C, Irving technique; D, Uchida technique; E, total salpingectomy; F, Falope ring; G, clip.

method that was used for sterilization; the less damage and shorter damage of the tubal segment the higher the success rate. It is claimed that the pregnancy rate after reversal of sterilization can approach 70%, but this is dependent on multiple factors including the woman's age, the location of the anastomosis, the final tubal length, and coexistent pelvic pathology.

IX. METHODS FOR TUBAL INVESTIGATIONS

The patency, course, and function of the tubes can be in part assessed by the following methods:

Hysterosalpingography: A radiopaque substance, such as lipoidal, or a water-soluble medium, such as urograffin, is injected into the uterus through a cervical canula. Imaging of the uterine cavity and tubes is obtained by fluoroscopy and X rays. The course and patency of the tube and potentially some peritubal adhesions can be assessed, the latter by intraperitoneal pooling of dye. Recently, sonography during intrauterine fluid instillation has been utilized to assess the uterine cavity. This method may also be applicable to assess tubal patency reliably, particularly as contrast agents which allow better visualization of tubal flow become more available.

Laparoscopy: Allows evaluation of the external tubal surface and peritubal pelvic adhesions, and by the addition of chromopertubation (the passage of colored dye) the patency of the tubes can be assessed. Additionally, the presence of a hydrosalpinx or tubal sacculations can be identified.

Salpingoscopy: Allows a panoramic view of the entire tubal mucosa, from the fimbria to the isthmus, and evaluation of intraluminal pathology. It may have a therapeutic role as well in lysing intratubal adhesions and/or treating other tubal pathology. Two types have been utilized in the past: the rigid and flexible salpingoscopes. Currently, narrow fibers are becoming available to allow viewing of the entire tube, thereby allowing examination of the cornual and isthmus segments which were not visible using rigid scopes. Currently, these devices are used in conjunction with laparoscopy or laparotomy to allow manipulation of the tube to facilitate passage of the scopes. If in the future it is possible to view the tubal lumen following transcervical passage without operative intervention, the clinical diagnostic value would increase dramatically.

Bibliography

Diamond, M. P. (1988). Surgical aspects of infertility. In *Gynecology and Obstetrics* (J. W. Sciarra, Ed.), pp. 1–23. Harper & Row, Philadelphia.

Hershlag, A., Diamond, M. P., and DeCherney, A. H. (1989). Tubal physiology: An appraisal. *J. Gynecol. Surg.* **5**, 2–25.

Hershlag, A., Seifer, D. B., Carcangiu, M. L., Patton, D. I., Diamond, M. P., and DeCherney, A. H. (1991). Salpingoscopy: Light microscopic and electron microscopic correlations. *Obstet. Gynecol.* **88**, 399–405.

Jansen, R. P. S. (1984). Endocrine response in the fallopian tube. *Endocr. Rev.* **2**, 525–549.

Lavy, G., Diamond, M. P., and DeCherney, A. H. (1987). Ectopic pregnancy: Its relationship to tubal reconstructive surgery. *Fertil. Steril.* **47**, 543–556.

Marshall, J. M. (1981). Effects of ovarian steroids and pregnancy on adrenergic nerves of uterus and oviduct. *Am. J. Physiol.* **240**, 165–174.

Penzias, A. S., Gutmann, J. N., and Diamond, M. P. (1993). Laparoscopic management of ectopic pregnancy. In *Extrauterine Pregnancy: Clinical Diagnosis and Management* (T. Stovall and F. King, Eds.), pp. 231–248. McGraw-Hill, New York.

Stovall, T. G., Ling, F. W., Gray, L. A., Carson, S. A., and Buster, J. E. (1991). Methotrexate treatment of unruptured ectopic pregnancy. *Obstet. Gynecol.* **77**, 749–753.

Thyrmond, A. S. (1994). Transcervical tubal cannulation in the diagnosis and treatment of tubal obstruction. In *The Fallopian Tube Clinical and Surgical Aspects* (J. G. Grudxinskas *et al.,* Eds.), pp. 151–156. Springer-Verlag, New York.

Westrom, L., Iosif, S., Svensson, L., and Mardh, P.-H. (1979). Infertility after acute salpingitis: Results of treatment with different antibiotics. *Curr. Ther. Res.* **26**, 752–764.

False Pregnancy

see Pseudocyesis

Family Planning

Victoria L. Dunning and Allan Rosenfield

Columbia University

GLOSSARY

contraceptive prevalence A measure, determined primarily by survey data, of the extent of contraceptive practice among a defined population group at a given point in time.

International Conference on Population and Development (ICPD) A conference convened every 10 years since 1974 (formerly called the World Population Conference) for governmental and nongovernmental leaders to address major issues related to population and development. In 1994, the population was held in Cairo, Egypt.

Planned Parenthood A network of family planning clinics that provide family planning and reproductive health services. International Planned Parenthood Federation (IPPF) is an umbrella organization for worldwide family planning affiliates. IPPF provides technical assistance and financial support to autonomous member family planning associations throughout the world. Planned Parenthood Federation of America (PPFA) is the national umbrella association for all U.S. Planned Parenthood affiliates. PPFA links member affiliates for support, advocacy, information exchange, technical support, and financial assistance.

population growth The proportion by which the population increases within 1 year. The rate measures the percentage change in population from one year to the next.

reproductive health The health related to human sexuality and reproduction. Reproductive health may include issues and services related to family planning, sexually transmitted disease, HIV/AIDS, pregnancy, childbirth, prenatal care, reproductive tract infections, reproductive cancers, infertil-

ity, female genital mutilation, menopause, sexuality, and sexual dysfunction.

Title X Federal funds allocated for family planning services for low-income women. It is the principal funding effort of the government for the reduction of unintended pregnancy.

total fertility rate The average number of children that would be born alive to a woman (or a group of women) during her lifetime if she were to pass through her child-bearing years conforming to age-specific fertility rates of a given year.

unmet need A measurement, determined largely by survey data, of women, married or living in union, who respond that they want to postpone or avoid child bearing but who are not using contraception.

unintended pregnancy Pregnancy that a woman states was either unwanted or mistimed at time of conception, irrespective of whether contraception was being used.

Family planning, the voluntary spacing or limiting of children by individuals and couples, contributes to the individual health and welfare of women, children, and families. Family planning has an impact on some communities by contributing to a reduction in both the birth rate and population growth, thus decreasing the burden on natural resources and government programs. Family planning programs provide education, counseling, care, and supplies to individuals to manage fertility and address reproductive health concerns.

I. INTRODUCTION

For most women, having children is a basic aspect of their lives. The process of pregnancy, childbirth, and child rearing greatly affect women's lives and the majority of this process is still largely their responsibility. Thus, the timing, spacing, and number of births is of great consequence for a woman's future. For young women and poor women, the implications of pregnancy and child rearing are particularly great. The ability to control the number and timing of the children she will bear is one of fundamental and personal importance to each woman. The implications of women's ability to control their fertility are global as well. Population growth can limit a society's economic and social growth. It can strain a communities' systems for feeding, educating, and caring for its young, thus limiting its potential. Family planning allows individuals to choose how many children they desire and when during a lifetime they will have them. The spacing and limiting of births that family planning affords has direct benefits on the health of women and children as well. Family planning helps avert unwanted pregnancy, thus reducing the abortions that may occur from unintended pregnancies and the numerous subsequent maternal deaths that occur when these abortions are performed in unsafe conditions. In short, family planning is basic to human progress and elemental to women's and children's health. Family planning programs provide the services and supplies necessary for women to control their fertility.

II. THE CASE FOR FAMILY PLANNING

The benefits of family planning are numerous and convincing. The spacing and limiting of births clearly has a beneficial impact on the health and welfare of women and children. Unintended pregnancy is prevalent in industrialized countries, with approximately half of all pregnancies in the United States categorized as unwanted or mistimed. Unmet contraceptive need is widespread in the developing world, with approximately 100 million or more women stating they do not wish to be pregnant but are not using contraception. High birth rates and population growth rates constitute an obstacle to social and economic development, contributing to a burden on natural resources and environmental degradation. Addressing these multifaceted implications provides

a clear and convincing case for comprehensive, accessible, quality family planning programs.

Women's and children's health directly benefits from family planning. The slogan, "too young, too old, too close, too many" sums up the risk factors for excessive or poorly timed pregnancy. Women whose pregnancy meets any of these criteria can endanger their health and threaten their life. For young women, under 18, pregnancy is more likely to result in preterm and low-birth-weight babies and increases the risk of infant mortality. The pregnancy and childbirth add additional risks for these women for complications which contribute to maternal morbidity and mortality. Child spacing of less than 2 years poses a health threat to the newborn as well as to the preceding child because of the increased risks for infant and child mortality associated with short birth intervals. High parity, over five births per woman, poses a health risk to mothers. Many births increase risks of both maternal mortality and morbidity.

In addition to the health risks from poorly timed and numerous births, unwanted pregnancies can have consequences for maternal morbidity and mortality as well. Women who, for a variety of reasons, do not wish to be pregnant, may seek unsafe or illegal abortions. Many maternal deaths are the result of botched abortions performed under unsanitary conditions.

Unintended pregnancies—pregnancies that are either mistimed or unwanted altogether—are prevalent throughout the world across all socioeconomic strata. In the United States, 50% of all pregnancies are unintended (Table 1). The consequences of such unintended pregnancies are serious. Women with unintended pregnancies are less likely to seek early prenatal care. They are more likely to expose the fetus to harmful risks, such as alcohol or tobacco. The unwanted pregnancy is more likely to result in preterm birth or low-birth-weight babies. Furthermore, an unwanted pregnancy may cause depression in the mother or cause strain in her relationship with her partner. Educational or career goals may be compromised due to the unintended pregnancy; economic strain that the parents were not prepared for may result.

TABLE 1
Estimated Proportion of Pregnancies (Excluding Miscarriages) by Outcome and Intention,
Percentage of Pregnancies Unintended, and Percentage of Pregnancies Ending in Abortion, 1994, by Marital Status, Age,
and Poverty Status at Interview

Demographic characteristics	Total pregnancies	Intended pregnancies ending in birth	Unintended pregnancies ending in birth	Abortions	Percentage of pregnancies unintended	Percentage of unintended pregnancies ending in abortion
Total	100.1	50.8	23.0	26.6	49.2	54
Marital status						
Currently married	100.0	69.3	19.3	11.3	30.7	37.0
Formerly married	100.0	37.5	21.8	40.7	62.5	65.1
Never married	100.0	22.3	31.0	46.7	77.7	60.1
Age						
15–19	100.0	22.0	42.7	35.3	78.0	45.3
20–24	100.0	41.5	26.2	32.3	58.5	55.2
25–29	100.0	60.3	17.2	22.5	39.7	56.7
30–34	100.0	66.9	14.6	18.4	33.1	55.7
35–39	100.0	59.2	17.9	23.0	40.8	56.3
≥40	100.0	49.3	17.9	32.8	50.7	64.7
Poverty status[a]						
<100%	100.0	38.6	31.3	30.1	61.4	49.0
100–199%	100.0	46.8	27.7	25.4	52.2	47.9
>200%	100.0	58.8	15.9	25.4	41.2	61.5

Note. Data modified from Henshaw, S.K. (1998). Unintended pregnancy in the United States. *Family Planning Perspectives, 30*, 1, 24–29.

[a] <100%, women whose household income fell below the poverty level; 100–199%, women with incomes between 100 and 200% of the poverty level; >200%, women with incomes which exceeded 200% of the poverty level.

Unintended pregnancy leads to approximately 1.4 million abortions per year in the United States. In 1994, there were an estimated 5.4 million pregnancies, of which about 2.7 million were unintended. Of these unintended pregnancies, approximately 1.4 million ended in an abortion, and the remaining resulted in a live birth. Only 2.7 million (approximately half) of all pregnancies in the United States were intended at the time of conception and resulted in a live birth. The reasons for unintended pregnancy are a complex combination of social, cultural, religious, medical, educational, and economic factors. Many of these unintended pregnancies resulted because of failure to use an effective method of contraception consistently and correctly, if at all. In some part, the unintended pregnancies resulted because of failure of the method itself. While the determinants of successful contraceptive use are vast and complex, we do know that education, counseling, access, and quality have a great impact on effective family planning use.

A similar indicator of family planning need, used more often in developing countries, is the measure of unmet contraceptive need. In the developing world, the number of unmet-need women who would not like to have a child at the present time or ever, but are not using contraception, is estimated to be at least 100 million. Of the unmet need in developing countries, India has the most women (31 million) in need of contraception for either spacing or limiting. Other countries with great shares of women with unmet need are Pakistan with 5.6 million, Indonesia and Bangladesh with 4.4 million each, and Nigeria with 3.9 million. Addressing the unmet need for contraception is essential in helping women meet their fertility goals.

Another argument for the importance of family planning is the effects of population growth on the environment and natural resources. This argument was the original rationale of many early "population" programs. Currently, the global population is over 5.9 billion and will reach 6.0 billion by 1998. By 2000, 800 million of the world's population will be teenagers, entering reproductive age. The child-bearing choices that this generation makes will profoundly affect how population continues to grow. The effects of such growth on natural resources, the environment, poverty, and hunger are hotly debated. While no resolution to this debate has been reached, it is clear that as a society grows, its ability to feed, educate, employ, and care for its people becomes more challenging. For the 800 million young people of reproductive age at the turn of the century, access to quality family planning services and education will be a critical point in determining the magnitude of future population growth. The practice of traditional methods of contraception such as periodic abstinence, even in the absence of family planning programs, illustrates the desire of women to control their fertility. Family planning can contribute to the ability of families to lead economically and socially satisfying, productive, and stable lives.

III. HISTORY

Population growth was originally low because of high mortality from a broad array of then untreatable infectious and other diseases. Traditional methods of family planning, including withdrawal, prolonged periods of breast-feeding, periodic abstinence, and attempted abortion, have been documented in preliterate societies. Recipes and prescriptions for traditional medicines which acted as spermicides, abortifacients, and sterility agents document efforts for fertility control as early as 1500 BC.

Prior to the development of modern contraceptive methods, Europe and North America went through a remarkable fertility decline in the nineteenth century, over a period of 100 or more years, in large part through the use of traditional methods. These trends of reduced fertility and subsequent decreased population growth were facilitated in the twentieth century both by the legalization of abortion and by the development of modern contraceptive technologies. In developing countries, on the other hand, there was little impact on death or birth rates until the late 1940s, when death rates began to fall over a much shorter period of time. This rapid decline led to a dramatic increase in rates of population growth and the implementation since the 1960s of international family planning efforts.

Before family planning programs became established in the United States, a network of private physicians and family planning advocates distributed contraceptives through covert and often illegal channels. As modern contraceptive technology became available in the 1960s with the development of oral contraceptives and IUDs, the need for programs and information to make them accessible became evident. The work of several individuals to bring family planning rights, supplies, and services to women, in the face of political and religious adversity, was elemental to the development and success of early family planning programs.

Margaret Sanger opened the first U.S. family planning clinic in Brooklyn in 1916 under popular and political adversity. In other countries, individuals such as Drs. Helena Wright and Marie Stopes of the United Kingdom, Elise Ottesen Jensen of Sweden, and Lady Rama Rau of India played similar pioneering and courageous roles in advocating for family planning services, supplies, and the reproductive rights of women. The efforts of these activists brought much needed, though limited, family planning services to women, particularly poor women.

Services were provided at first in urban areas, though later they expanded to meet the family planning needs of rural underserved communities. In 1952, the International Planned Parenthood Federation (IPPF) was formed as an umbrella organization for national affiliates for both developed and developing countries which provided family planning services, primarily to low-income and minority women. The Planned Parenthood Federation of America

(PPFA), the national office of the U.S. IPPF affiliate, provides technical support, advocacy, public information, and other services to its affiliates . The formation of IPPF and PPFA marked the beginning of the current family planning era.

IV. OVERVIEW OF CURRENT PROGRAMS

Currently, family planning services in the United States are provided through a variety of outlets. Women and men receive family planning education, counseling, services, and supplies through private-practice physicians, hospital clinics, public state and local clinics, and private voluntary agencies. While the majority of middle- and upper-income women are served through private physicians, government-supported clinics are the primary provider of family planning services for low-income women as well as for younger women under 20 years old.

Today, PPFA is the largest comprehensive network of private voluntary clinics in the United States, with approximately 900 Planned Parenthood clinics in 160 Planned Parenthood affiliates which offer family planning and reproductive health services to women. The majority of their client base is low-income, uninsured, and young women. PPFA clinics serve 30% of the low-income women who seek family planning services. State and local public clinics also provide family planning services to another 32% of these women. The remaining population is served through subsidized family planning services in hospital-based and independent clinics.

In the United States, funding for family planning services is provided through a variety of channels. Federal grants, state and local government funding, and Medicaid funds provide the bulk of revenues for subsidized family planning clinics. Private sources, such as foundations or individual donors, provide limited funds for family planning services, research, and evaluation. Client fees, in full payment or on a sliding scale, are a component of most family planning programs. For private providers of family planning services, third-party insurance payers cover the costs of services and supplies.

Title X is the principal funding effort of the federal government for programs whose goal is the reduction of unintended pregnancy. Established in 1970 by President Richard Nixon under the Public Health Service Act, these federal grants provide funds for family planning services and supplies for low-income women. In 1994, Title X funds provided $151 million to family planning clinics providing services to low-income women in the United States. Title X funding goes to the operation of different types of family planning clinics. In 1992, 52% of these funds went to clinics operated by state and local departments of health, 15% of the funds went to PPFA and its affiliates, 6% to hospital-based family planning services, and 27% to private voluntary organizations and other providers of subsidized family planning services. Title X funds provide family planning services and supplies, counseling and education, and other primary reproductive health care services, such as pregnancy testing, prenatal care, sexually transmitted disease diagnosis and treatment, and screening for cervical cancer, anemia, and high blood pressure. Title X has never provided funds for abortion or abortion-related services. The monies distributed help support family planning clinics in 50 states and two-thirds of U.S. counties which provide care for 6 million women annually. Eighty-eight percent of poor women of reproductive age live in a county with a Title X funded clinic. Many middle- and upper-income teenagers who may be insured seek family planning services through the subsidized clinic system, thus benefiting from the federal Title X grants. While these teenagers may have insurance which would pay for family planning services, they perceive the clinic services as more confidential than their family physician. Title V, better known as the maternal–child health block grant, provides limited federal funding to state government agencies, a portion of which is designated for family planning services. State and local governments provide a vital portion of family planning clinic funding. In 1994, $162 million in state monies was distributed for family planning services.

Medicaid, which ensures health and medical expenses for the eligible poor, provides the largest source of public funds for family planning services and supplies in the United States. Since 1972, family planning services have been included as one of Medicaid's mandated services, accounting for $332 million of Medicaid funds in 1994. Over 80% of subsidized family planning clinics received some of their revenue from Medicaid payments. Client fees are a part of most programs. For programs which serve low-income women, these fees are often determined on a sliding scale which ranges from free to full cost of supplies and services. For middle- and upper-income and insured women using private physicians for their family planning needs, costs are paid by insurers and client fees.

In the developing world, many family planning initiatives are supported through the collective efforts of international organizations and government bodies. Since 1964, a United Nations conference on population has been held every decade to discuss this collective effort. These conferences are attended by governmental and nongovernmental leaders central to the planning and implementation of family planning and reproductive health activities. The consensus reached at the conferences shapes the role and scope of family planning programs.

Family planning services in developing countries are provided through a similar patchwork structure of private voluntary clinics, government programs, and, to a much lesser extent than in the United States, private physicians. The initiatives of these programs are partially supported through financial and technical assistance of United Nations agencies, government agencies from industrialized donor countries, and nongovernmental organizations. In Asia, the majority of family planning programs are sponsored by the government. In Latin America, private voluntary organizations provide the majority of family planning services. Africa employs a combination of governmental and nongovernmental programs for its family planning programs.

Many family planning associations in the developing world are affiliates of IPPF, which provides technical support and limited financial assistance. In 1995, IPPF had 127 member family planning associations and worked in 150 countries. The member family planning associations provide family planning services at thousands of clinics and community service points in Latin America, Africa, Asia, and Eastern and Central Europe.

Family planning initiatives in developing countries are also funded through a variety of channels. Government, bilateral and multilateral international assistance, and private donations comprise the majority of international aid. Three-quarters of the costs of family planning programs are met by the individual country governments. The remaining 25% of costs are made up by the donor community through direct and indirect financial and technical assistance. Less than 1% of the U.S. budget is allocated to foreign assistance. Even with this small contribution to foreign aid, only .003 of 1% of the U.S. budget is spent on family planning activities.

Multilateral efforts are led by United Nations programs, primarily through the efforts of the United Nations Population Fund, the World Health Organization (WHO), UNICEF, and the World Bank. Bilateral initiatives are funded by industrialized country donors which provide monies to individual programs in developing countries. In the United States, bilateral efforts are primarily executed by the U.S. Agency for International Development (USAID). Nongovernmental organizations are often subcontracted by USAID to provide technical support and administration to programs. In addition to U.S. bilateral assistance, other bilateral efforts are initiated by European and Japanese donors. While the United States is the largest donor to family planning programs in terms of dollars, other countries, such as Norway and Sweden, allocate more money per capita to family planning programs in developing countries. Private donors, such as foundations, universities, and individuals, support national and international family planning initiatives with funding, technical assistance, and research.

Both in the United States and abroad, family planning programs and services are a public health success story. In the industrialized world, average family size declined gradually over a century to the current rate of about 1.6 births per women. In developing countries, the decline in fertility has happened more rapidly. In the three decades since the 1960s, the number of children per women in developing countries (including China) has decreased from an average of 6 to 3. Changes in fertility and child-bearing patterns have differed greatly by region and among individual countries. While some regions have experienced a great decline in fertility, others are experiencing a more modest one. In Latin American countries, fertility declined the most from 6.0 children per women in 1960 to approximately 3.0 in 1994. Asia has also experienced remarkable declines. In East Asia, total fertility rate has declined from 5.6 to 2.3. In Africa and the Middle East, fertility declines have been more modest. In sub-Saharan Africa, for example, fertility rates have declined to a modest 6.2 in 1994 from 6.6 in 1960. Nonetheless, more significant declines have begun to occur in some countries within this region.

Some countries have experienced remarkable successes in fertility decline (Fig. 1). Many of these declines can be attributed to widespread family planning program efforts. In Asia, Thailand achieved early success in significant fertility decline from 6 or 7 children per woman to approximately 2.0 in the relatively short period of time of about 25 years. In Bangladesh, which has extremely low social and economic indicators, a significant fertility decline has taken place from 6 or 7 children per woman in 1960 to 3.4 in 1990. In Latin America, Colombia stands out in fertility decline with a decrease in fertility rate from 6.8 in 1960 to 2.6 in 1990. Kenya, in sub-Saharan Africa, appears to be undergoing a marked fertility decline, from 8.1 in 1975 to 5.4 in 1993. Although the United States has reached a low fertility rate of 2.1, it has not succeeded as well as its European country counterparts in reducing unintended pregnancy, particularly among low-income women and teenagers.

Contraceptive prevalence, an indicator of current contraceptive use and a marker of the extent and success of family planning programs, varies greatly across regions and individual countries (Fig. 2). In sub-Saharan Africa, contraceptive prevalence remains at a modest 14% of women currently using a modern method of contraception. South Asia's rate of contraceptive prevalence is 44%, and South America boasts an average of 65% contraceptive prevalence. Once again, the contraceptive prevalence rates differ widely among individual countries. For example, despite sub-Saharan Africa's relatively low average contraceptive prevalence, the rate is 43% in Namibia. By contrast, other countries in the region, including Niger, Sierra Leone, and Somalia, report a contraceptive prevalence of less than 5%. Similar variations of contraceptive use exist in south Asia, where Sri Lanka and India have prevalence rates of 62 and 43% compared to Nepal and Pakistan at 23 and 12%. In South America, contraceptive prevalence rates vary from 30% in Bolivia to 66% in Colombia and 74% in Argentina. In the industrialized countries, contraceptive

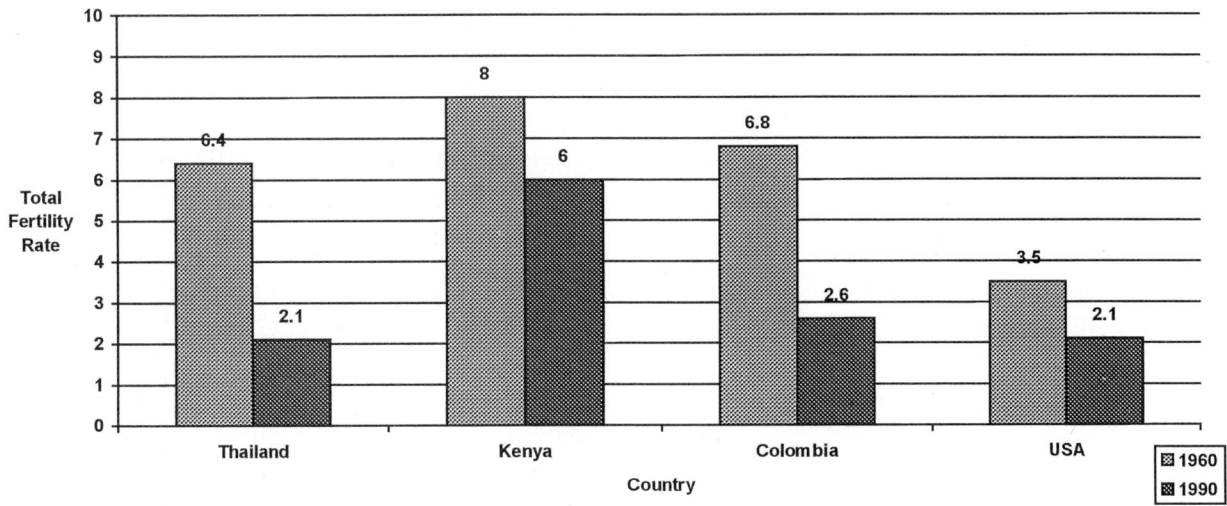

FIGURE 1 United Nations estimates of total fertility rates for selected countries, 1960 and 1990 (from UNICEF, 1993).

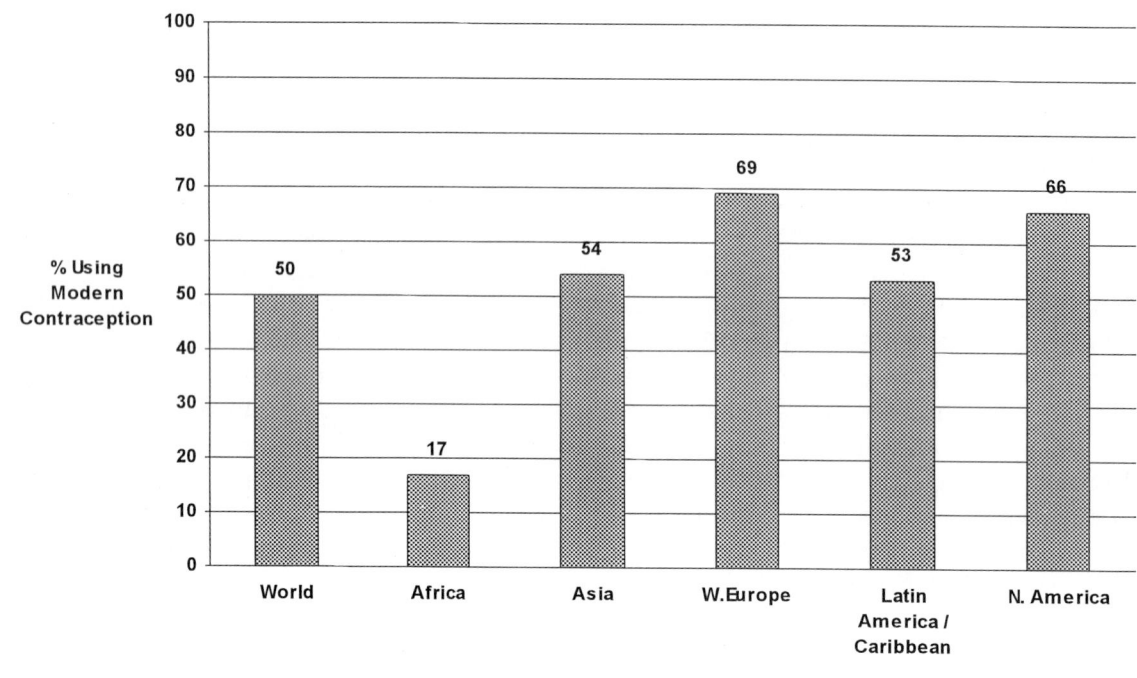

FIGURE 2   Regional estimates of contraceptive prevalence among married women (from Population Reference Bureau).

prevalence remains high with approximately two-thirds to three-quarters of women of reproductive age using contraception. Contraceptive prevalence in the United States is about 74%.

V. ESSENTIAL PROGRAM COMPONENTS

Family planning is an essential component of primary health care. Since the 1978 WHO- and UNICEF-sponsored Alma-Ata conference called for basic health services for all by the Year 2000, an emphasis has been placed on primary and preventive services in health care. The inclusion of family planning as an essential element of primary care recognizes its substantial contribution to the reduction of maternal and child morbidity and mortality. The most effective approach to the provision of family planning services as a primary and preventive health program is the subject of heated debate. While some argue that family planning is best integrated into

existing maternal–child health services, others fear this will dilute the programmatic impact and thus favor a vertical, single-purpose approach. Despite philosophical differences regarding the most appropriate approach, all family planning programs contain common elements.

Program requirements vary by the breadth and scope of services provided. Every program, however, will require basic components critical to success: supplies, staff, equipment, communication and education, logistical systems, and linkages to other services. Consistent stock and distribution of adequate supplies is necessary at all program points. Program staff require appropriate training and should possess adequate experience. Materials for information, education, and communication for the program, contraceptive method use, and client support should be available. Basic systems for client follow-up and record-keeping may vary in complexity by the program approach; however, these systems are important to establish quality consistent care. Lastly, at any program level, appropriate linkages to the complementary health care system, including lab facilities and referral for further care or services, helps establish a comprehensive program.

Despite the advantages of a clinical system of family planning services, barriers to access in developing countries have necessitated alternative approaches to family planning services. Community-based distribution programs, for example, allow trained individuals or small local organizations to provide selected contraceptive method education and distribution. Such programs are particularly effective in reducing the barriers to access in rural communities. Another alternative family planning service mechanism is social marketing programs which promote, distribute, and sell contraceptive products to consumers at subsidized prices. These programs are usually limited to distribution and sale of condoms and oral contraceptives. Social marketing programs can be particularly effective in reducing social and cultural barriers to family planning use.

As the setting of the programs vary, so does the team of family planning providers. Each program's family planning team will include a variety of health workers. Provider requirements depend in part on the program and the selection of services offered. Physicians' roles may be in direct service provision or as supervisors of other health workers. In many developing countries, a midwife or nurse plays a primary role in the provision of family planning services. In the United States, midwives and nurse practitioners play an increasingly visible role in the provision of family planning and reproductive health services. The use of the midlevel and community health workers is of particular importance in countries or regions where physicians are scarce.

Where community-based distribution is employed, often a respected community member is trained to provide basic services and supplies to individuals seeking family planning services. It is critical that such community health workers have adequate support and referral systems linking them to clinical settings. The use of such auxiliary health workers provides wider access in underserved rural and urban slum areas in developing countries. Several trained community health worker programs have been quite successful in providing family planning methods and education to communities. In a clinical setting, ancillary workers, such as counselors, health educators, and social workers, play a critical role in the education and counseling of family planning clients. Much greater attention in recent years has been directed to ensuring the quality of service programs.

VI. FUTURE DIRECTIONS

The International Conference on Population and Development (ICPD) in Cairo in 1994 broadened the family planning mandate. It calls for a comprehensive agenda to address and improve reproductive health. Reproductive health includes family planning services, but it also encompasses a broad array of issues, including STDs, HIV/AIDS, prenatal care, maternal mortality, female genital mutilation, reproductive cancers, and the overall empowerment of women. With this broadened agenda, we can expect both expansion and integration of current family planning programs and attention to the other aspects of reproductive health.

The ICPD indicated the links between reproductive health, development, and the achievement of gender equity. It called for social and economic empowerment of women. While a gender equity initiative is not defined in and of itself, we can expect attention to be focused not only on the outcomes of family planning and reproductive health programs but also on the process. The ICPD recognized the importance of men in the process of improving reproductive health. It called for male involvement to achieve gender equity and to garner the support of men for women's health needs. We can expect from this a wider variety of education and services which target men as partners in the family planning and reproductive health equation.

Quality of care has been determined to be a major factor in the satisfaction and continuation of contraceptive use. Variety of methods; waiting time; consistent supplies; adequate privacy; convenient hours; courteous, professional, and respectful staff; and culturally appropriate education and counseling all contribute to the quality of family planning programs throughout the world. Without these factors, family planning programs can fail. Quality will affect U.S. programs, where managed care and market forces put family planning programs in competition for clients. Quality assurance and monitoring will be essential to the success of U.S. programs and services. In countries in which family planning programs are new or recently established, quality must be incorporated from the beginning to ensure successful programs and satisfied users.

Unmet need will have to be addressed through targeted efforts of innovative family planning programs. Identification of barriers to providing services to women with unmet need is of great importance and the subsequent strategies to overcome them will need to be developed. Quality services, effective communication, male participation, and collaborative services with maternal–child health programs have been suggested as ways to help reduce unmet need in developing countries.

Much of the funding of family planning programs in the United States and the developing world is determined by the political climate and congressional budget approval. In recent years, family planning funding, both Title X and international family planning programs, have been under threat. Cuts, limitations, and harsh restrictions are some of the recent attacks on the management and operation of efficient and successful family planning programs. Response to this funding threat will remain critical for the continued existence, survival, and success of family planning programs.

In addition to political and funding threats, the U.S. health care system is experiencing great changes. As Americans search for new ways to offer efficient, effective, and less costly health care, family planning programs will have to determine their position within the emerging health care system. Managed care and health maintenance organizations in the United States will largely determine how, when, and where individuals receive family planning services.

VII. CONCLUSION

The benefits of family planning to individuals, families, and communities are clear. Family planning programs contribute greatly to the ability of women and couples to control their fertility and achieve their child-bearing goals. While it is not the sole solution, improved economic and social welfare, improved health indicators, and a decrease in the rate of population growth can be attributed at least in part to the success of voluntary family planning programs. Although access to voluntary and quality family planning services is of particular importance to poor women and young women, the benefits of accessible family planning directly and indirectly extend to all. Investment in voluntary, quality family planning programs provides a return in economic and social growth, health status, and the achievement of gender equality. The common goal, which family planning helps achieve, is that every pregnancy and every child is desired and cared for.

See Also the Following Articles

CONTRACEPTIVE METHODS AND DEVICES, FEMALE; CONTRACEPTIVE METHODS AND DEVICES, MALE

Bibliography

The Alan Guttmacher Institute (1995). *Hopes and Realities: Closing the Gap Between Women's Aspirations and Their Reproductive Experience.* The Alan Guttmacher Institute, New York.

Bruce, J. (1990, March/April). Fundamental elements of the quality of care: A simple framework. *Stud. Family Planning* **21**(2), 61–91.

Frost, J. (1996, May/June). Family planning clinic services in the United States, 1994. *Family Planning Perspect.* **28**(3), 92–100.

Hatcher, R. A., Trussell, J., Stewart, F., *et al.* (1994). *Contraceptive Technology,* 16th ed. Irvington, New York.

Henshaw, S. K., and Torres, A. (1994, March/April). Family planing agencies: Services, policies, and funding. *Family Planning Perspect.* **26**(2), 52–59.

Institute of Medicine (1995). *Best Intentions: Unintended Pregnancy and the Well-Being of Children and Families* (S. Brown and L. Eisenberg, Eds.). National Academy Press, Washington, DC.

Macro International, Inc. (1994). *DHS Program: Women's Lives and Experiences: A Decade of Research Findings from the Demographic and Health Surveys Program.* Macro International, Calverton, MD.

Maine, D. (1981). *Family Planning: It's Impact on the Health of Women and Children.* Center for Population and Family Health, Columbia University, New York.

Robey, B., Ross, J., and Bhushan, I. (1996, September). Meeting unmet need: New strategies. In *Population Reports*, Vol. 24, No. 1. Population Information Program, Center for Communications Programs, The Johns Hopkins University, Baltimore, MD.

The Rockefeller Foundation (1997). *High Stakes: The United States, Global Population, and Our Common Future.* The Rockefeller Foundation, New York.

Rosenfield, A., and Fathalla, M. (1990). Family planning programs and services. In *The FIGO Manual of Human Reproduction, Volume 2 Family Planning.* Parthenon.

Sherris, J. (Ed.) (1994, August). Women's reproductive health: The role of family planning programs. *Outlook* **12**(2).

Sollom, T., Benson Gold, R., and Saul, R. (1996, July/August). Public funding for contraceptive, sterilization, and abortion services, 1994. *Family Planning Perspect.* **28**(4), 166–173.

UNICEF (1994). *The Progress of Nations: The Nations of the World Ranked According to Their Achievements in Child Health, Nutrition, Education, Family Planning, and Progress for Women. 1994.* UNICEF House, New York.

Fecundity

see Fertility and Fecundity

Felidae

see Cats

Female Contraceptives

see Contraceptive Methods and Devices, Female

Female Reproductive Behavior

see Mating Behaviors

Female Reproductive Disorders, Overview

Robert W. Rebar

University of Cincinnati College of Medicine

GLOSSARY

follicle A structure in the ovary consisting of an oocyte or egg and the supporting cells, called granulosa and theca cells, which surround it; the cells of the follicle secrete androgens, estrogens, and progestins. As the follicle matures it develops a fluid-filled cavity and just before ovulation is termed a Graafian follicle.

follicle-stimulating hormone A peptide hormone produced by the anterior pituitary gland which stimulates the growth and development of Graafian follicles within the ovary and of spermatogenesis within the testis; hence, the term "gonadotropic" hormone is often applied.

gonadotropin-releasing hormone A peptide hormone composed of 10 amino acids that is produced in the hypothalamus and stimulates secretion of luteinizing hormone and follicle-stimulating hormone by the anterior pituitary gland.

hymen A membranous fold of tissue which partly occludes the external opening of the vagina in the female.

hypothalamus The ventral portion of the primitive diencephalon of the brain which activates, controls, and integrates peripheral autonomic nervous mechanisms, endocrine activity (including control of reproductive function), and several somatic functions, including water balance, body temperature, sleep, appetite, and food intake.

luteinizing hormone A peptide hormone produced by the anterior pituitary gland which, together with follicle-stimulat-ing hormone, stimulates estrogen secretion and triggers ovulation of the oocytes from mature follicles in the ovary and stimulates androgen production by the testis; luteinizing hormone is another gonadotropic hormone.

Müllerian ducts Paired ducts which develop into the Fallopian tubes, uterus, and the proximal portion of the vagina during fetal development in females and regress in males.

ovary The female gonad in which oocytes or eggs mature and are released on a cyclic basis each month during the reproductive years.

pituitary A gland at the base of the brain in a bony cavity termed the sella turcica which is attached by a stalk to the hypothalamus, from which it receives neural and hormonal input and with which it shares its blood supply. The pituitary gland is divided into two main lobes, the anterior and posterior pituitary. The anterior pituitary secretes several hormones, three of which (luteinizing hormone, follicle-stimulating hormone, and prolactin) are central to normal reproductive function.

prolactin A peptide hormone produced by the anterior pituitary gland that stimulates and sustains lactation (i.e., milk production) after delivery; prolactin also seems to have both stimulatory and inhibitory effects on both the ovary and the testis.

testis The male gonad in which spermatozoa required for reproduction develop.

vulva The external genitalia of the female, including the labia majora and labia minora (or so-called large and small lips), the vaginal opening, and a number of glands located in the region.

Wolffian ducts Paired ducts which develop before birth into the ducts which allow sperm to leave the testes during orgasm in males and which regress in females.

Normal reproductive function in women requires the complex interplay of the hypothalamus and higher brain centers, the anterior pituitary gland, the ovaries, and the reproductive tract, including the

Fallopian tubes, uterus, cervix, vagina, and external genitalia. Abnormalities that impact on any part of the reproductive system can adversely affect reproduction. Moreover, abnormalities can arise at any time in the life of a female—developmentally even before birth, prepubertally before maturation of the reproductive organs and development of secondary sexual characteristics (i.e., pubic and axillary hair and breasts), during the reproductive years between menarche (i.e., the first menstrual period) and menopause (i.e., technically defined as beginning with the last menstrual period), and during the postmenopausal years. It is also important to recognize that disorders of other bodily systems can impact reproductive function and pregnancy and that disorders of reproductive function and pregnancy can impact other systems in the body.

I. INTRODUCTION

There is no single accepted classification of all the various disorders of the reproductive system and reproductive functions but an attempt has made here to present an overview of the possible abnormalities in a logical fashion (Table 1). There has been no attempt to list all known disorders but rather to develop a framework for the abnormalities that exist. Another common approach is to consider the abnormalities on the basis of the signs and symptoms with which they present (e.g., abnormal uterine bleeding and pelvic pain).

II. DEVELOPMENTAL AND ANATOMIC ABNORMALITIES

A. Disorders of Sexual Differentiation

Early in development the gonads of both sexes are indifferent and bipotential and can develop into either testes or ovaries depending on the chromosomal complement of the individual. In gonadal dysgenesis the gonads are typically "dysgenetic," consisting of just "streaks" of fibrous tissue and are devoid of germ cells. Affected individuals commonly fail to develop secondary sexual characteristics at puberty (i.e., breasts and pubic and axillary hair)

and fail to menstruate (i.e., have amenorrhea). The term Turner's syndrome is most commonly applied to individuals with gonadal dysgenesis but is perhaps more appropriately used only for individuals with streak dysgenetic gonads, short stature (generally under 60 in. in height), and certain other physical abnormalities, most commonly including a "webbed" neck with extra skin on each side, low hair line at the neck, shield-like chest, poorly developed fingernails and toenails, abnormalities of the heart and aorta, and renal anomalies. Turner's syndrome is due most commonly to the absence of a single X chromosome so that the karyotype is 45,X (consisting of 45 rather than 46 chromosomes); this is the single most common chromosomal disorder in humans. More than 99% of all 45,X embryos are aborted, however, accounting for an incidence at birth of only about 1 per 10,000 females. Gonadal dysgenesis also may be associated with a number of different abnormalities of the X chromosome and with mosaicism involving either an X or a Y chromosome. Individuals who are mosaic have some cells with one karyotype (such as the normal 46,XX) and some with another (e.g., 45,X). True hermaphrodites, who are quite rare, by definition possess both ovarian and testicular tissue, either separately or, more commonly, together as one or more "ovotestes." Perhaps 50–70% of true hermaphrodites have a normal female 46,XX chromosomal complement, but others may be mosaic or even have a 46,XY karyotype.

Until about 8 weeks into pregnancy, the urogenital tracts and external genitalia of both males and females are indistinguishable. Initially each sex has two paired ductal systems, but normal sexual differentiation involves growth of one system of ducts coupled with regression of the other. Two hormones secreted by the developing testis, Müllerian inhibiting hormone (MIH) and testosterone, play major roles in this process; in the absence of these testicular hormones, the female ductal (or Müllerian) system develops. In the presence of these hormones the male ductal (or Wolffian) system develops. In contrast to the internal genitalia, the external genitalia develop in a continuous manner from an indifferent stage common to both sexes. However, here, too, testicular hormones appear necessary for the differentiation of male genitalia.

TABLE 1
Classification of Female Reproductive Disorders

Developmental and anatomic abnormalities	Cancers of the Reproductive System
Disorders of sexual differentiation	Uterus
Gonads	Cervix
Ductal abnormalities	Ovaries
External genitalia	Vulva
Disorders of the breasts	Vagina
Hypoplasia	Fallopian tubes
Supernumerary nipples	Gestational trophoblastic disease
Disorders of hypothalamic–pituitary system	Breast disorders
Kallmann's syndrome	Fibrocystic breast changes
Septooptic dysplasia	Nipple discharge
Hypopituitarism	Breast infection and abscess
Anatomic disorders of the reproductive tract	Breast cancer
Uterine fibroids	Disorders of pregnancy
Endometriosis	Abortion and stillbirth
Problems of pelvic support	Ectopic pregnancy
Endometrial polyps	Anemias
Anatomic abnormalities of the cervix and vagina	Rh incompatibility
Functional abnormalities	Abruptio placentae
Hypothalamic–pituitary dysfunction	Placenta previa
Hypothalamic chronic anovulation	Hyperemesis gravidarium
Dysfunction secondary to tumors and other anatomic lesions	Preeclampsia and eclampsia
Polycystic ovarian syndrome	Infertility
Premature menopause	Evaluation
Premenstrual syndrome	Treatment
Dysmenorrhea	Family planning
Benign ovarian masses	Natural family planning
Infectious abnormalities	Barrier contraception
Sexually transmitted diseases	Hormonal contraception
Lower tract disease	Intrauterine devices
Upper tract disease	Sterilization
Tuberculosis	Abortion
	Menopausal problems

Anomalies of the distal female genital tract represent a heterogeneous group of malformations that result from abnormal development of Müllerian ducts. Several of the less severe anomalies do not impact on reproductive function, whereas others may increase fetal losses during pregnancy and still others may preclude coital activity until corrected. Fertility seems unaffected in patients with uterine malformations only. The development of secondary sexual characteristics is virtually always normal, and affected women generally ovulate regularly. Although the incidence of these anomalies is unknown, estimates approximate 0.02% of the female population.

Abnormalities related to maternal ingestion of the synthetic estrogen diethylstilbestrol (DES) have been widely publicized in the lay press but are found in a relatively small number of women. Of the anomalies unrelated to drug ingestion, the septate uterus is most common, with the uterine cavity partially divided into two parts reflecting the embryological development of the uterus from the paired Müllerian ducts. An imperforate hymen, in which the entire ductal system is normal but there is no opening of the vagina externally, must be considered in the diagnosis of several anomalies of the ductal system. In this disorder, menstrual blood and mucus accumulate in the vagina or uterus, and the disorder is characterized by cyclic abnormal pain at puberty. Although not particularly rare, the disorder is generally easily corrected by surgical incision of the hymen. This disorder must be distinguished from vaginal atresia, in which no vagina ever developed, and transverse vaginal septa, in which a band or septum of vaginal tissue completely blocks the vagina. Failure of fusion of the paired Müllerian ducts may lead to a variety of disorders from the septate uterus already discussed to complete duplication of the reproductive tract including the vagina.

In pseudohermaphroditism, the individual has sex chromosomes and characteristics of one sex but external genitalia that are ambiguous or more characteristic of the opposite sex. In female pseudohermaphroditism, genital differentiation is not that expected for normal 46,XX individuals; however, ovaries are present. Affected individuals generally present with genital ambiguity at birth, but sometimes the disorder is not detected for several years.

Most commonly, female pseudohermaphroditism is due to a form of congenital adrenal hyperplasia (CAH) in which an adrenal enzyme deficiency results in decreased cortisol secretion with consequent increased pituitary corticotropin secretion and adrenal androgen overproduction. 21-Hydroxylase deficiency is the most common form of CAH, accounting for 95% of cases. All forms are transmitted as autosomal recessive traits. Replacement of the deficient cortisol will correct all the metabolic abnormalities, allow normal growth and development, and permit normal reproductive function. Rarely, ambiguous genitalia and female pseudohermaphroditism may result from maternal ingestion of various teratogens, typically androgenic steroids. The only steroid likely to result in androgenization when used at appropriate therapeutic doses is danazol, which may be used to treat endometriosis in women. There is no evidence that the inadvertent ingestion of combination oral contraceptives or of medroxyprogesterone acetate (Provera) during pregnancy will increase the risk of fetal anomalies.

In male pseudohermaphroditism, genital differentiation is not that expected for normal males even though the chromosomal complement is 46,XY and testes are present. Complete androgen insensitivity, also known as testicular feminization, is the classic

example. In this disorder, 46,XY individuals have bilateral testes, female external genitalia, a blindly ending vagina, and no Müllerian duct derivatives. Affected individuals undergo breast development and pubertal feminization, but they have little or no pubic and axillary hair. Elegant studies have documented that the disorder is due to a variety of molecular abnormalities, most commonly deficient or absent androgen receptor. Although testosterone is secreted in normal quantities by the testes, the individual is unable to respond to the androgens and is "insensitive" to them. A uterus and Fallopian tubes are not present because MIH is also secreted normally by the testes. Treatment involves removal of the testes after puberty, estrogen replacement, and emotional support and counseling for the individual and her family.

B. Disorders of the Breasts

There are several developmental disorders of the breast, including breast asymmetry, hypoplasia (i.e., underdevelopment) of the breast, breast hypertrophy (i.e., overdevelopment), nipple inversion, and development of additional (supernumerary) nipples (polythelia) and breasts (polymastia). Polythelia occurs in 1 or 2% of women and has a familial tendency. Polymastia may not be evident except during pregnancy when women complain of enlarging masses, most commonly in the axilla. Breast asymmetry is extremely common and is presumably due to differential sensitivity of the two breasts to gonadal steroids.

C. Disorders of the Hypothalamic–Pituitary System

Developmental abnormalities of the hypothalamus and anterior pituitary gland may manifest themselves by failure of girls to undergo pubertal development. Perhaps the classic disorder is Kallmann's syndrome in which the gonadotropin-releasing hormone (GnRH) neurons fail to migrate into the hypothalamic region during prenatal development. GnRH signals the secretion of luteinizing hormone (LH) and follicle-stimulating hormone (FSH) from the anterior pituitary gland; these hormones stimulate the ovaries to secrete estrogen and prepare an egg to be released at ovulation. The otic bulbs also may fail to develop completely so that hypogonadotropic hypogonadism (i.e., underdevelopment of the ovaries due to low levels of LH and FSH) occurs in association with hyposmia (defective sense of smell). Other central defects, including color blindness and cleft lip and palate, may occur in other family members or the affected individual. Although initially described in males, it is now clear that the disorder occurs in both sexes. In the extreme developmental abnormality the entire septooptic portion of the brain (i.e., central portion near the optic nerves) may fail to develop properly. Hypoplasia (underdevelopment) of the pituitary gland also may occur with resulting hypopituitarism (i.e., low secretion of all pituitary hormones). Such hypothalamic–pituitary disorders can be treated with estrogen and a progestin to induce pubertal development. Because the ovaries are normal in such individuals, ovulation can be induced with exogenous gonadotropin (LH and FSH) or GnRH stimulation.

D. Anatomic Disorders of the Reproductive Tract

A variety of anatomic abnormalities of the reproductive tract can impact reproductive function. These include uterine fibroids, endometriosis, uterine prolapse, endometrial polyps, and anatomic abnormalities of the cervix and vagina. Some benign tumors of the ovary may also impact reproductive function and will be discussed subsequently in considering other functioning abnormalities.

1. Uterine Fibroids

Fibroids are benign tumors of uterine smooth muscle. Fibroids occur in more than 20% of women over age 35 and are more common in black than in white women. Although the cause of fibroids is unknown, they often appear to grow more quickly in a high estrogenic milieu such as pregnancy and to shrink after menopause. They range in size from the microscopic to the very large, with each tumor being derived from a single uterine smooth muscle cell. Symptoms depend on the size of the fibroids, their number, and their location, but even women with large fibroids may not complain of any symptoms. Symptoms may include abnormal uterine bleeding; pain,

pressure, or heaviness in the pelvic region; increased abdominal girth; urinary urgency; and, rarely, infertility due to distortion of the uterine cavity or obstruction of the Fallopian tubes. The bleeding can be so heavy as to result in anemia (i.e., a reduced number of red blood cells). Most fibroids do not need any specific treatment, but removal of the tumors or even hysterectomy may be required for severe symptoms or rapid growth of the fibroids.

2. Endometriosis

Endometriosis is the occurrence of endometrial glands and stroma (i.e., the lining of the uterus) outside of the uterine cavity. Most commonly the misplaced tissue is found within the abdominal cavity and is believed to develop there as a result of flow of blood and tissue from the uterus back through the Fallopian tubes. Because the endometrial implants can respond to gonadal steroids, they may bleed during menses, causing pain and stimulating the development of scar tissue. As endometriosis progresses, the scar tissue may form fibrous bands (i.e., adhesions) that distort organs within the abdominal cavity and block the Fallopian tubes. Although the true incidence of endometriosis is unknown, it has been estimated to occur in 10–15% of menstruating women. As many as 25–50% of infertile women may have endometriosis, but any additional mechanisms by which endometriosis may cause infertility aside from physically obstructing fertilization are unknown. Although endometriosis may be suspected, it can be diagnosed only by direct visualization. Treatment is based on the extent of the disease, the woman's symptoms, her desires regarding pregnancy, and her age. Generally, treatment consists of surgical eradication of endometriosis and lysis of any adhesions, medical suppression of gonadal steroids (to suppress activity of the endometriosis), or a combination of medication and surgery. Endometriosis must be regarded as a chronic condition; only surgical removal of both ovaries will prevent endometriosis from recurring.

3. Disorders of Pelvic Support

Childbearing appears to be the cause for most problems of pelvic support, although age, hormonal status, nutrition, and inappropriate surgery can exacerbate any existing defect. Common problems include cystocele, rectocele, enterocele, uterine prolapse, and vaginal vault inversion.

Most patients complain of a bulging sensation in the vagina. In uterine prolapse, the cervical protrudes beyond the external genitalia, followed by the vagina and uterus. Cystoceles, which are herniations of the posterior bladder into the anterior vaginal wall, may be asymptomatic, but patients may note a protrusion into the vagina and may have urinary symptoms. Rectoceles, in which a portion of the anterior rectal wall bulges in the posterior vaginal wall with defecation, may cause a patient to place a finger in the vagina in order to expel feces from the rectum. Enteroceles are characterized by bulging of the posterior cul-de-sac (i.e., the area directly posterior to the cervix) into the rectovaginal septum (i.e., the space between the vagina and the rectum) and are typically associated with heaviness on standing which is absent when lying down. Pain with intercourse (i.e., dyspareunia) is rarely associated with genital relaxation. Both medical and surgical approaches to these problems are possible. Pessaries, which are devices placed in the vagina to support the uterus or rectum, are useful in women who are not candidates for surgical repair because of ill health.

4. Endometrial Polyps

Benign endometrial polyps or growths may develop in the glandular tissue lining the uterus. Such polyps may cause abnormal uterine bleeding or pain in association with increased menstrual bleeding. Polyps may be identified as intrauterine masses on transvaginal ultrasonography, and the diagnosis can be substantiated at hysteroscopy or by removal at dilatation and curettage (D&C). Because polyps may be missed on D&C, an argument can be made for performing hysteroscopy if a polyp is suspected. Treatment consists of removing any polyps.

5. Anatomic Abnormalities of the Cervix and Vagina

A number of nonneoplastic anatomic abnormalities can involve the cervix, including cervical polyps, condyloma acuminata (i.e., genital warts associated with human papilloma virus), nabothian cysts (i.e., epithelial retention cysts), and changes consistent with maternal DES exposure (such as a "cockscomb" cervix, which is a ridge of raised tissue over the anterior cervix; a cervical "collar," which consists of a head or rim surrounding most of the cervix; and cervical hypoplasia, in which there is essentially no vaginal portion to the cervix). Vulvovaginal abnormalities are not uncommon and can include a variety of benign lesions and skin disorders affecting the vulva. Cysts may develop in any of the several glands that are present in the area.

III. FUNCTIONAL ABNORMALITIES

A. Chronic Anovulation

Chronic anovulation is present in a variety of disorders in women of reproductive age in which the ovaries contain normal numbers of oocytes but the women fail to ovulate. This group accounts for the largest number of patients presenting with amenorrhea (i.e., the absence of menstrual periods) and includes individuals with hypothalamic and/or pituitary dysfunction, adrenal and thyroid disorders, and inappropriate steroid feedback as occur in polycystic ovarian syndrome (PCO).

Hypothalamic forms of anovulation and amenorrhea may be suspected in women with significant weight loss, anorexia, bulimia, or emotional disorders and in those engaged in endurance training. These forms of anovulation often must be diagnosed by exclusion because all hormonal values are typically in the normal range for women of reproductive age. Because circulating estrogen levels are typically at the lower end of the normal range, however, bone loss is often accelerated and affected individuals benefit from exogenous estrogen therapy. Ovulation usually can be induced easily with either clomiphene citrate or exogenous gonadotropin or GnRH.

PCO, also known as Stein–Leventhal syndrome, is a disorder in which the ovaries contain many fluid-filled cysts. In this disorder the pituitary gland commonly secretes larger amounts of LH than normal, stimulating the ovarian production of excess androgens (male hormones), sometimes causing the affected woman to develop masculine characteristics such as acne and coarse facial hair. Untreated, some of the androgens may be converted to estrogens, and the high levels of estrogens may increase the risk of cancer of the endometrium (i.e., uterine lining).

Patients are frequently obese and typically present with irregular bleeding or amenorrhea from puberty. No ideal treatment exists, and the choice of therapy depends on the signs and symptoms present, the woman's age, and her desires regarding pregnancy. Combination oral contraceptives, which are frequently used, effectively reduce circulating androgen levels and eliminate the increased risk of endometrial cancer while simultaneously providing effective contraception. Specific treatments exist to treat the hirsutism and to induce ovulation.

B. Premature Menopause

In premature menopause, also known as premature ovarian failure, the ovaries stop functioning and menstrual periods stop before age 40. Typically, circulating levels of estrogen are low, leading to an increase in the pituitary hormones that normally stimulate the ovaries, LH and especially FSH. The causes are varied and include autoimmune dysfunction (in which the immune system attacks its own ovaries), genetic disorders (such as Turner's syndrome), chemotherapy and/or irradiation for any of several cancers, and idiopathic (or unexplained) causes. Cigarette smoking is associated with menopause beginning several months early.

With the onset of premature menopause, affected women commonly complain of symptoms common to the menopause, including hot flashes, discomfort with intercourse, vaginal dryness, and mood swings. Because a small percentage of women with this disorder do in fact get pregnant after the diagnosis is made, it may be better to term the disorder "primary hypogonadism" rather than premature menopause. However, clinicians do not know how to induce ovulation effectively in affected women, and the use of donor oocytes with *in vitro* fertilization is clearly the most effective opportunity for pregnancy, with many programs reporting success rates in excess of 50%. Estrogen replacement therapy is warranted to prevent signs and symptoms of estrogen deficiency.

C. Premenstrual Syndrome

Premenstrual syndrome (PMS, premenstrual dysphoric disorder, and late luteal phase dysphoric disorder) is a condition in which several physical and emotional symptoms, including (but not limited to) irritability, anxiousness, depression, headaches, bloating, food cravings, and breast tenderness, recur in the several days prior to each menstrual period and are sufficiently severe so as to interfere with the normal life of the individual. The cause of the disorder is unknown, and treatment is largely aimed at reducing the worst symptoms. There is now clear evidence that a class of drugs affecting chemicals in the brain and known as serotonergic reuptake inhibitors, often used in larger doses to treat depression, are effective in a significant percentage of women with PMS.

D. Dysmenorrhea

Dysmenorrhea refers to uterine cramping during menstruation. Primary dysmenorrhea indicates that there is no underlying cause, and secondary dysmenorrhea is used when an underlying disorder explains the discomfort. Primary dysmenorrhea is quite common, probably occurring in more than 50% of women. Primary dysmenorrhea is believed to be due to uterine contractions caused by substances called prostaglandins contained in menstrual glands and blood. Secondary dysmenorrhea can have numerous causes, including most commonly endometriosis and uterine fibroids. The dysmenorrhea is treated most effectively by nonsteroidal antiinflammatory drugs, including ibuprofen and naproxen. If the pain continues, low-dose combination oral contraceptives are commonly added to suppress ovulation, especially because dysmenorrhea is commonly much worse in ovulatory menstrual cycles. If the pain continues, a search for a secondary cause by diagnostic laparoscopy, in which a fiberoptic scope is used to examine the abdominal cavity, is commonly the next step.

E. Benign Ovarian Masses

A variety of benign ovarian masses can develop within the ovary. Some of them affect reproductive function and ovulation either because the masses secrete abnormal quantities of ovarian hormones or because the masses alter normal hormonal secretion by compressing adjacent normal ovarian tissues. Although not all masses affect reproductive function, any benign mass can do so, and all will be considered here. Functional ovarian cysts, which do secrete hormones, are common and generally resolve without intervention over time. Benign tumors, most commonly benign cystic teratomas (also called dermoids), often must be removed surgically, if for no other reason than to eliminate the possibility of ovarian cancer.

IV. INFECTIOUS ABNORMALITIES

Most infections involving the reproductive tract are transmitted sexually by intercourse. Typically infections can be divided into those affecting the "lower" genital tract (i.e., vulva and vagina) and those affecting the "upper" genital tract (i.e., uterus and Fallopian tubes).

Vulvitis is inflammation of the vulva (the external genital organs), and vaginitis is inflammation of the lining of the vagina. Vulvovaginitis, involving both organs, is common. Common causes include infections and irritating substances. Poor hygiene can lead to infections that are not transmitted sexually. The most common symptom of vaginitis is an abnormal vaginal discharge. Vulvovaginitis is commonly more irritating than dangerous, with itching, soreness, and a foul odor being common symptoms. Common causes include *Candida* (yeast) and *Trichomonas vaginalis* (a protozoan). A variety of antimicrobial preparations are available to treat the various infectious agents commonly causing vaginitis. The drugs used to treat vulvitis depend on the cause and are the same as those used to treat vaginitis. In addition, cotton underwear may be recommended as well as the use of nonirritating soap and detergent for laundering.

Pelvic inflammatory disease (PID), also termed salpingitis, is inflammation of the Fallopian tubes and is commonly caused by sexually transmitted infections. The inflammation is usually caused by an "ascending" bacterial infection which enters through the vagina and moves up through the uterus into the Fallopian tubes. Such infections may also occur following vaginal delivery, a miscarriage, or an abortion. Both tubes are usually infected, and the infection can spread into the abdominal cavity and cause peritonitis. It is important to remember that pelvic tuberculosis remains a common cause of PID worldwide. Typically, affected individuals present with severe lower abdominal pain, fever, and nausea. Abscesses can develop in tubes, abdominal cavity, or ovaries. Patients should be treated with high-dose antibiotics and hospitalized if the infection is severe or if rapid improvement does not occur. The risk of complications, including infertility, increases as the severity of the infection increases.

V. CANCERS OF THE REPRODUCTIVE SYSTEM

Cancers can occur in any part of the female reproductive system. Cancer of the uterus more appropriately is termed endometrial cancer because it begins in the lining (i.e., endometrium) of the uterus. Endometrial cancer is the most common cancer of the female reproductive system and the fourth most common cancer in women. This cancer most commonly begins after menopause with abnormal uterine bleeding being the classic early symptom. Risk factors include irregular menstrual periods, obesity, diabetes, tamoxifen therapy (commonly used in women with breast cancer), and never having had children. Hysterectomy, the removal of the uterus, is the first line of treatment for early cancer and is almost always curative if the cancer has not spread beyond the uterus. When the disease has spread beyond the uterus, chemotherapy and progestins may be utilized. Cancer of the cervix, which forms the opening to the uterus and extends into the upper vagina, is the second most common cancer of the reproductive system in women and the most common in younger women. It most commonly affects women between 35 and 55 years of age. This cancer is closely linked to the human papillomavirus, which is transmitted by sexual intercourse, and may be caused by the virus. Early age at first coitus and multiple sexual partners increase the risk for cervical cancer. Postcoital spotting is the classic symptom, but most cases are diagnosed by abnormalities found on Pap smear. Treatment depends on how advanced the cancer is (i.e., its stage) and can consist of surgery and/or radiation, sometimes with chemotherapy for widespread disease. Early cervical cancer can be treated very effectively by surgical removal.

About 1 of 70 women will develop ovarian cancer,

most commonly after menopause. Although it is only the third most common cancer of the female reproductive system, more women die of ovarian cancer than of any other cancer of the reproductive system. There are many kinds of ovarian cancer, with the treatment and prognosis varying depending on the type of cancer. Most patients are asymptomatic until the cancer is advanced, and there is no good test for diagnosing ovarian cancer early.

The vulva is the term used to identify the external female reproductive organs. Vulvar cancer is almost always a skin cancer near or at the opening to the vagina. The key to diagnosis involves performing a biopsy on any suspicious lesion in the area. Excision of the tumor is the first line of therapy with radiation therapy sometimes added for advanced disease.

Cancers of the vagina and Fallopian tubes are rare. In contrast, gestational trophoblastic disease, which is tumorous growth of tissue from the placenta, occurs in perhaps 1 of every 2000 pregnancies in the United States. Benign tumors are commonly called hydatiform moles; usually arising from abnormal pregnancies, they may be termed molar pregnancies. Malignant trophoblastic disease is termed choriocarcinoma. The abnormal pregnancies associated with trophoblastic disease are typically characterized by abnormal uterine bleeding, and nausea and vomiting may be severe. The diagnosis can be made by scanning the uterus with ultrasound. Surgical evacuation of the contents of the uterus is indicated. After surgery human chorionic gonadotropin (hCG), the hormone that forms the basis for pregnancy tests and which is produced by both normal and abnormal placental tissue, is measured at intervals. Failure of the hCG levels to decrease to undetectable levels indicates persistent disease and indicates the need for chemotherapy.

VI. BREAST DISORDERS

Fibrocystic breast changes are quite common and include breast pain, formation of small cysts, and generalized benign "lumpiness" in the breast tissue. Persistent cysts can be aspirated with a needle, and women with such changes can be treated symptomatically. The major challenge is distinguishing fibro-cystic changes from breast cancer. Fibrocystic changes are not a disease although they are sometimes inappropriately called fibrocystic disease.

Nipple discharges are not uncommon but require evaluation. Secretion of milk (galactorrhea), which is defined as fluid containing fat regardless of color, may indicate hypothalamic–pituitary dysfunction and is commonly associated with amenorrhea. Galactorrhea also may be caused by hypothyroidism (i.e., underfunctioning of the thyroid gland) and is associated with use of tranquilizers and some other medications. A nipple discharge also may be caused by fibrocystic changes or a benign tumor in a milk duct (intraductal papilloma). Cancer is uncommon in women with a breast discharge.

Breast infections (mastitis) are uncommon except after childbirth or following an injury. Rarely, inflammatory breast cancer may appear similar to mastitis. Warm, red, swollen, and tender breasts typical of mastitis are treated with antibiotics.

There are different kinds of breast cancer, and prognosis is dependent on both the type and the extent of spread when diagnosed. There are several important risk factors for breast cancer; perhaps most important is a history of breast cancer in a first-degree relative (i.e., mother, sister, or daughter). Commonly, there are no symptoms, and breast cancer is first identified by the finding of a breast lump or on screening mammography. The finding of a lump or of a suspicious area on mammography requires biopsy. Treatment is complex and may include surgical excision, radiation therapy, chemotherapy, and use of estrogen antagonists (i.e., tamoxifen). Treatment of breast cancer is complicated by the fact that it must be regarded as a systemic disease, and recurrences may occur many years after treatment.

VII. DISORDERS OF PREGNANCY

A. Abortion and Stillbirth

In spontaneous abortion, or miscarriage, a fetus is lost from natural causes prior to the twentieth week of pregnancy. In a stillbirth the fetus dies after the twentieth week of pregnancy.

Perhaps one-third of pregnant women note some

bleeding and/or cramping in the first trimester (i.e., first third) of pregnancy. Any bleeding leads to use of the term "threatened abortion"; about half the women who experience bleeding subsequently abort. Most first-trimester abortions are genetic in that the chromosomal complement of the fetus is abnormal.

If all the contents of the uterus are expelled (i.e., a complete abortion), no further therapy may be necessary. Alternatively, a D&C may be required to dilate the cervical os (i.e., opening) and remove all the contents of the uterus. If the products of conception become infected, a D&C must be performed and systemic antibiotics utilized.

B. Ectopic Pregnancy

In ectopic pregnancy the pregnancy develops outside of the uterus, most commonly in the Fallopian tubes. Ectopic pregnancy generally occurs in women who have developed tubal and intraabdominal scarring from a previous infection or surgery. Sexually transmitted pelvic infections are probably the most common cause of scarring of the Fallopian tubes. Rarely, ectopic pregnancies occur in women with normal anatomy, perhaps because an egg released from one ovary is picked up and transported down the contralateral Fallopian tube. Ectopic pregnancies are potentially life-threatening; the pregnancies can be destroyed medically with drugs such as methotrexate if diagnosed early. Otherwise, they must be removed surgically.

C. Anemias

Because the blood volume of the mother increases by at least one-third in normal pregnancy and because the fetus must manufacture blood as well, more iron, which is needed to produce red blood cells, is required during pregnancy. The most common anemia, i.e., reduction in red blood cells or in the amount of hemoglobin (i.e., the protein which carries oxygen) in the red blood cells below normal, in pregnancy is iron-deficiency anemia. Commonly, the anemia is due to inadequate maternal ingestion of iron; in fact, excess iron, generally in vitamins, must be consumed by pregnant women in order to prevent iron-deficiency anemia. Less commonly the anemia

is due to folic acid deficiency, a B vitamin also required for the production of red blood cells.

D. Rh Incompatibility

In Rh incompatibility there is a mismatch between the Rh blood type of the pregnant woman and that of her fetus. As a consequence of the Rh incompatibility, the woman may produce antibodies against the fetus's red blood cells, causing them to rupture and leading to a "hemolytic" anemia. The Rh blood group includes molecules on the surface of the red blood cells that are unique to each individual. Problems occur in mothers who are Rh negative and who have babies that are Rh positive, with the Rh factor inherited from the father. There are rarely problems in a first pregnancy; because fetal blood rarely comes into contact with maternal blood, antibodies do not form in the absence of any contact. Small quantities of fetal blood commonly mix with maternal blood at delivery but may mix earlier if there is any fetal bleeding. Formation of maternal antibodies can be prevented by giving the mother Rh antibodies (in the form of immune globulin) to destroy any fetal cells in the maternal circulation. This is routinely done after delivery of an Rh-negative mother and at the time of any recognized fetal–maternal bleed. Rarely, hemolytic anemia may develop in the fetus that has an incompatibility of another major blood group.

E. Abruptio Placentae

In an abruption the placenta detaches from the uterine wall prematurely before delivery of the fetus. An abruption may be complete or incomplete. Women with high blood pressure, diabetes, multiple pregnancies, or who use cocaine are more apt to have abruptions. Severe abdominal pain and vaginal bleeding are the common symptoms. Significant blood loss by the mother and/or the fetus can result. Once the diagnosis is made, hospitalization is warranted. Treatment is delivery if there is evidence of continuing bleeding, fetal distress, or the pregnancy is near term. Otherwise, bed rest may stabilize the situation.

F. Placenta Previa

In placenta previa the placenta is implanted over the cervix at least in part. Individuals with placenta previa commonly present with painless, bright red vaginal bleeding in the last trimester of pregnancy. Delivery is warranted if the bleeding is profuse or will not stop. A cesarean section is almost always required for delivery.

G. Hyperemesis Gravidarum

Hyperemesis gravidarum is excessive nausea and vomiting, common in early pregnancy, so severe as to lead to dehydration and even starvation. Treatment is supportive and commonly involves hospitalization and administration of intravenous fluids. The cause is unknown.

H. Preeclampsia and Eclampsia

The diagnosis of preeclampsia is made when a pregnant woman develops two of the following signs in the last 20 weeks of pregnancy: high blood pressure, significant protein in the urine (proteinuria), and retention of fluid (edema), most commonly in the hands or face. Eclampsia is diagnosed when a woman with preeclampsia has seizures.

Preeclampsia occurs in about 5% of pregnancies. It is more common in women having their first pregnancies and in women with high blood pressure, diabetes, or multiple pregnancies. Treatment is bed rest if the preeclampsia is mild. If severe, delivery is warranted no matter how premature the infant. Magnesium sulfate is commonly given to prevent seizures. In the presence of preeclampsia, blood flow to the fetus is reduced, and fetal distress may occur. When severe, preeclampsia can result in the death of the mother (as well as of the fetus); eclampsia can lead to permanent impairment of the mother.

VIII. INFERTILITY

Infertility is commonly defined as the inability of a couple to achieve a pregnancy after repeated intercourse without contraception for 1 year. It is esti-mated that infertility affects one couple in five in the United States. The frequency appears to be increasing because individuals now commonly delay marriage and childbearing until later in life. The principal causes of infertility include problems with sperm, abnormal or absent ovulation, and obstruction of the Fallopian tubes. Diagnosis and treatment of infertile couples requires evaluation of both partners.

Treatment depends on the problems identified. Most readily treated are disorders of ovulation. If oocytes remain in the ovaries, ovulation can be induced with a number of properly selected endocrine preparations, but pregnancy results in only perhaps 75% of treated couples. Obstruction of the Fallopian tubes may be treated surgically in some cases. In all such cases, *in vitro* fertilization and embryo transfer [i.e., so-called assisted reproductive technologies (ART)] can be used to enable a couple to bear a child. Problems of sperm can be treated in a variety of ways, perhaps most successfully by ART with intracytoplasmic sperm injection, in which a single sperm is injected into the cytoplasm. Technologies to treat infertility are evolving rapidly.

IX. FAMILY PLANNING

Family planning refers to efforts to control the number and spacing of pregnancies. Couples may use contraception to prevent pregnancy temporarily or sterilization to prevent pregnancy permanently. Abortion may be used to terminate pregnancies when contraception has failed.

Many different forms of reversible contraception exist. Their success rates and any side effects vary greatly. Contraceptives are more apt to be used incorrectly by people who are younger, less educated, and/or less motivated to prevent pregnancy. Perhaps 5–15% of women using contraceptives intended for use at time of sexual intercourse (i.e., diaphragm, condom, contraceptive foam, or jelly) become pregnant during the first year of use. Such barrier methods are much less effective than "coitus independent" methods such as oral contraceptives, implants, injectable contraceptives, and intrauterine devices, which have typical failure rates of 0.1–0.3% in the first year of use.

So-called "natural family planning" depends on abstaining from sexual intercourse during the woman's fertile period and requires regular menstrual cycles on the part of the woman and motivation. On average, ovulation occurs 14 days before the beginning of the next menstrual period. Because sperm can survive at least 3 or 4 days in the female genital tract, pregnancy theoretically can result even when intercourse occurred several days before ovulation. Generally, eggs only survive about 24 hr if fertilization does not occur. Methods of natural family planning typically involve patients charting changes in their cervical mucus, basal body temperature, and menstrual cycle length in order to determine the "safe" days each month when intercourse is permissible. Body temperature generally increases after ovulation because of the hormone progesterone, which is secreted by the ovary in the second half of the normal ovulatory menstrual cycle. Under the influence of ovarian estrogen cervical mucus increases dramatically until the time of ovulation, becoming thin, slippery, and elastic, and then under the influence of progesterone becomes thick and cloudy. Estrogenic mucus assists sperm in traversing the female genital tract, whereas progestogenic mucus blocks migration of sperm into the uterus. Because of purposeful and accidental failure to follow appropriate guidelines associated with natural family planning, perhaps 20% of women conceive during the first year of use.

Perhaps the oldest form of contraception is coitus interruptus, in which the penis is withdrawn from the vagina before ejaculation of sperm during orgasm. The method is not totally reliable because some sperm may be released before orgasm. It obviously requires self-control and motivation on the part of the male partner.

Oral contraceptives, commonly called "birth control pills," contain either a combination of an estrogen and a progestin or a progestin alone. The synthetic steroids inhibit the secretion of hormones from the pituitary gland (LH and FSH) which stimulate the ovary to release an egg and keep the cervical mucus thick so that sperm cannot pass into the uterus easily. Implantable and injectable contraceptives have similar effects. In contrast, intrauterine devices, so-called IUDs, act by causing an inflammation inside the uterus and impede motility of both eggs and sperm, fertilization, and (should fertilization rarely occur) implantation. Although IUDs are used rarely by women in the United States, they are the most common form of reversible contraception used worldwide.

About one-third of all couples in the United States who use family planning methods choose permanent sterilization, which may consist of vasectomy (cutting the vasa deferentia, the tubes carrying sperm from the testes) in men or tubal ligation (cutting, cauterizing, or physically blocking the Fallopian tubes, which carry eggs from the ovaries to the uterus) in women. Surgical removal of the uterus (i.e., hysterectomy) is occasionally used to effect sterilization.

Abortion can be performed surgically, in which the contents of the uterus are evacuated generally through the vagina, and medically, in which drugs such as mifepristone (RU 486) and prostaglandins stimulate the uterus to contract and expel the products of conception.

X. MENOPAUSAL PROBLEMS

At menopause, technically defined as beginning following a woman's last menstrual period, estrogen production by the ovaries decreases dramatically. On average, menopause begins at age 51 and is abnormal before the age of 40. It is the cells surrounding the eggs in the ovaries (i.e., the granulosa cells) that secrete estrogen so that menopause commonly indicates the end of reproductive life and the loss of all oocytes. As estrogen levels fall in the perimenopausal period women may complain of symptoms such as hot flashes (occurring in approximately 80–85% of women), night sweats, insomnia, discomfort with intercourse, and emotional liability (including such nonspecific symptoms as fatigue, irritability, nervousness, and forgetfulness). Bone loss from the skeleton may accelerate, leading to osteoporosis, which is severe thinning of the bones, such that the risk of fractures increases. The risk of cardiovascular disease may increase after menopause in the presence of low circulating levels of estrogen.

Exogenous estrogen is generally effective in relieving these changes and symptoms in postmenopausal women. However, the risk of cancer of the lining of the uterus is increased dramatically unless an exogenous progestin is added as well. The addition of the progestin causes uterine bleeding in many women (after spontaneous menses ended) and the synthetic progestins may cause bloating and mood changes. Whether taking estrogen increases the risk of breast cancer is unclear, but there may well be an increased risk for women using estrogen for several ($\geqq 10$) years. However, protection against osteoporosis and cardiovascular disease requires continuous usage. Research in this area continues, especially in the large Women's Health Initiative study sponsored by the National Institutes of Health at 40 sites in the United States. Women who cannot or should not take estrogen can be treated symptomatically with a variety of drugs. Several new agents for treatment of osteoporosis are now available or soon will be and include bisphosphonates, calcitonin, fluoride, and selective estrogen receptor modulators.

See Also the Following Articles

CONGENITAL VIRILIZING ADRENAL HYPERPLASIA; BREAST CANCER; CERVICAL CANCER; ENDOMETRIOSIS; KALLMANN'S SYNDROME; LEIOMYOMA; OVARIAN CANCER; PELVIC INFLAMMATORY DISEASE (PID); PMS (PREMENSTRUAL SYNDROME); POLYCYSTIC OVARIAN SYNDROME; TURNER'S SYNDROME

Bibliography

Beckman, C. R. B., *et al.* (1992). *Fundamentals of Obstetrics and Gynecology*. Williams & Wilkins, Baltimore.

Lewis, B. J. (1992). Breast cancer. In *Cecil Textbook of Medicine* (J. B. Wyngaarden, L. H. Smith, Jr., and J. C. Bennett, Eds.), 19th ed., pp. 1381–1386. Saunders, Philadelphia.

Littman, B. A., Smotrich, D. B., and Stillman, R. J. (1995). Endometriosis. In *Principles and Practice of Endocrinology and Metabolism* (K. L. Becker *et al.*, Eds.), 2nd ed., pp. 906–909. Lippincott, Philadelphia.

Marchant, D. J. (1992). Nonmalignant disease of the breast. In *Cecil Textbook of Medicine* (J. B. Wyngaarden, L. H. Smith, Jr., and J. C. Bennett, Eds.), 19th ed., pp. 1378–1381. Saunders, Philadelphia.

Ory, S. J. (1995). The differential diagnosis of female infertility. In *Principles and Practice of Endocrinology and Metabolism* (K. L. Becker *et al.*, Eds.), 2nd ed., pp. 947–954. Lippincott, Philadelphia.

Rebar, R. W. (1992). The ovaries. In *Cecil Textbook of Medicine* (J. B. Wyngaarden, L. H. Smith, Jr., and J. C. Bennett, Eds.), 19th ed., pp. 1355–1376. Saunders, Philadelphia.

Rebar, R. W. (1995). Disorders of menstruation, ovulation, and sexual response. In *Principles and Practice of Endocrinology and Metabolism* (K. L. Becker *et al.*, Eds.), 2nd ed. Lippincott, Philadelphia.

Reid, R. L., and Fretts, R. C. (1995). Premenstrual syndrome. In *Principles and Practice of Endocrinology and Metabolism* (K. L. Becker *et al.*, Eds.), 2nd ed., pp. 909–914. Lippincott, Philadelphia.

Rittmaster, R. S. (1992). Hirsutism. In *Cecil Textbook of Medicine* (J. B. Wyngaarden, L. H. Smith, Jr., and J. C. Bennett, Eds.), 19th ed., pp. 1376–1378. Saunders, Philadelphia.

Schiff, I., and Walsh, B. (1995). Menopause. In *Principles and Practice of Endocrinology and Metabolism* (K. L. Becker *et al.*, Eds.), 2nd ed., pp. 915–924. Lippincott, Philadelphia.

Simpson, J. L., and Rebar, R. W. (1995). Normal and abnormal sexual differentiation and development. In *Principles and Practice of Endocrinology and Metabolism* (K. L. Becker *et al.*, Eds.), 2nd ed., pp. 778–822. Lippincott, Philadelphia.

Speroff, L., and Darney, P. (1996). *A Clinical Guide to Contraception*, 2nd ed. Williams & Wilkins, Baltimore.

Yeh, I.-T., Zalondek, C., and Kurman, R. J. (1995). Functioning tumors and tumor-like conditions of the ovary. In *Principles and Practice of Endocrinology and Metabolism* (K. L. Becker *et al.*, Eds.), 2nd ed., pp. 940–947. Lippincott, Philadelphia.

Female Reproductive System, Amphibians

Rakesh K. Rastogi and Luisa Iela

University of Naples Federico II

GLOSSARY

follicular atresia Degeneration of the ovarian follicle at any stage of its development.

folliculogenesis The process during which the oocyte grows and is ensheathed by granulosa and theca layers.

ovarian cycle The sequence of correlated intraovarian events leading to the release of a set of eggs.

ovarian follicle The oocyte with its surrounding layers, granulosa and theca.

oviduct The female genital tract, the Müllerian duct.

ovulation The process during which mature oocytes, devoid of their granulosa and theca layers, are expelled from the ovarian surface into the abdominal cavity.

spermathecae Sperm storage glands in the cloacal wall.

Amphibians include frogs and toads (anurans), newts and salamanders (urodeles), and the limbless gymnophiones (apodans or caecilians). The female reproductive system of amphibians is essentially composed of a pair of ovaries and a pair of oviducts (Müllerian ducts).

I. INTRODUCTION

As in other vertebrates, the amphibian ovary is both an endocrine gland producing sex hormones and a special organ producing female gametes. Each ovary is suspended by a mesovarium underneath the kidney. Whereas ovaries of frogs and toads are large and lobulated, and occupy most of the coelomic cavity when fully grown, those of newts and salamanders are elongated and lobate. Gymnophione ovaries are still more elongated, in line with the serpentiform body, and the growing ovarian follicles are arranged in a single row craniocaudally. In some species of gymnophiones the ovarian lobes appear to have a segmental arrangement. While the oviducts of frogs and toads become greatly elongated and convoluted during the breeding season, in urodeles they are slightly convoluted and in gymnophiones they remain more or less straight. In viviparous species, e.g., *Nectophrynoïdes occidentalis* (Anura), *Salamandra atra* (Urodela), and *Typhlonectes compressicauda* (Gymnophiona), they form uterus-like dilations where eggs are retained for gestation. The cranial extremity of each oviduct opens into the abdominal cavity by a ciliated ostium (= infundibulum). Caudally, each oviduct opens, as a rule, separately into the cloaca. However, in some species of frogs they have an unpaired opening, and in others the lower parts of both oviducts are themselves confluent. When ovulation occurs, oocytes are released into the abdominal cavity and are captured by the ostium. In all species of amphibians, the ovaries and oviducts undergo marked morphological and functional changes during each reproductive cycle. The clutch size can be in the thousands for most anurans, hundreds for most urodeles, and tens for most gymnophiones. The reproductive modes of amphibians include oviparity, ovoviviparity, and viviparity. The "placental" form of viviparity is not yet recognized in amphibians, and in most viviparous species the nutrition of the developing embryo is shown to be provided by oviductal secretions. While ovoviviparity and varying degrees of viviparity are common among urodeles and gymnophiones, the majority of anurans lay eggs and only a few of them are viviparous, e.g. *N. occidentalis*. External fertilization is most common among anurans, rather limited in urodeles, and totally unknown among gymnophiones. Among many salamandrids with internal fertilization, the female reproductive system also includes sperm storage glands in the cloaca called spermathecae. In the viviparous *S. atra*, the cloacal chamber has a specialized structure, the organ of Siebold, in which spermatophores are deposited during copulation. When the eggs have reached the uterine segment of the oviduct, the smooth musculature of the sperm reservoir works to squeeze spermatophores through the uterine openings converging into the cloacal chamber.

II. OVARY

A. Gross Morphology and Folliculogenesis

In terms of gross morphology, the amphibian ovary is like a folded sac with maturing oocytes protruding into both the abdominal cavity and the intraovarian cavity. The size of the ovary varies greatly during the reproductive cycle. The differences are due to clutch size and whether the species reproduces seasonally (annually or biennially) or aseasonally since these variables affect the number of vitellogenic follicles present at any one time. The amphibian ovary is not uniform in structure and is covered by a thin epithelium enclosing numerous oogonia and oocytes (ovarian follicles) embedded in a scarce connective stroma. The oogonia are usually embedded in the so-called germinal beds distributed haphazardly in the network of ovigerous folds of the germinal epithelium. The primary oocytes emerge from the same and are invested by a single layer of flattened granulosa/follicle cells as well as an external theca layer. The ultrastructural, histochemical, and functional characteristics of theca and granulosa layers are basically the same in all three groups of amphibians. The oocyte–granulosa–theca complex is commonly termed ovarian follicle. As a rule, amphibian oocytes

are mononucleated. There are, however, some exceptions. Each developing oocyte in the tailed frog, *Ascaphus truei*, has eight nuclei, whereas in the marsupial frog, *Flectonotus pygmaeus*, each may contain up to or more than 2000 nuclei. In both cases, all nuclei except one disappear before ovulation. In all amphibians, the oocyte growth (folliculogenesis) is accounted for mainly by the deposition of nutrients composed of phosphoproteins, lipids, and glycogen, collectively referred to as yolk. When yolk is deposited (a process termed vitellogenesis) in the cytoplasm of the developing oocytes in the form of small oval or spherical platelets, these follicles are usually called vitellogenic follicles. Previtellogenic ovarian follicles are those in which yolk deposition has not yet begun. On the basis of the time course of vitellogenesis or the amount of yolk deposited as well as their size, the vitellogenic follicles can be distinguished into early vitellogenic, advanced or late vitellogenic, and postvitellogenic follicles. Oocyte growth before vitellogenesis begins is sometimes referred to as early growth phase or first growth phase, followed by the late or second growth phase when vitellogenesis takes place. After vitellogenesis is terminated, the oocyte nucleus, often called the germinal vesicle, moves to a more peripheral position. Germinal vesicle breakdown is the event related to final oocyte maturation which generally takes place just prior to ovulation and before the oocyte can be fertilized. Hormonal and nutritional factors are involved in the recruitment of first growth phase oocytes for vitellogenesis. In fact, surgical removal of the pituitary gland totally inhibits vitellogenic growth of oocytes and many of them show signs of degeneration. Replacement therapy with either pituitary extract or purified pituitary gonadotropins reinstates oocyte growth. What determines the recruitment of a given ovarian follicle for vitellogenesis or for its final maturation and ovulation may depend on a change in responsiveness of the ovarian follicle to hormones such as the follicle-stimulating hormone (FSH), luteinizing hormone (LH), and other chemical factors such as ovarian opioid peptides, prostaglandins, and growth hormone. The molecular bases of these phenomena are little studied in amphibians, and we have no knowledge of how growth factors are involved in amphibian folliculogenesis.

B. Follicular Atresia

The phenomenon of degeneration of ovarian follicles has been described in a number of amphibians from a morphological and numerical standpoint. Under natural conditions, atretic removal of ovarian follicles is to some degree ubiquitous among amphibians, although the magnitude of its incidence varies widely both within and between the groups. It has long been recognized that, in amphibians, under adequate nutritional conditions the majority of ovarian follicles achieve ovulation, with approximately 5% or less undergoing atresia during the entire ovarian cycle. Some exceptions have been reported in which up to 50% of the ovarian follicles were found to undergo atresia but, in such cases, caution is warranted because it is imperative to know the nutritional conditions of animals. Thus, unlike mammals, in which <1% of oocytes reach final maturation, with the remaining being removed by atresia, atresia is not a significant phenomenon in amphibians. This contention gains strength from the fact that in amphibians each ovarian cycle generates a new subset of oocytes. Although atresia can occur at any stage of follicular development, it undoubtedly affects to a greater extent follicles that are undergoing vitellogenesis as well as mature follicles that have not been ovulated during the breeding season. In viviparous forms, the ovaries of pregnant females may exhibit a new cycle of folliculogenesis but all vitellogenic follicles undergo atresia at the time when ovulation should occur in nature. Follicles under atresia possess morphological characteristics that easily distinguish them from healthy/normal ovarian follicles. Morphological changes undergone by an atretic follicle include invasion of the ooplasm by granulosa cells (sometimes hypertrophic) which later become phagocytic, pycnosis and breakdown of the nucleus, luteinization of yolk when present, and a thickened theca layer which becomes highly vascularized. Little is known of the mechanisms and causes underlying this specific degenerative phenomenon. It has been pointed out that inadequate hormonal (brain and pituitary hormones, as well as intraovarian factors) and/or nutritional status and age of the animal as well as stage of the reproductive cycle, separately or in synchrony, might be responsible for the demise, in a variable percentage, of healthy ovarian follicles through atresia. Experimental manipulations, such as the removal of the pituitary gland and/or the fat bodies, and unusual environmental conditions, such as a long spell of harsh weather or poor feeding conditions, enormously augment the incidence of follicular atresia. Needless to say, we are still far from understanding specific atretogenic factors in Amphibia and we have no knowledge of why many ovarian follicles escape the fate of atretic demise.

C. Oocyte Maturation, Ovulation, and Postovulatory Follicles/Corpora Lutea

Germinal vesicle breakdown is the event which culminates in oocyte maturation and ovulation. Under "favorable" conditions a postvitellogenic follicle is induced by the maturation-inducing steroid, widely known as progesterone, to mature and ovulate. It has been confirmed that pituitary gonadotropins, mainly LH, induce oocyte maturation and ovulation through the secretion of progesterone by granulosa cells. It has been pointed out that progesterone reaches the oocyte via granulosa–oocyte heterocellular gap junctions and acts via high-affinity, low-capacity binding sites on the oocyte membrane. Progesterone-induced oocyte maturation may involve activation of protein kinases. This may, in fact, explain why many nonprogestational chemicals with protein kinase C-activating potential are capable of inducing oocyte maturation and ovulation. *In vitro* induction of germinal vesicle breakdown and/or ovulation has been obtained with substances as diverse as insulin, prostaglandins, prolactin, corticosteroids, and even androgens.

At the time of ovulation, the mature oocyte is physically expelled from the follicular layers and is released into the abdominal cavity. Ovulation may occur a few hours or even days ahead of mating; during the act of mating, when it is soon followed by ovodeposition; or several months after mating. In many species, fine red dots on the ovarian surface indicate the position of recently ovulated (evacuated) follicles, unanimously called postovulatory follicles.

In most oviparous amphibians, postovulatory follicles are transient. Following ovulation, the evacuated follicle collapses and a thickened and highly vascularized theca surrounds the mass of granulosa cells. The whole structure is rapidly removed within a few days presumably by phagocytic action of granulosa cells without forming a corpus luteum-like structure. In ovoviviparous and viviparous forms, however, all postovulatory follicles persist longer and transform into a secretory corpus luteum due to luteinized granulosa cells. It is thus obvious that the granulosa cells, like those in other vertebrates, are multifunctional in that they play a role in the uptake by the developing oocyte of circulating nutrients, secrete steroidogenic enzyme systems, and become phagocytic at the onset of atresia and after ovulation in oviparous forms. In some ovoviviparous and viviparous urodeles and gymnophiones the number of active corpora lutea in the two ovaries corresponds to the number of developing embryos in the uterine oviducts. Progesterone-secreting potential of corpora lutea is greater in pregnant females than in nonpregnant ones, and *in vitro* progesterone production by the ovary is higher during early pregnancy than after parturition. Since a new ovarian cycle is not usually initiated during pregnancy, it is likely that progesterone production by corpora lutea serves to keep the ovary quiescent. A new cycle of oocyte development in most viviparous amphibians begins only slightly before the end of gestation when corpora lutea are regressed—consequently, progesterone production should decline. This procedure may represent a mechanism by which the pregnant female can convey nutrients to the developing fetus rather than to the production of a new set of ovarian follicles.

D. Ovarian Hormones

Much of our current knowledge of ovarian hormones in amphibians comes from studies on anurans and urodeles. Inside the ovary, granulosa and theca layers undoubtedly represent the major, if not the only, source of steroids. This is supported by the histochemical localization in the granulosa and theca layers of cholesterol, lipids, and steroidogenic enzyme systems, mainly hydroxysteroid dehydrogenases, though they are variable in location and in stage of follicular development. Ultrastructural studies on the granulosa cells of some species have shown them to have organelles associated with steroid-

secreting cells. Studies *in vitro* involving incubation of ovarian fragments with labeled steroid precursors have demonstrated the ability of amphibian ovary to synthesize androgens, estrogens, and progestins. Indeed, chromatographic analysis and radioimmunoassay of ovarian tissue and/or plasma in several species confirm the presence of estradiol, estriol, estrone, testosterone, 5α-dihydrotestosterone, androstenedione, and progesterone. There is convincing evidence that the steroidogenic potential of ovarian follicles changes with the stage of follicular development and not the stage of the ovarian cycle. Not only the type of sex steroid but also its relative amount appear to be stage dependent. As a rule, the more advanced the stage of vitellogenesis the greater is the magnitude of steroidogenesis (e.g., *Pachymedusa dacnicolor* and *Rana dybowskii*). It is, in fact, the relative proportion of each stage of ovarian follicle that determines the overall steroidogenic potential of the ovary during a particular phase of the reproductive cycle. Which factors regulate the developmental stage-dependent changes in the steroidogenic potential of ovarian follicles is poorly understood. *In vitro* studies have shown that the postvitellogenic follicles of the African clawed frog, *Xenopus laevis*, year-round produce more testosterone than progesterone. Ovarian follicles of the frog, *R. nigromaculata*, during hibernation do the same, but if collected in the breeding season they produce more progesterone and very little testosterone.

While estrogens are well-known for their trophic action on the seasonal structural and functional modifications of the oviduct and of the female reproductive behavior, androgens are also known to induce oviduct development. Progesterone, instead, induces maturation of the amphibian oocyte through its action on the plasma membrane.

The secretion of estradiol from the ovary is necessary for the normal development and function of oviducts. In fact, surgical removal of both ovaries invariably leads to the regression of oviducts, whereas replacement therapy with an adequate amount of exogenous estradiol stimulates their development. In addition, in some species with unusual modes of reproduction, such as *Gastrotheca marsupiata*, estradiol injections can induce the development of marsupium-like structures in which, in the adult female, fertilized eggs are retained until their larval development is completed.

Surprisingly, in most species investigated, circulating androgen levels are higher than those of estradiol. It has been argued that androgens can be aromatized in granulosa cells, in specific brain regions to modulate reproductive behavior of the female, and in the oviduct to regulate its growth. Circulating and ovarian levels of sex steroids are strikingly different among species and in relation to the reproductive status of the female. However, in many species, the seasonal profile of circulating levels of estradiol and androgens is highly correlated with ovarian and oviductal weight changes. Only in some species has plasma progesterone peak been described in females during the breeding season (e.g., *P. dacnicolor*). In many species, plasma levels of progesterone may be highly variable during the entire reproductive cycle. Circulating levels of androgens and estrogens are usually closely related to the phases of the reproductive cycle, being relatively high prior to and/or during breeding and drastically lower for some time after breeding.

III. OVIDUCT

The oviduct is a filiform organ, often divided into regions, and has a lumen lined with secretory and ciliated cells. Among oviparous forms, the secretory nature of the oviduct indicates that it furnishes secretory products which are necessary in the formation of a jelly coat around the ova during their passage through oviducts. The multilayered gelatinous covering has been suggested to have mechanical and nutritional significance during cleavage and to somehow interact with spermatozoa in fertilization. The eggs can be stored in the slightly dilated caudalmost portion of the oviduct for variable lengths of time. In ovoviviparous species, the egg develops within the uterine oviduct, and in addition to an initial nutrient uptake from the stored yolk, a physiological exchange occurs with uterine tissue. In the viviparous toad (*N. occidentalis*), salamander (*S. atra*), and gymnophione (*T. compressicauda*), the internally fertilized eggs hatch precociously in the uterine segment of the oviduct where the larvae obtain their nutrition

from uterine secretions. Survival of the developing larvae in the uterine oviduct may also involve phenomena such as adelphophagy, as in the ovoviviparous *S. salamandra*. In ovoviviparous and viviparous species the uterine segment of the oviduct is well vascularized and has a well-developed smooth muscular covering. Changes in oviductal weight during the reproductive cycle are strictly related to the seasonal profile of circulating levels of estrogens as well as of ovarian weight. In strictly seasonal breeders the oviduct displays pronounced changes not only in size but also in its structural complexity and histochemical and biochemical characteristics. In all amphibians it shows full growth before ovulation. Growing ovarian follicles during vitellogenesis gradually increase their estrogen production to prepare the oviduct in the oviparous forms for the passage of ovulated eggs when gelatinous coatings are laid around them and in the viviparous forms when the eggs are fertilized and then retained inside the uterine segment of the oviduct for gestation. In some viviparous species, the uterine wall undergoes further modifications which are related to the stage of the embryronic development. Following ovodeposition or parturition, the oviduct becomes fully regressed and remains quiescent until a new reproductive cycle begins.

IV. FECUNDITY AND BREEDING CYCLE

The size and number of ova vary enormously among and within groups. Oviparous species in any amphibian group usually produce a greater number of eggs per female than their ovoviviparous and/or viviparous relatives. Many frogs and toads are the most fecund among amphibians, laying a few hundred to a few thousand small yolked eggs, ~1 or 2 mm in diameter. Among urodeles, oviparous species usually lay a few thousand eggs per season, whereas ovoviviparous and viviparous species lay tens of eggs. Of these, the ovoviviparous forms lay the biggest eggs (~5 mm diameter). Caecilians lay tens of eggs per season. Whereas oviparous species may lay eggs as big as 9 or 10 mm in diameter, those with viviparous mode of reproduction produce eggs that do not exceed 2 or 3 mm of diameter and that have a small amount of yolk.

The interval of time between two breeding seasons in sequence can represent the length of breeding cycle of a species. Most amphibians breed seasonally once a year, including temperate and tropical species. In many species, in fact, annual cyclic changes in the weight and histological composition of the ovary are consistent with the fact that they undergo a single reproductive period each year. The increase in ovarian weight is related to an increase in the number of vitellogenic follicles. However, morphological differences do occur among species. Thus, there may be variations in the periods of oogonial proliferation which may occur around the time of breeding, soon after breeding, or throughout the year. Similarly, vitellogenesis may take place throughout the year in many tropical species, autumn–spring in most temperate species, or during a surprisingly short amount of time before breeding, as in the Mexican leaf frog, *P. dacnicolor*. In explosive breeders, such as the spade-foot toad (*Scaphiopus couchii*), usually there is a short period of postovulatory ovarian quiescence followed by the period of vitellogenesis which terminates before overwintering so that the female is ready to lay eggs as soon as favorable environmental conditions, e.g., rainfall, occur. According to whether a species undergoes a prolonged or a short period of breeding, the ovarian cycle shows differential aspects in which there is either synchrony (explosive breeding) or asynchrony (prolonged breeding) in the recruitment of previtellogenic follicles for vitellogenesis and in the secretion of regulatory hormonal factors such as pituitary gonadotropins.

Among all amphibian groups, there are species which typically breed once a year, called seasonal/annual breeders, and others which exhibit biennial breeding. In many species, e.g., neotropical plethodontid salamanders, mating may occur throughout the year, but breeding is highly seasonal. As cold-blooded vertebrates, amphibians are well-known for their dependence on environmental correlates to lay eggs. Field studies have confirmed that most temperate zone species reproduce during spring and early summer; some species of frogs, such as *R. pipiens berlandieri*, breed in autumn, and some other temperate zone amphibians, such as *R. italica* and *Triturus*

italicus, even breed in late winter. It is not uncommon to find distinct patterns of the female reproductive cycle related to the geographical location of the population. While most temperate zone populations of some ranids are known to manifest one ovarian cycle per year, populations inhabiting warmer zones may show no seasonality and females can breed whenever there is sufficient rainfall. Many species inhabiting tropical, unpredictable, and aseasonal environments manifest opportunistic breeding at any time of the year. Thus, amphibians comprise temperate seasonal breeders and explosive seasonal or aseasonal breeders. Species with explosive breeding pattern are characterized by a fairly short (a few days) period of oviposition, and all gravid females of a population apparently oviposit synchronously. An asynchronous pattern of egg laying is typical of species with a prolonged breeding season. In many such species inhabiting temperate climate, each female can lay more than one clutch of eggs per breeding season, which may extend to a couple of months or more. In contrast, species inhabiting aseasonal climates may breed several times a year, usually every 2 or 3 months, and can easily be induced to breed in the laboratory, e.g., *X. laevis*.

Seasonality of reproduction in females is accounted for by an interaction of synchronizing environmental factors with an "endogenous" physiological rhythm of the ovary. The latter is reflected not only in seasonal modifications of ovarian follicular composition but also in sex steroid secretion as well as oviductal changes. The breeding cycle can be subdivided into particular phases indicating pre- and postovulatory events. In fact, the general trend is to classify the breeding cycle into periods of prespawning/preovulatory/prebreeding, spawning/ovulatory/breeding, postspawning/quiescence, and recovery/preparatory. Immediately after breeding, the female can be refractory to hormonal and/or environmental stimulation and this may represent the period of ovarian quiescence. The preparatory period, however, is the time when active vitellogenesis takes place. A transitory elevation of progesterone during periovulatory period, as seen in many species, suggests a possible preovulatory initiation role for progesterone. In fact, *in vitro* studies have indicated that progesterone is the natural ovulation-inducing hormone in amphibians. There are differences between species as to the circulating levels of estrogens and androgens in different periods of the reproductive cycle. For example, in species showing the explosive breeding pattern, plasma levels of androgens and estrogens remain at the baseline level during much of the year and rise rapidly shortly before breeding; they decline soon after egg laying is over. Species with a prolonged breeding season exhibit a gradual rise in circulating levels of sex steroids, androgens, and estrogens throughout autumn and winter with remarkable individual variation, and the annual peak is reached either in late winter or slightly before breeding.

V. REGULATORY MECHANISMS

Ovarian activity related to the breeding cycle is, as a rule, regulated by the hormones of the brain–pituitary axis through the secretion of pituitary gonadotropins, FSH and LH. Since breeding should occur during a period most appropriate for survival of eggs, embryos, and the young, there exist mechanisms through which brain–pituitary axis responds to specific environmental cues, of which temperature, rainfall, and food supply are the most well-known. Pituitary gonadotropin secretion is regulated by gonadotropin-releasing hormone (GnRH) produced in the brain. Pituitary hormones regulate ovarian activity: oogenesis, vitellogenesis, and steroidogenesis. Ovarian steroids, in turn, through their selective uptake in certain brain areas act on the brain tissue to modulate the timing and the magnitude of GnRH release. In some species, *in vitro* studies have shown that GnRH can regulate ovarian steroidogenesis directly. However, the molecular pathway which can convincingly document a direct effect of GnRH is far from clear. Sex steroid-binding proteins present in the blood plasma and their concentration may be related to the amount of sex steroid that can reach the target tissue. The ovarian steroids markedly affect the secretory activity of the pituitary through GnRH. In fact, it has been shown that removal of ovaries and replacement therapy with estradiol and testosterone, either alone or together, markedly affect the GnRH neuronal system in the brain. The cyclic ovarian steroidal secretion is manifest externally in behavioral profile. GnRH concentration in the brain tissue has been found to change in relation to the reproductive status of the female. The brain concentration is, in fact, highest in prebreeding females, e.g., *R. esculenta*, and declines drastically soon after breeding. Amphibian ovary has also been shown to synthesize prostaglandins and proopiomelanocortin-derived peptide, β-endorphin, which are considered to play a role in the ovarian activity. Like other ovarian parameters, these compounds also manifest reproductive cycle-related changes in the ovary. Thus, a sequential interaction of a multitude of external and internal factors is necessary for the successful completion of a breeding cycle culminating in the deposition of eggs and birth of the young.

See Also the Following Articles

Amphibian Ovarian Cycles; Amphibian Reproduction, Overview; Male Reproductive System, Amphibians

Bibliography

Bagnara, J. T., and Rastogi, R. K. (1992). Reproduction in the Mexican leaf frog, *Pachymedusa dacnicolor*. In *Reproductive Biology of South American Vertebrates* (W. C. Hamlett, Ed.), pp. 98–111. Springer-Verlag, New York.

Boujard, D., Charbonneau, and Joly, J. (1990). Hormonal control of oocyte maturation: Role of pituitary and follicle cells. In *Progress in Zoology: Biology and Physiology of Amphibians* (W. Hanke, Ed.), Vol. 38, pp. 85–102. Fischer-Verlag, Stuttgart.

Exbrayat, J.-M. (1993). Quelques aspects de la biologie de la reproduction chez *Typhlonectes compressicaudus* (Dumeril & Bibron, 1841), amphibien gymnophione. *Cahiers Univ. Cathol. Lyon Ser. Sci.* 7.

Kwon, H. B., and Ahn, R. S. (1994). Relative roles of theca and granulosa cells in ovarian follicular steroidogenesis in the amphibian, *Rana nigromaculata*. *Gen. Comp. Endocrinol.* **94**, 207–214.

Lofts, B. (1984). Reproduction. In *Marshall's Physiology of Reproduction: Reproductive Cycles of Vertebrates* (G. E. Lamming, Ed.), Vol. 1, pp. 127–205. Churchill Livingstone, Edinburgh, UK.

Rastogi, R. K., and Iela, L. (1994). Gonadotropin-releasing hormone: Present concepts, future directions. *Zool. Sci.* **11**, 363–373.

Saidapur, S. K. (1989). Reproductive cycles of amphibians. In *Reproductive Cycles of Indian Vertebrates* (S. K. Saidapur, Ed.), pp. 166–224. Allied, New Delhi.

Van Gansen, P. (1986). Ovogenèse des amphibiens. In *Traité de zoologie, tome XIV, fasc. 1B: Batraciens Ed.* Masson, Paris.

Wake, M. H. (1992). Reproduction in caecilians. In *Reproductive Biology of South American Vertebrates* (W. C. Hamlett, Ed.), pp. 112–122. Springer-Verlag, New York.

Wallace, R. A., and Selman, K. (1990). Ultrastructural aspects of oogenesis and oocyte growth in fish and amphibians. *J. Electron Microsc. Technol.* **16**, 175–201.

Xavier, F. (1987). Functional morphology and regulation of the corpus luteum. In *Hormones and Reproduction in Fishes, Amphibians and Reptiles* (D. O. Norris and R. E. Jones, Eds.), pp. 241–282. Plenum, New York.

Female Reproductive System, Birds

Patricia A. Johnson

Cornell University

I. Neuroendocrine System
II. Ovary
III. Oviduct

GLOSSARY

clutch A number of eggs laid sequentially and incubated as a group.

hierarchy The arrangement of follicles by size on the avian ovary, with the largest follicle destined to ovulate next, the second largest next, and so on.

oviduct The avian reproductive structure which receives the ovulated oocyte and in its specialized regions is responsible for the formation of the avian egg.

vitellogenesis The process of estrogen-stimulated yolk formation in the liver and receptor-mediated uptake of the yolk proteins by the oocyte.

One of the obvious differences between birds and mammals is that there is no estrous cycle or pregnancy. The embryo must have everything that it needs when the egg is laid. This fact dictates that the avian reproductive system be adapted to produce the completed egg. Because of the commercial importance of chickens, much of the basic investigation of the avian reproductive system has been performed using these domesticated birds. Most of the fundamental processes related to egg production in wild and domesticated species are similar, however, with exceptions noted for seasonal effects and level of productivity.

I. NEUROENDOCRINE SYSTEM

When the photoperiodic stimulus is adequate, the hypothalamus produces gonadotropin-releasing hormone (GnRH), which is secreted into the portal vessels and which stimulates the pituitary gland to release follicle-stimulating hormone (FSH) as well as luteinizing hormone (LH). In temperate-zoned birds in the wild, lengthening of the day provides the photoperiodic stimulus, whereas in domestic birds reproductive activity is stimulated by a long light phase. Among wild birds, behavioral signals, such as availability of a mate and nest materials, contribute to neuroendocrine stimulation. Interestingly, light is necessary as a stimulus to the reproductive system and also serves to set the phase during which preovulatory events occur within the ovulatory cycle. That is, among domesticated birds, when the photoperiodic stimulus is adequate, ovulation and oviposition

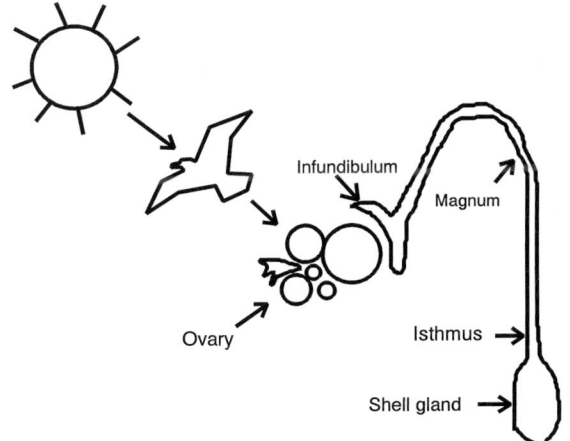

FIGURE 1 The influence of photoperiod on the avian female reproductive system. Light is perceived by the avian hypothalamus and pituitary gonadotropins result in the stimulation of the ovary. Ovarian hormones stimulate the growth and development of the oviduct, which consists of infundibulum, magnum, isthmus, and shell gland.

occur within a defined part of the light:dark cycle (Fig. 1).

Avian GnRH differs modestly from mammalian GnRH and exists in two forms in birds (I and II). GnRH-I is the form that is believed to exert physiological control over gonadotropic secretion. As is true in mammals, the pituitary gonadotropins, FSH and LH, are released in a pulsatile manner. The role of FSH in female birds is not well-defined, although it is believed to be involved in the process of follicular selection. There is an increased number of receptors for FSH in small follicles compared to large follicles, suggesting that these follicles may be more responsive at this stage of growth. LH is the hormone which most effectively stimulates steroid production from large follicles and which causes ovulation. The preovulatory surge of LH in birds is stimulated by increasing levels of progesterone originating from the largest follicle. Ovarian hormone negative feedback through the steroid hormones estrogen and progesterone, as well as the protein hormone inhibin, regulates the secretion of LH and FSH (Fig. 2).

II. OVARY

In most birds only the left ovary develops; the right ovary is formed embryologically but regresses under the influence of estrogen secretion from the left ovary. The development of the right gonad can be induced, however. If the left ovary is removed from a young hen, the right gonad develops testicular features. In contrast, following removal of the left ovary from a mature hen, a right ovotestis is formed. Consistent with gonadal lateralization, only the left oviduct/shell gland develops. There are some exceptions to this pattern of development; for example, some hawks and falcons have two well-developed ovaries and oviducts.

During follicular development, the avian ovary is a very dynamic and well-vascularized organ. The growing follicles are arranged in a hierarchy according to size and time to ovulation. This arrangement permits identification of the order in which follicles will ovulate. The number of large follicles present on the ovary is characteristic of the species and related to the number of eggs in a clutch. It is

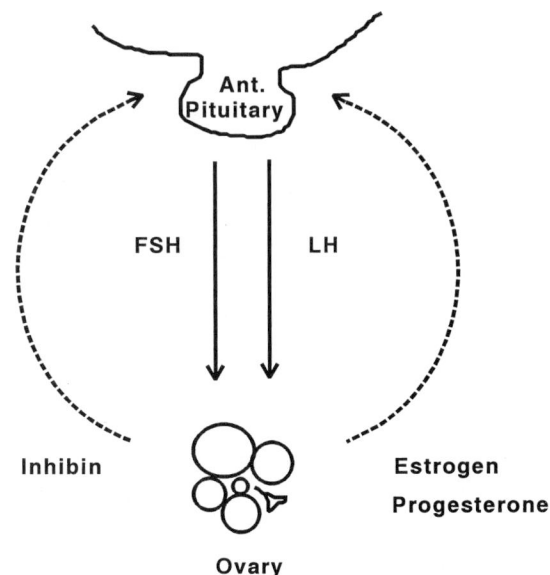

FIGURE 2 The feedback relationship between the pituitary gland and the ovary. Pituitary gonadotropins (FSH and LH) stimulate the ovary and are, in turn, regulated by ovarian hormones (inhibin, estrogen, and progesterone). Estrogen and progesterone also likely have a negative feedback effect at the hypothalamus.

not known what regulates this pattern of development but the follicles in the hierarchy have been characterized in terms of steroid secretion and other factors associated with development. Interestingly, the large avian follicles secrete an increasing amount of progesterone with development and, conversely, a decreasing amount of estrogen.

The cell types of the avian follicle are similar to those of the mammal. The oocyte is contained within the vitelline membrane, and surrounding this membrane is the avascular granulosa layer, which is the source of progesterone. A basement membrane separates the granulosa and theca cell layers. The theca/interstitial component has been classified as theca interna and theca externa; the theca interna produces androgens and the theca externa estrogens.

A hallmark of developing avian follicles is the accumulation of yolk. Vitellogenin synthesis occurs in the liver under the regulation of estrogen. In fact, the liver of a male bird can be induced to synthesize yolk components by estrogen treatment. In the process of vitellogenesis, yolk precursors are released into the blood from the liver and taken up by the growing follicles by a specific receptor-mediated process. Avian follicles acquire the majority of their yellow yolk in the final days prior to ovulation. As mentioned previously, the follicles are very well vascularized and this may facilitate the large transfer of yolk components prior to ovulation. As yolk is accumulated, it is deposited in concentric layers. Examination of a mature oocyte prior to ovulation reveals that the bulk of the volume is composed of yolk. The nuclear material and oocyte organelles, which comprise the germinal disk region, occupy a small opaque area on the surface of the oocyte. At ovulation, the oocyte is released through the stigma, a longitudinal slit in the follicular wall. After release of the oocyte, the follicle remnants gradually regress in the subsequent days. The postovulatory follicle does not reorganize and produce progesterone as in the corpus luteum of mammals. It does produce prostaglandins, however, and is involved in the timely oviposition of the ovulated oocyte.

Birds lay their eggs in a pattern called a clutch or sequence. A clutch generally refers to the number of eggs laid sequentially and then incubated as a group. In domesticated species, in which broodiness does not occur or is not permitted, the sequence refers to the number of eggs laid on successive days followed by a pause day. After the pause day, laying continues provided photoperiod is adequate. With time, there is some diminution of sequence length, most likely due to age or relative photorefractoriness. The typical number of eggs laid within a clutch or sequence varies with respect to species as well as age and resources within a species.

III. OVIDUCT

The avian oviduct is basically a tube or conduit with highly specialized regions that extends from the ovary to the cloaca. Each region of the oviduct is specialized for a particular function and the oocyte spends a unique amount of time in each section. The times indicated in this section are provided for comparison of relative times. They apply for the formation of an egg by a chicken laying at approximately 26-hr intervals. The infundibulum is the most anterior portion of the oviduct and the area that engulfs the oocyte after ovulation. Some sperm storage glands are located here and fertilization occurs in the infundibulum. The egg typically spends 15–30 min here. Regardless of whether fertilization has occurred, the egg progresses along the subsequent regions of the tract. The magnum is the largest portion of the oviduct and it is here that the egg white proteins are deposited. The principal egg white protein is ovalbumin and this accounts for more than one-half of the egg white protein in the egg. The synthesis of this protein is initiated by estrogen but secretion is somewhat more complicated. It is deposited on the egg as it moves along the magnum but it is also deposited on objects or irritants placed in the magnum. This indicates that some mechanical stimulus is involved in secretion. The egg spends about 2 or 3 hr traversing the magnum. The isthmus is a somewhat narrowed portion of the oviduct and the egg spends about $1\frac{1}{4}$ hr here. Inner and outer fibrous shell membranes are deposited on the layer of albumen on the gradually forming egg. From here, the egg moves to the shell gland, where it remains for approximately 20 hr and where calcification of the egg occurs. The plumping of the egg prior to calcification also occurs here. Plumping is the addition of water and salts to the egg white proteins. During calcification, calcium carbonate is deposited on the egg in a process that demands a high level of calcium mobilization by the female. Pigmentation of the shell also occurs in the shell gland. The uterovaginal junction is posterior to the shell gland. This is the main site of sperm storage glands. Sperm stored in these glands are likely released continuously and make their way to the infundibulum, the site of fertilization. Sperm deposited in the tract of a female are fertile for up to 7–14 days in chickens and 40–50 days in turkeys. The vagina in birds has no part in egg formation but is the portion of the oviduct leading to the cloaca.

See Also the Following Articles

FEMALE REPRODUCTIVE SYSTEM, AMPHIBIANS; AVIAN REPRODUCTION, OVERVIEW; AVIAN REPRODUCTIVE SYSTEM, DEVELOPMENTAL ENDOCRINOLOGY; FEMALE REPRODUCTIVE SYSTEM, NONHUMAN MAMMALS; MALE REPRODUCTIVE SYSTEM, BIRDS; FEMALE REPRODUCTIVE SYSTEM, REPTILES

Bibliography

Bakst, M. R. (1987). Anatomical basis of sperm-storage in the avian oviduct. *Scanning Microsc.* **1**, 1257–1266.

Bakst, M. R. (1993). Oviductal sperm storage in poultry: A review. *Reprod. Fertil. Dev.* **5**, 595–599.

Gilbert, H. D., Perry, M. M., and Gilbert, A. B. (1984). Yolk formation. In *Physiology and Biochemistry of the Domestic Fowl*(B. M. Freeman, Ed.). Academic Press, London.

Johnson, A. L. (1996). The avian ovarian hierarchy: A balance between follicle differentiation and atresia. *Poultry Avian Biol. Rev.* **7**(2/3), 99–110.

Morris, T. R. (1973). The effects of ahemeral light and dark cycles on egg production in the fowl. *Poultry Sci.* **52**(2), 423–445.

Muramatsu, T., and Sanders, M. M. (1995). Regulation of ovalbumin gene expression. *Poultry Avian Biol. Rev.* **6**(2), 107–123.

Murton, R. K., and Westwood, N. J. (1977). Avian breeding cycles. Oxford Univ. Press, Oxford, UK.

Female Reproductive System, Fish

Martin A. Connaughton
Washington College

Katsumi Aida
The University of Tokyo

GLOSSARY

final oocyte maturation The final stage of oocyte development prior to ovulation, characterized by germinal vesicle breakdown, the completion of the first meiotic division, and hydration, resulting in a mature oocyte capable of being fertilized. This process is triggered by a maturational surge of gonadotropins.

gonadotropin (GTH) Two different peptide hormones which stimulate gonadal development, GTH-I (follicle-stimulating hormone-like GTH) and GTH-II (luteinizing hormone-like GTH).

maturation-inducing hormone A C21 steroid produced in the follicular tissue which mediates final oocyte maturation of teleost oocytes.

maturation-promoting factor The final mediator of final oocyte maturation, a nonsteroidal substance produced within the oocyte in response to maturation-inducing hormone stimulation of the oocyte membrane.

ovulation Oocyte expulsion from the follicle and into the ovarian cavity or body cavity following final oocyte maturation.

Teleostei A classification of fishes that excludes jawless fishes, cartilaginous fishes, lobe-finned fishes, and sturgeons.

vitellogenesis Ovarian steroid-induced production and release of vitellogenin by the liver and accumulation of this material in the developing oocytes as yolk.

vitellogenin A glycolipoprotein produced in the liver and accumulated via micropinocytosis in the oocyte as yolk.

The female reproductive system of teleosts produces and supports the oocytes necessary for successful reproduction and in some cases supports the developing embryo for a period of time after fertilization. This system includes the hypothalamus, pituitary, ovary, and liver, which is responsible for the production of vitellogenin, a class of egg yolk proteins. The great diversity in Teleostei is reflected in a wide diversity of reproductive strategies, morphology, and physiology. In all species, however, oogonia proliferate in the ovary, become oocytes, undergo vitellogenesis, maturation, and ovulation, followed by either internal or external fertilization and internal or external development of the embryo. Endocrine controls of the system begin with hypothalamic responses to environmental cues and the consequent release of gonadotropin-releasing hormone to the pituitary, which responds with the release of gonadotropins (GTH) into the blood. Ovarian response to GTH release is the production of sex steroids and the initiation or continuation of oogenesis.

I. INTRODUCTION

Fishes are the largest and most diverse group of vertebrates. The variety of body shapes and life histories in this group has resulted in tremendous diversity in reproductive anatomy, physiology, and behavior. Though most fishes are oviparous, expressing external fertilization and development, a range of viviparous forms also exist, including a stunning array of adaptations for internal fertilization and development. Even among strictly oviparous forms, reproductive strategies can vary widely. Some species are oligotrophic, producing a large number of eggs

and sacrificing yolk content, whereas others are lecithotrophic, producing far fewer eggs but supplying them with a surplus of yolk and often substantial effort in the form of parental care. In addition, while most fish are iteroparous, spawning multiple times during their life history on some daily or annual schedule, semelparous species also exist, spawning only once and dying.

Sexual differentiation is also greatly variable among fishes. While most fish species are dioecious and gonochoristic (expressing two sexes, which are fixed at maturation), a number are hermaphroditic, either as sequential hermaphrodites (protandry or protogyny) or simultaneous hermaphrodites. In addition, some fish species express multiple forms of one sex, i.e., nesting and nonnesting male midshipman, whereas other species are parthenogenic, expressing only females. Even among gonochoristic species, the age at first reproduction can vary from a matter of weeks to as late as 15 years of age. One sex may reach maturity years before the other and one or both sexes may be neotenous, maturing precociously.

Even given the great variety of reproductive strategies observed among teleosts, there are common systems and processes in the female reproductive system, and it is on these central aspects that this article shall concentrate.

II. ENVIRONMENTAL PHYSIOLOGY

A. Rhythmicity and Environmental Cues

Seasonal rhythmicity of spawning results from an endogenous rhythm of gonad development entrained by environmental cues. Annual spawning is the result of evolutionary selection of a season when biotic and abiotic factors are optimal for the development and survival of the vulnerable early life history stages of fishes. These cycles are stronger in higher latitudes, which experience a greater annual range in abiotic factors, particularly photoperiod and temperature. Photoperiodic cues initiate and sustain gonadal recrudescence in many species of fish, whereas temperature is believed to be "permissive" in many species, allowing initial recrudescence to continue toward full maturation. Changes in both photoperiod and temperature cues have been observed to affect daily cycles of gonadotropin (GTH) release.

The timing of spawning is species specific and can fall into a number of general patterns, the most common of which are autumn–winter spawners (salmonids), spring–summer spawners (many coastal marine species), and summer spawners (sticklebacks). In lower latitudes, annual reproductive cycles are absent in some species (continuous spawners) but are very regular in other tropical species. Due to the constant climate in low latitudes spawning cycles are believed to be entrained by changes in food availability and water chemistry linked to seasonal changes in rainfall.

In both seasonal and continuously spawning species, daily, semilunar, or lunar cycles are often expressed once spawning has been initiated. Evening spawning is very common because it may decrease predation pressure on participants as well as fertilized eggs. Semilunar or lunar cycles link the reproductive activities of some species to the tides. This places the fertilized eggs out of reach of aquatic predators and/or avoids anoxic conditions.

B. Transduction of Environmental Cues

Environmental cues initiate oogenesis and ovarian growth and may initiate migration, nest building, and other reproductive behaviors. These changes are transduced primarily through increased output of gonadotropin-releasing hormone (GnRH) from the hypothalamus. In addition, dopamine, which acts as a gonadotropin-inhibiting factor in teleosts, and serotonin, argenine vasopressin, norepinephrin, and other substances are believed to play a role.

Environmental cues impinging upon visual, olfactory, temperature, or other sensory mechanisms are passed on to the hypothalamic–pituitary–ovarian axis (Fig. 1). Both the retina and the pineal gland are believed to play a role in the transduction of light cues because blinding and/or pinealectomy can cause gonadal regression in many species. The terminal

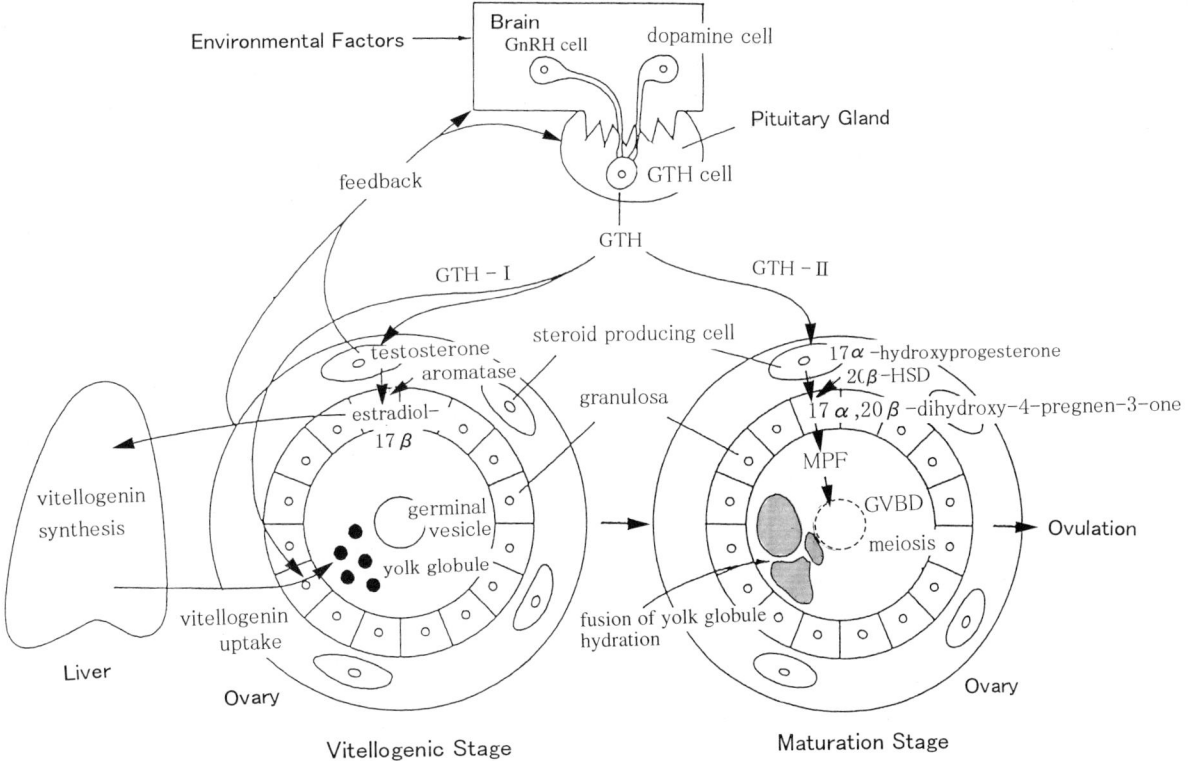

FIGURE 1 A schematic of the components of the female reproductive system in teleosts, including the organs, tissues, and endocrine pathways involved in the production of mature oocytes.

nerve in fishes is also believed to play a role in the transduction of seasonal cues because it is linked to the hypothalamus, pituitary, and olfactory systems and possibly to the retina and the pineal gland. The olfactory nerves in some species express GnRH and may affect the pituitary directly, without impinging upon the hypothalamus.

C. Proximal Cues and Pheromones

Proximal cues, such as substrate type, courtship behaviors, and pheromones, are often required for oocytes to reach full maturation, for ovulation, and for oviposition. The presence of the proper substrate for spawning or nest building has been observed to initiate the maturational GTH surge, which triggers final oocyte maturation and ovulation. Sound production by male fishes can attract females from a distance in several species and has been observed to induce oviposition of eggs in others.

Pheromones may act as "primers," producing long-term effects measured in days or hours, or as "releasers," resulting in nearly immediate changes in physiology or behavior. Pheromones produced by females, one class of which are steroid glucuronides released from the ovarian cavity during ovulation, may attract males from a distance and can increase sexual activity (courtship and nest building) in the male. The effects of male pheromones on females include attraction of the female to the producing male (releasing effect) and induction of ovulation and oviposition after brief exposure (priming effect).

III. CENTRAL ENDOCRINE CONTROL SYSTEMS

A. GnRH

The hypothalamus of teleosts directly innervates all portions of the pituitary rather than communicating with the anterior pituitary via a portal system. GnRH neurons are found in the forebrain, the ventral and preoptic telencephalon, and occasionally in the diencephalon, the terminal nerve, and the olfactory epithelium of some species. The location of GnRH neurons in the forebrain supports the hypothesis that

a migratory pathway from the olfactory plate to the forebrain may exist for teleost GnRH neurons, similar to that observed in mammals.

Nine distinct forms of the decapeptide GnRH molecule have been sequenced in vertebrates, all sharing at least 50% sequence similarity. Multiple GnRH molecules may be expressed simultaneously in teleosts. All fishes examined thus far express chicken-II GnRH and may also express one or more of the following forms: salmon GnRH, catfish GnRH, mammalian GnRH, or seabream GnRH. There is some evidence from localization studies that the multiple forms of GnRH may differ in function within a species. GnRH mRNA levels can be influenced by many factors, including the state of sexual maturation, photoperiodic conditions, and sex steroid levels, indicating that GnRH production is controlled by both internal and external factors. In teleosts, GnRH works in conjunction with dopamine (a gonadotropin-inhibiting factor) to control the secretion of GTH from the pituitary.

B. GTH

Two forms of GTH exist in fishes, GTH-I and GTH-II, which express follicle-stimulating hormone (FSH)- and luteinizing hormone (LH)-like effects, respectively. These two molecules are glycoprotein heterodimers consisting of two different noncovalently bonded peptides, the α and β subunits. The α subunit is common to both GTH-I and GTH-II, whereas each GTH has a distinct β subunit. The two GTH molecules are produced by two different types of gonadotropes in the proximal pars distalis. The presence and activity of these two cell types varies throughout the reproductive cycle, as does the activity of the two hormones. GTH-I is secreted for an extended period during the early phases of ovarian development, including vitellogenesis (Fig. 2). In response, the ovary produces estradiol-17β which stimulate the synthesis and release of vitellogenin by the liver. In circulation, ovarian steroids exert a negative feedback effect on the release of GTH-I indirectly via the hypothalamus and directly via the pituitary. GTH-I is also responsible for eliciting the production and storage of GTH-II in the pituitary. After the completion of vitellogenesis, a matura-

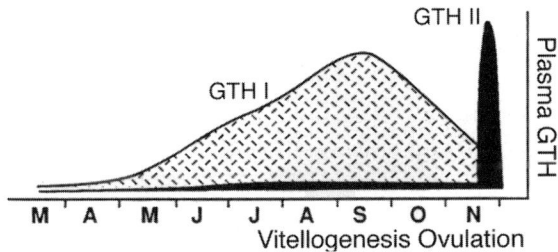

FIGURE 2 Diagrammatic representation of changes in plasma levels of GTH-I and GTH-II in coho salmon during reproductive maturation, reflecting the long-term presence of GTH-I in the plasma during vitellogenesis and the maturational surge of GTH-II just prior to ovulation (reproduced with permission from Dickoff and Swanson, 1990).

tional, preovulatory surge of GTH-II is triggered, initiating final oocyte maturation and ovulation (Fig. 2).

IV. THE OVARY

A. Development, Anatomy, and Histology

1. Origin/Differentiation

The ontogeny of the ovaries of teleosts is initially similar to that of other vertebrates. The bilateral ovaries arise from the germinal ridges located on either side of the dorsal mesentery and comprising the dorsolateral lining of the peritoneal cavity. Evidence suggests that primordial germ cell development is widespread, followed by migration of these cells into the germinal epithelium of the developing ovary. However, unlike other vertebrates, teleostean ovaries are derived from cortical tissue only, not from dual corticomedullary tissue. This lack of medullary tissue has been hypothesized as the reason for the frequency of hermaphroditic forms in fishes.

The timing of ovarian development is variable in fishes: It may be delayed many years or may occur very slowly, resulting in late sexual maturation. Differences in the age at maturation of male and females may also exist, and some fishes develop gonadal tissue precociously. In addition, though the ovaries always originate from bilateral primoridia, many species of fish have only one ovary at maturation. In these cases, the primordia may fuse or partially fuse during development, or one of the ovaries may fail to develop fully.

In simultaneously hermaphroditic or naturally sex-changing (sequentially hermaphroditic) species, both ovarian and testicular tissue exists simultaneously in the gonads (termed ovotestes). The appearance of testicular and ovarian tissue is species specific and depends on the form of hermaphrodism. In simultaneous hermaphrodism, as seen in some labrids, both tissues may exist and be fully functional concurrently. In cases of sequential hermaphrodism, both tissue types may be present with only one producing gametes, or one of the two tissues may be absent or regressed, depending on the species. For example, in a protandrous sparid, both testicular and ovarian tissues are present during the earlier, male, portion of the life cycle, but the oocytes in the ovarian tissue are arrested at the perinucleolus stage of oogenesis. However, after the change of sex to female, the male tissue regresses almost entirely, leaving a normal appearing ovary.

2. Anatomy/Histology

The ovary consists of (i) germinal epithelium, (ii) follicular tissue surrounding the developing oocytes, and (iii) associated connective and interstitial tissues, all of which undergo complex folding to form ovigerous folds (Fig. 3C). The outer surface of the ovary may be naked or partially or completely encased in the ovarian capsule, which develops from peritoneal folds. Teleost ovarian structure is classified into two forms: gymnovarian and cystovarian (Figs. 3A and 3B). The gymnovarian condition is possessed by all vertebrates, with the exception of some specialized teleosts (eels, salmonids, and smelts). Gymnovarian ovaries are characterized by ovigerous folds which lie exposed or partially exposed to the peritoneal cavity. Oocytes are ovulated into the coelom and moved to the oviduct and the cloaca by ciliated epithelium lining the wall of the peritoneal cavity. These ovaries lack an ovarian cavity and are either naked or partially encapsulate by an ovarian capsule.

Cystovarian ovaries are completely encased by the ovarian capsule, which extends posteriorly from the ovary and forms the oviducts in conjunction with the developing Müllerian ducts. The ovary bears a

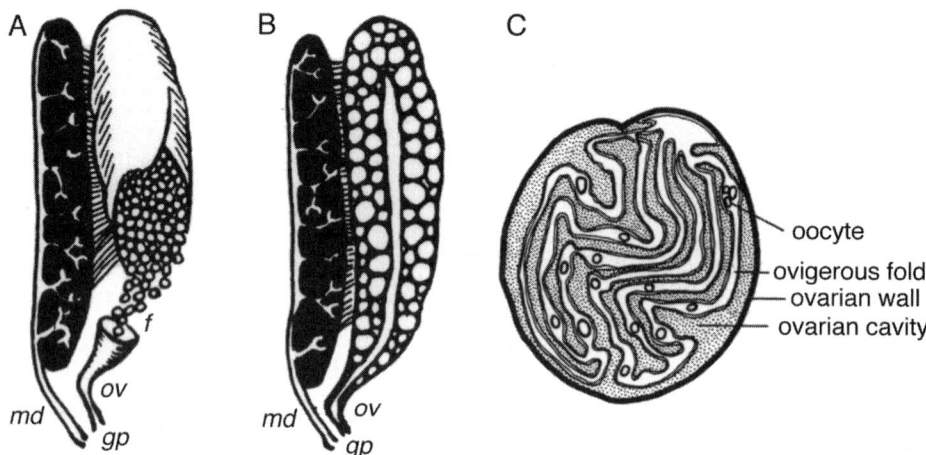

FIGURE 3 Types of ovaries in teleosts: (A) gymnovarian condition and (B) cystovarian condition. Black organs represent the mesonephric kidneys. f, open funnel of oviduct; gp, genital pore; md, mesonephric duct (adapted from Hoar, 1957). (C) A section though a cystovarian ovary showing the ovigerous folds and location of oocytes (adapted from Turner, 1983).

true ovarian cavity, into which the ovigerous folds project and may fill entirely. The oocytes are ovulated into the intraovarian space and moved down the oviduct to the cloaca without entering the peritoneal cavity. Two types of cystovarian ovaries can be differentiated by the mechanism of transport of the oocytes out of the ovarian cavity. In one type, ciliated epithelium lining the ovarian cavity move the oocyte out of the ovary, whereas in the other smooth muscles lining the ovarian capsule expel the ovulated oocytes by pulsatile contraction of the ovary.

3. The Follicle

The ovarian follicle consists of a developing oocyte and two surrounding cell layers separated by a basement membrane. The innermost of these consists of steroidogenic granulosa cells and the outer, thecal layer, consists of fibroblasts, collagen fibers, capillaries, and steroidogenic special theca cells (Fig. 4). Prior to and during vitellogenesis, a noncellular zona radiata will form between the granulosa cells and the oocyte, resulting from the growth of microvilli from both surfaces.

Atretic oocytes, or preovulatory corpora atretica, are common in the teleost ovary and result from atresia of the oocyte followed by hypertrophy of the granulosa cells. This process may occur at any stage of oocyte development or maturation. After the com-

plete reabsorption of the atretic oocyte by the granulosa cells, they collapse to form an irregular cell mass. Postovulatory follicular tissues, corpora lutea, are characterized by a highly vascularized thecal layer and hypertrophied granulosa cells and are believed

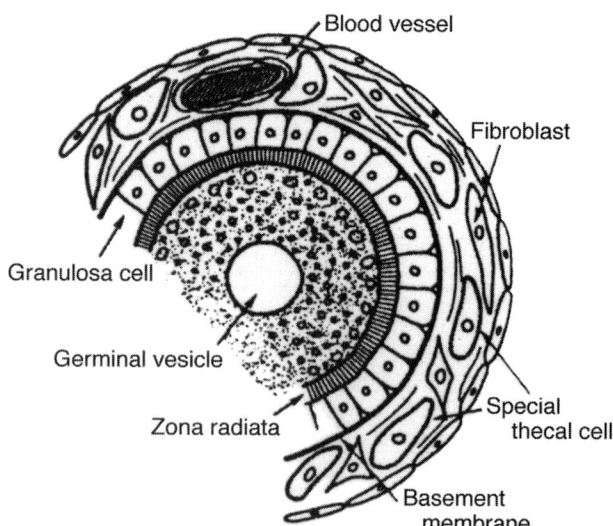

FIGURE 4 Diagrammatic representation of the follicle and oocyte during early vitellogenesis. The granulosa layer is separated from the thecal layer by a distinct basement membrane. The thecal layer is composed of fibroblasts, blood vessels, and steroidogenic special thecal cells (reproduced with permission from Nagahama, 1983).

to be the major site of postovulatory progesterone production in fishes.

B. Oocyte Recruitment

Three modes of oocyte development based on differences in the recruitment of new oocytes exits in teleosts: synchronous, group synchronous, and asynchronous. In fish showing synchronous oocyte development, all oocytes develop synchronously and ovulate at the same time. This is characteristic of semelparous species which spawn once and then die, including species of anadromous salmon and catadromous eels. In group-synchronous oocyte development, vitellogenic oocytes are divided into several synchronously developing groups, which are ovulated over the course of one breeding season; however, oocytes are not supplied from the nonvitellogenic pool during the spawning season. Therefore, in group-synchronous spawners, the total number of oocytes to be spawned in a season is predetermined by the number of vitellogenic oocytes. Group-synchronous spawners are multiple spawners but tend to have relatively short spawning seasons. In asynchronous oocyte development, oocytes at all developmental stages are present in the ovary and a clear grouping of oocytes into cohorts is difficult. Oocytes are continuously recruited from the nonvitellogenic pool in asynchronous ovaries. These species are multiple spawners exhibiting a long spawning season or continuous spawning.

V. OOGENESIS

A. Proliferation: Primary Growth

Oogonia proliferate by mitosis within the germinal epithelium of juveniles and adult females in most species. In general, proliferation of the oogonia occurs at a specific time of year in seasonal breeders, though in continuous or nonseasonal spawners proliferation may occur continuously or may be repeated within a short cycle. Oogenesis can be divide into a number of stages based on the size of the cell, appearance of the nucleus and nucleolus, and the type and location of yolk inclusions. These stages can be loosely grouped into primary growth and secondary growth phases. Primary growth refers to differentiation of the oogonia into oocytes and growth of the oocyte prior to yolk inclusion. Secondary grow refers to previtellogenic and vitellogenic inclusion of yolk into the oocyte and is dependent on the presence of GTH-I. In the following description of oogenesis and vitellogenesis, the sizes of goldfish oocytes will be used as an example of increasing size of the developing oocyte.

During the primary growth phase, oogonia become primary oocytes when meiosis is arrested at the diplotene stage of the first prophase. In the chromatin–nucleolus stage, the oocytes are smaller than 20 μm in diameter and are characterized by a large nucleolus associated with chromatin threads. During the early perinucleolus stage (Fig. 5) the oocyte and nucleus increase in size and multiple nucleoli form around the periphery of the nucleus. Further enlargement of the oocyte, up to 110–160 μm in the goldfish, characterizes the late perinucleolus stage. Also, irregular spaces develop between the granulosa cells and the oocyte, where microvilli develop from both surfaces.

B. Vitellogenesis: Secondary Growth

1. Cellular Processes

Secondary growth begins with the previtellogenic or yolk vesicle stage (Fig. 5). At this time, yolk vesicles appear in the ooplasm where they gradually increase their number and volume. These vesicles are endogenously produced at the Golgi apparatus and later migrate to the periphery where they become cortical alveoli. In most species, oil droplets also appear in the ooplasm during this phase. Microvilli continue to develop from the surface of the oocyte and the granulosa cells, and electron-dense substances begin to accumulate between both cells in the previtelline space.

The primary yolk stage, which follows the yolk vesicle stage, represents the initiation of the vitellogenic phase, during which vitellogenin produced in the liver is included in the oocyte. At this time, small yolk globules begin to appear among the yolk

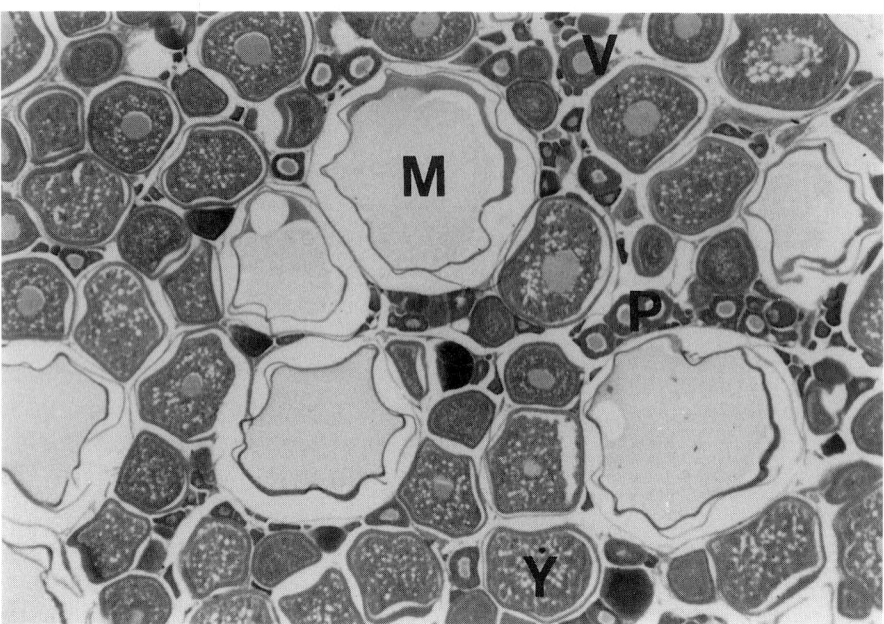

FIGURE 5 Light micrograph of mature ovary of the red seabream (magnification, ×40) containing oocytes at the perinucleolus stage (P), yolk vesicle stage (V), yolk stage (Y), and maturation stage (M).

vesicles, which continue to increase in number and volume. This process is initiated when the oocyte diameter reaches about 350 μm. Thereafter, oocytes grow rapidly with the accumulation and growth of yolk globules, which eventually occupy most of the cytoplasm (Fig. 5). Accumulation of yolk globules continues until the oocyte reaches a diameter of approximately 900 μm. The nucleus remains located in the center of the oocyte throughout vitellogenesis and, in general, the yolk globules maintain a granular structure. In many species, these yolk globules may fuse to form one large body of yolk and this fusion may lead to the transparency of eggs in some species.

Proteins, lipids, and polysaccharides are all detectable within yolk globules, which are surrounded by a limited membrane and show uniform internal structure. The main component of the yolk globules originates from vitellogenin, a glycolipoprotein whose molecular weight ranges from 326 to 600 kDa as examined by gel filtration and from 140 to 220 kDa based on SDS–PAGE. Tracer experiments using isotope-labeled vitellogenin have revealed that vitellogenin is incorporated into the oocyte by micropino-

cytosis after passing through the capillary, basal lamina, intercellular space of the granulosa cells, and the canaliculi of the developing chorion and binding to the vitellogenin receptor on the oocyte membrane.

During vitellogenesis, a vitelline envelope, consisting of two layers of glycoproteins, is formed between the granulosa layer and the oocyte. After fertilization, this envelope becomes the chorion, which is important in protecting the developing embryo physiologically and physically from the outer environment. The micropyle, through which sperm can make contact with oocyte surface during fertilization, forms at the animal pole of the vitelline envelope.

2. Hormonal Control

Vitellogenesis is initiated by estradiol-17β produced via a two-cell mechanism in the follicle. GTH-I initiates the production of testosterone from cholesterol in the steroidogenic special theca cell. Testosterone that does not go into circulation is transported to the granulosa cells where aromatization takes place, resulting in the production of estradiol-17β. It does not appear that the aromatization process in the granulosa cells requires the presence of GTH but rather only a sufficient level of androgen substrate. Estrogen-17β is subsequently released into the circulation where it binds at the liver, activating the vitellogenin gene (Fig. 1).

C. Final Oocyte Maturation

Fully vitellogenic oocytes are arrested in the first meiotic prophase and cannot be fertilized. In order to be ready for ovulation and fertilization, an oocyte must complete the first meiotic division, which involves germinal vesicle breakdown (GVBD), chromosome condensation, assembly of the first meiotic spindle, and extrusion of the first polar body. Shortly thereafter, meiosis is again arrested at the second metaphase, but at this point the oocyte can be fertilized because the final meiotic division does not take place until after fertilization. GVBD, the completion of the first meiotic division, and hydration, (the rapid uptake of water and increase in oocyte volume observed in many species) characterize final oocyte maturation. This process involves three levels of hormonal mediators, the first of which is a maturational surge of GTH-II from the pituitary, followed by steroidal substances produced by the follicular cells and nonsteroidal substances produced by the oocyte.

1. Primary Mediator: GTH-II

A rapid increase in blood levels of GTH-II is the event which triggers final oocyte maturation (Fig. 2), though the profile of this surge varies by species. The GTH-II surge is often initiated by some external cue such as photoperiod, temperature, adequate water speed, and/or the presence of the proper spawning substrate, which imply adequate environmental conditions for the survival of the larvae. The surge may also be elicited by pheromones, or visual or acoustic courtship behaviors by males, ensuring that the eggs will be fertilized if ovulated.

The oocytes of mature females have been induced to mature and ovulate by injection of GTH-II *in vivo*. Similarly, *in vitro* preparations of follicle-enclosed, postvitellogenic oocytes have been induced to undergo final oocyte maturation when incubated with GTH-II preparations. However, denuded oocytes (lacking follicular cells) could not be induced to undergo final oocyte maturation upon exposure to GTH-II, suggesting that at least one other mediator, apparently synthesized by the follicular cells, is necessary for final oocyte maturation.

2. Secondary Mediator: Maturation-Inducing Hormone

A variety of C21 steroids have been observed to be potent initiators of final oocyte maturation *in vitro*; including, 17α-hydroxyprogesterone, 17α,20β-dihydroxy-4-pregnen-3-one (17α,20β-DP), 17α,20β, 21-trihydroxy-4-pregnen-3-one (20β-S), cortisol, and deoxycorticosterone. Of these substances, however, only two (17α,20β-DP and 20β-S) have been identified as naturally occurring maturation-inducing hormones (MIHs) in fishes. MIH is synthesized in the follicular cells via a two-cell mechanism similar to that by which estradiol-17β is synthesized. For example, 17α-hydroxyprogesterone is synthesized by the special thecal cells and is transported to the granulosa cells, where it is converted to 17α,20β-DP by the enzyme 20β-hydroxysteroid dehydrogenase and released to the surface of the oocyte (Fig. 6). In this two-cell model, GTH has sites of action at both the thecal and the granulosa cells, unlike the model for estradiol-17β production.

3. Tertiary Mediator: Maturation-Promoting Factor

It has been determined that MIH acts on the surface of the oocyte because microinjection of MIH into the oocyte failed to elicit any changes in the cell. In addition, experiments with blockers have indicated that the mechanism of action of MIH is atypical for a steroid in that it is nongenomic. Receptors for MIH identified on the surface of the oocyte initiate the release of one or more cytoplasmic factors which mediate the actions of MIH inside the oocyte. These maturation-promoting factors (MPFs) appear to be very similar among all vertebrates and some invertebrates because MPF isolated from goldfish eggs was capable of producing maturation in frog and starfish ova. Frog and carp MPF involve two components, cdc2 kinase and cyclin B.

Thus, final oocyte maturation requires three levels of mediators (Fig. 7). The process is initiated by a

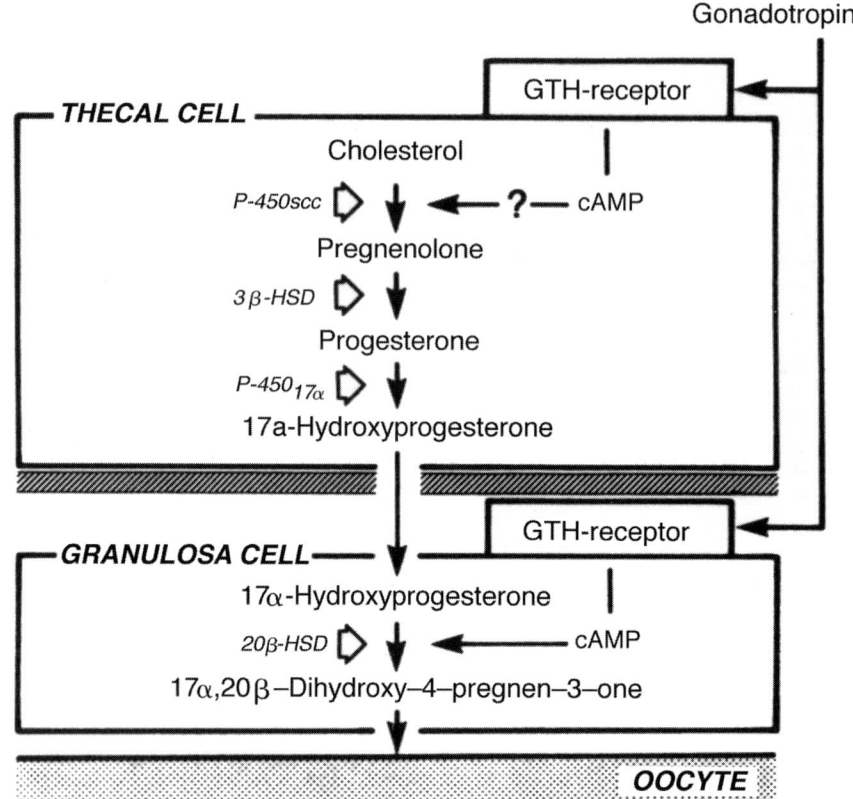

FIGURE 6 Two-cell type model for the synthesis of MIH, in this case, $17\alpha,20\beta$-dihyroxy-4-pregnen-3-one by salmonid postvitellogenic follicles. $P450_{scc}$, cholesterol side chain cleavage cytochrome P450; 3β-HSD, 3β-hydroxysteroid dehyrogenase isomerase; $P450_{17\alpha}$, P450 17α-hydroxylase; 20β-HSD, 20β-hydroxysteroid dehydrogenase (reproduced with permission from Nagahama *et al.*, 1994).

maturational surge of GTH-II, received by the thecal and granulosa cells and triggering the production of a C21 steroid MIH. MIH bonds to the oocyte membrane, initiating the intraoocyte production of the third and final mediator, the MPF, which triggers the cellular mechanisms of GVBD, the reinitiation of meiosis, and hydration. The resulting oocyte is ready for ovulation and fertilization.

D. Ovulation

Ovulation is the expulsion of the mature oocyte from the follicle. This event is preceded by follicular separation and the formation of a distinct hole in the follicle, through which the oocyte will pass. During follicular separation, the microvillar connections between the oocyte and the follicle retract from both sides, forming a wide space between the oocyte and

the granulosa cells. Proteolytic enzymes have been suggested to play a role in disrupting the follicle–oocyte connection and weakening the follicle wall. The weakening of the follicle wall prior to rupture appears to occur over a discrete section of the follicle because oocytes have been observed passing through discrete holes smaller in diameter than the oocyte.

Two mechanisms have been suggested for oocyte expulsion: (i) the buildup of pressure within the follicle due to oocyte hydration and (ii) the action of contractile elements in the follicle. In support of the second mechanism, microfilaments have been demonstrated in the thecal layer of some fish follicles. In addition, actin polymerization inhibitors and Ca^{2+} influx blockers can prevent ovulation *in vitro*.

Prostaglandins are believed to play a role in ovulation, perhaps in the stimulation of follicular contraction. Indomethacin, a prostaglandin synthesis inhibi-

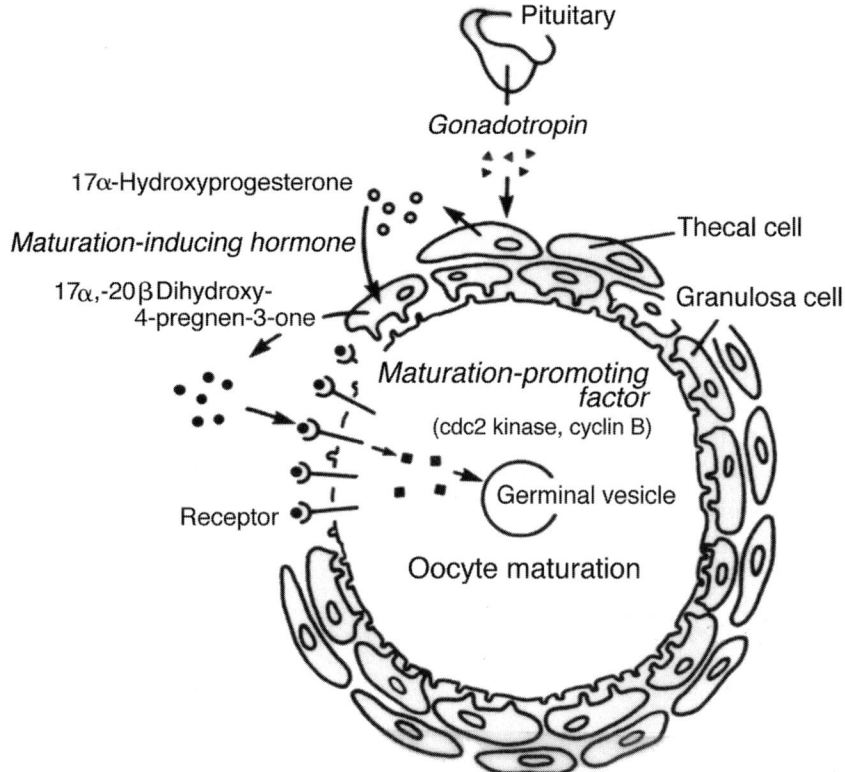

FIGURE 7 Hormonal regulation of oocyte maturation in teleosts. Three major mediators—gonadotropin, maturation-inducing hormone, and maturation-promoting factor—are involved (reproduced with permission from Nagahama *et al.*, 1994).

tor, can inhibit ovulation in goldfish exposed to warm temperatures, and prostaglandins E_2, E_1, and $F_{2\alpha}$ can induce ovulation in indomethacin-blocked females. Prostaglandin action has been suggested at the hypothalamic–pituitary level and at the ovarian level, and prostaglandins are produced by females prior to ovulation. Catecholamines can also elicit ovulation in fishes and may act at sympathetic nerve endings.

VI. MATERNAL–EMBRYONIC RELATIONSHIPS

A. Oviparity

Reproduction is teleosts is generally oviparous and fertilization external. Oviparity refers to external lecithotrophic development, indicating that the embryo relies solely on the yolk for nutrition during development. Oviparity is also characterized by oviposition of the egg prior to fertilization or the embryo prior to hatching, that is, while still encased by an egg envelope. Oviparity typically includes external fertilization, but some species undergo internal fertilization and even limited development of the embryo inside the female, although nutrition is still entirely lecithotrophic. The terms zygoparity and embryoparity have been coined to describe this process for embryos developing briefly and for longer periods, respectively, inside the female. These terms replace the less descriptive ovi- and ovoviviparity.

B. Viviparity

Viviparity (or live-bearing) describes internal fertilization of the eggs followed by an extended period of internal development and hatching from the egg envelope coincident with or preceding parturition. Almost all viviparous teleosts have a single fused, median ovary, and all are cystovarian. Gestation in fishes can take place either in the ovarian lumen (intraluminal) or within the ovarian follicle (intrafollicular). In this sense the ovary of the teleost is unique among vertebrates because it is the site of both egg production and gestation. Internal fertilization may take place inside the follicle prior to ovulation, or in the ovarian cavity following ovulation. Following intrafollicular fertilization, the embryo may be released from the follicle to continue gestation in the ovarian cavity, or gestation may continue within the follicle until parturition.

Facultative viviparity describes sporadic viviparity among species that are typically oviparous and involves only lecithotrophic nutrition, whereas obligate viviparity describes all species in which viviparity is the only mode of reproduction. Among the latter, embryonic nutrition ranges from strict lecithotrophy to extreme matrotrophy. Lecithotrophic viviparity is the most primitive form of viviparity and is trophically unspecialized. Matrotrophy refers to any maternal–embryonic trophic relationship in which nutrition is supplied from the maternal side during embryo development. Two major forms of matrophagy are observed in teleosts, including trophophagy, which involves the transfer of nutrients from a maternal epithelial surface across an open space to embryonic absorptive sites, and placentotrophy, which involves the transfer of nutrients across fused or intimately opposed maternal and embryonic tissues. In teleosts the nutritive maternal surface is derived from either the ovigerous folds of the ovary (in intraluminal gestation) or the follicle wall (in intrafollicular gestation). The absorptive tissue of the embryo might be derived from general body surfaces, finfolds, the yolk sac, buccal or branchial epithelium, or external projections derived from gut epithelium.

See Also the Following Articles

FEMALE REPRODUCTIVE SYSTEM, AMPHIBIANS; FEMALE REPRODUCTIVE SYSTEM, BIRDS; FEMALE REPRODUCTIVE SYSTEM, NONHUMAN MAMMALS, FEMALE REPRODUCTIVE SYSTEM, REPTILES; MALE REPRODUCTIVE SYSTEM, FISH; PHEROMONES, FISH; VITELLOGENINS AND VITELLOGENESIS

Bibliography

Chan, S. T. H., and Yeung, W. S. B. (1983). Sex control and sex reversal in fish under natural conditions. In *Fish Physiology* (W. S. Hoar, D. J. Randall, and E. M. Donaldson, Eds.), Vol. 9B, pp. 171–213. Academic Press, New York.

Dickoff, W. W., and Swanson, P. (1990). Functions of salmon pituitary glycoprotein hormones: "The maturational surge hypothesis." In *Progress in Comparative Endocrinology: Proceedings of the Eleventh International Symposium on Comparative Endocrinology* (A. Epple, C. G. Scanes, and M. H. Stetson, Eds.), pp. 349–356. Wiley-Liss, New York.

Goetz, F. W. (1983). Hormonal control of oocyte final maturation and ovulation in fishes. In *Fish Physiology* (W. S. Hoar, D. J. Randall, and E. M. Donaldson, Eds.), Vol. 9B, pp. 117–161. Academic Press, New York.

Hoar, W. S. (1957). The gonads and reproduction. In *The Physiology of Fishes* (M. E. Brown, Ed.), pp. 287–321. Academic Press, New York.

Liley, N. R., and Stacey, N. E. (1983). Hormones, pheromones and reproductive behavior in fish. In *Fish Physiology* (W. S. Hoar, D. J. Randall, and E. M. Donaldson, Eds.), Vol. 9B, pp. 1–49. Academic Press, New York.

Nagahama, Y. (1983). The functional morphology of teleost gonads. In *Fish Physiology* (W. S. Hoar, D. J. Randall, and E. M. Donaldson, Eds.), Vol. 9A, pp. 223–275. Academic Press, New York.

Nagahama, Y., Yoshikuni, M., Yamashita, M., and Tanaka, M. (1994). Regulation of oocyte maturation in fish. In *Fish Physiology* (W. S. Hoar, D. J. Randall, and A. P. Farrel, Ser. Eds.; N. M. Sherwood and C. L. Hew, Eds.), Vol. 13, pp. 393–472. Academic Press, New York.

Peter, R. E. (1983). The brain and neurohormones in teleost reproduction. In *Fish Physiology* (W. S. Hoar, D. J. Randall, and E. M. Donaldson, Eds.), Vol. 9A, pp. 97–127. Academic Press, New York.

Ravin, C. P. (1961). *Oogenesis.* Pergamon, Oxford, UK.

Selman, K., and Wallace, R. A. (1989). Cellular aspects of oocyte growth in teleosts. *Zool. Sci.* **6,** 211–231.

Sherwood, N. M., Parker, D. B., McRory, J. E., and Lescheid, D. W. (1994). Molecular evolution of growth hormone-releasing hormone and gonadotropin-releasing hormone. In *Fish Physiology* (W. S. Hoar, D. J. Randall, and A. P. Farrel, Ser. Eds.; N. M. Sherwood and C. L. Hew, Eds.), Vol. 13, pp. 3–66. Academic Press, New York.

Turner, C. L. (1983). Adaptations for viviparity in embryos and ovary of *Anableps anableps. J. Morphol.* **62,** 323–349.

Wourms, J. P., Grove, B. D., and Lombardi, J. (1988). The maternal–embryonic relationship in viviparous fishes. In *Fish Physiology* (W. S. Hoar and D. J. Randall, Eds.), Vol. 11B, pp. 1–34. Academic Press, New York.

Xiong, F., Suzuki, K., and Hew, C. L. (1994). Control of teleost gonadotropin gene expression. In *Fish Physiology* (W. S. Hoar, D. J. Randall, and A. P. Farrel, Ser. Eds.; N. M. Sherwood and C. L. Hew, Eds.), Vol. 13, pp. 135–158. Academic Press, New York.

Female Reproductive System, Humans

Bruce R. Carr

The University of Texas Southwestern Medical Center at Dallas

GLOSSARY

cervix The lowest part of the uterus which connects the vagina to the uterine cavity.

clitoris The erectile body of the female genitalia.

Fallopian tubes Tubes which extend from the uterine cavity to the ovary. They serve as a conduit for sperm, eggs, and the fertilized egg.

granulosa cells Small cells located inside the follicle and surrounding the ovum. Granulosa cells secrete estrogen.

theca cells Consist of two layers of spindle-like cells surrounding the ovarian follicle. Theca cells secrete androgen.

The female reproductive tract includes the ovary, Fallopian tubes, uterus, vagina, and external genitalia. Its function is to produce ova, receive sperm, and nurture an embryo and deliver a term infant.

I. INTRODUCTION

The female reproductive system includes primarily the ovary, which is the source of germ cells or ova, and hormones, which regulate sexual development. The Fallopian tubes transport the sperm to the egg and transport the fertilized ovum to the uterus. The uterus receives the sperm arriving from the vagina and external genitalia and is the site of the development of the fetus (Fig. 1). The precise coordination of the hormonal events of the hypothalamic–pituitary–ovarian axis leads to the formation of a single follicle which is ovulated and fertilized by a single sperm which has traversed the female genital tract. Pregnancy begins when the fertilized egg implants in the uterus and ends with the delivery of a newborn infant through the lower female reproductive tract.

II. OVARY

A. Gross Anatomy

The ovaries are paired, almond-shaped structures whose function is twofold: the production of the female gamete or ova and secretion of steroid hormones including estrogen and progesterone. The mature ovary is 2.5–5 cm in length, 1.5–3 cm in width, and 0.6–1.5 cm in thickness. Following menopause the ovarian size diminishes markedly. Normally the ovaries are situated in the upper part of the pelvic cavity near a slight depression in the lateral pelvic wall termed the ovarian fossa (Fig. 1). The ovary is attached to the broad ligament by the mesovarium. The uteroovarian ligament extends from the posterior portion of the uterus to the lower pole of the ovary. Infundibulopelvic or suspensory ligament extends from the upper pole of the ovary to the pelvic side wall. It is through this ligament that ovarian vessels and nerves reach the ovary.

The exterior surface of the ovary in the mature women is pitted and scarred by the process of ovulation. The ovary can be studied best in cross section as seen in Fig. 2. The ovary is divided into a cortex and medulla. The ova are located in the outer layer or cortex. The medulla or central portion of the

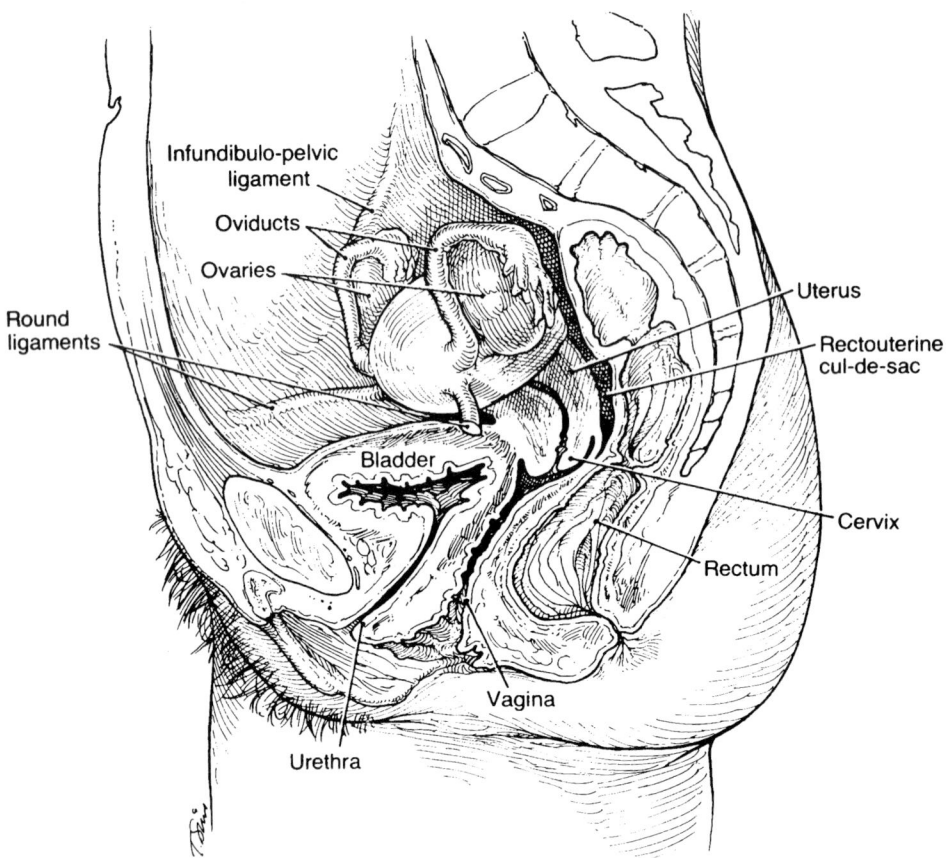

FIGURE 1 A sagittal section of the female pelvis demonstrating the structure and relationships of the female reproductive tract (used with permission from Cunningham *et al.*, 1997, p. 41).

ovary is composed of loose connective tissue that is connected to the mesovarium. The mesovarium contains a large number of arteries and veins.

B. Vasculature

The ovarian artery, a direct branch of the aorta, enters the broad ligament through the infundibular pelvic ligament and is the primary source of blood supply to the ovary (Fig. 3). At the ovarian hilum it is divided into a number of small branches which extend to the ovary, whereas the main artery transverses the entire length of broad ligament near the mesosalpinx and makes its way to the upper portion of the lateral source of the uterus. Here the ovarian artery branches and anastomoses with the uterine artery which is the secondary blood supply to the

ovary. The venous return includes that which leaves the ovarian vein through the infundibulopelvic ligament or the several veins within the broad ligament, known as the pampiniform plexus. The other source of drainage is via the uterine veins. On the right side, the ovarian vein enters the vena cava directly and on the left side the ovarian vein enters the left renal vein before entering the vena cava.

C. Innervation

The innervation of the ovaries accompanies the blood supply and is primarily sympathetic and parasympathetic nerves. The sympathetic nerves are derived in part from the ovarian plexus but accompany ovarian vessels and are derived from the plexuses that surround the ovarian branch of the uterine ar-

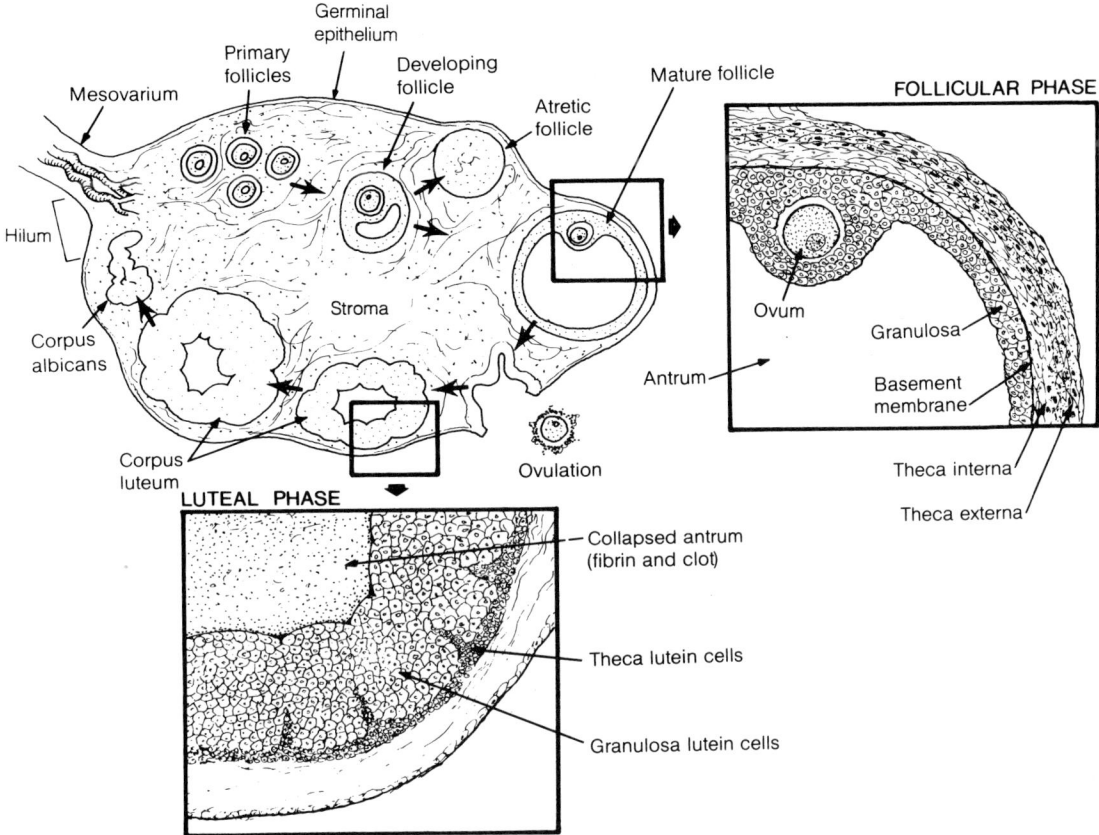

FIGURE 2 The structure of the ovary demonstrating the developmental changes of the adult ovary during the menstrual cycle (from B. R. Carr and J. D. Wilson, Disorders of the ovary and female reproductive tract, In *Harrison's Principles of Internal Medicine* (E, Braunwald, K. J. Isselbacher, R. G. Petersdorf, *et al.*, Eds.), 11th ed., pp. 1818–1837, McGraw-Hill, New York, 1987. Reproduced with permission of McGraw-Hill).

tery. The ovary is supplied with nonmyelinated nerve fibers, which for the most part accompany the blood vessels. Other autonomic nerves form wreaths around antral follicles and these give off many minute branches that reach up to but not through the granulosa membrane.

D. Lymphatics

The ovarian lymphatics also pass through the infundibulopelvic ligament along with the ovarian vessels to the lateral periaortic lymph nodes. On the left side primary nodes may be situated between the left ovary and the left renal veins. On the right side they may be found between the right renal vein and the

inferior vena cava. Shorter lymphatic pathways may also lead to the hypogastric nodes.

E. Microscopic Anatomy

The microscopic anatomy includes primarily connective tissue which is derived from atretic ovarian follicles (Fig. 2). Small primordial follicles, located near the cortex, comprise the majority of the ovarian follicles. During each menstrual cycle, from this pool of primordial follicles a cluster of primary follicles are recruited. One follicle becomes dominant and develops into a tertiary follicle with an antrum and finally a mature graafian follicle. Histologically the follicles are composed of an outer theca externa and

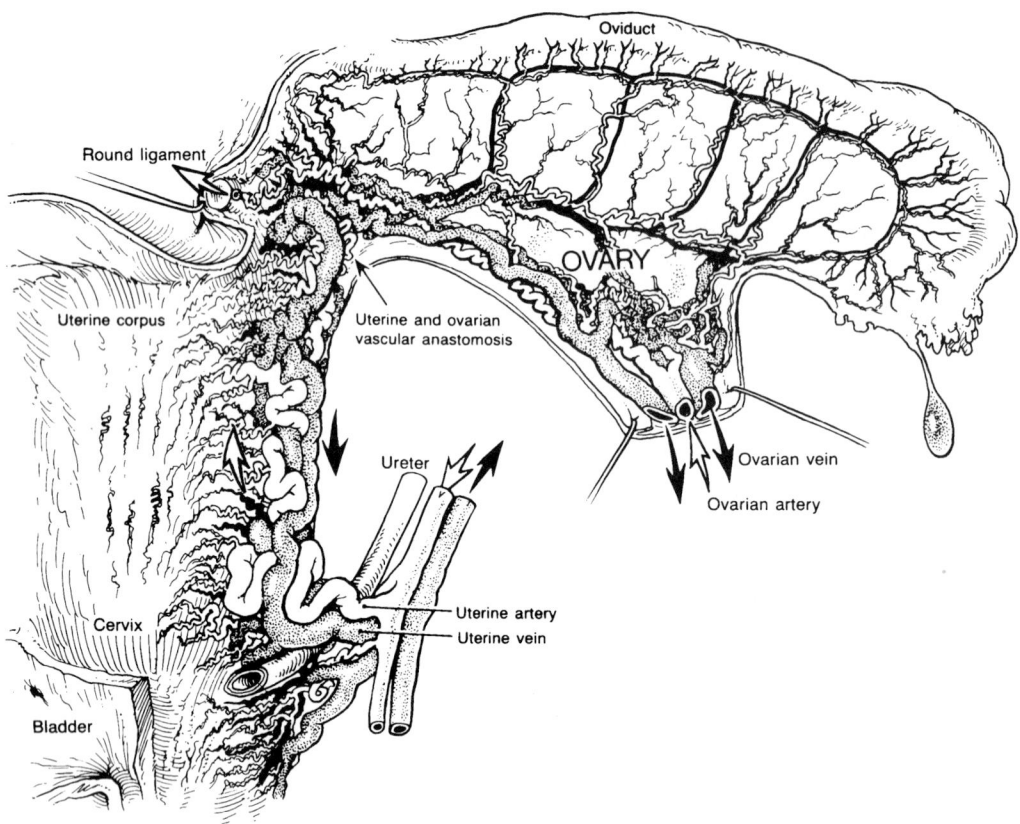

FIGURE 3 The blood supply to the left side of the ovary, Fallopian tube, and uterus. The ovarian and uterine vessels anastomose freely. Note the ureter passing below the uterine artery and vein near the cervix (used with permission from Cunningham *et al.*, 1997, p. 52).

theca interna and an inner granulosa layer separated by a basement membrane. The granulosa cells (which are not vascularized) also surround the developing ovum. Following ovulation capillaries enter the fragmented basement membrane and the granulosa becomes vascularized and transforms into a blood-filled corpus luteum. The theca externa and interna secrete primarily C19 steroids or androgens. The granulosa secretes primarily estrogen and the corpus luteum secretes primarily progesterone. The corpora albicantia represent old corpora lutea.

III. FALLOPIAN TUBES

A. Gross Anatomy

The Fallopian tubes or oviducts vary from 8 to 14 cm and usually average 12 cm in length. They are covered by peritoneum and the lumen is lined by a mucous membrane. The Fallopian tube is divided into three primary sections (Fig. 4):

1. An intramural section which traverses the uterine wall in more or less a straight fashion and is the narrowest part of the Fallopian tube.

2. The second portion or isthmus is not quite as narrow but adjoins the uterus and passes gradually into a wider or more lateral portion known as the ampullary section.

3. The ampullary section is very tortuous and gradually widens toward its outer end, which is composed of fimbriae also known as the infundibulum. One of the fimbria, the fimbria ovarica, is grooved and runs along the lateral border of the mesosalpinx to the ovary. The delicate fimbriae are believed to be important in ovum pickup following ovulation.

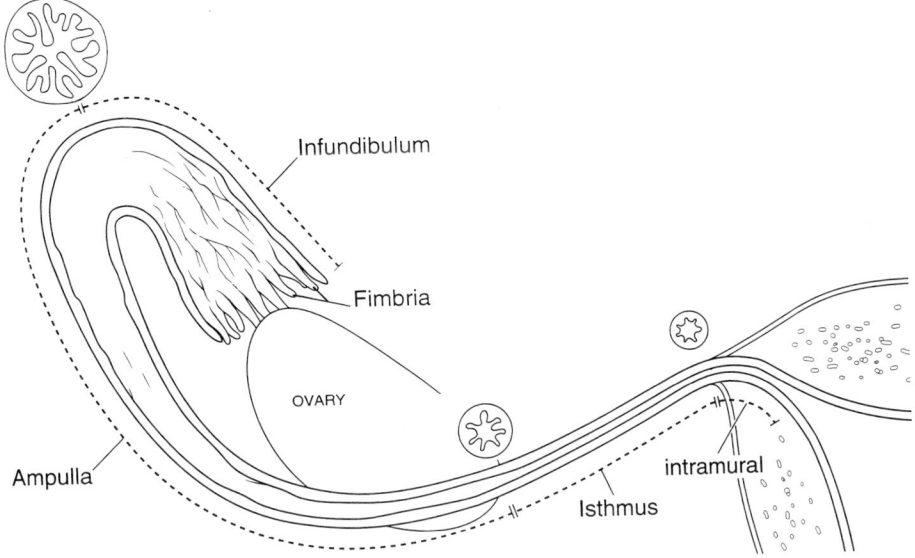

FIGURE 4 The anatomy of the Fallopian tube and cross section of the intramural, isthmic, and ampullary regions.

B. Vasculature

The blood supply to the Fallopian tubes is similar to that of the ovaries (Fig. 3).

C. Innervation

The nerves supplied to the Fallopian tubes are autonomic nerves and are similar to that for the ovary.

D. Lymphatics

The lymphatic drainage from the Fallopian tube is also similar to that of the ovaries. In addition, lymphatics drain to the common iliac nodes and into those of the sacral promontory.

E. Microscopic Anatomy

The wall of Fallopian tube consists of three layers: an outer or serosal coat, a muscular layer, and a mucosal lining. The muscular layer is composed of an inner circular and an outer longitudinal layer of smooth muscle fibers. At the point of the interstitium or intramural portion, there is an additional longitudinal muscle layer. The mucosal layer or endosalpinx consists of a single layer of columnar cells, some of which are ciliated, whereas others are secretory, and a third group noted to be peg-shaped cells. These peg-shaped cells contain dark nuclei which are found between the ciliated and secretory cells and may represent cells that are to be extruded into the Fallopian tube cavity.

IV. UTERUS

A. Gross Anatomy

The uterus is a pear-shaped and highly muscular organ (Fig. 5). The cavity of the uterus is lined by endometrium. During pregnancy, the uterus serves for reception, implantation, and nutrition of fetus and conceptus and expels the fetus during labor. The uterus is approximately 7.5 cm in length, 5 cm in width, and 2.5 cm in thickness. It maintains a horizontal position and orientation in between the bladder above and the rectum and the colon below (see Fig. 1). There are four parts of the uterus: the fundus, the body, the isthmus, and the cervix. The fundus of the uterus is the bluntly rounded upper portion of the uterus. It is the part above the line joining the points of entrance of the two Fallopian tubes. The body of the uterus narrows from the fundus down to the isthmus. The isthmus of the uterus is 1 cm

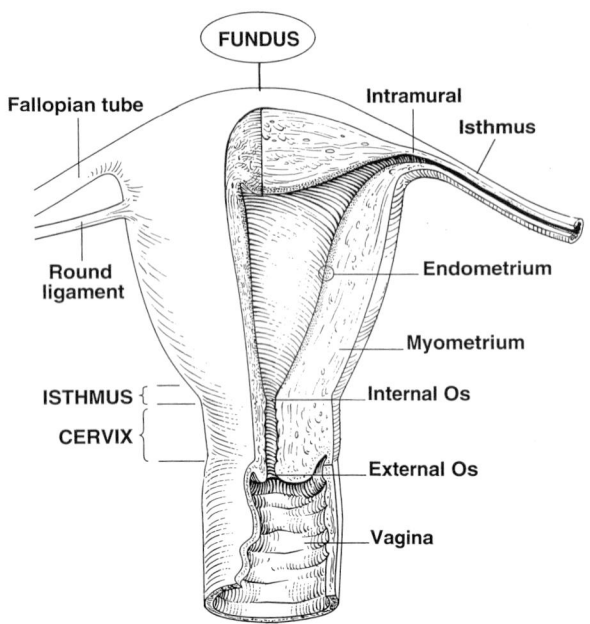

FIGURE 5 The anatomy of the uterus: left side, outside structures; right side, internal structures.

long and is a constricted region between the body and the cervix. The cavity of the uterine body is flattened and triangular in shape. The Fallopian tubes open into its basic angles and the apex is continuous with the cervical canal at the internal os. The cervix is cylindrical and slightly expanded in its middle and about 2 or 3 cm in length. Its canal is spindle shaped and opens into the vagina at the external os. The internal os is adjacent to the isthmus. The cervix is a specialized portion of the uterus. It is composed primarily of collagenous tissue plus elastic tissue and only a few smooth muscle fibers. In contrast, the body of the uterus is composed primarily of smooth muscle cells and includes serosal, muscular, and mucosal layers. The serosal layer is formed by the peritoneum that covers the uterus. The muscle layer, also known as the myometrium, makes up the major portion of the uterus and is composed of bundles of smooth muscle united by connective tissue. The inner lining of the uterus is known as the endometrium. It is a thin velvet-like membrane which has a large number of minute gland openings. The endometrium is composed of surface epithelium, endometrial glands, and stromal tissue. During the menstrual cycle, the endometrium grows in thickness and is asso-

ciated with an expansive growth in straight and spiral arterial blood supply.

B. Vasculature

The arteries of the uterus are the uterine, which is derived from the internal iliac, and the ovarian, which derives from the abdominal aorta (Fig. 3). The uterine artery approaches the organ from the lateral pelvic wall, where it is accompanied by veins, nerves, and lymphatics. One or 2 cm from the cervix, the ureter passing toward the bladder courses underneath the uterine artery. When it reaches the cervix, the uterine artery divides into a larger ascending branch which supplies the body of the fundus and smaller descending vaginal branch for the cervix and vagina. The larger branch has a tortuous course alongside the body of the uterus at the junctions of the ovarian ligament and Fallopian tube and anastomoses with branches of the ovarian artery. The veins follow the pattern of the arteries and drain into the iliac vein.

C. Innervation

The uterovaginal plexus located along the superior portion of the cervix, on either side of the uterus, is an extension of the inferior hypogastric plexus through which sympathetic and parasympathetic fibers pass.

D. Lymphatics

The lymphatics of the uterus are contained within three main networks of plexuses: one at the base of the endometrium, another in the myometrium, and a third subperitoneally. Lymphatics have not been detected in the endometrium. Drainage from the uterine body and from the cervix is similar, passing along the course of the uterine vessels.

E. Microscopic Anatomy

As described previously, the myometrium is composed of bundles and branches of smooth muscle cells united by connective tissue and the cervix is

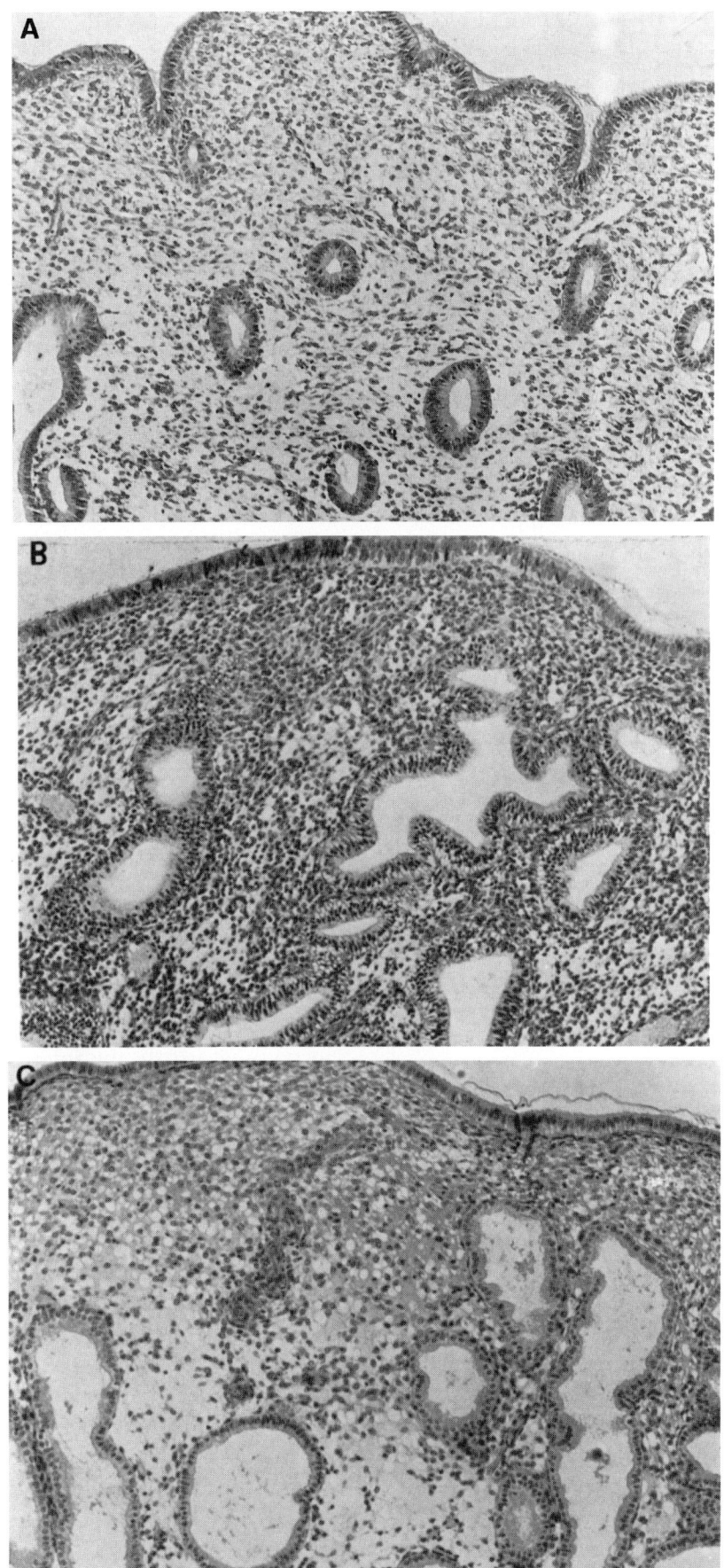

FIGURE 6 The histology of the endometrium during the menstrual cycle. A, Proliferative endometrium; B, 24–48 hr postovulatory endometrium; C, secretory endometrium.

composed primarily of collagenous tissue and elastic tissue. The most unique part of the microscopic anatomy of the uterus is the endometrium. The endometrium is composed of glandular cells and stroma. There are three principal hormonal changes that the endometrium undergoes during the menstrual cycle. These are proliferative changes during the follicular phase which consist of glandular mitoses, pseudo-stratification of the nuclei, and stromal mitoses (Fig. 6A). Following ovulation a second change of the endometrium is the development of basal vacuolization which occurs 6–48 hr after ovulation (Fig. 6B). During the luteal phase a secretory endometrium develops characterized by an increase in secretion and stromal edema (Fig. 6C). Just prior to menstruation, a pseudodecidual reaction takes place followed by necrosis and ending in sloughing of tissue and finally menstrual bleeding.

V. VAGINA

A. Gross Anatomy

The vagina is a thin-walled tube lined by squamous epithelium (Figs. 1 and 5). It is capable of dilation and constriction as the result of the interaction of supporting muscles and erectile tissue. It is the female organ of copulation. The vaginal space is only potential because the walls are usually in apposition. The vagina averages 8–10 cm in length and passes upward from the vestibule of the perineum to the cervix. The vagina has a mucous coat which consists of transverse elevations, the vagina rugae.

B. Vasculature

The vagina is supplied by the vaginal artery and vaginal branches of the uterine artery (Fig. 3). Additional blood supply comes from middle rectal and internal pudendal artery. The veins form a vaginal plexus along the lateral aspect of the vagina and the efferent vessels of the plexus are tributaries to the internal iliac veins.

C. Innervation

The nerves of the vagina are derived from the uterovaginal plexus. The vaginal nerves follow the vaginal arteries and end in the vaginal wall. The uterovaginal plexus is comparable to the prostatic plexus of the male.

D. Lymphatics

The lymphatics flow from the lower third of the vagina into the inguinal nodes, the middle into the internal iliac nodes, and the upper vagina drains into the iliac nodes.

E. Microscopic Anatomy

The vaginal mucosa is composed of noncornified stratified squamous epithelium. Below the epithelium there is a thin fibromuscular coat, usually consisting of an inner circular layer and an outer longitudinal layer of smooth muscle. There is a thin layer of connective tissue that overlies the mucosa and the muscularis containing blood vessels and lymph nodes. Under normal conditions glands are not present in the vagina.

VI. EXTERNAL GENITALIA

A. Gross Anatomy

The external genitalia is composed of the mons pubis, labia majora, labia minora, clitoris, vestibule, urethral opening, vestibular bulbs, vaginal opening, and hymen (Fig. 7). The mons pubis is a fat-filled portion that lies on the anterior surface of the symphysis pubis, and after puberty it is covered with hair. The labia majora consists of two rounded folds of adipose tissue that are covered with skin and extend downward and backward from the mons pubis. The labia majora is homologous with the male scrotum. The round ligaments from the uterus terminate at the upper border of the labia majora. The labia minora consist of two flat reddish folds of tissue that are visible when the labia majora are separated. These structures are joined at the upper extremity of the vulva. The labia minora vary greatly in length and shape. The labia minora converge superiorly where they fuse to form frenulum of the clitoris and posteriorly to form the posterior fourchette. The clitoris is

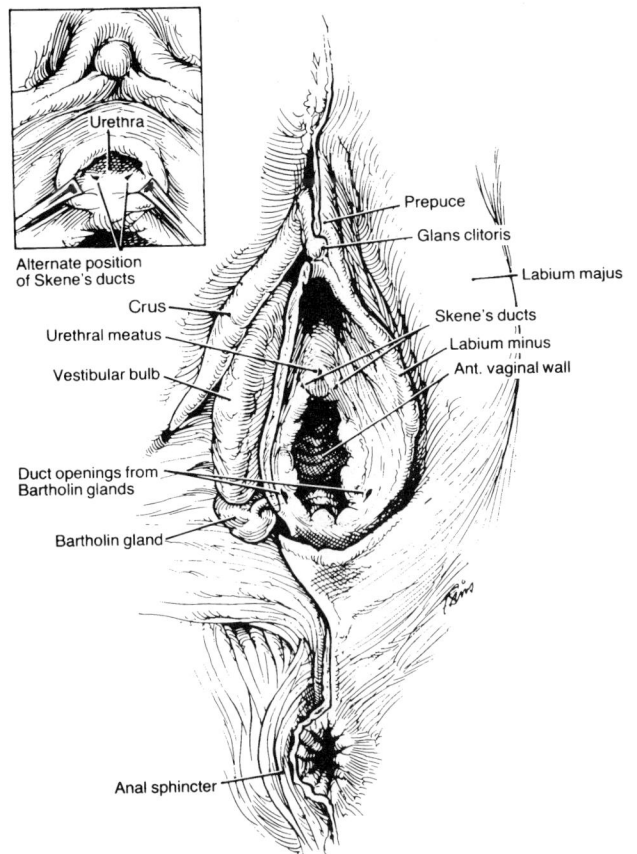

FIGURE 7 The external organs of reproduction of women. The lower anterior vaginal wall is visible through the labia minora. The insert illustrates the relationship of the urethra and Skene's ducts (used with permission from Cunningham *et al.*, 1997, p. 38).

homologous to the glans penis and is located near the superior extremity of the vulva. It is an erectile organ which projects downward between the branched extremities of the labia minora and it converges to form the prepuce in the frenulum of the clitoris. The clitoris is composed of a glans, a body or corpus, and two crura. The clitoris is usually <1–1.5 cm in length and rarely exceeds 0.5 cm in diameter.

The vestibule is an almond-shaped area that is enclosed by the labia minora laterally and extends from the clitoris to the posterior fourchette. The vestibule is a functional opening of the urogenital sinus of the embryo. In the mature state it is perforated by six openings: the urethra, the vagina, the two ducts of the Bartholin glands, and the two periurethral glands, also called Skene ducts. The posterior portion of the vestibule between the fourchette and vaginal opening is called the fossa navicularis. The Bartholin glands are a pair of small compound structures approximately 1 cm in diameter situated beneath the vestibule on either side of the vaginal openings and are the major vestibular glands. The lower two-thirds of the urethra lies immediately above the anterior vaginal wall. The urethral opening or meatus is at the midline of the vestibule approximately 1 cm below the pubic arch. Occasionally, two periurethral ducts known as Skene ducts open into the vestibule on either sides of the urethra. Beneath the mucous membranes of the vestibule on either side are the vestibular bulbs. The bulbs, which are a complex structure of veins, lie in close opposition to the ischial rami and are covered by the ischial cavernous, which are constrictual vaginal muscles. The vaginal opening is hidden by the overlapping labia and is usually appears almost completely closed by a membranous hymen. There are marked shapes and differences of the hymen, which is composed primarily of elastic collagenous tissue. The hymenal opening is usually crescent shaped or may be cribriform, septate, or fibrinated.

B. Vasculature

The blood supply to the external genitalia is complex. It includes the following vessels: the branches of the inferior gluteal artery, the internal pudendal artery, the inferior hemorrhoidal artery, the perineal artery, and the artery of the clitoris.

C. Innervation

The primary innervation of the external genitalia and perineum is by the pudendal nerve. The pudendal nerve has three branches: the inferior hemorrhoidal, the perineal, and the dorsal nerve. In addition, the following nerves contribute to the innervation of the perineal skin: (i) the anterior labial branches of the ilial inguinal nerve, (ii) the external spermatic branch of the genitofemoral nerve, and (iii) the perineal branches of the posterior femoral cutaneous nerve and branches of the perineal nerve.

D. Lymphatics

The lymphatics of the external genitalia include the inguinal lymph nodes, both superficial and deep. The superficial inguinal lymph nodes drain the external genitalia and lower third of the vagina. The deep inguinal lymph nodes receive the efferent lymphatics directly and indirectly from the superficial femoral lymphatics. The external genitalia and the lower third of the vagina are drained by networks of lymphatic anastomosis. From the region of the clitoris deeper lymphatics may pass directly to the deep femorals particular to Cloquet's in the femoral canal.

E. Microscopic Anatomy

The labia majora are supplied with numerous sebaceous glands. Beneath the skin there is a layer of dense connective tissue that is rich in elastic fibers and adipose tissue but is nearly void of muscular elements. The labia minora is covered by stratified squamous epithelium into which the papillae project. There are no hair follicles in the labia minora but there are sebaceous follicles encasing a few sweat glands. The anterior of the labia folds is composed of connective tissue with many vessels and smooth muscular fibers. The glans of the clitoris are made up of spindle-shaped cells and in the body there are two vascular corpora cavernosa, in the walls of which are smooth muscle fibers.

See Also the Following Article

MALE REPRODUCTIVE SYSTEM, HUMANS

Bibliography

Caldwell, W. E., and Moloy, H. C. (1933). Anatomical variations in the female pelvis and their effect in labor with a suggested classification. *Am. J. Obstet. Gynecol.* **26**, 479.

Cunningham, F. G., *et al.* (Eds.) (1997). *Williams Obstetrics,* 20th ed. Appleton & Lange, Stamford, CT.

DuBose, T. J., Hill, L. W., Hennigan, H. W., Jr., Nichols, D. H., Mezaraups, G. G., Porter, L., Marley, L., Butschek, C. M., Karnaze, G. C., and Walser, E. (1985). Sonography of arcuate uterine blood vessels. *J. Ultrasound Med.* **4**, 229.

Ferenczy, A., and Richart, R. M. (1974). *Female Reproductive System: Dynamics of Scanning and Transmission Electron Microscopy.* Wiley, New York.

Hodgson, B. J., and Eddy, C. A. (1975). The autonomic nervous system and its relationship to tubal ovum transport—A reappraisal. *Gynecol. Invest.* **6**, 161.

Krantz, K. E. (1958). Innervation of the human vulva and vagina. *Obstet. Gynecol.* **13**, 382.

Langlois, P. L. (1970). The size of the normal uterus. *J. Reprod. Med.* **4**, 220.

Moore, K. (1983). *The Developing Human.* Saunders, Philadelphia.

Spalteholz (1933). *Hand Atlas of Human Anatomy*, Vol 1. Lippincott, Philadelphia.

Woodruff, J. D., and Pauerstein, C. J. (1969). *The Fallopian Tube.* Williams & Wilkins, Baltimore.

Female Reproductive System, Insects

Erwin Huebner

University of Manitoba

GLOSSARY

aeropyle A small hole that traverses the egg chorion facilitating the passage of oxygen and carbon dioxide to and from the developing egg.

chorion The protective shell around oocytes produced by the follicular epithelium.

genital chamber Also called the *vagina* or *bursa copulatrix*; it is the last part of the female tract and receives the male genitalia in copulation.

hydropyle A specialized structure of the egg chorion in some insects that functions for water uptake.

micropyle An external opening that via the micropylar channel through the egg chorion allows the entry of sperm and fertilization to occur. The number of micropyles varies with species.

ootheca A protective casing around a group of oviposited eggs creating a packet; it is produced by female accessory glands.

ovariole The solid tubular functional unit of all insect ovaries.

panoistic ovariole An ovariole with a developing chain of oocytes each encased by follicle cells.

polytrophic ovarioles Ovarioles which have nurse cells associated with each oocyte and each unit is enclosed in a follicular epithelium. A chain of successive stages make up the ovariole.

spermatheca A diverticulum of the common oviduct or genital chamber.

sperm duct or ductus seminalis In Lepidoptera, a structure that connects the bursa copulatrix of the copulatory opening (which is separate from the ovipositioning opening) with the common oviduct so sperm can pass to the spermatheca.

telotrophic ovarioles Ovarioles which have all the nurse cells located in a common chamber, the tropharium, with oocytes connected by long bridges or trophic cords.

tunica propria The extracellular basal lamina that surrounds ovarioles.

The female insect reproductive system is responsible for producing viable oocytes, receipt and storage of sperm including facilitating fertilization, the transport of these eggs to the exterior, and for providing essential secretions with adhesive and/or protective functions for the laid eggs. Detailed study involves examination of the germ cells and somatic cells that comprise the various ovary types. Consideration of the growth and differentiation of oogonia into mature oocytes in the ovary is a central first step toward an understanding of the female reproductive system. Oogenesis results in mature eggs with protective chorions. These leave the ovary and pass down the oviducts into the rest of the tract where the timely release of sperm from the spermatheca fertilize them. Finally, as the eggs are oviposited secretions from the accessory glands are added. All these factors collectively are essential for the continuity of generations. The evolutionary success of insects is in large measure reflected in their well-designed and highly functional reproductive systems.

I. INTRODUCTION

Insects are masters among animals in capitalizing on their reproductive capacities. One need only recall the legendary reproductive prowess of queen termites and queen army ants. Well-engineered and efficient reproductive systems contribute significantly to the evolutionary success of insects. The reproductive tract consists of the manufacturing plant (the ovaries and sperm storage components), the packaging (ovaries and accessory glands), and delivery system (lateral and common oviducts, genital chamber or vagina or bursa copulatrix, and ovipositors). These ensure survival of the next generation and propagation of the species.

During the production of the egg cell in oogenesis a dowery of biochemical, molecular, and organelle components along with the developmental blueprint must be synthesized or accumulated and organized in preparation for fertilization as the eggs pass down the reproductive tract. Protective layers of the chorion are deposited around the egg during oogenesis and later additional secretions are added by the tract or accessory glands. These vary according to species and maximize survival after oviposition in terrestrial or aquatic environments. Simply put, insects are highly successful in reproduction because of well-designed reproductive tracts and the capacity to store sperm for extended periods.

Central to the reproductive process are the ovaries and their constituent ovarioles where the germline is housed and with the aid of associated cells generates eggs when needed. A number of basic cell-interacting themes of cells have evolved, resulting in three fundamental ovary types (panoistic, polytrophic meroistic, and telotrophic meroistic). In each, the coordinated orchestration of cell interactions under endocrine regulation produces mature oocytes usually encased in a chorion.

Besides the variations in ovarian cytoarchitecture, we must also consider the rest of the reproductive tract because it is also vital to reproductive success. The spermatheca stores sperm for extended periods and ensures fertilization and the initiation of embryonic development. The transportation of eggs from the ovary on their journey to the outside world involves the lateral oviducts, common oviduct, and genital chamber (often termed the vagina or bursa copulatrix). A diversity of accessory glands produce secretions that may include adhesives, foamy or gelatinous coverings, or tough tanned oothecal capsules. While the vast majority of insects are oviparous, there are special cases of viviparity in which the posterior part of the tract and accessory glands are modified to house and nourish the developing embryos or larvae, as in adenotrophic viviparity.

While reproduction serves to maintain the species rather than the individual and the process involves an integration of three levels (population, individual, and cellular process), the focus in this article is on the cellular, tissue, and organ level. Considering the huge array of insect diversity, an all-inclusive treatment of the subject is not possible here. The following attempts to provide an overview, the basic reproductive plan, some common variations, and some of the underlying functional mechanisms of the insect female reproductive system.

II. GENERAL MORPHOLOGY: BASIC ANATOMY AND EVOLUTIONARY VARIATIONS

Despite the enormous diversity in insect reproductive systems there are common elements and themes that apply to a broad spectrum of species. As described and depicted in the classical works of Snodgrass and Weber and repeated in many reviews in the intervening years, the female reproductive tract consists of paired ovaries, the lateral oviducts which usually connect to a common oviduct that in turn connects to the genital chamber referred to as a vagina by some authors or the bursa copulatrix by others. The bursa is where the male deposits the sperm, often in a spermatophore, during copulation. Figure 1 illustrates the basic plan, including the diverticulum that serves for sperm storage (the spermatheca) and the lower tract accessory glands. There are many variations on this theme. For an extensive coverage of the various insect orders the reader is referred to the books by Matsuda and Büning. Figures 2 and 3 provide some selected examples of tract and ovarian diversity.

The ovaries are usually encased in a mesodermally derived ovarian sheath which is rich in muscles and the tracheal supply. The complexity and degree of development of this sheath vary with the species. The ovaries are composed of their functional units the ovarioles. The ovarioles contain the germ tissue and mesodermally derived somatic tissues associated with the germ cells. An extracellular layer, the basal

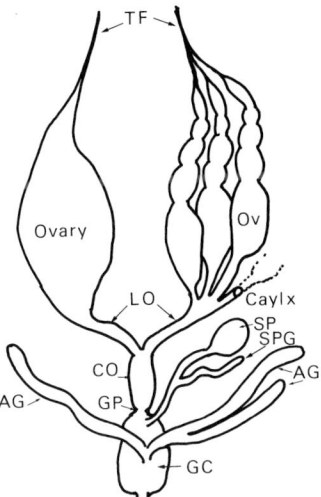

FIGURE 1 Generalized female insect reproductive tract. Components are terminal filaments (TF) and paired ovaries with ovarioles (Ov) attached to the calyx of lateral oviduct (LO) joined to common oviduct (CO), which opens via the gonopore (GP), the opening into the genital chamber (GC). Glands include the spermatheca (SP) with spermathecal gland (SPG), and accessory glands (AG) which may be single or have multiple tubules (redrawn and modified from Snodgrass, 1935).

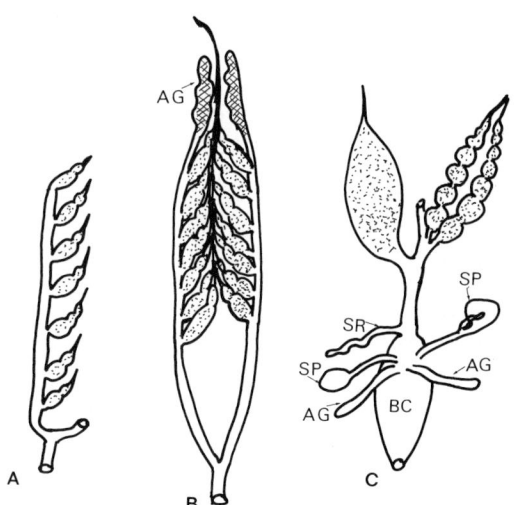

FIGURE 2 Ovarian variation. (A) The segmental arrangement panoistic ovarioles of the Dipluran *Catajapyx* (redrawn from Büning, 1994). (B) The ovary of the locust *Schistocerca* with an alignment of ovarioles with joined terminal filaments and the accessory gland (AG) at the tip of the lateral oviduct. (C) The *Drosophila* reproductive tract with consolidated ovaries; also included here is the rest of the tract with spermathecae (SP), seminal receptacle (SR), accessory glands (AG), and bursa copulatrix or vagina (BC).

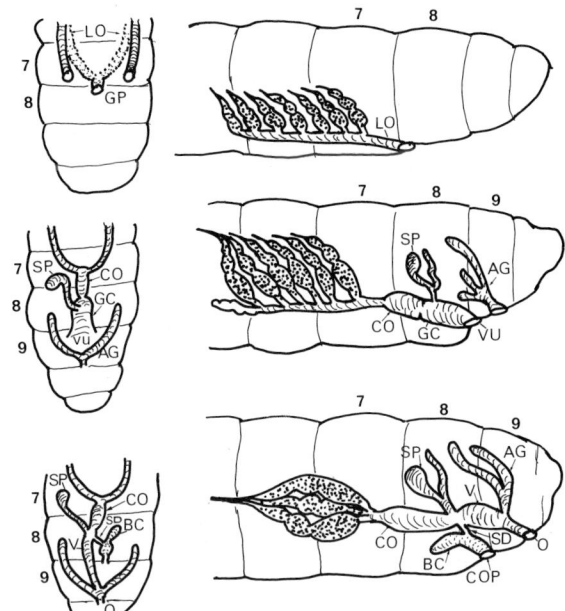

FIGURE 3 Dorsal and lateral views of the female reproductive tracts of three different insects. (Top) A simple primitive system as in the Ephemeroptera in which the lateral oviduct (LO) from each ovary opens independently. Also shown superimposed in the dorsal view is the situation in the Dermaptera in which the lateral oviducts fuse to a short common oviduct (dotted lines). (Middle) The generic Orthopteran tract in which the lateral oviducts from each ovary join the common oviduct (CO) in the seventh segment which connects to the genital chamber (GC) and its opening, the vulva (vu), with the spermathecal (SP) diverticulum. The accessory glands (AG) open in the ninth segment. (Bottom) The tract typical of many Lepidoptera in which there are two openings. The copulatory opening (COP) opens into a bursa copulatrix (BC) which connects via the sperm duct (SD) to the common oviduct (CO). Diverticula of the vagina (V) which opens in the ninth segment (oviparous or egg-laying opening, O) include the spermatheca (SP) and accessory gland (AG) (redrawn and modified from Snodgrass, 1935).

lamina or tunica propria, encases each ovariole and outside this are the ovariole sheaths. The cellular details and ovariole subtypes will be discussed later. Usually the ovarioles are packed together in a "consolidated" ovary but in some groups they may be spread out in a varying segmental fashion (Fig. 2). Basically, the ovarioles are solid tubular structures tapered at their apex, the germinal end, where they are attached to their suspensory terminal filaments. At their posterior or distal ends they usually have a tubular pedicel which connects to the calyx or

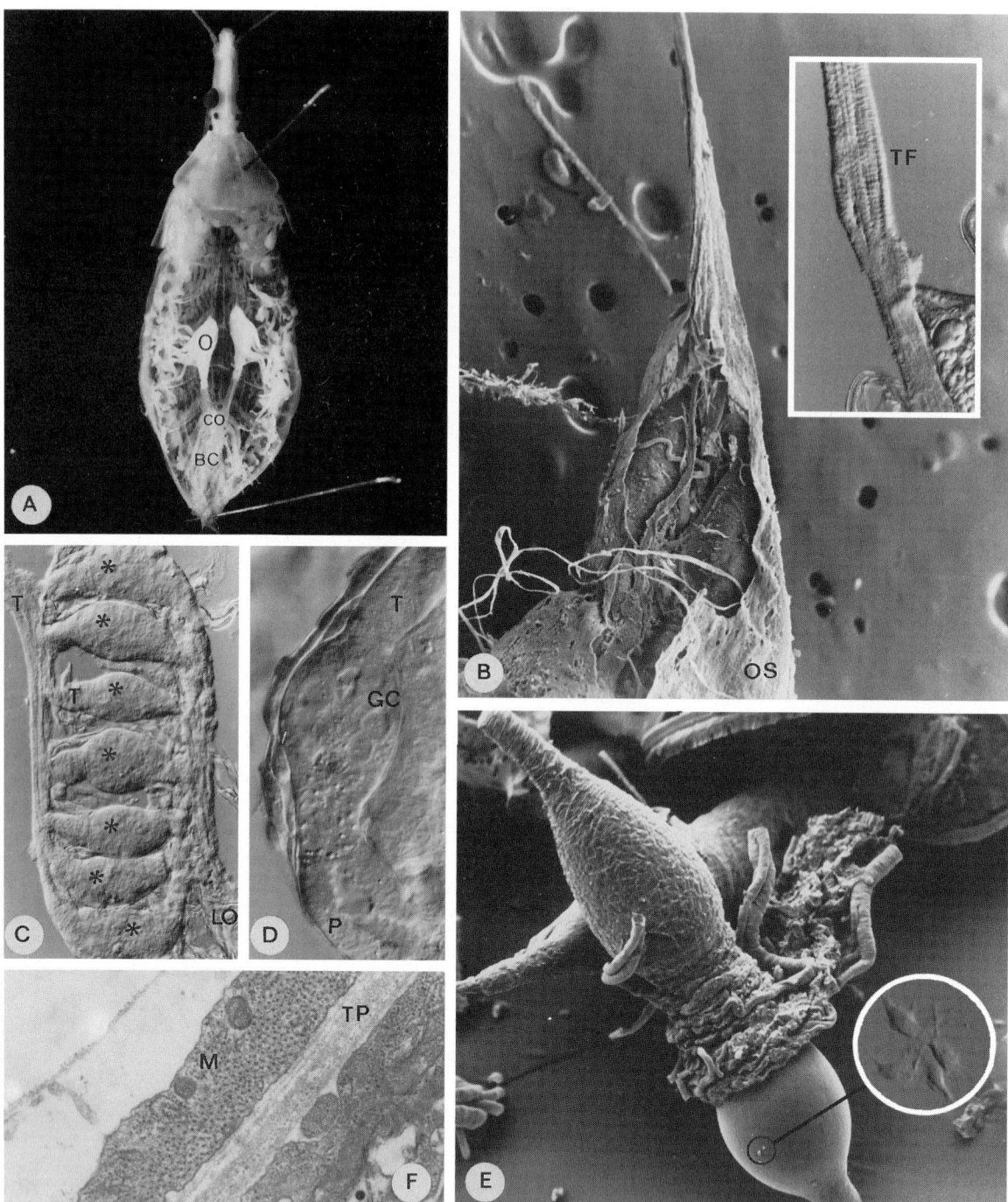

FIGURE 4 Female tract, ovarian sheaths, and ovarioles. (A) The female tract of *Rhodnius* with paired ovaries (O) connected via lateral oviducts to a common oviduct (CO) and bursa copulatrix (BC) and the accessory gland, the cement gland, just to the right of the bursa (reproduced with permission from D. Lutz and E. Huebner, *Tissue Cell* **12**, 773, 1980). The scanning EM of the *Rhodnius* ovary in B shows the ovarian sheath (OS) partially torn to reveal the ovarioles inside. The inset shows the muscular terminal filament (TF). (C and D) Nomarski differential interference microscopy views of the live developing ovary from second- and fourth-instar *Rhodnius*. Note the terminal filaments (T), cluster of germ cells (asterisks, GC), and pedicle (P). (E) An SEM of the functional unit the ovariole, with the ovariole sheath partially removed, revealing the smooth tunica propria (reproduced with permission from E. Huebner, *Tissue Cell* **13**, 105, 1981). The inset shows a DIC image of line ameboid cells seen on the tunica. The extracellular tunica propria (TP) of the beetle *Ips* is shown in F.

broadened end of the lateral oviduct. The number of ovarioles per ovary varies according to species and the ovariole type is characteristic and invariant within the taxonomic level of families or orders.

The somatic tissues of the ovaries and the lateral oviducts are of mesodermal origin. Embryonically invaginations of the seventh abdominal sternum give rise to the common oviduct and unpaired invaginations of the eighth and ninth segments, the vagina or bursa copulatrix. Thus, the common oviduct and the rest of the tract are of ectodermal origin and have a cuticular intima lining. In some higher insects the lateral oviducts may also be ectodermal.

There is a wide range of variation in the reproductive duct system from the simplistic one of the Ephemeroptera to the complex dual-opening system of many Lepidoptera. In the Ephemeroptera, the lateral oviducts of each ovary open independently to the exterior by attaching to short ectodermal invaginations of the seventh abdominal sternum forming a gonopore (Fig. 3). In most insects, however, the lateral oviducts join in the midline with the ectodermally derived invagination being the common oviduct which may open to the exterior as in the Dermaptera (Fig. 3) or as in most higher insects; the common oviduct links up with the genital chamber (bursa copulatrix or vagina), thereby forming a continuous tubular structure (Figs. 3 and 4). Other invaginations include the diverticula, the spermatheca, and accessory glands. As discussed later, these vary in location, structure, and function. In many Lepidoptera the posterior tract has two openings, one for egg laying and one for receiving the sperm in copulation which opens into a bursa copulatrix. A separate sperm duct, the ductus seminalis, connects the spermathecal duct with the egg passage so sperm can be stored in the spermatheca and subsequently fertilization can occur (Fig. 3).

III. OVARIES: BASIC MORPHOLOGY, TYPES, CELLULAR ARCHITECTURE, OOGENESIS, AND DEVELOPMENT

Since the first descriptions of the *Bombyx* ovary by Malpighi in 1661, a wealth of information has accumulated. The germline or primordial germ cells,

which constitute the lineage that perpetuates the species, developmentally are set aside already in the very early blastoderm embryo. This isolation of the germline from the rest of the embryo which will give rise to the somatic tissues was recognized in 1866 by Metshnikoff. During embryogenesis the primordial germ cells undergo a complex displacement and migration, arriving at the gonadal anlage in abdominal segment 5 in the case of *Drosophila*. Within the mesodermal gonad anlage the embryonic gonad begins to form with the interaction between the primordial germ cells and somatic tissue. This interaction establishes the foundation for the formation of the key components of the ovary and the partitioning into the functional units or ovarioles (Figs. 4C–4E). The details and timing relative to embryonic and postembryonic events vary according to group. For an indepth review of ovarian development see Huebner and Diehl-Jones.

Key components of the ovary that develop from the somatic tissue are the ovarian sheaths (Fig. 4B), the ovariole sheaths (Fig. 4E), the inner sheath of follicle cells, the terminal filaments (Fig. 4B), and the pedicels. Most insect ovarioles have pedicles. During either embryonic development as in the hemipteran *Rhodnius* or pupal development as in *Drosophila*, the cohort of primordial germ cells within the ovarian anlage are partitioned into identical repeating units the ovarioles (Figs. 4C and 4D). Each ovariole is composed of the germ cells and encasing somatic cells that will form some of the follicular cells of the germarium. A basal lamina forms around each ovariole, and apically a pad of presumptive terminal filament cells and basally the presumptive prefollicular and pedicel cells form (Fig. 4D). During postembryonic to adult metamorphosis the adult ovariolar morphology appears (Fig. 4E).

A. Ovariole Morphology
General Aspects

Each ovariole is covered by a relatively thick extracellular layer or basal lamina called the tunica propria (Fig. 4F). This is composed of a variety of extracellular matrix molecules, including collagen types I and IV, laminin, elastin, glycosaminoglycans, and entactin. It serves a structural role as well as acting as a sieve with chemical selectivity that determines the selection of molecules that can pass across into the ovariole. Usually the tunica propria continues over the tip of the ovariole where the muscular terminal filaments are attached. Where it passes between the ovariole tip and terminal filaments it is referred to as the transverse septum, which is lacking in the Plecoptera or stone flies. Outside the tunica propria are scattered ameboid wandering cells (Fig. 4E, inset) and usually a double-ovariole sheath consisting of a mesh of muscle cells, tracheoles, and connective tissue elements (Fig. 4E).

The number of ovarioles per ovary is variable according to the species; for example, 1 in certain Coleoptera (Coprinae), 2 in the Dipteran *Glossina*, usually 4 in Lepidoptera, 7 in the Hemipteran *Rhodnius*, 8 in the cockroach *Periplaneta*, 100 or more in Calliphora, 150–200 in queen honeybees, and more than 1000 in termite and army ant queens. Furthermore, in some species such as *Drosophila*, ovariole number can vary according to conditions and geography, varying from 10 to 30.

B. The Ovariole as a Functional Unit and Types

Among the arthropods, only the insect sister group, the Entognatha (Proturans, Collembolans, spring tails, and Dipturans) along with the insect taxa or Ectognatha have ovarioles. This unique and characteristic feature has recently prompted Büning to suggest a new classification as a monophylum "the Ovariolata." Putting aside taxonomic considerations, this highlights that a common and unique feature of insect ovaries (including the Entognatha) is ovarioles.

Alexander Brandt described the two major types of insect ovarioles 120 years ago; the panoistic and meroistic types. Around the turn of this century, Gross elegantly described the polytrophic and telotrophic subtypes of meroistic ovarioles.

During ovarian development primordial germ cells proliferate and associate with somatic cells. The different ovariole types arise due to different patterns of germ cell divisions. Of particular importance is if and when complete or incomplete cell division or cytokinesis of the germ stem cells occurs. For a de-

tailed discussion of the patterns of cell divisions within the evolutionary spectrum of ovaries the reader is referred to the books by Stys and Bilinski and Büning. If the stem cells in the germinal tip of the ovariole divide completely, a progression of oocytes are produced which each have their compliment of somatic follicle cells, thus forming a panoistic ovariole. On the other hand, if these germ cells undergo incomplete cytokinetic divisions and generate syncytial interconnected germ cell arrangements, these are polytrophic or telotrophic meroistic ovarioles. In all the ovariole types the oocytes are surrounded by a layer of simple epithelial cells, the follicle cells. It is generally believed that polytrophic ovaries evolved from panoistic ones and telotrophic from polytrophic ones. There is also evidence that in a few groups poytrophic ovaries have lost their nurse cells, becoming morphologically panoistic ovarioles (termed neopanoistic).

1. Panoistic Ovarioles

Panoistic ovarioles are found in a variety of insect orders, such as the Thysanura, Orthoptera, Odonata, Collembola, and Isoptera. Panoistic ovarioles (Figs. 5A–5C) have stem cells located in the germarium tip (Fig. 5B) of the ovariole. Under the appropriate hormonal and nutritional conditions these undergo complete cytokinesis, providing a steady source of oogonia. Each oogonium becomes encased by somatic prefollicular cells which proliferate generating a complete simple epithelium as the oogonia enlarge and differentiate, eventually forming mature oocytes. The initial growth phase, termed previtellogenesis, is due to the accumulation of ribosomes, a variety of mRNAs, proteins, and organelles. The oocyte nucleus or germinal vesicle is active in this synthesis in panoistic ovaries. Subsequently, yolk precursors vitellogenin and vitellin are incorporated via receptor-mediated endocytosis by the oocyte cell membrane or oolemma. These yolk precursors are synthesized outside the ovary by fat bodies and reach the oocyte by passing across the follicular epithelium via the spaces between adjacent cells. In chorionating eggs near the pedicel end of the ovariole, the follicle cells secrete the various layers of the chorion. The general cellular features of vitellogenesis and chorion formation also apply to many meroistic ovaries (Figs.

5N and 5O). Chorions may be endowed with aeropyles, hydropyles, and micropyles and have mechanical and protective properties. Along a panoistic ovariole there is a progression of developmental stages of follicles attached to each other by short stalks of interfollicular tissue (Figs. 5A and 5C).

In the panoistic ovariole the primary cell type intimately associated with the oocyte is the follicle cells. Coordination of activities of the follicle cells relative to the oocyte stage is mediated by gap junctions that couple oocytes with follicle cells and follicle cells to each other.

2. Meroistic Ovaries

In meroistic ovaries, division of the stem germ cells is dramatically different from that of panoistic ovaries. Postproliferative divisions characteristically have incomplete cytokinesis creating syncytial clones of germ cells. Subsequently, the majority of compartments in the syncytium differentiate into nurse cells or trophocytes, whereas a smaller number become oocytes. The compartments are interconnected by cytoplasmic bridges so the oocytes can receive nurse cell-produced products. The cellular architecture of the two types of meroistic ovarioles (polytrophic and telotrophic) differs considerably. The general stages of oogenesis, previtellogenesis, vitellogenesis, and chorion formation as mentioned earlier also occur in meroistic ovarioles; however, a major difference is that the oocyte nucleus relinquishes many of its synthetic functions to the nurse cells. These cells then produce most of the ooplasm during previtellogenesis.

i. Polytrophic Ovarioles The polytrophic ovarioles are found in orders such as the Adephaga coleopterans, Dermaptera, Neuroptera, Hymenoptera, Trichoptera, Diptera, and Lepidoptera. The general features are well exemplified in one of the most intensely studied insects, the fruit fly *Drosophila*.

In the germarium (Figs. 5D and 5E) of each ovariole, stem cells undergo proliferative mitotic divisions to give rise to cystoblasts. Cystoblasts in *Drosophila* then undergo three successive incomplete cytokinetic divisions creating syncytial clones of 16 cells interconnected by cytoplasmic channels or ring canals. The developmental molecular biology and functional morphology of the *Drosophila* ovary has been extensively studied. In other species with polytrophic ovarioles the number of cells differs depending on the number of synctial divisions. For example in the cecropia moth *Hyalophora* each clone consists of 8 cells (Fig. 5F). Figures 5D–5H show features of *Hyalophora*, *Drosophila*, and *Glossina* ovarioles.

Compartments of each clone differentiate such that one becomes the oocyte and remainder (15 in *Drosophila*) become nurse cells. Nurse cells in both polytrophic and telotrophic ovarioles have greatly enlarged polyploid nuclei with massive nucleoli and are rich in nuclear pores. In 1901, Giardina, using *Dytiscus*, recognized that nurse cells and oocytes have a common origin and are sister cells. This compartmentalization and the amplification of synthetic activity in the nurse cells allows for the rapid production of eggs. The oocyte is the recipient of the synthetic efforts of a number of nurse cells rather than relying on its own single nucleus. Molecular biology research has characterized the many messenger RNAs and proteins that *Drosophila* nurse cells manufacture and transport through the ring canals (Figs. 5G and 5H) to the oocyte. The structure, biochemistry, and functioning of ring canals is an area of intensive current research. The oocyte nucleus (Fig. 5H) is quiescent, relying on the efforts of the nurse cells to supply most of the nonyolk ooplasmic constituents. *Drosophila* is truly an egg-laying marvel. It has provided a superb experimental model for research on how embryonic cells are established and the foundation for embryogenesis is laid down during oogenesis.

Each unit consisting of the oocyte and its cohort of nurse cells is encased in a single layer of follicle cells. Follicle cells in some Dipterans contribute components to the yolk along with the usual extraovarian sources. Follicle cells also become regionally specialized and secrete the structures of the chorion. Interfollicular tissue connects adjacent polytrophic follicles so one finds a progression of stages along the ovariole length.

ii. Telotrophic Ovarioles The telotrophic (acrotrophic in the old literature) ovariole subtypes (Figs. 5I–5L) also establish syncytial germ tissue via incomplete cytokinesis. However, the timing and structural

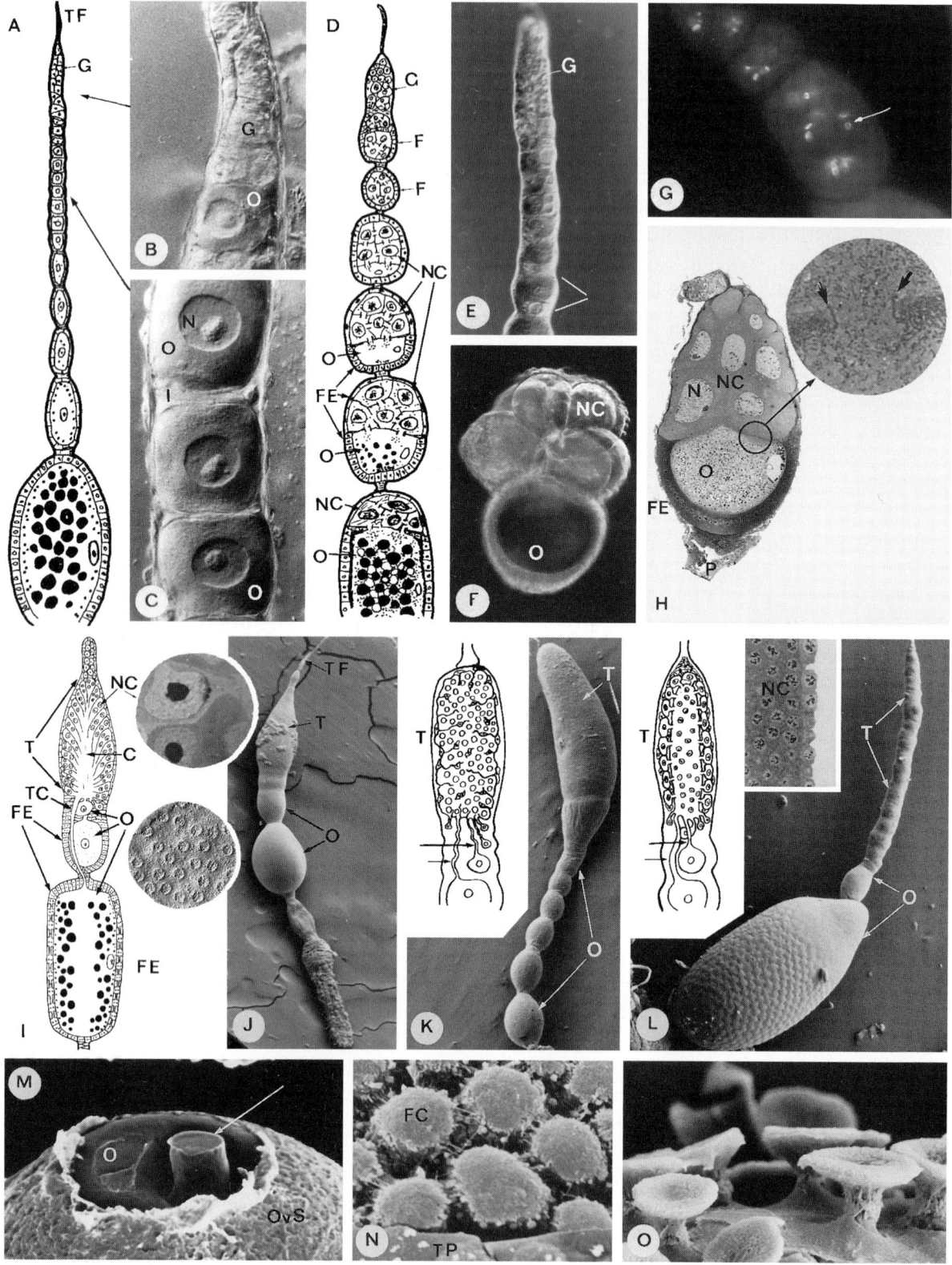

plan is totally different. The syncytial nature of the germ tissue is already established by incomplete cytokinesis during embryonic and postembryonic development of the ovary. During the last larval molt to the adult the syncytial germ tissue differentiates such that most of the compartments of a three-dimensionally complex syncytium become nurse cells, whereas a smaller number in the basal area become oocytes. There is a wide range of variation in the pattern of sibling cluster formation in taxa such as Hemiptera (Figs. 5I and 5J), some Coleoptera (Fig. 5K), and Ephemeroptera and Megaloptera (Fig. 5L). The nurse cells with enlarged nuclei rich in pores (Fig. 5I) for all the oocytes are collectively housed in a common chamber, the tropharium (Fig. 5M). Elongated cytoplasmic bridges or trophic cords connect the individual oocytes to the common nurse cell chamber. The structural organization of the tropharium varies between the Megalopterans, polyphage Coleopterans, and Hemipterans (Figs. 5J–5L). The Hemipteran tropharium has a nuclear-free core which contains an incredible array of microtubules. The nurse cells are housed in a lancelet or elongated structure (Fig. 5J). At the tropharium tip in some species, such as *Rhodnius,* there are mitotic cells. At the base of the tropharium are small oogonia which become encased by follicle cells derived from prefollicular tissue as the oocytes enlarge in previtellogen-esis. Nurse cell products are transported through elongated intercellular bridges, the trophic cords (Figs. 5I and 5K–5M). Upon growth, oocytes become displaced further from the tropharium with ever lengthening microtubule-rich trophic cords. Oocytes receive transport products throughout previtellogen-esis (e.g., *Oncopeltus*) and into early vitellogenesis (*Rhodnius*) when the cords close and vitellogenesis and chorion formation are completed. The number of growing follicles per ovariole varies considerably with species and may be highly regulated as in *Rhodnius*. The follicular epithelium in many telotrophic ovarioles become patent, creating extracellular spaces (Fig. 5N) so yolk precursors can reach the oocyte. These cells also secrete the often highly sculptured (Fig. 5N) chorions.

The functional interactions between nurse cells and oocytes relative to nurse cell functions and the mechanisms of nurse cell–oocyte transport have received much attention in recent years. Research efforts, which have focused intensely on a few model systems such as *Drosophila* and a few other common systems, have provided significant insights into how the various types of insect ovaries manufacture eggs. With the availability of new tools and current interests in evolutionary trends, new details about all three ovariole types are sure to emerge as more species are examined.

FIGURE 5 Ovariole types and features. These diagrams, light micrographs, and EMs depict a range of features of the panoistic, polytrophic, and telotrophic ovarioles. (A–C) Panoistic ovarioles: (A) the basic panoistic plan with the apical germarium (G) near the terminal filament (TF) and chain of developing follicles; (B and C) DIC light micrographs of Mantid ovaries depicting the germarial (G) area and early growing follicles with the large oocyte (O), nurse cell nucleus (N) with large nucleoli.(D–H) Polytrophic ovarioles: (D) a generic polytrophic ovariole with the germarial tip (G) and establishment of follicles (F) with follicle cells (FC) or cysts that have nurse cells (NC) and oocytes (O); (E) a DIC view of the germarium (G) of the cecropia moth and early follicles (white lines) (reproduced with permission from Huebner and Diehl-Jones, 1998); (F) a midstage follicle with nurse cells (NC) and oocyte (O); (G) staining of ring canals with immunostaining for the ring canal protein (P) hu-li-tai (hts) (photo by K. Yeow); (H) a section of the tsetse fly *Glossina austeni* (from E. Huebner *et al., Tissue Cell* 7, 535, 1975). Note the large nurse cell nuclei (N) and oocyte (O) connected by ring canal (arrow). (I–L) The three types of telotrophic ovarioles found in Hemiptera *Rhodnius* (I and J), polyphage Coleoptera *Ips* (K), and the Megaloptera *Sialis* (L) (J, reproduced with permission from E. Huebner, *Tissue Cell* **13**, 105, 1981; diagrams in K and L reproduced with permission from Huebner, 1984). Note the prominent nurse cell nuclei (I and inset) with extensive pores in the freeze fracture inset. NCs are in lobes around a central trophic core (C) of the tropharium (T). Trophic cords (TC) connect to the oocytes (O) (reproduced with permission from E. Huebner and H. Gutzeit, *Tissue Cell* **18**, 753, 1986). Eggs are surrounded by follicle cells. The insets in K and L illustrate the different cellular architectures of the three groups. (M) An SEM view of a trophic cord (arrow) entering the oocyte (O) (reproduced with permission from E. Huebner, *Tissue Cell* **13**, 105, 1981).(N) Follicle cells around a vitellogenic *Sialis* oocyte with the dark extracellular spaces. (O) The chorion that these cells eventually produce.

IV. TUBULAR TRACT: ANATOMICAL VARIATIONS, LATERAL OVIDUCTS, COMMON OVIDUCT, AND BURSA COPULATRIX

A. Anatomical Variation

As noted earlier, the ovarian somatic tissue and lateral oviducts are of mesodermal origin, whereas the remainder of the female tract, including glands, is ectodermal in origin. Comparatively there are three or four levels of complexity in the female tract. The simplest is in the Ephemeroptera, in which there are only lateral oviducts. Most insects have fusion of the lateral oviducts, with the common median oviduct derived from an invaginated pouch of the seventh abdominal segment. In these intermediate systems the common oviduct opens to the exterior via a gonopore. An example of this is the Dermaptera (Fig. 3). More complex and widespread systems have the common oviduct opening into a genital chamber (derived from invagination of the eighth and ninth segments) (Figs. 3 and 6A). A wide degree of variation exists in the structure of the genital chamber including its expansion or outpocketing into a bursa copulatrix. *Rhodnius* provides a basic example in which this chamber or vagina is the bursa copulatrix that serves as the female copulatory organ which receives the sperm-containing spermatophore (Figs. 6A and 6D). The opening of the female system at the end of the ninth segment is often called the vulva.

In some groups, primarily many Lepidoptera, cicadas, snake flies, and scorpion files, there are two openings to the female reproductive system (Fig. 3). One serves as a copulatory orifice and organ, whereas the other functions for oviposition. The posterior one serves to discharge the eggs and is the oviparous, whereas the anterior one is for copulation and opens into a bursa copulatrix which receives the penis and sperm. Sperm are carried from this tubular system by a narrow sperm duct to the common oviduct from where they can be passed into the spermatheca for storage (Fig. 3).

In viviparous insects such as the Dipteran *Glossina* and the roach *Diploptera,* the anterior part of the vagina or bursa can be greatly enlarged and modified into a uterus to house developing embryos and larvae.

B. Morphology and Function

Once eggs with their protective chorion have been made by the handy work of the ovary, the tract and associated organs are necessary for the storage, insemination, and laying of eggs. The initiation of the next generation begins with ovulation from the ovarioles into the lateral oviducts (Fig. 6A). In some species, such as unmated females of the Hemipteran *Rhodnius,* eggs can temporarily be stored in the pedicels and lateral oviducts, inhibiting further oogenesis until the female is mated. Normally, since mating includes the long-term storage of sperm, ovulated eggs are transported along and fertilized en route.

Morphologically, the lateral oviducts are lined by a simple cuboidal to columnar epithelium, rest on a

---

FIGURE 6   The tubular tract, spermatheca, and accessory glands. (A) The tract of *Rhodnius*; note the lateral oviduct (LO), the common oviduct (CO), the bursa copulatrix (BC) with the spermathecae (SP), and the accessory gland called the cement gland (CG) with its duct (arrows) opening at the posterior (reproduced with permission from Huebner, *J. Morphol.* **166**, 1, 1980). (B) Features of the common oviduct with the elaborate cuticle lining (arrows and inset), the epithelium (E), and extensive muscles (M). (C) The muscles and tracheoles in SEM. (D) The *Rhodnius* bursa copulatrix (BC) opened to reveal its interior and the cuticle lining which has small extensions (inset). (E and F) The muscular spermatheca and its duct (D) and a cracked view with sperm (arrow) in the lumen, the epithelial wall (E), and muscle (M) (reproduced with permission from Huebner, *J. Morphol.* **166**, 1, 1980). (G) The micropyles of the *Rhodnius* egg chorion (CH). In contrast to the *Rhodnius* tubular spermatheca, H shows the spherical spermathecae (arrows) of the dipteran *Glossina.*(I) The secretory and duct cell units of the *Rhodnius* cement gland in a light microscope section. In J, the cells have been digested away to show the cuticular lining of the duct cell (D) and end apparatus (EA) of the gland cell (I and J, reproduced with permission from D. Lococo and E. Huebner, *Tissue Cell* **12**, 557, 1980). (K) An egg during oviposition (reproduced with permission from G. Kelly and E. Huebner, *J. Morphol.* **199**, 175, 1989). The arrowhead highlights the cement gland adhesive applied to the ventral surface.

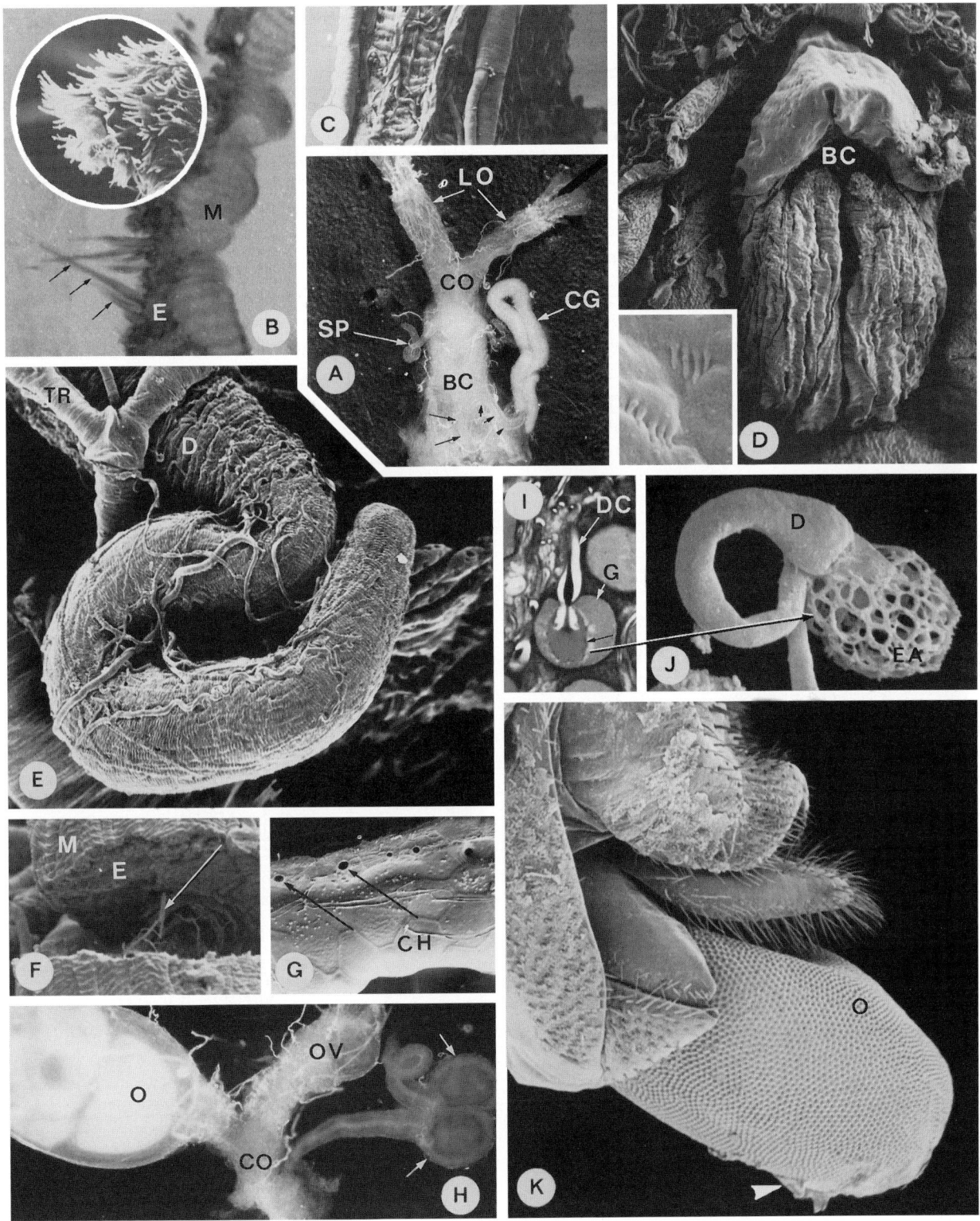

basal lamina, and small amount of connective tissue, and invested with a moderate layer of muscle tissue. The common oviduct is very similar, except it has a cuticular lining or intima reflective of its ectodermal origin (Fig. 6B). The muscle layer is more extensive as well (Figs. 6B and 6C). In many insects the spermathecal duct(s) opens at the level of the posterior area of the common oviduct, the juncture with the vagina, or the anterior dorsal area of the vagina (Figs. 6A and 6H). Eggs are fertilized and initiate early development as they enter the bursa copulatrix, which also has a cuticle lining (Fig. 5D) and is well endowed with muscles. Subsequently during oviposition the activities of the accessory glands come into play. The secretions may include gluing eggs down, encasing them in frothy or gelatinous secretions, and producing tough packets, the oothecae.

V. ACCESSORY GLANDS: TYPES, MORPHOLOGY, AND FUNCTIONS

As alluded to earlier, ectodermal diverticula of the female tract play important secretory functions and provide sperm storage. Both are essential to reproductive success. These ectodermal glands or ectadenes are usually considered in two categories; the spermatheca(e) and other accessory glands. The spermathecae function to store sperm and their ducts open into the vagina or, in a few cases, the common oviduct. The other ectodermal glands, the accessory glands, usually open in the ninth abdominal segment. Usually these are paired and called colleterial glands.

A. Spermatheca(e)

Spermatheca are found in all orders except Protura and Collembola. The number of spermathecae varies according to the group. For example, in the coleopteran *Blaps* there is one, in the hemipteran *Rhodnius* there are two, and in the dipteran *Drosophila* there are three. In the case of *Rhodnius* the spermathecae are not structurally homologous with other spermathecae but carry out the usual function of sperm storage.

Spermathecae can vary from a simple oval organ or pocket that acts as a spermatophore receptacle to a

pronounced cuticle-lined diverticular pouch or tube heavily invested with muscles (Figs. 6E and 6H). In *Rhodnius* the two finger-like diverticula open into the posterior common oviduct (Figs. 6A, 6E, and 6F). In Thysanura, Isoptera, and lower Orthoptera, the spermatheca opens just behind the juncture of the common oviduct and genital chamber, whereas in higher orders it often opens into the anterior dorsal part of the vagina or genital chamber. Detailed anatomical diagrams of the spectrum of insect orders are provided by Matsuda.

The spermathecae have a cuticular lining and vary from having a combined secretory and storage function in a tubular (Figs. 6A and 6E) or bulbous structure (Fig. 6H) to a more complex situation in which the secretory function is isolated into an associated spermathecal accessory gland. In *Rhodnius,* for example, the simple epithelial lining of the spermatheca (Fig. 7) also serves a secretory function, whereas in honeybees and most beetles there are distinct secretory glands with more complex epithelial structures.

FIGURE 7 This diagram summarizes the features of a typical type 1 secretory cell lining the spermatheca of *Rhodnius*. Note the cuticle lining (CU) and apical invagination (AI) (reproduced with permission from E. Huebner, *J. Morphol.* **166,** 1, 1980).

Because sperm can be stored for periods ranging from days and weeks to as long as 4 years in the case of queen honeybees, which can store up to 5×10^6 spermatozoa, the maintenance of sperm viability is an important factor. In general, the secretions of the spermatheca or gland presumably play a role in sperm storage. In *Rhodnius* the secretion is a mucoprotein providing a potential energy source.

In *Rhodnius* is the simplest form of epithelial structure termed a type I secretory cell (Fig. 7) according to a classification scheme proposed by Noirot and Quennedey in 1974. The columnar cells have an apical invagination where the secretion accumulates and then passes through the cuticular lining to the spermatheca's lumen, where the sperm are stored. A more complex morphology characterizes the structure of spermathecal glands as in bees and many beetles. Here there is also a cuticular lining which is produced by thin epithelial cells. This lining is perforated by many narrow cuticle-lined tubular invaginations that extend through the surface cells into enlarged secretory cells immediately below. These secretory cells have an apical invaginated area usually spherical in shape into which the cuticular tubules extend and may have an end apparatus. The secretory cells rich in RER and Golgi secrete into the invaginated cavity and the secretion passes into the cuticular tubule and out. There is an extensive literature on the ultrastructural features.

Coupled with the efficient and ready supply of sperm is the structural adaptation of the egg chorion to facilitate fertilization. As eggs pass down the tract, through the common oviduct, and into the bursa, the timely and controlled release of sperm occurs via extensive muscular contractions of the spermatheca. This region of the chorion has small channels or micropyles for sperm entry. The number and location of micropyles vary with species. In *Rhodnius* there are a number (10–20) of micropyles encircling the neck of the cap end (Fig. 6G). These are produced by retraction of a special group of elongate follicle cells that remain attached to the egg during chorion formation. In *Drosophila* there is a single micropyle, whereas in *Periplaneta* there are around 100.

Sperm reach the oolemma via the micropyle(s) and embryonic development is set in motion. The timely deposition of the eggs in an environment ap-

propriate for each species occurs next. Oviposition (Fig. 6K) is accompanied by the addition of products of accessory glands in most insects. These adhesives, coatings, casings, etc. help to maximize the survival of the embryos and/or larvae until hatching.

B. Other Nonspermathecal Accessory Glands

Nonspermathecal accessory glands vary greatly in structure, location, and function. Generally, they produce secretions which coat all or part of the egg surface. These can be proteins and mucoproteins that provide adhesion to substrates as in the cement glands of *Rhodnius* (Figs. 6A and 6I–K) and *Drosophila* colleterial glands. In groups that lay eggs in water (such as the Ephemeroptera, Trichoptera, and some Odonata) glands secrete gelatinous material that swells in water, providing a protective case. Insects such as the Mantidae encase their eggs in a frothy secretion that hardens, creating a hard spongy oothecal-like protective case. In the cockroach *Periplaneta* a hard, dark, tanned purse-like oothecal case is produced that protects the eggs. Adenotrophic viviparous species utilize their accessory glands to produce a protein-rich milk for the developing larvae. Many of the Hymenopterans have circumvented the reproductive function of their accessory glands into a defensive role producing poisons for their stinging apparatus.

Despite the variety of functions, many of the secretory cells of accessory or colleterial glands have a general morphology like the type 3 gland cell in the classification scheme of Noirot and Quennedy. By a complex developmental process a secretory unit is formed, with a secretory cell characterized by a large apical invagination. This cell is attached via septate desmosomal junctions to a tubular stalk cell that in turn inserts into the luminal epithelium. The unit has a cuticle lining which ends in an often enlarged end apparatus (Figs. 6I and 6J) that protrudes into the apical invagination of the secretory cell where the secretion is released. Figures 6A, 6I, 6J, and 8 provide views of this in the *Rhodnius* cement gland. This gland produces an adhesive protein applied to the ventral surface of the egg at ovulation (Fig. 6K).

The cockroach *Periplaneta* provides another valu-

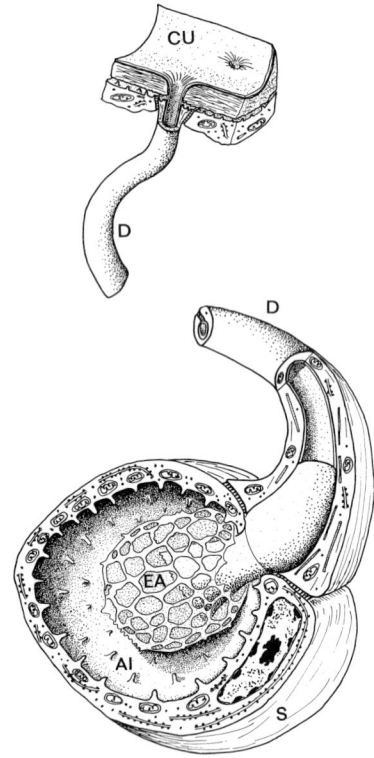

FIGURE 8 This diagram summarizes the features of the secretory cell duct cell units of the *Rhodnius* cement gland. Note the apical invagination (AI) of the secretory cell (S) with the end apparatus (EA) of the cuticle (CU) of the duct cell (D) (reproduced with permission from D. Lococo and E. Huebner, *Tissue Cell* **12**, 557, 1980).

able example. The right and left colleterial glands consist of many tubules and each side produces different secretions. The left has various secretory cell types that produce (i) the five structural oothecal proteins called oothecin, (ii) diphenoloxidase, (iii) the phenolic 4-0-β glucoside of protocatechuic acid, and (iv) structural protein and calcium oxalate. The right gland produces β-glucosidase. When eggs enter the posterior vagina and the right and left gland secretions are released and mix, a biochemical cascade results. The β-glucosidase liberates the protocatechuic acid, which is oxidized by the phenoloxidase to a quinone which cross links in the structural oothecins to produce the hardened, tanned ootheca which is deposited.

The orthopteran *Schistocerca* produces a foamy ootheca which interestingly is produced by secretory

gland cells located at the proximal end of the lateral oviduct and, unlike other accessory glands, is of mesodermal origin. The secretion comes not only from the gland at the anterior of the oviduct but also from the oviduct lining cells.

These examples draw attention to the diversity of locations and functional roles female accessory glands play. Each species has evolved a set of secretions which have enhanced their reproductive success and in turn promoted the survival of the species.

VI. SUMMARY

In summary, the accumulation and organization of the ooplasmic constituents during oogenesis in the ovary, the facilitation of internal fertilization, secretion of a protective egg chorion, and accessory gland secretions all contribute to reproductive success. The diverse spectrum of variation of each of these components of the female reproductive tract are major factors in the remarkable evolutionary success of insects.

See Also the Following Articles

Insect Reproduction, Overview; Male Reproductive System, Insects

Bibliography

Bate, M., and Martinez Arias, A. (Eds.) (1993). *The Development of Drosophila melanogaster*. Cold Spring Harbor Laboratory Press, Plainview, NY.

Büning, J. (1994). *The Insect Ovary: Ultrastructure, Previtellogenic Growth and Evolution*. Chapman & Hall, New York.

Davey, K. G. (1985). Reproduction. In *Comprehensive Insect Physiology, Biochemistry and Pharmacology* (G. A. Kerkut and L. I. Gilbert, Eds.), pp. 15–37. Pergamon, New York.

Huebner, E. (1984). The ultrastructure and development of the telotrophic ovary. In *Insect Ultrastructure* (R. C. King and H. Akai, Eds.), Vol. 2. pp. 3–48. Plenum, New York.

Huebner, E., and Diehl-Jones, W. (1998). Developmental biology of insect ovaries: Germ cells and nurse cell–oocyte polarity. In *Microscopic Anatomy of Invertebrates, Insecta* (F. W. Harrison and M. Locke, Eds.), Vol. 11C, pp. 957–993. Wiley-Liss, New York.

Mahajan-Miklos, S., and Cooley, L. (1994). Intercellular cytoplasmic transport during *Drosophila* oogenesis. *Dev. Biol.* **165**, 336–351.

Margaritis, L. H. (1985). Structure and physiology of the egg shell. In *Comprehensive Insect Physiology, Biochemistry and Pharmacology* (G. A. Kerkut and L. I. Gilbert, Eds.), pp. 153–213. Pergamon, New York.

Matsuda, R. (1976). *Morphology and Evolution of the Insect Abdomin.* Pergamon, New York.

Snodgrass, R. E. (1935). *Principles of Insect Morphology.* McGraw-Hill, New York.

Stys, P., and Bilinski, S. (1991). Ovariole types and the phylogeny of hexapods. *Biol. Rev.* **65**, 401–429.

Female Reproductive System, Nonhuman Mammals

Robert A. Dailey

West Virginia University

I. Introduction
II. Functions of the Reproductive System
III. Classes of Mammals Based on Reproductive Systems
IV. Development of the Reproductive System

GLOSSARY

cervix A structure considered the neck of the uterus; it effectively closes the uterine opening and functions as a sperm receptacle in some species.

corpus luteum The ovarian structure that forms after release of the egg and secretes progesterone.

follicle The ovarian structure that houses the oocyte and secretes primarily estrogen.

infundibulum A funnel-shaped structure at the tip of the oviduct that picks up the egg when released by the ovary.

ovary The primary sex organ of the female.

oviduct The tubal structure that is the passageway for the egg to the uterus and is the site of internal fertilization.

ovulation Release of the egg from the follicle.

ovum/egg A mature female gamete; almost all germ cells in the ovary are primary oocytes.

uterus The nutritive and protective organ for the fetus.

vagina The female copulatory organ and birth canal.

vulva The common passageway for the products of reproduction and for urine.

The female reproductive system consists of primary and secondary components essential for sexual reproduction. This system provides the female gamete, serves as the site for internal fertilization, provides for maternal nourishment of the embryo/fetus, and delivers the offspring.

I. INTRODUCTION

The female reproductive system is composed of cranial and pelvic portions. The cranial component includes the hypothalamus and pituitary gland. The hypothalamus integrates the internal and external messages that it receives and delivers an appropriate response through its neural secretions. One message is the hormone, gonadotropic hormone-releasing hormone, that acts on the anterior pituitary gland to promote secretion of luteinizing hormone (LH) and follicle-stimulating hormone (FSH). Another anterior pituitary hormone, prolactin, is under principally negative control by the hypothalamus. These hormones regulate most of the functions of the reproductive system. Another hypothalamic hormone, oxytocin, is secreted by cells in the hypothalamus and stored in the neural cell endings in the posterior pituitary. This hormone is released by the posterior pituitary and promotes contractions of smooth muscles of the reproductive tract, particularly for movement of gametes and delivery of the newborn, and the mammary glands for milk secretion. The pelvic portion consists of the primary sex organs, the female gonads (paired ovaries), and the secondary sex organs (the reproductive tract). The ovaries lie in the abdominal area in the lumbar region. Their positioning varies among species but in general is more caudal in species with a simplex uterus than in those with a more tortuous uterus. The reproductive tract opens through the pelvic area via the vulva in most species and is a common pathway for excretion of urine and for reproductive processes. This article addresses the anatomy of the pelvic portion of the female reproductive system.

II. FUNCTIONS OF THE REPRODUCTIVE SYSTEM

Gonads of both genders are cytogenetic organs. The ovaries are the sites of development and growth of the follicles, which house the female germ cells. The mature female germ cells, the ova, are released from the ovary at ovulation, completing the gametogenic or exocrine function of the ovary. In addition, the ovary is an endocrine organ and secretes the female hormones, progesterone and estrogen, in varying amounts throughout sexual development and reproductive cycles. These hormones are responsible for development of the female reproductive tract, maintenance of pregnancy, induction of the female mating response (estrus), and the development of the secondary characteristics that denote femaleness in a species, i.e., plumage, hair patterns, body size, and shape. The structure of the female reproductive tract is diverse among species of mammals, but its function is constant. The tubular portion consists of the oviduct, uterus, and vagina. The functions include transport of the ova and sperm in the tract, provision of the sites for fertilization and for development and growth of the embryo and fetus, and delivery of the fetus, and it serves as the organ for copulation and produces endocrine and exocrine products.

III. CLASSES OF MAMMALS BASED ON REPRODUCTIVE SYSTEMS

The class Mammalia can be divided on the basis of type of reproduction into two subclasses. Prototheria (Greek *Protos*, first; *therios*, beast) are egg-laying (oviparous) mammals. Belonging to the order Monotremata (Greek *monos*, single; *trema*, opening), these animals are so named because they have a cloaca, which is the single opening for excrement and reproduction. These mammals lay eggs with large yolks and tough, soft shells. They do not have a uterus or vagina, and their mammary glands are without teats or nipples. The duckbill platypus lays two eggs into a nest. After hatching, the young lap milk from mammary glands on the abdomen. The spiny anteater lays one egg, which is incubated in a pouch on the abdomen. Animals in the subclass Theria give birth to living young (viviparous) and belong to either Metatheria (Greek *meta*, between) or Eutheria (Greek *eu*, true) infraclasses. The order Marsupialia (Latin *Marsupium*, pouch) is in the Metatheria and includes the kangaroos as well as other marsupial species in Australia and the opossum in the Americas. These species have a short gestation and a poorly developed placenta and give birth to young that are in a very immature stage. The young migrate to the marsupium, an integumentary pouch, where they attach to a teat or nipple of the mammary gland. Eutheria (placental) is the largest group of mammals. These species also have separate anal and urogenital openings. Their eggs are microscopic with practically no yolk and no shell. They have embryonic membranes and well-developed placentas. The period of gestation varies from 20 days in the mouse to 22 months in the elephant. Their mammary glands are well developed with teats.

IV. DEVELOPMENT OF THE REPRODUCTIVE SYSTEM

A. The Ovary

1. Differentiation

The indifferent gonad forms as a bulge of tissue, the genital ridge, on the ventromedial aspect of each mesonephros. The germ cells arise as large cells [primordial gem cells (PGCs)] in the yolk sac and migrate to the genital ridge. These cells penetrate through the covering (germinal epithelium) into the medulla of the developing gonad, carrying with them associated mesenchymal cells, forming primary sex cords. If the gonadal tissue is genetically female, this medullary proliferation degenerates and is followed by a second proliferation of PGCs, which go only into the cortex (cortical proliferation). The PGCs continue to undergo mitosis (Fig. 1) until some time after entering the cortex. The female complement of germ cells is fixed at this point, which usually occurs before birth. As the second proliferation occurs, the cells enter in a cord-like fashion, giving rise to the term Pfleuger's egg cords. These cords break up, and many of the oogonia become connected by nexi, forming egg nests. Later, the egg nests break apart as mesenchymal cells differentiate and surround each oogonium.

The development of female germ cells from PGCs through complete ovum formation is oogenesis. The germ cells (oogonia) begin maturation and undergo meiosis through pachytene and diplotene (dictyate) stages of first meiosis. Most of the oocytes degenerate by the dictyate stage. The development becomes arrested at this stage, and meiosis will not continue until the inhibition is removed, which occurs with ovulation.

2. Follicles

The formation and development of follicles is folliculogenesis. The ovum with a single layer of flattened cells surrounding it is called a primordial follicle. The cells are surrounded by a basement membrane, which is not penetrated by blood vessels. Thus, nutrients and wastes travel to and from the ovum through diffusion processes. Development of follicles will proceed through several morphological stages (Fig. 2). In the first, the cells surrounding the ovum become cuboidal. This is the primary follicle because of the single layer of cells, which is called the granulosa. The primordial and primary follicles are the follicular reserve that will be used later in life. Development of the primary follicle into subsequent stages is termed recruitment, and the associated group of developing follicles is often called a cohort or crop of follicles. Recruitment from the reserve stage is thought to be under the influence of FSH. As mitosis occurs in the granulosa, the cell layers increase. In addition, stromal cells begin to organize around the outside of the basement membrane. These cells form the theca interna. Later, further differentiation of stromal cells forms the theca externa, which merges with the theca interna, without a clear demarcation of the two layers. This is the secondary follicle. In secondary follicles with multiple layers of granulosa, the oocyte is encased by the zona pellucida, a layer of mucopolysaccharides and proteins secreted by the granulosa cells. The area between the zona pellucida and the outer membrane (vitelline) of the oocyte is the perivitelline space. Development of follicles to this stage can occur in hypophysectomized (pituitary gland removed) animals and is termed gonadotropin independent. At this point in most species of mammals, the granulosa, under the influence of FSH, begin secreting fluid. As this fluid coalesces, lacunae form, and as these coalesce further, an antrum (cavity) is formed. The fluid is called follicular fluid. This type of follicle is the tertiary follicle, also referred to as antral or vesicular follicle. As the antrum forms, the oocyte is moved from its central position and toward the wall of the follicle (eccentric). Cells lining the wall of the follicle are mural or basal granulosa. Granulosa surrounding the developing ovum are called cumulus cells. The cumulus grades into the mural cells, and the area formed is the cumulus oophorous (egg-bearing hill). Follicles at this stage become dependent on gonadotropic support (gonadotropin dependent). Inadequate support will result in death of the follicle and ovum, a process referred to as atresia (failure to open). The vast majority (>90%) of follicles become atretic and undergo degeneration. Development of follicles into the gonadotropin-dependent stage is termed selection. As the follicles continue toward final maturation, some follicles are favored or exert selected growth over other growing follicles and promote atresia of those follicles. The selected follicles continue to grow and ovulate. Although growth of follicles from the reserve pools is continuous, the waxing and waning of the larger antral follicles appears in some species as waves. For example, the cow usually has two or three waves of

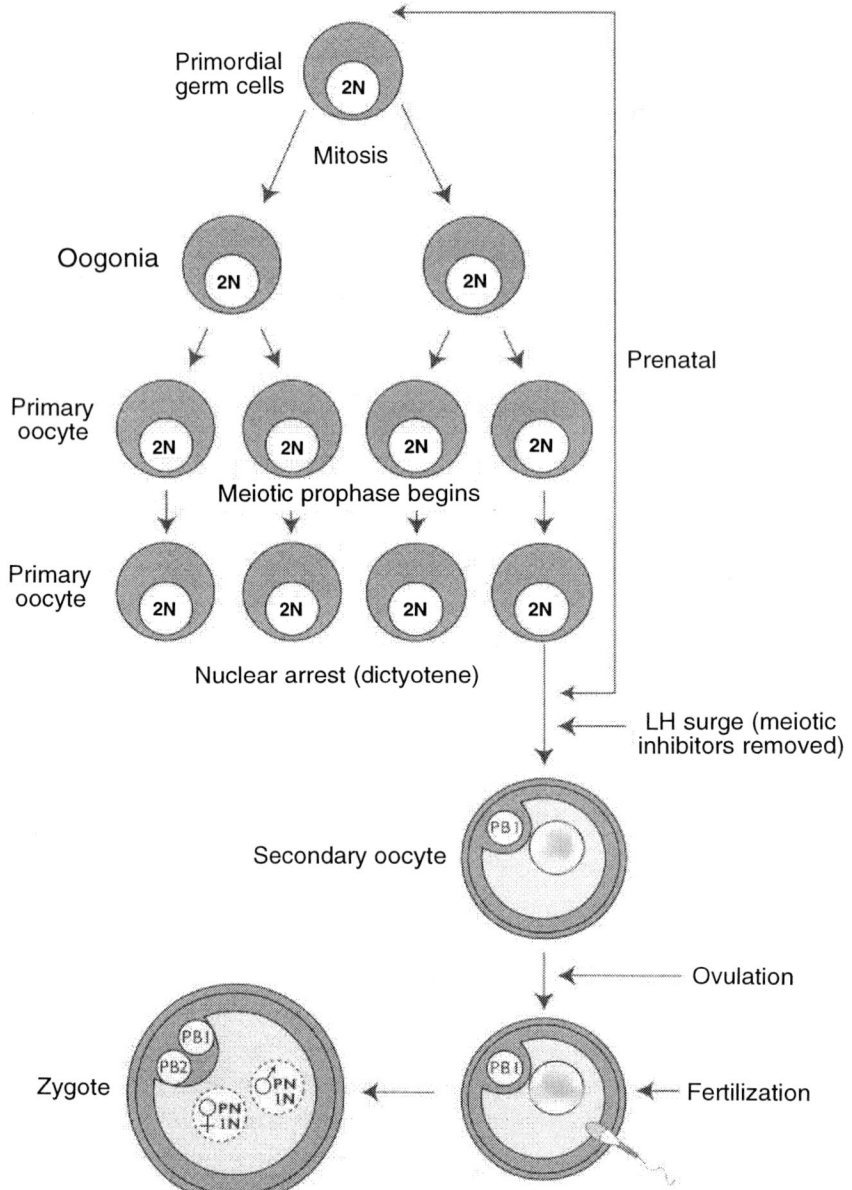

FIGURE 1 Schematic of oogenesis. Mitosis of oogonia yields primary oocytes until early follicular formation. No further mitosis occurs. Meiosis, germ cell division, begins in the primary oocyte and is arrested in the dictyotene (dictyate) stage. Resumption of meiosis, either after fertilization in some species or after the LH surge in others, results in division of the duplicated pairs of chromosomes (tetrads) between the secondary oocyte and the first polar body (PB1); the latter eventually degenerates. Division of the chromosomes between the ovum (ootid) and secondary polar body (PB2), which also degenerates, yields a haploid number of chromosomes and completes second meiotic division. Union of the male pronucleus with the female pronucleus at fertilization results in the normal complement of chromosomes (2N) in the zygote [from Senger, 1997 (graphic by Sonja Oei). Adapted with permission from Current Conceptions, Pullman, WA].

growth of a large follicle. These preovulatory follicles are called dominant at impending ovulation because of the clear separation, based on diameter, of the

follicle(s) that will ovulate from the subordinate follicles. Polytocous species give birth to multiple young and hence develop numerous follicles simultane-

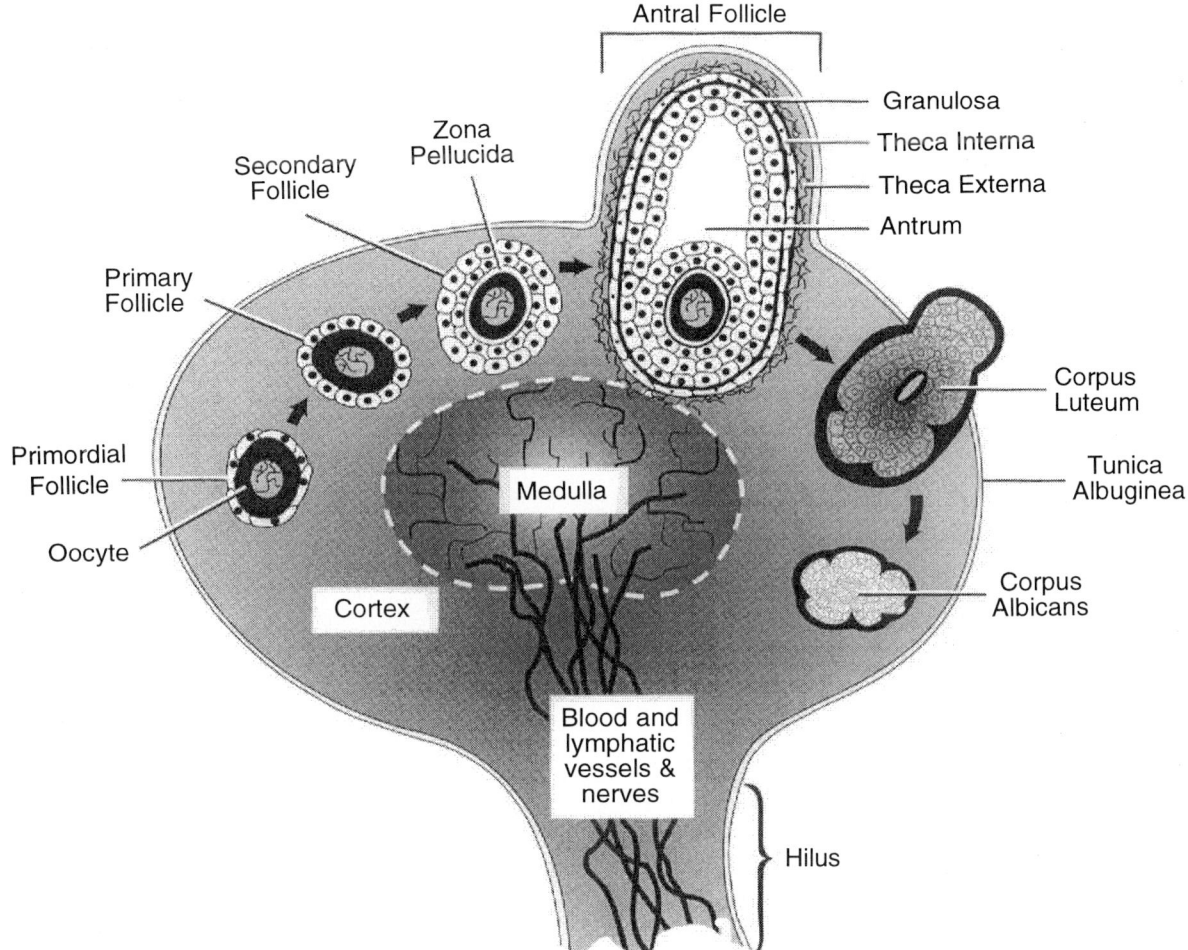

FIGURE 2 Schematic of the mammalian ovary. The stages, in order, that a follicle might attain are the primordial, primary, secondary, tertiary (antral), and ovulated. The latter is termed a corpus luteum when active and corpus albicans when no longer active. Potentially, all structures can be present at one time in the mature ovary; however, this relationship varies with reproductive stage and is species dependent [from Senger, 1997 (graphic by Sonja Oei). Adapted with permission from Current Conceptions, Pullman, WA].

ously, whereas monoecious species generally develop single follicles that ovulate. Not all mammalian species form antral follicles (e.g., Madagascar hedgehog). In others (rabbit), the antrum is incomplete with trabecular strands of cells connecting with the cumulus. Ovulation of follicles in the majority of species can occur anywhere on the ovary, except where the blood and nerve supplies enter at the hilus. In equids, however, ovulation occurs at the ovulation fossa because the ovary is inverted, with the cortex inside the medulla except at the ovulation fossa.

3. Corpus Luteum

After ovulation, the follicle is transformed first into a corpus hemorrhagicum (blood-filled body) and then into a corpus luteum (yellow body, although in many species the structure is red). The corpus luteum, except for the mare, can be seen on the surface of the ovary. The active corpus luteum secretes progesterone. Mammals are classed either as being spontaneous ovulators and responding to an endocrine stimulus or as induced ovulators and responding to neural afferents (mating stimulus). Likewise, function of the corpora lutea is either spontaneous or induced. The various combinations of types of ovulation and luteal function are species specific. The rabbit is an induced ovulator and has spontaneous luteal function; the mink is an induced ovulator and has seasonally induced luteal function; the domestic farm species are spontaneous ovulators and have spontaneous luteal function; and rodents are spontaneous ovulators with mating-induced luteal function. After function ceases, the corpus luteum regresses. The degenerated form is the corpus albicans (white body). The mature ovary can contain all classes of follicular and luteal structures. Thus, the ovary is a dynamic organ and its gross, observable form depends on the number and size of the structures visible on the surface of the ovary.

B. The Tubular System

1. Differentiation

The indifferent (i.e., before determination of sexual phenotype) stage is characterized by development of two tubular systems: the mesonephric and paramesonephric. The mesonephric ducts connect with the mesonephros. Each of the paired mesonephric kidneys is connected by several mesonephric ducts to the mesonephric tubules, which fuse with the urogenital sinus. At a comparable period, a second duct system forms near the mesonephric ducts as a pair of paramesonephric ducts, which also fuse with the urogenital sinus. Once sexual determination is made, one duct system proliferates while the other degenerates. In the female, the paramesonephric, or Müllerian, ducts develop, whereas the mesonephric, or Wolffian, ducts degenerate. The opposite occurs in the male. Noteworthy is the fact that it is the differentiation of the indifferent gonad into the testis that activates the embryonic testis to secrete testosterone and anti-Müllerian hormone, which results

---

FIGURE 3   Schematic of the types of uteri in mammals. The black areas represent the cervix of the uterus. The uterus is composed of the cervix, uterine body, uterine horns, and oviducts. The type of uterus reflects the amount of fusion of the components [from Senger, 1997 (graphic by Sonja Oei). Adapted with permission from Current Conceptions, Pullman, WA].

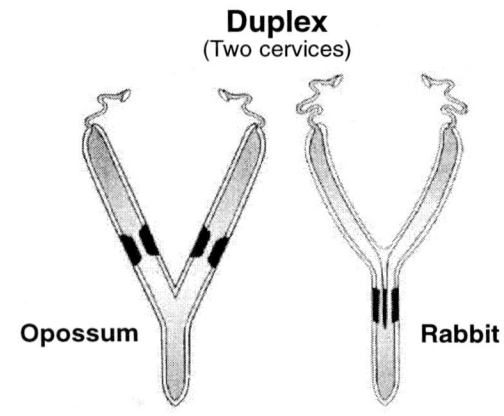

Duplex
(Two cervices)

Opossum   Rabbit

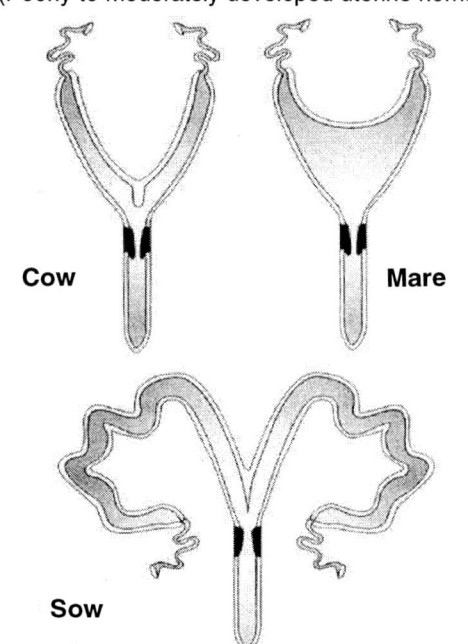

Bicornuate
(Poorly to moderately developed uterine horns)

Cow   Mare

Sow

Simplex
(No uterine horns)

Primate

in differentiation of the male system. Otherwise, the male system fails to develop and the female system is formed. In early experiments, rabbits that were gonadectomized before differentiation developed the female system. Placing a testis next to the ovary in female embryos resulted in regression of the female duct and development of the male duct, whereas placement of testosterone resulted in development of the male duct and the maintenance of the female duct. These results demonstrated the separate roles of testosterone and anti-Müllerian hormone. Remnants of the regressed mesosnephric system sometimes exist in appendage form and can be seen in the adult female. An example is Gartner's duct in the female pig, which lies parallel to the female tract and is embedded in the supporting connective tissue. An interesting anomaly (abnormality in development) occurs in female calves (Freemartin) born co-twin to a male. Because the twins share a blood supply, the anti-Müllerian hormone produced by the male promotes regression of the female system. Hence, the female reproductive system fails to develop beyond the urogenital stage, forming a blind vaginal cavity, whereas the male develops a proper duct system.

2. Oviduct

The duct system (Fig. 3) forms by canalization (hollowing out) of the tubules and fusion of the paired ducts. Variation in the amount of fusion of the ducts gives rise to the different types of female tracts. The Müllerian duct forms the oviduct, the uterus, and the anterior vagina. As with other hollow organs of the body, it is organized into four concentric layers: serosa, muscularis, submucosa, and mucosa. The serosa is the outer layer of squamous cells. The muscularis consists of an outer longitudinal and an inner circular layer of smooth muscle. The submucosa is the supporting layer for the mucosa and contains the nerves and circulatory vessels for the duct. The mucosa is secretory epithelium and varies in thickness from simple columnar in the oviduct to stratified squamous epithelium in the posterior vagina. The oviduct is paired and consists of the infundibulum, ampulla, and isthmus. The end nearest the developing ovary is open and shaped like a funnel,

hence the name infundibulum. The free end has microfilaments, called fimbriae, which make contact with the ovarian surface and pick up the ovum at ovulation. The ovum is transported by ciliary action and muscular contraction into the ostium of the ampulla (sac). The ampulla is the site of fertilization and early embryonic cleavage. The isthmus (narrow passage between two larger cavities) has a thicker muscularis and fewer folds of the mucosa. It connects to the uterus by the uterotubal junction. In some species (e.g., bat and mare), the ampullary–isthmus junction prevents unfertilized oocytes from entering the uterus, and the oocytes accumulate in the oviduct. The uterotubal junction might regulate the entry of embryos into the uterus.

3. Uterus

The amount of fusion of the remaining duct results in the simplex or duplex uterus. The simplex uterus results from complete fusion of both ducts except for the oviducts and is representative of the primates, the nine-banded armadillo, and the tamandua. Incomplete fusion of the ducts gives rise to the duplex (double) uterus. In this instance, the duct forms paired uterine ducts that connect with paired anterior vaginal ducts. The connecting portion is termed the cervix (neck) of the uterus. This type of uterus occurs in rodents. Partial fusion of the uterus results in a single cervix and uterine body (the fused portion) and paired uterine horns (the nonfused portions). This type of uterus is termed the bicornuate (two horns). This designation is classified further as bipartite, when residual tissue remained as a septum instead of complete separation, as occurs in ruminants. Because of confusion of these terms, bicornuate might be used for all classes with a single cervix and uterine body and paired uterine horns. The degree of fusion leads to a short uterine body and relatively long uterine horns (e.g., pig) or to a long uterine body and relatively short uterine horns (e.g., ruminants). Notably, species that are litter bearers have longer horns and more of the reproductive tract projects anterior to the pelvis. The names of the layers (Fig. 4) of the uterus (Greek *metrus*, uterus) are the perimetrium (serosa), myometrium (muscularis), and endometrium (mucosa and submucosa). The peri-

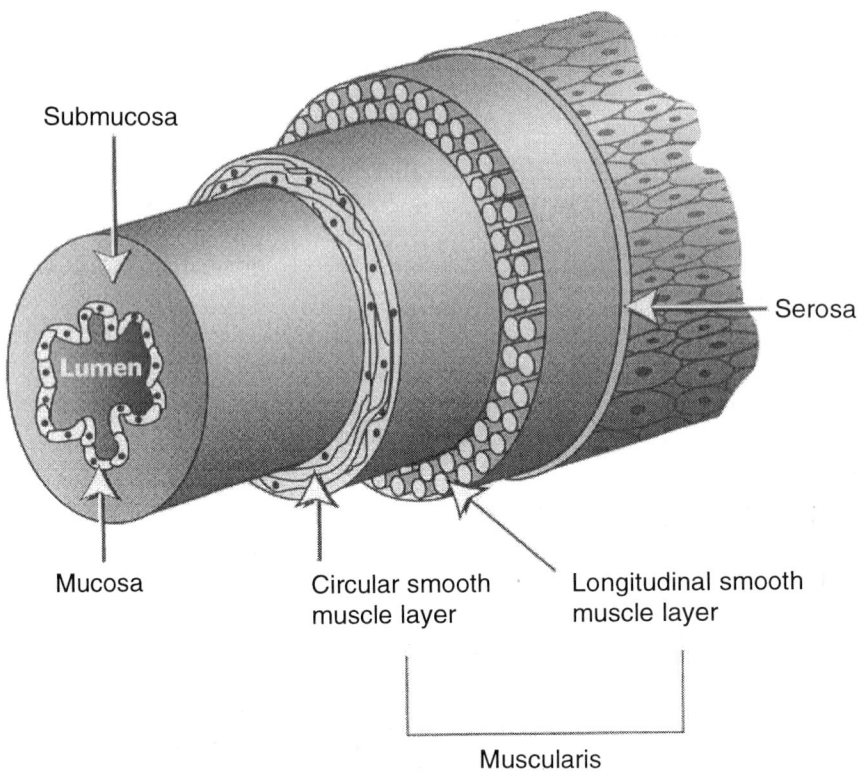

FIGURE 4 Schematic of the layers of the tubular portion of the reproductive tract. The lumen is lined by epithelial cells (mucosa). The outermost layer, serosa, is continuous with the peritoneum [from Senger, 1997 (graphic by Sonja Oei). Adapted with permission from Current Conceptions, Pullman, WA].

metrium is continuous with the serosa of the mesosalpinx. The myometrium has longitudinal and circular smooth muscle layers. It is responsible for uterine motility, tone, and expulsion of the fetus. The endometrium has uterine glands, which secrete substances to support the early embryo. The number of ciliated cells is reduced from that in the oviduct. The thickness of the endometrium varies with stages of the reproductive cycle; estrogen promotes development and progesterone promotes secretory activity. Unlike the primates that slough the endometrial lining at menses, nonprimates have minimal changes in the endometrium and no menstruation. Additionally, the endometrium of these mammals produces prostaglandin $F_{2\alpha}$, the hormone that causes regression of the corpus luteum in species with spontaneous luteal function. Ruminants have highly vascularized regions in the nonglandular areas, which protrude from the surface. These areas, the caruncles,

are the sites of the fetal attachments and in pregnancy form the cotyledons or placentomes. Other species (equine) develop endometrial cups during pregnancy that secrete a luteotropin (equine chorionic gonadotropin). Some species (swine and equine) have a diffuse fetal attachment, whereas others (cats and dogs) have a zonary attachment in which the chorionic villi are arranged in an equatorial belt. Rodents and primates have a discoidal arrangement of the chorionic villi.

The types of placental attachment are also named on the basis of the degree of erosion of the tissues separating the maternal and fetal blood. The layers of attachment on the maternal side are uterine vascular endothelium, connective tissue, and the epithelium; on the fetal side the layers are allantoic–chorionic epithelium, connective tissue, and vascular epithelium. In the epitheliochorial placenta, there is no erosion, and the uterine epithelium rests on the allanto–chorio epithelium as seen in the farm species. Felines and canines have an endotheliochorial placenta in which the maternal epithelium and connective tissue have been eroded. Primates have a hemochorial arrangement in which the maternal side has been completely eroded. Rodents have a hemoendothelial type, and only the fetal vascular lining separates the two blood supplies. A major modification of the posterior uterus occurs where it enters into the vagina. The tissue becomes less elastic and thick walled and is termed the cervix. The anterior end is continuous with the uterus, whereas the posterior protrudes into the vagina. In species with a duplex uterus, the cervix can be paired as in the rodents, but in the bicornuates it is fused into a single cervix. In some species the protrusion is gradual, as in the pig, whereas in ruminants, the cervix has an abrupt entrance resulting in formation of a fornix vaginae around the cervix. The cervix in canines and primates enters the vagina at an angle. The cervical canal takes different shapes. For example, ruminants have transverse annular rings; the pig has interdigiting pads, which conform to the corkscrew configuration of the boar's glans penis; and the mare has longitudinal folds. There are numerous mucus-secreting cells. Maximal cervical mucus is secreted prior (proestrus) to sexual receptivity (estrus). During pregnancy, the mucus thickens into a gel-like plug. In rodents, the mucus thickens after mating, forming a temporary cervical plug which is expelled within hours. Observance of the plug is a sign that mating occurred. The cervix is the neck of the uterus and serves primarily to keep the uterus from becoming contaminated. It also is a site of deposition of sperm in some species and has crypts that act as sperm reservoirs.

4. Vagina

The anterior vagina is the fused posterior portion of the Müllerian duct in most species. In marsupials, the vagina is separated into lateral vaginal canals, which accommodate the forked glans penis of the male, and a central birth canal. The vagina is the female copulatory organ and is lined with some ciliated cells and abundant mucus-secreting cells and some glands. The posterior vagina and vestibule are derived from the uorgenital sinus, which invaginates

from the exterior. Thus, the epithelium is stratified squamous and becomes keratinized as it continues to the external surface. The stratified epithelium cornifies under the influence of estrogen. Because of the changes in cell types, vaginal smears are used to follow the reproductive cycle of laboratory rodents. The vestibule is that portion of the reproductive tract that is common to the reproductive and urinary systems. The vestibule joins the vagina at the external urethral orifice. Vestibular glands produce lubricating mucus. The external portion of the vestibule is termed the vulva. It consists of inner labia minora, the homolog of the prepuce in the male, and outer labia majora, the homolog of the scrotum in the male. The clitoris, the homolog of the glans penis in the male, is erectile tissue located ventrally inside the labia. Rodents have separate urinary and reproductive openings and therefore do not have a vulva.

C. Supporting Tissue

1. Broad Ligament

The reproductive tract forms in a retroperitoneal position of the abdominal cavity (Fig. 5). As the tract grows it becomes enveloped in the peritoneum. The peritoneum fuses, and the tract is then suspended from the dorsolateral wall of the pelvis. This single sheet of connective tissue is called the broad ligament. The portion that suspends the ovaries is called the mesovarium. The mesovarium houses the nerves and blood vessels, which enter the ovary at the hilus. The mesosalpinx suspends the oviduct (salpinx). The mesometrium supports the remaining tract including the anterior vagina. The suspension of the reproductive tract results in the presence of two genital pouches. The rectogenital pouch is that portion of the body cavity between the rectum and the roof of the vagina. The vesicogenital pouch is that portion of the body cavity between the floor of the vagina and the bladder. These pouches are lined with peritoneum. The vagina and vulva are supported by loose connective tissue in the area caudal to the peritoneum, the retroperitoneal area. In large species, these tissue arrangements allow for rectal palpation of the reproductive tract or transrectal ultrasonic examination to monitor reproductive activity. Modification

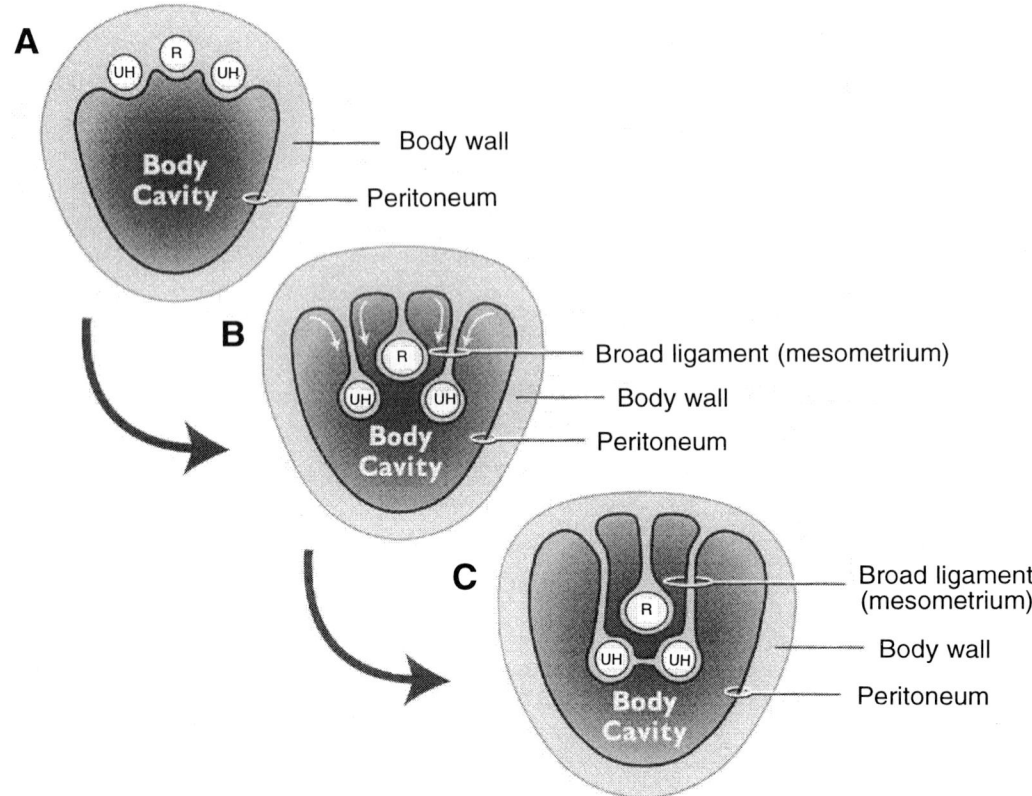

FIGURE 5 Schematic of the embryonic development of the broad ligament that supports the reproductive tract. The peritoneum is the connective tissue layer that surrounds the inner body wall and separates the body wall from the abdominal (body) cavity. As the rectum (R) and the uterine horns (UH) develop, each becomes enveloped by the peritoneum [from Senger, 1997 (graphic by Sonja Oei). Adapted with permission from Current Conceptions, Pullman, WA].

of the mesosalpinx and its position and the position of the oviduct result in the formation of a bursa around the ovary of most species. This membranous pouch may enclose part of or the complete ovary. The opening may be a slit, as in the rat, or a large opening, as in the cow.

2. Nervous and Vascular Tissue

Ovarian arteries or uteroovarian arteries branch from the dorsal aorta to supply the ovaries, oviducts, and anterior uterine horns. The middle uterine artery supplies the remaining uterine horn and the uterine body. The hypogastric artery supplies the cervix, vagina, and vulva. The uteroovarian vein drains the ovary, oviduct, and much of the uterus into the inferior vena cava. In species in which the uterine artery coils around the uteroovarian vein (e.g., ruminants), there is countercurrent exchange of $PGF_{2\alpha}$ from the

vein to the artery; this allows for local control of the life span of the corpus luteum. The pelvic nerve (parasympathetic) and the hypogastric nerve (sympathetic) innervate the reproductive tract.

See Also the Following Articles

Guinea Pig, Female; Marsupials; Parturition, Nonhuman Mammals

Bibliography

Asdell, S. A. (1964). *Patterns of Mammalian Reproduction*, 2d ed. Cornell Univ. Press, Ithaca, NY.

Bearden, H. J., and Fuquay, J. W. (1997). *Applied Animal Reproduction*. Prentice Hall, Upper Saddle River, NJ.

Bookhout, C. G. (1945). The development of the guinea pig

ovary from sexual differentiation to maturity. *J. Morphol.* 77, 233–263.

Mossman, H. W. (1953). The genital system and the fetal membranes as criteria for mammalian phylogeny and taxonomy. *J. Mammal.* 34, 289–298.

Mossman, H. W., and Duke, K. L. (1973). *Comparative Morphology of the Mammalian Ovary.* Univ. of Wisconsin Press, Madison.

Senger, P. L. (1997). *Pathways to Pregnancy and Parturition.* Current Conceptions, Pullman, WA.

Thibault, C., Levasseur, M., and Hunter, R. H. F. (1993). *Reproduction in Mammals and Man.* Ellipses, Paris.

Wodsedalek, J. E. (1963). *General Zoology.* W. C. Brown, Dubuque, IA.

Zuckerman, S. (1962). *The Ovary.* Academic Press, London.

Female Reproductive System, Reptiles

Valentine A. Lance

Center for Reproduction of Endangered Species

I. Ovary
II. Oviducts

GLOSSARY

atresia The process that occurs in ovarian follicles that fail to ovulate or fail to reach maturity. The basement membrane surrounding the oocyte breaks down, granulosa and macrophages invade, and the yolk is eventually resorbed.

germinal epithelium A layer or bed of germ cells in the ovary that divide mitotically and give rise to primary oocytes.

infundibulum The thin-walled anterior portion of the oviduct adjacent to the ovary into which oocytes are released at ovulation.

primordial follicles or *primary oocyte* Oogonia that have entered meiotic prophase I and are surrounded by rudimentary granulosa cells and connective tissue stromal cells.

pyriform cell A unique type of granulosa cell that forms an intercellular bridge with the oocyte. These cells occur only in the previtellogenic follicle of snakes and lizards and degenerate once vitellogenesis begins.

zona pellucida The region between the oocyte and the granulosa layer in preovulatory follicles; consists of a region of microvilli protruding from the surface of the oocyte.

The reproductive tract of the female reptile consists of a paired ovary and paired oviducts, which develop from the embryonic Müllerian ducts. In many species the right ovary is located slightly anterior to the left, is larger, and produces more eggs. In viviparous species the oviduct is modified for nutrient exchange.

I. OVARY

Reptile ovaries consist of a connective tissue stroma within which are found the developing oocytes, the germinal epithelium or germinal beds, blood vessels, and neural elements enclosed within the peritoneum. A connective tissue membrane, the mesovarium, connects the ovaries to the body wall. The ovarian stroma is thin and transparent with a generous blood supply via the ovarian artery. Venous drainage is into the inferior vena cava. The germinal epithelium divides mitotically, producing oogonia from which the primary follicles develop. The primary follicles consist of an enlarged oocyte surrounded by a single layer of granulosa cells. Ovaries of turtles, tortoises, and crocodilians are similar in appearance to those of birds in that the follicles are generally spherical and arranged in size classes, but unlike birds both a right and a left ovary develop. Each size class of follicle represents growth of 1 year;

thus, in the adult ovary there will be at least three and as many as five different size classes of follicles. The ovaries of snakes and some lizards are markedly different from those of crocodiles and turtles. The developing follicles are elongate rather than spherical and go through a growth stage sometimes referred to as hydration stage of growth or previtellogenic growth. At this stage they have a translucent or pearly white color. When vitellogenesis proper begins the follicles take on a yellow color.

A. Follicular Structure

The ovarian follicles of chelonians and crocodilians are similar to those of birds. Primary follicles develop when oogonia enter meiotic prophase I, move out of the germinal bed, and are surrounded by a layer of epithelial cells that will form the granulosa, a basement membrane, and a layer of connective tissue cells. The cytoplasm of the oocyte, sometimes referred to as ooplasm, is surrounded by a vitelline membrane, the zona pellucida, and the granulosa layer. The zona pellucida is formed by microvilli protruding from the surface of the oocyte (Fig. 1B). The granulosa is a single layer of cuboidal or flattened

epithelium resting on a basement membrane. Immediately outside the basement membrane is the theca. The theca appears to be composed of connective tissue cells several layers in thickness. Granulosa cells of both previtellogenic and vitellogenic follicles show a positive reaction for the enzyme, 3β-hydroxysteroid dehydrogenase (3β-HSD), whereas the theca

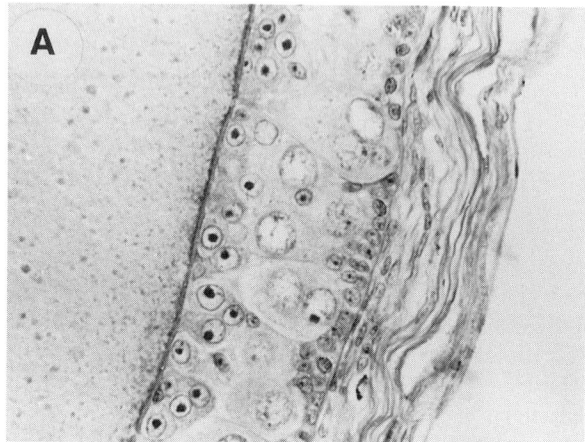

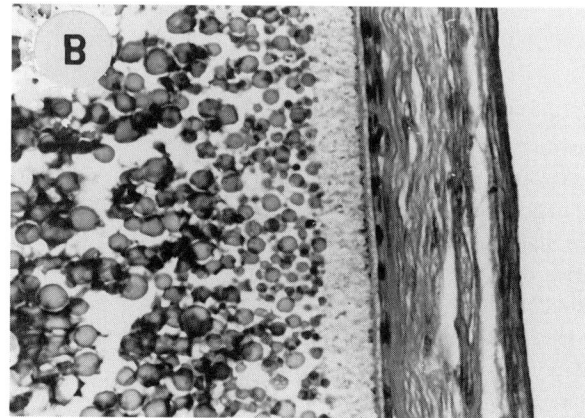

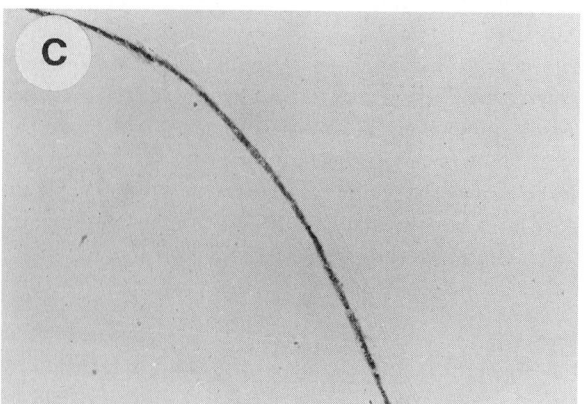

FIGURE 1 (A) Section of a previtellogenic follicle from a cobra ovary showing the three granulosa cell types: small cells against the basement membrane, intermediate cells adjacent to the oocyte, and the pyriform cells with their intercytoplasmic connection to the oocyte. The theca layer (right) surrounding the follicle is composed largely of connective tissue and blood vessels. Iron hematoxylin and orange G stain. Magnification, approximately ×500. (B) Section through a vitellogenic follicle from a cobra ovary. The pyriform cells have disintegrated and the granulosa is now composed of only a single cell type. Note the large, dark-staining yolk particles and the clear zona pellucida adjacent to a single flattened layer of granulosa. As in A, the theca is on the right and appears to be composed of connective tissue. Magnification, approximately ×350. (C) Histochemical demonstration of 3β-HSD in a frozen section through a vitellogenic follicle from a cobra ovary. The reaction is intense but restricted to the single layer of granulosa (B). No counterstain was used; hence, the yolk (left) and the theca (right) are negative for the enzyme and therefore invisible. Magnification, approximately ×50.

is consistently negative (Fig. 1C). These results and ultrastructural observations suggest that the granulosa layer is the source of ovarian steroid hormones. A number of preovulatory follicles undergo atresia. In these follicles the basement membrane breaks down and granulosa cells and macrophages invade the oocyte and resorb the yolk material. Following ovulation the granulosa cells luteinize and form the postovulatory follicle or "corpus luteum." The thecal cells do not form part of this structure and only the luteinized granulosa cells give a positive reaction for 3β-HSD. A correlation between the presence of this enzyme and elevated plasma progesterone suggests that these cells are the source of progesterone during the postovulatory period.

The main difference between the squamates and the other reptiles is in the structure of the granulosa layer of the developing follicle. The granulosa of the previtellogenic follicles of snakes and lizards consists of three cell types: small cells, pyriform or flask cells, and intermediate cells (Fig. 1A). The pyriform cell appears unique among the amniote vertebrates. In no other group do the cells of the granulosa layer form cytoplasmic bridges with the oocyte, but similar cells known as "nurse cells" are found in insect ovaries. The function of these cells remains speculative. Once vitellogenesis begins, the pyriform cells lose their connection with the oocyte and degenerate. The resulting granulosa consists of a single cuboidal epithelial layer similar in appearance to that of turtles and crocodilians (Fig. 1B).

II. OVIDUCTS

The oviducts are the structures within which the albumin layer and the eggshell are formed around the fertilized egg. Each region of the oviduct is specialized for its particular function. In the upper region fertilization takes place, in the middle regions albumin and the shell are deposited, and in the lower uterine region the eggs are held until oviposition. The oviducts of most reptiles are paired structures suspended from the body wall by the mesotubarium. In alligators, snakes, and a number of lizard species the right oviduct is larger and heavier than the left. In turtles and tortoises the paired oviducts appear similar in size, but a slight bias in favor of the right is sometimes seen. A number of known exceptions to this arrangement occur in some species of the tiny fossorial blind snakes in the families Typhlopidae and Leptotyphlopidae in which only a single right oviduct is present. However, in at least one species in each of these families, paired functional oviducts are present. In the limbless lizard, *Anniella*, the left oviduct is degenerate and nonfunctional. Because the anatomy of only a few of the more than 6000 extant squamates is known, more exceptions are likely to be discovered. There are no known variations in the Chelonia or Crocodilia, and the Tuatara is described as having a typical lizard-like reproductive tract. The blood supply is via a number of small arteries branching off the dorsal aorta and venous drainage is to the inferior vena cava. Each oviduct opens separately into the cloaca in most species examined, but they unite to form a single opening in some lizards. Based on its histological structure the oviduct of the tortoise may be divided into five major segments: infundibulum, uterine tube, isthmus, uterus, and vagina (Fig. 2). Squamate oviducts have been described as having only three segments: infundibulum, uterus, and vagina. Where more detailed microanatomical studies have been done, further subdivisions have been suggested. A recent description of the alligator oviduct lists seven separate regions. Each region is histologically distinct based on its function. As in the mammalian uterus, the tube is covered by a tough connective tissue membrane and an external longitudinal layer of smooth muscle and an inner circular layer of smooth muscle. The lining of the oviducts undergoes dramatic changes in response to estradiol and testosterone as the breeding season commences. All sections show changes, but the glandular region increases in size by several hundred-fold.

The infundibulum, or ostium, is a thin-walled section lacking multicellular glands but does have a thin bilaminar muscle layer. The lumen is thrown into long, complex folds or plicae and is lined with two or, in some species, three types of epithelium. These include a columnar ciliated cell with an apical nucleus, a nonciliated cuboidal or columnar cell, and a secretory cell with bleb-like cytoplasmic extensions. Other cell types have been identified based on both ultrastuctural characteristics and histochemical

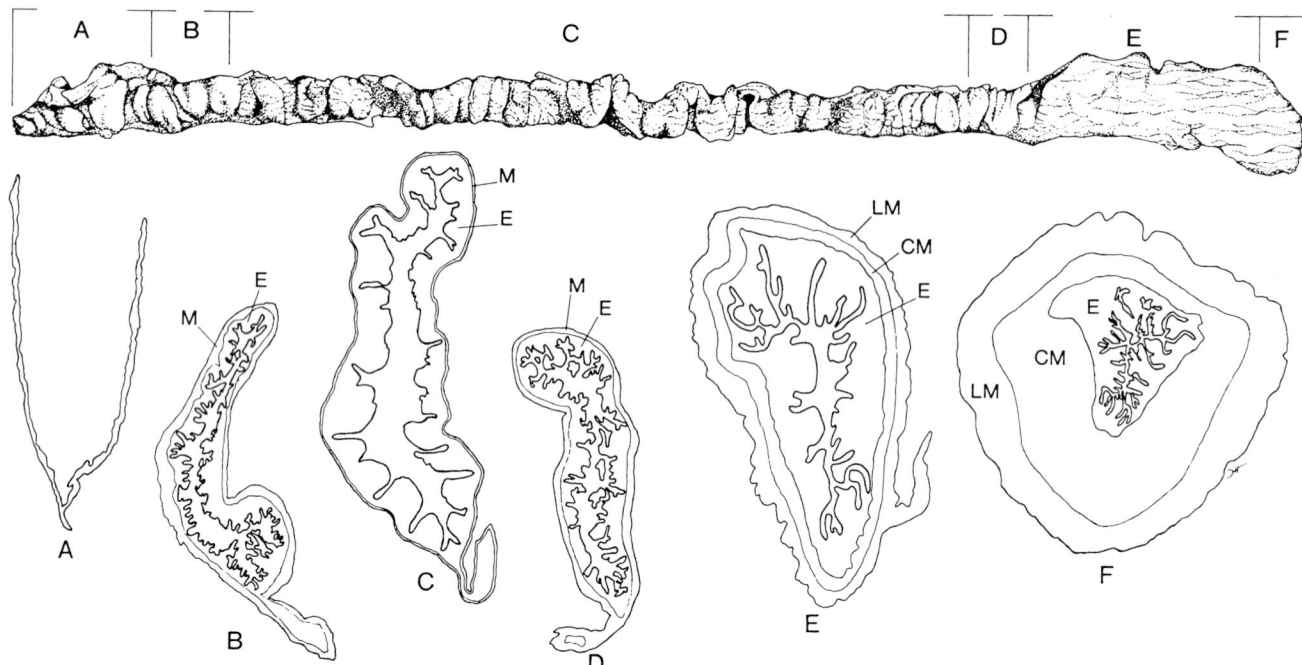

FIGURE 2 Morphology of the oviduct of the tortoise, *Gopherus polyphemus,* with representative cross sections of each region. A, Anterior infundibulum at the ostial opening; B, posterior infundibulum; C, uterine tube; D, isthmus; E, uterus; F, vagina. E, endometrium; CM, circular muscle; LM, longitudinal muscle; M, myometrium.

properties. During the breeding season the cells of the infundibular region increase in size and show signs of secretory activity. In some squamate species specialized crypts or sperm receptacles are found in the posterior region of the infundibulum

The uterus is a thick-walled tube with a submucosal layer containing numerous acinar glands, or endometrial glands, and, as mentioned previously, may be further subdivided based on the morphology and staining properties of the glandular cells (Fig. 2). The mucosal surface contains numerous ciliated cells interspersed among endometrial-like glands. Seasonal changes in the size and secretory activity of these glands that correlate with the seasonal changes in circulating ovarian steroids have been described. Injections of estradiol outside the breeding season result in a similar hypertrophy of the oviducts. Oviparous species have better developed uterine glands than viviparous species. There is considerable variation in the types of eggs produced by reptiles, from the hard-shelled, calcareous, avian-like eggs of the Crocodilia to the soft, thin, leathery-like eggs of snakes. There is also considerable variation in the

type of albumin produced. Interspecific differences in the histological structure of the uterine portion of the oviduct are therefore to be expected. In the tortoise (Fig. 2), the glandular portion is divided into an anterior uterine tube and a posterior uterus separated by an isthmus. The isthmus lacks acinar glands. In other species this region is referred to as the uterotubal junction. Evidence suggests that the uterine tube region secretes the albumen around the egg, which then passes through the nonglandular uterotubal junction into the uterus where the shell is deposited.

The vagina is a relatively short, thick-walled segment with a much tougher and thicker outer connective tissue layer and thicker outer longitudinal and inner circular muscle layers than the other regions of the oviduct. The lumen is lined with columnar ciliated epithelium interspersed with secretory cells. There are no acinar glands.

The reptile oviduct is a complex organ that contains a thickly ciliated lining that allows for the upward movement of the spermatozoa that are deposited in the vagina and secretes mucus for lubrication.

Various other proteins are secreted within this mucus coat that may provide protection against bacteria and fungi to both the tube and the freshly laid egg. Albumin, studied in few species, is a complex mixture of proteins produced by the oviduct. In addition, the oviduct secretes fibrous material and calcium carbonate for eggshell formation. Both prostaglandin $F_{2\alpha}$ and prostaglandin E_2 have been shown to be secreted by reptile uterine tissue *in vitro*. Estrogen and two different progesterone receptors have been characterized, and both α- and β-adrenergic receptors have been demonstrated. Histochemical identification of insulin-like growth factor-1 was reported for all regions of alligator oviduct. Egg laying can be induced in gravid uteri of reptiles by either arginine vasotocin, one of the endogenous posterior pituitary hormones of reptiles, or the mammalian posterior pituitary hormone, oxytocin, indicating that a receptor that recognizes these hormones is present.

See Also the Following Articles

FEMALE REPRODUCTIVE SYSTEM, AMPHIBIANS; FEMALE REPRODUCTIVE SYSTEM, BIRDS; FEMALE REPRODUCTIVE SYSTEM, FISH; MALE REPRODUCTIVE SYSTEM, REPTILES

Bibliography

Fox, H. (1977). The urogenital system of reptiles. In *Biology of the Reptilia Vol. 6. Morphology E* (C. Gans and T. S. Parsons, Eds.), pp. 1–157. Academic Press, New York.

Jones, R. E. (Ed.) (1978). *The Vertebrate Ovary*. Plenum, New York.

Klosterman, L. L. (1987). Ultrastructural and quantitative dynamics of the granulosa of ovarian follicles of the lizard *Gerrhonotus coeruleus* (Family Anguidae). *J. Morphol.* **192**, 125–144.

Palmer, B. D., and Guillette, L. J. (1988). Histology and functional morphology of the female reproductive tract of the tortoise *Gopherus polyphemus. Am. J. Anat.* **183**, 200–211.

Female Sterilization

Richard M. Soderstrom

University of Washington School of Medicine

I. Introduction
II. Nonlaparoscopic Methods
III. Laparoscopic Methods
IV. Sterilization Failures
V. Reversal of Sterilization

GLOSSARY

ampulla The outer third of the fallopian tube which is thicker than the isthmic portion, ending in a trumpet-shaped end known as the fimbrae.

bipolar Describing a method of electrocoagulation in which the radio waves pass through only the tissue grasped by a special bipolar instrument instead of through the entire body.

broad ligament A supporting membrane of the uterus and fallopian tubes.

catgut suture A dissolving suture.

colpotomy An incision made at the top of the vagina between the cervix of the uterus and the rectosigmoid colon.

cumulative pregnancy failure A measure of the failure rate for sterilization procedures, expressed as the number of pregnancies per thousand patients over a specific period of time, following a sterilization procedure.

cystotomy An incision into the wall of the bladder.

electrocoagulation The use of high-frequency radio waves to create heat within a living cell.

fimbriectomy The removal of the end of the fallopian tube.

isthmic The thinner portion of the fallopian tube attached to the uterus.

laparoscope A slender telescope, placed through a small incision, to view or operate within the abdominal contents.

laparotomy A large incision, usually more than 3 in. long, made to operate or view the abdominal contents.

myometrium The muscle wall of the uterus.

proctotomy An incision into the wall of the rectum.

random control trial A statistical method designed to eliminate potential bias by assigning an unproven treatment to one group of patients and a standard, proven treatment to another group through the use of a random assignment method.

serosa The thin, slippery coating which covers all organs within the abdominal cavity.

thermocoagulation The process of creating heat in an instrument which upon touching tissue will cause a burn.

unipolar Describing a method of electrocoagulation in which the radio waves pass through the entire body of the patient.

Sterilization remains the preferred method of birth control for couples who have completed their families or wish to have no children. For the female, the practicing gynecologist is confronted with a wide array of available techniques to recommend. Which technique is best, the easiest to perform, the most readily reversible, especially for young couples, or simply the most effective overall? Even better, is any technique highly efficacious and also readily reversible?

I. INTRODUCTION

For women, voluntary sterilization was difficult to obtain until May 1969. Prior to that date, the American College of Obstetricians and Gynecologists (ACOG) required, after approval of both the wife and husband, a three-person committee to review a written request from the patient's physician. Between May 1969 and October 1974, the ACOG recognized that most states had no law regarding sterilization and from a legal view the decision could be made between the physician and the patient. The ACOG did recommend, however, a consultation from two "senior" staff members following voluntary requests. In the 1974 *Standards for Obstetric–Gynecologic Services* the ACOG states, "If sterilization is requested by the patient and her physician agrees, consultation is not necessary." However, by 1970, most medical

centers had adopted their own guidelines similar to those of today. Today, when a medical condition is an indication for sterilization, a consultant skilled in that medical condition is still recommended. This new freedom of choice opened the door to use the laparoscope as a tool to perform female sterilization as an outpatient procedure. Today, it is one of the most common operations performed by gynecologic surgeons in the United States.

II. NONLAPAROSCOPIC METHODS

The tubes may be approached by conventional laparotomy, minilaparotomy, and colpotomy (an incision made transvaginally at the top of the vagina between the cervix of the uterus and the rectosigmoid colon). Practicing obstetrician–gynecologists are reluctant to perform a conventional laparotomy solely for tubal ligation except at cesarean section or during another major gynecologic or surgical procedure. In the immediate postpartum period, the uterus is enlarged up to the level of the umbilicus making it a desirable site for a small 1- or 2-cm incision which heals leaving only a small scar. Any of several techniques can be used.

A. Irving Technique

In 1924, Irving described a method to be used exclusively for sterilization coincidental with cesarean section. By burying the proximal stump within the myometrium and the distal portion between the leaves of the broad ligament, he was able to keep the ends of the severed tube away from each other. Sterilization failure has not been described with this technique, but because of episodes of heavy bleeding during the burying portion of the procedure it is seldom used.

B. Pomeroy Technique

Though never published by Pomeroy himself, this technique is the most widely used method for cesarean section patients or immediate postpartum sterilization. A knuckle or loop of the midportion of each tube is tied, not crushed, at its base with an absorbable suture, preferably plain or chromic catgut. Once the base is securely ligated, the knuckle of the tube (usually 2 cm in length) is excised and sent to pathology for confirmation.

C. Kroener Technique

This method is sometimes called a fimbriectomy, the removal of the end of the fallopian tube. Here, the end of the tube is completely removed. Because one of four patients will have a separate tongue of fimbria attached to the ovary, special care must be taken to remove that as well. Kroener emphasized the need to double tie the tube to reduce failures. This technique was meant to be used during colpotomy (vaginal incision), but the severe complications (about 1%) associated with vaginal tubal sterilization—cellulitis or pelvic abscess, hemorrhage, proctotomy, and cystotomy—may place too high a price on avoiding a small abdominal scar. One extensive review concluded: "At the moment, the serious nature of complications encountered after vaginal tubal sterilization would appear to prohibit wholehearted endorsement of this procedure as a quick, effective means of performing female sterilization."

D. Uchida Technique

From Japan, a method of tubal destruction claims a zero failure rate. Designed for the minilaparotomy approach, several centimeters of tube is stripped of its serosa, the denuded segment of tube is excised, and the proximal stump is buried in the serosal sleeve that remains. Data from other centers is not available.

III. LAPAROSCOPIC METHODS

Though Anderson, an American, performed the first laparoscopic tubal occlusion in 1947, it was not until the 1960s that this instrument's possibilities for female sterilization were appreciated. The new freedom of choice declared in 1969 made laparoscopy sterilization one of the most common operations performed by gynecologic surgeons in the United States.

During this transformation, one obstacle offered a new challenge to those intrigued with the laparoscope. The mechanics of the previously described sterilization methods are restricted by the remote control nature of laparoscopy. For instance, the Pomeroy technique, which ties a knuckle of the tube with a plain catgut suture followed by a segmental resection of the devitalized tube, was not possible with the laparoscopic techniques of that time without great ingenuity and instrument design. Therefore, in keeping with "a more simple approach," unipolar electrocoagulation, a tool of seasoned surgeons, seemed to meet this requirement. Since most sterilization techniques removed about 2 or 3 cm of fallopian tube, it seemed reasonable to coagulate the same amount of tube and if possible remove the coagulated specimen. To prove proper resection, a method to resect an uncoagulated specimen was designed in 1971 and was promoted as having a medicolegal advantage since the "coagulation first-resect after" methods frequently left the pathologist with coagulated tissue "consistent with but not diagnostic of fallopian tube."

It is no surprise that accidents, presumed to be due to aberrant unipolar electricity, soon became known to those involved with laparoscopy. Two physicians, Rioux and Corson, each working independently, designed a bipolar forceps which could harness the potential risk of aberrant electrical injury. Rioux's forceps never reached the marketplace because of a bureaucratic log jam within the Canadian patent system. Corson applied the bipolar principle to an existing forceps, the Frangenheim biopsy scissor. This instrument was difficult to manufacture and its ability to cut the tube proved to be unnecessary. The manufacturer of the Corson forceps (Cameron-Miller) stopped production in the late 1970s.

As early as 1974, several instrument companies had introduced their models of bipolar forceps; many have been modified as time and experience exposed design defects or bad outcome statistics (i.e., sterilization failures). Richard Kleppinger's bipolar forceps was designed as a system. Working with bioelectrical engineers, Kleppinger married his forceps design to the electrical output characteristics of a generator designed only for bipolar surgery. Today it would be considered the "gold" standard of bipolar systems.

Mechanical occlusion of the fallopian tube received considerable attention by several investigators. Lay and Yoon, independent of each other, introduced the silastic rubber ring method used by many today. The Agency for International Development chose this method for their sterilization training programs in developing countries.

In a prospective fashion, Hulka and Clemens created and evaluated a spring-loaded plastic clip which, with time and redesign, proved effective. Blier of Germany reported great success with a plastic clip which locked in place, but it fell prey to a high failure rate after its introduction to the United States. Filshie from England has developed a clip using a memory characteristic of silastic rubber to slowly occlude the tube. In 1996, after 13 years of experience abroad, it received FDA approval for sale in the United States.

Though the previous methods, as they were designed, discussed, and tested, received great attention during the mid-1970s, little was known of another electrical method—thermocoagulation. Instead of using the principle of electrodessication, low-voltage thermal energy was developed with two instruments.

Laparoscopy for sterilization offers the following advantages over other sterilization techniques:

1. Small incisions whose scars are barely visible
2. Same-day surgery
3. No vaginal drainage, i.e., colpotomy
4. No sexual restriction
5. Lower cost than other approaches
6. A panoramic view of the abdominal viscera

The disadvantages of laparoscopic sterilization are few. Though general anesthesia is used by most, regional and local techniques are favored by many, especially in developing countries, where the use of general anesthesia is infrequent. When compared to the minilaparotomy approach to sterilization, laparoscopy requires more training. Vasectomy, which may be preferable for some couples, is less expensive.

A. Electrical Methods

Electrical energy can be harnessed, shaped, and delivered in a variety of ways to accomplish a number of tasks. Equipment used to deliver this energy source must be designed with thought and knowledge about the principles of electricity or they will not perform as desired. If physicians do not understand the basics of electricity, even the best electrical system, used improperly, will fail. When using electricity for sterilization, two basic objectives must be met—obliteration of the vascular supply to, and complete disruption of, the living tissue. If one of these two objectives is incomplete, tubal patency may persist, which can lead to a sterilization failure.

1. Unipolar Techniques

i. Coagulation Alone   In the United States, electrocoagulation without transection or resection of the fallopian tube has the longest track record of all the methods of female sterilization. When laparoscopy sterilization began to grow in popularity, the simple application of unipolar electrical energy to the isthmic portion of the tube seem logical and efficient. By the mid-1970s, large "coagulation alone" studies were published with impressive outcome statistics. Each author coagulated 2 or 3 cm of the proximal third of each fallopian tube with a collective failure rate of 0.35%. Until 1996, these statistics stood as the standard by which all other methods were compared (Figs. 1 and 2).

ii. Coagulation with Transection   This method is practiced by many. However, the failure rate is not improved and the risk of a tear in the mesosalpinx

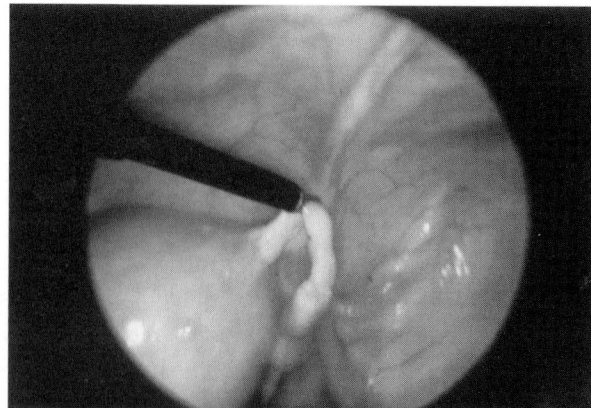

FIGURE 1   Using a laparoscopic grasping forceps, the right fallopian tube turns white as the electrosurgical energy destroys the tissue and seals off the blood supply.

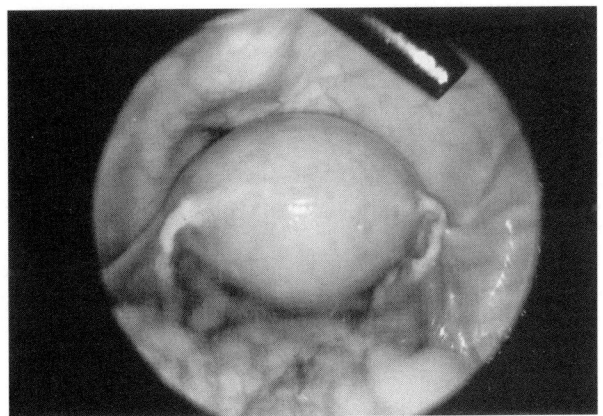

FIGURE 2 Both tubes have been coagulated near their attachment to the uterus.

with subsequent hemorrhage is increased. The advocates of this method argue that they prefer to see the fresh ends of the coagulated tube to assure themselves of a complete coagulation.

iii. Coagulation and Resection
This method of laparoscopic sterilization is more difficult to master and the failure rate is not improved. Bleeding from the mesosalpinx is a risk. Several instruments have been used to remove the specimen.

Where experience can say that all unipolar electrocoagulation methods are effective and acceptable, it also has shown that enough tissue is destroyed to occlude the tubal lumen and it obliterates the vascular tree to the coagulated segment promoting atrophy of that segment. In summary, unipolar electrocoagulation sterilization, and in particular the coagulation alone method, has stood the test of time and is much safer than once thought. Caution is still important, as with any electrosurgical procedure, when a unipolar method is employed.

2. Bipolar Methods
The principles of bipolar physics play a major role in the bipolar techniques of female sterilization. Though the gross appearance of the surface of a tube coagulated by bipolar energy may look the same as with unipolar energy, the depth of destruction is seldom the same. Because the lateral spread of destruction characteristic of unipolar energy does not occur with bipolar systems, the tube must be grasped and coagulated more times with the bipolar forceps than with the unipolar forceps. In reality the success of any electrocoagulation method, unipolar or bipolar, depends on the length of tube destroyed, not the number of applications. Most authorities feel that a 2-cm segment of coagulated tube is the minimum and 3 cm is preferable.

The system, designed by Richard Kleppinger, provides a continuous (undamped) waveform called the cut waveform. The waveform called coagulation is intermittent (damped); thus, less energy may be delivered to the tissue. His forceps is marketed with a warning to the purchaser not to adulterate the electrical cord so that it can be attached to another generator. His "bipolar system" has an additional feature worth noting. An ammeter displays the current flow during the process of electrocoagulation until the current ceases to flow. This technology removes the guesswork when one relies only on a visual endpoint. Recently, several companies have introduced similar generators.

3. Thermal Methods
Thermal coagulation or "true cautery" is available in the United States if one uses one of two instruments. The Waters' instrument is a wire hook heated much like an electric toaster wire, with the hook being sheathed in a heat-resistant plastic tube. The hook is extends beyond the sheath and hooks under the fallopian tube. The hook is then withdrawn into the protective sheath, the wire is heated, thermal cautery occurs, and the tube is severed by the wire hook when thermal damage is complete. A ground plate is not necessary but the tube should be elevated away from surrounding structures because the protective sheath can become hot. The amount of tissue destruction after using the Waters' instrument is less than 1 cm.

The second instrument, by Semm, looks like a grasping forceps but is coated in Teflon. Again, the instrument is heated (about 100°C) and direct cautery occurs. The area of damage is limited to the width of the forceps, so many applications are required to coagulate 2 or 3 cm of tube. Most of the experience with the Waters' instrument has been concentrated in only a few states and Semm's thermal forceps experience has come from Europe. Neither advocate has presented controlled nor convincing studies that this approach to female sterilization has an acceptable failure rate.

B. Mechanical

1. Band Method

Using a silastic rubber band or ring, Yoon successfully occluded the fallopian tube in the early 1970s. He adapted this technology to the operating laparoscope and to the two-puncture techniques. It has become a popular method for both laparoscopic and nonlaparoscopic approaches to sterilization.

The applicator is more complex than the instruments used for the electrical methods of sterilization. The applicator has two concentric cylinders with a grasping forceps within the inner cylinder which grasps and elevates a segment of fallopian tube. By means of a "ring loader" the silastic rubber ring is stretched over the distal end of the inner cylinder, and the outer cylinder retracted up the inner cylinder shaft about 5 mm. The handle of this applicator allows the surgeon to extend the grasping forceps beyond the inner cylinder, grasp the fallopian tube, and draw a knuckle of tube inside the inner cylinder. Once the knuckle of tube is secure within the inner

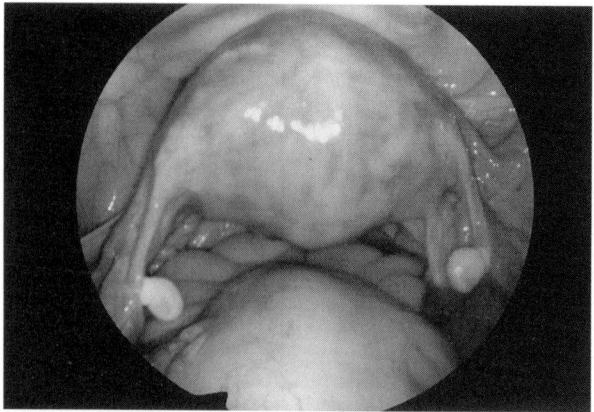

FIGURE 4 Both rings have been applied to each fallopian tube. Because of the tight application of the silastic rubber ring, the knuckle of tube turns whited due to the lack of blood supply. After several weeks, this knuckle of tube dies of atrophy.

cylinder, the handle of the applicator is "squeezed" against the outer cylinder causing this cylinder to push or "fire" the ring onto the secured tubal segment or "loop." The applicator is designed to draw into the inner cylinder a loop of tube which measures 2.6 cm before the band is fired (Figs. 3 and 4).

Because of the acute tissue necrosis which follows band application, postoperative discomfort can be a problem. Many laparoscopists apply sterile anesthetic jelly or solution on the tube when they perform

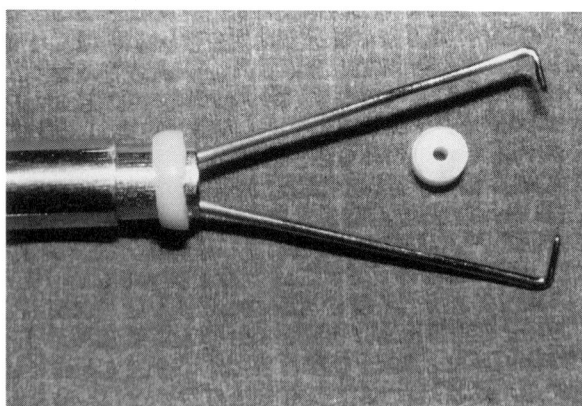

FIGURE 3 The silastic ring, in its natural state, is shown between the grasping tongs of the applicator. Another ring has been applied to the inner cannula which houses the grasping tongs. Once the tongs have grasp the fallopian tube, a knuckle of the tube is drawn into the inner cannula for about 1.3 cm. At that point, the outer sheath or cannula springs forward, pushing the ring onto the knuckle of tube.

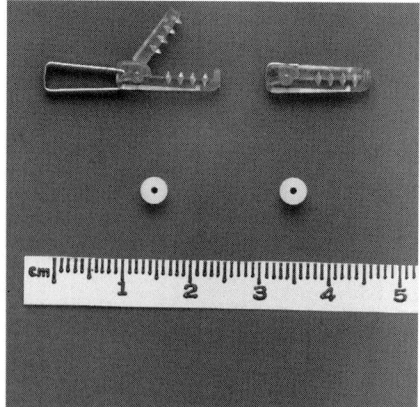

FIGURE 5 The Hulka clip in its open and closed states compared in size to the silastic rubber band. To close and secure the clip, the outer gold band is pushed by the applicator over the plastic jaws. The band, acting as a spring, closes the tube slowly over 2 days.

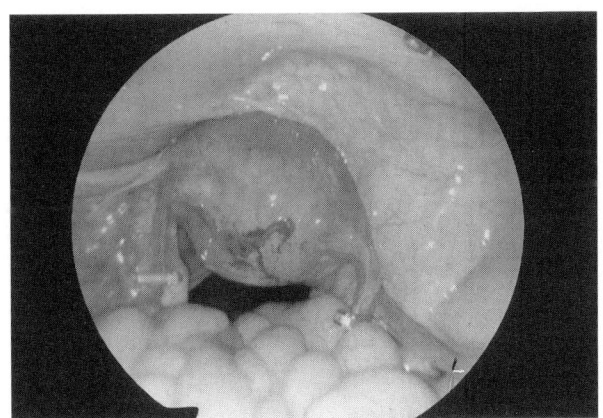

FIGURE 6 The Hulka clip has been applied to both fallopian tubes about 2 or 3 cm for the tube's attachment to the uterus.

ring sterilization. This helps in some patients but many will require intramuscular or intravenous narcotics to control their discomfort in the first few hours after surgery.

Unlike electrical methods, especially unipolar methods, the tissue destruction and atrophy is confined to the knuckle of tube involved in the initial procedure. Later, should a patient request a reversal operation, the chances for tubal restoration are better than those following electrocoagulation techniques.

2. Clip Sterilization

Tantalum clips used for vascular occlusion have been used for tubal occlusion but the failure rate is high because the clip bends over the muscular tube and occlusion is incomplete. Even multiple tantalum clip application has a high failure rate unless the vascular tree to a segment of tube is occluded and that segment dies.

Currently, two clips are available in the United States—the Hulka–Clemens clip and the Filshie clip. The Hulka–Clemens clip is made of 3-mm-wide Lexan plastic jaws hinged by a small metal spring which opens the jaws for application. Once the jaws are applied over the tube, a U-shaped gold-plated steel spring is pushed over and down the length of the jaws to hold them closed under constant pressure. In some patients it may take several days for this spring to squeeze the tube to complete occlusion. This latter fact protects the tube from an abrupt transection which could lead to bleeding or allow the clip to fall away from the tube (Figs. 5 and 6).

The Filshie clip is made of titanium with a silicon rubber strip embedded into the inner surface of each jaw. The clip locks into place with a latch to its end, the latch being locked under the fallopian tube at its mesosalpingeal attachment (Fig. 7). Both clips are to be applied approximately 2 or 3 cm from the cornua of the uterus at a 90° angle to the axis of the tube (Fig. 8).

The advantages of clip sterilization is that each destroy only 5 mm of tube; reanastomosis is always possible, and fear of inadvertent burns to surrounding structures is eliminated. Both have had extensive premarket testing and appraisal. There ap-

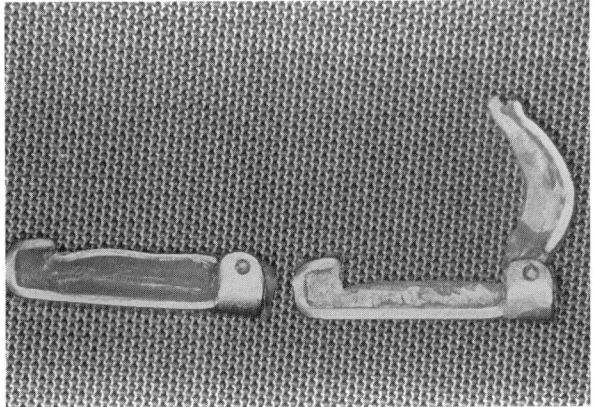

FIGURE 7 The Filshie clip in its closed and open states. Once the upper arm locks under the lower arm's jaw, the silastic rubber lining continues to compress the tube shut.

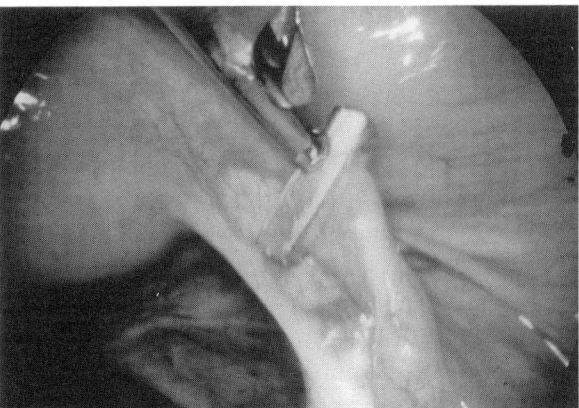

FIGURE 8 The Filshie clip has just been applied to the right fallopian tube with its special applicator.

pears to be less postoperative pain than for the band method. A disadvantage of the Hulka clip is that it requires more training than the band method (debatable), and the failure statistics of Hulka clip vary from study to study. The Filshie clip is easy to teach and, to date the failure rate data are excellent.

IV. STERILIZATION FAILURES

Unfortunately for patient and physician, sterilization failures occur. Each technique has inherent problems, and Mother Nature has a remarkable propensity to reconstitute the integrity of the normal fallopian tube. Statistical reports on the success and/ or failure of techniques are usually flawed because of anecdotal experience, retrospective review, and short-term follow-up. Too frequently, different techniques are "modified" by the individual surgeon, who in essence invents another method without the luxury of adequate statistics. The following comments are summarized from reports in the literature, anecdotal experiences of others, and the personal experience of this author.

Of the laparoscopic procedures described, unipolar electrocoagulation without transection has the lowest failure rate of all the laparoscopic techniques to date. When unipolar failures occur, a fistula is always the culprit. As previously mentioned, in the late 1970s, the presumed risks of unipolar electrocoagulation caused the laparoscopist to use other techniques presumed to be safer.

In recent years, problems with bipolar failure have become apparent in the literature. Incompatible equipment and inadequate coagulation appear to play a major role in these failures. Bipolar techniques that coagulate less than 2 cm of a tube or when the energy delivered to the tube is delivered below 20–25 W of power have an unacceptable failure rate. A cutting waveform should be used with the power setting at 25–35 W.

When bipolar failures are examined, a healthy endosalpinx is usually found with scarred, atrophic muscular tissue surrounding the patient's salpinx. These findings are consistent with incomplete electrocoagulation confirming that the visual endpoint of electrocoagulation relied on in unipolar methods

is not sufficient information to declare the tube coagulated. The use of an ammeter or optical meter, which tells the surgeon that all the energy that can be delivered has been delivered, should reduce the risk of incomplete coagulation.

In general, when ring failures occur, they are secondary to spontaneous reanastomosis. In these cases, once the loop contained above the application has atrophied, if the tube does not separate, as is the usual case, spontaneous anastomosis occurs. In such cases, the ring is frequently found perched on top of the anastomosis or adjacent to the anastomosis on the mesosalpinx.

Until 1996, the female sterilization failure rates were estimated from retrospective studies, each being subject to "lost to follow-up" bias. In April 1996, the Centers for Disease Control (CDC) published the only prospective study of the common methods of sterilization used in the United States. It took 14 years and over 10,000 patients to obtain a 10-year follow-up of each technique studied. More than a dozen teaching centers participated, making the power of the outcome statistics strong. The major weakness of this study was the exclusion of nonteaching centers which might have the more experienced surgeons. Still, an estimate of the probable failure rate of the different methods for female sterilization became more realistic; each method has a higher failure rate than previously believed. Two methods, the Hulka clip and the bipolar methods, had failure rates that were not only high when compared to the others but also seemed particularly high in patients who were under the age of 28. The cumulative pregnancy failure rate over 10 years was also higher with these two methods. The failure rates per thousand patients of the methods are shown in Table 1.

When those methods using electrosurgical techniques failed, the failure frequently presented as an ectopic pregnancy. This potentially life-threatening condition should be considered when such a patient presents, sometimes years later, with severe abdominal pain, especially with symptoms of an early pregnancy.

Frequently, failures associated with a Hulka clip follow placement of the clip too distal on the tube, placement at oblique angles to the axis of the tube,

TABLE 1

The Cumulative Failure Rate per Thousand
Patients of the Common Methods of Female
Sterilization over 10 Years

Method	Rate (confidence)
Unipolar	7.5 (1.1–13.9)
Postpartum "Pomeroy"	7.5 (2.7–12.3)
Silicone band	17.7 (10.1–25.3)
Bipolar	24.8 (16.2–33.3)
Spring clip	36.5 (25.3–47.7)

Note. From Peterson *et al.* (1996).

or the clip is misapplied to other structures. Also, when mechanical devices in medicine are used, occasional flaws in the individual device may render the operation unsuccessful.

During the FDA approval process for the Filshie clip, several smaller studies using a random control trial (RCT) study profile were reviewed. Though the patient numbers were smaller than with the CDC study and the period of study for the cumulative failure rate was only for 2 years, the failure rate was better than either the Hulka clip or the bipolar methods, being equal to the silastic band method.

V. REVERSAL OF STERILIZATION

The most common motivating factors are the so-called three D's: divorce, death, and disaster. Other factors include changes in lifestyle and economic status that may suddenly permit a couple to have a larger family. All candidates requesting a reversal should be thoroughly investigated, psychologically and physically. The operative report of the sterilization procedure should be available before any kind of surgery is contemplated, and an analysis of the partner's sperm should be obtained.

The end results of some methods of sterilization are frequently unpredictable. For example, a unipolar coagulation may destroy the whole tube, leaving barely a few fimbria behind. The Kroener technique can remove the fimbria and the entire ampulla. Conversely, if only the fimbria was excised, a complete ampulla may remain, permitting an adequate salpingoneostomy. A bipolar burn or Falope ring ligation can often be reversed by isthmic–ampullary anastomosis. This type of reversal, however, is more difficult than the isthmic–isthmic anastomosis performed in most cases of clip reversal.

In short, no method is 100% reversible. The clip methods using microsurgical techniques approach nearly 90% success. Other approaches have less success but are sufficient for some patients who request reversal, with a few accepting a success rate of only 10%.

See Also the Following Article

Family Planning

Bibliography

Gomel, V. (1978). Profile of women requesting reversal of sterilization. *Fertil. Steril.* **30**, 39.

Gomel, V, (1980). Microsurgical reversal of female sterilization: A reappraisal. *Fertil. Steril.* **33**, 587.

Peterson, H. B., Zhisen, X., Hughes, J. M., Wilcox, L. S., Tylor, L. R., and Trussel, J. (1996). The risk of pregnancy after tubal sterilization: Findings from the U.S. Collaborative Review of Sterilization. *Am. J. Obstet. Gynecol.* **174**(4), 1161–1170.

Soderstrom, R. M. (1985). Sterilization failures and their causes. *Am. J. Obstet. Gynecol.* **152**(4), 395–403.

Soderstrom, R. M., and Rioux, J. E. (1992). Laparoscopic sterilization. In *Operative Laparoscopy: The Masters' Techniques*. Raven Press, New York.

Soderstrom, R. M., and Yuzpe, A. A. (1979). Female sterilization. In *Clinics in Obstetrics and Gynecology*. Saunders, London.

Fertility and Fecundity

Philip J. Dziuk

University of Illinois

I. Introduction
II. Factors Affecting Fertility and Measures of Fertility

GLOSSARY

conception rate The proportion of a population that conceives, ranging from zero to 100%.
fecundity The number of offspring produced by a group per unit of time.
fertility The capability to reproduce.
litter size The number of young produced at one time.

We are all descended from two unbroken fertile lines of male and female ancestors. Reproduction is the mechanism for passing genes to the next generation. The number of offspring, and hence the number of genes an individual transfers to the next generation, is determined by the fertility and fecundity of the individuals. The proportion of the genes in the next generation that come from an individual determines the impact of that individual on the future of that species.

I. INTRODUCTION

Fertility is the ability to reproduce, in contrast to infertility or sterility. Infertility can be inferred to be temporary or relative to some degree of fertility, whereas sterility can be assumed to be permanent. The number and complexity of the interconnecting links necessary for production of offspring can lead to a degree of fertility if one or more links is weak.

Fertility is not an absolute. Not every mating or insemination results in a pregnancy and offspring; thus, fertility level is always relative to some standard or goal between zero and 100%. The transition from one level of fertility to another may take place gradually as with age from prepuberty to senescence.

Fecundity is limited by fertility in an absolute sense when only one offspring is produced for each pregnancy as is the usual case for humans, cows, and horses. In this case, fecundity is limited to fertility.

When more than one offspring is born in a litter, such as in pigs, rabbits, and mice, fecundity is affected by the number of young in the litter and their survival. Additional factors affecting fecundity are the proportion of nonreproducing members of the flock, herd, or group, interval between generations, intervals between successive births, number of potential pregnancies as influenced by the total fertile life of the individuals, and length of gestation. Each species has somewhat different factors impacting on the eventual fecundity of a population. Mice can mate at 35 days of age, deliver several young after a 21-day gestation, and mate again within 24 hr of giving birth. The life span of a mouse is measured in days. Humans and elephants are usually several years of age before they reproduce, produce one offspring after a gestation of many months, and can experience a period of infertility of several months if the young is nursed. Humans have a reproductive life of about 30–40 years. Both fertility and fecundity are quite variable and are influenced by many factors. Both fertility and fecundity are of considerable importance from several standpoints. Domestic livestock producers want to obtain maximum fertility and fecundity for their herds and flocks because reproductive efficiency is one of the most important factors in profitability of a livestock enterprise, whether it be chickens, pigs, sheep, horses, or laboratory animals. On the other hand, beautiful, useful, but also destructive animals, such as insects, wild rats, mice, birds, deer, raccoons, elk, coyotes, and geese, compete with humans for resources. Even the human race has come under scrutiny and criticism for its population growth. We need to be able to understand and deal with both the desirable and undesirable aspects of fertility and fecundity. The advent of assisted reproductive technologies has changed the ways that fertility can be achieved. Barriers to fertility can now be overcome by a variety of techniques.

II. FACTORS AFFECTING FERTILITY AND MEASURES OF FERTILITY

A. Anatomical

To achieve fertility, oocytes of the female must move from inside the ovarian follicle at ovulation, down the oviduct, meet the sperm, be fertilized, and then move down to the uterus and implant. Any physical barrier to this passage down the oviduct as a result of disease, surgery, or failure to develop embryologically will impede fertility. Any impediment to movement of the sperm through the male reproductive system and then to the site of fertilization in the upper oviduct will prevent fertility. Many of these obstacles can now be overcome by recovering the oocytes from the ovarian follicle and fertilizing them *in vitro* and transferring the fertilized embryo back into the oocyte donor or to another female, thus bypassing the obstacle. Sperm can be recovered from various segments of the male reproductive tract and used to fertilize oocytes *in vitro*, obviating the need to traverse the male and female reproductive tracts . Fertility can be achieved by heroic intervention under conditions that would prevent it normally even to the extent that an individual sperm can be injected directly into the oocyte, eliminating the need to pass through the male and female reproductive tract and the zona pellucida.

Fertility in the absolute sense as designated by conception rate is estimated by a number of means. The failure to return to estrus after mating or missed menstrual periods are an indication of pregnancy. These intervals from mating to failure to remate used to estimate fertility can vary from 1 or 2 months after mating to near term. The conception rate expressed as a percentage of animals not returning to estrus declines as pregnancy advances due to embryonic and fetal losses. Fertilization failure and prenatal losses often reduce fertility to a 50–70% conception rate. Much effort has been made to reduce prenatal losses. The complexity of the multitude of factors that must be integrated and progress perfectly to produce living young makes it difficult to ascribe any one cause to reduced fertility. It therefore follows that it has not been possible to prescribe a treatment or therapy that will improve fertility and fecundity in every case. The possibility of asynchrony between the rapidly developing embryo and the dynamic uterus has been proposed as one of the underlying causes. This cause could be brought about by one or more of many factors. In litter-bearing animals the number of offspring that develop fully can also be limited by available uterine space. Each fetus requires a minimum amount of space to survive and develop normally. A woman treated with fertility drugs can be pregnant with as many as five or six fetuses. Too many fetuses for the available space causes death, resorption, or stunting of the fetus.

There are a variety of tests for hormones of pregnancy, ranging from the presence of human chorionic gonadotropin in urine of women to estrone metabolites in pigs. Human chorionic gonadotropin has been detected by its ability to induce ovulation when injected into the rabbit or mouse (the rabbit or mouse does not die as is often depicted in television sit-coms; it only ovulates, which is not lethal). Currently, there are a number of simple colorimetric, immunochemical pregnancy tests available for home use by humans. Each fetal pig produces a certain amount of estrogen which can be detected in the mother's blood. The concentration of these estrogens indicate not only pregnancy but also the number of fetuses present. Progesterone is considered the hormone of pregnancy; its presence in blood, milk, and, in some cases, saliva, can be an indication of pregnancy. The sample must be taken at a stage when it would not be present except during pregnancy, such as during the follicular phase of estrual or menstrual cycles.

Currently, the determination of pregnancy is often done by ultrasonic analysis. There are a number of different methods varying from a simple echo analysis of the fluid surrounding the fetus to complex real-time scans that give a picture of a section of the fetus and its surroundings. These pictures can be used to detect gender of the fetus and stage and normality of development.

B. Endocrine and Age

Puberty is the stage at which an individual can reproduce. Puberty is under endocrine control. Prepubertal individuals are in an undeveloped state that does not yet permit gamete production and fertility. At the other end of the age scale, fertility declines until reproduction stops. Between puberty and cessation of fertility there is a period of maximum fertility for both males and females. In both wild and domestic animals, very young females and very old females are less likely to conceive and have smaller litters than females in the middle of the prime reproductive period. Males in early pubertal stages and old males produce fewer sperm and are less likely to impregnate a female than males in reproductive prime. Hormone production and reproductive tract morphology change with age, as does the fertility level.

Administration of hormones can either induce or reduce fertility depending on the particular hormone administered and the reproductive state of the animal. Gonadotropins can induce ovarian follicle growth of prepubertal females with subsequent ovulation and fertility, albeit a reduced fertility. The hormones of pregnancy, such as progesterone and estrogen, can be given to females beyond normal reproductive age, and by transferring embryos to these females, offspring can be produced. In these cases, fertility was induced either before or after it would occur normally.

C. Genetic

Both males and females have varying levels of fertility and fecundity. In extreme cases, chromosomal aberrations characterized by the absence of certain essential bits of chromosome (deletions) or additional bits of chromosomes (duplication) render the individual less fertile or perhaps infertile. The combinations of the X and Y chromosomes and modifica-

tions of segments within the chromosomes can have a profound effect on fertility. Although severe aberrations are to some extent self-limiting because the individuals cannot pass the defect on to any offspring, many genetic errors are passed on and reappear and others can arise spontaneously.

Certain strains of laboratory mice are much more prolific than others. In some cases the male is the primary cause of the difference in fertility between strains, and in other cases the female is a main cause. When two unrelated inbred strains are crossed, the resulting hybrid fetus is often more vigorous than an inbred fetus would be and thus survives more readily. A hybrid mother also imparts a survival advantage to her offspring. Incompatibility between mother and fetus due to blood group differences or other immune combinations can affect fertility. The variation in litter size between breeds of pigs is very striking, with wild pigs normally producing litters of 4 or 5 young, European domestic pigs producing about 10, and some Chinese pigs producing 14 or more pigs. The same variation exists between strains of laboratory mice, rabbits, sheep, and goats. Aside from the differences in fertility between breeds and strains, there are great differences among both male and female members of the same group. Bulls, rams, cocks, and boars used for artificial insemination of large numbers of females have a conception rate that is quite consistent and characteristic of that particular male. Those males with a relatively low conception rate are most often eliminated from breeding studs and artificial insemination centers unless they also have some extremely valuable and unique characteristic to compensate for the reduced fertility.

D. Environmental

The environment can include the physical surroundings and those things that can enter the body by the skin or intestine. Light has a profound effect on fertility, especially in certain species. Birds respond to increasing daylight by attracting and selecting mates and producing one or more broods of young. They become infertile in very late summer, fall, and winter. Sheep, elk, and deer, on the other hand, ordinarily mate only during the fall, and horses mate in the early summer. Fertility at other times is greatly reduced. Species have adapted the breeding period so that young are born during a period when food is more plentiful and climate is more favorable. Manipulating light:dark ratios can change reproductive patterns. There are many examples of species that migrate to a breeding ground for a congregation of potential pairs. Pairs or individuals that arrive too early or too late are less likely to raise young. Seals, turtles, penguins, and whales are extreme examples. Seasonal breeding animals are much more fertile and fecund during the height of the breeding season than at other times. Certain plants contain forms of hormones and toxins that can have profound effects on fertility. Sheep are prone to reduced fertility by ingestion of certain clovers that contain high levels of estrogenic compounds. Cattle that eat needles of specific pines are less fertile and can abort. Toxicants occur in nature and can also be the result of human activity. Chlorinated hydrocarbons, persistent insecticides and herbicides, and a myriad of other compounds all can have adverse effects on fertility. The classic example is that of the production of soft-shelled eggs by birds such as pelicans and eagles at the top of the food chain. The DDT that they accumulated from fish reduced the bird populations significantly. Fortunately, most species detoxify most introduced toxicants and those present in nature provided the concentration is not too high or too persistent. Not only does the light:dark ratio change with seasons but also the temperature changes. High temperatures interfere with production of normal sperm because the testes must be maintained at temperatures below body temperature. Anything that raises the temperature of the testes, such as a fever, high environmental temperatures, or restrictive clothing, will have a detrimental effect on sperm production and fertility of the male. These effects are often not seen for some time after the event because the interval from effect on the testes to detectable effect on fertility can be several months.

E. Nutrition

Very severe restriction of caloric intake—starvation—will often stop estrous and menstrual cycles and inhibit ovulation. A high caloric intake will often stimulate reproduction. A common practice of increasing the caloric intake before mating (flushing) has been used to enhance both fertility and fecundity in domestic animals. In some wild species the periodic abundance of food triggers a breeding period. High caloric intake promotes a greater number of ovulations so that fecundity is increased both by greater fertility and by inducing larger litters. The interval between successive births is influenced by availability of food. A plentiful supply shortens the interval between successive births.

In most free-ranging species, both males and females often have more than one mate. Even in those species considered to be monogamous, many individuals have been found to not always be faithful. This infidelity may be a form of progeny insurance in cases in which one of the pair is less fertile. When sperm from two or more males are present at the site of fertilization and compete for oocytes, most of the offspring will be sired by the more fertile male. The male with greater fertility will also have larger litters. The inherent advantage of one male disappears if his sperm arrive at the site of fertilization at an inappropriate time. Timing of insemination relative to ovulation is critical for optimum fertility. Gametes have finite lives measured in hours. Fertility is reduced when an insemination precedes ovulation by many hours. Insemination after ovulation also results in reduced fertility because of the very finite fertilizable life of the oocyte. Insemination prior to ovulation by sufficient time to allow for the period of sperm transport capacitation and penetration of oocyte membranes will give the greatest fertility. Avoidance of mating during the fertile period is a form of contraception. In most domestic and wild animals receptivity of the female occurs only during a fertile period.

F. Manipulations

In most cases, natural matings result in optimum fertility. When sperm are collected, cooled, stored, extended in selected media, frozen or in other ways handled, the fertility is usually less than when large numbers of sperm are deposited naturally. Extending the semen almost universally means the concentration of sperm is also reduced. Fertility of semen declines with storage period. The chemical makeup of the extender can influence the effectiveness of the extender; thus, there is an ongoing search for combinations of components that will best preserve the fertilizing ability of stored sperm. Sperm of bulls are now most often stored in a frozen state and there is a continuous attempt to devise procedures that will preserve fertilizing ability of sperm of many species both domestic and wild. Removing embryos from the mother and transferring them, freezing and storing them, and otherwise manipulating them reduces their chances of survival. Gamete and embryo manipulation can be a solution to an otherwise infertile condition. Sperm can be recovered from the male, oocytes can be recovered from the female, and fertilization can occur outside the body, and the embryo can be inserted back into a female and fertility is restored. Assisted reproductive technology has made it possible for infertile pairs to produce young under a wide variety of conditions.

See Also the Following Articles

INFERTILITY; NUTRITIONAL FACTORS AND REPRODUCTION; STERILITY; STRESS AND REPRODUCTION

Bibliography

Barnett, J. L., Batterham, E. S., Cronin, G. M., Hansen, D., Hemsworth, P. H., Hennessy, D. P., Hughes, P. E., Johnston, N. E., and King, R. H. (1987). *Manipulating Pig Production*. Australasian Pig Science Association, Alloury, New South Wales.

Cole, D. J. (1972). *Pig Production*. Butterworths, London.

Cupps, P. T. (1991). *Reproduction in Domestic Animals*. Academic Press, San Diego.

Dieleman, S. J., Colenbrander, B., Booman, P., and Van der Lende, T. (1992). *Clinical Trends and Basic Research in Animal Reproduction*. Elsevier, Amsterdam.

Dziuk, P. J. (1996). Factors that influence the proportion of offspring sired by a male following heterospermic insemination. *Anim. Reprod. Sci.* **43**, 65–88.

Foxcroft, G. R., Hunter, M. G., and Doberska, C. (1993). *Control of Pig Reproduction IV*. Journals of Reproduction and Fertility Ltd., Cochester, UK.

King, G. J. (1993). *Reproduction in Domestic Animals*. Elsevier, Amsterdam.

Sreenan, J. M., and Diskin, M. G. (1986). *Embryonic Mortality in Farm Animals*. Nijhoff, Dordrecht.

Zavy, M. T., and Geisert, R. D. (1994). *Embryonic Mortality in Domestic Species*. CRC Press, Boca Raton, FL.

Fertilization

Gerald Schatten

Oregon Regional Primate Research Center and Oregon Health Sciences University

I. Sperm Events
II. Sperm Incorporation
III. Metabolic Activation of the Egg
IV. Block(s) to Polyspermy
V. Pronucleus Formation and Nuclear Fusion
VI. Pronuclear Migrations and Genomic Union
VII. Implications and Future Directions

GLOSSARY

centrosome, sperm vs **zygote** The centrosome is the cell's microtubule organizing center that organizes the poles for mitotic and meiotic spindles, establishes the axis for cell divisions, and serves as the cell's internal compass endowing it with directionality for locomotion. During oogenesis, the maternal centrosome is destroyed and the sperm contributes the precursor centrosome (sperm centrosome). After the recruitment of maternal gene products and posttranslational modifications, the sperm centrosome is transformed into the zygote centrosome, capable of directing microtubule assembly and reproducing at each cell cycle.

cortical reaction Among the first microscopic events during fertilization, when tens of thousands of secretory granules (the cortical granules) subjacent to the egg's plasma membrane undergo a wave of exocytosis radiating from the site of sperm incorporation. The cortical reaction modifies the extracellular matrix of the egg (zona pellucida, vitelline layer, and fertilization coat) and renders it incapable of penetration by additional sperm as well as by bacteria and viruses.

cytoskeleton The cytoplasmic architectural system of fibers responsible for shape and motion. It is composed of three major networks—microfilaments (7 nm diameter), intermediate filaments (10 nm), and microtubules (25 nm)—along with a myriad of accessory proteins including motor proteins, stabilizing proteins, and cargo-docking proteins.

egg activation The cascade of events, involving elevations in calcium ion concentrations as well as other ionic and enzymatic changes, that leads to the metabolic activation of the dormant egg.

fertilization The process that culminates in the union of one, and only one, sperm nucleus with the egg nucleus within the activated egg cytoplasm.

fertilization cone The microfilament-containing eruption on the egg's surface that assembles around the entering, successful sperm. The fertilization cone is responsible for the physical incorporation of the sperm into the egg cytoplasm in most nonmammalian systems.

incorporation cone In mammals, the incorporation cone forms as a result of new microfilament and myosin II assembly at the cortex overlying the position of the incorporated sperm nucleus. Unlike the fertilization cone, the incorporation cone does not appear to be responsible for sperm incorporation but rather is the result of the sperm nucleus incorporation. This transient structure is disassembled at the time when the nuclear envelope fully encloses the incorporated sperm nucleus, transforming it into the male pronucleus.

microfilaments Seven-nanometer fibers composed primarily of actin. Microfilaments are involved in muscle contraction, blood clotting, nerve outgrowth, and cell motility. During fertilization, egg microfilaments are responsible for the formation of the polar bodies; sperm incorporation; adherence of the cortical granules and meiotic spindle to the egg cortex; microvillar organization as well as, later, cytokine-

sis. In some sperm, microfilaments are assembled or extended during the acrosome reaction to produce the acrosomal process, a harpoon-like structure (in some cases longer that the sperm tail) that establishes the initial contact between the gametes.

microtubules Twenty-five-nanometer fibers composed primarily of tubulins. Microtubules are responsible for chromosome separation during meiosis and mitosis; the swimming behavior of sperm tails, flagella, and cilia; the sperm aster that unites the male and female pronuclei during fertilization; as well as cellular motility.

oocyte Since most eggs are inseminated while arrested at a phase of meiosis, they are usually referred to as "oocytes" rather than mature eggs. In vertebrates, fertilization starts with the "egg" as an oocyte arrested in second meiotic metaphase.

polyspermy, block to Polyspermy is a lethal condition in which more than one sperm enters the egg, and the imbalances between the centrosomes and chromosomes result in aberrant cell divisions. The block(s) to polyspermy is the processes that prevent supernumerary sperm incorporations and usually involves the cortical reaction.

pronucleus After the sperm nucleus enters the egg cytoplasm, its DNA decondenses and it is invested with a maternal set of chromatin proteins and a nuclear envelope rich in nuclear pores and nuclear lamins. This decondensed, but still haploid, nucleus is referred to as the *male pronucleus*, and the complimentary structure derived from the maternal genome is called the *female pronucleus*.

sperm aster As the male and female pronucleus form, a new radially arrayed microtubule structure assembles within the egg, nucleated by the sperm centrosome. This sperm aster is responsible for pulling the female pronucleus to the male pronucleus as well as for moving both pronuclei toward the egg center; these motions are called the pronuclear migrations.

sperm incorporation The physical incorporation of the sperm into the egg cytoplasm.

tubulin, α-, β-, and γ- The 55-kDa proteins that assemble into microtubules. The majority of the microtubule is composed of a heterodimer of α-tubulins and β-tubulins. γ-Tubulin is a recently discovered rare, but ubiquitous, tubulin that serves as the nucleating site for each microtubule.

zygote nucleus The diploid nucleus formed during the first cell cycle by the fusion between the male and female pronuclei. Many systems, including mammals, do not have a zygote nucleus and instead the separate, but adjacent, male and female pronuclei undergo synchronous nuclear breakdown at the time of first mitosis. Fertilization is culminated as the parental chromosomes intermix at the equator of the first mitotic spindle.

Fertilization is the process that culminates in the union of one, and only one, sperm nucleus with the egg nucleus within the activated egg cytoplasm. For it to occur successfully, several events must occur, including (i) sperm activities; (ii) incorporation of the sperm into the egg cytoplasm, which includes the binding between and fusion of the egg and sperm's plasma membranes; (iii) egg activation and the initiation of the first cell cycles; (iv) formation of the sperm and egg nuclei (male and female pronuclei, respectively); (v) migrations of the pronuclei that lead to genomic union; and (vi) initiation of first division and early development. While understanding the cell biological basis of fertilization is central in creating a complete body of knowledge regarding reproduction, the study of fertilization itself has served as an intellectual foundation on which the entire field of cell biology has been constructed. With *in vitro* fertilization of mammals, including humans, now routine, it has been possible to solve many of the critically important problems of fertilization at the molecular cell biological level. This review will focus on the frontiers of knowledge regarding the cell biological basis of fertilization, especially in mammals.

I. SPERM EVENTS

The process of fertilization is initiated upon the recognition of the sperm that is in the egg's vicinity. This cellular recognition event leads to the sperm acrosome reaction that results in the externalization of the contents of the acrosomal vesicle. These contents include hydrolytic enzymes that permit the sperm to penetrate the outer investments of the egg (the zona pellucida in mammals and the egg jelly in most other species) as well as proteins and glycoproteins that bind the sperm to the egg, usually in a species-specific manner (e.g., bindin in sea urchins and several proteins in mammals including galactosyltransferase and fertilin). Capacitation is the prerequisite event to developing the ability to undergo the acrosome reaction in mammals.

The swimming of the sperm is also influenced by the egg. Perhaps the most extreme case is found in systems such as that of the horseshoe crab, *Limulus*,

in which the sperm is immotile until it senses that it is in the vicinity of the egg. In most other systems, egg proteins and glycoproteins affect sperm motility. In mammals, the sperm swims in long curvilinear tracks, probably to span the distances within the female reproductive tract, until it nears the eggs when the faster "homing" pattern called hyperactivation occurs. The figure-eight trajectory during hyperactivation is thought to participate in the sperm's precise location of the oocyte.

Because it is perhaps the most highly differentiated cell, the sperm's components are reduced to the barest minimum. In addition to the loss of most cytoplasmic organelles, except for the mitochondria essential for swimming, spermatogenesis results in the segregation and packaging of the haploid sperm nucleus into the compact, inactive nucleus with its lean nuclear envelope largely devoid of nuclear pore complexes and modified nuclear lamins. In addition, the sperm centrosome (the cell's microtubule organizing center) is reduced during spermatogenesis to that of a "precentrosome" capable of recruiting from maternal stores the essential gene products necessary for mitotic spindle pole formation during fertilization and early development.

For the egg, the sperm is essential for the contributing critical components: (i) the paternal haploid genome; (ii) the signal to initiate the metabolic activation of the egg; and (iii) the centrosome, which directs microtubule assembly within the inseminated egg and leads to the union of the sperm and egg nuclei in the activated egg cytoplasm as well as the formation of the mitotic spindles during development (Fig. 1 and 2).

II. SPERM INCORPORATION

The physical incorporation of the sperm into the egg cytoplasm occurs simultaneously with egg activation and the cortical reaction. In mammals, the sperm rests tangentially with the egg surface as it enters, whereas in lower systems the sperm is incorporated from a perpendicular orientation. Microfilaments, polymers assembled from maternal actin, assemble around the entering sperm to form the "fertilization cone" in nonmammalian systems or the "sperm incorporation cone" in mammals. Evidence that microfilaments are vital for sperm incorporation includes the correlation with new microfilament assembly specifically by the entering sperm as well as the inhibition of sperm incorporation (or in mammals, sperm tail incorporation) by drugs which prevent egg microfilament assembly.

III. METABOLIC ACTIVATION OF THE EGG

Oocytes and eggs are metabolically quiescent, and sperm–egg binding and fusion initiates a cascade of events that transforms the dormant egg into the dynamic, animated zygote. While there are still diverse opinions on the precise manner in which the sperm activates this cascade, it is clear in all fertilization systems that an elevation in intracellular calcium ion concentration is the central messenger in communicating the activating signal. Sperm incorporation in eggs in which the calcium increase is prevented do not activate, whereas the artificial increase of intracellular calcium ion concentrations leads to activation and parthenogenetic development.

At the beginning of this century the pioneering fertilization researcher Jacques Loeb asked the incredibly chauvinistic question: "How does the sperm save the life of the egg?" Strong evidence is provided by two often disparate teams of researchers. On the one hand, eggs possess receptors for sperm glycoproteins in their membranes, and compelling data indicate that the sperm might well behave as an honorary hormone or neurotransmitter activating the egg through well-understood signal transduction pathways. In this model, the binding of the sperm to the receptor in the egg alters the receptor so that it might trigger the G-protein cascade or activate tyrosine kinase pathways that then lead to the activation of phospholipase C and the production of inositol trisphosphate, which mobilizes internal calcium. In many eggs, the calcium essential for activation is internally sequestered, whereas in others, the opening of plasma membrane calcium channels leads to the influx of external calcium.

On the other hand, the sperm may activate the egg not by binding and modifying receptors in the egg membrane but by the direct injection into the egg cytoplasm of a sperm-derived protein or factor referred to as "oscillin." Likely candidates have been isolated from several species of sperm, and some stimulate egg activation across species lines. This theory helps explain the clinical success of intracytoplasmic sperm injection in which the sperm is introduced directly into the human egg cytoplasm, and it also avoids the danger of a sperm, incapable of being incorporated, activating the egg by binding to a receptor.

Still other researchers are convinced that fertilization might have evolved multiple overlapping pathways to ensure the success of egg activation, and that both the receptor pathway and the oscillin protein results are accurate and valid.

All eggs are arrested prior to sperm incorporation, but the precise stage during meiosis varies from species to species. In vertebrates, the eggs of most species arrest at metaphase of the second meiosis, whereas others arrest at other stages of egg maturation. Sea urchins are among the rare groups in which the eggs complete meiotic maturation and arrest with a haploid female pronucleus (egg nucleus). Because most eggs are actually at arrested stages of meiosis, they are usually referred to as oocytes. The cell cycle arrest in vertebrate eggs is due to the presence of a cytostatic factor (CSF), which is known to involve the c-Mos gene product as well as a MAP kinase cascade. The elevation in intracellular calcium at sperm incorporation destroys CSF and leads to the destruction of the cyclin portion of the M-phase promoting factor. With the biodegradation of cyclin, the metaphase-II arrested oocyte is capable of entering anaphase and the first cell cycle.

The molecular mechanism permitting the fusion between the plasma membranes of the sperm and egg is not well understood. Sperm proteins, such as the mammalian fertilin and the invertebrate bindin, are critical for the adhesion of sperm to eggs, usually in a species-specific fashion. While theories have been advanced that these or similar proteins might mediate fusion, this conclusion awaits further investigation.

IV. BLOCK(S) TO POLYSPERMY

Fertilization is defined as the fusion of one, and only one, sperm nucleus with the egg nucleus. Different eggs avoid the pathological condition of poly-spermy in various ways, though most rely on the secretion of cortical granules to modify the egg's extracellular matrix to ultimately prevent the incorporations of supernumerary sperm. The secretion of the cortical granules (the cortical reaction) prevents blocks to polyspermy by proteolytically digesting the sites for sperm binding and also by enzymatically modifying the egg's coat to toughen and harden it. In mammals, this results in the modification of the zona pellucida, whereas in many lower systems the vitelline layer that is closely adherent to the egg's plasma membrane is elevated and laminated to form the fertilization coat after the cortical reaction.

The initial block to polyspermy varies considerably. For example, in most bony fish, the extracellular matrix of the chorion has an opening precisely the size of a single sperm, thereby preventing multiple sperm entrances. At the other extreme, in large eggs of birds, tailed amphibians (urodeles such as newts and salamanders), and lower fish (such as shark and rays), multiple sperm enter the egg in a condition known as physiological polyspermy. While the supernumerary sperm decondense to form sperm nuclei (male pronuclei) and organize microtubule assembly into sperm asters (see Section VIII), only one sperm nucleus fuses with the egg nucleus prior to first mitosis; all other sperm nuclei and centrosomes degenerate. Perhaps the best understood fast block to polyspermy is the fast, electrical block of many invertebrates and some vertebrates. In this system, the successful sperm initiates the opening of ion channels, (e.g. Na^+ channels in sea urchins), resulting in the reversal of the membrane potential that prevents additional sperm fusion until the time expected for the establishment of the cortical reaction blockage.

V. PRONUCLEUS FORMATION AND NUCLEAR FUSION

Since the sperm enters the egg with a highly condensed, inactive nucleus, and since most eggs are actually arrested oocytes with condensed meiotic chromosomes, fertilization also leads to the decondensation and remodeling of the parental genomes (chromatin). In many systems, the sperm nucleus is modified during spermiogenesis so that the typical histone chromosome proteins are reduced by protamines, leading to a tighter packaging of the DNA from the usual nucleosome pattern. In all systems, the paternal DNA binds maternal histones and other proteins and is invested with a new nuclear envelope of maternal origin containing nuclear lamins and nuclear pore complexes. The completion of fertilization is signaled by genomic union. In most nonmammalian species, the decondensed male and female pronuclei migrate into close apposition (see Section VIII), and then the two nuclear membranes circumscribing each pronucleus fuse. This results in the formation of the diploid zygote nucleus. While data

FIGURE 1 Microtubule and DNA organization in normal inseminated human oocytes. (A) The meiotic spindle in mature, unfertilized human oocytes is anastral, oriented radially to the cell surface, and asymmetric, with a focused pole abutting the cortex and a broader pole facing the cytoplasm. No other microtubules are detected in the cytoplasm of the unfertilized human oocyte. (B–D) Shortly after sperm incorporation (3–6.5 hr postinsemination), sperm astral microtubules assemble around the base of the sperm head, as the inseminated oocytes complete second meiosis and extrude the second polar body. The close association of the meiotic midbody identifies the female pronucleus. Short, sparse disarrayed cytoplasmic microtubules can also be observed in the cytoplasm following confocal microscopic observations of these early activated oocytes (C). (E) As the male pronucleus continues to decondense in the cytoplasm (M), the microtubules of the sperm aster enlarge, circumscribing the male pronucleus. (F) By 15 hr postinsemination, the centrosome splits and organizes a bipolar microtubule array that emanates from the tightly apposed pronuclei. The sperm tail is associated with an aster (arrow). (G) At first mitotic prophase (16.5 hr postinsemination), the male and female chromosomes condense separately as a bipolar array of microtubules marks the developing first mitotic spindle poles. (H) By prometaphase, when the chromosomes begin to intermix on the metaphase equator, a barrel-shaped, anastral spindle forms in the cytoplasm. The sperm axoneme remains associated with a small aster found at one of the spindle poles (arrow). M, male pronucleus; F, female pronucleus. Scale bar = 10 μm (reprinted with permission from Simerly *et al.*, 1995).

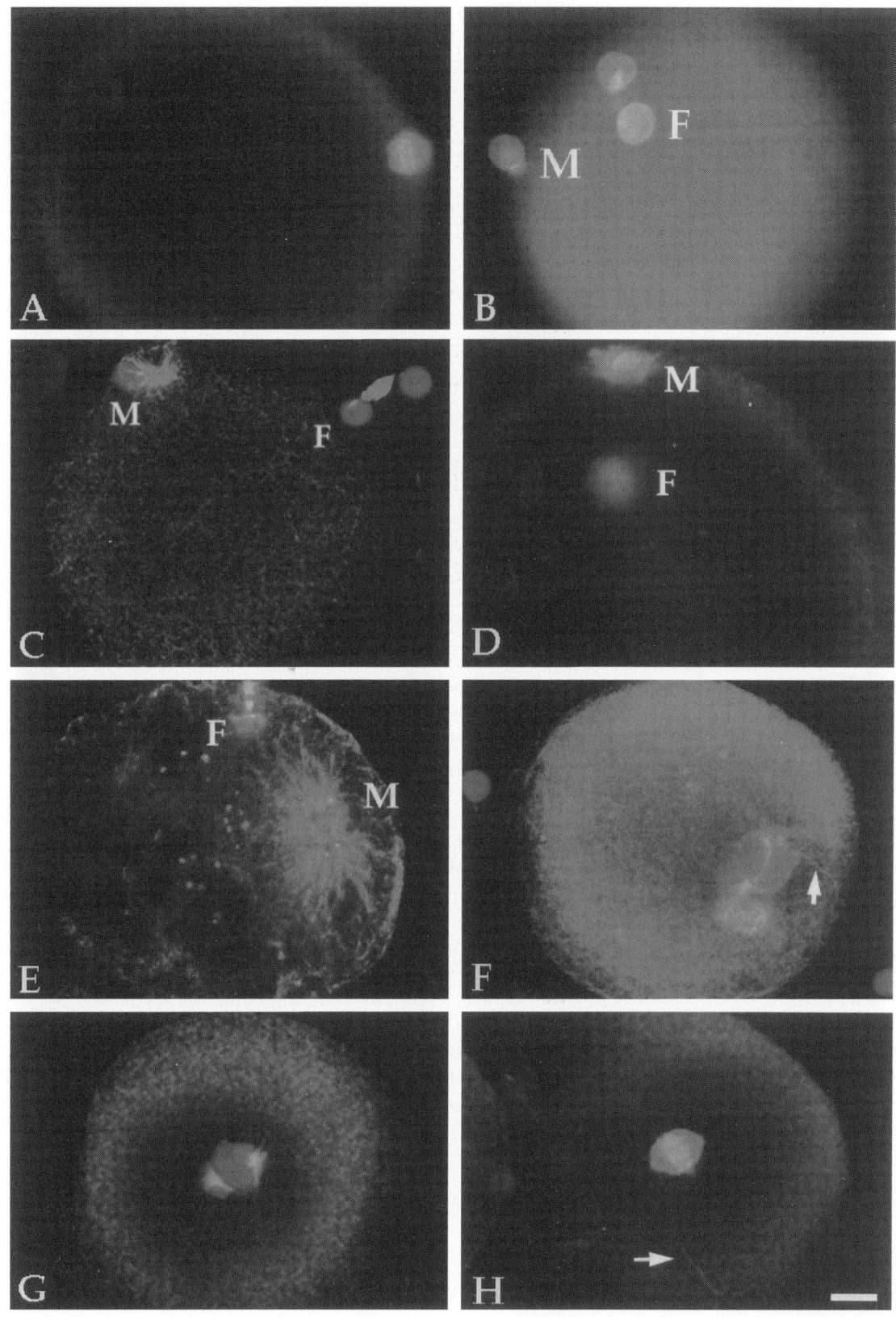

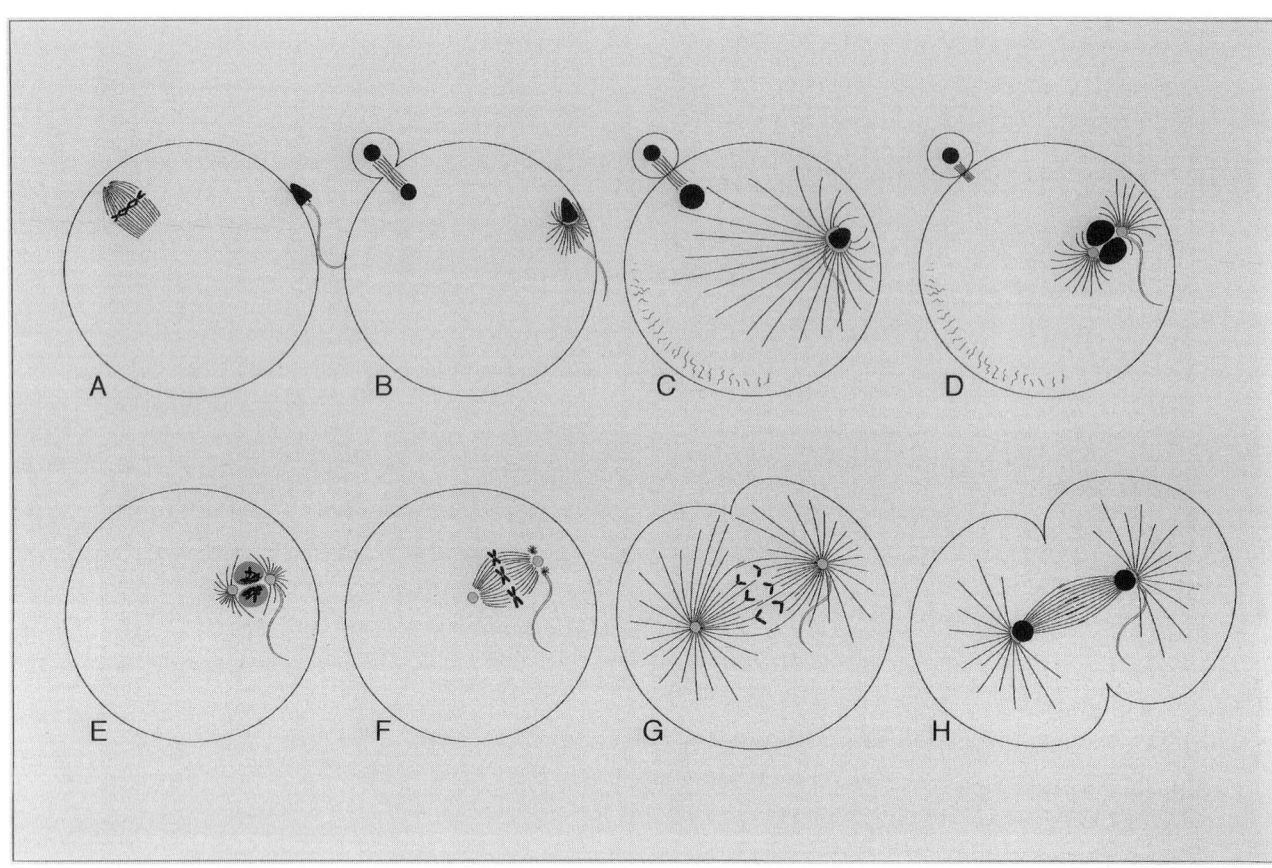

FIGURE 2 Microtubule organization and centrosome behavior during fertilization in humans. Unfertilized oocytes have microtubules only within the metaphase-arrested second meiotic spindle (A). After introduction of the sperm centrosome following sperm incorporation (B), new microtubules assemble to form the radial sperm aster adjacent to the decondensing male pronucleus (B). Centrosome position and shape is inferred from the observed pattern of microtubules. Microtubules are also found in the meiotic midbody and as disarrayed fibers in the cytoplasm distal to the sperm aster (B, C). As male and female pronuclei decondense (C), the sperm aster enlarges, and the male pronucleus is displaced centripetally. The female pronucleus moves toward the male (D) as the sperm aster becomes asymmetric and bipolar. Two microtubule tufts emanate from apposed pronuclei (D), suggesting that centrosome splitting occurs late in interphase. The still separate maternal and paternal chromosomes condense at prophase (E). They align during prometaphase so that by metaphase (F), parental genomes have united at the spindle equator. One or two small asters form at the metaphase spindle pole with the sperm tail (F). Pronuclei remain eccentric until first anaphase (G), when asters enlarge and preferentially interact with the adjacent cortical region; this displaces the spindle toward the zygote center. The cleavage furrow initially forms at this site, suggesting that sperm aster placement plays a role in specification of the first cleavage axis (H). The events past metaphase are inferred from rhesus zygotes (Wu *et al.*, 1996). Defects observed in oocytes from infertile patients include failures to complete (i) sperm incorporation, egg activation, and/or exiting of second meiosis (A → B); (ii) sperm aster nucleation (B); (iii) aster enlargement and pronuclear decondensation (C); (iv) coordinated centration of male pronucleus with sperm astral microtubules and sperm tail (B → C); (v) migration of female pronucleus (C → D); and (vi) cell cycle progression [meiosis to interphase (A → B), interphase to mitosis (E → H), or mitotic metaphase to interphase (F → H)] (reprinted with permission from Simerly *et al.*, 1995).

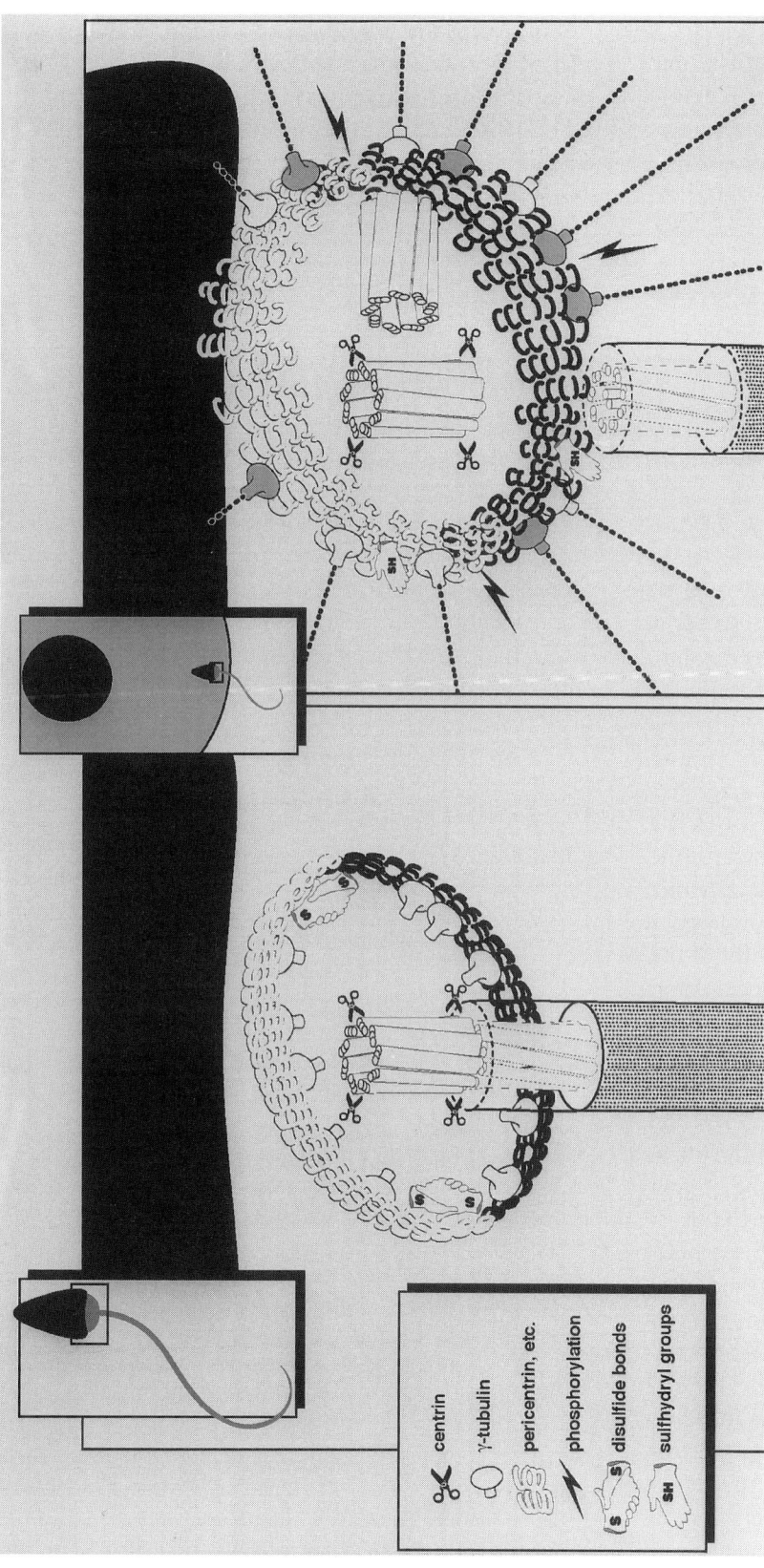

FIGURE 3 Molecular dissection of the human sperm centrosome and its reconstruction in the zygote. The sperm centrosome (left): The human sperm centrosome has centrin concentrated in one or two focal sites, corresponding to the centrioles. γ-Tubulin is not apparent in mature human sperm but becomes detectable after centrosomal priming of the sperm with disulfide reducing agents; this is a novel type of cytoplasmic capacitation. γ-Tubulin is also detectable on Western blots with intact or sonicated human sperm. The centrosome is not phosphorylated and the sperm tail microtubules extend from a centriole. The coiled-coil infrastructure of the centrosome probably anchors the centrosome to the sperm nucleus and regulates the exposure of, and binding sites for, γ-tubulin. The zygote's centrosome (right): After permeabilization and incubation in extracts from *Xenopus* oocytes, the human sperm becomes phosphorylated and heavily immunoreactive with antibodies to γ-tubulin. The γ-tubulin found in the human sperm is probably a combination of some paternal and largely maternal protein. The binding of calcium ions, released during the transient increase during egg activation, to centrin is predicted to result later in a centrin-induced severing of the doublet sperm tail microtubules from the triplet microtubules of the centriole. Perhaps the severing of the tail microtubules from the basal body frees the basal body complex so that it can bind additional γ-tubulin and undergo transformation into a centriole. In humans, calcium-mediated centrin excision does not lead to complete sperm tail dissociation from the centriole/centrosome complex. The coiled-coil domains of the centrosome are drawn as unraveling, expanding, and everting in the zygote; this exposes paternal γ-tubulin and also exposes binding sites for maternal γ-tubulin. The halo of γ-tubulin nucleates the microtubules which assemble into the sperm aster (Reprinted with permission from Schatten, 1994).

are limited on the molecular mechanisms involved in plasma membrane fusion, they are virtually completely missing in the case of nuclear fusion. In mammals, the pronuclei do not fuse during interphase: Instead, the adjacent pronuclei undergo separate nuclear breakdowns at first mitosis and the parental chromosomes intermix as they align at the metaphase plate of the first mitotic spindle. Consequently, the events of mammalian fertilization start during second meiosis and are not complete until first mitosis.

VI. PRONUCLEAR MIGRATIONS AND GENOMIC UNION

Following the successful incorporation of the sperm into the egg and egg activation, and coincident with pronuclear formation, a new motility system assembles within the egg and its organization is directed by the newly introduced sperm centrosome. The microtubule based sperm aster is a radially arrayed three-dimensional structure organized by the sperm centrosome, which is found adjacent to and affixed to the sperm nucleus. The sperm aster has three essential roles: (i) Its growth subjacent to the egg cortex pushes the sperm nucleus toward the egg center; (ii) when the distal ends of the dynamic microtubules contact the egg nucleus, it undergoes a sudden and swift translocation to the center of the sperm aster where it meets the sperm nucleus; and (iii) the continuing elongation of the sperm astral microtubules moves the now adjacent pronuclei toward the egg center.

The role of the sperm aster during fertilization cannot be minimized and certain forms of fertilization failures appear to be due to defects in the organization or functioning of the sperm aster. It is organized by the paternally inherited sperm centrosome, the structure of which is not well understood. The sperm centrosome (Fig. 3) attracts and binds numerous maternal proteins, especially γ-tubulin and nuclear mitotic apparatus protein (NuMA). These transform the inactive precursor sperm centrosome into the microtubule-nucleating, reproducing zygote centrosome. The pattern of the paternal inheritance of the centrosome appears universal, except for rodents and parthenogenetic systems.

Following the migration and union of the pronuclei, the centrosome duplicates and separates to become the future two poles for the first mitotic spindle. The DNA undergoes replication during interphase, and a new cascade of cell cycle-mediating events leads to the initiation of first division and early development.

VII. IMPLICATIONS AND FUTURE DIRECTIONS

Because fertilization bridges the fields of reproductive and developmental biology, a full understanding of the cellular and molecular events during fertilization is critical. In addition, many clinical problems including sophisticated infertility treatments and contraception hinge on a more complete knowledge underlying this crucial but, in many ways, mysterious and miraculous process.

Bibliography

Bedford, J. M., and Hoskins, D. D. (1990). The mammalian spermatozoon: Morphologyy, biochemistry and physiology. In *Marshall's Physiology of Reproduction* (G. E. Lamming, Ed.), Vol. 2, pp. 379–568. Longman, Harlow/New York.

Jaffe, L. A. (1990). First messengers at fertilization. *J. Reprod. Fertil.* **42**(Suppl.), 107–116.

Myles, D. G. (1993). Molecular mechanisms of sperm–egg membrane binding and fusion in mammals. *Dev. Biol.* **158**(1), 35–45.

Schatten, G. (1994). The centrosome and its mode of inheritance: The reduction of the centrosome during gametogenesis and its restoration during fertilization. *Dev. Biol.* **165**(2), 299–335.

Schatten, G., Hewitson, L., Simerly, C., Sutovsky, P., and Huszar, G. (1998). Cell and molecular biological challenges of ICSI: A.R.T. before science? In press.

Schultz, R. M., and Kopf, G. S. (1995). Molecular basis of mammalian egg activation. *Curr. Topics Dev. Biol.* **30**, 21–62.

Simerly, C., Wu, G. J., Zoran, S., Ord, T., Rawlins, R., Jones, J., Navara, C., Gerrity, M., Rinehart, J., Binor, Z., Asch, R., and Schatten, G. (1995). The paternal inheritance of the centrosome, the cell's microtubule-organizing center, in

humans, and the implications for infertility. *Nature Med.*
1(1), 47–52.

Swann, K., and Lai, F. A. (1997). A novel signalling mechanism for generating Ca^{2+} oscillations at fertilization in mammals. *Bioessays* 19(5), 371–378.

Trounson, A., and Bongso, A. (1996). Fertilization and development in humans. *Curr. Topics Dev. Biol.* 32, 59–101.

Yanagimachi, R. (1994). Mammalian fertilization. In *The Physiology of Reproduction* (E. Knobil *et al.*, Eds.), pp. 189–317. Raven Press, New York.

■

Fetal Adrenals

Sam Mesiano and Robert B. Jaffe

University of California, San Francisco

I. Introduction and Historical Perspective
II. Development and Growth of the Fetal Adrenals
III. Regulation of Fetal Adrenal Development and Function
IV. Function of the Fetal Adrenals

GLOSSARY

adrenal cortex The outer compartment of each adrenal gland composed of steroid hormone-producing endocrine cells.

adrenal glands Endocrine glands located on the top of each kidney which regulate carbohydrate and mineral metabolism and the body's response to stress.

adrenal medulla The central compartment of each adrenal gland composed of neuroendocrine chromaffin cells which secrete adrenalin.

definitive zone A narrow cortical zone which surrounds the fetal zone.

fetal zone The large central compartment of the primate fetal adrenal cortex present only during fetal life.

fetoplacental unit A functional relationship between the fetal adrenal cortices and the placenta such that androgens produced by the adrenals are utilized as essential substrates for placental estrogen production.

parturition The process of giving birth, involving the coordinated contractions of the uterus and dilation of the cervix.

T he adrenal glands are essential endocrine organs whose secretions are crucial for the maintenance of metabolic and fluid homeostasis and the physiological response to stress. Located at the cranial end of the kidneys, each adrenal is composed of two compartments: (i) the medulla and (ii) the cortex. The adrenal medulla produces adrenaline and noradrenaline whose secretions are directly regulated by neuronal interactions. The adrenal cortex is a steroid hormone-producing organ which can be divided into three morphologically and functionally distinct compartments: the outer glomerulosa, the central fasciculata, and the inner reticularis, which secrete mineralocorticoid (mainly aldosterone), glucocorticoid (mainly cortisol), and androgen [mainly dehydroepiandosterone (DHEA)], respectively. Growth and function of the adrenal cortex are regulated primarily by corticotropin (ACTH) secreted by the anterior lobe of the pituitary gland. ACTH is the principal regulator of cortisol production by fasciculata cells but also stimulates DHEA and aldosterone secretion from fasciculata and glomerulosa cells, respectively. Cortisol inhibits ACTH secretion resulting in a classic negative feedback endocrine loop. Aldosterone secretion by glomerulosa cells is also regulated by angiotensin II. In humans and higher primates, adrenal development is characterized by rapid cortical growth and a high level of steroidogenic activity during fetal life. The following discussion is concerned with the development, function, and regulation of the adrenal glands (and in particular the adrenal cortices) during fetal life.

I. INTRODUCTION AND HISTORICAL PERSPECTIVE

The fetal adrenal glands play a pivotal role in the regulation of fetal development and its preparation for extrauterine life. Development and function of the primate fetal adrenal cortex are unique among mammalian species. The unusual characteristics of the primate (particularly human) fetal adrenal were first realized in the early 1900s when its morphology was examined in detail and compared to that of other species. The distinct architecture of the human fetal adrenal cortex and its dramatic remodeling soon after birth captured the interest of developmental anatomists. Elegant experiments during the 1950s and 1960s demonstrated the central role of the primate fetal adrenal cortex in establishing the estrogenic milieu of pregnancy. Those findings were among the first indications of the function and physiological role of the human fetal adrenal cortex, and led Diczfalusy and co-workers to propose the concept of the "fetoplacental unit," in which the C19 steroid dehydroepiandrosterone sulfate (DHEA-S) produced by the fetal adrenal cortex is used by the placenta for estrogen synthesis.

During the 1960s, Liggins and colleagues demonstrated in the sheep that cortisol secreted by the fetal adrenal cortex late in gestation regulates maturation of the fetus and initiates the cascade of events leading to parturition. These pioneering experiments provided insight into the mechanism underlying the timing of parturition and therefore were of particular interest to obstetricians and perinatologists confronted with the problems of preterm labor. However, although cortisol emanating from the fetal adrenal cortex promotes fetal maturation in primates as it does in sheep, its role in the regulation of primate parturition, unlike that in sheep, has not been completely elucidated. Although fetal adrenal cortisol was found to promote fetal maturation in primates as it does in sheep, it was soon realized that primates and sheep have little in common regarding the role of the fetal adrenal cortex in the regulation of parturition. Recently, based on studies in the rhesus monkey, Nathanielsz and colleagues have proposed that the primate fetoplacental unit plays a role in regulating the timing of parturition in primates. These investigators provided evidence supporting the hypothesis that estrogens, produced by the placenta from androgens supplied by the fetal adrenal cortex, influence the timing of parturition.

II. DEVELOPMENT AND GROWTH OF THE FETAL ADRENALS

Fetal adrenal development in humans and higher primates differs qualitatively and quantitatively from that of other species and is characterized by extraordinarily rapid growth, high steroidogenic activity, and a unique morphologic appearance. Throughout gestation, the primate fetal adrenals are predominantly composed of cortical tissue; the medulla is essentially absent except for islands of chromaffin cells in the body of the cortex which do not coalesce to form a diecrete medulla until the fourth postnatal week. Unlike other species (e.g., ovine and rodent) in which the fetal adrenals are similar in structure to the adult adrenals, the morphology of the primate fetal adrenal cortex is unique and differs significantly from that of the adult adrenal cortex. In humans, the fetal adrenal cortex is made up of two morphologically distinct zones, the fetal zone and the definitive zone. The large inner fetal zone accounts for the bulk (80–90%) of the cortex and is the primary site of growth and steroidogenesis. The definitive zone (also referred to as the adult cortex, neocortex, or permanent zone) occupies the remainder of the cortex and is composed of a narrow band of tightly packed cells surrounding the fetal zone (Fig. 1).

The anlage of the human adrenal cortex are first identified at about the fourth week of gestation and form a thickening of the coelomic epithelium in the notch between the primitive urogenital ridge and the dorsal mesentry. By the fifth week, these primitive cells begin to migrate toward the cranial end of the mesonephros and by the eighth week of gestation they organize into anastamosing cords and eventually differentiate into large polyhedral cells, developing into the primordium of the fetal zone. The definitive zone is derived about a week later when a separate population of cells in the same area of coelomic epithelium migrates to the developing adrenal and surrounds the primordial fetal zone. At about

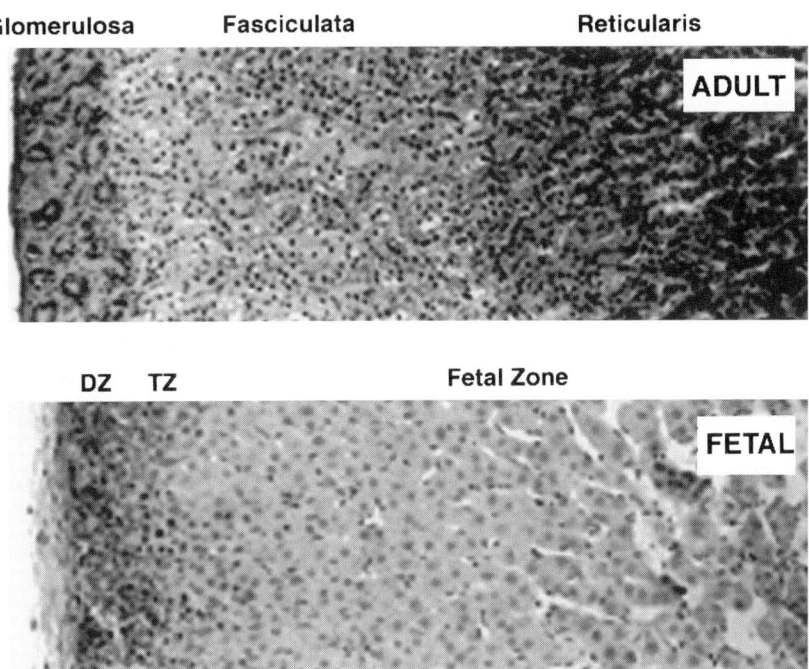

FIGURE 1 Morphology of the human fetal (midgestation) and adult adrenal cortex. The adult cortex is composed of three distinct zones: the glomerulosa, fasciculata, and reticularis (medulla not shown). In contrast, the fetal adrenal cortex is composed of two distinct zones: the outer definitive zone (DZ) and the large inner fetal zone. The transitional zone (TZ) comprises the outer edge of the fetal zone and forms a functionally distinct compartment between the fetal and definitive zones.

the ninth week of gestation, the adrenal is completely enclosed by a capsule, and an extensive network of sinusoidal capillaries develops between the cords of the fetal zone. This vasculature predominates in the central portion of the fetal zone and persists throughout fetal life. Consequently, the adrenal cortex is one of the most highly vascularized organs in the fetus. Abundant vascularization is likely required to facilitate access of hormonal products to the circulation.

After 10–12 weeks of gestation, the morphology of the human fetal adrenal cortex remains relatively constant. By midgestation (16–20 weeks), the fetal zone clearly dominates and is composed of large (20–50 mm) eosinophilic cells which exhibit ultrastructural characteristics typical of steroidogenesis. In the outer regions of the fetal zone, the cells are arranged in tightly packed cords, whereas the cells in the central portion are more widely spaced into a reticular pattern and separated by many vascular sinusoids. The definitive zone is composed of a narrow band of small (10–20 mm) tightly packed basophilic cells which exhibit structural characteristics

typical of cells in a proliferative state. The inner layers of the definitive zone form arched cords which send finger-like columns of cells into the outer rim of the fetal zone. Although definitive zone cells are lipid poor during midgestation, they accumulate some cytoplasmic lipid and begin to resemble steroidogenically active cells with increasing gestational age.

A feature which distinguishes primate fetal adrenals from those of other species is their rapid and disproportionate growth. In humans, this begins at approximately Week 10 of gestation and continues to term; growth is almost entirely due to enlargement of the fetal zone. By Week 20 of gestation, the human fetal adrenal glands are as large as their associated kidneys and the fetal zone accounts for 80–90% of the cortical volume.

As its name implies, the fetal zone exists only during fetal life; soon after birth it atrophies and the adrenal cortex is dramatically remodeled. This phenomenon appears to be unique to those primate species whose fetal adrenals include a fetal zone. The postnatal remodeling of the primate adrenal cortex involves a complex wave of differentiation such that the fetal zone involutes by a process involving apoptosis and the adult zonae glomerulosa and fasciculata develop from the persistent definitive zone. However, because there is no evidence of adrenal cortical insufficiency during the perinatal period, rudimentary but functional forms of the zonae glomerulosa and fasciculata are likely present prior to birth.

III. REGULATION OF FETAL ADRENAL DEVELOPMENT AND FUNCTION

It is not unexpected that the extraordinary growth and steroidogenic activity of the primate fetal adrenal cortex are dependent on an intact fetal pituitary gland, because it produces ACTH, the primary tropic regulator of the adrenal cortex postnatally. Several observations firmly establish the pivotal role of ACTH secreted by the fetal pituitary in primate fetal adrenal cortical regulation: (i) Disruption of hypothalamic–pituitary function in the human fetus (e.g., in anencephalics or associated with maternal glucocorticoid treatment) results in failure of the fetal zone to develop beyond the size attained at approximately the 15th week of gestation and is associated with dramatically reduced estrogen concentrations in the maternal circulation; (ii) administration of dexamethasone to normal human fetuses *in utero* reduces maternal estrogen levels, and dexamethasone treatment of pregnant rhesus monkeys during late gestation causes atrophy of the fetal zone and a marked decrease in maternal plasma estradiol and estrone levels and a parallel decline in fetal plasma cortisol levels. Similarly, experimental anencephaly (produced by fetal decapitation at around Day 80 of gestation; term = 165 ± 5 days) in the fetal rhesus monkey also caused marked atrophy of the fetal adrenal cortex and decreased maternal estrogen levels; (iii) *in utero* administration of ACTH to normal human fetuses increases maternal estrogen levels, and in anencephalics it partially restores adrenal size; (iv) in congenital adrenal hyperplasia, the fetal zone is markedly enlarged and adrenal androgen concentrations are elevated to virilizing levels; and (v) in fetal

rhesus monkeys, administration of ACTH increases fetal cortisol and maternal estrone concentrations, and increased endogenous ACTH secretion (induced by metyrapone treatment) increases fetal adrenal cortical growth and advances functional maturation, as assessed by 3β-hydrosteroid dehydrogenase/isomerase expression.

The mechanism by which ACTH regulates fetal adrenal cortical growth and function is not clearly understood. The actions of ACTH are mediated via its interaction with specific receptors on the cell surface of adrenal cortical cells. A human ACTH receptor has been cloned and characterized. This receptor is coupled to heterotrimeric guanine nucleotide-binding proteins that activate adenylate cyclase leading to an increase in intracellular cAMP, which activates protein kinase A and initiates the cascade of intracellular signaling events. Consequently, the actions of ACTH can be mimicked by cAMP analogs (e.g., 8-bromo-cAMP) or substances which increase adenylate cyclase activity (e.g., forskolin). The signaling events downstream from the activation of protein kinase A have not been characterized.

Several observations support the concept that human fetal adrenal growth and function are also influenced by factors which act independently from, or in conjunction with, ACTH. Evidence includes the following: (i) Human fetal adrenal cortical growth and steroidogenic activity are maximal during mid- to late gestation even though circulating ACTH concentrations in the fetus may be decreasing; (ii) prior to 10–15 weeks of gestation, adrenal development in anencephalic fetuses (presumably with markedly reduced ACTH) is normal, but thereafter the fetal zone fails to develop and does not exhibit its characteristic growth and steroidogenic activity, indicating that early in gestation fetal zone growth and function are not dependent on ACTH; (iii) in contrast, the definitive zone appears normal in anencephalic fetuses despite the absence of ACTH stimulation, suggesting that its growth is not dependent on ACTH at any stage in gestation, although its functional maturation appears to be regulated by ACTH; (iv) the fetal zone rapidly involutes once the newborn is delivered and separated from the placenta despite relatively unchanged exposure to ACTH, indicating that fetal zone growth and function are maintained by a factor(s) specific to intrauterine life—placental estrogens, chorionic gonadotropin, and corticotropin-releasing hormone have been implicated as placental-derived regulators of fetal adrenal cortical growth and function; and (v) ACTH is not a growth factor per se and is not a mitogen for adrenal cortical cells *in vitro*; however, it stimulates proliferation of adrenal cortical cells *in vivo*, suggesting that its proliferative actions may be mediated by locally produced growth factor(s).

Specific growth factors, including insulin-like growth factors-I and -II, basic fibroblast growth factor (bFGF), epidermal growth factor, activin, inhibin, and transforming growth factor-α and -β, acting in an autocrine and/or paracrine fashion, are likely candidates as mediators and/or modulators of the trophic actions of ACTH on the primate fetal adrenal cortex. This notion is not without precedent because several classical growth-promoting hormones are known to act through the stimulation of local autocrine/paracrine growth factors. In human and rhesus monkey fetal adrenal cortical cells, ACTH stimulates the expression of IGF-II and bFGF, both

of which are potent mitogens for these cells. Thus, fetal adrenal cortical growth and function are primarily regulated by ACTH, the actions of which are mediated and/or modulated by a cohort of locally expressed and placental-derived factors (Fig. 2).

IV. FUNCTION OF THE FETAL ADRENALS

The fetal adrenal gland is primarily a steroid-producing endocrine organ. Studies of perifused human fetal and definitive zone cells and the localization of steroid-metabolizing enzymes in the midgestation human and late-gestation rhesus monkey fetal adrenal cortex revealed that the fetal zone is the site of DHEA-S synthesis, whereas cortisol is produced only during late gestation (after 25–30 weeks in the human) by a third functionally distinct cortical zone, the transitional zone, located on the outer edge of the fetal zone. The definitive zone is not steroidogenically active *in vivo* until late gestation, when it is the likely site of mineralocorticoid synthesis.

The pioneering work of Liggins and colleagues in sheep first demonstrated that increased activity of the fetal hypothalamic–pituitary–adrenal axis is crucial in the regulation of parturition and the maturation of the fetal organ systems essential for extrauterine life. In this species, increased secretion of cortisol from the fetal adrenal glands during the final week of pregnancy initiates a cascade of events which culminates in the birth of a viable neonate. In most mammalian species, including humans, cortisol also stimulates events associated with preparation for extrauterine life, e.g., surfactant production by the fetal lungs, activity of enzyme systems in the fetal gut, retina, pancreas, thyroid and brain, and deposition of glycogen in the fetal liver. Clearly, perinatal survival is dependent on the timely initiation of labor when organ systems necessary for extrauterine life are sufficiently mature to allow the newborn to survive independent of the uterus and placenta. Thus, in a number of mammalian species, regulation of fetal maturation and the timing of parturition are controlled by a single hormone, cortisol, produced by the fetal adrenals, that appears to coordinate these processes such that fetal maturation proceeds appro-

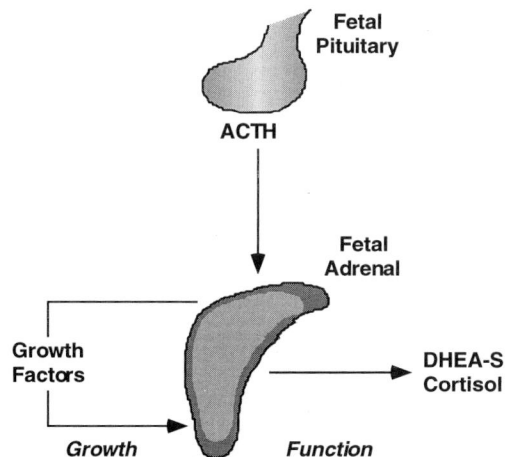

FIGURE 2 Mechanism of primate fetal adrenal cortical regulation. The principal tropic regulator of fetal adrenal growth and function is ACTH produced by the fetal pituitary gland. The growth stimulatory actions of ACTH are mediated by locally produced growth factors which act in an autocrine and/or paracrine fashion. ACTH directly regulates function of fetal adrenal cortical cells by augmenting steroid hormone (mainly DHEA-S and, later in gestation, cortisol) secretion by stimulating expression of steroidogenic enzymes.

priately prior to parturition. However, it was soon realized that fundamental differences exist between sheep and humans with regard to the regulation of parturition and that, although fetal adrenal cortisol clearly orchestrates the initiation of parturition in sheep, a similar role for cortisol in primates was not apparent.

In primates, the fetal adrenal cortex is essential in establishing the estrogenic milieu of pregnancy. Because the primate placenta lacks the steroid-metabolizing enzyme cytochrome P450 17α-hydroxylase 17/20 lyase (P450c17), it cannot produce C_{19} steroids from pregnenolone or progesterone and therefore is dependent on an exogenous supply of C_{19} steroid precursors for estrogen synthesis. In con-

trast to the placenta, the fetal adrenal cortex (particularly the fetal zone) expresses high levels of P450c17 and produces large amounts (approx 200 mg/day during late gestation) of the C_{19} steroid DHEA-S. The human placenta possesses high levels of aromatase activity and is efficient at converting DHEA-S provided by the fetal adrenal cortex and its 16-hydroxylated metabolite to estrogens. The combination of these two incomplete steroidogenic pathways in the two disparate organs results in a complete estrogen-synthesizing system. This strategy for estrogen formation in pregnancy is unique to primate species and is referred to as the fetoplacental unit.

In other species, the fetal adrenal cortex also influences placental estrogen production. The sheep

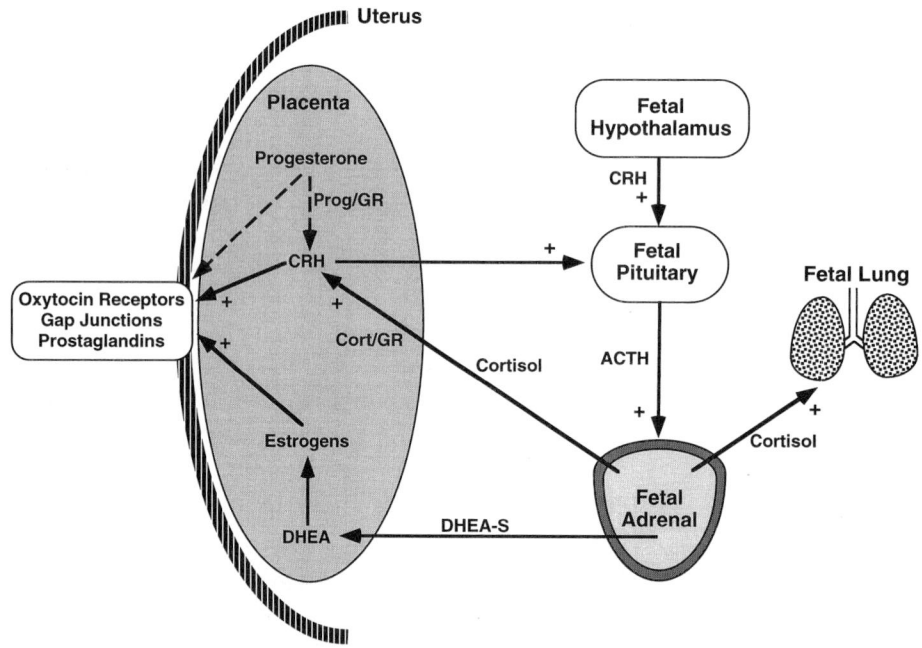

FIGURE 3 Schematic depiction of the theoretical model proposed by Karalis *et al.* (*Nature Med.* **2**, 556–560, 1996) describing the endocrine interaction between the fetal hypothalamic–pituitary–adrenal axis and the placenta in the regulation of human parturition. Late in gestation, cortisol produced by the fetal adrenal cortex blocks the inhibitory effects of progesterone on placental corticotropin-releasing hormone (CRH) production by competing with progesterone for the glucocorticoid receptor (GR). As a consequence, CRH secretion by the placenta into the fetal compartment increases, providing further stimulation to fetal pituitary ACTH production which in turn stimulates cortisol and DHEA-S production by the fetal adrenal cortex. Cortisol promotes maturation of fetal organ systems, such as the lungs, in preparation for extrauterine life and, by further enhancing placental CRH production, completes a positive feedback loop. DHEA-S is converted to estrogens by the placenta, which stimulate gap junction formation and oxytocin receptor expression by the myometrium and prostaglandin production by the aminon and decidua—events necessary to facilitate uterine contraction and labor. CRH also may directly enhance prostaglandin production and myometrial responsiveness to oxytocin. These events are inhibited by progesterone (adapted from Karalis *et al.*, *Nature Med.* **2**, 556–560, 1996).

placenta also lacks P450c17 for most of gestation and, because it is not supplied with C19 steroids, produces very little estrogen. However, late in gestation, cortisol secreted by the fetal adrenals induces expression of P450c17 in the placenta which results in the conversion of progesterone to androstenedione and subsequently to estrone and estradiol. The consequence of this is that during the final days before parturition, placental production of progesterone declines as its conversion to androstenedione and estrogen increases. In several species, an increase in the estrogen/progesterone ratio occurs at the end of pregnancy which is necessary for parturition. In primates, the placenta lacks P450c17 throughout gestation; therefore, progesterone production does not decline at the end of pregnancy as it does in sheep. However, toward the end of pregnancy, increased DHEA-S production by the fetal adrenal cortex results in increased substrate available to the placenta for estrogen production and a rise in maternal estrogen levels. Unlike the sheep, maternal estrogen concentrations in the pregnant woman, rhesus monkey, and baboon rise gradually over several weeks prepartum. Thus, in primates, androgen produced by the fetal adrenals as a source of aromatizable substrate for estrogen synthesis by the placenta may be the link between the fetus and mother in the initiation of parturition. Studies in the rhesus monkey have indicated that estrogens produced by the placenta from aromatizable C_{19} steroids may significantly influence the timing of primate parturition. The physiological roles of estrogens in primate pregnancy are diverse and beyond the scope of this discussion; however, it is clear that the fetal adrenal cortex is essential for the production of placental estrogens and in establishing the estrogenic milieu of primate pregnancy and possibly plays a major role in regulating the timing of parturition (Fig. 3).

Acknowledgment

This work was supported, in part, by NIH Grant R01 HD08478.

See Also the Following Articles

Adrenal Androgens; DHEA (Dehydroepiandrosterone); Fetal-Placental Unit

Bibliography

Diczfalusy, E. (1969). Steroid metabolism in the foeto-placental unit. In *The Foetal–Placental Unit* (A. Pecile and C. Finzi, Eds.), pp. 65–109. Excerpta Medica Foundation, Amsterdam.

Jirasek, J. (1980). *Human Fetal Endocrines*, pp. 69–82. Nijhoff, London.

Liggins, G. C. (1981). Endocrinology of parturition. In *Fetal Endocrinology* (M. J. Novy and J. A. Resko, Eds.), pp. 211–237. Academic Press, New York.

Liggins, G. C., Fairclough, R. J., Grieves, S. A., Kendall J. Z., and Knox, B. S. (1973). The mechanism of initiation of parturition in the ewe. *Rec. Prog. Horm. Res.* **29**, 111–159.

Mecenas, C. A., Giussani, D. A., Owiny, J. R., Jenkins, S. L., Wu, W. X., Honnebier, B. O., Lockwood, C. J., Kong, L., Guller, S., and Nathanielsz, P. W. (1996). Production of premature delivery in pregnant rhesus monkeys by androstenedione infusion. *Nature Med.* **2**, 443–448.

Mesiano, S., and Jaffe, R. B. (1996). Role of growth factors in the developmental and functional regulation of the human fetal adrenal cortex. *Steroids* **62**, 62–72.

Mesiano, S., and Jaffe, R. B. (1997). Developmental and functional biology of the primate fetal adrenal cortex. *Endocr. Rev.* **18**, 378–403.

Pepe, G. J., and Albrecht, E. D. (1995). Actions of placental and fetal adrenal steroid hormones in primate pregnancy. *Endocr. Rev.* **16**, 608–648.

Siiteri, P. K., and MacDonald, P. C. (1963). The utilization of circulating dehydroepiandrosterone sulfate for estrogen synthesis during human pregnancy. *Steroids* **2**, 713–730.

Fetal Alcohol Syndrome

David M. Stamilio and Nancy C. Rose

University of Pennsylvania Health System

GLOSSARY

acetaldehyde (aldehyde) A potentially toxic proximal metabolite in the catabolism of alcohol.

disulfiram (Antabuse) A drug that inhibits the metabolism of alcohol at the acetaldehyde dehydrogenase stage. It is used in alcohol rehabilitation since it produces a highly unpleasant reaction when combined with small amounts of alcohol.

ethanol (ethyl alcohol or alcohol) A central nervous system depressant commonly used socially or recreationally.

fetal alcohol effects Birth defects that may be attributed to maternal alcohol use but do not manifest as the complete fetal alcohol syndrome.

fetal alcohol syndrome A series of birth defects attributed to maternal alcohol use, including pre- and postnatal growth restriction, central nervous system abnormalities, and midface hypoplasia.

Fetal alcohol syndrome (FAS) is now the leading cause of mental retardation in the United States. Initially described by Jones and colleagues (1973), FAS is defined as a series of birth defects that are attributed to maternal use of alcohol during pregnancy. Its features include intrauterine and/or postnatal growth restriction, central nervous system (CNS) abnormalities, and facial hypoplasia. A spectrum of less severe findings, sometimes referred to as fetal alcohol effects (FAE), are attributed to the incomplete expression of FAS.

I. CLINICAL PRESENTATION AND DIAGNOSIS

A. Fetal Alcohol Syndrome

Using criteria established by the Fetal Alcohol Study Group of the Research Society on Alcoholism (1980), FAS is defined by the presence of one or more physical features in each of three principal categories (Fig. 1):

1. Intrauterine and/or postnatal growth retardation: defined as body weight, postnatal body length, or head circumference below the tenth percentile corrected for gestational or postnatal age;

2. Craniofacial abnormalities: at least two of the following must be present: (i) head circumference below the third percentile, (ii) microophthalmia or short palpebral fissures, or (iii) poorly developed philtrum, thin vermilion of the upper lip, and flattening of the maxilla;

3. CNS dysfunction, including neurologic abnormality, developmental delay, or intellectual impairment.

Other associated findings include skeletal or joint abnormalities, cardiac anomalies (most commonly ventricular and atrial septal defects), ptosis, cleft lip or palate, webbed or short neck, renal anomalies, cervical vertebral malformations, meningomyelocele, or hydrocephalus.

Sensory deficits, often secondary to physical anomalies, are also common in FAS. These deficits frequently involve vision, hearing, and speech. Common visual problems include strabismus, nystagmus, amblyopia, and optic nerve hypoplasia. Hearing loss can be conductive or sensorineural in nature. Approximately 93% of patients have recurrent episodes

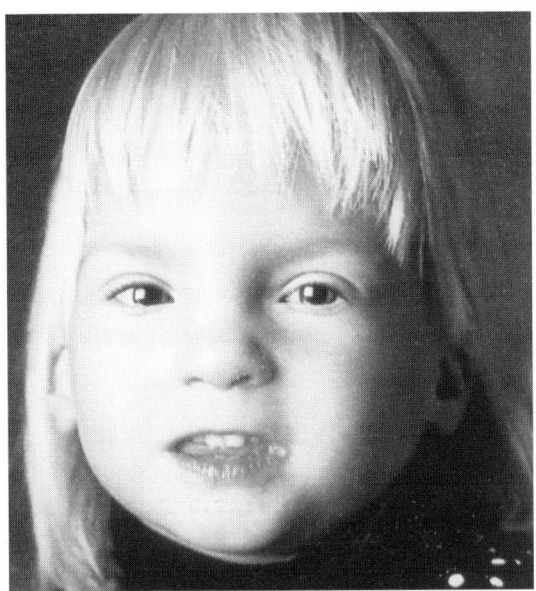

FIGURE 1 Characteristics of fetal alcohol syndrome.

of serous otitis media, contributing to hearing deficits.

The neonate with FAS may exhibit irritability or tremulousness. During childhood, attention deficit disorders are common. Cognitive deficits tend to be greater in verbal than in performance testing. Other studies have demonstrated that behavioral abnormalities, such as hyperkinesis, sleep disorders, and cognitive impairment, persist into adulthood. In the landmark report by Streissguth and co-workers (1985), a 10-year follow-up of the 11 children initially diagnosed with FAS revealed that 4 were mildly developmentally delayed, and 4 were severely handicapped. Affected females were noted to be obese at the time of puberty, whereas males were more likely to be thin.

B. Fetal Alcohol Effects (Alcohol-Related Birth Defects)

Although the effects of heavy antenatal alcohol use on the developing fetus are well established, the consequences of consuming lesser amounts of alcohol during pregnancy are less clear due to several factors. Animal studies are abundant, but great variation exists among designs and the results may not be generalizable to humans. It is difficult to quantify

the amount of maternal alcohol ingestion. Patients are unreliable in estimating the quantity and type of alcohol consumed. A 12-oz. beer, 4-oz. glass of wine, and a single shot glass of hard liquor all contain about 0.5 oz. of ethanol. Additionally, studies are often difficult to interpret and compare since definitions of light, moderate, and heavy alcohol use vary and frequently are vague. Drinking patterns are difficult to compare and quantitate for study designs. Given equal quantities of alcohol consumed, a binge drinking pattern will achieve higher acute blood alcohol levels and may be more deleterious to the fetus than if the quantity of alcohol were ingested over a period of several days. Furthermore, the population of heavy drinkers tends to be demographically distinct from nondrinkers. They are generally found to be older, of higher parity, less often married, more often smokers, larger consumers of coffee, and more often multidrug users. Although these characteristic differences could account for many of the poor pregnancy outcomes attributed to alcohol use, several studies have not controlled for such confounders. These limitations make it difficult to develop a dose–response curve and to identify a dose threshold for alcohol.

Harlap and Shiono (1981) and Kline and co-workers (1980) found an association between moderate alcohol consumption and spontaneous abortions or stillbirths at levels as low as 1 or 2 drinks per day, but this finding is not consistent in all studies. Little and colleagues (1986) found an increase in preterm delivery (PTD) rate in those patients having 1.3 drinks per day during but prior to their recognition of the pregnancy, and other studies report the PTD rate to be higher in mothers who are "heavy drinkers." However, other investigators have not been able to document this association. Multiple other factors (e.g., multidrug use) may be responsible for differences in PTD rates. Lower birth weight also seems to be associated with maternal alcohol use of at least 2 or 3 drinks per day. Many investigators have found an increase in congenital anomalies of varied types as mentioned previously with moderate or heavy alcohol use. Mills and colleagues (1987) performed a large prospective study of 32,870 pregnant women which investigated first-trimester moderate alcohol consumption. They found no difference in total fetal malformation rates in the offspring of women who consumed <1 alcoholic drink per day (77/1000) or 1 or 2 drinks per day (83/1000) compared to the rate in offspring of nondrinkers (78/1000).

II. PATHOPHYSIOLOGY

Although the exact pathophysiology and mechanism of teratogenicity in FAS is unknown, alcohol itself and/or its proximal metabolite, acetaldehyde, are commonly thought to be responsible for the disruption of fetal growth and development. The teratogenic mechanism is postulated to be multifactorial.

There appears to be critical developmental periods in which specific fetal tissues are more vulnerable to alcohol-related damage. In animal studies, facial dysmorphism has resulted from ethanol exposure in the late first-trimester equivalent, whereas exposure in the second and third trimesters has produced neuronal damage. In the third trimester during the particularly vulnerable "perinatal growth spurt," even 1 day of exposure in an animal model has been shown to decrease the number of cerebellar Purkinje cells.

A. Teratogenic Mechanisms

Alcohol and/or acetaldehyde have been shown to interfere with cellular and mitochondrial transport, induce local hypoxia possibly via vasoconstriction, promote the production of mutagenic free radicals via superoxide dismutase inhibition and aldehyde metabolism, and alter prostaglandins which are involved in the regulation of neurotransmitter release, cAMP formation, and cell differentiation in the CNS. Investigators have proposed that many of the adverse fetal effects attributed to alcohol are mediated by these alcohol-induced metabolic alterations. However, clinical and animal studies do not consistently show a cause and effect relationship between alcohol's metabolic effects and fetal outcome.

Alcohol may alter fetal growth and development by inducing a decrease in protein synthesis. This may be due to DNA hypomethylation which limits transcription and/or translation. Embryonic DNA is highly methylated. In the pregnant dam model, Garro

and co-workers (1991) showed that acute alcohol administration resulted in fetal DNA hypomethylation. The same investigators were able to show that aldehyde inhibits fetal DNA methylase activity. Alcohol has also been shown to interfere with O^6-methyl guanine transferase, an enzyme involved in DNA methylation and repair.

Furthermore, alcohol may induce some of its teratogenic effects via neurotransmitter alterations. Alcohol has been shown to alter ion currents mediated by glutamate, the major excitatory neurotransmitter found in high concentrations in the mammalian hippocampus. It has been hypothesized that these alcohol-initiated changes in glutamate neurotransmission result in delayed neuronal growth and/or neurotoxic effects which are responsible for alcohol's neurologic ill-effects. Alcohol may enhance neurotransmission mediated by γ-butyric acid and/or dopamine, but research to support these findings has been inconsistent and their relation to the pathophysiology of FAS is largely unknown.

B. General Mechanisms for Alcohol-Induced Neuropathology

An abundance of animal study data supports that the main ethanol target sites for neuronal damage are the hippocampus, regions of the cerebellum, and the cerebral cortex. There are at least four proposed mechanisms of neurologic developmental damage. First, alcohol alters the proliferation and migration of neurons. Second, ethanol, acetaldehyde or both cause direct (a.k.a., target site) death. Third, neurons surviving alcohol insult are altered. Alcohol has been shown to increase or decrease dendritic growth, change neuron morphology, and alter interneuron connectivity and synapses. Fourth, alcohol can alter nonneuronal elements in the brain, including causing astrogliosis, hindered microvasculature, and increased glial fibrillary acidic protein content.

C. Predisposing Factors for Alcohol Teratogenicity

Genetic variation among individuals in the formation and induction rate of the different enzymes needed to metabolize alcohol and its proximal by-product acetaldehyde plays an important part in the pathogenesis of FAS/FAE. A less efficient metabolic system will theoretically allow longer and higher fetal exposure. Genetic variation can also partially account for individual differences in the rate of trans-placental nutrient transport, uterine blood flow, and fetal neuronal sensitivity. Decreased nutrient transport and uterine blood flow could increase the fetus' risk of insult. Maternal polydrug use, especially cocaine and alcohol, and nutritional deficiency have also been documented as risk factors for FAS/FAE.

III. NUTRITIONAL DEFICITS IN RELATION TO ALCOHOL USE AND FETAL ALCOHOL SYNDROME

Malnutrition is common in alcoholics. Deficiencies in certain vitamins and minerals essential to growth and development may contribute to or exacerbate fetal alcohol effects. For example, vitamin B_6 deficiency, which is common in alcoholics, has been shown to cause fetal growth and development abnormalities. Folic acid, which is necessary for the synthesis and methylation of DNA during fetal and postnatal development, is commonly deficient in alcoholics. Alcohol has been shown to decrease the placental transport of zinc, an essential mineral in the synthesis of DNA, RNA, and protein in the developing fetal brain. Deficiency has been shown to be teratogenic, causing intrauterine growth retardation, brain mal-development, and cognitive dysfunction.

IV. EPIDEMIOLOGY

Recent reviews cited the incidence of FAS in the Western world to be 1 per 3000 live births and the worldwide incidence to be 0.97 per 1000 live births. Although 55–70% of U.S. women use alcohol recreationally and about 6% are considered problem drinkers, it has been estimated that only 1.4% report ethanol use during pregnancy. Since many patients do not report alcohol use and many physicians do not effectively elicit a history of alcohol abuse, estab-

lishing a population's true incidence of ethanol use during pregnancy is virtually impossible. In addition, the incidence of FAS varies widely as alcohol use varies. This variation occurs culturally and geographically. For example, there is variation among different European reports. The incidence of FAS in northern France is reported as 3.3 per 1000 live births, whereas Swedish data reveal an incidence of 1.7 per 1000 live births and estimate that prenatal alcohol accounts for 10% of the patients with IQs in the 50–80 range.

In the United States, the incidence of FAS varies between 3 per 10,000 and 3 per 1000 live births, resulting in approximately 1200 children a year born with FAS. The yearly cost to care for these patients is estimated at over $300 million. Jones and co-workers (1973), defining alcoholism as ≧8 drinks per day, reported a 30–40% prevalence of FAS in offspring of alcoholic women. The incidence of FAS in children born to heavy drinkers, defined variably, is 4.3%. Fetal alcohol syndrome is more prevalent in Native American and African American populations and appears to be associated with lower socioeconomic status. Fetal alcohol effects in the United States are estimated to be 5–10 times more common than FAS. However, rates of FAS and FAE in the offspring of mothers using lower doses of alcohol are less clear and a minimum dose at which below there is no increase in birth defects has not been established.

V. PREVENTION, INTERVENTION, AND MANAGEMENT

There is no cure or treatment for children affected by FAS or FAE. Therefore, the most important therapy for fetal alcohol effects is prevention. If possible, preconceptional counseling regarding prenatal alcohol use should be completed. All women of reproductive age who may potentially conceive should be warned not to drink during pregnancy and informed of its possible fetal effects. Maternal alcohol use should be a routine part of the preconceptional or prenatal history. An indirect, nonjudgmental approach is generally most effective. Some authors suggest inquiring about use of other drugs first, such as nicotine or caffeine, or eliciting a family history (including the father of the baby) of alcohol abuse initially. Another indirect measure of alcohol use is to assess the patient's tolerance to alcohol. If the patient drinks alcohol more than twice a week, has more than two or three drinks per occasion, or does not feel high when she has two or three drinks, this pattern may be indicative of current or developing alcohol dependency. Several methods to screen for alcohol use have been developed. The "CAGE" and "TACE" approaches, both consisting of four standard questions (forming the acronyms), have become popular since they are easy to use. The CAGE approach asks, Have you ever

1. Felt the need to *c*ut down drinking?
2. Felt *a*nnoyed by criticism of your drinking?
3. Had *g*uilty feelings about drinking?
4. Taken a morning *e*ye-opener?

Depending on the study cited, two or three positive answers are highly suggestive of alcoholism. The CAGE questionnaire has a high specificity (99.8%), but only a moderate sensitivity (51%) for detecting alcohol dependency; therefore, it only detects a little over half of alcohol-abusing patients. The TACE questionnaire is similar to CAGE in its content and efficacy, but asks about *t*olerance, getting *a*nnoyed, feeling the need to *c*ut down, and taking an *e*ye-opener. A score of $\geqq 2$ suggests that the patient has a problem with alcohol. Most authors agree that some screening tool for alcoholism should be used while obtaining a history. One screening method is not clearly better than the next. Biochemical markers, such as γ-glutamyl transferase and mean red cell volume, are being investigated as screening tests for alcohol abuse, but preliminary results are inconsistent. If a patient reports alcohol use, one should attempt to determine the pattern of drinking since different patterns may present the fetus with different risks.

Once alcohol abuse has been identified, the physician should reinforce the dangers of alcohol use during pregnancy. The patient should also be informed that no safe level of alcohol has been established because there is no known threshold for adverse fetal effects. Patients can be counseled that there is evidence that the cessation of heavy alcohol prior to

the third trimester can benefit the fetus. Once the risk of prenatal alcohol is identified, the prenatal evaluation should include frequent prenatal visits with persistent encouragement and supportive counseling, careful gestational dating, a detailed anatomy sonogram, and serial growth ultrasounds. Monitoring fetal and maternal well-being throughout the pregnancy is essential. An alcoholic patient refractory to counseling should be referred to a rehabilitation program or Alcoholics Anonymous. Pregnant patients with alcohol withdrawal can be treated with benzodiazepines, improved nutrition, and treatment of associated medical problems.

Disulfiram is contraindicated in pregnancy. Alcohol consumption concomitantly with disulfiram therapy will increase blood acetaldehyde concentration via aldehyde dehydrogenase (ADH) inhibition, which theoretically could exaggerate adverse fetal alcohol effects. Metronidazole (Flagyl) also inhibits ADH and should be avoided in pregnant patients that are known or suspected to use alcohol.

Lastly, the physician must remember that alcoholism and FAS/FAE are not short-term problems. The mother should have a long-term follow-up plan for rehabilitation, counseling, and support. The burden of physical and mental stresses carried by the mother and other family members should not be underestimated. They may harbor great feelings of guilt for the affected offspring. Care for the affected child requires extensive material and temporal resources, frequently taxing what is often an already stressed family structure. The FAS/FAE affected child could have any of the long-term CNS or non-CNS sequelae. These children frequently have dental caries requiring thorough dental care. They also more commonly need myringotomies, eyeglasses, and hearing aides. Affected children should also have IQ testing to help determine the need for special education programs.

VI. SUMMARY

Prenatal alcohol use and its adverse fetal effects continue to be major medical and public health problems. No safe threshold in which alcohol can be consumed without fetal defects has been established. The pathophysiology and mechanism involved in

FAS and FAE have not been elucidated, but research continues to slowly identify the multifaceted teratogenic mechanism. Since no cure exists for FAS/FAE, the best treatment to date is prevention. This requires early detection of the patient's alcohol problem and frequent and close patient contact. No single prevention program has been shown to be superior. Since investigations into FAS prevention have been suboptimal, future research should focus on optimizing prevention in addition to physiologic studies. Lastly, alcoholism and fetal alcohol effects are lifelong diseases and should be treated as such by all care providers involved.

See Also the Following Articles

Bibliography

Abel, E. L. (1995). An update on incidence of FAS: FAS is not an equal opportunity birth defect. *Neurotoxicol. Teratol.* **17**(4), 437–443.

Bonthius, D. J., and West, J. R. (1990). Alcohol-induced neuronal loss in developing rats: Increased brain damage with binge exposure. *Alcoholism Clin. Exp. Res.* **14**, 107–118.

Gardner, R. J., and Clarkson, J. E. (1981). A malformed child whose previously alcoholic mother had taken disulfiram. *N. Z. Med. J.* **93**(680), 184–186.

Garro *et al.* (1991). *Alcoholism Clin. Exp. Res.* **15**, 395–398.

Goodlett, C. R., Marcussen, B. L., and West, J. R. (1990). A single day of alcohol exposure during the brain growth spurt induces brain weight restriction and cerebellar Purkinje cell loss. *Alcohol* **7**, 107–114.

Harlap and Shiono (1981). *Lancet* **1**, 173–175.

Jones, K. L., *et al.* (1973). *Lancet* **1**, 1267–1271.

Jones, K. L. (1988). Fetal alcohol effects. In *Smith's Recognizable Patterns of Human Malformation* (L. Mills and M. Tanner, Eds.), 4th ed., pp. 491–494. Saunders, Philadelphia

Kline *et al.* (1980). *Lancet* **2**, 176–180.

Larroque, B. (1992). Alcohol and the fetus. *Int. J. Epidemiol.* **21**(4, Suppl. 1), S8–S16.

Ledig, M. L., M'Paria, J. R., Louis, J. C., Fried, R. , and Mandel, P. (1980). Effect of alcohol on superoxide dismutase activity in cultured neural cells. *Neurochem. Res.* **5**, 1155–1162.

Little *et al.* (1986). *Am. J. Epidemiol.* **123**, 270–278.

Mills, J. L., Graubard, B. I., Harley, E. E., Rhoades, G. G., and Berendes, H. W. (1984). Maternal alcohol consumption and birth weight. How much drinking during pregnancy is safe? *J. Am. Med. Assoc.* **252**, 1875–1879.

Mills, J. L., *et al.* (1987). *Pediatrics* **80**, 309–314.

Pierce, D. R., and West, J. R. (1986). Blood alcohol concentration: A critical factor for producing fetal alcohol effects. *Alcohol* **3**, 269–272.

Renolds, J. D., and Brien, J. F. (1995). Ethanol neurobehavioral teratogenesis and the role of L-glutamate in the fetal hippocampus. *Can. J. Physiol. Pharmacol.* **73**, 1209–1223.

Richardson, B. S., Patrick, J. E., Bousquet, J., Homan, J., and Brien, J. F. (1985). Cerebral metabolism in fetal lamb after maternal infusion of ethanol. *Am. J. Physiol.* **249**, R505–R509.

Sokol, R. J. (1980). Alcohol and spontaneous abortion. *Lancet* **2**, 1079.

Sokol, R. J., and Martier, S. (1996, December). High-risk pregnancy: Alcohol. *Contemp. OB/GYN*, 19–23.

Streissguth, A. P., Aase, J. M., Clarren, S. K., Randels, S. P., LaDue, R. A., and Smith, D. F. (1991). Fetal alcohol syndrome in adolescents and adults. *J. Am. Med. Assoc.* **265**, 1961–1967.

Streissguth, A. P., *et al.* (1985). *Lancet* **2**, 85–91).

West, J. R., Chen, W. A., and Pantazis, N. J. (1994). The vulnerability of the developing brain and possible mechanisms of damage. *Metab. Brain Dis.* **9**(4), 291–322.

Fetal Anomalies

E. Albert Reece

Temple University School of Medicine

Arnon Wiznitzer

Soroka Medical Center/Ben Gurion University School of Medicine

GLOSSARY

amniocentesis The passage of a needle through the abdomen into the amniotic sac to obtain a sample of amniotic fluid for genetic studies, usually performed between 10 and 18 weeks' gestation.

amniotic fluid Fluid surrounding a developing baby.

chorionic villus sampling The passage of a catheter through the cervix or a needle through the mother's abdomen into the developing placenta to obtain cells for genetic analysis; performed between 10 and 12 weeks' gestation.

chromosome One of the bodies (normally 46 in man) in the cell nucleus that is the bearer of genes.

Down's syndrome A chromosomal disorder which causes delays in physical and intellectual development.

neural tube defect An abnormal opening in the brain or spine.

teratogen An agent, such as a drug or virus, that causes birth defects.

Congenital malformations affect 3–5% of all pregnancies and are of the most significant factors associated with increased perinatal morbidity and mortality. Approximately 2% of newborns have a major congenital anomaly obvious at birth, and between 10 and 15% have minor congenital anomalies. The causes of congenital anomalies are variable and may be multifactorial, with some being genetic (10–20%), others due to environmental insults (10–20%), and a majority being caused by unknown factors (68%) by interactions of environmental factors with genetic predispositions.

I. INTRODUCTION

The effects of environmental agents on the developing fetus are both time and dose dependent and the period of susceptibility is variable. The growth rate of particular tissues at particular times or changes in metabolism of tissues during development may influence tissue susceptibility, which changes during development. Environmental influences on early development can be divided into the following categories:

1. Drugs and environmental chemicals
2. Maternal metabolic imbalance
3. Infections
4. Radiation
5. Programmed cell death
6. Growth factors

During the first period of embryonic development (from Day 18 to about Day 60 of gestation in the human), the embryo is most sensitive to the toxic effects of drugs and chemicals resulting in embryo lethality. However, the palate, central nervous system, and genital structures can be affected at a later stage of development.

II. ETIOLOGY

A. Drugs and Environmental Chemicals

Every teratogenic agent which has been tested shows a dose–response relationship and a threshold dose response. The dose to which the fetus is exposed is dependent on many factors, including metabolism of the substance, maternal pharmacokinetics, fetal and placental metabolism, and fetal distribution of the substance. Additionally, some drug-related disorders are related to fetal sex, for example, some hormones can affect normal development of either a male or female fetus.

Many drugs have been associated with an increased risk of reproductive toxicity. Even though an agent can produce one or more of these pathologic processes, exposure to such an agent does not guarantee that maldevelopment will occur. Furthermore, it is likely that a drug can have more than one effect on a pregnant woman and her developing conceptus; therefore, the nature of the drug or its biochemical or pharmacologic effects will not, in themselves, predict a human teratogenic effect. In fact, reports of human teratogens have come primarily from human epidemiological studies. Animal studies and *in vitro* studies can be very helpful in determining the mechanism of teratogenesis and the pharmacokinetics related to teratogenesis.

One of the most important teratogenic agents is alcohol. Alcohol will be discussed briefly because of its relatively large social impact. Fetal alcohol syndrome has been described in children with intrauterine growth restriction, mental retardation, cardiac abnormalities, microcephaly, reduction in the width of palpebral fissures, and developmental delay. The presence of any of these abnormalities in the offspring of an alcohol-using mother raises the question of causal relationship. Because these anomalies also occur in children who were not exposed to alcohol *in utero*, they can only be ascribed to maternal drinking when seen with the characteristic features of the alcohol syndrome.

The pathophysiologic mechanism for the adverse effects of alcohol on the developing fetus is poorly understood. Either alcohol itself or acetaldehyde are responsible for the disruption of fetal growth and development. It has been suggested that amniotic fluid may function as a reservoir, exposing the fetus to alcohol or ethanol for an extended time. This may be related to the fact that maternal enzymes such as alcohol dehydrogenase that are responsible for the metabolism of alcohol are lacking in the fetal liver. In addition, ethanol or acetaldehyde may interfere with placental transport of vital nutrients, leading to a form of fetal malnutrition.

B. Maternal Metabolic Imbalance

Extensive experimental and clinical work has been done examining the role of maternal metabolic imbalance on fetal development. It is now known that aberrant metabolic fuels during a critical period of organ formation act as a teratogen, inducing certain types of malformations. A typical example of this mechanism in pregnancy is poorly controlled diabetes. Neural tube defects are one of the most common serious malformations occurring in infants of diabetic women. In the general population, such defects occur in about 1 or 2 per 1000 live births in North America but at a much higher rate in the British Isles. Over the past quarter century, the rate has declined appreciably. Recently, folic acid has been shown to be a very effective dietary supplement used in the preconceptional period to decrease the incidence of neural tube defects. This finding has led to the recommendation of dietary intake of 0.4 mg of folic acid daily. Studies have been conducted in the United States and the United Kingdom documenting the significant benefit of folic acid supplementation. Such supplements have resulted in a significant reduction in the incidence of these anomalies. The Food and Drug Administration in the United States is strongly considering fortifying foods with folic acids as a means of ensuring that women in the reproductive age group receive adequate folic acid intake. Multivitamins, which contain some folic acid, have also been shown to reduce not only neural tube defects but also facial clefts.

C. Infections and Other Factors

The following maternal infections may induce fetal malformations: *Toxoplasma gondii*, rubella virus, cytomegalovirus, and herpes.

1. *Toxoplasma*

Toxoplasma gondii is a protozoan parasite. Its oocysts are found in cats that have ingested rodents infected with these cysts. Humans become infected if they eat uncooked or fresh meat from infected animals. When the disease is clinically apparent, symptoms include malaise, sore throat, fever, and lymph mode enlargement. The fetus may develop intrauterine growth restriction, nonimmune hydrops fetalis, and neural tube defects. In addition, the body may suffer from fever, jaundice, anemia thrombocylopenia, rash, and diarrhea. Long-term complications include mental retardation, visual deficits, and seizures.

2. *Rubella*

The rubella virus is spread by respiratory droplets in prolonged, closed exposure. Fetal infection requires maternal viremia. This type of malformation is gestational age specific. First-trimester rubella is believed to cause abortion. Deafness, mental retardation, heart lesions, and ophthalmologic abnormalities are associated with maternal viremia during early pregnancy. Second- and third-trimester rubella infections are not without clinical consequence. Growth retardation, developmental delay, hearing loss, and hair problems can occur. To date, no cases of congenital defects have been reported due to the rubella vaccine.

3. *Cytomegalovirus*

Cytomegalovirus (CMV) is the most common viral cause of congenital infection. Over 30,000 infants are born in the United States with congenital CMV every year. Infection in the mother is generally asymptomatic, whereas in some patients a flu-like disease may be present. Fetal infection may be associated with growth restriction, microcephaly, periventricular calcification, chorioretinitis, and hearing deficits. Approximately 5–15% of initially asymptomatic infants may subsequently develop learning disabilities and hearing deficits.

4. *Herpes Simplex Virus*

Transmission of genital herpes virus (HSV) to the baby requires intimate contact of infectious secretion with susceptible mucous membranes. Therefore, neonatal transmission occurs mainly intrapartum. Transplacental infection with HSV has been reported but is not well understood. In these cases HSV infection has been linked to early abortion, preterm labor, and congenital malformations. The spectrum of anomalies is similar to congenital rubella infection, i.e., microcephaly, chorioretinitis, intrauterine growth restriction, and periventricular calcification.

D. Radiation

External irradiation has been associated with several congenital abnormalities, such as microcephaly, intrauterine growth restriction, mental retardation, and eye anomalies. All of these malformations depend on dose and stage of exposure. No measurable risk is associated with exposure to 5 rads or less of X ray at any stage of pregnancy. For example, the dose required for a chest X ray is about 0.2 rad.

There are a number of other factors which are believed to be associated with developmental anomalies, such as aneuploidy, growth factors, programmed cell death, imprinting, cytoplasmic factors, and genetic and protein factors (Table 1).

E. Genetic Factors

Genetic causes for human malformations can be classified into two categories: abnormalities of chromosomes, such as Down's syndrome, and abnormalities caused by alterations in single genes, such as achondroplasia or osteogenesis imperfecta.

Down's syndrome is the most common chromosomal abnormality, occurring in approximately 1 of every 800 births. Obstetricians and perinatologists recommend that all women age 35 or older should be offered karyotypic analysis via amniocentesis or chorion villous sampling because the risk of aneu-

TABLE 1

Factors Associated with Developmental Anomalies

Genetic factors
Programmed cell death
Growth factors
Cytoplasmic, genetic, and protein factors

ploidy increases with maternal age. The chance that a 35-year-old woman in her second trimester is carrying a fetus with Down's syndrome is about 1 in 300. Some genetic diseases, such as cystic fibrosis and Tay–Sachs, occur primarily in families with a prior history of the disease. For example, as a couple begins planning for their first child, they may discover that each of them carries a defective or mutant copy of the gene responsible for causing, for example, cystic fibrosis. The couple's child has a one in four chance of inheriting cystic fibrosis, a disease in which affected individuals often find themselves struggling for breath throughout their lives, although sometimes a lung transplant will help to relieve their suffering.

Imprinting is a relatively new concept that is evolving and, in part, is serving to explain certain types of anomalies. This concept relates to the fact that there are certain alterations in the genetic information depending on whether this genetic information is of maternal or paternal origin. Depending on the source, the function may be different. Some human conditions, for example, Prader–Willi syndrome and Angelman syndrome, result from the same chromosomal defect, 15q12. However, they have different phenotypic manifestations believed to result from the deletion being of paternal or maternal origin. For example, in Prader–Willi, the deleted locus is of paternal origin, whereas in Angelman syndrome it is of maternal origin. The concept of imprinting has not been generalized to many different types of anomalies. However, its implications for other congenital anomalies are being examined.

F. Programmed Cell Death

There are many cells in the body in which growth and proliferation are carefully timed, and their death is also carefully timed or programmed. It is believed that in certain circumstances, a teratogen may increase the amount of programmed or predestined cell death, and if this occurs, it may induce certain types of anomalies.

G. Growth Factors

Fetal growth is promoted by a number of *in utero* growth factors, including insulin and insulin-like growth factors. The latter are believed to have a number of effects, both endocrine and paracrine, in stimulating both *in utero* and neonatal tissue growth. Therefore, if these growth factors are lacking or altered, tissue growth may be altered and developmental anomalies may result.

III. GENETIC COUNSELING

Genetic counseling is important for a couple at risk of delivering a fetus with anomalies. Counseling enables the couple to acquire basic medical information regarding fetal diagnosis, management, and possible outcomes; the role or lack thereof of genetic factors; the recurrence rate; and their reproductive options. This information is expected to allow the couple to make an informed decision regarding a course of action.

Families should be given basic information on the definitions of terms, significance, incidences, etc. They should know that anomalies may occur singly or may be in clusters. Single anomalies are usually multifactorial in origin, whereas multiple anomalies may represent a syndrome caused by teratogen, chromosomal anomaly, or a gene defect. The term syndrome usually designates events not occurring by chance but rather in some predictable association. Anomalies can be divided into at least three types: malformation, disruption, and deformation.

Malformation is a developmental abnormality in which the structure was not normally formed, such as failure of the neural tube to close. Disruption occurs when there is interruption or breakdown of a structure that was normally developed. Deformation occurs when a normally formed organ takes on abnormal shape, typically due to mechanical forces.

Live-born infants have about a 0.4% rate of chromosomal defects and an additional 0.2% rate of balanced chromosome rearrangements. Although the rate of structural malformations varies, it is estimated that about 3 or 4% of normal births are complicated by congenital malformations, mental retardation, or genetic disorders. This rate is believed to double by 7 or 8 years of age given later appearance and later diagnosis. This is important information of which parents should be aware so that they do not feel that congenital malformation of a fetus is something that they caused or something very unique to their family.

IV. PRENATAL DIAGNOSIS

Who should undergo prenatal diagnosis? Some have argued that the entire population should be screened, whereas others have suggested selective screening. Currently, screening of the entire pregnant population is not feasible, nor would it be cost-effective given the anticipated low yield. The latter is due primarily to the relatively low prevalence of fetal anomalies which are clinically recognized and the lack of a marker(s) sensitive enough to be applied universally. Most centers in the United States use selective screening based on the following factors:

1. Advanced maternal age (>34 years)
2. Previous offspring with chromosomal trisomy
3. A couple with a known history of having balanced translocation
4. Parents carrying the same deleterious autosomal recessive gene defect
5. Mother carrying a deleterious X-linked gene
6. Prior fetus with disorder of multifactorial inheritance but with know recurrence rate

Ultrasound has become a pivotal tool in both screening and diagnosis of congenital anomalies. This technology is useful in dating the pregnancy, in assessing fetal growth, and also in diagnosing most major structural malformations with a high degree of accuracy. Ultrasound has been the biophysical mainstay, applied extensively in pregnancy; usually applied in the first trimester for dating and the establishment of viability and in the second trimester for screening of congenital malformations. There is considerable debate as to whether routine ultrasound used in pregnancy does improve the perinatal mortality rate. Studies done in Europe including the United Kingdom demonstrate that ultrasound screening, when used for both high-risk and low-risk pregnancies, results in a high detection rate of anomalies and, as a result, reduced perinatal mortality rates

have been reported. On the other hand, other studies conclude that routine ultrasound scanning does not improve the outcome of pregnancy by reducing perinatal morbidity but may be effective for screening for malformations and lowering mortality.

The recent RADIUS (routine antenatal diagnosis of ultrasound) study in the United States, which was a multicenter study sponsored by the National Institutes of Health, demonstrated that ultrasound when used routinely in pregnancy does not reduce perinatal mortality. This is an extremely controversial study because, in fact, the detection rate of malformation was very low in the study; the population studied was homogenously relatively low risk with a variety of operators with varying experience. This study has been criticized heavily in that it does not represent the cross section of America, particularly those seen in large major urban centers where the risk is greatest. Therefore, the application of ultrasound in screening for congenital malformations and its impact on outcome is controversial.

With the improvement in ultrasound resolution as well as in various biochemical markers, significant strides have been made in the area of screening and prenatal diagnosis of congenital malformation. The primary modality of screening for congenital malformations are biochemical and biophysical methods. Some biochemical markers include the "triple screen," which includes a combination of α-fetoprotein, unconjugated estriol, and human chorionic gonadotropin (hCG). These three analyses, when measured between 15 and 18 weeks, result in a detection rate of Down's syndrome in the range of 60%, which is a significant increase over the detection rate using α-fetoprotein alone (20%). Other serum markers are evolving and they include pregnancy-associated plasma protein-A (PAPP-A) and the free β subunit of hCG (free β-hCG). These have been found to be effective before 15 weeks of pregnancy. There are other markers which are evolving but are not currently used clinically.

Nuchal thickness assessed by ultrasonography has been reported as a marker for Down's syndrome when measured in early pregnancy. There are new studies being performed in the first and second trimesters that use a translucency of 3 mm or more as being associated with an increased risk of chromosome anomaly, particularly trisomy 18, 21, and 13. However, its diagnostic accuracy has not been shown to improve when it is combined with PAPP-A or free β-hCG.

Use of a four-chamber view as a screening tool has also been applied in large population studies. It has been shown that implementation of this four-chamber view increases the detection rate of cardiac anomalies. Other studies have confirmed this utility and have determined that the four-chamber screening can be a very useful, even in level 1 centers, in identifying women who need to be referred to tertiary care centers for more definitive studies as well as karyotypic analyses.

However, many investigators believe that when ultrasound is applied during the course of pregnancy it is effective in detecting major structural anomalies with a relatively high degree of accuracy, and it can also identify a number of other aberrant growth patterns and, as a result, influence management and outcome.

Other techniques for prenatal diagnosis include chorionic villus sampling and, recently, embryofetoscopy. Amniocentesis has been performed throughout the world for many years in the midtrimester (15–18 weeks) and requires the use of a 20- to 22-gauge spinal needle. Under ultrasound guidance, the needle is inserted into the amniotic cavity. In the past, ultrasound was used to identify an appropriate pocket of fluid and during a separate setting the needle was passed into the amniotic cavity. This technique has been improved so that following the identification of the pocket of fluid, ultrasound monitors the simultaneous insertion of the needle so as to prevent any injury to the fetus. Once the needle enters the amniotic fluid, the stilette is removed, a syringe is then placed on the hub of the needle, and approximately 20 cc of amniotic fluid is withdrawn and sent for karyotypic analysis and other studies. Over many years, research has shown that the complication rate with this midtrimester amniocentesis is very low, with a reported incidence of fetal loss rate of about 0.5% or less. There are usually no other significant complications reported as ultrasound monitors the entire procedure.

In recent years, amniocentesis has been conducted in the first trimester and referred to as early amniocentesis. It is conducted in the same manner as the second-trimester amniocentesis, except the amount of fluid withdrawn is usually less. Typically, many operators tend to use the empiric rule that the number of cc's withdrawn is equivalent to the gestational age of the fetus at the time of this procedure. This is an attempt to reduce the amount of fluid withdrawn. However, the loss rate reported in many series is higher than that observed in the midtrimester. This rate seems to be about 2–2.5% above background. The amniocytes extracted from the fluid are reported to have a longer culture period and also a higher culture failure rate.

Chorionic villus sampling (CVS) is done in the first trimester, between 10 and 12 weeks. A tiny portion of the trophoblast is removed and sent for karyotypic or enzyme analyses. Technically, this procedure requires either a 20-gauge spinal needle passed into the trophoblast or a plastic catheter inserted through the cervix and passed along the site of the trophoblast. Using a 30-cc syringe, negative suction is exerted so that portions of the trophoblast can be removed. This procedure has been done successfully on hundreds of thousands of women. Based on two large multicenter studies in North America, the loss rates appear to be about 0.7% higher than the background loss rate for midtrimester amniocentesis based on singleton pregnancies. This results in an overall loss rate of about 1.2%. Obviously, the information derived from this procedure is very comparable to that obtained with amniocentesis. However, the loss rates are distinctly different: 1.2% for CVS versus 0.5% for midtrimester amniocentesis and about 2.5% or more for early amniocentesis.

As improvements in ultrasound technology have occurred, investigators have been fascinated by the live observations of fetal body limbs, breathing movements, heart motion, bladder filling and emptying, and the recording of fetal heart rate patterns. These improvements led, in the early 1970s, to the introduction of endoscopic visualization of the fetus performed in the second trimester using fairly large, rigid endoscopes and associated with fairly high (5–7%) fetal loss rates. In the mid-1980s a new technique was introduced called the percutaneous umbilical blood sampling. Here, accessibility to fetal circulation enhanced the ability to diagnose and treat several genetic diseases.

The previously described diagnostic and therapeutic techniques were limited to use during the second and third trimesters of pregnancy. The early conceptus has received only little diagnostic and relatively no therapeutic attention. The question is whether or not we are ready to take on the next step, which is the correction *in utero* of genetic diseases. It is now possible to identify the causative genes for certain congenital disorders as well as to treat some of these conditions with stem cell replacement therapy. In fact, this has been performed in the neonatal period. Congenital hematologic diseases, such as sickle cell anemia, β-thalassemia, and infantile malignant osteopetrosis, have been successfully treated by postnatal bone transplantation. However, human leukocyte antigen (HLA) identical donors are not often available. In addition, this type of transplantation requires concomitant use of immunosuppressive agents because the transplant is performed at a time when the infant is immunologically competent. By the time most of these neonates are considered for treatment, they have already been compromised by the disease. The questions that are intuitively raised are can we can provide this treatment earlier during the prenatal period and, if so, during what stage of pregnancy?

In cases in which a genetic disease is diagnosed prenatally, therapy using gene transfer must be initiated early enough to reduce or prevent a reversal of damage to the affected organ. Model systems for human gene delivery in the treatment of inherited diseases are being tested *in vitro* and in animals. Many available methods of gene transfer are used. Retroviruses have been utilized to transfer genes and have been shown to replicate and transmit genetic information. This technique has been found to be highly efficient in a broad range of host cells and has been performed without injury of the cell.

It is now clear that if gene therapy is to be undertaken in humans, the genes should be introduced in the fetal circulation in the first trimester. Since HLA antigens are laid down by approximately 14 weeks' gestation, any attempt to circumvent the consequence of immunosuppression would need to be initiated prior to this time. Until recently, the technol-

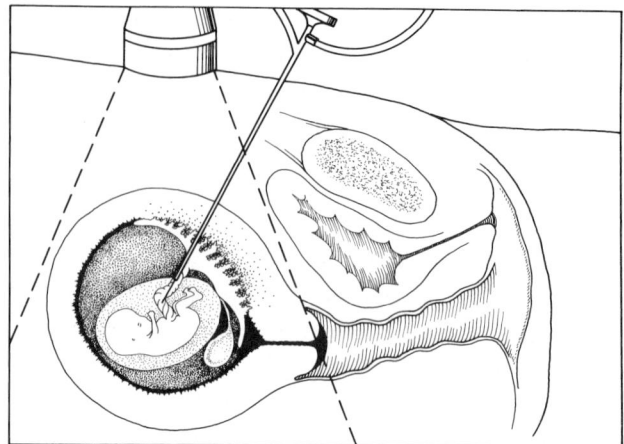

FIGURE 1   Schematic diagram of embryoscopy done by the abdominal route. (From E. A. Reece and C. Homko (1995). *Ultrasound Obstet. Gynecol.* 5, 282. © Partheon.)

ogy was not available to perform gene transfer and/ or stem cell therapy that early in gestation.

One of the newest techniques for prenatal diagnosis is embryoscopy or embryofetoscopy (Figs. 1 and 2) which provides access to the embryo–fetus during the first trimester when the fetus is still immunologically naive. Historically, transvaginal embryoscopy was introduced, but it was quickly abandoned because it required passage through the vagina, cervical canal, and chorion, all of which potentially increased the risk of infection during the procedure. Recently, our group modified the technique and began using a transabdominal approach. The needle is passed transabdominally under ultrasound guidance

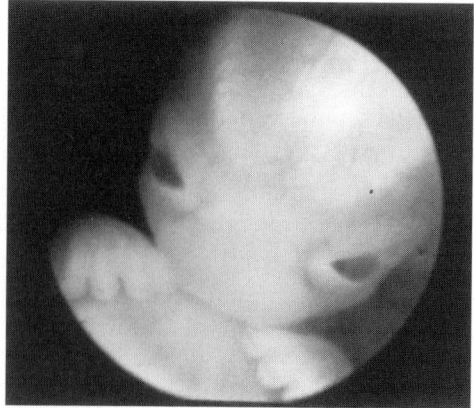

FIGURE 2   Direct visualization of fetus at 9 weeks by embryoscopy. (From E. A. Reece *et al.* (1992). *Am. J. Obstet. Gynec.* 166, 777. © Mosby.)

through the uterine wall and into the amniotic cavity very similar to amniocentesis. Fetal heart rate remains unaltered both during and following the procedure. Transabdominal embryoscopy has been used to observe the fetal morphology and also to diagnose certain disorders. Transabdominal embryoscopy will simply enhance our capability for early diagnosis and fetal therapy. Potential application includes first-trimester diagnosis of genetic diseases and embryonic blood sampling and/or tissue retrieval. It provides access to the embryo at a time when the embryo is immunologically naive and thus begins a new era for stem cell therapy before the lethal or disabling effects of disease are realized. Embryoscopy is expected to have a major impact on fetal medicine in the future, especially in the area of fetal therapy. Prenatal diagnosis will considerably change.

V. PRECONCEPTION COUNSELING AND CARE

There has been significant advancement in the impact of preconception care and counseling on the outcome of pregnancies. Many studies have demonstrated that women who are at high risk for various adverse outcomes to either themselves or the fetus can improve these outcomes by joining a preconception care program.

In these programs, patients are evaluated fully and any abnormality or chronic disease is treated or controlled prior to pregnancy. This has been shown to have a very positive impact in enhancing pregnancy outcomes.

See Also the Following Articles

AMNIOTIC FLUID; FETAL ALCOHOL SYNDROME; FETAL MONITORING AND TESTING; PRENATAL GENETIC SCREENING; TERATOGENS; ULTRASOUND

Bibliography

Brambati, B., Tului, L., Bonacchi, I., Shrimanker, K., Suzuki, Y., and Grudinskas, J. G. (1994). Serum PAPP-A and free beta hCG are first screening markers for Down's syndrome. *Prenatal Diagnosis* **14**, 1043–1047.

Crane, J. P., LeFevre, M. L., Winborn, R. C., Evans, J. K., Ewigman, B. G., Bain, R. P., Frigoletto, F. D., McNellis, D. (1994). A randomized trial of prenatal ultrasonographic screening: Impact on the detection, management, and outcome of anomolous fetuses. *Am. J. Obstet. Gynecol.* **171**, 392–398.

Dimmick, S. E., Kalousek, D. K. (1992). *The Embryo Fetus Textbook.* Lippincott-Raven.

Ewigman, B., Crane, J. P., Frigoletto, F. D., LeFevre, M. L., Bain, R. P., McNellis, D. (1993). Effect of the prenatal ultrasonographic screening on prenatal outcome. *N. Engl. J. Med.* **329**, 821–827.

Reece et al. (1992). *Medicine of the Fetus and Mother.* Lippincott-Raven.

Tegnander, E., Eik-Nes, S. H., Johansen, O. J., Linker, D. T. (1995). Prenatal detection of heart defects at the routine fetal examination at 18 weeks in a non-selected population. *Ultrasound Obstet.* **5**, 372–380.

Fetal Diagnosis, Invasive

Celeste Sheppard and Carl P. Weiner

University of Maryland Medical School

GLOSSARY

amniocentesis The withdrawal of amniotic fluid from the amniotic sac surrounding the fetus.

amniocytes Cells floating in amniotic fluid; the substrate for genetic testing by amniocentesis.

aneuploidy An abnormal number or arrangement of chromosomes.

chorionic villus sampling A technique of harvesting tissue of the chorion frondosum in the first trimester for genetic and biochemical testing.

chromosomes Discrete bodies of tightly coiled, condensed DNA or chromatin having characteristic size, shape, and staining patterns.

cordocentesis The sampling of fetal blood from the umbilical cord, undertaken for fetal diagnosis or therapy.

karyotype The representation of a complete set of chromosomes prepared from a photomicrograph of chromosomes within a dividing cell. A normal karyotype consists of 46 chromosomes: 22 pairs of autosomes and a pair of sex chromosomes.

maternal serum α-fetoprotein Used as a screening test for fetal neural tube defects. When combined with other serum markers, it can also identify pregnancies at risk for aneuploidy.

Antenatal diagnosis of a vast assortment of fetal conditions is available to pregnant women today. Rapid advancement of ultrasound technology has provided a virtual window through which the fetus can be observed in spectacular detail, even before the mother perceives fetal movement. Discoveries in the fields of molecular biology and genetics have identified the genetic basis of many inherited and chromosomal disorders, and prenatal diagnosis of many common genetic defects is possible. Families at risk for offspring affected with serious genetic disorders can now be offered prenatal testing early in pregnancy, giving them the opportunity to make informed, though often difficult, choices in their reproductive lives. The diagnosis of acquired medical disorders, such as fetal infection, is made possible by these techniques as well, advancing the frontier for fetal medicine.

I. INTRODUCTION

Any invasive procedure entails risk. Before accepting invasive fetal testing, the woman and her partner must understand the degree of risk to her and her fetus, the reliability of the information sought, and options for treatment or management of her pregnancy following the diagnosis. This information is usually discussed within the context of genetic counseling provided by an obstetrician, geneticist, or genetic counselor. Occasionally, the couple considering prenatal diagnosis may also consult with a medical ethicist, social worker, or member of the clergy for guidance and support. The patient's written informed consent is obtained before any invasive test is undertaken. Because greater operator experience and technical proficiency are associated with lower complication rates, these techniques should be performed only by clinicians trained to perform them.

The most common indication for invasive fetal diagnosis is advanced maternal age, but the list of indications is long and growing. As women age, their risk for delivering a child with a chromosomal abnormality, particularly Down's syndrome (trisomy 21), increases. Invasive testing is usually offered to women 35 years old or older because, at age 35, the risk of delivering a child with aneuploidy is the same or greater than the risk of pregnancy loss from an amniocentesis. Maternal serum screening tests, widely offered during the second trimester, identify pregnancies at increased risk for chromosomal or neural tube defects in the fetus. Amniocentesis can reveal the definitive diagnosis. Likewise, identification of a fetal malformation on a prenatal ultrasound examination usually prompts the offer of invasive diagnosis because of the frequent association of structural anomalies with chromosomal defects or fetal medical disorders. Because inherited disorders, such as sickle cell anemia, cystic fibrosis, muscular dystrophy, and many others, can be diagnosed by genetic analysis of fetal cells, testing is offered when there is a family history of hereditary genetic disease. Acquired medical conditions, such as anemia caused by red blood cell alloimmunization, can be diagnosed and often treated *in utero* when these techniques are employed (Table 1).

II. AMNIOCENTESIS

Amniocentesis is the withdrawal of amniotic fluid surrounding the fetus within the amniotic sac. It is

TABLE 1
Prenatal Diagnostic Procedures

Procedure	Recommended gestational age	Studies	Time to results (days)	Fetal loss rate (%)
Chorionic villus sampling	10–13 weeks	Karyotype	2–10	0.5–1
Amniocentesis	15 weeks to term	Karotype, AFP, AchE; lung maturity	2–14	≤0.5
Early amniocentesis	12–14 weeks	Karyotype	2–14	Unknown
Cordocentesis	18 weeks to term	Karyotype, hematology, chemistries	3–7	<1

traditionally performed at 15–18 weeks of gestation but it can be performed anytime after the first trimester. By the midtrimester, fetal urine is the chief source of amniotic fluid, and fetal swallowing and breathing movements contribute to the maintenance of a normal volume. The fluid contains many cells called amniocytes, exfoliated from the amnion, skin, and respiratory and urinary tracts; each of the living cells contains a complete genetic complement or "fingerprint" of the fetus. Cells filtered from the amniocentesis sample are cultured *in vitro*. A karyotype is prepared from dividing amniocytes when the DNA is condensed into discrete chromosomes. Chromosomal abnormalities, including aneuploidy, deletions, or rearrangements, are diagnosed by examination of the karyotype. Amniocyte DNA can be subjected to a variety of molecular studies to identify an aberrant gene responsible for a genetic disorder.

The chemical components of amniotic fluid can also supply valuable information about fetal health. When the maternal serum α-fetoprotein (MSAFP) is elevated, amniotic fluid levels of α-fetoprotein and acetylcholinesterase aid in the diagnosis of fetal structural defects. Hemolysis due to red blood cell alloimmunization results in the excretion of hemoglobin metabolites into the amniotic fluid which change the absorption of light at the wavelength 450 nm, termed OD450. If intraamniotic infection is suspected, amniotic fluid Gram stain, culture, and biochemical studies may be useful in making the diagnosis. In the near future, detection of microbial DNA by polymerase chain reaction may become the principal method of detecting intraamniotic infection. Assessment of fetal lung maturity later in gestation by measuring various components of surfactant

is an important indication for amniocentesis in the third trimester.

A. Technique

Amniocentesis is performed under continuous ultrasound guidance. When performed for genetic diagnosis, the procedure is preceded by a complete fetal anatomic survey to confirm fetal number and cardiac motion and to detect malformations and markers of aneuploidy. The location of the placenta and umbilical cord insertion are identified so that these structures may be avoided, if possible. A pocket of fluid that is accessible and free of hazards such as loops of cord or fetal extremities is identified and the approach planned. The maternal abdomen is then prepared with an antiseptic, such as a povidone iodine or alcohol, to reduce the risk of infection. Sterile drapes are used to isolate the sterile field, and the ultrasound transducer is covered with a sterile cover or glove. Sterile ultrasound gel is applied to the abdomen, and the pocket of fluid is again identified. A 22-gauge needle is inserted through the maternal abdominal wall into the fluid under direct, real-time ultrasound guidance. When the needle tip is observed within the fluid, the stylet is removed. Free return of amniotic fluid confirms proper placement. A syringe is affixed to the hub of the needle, and fluid is aspirated. The first 1 or 2 ml of amniotic fluid may be contaminated with maternal cells and is discarded if cytogenetic analysis is planned. For a karyotype, 20 ml of amniotic fluid (about 10–15% of the total volume at 16 weeks) is withdrawn. After the fluid collection is completed, the needle is withdrawn. Fetal heart activity is documented after the procedure, and the duration of bleeding from the site of needle insertion, if detected, is recorded.

In twin gestations, both fetal sacs are sampled. This usually requires a two-puncture technique in which the first sac is sampled and then injected with dye (usually indigo carmine). Successful amniocentesis of the second sac is confirmed by the withdrawal of clear fluid. The same technique is used for higher order multiple gestations.

B. Accuracy and Safety

The success rate for obtaining karyotype results is over 99%. Biochemical studies are also highly reliable, usually correlating well with the clinical condition of the fetus or neonate. Microbiological tests, such as amniotic fluid culture, are somewhat less accurate but can be helpful with clinical decision making.

The pregnancy loss rate attributable to midtrimester amniocentesis is approximately 0.5% or less. The first large, multicenter trials, performed in the 1970s, reported loss rates in pregnancies subjected to amniocentesis of 0.5–1.0% above the background rate. However, the common practice during that era often did not include fetal evaluation by ultrasound before the procedure, and real-time ultrasound guidance was not routinely practiced. Therefore, a proportion of the reported amniocentesis-associated losses may have included pregnancies that were not viable at the time of the procedure. In experienced centers, the pregnancy loss rate attributable to traditional amniocentesis is 1/300 or less. Up to 2% of women who undergo amniocentesis experience leakage of amniotic fluid. In the majority, the leak ceases, fluid reaccumulates, and the pregnancy continues without interruption. Identification of risk factors for pregnancy loss following amniocentesis, such as the presence of viral DNA in the amniotic fluid, is an active area of investigation. There appears to be no risk to the growth and development of children exposed to amniocentesis in the absence of immediate complications of the procedure.

C. Early Amniocentesis

In order to offer diagnosis earlier in gestation, amniocentesis performed between 11 and 15 weeks of gestation has been proposed. The first studies of this technique indicate that the pregnancy loss rate attributable to the procedure may be several times higher than that of traditional amniocentesis, perhaps as high as 2–5% or more. Furthermore, amniocentesis performed before 13 weeks is associated with an increased incidence of talipes equinovarus, or club-foot malformation. Studies are ongoing to define the earliest gestational age at which amniocentesis may be performed as safely and reliably as it is after 15 weeks.

III. CHORIONIC VILLUS SAMPLING

Chorionic villus sampling (CVS) is typically offered between 10 and 13 weeks of gestation for prenatal diagnosis, although it can be performed throughout the pregnancy. Villi within the rapidly growing chorion frondosum, the precursor of the placenta, contain cells which are cultured *in vitro* for karyotype, genetic, and biochemical testing. Karyotype determination from direct preparations of dividing trophoblast cells can yield results within a few hours, although *in vitro* culture techniques are usually employed for confirmation. The chief advantage of CVS over amniocentesis is the ability to perform the procedure several weeks earlier than amniocentesis. The sample also provides sufficient tissue for DNA analysis, often eliminating the need for culture.

A. Method

An ultrasound examination is performed to confirm a live gestation, fetal number, gestational age by crown–rump length measurement, and location of the chorion frondosum before CVS is attempted. The patient's history is reviewed, including indications for the procedure and risk factors for miscarriage. Written informed consent is obtained. Uterine bleeding, unexplained maternal fever, or other acute medical complaints are generally considered contraindications to the procedure.

There are two techniques commonly used to sample chorionic villi. Transcervical CVS involves aspirating a sample of the chorion frondosum into a cannula passed through the cervix. Transabdominal CVS is similar to amniocentesis except that the needle tip is directed into the chorion frondosum without entering the amniotic or extracoelomic cavities. The choice of technique depends primarily on patient and physician preference but can be influenced by the location and accessibility of the chorion frondosum, the position of the uterus, and the size of the patient. Success of both techniques depends on a skilled operator and trained assistant providing the ultrasound guidance.

The transcervical approach employs a thin, flexible cannula inserted through the cervix and into the chorion frondosum. With the patient in the lithotomy position, a speculum is placed in the vagina, and the cervix is prepared with an antiseptic solution. A thin, malleable cannula is introduced through the cervix and directed into the chorion frondosum as the operator observes the advancement of the cannula tip by ultrasound. Once the tip has traversed the chorion frondosum, the stylet is removed. A syringe is attached to the hub to provide gentle suction as the cannula is withdrawn, aspirating a sample of the gestational tissue.

The transabdominal route requires the use of a needle, usually 20 gauge, to aspirate villi. As with amniocentesis, the abdomen is prepared with an antiseptic solution, and the ultrasound transducer is wrapped in a sterile cover. The best approach is selected and the needle tip is introduced into the chorion frondosum using ultrasound guidance. With continuous negative pressure exerted by a syringe attached to the hub of the needle, several passes through the chorion frondosum are made to aspirate an adequate sample.

B. Accuracy and Safety

The operator inspects the tissue under a microscope to confirm the adequacy of the sample of branching villi. The sampling failure rate is extremely low in experienced hands, and the accuracy for genetic and biochemical diagnosis is comparable to that for midtrimester amniocentesis. Cytogenetic mosaicism, or the presence of two or more karyotypically distinct cell populations, is present in about 1% of villous samples. Although mosaicism is usually limited to the trophoblast, an amniocentesis may be offered to confirm the fetal karyotype.

The most common complication following CVS is bleeding, which occurs in up to one-third of patients following transcervical CVS. Immediate postprocedure bleeding is usually transient and of no consequence. The rates of infection and rupture of membranes are quite low—0.3% or less in experienced centers. Because of the potential for a small amount of fetal blood to enter the maternal circulation after CVS, Rh-negative women are counseled about the risk of Rh sensitization before accepting the procedure. To prevent sensitization, Rh immunoglobulin is administered to all unsensitized Rh-negative women who undergo CVS.

The total fetal loss rate following CVS, estimated by several large studies, is approximately 2%, or about 1% above the background loss rate in controls. Thus, the procedure-related risk is about 1% for CVS performed by either technique in experienced centers. This risk is comparable to the risk of traditional amniocentesis, and in most studies comparing the two procedures, no statistically significant difference in loss rate has been found. However, operator experience seems to influence loss rates considerably, and comparison of the outcomes of well-established and busy programs with centers with little experience in CVS suggests a significant influence of the learning curve.

Several reports of an increased incidence of limb reduction defects and facial defects in infants following CVS raised the concern that these rare defects might have been caused by the procedure. Large reviews of pregnancy outcomes of CVS have not confirmed an increase in the incidence of such defects. All of the defects reported occurred following CVS procedures performed before 10 weeks of gestation, suggesting that early gestational age at the time of sampling may confer risk. There is no apparent increase in the incidence of limb or facial malformations after CVS at or beyond 10 weeks. Today, most centers offering CVS perform the procedure after 10 completed weeks of gestation.

Placental biopsy, a method of transabdominal CVS performed after the first trimester, is used by some investigators as an alternative to cordocentesis for rapid fetal karyotype assessment. Success rates for obtaining a karyotype by either direct or culture techniques are comparable to those of conventional CVS, and results from direct preparations are usually available within a few days. Cultured preparations require 2 or 3 weeks, approximately the same time as required by amniocentesis. This method has not been studied as extensively as first-trimester CVS or cordocentesis, but reported rates of fetal loss are comparable.

IV. FETAL BLOOD SAMPLING

After initial reports in the late 1970s and early 1980s describing its feasibility, fetal blood sampling was enthusiastically developed in European and United States centers for prenatal diagnosis and therapy. Access to the fetal circulation was a seminal breakthrough in fetal medicine: A broad range of diagnostic tests could be performed with relative ease and a high degree of accuracy, physiological variables such as the umbilical intravascular pressures could be measured, and medical therapy could be administered directly to the fetus. The now well-established procedure, often called cordocentesis or percutaneous umbilical blood sampling, involves the ultrasound-guided insertion of a needle into the umbilical cord from which an aliquot of pure fetal blood is aspirated.

The most common indications for cordocentesis are rapid fetal karyotype assessment and hematologic studies in the management of red blood cell alloimmunization. The fetal karyotype is prepared from white blood cells by direct karyotype preparation, *in situ* hybridization techniques, or cell culture. Reliable results are typically available within 48–72 hr, making this procedure particularly appealing to parents and physicians when timely decisions would be influenced by knowledge of the fetal karyotype. In the management of red blood cell alloimmunization, the fetal blood type, complete blood count, Coomb's titer, and reticulocyte count are assessed reliably by this technique. Due to the safety and relative ease of cordocentesis, serial intravascular transfusion is the mainstay of therapy for affected fetuses until delivery near term. Other studies commonly performed include pH and blood gas analysis, infectious disease serology, renal and hepatic function tests, and other biochemical analyses.

A. Method

As with the other ultrasound-guided techniques described, a thorough sonographic examination is performed before proceeding. Disclosure of the information sought and the risks and benefits of the procedure precedes written informed consent. Because cordocentesis requires accurate placement of the needle tip into a relatively small target, the patient is made comfortable and encouraged to lie especially still. Maternal sedation and local anesthesia are sometimes used. Left uterine displacement is helpful to prevent supine hypotension after midgestation.

The maternal abdomen is prepared with an antiseptic solution, and the sterile field is isolated with drapes. Under direct, real-time ultrasound guidance keeping the needle tip in view, a 22-gauge needle is directed through the maternal skin and uterine wall into the umbilical vein. The placental insertion site of the cord is usually the preferred target because the cord is more stable at its placental attachment than a midsegment floating freely in the amniotic fluid. This site is often easy to identify, especially with the aid of color Doppler detection of blood flow, and in the majority of cases the cord root can be reached without obstruction by the fetus. Occasionally during blood sampling, and frequently during longer procedures such as fetal transfusion, fetal paralysis by intravascular or intramuscular injection of pancuronium is required to secure undisturbed access to the cord. The intrahepatic vein is occasionally the preferred target for fetal blood sampling because of its stability and ease of identification, especially in the presence of oligohydramnios. Rarely, a cardiac puncture is chosen when sampling at other sites has been unsuccessful.

Two methods of needle insertion have been developed. The "free hand" method allows the operator to make adjustments to the angle and direction of the advancing needle tip as it is passed through the abdominal and uterine walls toward the target. Motion of the maternal abdominal wall, as happens with a deep breath or tensing of the abdominal wall muscles, can shift the cord root in relation to the skin insertion site, requiring adjustment of the needle trajectory. However, the meandering needle tip may expose the uterine wall, membranes, cord, or placenta to more trauma than a direct, linear puncture. Some operators, therefore, advocate use of a needle guide. The device is affixed to the ultrasound transducer and an outline of the needle trajectory is superimposed onto the ultrasound screen. A needle passed through the fixed channel in the needle guide can be directed only in the fixed path: It can only be advanced or withdrawn along one line. This method requires careful placement of the trajectory template over the targeted site of the vein before insertion of the needle because neither maternal nor fetal motion can be compensated during a single needle pass. Occasionally, a second attempt must be made to reach the target, but because lateral movement of the needle is restricted, it is believed to be less traumatic to the cord or placenta than a freely mobile needle tip. The relative safety of techniques has not been compared prospectively.

B. Accuracy

A specimen of blood is obtained in nearly all cordocentesis procedures. Confirmation of fetal origin of the blood sample is the first step to ensure accuracy of the results. When the sample is obtained from a free loop of umbilical cord floating in amniotic fluid, there is no ambiguity about its origin. However, when a sample is obtained from the cord at its placental insertion, particularly if the needle approaches the cord root through the placenta, the possibility of aspiration of maternal blood requires accurate identification of the sample. Fetal blood differs from adult blood in several ways. The mean corpuscular volume (MCV) of fetal blood is generally higher than maternal blood (>100 vs <100 fl). If the maternal MCV is known, the MCV of the sample can be estimated rapidly by an automated cell counter, even before the termination of the procedure. The Apt test, an alkali elution technique, and the Kleihauer–Betke acid elution test can be used to identify the source of the cells in the sample and quantify the proportion of maternal cells in a mixed sample. Finally, sample identification by more sophisticated red blood cell antigen studies can be performed if necessary.

Fetal karyotype assessment from a proper blood sample is accurate. "Direct" preparations of the karyotype are so-called because they are made from photomicrographs of dividing white cells found in a blood smear prepared directly from the sample. Cultured cells can be induced to divide sooner than amniocytes, confirming the results within days rather than weeks. *In situ* hybridization techniques have been employed for immediate aneuploidy assessment, but they are not as reliable as direct and culture techniques and generally should not be the basis for critical decisions of pregnancy management.

Most other tests performed on fetal blood are as accurate as practicable in a clinical laboratory. Some tests, particularly immunological and infectious disease studies, are most reliable when they are performed by a laboratory specializing in fetal blood analysis. Accurate interpretation of the results from any laboratory, however, requires an understanding of fetal physiology and knowledge of the appropriate ranges of normal in the fetus.

C. Safety

The fetal loss rate from cordocentesis has been reported to be 1 or 2% in several large series. Estimation of the true loss rate is difficult, however, because many of the indications for cordocentesis are associated with high perinatal mortality. The majority of fetal losses occur following severe bradycardia or thrombosis of the cord during or immediately after the procedure or as a result of premature delivery due to obstetric complications, such as rupture of membranes or chorioamnionitis. Not surprisingly, fetuses in the poorest condition at the time of cordocentesis, especially hypoxemic fetuses, those with severe growth restriction or hydrops, are at greatest risk for procedure-related bradycardia and death. Factors other than fetal condition which appear to influence the loss rate include operator experience, gestational age at the time of the procedure, and technique. The best available information suggests that the loss rate of the otherwise healthy fetus approaches 1% using the free-hand technique and may be as low as 0.3% with the needle-guide technique.

Extravasation of blood at the site of cord puncture and transient slowing of the fetal heart rate (bradycardia) are the most common transient complications of cordocentesis. Bleeding at the site of cord puncture can be observed in up to 40% of cases and usually lasts only a few seconds and is well tolerated by most fetuses. Fetuses with thrombocytopenia may be at higher risk for bleeding than those with a normal clotting mechanism. Puncture of the umbilical artery is associated with longer duration of bleeding and a higher incidence of bradycardia. Therefore, the umbilical vein is the preferred target when it is accessible. Infection of the amniotic fluid or membranes is a rare but serious complication of any invasive procedure, so careful aseptic technique is strictly observed. Fetomaternal hemorrhage of small volume occurs commonly when the transplacental approach to cordocentesis is selected. Because of the potential for Rh isoimmunization, Rh-negative women should be treated with Rh immunoglobulin prophylaxis after cordocentesis.

V. OTHER PROCEDURES

Almost any identifiable collection of fluid in the fetus can be sampled using the ultrasound-guided techniques described. As in adult medicine, fetal medicine makes use of investigations of the urine or ascites or pleural fluid. A karyotype can usually be obtained from fluid aspirated from any of these unorthodox sites, and information on the chemistry of the fluid or the presence of infection may be useful for diagnosis and prognosis for the fetus. In the evaluation of obstructive uropathy, for example, fetal urinary chemistries contribute to the assessment of renal function and the options for fetal therapy.

In rare circumstances in which fetal genetic diagnosis requires study of the tissues likely to be affected, fetal biopsy may be possible. Usually performed in the second trimester and under direct ultrasound guidance, biopsy of the fetal skin, liver, or muscle can be obtained by passing a thin-gauge biopsy forceps or needle through a trocar introduced into the fetus through the maternal abdominal wall. Small samples of tissue thus obtained are amenable to histologic, genetic, or enzymatic analysis. These procedures are generally performed at about 18 weeks of gestation and carry essentially the same risks as amniocentesis.

VI. SUMMARY

Several invasive procedures are available for prenatal diagnosis of fetal genetic or chromosomal disorders in the first trimester and beyond. These procedures are performed under direct ultrasound guidance and strict aseptic technique by clinicians with special training in prenatal diagnosis. The method chosen depends on the age of gestation, diagnostic information sought, and the preference of the practitioner and the patient. Although each invasive procedure entails some risk, the minor complication rates are small, and the fetal loss rates are generally 1% or less. Because of the reliability and safety of these techniques, the horizons of fetal medicine are advancing rapidly.

See Also the Following Articles

Fetal Anomalies; Fetal Monitoring and Testing; Fetal Surgery; Prenatal Genetic Screening; Ultrasound

Bibliography

D'Alton, M. E. (1994, June). Prenatal diagnostic procedures. *Sem. Perinatol.* **18**(3), 140.

Evans, M. I., Johnson, M. P., and Moghissi, K. S. (Eds.) (1997). *Invasive Outpatient Procedures in Reproductive Medicine.* Lippincott-Raven, Philadelphia.

Simpson, J. L., and Elias, S. (Eds.) (1993). *Essentials of Prenatal Diagnosis.* Churchill Livingstone, New York.

Wapner, R. J. (1997, March). Chorionic villus sampling. *Obstet. Gynecol. Clin. North Am.* **24**(1), 83.

Weiner, C. P., and Okamura, K. (1996, May). Diagnostic fetal blood sampling-technique related losses. *Fetal Diagn. Ther.* **11**(3), 169–175.

Fetal Growth and Development

William W. Hay, Jr.

University of Colorado

GLOSSARY

AGA Appropriate for gestational age (weight >10th, <90th percentile for gestational age).

anthropometric Referring to measurement of body weight, length, circumference, and weight/length ratio.

autocrine Acting on cells that are the source of production or secretion.

conceptus The entire pregnant uterus, including the uterine tissues, the placenta and its membranes, and the fetus.

eutherian Describing mammals that have placentas during fetal development.

gestation The period of embryonic and fetal growth from conception to birth.

hyperplasia An increase in cell number.

hypertrophy An increase in cell size.

IUGR Intrauterine growth retardation.

LGA Large for gestational age (weight >90th percentile for gestational age).

macrosomia The condition of being fatter than normal (macrosomic fetuses also are LGA).

maternal constraint Nongenomic limitation of fetal growth by maternal (and thus uterine size) and related physiological factors such as uterine blood flow and uterine capacity to support placental growth and function.

paracrine Acting on cells adjacent to the cells that are the source of production or secretion.

ponderal index A ratio expressed as (weight, g)/(length, cm)^3.

polytocous Producing more than one offspring at a time.

SGA Small for gestational age (weight <10th percentile for gestational age).

Fetal growth and development refers to the growth of the fetus in size, as determined by anthropometric measurements of body weight, length, circumference, and weight/(length)^3 ratio, as well as changes in fetal body composition.

I. INTRODUCTION

The period of fetal growth is from the end of embryogenesis, at about the end of the first third of gestation, until term. Fetal growth occurs by increases in cell number and size. In the first third of gestation, during the embryonic period, growth occurs primarily by increased cell number (hyperplasia); in the middle third of gestation, cell size also increases (hypertrophy), while the rate of cell division becomes stable. In the last third of gestation, the rate of cell division declines, while cell size continues to increase. This pattern is found in all species, but there is considerable variation among species with respect to developmental changes in cell number and maturation for different organs; this phenomenon has been studied best in the brain (Fig. 1).

Many terms are used to describe variations in fetal growth. For example, human newborns are classified as having normal birth weight ($\geqq$2500 g), low birth weight (<2500 g), very low birth weight (<1500 g), or extremely low birth weight (<1000 g). Obviously, classification by weight alone says little about fetal growth rate because most infants with less than normal birth weights are the result of a shorter than normal gestation, i.e., they are preterm. Similarly, classifying newborns as preterm or term on the basis of birth weight also is erroneous because infants with intrauterine growth retardation are smaller than

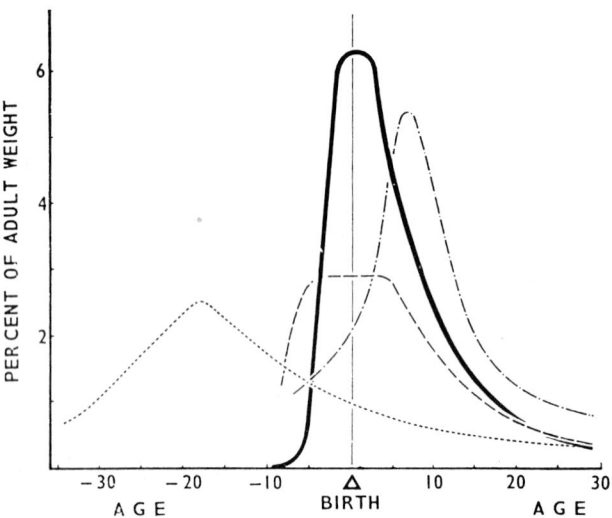

FIGURE 1 Velocity of human brain growth (wet weight) compared with that in other species. Prenatal and postnatal brain growth velocities are expressed as follows: human (——), in months; guinea pig (······), in days; pig (– – –), in weeks; rat (— · — · —), in days [reproduced with permission from J. Dobbing, Later development of the brain and its vulnerability, in *Scientific Foundations of Paediatrics* (J. A. Davis and J. Dobbing, Eds.), p. 567, Saunders, Philadelphia, 1974].

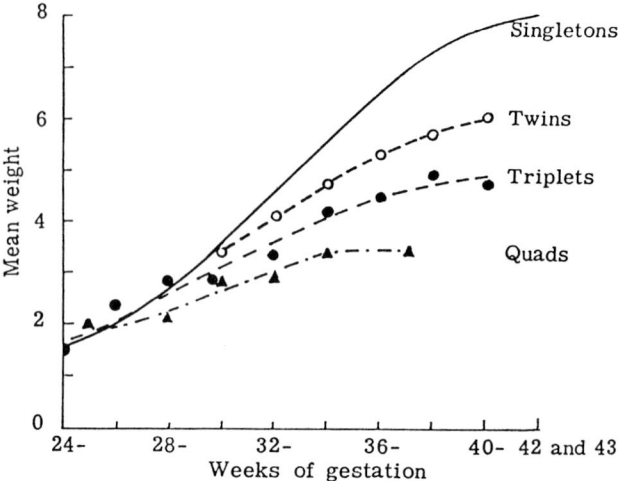

FIGURE 2 Mean birth weight of single and multiple fetuses related to duration of gestation [adapted from McKeown and Record, *J. Endocrinol.* 8, 386, 1952; reproduced with permission from M. Ounsted and C. Ounsted, *On Fetal Growth Rate*, Spastics International (Clinics in Developmental Medicine No. 46), p. 17, W. Heinemann Medical Books, London, 1973].

normal at any gestational age. Furthermore, it is inappropriate to label newborns as abnormally grown when their birth weight is less than some arbitrarily determined "normal" birth weight but their mother was quite small to begin with; such newborns are considered constitutionally small but not abnormal.

II. GROWTH OF FETAL SIZE

Under usual conditions, fetal growth follows its genetic potential, unless the mother is unusually small for the species and limits fetal growth by a variety of factors considered collectively to represent "maternal constraint." Maternal constraint results primarily from a limited uterine size and thus the capacity to support placental growth and nutrient supply to the fetus, not to any particular genetic factor. A clear example of maternal constraint is shown in Fig. 2, which documents the reduced rate of fetal growth of multiple fetuses in a species (human) which optimally supports only one fetus. Obviously, small fetuses of small parents or large fetuses of large parents do not reflect fetal growth retardation or fetal

overgrowth, respectively; in fact, their rates of growth are normal for their genome and for the size of the mother. Unless maternal constraint is particularly prominent, such fetuses would not grow faster or to a larger size if more nutrients were provided.

In most species, fetal weight tends to increase exponentially in the middle part of gestation, but there is considerable interspecies variation. In humans, the midgestational exponential increase in fetal weight tends to produce a typical S-shaped curve of fetal weight vs gestational age that is derived from cross-sectional evaluations of newborn weight vs gestational age (Fig. 3). In sheep, fractional weight gain of the fetus is greater in early gestation.

At the end of gestation, fetal mass varies considerably among mammals, reflecting differences in both fetal growth rate and gestational length. Including marsupials, the range in newborn weight from the smallest marsupial weight of 10 mg to the largest eutherian mammal—the blue whale, weighing 2000 kg—is 200-million fold. Among eutherian mammals, total newborn weight is inversely related to adult animal size and directly related to the length of gestation.

The length of gestation is more strongly related to the growth of neural tissue (range, 0.015–0.033

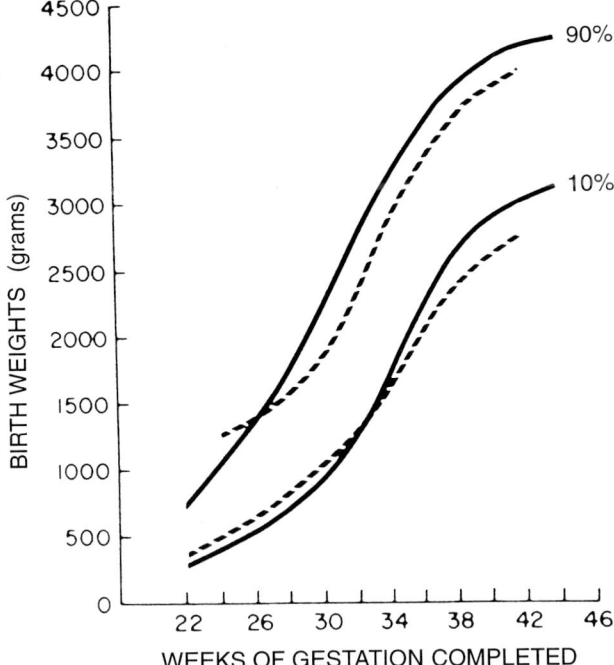

FIGURE 3 Birth weight percentiles for gestational age. Solid lines represent California total singleton live births (1970–1976); dotted lines represent Colorado General Hospital (Denver) live births (1948–1960) [reproduced with permission from R. K. Creasy and R. Resnik, Intrauterine growth retardation, in *Maternal–Fetal Medicine* (R. K. Creasy and R. Resnik, Eds.), 2nd ed., pp. 549–564, Saunders, New York, 1989].

$g^{1/3}$/day—a 2.2-fold range) than to the growth of the fetal body (range, 0.033–0.25 $g^{1/3}$/day—a 7.6-fold range). The physiological significance of this relationship is not known, but intrauterine development of a large brain/body mass ratio in humans is favored in a single fetus and is made possible by a slow rate of somatic growth. The latter allows a steady increase in fetal cerebral metabolic demand while the total metabolic demands of the conceptus are kept within limits that the mother can easily support without stress on her own metabolism.

III. DEVELOPMENTAL CHANGE OF FETAL BODY COMPOSITION

Cellular hypertrophy and the increased deposition of extracellular material in the fetus during the last third of gestation require large increases in nutrient supplies and appropriate utilization of these nutrients. Nutrient substrate supply is coupled with increased development of anabolic hormones and growth factors in fetal tissues and fetal plasma to produce increased nitrogen and carbon deposition in protein, carbohydrate deposition in glycogen, and fatty acid and triglyceride deposition in adipose tissue. Increases in these tissues gradually replace water in the fetal extracellular space.

Chemical composition studies of normal human infants are limited. Based on data from 15 studies accounting for 207 infants, nonfat dry weight and nitrogen content (predictors of protein content) show a linear relationship with fetal weight and an exponential relationship with gestational age (Fig. 4). At each gestational age, fetal weights are highly variable; however, larger fetuses grow faster than

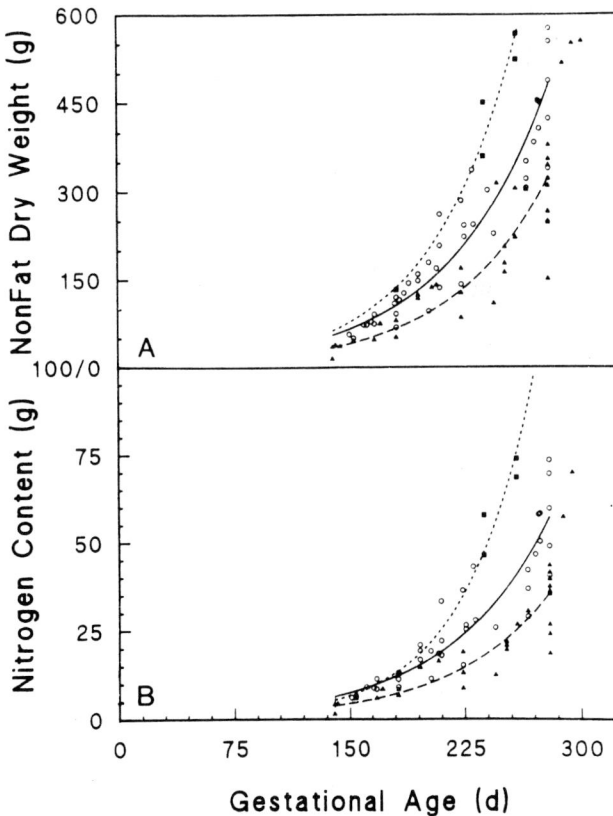

FIGURE 4 Nonfat dry weight (A) and nitrogen content (B) are plotted against gestational age for LGA (■), AGA (○), and SGA (▲) infants [reproduced with permission from J. W. Sparks, Intrauterine growth and nutrition, in *Fetal and Neonatal Physiology* (R. A. Polin and W. W. Fox, Eds.), p. 184, Saunders, Philadelphia, 1992].

TABLE 1

Increments per Day of Nutrients in the Fetal Body at Selected Intervals during Gestation

Fetal age range (weeks)	12–16	16–20	20–24	24–28	28–32	32–36	36–40
Weight range (kg)	0.02–0.1	0.1–0.3	0.3–0.75	0.75–1.35	1.35–2.0	2.0–2.7	2.7–3.4
Increments of nitrogen (N) and protein in fetal body/24 hr							
Total N	29	93	243	326	386	504	714
Protein (g) ($N \times 6.25$)	0.18	0.58	1.52	2.04	2.41	3.15	4.46
Increments of individual amino acids in fetal body (mg/day)							
ILE	6	26	53	71	82	109	148
LEU	13	43	111	151	174	231	330
LYS	13	41	107	145	167	222	313
MET	4	11	28	39	44	59	92
PHE	7	23	61	83	95	127	184
TYR	5	17	44	59	68	91	127
THR	7	23	61	83	95	127	184
VAL	8	27	70	94	109	145	210
ARG	14	43	114	154	177	236	340
HIS	5	15	39	53	61	81	112
ALA	13	41	107	145	167	222	319
ASP	17	52	136	183	211	281	392
GLU	23	74	195	263	303	403	568
GLY	21	68	177	240	276	367	513
PRO	15	48	125	168	194	258	300
SER	8	25	66	89	102	136	191

Note. From E. M. Widdowson, Chemical composition and nutritional needs of the fetus at different stages of gestation, in *Maternal Nutrition during Pregnancy and Lactation* (H. Aebi and R. Whitehead, Eds.). pp. 39–48, H. Huber, Berne, 1980.

smaller fetuses, and protein accretion follows accordingly. Table 1 shows nitrogen, protein, and selected amino acid composition and accretion rates for normal human fetuses. According to data from sheep and guinea pigs, however, only about 80% of the nitrogen content of the fetus is found in protein; the rest is found in urea, ammonia, and free amino acids. The data for human fetal protein content and accretion in Table 1 may be high because they are based solely on nitrogen content. Additional nitrogen requirements for urea excretion and for other possible nitrogen excretion products are not known for human fetuses.

A. Water

Fetal water content increases directly with body weight with increasing gestation, but not proportionally to body weight, as fetal body water, expressed as a fraction of body weight, decreases with increasing gestation. This process is greater in species such as human and guinea pig, in which the fetus produces relatively large amounts of adipose tissue that further dilutes concentrations of water in the fetus. Extracellular water, as a fraction of fetal body weight, also decreases more than intracellular water as gestation advances; this is mainly due to increasing cell number and increasing cell size rather than the intracellular concentration of osmotic substances.

B. Nonfat Dry Weight

Several items of comparative chemical and physical growth in six species are summarized in Table 2. Variation among certain parameters is considerable: Growth rate variation is 20-fold and weight-specific

TABLE 2

**Growth Characteristics and Chemical Composition at Term of Selected Mammals
and a Representative Human Fetus**

Characteristic	Human	Mammal				
		Monkey	Sheep	Pig	Rabbit	Rat
Gestation (days)	280	163	147	67	30	21.5
Number of fetuses	1	1	1	3–5	4–6	10–12
Growth rate (g/day/kg)	15	44	60	70	300	35
Fetal weight (g)	3500	500	4000	100	60	5
Dry weight (g/%)	1050/30	125/25	760/19	25/25	9/15	0.2/4
Nonfat dry weight (g/%)	490/14	—	640/16	14/14	—	—
Protein (g/%)	420/12	—	480/12	12/12	7.2/12	0.6/12

Note. From R. A. McCance and E. M. Widdowson, Glimpses of comparative growth and development, in *Human Growth* (F. Falkner and J. M. Tanner, Eds.), 2nd ed., Vol. 1, pp. 133–151, Plenum, New York, 1985.

fat content at term varies 16-fold, but nonfat dry weight and protein weight-specific contents (as percentages of total weight at term) are constant. Protein concentration is about 12% in all species at term. Limited chemical analyses at different gestational ages indicate that in other animals, as in the human, fetal protein content is linearly related to fetal weight. Thus, protein accretion in the fetal rat occurs about 23 times as fast as it does in the human. These species-related differences in growth rate are remarkable and require marked differences in the placental capacity to supply nutrients to the fetus.

C. Nitrogen Balance, Protein Turnover, and Protein Synthesis

A complete analysis of nitrogen balance in the fetus has been accomplished in only one species (the sheep) and for only one amino acid (leucine). Radioactive and stable isotopic tracers of selected amino acids, especially essential amino acids such as leucine and lysine, have been used to measure fetal protein synthesis, breakdown, and accretion. Protein accretion rate also has been measured in fetuses by whole carcass analysis at increasing gestational age. Figure 5 shows results of experiments in fetal sheep over the second half of gestation comparing fractional protein synthesis rates derived from tracer data and fractional body growth rates derived from carcass analysis data. Protein turnover per unit wet weight

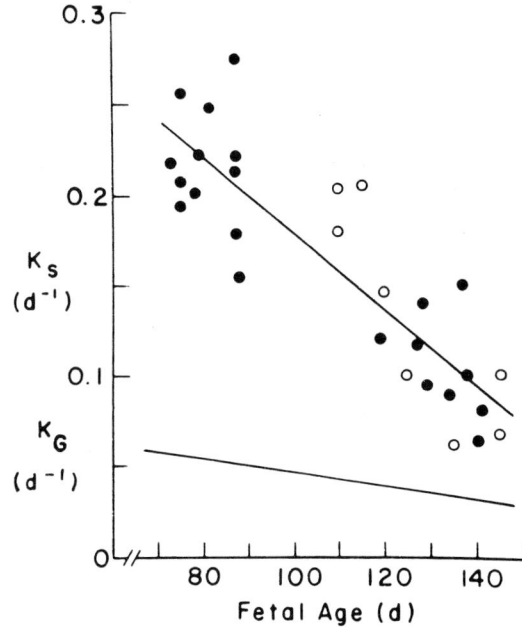

FIGURE 5 Fractional rate of protein synthesis (K_s) over gestation in fetal sheep studied with leucine (●) and lysine (○) radioactive tracers compared with the fractional rate of growth (K_G) in the lower portion of the figure (line) [reproduced with permission from W. W. Hay, Jr., Fetal requirements and placental transfer of nitrogenous compounds, in *Fetal and Neonatal Physiology* (R. A. Polin and W. W. Fox, Eds.), p. 439, Saunders, Philadelphia, 1992].

TABLE 3
Fetal Organ Weight as Percentage of Body Weight

Organ	Gestation		
	50%	*67%*	*90%*
Liver	6.5	5.1	3.1
Kidneys	1.6	1.2	0.7
Heart	0.9	0.8	0.8
Brain	3.4	2.9	1.7
Hindquarters	14.5	15.1	22.0

Note. From A. W. Bell *et al.*, *J. Nutr.* **117**, 1181–1186, 1987.

is higher in the early gestation fetus due primarily to increased rates of umbilical leucine uptake from the placenta (exogenous entry of leucine into the fetal circulation) and protein synthesis. This results in a 50% higher rate of net protein accretion.

In the growing fetus, net protein synthesis exceeds net protein breakdown, resulting in net protein accretion.

The mechanisms underlying the reduction in protein synthesis rate over gestation are not well understood, but they appear to be intrinsic to the fetus and not to a limitation of nutrient supply by the placenta. At least a partial explanation can be offered according to the changing proportion of body mass contributed by the major organs (Table 3). Based on the increased mass of skeletal muscle in the fetus with advancing gestation, whole body fractional synthesis rate should be lower because skeletal muscle has a relatively lower fractional protein synthetic rate in late gestation than in earlier gestation. A direct relationship between anabolic growth-promoting substances acting as principal regulators of fetal protein synthesis rate, and thus fetal growth rate, cannot be made, however, because plasma concentrations or secretion rates of these substances increase in the fetus as gestation proceeds, whereas protein synthetic rates decline.

D. Glycogen

Many tissues in the fetus, including brain, liver, lung, heart, and skeletal muscle, produce glycogen during the second half of gestation. Liver glycogen content, which increases with gestation (Fig. 6), is the most important store of carbohydrate for systemic glucose needs because only the liver contains sufficient glucose-6-phosphatase for release of glucose into the circulation. Skeletal muscle glycogen content increases during late gestation and forms a ready source of glucose for glycolysis within the myocytes. Lung glycogen content decreases in late gestation with change in cell type, leading to loss of glycogen-containing alveolar epithelium, development of type II pneumocytes, and onset of surfactant production. Cardiac glycogen concentration decreases with gestation owing to cellular hypertrophy, but cardiac glycogen appears essential for postnatal cardiac energy metabolism and function. Indeed, in an experimental animal model, neonatal survival following the onset of anoxia was dependent on cardiac glycogen content at birth. At term, fetal liver glycogen concentration in most species (80–120 mg/g) is at least twice the adult concentration. Glycogen synthesis rates are low (about 2 mg/day/g) in species with long gestations, such as human and sheep, and high (40–100 mg/day/g) in species with short gestations, such as the rat. These remarkably high glycogen synthesis rates require only 0.7–1.0 mg/min/kg of glucose, representing less than 20% of estimated glucose utilization rate in the fetal rat. In larger and slower growing fetuses (e.g., sheep, monkey, and human), glycogen synthesis by the liver accounts for an even smaller fraction of fetal glucose utilization.

The principal source of fetal glycogen is glucose derived from placental transport of glucose from the mother. Smaller fractions come from lactate and amino acids such as glutamine. Glycogen content of the fetus and selected fetal organs is directly related to the maternal and thus fetal plasma glucose concentrations. Thus, macrosomic fetuses of diabetic mothers have very high body and organ contents of glycogen, whereas intrauterine growth retardation (IUGR) fetuses that result from sustained maternal hypoglycemia or placental insufficiency and decreased placental glucose supply to the fetus have markedly decreased glycogen contents.

Net synthesis, degradation, and accumulation rates of fetal glycogen are controlled by the functional state

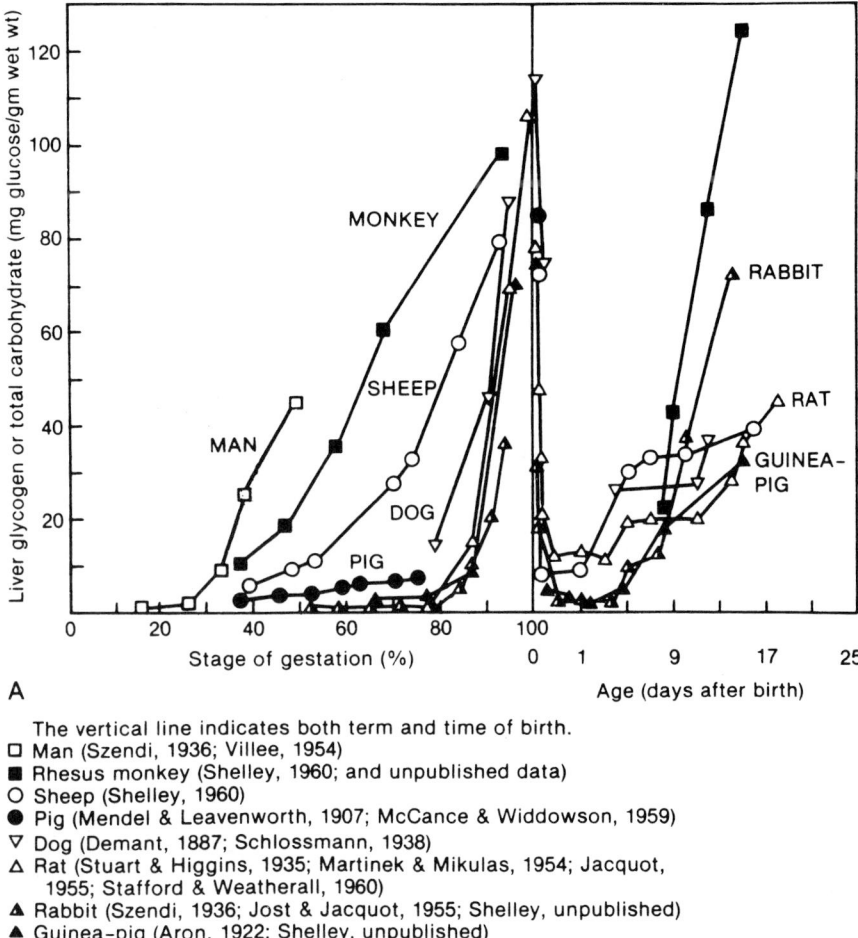

FIGURE 6 Change in liver glycogen content over gestation and following birth in several species (reproduced with permission from H. J. Shelley, *Br. Med. Bull.* 17, 137–143, 1961).

of two enzymes, glycogen synthase (which promotes glycogen formation) and glycogen phosphorylase (which promotes glycogen degradation). The total liver content of these two enzymes is relatively constant during gestation, but their functional states are regulated by hormone and substrate concentrations. Cortisol and glucose provide developmental regulation, whereas epinephrine (adrenaline) and glucagon provide acute and more variable regulation. Experimentally, cortisol infusion decreases glycogen content of the liver, whereas deficiencies in hypothalamic–pituitary regulation of the adrenal gland lead to cortisol deficiency and glycogen deficiency. Insulin acts synergistically with glucose to increase hepatic glycogen stores. Glucose also acts independently to activate glycogen phosphorylase and

glycogenolysis to keep hepatic glycogen content constant at higher glucose concentrations.

E. Fat

At term, fetal fat content, expressed as a fraction of fetal weight, varies markedly among species (Fig. 7). The fat content of newborns at term of almost all land mammals is 1–3% and is considerably less than that of the human (15–20%). Only the guinea pig, with 10–12% body fat at term, approaches the fat content of the human newborn. Differences in body fat content among species are due primarily to the capacity of the placenta to transfer fat to the fetus and to the capacity of the fetus to synthesize triglycerides and fat. Even in those species that take

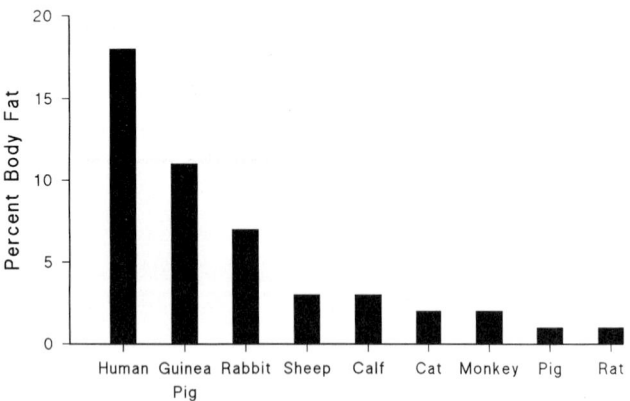

FIGURE 7   Fetal fat content at term as a percentage of fetal body weight among species [reproduced with permission from W. W. Hay, Jr., Nutrition and development of the fetus: Carbohydrate and lipid metabolism, in *Nutrition in Pediatrics* (W. A. Walker and J. B. Watkins, Eds.), 2nd ed., p. 376, Decker, Hamilton, Ontario, Canada, 1996].

up fat from the placenta and deposit fat in fetal tissues, the rate of fetal fatty acid oxidation is presumed low because plasma concentrations of fatty acids (and keto acid products, such as β-hydroxybutyrate and acetoacetate) are low and because the carnitine palmitoyl transferase enzyme system is not sufficiently developed to deliver long-chain fatty acids to the respiration pathway inside the mitochondria.

In the human fetus, calories produced by the complete oxidation of glucose and lactate can fully meet energy required for maintenance metabolism and for conversion of glucose and lactate to fatty acids. The portion of glucose converted into fat has been estimated to be approximately 23 kcal/day/kg. This would permit accumulation of 2.4 g/day/kg of fat. In the fetal sheep, acetate, glucose, and lactate provide 50, 17, and 33%, respectively, of the C_2 units for fatty acid synthesis in adipose tissue slices studied *in vitro* (at about 30 days prior to term or about 80% of gestation). Unlike glucose, the contributions of acetate and lactate depend directly on their concentrations, although fetal lactate concentrations are directly related to fetal glucose supply and plasma concentrations. Insulin promotes and glucagon inhibits lipogenesis in the fetus.

Fat accretion for the human fetus is shown in Fig. 8. Between 26 and 30 weeks gestation, nonfat and

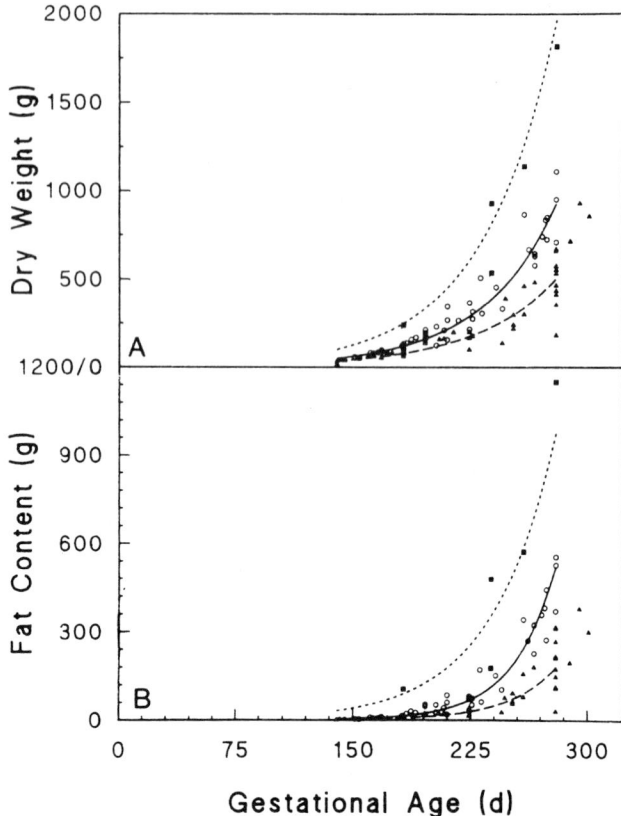

FIGURE 8   Dry weight (A) and fat content (B) plotted against gestation age in the same newborn human infants shown in Fig. 5b for LGA (■), AGA (○), and SGA (▲) infants [reproduced with permission from J. W. Sparks, Intrauterine growth and nutrition, in *Fetal and Neonatal Physiology* (R. A. Polin and W. W. Fox, Eds.), p. 184, Saunders, Philadelphia, 1992].

fat components contribute equally to the carbon content of the fetal body. After that period, fat accumulation considerably exceeds that of the nonfat components. At 36 weeks gestation, 1.9 g of fat accumulates for each gram of nonfat daily weight gain, and by term the deposition of fat accounts for over 90% of the carbon accumulated by the fetus. The rate of fat accretion is approximately linear between 36 and 40 weeks gestation, and by the end of gestation fat accretion ranges from 1.6 to 3.4 g/day/kg. At 28 weeks gestation, it is slightly less and ranges between 1.0 and 1.8 g/day/kg. By term, fat content of the human fetus is 15–20% of body weight, ranging from <10% in IUGR fetuses to 25% or more in macrosomic infants of diabetic mothers.

TABLE 4

Energy Value of Tissue Components

Tissue component	Energy value (kcal/g)[a]
H_2O	0
Fat	9.45
Nonfat dry weight	
Pig	4.0–4.6
Lamb	4.4–4.6
Guinea pig	4.6
Carbohydrate	4.15 (3.7–4.2)
Protein	5.65
In vivo catabolism	4.35

Note. Reproduced from F. C. Battaglia and G. Meschia, *An Introduction to Fetal Physiology*, p. 8, Academic Press, Orlando, FL, 1986.

[a] 1 kcal = 4190 J.

F. Caloric Accretion in the Fetus

The energy value of various tissue components is shown in Table 4. Fat has a high energy content (9.5 kcal/g) and a very high carbon content (approximately 78%). Thus, differences in fetal fat concentration among species lead to large differences in calculated caloric accretion rates and carbon requirements of the fetal tissues for growth. The caloric concentration of nonfat dry weight is fairly consistent across species and also within species at different developmental stages, indicating that the ratio of protein to nonprotein substrates in the tissues is relatively constant. Thus, caloric accretion rate of any fetus can be estimated from the growth curve of the fetus in question and the changing fat and water concentrations.

Data for caloric accretion and caloric distribution in the human fetus are shown in Table 5. Because growth of fat and nonfat (protein plus other) tissues is metabolically linked through energy supply that is used for protein synthesis and the production of anabolic hormones that promote positive protein, fat, and carbohydrate growth, restriction of nutrient supply is likely to produce growth deficits of all tissues, not just fat (i.e., growth retardation involves limitation of muscle growth as well as fat and glycogen). Indeed, chronic experimental selective caloric (glucose) restriction in the fetal sheep leads to increased protein breakdown as well as to lower rates of fetal growth and lipid content. In contrast, as shown by the growth curves in Fig. 8 from human infants born prematurely at different times during the last third of gestation, there is a bias toward thinner, small for gestational age (SGA) infants with less fat relative to nonfat weight and nitrogen content. This raises the possibility that in a species that does lay down considerable fetal fat during late gestation, differences in intrauterine growth rate may reflect fat deposition more than the growth of nonfat, protein-containing tissues.

G. Mineral Accretion in the Fetus

Fetal calcium content is best correlated with fetal body length; this is true for both appropriate for gestational age (AGA) and SGA infants. Using this index, fetal calcium content increases exponentially with a linear increase in length. Using this estimate, the human fetal rate of calcium accretion is about 85 mg/day/kg. Accretion of other minerals varies more

TABLE 5

Calculation of the Caloric Distribution in the Term Human Infant[a]

Wet weight	Distribution			
	Wet weight	Fat	Nonfat wet weight	Nonfat dry weight
Weight (g)	3450	386	3064	511
Total calories (kcal)	5950	3650	2300	2300
Caloric concentration (kcal/g)	1.73	9.45	0.75	4.5

[a] The values in this table are from E. E. Ziegler *et al.*, *Growth* **40**, 329–341, 1979.

TABLE 6
Chemical Composition of the Body of the Developing Fetus

Body weight (g)	Approx fetal age (weeks)	Per kilogram whole body		Per kilogram fat-free body										
		Water (g)	Fat (g)	Water (g)	N (g)	Ca (g)	P (g)	Mg (g)	Na (mEq)	K (mEq)	Cl (mEq)	Fe (mg)	Cu (mg)	Zn (mg)
30	13	900	5	906	10	3.0	2.0	0.10	20	40	81	—	—	—
100	15	890	5	894	10	3.0	2.0	0.10	100	40	70	50	—	—
200	17	885	5	889	14	4.0	3.0	0.15	100	40	70	50	3.5	18
500	23	880	6	885	14	4.4	3.0	0.20	100	44	66	56	3.5	18
1000	26	860	10	869	14	6.1	3.4	0.22	90	44	66	65	3.5	18
1500	31	847	23	867	17	6.8	3.8	0.24	85	44	66	68	3.8	18
2000	33	810	50	853	20	7.9	4.3	0.24	85	44	63	84	4.2	18
2500	35	776	74	838	21	9.0	4.8	0.25	85	48	56	95	4.3	18
3000	38	727	120	826	21	9.5	5.3	0.27	90	49	55	95	4.5	18
3500	40	686	160	816	21	10.2	5.8	0.27	95	51	54	95	4.8	18

Note. Reproduced with permission from E. M. Widdowson, Changes in body proportions and composition during growth, in *Scientific Foundations of Paediatrics* (J. A. Davis and J. Dobbing, Eds.), p. 155. Saunders, Philadelphia, 1974.

directly with body weight and according to the distribution of the minerals into extracellular (e.g., sodium) or intracellular (e.g., potassium) spaces (Table 6).

IV. REGULATION OF FETAL GROWTH

Fetal growth is the result of interaction among maternal, placental, and fetal factors, representing a mix of genetic mechanisms and environmental influences through which the genetic factors are expressed and modulated. The single most important environmental influence that affects fetal growth is the nutrition of the fetus. Nutrient supply to the fetus and the resulting increases in fetal tissue and plasma concentrations of anabolic hormones and growth factors are regulated by maternal health, maternal nutrition, uterine growth (including uterine blood flow and endometrial surface area), and placental growth and function.

A. Genetic Factors

Many genes contribute to fetal growth and birth weight of the normal term fetus. Maternal genotype is more important than fetal genotype in the overall regulation of fetal growth. Table 7 presents estimates of the quantitative contribution of fetal and parental factors to fetal growth and birth weight at term. The more modest regulation by the paternal genotype, acting through the fetal genotype, is essential for

TABLE 7
Factors Determining Variance in Birth Weight

	Percentage of total variance
Fetal	
Genotype	16
Sex	2
Total	18
Maternal	
Genotype	20
Maternal environment	24
Maternal age	1
Parity	7
Total	52
Unknown	30

Note. Derived from L. S. Penrose, *Proc. 9th Int. Cong. Genet.*, Part 1, 520, 1954. From R. D. G. Milner and P. D. Gluckman, Regulation of intrauterine growth, in Gluckman P. D., & Heymann MA (Editors), *Pediatrics and Perinatology* (P. D. Gluckman and M. A. Heymann, Eds.), 2nd ed., p. 285, Arnold, London, 1996.

trophoblast development. In fact, overexpression of the paternal genotype can produce trophoblast tumors. More specific gene targeting studies have shown the importance of genomic imprinting on fetal growth. For example, in mice normal fetal and placental growth require that the insulin-like growth factor IGF-II gene be paternal and the IGF-II receptor gene be maternal; paternal disomy producing IGF-II gene overexpression results in fetal overgrowth, whereas maternal disomy producing IGF-II underexpression results in fetal dwarfism. In humans, isopaternal inheritance of IGF-II alleles is associated with the Beckwith–Wiedemann syndrome that includes hyperinsulinism and fetal macrosomia.

B. Nongenetic Maternal Factors

There is a high correlation between birth weight of siblings which extends to cousins. The nongenetic, maternal nature of this effect is demonstrated by embryo transfer and cross-breeding experiments. For example, a small-breed embryo transplanted into a large-breed uterus will grow larger than a small-breed embryo remaining in a small-breed uterus. Furthermore, partial reduction in fetal number in a polytocous species such as the rat produces greater than normal birth weights in the remaining offspring. Conversely, embryo transfer of a large-breed embryo into a small-breed uterus will result in a newborn that is smaller than it would be in its natural large-breed environment. Such evidence demonstrates that fetal growth is normally constrained, and that this constraint comes from the maternal environment. This is a physiological process and includes the maternal-specific capacity of uterine size, placental implantation surface area of the uterus, and uterine circulation, which together support the growth of the placenta and its function.

C. Maternal Nutrition

Normal variations in maternal nutrition have relatively little effect on fetal growth because they do not markedly alter maternal plasma concentrations of nutrient substrates or the rate of uterine blood flow, the principal determinants of nutrient substrate delivery to and transport by the placenta. Human epidemiological data from conditions of prolonged starvation, as well as nutritional deprivation in experimental animals, indicate that even severe limitations in maternal nutrition limit fetal growth by only 10–20%. Restriction of caloric and protein intakes to less than 50% of normal for a considerable portion of gestation is needed before marked reductions in fetal growth are observed; such severe conditions often result in fetal loss before the impact on late-gestation fetal growth rate and fetal size at birth are manifested. Similarly, fetal macrosomia is common only in pregnancies complicated by gestational diabetes mellitus in which maternal plasma hyperglycemia and hypertriglyceridemia, plus fetal hyperinsulinemia, combine to produce excessive fetal adiposity.

D. The Placenta

The placenta exerts strong control over fetal growth by providing nutrients directly or in metabolically altered form and amount. Naturally and experimentally, placental growth precedes fetal growth, and failure of placental growth is directly associated with decreased fetal growth. There is considerable variation in this control. For example, experiments in sheep that limited placental growth did not result in proportionately reduced fetal weight, indicating that either the capacity of the smaller placenta to transport nutrients to the fetus increased adaptively or that the fetus developed increased capacity to extract nutrients from the placenta and direct those nutrients to growth. More characteristically, though, limitations in placental function to transfer nutrients to the fetus directly limit fetal growth. In fact, fetal growth retardation is seen as a natural and reproductively successful (though not perfect) adaptation to nutrient limitation. Figure 9 shows a direct relationship between fetal weight and placental weight in humans and indicates that large for gestational age (LGA), AGA, and SGA infants are directly associated with LGA, AGA, and SGA placentas. Clearly, placental size and fetal size are directly interrelated, although functional interrelationships between placenta and fetus are also important to fetal growth and development.

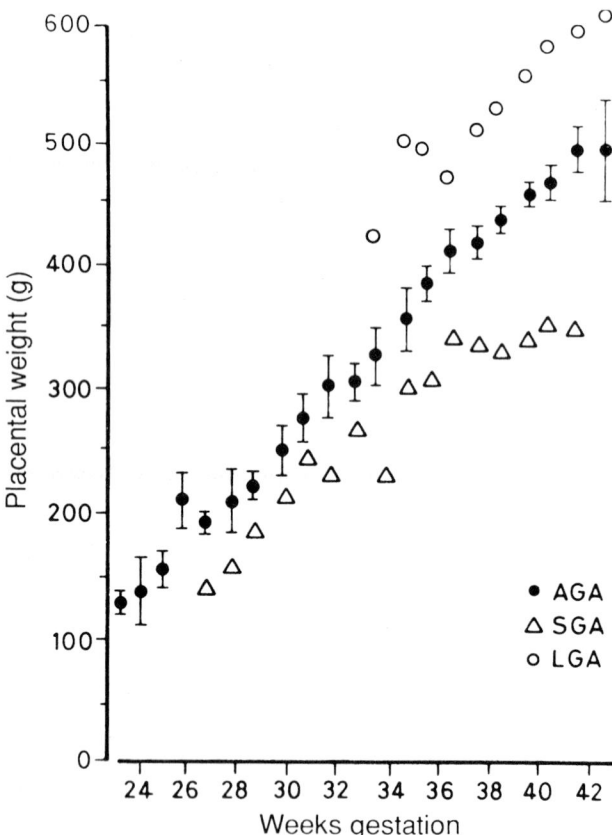

FIGURE 9 Mean placental weights for LGA (○), AGA (●), and SGA (△) human infants at each gestation age (±SEM given for AGA infants alone) (reproduced with permission in modified form from R. A. Molteni, *et al.*, *J Reprod. Med.* **21**, 327–334, 1978).

E. Maternal Endocrine Influences on Fetal Growth

Changes in maternal circulating growth hormone and growth hormone-like peptides such as placental lactogen which increase during pregnancy have combined effects that induce maternal insulin resistance and lead to higher circulating concentrations of glucose and lipids. These in turn are transported in increased amounts to the fetus where, combined with their stimulatory effects on fetal insulin and IGF-I and -II, they promote fetal adiposity (or macrosomia, as in the infant of the diabetic mother) and limit fetal protein breakdown, both of which promote fetal growth.

F. Fetal Endocrine and Autocrine/Paracrine-Acting Growth Factors Influences on Fetal Growth

Growth hormone, which classically acts as the major regulator of postnatal growth, has no demonstrable influence on fetal growth. Insulin produced by the fetus has important but not absolute influence over fetal growth. In sheep, fetal pancreatectomy in late gestation limits fetal growth rate by 20–30%. Pancreatic agenesis in humans produces severely growth retarded fetuses that are 30–50% less than normal weight near term. Insulin infusions into the fetus and excessive fetal insulin secretion enhance fetal glucose utilization and produce increased adiposity but only a 10–15% increase in fetal nonfat growth. Such hyperinsulinemic conditions also limit protein breakdown, which leads to increased protein accretion. The dominant effect on protein balance, however, is the plasma concentrations of amino acids. Insulin, by limiting protein breakdown, actually decreases plasma concentrations of amino acids, thereby limiting protein synthesis at the same time. It is not clear, therefore, how increased fetal insulin enhances, or its deficiency limits, protein accretion. The primary action of insulin may be to promote glucose and lipid utilization and, in turn, enhance protein accretion by providing more energy substrate to fuel protein synthesis and to substitute for amino acids to fuel oxidative metabolism. In support of this hypothesis, removal of insulin from the fetus increases fetal glucose concentration and the transfer of glucose from mother to fetus via the placenta, which reduces net fetal carbon accretion. Similar reductions in fetal growth, as occur with fetal insulin deficiency, result when glucose supply is restricted by chronic sustained or repetitive maternal hypoglycemia. Insulin also may enhance production of IGF-I by promoting intracellular nutrient availability, although hyperglycemia is a more direct stimulus of IGF-I secretion.

Both IGF-I and -II regulate fetal growth. Mice lacking the IGF-I gene have markedly reduced rates of fetal growth, principally in late gestation, and fetal size at term. IGF-II knockouts also have delayed fetal growth that is more pronounced in early to midgestation. IGF-I receptor knockout mice are more growth retarded than either IGF-I or -II knockouts alone. These IGF-I receptor knockouts are growth retarded to the same extent as mice in which both IGF-I and -II genes are deleted, confirming that receptor activation is the principal growth-regulating step in IGF-I and -II action. Infusions of IGF-I into fetal sheep demonstrate limited insulin-like effects on fetal glucose metabolism, but they do limit fetal protein breakdown, particularly when sustained hypoglycemia is present in the presence of increased proteolysis. IGFs may also regulate fetal growth by regulating placental growth. IGF-II gene knockout mice have small placentas and, in turn, lower IGF-I- and -II-binding proteins. IGF-binding proteins modulate effects of IGF-I and -II on fetal growth. Circulating IGF-II receptor may limit IGF-II effects by binding most of it in the circulation. Insulin-like growth factor-binding protein (IGFBP)-1 and -2 levels are relatively high in fetal plasma, perhaps limiting the effectiveness of IGF-I, whereas IGFBP-3 is low in the plasma of fetuses with IUGR, perhaps due to simultaneous insulin deficiency.

V. INTERPRETATION OF GROWTH CURVES

Cross-sectional growth curves have been developed from anthropometric measurements in populations of infants born at different gestational ages. Such curves have been used to estimate whether growth of an individual fetus or newborn is within or outside of the normal range of fetal growth, which is defined as between the 10th and 90th percentile, although what the curves actually show is simply how big a given fetus or newborn is relative to others at any given gestational age. Fetuses and newborns infants who are within the 10th and 90th percentiles for weight vs gestational age are considered AGA, those who are less than the 10th percentile are considered SGA, and those who are greater than the 90th percentile are considered LGA. In general, SGA infants come from small parents (particularly the

mother) and LGA infants come from large parents (again, particularly when the mother is big as well as the father).

Widely used curves of fetal growth are usually confined to the third trimester in humans. Each curve is based on local populations with variable composition of maternal age, parity, socioeconomic status, race, ethnic background, body size, degree of obesity or thinness, health, pregnancy-related problems, and nutrition, as well as the number of fetuses per mother, the number of infants included in the study, and how and how well measurements of body size and gestational age were made. Estimating gestational age is open to considerable error based on such factors as maternal postimplantation bleeding and irregular menses, wide variation in the development and presence of physical features of maturation in the infant, and interobserver variation in assessing an infant's developmental stage.

Mathematical analyses of various fetal growth curves have been used to determine growth rates over relatively short gestational periods or at discrete gestational ages. For example, the data used in the Lubchenco growth curves (Fig. 3) can be approximated by a simple exponential function showing fetal weight increasing at about 15 g/day/kg. This rate will vary from the smallest to the largest infants; however, there are only small differences among populations and studies of 1 or 2%.

Recently, fetal growth curves have been developed from serial ultrasound measurements, providing continuous rather than cross-sectional growth patterns. The growth of an individual fetus that is examined by ultrasound at different gestational ages can be better correlated with expected fetal growth rates than from cross-sectional, population-based fetal growth curves. Serial ultrasound measurements of fetal growth, in humans or in experimental animals, can also more accurately determine how environmental factors can inhibit (e.g., maternal undernutrition globally or hypoglycemia specifically) or enhance (e.g., maternal overnutrition globally or hyperglycemia specifically) fetal growth. Figure 10 shows ultrasound data for fetal growth curves in humans, including evidence for a particular fetus whose growth rate was clearly affected adversely by

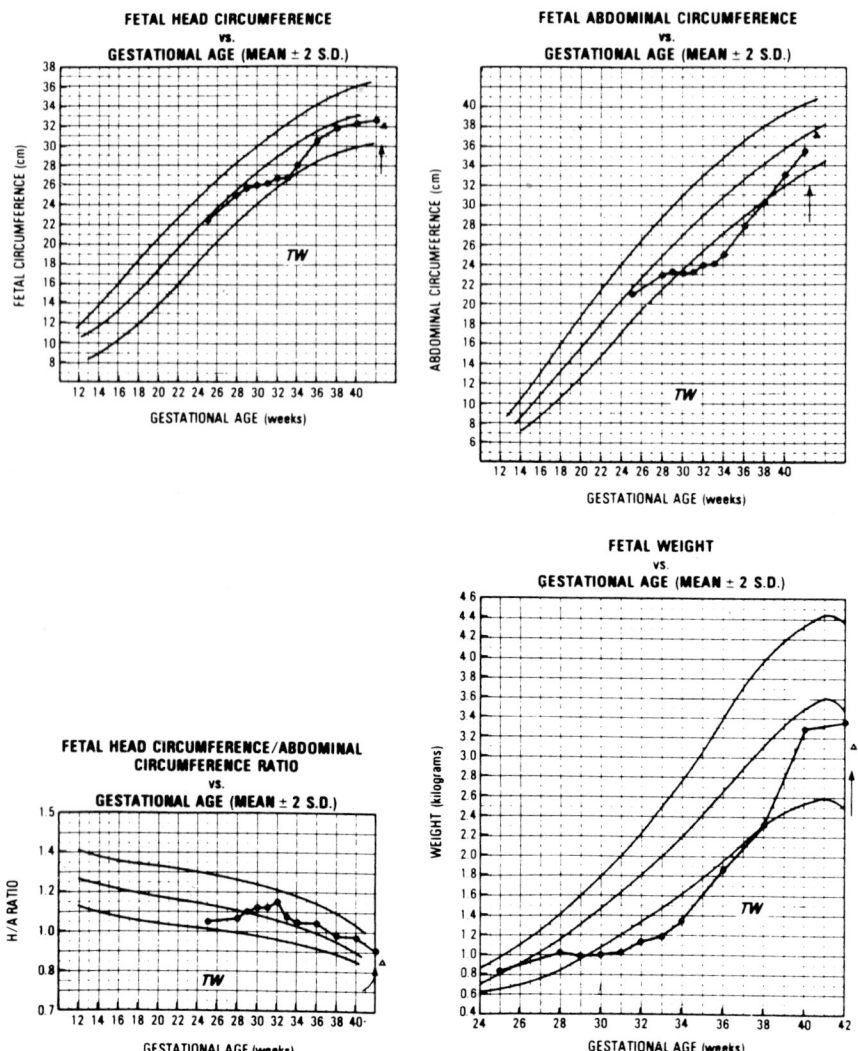

FIGURE 10   Serial fetal body measurements in a mother with severe ulcerative colitis. Note that fetal growth (———·———) begins to decrease markedly in midgestation but returns to normal following the initiation of central hyperalimentation at approximately 28–30 weeks. Solid lines represent the mean ±2 SD of a reference population of human fetuses with normal fetal growth rates according to ultrasound measurements [reproduced with permission from R. K. Creasy and R. Resnik, Intrauterine growth retardation, in *Maternal–Fetal Medicine* (R. K. Creasy and R. Resnik, Eds.), 2nd ed., p. 558, Saunders, New York, 1989].

poor maternal healthy and positively by improved maternal nutrition.

See Also the Following Articles

Placental Nutrient Transport; Ultrasonography

Bibliography

Barbera, A., Jones, O. W., III, Zerbe, G. O., Hobbins, J. C., Battaglia, F. C., and Meschia, G. (1995a). Ultrasonographic assessment of fetal growth: Comparison between human and ovine fetus. *Am. J. Obstet. Gynecol.* **173**, 1765–1769.

Barbera, A., Jones, O. W., III, Zerbe, G. O., Hobbins, J. C., Battaglia, F. C., and Meschia, G. (1995b). Early ultrasonographic detection of fetal growth retardation in an ovine model of placental insufficiency. *Am. J. Obstet. Gynecol.* **173**, 1071–1074.

Battaglia, F. C., and Meschia G. (1986). *An Introduction to Fetal Physiology*. Academic Press, Orlando, FL.

Davis, J. A., and Dobbing, J. (1974). *Scientific Foundations of Paediatrics*. Saunders, Philadelphia.

Hay, W. W., Jr. (1991). Glucose metabolism in the fetal–placental unit. In *Principles of Perinatal–Neonatal Metabolism* (R. M. Cowett, Ed.). Springer-Verlag, New York.

Hay, W. W., Jr. (1992). Fetal requirements and placental transfer of nitrogenous compounds. In *Fetal and Neonatal Physiology* (R. A. Polin and W. W. Fox, Eds.). Saunders, Philadelphia.

Hay, W. W., Jr. (1995). Current topic: Metabolic interrelationships of placenta and fetus. *Placenta* 16, 19–30.

Hay, W. W., Jr. (1996). Nutrition and development of the fetus: Carbohydrate and lipid metabolism. In *Nutrition in Pediatrics* (W. A. Walker and J. B. Watkins, Eds.), 2nd ed. Decker, Hamilton, Ontario, Canada.

Milner, R. D. G., and Gluckman, P. D. (1993). Regulation of intrauterine growth. In *Pediatrics & Perinatology, The Scientific Basis* (P. D. Gluckman and M. A. Heymann, Eds.), 2nd ed. Arnold, London.

Molteni, R. A., Stys, S. J., and Battaglia, F. C. (1978). Relationship of fetal and placental weight in human beings: Fetal/placental weight ratios at various gestational ages and birth weight distributions. *J. Reprod. Med.* 21, 327–334.

Nimrod, C. A. (1989). The biology of normal and deviant fetal growth. In *Maternal–Fetal Medicine* (R. K. Creasy and R. Resnick, Eds.), 2nd ed. Saunders, Philadelphia.

Ounsted, M., and Ounsted, C. (1973). *On Fetal Growth Rate*, Clinics in Developmental Medicine No. 46. Lippincott, Philadelphia.

Philipps, A. F. (1992). Carbohydrate metabolism of the fetus. General metabolism and glucose. In *Fetal and Neonatal Physiology* (R. A. Polin and W. W. Fox, Eds.). Saunders, Philadelphia.

Robinson, J. S., Owens, J. A., and Owens, P. C. (1994). Fetal growth and fetal growth retardation. In *Textbook of Fetal Physiology* (G. D. Thorburn and R. Harding, Eds.). Oxford Univ. Press, Oxford.

Sharp, F., Fraser, R. B., and Milner, R. D. G. (1989). *Fetal Growth*. Royal College of Obstetricians and Gynaecologists, London.

Shelley, H. J. (1961). Glycogen reserves and their changes at birth. *Br. Med. Bull.* 17, 137.

Sparks, J. W., and Cetin, I. (1992). Intrauterine growth and nutrition. In *Fetal and Neonatal Physiology* (R. A. Polin and W. W. Fox, Eds.). Saunders, Philadelphia.

Fetal Hormones

Theresa M. Siler-Khodr

The University of Texas Health Science Center at San Antonio

GLOSSARY

differentiation The specialization of a cell for a specific function.

feedback axis The interaction of hormones when the product of a hormone's action regulates the production of that hormone.

gestational age The fetal age as measured from the time of fertilization.

homeostasis The maintenance of internal balance by interacting stimuli and products.

hormone A substance that acts within a cell (autocrine action), on an adjacent cell (paracrine action), or during transport via blood to cells of another tissue (endocrine action) to regulate those cells' functions.

maternal–placental–fetal unit Mother, placenta, and fetal communication and interrelated production of hormones during pregnancy.

ontogenesis The development and changes of a system during its life cycle.

organogenesis The differentiation and development of an organ.

receptors Binding sites on a cell through which a hormone effects a specific action.

steroidogenesis The production of steroid hormones from cholesterol, catalyzed by specific enzymes.

This article will focus on the current knowledge of the role of fetal hormones in cellular differentiation, the ontogenesis of endocrine systems during gestation, their function in regulating system development and growth during gestation, and their interaction with the placental–maternal systems.

I. FETAL HORMONE ACTIONS

Certain chemical substances regulate and integrate cellular functions in all living organisms. In animal systems these factors are called hormones and may be of different chemical natures, such as peptides, proteins, steroids, and eicosanoids (Stein *et al.*, 1992b). Their cellular origins are multiple and their regulation, activity, and metabolism may be specific to the species or for the target cells. Initially, all hormones were thought to act by endocrine actions, i.e., the hormone is synthesized in a given tissue and then secreted into the circulation and transported via blood to the target tissue where it acts. Receptors for specific hormones, either on the target cell membrane or within the cell, through which the hormone may effect its action, have been identified in the target tissues. This concept was the basis of our understanding of hormonal actions until the mid-1970s. At that time our concept of hormonal actions was expanded by the finding that hormonal-like substances are produced not only in the endocrine glands but also locally, within or adjacent to the tissues in which they act. When a hormone acts on adjacent cells, the action is defined as its paracrine action, i.e., a hormone is produced in one cell and acts on another cell without being transported by blood. If a hormone is produced in the same cell in which it acts, then the action is called an autocrine action. This expanded concept of hormonal activities, i.e., the realization that hormones regulate physiology via endocrine (via blood), paracrine (cell to cell), and autocrine (within a cell) actions, has lead to a major expansion of endocrinology and further understanding of the regulation of physiologic systems.

This expanded concept has been especially significant to our understanding of the maintenance of pregnancy and the role of fetal hormones in fetal development. It has long been recognized that the ontogenesis of the endocrine system in the fetus begins during early gestation with differentiation of glandular tissues and functioning endocrine systems. This development of endocrine function is actually a continuum that occurs throughout life to differentiate and regulate tissue functions at each stage of a species' existence. However, only with the realization of paracrine and autocrine actions of hormones has the role of hormones in cellular differentiation been more fully appreciated. This is especially true in fetal development since autocrine and paracrine actions of hormones on cellular differentiation initiate and regulate organogenesis. Only after organogenesis do hormones assume their more classical role as regulators of tissue functions via endocrine actions. Thus, the initial mode of action of hormones in fetal life is that of their autocrine and paracrine functions. Then, autocrine, paracrine, and endocrine actions of hormones continue to regulate and maintain tissue functions and affect the dynamic differentiation of the evolving stages that each species transits in their life cycle.

This chapter will focus on the current knowledge of the role of fetal hormones in cellular differentiation, the ontogenesis of the endocrine system during gestation, their function in regulating system development and growth during gestation, and their interaction with the placental and maternal systems. What is known of the mechanism of action and function of fetal hormones and their ontogenesis from embryonic life throughout gestation will be reviewed. This will be followed by a discussion of the action and interactions of these hormones on the growth, metabolism, vasoregulation, respiration, and reproduction during fetal life and the long-term effects of the *in utero* actions of fetal hormones. Finally, the relationship of fetal hormones to placental and maternal factors will be discussed. This review will present the basic data existing on fetal hormones leading to the current theories of hormones in fetal development and their role in physiology of the fetus and future function of the organism.

II. ONTOGENESIS OF ENDOCRINE ORGANS AND THEIR HORMONAL PRODUCTION

A. Hypothalamic–Pituitary Axis

The pituitary in the human is made up of two lobes: the anterior pituitary, which is formed from a dorsal evagination of the oral ectoderm called Rathke's pouch, and the posterior lobe of the pituitary, which is actually neural tissue, formed from a ventral process of the diencephalon called the saccus infundibulum (Fig. 1). These two evaginations are

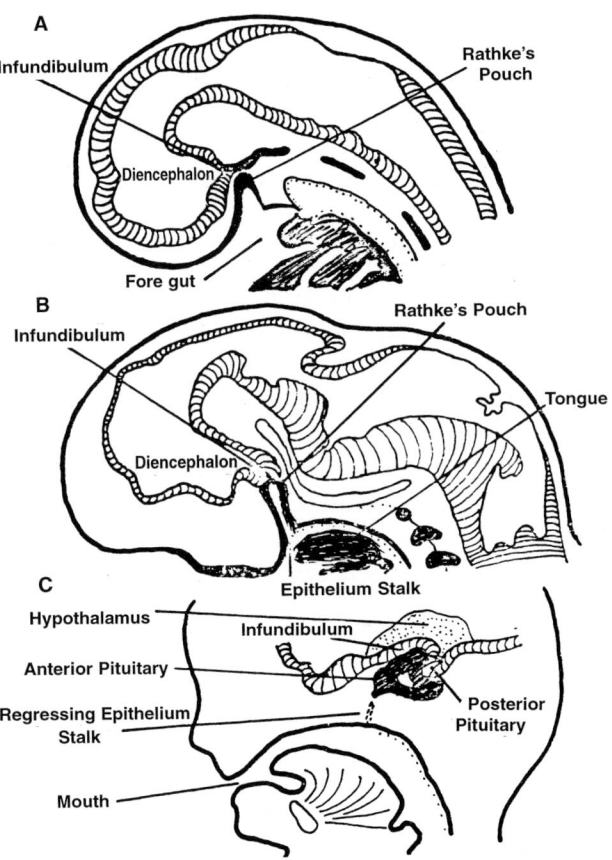

FIGURE 1 The development of the human anterior and posterior pituitary is shown at (A) 4 weeks of gestation, (B) 7 weeks of gestation, and (C) 8 weeks of gestation.

first evident in the human fetus at approximately 4 or 5 weeks of fetal life. By the seventh week of fetal life, Rathke's pouch has grown toward the diencephalon, making contact with the infundibulum. The tissue connecting Rathke's pouch to the oral epithelium degenerates by $7\frac{1}{2}$ weeks of fetal life, thus separating the anterior pituitary. The remaining tissue of the epithelial stalk forms the pharyngeal hypophysis. In some species, but not the human, the portion of the anterior pituitary in direct contact with the infundibulum forms a separate, intermediate lobe of the pituitary. In the human it is not a separate lobe but a differentiated area at the rear of the anterior pituitary. This early differentiation and organogenesis are mediated by autocrine and paracrine activities of the adjacent ectoderm and neural tissue of the diencephalon. By the seventh week of gestation the portal vessels have begun to form; thus, the possibility exists that endocrine actions may begin to play a role in fetal development. As ossification occurs in the fetus, the pituitary gland is enclosed on all but the dorsal side by the sella turcica, with the dorsal side of the anterior pituitary remaining directly connected to the diencephalon via the portal vessels. The human pituitary continues to grow rapidly prior to birth: From 3 to 9 months of gestation the human anterior pituitary increases in weight 200 times and the posterior lobe 33 times.

Very early in fetal life, autocrine and paracrine functions of the hypothalamus and the pituitary are present and induce the differentiation of this system and its endocrine competence. In the fetal pituitary growth hormone (GH), prolactin, luteinizing hormone, follicle-stimulating hormone (FSH), thyroid-stimulating hormone (TSH), adrenocorticotropic hormone (ACTH), α-melanocyte-stimulating hormone, (α-MSH), β-endorphin, and γ-lipotropin are produced. In some cases these have been shown at 5 weeks of fetal life (Adashi, 1996).

The diencephalon, which differentiates to form the hypothalamus and posterior pituitary, has also been shown to produce hormones from very early in gestation and throughout fetal life. Gonadotropin-releasing hormone (GnRH), thyrotropin-releasing hormone (TRH), corticotrophin-releasing hormone (CRH), somatostatin, and growth hormone-releasing hormone (GHRH) have all been demonstrated in the fetal hypothalamus in the human as early as 5 weeks of fetal life. Posterior pituitary hormones, oxytocin, vasopressin, vasotocin, and neurophysin have also been found. The origins of the neuronal cells are multiple. In the case of GnRH the cells migrate from the olfactory pit and this has been studied in detail (see review by Schwanzel-Fakuda and Pfaff, 1996). The regulation of this migration appears to be dependent on proteins expressed by these cells. After a nucleus of cells is established, the production of releasing and inhibiting factors in the hypothalamus begins. Their endocrine action on the anterior pituitary begins with the formation of portal vessels through which the hypothalamic-releasing and -inhibiting hormones can be transported to the pituitary. Hypothalamic nuclei for other releasing or inhibiting activities are similarly formed, as well as those for oxytocin and vasopressin. These latter hormones are stored primarily in the posterior pituitary. They are secreted from the posterior pituitary into the systemic circulation. Neurotransmitters and various hormones regulate hypothalamic hormone production and release through autocrine, paracrine, and endocrine actions.

B. Gonads

The differentiation of the gonad is initially similar in male and female fetuses. Very early in gestation the germ cells originate, possibly as early as the four- to eight-cell stage. Ectodermal cells in the epiblast of the inner cell mass of the blastocyst migrate from the extra embryonic site to the embryonic mesoderm of the primitive streak and then to the yolk sac. From there, they then migrate to the dorsal mesentery and then to the gonadal ridge along the mesonephros, where loose mesenchymal cells are covered by the coelomic epithelium (Byskov and Hoyer, 1994). In the 5-week fetus the germ cells number around 1000 and by 8 weeks have increased to 600,000. This massive proliferation is thought to be regulated by autocrine and paracrine factors such as mast cell factor, leukemia inhibitory factor, and basic fibroblast growth factor. The formation of the primitive gonad is dependent on the cells of the gonadal ridge, but without the germ cells formation of the gonad does not occur (Fig. 2; Byskov and Hoyer, 1994).

1. Differentiation of the Testis and Hormone Production

The coelomic epithelium is separated by a lamina from the germ cells where sex cords develop, and the tunica albuginea is formed from mesenchymal tissue. The formation of the testis is induced by a product(s) encoded on the short arm of the Y chromosome which is activated around 8 weeks of fetal life. Thus, chromosomal errors affecting X and Y chromosomes during early embryogenesis may interfere with gonadal formation and/or differentiation. Studies of steroidogenic enzymes in fetuses of 6–21 weeks, particularly for the production of testosterone, demonstrate activity from the eighth week of gestation, with a peak enzyme activity at 17–21 weeks. The testicular content and concentration of testosterone actually peaks around 14 weeks and then declines until a second peak around 28–32 weeks, followed by a decline by term. Increasing estrogen content with fetal age is also observed in the testis, although the concentration is actually highest in early gestation. Masculinization of the urogenital track is dependent on testosterone production. Mullerian-inhibiting factor, a peptide hormone produced by the testis, is responsible for inducing degeneration of the mullerian system which occurs in early gestation in the human fetus. Other hormones also known to be produced by the testis during fetal life include inhibin and activin.

2. Differentiation of the Ovary and Hormone Production

Differentiation of the definitive ovary may first be defined with the appearance of oogonial meiosis at 8 weeks of fetal life, leading to oocyte formation. No comparable event occurs in the testis until puberty. At this point, the oogonia continues undergoing mitosis, meiosis, or atresia. Mitosis stops by the seventh month of gestation in the human, and those oogonia that have not undergone meiosis will be lost to atresia. In the human, by 20 weeks of gestation the peak number of oogonia (7 million) is attained, but at the same time atresia results in rapid loss of the oogonia and by birth only 1 million remain. An autocrine and paracrine factor, meiosis-inducing factor, affects the initiation of meiosis. These oocytes become surrounded by migrating cells of mesonephric origin,

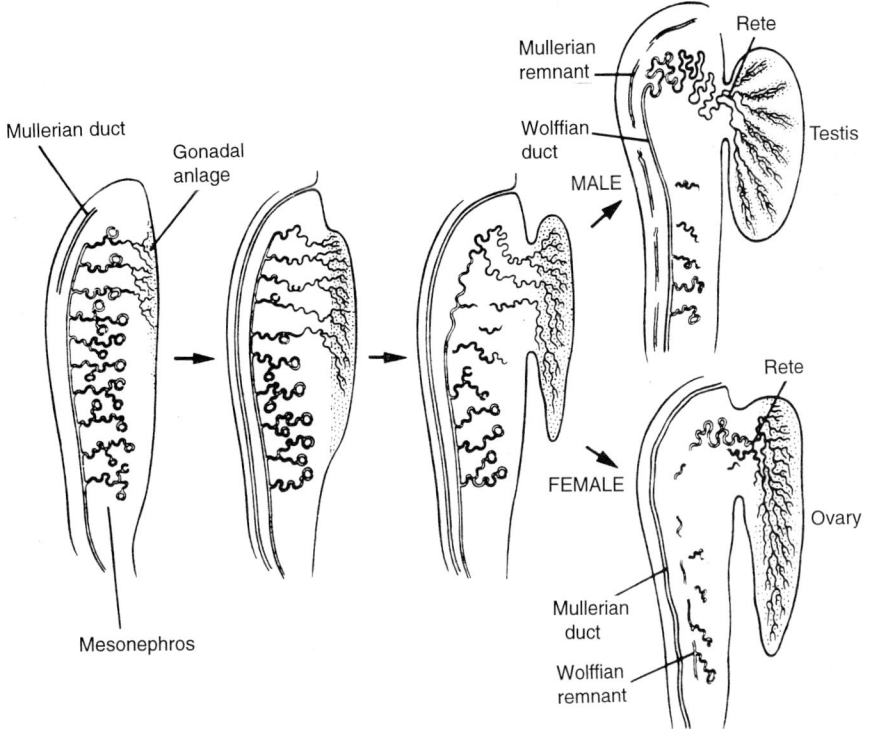

FIGURE 2 The transformation of the genital duct system as differentiating to testes and ovaries during early fetal life (reprinted with permission from A. Byskov, *in Germ Cells and Fertilization*, Cambridge Univ, Press, Cambridge, UK, 1982).

which differentiate to granulosa cells. These granulosa cells encircle the oocyte separated by a lamina and are thought to produce an oocyte maturation inhibitor, which acts as paracrine factor to arrest meiosis in the diplotene stage. In some species gonadotropins and growth factors seem to support follicular growth. The fetal follicle has been shown to produce FSH modulating factors such as inhibin. After 12 weeks steroidogenic cells can be seen, and biosynthetic competence has been demonstrated. However, the estrogen in the ovary is only detectable by 14–18 weeks of fetal age, attaining peak concentration from 20 to 24 weeks. Primordial follicles are present by 20 weeks and continue to increase throughout the remainder of gestation. Following the formation of primordial follicles in the fetal ovary, estradiol production falls to lower levels and then again increases by term. Testosterone is also produced by the fetal ovary, with the highest tissue total content at term, but the highest concentrations are observed in early gestation (10 weeks). Activin and inhibin are also produced by the fetal ovary.

C. Adrenal Cortex and Adrenal Medulla

Development of the fetal adrenal can first be seen at 4 weeks of gestation. This gland can be divided into two parts, the cortex and the medulla. The fetal adrenal cortex is derived from mesodermal cells, whereas the interior medulla is of neural origin. During fetal development the cortex grows very rapidly, attaining adult gland weight by term. The cortex in fetal life is composed of two zones the definitive outer cells and the inner cells which form an area called the fetal adrenal zone. Postnatally, this fetal zone will become the zona fasciculata. This well-differentiated fetal zone produces large amounts of steroids during fetal life. The predominant steroids are dihydroepiandrosterone (DHEA), aldosterone, corticosterone, and cortisol. Precursors for the steroidogenesis are fetal LDL-cholesterol and placental progesterone, and 17-OH pregnenolone, which is both fetal and placental in origin. Products of the fetal adrenal, primarily DHEA, which is a required precursor for placental estradiol production, thus provide a link and communication between the mother and fetus during pregnancy (Stein *et al.*, 1992a).

As early as 8 weeks of fetal life, the synthesis of these steroids occurs in the human adrenal and increases rapidly. Throughout midgestation, the dominant glucocorticoid is cortisone, shifting to cortisol by term and attaining levels of 60 ng/ml in cord blood. Much of the fetal cortisol is transferred to the mother. The shift to cortisol production is thought to be an important factor in the adaptation of the fetus for extrauterine life. Although the fetal adrenal probably works at maximal competence to maintain high cortisol production, the major product of the adrenal is DHEA. In addition, the fetal adrenal and liver have vast enzyme activities to hydoxylate and to sulfate this DHEA. The sulfated molecules are not biologically active. These weak androgens are transferred to the placenta, where these steroids serve as the precursor for estrogen production (Challis and Brooks, 1989; Jaffe *et al.*, 1981). The adrenal medulla, which produces epinephrine and norephinephrine, is also known to be functional in fetal life.

D. Thyroid and Parathyroid

By Day 16 or 17 of gestation, the anlagen of the thyroid is discernible. The thyroid anlagen is derived from the buccopharyngeal cavity. By 7 weeks of gestation it has migrated to the anterior neck, and by 10 weeks iodine accumulation in the gland has begun. Around 13 weeks colloids begin to form in the thyroid. Thyroglobulin, the glycoprotein that provides the tyrosyl residues that are the substrates for iodination, is the predominant protein of the colloid. The enzymes that catalyze thyroid hormone synthesis are active by 10 weeks fetal life, and this activity increases rapidly by 20 weeks. Interestingly, the predominant form of thyroid hormone that is produced in fetal life is reverse triiodothyroxine (rT_3), which is biologically inactive. In later gestation its production declines, allowing for biologically active tetraiodothyroxine (T_4) and triiodothyroxine (T_3) to be the dominant forms. As for the switch from cortisone to cortisol, this change in thyroxine is thought to be an important factor in the adaption of the fetus to

extrauterine life. Calcitonin-producing cells of the fetal thyroid, the thyroid C cells, differentiate around the 12th week of gestation in the human. Actual calcitonin production has been confirmed by direct assay of the gland, although studies throughout gestation are not available (Stein *et al.*, 1992a).

The parathyroid glands develop from the fifth buccopharyneal cavity and parathyroid hormone (PTH) is present in the gland by 10 weeks of gestation. Little is known about the synthesis, processing, secretion, and metabolism of PTH in fetal life. However, it is known to be secreted and the secretion is regulated (Stein *et al.*, 1992a).

E. Gastrointestinal Tissues

1. Pancreas

The pancreas is formed from diverticula of the endodermal cells of the foregut around 5 weeks of gestation. These diverticula will form the dorsal and ventral anlagen of the pancreas and by 7 weeks of gestation fuse to form the pancreas. The undifferentiated epithelial cells form into a lobular–tubular pattern by 9–12 weeks of gestation. A paracrine factor fetoacinar pancreatic protein (FAP protein) may affect differentiation. GH has also been shown to stimulate islet formation and secretion of insulin. Secretory enzymes are detectable by 14–16 weeks of gestation. Islets are first detectable around 12–16 weeks of age, but insulin is present from 8 weeks, although its secretion is first detected first at 9 weeks. Glucagon is detectable from 6 weeks of fetal life. Glucagon-producing A cells are discernible by 9 weeks, with increasing density thereafter. Glucagon is also released into circulation (Stein *et al.*, 1992a).

Somatostatin, as well as other gastric-regulating peptides, is also detectable in the early pancreas. TRH is also produced by the pancreas and present in high amounts by birth. GHRH is another hormone produced by the fetal pancreas, and CRH is also produced by fetal islet cells.

2. Gut and Lung

Regulatory peptides such as gastrin and somatostatin are present in the fetal stomach by 8 weeks of fetal life. Glucagon is found from 10 weeks of gestation. The gastrin and somatostatin content increases during gestation (Stein *et al.*, 1983). The fetal lung also produces hormones such as gastrin-releasing peptide, cytokines, and prostaglandins.

F. Vascular and Other Tissues

A number of vascular tissues have endocrine, paracrine, and autocrine functions in fetal life as well as during adulthood. The production of angiotensin II, through the activity of both liver and kidney enzymes catalyzing the conversion of angiotensinogen via renin and converting enzymes, is active in the fetus. Fetal vessels have been shown to express the endothelin gene. Prostaglandin production also occurs in fetal vascular tissue, and vasoactive intestinal peptide has been demonstrated in the blood of third-trimester fetuses.

There are multiple tissues in which growth factors have been identified in the fetus. Their cellular origin is mesenchymal and they are predominantly fibroblasts. Many different types of growth factors, such as insulin-like growth factors, fibroblastic growth factor, and epidermal growth factor, have been identified in tissues such as the bone, liver, and brain. In these sites they may exert both paracrine and endocrine functions during fetal life (Stein *et al.*, 1992a). Prostaglandins are also produced by many tissues and have endocrine function, such as the liver or the gonad. Another group of hormones previously thought to be limited to the immune system tissues are the cytokines. Cytokines are now known to occur in multiple tissue types beginning during fetal life.

III. FUNCTION AND INTERACTION OF FETAL HORMONES

For endocrine actions to occur, the target cell must have receptors, regardless whether the hormone acts in an autocrine, paracrine, or endocrine manner. Production of a hormone alone is not sufficient to produce its activity. Recent studies have shown that the presence of many specific receptors is expressed prenatally (Csaba, 1991). The specific receptors may be present on the cells' membrane, as is usually the case for protein and peptide hormones, or inside the cell, as is the case for steroid hormones. The regulation of receptor expression is also controlled by hormones in fetal and adult life. Specific errors in receptor expression may be lethal or lead to lifelong aberrations in endocrine or paracrine functions of the individual.

Endocrine actions of a hormone are dependent on the formation of a vascular system by which the hormones can be transported to their target tissue. The classical endocrine axis includes the hypothalamus–pituitary and some target tissue. Recent findings on the many sites of hormone production and the presence of receptors has led to the recognition of multiple axes not previously considered, such as the immune system and the adrenal gland or the reproductive system and vascular function. Multiple feedback interactions are known to function *in utero*. The following discussion will present some of the data on interacting tissues within a particular system. It should also be recognized that none of these systems operate in isolation and intersystem interactions also develop *in utero*.

A. Reproductive System

The classical reproductive system consists of the hypothalamus–pituitary–gonadal axis. Function of this system requires development of the vascular system such that hormones can work. Around the eighth week of fetal life, classical endocrine function begins. In the fetal testis Leydig cell differentiation and testosterone production are stimulated by gonadotrophin concentrations. The source of the gonadotropins in early pregnancy has been correlated to the peak of placental hCG production, which occurs at 7 or 8 weeks of fetal life. However, the early peak in fetal testosterone production does not occur until at least 12 weeks of fetal life. The hypothesis that the fetal pituitary is the source of gonadotropin is supported by the observation that fetuses with hypopituitary function have few Leydig cells and low testosterone production. On the other hand, in fetuses in which androgen receptors are defective, as in testicular feminization, Leydig cells are greatly increased as are the concentrations of gonadotropin. Circulating pituitary gonadotropins can also be correlated with testosterone production. Other studies have correlated a decrease in FSH production with increasing fetal inhibin production. These data demonstrate that the endocrine feedback systems develop and are functional in the fetus. Factors that interfere with these feedback axes will lead to abnormal development of the hypothalamic–pituitary–testicular axis.

Oocyte formation in the fetal ovary also occurs in early gestation, and steroidogenesis of androgens and then estrogens is present by 10–14 weeks of gestation. Peak estrogen production occurs at 20 weeks, just following peak hypothalamic GnRH concentrations and peak pituitary gonodotrophins in the pituitary and the circulation and maximal oogonia in the ovary (Fig. 3). After 20 weeks of gestation a negative feedback of the steroids leads to suppression of the hypothalamus and the pituitary. The finding that the fetal ovary in hypopituitary and anencephalic fetuses is undeveloped indicates that the hypothalamic–pituitary–ovarian axis plays a functional role in fetal life.

B. Growth and Metabolism

Growth hormone production in the fetus is regulated by the hypothalamus and is responsive to multiple stimuli (Kaplan *et al.*, 1972). For GH, such stimuli are hyperglycemia, arginine, and IGF-I. While GH may induce cell proliferation in early gestation, reports to date do not indicate that GH has a major role in somatic growth later in gestation. While many growth factors are produced during fetal life, little is yet known of their actions or feedback regulations. However, IGF-I is known to be affected by endocrine hormonal regulation. Receptors of IGFs have been documented in multiple fetal tissues. In addition, low IGF-I production in the fetus is associated with intrauterine growth retardation. Other hormones af-

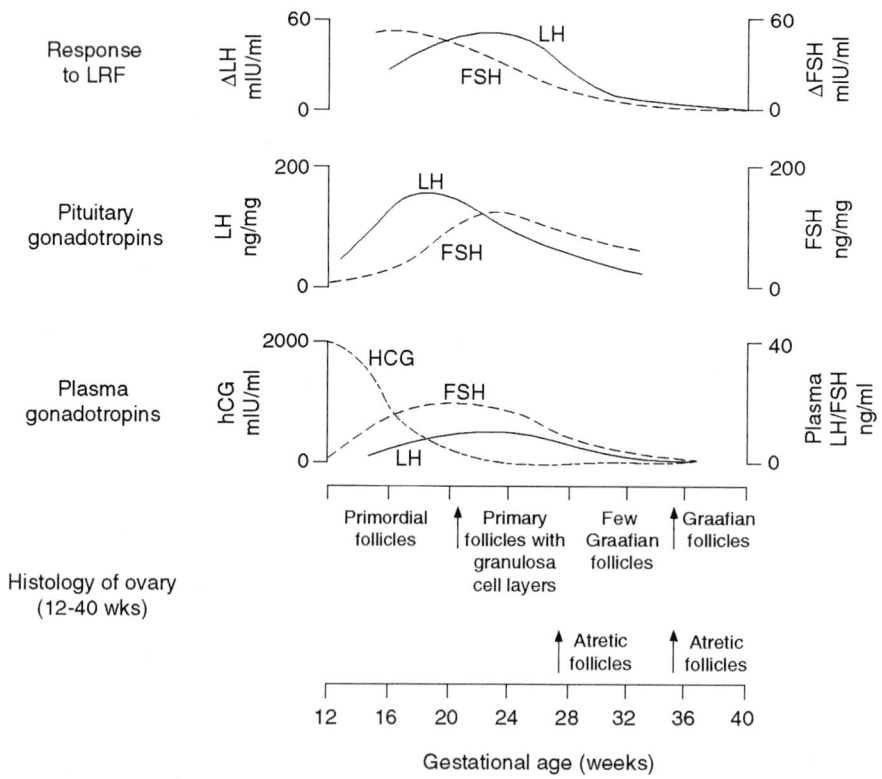

FIGURE 3 The correlation of the fetal hypothalamus and pituitary during ovarian development in the human (reprinted with permission from Tulclchinsky and Ryan, in Maternal–Fetal *Endocrinology*, Saunders, Philadelphia, 1980).

fecting fetal growth are calcitonin and parathyroid hormones, which have axes developed *in utero* regulating calcium metabolism of bone formation. A metabolic hormone, insulin, has feedback activity developed *in utero*. Fetal insulin production is responsive to both hyper- and hypoglycemia (Mayor and Cuezva, 1985).

Prolactin is affected by hypothalamic TRH, PIF, and dopaminergic regulation and osmolality. It may play a role in sodium homeostasis. Prolactin synergizes with other hormones to induce lung maturation. TRH induces a marked stimulation of TSH and thyroid hormones. In hypothyroid fetuses the lack of negative feedback results in an increase in TSH. Thyroid hormones are required for normal cellular growth, but since maternal thyroid hormones can cross the placenta, their absence does not result in irreversible damage until neonatal life. TRH has also been shown to enhance lung development when given with glucocorticoids.

The hypothalamic–pituitary–adrenal axis is highly developed *in utero*. Initially, α-MSH is produced by the pituitary and stimulates adrenal DHEA-S and DHEA production from the fetal adrenal zone (Goodyer and Branchaud, 1981). As gestation progresses, ACTH increases, stimulating cortisone production in the definitive zone of the adrenal using the DHEA as substrate. By late gestation, ACTH production, which is responsive to hypothalamic CRH stimulation, is greatly increased and the definitive zone production of cortisone is switched to cortisol, which stimulates lung development and prepares the fetus for delivery. The increased DHEA-S is the precursor for placental estrogens which ready the intrauterine tissues for delivery. The importance of the CRH–pituitary–adrenals axis to fetal development is supported by the observation that anencephalic fetuses have underdeveloped adrenals and are frequently born postterm, as are fetuses that cannot cleave DHEA-S to DHEA to be used for estrogen synthesis.

C. Other Endocrine Axes

Many other endocrine axes are known to exist *in utero*, such as the formation of angiotensin II, endothelins, and prostaglandin. Their regulation and

feedback systems have been indicated and are under study. The immune system (as well as other tissues) is active in cytokine production, which also affects many systems' function including the lung.

IV. Fetal–Placental–Maternal Interactions

The functions of fetal hormones are not limited to the fetal system. Communication between the fetus and the mother is critical for appropriate maintenance of pregnancy and the timely initiation of parturition. The placenta, a tissue of fetal trophoblastic origin, serves as an interface between mother and baby. Certain hormones of the placenta act to regulate placenta function and to modulate maternal functions. While placental hormones regulate and communicate with placenta, mother, and baby, only a few of these systems are now known.

The human placenta is very active in the production of steroids, pregnenolone, progesterone, and estrogens (Fig. 4). The former are transported in large amounts to the fetus, where they are sulfated and become biologically inactive and then transformed by the fetal adrenal to DHEA-S as discussed previously. This DHEA-S is a required precursor for the human placenta to synthesize estrogens. The sulfate is first cleaved from the DHEA-S to give DHEA, which is then used by the placenta to make active estrogens. This axis is of major significance for communication between mother, placenta, and baby. The human placenta also communicates with the fetus and mother via pregnenolone, progesterone, and CRH, which are produced in large quantities especially as term progresses (Fig. 5). CRH not only regulates placental ACTH, endorphins, prostaglandins, and cortisol production but is itself also sent to mother and fetus. Its role in the mother is open to speculation but may include interaction at the adrenal and the maternal hypothalamic–pituitary axis. In the fetus, the known functions of CRH are multiple. CRH can stimulate fetal pituitary ACTH, which leads to increased adrenal production of DHEA and cortisol. As described previously, DHEA is the precursor of placental estrogen production. Thus, in this fashion, CRH, together with a mature fetal pituitary, may

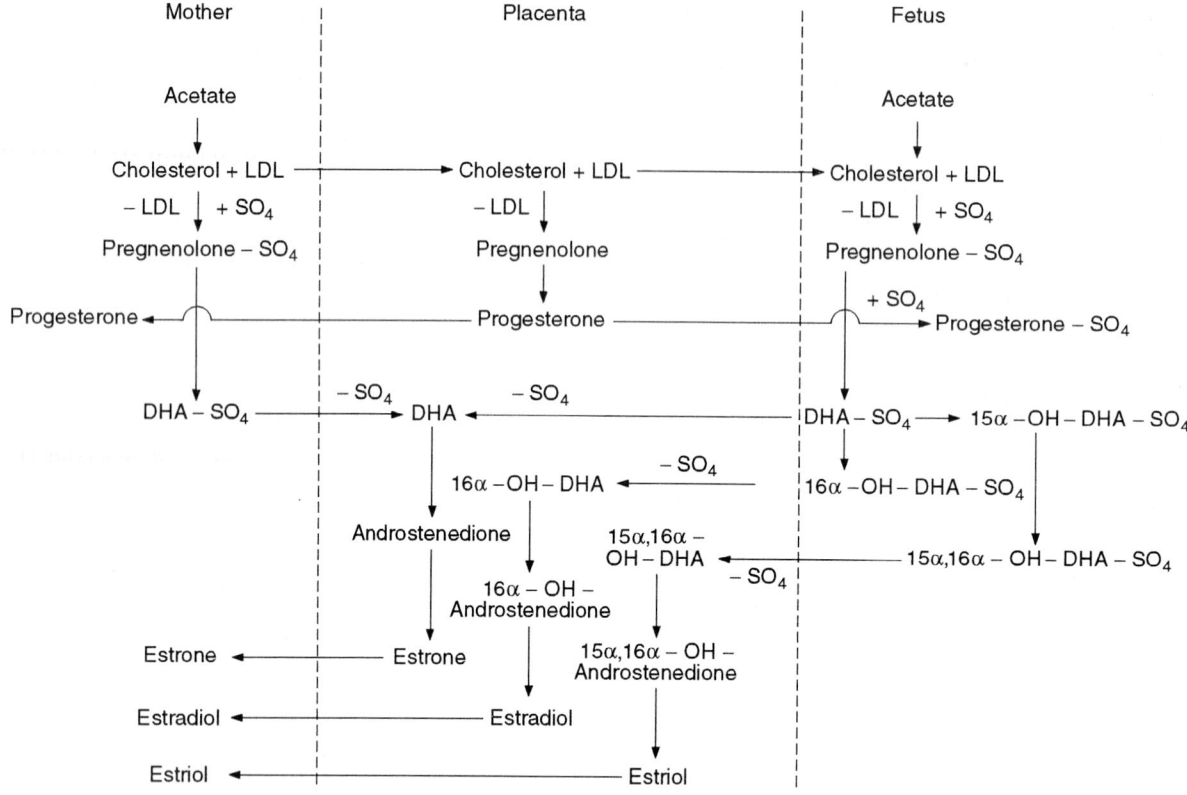

FIGURE 4   The steroidogenic pathways of the maternal–placental–fetal unit.

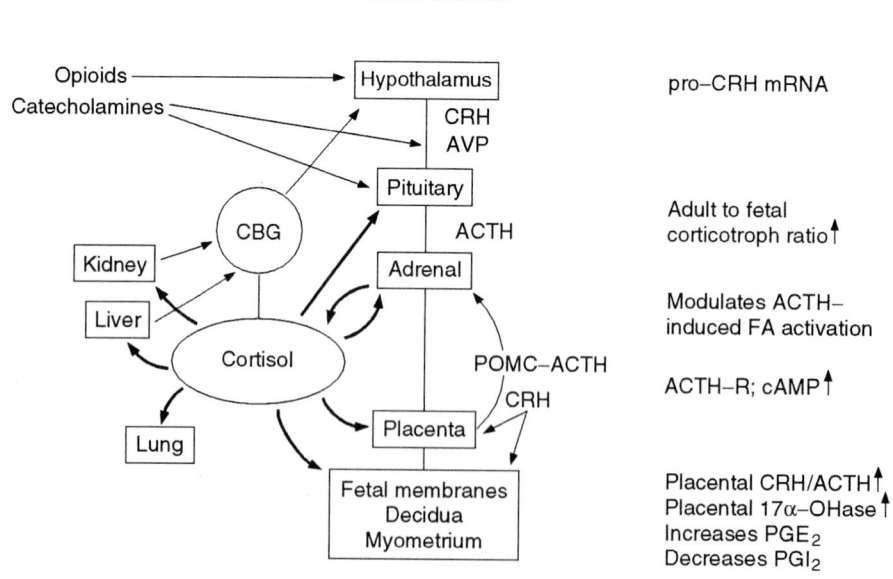

FIGURE 5   The maternal–placental–fetal interactions for CRH–ACTH and steroidogenesis (reprinted with permission from Challis and Brooks, 1989).

induce an estrogenic milieu facilitating parturition. In addition, CRH, via its stimulation of cortisol, may enhance lung maturation. It may also act directly on the fetal lung, stimulating prostaglandin production and hence lung maturation. CRH is prematurely increased in preeclampsia and certain cases of preterm labor, and thus it may be preparing the fetus for an untimely delivery. The role of maternal and fetal stress in the regulation placental CRH and fetal maturation is currently under intense study.

Many other placental paracrine axes exist and modulate CRH production. The enzyme that degrades GnRH in the placenta is stimulated by CRH and may be a mechanism to shift placental GnRH regulation of the placenta in early pregnancy to that of CRH production in later pregnancy. Numerous cytokines also increase placental prostaglandins and CRH which may lead to preparation for parturition. CRH also stimulates endorphins which correlate directly with CRH levels.

Insulin-like growth factors, especially IGF-I, of placental origin may also be of major significance to fetal growth. Not only can it inhibit placental vasoconstrictors and enhance placental growth and glucose transfer to the fetus but also it may be transferred to the fetus, where it can directly stimulate fetal growth. Placental IGF-I correlates with fetal size, whereas insulin and placental IGF-II do not.

V. SUMMARY

Autocrine and paracrine actions of hormones during fetal development initiate and regulate tissue differentiation. As the vascular system develops, the endocrine functions of hormones become operative. Essentially every tissue and system of the fetus is affected by hormones of fetal origin during gestation. The ontogenesis of the endocrine system occurs and in many species is irreversibly encoded. Hormonal regulations and feedback interactions are established and active during intrauterine life. The hypothalamic–pituitary–target tissue axes are functional and are critical for normal fetal development *in utero*. In addition, fetal hormones play an active role in communicating with the mother via the placenta regarding the status of fetal maturation and interact

in preparing the maternal system and the interuterine tissue for parturition. The coordinated functions of the maternal–placental–fetal unit and the fetal autocrine, paracrine, and endocrine activities ensure the normal development of the fetus and its timely delivery.

See Also the Following Article

Fetal–Placental Unit

Bibliography

Adashi, E. Y. (1996). The ovarian follicular apparatus (E. Y. Adashi, J. A. Rock, and Z. Rosenwaks, Eds.), pp. 17–40. Lippincott-Raven, Philadelphia.

Byskov, A. G., and Hoyer, P. E. (1994). The physiology of reproduction (E. Knobil and J. D. Neill, Eds.), 2nd ed., pp. 487–540. Raven Press, New York.

Challis, J. R. G., and Brooks, N. (1989). Maturation and activation of hypothalaic–pituitary–adrenal function in fetal sheep. *Endocrinol. Rev.* 10, 182–203.

Csaba, G. (1986). Receptor ontogeny and hormonal imprinting. *Experientia* 42, 750–759.

Goodyer, C. G., and Branchaud, L. (1981). Regulation of hormone production in the human feto-placental unit. *Ciba Foundation Symp.* 86, 89–123.

Jaffe, R. B., Seron-Ferre, M., Crickard, K., Koritnik, D., Mitchell, B. F., and Huhtaniemi, I. T. (1981). Regulation and function of the primate fetal adrenal gland and gonad. *Recent Prog. Horm. Res.* 37, 41–103.

Kaplan, S. L., Grumbach, M. M., and Shepard, T. H. (1972). The ontogenesis of human fetal hormones, I. Growth hormone and insulin. *J. Clin. Invest.* 51, 3080–3093.

Mayor, F., and Cuezva, J. M. (1985). Hormonal and metabolic changes in the perinatal period. *Biol. Neonat.* 48, 185–196.

Schwanzel-Fakuda, M. L., and Pfaff, P. W. (1996). The hypothalamic pituitary unit (E. Y. Adashi, J. A. Rock, and Z. Rosenwaks, Eds.)., pp 5–16. Lippincott-Raven, Philadelphia.

Stein, B. A., Buchan, A. M. J., Morris, J., and Polak, J. M. (1983). The ontogeny of regulatory peptide-containing cells in the human fetal stomach. *J. Histochem. Cytochem.* 31, 1117–1125.

Stein, B. A., Buchan, A. M. J., Morris, J., and Polak, J. M. (1992a). *The Placenta, Neonatal and Fetal Medicine* (R. A. Polin and W. W. Fox, Eds.). Saunders, Philadelphia.

Stein, B. A., Buchan, A. M. J., Morris, J., and Polak, J. M. (1992b). *Textbook of Endocrinology* (R. H. Williams, Ed.). Saunders, Philadelphia.

Fetal Lung Development

Susan H. Guttentag and Philip L. Ballard

University of Pennsylvania School of Medicine

GLOSSARY

alveolar–capillary membrane A respiratory membrane consisting of closely opposed endothelial cell membranes, interstitium, and alveolar type I cell membranes across which gas exchange occurs.

alveolus A blind-ended air sac of microscopic size which is the primary site of gas exchange in the lung.

angiogenesis The formation of new blood vessels from existent vessels.

branching morphogenesis Repetitive branching of epithelial structures into the surrounding mesenchyme governed by epithelial–mesenchymal interactions that establish the structural organization of the developing organ.

pulmonary acinus The tissue supplied with air by one terminal bronchiole.

septation Subdivision of the alveoli that increases alveolar number and surface area. Primary septa are produced by dichotomous branching of saccules. Secondary septa are formed when one of the two capillary layers of a primary septum rises up into the alveolar space creating a ridge.

surfactant A wetting agent: A substance, such as detergent, that reduces surface tension. Pulmonary surfactant is secreted by the type II pneumocytes lining the alveoli of the lungs to prevent the alveolar walls from collapsing together at end expiration.

vasculogenesis The formation of blood vessels from undifferentiated mesodermal cells.

The lung is a network of airways, alveoli, blood vessels, neural elements, and the supporting extracel-lular matrix with the primary function of accomplishing gas exchange. Lung organogenesis, the process by which these elements form and develop into mature structures, is separate from lung maturation, which involves morphologic and biochemical differentiation of precursor cells into specialized cells such as the type 1 and 2 alveolar cells with specific functions. Although the process of organogenesis and lung maturation share some common mechanisms, precision timing and the local milieu of cellular relationships as well as concentrations of growth factors and hormones play important roles in determining how developmental signals are interpreted by lung cells. Lung formation begins early in gestation and growth extends well into childhood, reaching completion by ~8 years of age. Organogenesis can be divided into five stages: the embryonic, pseudoglandular, canalicular, saccular, and alveolar stages. The timing of these stages in development and highlights of these periods appear in Fig. 1. It should be remembered that this scheme is an organizational tool and that considerable overlap may occur between stages.

I. EMBRYONIC STAGE OF LUNG DEVELOPMENT

The embryonic stage of lung organogenesis commences with the identification of the lung bud as an endodermal outpouching of the primitive foregut within a distinct mass of surrounding mesenchyme. The lung bud develops initially as a laryngotracheal groove of the ventral foregut at 26 days gestation and by 4.5 weeks that groove has closed such that the only remaining lumenal attachment to the foregut is in the region of the developing hypopharynx and larynx. Tracheoesophageal fistulae, tracheal atresia,

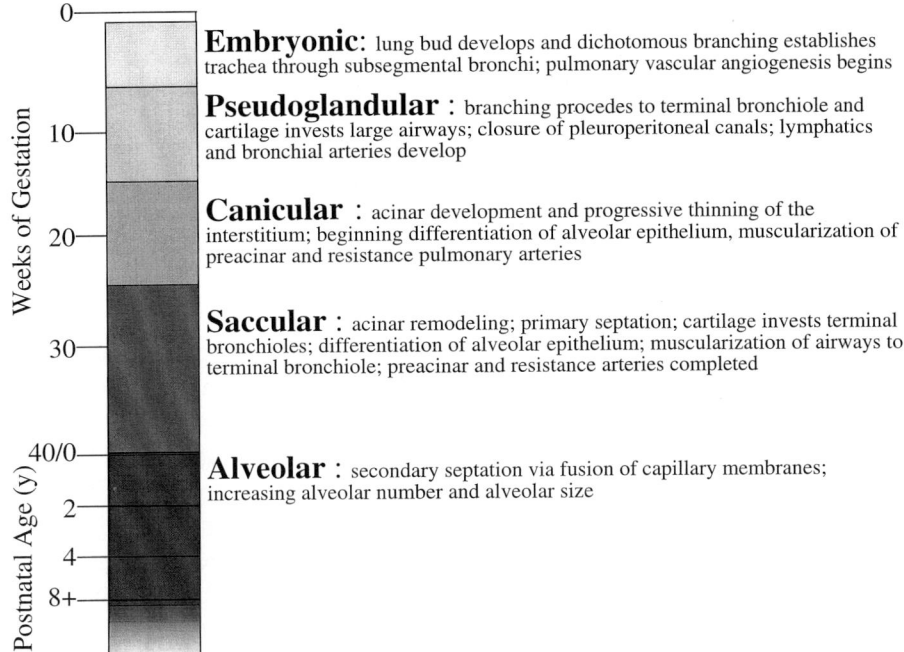

FIGURE 1 Time line of lung development. Lung development begins at 4.5 weeks gestation and extends through the postnatal period to roughly 8 years of age. Lung development can be organized into five stages: the embryonic, pseudoglandular, canalicular, saccular, and alveolar stages based on the structural organization of the most distal airspaces identifiable during that stage.

and stenosis are all developmental abnormalities of the embryonic stage. The lung bud then begins a series of dichotomous branchings that determine the five primordial lung lobes, two on the left and three on the right. Branching continues such that by the end of this period, the trachea and segmental and subsegmental bronchi are evident.

II. PSEUDOGLANDULAR STAGE OF LUNG DEVELOPMENT

The pseudoglandular stage of lung development (7–16 weeks) is marked by continuation of branching morphogenesis to establish the complete bronchial tree of conducting airways. Early studies indicated that by the end of 16 weeks all bronchial divisions are complete, resulting in a total of 23 divisions responsible for gas conduction. Others have since suggested that the pseudoglandular stage may also include the initial development of the rudiments of gas exchange (Fig. 2).

There is evidence of cell maturation within the airways during this period. The airway epithelium is tall and columnar, decreasing in height toward a more cuboidal appearance in the most distal segments. The epithelial lining cells of the conducting airways begin to differentiate in a centrifugal manner and ciliated, mucous-secreting, and basal cells are identifiable. Similarly, cartilaginous support of the bronchial tree begins and proceeds in a centrifugal fashion beginning in the primitive trachea at ~10 weeks and proceeding to the most distal terminal bronchioles by ~25 weeks.

Equally important, the pseudoglandular period is marked by development of the pulmonary vasculature. Branching of the pulmonary arteries follows branching of bronchi and bronchioli up to the acinar level, and for each airway branch the pulmonary artery branches are two or three times more numerous. Pulmonary venous development, which are as numerous as pulmonary artery divisions, occurs in the connective tissue septa between airway branches. The second vascular supply to the lung, the bronchial

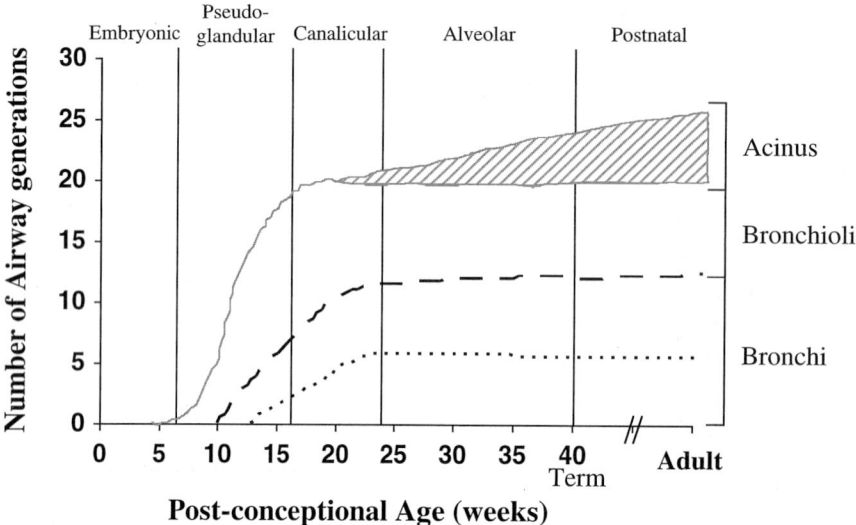

FIGURE 2 Preacinar versus acinar development. Preacinar airways include the trachea and segmental and subsegmental bronchi up to and including the terminal bronchiole. The acinus includes all airspaces derived from a single respiratory bronchiole. Preacinar airway development is completed during the canalicular stage of lung development and acinar development (shaded area) extends into childhood (adapted from Bucher and Reid, 1961).

circulation, arises first from the dorsal aorta in close proximity to the celiac axis (6 weeks) and is then replaced by definitive bronchial arteries arising from the aorta by 12 weeks. Persistence of rudimentary bronchial arteries is associated with developmental anomalies such as pulmonary sequestration.

Control of branching morphogenesis is a rapidly developing field of lung biology. Branching morphogenesis requires extensive communication between the epithelial cells lining the tubular lung bud and the invested mesenchyme. In studies with cultured embryonic lung, for example, removal of the mesenchyme or substitution of nonpulmonary mesenchyme prevent normal airway branching. The current theory of branching involves growth arrest of the epithelial cells at the branch point with continued or accelerated growth of epithelial cells lateral to the branch point. This is governed by gain and/or loss of expression of transcription factors such as thyroid transcription factor-1 and sonic hedgehog which control the expression of other genes in a complex array of both positive and negative feedback loops. Components of extracellular matrix are vital to branching morphogenesis and microvascular development because they influence cell adhesion, cell shape, and cytoskeletal organization; for example,

interference in the assembly of fibronectin matrix, which is thought to provide positional information for cell migrations, has been shown to inhibit branching in embryonic lungs.

The regulatory signals involved in branching morphogenesis are being explored in studies with cultured embryonic lung and *in vivo* after gene ablation. It is obvious from studies using lung organ culture, which is carried out in the absence of exogenous hormones and with low concentrations of serum, that circulating hormones are not essential for airway branching and lung growth. The physiological role of endogenous corticosteroids, thyroid hormones, and potentially other circulating hormones is limited to modulation of the rate of differentiation of both mesenchymal and epithelial cells after completion of airway branching. This is well illustrated for glucocorticoids by the arrest of lung structural development at about Fetal Day 16, after branching morphogenesis is complete, in transgenic mice with knockout of either corticotropin-releasing hormone or glucocorticoid receptor. Thus, signaling during formation of the lung must reside in endogenously produced factors which act in autocrine and paracrine fashion. A number of growth factors and cytokines have been studied as potential regulators of lung morphogenesis, and experimental evidence supports a role for epidermal growth factor, keratinocyte growth factor, vascular endothelial growth factor, retinoic acid, and possibly members of the insulin-like growth factor family.

Errors in the developmental program during the embryonic and/or pseudoglandular stages may result in a form of pulmonary hypoplasia that is characterized by decreased numbers of bronchial segments that contain an appropriate number of alveoli per bronchiole as indicated by radial alveolar counting. Closure of the pleuroperitoneal canals is another important event during the pseudoglandular stage. Lung hypoplasia is associated with congenital diaphragmatic hernia because branching and/or growth is inhibited by large amounts of abdominal viscera within the thoracic cavity.

III. CANALICULAR STAGE OF LUNG DEVELOPMENT

Development of the pulmonary acinus begins in the canalicular stage (16–24 weeks gestation). The acinus is the gas exchange unit of the lung and encompasses a respiratory bronchiole and all of its associated alveolar ducts and alveoli. A terminal bronchiole with all its associated acinar structures constitutes a lobule. Ultimately, the zone of gas exchange will attain a surface area of 50–100 M^2 and will be 2.5–3.0 liters in volume in the adult human. Branching of the most distal airways continues on a more limited basis while evolution of the relationships between the airspaces, capillaries, and mesenchyme takes on more significance. During Months 4–6 of gestation the epithelial cells lining the acini begin to differentiate further. The cuboidal epithelial cells accumulate large glycogen stores and develop small vesicles containing loose lamellae. In cells destined to become type 2 cells, lamellar bodies become larger, more numerous, and more densely packed with surfactant phospholipids and proteins, whereas those cells destined to become type 1 cells, upon losing their relationship to mesenchymal fibroblasts, lose the prelamellar vesicles and thin progressively, thereby adopting a phenotype more suitable to gas exchange. Type 1 and 2 cells are clearly evident by the end of

this stage. Table 1 lists common markers of differentiated lung epithelial cells including bronchiolar Clara and mucociliary cells as well as alveolar type 1 and 2 cells.

The basement membrane and interstitial extracellular matrix appear to play important roles in the differentiation of type 1 and 2 cells. The alveolar basement membrane includes laminin, entactin, type IV and V collagen, chondroitin sulfate and chondroitin sulfate proteoglycan, heparan sulfate proteoglycans, and possibly fibronectin. The local concentration of these components may regulate the location of type 2 cells (at the corners of alveoli) and relative ratio of type 1 and 2 alveolar cells. There is evidence that the type 2 cell basement membrane is discontinuous, allowing maintenance of direct cell–cell interactions with cells of the interstitium, specifically fi-

TABLE 1

Common Markers of Differentiated Lung Epithelial Cells

Type I cell markers	Clara cell markers
Cell membrane	Cell membrane
Ricinus comunis lectin	Nonciliated
Type I cell protein-α	Apocrine secretion
(T1-α)	Intracellular
Intercellular adhesion	Minimal glycogen
molecule-1 (ICAM-1)	Cytochrome P450
Carboxypeptidase-M	monooxygenases
Type II cell markers	Clara cell-specific protein (CC10, CCSP)
Cell membrane	Surfactant proteins A
Maclura pomifera lectin	and B
Pneumocin	Mucociliary cell markers
Glycoprotein 330	Plasticity of phenotype
Na,K ATPase (basolateral)	Basal, mucous-secreting, and ciliated
Intracellular	cells from same precursor cell
Surfactant proteins A–D	Cell membrane
Actin microfilaments	Ciliated vs nonciliated
Cytokeratin type 19	Intracellular
Enzymes (alkaline phosphatase, P450b, γ-glutamyl transferase, aminopeptidase N)	Substantial glycogen
	Viscoelastic mucous secretion
	Transforming growth factor-α and -β secretion

broblasts. In addition, *in vitro* studies have shown that extracellular matrix can have dramatic effects on maintenance of type 2 cell phenotype in isolated cell culture. Laminin and type 1 collagen may be important in maintaining the cuboidal appearance of type 2 cells in culture but only complex matrices have been shown to promote the synthesis and secretion of surfactant components in type 2 cell cultures. The developmental program governing extracellular matrix composition during alveolar epithelial cell differentiation is likely recapitulated after epithelial damage by airborne or blood-borne toxic agents. Since type 2 cells are the source of type 1 cells after injury, alterations in extracellular matrix composition are an essential component to reconstitute a functional, differentiated alveolar lining. Moreover, aberrant signaling during the repair process may lead to abnormalities in extracellular matrix deposition and reconstitution of the alveolar lining, thereby contributing to the pathobiology of lung diseases such as bronchopulmonary dysplasia.

The capillary network investing the acinus expands to match the increasing surface area of the developing acinar structures. Preacinar and resistance arteries which are present by 28 weeks of gestation develop through the process of angiogenesis whereby new vascular structures arise from existing vessels which send out endothelial cells that organize to form a central lumen, thereby extending the length of the existing vessel. The capillary network investing the developing acini develop by a process of vasculogenesis. Angioblasts within the mesenchyme surrounding the acini differentiate into endothelial cells, proliferate, and organize into clusters which send out endothelial cells that eventually interconnect and establish a lumen. This network then connects to the developing pulmonary arteries and veins. Alveolar capillary dysplasia, an unusual and fatal cause of respiratory failure in the term neonate, is the result of developmental errors preventing the appropriate interconnection of these vascular networks. During the canalicular stage, there is thinning of the interstitial mesenchyme such that by 28 weeks the alveolar–capillary membrane is similar in thickness to that in the adult (0.6 μm).

Muscularization of the preacinar and resistance arteries extends through the canalicular stage to term. The developing endothelial cells recruit smooth muscle and pericyte precursors, which are important in maintaining vascular integrity, through the action of growth factors such as angiopoietin-1. Muscularization is evident through the pulmonary arteries up to the level of the terminal bronchiole and decreases significantly in the vessels supplying respiratory bronchioles. Abnormal extension of smooth muscle along arterioles supplying acinar structures has been demonstrated in infants dying from persistent pulmonary hypertension of the newborn and is associated with severe bronchopulmonary dysplasia.

IV. SACCULAR STAGE OF LUNG DEVELOPMENT

Maturation of the pulmonary acinus is the hallmark of the saccular stage of pulmonary development (24 weeks to term). Lengthening and widening of the terminal sacs expands the surface area available for gas exchange. Each saccule will give rise to two or three alveolar ducts, further expanding the available surface area. The expansion of surface area and lumenal volume acts to compress the interstitium, bringing the capillary networks in close proximity to potential air spaces, thereby promoting alveolar development and later capillary bed fusion. At this stage there is a complete capillary network available for each acinus, effectively creating a double circulation between acini. Capillary networks investing separate acini will fuse as the interstitium separating alveoli thins toward birth, a process that is not complete until 18 months postnatal age.

Production of elastic fibers, composed of elastin and microfibrillar proteins, begins at midgestation and extends through the neonatal period. Elastin is a polymer of cross-linked tropoelastin precursor protein and constitutes ~90% of the mass of the fibers. A variety of glycoproteins comprise the microfibrils which serve as a scaffold for elastin deposition. Elastin is localized to the walls of alveoli and pulmonary vessels where its physical properties allow for repetitive deformations in the shape and size of these structures. Excess elastin or diminished content of elastin in diseases such as pulmonary hypertension and fibrotic conditions can severely compromise the strength and formation of lung vessels and alveoli.

Type 1 and 2 cells are readily identified early in the saccular stage (Fig. 3). Induction of the components of pulmonary surfactant results in the increased synthesis and storage of surfactant in the increasing numbers of lamellar bodies within type 2 cells. Secretion of sufficient amounts of surfactant to allow detection in amniotic fluid samples is evident by ~34 weeks gestation. During this stage the acini and the alveolocapillary membrane are sufficiently developed to support gas exchange but the limiting factor for successful air breathing becomes the maturation of the surfactant system. In the absence of adequate amounts of mature pulmonary surfactant, preterm infants born during this phase of lung maturation develop progressive atelectasis and respiratory distress syndrome. The rate of type 2 cell differentiation, and secondarily surfactant production by the fetal lung, is modulated by levels of endogenous corticosteroids and is accelerated by administration of glucocorticoid which is used clinically to prevent RDS in premature infants (Fig. 4). The response of the surfactant system to glucocorticoid involves all the lipid and protein components of surfactant and occurs primarily through increased gene expression, thus representing precocious maturation mimicking the normal developmental pattern and processes. Glucocorticoids likely also enhance maturation of other cell types as well as the extracellular matrix but these effects are less well characterized. Endogenous thyroid hormones, prolactin, prostaglandins, and catecholamines may also have stimulatory effects on type 2 cell maturation and clearance of lung fluid at birth. Certain cytokines (e.g., tumor necrosis factor-α and transforming growth factor-β) inhibit surfactant production in experimental systems and may downregulate surfactant in conditions such as sepsis and inflammation.

V. ALVEOLAR STAGE OF LUNG DEVELOPMENT

The alveolar stage of lung development begins at ~36 weeks and extends postnatally. Primitive saccules develop low ridges that subdivide the saccule

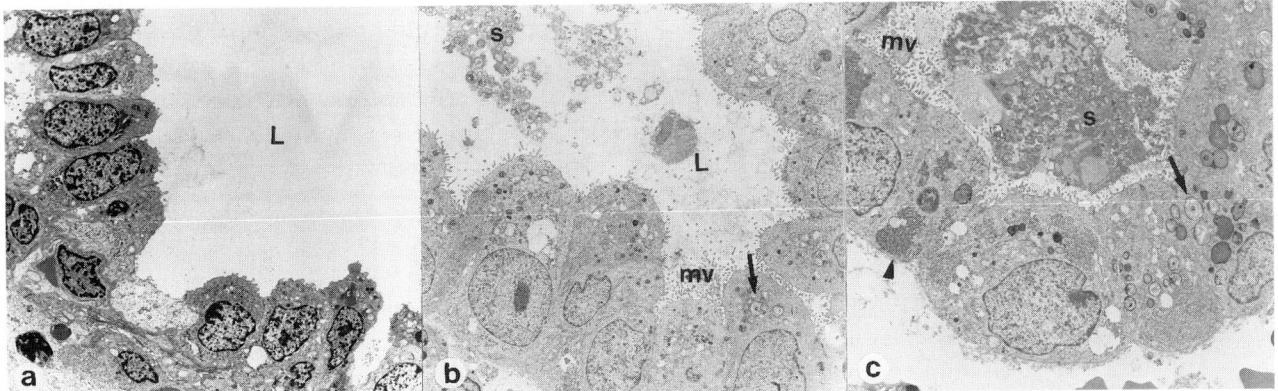

FIGURE 3 Transmission electron micrographs of human fetal lung epithelial cell differentiation *in vitro*. Explants were cultured for 6 days in the absence [control before (a) and after (b) culture] or the presence of dexamethasone-supplemented media (c). Small and dense, medium, and large lamellar bodies (LB) in control tissues are easily identified. Increased numbers and size of lamellar bodies as well as increased numbers of microvilli were noted after dexamethasone treatment. Cell shape and the presence or absence of glycogen zones indicate maturity of the type 2 cells. Maturation *in vitro* follows the events of type 2 cell maturation *in vivo* yet on a significantly compressed time frame. Arrow indicates LB; s, cellular debris and secreted material; arrowhead indicates glycogen; L, lumen; mv, microvilli (adapted from Chinoy *et al.*, *Am. J. Respir. Cell. Mol. Biol.*, 1995).

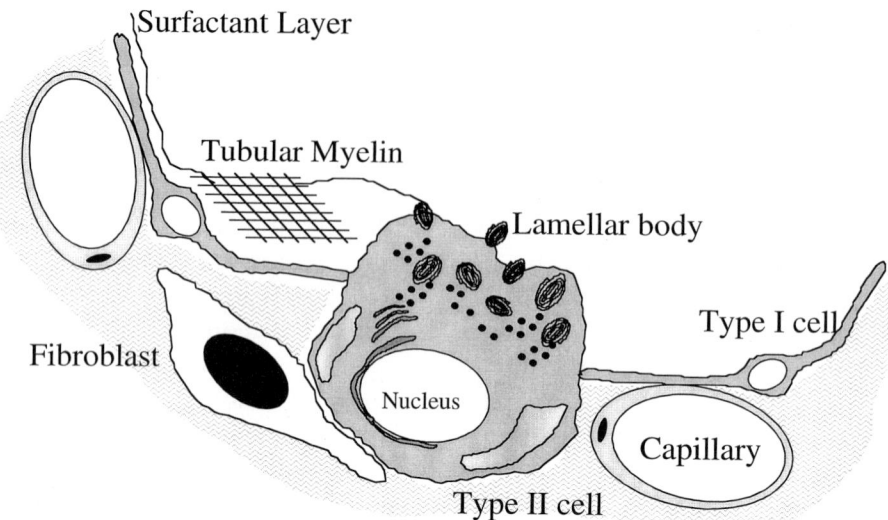

FIGURE 4 Idealized drawing of the mature alveolar epithelium. Lamellar bodies from type 2 cells are secreted into the alveolar space where pulmonary surfactant transitions through tubular myelin and then to the surfactant monolayer. Type 1 cells lie in close proximity to alveolar capilaries within the interstitium, thereby facilitating gas exchange.

into an alveolar duct (primary septa) and outpouchings between the ridges that will become alveoli (secondary septa). Regions destined for this secondary septation exhibit increased elastin deposition. Septa contain a connective tissue core separating two capillary membranes, suggesting that the septum is formed by the folding of the capillary on itself, which in general occurs in regions where the double capillary network is preserved between acini, or alternatively at pleural surfaces or adjacent to large blood vessels and conducting airways. In theory, septation limits the ability of the lung to repair damaged alveoli after the developmental processes of alveolarization and capillary network fusion are complete. Septation also leads to the development of the pores of Kohn, allowing gaseous continuity between acini. By term, alveolar type 1 and 2 cells are fully differentiated and there are sufficient levels of alveolar surfactant for normal respiration.

There is controversy regarding the completion of alveolarization. Current estimates suggest that the formation of new alveoli is complete by ~2 or 3 years postnatally, achieving a total of between 200 and 600 million. However, lung growth is not complete at this stage. Between 4 years of age and adulthood, there is extensive linear growth which includes expansion of the thoracic cavity. Increased intrathoracic volume increases the negative pressure generated by diaphragmatic excursions, thereby allowing three-dimensional stretch of alveoli which increases alveolar volume (Fig. 5). Lung volume increases 23-fold between birth and adulthood. The capillary bed expands at a greater pace, achieving a 34-fold expansion over the same period. Thus, compared to adult lung, the newborn lung has relatively less volume per kilogram of body weight, increased interstitial components relative to airspace, and a smaller capillary bed.

The physiological factors regulating growth of the lung during both fetal and postnatal life are not well understood. Circulating hormones of fetal origin are not essential since lung weight at term birth is not affected by fetal hypophysectomy or thyroidectomy. Treatment of animals with corticosteroids, depending on the dose, can reduce lung weight through direct effects on cell replication (e.g., in newborn rats) and indirectly in the fetus by accelerating placental senescence and impairing function. These corticosteroid effects are also observed in various other tissues and are reversible with catch-up growth. Lung

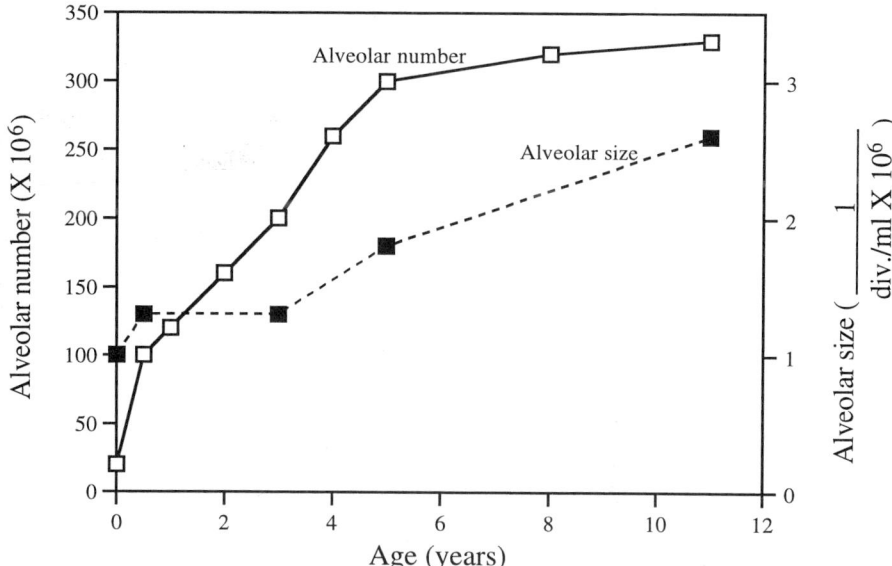

FIGURE 5 Alveolar development in human lung. Alveolar number increases from birth through approximately 8 years of age, whereas alveolar size does not increase significantly until after 4 years of age [adapted from P. Davies *et al.*, Structural methods in the study of development of the lung, In *Models of Lung Disease: Microscopy and Structural Methods* (J. Gil, Ed.), Lung Biology in Health and Disease, Vol. 47, pp. 409–472, Dekker, New York, and M. S. Dunnill, Quantitative methods in the study of pulmonary pathology, *Thorax* **17**, 320–328, 1962].

growth is stimulated after partial pneumonectomy but there is little information on mediating agents and mechanism of action.

The role of physical factors in lung size is well established: Normal growth requires adequate space in the chest cavity and appropriate distending forces. In fetal animals, lung growth is reduced by maneuvers such as cutting of the spinal cord (deenervating the diaphragm); removing a portion of the rib cage, thus reducing chest wall rigidity; chronic drainage of amniotic fluid; and tracheostomy, which allows free outflow of lung fluid. In contrast, tracheal ligation, which increases intraluminal pressure, increases lung dry weight. These and other experimental observations have led to the proposal that fetal breathing movements retard loss of lung liquid and maintain lung expansion in the face of decreased upper airway resistance, whereas during periods of nonbreathing there is tonic inflation secondary to laryngeal abduction. Both of these stretching events may stimulate production of growth-mediating factors such as IGF-II and platelet-derived growth factor

B by lung fibroblasts. The transduction mechanisms for increasing growth factors and subsequent stimulation of mitogenesis are largely unexplored.

VI. SUMMARY

Lung development follows a basic schema of airway branching, involving mesenchymal–epithelial interaction and a complex interplay of growth acceleration and growth arrest, followed temporally by maturation of cell types after the wave of branching passes more distal points in the lung. It remains unclear how these processes relate to each other and whether both completion of branching and continued growth are required for cellular differentiation. There is accumulating evidence that a complex array of growth factors and transcription factors present in epithelial and mesenchymal cells play key roles in these processes, and that effects of modulating levels of these factors can vary significantly depending on the developmental state of the cells.

See Also the Following Articles

Bibliography

Ballard, P. L. (1986). Hormones and lung maturation. In *Monographs on Endocrinology* (J. D. Baxter and G. G. Rousseau, Eds.), Vol. 28, pp. 24–341. Springer-Verlag, New York.

Brody, J. S., and Williams, M. C. (1992). Pulmonary alveolar cell differentiation. *Annu. Rev. Physiol.* 54, 351–371.

Bucher, U., and Reid, L. (1961). Development of the intrasegmental bronchial tree: The pattern of branching and development of cartilage at various stages of intra-uterine life. *Thorax* 16, 207–218.

Burri, P. (1997). Structural aspects of prenatal and postnatal development and growth of the lung. In *Lung Growth and Development* (J. McDonald, Ed.), Vol. 100, 1–36. Dekker, New York.

Cardoso, W. V. (1995). Transcription factors and pattern formation in the developing lung. *Am. J. Physiol.* 269(*Lung Cell. Mol. Physiol.* 13), L429–L442.

deMello, D. E., and Reid, L. M. (1995). Respiratory tract and lungs. In *Diseases of the Fetus and Newborn* (G. B. Reed, A. E. Claireaux, and F. Cockburn, Eds.), pp. 523–560. Chapman & Hall, London.

Dunsmore, S. E., and Rannels, D. E. (1996). Extracellular matrix biology in the lung. *Am. J. Physiol.* 270(*Lung Cell. Mol. Physiol.* 14), L3–L27.

Folkman, J., and D'Amore, P. A. (1996). Blood vessel formation: What is its molecular basis? *Cell* 87, 1153–1155.

Hogan, B. L. M. (1995). The TGFb-related signalling system in mouse development. *Sem. Dev. Biol.* 6, 257–265.

Kuroki, Y., and Voelker, D. R. (1994). Pulmonary surfactant proteins. *J. Biol. Chem.* 269(42), 25943–25946.

Mallampalli, R. K., Acarregui, M. J., *et al.* (1997). Differentiation of the alveolar epithelium in the fetal lung. In *Lung Growth and Development* (J. A. McDonald, Ed.), Vol. 100, pp. 119–149. Dekker, New York.

Mensah, E. A., Kumar, N. M., *et al.* (1996). Distribution of alveolar type II cells in neonatal and adult rat lung revealed by RT-PCR in situ. *Am. J. Physiol.* 271(*Lung Cell. Mol. Physiol.*, 15), L178–L185.

Plopper, C. G. (1997). Clara cells. In *Lung Growth and Development* (J. A. McDonald, Ed.), Vol. 100, pp. 181–210. Dekker, New York.

Roman, J. (1997). Cell–cell and cell–matrix interactions in development of the lung vasculature. In *Lung Growth and Development* (J. McDonald, Ed.), Vol. 100, pp. 365–400. Dekker, New York.

Fetal Membranes

Erdal Budak and Jerome F. Strauss, III

University of Pennsylvania Medical Center

I. Formation of the Fetal Membranes
II. Structure of the Fetal Membranes
III. Secretory and Metabolic Functions of the Fetal Membranes
IV. Rupture of the Fetal Membranes at Term
V. Premature Rupture of the Fetal Membranes

GLOSSARY

amnion The avascular membrane surrounding the fetus composed of a layer of amniotic epithelial cells resting on a basement membrane which overlies a dense collagen-rich matrix and fibroblasts. The human amnion is adherent to the chorion.

chorion laeve The chorion consisting of trophoblast cells embedded in a collagen matrix and adherent maternal decidua cells.

decidua The endometrium of pregnancy consisting of differentiated stromal cells embedded in a matrix rich in laminin.

Membranes envelope the amniotic cavity, which contains the fetus and amniotic fluid. In humans, these membranes consist of the amnion and the chorion. In lower species, there is an analogous set of membranes that encompass the developing embryo. The fetal membranes retain and absorb amniotic fluid, contribute to the formation of amniotic fluid early in pregnancy, secrete substances into the amniotic fluid as well as toward the uterus, and protect the fetus from infectious agents that can ascend up through the reproductive tract. These membranes normally break during labor, allowing the fetus to escape into the extrauterine world. However, if the membranes are breached prematurely, the fetal living space is compressed due to leakage of amniotic fluid, predisposing to abnormal fetal development, and there is a serious risk that the fetus and uterus can become infected, threatening both the fetus and the mother.

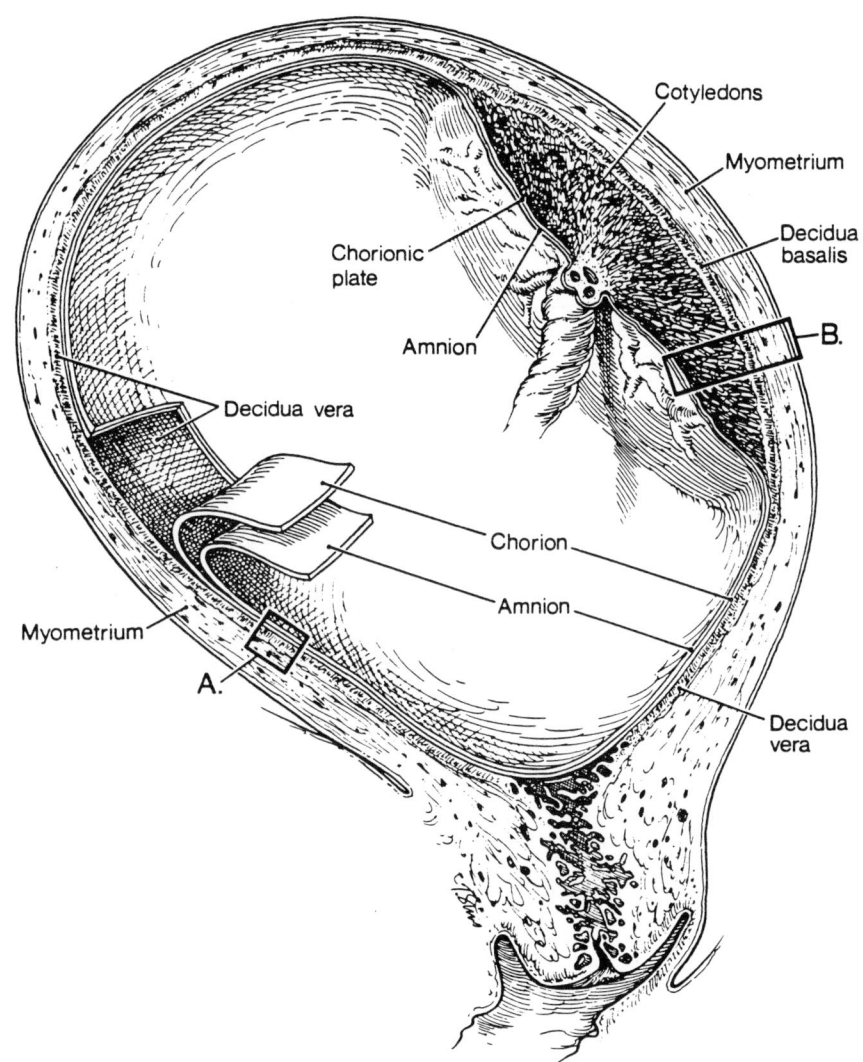

FIGURE 1 The human uterus, fetal membranes, and placenta at term [reprinted with permission from *Williams Obstetrics* (F. G. Cunningham, P. C. MacDonald, and N. F. Gant, Eds.), 18th ed., Appleton & Lange, Norwalk, CT, 1989].

I. FORMATION OF THE FETAL MEMBRANES

The amnion and amniotic cavity begin as a small cavity that forms near the embryonic pole between the embryonic disc and the trophoblast layer. As this cavity enlarges, it is surrounded and delimited by amnioblasts, specialized cells derived from the ectoderm of the embryonic disc. These amnioblasts organize to form the amnion (Fig. 1).

The chorion is formed from the trophectoderm and extraembryonic mesoderm. It surrounds the amniotic sac, the embryonic structures, and yolk sac. As the embryo grows and protrudes into the uterine cavity, the amniotic cavity and amnion enlarge but the overlying chorionic sac does not expand or grow at the same rate. Eventually, the chorionic cavity is obliterated so that the amnion and chorion come into close apposition around the 13th week of gestation.

The increase in amniotic fluid and the intraamniotic pressure compress the decidua capsularis and underlying chorionic villi leading to atrophic degeneration of these structures. The chorionic villi associated with the decidua capsularis become a thin layer, forming a structure called the chorion laeve. As pregnancy progresses, the chorion laeve becomes fixed to the uterine wall decidua.

In early development, the amnion is attached to the margins of the embryonic disc and its junction is located on the front or ventral surface of the embryo. As the embryo folds forward and the umbilical cord forms, the amnion entraps the entire embryo and also covers the umbilical cord. Ultimately, after the placenta develops and the chorionic cavity is obliterated, the amnion covers the umbilical cord and fetal surface of the placenta, and then extends outward from the margins of the placenta to attach and cover the inner surface of the chorion.

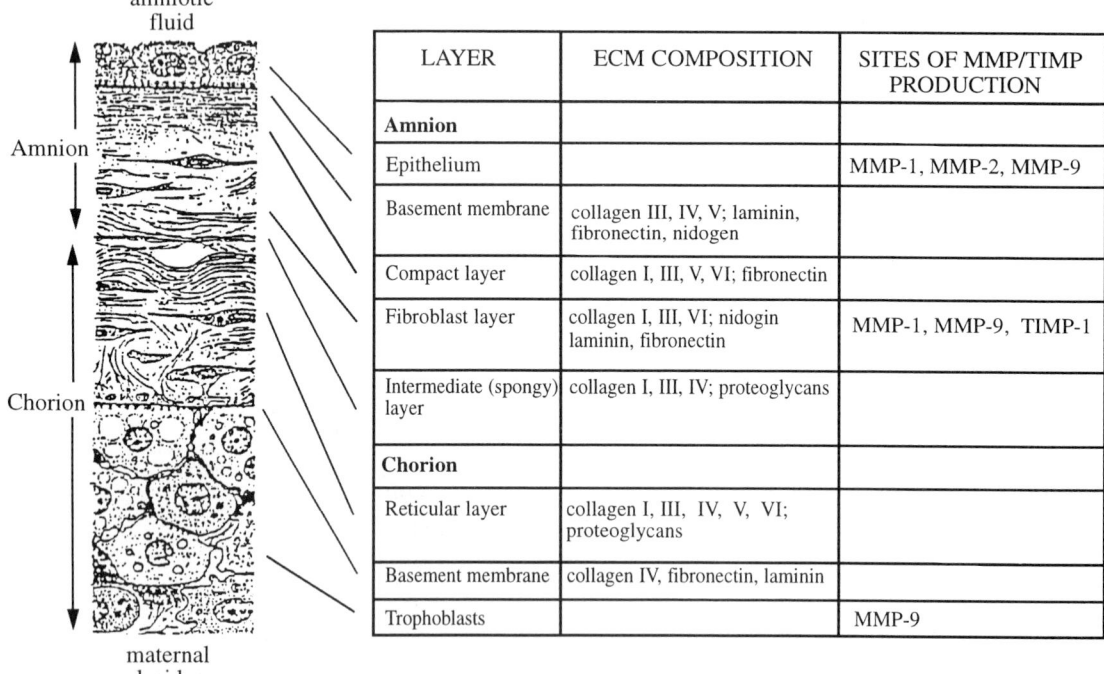

LAYER	ECM COMPOSITION	SITES OF MMP/TIMP PRODUCTION
Amnion		
Epithelium		MMP-1, MMP-2, MMP-9
Basement membrane	collagen III, IV, V; laminin, fibronectin, nidogen	
Compact layer	collagen I, III, V, VI; fibronectin	
Fibroblast layer	collagen I, III, VI; nidogin laminin, fibronectin	MMP-1, MMP-9, TIMP-1
Intermediate (spongy) layer	collagen I, III, IV; proteoglycans	
Chorion		
Reticular layer	collagen I, III, IV, V, VI; proteoglycans	
Basement membrane	collagen IV, fibronectin, laminin	
Trophoblasts		MMP-9

FIGURE 2 Schematic representation of the structure of the human fetal membranes at term. The primary extracellular matrix (ECM) components of each layer are identified and the sites of production of matrix metalloproteinases (MMPs) and endogenous MMP inhibitors, tissue inhibitors of metalloproteinases (TIMPs) are noted. The MMPs promote fetal membrane rupture by degrading the ECM.

II. STRUCTURE OF THE FETAL MEMBRANES

The human amnion contains five distinct layers (Fig. 2). It has no blood vessels or nerves and acquires oxygen and nutrients from the amniotic fluid. The innermost layer of the human amnion, facing the fetus, is a cuboidal amnion epithelium, which rests on a basement membrane containing collagen and various noncollagenous proteins and glycoproteins. Beneath the basement membrane is a compact layer of connective tissue, which is the main fibrous skeleton of the amnion. This compact layer contains fibrils of type I and type III collagens which impart tensile strength to the amnion.

Beneath the compact layer is the fibroblast layer, which is the thickest layer of the amnion and contains fibroblasts that produce the collagens of the compact zone and occasional macrophages. An intermediate or spongy layer lies between the amnion and chorion. It is composed mainly of hydrated proteoglycans and glycoproteins that give the "spongy" appearance to this layer in histological sections. The spongy layer also contains a nonfibrillar meshwork of collagen. It absorbs physical stresses by permitting the amnion to slide over the underlying chorion, which is anchored to the maternal decidua.

The chorion is thicker than the amnion, but has less tensile strength than the amnion. It has a polarity with its epithelial components (trophoblast cells) facing the maternal decidua. Beneath the trophoblast cells is a basement membrane and collagen fibrils.

The human fetal membranes show regional variation that distinguish the membranes overlying the placenta from the so-called reflected membranes that surround the fetus.

III. SECRETORY AND METABOLIC FUNCTIONS OF THE FETAL MEMBRANES

In addition to retaining the amniotic fluid and serving as a protective barrier, the fetal membranes secrete a number of protein hormones and prosta-glandins. They also actively metabolize steroid hormones and uterotonic substances. These secretory and metabolic activities of the membranes have been implicated in the maintenance of myometrial quiescence as well as in the process of labor.

Chorion oxytocinase and enkephalinase degrade the uterotonic peptides oxytocin and endothelin-1. Chorionic 15-hydroxyprostaglandin dehydrogenase catabolizes active prostanoids; platelet-activating factor 2-acylhydrolase inactivates platelet-activating factor, a potent stimulator of smooth muscle contraction. These metabolic activities prevent substances that promote contraction from acting on the myometrium. Although these catabolic activities do not decline during normal labor, reductions in their activities may promote preterm delivery. Thus, lower chorion laeve 15-hydroxyprostaglandin dehydrogenase activity is associated with preterm birth.

The fetal membranes metabolize steroid hormones. At term, the human chorion converts weaker estrogens to estradiol and catabolizes progesterone into 20α-hydroxyprogesterone, a biologically inert progestin. This effectively increases the estrogen/progesterone ratio, favoring myometrial activation.

The amnion is the primary site of fetal membrane prostaglandin synthesis and labor is accompanied by an increase in cyclooxyganse-2, the key enzyme involved in prostaglandin synthesis, in amnion epithelial cells. Although it has been postulated by some authors that prostaglandins produced by the amnion at term stimulate uterine contraction, a direct role for amnion-derived prostanoids in myometrial activation is questionable. However, these prostaglandins may stimulate matrix metalloproteinase activity in the fetal membranes and cervix, promoting degradation of extracellular matrix proteins (e.g., collagen) and thus facilitating fetal membrane rupture and cervical effacement and dilatation.

The protein products of the fetal membranes include corticotrophin-releasing factor, which may be involved in stimulating fetal membrane prostaglandin production; parathyroid hormone-related peptide, which could be involved in uterine smooth muscle relaxation and calcium homeostasis; and neuropeptide Y, which may participate in the regulation of uterine contractility.

IV. RUPTURE OF THE FETAL MEMBRANES AT TERM

The human fetal membranes generally rupture during labor. This event usually occurs spontaneously, causing a gush of fluid to be released through the cervix and vagina (breaking of the bag of waters). Strong and regular uterine contractions cause effacement and dilatation of the cervical orifice. The intrauterine pressure is lowest at this point, forcing the fetal membranes into the cervical canal, further accelerating cervical dilatation as a hydrostatic wedge. When the intraamniotic pressure exceeds the tensile strength of the membranes, they rupture. This usually happens when cervical dilatation has reached 4 or 5 cm.

Obstetricians may mechanically rupture the membranes if they do not break spontaneously, a procedure called amniotomy. This speeds labor and delivery by allowing the fetal head to come into direct contact with the cervix, facilitating further cervical effacement and dilatation.

The spontaneous rupture of the membranes at term probably results from biochemical changes in the structure of the membranes as well as the physical forces of uterine contraction. The biochemical changes include the increased production of collagenases (matrix metalloproteinases) that break down the amnion and chorion basement membrane (matrix metalloproteinase-9), causing amnion epithelial cells and chorionic trophoblast to lose there attachments to their basement membranes and subsequently undergo a programmed cell death or apoptosis, and degradation of the fibrillar collagens that give tensile strength to the membranes (interstitial collagenases). These structural changes probably occur in localized regions of the fetal membranes and are associated with loss of collagen and amnion and chorion cell death. The factors that increase the production of collagen-degrading enzymes right before or during labor are not well understand but may include various cytokines (tumor necrosis factor-α and interleukin-1) and prostaglandin E. Stretching of the membranes as a result of fetal growth and amniotic fluid accumulation and uterine contractions may induce the production of fetal membrane cytokines (interleukin-8), which in turn increases the production of cytokines that increase collagenase production by cells in the amnion and chorion.

V. PREMATURE RUPTURE OF THE FETAL MEMBRANES

Premature rupture of the membranes can occur at term before the onset of labor or preterm, which is considered membrane rupture before 37 weeks of completed gestation. Approximately 8–10% of pregnant women may have premature rupture of membranes at term, which increases the risk of intrauterine infection if the interval between membrane rupture and delivery is prolonged. One percent of women experience preterm premature rupture of membranes, which is a more serious complication of pregnancy. It is the main reason why women deliver premature infants.

Premature membrane rupture is usually announced by a sudden leakage of fluid from the vagina. The diagnosis can be made by testing the vaginal fluid pH. Vaginal secretions are usually acidic (pH 4.5–5.5), whereas amniotic fluid has a pH of 7.0–7.5. This is usually done with a paper impregnated with the pH-sensitive dye, nitrazine. In the presence of amniotic fluid with a pH of 7.0–7.5, the paper turns blue or blue-green. When dried on a slide, amniotic fluid forms a fern-like pattern as a result of salt and protein condensation. Significant loss of amniotic fluid is seen on ultrasound examination.

The causes of premature membrane rupture include defects in collagen synthesis which leave the membranes structurally weak. Premature fetal membrane rupture occurs when the fetus is affected with Ehlers–Danlos syndrome, a group of 11 genetic disorders in collagen synthesis. Nutritional deficiencies that affect collagen synthesis, including copper, which is required for the activity of the collagen cross-linking enzyme, lysyl oxidase, and ascorbic acid, required for the formation of the fibrillar collagen triple helix, may cause the development of weaker membranes. Cigarette smoking may predispose to premature membrane rupture by causing the induction of copper-binding proteins in the fetal membranes. Other factors that may contribute to premature membrane rupture include vaginal or uterine infection, which may increase local cytokine release and thus trigger production of collagenases that degrade fetal membrane collagens. Cortisol, which may increase in response to stress, inhibits the synthesis of fetal membrane proteins and could also promote weakening of the membranes.

See Also the Following Articles

Amniotic Fluid; Decidua; Fetal Anomalies; Labor and Delivery, Human; Preterm Labor and Delivery

Bibliography

Curtis, N. E., Ho, P. W., King, R. G., Farrugia, W., Moses, E. K., Gillespie, M. T., Moseley, J. M., Rice, G. E., and Wlodek, M. E. (1997). The expression of parathyroid hormone-related protein mRNA and immunoreactive protein in human amnion and choriodecidua at term compared with preterm gestation. *J. Endocrinol.* **154**, 103–112.

Germain, A. M., Smith, J., Casey, M. L., and MacDonald, P. C. (1994). Human fetal membrane contribution to the prevention of parturition: Uterotonin degradation. *J. Clin. Endocrinol. Metab.* **78**, 463–470.

Mitchell, B. F., and Wong, S. (1993). Changes in $17\beta,20\alpha$-hydroxsteroid dehydrogenase activity supporting an increase in the estrogen/progesterone ratio of human fetal membranes at parturition. *Am. J. Obstet. Gynecol.* **168**, 1377–1385.

Parry, S., and Strauss, J. F., III (1997). Premature rupture of the fetal membranes. *N. Engl. J. Med.* **338**, 663–670.

Polzin, W. J., and Brady, K. (1991). Mechanical factors in the etiology of premature rupture of the membranes. *Clin. Obstet. Gynecol.* **34**, 702–714.

Schmidt, W. (1992). The amniotic fluid compartment: The fetal habitat. *Adv. Anat. Embryol. Cell Biol.* **124**, 1–100.

Skinner, K. A., and Challis, J. R. (1985). Changes in the synthesis and metabolism of prostaglandins by human fetal membranes and decidua at labor. *Am. J. Obstet. Gynecol.* **151**, 519–523.

Warren, W. B., and Silverman, A. J. (1995). Cellular localization of corticotrophin releasing hormone in the placenta, fetal membranes and decidua. *Placenta* **16**, 147–156.

Fetal Monitoring and Testing

Iraj Forouzan

University of Pennsylvania

GLOSSARY

acidemia An increased concentration of hydrogen ions in blood.

acidosis An increased concentration of hydrogen ions in tissue.

Apgar scores A scoring system applied at 1 min and again at 5 min after birth to assess the condition of newborns. Parameters used are heart rate, respiratory effort, muscle tone, reflex irritability, and color.

asphyxia A decreased level of oxygen in tissue with metabolic acidosis.

false-negative rate The probability of a test being negative (normal) among diseased population.

false-positive rate The probability of a test being positive (abnormal) among nondiseased population.

hypoxemia Decreased oxygen content in blood.

hypoxia A decreased level of oxygen in tissue.

oligohydramnios A diminished volume of amniotic fluid.

oxytocin A nine-amino acid peptide secreted from posterior pituitary which can cause uterine contractions.

Perinatal mortality rate, defined as the number of fetal deaths plus neonatal deaths per 1000 total births, has declined steadily since 1965. The decline in perinatal mortality is mostly attributable to improved neonatal survival of low-birth-weight infants and improved perinatal care. Today, Intrapartum death is an unusual occurrence. Although not definitely proven, intrapartum fetal heart rate monitoring and antenatal testing may have contributed to the improvement in perinatal survival along with other changes in the perinatal obstetrics.

I. INTRAPARTUM FETAL MONITORING

The first description of the fetal heart in the literature appeared during the seventeenth century. In those early days physicians placed their ear on the maternal abdomen to hear fetal heart tone. In the beginning of the twentieth century stethoscopes were designed to listen to the fetal heart tones. Continuous recording of the fetal heart became possible in 1958. Empirical data accumulated since then by many observers throughout the world convinced clinicians that certain fetal heart rate patterns can predict poor perinatal outcome. The first commercially available continuous electronic fetal heart rate monitor was produced in 1968. Advances in technology since then have made fetal monitoring equipment more precise in recording real beat-to-beat variability in heart rate. Clinical data collected in the mid- and late 1970s suggested that electronic fetal heart rate monitoring reduces the intrapartum stillbirth rate and perinatal mortality. Certain fetal heart rate patterns were found to be associated with low Apgar scores and presumably fetal hypoxia. Some clinicians and nonobstetric professionals tried to associate some of these fetal heart rate patterns with outcomes such as infantile spastic palsies. This association, believed to exist by some, is not established by scientific data. Fetal hypoxia has not been proven to increase brain injury unless associated with severe asphyxia and metabolic acidosis. Even then, this intrapartum complication probably is a contributory factor in 15% of spastic palsies. Therefore, the value of continuous fetal electronic monitoring has recently come under question. During a normal labor transient fetal hypoxemia and even hypoxia frequently occurs; but the fetus generally tolerates these changes. Fetal acidemia occurs in all labors. This should not be confused with asphyxia, which is a severe form of hypoxia with metabolic acidosis.

The fetal heart rate at term ranges from 120 to 160 beats per minute (bpm). Normally there is variation in fetal heart rate with successive beats. This variability is normally a sign of fetal well-being. Decreased variability or flattening of the baseline fetal heart rate is nonreassuring. Frequently, there are periodic changes in fetal heart rate. Periodic increases in fetal heart rate (accelerations) are common and usually reassuring. Periodic slowings of fetal heart rate (decelerations) are of concern. Variable decelerations, which usually are associated with umbilical cord compression, are characterized by a slowing in fetal heart rate with abrupt onset and return. The depth, duration, and shape of these decelerations are variable (Figs. 1 and 2). When fetal oxygenation is reduced in response to uterine contractions, late decelerations occur. These decelerations are characterized by slowing of the fetal heart rate with gradual onset and gradual return. They are usually shallow; that is, the nadir of the deceleration is usually 10–30 bpm below the baseline. The timing is also late in relationship to uterine contractions (Fig. 3). Late decelerations are nonreassuring and when present require that the condition of the fetus and mother be carefully reviewed. In some cases, late decelerations may persist, despite the correction of contributing factors. In these situations direct sampling of the fetal scalp blood for gas analysis is indicated. Depending on the fetal and maternal condition, obstetricians may sometimes proceed with cesarean delivery.

II. ANTENATAL FETAL TESTING

Antenatal testing can identify those fetuses that are at high risk for intrauterine death. Pregnancies complicated by factors such as insulin-dependent diabetes, hypertension, fetal growth restriction, chronic renal diseases, collagen vascular diseases, hyperthyroidism, or previous stillbirth are usually

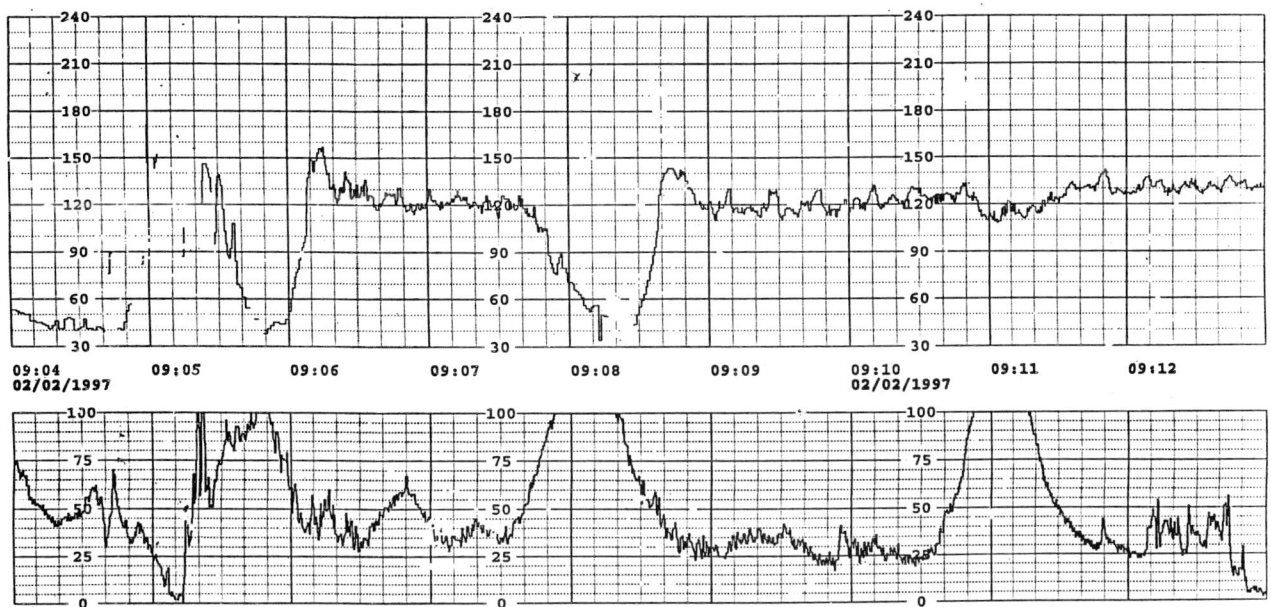

FIGURE 1 Variable decelerations. The speed of recording paper is 3 cm/min represented by horizontal line. Thick vertical lines are 1 min apart. The vertical lines in the top panel represent fetal heart rate from 30 to 240 bpm. Vertical lines of the bottom panel show uterine pressure recording in mm Hg. Note the fluctuations in the uterine pressure corresponding to the contractions.

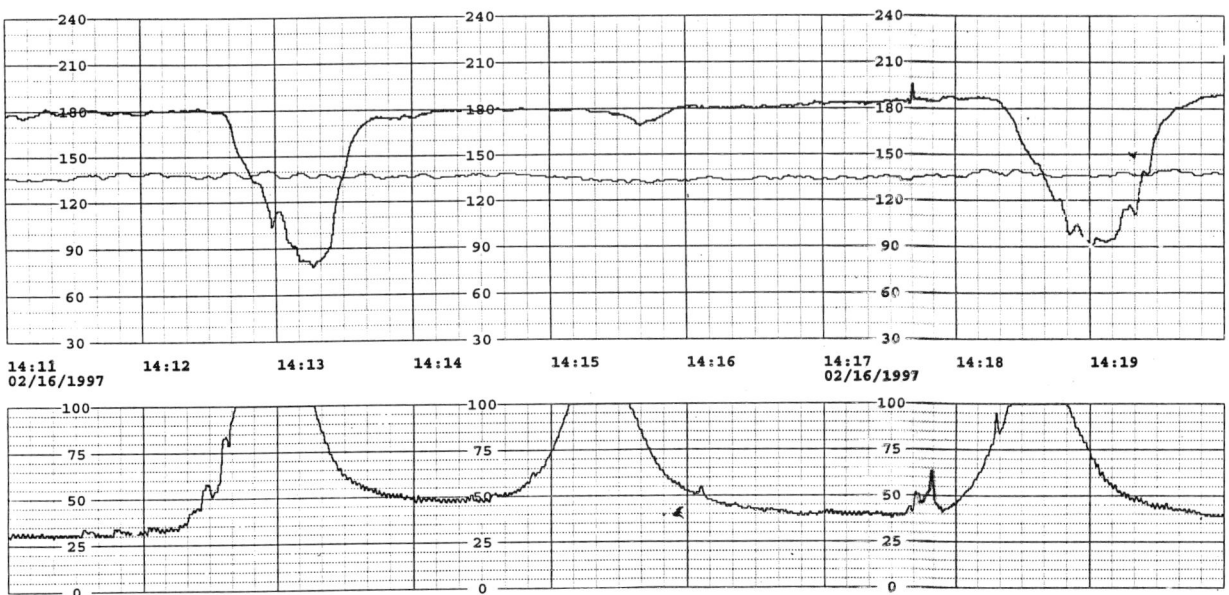

FIGURE 2 Variable decelerations in one twin only. (Top) Simultaneous fetal heart rate tracing of a twin gestation.

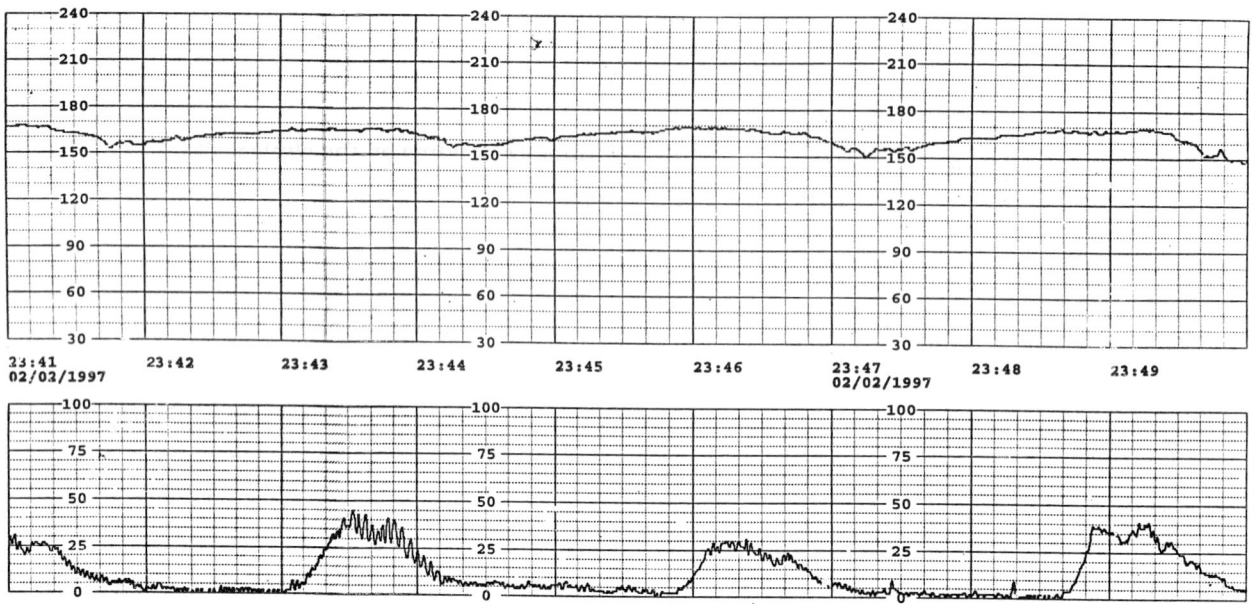

FIGURE 3 Late decelerations. (Bottom) Uterine contractions: There are four contractions in 9 min. (Top) Late decelerations with every uterine contraction. Decelerations are shallow and start after the beginning of the uterine contractions, nadir after the peak of the contractions, and end after the contractions end.

followed by one or a combination of these antenatal tests. Application of antenatal testing to these pregnancies may reduce the rate of fetal death. Since the overall incidence of fetal death is low, a large prospective study is needed to evaluate the impact of these tests on the number of deaths in the general population. Another problem with these tests is that intervention in response to an abnormal test is usually delivery of the fetus. These tests all have very high false-positive rates; in other words, a test may indicate a poor outcome but in reality the fetus is not at risk for demise.

A. Maternal Assessment of Fetal Activity

Real-time ultrasound has shown that the fetus spends 10% of its time making gross body movements—approximately 30 such movements per hour. The mother is able to appreciate about 70–80% of gross fetal movements. Fine movements such as hand grasping and sucking movements are not perceived by the mother. Different methods have been utilized for maternal assessment of fetal activity. One method is to count fetal activity for a specified duration (30–60 min) two or three times per day. If fewer than 3 movements occur in 1 hr further evaluation of the fetal condition should be undertaken. Another method is to count fetal movement for 12 hr while the mother is not interrupting her normal activity. With this method, if <10 movements are perceived in 12 hr further evaluation is indicated. There are some factors that may influence a mother's ability to perceive fetal movements, thus increasing the false-positive rate. Diurnal variation in fetal activity, maternal attention span and activity, placental position, maternal obesity, and volume of the amniotic fluid are some of the common factors. During the third trimester, the fetus spends most of its time in a sleep state. During active periods, which may last 40 min and are characterized by rapid eye movement (REM) sleep, the fetus exhibits intermittent abrupt movements of its limbs and complex movements of its head, limbs, and trunk. During quiet or non-REM sleep, which may last 20 min, fetal heart rate slows and the fetus may make startling movements.

The false-positive rate of daily fetal movement counts is 50% and false-negative rates is <10%. The majority of studies suggest that maternal assessment of fetal movements should be used as a universal screening test starting at 28 weeks gestation.

B. Contraction Stress Test

Analysis of intrapartum fetal heart rate monitoring demonstrated that a fetus with inadequate placental respiratory reserve will show late decelerations in response to hypoxia. To identify these fetuses prior to the actual labor, the contraction stress test is designed to simulate the contraction pattern of a normal labor. To perform the test the patient is placed in a semisitting position with a 30 to 45° angle and tilted slightly to the left. The reason for testing in this position is to avoid supine hypotension syndrome. Baseline fetal heart rate and uterine tone are recorded for 10–20 min. Uterine contractions are induced by intravenous infusion of oxytocin. In some patients adequate spontaneous contractions occur. An adequate contraction stress test requires three uterine contractions with moderate intensity lasting 40–60 sec. This pattern is approximately similar to the first stage of labor. Oxytocin is infused at a rate of 0.5 mU per minute, doubling every 20 min until the occurrence of adequate uterine contractions. The dosage of oxytocin usually does not exceed 10 mU per minute. Maternal nipple stimulation, which causes the release of endogenous oxytocin, can also be employed to induce uterine contractions. Nipple stimulation is easier to perform but it is accompanied by higher incidence of uterine hyperstimulation compared to the oxytocin induction. Hyperstimulation of the uterus has been reported in 2–10% of nipple stimulation cases. Interpretation of a contraction stress test is difficult in the presence of uterine hyperstimulation.

A negative contraction stress test is without decelerations. When late decelerations are present following 50% or more of uterine contractions, the test is interpreted as positive. The test is called suspicious or equivocal when there are intermittent or significant variable decelerations.

The incidence of perinatal death within 1 week of a negative contraction stress test is 0.4/1000. Positive contraction stress tests are associated with an increased incidence of intrauterine death, late decelerations in labor, low 5-min Apgar scores, fetal growth restriction, and meconium-stained amniotic fluid. The false-positive rate for contraction stress tests is 30%. Depending on the clinical situation of the fetus and mother, delivery may be indicated when the contraction stress test is positive. A suspicious or equivocal contraction stress test, and those that are unsatisfactory or show hyperstimulation, should be repeated in 24 hr.

C. Nonstress Test

Near-term fetuses spend 25% of their time in a quiet sleep state and the remaining time in an active state. The active sleep state is associated with REMs and fetuses show regular breathing movements and intermittent movements of head, limbs, and trunk. The controlling mechanisms for these periods are unknown. During the active sleep state fetal heart rate exhibits accelerations. Compared to the contraction stress test, this test is simple, less invasive, less time-consuming, less expensive, and can be conducted in an outpatient setting with less skilled personnel. It is also technically feasible for some women with polyhydramnios, marked obesity, or multiple gestation. The patient is seated in a reclining chair and fetal heart rate monitored using Doppler ultrasound transducer. Fetal movements are recorded either by the patient or by the person performing the test.

A reactive (normal) nonstress test should have at least two accelerations of the fetal heart rate within a 20-min period. Accelerations should peak at least 15 bpm above baseline and should last 15 sec from baseline to baseline (Fig. 4). If the criteria for reactivity are not met within 20 min of monitoring, some clinicians extend the test for an additional 20 min. If after 40 min of monitoring the fetal heart rate remains without adequate reactivity, the test is considered nonreactive and further action depending on the clinical situation should be undertaken. The period of fetal inactivity or sleep is the most common cause for nonreactivity. Twenty-five percent of patients with nonreactive nonstress test will show a positive contraction stress test. The false-positive rate

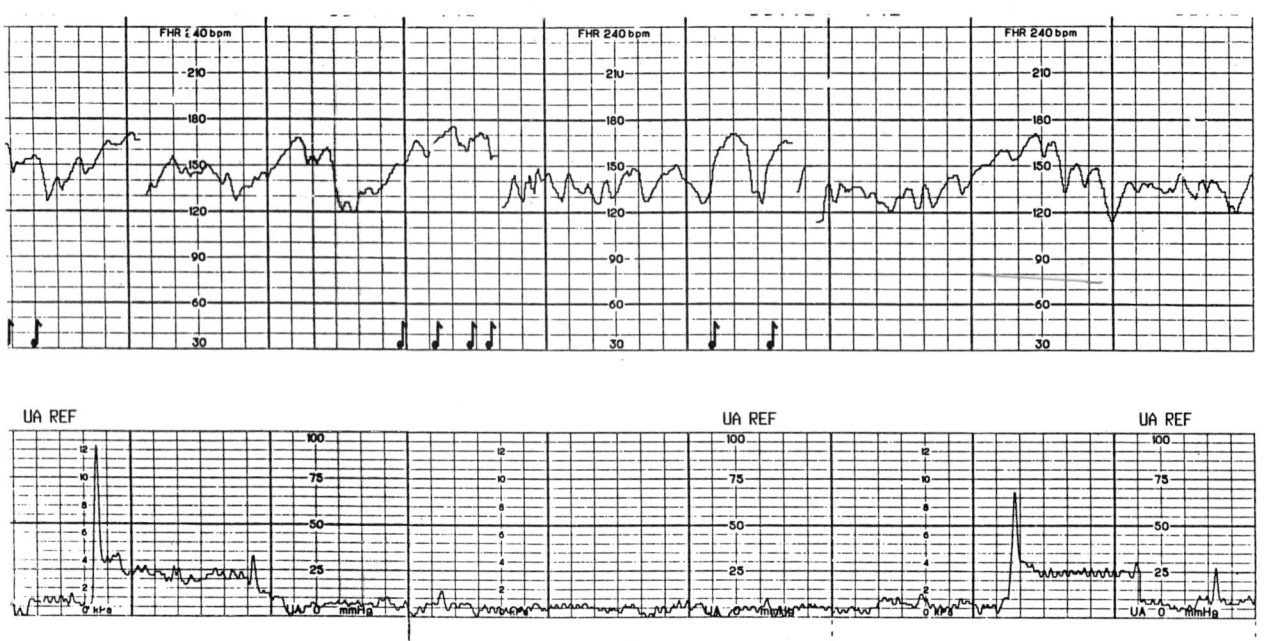

FIGURE 4 A reactive nonstress test. (Top) Accelerations of 15 bpm above the baseline lasting 15 sec from baseline to baseline.

for the nonstress test is estimated to be 50–75%. The false-negative rate is low, similar to the other antenatal tests, and estimated to be 1.4/1000.

Fetal movement in response to external sound stimulation was first reported in 1925. In the mid-1980s researchers observed alteration in fetal heart rate patterns after vibratory acoustic stimulation of term fetuses. Some clinicians use vibroacoustic stimulation to cut down on the number of false-positive nonstress tests and shorten the time required to perform the test.

D. Biophysical Profile

The fetal biophysical profile is a combination of the nonstress test plus four other parameters observed by ultrasound. The four parameters are fetal breathing movements, gross body movements, fetal tone, and quantitation of the amniotic fluid volume. The test is done by ultrasound observation of the fetus for a maximum of 30 min. During this 30 min a normal fetus will exhibit one or more episodes of rhythmic breathing movements of 30 sec or more duration (fetal breathing movements), three or more discrete body or limb movements (gross body movements),

and one or more episodes of extension of an extremity with return to flexion (fetal tone). There is no universal agreement with regard to quantitation of the amniotic fluid volume, but the majority of obstetricians consider the volume to be adequate when at least a single pocket of amniotic fluid measures in excess of 2 × 2 cm in two perpendicular planes. A score of 2 is given to each parameter when requirements for that parameter are met, and in the absence of these requirements, the given score is 0. Therefore; with each biophysical profile a score is generated between 0 and 10. Scores between 8 and 10 are good and usually no further action is needed. A score of 6 usually means that the test should be repeated within 24 hr or delivery should be contemplated if clinically appropriate. Scores of less than 4 are concerning and the fetus usually has to be delivered. Interestingly, those biophysical activities that are present earlier in fetal development are the last to disappear with fetal hypoxia. The fetal tone center in the cortex begins to function at 7.5–8.5 weeks. The fetal movement center in cortex–nuclei is functional at 9 weeks. Fetal breathing movements become regular at 20 or 21 weeks. Reactivity of the fetal heart rate, the basis for the nonstress test, which is an interplay between the sympathetic and the parasympathetic system, is established around 26 weeks. Amniotic fluid volume provides information about the presence of chronic fetal hypoxia. With a reduction in fetal oxygenation, oligohydramnios may result. The false-negative rate for the biophysical profile is 0.6/1000.

III. SUMMARY

Intrapartum fetal heart rate assessment is an integral aspect of the medical care of the fetus. Continuous electronic fetal heart rate monitoring may not be superior to intermittent auscultation. However, it does eliminate the need for one-to-one nursing. It is believed that antenatal testing reduces the incidence of fetal demise. An important benefit of these antenatal tests is that with a reassuring test the risk of fetal demise is extremely low in an otherwise high-risk pregnancy. These are not to be substituted for a thorough clinical evaluation of a pregnancy. Antenatal testing and intrapartum continuous electronic fetal heart rate monitoring are only tools to assist the clinician in identifying those fetuses in need of further attention.

See Also the Following Articles

APGAR SCORE; AMNIOCENTESIS; CESAREAN DELIVERY; FETAL DIAGNOSIS, INVASIVE; FETAL GROWTH AND DEVELOPMENT; FETAL SURGERY; OXYTOCIN

Bibliography

American College of Obstetricians and Gynecologists (ACOG) (1994). Antepartum fetal surveillance, ACOG Tech. Bull. No. 188. ACOG, Washington, DC.

American College of Obstetricians and Gynecologists (ACOG) (1995). Fetal heart rate patterns: Monitoring, interpretation, and management, ACOG Tech. Bull. No. 207. ACOG, Washington, DC.

Freeman, R. K., Anderson, G., and Dorchester, W. (1982). A prospective multi-institutional study of antepartum fetal heart rate monitoring. II. Contraction stress test versus nonstress test for primary surveillance. *Am. J. Obstet. Gynecol.* 144, 218–223.

Freeman, R. K., Garite, T. J., and Nageotte, M. P. (Eds.) (1991). *Fetal Heart Rate Monitoring*, 2nd ed.. Williams & Wilkins. Baltimore.

Gabbe, S. G., Niebyl, J. R., and Simpson, J. L. (1996). *Obstetrics Normal and Problem Pregnancies*, 3rd ed. Churchill Livingstone, New York.

Moore, T. R., and Piacquadio, K. (1989). A prospective evaluation of fetal movement screening to reduce the incidence of antepartum fetal death. *Am. J. Obstet. Gynecol.* 160, 1075.

Fetal–Placental Unit

Bruce R. Carr

The University of Texas Southwestern Medical Center at Dallas

GLOSSARY

decidua Specialized endometrial cells characteristic of pregnancy.

estriol An estrogen formed by the placenta from 16α-hydroxylated dehydroepiandrosterone sulfate.

fetal zone of the adrenal cortex The unique large zone of the fetal adrenal cortex which secretes androgens and regresses following birth.

human chorionic gonadotropin A protein hormone secreted by the trophoblast unique to pregnancy.

human placental lactogen A single-chain polypeptide secreted by the trophoblast which exhibits lactogenic and growth hormone activity.

low-density lipoprotein (cholesterol) The source of cholesterol utilized by the placenta and corpus luteum to produce progesterone and androgens by the fetal adrenal.

The fetal–placental unit demonstrates the importance of the interaction of the fetus and placenta in the formation of hormones which regulate fetal growth and maintenance of pregnancy.

I. INTRODUCTION

During human pregnancy there are marked hormonal as well as metabolic changes that occur. The placental unit, supplied with precursor hormones from both the maternal and fetal unit, synthesize large quantities of steroid hormones as well as a variety of protein and peptide hormones, and these products are in turn secreted into the fetal and maternal circulation. As a consequence of these events, steroid and protein hormones levels rise so that near term the mother and fetus are exposed to large quantities of estrogen, progesterone, mineralocorticoids, and glucocorticoids. The mother and, to a lesser extent, the fetus are also exposed to large quantities of protein hormones such as human placental lactogen, human chorionic gonadotropins (hCGs), prolactin, relaxin, and a variety of other peptide-releasing hormones. Implantation, the maintenance of pregnancy, parturition, and, finally, lactation are dependent on the complex interaction of these hormones in the maternal–fetal–placental unit.

II. PLACENTAL UNIT

In humans the placenta has evolved into a complex structure that is involved in delivery of nutrients to the fetus and production of numerous steroid and protein hormones as well as removal of waste products and metabolites from the fetus which are then transferred to the maternal compartment for excretion.

A. Progesterone

Progesterone is the principal steroid hormone secreted by the placenta during pregnancy. However, during early pregnancy the main source of progesterone is from the human corpus luteum. The corpus luteum of pregnancy, stimulated by hCG from the placenta, provides the major source of progesterone through 8–10 weeks of gestation. After this period of time the corpus luteum continues to secrete progesterone but the levels decline and are only a fraction of that secreted by the placenta. The increase in progesterone due to placental origin rises from 25 to about 150 ng/ml at term (Fig. 1). Most of the progesterone secreted by the placenta enters the maternal compartment.

The biosynthetic origin or precursor for progesterone production by the placenta appears to be independent of the fetal unit. That is, soon after fetal death *in utero*, pregnancy complicated by anencephaly, or in animal models such as the rhesus monkey, in which the umbilical cord has been ligated, progesterone levels do not decline significantly. In contrast, the major source of precursor hormone for placenta progesterone secretion is from the maternal unit which is derived from low-density lipoprotein (LDL) or LDL cholesterol. Thus, LDL cholesterol from the maternal plasma enters the placenta where it is rapidly converted into progesterone and then secreted back into the maternal compartment. The placenta by itself has very little capacity to secrete cholesterol from acetate. The role of progesterone appears to be maintenance of pregnancy and prevention of uterine contractions as well as prevention of rejection of the fetus by immunological mechanisms.

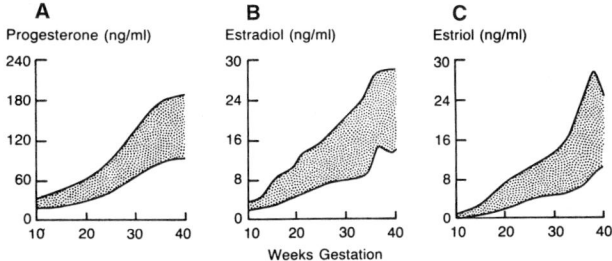

FIGURE 1 The range (mean ± 1 SD) of (A) progesterone, (B) estradiol-17β, and (C) estriol in plasma of normal pregnant women as a function of weeks of gestation [adapted from B. R. Carr, In *Principles and Practice of Endocrinology and Metabolism* (K. L. Becker, Ed.), pp. 887–898, Lippincott, Philadelphia, 1990].

B. Estrogen

Throughout pregnancy the rate of estrogen production and levels of circulating estrogen increase significantly (Fig. 1). Although the ovary may be the source of estrogen secretion during the early part of the first few weeks of pregnancy, thereafter nearly all the estrogen is formed by the trophoblasts of the placenta. The mechanism by which estrogen is produced by the placenta is unique and is the best example of the coordinated events of the fetal–placental-unit. The placenta cannot convert progesterone to estrogens because of deficiency of the enzyme 17α-hydroxylase. The placenta must rely on precursor androgens formed by the maternal and fetal adrenal glands (Fig. 2). As seen in Fig. 2 (right), the main source of cholesterol formation in the fetus is from the fetal liver. This is true because only approximately 20% of the fetal cholesterol is derived from the maternal compartment and since amniotic fluid (i.e., oral intake) cholesterol levels are negligible, the main source is the liver. The liver synthesizes cholesterol at a high rate, and if one considers the size of the fetal liver it can be computed that the fetal liver may supply sufficient amounts of cholesterol to maintain fetal adrenal steroidogenesis. Cholesterol from the fetal liver enters the fetal adrenal in the form of the LDL cholesterol where it is converted to dehydroepiandrosterone sulfate (DS). DS can then enter the placenta where it can be converted to estrone (E_1) and estradiol (E_2) or, more commonly, it enters the fetal liver where it undergoes 16α-hydroxylation and then enters the placenta where it is converted to estriol (E_3).

The measurement of estrogens, particularly estriol, was used at one time to monitor high-risk pregnancies. It is known that in certain pregnancies in which there is deficiency in estrogen production, such as sulfatase deficiency (in which there is deficiency in the sulfatase enzyme), aromatase deficiency (in which there is deficiency of the aromatase enzyme in the fetus which can lead to virilization of the fetus and mother), and anencephaly (in which there is a marked reduction in fetal adrenal androgen production): All these conditions are associated with prolonged pregnancies. The role of estrogen, in which human pregnancy exhibits the highest levels of all

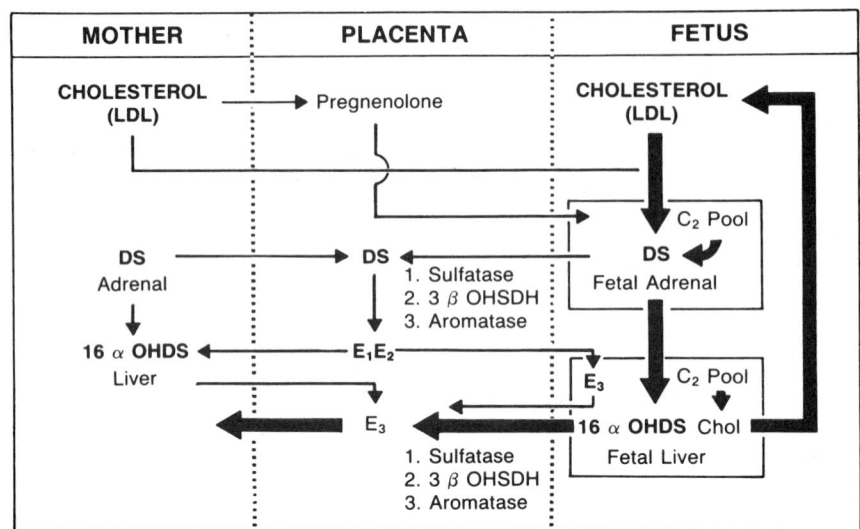

FIGURE 2   Sources of estrogen biosynthesis in the maternal–fetal–placental unit. LDL, low-density lipoprotein; Chol, cholesterol; C_2 pool, carbon–carbon unit; DS, dehydroepiandrosterone sulfate; E_1, estrone; E_2, estradiol-17β; E_3, estriol (adapted from B. R. Carr and N. F. Gant, The endocrinology of pregnancy-induced hypertension, *Clin. Perinatol.* **10**, 737, 1983).

animal species, is unclear. It is proposed that estrogen may regulate or fine-tune the events leading to parturition. Estrogen may play a role in stimulating uterine contractions as well as increasing uterine blood flow and maturation of fetal organ development

C. Protein and Peptide Hormones

The two principal hormones secreted by the placenta are hCG and human placental lactogen (hPL), also known as human somatomammotropin. hCG is secreted by the syncytiotrophoblasts and enters both the maternal and fetal compartments. When it enters the maternal compartment it stimulates progesterone secretion by the early corpus luteum. On the other hand, when it enters the fetal compartment it stimulates testosterone in the human fetal testis causing masculinization of the external genitalia. Placental levels of hCG rise in early pregnancy, doubling in concentration every 2 or 3 days until they reach a peak between 60 and 90 days. Thereafter, the levels of hCG decline, plateauing about 120 days before delivery (Fig. 3). In contrast, hPL, which has both lactogenic and growth hormone-like activity, increases progressively throughout pregnancy, reaching a peak at term (Fig. 3). The level of hPL correlates very well with the increasing placental weight. Its

half-life is so short that soon after delivery a hPL level is hard to detect. The role of hPL is unclear but it may play a role of sparing maternal glucose for the developing fetus. In this aspect both hCG and hPL play a role in regulating the fetal–placental unit. There are a variety of other peptide hormones, including all those made in the gut and in the hypo-

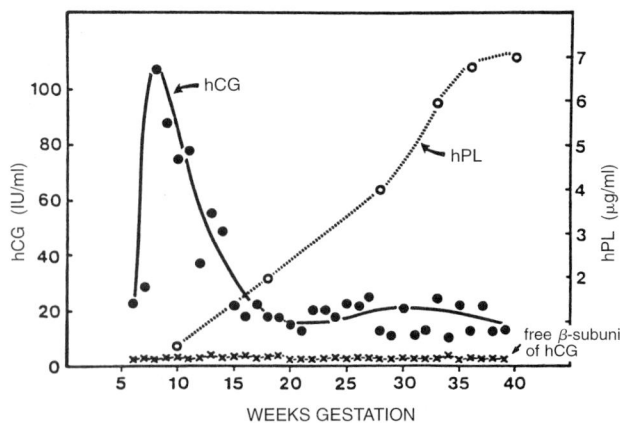

FIGURE 3   Mean concentration of chorionic gonadotropin (hCG) and placental lactogen (hPL) in sera of women throughout normal pregnancy (reproduced with permission from J. A. Pritchard, P. C. MacDonald, and N. F. Gant, *Williams Obstetrics*, 17th ed., p. 121, Appleton-Century-Crofts, Norwalk, CT, 1985).

thalamus including proopiomelanocortins, corticotropin-releasing hormone, growth hormone-releasing hormone, and a number of others. The role of these hormones has not been completely elucidated.

III. FETAL MEMBRANES AND DECIDUA

A. Fetal Membranes

The fetal membranes consist of an amnion and a chorion. The amnion is a unique structure: It is thin (0.02–0.5 mm) and contains no blood vessels or nerves, whereas the chorion is thick and well vascularized. Although fetal membranes do not synthesize hormones *de novo*, they exhibit extensive enzymatic capabilities for regulating hormone metabolism. In addition, the fetal membranes contain large quantities of arachidonic acid the obligate precursor of prostaglandins.

B. Decidua

The decidua is a complex structure of specialized endometrial stromal cells and thus is part of the maternal compartment. These cells develop in response to progesterone secreted during the luteal phase of the menstrual cycle and are further stimulated by progesterone secreted by placental trophoblasts. There is strong evidence to support that the decidua is a rich source of enzymatic activity and secretion of hormones. The decidua may be important in fetal homeostasis as well as maintenance of pregnancy since the decidua appears to communicate directly with the fetus via transport through the fetal membranes and into the amniotic fluid as well as directly by the myometrium by simple diffusion. Hormones that are primarily unique to the decidua include prolactin, relaxin, and prostaglandins. The prolactin that is secreted by the decidua is immunologically, structurally, and biologically similar to that of pituitary origin. However, the regulation of decidual prolactin secretion is much more complex. Bromocriptine treatment of pregnant women reduces the maternal and fetal plasma levels of prolactin but does not influence amniotic fluid levels of prolactin which reach a peak in midgestation. Prolactin secretion by decidual cells is not affected by treatment with dopamine, dopaminergic agonists, or thyroid-releasing hormone (TRH). Prolactin is thought to play a role in fluid osmolality. Relaxin, which is also secreted by the corpus luteum, is secreted in large amounts by the human decidua. It is thought that relaxin may play a role along with progesterone in reducing uterine activity as well as softening of pelvic tissues prior to parturition. The release of prostaglandins by the decidua is thought to play a role in parturition.

IV. FETAL UNIT

Understanding of the human fetal endocrine system has required the development of measurements of minute quantities of hormones. Using recent techniques in molecular biology, it has also been discovered that certain hormones are produced at very early stages of development in the human fetus. As the fetus develops, its endocrine system in part interplays with the placental and maternal unit, but near term certain aspects of the endocrine system are independent for preparation of extrauterine existence.

A. Fetal Hypothalamic–Pituitary Axis

The fetal hypothalamus differentiates from the forebrain during the first few weeks of fetal life, and by 12 weeks, hypothalamic development is well advanced. Hypothalamic-releasing hormones, including gonadotropin-releasing hormone, TRH, corticotropin-releasing hormone, growth hormone-releasing hormone, and their respective hypothalamic nuclei, have been identified early in fetal life in the hypothalamus.

With respect to the anterior pituitary gland, it develops from the outpouching of the gut called Rathke's pouch, which first appears in the human fetus at 4 weeks. The hormones of the anterior pituitary gland, including growth hormone, prolactin (Prl), follicle-stimulating hormone (FSH), luteinizing hormone (LH), thyroid-stimulating hormone (TSH), and adrenocorticotropin hormone (ACTH), can be detected as early as 7 weeks of fetal life.

To be fully functional, the connection in the hypothalamus–pituitary known as the hypothalamic pituitary portal system needs to complete development. This is not achieved until approximately 18 weeks of fetal life. However, there is evidence that the feedback system is effective as early as 6 weeks of fetal life. The levels of pituitary hormones, including growth hormone, LH, FSH, ACTH, TSH, and Prl, are depicted in Fig. 4. A few summary points can be made regarding Fig. 4. First, fetal prolactin levels are low in fetal blood but rise throughout pregnancy. Levels of ACTH and TSH are high at about midgestation and fall somewhat near term, but the most dramatic changes occur in both growth hormone and LH and FSH. These hormones are almost undetectable in early fetal life and increase to reach a peak about midgestation, and thereafter they fall and are barely detectable near term. This rapid decline in growth hormone, LH, and FSH is thought to be due to the maturation of the feedback mechanisms which occur at the level of the hypothalamic–pituitary axis.

B. Fetal Thyroid Gland

As opposed to interaction of some of the glands of the fetus such as the fetal adrenal, the fetal hypothalamic–pituitary–thyroid axis is independent of that of the maternal system as well as the placental unit. During early fetal life, the thyroid is able to concentrate iodine and synthesize iodothyronines. Iodine will easily cross the placenta, where it then enters the fetal compartment and converts to thyroid hormones and influences fetal TSH. The levels of thyroid hormones are relatively low in fetal blood in early pregnancy but thyroxin peaks at delivery. At the time of birth there is an abrupt rise in TSH, T_4, and T_3, with a rapid fall thereafter. This relative hyperthyroid state of the fetus appears to facilitate thermoregulatory mechanism for extrauterine life. Little if any maternal TSH, T_4, or T_3 crossed the placenta and enters the fetal unit.

C. Fetal Gonads

The fetal gonad develops under the influence of gene expression from the X or Y chromosome, thereby enabling a bipotential gonad to develop into the fetal testis or fetal ovary. The fetal gonad becomes a testis around 7 weeks of life. Placental hCG and fetal LH, which reach a peak around midgestation, cause fetal testosterone secretion to increase, which then causes masculinization of fetal external genitalia. The fetal testis, however, does not make sperm. In contrast, the fetal ovary, which develops at about 11 weeks of fetal life, is steroidogenically inactive. It is, however, very prolific in producing germ cells

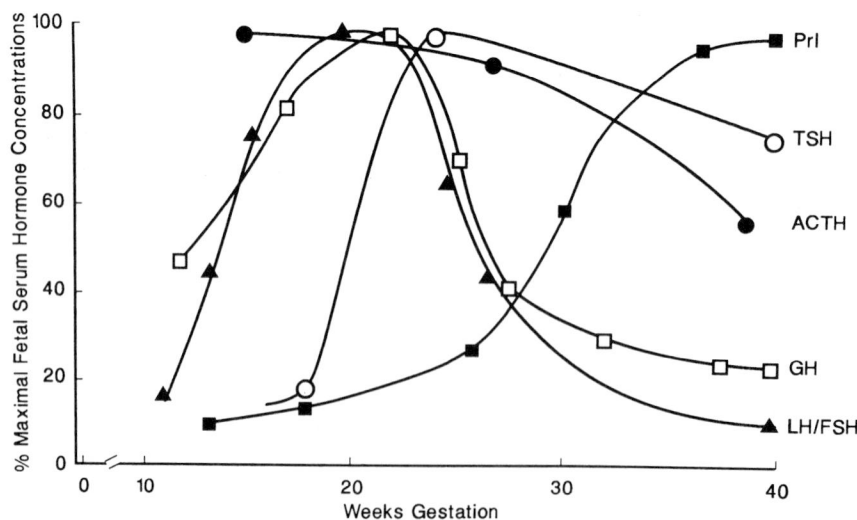

FIGURE 4 Ontogeny of pituitary hormone levels in human fetal sera. Prl, prolactin; TSH, thyroid-stimulating hormone; ACTH, corticotropin; GH, growth hormone; LH/FSH, luteinizing hormone/follicle-stimulating hormone (from Parker, 1993, p. 28).

and the number of ova in the fetal ovary reaches a peak at midgestation and thereafter decreases. For example, 8 million eggs form by midgestation but decline to 1 million eggs at delivery. This is due to a process of apoptosis known as ovarian germ cell atresia.

D. Fetal Adrenal Glands

The fetal adrenal glands are the endocrine glands which have elicited the most extensive research. The fetal adrenal gland secretes large quantities of C-19 steroids, particularly dehydroepiandrosterone sulfate. The fetal adrenal gland is unique and contains a section called the fetal zone which undergoes rapid growth and reaches a peak in size and secretion of C-19 steroids at term. Soon after delivery it undergoes involution and regression by apoptosis. The other section of the fetal adrenal gland, the neocortex or definitive zone, secretes small amounts cortisol and is relatively inactive until late gestation. The regulation of adrenal C-19 steroid section and cause of the tremendous growth of the fetal zone and its rapid regression after birth are unknown.

E. Fetal Parathyroid Gland and Calcium Homeostasis

The level of calcium in the fetus is regulated largely by the transfer of calcium from the maternal compartment across the placenta. The maternal compartment undergoes a number of adjustments that ultimately allow for a net transfer of a sufficient amount of calcium to the fetus to sustain fetal bone growth.

The changes in the maternal compartment that permit fetal accumulation of calcium include an increase in maternal dietary intake, an increase in circulating maternal 1,25-dihydroxyvitamin D [1,25-$(OH)_2D$], and an increase in circulating parathyroid hormone (PTH). No significant changes are observed in maternal calcitonin levels. The levels of total calcium and phosphorus decline in maternal serum, but ionized calcium remains unchanged. The "placental calcium pump" allows for a positive gradient of calcium and phosphorus to the fetus. Circulating fetal calcium and phosphorus levels increase steadily throughout gestation. Fetal levels of total and ionized

calcium as well as phosphorus exceed maternal levels at term.

The fetal parathyroid gland contains PTH, and the gland is capable of hormone secretion by 10–12 weeks of gestation. Fetal plasma levels of PTH reportedly are low but increase after delivery. The fetal thyroid contains calcitonin, and, in contrast to maternal plasma levels, calcitonin levels in the fetus are elevated. Since there is no transfer of PTH or calcitonin across the placenta, the consequences of the observed change in these hormones on fetal calcium are consistent with an adaption to conserve and stimulate bone growth within the fetus.

Plasma levels of the various forms of vitamin D are lower in the fetus than in the mother. The placenta and decidua are capable of 1α-hydroxylation and of formation of the active metabolite 1,25-$(OH)_2D$. However, the role, if any, of this hormone in the fetus is unknown since its major effect is on intestinal absorption of calcium.

After birth, serum calcium and phosphorus levels fall in the neonate. PTH levels begin to rise 48 hr after birth, and calcium and phosphorus levels gradually increase over the following several days, depending on dietary intake of milk.

F. Fetal Pancreas

The fetal pancreas appears during the fourth week of fetal life. The α cells contain glucagon and the γ cells containing somatostatin develop early, before β cell differentiation, although insulin can be recognized in the developing pancreas before apparent β cell differentiation. Total human pancreatic insulin and glucagon content increase with fetal age and are higher than the concentrations of the adult human pancreas.

In contrast to the pancreatic content of insulin, fetal insulin secretion is low and relatively unresponsive to acute changes in glucose in *in vitro* studies of pancreatic cells, in cord blood at delivery, and in blood samples obtained from the scalp of the fetus at term. In contrast, fetal insulin secretion *in vitro* is responsive to amino acids and glucagon as early as 14 weeks of gestation. Although the acute response to glucose is impaired in the fetal pancreas, β cells that are chronically exposed to elevated glucose levels, as may occur in maternal diabetes mellitus, undergo hypertrophy so that the rate of insulin secretion increases.

Glucagon has been detected in human fetal plasma as early as 15 weeks of gestation. Although secretion of glucagon is stimulated in late pregnancy by amino acids and catecholamines, acute changes in glucose appear to have little effect on fetal pancreatic glucagon secretion.

See Also the Following Articles

CHORIONIC GONADOTROPIN, HUMAN; CORPUS LUTEUM OF PREGNANCY; DECIDUA; FETAL ADRENALS; FETAL MEMBRANES; HUMAN PLACENTAL LACTOGEN; PROGESTERONE ACTIONS ON REPRODUCTIVE TRACT

Bibliography

Carr, B. R. (1995). The endocrinology of pregnancy: The maternal–fetal–placental unit. In *Principles and Practice of Endocrinology and Metabolism, 2nd ed.* (K. L. Becker, Ed.), pp. 987–1000. Lippincott, Philadelphia.

Carr, B. R., and Rainey, W. E. (1994). The adrenal. In *Infertility and Reproductive Medicine Clinics of North America* (J. P. Bruner, Ed.), Vol. 5, 749–764. Saunders, Philadelphia.

Carr, B. R., and Simpson, E. R. (1981). Lipoprotein utilization and cholesterol synthesis by the human fetal adrenal gland. *Endocr. Rev.* 2, 306–326.

Parker, C. R., Jr. (1993). The endocrinology of pregnancy. In *Textbook of Reproductive Medicine* (B. R. Carr and R. E. Blackwell, Eds.). Appleton & Lange, Norwalk, CT.

Tulchinsky, D., and Little, A. B. (Eds.) (1994). *Maternal–Fetal Endocrinology.* Saunders, Philadelphia.

Fetal Surgery

Theresa M. Quinn and N. Scott Adzick

The Children's Hospital of Philadelphia

GLOSSARY

bronchopulmonary sequestration A nonfunctional lung lesion that does not communicate with the tracheobronchial tree and receives its blood supply from a systemic rather than pulmonary vessel.

congenital cystic adenomatoid malformation A benign cystic, solid, or mixed lung mass that communicates with the tracheobronchial tree and receives a blood supply from the pulmonary circulation.

fetal surgery An operation performed on a fetus while the fetus remains inside the uterus.

fetoscopy An operative technique for fetal surgery using tiny uterine puncture holes through which are placed instruments and a telescope attached to a fiber-optic camera.

hydrops Fetal distress manifested by fluid collections around the heart and lungs, in the abdomen, or in the skin; can be caused by maternal antibodies cross-reacting with the fetus or related to fetal heart failure.

oligo/polyhydramnios Disorders of amniotic fluid volume (inadequate/excessive volume).

ultrasound An imaging modality using sound waves for prenatal screening.

Prenatal ultrasound provides a window into a previously inaccessible period of growth and differentiation: human fetal development. With the advent of sonography we have identified the typical features of a growing fetus and can recognize when the process goes awry.

I. INTRODUCTION

Most correctable malformations that can be diagnosed *in utero* are managed optimally after birth. The neonate can be delivered at term and appropriate medical and surgical therapy can be instituted at a center with neonatal critical care and pediatric surgical expertise. However, some congenital malformations can cause a fetus to die before these measures can be taken. For a highly selected group of cases, correction before birth may be possible. The goal of fetal therapy is to intervene in a life-threatening fetal disorder to change the otherwise devastating outcome. The major hurdles for the development of fetal surgery have been defining patient selection criteria, crafting effective fetal surgical techniques, devising fetal and uterine monitoring, generating new strategies for tocolytic management, and minimizing maternal and fetal risk.

II. HISTORY OF OPEN FETAL SURGERY

The first open fetal surgery, an exchange transfusion in a fetus with Rh isoimmunization, was attempted in the 1960s. Thereafter, with ultrasound guidance, less invasive means for fetal blood sampling and transfusion became possible. In the early 1980s, the era of open fetal surgery began with the correction of structural abnormalities. Harrison and colleagues performed the first open fetal operative procedure in 1982 by creating bilateral ureterostomies in a 21-week-gestation fetus with obstructive uropathy. Over the next 15 years, about 100 human fetuses with defined structural lesions and a poor prognosis have been treated with fetal surgery. The lesions have included urinary tract obstruction, congenital diaphragmatic hernia, cystic adenomatoid malformation, and sacrococcygeal teratoma.

Before fetal surgery could be performed in humans, fetal structural abnormalities needed to be modeled and corrected in animals. From the experience garnered from fetal sheep and nonhuman primate studies, operative, anesthetic, and tocolytic regimens were developed that would later be crucial to the success of human fetal surgery. The innovations included techniques for a quick and "bloodless" uterine opening and "water-tight" uterine closure, the creation of radiotelemetric fetal monitors, and the institution of aggressive tocolytic therapies for the inevitable uterine contractions provoked by fetal surgery.

Maternal safety remains the most important aspect of fetal surgery. The risks to the mother must be balanced with the benefits of the surgery for the severely compromised fetus. Maternal outcome is generally good: There have been few postoperative complications and no maternal deaths. The risks to the pregnant mother include those associated with general anesthesia, a laparotomy, and cesarean section—all of which will need to be repeated for the delivery of the operated fetus and for all future pregnancies because of the location of the uterine incision. However, the primary source of morbidity is preterm labor and its treatment, with tocolytic-induced pulmonary edema being the most prominent manifestation. The incidence of pulmonary edema has been minimized by careful fluid management and close patient monitoring. Amniotic fluid leak was notable in the early experience, but improved uterine closure techniques have reduced this problem. Fetal surgery does not appear to jeopardize future fertility and delivery, as evidenced by the more than 30 patients who have had subsequent uncomplicated pregnancies. Despite the risks, fetal surgery increases the choices for families when faced with an unborn baby with a life-threatening defect: Fetal surgery is an alternative to pregnancy termination, the likelihood of fetal loss, or the burden of raising a child with a severe malformation. The principal benefit is the prospect of a healthy baby.

III. PERIOPERATIVE CONSIDERATIONS

Maternal safety is the cardinal issue. To ensure her safety and maximize fetal benefit after surgery, specialized monitoring and a critical care approach have been developed. While the mother is monitored by arterial and central venous pressure recordings that begin in the operating room, the fetus is followed with continuous electrocardiogram and intrauterine pressure measurements collected by an implanted radiotelemeter. External Doppler fetal heart rate and tocodynamometers are also used. The critical care approach to the fetal patient entails careful fluid management, effective postoperative pain control, and an aggressive tocolytic regimen.

Beginning in the operating room, judicious fluid management is imperative. Hypovolemia results in poor uterine perfusion and provokes preterm labor, whereas hypervolemia increases the risk of maternal pulmonary edema associated with the use of magnesium sulfate and terbutaline.

The closer to term a fetus can be delivered, the more likely the fetus will benefit from the surgery and survive. However, incipient preterm labor remains the greatest challenge to the success of fetal surgery. To address this problem, preterm labor control begins in the operating room. In addition to providing anesthesia for both mother and fetus, halogenated inhaled anesthetics cause smooth muscle relaxation. Uterine muscle relaxation is important to maintain umbilical perfusion during the operation. After fetal surgery, preterm labor is treated with a combination of effective analgesia, prostaglandin inhibitors, magnesium sulfate, and β-mimetic agonists. Recent experimental studies suggest that endogenous nitric oxide mediates normal uterine relaxation during pregnancy. In rhesus monkeys, nitric oxide donors inhibit hysterotomy-induced uterine contractions. This experimental work has provided the impetus for novel approaches to labor control using nitric oxide donors such as nitroglycerin in human fetal surgery patients.

Despite the aggressive treatment of postoperative uterine contractions, most fetal surgery patients are delivered before 35 weeks' gestation. When labor can no longer be averted, a controlled delivery is performed using a modified cesarean section. Caring for the maternal–fetal patient in a specialized intensive care unit setting and ensuring the best outcome for the fetal patient require a multidisciplinary team with the collaboration of dedicated pediatric surgeons, perinatal obstetricians, sonographers, anesthesiologists, neonatologists, echocardiographers, and neonatal and obstetrical nurses.

IV. APPLICATIONS

Because of the maternal risks of fetal surgery and tocolysis, and the fetal risk of premature delivery, fetal surgery has been reserved for fetuses who have a poor prognosis without intervention. In particular, fetal structural malformations have been the most promising candidates for *in utero* repair. Using prenatal sonography, the structural abnormality has been identified, natural history elucidated, and pathophysiology modeled in animals. The malformations have included fetal lung masses, congenital diaphragmatic hernia, obstruction of the fetal urinary tract, and sacrococcygeal teratoma.

A. Fetal Lung Lesions

Ultrasound has been instrumental to the diagnosis and assessment of fetal thoracic malformations. Using prenatal ultrasound to define the natural history, it is evident that prenatally diagnosed congenital cystic adenomatoid malformation (CCAM) and bronchopulmonary sequestration (BPS) have a different outcome than their postnatal counterparts: The difference in mortality has been referred to as a "hidden mortality." The mortality of fetal lung lesions is underestimated because some affected fetuses die *in utero,* before diagnosis of the lesion, from the consequences of these space-occupying lesions. The masses can cause mediastinal compression and hydrops leading to intrauterine fetal demise. The mortality of fetal lung lesions is further underestimated when neonates, born with significant pulmonary hypoplasia from *in utero* lung compression, die prior to transfer to a tertiary care center. With improved prenatal detection of these malformations, we have begun to unravel the differences in pathophysiology between the lung lesions detected in the fetal period and those presenting postnatally. The prognostic indicators assist in perinatal management decisions.

1. *Congenital Cystic Adenomatoid Malformation*

CCAM is a benign cystic lung mass. The prognosis of a prenatally diagnosed CCAM relates to the size of the lesion and the physiologic derangement that results. A large CCAM compresses the fetal esophagus and decreases fetal swallowing of amniotic fluid, resulting in polyhydramnios. Compression of the lungs leads to pulmonary hypoplasia, whereas vena caval and cardiac compromise can lead to fetal hydrops and death (Fig. 1). Large fetal CCAMs associated with hydrops can have maternal consequences known as the "maternal mirror syndrome." This poorly understood hyperdynamic, preeclamptic state can be life threatening to the mother, "mirroring" the sick condition of the fetus. The mirror syndrome is also seen in sacrococcygeal teratoma and other fetal conditions that result in poor placental perfusion and likely endothelial cell injury.

Although percutaneous catheter decompression of the rare CCAM with a solitary large cyst has been performed, a hydropic fetus with a solid or multicystic CCAM diagnosed before 32 weeks of gestation can be considered for fetal lobectomy since prognosis is dismal without operation. After *in utero* CCAM resection, there is reversal of hydrops and sufficient lung growth during the remaining gestation to permit survival.

2. *Bronchopulmonary Sequestration*

BPSs are echodense, nonfunctional lung lesions with a systemic arterial blood supply arising from the aorta. The vascular supply can usually be demonstrated by color flow Doppler. The majority of prenatally diagnosed BPSs will decrease in size over gestation. A small number of lesions have an associated tension hydrothorax with secondary hydrops that necessitates fetal thoracentesis or thoracoamniotic shunting. A hydropic fetus with a BPS will die without operative treatment.

B. Congenital Diaphragmatic Hernia

Congenital diaphragmatic hernia (CDH), a failure of development of part of the diaphragm, results in pulmonary hypoplasia from compression of the lung by herniated abdominal viscera. Despite optimal postnatal care, including extracorporeal membrane oxygenation, fetuses with isolated CDH diagnosed by ultrasound prior to 25 weeks of gestation exhibit a mortality of 58%. Although sonographic predictors of survival have been difficult to define, it is now clear that liver herniation into the chest is a poor predictor of survival in fetal CDH. Sonographic detection of intrathoracic liver herniation is difficult, and recently we have begun using magnetic resonance imaging scanning to define this important sign. Other sonographic predictors of survival in CDH have included the ratio of right lung area (at the level of the four-chamber heart view) to head circum-

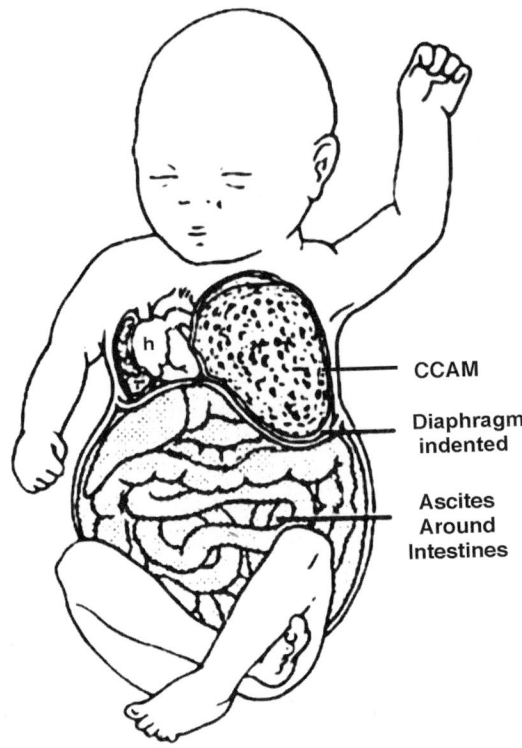

FIGURE 1 Pathophysiology of congenital cystic adenomatoid malformation (CCAM). A large fetal CCAM compresses the fetal esophagus, decreases fetal swallowing of amniotic fluid, and results in polyhydramnios. Compression of the lungs leads to pulmonary hypoplasia, whereas compression of the vena cava and heart (h) leads to fetal heart failure and death. Removal of the CCAM *in utero* reverses the fetal heart failure and allows for sufficient lung growth during the remaining gestation to allow survival.

Labels in figure: CCAM; Diaphragm indented; Ascites Around Intestines

ference as a measure of the degree of appropriate lung growth.

Two approaches to improving the survival of the fetus with congenital diaphragmatic hernia are based on accelerating fetal lung growth: primary repair of the diaphragmatic defect or occlusion of the trachea. There is significant lung growth that results from diaphragm replacement before birth. However, in fetuses with incarceration of the left lobe of the liver into the chest, reduction of the liver back into the abdomen kinks the umbilical vein, the fetal "lifeline" with blood from the placenta, and causes fetal demise. CDH fetuses without liver herniation into the chest have a reasonable prognosis when repaired after birth and are not considered candidates for *in utero* repair.

1. The Plug the Lung until It Grows Procedure

An interesting therapeutic strategy for fetal CDH, controlled tracheal obstruction to prevent the normal efflux of fetal lung fluid, was derived from animal models of CDH. Experimental tracheal occlusion increases the positive pressure in the developing lungs that is created by fetal lung fluid production. Increased intrapulmonary pressure enlarges hypoplastic lungs and reduces viscera into the abdomen (Fig. 2). After birth, the treated neonates show improved pulmonary ventilation and compliance com-

pared with untreated CDH controls. After extensive work in fetal sheep CDH models, the plug the lung until it grows (PLUG) procedure has been employed in human fetuses with CDH and liver herniation. Tracheal occlusion has been effected with intratracheal plugs and externally applied clips. It may be possible to perform tracheal occlusion using fetoscopic techniques to reduce the incidence of preterm labor that is noted after the hysterotomy used for open fetal surgery.

C. Urinary Tract Obstruction

Obstruction of the fetal urinary tract leads to renal developmental abnormalities that range from mild hydronephrosis to renal dysplasia. Severe urinary tract obstruction can also affect pulmonary development. The decrease in fetal urine output leads to oligohydramnios and subsequent fetal pulmonary hypoplasia. One of the most common causes of fetal urinary tract obstruction is posterior urethral valves, a malformation that leads to end-stage renal disease in 40% of patients by adolescence.

As with the previous fetal malformations, animal models were employed to develop methods of correction. Shunting the obstructed urinary tract before birth is effective to prevent renal and pulmonary damage. To determine when shunting was necessary, it was important to understand the natural history

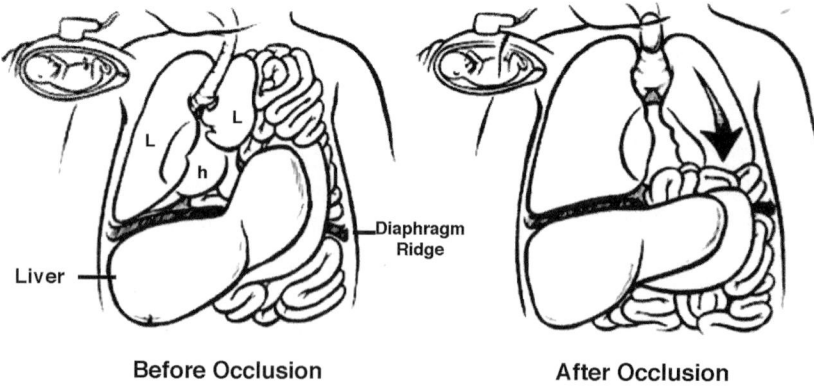

Before Occlusion   **After Occlusion**

FIGURE 2 Pathophysiology of fetal congenital diaphragmatic hernia. Failure of development of the diaphragm causes a defect through which the abdominal organs herniate. The herniated organs compress the lungs (L) and lead to pulmonary hypoplasia. Tracheal occlusion prevents the normal efflux of fetal lung fluid, enlarging the hypoplastic lungs and reducing the viscera into the abdomen. After birth, treated neonates show improved lung function compared with untreated congenital diaphragmatic hernia control patients. h, heart.

of fetal urinary obstruction. Human fetuses with urinary tract obstruction underwent ultrasound evaluation and urine sampling to formulate selection criteria for intervention. It was found that renal dysplasia can be predicted sonographically when renal cortical cysts or increased renal echogenicity are present. Poor renal function is also likely when fetal urinary electrolytes (sodium and chloride) and β 2-microglobulin are elevated. Applying these criteria, it is clear that most fetuses with obstructive uropathy do not require *in utero* treatment. Mild bilateral hydronephrosis without resultant amniotic fluid volume changes can be treated postnatally. The development of oligohydramnios warrants a complete evaluation of fetal renal function. If sonographic changes predicting renal dysplasia are present, *in utero* decompression is not warranted. When renal function is preserved, as assessed by electrolyte and sonographic criteria, early delivery and postnatal therapy can be instituted when lung maturity is present. If pulmonary maturity is not sufficient, *in utero* decompression can be considered.

The most widely employed method of shunting the fetal urinary tract has been a double pigtail catheter placed percutaneously under sonographic guidance; open fetal surgery vesicostomy to drain the bladder directly into the amniotic fluid cavity has also been performed. The effectiveness of shunting has been limited by obstruction or dislodgment of the catheters or inadequate drainage of the fetal urinary tract. A more novel approach has entailed antegrade YAG laser ablation of posterior urethral valves. Under ultrasound guidance, a technique known as fetoscopy is used to access the fetal bladder with multiple percutaneous ports through which a telescope and laser are threaded. Once the fetal urethral valves are identified, they are ablated with multiple pulses from the laser. Although there have been interesting technical advancements in the prenatal treatment of obstructive uropathy, it is still unclear whether fetal therapy can stop or reverse the renal dysplastic changes that result from severe urinary obstruction.

D. Sacrococcygeal Teratoma

While teratomas of the sacrococcygeal region represent the most common neonatal tumor, have a low malignant potential, and have a good prognosis when diagnosed after birth, the prenatal counterpart can behave very differently. A subgroup of fetuses with large teratomas will develop hydrops and die. Fetal death occurs because of high-output cardiac failure associated with a "vascular steal" to the tumor. The goal of *in utero* resection of sacrococcygeal teratoma (SCT) is to reverse high-output failure by ligating the blood flow to the tumor. We have performed fetal resection successfully in a 26-week-gestation fetus with a large SCT and high-output cardiac failure. There was no evidence of residual tumor after birth. Sonographic evaluation of fetal hemodynamics and placental dimensions may be useful to predict the development of fetal hydrops and need for surgery in an SCT fetus.

V. FETAL AIRWAY MANAGEMENT

Airway obstruction at birth due to a huge neck mass is life-threatening. In the course of treating fetuses with predictable airway obstruction, we have developed a systematic approach, the *ex utero* intrapartum treatment (EXIT) procedure, to secure the airway during delivery. The EXIT procedure is performed by using high doses of maternal inhaled halogenated anesthetic agents to facilitate uterine relaxation, performing a cesarean section, securing an airway with the placental circulation still intact, and then dividing the umbilical cord. This approach permits a variety of procedures to be performed, including bronchoscopy, endotracheal intubation, tracheostomy, and instillation of surfactant. The combination of intensive maternal–fetal monitoring, cesarean section with maintenance of fetoplacental circulation, and complete uterine relaxation provides a controlled environment for securing the airway in babies with prenatally diagnosed airway obstruction.

VI. FUTURE DIRECTIONS

A. Fetoscopy

Although fetoscopy has been used as a diagnostic tool since the 1960s, minimally invasive approaches to the fetus have progressed due to the recent advances in instrumentation and optics used for abdominal laparoscopy in children and adults. The small uterine puncture wounds required for fetoscopy may reduce the incidence of preterm labor compared to hysterotomy. Based on the benefits seen with laparoscopy, it is anticipated that the avoidance of a maternal laparotomy will reduce the maternal morbidity associated with fetal surgery.

Fetoscopy has been used in human fetuses for laser ablation of posterior urethral valves, umbilical cord ligation in twin gestations complicated by an acephalic/acardiac twin, YAG laser ablation of connecting placental vessels in twin–twin transfusion syndrome, and the PLUG procedure for tracheal occlusion in CDH. Despite its promise, fetoscopy suffers from several technical limitations. All of the instruments need to be reduced in size to fit in small trocars and reduce the risk of amniotic fluid leak and chorioamniotic separation. There is the additional challenge of uterine trocar site closure. A steep learning curve is apparent for the mastery of fetoscopic techniques. Despite these challenges, fetoscopy promises to be a significant addition to the fetal operative repertoire in the future.

B. Cellular Transplantation: Stem Cell Therapy

With the advent of prenatal bone marrow transplantation, the treatment of diseases before birth has recently been extended to include "cellular-deficiency" diseases. Flake reported the first successful treatment of a fetus with severe combined immunodeficiency syndrome using paternal, T cell-depleted bone marrow in early gestation to create hematopoietic chimerism—coexistence of both paternal and fetal bone marrow cells–resulting in a functional immune system. Fetal cellular therapy is an approach for diseases that would respond to engraftment of normal hematopoietic stem cells, including chronic granulomatous disease and some hemoglobinopathies. In addition, tolerance induction to a transplant in the developing fetal immune system may allow for later postnatal transplantation of partial or whole organs obtained from a living donor. In combination with cellular therapy, replacement

gene therapy during fetal life is an interesting consideration for fetuses with a diagnosis of a metabolic disease. Some diseases that may be amenable to fetal gene therapy include the genetic deficiencies of the urea cycle enzymes, such as ornithine transcarbamylase, cystic fibrosis, and surfactant protein B deficiency. Fetal therapy during early gestation may allow the gene and transfer vehicle to be recognized as self-antigens by introducing them prior to the development of immune competence.

VII. CONCLUSION

Using the novel technique of fetal surgery, some human fetal abnormalities may be corrected at a point in gestation before fetal injury or death have occurred. Prenatal operative intervention may become the more cost-effective and humane approach to a series of otherwise devastating fetal diseases. From the extensive research efforts to develop fetal surgery, the pathophysiology and natural history of a growing list of fetal anatomic abnormalities will be clarified for perinatologists, neonatologists, and pediatric surgical specialists.

See Also the Following Articles

FETAL ANOMALIES; FETAL MONITORING AND TESTING; TERATOGENS, OVERVIEW; TOCOLYTIC AGENTS; ULTRASOUND

Bibliography

Adzick, N. S. (1993). Fetal thoracic lesions. *Sem. Pediatr. Surg.* 2, 103–10/8.

Adzick, N. S., Harrison, M. R., Flake, A. W., Howell, L. J., Golbus, M. S., and Filly, R. A. (1993). Fetal surgery for cystic adenomatoid malformation of the lung. *J. Pediatr. Surg.* 28, 806–812.

Adzick, N. S., Crombleholme, T. M., Morgan, M. A., and Quinn, T. M. (1997). A rapidly growing fetal teratoma. *Lancet* 349, 538.

Crombleholme, T. M., Harrison, M. R., and Langer, J. C. (1988). Early experience with open fetal surgery for congenital hydronephrosis. *J. Pediatr. Surg.* 23, 1114–1118.

Flake, A. W., Harrison, M. R., Adzick, N. S., Laberge, J. M., and Warsof, S. L. (1986). Fetal sacrococcygeal teratoma. *J. Pediatr. Surg.* 21, 563–566.

Flake, A. W., Roncarlo, M., Puck, J. M., Almeida-Porada, G., Evans, M. I., Johnson, M. P., Abella, E. M., Harrison, D. D., and Zanjani, E. D. (1996). Treatment of X-linked severe combined immunodeficiency by in utero transplantation of paternal bone marrow. *N. Engl. J. Med.* **335**, 1806–1810.

Harrison, M. R., and Adzick, N. S. (1993). Fetal surgical techniques. *Sem. Pediatr. Surg.* **2**, 136–142. [Review]

Harrison, M. R., Adzick, N. S., Jennings, R. W., Duncan, B. W., Rosen, M. A., Filly, R. A., Goldberg, J. D., deLorimier, A. A., and Golbus, M. S. (1990). Antenatal intervention for congenital cystic adenomatoid malformation. *Lancet* **336**, 965–967.

Harrison, M. R., Adzick, N. S., Flake, A. W., Jennings, R. W., Estes, J. M., MacGillivray, T. E., Chueh, J. T., Goldberg, J. D., Filly, R. A., and Goldstein, R. B. (1993). Correction of congenital diaphragmatic hernia in utero: VI. Hard-earned lessons. *J. Pediatr. Surg.* **28**, 1411–1417.

Harrison, M. R., Adzick, N. S., Estes, J. M., and Howell, L. J. (1994). A prospective study of the outcome for fetuses with diaphragmatic hernia. *J. Am. Med. Assoc.* **271**, 382–384.

Harrison, M. R., Adzick, N. S., Flake, A. W., Vanderwall, K. J., Bealer, J. F., Howell, L. J., Farrell, J. A., Filly, R. A., Rosen, M. A., Sola, A., and Goldberg, J. D. (1996). Correction of congenital diagphragmatic hernia in utero: VIII. Response of the hypoplastic lung to tracheal occlusion. *J. Pediatr. Surg.* **31**, 1339–1348.

Hedrick, M. H., Estes, J. M., Sullivan, K. M., Bealer, J. F., Kitterman, J. A., Flake, A. W., Adzick, N. S., and Harrison, M. R. (1994). Plug the lung until it grows (PLUG): A new method to treat congenital diaphragmatic hernia in utero. *J. Pediatr. Surg.* **29**, 612–617.

Jennings, R. W., MacGillivray, T. E., and Harrison, M. R. (1993). Nitric oxide inhibits preterm labor in the rhesus monkey. *J. Maternal Fetal Med.* **2**, 170–175.

Metkus, A. P., Filly, R. A., Stringer, M. D., Harrison, M. R., and Adzick, N. S. (1996). Sonographic predictors of survival in fetal diaphragmatic hernia. *J. Pediatr. Surg.* **31**, 148–152.

Mychaliska, G. B., Bealer, J. F., Graf, J. L., Adzick, N. S., and Harrison, M. R. (1997). Operating on placental support: The ex utero intrapartum treatment (EXIT) procedure. *J. Pediatr. Surg.* **32**, 227–230.

<hr/>

α-*Fetoprotein and Triple Screening*

Jacob A. Canick
Brown University School of Medicine

I. Down Syndrome and Open Neural Tube Defects
II. Prenatal Screening
III. α-Fetoprotein and Prenatal Screening
IV. Triple Markers and Prenatal Screening
V. The Concept of the MoM
VI. Clinical Implementation of Triple Screening

GLOSSARY

analytes Biochemicals found in maternal serum whose concentrations are determined as part of triple screening.

Down syndrome (trisomy 21) The most common chromosomal abnormality and leading cause of severe mental retardation in the industrialized world, caused by the presence of an extra copy of the number 21 chromosome in each cell of the affected individual.

α-fetoprotein A glycoprotein synthesized by the embryonic yolk sac and fetal liver during human pregnancy. It is used clinically in second-trimester prenatal screening of fetal open neural tube defects and Down syndrome through measurement of its levels in maternal serum.

human chorionic gonadotropin A glycoprotein subunit hormone synthesized by the syncytiotrophoblast layer of the placenta during human pregnancy, used clinically in second-trimester prenatal screening of fetal Down syndrome through measurement of its levels in maternal serum.

open neural tube defects Congenital abnormalities caused by incomplete closure of the neural tube (spinal cord) during embryogenesis, the two most common forms being anencephaly (failure of the cephalic end of the neural tube to close) and open spina bifida (failure of inner portions of the neural tube to close).

screening Identification within the general population of individuals at increased risk of a medical condition that warrants treatment or preventive action.

triple screening A method of screening for fetal Down syndrome during the second trimester of human pregnancy through the measurement of α-fetoprotein, human chorionic gonadotropin, and unconjugated estriol.

unconjugated estriol Estriol is a steroid hormone synthesized by the placenta from steroid intermediates provided by the fetal adrenal and liver. Unconjugated refers to the absence of sulfate or glucuronide residues which are added to estriol through maternal metabolism to speed its clearance. Unconjugated estriol is used clinically in second-trimester prenatal screening of fetal Down syndrome through measurement of its levels in maternal serum.

Triple screening, entailing the measurement of α-fetoprotein, human chorionic gonadotropin, and unconjugated estriol, is an established clinical test for the identification of pregnant women at increased risk of carrying fetuses affected by either open neural tube defects or Down syndrome. Through the measurement of these secretory products of the fetoplacental unit in a sample of maternal serum, an estimate of a woman's risk of carrying an affected fetus can be made.

I. DOWN SYNDROME AND OPEN NEURAL TUBE DEFECTS

Down syndrome and open neural tube defects are two of the most common serious abnormalities identified after birth. The birth prevalence of Down syndrome is approximately 1 in 800 and the birth prevalence of open neural tube defects ranges from 0.5 to 2 per 1000, depending on the geographic region. The risk of giving birth to a baby with Down syndrome increases with the age of the mother so that a 20-year-old's risk is 1 in 1500 whereas a 30-year-old's risk is 1 in 900 and a 40-year-old's risk is 1 in 110. In contrast, the risk of giving birth to a baby with an open neural tube defect does not change with maternal age. Generally, Down syndrome and open neural tube defects are not inherited so that family history is usually not helpful in identifying pregnancies at high risk.

Down syndrome is caused by the presence of an extra copy of the number 21 chromosome or a portion of that chromosome in each cell of an affected individual, resulting in a karyotype of 47,XX,+21 or 47,XY,+21. Down syndrome is the most common chromosomal abnormality identified at birth and is the leading cause of severe mental retardation in the industrialized world. All affected individuals have mental retardation, moderate on average, a characteristic facial and body type, and a variety of medical abnormalities ranging from cardiac defects in 40% of cases to increased incidence of neonatal leukemia and congenital hypothyroidism. Individuals with Down syndrome who live into their fourth or fifth decade of life have a high probability of developing presenile Alzheimer's disease because of the extra dose of the amyloid precursor gene on the number 21 chromosome.

Open neural tube defects are disorders involving failure of the developing embryo's neural tube to close completely, a process which is usually completed 28 days after conception. When the cephalic (upper) end of the neural tube fails to close, the resulting condition is called anencephaly, a lethal defect in which the cranial vault is left almost completely empty. When the neural tube fails to close at some point below the anterior end, the resulting condition involves herniation of the meninges and a defect in the vertebrae and is called spina bifida. Eighty percent of spina bifida are open defects, with only a thin membrane covering the lesion. As many as 40% of infants with open spina bifida will not survive and most others have substantial problems which might include paralysis, incontinence, and hydrocephalus. Open neural defects are multifactorial in origin, although it is now known that folic acid deficiency may contribute to 75% of cases. Folic acid supplementation beginning before conception for the prevention of open neural tube defects is strongly recommended by professional societies and regulatory agencies.

II. PRENATAL SCREENING

Medical screening has been defined by Cuckle and Wald as the identification, among apparently healthy individuals, of people who are sufficiently at risk of a specific disorder to justify a subsequent diagnostic test or procedure or, in certain circumstances, direct preventive action. The same definition applied to prenatal screening for Down syndrome and open neural tube defects would be the identification of those pregnant women at increased risk of a serious fetal disorder to justify follow-up invasive and/or costly diagnostic tests such as genetic amniocentesis for chromosome analysis in fetal cells and high-resolution ultrasonography for abnormal fetal structures. The method of identification of high-risk pregnant women who are candidates for such diagnostic testing constitutes the prenatal screening test.

Before the discovery in the 1970s of a biochemical marker in maternal serum, prenatal screening for open neural tube defects was not possible. However, prenatal screening for Down syndrome was implemented before the advent of biochemical markers. The original screening test for Down syndrome was simply to determine the age of the mother based on the association between advancing maternal age and increased risk of Down syndrome pregnancy. Common practice was and continues to be to offer amniocentesis for genetic analysis to women during the early second trimester based on their maternal age. Those who will be 35 or older at the time of delivery, the so-called "advanced maternal age" group, generally comprise <5–10% of all pregnant women but give birth to about 30% of babies with Down syndrome. This enhancement in risk in older pregnant women makes them a high risk group. In contrast, women under the age of 35, while comprising more than 90% of all pregnant women, give birth to about 70% of all babies with Down syndrome Before triple screening, pregnant women only had their maternal age-related risk to guide them in deciding whether to have genetic amniocentesis.

III. α-FETOPROTEIN AND PRENATAL SCREENING

In 1956, a glycoprotein which migrated on electrophoresis between albumin and α1-globulin was found to be present in fetal but not adult blood. This so-called α-fetoprotein (AFP) is synthesized by the embryonic yolk sac and fetal liver and reaches concentrations in the fetal circulation as high as 2 or 3 mg per milliliter by the end of the first trimester of pregnancy. AFP serves as the functional equivalent of albumin in the fetal circulation, acting as an osmotic regulator and as a fatty acid carrier.

The potential clinical use of AFP in prenatal screening for open neural tube defects was discovered in the early 1970s in the United Kingdom, where in certain regions the incidence of anencephaly and open spina bifida was as high as 7 or 8 cases per 1000 babies born. For this reason, a test for these birth defects prenatally was a major public health goal. Researchers found that because of the very high concentrations of AFP normally present in the fetal circulation and cerebrospinal fluid, leakage of AFP from an open lesion into the amniotic fluid, as is found in anencephaly and open spina bifida, resulted in elevated levels of AFP measured in maternal blood. Collaborative studies in the United Kingdom during the late 1970s demonstrated the effectiveness of maternal serum measurement of AFP in screening for open neural tube defects between 15 and 20 gestational weeks. Eighty percent of open spina bifida and more than 90% of anencephaly were found in the 5% of the screened population who had the most elevated maternal serum AFP levels.

Amniotic fluid AFP levels are almost always elevated in cases of anencephaly and open spina bifida. Therefore, amniocentesis for the purpose of amniotic fluid AFP measurement serves as a diagnostic follow-up test for patients who are confirmed to have elevated maternal serum AFP levels. In addition, the amniotic fluid sample can be tested for the presence of the neuron-specific form of the enzyme acetylcholinesterase, which is almost always present when there is an open neural tube defect. As a noninvasive alternative to amniocentesis, targeted high-resolution ultrasound examination of the fetal head and spine is a common diagnostic method for identification of neural tube defects.

Maternal serum screening for open neural tube defects was widespread in the United Kingdom by 1980 but was in limited use in the United States until, in 1983, the U.S. Food and Drug Administration licensed the first AFP assay kits for clinical use. In 1985, the American College of Obstetricians and Gynecologists issued a "liability alert" to its membership recommending that all pregnant women be offered maternal serum AFP screening for the detection of open neural tube defects. By 1988, more than half of all pregnant women in the United States were having the AFP test.

In 1984, it was discovered that pregnancies affected with fetal Down syndrome had, on average, almost 30% lower levels of maternal serum AFP than unaffected pregnancies. This information could be used in combination with maternal age-associated risk of Down syndrome to calculate a new patient-specific risk. For the first time, it was possible to determine whether a woman under the age of 35, previously considered to be at relatively low risk, might be at sufficiently high risk to be offered genetic amniocentesis. When applied as a screening test to women under the age of 35, approximately 20% of all Down syndrome pregnancies could be found in the 5% of the screened population with the highest calculated risk.

IV. TRIPLE MARKERS AND PRENATAL SCREENING

With the discovery that maternal serum AFP levels were low in Down syndrome pregnancy, other products of the fetoplacental unit, measurable in maternal serum, were examined. In 1987, maternal serum concentrations of human chorionic gonadotropin (hCG), the placental hormone commonly measured in early pregnancy, were found to be on average two times higher, whereas concentrations of the fetoplacental steroid hormone, unconjugated estriol (uE3), were found to be about 30% lower in Down syndrome pregnancies than in unaffected pregnancies. The three analytes measured together in a sample of maternal serum between 15 and 22 gestational weeks (triple screening) were found to substantially enhance screening for fetal Down syndrome. When applied as a screening test to all pregnant women, approximately 60% of all Down syndrome pregnancies could be found in the 5% of the screened population with the highest calculated risk, about a three-fold improvement compared to screening with AFP alone.

The pathophysiologic basis of triple screening for Down syndrome is poorly understood. There appears to be a reciprocal relationship between maternal se-

rum levels of analytes whose origin is at least in part the fetus and analytes whose origin is the placenta alone. Thus, in the second trimester, maternal serum levels of AFP (a fetal liver product), uE3 (a placental product derived from precursors synthesized in the fetal adrenal and liver), and 16α-hydroxy DHEA sulfate (a fetal liver product) are all low in Down syndrome pregnancy. In contrast, a variety of pure placental products, including hCG, the free α and β subunits of hCG, inhibin-A, human placental lactogen, Schwangershaft protein, and progesterone, are all elevated in maternal serum from Down syndrome pregnancy in the second trimester. These findings indicate a possible generalized compensatory mechanism between a poorly functioning trisomy 21 fetus and a hyperfunctional trisomy 21 placenta.

Triple marker patterns in certain other chromosomal abnormalities have been characterized. The pattern in the usually lethal chromosomal abnormality trisomy 18 (Edwards syndrome) is best known; AFP and uE3 levels are even lower than they are in Down syndrome, and hCG and free β-hCG levels are very low instead of elevated. Because of these differences, cases of trisomy 18 will not generally be detected within a Down syndrome screening protocol. However, various screening strategies to identify women at greatly increased risk of trisomy 18 have been devised and implemented. Currently, trisomy 18 is the only chromosomal defect other than Down syndrome specifically targeted for screening with triple markers.

V. THE CONCEPT OF THE MoM

The method of prenatal screening using maternal serum analytes, exemplified by triple screening, was conceived by Wald for neural tube defect screening with AFP. Certain key insights were incorporated into the method. Maternal serum concentrations of AFP, as is true for all maternal serum analytes, are continually changing during the course of the early second trimester when screening is done. In addition, the levels of each screening analyte are measured by immunoassay, and different antibodies found in different assays will not always provide comparable results across assays. For this reason, Wald developed the concept of the MoM (multiple of the median) as a method of normalizing screening results. The median rather than the mean was chosen as a better estimate of the reference value because the distribution of analyte concentrations was right skewed, or log-Gaussian. By dividing the patient-specific concentration by the population median for the same gestational age, a simple yet powerful normalizing unit, the MoM, was created.

By definition, the most common value for an unaffected, singleton pregnancy is 1 MoM, and simple ratios of that value can define cutoffs for generation of screen-negative and screen-positive results. Patient-specific MoM values allow comparison of results from different laboratories, from different periods of gestation, and from different clinical circumstances. The MoM has become the currency used in prenatal screening throughout the world, and its value cannot be overstated.

VI. CLINICAL IMPLEMENTATION OF TRIPLE SCREENING

In screening for Down syndrome, the patient-specific MoM values calculated for each of the three

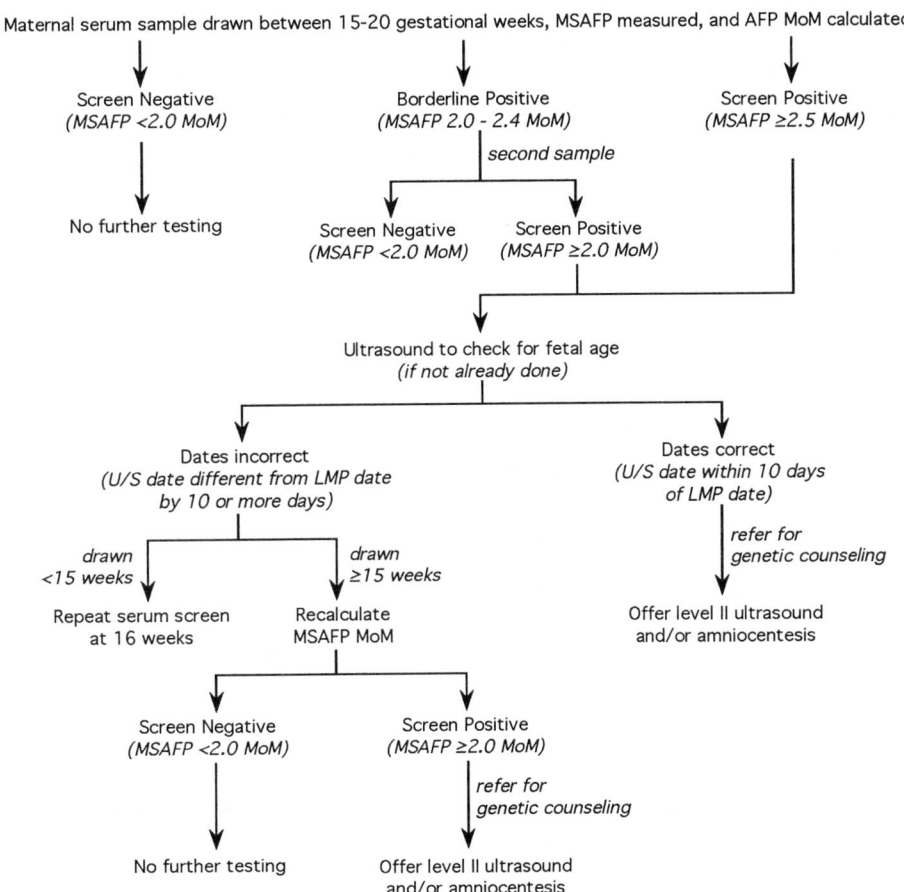

FIGURE 1 A clinical protocol used in maternal serum AFP screening for open neural tube defects. The choice of a cutoff in this protocol is 2.0 MoM, although 2.5 MoM is also commonly used. The extent of dating discrepancy between last menstrual period (LMP) and ultrasound (U/S) measurement which would trigger recalculation of the AFP MoM varies among screening programs but is usually between 10 and 17 days. Each screening program must decide on its own MoM cutoff and degree of dating discrepancy in maternal serum AFP screening for open neural tube defects. Level II ultrasound refers to targeted, high-resolution sonographic examination of the fetal head, spine, and abdominal wall as a follow-up diagnostic test for confirmed screen-positive patients.

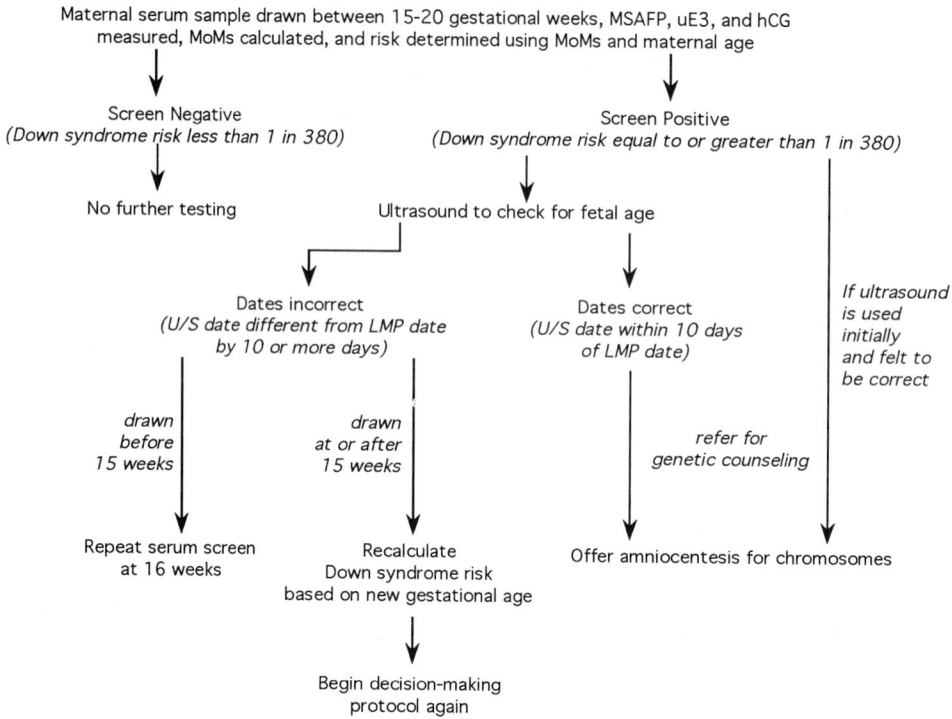

Maternal serum sample drawn between 15-20 gestational weeks, MSAFP, uE3, and hCG measured, MoMs calculated, and risk determined using MoMs and maternal age

Screen Negative
(Down syndrome risk less than 1 in 380)

No further testing

Screen Positive
(Down syndrome risk equal to or greater than 1 in 380)

Ultrasound to check for fetal age

Dates incorrect
(U/S date different from LMP date by 10 or more days)

drawn before 15 weeks

drawn at or after 15 weeks

Dates correct
(U/S date within 10 days of LMP date)

If ultrasound is used initially and felt to be correct

refer for genetic counseling

Repeat serum screen at 16 weeks

Recalculate Down syndrome risk based on new gestational age

Offer amniocentesis for chromosomes

Begin decision-making protocol again

FIGURE 2 A clinical protocol used in triple screening for Down syndrome. The choice of a risk cutoff in this protocol is 1 in 380, the term risk of a 35-year-old woman, although 1 in 250, the term risk of a 37-year-old, is also commonly used. The extent of dating discrepancy between last menstrual period (LMP) and ultrasound (U/S) measurement which would trigger readjustment of risk varies among screening programs but is usually between 10 and 17 days. Each screening program must decide on its own risk cutoff and degree of dating discrepancy in triple screening for Down syndrome.

analytes are combined in a statistical algorithm to modify that patient's a priori, maternal age-associated risk. The most commonly used method is called multivariate Gaussian distribution analysis, first applied univariately in screening with AFP. Other statistical methods, such as logistic regression and discriminant analysis, are less often used. The end result is that each person screened has a new risk based on a combination of results which better defines her actual risk of having a baby affected with Down syndrome. A screen-positive result is defined as a patient-specific risk sufficiently high that follow-up diagnostic testing should be offered. Most commonly, a risk cutoff of 1 in 380, the risk of a 35-year-old at term, or 1 in 250, the risk of a 37-year-old at term, is chosen to indicate screen positive vs screen negative.

A clinical program must integrate patient and pregnancy dating information with the laboratory results and then have in place genetic counseling, obstetrical follow-up procedures, and cytogenetic laboratory analysis. Once a screen-positive result is generated, rapid patient follow-up is essential because of the anxiety created in the patient and the limited time available for decision making if an abnormality is ultimately identified. Examples of actual protocols used in screening for open neural tube defects using AFP and Down syndrome using triple markers are shown in Figs. 1 and 2.

See Also the Following Articles

Fetal Monitoring and Testing; Teratogens, Overview

Bibliography

Canick, J. A., and Knight, G. J. (1992). Multiple-marker screening for fetal Down syndrome. *Contemp. Ob/Gyn* 36, 25–42.

Cuckle, H. S., and Wald, N. J. (1984). Principles of screening. In *Antenatal and Neonatal Screening* (N. J. Wald, Ed.), pp. 1–22. Oxford Univ. Press, Oxford, UK.

Haddow, J. E., and Palomaki, G. E. (1993). Prenatal screening for Down syndrome. In *Essentials of Prenatal Diagnosis* (J. L. Simpson and S. Elias, Eds.), pp. 185–220. Churchill Livingstone, New York.

Knight, G. J., and Palomaki, G. E. (1992). Maternal serum alpha-fetoprotein and the detection of open neural tube defects. In *Maternal Serum Screening for Fetal Genetic Disorders* (S. Elias and J. L. Simpson, Eds.), pp. 41–58. Churchill Livingstone, New York.

Saller, D. N., and Canick, J. A. (1996a). Maternal serum screening for fetal trisomy 21: Clinical aspects. *Clin. Obstet. Gynecol.* **39**, 783–792.

Saller, D. N., and Canick, J. A. (1996b). Maternal serum screening for fetal trisomy 21: The detection of other pathologies. *Clin. Obstet. Gynecol.* **39**, 793–800.

Wald, N. J. (1976). The detection of neural tube defects by screening maternal blood. *INSERM* **8**, 227–238.

Wald, N. J., and Cuckle, H. S. (1984). Open neural-tube defects. In *Antenatal and Neonatal Screening* (N. J. Wald, Ed.), pp. 25–73. Oxford Univ. Press, Oxford, UK.

Wald, N. J., Cuckle, H. S., Densem, J. W., Nanchahal, K., Royston, P., Chard, T., Haddow, J. E., Knight, G. J., Palomaki, G. E., and Canick, J. A. (1988). Maternal serum screening for Down's syndrome in early pregnancy. *Br. Med. J.* **297**, 883–887.

Fetus, Overview

Timothy A. Cudd

Texas A&M University

I. Introduction
II. Central Nervous System
III. Pulmonary System
IV. Cardiovascular System
V. Fluid Balance
VI. The Determination of the End of the Fetal Period
VII. Issues and Questions

GLOSSARY

apoptosis Programmed cell death.

ductus arteriosus A vascular connection in the fetus between the pulmonary artery and aorta that permits blood to bypass the lung and left heart.

ductus venosus A vascular connection in the fetus between the umbilical vein and the vena cava that permits blood to bypass the liver.

foramen ovale An opening between the vena cava and the left atrium in the fetus that permits blood to bypass the right ventricle and proceed directly to the left heart.

in utero Within the uterus.

meconium Fetal gastrointestinal excrement composed mainly of digested debris from swallowed amniotic fluid.

organification The process of original organ formation that occurs during the embryonic period.

PO₂ Partial pressure of oxygen.

postnatal Following birth.

surfactant A complex mixture of lipid (90%) and protein (10%) produced by type II pulmonary epithelial cells that is responsible for reducing surface tension in pulmonary alveoli.

urachus A structure connecting the fetal bladder to the allantoic cavity.

The fetal period of life is defined as the period of development beginning with the end of organification (the embryonic period) and ending with parturition and the beginning of postnatal life. In the human, the fetal period begins at 8 weeks of gestation. During the embryonic period, rudimentary formation of the organ systems is accomplished. However, these systems become better organized, increasingly functional, and much larger during the fetal period. The fetal period ends with the conceptus fully capable of existence outside of the womb. Postnatal life begins following a complex series of events organized around birth and highlighted by the beginning of respiration and the redirection of blood flow from the fetal to the postnatal pattern.

I. INTRODUCTION

The fetal period of development has attracted the attention of philosophers, physicians, and scientists since at least as early as writings are available. Pre-Socratic philosophers expressed views on human generation though these views survive only in fragmentary form. Hippocrates, who fit all these labels, provides the earliest surviving written treatise on the subject of fetal development and birth. Hippocrates discussed the period of organification and made the distinction between this period and the period of the most significant gestational growth of the conceptus, the fetal period. He related the termination of gestation with the end of the ability of the mother and placenta to adequately provide for the growing fetus:

The nutrient for growth which the mother's body provides is no longer sufficient for the child. Once these are no longer sufficient and the child is already big, in its desire for more nutriment than is there, it tosses about and so ruptures the membranes. By the same principle animals both domestic and wild give birth at the proper time for each species, and no later: for there must necessarily be a definite time for every species of animal, at which the food supply becomes insufficient.

Even though Hippocrates relied entirely on observation and reason, his conclusion, that the increase in fetal demand beyond supply is responsible for triggering parturition, is still championed in albeit refined versions of this hypothesis.

While the ancients were greatly interested in fetal life, they could only make remarks on the fetal condition based on postmortem observations of the fetus at different times throughout gestation. They were unable to critically test any hypothesis about fetal function, regulation, or development. The ability of investigators to monitor or manipulate the fetus to test hypotheses on how the fetus functions and devel-

ops was achieved much more recently than for postnatal animals; observing, instrumenting, and altering physiological functions within the womb presented investigators with a greater technical obstacle. Knowledge increased dramatically after techniques were developed to study the living fetus. Sir Joseph Barcroft and colleagues during the early twentieth century were the first to systematically study fetal physiological function. This was made possible by the development of experimental preparations that permitted the manipulation and sampling from the fetal circulation in the living fetus; the mother was anesthetized and the fetus was surgically exteriorized and the fetus survived through the preservation of umbilical blood flow. Geoffry Dawes and colleagues, after World War II, utilized these techniques to describe the alterations of fetal circulation at birth. However, the value of this technique was limited to the study of physiological processes that were not altered by anesthesia and that could be studied within a time span of a few hours.

The next important step in fetal experimentation, made by Donald Barron and co-workers, was the development of a chronic preparation that would allow the fetus to be studied in the womb free of the influence of anesthesia. Fetuses were surgically instrumented with catheters, returned to the womb, recovered, and studied without the effects of anesthesia. This technical development made it possible to study all fetal systems or dynamics that could not be assessed in anesthetized, exteriorized, and short-term preparations including the function of the endocrine, pulmonary, renal systems, cardiovascular reflex function, fluid balance, and metabolism. The sheep has contributed significantly to this knowledge because of the success investigators have had in instrumenting the ovine fetus both as an acute and as a chronic preparation. The study of fetal development and function continues to be an area of intense study that is being advanced by increasingly sophisticated scientific approaches employed by investigators around the world.

II. CENTRAL NERVOUS SYSTEM

During the embryonic period, the primitive nervous system arises from ectodermal cells interacting with surrounding mesoderm. Ectodermal tissue becomes committed to the formation of the neural plate upon initiation of neural induction. The neural plate gives rise to the formation of the neural tube, which finally becomes the brain and spinal cord. Soon after its formation, the neural tube becomes segmented into the forebrain, midbrain, hindbrain, and spinal cord. These regions then give rise to the telencephalon and diencephalon, the mesencephalon, the mesencephalon and myelencephalon, and spinal cord, respectively—regions of the brain that become distinguishable by the beginning of the fetal period. During the fetal period, there is tremendous proliferation of neurons and glia and differentiation of the wide variety of neuronal phenotypes. Neurons migrate to form the many billions of connections necessary for mature nervous system function. Brain growth and development occurs in a series of continuous stages of cell number increase, differentiation, migration, and selective apoptosis. These changes begin in the embryonic period and continue into the postnatal period.

The metabolic needs of fetal nervous tissues are high because of the high rate of growth and development. Therefore, reductions in nutrients or oxygen can have serious negative consequences on normal nervous system development. Short-term substantial or long-term but more modest reductions in oxygen delivery can result in fetal brain injury or developmental abnormalities. This period of vulnerability exists in large part because of the highly orchestrated nature of central nervous system development. Conditions that result in delayed growth and development will have permanent consequences; developmental events in the central nervous system must occur at the appointed time during fetal and early neonatal development or normal development is disturbed or the opportunity for sufficient cellular replication and neuronal migration is lost.

III. PULMONARY SYSTEM

One of the most obvious differences when comparing the fetus and postnatal animal is that the fetus does not require lungs for the oxygenation of blood until the moment of birth; oxygen is acquired from the placenta before birth. Though the lung does not play a role in fetal oxygenation, its functions *in utero* are essential for fetal life. By the end of the embryonic period, pulmonary development is rudimentary. The lungs at this stage of development are composed of the major airways and appear histologically as structures that are secretory in function (which they are). The pulmonary tree becomes increasingly branched and alveoli form during the fetal period. The air-filled structures of the lung after birth are fluid-filled during fetal life. Lung liquid secretion begins at least by midgestation in the fetal sheep. The production of fetal lung liquid is an active transport process performed by pulmonary epithelial cells resulting, in sheep, in the production of 1–4 ml/kg/hr liquid (increasing in rate with advancing gestation and pulmonary surface area). Fetal lung liquid is plasma-like in osmolality and in sodium and potassium concentration but is higher in chloride and lower in bicarbonate and protein content when compared to plasma. Once secreted by pulmonary epithelial cells, the fluid moves to the major airways, out of the trachea and larynx and into the pharynx, where it is either swallowed or passes into the amniotic cavity. The production of fetal lung liquid is essential for the normal development of lung architecture. During labor and delivery, lung liquid secretion ceases and pulmonary epithelial cells become resorptive in response to β-adrenergic receptor activation and vasopressin; plasma catecholamine and vasopressin concentrations increase to high concentrations at the time of labor and delivery and are essential in signaling lung epithelial cells to switch from secretory to absorptive function. Even though the fetal lung is not involved in oxygenation, fetal breathing begins before birth and is necessary for normal lung development. Fetal breathing movements are first detectable at 10 weeks in the human fetus and occur as intermittent bouts through the end of gestation.

Among the most clinically important disorders of the premature infant is respiratory insufficiency. While there are many causes of respiratory insufficiency, a predictable one in the premature infant is respiratory distress syndrome. As a result of pulmonary immaturity at the time of birth, the infant develops progressive pulmonary atelectasis. This is largely due to a deficiency of the surface-active material, surfactant. This material lines the alveoli and reduces surface tension resulting in the maintenance of alveolar stability at the low pressures of expiration. Its absence results in the progressive collapse of alveoli. The severity of this problem in premature infants has been greatly reduced by the administration of exogenous surfactants to premature infants especially if performed early in the postnatal period before the progression of atelectasis. The endogenous production of surfactant is promoted by glucocorticoids. Glucocorticoid concentrations rise at the end of gestation and especially at the time of delivery. Women expected to give birth prematurely are given exogenous glucocorticoids. Glucocorticoids readily cross the placenta and act on the fetal lung to accelerate fetal lung maturation and the formation of surfactant. If there is sufficient time between the initiation of glucocorticoid therapy and parturition, the infant will produce sufficient surfactant even if birth is significantly premature. Advances in the understanding of pulmonary development have been among the most important in terms of reducing neonatal morbidity and mortality.

IV. CARDIOVASCULAR SYSTEM

The fetal and maternal circulations are physically separated. Oxygenation of fetal tissues involves maternal circulation of blood to the uterus, fetal circulation of blood to the placenta, and exchange of oxygen across the placenta. Oxygen from maternal blood perfusing the uterus must diffuse to the fetal blood within the fetal placenta. The fetal cardiovascular system is then responsible for distributing oxygenated blood to fetal tissues. The placenta does not impede the transfer of oxygen from mother to fetus. However, the partial pressures of oxygen in fetal blood are considerably lower in the fetus than in the mother. The large difference in fetal and maternal partial pressure of oxygen (PO_2) is thought to be due to placental metabolism and oxygen consumption and to perfusion inequalities across the placenta. While fetal and maternal PO_2 values are very different, oxygen content is less so. Umbilical venous blood (more highly oxygenated than umbilical artery blood) and maternal arterial blood PO_2 values are 35 and 95 mmHg, respectively, and oxygen contents are

5.5 and 6.8 mM, respectively. The differences in these values narrow in part as a consequence of higher hemoglobin concentrations in the fetus. Contributing more than the higher hemoglobin concentration in fetal blood to the oxygen carrying capacity is the difference in hemoglobin affinity for oxygen. Fetal hemoglobin is said to be "shifted to the left" compared to maternal hemoglobin, meaning that fetal hemoglobin achieves a greater degree of saturation for a given PO_2 (Fig. 1). Nevertheless, the greater fetal affinity of fetal hemoglobin for oxygen does not present a problem for unloading oxygen to fetal tissues. This is because the operational range of the fetal hemoglobin saturation curve is very steep, meaning that for small changes in PO_2 there are large changes in saturation. Hence, there is effective unloading of fetal hemoglobin at the tissues.

Even before the beginning of the fetal period there is circulation of blood through a primitive vasculature and a four-chambered heart. However, important differences exist in the pattern of blood flow in the fetus when compared to the postnatal animal (Fig. 2). Cardiac output in postnatal mammals is the rate of blood flow through either the aortic valve or the pulmonary valve; the outputs from the left and right heart are equivalent. However, in the fetus, cardiac output is usually quantified as and referred

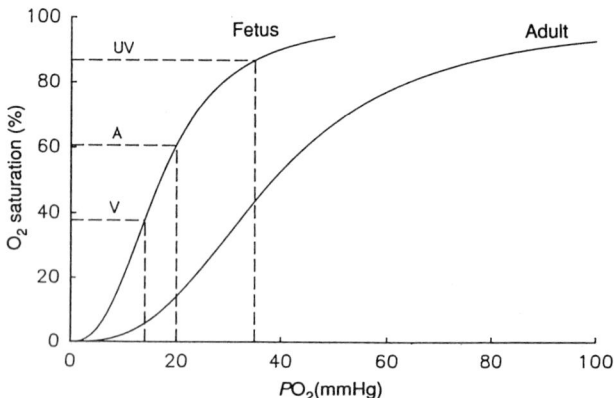

FIGURE 1 Oxygen dissociation curves for fetal and adult sheep. UV, A, and V denote the PO_2/O_2 saturation relationship for umbilical venous, umbilical arterial, and systemic venous blood, respectively [reproduced with permission from D. W. Rurak, In *Textbook of Fetal Physiology* (G. D. Thorburn and R. Harding, Eds.), Oxford Univ. Press, Oxford, UK, 1994].

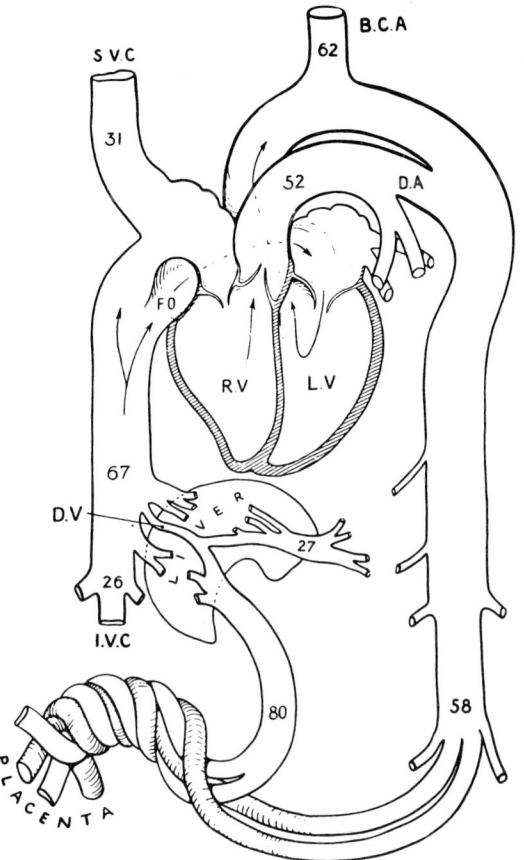

FIGURE 2 Fetal circulation. Values are O_2 saturation (%). RV, right ventricle; LV, left ventricle; SVC, superior vena cava; BCA, brachiocephalic artery; FO, foramen ovale; DA, ductus arteriosus; DV, ductus venosus (reproduced with permission from Born *et al.*, 1954).

to as "combined ventricular output" because outputs from the left and right heart in the fetus are not equivalent; 60–70% of the combined total is from the right heart. Venous blood returns to the heart and 34% of combined ventricular output passes directly to the left atria through a structure referred to as the foramen ovale. Because the left heart provides perfusion pressure for the circulation of the placenta (a low-resistance structure) as well as the fetal body, the left heart pressures are low and shunting of blood through the foramen ovale is right to left.

Most of the blood entering the right heart (55% of the combined ventricular output) passes from the pulmonary artery through the ductus arteriosus and into the aorta, whereas only about 10% passes through the lungs. Combined ventricular output is prodigious in the fetus: 462–548 ml/min in a 2–3 kg late-term fetal lamb compared to a cardiac output of 74 ml/kg/min in the nonpregnant adult female sheep. This is necessarily so in order to meet the high oxygen demands in the fetus. Oxygen delivery and oxygen consumption are much higher in the fetus compared to the mother (1100 compared to 650 μ and 340 compared to 195 μmol/min/kg in the fetal and maternal sheep, respectively). The high combined ventricular output together with high hemoglobin concentrations and the "left-shifted" fetal hemoglobin provide for the high oxygen demands of the fetus. Fetal blood is also shunted through the liver by the ductus venosus; adult liver, kidney, and respiratory functions are all largely performed by the placenta during fetal life.

At parturition, a process occurs by which the full complement of blood pumped by the right heart is directed through the lungs. This process involves the ventilation of the lungs, complex paracrine and endocrine actions not yet fully elucidated that result in pulmonary vasodilation, and closure of the ductus arteriosus. The umbilicus is broken and the remnants constrict. The ductus venosus also closes. Together, these changes result in an elevation in arterial and left heart pressures and decreases in pulmonary artery and right heart pressures, resulting in the end of right to left shunting of blood through the foramen ovale. The valve-like construction of this opening prevents left to right shunting. With the initiation of respiration and the successful orchestration of these changes in circulatory pattern, the fetus becomes a neonate, capable of maintaining sufficient tissue oxygenation outside of the womb.

The availability of oxygen to the fetus may be reduced by decreases in maternal perfusion of the uterus, a decrease in placental function, compression of the umbilical cord, or decreases in fetal combined ventricular output. The fetus employs protective strategies in response to reductions in oxygen availability. The fetal reflex responses to hypoxia, bradycardia, apnea, and redistribution of blood flow, are phylogenetically ancient because they are present in all classes of vertebrates. These same responses occur in aquatic animals such as ducks and seals and are referred to collectively as the "diving response," so called because these changes occur during dives. These reflex responses are lost in humans and other land mammals following parturition. Reductions in oxygen availability also result in cessation of all fetal body movements as well as breathing to reduce oxygen demand. Fetal blood flow is redistributed much like that in terrestrial mammals facing crises such as significant hemorrhage or like that of a diving sea mammal: the shunting of blood from tissues of lesser short-term importance (e.g., skin and gut), to heart, brain, and adrenals. While the mammalian fetus reduces oxygen consumption through behavioral means, lower animals such as turtles exposed to hypoxic conditions demonstrate the so-called "reverse Pasteur effect"; that is, glycolytic activity and energy demand by the tissues decrease with decreased availability of oxygen, a tremendously advantageous adaptation. There is no evidence that the mammalian fetus is capable of this type of adaptation.

V. FLUID BALANCE

Water content in the early conceptus is very high compared with the postnatal state. With advancing fetal development there is a decline in water content from 95% in the human fetus at 8 weeks' gestation (the beginning of the fetal period) to adult values of around 60%. While total body water declines overall with advancing gestation, extracellular water declines steeply while intracellular water increases. The fetus, like the postnatal animal, has intracellular, extracellular, intravascular, and interstitial fluid spaces as well as additional fluid compartments and fluid exchange interfaces not found in the postnatal animal (Figs. 3 and 4). The fetus develops within the fluid-filled amniotic cavity. Amniotic fluid is continually in contact with the skin of the fetus, the fetal respiratory tract through fetal lung liquid secretions, and the gastrointestinal tract. Amniotic fluid contains urine, squamous debris, meconium, and growth factors. Fetal swallowing begins by 11 weeks' gestation in humans and by 60 days in the sheep. The sheep fetus swallows 100–1000 ml/day. This fluid is composed of amniotic fluid and fetal lung liquid. The normal development of the gastrointestinal tract depends on the fetus swallowing amniotic fluid and lung liquid. Fetal urine production begins near the eighth week of gestation and is the largest contribution to amniotic fluid volume. The fetal urinary tract in sheep and other species is continuous with the allantoic cavity by means of the urachus. Urine may also pass into the amniotic cavity via the urethra.

Amniotic and allantoic compartments are in close apposition to one another and important transfers of fluids occur between the two compartments. Fluids in the fetus move between the compartments by intramembranous pathways (the exchange between the fetal blood and amniotic fluid via the placenta,

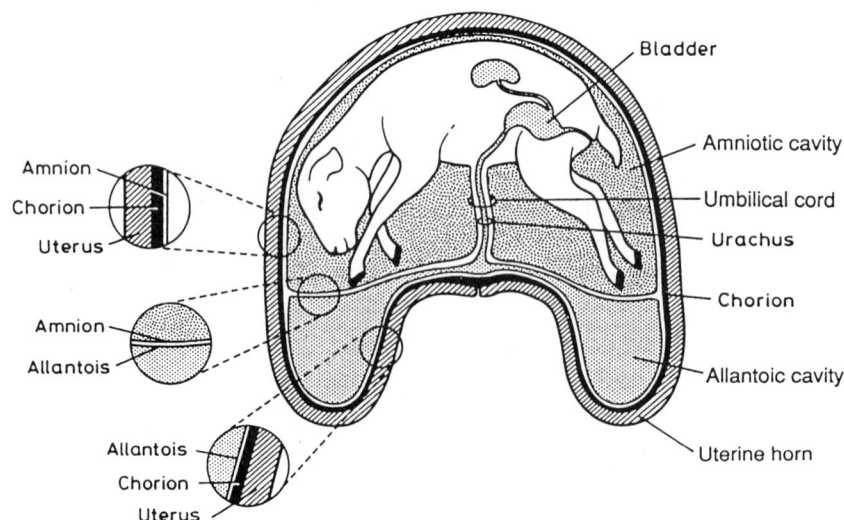

FIGURE 3 Schematic of ovine fetal membranes (from Gilbert, 1993. Reprinted with permission from Wiley-Liss).

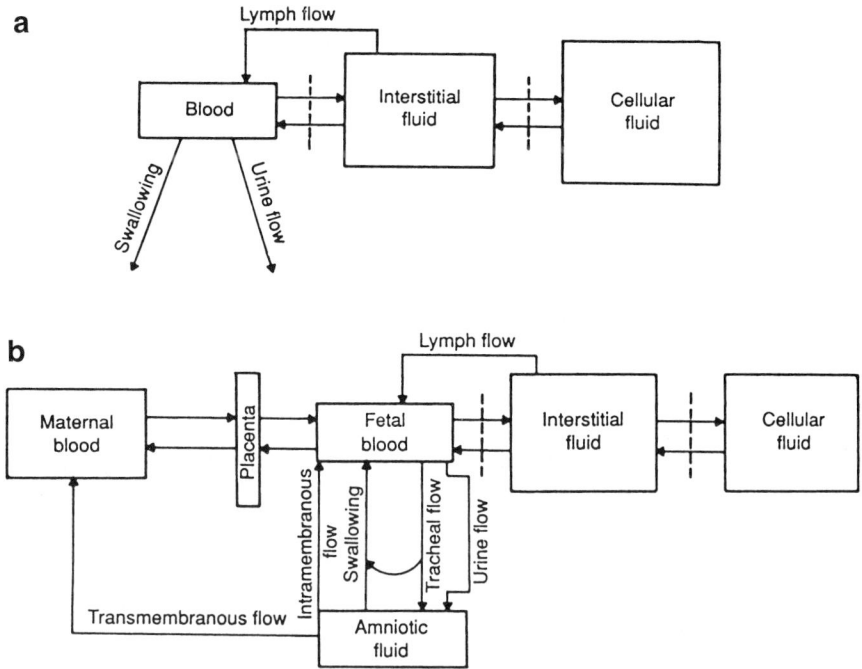

FIGURE 4 A comparison of adult (a) and fetal (b) fluid compartments and routes of fluid movement [reproduced with permission from R. A. Brace, In *Textbook of Fetal Physiology* (G. D. Thorburn and R. Harding, Eds.), Oxford Univ. Press, Oxford, UK, 1994].

fetal membranes, skin, and umbilicus) and by transmembranous pathways (the exchange between the amniotic fluid and maternal blood). Blood volume in the fetus is much higher than that in the adult(110–115 ml/kg compared to an average of 80 ml/kg in the adult). The fetus is able to regulate the movement between compartments to defend blood volume and blood pressure. Transcapillary and lymphatic flow is much greater in the fetus compared to the postnatal animals. As a result, the fetus can restitute blood volume following hemorrhage far faster than the adult. Conversely, increases in fetal blood pressure will reduce blood volume again because of the high conductance of the fetal capillaries.

VI. THE DETERMINATION OF THE END OF THE FETAL PERIOD

In the majority of cases, the fetus is delivered at an appropriate time, when maturation is complete enough to support life outside of the womb. Nevertheless, premature parturition occurs often enough to give it status as an important health problem.

In some cases, severe fetal malformation, death, or infection are the cause or rationale for premature birth. In other cases, a cause is not apparent. These cases have served to promote curiosity into what controls the timing of birth. A variety of hypotheses have been put forward to explain how the fetus might "know" when development is sufficient to permit successful birth. It has long been believed that the decision resides with the fetus. As early as the fifth century BC, Hippocrates believed that the fetus determined when the womb could no longer support the expanding needs of a growing conceptus and would then initiate the process of birth: "When it is time for the mother to give birth, what happens is that the child by the spasmodic movements of its hands and feet breaks the internal membranes." In the early twentieth century, Barcroft proposed that the fetus senses a diminished supply of oxygen and this leads to the initiation of parturition. Recently, Geoffry Thorburn proposed that a diminished supply of nutrients relative to demand results in an increase in circulating prostaglandins emanating from the placenta and that these prostaglandins act on the fetus to begin the process of parturiton. Others have focused on the possibility that a "clock" resides in the fetal brain that dictates when it is time for parturition to proceed. The basis for this hypothesis was an observation by an obstetrician, Percy Malpas, who reported that the length of gestation in anencephalic fetuses is significantly prolonged. A second observation, that gestation is greatly prolonged in sheep giving birth to cyclopian lambs, bolstered this hypothesis. Sheep raised in the western United States that consume the toxic plant, *Veratrum californicum*, early in pregnancy give birth to lambs that have defects in midline structure formation including cyclopia and agenesis of the pituitary and hypothalamus. Graham Liggins was the first to champion the hypothesis that these missing tissues, the fetal brain and pituitary, are responsible for controlling the timing of birth by increasing fetal cortisol production at the end of gestation. Liggins provided substantial experimental evidence to support this hypothesis by demonstrating that fetal adrenalectomy or hypophysectomy prolongs gestation, whereas the administration of adrenocorticotropin or cortisol promotes the initiation of parturition. In the sheep, it is well established that fetal hypothalamic drive to the fetal pituitary and successively to the fetal adrenals results in an increase in fetal cortisol production and cortisol-mediated induction of placental $P450_{c17}$. Increased expression of this enzyme results in an increase in estrogen production. Estrogen acts at the myometrium to set the stage for the initiation of parturition. What drives this system in the first place in unknown. How does the brain "know" that it is an appropriate time for parturition to take place? Is there an intrinsic clock as some have suggested or does the brain receive a signal from somewhere else as suggested by Thorburn? Is this the only mechanism by which the timing of parturition is controlled or are multiple signals integrated to create a final triggering signal? Blockade or removal of the hypothalamus–pituitary–adrenal axis does not prevent parturition but delays it, suggesting that removal reduces the precision of the mechanisms that control the timing of parturition.

It is likely that the fetus is important in determining the timing of parturition in all mammalian species. However, much less is known about how the timing of birth is controlled in species other than the sheep. The mechanisms controlling the timing of birth in human and nonhuman primates are known to be different from that of the sheep. Primate placental tissues do not express $P450_{c17}$ and this enzyme is not induced by fetal cortisol. Instead, fetal dehydroepiandrosterone (DHEA), a 17-hydroxylated adrenal product, is made available for the formation of estrogen. Fetal adrenal DHEA production increases at the end of gestation. What controls the increase in fetal DHEA production is not known. It is believed that this mechanism, though perhaps a contributing signal, is not the only one controlling the timing of parturition in primates. Others have proposed that the timing of parturition in primates is controlled by the placenta.

VII. ISSUES AND QUESTIONS

Many important questions remain to be answered. As discussed, the mechanisms controlling the timing of parturition remain to be elucidated. A complete understanding of the regulation of pulmonary vascular resistance during the transition from fetal to neonatal life has not yet been achieved. Regulation of fetal development and differentiation is poorly understood. While it is well appreciated that hormones and growth factors are involved in orchestrating fetal growth and differentiation, recent reports suggest that changes in maternal nutrition may alter this "programming." David Barker has presented considerable epidemiological evidence suggesting that conditions during the fetal period, specifically maternal and hence fetal nutrition, have a lifelong impact on the individual. For example, the predilection for an individual to develop cardiovascular disease as an adult is "programmed" for during the fetal period by the nutritional status during gestation. The possibility that programming for disease incidence in the adult occurs during the fetal period is an intriguing one that has drawn further attention to the fetal period. Finally, alcohol and other drugs are known to alter central nervous system and other tissue development. The societal consequences of this are large because these substances are frequently consumed by pregnant women, especially during the early gestation before the pregnancy has been recognized.

These questions, and many others, drive a continuing interest in the fetal period of life.

See Also the Following Articles

NUTRITIONAL FACTORS AND REPRODUCTION; PRETERM LABOR AND DELIVERY; RESPIRATORY DISTRESS SYNDROME

Bibliography

Barker, D. J. P., Gluckman, P. D., Godfrey, K. M., Harding, J. E., Owens, J. A., and Robinson, J. S. (1993). Fetal nutrition and cardiovascular disease in adult life. *Lancet* **341**, 938–941.

Battaglia, F. C., and Meschia, G. (1986). *An Introduction to Fetal Physiology*. Academic Press, New York.

Born, G. V. R., Dawes, G. S., Mott, J. C., and Widdicombe, J. G. (1954). Changes in the heart and lungs at birth. *Cold Spring Harbor Symp. Quant. Biol.* **19**, 102–108.

Dawes, G. S. (1968). *Foetal and Neonatal Physiology*. Year Book Med. Pub., Chicago.

Gilbert, W. M. (1993). Intramembranous absorption of water from the ovine allantoic cavity. *J. Maternal–Fetal Med.* **2**, 55–61.

Lonie, I. M. (1981). *The Hippocratic Treatises "On Generation," "On the Nature of the Child" and "Diseases IV."* de Gruyter, New York.

Nathanielsz, P. W. (1996). *Life Before Birth: Challenges of Fetal Development*. Freeman, New York.

Rosenfeld, C. R. (1977). Distribution of cardiac output in ovine pregnancy. *Am. J. Physiol.* **232**(3), H231–H235.

Thorburn, G. D., and Harding, R. (1994). *Textbook of Fetal Physiology*. Oxford Univ. Press, New York.

Fish, Modes of Reproduction in

Rudolf Reinboth

University of Mainz, Institute for Zoology

I. Introduction
II. Unisexuality
III. Main Types of Ambisexuality
IV. Mechanisms Controlling Sex Change
V. Germ Line
VI. Outlook

GLOSSARY

ambisexuality The normal occurrence of male and female function in one individual, either as simultaneous or as sequential hermaphroditism.

diandry The existence of two types of males which originate in two different ways.

gonadotropin A hormone that is produced in the pituitary gland and influences gonadal functions.

hypothalamus A part of the central nervous system of vertebrates that controls visceral functions.

meiosis A special kind of nuclear division in which the chromosome complement is exactly halved, leading to the formation of gametes.

neurosectory hormone A hormone that is produced by specialized neuronal cells in the hypothalamus.

protandry A type of ambisexuality in which the male sex matures first and is followed by female function.

protogyny A type of ambisexuality in which the female sex matures first and is followed by male function.

Teleost fishes display a large variety of reproductive modes. Among these the natural occurrence of ambisexuality (simultaneous or sequential functioning of the two sexes in one individual) is unique among vertebrates. Although a steadily growing number of research efforts have been devoted to this subject during the past five decades, our knowledge is still fragmentary. Beyond descriptive work on gonadal morphology, ethological observations emphasize that in many species social influences are most likely involved when an adult fish changes its sexual function. Physiological studies are still rare because most of the relevant species do not lend themselves very easily to experimental studies in the laboratory.

I. INTRODUCTION

In the large majority of vertebrates the production of the two kinds of sex cells (gametes) is partitioned between the sexes. Females produce eggs and males supply sperm. This division of sexual functions is known as gonochorism. For a normal vertebrate species, this condition seemed to be the rule. However, in 1932 the Amazon Molly (*Poecilia formosa*) became famous as the first unisexual vertebrate. Later on it was discovered that unisexuality occurs in other vertebrate groups (amphibia and reptiles).

Until the middle of this century, textbooks on zoology taught that in vertebrates there is never a functional hermaphrodite and that the sexes are functionally separate in every case. However, more than 2000 years ago Aristotle reported on a small sea bass from the Mediterranean as being hermaphroditic, and toward the end of the eighteenth century an Italian author gave an exact and masterly illustrated description of the reproductive system of that particular species (*Serranus cabrilla* L.). Apart from a few scattered papers on this topic, it took more than 150 years until it was generally recognized that simultaneous hermaphroditism (coexistence of functional ovarian and testicular tissue at the same time in one individual) and change of sex (sex inversion) in adult fishes is rather widespread among various groups of teleosts. New cases of teleost ambisexuality are now reported every year.

The gonads of teleost fishes are elongate in shape and paired. In males the testes are solid structures from which a main sperm duct, the vas deferens, arises and leads to the urogenital papilla. In females the ovary is usually a hollow paired organ with longitudinal lamellae protruding from the ovarian wall into the lumen. This cavity is continuous with the oviduct. Mature oocytes are discharged into the ovar-

ian cavity and move into the oviduct which ends in the urogenital papilla.

II. UNISEXUALITY

Among teleost fishes, about 10 species are known which have no males and in which females produce only female offspring either by gynogenesis or by hybridogenesis. In *Poecilia (Mollienesia) formosa* the females mate with males of other species of the same genus. Their sperm triggers the development of the egg nucleus but the male nucleus does not fuse with it (gynogenesis). In the genus *Poeciliopsis,* which also belongs to the killifishes, several unisexual species have been found which are all of hybrid origin. They depend on at least one parental species for sperm. As a result of gametic fusion, characteristics of the paternal genome are expressed. However, only the female genome is transmitted to the ovum because the entire paternal set of chromosomes is eliminated during or prior to meiosis (hybridogenesis). Several species of carp and goldfish have also developed unisexuality.

III. MAIN TYPES OF AMBISEXUALITY

Among ambisexual species the gonads on the whole do not differ fundamentally from those of gonochoristic fishes. However, there are considerable differences with regard to the arrangement of the germinal tissues in space and in time.

A. Simultaneous Hermaphroditism

In simultaneous hermaphrodites male and female sex products mature side by side more or less synchronously. The gonads contain well-defined ovarian and testicular areas. Quantitatively, the female part greatly outweighs the male zone. Both tissues lie within a common gonadal capsule (Fig. 1a). Quite like the situation in most of the females of gonochoristic teleosts, the ovary has an ovarian cavity which is continuous with the oviduct. Adjacent to the testicular part of the gonad a system of lacunae within

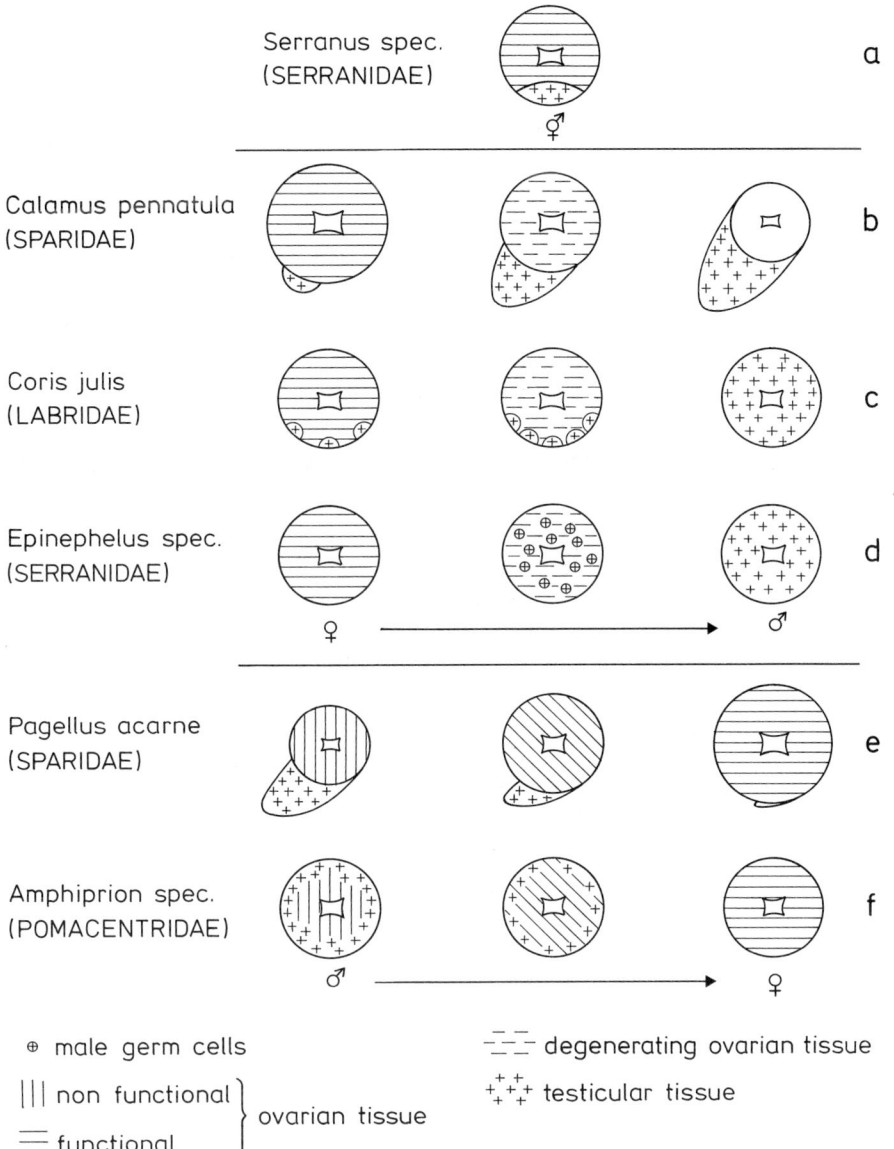

FIGURE 1 Schematic drawing of gonadal cross sections of fishes representing the main types of teleost ambisexuality: a, simultaneous hermaphroditism; b–d, protogyny; e and f, protandry.

the gonadal wall serves as a sperm duct. Near its caudal end it surrounds the oviduct nearly completely and finally the lacunae merge into a tube that opens to the exterior dorsally of the oviduct.

In the laboratory it was shown that self-fertilization is possible but in "typical" hermaphrodites (as described for several species of sea basses) pair spawning is the rule: Two individuals—one functioning as a female and the other one as a male—release in close contact to each other their sex products at the peak of a rapid spawning rush. Then, within seconds, the same fishes can change their sex roles. Each mate releases eggs several times during a spawning period and partners switch sexual function with each spawning. A unique case is represented by the small killifish *Rivulus marmoratus*, known from Florida and the Netherlands Antilles. It reproduces by internal self-fertilization. The eggs are incubated intraparentally for from <1 hr (one-cell stage) to as long as 2.5 days (stage of pectoral fin buds) and then develop extraparentally for about 14 days. In this fish self-fertilization produces an unbroken succession of uniparental generations of isogenous simultaneous hermaphrodites.

B. Protogyny

Protogyny is the most frequent form of consecutive (sequential) hermaphroditism. Both anatomical details and the course of sex change show considerable variation from species to species. In protogynous porgies (Sparidae) and in members of a related family, the juvenile gonad consists of two different zones lying side by side but separated from each other by connective tissue. The dorsal area develops like a typical ovary as known in gonochoristic species. The ventral region remains dormant and looks like a solid small band of tissue attached to the ovary. It consists mainly of gonocytes (spermatogonia) which, during the female phase, do not participate in the reproductive cycle. The onset of sex inversion (usually after a spawning period) is marked by a degeneration of the ovarian tissue. The oocytes break down and immune cells, such as granulocytes and lymphocytes, invade the ovarian lamellae. Concomitantly, the testicular part increases in size and sooner or later nests of ongoing spermatogenesis appear. The former ovarian lamellae disappear progressively and at the end of this process the strongly developed testis surrounds nearly completely the former ovary, which is now reduced to a small sac which retains its lumen and keeps some remnants of sterile somatic tissue (Fig. 1b).

In other species (e.g., in the labrid fish *Coris julis*) small islets of gonial cells within the ovarian wall are noticeable during the entire female stage (Fig. 1c). When the first symptoms of ovarian degeneration appear, the formation of the future testis starts from these peripheral "male nests." They display spermatogenetic activity and proliferate into neighboring ovarian lamellae. Even in the mature testis the arrangement of the germinal tissue is reminiscent of the lamellar organization of the former ovary. The persistence of its cavity characterizes the testis as a secondary structure of the gonad. The spermiduct develops by rupture of muscular and connective tissue layers in the old ovarian wall. In this way anastomozing sperm sinuses are formed which run the length of the two gonadal lobes and drain into a posteriorly located vas deferens.

In many protogynous groupers (e.g., certain members of the genus *Epinephelus*) male elements are not visible until the beginning of sex inversion. They first show up as spermatogenic nests amidst the female tissue in the ovarian lamellae. In such species the nonfunctional gonadal mixture of male and female characters can serve as a useful indicator of ongoing sex inversion (Fig. 1d). After resorption of the ovarian tissue usually at least some parts of the former ovarian cavity are maintained. Again, the spermiduct arises by a splitting of tissue layers in the former ovarian wall.

In many protogynous wrasses and parrot fishes, sexual organization is complicated by a very peculiar feature: the occurrence of two types of males with different origin (diandry). One kind differentiates as male from the very beginning (primary male), whereas the other passes through a functional female stage before transforming into male (secondary male). The anatomical organization of the testes of these two male types is completely different: The two lobes of the testis of the primary male are solid structures, whereas the testicular lobes of the secondary male (as described for *Coris*) retain the former ovarian cavity. The phenomenon of diandry is not well understood. It is even more puzzling that in such species two different types of spawning behavior are known. In pair spawnings one (territorial) male and a female spawn together. In group spawnings one female and several males participate in a spawning rush out of a larger group of fishes. The males acting as group spawners look like females, whereas the pair-spawning male displays a color pattern which is much more brilliant than that of females. This particular livery, the so-called terminal color pattern, is characteristic of larger (= older) males. The outer appearance of the smaller (younger) males, the initial color pattern, cannot be distinguished from that of females. In some species the transformation of the livery in males is closely related to change of sex (the terminal color pattern being restricted to secondary males), whereas in others the transformation of the livery from the initial to the terminal color pattern affects both types of males and seems to be a matter of size and age.

There are considerable differences with regard to the moment when sex inversion takes place. In some species all or at least the majority of females transform into males at a definite age, whereas in others sex change occurs over a broad range of ages and sizes. The description of a species as being protogynous does not automatically imply that in the natural habitat all females will change sex. The field data available are by far too scanty to draw such conclusions.

Data on the duration of gonadal transformation are scarce and partially conflicting. It is obvious that in fishes with an annual reproductive cycle it usually takes about a year before a female will function as a male. In tropical species that have continuous breeding throughout the year the lapse of time from the last spawning as a female to the first successful function as a male may be as short as 8 days. However, behaviorally a switch of sex may happen within a much shorter time as it has been observed under particular experimental conditions.

C. Protandry

In teleosts, protandry is much less common than protogyny. In protandric porgies the basic organization of the gonad is the same as in protogynous members of this family: the presence of the two heterologous germinal tissues in juvenile gonads. However, as opposed to the situation shown in Fig. 1b, the ventral part of the gonad, the testis, matures first, whereas the dorsal ovary with young oocytes stagnates. The most conspicuous sign for the onset of sex inversion is a striking change in the ovarian part: The number of the oocytes increases and they start to grow. Their enlargement is caused by the accumulation of yolk. At the same time the testis involutes, immune cells help break down the tissue, and, at the end of this process, a tiny thickening in the ventral of the ovary is the only morphological remnant of the fish's male past (Fig. 1e).

Clown fishes of the genus *Amphiprion* (Pomacentridae) represent another type of protandry. In these animals the functioning males possess an ovotestis in which mature testicular tissue and ovarian tissue are both present and not clearly delimited. The spermatogenic tissue is located around the periphery of the gonad. In the anterior part of the genital organ several layers of immature oocytes surround a central lumen (Fig. 1f). Posteriorly this lumen opens and the gonadal lobes become bands of germinal tissue with the male portion medially and the ovarian part laterally. The most obvious morphological symptom of ongoing sex change in these animals is the appearance of young oocytes in the peripheral area of the gonad. At this stage the amount of spermatogenic tissue is already markedly reduced and it disappears completely by the time sex change is complete. Information on the duration of the sex-changing process in protandric species is scarce and limited to observations on members of the genus *Amphiprion*. The duration varies from 45 days to 5 or 6 months. It appears that in such fishes the transformation from male to mature female takes much longer than the change from female to male in protogynous species. Gender change in *Amphiprion* is a facultative event. Its occurrence depends on particular social constellations.

IV. MECHANISMS CONTROLLING SEX CHANGE

The occurrence of sequential hermaphroditism has stimulated the search for factors which trigger the succession of the two heterologous sex phenotypes in one adult individual. For that purpose, not only environmental (extrinsic) influences must be taken into consideration but also processes working inside the fish (intrinsic factors).

A. Extrinsic Factors

The physical environment probably does not play any role in adult sex inversion. The large majority and greatest variety of ambisexual species live in tropical and subtropical marine waters and share this environment with species that are gonochoristic. Many of them reproduce year-round. Therefore, it is difficult to imagine how light or temperature could be of importance for sex change. However, during the past 25 years a steadily growing body of information has demonstrated that in many species social interactions affect sex inversion.

The protogynous cleaner wrasse *Labroides dimidiatus* represents one of the first examples for which a social control of sex change has been shown. This coral-reef fish lives in small groups consisting of a male with a harem of females, among which larger individuals dominate smaller ones. Apparently, the male in each harem suppresses the tendency of the females to change sex by actively dominating them. Death (or experimental removal) of the male releases this suppression and the dominant female of the harem changes sex immediately.

Conversely, in protandric clown fishes females are the largest and behaviorally dominant members of social groups, displaying frequent aggression toward their smaller mates. Typically, a reproductive pair inhabits a sea anemone with other, smaller, nonreproductive individuals. When the female disappears, the second-ranking individual, the reproductive male, suddenly becomes the largest member of the group. At this point, the gonad transforms and matures into a functional ovary. Concomitantly, that fish grows larger indicating that an apparently inhibitory influence by the dominant partner has been removed. It has been observed that in clown fishes mature female function can also be reached on a second pathway: If no larger individuals are present, juveniles may develop directly as females with no intervening functional male phase.

For other species (e.g., the protogynous sea bass *Anthias squamipinnis* and the Caribbean bluehead wrasse *Thalassoma bifasciatum*) it has been shown that social interactions may play a key role in the initiation of sex transition. It must be concluded that in all species (either protogynous or protandric) for which social control of sex change has been established, the occurrence of this event depends very much on a disturbance of the existing hierarchical order. Observations in the coral reef as well as laboratory experiments have clearly demonstrated that disruptions of interactions with the opposite sex and/or shifts in the interactions with other like-sexed group members may start gender change. On the other hand, nothing will happen in a stable social situation. Isolated fish usually do not change sex. There are reliable reports that pairs of clown fishes

which breed regularly and successfully in aquaria maintain their gender for years.

Until recently, it was generally believed that sex change in a fish happens only once in its life. However, laboratory experiments on a gobiid fish suggest that there may be exceptions to this rule. When the dominant male was removed from a social unit with several females, the largest one changed sex and became a male. Histological examination of the gonad, however, revealed that ovarian tissue was retained within a functional testis. When such males were put together with a larger male they changed back to the female sex.

From all studies that have been carried out so far it appears that, in addition to dominance and aggression, other parameters (e.g., the size of groups and their sex ratio) have to be taken into consideration. For a better comprehension of the way signals between conspecifics are transmitted, much more experimental work is needed. However, there are sufficient data to assume that besides behavioral traits (such as aggressive chases, fin displays, elaborate movements, etc.) chemical (olfactory and/or gustatory) and/or perhaps acoustic cues are involved. The perception of such stimuli and their transduction within the animal is the bridge that links environmental (extrinsic) and intrinsic factors, although research on that topic is, at best, in its very beginning.

B. Intrinsic Factors

1. Steroid Hormones

In all vertebrates, hormones play a decisive role in reproductive processes. Some of the steroid hormones, produced in the gonads, are related to the two sexes, as the terms "androgens" for the male sex hormones and "estrogens" for the female hormones indicate. The search for these compounds in some of the ambisexual teleosts has not yielded conclusive evidence that they may cause gonadal transformation. There are some reports that administration of androgens to functional females of protogynous fishes leads to an onset of spermatogenesis and ovarian degeneration. Other studies could not corroborate such findings. For a few species a correlation between gonadal status and hormone level has been observed (an increase of androgens toward the male phase in protogynous species, and a decrease of such hormones with the loss of male function in a protandric clown fish). However, it remains a matter of dispute whether and to what extent sex-specific steroids trigger gonadal changes or whether the appearance (or disappearance) of these hormones during sex change must be considered as the result of another primary event.

2. *Hypothalamo–Hypophysial Axis*

According to an older hypothesis, gonadotropin might be responsible for sex inversion. This idea is supported by the observation that injection of gonadotropin or gonadotropin-releasing hormone (produced as a neurosecretory substance in the hypothalamus) can induce sex inversion in females of protogynous fishes. This kind of treatment has been successful even in a species in which administration

of androgenic hormone proved to be ineffective. In all vertebrates the hypothalamo–hypophysial axis is recognized as the foremost point of intersection between the nervous and hormonal systems. Its importance for the control of reproductive processes is well established. Consequently, it appears quite plausible that the triggering factor is inside a sex-changing fish at the uppermost level of the hormonal hierarchy (Fig. 2). However, the definition of a general concept would be premature. Even if future research confirms that hypothalamic secretions (influenced by environmental cues) initiate gonadal transformation (including shifts in the ratio of gonadal steroids), many questions remain to be answered. For example, assuming that gonadotropin elicits gender change in females of protogynous fish, on what cell type in the gonad does it act? Does it act by inhibition or stimulation of some process? Moreover, currently no data are available on the effects of such hormones in males of protandric species.

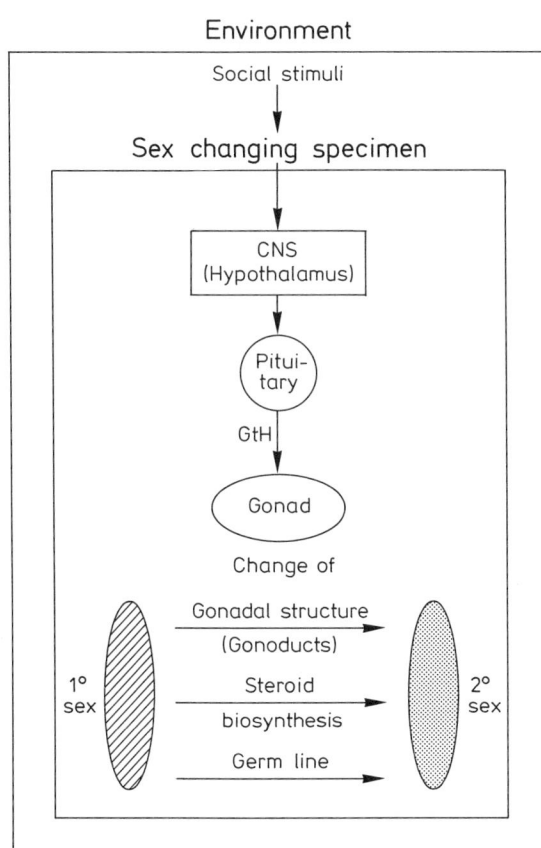

FIGURE 2 Hypothetical scheme on the control of sex inversion in an adult fish (see text for further explanation).

V. GERMLINE

When sex differentiation is taking place in the germline, sexually bipotent gonocytes turn, irreversibly, into either spermatogonia or oogonia from which spermatogenesis or oogenesis departs by meiotic divisions. This very general concept is valid for nearly all multicellular animals. Difficulties arise, however, when it is applied to species in which the sexual phenotype of the germinal tissue does not remain stable. Then, the problem of classical dignity is still unsolved: the fundamental questions (i) of the duration of the period during which a given gonocyte retains its sexual bipotency and (ii) of the factor(s) that orients this potency into the male or female direction. Unfortunately, no clear-cut criteria exist to discriminate between undifferentiated bipotent gonocytes, spermatogonia, and oogonia before the onset of meiosis. Therefore, it cannot be decided (e.g., in cases such as that shown in Fig. 1d) whether at the moment of sex inversion the alteration in the germline reflects differentiating influences on bipotential gonocytes or whether early differentiated spermatogonia are activated after they persist in a quiescent form through the first adult sex stage. Presumably, the factor which causes the loss of sexual bipotentiality in a germ cell is different from the stimulus which initiates spermatogonial proliferation and ensuing spermatogenesis. However, these extragerminal influences need to be identified.

VI. OUTLOOK

The number of teleosts for which one or the other type of ambisexuality has been reported has grown to over 100. Nearly all the relevant species inhabit marine waters, with most of them living in tropical and subtropical coral reefs. The vast majority of these fishes belong to the large order Perciformes (or Percomorphi). However, examples from other, taxonomically distant, orders are also known. There is ample evidence that teleost ambisexuality is a derived, specialized condition, the evolution of which has arisen repeatedly in unrelated groups of fishes.

For physiological work on ambisexual teleosts, their marine life is a handicap. All of them are oviparous and most of them produce tiny eggs that are released into the plankton. Therefore, it is difficult to rear such animals under laboratory conditions. For a few species, which are commercially valuable, methods to culture them in captivity have been developed. However, other factors restrict the possibilities of experimental research. It remains to be seen whether members of the genus *Amphiprion* might become a useful model for future studies. Clown fishes practice parental care and reproduce (relatively) easily in captivity. They display many traits of teleost ambisexuality which merit more physiologically oriented investigations. Therefore, developmental biologists, endocrinologists, and sensory physiologists should feel equally challenged to bring more light on the unique phenomenon of adult sex change among vertebrates.

See Also the Following Articles

FEMALE REPRODUCTIVE SYSTEM, FISH; MALE REPRODUCTIVE SYSTEM, FISH

Bibliography

Cardwell, J. R., and Liley, N. R. (1991). Hormonal control of sex and color change in the stoplight parrotfish, Sparisoma viride. *Gen. Comp. Endocrinol.* **81**, 7–20.

Chan, S. T. H., and Yeung, W. S. B. (1983). Sex control and sex reversal in fish under natural conditions. In *Fish Physiology* (W. S. Hoar, D. J. Randall, and E. M. Donaldson, Eds.), Vol IX b, pp. 171–222. Academic Press, New York.

Cole, K. S., and Shapiro, D. Y (1995). Social faciliation and sensory mediation of adult sex change in a cryptic, benthic marine goby. *J. Exp. Mar. Biol. Ecol.* **186**, 65–75.

Godwin, J. (1994). Behavioral aspects of protandrous sex change in the anemonefish, Amphiprion melanopus, and endocrine correlates. *Anim. Behav.* **48**, 551–567.

Reinboth, R. (Ed.) (1975). *Intersexuality in the Animal Kingdom.* Springer, Berlin.

Reinboth, R., and Bruslé-Sicard, S. (1997). Histological and ultrastructural studies on the effects of hCG on sex inversion in the protogynous teleost Coris julis (L.). *J. Fish Biol.* **51**, 738–749.

Ross, R. M. (1990). The evolution of sex-change mechanisms in fishes. *Environ. Biol. Fishes* **29**, 81–93.

Shapiro, D. Y. (1992). Plasticity of gonadal development and protandry in fishes. *J. Exp. Zool.* **261**, 194–203.

Sunobe, T., and Nakazono, A. (1993). Sex change in both directions by alteration of social dominance in Trimma okinawae (Pisces: Gobiidae). *Ethology* **94**, 339–345.

Warner, R. R., and Swearer, S. E. (1991). Social control of sex change in the bluehead wrasse, Thalassoma bifasciatum (Pisces: Labridae). *Biol. Bull.* **181**, 199–204.

Follicular Atresia

R. H. Oliver, G. D. Chen, and J. Yeh

Regions Hospital and University of Minnesota

GLOSSARY

atresia The degeneration of nonovulatory follicles within the ovary.

attrition The depletion of oogonia and primary oocytes during fetal development.

intraovarian factors Steroid hormones and growth factors synthesized and released within the ovary that act as paracrine or autocrine regulators of follicle development.

pseudomaturation Oocyte maturation during atresia.

pyknosis Darkened and irregularly stained nuclei.

I. THE SIGNIFICANCE OF FOLLICULAR ATRESIA

A. The Physiological Role of Atresia

Follicular atresia refers to the degeneration of nonovulatory follicles within the ovary. The literal meaning of atresia is the closure of a body orifice, referring to the shrinkage and eventual disappearance of the follicle. Follicular atresia is accompanied by loss of the oocyte and is a principal mechanism by which the finite endowment of oocytes in the ovary is depleted throughout life.

In humans, several million oogonia develop in the fetal ovary, but only a few hundred are ovulated during the course of reproductive life. Attrition, a process distinct from atresia, results in the depletion of oogonia and primary oocytes during fetal development. In humans, the number of developing oocytes peaks by midgestation to approximately 6 or 7 million but is reduced through attrition to less than 2 million at birth. Germ cell attrition differs from atresia in that it occurs during prenatal development and does not involve the follicle.

While the most certain physiological effect of atresia is the depletion of oocytes, there are indications that atretic follicles may also secrete ovarian steroids. Atretic follicles are the probable source for estrogen released from ovaries of premenarchal girls. Also, remnants of atretic follicles appear to serve as accessory sites of steroid production during the development of the primary follicle and in the luteal phase of the ovarian cycle.

B. Stages of Follicular Atresia

The functional unit of the postnatal ovary is the follicle. Atresia occurs continuously throughout the life of the female, and follicles at all stages of development undergo atresia. Notable instances of atresia occur prior to puberty, during the active reproductive cycle, and at an accelerated rate prior to menopause. Of the approximately 2 million healthy follicles present at birth, only about 300,000 remain at the onset of puberty. The rest undergo physiological death and degeneration from atresia. The 400 or 500 follicles that ovulate spontaneously in menstrual cycles are the only follicles that do not eventually undergo atresia. Follicles that mature during reproductive cycles but are not selected for ovulation display an increased incidence of atresia that suggests a heightened sensitivity to hormonal signals.

Many atretic follicles are found in the antral stage at a time when follicle-stimulating hormone (FSH) has already affected follicular development. This suggests these follicles have become FSH dependent and atresia could be triggered by declining levels of FSH insufficient to support their growth or that under the influence of the dominant follicle, they have become insensitive to endogenous levels of FSH. During the perimenopausal transition all follicles remaining in the ovary, including those still in the primordial stage, will succumb to atresia.

II. THE MORPHOLOGY OF ATRETIC FOLLICLES

Both the granulosa cells and the oocyte undergo changes in morphology during follicular atresia (Fig. 1). In initial stages, pyknotic nuclei are observed in some granulosa cells, especially those in contact with the antrum. Pyknosis, the appearance of darkened and irregularly stained nuclei, has long been recognized as an indication of impending cell death and is now thought to be due to fragmentation of the genetic material. A marked reduction in DNA synthesis occurs within the follicle, but some granulosa cells may still be proliferating. In later stages of atresia, granulosa cell proliferation is virtually nonexistent and cell death and fragmentation of granulosa cells is apparent. Debris originating from dead granulosa cells accumulates in the follicular fluid. Breakage in the basal membrane is apparent, and leukocytes begin to invade the follicle.

At this point, the oocyte may undergo germinal vesicle breakdown, similar to that which occurs in normal oocyte maturation. Oocyte maturation during atresia is called "pseudomaturation." Pseudomaturation probably arises as a consequence of the loss of contact between cumular granulosa cells and the oocyte that would normally keep the oocyte in meiotic arrest. However, pseudomatured oocytes can be fertilized *in vitro*.

In the final stages of atresia, follicular shrinkage progresses until the almost complete disappearance of the follicle. Very few granulosa cells remain in the follicle and most are detached from the basal membrane. Eventually, the oocyte degenerates and fragments. Remnants of atretic follicles persist as scar-like structures within the ovary.

III. REGULATION OF FOLLICULAR ATRESIA

A. Gonadotropins

Follicular development, ovulation, and atresia are hormonally regulated. The pituitary gonadotropins, FSH and luteinizing hormone (LH), are the principal mediators of follicular development. FSH, in particu-

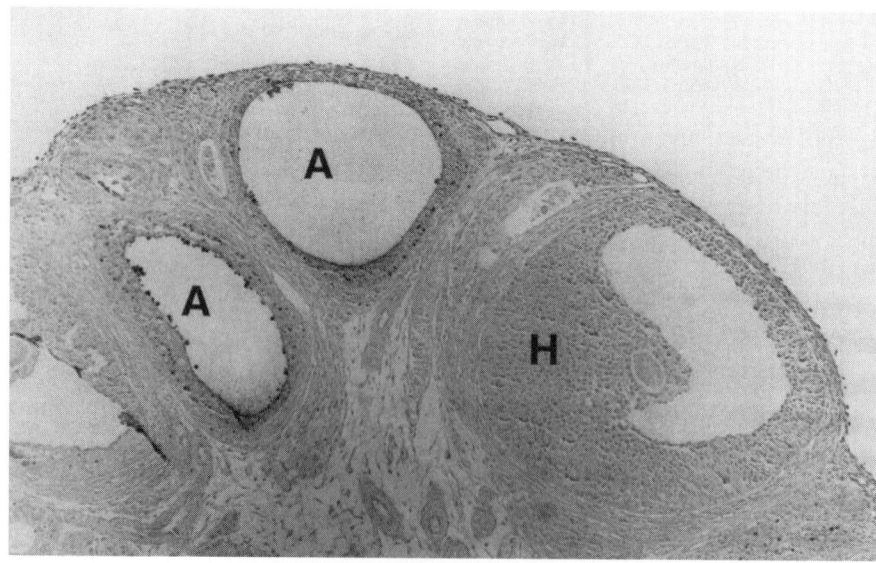

FIGURE 1 Photomicrograph of rat ovary showing atretic follicles. Fragmented DNA is stained in atretic follicles. H, healthy follicle; A, atretic follicle (reproduced with permission from Palumbo and Yeh, 1994).

lar, seems to be involved in regulation of atresia. Most postnatal follicles exist as primordial follicles with only a single layer of flattened granulosa cells surrounding the oocyte. FSH is required for the maturation of the follicle and for the development of the fluid-filled antrum, which is characteristic of the preovulatory or graafian follicle. Preantral follicles can develop in the absence of FSH, but follicles with antral compartments are dependent on FSH for survival. In this regard, FSH appears to confer an antiatretic effect on follicles and suppresses death and degeneration of the granulosa cells. Follicular degeneration during the prepubertal period appears to be due to the presence of threshold levels of FSH that are able to initiate antral development but insufficient to support follicular growth to term. Luteinizing hormone, human chorionic gonadotropin, and growth hormone inhibit granulosa cell death and thus also exhibit antiatretic effects.

B. Atresia as an Apoptotic Process

A recent trend in research on atresia focuses on the role of apoptotic processes in the mechanisms of follicular degeneration. Apoptosis was first related to follicular atresia as a result of the recognition that the changes in morphology associated with atresia were similar to those observed during programmed cell death. Numerous studies have now shown that atresia in vertebrate species, including humans, is mediated through evolutionarily conserved mechanisms of the apoptotic or "cell-suicide" pathways. Demonstrations of internucleosomal DNA cleavage, characteristic of apoptotic cell death, provide biochemical evidence of apoptosis during follicular atresia. Specific genes involved in apoptosis in invertebrates and in other vertebrate tissues are known to be expressed during follicular atresia. Among the genes investigated in this regard are members of the Bcl-2, Fas, and Caspase families. Involvement of these genes suggests physiological regulators of atre-

sia such as the gonadotropins and intraovarian factors may control the process of atresia by acting to induce or suppress cell death pathways linked with apoptosis.

C. Intraovarian Regulators

Other hormones and growth factors, several of which are intraovarian in origin, have roles in determining the fate of the follicle. Experimentally, these factors have been shown to either induce or prevent programmatic DNA degradation in follicular granulosa cells. Degradation of DNA is characteristic of apoptotic cell death and is an indication of the onset of atresia in ovarian follicles. Androgens, gonadotropin-releasing hormone, and tumor necrosis factor-α induce DNA degradation in granulosa cells. Estrogen, progesterone, epidermal growth factor, insulin-like growth factor-I, interleukin-1β, activin, and basic fibroblast growth factor prevent DNA degradation in granulosa cells.

See Also the Following Articles

Apoptosis (Cell Death); Follicular Development, Control of; FSH (Follicle-Stimulating Hormone); LH (Luteinizing Hormone)

Bibliography

Kapia, A., and Hsueh, A. J. W. (1997). Regulation of ovarian follicle atresia. *Annu. Rev. Physiol.* **59**, 349–363.

Palumbo, A., and Yeh, J. (1994). In situ localization of apoptosis in the rat ovary during follicular atresia. *Biol. Reprod.* **51**, 888–895.

Palumbo, A., and Yeh, J. (1995). Apoptosis as a basic mechanism in the ovarian cycle: Follicular atresia and luteal regression. *J. Soc. Gynecol. Invest.* **2**, 565–573.

Yeh, J., and Adashi, E. Y. (1998). The ovarian life cycle. In *Reproductive Endocrinology* (S. S. C. Yen, R. Jaffe, and R. Barbieri, Eds.). Saunders, Philadelphia.

Follicular Development

Bradley J. Van Voorhis

University of Iowa

GLOSSARY

aromatase A cytochrome P450 steroidogenic enzyme which converts androgens to estrogens. In the ovary, aromatase is localized in granulosa cells.

dominant follicle Of the cohort of follicles that begin final maturation, the follicles that are destined to ovulate.

follicular antrum The fluid-filled cavity of the growing ovarian follicle.

follicular atresia The process by which nondominant follicles are deleted from the population of ovarian follicles. The mechanism of atresia is by apoptosis or programmed cell death.

meiosis A process of germ cell division whereby each daughter nucleus receives half the number of chromosomes characteristic of the somatic cells of the species.

oogonia A premeiotic egg found in the fetal ovary. Oogonia increase in numbers by mitotic division.

Follicular development begins when follicles leave the pool of resting follicles and begin a process of growth, cell division, and cellular differentiation that culminates in ovulation. These vital processes occurring within the ovarian follicle have been known for many years to be controlled by exogenous endocrine factors, most notably gonadotropic hormones from the pituitary gland. Recently, a variety of intraovarian factors acting in both autocrine and paracrine fashions have been proposed to modulate some of the processes of follicular development. Although many follicles begin the process of follicular growth, very few actually fully develop into an ovulatory follicle. Most follicles become atretic and are eliminated by a process of atresia. The mechanism by which follicles become atretic is now known to be apoptosis or programmed cell death. This chapter discusses the important concepts and principles that have emerged from the study of follicular development.

I. FORMATION OF THE OVARIAN FOLLICLE

During embryonic development, the ovary is endowed with a finite number of germ cells. Primordial germ cells can first be identified in the yolk sac early in embryonic development, and these germ cells are then known to migrate to the genital ridge where they play an indispensable role in the induction of gonadal development. Indeed, no gonadal structure forms in the absence of germ cells. These premeiotic germ cells are referred to as oogonia and the number of oogonia increases in the ovary by mitotic activity. The majority of mammals restrict oogonial mitotic proliferation to prenatal development and shortly after birth. It is this process of mitotic division that leads to the finite number of oocytes available for the reproductive life span. Oogonia then depart the mitotic cycle and begin meiosis, becoming arrested in prophase of the first meiotic division. Oocytes at this stage are referred to as primary oocytes and they remain arrested in prophase of the first meiotic division throughout follicular development until the time of ovulation. The cellular mechanism behind meiotic arrest of the primary oocyte is unknown. Primary oocytes become surrounded by a squamous layer of pregranulosa cells and this combination forms the primordial follicles of the ovary (Fig. 1). Primordial follicles constitute the population of resting, nongrowing follicles which are progressively depleted during the reproductive life span. The depletion of primordial follicles occurs as the result of two processes: atresia or entry into the growth phase. In the human ovary, the percentage of atretic resting follicles has been estimated to be about 50% at birth and this percentage decreases from birth until approximately 30 years of age. After the age of 30, loss of resting follicles is mainly due to entrance of follicles into the growth phase. The mechanisms regulating resting follicle depletion by atresia are currently unknown.

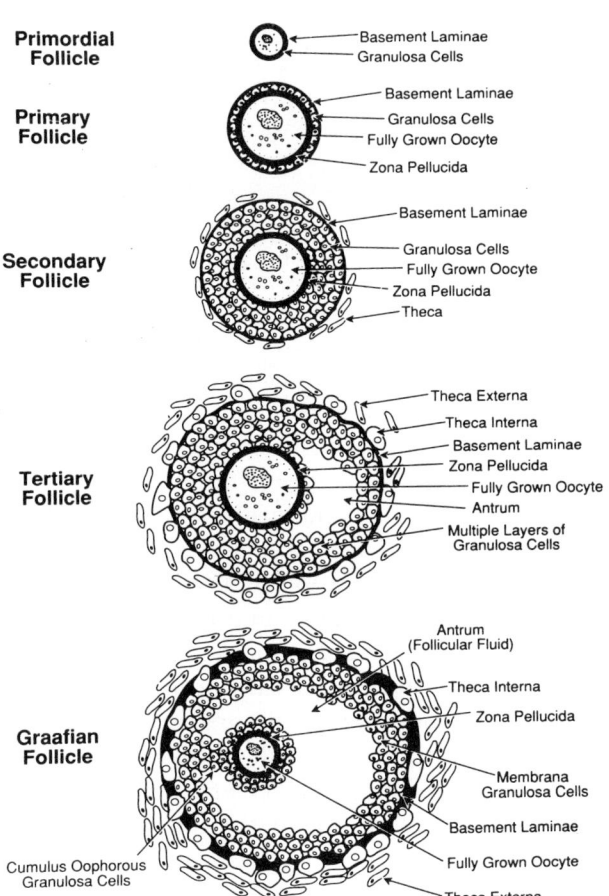

FIGURE 1 Depiction of follicular development from the primordial follicle to the preovulatory follicle (Graafian follicle) [reproduced with permission from *Textbook of Reproductive Medicine* (B. Carr and R. Blackwell, Eds.), Appleton & Lange, Stamford, CT].

II. DESCRIPTION OF FOLLICULAR GROWTH

Entry of primordial follicles into the growth phase occurs throughout the reproductive life span and is characterized by conversion of the flattened pregranulosa cells surrounding the oocyte into a single layer of cuboidal granulosa cells. This follicle is now termed a primary follicle (Figs. 1 and 2). The mechanism triggering initiation of follicular growth remains largely unknown. Recent research in this area has focused on growth factors and their receptors and protooncogenes which may be involved in the transition of resting follicles to growing follicles.

Once follicles enter the growth phase, they enlarge both by the proliferation of granulosa cells and by an increase in the size of the oocyte. In the growing follicle, a zona pellucida is formed around the oocyte. Whether the zona pellucida is secreted by the oocyte, surrounding granulosa cells, or both remains controversial and species differences may account for some of the conflicting results. Granulosa cell proliferation results in the oocyte becoming surrounded by several layers of granulosa cells. These follicles are termed secondary follicles (Figs. 1 and 2). Small growing follicles lack an independent blood supply but secondary follicles are served by one or two arterioles with an anastomotic network of capillaries just outside the basal lamina. This blood supply allows the follicle to be exposed to factors circulating in the blood stream which affect further follicular development. As secondary follicles enlarge, stromal cells near the basal lamina of the follicle become aligned parallel to each other and form a theca interna and a theca externa. The cells of the theca interna differentiate, become epitheliod in appearance, and acquire organelles characteristic of a steroid-secreting cell. The cells of the more peripheral theca externa retain their more spindle-shaped morphology and merge into the surrounding stroma. The cells of the theca layer play a critical role in follicular development and steroidogenesis by secreting androgens necessary as substrates for granulosa cell estradiol production (discussed later). As follicles continue to develop, small fluid-filled cavities or lacunae form between the granulosa cells. These lacunae are thought to aggregate and form the fluid-filled antrum

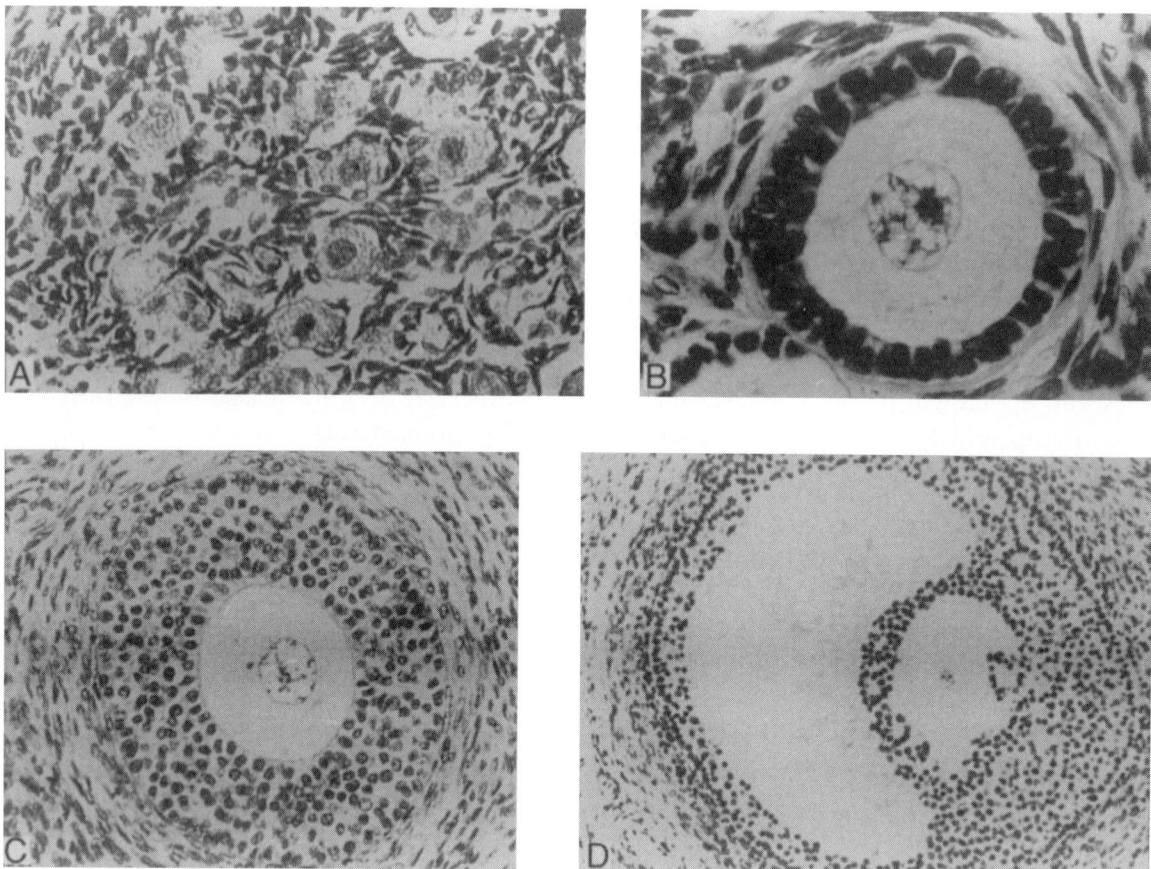

FIGURE 2 Follicular development in the human. A, Primordial follicles; B, primary follicle; C, secondary follicle; D, early antral follicle (reprinted with permission from the American Society for Reproductive Medicine *Fertility and Sterility,* 1996, Vol. 65, 238).

of the antral follicle (Fig. 1). In the antral follicle, the granulosa cells immediately surrounding the oocyte constitute the cumulus oophorus, whereas the other granulosa cells are referred to as either antral granulosa cells (cells closest to the antral cavity) or mural granulosa cells (cells closest to the basement membrane of the follicle). Granulosa cells in the follicle may not be homogeneous in activities and functions. For example, mural granulosa cells are more steroidogenically active than granulosa cells of the cumulus oophorus. The small antral follicle continues to rapidly grow by cellular multiplication and accumulation of fluid in the antrum until the ovulatory gonadotropin surge, which leads to the process of ovulation. During the final stages of follicular development, the preovulatory follicle becomes a highly vascularized structure and, just hours before ovulation, the granulosa wall which was avascular becomes bloody with invasion of blood vessels.

An interesting observation in the ovarian follicle is the presence of extensive intracellular gap junctions both between granulosa cells and between granulosa cells and the oocyte across the zona pellucida. The gap junctions are thought to allow intercellular communication and transport of small molecules between cells and allow granulosa cells of the follicle to function as a syncytium. Gap junctions across the zona pellucida to the oocyte may allow for control of the resumption of meiosis by cumulus granulosa cells.

Growth of the follicle does not occur in a linear fashion (Fig. 3). Rather, ovarian follicular growth

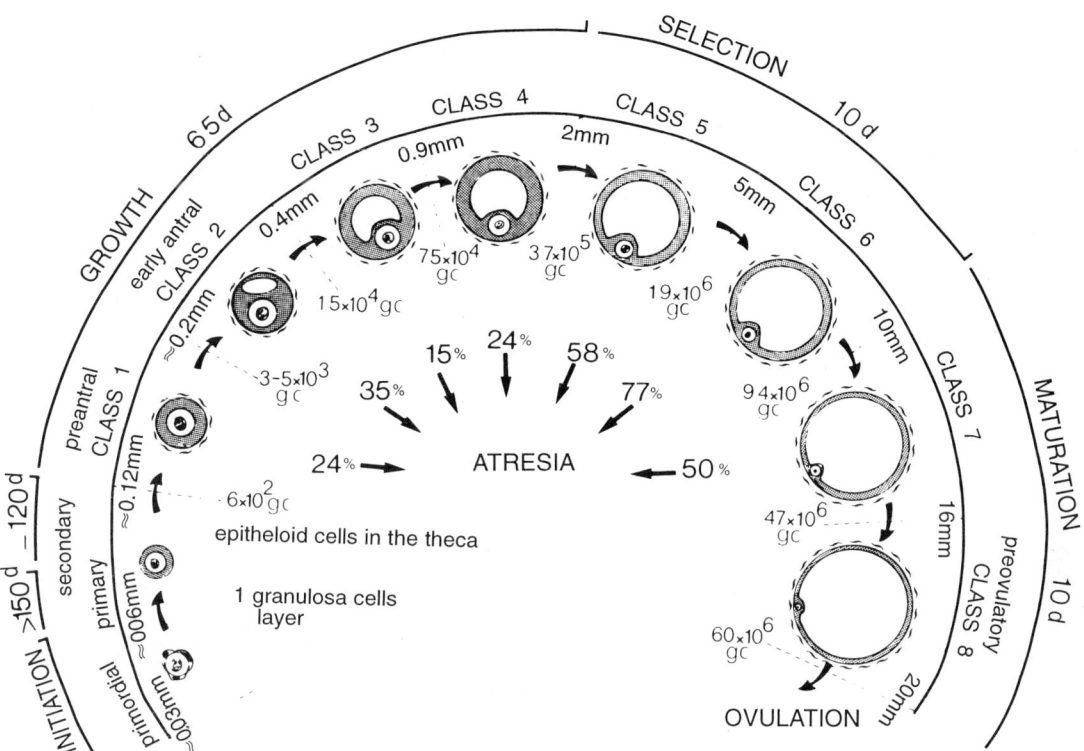

FIGURE 3 Time course of follicular development in the human ovary. Note the time course on the outer half-circle (reproduced from A. Gougeon, *Hum. Reprod. 1*, 82, 1986, by permission of Oxford University Press).

occurs at different rates and can be artificially divided into three phases: a preantral growth phase, a tonic growth phase, and an exponential growth phase. The preantral growth phase begins with the conversion of primordial follicles to primary follicles and ends with the formation of preantral follicles. In humans, the time for early growing follicles to attain the preantral stage is not known. However, based on data from monkeys and rats, in which the growth rate of small follicles is very slow, it is estimated that this time is probably several months in humans. The tonic growth phase is defined as the development of preantral follicles to small antral follicles or tertiary follicles (Fig. 1). In humans, this phase is estimated to occur over approximately 2 months and it ends when the small antral follicle is approximately 2–5 mm in diameter. These small antral follicles are then capable of entering an exponential growth with development to large, mature, ovulatory follicles. In humans, the exponential growth phase lasts approximately 21

days, beginning in the midluteal phase of one cycle and ending with ovulation of the follicle in the next cycle. Thus, during the exponential growth phase in humans, several early antral follicles are recruited to begin final maturation in the luteal phase of the preceding menstrual cycle. During the following follicular phase, a selection process occurs whereby a single follicle becomes dominant by the midfollicular phase and is the follicle that is destined to ovulate. Out of the follicles selected to begin the exponential growth phase, certain follicles achieve dominance and therefore ovulate, whereas others undergo atresia (Fig. 4). The numbers of dominant follicles out of the group of follicles that have been selected is species specific depending on the number of oocytes that are typically ovulated. Characteristics of dominant follicles include high mitotic activity in the granulosa cells, increased aromatase activity, a high estrogen to androgen ratio, and larger size than other nondominant follicles. Typically, the largest follicle

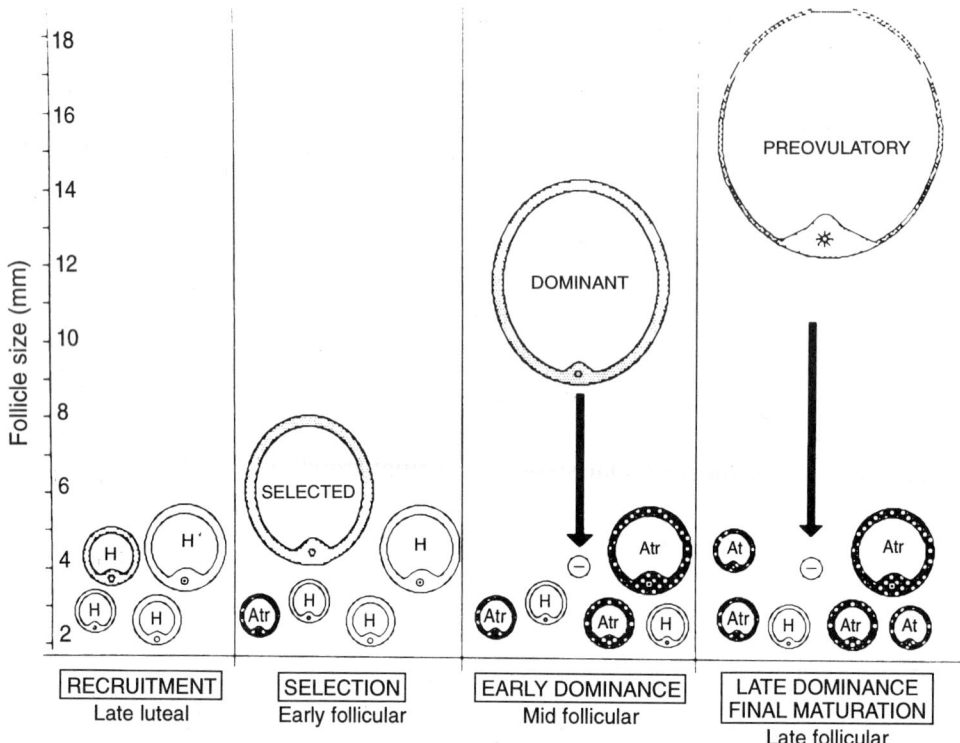

FIGURE 4 Development of a dominant follicle out of a cohort of growing follicles (courtesy of A. Gougeon).

at the midfollicular phase in humans is the dominant follicle which will continue to grow and will then ovulate.

III. REGULATION OF FOLLICULAR DEVELOPMENT

A. Gonadotropins

The primary importance of the two gonadotropic hormones, follicle-stimulating hormone (FSH) and luteinizing hormone (LH), in stimulating follicular development is unquestioned. FSH and LH are glycoprotein hormones secreted in a pulsatile fashion by the anterior pituitary gland under the influence of the hypothalamic hormone gonadotropin-releasing hormone (GnRH). Experimental induction of low circulating gonadotropin levels by either hypophysectomy or use of the GnRH agonist medications

leads to marked reductions in the number of developing follicles in a variety of species, including rat, mouse, hamster, guinea pig, and sheep. In addition, restoration of follicular growth can be observed by injecting these animals with gonadotropins. In women, a number of clinical conditions have also demonstrated the importance of gonadotropins to ovarian follicular development. In patients with Kallmann's syndrome, a condition characterized by the absent GnRH secretion and therefore very low levels of FSH and LH, follicular development past the primordial stage is rarely observed. In addition, women with ovarian failure due to gonadotropin-resistant ovary syndrome, a condition thought to be due to ovarian nonresponsiveness to gonadotropins, follicular development is limited to primordial and a few primary follicles.

Although the importance of the gonadotropin hormones is unquestioned in follicular development, it appears that there is a differential sensitivity of the ovarian follicle at various stages of development. Furthermore, the relative importance of FSH and LH in follicular development has been debated.

1. Role of Gonadotropins in Initiating Follicular Growth

Primordial follicles are almost certainly formed without the influence of gonadotropin stimulation since primordial follicles are observed in women with no gonadotropin activity due to either low circulating gonadotropin levels (Kallman's syndrome) or the absence of gonadotropin receptor activity (resistant ovary syndrome or ovarian dysgenesis due to FSH receptor mutations). One of the most crucial steps in folliculogenesis is the transformation of primordial follicles into primary follicles which signals entry into the population of growing follicles. It seems likely that transformation of the flattened pregranulosa cells of the primordial follicle into cuboidal granulosa cells may involve signals from the oocyte. Indeed, conversion of primordial follicles to primary follicles does not occur until the germinal vessicle oocyte grows to at least 20 μm in diameter. Although the precise stimulus for recruitment of primordial follicles to enter the growth phase is unknown, it is likely that even this earliest stage of follicular growth is dependent on at least basal levels of gonadotropin support. Evidence for this comes from clinical observations and experimental conditions in animals in which follicular development beyond the primordial stage is markedly reduced in the absence of gonadotropins.

2. Role of Gonadotropins in Follicular Growth to the Early Antral Stage

It is likely that at least basal levels of gonadotropins are necessary for development of the ovarian follicle from the primary follicle to the early antral follicle stage, although this is controversial. Ovaries from hypophysectomized animals (leading to low gonadotropin levels) have a few follicles that have matured to the early antral stage, leading some to believe that this growth is gonadotropin independent. However, from a quantitative standpoint, preantral follicular development is markedly reduced in all of these studies as well as in humans with low gonadotropin lev-

els. Good evidence for the importance of gonadotropins in preantral follicular development comes from experiments using isolated follicles from rats and mice. FSH alone in serum-free media can stimulate preantral follicles with two or three layers of granulosa cells and develop into normal-size antral follicles after 5 days of culture, whereas this development fails to occur in the absence of gonadotropins.

3. Role of Gonadotropins in Exponential Follicular Growth

In contrast to the tonic growth phase which is dependent on low levels of gonadotropins, the exponential growth phase of the ovarian follicle is heavily dependent on gonadotropins. The explosive growth of follicles from the early antral stage to an ovulatory follicle involves granulosa cell division and increasing follicular steroidogenesis, both of which are stimulated by gonadotropins. It is likely that gonadotropins act in synergy with steroid hormones and/or growth factors to mediate these effects on the follicle. Granulosa cells of selected early antral follicles exhibit a significant increase in the rate of proliferation which parallels the increase of circulating levels of FSH. In the newly selected follicles, by the midfollicular phase one follicle has produced relatively more estrogen than the other follicles in its cohort. The dramatic increase in estrogen by the dominant follicle leads to a feedback inhibition of FSH secretion by the pituitary gland. It is hypothesized that the resulting drop in circulating FSH leads to atresia of the nondominant follicles in the cohort of follicles that were selected. Thus, gonadotropins not only stimulate growth of the dominant follicle but also a relative lack of gonadotropin effect on nondominant follicles may be responsible for follicular atresia.

4. Role of Gonadotropins in Stimulating Steroidogenesis

A well-accepted principle of ovarian follicular development is the two-cell, two-gonadotropin theory of follicular steroidogenesis which states that both gonadotropins (FSH and LH) and both steroid-secreting cells of the ovarian follicle (granulosa cells and theca cells) are required for the production of estradiol (Fig. 5). Estradiol is synthesized by the enzyme

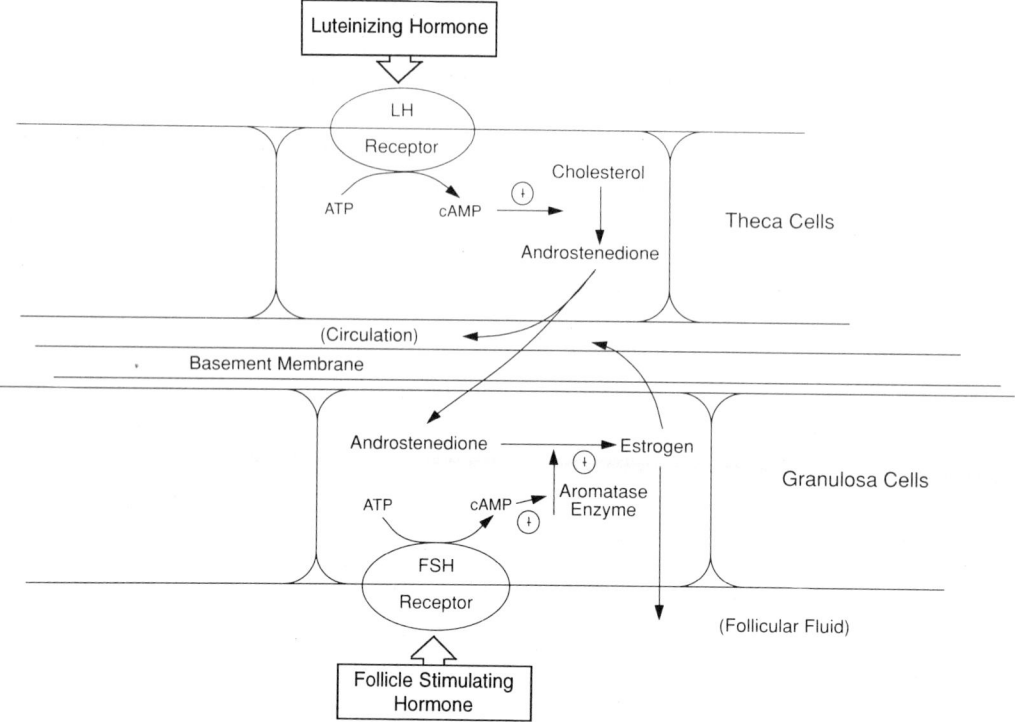

FIGURE 5 The two-cell, two-gonadotropin theory of follicular steroidogenesis [reproduced with permission from *Reproductive Endocrinology, Surgery and Technology* (Adashi, Rock, and Rosenwaks, Eds.), Lippincott-Raven, Philadelphia].

aromatase which is present in granulosa cells. Aromatase activity is stimulated by FSH binding to granulosa cells. Granulosa cells are incapable of producing androgens, a necessary substrate for the aromatase enzyme. However, theca cells can secrete androgens when stimulated by LH and this is the source of substrate for aromatase in the granulosa cell. Thus, estradiol production by the ovarian follicle requires the concerted actions of both theca cells (to synthesize androgens) and granulosa cells (to convert androgens to estradiol) under the influence of LH and FSH, respectively.

5. The Relative Importance of FSH and LH in Follicular Development

The availability of recombinant, pure FSH and LH has allowed exploration of the relative importance of these hormones in follicular development. In hypophysectamized rats and mice, recombinant FSH alone can stimulate follicular growth, ovulation, and even corpus luteum formation. In the absence of LH, estradiol secretion by the follicle remains low since

production of androgens by theca cells is limited. In similar experiments, monkeys treated with GnRH agonists to diminish endogenous FSH and LH can have follicular maturation induced by recombinant FSH alone. Likewise, in women with Kallmann's syndrome and hypogonadotropism, purified FSH induces follicular development, although estradiol levels are below what those seen when both FSH and LH are used clinically. Therefore, as its name suggests, follicle-stimulating hormone is the main promoter of follicular maturation. However, LH also plays an important, although perhaps secondary, role to FSH by stimulating androgen production by thecal cells and perhaps by synergizing with FSH in the more advanced phases of follicular development.

B. The Gonadotropin Receptors

Gonadotropins stimulate folliculogenesis by coupling with receptors localized on the cell surface of follicular cells. Gonadotropin receptors (FSH and LH receptors) are G-protein coupled receptors which activate adenylate cyclase and increase cAMP as the "second messenger" in follicular cells. FSH binds exclusively to the FSH receptor in granulosa cells, whereas LH binds to the LH receptor, which is initially localized to theca cells and then becomes expressed on granulosa cells during maturation of the preovulatory follicle (Fig. 6). FSH receptors have clearly been detected in human follicles from the secondary stage onward. The density of FSH receptors in granulosa cells is higher in antral than in preantral follicles and this may explain, in part, the increasing responsiveness of later follicular development to gonadotropin stimulation compared to earlier follicular growth. LH receptors are present on thecal cells at all stages of follicular development but only appear on granulosa cells in the preovulatory follicle. The induction of LH receptors on granulosa cells (Fig. 6) is mediated by the synergistic actions of both FSH and estradiol, leading to increasing gene transcription for this receptor. The induction of LH

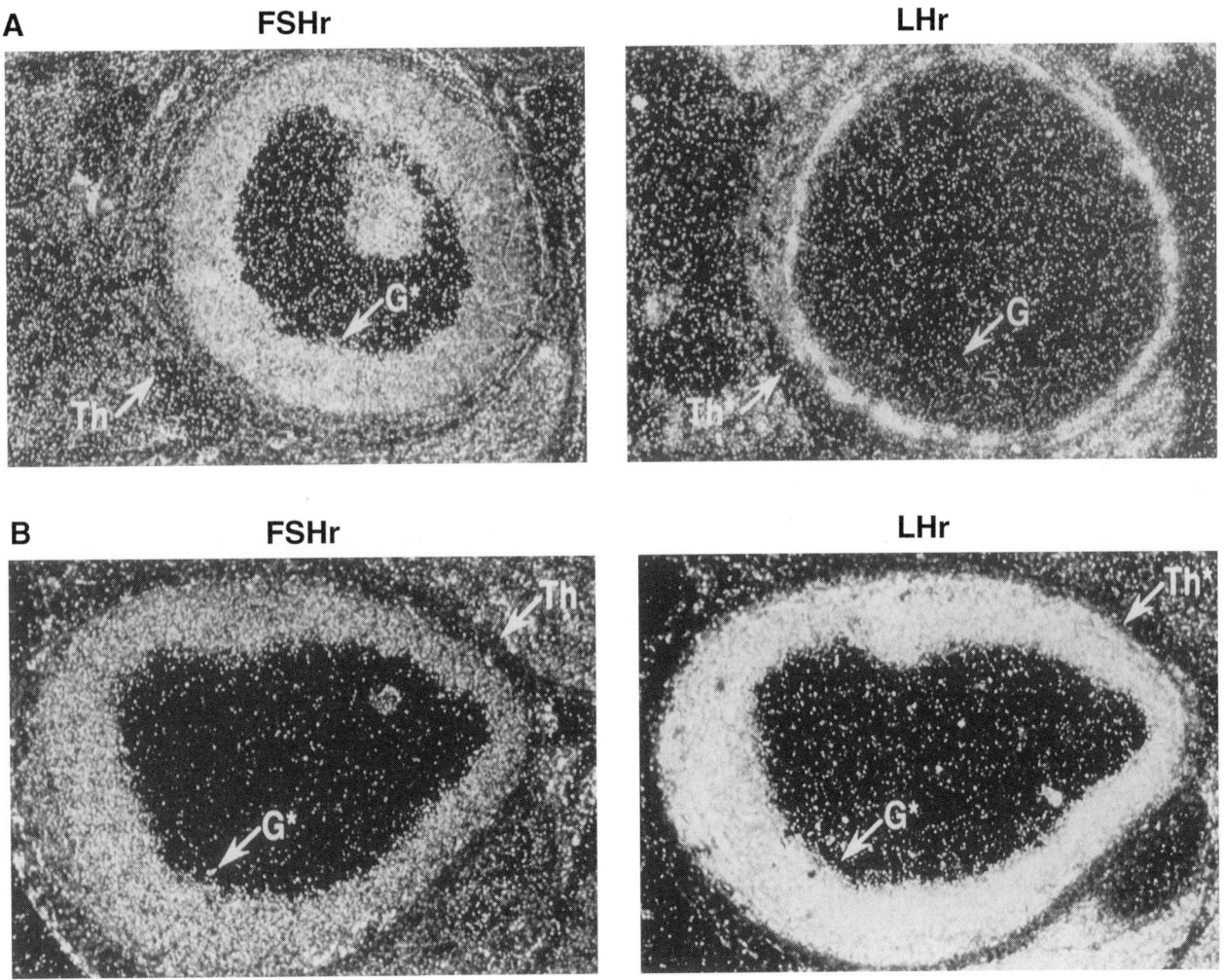

FIGURE 6 Cellular localization of FSH receptor (FSHr) and LH receptor (LHr) mRNA in the rat ovarian follicle by *in situ* hybridization. (A) Adjacent sections of a growing antral follicle. Note FSHr mRNA confined to granulosa cells of the follicle and LHr mRNA confined to thecal cells. (B) Adjacent sections of a preovulatory follicle. Note FSHr mRNA confined to granulosa cells of the follicle. LHr mRNA is now abundantly expressed in both granulosa and thecal cells. G, granulosa cells; Th, thecal cells. The asterisks indicate regions of positive hybridization (reproduced with permission from T. Camp, *Mol. Endocrinol.* 5, 1412, 1991. © The Endocrine Society).

receptors in granulosa cells is critical for final ovarian follicular maturation and ovulation. The results of the actions of luteinizing hormone on the mature granulosa cell include an increase in steroidogenesis, luteinization of granulosa cells, and ovulation of the oocyte.

The importance of the FSH receptor to follicular development has recently been demonstrated in women who have a familial type of premature ovarian failure termed hypergonadotropic ovarian dysgenesis. These women lack secondary sexual development and have ovarian dysgenesis with markedly elevated serum FSH values. Analysis of their FSH receptor gene has revealed a point mutation which leads to the complete absence of second messenger activity (cAMP generation) in granulosa cells. In essence, these women have an absence of FSH activity in the ovary. Ultrasonic and histologic examination of the ovaries of these women have revealed extremely small dysplastic ovaries containing predominantly primordial and some primary ovarian follicles. Although a few preantral and even one mature follicle have been seen in biopsy specimens, the number of maturing follicles was markedly reduced. Thus, when FSH action is abolished in the ovary due to a point mutation in the FSH receptor, ovarian follicle development is markedly impaired, demonstrating the importance of this gonadotropin receptor to follicular development.

C. Estradiol as a Regulator of Follicular Growth

In the rodent, estradiol synthesized by the ovarian follicle has not only endocrine effects on pituitary gonadotropin secretion but also local autocrine roles in the follicle. Estrogen receptors have been found in rodent granulosa cells. Estrogen has been shown to prevent follicular atresia and enhance granulosa cell growth, expression of steroidogenic enzymes, and expression of gonadotropin receptors. Recently, the importance of estrogen as an intraovarian regulator was confirmed in the mouse since transgenic mice with the estrogen receptor "knocked out" are infertile and have dysfunctional ovaries with poor follicular development beyond the secondary follicle stage of development.

Extension of these findings to the primate ovary has proven to be controversial. In contrast to many lower animals, estrogen receptors have not been consistently found in the primate granulosa cell. *In vitro* studies on the effects of exogenous estrogen on primate granulosa cell function have demonstrated that estrogens tend to reduce rather than stimulate steroidogenesis as seen in lower animals. In addition, *in vivo* studies in the monkey have demonstrated that exogenous estradiol leads to follicular atresia, in contrast to the opposite findings in lower animals. In addition, human metabolic diseases can be informative with regard to the role of estrogen in follicular development in higher species. Women with an autosomal recessive disorder in the steroidogenic enzyme 17-α-hydroxylase, 17-20-lyase have a markedly impaired ability to synthesize estradiol. However, apparently normal follicular development can be induced with gonadotropin injections in these patients despite the lack of estradiol synthesis. These observations suggest that follicular growth can occur even if granulosa cell estradiol synthesis is impaired and estradiol may not be a crucial intraovarian regulator of follicular development in primates as it is in many lower animals.

D. Putative Intraovarian Regulators of Follicular Development

Although there is no disputing the importance of gonadotropins and gonadal steroids in follicular development, there is much evidence that intraovarian factors may also play an important role in follicular development. It has been hypothesized that a variety of growth factors and cytokines as well as other substances may have important roles in "fine-tuning" follicular development. Many of the processes involved in follicular development cannot be fully explained merely by alterations in gonadotropin release. Examples include initiation and arrest of meiosis, initiation of early follicular growth, and the exponential growth and differentiation of the dominant follicle. Some of these processes are likely to be influenced not only by gonadotropins but also by intraovarian factors. A large effort has gone into establishing various growth factors and cytokines as bona fide regulators of follicular development. To qualify as an intraovarian regulator, the substance must demonstrate local production, local reception, and action within the ovarian follicle. In addition, some evidence for *in vivo* effects on ovarian function should be provided. The evidence for some of the better studied factors as regulators of follicular development is presented in the following sections.

1. Insulin-like Growth Factors

One family of growth factors that has received attention within the ovarian follicle are the group of insulin-like growth factors (IGF). The IGF system consists of IGF-1 and IGF-2 as well as IGF type I and IGF type II receptors. The IGF system is modulated by a family of binding proteins (IGF BPs) that bind IGFs and reduce their biological actions. In rodents, IGF-1 gene expression and protein have been detected predominantly in granulosa cells of developing follicles. Both granulosa and thecal cells possess receptors for IGF-1. IGF-1 seems to play a role in amplification of the gonadotropin hormonal action on the follicle. Both proliferation of follicular cells and stimulation of steroidogenesis have been demonstrated for IGF-1. The precise role of IGF-2 in rodent follicular development has not been thoroughly explored.

The importance of IGF-1 to rodent follicular development has been confirmed by knockout experiments in mice. Both sexes of adult mice that are homozygous for a targeted mutation in the IGF-1 gene are infertile dwarfs. The female mutants fail to ovulate even after administration of gonadotropins.

In humans it appears that IGF-2 may play the more important role as a local regulator of follicular development. IGF-2 protein has been demonstrated in both theca and granulosa cells of primate follicle and its receptors are also found in both of these cell types. In contrast to rodent species, IGF-1 protein expression has not been detected in the primate granulosa or theca cells. However, IGF-1 receptors are present on both granulosa and thecal cells and IGF-1 can be detected in follicular fluid of the primate follicle. The impact of IGFs on both granulosa and thecal cells argues in favor of the ovarian IGF-2 system playing the key autocrine role in human and primate ovarian follicular growth. In humans, IGF-1 is predominantly synthesized in the liver and it may play an endocrine role in modulating follicular development.

2. Epidermal Growth Factor in Transforming Growth Factor-α

Both epidermal growth factor (EGF) and transforming growth factor-α (TGF-α) are polypeptide growth factors which bind to the same receptor, the EGF receptor. The EGF receptor has been localized in granulosa cells in a variety of species. In human ovarian follicles, both EGF and TGF-α, as well as their common receptor, have been detected by immunohistochemistry. Intrafollicular concentrations of TGF-α and EGF decrease as follicles develop. *In vitro* experiments suggest that both EGF and TGF-α stimulate granulosa cell proliferation and inhibit estradiol production. Thus, TGF-α and EGF may function as inhibitors of gonadotropin-supported granulosa cell differentiation.

3. Transforming Growth Factor-β

Transforming growth factor-β (TGF-β) is a polypeptide with at least three isoforms (TGF-β1, TGF-β2, and TGF-β3). TGF-β is produced in both the theca and granulosa cells of rodent species. In humans, TGF-β production has been demonstrated by immunohistochemistry in growing follicles from the early antral stage onward. The actions of TGF-β include stimulation of granulosa cell proliferation. The importance of these growth factors in ovarian physiology remains unclear.

4. Basic Fibroblast Growth Factor

Basic fibroblast growth factor (FGF) is a peptide that has been isolated from many different tissues. It acts as a mitogen to mesodermally derived cells. In the ovary, FGF has been shown to be mitogenic to granulosa cells and it also inhibits thecal cell steroidogenesis. FGF may have an important role in stimulating angiogenesis within the ovary, particularly in the corpus luteum. The exact cellular localization for production of basic FGF in the ovary is still under investigation.

5. Inhibin, Activin, and Follistatin

Inhibin is a glycoprotein which is a heterodimer consisting of a common α subunit combined with one of two β subunits. Activins are formed by a combination of two β subunits. Follistatin is a high-affinity binding protein for both inhibin and activin. In the primate ovary, activin, inhibin, and follistatin have been predominantly detected in granulosa cells. In addition to the well-characterized endocrine effects of these compounds on pituitary gonadotropin secretion, it is also hypothesized that these compounds may play both autocrine and paracrine roles during follicular development. Activin production by granulosa cells declines during follicular growth, whereas inhibin and follistatin production increases in larger follicles. *In vitro* studies suggest that inhibin may act on theca cells to enhance androgen production, whereas activin may play an autocrine role in enhancing granulosa cell proliferation and aromatase expression. Follistatin's effects on follicular development are thought to be indirect by binding primarily to activin and inhibin to reduce their biological activity. The roles of inhibins and activins are being actively investigated using transgenic mouse models whereby the genes for these products are being knocked out. These studies should shed further light on the intraovarian role of these substances.

6. Interleukin-1

Recognition that white blood cells constitute an important cell type within the ovary has lead to the hypothesis that cytokines secreted by these cells may modulate ovarian function including ovarian follicular development. The most prominent cell type in the ovary is the macrophages, but mast cells, eosinophils, and T lymphocytes, among other cells of white cell lineage, have all been localized within the ovary. Interleukin-1 (IL-1) is a cytokine that is produced by activated macrophages and is known to have an important role as an immune mediator. IL-1 may affect follicular development since IL-1 at physiologic concentrations appears to have an antigonadotropic activity. Specifically, IL-1 has been shown to inhibit both FSH- and hCG-stimulated estradiol production in granulosa cells. Importantly, IL-1β and IL-1α mRNA have been detected in human preovulatory follicular aspirates. Whether or not IL-1 is produced by white blood cells present within the human follicular wall or by granulosa cells remains somewhat uncertain, although even after purifying the granulosa cells obtained from human follicular aspirates, IL-1 could still be detected. This suggests that granulosa cells may be the site of IL-1 production.

7. Tumor Necrosis Factor-α

Another potentially important cytokine in ovarian function is Tumor Necrosis Factor-α (TNF-α), which is another macrophage product which can exert cytostatic effects on many cells. TNF-α has been shown to attenuate the differentiation of cultured granulosa cells from the rat. In human granulosa cells obtained from IVF procedures, TNF-α inhibits hCG-induced progesterone production and also reduces FSH-induced estradiol production. Studies in the rat suggest that TNF-α may be produced in granulosa cells as well as in macrophages. The exact role that TNF-α is playing in follicular development is unclear, although one might speculate a role in follicular atresia.

8. Nitric Oxide

Another potential intraovarian regulator is nitric oxide (NO), which is a gas that has been shown to mediate a wide variety of physiologic responses. Its role in the ovary has recently been investigated. NO is produced in cells by a family of isozymes terms nitric oxide synthases (NOS). There are currently three recognized isoforms of NOS: inducible NOS, endothelial NOS, and neuronal NOS. NO mediates a wide variety of physiologic responses, including vasodilation, neurotransmission, and aspects of immune response of the macrophage. Both inducible and endothelial NOS isoforms have been localized in the rat and human ovarian follicle. Inducible NOS mRNA is highest in small secondary follicles and levels decrease in larger preantral and antral follicles. In contrast, endothelial NOS mRNA is stimulated by gonadotropin stimulation and peaks at the time of ovulation. Certainly, the exact cellular localization of each of these isozymes has varied somewhat from different reports, but both message and protein are consistently found within the ovarian follicle in all reports. NO has been proposed to play a role in granulosa cell steroidogenesis because it decreases estradiol secretion by human granulosa cells. NO may also play a role in inhibiting apoptosis in preovulatory rat follicles and in stimulating vasodilation and increased blood flow to the developing and preovulatory follicle.

E. Use of Transgenic Animals to Study Follicular Development

Many laboratories are producing transgenic knockout mice; this takes advantage of the pluripotential nature of embryonic stem cells. Gene targeting in embryonic stem cells involves disruption of specific genes in the cells, which ultimately allows one to generate null or "loss of function" mutations. These embryonic stem cells can be propagated as cell lines and used to produce chimeric mice and eventually mutant mice which lack the protein being studied. This knockout technique has demonstrated several new proteins as being important in follicular development as well as confirmed the importance of other proteins based on past experimental work. For example, the importance of estrogen and progesterone in follicular development has been demonstrated by mice with knockout mutations for the estrogen and progesterone receptors, respectively. Estrogen receptor knockout mice are infertile and histologic examination of ovaries reveals the presence of only primordial, primary, and secondary follicles. This study demonstrates the importance of estrogen in modulating the final stages of follicular development at least in the mouse. Mice with no functional progesterone receptors are also infertile due at least in part to an inability to ovulate. Histologic examination of the ovary revealed mature ovarian follicles which failed to ovulate and subsequently form a corpus luteum. Interestingly, the cumulus cells around the oocytes underwent expansion within the ovarian follicle. This demonstrates that progesterone may not be important for follicular development but may play a vital role in ovulation.

Other protein growth factors have also been shown to be vital for follicular development. Mice with knockouts for the IGF-1 gene are infertile dwarfs. Their ovaries demonstrate primordial, primary, secondary, and antral follicles, and only rarely are preovulatory follicles found in their ovaries. Granulosa cell steroidogenesis also appears to be impaired since serum estradiol levels are reduced by 50% compared to wild-type controls.

Several members of the TGF-β growth factor superfamily have been shown to be important in follicular development based on mouse knockout experiments. Growth differentiation factor-9, a member of the TGF-β superfamily, is an oocyte-derived growth factor and mice deficient in this growth factor are infertile. Ovarian histology reveals the presence of primordial and primary follicles but no growth beyond this stage. This model is interesting because it is a convincing demonstration that oocyte products may be important in directing follicular development beyond the primary follicle stage. Another member of the TGF-β superfamily is inhibin. The ovaries of inhibin-deficient mice form granulosa cell tumors and follicular development to the antral stage is markedly impaired. Despite superovulation with gonadotropins, ovulation of oocytes was severely limited and any oocytes collected were abnormal in appearance. This mouse model demonstrates the importance of inhibin in the final stages of follicular development and suggests the importance of an autocrine or paracrine role of inhibin in the ovary.

These examples demonstrate the importance of this new technique in the study of ovarian follicular development and ovulation. It is likely that many novel proteins affecting ovarian physiology will be discovered via this type of investigation.

IV. FOLLICULAR ATRESIA

No discussion of follicular development is complete without considering the process of atresia. Indeed, in mammals it has been recognized that >99% of follicles present at birth become atretic with only <1% achieving ovulation (Fig. 4). Therefore, it is a rare follicle that actually achieves ovulation; most follicles undergo atresia. Atresia can occur at any stage of follicular development, although careful analysis has revealed that atresia is not equally prevalent across all stages. A consistent finding in several species is that most follicular atresia occurs just before the final stages of follicular development. In human follicles, the highest rates of atresia are in early antral follicles. Similar findings have been revealed in cattle and in rodent species.

In primates, atresia rarely occurs in early growing follicles and the atresia that does occur is not dependent on cyclic changes in circulating hormones. In contrast, atresia of large follicles varies significantly throughout the cycle and is inversely related to the circulating FSH concentrations.

A. Characteristics of Atretic Follicles

Morphologically, atretic follicles are distinguished by oocyte involution, irregular shape of the follicle, increased intracellular spaces between granulosa cells, and nuclear pyknosis in the granulosa cell layer. In addition, atretic follicles in a variety of species have reduced vascularity in the thecal layer.

It has been recognized that atretic follicles have a more androgenic milieu in their follicular fluid compared to healthy follicles. Healthy growing follicles have a low androstenedione to estradiol ratio in their follicular fluid and the granulosa cells have abundant aromatase activity. In contrast, atretic follicles have a very high androstenedione to estradiol ratio and aromatase activity is either lacking or poorly expressed. As follicles degenerate, the granulosa cells lose their capacity to respond to FSH.

B. Follicular Atresia as an Apoptotic Process

It is now generally accepted that follicular atresia is a result of apoptosis or programmed cell death (Fig. 7). Apoptosis is an active energy-requiring process that allows for the selective deletion of cells. Apoptosis plays a role in a wide range of physiologic processes and is characterized morphologically by condensation of nuclear chromatin or pyknosis of the nucleus, a feature that is commonly seen in atretic ovarian follicles. In addition, apoptosis is characterized by fragmentation of genomic DNA into nucleosomal oligomers. A Ca^{2+}/Mg^{2+}-dependent endonuclease is responsible for cleaving DNA into roughly 180 base pair fragments which results in the ladderlike appearance when DNA is separated on an agarose gel. This DNA ladder appearance distinguishes poptotic cell death from necrosis which presents as a smeared pattern of DNA upon gel electrophoreses.

The evidence that follicular atresia is an apoptotic process includes characteristic morphologic findings, demonstration of DNA laddering in follicles known to be apoptotic, and the presence of endonucleases responsible for apoptosis in granulosa cells. Therefore, there is both histologic and biochemical evidence that apoptosis is the mechanism by which follicles undergo atresia.

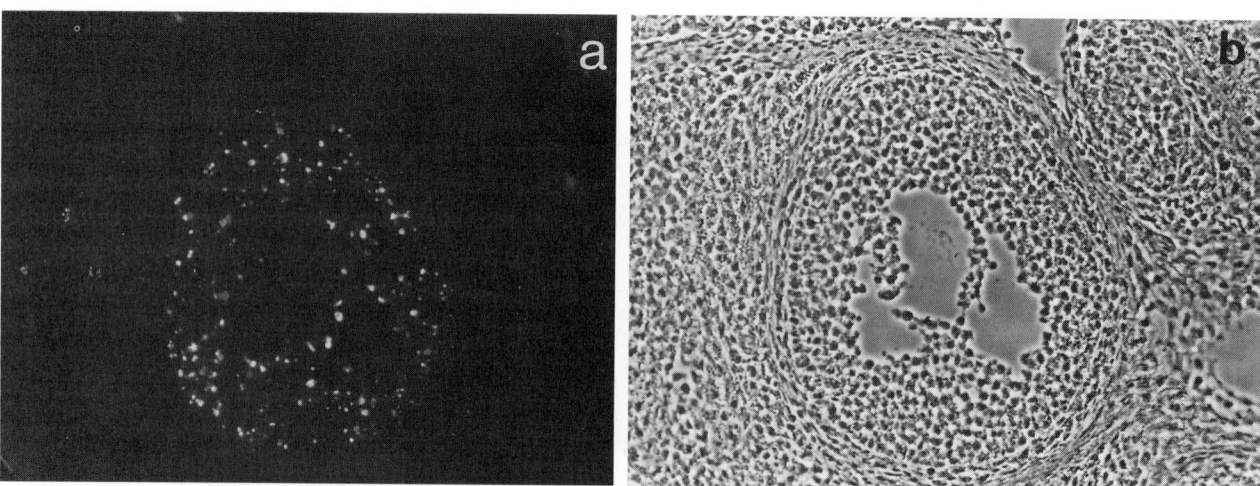

FIGURE 7 Apoptosis in an early antral follicle. (a) Apoptotic nucleii detected by an immunofluorescent *in situ* technique. (b) Phase-contrast view of the same follicle.

The factors controlling apoptosis within ovarian follicles are currently under intensive investigation. The importance of gonadotropins in preventing apoptosis in granulosa cells has been demonstrated in rat follicles both *in vivo* and *in vitro*. Therefore, insufficient gonadotropin stimulation may be pivotal in inducing atresia in ovarian follicles. Several growth factors, including EGF, TGF-α, and β-FGF have also been shown to reduce apoptosis in rat ovarian follicles cultured *in vivo*.

An important factor which may induce apoptosis in the follicle may be oxidative stress. With oxidative stress, free radicals are generated as a consequence of normal cellular respiratory metabolism. It has been shown that oxidative stress can induce granulosa cell apoptosis *in vitro*. In addition, the ability of gonadotropins to inhibit apoptosis can be mimicked by inhibitors of oxidative free radical formation, including superoxide dismutase, catalase, and ascorbic acid, when follicles are cultured *in vitro*. Little is known regarding the molecular mechanism behind the initiation of apoptosis by oxidative stress or other factors, but this is an area of active investigation.

See Also the Following Articles

APOPTOSIS; GONADOTROPIN RECEPTORS; STEROIDOGENESIS; TRANSGENIC ANIMALS

Bibliography

Aittomaki, K., Lucena, J. L. D., *et al.* (1995). Mutation in the follicle-stimulating hormone receptor gene causes hereditary hypergonadotropic ovarian failure. *Cell* **82**, 959–968.

Findlay, J. K. (1993). An update on the roles of inhibin, activin and follistatin as local regulators of folliculogenesis. *Biol. Reprod.* **48**, 15–23.

Fortune, J. E. (1994). Ovarian follicular growth and development in mammals. *Biol. Reprod.* **50**, 225–232.

Gougeon, A. (1996). Regulation of ovarian follicular development in primates: Facts and hypotheses. *Endocr. Rev.* **17**, 121–155.

Hillier, S. G. (1994). Current concepts of the roles of follicle stimulating hormone and luteinizing hormone in folliculogenesis. *Hum. Reprod.* **9**(2), 188–191.

Strauss, J. F., and Steinkampf, M. P. (1995). Pituitary–ovarian interactions during follicular maturation and ovulation. *Am. J. Obstet. Gynecol.* **172**, 726–735.

Follicular Steroidogenesis

Bradley J. Van Voorhis

University of Iowa

Glossary

androgen Any substance which possesses masculinizing activities—for example, the sex steroid testosterone.

cytochrome P450 enzymes A family of hemoprotein enzymes whose iron group, when complexed with carbon monoxide, shows a peak of light absorption at 450 nm. These enzymes are monoxygenases for the most part and many of the steroidogenic enzymes are in this family.

estrogen A generic term for estrus-producing steroid compounds—for example, the female sex steroid estradiol.

progestin A generic term for the group of steroids made chiefly in the placenta and corpus luteum whose function is to prepare the uterus for implantation of the embryo.

The ovarian follicle has two basic functions: to develop a mature oocyte and to secrete steroids. This article will concentrate on the key pathways of ovarian follicular steroidogenesis. Ovarian follicles secrete three classes of steroids: androgens, estrogens, and progestins. These steroids are then secreted into the blood steam and act as hormones affecting most notably the reproductive system but also the central nervous system, the musculoskeletal system, and the cardiovascular system, among others. Steroid secretion by the ovarian follicle controls reproduction by a feedback effect on the hypothalamic–pituitary axis, influencing gonadotropin secretion. Steroid secretion by the follicle also acts on tissues in the reproductive tract, including the endometrium, preparing it for implantation and pregnancy. In addition to acting on distant sites within the reproductive system, steroid secretion by the ovarian follicle also has autocrine and paracrine effects which directly influence follicular development (see Follicular Development, Control of). The importance of ovarian steroidogenesis is emphasized by women with genetic defects in aromatase; a key enzyme in estrogen synthesis. Several women have been noted to have point mutations in their aromatase gene which lead to a complete absence of aromatase activity and estradiol secretion. These women fail to undergo pubertal development and they have ambiguous external genitalia. In addition, they have multicystic ovaries and markedly elevated gonadotropin levels, emphasizing the importance of estradiol for proper follicular development and for feedback effects on gonadotropin secretion. As expected, these women are also infertile. Thus, ovarian follicular steroidogenesis is vitally important for propagation of our species.

I. STEROIDS SECRETED BY THE OVARIAN FOLLICLE

The three classes of steroids secreted by the ovarian follicle are progestins, androgens, and estrogens. Pro-gestins are 21-carbon compounds secreted predominantly by the corpus luteum following ovulation to prepare the endometrium for implantation. Progestins are also secreted by large preovulatory follicles. Progestins secreted by the follicle are pregnenolone, which is precursor to all steroid hormones, 17-hydroxy progesterone, and progesterone. Androgens are 19-carbon compounds which, in addition to having hormone effects at distant sites, serve as an immediate precursor to estrogen secretion in the follicle. The two most important androgens secreted by the ovarian follicle are androstenedione and testosterone. Estrogens are the most important follicular steroids secreted from a physiologic standpoint because of the key roles they play in female development and reproduction. The two major estrogens secreted by the ovary are estrone and estradiol. Estradiol is the most potent estrogen on a molar basis, being 10 times more potent in its biological activities than estrone. Indeed, estradiol is the most biologically active of any steroid made by the ovary.

II. STEROIDOGENIC PATHWAYS IN THE OVARIAN FOLLICLE

All steroids are derived from cholesterol, which is a 27-carbon compound. Sources of cholesterol in the ovary include circulating cholesterol, newly synthesized cholesterol from the 2-carbon precursor acetate, and existing stores of cellular cholesterol. Cellular cholesterol includes free cholesterol, cholesterol stored in the cell membrane, and cholesterol stored in cytoplasmic lipid droplets. Of these, it appears that circulating cholesterol is the most important precursor to ovarian follicular steroidogenesis, accounting for 90% or more of the cholesterol destined for steroid production. There is considerable evidence that circulating cholesterol is taken into the ovarian follicular cells via binding of lipoproteins to the cell surface membrane. The protein component of lipoproteins is termed apoproteins, and these constitute the sites which bind lipoprotein receptor molecules on the cell surface. The lipoproteins which contain cholesterol are then internalized into the cell and free cholesterol is liberated by lysosomal esterases. Cholesterol is then available to be further

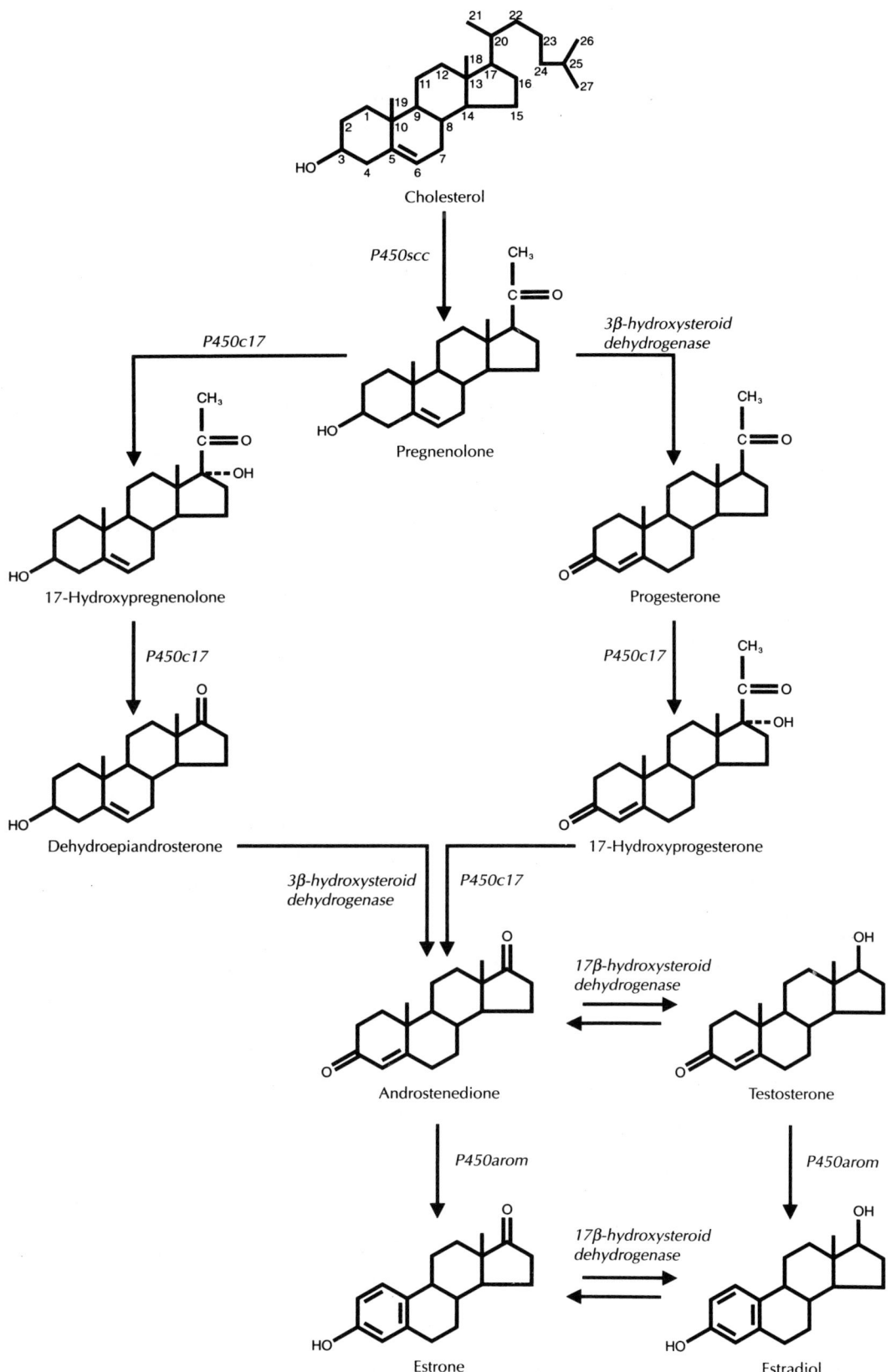

FIGURE 1 Pathways of ovarian follicular steroidogenesis [reproduced with permission from *Clinical Gynecologic Endocrinology and Infertility* (Speroff, Glass, and Kase, Eds.), 5th ed.].

metabolized by steroidogenic enzymes. The relative importance of various types of lipoproteins to ovarian steroidogenesis has been investigated. Both low-density lipoproteins (LDL cholesterol) and high-density lipoproteins (HDL cholesterol) have been shown to be the sources of precursor cholesterol in the ovary. There seems to be some species differences regarding the relative importance of these two lipoproteins for follicular steroidogenesis, with HDL cholesterol being more important in rodents and LDL cholesterol being more important in other species, including humans. Mouse gene knockout experiments have identified the specific apoprotein termed apoprotein A-1 as being crucial for HDL cholesterol uptake into steroidogenic cells in the rodent.

Once cholesterol has entered the cell, it undergoes a series of enzymatic conversions to either 21-carbon progestins, 19-carbon androgens, or 18-carbon estrogens (Fig. 1). The first step in the steroidogenic pathway is the conversion of cholesterol to pregnenolone via cleavage of the carbon 20,22 bond. This cleavage reaction occurs in the mitochondria of the ovarian follicular cell and is believed to be the rate-limiting step in steroidogenesis. The enzyme that catalyzes this reaction is termed the cytochrome P450 side chain cleavage enzyme (P450scc).

Recent studies have identified an important protein termed the steroidogenic acute regulatory protein (StAR) which is important in transporting cholesterol into the mitochondria, where it can be metabolized by P450scc. The StAR protein appears to mediate the acute trophic regulation of steroid hormone synthesis. Individuals deficient in this protein due to an autosomal recessive disorder termed congenital lipoid adrenal hyperplasia have deficient adrenal and gonadal steroidogenesis. Thus, steroidogenesis is regulated not only by steroidogenic enzymes but also by transport of cholesterol within the cell.

After pregnenolone is formed, it can be further metabolized by two different enzymes (Fig. 1). The first pathway is via the 3β-hydroxysteroid dehydrogenase enzyme, which converts pregnenolone to progesterone. The 3β-hydroxysteroid dehydrogenase enzyme is located in the endoplasmic reticulum and is responsible for the formation of progesterone, a particularly important ovarian hormone during preg-

nancy and the luteal phase of the cycle. The second pathway of metabolism for pregnenolone is via the cytochrome P450 enzyme, 17-hydroxylase, 17,20-lyase (P450c17). P450c17 is located in the endoplasmic reticulum of steroidogenic cells and is responsible for the conversion of pregnenolone to dehydroepiandrosterone. P450c17 can also convert progesterone to androstenedione. These two alternative pathways for the metabolism of pregnenolone are called the Δ-4 pathway or the Δ-5 pathway, referring to the number of the first carbon having a double bond in either the first (Δ-4) or second ring (Δ-5) of the steroid molecule.

Once androstenedione is formed within the ovarian follicular cells, it can either be converted to testosterone via a 17β-hydroxysteroid dehydrogenase or be converted to an estrogen by aromatase (Fig. 1). The 17β-hydroxysteroid dehydrogenase enzyme is a non-cytochrome P450 enzyme found in the endoplasmic reticulum. Aromatase (P450arom) is a cytochrome P450 enzyme located in the endoplasmic reticulum. This enzyme is responsible for the conversion of androstenedione to estrone or the conversion of testosterone to estradiol 17β. Aromatase is a single enzyme encoded by a single gene, and aromatiase has been localized not only within the ovary but also in other sites, most notably the placenta, peripheral fat, and the brain.

III. TWO-CELL, TWO-GONADOTROPIN THEORY OF OVARIAN STEROIDOGENESIS

The biosynthetic pathways of estradiol production in the ovary outlined in the previous section do not occur within a single cell type. Rather, estradiol secretion by the ovarian follicle requires a unique cooperation between cell types allowing follicular steroidogenesis (Fig. 2). A well-accepted theory of ovarian estradiol secretion holds that theca cells surrounding the follicle are responsive to luteinizing hormone (LH) and secrete predominantly androgens. These androgens diffuse across the follicular basement membrane to granulosa cells, which are responsive to the gonadotropin hormone, follicle-stimulating hormone (FSH). FSH induces aromatase

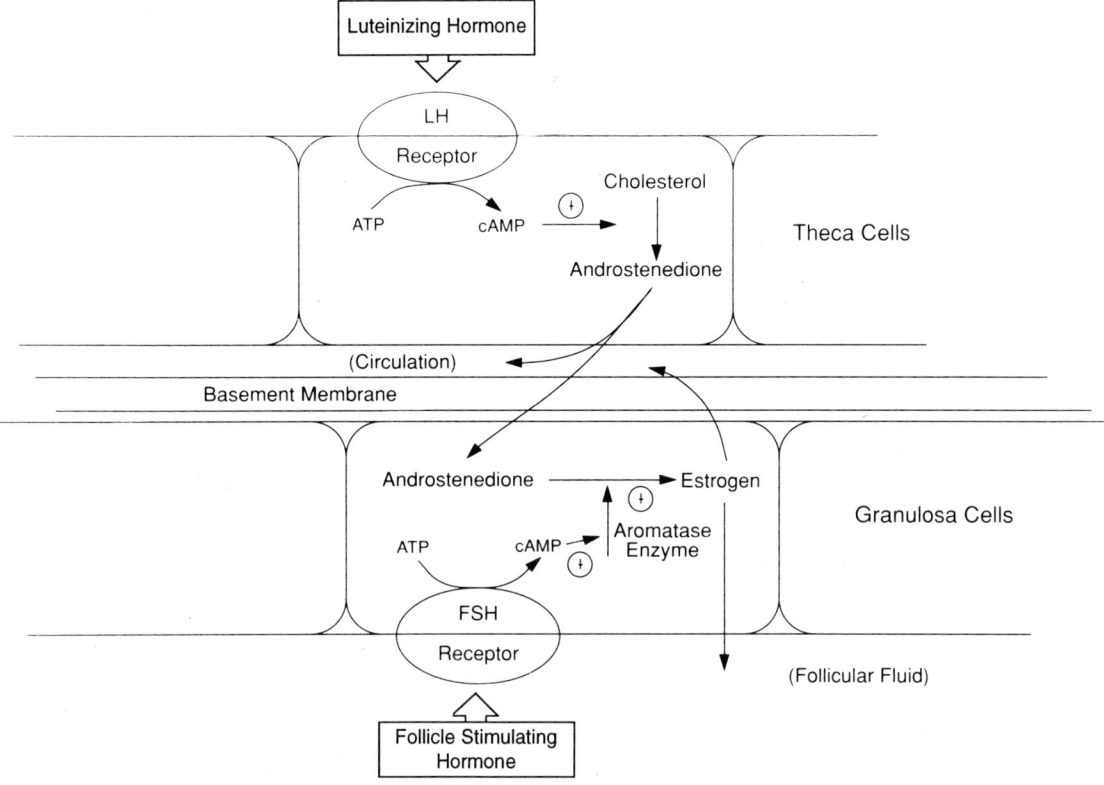

FIGURE 2 The two-cell, two-gonadotropin theory of follicular steroidogenesis [reproduced with permission from *Reproductive Endocrinology, Surgery and Technology* (Adashi, Rock, and Rosenwaks, Eds.), p. 214, Lippincott-Raven, Philadelphia].

within granulosa cells; aromatase is the enzyme responsible for conversion of androgens to estrogens. Granulosa cells are not capable of producing the substrate androgens for the aromatase enzyme because they lack P450c17 and thus are dependent on theca cells for this substrate. Therefore, the follicle is dependent on two cells (theca cells and granulosa cells) and two gonadotropins (LH and FSH) for the secretion of estradiol.

IV. THECA CELL STEROIDOGENESIS

The most abundant steroid products of both theca and interstitial cells of the ovary are androgens. Following formation of pregnenolone by P450scc, the preferred metabolic pathway in theca cells is via the Δ-5 pathway to dehydroepiandrosterone. This androgenic steroid is then metabolized via 3β-hydroxy-

steroid dehydrogenase to androstenedione, which is the primary androgen produced by the ovarian follicle. The gonadotropin hormone which stimulates thecal cell steroidogenesis is LH. LH receptors are present on thecal cells at all stages of follicular development. LH binds to these receptors and acts via cyclic AMP as its second messenger to stimulate androgen production in the follicle predominantly by increasing steroidogenic enzyme activity. LH has been shown to increase both P450scc and P450c17 activity within the theca cells. As a result of these enzyme activities, androstenedione is the principal aromatizable steroid produced by theca cells. Androstenedione then diffuses to granulosa cells, which possess abundant aromatase activity. There is some evidence that theca cells may also aromatize androgens and therefore contribute directly to ovarian estrogen secretion. However, the amount of aromatase in theca cells is significantly less than what is found in granulosa cells and therefore theca cell production of estrogens is very minor compared to the estrogen production by granulosa cells.

V. GRANULOSA CELL STEROIDOGENESIS

Whereas the steroidogenic pathways in theca cells are organized for androgen production, the primary steroidogenic pathways in granulosa cells are organized for the production of estrogens. Estradiol production is the result of cytochrome P450 aromatase, which metabolizes androgens to estrogens. FSH is the gonadotropin which stimulates granulosa cell steroidogenesis. FSH receptors are present on granulosa cells from an early point in follicular development and the actions of FSH via the FSH receptor have been shown to stimulate aromatase transcription and activity in granulosa cells. Granulosa cells also have abundant 17β-hydroxysteroid dehydrogenase enzyme activity. This enzyme is responsible for converting androstenedione to testosterone and estrone to estradiol.

In addition to secreting estradiol, granulosa cells are also capable of secreting progesterone. The major source of progesterone is the corpus luteum following ovulation. In the ovarian follicle, progesterone is predominantly secreted in large antral and immediately preovulatory follicles. Both FSH and LH stimulate progesterone secretion by granulosa cells. LH can act on granulosa cells of large preovulatory follicles because these mature granulosa cells have acquired LH receptors which are not present in more immature granulosa cells in smaller follicles. The source of cholesterol for progesterone production in the granulosa cells is thought to be *de novo* synthesis of cholesterol from acetate. Circulating cholesterol in lipoproteins is not available to the granulosa cells since the developing follicle is avascular and the large lipoproteins are not able to penetrate to the granulosa cell layer. Following the LH surge, circulating lipoprotein is likely to be the major source of cholesterol for steroidogenesis in the granulosa cell since blood vessels invade the follicle at this stage of folliculogenesis. Both FSH and LH greatly stimulate 3β-hydroxysteroid dehydrogenase enzyme activity, which is responsible for the conversion of pregnenolone to progesterone.

VI. STEROIDOGENESIS AND THE GROWING OVARIAN FOLLICLE

Steroidogenic capabilities vary not only according to cell types within the follicle but also according to the stage of follicular development. As ovarian follicles grow and develop, steroidogenesis markedly increases. Follicles developing from the primary follicle to preantral follicles have limited steroidogenic capabilities. Although granulosa cells have FSH receptors in these early follicles, steroidogenesis in these follicles is extremely limited. Experimental *in vitro* studies have demonstrated that large preantral follicles will respond to gonadotropins with increased steroidogenesis; however, it is uncertain to what extent the preantral follicles actually secrete steroids *in vivo* in response to circulating levels of gonadotropins. Steroidogenic capability increases dramatically as follicles develop into the antral and preovulatory stage. The dominant follicle (the follicle destined to ovulate) developing from the midfollicular phase onward is clearly the major source of estradiol in the human reproductive cycle. Dominant follicles in humans are distinguished by very high follicular fluid levels of estradiol that often exceed 2000 ng/ml. In conjunction with the high estradiol levels, FSH values in follicular fluid are also elevated, supporting the essential role of FSH in stimulating steroidogenesis. The dominant follicle is also endowed with larger numbers of granulosa cells producing estradiol when compared to nonovulatory or atretic follicles. Over 90% of circulating estradiol is thought to originate from the dominant follicle. This has been experimentally demonstrated by enucleating the dominant follicle from the ovary and concomitantly demonstrating a marked decrease in plasma estradiol levels which cannot be compensated for by the remaining nondominant follicles.

In addition to secretion of estrogen by the antral and preovulatory follicles, androgens are also secreted by these follicles. Androstenedione levels in follicular fluid are approximately 100 times those found in peripheral plasma. The highest levels of androgens are found in smaller follicles compared to the dominant follicle in the late follicular phase. As the follicle matures, the concentrations of androgens and estrogens are inversely related, with estrogen levels substantially increasing while androgen levels decrease. The androgen levels may be less in the large dominant follicle compared to atretic follicles because of the greater aromatase activity converting androgens to estrogens within granulosa cells of dominant follicles.

Progestins are also produced in a human follicle and concentrations are higher in larger antral follicles than in smaller ones. The largest increase in progesterone occurs shortly before ovulation, in response to the LH surge. Another progestin, pregnenolone, also shows an increase in follicular fluid from large antral follicles beginning in the mid- to late follicular phases.

Steroidogenesis changes markedly in the preovulatory follicle following the LH surge at midcycle. With the LH surge, there is a transient increase in estrogen secretion which lasts only for a few hours followed by large decreases in androgens and estrogens in follicular fluid and serum. The major cause of this rapid decline in estrogen secretion after the LH surge appears to be limited availability of aromatizable androgens due to decreased activities of cytochromes P450scc and P450c17 in theca cells. Following the LH surge, the granulosa cells undergo luteinization and the predominant steroidogenic pathways are directed toward production of progesterone in the corpus luteum. Therefore, it is apparent that steroidogenesis in the follicle is highly dependent on the stage of development and the influences of gonadotropins, particularly LH, at the time of the midcycle LH surge.

In summary, ovarian follicular steroidogenesis is a vital and complex process necessary for sexual development and reproduction. The key concept of steroidogenesis is the two-cell, two-gonadotropin theory of steroidogenesis. Ovarian follicular steroidogenesis is under the control of gonadotropin hormones which alter expression of steroidogenic enzymes in theca and granulosa cells. In turn, the steroids produced have feedback effects on gonadotropin secretion, allowing for the cyclic nature of the reproductive process.

See Also the Following Articles

GONADOTROPIN BIOSYNTHESIS; GRANULOSA CELLS

Bibliography

Adashi, E. Y. (1994). Endocrinology of the ovary. *Hum. Reprod.* 9(5), 815–827.

Bulun, S. E. (1996). Aromatase deficiency in women and men: Would you have predicted the phenotypes? *J. Clin. Endocrinol. Metab.* 81(3), 867–871.

Hillier, S. G., Whitelaw, P. F., and Smyth, C. D. (1994). Follicular oestrogen synthesis: The "two-cell, two-gonadotrophin" model revisited. *Mol. Cell. Endocrinol.* 100, 51–54.

Lin, D., Sugawara, T., Strauss, J. F., Clark, B. J., Stocco, D. M., Saenger, P., Rogol, A., and Miller, W. L. (1995). Role of steroidogenic acute regulatory protein in adrenal and gonadal steroidogenesis. *Science* 267, 1828–1831.

Miller, W. L. (1988). Molecular biology of steroid hormone synthesis. *Endocr. Rev.* 9, 295.

Plump, A. S., Erickson, S. K., Weng, W., Partin, J. S., Breslow, J. L., and Williams, D. L. (1996). Apolipoprotein A-1 is required for cholesteryl ester accumulation in steroidogenic cells and for normal adrenal steroid production. *J. Clin. Invest.* 97(11), 2660–2671.

Simpson, E. R., Mahendroo, M. A., Means, G. C., Kilgore, M. W., Hinshelwood, M. M., Graham-Lorence, S., Amarneh, B., Ito, Y., Fisher, C. R., Michael, M. D., Mendelson, C. R., and Bulun, S. E. (1994). Aromatase cytochrome P450, the enzyme responsible for estrogen biosynthesis. *Endocr. Rev.* 15(3), 342–355.

Folliculogenesis

see Follicular Development

Follistatin

Christopher Gilfillan and David Mark Robertson

Prince Henry's Institute of Medical Research

GLOSSARY

activin A protein hormone (inhibin βA–βA, βA–βB, or βB–βB subunit dimers) produced widely with effects on development, the pituitary (stimulates follicle-stimulating hormone secretion), gonads, and other organs.

follicle-stimulating hormone and *luteinizing hormone* Pituitary glycoprotein hormones which regulate ovarian and testicular function, including gametogenesis.

follistatin A single-chain glycoprotein that acts as a binding protein for activin and neutralizes its diverse functions.

gonadotropin-releasing hormone A peptide hormone secreted from the hypothalamus to stimulate pituitary follicle-stimulating hormone release.

inhibin A gonadal hormone (α-βA or α-βB subunit dimer) which inhibits follicle-stimulating hormone secretion by the pituitary.

transforming growth factor β (TGFβ) A growth factor with cell regulatory activity representing a family of proteins with sequence homology and dimeric subunit structure. Other members consist of inhibin, activin, bone morphogenetic protein, and Mullerian inhibitory substance.

I. INTRODUCTION

Follistatin was originally isolated from side fractions obtained during the purification of inhibin and activin from ovarian follicular fluid on the basis of its ability to suppress follicle-stimulating hormone (FSH) secretion by pituitary cells in culture. Compared to inhibin, follistatin's bioactivity was low, suggesting that it was probably a weak inhibin agonist.

However, the observation that follistatin was also an activin-binding protein which neutralized activin bioactivity has implicated follistatin as a regulator of activin activity. Thus, follistatin is important in many biological systems in which activin plays an major role; for example, follistatin antagonizes activin-induced stimulation of FSH release from pituitary cells, activin-induced erythroid differentiation, activin-stimulated induction of mesoderm, and activin-enhanced FSH-induced differentiation of ovarian granulosa cells.

II. ISOLATION OF FOLLISTATIN AND CLONING THE GENE

Isolation of follistatin from bovine and porcine ovarian follicular fluid revealed a single-chain polypeptide of six or more isoforms (31–45 kDa). These isoforms have identical N-terminal sequences, are cysteine rich, and are glycosylated. They showed no marked sequence homology with other proteins (e.g., inhibin and activin) but did show limited sequence homology with epidermal growth factor and human pancreatic secretory trypsin inhibitor. There is some controversy over the *in vitro* bioactivity of the various isoforms, with some studies showing similar activities, whereas others suggest that the lower molecular weight forms may be the more active.

Northern blotting of polyadenylated mRNA from rat, bovine, and ovine tissues has confirmed the presence of two main bands of approximately 1.6 and 2.5 kb. In all cases, the larger mRNA transcript was the more abundant and represents the mRNA for both the 344- and 317-amino acid precursors, which are indistinguishable on Northern blots. The role of the smaller transcript is uncertain. Sequence analysis

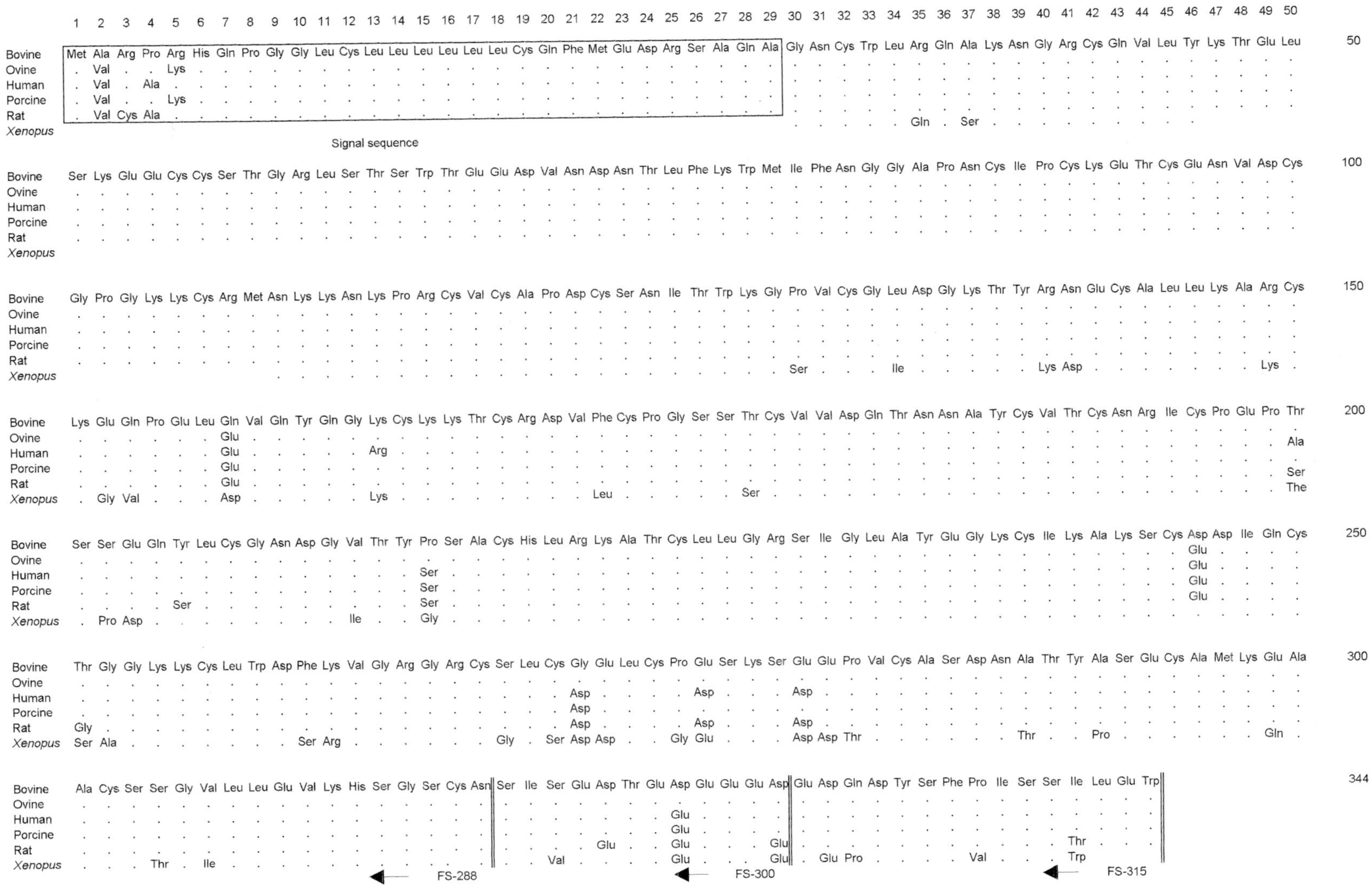

FIGURE 1 A comparison of the amino acid sequence of bovine follistatin with the other known follistatins. The very high sequence homology is evident with the first 127 amino acids of the mature protein identical in all mammalian species. There are only two unique amino acid substitutions in the human protein at position 163 (arginine for lysine) and 200 (alanine for threonine). The *Xenopus* amino acid sequence is incomplete.

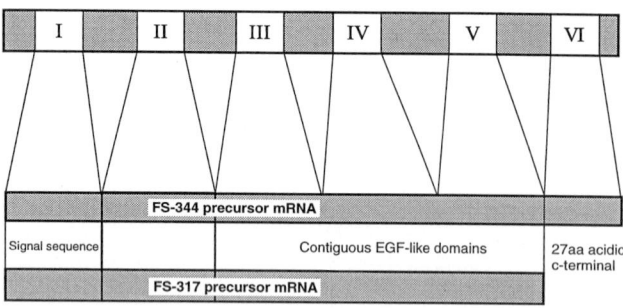

FIGURE 2 The structure of the follistatin gene and the alternative splicing of mRNA to produce two transcripts, one encoding a 317-amino acid precursor and the other a 344-amino acid precursor.

suggested a 29-amino acid signal sequence and, therefore, mature proteins of either 288 or 315 amino acids were predicted. The amino acid sequences of follistatin from pig, human, rat, sheep, chick, and cow have been published and have a shared homology of 97%. A similar protein in *Xenopus* with 87% homology has been identified. These sequences are shown in Fig. 1.

The six isoforms of follistatin isolated from porcine ovarian homogenates consist of follistatin-315 (FS-315) glycosylated at Asn259; three forms of a truncated follistatin-303 (FS-303) glycosylated at both Asn95 and Asn259, Asn259 alone, and nonglycosylated at either site; and follistatin-288 (FS-288) glycosylated at Asn259 and nonglycosylated. Approximately 75% of the recovered protein was in the form of FS-303, which is truncated at the termination of the acidic stretch Glu292–Asp304. It appears, therefore, that the predominate form of follistatin in follicular fluid is a proteolytically cleaved form of the larger precursor (FS-315) with FS-288 in relatively low abundance. Further analysis of porcine follistatin, which was affinity purified from porcine follicular fluid, revealed two major bands at 31 and 32 kDa following deglycosylation with *N*-glucosidase-F. The 31-kDa band can be attributed to deglycosylated FS-288, but the more abundant 32-kDa band represents FS-315 that has been proteolytically cleaved at Asp300. This confirms that the major forms of native follistatin in porcine follicular fluid are a mixture of glycosylated forms of a 300-amino

acid core protein derived from a mRNA encoding the 344-amino acid prefollistatin molecule.

A. The Follistatin Gene

The single follistatin gene, consisting of six exons, has been shown to have high structural homology between species (pig, cow, rat, and human). The first exon encodes the 29-amino acid signal sequence. Exons II–V encode four contiguous domains, three of which are highly homologous (III–V). The sixth exon encodes the 27-amino acid C-terminal extension included in the longer of the two mRNAs by an alternative splicing event. This sixth domain was found to contain an extremely acid amino acid sequence (Glu-Asp-Thr-Glu-Glu-Glu-Glu-Glu-Asp-Glu-Asp-Gln-Asp) which may have physiological significance. The structure of the follistatin gene and the alternative mRNA processing are illustrated schematically in Fig. 2.

The rat follistatin promoter region has three distinct transcription start sites (cap sites) termed α, β, and γ, each about 30 base pairs downstream from a potential TATA box. Potential response elements for the transcriptional regulators, AP-1, SP-1, and AP-2, and a cyclic AMP response element are also found in this region. In addition, two regions with inhibitory activity have been identified (-530 to -366 and -376 to -309).

III. FOLLISTATIN AND ITS ACTIVIN-BINDING ACTIVITY

A. Follistatin as an Activin-Binding Protein

Soon after the purification of follistatin was reported, binding proteins for activin from rat ovaries and bovine pituitaries were isolated that were structurally and functionally identical to follistatin. The stoichiometry of the binding interaction between activin and follistatin was established by observing the binding of labeled activin or follistatin to either protein adsorbed onto nitrocellulose membranes. These

TABLE 1
Comparison of Activin and Follistatin Actions in Various Species, Tissues, and Cells

Species	Tissue/cell type	Activin action	Follistatin action
Xenopus	Embryonic ectoderm	Mesoderm formation	Prevents mesoderm formation
		Inhibition of neural differentiation	Allows for neural differentiation in the presence of mesoderm
Rat	Ovarian granulosa cells		
	Early estrus	Increases aromatase activity	Inhibits activin effect
		Increases LH receptor expression	Inhibits activin effect
		Increases FSH receptor expression	Inhibits activin effect
		Increases progesterone production	Inhibits activin effect
		Increases inhibin subunit expression	Inhibits activin effect
	Late estrus	Promotes atresia	Inhibits activin effect
		Reduces progesterone production	Inhibits activin effect
	Oocytes	Promotes meiotic maturation	Inhibits meiotic maturation
Bovine	Ovarian granulosa cells	Inhibits progesterone production	Inhibits activin effect
		Inhibits oxytoxin production	Inhibits activin effect
	Theca cells	Inhibits progesterone production	Increases progesterone production even in the absence of activin
Human	Ovarian granulosa cells	Inhibits progesterone production	Increases progesterone production
		Inhibits aromatase activity	Increases aromatase activity
Rat	Leydig cell	Increases testosterone production	
	Spermatogonia	Increases proliferation	Inhibits proliferation
Rat	Gonadotroph	Increases FSH β expression and FSH secretion	Reduces FSH β expression and FSH secretion
	In vivo	Increased FSH pulse amplitude	Decreases FSH pulse amplitude
	Somatotroph	Decreases proliferation	
		Decreases GH secretion	Increases GH secretion
	Corticotroph	Decreases ACTH secretion	Increases ACTH secretion
	Lactotroph	Decreases prolactin secretion	Increases prolactin secretion
Human	Placental cells	Increases progesterone secretion	Inhibits activin effect
		Increases hCG secretion	Inhibits activin effect
Rat	Pancreatic β cells	Stimulates insulin secretion	
Human	Fetal adrenal cells	Inhibits proliferation	
		Inhibits steroidogenesis	Partially inhibits activin effect
Porcine	Thyroid follicular cells	Increases proliferation	Decreases proliferation
		Decreases TSH-stimulated iodine uptake and cyclic AMP accumulation	
Rat	Liver	Decreases proliferation postresection	Increases proliferation postresection
Rat	Osteoblasts	Increases proliferation	
		Decreases differentiated function (ALP production)	Increases ALP production
Rat	Endothelial cells	Inhibits proliferation (angiogenesis)	
Rat	Vascular smooth muscle	Inhibits proliferation at low dose	Increases PDGF-stimulated proliferation
		Stimulates proliferation at high dose	

experiments suggested that activin has two binding sites for follistatin (one for each subunit), but that the follistatin has only one binding site for activin. Inhibin is also able to bind weakly to follistatin but appears to have only one binding site.

Further characterization of follistatin–activin binding came from the production of mutated forms of follistatin. Site-specific mutagenesis of human FS-315 clones produced a series of mutated recombinant human follistatins with either or both of the two potential Asn-linked glycosylation sites inactivated, however, activin binding was unaffected suggesting that glycosylation of follistatin was not required for activin binding. There was no activin binding when the protein was unfolded under reducing conditions. Furthermore, mutants with a modified amino-terminal sequence failed to bind activin. The activin binding activities of the six forms of native follistatin isolated from porcine ovaries (discussed previously) were similar (k_{dis} 0.54–0.68 nM) and unaffected by deglycosylation. This suggests that the carboxy-terminal and/or carbohydrate moieties are not important in follistatin–activin binding interactions. The affinity of derivatives of FS-288 for activin may be enhanced by virtue of an exposed heparin binding site.

In order to identify the follistatin-binding region of activin, synthetic peptide fragments of the inhibin βA subunit were tested for their ability to compete with activin for binding to follistatin. Three activin peptides had inhibitory activity, one near the amino-terminal (15–29) and two overlapping sequences near the carboxy terminus (99–133 and 102–116), suggesting that there are at least two regions of the molecule involved in activin binding.

Activin and inhibin binding activity in human serum was detected by incubating iodinated activin/inhibin with serum and separating complexes by size-exclusion high-performance liquid chromatography (HPLC). Labeled activin elutes with apparent molecular mass of 60–80 kDa (corresponding to a follistatin/activin complex) and 800 kDa (probably an activin/α2-macroglobulin complex). Inhibin was also associated with α2-macroglobulin in serum. A follistatin/inhibin complex in serum was not identi-

fied. It would appear that in serum, activin and inhibin are bound to both follistatin and α2-macroglobulin, with α2-macroglobulin the major binding protein. In follicular fluid, the situation is reversed, with follistatin the major binding protein for activin and inhibin and α2-macroglobulin playing a lesser role. Recently, it has been shown that the apparent molecular weight of the activin/follistatin complex in serum is larger than that in follicular fluid, suggesting that the larger molecular weight forms of follistatin (follistatin-315-derived proteins) predominate in serum.

The nature of the follistatin/activin–inhibin interaction *in vivo* has been addressed in two studies. Radiolabeled inhibin A and activin A were injected into rats. High-molecular-weight radioactive-bound material (possibly a complex with α2-macroglobulin) persisted longer than free inhibin or activin in the plasma. A follistatin–activin/inhibin complex was not seen, suggesting that the major activin/inhibin binding protein in serum is α2-macroglobulin. The clearance pattern of activin/inhibin-binding protein complexes *in vivo* in serum was examined by size-exclusion HPLC following intravenous injection of radiolabeled inhibin/activin in rats. In these experiments, complexes corresponding to inhibin/activin–α2-macroglobulin (470 kDa) and inhibin/activin–follistatin (125 kDa) were formed within 2 min of injection and persisted beyond the disappearance of unbound inhibin/activin at 4 hr. The inhibin/activin–follistatin complex was displaceable by unlabeled inhibin/activin, whereas the inhibin/activin–α2-macroglobulin complex was not, consistent with the view that the inhibin/activin–α2-macroglobulin complex is a low-affinity, high-capacity binding system and that the inhibin/activin–follistatin complex is a low-capacity, high-affinity system.

There is currently no clear evidence that the levels of either activin or follistatin in the circulation are specifically regulated or that they perform any endocrine role. Investigation of this area, however, has been limited by the inability to reliably measure either activin or follistatin in the serum or to distinguish between the free and bound fractions of these proteins (see Section IV).

B. Follistatin–Heparin Interactions

In addition to binding activin and inhibin, follistatin has affinity for sulfated polysaccharides, particularly dextran sulfate and heparin. This property of follistatin may have an important role in locating the molecule in the extracellular matrix close to its potential sites of action. Follistatin, affinity labeled with [^{125}I]activin, localizes to the surface of granulosa cells and this binding displays saturation kinetics consistent with a specific binding site. This binding is inhibited by heparin and heparan sulfate but not by other matrix glycoaminoglycans, such as chondroitin, keratan, and dermatan sulfates. The binding can also be eliminated by heparinase pretreatment of the cells. Granulosa cells treated in this way retain full activin responsiveness suggesting that the cell surface binding of follistatin is not a prerequisite for activin/activin receptor interactions. A synthetic fragment of the basic amino acid-rich region of follistatin (residues 75–86) was found to bind heparin and to compete with both recombinant human FS-288 and recombinant human FS-315 for sulfate cellufine binding. This fragment may form a helix that places basic residues Lys75, Lys82, and Arg86 in alignment, enabling binding to the negatively charged sulfate residues of heparin. Recombinant human FS-315 has a lower affinity for heparin possibly related to interference by the extra acidic amino acid carboxy-terminal "tail" in this molecule. The heparin binding activity of six molecular forms of native porcine follistatin is dependent on the carboxy-terminal structure of the isoforms. The short forms ending at Asn288 have high affinity and the intermediate-sized forms ending at Gln303 have intermediate affinity. The longer forms of follistatin ending at Trp315 had no affinity for the granulosa cell surface.

IV. BIOLOGY OF FOLLISTATIN

A. Follistatin and Activin in Embryogenesis

Experiments largely in *Xenopus* embryos have established the important role of activin as an inducer of mesoderm. Activin induces mesoderm formation from animal cap ectoderm, whereas transfection with a truncated activin receptor acts as a dominant negative mutation and eliminates activin action, thus preventing the development of mesoderm. It appears that different concentrations of activin can dictate a range of fates for embryonic ectoderm. High-level exposure leads to mesoderm formation, low-level exposure leads to epidermal differentiation, and the absence of activin leads to a default response which appears to lead to neural differentiation. Neural differentiation is normally dependent on the formation of adjacent mesoderm by activin and so the default neural state can only develop if the local effects of activin can be limited to developing mesoderm and epidermis.

Xenopus follistatin was first demonstrated by screening a *Xenopus* ovarian cDNA library. The predicted sequence showed 84% homology with porcine and human follistatin. The embryonic mRNA was first expressed in the early gastrula stage, just after the expression of the inhibin βB subunit mRNA, which was expressed in the late blastula stage. Follistatin can prevent the mesoderm inducing activity of activin. Follistatin is expressed in the dorsal midline mesoderm or "organizer" region that is the source of neuralizing signals, and this expression is regulated by activin. It has been shown that ectopic expression of follistatin in the absence of mesoderm is capable of directly neuralizing embryonic explants. This suggests that follistatin, by blocking activin signaling via the activin type II receptor, induces neural differentiation in *Xenopus* embryogenesis. The presence of other independent neural inducers (such as Noggin) and several members of the transforming growth factor-β (TGF-β) superfamily that can induce mesoderm (TGF-β, Vg-1, bone morphogenetic protein-3 and -4, and activin) suggests that there is considerable redundancy in these morphogenetic signaling systems.

Activin also induces mesoderm formation in the chicken. As in the frog, the inhibin βB subunit is expressed early in pregastrulation embryos and so activin B is one of the likely physiological mesoderm inducers. In the mouse, activin induces mesodermal differentiation in an embryonal carcinoma (EC) cell line (P19) and is an inhibitor of neural differentiation. Follistatin expression was seen in undifferentiated P19 EC and embryonic stem cells and in whole embryos and may modulate activin action after early murine development.

B. The Follistatin Knockout Mouse

Follistatin-deficient homozygous transgenic mice were produced by targeted deletion of the follistatin gene in embryonic stem cells. At birth, homozygous mice were present in proportion to the number expected by Mendelian inheritance, suggesting that these mice survived normally to birth. However, all follistatin-deficient mice died soon after birth because they failed to breath properly. The homozygous mice were small, growth retarded, and had a tight skin with increased thickness of the stratum corneum reminiscent of changes seen in transgenic mice over-expressing TGF-β. They also had atrophic diaphragmatic and intercostal muscles, defects in the 13th pair of ribs, and a reduced number of lumbar vertebrae. There was also abnormal dental development with absent incisors. Abnormal whisker development and hard palate abnormalities were similar to those seen in the activin βA subunit-deficient mouse suggesting that follistatin may promote activin action at these sites. It is of interest that despite the evidence implicating follistatin as a neural inducer in early embryonic development, the neural development of follistatin-deficient mice was normal from a gross anatomical and histological perspective.

C. The Physiology of Follistatin in the Ovary

In the rat ovary, follistatin mRNA expression is limited to the granulosa and early luteal cells, with the intensity of signal increasing throughout follicular maturation and declining in the preovulatory Graafian follicle. Following ovulation, the signal declines further with advancing luteinization and disappears with luteolysis. The signal also declines sharply following atresia in nondominant follicles. Other ovarian cell types, such as thecal and stromal cells, do not express follistatin. Follistatin protein, as detected by immunohistochemistry, is expressed only in tertiary follicles destined to ovulate during late estrus. The signal increases in intensity, with follicular development peaking in the mature preovulatory follicle. Expression then declines after ovulation and becomes undetectable in old corpora lutea. There is no expression of follistatin protein detectable in atretic follicles.

The regulation of follistatin mRNA expression in granulosa cells has been studied in the rat, porcine, and human species. In the rat, follistatin mRNA expression in granulosa cells obtained from diethylstilbestrol (DES)-treated immature rats was stimulated by FSH and activin, individually and synergistically. Epidermal growth factor inhibited FSH-stimulated follistatin mRNA accumulation, whereas LH stimulation had little effect.

Follistatin produced locally in the ovary has the potential to interact with other locally produced factors, in particular activin, in the control of follicular function. In granulosa cells obtained from DES-treated immature rats, follistatin has no effect on its own but has the opposite effect to activin in the presence of FSH, inhibiting activin-induced aromatase activity and inhibin production. Interestingly, in some studies, follistatin further stimulated activin-induced progesterone production. In other studies, follistatin dose-dependently inhibited the activin-induced increase in progesterone and inhibin production and aromatase activity in culture of rat granulosa cells, and it also prevented the activin-induced morphological transformation of the cultured cells. Follistatin-inhibited activin stimulated granulosa cell inhibin production in a manner that suggested the formation of inactive follistatin/activin complexes with a k_{dis} of 0.13 nM and a follistatin to activin molar ratio of 2:1. Follistatin prevented activin-induced increases in granulosa cell FSH and LH receptor number and responsiveness, but it had no effect on the FSH responsiveness of granulosa cells when given alone. Rat granulosa cell/oocyte complexes can be induced to proliferate and form follicle-like structures in culture and these changes can be prevented by follistatin. In immature bovine granulosa cells, follistatin enhances the secretion of oxytocin and progesterone induced by bovine LH. The effect of activin in this system is to inhibit the secretion of both these markers of differentiated corpus luteum function. Follistatin protein production from bovine granulosa cells in primary culture is increased in the presence of FSH in a dose-dependent manner. The effect of FSH is mimicked by 8-bromo-cAMP but not by LH.

Follistatin may also have effects on oocytes. Inhibin A subunit mRNA and activin type IIa receptor mRNA are found in rat oocytes. Activin A stimulates meiotic maturation in rat oocytes and this effect is antagonized by follistatin. Follistatin alone exerts an inhibitory effect on oocyte maturation suggesting that it is antagonizing endogenously produced activin, which may act in an autocrine manner.

D. The Physiology of Follistatin in the Testes

Follistatin protein is detectable by immunohistochemistry in adult rat testes. Within the seminiferous tubule, only the spermatogenic cells show evidence of immunostaining, with no staining visible on the Sertoli cells. This distribution is similar to the reported distribution of inhibin βA subunit protein. Follistatin mRNA expression also parallels that of inhibin βA mRNA in the seminiferous tubule. *In situ* hybridization localizes follistatin mRNA expression to the Sertoli cell cytoplasm, with signal increasing from stage VIII through stages IX–XI of spermatogenesis. The expression of the activin A receptor mRNA was located primarily in the developing spermatogenic cells following the expression of follistatin and inhibin βA subunit in neighboring Sertoli cells. Therefore, it would appear that follistatin is synthesized by Sertoli cells and is localized to dividing germ cells possibly by binding to cell surface proteoglycans.

Sertoli cell-enriched cultures express follistatin mRNA. Expression can be increased by epidermal growth factor and by the protein kinase C activator, PMA. FSH, testosterone, activin, glucocorticoid, and retinoic acid fail to stimulate expression.

Activin receptors are located on late pachytene spermatocytes and round spermatids (ActR-II) and Sertoli cells and spermatogonia in stages IX–XI (ActR-IIB). Activin also has profound effects on spermatogenesis, causing reaggregation of Sertoli cells and germ cells in culture. Spermatogonial proliferation is also increased, and follistatin can antagonize

this effect. This suggests that activin may have a role during testicular development in directing tubular structure. This effect may be tempered in later life by follistatin.

E. The Physiology of Follistatin in the Pituitary

Follistatin mRNA has been detected in the rat pituitary and specifically located in the gonadotrophs (70% LH cells; 20% FSH cells) and folliculostellate cells. All cell types, with the exception of corticotrophs, express follistatin mRNA in culture, and the most abundant is the somatotroph. Inhibin α and βB subunits are also expressed in the pituitary and it is likely that interaction between locally produced follistatin and activin B, and possibly inhibin B, forms an important regulatory system for controlling FSH release.

Rat pituitary follistatin mRNA increases following gonadectomy in male and females. The increase occurs within 24 hr and may precede the increase in FSH β subunit mRNA. Gonadectomy increases gonadotrophin-releasing hormone (GnRH) secretion and the increase in follistatin mRNA expression following gonadectomy in male rats can be prevented by administration of either testosterone or GnRH antagonists. If gonadectomized, testosterone-replaced male rats are pulsed with GnRH at various frequencies, follistatin mRNA levels peak at intervals of 8 min, and FSH β subunit mRNA levels begin to rise at GnRH pulse intervals of 30 min. When the time intervals have lengthened beyond 120 min, follistatin mRNA expression is downregulated, whereas gonadotrophin subunit mRNA expression continues to be stimulated. This provides a mechanism whereby alterations in GnRH pulse frequency can selectively modulate FSH and LH secretion. In long-term ovariectomized rats, a GnRH antagonist is unable to reduce the elevated follistatin mRNA expression but administration of follistatin completely reverses the ovariectomy-induced increase in follistatin gene expression, presumably by binding locally produced activin. Administration of porcine follistatin to ovariectomized rats suppressed mean FSH levels due to reduced FSH pulse amplitude but not frequency. Pulsatile LH secretion was not affected. The effects of follistatin were additive with estradiol benzoate. It was concluded that ovariectomy leads to increased activin B production, which stimulates FSH β subunit expression, FSH secretion, and follistatin mRNA levels. Follistatin thus generated may act as a short loop feedback inhibitor of activin action.

Follistatin mRNA can be detected by *in situ* hybridization in somatotrophs, lactotrophs, and thyrotrophs. Activin has diverse actions on these cell types, including inhibition of somatotroph proliferation and GH secretion in response to growth hormone-releasing hormone and inhibition of ACTH and prolactin secretion. Therefore, it is likely that the presence of activin favors FSH secretion and fertility and the presence of follistatin inhibits activin, which favors the "stress" hormones and downregulated fertility. The interplay of activin and follistatin at the local level in the pituitary could potentially have profound effects on the functional state of the whole organism.

F. The Physiology of Follistatin in the Decidua and Placenta

The colocalization of inhibin βA subunit and follistatin mRNAs in the decidua and placenta of the rat, mouse, and human suggests that there is a local autocrine regulatory system operating in the placenta akin to that in the ovary and the pituitary. This system may impact the process of placentation as well as the control of hCG and progesterone production. The clear importance of activin/follistatin in embryonic development also suggests that decidual production of these factors may impact early development.

G. The Physiology of Follistatin in the Pancreas

Follistatin expression has been detected along with inhibin βA, βB, and α subunits in the pancreatic islets. Immunostaining using an antisera directed against residues 123–134 revealed follistatin protein only in the insulin-producing pancreatic B cells. Interestingly, if an antiserum raised against the peptide (300–315) was used, which should be specific for FS-315, no staining was seen. Inhibin βA subunit

staining was found in the A or B cells depending on the antisera used, and inhibin βB subunit staining was most intense in the A cells. Immunostaining for follistatin and inhibin β subunits was diminished by pretreatment of rats with the B cell toxin, streptozotocin. Activin stimulated insulin production from rat islets in culture and it is therefore likely that activin/inhibin and follistatin play a role in the local regulation of islet cell function.

H. The Physiology of Follistatin in the Adrenal Cortex

Follistatin and inhibin α and β subunit mRNAs were detected in the rat adrenal, primarily in the zona glomerulosa and reticularis. Activin A can inhibit the proliferation of human fetal adrenal cells *in vitro* and inhibit steroidogenesis in cultured ACTH-stimulated bovine adrenocortical cells. The effect was specific for inhibition of 17α-hydroxylase, a property shared with TGF-β. This inhibitory effect can be partially reversed with follistatin. These findings argue strongly for a paracrine or autocrine role for activin and follistatin in the adrenal cortex.

I. The Physiology of Follistatin in the Thyroid

Activin A has been identified by immunohistochemistry in thyroid follicular cells. Activin A induces increased [^{3}H]thymidine incorporation in porcine thyroid cells in culture but inhibits thyroid-stimulating hormone-induced iodine uptake and cAMP accumulation. The effect of activin on DNA synthesis was blocked by follistatin. Epidermal growth factor also stimulated [^{3}H]thymidine incorporation but this effect was not inhibited by follistatin.

J. The Physiology of Follistatin in the Liver

In the liver, intense follistatin immunoactivity was detected over hepatocytes, and follistatin is produced by cultured rat hepatocytes. Activin is also expressed in hepatocytes, and expression is increased dramatically after partial hepatectomy. The effect of activin on hepatocytes *in vitro* is to inhibit proliferation, and so activin is an autocrine negative regulator of hepatocyte growth and plays a role in restricting the regrowth of hepatic tissue following partial hepatectomy. Follistatin can block this inhibitory effect of activin both *in vitro* and *in vivo*. A single intraportal administration of recombinant human follistatin following 70% hepatectomy in rats resulted in a significantly increased rate of liver regeneration and a greater final weight of liver. Hepatoma cell lines also produce follistatin, and this locally produced follistatin may be responsible for attenuation of the growth-inhibitory effects of activin at high cell density even in cell lines that possess activin receptors.

K. The Physiology of Follistatin in Bone

The TGF-β family of peptides are involved in bone cell physiology, and activin and follistatin are no exception. Studies have been performed on osteoblasts isolated from rat parietal bones and in osteoblast cell lines. Although activin does not seem to be expressed by these cells, they have abundant high-affinity binding sites for activin, and activin has been shown to have TGF-β-like effects, stimulating proliferation and inhibiting differentiated function (alkaline phosphatase production) induced by retinoic acid. Exogenous follistatin inhibits this suppressive effect of activin on alkaline phosphatase production, and follistatin is expressed endogenously by these cells and secreted into the medium. Follistatin expression is inhibited by the differentiating influence of retinoic acid. Therefore, it appears that follistatin is expressed locally by osteoblasts to enable the continuation of differentiated function in the face of exogenous activin.

L. The Physiology of Follistatin in the Brain and Nervous Tissues

Follistatin mRNA has been detected in homogenates of rat cerebral cortex by Northern analysis, whereas *in situ* hybridization gave a weak signal over the olfactory tubercle and layer II of the cortex. However, it is in embryonic development where follistatin expression in the developing nervous system has received most attention. In *Xenopus* embryology, follistatin expression occurs at the gastrula stage in tissues with potent neural-inducing capacity such as anterior mesoderm and prospective notochord. Following closure of the neural tube, expression can be detected in two stripes encompassing the midbrain, hindbrain, and, later, the forebrain. It is thought that follistatin expression in these tissues leads to neural differentiation by neutralizing the inhibitory effect of activin which normally favors epidermal differentiation. In the developing mouse embryo, follistatin expression occurs in the developing hindbrain in rhombomeres 2, 4, and 6. In the human fetus, at 15–17 weeks, follistatin expression was detected by Northern blotting in all tissues examined (cerebrum, cerebellum, and spinal cord).

M. The Measurement of Follistatin in Serum

There is no current consensus on the changes in plasma follistatin under various physiological conditions. This probably relates to the limited knowledge available of the forms of follistatin present in plasma. In this situation the appropriateness of the various assay methods cannot be established. In general it appears that circulating follistatin in the sheep and human varies little in a variety of reproductive circumstances and is therefore unlikely to be involved in the endocrine regulation of FSH secretion. It seems likely that serum follistatin rises during human pregnancy and may be differentially regulated by GnRH in men and women.

See Also the Following Articles

Fetal Adrenals; FSH (Follicle-Stimulating Hormone; LH (Luteinizing Hormone)

Bibliography

Hemmati-Brivanlou, A., Kelly, O. G., and Melton, D. A. (1994). Follistatin, an antagonist of activin, is expressed in the Spemann organizer and displays direct neuralizing activity. *Cell* 77, 283–295.

Inouye, S., Guo, Y., Ling, N., and Shimasaki, S. (1991). Site-specific mutagenesis of human follistatin. *Biochem. Biophys. Res. Commun.* 179, 352–358.

Krummen, L. A., Woodruff, T. K., De Guzman, G., Cox, E. T., Baly, D. L., Mann, E., Garg, S., Wong, W. L., Cossum, P., and Mather, J. P. (1993). Identification and characterization of binding proteins for inhibin and activin in human serum and follicular fluids. *Endocrinology* **132**, 431–443.

Matzuk, M. M., Lu, N., Vogel, H., Sellheyer, K., Roop, D. R., and Bradley, A. (1995). Multiple defects and perinatal death in mice deficient in follistatin. *Nature* **374**, 360–363.

Nakamura, T., Sugino, K., Titani, K., and Sugino, H. (1991). Follistatin, an activin-binding protein, associates with heparan sulfate chains of proteoglycans on follicular granulosa cells. *J. Biol. Chem.* **266**, 19432–19437.

Robertson, D. M., Klein, R., de Vos, F. L., McLachlan, R. I., Wettenhall, R. E. H., Hearn M. T. W., Burger, H. G., and de Kretser, D. M. (1987). The isolation of polypeptides with FSH suppressing activity from bovine follicular fluid which are structurally different to inhibin. *Biochem. Biophys. Res. Commun.* **149**, 744–749.

Shimasaki, S., Koga, M., Esch, F., Mercado, M., Cooksey, K., Koba, A., and Ling, N. (1988). Porcine follistatin gene structure supports two forms of mature follistatin produced by alternative splicing. *Biochem. Biophys. Res. Commun.* **152**, 717–723.

Sugino, K., Kurosawa, N., Nakamura, T., Takio, K., Shimasaki, S., Ling, N., Titani, K., and Sugino, H. (1993). Molecular heterogeneity of follistatin, an activin-binding protein. Higher affinity of the carboxyl-terminal truncated forms for heparan sulfate proteoglycans on the ovarian granulosa cell. *J. Biol. Chem.* **268**, 15579–15587.

Ueno, N., Ling, N., Ying, S. Y., Esch, F., Shimasaki, S., and Guillemin, R. (1987). Isolation and partial characterization of follistatin: A single chain Mr 35000 monomeric protein that inhibits the release of FSH. *Proc. Natl. Acad. Sci. USA* **84**, 8282–8286.

Vale, W. W., Hseuh, A., Rivier, C., and Yu, J. (1990). The inhibin/activin family of hormones and growth factors. In *Peptide Growth Factors and Their Receptors: Handbook of Experimental Physiology* (M. A. Sporn and A. B. Roberts, Eds.), Vol. 95, pp. 211–248. Springer-Verlag, Berlin.

Freemartin

Sherrill E. Echternkamp
Roman L. Hruska U.S. Meat Animal Research Center

GLOSSARY

anti-Müllerian hormone or *Müllerian-inhibiting substance* or *factor* A dimeric glycoprotein hormone of the transforming growth factor-β multigene family that causes regression of the Müllerian ducts.

chimera An animal containing cells or tissues of diverse genetic origin.

freemartin An intersex or hermaphroditic genotypic female born cotwin to a male.

Müllerian ducts Anlagen (embryonic precursor) for the oviducts (or Fallopian tubes), uterus, cervix, and anterior vagina of the female reproductive tract.

retained fetal membranes (placenta) Natural expulsion of placenta is delayed for 48 hr or longer after parturition.

Wolffian ducts Anlagen for the epididymis, vas deferens, seminal vesicles, and ejaculatory duct of the male reproductive tract.

The freemartin syndrome is an intersexed female born cotwin to a male. Freemartinism results from fusion of the placental blood vasculature which allows an interchange of primordial cells and biochemical mediators of sexual differentiation between the twins early in gestation. Because sexual development occurs earlier in male than in female fetuses, the reproductive system of the female twin is masculinized by factors controlling development of the male reproductive system.

I. INTRODUCTION

The freemartin syndrome is a distinctive form of intersexuality, found primarily in cattle, in which a genetically female fetus is masculinized *in utero* by the presence of a male fetus. The intersexuality results from fusion of the placental chorion and anastomosis of chorioallantoic arterial blood vessels between adjacent twin or multiple fetuses (Fig. 1). The consequence of the vascular anastomosis is XX/XY chimerism in both the male and female fetus due to exchange of hemopoietic stem cells, and possibly germ cells, and to masculinization of the reproductive system of the female twin fetus by hormones (e.g., androgens and anti-Müllerian hormone) controlling development of the reproductive system of the cotwin male fetus; the chorioallantoic anastomosis occurs just prior to the critical period for sexual differentiation. In the absence of the testis-determining gene linked to the Y chromosome, the undifferentiated gonad becomes a female ovary. Likewise, the brain centers for sexual behavior and gonadotropin secretion remain female (feminized) in the absence of testicular androgens during critical periods of fetal development. Although isolated cases of the freemartin syndrome have been reported in pigs, goats, and sheep, the highest incidence is found in cattle; for example, 92–96% of the bovine females born cotwin to a male express the freemartin syndrome and, thus, are sterile. In contrast, the male cotwin is generally classified as a normal virile, fertile male. However, some reports have suggested a higher incidence, or precocious onset, of testicular degeneration and/or infertility in young bulls born cotwin to a freemartin compared to single-born bulls.

II. TWINNING IN DOMESTIC LIVESTOCK

Based on their genetic origin, natural twins are classified as either dizygotic (fraternal) or monozygotic (identical) twins. Dizygotic twins originate from two separate female gametes or eggs that are released from two separate ovarian follicles and are fertilized individually, whereas monozygotic twins originate from the same zygote; i.e., the zygote splits into two zygotes early in embryonic development. Identical twins are less common, constituting about 10% of the twin births, and occur randomly. An exception is the artificial production of identical animals by bisecting or cloning of embryos. Because of their common origin, identical twins are of the same sex, whereas fraternal twins are of independent genetic origin, so their sex ratio is governed by the principles of probabilities; 25% male cotwins, 25% female cotwins, and 50% heterosexual (unlike-sex) twins.

A. Cattle

The frequency of twin or multiple births varies widely among species. Cattle are primarily monotocous (single bearing) so the frequency of twin or multiple births is low but does vary with breed of cattle, age of dam, and season or month of the year.

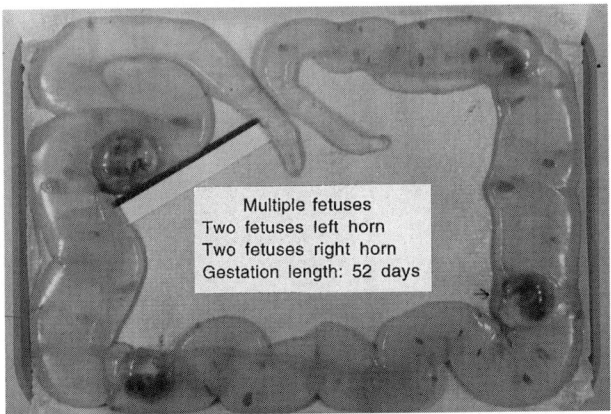

Multiple fetuses
Two fetuses left horn
Two fetuses right horn
Gestation length: 52 days

FIGURE 1 Illustration of fusion of placental chorion among multiple, 52-day-old bovine fetuses; quadruplet fetuses (two per uterine horn) from a cow treated with follicle-stimulating hormone to induce multiple ovulations.

Twinning rate is highest for the Holstein breed (4–10%), intermediate for the other dairy breeds and for dual-purpose continental breeds (2 or 3%), and lowest for the British beef breeds (0.3–0.5%). The frequency of triplet births in cattle is about 1% the frequency of twin births, and the largest number of live calves reportedly born to a cow is quintuplets. Frequency of twin ovulations, and consequently dizygotic twin births, increases with age up to 4 years and, in the Northern Hemisphere, is highest for females bred during the fall months (i.e., October–December) and lowest in the spring months (i.e., March–May). Although heritability estimates for twin births or ovulations are low ($h^2 = 0.07$–0.11), selection for twin ovulations has increased the annual twinning rate in a composite herd of cattle at the Roman L. Hruska U.S. Meat Animal Research Center from 4 to 40% in three generations. Current results in cattle provide no evidence for a major gene controlling ovulation rate or litter size but suggest instead a number of genes having an effect on ovulation rate, i.e., polygenic inheritance as a quantitative trait. In addition to the usual 20–30% early embryonic mortality (Days 5–15 of gestation) associated with single bovine embryos, bovine twin fetuses have increased mortality at 7–9 weeks of gestation (6.3% for twins vs 2.7% for singles) and late-term (7+ months of gestation) abortions (6.1 vs 0.8%, respectively). The slow reabsorption or delayed expulsion of fetuses dying at 7–9 weeks of gestation delays return to estrus 10–14 days or longer. Furthermore, associated with twin births in cattle is a high incidence (37%) of abnormal presentations or breech positioning of either one or both fetuses at parturition. Without timely obstetrical assistance, such deliveries result in stillborn calves or death shortly after birth. Gestation length is about 7 days shorter for twin pregnancies, which may contribute to the increased percentage of cows (27.9% for twins vs 1.9% for single births) that pathologically retain the fetal membranes or placenta for longer than 48 hr, frequently requiring therapeutic treatment.

B. Sheep and Goats

Breeds of sheep vary widely in litter size, with the majority of breeds averaging 1.3–1.8 lambs per ewe.

Within the past few decades, genetic selection for multiple births in sheep or the expanded use of prolific sheep breeds in interbreed matings has converted ewes from being primarily monotocous to being polytocous (litter-bearing). For example, mature ewes of the Romanov breed have a litter size of up to 8 lambs; however, pubertal or yearling Romanov ewes produce primarily single births, again illustrating the effect of age on ovulation rate. In some breeds or populations of sheep, ovulation rate is controlled by a major gene, e.g., the Booroola fecundity gene or locus (*FecB*). Each copy of the *FecB* allele increases ovulation rate by about 1.5 ova. The *FecB* locus has been mapped to sheep chromosome 6. In contrast, in other sheep breeds (e.g., Finnsheep and Romanov), the ovulation trait is polygenic and under quantitative genetic control. Twin or multiple births are also relatively common in goats.

C. Horses

Despite the high incidence of multiple ovulations (4–43% twin ovulations) in horses, only about 1% of bred mares produce live twin foals, which are primarily dizygotic. As in other species, frequency of multiple ovulations increases with age and varies with breed; it is highest for draft mares (31.6% multiple ovulations), intermediate for race or saddle breeds (20.3%), and lowest for pony mares (10.7%). Of the multiple ovulations, 94.1% were twin, 5.7% were triplet, and 0.2% were quadruplet ovulations. In addition to equine husbandry practices to prevent twin pregnancies, a natural reduction in twin pregnancies results from (i) a proposed embryo-reducing mechanism present between Days 7 and 40 postovulation and (ii) a high rate of fetal mortality (either one or both) in twin pregnancies either in the first trimester or as late-term abortions. About 70% of the equine twin embryos implant in the same uterine horn (unilateral twins), and about 85% of these unilateral twins undergo embryo reduction to eliminate one of the twin embryos by Day 40 of gestation. Also, elimination of a twin embryo by the proposed embryo-reducing mechanism seems to occur more frequently for twin embryos resulting from synchronous twin ovulations. However, only about 50% of the twin ovulations in mares are synchronous (i.e., both ovulations occur within 24 hr), whereas the interval between asynchronous twin ovulations may range from 1 to 10 days: 1–4 days of asynchrony, 75.2%; 5–8 days, 18.2%; and >8 days, 6.2%. Diestrous ovulations are common in mares and do produce fertilizable ova. Also, sperm have a long survival time in the reproductive tract of the mare, with pregnancies occurring in mares bred up to 6 days before ovulation. Although most bilateral twins survive the embryo-reducing mechanism in mares, the majority of these twins are lost to abortion (70%) or prenatal or perinatal mortality of one or both fetuses. A high incidence of fetal mortality has been observed in pony mares shortly after fusion of the chorioallantois between bilateral twins (i.e., Day 46 of gestation), which may account for the lack of equine freemartins. Collectively, <20% of mares diagnosed with twin embryos produce viable single or twin foals.

III. ANATOMICAL CHARACTERISTICS, SYMPTOMS, AND DIAGNOSIS OF THE FREEMARTIN

A. Anatomy and Symptoms

The external genitalia of the intersexed bovine freemartin is generally phenotypically female but may include abnormalities in development of the urogenital structures (e.g., enlarged protruding clito-

ris, tuft of coarse hairs arising from the inferior vulval commissure, increased distance between the anogenital openings, and a rudimentary penis, prepuce, and scrotal sac). Masculinization of the external genitalia is rare. Mammillary teats are rudimentary, too. In contrast, the internal genitalia of the freemartin (Fig. 2) is characterized by gross anatomical anomalies indicative of partial or total inhibition of the anlagen of the female reproductive system during embryonic development, resulting in gonadal hypoplasia, repression of the Müllerian duct derivatives, and variation in the stimulation or masculinization of the Wolffian duct system. The small freemartin gonads (0.2–0.4 g) are located at the normal site of the ovaries. Although the gonads are composed of both ovarian and testicular tissue, most ovarian characteristics are lacking (e.g., no primordial or antral follicles) and replaced by a prominence of testicular differentiation (Fig. 3). Transformation of the female gonads varies among freemartins and includes either (i) small gonads composed primarily of fibrous stroma and interstitium with rosettes of Sertoli cells or poorly organized seminiferous cords, a disproportional increase in rete ovarii or rete testis, and a thickened albuginea; (ii) an intermediate degree of testicular development in which the gonads are larger, have the gross morphology of an immature testis, and contain regions in which the Sertoli cells are organized into seminiferous tubules, but lack germinal epithelium, that connect with the rete testis and subsequently a small epididymis; or (iii) extreme

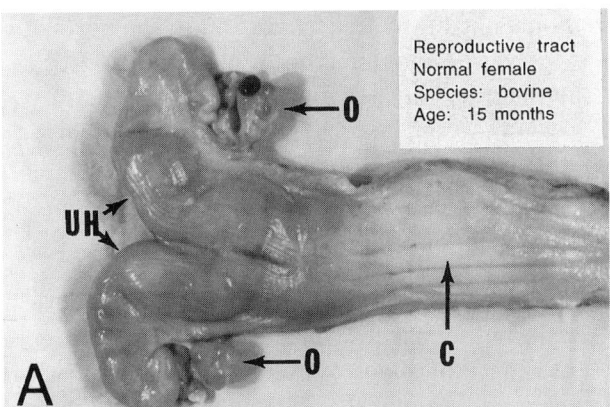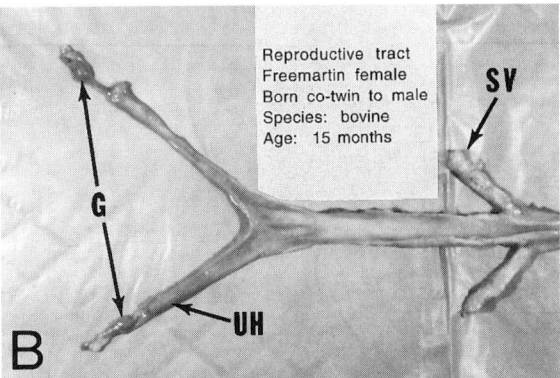

FIGURE 2 Comparison of the genital tract from a normal cyclic heifer (A) and a freemartin (B) at 15 months of age. Note the absence of uterine body, cervix, and anterior vagina and the presence of small seminal vesicles in the freemartin genital tract. O, ovary; G, gonad; UH, uterine horns and body; C, cervix; SV, seminal vesicle.

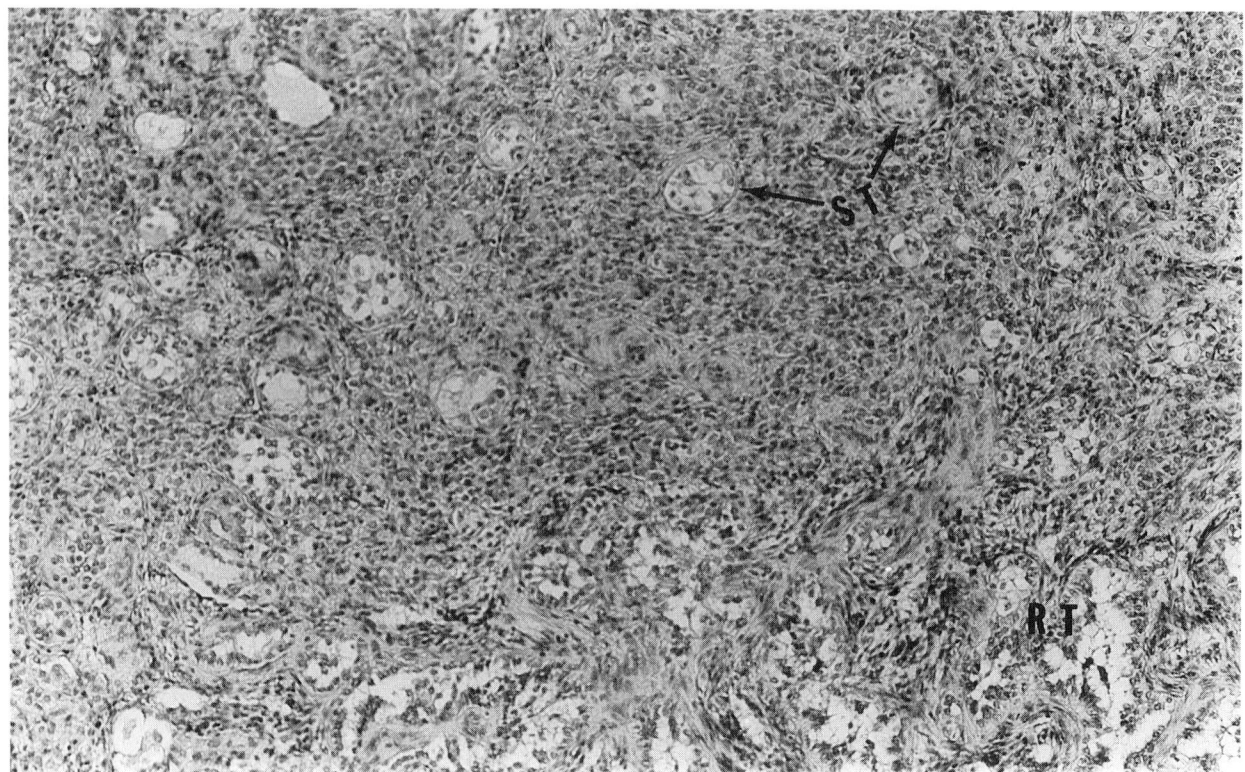

FIGURE 3   Histological section through the gonad from a 15-month-old freemartin showing organization of Sertoli cells into seminiferous tubules. Interstitial cells (Leydig cells; arrows) are interspersed among the tubules, and rete tubules (RT) are present in the lower right corner.

cases in which one or both gonads resemble a testis with an epididymis, pampiniform plexus, and vas deferens attached and are encapsulated in the tunica vaginalis with an extending spermatic cord and gubernaculum, and one or both testis descend into the inguinal canal. Consequently, variations occur postnatally among freemartins in the expression of female and male secondary sex characteristics (e.g., thickened neck with crest and coarse hair on the vulva), which are indicative of gonadal composition (masculinization) and steroid production. Deviations in the development of the Müllerian ducts (Fig. 2) include varying degrees of development of the oviducts and uterus (uterine horns ranging from total absence to almost normal length, but generally they are a few centimeters in length); undifferentiated cords in place of the uterine body, cervix, and anterior vagina; and a short posterior vagina (7–11 cm in length in yearling freemartins) without an anterior orifice. Consequently, freemartin females are sterile.

Seminal vesicles are generally present, with their developmental status being reflective of androgen production by the gonads. Vas deferens may also be present. Transformation of the ovaries and regression of the Müllerian duct system occur at the same gestational age in the freemartin and in the normal male, whereas development of the Wolffian structures and of the masculine secondary sex characteristics occurs several weeks later in the freemartin than in the normal male.

Although the frequency of twin or multiple births is relatively high in sheep and goats, unlike cattle, the frequency of freemartins is very low (i.e., 0.8–1.2%). However, frequencies as high as 5–10% have been reported for sheep which concurred with a reported incidence of about 5% red blood cell chimerism in ovine multiple births. If fusion of the chorions does occur in sheep or goats, only a small portion of the pregnancies show vascular anastomosis because the fused areas have a fibrous thickening or formation of fibrous bands or sutures at the fusion site to prevent the vascular anastomosis. Ovine freemartins show extreme masculinization of the internal reproductive tract with less variation among animals and usually show no remnants of the Müllerian ducts. A major gene (i.e., Inverdale gene) has been found on the X chromosome in the Romney breed of sheep that in the homozygous condition results in small nonfunctional "streak ovaries." The ovaries are one-eighth of normal size with a flattened streak-liked organ morphology and are void of follicular development and, thus, infertile. Conversely, one copy of the Inverdale gene increases ovulation rate by one ovum.

The incidence of freemartinism in goats is thought to be very low, as is the incidence of placental anastomosis. However, other types of intersex conditions are common in goats. A hereditary pseudohermaphroditic condition phenotypically similar to freemartinism has been identified in some breeds of polled goats (e.g., 6.0% in Toggenburgs, 11.1% in Saanens, and 2.0% in French-Alpines). Unlike the XX/XY freemartin, the majority of these intersexes are XX female as well as homozygous for the polled gene (horn gene is recessive in these breeds). These intersexes have an enlarged clitoris, often penile and sometimes with hypospadias, and a vulval opening that varies in distance from the anus. The gonads are of the testicular type with varying degrees of testicular hypoplasia. As with the freemartin, XX germ cells are present in medullary tissue of embryonic testes but gone at birth; thus, spermatogenesis is absent. The XX germ cells either degenerate or undergo a very limited number of mitotic divisions before meiosis and are difficult to find. However, both polled and horned goats with XX/XY blood chimerism and a phenotype indistinguishable from the XX intersex have been identified and are likely freemartins.

The early appearance of avascular necrotic tips in the individual chorionic sacs in swine placentae may prevent vascular anastomoses among the multiple porcine fetuses and, thus, the rare occurrence of freemartinism in swine, whereas side-to-side fusion of fetal chorionic membranes between porcine fetuses occurs later in embryonic development after sex determination. The fibrous bands or avascular necrotic tips in ovine or porcine placentae are seldom found in placentae of bovine twins. Several forms of sexual anomalies occur in horses. As noted earlier, a high incidence of fetal mortality has been observed in pony mares shortly after fusion of the chorioallantois between bilateral twins (i.e., Day 46 of gestation), which would reduce the opportunity for a freemartin. A genetic XY sex-reversal syndrome has been identified in horses in which XY males have anatomical characteristics ranging from a phenotypic mare with normal ovaries and fertility to virilized intersexes with elevated testosterone secretion, gonadal dysgenesis, abnormal Müllerian duct development, and an enlarged clitoris.

B. Diagnosis

Symptoms and methods for diagnosis of the freemartin syndrome include clinical genital anomalies or aberrations, sex chromosome chimerism (XX/XY) in hemopoietic cells, the presence of sex chromatin bodies in blood leukocytes of the male cotwin, blood typing of erythrocytes, skin grafting between twins, and molecular probes to the Y chromosome. Diagnosis in cattle is commonly made by measurement of vaginal length (5 or 6 cm in freemartin vs >12 cm in normal heifer); by visual evaluation of the external genital, secondary sex characteristics, and sexual behavior; by palpation or ultrasound of the reproductive tract; and/or by chromosomal analysis of lymphocytes for sex chromosome chimerism. Because of the wide range in XX:XY ratios among freemartins, accurate diagnosis of a chimera requires karotyping a large number of metaphase smears at an added economic expense. The recent development of molecular probes to DNA sequences of the Y chromosome combined with polymerase chain reaction has enhanced sensitivity and diagnostic accuracy for detection of sex chromosome chimerism in white blood cells. With either method, economic cost limits chromosomal analysis to those cases that cannot be confirmed anatomically. However, XX/XY chimerism does not necessitate sterility because some of the 4–8% fertile females born cotwin to a male were found to be chimeric. Other possible explanations for these fertile females were that placental anastomosis occurred after sexual differentiation or did not occur (i.e., not chimeric). The occurrence of a single-born freemartin in cattle is very rare (1 or 2% of single-born females in cattle population with 35% twinning), suggesting that death of one bovine twin fetus after placental vascular anastomosis generally causes death of both fetuses.

IV. EMBRYOLOGICAL DEVELOPMENT AND ONTOGENY OF THE FREEMARTIN SYNDROME

Anatomical anomalies associated with the freemartin syndrome are suggestive of masculinization of the female reproductive tract during embryonic development. Thus, a comparative, chronological review of normal organogenesis between male and female fetuses, primarily bovine, will enhance the understanding of the origin and cause(s) for the intersexual traits. In mammals, the primordium of the ambisexual reproductive tract is composed of two germinal ridges (undifferentiated gonads) ventral–medial to the mesonephros and of two duct systems, the Müllerian and the Wolffian ducts, that extend from the undifferentiated gonads to the urogenital sinus in parallel with the metanephric ducts. The primordial gonad consists of an inner medullary region surrounded by the cortex. In the male gonad, the primary sex cords contained within the medulla stimulate development of the medullary region and suppress development of the cortex, whereas the developmental priority is reversed in the female gonad with stimulation occurring in the cortex rather than in the medulla. In the female, the Müllerian ducts are the anlagen of the oviducts (or Fallopian tubes), uterus, cervix, and anterior vagina, and the Wolffian ducts remain rudimentary. Conversely, in the male, the Wolffian ducts are the anlagen of the epididymis, vas deferens, seminal vesicles, and ejaculatory duct, and the Müllerian ducts remain rudimentary.

Very early in embryonic development, primordial germ cells (either male or female) are attracted to migrate from their extragonadal site in the yolk sac endoderm to the bipotential germinal ridges or gonadal primordia. These germinal ridges are composed of somatic cells derived from the mesonephros and coelomic epithelium. Association with the somatic cells of the gonad appears to be required for optimal growth and multiplication of the primordial germ cells. Normally, male germ cells, or gonocytes, migrate into the medulla of the undifferentiated gonads and become surrounded by specialized somatic cells with an abundant clear cytoplasm (i.e., differentiating Sertoli cells). As these Sertoli cells replicate and differentiate, they, along with the germ cells, align themselves into sex cords. With the formation of a central lumen and basement membrane, these sex cords become the seminiferous tubules and the future site of spermatogenesis. However, the presence of the germ cells does not seem essential for differentiation of the Sertoli cells and formation of the seminiferous cords. An ingrowth of mesonephric cell cords forms the rete testes, which connect the seminiferous tubules to the efferent ducts. Differentiation of the Leydig cells (steroid-secreting cells) within the testicular interstitium occurs several days later; initiation and maintenance of steroidogenesis by the Leydig cells are influenced by products (e.g., inhibin and activin) of the Sertoli cells.

Normally, migrating female germ cells become enclosed either within elongated cords of cells from the coelomic epithelium or within the less organized somatic tissue of the germinal ridge; both somatic tissues originate from the mesonephros. The cords of mesonephric cells promote ovarian remodeling by colonizing in the central part (medulla) of the presumptive gonad and then displacing the germ cells toward the periphery or cortex of the gonad. Germ cells remaining in the medulla usually degenerate, whereas those in the cortex continue to proliferate and form nests of germ cells (oogonia). The subsequent differentiation of oogonia into primary oocytes is usually completed by birth. The tunica albuginea covering the ovary forms from mesenchymatous cells and is very thin compared to the thicken albuginea found on the testis or freemartin gonad. When the oocytes reach the diplotene stage of the first meiotic prophase, they initiate formation of primordial follicles, which are composed of a single layer of granulosa cells surrounding the oocyte. Granulosa cells develop from rete cells, which also originate from the mesonephros. Follicular formation can occur both pre- and postnatally depending on the species. Unlike the testes, development of the ovaries and maintenance of the follicles depends on the presence of germ cells or oocytes. Without germ cells, the cortex of the ovary fails to develop and a gonadal streak results. By comparison, loss of the oocytes after follicle formation causes follicular cells to differentiate into Sertoli cells and form seminiferous cords or tubules characteristic of males. Conversely, the presence of an oocyte will transform Sertoli cells into follicular cells. In aging rats, the loss of oocytes after follicle formation also causes follicular cells to differentiate into XX Leydig cells. In both mammals and birds, gonocytes of one sex do not complete gametogenesis in the somatic tissues of the other sex.

In mammals, germ cells are predisposed to become oocytes and undifferentiated gonads become ovaries. Likewise, the centers of the mammalian brain responsible for sexual behavior and for gonadotropin secretion are inherently female and are subsequently sexually organized or programmed by steroids (i.e., androgens and estrogens) from the developing gonads. Differentiation of the male testis from the bipotential germinal ridge is generally believed to be induced by the testis-determining factor, a transcriptional regulatory protein encoded by the sex-determining region (SRY) located on the short arm of the Y chromosome and expressed in the somatic portion of the germinal ridge. Testis-determining factor, or SRY, is a DNA-binding protein that is thought to act as a switch to initiate transcription of a cascade of other genes, which contribute to testicular development. Included in these events is the initiation of production of two separate testicular hormones which direct the development of the male phenotype: (i) androgen steroids, which are produced in the fetal Leydig cells and masculinize the male genitalia, and (ii) anti-Müllerian hormone (or Müllerian-inhibiting substance), which is produced in fetal Sertoli cells and inhibits development of the Müllerian ducts. The dimorphic action of the testis-determining factor on male sexual differentiation is mediated through expression of the nuclear receptor steroidogenic factor-1 gene. Expression of steroidogenic factor-1 was first identified in Leydig cells of the testis as an important regulator of steroid hydroxylase gene expression and of androgen production in Leydig cells (e.g., 3β-hydroxysteroid dehydrogenase). Recently, steroidogenic factor-1 expression has been identified in the primary Sertoli cells of the embryonic mouse testis, where it activates expression of the anti-Müllerian hormone gene and its regression of the Müllerian ducts in the male. However, activation of the anti-Müllerian hormone gene by steroidogenic factor-1 may involve an additional ligand(s) or cofactor(s) of autosomal and/or sex chromosomal gene origin. During differentiation of the sexually dimorphic germinal ridges into testes, immature Sertoli cells originating from the germinal epithelium grow into the mesenchyme and form cord-like masses, which differentiate into seminiferous tubules. Formation of these testicular cords or seminiferous tubules in the early ovine testis is preceded by the presence of anti-Müllerian hormone and 3β-hydroxysteroid dehydrogenase immunopositive cells. Presumably these immunopositive cells were Sertoli and Leydig cells, respectively, because the subsequent production of anti-Müllerian hormone and 3β-hydroxysteroid dehydrogenase was associated with immature Sertoli cells present in seminiferous cords and with Leydig cells present in the interstitium, respectively. Initiation of androgen production by the bovine fetal testis begins by Day 42 of gestation.

Chorionic vascular connections between bovine twin fetuses are well established by 39 days of gestation and occur in advance of, or in conjunction with, testicular organogenesis in male fetuses. However, timing of the placental anastomosis between bovine twins can vary (i.e., ranges from 30 to 50 days of gestation), being influenced by factors such as whether the twin fetuses are located in the same or separate uterine horns (i.e., occurring later in bilateral twin pregnancies). This difference in timing of placental anastomosis may account for some of the observed variation in degree of anatomical anomalies or masculinization among freemartins. Testes become histologically recognizable on Days 39 or 40 of gestation in the bovine male, whereas the tunica albuginea and the testicular interstitial cells are identifiable at 43 and 45 days of gestation, respectively. In contrast, development of the female ovaries occurs several weeks later. For example, the beginning of the slow, prolonged thickening of the superficial ovarian layers and appearance of the Pfluger's cords (i.e., clusters of steroidogenic cells in the medulla of the gonad, which are precursors of the follicular cells found later in the ovarian cortex) occurs on Days

47–49 of gestation, whereas the appearance of the first premeiotic figures occurs by Day 75 and inclusion of oocytes into primordial follicles by Day 130 of gestation. In contrast to the increase in gonadal volume or size observed in normal females, gonadal development in the freemartin stops at Days 48 or 49 of gestation, and the gonads remain stunted for 6 weeks or more, except for development of the tunica albuginea. Since this period of delayed gonadal development corresponds to the period of gonadal growth and oogonial proliferation in the normal female gonad, the freemartin gonad becomes severely inhibited and devoid of germ cells. As a result of the gonadal inhibition and loss of oocytes, cords of Sertoli cells begin to develop (or medullary cords persist) adjacent to the rete ovarii and encompass any remaining germ cells. However, these XX germ cells do not survive postnatally in this pseudotesticular environment. Leydig cells begin to appear in the freemartin gonad between 70 and 100 days of gestation. The freemartin fetal gonad secretes androstenedione instead of estradiol with elevated blood levels of androstenedione and testosterone after 75 days of gestation; the elevation in androgens enhances stimulation of the Wolffian duct system in the freemartin. Since anti-Müllerian hormone inhibits aromatase, the decreased aromatase activity in the freemartin gonad is attributed to its inhibition by anti-Müllerian hormone from the male cotwin.

Differentiation of the male internal and external genitalia proceeds rapidly between Days 47 and 60 of gestation in cattle. An increase in the anogenital distance in male fetuses is identifiable by Day 47 of gestation and continues until the penis comes to open below the umbilicus on Day 58. The scrotum is identifiable in the male by Day 60 of gestation. Initially, the Müllerian ducts begin to grow in both sexes but begin to regress in the male fetuses at about 50 days and disappear by about 70 days of gestation. Likewise, from Day 52 onward, the diameter of the upper Müllerian ducts is smaller in freemartins than in normal females. Regression of the Müllerian ducts in the freemartin parallels that in the normal male fetus but at a slower rate so that the reduction is somewhat less pronounced or more variable in freemartins than in males of the same age. Regression of the Müllerian ducts within the freemartin is correlated positively with a reduction in gonadal volume.

Sex differences in development of the prostatic urethra and of connections between Wolffian ducts and the urogenital sinus in males verses development of the vagina and vestibule in females appear on about Day 56; freemartins primarily display vaginal organogenesis. Also, growth of the urethra proper between Days 55 and 70 occurs more rapidly in female than in male fetuses, resulting in increased length in the normal female on Day 70; the freemartin tends to be intermediate between the normal female and male. Prostate buds begin to appear on the prostatic urethra on Day 56 and increase in number and size with age. Conversely, prostatic buds, or a prostate, in freemartins are rarely if ever present. Seminal vesicles appear in normal male fetuses on Day 56 as an outgrowth of the lower Wolffian duct. The primordium of a male-type seminal vesicle has been found on Days 61 or 62 of gestation in freemartin fetuses with ovarian and Müllerian duct inhibition. Presumably, these structures contribute to the poorly developed seminal vesicles found in postnatal or older freemartins. In normal female fetuses, slow regression of the anterior Wolffian ducts begins by Day 70, whereas the posterior part of the ducts persists for a longer period.

V. HORMONAL VS CELLULAR INDUCTION OF FREEMARTINISM

A. Hormonal Theory

1. Androgens

As noted previously, the first morphological sign of sexual dimorphism between the testis and the ovary is the development of primordial Sertoli cells and their alignment into histologically visible cord-like structures and seminiferous tubules; their secretion of anti-Müllerian hormone; and development of androgen-producing Leydig cells in the interstitium. Attempts to reproduce the freemartin syndrome experimentally by administering androgens to pregnant cows early in gestation have failed to influence gonadal differentiation or to inhibit transformation of the Müllerian ducts; oviducts or Fallopian tubes, uterus, and ovaries were preserved in female fetuses of androgen-treated cows. However, these androgen treatments caused partial to complete masculinization of the Wolffian duct system and of the external genitalia of female fetuses. Thus, some of the anomalies found in the internal and external genitalia of the freemartin are indicative of androgenization of the Wolffian duct system and/or the urogenital sinus. Androgenization of the freemartin genital tract results from transfer of androgens from the male fetal testis and/or androgen production by the masculinized freemartin gonad. Higher androgen and lower estrogen concentrations were found in freemartin than in normal female fetuses by 40 days of gestation.

2. Anti-Müllerian Hormone

Anti-Müllerian hormone is a transforming growth factor-β-like glycoprotein hormone which, as the name implies, causes morphological cell death and regression of the Müllerian ducts in mammalian male embryos. Production of anti-Müllerian hormone by immature Sertoli cells is detectable in incubated testicular tissue from normal male bovine fetus by Day 42 of gestation, which coincides with the first signs of sexual differentiation of the male reproductive system and organization of the Sertoli cells into seminiferous tubules within the bovine testis. However, cells immunopositive for anti-Müllerian hormone were detected in the ovine fetal testis prior to cord organization (Day 30 of gestation). Likewise, intense immunostaining for anti-Müllerian hormone was associated with Sertoli cells in the seminiferous tubules of the ovine testis at Day 40 of gestation. Peak production of anti-Müllerian hormone by fetal testes of singleton bulls, as well as fetal testes of bulls cotwin to a freemartin, occurs between 50 and 80 days of gestation. Thus, elevated serum concentrations of anti-Müllerian hormone in bovine male fetuses coincide with the period of physiological regression of the Müllerian ducts in the normal male fetus (i.e., Days 50–80 of gestation). Production of anti-Müllerian hormone by the fetal testis continues for the duration of gestation, which is followed by a postnatal decline in activity associated with testicular maturation. In the absence of anti-Müllerian hormone, Müllerian ducts give rise to the oviducts (Fallopian tubes), uterus, cervix, and anterior vagina in either female or male fetuses.

A role for anti-Müllerian hormone in the induction of the freemartin intersexual traits is supported by results from the treatment of rat ovarian tissue with anti-Müllerian hormone or from ovarian grafts in mice. Treatment of fetal rat ovaries with anti-Müllerian hormone consistently induced formation of cord-like structures in the gonadal blastema, which resembled seminiferous tubules. The somatic cells contained within these tubules assumed an epithelial appearance and resembled differentiating Sertoli cells, became polarized against a discontinuous basal membrane, and produced anti-Müllerian hormone. In addition, anti-Müllerian hormone inhibited XX germ cell proliferation in rat ovaries but had no effect on XY germ cells in male gonads. Without germ cells, fetal rat ovaries develop cord-like structures containing anti-Müllerian hormone activity. Likewise, when mouse XX gonadal primordium was grafted beneath the renal capsule of the adult female mouse, it developed seminiferous-like tubules composed of Sertoli cells secreting Müllerian hormone, again suggesting that XX cells differentiate into Sertoli cells as a consequence of oocyte loss in the gonadal graft.

In the bovine freemartin, inhibition of ovarian development and regression of the Müllerian ducts reportedly occur simultaneously in the freemartin beginning on Days 48 or 49 of gestation, which also parallels regression of the Müllerian ducts in the male cotwin. Because production of anti-Müllerian hormone by immature Sertoli cells is detectable in incubated testicular tissue from normal bovine fetuses at 42 days of gestation, it is hypothesized that anti-Müllerian hormone from the male fetal testis is transferred to the cotwin female fetus via existing anastomosed chorioallantoic arterial blood vessels to cause morphological stunting, loss of ovarian germ cells, and inhibition of aromatase activity and estrogen secretion by the gonad early in ovarian development and to induce regression of the Müllerian ducts. Conversely, interruption of the placental anastomosis between the male and female twins before 45 days of gestation prevents Müllerian regression and ovarian stunting in the female. Furthermore, correlated ($r = .97$) high concentrations of anti-Müllerian hormone are found in the serum of both the male and freemartin fetus from Days 50 to 110 of gestation. However, freemartin gonads only rarely produce anti-Müllerian hormone between 50 and 65 days of gestation and then in minute amounts compared to the active production of anti-Müllerian hormone by fetal testes of singleton bulls and bulls twin to a freemartin, suggesting that the anti-Müllerian hormone in the early freemartin is of male origin. After 80 days of gestation, anti-Müllerian hormone production by freemartin fetal gonads was found in one-half of the freemartins and usually associated with the presence of seminiferous tubules; however, the presence of seminiferous tubules within the gonad of a freemartin had no apparent relationship to the concentration of anti-Müllerian hormone in its serum. Fetal gonads of the normal female produce no anti-Müllerian hormone, but ovarian follicular cells do produce the hormone postnatally.

B. Cellular Theory

Because the interchange of both germ cells and somatic cells has been reported between the freemartin and its male cotwin, some investigators have proposed that ovarian stunting and sex reversal in the freemartin results from transplacental migration of male germ cells to the undifferentiated gonadal ridge of the female twin where they induce masculinization of the developing ovaries. Similarly, some investigators have reported a positive correlation between the degree of masculinization and the percentage of XY blood chimerism within the freemartin. However, current evidence indicates that male germ cells do not mediate structural development of the testis but rather that testicular development is mediated by the sex and/or autosomal chromosomes of the nongerminal tissue of the germinal ridge. Also, blood marrow chimerism is common between marmoset twins of unlike sex, but development of the female is unimpaired. Likewise, normal sexual development occurs in human twins of the opposite sex in the presence of XX/XY chimerism.

VI. FERTILITY OF MALE COTWIN

The effect of placental anastomosis on sexual development in the male twin of a heterosexual pregnancy is less pronounced. Unlike the freemartin, the chimeric cotwin bull is a phenotypically normal male with few reported variations or aberrations in his reproductive tract and is generally assumed to possess normal fertility. However, some evaluations have indicated a higher percentage of males born cotwin to a freemartin being culled for poor reproductive performance in comparison to single-born bulls. Reasons for culling include poor libido, no semen, low sperm output or sterility, high concentration of abnormal sperm (e.g., cytoplasmic droplets, abaxial midpieces, or abnormalities of the acrosome), and/or low sperm motility—all factors that contribute to lower conception rates. The reduced fertility reported for some chimeric bulls has been associated with an increased incidence or amount of testicular degeneration. Also, some chimeric bulls have abnormally low testosterone concentrations both in testicular tissue and in spermatic vein blood and fail to give a testosterone response to gonadotropin stimulation. Because of the wide variation in semen quality and production among chimeric bulls, evaluations of libido and semen characteristics are indicated before retaining bulls born cotwin to a freemartin for breeding purposes.

Just as in the freemartin, estimates of the XX:XY ratio among lymphocytic or erythrocytic chimeric bulls are highly variable, ranging from 5 to 95% XX. Likewise, no association has been found between percentage of XX blood cells and reproductive performance of a chimeric bull. In addition to the exchange of primordial hemopoietic and lymphopoietic cells, evidence exists for the interchange of primordial germ cells between the freemartin and its cotwin male because XX germ cells have been found in the gonads of male cotwin fetuses and XY cells in gonads of freemartin fetuses. Similarly, XX spermatogonia and primary spermatocytes were detected in testicular specimens from postnatal or pubertal bulls. However, several other studies have failed to find XX germ cells in testes of chimeric bulls ranging in age from 1 to 9 years. Thus, these transplanted XX germ cells may only persist until sexual maturity, and some investigators have speculated that the increased testicular degeneration reported for chimeric bulls with lower fertility is an autoimmune response to the XX germ cells. Regardless, evidence from cattle, goats, and mice indicates that most XX germ cells fail to survive or to undergo spermatogenesis when placed in a testicular environment. Conversely, if the XX germ cells did express normal spermatogenesis, then an increase in X-bearing spermatozoa and a consequential increase in the proportion of female progeny would be anticipated. Most studies have failed to demonstrate a deviation from the normal 1:1 sex ratio in progeny of chimeric bulls, but a few examples of chimeric bulls siring disproportionally more female progeny have been reported. Possibly, the possession of more than one X chromosome is incompatible with further differentiation of testicular primordial germ cells to spermatogonia.

VII. ECONOMIC VALUE

For commercial beef herds, the 25% female–female twin pairs results in the same number of replacement females per year as in herds producing only singletons. However, the freemartin syndrome may pose unique considerations in the production and sale of foundation animals. Because of the infertility associated with freemartinism, the economic value of the freemartin is in meat production. As a result of gonadal hypoplasia and minimal gonadal steroidogenesis found in most freemartins, growth rate and carcass characteristics of the freemartin reflect those of an early castrate. Birth weight and weaning weight are similar for freemartin and gonadally intact twin-born females but 10–15% lighter than those for single-born females. Growth rate is similar for genotypic females regardless of type of birth. The economic loss due to the smaller area of the longissimus dorsi muscle is compensated for by the increased intramuscular deposit of fat (i.e., marbling) in the muscle of the freemartin, which results in about 90% of the freemartin carcasses having a USDA grade of choice or prime.

See Also the Following Articles

Cattle (Bovidae); Horses (Equidae); Intersexuality; Sheep and Goats; Twinning

Bibliography

Dominguez, M. M., Liptrap, R. M., Croy, B. A., and Basrur, P. K. (1990). Hormonal correlates of ovarian alterations in bovine freemartin fetuses. *Anim. Reprod. Sci.* 22, 181–201.

Echternkamp, S. E. (1992). Fetal development in cattle with multiple ovulations. *J. Anim. Sci.* 70, 2309–2321.

Echternkamp, S. E., Gregory, K. E., Dickerson, G. E., Cundiff, L. V., Koch, R. M., and Van Vleck, L. D. (1990). Twinning in cattle: II. Genetic and environmental effects on ovulation rate in puberal heifers and postpartum cows and the effects of ovulation rate on embryonic survival. *J. Anim. Sci.* 68, 1877–1888.

Ginther, O. J., and Griffin, P. G. (1994). Natural outcome and ultrasonic identification of equine fetal twins. *Theriogenology* 41, 1193–1199.

Gregory, K. E., Echternkamp, S. E., and Cundiff, L. V. (1996). Effects of twinning on dystocia, calf survival, calf growth, carcass traits, and cow productivity. *J. Anim. Sci.* 74, 1223–1233.

Guerrier, D., Boussin, L., Mader, S., Joss, N., Kahn, A., and Picard, J. Y. (1990). Expression of the gene for anti-Müllerian hormone. *J. Reprod. Fertil.* 88, 695–706.

Hunter, R. H. F. (1995). Anomalous sexual development in domestic species. In *Sexual Determination, Differentiation and Intersexuality in Placental Mammals.* Cambridge Univ. Press, New York.

Jost, A., and Magre, S. (1993). Sexual differentiation. In *Reproduction in Mammals and Man.* Ellipses, Paris.

Jost, A., Vigier, B., and Prepin, J. (1972). Freemartins in cattle: The first steps of sexual organogenesis. *J. Reprod. Fertil.* 29, 349–379.

Kastli, F., and Hall, J. G. (1978). Cattle twins and freemartin diagnosis. *Vet. Rec.* 102, 80–83.

Long, S. E. (1979). The fertility of bulls born twin to freemartins: A review. *Vet. Rec.* 104, 211–213.

Lyet, L., Louis, F., Forest, M. G., Josso, N., Behringer, R. R., and Vigier, B. (1995). Ontogeny of reproductive abnormalities induced by deregulation of anti-Müllerian hormone expression in transgenic mice. *Biol. Reprod.* 52, 444–454.

Shen, W.-H., Moore, C. C. D., Ikeda, Y., Parker, K. L., and Ingraham, H. A. (1994). Nuclear receptor steroidogenic factor 1 regulates the Müllerian inhibiting substance gene: A link to the sex determination cascade. *Cell* 77, 651–661.

Sweeney, T., Saunders, P. T. K., Millar, M. R., and Brooks, A. N. (1997). Ontogeny of anti-Mullerian hormone, 3β-hydroxysteroid dehydrogenase and androgen receptor expression during ovine fetal gonadal development. *J. Endocrinol.* 153, 27–32.

Taketo, T., Saeed, J., Manganaro, T., Takahashi, M., and Donahoe, P. K. (1993). Müllerian inhibiting substance production associated with loss of oocytes and testicular differentiation in the transplanted mouse XX gonadal primordium. *Biol. Reprod.* 49, 13–23.

Vigier, B., Tran, D., Legeai, L., Bézard, J., and Joss, N. (1984). Origin of anti-Müllerian hormone in bovine freemartin fetuses. *J. Reprod. Fertil.* 70, 473–479.

Wilkes, P. R., Wijeratne, W. V. S., and Munro, I. B. (1981). Reproductive anatomy and cytogenetics of freemartin heifers. *Vet. Rec.* 108, 349–353.

FSH (Follicle-Stimulating Hormone)

Leo E. Reichert, Jr.

Albany Medical College

GLOSSARY

glycoprotein A protein hormone with covalently associated carbohydrate.

gonadotropes Cells in the pituitary gland wherein gonadotropic hormones are synthesized.

gonadotropic hormones (gonadotropins) Anterior pituitary hormones that have their target cells in the testis or the ovary.

heterodimers Molecules composed of two dissimilar subunits.

oligosaccharides Chains composed of variable numbers and types of sugars.

polyclonal/monoclonal antibodies Either single (mono-) or multiple (poly-) types of immunoglobulins synthesized in response to antigen stimulation.

Follicle-stimulating hormone (FSH), also called follitropin, is a hormone synthesized in the pituitary gland of mammals and most lower vertebrate species. It is one of three structurally related glycoprotein hormones secreted by the pituitary gland, the others being luteinizing hormone and thyroid-stimulating hormone.

I. INTRODUCTION

The follicle-stimulating hormone (FSH) content of the pituitary gland is quite low, about 1/20 that of other pituitary hormones such as growth hormone or luteinizing hormone (LH). FSH is synthesized in specialized cells (gonadotropes) of the anterior pituitary gland and released upon stimulation of the gland by the hypothalamic decapeptide, gonadotropin-releasing hormone (gonadoliberin, GnRH). The biological effect of FSH is stimulation of synthesis of the steroid hormones progesterone and estrogen. In the female, peak serum levels of FSH, as well as of LH, occur immediately prior to and are essential for the process of ovulation. FSH is also synthesized in the male pituitary gland. The designation "FSH" is used to identify this hormone in either sex, and this also applies to LH. FSH, along with LH and testosterone, is required for initiation and maintenance of normal spermatogenesis in the prepubertal male and for maintenance of qualitatively competent spermatogenesis in the adult male. In each sex, FSH also stimulates synthesis and activation of an enzyme, aromatase, which is responsible for conversion of testosterone to estrogen in gonadal tissue as well as peripherally.

Circulating FSH binds to specific receptors on the outer membrane of ovarian granulosa cells and testicular Sertoli cells. Biological effects of FSH are the result of its high-affinity binding to membrane receptors, resulting in generation of a "second messenger," the nucleotide cyclic adenosine monophosphate (cAMP). cAMP dissociates a regulatory component from the critical enzyme, protein kinase A, thus allowing it to act on appropriate substrates. In the gonads, protein kinase A phosphorylates and thereby activates key enzymes required for steroidogenesis, such as the enzyme aromatase, as well as other rate-limiting enzymes involved in the process of steroidogenesis from the obligatory precursor, intercellular-free (nonesterified) cholesterol. Protein kinase A also phosphorylates and thereby activates cytoplasmic proteins, termed transcription factors, which are responsible for a more delayed, but longer lasting response to FSH. After phosphorylation, transcription factors move to the nucleus, where they bind as dimers to regions on nuclear DNA termed response elements, leading to RNA polymerase activation, synthesis of messenger RNA, and, ultimately, synthesis of key enzymes necessary for steroidogenesis and maintenance of FSH biological effects.

II. STRUCTURE

Although there are some variations in the details of structure, pituitary FSH of most mammalian species has a consensus molecular weight of about 30,000 and is composed of two noncovalently associated subunits, each with a molecular weight of about 15,000 but with different amino acid and carbohydrate compositions. One subunit is designated the α subunit and is structurally very similar to the α subunit of LH, thyroid-stimulating hormone (TSH), and the placental hormone, human chorionic gonadotropin (hCG). Therefore, the FSH α subunit is more appropriately termed glycoprotein hormone α subunit (GH-α). Because of its greater abundance in pituitary tissue, it has been possible to isolate and determine the structure of GH-α from a greater variety of species than has been possible for FSH β subunit, including ovine, bovine, porcine, rabbit, equine, mouse, rat, human, whale, rhesus monkey, camel, hamster, chicken, turkey, bullfrog, carp, salmon, and eel.

The other subunit, designated the β subunit, is specific for FSH in the same manner that the LH β subunit is specific for LH and the TSH β subunit is specific for TSH. As used here, the term "specific" refers to the ability of the β subunit, when combined with the GH-α subunit, to form a noncovalently associated heterodimer having a conformation appropriate for high-affinity interaction with one, and not all, membrane receptors. It has been possible to isolate and determine the structure of the β subunit from equine, human, ovine, bovine, porcine, rhesus, and bullfrog FSH. It is anticipated that with the advent of techniques of molecular biology, the structure of the β subunits of many other species will soon become available. The biologic activity of glycoprotein hormone α/β heterodimers depends on the nature of the β subunit. Thus, artificially prepared molecules composed of GH-α + FSH β have FSH activity, whereas a molecule composed of GH-α + LH β has LH activity. Individual α and β subunits are without significant biological activity, but when combined they recover nearly all that present in the native hormone. Some reports attribute various types of biologic activity to isolated glycoprotein hormone subunits, but these remain controversial and their significance is unclear.

The gonadotropin α and β subunits of most mammalian species are the products of separate genes. The genomic FSH genes among currently characterized species seem evolutionarily conserved and contain two exons and three introns. The FSH β subunit gene is unique among those for pituitary hormones by virtue of an unusually long 3' untranslated region. However, for lower primates and some teleosts (such as carp, salmon, and eel), a single gene product for a primordial gonadotropin having both FSH- and LH-like activities, termed gonadotropic hormone (GTH), has been reported. Some structural diversity also exists for GTH among and within different species. For example, the primary structure (amino acid sequence) of one such form, sGTH-1β from chum salmon, has greatest homology to FSH β, whereas another form, sGTH-2β, from chinook salmon, bears the closest homology to LH β. Although details of structure for many different species remain to be elucidated, it seems clear that a FSH represents a true primordial reproductive hormone.

III. COMPOSITION

A. Amino Acid Sequence

The consensus primary amino acid sequence of GH-α and FSH β subunits are given in Table 1. For FSH (glycoprotein hormone) α subunit, the primary structure is strongly conserved phylogenetically, with a minimum of 80% sequence homology in 25 of 35 species studied. For the hormone-specific FSH β subunit, homology ranges from 78 to 92% between FSH of different species. Sequence homology among

TABLE 1

Consensus Amino Acid Sequence of Glycoprotein Hormone α Subunit and the β Subunit of Follicle-Stimulating Hormone

Consensus sequence of the α subunit

```
1      5       10      15      20      25      30
f P d g e   f . . q g   C p E C k   l k e N k   . F s k p   g a p I y
31     35      40      45      50      55      60
Q C m G C   C f S R A   Y P T P .   R S k k T   M L V p K   n I T S E
61     65      70      75      80      85      90
a t C C V   A k a f t   k . t v m   g n . k v   e N H T .   C h C S T
       95
C y y H   k s
```

Consensus sequence of the FSH β subunit

```
1      5       10      15      20      25
. S C E L   t N I T I   . v E K E   e C . f C   I s I N t
26     30      35      40      45      50
T W C a G   Y C Y T r   D l v y k   d P a r p   n I Q k t
51     55      60      65      70      75
C T f k E   l v Y E T   v k v P G   C A h h a   d s l Y T
76     80      85      90      95      100
Y P V A t   e C H C g   k C d s d   s T D C T   V R g L G
101    105     110
P s Y C S   f . e m k   E
```

Note. The amino acid single-letter code has been used in this compilation. A, alanine; V, valine; L, leucine; I, isoleucine; P, proline; M, methionine; F, phenylalanine; W, tryptophan; G, glycine; S, serine; T, threonine; C, cysteine; Y, tyrosine; N, asparagine; Q, glutamine; D, aspartic acid; E, glutamic acid; K, lysine; R, arginine; H, histidine. Capital letters indicate unanimous residues. This means the same amino acid is present in all species studied. Lowercase letters indicate majority residues. This often reflects a difference of only one or two amino acids in the group studied, often conservative substitutions. Periods (·) reflect no majority and considerable variation in the amino acid identified in that position. The structural homology between the FSH α subunit and other glycoprotein hormone α subunits ranges from 72% (human vs whale or equine) to over 97% (human vs rhesus). On the basis of 210 available comparisons, homology between the FSH β subunit and β subunits of other glycoprotein hormones ranged from 31% (TSH) to 43% (LH).

β subunits of FSH, LH, and TSH are predictably much less, as low as 30% in some instances. Such structural differences allow for the diversity and functional specificities that exist among the hormone-specific β subunits.

Since each subunit is rich in cysteine content, an amino acid having a free sulfhydryl (-SH) group, there is the potential for significant disulfide (S–S) bond formation within the individual subunits. The most significant of these appear to be between positions 32 and 64 in the α subunit and between 93 and 100 in the β subunit. The elucidation of three-dimensional structural features of the intact molecule, strongly influenced by the positioning of intra- and interchain disulfide bonds, awaits solving of its crystal structure. However, the crystal structure of the closely related glycoprotein gonadotropic hormone hCG has been determined, and it is assumed that the crystal structure of FSH will prove to be similar.

B. Carbohydrates

Each FSH subunits contains two oliogsaccharide chains, linked to a free amide group of the amino acid asparagine contained within its structure. In the α subunit, N-linked oligosaccharide chains are at positions 56 and 82. In the β subunit, these are at positions 6 or 7 and 23 or 24. The exact positioning is not certain. The carbohydrates associated with FSH are needed for optimal expression of *in vivo* activity. Circulating FSH appears to represent a family of variably glycosylated FSH molecules having differing degrees of biologic activity. Chemically or enzymatically deglycosylated FSH still binds to receptors on gonadal tissue (granulosa cells in the female and Sertoli cells in the male) with the same or greater affinity than does native FSH, but it lacks the ability to induce postreceptor-binding events such as steroidogenesis. As a result, deglycosylated FSH has been proposed as a contraceptive, putatively blocking the action of circulating, fully glycosylated FSH. FSH-associated carbohydrate is also important for maintenance of the circulatory survival time of FSH, particularly the terminal sialic acid residues on the associated oligosaccharides. Once removed, the exposed galactosyl residues facilitate uptake of desialylated FSH by receptors on the liver, its removal from the circulation, and subsequent degradation by liver cells.

IV. MEASUREMENT

A problem in measuring the activity of FSH is that of interference by LH in any *in vivo* response. This is because FSH and LH generally act synergistically in biological systems. The classical, and the only, specific assay for measurement of FSH activity in intact animals is that developed by Steelman and Pohley utilizing immature female rats. In that assay, maximum ovarian growth due to LH is achieved by administration of massive doses of hCG, which is an easily obtained glycoprotein hormone secreted by the placenta during pregnancy and has biologic activity comparable to pituitary LH. Any ovarian weight gain that occurs in excess of that caused by hCG (assessed in a separate control group) is assumed to be due to FSH. Although an extremely useful classical assay for research endocrinologists, its lack of sensitivity (in the microgram range) was a serious handicap to its clinical application in, for example, measuring FSH activity in blood.

Small amounts (picograms) of FSH can be detected through use of immunologic techniques, utilizing polyclonal or monoclonal antibodies raised against FSH. The technology of immunoassay has advanced steadily so that now FSH levels can be quantitated in a few microliters of blood, either fresh or after elution from a solid medium, such as filter paper, which can be sent to central laboratories for analysis through the mail. The subunit nature of FSH and its structural homologies to LH and TSH have complicated interpretation of immunoassay results utilizing polyclonal antisera to intact FSH. Modern methods utilize monoclonal antibodies raised to hormone-specific epitopes, such as those uniquely present on the FSH β subunit. It should be noted, however, that these results assess immunologic activity, not biologic activity, and may be responsive to inactive hormone or circulating hormonal fragments. FSH can also be measured in an assay technically similar to immunoassay, wherein receptors present on membranes of gonadal cells, rather than antibodies, are utilized to measure the concentration of FSH. This assay measures receptor affinity of FSH, but not necessarily its biologic activity in terms of postbinding events. Recall that deglycosylated FSH also binds to receptor. Another approach involves incubation of FSH or serum with cultured gonadal cells (granulosa cells or Sertoli cells) and measurement of resulting synthesis of specific steroid hormones by immunoassay. This is mostly an experimental procedure due to questions of sensitivity and interference by nonspecific serum factors. The defining measurement of FSH biologic activity, however, remains its ability to effect gonadal function in the intact animal model.

V. USES

In addition to being present in only relatively small amounts, highly purified FSH is particularly unstable, and total FSH activity is lost as purification proceeds. This property complicated preparation of FSH from pituitaries in limited supply, such as human, primate, or rodent pituitary glands. This was a serious impediment to general progress in studies on basic aspects of mammalian reproduction. Clinically, a program sponsored by the National Institutes of Health (NIH) supported collection of sufficient numbers of human and farm animal pituitary glands to allow isolation of FSH (as well as the other pituitary hormones) for structural characterization and rigorous biological evaluation. The resulting FSH was distributed gratis by NIH and proved invaluable for use in human clinical and experimental whole animal studies as a reliable and readily available reference preparation for quantitative assessment of FSH (allowing laboratories worldwide to compare results in a meaningful way) and as a primary antigen for generation of antibodies and development of radioimmunassays sufficiently sensitive for measurement of FSH in tissue extracts, cell culture media, and, perhaps of greatest significance, in small amounts of human blood. Recently, the problem of supply has been alleviated by production of genetically engineered FSH through use of techniques of molecular biology; this FSH has properties indistinguishable from native FSH and, in some cases, it is improved over that of the natural hormone. Currently, FSH is used clinically for induction of ovulation and stimulation of spermatogenesis in humans and in animal husbandry for superovulation of farm animal. Active immunization with FSH or its β subunit is also being studied as a possible approach to contraception. Finally, measurement of serum FSH levels in humans is an important clinical parameter for diagnosis and treatment of numerous fertility problems in men and women.

See Also the Following Articles

LH (LUTEINIZING HORMONE); PITUITARY GLAND, OVERVIEW; SPERMATOGENESIS, HORMONAL CONTROL OF

Bibliography

Bousefield, G. R., Perry, W. M., and Ward, D. N. (1994). Gonadotropins: Chemistry and biosynthesis. In *The Physiology of Reproduction* (E. Knobil and J. D. Neill, Eds.), pp. 1749–1792. Raven Press, New York.

Campbell, R. K., Dean-Emig, D. M., and Boyle, W. R. (1991). Conversion of human choriogonadotropin into follitropin by protein engineering. *Proc. Natl. Acad. Sci. USA* **88**, 760–764.

Moudgal, A. R., Murthy, G. S., and Saxena, B. N. (1997). Responsiveness of human male volunteers to immunization with ovine follicle stimulating hormone vaccine: Results of a pilot study. *Hum. Reprod.* **12**, 457–463.

Steelman, S. L., and Pohley, K. (1953). Assay of follicle stimulating hormone based on augmentation with human chorionic gonadotropin. *Endocrinology* **53**, 604–610.

Worthman, C. M., and Stallings, J. F. (1997). Hormone measures in finger-prick blood spot samples. New fields for reproductive endocrinology. *Am. J. Phys. Anthropol.* **104**, 1–21.

Galactorrhea

Howard A. Zacur

Johns Hopkins Medical Institutions

GLOSSARY

duct ectasia Dilation of the duct.
euprolactinemic Having a serum or plasma prolactin concentration within the normal range.
galactopoiesis Maintenance of milk secretion.
galactorrhea Secretion of milk from the nipples of the breast unrelated to pregnancy or postpartum time periods.
hyperprolactinemia Elevation of the serum or plasma prolactin concentration above the normal range.
lactogenesis Initiation of postpartum milk secretion.
mammogenesis Growth and differentiation of the mammary gland.
nipple discharge Spontaneous escape of fluid from the nipple.
nipple secretion Fluid within the ducts of the breast collected by nipple aspiration or breast massage.

Galactorrhea is derived from the Greek, Latin, and French words for milk, i.e., *galact*, and flow, i.e., *rrhea*. Galactorrhea is the discharge or secretion of milk from the nipples of the breast unrelated to pregnancy or suckling postpartum. Demonstration of its existence requires that it be distinguished from other types of breast discharges or secretions and that an evaluation be undertaken to determine its cause.

I. INTRODUCTION

A relationship between the presence of breast milk unrelated to pregnancy and its effect on female reproduction has been known since "ancient" time. Some authors have mentioned that Hippocrates once wrote that "if a woman who is not with child, nor has brought forth, has milk, her menses are obstructed." Chiari in 1855 and Frommel in 1882 described persistence of galactorrhea associated with amenorrhea in postpartum women. In 1932 Ahumada and del Castillo described the presence of galactorrhea associated with amenorrhea in women not recently pregnant. In 1954 Forbes and colleagues reported that women with galactorrhea and amenorrhea unrelated to pregnancy could often be found to have a pituitary tumor. These authors also postulated that the syndrome of galactorrhea and amenorrhea was caused by secretion of a lactogenic hormone from the pituitary.

In 1971 the existence of a human pituitary prolactin molecule was demonstrated by Hwang *et al.* working in the laboratory of Henry Friesen. Their efforts resulted in development of a radioimmunoassay for this hormone. Once created, this assay was used to measure the concentration of prolactin in the plasma of human subjects complaining of a variety of medical conditions. It was soon discovered that in many instances women who were amenorrheic or who complained of galactorrhea were found to have an elevation of prolactin in their peripheral plasma. In some of these cases the presence of a pituitary tumor was demonstrated as well.

This article describes the evaluation and management of the patient with galactorrhea. A review of breast anatomy and the physiology of lactation will be described followed by a discussion of various types of nipple discharges and secretions. Specific management of the patient with galactorrhea will follow.

II. BREAST ANATOMY AND PHYSIOLOGY OF LACTATION

The human breast consists of ducts, glands, supportive connective tissue, and fat. Nipples of the breasts are composed in part of a muscular–elastic fiber system designed to cause nipple erection and to permit suckling. Circular pigmented areas surrounding the nipples of the breasts are called areolae. Small ducts emanating from Montgomery glands open onto the surface of the areolae. These glands are in part sebaceous glands and in part small milk glands and may provide lubrication for the surface of the nipple during suckling.

Each nipple is composed of 15–25 milk ducts which descend beneath the surface of the areola and then expand to form lactiferous sinuses in which milk may be stored (Fig. 1). Lactiferous sinuses receive milk from mammary ducts emanating from 15–25 breast lobes. Breast lobes are separated from each other by connective tissue and fat. Each mammary duct that ends at the nipple will drain fluid from a single breast lobe. Each breast lobe is composed of many lobules which contain grape-like glandular structures known as alveoli or acini. Alveoli are lined by dark luminal cells called A or "foam" cells and by clear basal or B cells also known as "chief" cells. Chief cells are responsible for the development of the A cells, which are responsible for the synthesis of milk proteins. Surrounding the alveoli and its ducts are contractile epithelial cells known as myoepithelial cells. Contractions of myoepithelial cells induced by oxytocin released by the pituitary forces milk out of the alveoli and into the ducts leading toward the nipple. This process is known as the "letdown reflex."

After appropriate hormonal stimulation, lipid droplets may be seen to form in the cells of the alveoli that produce milk. Eventually, each of these lipid droplets becomes surrounded by the aveolar cell

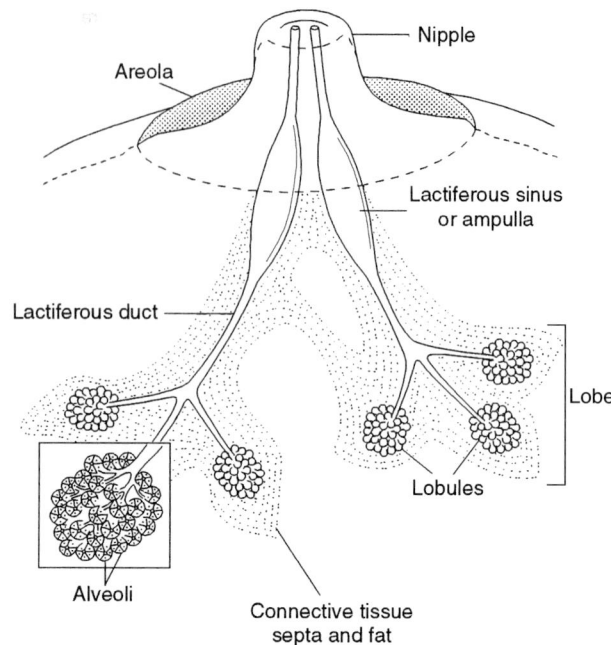

FIGURE 1 Ducts, sinuses, lobes, lobules, and aveoli of the human breast depicted schematically. Each lactiferous sinus or ampulla drains one of 15–25 breast lobes.

membrane which eventually results in the separation of a membrane-bound fat globule from the milk-producing cell. This is a secretory process which has been described by Urosumi *et al.* as being unique to the mammary gland.

Milk secretion during nursing results from the following four transcellular mechanisms which have been described by Neville: (i) the exocytotic secretion of aqueous components of milk, i.e., lactose, casein, calcium, citrate, and phosphate; (ii) the unique secretion of milk fat as described previously; (iii) the release of small molecules, such as sodium, potassium, chloride, and glucose, from milk-producing cells via pumps, and (iv) the secretion of large molecules, e.g., immunoglobulins utilizing coated pits. Secretion of immunoglobulins involves extracting these molecules from the plasma by use of a "coated pit" within the membrane of the milk-producing cell at its basal end. Once trapped, these proteins are transported within the cytoplasm of this cell to its apical surface, which faces the alveolar lumen. Once at this location these proteins are released into the alveolar lumen (Fig. 2).

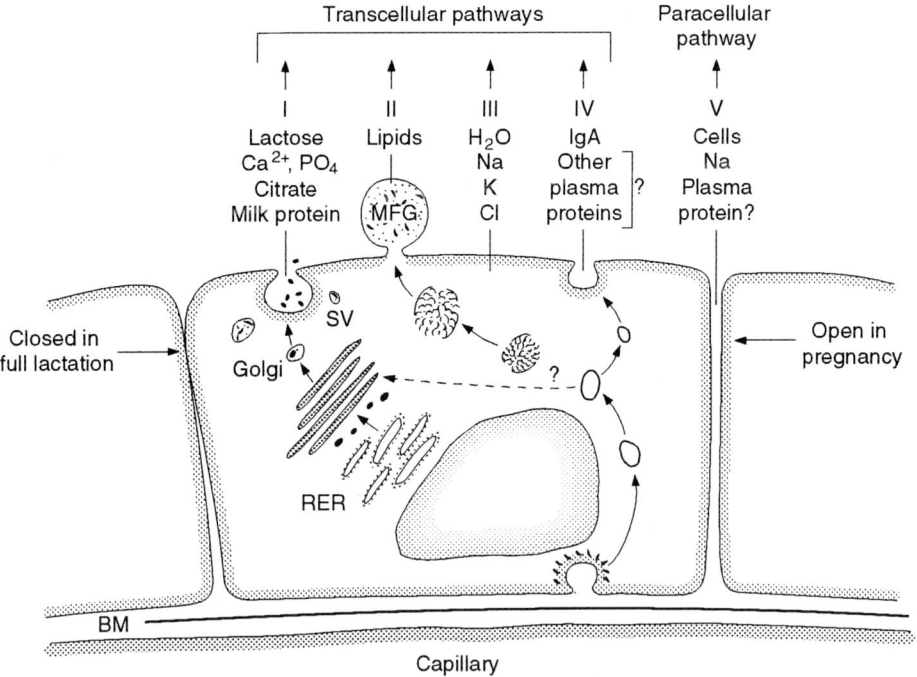

FIGURE 2 Four different pathways for secreting milk components are depicted (modified from Neville, 1990).

Alveolar milk-producing cells are connected by tight junctions. During active nursing these junctions can become loose, allowing additional plasma constituents, e.g., sodium and chloride, to enter the alveolar lumen via a paracellular route. As a consequence, the ionic composition of breast milk may be altered during active nursing.

Breast milk synthesis and secretion are physiological processes responsive to appropriate hormonal stimulation. Hormonal stimulation affects mammogenesis or the growth and differentiation of the mammary gland as well as lactogenesis or the initiation of postpartum milk secretion. Maintenance of milk secretion or galactopoiesis in the postpartum period also requires a hormonal influence.

The type and amount of hormonal stimulation involved in lactation varies depending on whether mammogenesis, lactogenesis, or galactopoiesis is occurring. Prolactin, growth hormone, thyroxine, adrenal glucocorticoids, insulin, estrogen, and progesterone have all been shown to be involved in mammogenesis. Priming of breast tissue by increased amounts of estrogen and progesterone in conjunction with persistently elevated prolactin levels precedes lactogenesis. Following the initiation of lacta-

tion, breast milk secretion can be maintained by concentrations of prolactin within "normal" limits as long as nipple stimulation is maintained.

III. NIPPLE DISCHARGES AND SECRETIONS

A knowledge of mammary gland anatomy and physiology is required to understand both the significance and the various differences in breast discharges and secretions. The spontaneous escape of fluid from the nipple is termed a "nipple discharge." Fluid in breast ducts that must be collected by nipple aspiration or breast massage is termed a "nipple secretion."

During nonpregnancy time periods, fluid is continuously released into the duct system of the breasts and then reabsorbed into lymphatics and blood, creating a balance between breast duct fluid production and reabsorption. This may be altered during active states of fluid secretion into the ducts, e.g., lactation. Fluid normally present within the duct system of the breast may be revealed as a consequence of an active attempt to elicit it, e.g., use of a breast pump. When an active attempt is made to express nipple fluid and none is found, the lactiferous ducts of the nipple may be blocked by keratotic plugs or fluid may not be present within the ducts. Nipple secretion in non-lactating women is infrequently observed because of the presence of keratotic plugs.

Occasionally, a pseudonipple discharge may be seen. Pseudonipple discharge represents breast fluid that is not produced within the alveoli, mammary ducts, or lactiferous sinuses of the breast. Infections resulting in perialveolar abscesses with fistulae formation may produce pseudonipple discharges. Trauma, e.g., jogger's nipples, or skin lesions, e.g., an eczematoid lesion, may also produce a pseudonipple discharge that should be differentiated from true nipple discharge. Paget's disease should always be excluded if an eczematoid lesion of the nipple does not heal.

Leis has classified fluid arising from the nipple of the breast either spontaneously or by aspiration into one of the following seven basic types: milky, multicolored and sticky, purulent, clear or watery, yellow or serous, pink or serosanguineous, and bloody.

Purulent nipple discharge is indicative of infection. Multicolored discharge which is sticky and emanates from one or more ducts may be the result of breast duct dilation or ectasia which is a benign condition. Colors of nipple fluids which are frequently encountered include green, yellow, white, brown, gray, reddish brown, or even black. Darker colors have been found to be associated with higher concentrations of cholesterol, estrogen, and lipid peroxides within the fluid by Petrakis and colleagues. Petrakis' research group has also speculated that substances within mammary duct fluid may be responsible for causing malignancy. Spontaneous nipple discharges which are clear, yellow or serous, pink, or bloody are more likely to be associated with an underlying breast malignancy. Milky discharge is called galactorrhea, which is almost always benign.

IV. EVALUATION OF NIPPLE DISCHARGE

Spontaneous discharge or elicited secretion from multiple ducts of the nipple of milky fluid is almost always galactorrhea. Placement of a drop of the milky nipple secretion beneath the coverslip of a microscope slide will reveal the presence of numerous fat globules under low microscope magnification thus verifying that the fluid is galactorrhea. It is not routinely necessary to send this type of nipple fluid for cytologic examination or to obtain breast imaging studies if the nipple discharge/secretion is identified as galactorrhea. Measurement of the serum/plasma prolactin level is a helpful diagnostic test when galactorrhea is detected. Measurement of a serum/plasma insulin growth factor-1 (IGF-1) level may be helpful as well.

Nipple discharge/secretion that is multicolored and sticky and arising from multiple nipple ducts is almost always benign, particularly when it is elicited by breast exam or pump. Cytologic examination is rarely positive and not believed to be useful as a routine screen.

If the nipple discharge is purulent, a search for an infection or abscess of the breast is indicated. Surgical drainage and antibiotic treatment are provided as necessary. If an abscess is identified histological evaluation of the abscess wall should be performed to exclude a malignancy.

Nipple discharges which are spontaneous and are sanguineous, serosanguineous, or watery are of the greatest concern since an associated breast malignancy may be found in 6, 13, 28, and 33% of these cases, respectively. Cytology and mammography along with a very careful breast exam are indicated. Galactography, i.e., contrast medium injected into the discharging duct, may also be helpful. Surgical excisional biopsy of the discharging ducts may also be performed. Reliance only on a negative cytological exam of the nipple discharge as a reason not to proceed with further evaluation is not recommended because several studies have shown that carcinomas are not routinely identified by this method. This may occur because breast tumors may block the outflow of fluid from the ducts of the breast.

V. EVALUATION OF GALACTORRHEA

Once a diagnosis of galactorrhea has been made an evaluation as to its cause is required. Kleinberg and colleagues, in a classic study in 1977, reported on 235 cases of galactorrhea which included 12 men. This study showed that for 76 women with galactorrhea who were having menses, 65 (85%) had a normal serum prolactin concentration. This percentage included 17 women who were also having irregular menses. This is surprising since menstrual cycle disturbances, i.e. ,skipped or shortened menstrual cycle intervals, are often observed in women as the serum prolactin level rises. Subjects with regular menses and galactorrhea would therefore be expected to have a normal serum or plasma prolactin level >85% of the time. For the 20 women with galactorrhea who were amenorrheic in Kleinburg's study, 18 (90%) had an elevated prolactin level.

There was a 20% chance of detecting a pituitary abnormality radiographically in the 235 patients with galactorrhea. This incidence of pituitary abnormality is similar to that reported for "normal" patients undergoing radiographic imaging of the pituitary gland. The risk of detecting a pituitary abnormality increased to 34% in women who were amenorrheic and also exhibited galactorrhea.

Identification of a hyperprolactinemic state requires that serum/plasma samples obtained for prolactin measurement be taken under rigorously controlled conditions on more than one occasion. If a diagnosis of hyperprolactinemia is made, a thorough evaluation to determine the cause of the hyperprolactinemia is then performed.

Individuals with galactorrhea whose serum or plasma prolactin concentration is within the normal range do not require hypothalamic/pituitary imaging either by computed axial tomography or by magnetic resonance imaging. Some investigators have indicated that individuals with galactorrhea persisting over time may eventually become hyperprolactinemic. This must be viewed as speculation since strong clinical evidence has not been published to support this conclusion. Consequently, it is difficult to recommend, as others have, that pituitary imaging studies should be performed for individuals with euprolactinemic galactorrhea. It is not difficult to agree with the recommendation that serial prolactin measurements be performed, e.g., at yearly or longer intervals, in individuals with galactorrhea and normal prolactin concentrations.

Euprolactinemic individuals with galactorrhea as well as 30% of individuals with hyperprolactinemia and galactorrhea have been shown to have acromegaly. Due to the very severe consequences from undiagnosed and untreated acromegaly, screening for growth hormone should be considered for individuals with euprolactinemic galactorrhea. Hyperhidrosis, glucose intolerance, neuropathy, arthritis, hypertension, and heart disease are some of the symptoms caused by excess growth hormone in addition to the well-known symptoms of acral and soft tissue overgrowth. It has been estimated that on average the disease of acromegaly may be present for 6–10 years before a diagnosis is made. Screening for acromegaly can be performed by requesting a serum/plasma somatomedin C or IGF-1 concentration.

An explanation for galactorrhea occurring in an individual who is euprolactinemic remains unclear. It is well-known that women who breast-feed for many months following delivery of their children will produce copious amounts of breast milk despite having a serum plasma prolactin concentration within the normal range. These women may also continue to breast-feed while ovulating and menstruating. Euprolactinemic galactorrhea may also be seen in individuals using the combination oral contraception pill or in women with anterior chest wall irritation.

Euprolactinemic galactorrhea cases have traditionally been "explained" by postulating the existence of increased sensitivity of the mammary gland prolactin receptor to the prolactin hormone or by the simultaneous secretion of another hormone capable of amplifying the biological activity of the prolactin molecule. An example of the latter would be the secretion of "synlactin" by the liver. This molecule has been reported to amplify the biological activity of the prolactin molecule several-fold in rats and pigeons. Another explanation is that the "integrated" daily or nightly prolactin concentration may be elevated even though an individual serum value may not be elevated. This condition would therefore not be diagnosable on the basis of a single serum/plasma prolactin determination. Serial measurements of the prolactin hormone concentration taken throughout the day or night or following a challenge by thyrotropin-releasing hormone have been recommended to detect less obvious or "subclinical" states of hyperprolactinemia. It has also been argued that unless the breast has been appropriately hormonally primed as normally occurs during pregnancy, breast milk secretion will not occur even when the prolactin concentration is elevated. Use of the combination oral contraceptive pill may provide sufficient hormonal stimulation to ducts and alveoli of the breast to allow the breast to respond to endogenous basal prolactin levels in some women who have never been pregnant.

VI. TREATMENT OF GALACTORRHEA

Treatment of the patient with galactorrhea will depend on the serum plasma prolactin and growth hormone (IGF-1) measurements as well as the existence of associated clinical problems, e.g., breast augmentation. Individuals with an elevation of the serum/plasma prolactin concentration require a specialized evaluation.

Treatment of euprolactinemic galactorrhea is strongly influenced by the inclination of the patient. For individuals not wishing to take medication, no treatment with continued observation may be offered. Patients should be advised that spontaneous remission of this condition may occur. Repeated attempts to detect nipple secretion using breast tissue compression are to be minimized because this type of breast stimulation could perpetuate the galactorrhea.

Euprolactinemic galactorrhea may be treated using dopamine agonist drug therapy to lower the basal serum/plasma prolactin concentration. Two dopamine agonists exist in the drug formulary which are currently approved by the United States Food and Drug Administration (FDA) for the treatment of hyperprolactinemia: Parlodel (bromocriptine) and Dostinex (cabergoline). When either of these drugs is given to individuals with hyperprolactinemia minor side effects may occur in 50% or more of the subjects treated. Side effects include orthostatic hypotension, lethargy, nausea, and nasal stuffiness. The chance of experiencing one of these side effects may be even greater for individuals given dopamine agonist ther-

apy who have a serum/plasma prolactin level within the normal range. Dopamine agonist drug therapy is not inexpensive and patients with euprolactinemic galactorrhea should be made aware of this before beginning treatment.

Following the initiation of drug therapy, lowering of the serum/plasma prolactin concentration is almost immediate and cessation of galactorrhea should be expected within weeks. Duration of drug therapy may be individualized, with drug treatment provided only until the galactorrhea is no longer detected, after which time therapy may be stopped. This therapeutic approach is similar to that used for suppression of lactation postpartum. If galactorrhea recurs, administration of the medication for a longer time period before stopping may be attempted. It should be mentioned that cabergoline is not approved by the FDA for postpartum lactation suppression and Sandoz Pharmaceuticals has voluntarily withdrawn its previously approved FDA indication for the use of Parlodel in the treatment of lactation suppression. Voluntary withdrawal of this indication by Sandoz occurred after it was reported that several patients who were given bromocriptine postpartum to suppress lactation experienced hypertensive crises and cerebrovascular accidents. A causal relationship between these adverse events and the use of dopamine agonists has never been proven. Other well-known medical disturbances of pregnancy, e.g., preeclampsia and eclampsia, are known to cause hypertension and cerebrovascular accidents and are suspected of having caused some of these unfortunate events. Patients given dopamine agonists for suppression of euprolactinemic galactorrhea should be informed of the fact that these drugs do not have FDA approval for this indication and that the potential for serious side effects may exist.

Surgery may be considered for patients unwilling to try or unable to tolerate dopamine agonist therapy for galactorrhea. When surgery is used as therapy, a circumareolar skin incision is made and the duct system beneath the nipple is dissected free and identified. A portion of the ducts is excised followed by ligation of the proximal and distal ends of the ducts.

Surgery is provided to women whose nipple discharge is caused by papillomas of the ducts and not by galactorrhea. Complete duct excision should not be performed in any woman contemplating future pregnancy for obvious reasons.

See Also the Following Articles

HYPERPROLACTINEMIA; LACTATION, HUMAN; LACTOGENESIS; PSEUDOCYESIS

Bibliography

Ahumada, J. C., and Del Castillo, E. B. (1932). Amenorrhea y galactorrhea. *Bol. Soc. Obstet. Ginecol.* **11**, 64–66.

Chiari, J., Braun, C., and Spèth, J. (1855). *Klinik, der Geburtshife und Gynèkologie*, pp. 362–415. Erlangen, Verglag Von Ferdinand Enke.

De Pablo, F., Eastman, R. C., Roth, J., and Gorden, P. (1983). Plasma prolactin in acromegaly before and after treatment. *J. Clin. Endocrinol. Metab.* **53**, 344–352.

Forbes, A. P., Henneman, P. H., Griswold, G. C., and Albright, F. (1954). Syndrome characterized by galactorrhea, amenorrhea and low urinary FSH: Comparison with acromegaly and normal lactation. *J. Clin. Endocrinol. Metab.* **14**, 265–271.

Frommel, R. (1882). Über puerperal: Atrophie des Uterus. *Z. Geburtshife* **7**, 305–313.

Hwang, P., Guyda, H., and Friesen, H. (1971). A radioimmunoassay for human prolactin. *Proc. Natl. Acad. Sci. USA* **68**, 1902–1906.

King, E. B., and Goodman, W. H. (1991). Discharges and secretions of the nipple. In *The Breast: Comprehensive Management of Benign and Malignant Disease* (K. I. Bland and E. M. Copeland, III, Eds.), pp. 46–47. Saunders, Philadelphia.

Kleinberg, D. L., Noel, G. L., and Frantz, A. G. (1977). Galactorrhea: A study of 235 cases, including 48 with pituitary tumors. *N. Engl. J. Med.* **296**, 589–600.

Leis, H. P. (1989). Management of nipple discharge. *World J. Surg.* **13**, 736–742.

Mick, C. C. W., and Nicoll, C. S. (1985). Prolactin directly stimulates the liver in vivo to secrete a factor (synlactin) which acts synergistically with the hormone. *Endocrinology* **116**, 2049–2053.

Neville, M. C. (1990). The physiological basis of milk secretion. *Ann. N. Y. Acad. Sci.* **586**, 1–11.

Petrakis, N. L. (1993). Nipple aspirate fluid in epidemiologic studies of breast disease. *Epidemiol. Rev.* **15**, 188–195.

Turton, D. B., Mohamed Shakir, K. M. (1995). Galactorrhea caused by esophagitis. *Am. J. Obstet. Gyencol.* **173**, 1629–1630.

Urosumi, K., Kobayashi, Y., and Baba, N. (1968). The fine structure of mammary glands of lactating rats, with special reference to the apocrine secretion. *Exp. Cell Res.* **50**, 177–192.

Gamete Intrafallopian Transfer (GIFT)

see Reproductive Technologies

Gametes, Overview

James M. Robl and Rafael A. Fissore

University of Massachusetts

I. Gamete Morphology
II. Gamete Production
III. Fertilization and Oocyte Activation

GLOSSARY

axoneme A collection of microtubules that consists of an outer ring of nine microtubule doublets and an inner triplet. The microtubules in the doublets slide along one another to cause the axoneme to bend.

genetic recombination The mixing of DNA, in the form of chromosomes, during development of the gametes.

gonads The reproductive organs of males (testes) and females (ovaries).

microtubules Long filamentous skeletal structures in cells that help to maintain the shape of the cell and also move the DNA during the cell division cycle.

polyspermy Fertilization of an oocyte with more than one sperm.

Gametes (germ cells) are cells from the male and female gonads (testes and ovary) that are highly specialized and have the capacity, in nature, to combine and give rise to an embryo. The male gamete (sperm and spermatozoa) is one of the smallest cells in the body and, to facilitate interaction with the egg, is motile. It functions to physically transport the paternal complement of DNA to the female gamete and to initiate development. In contrast, the female gamete (secondary ooctye, mature oocyte, and egg) is one of the largest cells in the body because it provides stored cellular components to facilitate early embryo development. The gametes execute the primary function of the two sexes; which is to provide a mechanism for genetic recombination between individuals of a species to produce a new, genetically unique individual. Producing unique genetic variants allows a species to readily adapt to a changing environment.

I. GAMETE MORPHOLOGY

The male and female gametes are two of the most unique cells in the body. The structure of these cells relates directly to their specialized functions.

A. Sperm

The sperm consists of a head and a tail. The head contains the tightly compacted haploid male DNA and the acrosome, a vesicle containing enzymes that aid in passage of the sperm to the surface of the egg. The tail, which propels the sperm to the surface of the egg, consists of a neck, middle piece, and principle piece.

1. Sperm Head

The largest component of the sperm head is the nucleus. The nucleus in the sperm is different from other cells in that the DNA is tightly compacted and inactive. Condensation of the nuclear DNA is the result of the loss of specific DNA-binding proteins normally present in cells (histones) and their replacement by another type of DNA-binding protein which, in mammals, is called a protamine. The shape of the sperm head varies among species. For example, bull sperm heads are paddle shaped (Fig. 1), mouse sperm heads are hook shaped, chicken sperm heads are spindle shaped, and zebrafish sperm heads are round. The shape of the sperm head is determined by a rigid cytoskeleton lying beneath the plasma membrane referred to as the perinuclear theca.

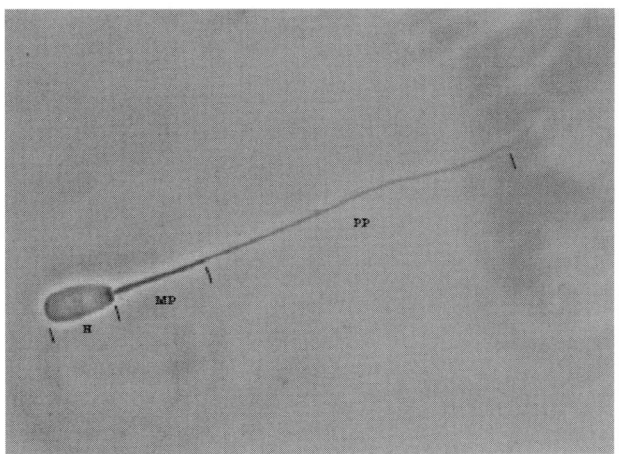

FIGURE 1 Bull sperm showing the sperm head (H) and the midpiece (MP) and the principal piece (PP), which are the major components of the tail.

The second most prominent component of the sperm head is the acrosomal cap. The acrosomal cap surrounds the anterior portion of the sperm nucleus. It is a vesicular structure derived from the Golgi complex and contains a wide variety of enzymes. These enzymes are released at the time of fertilization as a result of fusion between the outer acrosomal membrane and the plasma membrane (acrosome reaction) which lie in close contact to one another.

The plasma membrane of the sperm head can be divided into three regions. The acrosomal region contains receptors for initial interaction with the egg's extracellular coat to signal the sperm to undergo the acrosome reaction. The equatorial segment is a broad band located at the posterior region of the acrosome cap and functions in fusion of the sperm with the egg. The final region of the plasma membrane is the postacrosomal region, which extends from the acrosome to the base of the sperm head.

2. Sperm Tail

The anterior portion of the sperm tail is the connecting piece (neck), which attaches the tail to the head. The connecting piece contains a centriole, which initiates polymerization of the microtubules that form the tail axoneme. The anterior portion of the neck contains the capitulum, which attaches to the basal plate of the sperm head. Surrounding the centriole are the segmented columns. The function of the segmented columns is probably structural; providing a tough, flexible skeleton for the sperm neck.

Posterior to the connecting piece is the midpiece. The midpiece is the region containing the mitochondrial sheath. The mitochondria provide energy to drive sperm motility. The outer dense fibers lie longitudinally beneath the mitochondrial sheath and adjacent to the sperm axoneme and extend from the connecting piece to the end of the tail. The outer dense fibers appear to give the tail an elastic rigidity that is necessary for the recoil of the tail following axoneme-induced bending. The axoneme is in the core of the tail and is composed of nine outer doublets and a single inner doublet of microtubules similar to other flagella. It extends from the connecting piece to the tip of the tail.

At the boundary of the midpiece and the principal piece is the annulus. It forms the distal margin of the mitochondrial sheath. Below the annulus is the principal piece, which is composed of the central axoneme surrounded by seven outer dense fibers with the two remaining outer dense fibers modified to form two longitudinal columns. In addition, the fibrous sheath contains a circumferential sheath which spirals down the length of the principal piece. As with the outer dense fibers of the midpiece, the fibrous sheath provides elastic rigidity to the sperm tail.

B. Oocyte

The unique features of the oocyte, when compared to other cells, are its large size, the presence of an acellular coat surrounding the oocyte, and the outer cortical region of the cytoplasm, which contains a unique type of granule.

1. Oocyte Size

Oocytes are invariably large cells because they must store and accumulate the material necessary for growth and development of the early embryo (Fig. 2). The amount of accumulated yolk, a nutritious material rich in proteins and lipids, determines in part the size of oocytes. Mammalian oocytes contain very small amounts of yolk and they are relatively small in size, ranging from 50 to 250 μm. Conversely,

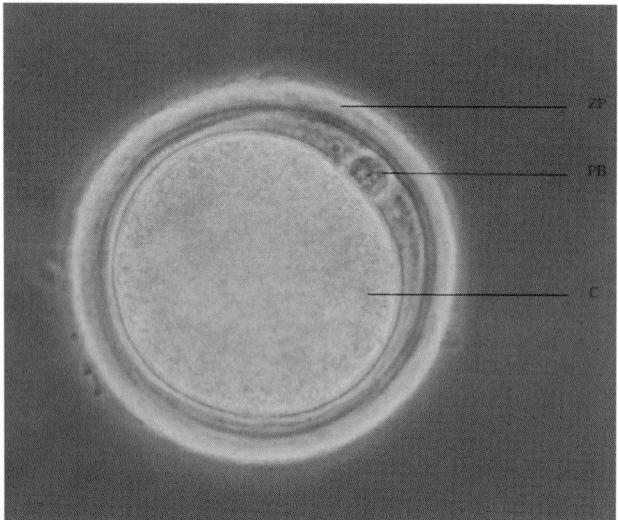

FIGURE 2 Cow oocyte showing the zona pellucida (ZP), the first polar body (PB), and the cytoplasm of the oocyte (C).

frog or hen eggs contain very large amounts of yolk and they are considerably larger—from millimeters to several centimeters, respectively.

2. Extracellular Coat

The oocyte is surrounded by an acellular coat formed by glycoproteins. In mammals this layer is called the zona pellucida, and in many non-mammalian oocytes it is called the vitelline envelope. The role of this coat is probably to keep the cells of the embryo from coming apart until they have developed sufficient junctional attachments to stay adhered together on their own. It is also involved in species-specific recognition by sperm and in blocking further sperm penetration after the first sperm has contacted the egg (block to polyspermy). In addition, many mammalian oocytes are surrounded by a layer of cells—the cumulus cells. In the oviduct, oocytes of some species (rabbits, hares, and mares) acquire a mucoid coat. Oviductal cells also produce a gelatinous material that surrounds frog eggs and, in the hen, they secrete proteins that form the white and shell.

3. Oocyte Cortex

Beneath the plasma membrane, oocytes of mammals, frogs, and fish contain a specialized organelle, the cortical granule. These vesicles contain proteo-

lytic enzymes, mucopolysaccharides, and other undefined components. Cortical granules fuse to the oocyte's plasma membrane upon fertilization and release their contents. In mammals, the contents of the cortical granules modify proteins of the zona pellucida, which then acts as a block to polyspermy.

II. GAMETE PRODUCTION

There are several common features of sperm and oocyte production. Both male and female germ cells begin as "gonia" (spermatogonia and oogonia) which divide mitotically to increase germ cell numbers. They then enter a process of genetic recombination and reduction, called meiosis, that occurs in two unique cell divisions. Upon entry into the first division of meiosis they are referred to as primary spermatocytes and primary oocytes. While progressing through the second division of meiosis they are called secondary spermatocytes and secondary oocytes. Finally, male germ cells that are undergoing the morphological changes to become a mature sperm are called spermatids. Mature sperm and oocytes are produced only following puberty.

A. Sperm

Following puberty, sperm, in contrast to oocytes, are produced continuously for the remainder of the life of most males by a process called spermatogenesis. Spermatogenesis consists of two processes: (i) spermatocytogenesis, which consists of the two meiotic divisions, and (ii) spermiogenesis, which includes the metamorphosis of the round spermatid into the mature, motile sperm cell. These two processes occur in the seminiferous tubules of the testis in direct association with the Sertoli cells.

1. Spermatocytogenesis

Entry of spermatogonia into meiosis is coincident with the movement of the germ cell from the periphery of the seminiferous tubule toward the central lumen. Prior to the first division of meiosis the genetic material undergoes recombination, which is the mixing of genes derived from each of the parents of the male. The second division, which occurs as the cells are moving toward the lumen of the seminiferous tubule, is a reduction division and results in the germ cell containing only half the number of chromosomes normally present in cells.

2. Spermiogenesis

During spermiogenesis, formation of the acrosome, condensation of the sperm head, and elongation of the tail occur simultaneously. The acrosome forms by the association and attachment of an acrosomal vesicle with the nucleus. The acrosomal vesicle arises from the Golgi apparatus and spreads over the surface of the nucleus, forming a caplike structure that covers nearly half of the nucleus, depending on the species. Condensation of the chromatin occurs as the final shape of the sperm nucleus and head take form. Coincident with the association of the acrosomal vesicle with the nucleus, the two centrioles become associated with the opposite pole of the nucleus and begin to polymerize the microtubules that form the axoneme. As the axoneme extends, the proteins that form the outer dense fibers begin to collect along the microtubule doublets. Finally, the mitochondria associate and collect in a spiral organization along the midpiece. The final process of spermiogenesis is the shedding of the excess cell cytoplasm from the sperm and release of the fully formed, but not yet motile, sperm into the lumen of the seminiferous tubule. Final maturation of the sperm takes place as they move out of the testis and through the epididymis to reside in the final segment of the epididymis to await ejaculation.

B. Oocyte

In contrast to the sperm, all oogonia enter meiosis at approximately the same time during fetal development and arrest at a point following genetic recombination but prior to the first meiotic division. From birth, ovaries are populated by a large number of these primary or immature oocytes. These oocytes are surrounded by follicular cells which nurse the oocyte through a substantial growth phase before the oocyte is (i) competent to resume meiosis and (ii) able to support embryonic development following fertilization. Following growth of the oocyte and at-

tainment of puberty in the female, a few of the oocytes receive a signal that causes them to mature, which consists of progression through the first division of meiosis and immediately on to the second. However, the oocyte arrests before the second meiotic division is complete, which occurs only following fertilization.

1. Oocyte Growth

Cells surrounding the oocyte play a significant role in oocyte growth. In some invertebrates, these cells, called nurse cells, contain cytoplasmic bridges through which they transfer yolk to the oocyte. In mammalian oocytes, these cells are called cumulus cells. They provide oocytes with precursors for macromolecular synthesis. Cumulus cells also communicate with the oocyte and relay the female's hormonal signal to the oocyte, initiating maturation.

Oocyte growth is a continuous event that lasts 2 or 3 weeks in mouse oocytes but takes several months in large domestic mammals. Oocyte growth is accompanied by the appearance of cortical granules and the zona pellucida and by a marked change in size and shape of other organelles. These changes, indicative of high metabolic activity, result in the accumulation of RNA and proteins. Among the latter group are likely to be proteins that regulate resumption, progression, and completion of meiosis and initiation of development.

2. Oocyte Maturation

Oocyte maturation is the progression of the oocyte through meiosis to the second meiotic division where it becomes arrested until activated by the sperm. The oocyte acquires this ability only after completing the growth phase and receiving a signal from the female. Resumption of meiosis entails dissolution of the nuclear envelope, characteristically referred to as germinal vesicle breakdown; condensation of chromatin; and extrusion of the first polar body. The polar body is the result of an unequal cell division that leaves most of the cytoplasm with half of the DNA to form the oocyte and serves to dispose of the other half of the DNA in the small polar body. Following the release of the first polar body, oocytes enter the second mitosis of meiosis and are often referred to as eggs. Following maturation the egg is released from the ovary and is transported about half way down the oviduct to wait for a fertilizing sperm.

III. FERTILIZATION AND OOCYTE ACTIVATION

During copulation the sperm are ejaculated into the female reproductive tract. In some species sperm are deposited in the vagina, and in other species they are deposited in the cervix or uterus. The sperm are then transported to the eggs by a combination of contractions in the female reproductive tract and sperm motility. When the sperm reaches the egg, it must first pass between the cumulus cells that surround the egg before it reaches the zona pellucida. The plasma membrane of the sperm attaches to the zona pellucida and this interaction induces the acrosome reaction, exposing the zona pellucida to the inner acrosomal membrane and a collection of enzymes released from the acrosome. The sperm then burrows through the zona pellucida using a combination of enzymatic digestion and mechanical force. Following passage through the zona pellucida, the sperm attaches to the egg plasma membrane by re-

ceptors in the equatorial segment. The sperm and egg membranes fuse and the sperm is engulfed by the egg. Upon entry into the egg the sperm induces a series of increases in intracellular calcium concentration which cause the egg to complete meiosis. With the completion of meiosis, the second polar body is extruded and the two haploid sets of DNA, derived from the sperm and egg, decondense, form pronuclei, and migrate to become tightly opposed to one another in the center of the egg.

Bibliography

Alberts, B., Bray, D., Lewis, J., Raff, M., Roberts, K., and Watson, J. D. (1994). *Molecular Biology of the Cell.* Garland, New York.

Austin, C. R., and Short, R. V. (1976). *Reproduction in Mammals: Book 6: The Evolution of Reproduction.* Cambridge Univ. Press, New York.

Gilbert, S. (1994). *Developmental Biology.* Sinauer, Sunderland, MA.

Knobil, E., Neill, J. D., Greenwald, G. S., Markert, C. L., and Pfaff, D. W. (1994). *The Physiology of Reproduction.* Raven Press, New York.

Moore, K. L. (1977)/ *The Developing Human: Clinically Oriented Embryology.* Saunders, Philadelphia.

Gastrulation

see Embryogenesis, Mammalian

Genetic Counseling

see Prenatal Genetic Screening; Fetal Anomalies

Gene Transfer, Sperm-Mediated

Jagdeece J. Ramsoondar and Jorge A. Piedrahita

Texas A&M University

GLOSSARY

episomal integration "Foreign" DNA that enters the cell but is not integrated into the host genome. The extrachromosomal DNA may replicate independently.

exogenous DNA "Foreign" DNA sequences that are not homologous to the host genome.

gene transfer A collective term for the methods used for carrying new or altered genes into cells.

genome The genetic makeup (the set of all genes and gene signals) of cells and viruses. It represents the total number of chromosomes which is the diploid number ($2n$) in somatic cells and the haploid number (n) in gametes.

germline transmission The passage of transgenes from the transgenic founder animals through their gametes to the offsprings.

recombinant DNA technology The procedures used to insert (clone) DNA segments (genes) of interest into DNA molecules that are able to replicate (cloning vectors) when placed within the environment of a cell that permits replication and expression of the cloned gene.

transgenic animal An individual whose genetic composition (genome) has been artificially modified by the addition or deletion of a specific gene sequence. The transgenic animal called the *founder* may transmit the genetic mutation to its progeny in subsequent generations.

Sperm-mediated gene transfer refers to the artificial process whereby a foreign gene in the form of recombinant DNA is integrated into spermatozoa, which then carry the exogenous DNA into the egg during the process of fertilization.

I. INTRODUCTION

Ideally, the foreign gene becomes stably integrated into the embryonic host genome, expressed appropriately in the adult host tissue, and transmitted through the germline to be inherited in a Mendellian fashion by progeny.

In theory, this is a simple and convenient method for generating transgenic animals as spermatozoa, nature's vectors for carrying foreign DNA into the egg's genome which may be utilized to deliver recombinant DNA. In practice, however, it is difficult to obtain the incorporation of recombinant DNA into the zygote's genome by this method. Furthermore, if exogenous DNA incorporation does occur, the recombinant DNA usually becomes rearranged. Sperm-mediated gene transfer is a controversial method that is currently receiving critical attention.

Available technologies for gene transfer are technically difficult, expensive, and inefficient. Thus, the use of sperm cells as vectors for gene transfer is an attractive possibility. This simple concept was demonstrated about 25 years ago in the rabbit and not followed up, only to be revisited almost 20 years later when it was reported that a 30% frequency of transgenic animals was obtained by the simple preincubation of mouse sperm with naked DNA prior to fertilization. The renewed interest generated from this study stimulated attempts in a number of labs to reproduce these results. The technology for sperm-mediated gene transfer and a brief account of recent progress are presented in the following sections.

II. METHODS FOR THE LOADING OF EXOGENOUS DNA INTO SPERMATOZOA

The production of genetically transformed sperm cells by their association with exogenous DNA is the first and probably the most crucial step in the generation of transgenic animals by sperm-mediated gene transfer. In theory, the exogenous DNA must be internalized within the nucleus of the sperm before it can be stably integrated into the zygote's genome. If not, copies of the recombinant plasmid may be deposited into the egg cytoplasm which may be detected at the early embryonic stages but are not stably incorporated into the chromosomes. Alternatively, foreign DNA internalized in the sperm nucleus may not become integrated in the host's genome and may exist extrachromosomally as episomal DNA. The following methods are utilized to load exogenous DNA into spermatozoa:

A. Spontaneous Binding and Uptake of Exogenous DNA

Incubating mouse epididymal sperm with DNA containing the marker gene chloramphenicol acetyltransferase (CAT) for 15 min before fertilization results in a high frequency (30%) of transgenic animals. These results led to investigations of DNA-binding ability of spermatozoa. Spermatozoa from a variety of species spontaneously bind exogenous DNA, including sea urchin, insect, fish, bird, sheep, goat, cattle, pig, and human. The binding and uptake of DNA is rapid. Radiolabeled DNA can be found in 45% of mouse sperm nuclei by 10 min and in 65% after 2 hr of sperm/DNA coincubation. Exogenous DNA specifically localizes to the posterior region of the sperm head, known as the postacrosomal segment (Fig. 1). Incidentally, the plasma membrane of this region is involved in sperm–egg membrane fusion during fertilization and appears to be easily penetrated by foreign DNA.

About 15–22% of the sperm-bound foreign DNA is internalized within the nucleus and the remainder localized at the periphery of the nucleus between the plasma and nuclear membranes of sperm. Viable sperm take up significant amounts of foreign DNA,

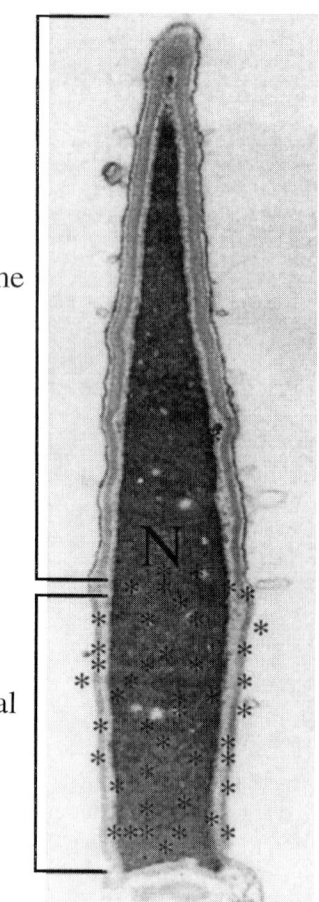

acrosome

post acrosomal region

FIGURE 1 Transmission electron micrograph (provided by Dr. Larry Johnson, Texas A&M University) of a section edge-on through the head of a sperm to illustrate (*) the region of the sperm head (postacrosomal region) that spontaneously binds and internalizes exogenous DNA. The anterior region of the sperm nucleus (N) is covered by the acrosome.

whereas damaged or killed (glutaryaldehyde-fixed) sperm do not take up DNA, and not all live sperm bind exogenous DNA. Greater numbers of mature epididymal sperm bind exogenous DNA compared to that of washed ejaculated spermatozoa. The binding capacity of foreign DNA to sperm is high and a single mouse sperm can interact with as many as 4000 DNA molecules within 30 min.

Evidence from investigations into the molecular basis of DNA binding and internalization by sperm suggest that rather than being a random event, it is an active process mediated by specific protein factors. Three classes of DNA-binding proteins have been identified to date (50, 30–35, and <20 kDa). The 30- to 35-kDa class is conserved among species and is the only one in intact sperm that is accessible to exogenous DNA. Purified 30- to 35-kDa proteins also binds DNA *in vitro*. The protein–DNA interaction is ionic and can be inhibited by nonradioactive competitor DNA or polyanions such as heparin and dextran sulfate. Governed by ionic charge interactions, larger DNA fragments bind more readily than smaller fragments.

Seminal plasma is a potent inhibitor of binding of foreign DNA to spermatozoa. This explains why after extensive washing, ejaculated spermatozoa still bind DNA less efficiently than mature epididymial sperm. A protein factor that blocks the binding of exogenous DNA to sperm is present in seminal plasma. The inhibitory factor specifically binds to the 30- to 35-kDa protein on sperm, preventing its interaction with foreign DNA. This factor is not species specific, has been detected in the seminal plasma of mammals and on sea urchin sperm, and appears to be a 37-kDa glycosaminoglycan. Although the mechanism of nuclear internalization of foreign DNA is not known, preliminary evidence suggests that specific proteins are also involved. This has been hypothesized to be nature's way of ensuring that the genome of animals is not altered by challenges from exogenous DNA.

B. Electroporation of Exogenous DNA into Spermatozoa

Electroporation is widely used for transfecting cells. A high electric field is applied in brief pulses to cell suspensions creating temporary pores in the cell membrane that allow the passage of molecules into the cell. This procedure enhances the entry of exogenous DNA into cells. Electroporation also enhances foreign DNA uptake by sperm cells compared to simple coincubation with DNA. Both the number of cells incorporating DNA and the number of DNA molecules per cell are increased by electroporation.

One adverse consequence of electroporation of sperm with DNA is a decrease in fertility which may be attributable to damage and premature breakdown of the sperm acrosomes and/or reduced sperm motility. The efficiency of transfection by electroporation is dependent on the cell type and the parameters of electroporation. This holds true for spermatozoa. Altering the electric field strength or pulse length does not significantly change DNA uptake by bovine sperm. The opposite is true for fish sperm up to a certain threshold.

C. Liposome-Mediated DNA Uptake by Spermatozoa

Liposomes are artificial fluid-filled vesicles surrounded by a lipid bilayer not unlike that found in cell membranes. They are used both *in vivo* and *in vitro* to take encapsulated compounds into cells. Special liposomes containing cationic lipids have been produced to transfer DNA into sperm cells without compromising their fertilizing ability. The liposome-encapsulated DNA is protected from degradation by enzymes present in seminal plasma and the female reproductive tract. The cationic lipid interacts with and coats the negatively charged DNA molecules. The positively charged outer surface of the cationic liposomes fuses with the negatively charged sperm cell membranes in the postacrosomal region. This is a very efficient method for transfer of foreign DNA into sperm cells. Confocal microscopy revealed that 80% of liposome-treated mouse sperm internalized foreign DNA ranging in size from 1.8 to 14 kb. About five DNA molecules per sperm cell are internalized by this method in poultry. Studies in the mouse, rabbit, and chicken using liposome-encapsulated DNA showed enhanced foreign DNA uptake by spermatozoa.

III. THE TRANSFER OF EXOGENOUS DNA INTO THE EGG BY SPERMATOZOA DURING FERTILIZATION

Following treatment to allow binding and internalization of exogenous DNA, spermatozoa are used to fertilize recently ovulated ova either *in vivo* by artificial insemination or by *in vitro* fertilization. Ideally, the sperm carries the exogenous DNA into the egg which becomes stably integrated into the genome. A transgenic animal is produced which expresses the transgene and passes it down to its progeny in a Mendellian fashion. Currently, the status of sperm-mediated gene transfer is far from ideal.

The original report indicated that mouse spermatozoa transferred a 5-kb plasmid containing the chloramphenicol acetyltransferase (CAT) reporter gene (*pSV2CAT*) into mouse ova *in vitro* and this gene became stably integrated in 30% of the resulting transgenic animals. Southern blot analysis showed that the foreign DNA was inherited by the F1 progeny in a Mendellian fashion, demonstrating that the founder animals were truly transgenic. In addition, high expression of the CAT gene was detected in tissues of F1 individuals. Within the next 2 years eight different labs failed to reproduce these results in mice, some using the exact material as that of the original study. Upwards of 1300 mice were produced using this method; however, none was transgenic. This prompted the conclusion that it is not as easy as first reported to produce transgenic animals by sperm-mediated gene transfer. The most recent results are that exogenous DNA expression could not be detected in mouse embryos fertilized *in vitro* by liposome/DNA transfected epididymal sperm but transfected testicular stem cells did pass the exogenous DNA to developing spermatocytes. Whether these sperm successfully transferred the exogenous DNA into the egg's genome is currently not known.

In other reports, sperm-transferred DNA has been detected in sea urchin embryos, blastocysts of mice using sperm incubated with either naked or liposome-enclosed exogenous DNA, and in bovine blastocysts using sperm transfected by electroporation or simple coincubation. Fertilization of fish eggs with spermatozoa electroporated with exogenous DNA resulted in the development of offsprings containing the transgene. However, the foreign DNA was only detected in DNA preparations from offspring for the next two generations. In chickens, liposome-treated or electroporated spermatozoa transferred the exogenous DNA into the embryo and the foreign DNA was detected in newly hatched offsprings. In all these reports the foreign DNA was not integrated into the embryonic genome but existed extrachromosomally (episomal integration).

Encouraging results were recently obtained from studies of pigs. Southern blot analysis of genomic tail DNA showed that 5 of 82 offspring were transgenic after artificial insemination with sperm loaded by simple incubation with the same plasmid (*pSV2CAT*) DNA used in the original mouse study. However, the plasmid DNA in all five transgenic animals was rearranged. There also appeared to be some sequence selection exerted on the foreign DNA. Certain sequences within the plasmid were more frequently involved in rearrangement. DNA rearrangement appears to be a common feature of sperm-mediated DNA transfer in swine, cattle, mice, and fish. Studies have also shown that certain plasmids were more efficient than others in effecting sperm-mediated foreign DNA transfer. This underscores the need for a better understanding, at the molecular level, of processes involved in sperm-mediated DNA transfer. Despite the problems encountered, recent results are encouraging enough to continue research on this simple and convenient means of producing transgenic animals.

See Also the Following Articles

Cloning; Transgenic Animals

Bibliography

Bachiller, D., Schellander, K., Peli, J., and Ruther, U. (1991). Liposome-mediated DNA uptake by sperm cells. *Mol. Reprod. Dev.* 30, 194–200.

Birnstiel, M. L., and Busslinger, M. (1989). Dangerous laisons: Spermatozoa as natural vectors for foreign DNA? *Cell* 57, 701–702.

Brackett, B. G., Baranska, W., Sawicki, W., and Koprowski, H. (1971). Uptake of heterologous genome by mammalian spermatozoa and its transfer to ova through fertilization. *Proc. Natl. Acad. Sci. USA* 68, 353–357.

Brinster, R. L., Sandgren, E. P., Behringer, R. R., and Palmiter, R. D. (1989). No simple solution for making transgenic mice. *Cell* 59, 239–241.

Francolini, M., Lavitrano, M., Lamia, C. L., French, D., Frati, L., Cotelli, F., and Spadafora, C. (1993). Evidence for nuclear internalization of exogenous DNA into mammalian sperm cells. *Mol. Reprod. Dev.* 34, 133–139.

Kim, J.-H., Jung-Ha, H.-S., Lee, H.-T., and Chung, K.-S. (1997). Development of a positive method for male stem cell-mediated gene transfer in mouse and pig. *Mol. Reprod. Dev.* **46**, 515–526.

Koo, H.-W., Ang, L.-H., Lim, H.-B., and Wong, K.-Y. (1992). Sperm cells as vectors for introducing foreign DNA into zebrafish. *Aquaculture* **107**, 1–19.

Lauria, A., and Gandolfi, F. (1993). Recent advances in sperm cell mediated gene transfer. *Mol. Reprod. Dev.* **36**, 255–257.

Lavitrano, M., Camaioni, A., Fazio, V. M., Dolci, S., Farace, M. G., and Spadafora, C. (1989). Sperm cells as vectors for introducing foreign DNA into eggs: Genetic transformation of mice. *Cell* **57**, 717–723.

Nakanishi, A., and Iritani, A. (1993). Gene transfer in the chicken by sperm-mediated methods. *Mol. Reprod. Dev.* **36**, 258–261.

Niemann, H. (1996). A survey of sperm mediated DNA-transfer in farm animals. *Reprod. Domestic Anim.* **31**, 211–216.

Sin, F. Y. T., Bartley, A. L., Walker, S. P., Sin, I. L., Symonds, J. E., Hawke, L., and Hopkins, C. L. (1993). Gene transfer in chinook salmon (*Oncorhynchus tshawytscha*) by electroporating sperm in the presence of pRSV-lacZ DNA. *Aquaculture* **117**, 57–69.

Sperandio, S., Lulli, V., Bacci, M. L., Forni, M., Maione, B., Spadafora, C., and Lavitrano, M. (1996). Sperm-mediated DNA transfer in bovine and swine species. *Anim. Biotechnol.* **7**, 59–77.

Zani, M., Lavitrano, M., French, D., Lulli, V., Maione, B., Sperandio, S., and Spadafora, C. (1995). The mechanism of binding of exogenous DNA to sperm cells: Factors controlling the DNA uptake. *Exp. Cell Res.* **217**, 57–64.

Genitalia

Mary Min-chin Lee

Massachusetts General Hospital

I. Bipotential Genital Precursors
II. Female Genitalia
III. Male Genitalia

GLOSSARY

mesonephros The second of three sets of excretory organs that develop sequentially during human embryogenesis. The mesonephros function as interim kidneys until the permanent kidneys evolve, then partially regress. The glomeruli and tubules of the mesonephros connect to the mesonephric (Wolffian) duct and contribute to the epididymis and paraepididymis.

Müllerian duct The paramesonephric duct; the anlagen of the female internal reproductive structures.

oocytes Oogonia which have entered meiotic division in the differentiating ovary. This process starts at 11 or 12 weeks gestation and is completed by the end of gestation.

oogonia Mitotically dividing immature germ cells within the follicles of the fetal ovary.

spermatocytes Differentiated spermatogonia which have entered the first meiotic division of the spermatogenic cycle and moved into the adluminal compartment of the tubule.

spermatogonia Mitotically dividing immature germ cells in the testis that reside in the basal compartment of the seminiferous tubule.

Wolffian duct The mesonephric duct; the anlagen of the male internal accessory reproductive organs.

Genitalia is an inclusive term for the organs of the reproductive system, which includes the internal reproductive tract, the external genital structures, and the gonads. The mammalian reproductive system develops in three stages; the initial formation of the bipotential precursors of the genitalia followed by their primary sexual differentiation and subsequent secondary sexual maturation with the attainment of adult reproductive function. Gonadal sex determination and differentiation of the internal and external

Encyclopedia of Reproduction VOLUME 2

genitalia are dependent on the sex chromosome endowment and the expression of the testis-determining gene, *SRY*. The following discussion focuses on development of the genitalia in humans.

I. BIPOTENTIAL GENITAL PRECURSORS

A. Gonads

The coelomic germinal epithelium and the underlying mesenchyme of the mesonephros proliferate to form the undifferentiated gonadal ridges, the primordia of the internal genitalia. Protrusion of the epithelium into the mesenchyme produces the primary sex cords containing the precursors of the somatic cells of the gonad. During early embryogenesis, primordial germ cells in the developing allantois migrate through the yolk sac, along the hindgut, to the gonadal ridges and proliferate in the primary sex cords. The undifferentiated embryonic gonads, consisting of primary sex cords containing primordial germ cells, are initially identical in males and females. Subsequent sexual differentiation of the bipotential gonads is contingent on the genetic sex of the somatic cells, i.e., whether *SRY* and other genes in the sex determination pathway are expressed to induce testicular differentiation.

B. Genital Ducts

The Wolffian and Müllerian urogenital ducts are the anlagen of the male and female internal reproductive tracts, respectively (Fig. 1). The Wolffian (mesonephric) ducts arise first as the longitudinal excretory tubule of the mesonephros. By 6 weeks gestation, the Müllerian (paramesonephric) ducts develop from the invagination of the coelomic epithelium lateral to the Wolffian ducts. Caudally, the Müllerian ducts cross medially over the Wolffian ducts and both pairs of ducts extend to the urogenital sinus. Before the seventh week of gestation, both sets of urogenital ducts consist of straight tubules lined with epithelium that are indistinguishable between males and females. Ensuing primary differentiation of the genital ducts to sex-specific internal reproductive organs

is regulated by the absence or presence of specific hormones produced by the gonads. In males, the secretion of testosterone by the Leydig cells and Müllerian inhibiting substance (MIS) by the Sertoli cells promotes the differentiation of the Wolffian ducts to the male internal accessory organs and the involution of the Müllerian ducts, respectively. Conversely, in females, the Wolffian ducts regress without androgens and the Müllerian ducts autonomously differentiate into the uterus, Fallopian tubes, and upper vagina in the absence of the inhibitory effects of MIS.

C. Undifferentiated External Genitalia

The primordia of the external genitalia are also identical in male and female fetuses during the indifferent or bipotential stage of sexual differentiation (Fig. 1). The undifferentiated genital tubercle at the cranial end of the cloacal membrane elongates to form a phallus. Labioscrotal (genital) swellings and urogenital (urethral) folds develop bilaterally along the cloacal membrane, which subsequently divides into a ventral urogenital and a dorsal anal membrane. The common genital precursors further differentiate as either male external genitalia under the stimulation of androgens secreted by the developing testis or as female structures in the absence of male hormones.

II. FEMALE GENITALIA

A. Ovary

The onset of ovarian organogenesis in females lags behind *SRY* induction of the testicular pathway in males by 2 or 3 weeks; thus, in the 6- to 10-week-old human fetus, the lack of testicular cords is consistent with eventual development of the gonad as an ovary. Ovarian morphogenesis proceeds with the formation of secondary sex cords as the epithelium extends into the underlying mesenchyme to surround the primordial germ cells which proliferate rapidly to peak numbers of 6 or 7 million at 5 months gestational age. The centrally located mesonephric cells push the oogonia to the periphery, creating the pattern of a cortex and medulla. The secondary sex cords break down to form the primordial follicles

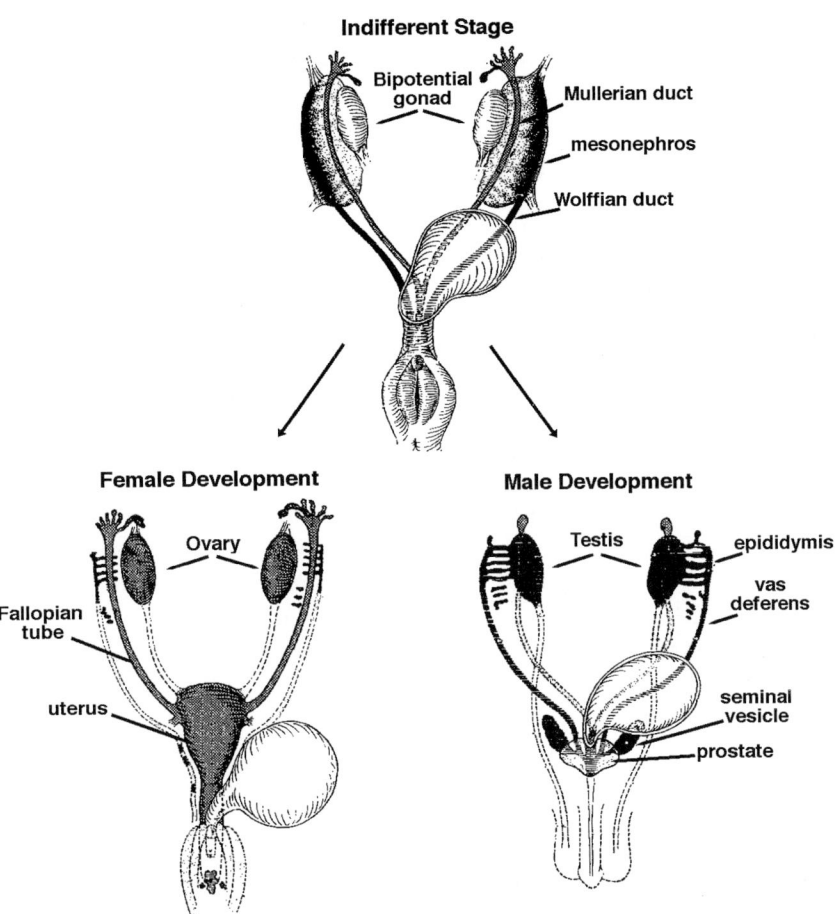

FIGURE 1 Embryonic differentiation of the internal reproductive structures. Both Wolffian and Müllerian urogenital ducts are present in male and female embryos during the indifferent stage of sexual differentiation. Subsequent differentiation of the genital ducts to sex-specific internal reproductive tracts is governed by gonadal determination. In females, the Müllerian ducts differentiate into the female internal reproductive structures in the absence of MIS, whereas in males, the Wolffian ducts differentiate into the male internal reproductive structures under the stimulatory influence of testosterone (adapted with permission from Grumbach and Conte, 1992).

within which the oogonia initiate the first meiotic division to become oocytes. Enclosure of the oogonia within follicles is essential for their survival and further differentiation. The transformation of oogonia to oocytes starts at 11 or 12 weeks gestation and is completed shortly after birth. During the process of folliculogenesis and oogenesis, oocyte loss by degeneration results in a decline in numbers to 1 or 2 million at birth. In the postnatal ovary, the oocytes remain arrested at late prophase of meiosis within primordial follicles until follicular recruitment and growth start at puberty. During each menstrual cycle, several follicles grow and develop, but only one is

selected as the dominant follicle and proceeds to ovulation; the rest undergo atresia. In the sexually mature ovary, the coordinated action of a number of hormones and local growth factors such as follicle-stimulating hormone, luteinizing hormone, inhibin, meiosis-inducing substance, and estradiol, as well as the expression of specific genes on both X chromosomes are necessary to achieve normal oogenesis. The absence or deletion of one X chromosome is associated with apparently normal early migration and proliferation of the germ cells, followed by excessive degeneration and massive loss of oogonia by birth, as typified by patients with 45,X gonadal dysgenesis (Turner's syndrome). More subtle ovulatory defects might result from endocrine dysfunction of the granulosa and theca cells or from disturbances of the hypothalamic–pituitary–gonadal axis.

In nongrowing, primordial follicles, granulosa cells surround the oocyte as a single layer and produce local factors such as oocyte maturation inhibitor and MIS to help maintain the oocyte in meiotic arrest. As recruited follicles grow and develop, the granulosa cells proliferate to form multiple layers around the oocyte, enclosed by layers of thecal cells. The granulosa cells, derived from the coelomic epithelium, and the theca cells, derived from the mesenchyme, are the principal steroidogenic cells of the ovary. Although the fetal ovary has the capacity to synthesize androgens and estrogens, steroidogenesis in the developing ovary is negligible and appears to be nonessential for differentiation of the genitalia. In the mature ovary, estrogen synthesis requires the coordinated actions of both theca and granulosa cells because the various steroidogenic enzymes are differentially localized in the two cell types. The theca cells produce primarily androgens which are aromatized in the granulosa cells to estrogens.

B. Internal Reproductive Structures

The absence of MIS expression in female embryos enables the Müllerian ducts to start differentiating into the female internal reproductive tract structures at 6 or 7 weeks gestation. The Fallopian tubes, or oviducts, evolve from the cranial unfused portions of the Müllerian ducts and acquire the distinct characteristics of the ampulla, infundibulum, and isthmus. Simultaneously, the caudal fused sections of the Müllerian ducts differentiate as the uterus and upper vagina. The corpus and cervix of the uterus first become identifiable and then are distinguishable from the vagina as the vaginal fornices develop. The lower two-thirds of the vagina arises from the urogenital sinus rather than the Müllerian anlagen. The endometrium and myometrium of the uterus evolve from splanchnic mesenchyme and the broad ligaments are developed from the peritoneal folds that are formed when the Müllerian ducts cross over the Wolffian ducts.

In female embryos, the Wolffian ducts degenerate and involute soon after gonadal sex determination in the absence of local androgens. The residual caudal part of the duct becomes the ureteric bud and forms part of the bladder and urethra. The process of female reproductive tract development, i.e., differentiation of the Müllerian structures and regression of the Wolffian ducts, occurs independently of fetal gonadotropins, sex steroids, or gonadal growth factors. The role of maternal estrogens in female sexual differentiation remains to be resolved. The development of abnormal gonads, such as ovotestes, that cause the female embryo to be exposed to androgens and MIS during the critical period of sexual differentiation will virilize the Wolffian duct and inhibit normal formation of the female reproductive structures. Congenital malformations of the uterus, oviducts, and vagina can also arise during morphogenesis of the Müllerian ducts despite normal ovarian development. Moreover, the mesonephros has a critical role in the early formation of the Müllerian structures; thus, structural defects of the reproductive tract are often found in association with renal anomolies.

C. External Genitalia

Feminization of the external genitalia also occurs without requiring any known hormonal stimulation. In the absence of testosterone, the phallus stops elongating and the urogenital folds fail to fuse, except posteriorly where they form the frenulum of the labia minora (Fig. 2). The urogenital folds develop as the labia minora and the labioscrotal folds as the labia majora. The labioscrotal folds fuse posteriorly and anteriorly to form the labial commissures. Infants with gonadal agenesis or early gonadal loss will develop female external genitalia, whereas exposure of the undifferentiated genital anlagen to androgens causes masculinization of the genitalia. The degree of genital ambiguity correlates with the timing and extent of the exposure to androgens. The most common etiology of virilized genitalia in 46,XX girls is congenital adrenal hyperplasia. In this disorder, fetuses are exposed to excessive adrenal androgens secondary to an enzymatic defect in cortisol biosynthesis that results in overproduction of androgens.

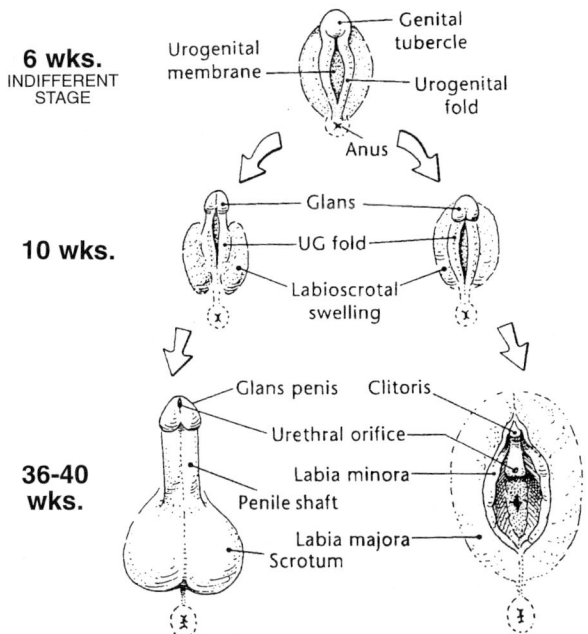

6 wks.
INDIFFERENT
STAGE

Genital tubercle
Urogenital membrane
Urogenital fold
Anus

10 wks.

Glans
UG fold
Labioscrotal swelling

36-40 wks.

Glans penis
Urethral orifice
Penile shaft
Scrotum

Clitoris
Labia minora
Labia majora

FIGURE 2 Differentiation of the external genitalia. The external genitalia originate as identical bipotential anlagen that differentiate as male genitalia in response to androgen secretion by the testis and as female genitalia in the absence of androgens (from Gustafson and Donahoe, 1994, with permission from Annual Reviews, Inc.).

III. MALE GENITALIA

A. Testis

The initial event in testicular determination of the bipotential gonad is the expression of *SRY*, the testis determining gene, followed by differentiation of the Sertoli cells, formation of testicular cords, and vascularization of the testis. Sertoli cells are first identified by their expression of MIS and their localization within the testicular cords, a convoluted network of cells surrounding the germ cells. Any remaining germ cells that are not encompassed within these testicular cords degenerate. The Sertoli cells arise from mesenchymal and/or epithelial cells and proliferate rapidly in the fetus and neonate, then cease dividing at sexual maturity. During embryogenesis, their primary secretory product is MIS, a hormone which has an essential paracrine role in male sexual differentiation. The testicular cords, which are aligned perpendicular to the long axis of the testis,

are enclosed by a basal lamina formed by Sertoli and peritubular cells that partitions the testis into intratubular and interstitial compartments. The fetal Leydig cells differentiate from mesenchymal or mesonephric cells after testicular cords are recognizable. They arise in the interstitial spaces between the testicular cords and have the abundant smooth endoplasmic reticulum and mitochondria with tubular cristae that are characteristic of steroid-producing cells. The testosterone produced by the fetal Leydig cells is essential for masculinization of the Wolffian ducts into the male reproductive structures and virilization of the external genitalia. The fetal Leydig cells proliferate slowly and are steroidogenically active in the fetus and then dedifferentiate and/or degenerate after the neonatal period.

At the onset of secondary sexual maturation, a second population of Leydig cells arises from mesenchymal precursors, proliferates, and gains adult steroidogenic capacity. Conversely, the Sertoli cells become mitotically inactive as they achieve their adult population and acquire differentiated function to supply nutrients to the germ cells and support spermatogenesis. The Sertoli cells establish tight junctions that separate the seminiferous tubule into basal and adluminal compartments and produce a number of products, including androgen-binding protein, proteases, extracellular matrix components, inhibin, and MIS. The earliest signs of pubertal maturation of the seminiferous tubule are the acquisition of a lumen and the onset of spermatogenesis.

B. Internal Reproductive Structures

The growth and differentiation of the Wolffian ducts are dependent on the normal expression and signaling of testosterone through the androgen receptors during a critical period of sexual differentiation at 7–10 weeks gestation. The cranial part of the duct lengthens and becomes convoluted to form the epididymis. The outer muscular layer of the epididymis is derived from mesenchymal cells surrounding the duct. The central section of the Wolffian duct develops as the vas deferens and gains a thick muscular wall originating from the mesenchyme. The caudal Wolffian duct becomes the ejaculatory duct and the seminal vesicle. The section of the mesonephric tubules that is adjacent to the testes becomes the efferent ductules of the epididymis, whereas the cranial end degenerates and the caudal part develops as the paradidymis. The mechanism of Wolffian duct differentiation is not well understood, but the process is stimulated by testosterone and requires the synthesis of prostaglandins and the actions of basic fibroblast growth factor. Abnormalities in either testosterone synthesis or molecular defects of the androgen receptors can cause inadequate virilization of the male reproductive tract structures.

MIS secreted by Sertoli cells promotes involution of the adjacent Müllerian duct in males by signaling through a heteromeric complex of transmembrane serine–threonine kinase receptors. The MIS receptor is expressed in the urogenital ridges of both male and female embryos at the time of sex determination; thus, the sex specificity of Müllerian duct regression is mediated by the expression of MIS. Full regression of the Müllerian duct requires exposure to MIS during a critical period of embryogenesis following gonadal differentiation. The Müllerian duct regresses by condensation of the mesenchyme and degradation of the extracellular matrix of the duct. Molecular defects in either MIS or its receptor lead to persistence of the Müllerian ducts in males who are otherwise normally virilized.

C. External Genitalia

The conversion of testosterone to its more active metabolite, dihydrotestosterone, by 5α-reductase enzyme in genital skin is necessary for masculinization of the external genitalia. During the second and third trimesters, the genital tubercle elongates to form the phallus and brings the urethral folds forward to constitute the walls of the urethral groove (Fig. 2). The penile urethra is created by the fusion of the two urethral folds, whereas the distal urethra is formed by invagination and canalization of the ectodermal cells at the tip of the glans. The body of the penis arises from the mesenchyme. The genital swellings in the inguinal region move caudally and fuse in the midline to form the scrotum. Disorders of androgen synthesis or action lead to undervirilization of the external genitalia and genital ambiguity and to abnormalities in positioning of the urethral meatus (hypospadias).

During mammalian sexual differentiation, gonadal sex determination is the initial step in directing the phenotypic development of the reproductive system. Differentiation of the external genitalia and internal reproductive tracts in females does not require ovarian secretion of hormones or growth factors; therefore, female sexual differentiation progresses normally even if the ovaries are dysgenetic. Normal testicular morphogenesis and adequate Sertoli and Leydig cell function, however, are necessary for differentiation of the male reproductive tract and external genitalia. Thus, molecular defects in either MIS or testosterone or their pathways of action as well as abnormalities in testicular formation will disrupt the development of male sexual characteristics and perturb normal virilization.

Bibliography

Byskov, A. G. (1986). Differentiation of mammalian embryonic gonad. *Physiol. Rev.* **66,** 71–117.

Byskov, A. G., and Hoyer, P. E. (1994). Embryology of mammalian gonads and ducts. In *The Physiology of Reproduction* (E. Knobil and J. D. Neill, Eds.), 2nd ed., pp. 487–540. Raven Press, New York .

Grumbach, M. M., and Conte, F. A. (1992). Disorders of sex differentiation. In *Williams Testbook of Endocrinology* (J. D. Wilson and D. W. Foster, Eds.), 8th ed., pp. 853–952. Saunders, Philadelphia.

Gustafson, M. L., and Donahoe, P. K. (1994). Male sex determination: Current concepts of male sexual differentiation. *Annu. Rev. Med.* **45,** 505–524.

Lee, M. M., and Donahoe, P. K. (1997). The infant with ambiquous genitalia. In *Current Therapy in Endocrinology and Metabolism* (C. W. Bardin, Ed.), 6th ed., pp. 216–223. Mosby/Year Book, St. Louis, MO.

Moore, K. L., and Persaud, T. V. N. (1993). *The Developing Human,* 5th ed. Saunders, Philadelphia.

Schafer A. J., and Goodfellow, P. N. (1996). Sex determination in humans. *Bioessays* **18,** 955–963.

Germ Layers

Jonathan J. H. Pearce

Mount Sinai Hospital, Toronto

GLOSSARY

ectoderm The outer layer, which forms the skin and nervous system; from the Greek words *ektos* (outside) and *derma* (skin).

endoderm The inner layer, which forms the respiratory and gastric systems; from the Greek word *endon* (within).

germ layer A morphologically distinguishable sheet of cells present in the early embryo that will form one of the body's three principal organ systems.

mesoderm The middle layer, which forms the skeletal, muscle, and vascular systems; from the Greek word *mesos* (middle).

Germ layers are the principal subdivisions of the early embryo. They are formed at the time of gastrulation by morphogenetic movements that convert a bilaminar blastula stage embryo into a trilaminar gastrula. Recent genetic and molecular studies have identified factors that are required for the specification of these layers. The nature of these factors and their roles in amphibians, chicks, and mice will be discussed. Once formed, inductive events between the germ layers drive the development of the body organs and appendages. The molecular models of germ layer formation in vertebrates have allowed a reappraisal of the extent of homology between these layers and those found in other animals.

I. A BRIEF HISTORY OF GERM LAYERS

In the eighteenth century, the prevailing models of animal development postulated that a miniature version of the animal was present in the fertilized egg and that development was simply the growth of this preformed body. Working at this time, Caspar Friedrich Wolff described the development of flower organs from simple common progenitors. To establish whether his findings in plants had broader application, Wolff performed a study of the early development of the chick. He ascertained that systems as diverse as the gastric, nervous, and vascular all arise from visibly similar and simpler precursors, thereby disproving the preformation theory.

The nature of these organ precursors was investigated further by Christian Heinrich Pander. In 1817, he published the first description of separate layers in an embryo. Pander proposed that there were two primary layers and that a third layer formed between these. A colleague of Pander's, Karl Ernst von Baer, undertook a careful analysis of the metamorphosis of these layers into organs and described similar layers in the embryos of other vertebrates. von Baer called the upper of the two primary layers the animal and the lower the vegetative, because he believed that it was this layer that gave rise to the gut and vasculature. These terms are still used to describe the two different poles of an egg, the animal and vegetal. The common names for the three germ layers—the ectoderm, mesoderm, and endoderm— were proposed by Robert Remak in the midnineteenth century. The names reflect the relative positions of the germ layers in the embryo: *ecto* (outer), *meso* (middle), and *endo* (inner).

II. THE FORMATION OF THE GERM LAYERS

The views of the pioneers of germ layer theory differed significantly on the origins of the layers. A resolution of these differences had to await the refinement of techniques that allowed investigators to establish the developmental fates of subdomains of the early embryo. When a permanent stain is applied to a small area of an embryo, the fate of this region can be assessed by analyzing the distribution of the stain later in development. By collating the data of many such markings, the origin of later formed structures, such as the germ layers, can be mapped out on the surface of the embryo. The first maps of embryonic fate were produced by the application of vital dyes or carbon particles; more modern protocols use florescent carbocyanine dyes. The early strategies all marked clumps of cells. By marking individual cells, the resolution of the maps is increased and, importantly, the maps also show at what stage cells become committed to individual structures.

Fate mapping has shown that the three definitive germ layers are laid down at gastrulation. Prior to gastrulation, individual cells can contribute progeny to more than one germ layer but during gastrulation cells become committed to a single layer. A notable exception to this is the neural crest, an ectodermal derivative that contributes to mesodermal structures.

The start of vertebrate gastrulation is generally marked by cells leaving the primary ectodermal layer. The departing cells travel below the ectoderm and form both the endodermal and mesodermal germ layers (Fig. 1). The zone of transition is called the blastopore lip in amphibians, the germ ring in teleosts, and the primitive streak in amniotes. The first cells to leave the ectoderm contribute to both lower layers, whereas later departing cells form just mesoderm. The complex morphogenetic movements of gastrulation simultaneously form the major body axes and the definitive germ layers. Although vertebrates differ significantly in the nature of these movements, the similarity of the fate maps of pregastrula embryos suggests that a common mechanism acts to specify the germ layers in all vertebrates (Fig. 2).

III. THE SPECIFICATION OF THE GERM LAYERS

A. Amphibians

Due to its large size and accessibility, the development of the amphibian egg has been extensively studied. Seminal work by Spemann and Mangold identified a region of the blastopore lip, the so-called Spemann organizer, that is able to instigate the formation of an entire duplicate body axis when transplanted to an ectopic site on a host blastopore. As the organizer coopts host tissues into the secondary axis, it is believed to emit inductive signals.

The success of Spemann's investigations led to the identification of other inductive sources in the early amphibian embryo. Nieuwkoop demonstrated that a signal emanating from the vegetal endoderm is able to induce mesoderm formation in animal caps, a tissue normally fated to form ectoderm. Using animal cap inductions as an assay, screens have identified a number of mesoderm-inducing factors. The first identified factor was activin A, a member of the transforming growth factor-β (TGF-β) superfamily. A second factor, basic fibroblast growth factor (bFGF), was reported shortly after.

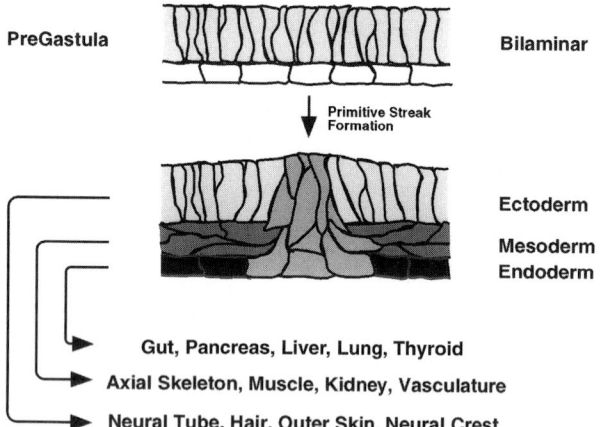

FIGURE 1 Primitive streak formation. In gastrulating chick and mouse embryos, ectodermal cells undergo an epithelial to mesenchymal transformation and migrate below the ectoderm to form both the mesoderm and the endoderm. The definitive germ layers then form the body's appendages and organs.

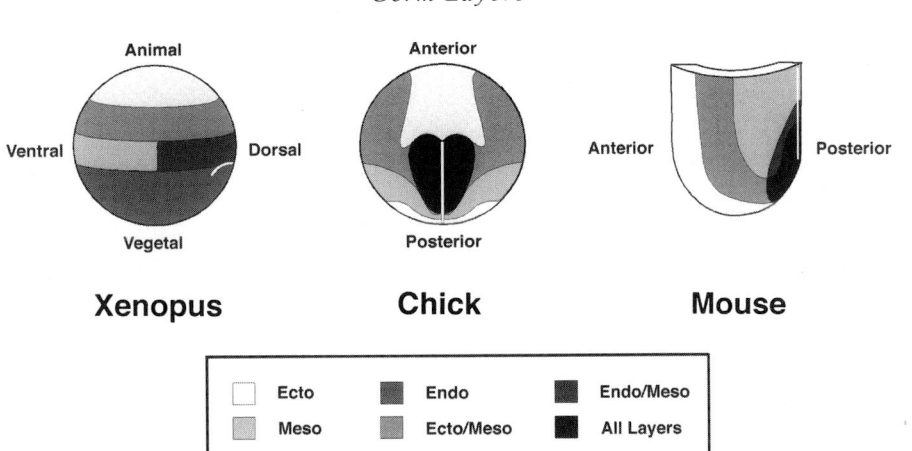

FIGURE 2 Pregastrula fate maps. These maps show which regions of the early embryo will contribute to each of the germ layers. The similarity in the distribution of the regions relative to the blastopore lip (in *Xenopus*) and the primitive streak (in the chick and mouse) implies that the germ layers are specified in similar ways in all these animals. White lines demarcate the future positions of the blastopore lip and primitive streak.

Factors identified by induction screens may be endogenous signals or they may merely mimic the effects of those signals. To differentiate between these possibilities, a standard set of tests is applied. First, a true candidate will be expressed in an active form at the right time and in the right place. Second, it will be sufficient for the induction, and third, often the hardest to prove, it will be necessary for the induction. Both activin A and bFGF fulfill the second requirement. However, they both fail the first. Vegetal endoderm dissected from *Xenopus* embryos prior to zygotic gene activation is capable of inducing mesoderm in animal caps. The endogenous factor or factors are therefore maternally supplied and present in the early embryo. There is, however, no detectable activin A mRNA in the early embryo. There are very low levels of protein but it is unknown whether these levels can induce mesoderm. Both bFGF mRNA and protein are found in the early embryo but the protein product, unusually for an FGF, is not secreted. It has been proposed that bFGF might act during wound repair because it could be released from damaged cells.

Although neither activin A nor bFGF fulfill all the necessary criteria, close relatives of these genes may. Screens of the FGF family have identified genes that are active during endogenous mesoderm induction, such as eFGF. Likewise, other members of the TGF-β superfamily, the bone morphogentic proteins (BMP) -2, -4, and -7, can all induce mesoderm and are present at the right time. The mesoderm induced by the BMPs is of a ventral type: blood. In contrast, another member of the TGF-β superfamily, Vg1, is capable, in special circumstances, of inducing dorsal mesoderm: muscle and notochord. *Vg1* mRNA is expressed during oogenesis and is localized to the vegetal cortex. The coincidence of *Vg1* mRNA localization and mesoderm-forming activity has long made Vg1 a favored candidate for an endogenous signal. However, numerous studies have failed to show that Vg1 protein is processed to an active form. The demonstrations of Vg1 activity have relied on the production of processable chimeric BMP–Vg1 proteins. Nevertheless, a model in which BMPs specify ventral mesodermal fates and Vg1 specifies dorsal fates remains attractive.

The third and most stringent test of inducing genes, the necessity criteria, is commonly performed by mutational analysis. Amphibians, due to their complex genetics and long life cycles, are not well suited for this. An alternative strategy has therefore been adopted. A construct, capable of suppressing the activity of the candidate gene, is designed and injected in excess. These so-called dominant-negative constructs have been used to test the necessity of both TGF-β and FGF signaling. Instead of targeting individual ligands, most dominant-negative tests have inactivated the ligand's receptor. It was believed that, by targeting specific receptors, the role of particular ligands could be addressed. Further analysis has found, though, that many of the receptors targeted in this fashion are more promiscuous than expected. Rather than disrupting signaling by individual ligands, these strategies have tended to disrupt signaling by whole families of ligands. These experiments, however, are not without merit because they have elucidated which signaling pathways, if not which specific ligands, are required. When a dominant-negative FGF receptor-1, XFD, is introduced into the early *Xenopus* embryo, mesoderm formation, as revealed by the expression of the mesoderm-specific *Xbrachyury* (*Xbra*) gene, is initiated but not maintained. In contrast, a dominant-negative activin receptor-IIB, XAR1, is able to block mesoderm formation. Because the dominant-negative activin receptor is able to interfere with at least activin and Vg1 signaling, this result does not distinguish which, if either, of these ligands is necessary but it implies that TGF-β signaling is critical for mesoderm formation.

In addition to its roles in mesodermal patterning, TGF-β signaling has been implicated in endoderm formation. When XAR1 is injected into the vegetal pole, the expression of the endoderm-specific marker *XlHbox-8* is blocked and mesodermal and ectodermal markers are activated. In accord with this, activated Vg1 is able to induce the expression of *XlHbox-8* in animal caps.

An additional and surprising finding of the XAR1 studies was that animal caps expressing XAR1 express neural markers at the expense of epidermal markers. This helped to explain an earlier observation that dissociated *Xenopus* embryos fail to turn on mesodermal markers but do activate neural markers. These data suggest that neurectoderm is the embryonic default state and that TGF-β signaling is able to alter this to more ventral and mesodermal fates.

TGF-β signaling may then globally pattern the early *Xenopus* embryo. Neurectoderm, the default state, is formed in the absence of signaling. Mesoderm is specified at low levels and endoderm at higher levels of signal. This model is likely incomplete. FGFs may still play a role because a dominant-negative FGF receptor blocks *XlHbox-8* expression and FGFs induce endodermal markers in animal caps.

To exert their effects, inductive signals are transduced from cell surface receptors, via cytoplasmic intermediates, to transcriptional regulators. Genetic studies have identified genes that are required for TGF-β and FGF signaling, such as the *smad* and *ras* genes. The disruption of these necessary components in *Xenopus* embryos has effects on germ layer specification that are similar to those caused by the disruption of the upstream ligands and receptors. However, only those components that are known to have functions specific to germ layer patterning will be discussed.

Recent studies have identified a number of transcription factors involved in germ layer specification. Some of these are the nuclear mediators of the signal transduction pathways, whereas others are the targets of these pathways. The target genes are distinguishable from more downstream genes because they alone are induced in the presence of cycloheximide, an inhibitor of protein synthesis. One such gene, *Mix.2*, a homeobox transcription factor, is expressed broadly in the vegetal and equatorial zones where, based on its homology with *Mix.1*, it may play a role in ventral mesoderm specification. *Mix.2* is induced by activin and other members of the TGF-β family but not by the FGFs. Activin induction is mediated by an activin responsive element (ARE) in the *Mix.2* promoter. It has recently been shown that a winged helix transcription factor, forkhead activin signal transducer-1 (FAST-1), binds to the ARE as part of a complex. Supershift analysis has found that this complex also contains the XMAD2 and XMAD4 proteins, two necessary components of activin signaling. The XMAD/FAST-1 complex appears to deftly link signal transduction and gene activation. AREs have been identified in two other activin target genes, *goosecoid* (*gsc*) and *XFD-1'*. However, since none of the known AREs are homologous, activin, or activin-like, signaling is likely to be transduced in a number of ways.

Brachyury is the founding member of a family of genes, the *T-box* (*Tbx*) family, that play important roles in mesendodermal patterning. *Xbra*, the *Xenopus* ortholog, is an activin-inducible target gene that, when expressed in animal cap cells, causes mesodermal differentiation. Work from a number of groups has led to a model of *Xbra* regulation in which expression is initiated by activin-like signals and maintained by a regulatory loop linking expression and FGF signaling.

Two newly identified members of the *Xenopus Tbx* family, *Xombi/Brat* and *Eomesodermin* (*Eomes*), can, like *Xbra*, induce mesoderm formation in animal caps. *Xombi* has additionally been shown to induce the endodermal marker *XlHbox-8*. *Xombi* is expressed from oogenesis through gastrulation; *Eomes* expression is not seen until the start of gastrulation (stage 10); and *Xbra* is not activated until stages 10 or 11. Because both *Eomes* and *Xbra* induction are cycloheximide independent, it is unlikely that the temporal cascade reflects a regulatory cascade, although ectopic expression of *Eomes* in animal caps can induce *Xbra* but not visa versa. For the same reasons that dominant-negative approaches were adopted for assessing the function of the inducing signals, the effects of disrupting *Tbx* gene function have been studied, not by genetics, but by the creation of blocking molecules. By replacing the transcription activation domain of a Tbx protein with the repressing domain of the *Drosophila* engrailed protein (En^R) and injecting this into early embryos, it is possible to block *Tbx* function *in vivo*. This has been done for all three Tbx proteins. Xombi-En^R blocks the expression of both *Xbra* and *gsc*, suggesting that it is required panmesodermally. Eomes-En^R causes an arrest during gastrulation and a marked reduction in the expression of the mesoderm marker muscle actin. Neural markers, on the other hand, were either unchanged or increased, consistent with the proposal that neurectodermal fate is an inductive signaling independent default state. Xbra-En^R has the weakest effect. Embryos arrest during gastrulation, at a later stage than those injected with Eomes-En^R. Anterior muscle and notochord are formed but posterior structures are missing. Because they fulfill the three criteria of expression, sufficiency, and requirement, the *Tbx* genes appear to play a central role in the specification of mesendodermal fate. It will be inter-

esting to determine how these genes are controlled by the TGF-β signaling pathway.

Amphibian studies have identified a number of genes that are required for correct germ layer specification. These genes define signaling pathways and gene families, orthologs of which have been sought in other vertebrates.

B. Chick

Mesoderm induction in the chick is initiated just prior to overt primitive streak formation, in the posterior marginal zone (PMZ) at about stage XIII. Chicken *Vg1* (*CVg1*) is expressed from stage X in the PMZ and throughout the nascent primitive streak. Although CVg1 can be processed by *Xenopus* oocytes, CVg1, like its *Xenopus* otholog, is not secreted from tissue culture cells. Misexpression studies have therefore used a secretable dorsalin–CVg1 chimeric protein. This protein can induce mesoderm in both *Xenopus* animal caps and chick extraembryonic epiblast. In addition, when misexpressed in the marginal zone, the dorsalin–CVg1 protein can induce an ectopic primitive streak. CVg1 is therefore a likely PMZ signal that directs mesoderm and primitive streak formation.

Chicken activin, like CVg1, is able to induce mesoderm and ectopic primitive streaks. Activin B is present from stage XIII but the two known activin type II receptors are not expressed until stages 3 or 4, after primitive streak formation. Therefore, if activin is an endogenous chicken mesoderm inducer, it must act through other, as yet unidentified, type II receptors.

The T box family members *Ch-T*, *Ch-TbxT*, and *Ch-Tbx6L* are all expressed during mesendoderm formation in the chick. *Ch-Tbx6L* is the earliest activated of these, being expressed in the PMZ from stage X. *Ch-T* expression is initiated at stage XIII in the PMZ epiblast and in a domain just anterior to this. *Ch-TbxT* is not expressed until after primitive streak formation. *Ch-Tbx6L* and *Ch-T* may therefore, as in *Xenopus*, be direct targets of mesoderm-inducing factors. Both genes are rapidly activated in dissociated blastoderm cells by either FGF4 or activin but surprisingly they are both cycloheximide sensitive. It is not known whether the difference in cycloheximide sensitivity between the chick and *Xenopus* reflects a true species difference or is due to dissimilar experimental designs.

C. Mouse

Due to its small size and inaccessibility, manipulating the mouse embryo is difficult. The mouse genome, though, in contrast to those of amphibians and chicks, can be readily altered. A large collection of spontaneous mutants has recently been augmented by mutants from reverse genetic screens. By selecting rare homologous recombination events in mouse embryonic stem cells, a gene can be disrupted at will. This power has been utilized to study the role of mouse orthologs of the *Xenopus* genes.

Although neither the activin-βA or -βB subunits are expressed in the gastrulating mouse embryo, activinβA is expressed in the surrounding decidua. Mesoderm formation, however, is unaffected by the disruption of either subunit alone or in combination. Similarly, no early defect in mesoderm induction is observed in mice null for the activin type IIB receptor. In contrast, a proportion of embryos lacking BMP4 and all those missing the BMP receptor-I fail to initiate primitive streak, and hence mesendoderm, formation. The interpretation of defects in early mouse development is complicated by the fact that the mouse embryo must attain a minimum size before it can gastrulate. Because BMP receptor-I embryos are smaller than littermates and this defect is apparent some time prior to gastrulation, it is possible that primitive streak failure is a secondary consequence of reduced epiblast proliferation.

An alteration in the founding member of the *Tbx* gene family, *Brachyury*, is responsible for the classical mouse mutant *T*. In homozygous *T* embryos, both the posterior mesoderm and notochord are either absent or disorganized. The similarity between these defects and those observed in *Xenopus* Xbra-En^R embryos is striking. In addition to *Brachyury*, there are seven other known murine *Tbx* genes. One of these is expressed in the endoderm and at least five others are expressed in different mesodermal populations of the primitive streak. Given their known expression patterns and functions, it seems likely that the *Tbx*

gene family plays a conserved role in the establishment of vertebrate mesendodermal patterning.

Due in part to the lack of appropriate markers, the formation of the endoderm is less well characterized than that of the mesoderm. One gene, though, is known to be critical for endoderm formation. *Hepatic nuclear factor 3β (HNF-3β)*, a winged helix transcription factor, is expressed in the anterior primitive streak, from its inception. Later, *HNF-3β* is expressed in the gut endoderm and the midline structures, the prechordal mesendoderm, and the notochord and the floor plate. Fate maps show that all these tissues are derived from the anterior primitive streak. In homozygous *HNF-3$\beta^{-/-}$* embryos, both the midline structures and the anterior endoderm are missing. *HNF-3β* orthologs, expressed in similar patterns to the mouse gene, have been identified in *Xenopus* and chicks. The role of these genes in endoderm formation is unknown, however.

The use of a variety of model organisms, each with their own particular strengths, has aided the identification and analysis of genes important for germ layer specification. The TGF-β superfamily members Vg1, BMP-2, -4, and -7, along with an as yet unidentified activin-like signal, are able to induce mesoderm and endoderm from a neurectodermal default state. These signals activate target genes that include the *Tbx* family, *Mix.2, gsc,* and *HNF-3β*. The target genes in turn control batteries of downstream genes that specify the different behaviors of the germ layers.

To complement the approaches described here, large-scale mutagenesis screens of the zebrafish have recently been undertaken. These screens have already isolated a number of mutants, such as one-eyed pinhead, a mutant that fails to form prechordal mesoderm and endoderm, that will undoubtedly enhance our understanding of germ layer specification.

IV. ORGANOGENESIS

Once formed, the germ layers act as the basic building blocks of future development. Some processes occur within germ layers, such as the formation of somites, but many involve interactions between layers. These interactions drive the formation of appendages and organs. A signal emanating from the mesoderm, perhaps FGF-8, is believed to be responsible for the induction of the limb apical ectodermal ridge (AER). Limb outgrowth and patterning then depends on interactions between the AER and the underlying mesoderm. Similarly, a signal from the endoderm, perhaps established by the notochord, determines the position of the pancreas. As in the limb, an interactive loop between the germ layers is then required for continued pancreatic development.

V. BEYOND VERTEBRATES

The identification of genes necessary for vertebrate germ layer specification has allowed a reevaluation of the homology between these layers and those of other animals. Homology is not simply a similarity of form but rather a similarity due to common descent.

The discovery that a common set of genes is involved in the specification of the germ layers of all vertebrates suggests that their common ancestor used the same set of genes during its own development. The ascidians and amphioxus are extant protochordates and both possess *HNF-3β* and *Tbx* genes. These genes, like their vertebrate counterparts, are expressed in the developing germ layers. Antisense procedures have demonstrated a role for the *HNF-3β* homolog *MocuFH1* in ascidian gastrulation, and bFGF, but not activin, can induce the expression of the ascidian T homolog, *As-T*.

Because it seems probable that *HNF-3β* and the *Tbx* family were present in the common vertebrate ancestor, perhaps they played similar roles even earlier in evolution. To address this, a much more distantly related insect, *Drosophila*, has been examined. *Drosophila* also has *HNF-3β* and *Brachyury* homologs, the genes *forkhead* and *Brachyentron*. Notably, mutations in the *forkhead* gene result in a loss of endoderm structures, suggesting that an *HNF-3β/forkhead* gene may have acted to specify endoderm in the common ancestor of insects and vertebrates. *Brachyentron* mutants have defects in hindgut and anal pad formation. Whether these findings suggest that the vertebrate notochord and the insect gut arose from a common precursor or whether these organisms use the same gene for different purposes is unknown.

Once formed, a common set of genes pattern the mesoderm of *Drosophila* and vertebrates. BMPs specify dorsal–ventral differences in both groups and the patterning of the heart and muscles is under the control of homologous genes, the *Nkx/tinman* and *MyoD/nautilus* genes, respectively. These findings suggest that the mesoderm of insects and vertebrates is indeed homologous. However, there is no strong evidence that these layers are specified in similar ways. *Drosophila* mesoderm is specified by the *dorsal* gene, a gene whose closest vertebrate homolog, NF-κB, is involved in the immune response. That *Drosophila* and vertebrates differ in specification but not patterning is not without precedent. Both *Drosophila* and vertebrates use a complex of genes, the *Hox* complex, to pattern their anterioposterior axes. However, they use different pathways to activate these complexes. These examples suggest that diverse animals share common developmental programs, such as germ layers, inherited from a common ancestor, but that they activate these in different ways.

Bibliography

Heasman, J. (1997). Patterning the Xenopus blastula. *Development* **124**, 4179–4191.

Hertwig, O. (1905). *Text-Book of the Embryology of Man and Mammals*. Macmillan, New York.

Holley, S. A., Jackson, P. D., Sasai, Y., Lu, B., De Robertis, E. M., Hoffmann, F. M., and Ferguson, E. L. (1995). A conserved system for dorsal–ventral patterning in insects and vertebrates involving sog and chordin. *Nature* **376**, 249–253.

Krumlauf, R. (1992). Evolution of the vertebrate Hox homeobox genes. *Bioessays* **14**, 245–252.

Lawson, K. A., and Pedersen, R. A. (1992). Clonal analysis of cell fate during gastrulation and early neurulation in the mouse. In *Postimplantation Development in the Mouse*, Ciba Foundation Symposium No. 165. Wiley, Chichester, UK.

Smith, J. (1997). Brachyury and the T-box genes. *Curr. Opin. Genet. Dev.* **7**, 474–480.

Stern, C. D. (1992). Mesoderm induction and development of the embryonic axis in amniotes. *Trends Genet.*. **8**, 158–163.

Tam, P. P. L., and Behringer, R. R. (1997). Mouse gastrulation: The formation of a mammalian body plan. *Mech. Dev.* **68**, 3–25.

Tam, P. P. L., and Quinlan, G. A. (1996). Mapping vertebrate embryos. *Curr. Biol.* **6**, 104–106.

Global Zones and Reproduction

F. H. Bronson
University of Texas at Austin

I. Climate and the Global Distribution of Biomes
II. The Major Biomes
III. Global Patterns of Human Reproduction

GLOSSARY

biome A region of the earth having a distinct form of vegetation, e.g., a tropical forest.

food chain The feeding pattern of various groups of organisms in a given habitat, e.g., plants are consumed by plant-eating mammals, which in turn are eaten by other, carnivorous mammals.

photoperiod The period of time within a day in which an organism is exposed to light, varying with the season of the year.

photosensitive or *photoresponsive* Having patterns of reproductive behavior that are affected by changes in the amount of available daylight.

seasonality The fact of varying with the season of the year; specifically, the characteristic in some species of having reproductive behavior that varies according to season, such as by not reproducing in winter.

taiga A region of the Northern Hemisphere, below the arctic tundra and above temperate forests, characterized by a layer of tall evergreen trees and minimal growth of shrubs and ground cover.

trophic level One aspect of a food chain, e.g., large predatory mammals such as the wolf and wolverine can be described as being at the same trophic level.

Mammals survive and reproduce successfully in environments as diverse as equatorial rain forests, arctic tundra, hot deserts, and the upper strata of deep oceans. Most of these habitats are characterized by strong seasonal change, including those in the tropics. Thus, the total amount of environmental variation exploited by mammals is immense. This article discusses the strategies that allow them to reproduce under such a broad range of conditions. On a global scale, these strategies are probably best visualized within the context of the world's major biomes. A biome is defined as a geographical region characterized by a particular kind of vegetation. It is a direct reflection of the region's climate and it characterizes the base of the region's food chains. For purposes here, it is reasonable to consider the marine environment as a unit and divide the rest of the world into eight terrestrial biomes: arctic tundra, taiga, temperate forests, temperate grasslands, desert/semidesert, savanna/thorn forest, tropical forests, and Mediterranean-type shrublands.

I. CLIMATE AND THE GLOBAL DISTRIBUTION OF BIOMES

Climate is determined in large part by two factors: the intensity of the sun's rays and the pattern of global air circulation. The sun's rays are more spread out at the higher latitudes and the only biome above 75° of latitude is arctic tundra, defined as a collection of plants adapted to live over permafrost. Sunlight is more intense in the equatorial region, making it warmer, which in turn causes the air to rise and move toward the poles. The curvature of the earth and its rotation on its axis interact with this flow of air to produce three great circulation cells in each of the Northern and Southern Hemispheres at 30° intervals. Thus, air rises at 60° as well as at the equator, making these regions generally rainy, and it falls at 30° and at the poles, making these regions much drier. As a result, there is a broad belt of coniferous forest (taiga) in North America, Europe, and Asia between 60° and 70° of latitude, another band of forests of one kind or another in the equatorial regions of South America, Africa, and Asia, and many of the world's great deserts are located around 30° of latitude.

A variety of factors produce variation in these general patterns. Wind direction varies in relation to the six circulation cells and this in turn interacts with ocean currents of different temperatures, the shape of land masses, and the presence of mountain ranges to produce great regional variation in rainfall on the continents. Thus, desert/semidesert, grassland, forest, or Mediterranean-type shrubland can be found scattered throughout the temperate zone, depending on the amount and seasonal pattern of precipitation.

The earth rotates around the sun with its axis slanted 23.5° in relation to the sun. This results in an annual cycle of temperature at the higher latitudes and seasonal variation in rainfall in many parts of the tropics. Deciduous forest of one kind or another is found in the tropics where rainfall is seasonal and relatively high; savanna/thorn forest is found where rainfall is lower but still adequate for grass and shrubs, whether seasonal or not. Tropical rain forest occurs only where rainfall is both high and nonseasonal.

THE MAJOR BIOMES

A. Arctic Tundra

This is the most seasonal of the biomes since the sun never sets in midsummer and never rises in midwinter. Thus, photosynthesis and hence ecosystem productivity is continuous during the summer and absent during the winter. At the base of food chains is a collection of plants that are adapted to live in moist conditions over ice, mostly sedges, heaths, mosses, and willows.

The mammalian herbivores of the arctic tundra include rodents, such as lemmings (e.g., *Lemmus* sp.) and the arctic hare (*Lepus arcticus*), and a few large ungulates, such as the musk ox (*Ovibos moschatus*) and caribou or reindeer (*Rangifer tarandus*). At the next higher trophic level are several carnivores, including some of the weasels (*Mustela* sp.), foxes (*Alopex* and *Vulpes* sp.), and the gray wolf (*Canis lupus*).

When food availability permits, there are two basic patterns of reproduction among the tundra mammals: continuous breeding, even under the snow, and rigid seasonality in which one litter of offspring is produced each year.

Many lemmings and weasels are insensitive to variation in photoperiod. They have the capacity to reproduce throughout the winter, energetic conditions permitting. Dramatic 3- or 4-year cycles of population density, whose cause is poorly understood, are seen in some lemming populations. Weasel reproduction is heavily dependent on the number of lemmings. Thus, with a delay of several months, the number of weasels closely tracks the lemming cycles.

Rigid seasonality in the production of litters characterizes the reproduction of the hares, ungulates, and carnivores, all of which are monoestrous. Births are timed to take advantage of the short, explosively productive arctic summer, and undoubtedly regulated routinely by photoperiod.

There is no equivalent of the arctic tundra in the Southern Hemisphere (Fig. 1).

B. Taiga

South of the arctic tundra lies a broad band of coniferous forest known either as taiga or as northern boreal forest. Trees with water-conserving, waxy, needle-like leaves, such as the spruces and firs, are the dominant plants of this biome. Little light reaches the ground in thick conifer groves even at midsummer; thus, there is little ground vegetation and essentially no mammalian food chain. Where the forest is broken open by lakes, bogs, streams, or human interference, pines, alders, aspens, and mammals all become more common.

Rodents such as voles (*Microtus* sp.) and the red squirrel (*Tamiasciurus hudsonicus*) are the most common herbivores of the taiga. Also feeding at this trophic level are a variety of hares (*Lepus* sp) and the ubiquitous *Alces alces*, called the moose in North America and the elk in Europe and Asia. At the next highest trophic level are a variety of insectivores such as the widespread hedgehog (*Erinaceus* sp) and short-tailed shrews (*Sorex* sp). The carnivores include mustelids such as the wolverine (*Gulo gulo*) and sable (*Martes zibellina*), canids such as some of

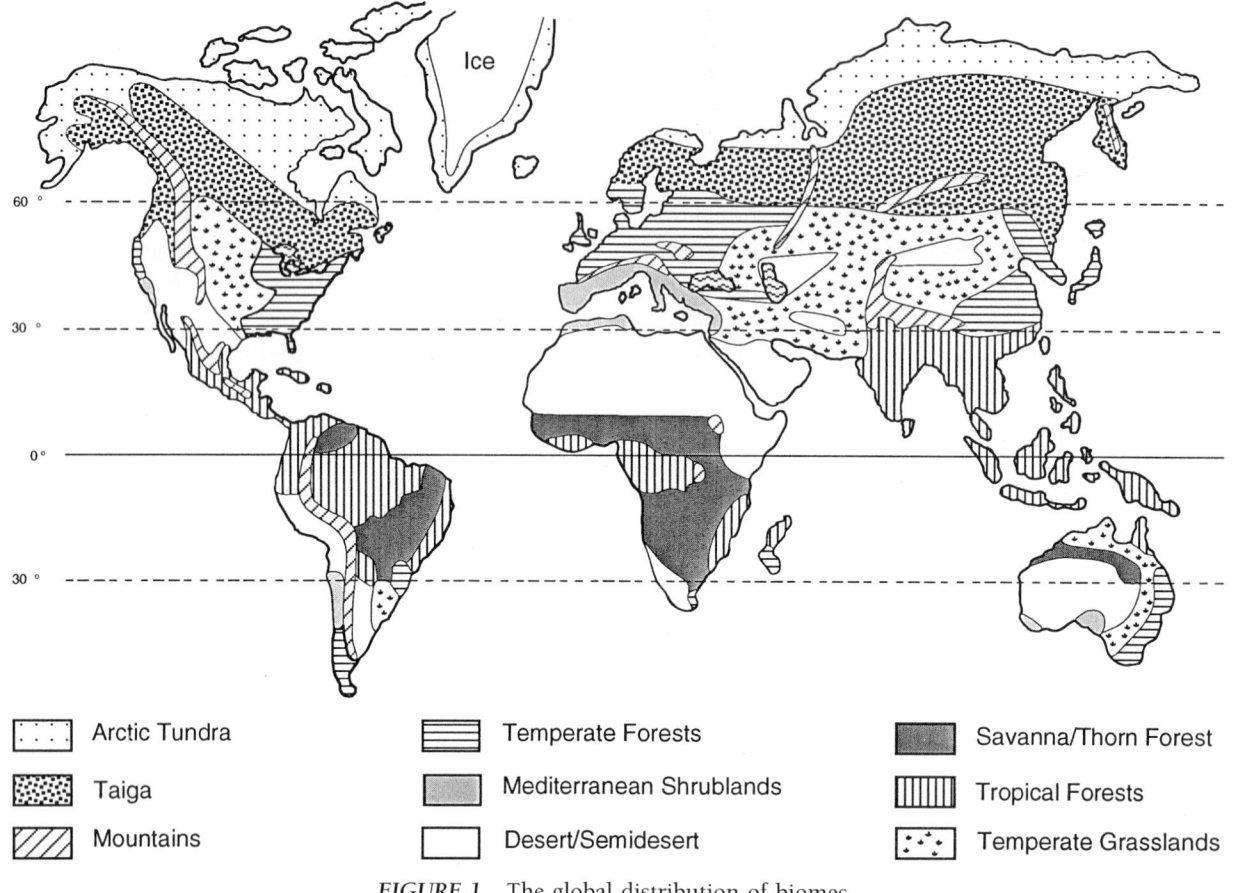

Arctic Tundra   Temperate Forests   Savanna/Thorn Forest

Taiga   Mediterranean Shrublands   Tropical Forests

Mountains   Desert/Semidesert   Temperate Grasslands

FIGURE 1 The global distribution of biomes.

the foxes (*Vulpes* sp.), and felids such as the lynx (*Felis lynx*).

Photoperiodic regulation and thus rigid seasonality is the common pattern of reproduction among most of the mammals of the taiga, sometimes associated with hibernation, as in the case of the hedgehog. Births are timed to take advantage of maximum food availability and warming temperatures: late spring, early summer for herbivores, and often somewhat later for carnivores. Some taiga vole populations include both photoresponsive and unresponsive individuals and thus some individuals can reproduce during the winter even at the higher latitudes of this biome, energetic conditions permitting. The linked 10-year cycles characterizing the snowshoe hare (*Lepus americanus*) and lynx populations of this biome are cited routinely in biology texts as an example of classic predator–prey cycles.

There is no equivalent of taiga in the Southern Hemisphere (Fig. 1).

C. Temperate Forests

Included in this category are the classic deciduous forests and mixed hardwood and coniferous forests of the temperate zone, most notably those in eastern North America, Europe, eastern Asia, and the east coast of Australia. The longer growing season at the midlatitudes of the temperate zone promotes the evolution of trees that drop their leaves during the winter to conserve water and then regrow them every spring. Species diversity is low compared to the tropical biomes but higher than that of either the tundra or taiga. Except in Australia, the mammalian fauna of the temperate forests are mostly rodents, shrews, bats, and carnivores, with lesser numbers of primates and ungulates.

As is typical of most biomes, the most common mammals are rodents, such as jumping mice (*Zapus hudsonicus*), various species of *Peromyscus*, and the gray squirrel (*Sciurus carolinensis*) in North America and several species of Old World field mice (*Apodemus* sp) and bank voles (*Arvicola* sp.) in Europe and Asia. Ungulates feeding at this level include the white-tailed deer (*Odocoileus virginianus*) and red deer (*Cervus elaphus*).

The bats of the temperate forests are all small and insectivorous, and they routinely avoid winter by either hibernating or migrating. Most are monestrous. Examples include the little brown bats (*Myotis* sp.) and big brown bats (*Eptesicus* sp.). Also feeding on insects are moles (e.g., *Talpa* sp. and *Scalopus* sp.) and shrews (e.g., *Blarina* sp. and *Soriculus* sp.).

There are a variety of carnivores in the temperate forests, including foxes (*Vulpes* sp.), bears (*Ursus* sp.), mustellids such as the stoat (*Mustela erminea*), small cats of the genus *Felis* and, less commonly, big cats of the genus *Panthera*.

The primates inhabiting this biome include lemurs, such as the ring-tailed lemur (*Lemur catta*), and some of the macaques. The common rhesus monkey (*Macaca mullata*) penetrates this biome in eastern Asia.

With the exception of some of the smaller rodents, and perhaps some of the insectivores, strict seasonality, regulated by photoperiod, is probably the routine seasonal strategy employed in the temperate forests. As in the taiga, births are timed to take advantage of maximum food availability and warming temperatures. In many mammals in these forests, the reproductive cycle is timed in relation to hibernation, as it is in some of the bats, bears, and even an occasional canid (e.g., the raccoon dog, *Nyctereutes procyonides*). Associated with hibernation in the bats are such interesting adaptations as delayed ovulation, delayed implantation, and sperm storage in female and dissociation of the timing of hormone production and sexual behavior in males. Some of the mustellids, such as the stoat, employ delayed implantation to adjust their basic cycle to the seasons of these latitudes. Seasonal reproduction, often in relation to the pattern of rainfall, is common among the primates. As at the higher latitudes, heterogeneity of reproductive photoresponsiveness characterizes some and possibly many populations of the smaller, shorter lived rodents in these forests, and thus some individuals can reproduce opportunistically during the winter.

A variety of small mouse-like marsupials, including those of the genera *Antechinus*, are found in the temperate forests of eastern Australia. Larger marsupials residing here include some of the long-nosed bandicoots (*Perameles* sp.), gliders (*Petaurus* sp.), the koala (*Phascolarctos cinereus*), and the common wombat (*Vombatus ursinus*). Seasonal reproduction, probably routinely regulated by photoperiod, is common in this part of Australia. *Antechinus* show a spectacular reproductive strategy. Despite their small size, these animals reproduce only once each year during a well-timed 2-week breeding season, at the end of which all the females ovulate and all the males die of stress-related conditions. The only males left alive are those in the uterus.

D. Temperate Grasslands

Called steppes in Asia and pampas in South America, the temperate grassland biome, as defined here, includes arid short-grass prairies that merge into semidesert, long-grass prairies that border on deciduous forest, and the highly seasonal, monsoonal grasslands of southern Asia. Much of the interior of North America and Asia is grassland, as is most of Argentina. There is also a broad band of grassland in eastern and northern Australia.

The dominant mammals of this biome are often livestock and many of the world's grasslands have been altered dramatically by overgrazing. In less disturbed areas are a wide variety of rodents. In North America these include voles (*Microtus* sp.), deermice (*Peromyscus maniculatus*), and several ground squirrels (*Citellus* sp.). In South America are field mice of the genus *Akodon*, cavies (*Cavia* and *Galea* sp.), and the plains viscacha (*Lagostomus maximus*). Depending on aridity, latitude, and altitude, lemmings of the genus *Lagurus*, voles (*Microtus* sp.), gerbils of several genera, and/or jerboas of the family Dipodidae may predominate in Asia. Also common in North America and Asia are hares and jackrabbits (*Lepus* sp.). The most visible herbivores in all undisturbed grasslands are ungulates such as the pronghorn (*Antilocapra americana*) in North America, the pampas deer (*Ozotoceros bezoarticus*) in South America, and a variety of gazelles (*Procapra* sp) in Asia. The first carnivore level is occupied by canids such as the coyote (*Canis latrans*), felids such as South America's gato pajero (*Felis colocolo*), weasels, including Old

World and New World badgers (*Meles meles* and *Taxidea taxus*) , and a variety of shrews and armadillos.

Seasonal patterns of reproduction vary greatly in the grasslands depending on latitude, seasonality of rainfall, and the reproductive adaptations of the mammal under consideration. At the higher latitudes, strict seasonality enforced by photoperiod is probably routine among the larger rodents that hibernate as well as the ungulates and carnivores. This is probably true for many of the smaller rodents as well, but mixed populations of photoresponsive and unresponsive individuals undoubtedly occur here quite commonly. At the lower latitudes of North America and in the monsoonal grasslands of southern Asia, the small mammals often reproduce opportunistically in relation to rainfall and hence food availability without response to variation in day length. The larger mammals also reproduce in relation to the monsoonal rains, but how this is timed in advance is unknown.

The Australian grasslands have a rich fauna of marsupials, including perhaps the smallest mammal, the long-tailed planigale (*Planigale ingrami*). Another common marsupial of small size is the fat-tailed dunnert (*Sminthopsis crassicaudata*). Many of the large macropods, such as the common wallaroo (*Macropus robustus*), move from forested areas into grassland to feed, and some, such as the red kangaroo (*M. rufus*) spend much of their time there. Most of these species have the capacity to reproduce continuously, depending on rainfall and hence food availability, although the reproduction of the smaller marsupials is probably regulated to some degree by photoperiod at the higher latitudes. Embryonic diapause allows the macropods to have an embryo awaiting development while the female is suckling a joey in her pouch.

E. Desert/Semidesert

More than one-third of the earth's land masses are arid or semiarid, without enough rainfall to support crops. Deserts occur where rainfall is low and evaporation exceptionally high. Vegetation is rare or absent in many parts of the world's great deserts, such as the Sahara, Gobi, Kyzyl-Kum, and Great Victoria. As rainfall increases slightly, one begins to see deep-rooted evergreen shrubs and highly specialized plants such as the cacti. Rainfall often is episodic and unpredictable in deserts and semideserts and the result of a rain can be a dramatic opportunistic flowering of annuals and perennials.

Rodents predominate in this biome, more so than in any other. Most visible in North America are the kangaroo rats of the genus *Dipodomys* and some of the antelope ground squirrels (*Ammospermophilus* sp.). Field mice of the genus *Akodon* are common in the Patagonian desert/semidesert of South America, as is the rabbit-like mara (*Dolichotis patagonum*). The deserts of Asia and Africa are populated by many species of jirds (*Meriones* sp.), gerbils of the genera *Gerbillus and Rhompomys,* and jerboas of the genera *Jaculus* and *Dipus.* Two other notable rodents of the African deserts/semideserts are the rock hyrax (*Procavia capensis*) and the fat sand rat (*Psammomys obesus*). The ungulates include a wide variety of gazelles of the genus *Gazella*, feral asses (*Equus asinus*), and the handsome saiga antelope (*Saiga tatarica*) in Asia, the guanaco (*Lama guanicoe*) in Patagonia, and the ibex (*Capra ibex*) in the near eastern deserts. Carnivores include a variety of felids such as the spotted wildcat (*Felis ocreata*) in central Asia, foxes such as *Dusicyon culpaeus* in Patagonia, and the fennec (*Fennecus zerda*) and caracal lynx (*Felis caracal*) in Africa.

In temperate deserts, mammals face the complex problem of reproducing in relation to both an annual cycle of temperature and an unpredictable pattern of rainfall. Many have solved this problem by using multiple cues to regulate their reproduction: photoperiod, immediate energetic conditions (temperature and food availability), and, in some cases, secondary plant compounds. The shorter-lived mammals rely more on the short-term cues, probably most often the immediate energetic conditions, and less on photoperiod. The longer-lived ungulates and carnivores rely more on photoperiod. Photoperiod cannot be used as a cue in the equatorial regions and reproduction in the equatorial deserts seems to be strictly opportunistic, although the precise controlling mechanisms are largely unknown.

The mouse-sized marsupials inhabiting the Australian deserts/semideserts include several species of dunnarts (*Sminthopsis* sp.), the kultarr (*Antechinomys laniger*), and native murids such as the spinifex hopping mouse (*Notomys alexis*). At the first carnivore level is the hamster-sized mulgara (*Dasycercus cristicauda*), which usually eats insects and small vertebrates but feasts on house mice during plagues. Carnivorous quolls (*Dasyurus* sp.) are also occasionally seen in the desert and red kangaroos can be common The potential for continuous reproduction characterizes the mammals of the Australian deserts/semi-deserts, with success being determined by rainfall and hence food availability. In prolonged drought, for example, both male and female kangaroos become infertile and this is probably true of the smaller marsupials as well.

F. Savanna/Thorn Forest

Rainfall is often highly seasonal in the tropics and savanna is typically defined as tall grassland with occasional stands of trees that leaf out during the rainy season. As the amount of rainfall increases, the thorny acacia begins to dominate until it is thick enough to be described as thorn forest. Much of tropical Africa is classified as savanna/thorn forest. This biome is also found in northern Australia, Brazil, and Venezuela. It is of minor importance in Asia and absent in Europe and North America.

To most people the term "savanna" evokes the image of vast herds of grazing ungulates and their large felid predators. This can still be seen in a few parks in east and southern Africa, but much of this is gone and, indeed, it was never typical of savanna in other parts of the world. In most places, the savanna/thorn forest is dominated by rodents, ungulates, bats, and sometimes marsupials or primates.

On the African savanna the common rodents include the multimammate rats (*Mastomys* sp.), soft-furred rats (*Praomys* sp.), and striped grass rats (*Lemniscomys* sp.). In South America are vesper mice (*Calomys* sp.), cane mice (*Zygodontomys* sp.), and South American field mice (*Akodon* sp.). A variety of hares of the genus *Lepus* occur on the African savanna. The African ungulates are often those commonly seen in zoos: the black rhino (*Diceros bicornis*), giraffe (*Giraffa camelopardalis*), and a large number of antelopes, including the springbok (*Anti-*

dorcas marsupialis). Their counterparts in South America include tapirs (*Tapirus terrestris*), peccaries (*Tayassu* sp.), and deer of several genera. Insectivorous bats are particularly common on the savanna, including the often seen tomb bats (*Taphozous* sp.) and slit-faced bats (*Nycteris* sp.) in Africa. Primates can be seen the African savanna/thorn forest biome, including the bush baby (*Galago senegalensis*) and a variety of baboons (*Papio* sp.). The small marsupials inhabiting the savanna/thorn forest of Australia include some of the thin-tailed dunnarts (*Sminthopsis* sp.) and *Antechinus bellus*. Larger grazing marsupials include the brush-tailed possum (*Trichosurus vulpecula*). Insectivorous marsupials include some of the bilbies (*Macrotis* sp.). Along with the ubiquitous red kangaroo, the antilopine wallaroo (*Macropus antilopinus*) frequents this biome.

Typically, the smaller mammals with short life expectancies and short gestations that live in this biome—rodents, lagomorphs, and some of the marsupials—try to reproduce continuously, year-round, but succeed or fail depending on food availability and hence rainfall. Where there is slight seasonality of rainfall there is a slight reduction in reproductive success during the dry season. Where the dry season is harsh and prolonged, reproductive success is limited entirely to the wet season. In regions where there are two rainy seasons, the smaller mammals reproduce twice annually. There are many exceptions to this generality, including *Antechinus bellus*, which has a restricted breeding season that is probably regulated by photoperiod.

Year-round breeding with seasonal peaks of births during the wet season has been noted in Africa for many of the larger, longer-lived mammals. An interesting question relates to the way births are correlated with the wet season in these animals since their gestation periods range from several months to almost 2 years. Seasonality of reproduction with births tending to occur most often in the wet season also characterizes the ungulates living in the savanna/thornbush regions of South America. Opportunistic reproduction in relation to rainfall characterizes the large grazing marsupials of the Australian savanna/thorn forest, but this is reversed in at least one case. Seasonal rains, monsoonal in origin, can flood vast areas of northern Australia and this depresses reproductive success in the agile wallaby (*Macropus agilis*).

With their food supply being less directly dependent on rainfall, the carnivores that prey on the herbivores tend to reproduce year-round in the savanna/thornbush biomes with much less seasonality of births. This is also true for the primates inhabiting this biome and the insectivorous bats.

G. Tropical Forests

Much of Central America, tropical South America, and southern Asia is heavily forested, as are most of the islands between Southeast Asia and Australia. Classic rain forest is found where rainfall is both high and nonseasonal, as in the Amazon basin and in a large band stretching from eastern India down the Malaysian peninsula and on to the York peninsula of northern Australia. There is a relatively narrow band of rain forest running east and west and surrounded by savanna/thorn forest in Africa. Deciduous forest is found in the tropics, where rainfall is relatively high but seasonal, as it is in southern China. These trees drop their leaves during the dry season, making it the equivalent of winter at the higher latitudes. Tropical scrub forest is found where the wet season is routinely short, as in parts of Mexico, western India, and central South America.

Species diversity is almost unbelievably high in tropical forests, particularly the rain forests, and it is impossible to discuss the mammal composition of this biome with any brevity. Two orders of mammals stand out here, however—the bats and primates. Bats account for one-quarter of all species of mammals, and most of them are found in the tropical forests where they have diversified to feed upon prey or substances as diverse as insects, fish, blood, or fruit. Some of the most common fruit-eating bats in Central and South America are the short-tailed leaf-nosed bats (*Carollia* sp.) and the neotropical fruit bats (*Artibius* sp). In Africa the most common fruit-eating bats are the common rousette fruit bat (*Rousettus aegyptiacus*) and straw-colored fruit bat (*Eidolon helvum*). The common fruit bats of Asia and northern Australia include the huge flying foxes of the genus *Pteropus* and the dog-faced fruit bats (*Cynopterus* sp.). The long-nosed bat (*Rhynochonycteris naso*) and

some of the mastiff bats of the genus *Molossus* are common insectivorous bats in South America. Their counterparts in Africa are pipistrelles (*Pipistrellus* sp.) and in Asia and northern Australia the long-winged bats (*Miniopterus* sp.).

South American primates include tamarins of many genera, squirrel monkeys (*Saimiri* sp.), howlers (*Alouatta* sp.), and spider monkeys (*Ateles* sp.). In Africa, including Madagascar, are lemurs of a variety of genera, colobus monkeys (*Colobus* sp.), gorillas (*Gorilla gorilla*), and both species of chimpanzee (*Pan* sp.). In Asia and the South Pacific are many macaques (*Macaca* sp.), langurs of the genus *Presbytis* or *Semnopthecus*, tarsiers (*Tarsius* sp.), and the orangutan (*Pongo pygmaeus*).

As was discussed in relation to the savanna/thorn forest biome, most mammals living in the tropics have the capacity to reproduce continuously depending on food availability and hence rainfall. Little seasonality of reproduction is seen in lush rain forests where rainfall is both heavy and absolutely nonseasonal. Seasonal trends in reproduction are common in tropical deciduous forests, however, and sometimes there are highly restricted seasons for mating and birthing. Two species of fruit bats living side by side may feed on fruits that appear at different seasons and thus their birthing seasons might differ by several months. Nectar-feeding bats time their births in relation to the time of flowering, which can also vary seasonally depending on the species of plant. An exceptionally interesting question relates to the unknown nature of the predictive cue used by seasonal breeders to trigger gonadal development in preparation for mating when they live close to the equator. At least some bats in these forests have a circannual rhythm that must be reset each year by this cue.

The term opportunist could probably be applied to all the tropical primates even though it seems to offer a poor fit for mammals with such long gestations. In fact, the primates at these latitudes all seem to have the capacity to reproduce continuously and the relatively low energetic demands of lactation seem to ensure that they do reproduce more or less continuously in modestly seasonal environments. In strongly seasonal forests, they reproduce seasonally.

H. Mediterranean-Type Shrublands

This is the smallest biome in terms of land area. Most of it occurs as a partial ring around the Mediterranean Sea, but it can be seen in several other parts of the world, including southern and southeastern Australia, where it is called mallee, and California, where it is called chaparral. Climatically, this biome is characterized by wet winters and prolonged dry summers and temperature variation typical of the lower temperate zone. The dominant vegetation is woody shrubs with thick, water-conserving leaves that are not shed seasonally.

Small native murids of the genus *Pseudomys* are quite common in the Australian mallee, as are many marsupials such as the small ningauis (*Ningauis* sp.), the large western gray kangaroo (*Macropus fuliginosus*), and, as in many parts of Australia, the introduced European rabbit (*Oryctalagus cunniculus*). Two of the common small mammals in the California chaparral are the California pocket mouse (*Perognathus californicus*) and the Pacific kangaroo rat (*Dipodomys agilis*). Brush rabbits (*Sylvilagus bachmani*), mule deer (*Odocoileus hemionus*), and coyotes (*Canis latrans*) are also common in the chaparral. Typical mammals in Israel include the jird (*Meriones tristrami*), one of the blind mole rats (*Spalax ehernbergi*), and the Egyptian mongoose (*Herpestes ichneumon*).

An interesting aspect of this biome is the fact that most of its mammals reproduce during the spring and early summer, probably routinely enforced by photoperiod, despite the prolonged summer drought. There are exceptions, however. The California vole (*Microtus californicus*) routinely reproduces during the winter rainy season despite the fact that it reacts like a long-day breeder in the laboratory.

I. Marine Mammals

Three orders of mammals exploit marine or freshwater habitats: the seals, sea lions, and the walrus of the order Pinnipedia; the whales, dolphins, and porpoises of the order Cetacea; and the manatees and dugong of the order Sirenia. Almost all the pinnipeds show a marked seasonal cycle, often associated with feeding migrations and molting, in which they "haul out" onto land or ice once a year to give birth and mate. Photoregulated, delayed implantation allows them to lengthen the period between mating and birth in order to accommodate the annual cycle of their environment. Hundreds of females may be birthing and mating in a restricted locale in a short period of time. Most pinnipeds are polygynous, and fighting between the older males for females for their harems can be fierce at haul out.

Most cetaceans reproduce with at least some seasonal tendency. In equatorial regions many dolphins either live in rivers and estuaries or migrate there to reproduce; peak periods often follow the rainy season and the resulting flooding. In colder waters, killer whales (*Orcinus orca*) reproduce year-round, but again with seasonal peaks associated with feeding migrations. The other whales also often reproduce seasonally, and often in relation to annual migrations to higher latitudes and colder water to feed and then back to the warmer lower latitudes to mate and give birth. The environmental control over this cycle, and in particular the role of photoperiod in regulating it, is unknown.

The sirenians inhabit the warm coastal waters of the tropics where they have the capacity to reproduce year-round but sometimes they show seasonal tendencies in mating and births.

III. GLOBAL PATTERNS OF HUMAN REPRODUCTION

Almost all human populations show seasonal variation in births, owing mostly to seasonal variation in conception. Many of these annual cycles are robust and highly repeatable over decades and even centuries. In many and probably most instances, these cycles reflect cultural variation with increased frequency of conception associated with holidays or festivals, for example, but this is not always the case. Climactic variation can influence conception directly in at least two and possibly three ways.

First, there is good evidence that ovulation can be regulated seasonally in human populations experiencing seasonal variation in available energy. Inadequate food intake, increased energy expenditure, or both can delay menarche, suppress the frequency of ovulation in nonlactating females, and prolong lactational amenorrhea, all on a seasonal basis. This is best seen in tropical subsistence societies in which food availability varies greatly due to a strongly seasonal pattern of rainfall. It may also account for some of the potent seasonal variation seen in arctic Inuit populations, but this is much more controversial.

Second, it seems likely that hot summer temperatures can suppress spermatogenesis in some populations at the mid- and lower latitudes where air-conditioning is absent. This has been documented most thoroughly in the southern United States. Unfortunately, little is known about this possibility in human populations in the tropics.

Third, correlational studies argue that there may be some regulation by photoperiod at the higher latitudes in humans. This is the most controversial possibility and awaits further study.

See Also the Following Articles

Circadian Rhythms; Circannual Rhythms; Nutritional Factors and Reproduction; Photoperiodism, Vertebrates; Seasonal Reproduction, Mammals

Bibliography

Bronson, F. H. (1989). *Mammalian Reproductive Biology*. Univ. Chicago Press, Chicago.

Bronson, F. H. (1995). Seasonal variation in human reproduction: Environmental factors. *Q. Rev. Biol.* 70, 141–164.

Goodall, D. W. (Ed.) (1983–1992). *Ecosystems of the World*, Vols. 1–16. Elsevier, Amsterdam.

Hayssen, V., van Tienhoven, A., and van Tienhoven, A. (1993). *Asdell's Patterns of Mammalian Reproduction*. Cornell Univ. Press, Ithaca, NY.

Nowak, R. M. (1991). *Walker's Mammals of the World*, Vols. I and II, 5th ed. Johns Hopkins Univ. Press, Baltimore.

Strahan, R. (Ed.) (1995). *Mammals of Australia*. Smithsonian Inst., Washington DC.

Glucocorticoids

see CBG

Gnathostomulida

Wolfgang Sterrer
Bermuda Natural History Museum

GLOSSARY

bursa (copulatrix) The organ in which sperm are stored after copulation and prior to fertilization.
conulus A type of aflagellate, cone-shaped sperm found in Gnathostomulida Conophoralia.
monociliated epidermis Each epidermal cell is provided with only one cilium.

Gnathostomulida are microscopic, free-living, unsegmented marine worms of enigmatic phylogenetic affiliation.

I. INTRODUCTION

Initially described as an aberrant taxon of turbellarian flatworms, Gnathostomulida were recognized as a distinct phylum by Riedl and Sterrer, who proposed the currently used classification. Found exclusively in the interstices of shallow marine sandy bottoms, the phylum currently comprises fewer than 100 species, of which many have worldwide distribution. Worm-shaped and ranging from 0.3 to 3 mm in length, Gnathostomulida differ from all known invertebrates in having a monociliated epidermis, i.e., each epidermal cell carries only a single cilium. They are further distinguished by the possession of a bilaterally symmetric pharynx that usually contains complex cuticular mouth parts consisting of paired jaws and an unpaired basal plate. The phylum comprises two orders, Filospermoidea and Burso-vaginoidea, the latter divided is into two suborders, Scleroperalia and Conophoralia. Gnathostomulid evolution most likely progressed from the filospermoid condition (sperm with one 9+2 flagellum, no bursa, and a simple male pore) to the bursovaginoid condition (sperm aflagellate, bursa system present, and a complex copulatory organ).

Ax proposed to unite Gnathostomulida and Platyhelminthes as sister groups in the taxon Platyhelminthomorpha, at the basis of Bilateria, whereas Sterrer *et al.* and Rieger and Tyler pointed out possible relationships with aschelminths such as Gastrotricha and Rotifera. The only available molecular evidence, based on an 18S rRNA gene sequence, supports neither close platyhelminth nor aschelminth affinities but suggests Chaetognatha as a potential sister taxon.

Although Gnathostomulida are often the dominant invertebrate taxon in the detritus-rich, oxygen-poor sands in which they typically occur, our knowledge of their biology is still very scanty. It is assumed that they feed by grazing on the microflora (bacteria and fungal hyphae) that coats sand grains. Of their reproduction, very little is known beyond overall anatomy, gamete structure, and gametogenesis.

II. SEXUAL REPRODUCTION

Gnathostomulida are hermaphrodites (Fig. 1), and there is no evidence for asexual reproduction or parthenogenesis. The tubular or follicular testes are located in the posterior third of the body, between the gut and the epidermis. They are paired in Filospermoidea Haplognathiidae and all Bursovaginoidea, but unpaired in Filospermoidea Pterognathiidae and all Conophoralia, in the latter arising from the fusion of paired anlagen. There are three types of sperm;

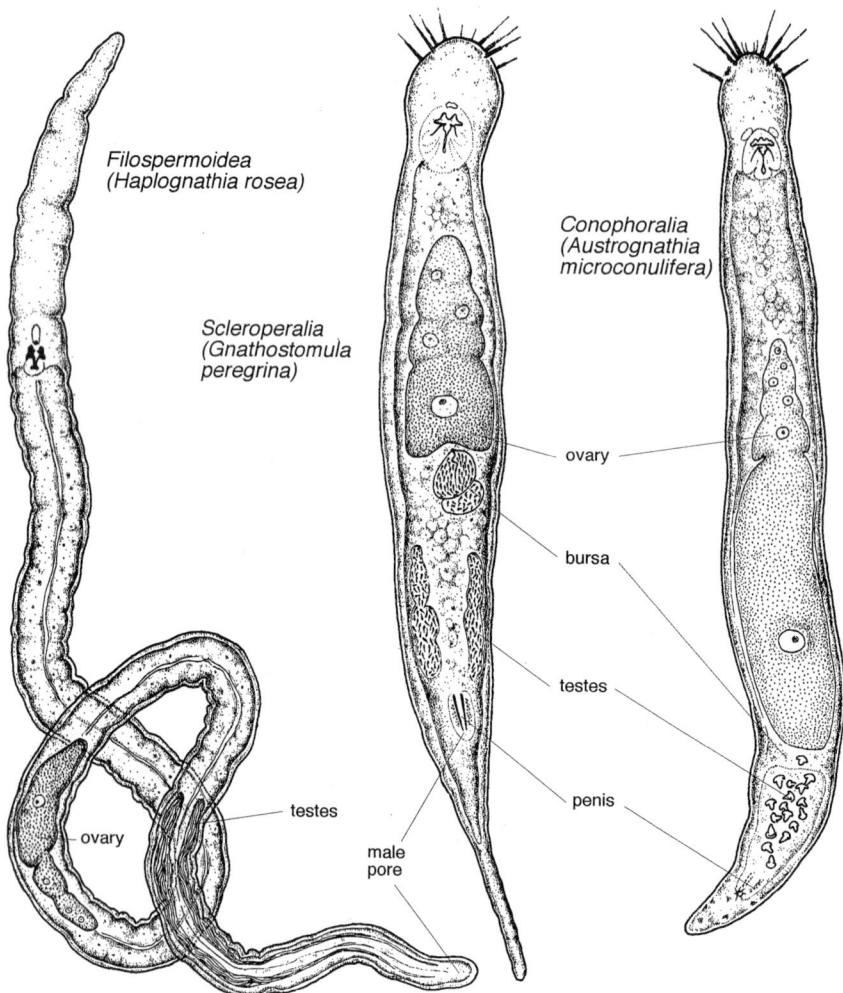

Filospermoidea
(Haplognathia rosea)

Scleroperalia
(Gnathostomula
peregrina)

Conophoralia
(Austrognathia
microconulifera)

ovary

bursa

testes

penis

ovary

testes

male
pore

FIGURE 1   Representatives of the three major taxa of Gnathostomulida.

their homology is still uncertain (Fig. 2). The Filospermoidea have filiform sperm to 100 μm long (Figs. 2A), which consists of a spiral nucleus topped by an acrosome, a middle piece with mitochondrial derivatives, and a flagellum with one $9+2$ axoneme. Scleroperalia have aflagellate "dwarf" sperm of only 2–8 μm length (Fig. 2B), which may either be round or droplet shaped and is usually provided with short projections (micropodia). A third sperm type, called conulus, is found in Conophoralia only (Fig. 2C); it is aflagellate, cone-shaped, and up to 70 μm long. Filiform sperm is able to move and rotate like a corkscrew, whereas dwarf sperm and conuli are immotile.

Filospermoidea, and probably Conophoralia, have a simple, rosette- or funnel-shaped penis that is

weakly muscular but richly glandular. Located posteriorly to the testes, it empties into a subterminal ventral pore. Scleroperalia are characterized by a bulbous, muscular penis which usually surrounds a tubular penis stylet made up of 8–10 rod-shaped cell extensions.

The unpaired, pear-shaped ovary lies dorsally, between the gut and the epidermis, extending from behind the pharynx to about the midbody region. The germinal zone is located anteriorly, and size, maturity, and yolk content of oocytes increase caudally during oogenesis. Each growing oocyte is surrounded by one or more accessory cells. A single mature egg usually takes up the posterior-most half or two-thirds of the ovary.

Sperm transfer is by copulation. Filospermoidea

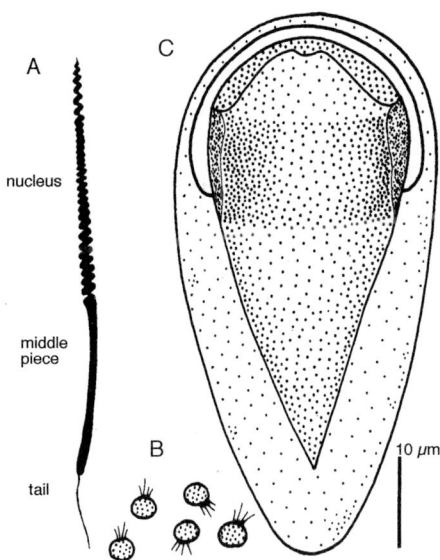

FIGURE 2 Sperm types of Gnathostomulida. (A) *Haplognathia simplex* (Filospermoidea), (B) *Gnathostomula jenneri* (Scleroperalia), and (C) *Austrognathia riedli* (Conophoralia).

lack a vagina and a bursa for sperm storage; apparently, sperm is injected under the epidermis and then distributed throughout the body where it is stored prior to use in fertilization. Most Conophoralia have a permanent vagina situated dorsally behind the ovary; the vagina leads into a pouch-shaped bursa in which usually only one or two sperm are stored. Scleroperalia lack a permanent vagina but are characterized by a bursa system consisting of a caudal, rounded prebursa which connects anteriorly to a conical bursa. The wall of the bursa is composed of flattened cells which meet laterally to form crests and anteriorly to form a perforated mouthpiece through which stored sperm is channeled to the mature egg.

III. DEVELOPMENT

Oviposition, at least in Scleroperalia, is by rupture of the dorsal epidermis behind the ovary and bursa, at the spot where a vagina may be located. The egg then becomes spherical, about 55 μm in diameter, and sticks to sand grains. Development is direct. Cleavage, still insufficiently known, seems to be of the spiralian type, with mesolecithal–epibolic gastru-

lation, resulting in a juvenile which hatches at a length of 100 μm. The juvenile lacks jaws but has a rudimentary pharynx.

Nothing is known about reproductive physiology and endocrine control.

Bibliography

Alvestad-Graebner, I., and Adam, H. (1983). Gnathostomulida. In *Spermatogenesis and Sperm Function* (K. G. Adiyodi and R. G. Adiyodi, Eds.), pp. 171–180. Wiley, New York.

Ax, P. (1956). Die Gnathostomulida, eine rätselhafte Wurmgruppe aus dem Meeressand. *Abhandl. Akad. Wiss. Lit. Mainz, math.—Naturwiss. Kl* **8**, 1–32.

Ax, P. (1985). The position of the Gnathostomulida and Platyhelminthes in the phylogenetic system of the Bilateria. In *The Origins and Relationships of Lower Invertebrates* (S. Conway Morris, J. D. George, R. Gibson, and H. M. Platt, Eds.), pp. 168–180. Clarendon, Oxford, UK.

Falleni, A. (1993). Ultrastructure of growing oocytes and accessory cells in *Austrognathia* (Gnathostomulida, Bursovaginoidea). *Tissue Cell* **25**, 777–790.

Knauss, E. B., and Rieger, R. M. (1979). Fine structure of the male reproductive system in two species of *Haplognathia* Sterrer (Gnathostomulida, Filospermoidea). *Zoomorphologie* **94**, 33–48.

Littlewood, D. T. J., Telford, M. J., Clough, K. A., and Rohde, K. (1998). Gnathostomulida—An enigmatic metazoan phylum from both morphological and molecular perspectives. *Mol. Phylogenet. Evol.* **9**, 72–79.

Mainitz, M. (1979). The fine structure of gnathostomulid reproductive organs. I. New characters in the male copulatory organ of Scleroperalia. *Zoomorphologie* **92**, 241–272.

Riedl, R. J. (1969). Gnathostomulida from America. *Science* **163**, 442–445.

Rieger, R. M., and Tyler, S. (1995). Sister-group relationship of Gnathostomulida and Rotifera-Acanthocephala. *Invertebr. Biol.* **114**(2), 186–188.

Sterrer, W. (1972). Systematics and evolution within the Gnathostomulida. *Syst. Zool.* **21**(2), 151–173.

Sterrer, W. (1974). Gnathostomulida. In *Reproduction of Marine Invertebrates, Vol. 1. Acoelomate and Pseudocoelomate Metazoans* (A. C. Giese and J. S. Pearse, Eds.), pp. 345–357. Academic Press, New York.

Sterrer, W., Mainitz, M., and Rieger, R. M. (1985). Gnathostomulida: Enigmatic as ever. In *The Origins and Relationships of Lower Invertebrates* (S. Conway Morris, J. D. George, R. Gibson, and H. M. Platt, Eds.), pp. 181–199. Clarendon, Oxford, UK.

GnRH (Gonadotropin-Releasing Hormone)

P. Michael Conn

Oregon Regional Primate Research Center and Oregon Health Sciences University

L. Jennes

University of Kentucky

J. A. Janovick

Oregon Regional Primate Research Center

GLOSSARY

analog A chemical compound which is chemically similar to another compound.

agonist A chemical compound which evokes the same or very similar actions in a target cell in comparison to another.

antagonist A chemical compound which blocks the actions of another chemical.

gonadotropin-releasing hormone A peptide hormone which stimulates the release of the gonadotropins from the pituitary and which has other central nervous system actions.

gonadotropins Collectively, luteinizing hormone and follicle-stimulating hormone. These regulate sex hormone steroidogenesis as well as production of ova and sperm.

I. INTRODUCTION

Gonadotropin-releasing hormone (GnRH: pyro-Glu1-His2-Trp3-Ser4-Tyr5-Gly6-Leu7-Arg8-Pro9-Gly-amide10) is synthesized, then stored, in the medial basal hypothalamus. In response to neural signals, pulses of GnRH are released into the hypophysial portal system and then conducted to the anterior pituitary where they stimulate release of the gonadotropins, luteinizing hormone (LH), and follicle-stimulating hormone (FSH). Accordingly, GnRH should be viewed as providing a humoral link between the neural and endocrine systems. Pituitary gonadotropins are released into the systemic circulation and regulate gonadal steroidogenesis and gamete maturation; LH regulates ovulation and corpus luteum formation in females and androgen secretion in males, whereas FSH stimulates the growth and maturation of ovarian follicles in females and spermatogenesis in males. GnRH and its analogs also have direct gonadal actions in some species by inhibiting steroidogenesis in males and ovulation in females. Hypothalamic-derived GnRH is unlikely to be responsible for these actions (because of the very low concentrations in the peripheral circulations), although the possibility remains that a different, locally produced molecule may bind a gonadal "GnRH receptor" *in vivo*. Identification of such a molecule has not been forthcoming, however.

The ability of GnRH to stimulate reproductive functions (at pulsatile low doses) or suppress them (at high doses) has been applied for various purposes, including the induction of ovulation and spermatogenesis, contraception, and the treatment of precocious puberty, endometriosis, and steroid-dependent tumors (i.e., prostate and breast cancer), as well as for a wide range of veterinary purposes.

Probes for immunohistochemistry and for *in situ* hybridization have provided data that suggest that GnRH and its receptor may have a broader distribution in the central nervous system than originally believed. This result presents the possibility of releasing hormone actions that have not yet been described.

An understanding of the basic mechanisms by which GnRH alters physiological function has been a valuable resource for the design of strategies for clinical and veterinary interventions which have provided a powerful incentive for work in this area. This review describes current knowledge about the means by which the releasing hormone interacts with its receptor and subsequently activates the responses of the target cell.

II. CELL BIOLOGY AND DISTRIBUTION OF THE GnRH RECEPTOR

A. Ligand Binding and Physical Characteristics

The first step in GnRH action is recognition of the ligand by its plasma membrane receptor. A great deal is known about the binding step because of the availability of literally thousands of analogs. Highly effective radioligands can be prepared by using high-affinity, metabolically stable agonists. Such synthetic agonists frequently share the presence of a D-amino acid in the sixth position, which inhibits vicinal enzymatic degradation, and the further substitution in which an ethylamide is substituted for the C-terminal Gly10-amide. These substitutions, when present, enhance receptor binding affinity. Changes in GnRH receptor number (but not binding affinity) have been reported to occur during the rat estrous cycle, lactation, castration, and aging, as well as in other endocrine states. In a general way, the frequency of the receptors is predictive of the responsiveness of the gonadotrope cell to GnRH. However, conditions can be devised in cell cultures and *in vivo* that lead to diminished cellular responsiveness, even in the face of elevated receptor numbers. Thus, regulation of receptors is not the sole determinant of cellular responsiveness; this concept becomes especially important in consideration of the mechanism of development of the refractory state in desensitization.

Before genetic approaches were available, many physical techniques were used to study the receptor. Target size analysis, which provides a measure of the functional molecular weight of a particular moiety *in situ* (including any other intimately associated molecules), yields a molecular weight of 136,346 ± 8,120. A lower molecular weight (60,000) has been reported for photoaffinity-labeled (PAL) receptor that is analyzed under denaturing conditions. Conceivably, this moiety is the hormone-binding component of a larger (functional) complex described in the preceding study. Such a complex could be composed of multimers of receptors or of receptor–effector complexes.

In 1992 the first report appeared describing the cloning and functional expression of the mouse GnRH receptor. Since that time, at least five more reports have appeared for the same and other species. In the first report, a 327-amino acid protein is described with a molecular weight of 37,683 and three consensus sites for N-linked glycosylation (likely explaining the disparity in molecular weights for the cloned form and the PAL receptor). The overall sequence of the receptor suggests that it is a member of the seven-transmembrane segment (7-TMS) class, characteristic of G-protein-linked receptors. The highly conserved aspartate or glutamate found in TMS-2 of many receptors (which is essential for function) is replaced in the GnRH receptor by asparagine. The GnRH receptor lacks the polar cytoplasmic C-terminal region frequently observed in this group of molecules and, in some cases, believed to be associated with biological function. Another unique feature of the molecule is the substitution of serine for tyrosine adjacent to TMS-3, thus creating a potential phosphorylation site. A highly basic site, flanked by two consensus phosphorylation sites, is located at the first internal loop after TMS-1. A recent review compares the GnRH receptor structure with other well-characterized receptors (Ulloa-Aguirre and Conn, 1998).

III. GnRH RECEPTORS IN THE BRAIN

In addition to regulating LH and FSH release from the anterior pituitary gonadotropes, GnRH exerts a variety of effects inside the central nervous system. Thus, lordosis and mounting behaviors are facilitated by administration of GnRH into the hypothalamic ventromedial nucleus and the central gray as well as by intraventricular, subarachnoid, and even peripheral injections of GnRH. These effects appear to be specific for GnRH since they can be prevented by coadministration of antibodies against GnRH or by GnRH antagonists. This high degree of specificity as well as the observations that GnRH can change the firing patterns of certain neurons and that GnRH is present in presynaptic nerve terminals led to the view that GnRH actions in the brain are mediated by specific GnRH receptors. This view has been substantiated by the identification and characterization of specific intracerebral GnRH binding sites with *in vitro* autoradiography as well as by *in situ* hybridization, which localizes the GnRH receptor mRNA to select neurons in the brain. A detailed review relating GnRH peptide, receptor, and receptor mRNA has been published (Jennes *et al.*, 1994).

A. Distribution of GnRH Receptors in Rat Brain

Incubation of unfixed frozen brain sections with [^{125}I]Buserelin, [^{125}I-]D-Ala6-NaMeLeu7-Pro9-NHEt-GnRH, or [^{125}I]-D-Ala6-desGly10-NHEt-GnRH results in specific binding of the ligands to selected regions of the central nervous system. Beginning rostrally, GnRH agonists bind to the laminae glomerulose and plexiformis externa of the olfactory bulb as well as to the nucleus olfactorius anterior, pars externa. The piriform cortex labels consistently over its pyramidal cell layer up the sulcus rhinalis, which exhibits GnRH binding throughout its rostrocaudal extent. In the septum, GnRH binding sites are concentrated in the dorsolateral portion of the lateral septum, whereas the medial septum contains only a small number of binding sites. More caudally, GnRH agonists bind to neurons in the hypothalamic ventromedial nucleus and, to a lesser extent, in the arcuate nucleus, whereas binding sites have not yet been identified in the preoptic region. The interpeduncular nucleus and the central gray are moderately labeled without any apparent differences in the labeling intensity according to their anatomical subunits. The major targets for GnRH in the brain are the amygdala and the hippocampus. In the amygdala, the medial, lateral, and cortical nuclei exhibit moderate labeling, whereas in the hippocampus most labeling is observed over the strata oriens and radiatum of areas CA1–4. The perikarya of the pyramidal cells

and the granule cells of the dentate gyrus are not labeled with *in vitro* autoradiography.

B. Distribution of GnRH Receptor mRNA in Rat Brain

The discovery of the nucleotide sequence encoding the pituitary GnRH receptor permitted the generation of high specific activity cRNA probes that have been used to localize the GnRH receptor mRNA in the rat central nervous system. The distribution of the neurons containing the mRNA encoding the GnRH receptor as assessed by *in situ* hybridization shows an extensive parallelism compared with the distribution of the GnRH receptor protein as studied by *in vitro* autoradiography. Thus, the mRNA is present in neurons of the nucleus olfactorius anterior, the pyramidal neurons of the piriform cortex, and the sulcus rhinalis, as well as in the hypothalamic ventromedial and arcuate nuclei. However, the mRNA is not detected in the septum, interpeduncular nucleus, or central gray, but is detected in the habenular complex of the epithalamus. In the amygdala, GnRH receptor mRNA is found in the caudal aspects of the medial and cortical nuclei; in the hippocampus, the mRNA is present in high concentrations in the pyramidal cells of areas CA1–4 and in the granule cells of the dentate gyrus. Based on current knowledge of the anatomical projections, it is suggested that (i) the neurons in the habenula synthesize the GnRH receptor and transport the protein intraaxonally to their nerve terminals in the interpeduncular nucleus, (ii) the GnRH receptors produced in the ventromedial nucleus are transported to the central gray, and (iii) the pyramidal and granule neurons of the hippocampus transport the GnRH receptor protein to their dendrites, which are located in the strata oriens and radiatum of the hippocampus.

C. Ligand Specificity of the Hippocampal GnRH Receptor

The structure-binding relations of the brain GnRH receptor have been characterized extensively by radioreceptor assays and *in vitro* autoradiography. The results of quantitative *in vitro* autoradiographic analyses of GnRH analog binding in the presence or absence of increasing amounts of unlabeled GnRH show that most GnRH analogs that bind to the GnRH receptor in the anterior pituitary exhibit similar affinities for the GnRH receptor in the brain. Binding of ^{125}I-labeled GnRH analogs to the hippocampus reaches its maximum after 1 hr at 4°C; it is saturable and specific. Thus, the GnRH decapeptide, the agonists D-Ala6-GnRH, desGly10-GnRH ethylamide, and Buserelin, as well as the antagonist N-Ac-D-(pyro)-Cl-Phe1,2-D-Trp3-D-Lys6-D-Ala10-GnRH compete with ^{125}I-labeled Buserelin for binding with high efficacies (EC$_{50}$ 5 nM, 100 pM, 5 nM, 0.6 nM, and 100 pM, respectively). Conversely, the GnRH fragments des1-GnRH, des2-GnRH, [4–10]GnRH, [5–10]GnRH, and Ac-[5–10]GnRH, which do not bind to the pituitary GnRH receptor, also do not recognize the brain GnRH receptor. In addition, GnRH sequences typical for other animal species, such as Tyr3-Leu5-Glu6-Trp7-Lys8-GnRH (lamprey), Gln8-GnRH (chicken I), Leu7-Gln8-GnRH (chicken II), and Trp7-Leu8-GnRH (salmon), which have no or only very low affinity for the pituitary GnRH receptor, also do not recognize the brain GnRH receptor. Although these studies show that the binding characteristics of the brain and pituitary GnRH receptors are very similar, a difference in the efficacy has been reported for the antagonist D-*p*-Glu1-D-Phe2-D-Trp3,6-GnRH. This antagonist is very potent in blocking the GnRH receptor in the anterior pituitary but ineffective in competing with ^{125}I-labeled Buserelin for binding to the brain GnRH receptor. The reason for this difference in the binding characteristics is not clear. However, since more hydrophobic antagonists such as N-Ac-D-(pyro)-Cl-Phe1,2-D-Trp3-D-Lys6-D-Ala10-GnRH block the brain GnRH receptor, the lipid environment of the neuronal plasma membrane appears unlikely to be responsible for the ineffectiveness of D-*p*-Glu1-D-Phe2-D-Trp3,6-GnRH to interact with the brain GnRH receptor.

Similar ligand specificities and affinities have been obtained when crude membrane fraction of the hippocampus were exposed to ^{125}I-labeled Buserelin. The binding is specific and saturable and reaches its maximum after 1 hr at 4°C. Scatchard analyses of satura-tion experiments indicate the presence of a single class of high-affinity binding sites that has a K_a of 6.2×10^9 M^{-1}. ^{125}I-Labeled Buserelin is displaced by D-Ala6-GnRH with an EC$_{50}$ of 400 pM and by GnRH with an EC$_{50}$ of 5 nM, whereas GnRH and D-Ala6-GnRH compete for binding with an EC$_{50}$ of 23 and 0.7 nM, respectively. The fragments [4–10]GnRH, [5–10]GnRH, and Ac[5–10]GnRH are ineffective, whereas micromolar concentrations of chicken I and II, lamprey, or salmon GnRHs are needed to displace 50% of ^{125}I-labeled Buserelin binding.

Photoaffinity labeling of hippocampal membranes with ^{125}I-labeled azido-benzoyl-D-Lys6-GnRH followed by one-dimensional gel electrophoresis and exposure to X-ray film reveals two major families of specifically labeled proteins: one single band with an apparent molecular mass of 29 kDa and two groups of proteins with an apparent molecular mass between 50 and 60 kDa. The molecular weight range of these proteins is similar to those of the photoaffinity-labeled proteins of pituitary membranes. The labeling of both the hippocampal and the pituitary membranes appears to be specific since it can be prevented by coincubation of ^{125}I-labeled azido-benzoyl-D-Lys6-GnRH with excess unlabeled GnRH.

D. Regulation of GnRH Receptor Expression in the Brain

The content of GnRH receptor mRNA and protein in the brain appears to be regulated by the hormonal conditions of the animals. Thus, ovariectomy or orchidectomy cause a twofold increase in the number of GnRH binding sites in the hippocampus which is paralleled by a similar increase in the amount of GnRH receptor mRNA. This increase is reversible by the administration of gonadal steroid hormones. Similarly, the GnRH receptor mRNA levels in the mediobasal hypothalamus vary during the estrous cycle as determined by *in situ* hybridization followed by autoradiography and image analysis. The results of these studies show that GnRH receptor mRNA levels in the arcuate and ventromedial nuclei are relatively low during diestrus I and II, and then they rise to the highest levels on the morning of proestrus. This increase is followed by a sharp decline in the relative content of GnRH receptor mRNA during the afternoon of proestrus. The mRNA levels remain low through the morning of estrus before they return to baseline levels by the afternoon of diestrus I. The timing of the changes in the GnRH receptor mRNA levels shows that the GnRH receptor mRNA content in the mediobasal hypothalamus is increased prior to and not as a consequence of the GnRH-mediated preovulatory LH surge. The data suggest that rising levels of circulating estradiol induce the early rise during the morning of proestrus, whereas other factors such as homologous upregulation may not be involved since GnRH release is relatively low at this time point. On the other hand, the steep decline during the afternoon of proestrus may be caused by declining estradiol levels or by homologous down-regulation since GnRH release from the mediobasal hypothalamus is high. However, the regulation of the GnRH receptor mRNA during the estrous cycle is not found uniformly throughout the central nervous system. For instance, the relative amounts of the GnRH receptor mRNA in region CA3 of the hippo-campus do not change under these physiological con-ditions; the mRNA content in the pyramidal neurons of area CA1 and the dentate gyrus are high during the afternoon of diestrus II and the morning of proestrus before they decline to their lowest level by the after-noon of proestrus. However, none of the changes in the hippocampus reach statistical significance. While it is clear that the gonadal steroids affect the ultra-structure, axonal sprouting, and synaptic receptor density of hippocampal neurons, the role of these neurons in the behavioral or hormonal regulation of the estrous cycle remains to be clarified.

E. Second Messenger Systems

Very limited data are available on the signal trans-duction mechanisms which are activated by GnRH binding to its receptors in the brain. In order to begin to determine the identity of the second messengers, hippocampal slices were preloaded with L-myo-[1,2-^{3}H](N)-inositol, exposed to GnRH in the presence or absence of the GnRH antagonist Nal-Glu-GnRH, and total [^{3}H]inositol phosphates were measured. The results show that GnRH stimulates the produc-tion of total [^{3}H]inositol phosphates in a dose-depen-dent manner and that this effect is prevented in the presence of the GnRH antagonist. When [^{3}H]inositol monophosphate, [^{3}H]inositol bisphosphate, and [^{3}H]inositol trisphosphate were eluted sequentially by ion exchange chromatography, it was shown that at low physiological concentrations of GnRH (100 pM–10 nM), all three forms of inositol phosphates were formed at similar rates. However, at higher concentrations of GnRH (100 nM–1 μM) mostly inositol monophosphate and inositol bisphosphate were produced. In order to determine if the gonadal steroids can alter the rate of inositol production in response to GnRH, hippocampal slices of ovariecto-mized or estradiol–progesterone-treated animals were used. The results show that low doses of GnRH (100 pM–1 nM) cause a twofold larger increase in [^{3}H]inositol phosphate synthesis in the estradiol–progesterone-treated animals when compared to the ovariectomized animals. However, at higher concen-trations of GnRH (10 nM–1 μM), no difference be-tween these two animal models was apparent. These results suggest that the response to GnRH in hippo-campal pyramidal cells is modulated by gonadal ste-roids such that the neurons appear to be sensitized by low physiological concentrations of the hormone. If it can be shown that the gonadal steroids have a similar effect on other GnRH target neurons in the brain, this sensitization may be an important mecha-nism for the regulation of other intracerebral effects of GnRH, such as the facilitation of reproductive be-haviors.

IV. RECEPTOR–EFFECTOR COUPLING

A. Microaggregation in Early Receptor Actions

The action of GnRH is mediated via a 7-TMS G-protein-coupled receptor. It has been shown that a pure GnRH antagonist is converted to an agonist if it is bestowed with the ability to dimerize receptors. Similarly, dimerization enhances the potency of GnRH agonists. Although these observations made it attractive to consider receptor dimerization as an early event in hormone action, there has not been a sufficient technique to allow demonstration of ligand-induced receptor–receptor interactions for this receptor or for any other member of the ubiquitous 7-TMS class.

A GnRH analog (agonist) was coupled to lactoperoxidase. Primary pituitary cell cultures (approximately 9×10^7 cells) were incubated (1.7 ml, 4°C) with 10^{-7} M GnRHa-LPO. At this temperature, agonists bind to the receptor but do not produce any measurable response (activation of second messenger systems or LH release). After 2 hr, the cell suspension was mixed with 20 ml medium 199 containing 10 mM Hepes buffer, pH 7.4, and 0.3% bovine serum albumin; cells were collected by centrifugation and resuspended in 1 or 2 ml medium. The cells were aliquotted, then warmed in a water bath to 23°C (chosen because cells respond to GnRH with LH release, although the membrane fluidity, desensitization, and ligand dissociation are markedly decreased) for 0, 1, 3, 5, 10, 15, 30, or 45 min. The cells were chilled to 4°C. The cell suspension was then mixed with 100 μCi ^{125}I and 0.015% H_2O_2 (final concentration) and warmed to 23°C for 1 min. The reaction was terminated and stabilized by the addition of 20 μl of a saturated solution of potassium iodide and sodium azide. The cells were solubilized in lysis buffer (0.15 M NaCl, 20 mM Hepes, 1 mM EDTA, 0.5% Triton X-100, 1 μM leupeptin, and 1 μM pepstatin A); receptor was partially purified on a 200-μl agarose-bound wheat germ agglutinin column which was eluted in 1 mM N, N', N''-triacetylchitotriose and then it was electrophoresed on a 7.5% SDS–PAGE system. The gel was dried and autoradiographed. For competition binding studies only the 5-min warming time point was examined.

The iodinated cell profile shows a number of bands, including a dark band in the lower half of the gel which appears even in the absence of LPO and provides a convenient internal marker. Beneath this band is a diffuse band which (i) increases in density with a warming period, (ii) comigrates with photoaffinity-labeled receptor, and (iii) is inhibited by competition with GnRH agonists and antagonists but not by a chemically related analog (des-pyroGlu1-GnRH) that does not bind to the receptor, other hypothalamic-releasing hormones [thyrotropin-releasing hormone (TRH), CRF, and GHRH], or luteinizing hormone. Likewise, this band is absent when the GnRHa-LPO conjugate is omitted from or added without a preincubation for binding and does not appear when the corresponding LPO conjugate is prepared from a GnRH antagonist (D-p-Glu1-D-Phe2-D-Trp3-D-Lys6-GnRH). For lanes derived from cells maintained completely in the cold prior to radioiodination (i.e., time zero), there is virtually no detectable receptor iodination, whereas increasing iodination is measured following even brief warming periods (1 min). This suggests that self-iodination of receptor is not occurring; the warming period is likely permissive of lateral mobility (microaggregation) of the receptor. There is an increase in background progressive with time in the gels of partially purified receptor. No detectable increase is seen in gels derived from the total cell mixture (i.e., prior to the agarose–lectin column, in which the receptor iodination is not visible; not shown). It is conceivable, therefore, to recognize that the increased background results from degraded receptor and the presence of other proteins present in the neighborhood of the receptor which are copurified. Receptor iodination is also not likely a result of receptor patching and capping which occurs at substantially later times (20 min even at 37°C) and is blocked by 100 μM vinblastine, unlike LH release and the presently described process.

The observation that substitution of receptor Lys121 (located near the cytoplasmic face of the third TMS) decreases agonist binding affinity and that Glu301 (located at the cytoplasmic face on the seventh TMS) apparently interacts with Arg8 in the ligand suggests that the D-Lys6 in the releasing hormone analog is oriented toward the outside of the cell. Likewise, the ability to form biologically active GnRH analogs with bulky substitutions (lactoperoxidase, ferritin, colloidal gold, rhodamine, and t-butyl-D-Ser) at the sixth position (or radioiodine at Tyr5) suggests that these positions are not buried in the lateral plane of the membrane but rather project outward from the cell. Also, in the present study, the LPO is separated from the releasing hormone analog by an 11.6 Å spacer arm allowing considerable rotational freedom. These observations and the recognition that allosteric changes occur virtually instantaneously with binding, unlike the slight lag noted in the time course here, suggest that self-iodination of a receptor as a result of a temperature-requiring agonist-dependent allosteric change is an unlikely explanation of the current data. The likely explanation is that LPO molecules are iodinating tyrosyl residues on neighboring receptors. External (i.e., accessible to iodination) Tyr residues are located at positions 108 (second external loop) and 182 (third external loop).

Receptor iodination can be measured at the earliest time point (1 min) after warming of agonist-occupied receptor. This precedes detectable gonadotropin release from the cells (20 min at 37°C) and it is consonant with the earliest detectable responses to the releasing hormone, including calcium flux and enhanced IP metabolism. In addition, the time course of development of receptor–receptor interactions measured by this method suggests that it is associated with the onset of endocrine response rather than patching, capping, and internalization of receptor, which is scarcely measurable at 15 min and develops at 30 min. This technique, which should be generally applicable to other receptor systems, provides evidence that GnRH agonists provoke microaggregation of the 7-TMS receptor as an early mechanistic event in hormone action—the first measurable step after hormone binding to its receptor.

B. Calcium Dependence of Gonadotropin Release

Clearly, GnRH requires ionic calcium (Ca^{2+}) for many of its actions. Omission (or chelation) of this ion in the extracellular medium bathing the cells inhibits both depolarization and hypothalamic extract-stimulated LH release from pituitary tissue; this observation provided an early indication of the significance of calcium. This ion was among the first components of the GnRH response system to be systematically evaluated when synthetic releasing hormone became generally available. Agents that increase the levels of intracellular Ca^{2+}, including ionophores A23187, X537A, and ionomycin, depolarizing agents KCl and veratridine, calcium channel activators, and calcium-loaded liposomes, release LH with efficacies similar to that of GnRH. Other cations do not substitute for calcium in this activity. An increase in transmembrane Ca^{2+} flux in response to

GnRH was demonstrated by measuring efflux from $^{45}Ca^{2+}$-loaded cells. The use of Ca^{2+}-sensitive fluorophores has provided evidence that intracellular free Ca^{2+} concentrations rise transiently following exposure of target cells or GnRH.

A role for activation of specific plasma membrane calcium channels in response to GnRH is also indicated by the ability of Ca^{2+} channel-blocking agents (verapamil or D600) to inhibit stimulated LH release, both in isolated cells and in human subjects, and by the LH-releasing action of maitotoxin, a Ca^{2+} channel-activating agent. In addition, patch-clamp studies of isolated gonadotropes have shown that although these cells do not depolarize in response to GnRH, the releasing hormone is capable of activating a plasma membrane Ca^{2+} current. This finding was consistent with the observation that tetrodotoxin, a potent inhibitor of the sodium channel, inhibited LH release in response to agents that provoked depolarization (Kcl and veratridine) but not in response to GnRH itself. Elevation of intracellular Ca^{2+}, delivered in part via GnRH-activated plasma membrane Ca^{2+} channels, is likely to be an important signal for the stimulation of LH release by GnRH.

A potential intracellular mediator of the Ca^{2+} signal in gonadotropes is calmodulin, which, after binding Ca^{2+}, alters the activity of several enzymes and cytoskeletal proteins that have been implicated in the secretory process. Calmodulin redistributes from the soluble to the particulate fraction (most likely plasma membrane) of the pituitary following administration of GnRH to rats. In immunohistochemical studies, calmodulin localizes in association with the GnRH receptor "patch" following GnRH treatment. In addition, pharmacological agents that inhibit the activity of Ca^{2+}-calmodulin also block LH release in response to GnRH of Ca^{2+} ionophores. These agents include pimozide, penfluridol, and the naphthalene sulfonamide "W" compounds.

A systematic study was undertaken to characterize and identify the calmodulin-binding components of the gonadotrope. In one study relying on a [^{125}I]iodocalmodulin gel overlay assay (Midwestern blot) we showed tissue-specific and Ca^{2+}-dependent patterns of ^{125}I-labeled calmodulin binding. Five major [^{125}I]iodocalmodulin-labeled components of subunit M_r = 205,000, 200,000, 135,000, 60,000, and 52,000 were identified. Binding of calmodulin was abolished in the presence of 1 mM EGTA (a calcium chelator) or calmodulin antagonists such as penfluridol (1 μM) or pimozide (1 μM). Subcellular fractionation revealed that the Ca^{2+}-dependent calmodulin-binding components are localized primarily in the cytosolic fraction. Separation of dispersed anterior pituitary cells by a linear metrizamide gradient yielded gonadotrope-enriched fractions that contained all five [^{125}I]iodocalmodulin binding components corresponding to the major bands in the pituitary homogenate. Studies with ovariectomized or ovariectomized with steroid-replaced animals indicated that the tissue content of calmodulin binding components, like calmodulin itself, did not appear to be differentially regulated by steroids. A comparison of rat and bovine pituitary tissue homogenates revealed that binding components migrating at the same apparent molecular weights were found for four of the components. Several of the gonadotrope proteins that are regulated by Ca^{2+}-calmodulin have been identified as spectrin, caldesmon, and calcineurin based on their molecular weight, immunological character, calmodulin binding activity, and other physiological criteria. Recently, studies in which specific antiserum to caldesmon has been inserted into living cells have revealed potentiation of LH release in response to GnRH. A model has been proposed in which Ca^{2+} regulates gonadotropin release by control of the gel-sol state of the cytosol.

C. Inositol Phospholipids and Ca^{2+}-Mobilizing Signals

Interest in the relation between Ca^{2+} mobilization and inositol phospholipid metabolism grew out of studies showing that stimulated metabolism of inositol-containing phospholipids was involved in the gating (regulated entry) of Ca^{2+} in response to extracellular signals. This hypothesis provoked studies that linked GnRH to increased rates of metabolism of phosphatidic acid and phosphatidylinositol. The observation of rapid Ca^{2+}-independent phospholipase C-type hydrolysis of phosphatidylinositol 4,5-bisphosphate (PIP$_2$), which constitutes only a small fraction of the total inositol phospholipid pool, led to the discovery that a product of this cleavage, inositol-1,4,5-triphosphate (IP$_3$), releases Ca^{2+} sequestered in a nonmitochondrial storage site. Recently, IP$_3$ has

been implicated as a regulator of Ca^{2+} gating at the plasma membrane.

In GnRH-stimulated gonadotropes, an increase in phospholipid metabolism, accumulation of inositol phosphates, and an increase in the concentration on intracellular calcium have been observed.

GnRH-stimulated production of inositol phosphates has been measured in pituitary tissue slices and in dispersed cells. Inositol phosphates produced include IP$_3$; this isomer has been reported to release Ca^{2+} from subcellular organelles derived from the pituitary. In view of previous studies that indicated the dependence of LH release on continual access to extracellular Ca^{2+}, the extent to which Ca^{2+} derived from intracellular sources contributes to Ca^{2+}-dependent LH release remains unclear. Still, studies of perifused pituitary cells have shown that chelation of Ca^{2+} in the extracellular medium does not totally extinguish the earliest detectable release of LH after a pulse of GnRH, but that conditions that deplete cellular Ca^{2+} (EGTA + ionophore A23187) eliminate this early release. Collectively, these results oblige consideration of a potential role for inositol phosphates in gonadotrope Ca^{2+} mobilization.

To be considered a second messenger for GnRH-stimulated LH release, three requirements must be met: (i) GnRH should stimulate increased inositol phosphate turnover; (ii) inositol phosphate turnover, stimulated by any means, should provoke LH release; and (iii) inhibition of inositol phosphate turnover should block GnRH-stimulated LH release. The first requirement has been fulfilled. GnRH does stimulate IP turnover. Although GTP analogs and sodium fluoride stimulate both IP turnover and LH release, this behavior does not fulfill the second prerequisite because G-protein activation may provoke cellular effects in addition to IP turnover that could be responsible for LH release. The third criterion has not been demonstrated and, notably, one study reported that the Ca^{2+} channel blocker D600 and the calmodulin antagonist pimozide decreased GnRH-stimulated LH release without inhibiting IP turnover. The results of that study question the relation between GnRH-stimulated IP turnover and LH release.

To study the dependence of GnRH-stimulated LH release on IP turnover, this study utilized an inhibitor of phospholipase C activity, 1-(6{[17β-3-methoxyestra-1,3,5(10)-triene-17-yl]amino}hexyl)-

1H-pyrrole-dione (U-73122) and an inactive analog, 1-(6{[17β-3-methoxyestra-1,3,5(10)-triene-17yl] amino}hexyl])-2,5-pyrrolidinedione (U-73343). U-73122 (10 μM) decreased GnRH-provoked (1 μM, 45 min) IP accumulation from 873 ± 61 to 365 ± 50 dpm (basal accumulation also was decreased from 420 ± 18 to 207 ± 16 dpm), whereas LH release was not inhibited ($30.2 \pm 1.4\%$ of cellular LH in control compared with $30.3 \pm 1.1\%$ in U-73122-pretreated cells). GnRH provoked increased IP_3 accumulation (123% of basal) after 15 sec of stimulation, IP_2 accumulation (131% of basal) after 30 sec, and IP_1 (121% of basal) after 1 min. Pretreatment with U-73122 blocked accumulation of IPs at these early time points. Sodium fluoride-stimulated IP accumulation was also inhibited by U-73122 (from 1539 ± 132 to 414 ± 21 dpm), whereas LH release increased from $22.9 \pm 1.4\%$ total cellular LH to $28.0 \pm 2.2\%$. In contrast, GnRH- and sodium fluoride-stimulated IP accumulation were not significantly decreased in U-73343-pretreated cells (GnRH: 817 ± 43 dpm compared with 873 ± 61 dpm in control; sodium fluoride: 1133 ± 74 dpm compared with 1539 ± 132 dpm in control cells). Results of a perifusion study showed that U-73122 did not block the initial phase of GnRH-stimulated LH release or interfere with the development of desensitization to the releasing hormone. In addition, GnRH-stimulated intracellular Ca^{2+} fluctuations were similar in magnitude and duration in U-73122-pretreated cells and U-73343-pretreated cells. These results demonstrate that GnRH- as well as sodium fluoride-stimulated LH release can be uncoupled from IP turnover, questioning the role of IP_3 as a second messenger for GnRH-stimulated LH release.

U-73122 has also been used to show that IP accumulation and gonadotrope desensitization can be uncoupled.

D. Receptor Uncoupling from Phosphoinositide Hydrolysis

GnRH regulates pituitary gonadotropin release by a Ca^{2+}-dependent mechanism involving receptor-mediated phosphoinositide hydrolysis. Previous studies indicated that activation of pituitary PKC, although not required for acute gonadotropin release

in response to GnRH, is likely to be involved in the chronic regulation of gonadotrope responsiveness, and that activation of PKC by phorbol esters produces both the uncoupling of GnRH-stimulated phosphoinositide hydrolysis and the selective enhancement of GnRH agonist binding in pituitary cell cultures. We have examined the possibility that these processes are mechanistically related. Dissociation of bound agonist radioligand at 23°C was found to be reduced in the presence of phorbol esters; ligand bound in the presence of phorbol ester was resistant to displacement by competing ligands at 4°C. However, agonist bound in the presence of phorbol ester was dissociated by subsequently washing cells at pH 3. Receptor photoaffinity-labeling studies confirmed that agonist association with a membrane component identified as the GnRH receptor was increased in the presence of phorbol ester. These results suggest that in the presence of a phorbol ester PKC activator, agonist-occupied GnRH receptors remain at the cell surface but are sequestered in some manner. In other experiments, cells preloaded with [³H]inositol were treated with GnRH agonist ligand and phorbol ester at 4°C to form a pool of sequestered agonist-occupied receptors; then displaceable (nonsequestered) agonist was removed by incubation with antagonist ligand. After addition of LiCl and warming to 37°C, [³H]IP production (an index of phosphoinositide hydrolysis) in phorbol ester-treated cells was reduced to 67% of vehicle control, although residual specific agonist binding had been increased to >300% of control. The appearance of sequestered receptors and inhibition of [³H]IP production had similar phorbol ester concentration dependencies. These results suggest that the same agonist-occupied GnRH receptors sequestered as a result of PKC activation are also preferentially uncoupled from phosphoinositide hydrolysis.

E. G-Proteins

Many laboratories have focused on the possible involvement of GTP-binding proteins linking activated receptor to formation of IPs and diacylglycerol. This possibility was first suggested when researchers found that receptors linked to phosphoinositide hydrolysis altered affinities for agonist in the presence of guanine nucleotides. Thus, phospholipase C (PLC) may be regulated by a single-transduction G-protein in a manner analogous to adenylate cyclase or phosphodiesterase. In support of this hypothesis, activation of PLC occurs with GTP and with nonhydrolyzable GTP analogs and is inhibited by GDPβS (competitive inhibitor of G-proteins). In addition, hormonal activation of PLC is dependent on guanine nucleotides and occurs simultaneously with an increase in high-affinity GTPase activity.

Several studies suggested the involvement of a G-protein coupled to the GnRH receptor that, when activated, provoked LH release and IP production. Addition of GTP or a metabolically stable analog (guanylimidodiphosphate) to permeabilized pituitary cells stimulated a time- and concentration-dependent increase in inositol phosphate accumulation and LH release. These responses were insensitive to both pertussis toxin and cholera toxin, indicating that the putative G-protein-mediating GnRH actions, such as that hypothesized to mediate TRH in GH$_3$ pituitary cells, had properties different from G$_s\alpha$ and G$_i\alpha$. Further evidence consistent with an association of the GnRH receptor and a G-protein was provided in 1989 by the demonstration of decreased receptor affinity for a GnRH agonist in the presence of guanine nucleotides.

In other studies, sodium fluoride, an exogenous activator of G-proteins (when complexed with aluminum ions present as a contaminant in most solutions), was used to investigate the possibility of a G-protein link between GnRH receptor activation, PLC activity, and LH release. Treatment of primary pituitary cell cultures from immature female rats with sodium fluoride stimulated release of 20% of total cellular LH (compared with 40–50% at maximal GnRH concentrations) and increased IP accumulation. Sodium fluoride-stimulated LH release was insensitive to cholera toxin and pertussis toxin. Sodium fluoride-stimulated LH release was additive, with a maximally effective concentration of PMA, and was not inhibited by depletion of cellular PKC, suggesting that PKC does not mediate sodium fluoride effects. Treatment of cultures with 3 mM EGTA and 10 nM GnRH for 5 or 16 hr reduced pituitary responsiveness to subsequent treatment with GnRH but had no effect on sodium fluoride-stimulated LH release. Although

the precise mechanism of sodium fluoride-stimulated LH release remains to be described, these studies support a role for a G-protein in regulation of LH release by the releasing hormone.

Because a G-protein appears to be activated following GnRH stimulation of the gonadotrope, a role for this moiety in GnRH-stimulated alterations in gonadotrope responsiveness was also assessed. A 3-hr pretreatment of pituitary cell cultures with 10 mM NaF (a G-protein activator) resulted in decreased gonadotrope responsiveness to subsequent GnRH treatment (3 hr, 100 nM; 34.3 ± 1.6 vs 23.4 ± 1.5% of total cellular LH). Sodium fluoride-provoked gonadotrope desensitization to GnRH also occurred in the presence of 3 mM EGTA and in cells that had been depleted of PKC. Desensitization to GnRH did not occur in response to pretreatment with dibutyryl cAMP (dBcAMP; 8 hr, 1 mM). In addition, neither GnRH nor sodium fluoride stimulated IP accumulation above basal levels following the sodium fluoride pretreatment. GnRH receptor binding also decreased by 30% with sodium fluoride pretreatment. In contrast, a 3-hr sodium fluoride (10 mM) pretreatment enhanced responsiveness of the gonadotrope to the Ca^{2+} ionophore A23187 in a PKC- and cAMP-dependent manner. Responsiveness to the phorbol ester PMA was also increased, whereas responsiveness to the Ca^{2+} channel activator maitotoxin was unchanged. These data suggest that G-protein activation by sodium fluoride provokes gonadotrope desensitization to GnRH stimulation both by decreasing receptor numbers and by uncoupling of the receptors from IP turnover. In addition, a distinct G-protein action appears to be involved in sensitizing the gonadotrope to A23187 and PMA. Studies using cholera and pertussis toxins have provided additional support of the view that multiple G-proteins are involved in GnRH action.

Our laboratory has assessed the potential for crosstalk between a cholera toxin-sensitive G-protein and PKC. In these studies cells were depleted of PKC with 1 μM PMA (12 hr, followed by M199/BSA for 6 hr) prior to treating with vehicle, pertussis toxin (PTX), cholera toxin (CTX), or dibutyryl cyclic AMP (dBcAMP) for 18 hr. PTX (10 ng/ml) significantly decreased GnRH-stimulated IP production over a range of 10^{-8}–10^{-6} M GnRH. The degree of this inhibition was the same in control cells and PKC-depleted cells. Pretreatment with CTX (0.5 μg/ml) significantly decreased GnRH-stimulated IP production over a range of 10^{-9}–10^{-6} M GnRH in PKC-depleted cells. This effect was mimicked by pretreatment with 3 mM dBcAMP. Although CTX and dBcAMP both decreased GnRH-stimulated IP production in control cells, this effect was enhanced in PKC-depleted cells. CTX (0.1 μg/ml) and dBcAMP (3 mM) both enhanced GnRH-stimulated LH release, whereas PTX (100 ng/ml) had no effect. This effect is observed in control cells as well as in PKC-depleted cells.

Both PKA and PKC are capable of regulating IP turnover by phosphorylating PLC at distinct sites. CTX activates a G-protein that increases cAMP. Cyclic AMP can then activate PKA. In PKC-depleted cells, CTX inhibits GnRH-stimulated IP production. This effect is mimicked by dBcAMP, which suggests a role for PKA in the gonadotrope. The results of this study provide evidence for cross-talk between a CTX-sensitive G-protein and PKC.

We have used three different methods to assess GnRH receptor regulation of $G_{q/11}\alpha$ subunits ($G_{q/11}\alpha$). First, we used GnRH-stimulated palmitoylation of $G_{q/11}\alpha$ to identify their involvement in GnRH receptor-mediated signal transduction. Dispersed rat pituitary cell cultures were labeled with [9,10-^{3}H(N)]palmitic acid and immunoprecipitated with rabbit polyclonal antiserum made against the C-terminal sequence of $G_{q/11}\alpha$. The immunoprecipitates were resolved by 10% SDS–PAGE and quantified. Treatment with GnRH resulted in time-dependent (0–120 min) labeling of $G_{q/11}\alpha$. GnRH (10^{-12}, 10^{-10}, 10^{-8}, or 10^{-6} M) for 40 min resulted in dose-dependent labeling of $G_{q/11}\alpha$ compared to controls. Cholera toxin 5 μg/ml; activator of $G_s\alpha$), pertussis toxin (100 ng/ml; inhibitor of $G_i\alpha$ actions), and Antide (50 nM; GnRH antagonist) did not stimulate palmitoylation of $G_{q/11}\alpha$ above basal levels. However, phorbol myristic acid (100 ng/ml; PKC activator) stimulated the palmitoylation of $G_{q/11}\alpha$ above basal levels, but not to the same extent as 10^{-6} M GnRH. Second, we used the ability of the third intracellular loop (3_i) of other 7-TMS receptors that couple to specific G-proteins to antagonize GnRH receptor-stimulated signal transduction, and therefore act as an intracellular inhibitor. Because the third intracellular loop of α_{1B}-

adrenergic receptor ($\alpha_{1B}3_i$) couples to $G_{q/11}\alpha$, it can inhibit $G_{q/11}\alpha$-mediated stimulation of IP turnover by interfering with receptor coupling to $G_{q/11}\alpha$. Transfection (efficiency, 5–7%) with $\alpha_{1B}3_i$ cDNA, but not the third intracellular loop of M_1-acetylcholine receptor (which also couples to $G_{q/11}\alpha$), resulted in 10–12% inhibition of maximal GnRH-evoked IP turnover compared to vector-transfected GnRH-stimulated IP turnover. The third intracellular loop of α_{2A}-adrenergic receptor, M_2-acetylcholine receptor (both couple to $G_i\alpha$) and D_{1A} receptor (couples to $G_s\alpha$) did not inhibit IP turnover significantly from control values. GnRH-stimulated LH release was not affected by the expression of these peptides. Third, we assessed GnRH receptor regulation of $G_{q/11}\alpha$ in a prolactin-secreting adenoma cell line (GGH$_3$1′) expressing the GnRH receptor. Stimulation of GGH$_3$1′ cells with 0.1 μg/ml Buserelin resulted in a 15–20% decrease in total $G_{q/11}\alpha$ at 24 hr following agonist treatment compared to control levels; this action of the agonist was blocked by GnRH antagonist, Antide (10^{-6} g/ml). Neither Antide (10^{-6} g/ml, 24 hr) alone nor phorbol myristic acid (0.33–100 ng/ml, 24 hr) mimicked the action of GnRH agonist on loss of $G_{q/11}\alpha$ immunoreactivity. The loss of $G_{q/11}\alpha$ immunoreactivity was not due to an effect of Buserelin on cell-doubling times. These studies provide the first direct evidence for regulation of $G_{q/11}\alpha$ by the GnRH receptor in primary pituitary cultures and in GGH$_3$ cells.

F. Gonadotropin Biosynthesis

The gonadotropins LH and FSH consist of two heterogeneous subunits, α and β. The α subunit is common between these and another pituitary hormone, thyrotropin.

The availability of radiolabeled oligonucleotide probes for gonadotropin α and β subunit mRNAs has made it possible to investigate gonadotrope regulation of gene expression. Papavasiliou and colleagues reported in 1986 that mRNAs for α and β LH subunits in intact pituitaries are increased twofold after 48 hr of pulsatile exposure to GnRH. Recently, the ability of GnRH and phorbol esters to increase LH β mRNA levels in cell cultures has been shown. Moreover, depletion of PKC activity inhibits the ability of GnRH to increase mRNA levels. Therefore, in contrast to GnRH-stimulated LH release, receptor downregulation, and desensitization, the action of GnRH on LH mRNA levels appears to require PKC. Whether the rates of mRNA synthesis and degradation are both altered by activated PKC is not currently clear. The action of this enzyme on events at the transcriptional level has been implicated in a variety of systems and is certainly consistent with the long-appreciated effects of phorbol esters on cellular differentiation and transformation.

Pulsatile exposure to GnRH is also believed to be important for the maintenance of cellular LH pools. In this regard, GnRH has been reported to stimulate gonadotropin polypeptide biosynthesis and glycosylation in cell cultures, as measured by incorporation of labeled amino acids and monosaccharides into immunoreactive LH. Similarly, activators of PKC have been found to increase the rates of LH polypeptide biosynthesis and glycosylation. Whether these increases reflect an effect of PKC at the translational level or are secondary to increases in gonadotropin mRNA levels is not known, nor have researchers determined whether PKC is required for the stimulation of gonadotropin biosynthesis by GnRH.

V. DESENSITIZATION OF THE RESPONSE SYSTEM

A. Gonadotrope Desensitization

Diminished gonadotropin release as a result of prior exposure to GnRH-homologous desensitization occurs in the pituitary gonadotrope and is the basis of the clinical efficacy of this hormone when used to treat disease states benefiting from a functional castration. *In vitro* and *in vivo* studies show that pulsatile administration (pulse/60 min or pulse/90 min) of GnRH is associated with maintenance of gonadotropin release, whereas constant or rapidly pulsed GnRH (pulse/15 min) results in diminished gonadotropin release. Loss of responsiveness due to constant high-frequency stimulation can be restored by resetting the pulse frequency (pulse/90 min). The underlying molecular mechanism of desensitization has not been completely defined but appears to be associated with multiple changes in the gonadotrope.

B. GnRH Regulates Receptor Numbers

When pituitary cell cultures were incubated with 1 nM GnRH continuously, the receptor number underwent biphasic regulation. At initial times (0–3 hr), receptor numbers decreased below control levels but recovered (3–5 hr) and overshot the control values (6–9 hr). The initial decrease in receptor number is not likely to be due to occupancy of the GnRH receptor by the homologous hormone since the cells are washed under conditions that would be expected to elute the GnRH ($t_{1/2}$ for dissociation is a few minutes) and the 2-hr binding assay provides sufficient time for exchange of any remaining GnRH with the higher affinity radioligand.

In studies in which antiserum AB-9113 [which binds GnRH, but not the (analog) radioligand] was included during the binding incubation (titer, 1 : 100), no difference was seen in numbers of receptors compared with measurements made in the presence of preimmune serum.

C. The Functional State of the Receptor-Linked Ca^{2+} Ion Channel Changes in Desensitization

The time course and efficacy of LH release in response to GnRH and to maitotoxin (an activator of the GnRH receptor-linked Ca^{2+} ion channel) are similar; both secretagogues require extracellular Ca^{2+} and are inhibited by the selective Ca^{2+} ion channel antagonist methoxyverapamil (D600). LH release in response to either GnRH or maitotoxin is not measurably inhibited by two other chemical classes of Ca^{2+} ion channel inhibitor represented by nifedipine and diltiazem. The two secretagogues are nonadditive in their action on LH release when presented at high doses; prior studies indicate that maitotoxin has no endogenous ionophoretic activity. These observations indicate that maitotoxin is likely to stimulate LH release by activation of the GnRH receptor-associated Ca^{2+} ion channel in the gonadotrope.

We have therefore assessed the functional state of this channel during the development of homologous desensitization of the gonadotrope to GnRH by measuring the ability of maitotoxin to stimulate LH release. Cells were desensitized with GnRH in the presence of 3 mM EGTA. Under these conditions, the cells become refractory to GnRH in the absence of gonadotropin release since the latter process, but not the former, requires extracellular Ca^{2+}. Accordingly, this approach allows assessment of the degree of desensitization in the absence of the influence of gonadotropin depletion. Such desensitized cells are less responsive to GnRH. Desensitized pituitary cells also respond with diminished efficacy and potency to maitotoxin 3 hr or more after GnRH treatment but not at an earlier time (1 hr) when GnRH receptors are diminished. These data are consistent with a model in which homologous desensitization is viewed as developing in two phases. Initially, loss of responsiveness is due to receptor loss; subsequently it is maintained by loss of functional activity of the Ca^{2+} ion channel.

D. A CTX-Sensitive Guanyl Nucleotide Binding Protein Mediates the Movement of Pituitary Luteinizing Hormone into a Releasable Pool: Loss of This Event is Associated with the Onset of Homologous Desensitization to GnRH

Experiments were conducted in which perifused cell cultures were pretreated (18 hr) with CTX (5 μg/ml), PTX (100 ng/ml), or medium alone (control). The cells were then treated with 5 nM GnRH either continuously or pulsed (5 min) every 15, 30, 60, or 90 min. In all cases the initial response was 2- to 2.5-fold higher in CTX-pretreated cells than in PTX or control cells. In cultures that were exposed to continuous administration of GnRH, the difference in LH release between CTX-pretreated cells and controls or PTX-pretreated cells was gradually lost and the response level was virtually indistinguishable by 120 min (90 min after the initiation of GnRH). By this time, overall LH release dropped markedly from the original response levels and was essentially basal. In response to 1 μM GnRH, which circumvents desensitization, the CTX-pretreated cells produced a 1.8-fold higher response than the control or PTX-pretreated cells.

When perifused cultures were pulsed with 5 nM GnRH every 15 min, the difference in response be-

tween the CTX-pretreated and control cells was lost by the ninth pulse; pulsatile LH release, albeit substantially desensitized, could be seen throughout the 20 pulses given. As observed for continuous administration study, 1 μM GnRH results in an enhanced LH release from the CTX-pretreated cells compared with PTX-pretreated or control cells. A GnRH pulse interval of 30 min is also sufficient to provoke desensitization and, concomitantly, a loss of difference in response level between CTX-pretreated and control cells. In this experiment, 60-min pulse intervals of 5 nM GnRH only modestly desensitize the cells to GnRH and partially decrease the difference between CTX-pretreated and control cells. A pulse interval of 90 min appears sufficient to maintain LH release and the differences between response levels of CTX-pretreated and control cells.

The action of CTX pretreatment could be mimicked by pretreatment with 1 mM dBcAMP. When pretreated cells were pulsed (pulse/30 min) with 5 nM GnRH, the response in dBcAMP- or CTX-pretreated cells was initially 2- to 2.5-fold higher than in control cells. As pulses were continued and the cells desensitized, the LH release levels in the pretreated cells approached that of the controls; by pulse 7 or 8, the release levels were indistinguishable. When cells were rapidly pulsed (pulse/15 min or pulse/30 min) until CTX-pretreated cells and control cells had nearly the same response levels (12 and 6 pulses, respectively) and then pulsed at pulse/90 min, the enhanced level of release in CTX-pretreated cells was restored to higher than control levels. When a 60-min pulse period was used, only slightly desensitizing the cells, the administration of pulses at 90 min restored the difference in CTX-pretreated and control LH release and did not further desensitize the cells. The SEM total (i.e., released plus unreleased) LH for each treatment group varied by less than 4% of the mean; the percentage of LH released was uniformly higher in CTX-pretreated cells. In no experiment was the total amount of LH higher in CTX-pretreated groups than in control or PTX-pretreated cells.

These studies show that (i) CTX mediates the movement of LH from a nonreleasable to a releasable pool, (ii) the mechanism of this event is consistent with activation of a G-protein that is coupled to production of cyclic AMP because CTX stimulates the production of cyclic AMP and 1 mM dBcAMP mimics the action of CTX, and (iii) this action of CTX is lost progressively as cells become desensitized by continuous or rapidly pulsed (pulse/15 min) administration of physiological levels of GnRH (5 nM) and is restored by reduction of the pulse frequency to a physiological rate (pulse/60 min or pulse/90 min) or by administration of a high concentration (1 μM) of GnRH, both of which circumvent desensitization and result in elevated levels of LH release. PTX does not have the same action, although we have shown previously that the concentration used is sufficient to alter the production of inositol lipid in response to GnRH. The elevated response to GnRH seen in CTX-pretreated cells is also seen in response to the Ca^{2+} ionophore A23187, indicating that this effect is distal to Ca^{2+} mobilization; this result is consistent with the view that this effect results from movement from a nonreleasable to a releasable pool. Prior to the elucidation of the action of CTX on G-proteins, a report appeared indicating that pretreatment with CTX increased LH released from pituitary cultures. The mechanism by which unreleasable gonadotropin becomes releasable is unclear, although clearly depletion of releasable LH results in a time-dependent movement of the unreleasable gonadotropin into a releasable pool. The further observation that tunicamycin, but not cycloheximide, interferes with this event makes it attractive to consider that glycosylation but not protein synthesis is associated with this event. Since first described in an *in vivo* primate model, the phenomenon of gonadotrope desensitization has been demonstrated in multiple cell culture models, suggesting that the process is an intrinsic property of the gonadotrope cell, allowing it to respond to the frequency of GnRH administration. Initially, a loss of measurable receptor binding occurs in response to the releasing hormone; however, as binding returns and exceeds predesensitization levels, the cells remain refractory to the releasing hormone. This observation has catalyzed the search for other lesions mediating the loss of responsiveness. The use of maitotoxin has shown that as receptors

recover, the function of the receptor-linked Ca^{2+} channel is decreased and the refractory state is maintained. These studies indicate that homologous desensitization to GnRH is associated with a failure of LH to distribute into a releasable pool in the gonadotrope.

VI. CONCLUSION

This article summarizes the state of our understanding of the molecular and cellular mechanism of GnRH action. An interest in this area is warranted by the importance of this releasing hormone and its synthetic analogs in therapeutic applications.

Acknowledgments

Work described from our laboratories was supported by Grant Nos. HD19899 and HD24697.

See Also the Following Articles

Gonadotropes; Hypothalamic—Hypophysial Complex

Bibliography

Conn, P. M., and Crowley, W. F. (1991). Gonadotropin-releasing hormone and its analogues. *N. Engl. J. Med.* **324**, 93–103.

Jennes, L., Eyigor, O., Janovick, J. A., and Conn, P. M. (1997). Brain gonadotropin releasing hormone receptors: Localization and regulation. *Recent Prog. Horm. Res.* **52**, 475–493.

Sealfon, S. C., Weinstein, H., and Millar, R. P. (1997). Molecular mechanisms of ligand interaction with the GnRH receptor. *Endocrine Rev.* **18**(2), 180–205.

Stojikovic, S. S. (1998). Calcium signaling. In *Handbook of Physiology: Molecular and Cellular Endocrinology.* Oxford Press, New York.

Ulloa-Aguirre, A., and Conn, P. M. (1998). G protein-coupled receptors and the G protein family. In *Handbook of Physiology: Molecular and Cellular Endocrinology.* Oxford Press, New York.

GnRH Pulse Generator

Jon E. Levine

Northwestern University

GLOSSARY

GnRH Gonadotropin-releasing hormone; a decapeptide which is synthesized in hypothalamic neurons, released into the hypophysial portal vasculature, and conveyed to the anterior pituitary gland, where it stimulates the synthesis and secretion of the gonadotropins, luteinizing hormone, and follicle-stimulating hormone; also referred to as *luteinizing hormone-releasing hormone.*

hypothalamic–hypophysial portal vessels The specialized vasculature which forms a plexus in the median eminence of the hypothalamus and collects into larger vessels which extend into the hypophysis (pituitary gland); a major function of this vascular network is to convey substances which are neurosecreted from neurovascular junctions in the median eminence to the anterior pituitary gland, where these factors can regulate hormone secretions.

median eminence The neural tissue at the base of the hypothalamus which contains neurovascular junctions, from which releasing factors such as GnRH are released into the hypothalamic–hypophysial portal vessels.

pulsatile neurosecretion Rhythmic releases of neurohormone pulses from neurovascular junctions in the median eminence into the hypothalamic–hypophysial portal vessels.

pulse generator The cellular and molecular mechanisms which produce and sustain pulsatile neurosecretion; these mechanisms are currently unknown.

The central nervous system regulates reproductive hormone secretions primarily through the neurosecretion of gonadotropin-releasing hormone (GnRH). The GnRH decapeptide is synthesized in neurons of the basal forebrain, transported intraneuronally to neurovascular junctions in the median eminence of the hypothalamus, and released into the hypothalamic–hypophysial portal vasculature. Following transport to the anterior pituitary, GnRH can bind receptors located in the plasma membranes of gonadotropes and activate their associated signal transduction pathways. The major actions of GnRH include stimulation of the secretion and synthesis of luteinizing hormone (LH) and follicle-stimulating hormone (FSH). In the absence of GnRH release, secretions and synthesis of LH are virtually halted, FSH production is diminished, and gonadal activities are either impaired or completely arrested.

Neurosecretion of GnRH is almost invariably intermittent, consisting of pulses of secretion which occur regularly at intervals of 20 min or more, depending on the species and physiological circumstances. This pulsatile release pattern, moreover, is obligatory for sustaining normal gonadotropin secretion and synthesis. While the pulsatile infusion of the decapeptide can restore gonadotropin secretions in GnRH-deficient subjects, continuous administration of GnRH fails to sustain gonadotropin release. The pulsatile GnRH release pattern is thus considered a critical feature of the cascade of hormone secretions that constitute the reproductive axis. Those neurons and their processes which function to release the neuropeptide GnRH in this rhythmic manner are collectively referred to as the GnRH pulse generator.

I. GnRH PULSE GENERATOR ACTIVITY

The existence of a neural GnRH pulse generator was originally inferred from the observations that LH secretions are pulsatile in a wide variety of animal

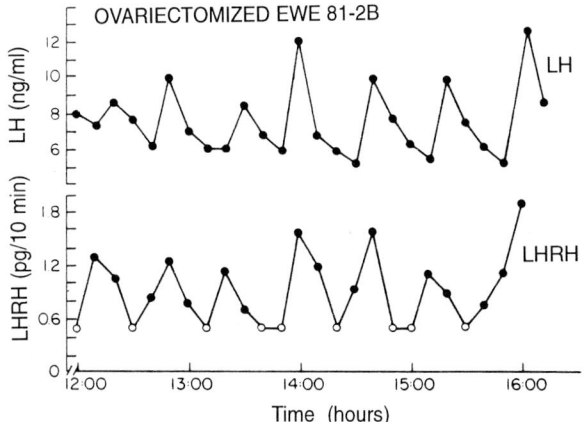

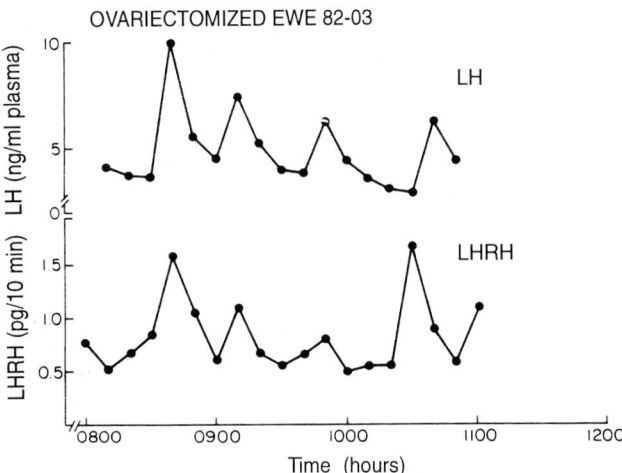

FIGURE 1 Simultaneous measurement of pulsatile GnRH (LHRH) and LH release patterns in two individual, ovariectomed ewes. Median eminence perfusates were collected by push–pull cannulae, and peripheral blood samples were obtained via jugular catheters. The GnRH levels in perfusates and LH levels in serum were determined by GnRH and LH radioimmunoassays, respectively (from Levine *et al.*, *Endocrinology* **111**, 1449–1455, 1982, with permission).

mented, with GnRH pulses found to directly precede or accompany peripheral LH pulses.

The frequency of GnRH pulses can vary among species, sex, age, and physiological circumstance. The GnRH pulse frequencies may range from as fast as one pulse/15 min in gonadectomized rodents, to as slow as one pulse/6 hr or more in anestrous sheep. In general, individual pulses in all species consist of a burst of neurosecretion lasting 0.5–5 min, after which the local concentration of GnRH in the median eminence decays according to the rate at which it is transported through the portal vasculature. The degradation rate in the extracellular fluid of the median eminence may also impact the waveform of an individual pulse. Figure 1 depicts the patterns of GnRH and LH release as measured in two ovariectomized sheep.

II. NEUROENDOCRINE REGULATION OF GnRH PULSE GENERATION

A. Ontogeny of GnRH Pulse Generation

In the vast majority of studies, patterns of GnRH release have not been measured directly but have instead been inferred from analysis of pulsatile LH measurements. Such measurements have indicated that the GnRH pulse generator is functional in the late gestational fetus and in infants in a variety of species. This early period of pulsatility appears to subside and is followed by a prolonged period of apulsatitity or diminished pulsatility during the juvenile phase of development. The GnRH pulse generator "reawakens" during early puberty, with the frequency and amplitude of pulses increasing throughout the pubertal process. During the pubertal transition in humans, a diurnal pattern of GnRH secretion emerges in which nocturnal increases in GnRH pulsatility are observed. These nighttime episodes of GnRH pulsatility disappear with further sexual maturation. It is not known how GnRH pulse generation is regulated throughout pubertal development, but studies suggest that peripheral cues and the actions of several growth factors may be of critical

species. The pulsatile LH secretions can be completely abolished, moreover, by treatments which block the actions of GnRH on the anterior pituitary gland. That a neural GnRH pulse generator exists was thereafter directly confirmed by simultaneous measurements of GnRH and LH release in rodents, domestic farm animals, and nonhuman primates. In all cases, pulsatile GnRH release has been docu-

importance in mediating the pubertal reactivation of GnRH pulsatility.

The GnRH pulse generator operates in an undiminished manner throughout the adult reproductive life span of male and female mammals. In reproductive senescence, the physiological role of the GnRH pulse generator is not clear, and it may vary among species. It is possible that an attenuation of GnRH pulsatility mediates reproductive senescence in some species, such as the rat. It has also been argued that the GnRH pulse generator may only exhibit changed activity in reproductive senescence in response to altered gonadal function; in menopausal women, for example, it is believed that ovarian failure precipitates subsequent activation of GnRH pulsatility as a result of withdrawal of the negative feedback actions of gonadal hormones.

B. Homeostatic Inhibition of the GnRH Pulse Generator

In the absence of any endocrine regulation, GnRH pulse generation in adult animals can occur at frequencies approaching one pulse/10 min. In all normal physiological situations, however, the frequency and amplitude of GnRH pulses are continuously subject to feedback regulation by gonadal hormones. In male animals, the major feedback mechanism is mediated by testicular androgens. The GnRH pulses evoke LH pulses, which stimulate episodic testosterone secretion. Testosterone and possibly other gonadal hormones exert negative feedback influences at the hypothalamic level through the restraint of GnRH pulse generator activity. Inhibition is also exerted at the pituitary level through suppression of gonadotropin responses to GnRH pulses.

In females the negative feedback actions of estrogen and progesterone (as well as inhibin and other gonadal peptides) are also of primary importance in restraining GnRH pulse generation and in suppressing responsiveness of the pituitary gland. While the relative influence of the two ovarian steroids may vary among species and throughout different physiological circumstances, it is likely that the appropriate homeostatic control over basal GnRH pulsatility requires the actions of both steroid hormones. In spontaneous ovulators, the negative feed-

back actions of gonadal hormones, along with their positive feedback effects during the preovulatory period, are major determinants of the frequency and amplitude of pulsatile GnRH release during the ovulatory cycle.

C. GnRH Pulse Generation during the Ovulatory Cycle

In female rats, sheep, primates, and a variety of other animals the characteristics of pulsatile GnRH release are known to change throughout the course of the ovulatory cycle. During the follicular phase of the ewe, for example, pulsatile GnRH release is of a relatively low-amplitude and high-frequency pattern. By contrast, the GnRH pulse frequency is reduced and the amplitude of GnRH pulses is increased during the luteal phase of the cycle. It is believed that estrogen and progesterone may exert negative feedback actions through divergent routes, and that the relative magnitude of their individual effects during the follicular or luteal phases of the cycle may determine these characteristic patterns of GnRH pulsatility during these portions of the ovulatory cycle.

The major change in pulsatile GnRH release occurs at midcycle, during the initiation of preovulatory gonadotropin surges. Under the influence of a rising tide of estrogen secreted by the ripening follicle(s), a surge of GnRH is released which evokes the secretion of the preovulatory LH and FSH surges, the former principally responsible for triggering ovulation. In some species, a secondary GnRH surge may occur which facilitates sustained release of FSH, viz. the secondary FSH surge. The latter functions to recruit ovarian follicles for the subsequent cycle.

The cellular and molecular mechanisms through which estrogen facilitates GnRH surges are not well understood, and much work has been directed toward an understanding of this important neuroendocrine control mechanism. Preovulatory luteinizing hormone-releasing hormone surges have been identified in a variety of species, including rats, sheep, and monkeys, and indirect measurements suggest that they may occur in women. Detailed analysis of GnRH patterns during surges suggest that GnRH pulse amplitude, as well as a basal GnRH secretory component, may be acutely increased as a major portion of this surge release. Mechanisms by which estrogen may exert these positive feedback actions include increased GnRH gene expression, activation of synaptic inputs to pulsing GnRH neurons, alterations in neurotransmitter receptor synthesis, and/or alterations in the expression of postreceptor signaling molecules.

D. Exteroceptive and Interoceptive Signals Which Regulate Pulsatility

The GnRH pulse generator is also regulated by numerous physiological signals from nongonadal sources. These include sensory stimuli, information retrieved from memory, and circulating hormonal and metabolic factors. Perhaps the best studied of these regulatory inputs are photoperiodic stimuli. In seasonally breeding animals, day length is registered through photic pathways which ultimately regulate the pattern of pineal melatonin secretions. Changes in the duration of pineal melatonin secretion, in turn, activate or depress GnRH pulse generation and thereby sustain or depress activity in the reproductive axis. In hamsters, long days are photostimulatory and short days are photoinhibitory for the reproductive axis, and these responses are presumably mediated by activation and inhibition of the GnRH pulse generator, respectively. In sheep, exposure to short days leads to a stimulation of GnRH pulsatility, whereas long days are photoinhibitory; this mechanism dictates the birth of young in spring, when chances of survival for offspring are maximal.

There are numerous other examples of reproductively relevant sensory cues which prompt adaptive alterations in GnRH pulse amplitude or frequency: GnRH pulses are acutely stimulated in ewes following visual exposure to rams; coitus induces an ovulatory GnRH surge in rabbits through activation of sensory pathways leading from the cervix to the hypothalamus; and pheromonal cues from male mice are critically important for release of GnRH surges in female mice. In all these circumstances, reflex pathways convey information regarding the presence of a sexual partner, and the GnRH pulse generator is activated to prepare the gonads for fertilization.

Chronic and acute stress can also alter the pattern of pulsatility, often in an inhibitory manner so as to favor the expenditure of metabolic energy on adaptive stress responses rather than on reproduction. Metabolic and hormonal cues, such as glucose, leptin, thyroid hormones, or glucocorticoids, can also be important in mediating adaptive physiological responses of the GnRH pulse generator. Food deprivation, for example, is a state which is not compatible with reproductive success; it is not surprising, then, that it is accompanied by suppression of GnRH pulsatility in many species. Again, this regulatory mechanism may spare energy stores from being depleted during attempts to reproduce in adverse environmental conditions. In disease states, GnRH pulsatility may also be suppressed. Immune stress and inflammation responses may be accompanied by suppression of the GnRH pulse generator, possibly through the actions of cytokines that are known to be produced in these conditions.

III. PHYSIOLOGICAL SIGNIFICANCE OF GnRH PULSE GENERATOR ACTIVITY

The pulsatile pattern of GnRH release is of critical physiological significance for the normal operation of the reproductive axis. Many studies have demonstrated that pulsatile administration of GnRH to experimental subjects *in vivo*, or to isolated pituitary tissues *in vitro*, can most effectively stimulate the synthesis and secretion of the gonadotropins LH and FSH. A continuous presentation of GnRH evokes far less synthesis or secretion and in some cases is known to be inhibitory in this regard. It is likely that continuous GnRH stimulus leads to downregulation of GnRH receptor binding and/or postreceptor signaling by intracellular messengers. The pulsatile signal, by contrast, appears to permit replenishment of the signaling capacity during the interpulse intervals.

There is evidence that the frequency of GnRH pulsatility may be physiologically regulated to allow for the differential release of LH and FSH. Higher frequencies of pulsatile GnRH release generally favor the secretion of LH over FSH, whereas lower GnRH frequencies are more often associated with preferential release of FSH. It has been proposed that the GnRH pulse generator may be regulated to produce frequencies which, in turn, drive FSH secretion selectively in several important physiological circumstances; these may include periods of FSH secretion in the early follicular phase of the ovulatory cycle, during the secondary surge of FSH secretion, and in the early stages of photo stimulation of the reproductive axis in seasonally breeding animals.

The pulsatile pattern of GnRH release has also been found to be of profound clinical importance. In men and women with insufficient or inappropriate GnRH secretion, a pulsatile GnRH treatment regimen has been found to be an absolute requirement for successful restoration of gonadotropin secretion. The use of portable infusion pumps to deliver regular GnRH pulses has proven extremely effective in restoration of fertility in both men and women with various forms of hypothalamic, hypogonadotropic hypogonadism.

IV. CELLULAR BASIS OF GnRH PULSE GENERATION

A. Electrophysiology

Electrophysiological correlates of GnRH pulse generator activity have been characterized using hypothalamic electrode recordings in monkeys, sheep, goats, and rats. Such studies have demonstrated that the pulse generator consists of some set of neurons which periodically fire in unison a high-frequency volley of action potentials which eventuate in the neurosecretion of a GnRH pulse into the hypophysial portal vessels. These recordings have also underscored the likely existence of at least two important elements of the GnRH pulse generating process: a "pacemaker" or pseudo-pacemaking mechanism (see Section IV,B) and a mechanism for electrophysiological synchronization among neurosecretory cells.

B. Cellular Models of Pulsatility

One simple model for pulsatility holds that pacemaking activity occurs within one or more GnRH neurons themselves and that the activity of "slave" GnRH neurons is entrained to the rhythm of a dominant pacemaker within the population. This idea

is supported by the recent observations that GnRH release from immortalized GnRH-producing tumor cells in culture is indeed pulsatile. A variation of this idea (pseudo-pacemaking) holds that any neuron within the pulsing network can initiate a pulse, and that the pulsatile rhythm follows from the continued repetition of (i) random autoactivation of a neuron within the interconnected GnRH network; (ii) stimulation of other neurons in the network by the excited cell; (iii) a refractory period, during which no neurons fire; and (iv) autoactivation of another neuron in the network. This stochastic model does not depend on the activity of a distinct pacemaker cell but is instead driven by the emergent, random activity of an interconnected network of cells.

In either model, some mechanism must operate to synchronize the pulsatile release activity among GnRH neurons. The simplest scenario is one in which synchronicity is achieved through intercellular signaling via GnRH–GnRH synaptic contacts or electrical coupling among pulsing neurons. Other possibilities include the production of other intercellular, synchronizing factors, such as nitric oxide or other "volume" neurotransmitters which can exert effects over a relatively wide area. The mechanism which mediates synchronization among pulsing cells and the cellular basis of pulsatility itself remain critically important scientific issues that are currently being addressed in a variety of *in vitro* and *in vivo* studies.

See Also the Following Articles

Hypothalamic–Hypophysial Complex; Neuroendocrine Systems

Bibliography

Crowley, W. F., Whitcomb, R. W., Jameson, J. L., Weiss, J., Finkelstein, J. S., and O'Dea, L. S. L. Neuroendocrine control of human reproduction in the male. *Recent Prog. Horm. Res.* 47, 27–62.

Knobil, E. (1981). Patterns of hypophysiotropic signals and gonadotropin secretion in the rhesus monkey. *Biol. Reprod.* 24, 44–49.

Knobil, E., and Hotchkiss, J. (1988). The neuroendocrine control of the menstrual cycle. In *The Physiology of Reproduction* (E. Knobil and J. D. Neill, Eds.), pp. 1971–1994. Raven Press, New York.

Leng, G. (1988). *Pulsatility in Neuroendocrine Systems*. CRC Press, Boca Raton, FL.

Levine, J. E. (1994). *Pulsatility in Neuroendocrine Systems,* Methods in Neurosciences Series, Vol. 20. Academic Press, San Diego 1994.

Levine, J. E., Bauer-Dantoin, A. C., Besecke, L. M., Conaghan, L. A., Legan, S. J., Meredith, J. M., Strobl, F. J., Urban, J. H., Vogelsong, K. M., and Wolfe, A. M. Neuroendocrine regulation of the luteinizing hormone-releasing hormone pulse generator in the rat. *Recent Prog. Horm. Res.* **47**, 97–153.

Levine, J. E., Chappell, P., Besecke, L. M., Bauer-Dantoin, A. C., Wolfe, A. M., Porkka-Heiskanen, T., and Urban, J. H. (1995). Amplitude and frequency modulation of pulsatile luteinizing hormone-releasing hormone. *Cell. Mol. Neurobiol.* **15**, 117–139.

Mellon, P. L., Wetsel, W. C., Windle, J. J., Vanenca, M. M., Goldsmith, P. C., Whyte, D. B., Eraly, S. A., Negro-Vilar, A., and Weiner, R. I. (1992). Immortalized hypothalamic gonadotropin-releasing hormone neurons. *Ciba Foundation Symp.* **168**, 104–126.

Gonadogenesis, Female

Jeff A. Parrott and Michael K. Skinner

Washington State University

I. Introduction
II. Early Development of the Ovary
III. Primordial Follicle Development
IV. Recent Progress: kit-Ligand Induces Primordial Follicle Development

GLOSSARY

female gametogenesis (oogenesis) A process that includes migration of primordial germ cells into the developing ovary, and continuous development of ovarian follicles in the adult as primordial follicles are recruited to develop.

granulosa cell A female somatic cell type in the ovarian follicle that surrounds the oocyte providing the required microenvironment for its maturation.

oocyte A female germ cell that develops within ovarian follicles until released from the ovary at ovulation.

ovary The female gonad; a structurally dynamic organ that supports oocyte maturation within developing follicles.

ovulation A process in which the mature Graafian ovarian follicle wall is ruptured and an oocyte is released.

primordial follicle A dormant ovarian follicle that consists of a single oocyte surrounded by a single or partial layer of squamous "pregranulosa cells" and will initiate follicular development and ovulate or degenerate. The pool of primordial follicles contains the available oocytes that a female will ever have.

theca cell A female somatic cell type that surrounds and provides structural integrity for the ovarian follicle. Theca cell–granulosa cell interactions are essential for ovarian follicular development.

Mammalian reproduction is dependent on the maturation of germ cells in the gonad. In the female, germ cells (i.e., oocytes) are contained in the ovary, which is a structurally dynamic organ that supports maturation of oocytes within developing follicles. Ovarian follicular development begins with the recruitment of primordial follicles to initiate development and continues through ovulation. After the oocyte is released at ovulation, the remaining theca cells and granulosa cells in the follicle develop into the corpus luteum. In humans, one dominant follicle ovulates from a single ovary during each menstrual cycle. Follicles that do not ovulate become atretic and degenerate. A great deal of research during the past 100 years has provided an understanding of some of the factors that control ovarian follicular development. However, two important aspects of ovarian folliculogenesis are still largely unknown. These processes are (i) the initiation of primordial follicle development and (ii) the selection of the dominant follicle.

I. INTRODUCTION

During embryonic development, the process of ovarian development in the embryo establishes the tissue structures and cell populations that are necessary for follicular development in the adult. Most important, the germ cell population is established during early ovarian development. Ovarian organogenesis involves the coordinated movements and induction of several dynamic cell populations. The sequence of events at specific developmental stages is conserved from mice to humans (Table 1). Although the primary function of the mammalian ovary is to support germ cell (i.e., oocyte) development, all germ cells initially develop outside of the genital ridge. Interestingly, primordial germ cells follow a specific migratory pathway through the embryo into the gonad.

II. EARLY DEVELOPMENT OF THE OVARY

In the mammalian embryo, gonads develop from (i) the coelomic epithelium, (ii) the mesenchyme of the mesonephric ridge, and (iii) the primordial germ cells. The development of gonads is unique in that the gonadal rudiment can differentiate into an ovary or a testis. This developmental decision determines the subsequent sexual development of the organism. Before this decision is made the mammalian gonad first develops through an indifferent stage. During this stage the gonadal rudiment is first recognizable as a thickening of the coelomic epithelium and budding mesonephric mesenchyme on the medial side of the mesonephric ridge. As the thickening tissue buds out from the mesonephros, it forms the genital ridge (Fig. 1A) during Embryonic Days 31–35 in humans and on Embryonic Day 9 in mice. The coelomic epithelial cells of the genital ridge proliferate and migrate into the adjacent loose mesenchymal tissue in the budding indifferent gonad (Fig. 1B), recognizable during Embryonic Week 6 in humans and Day 11 in mice.

Two important paired ducts are also present in the mesonephros during early embryonic development.

TABLE 1
Early Development of Germ Cells and the Ovary

Age of embryo		Developmental process
Human	*Mouse*	
Day 28	Day 7.5	First primordial germ cells recognizable
Weeks 4–7	Days 8–12.5	Germ cells proliferate during migration
Week 4 (Days 25–30)	Day 9	Mesonephric tubules form
Weeks 4–5 (Days 28–30)	Day 9	Germ cells migrate to hindgut
Weeks 4–5 (Days 28–30)	Days 9–10	Wolffian ducts form
Week 5 (Days 31–35)	Day 9	Coelomic epithelium thickens to form genital ridge
Week 6 (Days 35–42)	Days 10–11	Germ cells migrate to dorsal mesentery
Week 6 (Days 38–42)	Day 11	Budding indifferent gonad is recognizable
Week 6 (Days 40–42)	Day 11	Müllerian duct forms
Week 7 (Days 42–48)	Days 11–12	Germ cells migrate to indifferent gonad
Week 7 (Day 49)	Day 12.5	Germ cell migration to gonads is complete
Weeks 7–8	Day 13	Gonad is recognizable as ovary
Months 2–7	Days 12–13	Oogonia proliferate within ovary
Months 2.5–7	Days 13–16	Oocytes initiate then arrest in meiosis (oogenesis)

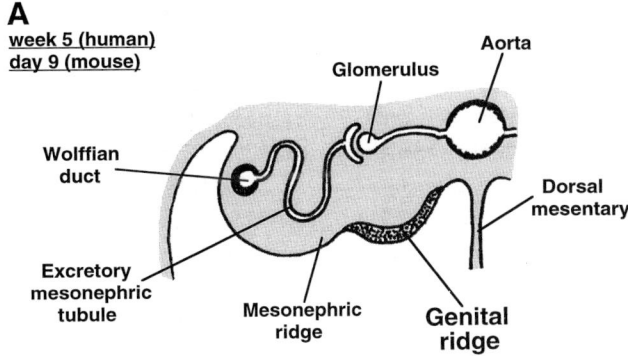

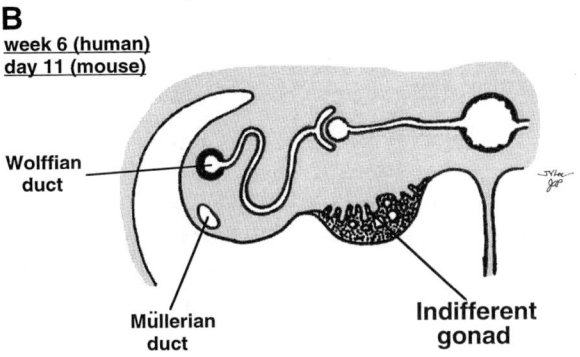

FIGURE 1 Early development of the indifferent gonad. The indifferent gonad can develop into an ovary or a testis. (A) During this stage the gonadal rudiment begins to bud out from the genital ridge to form the indifferent gonad. This is first observed as a thickening of the coelomic epithelium. (B) As the gonad grows, the coelomic epithelium of the genital ridge proliferates and migrates into the adjacent loose mesenchymal tissue. Germ cells are not present when the indifferent gonad buds out from the genital ridge. Germ cells migrate into the gonad during Embryonic Weeks 6 and 7 in humans and on Embryonic Days 11–12.5 in mice.

The Wolffian duct appears during Embryonic Week 4 in humans and on Day 10 in mice (Fig. 1A). The Müllerian duct is first seen during Embryonic Week 6 in humans and on Day 11 in mice (Fig. 1B). The Wolffian and Müllerian ducts will eventually develop into reproductive tracts of the male or female, respectively. Later in male development, the Müllerian duct will regress due to the production of mullerian-inhibiting substance by Sertoli cells in the testis. Androgen production by Leydig cells in the developing testis will support development of the Wolffian duct into the epididymus, vas deferens, and seminal vesicles. Later in female development, the Wolffian duct will regress due to the lack of high androgen production

by the gonad (i.e., ovary). The Müllerian duct will develop into the oviduct, uterus, and upper part of the vagina. The development of the female reproductive tract from the Müllerian duct is not dependent on estrogen. For example, targeted disruption of the estrogen receptor gene in mice (i.e., the ERKO mouse) has no effect on prenatal development of the reproductive tract.

In essentially all mammals, germ cells are not present in the indifferent gonad when the genital ridge begins to bud out from the mesonephros. Primordial germ cells migrate along a characteristic pathway and enter the indifferent gonad when the coelomic epithelium is invading the loose mesenchyme. Primordial germ cells begin to enter the developing ovary during Embryonic Weeks 6 and 7 in humans and on Embryonic Day 11 in mice. This process of germ cell migration is highly conserved from flies to frogs to humans and is thought to be the result of a specific evolutionary process. In particular, primitive Metazoa had germ cells but no gonads to harbor them. As higher animals acquired gonads, biological strategies evolved to support germ cell migration to developing indifferent gonads. Migration of primordial germ cells is controlled by short-range cell-to-cell contacts and long-range effects of the genital ridge.

Analysis of germ cell migration during embryonic development has been an area of research for several decades. Most of the initial work was performed in mice. A significant advance in mapping mammalian germ cell migration was achieved with the observation that primordial germ cells contained extremely high levels of alkaline phosphatase that distinguished embryonic cells. Using this staining procedure, primordial germ cells can first be seen in the allantois on Embryonic Day 28 in humans and Embryonic Day 7.5 in mice (Table 1). The germ cells then begin to migrate into the yolk sac at the base of the allantois (Fig. 2A). By this time, the germ cells have split into two populations that will eventually migrate into the left or right gonad. Migration continues from the yolk sac through the newly formed hindgut and up the dorsal mesentery into the genital ridge (Fig. 2B). Germ cells migrate by ameboid movements and migration into the gonads is complete by the end of Embryonic Week 7 in humans and Embryonic Day 12.5 in mice (Table 1).

A

<u>day 28 (human)</u>
<u>day 7.5 (mouse)</u>

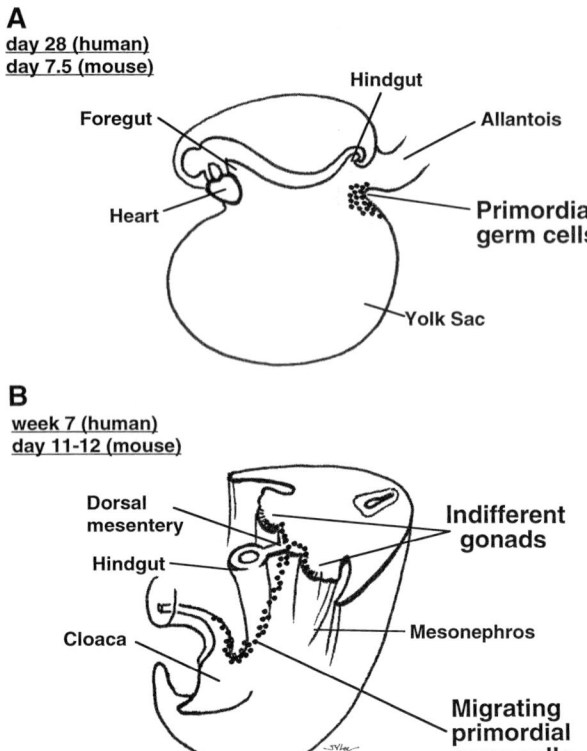

B

<u>week 7 (human)</u>
<u>day 11-12 (mouse)</u>

FIGURE 2 Germ cell migration in mammals. Primordrial germ cells are detected by staining for alkaline phosphatase. (A) Primordial germ cells can first be seen in the allantois on Embryonic Day 28 in humans and Embryonic Day 7.5 in mice. Germ cells then begin to migrate into the yolk sac at the base of the allantois. By this time, the germ cells have split into two populations that will migrate into the left or right gonad. (B) Migration continues caudally from the yolk sac through the newly formed hindgut and up the dorsal mesentery into the genital ridge. Primordial germ cell migration into the gonads is complete by the end of Embryonic Week 7 in humans and Embryonic Day 12.5 in mice. The overall process of germ cell migration is common from flies to frogs to humans.

The number of primordial germ cells is relatively small before germ cell migration takes place. In 70% of early vertebrate embryos, 20–100 primordial germ cells are present. The degree of consistency of early germ cell numbers in vertebrates suggests that some common processes are involved in the induction of the germ cell line. Primordial germ cells dramatically increase in number by proliferating. During germ cell migration in the human (Embryonic Weeks 4–7) and mouse (Embryonic Days 8–12.5), the number

of primordial germ cells increases from about 100 to 2500–5000. After migration to the gonad is complete, primordial germ cells (i.e., oogonia) organize into clusters and continue to proliferate. Oogonia continue to proliferate until Embryonic Month 7 in humans and Day 13 in mice (Table 1). In humans the number of oogonia increases to 6 or 7 million.

Migration and proliferation of primordial germ cells is critical for fertility. White Spotting (W) mutations in mice caused sterility because the W locus encodes the c-*kit* gene that is necessary for germ cell migration in the embryo. The c-*kit* gene is a protooncogene receptor that interacts with the kit ligand (KL) (described in Section IV). Both KL and c-*kit* are essential for germ cell migration and proliferation during embryonic development. The somatic cells that line the migratory pathway express KL, whereas migrating primordial germ cells express c-*kit*. In the absence of KL or c-*kit*, embryonic germ cell migration and proliferation is inhibited. Subsequent studies have demonstrated that KL and c-*kit* are important for several stages of ovarian folliculogenesis in the adult. One exciting function of KL/c-*kit* during recruitment of primordial follicles is currently being investigated (described in Section IV).

Indifferent gonads are first recognizable as ovaries when primordial germ cells (i.e., oogonia) stop proliferating and enter meiotic prophase. This process occurs during Embryonic Weeks 7 and 8 in humans and on Day 13 in mice (Table 1). Oogonia that have entered meiosis are called oocytes. Initiation of oocyte meiosis in the ovary marks the initiation of oogenesis. However, not all oogonia enter meiosis at the same time. Therefore, proliferating oogonia and meiotic oocytes are present in young ovaries at the same time. As shown in Table 1, oogenesis occurs during Embryonic Months 2.5–7 in humans and on Embryonic Days 13–16 in mice. Meiosis continues through the first meiotic prophase until arresting at the diplotene stage. The ability of germ cells to develop from primordial germ cells to meiotic oocytes does not require the presence of the gonad. However, the gonad is necessary for oocytes to develop beyond this stage. A single layer of presumptive granulosa cells organizes around each oocyte. These oocyte–pregranulosa complexes are called primordial follicles. Oocytes in primordial follicles remain arrested in meiosis until later in life when they resume development. In humans some oocytes are arrested in meiotic prophase in the form of primordial follicles for up to 50 years. The factors that control recruitment of primordial follicles to develop are not known.

Not all oocytes in the ovary survive. Proliferation of oogonia in the embryo establishes an abundance of germ cells in females. After this proliferative phase is complete, the number of germ cells in the ovary decreases progressively throughout life. Germ cells degenerate during embryonic ovarian development or during follicular development in the adult. The majority of germ cells are eliminated in females, perhaps to select germ cells with stable, viable genomes. In humans the number of germ cells is reduced from 6 or 7 million in the 7-month-old embryo to 2 million at birth and 400,000 at puberty.

III. PRIMORDIAL FOLLICLE DEVELOPMENT

A crucial event for ovarian development is the enclosure of oocytes by somatic cells into individual "compartments." These cellular structures are called primordial follicles. Survival and differentiation of germ cells depend on their organization into primordial follicles. Primordial follicles are formed when mesonephric-derived cells within the ovary contact and organize into a single layer of pregranulosa cells around the oocyte (Figs. 3A and 4A). These cell–cell

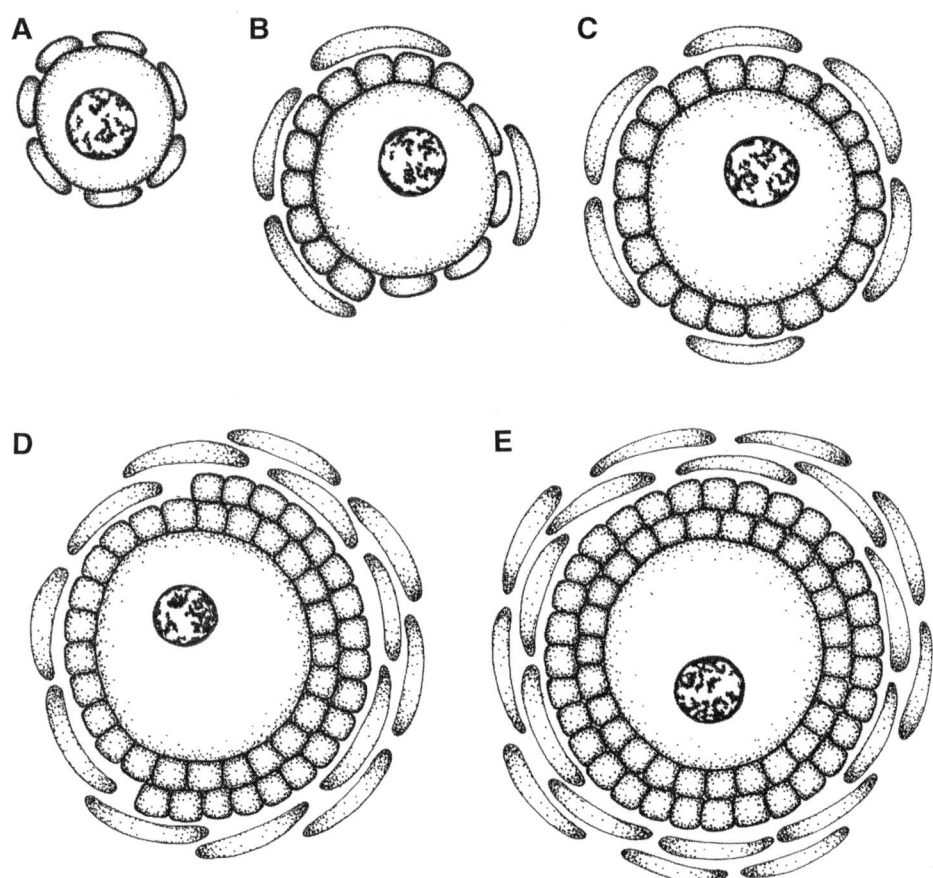

FIGURE 3 Schematic of primordial follicle development in the ovary. (A) Primordial follicles (stage 0) consist of an oocyte surrounded by a full or partial layer of flattened pregranulosa cells. (B) Early primary follicles (stage 1) have initiated development and contain some columnar (enlarged) granulosa cells. Theca cells are being recruited and are starting to organize around the follicle. (C) Primary follicles (stage 2) are surrounded by a single layer of cuboidal granulosa cells around the oocyte. (D) Transitional follicles (stage 3) contain 1 or 2 layers of columnar granulosa cells. Formation of theca cell layers continues. (E) Preantral follicles (stage 4) have two or more layers of columnar granulosa cells. Theca cells are well organized around the follicle.

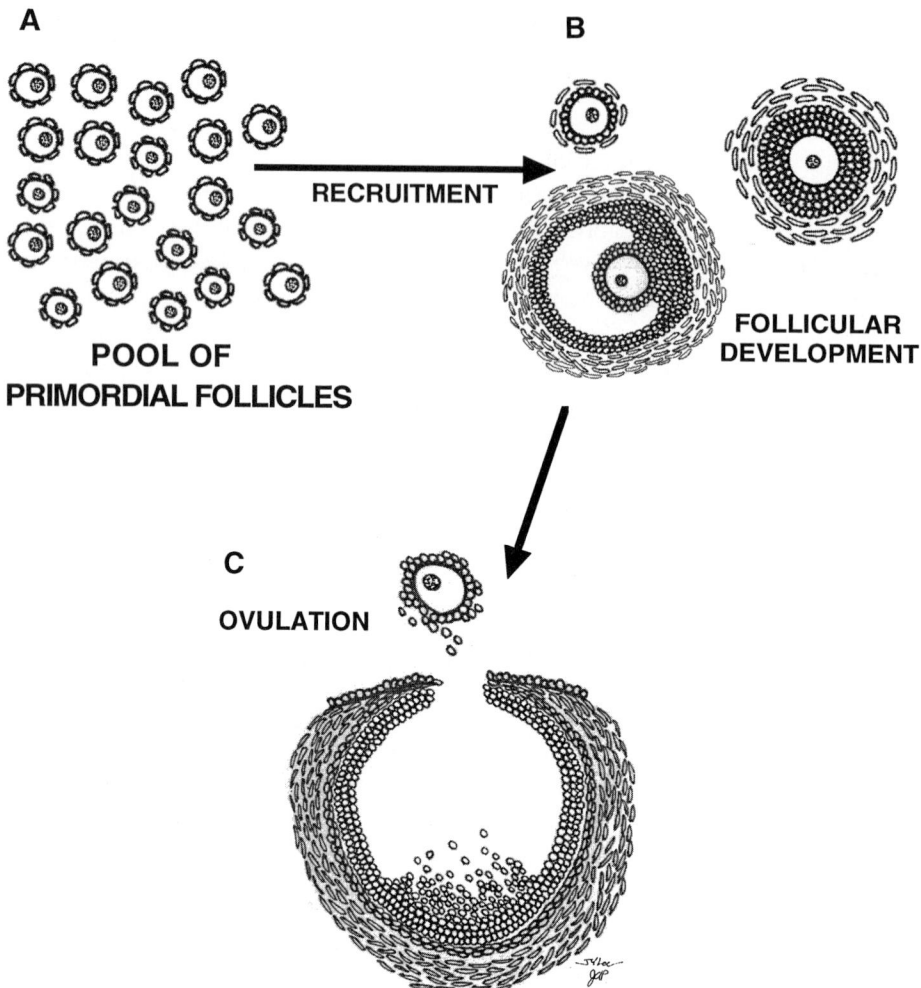

FIGURE 4   Primordial follicles are recruited to develop later in life. (A) A single layer of presumptive granulosa cells organizes around oocytes to form the pool of primordial follicles. Primordial follicles represent the complete pool from which all ovarian follicles can develop. Oocytes arrest at the diplotene stage of the first meiotic prophase. Oocytes remain dormant in the form of primordial follicles until later in life when they are recruited to develop. (B) Follicular development involves follicular expansion due to theca cell and granulosa cell proliferation as well as the formation of an antrum. Oocytes remain arrested in meiosis. (C) After the luteinizing hormone surge at ovulation, oocyte meiosis is resumed. In humans some oocytes are arrested in meiotic prophase in the form of primordial follicles for up to 50 years.

contacts may act as a trigger for the onset of meiosis in germ cells. After arrest in prophase of meiosis, oocytes remain dormant in the form of primordial follicles until recruited for follicular development later in life (Fig. 4).

Extensive research (i.e., histological, histochemical, and autoradiographic techniques) has established that primordial follicles represent the pool from which all developing follicles will emerge. Wal-

deyer correctly proposed in 1870 that the germ cells observed in the embryonic ovary developed sequentially into mature ova throughout life. In 1887 and 1903, Paladino claimed there was a continuous destruction of follicles with constant formation of new egg tubes and ova, with the process going on from birth to old age in mammals. Subsequent studies showed that *de novo* formation of oocytes does not occur in ovaries depleted of germ cells. Today it is well accepted that the oocytes established in the pool of primordial follicles are the only oocytes a female animal will ever have.

The large pool of primordial follicles in mammalian ovaries is progressively reduced through follicular development. Throughout life, a subset of available primordial follicles is continuously recruited to develop (Figs. 4 and 5). In embryonic and prepubertal ovaries, all developing follicles degenerate. In pubertal and adult ovaries, some follicles will continue development and successfully ovulate (Fig. 4C). All primordial follicles that initiate follicular development are destined to ovulate or degenerate through

atresia. Therefore, factors that control recruitment and initiation of primordial follicle development ultimately determine the number of available follicles in the ovary.

The factors that control initiation of primordial follicle development in the ovary are not known. Identification of such factors has been an extensive area of research for more that 100 years. During this time, no hormone (including gonadotropins and estrogen) has been identified that influences this process. However, there is evidence that the total number of primordial follicles (i.e., the pool) in the ovary may influence the number of developing follicles.

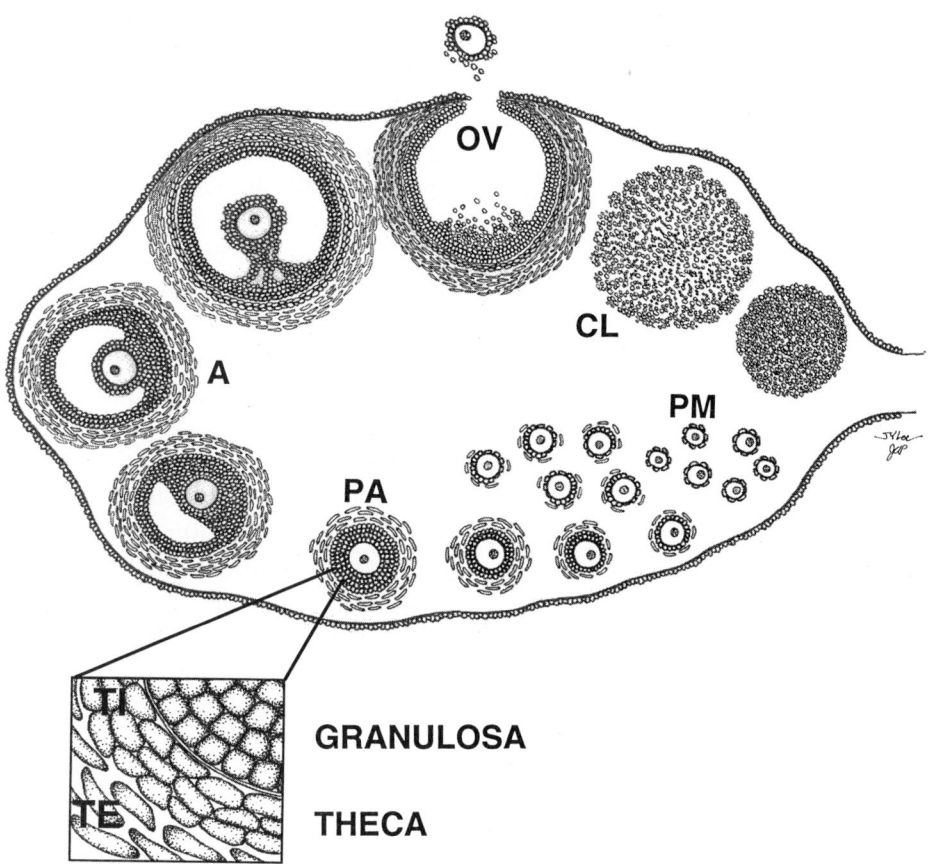

FIGURE 5 Schematic of cycling adult ovary. The basic functional unit is the ovarian follicle. Follicles are recruited to initiate development from the pool of primordial follicles (PM). Developing follicles consist of a single oocyte and two somatic cell types (enlarged box). Granulosa cells surround the oocyte and provide the critical microenvrionment for oocyte maturation. Surrounding the granulosa cells, theca cells provide structural integrity for the follicle. Theca cells differentiate into two distinct layers during folliculogenesis. Theca interna (TI) are adjacent to granulosa cells and theca externa (TE) are further outside. Theca cells and granulosa cells proliferate and differentiate to form preantral follicles (PA) and fluid-filled antral follicles (A). A small number of follicle continue development and eventually ovulate (OV). After ovulation the remaining theca cells and granulosa cells organize into the corpus luteum (CL).

Nevertheless, initiation of primordial follicle development is proposed to be controlled by a hormone or soluble factor. The mechanism of how only a subset of primordial follicles can be recruited to initiate development was being considered by Hargitt in 1930. Hargitt argues that

If an assumption be made that many latent primary follicles are stored in the mammalian ovary, a serious difficulty is encountered in explaining the stimulation of a few to renewed development, while others remain quiescent. Likely enough, some hormone might stimulate renewed activity, but why of a few not all of the same age?

This question has led to the proposal that initiation of primordial follicle development is controlled locally by a factor(s) produced in the ovary.

Peters' discussion of follicular development in immature mouse ovaries suggested that initiation of primordial follicle development may be influenced by a factor(s) produced by developing follicles. Such factors may induce or inhibit recruitment of primordial follicles. The local presence of stimulatory and inhibitory factors for this process may determine how a subset of primordial follicles begins to develop. The influence of antral follicles on initiation of primordial follicle development was tested by injecting follicular fluid (bovine) into neonatal mice for 4 days. Within a few days, the number of primordial follicles that initiated development was reduced by 35%. These results suggest that developing antral follicles produce an unknown inhibitory factor(s) for primordial follicle development. It has also been suggested that granulosa cells in early developing follicles produce an unknown factor that acts on theca cells to promote early follicular development. The identity of these putative factors is not known despite their importance for control of primordial follicle development.

IV. RECENT PROGRESS: KIT-LIGAND INDUCES PRIMORDIAL FOLLICLE DEVELOPMENT

KL and its receptor c-*kit* are important for gametogenesis, melanogenesis, and hematopoiesis. During embryonic development, KL and c-*kit* are essential for germ cell migration. In the adult ovary, granulosa cells produce KL that may be important for oocyte function during follicular development (e.g., oocyte development and meiotic arrest). Most of our understanding about the function of KL in the ovary is based on observations of KL actions on oocytes. However, granulosa cell-derived KL also influences theca cell growth and functional differentiation (i.e., steroidogenesis). Expression of the receptor c-*kit* has been observed in stromal cells, theca cells, and oocytes in very early developing follicles. Based on the variety of KL actions and the expression patterns of KL and c-*kit* in the ovary, KL has been proposed to induce initiation of ovarian primordial follicle development.

Ovaries from 4-day-old rats contain large numbers of primordial follicles that can initiate development. After 5 days in ovary organ culture, some of these primordial follicles spontaneously initiated development and KL induced primordial follicles to initiate development. When control ovaries were cultured with ACK-2, a c-*kit* antibody that strongly inhibits KL actions, spontaneous primordial follicle development was completely inhibited. These results suggested that KL was necessary and sufficient to induce initiation of primordial follicle development in ovarian organ cultures.

Initiation of primordial follicle development is one of the most basic aspects of ovarian folliculogenesis. KL is the first identified factor that is involved in initiation of primordial follicle development. The mechanism of how KL induces primordial follicle development is not known. It is possible that KL has important actions on oocytes, theca cells, and adjacent stromal–interstitial cells during early follicle development.

Acknowledgment

The figures were provided by my friend and colleague, Janet Y. Lee. Thank you for your dedication and kindness.

See Also the Following Articles

Corpus Luteum; Follicular Development, Control of; Ovary, Overview; Theca Cells

Bibliography

Besmer, P., Manova, K., Duttlinger, R., Huang, E. J., Packer, A., Gyssler, C., and Bachvarova, R. F. (1993). The kit-ligand (steel factor) and its receptor c-kit/W: Pleiotropic roles in gametogenesis and melanogenesis. *Dev. Suppl.*, 125–137.

Brambell, F. (1927). The development and morphology of the gonads of the mouse. Part I. The morphogenesis of the indifferent gonad and of the ovary. *Proc. R. Soc. London B Biol. Sci.* **101**, 391.

Buehr, M. (1997). The primordial germ cells of mammals: Some current perspectives. *Exp. Cell Res.* **232**, 194–207.

Byskov, A. G. (1986). Differentiation of mammalian embryonic gonad. *Physiol. Rev.* **66**, 71–117.

Cooley, L. (1995). Oogenesis: Variations on a theme. *Dev. Genet.* **16**, 1–5.

Hargitt, G. T. (1930). The formation of the sex glands and germ cells of mammals. *J. Morphol. Physiol.* **49**, 333–353.

Korach, K. S. (1994). Insights from the study of animals lacking functional estrogen receptor. *Science* **266**, 1524–1527.

Langman, J. (1981). *Medical Embryology*. Williams & Wilkins, Baltimore.

Medvedev, Z. A. (1981). On the immortality of the germ line: Genetic and biochemical mechanism. A review. *Mech. Ageing Dev.* **17**, 331–359.

Mintz, B. (1957). Embryological development of primordial germ cells in the mouse: Influence of a new mutation W. *J. Embryol. Exp. Morphol.* **5**, 396–403.

Mintz, B., and Russell, E. S. (1955). Developmental modifications of primordial germ cells induced by the W-series in the mouse embryo. *Anat. Rec.* **122**, 443.

Paladino, G. (1887). Ulteriori ricerche sulla distinzione e rinnovamento continuo del parenchima ovarico dei mammiferi. *Anat. Anz.* **2**, 835–842.

Paladino, G. (1903). Sulla rigenerazione del parenchyma ovarico e sul tipo di struttura dell' ovaja di Delfina. *Rend. R. Acad. Sci. Fisiche Matematiche Napoli* **12**, 289–298.

Parrott, J. A., and Skinner, M. K. (1997a). Direct actions of KL on theca cell growth and differentiation during follicle development. *Endocrinology* **138**, 3819–3827.

Parrott, J. A., and Skinner, M. K. (1997b). Kit ligand/stem cell factor induces initiation of primordial follicle development in the ovary. Submitted for publication.

Peters, H. (1979). Some aspects of early follicular development. In *Ovarian Follicular Development and Function* (A. R. M. a. W. A. Sadler, Ed.), pp. 1–13. Raven Press, New York.

Peters, H., and McNatty, K. P. (1980). *The Ovary. A Correlation of Structure and Function in Mammals*. Univ. of California Press, Berkeley.

Peters, H., Byskov, A. G., and Faber, M. (1973). Intraovarian regulation of follicle growth in the immature mouse. In *The Development and Maturation of the Ovary and Its Functions* (H. Peters, Ed.), Int. Cong. Ser. No. 267, Excerpta Medica, pp. 20–23. Amsterdam.

Peters, H., Byskov, A. G., Himelstein-Braw, R., and Faber, M. (1975). Follicular growth: The basic event in the mouse and human ovary. *J. Reprod. Fertil.* **45**, 559–566.

Rajah, R., Glaser, E. M., and Hirshfield, A. N. (1992). The changing architecture of the neonatal rat ovary during histogenesis. *Dev. Dyn.* **194**, 177–192.

Waldeyer, W. (1870). *Eierstock und Ei*. Engelmann, Leipzig.

Witschi, E. (1948). Migration of the germ cells of human embryos from the yolk sac to the primitive gonadal fold. *Contrib. Embryol.* **32**, 67–80.

Gonadogenesis, Male

Jose Teixeira and Mary M. Lee

Massachusetts General Hospital and Harvard Medical School

GLOSSARY

androgens Sex steroid hormones produced by Leydig cells of the testis, required for normal virilization of internal and external reproductive tract structures, spermatogenesis, and male sexual behavior.

germ cells or **gonocytes** Stem cells which in males are actively dividing and are precursors to sperm.

Leydig cells or **interstitial cells** Steroidogenic cells which produce the sex steroid hormones in the testis.

mesonephros The embryonic kidney, from which the somatic cells of the gonad arise.

orphan nuclear receptor A hormone-binding protein that is expressed in the nucleus with no known ligand.

Sertoli cells Cells in the seminiferous tubules of the testis which support the maturation of spermatocytes.

SRY The sex determining region of the Y chromosome; the testis determining factor gene that initiates development of the male phenotype.

urogenital ridge The gonadal primordium formed by enlargement of the coelomic epithelium in the early embryo and from which the gonad and internal reproductive tract ducts are formed.

Organogenesis of the testis has been described by descriptive experiments designed to elucidate the gross morphological changes involved. New insights into the molecular aspects of testicular development have been derived from experiments using homologous recombination in mice to "knock out" sex-determining genes and by genetic analysis of natural mutations in intersex patients. Development of the testis can be divided into four distinct stages, each required for male reproductive competence. The gonadal ridge develops early as an outcropping of the mesonephros and is populated by primordial germ cells. Soon thereafter, a molecular switch toward male-specific development of the gonad is initiated by the expression of *SRY*. Then, the Sertoli and Leydig cells necessary for sperm maturation and for virilization differentiate. Finally, production of mature sperm begins at puberty. Testicular development is described by changes in gross morphology and in some of the purported molecular details.

I. DEVELOPMENT OF THE BIPOTENTIAL GONAD

In the first stage of gonadal development, during the fifth week of human development, the bilateral urogenital ridges arise from the coelomic epithelium and underlying mesenchyme and are colonized by germ cells, producing the bipotential gonad primordium (Fig. 1). This process is characterized by morphologically discrete steps. The cells of the coelomic epithelium begin dividing rapidly, thickening the cell layers medial to the mesonephros and at the root of the dorsal mesentery. There is no basement membrane below the outer epithelial layer to restrain the proliferating inner epithelial cells. These cell layers will later form the underlying mesoderm from which the primary sex cords are derived.

Meanwhile, primordial germ cells, or gonocytes, which can be detected in the wall of the yolk sac proximal to the end of the allantois at the end of

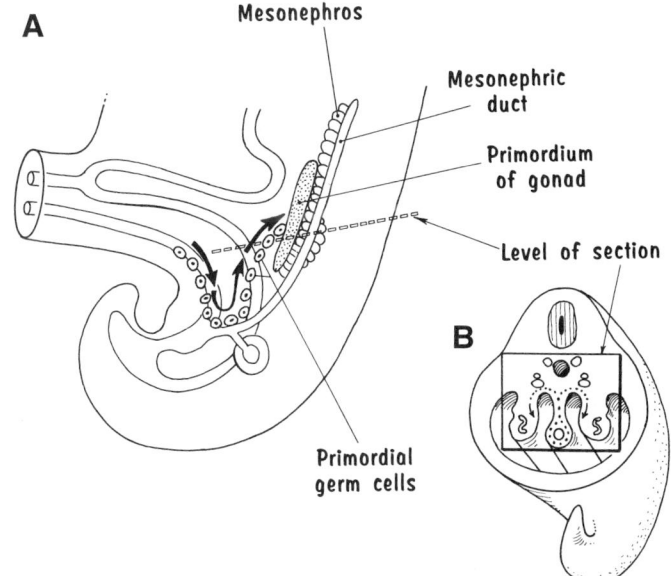

A

Mesonephros

Mesonephric duct

Primordium of gonad

Level of section

B

Primordial germ cells

FIGURE 1 (A) Sagittal section of a 5-week-old embryo showing migration of cells from the yolk sac to the gonadal ridge. (B) Transverse section (shown in more detail in Fig. 2) [adapted with permission from *The Developing Human* (K. L. Moore and T. V. N. Persaud, Eds.), 5th ed., Chap. 13, Fig. 20, Saunders, Philadelphia, 1993].

gastrulation by alkaline phosphatase staining, begin their migration at the four-somite stage early in the fourth week of human development. The germ cells leave the gut endoderm and dorsal yolk sac and migrate along the dorsal mesentery by ameboid movement with pseudopodial processes, proliferating along the way and increasing their number about 20-fold, from an initial 50 to approximately 1000. When the germ cells finish their migration, the urogenital ridge has differentiated from a thickened epithelial layer into a clearly discernible blastema posterior to the mesonephros. The germ cells continue to proliferate in the gonadal ridge to 600,000 by 8 weeks gestation in humans. The primordia of both male and female internal reproductive structures, the Wolffian ducts, which can differentiate into the epididymides, vas deferens, and seminal vesicles, and the Müllerian ducts, which are the progenitors of the uterus, fallopian tubes, and upper vagina, are present. In the male, the paramesonephros or Müllerian ducts will degenerate and the mesonephric or Wolffian tubules will form the efferent ductules of the testes.

II. GONADAL SEX DETERMINATION

Chromosomal or karyotypic sex is determined at fertilization by the presence of the Y chromosome. Establishment of male phenotypic sex involves development of the bipotential or undifferentiated gonads as testes and the subsequent hormonally mediated male differentiation of the internal reproductive ducts and external genitalia. Although the morphological changes have been well described (Burger and DeKrester, 1989; Fig. 2), the molecular mechanisms underlying those changes are only now being studied because of recent advances in molecular and cytological techniques. Significant progress has been made in the identification of some of the genes involved in gonadogenesis and sex determination but their roles in development are largely unknown.

Fifty years ago, it was observed that sex determination of the bipotential or indifferent gonad was independent of external influences and that the Y chromosome was required for development of the testis. Testosterone, secreted by the Leydig cells of the developing testis, actively induces differentiation of the Wolffian duct into the male-specific internal accessory structures, whereas Müllerian-inhibiting substance (MIS), a transforming growth factor-β family hormone, which is produced by Sertoli cells, promotes regression of the Müllerian ducts. Ovarian differentiation is often called the autonomous pathway because, shortly after testes would have developed, in the absence of testosterone and MIS the bipotential gonad develops into an ovary. Therefore, structural organization of the testes and regression of the female reproductive duct primordia are the first sexually dimorphic events in development.

The genetic basis for testicular organogenesis and subsequent male phenotypic development has only recently been described. Using Y chromosomal translocations in XX males and Y chromosome deletions in XY females that resulted in a sex-reversed phenotype, researchers were able to isolate the *SRY* gene. This gene is located on the short arm of the Y chromosome in the most distal portion of the Y-unique region, adjacent to the pseudoautosomal boundary. *SRY* encodes a male-specific gene that is evolutionarily conserved and contains a single open reading frame. Expression of *SRY* has been detected in all fetal male

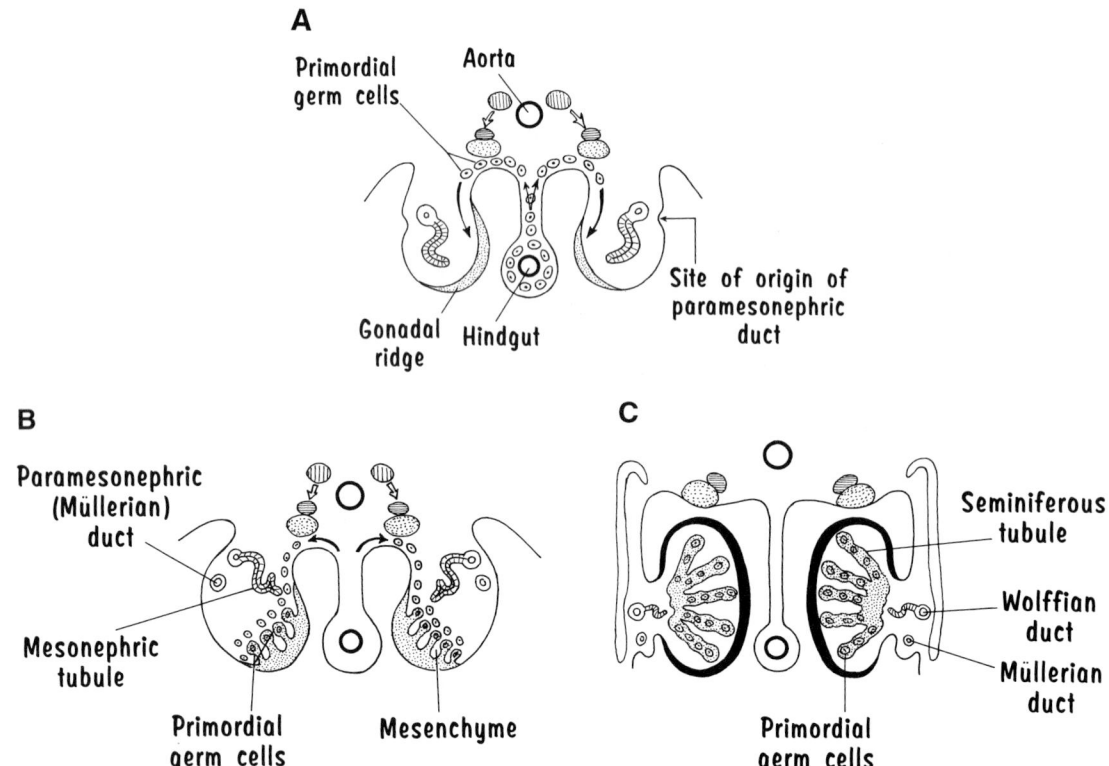

FIGURE 2   Human testicular differentiation. (A) In a 5-week-old embryo, the coelomic epithelium thickens to form the gonadal ridge and invaginates to form the paramesonephric or Müllerian duct. (B) After 6 weeks, primary sex cords start to develop under the influence of *SRY*. The undifferentiated Müllerian and Wolffian (mesonephric) ducts are completely formed and germ cells have finished migrating. (C) At 7 weeks, testicular cords (precursors of the seminiferous tubules) become apparent and the Müllerian duct starts degenerating under the influence of MIS. Testosterone has stimulated the mesonephric (or Wolffian duct) to undergo differentiation into the vas deferens, epididymis, and seminal vesicles [adapted with permission from *The Developing Human* (K. L. Moore and T. V. N. Persaud, Eds.), 5th ed., Chap. 13, Fig. 20, Saunders, Philadelphia, 1993].

tissues, and in mice expression is consistent temporally with induction of testicular differentiation from the indifferent gonadal ridge. The most convincing evidence that *SRY* expression is sufficient for testicular morphogenesis is that XX mice containing a transgene for *SRY* develop as normally virilized males but with abnormal spermatogenesis. Thus, transformation of the somatic cells to Sertoli and Leydig cells was successfully induced by *SRY*, although other Y-specific factors are apparently required for sperm maturation.

The predicted SRY protein contains 223 amino acids and a highly conserved HMG (high-mobility group protein) box domain of 79 amino acids. HMG-box proteins are chromatin-associated or DNA-binding proteins and transcription factors. SRY might

initiate testicular development by either transcriptional regulation of male-specific genes required for male phenotypic development or repression of genes suppressing the male pathway of gonadal development. Additionally, SRY expression had been localized to the nucleus and mouse Sry contains a transcriptional activation domain. SRY binds to the HMG proteins' cognate binding site, AACAAAG, and to an optimal A/TAACAAT site *in vitro*; however, correlation to *in vivo* binding has not been made. Therefore, the preponderance of data suggests that SRY is the testis determining factor, but the molecular mechanisms of SRY-induced male development remain unknown.

Although *SRY* is sufficient for initiation of male gonadal differentiation, autosomal genes are also involved. Some 46,XY human sex-reversed patients lack mutations in *SRY*, or on the Y chromosome, and deletional mutagenesis of certain autosomal genes results in murine sex reversal. An orphan nuclear receptor called steroidogenic factor-1 (*SF-1*) is a transcription factor involved in the production of sex hormones in steroidogenic tissues. *SF-1* expression has been observed at the earliest stages of organogenesis in both male and female mice, with levels decreasing in females after the sexual pathway of the gonad has been decided. Knockout of the *SF-1* gene in mice by homologous recombination resulted in gonadal and adrenal agenesis. The gonadal ridges develop normally in *SF-1* null mice until Embryonic Day 10.5, the onset of sexual dimorphism; they are resorbed after that point. Mice homozygous for null mutations of the *SF-1* gene die soon after birth from adrenal insufficiency. While a role for *SF-1* in primary sex determination seems unlikely, it is required for both male and female gonadal morphogenesis, probably by regulating expression of steroid hormone-producing enzymes, and the absence of *SF-1* is incompatible with life.

WT1 is a gene that is expressed early, at Embryonic Day 9, in the mouse urogenital ridge. The WT1 protein is a zinc finger-containing transcription factor that is important for early gonadal and kidney development. *WT1* probably plays no role in sex determination since it is expressed in both male and female ridges, but *WT1* is required for gonadal formation at an earlier stage than SF-1 and SRY. *WT1* was identified as the gene that causes Wilms' tumor, an early childhood cancer of the kidney, and its role in gonadal development was recognized when the gene was deleted by homologous recombination in mice. Homozygous null mutant mice had a rudimentary thickening of the coelomic epithelium of the urogenital ridge, but there was no apparent differentiation of the gonad at Embryonic Day 11.

Sox9 is the gene that has been associated with campomelic dysplasia, a syndrome characterized by congenital bone and cartilage malformation and sex reversal in 46,XY males. *Sox9* shares homology with *SRY* in its DNA-binding domain (the HMG box domain) which suggests it also functions as a transcription factor. *In vitro* experiments show *Sox9* binding to the same target DNA sequences as does *SRY*. Its

expression has been observed in most tissues, and its role in testis development is not clear. *Sox9* manifests a phenotype when one of the *Sox9* alleles is mutated, suggesting haploinsufficiency might also cause XY sex reversal.

46,XY gonadal dysgenesis has also been described with an apparent X-linked mode of inheritance. Overexpression of a dosage-sensitive gene because of duplication of an Xp locus resulted in XY sex reversal. Genetic studies of patients with this phenotype indicated that the duplicated gene might be *DAX-1*, the nuclear hormone receptor gene responsible for congenital adrenal hypoplasia. Little is known about the role of *DAX-1* in gonadal development.

III. SERTOLI CELLS AND LEYDIG CELLS

Differentiation of the testicular somatic cells follows soon after the initiation of *SRY* expression (Fig. 3). Four cell types define the testis in response to *SRY* expression. Sertoli cells provide scaffolding and nutritional support for spermatogenesis and produce MIS. Leydig or interstitial cells are the steroid hormone-producing cells, and spermatogonia or germ cells will be discussed in the next section. Peritubular cells, which help form the blood–testis barrier basement membrane of the avascular seminiferous tubules and maintain its structural integrity, will not be discussed.

The primary role of the testis is spermatogenesis, which is hormonally controlled by anterior pituitary gonadotropins, luteinizing hormone (LH) and follicle-stimulating hormone (FSH). LH stimulates steroid hormone synthesis, particularly testosterone, by Leydig cells. FSH stimulates the Sertoli cells to produce spermatogenic proteins thought to nurture and coordinate sperm maturation in the seminiferous tubule. Experimental evidence for paracrine communication between Sertoli cells and Leydig cells, which are separated by the blood–testis barrier, is provided by the finding that FSH indirectly induces Leydig cell hypertrophy and hyperplasia by increasing LH receptor numbers and steroidogenic responsiveness. Coculture of Sertoli and Leydig cells, as well as cul-

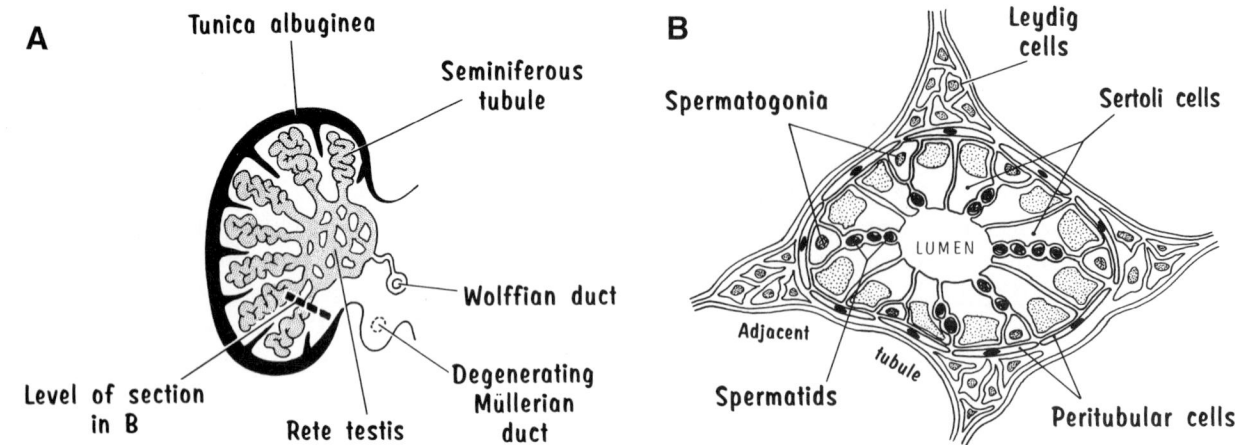

FIGURE 3 (A) Testis at 20 weeks showing seminiferous tubule formation. The Wolffian duct differentiates into the epididymus and efferent ductules and the Müllerian duct is completely regressed. (B) Transverse section through a testis showing a seminiferous tubule surrounded by the peritubular cells that provide structure and form the blood–testis barrier. The steroidogenic Leydig cells are located between the tubules, in the interstitial space. The tubule is composed of spermatogonia encompassed by the Sertoli cells which extend into the lumen of the tubule, sustaining and nurturing sperm maturation. During spermatogenesis the spermatocytes move toward the lumen and mature sperm are eventually released into the lumen [adapted with permission from *The Developing Human* (K. L. Moore and T. V. N. Persaud, Eds.), 5th ed., Chap. 13, Fig. 20, Saunders, Philadelphia, 1993].

ture of Leydig cells with Sertoli cell-conditioned media, enhances the LH and cAMP-stimulated increase in steroidogenesis in Leydig cells.

Sertoli cells, first described in the mid-nineteenth century, are the somatic epithelial cells of the semimiferous tubules in the gonad which are oriented radially around the semiferous tubule and extend from the basal lamina to a central point within the lumen. In the adult, Sertoli cells are irregularly columnar and have been described as tree-like in shape, encompassing the developing germ cells. Sertoli cells change in shape and in protein expression in conjunction with the stages of the spermatogenic cycle.

The origin of Sertoli cells has not been definitively determined except as somatic cells of the urogenital ridge. Their differentiation is intimately involved with differentiation of the testicular cords, following population of the ridge by germ cells. SRY expression is necessary for sertoli cell differentiation, whereas germ cells are not needed. Additionally, fetal and adult Sertoli cells are required for germ cell survival and maturation since ectopic germ cells degenerate. The Sertoli cells organize as one- to three-cell-thick cords with irregular outlines and become separated from the interstitium by a basement membrane

formed by Sertoli and peritubular cells. Sertoli cell proliferation is controlled by FSH from the pituitary and by local Sertoli and Leydig cell factors and ceases at puberty when spermatogenesis begins and spermatogonia are released from G_1 arrest.

Leydig cells are found in the interstitial spaces between the seminiferous tubules and develop from two generations of cells, a fetal and a pubertal/adult population. Fetal Leydig cells differentiate from mesenchymal somatic cells after sex cord formation and Sertoli cell differentiation. The initial signal triggering Leydig cell differentiation is unknown, but since Leydig cells develop after Sertoli cells, it might be due to Sertoli cell paracrine hormones. Leydig cells undergo a proliferative phase as androgen steroidogenesis greatly increases in response to hCG and fetal LH. After birth, the fetal Leydig cells begin regressing and androgen production declines, probably from loss of LH/hCG stimulation. Adult Leydig cells start to arise and proliferate from the mesenchymal presursors at puberty and achieve their adult population soon after sexual maturity. Androgen production by Leydig cells is essential for fetal sexual differentiation, secondary sexual maturation at puberty, spermatogenesis, and libido.

IV. SPERMATOGENESIS

Upon reaching the gonadal ridge, the primordial germinal cells continue to proliferate and are encompassed within testicular cords formed by Sertoli cell precursors. Germ cells that remain extragonadal or outside the testicular cords degenerate, whereas those within the tubules, termed prespermatogonia, undergo several cycles of mitotic division before entering a period of mitotic arrest. The duration of time until mitosis resumes varies from species to species but lasts throughout the prepubescent period in man. Further maturation of the testis occurs under the regulation of the pituitary gonadotropins LH and FSH as well as local growth factors and cytokines. Changes in both Sertoli and Leydig cell-differentiated function are essential for the initiation of spermatogenesis. The onset of puberty is characterized by the resumption of spermatogonial division, the formation of a lumen in the seminiferous tubules, and the establishment of tight junctions between the Sertoli cells that separate the adluminal surface from the interstitial space. These tight junctions create a blood–testis barrier restricting the passage of macromolecules and fluids between the two compartments which consequently differ in their ionic and protein composition.

In the mature testis, spermatogenesis occurs in cycles with well-defined stages. Each stage has specific germ cell associations with corresponding Sertoli cell-differentiated function and lasts for fixed periods of time. Spermatogonia mature by progressing through successive stages of the seminiferous tubule cycle multiple times before completion of spermatogenesis. Each spermatogonia divides several times before entering a prolonged meiotic phase to become preleptotene spermatocytes, which are sequestered in the adluminal surface. Except for the spermatogonia in the basal compartment, all the maturing spermatogonia and spermatocytes are dependent on Sertoli cells for nutrients and hormones. The preleptotene spermatocytes then complete meiosis and undergo another mitotic division to yield the haploid spermatids. This process of spermatogenesis is accompanied by varying degrees of germ cell degeneration. Spermiogenesis, the differentiation of the spermatids to mature spermatozoa, is characterized by morphological changes that include the development of an acrosome, nuclear condensation and elongation of the flagellum, and further reduction of the cytoplasm and release of the mature spermatid from the apex of the Sertoli cell. The initiation and maintenance of spermatogenesis requires the hormonal action of FSH and testosterone and is dependent on the supportive and paracrine actions of the Sertoli cells and on androgen production by the Leydig cells.

In summary, male gonadogenesis is a four-part process. The bipotential or sexually indifferent gonad develops from the coelomic epithelium and is populated by germ cells. Testicular determination is established by the expression of *SRY*, which initiates the development of the testicular cords and Sertoli and Leydig cells. Finally, germ cells undergo maturation into sperm, a cyclical process which requires the differentiated functions of Sertoli and Leydig cells.

See Also the Following Articles

Interstitial Cells of Leydig; Spermatogenesis

Bibliography

Burger, H., and DeKretser, D. (1989). *The Testis,* 2nd ed. Raven Press, New York.
Harley, V. R., and Goodfellow, P. N. (1994). The biochemical role of SRY in sex determination. *Mol. Reprod. Dev.* 39, 184–193.
Gustafson, M. L., and Donahoe, P. K. (1994). Male sex determination: Current concepts of male sexual differentiation. *Annu. Rev. Med.* 15, 505–524.
Kierszenbaum, A. L. (1994). Mammalian spermatogenesis in vivo and in vitro: A partnership of spermatogenic and somatic cell lineages. *Endocr. Rev.* 15, 116–134.
Martinez-Mora, J. (1994). *Intersexual States, Disorders of Sex Differentiation.* Ediciones Doyma, Barcelona, Spain.
Russell, L. D., and Griswold, M. D. (1993). *The Sertoli Cell.* Cache River Press, Clearwater, FL.
Saez, J. M. (1994). Leydig cells: Endocrine, paracrine, and autocrine regulation. *Endocr. Rev.* 15, 574–626.
Sharpe, R. M. (1994). Regulation of spermatogenesis. In *The Physiology of Reproduction* (E. Knobil and J. D. Neill, Eds.), 2nd ed. Raven Press, New York.

Gonadotropes

Gwen V. Childs

University of Texas Medical Branch

GLOSSARY

anterior lobe The largest glandular region in the adenohypophysis of the pituitary. Its principal products include six different hormones that stimulate various target organs throughout the body: growth hormone, adrenocorticotropin, thyroid-stimulating hormone, luteinizing hormone, prolactin, and follicle-stimulating hormone.

autocrine regulation Secretory product from one type of cell regulates the same set of cells.

axons (axonal) The cellular process projecting from a neuron that carries electrical or chemical regulatory information to another neuron/target organ/blood vessel.

bihormonal Containing two hormones. In the case of gonadotropes, a cell that contains both luteinizing hormone and follicle-stimulating hormone. Identified by cytochemical labeling which can distinguish two or more cellular products (dual cytochemistry) in the same cell.

castration/ovariectomy Surgical removal of the gonads (ovaries or testes).

endocrine secretion Secretion of a hormone into the bloodstream, which provides the major highway for transport of the hormone.

Golgi complex The cellular organelle responsible for condensation and packaging of products destined for secretion. It consists of flattened sacs arranged in a semicircular conformation. Material is moved by vesicles or connections between the sacs. Newly formed storage granules are seen in the innermost region of the Golgi complex.

hypothalamus A region at the base of the thalamus and around the third ventricle of the brain which contains groups of neurons and fiber tracts, many of which control the pituitary gland.

immunocytochemistry The process by which a product is identified in or on a cell or group of cells. The cytochemical process uses specific antibodies that react with that product. The antibodies themselves are either labeled or identified by other labeled antibodies or probes. The technology allows one to identify the site of production or action of a hormone, growth factor, or other regulatory product.

infundibular stalk The region connecting the pituitary to the floor of the hypothalamus. It is the extension of the neurohypophysis. It is also a downgrowth of nerve fibers from the hypothalamus and contains a large nerve fiber tract leading to the posterior lobe of the pituitary. It also contains a plexus of capillaries that receive fiber input from the neurons in the hypothalamus. These fibers contain hormones that regulate pituitary cells. Since they are secreted into the bloodstream, they are called neuroendocrine cells.

median eminence The portion of the neurohypophysis that actually connects with the hypothalamus (the top of the infundibular stalk). It contains capillaries that receive nerve fibers from the cells in the hypothalamus.

messenger ribonucleic acid (mRNA) A strand of nucleotide bases that code for particular proteins made by the cells. It is produced in the nucleus by the transcription of the genetic code on DNA molecules. mRNA travels to the cytoplasm where it is decoded by ribosomes and transfer RNA.

monohormonal Containing only one hormone. In the case of gonadotropes, a cell that contains only luteinizing hormone or follicle-stimulating hormone following dual-labeling cytochemistry.

negative feedback An effect by which the secretory products from a target organ prevent or limit further stimulation of that gland. They may do this by inhibiting the stimulatory hormone(s) from their regulatory gland(s).

neuroendocrine secretion Secretion of a hormone, from a neuron, into the bloodstream.

ovulation The release of the mature ovum from the ovarian follicle into the peritoneal cavity where it is picked up by the Fallopian tube.

paracrine regulation Secretory product from one type of cell regulates other cell types in the same organ.

periodic acid Schiff stain A stain that differentiates cells containing glycoproteins.

pituitary basophils Cells in the adenohypophysis that stain blue or purple with classical trichrome (triple color) stains used to study the pituitary. This term is actually a misnomer because none of the trichrome stains include basic dyes.

pituitary gland A hormone-producing gland that lies in a depression at the base of the skull. It is subdivided into the adenohypophysis and neurohypophysis based on embryologic origin. The neurohypophysis originated as a downgrowth from the brain and remains connected to the hypothalamus by an infundibular stalk. The adenohypophysis originated from the roof of the mouth.

positive feedback An effect by which secretory products from a target gland enhance secretions from their regulatory gland.

reproductive cycle In female mammals, the period during which a group of eggs or oocytes mature in cellular sacs called follicles leading to ovulation. At the same time, the uterine lining is prepared for a possible pregnancy. During most normal cycles, one or more oocytes reach maturation about half way in the cycle. They are then released (ovulation) to be picked up by the Fallopian tube of the uterus. The cycle requires carefully timed secretion of follicle-stimulating hormone to stimulate the growing follicles and a surge of luteinizing hormone just before ovulation. After ovulation, the oocyte may or may not be fertilized. The midcycle surge of luteinizing hormone also stimulates the cells from the ovulated follicle to form a corpus luteum to produce hormones needed to maintain the uterine lining for the eventual pregnancy. The increased levels of follicle-stimulating hormone also stimulate ovulation and maturation of follicles for a future cycle.

ribosomes The site where the message on the messenger ribonucleic acid is read and decoded to produce proteins. If the proteins are destined for secretion, the ribosomes produce a signal peptide which causes a complex series of reactions ending in attachment to the outer surface of endoplasmic reticulum and transfer of the new protein to the inside of the rough endoplasmic reticulum sac.

rough endoplasmic reticulum A cellular organelle arranged in sacs or vacuoles that may be flattened or dilated. The outside surface is covered with ribosomes which help to organize the synthesis of proteins. Newly synthesized proteins are inserted through a groove in the ribosome and stored in the rough endoplasmic reticulum.

secretion The process by which the hormone product is sent into the extracellular space. It usually involves the movement of storage granules to the plasma membrane, the joining of the membrane to that of the storage granule, and the release of the material inside the granule. The product may move to blood vessels (endocrine secretion) or it may move in the extracellular space to a neighboring cell (paracrine secretion).

storage granules Hormones or other products destined for secretion are sequestered first in the rough endoplasmic reticulum. Then, they are transported to the Golgi complex where the final package includes a Golgi complex membrane wrapped around the stored product. They may be secreted from the cell or they may be stored in the cytoplasm. Sometimes the size and shape of storage granules are used to identify a particular cell type. These granules may also be called *secretory granules*.

target organ The gland that is stimulated by a particular endocrine gland.

transport vesicles Small vesicles of endoplasmic reticulum membrane that bud from this organelle and carry newly synthesized protein to the Golgi complex.

Gonadotropes are a specific type of hormone-producing cell found in the anterior lobe of the pituitary gland. Their major function is the production and secretion of two hormones that regulate the ovaries and testes. One of these is luteinizing hormone (LH), which stimulates ovulation in the ovary and the synthesis and secretion of testosterone in the testes. In the male, this hormone has also been called interstitial cell-stimulating hormone. The other hormone is follicle-stimulating hormone (FSH), which stimulates the development of the follicle and ovum in the ovary and the development of the sperm in the testes. The hormones are called gonadotropins and they are glycoprotein molecules. Each gonadotropin molecule has two subunits, called the α and β subunits. LH and FSH molecules each contain the same α subunit. However, the β subunit is unique to each gonadotropin and confers biological and immunological specificity to the LH or FSH molecules.

I. STRUCTURE OF
THE GONADOTROPE

Gonadotropes are round or ovoid cells, about 10–12 μm in diameter. They often lie near blood vessels (Fig. 1). Classical studies at the light microscopic level have identified them by their specific reactivity to combinations of stains. Gonadotropes stain positively for the periodic acid Schiff reaction because of their glycoprotein hormone content. They also stain blue with most of the classical trichrome stains. Because of their particular staining reactions, they are called "basophils." Most researchers believe that gonadotropes are the Δ basophils that were named in the classification scheme developed in the early to mid-1900s.

When seen with the transmission electron microscope (TEM), an actively synthesizing gonadotrope contains rough endoplasmic reticulum which is arranged in flattened or dilated sacs scattered throughout the cell (Fig. 2). This is the site for the production of the glycoprotein hormones, luteinizing hormone (LH) and follicle-stimulating hormone (FSH). The sacs become more dilated as they are filled with proteins. Some of the carbohydrate groups on the gonadotropin molecules are also added in the rough endoplasmic reticulum. The rest are added in the Golgi complex.

After synthesis, the hormones are transported to the Golgi complex in small vesicles. Immature storage granules are seen in the Golgi complex packaging zone as dense patches in the sacs and large vesicles (Fig. 3). The mature storage granules then accumulate in the cytoplasm. Gonadotropes typically contain abundant storage granules scattered throughout the cell cytoplasm. Some of the gonadotropes produce granules that are of a uniform size (0.2–0.25 μm in diameter). Other gonadotropes produce two populations of granules, a small set that is 0.1–0.2 μm in diameter and a less abundant larger set that is 0.4–0.6 μm in diameter (Figs. 2 and 3). The significance of the two populations of granules is not known because both types may contain LH and FSH.

The hormone content of the gonadotropes can be identified at the light microscopic level by immunocytochemistry. This technique uses specific antibodies to cytochemically identify the β subunit on molecules of LH and FSH at sites of storage or synthesis. Figure 1 illustrates a field labeled for LHβ subunit. Figure 3 shows an immunocytochemically labeled field with the use of colloidal gold probes for LHβ and FSHβ. This dual-labeling protocol shows differential storage of the gonadotropin molecules. In Fig. 3, one vacuole in the Golgi region contains only smaller gold markers for FSHβ. Nearby is a vesicle that contains gold markers for both LHβ and FSHβ.

When the gonadotrope is stimulated to secrete, the storage granules move to the cell membrane and the membrane of the storage granule joins with the cell membrane. This creates an opening through which the hormone content of the granule can pass into the extracellular space. The hormone is then able to pass across the blood vessel wall into the bloodstream, where it is then sent to the ovary and testes. Immunocytochemical labeling viewed with the TEM shows that different amounts of LH and FSH can be found in the storage granules of bihormonal gonadotropes (Fig. 3). Over half of the granules may contain both LH and FSH, which allows the two hormones to be released together. However, some granules in bihormonal gonadotropes contain only LH or FSH (Figs. 3 and 4). This suggests that the Golgi complex may preferentially package one of the

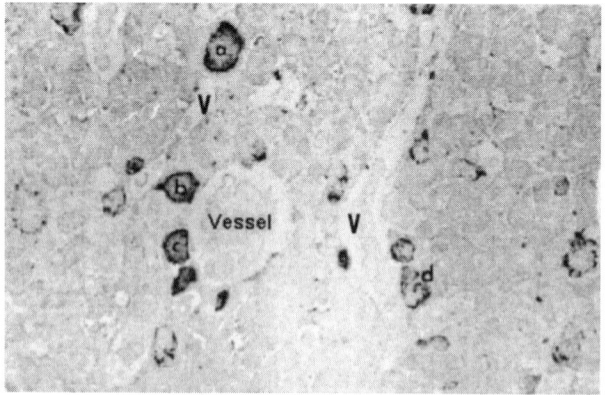

FIGURE 1 Semithin, 1-mm plastic section of the anterior pituitary. The field was immunocytochemically stained for the LHβ subunit. The LH-containing gonadotropes are the dark cells (a–d) in the field. They are mostly ovoid; however, one or two processes can also be seen. They also vary in size. Note that almost all of them are clustered around a blood vessel (v). Magnification, ×1100.

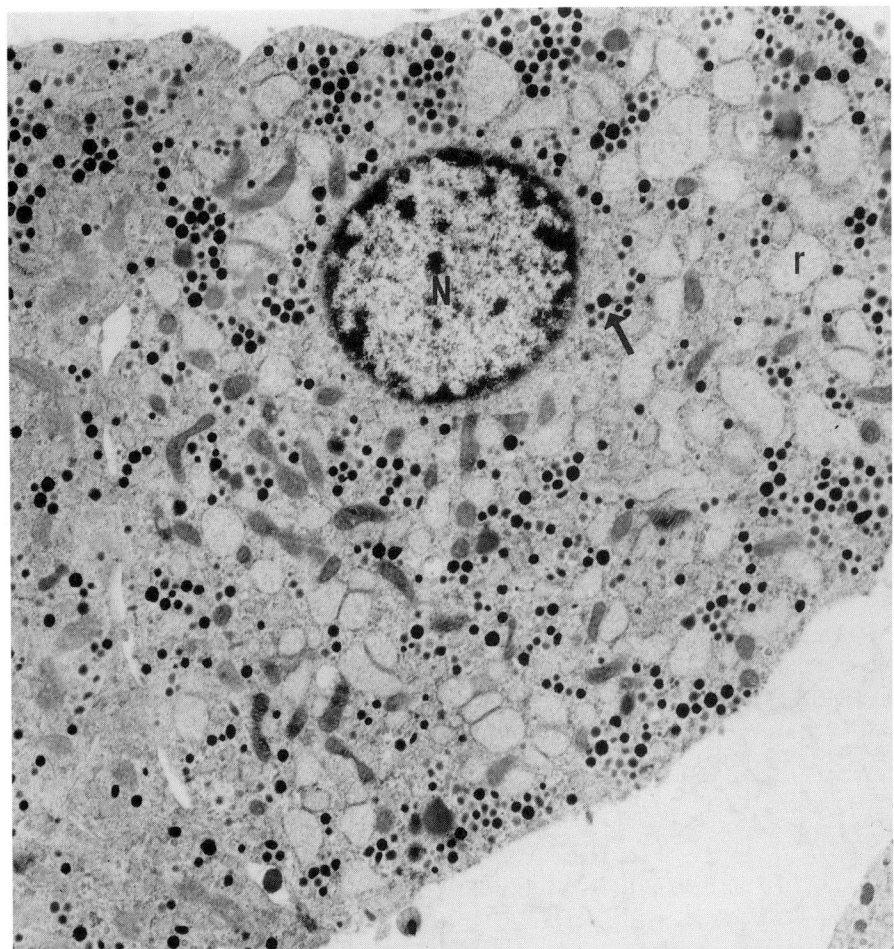

FIGURE 2 Transmission electron micrograph of a gonadotrope showing a section through the nucleus (N) and cytoplasm which contains sacs of rough endoplasmic reticulum (r) and storage granules. The arrow points to a cluster of storage granules. Note that two sizes are evident, which is a distinguishing feature for a subtype of gonadotrope. Magnification, ×12,500.

gonadotropins to secrete more of that gonadotropin. Figure 3 illustrates preferential packaging of FSHβ stores.

Analysis of the storage patterns in a number of mammalian species shows that from 50 to 100% of gonadotropes store both LH and FSH in the same cell. These are called bihormonal gonadotropes. The bihormonal state is found most frequently during periods of high secretory activity. However, some species also have gonadotropes that store only LH or FSH in each cell. These cells are called monohormonal gonadotropes. They are abundant among the smallest gonadotropes. Figure 5 illustrates a serially sectioned monohormonal gonadotrope that is immunocytochemically labeled for FSHβ but not LHβ.

II. GONADOTROPE RESPONSES TO GONADOTROPIN-RELEASING HORMONES

Gonadotropes produce receptors for key regulatory hormones. This allows them to be stimulated or inhibited in synchrony with the rest of the reproductive system. The production of these receptors is timed carefully during the reproductive cycle to enable the gonadotrope to be responsive at the appropriate time.

One of the most important regulatory factors is gonadotropin-releasing hormone (GnRH). This peptide hormone is produced by neurons in the hypothalamus of the brain. These neurons send axons to

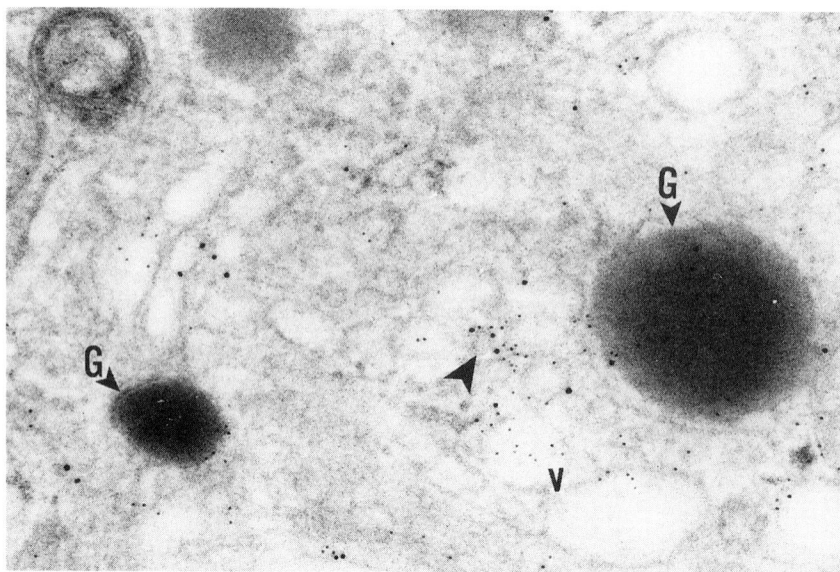

FIGURE 3 Transmission electron micrograph showing the Golgi complex of a gonadotrope. Storage granules are seen in this region (G). The field was immunocytochemically labeled for LHβ (small gold markers) and FSHβ (large gold markers). The density of the stores in the granules occludes the gold markers; however, underexposed photographs show that the larger granule contains both LHβ and FSHβ. On the nearby vesicles and vacuoles, the gold markers are seen more easily. One vacuole (v) contains label for LHβ only. Another (arrowhead) contains both LHβ and FSHβ. This suggests that the Golgi complex can dissociate the packaging of LH and FSH and produce monohormonal storage granules. Magnification, ×125,000.

end at sites along the capillary plexus in the infundibular stalk and median eminence. The capillaries connect with the portal vessel system supplying the anterior lobe of the pituitary. GnRH is stored in storage granules and, during times of stimulation, it is released from these axonal endings. It travels to sites in the anterior lobe where gonadotropes are located and binds to specific receptor molecules in the gonadotrope plasma membrane. After binding, GnRH stimulates both the synthesis and the release of LH and FSH. Pulsatile secretion of GnRH is necessary before the gonadotrope population can orchestrate a surge of LH. This surge of LH, along with a rise in FSH, is vital to ovulation.

Early in the reproductive cycle, GnRH may also stimulate developmental changes in gonadotropes that enable them to coordinate and support the gonadotropin secretory activity. GnRH may cause a monohormonal gonadotrope to become bihormonal by stimulating the production of the other hormone. GnRH-stimulated gonadotropes become angular or more irregularly shaped as they extend processes to blood vessels. When GnRH is added to gonadotropes in tissue culture it causes the extension of cellular processes that become filled with stores of gonadotropins. An example is shown in Fig. 4. The GnRH molecules themselves can also be cytochemically detected on the hormone-filled cellular process (Fig. 4). Finally, as it stimulates synthesis, GnRH also causes the dilation of rough endoplasmic reticulum. In fact, GnRH actually stimulates the gonadotropes to grow larger to accommodate the increased synthetic and secretory activity.

Finally, the past two decades have brought forth evidence of a FSH releasing factor (FSHRF) that is separate from GnRH and stored in separate regions of the hypothalamus. Evidence suggests it may be a variant (analog) of mammalian GnRH. This long-sought FSHRF may selectively stimulate secretion from monohormonal FSH cells or granules. Studies are underway to characterize the origin and functions of the candidate FSHRF analog.

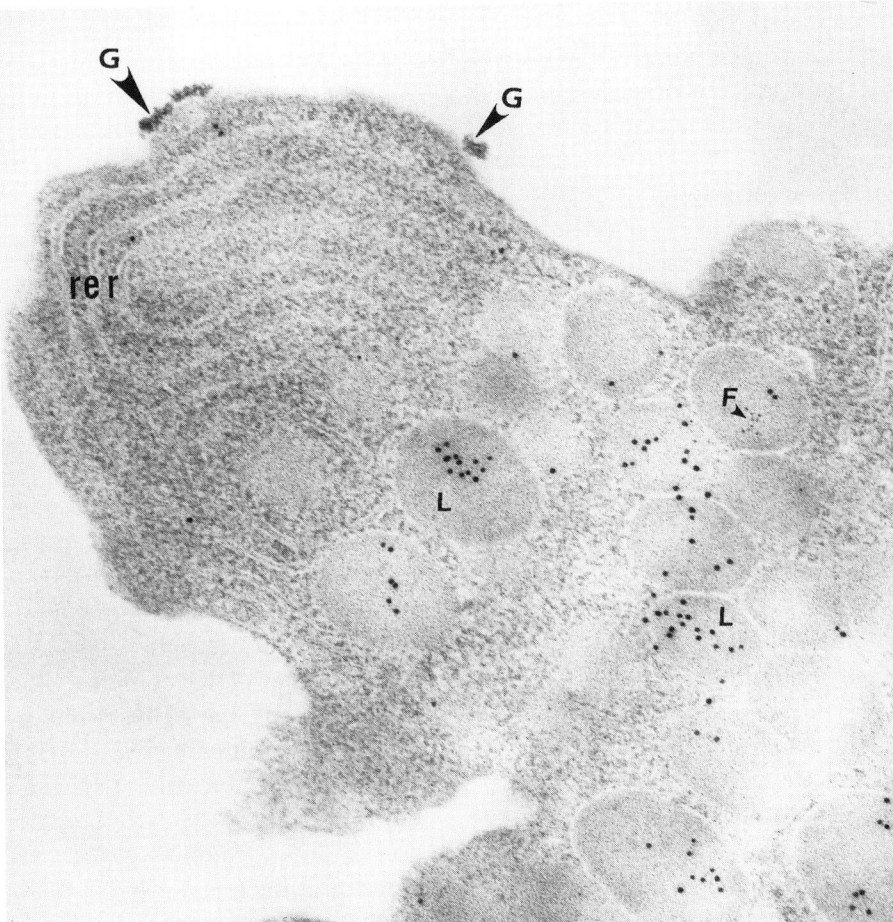

FIGURE 4 Cellular process of a gonadotrope that was first stimulated with an biotinylated analog of GnRH that was then detected cytochemically. Patches of label for the GnRH binding sites are noted (G). The process is filled with flattened rough endoplasmic reticulum (rer) and storage granules that are immunocytochemically labeled for LHβ [large gold markers (L)] or FSHβ [small gold markers (F)]. The field contains a number of storage granules labeled only for LHβ (monohormonal granules). One of the storage granules contains both LHβ and FSHβ (F). Magnification, $\times$125,000.

III. GONADOTROPE RESPONSES TO NEGATIVE FEEDBACK EFFECTS OF GONADAL STEROIDS

Another group of regulatory hormones important to gonadotropes are the steroids secreted by the ovary and testes. These include estrogens, testosterone, and progesterone. Most of the time, steroids inhibit gonadotropin secretion, providing feedback about the need for further stimulation (negative feedback). This is especially evident after a period of surge secretion (after ovulation). This type of feedback is vital to regulate the timing of the LH surge needed for ovulation and the maintenance of the uterine lining.

In classical studies, gonadotropes were first discovered and identified by their responses to the removal of this negative feedback. After surgical removal of the testes or ovaries, or after menopause, the gonadotropes respond by enlarging and increasing their secretion of LH and FSH. These cells are called castration cells or ovariectomy cells. They become filled with large vacuoles of dilated rough endoplasmic reticulum and begin to take on a "Swiss cheese-like" appearance. Eventually, some of the cells contain one large vacuole and little bordering cytoplasm, giving them the appearance of a "signet ring." After castration or ovariectomy, the cells become nearly 100% bihormonal and nearly three times more numerous than in intact animals. Within a few weeks, they are the largest cells in the pituitary. When studied at the electron microscopic level, it appears that secretion may outpace storage, resulting in fewer storage granules. Sometimes the rough endoplasmic reticulum sacs contain dense regions of hormone stores. Eventually (after 6 months) many of the largest castration cells are replaced by numerous small, densely granulated gonadotropes.

IV. POSITIVE FEEDBACK BY GONADAL STEROIDS

Positive feedback mechanisms are also seen during the reproductive cycle. Early in the reproductive cycle of mammals, FSH secretion is high in order to stimulate a developing crop of ovarian follicles. Estradiol is produced and secreted by these developing follicles. Estradiol then stimulates gonadotropes to produce more receptors for GnRH. This in turn promotes increases in gonadotropin synthesis and release. This stimulation culminates in a surge of LH which regulates ovulation. It also results in high levels of FSH after ovulation which stimulates the next group of developing follicles.

V. AUTOCRINE OR PARACRINE REGULATORY FACTORS

Two additional regulatory peptides important to gonadotropes are activin and inhibin. These hormones were originally discovered in ovarian follicular fluid. Activin stimulates and inhibin inhibits the synthesis and release of FSH and the expression of GnRH receptors. Some studies have also reported effects on LH release; however, these may be secondary to the effects on the expression of GnRH receptors. The subunits of activin and inhibin are found in the pituitary gonadotrope granules. This may indicate local or autocrine control of FSH synthesis and release by gonadotropes.

A third regulatory factor, also produced by pituitary cells, is follistatin. This factor is found in a number of cell types; the most numerous include LH and growth hormone (GH) cells. Follistatin binds activin and thereby limits its stimulatory effects on FSH. Thus, follistatin may be part of the autocrine or paracrine circuitry that regulates FSH synthesis and release.

VI. MATURATION OF GONADOTROPES

Gonadotropes mature during fetal and neonatal development to be ready for puberty. Immature gonadotropes are small and not well granulated. In fact, they are difficult to distinguish by morphology alone. They contain a few scattered granules that may vary in shape and size. Often the granules are concentrated in the Golgi complex; however, sometimes they lie along the cell periphery in a single row. In the immature rat, the developing gonadotrope population produces LH first and then FSH several days later. Thus, the early developing population of gonadotropes contains some bihormonal gonadotropes and the remainder are mostly monohormonal LH cells. In the neonatal rat, monohormonal FSH cells are not abundant until several days after birth.

Gonadotropes that resemble the immature subtype exist in the adult animal (Fig. 5). They are mostly found among the smallest of the pituitary cells. It has been postulated that these cells provide a source of reserve cells that can support the reproductive processes. Several findings support this hypothesis. First, newly synthesized messenger ribonucleic acid (RNA) for LH and FSH is found primarily in small, immature gonadotropes early in the cycle. Second, most of the small gonadotropes are also monohormonal in that they contain and express the messenger RNA (mRNA) or protein for only one gonadotropin. Third, their expression of GnRH receptors increases if they are treated with estradiol. Furthermore, GnRH stimulation will cause the small monohormonal cells to become bihormonal. Just before midcycle, many of the monohormonal cells begin to synthesize the other gonadotropin as their hormone stores increase. This produces a population of gonadotropes that is

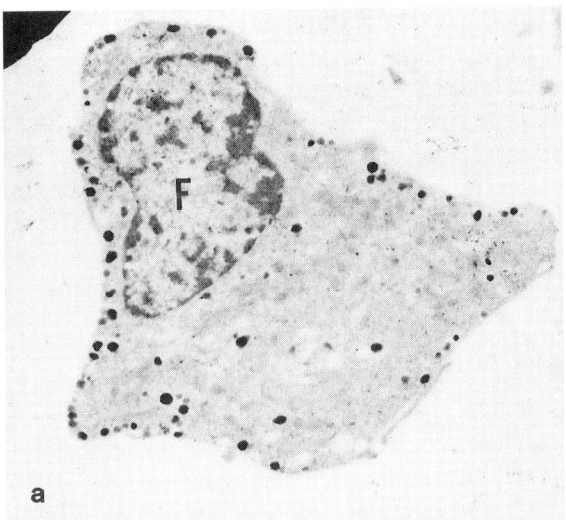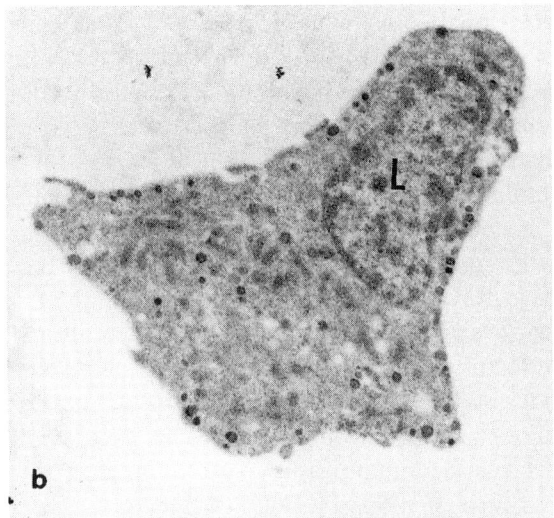

FIGURE 5 Transmission electron micrograph showing a serial section through a gonadotrope. This cell was in the fraction containing the smallest pituitary cells (the average size is a little larger than a red blood cell). (a) A field immunocytochemically labeled for FSHβ. The labeling method produced a dense black reaction product on the storage granules. Note that most are at the periphery of the cell and labeled for FSHβ. (b) A serially sectioned field immunocytochemically labeled for LHβ. The dense gray storage granules remain mostly unlabeled for LHβ. Thus, this gonadotrope is classified as a monohormonal FSH cell. It has the potential to produce LHβ and become bihormonal, however. Magnification, ×12,400.

eventually over 75% bihormonal just before ovulation.

VII. RELATIONSHIP TO OTHER PITUITARY CELL TYPES

The gonadotropes are a part of a cell population in the anterior pituitary containing at least six cell types. These different types may be differentiated by their storage of one principle hormone. The hormones include growth hormone, adrenocorticotropin, prolactin, and thyroid-stimulating hormone. Whereas each hormone is produced primarily by one cell type, cells and tumors that produce different combinations of these hormones have also been found. Thus, it has been postulated that certain pituitary cells may be multifunctional and capable of supporting other populations. Current research has begun to focus on transitional patterns of differentiation among pituitary cells.

Gonadotropes in particular are the subject of some of these studies because subsets may contain prolactin, growth hormone, thyroid-stimulating hormone, or adrenocorticotropin depending on the physiological state. The abundance of gonadotropes producing adrenocorticotropin during neonatal development suggests that they may stem from a common stem cell.

In addition, N-terminal fragments of the proopiomelanocortin gene (that produces ACTH) are secreted by gonadotropes to stimulate the development and differentiation of prolactin cells. Recently, the coexpression of GH and gonadotropins during the reproductive cycle has been correlated with the augmentation in the gonadotrope population just before ovulation. mRNA for gonadotropins has been found in subsets of growth hormone cells. This supports the hypothesis that some growth hormone cells may contribute to the high LH secretion just before ovulation or the high FSH secretion that promotes the development of the ovarian follicle. This hypothesis is further supported by findings that show GnRH receptors on growth hormone cells as well as growth hormone-releasing hormone receptors on gonadotropes during this time. Hence, it is possible that the gonadotrope population is actually augmented by these cell types. The regulatory process that causes the transitional conversion of some of the GH cells to gonadotropes remains to be determined.

Finally, several of the hormone-bearing cell types also produce growth factors and other regulatory peptides that may regulate certain neighboring cells. As stated earlier, gonadotropes produce activin and inhibin subunits which may serve as autocrine or paracrine regulators. Gonadotropes also produce regulatory factors belonging to the renin–angiotensin family as well as growth factors such as epidermal growth factor and nerve growth factor. These products may modulate secretion and synthesis in other pituitary cell types. The findings have led to research that focuses on communication patterns between different groups of pituitary cells, especially with respect to paracrine regulation.

See Also the Following Articles

Activin and Activin Receptors; Follistatin; FSH (Follicle-Stimulating Hormone); GnRH (Gonadotropin-Releasing Hormone); Gonadotropins, Overview; Gonadotropin Receptors; Gonadotropin Secretion, Control of; Inhibins; LH (Luteinizing Hormone)

Bibliography

Albanese, C., Colin, I. M., Crowley, W. F., Ito, M., Pestell, R. G., Weiss, J., and Jameson, J. L. (1996). The gonadotropin genes: Evolution of distinct mechanisms for hormonal control. *Rec. Prog. Horm. Res.* **51**, 23–58.

Catt, K. J., Loumaye, E., Kalikineni, M., Hyde, C. L., Childs, G. V., Amsterdam, A., and Naor, K. (1983). Receptor and actions of GnRH on pituitary gonadotropes. In *Role of Peptides in Control of Reproduction* (D. McCann and Dhindsa, Eds.), pp. 33–61. Elsevier, New York.

Childs, G. V. (1986). Functional ultrastructure of gonadotropes: A review. In *Current Topics in Neuroendocrinology* (D. Pfaff, Ed.), Vol. 7, pp. 49–97. Springer-Verlag, New York.

Childs, G. V. (1994). Division of labor among gonadotropes. *Vitamins Horm.* **50**, 217–283.

Childs, G. V., Unabia, G., and Rougeau, D. (1994). Cells that express luteinizing hormone (LH) and follicle stimulating hormone (FSH) beta (β) subunit mRNAs during the estrous cycle: The major contributors contain LHβ, FSHβ and/or growth hormone. *Endocrinology* **134**, 990–997.

Conn, P. M. (1986). The molecular basis of gonadotropin-releasing hormone action. *Endocr. Rev.* **7**, 3–10.

Gharib, S. D., Wierman, M. E., Shupnik, M. A., and Chin, W. W. (1990). Molecular biology of the pituitary gonadotropins. *Endocr. Rev.* **11**, 177–199.

Halmi, N. S., and Moriarty, G. C. (1977). The hypophysis (pituitary gland). In *Histology* (Greep and Weiss, Eds.), 4th ed., pp. 1093–1967.

Ibrahim, S. N., Moussa S. M., and Childs G. V. (1986). Morphometric studies of rat anterior pituitary cells after gonadectomy: Correlation of changes in gonadotropes with the serum levels of gonadotropins. *Endocrinology* **119**, 629–637.

Kaiser, U. B., Conn P. M., and Chin, W. W. (1997). Studies of gonadotropin-releasing hormone (GnRH) action using GnRH receptor-expressing pituitary cell lines. *Endocr. Rev.* **18**, 46–70.

Knight, P. G. (1996). Roles of inhibins, activins, and follistatin in the female reproductive system. *Front. Neuroendocrinol.* **17**, 476–509.

Lloyd, J. M., and Childs, G. V. (1988). Changes in the number of GnRH-receptive cells during the rat estrous cycle: Biphasic effects of estradiol. *Neuroendocrinology* **48**, 138–146.

Pierce J. G., and Parsons T. F. (1981). Glycoprotein hormones: Structure and function. *Annu. Rev. Biochem.* **50**, 465–495.

Schwartz, N. B. (1995). The 1994 Stevenson Award Lecture. Follicle-stimulating hormone and luteinizing hormone: A tale of two gonadotropins. *Can. J. Physiol. Pharmacol.* **73**, 675–684.

Turgeon, J. L., Kimura, Y., Waring D. W., and Mellon, P. L. (1996). Steroid and pulsatile gonadotropin-releasing hormone (GnRH) regulation of luteinizing hormone and GnRH receptor in a novel gonadotrope cell line. *Mol. Endocrinol.* **10**, 439–450.

Wen, H. Y., Karanth, S., Walczewska, A., Sower, S. A. and McCann, S. M. (1997). A hypothalamic follicle-stimulating hormone-releasing decapeptide in the rat. *PNAS* **94**, 9499–9503.

Gonadotropin Biosynthesis

Raymond Counis

Université Pierre et Marie Curie

GLOSSARY

Golgi apparatus A complex cellular organelle in close juxtaposition with the nucleus and rough endoplasmic reticulum (RER). It appears composed of parallel, flattened saccules, vesicles, and vacuoles, organized into three functionally distinct regions, the *cis*-Golgi network, the Golgi stack (containing several distinct subcompartments), and the *trans*-Golgi network (TGN). The *cis*- and *trans*-Golgi networks are the entry and exit faces of the stack. Proteins transit from RER to TGN and are subject to a progressive maturation.

intron A sequence in the transcriptional unit of a gene which is excised from the primary transcript during mRNA maturation within the nucleus.

mRNA An oriented ribonucleotidic sequence consisting of a coding region flanked by a 5' (upstream) and a 3' (downstream) untranslated region, the latter being extended by a stretch of 100–150 adenines (poly-A). Both initiation and termination of translation, as well as polyadenylation, are defined by specific signals on the mRNA. The mRNA is capable of interaction with several structures, including ribosomes and proteins.

polysome (or **polyribosome**) A structure generally attached to the endoplasmic reticulum to form the RER and constituted by several ribosomes linked to a specific mRNA. Each ribosome synthesizes a protein by translation using the information provided by the nucleotide sequence of mRNA according to the classical codon usage. Taking into account that amino acids are incorporated into polypeptides at an average rate of 20/sec, the synthesis of one gonadotropin subunit (the polypeptidic precursor) should take about 6–8 sec. Ribosomes are distant from about 60 nucleotides each on a specific mRNA and move along the mRNA from the 5' to the 3' end to synthesize one protein each. They detach thereafter and dissociate into subunits.

promoter A sequence of the gene upstream of the region corresponding to the mRNA (delimited by the transcription initiation site). This region contains information ("cis-elements") pertaining to the attachment of the RNA polymerase complex transcription, tissue-specific expression, and regulation of RNA polymerase complex fixation and activity. A combination of several elements are involved in this process, which may vary by gene depending on the cell type and transcriptional factors and regulators present.

signaling A cascade reaction by which extracellular signals exert their action(s) on a specific intracellular target(s). This process involves specific recognition of the signal by a specific receptor, transduction into the cell (via different mechanisms), and activation or induction of cellular and/or nuclear proteins. Phosphorylation–dephosphorylation represents an important mechanism in these processes.

signal peptide (or **leader sequence**) A highly hydrophobic sequence of 15–30 amino acids present upstream (representing the N terminus) of a newly synthesized protein (a precursor) that allows the protein to cross the membrane and enter the lumen of RER. This sequence is enzymatically cleaved on the internal face of the RER membranes.

Gonadotropins are glycoproteins secreted under the control of a variety of factors, implying an appropriate adaptation of the biosynthetic processes. These hormones, which include the pituitary luteinizing hormone (LH) and follicle-stimulating hormone (FSH) and, in certain species, the placental choriogonadotropin (CG), are heterodimers resulting from the noncovalent association of a common α and a specific β subunit. In contrast to CG, which is expressed only during early pregnancy, LH and FSH are produced and released in different degrees throughout life in a manner implying a physiological regulation. The complexity of gonadotropin structure and the absence of appropriate models or technology has long been an obstacle for the elucidation of the mechanisms of biogenesis and its regulation; however, much progress has been accomplished in these domains over the past 15 years.

I. INTRODUCTION

Two pituitary gonadotropins control gonadal function in higher vertebrates: the luteinizing hormone (LH) and the follicle-stimulating hormone (FSH). Both hormones are produced by specific cells, the gonadotropes, and are released primarily under the control of a hypothalamic neurohormone, the gonadotropin-releasing hormone (GnRH). In the male, LH stimulates testosterone production by Leydig cells of the testis, and in the female, LH stimulates steroidogenesis in maturing follicles of the ovary and also induces ovulation at the midcycle surge. FSH promotes follicular development in the ovary and sperm production in the testis. A third gonadotropin, choriogonadotropin (CG), is produced by the placenta of human and equines exclusively and is required for the maintenance of pregnancy. CG binds to LH/CG receptors on cells in the corpus luteum to induce progesterone secretion during the first trimester of pregnancy. The three gonadotropins are members of the structurally complex glycoprotein hormones family, which also includes the pituitary thyrotropin-secreting hormone (TSH).

Several attempts to isolate the *de novo* synthesized gonadotropins in pituitary or placental tissue were made as early as the 1970s, with the availability of antisera for the specific immunoextraction of radiolabeled hormones. However, given the methodology employed (incubation of tissue or cultured cells and the *in vitro* incorporation of radioactive amino acids or sugars) and the specificity of the antibodies used (directed against native subunits or complete, mature hormones), this procedure did not allow for the assessment of the very initial steps of subunit production. By the 1980s, the cell-free translation of pituitary or placental messenger RNA (mRNA) using heterologous systems (wheat germ extract and reticulocyte lysate), unable to induce the maturation of the primary translated products, provided the first evidence that gonadotropin subunits were synthesized as independent precursors, encoded by distinct mRNAs. Furthermore, only sequential antibodies were able to recognize each precursor, revealing conformational differences between the primary and processed subunit forms. These studies were subsequently confirmed by the isolation of cDNAs and genes (1980–1990). Other investigations also include the characterization of the intermediate forms of the subunit biosynthesis, especially with respect to glycosylation, and the association of subunits into heterodimers. In addition to the elucidation of the different steps involved in gonadotropin biogenesis, these studies have provided a variety of means and tools to explore how the biosynthetic processes could be regulated, in particular by the hormones already known to control gonadotropin release.

II. THE DIFFERENT STAGES OF GONADOTROPIN SYNTHESIS: RELATIONSHIP TO THE SECRETORY PATHWAY

A. Organization of Gonadotropin Subunit Genes and Characterization of mRNA Transcripts

Figure 1 depicts the most important specific structural characteristics of typical mammalian α or β gonadotropin subunits together with the corresponding precursor, mRNA, and gene. Gonadotropin subunit genes have been characterized in several mammalian species, including human, mouse, rat, bovine, ovine, and monkey. In each species, only one gene encodes for α, LHβ, and FSHβ. In humans, the LHβ gene is located in a cluster of about 50 kilobases (kb) which also contains six copies of the CGβ in an array of tandem and inverted sequences (Fig. 1A, inset), of which only three (designed as 5, 3, and 8) are functional. In contrast to humans, a single gene encodes for equine LHβ and CGβ expressed in pituitary gonadotropes and in the placenta, respectively.

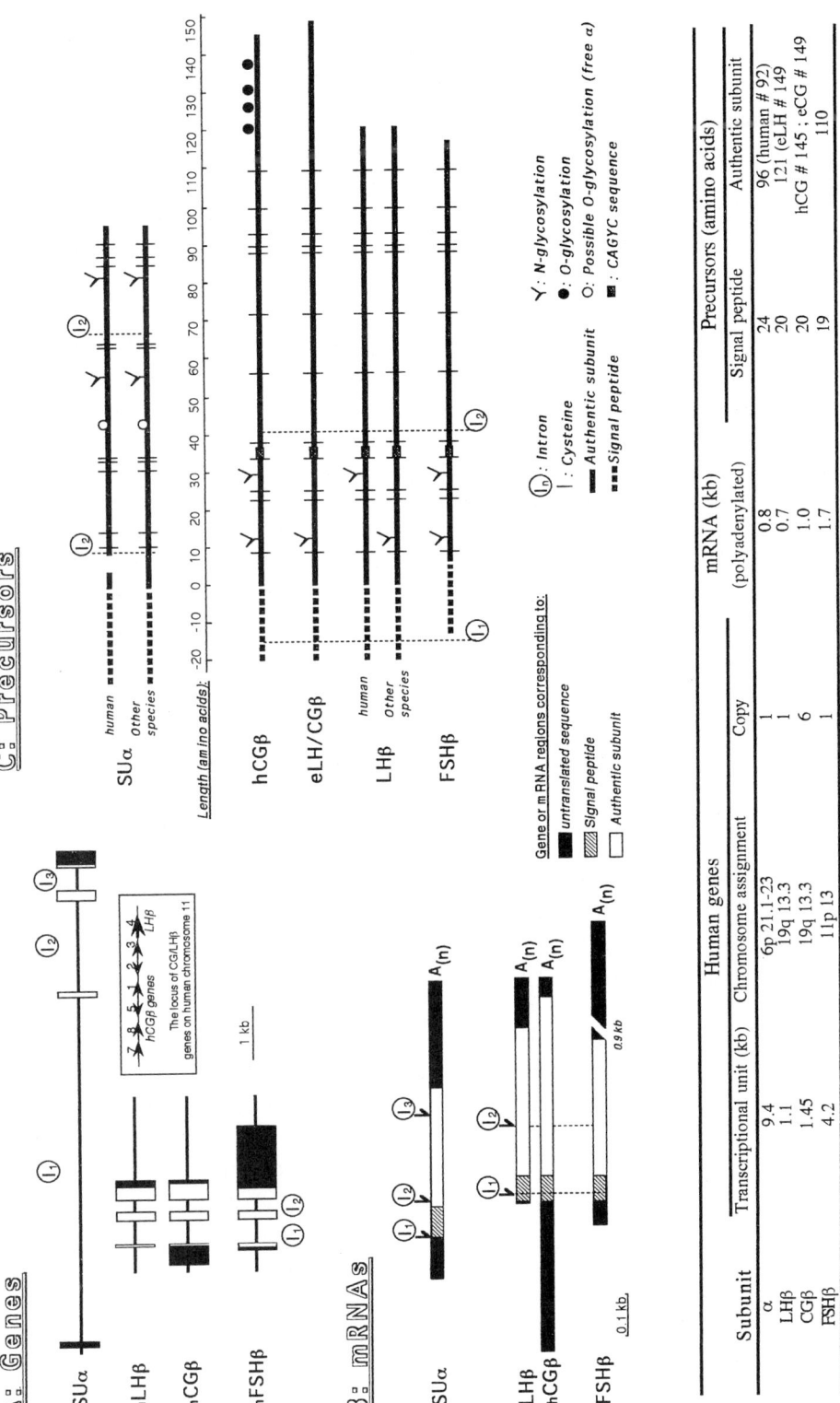

FIGURE 1 Characteristics of the gonadotropin genes, mRNA transcripts, and precursors. Human gonadotropins (which include hCG) are used as models for the diagrammatic represenation of genes (A) and mRNA (B). (C) Precursors of the typical gonadotropin subunits (adapted with permission from R. Counis and M. Jutisz, Regulation of pituitary gonadotropin gene expression. Outline of intracellular signaling pathways, *Trends Endocrinol. Metab.* **2**, 181–187, 1991).

From Fig. 1, it is apparent that a high degree of conservation is present in the organization of the α gene, on the one hand, and the β subunit gene on the other hand. For the α gene, the major differences between species include, in addition to some variation in nucleotide sequence, the size of introns, especially the first intron (13.5 kb in bovine, 5.4 kb in rat, and 6.4 kb in human), and a deletion in the human exon 2 (which encodes for the signal peptide plus the N terminus) resulting in a subunit reduced by 4 amino acids compared to the other species. With this unique exception, the amino acid sequence of the α subunit (and in particular the number and position of cysteines) appears considerably well conserved during phylogeny, emphasizing the important role of this subunit in the functional activity of the dimeric hormone. The high degree of conservation in the structure of the β subunit gene as well as in the characteristic features of the coding region (particularly the perfect alignment of the 12 cysteines and the position of N-linked carbohydrate chains) reflect evidence of a common ancestral origin. It is now well established that the CGβ and FSHβ have evolved from the LHβ gene by duplication modifications. Although the human LHβ and CGβ genes are highly homologous (95%), they exhibit differences in the transcriptional start sites which result in mRNA with variable 5′ untranslated extensions. In addition, the human CGβ mRNA encodes for a polypeptide with a 24-amino acid carboxy-terminal extension caused by a read-through mutation of the translational stop codon present in the LHβ gene. Compared to the other subunits, the FSHβ mRNA contains a long 3′ untranslated region ($\sim$1.3 kb) which may account for the high sensitivity to degradation (half-life $\sim$1–3 hr).

B. Subunit Production, Maturation, and Association in Relation to the Secretory Pathway

The synthesis of gonadotropin subunits and formation of mature hormone follow the general pathway of protein secretion in eukaryotes, which is tightly linked to the subcellular organization of cells. Typically, newly synthesized proteins undergo an intracellular transit analogous to that primarily estab-lished for the exocrine pancreas by Palade, which involves the following steps:

1. Synthesis on attached polysomes and transfer into the rough endoplasmic reticulum (RER)
2. Transport within the Golgi apparatus from the *cis* face to the *trans* face
3. Concentration and aggregation into electron-dense material and packaging into membrane-bound immature secretory granules in the *trans*-most Golgi saccule
4. Storage in mature secretory granules
5. Extracellular discharge of secretory granules by exocytosis or, alternatively, intracellular degradation of excess secretory granules, following their fusion with lysosomes

Compared to other proteins, the synthesis of gonadotropins presents some particularities especially with regard to the heterodimeric structure, the role, and the specific sugar type of N glycosylation. Moreover, in the pituitary, the two gonadotropins are not stored in the same granules, implying distinct addressing signals.

Consistent with the previously described general scenario for secreted proteins, spliced and 3′ poly-adenylated mRNAs derived from the transcription of gonadotropin genes are translocated into the cytoplasm and translation occurs on polysomes as for other proteins. For every subunit, the nascent polypeptide (precursor) enters the internal cisternae of RER and processing begins cotranslationally in the lumen of this structure, consisting of the cleavage and excision of the signal peptide and glycosylation. Both have been shown to be crucial for the synthesis of authentic subunits and their appropriate association into a dimeric hormone which is stable in the general circulation and has biological activity on target cells. These properties are conferred on gonadotropins by two types of carbohydrate units: those linked to asparagine (Asn) residues present in the specific sequence, Asn-X-serine or threonine (N glycosylation), and, in some cases, those linked to a serine or threonine residue (O glycosylation). N glycosylation is essential for the acquisition of correct subunit structure and for trafficking, whereas both N and O glycosylation play a role in hormonal stability and solubility in blood. Blocking of N glycosylation results in the accumulation of immature subunits in the cells and even in the arrest of translation. Several methodological approaches have shown that combination of complementary α and β subunits begins in the RER, as early as after cleavage of the signal peptide and initiation of N glycosylation. However, the final, stabilized structure is acquired later with the progression and maturation of dimers in the subcellular structures. As schematically represented in Fig. 2, N glycosylation is initiated in RER by the in-block branching of a mannose-rich structure, which will then undergo a series of complex sequential modifications (eliminations and additions of sugar residues) according to the specificity of glycosidases and glycosyltransferases present as the protein subunit transits along the different cisternae of the saccule network. The general structure of carbohydrate units diverges essentially in the late compartments of the Golgi (the *trans*-Golgi and the *trans*-Golgi network) leading to the formation of multiple forms of sugar chains with respect to the degree of ramification (mono-, bi-, or triantennae) and termini, accounting for the established microheterogeneity inherent to gonadotropins.

Thus, while the most common, ultimate step in the processing of carbohydrate chains consists of the addition of a terminal sialic acid on a galactose residue (the general situation for CG in placenta), an original feature for the pituitary consists of the possible addition of a sulfated *N*-acetylgalactosamine as demonstrated by Baenziger and Coll. This results from the exclusive expression in the pituitary of two highly specific enzymes: a *N*-acetylgalactosamine (GalNAc)-transferase, which catalyzes the transfer of a GalNAc residue onto a *N*-acetylglucosamine (GlcNAc), and a sulfotransferase, which catalyzes the transfer of a SO_4 radical onto GalNAc termini (in the specific terminal sequence: GalNAcβ1,4GlcNAcβ1,2Manα→Asn). Interestingly, the GalNAc-transferase is capable of discriminating between carbohydrate chains present on either LH or FSH subunits which results in the synthesis, within the same cell (pituitary gonadotrope), of two parent hormones that are differently glycosylated and, consequently, differently charged. For instance, in humans, LH contains carbohydrate chains with sul-

fated or sialylated termini in a similar proportion (sialylated:sulfated ratio, 1:18), whereas sialylated ends are by far predominant in FSH (sialylated:sulfated ratio, 12:58). In variance, bovine LH contains sulfated ends exclusively, illustrating that the ultimate steps in N glycosylation of pituitary gonadotropins are also species dependent. It is expected that the association of α and β subunits induces a conformational determinant essential for GalNAc-transferase recognition and action; however, the true motifs involved have, to date, not been identified. Since this process occurs only in the pituitary, in which two gonadotropins are produced and stored in distinct granules, it is attractive to speculate that such hormone-specific motifs could be involved in the addressing of each gonadotropin to a specific granule.

C. Differential Sorting of Gonadotropins and Free α Subunits

The subunit folding progressively acquired during trafficking of gonadotropins and their maturation along the Golgi saccules, due in part to glycosylation (as noted previously), is maintained by the step-by-step establishment of appropriate intrachain disulfide bonds. Completion of intramolecular disulfide bonds in the β subunit appears to be the rate-limiting step in the $\alpha\beta$ combination. Interestingly, the X-ray analysis of human chorionic gonadotropin (hCG) crystals has revealed the stabilization of the heterodimer by a C-terminal segment of the β subunit which enfolds the α subunit and is covalently linked, like a seat belt, by a disulfide bond (between cysteines 26 and 110). The final step in the production of gonadotropins consists in the distribution and packing of the hormone(s) into secretory granules. This process occurs in the *trans*-Golgi network and represents a link between the two fundamental processes in hormone secretion, i.e., synthesis and release. In the placental syncytiotrophoblasts, only one gonadotropin (CG) is produced and its abundant release probably occurs via a constitutive (nonregulated) pathway. In contrast, in the pituitary gonadotropes, the two gonadotropins, LH and FSH, are packed into distinct secretory granules and are essentially hormone dependent for their exocytotic release.

In addition to the complete heterodimeric hor-

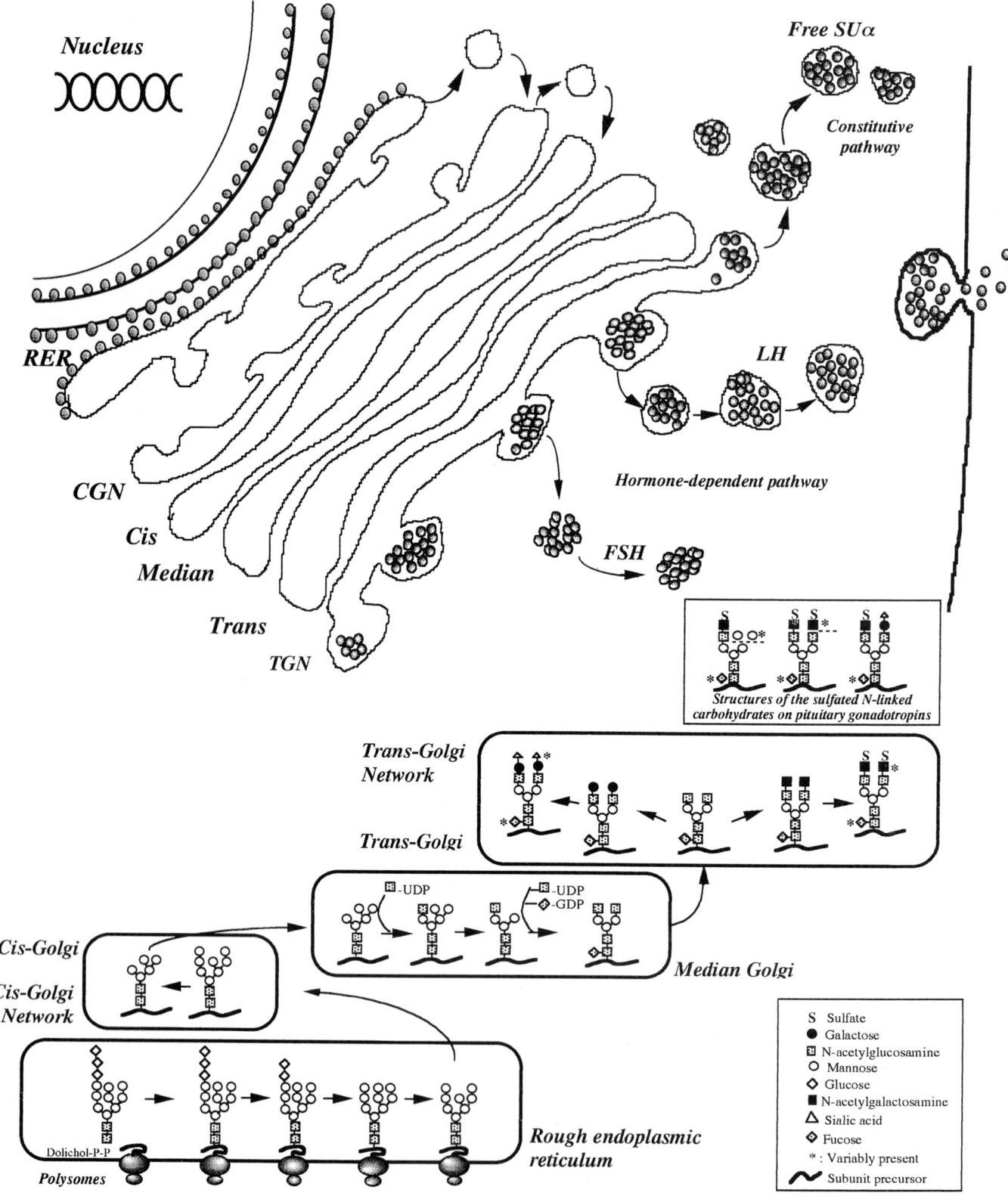

Structures of the sulfated N-linked carbohydrates on pituitary gonadotropins

Nucleus

RER

CGN

Cis

Median

Trans

TGN

Free SUα

Constitutive pathway

LH

Hormone-dependent pathway

FSH

Trans-Golgi Network

Trans-Golgi

Median Golgi

Cis-Golgi

Cis-Golgi Network

Rough endoplasmic reticulum

Dolichol-P-P

Polysomes

S   Sulfate
●   Galactose
⊞   N-acetylglucosamine
○   Mannose
◇   Glucose
■   N-acetylgalactosamine
△   Sialic acid
◈   Fucose
* : Variably present
〰   Subunit precursor

mones, large amounts of free α, which may represent about 50% of the total subunit synthesized, are released as such from the pituitary (as well as from the placenta in the appropriate circumstance). This reflects the excessive expression of the α subunit gene in these tissues. Interestingly, the glycosylation of the free α differs from that of the combined subunit in two ways: (i) The N-linked carbohydrates are much more complex in terms of ramification, suggesting that such a multiple ramification is impossible or hindered after the association of α with its β subunit counterpart, and (ii) an O-linked carbohydrate can be present on threonyl residue 43 (39 in human). The O glycosylation begins in the early Golgi saccules (*cis*-Golgi), and thus somewhat later than N glycosylation; however, compared to the latter, it is far less complex and displays no other known role besides enhanced protection of gonadotropins against degradation in the blood. Particularly, the genetically engineered production of LH, FSH, or even TSH with a β subunit bearing the O-linked carbohydrate-rich peptidic extension of hCGβ on its C terminus is proven to enhance *in vivo* half-life. With regards to the α subunit, it is clear that O glycosylation prevents association with a β subunit.

Several hypotheses have been elaborated to explain the occurrence and the role of O glycosylation in addition to N glycosylation of free α. The most persuasive are based on the presence of a potential O glycosylation site on α, inaccessible to glycosyltransferases or altered after even partial interaction of α

with a β subunit. Whatever the mechanism involved, the free α does not appear to be packed into the same secretory granules with LH or FSH, and free α-containing granules essentially follow a constitutive release. It is well established that only complete (dimeric) gonadotropins are biologically active and a possible role for free α is essentially speculative. Finally, constitutive secretion and arrival in a dense secretory granule have been estimated to occur with an approximate $t_{1/2}$ value of 1–1.8 hr. An additional period of about 1.5 hr appears necessary before a LH-containing granule becomes sensitive to release by GnRH. As a result, the $t_{1/2}$ for exocytosis under GnRH stimulation of a granule containing newly synthesized LH would be approximately 3–3.5 hr.

III. REGULATION OF GONADOTROPIN SYNTHESIS

The characterization of the different stages involved in the biosynthetic pathway of gonadotropins has provided means for the investigation of hormonal control in processes associated with the control of gonadotropin release. While the physiological paracrine regulation of CG in placenta is still poorly understood, a considerable amount of information has accumulated over decades concerning the regulation of pituitary gonadotropin release. The most important hormones and factors controlling gonadotropin release, namely, GnRH, gonadal steroids, activin,

FIGURE 2 Biosynthetic pathway for gonadotropins. Nascent subunits (precursors) enter the RER, where processing begins cotranslationally, consisting of the cleavage of the signal peptide and initiation of N glycosylation. This results in structural changes and subunit association. (Bottom) The different steps of N glycosylation which occur progressively during the transport of gonadotropin subunits through the different components of the Golgi saccules. Rough endoplasmic reticulum: transfer of a dolichol-linked glucose (Glc)₃/mannose (Man)₉/N-acetylglucosamine (GlcNAc)₂ unit to asparagine (Asn) at a N glycosylation site; Asn-X-serine/threonine and removal of Glc by glucosidases I and II and one Man by α-1,2-mannosidase. *cis*-Golgi network (CGN) and *cis*-Golgi: removal of another three Man. Median Golgi: addition of GlcNAc by GlcNAc-transferase I [on one or two of the branches of the carbohydrate unit, a process which determines the formation of mono-, bi-, or triantennary carbohydrates (not detailed)] and possible transfer of a fucosyl residue to the Asn-linked GlcNAc. *trans*-Golgi: the presence of a GlcNAc induces the addition of either a galactosyl (Gal) residue, in both the pituitary and placenta, or a N-acetylgalactosaminyl (GalNAc) residue in the pituitary only. *trans*-Golgi network (TGN): sialylation of GlcNAc by sialyltransferase or sulfation of GalNAc by the pituitary-specific sulfotransferase. The structures of the sulfated Asn-linked oligosaccharides are schematically illustrated in the inset, with variably present moieties annotated with an asterisk. In TGN, terminal stabilization of αβ dimers (via the ultimate disulfide bonding in the β subunit) and the distribution of hormones into secretory granules also occur. (Top) The differential sorting of LH, FSH, and free α molecules in the pituitary.

inhibin, and follistatin, appear to also affect the synthesis of their constitutive subunits. From the schematic representation in Fig. 2, it can be expected that the biosynthetic pathway of these subunits has potential for multiple control points. To date, much attention has been given to the regulation at the pretranslational level; nevertheless, there is some argument in favor of the existence of translational and posttranslational regulation.

A. Hormonal Regulation of LH and FSH Production at the Pretranslational Level

In a variety of animal models and under different physiological or experimental conditions, the levels of α, LHβ, and FSHβ mRNAs have appeared to be regulated either coordinately or differentially. Early investigations for the quantification of mRNA used an indirect approach consisting of the cell-free translation of pituitary mRNAs followed by the estimation of subunit precursor production. The availability of cDNAs allowed the more direct and classical assay using hybridization to radioactive probes. In fact, both methods provided convergent information; however, the latter was more sensitive and precise. An illustration of the complex physiological regulation and individual response of each member of this group of genes is provided by the analysis of subunit mRNA levels during the different phases of the rat estrus cycle. A differential control of gonadotropin mRNAs is evident by the fact that α and LHβ mRNA levels increase several hours prior to the preovulatory surge of LH, whereas an increase in FSHβ mRNA is observed several hours later. In addition, a second rise in FSHβ mRNA occurs at diestrus when LHβ mRNA levels are low. These changes in gene expression are consistent with functional changes in the pituitary activity. It is apparent that they reflect an interconnected influence of a variety of regulatory signals from the brain, the pituitary, and the gonads and deserve further study at the molecular and cellular level.

Another example of hormone-regulated gonadotropin gene expression is provided by castration. Surprisingly, this experimental model, consisting of the eradication of gonadal factors (such as inhibin, activin, and steroids), has emphasized the determinant role of hypothalamic GnRH on gonadotropin subunit gene expression. Indeed, a large, time-dependent increase in all three subunit mRNAs is observed postcastration which returns to normal after the replacement of gonadal steroids, suggesting an essential role for these hormones in the observed changes. In fact, the stimulatory effect of castration on the expression of gonadotropin genes does not occur after administration of a potent GnRH antagonist, indicating that steroids do not act through a direct negative action at the pituitary level but rather via a primarily action on the hypothalamus to decrease GnRH secretion and pituitary stimulation (Fig. 3). Actually, GnRH injection into castrated rats treated with testosterone or phenoxybenzamine (an α-adrenergic blocker) to neutralize the secretion of endogenous GnRH stimulated gonadotropin gene expression. Similarly, as after the administration of a GnRH antagonist, in sheep the removal of endogenous GnRH influence on the pituitary by surgical hypothalamic–pituitary disconnection induced a dramatic depletion in LH and FSH subunit mRNA levels, which were restored to normal levels by the administration of GnRH. Finally, an elegant demonstration of such a GnRH

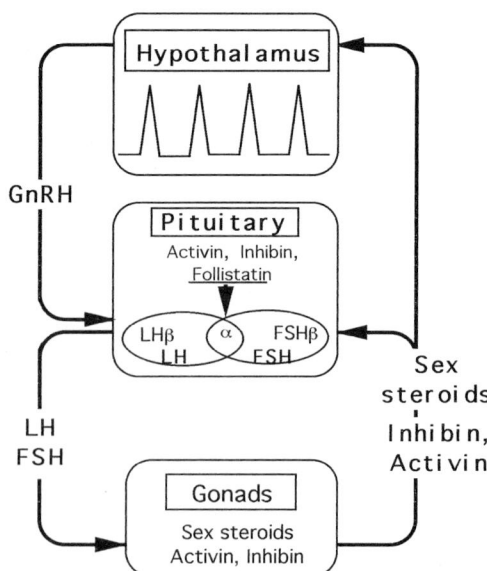

FIGURE 3 Autoregulation of the hypothalamic–pituitary–gonadal axis.

role has been provided by the use of female mice that lack a functional gene for GnRH (*hpg* mice). These mice do not produce gonadotropins; however, when a normal GnRH gene is introduced into these animals by transgenic technology or when a GnRH-secreting cell line (the GT-1 neuronal cells) is implanted, normal gonadotropin synthesis is restored. *In vitro* and *in vivo* studies have shown that GnRH altered the transcriptional activity of the α and gonadotropin β subunit genes establishing a means of augmenting steady-state mRNA levels. In addition, GnRH also increases the length of the 3′-polyadenylated tract of mRNAs and, perhaps in conjunction with this function, the stability of mRNAs.

Whether steroids could regulate the expression of gonadotropin genes directly, at the pituitary level, has been extensively investigated, especially since gonadotropes contain receptors for estradiol, testosterone, and progesterone. Interestingly, a great variation exists within vertebrate classes and subclasses regarding the nature and occurrence of such direct effects of steroids on the pituitary. In mammals, estradiol at low concentrations (picomoles) and/or progesterone stimulate LH and FSH production *in vitro* (tissue or cultured pituitary cells). In contrast, estradiol invariably inhibits LH and FSH production in amphibian pituitary, whereas in teleost fish, it stimulates the production of the LH-related gonadotropin hormone 2 (GTH2) in salmonids and has no effect in eel GTH2 which responds positively to androgens. While there is some evidence for a modulation of mammalian LH or teleost GTH2 via an alteration of the expression of LHβ or GTH2β, respectively, it is uncertain that an increased expression of FSHβ regulates mammalian FSH synthesis. Finally, testosterone represents an example of an intricate action of different hormones in mammals since *in vitro* the androgen readily stimulates expression of FSHβ, whereas *in vivo* it is inhibitory except when the endogenous GnRH action is blocked at the pituitary receptor level by a GnRH antagonist.

Another layer of complexity is introduced by the actions of activin, inhibin, and follistatin. Activin and inhibin have long been considered proteins exclusively produced by gonads (follicular fluid) but have recently appeared in fact to be produced in many other tissues including the pituitary. Activin and inhibin both act via specific receptors. In contrast, follistatin binds and thereby neutralizes activin. In rats, activin stimulates FSH production, whereas inhibin and follistatin suppress both its production and its release. The most recent data suggest that the activin/inhibin/follistatin system functions with activin as a powerful stimulator of FSHβ gene expression and both follistatin and inhibin acting as inhibitors by negating the stimulatory action of activin. Since these factors are produced in both the pituitary and the gonads, it could be predicted that they act as autocrine/paracrine factors at either level and it thus becomes necessary to identify their respective tissue origin and incidence *in vivo* (Fig. 3).

As was observed with GnRH, the regulation by steroids or activin/inhibin affects the transcriptional activity of gonadotropin gene(s) and/or, in some cases, the stability and polyadenylation of mRNA; an increased stability is associated with an extended 3′-polyadenylated tail. While GnRH augments the production of all three gonadotropin mRNAs having extended polyadenylated tail, activin induces similar effects only on FSHβ mRNA. In contrast, progesterone induced shortening of mRNA polyadenylated tails in both rat and ovine but only reduced transcription in ovine. The mechanisms involved in the stabilization of specific mRNAs are not well understood and whether they could be hormone regulated is not known. Polyadenylation of mRNAs normally occurs in the nucleus; however, there is some data in favor of a possible reversible alteration of the polydenylated tract in the cytoplasm of gonadotropes. On the other hand, because the polyadenylated tail is protected against degradation by specific polyadenylyl-binding proteins, any length reduction first requires destabilization and dissociation of the mRNA–protein complex. It has been shown for other mRNAs that this destabilizing process could be initiated by the interaction of specific regulatory proteins with typical motifs on mRNAs. The existence of such proteins and the possible hormonal regulation have not been examined to date but would be a fascinating development in the research of posttranscriptional controls of gonadotropin production in gonadotropes.

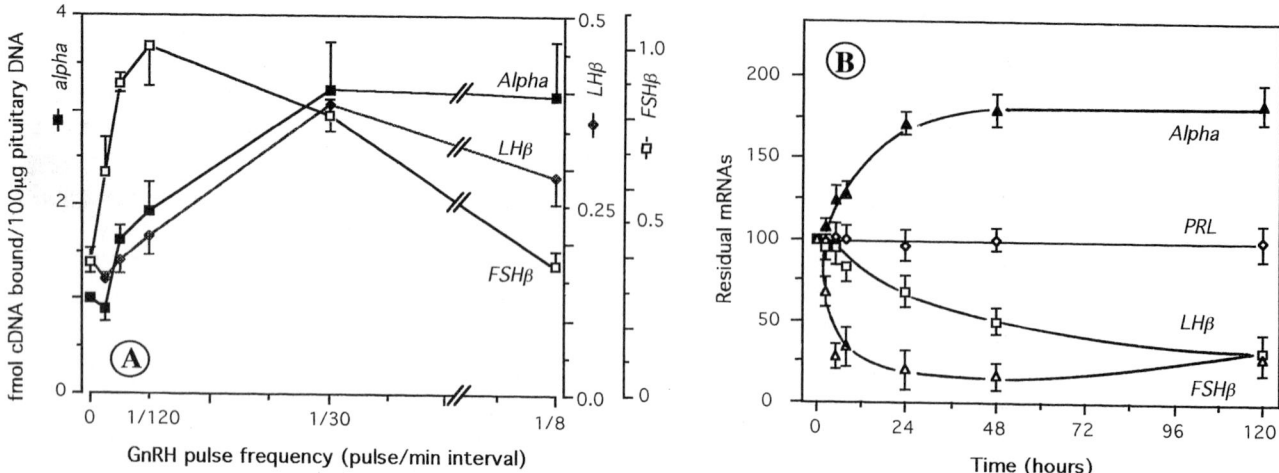

FIGURE 4 Effects of GnRH pulse frequency and permanent GnRH stimulation on the expression of pituitary gonadotropin subunit genes. (A) Orchidectomized testosterone-replaced rats were treated with exogenous GnRH at pulse intervals of 480, 240, 120, 30, and 8 min for 24 hr. Dose per pulse of GnRH was 25 ng. (B) Intact male rats were treated with a single injection of a long-acting GnRH analog (triptorelin) and pituitary gonadotropin mRNA was assayed after the indicated periods of time. PRL, prolactin mRNA (used as reference) (adapted with permission from (A) A. C. Dalkin, D. J. Haisenleder, G. A. Ortolano, T. Ellis, and J. C. Marshall, The frequency of gonadotropin-releasing hormone (GnRH) stimulation differentially regulates gonadotropin subunit mRNA expression, *Endocrinology* **125**, 917–924, 1989; and (B) Y. Lerrant, M. L. Kottler, F. Bergametti, M. Moumni, J. Blumberg-Tick, and R. Counis, Expression of gonadotropin-releasing hormone (GnRH) receptor gene is altered by GnRH agonist desensitization in a manner similar to that of gonadotropin β-subunit genes in normal and castrated rat pituitary, *Endocrinology* **136**, 2803–2808, 1995. Copyright The Endocrine Society).

B. Differential Regulation of Pituitary Gonadotropin Subunit Genes According to the Pattern of GnRH Stimulation

An intriguing feature of GnRH secretion and action is pulsatility. This phenomenon was established by E. Knobil and has been the subject of many studies. In particular, like gonadotropin release, the pattern of GnRH release may influence the response of gonadotropin subunit genes to GnRH stimulation. As shown in Fig. 4A, the individual gonadotropin genes respond differentially to the frequency of GnRH pulses. When GnRH is administered at rapid frequencies, LHβ rises preferentially. When the neurohormone is present at lower frequencies, FSHβ increases to a greater extent than LHβ mRNA. The α subunit gene responds well to GnRH in a broad range of pulse frequencies and particularly to a permanent presence of GnRH (in contact with pituitary cells) which corresponds to a *pulsatile* stimulation of these cells but with pulses at very high frequency. Under similar conditions, the β subunit genes appear to

become inactivated. The reasons for such differences between the response of the genes and particularly of individual genes expressed in a single cell type are unclear. Figure 4B shows as a function of time the effects on pituitary mRNA levels of a permanent exposure to GnRH in male rats using the injection of a potent long-acting GnRH agonist. Consistent with data in Fig. 4A, a rapid decrease in the pituitary content of LHβ and FSHβ mRNAs is observed with apparent half-lives (44 and 3 hr, respectively) compatible with an extremely rapid arrest of gene transcription, whereas the α mRNAs increased. Treatment had a marked effect on gonadotropin release since, after a rapid and important increase in response to stimulation, serum LH or FSH levels were considerably depressed. This phenomenon corresponds to the so-called "desensitization" which is interpreted as a loss of the pituitary responsiveness to GnRH and involves a multitude of membrane and postmembrane events within gonadotropes which, although extensively investigated, remain to be clarified. This paradoxical property of the gonadotropes has been exploited in animal reproduction and treatment of human disorders since the depletion of serum gonadotropins induced by GnRH agonists causes the eradication of circulating sex steroids. Interestingly, in amphibians or teleosts, constant GnRH stimulation does not desensitize but rather may enhance the response of pituitary gonadotrophs, providing an additional example of differences among vertebrates in the regulation of the gonadotropic function.

C. Translational and Posttranslational Regulation of Gonadotropin Synthesis: Adaptation of the Structural Organization of Gonadotrope Cells to Stimuli

Much less attention has been given to translational and posttranslational regulation of gonadotropin synthesis. Nevertheless, on several occasions the synthesis of gonadotropins has been shown to be altered with no detectable change in subunit gene expression. With respect to its complexity, it can be assumed that several steps of the biosynthetic pathway could be subject to regulation, including the translational efficiency of mRNA(s), glycosylation (and particularly N glycosylation), transport through the different compartments as depicted in Fig. 2, intrachain disulfide bond formation and heterodimerization, and sorting. As mentioned, prevention or interruption of N glycosylation can alter gonadotropin trafficking or even inhibit the synthesis of polypeptidic chains. Evidence for the occurrence of nongenomic regulation is fragmented or poorly established for gonadotropins, especially because the appropriate model or technology is lacking. For instance, LHβ mRNA with a shortened polyadenylated sequence has a reduced translational capacity in a heterologous cell-free system, but this is difficult to test in gonadotropes. In contrast, several authors have described regulation of N glycosylation on gonadotropin subunits by either GnRH or estradiol. This regulation occurs at the late stage of the N glycosylation process. Such an occurrence is strengthened by the fact that expression of both the GalNAc-transferase and sulfotransferase is under the control of estradiol. A role of GnRH for these enzymes is unknown.

Morphological changes observed in gonadotropes under the various experimental or physiological conditions examined (i.e., castration, aging in both male and female, estrus cycle in the female, or steroid treatments) support the idea of an effective occurrence of altered synthetic activity in these cells and the participation of the secretory machinery in response to hormone stimuli. For instance, castration is characterized by the appearance of the so-called "castration cells" or "signet-ring cells," which correspond to hypertrophied gonadotropes. These cells, which are under the sustained stimulation of endogenous GnRH and lack gonadal hormones, exhibit a dilated endoplasmic reticulum resulting in round vacuoles or irregularly shaped and dilated cisternae resembling filigree or a very large vacuole after the gradual expansion and confluence of the endoplasmic reticulum cisternae. Consistent with an increased synthetic activity, a larger proportion of LH- and/or FSH-producing cells is detectable. Conversely, immunization against GnRH or the administration of a GnRH antagonist counteracts these effects of castration.

Similarly, as during the early stages postcastration, a rapid and obvious extrusion of secretory granules via exocytosis occurs within minutes in response to GnRH in gonadotropes of intact rats, corresponding to the release of stored gonadotropins. The diameter and electron density of the secretory granules decrease rapidly after GnRH infusion. Moreover, along with granule exocytosis, a progressive extension of the RER and the Golgi zone occurs reflecting the delayed increase in gonadotropin synthesis. All these observations suggest that the acute hormonal release under a specific secretagogue is rapidly correlated with modifications of the entire ultrastructural organization of the cell. In addition, the subcellular structures of the cell may influence, and in any case contribute to or hinder, the synthetic process for gonadotropin production within the cell.

D. Multiple Signaling Pathways Are Involved in the Control of Gonadotropin Synthesis

There is increasing evidence for the presence of an array of intracellular signaling pathways in gonad-

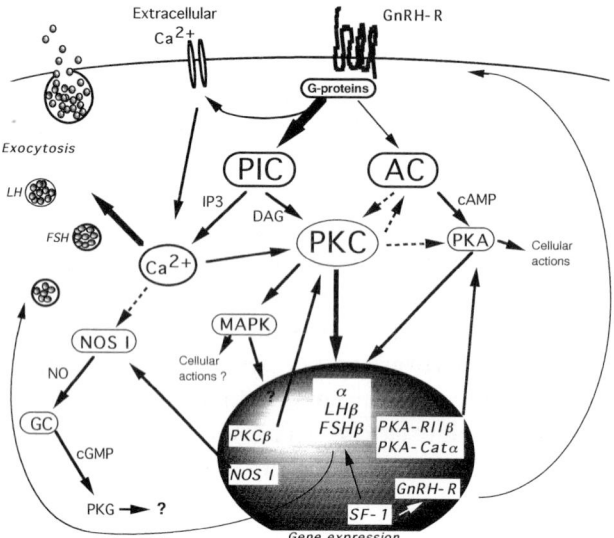

FIGURE 5 Major signaling pathways in gonadotropes and the integrated control of cellular function. Several signaling pathways and second messengers in gonadotropes that can be activated by GnRH, directly or indirectly, to regulate in an interconnected manner the release and synthesis of gonadotropins are shown. In addition, several elements of the signaling pathways which are under the control of GnRH are shown. For clarity, a possible interplay with the signaling of another hormone has not been represented. Abbreviations used are as in the text; GC, guanylyl cyclase; PKG, cGMP-dependent protein kinase). Dashed arrows represent observed actions or interactions whose mechanisms remain to be elucidated.

otropes (Fig. 5) which can be triggered by GnRH and/or other hormones such as pituitary adenylate cyclase-activating polypeptide (PACAP, a novel hypothalamic hypophysiotropic neurohormone). It is now well documented that GnRH interacts at the pituitary level via specific, seven-transmembrane G-protein-coupled receptors located at the surface of gonadotropes and stimulates the release of gonadotropins via the activation of the phosphoinositidase C (PIC) with the ensuing production of diacylglycerol (DAG) and inositol trisphosphate, responsible for the activation of protein kinase C (PKC) and the mobilization of intracellular Ca^{2+}, respectively. GnRH also induces the activation of phospholipases D and A_2, production of cyclic AMP (cAMP) and cyclic GMP (cGMP), and, in certain circumstances, activation of tyrosine kinases and the mitogen-activated protein kinase (MAPK) cascade. Similarly, the

PACAP receptor is capable of multiple signaling in gonadotropes through its efficient coupling to both the PIC and adenylate cyclase pathways.

A role of cAMP and cGMP as second messengers of GnRH in the acute release of gonadotropins has been dismissed. However, a clear, time- and dose-dependent stimulatory action of cAMP on LH subunit biosynthesis has been shown using anterior pituitary cells incubated in the presence of labeled amino acids followed by immunoextraction and purification by gel electrophoresis. In addition to the stimulation of synthesis, cAMP increases the release of the newly synthesized subunits, suggesting a possible contribution in the cAMP-induced delayed release of LH. Not surprisingly, DAG or phorbol esters (the potent direct activators of PKC) are also capable of causing an increased production of LH subunits, suggesting that activation of the PIC pathway not only stimulated acute release of gonadotropins but also played a role in the reconstitution of the intracellular stores. In fact, there are several reports indicating a stimulatory action of either second messenger on gonadotropin subunit gene expression; nevertheless, whether cAMP effectively mediates in these biosynthetic processes the action of GnRH or another hormone (such as PACAP) is still in question. Another possible candidate as a contributor to the mechanisms of hormonal regulation is calcium, which has been shown to be capable of stimulating the transcriptional activity of the α gene in the gonadotrope-derived cell line αT3-1. In these cells GnRH can activate the MAPK cascade via PKC and consequently increase the transcriptional activity of the α gene. Interestingly, MAPK is present in the rat anterior pituitary but is activated by GnRH exclusively during proestrus in females, suggesting the existence, at this stage, of a specific switch to establish the link between the GnRH receptor and MAPK. Hormonal signaling in pituitary gonadotropes thus involves multiple intracellular mediators and pathways that may operate in an interconnected manner according to a dynamic regulation.

In the human placenta, a clear action of cAMP has been shown on the expression of α and CGβ. This has led to an analysis of the promoters but the essential information to date has been obtained for the α subunit because all gonadotropin β promoters are highly cell specific and expressed very poorly in the available cell lines. Consistent with expression in different cell types (gonadotropes, thyreotropes, and syncytiotrophoblasts), multiple elements, including regulatory and tissue-specific elements, have been identified in the human α subunit promoter that function with close interdependence. Two identical motifs which confer responsiveness to the activation via cAMP [called cAMP-responsive element (CRE)] have been identified, and its corresponding transactivator has been isolated and characterized (cAMP-responsive element-binding protein). The CRE elements are truncated in the α promoter of all species other than human and absent in the CGβ promoter, suggesting a distinct mechanism of action. Surprisingly, current studies suggest that the α and CGβ genes share relatively few regulatory elements in common. Nevertheless, a variety of cell signaling pathways appear to converge upon relatively limited promoter regions that are initially defined based on responsiveness to cAMP. In the α and LHβ subunit promoters, a region required for transcriptional responsiveness to GnRH has been found which appears to function in correlation with a steroidogenic factor-1 (SF-1) motif(s) initially characterized as essential for gonadotrope-specific expression. Whether the GnRH-responsive region contains specific elements conferring response to DAG and/or cAMP is still under investigation. PKC seems to play a central role in pituitary gonadotropes since it mediates the hormone-induced activation of the MAPK cascade, and it may alter levels of PKA subunits.

IV. SIGNALING AND THE INTEGRATED CONTROL OF GONADOTROPIN FUNCTION

Much data have accumulated over the past 5 years indicating that hormones that control gonadotropin synthesis and release in pituitary gonadotropes also regulate levels and/or activity of several proteins that represent a determinant for the cellular function, especially for the secretory processes (Fig. 5). For instance, pulsatile GnRH increases, and desensitizing GnRH decreases, the number of GnRH receptors via pre- and posttranscriptional actions. GnRH can also

regulate levels and gene expression of PKCβ (a PKC isoform), several isoforms of PKA regulatory and catalytic subunits, and the transcriptional factor SF-1 which is required for cell-specific expression and regulation of several important gonadotropic proteins including the α, LHβ, and GnRH receptor genes. Finally, it has recently been shown that GnRH also stimulates the expression, in gonadotropes, of the neuronal type of nitric oxide synthase (NOS I), an enzyme that catalyzes the intracellular production of NO and appears to be responsible for the GnRH-induced increase in cGMP mentioned previously (and described about 20 years ago).

Collectively, this concert of regulations emphasizes the important role of GnRH in the control of not only the synthesis and release of gonadotropins but also the whole secretory function of gonadotropes. Taking into consideration its effects on its receptor and other signaling or transcriptional elements, it is evident that GnRH can regulate its own influence on the cell function. Other hormones such as PACAP or steroids, which can interfere with intracellular pathways at various levels including the production of messengers and/or the expression of genes, appear to be potent modulators of the GnRH action. Subtle intracellular rearrangements may thus account for the fine-tuned regulation and adaptation of gonadotrope function to physiological status.

V. CONCLUSIONS AND PERSPECTIVES

Studies during the past 15 years have improved our understanding of how the structurally complex gonadotropins are produced. Most of the different steps and mechanisms involved in this process have been clarified: Subunit genes have been isolated and their transcripts as well as the primary products of their translation characterized. In addition, the conditions for the biogenesis of mature heterodimeric hormone have been established, especially with respect to the secretory process and structural organization of the cells, which reconcile molecular and cellular aspects. Moreover, and more important, there is evidence that the production of gonadotropins is regulated, especially in pituitary gonadotropes, by hormones already known to control their release. Although intracellular systems that play a role in the transduction of the GnRH signal to the nucleus and the ultimate steps in the activation of gene expression are still unclear, there is now a considerable body of evidence suggesting that synthesis and release are intimately interconnected within gonadotropes.

Despite these considerable advances, it is apparent that a number of areas warrant further investigation. One of the important challenges for the future is to expand our knowledge of the promoter regulation. Much of our current knowledge of the α genes is derived from experiments using transient expression assays in cultured cells. With this approach very little information has been obtained for the gonadotropin β genes, and appropriate host cells are necessary for the further dissection of the LHβ and FSHβ promoters. The use of such methodology also requires complementary *in vivo* analysis to test the real impact of specific regulatory elements and, thus, it is probable that transgenic experiments will be useful in the future.

Signaling of gonadotropes and especially pituitary gonadotropes represents a fascinating subject of investigation. First, it is crucial to increase our understanding of the role of each of the potential signaling pathways and intracellular messengers on every member of the gonadotropin gene family and, when such a role exists, on posttranscriptional events. Second, it is extremely important to extend our investigation to the possible coregulation of receptor(s), enzyme(s), and other elements of the signaling cascade(s) in gonadotropes since this introduces a second level of complexity, the elucidation of which may help to deepen our understanding of the entire gonadotropic function at the molecular and cellular level and, by extension, the physiology of reproduction.

Finally, these studies will offer a variety of opportunities to apply the information obtained to pathologic situations. Such a strategy has been successfully applied in the case of GnRH agonists and antagonists to suppress gonadotropins and, consequently, circulating gonadal steroids. We can thus expect an application to any other specific target that may be susceptible to help in a definite manipulation of gonadotropic function.

See Also the Following Articles

Chorionic Gonadotropin, Human; FSH (Follicle-Stimulating Hormone); GnRH (Gonadotropin-Releasing Hormone); GnRH Pulse Generator; Gonadotropes; Gonadotropin Receptors; Gonadotropin Secretion, Control of; LH (Luteinizing Hormone); Pituitary Gland, Overview

Bibliography

Baenziger, J. U., and Green, E. D. (1988). Pituitary glycoprotein hormone oligosaccharides: Structure, synthesis and function of the asparagine-linked oligosaccharides on lutropin, follitropin and thyrotropin. *Biochim. Biophys. Acta* **947**, 287–306.

Barbieri, R. L. (1992). Clinical applications of GnRH and its analogues. *Trends Endocrinol. Metab.* **3**, 30–34.

Gharib, S. D., Wierman, M. E., Shupnik, M. A., and Chin, W. W. (1990). Molecular biology of the pituitary gonadotropins. *Endocr. Rev.* **11**, 177–199.

Jameson, J. L., and Hollenberg, A. N. (1993). Regulation of chorionic gonadotropin gene expression. *Endocr. Rev.* **14**, 203–221.

Kornfeld, R., and Kornfeld, S. (1985). Assembly of asparagine-linked oligosaccharides. *Annu. Rev. Biochem.* **54**, 631–664.

Krsmanovic, L. Z., Stojilkovic, S. S., and Catt, K. J. (1996). Pulsatile gonadotropin-releasing hormone release and its regulation. *Trends Endocrinol. Metab.* **7**, 56–59.

Mather, J. P., Woodruff, T. K., and Krummen, L. A. (1992). Paracrine regulation of reproductive function by inhibin and activin. *Proc. Soc. Exp. Biol. Med.* **201**, 1–15.

McArdle, C. A., and Counis, R. (1996). GnRH and PACAP action in gonadotropes. Cross-talk between phosphoinositidase C and adenylyl cyclase mediated signaling pathways. *Trends Endocrinol. Metab.* **7**, 168–175.

Quérat, B. (1994). Molecular evolution of the glycoprotein hormones in vertebrates. In *Perspectives in Comparative Endocrinology* (K. G. Davey, R. E. Peter, and S. S. Tobe, Eds.), pp. 27–35. National Research Council of Canada, Ottawa.

Rothman, J. E., and Orci, L. (1992). Molecular dissection of the secretory pathway. *Nature* **355**, 409–415.

Stockell Hartree, A., and Renwick, A. G. C. (1992). Molecular structures of glycoprotein hormones and functions of their carbohydrate components. *Biochem. J.* **287**, 665–679.

Stojilkovic, S. S., Reinhardt, J., and Catt, K. J. (1994). Gonadotropin-releasing hormone receptors: Structure and signal transduction pathways. *Endocr. Rev.* **15**, 462–499.

Tougard, C., and Tixier-Vidal, A. (1994). Lactotropes and gonadotropes. In *The Physiology of Reproduction* (E. Knobil and J. D. Neill, Eds.), 2nd ed., pp. 1711–1747. Raven Press, New York.

Gonadotropin Receptors

David Puett

University of Georgia

GLOSSARY

hormones Molecules with diverse chemical structures that regulate a variety of cellular functions. In classical endocrine function the term was used to denote a compound biosynthesized and secreted by one cell type that traveled through the bloodstream and acted on another cell type. It is now recognized that hormones can act on neighboring cells within the same tissue, i.e., paracrine function, or even on or within the same cell in which it is biosynthesized, i.e., autocrine function. Cell responsiveness to a hormone depends on the presence of a specific receptor, and the response can be either stimulatory or inhibitory to a biochemical process.

ligands Molecules that bind to a receptor and can be either a naturally occurring hormone or synthetic compound that behaves biologically like the natural hormone (agonist function) or blocks the action(s) of the natural hormone (antagonist function).

receptors Protein molecules, often posttranslationally modified, that serve as essential mediators in hormone action by recognizing the appropriate hormone(s) and forming a specific and high-affinity, but noncovalent, interaction. Upon binding its cognate hormone, the receptor undergoes a conformational change that initiates a series of cellular responses that can include signaling via one or more pathways or specific gene activation, i.e., enhanced transcription, followed by mRNA processing and translation to function protein(s). Thus, receptors, which can be localized on the cell surface or within the cell, act as transducers to translate the message of hormone binding into a language or form that can be understood by the cell.

signaling An intermediate series of biochemical reactions in response to hormone–cell surface-associated receptor complex formation ultimately leading to the appropriate cellular response. Several forms of intracellular signaling are operative and all, at one or more points, involve enzymatic amplification of the initial hormone–receptor complex such that a significant response can occur with the formation of only a limited number of binding events.

Sexual reproduction involves numerous biochemical processes, each of which is highly regulated. The gonadotropins and their receptors are essential for gonadal steroidogenesis and gametogenesis. Thus, they are key players in the development and maintenance of fertility; in addition, they are necessary for male sexual differentiation *in utero*. Aberrations in gonadotropin or receptor structure can lead to developmental abnormalities or to infertility.

I. INTRODUCTION

The reproductive axis involves a coordinated interplay between the hypothalamus, the adenohypophysis (or anterior pituitary), and the gonads. The potent decapeptide, gonadotropin-releasing hormone (GnRH), is biosynthesized in the median eminence as a larger, inactive precursor which is posttranslationally modified via several enzymes and travels through the hypophyseal portal system to the pituitary gland, where it binds to a specific, high-affinity G protein-coupled receptor (GPCR) localized on gonadotrophs in the adenohypophysis. In response to GnRH binding, gonadotrophs release two gonadotropins, lutropin [luteinizing hormone

521

(LH)] and follitropin [follicle-stimulating hormone (FSH)], into the bloodstream where they bind to distinct GPCRs, the LH receptor (LHR) and the FSH receptor (FSHR) that are present on different cell types in the testis and ovary.

The activation of these gonadal receptors to elicit cellular responses, along with other cellular events, leads to sex steroid synthesis and the production of gametes. Other tissues, such as uterus, breast, prostate, muscle, and brain, are responsive to the three major classes of sex steroids: progestins, estrogens, and androgens. Communication between the brain, pituitary, and gonads involves not just GnRH, LH, and FSH acting in a unidirectional manner; gonadally derived sex steroids and peptide hormones serve as messengers to the brain and pituitary to complete the cycle, thus ensuring proper regulation of each step of the reproductive process in females and males.

If fertilization and implantation occur, the placenta joins in the arena of reproductive tissues as both a producer of and responder to key hormones. One placental-derived hormone in primates and equids is a gonadotropin, choriogonadotropin [chorionic gonadotropin (CG)], that is necessary for the maintenance of early pregnancy. The structure of CG is sufficiently similar to that of LH that both hormones act through a common GPCR, LHR.

In summary, the three human gonadotropins, LH, FSH, and CG, and their two GPCRs, LHR and FSHR, function in various aspects of development and reproduction. This article deals primarily with the gonadotropin receptors, but a short description of their ligands will be presented.

II. THE LIGANDS OF GONADOTROPIN RECEPTORS

The three gonadotropins, along with another structurally related hormone, thyrotropin [thyroid-stimulating hormone (TSH)], constitute the glycoprotein hormone family. The four glycoprotein hormones are heterodimers, utilizing a common α subunit and a hormone-specific β subunit. The β subunits themselves are homologous; indeed, as mentioned earlier, LHβ and CGβ are sufficiently similar in structure that the two intact hormones bind to and activate the same receptor, LHR. Depending on the species, the α and β subunits contain about 90–100 and 105–120 amino acid residues, respectively; the exceptions are the primate and equid CGβs as well as equid LHβ, which contain an additional "extension" at the carboxy terminus of some 25 amino acid residues. Each of the subunits is glycosylated, and association to form an active heterodimer is reversible, i.e., no covalent intermolecular bonds are formed.

The crystal structure of only one glycoprotein hormone, human CG, has been solved, with most of its carbohydrate removed to facilitate crystallization. The structure is characterized by several unusual features. It is a rather elongated molecule with the two subunits being intertwined one with another. Of considerable interest was the finding that the α and β chains, although displaying no obvious amino acid sequence similarity, exhibited similar three-dimensional folding patterns reminiscent of several growth factors. Results of a considerable amount of research from many laboratories using a variety of experimental approaches, including site-directed mutagenesis, protein engineering, chemical modifications, selective proteolysis, synthetic peptides, and specific antibodies, have delineated a number of amino acid residues on the α and β subunits that are likely candidates for forming contact sites with LHR.

III. STRUCTURAL ASPECTS OF GONADOTROPIN RECEPTORS

Being integral membrane proteins, the gonadotropin receptors, like other members of the GPCR family, proved difficult to purify and characterize. The first cDNA clones were reported on LHR from two species in 1989; these were quickly followed by studies on LHR from other species and on FSHR and TSHR from several species. The three glycoprotein hormone receptors, like their cognate ligands, are homologous. Containing the typical features associated with this family of serpentine or heptahelical receptors, for example, three extracellular loops, three intracellular loops, seven transmembrane helices, and a carboxy-terminal cytoplasmic tail, these

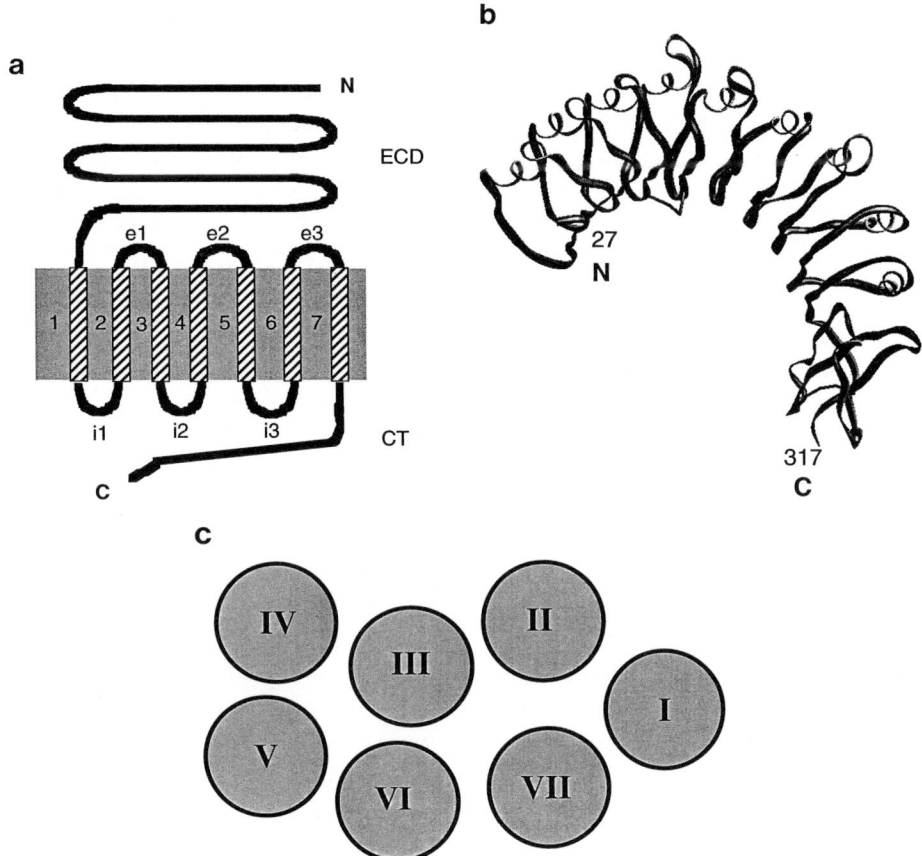

FIGURE 1 (a) Schematic structure of a gonadotropin receptor showing the amino-terminal (N) large glycosylated extracellular domain (ECD), the seven transmembrane helices (1–7), the three extracellular loops (e1–3), the three intracellular loops (i1–3), and the carboxy-terminal (C) cytoplasmic tail (CT) (modified from K. C. McFarland, R. Sprengel, H. S. Phillips, M. Kohler, N. Rosemblit, K. Nikolics, D. L. Segaloff, and P. H. Seeburg, Lutropin–choriogonadotropin receptor: An unusual member of the G protein-coupled receptor family, *Science* **245**, 494–499, 1989). (b) A model of a portion of the extracellular domain of rat LHR, encompassing amino acid residues 27–317, based on nine pseudoleucine-rich repeats (modified from N. Bhowmick, J. Huang, D. Puett, N. W. Isaacs, and A. J. Lapthorn, Determination of residues important in hormone binding to the extracellular domain of the luteinizing hormone/chorionic gonadotropin receptor by site-directed mutagenesis and modeling, *Mol. Endocrinol.* **10**, 1147–1159, 1996). (c) Model of the arrangements of transmembrane helices in GPCRs (modified from J. M. Baldwin, The probable arrangement of the helices in G protein-coupled receptors, *EMBO J.* **12**, 1693–1703, 1993).

receptors also have a relatively large amino-terminal domain on the extracellular face of the membrane that is now known to be responsible for most of the high-affinity hormone binding (see Fig. 1a for a schematic of the receptor structure).

Low-resolution structural information is available on two serpentine receptors: bacteriorhodopsin, a light-driven proton pump from *Halobacterium halobium* (not a GPCR), and bovine rhodopsin (a GPCR). Neither of these molecules contains the large extra-

cellular domain characteristic of LHR and FSHR. A large portion of the extracellular domain is composed of pseudoleucine-rich repeats, and several laboratories have reported models of this region of LHR based on comparisons with a leucine-rich protein of known three-dimensional structure. Figure 1b shows one such proposed model of this region of LHR. The heptahelical regions of the LHR and FSHR are presumed to be similar to those of other GPCRs, and a possible arrangement is shown in Fig. 1c based on

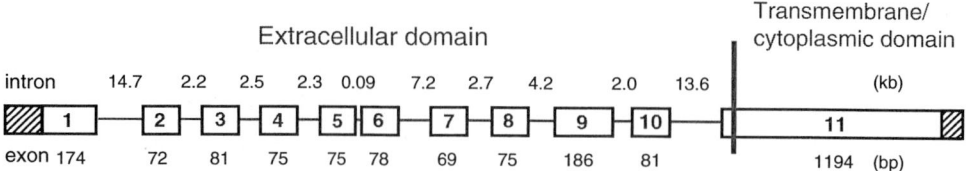

FIGURE 2 Genomic organization of the LHR gene showing the 11 exons (open boxes) and intervening sequences (introns, solid lines); exons and introns are not drawn to scale. Exons 1–10 encode most of the extracellular portion of LHR; exon 11 encodes a small portion of the extracellular domain, the seven transmembrane helices, the six connecting loops, and the cytoplasmic tail. The number of base pairs (bp) in each exon and the number of kilobases (kb) in each intron are indicated (modified from Y. B. Koo, I. Ji, R. G. Slaughter, and T. H. Ji, Structure of the luteinizing hormone receptor gene and multiple exons of the coding sequence, *Endocrinology* **128**, 2297–2308, 1991; and C. H. Tsai-Morris, E. Buczko, W. Wang, X.-Z. Xie, and M. L. Dufau, Structural organization of the rat luteinizing hormone (LH) receptor gene, *J. Biol. Chem.* **266**, 11355–11359, 1991).

modeling studies and low-resolution diffraction data. There are no structural data or modeling results on the connecting loops and the intracellular carboxy-terminal tail.

The genomic structure of LHR has been determined. Eleven exons encode the full-length receptor; exons 1–10 are responsible most of the extracellular amino-terminal domain, and exon 11 encodes a small portion of the extracellular domain, the seven transmembrane helices, the six connecting loops, and the carboxy-terminal cytoplasmic tail (Fig. 2). Elucidation of the genomic structure of this GPCR and a comparison with other GPCRs revealed an interesting feature: Exon 11 of the LHR gene is similar to the intronless adrenergic receptor genes; hence, the LHR gene appears to have originated by the addition of 10 exons, encoding most of the extracellular domain, to an intronless gene of a GPCR.

IV. MECHANISTIC ASPECTS OF GONADOTROPIN RECEPTORS

The GPCRs, unlike growth factor receptors, have no known enzymatic activity. Their intracellular signaling pathways involve association with a heterotrimeric ($\alpha\beta\gamma$) G protein, which in turn interacts with an enzyme such as adenylyl cyclase or phospholipase C (PLC). The major signaling pathway of LHR and FSHR, in response to their respective cognate ligand binding, LH (or CG), and FSH, is via G_s. The scenario that has developed for this pathway is as follows: A GDP-α_s complex binds to $\beta\gamma$, and the inactive heterotrimer associates with the GPCR. Ligand binding to the receptor leads to a conformational change that in turn alters the conformation of the α_s subunit such that GDP is released and GTP is bound. The GTP-α_s complex dissociates from the GPCR and from $\beta\gamma$ and associates with and activates adenylyl cyclase. Being a GTPase, the α_s subunit will eventually hydrolyze the bound GTP to GDP and reenter its inactive state, i.e., GDP-α_s, then reassociate with $\beta\gamma$, and finally bind to receptor to begin a new cycle.

When GTP-α_s is associated with the enzyme adenylyl cyclase, catalytic activity is increased and the substrate ATP is converted to the product cyclic AMP. Often referred to as a second messenger (with LH, CG, or FSH representing the primary messenger), cAMP binds to the regulatory subunit of a serine/threonine kinase, cAMP-dependent protein kinase, or protein kinase A (PKA). Inactive PKA consists of two regulatory subunits (R) and two catalytic subunits (C); binding of cAMP to R, at a stoichiometry of 2:1, promotes the dissociation of PKA to 2C and 2R(cAMP). Since R is inhibitory to C, its removal permits the full catalytic activity of C to be realized. With ATP as a phosphate donor, the catalytic subunit phosphorylates certain intracellular proteins at serine or threonine residues, including a transcription factor(s) that enhances gene activation. The enzyme, cyclic nucleotide phosphodiesterase, rapidly degrades cAMP, and as the intracellular concentration of the cyclic nucleotide diminishes, the bound cAMP dissociates from the regulatory subunit and the inactive form of PKA, i.e., R_2C_2, is again formed. The phosphorylated cytoplasmic and nuclear proteins continue their functions until the serine- or threonine-bound phosphate is removed by a protein phosphatase.

As can be ascertained from the previous description, the act of gonadotropin binding to its receptor is considerably amplified by virtue of enzyme activation and is extended temporally by the signaling cascade which, in itself, involves enhanced gene transcription, mRNA processing, and protein translation.

There is compelling evidence that in certain cell types and certain species gonadotropin-receptor activation involves another signaling pathway, namely, G_q, which utilizes a different α subunit, α_q. The overall regulation and mechanism of action of G_q are similar to those of Gs, except that the activated form of α_q, i.e., GTP-αq, binds to the enzyme PLC rather than adenylyl cyclase. PLC hydrolyzes the minor membrane lipid, phosphatidylinositol 4,5-bisphosphate, to yield two bioactive products, inositol 1,4,5-trisphosphate (IP_3) and 1,2-diacylglycerol. IP_3 leads to the intracellular release of stored calcium ions from the endoplasmic reticulum, and the combination of calcium ions and 1,2-diacylglycerol increases the catalytic activity of one or more isoforms of PKC, another serine/threonine protein kinase but distinct from PKA. Some cells will respond to gonadotropins with increased activity of both PKA and PKC; however, the dominant pathway in most cells appears to be the cAMP pathway, i.e., the G_s–adenylyl cyclase–PKA sequence.

It is noteworthy that the heterodimer $\beta\gamma$, released from GTP-α_s or GTP-α_q, can exhibit regulatory activities as well. Unfortunately, there is little information available on the potential roles of $\beta\gamma$ in cells expressing LHR or FSHR.

Once formed, the hormone–receptor complex can undergo one of several fates after α_s or α_q activation. The complex, with the receptor now in its activated form, may, in principle, bind another inactive G protein heterotrimer, or perhaps even GDP-α itself, and begin another round of activation. However, it is now known that even in the presence of ligand, gonadotropin receptors become desensitized, probably from phosphorylation of one or more serine or threonine residues located on the intracellular loops or the cytoplasmic tail.

Another possibility is that the hormone dissociates from the receptor, returning the latter to its inactive form which can then bind another hormone. Here again, however, the receptor may be in a desensitized form if intracellular phosphatases have not yet removed the phosphate(s). It is noteworthy that gonadotropin receptor binding is of high affinity, e.g., K_Ds on the order of 0.1–1 nM. The kinetics of binding are essentially those of diffusion-controlled reactions, whereas the kinetic dissociation constants are extremely low. Thus, the hormone–receptor complex is, in a thermodynamic sense, quite stable and unlikely to dissociate within the time frame of hormone action.

Another probable fate is that the hormone–receptor complex is internalized into the cell via an endosome. While there may be some as yet unknown biological activity associated with the internalized complex, it is recognized that fusion of the endosome with a primary lysosome can lead to degradation of both hormone and receptor; alternatively, some of the receptors may be recycled to the plasma membrane. This process is known as ligand-mediated receptor downregulation and, along with the process of desensitization, ensures that the hormone–receptor complex is uncoupled from signaling after a finite time. Teleologically, one could argue that these mechanisms prevent a given target cell from becoming overstimulated in response to an external signal.

This brief overview has shown that gonadotropin receptor activation is a dynamic process setting into motion a series of intracellular events that acutely result in sex steroid biosynthesis and chronically participate in the growth and differentiation of selected gonadal cells. A number of reports have appeared indicating the presence of LHR in many nongonadal cells, but the physiological importance of these receptors is yet to be determined.

V. ETIOLOGY OF DEVELOPMENTAL AND REPRODUCTIVE DISORDERS ATTRIBUTABLE TO GONADOTROPIN RECEPTORS

Gonadal expression of the two gonadotropin receptors is limited to a few cell types. In both females and males, a two-cell model has been developed to explain the actions of pituitary gonadotropins. While not complete, the model is very useful in delineating the actions of the two gonadotropins, LH and FSH.

Follicular development and maturation in the sexually mature female is based on two cell types: (i) theca cells that express LHR and respond to LH via increased androgen production, and (ii) granulosa cells that express FSHR and respond to FSH via proliferation and increased cytochrome P450 aromatase, an enzyme that converts the theca-produced androgens to estrogens. The major ovarian estrogen is 17β- estradiol, which exhibits effects via the estrogen receptor in many tissues, including an autocrine effect on granulosa cells. As the dominant follicle(s) matures in response to an increasing estrogen, androgen milieu, granulosa cells begin to express LHR, and the LH surge required for ovulation promotes differentiation of the granulosa cells into luteal cells that produce primarily progesterone in response to LH.

In a nonfertile cycle, the corpus luteum will regress since LH secretion is inhibited by the increasing concentration of progesterone in the luteal phase. In a fertile cycle, the corpus luteum is rescued by a luteotropic factor, e.g., CG in women, which maintains progesterone production that is required to sustain pregnancy. As the placenta grows, it develops the capacity to produce sufficient progesterone to maintain pregnancy, and, concomitantly, human CG production diminishes.

In sexually mature males, the two-cell model applies to (i) Leydig, or interstitial, cells that express LHR and respond to LH via increased androgen, mainly testosterone, production that is required for spermatogenesis and the maintenance of secondary sex characteristics, and (ii) Sertoli cells that express FSHR and respond to FSH via increased protein production required for spermatogenesis.

The gonadotropin receptors, particularly LHR, are also important in male sexual differentiation *in utero*. In eutherian mammals, the paradigm, chromosomal sex → gonadal sex → phenotypic sex, is followed. At birth, the reproductive tract is gender indifferent and by default will develop into a female phenotype unless other factors intervene.

Male sexual differentiation occurs between 6 and 8 weeks postconception in humans, and one factor required for differentiation to occur is expression of the *SRY* (sex-determining region of the Y chromosome) gene that encodes a transcriptional effector. This protein, probably acting in concert with others, leads to regression of the Müllerian ducts and differentiation of the Wolffian ducts into vas deferens. The indifferent gonads differentiate into testes that respond to human CG via LHR to produce testosterone, which is required for development of secondary sex glands such as the prostate and seminal vesicles. (These tissues have the capability of converting the potent androgen, testosterone, into an even more potent androgen, dihydrotestosterone via the enzyme 5α-reductase.) In time, the fetal pituitary will develop sufficiently to produce LH and FSH, which are both required for normal testicular growth and development.

After birth, young 46,XY children, i.e., genetic boys, produce testosterone that reaches adult levels for a brief period. Afterwards, the reproductive axis is rather quiescent in both young boys and girls until puberty begins. It is noteworthy, however, that pituitary gonadotropins and their gonadal receptors are present and can be further biosynthesized and activated with the proper stimulus. In a highly simplistic sense, the development of puberty or sexual maturation can be defined as an activation of the GnRH pulse generator.

Thus, functional gonadotropin receptors are essential for reproduction in sexually mature females and males; moreover, they are involved in male sexual differentiation and maturation. Not surprisingly, receptor dysfunction can lead to abnormal differentiation, development, and reproductive function.

Analysis of the LHR gene, with emphasis on exon 11, from patients presenting with developmental and reproductive disorders, such as familial male precocious puberty, male pseudohermaphroditism, hypogonadism, and amenorrhea, has revealed several naturally occurring mutations that result in constitutive receptor activation, i.e., in the absence of ligand, or receptor inactivation, i.e., loss of function even in the presence of ligand. Detailed structure–function studies of these errors of nature have provided considerable insight into the mechanism of transmembrane signaling and have complemented other molecular and cellular studies based on site-directed mutagenesis and protein engineering of LHR.

Familial male precocious puberty, or testotoxicosis, is a disorder characterized by the beginning of pubertal development in young boys, generally appearing at 3 or 4 years of age. In boys presenting with this disorder, serum testosterone levels are elevated, and testicular biopsies reveal Leydig cell hyperplasia; although sporadic cases occur, the inheritance is autosomal dominant. Only a limited number of cases have been so analyzed, and the majority of the mutations occur in transmembrane helix 6. Activating mutations have also been noted in transmembrane helices 2 and 5, as well as in intracellular loop 3. By repeating the experiments of nature in the laboratory, it has been possible to establish that each of these point mutations in exon 11 of the LHR gene leads to a form of LHR that is activated even in the absence of ligand, thus explaining the clinical phenotype. In addition to constitutive receptor activation, the mutant forms of LHR can also respond to exogenous ligand; also of interest is the clinical observation that the onset of puberty does not appear much before 3 years of age. That young girls are not affected by these activating mutations can be rationalized on the basis that recruitment and maturation of primordial follicles require FSH and factors other than gonadotropins. At the other end of the clinical spectrum are patients presenting with disorders in sexual differentiation and reproductive function. From the earlier discussion on male sexual differentiation *in utero*, it is apparent that a nonfunctional or only partially functional LHR would lead to feminization or pseudohermaphrodism in a 46,XY fetus. Depending on the severity of LHR dysfunction, pubertal development and reproductive function in the adult male would be severely affected. Whereas sexual differentiation would not be impaired in a 46,XX fetus, since the female phenotype will develop unless the indifferent reproductive tract is altered by SRY and other factors, pubertal development and adult reproductive function will be greatly compromised depending on the severity of LHR dysfunction. Analysis of the LHR gene from patients presenting

with these various disorders has revealed point mutations in exon 11 and a deletion of exon 8 that resulted in decreased receptor expression and responsiveness to ligand. Two cases were also identified in which premature stop codons were introduced by the mutations, one in transmembrane helix 5 and one in intracellular loop 3; for these, the prematurely truncated forms of LHR probably fail to fold properly and intracellular degradation of the aberrant receptor would likely occur. A similar argument may apply to the mutant LHR lacking the amino acid residues encoded by exon 8. The point mutations identified that result in loss of function LHR mutants were present in the extracellular domain, near transmembrane helix 1, and in transmembrane helices 6 and 7. For these, the level of mutant receptor expression was reduced relative to wild type, as was responsiveness to ligand.

Compared to the human LHR, far less has been determined of the role of human FSHR in reproductive disorders. The FSHR gene was studied in a group of Finnish women with normal XX karotype who presented with hypergonadotropic ovarian dysgenesis, a condition characterized by streak ovaries and elevated levels of LH and FSH. In this extended kindred, a point mutation was identified in exon 7 of the FSHR gene. When analyzed in the laboratory, it was found that this single amino acid residue replacement reduced the level of FSHR expression and signaling compared to the wild-type receptor. This finding explains the clinical phenotype in that, even in the presence of FSH, the FSHR is unable to respond appropriately to hormone and the prepubertal ovary is unable to develop into a mature, cycling ovary. Thus, one of the essential roles of FSH and FSHR is to rescue developing follicles from atresia, which is the normal event in the prepubertal girl.

A number of males from the same Finnish family were evaluated, and several were identified who were either heterozygous or homozygous for the FSHR mutant in exon 7. Surprisingly, these individuals exhibited varying degrees of spermatogenic failure, but none were azoospermic. This observation is consistent with a recent report that transgenic "knockout" mice missing FSH were fertile. Thus, it appears that FSH and FSHR are not as critical in male as in female sexual development and fertility.

Another loss of function mutation in FSHR was recently found in a high percentage of human ovarian sex cord tumors. This point mutation, which led to a change in one amino acid residue in transmembrane helix 6, reduced the level of FSHR expression and responsiveness to FSH. It is unclear how, or even if, this inactivating mutation in FSHR is contributory to neoplasia; it may be one of several factors that predisposes sex cord stromal tumors and ovarian small cell carcinomas.

In summary, proper functioning of the gonadotropin receptors is required for normal development and fertility. Moreover, dysfunctional receptors appear to be associated with tumors of the reproductive tract, although much more work is required to establish causal effects.

Acknowledgment

The National Institutes of Health (Grant No. DK33973) is gratefully acknowledged for supporting the author's research in gonadotropins and their receptors.

See Also the Following Articles

Chorionic Gonadotropin, Human; FSH (Follicle-Stimulating Hormone); GnRH (Gonadotropin-Releasing Hormone); Gonadotropin Biosynthesis; Gonadotropin Secretion, Control of; Hormone Receptors, Overview; LH (Luteinizing Hormone); *SRY* Gene

Bibliography

Bourne, H. R. (1997). How receptors talk to trimeric G proteins. *Curr. Opin. Cell. Biol.* **9**, 134–142.

Ji, T. H., Murdoch, W. J., and Ji, I. (1995). Activation of membrane receptors. *Endocrine* **3**, 187–194.

Puett, D., Bhowmick, N., Fernandez, L. M., Huang, J., Wu, C., and Narayan, P. (1996). hCG-receptor binding and transmembrane signaling. *Mol. Cell. Endocrinol.* **125**, 55–64.

Segaloff, D. L., and Ascoli, M. (1993). The lutropin/choriogonadotropin receptor . . . 4 years later. *Endocr. Rev.* **14**, 324–347.

Themmen, A. P. N., Martens, J. W. M., and Brunner, H. G. (1997). Gonadotropin receptor mutations. *J. Endocrinol.* **153**, 179–183.

Gonadotropin Secretion, Control of

Charles A. Blake

University of South Carolina School of Medicine

GLOSSARY

androgen A chemical that stimulates the accessory sex organs of the male such as the prostate gland. Endogenous androgens in animals include testosterone and other structurally similar steroids.

estrogen A chemical that induces estrus or heat; i.e., willingness of a female nonprimate mammal to accept the male sexually. Endogenous estrogens in animals include estradiol 17-β and other structurally similar steroids.

false-positive feedback With respect to the gonadotropins, an effect by which gonadotropin-induced decreased release of a gonadal target cell hormone results in increased release of the stimulating gonadotropin.

gonadotropin A hormone that acts primarily or solely at the gonads.

half-time With respect to hormones in blood, the time required for half of any starting amount of hormone to disappear.

hormone A chemical secreted by a cell that enters the blood directly for transport to another part of the body where it exerts a regulatory effect. In the classical sense, hormone-secreting cells are found in circumscribed areas called ductless or endocrine glands. This definition has been expanded to include diffusely localized cells or groups of cells in various tissues including cells in the nervous system that secrete neurohormones.

negative feedback With respect to the gonadotropins, an effect by which gonadotropin-induced release of a gonadal target cell hormone decreases the release of the stimulating gonadotropin.

orchidectomy The removal of the testes from an animal.

ovariectomy The removal of the ovaries from an animal.

positive feedback With respect to the gonadotropins, an effect by which gonadotropin-induced release of a gonadal target cell hormone increases the release of the stimulating gonadotropin.

secretion A process by which cells use energy to convert precursor molecules into more complex molecules and then release them. The term often is used interchangeably with *release* and this is appropriate when describing the effects of a chemical on cellular release as long as the chemical alters synthesis and release in the same direction. Otherwise, the component terms, *synthesis and release*, should be employed.

Gonadotropins are hormones that have an affinity for the gonads. Some invertebrates have a gland that secretes gonadotropin, for example, the optic gland of the octopus. In most, if not all, male and female vertebrates there are two gonadotropic hormones secreted by the pituitary gland. They are luteinizing hormone (LH) and follicle-stimulating hormone (FSH). Individual cells called gonadotropes usually contain both LH and FSH. In mammals, gonadotropes located in the anterior pituitary gland (APG) are responsible for secreting the LH and FSH that act on the gonads. Gonadotropes also may be present in the pars tuberalis, tissue that migrated out

of the developing APG to attach to the infundibular stalk or underlie the median eminence, but these cells apparently do not contribute significantly to the pool of LH or FSH in the systemic circulation. In pregnant primates and equids, there also is a nonpituitary gland source of a gonadotropic hormone. Chorionic gonadotropin (CG) is secreted by syncytio-trophoblasts in the primate placenta. In the horse and other equids, CG is secreted by chorion-derived, uterine endometrial cells. Although additional hormones act on the gonads, they are not normally referred to as gonadotropic hormones. Thyroid hormone is an example of a hormone that acts on most of the cells in the body including gonadal cells. Prolactin, another APG hormone, has principal actions on nongonadal tissue in vertebrates. However, it also has important actions on the gonads in some mammals. For that reason, it is sometimes referred to as a gonadotropic hormone. This article addresses the control of APG LH and FSH secretion in mammals.

I. INTRODUCTION

Through their actions on the testes in the male and the ovaries in the female, luteinizing hormone (LH) and follicle-stimulating hormone (FSH) regulate the reproductive system. Such actions include stimulating the secretion of testicular and ovarian steroids and, at least in mammals, one or two gonadal protein hormones called inhibin. The actions of steroids within the gonads are essential for sperm production in the male and follicular growth in the female. In addition to influencing the function of many cell types within the body, including those of the reproductive tract, gonadal steroids circulating in blood act to decrease or increase gonadotropin secretion and thereby exert negative or positive feedback on LH and FSH secretion. Feedback is exerted at the brain to alter the secretion of luteinizing hormone-releasing hormone (LHRH), a brain hormone that controls anterior pituitary gland (APG) gonadotropin secretion. It also is exerted directly at the APG to regulate gonadotropin secretion in the absence or presence of LHRH. Inhibin also plays roles in influencing gonadal processes and in controlling gonadotropin secretion. Regulation of gonadotropin secretion is mandatory for the maturational changes that occur in the reproductive system of immature vertebrates and for the maintenance of reproductive fertility in the adult.

II. NONCYCLIC AND CYCLIC GONADOTROPIN RELEASE

The brains of male and female mammals initially possess the capacity to develop mechanisms that result in the stimulation of cyclic gonadotropin release when adulthood is reached. This cyclic release of LH and FSH occurs in female mammals during the estrous or menstrual cycle. The pattern of gonadotropin secretion in the adult male is noncyclic. In some species, sexual differentiation of the brain to a male-type brain occurs in genetic males to render them incapable of exhibiting cyclic gonadotropin release as an adult. This particular masculinizing effect is induced by testicular androgen secretion before birth or shortly thereafter. In the rat, the androgens involved are likely testosterone and its precursor androstenedione. Androgens mediate this effect by being converted to estrogen by the enzyme aromatase, which is present in brain.

The cyclic release of LH and FSH in adult females is characterized by a marked increase or surge of gonadotropins in the circulation prior to ovulation. This surge is stimulated by rising concentrations of estrogen in the blood. Experimental rats in which androgen did not masculinize the brain during the neonatal period will respond, when adults, to exogenous estrogen with a surge of gonadotropin release. This response to estrogen occurs in adult male rats orchidectomized at birth and in females ovariectomized as adults but does not occur in adult females or in females ovariectomized as adults if they were administered testosterone or estrogen shortly after birth. Interestingly, the primate testes secrete androgen during development, but it does not exert this particular masculinizing effect on brain. The adult male primate does not exhibit a surge of gonadotropin release but is capable of doing so if orchidectomized and administered estrogen. The adult female primate exposed to androgen during development also can respond to estrogen with a surge of gonadotropin release.

III. CONTROL OF GONADOTROPIN SECRETION BY LHRH

The primary factor that controls gonadotropin secretion is LHRH. It is synthesized by nerve cells, the cell bodies of which are located primarily in the hypothalamus of the brain, and released from axon terminals in the median eminence from where it is transported to the APG via hypophysial portal veins. This is significant because the LHRH is not diluted in the systemic circulation before it reaches the gonadotropes, providing an economical use of LHRH and a rapid response of gonadotropes to hormonal control by the brain.

There is evidence to suggest that LHRH secreted by the mammalian fetus stimulates some of the cells in the developing APG to differentiate into gonadotropes. Thereafter, these cells are able to synthesize some LH in the absence of concurrent LHRH stimulation. However, LHRH likely acts later during fetal life in some species and does act during the perinatal period in others to stimulate the development of the full complement of gonadotropes and to initiate the synthesis of significant amounts of FSH within them. The long-term maintenance of LH and FSH synthesis is dependent on LHRH. Additionally, any significant release of LH by gonadotropes is dependent on LHRH stimulation. By contrast, FSH release, which also is stimulated by LHRH, can be substantial on a temporary basis in the absence of concurrent LHRH stimulation of gonadotropes. Thus, although both gonadotropic hormones are synthesized by the same cells, FSH can be released selectively. As discussed in Section VII, LH can be released preferentially. Because LHRH causes both LH and FSH secretion, it also is called gonadotropin-releasing hormone (GnRH).

The LHRH neurons synchronize to release LHRH episodically in pulses, and the amplitude and frequency of these pulses change during maturation to the adult reproductive state. During intervals between LHRH pulses, gonadotropes are not influenced by significant amounts of LHRH (Fig. 1). The LHRH which does not bind to LHRH receptors attached to the plasma membrane of gonadotropes is diluted in the systemic blood where it is rapidly cleared from the circulation presumably by the liver and/or kidney. The half-time for disappearance of circulating LHRH is brief and has been reported to be 6.7 min

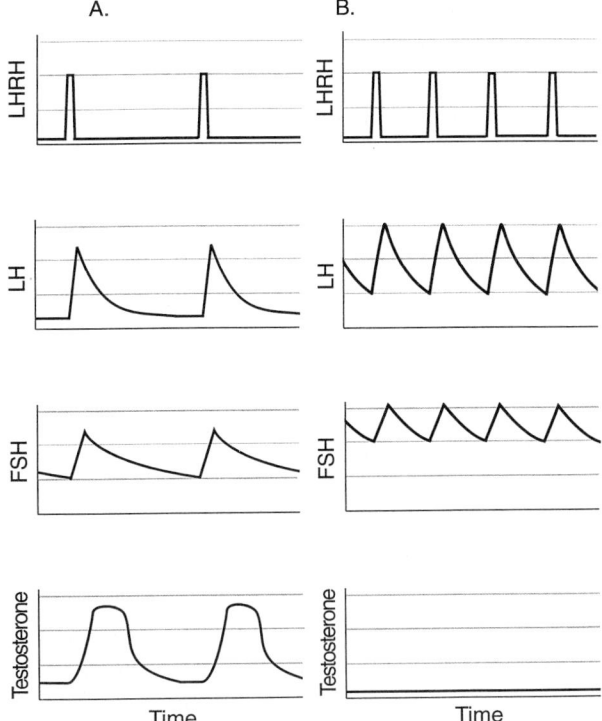

FIGURE 1 This diagrammatic representation of hormone patterns in normal mammals with intact testes (A) and in orchidectomized mammals (B) is based on data from a number of species and is designed to represent general principles. Actual fluctuations in hormone concentrations and patterns vary between species. In this diagram, there are no silent LHRH pulses, all FSH pulses are associated with LHRH pulses, and a single LH pulse precedes each episodic increase in testosterone. Patterns are illustrated for LHRH in hypophysial portal vein blood and for the other hormones in the systemic circulation.

collecting hypophysial portal vein blood or extracellular fluid in the median eminence or APG from an unanesthetized mammal for the measurement of LHRH, increases in LH concentration in the systemic circulation are considered to be representative of increases in LHRH release. Occasionally, it has been reported that there are LHRH pulses without accompanying LH pulses interspersed between LHRH-induced LH pulses in orchidectomized and ovariectomized mammals as well as in mammals with intact gonads. These silent LHRH pulses may be due to an artifact resulting from the collection procedure. However, it is known that release of a number of hormones occurs episodically in pulses, and continuous or excessive release of a hormone can decrease the number of active receptors for that hormone on the surface of its target cells. This process, called downregulation, also could explain the silent pulses. In mammals with intact gonads, silent LHRH pulses also could be due to gonadal secretions inhibiting LHRH action at gonadotropes (see Section V).

Pulsatile LH release is characterized by a marked rise in circulating LH concentration that occurs over a period of approximately 5 min in response to near concurrent pulsatile LHRH release (Fig. 1). Circulating concentrations of LH then decline exponentially with a half-time for disappearance that is similar to that observed for LH in the circulation immediately after the pituitary gland has been surgically removed. Thus, LH is released in pulses with intervening periods of little or no LH release.

B. Pulsatile FSH Release

Pulsatile increases in FSH release likely exist in all species, but they have been measured reliably only in a few (Fig. 1). The pulses have been more difficult to detect and study than those of LH due to the slow rate of clearance of FSH from blood. Unless FSH is measured with great precision in blood samples collected serially over short time periods to delineate the slow declining phase, it is difficult to identify an FSH pulse. In addition, an increase in FSH or LH pulsation frequency, such as occurs in orchidectomized mammals, functions to mask pulse detection unless there is an accompanying increase in the amplitude of the pulses. This is due to circulating gonadotropin concentrations rising to elevated levels

in sheep. Moreover, as determined in rats, <0.05% of the systemic blood reaches the APG each minute. These conditions ensure that secreted LHRH that escapes binding to gonadotropes and enters the systemic circulation has a minuscule chance of ever binding to gonadotropes and stimulating gonadotropin secretion.

A. Pulsatile LH Release

Pulsatile increases in LHRH concentration in hypophysial portal vein blood usually cause pulsatile increases in LH concentration in the systemic circulation (Fig. 1). Because of the difficulty involved in

in association with the decrease in time for gonadotropins to be cleared from the circulation between pulses. An increase in the frequency of a pulse of unvarying magnitude would mask FSH pulse detection more than LH pulse detection due to the longer half-time for clearance of FSH than of LH from blood.

In humans, FSH pulses are normally synchronized with LH pulses and administration of LHRH antagonists blocks both LH and FSH pulses. By contrast, in orchidectomized rat and ovariectomized hamster there often is a lack of synchrony between the LH and FSH pulses. Additionally, administration to these rodents of an LHRH antagonist to block the action of LHRH or antiserum to LHRH to inactivate the biological activity of LHRH blocks LH but not FSH pulses. In one study of ovariectomized sheep, hypophysial portal vein blood was collected on the surface of the APG, and LHRH and both gonadotropins were measured in each sample collected. Each pulse of LHRH was associated with a pulse of LH and FSH. However, there were additional FSH pulses not associated with LHRH pulses. Collectively, these observations in rodents and sheep have been interpreted as evidence that some FSH pulses are not caused by LHRH pulses. A number of investigators have postulated the existence of an FSH-releasing hormone that causes only FSH pulses.

IV. LH AND FSH CLEARANCE

The gonadotropins control the release of gonadal hormones that exert feedback regulation on LH and FSH secretion. Thus, the length of time that gonadotropins are elevated in the circulation plays a role in determining the timing of subsequent gonadotropin release.

The rate of decline of LH and FSH concentrations in the circulation is related primarily to the type of terminal residues on the carbohydrate side chains. Sialic acid termini markedly extend the life of gonadotropins in the circulation. Gonadotropins that are heavily sialylated are cleared slowly from the circulation. This occurs to some extent by glomerular filtration within and subsequent excretion by the kidney. In fact, intact gonadotropin can be collected from urine. In women, the presence of CG in urine forms

the basis for some pregnancy tests. Alternatively, sialic acid termini eventually are removed by neuraminidase, an enzyme that is ubiquitous within the body and especially concentrated in kidney. Gonadotropin molecules that are sufficiently desialylated are rapidly metabolized via binding to the asialoglycoprotein receptor in liver hepatocytes. Sulfated terminals bind to another receptor in endothelial and Kupffer cells in the liver that mediates rapid metabolism of the gonadotropin molecule. However, the clearance of sulfated gonadotropins from the circulation is not nearly as rapid as that for gonadotropins with nonsulfated galactosamine or nonsialylated galactose termini on the carbohydrate side chains that bind to the asialoglycoprotein receptor. Thus, sulfation or sialylation of carbohydrate terminals acts to extend the life of gonadotropin in the blood by preventing binding to the asialoglycoprotein receptor. Although all gonadotropins may undergo glomerular filtration, those gonadotropins that are heavily sulfated are metabolized fairly rapidly by the liver. Gonadotropins that are heavily sialylated can be metabolized by liver when given time for sufficient desialylation to occur.

Secreted LH in mammals has a high percentage of sulfated terminals. The relatively rapid clearance of these sulfated LH molecules contributes significantly to the pulsatile mode of circulating LH concentration. Isoforms of secreted FSH normally contain more sialic acid residues than do isoforms of secreted LH. This results in a longer half-time for disappearance of FSH from the circulation than that for LH. For example, in the orchidectomized rat the half-time for FSH is approximately 2 hr, whereas that for LH is approximately 25 min.

V. CONTROL OF GONADOTROPIN SECRETION IN THE MALE MAMMAL

Luteinizing hormone acts to promote testosterone secretion after binding to LH receptors attached to the plasma membrane of Leydig cells, which are located in the interstitial spaces between the seminiferous tubules in the testes. Some of the testosterone is utilized within the testes, where it supports sperm production, and some of it enters the systemic circulation to function as a hormone.

In most mammals studied, there is a close coupling between single LH pulses in the blood and single elevations in circulating testosterone concentrations (Fig. 1). In rat and man, two or more LH pulses usually occur before a rise in circulating testosterone occurs. In contrast to the circulating pulsatile LH concentrations that are characterized by a rapid rise in LH followed by the onset of a near immediate decline, circulating testosterone concentrations rise more gradually and often remain elevated for a considerable length of time. This is due to a long secretion time because the half-time for the disappearance of testosterone from blood is relatively brief (10–20 min in man). In the rat the elevation in circulating testosterone lasts a very long time (3–6 hr), and there usually are no LH pulses from the time circulating testosterone concentration reaches its peak until some time after testosterone has returned to low levels due to inactivation by the liver.

A. Negative Feedback on LH

Increased circulating testosterone acts to alter the function of many cells within the body and to suppress LH secretion in a negative feedback fashion. An enormous amount of experimentation has been conducted for more than three decades to determine whether the negative feedback effect of testosterone on LH release is exerted in the brain or pituitary gland. However, testosterone-mediated feedback on LH release is still poorly understood. The following discussion highlights just a few of the experimental approaches that have been employed and some of the difficulties encountered in trying to elucidate the effects of testosterone on LH release.

1. Orchidectomy and Effects of Testosterone on LH Release

The effects of testosterone on LH pulse frequency and amplitude have provided clues. For example, when the principal source of testosterone in the circulation is eliminated experimentally by orchidectomy, the frequency of LH pulses becomes more regular and increases to varying degrees depending on the species and length of time after orchidectomy

(Fig. 1). In some mammals such as rat, the magnitude of the LH pulses increases as well, but this parameter becomes more difficult to assess as the frequency of LH pulses increases causing an elevated baseline of circulating LH. The net effect in all mammals is an increase in mean circulating LH concentration. Administration of testosterone to orchidectomized mammals lowers mean circulating LH levels by suppressing the frequency of LH pulses and in species such as rat also lowers the amplitude of the LH pulses. These results clearly support the view that testosterone exerts negative feedback on LH release. However, such results are equivocal with respect to determining the site(s) of testosterone action. The steroid-mediated decrease in LH pulse frequency could be due to an action exerted at brain to suppress LHRH pulse frequency or at the APG to block LH release in response to some of the LHRH pulses. Testosterone's suppressive effect on LH pulse amplitude could be due to action exerted at brain to diminish LHRH pulse amplitude or at the APG to suppress LHRH-induced LH release. It also could be due to reduction in the synthesis of a readily releasable pool of LH as a result of testosterone altering LHRH secretion.

2. LH Release in Response to Exogenous LHRH

Other approaches have centered on testing the effects of testosterone on the ability of the APG to release LH in response to exogenous LHRH. Studies have been conducted *in vivo* using orchidectomized mammals. In most species, testosterone is very effective in suppressing LH release in response to synthetic LHRH. This indicates that testosterone can suppress LHRH-induced LH release. This does not necessarily mean that testosterone acts solely at the APG directly under physiological conditions. In these studies, testosterone was administered to maintain elevated levels of testosterone in blood rather than to simulate episodic increases that are present in the circulation of normal mammals with intact testis. This could result in testosterone exerting nonphysiological actions directly on gonadotropes. Moreover, testosterone alters endogenous LHRH release which can alter LH synthesis and release. With regard to the previously described results, false interpretations might be made if an endogenous LHRH pulse coincided with an exogenous LHRH pulse. Some investigators have circumvented the latter criticism by eliminating endogenous LHRH from the picture via one of several experimental approaches such as lesioning LHRH neurons in the hypothalamus. The experimental animals are maintained on a fixed frequency and dose of exogenous LHRH and then treated with testosterone. One drawback to such experiments is the possibility that procedures employed to eliminate endogenous LHRH may introduce new variables. Another concern is that the APG was not exposed to physiological episodic increases in circulating testosterone in these studies. A modification of this approach has been to examine the effects of unvarying testosterone concentrations on the response of the isolated APG to LHRH *in vitro*. All of these techniques have yielded important information regarding the ability of testosterone to exert negative feedback at the APG, but the results may not reflect the normal physiological situation.

3. Studies in Sheep

Further clarification of the site(s) of testosterone negative feedback emerged with techniques that enabled the collection of secreted LHRH over time. Sheep have proven to be especially suited for the collection of serial hypophysial portal vein blood samples for assay of LHRH. Data collected in a number of studies indicate the following results. Each LH pulse in rams with intact testes causes an episode of increased circulating testosterone, and subsequent LH pulses occur only after testosterone has returned to low levels in blood (Fig. 1). Orchidectomy causes an increase in frequency but not in amplitude of LHRH and LH pulses. Administration of testosterone to orchidectomized rams to restore it to unvarying magnitudes similar to those present between trough and peak levels in rams with intact testes reduces the frequency of LHRH and LH pulses but does not affect their amplitude. In all rams, virtually every LHRH pulse in hypophysial portal vein blood is followed by an LH pulse in the systemic circulation. These data on LHRH and LH pulse frequency are straightforward. However, the pulse amplitude of LHRH can vary depending on the site of hypophysial portal vein blood collection and removal of LHRH

from hypophysial portal vein blood can reduce LH pulse amplitude. Thus, comparisons of LHRH pulse amplitudes measured in different animals or of LHRH and LH pulse amplitudes in the same animal must be viewed with caution. In studies employing orchidectomized rams in which endogenous LHRH influence was eliminated, testosterone had no effect on LH release in response to exogenous LHRH pulses of fixed frequency and amplitude. Collectively, these results could be interpreted to mean that testosterone exerts negative feedback on LH release solely by suppressing the frequency of LHRH pulses. The site of this action of testosterone is in the hypothalamus or median eminence, and evidence has been obtained to indicate that it is likely not due to a direct action on LHRH neurons.

This scenario precludes the need for testosterone to act on gonadotropes. However, the fact that testosterone has been reported to suppress LH release in response to exogenous LHRH *in vitro* and in some experiments *in vivo* obfuscates matters. Nevertheless, these observations should not be interpreted to mean that testosterone exerts negative feedback at the APG to affect pulsatile LH release on a moment-to-moment basis. The mammalian testes secrete low amounts of testosterone in the absence of gonadotropin stimulation as well as between episodes of increased testosterone release in the presence of low circulating LH levels. It is important to remember that LH pulses occur when circulating testosterone is at basal levels and not during episodes of increased circulating testosterone. Interestingly, restoration of circulating testosterone in orchidectomized rams to unvarying levels comparable to the basal levels which are present between episodes of increased circulating testosterone in rams with intact testes did not alter LH pulse frequency but increased LH pulse amplitude. When testosterone was elevated to unvarying levels that approximated one-fourth of the peak levels observed during episodic increases in circulating testosterone in rams with intact testes, the frequency and amplitude of the LH pulses decreased and were similar to those observed in rams with intact testes. These observations suggest that LH release may be enhanced in response to an endogenous LHRH pulse which occurs in the presence of low basal circulating levels of testosterone.

Study of the relationships between LHRH and LH pulses and episodes of increased blood testosterone levels at different seasons has further clarified the relationships between these hormones. Between the transition period from breeding to nonbreeding season in March and the start of the breeding season in September in Hampshire rams, LHRH pulse frequency and amplitude remain unchanged. However, LH pulse amplitude and LH release in response to exogenous LHRH decrease. This is associated with increased testosterone release in response to LH pulses, despite the smaller amplitude of the LH pulses, and an elevation in basal circulating testosterone concentrations between LH pulses. These observations and those of the aforementioned studies indicate the following results. During the nonbreeding and breeding season, episodic increases in circulating testosterone control LH release on a moment-to-moment basis by suppressing LHRH pulse frequency. When basal circulating testosterone levels are very low between LHRH pulses during the nonbreeding season, testosterone may act at the APG to enhance LHRH-induced LH release. By contrast, when basal circulating testosterone levels are higher between LHRH pulses during the breeding season, testosterone likely acts at the APG to continuously suppress the amplitude of LH pulses in response to LHRH. The reduced magnitude of the LH pulses may function to limit testosterone secretion, which is highly responsive to LH stimulation during the breeding season. Thus, testosterone may act at the APG in a physiological setting to alter LH pulse amplitude on a seasonal basis.

4. Studies in Rats

Testosterone may exert negative feedback on LH release in rats in a different manner than it does in rams (Fig. 2). In orchidectomized males of both species, testosterone acts to decrease the frequency of LHRH and LH pulses. This supports the view that testosterone suppresses LHRH pulse frequency on a chronic basis. However, individual episodic increases in circulating testosterone may not function to decrease LHRH pulse frequency in rats with intact testes. Based on LHRH measurements in extracellular fluid collected from the median eminence, there was an appreciable number of LHRH pulses not accompa-

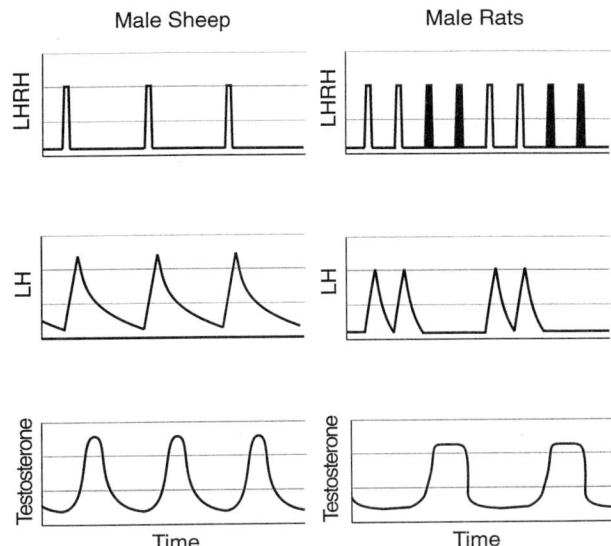

FIGURE 2 Diagrammatic representation of hormone patterns in normal male sheep and rats. The relationships between LHRH and LH pulses and episodes of increased testosterone suggest species differences with regard to testosterone exerting negative feedback on pulsatile LH release. In sheep, there is a one-to-one relationship between LHRH and LH pulses. Testosterone acts acutely to inhibit LHRH pulse frequency and therefore LH pulse frequency. In rat, the observation that there are many silent LHRH pulses (four are illustrated as darkened pulses) suggests that testosterone inhibits LH pulsatility by blocking LHRH-induced LH release. Each episode of increased testosterone is normally preceded by one LH pulse in rams and two or more LH pulses in rats. Patterns are illustrated for LHRH release profiles and for LH and testosterone in the systemic circulation.

nied by LH pulses in the peripheral circulation. It has been suggested that testosterone pulses block LH pulses on a moment-to-moment basis by blocking LHRH-induced LH release at the APG. This contrasts with rams with intact testes in which LHRH pulses are blocked by testosterone pulses and each LHRH pulse is almost always followed by an LH pulse. It is unclear whether the silent LHRH pulses in rats represent true species differences or an artifact of the collection procedure. Further research is needed to ascertain whether testosterone exerts moment-to-moment negative feedback on LH release in most mammals by suppressing LHRH pulse frequency, inhibiting gonadotrope responsiveness to LHRH, or both. Negative feedback control of LH release in males is diagrammed in Fig. 3.

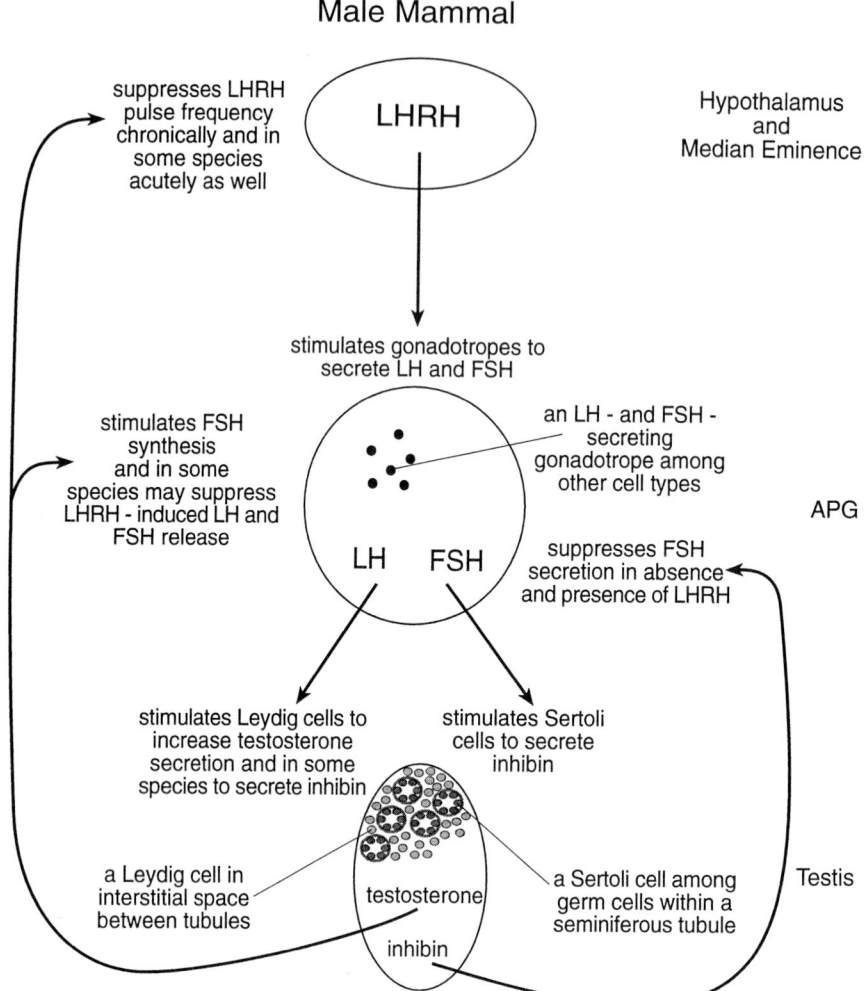

FIGURE 3 This diagram shows established interrelationships between LHRH, LH, FSH, testosterone, and inhibin in some species of adult male mammals. The importance of inhibin in exerting negative feedback on FSH secretion varies between species. The diagram does not depict Leydig cell DHT and estrogen release, Sertoli cell estrogen release, or conversion of testosterone to DHT or estrogen at target tissues; all of these factors contribute to the feedback regulation of LH and FSH secretion to varying degrees in different species.

5. Testosterone as a Prohormone

To further complicate our understanding of negative feedback on LH release in the male, it is unclear whether testosterone itself exerts negative feedback on LH release or acts as a prohormone. The brain and APG contain the enzyme 5α-reductase, which converts testosterone to another androgen, 5α-dihydrotestosterone (DHT). In males, circulating DHT concentrations are approximately one-ninth those of testosterone and result from testicular secretion and conversion of testosterone to DHT in peripheral tissues containing 5α-reductase. There is a single androgen receptor that binds testosterone and DHT, and DHT is more potent than testosterone after binding to the androgen receptor. Thus, conversion of testosterone to DHT at target tissues is a means of amplifying the testosterone signal. In orchidectomized males of most species, administration of testosterone or DHT inhibits LH release. If the negative feedback in the physiological setting is due to androgenic action, it is very likely that DHT may be more important than testosterone in this regard.

Testosterone also may exert negative feedback on LH release by conversion to estrogen. Circulating concentrations of estrogen are low in males. They arise by secretion by Leydig cells as well as by testicular Sertoli cells (at least in immature mammals and to some extent in adults of some species) by aromatization of androgen provided as substrate to the Sertoli cells by the Leydig cells. The cortex of the adrenal glands also secretes some estrogen in some species, and circulating concentrations of estrogen can increase as a result of conversion of androgen to estrogen by aromatase located in the smooth endoplasmic reticulum in cells of some peripheral tissues that include hypothalamus but not APG. Estrogen concentrations in the peripheral circulation by themselves are probably too low to account for negative feedback on LH release. However, administration of sufficient doses of estrogen will inhibit LH release in orchidectomized mammals. This could reflect a physiological process by which estrogen suppresses LHRH pulse frequency in brain after aromatization of testosterone. It also may reflect a physiological suppression of LHRH-induced LH release by estrogen at the level of the APG. The APG in the male contains estrogen receptors, and exogenous estrogen will suppress LHRH-induced LH release. The hypothalamus produces considerable amounts of estrogen by aromatization of androgen, and hypophysial portal vein estrogen concentrations may be elevated above those present in the systemic blood. Thus, estrogen of hypothalamic origin combined with estrogen of systemic origin may act directly on gonadotropes to suppress LHRH-induced LH release. However, because the APG does not contain aromatase (at least in species that have been studied), any direct effect of testosterone at the level of the gonadotrope must be due to androgenic rather than estrogenic action.

It is highly likely that the entire complement of circulating androgens and estrogens function to exert negative feedback control on LH release in males of all mammalian species. However, there is evidence that negative feedback on LH release is due primarily to androgenic action in some species and to estrogenic action in others. Two experimental designs have helped elucidate whether androgenic or estrogenic action is involved. First, the effects of DHT on LH release have been studied in orchidectomized mammals. This androgen cannot be aromatized to estrogen, and the effects of DHT can be ascribed to androgenic action. Second, aromatase inhibitors have been administered to normal males to ascertain whether suppression of the conversion of testosterone and other aromatizable androgens to estrogen interferes with negative feedback control of LH release. Collectively, the results suggest that androgens are the primary regulators of LH negative feedback in male rats, androgens and estrogen are involved in LH negative feedback in men, and estrogen is an important regulator of LH negative feedback in the male dog.

B. Negative Feedback on FSH

1. Inhibins

Follicle-stimulating hormone binds to FSH receptors attached to the plasma membrane of Sertoli cells, which are located along with germ cells within the seminiferous tubules in the testes. FSH acts to promote synthesis of aromatase and other proteins by the Sertoli cells. In a classical negative feedback control system, one would predict a Sertoli cell secretion to inhibit FSH release. This does occur, but its importance in exerting negative feedback on FSH release depends on the circumstances.

Among the Sertoli cell proteins secreted are two inhibins and three activins. Administration of exogenous inhibin suppresses FSH secretion and administration of exogenous activin increases FSH secretion in adult male mammals. However, it has been difficult to assess the physiological roles of testicular inhibin and activin secretion on the control of FSH release for four reasons. First, it has been difficult to measure the different inhibin and activin dimeric molecules due to similarities in their structures. Second, the different forms of inhibin or activin may bind to different receptors and exert similar biological effects through different mechanisms. Third, the APG secretes inhibin and activin. Fourth, in addition to FSH stimulating inhibin secretion by Sertoli cells, LH also stimulates inhibin secretion by Leydig cells in some species. Despite these factors, which complicate experimental design and interpretation of results, there is convincing evidence that FSH-induced Sertoli cell inhibin secretion exerts negative feedback on FSH release. This inhibitory action is exerted at the level of the APG to suppress FSH release in the absence or presence of LHRH stimulation. It is specific in the sense that physiological levels of inhibin in most species studied do not alter LH release. Additionally, there is evidence that activin functions to control FSH release. However, it is not clear whether circulating activin has any significant effect. Activin may be important as an APG synthesized product exerting paracrine or autocrine regulation of FSH secretion. Inhibin secreted by the APG also may exert local regulatory effects. Interestingly, inhibin can block the stimulatory action of activin on FSH release.

The importance of inhibin in exerting negative feedback on FSH release varies with age and between species. For example, in male rat, circulating inhibin concentration increases until about the fourth week after birth and plays an important role in the negative feedback control of FSH release at that time. Thereafter, circulating inhibin levels decline, and inhibin is not the primary testicular regulator of FSH negative feedback in the adult. In fact, administration of antiserum to inhibin to adult male rats to inactivate the biological activity of inhibin has been reported not to affect circulating FSH concentrations. By contrast, similar experiments conducted in rhesus monkeys have shown that circulating FSH levels rise, suggesting that inhibin plays an important role in exerting negative feedback on FSH release in the adult male monkey.

2. *Testosterone*

Testosterone also exerts negative feedback control on FSH release in male mammals. This should not be surprising because testosterone suppresses the frequency of LHRH pulses, and LHRH stimulates FSH secretion. The actions of testosterone at the level of the APG on FSH release are complex. Testosterone functions to suppress LHRH-induced FSH release at least under experimental conditions, but it also can act on the APG to enhance FSH secretion as has been observed in the temporary absence of LHRH stimulation. Testosterone is the primary regulator of FSH negative feedback in the adult male rat, and it undoubtedly plays an important role in other adult male mammals as well. In this sense, testosterone-

mediated negative feedback on FSH release deviates from a classical negative feedback system because FSH does not cause testosterone release. Rather, LHRH release causes both LH and FSH release, and testosterone release induced by LH exerts negative feedback on FSH release. The issue of whether or not testosterone acts as a prohormone to exert negative feedback on FSH release has already been addressed with regard to its action on brain to affect LHRH release (*vide supra*). There is no evidence to indicate that testosterone-mediated inhibition of LHRH-induced FSH release does not involve the same mechanisms involved in testosterone-mediated inhibition of LHRH-induced LH release. In summary, testosterone, through actions on brain and pituitary gland, and inhibin, through actions on pituitary gland, act to control FSH secretion in the male, and the importance of each hormone varies with regard to age and species. Negative feedback control of FSH release in males is diagrammed in Fig. 3.

C. Gonadotropin Isoforms

Testosterone and DHT act to extend the half-time for disappearance of gonadotropins from the circulation in adult rats. Estrogen has the opposite effect in adult rodents and rhesus monkeys. Steroids likely exert these effects by modulating sialylation of the gonadotropins that are secreted. It is not known whether steroids affect sialylation through actions exerted directly at gonadotropes or indirectly via alteration of LHRH release. Changes in LHRH release could affect sialylation directly by acting within gonadotropes or indirectly via alteration of FSH-mediated changes in inhibin secretion. It is established that less sialylated forms of LH and FSH bind better to gonadotropin receptors and exert greater biological activity than the more heavily sialylated forms. The net effect of sialylation is to maintain moderate biological activity over a relatively long period of time. By contrast, the same amount of secreted gonadotropin containing less sialic acid residues would exert a higher degree of biological activity over a shorter period of time. Because circulating LH and FSH concentrations are relatively low in normal males with intact testes, it has not been feasible to study the half-times for disappearance of circulating endogenous gonadotropins. Thus, it is not clear how the total complement of circulating steroids in normal males alters the rate of clearance of gonadotropins from blood. However, it is clear that gonadotropins synthesized by the APG in adult males of some species are heavily sialylated, presumably because of the dominance of androgen in the circulation of males. Thus, secreted LH and FSH would be expected to exert moderate biological activity over a relatively long time to maintain Leydig and Sertoli cell functions, which do not vary on a daily basis. Unlike the adult female, the male has no need to maximize gonadotropin bioactivity over a brief period of time. Sialylation acts to conserve the production of gonadotropin.

VI. CONTROL OF GONADOTROPIN SECRETION IN THE FEMALE MAMMAL

Control of gonadotropin secretion in the female mammal is more complex than that in the male. In both sexes, gonadal secretions exert negative feedback control on gonadotropin secretion. However, ovarian steroids also exert positive feedback effects on LH and FSH secretion that result in preovulatory increases or surges of these gonadotropins in blood during the estrous or menstrual cycle. The most important ovarian secretion that regulates gonadotropin secretion is estrogen. Changes in circulating estrogen concentrations which effect major changes in gonadotropin secretion are mirrored by changes in the cellular makeup of the ovaries. For that reason, it is helpful to examine gonadotropin secretion in three stages: the follicular phase of the ovarian cycle, the preovulatory period, and the luteal phase of the ovarian cycle. It should be noted that there are differences between species with regard to the effects of gonadotropin on the ovary and the apparent relative importance of the different ovarian secretions on LH and FSH secretion. The following description emphasizes some general principles that likely apply to many mammalian species, although the experimental evidence to support such principles may have been obtained in only one or a few species.

A. The Follicular Phase

Circulating estrogen levels increase during the follicular phase of the ovarian cycle when one or more follicles containing an ovum grow in preparation for ovulation. This growth includes proliferation of granulosa cells adjacent to the ovum and differentiation of mesenchymal cells in the ovarian stroma to form the theca interna which surrounds the follicle. Initially, FSH induces the granulosa cells to synthesize aromatase. Subsequently, theca interna cells synthesize androstenedione and, to a lesser extent, testosterone in response to LH. The androgens diffuse to the granulosa cells, where they are aromatized to estrogen. Estrogen is secreted into the follicular fluid located inside the antrum or cavity within the follicle, where it acts in high concentration in concert with other follicular factors to regulate growth of the follicle. Some of this estrogen also enters the blood. Although a principal function of the theca interna cells is to provide androgen as substrate for estrogen production by the granulosa cells, the theca interna cells do synthesize some aromatase and secrete some estrogen as well. Theca interna cells from follicles that failed to mature fully and underwent atresia in previous cycles collect in the stroma of the ovary as interstitial cells, forming pockets of cells called interstitial glands. These cells also secrete androgen and estrogen in response to LH. Estrogen release does increase episodically in response to LH pulses. Estradiol-17β is the principal estrogen secreted, and it is inactivated within the liver. In women, it has a half-time for disappearance from blood of approximately 30 min.

These episodic increases in circulating estrogen during all but the last portion of the follicular phase of the cycle exert negative feedback on LH and FSH release by actions exerted at brain and APG. When mammals are ovariectomized during the follicular phase, there is an increase in the amplitude of LH and FSH pulses without any appreciable change in frequency and an increase in mean circulating levels of both gonadotropins. Estrogen replacement in short-term ovariectomized mammals suppresses the magnitude of the gonadotropin pulses, restores circulating LH to low levels, and partially lowers circulating FSH. As demonstrated in sheep, estrogen acts to suppress the amplitude of LHRH pulses. Currently, there is no evidence to indicate that any action of estrogen at brain to alter LHRH release is due to a direct action on LHRH neurons. As demonstrated in sheep and several other mammals, estrogen also suppresses the LH and FSH responses of gonadotropes to LHRH by direct action at the APG. In sheep, high-frequency, low-amplitude LHRH pulses characterize the follicular phase of the cycle. This pattern of LHRH secretion coupled with estrogen negative feedback action on gonadotropes functions to hold

gonadotropin secretion in check while estrogen concentration rises and follicular growth proceeds. The failure of estrogen to suppress FSH as effectively as it suppresses LH and the fact that circulating FSH levels rise sooner after ovariectomy than do those of LH can be attributed to the loss of ovarian inhibin secretion (see Section VI,D). Ovarian feedback control of LH and FSH release during the follicular phase is diagrammed in Fig. 4.

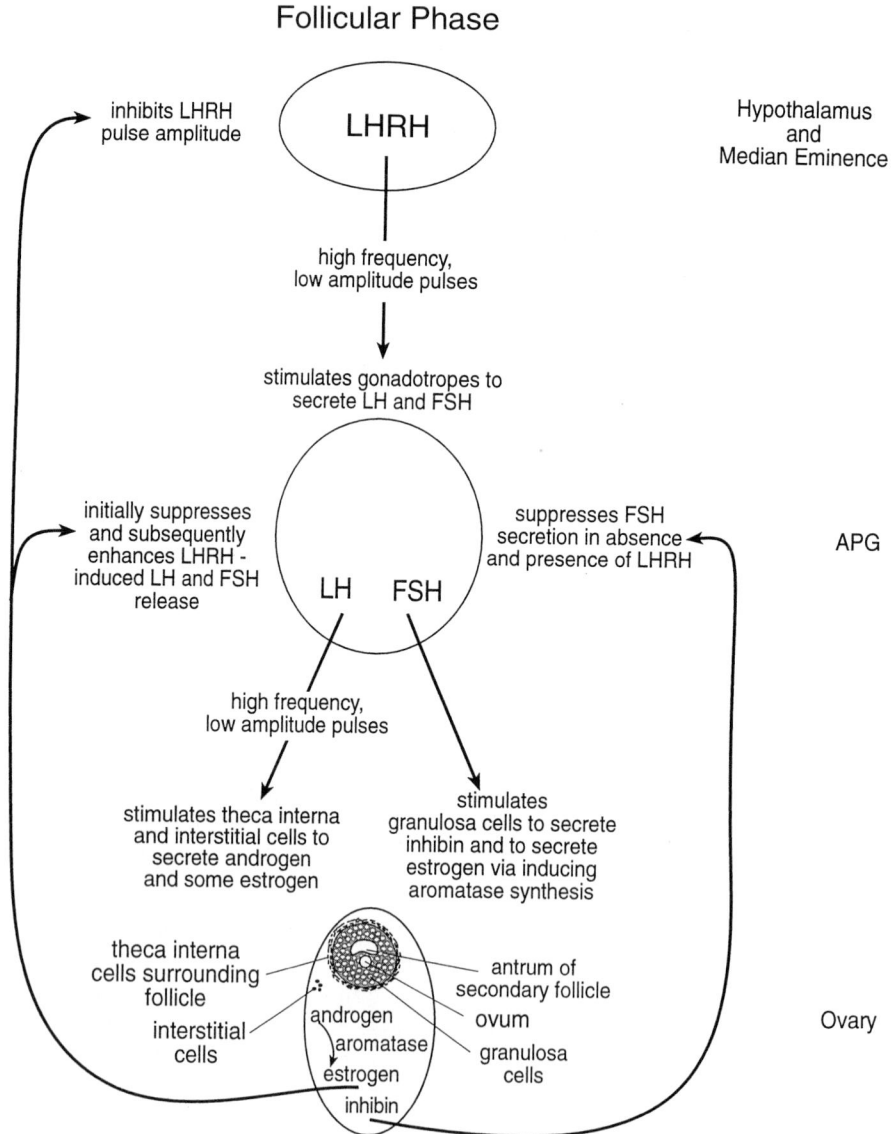

FIGURE 4   This diagram shows established interrelationships between LHRH, LH, FSH, estrogen, and inhibin in some species during the follicular phase of the estrous or menstrual cycle.

B. The Preovulatory Period

Near the end of the follicular phase of the cycle, there is a marked elevation in circulating estrogen concentration as a result of increased secretion by granulosa cells in the large mature follicle(s) destined to ovulate. In some species, the theca interna cells likely contribute to this increase in circulating estrogen. When estrogen reaches sufficiently high levels in the blood that are maintained for a sufficient length of time of about one and a half days, the steroid initiates positive feedback effects on gonadotropin release that result in marked increases in circulating LH and FSH. These surges cause ovulation and corpus luteum formation.

1. Positive Feedback at the Brain

Estrogen exerts positive feedback on gonadotropin release by acting at the brain to mediate increased LHRH release. This initiates the LH surge and accompanying FSH surge. There may be species differences with regard to whether this increase in LHRH release takes on the form of an increase in frequency of LHRH pulses of similar magnitude, an increase in frequency of LHRH pulses of increased magnitude, an increase in amplitude of LHRH pulses with no change in frequency, or continuous release of LHRH with or without accompanying LHRH pulses. Actual measurements of portal vein LHRH concentrations in sheep during the entire LH surge support the latter explanation. The release of LHRH is continuous as determined by assay of LHRH in consecutive samples collected continuously over 30-sec periods. Some fluctuations in LHRH levels were detected that could reflect desynchronization of LHRH neurons or superimposition of pulsatile LHRH release on continual LHRH release.

2. Positive Feedback at the APG

Estrogen also exerts positive feedback effects on gonadotropin release by acting at the level of the APG. Circulating levels of estrogen increase during the follicular phase in mammals, and there is an eventual loss of estrogen negative feedback at the APG to suppress LHRH-induced LH release. During the progression of the follicular phase an increase in the LH released following a single injection of LHRH

occurs. There is abundant evidence to indicate that elevated levels in circulating estrogen during the latter part of the follicular phase of the cycle act at the APG initially to cause a disappearance of estrogen's effect to suppress LHRH-induced LH release and subsequently to increase the responsiveness of the gland to LHRH. Furthermore, after the onset of the endogenous LHRH surge, there is a dramatic increase in LH release in response to a single injection of LHRH. Although the positive effect of estrogen to enhance LHRH-induced gonadotropin release occurs prior to the onset of the gonadotropin surges, circulating gonadotropin concentrations do not increase because the increase in circulating estrogen acts to suppress LHRH pulse amplitude. Ovarian feedback control of LH and FSH release during the preovulatory period is diagrammed in Fig. 5.

The marked increased responsiveness of the APG to release gonadotropin in response to LHRH after the onset of the preovulatory gonadotropin surges has been termed the LHRH self-priming response. It has been demonstrated in rats in which the preovulatory LHRH surge has been blocked with central nervous system-acting drugs. When exogenous LHRH of fixed dose is pulsed at 60-min intervals, the gonadotropes release a considerably greater amount of LH in response to the second pulse of LHRH. Similarly, continuous, constant-rate LHRH infusion causes some LH and FSH release for a period of time. This is followed by a marked increase in plasma LH and a smaller increase in FSH concentrations starting approximately 40 min after the onset of infusion. The doses of LHRH used to demonstrate LHRH self-priming caused ovulation and resulted in increased stimulation of the gonadotropes as occurs spontaneously during the preovulatory LH and FSH surges. The time delay for expression of LHRH self-priming after initial exposure to increased LHRH is compatible with the time needed to induce protein synthesis. Moreover, it is well established that LHRH stimulates LH and FSH synthesis, and estrogen facilitation of LHRH-induced LH release in culture is blocked by a protein synthesis inhibitor. Thus, the self-priming mechanism may represent a marked increase in newly synthesized gonadotropin that is readily releasable. Alternatively, it may represent a marked increase in synthesis of protein involved in release of

Preovulatory Period

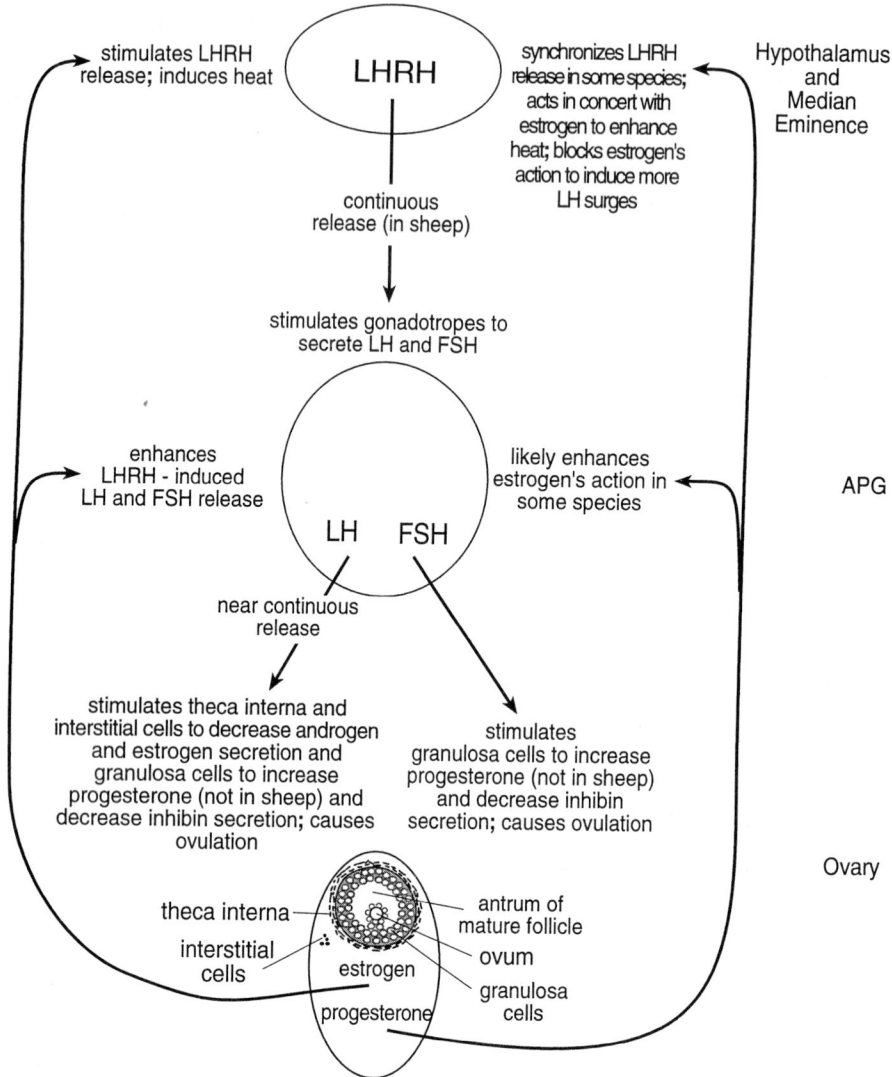

FIGURE 5 This diagram shows established interrelationships between LHRH, LH, FSH, estrogen, and progesterone in some species during the preovulatory period of the estrous or menstrual cycle. The listed actions of progesterone on hypothalamus and median eminence probably are exerted within those areas and not elsewhere in the brain.

either newly synthesized gonadotropin or the ample stores that are present within the APG throughout the estrous cycle. Although LHRH self-priming can be demonstrated on other days of the estrous cycle, the magnitude of the response is greatest near the end of the follicular phase on the day the preovulatory gonadotropin surges normally occur and this enhancement of the LHRH self-priming response has been shown to be related to increased circulating

levels of estrogen. Therefore, the preovulatory LH and FSH surges likely begin solely as the result of increased LHRH release that acts on gonadotropes already sensitized by estrogen to respond to this increase in neural signaling. Presensitization of gonadotropes by estrogen results initially in some amplification of gonadotropin release in response to increased LHRH. It also results in markedly more gonadotropin release in response to continued increased LHRH release via amplification of the LHRH self-priming mechanism.

3. *Reflex Ovulators*

Some mammals, such as cat, ferret, and rabbit, are reflex ovulators. Their estrous cycles spontaneously arrest at the end of the follicular phase because they do not have spontaneous gonadotropin surges. Circulating estrogen levels are high at this time and their action within brain induces estrus or heat which can be maintained for a considerable length of time in the presence of elevated circulating estrogen. Olfactory, visual, and auditory sensory stimulation associated with the presence of a male and, more important, tactile stimulation associated with mating activate neural pathways that ultimately stimulate LHRH release. This causes the midcycle gonadotropin surges and resumption of the estrous cycle.

4. *Timing of the Gonadotropin Surges*

The surges of LH and FSH start at the same time in response to LHRH and in some species, including sheep and human, end at the same time. By contrast, in other species with a short estrous cycle, such as rat and hamster, there is either an extended period of enhanced FSH release or a separate FSH surge immediately after the end of the LH surge. In the ewe there is a secondary FSH surge that starts about one day after the end of the first surge. Although the LH surge and the accompanying FSH surge depend on the immediate presence of LHRH stimulation, the secondary FSH surge is not dependent on the immediate presence of LHRH. Women have a separate FSH surge during the transition from the luteal to the follicular phase of the cycle. The secondary or separate FSH surges function to stimulate and thus signal a new cohort of follicles to start developing in the next cycle.

In sheep, the declining phase of the LH surge is associated with a loss of sensitivity of gonadotropes to LHRH. This is evidenced by the maintenance of continuously high hypophysial portal vein LHRH concentrations for a considerable period of time after circulating LH concentrations have returned to presurge levels. This may represent downregulation of LHRH receptors. In rats, the declining phase of the LH surge is due in part to decreased sensitivity of

gonadotropes to LHRH. Additionally, the declining phase in rats and hamsters is due to a shutdown of the brain conditions that altered LHRH secretion in the first place to stimulate the LH surge. It is not known whether a brain mechanism becomes active to stimulate LHRH release or inactive to allow its release to occur.

In rodents there is a potential activation period for the LH surge that is linked to the lighting schedule. The beginning of this period or the time that the preovulatory gonadotropin surges normally start during the afternoon of proestrus is set by the light:dark schedule in which rodents are maintained. By administering central nervous system-acting drugs such as nicotine, one can delay the onset of the LH surge 1 or more hours into the evening of proestrus. With progressive delay in the time of onset of the surge, there is a progressive attenuation of the length of time the surge persists and in all situations it ends at the same time during the late evening of proestrus. When drugs are utilized to block the LH surge during the entire potential activation period, the LH surge is delayed 24 hr from the time it normally would have started. The mechanism involved in this 24-hr rhythmicity is not known, but it likely involves a reversal of conditions that ended the potential activation period. Estrogen can stimulate daily LH surges during this window or potential activation period in ovariectomized rodents. However, the preovulatory surge in progesterone secretion (*vide infra*) prevents this from happening during the estrous cycle even if circulating estrogen concentrations are maintained at artificially elevated levels. Although rodents and other species of mammals may share similar brain mechanisms that result in termination of the LH surge, the onset of preovulatory gonadotropin surges is not linked to the time of day in most mammals.

5. *Gonadotropin Isoforms*

Increased circulating estrogen levels during the follicular phase of the cycle likely alter the isoforms of gonadotropin secreted during the surges. The half-times for disappearance of circulating LH and FSH are lengthened after ovariectomy. Estrogen replacement shortens the half-times and decreases sialylation of these gonadotropins in the APG of rodents and rhesus monkey. This appears to be of physiological significance during the estrous and menstrual cycle. During the beginning of the follicular phase, when estrogen levels are low, gonadotropes synthesize more heavily sialylated isoforms which likely function to exert long-term effects on developing follicles. As follicular estrogen secretion increases, less sialylated isoforms are synthesized which likely function to exert more potent biological activity. Massive amounts of poorly sialylated gonadotropin secreted during the preovulatory period would serve to maximize the biological effect of gonadotropin on the ovary to expedite luteinization and ovulation.

6. Gonadotropin Surges and Steroid Secretion

As its name implies, LH acts to cause luteinization. With onset of the preovulatory LH surge, LH initiates changes in both granulosa and theca interna cells that lead to corpus luteum formation. These changes include altered steroid secretion by these cells. During the follicular phase of the cycle granulosa cells secrete small amounts of progesterone in response to FSH. During the latter part of the follicular phase of the cycle, granulosa cells acquire LH receptors, but progesterone secretion changes little. Estrogen functions within the ovary to suppress progesterone secretion during the follicular phase. However, virtually immediately after the start of the LH surge, there is a marked increase in granulosa cell progesterone release that continues for several hours in many species, including rat and human but not sheep. Concomitant with the initial rise in circulating progesterone, there are brief increases in circulating testosterone and estrogen concentrations followed by declines to low levels. The decline in estrogen secretion is due to a decline in androgen synthesis.

Some actions of progesterone amplify or refine the effects of estrogen regardless of whether they are negative or positive. In rodents, there probably are at least five functions of this preovulatory increase in circulating progesterone. First, progesterone acts on brain to synchronize the effects of estrogen that cause the LHRH surge (Fig. 5). Appropriately timed injections of progesterone can advance the onset of the LH surge by about 1 hr or by 1 day. Progesterone

action shortly after the onset of the LH surge may result in an increased magnitude of the gonadotropin surges by synchronizing LHRH release. Second, in some species progesterone likely acts at the pituitary gland to enhance estrogen-mediated increased sensitivity to LHRH. Such action also would function to magnify the rise in circulating LH and FSH. Third, it acts on the estrogen-primed brain to enhance estrus. The female rat will mate with a male approximately 3 hr after the onset of the progesterone surge. Fourth, it is known that progesterone acts to decrease the frequency of LHRH pulses and by inference it may act to suppress the frequency of LHRH pulses to cause the decrease in frequency of LH pulses observed after the end of the LH surge. Fifth, as mentioned previously, it acts at the brain to prevent any further estrogen-induced daily LH surges. This effect may be mediated by its action to suppress the frequency of LHRH pulses.

C. The Luteal Phase

The LH surge and associated FSH surge likely act together to cause ovulation although either hormone alone is effective. After ovulation, the granulosa and theca interna cells remaining in the lining of the follicle complete their transformation into lutein cells and proliferate to form the corpus luteum. However, it should be noted that corpus luteum formation will occur in response to the gonadotropin surges even if ovulation is experimentally blocked. During the luteal phase of the cycle circulating concentrations of estrogen normally are low but sufficient in combination with progesterone to exert negative feedback on gonadotropin release. Primates are an exception in that the corpus luteum does secrete substantial amounts of estrogen. Those high levels of estrogen also function in combination with progesterone to exert negative feedback on gonadotropin release.

The corpus luteum in all mammals secretes large amounts of progesterone and in some species this occurs in the absence of concurrent gonadotropin stimulation. This progesterone functions to prepare the uterus for implantation. It also acts in the presence of low levels of circulating estrogen to suppress the frequency and increase the amplitude of LHRH pulses. The increase in amplitude of the LHRH pulses may result from increased accumulation of LHRH in neurons during the longer interpulse interval. In sheep, exogenous estrogen cannot induce an LH surge under these conditions. These changes in LHRH secretion are reflected in comparable changes in LH and FSH secretion. For example, in women LH and FSH pulses are of low frequency and high amplitude during the luteal phase of the cycle compared to those observed during the follicular phase.

Ovarian feedback control of LH and FSH release during the luteal phase is diagrammed in Fig. 6.

If pregnancy does not occur, the mammalian corpus luteum stops producing progesterone which had been functioning to suppress the frequency of LHRH pulses. This signals the onset of the follicular phase, with higher frequency, lower amplitude LH pulses occurring until the time of the preovulatory gonadotropin surges. If pregnancy occurs, there is stimula-

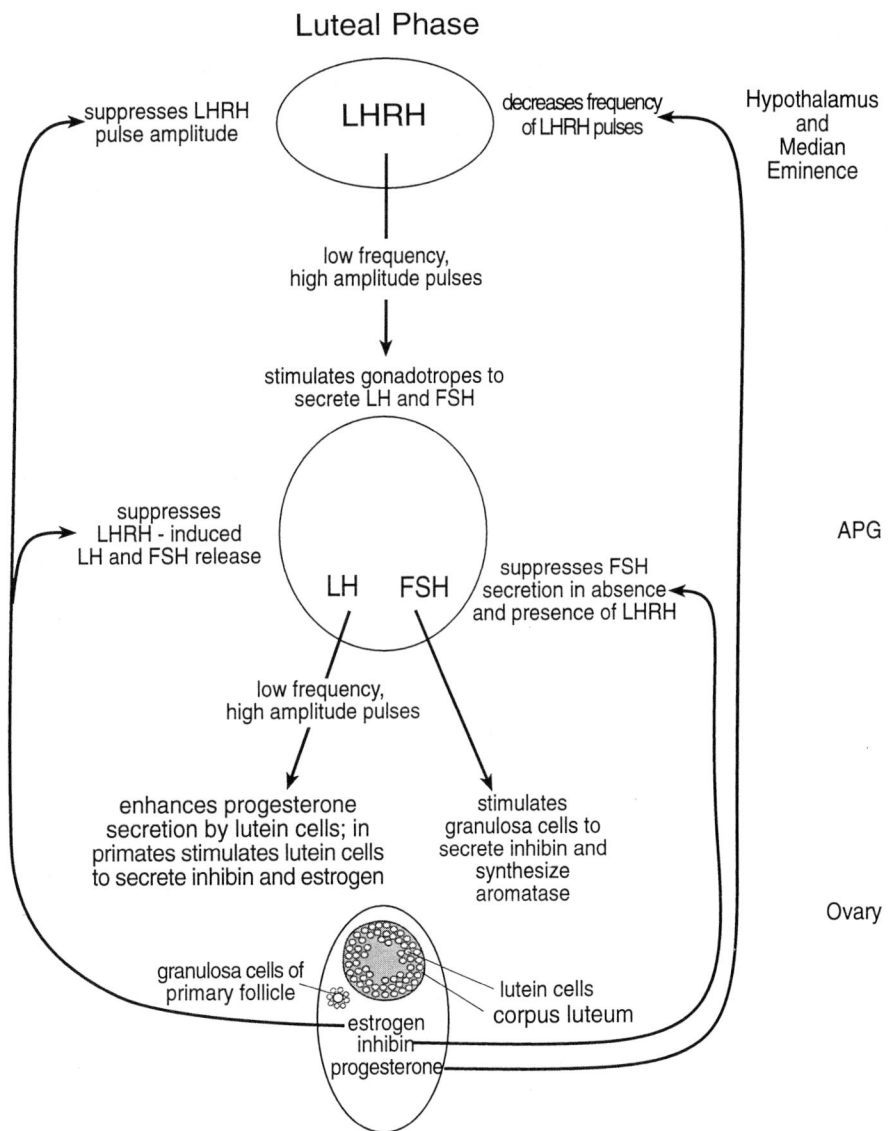

FIGURE 6 This diagram shows established interrelationships between LHRH, LH, FSH, estrogen, progesterone, and inhibin in some species during the luteal phase of the estrous or menstrual cycle.

tion of the corpus luteum to prolong its life span. In primates and equids this stimulating factor is chorionic gonadotropin (CG).

D. Inhibin

Inhibin functions in females as it does in males—it inhibits LHRH-dependent and LHRH-independent FSH secretion. It is secreted by granulosa cells in response to FSH and LH after the granulosa cells acquire LH receptors during the latter part of the follicular phase of the cycle. Inhibin secreted in response to FSH functions in a classical negative feedback system to suppress FSH secretion and thereby control follicular growth. Inhibin also acts within the ovary on theca interna cells to enhance androgen secretion in response to LH. Activin exerts the opposite effect.

In some species, circulating inhibin levels drop dramatically after the start of the gonadotropin surges in response to either LH or FSH. This effect of FSH to decrease rather than increase inhibin secretion, which results in an increase in FSH secretion, is an example of false-positive feedback. It may represent "disappearance" of the cells that secrete inhibin as granulosa cells change to become granulosa–lutein cells. The corpus lutea in these species secrete little or no inhibin. On the other hand, although primates also secrete less inhibin during the preovulatory period, they do secrete inhibin from the corpus luteum under LH stimulation. In women, inhibin levels are highest during the luteal phase of the cycle, which may contribute to the very low circulating FSH concentrations during this period.

VII. PREFERENTIAL LH OR FSH SECRETION

In general, changes in circulating LH and FSH concentrations parallel each other. This is not surprising because LH release is LHRH dependent, and LHRH stimulates both LH and FSH secretion. However, there are situations in which there is preferential LH or FSH release. For example, gonadotropin secretion is virtually selective for FSH release during the secondary phase of the preovulatory FSH surge in rodents. The decline in inhibin secretion mediated by either the LH surge or accompanying FSH surge results in a decline in inhibin-mediated negative feedback control of FSH release at the level of the gonadotrope. Subsequently, FSH release increases even if LHRH is neutralized prior to the onset of the secondary FSH surge by the administration of LHRH antiserum. This also has been demonstrated *in vitro*. When APGs are removed prior to the onset of the secondary FSH surge, they release markedly increased amounts of FSH in the absence of LHRH stimulation than do APGs removed prior to the time of the primary FSH surge. Interestingly, administration of activin antiserum between the two FSH surges reduces the magnitude of the secondary FSH surge. This observation, coupled with the fact that increased FSH release occurs *in vitro*, suggest a role for pituitary activin rather than circulating activin in causing the secondary FSH surge.

Other mechanisms could also be involved in causing preferential gonadotropin release. An acceleration of LHRH pulse frequency has been shown to favor LH secretion and a slowing of LHRH pulse frequency is known to favor FSH secretion under some conditions. It also should be kept in mind that failure to identify and isolate an FSH-releasing hormone should not be interpreted to mean that such a factor does not exist.

Divergences in circulating LH and FSH concentrations are not always due to the differential release of LH and FSH. A longer time for clearance of FSH than LH from the circulation can increase the circulating FSH:LH ratio in association with an increase in LHRH pulse frequency. For example, during most of the follicular phase of the human menstrual cycle the ratio of circulating FSH to LH concentrations is greater than that which occurs during the luteal phase of the cycle. This can be attributed, at least in part, to the increased frequency of LHRH-induced LH and FSH pulses during the follicular phase. Circulating FSH concentrations increase more than those of LH due to the decreased interpulse interval and a longer half-time for clearance of FSH than LH. The increased LH:FSH ratio during the luteal phase also may be explained, in part, by changes in circulating levels of inhibin. The increase in inhibin secretion during the luteal phase may increase negative feedback on FSH release.

VIII. AGE, REPRODUCTIVE STATUS, AND THE ENVIRONMENT

A. Immature Mammals

After birth there is a period during which the testes and ovaries are essentially inactive until pituitary gonadotropin secretion increases and puberty, or initial reproductive capacity, is achieved. The period of gonadal inactivity is brief in species such as rat that reach reproductive competence within 2 months after birth. By contrast, the period lasts years in primates. The lack of gonadotropin secretion is due to failure of the hypothalamus to release appreciable amounts of LHRH in a synchronous fashion. As demonstrated in nonhuman primates, the hypothalamus contains substantial amounts of LHRH, and exogenous LHRH will stimulate LH release during this period of gonadal inactivity. These data, coupled with the fact that pulsatile LH release can be demonstrated before birth, suggest that LHRH release is inhibited by some mechanism. The factors involved in arresting and in reinitiating LHRH release are unresolved.

B. Aging

In aging male mammals, there is no marked depletion of germ or endocrine cells. Testicular function declines gradually with age, but circulating testosterone concentration is reasonably well maintained. For example, in aged mice there is a slowing of LH pulse frequency and a reduction in LH pulse amplitude. This is associated with reduced frequency of episodes of increased testosterone and peak testosterone concentrations. These changes likely are due primarily to a reduction in LHRH pulse frequency. Neural aging also could explain the reduction in sexual activity that occurs in some old male mice because such decreased activity is not correlated with decreased circulating testosterone.

Depletion of ovarian follicles is a primary determinant of declining reproduction in aging females. In the female primate, menstrual cycles become irregular and cease during menopause. The ovaries depleted of their endocrine cells do not release estrogen or progesterone in response to gonadotropin. Removal of the negative feedback action of estrogen and progesterone and possibly of inhibin results in high-frequency, high-amplitude pulses of circulating gonadotropins that are more heavily sialylated. Similarly, estrous cyclicity often becomes irregular and eventually ceases in nonprimates. However, the term menopause is normally used only when referring to primates.

C. Pregnancy and Lactation

1. Pregnancy

Relatively little is known about the control of gonadotropin secretion during pregnancy. Elevated circulating levels of progesterone or structurally similar steroids function in concert with estrogen to maintain pregnancy and suppress gonadotropin secretion. Maternal recognition of pregnancy and maintenance of luteal progesterone secretion to curtail estrous or menstrual cyclicity involves different mechanisms in different species. In those with a short estrous cycle, in which the luteal phase ends before implantation takes place approximately 4 or 5 days after ovulation, mating functions to maintain luteal function. In rat, mating activates pelvic nerves which act via neural pathways to cause increased prolactin secretion that lasts for approximately 2 days. It also causes changes in the brain that are reflected by twice daily surges of circulating prolactin during early pregnancy. In rat, prolactin functions in concert with low circulating levels of LH at the corpora lutea to maintain progesterone secretion. During the latter half of pregnancy, the rat placenta secretes a prolactin-like hormone that substitutes for APG prolactin, and LH is no longer needed to maintain the corpora lutea. In other species, the luteal phase of the cycle is longer than the time required for implantation to occur and the placenta either secretes hormones that maintain luteal function or secretes progesterone and estrogen. In women, CG initially stimulates the corpus luteum to secrete both progesterone and estrogen. Placental steroids maintain pregnancy thereafter. In some species, removal of luteal tissue by ovariectomy at any time during pregnancy terminates pregnancy. In other species, ovariectomy during midpregnancy does not terminate pregnancy. In women, corpus luteum steroid secretion starts to decrease about 8 weeks after fertilization and ovariectomy at 6 weeks after fertilization does not terminate pregnancy. Thus, there are considerable differences between species with regard to stimuli that maintain luteal function and the sources of progesterone and estrogen at different times during pregnancy.

As in cyclic females, estrogen acts in pregnant mammals to exert negative feedback on APG gonadotropin secretion and induce the formation of progesterone receptors. Thus, progesterone acts in concert with estrogen at the uterus to maintain pregnancy and at the brain and possibly the APG to suppress gonadotropin secretion. Progesterone likely acts as it does during the luteal phase of the cycle to suppress LHRH pulse frequency. It may act at the APG in concert with estrogen to suppress LHRH-induced gonadotropin release because gonadotropin release in response to exogenous LHRH is severely blunted during pregnancy. However, this also could be due to decreased gonadotropin synthesis in response to suppressed LHRH release. Interestingly, pregnant horses secrete sufficient APG FSH during early pregnancy to stimulate follicular development. With increased CG secretion, ovulation occurs and additional corpora lutea are formed. These secondary corpora lutea function like the primary ones to release progesterone in response to CG.

Circulating concentrations of estrogen and progesterone decrease markedly prior to birth or, in the case of some primates, at birth with the shedding of the placenta. The decrease in estrogen concentration functions to increase milk production because estrogen no longer suppresses the stimulatory effect of prolactin on the mammary glands. For this reason, estrogen is administered to suppress milk production in women who do not wish to nurse. The decrease in circulating estrogen and progesterone results in a loss of negative feedback on gonadotropin secretion. Although suckling and the maintenance of lactation has no apparent effect on resumption of estrous cyclicity in some species such as the marmoset monkey, it influences cyclicity in other species in various ways.

2. Lactation

A preovulatory surge of LH and FSH occurs in the adult female rat within 1 day after delivery. It is linked to the time of birth rather than to photoperiod. If fertilization takes place, suckling delays implantation until circulating estrogen concentration rises during midlactation in response to LH. Regardless of whether fertilization occurs, estrous cyclicity ceases in response to suckling during lactation. The strength of the suckling stimulus determines to a large extent how long estrous or menstrual cyclicity is suppressed. Although nursing is reasonably effective as a birth control in women, particularly during early lactation when suckling frequency is greatest, it should be noted that a small percentage (<10%) of women who engage in intercourse during the first 6 months of nursing become pregnant.

The mechanisms and principles involved in the control of gonadotropin secretion during lactation in the rat may be important in many other species as well. Suckling functions to suppress estrous cyclicity for approximately 20 days in lactating rats suckling the average number of 9 or 10 pups. Suckling frequency decreases around midlactation, and this permits changes in gonadotropin secretion that will result in the resumption of estrous cyclicity. Suckling suppresses estrous cyclicity by suppressing gonadotropin secretion. During the first half of lactation, suckling acts via neural pathways to virtually block LH secretion and suppress FSH secretion. This action is exerted on brain to inhibit LHRH release. Lactating rats maintain the capacity to respond to estrogen with an LH surge and to release LH in response to exogenous LHRH although both these responses are markedly suppressed. These suppressed responses likely are due to decreased gonadotropin synthesis in the absence of LHRH stimulation. The inhibitory effect of suckling on LHRH release does not require suckling-induced prolactin release or prolactin-stimulated progesterone release from the corpora lutea during the first half of lactation.

As suckling frequency decreases at midlactation in rats, LH secretion increases to a small extent and it replaces the function of prolactin on maintaining the corpora lutea. However, LH secretion remains suppressed until late lactation, and there is evidence that suckling suppresses LH secretion during this time primarily by increasing circulating prolactin and progesterone levels. Whereas FSH secretion is suppressed during early lactation presumably because LHRH does not maintain FSH synthesis, circulating FSH concentrations are not suppressed during the second half of lactation when some LHRH is released and circulating inhibin levels are similar to those that are present during the luteal phase of the estrous cycle.

D. Seasonal Breeders

Many mammals are seasonal breeders. In fact, most species including human show at least some subtle changes in gonadal activity throughout the year. Although mammals in the wild usually synchronize their reproductive activity to changes in temperature, photoperiod, or food supply, other factors also may be involved. There are seasonal changes in the reproductive capacity of males as well as females. Seasonal breeding is an adaptation that ensures that birth occurs at a particular time of year that is compatible with raising young in the wild.

1. Males

The extent of seasonal change in testicular activity in males varies markedly between species. For example, it is dramatic in the pipistrelle bat in which testes size is decreased more than 90% in the inactive phase. Such changes in male mammals are associated with decreased gonadotropin secretion, which results in decreased testosterone production. By contrast, as demonstrated in Europe, seasonal changes in men are minimal and reflected by tendencies for slight decreases in testosterone secretion, sperm ejaculate number and volume, and frequency of intercourse in winter and spring. Changes in the Soay sheep are intermediate but sufficient to limit breeding to a particular time of year. The mechanisms involved in suppressed gonadotropin secretion in this species are fairly well understood.

In autumn, during the breeding season in Soay sheep, high-frequency LHRH pulses (about one every 2 hr) occur. These stimulate increased LH and FSH release which maintain testicular function and sex behavior. With increased daylight during the photoperiod in the transition from winter to spring, LHRH pulse frequency decreases to approximately one every day. This results in a decreased frequency of gonadotropin pulses, an increase in LH pulse amplitude, low testosterone concentrations, low sperm production, and a marked decrease in libido. Moreover, the response of the testes to release testosterone in response to a fixed amount of LH is markedly suppressed when the testes are regressed. In Soay sheep, the primary environmental stimulus for change in testicular function is a change in photoperiod, and this is mediated via neural pathways from the retina of the eye that alter melatonin secretion by the pineal gland. Light inhibits melatonin secretion. In Soay sheep, short photoperiod during autumn and an extended daily period of increased melatonin secretion are associated with high-frequency LHRH pulses. There is some evidence that melatonin may act within the medial basal hypothalamus to mediate its effects. Although species differences exist with regard to changing environmental conditions which impair reproductive function in seasonal breeders, a known effect is decreased LH pulse frequency and FSH secretion which ultimately alter testicular function. This probably occurs as a result of decreased LHRH pulse frequency in most, if not all, seasonal breeders.

2. Females

Within a species, mechanisms causing seasonal changes in gonadotropin secretion in females appear to be similar to those in males. For example, in ewes, LHRH and LH pulse frequency is low and LH pulse amplitude is high during anestrus in spring compared to that observed during the breeding season in autumn. Like in rams, these changes are the result of photoperiod-induced changes in pineal gland melatonin secretion. Evidence suggests that the pattern of melatonin secretion in anestrus, but not that during the breeding season, acts to maintain an estrogen-sensitive neural pathway that inhibits LHRH pulse frequency. Estrogen is not effective at suppressing LHRH pulse frequency during the breeding season. It is possible that melatonin also maintains a comparable inhibitory neural pathway in rams during the nonbreeding season that is activated by testosterone directly or via its aromatization to estrogen.

Interestingly, social cues and male pheromones

can stimulate LH release in some mammals, including seasonal breeders in anestrus. For example, exposure of an anestrous ewe to a new ram can increase LHRH pulse frequency which initiates the follicular phase of the estrous cycle. The pheromone(s) involved is a chemical secreted by sweat glands in skin in response to testosterone stimulation. It activates olfactory neural pathways in ewes that cause an increase in LH pulse frequency within minutes. This may involve inactivation of the estrogen-sensitive neural pathway that inhibits LHRH pulse frequency during anestrus.

E. Exercise, Food Consumption, and Stress

In addition to regularly occurring environmental influences that cause seasonal changes in gonadotropin secretion, environmental conditions can suppress gonadotropin secretion at any time of year in species that are continuously reproductively active. For example, although LH release in laboratory rats is unaffected by a variety of stressors that are associated with physical distress, immobilization which may be perceived by rats as especially stressful can temporarily block LH release. Of particular interest, humans are susceptible to developing impaired gonadotropin secretion as a result of lifestyle and societal influence. Chronic strenuous exercise, excessive dieting, and psychogenic stress can markedly impair gonadotropin secretion and compromise reproduction. The effects are not nearly as dramatic in men as they can be in women and appear to involve neural pathways that suppress LHRH pulse frequency.

In men, gonadotropin secretion usually is sufficiently maintained to support testicular function in response to adverse environmental or psychological conditions. Some regimens of endurance exercise, fasting, and some types of emotional stress reduce LH pulse frequency and circulating testosterone concentrations. These conditions may lead to impotence which is reversible. Otherwise, reproductive processes appear to be maintained at basal levels, although a prolonged period of impaired LH secretion due to chronic food deprivation can cause gonadal atrophy and sterility. These effects of food deprivation on impaired testicular function resulting from decreased LH secretion are separate from effects attributable to lack of vitamin A or other dietary factors that are utilized by the testis to maintain sperm production. There is evidence to suggest that the impaired LH secretion that occurs in all these situations is the result of decreased LHRH pulse frequency. Chronic undernutrition in boys also may delay the pubertal increase in LHRH release and sexual maturation. On the other hand, participation in sports does not alter the onset of sexual maturation, and some activities actually increase testosterone secretion in adults. Short-duration exercise conducted at maximal intensity, such as that which occurs with lifting heavy weights repetitively, increases circulating testosterone concentrations within minutes. This latter effect declines with age and appears to be absent in men in their seventies.

The mechanisms involved in suppressing LHRH release in response to endurance exercise, fasting, or stress are not known. With endurance exercise and fasting, a decrease in the basal metabolic rate in association with a depletion of energy reserves may be a trigger. With emotional stress, increased thought processes may activate neural circuitry inhibitory to LHRH release. Additional mechanisms also may suppress LHRH release. One possibility is an increase in circulating levels of prolactin.

Strenuous exercise, low blood sugar, or stress cause increased circulating prolactin concentrations. This may lead to impotence, which is readily reversible after prolactin returns to normal levels. Continuous elevation of circulating prolactin at high levels, which occurs in men with APG prolactin-secreting tumors, acts to cause impotence, lack of libido, suppressed gonadotropin secretion, and impaired fertility or even sterility. The loss of libido is not due solely to impaired testosterone secretion and likely represents a direct action of prolactin on the brain. As demonstrated in rats, hyperprolactinemia suppresses gonadotropin secretion by suppressing LHRH pulse frequency. These effects of hyperprolactinemia are reversed if blood prolactin levels are lowered, and it is probably rare that exercise, dieting, or stress elevate circulating prolactin levels over a sufficient length of time to impair male reproduction to the extent observed in men with prolactin-secreting tumors.

In women, exercise, insufficient food consumption, or stress can shorten the luteal phase of the menstrual cycle, cause anovulatory menstrual cycles, or even suppress gonadotropin secretion to the extent that menstrual cyclicity does not occur. In girls, chronic extensive exercise may cause temporary primary amenorrhea, i.e., a delay in menarche (the first menstrual period). In women, it can cause secondary amenorrhea (the interruption of menstrual periods after menarche). These effects do not appear to be due to exercise-induced increases in pituitary gland secretion of prolactin or β-endorphin nor to the amount of body fat present at the onset of exercise. Although the incidence of short luteal phases is high in response to extensive exercise, that of amenorrhea is relatively low unless the exercise is associated with body weight loss and a decrease in the basal metabolic rate. The extent that exercise is metabolically draining may be a critical factor in determining the extent of menstrual cycle alteration. Excessive weight loss associated with dieting also can lead to amenorrhea. Women with anorexia nervosa who have severely reduced food intake usually do not have menstrual cyclicity. Situations in which there is a reduction in the basal metabolic rate and lack of reproductive cyclicity may reflect physiological adaptations rather than physiological dysfunctions. Metabolism may be shifted from some normal body processes including reproduction to the processes involved in adaptation to exercise or to the maintenance of bodily functions necessary for survival. The amenorrhea associated with exercise and low food consumption is likely due to a reduction in LHRH pulse frequency. Psychogenic stress also can cause amenorrhea by interfering with LHRH release. In new stressful situations, the neural pathways likely involve the creation of stressful thoughts in the frontal cortex, their amplification to extremely disturbing emotional levels in the limbic system, and transformation of these neuronal perceptions as a physical response by the hypothalamus which suppresses LHRH release. The alterations in menstrual cyclicity that result from exercise, low food consumption, or psychogenic stress appear to be reversible if the causative condition is eliminated. However, other effects resulting from decreased estrogen secretion in response to compromised gonadotropin secretion, such as the weakening of bone structure that occurs with osteoporosis, may be permanent.

See Also the Following Articles

Anterior Pituitary; Critical Period, Estrous Cycle; FSH (Follicle-Stimulating Hormone); GnRH (Gonadotropin-Releasing Hormone); GnRH Pulse Generator; Gonadotropin Biosynthesis; LH (Luteinizing Hormone); Pituitary Gland, Overview; Testosterone Biosynthesis

Bibliography

Blake, C. A. (1974). Differentiation between the "critical period," the "activation period" and the "potential activation period" for neurohumoral stimulation of LH release in proestrous rats. *Endocrinology* 95, 572–578.

Bonen, A. (1994). Exercise-induced menstrual cycle changes: A functional, temporary adaption to metabolic stress. *Sports Med.* 17, 373–392.

Bousfield, G. R., Perry, W. M., and Ward, D. N. (1994). Gonadotropins: Chemistry and biosynthesis. In *The Physiology of Reproduction* (E. Knobil and J. D. Neill, Eds.), 2nd ed., Vol. 1, pp. 1749–1792. Raven Press, New York.

Chappel, S. C., Ulloa-Aguirre, A., and Coutifaris, C. (1983). Biosynthesis and secretion of follicle-stimulating hormone. *Endocr. Rev.* 4, 179–211.

Crowley, W. F., Jr., Whitcomb, R. W., Jameson, J. L., Weiss, J., Finkelstein, J. S., and O'Dea, L. St. (1991). Neuroendocrine control of human reproduction in the male. *Rec. Prog. Horm. Res.* 47, 27–67.

DePaolo, L., Mercado, M., Guo, Y.-L., and Ling, N. (1994). Follicle-stimulating hormone: Control from within. In *Frontiers in Endocrinology: Volume 3, Inhibin and Inhibin-Related Proteins* (H. Burger, J. Findlay, D. Robertson, D. deKretser, and F. Petraglia, Eds.), Ares-Serono Symposia, Rome, pp. 115–126.

D'Occhio, M. J., Schanbacher, B. D., and Kinder, J. E. (1982). Relationship between serum testosterone concentration and patterns of luteinizing hormone secretion in male sheep. *Endocrinology* 110, 1547–1554.

Elias, K. A., and Blake, C. A. (1981). A detailed *in vitro* characterization of the basal follicle-stimulating hormone and luteinizing hormone secretion rates during the rat four-day estrous cycle. *Endocrinology* 109, 708–713.

Ellis, G. B., and Desjardins, C. (1982). Male rats secrete luteinizing hormone and testosterone episodically. *Endocrinology* 110, 1618–1627.

Freeman, M. E. (1994). The neuroendocrine control of the ovarian cycle of the rat. In *The Physiology of Reproduction*

(E. Knobil and J. D. Neill, Eds.), 2nd ed., Vol. 2, pp. 613–658. Raven Press, New York.

Goodman, R. L. (1994). Neuroendocrine control of the ovine estrous cycle. In *The Physiology of Reproduction* (E. Knobil and J. D. Neill, Eds.), 2nd ed., Vol. 2, pp. 659–709. Raven Press, New York.

Gorski, R. A., Harlan, R. E., and Christensen, L. W. (1977). Perinatal hormonal exposure and the development of neuroendocrine regulatory processes. *J. Toxicol. Environ. Health* 3, 97–121.

Hotchkiss, J., and Knobil, E. (1994). The menstrual cycle and its neuroendocrine control. In *The Physiology of Reproduction* (E. Knobil and J. D. Neill, Eds.), 2nd ed., Vol. 2, pp. 711–749. Raven Press, New York.

Knight, P. G. (1996). Roles of inhibins, activins, and follistatin in the female reproductive system. *Frontiers Neuroendocrinol.* 17, 476–509.

Levine, J. E., Bauer-Dantoin, A. C., Besecke, L. M., Conaghan, L. A., Legan, S. J., Meredith, J. M., Strobl, F. J., Urban, J. H., Vogelsong, K. M., and Wolfe, A. M. (1991). Neuroendocrine regulation of the luteinizing hormone-releasing hormone pulse generator in the rat. *Rec. Prog. Horm. Res.* 47, 97–153.

McNeilly, A. S. (1994). Suckling and the control of gonadotropin secretion. In *The Physiology of Reproduction* (E. Knobil and J. D. Neill, Eds.), 2nd ed., Vol. 2, pp. 1179–1212. Raven Press, New York.

Rhim, T.-J., Kuehl, D., and Jackson, G. L. (1993). Seasonal changes in the relationships between secretion of gonadotropin-releasing hormone, luteinizing hormone, and testosterone in the ram. *Biol. Reprod.* 48, 197–204.

Turner, C. D., and Bagnara, J. T. (1976). *General Endocrinology*, 6th ed. Saunders, Philadelphia.

Gonads

see Ovary, Overview; Testis, Overview

Gonadotropins, Overview

M. Ram Sairam

Clinical Research Institute of Montreal

I. Introduction and Nomenclature
II. Sources of Gonadotropins
III. Structure and Function
IV. Molecular Biology of Gonadotropins: Mutations and Animal Models
V. Signal Transduction and Mechanism of Action
VI. Medical and Veterinary Use of Gonadotropins
VII. Perspectives and Future Directions

GLOSSARY

gametogenesis The development of the ovum in the ovary and sperm production by the testis.

glycoproteins Complex proteins with covalently attached sugars on the surface of the molecule; they perform a wide variety of functions.

gonadotropins Hormones secreted by the pituitary and placenta (in some animals) to sustain reproductive function.

pituitary and *placenta* Two important sources of gonadotropins and considered to be master glands of the endocrine orchestra.

recombinant protein Products engineered for large-scale production by manipulating genes.

signal transduction The process by which the chemical message contained in a molecule is transmitted to function inside a living cell.

steroidogenesis The process of synthesis and secretion of sex steroid hormones by appropriate cells in the ovary and testis following stimulation by gonadotropins.

Gonadotropin is derived from the Greek term "tropos" (meaning turning on) to indicate an agent that initiates and sustains the function of gonads, the ovary and testis, which are the principal sex organs in the body. These are complex proteins secreted by the pituitary and placenta into the bloodstream to stimulate the gonads in the lower part of the abdomen. While they are not essential for life, their concerted actions or sometimes sequential effects are critical for propagation of species. Many disorders of infertility in humans are due to insufficient or inappropriate secretion or action of gonadotropins and thus requiring therapeutic intervention. The field of study on gonadotropins is vast, involving various disciplines in life sciences extending from biochemistry to physiology, pharmacology, cell biology, molecular biology, medicine, veterinary science, and manufacturing. These studies integrate knowledge derived from biochemical methods of purification of the hormones, structure, signal transduction, and molecular biology into useful therapeutic application to humans and animals. This vast knowledge has contributed to numerous advances in treatment of infertility and development of novel contraceptive methodologies and fostered a multimillion dollar enterprise for manufacturing fertility agents.

I. INTRODUCTION AND NOMENCLATURE

The anterior pituitary gland secretes in a cyclic manner a number of hormones into the blood, of which two are required to stimulate the growth and activity of the gonads. These are called gonadotropic hormones or gonadotropins to designate a family of agents. In human and other primates and some other species, during pregnancy the placenta also secretes a powerful gonadotropin to sustain pregnancy and protect the fetus. The pituitary gonadotropins are usually referred to by their common names, such as follicle-stimulating hormone (FSH) and luteinizing hormone (LH). The names follitropin and lutropin respectively are also employed in the scientific literature to describe the previously mentioned hormones. Other designations or trade names are also used to indicate different pharmaceutical preparations exhibiting these activities. A third hormone of the pituitary, sometimes called luteotropic hormone, is also included in this family but because its actions are more diverse, this entity will not be considered in this article.

The hormone FSH stimulates the growth of one or more ovarian follicles, depending on the species and physiological context. For example, during the normal menstrual cycle in women, FSH stimulates the growth of only one dominant follicle in the ovary, thus allowing ovulation of only one egg, whereas during treatment for infertility the same hormone might cause multiple follicles to grow and ovulate. These follicles also produce estrogen, which is one of two sex hormones in women.

In other animals, such as pigs and bitches, FSH always produces multiple follicular growth. In the male, FSH acts on the gametogenic compartment of the testis to ensure the optimum production of sperms in a process called spermatogenesis. The second gonadotropin, LH, is responsible for acting on the fully grown mature ovarian follicle to cause ovulation. It is this rupture of the follicular bag that liberates the ovum into the peritoneal cavity to be collected by the oviducts, also called the Fallopian tube. The same hormone converts the ruptured follicle into a new structure called the corpus luteum, which now begins to produce the sex steroid progesterone. Together, the production of estrogen and progesterone maintains the menstrual (estrous cycle in other animals) cycle in women during their reproductive years.

In men LH stimulates steroidogenesis to cause secretion of the male hormone (testosterone) from the

Leydig cells of the testis. This androgen, together with actions of FSH, maintains spermatogenesis and sex accessory glands and induces male behavior.

The placenta in women during the first trimester of pregnancy produces copious amounts of a gonadotropin called human chorionic gonadotropin (hCG). Its production begins soon after implantation of the fertilized ovum in the uterus to such an extent that the detection of the hCG in urine forms the basis of all pregnancy tests. Currently, these tests are easy to use and cheap, and results from self-administered tests are usually available within an hour. Pregnancy urine is also a rich and commercially viable source for the manufacture of clinical-grade gonadotropin. hCG is structurally and functionally similar to pituitary LH and as such is used as a convenient surrogate for LH in clinical treatment. Other animals such as the pregnant mare also secrete very high amounts of gonadotropin into the bloodstream. This hormone, called equine chorionic gonadotropin [eCG or pregnant mare serum gonadotropin (PMSG)], is mainly used as a substitute for FSH in animals.

During the past 25 years, extraordinary and rapid advances regarding the biochemistry and molecular biology of gonadotropins and their receptors have given us remarkable insights into their structural architecture and mechanisms of action. Investigations on the structure–function relationships of hCG and other gonadotropins have led to determination of the three-dimensional (X-ray) structure of the molecule to create models of hormone–receptor interaction. Molecular cloning of the genes for the gonadotropin subunits, the constituent parts of these hormones, has resulted in their production by recombinant DNA techniques. Application of the advances in immunology such as hybridoma technology has revolutionized detection methods of these hormones in clinical tests and in at-home pregnancy tests. Cloning methods have also given us more detailed information on the genes that code for the receptors in the ovary and testis. These interdisciplinary advances are expected to herald a new era in treatment of infertility, contraceptive development, and animal management and welfare.

II. SOURCES OF GONADOTROPINS

A survey of the current sources of gonadotropins is given in Table 1 with emphasis on human preparations. The hormones are found in the pituitary (placenta, blood, and urine). Materials isolated from the pituitary tissue (e.g., human/animal glands) are called pituitary gonadotropins, whereas the hormone prepared from human urine is called urinary gonadotropin. Extensive research performed on animal gonadotropins extracted from large collections of glands of domestic animals, such as pigs, cattle, and sheep pooled from slaughter houses, has provided valuable lessons for designing appropriate methods for preparing human hormones. Thousands of glands were required to obtain small amounts of pure hormone necessary for treatment, but recombinant DNA technology has overcome this problem. A practical solution and alternative to the human pituitary gonadotropins has resulted from the discovery that they are excreted daily in urine. In certain physiological states, such as menopause, pregnancy, and other conditions such as eunuchoidism (no functional gonads), large amounts of the hormones are excreted in the urine. Thus, collection of urine on a commercial scale has provided a valuable source for manufac-

TABLE 1
Sources of Gonadotropins

Hormone	Origin	Excretion/transport	Manufacture
FSH	Pituitary [human, animal (pig, sheep, cow, horse)]	Human urine	Postmenopausal urine
LH	Pituitary [human, animal (pig, sheep, cow, horse)]	Human urine	Postmenopausal urine
hCG	Trophoblast tissue (pregnancy human)	Urine	Pregnancy urine
PMSG (eCG)	Endometrial cups (pregnant mare)	Blood	Mare serum
FSH, LH, hCG	Recombinant DNA	Cell culture	CHO cells or other systems

ture of pharmaceutical preparations containing FSH and hCG activity.

Animal gonadotropins, despite their high purity and availability, cannot be utilized for human treatment because of immune reactions that set in as a result of injection of foreign protein into the body. They are useful in the domestic livestock industry for embryotransfer technology and animal breeding. However, specific antibodies are very useful in developing immunologic tests and vaccine technologies for contraception.

Currently, most of the gonadotropins in clinical use for humans come from urine of postmenopausal subjects and pregnant women. Urine from postmenopausal women represents a rich source of FSH, whereas that from pregnant women as hCG offers a surrogate for LH. These urinary preparations are believed to be some form of metabolic products of the pituitary (FSH) and placental hormones (hCG) which have retained biological activity. In addition to the active materials, there are also other subunit forms or fragments that are excreted in the urine, but their significance is not well understood.

III. STRUCTURE AND FUNCTION

Systematic investigations into the existence of gonadotropins began in the 1920s when surgical techniques of hypophysectomy (removal of the pituitary gland) in animals led to gonadal regression and loss of reproductive function. This could be corrected by administration of suitably prepared water-soluble extracts of the pituitary gland. The existence of an important pregnancy hormone, hCG, was also discovered during the same period. During the next 50 years important advances were made in recognizing two gonadotropin molecules with different properties, and fractionation methods were developed to isolate the two active principles, referred to as FSH and LH (and hCG), as highly purified biochemical entities. The discovery of their subunit structure in the late 1960s and elucidation of the complete amino acid sequence in the early 1970s have laid the foundation for our current understanding of these important hormones.

A. Detection Methods

For the isolation of any biologically active principle, detection methods called bioassays are very crucial in their identification. Traditionally, these methods utilized living animal models such as sexually immature female or male rats into which extracts were injected. A few days after treatment the increases in weight of certain tissues, such as the ovaries, uteri, or ventral prostate, were taken as indices of biological action. With time and greater understanding of their biochemistry and physiology, *in vitro* test procedures using cells in culture or immunoassays or receptor-based assays largely replaced animal models. These are rapid and economical. However, they cannot replace *in vivo* assays to assess the final pharmacological properties. Such tests are required for new gonadotropin products. Currently, cells derived from certain gonadal tumors or those bearing cloned gonadotropin receptor genes (cDNA-complementary DNA) are available for routine use. For simplicity, we may classify the various tests in the following order of acceptance: immunoassays, receptor assays, *in vitro* cell bioassays, and *in vivo* bioassays.

B. The Glycoprotein Hormone Family: Heterodimer Structure

All the gonadotropins (FSH, LH, hCG, and eCG), together with another pituitary hormone called thyrotropin [thyroid-stimulating hormone (TSH)], belong to a family commonly designated glycoprotein hormones. They are so named because in addition to their protein backbone these molecules also have a considerable amount of carbohydrate (sugar) residues attached to the amino acids at certain key positions. Depending on the source (pituitary or placenta/urine), these glygoprotein hormones contain from 15 to 35% carbohydrate by weight. The presence of these sugar residues is important for biological function. In the field of endocrinology, although the term glycoprotein hormone has been used in a restricted manner to refer to gonadotropins and thyrotropin, it should be understood that there are other molecules, such as erythropoietin, which are also glycoprotein hormones. Such molecules, however, are excluded from this discussion.

Members of this glycoprotein hormone family are heterodimers, meaning their structure is composed of two nonidentical subunits or halves, designated α and β. Such a structure is found for virtually all gonadotropins examined to date, including human, animals, amphibians, fishes, birds, marsupials, and reptiles. This suggests that the dimeric structure designed to stimulate gonads and reproduction has been conserved during evolution.

When the three pituitary glycoprotein hormones (FSH, LH, and TSH) and pregnancy gonadotropin hCG were first studied, it became apparent that each molecule contained one α and one β subunit, which could be physically separated by a technique called dissociation. Remarkably, within a given species (e.g., human), the α subunit of the four glycoprotein hormones is identical in structure; thus, it is also called the common α subunit. The β subunit is hormone specific and contributes to the specificity of the hormones action.

A diagrammatic representation of the glycoprotein hormone model is shown in Fig. 1. As visualized in this model, the same common α subunit will combine with a different hormone-specific β subunit in a manner that produces a unique three-dimensional configuration, which can efficiently interact with the hormone–receptor system in the target cell. The hormone–receptor system functions in much the same way as electronic signals in the atmosphere which are received and interpreted by the television/radio receiver. The hormone circulating in the blood is the signal that is received by the receptor on the cell to perform its designated function.

According to this model, one can experimentally create any desired hormone by combining the α subunit with a given β subunit of required specificity. Thus, interspecies combinations are also feasible, and such work has contributed much to our understanding of the hormone's functions in different species by identifying potential similarities and sites of recombination.

In the previous model, each hormone (α–β complex) is depicted as interacting with a unique receptor. Notice that their shapes become different. However, human LH and hCG interact with the same receptor and perform the same function because their β subunits are very similar in structure. Thus, they are considered structural and functional homologs, meaning one can be used in place of the other if required for certain medical conditions. However, it must be remembered that the *in vivo* circulating half-life of human LH (1 hr) is much shorter than that of hCG (about 24 hr) due to the presence of a large extracarbohyrate-rich additional peptide segment at the end of hCG β subunit.

Understanding the structure of a glycoprotein hormones requires the determination of not only the sequence of amino acids but also the arrangement of carbohydrate residues that form an integral part of the molecule. The individual amino acids are like beads that can provide a distinct pattern to a necklace. In each subunit the constituent amino acids are arranged in a predetermined order dictated by genetic makeup of the species or individual, and unraveling this constitutes the amino acid sequence. These sequences were painstakingly determined by chemical means for most of the gonadotropins well before modern-day gene cloning procedures became available. Knowledge of their genes further clarified discrepancies that had been reported by chemical determinations and paved the way for production of recombinant gonadotropins.

The two gonadotropins produced in cells called gonadotrophs of the pituitary gland have a molecular mass of 28–30 kDa, whereas the placental gonadotropin produced during pregnancy by trophoblast cells or endometrial cups (e.g., women and mare) is about 37 kDa. Although the amino acid sequences of gonadotropins from about 25 species are known, we will consider here the human hormones as an example. The α subunit consists of 92 amino acids with two sites of attachment for carbohydrate chains at positions asparagines 52 and 78. Why only these two positions? Because the enzymatic mechanisms that operate within the mammalian cell recognize unique sets of amino acid sequences as a code within the polypeptide chain and attach preassembled carbohydrate units only at these locations. The presence or absence of these sugar chains modifies many biological properties of gonadotropins.

The hormone-specific β subunit is slightly variable depending on the hormone. Among the three, human FSH β subunit is the shortest, with 111 amino acids, LH β has 121 amino acids, and hCG β subunit is

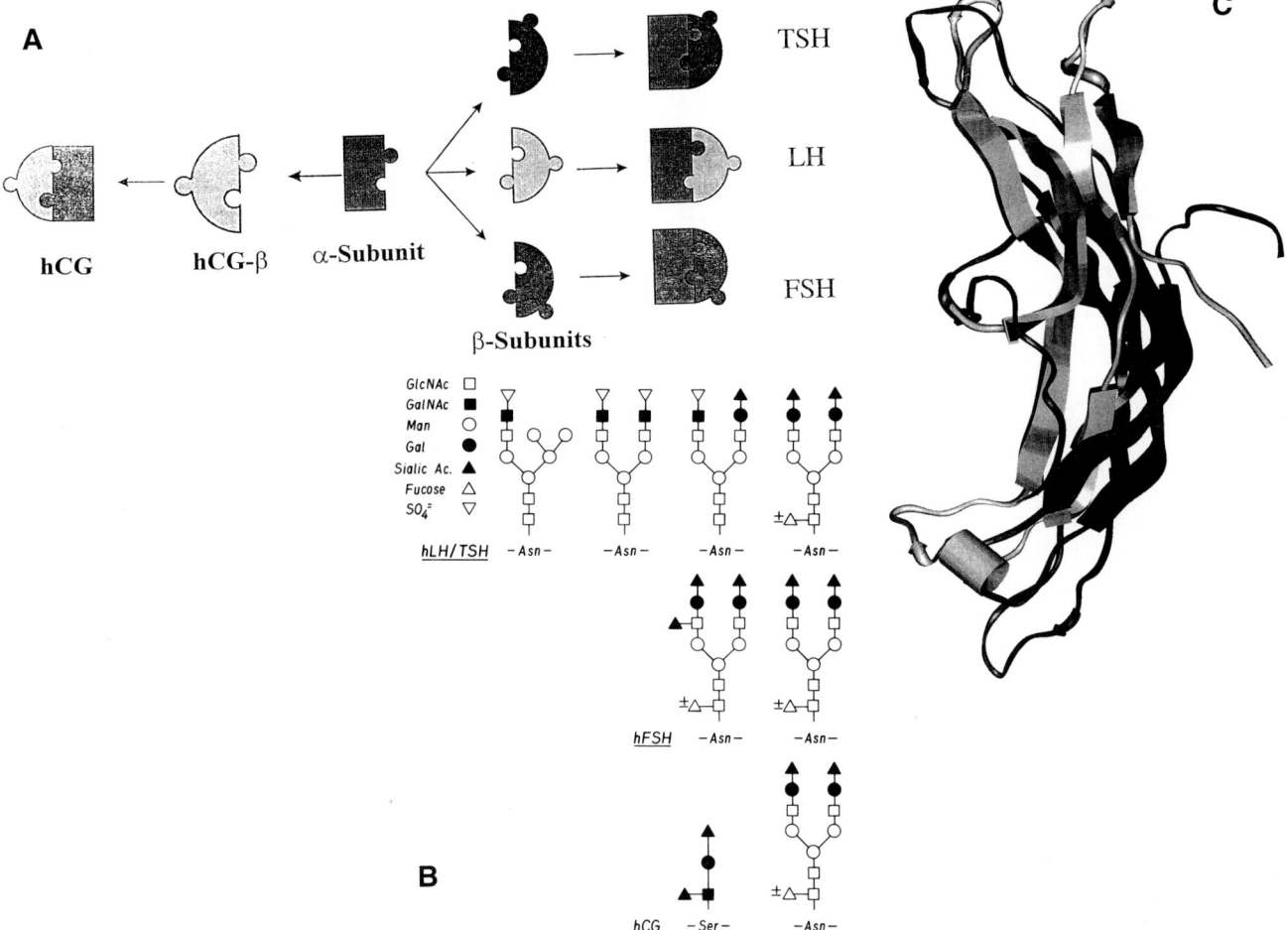

FIGURE 1 (A) The subunit model of glycoprotein hormones. In all animals (human in this figure) a common α subunit unites with a chosen β subunit to produce a different hormone. Note that final shape of the hormone becomes different in each case so that it may interact with its own receptor. Because LH β and hCG β subunits are very similar, the hormones LH and hCG, although produced in different tissues (pituitary and placenta, respectively), interact with the same receptor producing the same responses. The size of hCG β subunit as shown is larger but the final shape of the hCG fits in much the same way as LH does to the receptor. (B) The main assembly of sugar units (carbohydrate chains) that are present in human glycoprotein hormones. Each symbol represents a different type of sugar. GlcNAc, glucosamine; GalNAc, galactosamine; Man, mannose; Gal, galactose; Sialic Ac, sialic acid or *N*-acetyl-neuraminic acid; SO$_4$, sulfate group. $-$Asn$-$ indicates that the sugar linkage is of N-glycosylation type; $-$Ser$-$ shows O-glycosylation and is predominant in gonadotropins of placental origin. (C) A model of hCG dimer structure derived from X-ray crystallographic study of a deglycosylated hormone. The union of α and β subunits is stabilized by certain key $-$S$-$S$-$ bridges which are described to form a seat belt arrangement. The two chains shown here as ribbons in different shades contain α helical (barrel), β sheet (flat), and turn (thin lines) structures. Belonging to the same family, LH, FSH, and TSH also have similar configurations.

the longest, with 145 amino acids. The amino acid sequences of FSH and LH β subunits differ greatly, thus accounting for their action on different cells. However, the LH β and hCG β subunits are very similar, although the two are slightly different in

length and sugar composition. This explains why LH and hCG bind to the same receptor system in the gonads and stimulate similar mechanisms. The sites of carbohydrate attachment, also called glycosylation sites, in the β subunit are different: LH β has one at position 30 [asparagine (Asn)], FSH β has sites at Asn7 and Asn24, and hCG β undergoes such a modification at six sites. In the latter, there are two types of linkages: N-glycosylation, representing linkages to Asn at positions 13 and 30 (note that position 13 is also glycosylated in hLH β subunit), and O-glycosylation. At sites of O-glycosylations, of which there are four in the extreme tail end (carboxy-terminal portion) of the molecule, the sugar chains are linked to serine residues (amino acids) at positions 121, 127, 132, and 138.

C. Carbohydrate Structure and Biological Function

Sugar residues in their various forms constitute an important part of the glycoprotein hormones. To offer a simple comparison, the presence of sugars may be similar to the icing decorations laid out artfully on a cake. These patterns vary depending on the source of the glycoprotein hormone as shown in Fig. 1B. It can be readily discerned from this representation that the sugar chains vary slightly at the ends, with some terminating in acidic sugar called sialic acid and others have sulfate. Some of these differences are attributable to the presence of different enzymes in the pituitary and placenta, which produce the gonadotropins.

Owing to their high content in glycoprotein hormones, it was long suspected that the sugars may play an important role in biological function. The use of various techniques, including careful stripping off (denuding) of the peripheral sugars or gene modifications, has served to establish the critical role of sugars (such as sialic acid) in maintaining the hormone in circulation. hCG, which has a high content of sialic acid, remains in circulation for about 24 hr, whereas an equivalent amount of hLH of pituitary origin disappears quickly 1 hr following an intravenous injection. Also, sugar residues of the hormone, particularly those found on the α subunit, are critical in signal transduction. Thus, a hormone devoid of glycosylation at a unique site(s) may bind to a receptor but is unable to activate the target cell. Such a modified hormone would display a change in biological profile, now showing antagonistic properties (in-

hibitory properties *in vitro* and *in vivo* under appropriate conditions). Knowledge derived from such experiments has permitted the design of analogs with desired hormonal properties (e.g., antagonist, long-acting agonists, and short half-life molecules).

The presence of variable amounts of sugars has also conferred a considerable degree of polymorphism on gonadotropins. For example, even a highly active and pure preparation of gonadotropin may reveal the presence of closely related molecules on closer examination. This is in large part due to differences in the amount and variety of terminal sugars. In addition, there may be internal proteolytic snips introduced by the action of various enzymes in the circulation or tissues. Some of these products have been recognized in various pathological or physiological conditions. These are believed to be part of the complex signaling system that operates to maintain the reproductive cycle.

Because of the requirement of sugars for maintaining biological activity, complex proteins such as gonadotropins cannot be produced in microorganisms such as bacteria (*Escherichia coli*), which have been so successfully employed in the large-scale manufacture of other hormones such as insulin and human growth hormone. This is due to the fact that bacterial cells are unable to perform complex molecular maneuvers required for glycosylation and attachment to the protein backbone.

D. The Three-Dimensional Structures of Gonadotropins

As discussed in Section III,B, the dimeric structure of gonadotropins is known (Fig. 1A). The union of the α and β subunits in the correct order and configuration is critical for such a molecule to interact with its receptor for cellular activation. Once the subunits are physically separated, they no longer display the biological function of the parent molecule. When these two subunits are brought together under appropriate conditions in a process called reassociation, not only is the structural integrity of the hormone fully restored but also its complete biological activity is regained. Thus, this process is completely reversible. These observations therefore imply that structural changes in the form of different orientations occur during the process of dissociation–association.

To understand the precise details of the molecular changes contributed by different parts of the subunits, we need information on the crystal structure of the hormone. For many years model building of hCG, which could be prepared in large amounts, was unsuccessful because of erroneous disulfide assignments or incorrect homologies with other known structures. The disulfide pairings in a protein molecule represent critical bridges formed as –S–S– covalent links between two different cysteine (amino acids) residues present at different positions in the polypeptide chains. Currently, determination of the crystal structure of a molecule by X-ray diffraction offers the best hope of decoding its molecular configuration. Because hCG is a highly glycosylated molecule containing over 30% carbohydrate, good crystals of the hormone amenable for X-ray analysis could not be secured. Fortunately, one of the techniques (chemical deglycosylation) that we had applied many years ago as part of our structure–function analysis provided a hormone derivative with less sugar content suitable for crystallization. Using such methodology, the crystal structure of deglycosylated hCG, the best approximation of the active hormone that is currently available, was determined in 1994. A computer-projected image of the hormone structure is shown in Fig. 1C. The lightly shaded α subunit could attain different foldings upon interaction with the darker shaded β subunit.

Remarkably, this work has revealed striking features of gonadotropins that could not be predicted from their primary structures, namely, the sequence of amino acids in the α and β subunits of hCG. First, even though the α and β subunits are dissimilar in sequence, they each portray a similar topography. Second, their folding motif or pattern is typical of a structural organization present in other types of molecules which exhibit growth factor activity in other cells. Thus, hCG, and by implication the other gonadotropins, also belongs to the "cystine knot growth factor" family, which includes proteins such as transforming growth factor-β, platelet-derived growth factor, nerve growth factor, and vascular en-

dothelial growth factor. In hCG, the α and β subunits are believed to be engaged in a clasped embrace involving the cystine knots. Thus, the two subunits of hCG are highly intertwined with the surfaces of subunits contacting each other, forming a tight association. What is truly remarkable is that the structure shows how the same α subunit can associate with a different β subunit to display a configuration that will allow it to interact with its own receptor but not with another. The structure of hCG brings us one step closer to revealing the amino acid(s) that makes contact with the receptor sites to allow signal transduction to proceed. This has also formed the basis for many studies on mutagenesis and recombinant hormone design.

Third, the growth factor-related structure of the hCG and its subunits (as well as the other hormones) indicate that individual free subunits present in tissue or circulation could by themselves display new forms of biological activities. The discovery of such novel biological activities and their potential receptor mechanisms represent new challenges for further investigation. These may have important practical implications when we consider the fact that many cancer cells overproduce one or the other hCG subunit or fragment. If these cancer cells have learned to utilize such molecules to sustain their relentless growth properties, how can their production or action be stopped to achieve inhibition of tumor growth? Understanding these mechanisms may herald new approaches to management of certain cancers.

IV. MOLECULAR BIOLOGY OF GONADOTROPINS: MUTATIONS AND ANIMAL MODELS

The α and β subunits of glycoprotein hormones are products encoded by separate genes. These genes have been cloned from numerous species including man, and these advances have greatly enhanced our understanding of structure–function relationships. The advantage of molecular cloning technique is that the structure of the subunits can be deduced without actual isolation of the hormone from a given species.

		Gene Length (Kb)	mRNA (bp)	Chromosome Location	Gene Copy#	Amino acids
α		9.4	800	6p21.1-23	1	92
LHβ		1.1	700	19q13.3	1	121
CGβ		1.5	1000	19q13.3	6	145
FSHβ		4.2	1800	11p13	1	111
LH-R		~ 80	~ 2600	6p21	1	701
FSH-R		~ 80	~ 2400	6p21	1	695

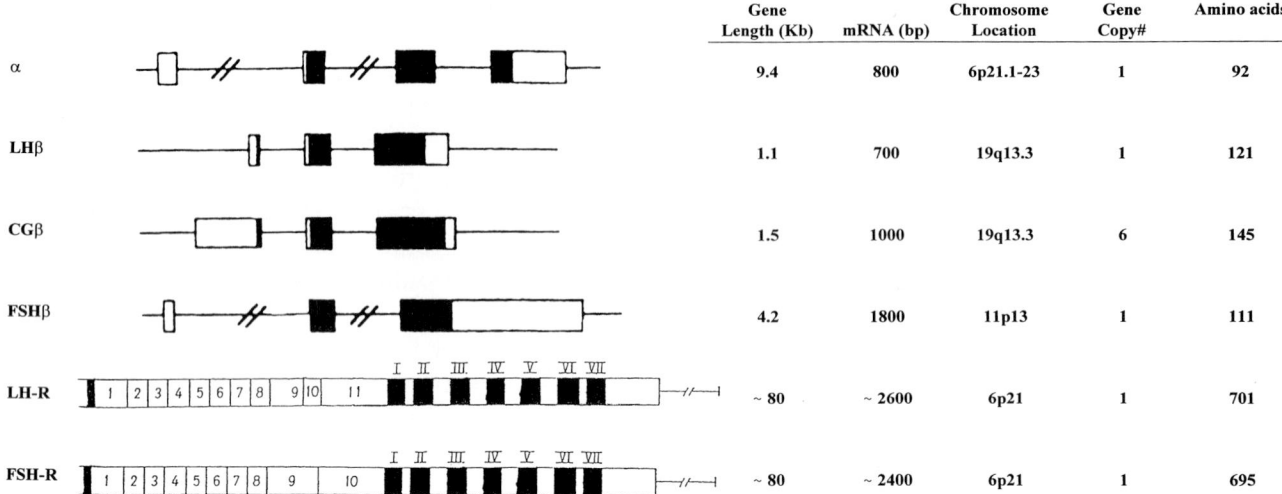

FIGURE 2 Summary of the structures of gonadotropin and receptor genes. The α subunit gene consists of 4 exons, whereas there are 3 exons in LH/CG β as well as FSH β. Compared to the hormones, the receptor genes are more complex and much larger. The presence of 11 and 10 (or 11) exons in LH and FSH receptor genes contributes much to the phenomenon of alternate splicing that gives rise to many gonadotropin receptor variants, adding to the complexity of hormone signal transduction and regulation. The boxes identified as exons contain the coding region of the gene used for assembly of the respective proteins.

The current information on human gonadotropin genes is summarized in Fig. 2 along with relevant data on respective receptors. There is a single gene for the α subunit as well as the β subunit of LH and FSH. Unlike the LH β gene, there is a cluster of six genes for the hCG β which is thought to have originated from a duplication of the former. In each gene the coding region (exon) of the DNA is interrupted by intervening sequences called introns. The β subunit genes share similar gene structures and are highly homologous. The FSH β subunit gene is about four times longer than the LH/CG β gene due to an extended 3' untranslated segment which is thought to alter its stability and account for the rapid turnover of the FSH β messenger RNA (mRNA).

Mutations in human pituitary gonadotropins genes are rare because such deleterious alteration would be detrimental to reproductive function. There are no known mutations of the α subunit. Any inactivating mutation of this gene would render the subject/animal hypothyroid and hypogonadotropic. Such animals have been designed for laboratory studies. A single base substitution in the hLH β gene resulting in change of amino acid glutamine at position 54 to

arginine inactivated the hormone, with infertility in the homozygous individual and drastic reduction in fertility for the male heterozygote. However, female heterozygotes had normal sexual development and fertility.

In men selective FSH deficiency apparently causes defective sperm production and infertility. When the FSH β gene underwent mutation in one individual, a shorter version of the protein could not form a dimer with the α subunit. Consequently, there was no hormonal activity.

In contrast to outright mutations, gene polymorphisms of gonadotropins in animals and humans that do not affect fertility are known. As opposed to natural mutations mentioned previously, alterations or total deletion of gene function can be induced in animal models, providing novel insights into pathological situations in humans. Thus, transgenic or knockout animals (mice) for LH and FSH have been created for laboratory study and may some day be used to enhance breeding performance.

As a corollary to mutations in the gonadotropin subunit genes, several genetic abnormalities that affect the LH and FSH receptors have also been described. Depending on the type of mutation, there could be deficiency of gonadal function or over-activity.

V. SIGNAL TRANSDUCTION AND MECHANISM OF ACTION

For a signaling molecule, the gonadotropins are rather large in size and complex, like many other protein hormones. How do numerous entities such as these present in the same circulation know where and when to act at selected sites in the body? This requirement and specificity of action is achieved by the evolution of a system called the hormone (ligand)–receptor interaction. The presence of receiver molecules called receptors at the target site allows a particular hormone to become attracted or trapped by a cell bearing receptors on the cell surface, which is accessible to the circulating blood. From this concept it follows that gonadotropins interact with specific receptors located on certain cells in the ovary and testis.

Large molecules, such as LH (hCG) and FSH, have the potential to display numerous sites of interaction on their surface which are created by the juxtaposition of distant amino acids brought near each other by the twisting and turning (folding) of the two subunits. These surfaces interact with unique pockets that are present in the gonadotropin receptors.

The gonadotropin receptors are also complex glycoproteins assembled on the surface of certain ovarian and testicular cells. There are two receptors, each specific for LH (hCG) and FSH. In the ovary FSH and LH receptors appear in succession on the granulosa cell as required at different times to coordinate the development (enlargement) of the follicle in which the ovum is growing toward its maturity. Up to this point, these and other cells in the follicle secrete the sex steroids (estrogen and progesterone) under hormonal stimulation to contribute to the cyclic changes that occur during the menstrual cycle. Once ovulation has occurred, FSH receptors are shut down and the new structure (the corpus luteum) has only LH receptors that will cause predominantly the secretion of progesterone. These cyclic changes are responsible for regular menstrual cycle in women.

In the testis of the male, however, FSH and LH receptors are present in two different compartments from the beginning of development. The FSH receptors for the most part are localized on the Sertoli cells present inside the seminiferous tubule where sperm production occurs. The receptors for LH are situated on cells that live in the space between the tubules. They are called Leydig cells, the main function of which is to produce testosterone, the male sex hormone.

As a result of rapid advances in molecular cloning technology, we now know the nature of the genes and have a better understanding of the structure of gonadotropin receptors. The two receptors are coded by separate genes whose total length is about 80,000–100,000 base pairs. These are ultimately translated into glycoproteins of approximately 700 amino acids and inserted into the membrane. Figure 3 depicts a proposed model for the structure of the LH/hCG receptor. As mentioned earlier these two hormones, which are structurally very similar, interact with the same receptor. The FSH receptor, which is very specific only to this hormone, also has a similar topography. We can visualize them as having three parts. A long segment composed of 350 amino acids, known as the extracellular domain, remains dangling outside the cell like an antenna to trap the hormone that it meets from circulation. The next part of the receptor, called the transmembrane region, is composed of seven barrel-type segments that serve to firmly anchor the molecule on the cell surface. The last part of the receptor is more or less the tail, which is tucked into the cells interior where it can establish contact with the inner machinery of the cell to bring about hormonal response. This model is called a heptahelical or serpentine structure and members of this group belong to a family designated G-protein-coupled receptors. In addition to such a predicted structure, there are other receptor units called alternately spliced products whose significance is not understood.

Contacts between the gonadotropin and receptors are initiated by numerous forces of intermolecular attraction to such an extent that minute quantities of the hormone are sufficient to evoke cellular response. Amounts as small as billionths of a gram (ng/ml) initiate a cascade of events that amplifies the signal

Human LH / hCG Receptor

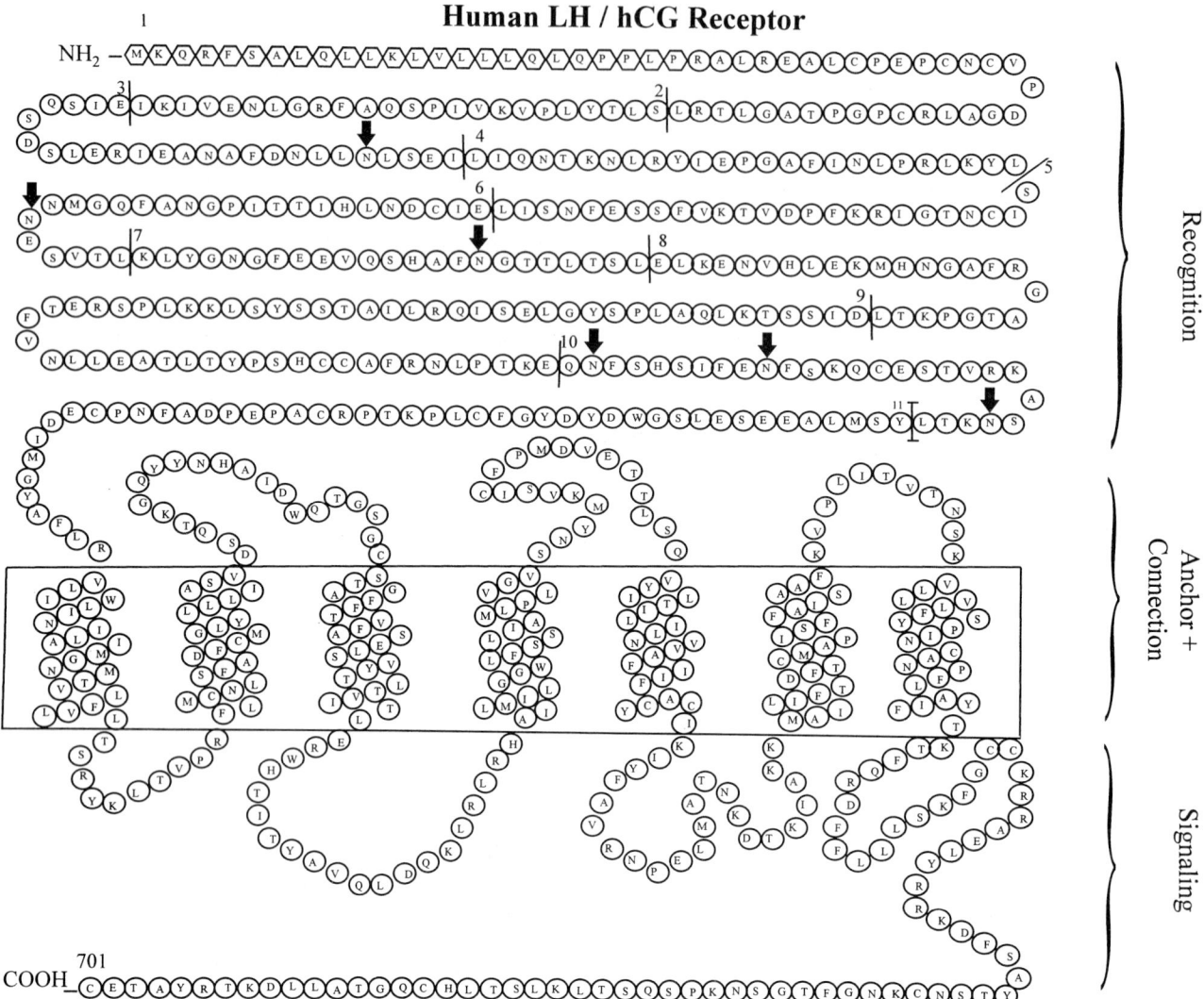

FIGURE 3 Schematic representation of the LH/hCG receptor molecule that controls hormone action in the ovary and testis. The long polypeptide chain is organized to perform three functions. The extracellular domain interacts with the hormone. The second part, consisting of seven barrels, anchors the receptor on the cell membrane, and the last segment, the intracellular tail, communicates with components inside the cell to produce appropriate signals. The first 24 amino acids that represent the signal peptide required for biosynthesis of the receptor are eventually removed from the mature protein. Note that the extracellular domain, although shown here as a linear chain, organizes itself into defined twists and turns to provide docking sites for specific recognition of the shape of each hormone molecule as present in the heterodimer (see Fig. 1A). The numbers and thick lines specify the exon and boundaries. The circled letters represent amino acids. Putative N-glycosylation sites are shown by arrows. The receptors for gonadotropins are present mainly in certain cells of the ovary and testis.

at each step of the reaction. Once hormone binding occurs with the receptor, there is an intramolecular change in the configuration of the complex to allow generation of other molecules which serve as second messengers. There are several small molecules or ions which can fulfill this function: The most common include cyclic adenosine monophosphate, calcium ions, diacyl glycerol, and inositol phosphate. These agents then go on to activate other enzymes which catalyze a host of diverse reactions. All these chain reactions occur in rapid succession, such that within a few minutes or hours, hormonal response such as sex hormone production can be observed. The cumulative effects of all these reactions may lead to cellular growth, DNA synthesis, and hormonal effect. For example, ovulation can occur within 24 hr following a timely injection of hCG in a patient who has been pretreated with FSH for *in vitro* fertilization. Thus, precise timing is required to coordinate the actions of the two gonadotropins during treatment.

In both the ovary and the testis, the combined action of both gonadotropins/receptor units is essential to ensure optimal functioning of the reproductive system. Several mutations of the LH as well as the FSH receptor have been recognized as a direct result of molecular cloning analysis. These mutations change the receptor in such a way that may lead to either inactivation or hyperactivation in the absence of a ligand. In certain cases of precocious puberty, a mutant LH receptor becomes hyperactive and functions without the need for hormone. In other individuals, lack of FSH receptor function leads to ovarian failure as well as decreased sperm production. Animal models defective in receptor function are expected to provide new insights into the workings of the signal transduction process and suggest strategies for overcoming associated pathologies or provide the means for their manipulation in useful ways.

Until recently, it was believed that gonadotropin receptors were localized exclusively in cells of the ovary and testis. With the availability of more sensitive and sophisticated DNA detection methods, the expression of these genes has been recognized in disparate tissues such as the brain, blood vessel, mammary gland, and prostate tissue. There are also some reports of detection in cancer cells. For the most part, these findings are confined to detection of so-called mRNA transcripts and/or immunoreactive LH/hCG receptor protein domains. The validity and physiological significance of these novel observations remain to be confirmed.

VI. MEDICAL AND VETERINARY USE OF GONADOTROPINS

As their names suggest, gonadotropins (FSH, LH, and hCG) are molecules that stimulate the growth, development, and function of the gonads. Naturally,

it follows that they can be used to treat patients or animals that are deficient in these hormones. Their use has spawned a multimillion dollar industry which is expected to grow as the incidence of infertility rises.

For nearly 50 years, gonadotropins have been employed in human medicine either as a therapeutic agent or as a diagnostic tool. The advent of assisted reproductive technology or *in vitro* fertilization procedures to alleviate problems related to infertility in childless couples has accelerated not only the quantity of human gonadotropins but also the quality of hormones required for controlled ovarian stimulation.

Human FSH and hCG are used to treat infertility in women who fail to ovulate due to insufficient or inappropriate secretion from their own pituitary. In such individuals, the presence of a functional receptor system is essential for successful gonadotropin therapy. In children, the use of hCG will correct infantile cryptorchidism due to undescended testis. In men, the same hormones are used to stimulate the testis to augment the production of sperm and the male hormone testosterone. Arising from the latter property is the potential misuse by athletes to stimulate their androgen production in sports medicine. This has prompted the International Olympic Committee to ban such inappropriate use of a natural hormone.

Hormones extracted from human pituitary glands are no longer used in clinical treatment to avoid latent effects due to viral contaminations. Currently, human FSH prepared from the urine of menopausal women and extracted in large quantity by two companies represents the major source of the hormone for infertility treatment. Likewise, hCG, the second hormone required for infertility treatment, is also a urinary product but is derived from pregnant women. Recombinant human FSH produced by genetic engineering in Chinese hamster ovarian (CHO) cells has undergone extensive clinical trials and successful live births have been recorded by women who have received treatment. It has been approved for use in Europe and North America and may eventually replace the current urinary FSH preparations. Although recombinant hCG can also be made, I believe that it is economically cheaper to extract hCG from pregnancy urine. This source will continue to find major application in human treatment.

Besides therapeutic treatment, gonadotropin antibodies developed over a period of 25 years are widely used as diagnostics in clinical medicine and family practice. Highly sensitive and rapid tests, including those that can be done at home, to detect ovulation and pregnancy are widely available. The pregnancy tests may also be used to monitor pregnancy maintenance. A negative test that cannot confirm previous positive reaction is an indication of early pregnancy loss. Because many tumors are known to secrete hCG (normally produced in sufficient amounts only during pregnancy), methods to monitor its subunits or degradation products in circulation are utilized to detect the presence of cancer (in the absence of pregnancy) and follow the course of treatment by chemotherapy, radiation, or other procedures as well as remission.

In veterinary medicine, gonadotropins are extensively used to induce superovulation. Superovulation is the ability to induce multiple ovulations in an animal. Following fertilization, embryos may be transferred into recipient animals. Cattle breeders, pig farmers, and horse breeders use such technologies to improve their stock quickly. For veterinary applications, FSH and LH extracted from the pituitaries of animals collected at slaughterhouses are widely used. In addition, a hormone (similar to hCG) extracted from the blood of pregnant mare (PMSG) is also used. These techniques are expected to gain more popularity with the advent of transgenic animal approaches to produce rare human proteins (pharmaceuticals) in animals. Another application of gonadotropins worthy of mention is their utility in the aquaculture industry of fish/marine farming to augment food production.

VII. PERSPECTIVES AND FUTURE DIRECTIONS

The last quarter of this century has witnessed remarkable advances in the chemistry, biology, immunology, and molecular biology of the gonadotropins and receptors. In addition to knowing the complete sequences of these complex proteins, we are also beginning to discern their three-dimensional structures. The information derived from model building and the ability to manipulate gonadotropin genes for production on a large scale in various expression systems will make therapeutic application much easier and more accessible. This will surely pave the way for designer gonadotropins with desired pharmacological properties to suit various applications. Various strategies for better (delivery) administration of these protein pharmaceuticals will also develop. Currently, such hormones are effective only in the form of injections. Owing to easy destruction in the digestive system, they cannot be taken orally. However, one may expect to witness significant developments along these lines in the early part of the next century. With a precise knowledge of the contact points between the gonadotropins and their respective receptors, as well as new exploratory techniques such as screening of combinatorial libraries, it may be possible to develop small synthetic compounds. These could be fabricated in the laboratory from simple starting materials. Because they are small, they could be designed as orally active formulations. Such new compounds, which would be classified as gonadotropin mimetics, may find application in treating infertility as well aid in the discovery of novel and better methods of contraception. Both types of uses will serve human and animal welfare immensely.

The increase in knowledge of the various forms of gonadotropins, their subunits, and potentially new receptor structures and variants could lead to new applications of gonadotropins in various fields, including cancer biology and immunology.

Gonadotropin (hCG) and receptor-like structures have also been discovered in other living organisms, such as certain species of bacteria and yeast. Understanding their functional role in the life cycle of these systems might facilitate development of novel pharmaceutical compounds.

See Also the Following Articles

Chorionic Gonadotropins, Human; FSH (Follicle-Stimulating Hormone); GnRH (Gonadotropin-Releasing Hormone); Gonadotropin Biosynthesis; Gonadotropin Receptors; Gonadotropin Secretion, Control of; LH (Luteinizing Hormone); Steroidogenesis, Overview

Bibliography

Combarnous, Y. (1992). Molecular basis of the specificity of binding of glycoprotein hormones to their receptors. *Endocr. Rev.* **13**, 670–691.

Dufau, M. (1995). The luteinizing hormone receptor. *Curr. Opin. Endocrinol. Diabetes* **2**, 365–374.

Ferin, M., Jewelewicz, R., and Warren, M. (1993). *The Menstrual Cycle.* Oxford Univ. Press, New York.

Howles, C. M. (1996). Genetic engineering of human FSH (gonal-F). *Hum. Reprod. Update* **2**, 172–191.

International Conference on Gonadotropins and Receptors (1996). Structure, function and molecular forms. *Mol. Cell. Endocrinol.* **125**, 1–182. [Special issue]

Lapthorn, A. J., Harris, D. C., Littlejohn, A., Lustbader, J. W., Canfield, R. E., Machin, K. J., Morgan, F. J., and Isaacs, N. W. (1994). Crystal structure of human chorionic gonadotropin. *Nature* **369**, 455–461.

Sairam, M. R. (1989). Role of carbohydrates in glycoprotein hormone signal transduction. *FASEB J.* **3**, 1915–1926.

Thotakura, N. R., and Blithe, D. L. (1995). Glycoprotein hormones: Glycobiology of gonadotrophins, thyrotrophin and free subunit. *Glycobiology* **5**, 3–10.

Gorillas

see Primates, Nonhuman

Graafian Follicle

R. H. Oliver, G. D. Chen, and J. Yeh

Regions Hospital and University of Minnesota

I. Introduction
II. Structure and Function of Follicle Components
III. Regulation of Follicular Maturation

GLOSSARY

antrum The fluid-filled compartment of the mature follicle.

basal (basement) membrane A noncellular structure that isolates and contains the follicle.

Call–Exner bodies Fluid-filled spaces between follicular granulosa cells that coalesce to form the antrum.

cumulus/oocyte complex (cumulus oophorus) The subpopulation of granulosa cells that surround the oocyte.

graafian follicle The mature, preovulatory ovarian follicle.

granulosa cells The principal cell type found in the follicle. Granulosa cells produce steroids and growth factors that regulate follicular development.

I. INTRODUCTION

The mature follicle is known as the graafian follicle, named after the Dutch anatomist, Regnier de Graaf (1641–1673). Its identifying characteristic is the presence of a central, fluid-filled cavity, the antrum (Fig. 1). The graafian follicle functions to support maturation and eventual ovulation of the oocyte. After ovulation, the follicle undergoes transformation to become the corpus luteum that secretes hor-

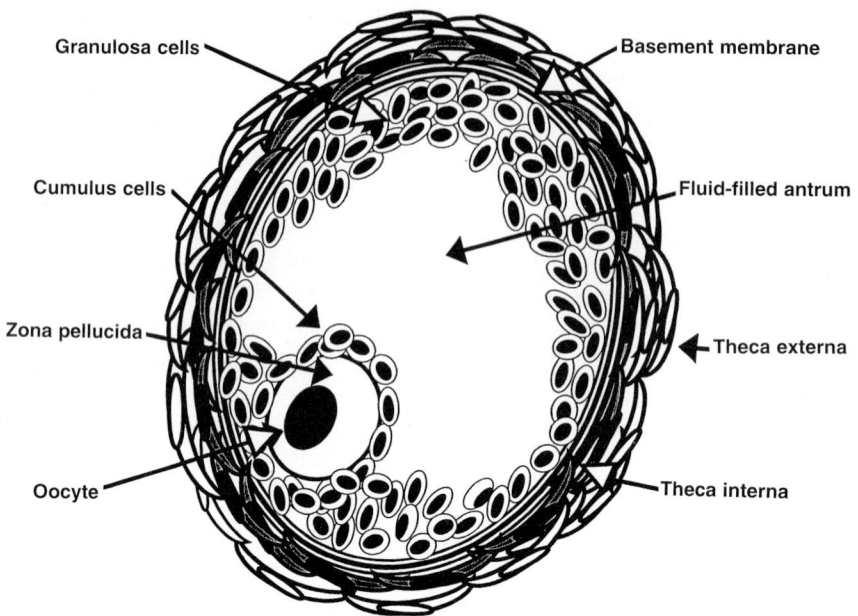

FIGURE 1 Diagram of the mature graafian follicle. The large antral compartment (antrum) is the distinctive characteristic of the mature graafian follicle. Cellular components include the thecal layers (theca externa and interna), the oocyte, granulosa, and cumulus cells. The basement membrane is a noncellular structure that isolates the granulosa cells, oocyte, and follicular fluid from direct contact with thecal layers and the circulatory system.

mones essential for the maintenance of pregnancy. The various cell types and structures of the follicle interact to control the hormonal milieu to which the developing oocyte is exposed.

Follicles pass through different stages as they mature in the ovary. In humans only one dominant graafian follicle ovulates during the menstrual cycle. Follicular development is regulated by gonadotropic hormones from the hypothalamic–pituitary system and by local factors produced within the ovary. The terms antral, tertiary, vesicular, and preovulatory are also used to designate the graafian follicle.

II. STRUCTURE AND FUNCTION OF FOLLICLE COMPONENTS

A. Theca Externa and Interna

The theca externa is the outermost layer of the follicle (Fig. 1). The theca externa is a structural tissue composed of cells resembling smooth muscle that may have some contractile function. The theca externa derives from pluripotent cells of the ovarian stromal tissue. It serves to compartmentalize the follicle within the ovary and is possibly involved in mechanisms of ovulation.

The theca interna, internal to the theca externa, also develops from stromal precursor cells (Fig. 1). The theca interna is well vascularized and can be distinguished from the externa due to its pink or reddish appearance. The blood vessels of the theca interna transport metabolic precursors and by-products to and from the follicle and serve as a plexus for the exchange of hormones between the follicle and systemic circulation. Individual theca interna cells exhibit a characteristic spindle-shaped morphology. The theca interna is a true glandular tissue and is the principal site of synthesis of androgens later converted to estrogens and progesterone by other components of the follicle. The mature, androgen-secreting thecal interna is often referred to as the interstitial gland.

B. Basal Membrane

The basal, or basement, membrane, a noncellular structure, separates the theca interna from the inner parts of the follicle (Fig. 1). It isolates the internal follicular components from blood vessels and provides a matrix for attachment of the granulosa cells. Nutrients and steroid precursors pass through the basement membrane. At the onset of luteinization, after ovulation, blood vessels enter through the basement membrane and establish an active blood supply to the follicle.

C. Granulosa Cells and the Cumulus/Oocyte Complex

Granulosa cells are the principal cell type found within the follicle (Fig. 1). Granulosa cell layers are avascular and appear pale in contrast to the adjacent thecal layer. Individual cells are spherical and contain numerous eosinophillic granules. Different populations of granulosa cells are distinguished by their proximity to the basal membrane, to the antrum, or to the developing oocyte. The membrana granulosa consists of mural granulosa cells that attach to the basement membrane and antral granulosa cells that are adjacent to the antrum. Cumulus cells are a subpopulation of granulosa cells that enclose and make intimate contact with the developing oocyte (Fig. 1). The primary steroidogenic function of granulosa cells in the graafian follicle is to convert thecal-produced androgens into estrogens. In addition, granulosa cells produce autocrine and paracrine factors that contribute to regulation of follicular development and oocyte maturation.

The oocyte occupies an eccentric position in the follicle. The zona pellucida, a glycoprotein coat involved in regulation of sperm penetration during fertilization, encloses the oocyte (Fig. 1). Layers of cumulus cells that surround and are contiguous with the oocyte are known collectively as the cumulus oophorus. Cells of the cumulus oophorus communicate and interact with the oocyte by means of intercellular gap junctions and secretion of paracrine growth factors. Cumulus cells seem to be involved in regulation of the meiotic arrest that is characteristic of oocyte maturation in vertebrate ovaries. At ovulation, cumulus cells are ejected along with the oocyte and may facilitate transport of the oocyte to the Fallopian tubes.

D. The Antrum

The fluid-filled antrum is the distinguishing morphological characteristic of the graafian follicle. In the mature graafian follicle the antrum occupies most of the space within the follicle (Fig. 1). During the development of the follicle, fluids secreted by granulosa cells accumulate between cells to form fluid-filled spaces, known as Call–Exner bodies. The Call–Exner bodies eventually coalesce to form the antrum. The follicular fluid within the antrum contains proteins, steroids, peptide growth factors, mucopolysaccharides, and electrolytes. In addition to serving as a reservoir for metabolites and medium for the exchange of hormones and growth factors, antral fluid may act to increase the pressure within the follicular wall and aid in the extrusion of the oocyte at ovulation.

III. REGULATION OF FOLLICULAR MATURATION

A. Gonadotropins and Intraovarian Factors

After puberty, follicles at all stages of development are present in the ovary. Follicular development occurs continuously throughout reproductive life and only ceases with menopause. The development of the follicle is regulated by the interaction of gonadotropins, steroid hormones, and growth factors. Follicular growth from the primordial to the preantral stage does not appear to require gonadotropic signals. However, antral development is dependent on gonadotropins and requires the presence of follicle-stimulating hormone (FSH). Follicular maturation appears to be largely a consequence of growth and differentiation of granulosa cells. The granulosa cells are the only cells known to possess functional FSH receptors. From early graafian stages to the preovulatory stage, granulosa cells actively proliferate under the stimulation of FSH. FSH induces follicular growth by increasing the number of granulosa cells as well as the amount of follicular fluid. At advanced stages of differentiation, granulosa cells achieve responsiveness to luteinizing hormone (LH). LH supports granulosa cell steroidogenesis and induces ovulation at the term of follicular development.

Intraovarian factors act through autocrine/paracrine pathways and cell-to-cell contact to orchestrate responses to gonadotropins and to regulate other aspects of follicular development. Growth factors, steroid hormones, and cytokines present in the ovary are known to affect follicular development. Important aspects of follicular development thought to be regulated by intraovarian factors include selection of dominant follicles, induction and prevention of atresia, and alterations in steroidogenesis.

B. The Dominant Follicle

The mechanism of dominance, by which only one or at most a few follicles are selected to complete maturation and to ovulate, is not well understood. It is clear that a significant aspect of this mechanism is the cyclical rise in FSH occurring at the beginning of the ovulatory cycle. In humans, this rise in FSH accelerates the growth of a cohort of follicles, but only one will generally survive and reach maturity. It is possible that certain of these follicles are more responsive to FSH than the others and become highly steroidogenic, generating large amounts of estradiol and inhibin that downregulates the secretion of FSH. As a result of declines in FSH, the less responsive follicles degenerate through a process known as atre-

sia, whereas the surviving follicle continues to grow. Granulosa cells from the dominant follicles become responsive to LH, which provides additional stimulus for steroidogensis as well as survival. A surge of LH, induced by increases in the circulating concentration of estradiol, takes place that induces the series of events that culminate in ovulation. After ovulation, the graafian follicle is converted into the corpus luteum and its function is altered from maintenance of the oocyte to maintenance of pregnancy.

See Also the Following Articles

Follicular Development, Control of; FSH (Follicle-Stimulating Hormone); Granulosa Cells; LH (Luteinizing Hormone) Ovarian Cycle, Mammals

Bibliography

Gougeon, A. (1986). Dynamics of follicular growth in the human: A model from preliminary results. *Hum. Reprod.* **1**, 81–93.

Richards, J. S., Fitzpatrick, S. L., Clemens, J. W., Morris, J. K., Alliston, T., and Sirois, J. (1995). Ovarian cell differentiation: A cascade of multiple hormone, cellular signals, and regulated genes. *Recent Prog. Horm. Res.* **50**, 223–254.

Yeh, J., and Adashi, E. Y. (1998). The ovarian life cycle. In *Reproductive Endocrinology* (S. C. C. Yen, R. Jaffe, and R. Barbieri, Eds.). Saunders, Philadelphia.

Granulosa Cells

Kenneth H. H. Wong and Eli Y. Adashi

University of Utah School of Medicine

I. Morphologic Development
II. Differentiation
III. Steroidogenesis

GLOSSARY

differentiation The process of acquiring character or function different from that of the original type.

follicle-stimulating hormone The gonadotropin responsible for the recruitment of follicles and the facilitation of their growth.

folliculogenesis The process by which a primordial follicle grows and develops into a specialized follicle with the potential to ovulate its egg or to die by atresia.

luteinization The transformation of the mature ovarian follicle into a corpus luteum after ovulation.

luteinizing hormone The gonadotropin responsible for promoting ovarian steroid secretion and inducing ovulation.

Granulosa cells (GCs) are a major component of the basic functional unit of the ovary, the ovarian follicle. GCs play a vital role in ovarian hormonal steroidogenesis and create conditions necessary for resumed oogenesis, ovulation, fertilization, and implantation.

I. MORPHOLOGIC DEVELOPMENT

The mammalian ovary is composed of an outer cortical and inner medullary layer. The ovarian follicle, the basic functional unit of the ovary, is located in the cortex and is composed of the oocyte and layers of GCs and theca cells. The GCs surrounding the oocyte are committed to undergo proliferation and steroidogenic and morphological differentiation. This differentiation in response to endocrine and paracrine stimuli plays a key role in the functional maturation of the entire follicle. The morphologic features of GCs vary depending on which particular follicular stage the GCs are associated with.

A. Embryonic Origin

During early development of the female gonad, germ cells migrate from the embryonic hindgut and endoderm of the yolk sac to the gonadal ridge. The exact origin of GCs is unknown, although there is speculation that GCs may arise from the rete ovarii (remnants of the mesonephric duct) which is mesothelial in character. These mesenchymal cells may then organize around the primary oocytes, differentiate into GCs, and ultimately organize into follicles.

B. Primordial Follicles

The primordial follicles represent a pool of nongrowing follicles from which all preovulatory follicles are selected. Each primordial follicle is composed of a small, immature oocyte and an outer single layer of squamous-appearing GCs.

C. Developing Follicle

As the follicle develops, GCs proliferate, forming 3–5 concentric layers around the oocyte forming the preantral follicle. As seen through light microscopy, the GCs within the maturing follicles are polyhedral in shape and individually measure 5–7 mm in diameter in the human. The large, healthy follicle destined to ovulate shows prominent GC proliferation with abundant mitoses. In the antral follicle, there is secretion of mucopolysaccharide-rich fluid which coalesces to form a single large cavity or antrum lined by several layers of GCs. As the oocyte reaches its definitive size, the GCs proliferate to form the

Copyright © 1999 by Academic Press.
All rights of reproduction in any form reserved.

cumulus oophorus that surrounds the oocyte (mature follicle). The membrana granulosa consists of 8–18 cell layers in its broadest part, and the cumulus oophorus also consists of 8–10 cell layers of GCs around the oocyte. Close to the time of follicular rupture, the parietal GCs become larger, reaching a diameter of 12–18 mm in the human. After ovulation, the GC nuclear volume reaches 125 mm³.

Histologically, GCs have scant pale frothy cytoplasm, indistinct cell borders, and small, round to oval hyperchromatic nucelii. GCs typically surround small cavities called Call–Exner bodies, which contain material consisting of excess basal lamina. Although there is abundant ultrastructural evidence of active protein synthesis, features indicative of steroidal biosynthesis are absent until shortly prior to ovulation. Gap junctions appear during the late developmental stage of the follicle concomitantly with the appearance of LH receptors. These junctions may serve as structures through which intracellular communication and local coordination of follicular maturation can be established. Through these gap junctions, cAMP may enter the oocyte and hold the oocyte in meiotic arrest. Also, the coordination of steroidogenic stimulation of GCs during transformation into luteal cells may occur through these junctions. *In vitro* studies conducted in primary cultures of rat, pig, and human GC demonstrate stimulation of gap junction formation by estrogen, follicle-stimulating hormone (FSH), insulin, and basement membrane components.

D. Corpus Luteum

Luteinized GCs in the human are large, measuring 30–35 mm in diameter, and are polygonal in shape. They contain abundant lipid droplets with histochemical features typical of steroidogenic cells. In the pregnant primate's corpus luteum, the granulosa–lutein cells enlarge, reaching a maximum size of 50–60 mm by 8 or 9 week's gestation. The GCs are round or polyhedral and contain one or two prominent nucleoli. GCs also contain large vacuoles that occupy almost the entire cell, which displaces the nucleus. In addition, diffusely scattered lipid droplets as well as abundant smooth and rough endoplasmic reticulum are commonly found in GCs.

E. Atretic Follicles

As degeneration of the oocyte begins, degeneration of the GC soon follows, with the eventual disappearance of the follicle. The earliest evidence of this process is a decrease in the number and mitotic activity of GCs, with resulting thinning of the GC layer. Eventually, there is invasion by the vascular connective tissue and filling of the central cavity.

II. DIFFERENTIATION

GCs display a high degree of structural change and play a key role in the functional maturation of the follicle. A finely tuned process of differentiation occurs in all constituents of the follicle during folliculogenesis. The GCs surrounding the oocyte are committed to undergo structural and functional differentiation to create the conditions necessary for resumed oogenesis, ovulation, fertilization, and implantation.

A. Developing Follicle

The understanding of GC differentiation has dramatically improved through the extensive use of cell culture approaches. Differentiation of GCs begins when the primordial follicle is recruited to initiate growth. There is early acquisition and binding of FSH receptors at or about the primary follicle stage (when the oocyte ceases to grow). With binding of the FSH receptors, profound alterations in GCs occur concurrent with morphological transformation. The once-flattened epitheliod GC assumes a cuboidal shape presumably as a result of cytoskeletal alterations. In addition, GC cells are shifted into prominent mitoses.

At the secondary or early tertiary follicular stage, differentiation of GCs become FSH dependent, in contrast to pituitary-independent differentiation in earlier stages. FSH binds to G-protein-linked FSH receptors in the GCs, which results in adenylate cyclase stimulation and increased levels of cAMP. In turn, cAMP binds to cAMP-dependent protein A kinases, which ultimately phosphorylate regulatory proteins and initiate the various GC-differentiated processes.

FSH binding causes significant changes in the intracellular organelles of GCs. Furthermore, there are rearrangements in three major cytoskeletal elements (microtubules, microfilaments, and intermediate filaments). These rearrangements are essential for the clustering of lipid droplets, mitochondria, lysosomes, and smooth endoplasmic reticulum, cellular machinery indicative of steroid production. All these changes are critical for the steroidogenic role of the differentiated GC demonstrated by increasing levels of aromatase activity and, ultimately, estrogen biosynthesis.

B. Luteinization

A major requirement for the maintenance of a pregnancy is the steroidogenic organ, the corpus luteum. The corpus luteum is derived from cells of the ovulatory follicle after the midcycle luteinizing hormone (LH) surge. Formation of the corpus luteum is dependent on the conversion of GCs into luteal cells, a differential process called luteinization.

At the time of antrum formation in the follicle, LH receptors appear in GCs. These LH receptors are found only in GCs from large preovulatory follicles. Most of the newly synthesized LH receptors are located over the numerous microvilli around the GC circumference. After receptor activation by ligand binding, there is a generation of intracellular signals through the cAMP and phosphotidylinositol cascades. These cellular responses to LH stimulation are characterized by biochemical and morphological aspects of luteinization (Fig. 1).

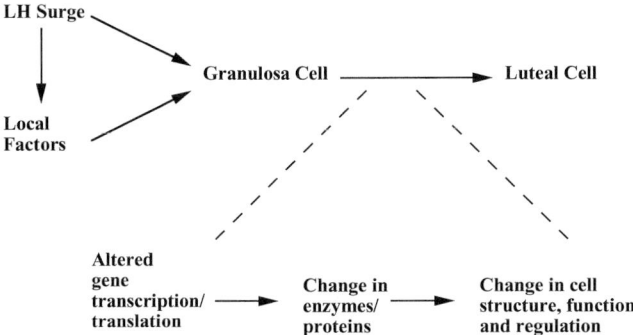

FIGURE 1 Biochemical and morphological aspects of luteinization.

In general, luteinization is a process associated with a decline in proliferation of GCs and an increase in enzymes involved with progesterone synthesis. The GC mitotic index falls after LH exposure while there is an increase in cell size, cytoplasm volume, and nuclear to cytoplasmic ratio. LH also causes a massive synthesis of progesterone mainly by the induction of cholesterol side chain cleavage enzyme. A concurrent morphologic change occurs in the GCs, with an increase in lipid inclusions and the mitochondria attaining increasing complexity with their cristae.

C. Atresia

Follicular atresia is a universal phenomenon and occurs at any stage of follicular development. Follicles in the growing or nongrowing pools may both become atretic. During atresia, GC alterations occur in two patterns: (i) prominent necrotic changes in the oocyte with secondary alterations in the GCs or (ii) degenerative changes in the granulosa with an almost unchanged oocyte.

The degeneration of GCs has all the characteristics of apoptotic cell death. Cells demonstrate nuclear condensation (pyknosis) and marked dilatation of cytoplasmic organelles. There is a loss of microvilli corresponding with loss of LH receptors. GCs flatten and show DNA fragmentation. Eventually, there is follicular basement membrane breakdown and clearance of GCs by macrophages and scavenger-type cells.

III. STEROIDOGENESIS

GCs are the cellular source of estrogens and progestins, the two most important ovarian steroids.

A. Estrogens

The principal site of aromatase activity and estrogen synthesis is the GCs of large antral and preovulatory follicles. Regulation of androgen aromatization in the GCs appears to be by the action of FSH. The effects of FSH on aromatase activity are to enhance both synthesis of aromatase enzymes and their activation.

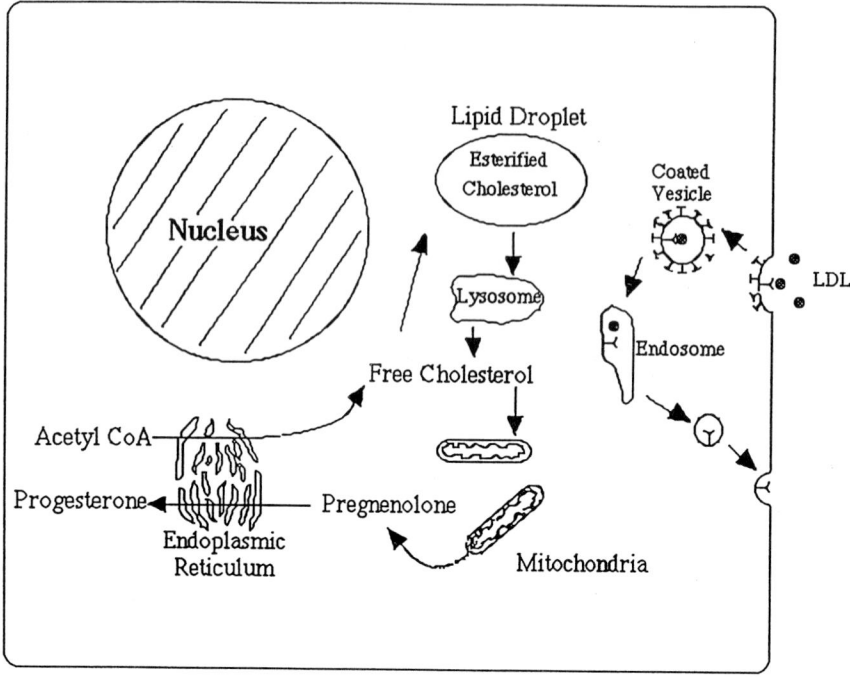

FIGURE 2 The steroidogenic granulosa cell. Free cholesterol is the precursor for pregnenolone and ultimately progesterone. The free cholesterol can be derived from one of three possible sources, depending on the follicular stage: circulating low-density lipoproteins (LDL), hydrolysis of cholesterol esters, or cholesterol synthesized *de novo*.

In addition to this FSH action, estrogen biosynthesis requires cooperation between the GCs and their thecal neighbors. The LH-dependent theca cells supply the androgen substrates to be acted on by the aromatase of the FSH-inducible GCs.

B. Progestins

The production of progesterone and its metabolites is one of the major biosynthetic activities of GCs. Typical of all steroidogenic cells, GCs convert cholesterol into pregnenolone and then to progesterone (Fig. 2). Central to this biosynthetic pathway, however, is the availability of cholesterol. In the GCs of antral and preovulatory follicles, intracellular cholesterol is probably derived from either *de novo* synthesis or endogenous stores. After ovulation, extracellular low-density lipoprotein will become the major source of cholesterol for steroid biosynthesis in the luteal cells.

Progesterone biosynthesis by GCs increases dramatically after the LH surge. The main action of LH appears to be in the conversion of cholesterol to pregnenolone, the rate-limiting step in progesterone biosynthesis. This reaction is mediated by the cholesterol side chain cleavage enzyme, P450$_{SCC}$. LH appears to both activate P450$_{SCC}$ and stimulate cholesterol transport.

See Also the Following Articles

CORPUS LUTEUM; FOLLICULAR DEVELOPMENT, CONTROL OF; STEROIDOGENESIS

Bibliography

Adashi, E. Y., and Leung, P. C. K. (Eds.) (1993). *The Ovary.* Raven Press, New York.

Amsterdam, A., and Rotmensch, S. (1987). Structure–function relationships during granulosa cell differentiation. *Endocr. Rev.* 8, 309–337.

Jones, R. E. (Ed.) (1978). *The Vertebrate Ovary.* Plenum Press, New York.

Knobil, E., and Neill, J. D. (Eds.) (1994). *The Physiology of Reproduction,* 2nd ed. Raven Press, New York.

Paton, A. C., and Collins, W. P. (1992). Differentiation processes of granulosa cells. In *Oxford Reviews of Reproductive Biology* (S. R. Milligan, Ed.), Vol. 14, pp. 168–223. Oxford Univ. Press, New York.

■

Great Apes

see Primates, Nonhuman

■

Growth Factors

J. Julie Kim and Asgerally T. Fazleabas

University of Illinois College of Medicine at Chicago

  I. Introduction
 II. Epidermal Growth Factor
III. Insulin-like Growth Factors
 IV. Transforming Growth Factor-β
  V. Vascular Endothelial Growth Factor
 VI. Fibroblast Growth Factor
VII. Platelet-Derived Growth Factor
VIII. Hepatocyte Growth Factor
 IX. Summary

GLOSSARY

autocrine Referring to a product which acts on the same cell that produces it.

growth factors Peptides that are produced locally within tissues and act via autocrine, paracrine, or juxtracrine mechanisms.

juxtacrine Referring to a product produced by one cell which acts on an adjacent cell with which it has contact.

paracrine Referring to a product produced by one cell which then acts on an adjacent cell.

signal transduction The biochemical pathway by which a growth factor alters cell function(s) when the growth factor binds its receptor and transduces a signal to the nucleus.

The reproductive tract undergoes dynamic changes during the fertile life span of both males and females. Historically, these changes were thought to be modulated by gonadotropins and steroid hormones. It is now apparent that other molecules, such as the ever growing family of growth factors, play an important role in regulating the hormonal responsiveness and functions of the reproductive tract.

I. INTRODUCTION

Growth factors interact with cell surface receptors and initiate a cascade of events which influence cell growth, metabolism, differentiation, and death. Significant progress has been made in characterizing the structure and function of growth factors, their receptors, and binding proteins. Much of this progress has resulted from the development of recombinant growth factors, molecular cloning techniques, and transgenic animals. Despite the immense increase in knowledge of growth factor biology, an absolute physiological role in reproductive tissues for any of the growth factors has not been convincingly demonstrated. This article briefly describes the growth factors and their receptors that have been localized to the ovary, testes, uterus, and placenta and highlights the postulated functions of these molecules in reproductive processes.

II. EPIDERMAL GROWTH FACTOR

A. Properties

The epidermal growth factor (EGF) family consists of EGF, transforming growth factor-α (TGF-α), heparin-binding EGF (HB-EGF), amphiregulin, betacellulin, vaccinia virus growth factor, and the heregulin subfamily. The receptor family belongs to the erb B class of oncogenes with erbB1 being the best characterized EGF receptor (EGF-R). This overview focuses primarily on the expression of EGF, TGF-α, and EGF-R in reproductive tissues.

EGF was first isolated in 1961 from submandibular gland extracts of newborn mice. EGF is a 6-kDa single-chain polypeptide of 53 amino acids with three disulfide bonds which are responsible for the tertiary structure of the molecule and its biological activity. EGF is synthesized by proteolytic cleavage of a large precursor molecule of 128 kDa, comprising 1217 amino acids in murine species and 1207 amino acids in the human. The human EGF precursor gene is about 110 kb containing 24 exons and is located on chromosome 4. Studies have demonstrated that this precursor can be biologically active and can activate the EGF-R on adjacent cells.

TGF-α is a 50-amino acid, single-chain polypeptide that is structurally similar to EGF. It is synthesized as a large precursor molecule which is cleaved by an elastase-like protease to a mature protein of 5.6 kDa. TGF-α can bind to EGF-R with equal affinity to EGF and activate the receptor. The gene for TGF-α is located on chromosome 2, distinct from EGF.

The EGF receptor, erbB1 is a 170-kDa monomeric protein of 1186 amino acids composed of an extracellular ligand-binding domain, a short 20- to 22-amino acid transmembrane-spanning region, and an intracellular cytoplasmic domain containing tyrosine kinase and autophosphorylation sites. A truncated form that only encodes the tyrosine kinase domain of the receptor has also been identified. Ligand binding results in dimerization or clustering of the receptor–ligand complex and activation of the receptor tyrosine kinase, which subsequently initiates a cascade of signaling events and a multitude of biological responses. Following activation, the EGF receptor–ligand complex is internalized and degraded in lysosomes.

B. Localization and Postulated Functions in Reproductive Tissues

The biological actions of EGF and its related peptides are extremely diverse. EGF is involved in cellular proliferation, regulation, and differentiation. At the cellular level, EGF stimulates ion transport, enhances endogenous protein phosphorylation, causes alterations in cell morphology, and stimulates DNA synthesis.

1. Ovary

EGF and EGF-R are expressed in follicular and stromal cells of the human ovary. Although the precise role of EGF in the ovary remains to be clarified, some studies have demonstrated effects of EGF on ovarian tissue. In the rat, EGF can attenuate follicle-stimulating hormone (FSH)-mediated differentiation of granulosa cells. In the perfused rabbit ovary, EGF can inhibit ovulation and human chorionic gonadotropin (hCG)-induced estradiol and progesterone secretion. EGF in conjuction with transforming growth factor-β (TGF-β) decreases the ability of granulosa cells to express cAMP-induced luteinizing hormone (LH) receptors *in vitro*.

2. Testes

EGF has been suggested to play a role in spermatogenesis, differentiation of Sertoli cells, and testosterone-dependent differentiation of the Wolffian duct. It may also be involved in the development of prostatic hyperplasia. The effects of EGF on Leydig cells depend in part on the presence of LH.

3. Oviduct

EGF, TGF-α, and EGF-R are present in both the human and the baboon oviduct. Results from human studies suggest that expression of these growth factors varies with cell type and reproductive stage. In the baboon, EGF and EGF-R are present in all tissue compartments, especially during the follicular stage. TGF-α is localized only in secretory epithelial cells.

4. Uterus

In the primate, EGF expression does not appear to be hormonally regulated during the menstrual cycle. TGF-α is localized in stromal cells of the human endometrium during the proliferative phase and in both surface and glandular epithelial cells during the secretory phase of the menstrual cycle. In contrast, TGF-α is localized exclusively to uterine epithelial cells in the baboon. EGF-R has been immunolocalized to both the glandular epithelium and the stromal cells, and binding studies demonstrated that EGF-R sites increase in number in proliferative stage endometrium and peak just prior to ovulation.

EGF, TGF-α, HB-EGF, and amphiregulin are expressed in the mouse uterus. EGF is restricted to the luminal and glandular epithelia and is associated with cell proliferation. EGF expression is upregulated by estrogen and it can independently mimic estrogenic effects in the mouse uterus. TGF-α is expressed in the luminal epithelium of the receptive uterus prior to implantation. Since implantation is normal in TGF-α knockout mice, it does not appear to be critical for this process. HB-EGF and amphiregulin are found in the luminal epithelium surrounding the activated mouse blastocyst and may be involved in the early processes of implantation. EGF-R is expressed in all uterine cell types, whereas the truncated form of EGF-R is limited to epithelial cells. Transgene knockouts of EGF-R have shown different responses depending on the genetic strain of mouse in which the gene was mutated.

5. Placenta and Decidua

The primate placenta and decidua express EGF, TGF-α, and EGF-R and their expression appears to be maximal in earlier rather than later stages of gestation. These ligands have been implicated to play a role in trophoblast invasion, proliferation, and differentiation. In the mouse, EGF-R is expressed on the blastocyst, which may interact specifically with HB-EGF, and amphiregulin that are expressed on the luminal epithelium during the initiation of decidualization.

III. INSULIN-LIKE GROWTH FACTORS

A. Properties

The insulin-like growth factors (IGFs) were first discovered as a sulfation factor mediating the effects of growth hormone on somatic growth and were therefore initially named somatomedins.

IGFs are single-chain polypeptide hormones that exhibit close homologies to insulin. IGF-I and IGF-II are low-molecular-weight peptides (7 kDa) that share 62% amino acid sequence homology, possess different structures, and are recognized by distinct receptors on the cell surface.

The human IGF genes consist of complex exon–intron structures and have multiple promoter elements. The human IGF-I gene spans a region of more than 95 kb of chromosomal DNA on chromosome 12 and contains five exons interrupted by four introns. The human IGF-II gene is located on chromosome 11 and spans 30 kb of chromosomal DNA and contains nine exons.

1. The IGF Receptors

IGFs bind to IGF-I and IGF-II/mannose-6-phosphate (M-6-P) receptors as well as to the insulin receptor, but with lower affinity. The type I IGF receptor, like the insulin receptor, is a heterotetramer consisting of two extracellular α subunits and two transmembrane β subunits. It binds IGF-I with higher affinity than IGF-II and insulin. The α subunits and some extracellular portions of the β subunits confer ligand binding, whereas the intracellular β subunits possess a tyrosine kinase catalytic domain. It has been shown that the tyrosine kinase of each β subunit transphosphorylates the opposite β subunit. The actions of IGF-II/M-6-P receptor in IGF-II signaling are less clear, but this receptor may play a critical role in clearing IGF-II from the circulation.

The biological actions of IGFs are modulated by a family of at least six distinct IGF-binding proteins (IGFBPs) that are produced by most tissues and are found in the circulation and in extracellular compartments. IGFBPs modulate the action of IGF in various ways. They act as transport proteins in plasma and control the efflux of IGFs from the vascular space, prolong the half-lives of the IGFs and regulate their metabolic clearance, provide a means of tissue and cell type-specific localization, and directly modulate the interaction of the IGFs with their receptors, thereby controlling their biological actions.

B. Localization and Postulated Functions in Reproductive Tissues

IGFs are synthesized by a number of cell types and may act in an autocrine, paracrine, or endocrine fashion. *In vitro*, IGF actions include cell cycle progression, proliferation, differentiation, regulation of cell function, and cell death.

Signaling via the IGF receptors is similar to that in other growth factors in that ligand binding autophosphorylates the IGF receptor, which subsequently phosphorylates other tyrosine-containing substrates. The predominant substrate of the IGF-I receptor, the insulin receptor substrate-1 (IRS-1), weighs 185 kDa and contains 21 potential tyrosine phosphorylation sites, of which 6 are recognition sites for binding proteins containing src homology 2 (SH2) domains. When IRS-1 is phosphorylated by either the insulin or IGF-I receptor, it can bind to at least two SH2-containing proteins. The p85 regulatory subunit of phosphotidylinositol-3 kinase and growth factor receptor-bound protein-2 (Gfb2) are proteins with SH2 domains.

1. Ovary

The mRNAs for IGFs and their receptors are found in human follicles at successive stages of their development. Specifically, IGF-I protein and mRNA are detected in granulosa cells, whereas their presence in theca cells is less clear. In the rat ovary, IGF-I gene expression is restricted to granulosa cells and type I and II IGF receptors exist on both granulosa and theca cells. IGFs are able to stimulate cell proliferation and steroidogenesis in human granulosa and theca interna cells from immature to preovulatory follicles. These effects are strongly enhanced by the presence of gonadotrophins.

2. Testes

IGF-I is produced by Sertoli cells and its receptor is present on Leydig, Sertoli, and germinal cells. IGFs have been shown to induce cell proliferation in spermatogonia and Sertoli cells, increase hormone secretion from Leydig cells, and stimulate Sertoli cell transferrin production.

3. Uterus

Immunoreactive IGF-I and -II protein or their mRNA transcripts have been demonstrated in uterine luminal fluid or tissue extracts from rat, mouse, pig, cow, nonhuman primates, and humans. IGFs participate in the mitotic and differentiative events in the endometrium during the menstrual cycle and in early pregnancy. In the rat, IGFs have been localized in endometrial epithelium and stroma as well as in uterine myometrium. Estrogen can increase uterine IGF-I and IGF-II mRNA expression. While endometrial IGF-I receptor numbers are also estrous cycle dependent with changes paralleling circulating estradiol levels, IGF-II receptors do not vary. In the mouse, estrogen induces IGF-I gene expression in the endometrial epithelium, whereas progesterone induces it in the endometrial stroma. In pigs, estrogen and/or progesterone increase endometrial IGF-I mRNA and protein, whereas these steroids have no effect on IGF-II gene expression in this tissue. IGF-II mRNA is most abundantly expressed in late gestation in porcine as well as bovine endometrial tissues.

In humans, the IGF-I gene is expressed primarily in proliferative and early secretory endometrium, whereas IGF-II is expressed abundantly in mid- to late secretory endometrium and early pregnancy decidua. Since IGF-II mRNA is expressed during differentiation of the endometrium, this growth factor may be involved in endometrial tissue shedding during each menstrual cycle or endometrial remodeling during early pregnancy. Circulating IGF-I levels increase as pregnancy advances, whereas IGF-II levels do not change or increase only slightly. This is in contrast to the baboon and rhesus monkey in which IGF-II and not IGF-I levels increase during gestation. After delivery, maternal circulating IGF-I and -II levels rapidly decline, suggesting that the major source of these growth factors is the placenta and/or decidua.

IGF receptors are also present in the endometrium and their expression is steroid dependent. Type I and II IGF receptor mRNAs are more abundantly expressed in the secretory phase and during early pregnancy in human endometrium. In the baboon, type I IGF receptor is localized to the proliferating glandular epithelium and is most evident during the follicular and early luteal stages of the menstrual cycle. The decrease during the mid- to late luteal phase corresponds to the decrease in epithelial mitotic activity and differentiation into a secretory endometrium. Type I IGF receptor is also expressed in decidualizing stromal cells at the site of implantation in the baboon.

4. Placenta

IGF-I and -II are produced in human placenta and are expressed as early as 9 weeks of gestation. IGF-I mRNA levels are highest in first trimester and decrease thereafter, whereas IGF-II mRNA expression is higher in second-trimester placentas than in first-trimester or term placental tissues. IGF-I mRNA is expressed by syncytiotrophoblast, whereas IGF-II mRNA appears in placental fibroblasts and in the cytotrophoblast. IGF-I and -II are present in the chorion, amnion, and decidua both in first-trimester and term placentas. Human placental cells express mRNA for IGF-I receptors in placental vessels, stroma, syncytiotrophoblast, and cytotrophoblast.

Gene targeting experiments demonstrate that IGF-I-deficient mice are smaller at birth and grow poorly postnatally. Defects are evident in skeletal muscle, skin, and bone maturation. IGF-II-deficient mice are small at birth but grow normally after birth with no signs of impairment. In IGF-I receptor knockout mice, growth is significantly impaired, even more so than in IGF-I-deficient mice. Most of these mice die shortly after birth. The IGF-II receptor-deficient mice are large at birth and have increased levels of circulating IGF-II. Mice with targeted disruption of IGFBP-2 are normal, whereas trangenic overexpression of IGFBP-1 impairs the action of estrogen in the uterus.

IV. TRANSFORMING GROWTH FACTOR-β

A. Properties

TGF-β is a 25-kDa dimeric protein composed of two identical disulfide-linked chains. The monomers of 112 amino acids are derived from a larger, biologically inactive 390-amino acid precursor by proteolytic digestion. Five forms of TGF-β have been cloned: TGF-β1–TGF-β5. All monomers are 70–80% identical and TGF-β1, -β2, -β3, and -β5 interact with the same receptor system.

There are three types of receptors: type I (65 kDa) and type II (85–95 kDa), which bind TGF-β1 with an affinity 10 times higher than TGF-β2, and type III, which has a similar affinity for TGF-β1 and TGF-β2.

B. Localization and Postulated Functions in Reproductive Tissues

TGF-β has widespread biological actions which can depend on cell type, its state of differentiation, growth conditions, and other growth factors present. TGF-β may have stimulatory, inhibitory, biphasic, or no effects on cell proliferation. TGF-β inhibits growth of most epithelial cells in culture with the exception of human mesothelial cells and rat uterine and vaginal epithelial cells, which appear to proliferate in response to TGB-β. Treatment of mesenchymally derived cells with TGF-β generally results in an increase in proliferation, although these effects can be modified by the presence of other factors. While inducing differentiation in some cells, TGF-β inhibits this process in adipocytes and myocytes. TGF-β has also been reported to modulate the differentiated state of granulosa and Leydig cells.

1. Ovary

Granulosa and thecal cells secrete TGF-β and both cell types are responsive to TGF-β. Maturation of ovarian granulosa cells involves cAMP production, steroidogenesis, and LH receptor formation and it appears that TGF-β has a dual effect on these processes. TGF-β enhances the stimulatory actions of low levels of FSH but selectively inhibits development of granulosa cells and induction of LH receptors at higher levels of FSH. TGF-β can also enhance the induction of EGF receptors by FSH and regulate the inhibitory action of EGF during granulosa cell differentiation. The role of TGF-β in ovarian granulosa cell differentiation *in vivo* is unknown. TGF-β is involved in regulation of growth and steroidogenesis during follicular development, and it is a potent *in vitro* stimulator of oocyte maturation in the rat.

2. Testes

Sertoli cells produce TGF-β which decreases with the addition of FSH. TGF-β produced by peritubular and Sertoli cells has been implicated in paracrine control of steroidogenesis in Leydig cells. This growth factor does not appear to affect proliferation of any of these cells, nor does it seem to affect their differentiation. It has been shown that TGF-β reduces expression of hCG receptors on porcine Leydig cells by almost 70%.

3. Uterus

TGF-β1 is expressed in the normal human endometrium and in endometrial cancer cells. In the reproductive tract, angiogenesis is especially important for implantation. Since TGF-β effects so many cell types, it is not clear if vascularization results from a direct effect of TGF-β on endothelial cells or from indirect stimulation of endothelial cells by other growth factors that are induced by TGF-β.

The preimplantation uterus of a pregnant mouse synthesizes TGF-β in the glandular and luminal epithelium, whereas TGF-β protein accumulates in the extracellular matrix of the stroma surrounding the epithelium. After implantation, the primary decidual zone appears to synthesize TGF-β that accumulates in the matrix of the secondary decidual zone and decidua capsularis. In addition, levels of TGF-β in the uterus may be regulated by estrogen and progesterone. This, in turn, may affect angiogenesis, extracellular matrix production, cell growth, and cell differentiation in the uterus.

4. Embryogenesis and Placentation

In amphibians, TGF-β3 is able to induce formation of mesoderm from ectoderm *in vitro*. In 11- to 18-day mouse embryos, TGF-β protein is localized to areas of critical epithelial–mesenchymal interactions, especially regions that show extensive tissue remodeling such as in mesenchyme underlying developing digits, heart valves, and palate.

TGF-β1 has been localized in human placenta and in trophoblast, mainly in the periphery of placental villi, i.e., the syncytiotrophoblastic layer. The highest levels of TGF-β1 mRNA expression are found between 17 and 34 weeks of gestation in humans.

5. Extracellular Matrix

One of the best characterized biological actions of TGF-β is its ability to increase the accumulation and response of cells to extracellular matrix proteins. Synthesis of matrix proteins by cells is enhanced and degradation of these proteins by proteases is inhibited by TGF-β. Matrix proteins induced by TGF-β are fibronectin, types I, III, IV, and V collagens, thrombospondin, osteopontin, tenascin, osteonectin, and chondritin/dermatan sulfate proteoglycans. The production of these proteins results from both activation of promoter elements of some matrix proteins and stabilization of their mRNAs. TGF-β also alters mRNA levels of genes that regulate matrix degradation. It decreases synthesis and secretion of proteases, such as plasminogen activator, collagenase, elastase, and transin, while increasing synthesis and secretion of protease inhibitors, such as plasminogen activator inhibitor and tissue inhibitor of metalloproteases. TGF-β also promotes interaction of cells with the extracellular matrix proteins by increasing expression of cell membrane receptors for cell adhesion proteins, such as integrins. Regulation of expression of extracellular matrix proteins is particularly important in the female reproductive tract.

6. Immune System

The effect of TGF-β on immune cell function is mainly inhibitory. It inhibits interleukin-dependent T and B lymphocyte proliferation *in vitro* and the generation and function of lymphokine-activated cells, and it suppresses immunoglobulin secretions by B lymphocytes and upregulates IL-2 and transferrin receptors in T lymphocytes. Immunoreactive TGF-β1 is present in both pre- and postimplantation mouse embryos. It has been postulated that the immune suppressive effects of TGF-β could be involved in preventing rejection of fetal–placental tissues by the mother during pregnancy.

V. VASCULAR ENDOTHELIAL GROWTH FACTOR

A. Properties

The human vascular endothelial growth factor (VEGF) gene is encoded by a region that spans approximately 14 kb of chromosomal DNA. Alternate exon splicing of a single VEGF gene results in the synthesis of four polypeptides encoded for by 121, 165, 189, and 206 amino acids. VEGF165, a basic heparin-binding homodimeric glycoprotein of 45 kDa, is the major isoform. VEGF121 is a weakly acidic polypeptide that fails to bind heparin. VEGF189 and VEGF206 are more basic and bind to heparin with greater affinity than VEGF165. VEGF121 is a freely soluble protein, whereas VEGF165 can be secreted or found bound to the cell surface or extracellular matrix. VEGF189 and VEGF206 are also found in the extracellular matrix.

Three cell surface receptors for VEGF have been identified: flt-1 (fms-like tyrosine kinase), flt-4, and KDR (kinase domain region)/flk-1 (fetal liver kinase-I, which is the murine homolog). Receptors have seven immunoglobulin-like domains in the extracellular domain, a single transmembrane region, and a consensus tyrosine kinase sequence that is interrupted by a kinase inert domain. The signal transduction mechanisms of VEGF receptors remain unclear; however, VEGF acting via endothelial tyrosine kinase receptors can induce phosphorylation of many proteins.

B. Localization and Postulated Function in Reproductive Tissues

VEGF is a potent mitogen that has been associated with regulation of angiogenesis. It can induce the expression of factors that promote cellular invasion and tissue remodeling, such as plasminogen activators, plasminogen activator inhibitor-1, and metalloproteinases.

1. Ovary

VEGF is localized in the thecal cell layer of healthy follicles with minimal VEGF peptide detected in the granulosa cell layer. In cultured bovine ovarian granulosa cells, LH can induce VEGF mRNA expression via the adenylate cyclase pathway. VEGF mRNA and protein are expressed in human ovarian granulosa and theca cells during the latter stages of follicular development and are not expressed in atretic follicles. VEGF may be involved in angiogenesis that occurs during growth and regression of the corpus luteum. It is expressed in luteal tissue throughout the ovine estrous cycle with increased expression during rapid luteal development when luteal vascular growth is maximum.

2. Uterus

Angiogenesis is an essential component of endometrial regeneration after menses and in preparation for implantation. VEGF is detected in both glandular epithelium and stroma in human endometrium. During the early proliferative phase, epithelial cells exhibit VEGF staining, whereas staining in stromal cells is limited. During mid- to late proliferative phase, stromal cell staining of VEGF is intense, whereas in late secretory endometrium, only the glands stain positively. The increase in VEGF parallels the increase in estradiol levels in the proliferative phase of the menstrual cycle and addition of estradiol to isolated human endometrial cells increases VEGF mRNA expression. After initial attachment of the mouse embryo, luminal epithelial and stromal cells of the endometrium immediately surrounding the blastocyst exhibit accumulation of VEGF mRNA. Human myometrium also expresses mRNA for VEGF, which appears to be higher in the secretory phase than in the proliferative phase of the cycle.

Expression of the VEGF receptors flt-1 and KDR is not restricted to endothelial cells. The mRNAs for flt-1 and KDR have been detected in the smooth muscle cells of the uterus and in uterine decidua at both the mesometrial and antimesometrial poles.

3. Placenta

The placenta displays a vascular network critical to its development. VEGF immunoreactive protein and flt-1 protein are present in the cytotrophoblast in first-trimester placenta, in Hofbauer cells within the mesenchyme, in macrophages, and in maternal decidual cells. In near-term ovine placenta, VEGF mRNA is detected in cotyledon, chorion, and amnion. KDR is found in endothelial cells and hemangioblasts of human placenta and in the endothelial cells of the villous stroma adjacent to areas of trophoblast proliferation.

Placenta growth factor, a new member of the VEGF family, has been found in abundance in human placental and maternal decidual cells. Evidence suggests that flt-1 and not KDR is the receptor for this growth factor.

Targeted disruption of the VEGF gene in mice results in embryonic lethality presumably due to the absence of vasculogenesis and angiogenesis. Targeted disruption of VEGF receptors flt-1 or KDR/flk-1 results in embryonic mortality during early development.

VI. FIBROBLAST GROWTH FACTOR

A. Properties

There are nine identified members of the FGF family, eight of which are expressed in the human (FGF-1 through -7 and -9), with FGF-8 identified only in mice. Alternative names have been given to the different FGFs: FGF-1, acidic FGF; FGF-2, basic FGF (bFGF); FGF-3, Int-2; FGF-4, k-FGF and Kaposi's-GF; FGF-5; FGF-6, Hst-2; FGF-7, keratinocyte growth factor (KGF); FGF-8, androgen-induced GF; and FGF-9, glia activating factors. They range from 17 to 38 kDa in molecular weight, with only 14% nucleotide sequence homology in the coding region among the eight human FGF genes.

Of all the FGFs, FGF-2 or bFGF has the widest tissue distribution and the broadest spectrum of biological activities. It was first identified in 1974 as a 146-amino acid protein that could stimulate proliferation of 3T3 mouse fibroblast cells.

The 34-kb bFGF gene is localized on human chromosome 4. It has three coding regions separated by two large noncoding regions. The 5′ end does not encode the conventional signal peptide sequence necessary for directed secretion of the protein to the cell exterior. Both longer and shorter forms of FGF-2 than that predicted from the cDNA sequence exist in guinea pig brain, rat brain and liver, human placenta, and several types of cultured cells due to the use of different translation initiation sites. Since the bFGF gene predicts a protein of 155 amino acids, these N-terminally shortened versions may result from proteolytic cleavage.

Four distinct genes code for receptors of FGF (FGFR-1, FGFR-2, FGFR-3, and FGFR-4). These receptors have an extracellular ligand-binding domain, a transmembrane domain, and a cytoplasmic tyrosine kinase domain. FGFs can also bind with low affinity to glycosaminoglycans, including heparin and heparan sulfate.

B. Localization and Postulated Functions in Reproductive Tissues

1. FGF-2

The FGFs play major roles in development, wound healing, hematopoiesis, and tumorogenesis. Their action is not restricted to fibroblasts per se and can induce mitogenic, chemotactic, neurotrophic, and angiogenic activities in a number of cell types.

Newly synthesized FGF is mainly secreted, but some protein translocates to the nucleus. Nuclear localization can occur following cytoplasmic synthesis or cell surface binding. Exogenous FGF can also be internalized and translocated to the nucleus of cells. The mitogenic effect of FGF-1 has been directly correlated with its appearance in the nucleus and FGF-2 has been proposed to activate nucleolar protein kinases and to regulate ribosomal gene transcription.

i. Ovary   FGF-2 has been immunolocalized in bovine follicles, specifically in the cytoplasm of oocytes from primordial and primary follicles. Strong staining is also evident in corpora lutea, surface epithelium of the ovary, smooth muscle cells surrounding blood vessels, and theca interna cells. In granulosa cells, the layers closest to the most basal region exhibit FGF-2 immunostaining. mRNAs for FGF-2 and its receptor have been observed in the fetal ovary and in luteinized granulosa cells obtained following ovarian hyperstimulation. The possible role of FGF-2 in human follicular function remains unclear. It is mitogenic for granulosa cells, inhibitory for granulosa cell differentiation, and can regulate steroidogenesis in thecal cells. When FGF was infused into the ovarian artery of ewes, the secretion of estradiol and androstenedione was suppressed, although there were no detectable effects on progesterone secretion or pulsatile LH secretion.

ii. Testes   FGF-2 may regulate gonadal function in the male. It is mitogenic for immature porcine Sertoli cells, which are thought to play a role in prostatic cell growth, and modulates basal and LH/hCG-stimulated Leydig cell function. Fast-growing prostatic tumors exhibit high FGF-2 expression and several spliced variants of FGF receptor. FGFR-1 transcripts have been found in Leydig cell-enriched fractions, peritubular cells, Sertoli cells, and, to a lesser extent, germ cells.

iii. Uterus FGF-2 is present in high concentrations in the normal endometrium of cycling women, whereas there is little evidence for FGF receptor expression in this tissue. FGF-2 stimulates tissue regeneration by inducing proliferation and differentiation of virtually all cell types. Thus, FGF-2 could play a pivotal role in regeneration and neovascularization of the endometrium during each menstrual cycle in preparation for implantation. In mouse endometrial epithelial cells, FGF-2 and two forms of FGFR-1 mRNA have been detected. FGF-2 is present in uterine luminal fluid and decidual cells in the mouse during early pregnancy.

iv. Placenta In humans, FGF-2 is present in cytotrophoblast cells, extravillous trophoblast, endothelial cells, and decidua during the first trimester. At term, FGF-2 staining was in the syncytiotrophoblast and fetal membranes. This growth factor may be critical for ensuring vascularization during placental development.

2. KGF

KGF (FGF-7) is a single-chain polypeptide of 28 kDa with mitogenic activity specific for epithelial cells. KGF is normally expressed by stromal fibroblasts and acts on epithelial cells in a paracrine fashion. A large induction of KGF expression occurs in the dermis during wound healing.

i. Ovary Bovine thecal cells express KGF mRNA and protein, whereas granulosa cells have no detectable KGF. Recombinant KGF stimulates the proliferation of granulosa cells but not thecal cells.

ii. Uterus KGF mRNA expression is high in the progesterone-dominated late secretory stage human endometrium and it has been postulated that KGF is the stromally derived growth factor that regulates epithelial cell function via the KGF receptor during the luteal phase. High levels of KGF receptor are evident when the endometrium is under estrogen dominance and it has been shown in the rhesus monkey that KGF mRNA is highest in the luteal phase or after progesterone treatment.

iii. Placenta KGF mRNA is strongly expressed in placental mesenchymal cells in rhesus monkey in early gestation and then decreases as pregnancy proceeds. KGF receptor mRNA is expressed in the adjacent trophoblastic epithelium.

Overexpression of the KGF gene in transgenic animals results in hyperplasia in the male genital tract, including the seminal vesicles, vas deferens, and the prostate. In female transgenic animals, the mammary epithelium undergoes hyperplasia.

VII. PLATELET-DERIVED GROWTH FACTOR

A. Properties

Platelet-derived growth factor (PDGF) consists of two distinct, but closely related, polypeptide chains of 30 kDa. Depending on the cell type, PDGF can exist as AA or BB homodimers or as AB heterodimers with potencies differing in various cells. Some cell types preferentially or exclusively express one chain over the other.

PDGF binds to two distinct, but closely related, PDGF receptors, termed α and β receptors, which have significant amino acid sequence homology.

The interaction of PDGF with its receptor activates the receptor tyrosine kinase domain resulting in its autophosphorylation, which in turn activates additional tyrosine kinases, serine/threonine kinases, and respective phosphatases.

B. Localization and Postulated Functions in Reproductive Tissues

PDGF is undetectable in the circulation, but it is locally produced. It has a biological half-life of <2 min and usually acts in an autocrine or paracrine fashion. Many different cell types synthesize and secrete PDGF-like molecules.

1. Testes

High levels of PDGF and PDGF receptor mRNA and protein are found in the rat testis. PDGF is predominantly expressed in Sertoli and peritubular myoid cells during prenatal and early postnatal periods. During fetal development in the rat, PDGF is chemotactic for peritubular myoid cells. Rat gonocytes, which are precursor cells of spermatogonia, express the PDGF receptor and treatment of these cells with PDGF induces cell proliferation.

2. Uterus

In the human endometrium only PDGF-BB mRNA has been detected. Endometrial stromal cells are a source of PDGF, which may act in an autocrine and/or paracrine fashion. PDGF is a potent mitogen for stromal and decidual cells and it inhibits prolactin production by decidual cells.

3. Placenta

The expression of PDGF-B chain transcript is tightly regulated during the first trimester of pregnancy. The extravillous cytotrophoblast cells that invade the endometrium to form the cytotrophoblast shell express the PDGF-B gene, and a subpopulation of these cells express the PDGF receptor genes. In contrast, neither the PDGF-B gene nor the PDGF receptor are expressed in villous trophoblast of first-trimester placenta.

VIII. HEPATOCYTE GROWTH FACTOR

A. Properties

Hepatocyte growth factor (HGF) was initially purified from the serum of hepatectomized rats. It is derived from a precursor of 728 amino acids which is proteolytically processed to form a mature protein of 87 kDa composed of an α subunit and a β subunit linked by a disulfide bond. HGF binds to a heterodimeric tyrosine kinase receptor that is encoded by the c-met oncogene.

B. Localization and Postulated Function in Reproductive Tissues

HGF is expressed in a variety of tissues and cell types. It induces cell proliferation, cell motility, and invasiveness.

1. Ovary

The Met/HGF receptor is detected on the surface epithelium of the human ovary and present at high levels in primary tumors of the ovaries. It is also expressed by mouse granulosa cells of developing and mature oocytes.

Bovine theca interna cells express HGF mRNA and protein. HGF mRNA is also expressed in fetal and neonatal rat tissues, specifically in both theca interna and granulosa cells. It has been demonstrated that HGF can impair LH-stimulated androgen production by rat thecal cells and delay the onset of apoptosis in rat antral follicles. Murine ovaries express increasing levels of HGF mRNA in response to estrogen. Despite these extensive *in vitro* studies, very little is known about the *in vivo* role of HGF on ovarian function.

2. Uterus

HGF mRNA is expressed by secretory phase endometrium and during menstruation. Its receptor is present in endometrial epithelial and decidual cells. The addition of recombinant HGF can promote proliferation of these cells.

3. Placenta

The human placenta is a rich source of HGF. It is expressed in the villous syncytium, extravillous trophoblast, and amnionic epithelium. HGF appears to promote growth of the cytotrophoblast via a paracrine mechanism.

IX. SUMMARY

It is apparent that growth factors play an important role in regulating reproductive function in a variety of species. The responses to growth factors are diverse and depend on the tissue, cell type, status of the cell, and hormonal environment. The actions of growth factors on many cell types in the ovary, testes, uterus, and placenta appear to influence cell physiology during the estrous or menstrual cycle or developmental biology *in utero*. It is apparent that growth factors act individually or synergistically with one another to regulate cellular and molecular events that lead to cell growth, remodeling, metabolism, differentiation, and death. Future studies will provide a better understanding of the role of growth factors in the complex events involved in reproductive processes and the potential to impact fertility and infertility therapies.

See Also the Following Articles

FOLLICULAR STEROIDOGENESIS; IGF (INSULIN-LIKE GROWTH FACTORS); SERTOLI CELLS

Bibliography

Bikfalvi, A., Klein, S., Pintucci, G., and Rifkin, D. B. (1997). Biological roles of fibroblast growth factor-2. *Endocr. Rev.* **18,** 26–45.

Ferrara, N., and Davis-Smyth, T. (1997). The biology of vascular endothelial growth factor. *Endocr. Rev.* **18,** 4–25.

Guidice, L. C. (1994). Growth factors and growth modulators in human uterine endometrium: Their potential relevance to reproductive medicine. *Fertil. Steril.* **61,** 1–17.

Jones, J. I., and Clemmons, D. R. (1995). Insulin-like growth factors and their binding proteins: Biological actions. *Endocr. Rev.* **16,** 3–34.

Kingsley, D. M. (1994). The TGF-β superfamily: New members, new receptors, and new genetic tests of function in different organisms. *Genes Dev.* **8,** 133.

Pimental, E. (1994). *Handbook of Growth Factors, Volume II: Peptide Growth Factors.* CRC Press, Boca Raton, FL.

Rotwein, P. (1997). Peptide growth factors. In *Endocrinology, Basic and Clinical Principles* (Conn and Melmed, Eds.), pp. 79–99. Humana Press, Totowa, NJ.

Shull, M. M., and Doetschman, T. (1994). Transforming growth factor-β1 in reproduction and development. *Mol. Reprod. Dev.* **39,** 239–246.

Guinea Pig, Female

Amanda L. Trewin and Reinhold J. Hutz
University of Wisconsin–Milwaukee

I. Ovarian Morphology and Function
II. Accessory Reproductive Organs
III. Physiology and Pathophysiology of Pregnancy and Lactation
IV. Uses in Biomedical Research

GLOSSARY

apoptosis Programmed cell death.
atresia A process by which ovarian follicles degenerate; the mechanism causing atresia is thought to be apoptosis.

cornification A process by which cells become flattened and jagged edged, lose their nuclei, and are filled with keratin.
dystocia A difficult birth.
lordosis A reflexive stance which facilitates copulation where the female arches and straightens the back while elevating the entire posterior region with dilation of the pudendum.
metritis Inflammation of the uterus.
precocious Advanced in early development.
pudendum The external genital organs of a female.
rete A network of blood vessels, nerve fibers, or other strands of interlacing tissue in an organ.
teratogen A substance that causes developmental abnormalities in a fetus.

The laboratory guinea pig is a member of the order Rodentia, suborder Hystricomorpha, family Caviidae, genus *Cavia*, and species *porcellus*. *Cavia porcellus* is a widely used animal model because of its docile nature and reproductive similarities to women.

I. OVARIAN MORPHOLOGY AND FUNCTION

A. Ovary

The ovaries are located in the lumbar region of the abdominal cavity caudal and lateral to the kidneys. They are amygdaloidal in shape without any lobes or fissures. The ovary is 6–8 mm in length and 4 or 5 mm in diameter. A thin tunica albuginea surrounds the ovary. The ovary has a large rete which is often cystic. The remainder of the ovary contains the developing follicles, corpora lutea, and interstitial tissue.

The follicles contained within the ovary are often at different stages of development. At any time during that development, a follicle can undergo atresia. There are two types of atresia that occur in the guinea pig ovary. Atresia of small follicles begins with connective tissue infiltration. This type of atresia is prevalent between Days 2 and 5 of the estrous cycle. Atresia of antral follicles, which occurs later in the cycle, begins with the appearance of pyknotic nuclei in the antrum; this is characterized by apoptosis. As the mural granulosa cells degenerate, the ovum degenerates and the overall follicle shrinks. The theca interna, on the other hand, continues to proliferate and becomes fibrolytic in appearance. The zona pellucida, although shriveled, also persists within the guinea pig ovary. As a result, the interstitial tissue in the guinea pig has a nodular appearance.

A preovulatory follicle in the guinea pig is approximately 0.8 mm in diameter. It has a large antrum, several layers of cumulus and mural granulosa cells, and a thick layer of theca. Growth of ovarian follicles in the guinea pig is slow but steady. The rate of growth rapidly increases at estrus to approximately nine times the usual rate. The growth spurt culminates with ovulation. Approximately three preovula-

tory follicles are produced per estrous cycle. Normally, the preovulatory follicles are divided equally between ovaries; however, if a guinea pig is hemiovariectomized the remaining ovary will compensate by producing the normal quota of follicles.

After ovulation, the corpora lutea become the functional structures within the ovary. Granulosa and theca interna cells become luteinized to form a functional corpus luteum. After 2 or 3 days, the corpus luteum exists as a narrow band of luteal cells with a central cavity. The size of a corpus luteum increases by hypertrophy and hyperplasia until it reaches a volume of approximately 2 mm³. During a nonpregnant cycle, the corpora lutea will persist until approximately Day 10 of the following cycle and then gradually regress. During a pregnancy, the corpora lutea persist through early lactation, reaching their greatest diameter at approximately 30 days of gestation.

There is no evidence to indicate that one ovary is superior to the other in reproductive function since the mean frequency of ovulation is similar between ovaries, each ovary has equal numbers of corpora lutea and total follicles, and the ovaries have similar weights.

B. Estrous Cycle

The guinea pig has an estrous cycle of approximately 16 days, 14 days of which constitute luteal phase; however, a great deal of variation appears to be normal. It does not appear to be abnormal for an occasional cycle to be as short as 12 days or as long as 20 days. The estrous cycle can be divided into four phases: proestrus, estrus, metestrus, and diestrus. Proestrus lasts for 1–5 days. During proestrus, the female is more active and pursues her cagemates. She typically sways her hindquarters while making a throaty rasping sound. She may display mounting behavior. Proestrus is followed by estrus. The animal is in estrus for 24–48 hr. Toward the end of estrus, she displays lordosis. Ovulation normally occurs approximately 10 hr after the onset of estrus. Following ovulation, the guinea pig will enter metestrus, which lasts 2.5–3 days. The remainder of the cycle is diestrus. Animals that are housed together tend to show similar estrous cycles. A researcher can determine when the female is exhibiting estrus by monitoring the vaginal membrane for perforation or analyzing vaginal smears for cornified epithelial cells. The day that the vaginal membrane is open and maximal cornification is observed is designated as Day 0, the day of ovulation. Days of the estrous cycle follow thereafter consecutively.

C. Hormonal Profile of the Estrous Cycle

Although atresia occurs continuously throughout the estrous cycle of the guinea pig, there appears to be a biphasic pattern of follicular growth, more similar to that exhibited by domestic species (e.g., cow and ewe). The first cohort of follicles reaches its greatest diameter on Day 10 or 11 and then becomes atretic. The corpora lutea from the previous cycle regress when the first cohort of follicles regresses. It has been hypothesized that the estrogen produced by the first cohort of follicles causes the release of prostagladin$_{F2alpha}$ from the uterus which then causes the regression of the corpus luteum. This hypothesis is supported by data that show that if a guinea pig undergoes a hysterectomy before Day 11, the corpora lutea will be maintained for at least 50 days. The second cohort of follicles then matures and ovulates on about Day 16. Ovulation is followed by a wave of atresia; thus, no antral follicles are present within the ovary following ovulation. This biphasic pattern of follicular growth has been confirmed with ovarian histology and estrogen synthesis; estrogen concentrations peaked on Days 10 and 0 and fell to their nadir on Day 12.

Plasma progesterone rises rapidly after ovulation due to the development of corpora lutea; peaking at approximately 2.8 ng/ml on Day 5. As the corpora lutea regress, serum progesterone concentrations decline to their nadir of approximately 0.5 ng/ml by Day 12. This drop in progesterone allows a rise in the release of follicle-stimulating hormone (FSH) from the anterior pituitary on Day 13, thus facilitating folliculogenesis and estrogen synthesis. Preovulatory follicles become ripe approximately 2 days prior to ovulation. Although ovulation can be induced with human chorionic gonadotropin (hCG), the endogenous surge of luteinizing hormone will not occur until Day 16.

Although ovulation can be induced with hCG, there is no evidence to indicate the ability to superovulate follicles regardless of the gonadotropin schedule used. This is supported by data that show that treatment with hFSH has no significant effect on follicular size or steroid output in the guinea pig.

D. Behavioral Estrus

As with most species exhibiting a spontaneous ovulation, the female goes into heat due to the effects of ovarian hormones. Progesterone alone is not enough to induce receptivity; the female must be primed with estrogen before exposure to progesterone. During estrus, there is a 6- to 11-hr time frame when the female accepts a male. If a light/dark cycle is maintained, the female is more likely to display behavioral estrus between 5 PM and 5 AM. During behavioral estrus, the female displays lordosis and mounts other females. There is a positive correlation between the number of follicles ovulated and the aggressiveness of the female during estrus presumably due to increased estrogen secretion.

E. Puberty

The female guinea pig normally exhibits first estrus between 49 and 63 days of age; however, she seldom exhibits first vaginal opening before she achieves a mass of 320 g. The total duration of vaginal opening is longer during the first estrous cycle. The hormonal profile of first estrous cycle is similar to that of the adult except that progesterone concentrations are slightly lower; this is most likely due to a lack of corpora lutea from a previous cycle.

II. ACCESSORY REPRODUCTIVE ORGANS

A. Oviduct

The infundibulum of the oviduct brushes against the dorsolateral surface of the ovary. The isthmus penetrates the uterine wall. At ovulation, the ova surrounded by a dense mass of cumulus cells are expelled from the ovary and move into the ampulla of the oviduct. The ova move slowly through the ampulla, taking 1 or 2 days. If sperm are present, fertilization normally occurs within the oviduct. Once the embryos reach the isthmus of the oviduct, they move rapidly toward the uterus. By Day 3 of the estrous cycle, the embryos arrive at the uterus. For pregnancy to occur, the ova must be fertilized within 20 hr of ovulation. Aging of the ova becomes evident as early as 8 hr after ovulation. Normally, there is a high fertilization rate of guinea pig ova. Polyspermy is rare, occurring in only 1.8% of all fertilized ova.

B. Uterus

The guinea pig has a bicornuate uterus. The two horns are each approximately 45 mm in length. The horns fuse at the cervices and become only one os cervix. The body of the uterus is approximately 12 × 10 mm. Implantation usually occurs between 6 and 7.5 days for normal pregnancy to follow.

C. Placenta

The guinea pig has a labyrinthine hemochorial placenta where a single layer of fetal trophoblasts contact the maternal blood. There is a visceral yolk sac which allows maternal transfer of antitoxin, antibodies, and serum proteins. Countercurrent blood flow allows for more efficient transport of oxygen across the fetoplacental unit.

D. Vagina

During most of the estrous cycle, the female guinea pig has an intact epithelial closure membrane. This membrane will becomes perforated during estrus and at parturition. The vaginal membrane is normally perforated for at least 1 day prior to ovulation. The vaginal vault is lined with epithelium. During most of the estrous cycle, the epithelium is intact. The individual cells are round and nucleated. During proestrus, round epithelial cells, some cornified epithelial cells, and leukocytes are present in the vaginal smear. During estrus, the vaginal membrane is perforated. The epithelium becomes cornified and sloughs off in sheets. The day of maximal cornification is designated as Day 0. Leukocytic invasion is characteristic of Day 1 of the estrous cycle. This return of leukocytes indicates that ovulation has occurred. It is by monitoring the status of this membrane and by analyzing vaginal smears that the researcher is able to determine the cyclic pattern of estrus in an individual female.

III. PHYSIOLOGY AND PATHOPHYSIOLOGY OF PREGNANCY AND LACTATION

A. Pregnancy

Breeding can begin at 2 or 3 months of age or when a mass of 350–400 g is reached. Most laboratory guinea pigs are bred between 18 months and 4 years; however, females older than 27 months generally have smaller litters (one or two). A female can be expected to have four or five litters. Normally, a boar is placed with 1–10 sows for breeding. More than one boar should not be placed together because males tend to be aggressive toward one another. Because the boar is housed with the sows for an extended time period, it is not necessary to detect estrus unless the timing of pregnancy is important. If the timing of pregnancy onset is of interest, the estrous cycle should be monitored. The timing of pregnancy can then be confirmed by the presence of sperm or a vaginal plug within the vaginal vault. Sperm can be observed in the vaginal smear for a few hours postcoitus. A vaginal plug is made of secretions from the vesicular and coagulating glands of the male. Epithelial cells from the vagina surround these secretions. The vaginal plug persists for a few hours, at which time it is dislodged and falls into the cage. Observation of the vaginal plug either within the vaginal vault or in the cage is indicative of impregnation.

Gestation in the guinea pig is 59–72 days with an average length of 63–68 days. The optimal length for gestation is 69 days. The litter size varies between one and six, with an average litter of three young. The length of gestation is correlated with the size of the litter, with a larger litter producing a shorter gestation. The size of the litter is also positively correlated with size of the sow. There is also a positive correlation between the interval since last gestation and litter size. The gestation can be divided into three trimesters of approximately 3 weeks each. A functional ovary is necessary for the maintenance of pregnancy through the first trimester or Day 25.

Fetuses are palpable by the fourth or fifth week of gestation. Under the influence of relaxin, the pubic symphysis will begin to separate during the last week of gestation. A pubic symphysis separated to 15 mm indicates that the pups will be born within 48 hr. At birth, the pubic symphysis should be separated to 22 mm. If the first breeding occurs after 7 months, the pubic symphysis will not separate as easily and fat pads may occlude the birthing canal; thus, there may be problems with dystocia and/or death. Parturition normally lasts 10–30 min, with approximately 7.4 min reported between each birth.

A sow can be expected to double her weight during pregnancy. Newborns normally weigh 60–100 g. There does not appear to be a difference in weight between the sexes at birth. There is a correlation between birth weight and the litter size; as the litter size increases, the weight of the pups decreases. Young born weighing less than 50 g rarely survive. Although newborns are precocious, they do require nursing bouts and licking from the mother for at least the first few days. Normally, weaning does not occur until Day 21.

Abandonment may occur if the mother has mastitis or is inexperienced. Environmental disturbances may also lead to abandonment. Spontaneous abortions are common in the guinea pig. Abortions occur in approximately 8.3% of all pregnancies and are not related to litter size. Stillborns are also common in the guinea pig and are correlated positively with litter size. Fetal phagia is not typical in the guinea pig.

Although folliculogenesis occurs throughout the reproductive life span of the guinea pig, spontaneous ovulation does not occur during pregnancy. Females exhibit estrus within 10 hr postpartum. If mated postpartum, 60–80% become pregnant. Postpartum estrus is similar to normal estrus except that it is shorter, lasting approximately 3.5 hr. Ovulation occurs at 12–15 hr postpartum. If she does not become pregnant during the postpartum estrus, she displays a normal estrous cycle regardless of lactation. Nor-

mally, there is no change in length of the estrous cycle due to sterile mating; therefore, pseudopregnancies are rare in the guinea pig. When pseudopregnancy does occur, it lasts approximately 17 days.

B. Mammary Gland

Although both sexes have two nipples located on the lower abdomen, only the female has mammary glands. The mammary gland in the guinea pig is an apocrine gland. The milk is secreted through a single large glactophore in each teat. A female guinea pig can successfully nurse more than two pups. The female will allow young other than her own to nurse from her. If allowing other pups to foster, it is important to realize that larger pups may monopolize the nipple. Weaning occurs in the guinea pig between 14 and 21 days of age and mass of 150–200 g. Early weaning is related to a higher mortality rate among the pups.

Lactation milk volume increases through Day 5, peaks on Days 5–8 , and decreases to its nadir by Days 18–23. The rate at which the pups gain weight is directly related to milk yield. A larger litter size is also correlated with increased yield. Guinea pig milk contains 3.9% fat, 8.1% protein, 3.0% lactose, and 83.6% water. The protein content increases and the lactose content decreases during the second half of lactation. Na^+ concentrations increase, whereas K^+ concentrations decrease throughout lactation. Guinea pig milk contains approximately 77 calories / 100 g. The female will normally lose 54–163 g due to lactation.

C. Infertility

There are many causes of infertility in the guinea pig. Increased age of the female can cause declining reproductive ability. In addition to a decreased ability to become pregnant, an older female will generally have smaller litters. Guinea pigs are also sensitive to environmental stress. Use of wire floors, an elevated ambient temperature (optimal 21°C), and bedding adhering to the genitalia are examples of environmental stress that reduce reproductive ability. The overall health of the guinea pig also affects reproduction. Guinea pigs with a nutritional deficiency or metritis are less likely to become pregnant. Estrogen within the diet of the guinea pig also decreases pregnancy rates. Finally, the age and size of the boar will affect pregnancy outcome. The boar should be older and larger than the sow in order to ensure a successful mating.

IV. USES IN BIOMEDICAL RESEARCH

The laboratory guinea pig, *C. porcellus*, is an animal often utilized by reproductive scientists. It is a preferred model for studies of ovarian function because it exhibits a functional luteal phase; for studies of ova transport since the progression of the ova through the oviduct is similar to that of humans; and for studies of endocrine control of pregnancy since this is also similar to humans, nonhuman primates, and horses. The guinea pig is also used as a model to study atresia since it has a wave of follicles that are expected to undergo atresia naturally during each cycle. Finally, because the guinea pig has a long gestation and a placenta similar to that of the primate, it is a good model to study the effects of various teratogens.

See Also the Following Articles

ESTROUS CYCLE; FOLLICULAR ATRESIA; RODENTIA

Bibliography

Bland, K. P. (1980). Biphasic follicular growth in the guinea pig oestrous cycle. *J. Reprod. Fertil.* **60**, 73–76.

Donovan, B. T., and Traczyk, W. (1962). The effect of uterine distension on the oestous cycle of the guinea pig. *J. Physiol.* **161**, 227–236.

Harkness, J. E., and Wagner, J. E. (1989). The guinea pig. In *The Biology and Medicine of Rabbits and Rodents*, 3rd ed. Lea & Febiger, Philadelphia.

Hugo, M., Salinas, L. A., Fernandez, E., and Pauerstein, C. J. (1977). Time course of ovum transport in guinea pigs. *Fertil. Steril.* **28**(8), 863–865.

Hutz, R. J., Bejvan, S. M., Durning, M., and Dierschke, D. J. (1990). Changes in follicular populations, in serum estrogen and progesterone, and in ovarian steroid secretion *in vitro* during the guinea pig estrous cycle. *Biol. Reprod.* **42**, 266–272.

Joslyn, W. D., Wallen, K., and Goy, R. W. (1971). Cyclic changes in sexual response to exogenous progesterone in female guinea pigs. *Physiol. Behav.* **7**, 915–917.

Mossman, H. W., and Duke, K. L. (1973). *Comparative Morphology of the Mammalian Ovary.* Univ. of Wisconsin Press, Madison.

Rawson, J. M. R., Galey, C. I., Weinberg, L. C., and Hodgson, B. J. (1979). Effects of gonadotropins on follicular development, ovulation, and atresia in the mature guinea pig. *Horm. Res.* **10**, 25–36.

Sisk, D. B., Wagner, J. E., and Manning, P. J. (1976). *Reproductive System. The Biology of the Guinea Pig.* Academic Press, New York.

Westfahl, P. K., and Vekasy, M. S. (1988). Changes in serum and ovarian steroids during reproductive development in the female guinea pig. *Biol. Reprod.* **39**, 1086–1092.

Gynecomastia

Samuel Smith

Harbor Hospital Center, Baltimore

GLOSSARY

androgens Sex steroids derived from the 19-carbon androstane ring responsible for male secondary sexual development. Testosterone is the primary androgen produced in the testes in men. Androstenedione and dehydroepiandrosterone are the principal androgens produced in the adrenal gland.

aromatase The enzyme that converts 19-carbon androgens to 18-carbon estrogens by aromatizing the A ring of the steroid nucleus.

estrogens Sex steroids derived from the 18-carbon estrane ring. Estradiol and estrone are two naturally occurring estrogens responsible for breast development and feminizing changes in females.

follicle-stimulating hormone A pituitary gland hormone responsible for the initiation of spermatogenesis during puberty and largely responsible for Sertoli cell function in the testes.

gynecomastia The enlargement of glandular breast tissue in males.

hypogonadism A reproductive condition characterized by impaired testicular production of testosterone or reduced or absent spermatogenesis. Primary hypogonadal states lead to lack of secondary sexual development in boys. Secondary hypogonadal states are due to conditions acquired after puberty has occurred. Hypogonadal states may be caused by early failure or destruction of the testes, by insufficient/inappropriate stimulation of the testes by hypothalalmic–

pituitary gland hormones, and by late-onset deficiencies of testosterone biosynthesis.

lipomastia A condition of excessive adipose (fat) tissue in the breast, commonly seen in obese men.

luteinizing hormone A pituitary gland hormone primarily responsible for stimulating the Leydig cells of the testes to produce testosterone.

macrogynecomastia A dome-shaped breast enlargement $\geqq 5$ cm in diameter that resembles the middle and late stages of normal female development.

I. INTRODUCTION

Gynecomastia is the benign enlargement of the male breast due to proliferation of the glandular tissue within the breast. Gynecomastia may reflect a transient or a permanent disturbance of male endocrine physiology. Estrogens stimulate and androgens inhibit glandular breast development. Gynecomastia occurs when there is a decreased ratio of androgen to estrogen in the blood or within breast tissue.

Gynecomastia may present as unilateral or bilateral breast tenderness, progressive painless enlargement of the male breasts, or may simply be an incidental finding during a routine physical examination. The majority of cases are transient and are not associated with serious conditions. A patient's main concerns are the discomfort and pain, cosmetic appearance, and risk of breast cancer. The physician needs to consider that the gynecomastia may be a presenting symptom of a malignancy or a sign of a serious underlying illness or endocrinopathy. Diagnostic tests are performed to identify those men who have serious conditions that require therapy.

II. PREVALENCE

Gynecomastia is very prevalent and three distinct peaks in age distribution occur, the first occurring at birth. Two-thirds of newborn babies may demonstrate palpable breast tissue that developed in response to placental and fetal hormones. There is a second peak in prevalence during puberty at which time up to 50% of boys may develop palpable breast tissue. Newborn and pubertal gynecomastia are usually transient phenomena. The third peak occurs in adult males aged 50–80, and in this age group gynecomastia is much less likely to be transient. Prevalence studies have not demonstrated any racial or ethnic differences for gynecomastia.

III. HISTOPATHOLOGY

Gynecomastia is associated with palpable disks of subareolar breast tissue that are freely mobile and nonadherent to the skin or underlying tissues. The tissue feels like coiled rope and should not be confused with lipomastia, the soft subcutaneous fat tissue frequently associated with obesity. The histologic features of gynecomastia correlate strongly with the duration of the process rather than its etiology. Characteristic changes associated with gynecomastia of less than 2 years duration include proliferation, elongation, and budding of the duct system and proliferation of the fibroblastic stroma. At this stage, gynecomastia will regress completely if the hormonal cause is eliminated. With longer durations there is progressive fibrosis and hyalinization associated with regression of the epithelial proliferation. Eventually, the number of ducts regresses. Resolution of gynecomastia is characterized by a reduction in size and glandular content with a gradual replacement by hyaline bands which will eventually disappear. If gynecomastia has been present for more than 4 years, fibrosis and hyalinization may be very extensive, and complete resolution of the breast mass may not occur. Surgery may be the only effective treatment option when gynecomastia has been present for many years.

Gynecomastia commonly develops in an asymmetric manner. One breast may enlarge and become tender months or years before the contralateral breast demonstrates any abnormality. Since bilateral changes usually develop, unilateral gynecomastia is considered an early stage in the development of bilateral gynecomastia.

IV. PHYSIOLOGIC CAUSES

A. Newborn Males

Transient breast enlargement is observed in 50% or more of newborn infants. Estrogens produced by the fetal–placental unit induce ductal hypertrophy; less commonly prolactin contributes to the development of a clear or milky nipple secretion once called witch milk. Neonatal gynecomastia usually resolves spontaneously within several weeks. Gynecomastia associated with bloody nipple discharge or persisting for more than several months suggests a pathologic process, such as intraductal papilloma or hormonal abnormality.

B. Pubertal Gynecomastia

Clinically palpable gynecomastia is very common during puberty and typically occurs in otherwise healthy adolescent boys. Its frequency ranges from 20 to 75%, and on average approximately 40% of boys between 10 and 16 years develop transient gynecomastia. Pubertal gynecomastia is most commonly observed at age 14 when the incidence peaks. It spontaneously resolves in 75% of boys within 2 years and in 90% of boys within 3 years. Studies variably define gynecomastia and this may contribute to the range of prevalence. Breast enlargement ≥ 0.5, ≥ 1, and ≥ 2 cm, firm subareolar tissue, and palpable glandular tissue have all been used as criteria. Visually obvious gynecomastia occurs in less than 10% of pubertal boys. A recent report found palpable gynecomastia exceeding 2 cm in 40.5% of 954 healthy unselected young Greek men aged 18–26 years. This unexpectedly high prevalence conflicts with the transient nature of pubertal gynecomastia and suggests a new wave of gynecomastia may occur in young adulthood. Alternately, they may be due to late-onset enzyme deficiencies in testosterone biosynthesis, or common environmental exposures to estrogens in plants and meats.

Pubertal gynecomastia is usually <4 cm in diameter and resembles early pubertal breast development in girls. The nipple and areola do not usually form a secondary mound above the breasts, and the process is usually bilateral. Signs of male secondary sexual development usually precede the onset of gynecomastia. Pubertal macrogynecomastia refers to dome-shaped breast enlargement ≧5 cm in diameter that resembles the middle and late stages of normal female development. The nipple and areola often form a secondary mound above the breast. Since macromastia generally does not resolve spontaneously, the condition should be evaluated and treated.

1. Etiology

Sustained elevations of estradiol and estrone have not been found in boys with pubertal gynecomastia. In early puberty the pituitary gland stimulates the testes to secrete testosterone during the nighttime hours, and only in later puberty does nocturnal and daytime stimulation of the testes occur. Serum estrogens, in contrast, rise during the day and night in early puberty. Estrogens are formed by the extraglandular aromatization of adrenal gland androgens such as androstenedione and dehydroepiandrosterone. Since testosterone suppresses the action of estrogen on breast tissue most effectively at night, the stage may be set for estrogens to stimulate breast changes during the daytime in early puberty if there is a relative imbalance between adrenal androgen and estrogen. The decrease in the adrenal androgen : estrogen ratio observed in many boys with pubertal gynecomastia has been linked to increased aromatase activity in breast tissue. Thus, an increased conversion of adrenal androgens to estrogens within breast tissue and low daytime testicular testosterone secretion contribute to the pathogenesis of pubertal gynecomastia. In some boys with pubertal gynecomastia, low levels of free testosterone have been noted in association with increased levels of sex hormone-binding globulin (SHBG).

2. Pathologic versus Physiologic

Pathologic conditions should be considered when gynecomastia develops prior to age 10, if puberty occurs precociously or after the onset of gynecomastia, if gynecomastia fails to spontaneously resolve, or if it occurs in a boy with genital tract disease or undermasculinizaton. A search for causative drugs, chronic illness, endocrine disorders, neoplasias, hypogonadism, and androgen resistance disorders must be conducted in this context.

C. Gynecomastia of Aging

Gynecomastia is observed in many healthy men aged 50–80. Autopsy data show that up to 40% of elderly men have gynecomastia. Alterations in estrogen and androgen physiology observed in the elderly that may contribute to the development of gynecomastia include decreased free and total testosterone, increased peripheral aromatization of androgen to estrogen, and a decreased androgen:estrogen ratio. Since gynecomastia may be a sign of an underlying disorder, seemingly healthy men should be evaluated for medical conditions that may cause gynecomastia. Mammography is useful in differentiating breast enlargement due to fat and true gynecomastia.

V. PATHOLOGIC GYNECOMASTIA

A. Endocrine Disorders

1. Hypogonadism

Congenital and acquired conditions that lead to male hypogonadism are associated with the onset of gynecomastia in the teenage years. When gynecomastia is associated with a failure of gonadal testosterone biosynthesis, pituitary gonadotropin [follicle-stimulating hormone (FSH) and luteinizing hormone (LH)] levels are generally elevated. Even though the testicles fail to produce testosterone, adrenarche occurs and adrenal androgens are peripherally converted to estrogens by tissue aromatases. Insufficient testosterone is produced to inhibit the effect of these estrogens on the breast tissue and gynecomastia develops.

Congenital anorchia is a rare disorder in which testes are absent in otherwise phenotypically normal 46,XY boys. At birth this condition may be confused with bilateral cryptorchidism. Since the external genitalia and penis otherwise appear normal at birth, it is presumed that the testicles regressed during fetal life after completion of male external genital development, which is dependent on testosterone and dihydrotestosterone. Approximately half of reported cases develop gynecomastia. The cause for the condition is unknown.

Klinefelter's syndrome is associated with primary hypogonadism occurring in men with a 47,XXY karyotype. The testes are small, virtually devoid of seminiferous tubules, and demonstrate adenomatous hyperplasia of Leydig cells. Serum FSH and LH levels are usually elevated, and serum testosterone levels are low, although many men will have low-normal testosterone levels. Testicular production of estradiol is frequently elevated because of enhanced Leydig cell aromatase activity. Approximately half of men with 47,XXY karyotypes and a third of men with Klinefelter's mosaicism develop gynecomastia at the time of expected puberty.

A variety of defects in enzymes leading to testosterone biosynthesis are associated with gynecomastia. Deficiencies in cytochrome P450c17 (17-ketosteroid reductase) and 3β-hydroxysteroid dehydrogenase inhibit testosterone synthesis. In congenital onset forms of these disorders, the external genitalia usually demonstrate some degree of microphallus and ambiguity; cryptorchidism, hypospadias, and bifid scrotum may be associated. Complete enzymatic blocks in testosterone biosynthesis lead to a 46,XY individual looking entirely like a normal girl who experiences primary amenorrhea. Milder late-onset forms of these two enzyme deficiencies occur and are characterized by gynecomastia, with or without hypogonadism, in boys and young men.

Defects in the structure or function of the androgen receptor lead to androgen-insensitivity syndromes. Complete androgen insensitivity, or testicular feminization syndrome, presents as primary amenorrhea in a female who has developed breasts but is lacking a uterus and has scant pubic and axillary hair. Her karyotype is 46,XY, and she has testicles located intraabdominally or in the inguinal region. Incomplete forms of androgen insensitivity present as genital ambiguity at birth and gynecomastia during puberty. The mildest forms of incomplete androgen insensitivity present as normal-appearing men with severe oligospermia and gynecomastia. In each instance gynecomastia occurs because estrogen action on breast tissue is insufficiently opposed by testosterone due to androgen receptor abnormalities. Assays for pubic skin fibroblast androgen receptors are used to confirm the diagnosis of incomplete androgen insensitivity syndrome.

Patients with true hermaphoditism have both testicular and ovarian components within their gonads. In these individuals breast development reflects either increased estrogen production or reduced testosterone production by the gonads.

Adult males can acquire or develop hypogonadism for a variety of reasons. One of the most common is cancer chemotherapy medications. Testicular damage can also occur due to radiation therapy/exposure, bilateral injury, viral orchitis in adolescence, and granulomatous disease. Viral orchitis is the most common cause of postpubertal testicular failure, and mumps is the most important pathogen that can lead to testicular atrophy. Mumps orchitis is associated with very low testosterone production, normal estrogen production owing to peripheral conversion of adrenal androgens, and a marked reduction in the testosterone:estrogen ratio. Trauma is the second most common cause of acquired hypogonadism. Gynecomastia also frequently develops after castration.

Primary hypogonadism accounts for approximately 8% of cases of gynecomastia. Secondary (acquired) hypogonadism accounts for approximately 2% of cases.

2. Hyperthyroidism

Thyrotoxicosis is often associated with increased circulating estradiol levels, and approximately 30% of young men with hyperthyroidism demonstrate gynecomastia. SHBG levels are increased in hyperthyroidism, and the differential binding affinity of SHBG for testosterone and estradiol leads to a lower free testosterone:estradiol ratio. In addition, hyperthyroid men produce excess amounts of androstenedione, which serves as a substrate for peripheral conversion to estrogens.

3. Adrenal Disorders

Adrenal insufficiency due to isolated pituitary ACTH deficiency and congenital adrenal hyperplasia due to 11-hydroxylase deficiency are rare causes of gynecomastia.

B. Tumors

1. Testicular

Testicular tumors account for approximately 3% of all gynecomastia, and it is not uncommon for gynecomastia to be the presenting symptom for testicular and other neoplasias. Testicular Leydig cell and Sertoli cell tumors may secrete estrogen or human chorionic gonadotropin (hCG). Estrogen-producing Leydig cell tumors are generally very small and nonmalignant. They can be detected by ultrasound examinations prior to enlarging enough to be palpable. Germ cell tumors may also secrete hCG, which enhances estradiol and testosterone production by the testes; they may occasionally demonstrate increased aromatase activity and actively convert androgens to estrogens.

2. Adrenal

Estrogen-producing (feminizing) adrenal gland tumors are frequently large and malignant. Adrenal tumors may also secrete androstenedione, which is then peripherally aromatized to estrone. Feminizing adrenal tumors are associated with advanced bone age as well as gynecomastia in pubertal boys.

3. Ectopic Production of Human Chorionic Gonadotropin

Hepatic, lung, mediastinal, and kidney cancers may secrete hCG, which potently stimulates testicular aromatase activity and estrogen production. Testicular choriocarcinoma is the most common tumor associated with ectopic hCG production.

C. Chronic Disease

1. Liver Disease

In cirrhosis, decreased hepatic extraction of androstenedione leads to increased substrate for peripheral aromatization to estrone. The cirrhotic liver's inability to inactivate and metabolize estrogens also contributes to a rise in circulating estrogen levels. Excess estrogens, however, are not produced by the liver.

2. Malnutrition

Starvation and malnutrition resulting from chronic debilitating illness can cause gynecomastia. Breast enlargement associated with malnutrition frequently occurs after caloric intake increases and is called refeeding gynecomastia. Presumably, hepatic dysfunction impairs the metabolism of estrogens and androgens, thereby facilitating the extraglandular conversion of androgens to estrogens. Gynecomastia typically resolves after hepatic function returns to normal.

3. Renal Failure

Transient gynecomastia develops in approximately half of men with renal failure who undergo hemodialysis. Chronic uremia induces testicular damage, reducing testosterone production and spermatogenesis.

4. Nervous System Disease

Decreased testicular function is observed in some men with paraplegia from trauma, myotonia atrophica, and other neurologic conditions. Breast enlargement is related to reduced testosterone:estrogen ratios.

D. Familial Excessive Peripheral Aromatase Activity

Excessive estrogen production occurs in some boys due to increased activity of tissue aromatase enzymes. Several families have been described in which gynecomastia was associated with high estradiol to testosterone ratios. Affected boys developed gynecomastia in early puberty. Rates of peripheral conversion of androgens to estrogens 10–50 times higher than normal have been identified in these boys. Statues of King Tutankhamun, his father, and his brothers from ancient Egypt suggest that they were affected by familial gynecomastia. The genetic basis of this abnormality has not been determined.

E. Drugs

A wide variety of drugs can cause gynecomastia (Table 1). There are several mechanisms implicated in drug-induced gynecomastia. Estrogens and drugs

TABLE 1
Drugs Associated with Gynecomastia

Category	Drug
Antiandrogens and inhibitors of androgen biosynthesis	Spironolactone, ketoconazole, cyproterone, finasteride, flutamide
Antibiotics	Ketoconazole, isoniazid, metronidazole minocycline
Antiseizure medications	Phenytoin
Antiulcer medications	Cimetidine, omeprazole, ranitidine
Cancer chemotherapeutics	Alkylating agents, methotrexate
Cardiovascular drugs	Digoxin, verapamil, nifedipine, captopril, amiodarone, methyldopa, reserpine, diltiazem
Hormones	Anabolic steroids, hCG, diethylstilbestrol, estrogens, growth hormone, oral contraceptives
Neuroleptics	Diazepam, haloperidol, phenothiazines, tricyclic antidepressants
Drugs of abuse	Alcohol, marijuana, amphetamines, heroin
Miscellaneous	Penicillamine, cyclosporine

Note. Table derived from Braunstein (1993) and Thompson and Carter (1993).

with estrogenic activity can directly induce severe gynecomastia. Other drugs act by enhancing endogenous estrogen production, inhibiting testosterone synthesis, or inhibiting the action of testosterone. Most drugs induce gynecomastia by unknown mechanisms. A large number of drugs have case reports temporally linking them with gynecomastia.

1. Drugs with Estrogenic Activity or Which Increase Estrogen Production

Approximately 10% of men who take digoxin for 1 year or longer develop gynecomastia. Initially gynecomastia was attributed to the drug's estrogen-like effects or the fact that digoxin may be an estrogen precursor. Digoxin may also produce gynecomastia by a refeeding mechanism in chronically debilitated men. Human chorionic gonadotropin administration to men and boys may result in gynecomastia due to hCG's direct effect on Leydig cell estradiol production. Anabolic steroids taken to increase muscle mass in athletes and body builders may be aromatized to estrogens and lead to gynecomastia; alternately they may inhibit the hypothalamic–pituitary unit leading to gonadal suppression. Gynecomastia has also been induced by ingesting estrogenic compounds in medications (e.g., diethylstilbestrol), meat, and milk and by skin contact with estrogen-containing lotions and oils.

2. Inhibitors of the Androgen Receptor

The antiulcer medicine cimetidine is a weak androgen receptor antagonist. Long-term administration is commonly associated with gynecomastia due to its competitive binding to the androgen receptor. Spironolactone, cyproterone acetate, zanoterone, and flutamide may cause gynecomastia by a similar mechanism.

3. Inhibition of Testosterone or Dihydrotestosterone Biosynthesis

Spironolactone and ketoconazole cause dose-dependent inhibition of cytochrome P450c17, the enzyme that converts progestogens to androgens. As a result, there is a lowering of the androgen:estrogen ratio and an increased chance of developing gynecomastia. Finasteride, an inhibitor of 5α-reductase used to treat benign prostatic hypertrohy, has been associated with gynecomastia which resolves upon discontinuation of the medication. Cancer chemotherapeutic agents may lead to the destruction of testicular Leydig cells, thereby reducing testosterone levels and the androgen:estrogen ratio. Leydig cell damage may occur regardless of the cancer being treated. Anabolic steroids, such as oxandrolone and stanozolol, may cause suppression of testosterone biosynthesis through inhibition of pituitary LH secretion.

4. Drugs Acting by an Unknown Mechanism of Action

Most medications listed in Table 1 have a small number of case reports temporally linking them to gynecomastia. If a true association exists, their mechanisms of action have yet to be elucidated.

F. Idiopathic Gynecomastia

Up to 25% of cases of gynecomastia may have no identifiable etiology and are therefore idiopathic. Senescent gynecomastia is frequently idiopathic. Some of these cases may be related to unrecognized environmental exposures to estrogenic or antiandrogenic substances.

VI. PATIENT EVALUATION

When a patient presents with a complaint of breast enlargement, a careful history should be obtained regarding the onset, timing, duration, and progression of the breast enlargement; the presence or absence of pain; use of any drugs, alcohol, or medications; and a detailed medical history including an evaluation of any potential environmental or inadvertent exposures to hormonally active substances. Information should be obtained regarding pubertal onset and progression, family history of gynecomastia, symptoms of hypogonadism (delayed puberty, erectile difficulty, decreased libido, and oligospermia), and the symptoms of cirrhosis, renal failure, hyperthyroidism, malnutrition, or other illness. Bodybuilders and competitive athletes should be directly questioned about the use of anabolic steroids.

Physical examination needs to differentiate gynecomastia from fatty enlargement of the breasts without gynecomastia. The examination should describe whether the breast enlargement is unilateral or bilateral, tender or nontender, and whether there is a nipple discharge. Breast tissue is grasped between the thumb and forefinger and gradually squeezed together. Gynecomastia feels like a firm or rubbery, mobile, disk-like mound of tissue beneath the nipple/areola complex. Adipose tissue does not have the same texture and feel. If a hard, irregular, eccentric, ulcerated, or immobile unilateral breast mass is palpated, the patient may have cancer and a breast biopsy is indicated. The presence and extent of secondary sexual characteristics, size of the testes, and the presence/absence of ambiguous genitalia are described. The abdomen and testes are palpated for tumors. Signs of underlying medical illness and malnutrition are sought.

Additional testing should be individualized. Painful, tender gynecomastia of recent onset during mid- to late puberty in an otherwise normal boy generally requires only history, physical examination, and observation. If the condition resolves within 1 year, no further evaluation is required. A malnourished pubertal boy should have chronic illness or malignancy excluded. An undermasculinized boy with small testes needs to be evaluated for hypogonadism, as do an adults with small testes. Asymmetric testicular size is a clue to a testes tumor. Liver, thyroid, and renal function tests should be performed when indicated by history and examination. Endocrine hormones frequently assayed include FSH, LH, estradiol, testosterone, dehydroepiandrosterone sulfate (DHEAS), and hCG. Prolactin is measured if galactorrhea is also present. If prolactin is elevated, MRI or CT scans can evaluate for pituitary adenomas. High serum LH or FSH and low testosterone suggests primary testicular failure. Karyotype analysis is indicated when there is evidence of hypogonadism, particularly Klinefelter's syndrome. High LH and normal serum testosterone suggests androgen-resistance disorders, although thyrotoxicosis is also associated with high LH and testosterone. A low LH and low testosterone suggests hypopituitarism. An elevated level of DHEAS warrants evaluation for an adrenal gland tumor by CT scan or ultrasound. High serum estradiol levels require that feminizing tumors of the liver, adrenal gland, and testes be excluded. If testicular ultrasound is normal, then abdominal CT or MRI scan is indicated. If these are all normal, increased extraglandular aromatase activity probably accounts for the elevated serum estradiol. Elevated serum hCG requires a search for hCG-producing tumors of the chest, abdomen, testes, and brain. Testicular ultrasound examinations can identify testes tumors even before they are large enough to be palpated. Chest X-ray and appropriate CT or MRI scans are used to exclude the other tumors.

If a patient presents with progressive gynecomastia and is receiving a medication associated with gynecomastia, the medication should be discontinued. The patient can be reevaluated in 1 or 2 months. Drug-induced gynecomastia should show some reduction in size or tenderness upon discontinuation of the medication. Marijuana and other drugs of abuse should be stopped to see if gynecomastia regresses.

When no abnormalities are identified after initial exam and hormonal evaluation, the patient can be reevaluated every 6 months for progression or resolution of the gynecomastia. In many cases, no specific etiology will be identified.

VII. TREATMENT

Gynecomastia is frequently a transient physiologic phenomenon. History and physical examination usually dictate whether breast enlargement needs to be immediately evaluated or whether patient reassurance and observation should be sufficient.

A small amount of gynecomastia in a phenotypically normal newborn boy without a positive family history should resolve spontaneously. The same amount of gynecomastia in a newborn with ambiguous genitalia should prompt a search for abnormalities in testosterone biosynthesis or androgen receptor function. Gynecomastia developing in a young prepubertal boy should prompt a search for feminizing tumors and testicular tumors.

Adolescent boys with gynecomastia comparable to that observed with breast budding in girls, usually <4 cm in diameter, can be reassured that their gynecomastia is 90% likely to disappear within 3 years without treatment. For boys who are embarrassed by their appearance or who have very tender breasts, medical therapy will induce regression in over half of cases within 6 months. Pain relief and resolution can frequently be accelerated by antiestrogens such

as tamoxifen and clomiphene citrate or by aromatase inhibitors, such as testolactone. Tamoxifen and clomiphene work by competing with estradiol for the estrogen binding sites in breast tissue. Tamoxifen 10–20 mg twice daily and clomiphene citrate 100 mg daily may cause mild gastrointestinal side effects. Testolactone inhibits the aromatization of adrenal and testicular androgen to estrogen; it is safe and effective at 150 mg three times daily. Neither medical option impairs pubertal growth and development. A 3-month therapeutic trial can be offered to boys with gynecomastia of recent onset without symptoms of hypogonadism or chronic illness. The decision to treat gynecomastia medically is influenced by its duration, pain, cosmetic appearance, and psychological reactions. Medical therapy is most effective in the early, proliferative stages of gynecomastia. Macrogynecomastia up to 6 cm in diameter frequently responds if it is treated early. Medical management is less likely to be effective after 2 years, when breast tissue begins to hyalinize and epithelial proliferation slows down. After 4 years, when fibrosis predominates and the breast tissue is very firm, surgical management is necessary. Periareolar or transareolar reduction mammoplasty produces good cosmetic results. In obese boys and men excess adipose tissue can be removed with suction lipectomy and the residual gynecomastia is then excised. Gynecomastia that recurs postoperatively is usually treated medically.

Pathologic conditions that cause pubertal and adult-onset gynecomastia should be treated appropriately. Pubertal gynecomastia in hypogonadal boys is treated with testosterone injections. Klinefelter's syndrome and androgen-insensitivity syndromes are less likely to respond to testosterone supplementation; antiestrogens and aromatase inhibitors may be used as well as reduction mammoplasty. Testicular, adrenal, and other tumors are treated surgically; metastatic disease is treated with adjuvant chemotherapy or radiation therapy. Correction of underlying medical illness generally results in resolution of gynecomastia over time. In patients with chronic renal failure, renal transplantation is more likely to reverse the gynecomastia and hypogonadism than dialysis. Patients with idiopathic gynecomastia respond to aromatase inhibitors.

Gynecomastia in the elderly may be due to malnutrition, cachexia, liver disease, cancers, and treatable systemic medical illnesses. Physiologic gynecomastia of the elderly is usually long-standing and fibrotic. It is best managed with reassurance, although some patients desire surgical correction.

When gynecomastia is thought to be induced by a medication, the drug should be withdrawn and replaced with a suitable alternative. Gynecomastia will usually resolve if the drug was responsible. Adolescent physiologic gynecomastia may temporally coincide with medication use and appear to be drug induced.

Two medications merit special discussion—growth hormone and anabolic steroids. Growth hormone (GH) is used to treat GH deficiency (GHD) in prepubertal boys and some cases of non-GHD short stature. Gynecomastia has been reported in boys who received GH therapy. In some boys it resolved even though therapy was continued. Other cases resolved when the GH was withdrawn. There is also ongoing research into the effects of growth hormone treatment on aging. Elderly men with GH-associated gynecomastia generally respond to discontinuing the hormone.

Bodybuilders and other athletes take anabolic steroids to increase muscle mass. They are widely used and may be an overlooked cause of gynecomastia in athletes who complain of gynecomastia. Fine-needle aspiration of breast enlargement that shows apocrine metaplasia in association with gynecomastia is very suspicious for anabolic steroid use. Some bodybuilders use tamoxifen to counteract the gynecomastia. Mild gynecomastia is suitable for surgical excision and good cosmetic results follow. Gynecomastia exceeding 6 cm in diameter is less likely to have good surgical results. Anyone confirmed to be using anabolic steroids, particularly in a polypharmacy manner, should be evaluated for their main toxic side effects—reproductive dysfunction, liver disease and tumors, lipoprotein abnormalities, coronary artery disease, polycythemia, and HIV due to sharing needles.

See Also the Following Articles

Hypogonadism; Klinefelter's Syndrome; Testosterone Biosynthesis

Bibliography

Biro, F. M., Lucky, A. W., Huster, G. A., and Morrison, J. A. (1990). Hormonal studies and physical maturation in adolescent gynecomastia. *J. Pediatr.* **116**, 450–455.

Braunstein, G. D. (1993). Gynecomastia. *N. Engl. J. Med.* **328**, 490–495.

Courtliss, E. H. (1987). Gynecomastia: Analysis of 159 patients and current recommendations for treatment. *Plastic Reconstr. Surg.* **79**, 740–750.

Dixon, J. M., and Mansel, R. E. (1994). Congenital problems and aberrations of normal breast development and involution. *Biol. Med. J.* **309**, 797–800.

Georgiadis, E., Papandreou, L., Evangelopoulou, C., Aliferis, C., Lymberis, C., Panitsa, C., and Batrinos, M. (1994). Incidence of gynecomastia in 954 young males and its relationship to somatometric parameters. *Ann. Hum. Biol.* **6**, 579–587.

Glass, A. R. (1994). Gynecomastia. *Endocrinol. Metab. Clin. North Am.* **23**, 825–837.

Mahoney, C. P. (1990). Adolescent gynecomastia: Differential diagnosis and management. *Pediatr. Clin. North Am.* **37**, 1389–1404.

Reyes, R. J., Zicchi, S., Hamed, H., Chaudary, M. A., and Fentiman, I. S. (1995). Surgical correction of gynecomastia in bodybuilders. *BJCP.* **49**, 177–179.

Samdal, F., Kleppe, G., Amland, P. F., and Abyholm, F. (1994). Surgical treatment of gynecomastia. Five years' experience with liposuction. *Scand. J. Plastic Reconstr. Surg. Hand Surg.* **28**, 123–130.

Thompson, D. F., and Carter, J. R. (1993). Drug-induced gynecomastia. *Pharmacotherapy* **13**, 37–45.

Hemichordata

Gary M. King

University of Maine

GLOSSARY

bromoaromatic Referring to various mono- and multisubstituted bromophenols, -indoles, and -pyrroles and their chloro- analogs.

stolon A basal tube-like structure of pterobranch colonies that interconnects zooids and serves to anchor colonies to the growth substrate.

tornaria A ciliated, motile, free-living planktonic larval stage characteristic of echinoderms and enteropneusts.

vermiform Worm-like.

The phylum Hemichordata includes two classes of strictly marine invertebrates closely allied phylogenetically to the echinoderms, with a limited number of traits suggesting weak affinities with chordates, especially the urochordates (tunicates).Hemichordates occur globally, ranging from polar to tropical regions and from shallow to abyssal depths. Only a small number of species (<100) have been formally described, and of these few are well-known. The class Enteropneusta consists of vermiform solitary individuals that typically burrow in sediments, whereas the Pterobranchia consists of colonial forms typically attached to suprasediment surfaces. The enteropneusts are also well-known for containing high concentrations of various bromo- and chloroaromatics, the functions of which are uncertain.

I. REPRODUCTION IN THE ENTEROPNEUSTA

Three clearly distinct regions characterize the bilaterally symmetrical enteropneusts (Fig. 1): the anterior proboscis, containing heart, glomerulus, and buccal diverticulum, which formerly and incorrectly was considered homologous to the chordate notochord; the collar, with buccal tube and pharynx; and the trunk, with a branchial region containing ciliated gill slits and a hepatic region containing a straight simple gut with hepatic diverticula or, in some cases, hepatic sacculations. Reproductive tissues also occur in the trunk extending variably from the posterior hepatic to the branchial region.

Hermaphroditism is unknown in enteropneusts, all of which have been described with separate sexes. Sexual dimorphism has been recorded in some cases, but more typically males and females are indistinguishable based on external characters except in the case of sexually mature individuals, the gonads of which can sometimes be discerned through the body wall of the trunk. When visible, eggs and sperm sacs are usually differentially colored within a species (Fig. 2).

In addition to their capacity for sexual reproduction, asexual reproduction has been described for several species of the family Ptychoderidae (Table 1). This process occurs after fragmentation of individuals anterior to the hepatic region. Up to 14 fragments have been reported to produce viable individuals after tissue reorganization and differentiation. Factors leading to the initiation of fragmentation are largely unknown, as is the relative importance of

FIGURE 1   *Protoglossus graveolens* (Giray & King) ripe male with prominent proboscis, collar, and trunk; sperm sacs occur as numerous white folds visible in the posterior trunk (from Giray and King, 1996).

asexual versus sexual reproduction. Since enteropneusts are relatively fragile, fragmentation could occur as a result of various forms of physical damage. Alternatively, asexual reproduction may ensue as a result of changes in environmental conditions, e.g., nutrient limitation. There are several consequences of sexual versus asexual reproduction in enteropneusts, one of which concerns dispersal and recruitment. Asexual reproduction ensures limited dispersal and probably high local recruitment, whereas sexual reproduction involving planktonic larvae ensures a relatively high degree of dispersal and perhaps

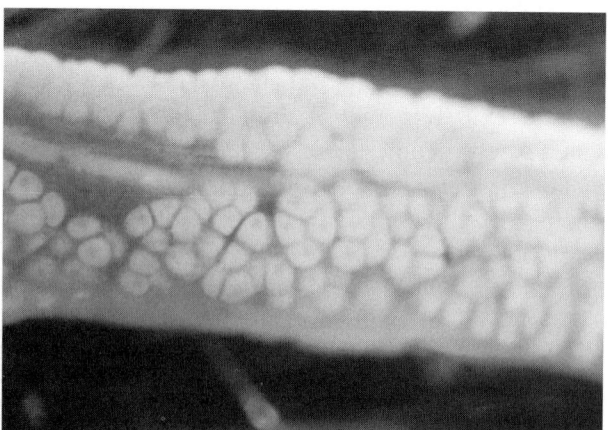

FIGURE 2   The posterior trunk of an adult female *Protoglossus graveolens* (Giray & King); numerous grape-like clusters are egg sacs visible through trunk epithelium (from Giray and King, 1996).

TABLE 1
Enteropneusts Reported to Reproduce Asexually or to Substantially Regenerate Tissues

Species	Asexual reproduction	Regeneration
Balanoglossus capensis	+	
Balanoglossus clavigerus		+
Glossobalanus minutus	+	+
Glossobalanus crozieri	+	
Ptychodera flava		+
Saccoglossus kowalevskii		+
Saccoglossus ruber		+

lower recruitment. Obviously, asexual reproduction also has profound implications for the genetic structure of local populations.

Although various aspects of sexual reproduction have been documented for a number of species, the most detailed information exists for several saccoglossids (e.g., *Saccoglossus horsti*, *S. pusillus*, *S. kowalevskii*, and *S. otagoensis*). For most, the gonads occur in the trunk coelom and are covered with coelomic epithelium. Apparently, gonads arise from mesenchymal cellular aggregation; they are typically composed of numerous ovoid sacs with a germinal epithelium layer and a highly vascularized basement membrane, the whole of which is enclosed by coelomic epithelium.

Enteropneust sperm are generally considered morphologically primitive. The exact structure varies among species, but a rounded head with an apical tip, four mitochondrial bodies, and a slender tail with an overall length of about 50 μm are common characteristics. Ova are variably associated with yolk cells; in the family Harrimaniidae, yolk cells are entirely absent (except for saccoglossids), whereas in the Ptychoderidae, yolk cells are found in all species examined thus far. When present, yolk disappears during oocyte growth and is not apparent outside of the eggs of ripe animals. Oocytes tend to be smaller in yolk-containing species. Egg sizes tend to vary more than sperm, ranging from <200 to >400 μm for mature eggs.

Gametogenesis can occur throughout the year in some taxa, though formation of fully ripe sperm and eggs is often temporally restricted, as is spawning. For example, in *S. horsti*, spawning occurs between May and June. In *S. horsti*, reproduction and spawning appear synchronized, with release of sperm following that of eggs in a given locale. The cues for synchronization are unknown but likely involve chemical signals or pheromones. In general, controls of gametogenesis and gametogenic cycles are unknown but likely involve nutrition as a major factor.

Four modes of spawning have been documented: (i) release of sperm and eggs by animals entirely or partially within burrows, with a flow of gametes out of burrows to surface sediments or water column (e.g., *S. horsti* and *Ptychodera flava*); (ii) release, fertilization, and development of gametes within burrows (*S. pusillus*); (iii) release of gametes external to burrows by shedding of a mucous cocoon (e.g., *S. otagoensis*); and (iv) possible planktonic release and fertilization of gametes (*Glossobalanus* spp. from Japan).

The process of fertilization and larval development has been well documented for *S. kowalevskii*. In this species, the acrosome is activated when sperm contacts an egg. The plasma membrane at the sperm tip degenerates, and an acrosomal tube penetrates the jelly external to the egg. The acrosomal tube and oocyte fuse, and the nucleus, mitochondria, and centrioles of the sperm are transferred to the egg, after which the fertilization membrane develops. Embryological development proceeds rapidly, with the first cleavage within a few hours and the seventh cleavage in <10 hr. Cleavage in the enteropneusta and pterobranchia is radial and holoblastic during early stages. Data for a number of species indicate that blastula and gastrula form within 6–15 and 10–24 hr, respectively. Hatching occurs within 30 hr to 7 days; tornarial larvae, which are characteristic of the Ptychoderidae, form within a few days.

Two modes of development are known for enteropneusts: direct and indirect. Species with small eggs (<150 μm) often exhibit indirect development, with a planktonic tornaria as is typical for echinoderms. Direct development, with or without a planktonic stage, is typical of those taxa with large eggs (>250 μm) that lack yolk cells. During indirect development, hatched tornaria assume a planktonic existence. Fully mature tornaria occur from 3 or 4 days to weeks after hatching. Tornaria feed in the plankton and slowly undergo a true metamorphosis. Coelomic development proceeds initially with the formation of a protocoel from a swelling of the archenteron; this becomes the proboscis coelom. The mechanism for formation of the remaining coeloms (those of the collar and trunk) varies among taxa, with up to five patterns recognized. In some cases, individual coelomic compartments arise separately from the archenteron wall, whereas in other cases the collar and trunk coeloms arise as paired archenteron evaginations. After development of proboscis, collar, and trunk, the tornaria assume a benthic existence following a course similar to that of the direct-developing larvae.

II. REPRODUCTION IN THE PTEROBRANCHIA

Relative to the enteropneusts, details of ptero-branch reproduction are poorly known. In large part this is due to the inaccessibility of most pterobrach taxa, which reside in the deep sea; logistical constraints have resulted in only a limited number of collections to date, few of which involve multiple observations of a given taxon or sufficient material for a complete characterization of reproductive biology.

The characteristics of *Rhabdopleura normani* have been described most extensively because it occurs in shallow water (<3 m depth) around Bermuda, where it is readily accessible on the underside of coral rubble. While the reproductive traits of *R. normani* are emphasized here, it must be understood that other lesser known taxa are likely to differ to some extent. *Rhabdopleura normani* is a representative of one of the two main pterobranch genera (*Cephalodiscus* is the other). *Rhabdopleura normani*, like virtually all pterobranchs, occurs in colonies consisting of zooids differentiated into three regions, homologous to the proboscis, collar, and trunk of enteropneusts. Individual *R. normani* zooids occupy a translucent tube cemented to coral substrate at the base; all zooids in a given colony are interconnected by a stolon running through the tube bases. Sexes are separate, and as in enteropneusts, reproductive structures are confined to the trunk. Colonies may consist of males or females, but both are usually present. Some evidence suggests that all zooids in a colony share a common parentage, and that mixed male and female zooids can arise asexually from an individual sexually produced zooid.

Asexual reproduction is common among the ptero-branchs, more so than among the enteropneusts. The process involves budding occurring at the zooid base. In *Cephalodiscus*, budding begins when a bulge arises from the trunk metacoel. After elongation, the anterior end of the bud swells, forming the cephalic shield (proboscis). This is followed by differentiation of the coelom and then formation of the gut, perhaps from the ectoderm. After further differentiation, the bud detaches from the parent and takes up existence as an individual within the colony. In *Rhabdopleura*, individual zooids remain connected by the stolon, from which buds arise as an evagination that differentiates subsequently as in *Cephalodiscus*. Factors controlling asexual versus sexual reproduction are unknown, although some evidence for environmentally sensitive cyclic patterns has been noted.

In contrast to enteropneusts, some pterobranchs exist as hermaphrodites or neuters (e.g., *Cephalodiscus*), with little sexual dimorphism. In others, such as *Rhabdopleura*, the sexes are separate. The hermaphrodites are generally characterized by two gonads, a testis and an ovary. In *Rhabdopleura*, gonads occur singularly as either testis or ovary. In both genera, gonopores occur between the anus and collar groove and are connected to the gonads by short ducts.

The testis of *R. normani* can extend the length of the trunk metacoel parallel to the gut. A median constriction differentiates the testis into an anterior and posterior region: The former is connected to the gonoduct and functions as a seminal vesicle, whereas the latter region is the site of spermatogenesis. For both *Cephalodiscus* and *Rhabdopleura*, sperm are uniflagellate with an elongated, tapering head lacking an acrosome. Relative to the pterobranchs, enteropneust sperm appear primitive. The "derived" nature of pterobranch sperm may be indicative of modified methods of fertilization compared to the enteropneusts. However, the mechanisms for sperm release are generally unknown. It has been speculated that *R. normani* is a "broadcast spawner," releasing sperm into the seawater around a zooid.

Pterobranch ovaries occur as simple, sometimes conical masses in the trunk coelom. In ripe zooids of *R. normani*, the ovary occupies much of the trunk volume and displaces the gut laterally. Both yolk cells and follicular elements occur in *Cephalodiscus* but not in *Rhabdopleura*. Oocytes form in the anterior portion of the ovary, with maturing oocytes occurring in the posterior region. Mature oocytes in pterobranchs (200 μm diameter in *R. normani*) often exceed the diameter of the gonoduct, giving rise to speculation that fertilization is internal or that oocytes are released after dehiscence of the body wall or disintegration of the zooid. For *R. normani*, some evidence suggests external fertilization in the brood chamber, though this has not been confirmed.

Fertilized eggs in pterobranchs appear to develop in two different locations, each characteristic of specific groups. In some *Cephalodiscus* spp., encapsulated eggs are attached to a stalk contiguous with the parent coenecium; in other *Cephalodiscus* and *Rhabdopleura*, eggs occur without a stalk underneath the parent in what amounts to a brood chamber. In *R. normani*, the tube in which female zooids reside is distinct from that of males. In particular, the female tube is coiled about 360° or more at the basal end. Eggs and developing embryos (up to seven) are aligned within this portion of the female tube. When embryos reach a motile stage, they swim out of the tube ostium and subsequently seek a suitable site for settling.

Emybronic development is not well described for pterobranchs. In *R. normani*, cleavage of the zygote gives rise first to a spherical cell mass that elongates to form a swimming larva approximately 400 μm in length 4–7 days postfertilization. The larvae are lecithotrophic (nonfeeding), depending on stored yolk. The anterior tip of the larvae has been postulated as a sensory organ, whereas the posterior region serves for temporary attachment during the search for a suitable substrate for settlement. Settling usually occurs within 24 hr under *in vitro* conditions.

See Also the Following Articles

ECHINODERMATA; HERMAPHRODITISM; TUNICATA

Bibliography

Burdon-Jones, C. (1951). Observations on the spawning behaviour of *Saccoglossus horsti* Brambell & Goodhart, and of other enteropneusta. *J. Mar. Biol. Assoc. UK* **29**, 625–637.

Burdon-Jones, C. (1952). Development and biology of the larva of *Saccoglossus horsti* (Enteropneusta). *Philos. Trans. R. Soc. London B* **235**, 553–589.

Colwin, A. L., and Colwin, L. H. (1953). The normal embryology of *Saccoglossus kowalevskii* (Enteropneusta). *J. Morphol.* **92**, 401–453.

Giray, C., and King, G. M. (1996). *Protoglossus graveolens*, a new hemichordate (Hemichordata: Enteropneusta: Harrimanidae) from the northwest Atlantic. *Proc. Biol. Soc. Washington* **109**, 430–445.

Hadfield, M. G. (1975). Hemichordata.. In *Reproduction of Marine Invertebrates. Vol. II. Entoprocts and Lesser Coelomates* (A. C. Giese and J. S. Pearse, Eds.), pp. 185–240. Academic Press, New York.

Lester, S. M. (1988). Ultrastructure of adult gonads and development and structure of the larva of *Rhabdopleura normani* (Hemichordata: Pterobranchia). *Acta Zool. (Stockholm)* **69**, 95–109.

Hemochorial Placentation

Erdal Budak and Jerome F. Strauss, III

University of Pennsylvania Medical Center

I. Introduction
II. Types of Hemochorial Placenta
III. The Circulatory Systems in the Hemochorial Placenta
IV. Advantages and Liabilities of Hemochorial Placentation

GLOSSARY

blastocyst The stage of embryonic development preceding implantation. The blastocyst consists of an outer layer of trophectoderm which is destined to become the placenta, a fluid-filled cavity or blastocoel, and an inner cell mass which will give rise to the fetus.

chorionic villi The primate placenta is organized as a tree-like tubular structure. The branches, or villi, become progressively smaller with each branching, ending in terminal villi where gas and nutrient exchange occurs between maternal and fetal blood.

decidua/decidualization The decidua is the endometrium of pregnancy. Decidual cells form from the differentiation of uterine stromal cells in response to hormones produced by the ovaries and placenta.

implantation   The process by which the blastocyst makes contact with the endometrium.

trophoblast/trophectoderm   The outer layer of cells of the blastocyst is the trophectoderm. The trophectoderm gives rise to trophoblast cells which form the placenta.

Prior to implantation, the free-living blastocyst acquires metabolic fuels and oxygen from the uterine secretions in which it is bathed. However, the continued growth of the embryo ultimately results in a size and metabolic rate that can no longer be sustained in the free-living state, requiring the embryo to establish a closer and more integrated relationship with the maternal host through the placenta. The form and structure of this interaction varies widely among and even within species.

I. INTRODUCTION

In hemochorial placentation, extraembryonic fetal cells (trophoblast cells) come into direct contact with maternal blood. This type of fetal–maternal interface is distinguished from endotheliochorial placentation, which occurs in the dog and cat in which maternal capillaries contact fetal trophoblast cells, and epitheliochorial placentation, which is characteristic of domestic animals in which uterine epithelial cells contact fetal trophoblast.

In humans, higher primates such as the baboon and rhesus monkey, mice, rats, guinea pigs, and rabbits, implantation of the blastocyst is invasive; the trophectodem penetrates the uterine epithelium and extends to various depths into the uterus. The penetration of trophoblast cells into the human uterus is extensive, so-called interstitial implantation, with the embryo burrowing deep into the uterine lining and ultimately being covered over by uterine tissue. The invading human trophoblast cells reach the uterine spiral arteries where they displace endothelial cells and remodel the arterial smooth muscle walls, depositing a fibrinoid substance, to create distensible, low-pressure channels by which maternal blood is brought into the emerging placenta. This remodeling is completed by the 20th week of gestation. The maternal and fetal circulations remain distinct in the hemochorial placenta with trophoblast cells serving as the barrier between the two systems.

II. TYPES OF HEMOCHORIAL PLACENTA

In the human, baboon, rhesus monkey, and armadillo, the placenta is a hemomonochorial structure (Table 1). The human placenta is organized into a branching tree-like structure, the chorionic villi, with an outer layer of syncytiotrophoblast surrounding a core of stroma and fetal vessels, ultimately connected to the fetal umbilical arteries and vein (Figs. 1 and 2). By progressive branching these tubular structures form smaller and smaller ramie, ending in terminal villi. The terminal villi are the major sites of exchange between the maternal and fetal circulations. They consist of an outer layer of multinucleated syncytiotrophoblast that rests on a basement membrane. The nuclei are usually clustered together, leaving a thin

TABLE 1
Types of Hemochorial Placenta

Name of species	Type of implantation	Depth of invasion	Shape of placenta	Material tissue in contact	Placental classification
Human	Invasive	Interstitial	Discoid	Blood	Hemomonochorial
Rhesus monkey	Invasive	Modest	Bidiscoid	Blood	Hemomonochorial
Rabbit	Invasive	Modest	Discoid	Blood	Hemodichorial
Mouse/rat	Invasive	Modest	Discoid	Blood	Hemotrichorial

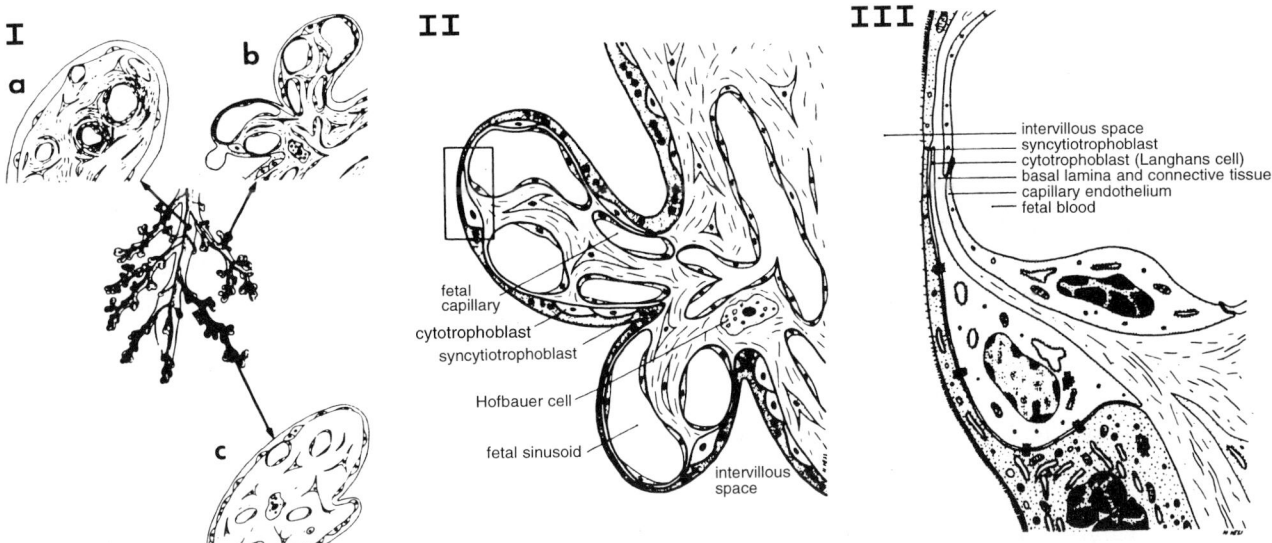

FIGURE 1 Schematic drawings of the human villous tree. (I) a, stem villus, one of the larger branches; b, a terminal villus, the smallest of the branches; c, an intermediate villus. (II) Intermediate power view of a terminal villus. (III) High-power view of a terminal villus. Hofbauer cells are fetal macrophages. Langhans cells are another name for cytotrophoblast cells (reprinted with permission from G. E. Ringler and J. F. Strauss, III, *In vitro* systems for the study of human placental endocrine function, *Endocr. Rev.* **11,** 105–123, 1990).

layer of cytoplasm to cover most of the villus surface. Fetal capillaries come into close apposition to the basement membrane, minimizing the distance between the maternal and fetal blood. The syncytiotrophoblast is a polarized cell type with a rich microvillous apical surface that greatly increases the absorptive area in contact with maternal blood. Transport molecules and ion channels are also polarized to facilitate uptake of maternal nutrients at the apical

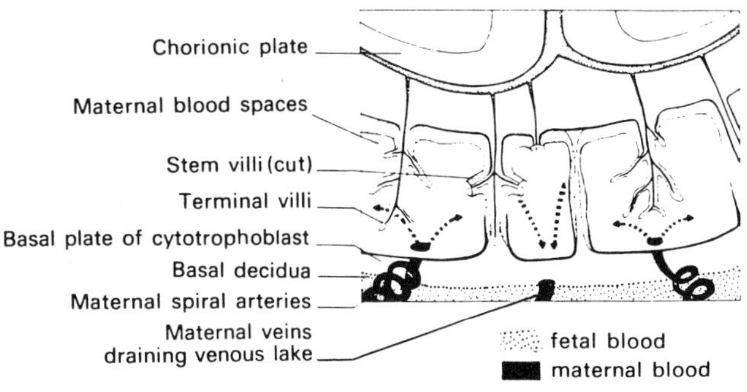

FIGURE 2 Schematic representation of terminal villi, the lacunae of the intervillous space, and maternal blood vessels [reprinted with permission from *Essential Reproduction* (M. Johnson and B. Everitt, Eds.), 3rd ed., p. 246, Blackwell Scientific, Oxford, UK, 1988].

surface and discharge them at the basal surface where they can diffuse into fetal capillaries. In the mature placenta, maternal arterial blood enters lacunae or lakes to bathe the chorionic villi. Veinules drain blood from the lacunae.

In the rabbit and the rat and mouse, two or three trophoblast layers, respectively, separate maternal blood from fetal capillaries forming hemodiochorial and hemotrichorial placentae (Table 1). There is little reason to believe that the number of trophoblast layers that separate the maternal and fetal circulations affects in a major way the efficiency of placental nutrient uptake and fetal waste discharge. The forces that sponsor uptake and exchange including active transport and the development of metabolic gradients that drive diffusion and carrier-mediated exchange depend largely on the rate of blood flows in the two circulations.

III. THE CIRCULATORY SYSTEMS IN THE HEMOCHORIAL PLACENTA

Maternal blood reaches the placenta through uterine and ovarian vessels. About 10% of cardiac output reaches the placenta at the end of pregnancy. The perfusion pressure within the intervillous space is normally low (4–10 mm Hg) as a result of the vascular remodeling of maternal arteries by invading trophoblast cells. In addition, the circulation of maternal blood in the intervillous space in primates is relatively sluggish because of the convoluted structure of the maternal terminal endometrial arteries. The low pressure and low velocity of flow may prevent the early conceptus from being dislodged; it also facilitates effective nutrient exchange. The velocity of fetal blood flow in the terminal villi is correspondingly reduced due to the marked increase in the total vascular cross-sectional area resulting from the extensive network of terminal capillaries. If remodeling of the maternal uterine arteries by invading trophoblast cells does not take place, the vessel muscular walls remain responsive to sympathetic stimulation and can constrict, limiting blood flow to the placenta and increasing maternal blood pressure. This phenomenon is believed to be a cause of pregnancy-induced hypertension or preeclampsia.

Reduced uteroplacental perfusion results from inadequate remodeling of the maternal uterine vessels by trophoblast cells and vasoconstriction of maternal vessels. The consequence of reduced blood flow to the intervillous space is reduced nutrient uptake and intrauterine fetal growth retardation.

IV. ADVANTAGES AND LIABILITIES OF HEMOCHORIAL PLACENTATION

The advantages of hemochorial placentation to the fetus are direct access to maternal nutrients and gasses and easy disposal of fetal waste products and placental hormones into the maternal circulation. The liabilities of hemochorial placentation include the fact that invasive implantation is required, necessitating the maternal host to develop effective barriers to limit the trophoblast invasion. Differentiation of uterine stromal cells into decidual cells, a process called decidualization, appears to be the primary mechanism by which trophoblast invasion is re-

strained. Decidualization is stimulated by progesterone as well as other hormones including relaxin. If this maternal barrier is not in place, trophoblast invasion can be excessive as occurs when implantation takes place outside of the uterus. Ectopic pregnancy in the Fallopian tube, whose stroma cannot decidualize, may lead to trophoblast erosion through the tube and its blood vessels and serious hemorrhage. As noted previously, failure to establish a low-pressure linkage between the maternal arteries and placental bed can lead to reduced uteroplacental perfusion and nutrient restriction of the fetus.

Contact between maternal blood and fetal extraembryonic tissues bearing paternal antigens requires the development of a mechanism to suppress the maternal immune response to the fetal semiallograft. In addition, the risk of immunization of the mother against fetal antigens spilled into maternal blood is also increased because of the intimate relationships of the two circulations.

See Also the Following Articles

Blastocyst; Decidua; Fetal Anomalies; Fetal-Placental Unit; Implantation; Preeclampsia/Eclampsia; Trophoblasts to Human Placenta

Bibliography

Enders, A. C., and Welsh, A. O. (1993). Structural interactions of trophoblast and uterus during hemochorial placenta formation. *J. Exp. Zool.* **266**, 578–587.

Gerretsen, G., Huisjes, H. J., Hardonk, M. J., and Elema, J. D. (1983). Trophoblast alterations in the placental bed in relation to physiological changes in the spiral arteries. *Br. J. Obstet. Gynaecol.* **90**, 35–39.

Moore, K. L., and Persaud, T. V. N. (1993). *The Developing Human. Clinically Oriented Embryology*, 5th ed. Saunders, Philadelphia.

Poston, L. (1997). The control of blood flow to the placenta. *Exp. Physiol.* **82**, 377–387.

Redline, R. W. (1997). The structural basis of maternal–fetal immune interactions in the human placenta. *Curr. Topics Microbiol. Immunol.* **222**, 25–44.

Steven, D. H. (Ed.) (1975). *Comparative Placentation*. Academic Press, New York.

Hemocoelic Insemination

K. G. Davey

York University

I. Hemocoelic Insemination in *Cimex lectularius*
II. Evolution of Hemocoelic Insemination

GLOSSARY

aedeagus The extensible portion of the intromittent organ or penis of most insects.

bursa copulatrix The portion of the female tract of the insect which receives the male genitalia, analogous to the vagina of mammals.

chorion The eggshell surrounding the insect egg and a product of the follicular epithelium which surrounds each developing egg.

hemocoel The body cavity of arthropods within which the hemolymph, or blood, circulates freely.

gonostylus A specialized structure on the ninth abdominal segment of male insects used to grasp the female during copulation.

Onychophora A phylum of worm-like animals with characteristics of both the annelids and the arthropods.

spermalege A specialized structure found on the abdominal wall only in cimicoid bugs through which the male inserts the spermatozoa. An **ectospermalege** is a specialized area on the cuticle for the reception of the sharp gonostylus of the male. A **mesospermalege**, a pad of tissue beneath the ectospermalege to receive the semen, may also be present.

spermatophore A structure produced by the males of many animals which encloses the semen while it is transferred to the female.

Hemocoelic insemination refers to a mode of insemination in certain insects whereby the semen is deposited into the hemocoel or body cavity of the female, typically through a wound in the cuticle of the female produced by a specialized "clasper" (gonostylus) of the male. In insects, the phenomenon is seen only in members of the superfamily Cimicoidea of the order Heteroptera. Hemocoelic insemination is known in the families Nabidae, Cimicidae, Plokiophilidae, and Polyctenidae of the superfamily.

Spermatozoa are also injected through the body wall in a variety of other invertebrates. In a number of species widely scattered through the phylum Platyhelminthes, spermatozoa are injected through the body surface, usually via a heavily armored penis. Examples can be found in the Turbellaria and the Cestoda. A similar phenomenon occurs in the majority of rotifers. This mode of insemination is usually referred to as hypodermic impregnation, does not, of course, involve a hemocoel, and will not be considered further in this entry.

In some leeches, such as *Placobdella*, and in some Onychophora, spermatophores are deposited on the body surface which then erodes beneath the spermatophore so as to allow the spermatozoa to enter the body cavity. This interesting phenomenon is also of a different nature from hemocoelic insemination and will not be considered further here.

I. HEMOCOELIC INSEMINATION IN *CIMEX LECTULARIUS*

Hemocoelic insemination has been described in some detail in the common bed bug, *C. lectularius*. This is a blood-feeding heteropteran, well-known for its infestations in dwellings.

The genitalia of the male are modified only slightly for hemocoelic insemination. In most other Heteroptera, the aedeagus is contained within a bulbous modification of the ninth abdominal segment, the genital capsule, and the capsule is flanked by two claspers (gonostyli) which serve to hold the female during copulation. In *C. lectularius*, the right gonostylus has been lost, the left gonostylus is sharpened at its tip, and a channel or groove runs from the base of the gonostylus almost to its tip. The aedeagus, when extended from the reduced genital capsule, enters the groove of the modified gonostylus.

The sharp gonostylus becomes the copulatory organ of the male. During copulation, the gonostylus is inserted into a sclerotized pocket on the right ventral abdomen of the female between the fifth and sixth segments. This is the ectospermalege (sometimes called the Organ of Ribaga), beneath which lies a pad of mesodermal tissue, the mesospermalege (called the Organ of Berlese by some authors). The gonostylus penetrates the cuticle of the ectospermalege, the aedeagus is extended along the groove in the gonostylus, enters the mesospermalege and the spermatozoa, and associated secretions are poured into the mesospermalege. The spermatozoa are placed in a single mass centrally in the tissue which forms the mesospermalege, and during the next few hours, this mass becomes diffuse and the spermatozoa enter the hemolymph.

The spermatozoa find their way through the hemolymph to specialized sperm storage organs, one at the base of each of the paired lateral oviducts. These structures consist of a mass of loose cells similar to those in the mesospermalege covered by a single layer of epithelial cells. The spermatozoa enter the sperm storage organs, and by 4 hr after copulation, they are present among the cells of the sperm storage organ in great numbers. Within a few days, the spermatozoa are gathered into a number of discrete bundles in which the individual sperm cells are aligned with one another.

In an unfed female, the spermatozoa, together with those from subsequent matings, remain in the storage organ. Any eggs which develop in the ovary do not enter vitellogenesis and eventually degenerate. When the female feeds, however, some of the sperm leave the storage organs as single cells and enter the walls of the lateral oviducts, where they migrate upwards to the ovary. This migration is accomplished within specialized branched cells which form an anastomosing network in the wall of the lateral oviduct. While fertilization has not been observed, spermatozoa have been seen within the developing follicle and fertilization occurs well before the chorion is formed. Not all the spermatozoa leave the sperm storage organ after a single feed. Some remain behind to be used at subsequent feeds, and a single copulation will suffice for six bouts of egg production. After each feed, a wave of spermatozoa leaves the storage organ and migrates upward to the ovary. The decline in fertility after the sixth bout of egg production from a single mating is apparently a result of a decay in the remaining spermatozoa rather than an inadequate supply.

II. EVOLUTION OF HEMOCOELIC INSEMINATION

Hemocoelic insemination as exemplified in *C. lectularius* involves a number of bizarre adaptations. The dangers implicit in the wounding of the female by the male, the development of a novel route for the transport of the spermatozoa to the eggs, and their fertilization before the chorion is laid down all raise questions about the evolution of this novel mode of insemination. It is not at first easy to see what advantages are conferred by the adoption of such an unusual reproductive strategy.

The work of the French entomologist Jacques Carayon in the 1950s (summarized in Hinton, 1964) on the reproduction of various Cimicoidea throws some light on these questions. The evolutionary steps leading to the situation in *C. lectularius* are suggested to be as follows. In some members of the family Nabidae, such as *Alloeorhynchus plebejus*, spermatozoa are deposited by the male into the bursa copulatrix of the female, but the spermatozoa are not transported to spermathecae as in other Heteroptera (see *Rhodnius prolixus*). Instead, they enter the lumen of the oviducts and migrate to the ovary. In the process, some spermatozoa become "lost" and enter the hemocoel, where they are phagocytosed by hemocytes. In several other nabids and a few cimicids, the male actually penetrates the wall of the bursa and deposits semen directly into the hemocoel. While many are phagocytosed by hemocytes, the number injected is large, and many survive to reenter the oviducts or ovary. In a few cases, specialized accumulations of cells derived from hemocytes or fat body cells develop as organs which apparently phagocytose the spermatozoa. This is seen as the precursor of the mesospermalege.

A further step is exemplified by *Primicimex cavernius*, a bed bug associated with bat caves. In this species, the spermatozoa are injected by the male through the general body surface using a modified gonostylus as in *C. lectularius*, but neither ecto- nor mesospermalege are present. Copulations can be detected by the scars left on the female by the male. Spermatozoa become widely dispersed in the body by the circulation but eventually accumulate in sperm storage organs such as those in *C. lectularius*.

The ectospermalege appears as a localized protective pocket to receive the gonostylus of the male, and this is accompanied by the development of the mesospermalege as in *C. lectularius*. From the work of Davis (1956, 1965), it seems probable that the mesospermalege functions to protect the spermatozoa from phagocytosis and to provide an environment for the seminal fluid to activate the spermatozoa.

Another development, exemplified in several cimicids such as *Strictocimex brevispinosus* or many species within the subfamily Anthocorinae, involves the extension of the mesospermalege to form a tissue bridge directly to the sperm storage organs. By this means the spermatozoa are protected from phagocytosis by the hemocytes.

According to this scheme, the original advantage derived from hemocoelic insemination has to do with the nutrition gained by the female from the spermatozoa. The nutritional gain my be very modest, but many of the cimicoids are opportunistic blood feeders, and the provision of even a tiny nutritional increment may be important in facilitating survival until a blood meal becomes available. More direct, if bizarre, support for this view is to be found in the genus *Afrocimex*, members of which inhabit bat caves in Africa. The females have two ectospermalege, both on the left side of the abdomen, associated with a single mesospermalege. Surprisingly, males also have two ectospermalege, but without the accompanying mesospermalege. Males have been observed to copulate with other males, and apparently semen is transferred because the same scars appear in their ectospermalege as appear in the ectospermalege of females. The benefit conferred on the recipient male is the protein and nucleic acid-rich nutrition derived from the transfer of semen which will be phagocytosed because of the absence of a mesospermalege. A male which has been inseminated will survive longer in the absence of its host. This appears to be a rare case in which homosexual copulation is important to the survival of the species.

See Also the Following Articles

Egg Coverings, Insects; Insect Reproduction; Male Reproductive System, Insects

Bibliography

Davey, K. G. (1965). *Reproduction in the Insects.* Oliver & Boyd, Edinburgh, UK.

Davis, N. T. (1956). The morphology and functional anatomy of the male and female reproductive systems of *Cimex lectularius* L. (Heteroptera, Cimicidae). *Ann. Ent. Soc. Am.* **49**, 466–493.

Davis, N. T. (1965). Studies of the reproductive physiology of the Cimicidae (Hemiptera)—III. The seminal stimulus. *J. Insect Physiol.* **11**, 199–1211.

Hinton, H. E. (1964). Sperm transfer in insects and the evolution of haemocoelic insemination. In *Insect Reproduction* (K. C. Highnam, Ed.). Royal Entomological Society, London.

Hermaphroditism

Kerstin Wasson

Humboldt State University

GLOSSARY

clonal animal An animal in which the genet is composed of multiple ramets that are physiologically separated and able to function independently.

colonial animal Animal in which the ramet is composed of multiple, physiologically interconnected modules.

genet A genetic individual; the entire mitotic product of one zygote.

gonochoric Single-sexed: either male or female; making only one type of gametes.

module The fundamental functional unit of construction that is repeated in the development of a colony or clone.

ramet A somatic individual; a coherent "body" able to function independently.

sequentially hermaphroditic Having temporally separate male and female phases; both types of gametes are produced by the same body but not at the same time.

sex allocation The allocation of resources (especially time and energy) to male vs female function.

simultaneously hermaphroditic Simultaneously male and female; both types of gametes are produced by the same body at roughly the same time.

unitary animal An animal that does not undergo asexual reproduction; with a single ramet per genet.

Hermaphroditism is the expression of male and female functions by a single genet. While the majority of animal species are probably gonochoric (single sexed), hermaphroditism is very common in the ani-mal kingdom. Hermaphroditism occurs in many different lineages of animals, apparently as the ancestral sexual mode in some taxa and as a derived state in others. The expression of hermaphroditism varies greatly among different animals; any system for understanding hermaphroditism should encompass temporal and spatial variation in its expression. The evolution of hermaphroditism, and variations of hermaphroditism, can best be understood by drawing on the predictions of sex allocation theory.

I. TYPES OF HERMAPHRODITISM

Animals are either gonochoric or hermaphroditic. Hermaphroditism can usefully be further subdivided, based on whether both types of gametes are produced simultaneously.

Spatial patterns of hermaphroditism may also vary; in some animals, a single gonad makes both sperm and eggs, whereas in others there are separate ovaries and testes. However, despite temporal and spatial variation in patterns of hermaphroditism, its definition is clear and distinct from that of gonochorism: A hermaphrodite expresses both male and female functions by producing both types of gametes.

II. SEXUAL MODES OF UNITARY, CLONAL, AND COLONIAL ANIMALS

The sexual modes of unitary animals (animals with a single body per genetic individual) can easily be classified. Following from the definitions given previously, a unitary animal may be gonochoric (G), sequentially hermaphroditic (S), or simultaneously hermaphroditic (H). (A very few unitary animals have more bizarre sexual modes, such as gynodioecy;

these are very rare and are here lumped under hermaphroditism since individuals have the potential to express both male and female functions.) The three sexual modes of unitary animals are shown in the innermost circle in Fig. 1.

Clonal animals have two structural levels at which sexual mode can be assessed—the genet and the module—and so a two-letter code is needed to specify mode of sex. In this system, the first letter refers to the genet and the second to the modules. Mode HS is thus a simultaneously hermaphroditic genet with sequentially hermaphroditic modules. The six potential sexual modes of clonal animals are represented by the middle ring in Fig. 1.

Colonial animals have three structural levels (genet, colony, and module), and so three-letter codes must be employed to describe their three-level structural composition. The first letter of the code refers to the genet, the second to the colony, and the third to the modules. For instance, HsG is the code for a simultaneously hermaphroditic genet with sequentially hermaphroditic colonies composed of gonochoric modules. There are 10 potential sexual modes for colonial animals, as shown in the outermost ring in Fig. 1. While the 3 sexual modes of unitary animals are familiar and easily understood,

the sexual modes of clonal and colonial animals are more complicated. Examples and characterizations of these sexual modes are provided by Wasson and Newberry.

All 3 sexual modes of unitary animals are common, but only 4 of the 6 clonal modes and 5 of the 10 colonial modes have known examples (modes with no known examples are bracketed in Fig. 1). Some of the logically possible but biologically unknown sexual modes may be very difficult to achieve (e.g., coordinating sex change of all clonal modules of a genet in mode SS), whereas others may be selectively disadvantageous.

III. TAXONOMIC DISTRIBUTION OF HERMAPHRODITISM AMONG ANIMALS

The distribution and frequency of all the sexual modes among animals are shown in Tables 1 and 2. Hermaphroditism is clearly widespread, occurring in many animal taxa. Hermaphroditism appears to have arisen from gonochorism many times, although in some lineages hermaphroditism may be ancestral.

A comparison at the genet level (Fig. 2a) reveals that hermaphroditism is more common than gonochorism in both clonal and colonial animals, but the reverse is true for unitary animals. (Note that Fig. 2 is based on the number of occurrences of each sexual mode among the taxa listed in Tables 1 and 2. Counting number of species, rather than number of higher taxa, might yield a somewhat different result.) The prevalence of hermaphroditism among clonal and colonial animals is accounted for almost entirely by genet simultaneous hermaphrodites; no clonal genets are sequentially hermaphroditic, and very few colonial ones are sequentially hermaphroditic. In contrast, a moderate number of unitary animal taxa are sequentially hermaphroditic. The sheer difficulty of synchronizing sex change of all component modules and ramets of clonal and colonial animals may explain the rareness of sequential hermaphroditism in these animals relative to their unitary counterparts.

A comparison of sexual modes at the module/body level (Fig. 2b) yields a completely different pattern. Instead of a trend of increasing simultaneous her-

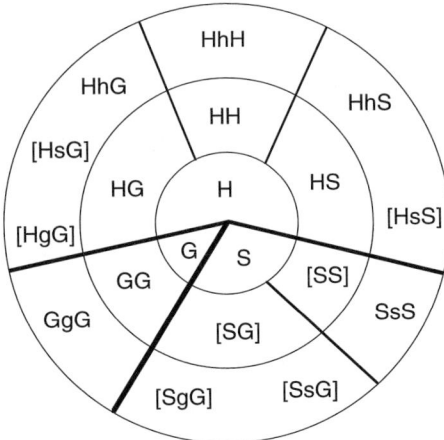

FIGURE 1 Schematic representation of the sexual modes of animals. The 3 sexual modes of unitary animals are shown in the inner circle, the 6 of clonal animals are shown in the middle ring, and the 10 of colonial animals are shown in the outermost ring. Bracketed modes are logically possible but biologically unknown. See text for abbreviations for each mode (modified from Wasson and Newberry, 1997).

TABLE 1

Frequency of Gonochorism (G), Sequential Hermaphroditism (S), and Simultaneous Hermaphroditism (H) among Unitary Members of Animal Taxa

Taxon	Sexual mode		
	G	S	H
Cnidaria			
Hydrozoa	X		
Scyphozoa	XXX		
Anthozoa	XX	X	X
Ctenophora			XXX
Platyhelminthes			
Turbellaria	X	X	XXX
Cestoda	X		XXX
Gnathostomulida		X	XXX
Gastrotricha		XX	XX
Kinorhyncha	XXX		
Nematoda	XXX	X	X
Nematomorpha	XXX		
Rotifera	XXX		
Acanthocephala	XXX		
Loricifera	XXX		
Nemertea	XXX	X	X
Annelida			
Polychaeta	XXX	X	X
Oligochaeta			XXX
Hirudinea			XXX
Sipunculida	XXX	X	X
Pogonophora	XXX		
Vestimentifera	XXX		
Echiura	XXX		
Priapulida	XXX		
Mollusca			
Aplacophora	XX		XX
Polyplacophora	XXX		X
Monoplacophora	XXX		
Scaphopoda	XXX		
Pelycepoda	XXX	X	X
Gastropoda	XX	X	XX
Cephalopoda	XXX		
Arthropoda			
Insecta	XXX	X	X
Crustacea	XX	X	XX
Chelicerata	XXX		
Onychophora	XXX		
Tardigrada	XXX		X
Pentastomida	XXX		
Phoronida	XX		XX
Brachiopoda	XXX	X	X

Taxon	Sexual mode		
	G	S	H
Echinodermata			
Crinoidea	XXX		
Asteroidea	XX	X	X
Echinoidea	XXX		
Ophiuroidea	XX	X	X
Holothuroidea	XX		X
Concentricycloidea	XXX		
Chaetognatha			XXX
Hemichordata			
Enteropneusta	XXX		
Chordata			
Urochordata			XXX
Cephalochordata	XXX		
Vertebrata	XXX	X	X
Total number of taxa displaying mode	42	16	27

Note. X, some members display this mode; XX, many members do; XXX, most or all do (modified from Wasson and Newberry, 1997).

maphroditism and decreasing gonochorism as seen in the genet-level comparison, the frequencies for all three animal groups are remarkably similar. Of course, the graphs are identical for unitary animals (Figs. 2a vs 2b) because the genet and the body are synonymous in this animal group. However, the graphs for clonal and colonial animals change drastically when modes that end with the same letter are grouped (recall that the final letter in the code represents the module's sexual mode) rather than modes that start with the same letter (recall that the first letter represents the genet's sexual mode). The module/body-level comparison reveals that across all three groups, about 40–45% are gonochoric, 20–25% are sequentially hermaphroditic, and 30% are simultaneously hermaphroditic. This similarity between unitary, clonal, and colonial animals suggests that the bodies may face the same sorts of selective pressures, and with similar frequency, despite the different structural organization of the genets which they comprise. Selection for different sexual modes must be understood at the level of both the bodies and the whole genets.

TABLE 2
Frequency of 6 Clonal and 10 Colonial Modes of Sex among Animal Taxa

Taxon		Sexual mode															
	Genets	Gonochoric		Sequentially hermaphroditic					Simultaneously hermaphroditic								
	Modules	Gonochoric		Gonochoric			Seq. herm		Gonochoric				Seq. herm			Sim. herm.	
	Mode	GG	GgG	SG	SgG	SsG	SS	SsS	HG	HgG	HsG	HhG	HS	HsS	HhS	HH	HhH
Porifera		X											XX			XX	
Cnidaria																	
Hydrozoa		XX	XX						X			X	X		X	XX	X
Scyphozoa		XXX											X			X	
Cubozoa		XXX															
Anthozoa		XX	XX									X	X		X	XX	XX
Ctenophora																	
Platyctenea													X			XX	
Mesozoa																	
Dicyemida																XXX	
Orthonectida		XXX														X	
Rotifera									XXX								
Platyhelminthes																	
Turbellaria													X			XXX	
Cestoda		X														XXX	
Trematoda		X														XXX	
Nemertea		XXX											X				
Kamptozoa																	
Solitaria													XXX				
Stolonata			X									X					X
Annelida																	
Polychaeta		XXX											X			X	
Oligochaeta																XXX	
Sipunculida		XXX															
Pogonophora		XXX															
Arthropoda																	
Insecta		XX							XX								
Crustacea									XXX								
Bryozoa													XX		X		XX
Phoronida		XX														XX	
Echinodermata																	
Asteroidea		XX											X			X	
Ophiuroidea		XX														X	
Holothuroidea		XXX															
Hemichordata																	
Enteropneusta		XXX															
Pterobranchia												XX					X
Chordata																	
Urochordata			X					X							X	X	XXX
Vertebrata		XXX															
Total number of taxa displaying mode		19	4	0	0	0	0	1	4	0	0	5	10	0	4	16	6

Note. X, some members display this mode; XX, many members do; XXX, most or all do (modified from Wasson and Newberry, 1997).

a) genet-level comparison

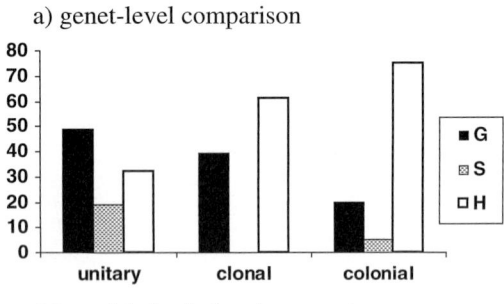

b) module/body-level comparison

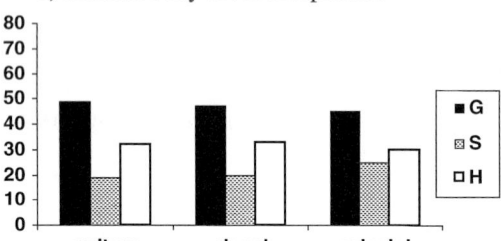

FIGURE 2 Comparison of the frequency of sexual modes among animal groups. (a) The percentages of unitary, clonal, and colonial taxa that have gonochoric (G), sequentially hermaphroditic (S), and simultaneously hermaphroditic (H) genets are shown; these percentages were calculated by counting the number of taxa displaying each mode, shown at the bottom of Tables 1 and 2. (b) The percentages of unitary, clonal, and colonial taxa that have G, S, and H modules/bodies are shown based on similar calculations.

IV. THE EVOLUTION OF HERMAPHRODITISM

A. Sex Allocation Theory

Currently, the most powerful conceptual understanding of the evolution of hermaphroditism comes from sex allocation theory. This theory invokes natural selection in the face of reproductive trade-offs and emphasizes that the sexual mode of a species is determined by selection acting on the number of offspring produced by individuals of different types. Sex allocation theory examines ultimate causes ("why" questions), not proximate mechanisms ("how" questions). The following are the kinds of questions addressed by theorists: When is hermaphroditism favored over gonochorism? In sequential hermaphrodites, when should sex change occur and which sex should be expressed first? In sequential

or simultaneous hermaphrodites, what proportion of resources should be allocated to male vs female function?

This latter question has proved particularly provocative because in many hermaphroditic species individuals can vary in their extent of paternal vs maternal investment. This variation lends itself to hypothesis testing; for instance, by looking for predicted changes in sex allocation in response to the physical or biological environment of individuals. To assess variation between individuals of a species, which is often quantitative rather than qualitative, Lloyd introduced quantitative measures of phenotypic gender (parental investment in male vs female function) and functional gender (realized reproductive success through male vs female function). Although this approach was developed for plants, it applies equally to hermaphroditic animals. The guiding theoretical principle when examining such intraspecific variation is that the evolutionarily stable strategy (ESS *sensu* Maynard Smith) of resource allocation to male vs female function is the one which maximizes the product of fitness gained through male function and that gained through female function.

Examining variation of sexual modes between species or higher taxa is more challenging than studying intraspecific variation. Sex allocation theory is still illuminating, but it is more difficult to determine the selective pressures which have driven the evolution of sexual modes. Sex allocation theory can successfully explain the sexual modes of a number of species. However, lest these cases be dismissed as "just-so" stories, the goal for the future should be the ability to accurately predict the sexual modes of species, given sufficient knowledge about their ecology and history.

I will summarize theoretical predictions about the conditions under which sequential hermaphroditism and simultaneous hermaphroditism should evolve, drawing mostly on sex allocation theory. This discussion should help to clarify the circumstances in which these sexual modes might be favored. Nevertheless, there are plenty of gonochoric species which seem to meet these criteria, so our understanding of the evolution of hermaphroditism is still far from complete.

B. Sequential Hermaphroditism

Our best explanation for the evolution of sequential hermaphroditism is the size-advantage hypothesis, first articulated by Ghiselin. He suggested that sequential hermaphroditism is favored if the relationship between reproductive success and size differs between sexes. Typically, this means that the reproductive success of one sex benefits more from large size than does the other. For instance, protogyny may be favored in species in which male success increases with experience and territory that can only be gained by large individuals. On the other hand, protandry may be favored in species in which female fecundity increases sharply with size.

Warner formalized and expanded the size-advantage hypothesis, demonstrating mathematically that sequential hermaphroditism is favored whenever one sex benefits from being larger or smaller than the other. Charnov noted that differences in growth or mortality rates between the sexes could drive the evolution of sequential hermaphroditism as well as differences in size-specific reproductive success. If one sex has a higher growth rate, it is advantageous to be that sex first. If one sex has a higher mortality rate, it is advantageous to be the other sex first.

The size-advantage hypothesis (and subsequent variations) has proved to be a powerful tool for understanding why some species are sequential hermaphrodites. However, Policansky notes that many organisms that would seem to benefit from sequential hermaphroditism do not display this sexual mode. So why aren't more animals sequentially hermaphroditic? This question will continue to motivate research in the future because we have only a few tantalizing hints toward the answer(s). Sequential hermaphroditism appears to be significantly more common in animals with indeterminate growth. Perhaps those with determinate growth do not generally have the developmental flexibility to express both sexes. Sequential hermaphroditism is more common in marine than freshwater animals; perhaps only certain types of habitats or environmental pressures favor this sexual mode. Finally, the cost of sex change, which has been mostly overlooked in theoretical models, may help to explain why more animals are not sequentially hermaphroditic.

Another fruitful area for exploration is the examination of the variation in timing of sex change within individuals of a species or between closely related species. There are theoretical models that predict when sex change should occur, but these predictions have only been tested in a few sequential hermaphroditic taxa. The sex ratio of a sequentially hermaphroditic species is generally skewed toward the first sex (female in protogyny and male in protandry). We have yet to adequately explain why more time is usually spent in the first than the second sex.

C. Simultaneous Hermaphroditism

Simultaneous hermaphroditism is favored when it is hard to find a mate—either because population densities are low or because motility is limited—because a simultaneous hermaphrodite can mate with any sexually mature conspecific, whereas a gonchore or sequential hermaphrodite can mate only with those of the opposite sex. Simultaneous hermaphroditism thus doubles the chance of encountering a mate. Simultaneous hermaphroditism is disproportionately common among sessile, parasitic, deep-sea, and planktonic animals, perhaps because all of these ecological circumstances can result in rare access to conspecifics.

Other factors may also drive the evolution of simultaneous hermaphroditism. The potential for self-fertilization in the absence of conspecifics may be important for some species, though many hermaphroditic species appear to have mechanisms to minimize or prevent selfing. Small animals may make the most of limited body space by brooding a few yolky eggs; they cannot produce enough free-spawned eggs to overcome the high risk of mortality prior to settlement. Such a small brooder can do little to increase its reproductive success through female function since brood space is limiting, but it might increase reproductive success through male function by producing a bit of sperm. This type of scenario may help to explain the observed correlation in many animals of small size, brooding, and hermaphroditism.

Most theory attempts to elucidate the advantages of simultaneous hermaphroditism over gonochorism. A necessary expansion of this approach is to determine the conditions under which simultaneous hermaphroditism is favored over sequential hermaphroditism. Also, for clonal and colonial animals, there are no theoretical predictions as to when each of the many pathways to genet simultaneous hermaphroditism (Fig. 1) are favored over others. Hermaphroditism is a fertile ground for scientific inquiry and will continue to be a rich source for future discoveries.

See Also the Following Articles

Sex Chromosomes; Sex Determination, Genetic; Sex Ratios

Bibliography

Charnov, E. L. (1982). *The Theory of Sex Allocation*. Princeton Univ. Press, Princeton, NJ.

Clark, W. C. (1978). Hermaphroditism as a reproductive strategy for metazoans; Some correlated benefits. *N. Z. J. Zool.* 5, 769–780.

Ghiselin, M. T. (1969). The evolution of hermaphroditism among animals. *Q. Rev. Biol.* 44, 189–208.

Lloyd, D. G. (1980). Sexual strategies in plants III. A quantitative method for describing the gender of plants. *N. Z. J. Bot.* 18, 103–108.

Maynard Smith, J. (1978). *The Evolution of Sex*. Cambridge Univ. Press, Cambridge, UK.

Policansky, D. (1982). Sex change in plants and animals. *Annu. Rev. Ecol. Syst.* 13, 471–495.

Tomlinson, J. (1966). The advantages of hermaphroditism and parthenogenesis. *J. Theor. Biol.* 11, 54–58.

Warner, R. R. (1975). The adaptive significance of sequential hermaphroditism in animals. *Am. Nat.* 109, 61–82.

Wasson, K., and Newberry. A. T. (1997). Modular metazoans: Gonochoric, hermaphroditic, or both at once? *Invertebr. Reprod. Dev.* 31, 159–175.

Hibernation

see Seasonal Reproduction

Hirsutism

William J. Butler

Medical University of South Carolina

I. Androgen Metabolism
II. Differential Diagnosis
III. Workup
IV. Other Diagnostic Modalities
V. Treatment

GLOSSARY

acanthosis nigricans A dermatologic condition with velvety, verrucus hyperpigmentation in skin creases.

androgen A class of C-19 steroids derived from cholesterol and secreted by the adrenal cortex and gonads. Androgens act through a specific receptor.

cytochrome P450 17-hydroxylase/70-20 lyase A steroidogenic enzyme catalyzing 17α-hydroxylation of C-21 steroid precursors and then scission of the bond between carbons 17 and 20 to form C-19 androgens.

dehydroepiandrosterone sulfate The primary circulating adrenal androgen.

17α-hydroxyprogesterone The androgen marker for deficient 21-hydroxylase activity.

testosterone A steroid hormone derived from ovary (25%), adrenal gland (25%), and peripheral conversion (50%). It is the most potent circulating androgen in females.

Hirsutism is excess terminal hair growth in women. Most commonly this is secondary to increased androgen synthesis and/or increased responsiveness of the skin/hair follicle to androgens. Androgen-dependent hirsutism occurs in a masculine-type hair distribution with increased facial, chest, and abdominal hair. Associated features are acne, menstrual dysfunction, and increased weight. Virilization is secondary to more extreme levels of hyperandrogenism. It may involve deepening of the voice, breast atrophy, temporal balding, and clitoromegaly.

I. ANDROGEN METABOLISM

Androgen metabolism involves the secretion, transport, peripheral conversion, metabolic degradation, and excretion of C-19 steroids. A brief review is appropriate in the discussion of hirsutism. The major circulating androgens in the adult woman are testosterone, androstenedione, dehydroepiandrosterone (DHEA), and dehydroepiandrosterone sulfate (DHEAS). Testosterone is secreted by both the ovary and the adrenal gland but 50% of it is derived from peripheral conversion of androstenedione. Both testosterone and androstenedione can be metabolized to dihydrotesterone (DHT) by 5-α-reductase. Androstenedione is produced in equal amounts from the ovary and the adrenal cortex. The adrenal gland produces 90% of circulating DHEA and 99% of circulating DHEAS. DHEAS is the most abundant androgen in circulation but has a very weak androgenic action compared to testosterone and DHT.

Androgen biosynthesis includes many enzymatically catalyzed steps involving a series of enzymes, many of which are from the cytochrome P450 family. These metabolic pathways are shared with the synthesis of other steroids such as glucocorticoids and estrogens. Congenital deficiencies in the activity of any of these enzymes lead to accumulation of precursors and the potential for a clinical condition of hyperandrogenism.

Most testosterone and DHT in the circulation are tightly bound to a hepatic protein termed sex hormone-binding globulin (SHBG) and, to a lesser extent, albumin. Only the free steroid fraction can bind to the androgen receptor and induce metabolic activity. The percentage bound hormone also affects metabolic clearance. For example, in obesity there is a decrease in SHBG leading to increased androgen clearance. Androgens can also be peripherally metabolized to estrogens by the enzyme aromatase which is present in adipose tissue.

II. DIFFERENTIAL DIAGNOSIS OF HIRSUTISM

A. Polycystic Ovarian Syndrome

Between 70 and 80% of hyperandrogenic women with hirsutism have Polycystic Ovarian Syndrome (PCOS). The diagnosis is one of exclusion after elimination of other known causes of hyperandrogenemia. The clinical features of PCOS include perimenarchal onset of hirsutism, oligoamenorrhea secondary to oligoanovulation, and obesity. This triad was described as the Stein–Leventhal syndrome but it should be understood that individual presentations may vary and not all patients have all features. Insulin resistance is commonly associated with PCOS and may be involved in the pathogenesis of the syndrome. The clinical finding of polycystic ovaries gives the syndrome its name. These ovaries are enlarged with a thickened hyalinized capsule with an underlying layer of atretic follicular cysts <1 cm in diameter. There is hyperplasia of the theca cell layer around these follicles that is the source of at least part of the excess androgen production. A common complaint is infertility secondary to the chronic anovulatory state.

1. Endocrine Evaluation

One theory of the pathogenesis of PCOS is an alteration in gonadotropin-releasing hormone (GnRH) production from the hypothalamus. An increased pulsatile frequency of GnRH release could lead to an imbalance in the synthesis and release of follicle-stimulating hormone (FSH) and luteinizing hormone (LH) from the anterior pituitary. LH concentrations are commonly elevated and FSH levels normal or low in PCOS. This alteration in ratio of LH to FSH has been described to assist in the diagnosis, with a ratio $\geqq 3:1$ being consistent with the syndrome. The increased LH secretion can contribute to the excess androgen production by its effect on ovarian thecal cells to stimulate androgen synthesis.

The source of the excess androgen production in PCOS reflects both ovarian and adrenal contributions. Testosterone levels are high-normal to elevated as are androstenedione levels. GnRH agonist suppression studies have demonstrated that the source for these androgens is the ovary. However, both DHEA and DHEAS levels can also be elevated in PCOS and these are not affected by GnRH agonist suppression of ovarian steroidogenesis. The precise mechanism for increased adrenal androgen production has not been demonstrated. An increased sensitivity to ACTH has been proposed as has dysregulation of the cytochrome P450 17-hydroxylase/17-20 lyase enzyme (P450 17). Dysregulation of P450 17 may be secondary to an intrinsic genetic defect in some patients or potentially induced by extraadrenal factors such as altered ovarian steroidogenesis or hyperinsulinemia secondary to insulin resistance. Estrogen has been demonstrated *in vitro* to increase P450 17 activity and estrogen biosynthesis is certainly altered in patients with PCOS. Not only is estradiol production relatively deficient due to inadequate follicular growth and maturation but also increased peripheral conversion of androstenedione to estrone gives an altered hormonal milieu. Estrone feedback effects on the pituitary may further alter the relative synthesis and release of FSH and LH. Prolactin has been noted to be elevated in 20–30% of women with PCOS. This may be secondary to estrone feedback on pituitary lactotropes or may be due to altered dopamine activity that also has been

postulated to be the etiology for altered GnRH pulse frequencies. Another theory is a paracrine effect of GnRH between gonadotrope and lactotrope cells in the anterior pituitary. Data to support any specific theory are lacking.

2. Insulin Resistance

Hyperinsulinemic insulin resistance is characteristic of most PCOS patients. This is independent of the presence of obesity and appears to be due to a different mechanism than the insulin resistance known to be associated with obesity. Because the insulin resistance persists despite surgical removal of the ovaries or GnRH agonist suppression of androgen synthesis, it is unlikely that the insulin resistance is secondary to the hyperandrogenemia. Instead, the hyperinsulinemia may serve an important pathogenic role in the initiation of PCOS. Insulin resistance is due to a postreceptor block the exact nature of which has not been determined. The effect of hyperinsulinemia to induce androgen synthesis may be direct or may involve an as yet undetermined third factor such as a PCOS gene. Insulin has also been shown to directly lower circulating SHBG levels, resulting in increased free bioavailable androgens. Growth factors such as insulin-like growth factor-1 (IGF-1) and their transport proteins may also play a role in pathogenesis of PCOS. Serum insulin-like growth factor binding protein-1 (IGFBP-1) levels are lower in PCOS patients than normal controls. IGFBP-1 can act as an FSH antagonist and therefore play a role in inhibiting follicular development. Serum-free IGF-1 levels are increased and this may further stimulate thecal androgen production.

3. Hyperthecosis

Patients with hyperthecosis are often virilized and have high levels of circulating testosterone. They may represent the extreme of the continuum of PCOS patients. The diagnosis is made on pathological evaluation of the ovary demonstrating extensive hyperplasia of luteinized thecal cells with relatively few atretic follicles.

These patients may also have HAIR-AN syndrome (hyperandrogenemia, insulin resistance, and acanthosis nigricans). HAIR-AN patients have more severe insulin resistance of the Kahn type A (due to genetic mutations in the insulin receptor) or Kahn type B (due to autoantibodies to the insulin receptor). They may also have associated lipoatrophic diabetes and significant dyslipidemia with a high risk of hypertension and cardiovascular disease.

B. Nonclassical Adrenal Hyperplasia

Nonclassical adrenal hyperplasia (NCAH; also known as late onset or attenuated adrenal hyperplasia) presents a clinical picture essentially indistinguishable from PCOS. There is perimenarchal onset of menstrual irregularity and hirsutism secondary to adrenal hyperandrogenemia. The pathogenesis is due to a genetic defect in one of the enzymes catalyzing steroidogenesis. The most commonly affected enzyme is a partial deficiency of cytochrome P450 21-hydroxylase (P450 21) which affects 1 or 2% of androgenic women. Certain ethnic groups, such as Ashkenazi Jews and Hispanics, report a higher frequency. In contrast to patients with congenital adrenal hyperplasia who have a complete absence of P450 21 activity, NCAH patients have a partial defect that allows compensation with adequate glucocorticoid synthesis but at the expense of increased ACTH stimulation and accumulation of androgenic precursors resulting in hyperandrogenemia. The androgen profile is not differentiable from PCOS patients and therefore diagnosis is dependent on measurement of 17α-hydroxyprogesterone (17-OHP) levels. Individuals with a basal level of 17-OHP $\geqq 200$ ng/dl should undergo an ACTH stimulation test to rule out NCAH.

NCAH is an autosomal recessive condition with the gene identified on the short arm of chromosome 6 in the HLA region. The P450 21 gene lies in close proximity to a pseudogene which allows for gene conversions or deletions during meiosis. There are numerous described mutations and the combined carrier frequency for a single mutated gene in the general population has been estimated at 5%.

Several other steroidogenic enzymes have been described in lesser frequency to cause NCAH. Both the cytochrome P450 11 hydroxylase gene and 3β-hydroxysteroid dehydrogenase gene have been deter-

mined to have deficient activity in some patients. Diagnosis of these variants requires more extensive analysis of steroid precursors in both the basal and stimulated states.

C. Androgen-Producing Neoplasms

Although rare, neoplastic causes of hyperandrogenism must be considered in the differential diagnosis of hirsute patients, particularly in the presence of signs of virilization or severe hyperandrogenemia. A clinical marker is the rapid onset of hirsutism and/or virilization with menstrual irregularity in a previously asymptomatic patient. Ovarian neoplasms are the most common source of neoplastic androgen production. As a group, the granulosa–theca cell tumors are most frequent and account for 15–20% of solid ovarian tumors. They are derived from gonadal stroma and may be hormonally active, producing predominantly estrogens but also androgens. They occur in patients of any age and may be bilateral in 10–15% of cases. Up to 20% are thought to have a malignant potential.

Gonadal stromal tumors account for 1% of solid ovarian tumors. These include tumors previously described as arrhenoblastomas and Sertoli–Leydig cell tumors. They occur in women of reproductive age, are usually unilateral, and have a lower risk of malignancy of between 5 and 20%. Another variant is the hilus cell tumor, which is a small tumor of luteinized cells often found in the hilar region of the ovary in peri- and postmenopausal women.

Almost any ovarian neoplasm has the potential to be steroidogenically active, including germ cell tumors, epithelial neoplasms, and even metastatic cancer. The mechanism appears to be stimulation of the ovarian stroma by the presence of the neoplastic cells. Both estrogen and androgen synthesis have been described.

Androgen-producing adrenal tumors are relatively rare. Ninety percent of these are benign functioning adenomas. They typically produce DHEAS but rarely produce testosterone. The DHEAS-producing adenomas have a peak age range of 20–40 years with the testosterone-secreting tumors more commonly found in menopausal women. They are almost uniformly small and unilateral and can be detected only by sophisticated imaging studies of the adrenal gland.

D. Other Disorders Associated with Hirsutism

1. Cushing's Syndrome

Both menstrual irregularity and hirsutism are common presenting complaints of patients with Cushing's syndrome. The hirsutism is relatively diffuse and mild and accompanied by hypertrichosis. In Cushing's disease a pituitary tumor produces ACTH, leading to excessive adrenal androgen synthesis and secretion. Cushing's syndrome may also result from ectopic ACTH-producing tumors or a glucocorticoid-producing adrenal adenoma. Although presenting with a similar clinical picture including insulin resistance, Cushing's syndrome is a rare cause of hirsutism. Evaluation should be limited to those patients who present with other signs and symptoms, including Cushingoid features, generalized weakness, hypertension, and easy bruisability.

2. Exogenous Androgens

Androgenic drugs are another rare cause of hirsutism. Danazol, a drug commonly used to treat endometriosis, is a synthetic steroid that may have endogenous androgenic potency as well as suppressing SHBG levels with elevation of free bioavailable testosterone. Other anabolic steroids that may be used by athletes and bodybuilders may also induce hirsutism.

3. Idiopathic Hirsutism

This is a diagnosis of exclusion and may be made only when circulating androgen levels are found to be consistently normal in the presence of obvious hirsutism. Between 5 and 25% of hirsute women have been described to have otherwise normal endocrine evaluations. Some of these women may have excessive 5-α-reductase activity in the hair follicles resulting in increased metabolic conversion of testosterone to DHT. Unfortunately, measurements of 3-α-androstanedione glucuronide, the primary metabolite of DHT, have not proven to be useful in

making this diagnosis. Many cases of so-called hirsutism may merely be genetic or ethnically based hypertrichosis with heavy vellus hair growth in a normal female pattern and therefore not true hirsutism.

III. WORKUP

A. History

The menstrual history, including age of menarche and course of pubertal development, is very important given the perimenarchal onset of benign forms of hirsutism. The age of onset of hirsutism and rapidity of its development should be ascertained. Any associated complaints, including obesity, acne, or signs of virilization such as hair loss and deepening of the voice, should be noted. Acute onset or signs of virilization raise the possibility of Cushing's syndrome or an androgen-secreting neoplasm.

A complete medication history should be obtained, although female athletes may be reluctant to admit anabolic steroid use. A positive family history suggests a benign etiology. Patients with NCAH will have a negative family history because it is an autosomal recessive disorder and heterozygotes do not have hirsutism. Documentation should be obtained regarding methods of self-treatment such as shaving, plucking, or dipilatory use because this may interfere with accurate evaluation on physical exam.

B. Physical Examination

This is the basis for diagnosis because the hair follicle is an *in vivo* bioassay for hyperandrogenemia. If there is terminal hair growth in a male-type distribution, the diagnosis of androgen-dependent hirsutism has been established. The amount and distribution of hair can be qualitatively described or semiquantitated using one of several scoring techniques.

Signs of virilization, such as temporal balding, breast atrophy, increased muscle mass, and clitoromegaly, should be noted. The skin should be carefully inspected particularly around the neck for the presence of acanthosis nigricans, a marker for significant insulin resistance. Features suggestive of Cushing's syndrome include central obesity, thin skin with violaceous striae on the abdomen, facial rounding, and proximal muscle weakness. On pelvic examination, clitoromegaly is a marker for more severe hyperandrogenemia and a palpable adnexal mass requires evaluation to rule out an ovarian neoplasm.

C. Laboratory Evaluation

As already noted, androgen-dependent hirsutism can be diagnosed on physical examination. The laboratory evaluation is directed toward excluding more serious diagnoses such as a neoplasm. Baseline studies should include a serum total testosterone with repetitive values $\geqq 150-200$ ng/dl worrisome for an androgen-secreting neoplasm. DHEAS values $\geqq 7000$ ng/ml call for evaluation to rule out an adrenal tumor. FSH and LH values can be obtained to help confirm a diagnosis of PCOS. An absolute elevation of LH $\geqq 20$ mIU/ml or a ratio of LH : FSH $\geqq 3:1$ are consistent with the diagnosis but are not seen in every patient. Serum prolactin may be obtained as part of the appropriate evaluation for any patient with menstrual irregularity to rule out a prolactinoma. Approximately 30% of patients with PCOS will have mild elevations. The diagnosis of NCAH is made by obtaining a single baseline screening test for 17-hydroxyprogesterone. This should be obtained in the follicular phase of an ovulatory patient but because most of these hirsute patients are anovulatory it can be obtained at random. A value <200 ng/dl essentially rules out NCAH. A value $\geqq 200$ ng/dl requires an ACTH stimulation test. This is performed by injecting 250 mg of Cortrosyn (1-24-ACTH) intravenously and repeating 17-hydroxyprogesterone levels at 30 or 60 min. If the poststimulation 17-hydroxyprogesterone level is $\geqq 1000$ ng/dl, this confirms the diagnosis of P450 21-deficient NCAH. Other steroid precursors can be measured to detect the rarer forms of NCAH.

If Cushing's syndrome is suspected, a dexamethasone suppression test should be obtained. One milligram of dexamethasone is given at 11:00 PM, and at 8:00 AM a serum cortisol is drawn. A value <5 mg/dl rules out Cushing's syndrome. False-positive results may be obtained in obese and depressed patients and then a 24-hr urinary-free cortisol should be obtained with a value <100 mg/dl per 24 hours consistent with a negative diagnosis.

Insulin resistance may be determined by obtaining a basal insulin level. Values $\geqq 80$ mU/ml in the fasting state are consistent with significant insulin resistance. An oral glucose tolerance test should be obtained to rule out frank diabetes. Insulin levels measured after an oral glucose load that exceeds 300 mU/ml are also consistent with severe insulin resistance such as occurs in the HAIR-AN syndrome.

IV. OTHER DIAGNOSTIC MODALITIES

Vaginal ultrasound can serve an important role in the differential diagnosis of androgen-dependent hirsutism. The characteristic appearance of polycystic ovaries, while not pathognomonic, is very supportive of a diagnosis of PCOS. The ovaries are enlarged with a thickened echogenic capsule and the classical string of pearls appearance with at least 8–10 <1-cm-diameter subcapsular follicular cysts. This appearance has been described in normally ovulatory patients and therefore is not alone sufficient for diagnosis. Vaginal ultrasound may also detect ovarian neoplasms but its sensitivity in detecting nonpalpable tumors has not been accurately determined. Doppler flow ultrasound looking at vascular flow in the ovary may be of additional benefit.

Adrenal visualization requires either computed tomography or magnetic resonance imaging. Selective ovarian and adrenal vein catheterization to detect the source of increased androgen production has been useful in the past but is less frequently utilized now because of improved imaging techniques. Lateralizing of the elevated androgens to one ovary or one adrenal is highly diagnostic of the presence of a neoplasm. Serum testosterone levels in the neoplastic range with failure of lateralization on selective ovarian vein catheterization is consistent with hyperthecosis.

V. TREATMENT

A. Medical Treatment

1. *Oral Contraceptives*

The primary treatment modality for androgen-dependent hirsutism is oral contraceptives. Although designed to prevent ovulation, the inhibition of gonadotropin secretion results in decreased ovarian steroidogenesis. A secondary effect is induction of SHBG synthesis from the liver. Theoretically, a pill containing a new generation non-19-nortestosterone derivative progestin would be less inherently androgenic and more effective. However, a more potent higher dose progestin may better suppress LH and further reduce ovarian androgen synthesis. There is no direct clinical data available comparing different formulations of oral contraceptives as to efficacy.

2. *GnRH Analogs*

Gonadotropin-releasing hormone analogs are superagonists that downregulate pituitary gonadotropes and suppress LH and FSH synthesis and release. This has been termed a medical castration because of the profound inhibition of ovarian steroidogenesis that results. These drugs are very effective in suppressing ovarian androgen production but have major clinical disadvantages. They are expensive, must be given parenterally, and are associated with significant hypoestrogenic side effects including hot flashes and loss of bone mineral mass. Long-term treatment would require an add-back regimen of low-dose estrogen and progestin to alleviate these adverse effects.

3. *Antiandrogens*

The prototypical antiandrogen available is spironolactone. It is an aldosterone antagonist and diuretic but also has antiandrogenic effects by competitive inhibition of androgen receptor binding and is a weak inhibitor of testosterone biosynthesis. The usual dosage is 25–100 mg given orally twice a day. Side effects include menstrual irregularity, nausea, and fatigue. Some studies have shown a synergistic effect when used in combination with oral contraceptives, with an excellent overall response rate of 75–80%.

Flutamide is an antiandrogen used to treat prostate cancer. It has less affinity for the androgen receptor than spironolactone and is used in a higher dose of 125–250 mg twice a day. It has not been shown to be more effective than spironolactone and the major side effect is the risk of drug-induced hepatitis which may be fatal.

Finasteride is a 5-α-reductase inhibitor used for the treatment of benign prostatic hyperplasia. A dose of 5 mg daily has been shown to be as effective as 100 mg of spironolactone in the treatment of hirsutism. The safety and side effect profiles are excellent, but it may cause ambiguous genitalia in male offspring and therefore should be used with caution in patients at risk for pregnancy.

Cimetidine is a weak antiandrogen and has not proved effective for the treatment of hirsutism. Ketoconazol is a potent antifungal agent and inhibits testosterone synthesis but also inhibits cortisol biosynthesis and its potential hepatotoxicity mitigates against its use for treatment of hirsutism.

4. *Adrenal Suppression by Glucocorticoids*

Patients with NCAH with a primary adrenal source for their hyperandrogenemia can be treated with exogenous glucocorticoids to suppress ACTH and decrease adrenal stimulation, thereby reducing the production of androgenic precursors. Although any glucocorticoid derivatives can theoretically be used, a potent steroid such as dexamethasone given in a low dose of 0.25–0.5 mg qd is usually utilized. This reduces the risks of glucocorticoid excess while potentially maximizing suppression of androgen synthesis.

B. Other Therapies

1. *Surgical Therapy*

The treatment of choice for adrenal or ovarian neoplasms is surgical excision. In the absence of metastatic disease this results in a prompt fall in circulating androgen levels and significant improvement in hirsutism and other symptomatology.

Patients with the more severe forms of PCOS, and particularly those with hyperthecosis or the HAIR-AN syndrome, may not respond adequately to medical therapy. Another option for their treatment is the ovarian wedge resection. Classically performed via laparotomy, this can now be performed laparoscopically either with resection of the ovary or by using a laser or cautery to drill holes in the ovarian cortex and thus destroying androgen-producing tissue. The rationale is to remove a part of the thecal cell mass, thereby reducing overall androgen synthesis. The original series by Stein and Leventhal reported long-term improvement in both hirsutism and ovulatory function without recurrence of the ovarian pathology. Recent reports have noted short-term resumption of ovulation with eventual recurrence of hyperandrogenemia and anovulation. Patients who are candidates for this therapy include those with hyperthecosis and those with PCOS who desire pregnancy and are unresponsive to standard ovulation induction medications. They run the risks of surgery and in particular a risk of pelvic adhesion formation which may further compromise their fertility.

2. Mechanical Treatment

Mechanical methods of hair removal treat the superficial problem but do not change the underlying hyperandrogenemic state. Bleaching can lighten the visible hair but skin irritation is a problem. Shaving is both easy and safe. Contrary to popular opinion, it does not increase the rate or coarseness of hair growth. It may be used as an adjunct to electrolysis. Plucking or waxing can be uncomfortable and there is a risk infection and scarring. Depilatory creams are also effective but can cause a problem with skin irritation. All these treatments are temporary and must be repeated as hair regrows.

Electrolysis is claimed to be a permanent method of hair removal but regrowth may occur in some patients. It involves electrocoagulation of the hair follicle using shortwave current. There is usually minimal discomfort and no scarring, but the technical skill of the operator can be highly variable in states not having licensing procedures. Treatment can be time-consuming and very expensive.

See Also the Following Articles

Androgen Inhibitors; DHEA (Dehydroepiandrosterone); Polycystic Ovary Syndrome

Bibliography

Faber, K., and Hughes C. L., Jr. (1991). Laboratory evaluation of hyperandrogenic conditions. In *Infertility and Reproductive Medicine Clinics of North America* (D. E. Pitaway, Ed.), Vol. 2, pp. 495–509. Saunders, Philadelphia.

Gonzalez, F. (1997). Adrenal involvement in polycystic ovarian syndrome. *Sem. Reprod. Endocrinol.* **15**, 137–158.

Morris, D. (1995). The etiology of hyperandrogenism in women. *Curr. Opin. Obstet. Gynecol.* **7**, 224–226.

Nessler, J. E. (1997). The role of hyperinsulinemia in the pathogenesis of the polycystic ovary syndrome and its clinical implications. *Sem. Reprod. Endocrinol.* **15**, 111–122.

Pitaway, D. E. (1991). Neoplastic causes of hyperandrogenism. In *Infertility and Reproductive Medicine Clinics of North America* (D. E. Pitaway, Ed.), Vol. 2, pp. 531–546. Saunders, Philadelphia.

Rittmaster, R. S. (1995). Medical treatment of androgen dependent hirsutism. *JCEM* **80**, 2559–2563.

Scherzer, W. J., and Adashi, E. Y. (1991). Adrenal hyperandrogenism. In *Infertility and Reproductive Medicine Clinics of North America* (D. E. Pitaway, Ed.), Vol. 2, pp. 479–494. Saunders, Philadelphia.

Shanti, A., and Murphy, A. A. (1997). Surgical approaches to ovulation induction. *Sem. Reprod. Endocrinol.* **15**, 183–192.

HIV Infection and AIDS

Kearston Schmidt and Penelope J. Hitchcock

National Institutes of Health

GLOSSARY

cytoplasm The substance of a cell, exclusive of the nucleus, which contains various organelles.

cytotoxic Toxic to the cells.

epidemiology The study of the distribution and determinants of health-related events in specified populations and the application of this study to control health problems.

epithelium The tissue cells comprising the skin; also the cells lining all the passages of the hollow organs of the respiratory, digestive, reproductive, and urinary systems.

Food and Drug Administration (FDA) An agency of the U.S. Department of Health and Human Services that regulates the testing and approval of experimental drugs, vaccines, and devices based on evidence of safety and efficacy.

genome All the hereditary factors present in the organism or virus.

helper T lymphocytes Part of the cellular immune response to fight infections.

inhibitor An agent that restrains or retards physiologic, chemical, or enzymatic action.

monocyte/macrophage A large white blood cell active in destroying and devouring bacteria or foreign matter that has entered the body (monocyte). A macrophage is a monocyte that has left the bloodstream and resides in the tissues.

polymerase chain reaction (PCR) An enzymatic process that allows the amplification and identification of specific DNA or RNA sequences; this is the basis for diagnosis of infection.

retrovirus An RNA virus that carries an enzyme which makes a DNA copy of the viral genetic information.

shedding The movement of virus across an epithelial surface into body secretions.

suppressor T lymphocytes which dampen or shut down the immune response to an infection or other foreign body.

Acquired immunodeficiency syndrome (AIDS) was first recognized as a disease in the early 1980s. AIDS is virtually always fatal. The cause, human immunodeficiency virus (HIV), is sexually transmitted by vaginal, anal, and rarely by oral intercourse. It can also be transmitted by blood through transfusion or use of contaminated needles. Finally, it can be transmitted from infected mother to her baby three ways. The virus can infect the baby in the uterus, in the birth canal, or through breast milk.

Effective methods are available to prevent HIV transmission by all routes. Sexual transmission can be prevented by consistent and correct use of the male (latex) condom. Blood transmission can be prevented by proper processing of blood and blood products and by use of sterile needles. Maternal transmission can be prevented by drugs that reduce the amount of virus in the mother's blood and by substituting commercial formula for breast milk.

I. RISK FACTORS DRIVING THE HUMAN IMMUNODEFICIENCY VIRUS EPIDEMIC

Compared to other sexually transmitted viruses, such as the genital wart virus, human papillomavirus

(HPV), the genital herpes virus, or hepatitis B virus, human immunodeficiency virus (HIV) is relatively difficult to transmit. It is estimated that the chance of becoming infected if exposed ranges between 1/500 and 1/1000. By contrast, the chance of infection with HPV is estimated to be 1/4 if exposed.

Three aspects of acquired immunodeficiency syndrome (AIDS) have favored rapid spread of HIV throughout the world: (i) the sexual transmission of the virus, (ii) the long silent period of infectiousness before AIDS develops, and (iii) the destruction of the body's defense system. Sexual intercourse is fundamental to the biology of our species: It enables us to have children. Given that there are no tools to prevent infection that permit pregnancy, one cannot get pregnant without risking infection. The length of the incubation period between infection and disease means that people are contagious before they recognize they are ill. In the absence of routine screening for HIV infection, the virus is silently spread, often without the knowledge of either partner. Finally, the virus targets the body's defense system, which means that the body cannot fight HIV, and that ultimately it cannot fight other infections either. Most AIDS patients die of opportunistic infections or cancer.

Three other behaviors or conditions have increased the risk of infection in certain populations: (i) anal intercourse, (ii) intravenous drug use, and (iii) other sexually transmitted diseases (STDs). Increased risk of HIV infection occurs if people engage in anal intercourse. This is a very efficient method to transmit HIV, in part because of the trauma which allows the virus to come in contact with the blood. Importantly, rectal cells have been shown to possess receptors for HIV (CD4-like receptors) making viral entry into rectal epithelial cells quite efficient. In the United States and other areas of the world where HIV infection is common in gay men, this has been an important risk factor. The Centers for Disease Control recently reported that more than half of heterosexual men with AIDS (58.8%) reported unprotected vaginal or anal sex in the past year. More than one-third of homosexual men (35.0%) reported unprotected anal sex in the past year. Tragically, it has become recognized that adolescents are engaging in anal intercourse, both for prevention of pregnancy and (erroneously) for prevention of infection.

With intravenous exposure, whether through tainted blood products or through shared needles, direct inoculation into the bloodstream (which contains many target cells) has a transmission efficiency of approximately 100%. Efforts to protect the blood supply by screening donors for HIV infection and by heat treating blood have virtually eliminated transmission through blood transfusion. Use of needle exchange (as opposed to the bleaching or disinfection of needles, which is not always effective) has proven to be extremely effective. HIV infection rates are decreased and intravenous drug use rates are unchanged in areas that have well-run needle exchange programs.

The sexually transmitted discharge diseases, such as gonorrhea and chlamydial infections, and ulcerative STDs, such as syphilis, chancroid, and herpes infection, increase the risk of HIV infection between 3- to 5-fold and 5- to 10-fold for the discharge and ulcerative diseases, respectively. With STDs, the natural defense mechanisms are compromised, the mucosal surface is damaged, and cells with HIV surface receptors are recruited to the reproductive tract to fight these other infections. In addition, an HIV-infected person who has another STD is more infectious. For example, treatment for gonorrhea will greatly reduce HIV shedding in the semen of coinfected men since the "window" of viral exit—the damaged tissue—is closed when the tissue heals.

II. EPIDEMIOLOGY OF AIDS AND HIV INFECTION

It is currently estimated that over 20 million people are HIV infected worldwide. The rates of infection in Asia are extremely high; as a result, projections for the number of persons who will be HIV infected in the year 2000 are being revised upwards at regular intervals. Given that there is no surveillance system in place to accurately measure HIV infections, we must rely on results from research studies to make the estimates.

A. Developing Countries

Over 90% of the world's HIV-infected population lives in developing countries. Here this disease is spread through heterosexual vaginal intercourse. The high prevalence of other sexually transmitted diseases, practices of vaginal douching and drying, and (perhaps) exposure to strains of viruses that are more easily transmitted through vaginal intercourse have contributed to high HIV rates. The separation of families and the economic hardships that give rise to seeking and exchanging sex for money, for food, or for other essential needs have resulted in the predictable distribution of high prevalence populations along trucking routes and in the inner cities.

B. United States and Other Developed Countries

In the United States, some countries in northern Europe, and in a few Asian countries, HIV spread rapidly through populations of gay men and/or intravenous drug users. In these areas, secondary spread is now occurring in the heterosexual partners of bisexual men and intravenous drug users in which transmission during vaginal intercourse is likely facilitated by other STDs (Fig. 1).

Indeed, in the United States the profile of the AIDS epidemic is changing; rates have decreased among men who have sex with men, remained unchanged among intravenous drug users, and increased dramatically among women and among others with "no known risk factors." African Americans and Hispanics bear an enormous burden of disease. Furthermore, the average age of infection with HIV has decreased dramatically in the past two decades from 26 to 20; one of five people living today with AIDS was infected as a teenager.

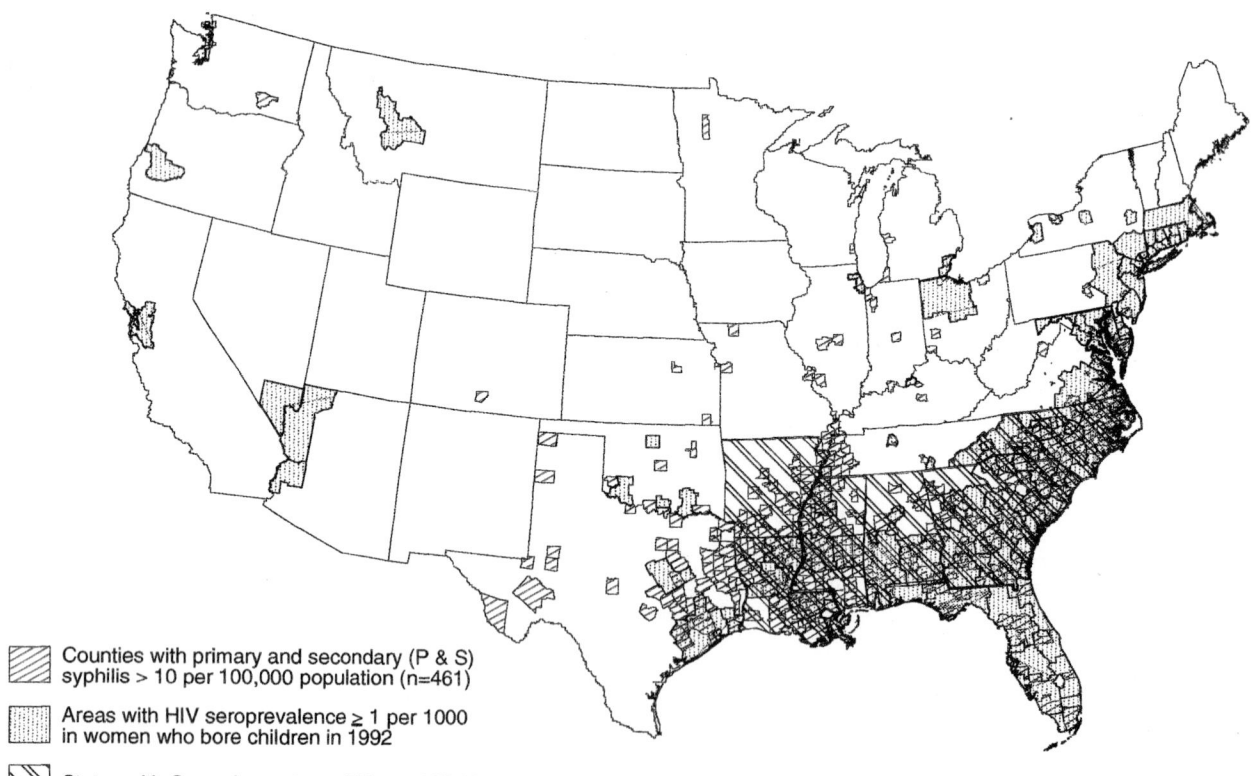

FIGURE 1 HIV seroprevalence. Areas with the highest HIV seroprevalence in women who bore children in 1992, counties reporting the highest primary and secondary syphilis rates (1993), and states reporting the highest gonorrhea rates (1993), United States.

III. THE VIRUS

The virus is made up of an outer membrane, genetic material (RNA), and proteins. The membrane, which is actually taken from the host cell as the virus exits the cell, and the protein shell serve to protect the genetic material. It is the genetic material that orchestrates the production of new virus in the cell. HIV is a retrovirus, a name given to a group of viruses that share a "reverse mechanism" for making new virus. Instead of genes made of DNA, the virus contains RNA copies of the genes and a protein, reverse transcriptase, that can make a DNA copy of the RNA. The DNA copy, since it is made in the host, is recognized as "self" and integrated into the host's genetic material. This DNA copy of the virus is called a provirus. It becomes a permanent part of the host's genome and can serve as source of new, infectious virus at any time (Fig. 2).

There are two virus types, HIV-1 and HIV-2. Throughout the world, HIV-1 causes most cases of AIDS; however, infection with HIV-2 can and does cause the fatal disease AIDS. Through HIV-2, we can understand the animal origin of these two AIDS viruses. HIV-2 is very similar to the monkey virus, simian immunodeficiency virus (SIV), and it is likely that HIV moved from monkeys to humans in Africa some time during the first half of this century. In the course of studying SIV infection in monkeys, rare laboratory accidents have resulted in the transmission of SIV to laboratory workers. Also, in Africa analysis of cases due to HIV-2 infection have confirmed that HIV-2 and SIV are virtually the same. SIV-infected and HIV-2-infected people have subsequently developed AIDS.

Movement of a virus from one species to another is probably an uncommon event; however, it has occurred many times in the past. For example, in the 1970s, the virus that causes cat distemper underwent a genetic alteration that enabled it to survive (and cause disease) in dogs; it causes a serious intestinal infection, quite similar to that caused in its original host, the cat. In terms of retroviruses, it is not surprising that the cat and the mouse viruses are

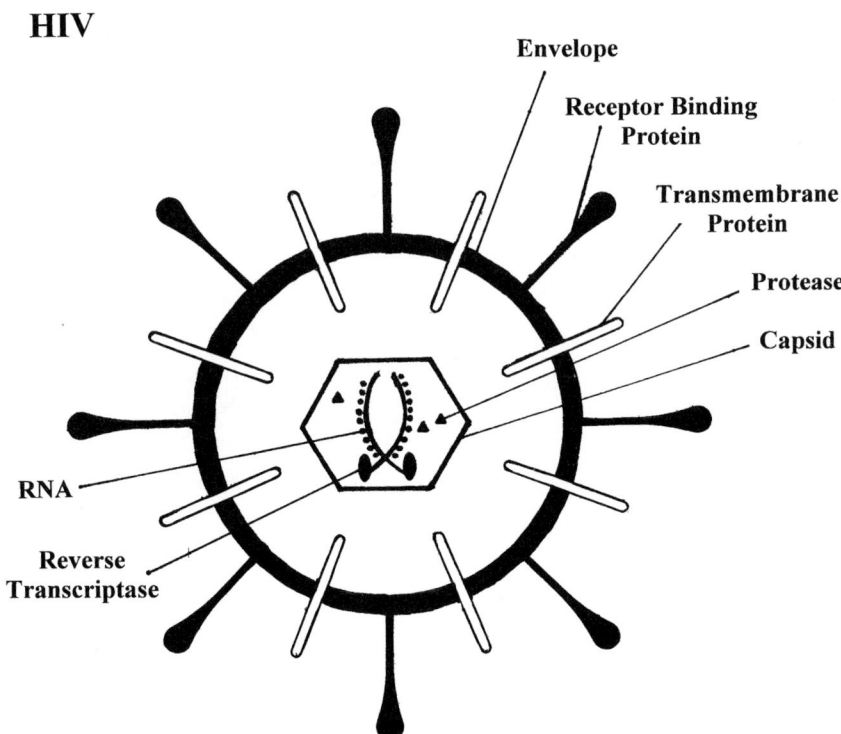

FIGURE 2 HIV structure. Schematic drawing showing the viron proteins, genetic material, and envelope.

quite similar, suggesting that the virus of the prey (the mouse) underwent genetic change that enabled it to infect the predator (the cat).

IV. EARLY STEPS IN THE INFECTIOUS PROCESS

A. Attachment to Host Cells

The virus can bind to cells with CD4 receptors, helper T lymphocytes and macrophages, and cells with CD4-like receptors, rectal epithelial cells. It binds to the receptor with its envelope glycoprotein (this is a sugar containing viral protein that sticks through the envelope). Recently, other host cell receptors have been shown to be necessary for secondary attachment steps. Antibodies to the glycoprotein will block attachment (Fig. 3).

B. Membrane Fusion

Once the virus attaches, the viral envelope fuses with the target cell's membrane. Because the virus obtained the envelope when it escaped from its host cell, the membranes are very similar and fusion occurs readily. In the process of fusion, the contents of the virus, the RNA and proteins, are released into the cytoplasm of the host cell.

C. Synthesis of DNA

The viral enzyme, reverse transcriptase (RT), makes a copy of the viral RNA and turns the information into DNA, first single stranded and then double stranded.

D. Integration

Although the information in the DNA is completely viral, from the host cell's perspective, this double-stranded DNA "looks like" its own set of genes. The DNA is transported to the nucleus of the cell. There, it is integrated into the host genome with a viral enzyme called "integrase." The copy of the viral genome is called a provirus ("before" virus). The provirus has signals that can instruct the cell to make more copies of the virus. In the meantime, the provirus, like the host cell's genome,

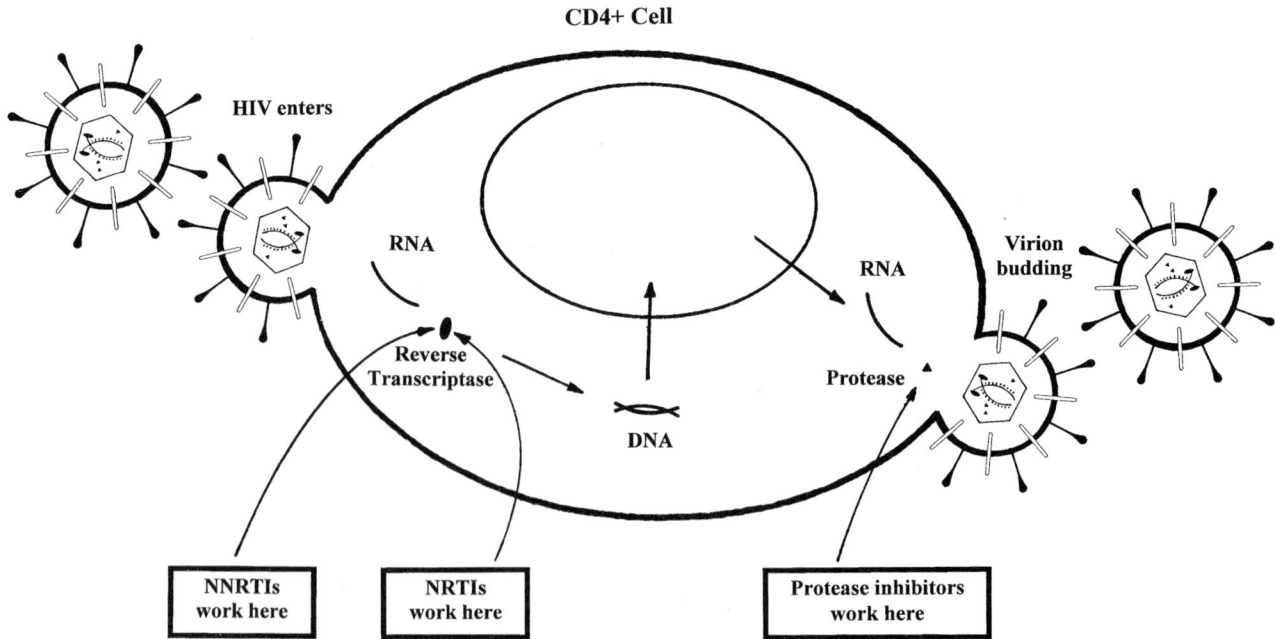

FIGURE 3   Schematic drawing showing the life cycle of HIV and a representation of the mechanism of action of each class of antiviral drug.

is passed to daughter cells every time the cell divides.

E. Syntheses of New Viruses

In order to make new viruses, the DNA copies of the viral genome are changed back into messenger RNA by the host cell. The host enzyme RNA polymerase recognizes specific directions found in a viral sequence, the long terminal repeat. The viral RNAs are used to make proteins such as RT or capsid proteins; some RNAs are put into the new virus particles to serve as the viral genome. The capsid proteins, the genome, and the enzymes move to the cell surface and undergo assembly. Once assembled, virus particles move out of the cell and acquire their envelope.

F. Latency and Eclipse Period

HIV may not synthesize new viruses after becoming a provirus. It may sit quietly in the host genome for an indeterminate time period. This is the latent phase. In the absence of new viruses, the body's immune system may not recognize the infection. This eclipse period is usually less than a few weeks in length, although in rare individuals it could be as long as 6 months. During this time, the person could have a negative HIV test result even though he or she was infected. In the absence of virus production, the person may not be infectious but this is unknown. The process whereby latent provirus becomes active may involve some viral genes such as *tat* and *rev*. Other stimuli including other viral infections in the same cell, e.g., genital herpes virus, may affect these processes and induce latency or increase expression of HIV genes leading to explosive production of HIV particles.

V. THE IMMUNE RESPONSE

The virus causes a dramatic response to infection. Within a few weeks the host will produce virus-specific antibodies. However, these antibodies are unable to completely shut down virus production. The virus escapes the immune response by not marking infected cell surfaces during the nonproductive or latent phase and by changing the viral surface so that the response made to today's virus will not "recognize" tomorrow's virus. In fact, the ability of the virus to undergo mutation is quite extraordinary. The envelope gene is particularly likely to change, leading to changes in the surface protein and avoidance of the host's antibodies. Interestingly, the part of the protein that binds to the host receptor is stable; the other changeable parts evoke an antibody response that is obsolete almost as soon as it is made by the host. In addition, host enzymes add sugars to the proteins; this coat of carbohydrate acts as another type of camouflage.

VI. DETERIORATION OF THE BODY'S DEFENSE SYSTEM

HIV infects two important CD4-bearing immune cell types: helper T cells and monocytes/macrophages. The virus kills T-helper cells. As the name implies, these cells are key in helping the body respond to infections. In addition to a loss of T-helper cells, numbers of the complementary cell, suppressor/cytotoxic cells, are relatively increased. Loss of the T-helper cell has many consequences because this cell serves to coordinate the immune response. The monocytes/macrophages are important in processing and presenting infectious agents to the immune system. The macrophage also kills other viruses and bacteria that invade the body; this function is impaired in HIV-infected macrophages. Recently, a special type of macrophage, the Langerhans cell, has been shown to possess the receptor for HIV. This may be important since Langerhans cells are found in the epithelium and could be the entry site for the virus. Although HIV-infected patients do make antibodies, there may be defects in the antibody response to certain types of antigens, especially carbohydrates. This is interesting since these carbohydrate responses are not thought to depend on T-helper cells.

VII. DISEASE PROGRESSION

Early HIV infection (group I) may be associated with a flu or mononucleosis-type illness approximately 1 or 2 weeks after infection. Signs and symptoms may include fever, swollen lymph nodes, muscle soreness, and meningitis. This period usually marks the initial replication of the virus. High numbers of viral particles are made in infected cells. Viral antigens (P24) can be detected in blood samples within 1 or 2 weeks; this is followed by HIV-specific antibodies (Fig. 4). These antibodies form the basis of the HIV test. Most people become test-positive within weeks or months of infection. During this period the patient may be extremely contagious even though an HIV test may not yet reveal infection. The antibody production coincides with a decrease in detectable virus and a disappearance of symptoms. Viral replication continues at low levels (group II). In some people, the initial short period of symptomatic disease may be followed by persistent swollen lymph nodes (group III). Pre-AIDS may manifest with diarrhea, oral and vaginal yeast infection, weight loss, and fever.

AIDS (group IV), more than anything, reflects extensive destruction of the body's immune system. Without a functional defense system the AIDS patient is susceptible to infection and cancer. Opportunistic infections include fungal and parasitic diseases. The cancer common in gay men is Kaposi's sarcoma (also caused by a virus, human herpesvirus 8). B cell lymphoma and cervical/anal cancers are also seen in AIDS patients. Most terminal patients have central nervous system disease as well.

VIII. HIV AND PREGNANCY

HIV can be transmitted from infected mothers to the fetus/infant three ways: (i) prenatal—within the uterus the fetus is infected when the virus crosses the placental membranes; (ii) conatal—during the birth process; and (iii) postnatal—through breast milk. The evidence for infection in the uterus comes from studies that have identified the virus in fetal tissue and in the neonate. Results of studies from investigators in Uganda suggest that HIV-infected women may be less fertile (up to 25%), i.e., less likely to become pregnant than women who are uninfected;

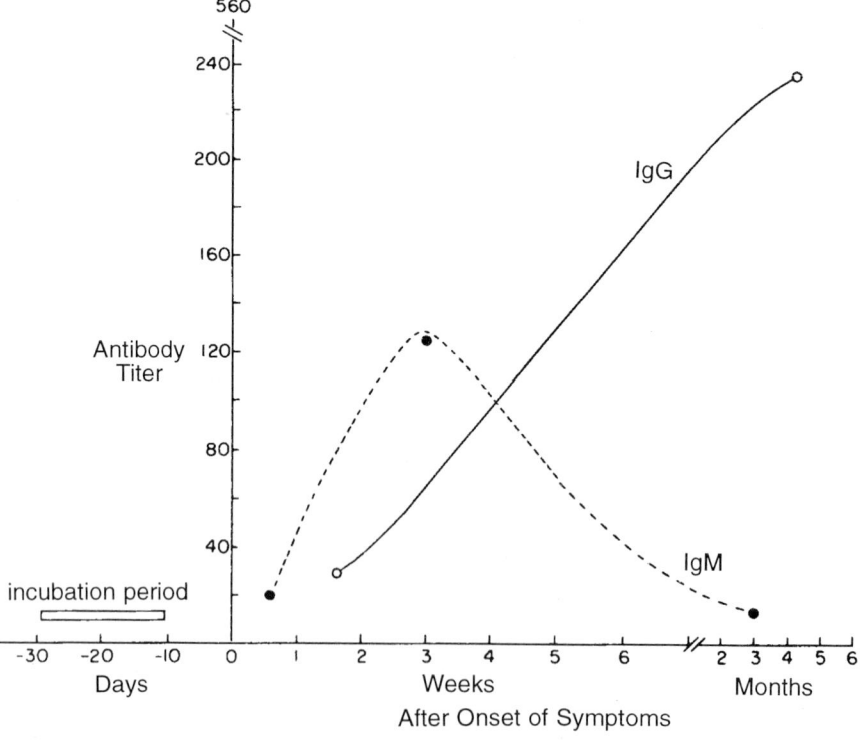

FIGURE 4 Chronology of viral and antibody production. This graph represents the average time line of virus isolation and HIV antibody production during the course of an infection.

this means that the adverse impact of HIV infection on pregnancy may be significantly greater than is currently appreciated. Perinatal transmission reflects the exposure of the infant to HIV in maternal secretions and blood as it passes through the birth canal. Breast milk as the source of neonatal infection reflects the isolation of HIV from breast milk, transmission of HIV to infants by breast-feeding mothers who were infected after the infant was born, and prospective studies that demonstrate significant risk of HIV transmission associated with breast-feeding. The risks associated with these methods of transmission vary depending on the birth practices (e.g., use of scalp electrodes) and other factors (e.g., nutritional status and CD4 count of the mother at the time of birth).

Prevention of transmission from mother to infant has been demonstrated by two approaches: antiviral therapy to reduce maternal viral load and avoiding infant exposure to breast milk. Landmark studies in the developed and the developing world have demonstrated that the use of azidothymidine (AZT) during pregnancy with or without AZT treatment of the infant will dramatically reduce the risk of HIV transmission that occurs late in the pregnancy or at the time of birth. Other studies have demonstrated that replacement of breast milk with commercial formula will reduce the risk of postnatal transmission.

IX. DIAGNOSIS

Diagnosis of HIV infection requires sensitive and specific tests to identify a unique response to infection (antibody or serology) or detect a part of the virus (antigen). There are two test characteristics that are important: specificity and sensitivity. Specificity is a measure of the test to correctly identify a particular infection from any other. In other words, the HIV test result is only positive when the patient has HIV infection. Sensitivity is a measure of how efficient the test is in detecting the infection. For instance, if a test is 50% specific, it means that the test result will be positive for an infection that is not HIV, e.g., genital herpes. If a test is 50% sensitive it means that if two patients have a particular infection, only one will have a positive test result. The test will not "see" the other infection. Because of the social stigma of HIV infection and the negative legal and/or economic consequences of HIV infection, an extremely high specificity is necessary. In addition, since early diagnosis and treatment reduces the chances of spreading the infection and improves the health of the infected person, a high degree of sensitivity is essential. In other words, we expect the test to identify everyone who is infected and no one who is not infected. The most extreme example of these standards has been achieved with HIV diagnostic tests in which specificity is >99% and sensitivity is also >99%.

A. Serology

Most HIV tests are performed on serum. Serum is the straw-colored fluid that is left when red blood cells are separated from the blood sample. The serum contains many proteins, including antibodies. Antibodies are proteins that develop as part of the immune response to infection. HIV tests attempt to identify two components that can be found in the serum: HIV antibodies that have been made in response to the virus and HIV antigens, parts of the virus itself. There are two types of HIV test that utilize serum: those that test for antibodies and those that test for antigens.

1. Enzyme-Linked Immunosorbent Assay

The enzyme linked immunosorbent assay (ELISA) test can detect antibodies to the HIV virus within 3 months of becoming infected. Serum is collected, mixed with purified HIV antigen, and allowed to react. An indicator, which gives a color range, is then added to the mixture. If the serum contains HIV-specific antibodies bound to the antigen, the indicator will bind to the complex. The attachment of the indicator to the complex produces a color. If the color is bright enough (according to an electric photo eye), the test is said to be positive. If this test is positive, it will be repeated. If the second ELISA is positive it will be confirmed by another test, either the Western blot or an immunofluorescence assay.

2. Western Blot

Western blot detects antibodies to specific viral proteins. Serum is reacted with purified HIV antigens that have been electrically separated and electrically transferred to a special membrane. If HIV-specific antibodies are present, these will bind to the antigen. These complexes are detected and the pattern of bands is compared to the test standard. If it is the same, the test is positive for HIV antibodies.

3. Immunofluorescence Assay

Immunofluorescence assay (IFA) also detects antibodies and is a less expensive alternative to the Western blot method. For this method serum is placed on microscope slides containing HIV infected. If the serum contains antibodies to HIV, complexes of virus and antibody form. The slide is then washed to remove unbound antibodies. A fluorescent "tag" which adheres to human antibody is then added to the slide. If the antibody is present (bound to virus), cells will appear fluorescent when viewed under an ultraviolet microscope.

4. HIV p24 Antigen Assay

Another serum test, HIV p24 antigen assay, has been Food and Drug Administration (FDA) approved for screening blood bank donations. The p24 test detects a viral protein that is part of the viral outer membrane. This test is unlike other HIV tests in that it does not detect antibodies that are produced in response to infection. This test is limited because it is effective only early in the viral infection cycle but it is able to detect infection, before the immune system makes antibodies to HIV. The protein, p24, can be detected in blood serum and in cerebral spinal fluid.

5. Saliva-Based Tests

HIV testing can be done by collecting a saliva sample rather than a blood sample since saliva contains the same antibodies that are in blood. Collection for saliva-based tests is fast and easy: A cotton pad is placed in between the cheek and gum and an oral specimen is obtained. An indicator line appears on the collection pad once enough fluid is collected; the collection device is then placed in a special vial and sent to the lab. In the lab, the sample is tested by an ELISA test. A second saliva sample is taken if the first test is positive and is analyzed again. As with positive serum samples, if the second result is posi-

tive further testing by Western blot or IFA is performed.

6. Urine-Based Tests

Urine samples are used as a source for antibodies in the initial screening test, the ELISA test, or in the confirmatory Western Blot assay. This test has the advantage of not requiring an invasive blood sample and it appears that test results fall into the "indeterminate" category less often than with the serum-based tests.

7. Home-Based Tests

Blood samples for anonymous HIV testing may now be collected at home. Home Access Express HIV-1 Test System is currently available by telephone order and over-the-counter at pharmacies. Home-based tests require a blood sample from a finger stick. The blood drops are placed on special paper and air-dried. The specimen is mailed to a laboratory where it is tested. Like other HIV diagnostic tests, the home-based test detects HIV antibodies using the ELISA method. If the test is positive, a new sample is taken and tested by the same method. If a second positive is found, it is recommended that the test be confirmed by Western blot or IFA using a serum specimen. Each test kit has a personal identification number to access results by phone. Counseling services are available, three sessions are offered to HIV− persons and six to those that have HIV+ results.

B. Testing Policies

Two types of HIV testing services are offered: confidential (the patient is identified but the test results are not available to anyone but the patient) and anonymous (the patient's name is not given, a number identifies the sample, and the test results are given only to the person presenting the identification number). In the United States HIV testing is voluntary; compulsory testing, in which people are required to be tested, is not practiced in the United States.

X. TREATMENT

The main objective of treating HIV infection is to decrease or stop replication of the virus and consequently stop the destruction of the immune system. This is a daunting task because billions of virus particles are produced daily and the HIV virus is almost continuously mutating. The high frequency of mutation results in drug-resistant strains of virus. With this in mind, three classes of antiretroviral drugs have been developed to target specific viral sites (see Fig. 3). When drugs from two or three of these classes are used in combination, this regimen achieves maximum suppression of viral replication. Three classes of drugs are currently used to treat HIV infection: nucleoside reverse-transcriptase inhibitors (NRTI), protease inhibitors (PI), and nonnucloside reverse-transcriptase inhibitors (NNRTI).

A. NRTI

This class of drugs, which includes AZT, is central to combination therapy. It functions to block the HIV enzyme, reverse transcriptase (RT), that uses the RNA sequence to make a DNA copy. If the HIV virus is unable to make a DNA copy, new RNA molecules cannot be transcribed and the infection is aborted. Currently, there are five nucleoside RT inhibitors approved by the FDA and a number of other agents are in development.

Use of NRTI for HIV infection may cause some side effects, ranging form nausea, diarrhea, and oral ulcers to inflammation of the pancreas (pancreatitis), low red or white blood cell counts (anemia and neutropenia, respectively), and numbness in the limbs and appendages (peripheral neuropathy). As with most drug use, side effects vary from person to person in degree of severity.

B. Protease Inhibitors

The HIV virus makes an enzyme called protease that is responsible for cleaving or cutting long protein chains into functional individual protein products. These protein products include such enzymes as RT and core proteins that make up the viral capsid. Protease inhibitors act late in the viral cycle and prevent the enzyme, protease, from cutting the long-chain protein into functional protein products. This effectively blocks the formation of enzymes and the assembly processes that are essential for the production of new viruses.

As with NRTI, several PIs exist and have been approved by the FDA, whereas others are in development. As with any treatment, side effects vary considerably in severity. Like most drugs, PIs are processed by the liver. Sometimes chemical interactions between PIs and certain other medications take place and may cause liver problems. Protease inhibitors may interact with antibiotics, antidepressants, antifungals, and antituberculosis drugs. These drugs should be avoided when treatment with PIs is recommended. Another side effect of PI therapy may be the onset of diabetes. Recently, reports of abnormal depletion of facial fat have been published; the mechanism of action is unknown.

C. NNRTI

NNRTIs block the RT enzyme directly. They are used in combination with NRTIs and PIs and have very few side effects with the exception of skin rash. Currently, NNRTIs are used only as the third or fourth drug in dual therapy because viruses quickly develop resistance to them if they are used alone.

D. Highly Active Antiretroviral Therapy (HAART)

Use of a combination of drugs is currently the standard treatment for HIV infection. In combination, these drugs limit viral replication and inhibit HIV mutation rates which cause drug resistance. Treatment regimens are tailored to each individual to ensure the best possible therapy. As stated previously, the fundamental component of combination therapy is the NRTI. Treatment usually consists of two NRTIs, which function at different sites. The first group works in the cells that are actively replicating viral genetic material. The second group works in cells that are in a resting state and are not actively replicating. Because these drugs target different cell populations, there is little competition between them for active sites so that they are able to work together to slow the HIV replication cycle.

In addition to NRTIs, HAART includes one PI and may include a NNRTI. Triple drug therapy includes two NRTIs and one PI or NNRTI.

XI. ADOLESCENTS AND HIV/STDs

In the United States, the majority of young people have sexual intercourse for the first time in their teens. A recent report indicates that more than half of women and almost three-fourths of men have had sexual intercourse by their 18th birthday. Two recent studies have provided data indicating that sexual activity among teenagers is increasing. Other studies have also documented that adolescents and young adults are inconsistent users of male condoms, although there is evidence to suggest the trend is improving.

The significance of lifetime partners can be best appreciated with a simple arithmetic problem. Keeping in mind the concept of incurable viral STDs accumulating in the population and two assumptions— (i) adolescents practice serial monogamy (i.e., one partner at a time in series) and (ii) all adolescents in a given population have the same partner acquisition rate—how many exposures real and phantom (partners of partners) are represented in an adolescent's life between ages 15 and 24 if

The adolescent has sexual intercourse for the first time at 15;
Each relationship/partner lasts 1 year; and
They have a new partner each year.

In the nine years they will have nine real partners. In terms of incurable viral STDs, where both the real and the phantom partners are relevant, they have been exposed to over 500 partners.

Given these behavioral factors, the implications for risk of STDs, including HIV infection, are undeniable and are supported by the data trends from studies in teen populations. The cumulative number of AIDS cases has increased steadily since the beginning of the epidemic. Furthermore, if one assumes an incubation time of 8 years from the time of HIV infection to AIDS, 20% of all persons with AIDS were infected as teenagers. A closer look at the demographics of the new AIDS cases reveals that gay teens account for most of these. Arguably, this represents cases among new susceptibles in a larger population of gay men that has been at high risk for HIV infection since the beginning of the AIDS epidemic.

If other STDs are considered, however, a larger picture of "high-risk adolescent populations" emerges. In the United States it is estimated that 12 million incident cases of STDs occurred in 1997; the majority of these cases occur in young adults, with an estimated 3 million occurring in teenagers. Age rates for two bacterial STDs, gonorrhea and chlamydial infection, are highest among youth. Gonorrhea cases are found among populations of low socioeconomic status in urban and rural areas of the United States. The highest rates are seen in teenage girls 15–19 years of age and young men 20–24 years of age. With respect to chlamydial infection, two trends are clear: The number of cases is increasing due to better diagnostic methods and the rates are highest among 15- to 19-year-old women regardless of socioeconomic status.

For the other viral STDs, the situation is similar although the trends have more serious implications for transmission. Because these viral infections are incurable, each incident case becomes a prevalent case; in other words, the total number of cases in the population increases over time. For human papillomavirus infection, the cause of genital warts and cervical cancer, the attack rate is 25% in the first year of sexual activity with fewer than two sexual partners. Similarly, the rates of genital herpes begin to increase in adolescents, concordant with the onset of sexual behavior. Arguably, the incidence rates of HIV infection could increase in any sexually active teenage population, especially if young women choose partners that are older and are from populations in which HIV prevalence is high. Recent studies note high HIV infection rates among gay teens, most of whom engage in bisexual behavior prior to recognizing that they are gay men. These bisexual teens represent a high-risk population that is juxtaposed to adolescents who might otherwise be at low risk for HIV infection.

XII. PREVENTION OF INFECTION

We are now in the second decade of the AIDS epidemic, and our hopes and dreams for a vaccine or a microbicide to prevent HIV infection and a drug to cure AIDS are still just that. Arguably, we are closer to developing a vaccine, a microbicide, or a cure and we are smarter about this virus and the disease it causes. However, while we are waiting for the scientific breakthroughs, is there more that we can do today to prevent HIV/AIDS? Current recommendations include (i) abstinence; (ii) sexual relationships between mutually monogamous, uninfected partners; (iii) correct and consistent use of the male condom; (iv) use of clean needles; (v) prevention and control of other STDs so as to decrease susceptibility to HIV infection and to reduce infectiousness of HIV-infected people; (vi) use of antiviral therapy to reduce transmission to infants; and (vii) substitution of breast milk for commercial formula. Other behavior changes may reduce risk of infection; these include (i) delay sexual debut as long as possible (the earlier the age of first intercourse the more susceptible to HIV infection and other STDs and the greater the number of lifetime partners); (ii) avoid sex during menses (HIV-infected women are probably more infectious at this time and HIV-uninfected women are probably more susceptible, given the direct access to blood that is created by the sloughing of the uterine lining); (iii) decline opportunities for sexual intercourse that is not associated with relationships that are intellectually and emotionally intimate; (iv) avoid anal intercourse, and if practiced always use a condom; (v) avoid douching—this practice is associated with loss of protective normal bacteria in the vagina and increased risk of some STDs including HIV infection; and (vi) seek screening and treatment of STDs based on subtle symptoms and/or sex with a new partner.

XIII. CONCLUSIONS

Although HIV infections in the United States are currently clustered in high-risk populations, there is no reason why they will remain clustered. HIV has a number of characteristics that contribute to the sustained presence within a population:

It is transmitted sexually, and sex is required to reproduce our species.
It causes an asymptomatic disease for many years while the host is infectious.
It integrates into the body and can remain "invisible" for a long time.
It lives within the body's disease fighting system, making vaccine development efforts extremely difficult.

Unfortunately, the adolescent populations and the STDs (chlamydial infection and genital herpes) will provide the messenger and the opportunity for HIV infection to become more broadly distributed in the population. The possibility that this is occurring clandestinely, during this apparent plateau in AIDS case rates, is raised in studies of adolescent populations geographically and socially juxtaposed to high-risk populations. Whether we will be able to aggressively and effectively address the implicit research and program needs and opportunities that must be met in order to prevent the spread of HIV infection into adolescent populations remains to be determined.

See Also the Following Article

Sexually Transmitted Diseases

Bibliography

Centers for Disease Control and Prevention (1998). Public Health Service Task Force recommendations for the use of antiretroviral drugs in pregnant females infected with HIV-1 for maternal health and for reducing perinatal HIV-1 transmission in the United States. *Morbidity Mortality Weekly Rep.* 47(RR-2).

Cohen, P. T., Sande, M. A., and Volberding, P. A. (1990). *The Aids Knowledge Base*. Medical Publishing Group, Waltham, MA.

Eng, T., and Butler, W. (1997). *The Hidden Epidemic*. National Academy Press, Washington, DC.

Fishbein, M., Guenther-Grey, C., Higgins, D., Johnson, W., and Moseley, R. (1996). Community-level prevention of human immunodeficiency virus infection, among high-risk populations: The AIDS Community Demonstration Projects. *Morbidity Mortality Weekly Rep. 45*

Holmes, K. Mardh, P., Sparling, P., and Wiesner, P. (1990). *Sexually Transmitted Diseases,* 2nd ed. McGraw-Hill, New York. [Refer to units 28–32, 3rd edition]

Kelly, J. (1995). *Changing HIV Risk Behavior.* Guilford, New York.

Morse, S. A., Moreland, A. A., and Thompson, S. E. (1992). *Atlas of Sexually Transmitted Diseases.* Gower Medical, New York. [Refer to unit 8]

Normand, J., Vlahov, D., and Moses, L. (1995). Preventing HIV transmission. In *The Role of Sterile Needles and Bleach.* National Academy Press, Washington, DC.

Homosexuality

Vivienne Cass

Corporate Psychology Consultants, South Perth, Western Australia

GLOSSARY

bisexuality Sexual and/or romantic attraction to both same- and opposite-sex individuals, either at the same time or at different times.

essentialism A framework of thinking which holds that a common "essence" or inner core exists in regard to a particular human quality and that this essence occurs across cultural and historical context.

gender The cultural expression of biological sex in terms of masculine or feminine behavior; beginning to be used interchangeably with "sex."

heterosexuality Sexual and/or romantic attraction to individuals of the opposite sex to oneself.

homophobia Conscious and unconscious prejudice against lesbians, gay men, and bisexuals.

homosexual identity The belief and sense of self as "a homosexual."

homosexuality Sexual and/or romantic attraction to individuals of the same sex as oneself.

sex The biological sex of an individual (i.e., external and internal anatomy and chromosomal makeup).

sexual orientation A pattern of persistent and consistent sexual and romantic attraction.

social constructionism A framework of thinking which holds that human behavior is created from (i.e., a product of) the sociocultural environment in which it occurs.

Homosexuality refers to the expression of sexual–romantic desires and activities occurring between individuals of the same sex. Homosexual experience may be expressed covertly (e.g., sexual desire and fantasies) and/or overtly (e.g., sexual contact and expressions of romantic love). Although most definitions of homosexuality include reference to both behavioral and psychological components, a distinction is often made between homosexual sexual behavior which does not necessarily involve sexual desire or love (e.g., in prisons) and homosexual attraction which is considered to include one or both of these variables. The term homosexual was coined in the mid-nineteenth century to describe individuals who experience sexual–romantic desires for someone of the same sex as a persistent and consistent attraction. In Western cultures such a pattern of sexual–romantic attraction is described as a homosexual

sexual orientation (superseding the now outdated sexual preference). In the twentieth century the concept of sexual orientation has been extended to include self-identity. Homosexual identity, the self-perception individuals have of themselves as "a lesbian" or "a gay man," is increasingly used to identify and categorize people as homosexual. The study of homosexuality is complex. The concepts of sexual orientation, sexual identity, homosexual, and even "sexual" are not universal to all cultures and historical periods. Hence, an examination of historical, anthropological, psychological, biological, and sociocultural factors is required to fully understand this multifaceted subject. The current literature on homosexuality focuses on traditional areas such as causation as well as the recent issues of homophobia, parenting, and identity formation. The past decade has also given rise to a significant literature offering critical analysis and review of the cultural meanings of sexuality and gender.

I. INTRODUCTION

Individuals in most cultures and throughout history have engaged in sexual and/or romantic behavior and relationships with those of the same sex. In Western cultures, such behaviors came to be seen in a negative light, the result of early Christian teachings on sexuality which deemed sinful any sexual activity other than that engaged in for reproduction. Development of the early legal code incorporated this moral code, leading to the classification of same-sex sexual behaviors as deviant and criminal. The legal code, in turn, became the basis for the medical classificatory systems developed in the nineteenth century which saw same-sex sexual and romantic activity, now described by the term "homosexuality," as a pathological and medical condition.

During the second half of the nineteenth century the focus shifted from sexual behaviors occurring between individuals of the same sex to the concept of "homosexuals" as a separate type of person. Homosexuality was described as a particular medical condition characterized by sexual inversion (i.e., being like the opposite gender). The genesis of this condition was considered biological pathology, although

some dissenting voices claimed that inversion was an innate and natural variation of human nature. The concept and label of heterosexuality appeared only in the last decade of the nineteenth century and eventually came to signify an opposite state to homosexuality, one characterized as normality. As such, it was not seen to require study. These historical developments resulted in the construction of concepts that now form the basis of Western thinking on sexual–romantic orientations.

The twentieth century saw these concepts and terminology adopted by the individuals to whom they referred. By the mid-1970s self-identification as "lesbian" or "gay" was a well-established practice and soon extended to group identification as the notion of homosexuals as a minority group emerged and solidified. Within a short period the minority group concept had entered the public domain. Societal and professional attitudes toward homosexuality and homosexuals were also changing in this period, culminating in 1974 with the removal of homosexual attractions as a mental disorder from the standard classification system of mental disorders (the *DSM*) used throughout the industrialized world.

II. CONCEPTIONS OF HOMOSEXUALITY

A. Western Assumptions

The Western concept of homosexuality which has evolved from the previous century until present times incorporates several assumptions that are widely held within scientific as well as general circles. The first is that the expression of sexual and/or romantic desire for others of the same sex is a strong indication of an underlying predisposition or homosexual sexual orientation. Although it is known homosexual behavior can be distinguished from orientation, anyone engaging in homosexual behavior who asserts to being heterosexual is viewed with suspicion. The second assumption is that once such an orientation shows itself, it will remain the same throughout an individual's lifetime. Related to this is the assumption that homosexual orientation is created early in life and can lie dormant until the individual concerned becomes aware of its existence.

The defining characteristic of sexual orientation is entirely that of gender of sexual partners. On this basis it is considered that there are two primary sexual orientations, heterosexuality and homosexuality, with bisexuality being a third and somewhat secondary orientation to the other two. All three orientations are seen to be discrete entities, an individual being classified either as homosexual, heterosexual, or bisexual (although bisexuality is sometimes considered a holding position while deciding which of the other two is appropriate). Finally, homosexual orientation is also assumed to be linked to cross-gender behavior, with gay males perceived as being female-like and lesbians as male-like.

B. Criticism of Western Assumptions

Evidence suggests that Western conceptions of homosexuality and sexual orientation do not fit the actual experiences of human beings. The famous early work of Kinsey and associates, and that of others since, indicates that it is more accurate to think of individuals experiencing degrees of attraction to those of the same and opposite sex along a number of different dimensions, including sexual fantasy, emotional bonding, sexual activity, and identity. Many people experience varying degrees of attraction to both sexes at different times in their lives. Individuals can also relate simultaneously to each sex on different dimensions; for example, a man who feels sexually or romantically attracted to his female partner and also fantasizes about men. It has been concluded that a simple division into the categories of heterosexual, homosexual, and bisexual does not capture the flexibility and complexity of human experience which often sees individuals move from same-sex to opposite-sex attractions and back again and vice versa. It has also been found that women's sexuality frequently presents differently to men's, with many women, for example, developing an attraction for other women in later life where previously none existed.

Nevertheless, studies indicate that the majority of people do exhibit heterosexual attractions. From this, biological scientists claim that the evidence from Western populations suggests individuals tend toward sexual orientations of either homosexuality or heterosexuality. They identify a bimodal, J-curve model of sexual orientation as descriptive of the findings, with heterosexuality representative of the majority of people. Social scientists propose, on the other hand, that strong directives from Western (or Western-influenced) cultures, which promote heterosexuality in a more positive light than homosexuality and set up expectations that individuals will fit themselves into one of the sexual orientation categories, are at least partially responsible for a bimodal distribution. There is also strong clinical evidence that individuals feel more comfortable if they can claim to belong to one of the sexual orientation groups rather than remain unlabeled.

III. HOMOSEXUAL BEHAVIOR IN NON-WESTERN CULTURES

Anthropological and historical research in the 1980s and 1990s into same-sex sexual and romantic desires in non-Western cultures provides evidence of cultural variations in the way such feelings and behaviors are defined and given expression. For example, in some Melanesian societies, all young boys are taken away from contact with females until they are of marrying age. As part of male initiation rites, oral (and sometimes anal) sex takes place regularly between older males and the boys, the ingestion of semen by the latter being considered the path to masculinity. These sexual acts are also sources of sexual enjoyment. Upon reaching marrying age, these boys engage in sexual activity with their wives and usually cease sexual activity with males except to later return in the role of older male to a new group of young boys. Societies such as these do not categorize people according to the gender of sexual partners and there is no belief that the boys should or will stay focused on male partners, despite having lived in an all-male environment and engaged in same-sex activities for much of their young lives.

Similarly, in some traditional South American and African societies, women enter sexual–romantic relationships with other women while at the same time having relationships with men. Such behavior is considered acceptable and normal. In Latin American culture a different form of homosexuality exists in which it is the type of sexual act, rather than the gender of partner, that is a significant and defining variable. Sex with other men is acceptable provided a masculine "active" role is taken but roles taken will vary from situation to situation according to whom is seen to be the more masculine. Traditional North American Indian tribes recognized another version of same-sex relations. They acknowledged the importance of the spiritual berdache, an individual who dressed and acted as someone of the opposite sex and was considered to be neither male or female but belonging to a third category which combined both male and female qualities. The partners of berdache could be same- or opposite-sex individuals or both.

Other variations of same-sex activities and relations have been documented historically in ancient societies in which social power and status are the key variables in governing such behavior. In such societies, socially powerful men expressed sexual desire and emotions with others who were their social inferior regardless of sex, race, or age. In Greece, older men engaged in relationships with young males to whom they acted as mentors while simultaneously maintaining a relationship with wives.

From the anthropological and historical literature social scientists have concluded that Western notions of homosexuality and sexual orientation are not universally held and sexual attractions are more variable than Western views allow for. They urge caution against treating homosexuality as a unitary phenomenon and against assuming that persistent patterns of same-sex sexual–romantic behavior, desire, and attraction are automatically indicative of a fixed homosexual sexual orientation as defined by Western thinking. Hence, an understanding of homosexuality requires a sound knowledge of the cultural context within which such behaviors occur.

IV. MULTIPLE HOMOSEXUALITIES

A. Expressions of Homosexuality

On the basis of the cultural and historical data, various forms of homosexuality are evident, including (i) homosexuality, occurring between older and younger individuals, which is seen to be a normal part of development for the younger person. Marriage and parental responsibilities are expected to occur in adulthood. This may lead to a cessation of homosexual activity or continuation alongside heterosexual relations; (ii) homosexuality occurring between an individual who acts and/or dresses as someone of the opposite gender and a partner who presents according to biological sex; (iii) homosexuality occurring in an egalitarian setting in which partners are social equals. This behavior may be common in adolescence and cease after marriage, but this is not always the case; (iv) homosexuality occurring between partners drawn from different economic or social classes or where there is a power difference, with the superior individual defining the nature of the relationship; (v) romantic same-sex relationships in which an overt erotic component is not necessarily present—more common to women than men; and (vi) modern homosexuality found in Western or Western-influenced cultures which involves a conscious adoption of a lesbian or gay identity signifying membership in a homosexual social group and subculture. Less apparent is a parallel development among those identifying as bisexual. Each of these forms of homosexuality or bisexuality is complex and holds meanings and cultural functions deeply rooted in the culture and times in which it occurs.

B. Homosexuality versus Sexual Orientation

Scientists have been divided on what these different forms of homosexuality mean for the study of same-sex attractions. One group (consisting largely of biological scientists) distinguishes homosexuality from sexual orientation. The latter is considered a trait with roots in the genetic and endocrinological makeup of the individual (and sometimes the postnatal environment) which predisposes that person to develop into a homosexual. Individuals who do not show clear indications of having a strong, fixed pattern of sexual–romantic attractions for others of the same sex are not considered to be "true" homosexuals.

This approach to the study of homosexuality incorporates the essentialist approach which assumes that a human propensity for homosexual sexual orientation exists regardless of cultural context. Such an orientation represents an essential core of human nature that signifies the existence of universal inner mechanisms of development within individuals (although its expression may vary according to the culture in which it occurs). An essentialist approach assumes that sexual orientations are natural realities and that the terms homosexual, bisexual, and heterosexual simply reflect such realities. Claims of biological etiology usually accompany an essentialist approach but psychological factors have also been proposed. Evidence for these views come from (i) the literature on conversion therapy which documents repeatedly unsuccessful attempts to change homosexuals into heterosexuals, (ii) the anthropological and historical literature which describes a small group of individuals in different societies who appear to maintain same-sex attractions despite cultural directives to shift to heterosexuality, and (iii) various surveys of Western sexual practices which suggest that the numbers of people identifying as exclusively or nearly exclusively homosexual has remained relatively constant at 4–6% for men and 2 or 3% for women.

Another group of scientists (primarily social scientists) has taken a different approach to the anthropological and historical data, concluding that the essentialist distinction between homosexual behaviors and sexual orientation is too simplistic and narrow. The social constructionist approach holds that an understanding of homosexual expression should give weight to all the different variations of that expression without nominating one form (exclusive homosexuality) as particularly distinctive. Recognition is given to the interaction between biological, psychological, and sociocultural variables in producing sexual behaviors and creating different sexual realities for the members of any society. Social constructionist thinking perceives sexual orientations to be cultural entities and the categories of homosexual, bisexual, and heterosexual as culturally imposed divisions on what is really a continuum of sexual experience. History shows these categories of sexuality to have arisen in Western culture during the twentieth century when the concept of "sexuality" was first identified as a separate category of human existence.

The significant debate that developed between the essentialist and social constructionist approaches in the 1980s and 1990s, at times fierce, has encouraged researchers from all disciplines to critically examine the assumptions underlying their work. Currently, there is a shift from the more extreme positions taken previously in the debate, with constructionists acknowledging the existence across cultures of a small number of individuals whose strong same-sex attractions seem less influenced by cultural influences, and essentialists recognizing that sociocultural parameters such as gender and social class play a part in constructing and reinforcing sexual–romantic attractions.

V. INCIDENCE OF HOMOSEXUALITY

Surveys of homosexual incidence carried out on Western populations indicate that homosexual expression is found in a small but significant proportion of the population. Such surveys reveal that it is inaccurate to use a single incident statistic to describe the prevalence of homosexuality since factors such as the particular aspect of homosexual expression being studied, age, gender, place of residence, and education can heavily influence its distribution in a community.

A recent study of Americans found that 9.1% of men and 4.3% of women had engaged in some specific same-gender sexual activity since puberty. However, questions about same-gender partners since puberty elicited frequencies of 7.1% of men and 3.8% of women. When participants were asked about same-gender desire, 7.7% of men and 7.5% of women reported feeling attracted to same-gender people or found the idea of sex with a same-gender partner appealing. Only 2.8% of men and 1.4% of women in the study considered themselves homosexual or bisexual. Combining the three elements of homosexual behavior, desire, and identity, 10.1% of men and 8.6% of women were found to report same-gender sexuality since 18 years. Place of residence also indicated the complexity of measuring homosexuality. Incidence of city dwellers who engaged in same-gender sexual activity since puberty was 15.8% for men and 4.6% for women compared with 2.7 and 2.2% for rural dwellers, respectively. The complex and multidimensional character of Western homosexuality, evident in these data, points to the difficulty in defining what is meant by the terms homosexual and homosexuality.

VI. CAUSALITY

A. Early Research Interests

Interest in the etiology of homosexuality and bisexuality has existed since the nineteenth century when the medical profession became involved in these areas. Early investigation of homosexuality as a medical/psychiatric problem attempted to identify a cure for what was considered a pathological state. Early theories of homosexuality, based primarily on the work of Ulrich, Hirschfeld, Ellis, and Kraft-Ebbing, perceived homosexuality to be inborn although distinctions were frequently made between those for whom homosexuality was seen to be congenital and those for whom it was not. The predominant underlying assumption was that homosexuals were "inverts," exhibiting characteristics of the opposite sex, but some theories depicted homosexuality as a third sex. Biological factors were thought to be responsible for the so-called femininity of male homosexuals and so-called masculinity of female homosexuals.

Freud's psychoanalytic theories provided the first departure from the pathology/biological focus and suggested that the expression of homosexuality was a natural process in the developing individual. Some individuals did not achieve the final phase of development which was heterosexuality, and Freud saw this as acceptable and without any indication of mental pathology. Freud's theories led to a greater emphasis on underlying psychological factors, but while the research focus shifted to the examination of familial relations and upbringing, many of the studies continued to adopt the pathological model of homosexuality. Until the 1970s much of this research addressed "maladaptive" child-rearing patterns thought to be responsible for the development of homosexuality. However, a consistent lack of evidence for such a hypothesis has led to the abandonment of this early research approach. Little modern psychological research into causation has replaced this earlier work,

despite suggestions that variables such as homosexual identity may play a part in guiding the development of lesbian or gay sexual attractions.

B. Biology and Predisposing Factors

Biological research, focusing on the individual who is more or less exclusively homosexual in sexual orientation, has provided data suggesting that biological factors may play a part in predisposing someone to this form of homosexuality. Such findings must be taken with caution, however, since there has been significant criticism of the methodological and theoretical approaches of these studies. Several lines of research have been taken, some more fruitful than others.

1. Homosexuality as Sexual Inversion

The idea that homosexual men are female-like and lesbians are male-like has dominated the literature since the nineteenth century. While there is considerable evidence that sexual inversion is not the norm among homosexuals, it seems clear that a small percentage fits this description. Research into this group indicates that about 75% of feminine men develop a homosexual orientation, suggesting that a link between cross-gender behavior and homosexuality may exist.

Studies pursuing this link have examined the hypothesis that homosexuality is the result of hormonal abnormalities. Investigations of male and female homosexuals have found them to be indistinguishable from heterosexuals on adult hormonal levels. Another line of research has examined the possibility of hormonal imbalances occurring during prenatal development which would lead to an inversion of "normal" heterosexual development. According to this approach inappropriate androgenization (i.e., low levels) of the brain in males is believed to lead to retention of a female pattern of brain organization and hence potential for attraction to men. Female homosexuals are believed to result from prenatal exposure to inappropriately high levels of androgens producing a masculinized brain and a subsequent attraction to women. Different theories posit the cause of these hormonal irregularities as either genetic conditions (e.g., autosomal recessive genetic conditions or pseudohermaphodism) or environmental triggers (e.g., maternal stress) which work upon a genetic vulnerability to androgens (e.g., androgen sensitivity syndrome).

Results have been inconsistent and inconclusive in regard to a connection between hormonal levels and sexual orientation. Data suggest that socialization and cognitive factors play a significant role in those with chromosomal prenatal defects, with sexual orientation more likely to be related to gender allocated at birth than internal physiology. Several aspects of the research into hormonal levels have been severely criticized, including reliance on the assumption that brain differences between men and women are prenatally determined.

2. Homosexuality as Anatomical Difference

Another line of research has investigated possible anatomical differences between the brains of homosexual and heterosexual men (based primarily on the assumption that these differences were caused by hormonal imbalances during fetal development). Some preliminary studies have concluded that anatomical differences exist which show male homosexuals as having certain brain structures more like those of women than men. Parts of the hypothalamus (the suprachiasmatic nucleus, the third interstitial nuclei of the anterior hypothalamus, and the anterior commissure) have been implicated in sexual orientation toward males. Studies of brain structures linked to cognitive and visuospatial abilities have investigated possible differences between homosexual and heterosexual men (based on supposed differences between men and women) and also concluded that some differences between these groups were observed. However, the quite considerable criticism made of anatomical studies cautions against drawing any firm conclusions about the existence of brain dimorphism between homosexual and heterosexual men. Knowledge about the role of these brain structures is still limited and prevents researchers from concluding whether any differences found are indications of a cause or effect of sexual orientation.

3. Homosexuality as Genetic Variation

Familial studies examining twins of both genders have found evidence for a hereditary component of homosexual sexual orientation. Identical twins have been found to show a 50–70% likelihood of both twins having a homosexual orientation, more than double the rate of nonidentical twins and five times the rate for adopted brothers or sisters, as would be expected if hereditary factors are implicated. Since a large proportion of identical twins were not concordant for homosexuality, this research also points to the need to examine environmental factors more closely. Critics of the familial research point to the difficulty of partitioning out environmental effects and of claiming genetics as the sole and direct cause of homosexuality.

Familial research also reveals higher rates of male homosexuality in particular families, with one study finding 22% of brothers of homosexual twins (concordant for homosexuality) also being homosexual compared with 4% of brothers of mixed sexual-orientation twins. These findings have spurred studies searching to find more direct links to genes.

Preliminary results of genetic studies, although still tentative, have found that gay brothers were highly likely (67–83%) to show concordance for DNA markers on the X chromosome in the area of Xq28. Although markers are not themselves genes, they point to the probability of genetic factors. It has been concluded that genes probably play a role in 5–30% of male homosexuals and this influence is likely to come from a complex interaction between a number of genes which act together to predispose the individual to homosexuality. It has been suggested that the results, if valid, highlight the presence of different types of homosexualities. Genetic researchers now recognize that their studies probably investigate a particular type of male homosexual. It has also been suggested, on the basis of findings to date, that lesbian orientation may be a different phenomenon to male homosexuality.

However, the genetic research has also been criticized in the literature and some replication studies have found no evidence for X-chromosome inheritance. Studies of lesbians also show no link between homosexuality and Xq28.

4. Homosexuality as Evolutionary Advantage

The finding that some forms of homosexuality may be genetically predisposed has led recently to the consideration of the evolutionary purpose or reproductive advantage that homosexual genes may confer on the heterosexual who is likely to be genetically dominant but could also inherit one copy or allele of the genes related to homosexuality. It has been proposed, for example, that homosexual genes may be linked to enhanced sexual drive and performance, to personable attractiveness, or to social altruism, qualities that would provide an advantage to heterosexual reproduction.

C. Social Science Theory and Causation

Although modern psychosocial theories of exclusive homosexuality have not yet been developed, various social science theories (e.g., scripting theory, social learning theory, and choice or decision-making theory) have been adapted to sexual behavior and are relevant to understanding homosexuality. These theories emphasize the significant role sociocultural processes, relationships, emotional experiences, and cognitive development play in determining the way sexual thoughts, emotions, and actions are interpreted and acted upon as well as what is defined as sexual. Individuals are not passive creatures but rather have the capacity to react intentionally to their environments and hence to influence the outcome. Sexual behavior is not simply an internal quality but involves significant social, public, and relational dimensions. Biological factors of homosexuality, should they be found to exist, will play only a small role in determining specific sexual and romantic behavior in adults, and it would be expected that the effects of any genetic potential would be expressed differently in different sociocultural contexts. Currently, there is little research into the impact of psychological and relational development, sexual scripts, or decision making on the development of homosexuality.

VII. CONCLUSIONS

Homosexuality is not a simple unidimensional behavior or concept. It is a complex phenomenon that is best recognized by the notion of plural homosexualities and hence different causes and developmental paths. Even the notion of exclusive homosexuality is not a simple one but rather involves those who feel homosexual attractions well before puberty and those who do so only in late adolescence or adulthood. Researchers are beginning to recognize the need to examine biological, psychological, and sociocultural factors in the development of exclusive and other forms of homosexuality. Although the relative simplicity of a biological causal model is appealing to many, it is highly unlikely that any one factor will be able to account for the complex multidimensional components that make up homosexuality. Rather, it is likely to be the reciprocal interaction between these variables that leads to homosexuality and sexual orientation, an understanding of which has yet to be translated into a theoretical model or research strategies.

Bibliography

Blackwood, E. (Ed.) (1986). *Anthropology and Homosexual Behavior.* Haworth, New York.

Cabaj, R., and Stein, T. (Eds.) (1996). *Textbook of Homosexuality and Mental Health.* American Psychiatric Press, Washington, DC.

DeCecco, J., and Parker, D. (Eds.) (1995). *Sex, Cells and Same-Sex Desire.* Haworth, New York.

Faderman, L. (1982). *Surpassing the Love of Men.* Women's Press, London..

Greenberg, D. (1990). *The Construction of Homosexuality.* Univ. of Chicago Press, Chicago.

Hamer, D., and Copeland, P. (1994). *The Science of Desire.* Simon & Schuster, New York.

Herdt, G. (Ed.) (1993). *Ritualized Homosexuality in Melanesia.* Univ. of California Press, Berkeley.

Laumann, E., Gagnon, J., Michael, R., and Michaels, S. (1994). *The Social Organisation of Sexuality: Sexual Practices in the United States.* Univ. of Chicago Press, Chicago.

McKnight, J. (1997). *Straight Science? Homosexuality, Evolution and Adaptation.* Routledge, London.

McWhirter, D., Sanders, S., and Reinisch, R. (Eds.) (1990). *Homosexuality/Heterosexuality: Concepts of Sexual Orientation.* Oxford Univ. Press, Oxford.

Rosario, V. (Ed.) (1997). *Science and Homosexualities.* Routledge, London.

Williams, W. (1992). *The Spirit and the Flesh: Sexual Diversity in American Indian Cultures.* Beacon, Boston.

Hormonal Contraception

Malcolm Potts and Claire Norris

University of California, Berkeley

GLOSSARY

azoospermia A lack of spermatazoa in the semen; the failure to form spermatozoa.
COCs Combined oral contraceptives; contraceptive pills that combine an artificial progestin and an artificial estrogen.
contraception The fact or process of preventing pregnancy, whether by the use of certain physical agents or devices, or by certain procedures.
Pill, the An informal collective term for oral contraceptives.
POPs Progesterone-only pills; contraceptive pills that contain only a progestin.

Hormonal methods of contraception inhibit ovulation and bring about other changes in a woman's reproductive system that prevent pregnancy. Hormones can be given by mouth, injected into a muscle, or implanted. The introduction of the Pill revolutionized family planning because use of the method is unassociated with sexual intercourse and not even anatomically related to the reproductive tract. Hormonal contraceptives have effects that are unrelated to contraception; some are harmful, such as changes in the cardiovascular system, whereas others are beneficial, such as a marked reduction in the incidence of two important female cancers. The study of the long-term effects of hormonal contraceptives is difficult and much of the most useful information can only be obtained after the method as been

on the market for some years and used by many women. After 30 years of use, the Pill is one of the best understood drugs in medicine and for nearly all women the benefits of use outweigh the risks by a wide margin.

I. HISTORY OF HORMONAL METHODS

Folk medicine, back to classical times, sought plants with contraceptive properties, and when oral contraceptives were first developed in the 1950s they depended on plant sources for the hormones used.

As soon as pregnancy occurs, hormones produced in the afterbirth imitate those made in the brain and ovaries. The developing pregnancy hijacks the woman's endocrine system and prevents the next ovulation. As early as the 1920s, Ludwig Haberlandt in Austria understood that this mechanism could form the basis of artificial contraception, but it was not until 1940 when Russel Marker was able to synthesize steroid hormones from plant sources (in this case, Mexican wild yams) cheaply, making the Pill possible. Even so, it took the drive of Margaret Sanger, the American family planning pioneer then in her seventies, to persuade scientists to bring a usable oral contraceptive to fruition. Real work began in 1952, and it was carried out at the Worcester Foundation for Experimental Biology in Shrewsbury, Massachusetts, and in Boston by a Roman Catholic obstetrician, John Rock, a Jewish reproductive physiologist, Gregory Pincus, and a Chinese reproductive physiologist, M. C. Chang. An orally active progesterone analog was used to inhibit ovulation. By accident the first oral contraceptives also contained a small amount of an artificial estrogen, and it was discovered this made a better product, giving an excellent control of uterine bleeding.

Contraception was illegal in Massachusetts in the 1950s, which is why the first large-scale oral contraceptive trials were conducted in Puerto Rico. Abortion was unthinkable and to ensure absolutely that no accidental pregnancies occurred, the early oral contraceptives used exceedingly high doses of hormones. These high doses produced many short-term and a few serious long-term adverse side effects.

No one foresaw the revolution the Pill would bring about and it was marketed somewhat reluctantly by what was at the time a relatively small company. G. D. Searle & Company received U.S. Food and Drug Administration approval to sell Enovid in 1957. Injectable contraceptives, beginning with a brand marketed by Upjohn under the name DepoProvera, were introduced in the 1960s. Initially they were approved to treat menstrual disorders and uterine cancer. Despite sound scientific evidence that DepoProvera was also a safe and effective contraceptive, approval for its use in the United States was delayed until 1992.

II. ORAL CONTRACEPTIVES, INJECTABLES, AND IMPLANTS

Oral contraceptives are of two types. Combined oral contraceptives (COCs) use an artificial progestin and an artificial estrogen taken for 21 days out of every 28. COCs prevent ovulation, make the cervical mucus hostile to sperm, and alter the lining of the womb so that if by chance sperm and ovum do meet, then the fertilized egg cannot attach to the uterus. The second type of oral contraceptives are progesterone-only pills (POPs) that only contain a progestin. Ovulation often occurs using this method and the primary contraceptive effect of POPs is mediated through changes in the cervical mucus and the uterine endometrium.

Injectables and implants contain artificial progestins that are released slowly over long periods into the bloodstream, where their actions and side effects are similar to those of POPs. Injectables can be given in the shoulders or buttocks. DepoProvera uses a crystalline progestin analog that is released over a minimum of 3 months. Another depot injectable, norethisterone enanthate, must be given every 2 months. While using injectables, a woman's periods are usually irregular, and they may stop altogether for many months at a time. There is no harm to this, providing the user is warned and understands what is happening. Newly introduced monthly injectables, containing both a progestin and an estrogen, however, provide good contraception and regular menstrual cycles.

Contraceptive pills have now been in use for 30 years and are among the most intensively studied drugs in the history of medicine. Since they were first introduced, the hormone content has been lowered to the minimum necessary to prevent pregnancy. Correctly used, oral contraceptives have a failure rate of less then 1 per 100 users per year. In the real world, however, when tablets are forgotten, failure rates are often much higher. POPs are less tolerant of mistakes in use than COCs, and in clinical trails and in everyday use POPs have a higher failure rate. Injectables and implants, once given, cannot be forgotten and are among the most effective contraceptive methods available.

Women who use COCs experience uterine bleeding during the 7 days they are not using the hormone tablets (some formulations add dummy, or placebo, tablets to help keep track of the days when the active tablets are not being used). The imitation menstrual period experienced by a woman on COCs involves less blood loss than a normal one. Once a women is using COCs, she could have any pattern of uterine bleeding she chooses, and some women use COCs for more than 28 days or to shift their cycles a few days to meet their own convenience (e.g., in order not to menstruate at the weekend or during an athletic competition, to avoid the time of an examination, or to plan for a honeymoon). A woman's background menstruation continues when using POPs, and her periods are often irregular.

Implants use a progestin called levonorgestrel, which has a long history of use in oral contraceptives. The hormone is contained in six silastic capsules that are implanted just under the skin on the inside of the upper arm using a local anesthetic. The hormone diffuses through the silastic capsules at a constant low rate over 5 years. At the end of this time, or if the woman requests it earlier, the capsules are removed. Removal is often more difficult than insertion, and it can be painful and tiresome. Newer implants containing levonorgestrel or other progestins that require only one or two rods to be implanted are under clinical trial and may soon supplant the original six-rod system.

III. USE AND SHORT-TERM SIDE EFFECTS

There are certain well-defined groups of women for whom oral contraceptives may not be a first choice or should not be used at all (Table 1). The largest number consists of women over age 35 who also smoke heavily. If a woman is not a smoker then there is no upper limit to the age when pills can be used. If a young women is at risk of pregnancy then she can almost always use pills if she wants to. COCs provide some protection against the spread of pelvic infections, but they do not protect the user in any way against catching a sexually transmitted disease or HIV.

Users should begin taking active tablets on the first day of menstrual bleeding. If they forget one pill, they should take two the next day. As noted, with COCs patterns of menstruation will be regular and if the woman has been having period pains these usually (although not always) disappear. The user may feel some nausea and breast tenderness, similar to those experienced in early pregnancy. These symptoms almost always disappear after a couple of cycles. All hormonal contraceptive are associated with a 100% return of fertility. (Although if a women who has never had a child takes the Pill for many years, she will be older when she wants to start a family and perhaps less likely to get pregnant than when she was younger.) The artificial hormones used in COCs and POPs are nearly all excreted within 24 hr of taking the last active tablet, but injectables can remain in the body much longer and users should be advised that it may take some time before fertility is fully restored.

IV. EMERGENCY CONTRACEPTION

It has been known since the 1960s that changing the balance of ovarian hormones can prevent a fertilized egg from implanting in the uterus. If a couple takes risks and has unprotected intercourse, the condom breaks, or the tragedy of rape occurs, then the oral ingestion of steroid hormones greatly reduces the chances of pregnancy occurring. A women can take four low-dose oral contraceptive pills as soon

TABLE 1
Women Who Should Not Use Low-Dose Oral Contraceptives

Risks of use usually outweigh benefits	*Method should not be used*
	Pregnancy
Breast-feeding and 6 months or less since delivery	Breast-feeding and 6 weeks or less since delivery
Age 35 and over and smoke less than 20 cigarettes/day	Age 35 and over and smoke 20 cigarettes or more/day
High blood pressure (140–159/90–99)	Essential hypertension 160+/100+
	Diabetes of 20 years duration or more or with signs of vascular damage
	History of deep venous thrombosis
	Major surgery with prolonged immobilization
	Angina or history of heart attack
	Stroke (or very severe headaches with visual or other physical changes)
Gall bladder disease	Viral hepatitis, liver tumors, or severe cirrhosis of the liver
Past history of breast cancer with no evidence of recurrence for 5 years	Cancer of the breast in the past 5 years or existing cancer of the uterus

as she realizes she is exposed to the risk of pregnancy and four more 12 hr later. The easiest source of eight tablets for emergency contraception is sometimes a sister's or best friend's regular contraceptive pills. Any COC user who donates tablets in this way needs to know that the simplest thing to do is to use the remainder of the cycle of pills from which they donated the eight and then go directly to the first active tablet in the next packet. In this way they will have a longer than usual interval between two menstrual bleeds but run no risk of getting pregnant themselves. An IUD can also be inserted to act as postcoital or "morning-after" contraceptive. It can be inserted at any time during the menstrual cycle.

V. HORMONAL CONTRACEPTIVES, HEALTH, AND CANCER

Like natural hormones, hormonal contraceptives produce many changes in a woman's body. They imitate the hormonal environment of pregnancy and lactation more than that of a woman who is having regular menstrual cycles. Some cancers are influenced by a woman's pattern of child bearing (which alters her hormone pattern) and important and valid questions have been asked about the possible impact of hormonal contraceptives on a woman's risk of cancer.

Hormonal contraceptives can be tested on rats, dogs, or monkeys, but the reproductive systems of other animals often differ from that of humans (dogs, for example, only ovulate two or three times a year, not every month like humans). In the end, the use of any new hormonal contraceptive is an experiment on our species. In other words, safety can never be fully proved in advance of widespread human use. COCs were almost withdrawn from the market in the 1960s because of unforeseen effects on the cardiovascular system, and, as noted, injectables were kept out of the U.S. market for nearly 30 years as a result of fears of cancer. Even today, women all over the world have a more negative picture of hormonal contraceptives than large-scale epidemiological studies justify, and only a minority of people know that the Pill prevents two important cancers.

Following animal tests and small-scale human tests of safety, the U.S. Food and Drug Administration requires about 600 women-years of clinical testing to license a new hormonal contraceptive. Such trials give important information on failure rates and on short-term side effects, but they provide little or no information about possible rare, long-term effects. If, for example, the pill causes heart disease or stroke in 1 or 2 women in every 100,000 users, it may take 1 million users and several years exposure to the drug to demonstrate the problem. If, as also occurs, pill use prevents 1 or 2 deaths in 100,000, then it can be even more difficult to demonstrate.

There are two chief ways of studying any drug once it has been licensed for sale. The first is a prospective study, in which many thousands of users and nonusers are followed for several, sometimes tens, of years. Second, in a retrospective study a suspected side effect, for example, heart disease or stroke, is identified and data are collected to show whether women using hormonal contraceptives are more likely than average to suffer from the condition. For prospective studies as few as 150 cases may be enough to detect a difference. Both types of study are fraught with problems and it is often rash to decide on a course of action based on a single study. In the case of gall bladder disease, for many years there was thought to be an adverse relationship with COC use, but recent studies have shown this not to be the case.

Numerous retrospective and a few prospective studies of COCs have been conducted in America and Europe. The most important adverse effects are on the cardiovascular system, in which high-dose pills increase the risk of heart attack, stroke, and blood clots in the legs, which in rare cases can break off and block the lungs, causing thrombolism. The risk of cardiovascular illness among COC users rises with the age of the women and is greatly magnified if the women smoke. Because prospective and retrospective studies can only be done when the pill has been on the market for some time, much of the data being used today relates to high-dose pills no longer in use. The use of modern low-dose COCs has significantly reduced the cardiovascular risks and in some cases may have eliminated these risks altogether. There may be subtle differences in cardiovascular risks according to the exact type of progestin used.

The most important long-term beneficial effects are on reducing ovarian and uterine cancer. If a woman takes COCs for several years, she halves her risk of developing these cancers and the protection persists for 10 years or more after stopping use of COCs. Women using COCs have less ovarian cysts and less benign tumors of the breast. Extensive studies of breast cancer among COC users show no overall adverse or beneficial effects. Among women who begin COCs before age 20 there is a small elevation in the risk of breast cancer over the next 5 years. Fortunately, the background level of cancer in young women of this age group is very low and the absolute risk of a young user developing breast cancer remains very small. In general, the protective effects against uterine and ovarian cancer occur in older women when these cancers are relatively more common.

In short, using the Pill changes the profile of disease to which a woman is subject. When the whole equation of risk and benefit is put together, even ignoring the prevention of unintended pregnancy (which carries a measurable risk of death and illness) and abortions, for nearly all women the benefits of use outweigh the risks. There is, however, a group of women who should choose an alternative method (Table 1).

VI. NEW METHODS

The hormones controlling ovulation can be interrupted in other ways besides using progesterone and estrogen analogs. A group of hormones called gonadotrophin-releasing factors are well understood. They can be delivered, for example, by a nasal spray.

Women who have many children and breast-feed for long intervals have less ovarian, uterine, and breast cancer than women who have few or no children. The fact that the Pill (which imitates pregnancy and breast-feeding) reduces the risk of uterine and ovarian cancer is biologically explicable. Could a hormonal contraceptive be developed which would also reduce the risk of breast cancer? Malcolm Pike and colleagues in Los Angeles have begun small-scale clinical studies aimed at developing a combination of releasing hormones and progestins designed to meet this goal. Developing and testing such a formulation and demonstrating if it actually reduces breast cancer will take time and patience. However, the goal for the twenty-first century should be to develop a method of contraception that permits women to have the children they want when they want them and that also ensures they reach the menopause as healthy individuals with a low risk of reproductive cancers, which now attack 1 in 11 American women.

Is a male pill using hormones possible? Both sperm and the male hormone testosterone are produced at closely linked sites in the testis. Sperm can only develop bathed in high concentrations of testosterone, whereas the other effects of the male hormones are brought about by a relatively much lower concentration at other sites such as the brain, bones, and muscles. The World Health Organization has completed preliminary trials using injections of testosterone as a male contraceptive. In this method, the injected testosterone deceives the brain into reducing the natural production of testosterone in the testes. The injected testosterone replaces the natural production everywhere except in the testes where the levels are too low to maintain sperm production. Because it takes about 70–80 days for sperm to undergo spermatogenesis and reach maturity in the epididymis, the method is not immediately effective and fertility does not return as soon as the injections stop. When testosterone is given alone, not all men reach azoospermornia: About 58% reached this state in 120 days on average. The combination of testosterone and a progestin leads to more rapid onset and a greater degree of azoospermia and may thus prove to be more practical.

See Also the Following Articles

Contraceptive Methods and Devices; Ovulation; Progestins

Bibliography

Asbell, B. (1995). *The Pill: A Biography of the Drug That Changed the World*. Random House, New York.

Committee on the Relationship between Oral Contraceptives and Breast Cancer, Institute of Medicine (1991). *Oral Contraceptives and Breast Cancer*. National Academy Press, Washington, DC.

Collaborative Group on Hormonal Factors in Breast Cancer (1996). Breast cancer and hormonal contraceptives: Collaborative reanalysis of individual data on 53,297 women with breast cancer and 100,239 women without breast cancer from 54 epidemiological studies. *Lancet* **347**, 1713–1727.

Fortney, J., Harper, J. M., and Potts, M. (1987). Oral contraceptives and life expectancy. *Stud. Family Planning* **17**, 117–125

Guillebaud, J. (1987). *The Pill*, 3rd ed. Oxford Univ. Press, Oxford, UK.

Hatcher, R., Trussel, J., Stewart, F., *et al.* (1997). *Contraceptive Technology 1997*. Irvington, New York. (This text is updated annually)

Petitti, D. B., Sidney, S., Bernstein, A., and Wolf, A. (1996). Stroke in users of low-dose oral contraceptives. *N. Engl. J. Med.* **335**, 8–15.

Ory, H. (1982). The noncontraceptive health benefits from oral contraceptive use. *Family Planning Perspect.* **14**, 182–184.

Samules, S. E., and Smith, M. D. (Eds.) (1994). *The Pill: From Prescription to Over the Counter*. Kaiser Foundation, Menlo Park, CA.

Thomas, D. B. T., and Ray, R. M. (1996). Oral contraceptives and invasive adenocarcinoma and adenosquamous carcinomas of the uterine cervix. The World Health Organization Collaborative Study of Neoplasia and Steroidal Contraceptives. *Am. J. Epidemiol.* **24**, 281–289.

Thorogood, M., Mann, J., Murphy, M., *et al.* (1991). Is oral contraceptive use still associated with an increased risk of fatal myocardial infarction? Report of a case control study. *Br. J. Obstet. Gynaecol.* **98**, 1245–1253.

World Health Organization (WHO) (1992). *Oral Contraceptives and Neoplasia*, WHO Technical Report Series. WHO, Geneva.

World Health Organization (WHO) (1996). *Improving Access to Quality Care in Family Planning: Medical Eligibility Criteria for Initiating and Continuing Use of Contraceptive Methods*. WHO, Geneva.

Hormonal Control of the Reproductive Tract, Subavian Vertebrates

Ian P. Callard, Vicki Abrams-Motz, Georgia Giannoukos, and Lisa A. Sorbera

Boston University

I. The Oviduct
II. Hormones Acting on the Oviduct
III. The Biological Actions of the Hormones on Müllerian Duct Derivatives
IV. Summary

GLOSSARY

corpus luteum An ovarian structure formed from the ruptured follicle after ovulation; synthesizes progesterone and estradiol.

elasmobranchs Sharks and rays, the most ancient jawed vertebrates, originating about 400 million years ago.

estradiol An 18-carbon steroid derivative of cholesterol; primary hormone of the ovarian follicle responsible for reproductive tract growth and function.

follicle The ovarian structure responsible for nurturing the oocyte prior to ovulation and the synthesis of estradiol.

homologous Having a common evolutionary origin.

Müllerian duct Duct of the embryonic kidney (mesonephros); develops into reproductive tract of female; disappears in male.

neurohypophysial hormones Hormones synthesized in the brain and stored in a ventral outpocketing of the brain called the neurohyphysis or neural lobe; control water balance and smooth muscle contraction (e.g., uterus).

oviparous A mode of reproduction in which the female organism lays eggs (e.g., bird and turtle).

oviposition The act of egg laying.

parturition The birth of young developed internally in the maternal organism.

progesterone A 21-carbon steroid derivative of cholesterol;

primary hormone of the corpus luteum; maintains pregnancy.

relaxin A peptide hormone of the ovary and placenta, similar to insulin in structure; participates in the regulation of reproductive tract smooth muscle and connective tissue remodeling.

vasotocin The most ubiquitous of the nonmammalian neurohypophysial hormones; causes contraction of smooth muscle of the oviduct.

viviparous Referring to a mode of reproduction in which the female organism nurtures embryos internally (e.g., some lizards, snakes, sharks, and rays).

I. THE OVIDUCT

The vertebrate oviduct is a complex structure derived from the Müllerian duct during sexual differentiation in most vertebrates. Its function is the conveyance of gonadal products to the exterior, the provision of additional nutritive and protective (shell) coverings for the egg, and to provide a protected site for development in the case of viviparous species. Its homologies are uncertain. It is absent in the extant jawless vertebrates (hagfishes and lampreys). In elasmobranchs, the well-developed Müllerian ducts arise early in ontogeny from the pronephros and its duct, an origin somewhat different from that of tetrapods in which the funnel is believed to be of pronephric derivation and the duct of Wolffian derivation. The homologies of the teleost oviduct are also unclear. Although teleosts do not have a "typical" Müllerian funnel or oviduct, they have either a truncated oviduct (e.g., some salmonids and cyprinids) or, in the majority of teleosts, the duct is formed by backward growth of peritoneal folds which enclose the oviduct during development. In this condition, the lumen of the oviduct and ovary ("closed ovaries") are contiguous. Due to this arrangement, it is only in the teleosts and sturgeons that ovarian products are not exposed to the general body cavity. For amniotes (reptiles, birds, and mammals), at least, it appears to be agreed that the Müllerian duct is derived from the coelomic epithelium and mesonephros.

Although the homologies and evolutionary origins of the Müllerian duct and its component parts are unclear, there is considerable structural (Fig. 1) and functional similarity. A suite of primary regulatory hormones appears to have been established as early as

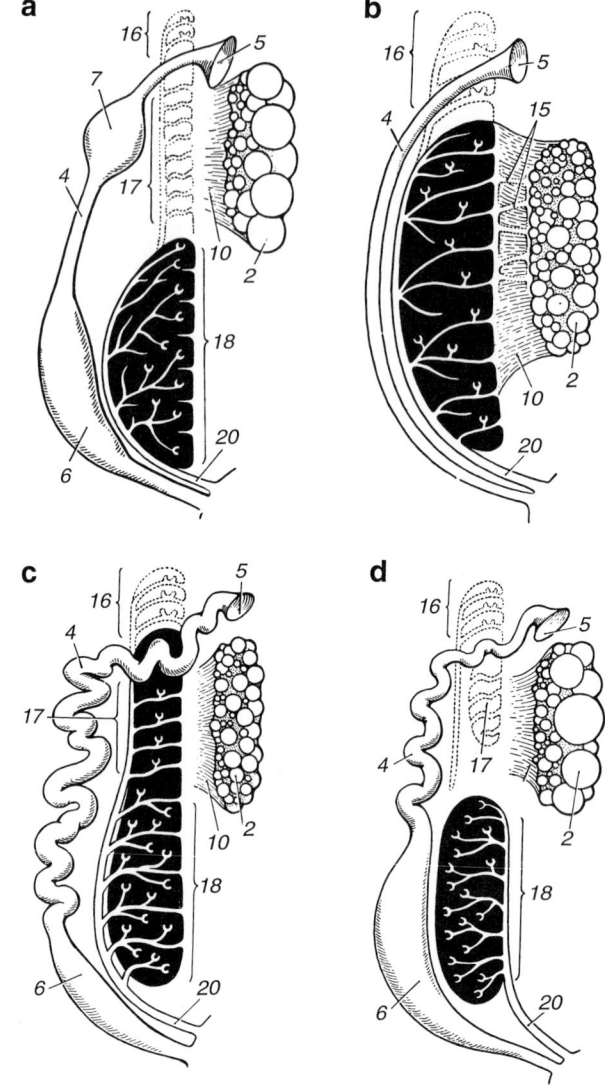

FIGURE 1 The Müllerian duct and its derivatives in vertebrates. (a) Elasmobranch (shark). (b) Primitive teleost (sturgeon). (c) Amphibian (urodele). (d) Reptile/bird. (Reproduced from Hoar, W. S., and Randall, D. J. (1969). "Reproduction." In *Fish Physiology* Academic Press and Portman, A. (1969). *Einführung in die vergleichende morphologie der wirbeltiere.* Schwabe and Co. AG. Verlag. Basel.) Key: 2 = ovary; 4 = oviduct, Fallopian tube; 5 = funnel; 6 = uterus; 7 = shell gland; 10 = mesovarium; 15 = renal remnant; 16 = pronephros; 17 = opisthonephros, gonadal region; 18 = functional opisthonephros; 20 = ureter.

the elasmobranchs and to have been retained through the vertebrates. With regard to the mammalian uterus, a derivative of the Müllerian duct, Heller suggested that it is the target of more hormones than any other single organ. As the following material will show, at least some of these are held in common throughout vertebrates.

II. HORMONES ACTING ON THE OVIDUCT

The principal hormones involved in the development and regulation of the vertebrate reproductive tract, egg transport, gestation, oviposition, and parturition are the gonadal steroids, progesterone and estradiol, the peptide relaxin, and small peptides of neurohypophysial origin, which include vasotocin and oxytocin (see Fig. 2). These hormones are the principal focus in this article. In addition, control of reproductive tract smooth muscle and glandular function is provided by direct neural (adrenergic/cholinergic) input, circulating epinephrine, and prostaglandins from local or distant sources. There is evidence that these bioactive agents are present in and are involved in the control of the reproductive tract of subavian species, but they will not be dealt with here.

A. Progesterone and Estradiol

Adequate evidence for the synthesis and secretion of these two hormones in all vertebrate groups by pathways identical to those found in the mammalian gonads exists. In addition, in all vertebrates, the pattern of secretion of these hormones is well correlated with imputed physiologic regulation of the tissues of the reproductive tract and its functions. For some subavian species at least, regressive changes in the reproductive tract following gonadal removal indicate the gonad as a primary source of trophic hormones, and experiments involving hormone injection further support this conclusion. The presence of classical progesterone and estradiol receptors in the oviduct of subavian species also points to steroid regulation of the reproductive tract of these vertebrates.

B. Peptides: Relaxin and Neurohypophysial Peptides

In the subavian vertebrates, relaxin has only been chemically isolated and its structure determined for elasmobranchs. However, in both elasmobranchs and reptiles it has effects on the reproductive tract, which suggest it is part of the suite of hormones which control the reproductive tract of all vertebrates.

Small peptides (nine amino acids), with slight variations in amino acid sequence, have been isolated from the posterior lobe/hypothalamus of all jawed vertebrate groups. Furthermore, adequate evidence exists to allow the assertion that these neurohormones are involved in the regulation of oviduct smooth muscle contraction and the expulsion of eggs or developed young in all vertebrates with the exception of cyclostomes.

III. THE BIOLOGICAL ACTIONS OF THE HORMONES ON MÜLLERIAN DUCT DERIVATIVES

A. Steroids

1. Estrogens

It is accurate to say that estrogens are the primary trophic hormones of the subavian reproductive tract; in all groups ovariectomy results in atrophy of the oviduct, and estrogen treatment stimulates oviduct growth. However, there are relatively few observations of this nature in elasmobranchs, teleosts, and amphibians and somewhat more for reptiles. The cellular basis for the action of estrogenic steroids is undoubtedly the estrogen receptor(s) which has been at least partially characterized in the reproductive tract of all groups except teleosts. More detailed studies of the effects of estrogens on the oviduct with regard to the processes of oviposition and parturition are rare and mainly encompass work on the reptiles and elasmobranchs. In both groups, estradiol, known to be secreted by the ovarian follicle, has important effects on myometrial contractility, connective tissue and glandular elements which are associated with transport, accommodation of the egg/embryo, and provision of secondary and tertiary envelopes for the egg. In the turtle, oviduct contractility increases with estradiol titer and with estradiol treatment, and changes in tissue extensibility (connective tissue) and epithelial glandular secretory activity are consistent with important physiologic actions in the control of oviposition and related events.

2. Progesterone

Although the role of progesterone as a trophic stimulus for the oviduct of nonmammalian species is less clear in subavian species, the identification of progesterone receptors in elasmobranch, amphibian, and reptilian (turtle and snake) suggests a physiological role. This is supported by the immuncytochemical localization of progesterone receptors in the smooth muscle and glandular regions of the turtle oviduct. In reptiles and elasmobranchs, progesterone is secreted by both the follicle and the corpus luteum, and plasma levels of the hormone in both viviparous elasmobranchs and reptiles are well correlated with the presence of the corpus luteum. The available evidence does not support a role for progesterone in the maintenance of pregnancy in viviparous non-mammals. Removal of ovaries or corpora lutea during gestation in reptiles has been reported to cause variable effects, and exogenous progesterone does not prevent abortion in ovariectomized water snakes but may prolong gestation and parturition in intact lizards. Removal of corpora lutea from gravid egg-laying (oviparous) lizards causes early oviposition, suggesting a role for progesterone in egg retention. In other lizards, spontaneous contractility increases during gestation and falls after parturition. Furthermore, progesterone significantly inhibits normal and estrogen-stimulated myometrial contractions and can prevent egg laying in turtles treated with a dose of vasotocin adequate to induce egg laying in the absence of progesterone. In pregnant lizards, progesterone treatment prolongs pregnancy by preventing parturition, and removal of the corpora lutea results in premature oviposition in egg-laying lizards. In the spiny dogfish, progesterone has been reported to inhibit the action of relaxin on the oviduct, and in the little skate it reduced the duration of stay for egg capsules in the oviduct.

B. Peptide Hormones

1. Relaxin

In the spiny dogfish, this hormone, when injected after estrogen pretreatment into late-stage pregnancy, induces premature release of the free-swimming "pups." This relaxin effect appears to be due to a change in the extensibility of the connective tissue of the narrow "cervical" region of the oviduct which is normally responsible for the retention of the pups. This effect is very similar to the effect that the hormone has on connective tissue of the mammalian reproductive tract. Furthermore, since the action of relaxin is only seen after estradiol treatment, it appears that estradiol may induce relaxin receptors in the small shark, as in mammals. Both *in vivo* and *in vitro*, relaxin treatment significantly slows the contractile rhythm of the oviduct smooth muscle. Furthermore, as indicated previously, progesterone inhibits this action of relaxin. Relaxin has a similar effect on the reproductive tract of the turtle.

C. Neurohypophyseal Hormones

Gravid lizards respond to injections of oxytocin or mammalian neurohyophyseal extract by commencing parturition within 2 hr, and snakes also respond by delivering young in response to homologous neural lobe extract. In *in vitro* studies of isolated painted turtle oviduct, arginine vasotocin, the primary avian and reptilian neural lobe hormone, was found to be much more potent than the mammalian hormone, oxytocin, in stimulating contraction of the oviducal smooth muscle. Similar observations were made on lizards. In the painted turtle, *in vivo* oviduct contractions have been monitored in gravid animals after vasotocin injections, and a definitive pattern of contractile spikes accompanies the expulsion of individual eggs; these spikes which can be inhibited by progesterone.

A variety of amphibians respond to several neural lobe hormones and extracts by contractions of the caudal regions of the oviduct, and several species show a marked seasonal variation in responsivity, suggesting an important role for steroids. In teleosts, although there is considerable variability of response, it is fair to say that both the ovary, which contains variable amounts of smooth muscle, and the oviduct, respond to neural lobe preparations and hormones, particularly isotocin. As in amphibia, sensitivity varies according to follicular development and suggests a role for ovarian hormones in the response. Early work on a variety of elasmobranchs produced modest

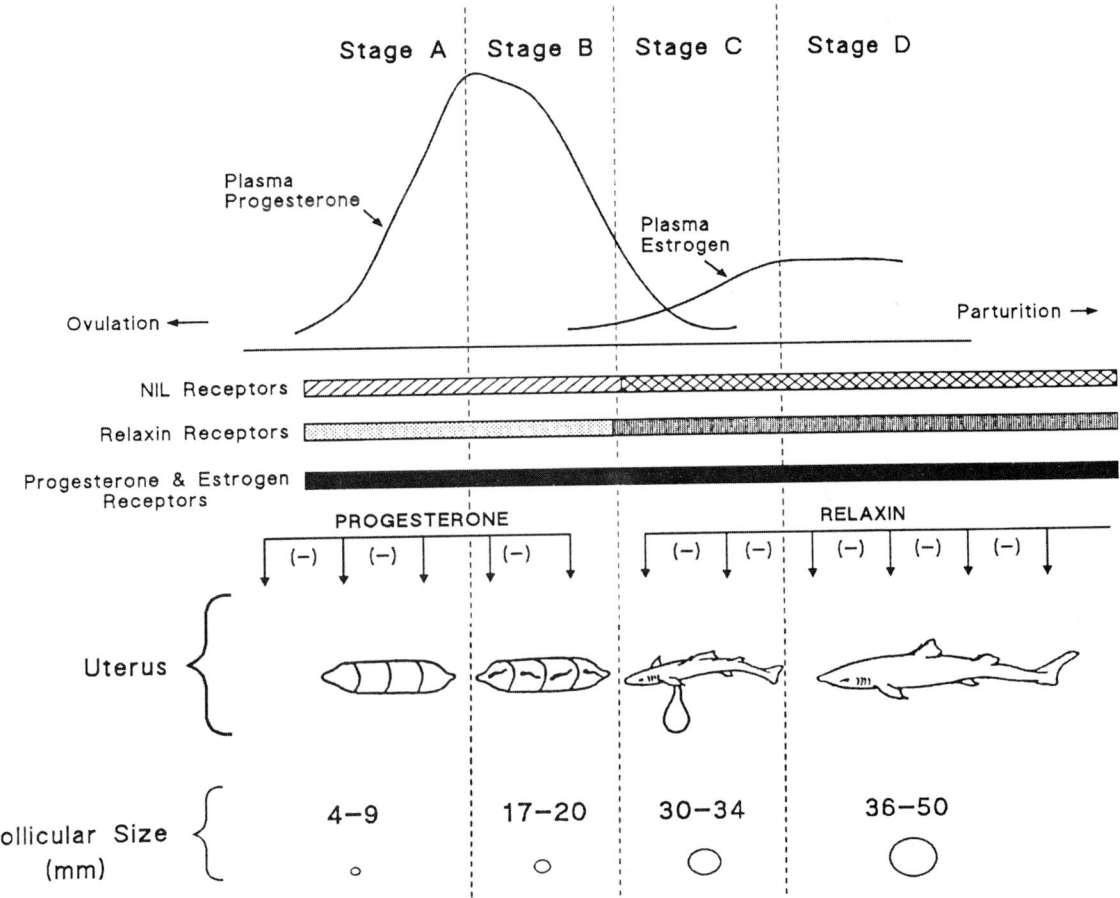

FIGURE 2 Model of the interactions of steroid and peptide hormones in the control of the spiny dogfish reproductive tract. The presence of relaxin and neuohypophysial hormone receptors is inferred from the biological actions of the hormones *in vivo* and *in vitro*. Estrogen and progesterone receptors have been identified in elasmobranch reproductive tracts (reproduced with permission from Sorbera and Callard, 1995).

and variable responses to neural lobe extracts, but recent studies have indicated definite *in vivo* and *in vitro* responses of the pregnant spiny dogfish oviduct to homologous neural lobe extracts.

IV. SUMMARY

The existing body of information indicates that the suite of hormones and blueprint for the control of the vertebrate reproductive tract were established at the beginning of the vertebrate era, approximately 450 million years ago. These same hormones, established in the ancestors of present-day elasmobranchs,

are used in the control of human reproductive tract and the reproductive process.

See Also the Following Articles

Elasmobranch Reproduction; Teleosts, Viviparity; Viviparity and Oviparity

Bibliography

Callard, I. P., and Koob, T. J. (1993). Endocrine regulation of the elasmobranch reproductive tract. *J. Exp. Zool.* **266** 368–377.
Giannoukos, G., and Callard, I. P. (1996). Radioligand and

immunochemical studies of turtle oviduct progesterone and estrogen receptors: Correlations with hormone treatment and oviduct contractility. *Gen. Comp. Endocrinol.* **101**, 63–75.

Handbook of Physiology Section 7: Endocrinology (1974). Volume IV: The pituitary gland and its neuroendocrine control, Chaps. 6–20.

La Pointe, J. (1977). Comparative physiology of neurohypophysial hormone action on the vertebrate oviduct–uterus. *Am. Zool.* **17**, 763–773.

Sorbera, L. A., and Callard, I. P. (1995). Myometrium of the spiny dogfish *Squalus acanthias*: Peptide and steroid regulation. *Am. J. Physiol.* **269** (*Regul. Integr. Comp. Physiol.* **38**), R389–R397.

Hormone Receptors, Overview

Adviye Ergul and David Puett

University of Georgia

I. Introduction
II. Membrane-Associated Cell Surface Receptors
III. Intracellular Receptors

GLOSSARY

cytokines A large family of proteins that serve as messengers between cells of the immune system and thus regulate many aspects of the immune response.

growth factors Generically any compound that facilitates or is required for cell division or whole animal growth, although the term generally applies to protein molecules that bind to specific cellular receptors and promote cell division.

hormones Molecules with diverse chemical structures that regulate a variety of cellular functions in target cells characterized by a receptor to the particular hormone.

ion channels Integral membrane proteins that form a core in the membrane to allow passage of ions upon appropriate cell activation.

ligands Naturally occurring or synthetic molecules that bind to a receptor and either activate the receptor (agonist function) or prevent the receptor from binding and responding to an activating molecule (antagonist function).

receptors Proteins, located intracellularly or localized on the cell surface, that recognize and bind ligands such as hormones, cytokines, and growth factors with high affinity and specificity and initiate a cellular response via transcriptional activation or one or more signaling pathways.

From one perspective, reproduction can be viewed simplistically as an orchestrated series of processes, each of which depends on appropriate intramolecular and intermolecular recognition and response. In a variety of tissues, including the brain, adenohypophysis, gonads, and secondary sex organs, hormone receptors are central to the concepts of recognition and response. Without exception, hormone receptors are proteins which, in many cases, are posttranslationally modified, e.g., via glycosylation, phosphorylation, and myristoylation. They can be localized either as integral membrane proteins on the cell surface or within a cell; in at least one instance a portion of a cell surface receptor circulates and acts as a binding protein of the hormone.

I. INTRODUCTION

Hormone receptors have two major functions: (i) They must recognize their cognate ligand(s), which is normally present at low concentrations, and discriminate against other contenders, and (ii) once recognition and high-affinity hormone binding have occurred, the receptor must act as a transducer and trigger one or more intracellular processes.

In the case of many cell surface receptors, ligand binding occurs to a portion of the receptor in the extracellular millieu or to a pocket formed by receptor helices within the transmembrane domain. In addition to this hormone recognition domain, there is another recognition domain on the intracellular side of the cell membrane that can activate other proteins or serve in an enzymatic role. In all cases, however, the ligand-occupied receptor can activate one or more intracellular signaling cascades; within each cascade there exists one or more steps of amplification. Moreover, it is now recognized that often the separate cascades are not independent. There may be cross talk between the different signaling pathways or cascades, and conceivably a temporal switch may occur from one pathway to another. Frequently, hormones acting on cell surface receptors will elicit rapid responses since the necessary cellular machinery is in place and the formation of a hormone–receptor complex is analogous to the act of turning a light switch "on." In addition to these acute cellular responses, which may simply be the opening of an ion channel, there is often a delayed or chronic phase involved in hormone activation arising, generally, from gene activation via one of the signaling pathways.

Intracellular receptors, unlike their counterparts located on the plasma membrane, do not activate specific signaling pathways. Rather, these receptors are, in effect, transcriptional regulators, generally activators, that themselves are regulated by ligand binding, i.e., the formation of a hormone–receptor complex. Thus, all intracellular receptors have, in addition to a hormone-binding domain, a DNA-binding domain composed of two zinc finger recognition motifs.

It is now recognized that hormone receptors are not static cellular entries. In response to hormone binding, they can undergo conformational changes; they can self-associate or associate with other cellular proteins; they can translocate from one cellular compartment to another; and they can, in concert with other cellular proteins, regulate the duration of activation, i.e., some, after a finite time, become uncoupled such that no additional signaling occurs even in the presence of hormone.

This article provides an overview of hormone receptors and treats the major classes known to date. Because of space limitations, it is not possible to discuss all the different hormones that act through these receptors; rather, we have chosen to present a succinct review of the various receptor types with specific examples given throughout the article.

II. MEMBRANE-ASSOCIATED CELL SURFACE RECEPTORS

Various ligands, ranging from small biologic amines to peptides and large glycoprotein hormones, mediate their effects by binding to their cognate receptors present on the plasma membrane of target cells. There are three major classes of membrane-associated receptors: single membrane-spanning receptors with or without tyrosine kinase activity, seven transmembrane-spanning receptors coupled to G-proteins, and ion channels (Table 1). Once the ligand, which may be a neurotransmitter, hormone, cytokine, etc., binds to its receptor, this complex can activate various signal transduction pathways either on the cytoplasmic face of the membrane or after internalization, and the second messengers involved in this process are summarized in Table 2.

A. Single Membrane-Spanning Receptors

In addition to the single membrane-spanning domain, receptors in this category have an extracellular domain characterized by the ligand binding site and an intracellular domain characterized by a specific region that couples to a different signaling pathway depending on the type of receptor. A cysteine-rich domain or highly conserved cysteine residues are present in the intracellular domain of receptors of this class, and they can be classified in two subgroups based on their mode of subunit assembly on the plasma membrane (Figs. 1A and 1B).

1. Single Polypeptide Chain Receptors

This subgroup includes hormone receptors such as growth hormone receptor, prolactin receptor, growth factor receptors such as platelet-derived growth factor receptor, epidermal growth factor receptor

TABLE 1
Membrane-Spanning Hormone Receptors[a]

Single membrane-spanning receptors		Seven transmembrane-spanning receptors coupled to G proteins	Hormone-gated ion channels
With tryosine kinase activity	*Without tyrosine kinase activity*		
Insulin	Growth hormone	Glycoprotein hormone receptors	Acetylcholine
Interleukin-2	Nerve growth factor	LH/hCG	Glutamate
Platelet-derived growth factor	Prolactin	FSH	γ-Aminobutyric acid
Epidermal growth factor		TSH	Glycine
Macrophage colony-stimulating factor		Adrenergic receptors	Serotonin receptor-3
		Parathyroid hormone	
		Growth hormone-releasing hormone	
		Calcitonin	
		Histamine	
		Gonadotropin-releasing hormone	

[a] Partial list.

(EGF-R), and some of the cytokine receptors including interleukins (IL-1-R and IL-7-R) and granulocyte colony-stimulating factor receptor (Fig. 1A).

The EGF family of receptors, also known as erbB or heregulin receptors, have four members and exhibit interesting properties. EGF and transforming growth factor are the natural ligands for EGF-R (erbB), and, upon binding, other members of EGF receptors are recruited to form homodimers or heterodimers. Neu

(erbB-2) receptors do not appear to have a natural ligand and seem to be involved in EGF-induced heterodimerization with EGF-R (erbB), a necessary event for cell proliferation. ErbB-3 receptors do not signal but form heterodimers with other EGF receptors. Heregulin binds to erbB-4 receptors and is involved in cell differentiation. It is believed that the relative distribution of EGF-R (erbB), erbB-2, and erbB-4 is important to maintain a balance between

TABLE 2
Hormones That Act via Various Signaling Systems[a]

Via cAMP		Via cGMP	Via IP$_3$	Via ion channels
Excitatory	*Inhibitory*			
Calcitonin	Somatostatin	Atrial natriuretic peptide	Epinephrine	Acetylcholine
Glucagon	Norepinephrine (α-adrenergic)		Endothelin	GABA
VIP	Bradykinin		Angiotensin II	Nitric oxide
LH			TRH	Glutamate
FSH			Histamine (H1)	Serotonin (5HT$_3$)
TSH			Thromboxanes	
ACTH			GnRH	
Norepinephrine (β-adrenergic)				
Histamine (H2)				
Vasopressin				

[a] Only a partial list of the major accepted pathways is given.

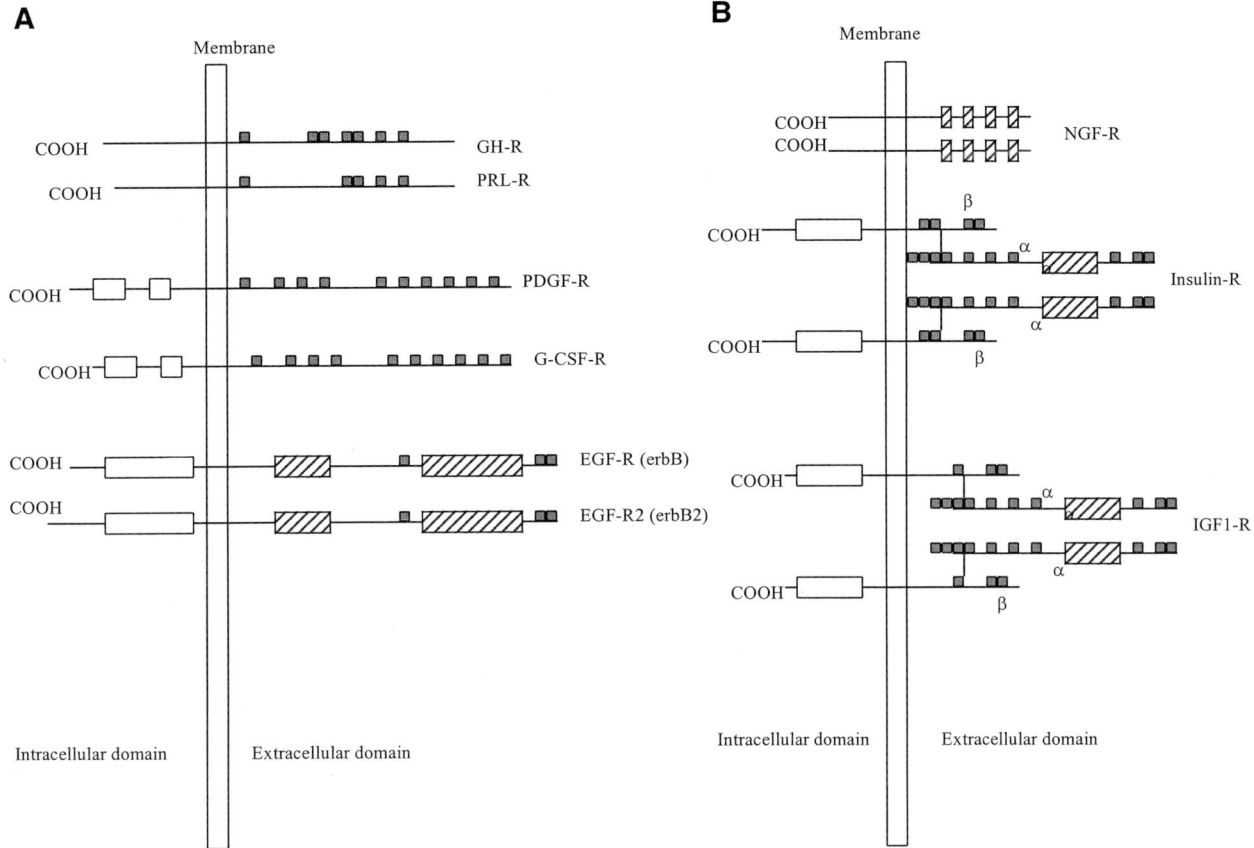

FIGURE 1 Single membrane-spanning receptors consist of single polypeptide chains (A) and multiple subunits (B). Open boxes, hatched boxes, and small closed squares represent tyrosine kinase domains, cysteine-rich domains, and conserved cysteines, respectively. All proteins are drawn to the same scale.

cell proliferation and differentiation in responsive tissues. Overexpression of erbB-2 may lead to tumor formation as seen in breast cancer.

2. Multiple Subunit Receptors

Receptors formed by the assembly of two or more polypeptide chains are illustrated in Fig. 2B. Receptors for tumor necrosis factor (TNF-R) and nerve growth factor are composed of two noncovalently attached identical peptide chains. A recently discovered member of the TNF-R superfamily, the fas receptor (fas-R), is involved in cell death and it is found as a single polypeptide chain receptor or dimer. Interestingly, TNFp53-R and fas-R do not exhibit any kinase activity and are not associated with any known signaling pathway. In addition to the extracellular ligand-binding domain, these receptors contain a cy-

toplasmic death domain. Upon ligand binding, adapter proteins are recruited to the plasma membrane and a series of events, which involves protein–protein interactions leading to the activation of various interleukin-converting enzymes, is triggered. These activated cysteine proteases ultimately cause cell death via apoptosis.

Some of the cytokine receptors, such as those for IL-2-R, IL-3-R, IL-4-R, IL-5-R, IL-6-R, and granulocyte-macrophage colony-stimulating factor, belong to this category and are composed of multiple subunits. Each interleukin receptor has specific α and β subunits, whereas three subunits, known as gp130, KH97, and IL-2-Rγ, are common to several interleukins. The composition of the subunits determines the affinity of the receptor for its respective ligand. For example, in the case of IL-2-R, there are three

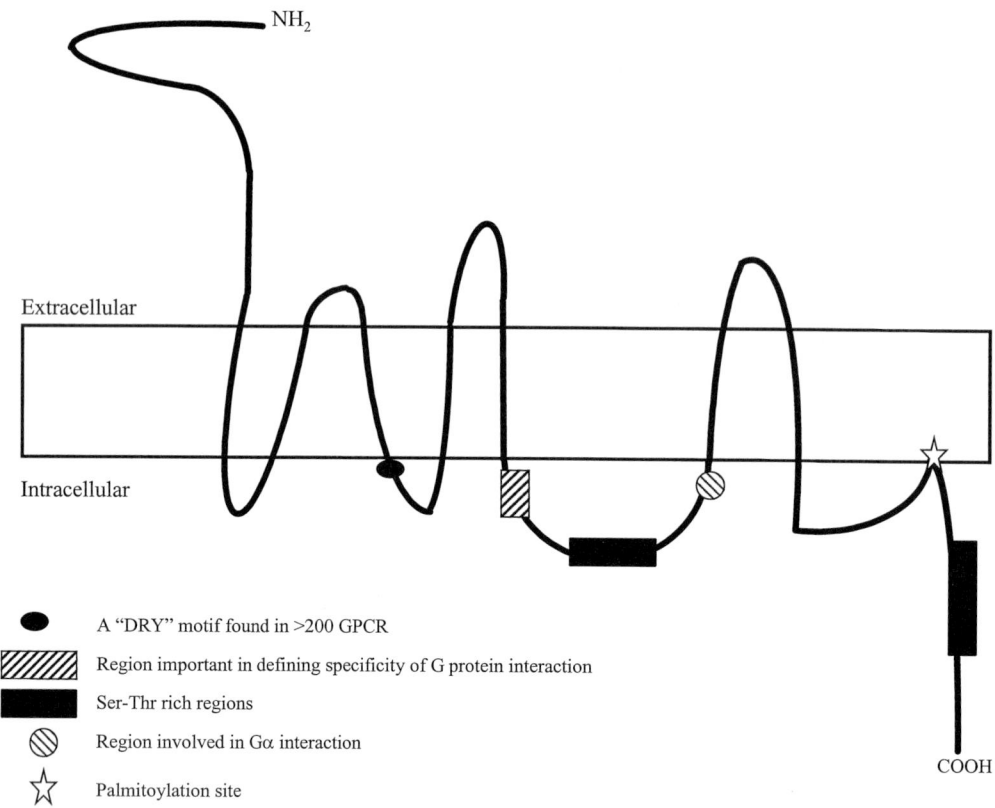

NH₂

Extracellular

Intracellular

● A "DRY" motif found in >200 GPCR

▨ Region important in defining specificity of G protein interaction

■ Ser-Thr rich regions

⊘ Region involved in Gα interaction

☆ Palmitoylation site

COOH

FIGURE 2   Structural elements of G protein-coupled receptors (modified from Birnbaumer, 1995).

forms expressed on normal cells. The low-affinity receptor, consisting of just the α chain, is not capable of signaling. The intermediate affinity receptor contains β and γ chains, whereas the high-affinity receptor, which is responsible for the major mitogenic responses of IL-2, consists of α, β, and γ subunits.

The majority of the cytokine receptors, including the single polypeptide chain receptors IL-1-R and IL-7-R, are also found in a soluble form in the serum. These soluble forms are generated either from proteolytic cleavage of the extracellular domain or from the alternative splicing of the mRNA encoding these receptors. The role of soluble receptors is quite complex, and they have been postulated to both enhance and inhibit the effects of cytokines. The circulating cytokine–receptor complexes have been suggested to interact with the common subunits on cells that normally do not express cytokine receptors, thus leading to an enhancement of cytokine action. On

the other hand, soluble receptors may also compete with the membrane-bound receptors by preventing the cytokines from reaching a sufficient concentration to generate biological effects.

Insulin (I) and insulin-like growth factor receptors are also included in this category. These receptors, however, are composed of four polypeptide chains linked by three disulfide bonds. Two of the subunits that span the membrane contain a tyrosine kinase domain, whereas subunits on the outside of the cell are mainly involved in ligand binding.

B. Seven Transmembrane-Spanning Receptors

The receptors in this class, also known as heptahelical G protein-coupled receptors (GPCR), share structural homology despite the chemically diverse range of their ligands. The overall structural characteristics of the GPCR family are highly conserved. In addition to the seven hydrophobic transmembrane helices, these receptors have an extracellular N-terminal domain, which can be quite variable in size, three intracellular and three extracellular connecting loops, and, with one exception, an intracellular C-terminal tail. Most, but not all, of these receptors also have a palmitoylated cysteine residue in the C-terminal tail that anchors to the membrane creating a fourth intracellular loop.

An analysis of GPCRs identified to date, encompassing over 250 receptors, shows that the majority of the primary sequence homology is found within the transmembrane helices, whereas the hydrophilic loop regions are more divergent. While the transmembrane core, extracellular loops, and the first two intracellular loops are similar in size, the length of the third intracellular loop and the N and C termini varies. For instance, the extracellular N-terminal domains of glycoprotein hormone receptors [luteinizing hormone receptor (LH-R), follicle-stimulating hormone receptor (FSH-R), and thyroid-stimulating hormone receptor (TSH-R)] and adrenergic receptors are about 350 and 35 amino acid residues long, respectively; the C-terminal domains of α1B adrenergic and gonadotropin hormone-releasing hormone are about 165 and 2 amino acid residues long, respectively.

Based on a comparison of amino acid sequences, the GPCRs have been classified into three subfamilies: the rhodopsin/β-adrenergic, secretin/vasointestinal peptide (VIP), and metabotropic glutamate (mGlu) receptors. The latter will be discussed under Section II,C.

1. The Rhodopsin/β-Adrenergic Receptor Subfamily

This group of receptors constitutes the vast majority of the GPCRs. The ligands range from small biologic amines, such as histamine, to small peptides, such as endothelin-1, and to larger hormones, such as erythropoietin and LH. Small molecules such as the endogenous catecholamines have been reported to bind within the transmembrane domains. The binding of peptide ligand and glycoprotein hormone receptors, however, involves both the extracellular and the transmembrane domains. Upon ligand binding, receptors undergo a conformational change leading to the activation of signaling mechanisms.

2. The Secretin/VIP Receptor Subfamily

The ligands for this subfamily include secretin, glucagon, growth hormone-releasing hormone, parathyroid hormone, and calcitonin. These receptors show little sequence homology to the rhodopsin/β-adrenergic receptor family, and all contain a hydrophobic signal sequence at the N terminal. A hydrophilic, cysteine-rich stretch of 115–160 amino acid residues is located between the signal sequence and the first transmembrane domain. This subfamily also contains potential glycosylation sites in the extracellular region and one highly conserved cysteine residue in each of the first two extracellular loops. Originally, these receptors were reported to couple to G_s leading to increased levels of cAMP, but recent findings indicate that they can signal through multiple pathways.

C. Ion Channels

Ion channels are composed of various combinations of either one or more protein subunits that associate to form a core in the membrane, and upon stimuli, also known as gating, they allow a rapid flow of ions across the membrane (Fig. 3). There are three subgroups: voltage-gated ion channels, such as Na^+ and K^+ channels; neurotransmitter-gated ion channels, including acetylcholine and γ-aminobutyric acid (GABA) receptors; and ligand-gated channels that are activated by Ca^{2+} ions and ATP. Since this article is mainly focused on hormone receptors, only neurotransmitter-gated ion channels will be discussed in greater detail.

Neurotransmitter-gated channels encompass the acetylcholine receptor, GABA and glycine receptor, and glutamate receptor subfamilies. It should be noted that GABA, 5-hydroxytryptamine (5-HT), and glutamate receptors can also produce much slower synaptic responses by acting via different receptors that are coupled to G proteins such as $GABA_B$, 5-HT_1, and 5-HT_4; these, however, do not contain intrinsic ion channels. Neurotransmitter-gated ion re-

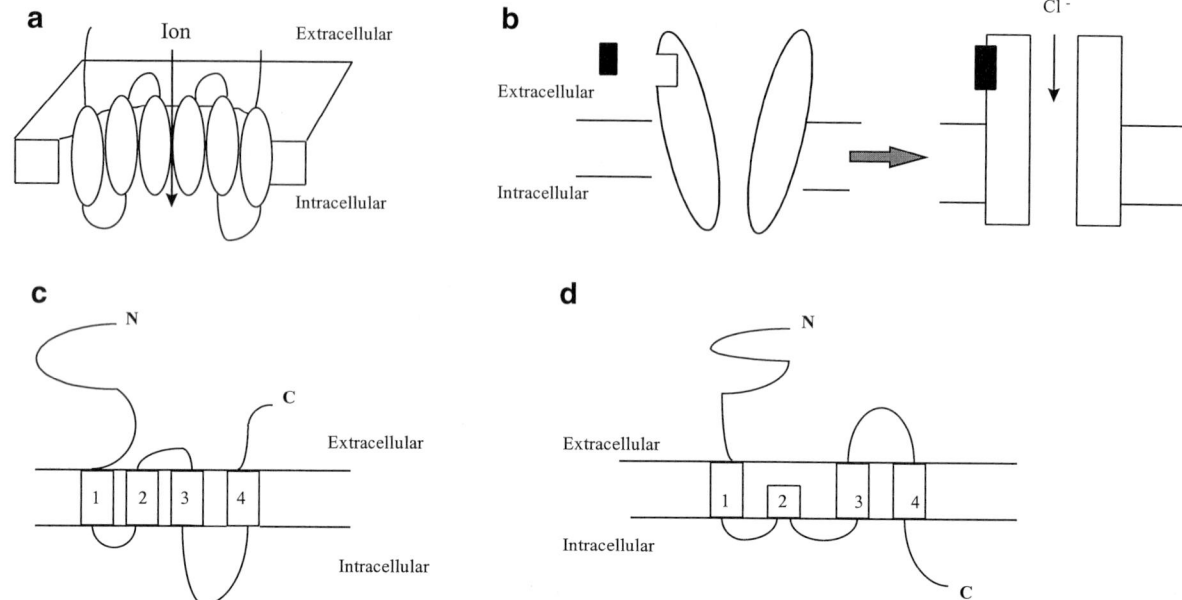

FIGURE 3 Neurotransmitter-gated ion channels: Schematic representations of a membrane-spanning ion channel on the cell membrane (a), gating of the channel upon ligand binding (b), and membrane topology of acetylcholine (c) and glutamate (d) receptors. Numbers 1–4 in (c) and (d) indicate transmembrane or partial transmembrane segments.

ceptors span the membrane four times, and each has a large extracellular N-terminal domain and a small extracellular C-terminal tail. In glutamate receptors, the C-terminal domain is intracellular.

1. Acetylcholine Receptors

Nicotinic acetylcholine receptors consist of five subunits termed α, β, γ, and δ;. In each receptor, there are two α chains, which are involved in ligand binding, and one each of the others. The membrane topology of these receptors is shown in Fig. 3. The muscarinic acetylcholine receptor, however, is a member of the GPCR family.

2. GABA and Glycine Receptors

GABA$_A$ receptors are widely distributed throughout the brain, whereas glycine receptors are mainly found in the brain stem and spinal cord. Both receptors regulate the opening of Cl$^-$-selective channels leading to rapid inhibitory transmission in the mammalian central nervous system. Similar to nicotinic acetylcholine receptors, they also consist of five subunits. The glycine receptors, for example, form a pentamer in the ratio $\alpha_3\beta_2$.

3. Glutamate Receptors

The ligand for these receptors is glutamate (Glu), which is the major neurotransmitter involved in fast excitatory transmission in the central nervous system. This type of transmission is important in some of the long-lasting changes in the synaptic efficiency which involve learning and memory. This family of receptors can be further divided into two subfamilies: (i) receptors with intrinsic ion channels, such as AMPA, kainate, and NMDA receptors, which are named according to the synthetic agonists that activate them, and (ii) metabotropic glutamate (mGlu) receptors which belong to the GPCR family. There are at least six closely related mGlu receptor subtypes reported to date and all are found in the central and peripheral nervous system. The members of this family are quite large in comparison to other GPRCs; mGlu-R1, for example, contains almost 1200 amino acid residues. Similar to the secretin/VIP receptor family, there are highly conserved cysteine residues in the extracellular domain. Receptors with intrinsic ion channels have a long extracellular domain and four transmembrane domains. Recent reports have

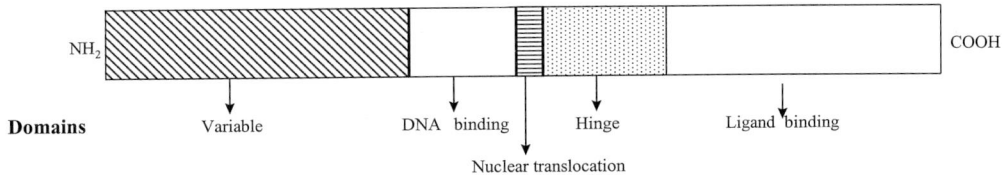

FIGURE 4 Schematic representation of major domains found in the intracellular hormone receptors.

shown that the second transmembrane domain does not span the membrane entirely (Fig. 3).

III. INTRACELLULAR RECEPTORS

Intracellular receptors, also termed the nuclear hormone receptor superfamily and steroid hormone receptors, are structurally related DNA-binding proteins that regulate transcription of target genes and involve receptors for glucocorticoids, androgens, mineralocortocoids, progestins, estrogens, thyroid hormones, vitamin D, retinoic acid, and orphan receptors which do not have a known ligand.

The members of this family are single polypeptide chains with three main domains as depicted in Fig. 4. The N-terminal domain is highly variable and contains transcriptional activation function that can be hormone independent. The DNA-binding domain contains two zinc finger elements and confers specificity for binding to hormone response elements founds on DNA. The ligand-binding domain determines the ligand specificity and has a number of functions involving receptor release from the heat shock complex, translocation to the nucleus, homodimerization, heterodimerization, and transcriptional activation. In addition to these three domains, there is also a small domain involved in nuclear translocation distal to the DNA-binding domain and a hinge region proximal to the ligand-binding domain.

The subcellular localization of intracellular receptors varies (Table 3). For example, thyroid hormone receptors are found exclusively in the nucleus, whereas the majority of the steroid hormone receptors partition between the cytoplasm and the nucleus. Unliganded intracellular receptors, with the exception of those for thyroid hormones and retinoic acid, form a complex with the heat shock proteins (hsp), including hsp90 and hsp70. The majority of receptor–hsp complexes are localized in the cytoplasm; however, estrogen receptor complexes are present in the nucleus. Hormone binding leads to the dissociation of receptors from these complexes, and the hormone–receptor complex then forms homodimers which bind to hormone response elements in the promoter regions of target genes.

TABLE 3
Intracellular Receptors

Hormone	Localization of free receptors (%)	Size (No. of amino acids)
Thyroid hormones	Nucleus (100)	395
Retinoic acid	Nucleus (95)	432
Estrogen	Nucleus (95)	595
Progesterone	Nucleus (95)	933
Androgen	Nucleus (90)	917
Vitamin D	Nucleus (75)	427
Aldosterone	Nucleus (40)	984
Glucocorticoids	Cytosol (90)	777

See Also the Following Articles

Bibliography

Aidley, D. J., and Stanfield, R. P. (1996). *Ion Channels: Molecules in Action.* Cambridge Univ. Press, Cambridge, UK.

Birnbaumer, M. (1995). Mutations and diseases of G protein coupled receptors. *J. Recept. Signal Transduct. Res.* **15**, 131–160.

Castellino, A. M., and Chao, M. V. (1996). Trans-signalling by cytokine and growth factor receptors. *Cytokine Growth Factor Rev.* **7**, 297–302.

Falus, A. (1995). Cytokine receptor architecture, structure and genetic assembly. *Immunol. Lett.* **44**, 221–223.

Hamblin, A. S. (1993). *Cytokines and Cytokine Receptors.* Oxford Univ. Press, New York.

Norman, A. W., and Litwack, G. (1997). *Hormones.* Academic Press, New York.

Riberio, R. C. J., Kushner, P. J., and Baxter, J. D. (1995). The nuclear hormone receptor gene superfamily. *Annu. Rev. Med.* **46**, 443–453.

Strader, C. D., Fong, T. M., Graziano, M. P., and Tota, M. R. (1995). The family of G-protein-coupled receptors. *FASEB J.* **9**, 745–754.

Hormones and Reproductive Behaviors, Fish

Kent D. Dunlap and Harold H. Zakon

University of Texas, Austin

GLOSSARY

electric organ discharge The electrical signal produced by modified neural and muscle structures in some fish. In some species, the discharge is sexually dimorphic and is used in gender recognition.

electrocytes The cells of the electric organ of electric fish which generate the electric organ discharge.

11-ketotestosterone The principal androgen regulating male reproductive behavior in fish.

lateral line organ A sensory organ in many fish that detects low-frequency vibrations in the water. It is sometimes used in communication during courtship.

sonic muscles Muscles found in some teleosts which contract at very high frequency and cause the swim bladder to vibrate. The vibrations produce a sound that is used as a courtship call.

Among their ~20,000 species, teleost fish display an enormous diversity of reproductive modes and mating behaviors. For example, some species change sex during their lifetime, some are simultaneously both male and female, and some are a single sex throughout their life. Some species mate with multiple partners in a single season, whereas others are monogamous. Many species have external fertilization, whereas others have internal fertilization. Underlying such diversity is also a remarkable variety of neuroendocrine mechanisms coordinating reproductive physiology and behavior. Fish do not show the common vertebrate pattern whereby behavioral patterns are organized by steroids early in life and activated by steroids at the initiation of the breeding season. Rather, fish have a diversity of hormone–behavior interactions that reflect particular mating systems. Thus, species range from those whose behavior is closely regulated by circulating hormones to those whose behavior is hormone independent.

I. INTRODUCTION

Despite the diversity of reproductive patterns, all forms of sexual reproduction in teleosts require communication between the sexes. At the very least, individuals must broadcast and perceive signals that convey information about species identity, gender, and reproductive state. Often, the same hormones (and other signals) which stimulate production of gametes and mating behaviors also act on the brain and peripheral organs to influence the generation and reception of reproductive signals.

During communication, "senders" produce signals, such as acoustic signals or visual displays, that pass through the environment to "receivers," which use specialized sense organs to collect and transduce the information. Most of us are familiar with how communication signals pass through the terrestrial environment, but the aquatic medium in which fish live transmits signals in unique and often nonintuitive ways. In addition, the structures that fish use to produce and perceive communication signals are often quite different than those used by terrestrial vertebrates.

In this article, we give an overview of hormonally regulated reproductive behaviors in fish, outline the major communication modes used by fish in reproductive signaling, and discuss how these signals are transmitted though the water medium. We then present two case studies which demonstrate how hormones influence some of the unusual communication modes used by fish in reproduction.

II. HORMONAL REGULATION OF REPRODUCTIVE BEHAVIOR

In fish, as in other vertebrates, gonadal steroids participate in the regulation of sexual behavior. However, the extent and exact nature of steroidal control differs widely among species and often correlates with reproductive mode or mating system.

A. Females

The relative importance of estrogen in stimulating female reproductive behavior appears to depend on the mode of fertilization. Internally fertilizing species, such as guppies (*Poecilia reticulata*), require the presence of an ovary to consistently express sexual receptivity. Ovariectomized females are sexually receptive for a few days, but they quickly lose their responsiveness to male courtship display; estrogen injections reinstate receptivity.

Estrogen is comparatively less important in externally fertilizing species such as goldfish (*Carassius auratus*). Although estrogen probably plays a role in priming the neural structures stimulating sexual behavior, cyclical fluctuations of estrogen associated with ovulation are not critical for activating female sexual activity. Instead, spawning behavior in goldfish is activated by prostaglandins that are released by the ovarian follicle and the reproductive tract at the time of ovulation. Injection of prostaglandin $F_{2\alpha}$ induces female sexual activity within minutes, even in ovariectomized females. Prostaglandin has greatest efficacy if delivered to the ventricles of the brain, suggesting that it acts directly on neural circuits controlling behavior. This tight coupling of spawning behavior to ovulation via prostaglandins helps to ensure that mating occurs at a time that is most conducive to fertilization. Moreover, prostaglandins released into the water by the female act as pheromones to stimulate male sexual behavior, ensuring that the males' behavior is coordinated with ovulation as well.

B. Males

Male reproductive behaviors are generally regulated by androgens. In many species, castration eliminates and androgen replacement restores a broad range of behaviors associated with reproduction. For example, in blue gouramis (*Trichogaster trichopterus*), androgens control prespawning behaviors such as nest building as well as mating behaviors per se. However, in some species, postspawning reproductive behaviors occur only when androgen levels are low. In garabaldi (*Hypsypops rubicundus*) and bluegill sunfish (*Lepomis marochirus*), androgens rise during courtship but fall precipitously during the egg brooding phases of the reproductive season. This negative relation between androgen levels and paternal care is analogous to that observed in several avian species.

11-ketotestosterone, rather than testosterone, is usually the most potent androgen in stimulating male sexual behavior. Testosterone levels are often as high or higher in females as in males, but 11-ketotestosterone levels are much higher in males, especially during the breeding season. The predominance of 11-ketotestosterone in male courtship behavior is also supported by species with multiple male morphs in which one morph actively courts females and the other does not (see Case Study I). In a variety of such species, both male types have similar testosterone levels, but courting males have higher levels of 11-ketotestosterone. Recent studies indicate there may be multiple androgen receptor forms in fish, one that has greater affinity for testosterone and another that preferentially binds 11-ketotestosterone and dihydrotestosterone. This may explain how several reproductive processes (e.g., spermiation and courtship behavior) can be differentially responsive to various androgen forms.

C. Bipotential Sexuality

Some fish have a remarkable potential to exhibit sexual behavior typical of either sex when treated with different hormones. Electric fish emit sexually dimorphic electric signals that are used in gender recognition (see Case Study II). The nature of their discharge depends heavily on their steroidal condition and not on their gonadal sex. In several species, treatment of intact or gonadectomized individuals of either sex with androgens shifts the discharge in the male-typical direction; treatment with estrogens shifts the discharge in the female-typical direction.

Such bipotential sexuality reaches an extreme form in the simultaneous display of male and female behavior exhibited by hormone-treated goldfish. Individuals treated with both 11-ketotestosterone and prostaglandin simultaneously exhibit female-typical courtship behaviors toward males and male-typical behaviors toward females. Considering this bipotentiality of sexual behavior in gonochoristic species (species in which individuals are either male or female throughout their life), it is easy to imagine evolutionary transitions between gonochorism and hermaphrodism (species which are simultaneously or sequentially both male and female).

D. Steroid-Independent Reproductive Behavior

Although gonadal steroids have potent effects on reproductive behavior in many fish species, gonads are not critical for the expression of sexual behavior in all species. Steroids typically exert their effects over the time course of hours to days, yet some fish exhibit changes in sexual behavior within seconds to minutes after the social environment changes. In the blue head wrasse, *Thalassoma bifasciatum*, females rapidly begin to show male-typical spawning displays within minutes after the dominant male is removed from the group. This drastic and rapid change in courtship behavior is displayed even in females that have been gonadectomized before the removal of the dominant male, indicating that gonadal steroids are not necessary for this behavioral change.

E. Behavioral Influences on the Endocrine System

It is important to recognize that in fish, as in other vertebrates, hormones and behavior act reciprocally. Behaviors affect hormones and vice versa. One well-studied example is the social regulation of neuroendocrine cells in a cichlid, *Haplochromis burtoni*. In this species, a small fraction of the males (~10%) are highly territorial and monopolize breeding opportunities. Compared to nonterritorial males, territorial males have higher androgen levels, larger testes, and brighter sexual coloration. They also have larger gonadotropin-releasing hormone (GnRH) neurons in the preoptic area of the hypothalamus. (These neurons regulate gonadal activity through their stimulation of gonadotropin secretion from the pituitary.) When a territorial male is removed, one of the nonterritorial males rapidly changes its behavior and coloration. After 4 weeks, the previously nonterritorial male will have testes and GnRH-containing neurons equivalent in size to those of other territorial males. Researchers hypothesized that behavioral interactions (i.e., aggression) are perceived by the senses and integrated by higher brain centers, which then send projections to the GnRH-containing neurons of the hypothalamus. These neurons then regulate the hormonal and gonadal state of the fish.

III. COMMUNICATION MODES IN FISH REPRODUCTIVE BEHAVIOR

A. Chemical Signaling

One common form of communication used by fish in reproductive behaviors occurs through chemical signals, also known as pheromones. Because chemicals do not diffuse very rapidly in water (compared to in air), chemical communication is typically limited to short distances. Unlike in terrestrial environments, chemical messengers in aquatic environments must dissolve in water. Thus, solubility rather than volatility characterizes waterborne chemical signals.

Chemical signaling may be the most direct form of communication because often the signaling molecules themselves are closely related to hormones (i.e., steroids and prostaglandins) that regulate the reproductive physiology of the sender. Both the production of pheromones and associated olfactory structures are sexually dimorphic and steroid sensitive in many species.

B. Acoustic Signaling

Many fish use sound to communicate their reproductive state to conspecifics. Sound travels much farther and faster in water than in air. In addition, because most objects (e.g., other animals and plants) in the water are relatively close to the same density

as water, they are almost "acoustically transparent" and do not greatly impede the transmission of acoustic signals. In deep water (>10 m), acoustic signals can travel as much as hundreds to thousands of kilometers with very little attenuation. Consequently, sound can be used as a long-distance form of communication to attract and court mates. Sound is also used for communication in shallow water, but interactions of sound waves with the substrate and air–water interface cause substantial refraction and absorption and thereby limit the range of frequencies that propagate effectively.

Fish use a diversity of structures to produce sound. For example, some species grind hard parts (e.g., pharyngeal teeth and spine-bearing fins) against each other to create sounds called stridulations. Other species employ various anatomical structures, which are sometimes sexually dimorphic and steroid sensitive, to cause the swim bladder to vibrate (see Case Study I). Many fish have remarkably sensitive hearing, usually with the greatest sensitivity in the mid-range of tones (200–1000 Hz). Some species, such as goldfish and other Ostariophysan fish, have bony structures (Weberian ossicles) that connect the inner ear with the swim bladder and, as a consequence, are sensitive to high-frequency sounds (up to ~6000 Hz).

C. Visual Signaling

Many fish have sexually dimorphic size, shape, coloration, and movement patterns that signal to conspecifics the sex and reproductive state of an individual. Such information is transmitted by light to the visual system of the receiver. Light is scattered and absorbed by water and suspended particles, and thus vision is rarely used for distances longer than 100 m, even in the clearest water. Moreover, water scatters and absorbs light in ways that depend on the color (wavelength) of light. Thus, what appears as a red eyespot on land may look very different to a fish 10 m underwater.

Visual signals can be permanent, seasonal, or highly dynamic. For example, some fish are permanently sexually dimorphic in size; some show bright sexual coloration or shape changes only during the breeding season, and some have very rapid color changes (within seconds) that signify the motivational state of the fish. Hormones participate in the generation of dimorphic color and morphology, even though such signals are expressed on widely different time scales. Androgens induce seasonal development of male nuptial coloration in several brightly colored species (e.g., sticklebacks, minnows, and gouramis). Also, brain regions which regulate rapid color changes contain sex steroid receptors and GnRH-containing neurons.

Visual signals are also conveyed through stereotyped body movements during courtship. These movements are often sex specific and frequently accentuate dimorphic body markings. For example, the male minnow performs a "waggle dance" in which his snout is pointed downward to expose his turquoise flank patches, which signify readiness to mate, and to hide his red belly, which signifies aggression.

D. Vibrational Signaling

Fish have mechanoreceptors (neuromasts) located in rows along the face and body axis. This sensory organ, the lateral line organ, is unique to fish and some amphibians, and it serves to sense low-frequency (10–200 Hz) vibrations in the water. Such vibrations dissipate rapidly in water and thus are only biologically relevant over short distances (up to 1 m).

Although the lateral line usually functions in the detection of movement produced by predators and prey, some species produce rhythmic movements during courtship that serve as communication signals. For example, male and female himé salmon, *Onocorhynchus nerta*, coordinate their spawning through an elaborate exchange of vibrational signals generated by rapid contractions (10–30 Hz) of the tail musculature. These movements may have originally served to eject gametes from the body cavity but were subsequently incorporated into the ritualized courtship sequence.

E. Electric Signaling

The aquatic environment conducts electricity much better than air. All fish produce small electrical fields from muscular activity associated with swimming, breathing, and heart contraction, and many fish species can sense the weak fields produced by their predators and prey. In two groups of freshwater teleost fish, one in Africa (Mormyriformes) and one in South America (Gymnotiformes), muscular and neural structures have been modified to produce highly stereotyped electric signals used in reproductive signaling. The range of electrocommunication signals is relatively short (~0.5–1 m), but this form of communication is effective in dark, murky, or noisy aquatic environments in which visual or acoustic signals are difficult to detect.

The brains of electric fish have pacemaker neurons which stimulate the activity of the electric organ located in the fish's tail. The electric field generated by the electric organ is perceived by other electric fish through specialized electroreceptors located on the fish's face and body. The pacemaker in the brain, the electric organ, and the electroreceptors are all sexually dimorphic and sensitive to sex steroids. Consequently, the electric discharge can be used to signal the gender and reproductive state of an individual.

F. Multiple Communication Modes

Although the previous discussion emphasizes discrete classes of communication signals, most fish probably use multiple communication modes at the same time or in different phases of reproductive behavior. For example, during courtship behavior, a male haddock initially rests on the substrate and displays a characteristic courtship visual signal by altering its pattern of pigmentation. When a female appears, the male emits a train of continuous sound pulses while approaching the female. When very close to the female, the male flicks his tail in broad strokes, producing waves that stimulate her lateral line. In several species it has been demonstrated that only a combination of such communication signals is sufficient to elicit the complete spawning sequence.

In some cases, a single courtship display stimulates multiple sensory modalities. For example, because pheromones diffuse slowly in water, fish frequently create waves that transport pheromones to the olfactory surfaces of their mate. This kind of wave likely also stimulates the receiver's lateral line. Similarly, some fish produce communication signals at ~50–

100 Hz which are perceptible by both the auditory and lateral line systems of the recipient fish.

Sometimes there is an inverse relationship between the use of various communication modes in reproductive signaling. Individuals of the mormyrid species, *Pollimyrus isidori*, have the capacity to produce both electric and acoustic signals, but, during courtship, they silence their electrical discharge when producing "grunt" and "moan" acoustic signals. A similar inverse relation is found interspecifically among cichlids: Species which display the greatest number of sound patterns have the fewest number of courtship color patterns.

IV. CASE STUDY I: VOCAL COMMUNICATION IN MIDSHIPMEN

Plainfin midshipmen, *Porichthys notatus*, use vocal communication to coordinate the sexual behavior of males and females. These fish, also known as California singing fish, live intertidally along the northwest coast of North America. They have a mating system that is relatively common among fish in which males exhibit two distinct phenotypes. One male type is typically large and actively courts females and defends territories, and the other male type is small and reproduces by dashing in and fertilizing the eggs of females that have been courted by a large male.

During the breeding season, one male type—termed the type I male—builds nests under rocks and emits acoustic hums that last as long as 15 min (Fig. 1). These hums attract females to the nest. When a gravid female enters the nest, the male stops humming. A female lays ~200 eggs and then leaves the nest, and the male guards the nest until hatching.

The sound produced by type I males is generated by muscles that attach to the swim bladder and contract very rapidly (~100 times/sec), causing the swim bladder to resonate like a drum. These sonic muscles

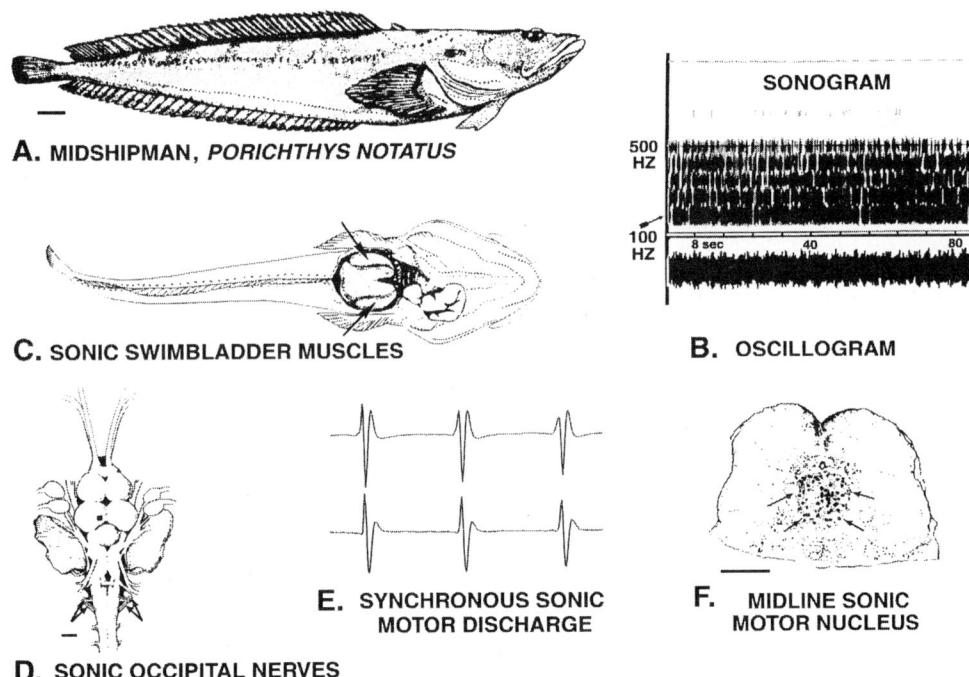

FIGURE 1 The sonic motor system of the midshipman. (A) Side view of a type I male midshipman (scale bar = 1 cm). (B) Vocalization of the type I male midshipman. The call is composed of a fundamental frequency at 100 Hz and a series of higher harmonic components. (C) The call is made as the male rapidly vibrates the swim bladder with a pair of attached muscles (arrows). (D) The brain of a midshipman. Arrows denote the nerves that innervate the swim bladder muscles. (E) At a fast time scale individual nerve volleys can be recorded from the occipital nerves. (F) The occipital nerves emanate from a midline nucleus located in the medulla called the sonic motor nucleus (scale bar = 500 μm). During breeding season the motoneurons and swim bladder muscles grow larger in type I males under the influence of 11-ketotestosterone (from Bass and Baker, 1990).

are stimulated by clusters of motoneurons in the brain, whose firing, in turn, is coordinated by another set of pacemaker neurons.

In contrast to type I males, type II males neither vocalize nor court females. Rather, they occupy the nests of type I males, wait for a female to enter the nest, and spawn rapidly when the eggs are laid. Associated with this difference in reproductive behavior are marked differences in the timing of sexual maturation, the morphology of the gonads and vocal apparatus, and the circulating level of androgens.

Both types show similar developmental pathways in brain regions that produce GnRH to initiate sexual maturation, but the timing of the hormonal cascade leading to maturation is different. As a consequence, type I males mature after 2 years, whereas type II males mature in ~7 months.

Because type II males begin sexual maturation earlier and show truncated somatic growth, they have much larger gonads (relative to body weight). However, type I males have larger vocal muscles and larger motor neurons and pacemaker neurons that stimulate vocal sounds. Experimental studies showed that androgens regulate the size and ultrastructure of vocal muscles, which suggested that these male types differ hormonally. Indeed, subsequent studies showed that, although both male types have similar levels of testosterone, type I males have much higher levels of 11-ketotestosterone than do type II males.

V. CASE STUDY 2: ELECTROCOMMUNICATION IN ELECTRIC FISH

The use of weakly electric field in reproductive signaling has evolved independently in two orders of freshwater teleosts: Gymnotiformes of South America and the Mormyriformes of Africa. Comparisons of species between and within these orders show that (i) analogous behaviors can be regulated by hormones in very different ways, and (ii) evolutionary changes in reproductive signals can be accompanied by changes in sensitivity to steroids.

Electric fish are generally active only at night, and in many cases, they inhabit waters that are muddy and/or shallow. Visual and acoustic information does

not travel effectively under these conditions. Throughout their life, these fish produce weak voltages (in the range of millivolts) that create an electric field around them. One principal function of the electric organ discharge (EOD) is to aid in localizing objects around them, similar to the way bats use sonar. During the breeding season, steroid hormones secreted from the maturing gonads modify the form of the EOD into a signal that communicates an individual's sex and reproductive condition.

The EOD can be characterized by its frequency and pulse duration. The frequency is determined by pacemaker neurons in the medulla of the brain (Fig. 2). These neurons connect onto spinal neurons, which then synapse onto electrocytes in the electric organ. The pulse duration is determined by electrical properties of the electrocytes. In African electric fish, such as *Brienomyrus*, or South American fish, such as *Hypopomus*, males and females differ in EOD pulse duration. Males have longer pulses than females, and androgens cause a broadening of the EOD pulse (Fig. 3). Androgens exert their effect by increasing the

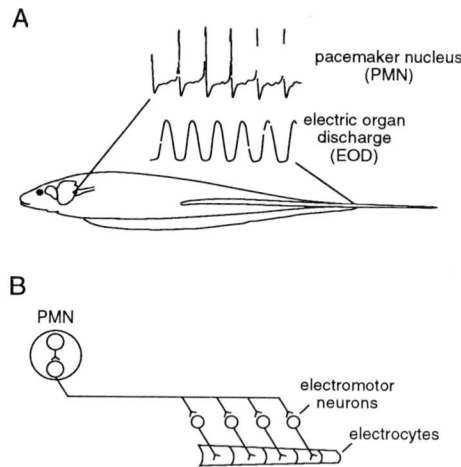

FIGURE 2   Control circuitry of the electric organ discharge in weakly electric fish. (A) Electrical recordings made from the neurons in the pacemaker nucleus (PMN) and the electric organ discharge (EOD) recorded with electrodes placed next to the fish. Note that every impulse in the pacemaker is followed by a discharge from the electric organ. (B) Schematic illustration of the circuit controlling the discharge. Pacemaker neurons within the PMN synapse on a second type of neuron that connects with electromotor neurons in the spinal cord. They synchronously activate the electrocytes, the cells of the electric organ, to produce each discharge pulse (from Zakon, 1987).

A

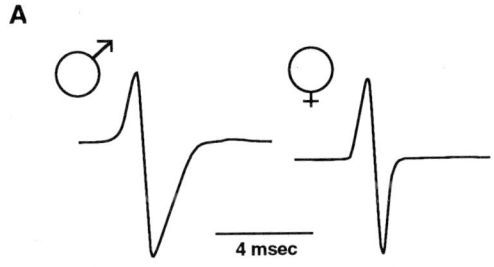

4 msec

B

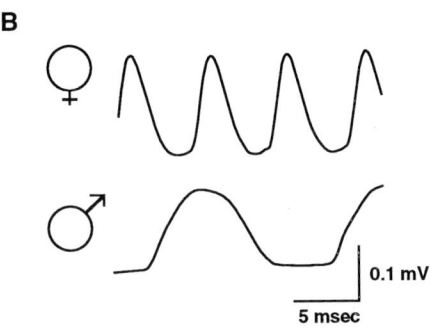

0.1 mV

5 msec

FIGURE 3 Sexually dimorphic discharges of South American electric fish. (A) In *Hypopomus*, both sexes produce pulses with a positive and negative phase, but the negative phase of the male discharge is of longer duration than that of the female. (B) In *Sternopygus*, the males discharge at a lower frequency than females. In both species treatment of fish with androgens causes the discharge to become more male-like (from Zakon, 1993).

membrane surface area in the electrocytes. This slows down the rate at which the electrocytes depolarize and thereby increases the duration of the pulse.

In South American electric fish that emit a sinusoidal discharge (e.g., *Sternopygus*), steroid hormones modify the activity of both the pacemaker neurons and electrocytes. Thus, sexual dimorphism is evident in both EOD frequency and pulse duration. Estrogen causes the frequency and pulse duration to shift in the female-typical direction; androgens cause a shift in the male-typical direction. Estrogens and androgens both have receptors in the electrocytes, and they induce their opposite effects on EOD pulse duration by acting in opposite directions on the rates at which the electrocyte sodium current turns on and off. Rather than affect electrocyte morphology, as in the African *Brienomyrus*, steroids appear to affect the electrical excitability of the *Sternopygus* electrocyte. This example demonstrates how evolution can pro-

duce convergent reproductive signaling systems through different cellular responses to steroids.

In addition to emitting a continuous EOD, many species in the genus *Apteronotus* produce rapid increases in frequency known as chirps. Chirps are exhibited only in social contexts, i.e., during courtship and aggression. In an ancestral species, *Apteronotus leptorhynchus*, chirping is both sexually dimorphic and androgen dependent. Males chirp much more than females, but if females are treated with androgen, they chirp more frequently. By contrast, in a more derived species, *A. albifrons*, chirping is neither sexually dimorphic nor inducible by androgens. Thus, it appears that evolutionary loss of sexual dimorphism in chirping was accompanied by, and perhaps caused by, a parallel loss in steroid sensitivity.

See Also the Following Article

Fish, Modes of Reproduction in

Bibliography

Bass, A. H. (1996). Shaping brain sexuality. *Am. Sci.* **84**, 352–363.

Bass, A. H., and Baker, R. (1990). Sexual dimorphisms in the vocal control system of a teleost fish: Morphology of physiologically identified neurons. *J. Neurobiol.* **21**, 1155–1168.

Francis, R. C, Soma, K., and Fernald, R. D. (1993). Social regulation of the brain–pituitary–gonadal axis. *Proc. Natl. Acad. Sci. USA* **90**, 7794–7798.

Hopkins, C. D. (1988). Social communication in the aquatic environment. In *Sensory Biology of Aquatic Vertebrates* (J. Atema, R. Fay, A. Popper, and W. Tavolga, Eds.), pp. 233–268. Springer-Verlag, Berlin.

Stacy, N. E. (1987). Roles of hormones and pheromones in fish reproductive behavior. In *Psychobiology of Reproductive Behavior: An Evolutionary Perspective* (D. Crews, Ed.), pp. 28–60. Prentice Hall, Englewood Cliffs, NJ.

Stacy, N. E., and Kobayashi, M. (1996). Androgen induction of male sexual behaviors in female goldfish. *Horm. Behav.* **30**, 434–445.

Zakon, H. H. (1987). Hormone-mediated plasticity in the electrosensory system of weakly electric fish. *Trends Neurosci.* **10**, 416–421.

Zakon, H. H. (1993). Weakly electric fish as model systems for studying long term actions of steroids on neural circuits. *Brain Behav. Evol.* **42**, 242–251.

Hormones of Pregnancy

Glen E. Hofmann

Bethesda Hospital

GLOSSARY

blastocyst The highly differentiated fertilized egg that will implant.

endometrium The lining of the uterine cavity where implantation of the fertilized egg takes place.

implantation The process by which the fertilized egg adheres to and invades the hormonally prepared endometrium.

protein hormone Any of the various hormones made from amino acids which usually but not always have sugar molecules attached. Protein hormones act by binding to receptors on the cell surface or within the cell binding to receptors in the membranes of intracellular organelles including the nucleus.

steroid hormone Any of the various hormones synthesized from the 27-carbon compound cholesterol. Steroid hormones elicit their biological response by binding to intracellular but mostly nuclear receptors.

Human pregnancy requires a magnificent orchestration of maternal hormones prior to implantation, maternal and embryonic communication prior to, during, and after implantation, and continued maternal–fetal communication during pregnancy up to the moment of the initiation of labor and finally delivery. The process by which all this communication is successfully achieved is one of the most remarkable events of nature and is the subject of this article.

I. PREIMPLANTATION MATERNAL–UTERINE COMMUNICATION

Prior to ovulation, fertilization of the human egg (oocyte), and the initiation of pregnancy (implantation), the uterus must be hormonally prepared for implantation. The hormonal requirements necessary to prepare the endometrium for implantation are elegantly defined by the ovum donation model. This clinical model uses eggs from young healthy egg donors which are fertilized by the sperm of the husband of the recipient woman. The recipient woman is often much older and incapable of efficiently producing healthy eggs of her own, or she may be frankly menopausal. The endometrium of the recipient woman is hormonally prepared for implantation by the administration of estradiol (E_2), either orally or transdermally, and progesterone (P), usually by intramuscular injection (Fig. 1). These hormones are normally produced by the ovary. When administered properly, these two hormones alone can prepare the endometrium to be receptive to implantation and the initiation of pregnancy. This simplicity does not deny the complex and poorly understood response of the endometrium to these two hormones. The response of the endometrium is a beautifully orchestrated proliferative response initially to E_2, followed by secretory transformation, differentiation, and decidualization of the endometrium in response to the addition of P to E_2 treatment. Additionally, in response to E_2 and P, the endometrium produces many hormones that act within the same cell of origin (intracrine hormone action), on the surface of the cell of origin (autocrine hormone action), and/or on neighboring cells (paracrine hormone action) in order to become prepared for implantation. These hormones are known as growth factors, oncogenes, and cytokines.

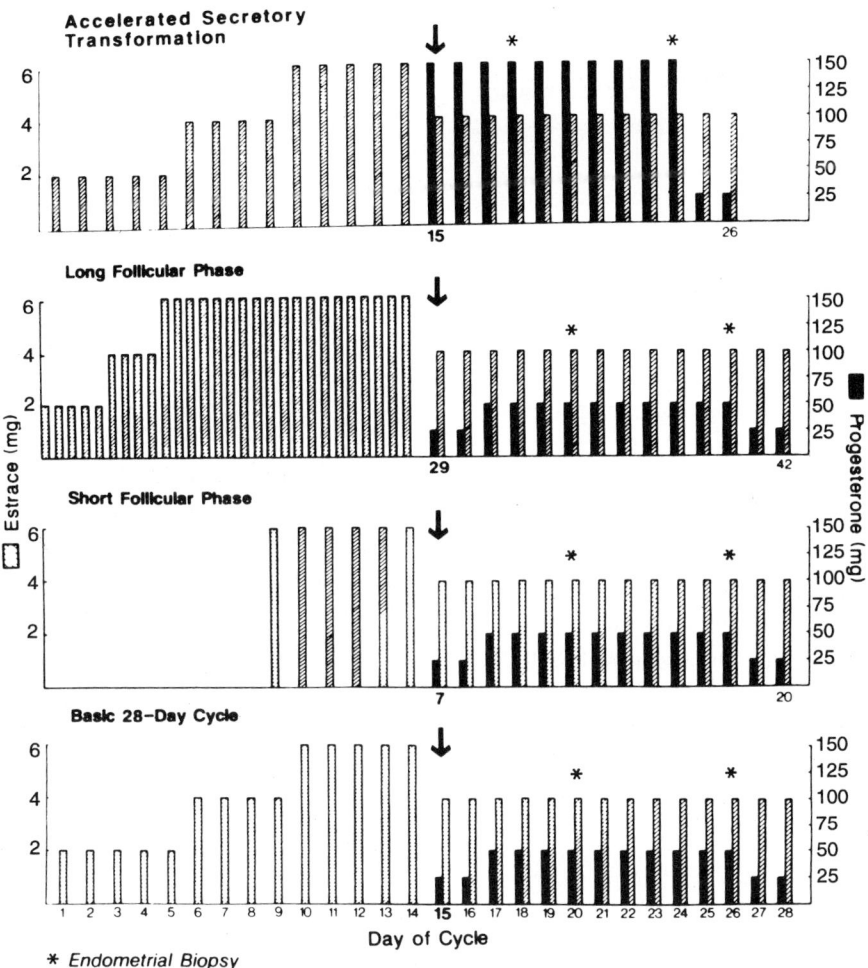

FIGURE 1 Various hormone replacement protocols using estradiol and progesterone in oil for endometrial proliferation and secretory transformation prior to embryo transfer in women undergoing ovum donation (reprinted with permission by The Endocrine Society from Navot *et al.*, 1989).

Additionally, a class of proteins known as integrins appears to be necessary for implantation, and a specific integrin is expressed only during the window of implantation when the endometrium is receptive to implantation. Further understanding of the role of various growth factors and integrins will be necessary for a complete understanding of the hormonal events required for implantation.

As the fertilized egg transverses the Fallopian tube on the way to the uterine cavity, it continues to grow and differentiate into a blastocyst. The blastocyst, now in the uterine cavity, prepares to come into apposition with the uterine lining and undergoes the process of hatching so that the blastocyst can escape the zona pellucida and begin the implantation process. As demonstrated by culturing human embryos during *in vitro* fertilization (IVF) procedures, the blastocyst starts to signal the endometrium of its presence prior to hatching. The first embryonic signal is the pregnancy hormone measured in all pregnancy tests known as human chorionic gonadotropin (hCG). The hCG is produced by the trophectoderm (the developing placenta) which surrounds the blastocyst and can be detected in IVF growth media prior to hatching of the blastocyst. In addition to the production of hCG, the trophectoderm produces multiple growth hormones (epidermal growth factor and its receptor, transforming growth factor-α and -β, and insulin-like growth factor-1 and its binding proteins) and multiple other cytokines and growth regulators.

Upon hatching, protein hormones and protein hormone receptors on the surface of the trophectoderm recognize their counterparts on the luminal surface of the endometrium, and the embryo attaches to and starts to invade the maternal decidualized (progesterone differentiated) stromal cell compartment of the endometrium. Multiple growth factors, proteases, angiogenic factors, and hemostatic factors are necessary for the controlled invasion of the decidua to allow the early placenta to breech the maternal vasculature to establish maternal–fetal communication.

During the time of embryo transport through the Fallopian tube and during and after implantation, the ovary continues to produce E_2 and P. In the absence of pregnancy, the production of these two hormones declines and, when they reach very low levels, menstruation begins. In the presence of a pregnancy, the hCG produced rescues the corpus luteum to ensure continued production of E_2 and P to support the pregnancy until the placenta is capable of adequate hormone production to support itself. The luteal–placental shift (the time when the corpus luteum is no longer necessary to support the pregnancy), originally thought to occur after the 12th week of pregnancy, was shown through the ovum donation model to occur much earlier. The developing placenta is now known to be capable of adequate hormone synthesis to support the pregnancy by the end of the 7th week of pregnancy.

Once implantation has occurred, maternal–fetal communication has been established. The placenta acts as a mechanism by which nutrients reach the fetus from the mother, and the fetal waste is transported to the maternal circulation. The placenta also acts a barrier to protect the fetus from many potential toxins (both chemical and infectious) and is a virtual factory for the synthesis of multiple hormones required to support the pregnancy and enhance fetal growth. The hormones are steroidal and nonsteroidal and protein or peptide in nature.

II. PLACENTAL PROTEIN HORMONES

A. hCG

The first identified and most well-known protein hormone of placental origin is hCG, produced by the syncytiotrophoblast and intermediate trophoblast. hCG is a dimeric glycoprotein composed of an α subunit, which it shares in common with follicle-stimulating hormone (FSH), luteinizing hormone (LH), and thyroid-stimulating hormone (TSH), and a β subunit, which is unique to hCG and makes it biologically and immunologically distinct from FSH, LH, and TSH. hCG, as well as the other protein hormones just mentioned, is glycosylated. The extent of glycosylation and the amount of sialic acid confers the bioactivity and half-life to the hormone. First detectable a few days prior to the missed period, hCG levels rise rapidly during the first 2 weeks of pregnancy, reaching maximal levels by 10–12 weeks gestation (Fig. 2). After this time, the levels slowly fall but do not reach nondetectable levels until after delivery. Early in pregnancy, hCG levels at least double every 48 hr, thus indicating a healthy pregnancy. The high hCG levels are believed to be responsible for the morning sickness in early pregnancy which sometimes leads to hyperemesis gravidarum. The only certain function of hCG is maintenance of the corpus luteum, but there are data suggesting a role for hCG in adrenal as well as gonadal development. In particular, it is believed but not proven that hCG stimulates androgen (male hormone) production by the fetal testicle, leading to masculinization of the male fetus. It is speculated that hCG may also stimulate the inner zone of the adrenal cortex, leading to adrenal androgen production. Additionally, biologically active receptors for hCG have been identified in the uterus, which may imply a role for hCG in uterine function during pregnancy, and in the brain, also suggesting a role for hCG in neuronal development.

hCG is produced abnormally in abnormal pregnancies. For example, in ectopic pregnancies, hCG levels rise slowly, indicating the abnormality of the pregnancy. Thus, the abnormal hCG levels lead to

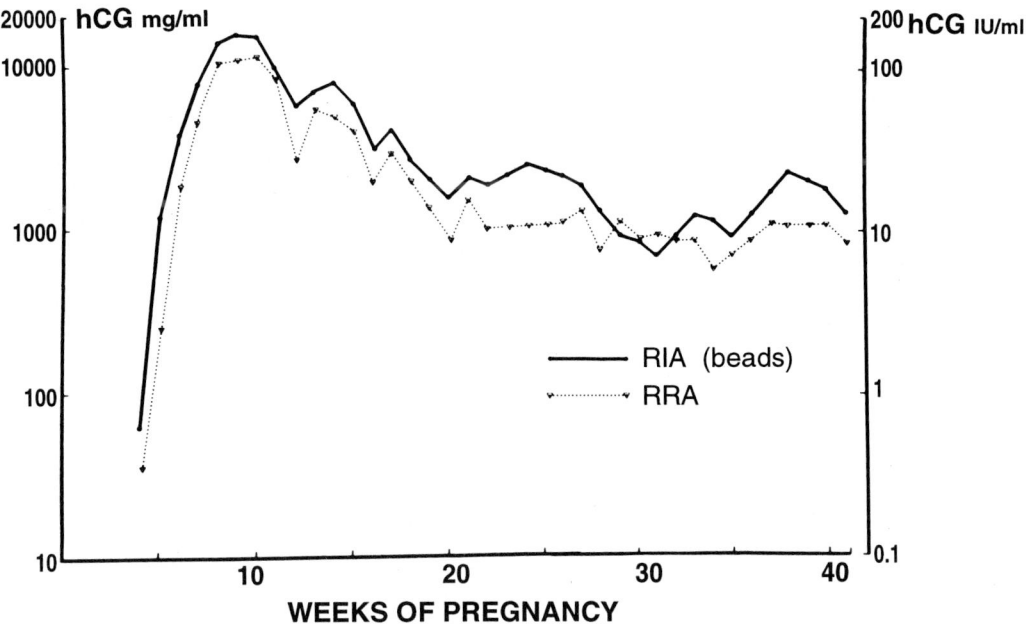

FIGURE 2 Serum concentrations of hCG during normal pregnancy (reprinted with permission from B. B. Saxena, Human chorionic gonadotropin, In *Endocrinology of Pregnancy* (F. Fuchs and A. Klopper, Eds.), 3rd ed., p. 53, Harper & Row, Philadelphia, 1983).

the diagnosis of a miscarriage or tubal pregnancy. In gestational trophoblastic disease (molar pregnancies), there is no fetus but rather an abnormal placenta, and the hCG levels are very high for the gestational age of the pregnancy. Thus, monitoring serum hCG levels can be used to document pregnancy and evaluate the normalcy (health) and location of the pregnancy, the presence of trophoblastic disease, and the fate of the abnormal pregnancy after appropriate diagnosis and treatment.

B. Human Placental Lactogen

The second major placental hormone is human placental lactogen (hPL; also known as human somatomammotropin). hPL is produced only by the syncytiotrophoblast. hPL is a single-chain polypeptide with ~85% sequence homology to human growth hormone (hGH) but has only ~3% of the somatotropic activity of hGH. hPL also has lactogenic activity in some assays, but its lactogenic role in humans remains doubtful. hPL serum levels are dependent on placental weight and are significantly elevated in

multiple births compared to singleton gestations (Fig. 3). The major role of hPL during pregnancy appears to be involved with mobilization and utilization of energy stores. Specifically, hPL mobilizes lipids in the form of free fatty acids, which are the body's greatest energy source. In normal circumstances, there is adequate glucose to meet most energy needs of the body. However, during pregnancy, there is a relative disruption of normal maternal glycemic control. This is due to two influences: first, both the mother and fetus require glucose as the major but not sole source of metabolic energy. Thus, there is a constant drain on the maternal circulating glucose levels giving rise to the relative hypoglycemia. Second, the placental hormones E_2 and P, but especially hPL, interfere with the action of maternal insulin. Thus, the placental hormones are diabetogenic. While the concentrations of all three of these hormones rise during pregnancy, hPL levels increase significantly during the second trimester of pregnancy and this is a major force in the propensity for pregnant women to demonstrate diabetes during pregnancy (gestational diabetes). This effect results

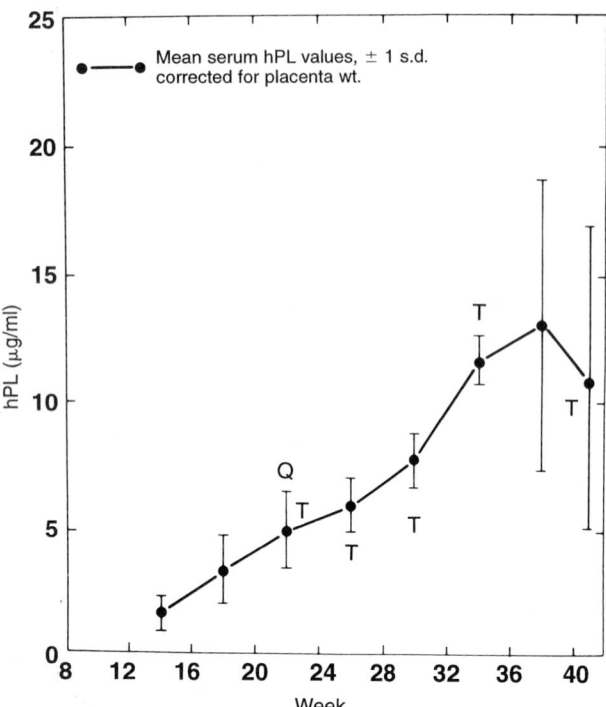

FIGURE 3 Serum concentrations of hPL during normal pregnancy after normalization for multiple gestations and placental weight at delivery (T, twins; Q, quintuplets) (reprinted with permission from J. B. Jasimovich, Placental lactogen and pituitary prolactin, in *Endocrinology of Pregnancy* (F. Fuchs and A. Klopper, Eds.), 3rd ed., p. 148, Harper & Row, Philadelphia, 1983).

in increased maternal insulin levels which are driven by a decreased cellular response to insulin caused by hPL. This mechanism provides for adequate levels of glucose to supply the fetus. In times of fasting, glucose levels decline, and in response hPL levels rise, causing lipolysis and the generation of free fatty acids. With prolonged fasting, the fatty acids generated in response to hPL are inadequate, and maternal fat and amino acids are mobilized and in certain instances result in a rise in maternal ketones. When this occurs, the ketones are detectable in maternal urine and signify the extent of the fasting state. As one might expect, hPL is also involved with regulation of maternal levels of triglycerides, cholesterol, phospholipids, and lipoproteins. Clinical uses of hPL for monitoring the health of a pregnancy have not been shown to be of value.

C. Human Chorionic Thyrotropin

Human chorionic thyrotropin (hCT) is a placental thyrotropic hormone similar to pituitary TSH. It is one of two thyrotropic substances of placental origin. However, the thyrotropic activity of hCT compared to TSH is very small. Some women with trophoblastic disease become severely hyperthyroid, and it has been postulated but not proven that hCT may play a role in hyperthyroidism associated with gestational trophoblastic disease.

Pregnancy is a time of relative hyperthyroidism. Maternal levels of thyroxine (T_4) are in the high normal range, and thyroid hormone replacement in hypothyroid mothers attempts to keep the T_4 in the high normal range. Most of the increase in maternal thyroid function is due to the increased metabolic needs of the mother and the high estrogen levels which increase thyroid binding globulin, decreasing bioavailable T_4. The fetus and placenta contribute little to the maternal thyroid state. There is very little placental transfer of any thyroid hormones in either direction. Thus, the maternal and fetal compartments are relatively isolated from each other, and, other than hCT, the placenta contributes little thyroid hormone function to either the mother or the fetus. Fetal TSH and T_4 appear in fetal circulation by 10 weeks gestation and rise dramatically by 20 weeks. Fetal TSH levels plateau at 28 weeks gestation and remain at these relatively high levels until term. Fetal T_4 levels also exceed maternal levels at term, giving rise to modest fetal hyperthyroidism at delivery. Fetal thyroid function returns to normal soon after parturition.

D. Prolactin

Prolactin is one of the major determinants of breast development in the pubertal female and plays a central role in breast physiology during pregnancy, preparing the breast to lactate. Maternal prolactin levels begin to rise immediately after the initiation of pregnancy and continue to rise during gestation, reaching levels five or six times greater than those in the nonpregnant state. However, the maternal pituitary is not the only source of prolactin during pregnancy.

The uterus and fetal pituitary are both capable of producing significant amounts of prolactin. The endometrium is a major source of prolactin during pregnancy, with the decidualized stromal cells producing large quantities of prolactin. While decidualization (the result of estradiol priming and progesterone secretory transformation) of the endometrium is required for prolactin synthesis, the control of prolactin production by the decidua during pregnancy is unknown. The decidua is the source of prolactin found in amniotic fluid. That the maternal and decidual compartments are regulated differently is demonstrated by the normalization of maternal serum prolactin levels in hyperprolactinemic women by the medication bromocriptine, which has no effect on amniotic fluid levels of prolactin. Amniotic fluid levels of prolactin parallel maternal levels until about the 10th week, rise dramatically until the 20th week, and then decrease until term. Decidual prolactin may be important in fluid and electrolyte regulation in amniotic fluid.

E. Corticotropin-Releasing Hormone

Human corticotropin-releasing hormone (CRH) (factor) is a 41-amino acid hypothalamic peptide that stimulates the pituitary corticotropes to release ACTH and other peptides derived from the proopiomelanocortin (POMC) peptide. ACTH, in addition to the other peptides released (most notably β-endorphin, β-lipotrophic hormone, and α-melanocyte-stimulating hormone from the common precursor, POMC), stimulates the adrenal gland initiating corticosteroid synthesis. A placental CRH has been identified that is similar in size and biological activity and immunologically identical to hypothalamic CRH. Message and protein product for CRH has been identified in placenta (cytotrophoblasts more than syncytiotrophoblasts), amnion, and decidua, and in the fetal circulation. However, CRH gene regulation in the placenta appears to be regulated differently from that of the hypothalamus. Specifically, hypothalamic mRNA for CRH is low and negatively modulated by glucocorticoids, whereas mRNA for placental CRH increases during pregnancy and is stimulated by glucocorticoids. Maternal concentrations of CRH are low during early pregnancy, rise significantly in midpregnancy, and rise further to term, falling precipitously after delivery (maternal levels are independent of whether the mother is in labor or not or the route of delivery), suggesting that the maternal CRH is of placental origin. Amniotic fluid concentrations of CRH also rise throughout pregnancy, and CRH can be identified in fetal blood at cordocentesis or at delivery. Binding proteins for CRH (CRH-BP) have also been identified and studied. These specific BPs are elevated in amniotic fluid in midpregnancy but decline toward term as amniotic fluid levels of CRH continue to rise. Thus, at term, the amount of CRH that is biologically available increases as does the relative amount present in amniotic fluid. The elevated CRH in maternal serum stimulates maternal ACTH release, which in turns causes increased maternal glucocorticoid synthesis. The role of CRH in the fetus is less well defined. It is known that the fetal pituitary can respond to CRH as early as the 14th week of gestation. The fetal adrenal response to the ACTH released from the fetal pituitary may be responsible for the development and maintenance of the fetal zone of the adrenal gland. Later in pregnancy, the role of CRH may be to stimulate the fetal adrenal gland to produce the glucocorticoids necessary for organ maturation, particularly the fetal lung production of surfactant required for normal lung function after birth. Finally, alterations of CRH concentrations and its associated BPs have been associated with pregnancy-induced hypertension. Further research is required to fully understand the mechanism(s) of action of CRH and CRH-BPs.

III. PLACENTAL STEROID HORMONES

The placenta, working with the mother and fetus, is involved in the production of multiple steroid hormones during pregnancy. Steroid hormones are of gonadal (both maternal and fetal), placental, and adrenal (both maternal and fetal) origin. The role of E_2 and P in the proliferation and secretory transformation of the endometrium in anticipation of the initiation of pregnancy has already been discussed.

The placenta is capable of synthesizing many other steroids independent of the mother and/or fetus, but in many cases precursor hormones are shuttled back and forth to complete hormone synthesis.

Steroid hormones are all derivatives of the 27-carbon compound cholesterol (Fig. 4). Cholesterol is carried to the placenta via protein carriers know as lipoproteins. The levels of these proteins are regulated by genetics, diet, and metabolic need. These carrier proteins are required to deliver the precursor cholesterol to all tissues. The placenta has specific receptors for the major delivery protein known as low-density lipoprotein (LDL). Upon binding to its receptor, the LDL molecule, which contains protein, lipids, and cholesterol, is internalized into the cell, and through multiple, highly orchestrated enzymatic reactions, cholesterol is released from LDL and becomes available for steroid hormone synthesis. When discussing steroid hormone synthesis, it is helpful to visualize the process in terms of compartmentalization into maternal, fetal, and placental areas (Fig. 5). For structures of the individual steroid hormones, the reader should refer to Fig. 4.

Progesterone is synthesized by the placenta by the seventh week of pregnancy. Prior to 7 weeks, the corpus luteum is responsible for the majority of progesterone synthesis. Progesterone is required for the maintenance of the pregnancy but also serves as a critical precursor for the synthesis of fetal adrenal gluco- and mineral corticosteroids and gonadal steroids. The fetus cannot synthesize the important corticosteroids directly from cholesterol, as can the mother, because the fetal adrenal gland lacks a necessary enzyme (3β-hydroxysteroid dehydrogenase, β-4,5-isomerase) required for the synthesis of progesterone. Thus, the placental progesterone is

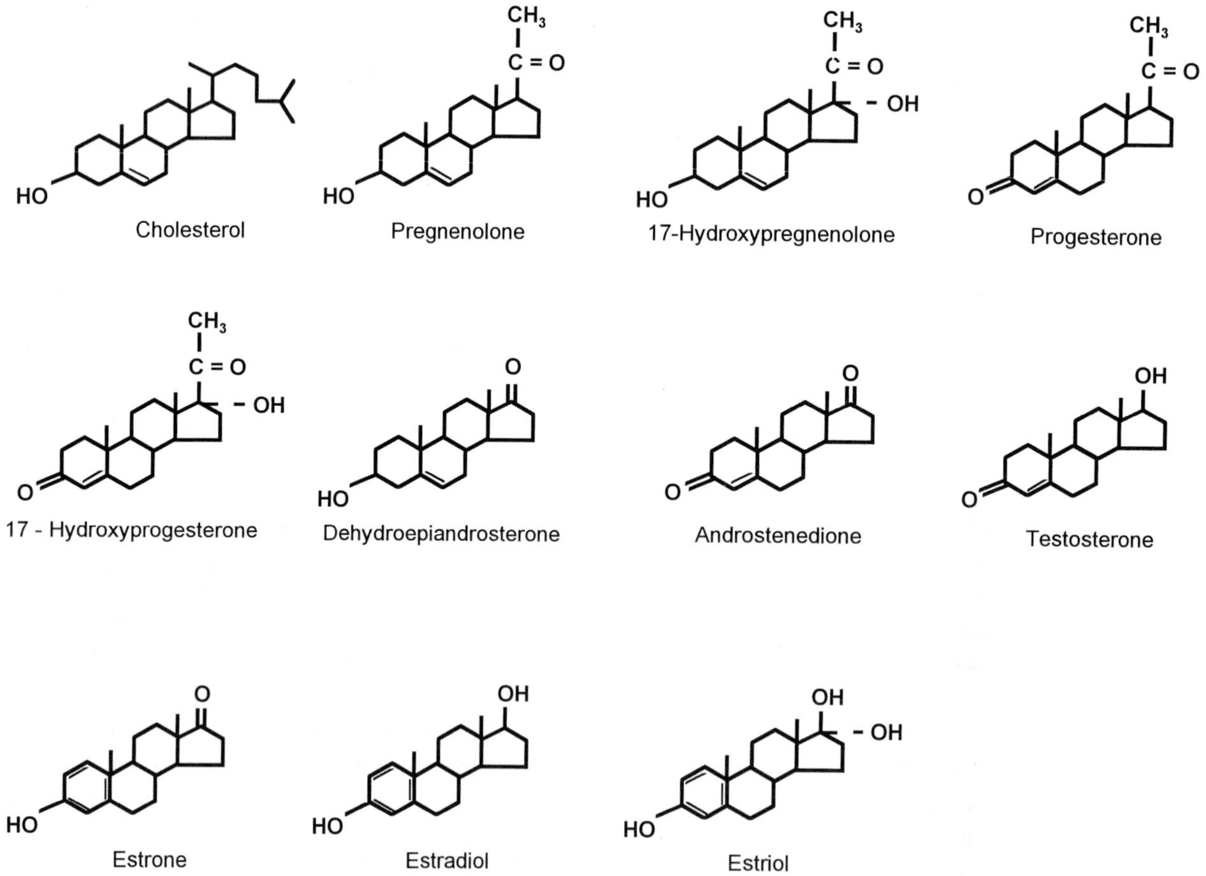

FIGURE 4   Chemical structures of steroid hormones synthesized by the maternal–placental–fetal unit.

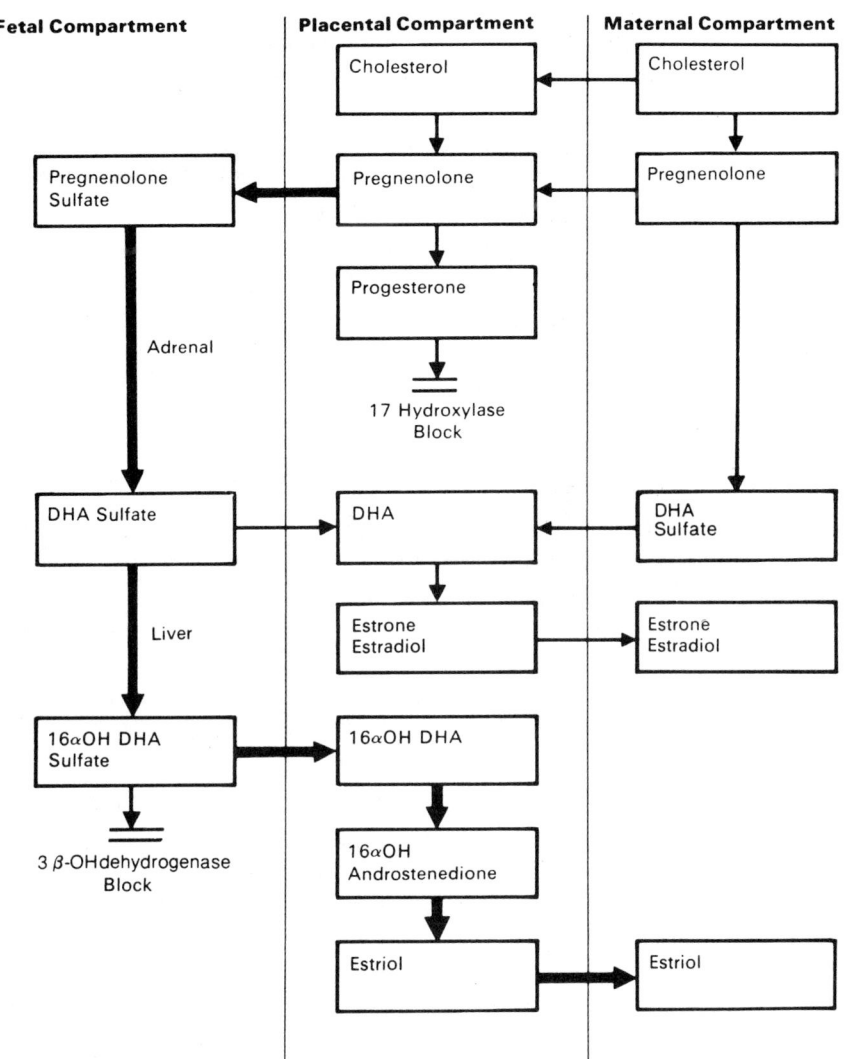

FIGURE 5 Compartmentalization of steroid hormone synthesis by the maternal–placental–fetal unit (reprinted with permission from Speroff *et al.,* p. 322).

transported to the fetus and then converted into the corticosteroids by the fetal adrenal gland. Therefore, the progesterone produced by the placenta is critical because it is transported to the fetus to serve as the precursor for fetal gluco- and mineral corticosteroid synthesis and through direct synthesis gonadal steroids by the fetal gonads.

Levels of progesterone rise throughout pregnancy (Fig. 6). The exact physiologic role for progesterone in later pregnancy is unknown, but progesterone is thought to play a role in suppressing the maternal immune system to allow placental invasion of mater-

nal tissues without rejection and to maintain uterine quiescence prior to the initiation of labor.

The levels of estrogens rise during pregnancy (Fig. 7). There are three different estrogens: estrone, estradiol, and estriol. While differing only in the number of hydroxyl groups on the steroid molecule and placement of double bonds (Fig. 4), the various estrogens differ significantly in biologic activity, with estradiol being the most biologically active. The precursors of all estrogens are the 19-carbon androgens, androstenedione and testosterone. The maternal and fetal (if the fetus is a female) ovaries are capable of

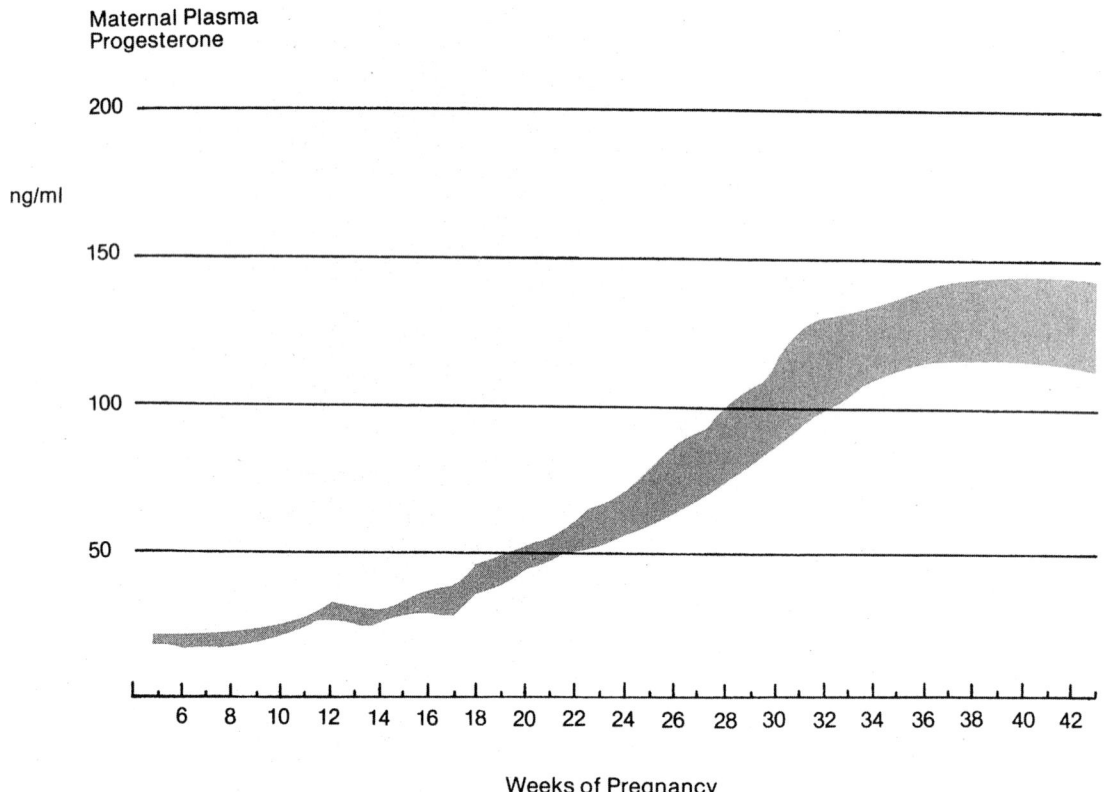

Maternal Plasma
Progesterone

FIGURE 6   Maternal plasma concentrations of progesterone during pregnancy (reprinted with permission from Speroff *et al.*, p. 319).

synthesizing estradiol *de novo* from cholesterol. The placenta, however, cannot because it lacks an enzyme (17-hydroxylase) critical for metabolizing progesterone to estrogens. To produce estrogens, the placenta shuttles progesterone to the fetus where the fetal adrenal gland metabolizes the progesterone to dehydroepiandrosterone sulfate (DHAS). However, the placenta cannot metabolize the sulfated steroid. Thus, the fetus must remove the sulfate prior to transport of the resulting DHA to the placenta. The ability of the fetus to conjugate biologically potent steroids is an important and critical function and may act as a protective mechanism for the fetus against high concentrations of highly biologically active steroids. The DHA produced by the fetus then diffuses to the placenta, where it is converted into androstenedione and eventually into estradiol, estrone, and estriol. Because maternal tissues synthesize estradiol and estrone normally in the pregnant and nonpregnant states, their concentrations during

pregnancy have not been useful to determine fetal well-being. However, maternal synthesis of estriol in the nonpregnant state is minimal, and the levels of maternal circulating estriol have been used extensively in the past to determine fetal–placental well-being, especially in diabetic pregnancies. This is because the synthesis of estriol requires the normal functioning of the placenta and fetal adrenal gland and normal placental transport. Hence, normal levels of estriol were reassuring of normal fetal–placental function. This method of screening pregnancies for fetal health has been replaced by monitoring of fetal heart rates (nonstress tests and oxytocin challenge tests) and ultrasound monitoring (biophysical profile).

Estradiol is required for endometrial preparation prior to implantation. Throughout pregnancy, however, the role of estrogen in placental function and uterine physiology is less well-known. However, estrogens, particularly estradiol and estrone, are capa-

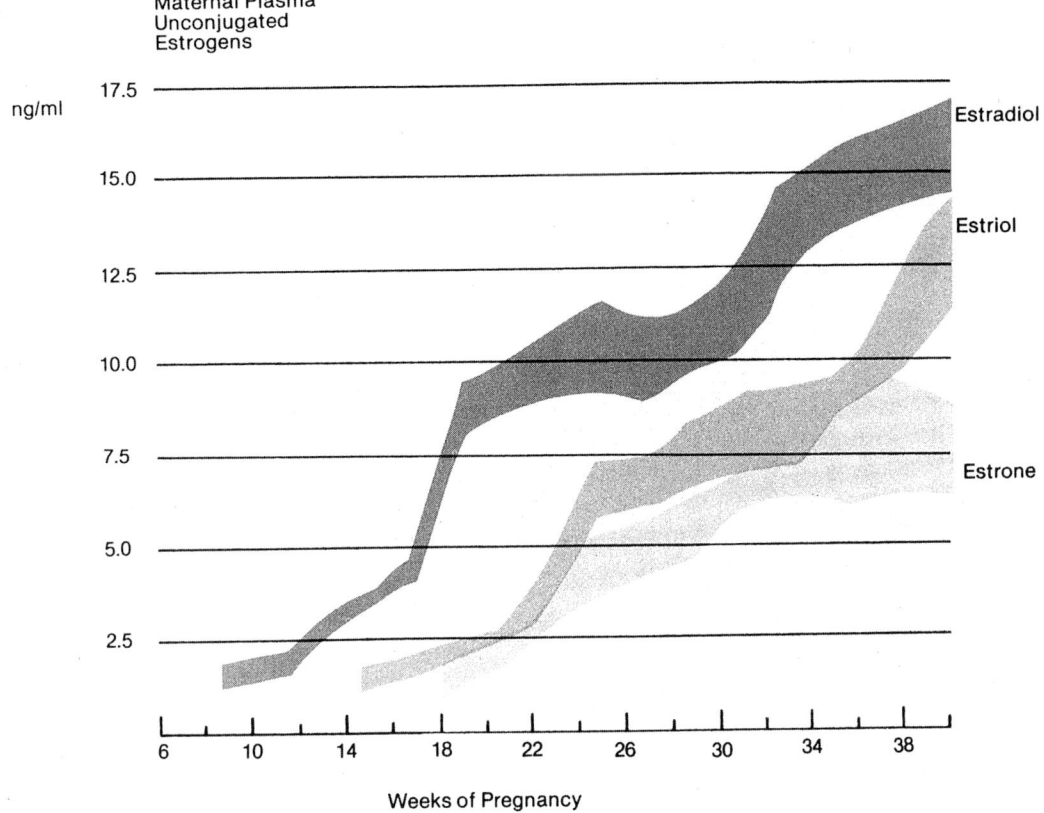

**Maternal Plasma
Unconjugated
Estrogens**

FIGURE 7   Maternal plasma concentrations of unconjugated estrogens throughout pregnancy (reprinted with permission from Speroff *et al.*, p. 323).

ble of influencing fetal adrenal physiology. The fetal adrenal is exposed to a high-estrogen environment. Indeed, it is believed that it is the high-estrogen tone and not the pituitary peptides (ACTH) that is responsible for fetal adrenal physiology. Estrogens at high concentrations (10–100 ng/ml) are capable of inhibiting the adrenal enzyme 3β-hydroxysteroid dehydrogenase isomerase and, in the presence of ACTH, enhance secretion of DHA which is required for placental estrogen synthesis. At these high estrogen levels, which are easily obtained during pregnancy, cortisol production is inhibited, giving rise to the high levels of ACTH which result in the hyperplasia of the adrenal cortex. With delivery and normalization of the estrogen levels, the fetal adrenal cortex quickly changes to the normal adrenal gland found in adult life. Thus, the fetal adrenal glands major function may be to provide adequate DHAS for placental estrogen production with the resulting

estrogen feeding back on the fetal adrenal to ensure an adequate supply of DHAS. No other function for DHAS in pregnancy is currently known.

IV. OTHER PREGNANCY-RELATED FACTORS

There are many other peptides and hormones that are known to change concentrations during pregnancy, particularly the renin–angiotensin system, growth factors, oncogenes and cytokines, and the family of insulin-like growth factors (I and II) and their carrier proteins. There is an increasing interest in the family of related peptides that include the transforming growth factor-β family, inhibins, and activin. Their roles, if any, in pregnancy and fetal growth and development are poorly defined and are the subject of intensive research.

V. SUMMARY

Pregnancy involves the elegant communication between maternal, placental, and fetal compartments from the moment of implantation to the onset of labor, requiring the careful orchestration of steroid, nonsteroid, and protein hormones. Clearly, the fetal and placental compartments are intimately involved in the initiation of labor, culminating in parturition. The mechanisms by which the onset of labor is controlled are beyond the scope of this article.

See Also the Following Articles

BLASTOCYST; ENDOMETRIUM; HUMAN PLACENTAL LACTOGEN; PLACENTAL AND DECIDUAL PROTEIN HORMONES; STEROID HORMONES

Bibliography

Bergh, P. A., and Navot, D. (1992). The impact of embryonic development and endometrial maturity on the timing of implantation. *Fertil. Steril.* **58**, 537–542.

Fuchs, F., and Klopper, A. (1983). *Endocrinology of Pregnancy*, 3rd ed. Harper & Row, Philadelphia.

Giudice, L. C. (Ed.) (1996). Growth factors and reproduction. *Sem. Reprod. Endocrinol.* **14**, 179–282.

Giudice, L. C., and Polan, M. L. (Eds.) (1995). Growth factors in the endometrium. *Sem. Reprod. Endocrinol.* **13**, 93–171.

Navot, D., Anderson, T. A., Droesch, K., Scott, R. T., Kreiner, D., and Rosenwaks, Z. (1989). Hormonal manipulation of endometrial maturation. *J. Clin. Endocrinol. Metab.* **68**, 801–807.

Scott, R. T., Navot, D., Liu, H.-C., and Rosenwaks, Z. (1991). A human in vivo model for the luteal–placental shift. *Fertil. Steril.* **56**, 481–484.

Speroff, L. (Ed.) (1992). Pregnancy proteins: Basic concepts and clinical applications. *Sem. Reprod. Endocrinol.* **10**, 61–182.

Speroff, L., Glass, R. H., and Kase, N. G. (Eds.) *Clinical Gynecologic Endocrinology and Infertility*, 4th ed. Williams & Wilkins, Baltimore.

Sueoka, K., Shiokawa, S., Miyazaki, T., Kuji, N., Tanaka, M., and Yoshimura, Y. (1997). Integrins and reproductive physiology: Expression and modulation in fertilization, embryogenesis and implantation. *Fertil. Steril.* **67**, 799–811.

Horses

Dan C. Sharp
University of Florida

GLOSSARY

capsule An acellular covering of entire equine conceptus from blastocyst stage to Days 21–25. The capsule consists of mucin-like glycoproteins, including sialic acid. The function of the capsule is debated but it likely provides structural integrity for the spherical equine conceptus and is postulated to act as a shield against bacterial and viral invasion.

chorionic girdle An avascular band of cells circling the spherical conceptus in the area between the advancing allantoic cavity and receding vitelline cavity. The trophectoderm of the chorionic girdle is underlain by exocoelom but is at the juncture of the developing allantochorion which contains a third tissue layer, the mesoderm. It is postulated that mitogens from the allantochorionic mesoderm participate in the development of this unique tissue. Cells of the chorionic girdle contain the gene for production of equine chorionic gonadotropin and can produce equine chorionic gonadotropin *in vitro* prior to endometrial invasion.

endometrial cups Specialized and apparently unique structures usually found in the caudal portion of the gravid uterine horn. The endometrial cups are placental in origin, derived from specialized invasive cells of the chorionic girdle and colonizing endometrial epithelia beginning about Days 35–40 of pregnancy. Because of the circular structure of the girdle, endometrial cups are found in a ring or horseshoe shape. Endometrial cups are the source of equine chorionic gonadotropin.

episodic secretion Characterized by a series of notable secretory events, not necessarily regular in occurrence or predictable based on observation of preceding events.

microcotyledon A placental exchange unit created by interdigitation of trophoblastic microvilli and apposing endometrial epithelium. Its structural appearance resembles ruminant placentomes, except that equine microcotyledons are considerably smaller (≈ 1 mm). The microcotyledons develop gradually from an ill-defined initial attachment beginning around Day 40 and not considered complete until Day 150.

ovulation fossa The cavity formed following inversion of ovarian cortical and medullary tissues embryologically. The ovulation fossa is located on the ventral surface of the ovary and is the only location of surface germinal epithelium. The presence of the germinal epithelium at the fossa restricts follicular rupture to that site.

pulsatile secretion Characterized by a series of regularly recurring, predictable secretory events. Observation of a given sequence of secretory events would permit prediction of the next sequential event.

seasonality The exhibition of periodic reproductive activity and inactivity associated with environmental factors that vary seasonally.

yolk sac The first functional placenta formed by the equine conceptus; consists of outer trophectoderm and a tightly adhered inner layer of endoderm. The endoderm emanates from the area of the inner cell mass and develops in an "abembryonic" direction, completing the inner layer between Days 10 and 12.

Reproductive patterns in Equidae often represent a fascinating departure from established dogma as learned from studies of other domestic, laboratory, or companion animals. The reasons for these departures are not known, but study of reproduction in Equids tends to provide a healthy open-minded attitude toward reproductive biology generally. The reader of this article is urged to consider that the reproductive challenges faced by most species are reasonably similar, but the evolutionary approaches to surmounting those challenges provide a fascinating glimpse at biological diversity, illustrated often by horses.

I. HORSES AS SEASONAL BREEDERS

The fossil record provides ample evidence that horses evolved in a temperate environment, characterized by distinct seasons of alternating warmth and cold, driven by annual changes in day length and by solar radiation. As with most species that evolved in such a climate, survival advantage for evolving Equidae lay in timing parturition at a time of year that would best provide appropriate temperatures and nutrition for the young to survive, i.e., spring and summer. It is important to remember that the key to seasonal breeding patterns is the season for birthing, not the season for mating. Therefore, the breeding season should be timed a gestation length ahead of parturition. Because mares have an approximately 11 month gestation, the breeding season also takes place during the spring and summer. The environmental factor capable of such precise timing in temperate climates is photoperiod. When one considers that the annual reproductive cycle must be timed so far in advance, it seems unlikely that changes in temperature, rainfall, vegetative growth, or other more acutely regulated factors would provide the long-term timing required to satisfy the need for regulating foaling.

A. Mares

The annual reproductive cycle consists primarily of four phases or periods: (i) anestrus, (ii) vernal transition, (iii) the breeding season, and (iv) autumnal transition (Fig. 1). Timing and precision of these stages vary, and the anestrus and vernal transition stages likely have the greatest economic impact. These four stages are characterized by the whole range of reproductive functions from reproductive

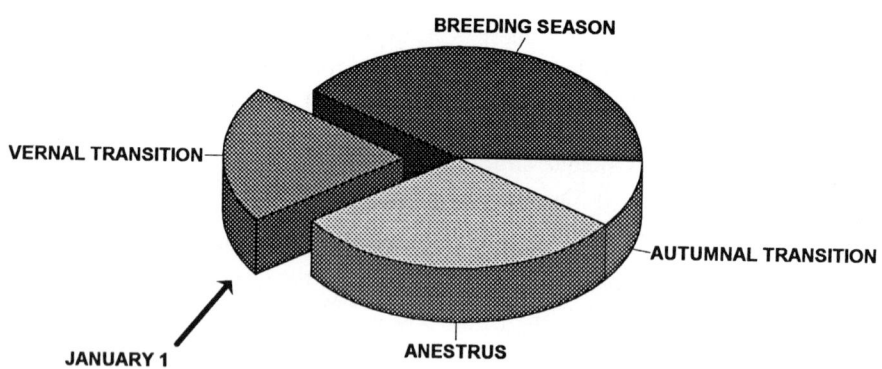

FIGURE 1 Phases of the annual reproductive cycle of mares in the Northern Hemisphere. The vernal transition is accentuated because of its economic and research importance.

competence (the breeding season) to complete reproductive quiescence (anestrus) with the two transitional stages reflecting the processes of reproductive renewal (vernal transition) or decline (autumnal transition). The months shown for the four stages are approximations only because there is extreme individual variation during certain stages (e.g., autumnal transition), and these times should be viewed as guidelines only.

1. Anestrus

Anestrus (November–early January in the Northern Hemisphere) is characterized by almost complete reproductive quiescence, reflecting drastically reduced hypothalamic gonadotropin-releasing hormone (GnRH) content and secretion and reduced pituitary gonadotropin [luteinizing hormone (LH) and follicle-stimulating hormone (FSH)]. As a result of loss of gonadotropin secretion, ovaries of mares become inactive and atrophic morphologically. There are few ovarian follicles larger than 5–7 mm in diameter and certainly no corpora lutea. The lack of ovarian activity results in cessation of sex steroid production, and estrogen and progesterone are essentially undetectable during anestrus. Behaviorally, mares often become indifferently tolerant to a stallion's advances, apparently lacking the guiding action of sex steroids. There is a marked pelage change during anestrus in mares that has a close association with reproductive function. During anestrus, the hair coat becomes long and firmly attached. Shedding of the long hair coat in the springtime is closely associ-

ated with reproductive renewal, likely through photic mechanisms.

The mechanisms for this wintertime reproductive collapse are not completely understood but likely represent a cascade of failures beginning in the hypothalamus. Hypothalamic GnRH content and secretion fall to minimal levels during anestrus. GnRH secretory patterns were determined by push–pull perfusion, a technique in which hypothalamic or pituitary extracellular fluids are sampled directly via an indwelling stainless-steel catheter. This technique is necessitated by the fact that GnRH secreted from the hypothalamus to the pituitary does not reach the general circulation in sufficient concentrations to be detected due to dilution and proteolytic degradation (Fig. 2). During anestrus three of four mares did not have detectable GnRH in push–pull perfusates. These data agree well with the low tissue content measured within the hypothalamus. Thus, it appears that a key factor in the reproductive quiescence of mares during the winter is a reduction in hypothalamic GnRH secretion. Probably as a result of this GnRH secretory reduction, LH and FSH concentrations in blood decline drastically. Additionally, pituitary content of LH is dramatically reduced during anestrus, although pituitary FSH content does not decline. Analysis of pituitary ribonucleic acid (RNA) for specific messenger RNA (mRNA) encoding for the α and β subunits of LH indicated that the gene regulating expression of these subunits is essentially turned off during anestrus. As will be discussed later, reactivation of this gene during vernal transition is

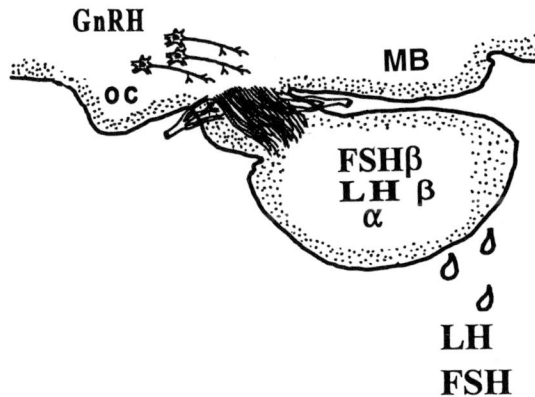

FIGURE 2 Schematic representation of hypothalamic–pituitary interaction. GnRH is synthesized in neuron cell bodies within the hypothalamus and secreted into capillaries of a portal system in the median eminence. The portal system delivers GnRH to the anterior pituitary gland, where it stimulates synthesis of gonadotropin (LH and FSH) subunits and secretion of mature hormone into the peripheral circulation.

the crux of reproductive recrudescence. Thus, anestrus represents ovarian inactivity, both morphologically (no follicular development and no corpora lutea) and hormonally (undetectable estrogen and progesterone), which reflects, in turn, the lack of pituitary gonadotropin secretion. Of interest, failure of LH and FSH secretion during anestrus may reflect different regulatory mechanisms. Because pituitary FSH content does not change throughout the year, failure of FSH secretion during anestrus probably reflects the failure of a hypothalamic signal via GnRH. Failure of LH secretion, on the other hand, reflects the lack of GnRH secretion but also reflects failure of expression of the gene encoding LH subunit production.

2. Vernal Transition

Of the four stages of the annual reproductive cycle, vernal transition [mid-January to early April (horses) or early May (ponies) in the Northern Hemisphere] is the most troublesome economically due to disparity between the physiological breeding season and human management concerns. Many horse breeders prefer to have foals born as early in the year as possible, setting up a conflict between economic goals and physiology. Furthermore, as will be seen, vernal transition is fraught with ambiguous signs and

events which may seem confusing to horse breeders and lead to inappropriate breeding practices. In fact, the vernal transition is a beautifully orchestrated series of events, which appear to be obligatory, leading to the first ovulation of the year. By definition, the first ovulation of the year indicates successful transfer from vernal transition into the breeding season. Timing of the first ovulation of the year, and the events which lead up to it, is surprisingly precise, permitting relatively easy study of this phase. The following is a chronological account of the events characterizing vernal transition.

i. Renewal of GnRH Secretion Shortly after the winter solstice (December 21 in the Northern Hemisphere) hypothalamic GnRH content begins to increase through mechanisms that are not understood. This increase in GnRH neuronal activity is also reflected in increased secretion. Whether the increase in GnRH content and secretion reflects some acute response to the change in day length or whether it reflects a refractoriness to short days remains to be determined. It is clear that exposure of anestrous mares to an artificially lengthened day prior to the winter solstice can stimulate GnRH secretion substantially. Within 2 weeks of exposing anestrous mares to 16 hr of light daily, GnRH secretion was increased more than fivefold. Stimulation of GnRH secretion with artificial lighting systems may not necessarily reflect the same mechanism that drives the increase in GnRH by the first of the year, however, because natural day length changes slowly around the winter solstice. There is essentially no change in day length for a week before and after December 21 at 30°N latitude.

ii. FSH Secretion Presumably in response to increased GnRH secretion in early to mid-January FSH concentrations increase in the peripheral blood. The pattern of FSH secretion during vernal transition can be described as high, variable concentrations. The extreme variation in mean samples likely indicates dynamic secretion.

iii. LH Secretion Throughout the vernal transition, there is very little change in LH secretion until shortly before the first ovulation of the year. There is discrepancy in the literature as to whether or not LH increases pulse frequency as the first ovulation approaches. Regardless, mean LH secretion does not increase in a preovulatory-like surge until 5–7 days prior to the first ovulation of the year. This unusual environment, high FSH:low LH, contributes to some of the ambiguous events that mark vernal transition (see Section I,A,2,iv). The lack of LH secretion, despite increased GnRH secretion, and increased FSH secretion probably reflects failure to reestablish the gene encoding LH subunit synthesis. As mentioned previously, this is the crux of vernal transition, and reestablishment of the LH gene is required for full reproductive function.

iv. Ovarian Follicular Development

Shortly after the increase in serum concentration of FSH, a small increase in ovarian follicular development is observed. This is characterized by increased numbers of small, intermediate, and large follicles. Ovarian size also increases, reflecting the increased follicular development, and at ultrasound examination, mare's ovaries during midvernal transition often exhibit multifollicular or "grape-cluster" appearance. Figure 3 demonstrates the progressive changes in follicular development throughout vernal transition. The unusual aspect of this renewal of follicular activity is

that large, preovulatory-like follicles develop but fail to ovulate. In some studies, it has been shown that an average of 3.7 ± 0.9 follicles develop to a diameter of >30 mm and then regress without ovulating. The development of large, but anovulatory, follicles is one of the ambiguous signs leading to inappropriate breeding and mismanagement during vernal transition. Breeders, anxious to accomplish their goals of producing foals early in the subsequent year, may schedule breeding of mares during the vernal transition based on clinical detection of these follicles, which contributes to reproductive inefficiency.

Studies of the steroidogenic function of vernal transition follicles have demonstrated that they are poorly steroidogenic, with little production of either androgens or estrogens until just before the first ovulation of the year. Seamans and Sharp first demonstrated that vernal transition follicles do not produce estrogen, and Tucker *et al.* further demonstrated that this reflects a lack of the steroid enzymes which convert progesterone to major steroid metabolites further along the steroidogenic pathway. Specifically, the enzyme, 17β hydroxylase, 20 lyase, which converts progesterone to 17β hydroxyprogesterone, is missing from granulosa and theca cells from early vernal transition follicles. Of interest, aromatase and 3β hydroxysteroid dehydrogenase enzymes appear to be present. It is not known what limits the presence of the 17β hydroxylase, 20 lyase enzyme, but the high FSH:LH environment is a likely candidate. Furthermore, it is not known what brings about the induction of this enzyme to create a steroidogenically functional follicle. The slight increase in mean LH secretion observed during the 2 or 3 weeks prior to the first ovulation of the year, but prior to the ovulatory LH surge, is a likely candidate. The ability of a given follicle to produce androgens and estrogens apparently is accompanied by the ability to enter into feedback interactions with the hypothalamic–pituitary axis because it is associated with an LH "surge" and ovulation. The intricacies of these interactions remain to be understood.

The early vernal transition follicles are incompetent in ways other than steroidogenesis, probably indicating the inappropriate gonadotropin environment in which they develop. Although vernal transition follicles achieve preovulatory-like size, their

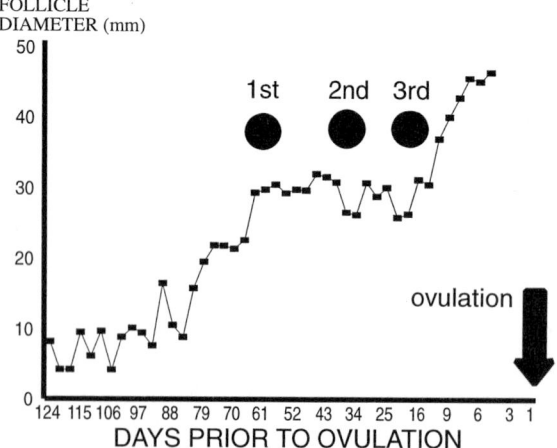

FIGURE 3 Diameter of the largest ovarian follicle during the vernal transition in mares. The first ovulation of the year is designated as "ovulation." An average of 3.7 large, anovulatory follicles, designated by ●, usually precede development of the first ovulatory follicle.

growth rate is considerably slower than that of follicles during the breeding season, and monitoring the time it takes vernal transition follicles to reach a diameter of 30 mm may be useful in attempting to determine their status. The granulosa cell investment is also significantly less in vernal transition follicles, indicating that growth of a given follicle does not necessarily represent functional status reliably. Early vernal transition follicles are poorly vascularized as well, further pointing to the inappropriate environment in which they develop. Follicular fluid from normally functioning follicles contains significantly more insulin-like growth factor-I than do their vernal transition predecessors. Further research will likely reveal a host of components either lacking or present in insufficient amount to permit normal follicular function during vernal transition in mares.

The vernal transition represents a phase of reproductive renewal that is relatively repeatable and consists of sequential interactions between the gonads and the hypothalamic–pituitary axis. The primary failure in anestrus may be loss of GnRH in the hypothalamus, resulting in loss of reproductive response all the way downstream to the ovaries, as illustrated in Fig. 4. On the other hand, the crux of reproductive recrudescence appears to be reestablishment of the LH gene (Fig. 4). The contribution of the ovaries to the process of reproductive quiescence and recrudes-

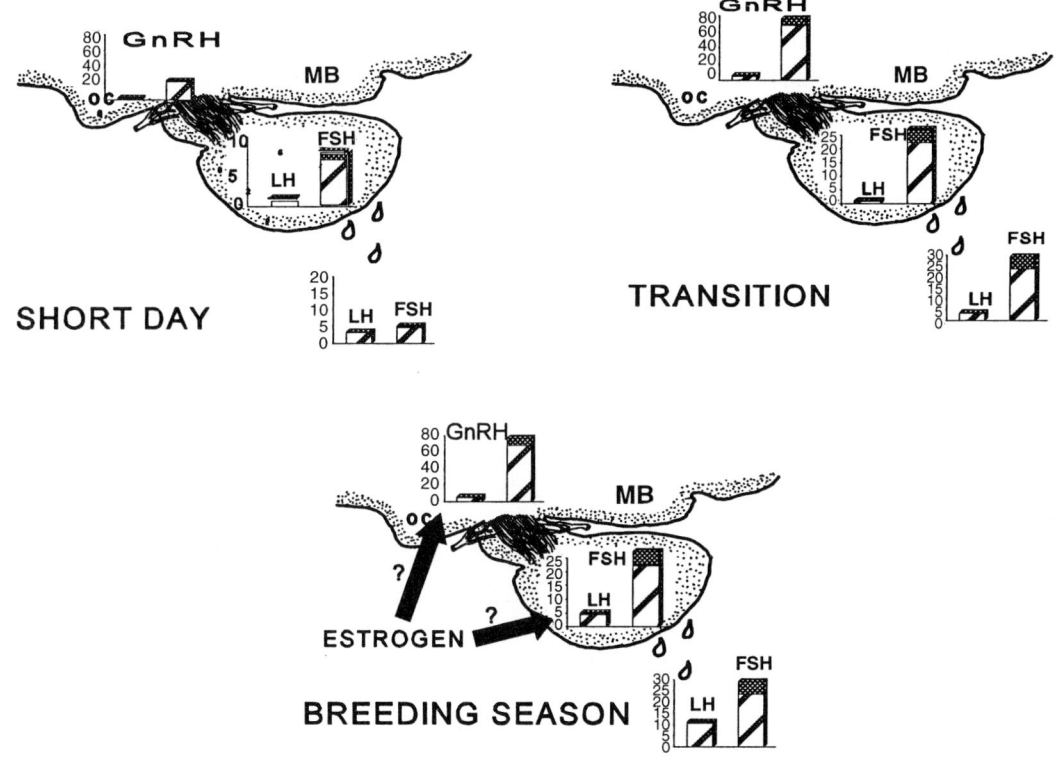

FIGURE 4 Schematic model of changes in the hypothalamic–pituitary axis during vernal transition. During anestrus (short day), GnRH content in the hypothalamus and secretion are low, pituitary LH content is low due to reduced expression of the LH gene, and pituitary FSH content remains unchanged. Peripheral concentrations of LH and FSH are low due to loss of gene expression (LH) and reduced hypothalamic GnRH stimulation (LH and FSH). During vernal transition, GnRH secretion increases, leading to increased peripheral FSH concentration but not LH due to continued lack of gene expression. The ovarian response to high FSH/low LH is development of large, anovulatory follicles. Through mechanisms unknown, but possibly involving estrogen from the first steroidogenically competent follicle of the year, LH gene expression is restored and the ovulatory surge of LH occurs.

cence appears to be very little, although ovarian modulation of the final events of reproductive recrudescence may occur.

3. The Breeding Season

This phase of the annual reproductive cycle is characterized by repeated estrous cycles in the absence of breeding, as described in Section II.

4. Autumnal Transition

Transition from the breeding season to anestrus is perhaps the least understood phase of the annual reproductive cycle. This is partly due to the relative lack of economic interest in this phase, because few breeders wish to breed their mares at this time, and partly due to the fact that variation among mares entering autumnal transition is extreme, making scientific studies difficult. Whereas variation about the date of the first ovulation of the year may be small—of the order of a few days to a week—variation in the timing of the last ovulation of the year may extend to several weeks or months. Because the mechanisms are unknown, speculation as to how the reproductive collapse occurs may be inappropriate. One feature that appears to be borne out by several investigations is that autumnal transition is characterized, eventually, by failure of ovulation and reduced gonadotropin secretion.

B. Stallions

Stallions also exhibit seasonal reproductive patterns. In contrast to mares, whose anestrus is absolute, stallions show a marked reduction in reproductive function but not a complete loss. That is, semen quality and stallion reproductive behavior are reduced in the wintertime but are not completely ablated. Because cessation of reproductive function in mares is sufficient to achieve the goals of seasonal reproductive control, one wonders what purpose the reduction in stallion reproduction serves.

Seminal quality, number of progressively motile sperm, and volume of gel-free semen all decrease during the winter months as do behavioral traits such as number of mounts per ejaculate and latency time to mount. However, the diminution of these repro-

ductive events in stallions does not necessarily mean that stallions become sterile, only reduced. Perhaps the wintertime reduction in reproductive capability in stallions is more a form of conservation than contraception.

II. THE ESTROUS CYCLE OF MARES

Once a mare has experienced the first ovulation of the year, she has entered, by definition, the breeding season. This time of the year is characterized by repeated opportunities to conceive (i.e., by repeated estrous cycles) unless pregnancy occurs. Numbering the days of the estrous cycle is problematic in horses compared with other domestic stock because of the relatively prolonged period of estrus. Numbering the days of the estrous cycle consecutively beginning on the first day of estrus is one typical convention, but it does not convey much information about the length of estrus, leading some to number estrus and diestrus separately. Yet another convention, most often employed where precision is required, is to number the consecutive days after ovulation. Ovulation is the event that sets the timing for subsequent events; therefore, this is one of the most precise conventions.

A. Behavioral Aspects

The behavioral displays of mares are windows into the driving forces of morphological and hormonal changes of the estrous cycle. Reproductive management of mares utilizes this information through organized encounters with stallions. This process is called "teasing" and is the cornerstone of a good reproductive management program because it outlines not only the current behaviors of a mare but also provides a history of aberrant or irregular cycles. Estrus, the period of sexual receptivity, is 7 days on average but may be as short as 2 or 3 days and as long as 10 days. Sexual receptivity is exhibited through a variety of visual displays, including elevated tail, squatting or "posturing," and general calmness or placidity in the presence of a stallion. Perhaps one of the more unique behavioral signs of sexual receptivity is eversion of the clitoral fossa, or "winking" in the presence of a stallion. The stallion's response to these behavioral signs is generally one of increased interest.

Shortly after ovulation, mares become sexually nonreceptive, launching the major portion of the estrous cycle, diestrus. Diestrus is a less variable phase of the estrous cycle, with most mares experiencing a diestrus length of 14–16 days. Diestrus is less variable because it is regulated by corpus luteum function and progesterone secretion.

B. Morphological Aspects

Estrus is the phase of follicular development and is therefore characterized by development of several (one to four) ovarian follicles to a substantial size. Usually one of these follicles is selected through unknown mechanisms to become the dominant follicle, although some horse breeds have a tendency to produce two follicles that eventually ovulate. Specifically, the incidence of multiple ovulations are as follows: thoroughbred > standardbred > quarter horse > Arabian > ponies. One of the unique features of equine reproduction is that the ovary is "inside out" relative to most other species. Ovaries of most mammalian species consist of a central medullary structure containing support elements, such as connective tissue, blood vessels and nerves, and peripheral cortical tissue, which contains the germ cells. Furthermore, the outer surface of the ovary of most species is covered with surface germinal epithelium, which presumably permits ovulation to occur wherever a given follicle develops. In mares, however, the cortical tissue bearing primordial follicles is in the center of the ovary, with medullary elements surrounding it. Furthermore, the ovary of the mare is almost completely surrounded with thick mesothelial tissue, precluding the possibility of follicle rupture. Ovulation can take place only at a specified place known as the ovulation fossa, where there is no mesothelial covering but rather an investment of germinal epithelium. This unusual inside-out structure may well contribute to some of the unique aspects of mare reproduction. For instance, one can speculate that the relatively lengthy follicular development phase and prolonged estrus represent necessary evolutionary adaptations to allow time for the

centrally located developing follicle to come into contact with the ovulation fossa, the only place that ovulation can take place. Further speculation includes the possibility that the prolonged LH peak (5–7 days) characteristic of this species is an evolutionary necessity to maintain some sort of stimulus until the follicle contacts the ovulation fossa. Such a speculation, while just that, would also include the intriguing possibility that evolutionary pressure to maintain a prolonged LH peak might have forced the development of hypothalamic–pituitary interactions requiring special tolerance to prolonged hormone secretion. Most domestic species have an acutely regulated hypothalamic–pituitary interaction in which GnRH, of necessity, is secreted in an intermittent, or pulsatile manner to prevent downregulation of pituitary GnRH receptors and extinction of the hypothalamic tropic signal. Many studies have demonstrated in horses, however, that GnRH can be administered in continuous fashion, and in surprisingly high doses, without leading to extinction of the LH response. While there is no experimental design that can test the hypothesis that these strange regulatory interactions evolved in response to these pressures, the ovary's inside-out structure certainly lends strength to such speculation.

The corpus luteum of the mare does not seem to present any such bizarre morphological characteristics, although there does not seem to be the mixture of small and large luteal cells so prevalent in cattle and sheep. Also, due to the necessity of ovulating at the ovulation fossa, the corpus luteum usually retains a triangular or "guitar-pic" shape, with the apex of the triangle located in the fossa.

C. Hormonal Aspects

As pointed out previously, one of the unique aspects of mare reproduction is the prolonged LH "peak" lasting from 5 to 7 days, and indeed reaching highest concentrations of LH often after ovulation has occurred. It is not difficult to appreciate, after viewing the ovarian structure of mares, that an acute short-lived ovulatory surge of LH such as seen in ruminants might not accomplish whatever tissue remodeling is required to bring an ovarian follicle to the fossa where it can rupture. This prolonged procedure is also accompanied by relatively prolonged estrogen secretion from the developing follicle(s), although not of the same duration as LH. Estradiol concentrations rise from nearly undetectable concentrations (<5 pg/ml) by midway through estrus to 50–60 pg/ml in the peripheral circulation. Changes in FSH are less well understood, although FSH is elevated during estrus as is LH. There is a second peak in FSH, however, observed in middiestrus that appears to have a seasonal component. During the breeding season proper, mares exhibit two FSH peaks during an estrous cycle, but as autumnal transition nears only the estrus-related peak, and not the middiestrus peak, occurs. The physiological significance of this is not well understood.

After ovulation, the corpus luteum begins to form rapidly, and concentrations of progesterone rise rapidly in the peripheral circulation such that they may exceed 1 ng/ml by the first or second day after ovulation. Not coincidentally, the rising progesterone concentration is associated with loss of sexual receptivity and onset of diestrus. This remarkable effect on behavior is often utilized by managers of performance horses. It is generally thought that the behaviors associated with estrus are distracting either to the mare herself or to nearby stallions. To counter this perceived problem, progesterone is often administered to performance mares to maintain a prolonged diestrus-like state. Once the corpus luteum is formed, progesterone concentrations rise rapidly to peak concentrations (7–10 ng/ml) by 7 or 8 days after ovulation. As the name suggests, progesterone is the hormone of pregnancy, and diestrus is a time of intrauterine histotroph accumulation. Histotrophe is defined as the sum total of nutrient substances delivered to the embryo of viviparous animals from sources other than the mother's blood. Many intrauterine proteins that have been identified have an absolute requirement for progesterone for their synthesis and secretion. The process of progesterone induction of intrauterine proteins takes place every estrous cycle, regardless of whether or not the mare was bred. By 14–16 days after ovulation, however, the endometrium is also primed to produce a luteolysin, prostaglandin $F_{2\alpha}$ (PGF), unless secretion of

this luteolysin is blocked by some conceptus-mediated process.

III. ESTABLISHMENT OF PREGNANCY

The process by which the luteolytic secretion of PGF is averted, permitting continued corpus luteum function, is known as the maternal recognition of pregnancy and is an important milestone in the establishment of pregnancy. It is pertinent to point out that failure of pregnancy establishment at this time (first 3 weeks after ovulation) is relatively common (20–25% embryo loss), indicating that this is a critical time in the reproductive life of the mare and her progeny. Because pregnancy appears to be established by removal (or prevention of secretion) of the luteolytic substance, it has been called a "luteostatic" process, implying that the corpus luteum will continue to function if the luteolytic process is averted. Prostaglandin $F_{2\alpha}$ is released from the nonpregnant uterus at 14 days after ovulation and results in luteal regression in nonpregnant mares. Thus, secretion of PGF can be considered as a critical deadline for pregnancy establishment because whatever the mechanism evoked by the developing conceptus, it must take place by Day 14 or suffer the consequences of pregnancy loss. The conceptus factor that blocks PGF secretion in horses is still unknown, but it appears to be different from conceptus products of other domestic species. For instance, it is widely accepted that interferon-τ (IFN-τ) from ruminant conceptuses plays a major role in maternal recognition of pregnancy in those species. To date, there is little or no evidence that an IFN-τ is present at all in mare reproductive tracts. In swine, the secretion of prostaglandin is converted from endocrine to exocrine secretion by the action of conceptus estrogens. This has the effect of sequestering prostaglandin within the uterine lumen, thus preventing luteolysis. Because equine conceptuses also produce prodigious amounts of estrogen, it is easy to speculate that similar mechanisms exist in this species. However, there has been no convincing study in which estrogen administration to nonpregnant mares prolonged luteal maintenance as it does in swine. Furthermore, the endocrine to exocrine shift does not occur when mare endometria are tested in an *in vivo* system, which demonstrates sidedness (endocrine versus exocrine) of prostaglandin secretion. Therefore, the conceptus signal for maternal recognition of pregnancy in horses remains elusive. Release of prostaglandin from the uterus involves a cascade of events which includes oxytocin binding to endometrial receptors as an initial part of the cascade. Recent studies in mares have indicated that the conceptus interrupts this oxytocin–prostaglandin cascade and thereby blocks luteolytic release of PGF. This interruption of the luteolytic release of prostaglandin may reflect two mechanisms for which evidence exists: (i) There appears to be a decrease in oxytocin receptor number and affinity in pregnancy, and (ii) cervical stimulation, which leads to oxytocin and PGF release in nonpregnant mares, leads only to oxytocin release in pregnant mares. Thus, whatever the conceptus factor, it acts to uncouple the normal luteolytic mechanisms involving the oxytocin–prostaglandin cascade.

IV. PREGNANCY

Perhaps in pregnancy more than any other aspect of horse reproduction the word "unique" applies. Pregnancy in mares involves utilization of a functional yolk sac placenta that later switches to a more conventional chorioallantoic form of placenta; development of specialized endometrial structures, of fetal origin, that secrete a chorionic gonadotropin; development of multiple ovulations in early pregnancy to form accessory corpora lutea; and development of highly steroidogenic, functional fetal gonads that constitute the fetal half of a feto–placental unit.

A. Embryogenesis and Placentation

The most notable departure of equine embryonic development from that of other livestock species is the fact that the equine conceptus remains spherical throughout the first 2 months of gestation. This spherical configuration is associated with morpho-

logical and physiological characteristics that are unusual and apparently unique to Equids. Presumably, the elongation of conceptuses from other livestock species allows for endometrial–conceptus communication to some extent prior to intimate maternal–fetal connection. Because equine conceptuses do not elongate, however, it is reasonable to speculate that they achieve a similar form of communication opportunity by moving throughout the uterus during the time of maternal recognition of pregnancy. Ginther reported that equine conceptuses move about the uterus freely, covering the equivalent distance of several football fields during the 5- or 6-day mobility phase. The cause of this mobility is still debated, but close observation with diagnostic ultrasound suggests that a strong component of the conceptus mobility lies in maternal uterine contractions. Whether or not the conceptus plays an active role in stimulating such contractility has not been fully resolved. Conceptus mobility is important for maternal recognition of pregnancy, however, because restricting equine conceptus mobility prior to pregnancy establishment leads to corpus luteum regression and embryonic loss. Perhaps not coincidentally, the conceptus mobility phase stops at about Day 16, after the critical deadline for maternal recognition of pregnancy, as the conceptus becomes physically lodged or "fixated" at the bifurcation of one of the uterine horns. Fixation, as termed by Ginther, likely results from the combination of tonic uterine contractions, typical of early equine pregnancy, and increasing spherical enlargement of the conceptus, filling the uterine lumen. Further contribution to the cessation in mobility by tissue adhesion is unknown but may be a reasonable consideration. Therefore, it is speculated that the mobility of equine conceptuses from Days 10 to 16 plays an important role in maternal recognition of pregnancy by interacting with the entire maternal endometrium to block PGF secretion and perhaps to gain access to histotrophe.

The spherical nature of the equine conceptus may impart structural requirements that are at least partially fulfilled by the acellular covering of the conceptus known as the capsule. The capsule is known to be present as early as Day 7, and it persists at least until about Day 25. The function of the capsule is not completely understood, but it likely provides structural integrity to the otherwise diaphenous extraembryonic membranes. Uterine tone increases markedly during the period of conceptus mobility, and by witnessing, via ultrasound examination, the continuous deformation of the conceptus by uterine contractions it is easy to imagine that the capsule is an important structural feature. However, it is probably naive to think that the capsule functions only structurally because there is apparently complex biochemical change taking place as well.

The equine conceptus begins life with a functional choriovitelline placenta. Endodermal cells line the inner surface of the trophectoderm around Day 11, forming the yolk sac. The Day 11 conceptus has a diameter of approximately 6 mm and contains approximately 0.04 ml of yolk sac fluid. The yolk sac fluid accumulates rapidly through unknown mechanisms, however, and by Day 18, the conceptus diameter may be as much as 25 mm and contain approximately 10 ml of yolk sac fluid. Up until approximately Day 14, the equine conceptus consists primarily of a simple bilaminar structure, outer trophectoderm, and inner endoderm, surrounded by the acellular capsule. Sometime on Day 14 after ovulation, however, a third layer of cells, the mesoderm, begins to emanate from the region of the embryonic disc and becomes interposed between ectoderm and endoderm layers. Temporally coincident with this new developing layer, the array of proteins synthesized by the conceptus changes from a relatively simple array of three to five proteins to a complex array consisting of dozens of proteins. This suggests that growth of the mesoderm may be associated with induction of a number of genes. Growth of the mesoderm is also associated with development of primitive blood islets and the beginnings of hematopoiesis. One of the prominent uterine proteins observed at this time is called "uteroferrin," an iron-containing glycoprotein that may serve to transport iron to the developing embryo.

The avascular junction between advancing allantois and receding yolk sac becomes known as the "chorionic girdle" as it encircles the still spherical conceptus. Cells of the chorionic girdle invade the maternal endometrial epithelium 35–40 days after ovulation, setting up a maternal structure known as the "endometrial cups." Named because they were originally thought to be strictly maternal in origin, it is now clear that the endometrial cups are the result of girdle cell invasion and colonization. These cells contain the gene for equine chorionic gonadotropin (eCG) synthesis and secretion. eCG becomes evident in the peripheral circulation about this time, reaching peak concentrations about Day 60, then declines steadily until approximately Day 120. The purpose of the endometrial cups and eCG secretion is still not understood, but it is clear that this initial intimate maternal–fetal connection serves some purpose other than nutritive. Perhaps this strange reaction represents an immunologic "testing of the water" in advance of the more intimate chorioallantoic–endometrial interdigitation that begins to take place soon thereafter. On approximately Days 40–60, chorioallantoic macrovilli begin to interdigitate into maternal endometrial crypts to form the microcotyledons. These structures, miniature versions of the placentomes of ruminants, begin forming about Day 60 but do not become fully formed until approximately Day 150.

B. Endocrinology of Pregnancy

Early pregnancy (20–60 days) in mares is characterized by very low circulating LH, due to the negative feedback of progesterone but continuing FSH secretion. This results in development of large preovulatory-like follicles, but ovulation does not occur. However, at about 45–50 days after ovulation, eCG begins to be secreted from the endometrial cups. Because eCG is structurally identical to LH, the follicles luteinize and/or ovulate, forming multiple accessory corpora lutea (CL). Progesterone rises with the formation of these accessory CL. The presence of the accessory CL has raised considerable debate because a potential physiological role for such structures is unclear. Misconceptions regarding whether or not the primary corpus luteum (developing from the ovulation that leads to fertilization) regressed and was replaced by the accessory corpora lutea has led horse breeders to administer progesterone in hopes of preventing pregnancy loss due to progesterone insufficiency. This misconception was clarified and it is now well-known that the primary corpus luteum does not regress; however, administration of progesterone continues to be a popular treatment for pregnant mares despite the lack of clear proof that it is helpful.

Contrary to human chorionic gonadotropin (hCG), which is expressed from a gene distinct from the human pituitary LH gene, the gene that expresses eCG is the same gene that expresses equine pituitary LH. It is curious that expression of the eCG gene occurs only in cells of the chorionic girdle, which raises some interesting questions about tissue-specific gene regulation. Also of interest, the quantity of eCG secreted reflects, to some degree, the genetic contribution of the sire. What role this genetic contribution plays is unclear, but the information could prove useful in providing information about what traits the stallion contributes to successful pregnancy. The long circulatory half-life of eCG is due to the complex glycosylation, making it a very useful tool in stimulating ovarian function in a variety of laboratory and livestock species.

About the time the microcotyledonary attachment begins (Day 60), concentrations of estrogens rise in the peripheral circulation. Unlike other species, in horses estrogens increase tremendously during midpregnancy but begin to decline after about Day 250. The source of these estrogens is thought to be the fetal gonad, which enlarges tremendously during midpregnancy and contributes androgens to a fetal–placental steroidogenic unit. Included in the mix of estrogens secreted by the pregnant mare are two unique estrogens, equilin and equilenin, which are synthesized in a pathway that diverges from the *de novo* estrogen synthetic pathway before cholesterol. Thus, these unique equine estrogens are not produced from cholesterol, unlike progesterone, testosterone, and other sex steroids.

The accessory corpora lutea, as well as the primary corpus luteum, regress around Days 120–150, and progesterone concentrations in blood become essentially undetectable after this time. However, there is a slight increase in 5α-pregnanes in mid- to late pregnancy, followed by a sharp rise in these progestins in the final 2 or 3 weeks of pregnancy. Maternal ovarian contribution to maintenance of pregnancy is questionable because ovariectomy as early as Day 70 results in normal gestation and parturition.

V. PARTURITION

Little is known about the mechanisms of parturition initiation in mares. The role of the hypothalamic–pituitary–adrenal axis in parturition initiation in most livestock species is well accepted, but there is little evidence to suggest that similar mechanisms exist in mares. Levels of cortisol and ACTH are much lower in mares than in ruminants, and the rise in equine cortisol just prior to parturition occurs much more abruptly. Furthermore, the rise in fetal cortisol concentrations prior to parturition in equids suggests some involvement of the equine fetal adrenal axis, but parturition cannot be unambiguously induced by administering cortical steroids to the mare as it can in ruminants. More work is clearly needed to determine the signal for parturition induction in mares.

VI. REPRODUCTIVE MANAGEMENT

A. The Annual Reproductive Cycle of Mares

The disparate timing of the physiological breeding season and the "man-made," economically based breeding season has prompted many attempts to alter the timing of the physiological breeding season. To date, none has been more effective than acceleration of the onset of the breeding season by exposure to artificially lengthened day. Exposing mares to increasing day length prior to the winter solstice results in earlier onset of vernal transition and, hence, earlier onset of the breeding season as judged by timing of the first ovulation of the year. Importantly, events of the vernal transition do not appear to be truncated or omitted. Rather, the entire process appears to be shifted to an earlier start date. Thus, it is very important to begin such artificial lighting systems well before the solstice. The mode of exposure to artificially lengthened day does not appear to be critical. That is, exposing anestrus mares to 16 hr of artificial light daily, beginning sometime in November, in the Northern Hemisphere, appears to be as effective as exposing them to only a few hours of artificial light at the time of sunset in combination with exposure to natural sunlight during the day. Use of artificial photoperiod, while effective, requires forethought and often requires construction or modification of housing facilities. Therefore, alternative means of modifying onset of the breeding season are desirable but unavailable.

One such alternative, in theory, would be to administer GnRH because the limitation in GnRH secretion appears to be an essential reason for anestrus. However, current understanding of the mechanisms controlling GnRH synthesis, secretion, and pituitary gonadotrope stimulation is incomplete, making widespread use of GnRH unjustifiable. As pointed out, reestablishing the LH gene in the pituitary appears to be the crux of vernal transition and GnRH administration alone may not accomplish that.

B. The Estrous Cycle of Mares

Management of the estrous cycle relies mainly on monitoring sexual behavioral changes in mares as well as monitoring ovarian follicular dynamics by transrectal manual palpation or ultrasound examination. Two of the most critical management needs in the horse breeding industry are (i) ovulation prediction and induction and (ii) corpus luteum control. The latter is efficiently accomplished by administration of PGF at any time after corpus luteum maturation (Day 5 postovulation). Regression of the corpus luteum with PGF results in return to estrus in 2–4 days, providing the opportunity to schedule mares for breeding earlier than would occur otherwise. Predicting or inducing ovulation is a management need that remains to be perfected. In most horse breeding management systems, the mare and stallion are housed separately and may get the opportunity to mate only a few times during estrus. This means the management decision to schedule mating must be appropriately timed to occur near (24–38 hr prior to) ovulation. Because of the prolonged estrus, management errors in this scheduling can reduce reproductive efficiency. Prediction of ovulation time by transrectal palpation or ultrasound is not completely reliable. Therefore, hCG is often administered at the time of mating to hasten or ensure ovulation within 48 hr. Human chorionic gonadotropin is LH-like and is thought to hasten the ovulatory process in mares. Because it is a foreign protein, however, antibody

production can lead to reduced efficacy with overuse. Thus, reliable management tools to predict or to stimulate the ovulatory process are not yet readily available.

See Also the Following Articles

CATTLE (BOVIDAE); EPITHELIOCHORIAL PLACENTATION; PREGNANCY, MATERNAL RECOGNITION OF; SEASONAL REPRODUCTION, MAMMALS

Bibliography

Baker, C. B., McDowell, K. J., Rabb, M. H., and Bailey, E. (1992). Lack of expression of interferon genes by equine embryos. *J. Reprod. Fertil.* **44**(Suppl.).

Ball, B. A., and Woods, G. L. (1987). Embryonic loss and early pregnancy loss in the mare. *Compendium Equine* **9**, 459–469.

Ball, B. A., Little, T. V., Hillman, R. B., and Woods, G. L. (1986). Pregnancy rates at days 2 and 14 and estimated embryonic loss rates prior to day 14 in normal and subfertile mares. *Theriogenology* **26**, 611–619.

Bazer, F. W., and Thatcher, W. W. (1977). Theory of maternal recognition of pregnancy in swine based on estrogen controlled endocrine versus exocrine secretions of prostaglandin F2α by the uterine endometrium. *Prostaglandins* **14**, 397–401.

Fitzgerald, B. P., Affleck, K. J., Barrows, S. P., Murdoch, W. L., Barker, K. B., and Loy, R. G. (1987). Changes in LH pulse frequency and amplitude in intact mares during the transition into the breeding season. *J. Reprod. Fertil.* **79**, 485–493.

Franklin, K. J., Gross, T. S., Dubois, D. H., and Sharp, D. C. (1989). In vitro prostaglandin secretion from luminal and myometrial sides of endometrium from cyclic and pregnant mares at day 14 post-estrus. *Biol. Reprod.* **40**(Suppl. 1), 199.

Friedman, L. J., Garcia, M. C., and Ginther, O. J. (1979). Influence of photoperiod and ovaries on seasonal reproductive activity in mares. *Biol. Reprod.* **20**, 567–574.

Ginther, O. J. (1985). Dynamic physical interactions between the equine embryo and uterus. *Equine Vet. J. Suppl.* **3**, 41–47.

Ginther, O. J. (1993). *Reproductive Biology of the Mare.* Equiservices, Cross Plains, WI.

Hart, P. J., Squires, E. L., Imel, K. J., and Nett, T. M. (1984). Seasonal variation in hypothalamic content of gonadotropin-releasing hormone (GnRH), pituitary receptors for GnRH, and pituitary content of luteinizing hormone and follicle stimulating hormone in the mare. *Biol. Reprod.* **30**, 1055–1062.

Holtan, D. W., Squires, E. L., and Ginther, O. J. (1979). Effects of ovariectomy on pregnancy in mares. *J. Anim. Sci.* **41**, 359.

McDowell, K. J., Sharp, D. C., Grubaugh, W. R., Thatcher, W. W., and Wilcox, C. J. (1988). Restricted conceptus mobility results in failure of pregnancy maintenance. *Biol. Reprod.* **39**, 340–348.

Nett, T. M., Holtan, D. W., and Estergreen, V. L. (1975). Oestrogens, LH, PMSG, and prolactin in serum of pregnant mares. *J. Reprod. Fertil. Suppl.* **23**, 457–462.

Oriol, J. G., Sharom, F. J., and Betteridge, K. J. (1993). Developmentally regulated changes in the glycoproteins of the equine embryonic capsule. *J. Reprod. Fertil.* **99**, 653–664.

Peltier, M. R., Robinson, G., and Sharp, D. C. (1997a). Effects of melatonin implantation in Pony mares: 1. Acute effects. *Theriogenology*, in press.

Peltier, M. R., Robinson, G., and Sharp, D. C. (1997b). Effects of melatonin implantation in Pony mares: 2. Long term effects. *Theriogenology*, in press.

Seamans, K. W., and Sharp, D. C. (1982). Changes in equine follicular aromatase activity during sexual recrudescence. *J. Reprod. Fertil. Suppl.* **32**, 225–233.

Sharp, D. C., and Davis, S. D. (1993). Vernal transition. In *Equine Reproduction* (J. L. Voss and A. O. McKinnen, Eds.), pp. 133–144. Lea & Febiger, Philadelphia.

Sharp, D. C., and Ginther, O. J. (1975). Induction of ovarian activity and estrous behavior in anestrous mares with light and temperature. *J. Anim. Sci.* **41**, 1368–1372.

Sharp, D. C., and Grubaugh, W. R. (1987). Use of push-pull perfusion techniques in studies of gonadotrophin-releasing hormone secretion in mares. *J. Reprod. Fertil. Suppl.* **35**, 289–296.

Sharp, D. C., Grubaugh, W. R., Weithenauer, J., and Sheerin, P. (1988). Exposure to long photoperiod results in increased GnRH secretion in anestrous pony mares. *Biol. Reprod. Suppl.* **1**, 39.

Sharp, D. C., McDowell, K. J., Weithenauer, J., Franklin, K., Mirando, M., and Bazer, F. W. (1989). Is an interferon-like protein involved in the maternal recognition of pregnancy in mares? *Equine Vet. J. Suppl.* **9**, 7–9.

Sharp, D. C., Cleaver, B. D., and Davis, S. D. (1993). Photoperiod. In *Equine Reproduction* (J. L. Voss and A. O. McKinnon, Eds.), pp. 179–186. Lea & Febiger, Philadelphia.

Sharp, D. C., Thatcher, M.-J.., Salute, M. E., and Fuchs, A.-R. (1997). Relationship between endometrial oxytocin receptors and oxytocin-induced prostaglandin F2α release during the oestrous cycle and early pregnancy in pony mares. *J. Reprod. Fertil.* **109**, 137–144.

Sherman, G. B., Wolfe, M. W., Farmerie, T. A., Clay, C. M., Threadgill, D. S., Sharp, D. C., and Nilson, J. H. (1992). A single gene encodes the β subunits of equine luteinizing hormone and chorionic gonadotropin. *Mol. Endocrinol.* **6**, 951–959.

Squires, E. L., Douglas, R. H., Steffenhagen, W. P., and Ginther, O. J. (1974). Ovarian changes during the estrous cycle and pregnancy in mares. *J. Anim. Sci.* **38**, 330–338.

Strauss, S. S., Chen, C. L., Kalra, S. P., and Sharp, D. C. (1979). Localization of gonadotropin releasing hormone in the hypothalamus of ovariectomized pony mares. *J. Reprod. Fertil. Suppl.* **27**, 123–129.

Tucker, K. E., Cleaver, B. D., and Sharp, D. C. (1993). Does resumption of follicular estradiol synthesis during vernal transition in mares involve a shift in steroidogenic pathways? *Biol. Reprod.* **48**(Suppl. 1),188.

Villahoz, M. D., Squires, E. L., Voss, J. L., and Shideler, R. K. (1985). Some observations on early embryonic death in mares. *Theriogenology* **23**, 915–917.

HPL

see Human Placental Lactogen

Human Chorionic Gonadotropin

see Chorionic Gonadotropin, Human

Human Placental Lactogen
(Human Chorionic Somatomammotropin)

Michael Freemark

Duke University Medical Center

GLOSSARY

hGH and human prolactin receptors Single-chain polypeptides that mediate the biological actions of the somatogenic and lactogenic hormones.

human growth hormone (hGH) gene cluster The region on the long arm of chromosome 17 that encodes the genes for pituitary and placental growth hormones and human placental lactogen (hPL).

lactogens The family of polypeptide hormones that includes prolactin, the placental lactogens (chorionic somatomammotropins), and the primate growth hormones.

syncytiotrophoblasts Multinucleated cells of the placenta that synthesize and secrete hPL, placental growth hormone, human chorionic gonadotropin, and other hormones and growth factors.

Human placental lactogen (hPL; also called human chorionic somatomammotropin) is the major secretory protein product of the human placenta. It is a polypeptide hormone that has structural and functional similarities to human growth hormone and human prolactin. With diverse effects on nutri-

ent metabolism, hormone secretion, growth factor production, and mammary development, hPL plays a major role in maternal metabolism and fetal growth.

I. INTRODUCTION

A successful pregnancy requires support, sustenance, and immunological tolerance of the developing embryo and the provision and utilization of substrates for fetal growth. Adaptations in maternal metabolism must ensure the delivery of nutrients and growth factors to the fetus and must prepare the mother to feed and care for her newborn young. In these critical functions of pregnancy, the lactogenic hormones of the pituitary and placenta appear to play a central role.

The lactogenic hormones comprise a family of polypeptides that include prolactin, the placental lactogens, and the primate growth hormones. The lactogens maintain luteal function in early pregnancy in the rat and mouse, increase food intake and pancreatic insulin secretion, stimulate mammary growth and milk protein production, and promote maternal caretaking activity. These lactogenic effects ensure the adequacy of the uterine environment for embryonic implantation, bestow a continuous supply of nutrients for fetal development, and provide food, warmth, shelter, and protection for the newborn infant. Through effects on lymphocyte function, lactogens may modulate the maternal immune response to the fetus. Also, through actions on the renal tubule, gastrointestinal mucosae, and chorionic membranes, lactogens may play a role in maternal and amniotic fluid and electrolyte balance. Secreted directly into the fetal circulation and binding with high affinity to fetal tissues, the lactogens also exert direct effects on fetal metabolism and may modulate fetal growth.

II. HUMAN PLACENTAL LACTOGEN GENE EXPRESSION AND PLASMA CONCENTRATIONS IN THE MOTHER AND FETUS

The major lactogen of human pregnancy is called human placental lactogen (hPL) or human chorionic somatomammotropin. It is a 22.2-kDa nonglycosylated polypeptide containing 191 amino acids and two intramolecular disulfide bridges. The hormone is synthesized as a precursor (25 kDa) containing a 26-amino acid signal peptide. Cleavage of the signal peptide by membrane peptidases produces the mature 22.3 kDa secreted peptide. The mature protein has striking structural similarities to pituitary human growth hormone (hGH) and, to a lesser extent, human prolactin. The amino acid sequence of hPL is 85% identical to that of hGH and 27% identical to that of human prolactin. The genes encoding hPL reside on chromosome 17, band q22–24, within the 66-kb hGH gene cluster (Fig. 1). The five genes contained within this cluster are thought to have evolved from a common ancestral gene by duplication and nonreciprocal crossover events among nonallelic repetitive Alu elements in the intergenic regions. Each of the genes approximates 2.6 kb in length and contains five exons and four introns. The hGH-N gene is expressed in the anterior pituitary gland and in lymphoid and hematopoietic cells, whereas the four other genes are expressed in the placental syncytiotrophoblasts. The latter include the hGH-V gene, which encodes a 22-kDa protein that differs from pituitary GH by 13 amino acids; the hPL-A and hPL-B genes, which encode identical mature proteins that differ by a single amino acid in their

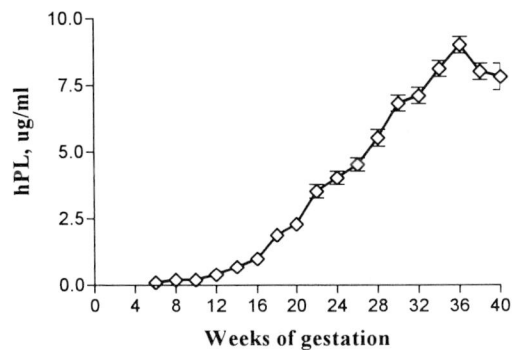

FIGURE 2 Serum concentrations of hPL in the mother during pregnancy (redrawn with permission from F. Frankenne *et al.*, The physiology of growth hormones in pregnant women and partial characterization of the placental GH variant, *J. Clin. Endocrinol. Metab.* **66**, 1171–1180, 1988).

signal peptides; and the hPL-L gene. Initially thought to be a pseudogene, the hPL-L gene appears to be expressed in minute amounts during normal pregnancy; however, the protein encoded by this gene has not yet been identified. Alternative splicing of the primary hPL-A and hPL-B transcripts may occur to a limited extent, though the proteins encoded by the alternatively spliced mRNAs have not yet been characterized.

The mRNAs encoding hPL are expressed uniformly in syncytiotrophoblast cells. Immunoreactive hPL can be identified in syncytiotrophoblasts within 5–10 days after implantation of the fertilized ovum, and by the third week postconception hPL can be detected by radioimmunoassay in the maternal circulation. Thereafter, maternal plasma hPL levels rise progressively to reach a peak of 5000–15,000 ng/ml several weeks prior to term (Fig. 2). During the third trimester, synthesis and release of hPL account for 10% of placental protein production, making hPL the most abundant peptide hormone produced in primates. A modest reduction in maternal plasma hPL concentrations precedes the onset of labor. Following delivery hPL disappears rapidly from maternal plasma, with a half-life of 10–30 min.

In amniotic fluid, the levels of hPL range from 300 to 400 ng/ml in early to midgestation. Concentrations in amniotic fluid rise to maximal levels (750–850 ng/ml) at 34–36 weeks and decline prior to term. hPL appears to be secreted directly into the amniotic

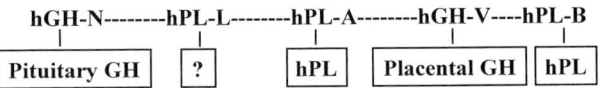

FIGURE 1 Schematic diagram of the organization of the genes encoded in the hGH gene cluster.

fluid because there is no transfer of the hormone from the maternal or fetal blood.

hPL is found in extraembryonic coelomic fluid and is detected in fetal blood as early as 8 weeks of gestation. The concentrations of hPL in fetal serum vary independently of maternal concentrations, owing to the failure of the somatogenic and lactogenic hormones to traverse the placental/decidual barrier and to differential secretion and regulation within the fetal and maternal compartments. Studies in human abortuses suggest a striking rise in fetal serum hPL concentrations between 12 and 20 weeks of pregnancy, with peak levels approximating 50–400 ng/ml, followed by a progressive decline in fetal PL levels to a mean of 15–30 ng/ml in term cord blood. However, direct fetal periumbilical cord blood sampling at various times during pregnancy reveals a different pattern of hormone expression. In the intact human fetus, hPL levels rise from a mean of 5 ng/ml at 20 weeks of gestation to a mean of 20–30 ng/ml at birth. Differences between the levels of hPL measured in abortuses and those measured in intact fetuses may reflect the stress of the abortion and/or the disruption of the normal fetal/placental circulation.

III. REGULATION OF PLACENTAL hPL SECRETION AND GENE EXPRESSION

For reasons related to the dangers of human, and particularly maternal and fetal, experimentation, it has been difficult to study the factors that regulate plasma PL concentrations in the mother or fetus. *In vivo*, the major factor controlling maternal plasma PL concentrations appears to be the relative mass of the placenta. The concentration of hPL mRNA in individual syncytiotrophoblast cells remains relatively constant throughout gestation, and the striking increases in placental hPL mRNA and maternal hPL concentrations during the latter half of pregnancy correlate closely with progressive increases in placental weight and syncytial mass. Consequently, maternal plasma hPL concentrations are commonly elevated in twin and triplicate pregnancies and are reduced in women with placental insufficiency.

A number of studies suggest that PL concentrations in the pregnant mother are regulated by nutrient availability; for example, prolonged (48–60 hr) fasting stimulates a rise in maternal hPL concentrations. However, short-term changes in plasma glucose, insulin, triglycerides, fatty acids, or amino acids have inconsistent or no effects on maternal plasma hPL concentrations. Similarly, estrogens, progestogens, glucocorticoids, thyrotropin-releasing hormone, somatostatin, and dopaminergic agonists have little or no effect on maternal plasma hPL concentrations *in vivo*. Corresponding changes in fetal hPL concentrations have never been assessed. Studies in intact and aborted fetuses at mid- and late gestation and in newborn infants at term have demonstrated positive correlations between fetal plasma hPL concentrations and fetal weight and fetal plasma insulin-like growth factor I (IGF-I) and IGF-II concentrations. However, studies of cord blood hPL levels in growth-retarded neonates and in infants of diabetic mothers have yielded inconsistent results, possibly reflecting the heterogeneity of the underlying diseases.

In contrast to these studies demonstrating lack of regulation of hPL expression *in vivo*, a number of *in vitro* studies suggest roles for hormones, cytokines, and growth factors in the regulation of hPL gene expression and placental hPL secretion. For example, hPL gene transcription, expression of hPL mRNA, and release of hPL in placental trophoblast cells in culture are induced by thyroid hormone, retinoic acid, 1,25-dihydroxy vitamin D_3, interleukin-6, interleukin-1β, and high-density lipoproteins. The effects of thyroid hormone, retinoic acid, and vitamin D_3 are exerted through binding to specific receptors that are expressed in trophoblast cells and are mediated through the interactions of the thyroid hormone and retinoic acid receptors with steroid response elements situated at nucleotides −954 to −1170 of the hPL promoter. The physiological significance of the response to thyroid hormone is unclear because concentrations of free T3 and free T4 in maternal serum and placental expression of thyroid hormone receptors decline between early and late pregnancy. Similarly, induction of hPL expression by IL-6 and IL-1β is difficult to interpret because the expression of these cytokines declines markedly as cytotrophoblasts differentiate into syncytiotrophoblasts. On the other hand, the effect of high-density lipoproteins (HDLs) on hPL expression is of interest because concentrations of HDLs in maternal serum increase in parallel with the rise in maternal hPL concentrations. The effect of HDLs appears to be mediated through apolipoprotein A1, which is expressed in the placenta as a truncated 22.7-kDa form. Interestingly, the administration of HDLs to the pregnant ewe stimulates an increase in plasma ovine PL concentrations, suggesting that HDLs may play a physiological role in the control of PL secretion *in vivo*.

IV. RECEPTORS FOR hPL AND THE LACTOGENIC HORMONES IN MATERNAL, FETAL, AND UTEROPLACENTAL TISSUES

A. The hGH and Human Prolactin Receptor Family

The biological actions of hGH, prolactin, and hPL are mediated through binding to specific, high-affinity receptors localized to plasma membranes of target tissues. Two receptors for the human somatogenic and lactogenic hormones have been characterized: the hGH receptor (hGHR) and the human prolactin receptor (hPRLR). The GH and PRL receptors are encoded by genes located on chromosome 5 (5p13–14). Each receptor is a single-chain polypeptide comprising extracellular, transmembrane, and cytoplasmic domains. The extracellular domain functions as a hormone-binding unit. Two pairs of cysteine residues situated near the amino terminus of each receptor are critical for protein folding. In the PRLR, a tryptophan-serine-residue-tryptophan-serine (WSXWS) motif lying just proximal to the transmembrane region appears essential for protein–protein interactions, homodimerization of the receptor, and cellular trafficking. Homodimerization of the GH as well as the PRL receptors appears necessary for efficient signaling. Within the proximal cytoplasmic domain of the GH and PRL receptors lies a proline-rich motif that interacts with a receptor-associated tyrosine kinase termed Janus kinase 2 (JAK2). The proline-rich motif is critical for receptor-mediated signaling. Binding of ligand to the full-length hGH or hPRL receptors is accompanied by activation of JAK2 and tyrosine phosphorylation of the receptor; activation of other cellular kinases including the Fyn, Raf-1, and mitogen-activator protein kinases; and tyrosine phosphorylation of various cytosolic proteins including insulin receptor substrate-1 and signal transducers and activators of transcription 1, 3, and 5.

Despite the striking homologies between hPL and hGH in gene structure and amino acid sequence, hPL binds with only low affinity (k_d 770 nM) to the hGH receptor. In contrast, hPL binds with very high affinity (k_d 46 pM) to the hPRLR, suggesting that hPL functions predominantly as a lactogen rather than as a somatogen in the pregnant mother and fetus. Limited evidence suggests that hPL may also bind to a distinct hPL receptor, expressed in human fetal skeletal muscle but as yet poorly characterized. Interestingly, the binding of hPL (and of hGH) to the hPRLR requires zinc, whereas the binding of hPRL to the hPRLR does not. The concentrations of zinc required for optimal binding of hPL or hGH (10–50 μM) are comparable to the concentrations of zinc in human fetal and maternal serum (9–12 μM), suggesting that zinc availability during pregnancy may modulate the fetal tissue response to the various lactogenic hormones.

B. PRL Receptors in Maternal, Fetal, and Uteroplacental Tissues

Ethical and practical considerations preclude the study of the expression of PRLRs in the mother during human pregnancy. In experimental animals such as the rat and mouse, the PRLR is expressed widely in maternal tissues, including the liver, ovary, adrenal, kidney, breast, pancreas, thymus, small intestine, choroid plexus, and hypothalamus. The PRLR is also expressed in the amnion, chorion, decidua, and placental trophoblast in humans and experimental animals.

In human abortuses, hPL binding activity and messenger RNA encoding the PRLR are expressed in diverse tissues, including the liver, small intestine, adrenal, heart, lung, pancreas, thymus, kidney, brain, testis, and skeletal muscle. In early gestation (7.5–10 weeks), PRLR immunoreactivity is detected in tissues derived from embryonic mesoderm, including the periadrenal and perinephric mesenchyme, the pulmonary and duodenal mesenchyme, the skeletal and cardiac myocytes, and the mesenchymal precartilage and maturing chondrocytes of the endochondral craniofacial and long bones, vertebrae, and ribs. Subsequently, there are striking changes in cellular distribution and magnitude of expression of PRLRs in a number of tissues. In the fetal adrenal the initial mesenchymal PRLR expression is succeeded (by 14 weeks gestation) by emergence of PRLR immunoreactivity in deeper fetal cortical cell layers. In the fetal kidney and lung, invagination of cortical mesenchyme is accompanied by progressive PRLR immunoreactivity in renal tubular and bronchial airway epithelial cells. In renal tubules and collecting ducts, the PRLR is expressed primarily on luminal surfaces, consistent with the effects of lactogenic hormones on tubular fluid and electrolyte transport in postnatal animals. In pulmonary airways, the PRLR is expressed initially in basal epithelial cells and subsequently in surface epithelial cells lining the bronchiolar lumen. In pancreas, the PRLR is detected in acinar cells and ducts in early gestation; in late gestation and in the postnatal period, the PRLR is expressed predominantly in pancreatic islets, colocalizing with insulin and glucagon. The widespread expression of the PRLR in the human fetus in early and midgestation, and the changing distribution of PRLRs in fetal tissues during ontogeny, suggest roles for hPL and other lactogenic hormones in tissue differentiation and development.

V. BIOLOGICAL ACTIONS OF hPL IN THE MOTHER AND FETUS AND THE ROLE OF hPL IN PREGNANCY

The mother undergoes striking changes in intermediary metabolism during the course of normal pregnancy. In early gestation, body fat accumulates in a centripetal distribution. In mid- to late gestation, the sensitivity to insulin is reduced and the mother develops postprandial hyperglycemia, hypertriglyceridemia, and hyperinsulinemia. Prolonged fasting (12–18 hr), on the other hand, leads to exaggerated production of free fatty acids and ketone bodies. These adaptations are thought to ensure continuous supply of glucose and amino acids to the fetus. A role for hPL in the metabolic adaptations to pregnancy is suggested by investigations in humans and experimental animals. For example, hPL and other lactogens increase food intake and stimulate glucose uptake, glucose oxidation, and incorporation of glucose into glycogen, glycerol, and fatty acids in isolated rat adipocytes. These effects may facilitate adipose accumulation in the mother in early pregnancy and during the fed state. hPL reduces insulin sensitivity and induces carbohydrate intolerance *in vivo* and stimulates [³H]thymidine incorporation, insulin gene transcription, insulin production, and glucose-dependent insulin secretion in pancreatic islet cells; these effects may contribute to postprandial hyperglycemia and hyperinsulinemia in the pregnant mother in mid- to late pregnancy. hPL also increases the basal rates of lipolysis in adipocytes *in vitro* and the plasma concentrations of nonesterified fatty acids, ketones, and glycerol *in vivo*. Thus, the increase in plasma hPL concentrations during fasting may facilitate mobilization and utilization of maternal free fatty acids for energy and thereby spare maternal glucose for the fetus. However, hPL is unlikely to regulate metabolism of carbohydrates and lipids in response to short-term fluctuations in nutrient availability because glucose, amino acids, and fatty acids exert little or no effect on maternal plasma hPL levels. Long-term effects of hPL on maternal metabolism are likely coordinated with metabolic effects of other maternal hormones such as placental growth hormone, prolactin, and sex steroids.

Secreted directly into the fetal as well as the maternal circulations, hPL has anabolic effects on fetal metabolism that may promote fetal growth. For example, hPL stimulates amino acid uptake, DNA synthesis, and IGF-I production in human fetal myoblasts, fibroblasts, and hepatocytes in culture. The effects of hPL on [³H]thymidine incorporation and amino acid transport are blunted though not abolished by an antiserum to IGF-I, suggesting that the action of hPL is mediated in part through paracrine release of IGF-I. hPL also stimulates DNA synthesis and insulin production in fetal pancreatic explants and promotes formation of islet-like cell clusters in cultures of human fetal pancreas. These findings suggest roles for hPL in the induction of islet cell growth and insulin production during late gestation.

A role for lactogenic hormones, such as hPL and prolactin, in the production of fetal adrenocortical steroid hormones is suggested by studies in the chronically catheterized baboon fetus. Intravenous administration of prolactin to the baboon fetus at midgestation stimulates an increase in fetal plasma dehydroepiandrosterone concentrations. Similar effects of lactogenic hormones have been recorded in some, but not all, studies using human fetal adrenal slices, baboon fetal adrenocortical cells, and isolated adrenocortical cells from human adults. The induction of fetal dehydroepiandrosterone secretion suggests a mechanism by which fetal hPL may modulate placental estrogen production.

A role for lactogenic hormones in the development of the human fetal lung is suggested by studies in pulmonary explants from human abortuses at midgestation. In concert with glucocorticoids, prolactin stimulated a two- or threefold increase in pulmonary surfactant production and lamellar body secretion. On the other hand, cortisol, prolactin, and hPL reduced the production of IGF-I in lung explant cultures. These observations suggest a role for lactogens in pulmonary maturation—a hypothesis supported by some, but not all, clinical studies in premature infants demonstrating a negative correlation between umbilical cord prolactin levels and the incidence of respiratory distress syndrome.

Care of a newborn baby requires induction of maternal mammary growth and milk production as well as changes in maternal behavior that facilitate maternal–infant interaction. hPL stimulates DNA synthesis in human mammary epithelial cells and growth of the ductal epithelium, suggesting that the hormone may facilitate mammary development prior to delivery. Recent studies demonstrate that administration of hPL to virgin rats stimulates maternal caretaking and nesting behavior; this effect is mediated through a direct action of the hormone on the medial preoptic area of the hypothalamus. Since hPL is detected in cerebrospinal fluid of pregnant women, the hormone may help to prepare the mother to nurture and care for her newborn baby after birth.

Despite abundant evidence that hPL plays important roles in maternal metabolism and fetal development, the virtual absence or very low levels of hPL in pregnant women with deletions of the hPL A and B genes is generally not accompanied by metabolic,

immunologic, or reproductive abnormalities in either the mother or the fetus. In some cases the gene deletion is associated with intrauterine growth retardation. Presumably, overlapping biological actions of various somatogenic and lactogenic hormones may provide a redundancy of function that may protect the mother and fetus against the loss of a single hormone. In a number of experimental systems, for example, the biological actions of human GH and prolactin are similar to those of hPL. Thus, the expression of hGH and/or prolactin may compensate for reductions in PL in patients with deletions of one or more of the hPL genes. There is currently no experimental evidence to confirm or refute such speculation. However, preliminary studies suggest that variant genes that are normally expressed at negligible levels (e.g., the hPL-L gene) may be expressed in pregnant patients bearing deletions of the hPL-A and B genes.

The roles of lactogenic hormones in maternal metabolism and fetal development may be clarified by studies of "knockout" mice with deletions of the mouse PRLR. Deletions of the PRLR produce a resistance to the actions of all lactogenic hormones including prolactin and the placental lactogens. Initial studies reveal striking reproductive abnormalities in PRLR-deficient mice: Homozygous females are sterile owing to defects in implantation and ovarian function, and heterozygous females have delayed mammary development and lactate poorly after the first pregnancy, and maternal behavior is aberrant. Homozygous males may also show reduced fertility. The effects of the PRLR knockout on mouse embryos have not yet been examined, though homozygous mutants survive to adulthood. No human model of tissue resistance to lactogenic hormones has yet been described.

VI. CONCLUSIONS AND FUTURE DIRECTIONS

Maternal and fetal metabolism and growth are complex processes subject to regulation by a multitude of genetic, nutritional, environmental, and hormonal factors. While hPL may not play an indispensable role in maternal and fetal metabolism, the hormone likely serves important functions in embryonic and fetal development. Studies of ontogenesis of prolactin receptors and biological actions of lactogenic hormones in maternal and fetal tissues suggest roles for hPL in maternal carbohydrate and lipid metabolism, mammary development, and maternal behavior. In the fetus, hPL may play roles in adrenocortical function, pulmonary maturation, hepatic growth factor production, and pancreatic development and insulin secretion. The effects of the lactogens on fetal skeletal growth remain speculative, though expression of prolactin receptors in human fetal chondrocytes and skeletal myocytes and direct anabolic effects of hPL in fetal myoblasts suggest roles for lactogenic hormones in cartilage and muscle growth and development.

Future research will likely explore differences in the mechanisms by which hGH, human prolactin, and hPL interact with the GH and PRL receptors and should probe the existence of a distinct hPL receptor in maternal and fetal tissues. The roles of hPL and other lactogens in tissue differentiation and organ development should be clarified, and the effects of amniotic fluid lactogens on fetal metabolism and growth should be elucidated. Finally, investigations should examine the mechanisms by which nutritional factors and trace mineral (e.g., zinc) availability in the pregnant mother modulate the fetal tissue response to lactogenic and somatogenic hormones.

See Also the Following Articles

Growth Factors; Hormones of Pregnancy; Prolactin, Actions of

Bibliography

Brelje, T. C., Scharp, D. W., Lacy, P. E., Ogren, L., Talamantes, F., Robertson, M., Friesen, H. G., and Sorenson, R. (1993). Effect of homologous placental lactogen, prolactins, and growth hormones on islet beta-cell division and insulin secretion in rat, mouse, and human islets: Implication for placental lactogen regulation of islet function during pregnancy. *Endocrinology* **132**, 879–887.

Chen, E. Y., Liao, Y.-C., Smith, D. H., Barrera-Saldana, H. A., Gelinas, R. E., and Seeburg, P. H. (1989). The human growth hormone locus: Nucleotide sequence, biology and evolution. *Genomics* **4**, 479–497.

Freemark, M. (1998). The roles of growth hormone, prolactin and placental lactogen in human fetal development: Critical analysis of molecular, cellular and clinical investigations. In *The Molecular and Cellular Basis of Pediatric Endocrinology* (S. Handwerger, Ed.). Humana Press, Totowa, NJ.

Freemark, M., Driscoll, P., Maaskant, R., Petryk, A., and Kelly, P. A. (1997). Ontogenesis of prolactin receptors in the human fetus in early gestation: Implications for tissue differentiation and development. *J. Clin. Invest.* **99**, 1107–1117.

Goffin, V., Shiverick, K. T., Kelly, P. A., and Martial, J. A. (1996). Sequence–function relationships within the expanding family of prolactin, growth hormone, placental lactogen, and related proteins in mammals. *Endocr. Rev.* **17**, 385–410.

Handwerger, S. (1998). Regulation of human placental lactogen gene expression. In *The Molecular and Cellular Basis of Pediatric Endocrinology* (S. Handwerger, Ed.). Humana Press, Totowa, NJ.

Hill, D. J., Crace, C. J., Strain, A. J., and Milner, R. D. G. (1986). Regulation of amino acid uptake and deoxyribonucleic acid synthesis in isolated human fetal fibroblasts and myoblasts: Effect of human placental lactogen, somatomedin-C, multiplication stimulating activity and insulin. *J. Clin. Endocrinol. Metab.* **62**, 753–760.

Hill, D. J., Freemark, M., Strain, A. J., Handwerger, S., and Milner, R. D. G. (1988). Placental lactogen and growth hormone receptors in human fetal tissues: Relationship to fetal plasma human placental lactogen concentrations and fetal growth. *J. Clin. Endocrinol. Metab.* **66**, 1283–1290.

Lowman, H. B., Cunningham, B. C., and Wells, J. A. (1991). Mutational analysis and protein engineering of receptor-binding determinants in human placental lactogen. *J. Biol. Chem.* **266**, 10982–10988.

Ormandy, C. J., Binart, N., Camus, A., Lucas, B., Buteau, H., Barra, J., Babinet, C., Edery, M., and Kelly, P. A. (1997). Null mutation of the prolactin receptor gene produces multiple reproductive defects in the mouse. *Genes Dev.* **11**, 167–178.

Rygaard, K., Revol, A., Esquivel-Escobedo, D., Beck, B. L., and Barrera-Saldana, H. A. (1998). Absence of human placental lactogen and placental growth hormone (hGH-V) during pregnancy: PCR analysis of the deletion. *Human Genetics* **102**, 87–92.

Talamantes, F., and Ogren, L. (1996). Human placental lactogen. In *Reproductive Endocrinology, Surgery and Technology* (E. Adashi, J. A. Rock, and Z. Rosenwaks, Eds.), pp. 769–781. Lippincott-Raven, Philadelphia.

Walker, W. H., Fitzpatrick, S. L., Barrera-Saldana, H. A., Resendez-Perez, D., and Saunders, G. F. (1991). The human placental lactogen genes: Structure, function, evolution and transcriptional regulation. *Endocr. Rev.* **12**, 316–328.

Hybridization

Michael L. Arnold

University of Georgia

GLOSSARY

endogenous selection Natural selection that acts against genotypes regardless of the environment.

exogenous selection Natural selection that acts for or against genotypes depending on environment X genotype interactions.

fitness Differential viability and fecundity of genotypes leading to differences in offspring numbers.

gene flow Transfer of DNA resulting from matings between individuals belonging to different populations.

hybrid zone A region in which two populations of individuals that are distinguishable on the basis of one or more heritable characters overlap spatially and temporally, mate, and form viable (and at least partially fertile) offspring.

introgressive hybridization The transfer of DNA between genetically distinguishable populations or groups of populations through repeated matings of hybrid offspring with parental individuals.

natural hybridization Successful (i.e., those that produce viable and at least partially fertile offspring) matings in nature between individuals from two populations, or groups of populations, which are distinguishable on the basis of one or more heritable characters.

natural selection A process whereby certain genotypes are favored or disfavored leading to differential survivorship and reproduction and the increase or decrease in the frequency of different alleles.

Natural hybridization has been investigated in three different types of evolutionary analyses. First, the systematics (i.e., the taxonomic or evolutionary relationships) of a particular group of organisms has been inferred from the presence/absence of crosses in nature. A second type of analysis has involved examining natural hybridization to detect barriers for gene flow and thus understand one aspect of the process of speciation—the development of reproductive isolation. The third category of study assumed that natural hybridization was an important evolutionary process per se. Its importance resided in the ability to generate novel genotypes that could lead to adaptive evolution and/or the founding of new evolutionary lineages (e.g., new species and genera). This article will focus on natural hybridization rather than artificial hybridization due to interest in understanding those processes that occur in nature that affect evolutionary change.

I. MODELS THAT PREDICT THE EVOLUTIONARY CONSEQUENCES OF NATURAL HYBRIDIZATION

Several models have been proposed that predict the outcome of natural hybridization. Common components of these models are natural selection and dispersal. The relative roles of these two processes vary greatly between models. Inferring which of the several models best explains a majority of instances of natural hybridization will indicate the relative effects of dispersal and selection on hybrid zone evolution and the evolutionary importance of natural hybridization.

Studies emphasizing the population-level processes of gene flow, natural selection, etc. have usually incorporated a conceptual framework based on selection against all hybrid individuals balanced by dispersal of organisms with parental genotypes into the hybrid zone. It is important to note that this model assumes that selection against hybrids is independent of environment (i.e., endogenous). Endogenous selection results from the disruption of parental gene arrays by recombination.

A second conceptual framework assumes that natural selection in hybrid zones involves interactions between specific genotypes and defined environments. These so-called environment-dependent models incorporate the effect of selection resulting from environmental variability (i.e., exogenous selection). Certain environment-dependent models incorporate dispersal as a major factor, whereas others do not. One model assumes that parental individuals are more fit in their respective habitats, whereas hybrids are assumed to be uniformly unfit. Another set of environment-dependent models holds that hybrids may be more fit than their parents in certain environments.

A final characteristic used to differentiate between models of hybrid zone evolution is stability. Many evolutionary biologists have argued that hybrid populations are ephemeral. This viewpoint is based on the assumption that there are only two possible outcomes for natural hybridization—reinforcement of reproductive barriers or genetic fusion of the hybridizing populations. Gene flow is thus viewed as capable of amalgamating the hybridizing populations. Alternatively, it has been hypothesized that, given sufficiently low fitnesses for the hybrid offspring, selection will finalize the development of reproductive isolation. Findings from empirical and theoretical (i.e., mathematical) analyses that indicate stability over long periods of time do not support ephemeral models.

A. Tension Zone Model

The assumptions of the tension zone model include selection against all hybrid genotypes across all environments. That such hybrid zones can be relatively stable is viewed as a consequence of continued immigration of parental individuals into the zone. This immigration leads to further hybridization and thus renews the hybrid genotypes depleted by negative selection; selection against hybrid individuals results from lowered viability and/or fertility. The interplay between selection against hybrid genotypes

and continued dispersal of parental individuals into the area of geographic overlap results in smooth clinal transitions. The slopes of these clinal transitions are determined by the strength of selection against the hybrid genotypes. Furthermore, the position (spatial) of the transitions results not from habitat preferences but rather from where the parents happen to coincide in space and the population densities of the two parents (i.e., the hybrid zone will tend to move toward areas of lower density).

B. Mosaic Model

Studies that have included an analysis of the genotypic (or morphologic) variation and habitat distributions have normally discovered correlations between the genetic and environmental characters. Indeed, a frequently made observation is that hybrid zones are "mosaics" of genotypes rather than smooth transitions from one parental form through an area of hybrid genotypes into the other parent. Figure 1

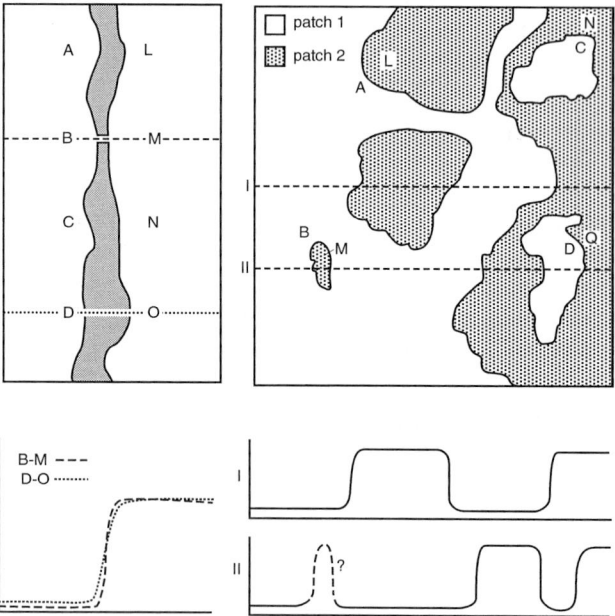

FIGURE 1 Two models of hybrid zone structure. (Left) Long, narrow hybrid zone with many characters showing concordant, coincident clinal variation. (Right) Patchy distribution of different genotypes due to habitat/genotype associations leading to fluctuations in genetic variation associated with different forms (reproduced with permission from Harrison, 1990). Letters indicate hypothetical sampling sites.

illustrates the spatial and genetic structure of a clinal and a mosaic hybrid zone. Mosaic zones are assumed to reflect the adaptation of the two parents to different, patchily distributed habitats. Although the mosaic model is environment dependent for the parental genotypes, hybrids are considered to be less fit in all habitats.

C. Bounded Hybrid Superiority Model

Under the bounded hybrid superiority (or hybrid-superiority) model some hybrid genotypes are assumed to be more fit than their parents in certain habitats. This model is thus environment dependent. The term "bounded" reflects the observation that hybrid zones between vertebrate species were sometimes narrow and located in ecotonal (i.e., transitional habitat) regions. Because hybrid zones were localized to areas of habitat transition it was concluded that the parental taxa were less well adapted to these regions relative to certain hybrid genotypes.

A second assumption is also generally included in this model: Hybrids are less fit than their parents in the parental habitats. This conclusion is not supported by the numerous instances of introgressive hybridization in which hybrid genotypes occupy parental environments. This latter observation suggests that hybrids may displace parental individuals from parental habitats.

D. Evolutionary Novelty Model

The evolutionary novelty model reflects the viewpoint that environment X genotype interactions are of fundamental importance for determining the outcome of hybridization episodes. Furthermore, the conceptual basis for this model assumes that hybrid genotypes are not always less fit than their parents (Fig. 2), even in the habitats normally occupied by the parental form(s). Under the evolutionary novelty model endogenous selection is proposed to act against certain hybrid genotypes through lowered viability or fertility. This selection results from chromosome structural differences and/or recombination that leads to the disruption of the coadapted parental genomes. Exogenous selection, on the other hand,

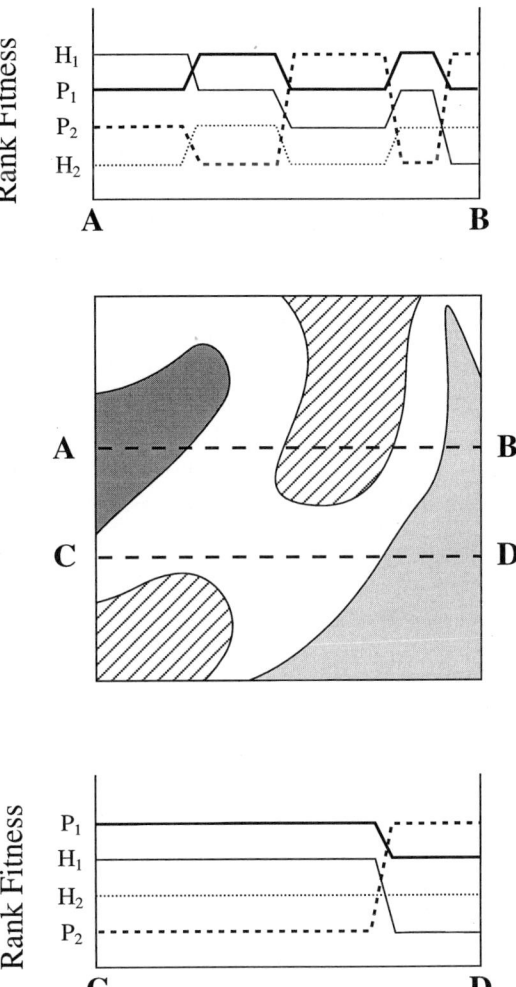

FIGURE 2 An illustration of the ecological component of the evolutionary novelty model. The environment in the central box consists of four habitats, indicated by the shading pattern. A transect across A–B would reveal a pattern of changing rank fitnesses among two parental species (P_1 and P_2) and two hybrid genotypes (H_1 and H_2), as shown in the top graph. The two parental species and one of the hybrid genotypes (H_1) each has the highest rank fitness in at least one habitat. Hybrid genotype H_2 has low rank fitness in all habitats. A transect across C–D would sample only two of the habitats and would show that both hybrid genotypes had either intermediate fitness or lower fitness than both parental species, as shown in the bottom graph. Many other patterns of changing rank fitness are, of course, possible.

is viewed as either positive or negative and affects the frequency of both hybrid and parental individuals. The assumptions of the evolutionary novelty model include the following: (i) Hybrid zones are environmentally complex; (ii) the initial hybrid generation (i.e., F_1) may be relatively difficult to form, resulting in hybrid zones being located in areas where sympatry is most likely to be long-lived (e.g., ecotones or disturbed habitats); (iii) exogenous selection results in different hybrid genotypes possessing fitnesses greater than, equivalent to, or less than those of their parental taxa in various habitats, including those occupied by the parental taxa; and (iv) endogenous selection results in some hybrid genotypes that possess uniformly lower fitness than their progenitors.

E. Natural Selection and the Term "Hybrid"

Every model of hybrid zone evolution assumes something about the roles of natural selection and/ or dispersal. However, deciding which of the models is applicable to a specific case of natural hybridization depends on deciphering the form of selection. In contrast, it is not possible to strongly infer whether a hybrid zone is environment dependent or independent based solely on estimates of dispersal distances for the parents. By the "form" of selection, it is meant whether hybrids are always selected against—or sometimes selected for—across environments.

Inferences concerning the relative fitness of hybrids are the basis for choosing between the alternative frameworks. However, an inherent problem exists with the use of the term hybrid to discuss the fitness of recombinant individuals (i.e., B_1 and F_2). In particular, placing all recombinant individuals within a single class (i.e., hybrid) can lead to incorrect conclusions concerning their relative fitnesses. For example, a general assumption has been that hybrids are uniformly unfit. Almost invariably the group of individuals referred to are genotypically heterogeneous. A similar conclusion might be reached if individuals belonging to two species were placed into a single class and then used for estimates of fitness across environments. In contrast, when hybrids are subdivided into well-defined genotypes, fitness estimates for these classes often range from less than to greater than the fitness of the parental forms.

II. BARRIERS TO NATURAL HYBRIDIZATION

Natural hybridization has been detected in literally thousands of species of plants and animals. One conclusion drawn from this observation is that natural hybridization may have a major evolutionary effect through the production of new genotypes that are more fit than their progenitors in certain environments. Paradoxically, there are numerous barriers to the production of hybrid individuals. Processes that act as barriers determine the frequency of hybridization events and the genotypes produced and thus affect the evolutionary outcome of particular hybridization episodes. Furthermore, as discussed previously, if certain hybrid generations (e.g., F_1 individuals) are difficult to produce, this can affect (i) where hybridization is most likely to occur (i.e., ecotones or disturbed habitats) and (ii) the types of advanced-generation hybrids formed (e.g., predominance of B_1 instead of F_2 individuals). Such limitations will determine which hybrid genotypes will be available for natural selection.

This paradox can be reconciled by a model incorporating positive selection favoring certain advanced-generation hybrid genotypes. Positive selection may indeed play a major role in the origin of new evolutionary lineages consisting of hybrid genotypes. This would occur if hybrids are favored over parents in certain environments. In addition, barriers to reproduction are also important for the founding of new evolutionary lineages (hybrid or not). Thus, processes that first limit natural hybridization may ultimately serve to isolate and protect the new hybrid lineage from amalgamation with other hybrids or the parental forms.

Reproductive barriers can be characterized in various ways (Fig. 3), such as life history stages and behavioral interactions (e.g., mating calls of different males). Such stages and interactions form a continuum of biological phenomena. However, using these processes as a framework for discussion permits the identification of definable phenomena that affect hybridization. Various processes have been identified

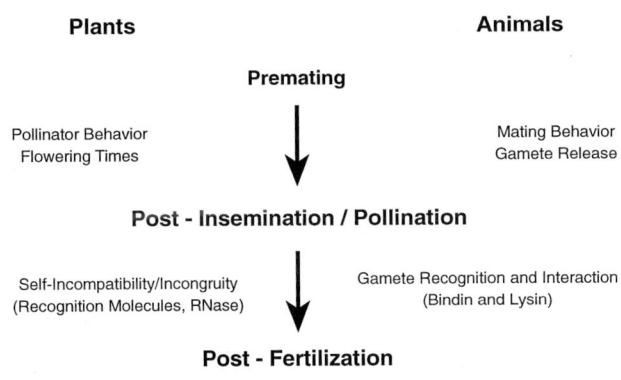

Plants   **Animals**

Premating

Pollinator Behavior   Mating Behavior
Flowering Times   Gamete Release

Post - Insemination / Pollination

Self-Incompatibility/Incongruity   Gamete Recognition and Interaction
(Recognition Molecules, RNase)   (Bindin and Lysin)

Post - Fertilization

Exogenous and Endogenous Selection   Exogenous and Endogenous Selection
(Viability, Survivorship, Fertility)   (Viability, Survivorship, Fertility)

FIGURE 3 Some barriers to hybridization for plant and animal taxa.

as important in limiting gene flow between divergent individuals and populations. These processes include premating behavior, gamete competition, embryo abortion, hybrid establishment, and hybrid fertility. It is interesting that studies of these barriers have been limited by certain conceptual frameworks that arose during the neo-Darwinian synthesis. An example of this has involved the examination of barriers at the postinsemination, prefertilization stage of reproduction (Fig. 4).

Barriers to natural hybridization that occur prior to fertilization have been recognized for centuries. For example, Darwin recognized that when gametes from different species were mixed together and then placed on the female reproductive structure, the most likely progeny were conspecific. Darwin's recogni-

tion of the importance of gamete competition does not reflect a widespread appreciation for this type of reproductive barrier. Indeed, this may be one of the strongest barriers to gene exchange between differentiated populations. There are at least two reasons for the underestimation of the affect of gamete competition on reproductive isolation. First, the goal of most crossing experiments is to determine the difficulty of hybrid formation. However, these experiments almost always use only one type of gamete (i.e., either conspecific or heterospecific) in crosses with a given female. For example, a female is normally placed into a population cage with numerous conspecific or heterospecific males, but not both. This design is assumed to give an accurate estimate of the barriers to hybridization. Instead, this design will most likely yield an underestimate of the strength of the barriers to hybridization. The competitive interactions between the differentiated gametes often result in the formation of conspecific rather than hybrid progeny (Fig. 5).

As highlighted previously, the presence of numerous, highly limiting reproductive barriers is paradoxical in light of the widespread occurrence of natural hybridization. Deciphering this paradox rests with the same conceptual framework used to explain evolution per se. The successful establishment of new genotypes (hybrid or otherwise) is thus determined by relatively rare events. In particular, the formation of hybrid progeny and their establishment in natural populations may initially be nonadaptive, or even

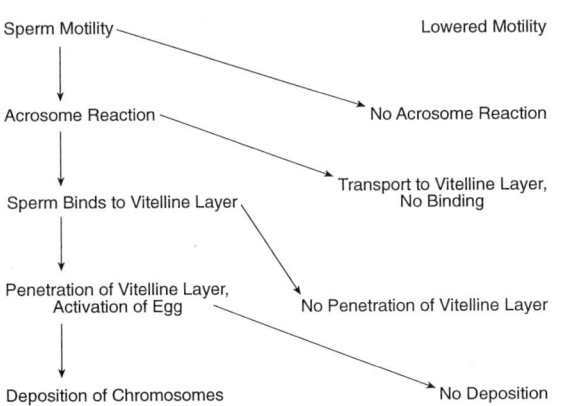

Sperm Motility   Lowered Motility

Acrosome Reaction   No Acrosome Reaction

Sperm Binds to Vitelline Layer   Transport to Vitelline Layer,
                                  No Binding

Penetration of Vitelline Layer,   No Penetration of Vitelline Layer
Activation of Egg

Deposition of Chromosomes   No Deposition

FIGURE 4 Postinsemination processes that may limit natural hybridization.

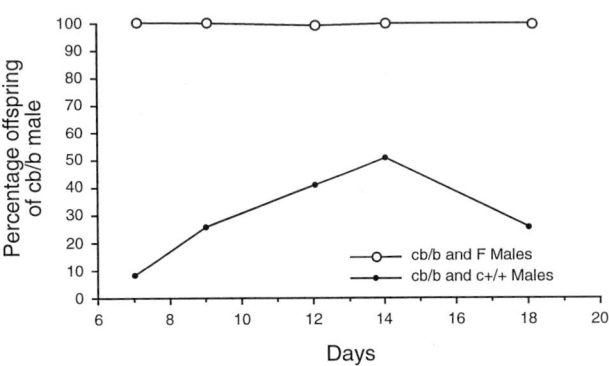

FIGURE 5 Percentage of offspring sired by a *Tribolium* (flour beetle) cb/b (i.e., conspecific) male produced by *Tribolium castaneum* females mated to the cb/b male and a *T. freemani* male (○) or the cb/b male and another conspecific male (●) (reproduced with permission from Robinson *et al.*, 1994).

maladaptive for the hybridizing individuals. However, these crosses may lead to adaptive evolution through the production of hybrid genotypes that are more fit than their parents in parental or novel habitats. Furthermore, barriers to natural hybridization may subsequently isolate the hybrid individuals and lead to divergent evolution.

III. OUTCOMES OF NATURAL HYBRIDIZATION

The potential outcomes of natural hybridization may include introgressive hybridization and both polyploid and diploid speciation. When introgressive hybridization occurs there is the opportunity for positive selection for alleles or genotypes in the introgressed form. This selection may involve either (or both) the nuclear or cytoplasmic genetic elements. New species may arise at the diploid level as a result of processes such as recombinational speciation (reorganization of novel parental chromosomal rearrangements). Polyploid "species," on the other hand, may have multiple origins from hybridization between different populations.

An increasingly important aspect of natural hybridization is its effect on biodiversity. In this regard, natural hybridization has largely been discussed as a negative influence on rare and endangered taxa. For example, the danger of genetic assimilation of rare forms by a more common taxon has been exemplified in numerous animal and plant species complexes. The conservation issues may, however, be more complex. For example, complexity arises if natural hybridization results in the salvaging of maximum genetic diversity in a rapidly changing environment. Genetic assimilation of a rare form by a more common taxon may thus be due to the numerical superiority of the latter. Alternatively, elevated hybrid fitness might result in the same pattern of assimilation. Similarly, introgression into the rare taxon may increase its fitness and facilitate an increase in the frequency of individuals belonging to this form (although now technically a hybrid). Finally, it is possible that hybrid genotypes may serve as vessels for the parental genotypes/phenotypes that may then be reconstituted via natural or artificial selection.

Regardless of the specific outcome of a given instance of natural hybridization, this process is now recognized as widespread. Furthermore, the effects of natural hybridization are likely to include adaptive and taxonomic diversification.

Bibliography

Anderson, E. (1949). *Introgressive Hybridization*. Wiley, New York.

Arnold, M. L. (1997). *Natural Hybridization and Evolution*. Oxford Univ. Press, Oxford.

Arnold, M. L., and Hodges, S. A. (1995). Are natural hybrids fit or unfit relative to their parents? *Trends Ecol. Evol.* **10**, 67–71.

Avise, J. C. (1994). *Molecular Markers, Natural History and Evolution*. Chapman & Hall, New York.

Barton, N. H., and Hewitt, G. M. (1985). Analysis of hybrid zones. *Annu. Rev. Ecol. Syst.* **16**, 113–148.

Dobzhansky, Th. (1940). Speciation as a stage in evolutionary divergence. *Am. Nat.* **74**, 312–321.

Endler, J. A. (1977). *Geographic Variation, Speciation, and Clines*. Princeton Univ. Press, Princeton, NJ.

Grant, V. (1986). *Plant Speciation*. Columbia Univ. Press, New York.

Harrison, R. G. (1986). Pattern and process in a narrow hybrid zone. *Heredity* **56**, 337–349.

Harrison, R. G. (1990). Hybrid zones: Windows on evolutionary process. *Oxford Surv. Evol. Biol.* **7**, 69–128.

Howard, D. J. (1986). A zone of overlap and hybridization between two ground cricket species. *Evolution* **40**, 34–43.

Lewontin, R. C., and Birch, L. C. (1966). Hybridization as a source of variation for adaptation to new environments. *Evolution* **20**, 315–336.

Moore, W. S. (1977). An evaluation of narrow hybrid zones in vertebrates. *Q. Rev. Biol.* **52**, 263–277.

Remington, C. L. (1968). Suture-zones of hybrid interaction between recently joined biotas. *Evol. Biol.* **2**, 321–428.

Rieseberg, L. H., and Wendel, J. F. (1993). Introgression and its consequences in plants. In *Hybrid Zones and the Evolutionary Process* (R. G. Harrison, Ed.), pp. 70–109. Oxford Univ. Press, Oxford, UK.

Robinson, T., Johnson, N. A., and Wade, M. J. (1994). Postcopulatory, prezygotic isolation: Intraspecific and interspecific sperm precedence in *Tribolium* spp., flour beetles. *Heredity* **73**, 155–159.

Stebbins, G. L., Jr. (1959). The role of hybridization in evolution. *Proc. Am. Philos. Soc.* **103**, 231–251.

Wagner, W. H., Jr. (1970). Biosystematics and evolutionary noise. *Taxon* **19**, 146–151.

Hydra

Vicki J. Martin

University of Notre Dame

GLOSSARY

budding An asexual form of reproduction that generates genetically identical individuals.

Cnidaria A phylum of diverse animals including the hydras, jellyfish, sea anemones, corals, hydroids, sea fans, siphonophores, and zoanthids.

diploblastic Consisting of two germ layers, an ectoderm and an endoderm, which give rise to two adult epithelia, the epidermis and gastrodermis, respectively.

epidermis The outer epithelium of the body wall.

gastrodermis The inner epithelium of the body wall that surrounds the gastrovascular cavity.

interstitial cells Multipotent stem cells that are lodged in the interstices among the epithelial cells.

mesoglea An extracellular basement membrane that separates the epidermis and gastrodermis.

morphogen A substance existing in a concentration gradient along the body axis that specifies position along the body axis.

nematocytes Stinging cells of cnidarians.

polyp An adult morphology with the shape of an elongated cylinder; the aboral end, the foot (basal disc), is attached to the substrate, whereas the opposite free end, the oral end, bears a mouth and tentacles.

radial symmetry Symmetry in which the body parts are arranged radially around a central oral–aboral axis.

stem cell A cell capable of extensive proliferation, creating more stem cells (self-renewal), as well as differentiation into cellular progeny.

T he freshwater hydra is a member of the Cnidaria, a group of animals that arose early in metazoan evolution. Because of their anatomical simplicity, hydra lend themselves well to studies in cell growth, morphogenesis, budding, regeneration, patterning, differentiation, embryogenesis, and neurobiology. A number of reasons account for hydra's popularity. It has a simple body plan consisting of a tube with a head and a foot at opposite ends. It is one of the simplest organisms, composed of 20–25 cell types with a total of 30,000–200,000 cells depending on the species. These cells are arranged in two concentric germ layers, an ectoderm and an endoderm, separated by a mesoglea, an extracellular matrix similar to a basement membrane. Both germ layers consist of epithelial cells and the interstitial cell lineage, a population of multipotent stem cells, and their differentiation products: secretory cells, nematocytes, neurons, and gametes. Hydra have distinct tissues but no organs. They possess digestive, muscular, and nervous systems in a primitive stage of development. For example, the nervous system is a simple nerve net composed of naked polar and nonpolar neurons. The muscle system is formed by epitheliomuscular cells, cells that combine features of epithelial cells and muscle cells. These are the most primitive muscle cells found in the metazoans. There are no distinct respiratory, excretory, circulatory, or reproductive systems. Gonads consist merely of aggregations of sex cells. Animals are easy and inexpensive to grow in the laboratory and have a short doubling time of 2 or 3 days. They can be asexually propagated forever via budding, and animals produced by budding are genetically identical. Hydra are susceptible to experimental manipulations such as excision and grafting, cell dissociation and reaggregation to form intact animals, and altering the cell composition, such as completely eliminating the entire interstitial cell lineage. They exhibit an amazing capacity for regeneration, whether from small pieces of excised tissue or from a reaggregate of dissociated cells. Finally, studying the development of hydra has become important because diploblastic hydra occupy a strategic position for examining the evolution of developmental mechanisms. A comparison of the development of hydra with those of the more complex triploblastic metazoans will reveal those features that are common to all animals. Furthermore, an investigation of these common features will uncover basic genetic mechanisms underlying development.

I. INTRODUCTION

Hydra are the most thoroughly studied cnidarian. They are small freshwater polyps ranging from 1 to 20 mm in body length. There is no medusa. Each polyp is an elongated cylinder. The aboral end, the foot or basal disc, is attached to the substrate, whereas the opposite free end, the oral end, bears the mouth and tentacles (Figs. 1 and 2). Tentacles are slender evaginations of the body wall that encircle the oral end and function in defense and the capture of food. Usually the tentacles are the same length as the body; however, in some species tentacles may extend 1.5 to 5 times the length of the hydra body. Tentacles are armed with stinging cells called nematocytes. Each nematocyte has a capsule with a poisonous thread called a nematocyst. Hydra exhibit radially symmetry. There is one main axis of symmetry, the oral–aboral axis, that extends from mouth to base. All parts are arranged concentrically around this axis. Reproduction is by budding or by production of gametes in the ectoderm. A fertilized egg develops into a tiny polyp. There is no larval stage. Both gonochoristic (dioecious) and hermaphroditic individuals exist.

Hydra are found in bodies of fresh water from the Amazon to Alaska. They occur at depths from shallow water to 60 m or more and at temperatures from near freezing to 25°C. Many hydra species coexist in a single pond. Frequently, small green hydra containing symbiotic algae and brown hydra are found together. Animals attach to any reasonable hard surface. Favorite substrates include water lily stems, dead leaves, sticks, and stones. Hydra often detach and reattach to substrates. They can somer-

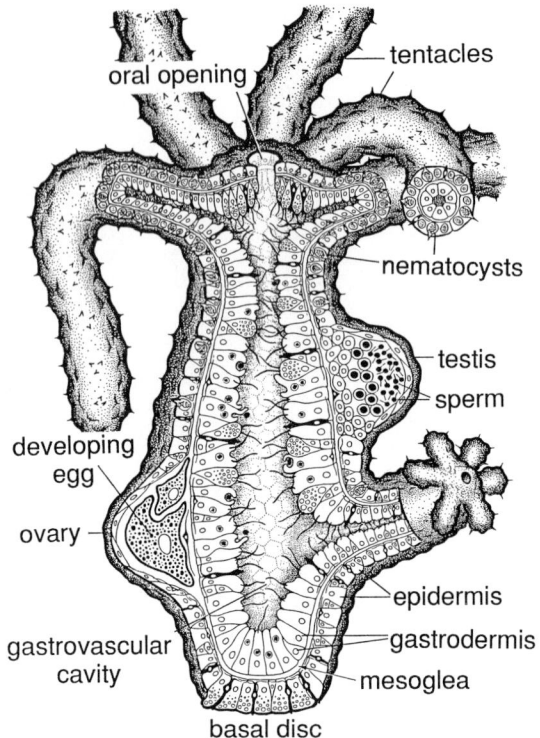

FIGURE 1 Polyp of the freshwater hydra. The anterior oral end bears the mouth and tentacles, whereas the posterior aboral end, the basal disc (foot), attaches to a substrate. The body wall is composed of an outer epidermis, a middle mesoglea, and an inner gastrodermis that surrounds a gastrovascular cavity. Hydra reproduces asexually by budding and sexually by producing eggs or sperm. [Reproduced with permission from Pearse *et al.* (1987) *Living Invertebrates,* p. 113. Boxwood Press, Pacific Grove, CA.]

sault through the water and frequently are found floating upside down in the water surface film with their tentacles hanging down.

II. STRUCTURE OF HYDRA

A. Body Plan

Hydra has a simple body plan with a head and a foot at opposite ends (Figs. 1 and 2). The body wall contains two epithelial layers, an outer ectoderm (the epidermis) and an inner endoderm (the gastrodermis), which surrounds a central digestive cavity, the gastrovascular cavity. This cavity has a single opening, the mouth, at the oral end of the polyp. This

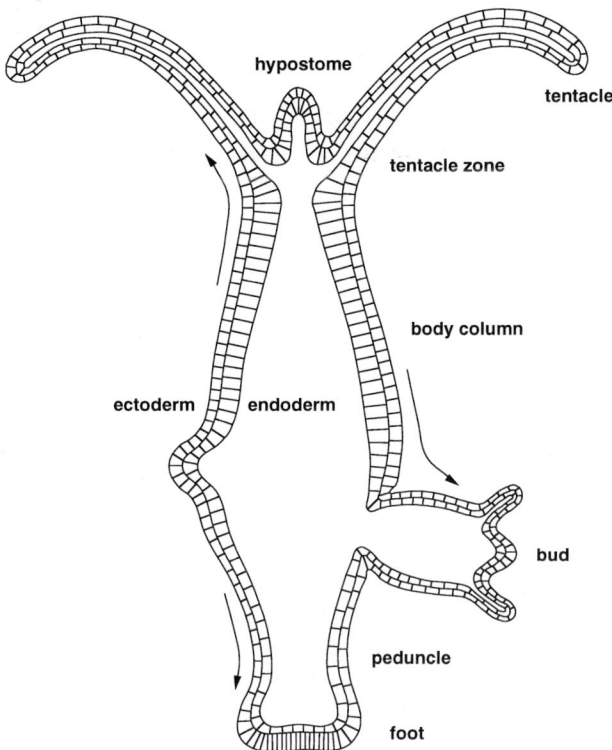

FIGURE 2 Regions of a hydra polyp indicating directions of tissue displacement along the body column (reproduced with permission from Bode, 1996).

oral opening functions as both mouth and anus, serving for food intake and ejection of indigestible material. The cavity is often referred to as a coelenteron because it functions both for digestion and for circulation. Between the two epithelia is an extracellular layer, the mesoglea. In hydra, the mesoglea is a thin, noncellular layer composed of type IV collagen, fibronectin, heparan sulfate proteoglycan, and laminin.

The body regions of hydra are few (Fig. 2). The head contains two parts: a hypostome and tentacle zone. The hypostome is a cone-shaped region bearing the mouth. Below the hypostome emerges a tentacle zone, bearing five to seven evenly spaced tentacles. Below the head region is the body column. This column contains three regions: an upper gastric region, a middle budding zone where asexual progeny form, and a lower peduncle. Below the peduncle, at the basal end of the animal, is the foot (basal

disc), which functions to attach the animal to a substrate.

B. Cell Types

The bilayered epithelial tissue arrangement of the body wall is the same throughout the animal (Fig. 2). The epithelial cells of the ectoderm are epitheliomuscular cells, whereas the epithelial cells of the endoderm are nutritive muscle cells. Lodged in the interstitial spaces among the epithelial cells of both layers are cells of the interstitial cell lineage (Figs. 3–5A). This lineage consists of a population of multipotent stem cells, the interstitial cells, three classes of somatic differentiation products (secretory cells, nematocytes, and neurons), and both types of gametes (eggs and sperm). There are two types of secretory cells (gland cells and mucous cells), four types of nematocytes (desmonemes, atrichous isorhizas, holotrichous isorhizas, and stenoteles), at least three types of sensory neurons, and an undefined number of types of ganglion neurons. Cells of the interstitial lineage comprise about 80% of all cells in the adult. Interstitial cells and differentiation intermediates of the three classes of somatic cells are found in the body column and parts of the head but are few or absent in the tentacles, foot, and apex of the hypostome. Furthermore, nematocyte differentiation is confined to the ectoderm of the body column, whereas secretory cell differentiation is confined to the endoderm of the head and body column. Upon completion of differentiation, the majority of the nematocytes migrate into the ectoderm of the tentacles. Neuron differentiation occurs throughout the body column of both layers as well as in the tentacle zone and the base of the hypostome. These neurons form a continuous nerve net throughout both layers of the animal (Fig. 5B).

C. Tissue Dynamics

The epithelial cells of hydra are in a steady state of production and loss. In both layers of the body column, the epithelial cells continuously divide, with a cell cycle time of 3 or 4 days. Thus, the tissue mass of the animal doubles in 3 or 4 days; however, the animal remains constant in size. As the epithelial

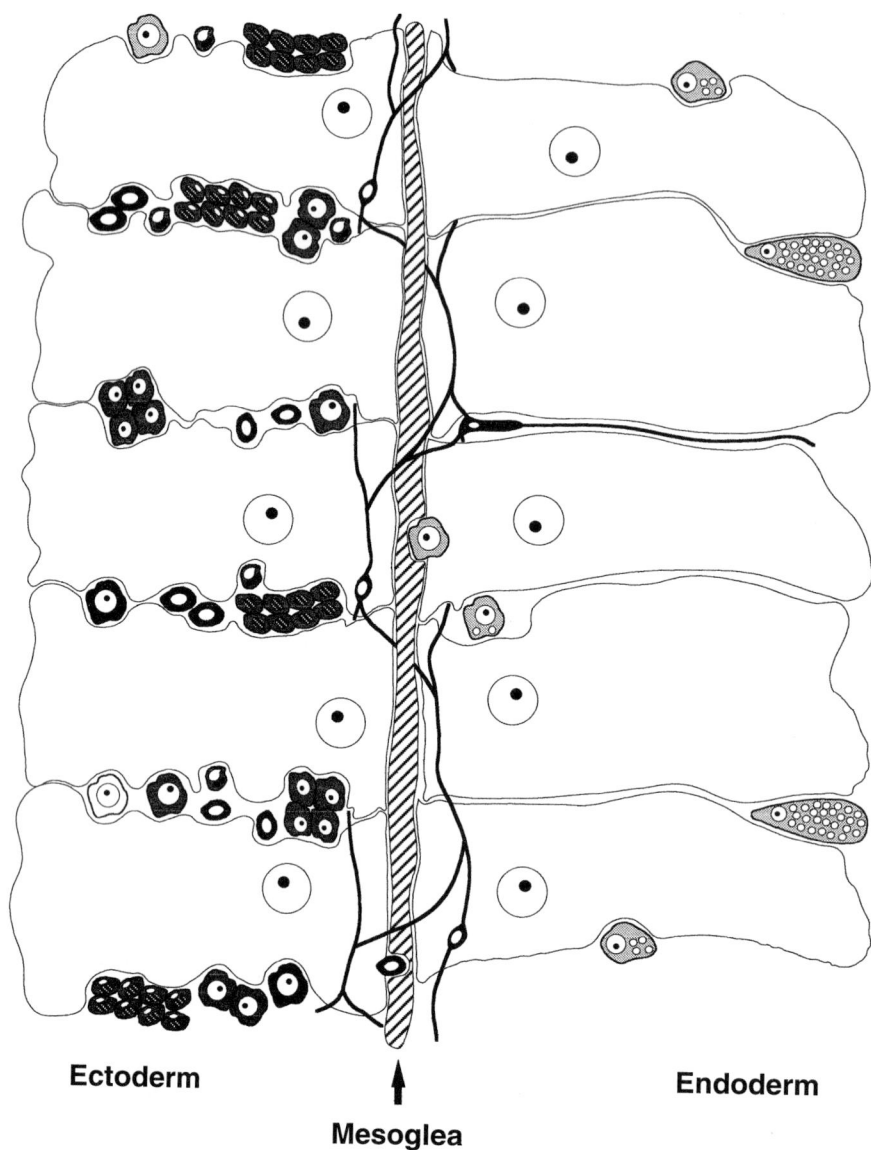

Ectoderm   ↑   **Endoderm**

Mesoglea

FIGURE 3   Section of the body column showing the location of cells of the interstitial cell lineage among the interstices of the epithelial cells. Interstitial lineage cell shades are as in Fig. 4 (reproduced with permission from Bode, 1996).

cells divide, they are continuously displaced along the body column toward an extremity (tips of tentacles, apex of hypostome, and base of foot) (Fig. 2). Cells in the upper part of the body column are displaced into the head, onto the tentacles, and eventually lost from the animal by sloughing at the tips of the tentacles. Cells in the middle or lower end of the body column are displaced down the body column and lost into developing buds or sloughed at the foot. Cells of the interstitial cell lineage are also con-

tinuously lost, as they too are displaced into an extremity or into a bud. Consequently, the interstitial stem cells in the body column must also continuously proliferate and differentiate. Thus, hydra consist of three stem cell systems: two epithelial cell lineages and the interstitial cell lineage. Under normal conditions these self-renewing cells are in a steady state in which the production of cells is balanced by the loss of these cells. Hence, hydra never grow old and are potentially immortal.

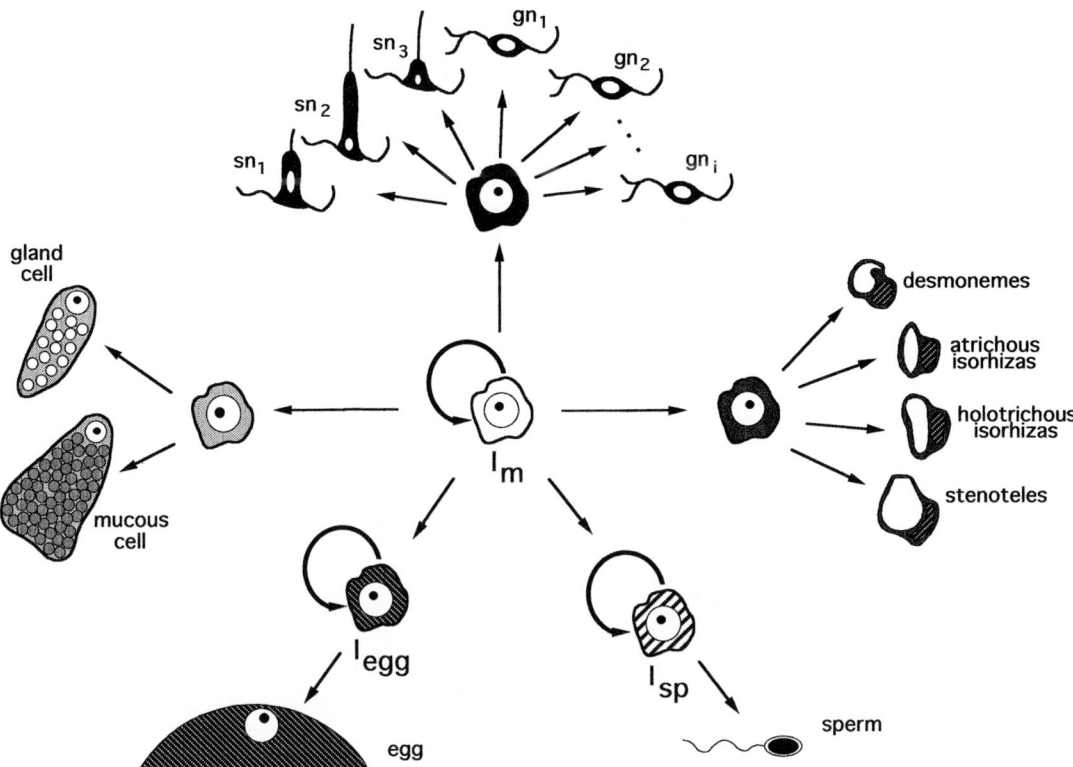

FIGURE 4   The interstitial cell lineage of hydra. This lineage includes multipotent stem cells (I_m), unipotent stem cells restricted to egg (I_{egg}) and sperm (I_{sp}) differentiation, four types of nematocytes, two types of secretory cells, sensory neurons (sn), ganglion cells (gn), and two types of gametes (eggs and sperm) (reproduced with permission from Bode, 1996).

III. REPRODUCTION OF HYDRA

A. Asexual Reproduction

Budding is the chief means of reproduction in hydra (Fig. 6A). All buds are genetically identical to their parent. A bud develops as a simple evagination of the body wall from a region of the body column designated the budding zone.

This zone is located at the lower end of the gastric region above the foot. Each bud contains an extension of the parent gastrovascular cavity and both epithelia. The bud rapidly elongates and forms a mouth and tentacles at its distal end. Eventually, the bud detaches from the parent and becomes an independent hydra. The entire process requires 2 or 3 days, and well-fed hydra bud every 3 or 4 days. At times multiple buds may arise simultaneously from the budding zone. Also, buds and gonads may occur concurrently in the same individual (Fig. 6A).

B. Sexual Reproduction

In some hydra species, e.g., *Hydra vulgaris* and *Hydra utahensis*, sexual reproduction occurs spontaneously and no specific environmental cues have been identified. In *Hydra oligactis* and *Hydra hymanae* sexual reproduction can be induced by lowering the temperature and in *Hydra magnipapillata* by decreasing the food supply.

1. Gametogenesis

Gametes arise from a population of interstitial cells that are dispersed within intercellular spaces of the epitheliomuscular cells (Fig. 1). During gametogenesis interstitial cells aggregate in specific ectodermal regions of the body column to form gonads. Each gonad consists of clusters of interstitial cells that cause the overlying ectodermal epithelium to bulge. Each bulge is supported by the overlying stretched ectodermal cells (Fig. 1).

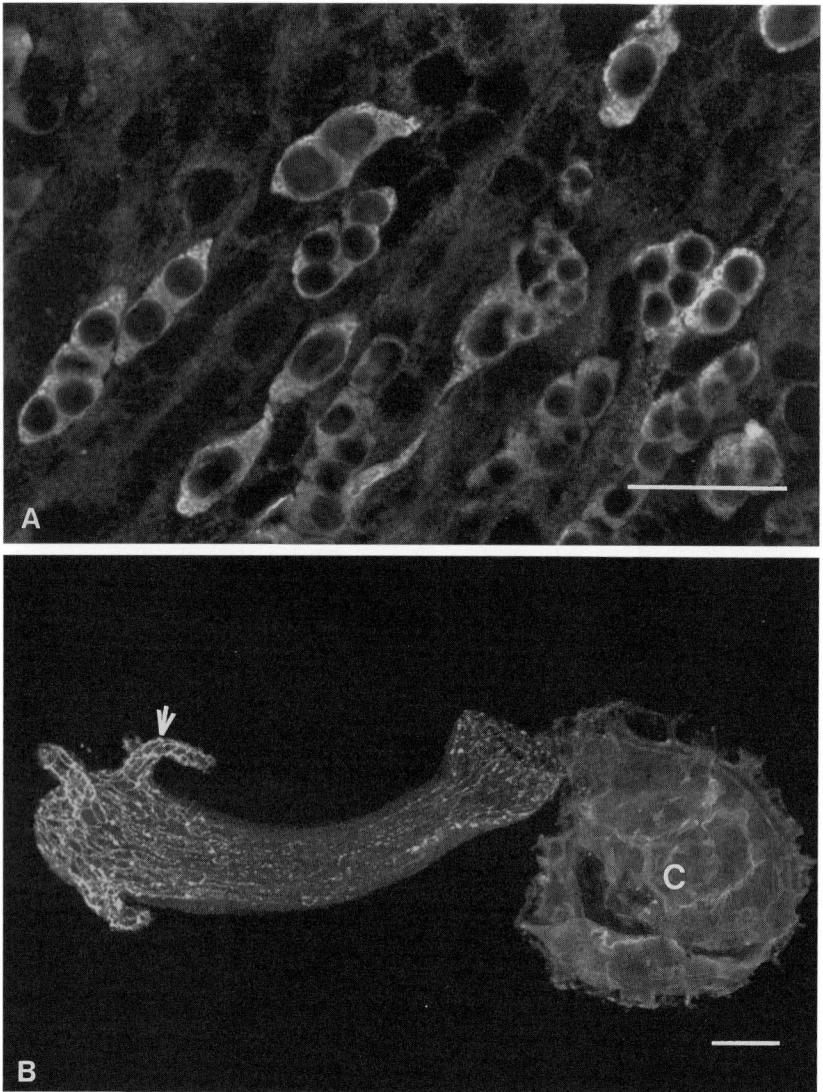

FIGURE 5 (A) Body column of hydra showing interstitial cells and their differentiating intermediates stained with a CP4 monoclonal antibody. Scale bar =25 μm. (B) Hydra hatching out of its cuticle (C). The young hatchling is forming a RFamide-positive nerve net (arrow) composed of sensory neurons and ganglion cells. Scale bar =100 μm.

Testes are conical protrusions with an opening, the nipple, through which sperm exit. Testes are found along the entire upper body column from beneath the head to the budding zone (Fig. 6B). Ovaries are large ectodermal bulges in the lower half of the body column near the budding region (Fig. 6C). Each ovary houses one to several developing eggs covered by a thin layer of unspecialized flattened epitheliomuscular cells.

The events of spermatogenesis in hydra are similar to those in higher metazoans, including humans. Interstitial cells committed to the sperm lineage proliferate throughout the entire body column, forming nests containing 4–32 cells. Each nest contains a syncytium of interstitial cells connected by intercellular bridges. These nests of interstitial cells migrate to testes-forming areas where they undergo a morphological differentiation into sperm intermediates.

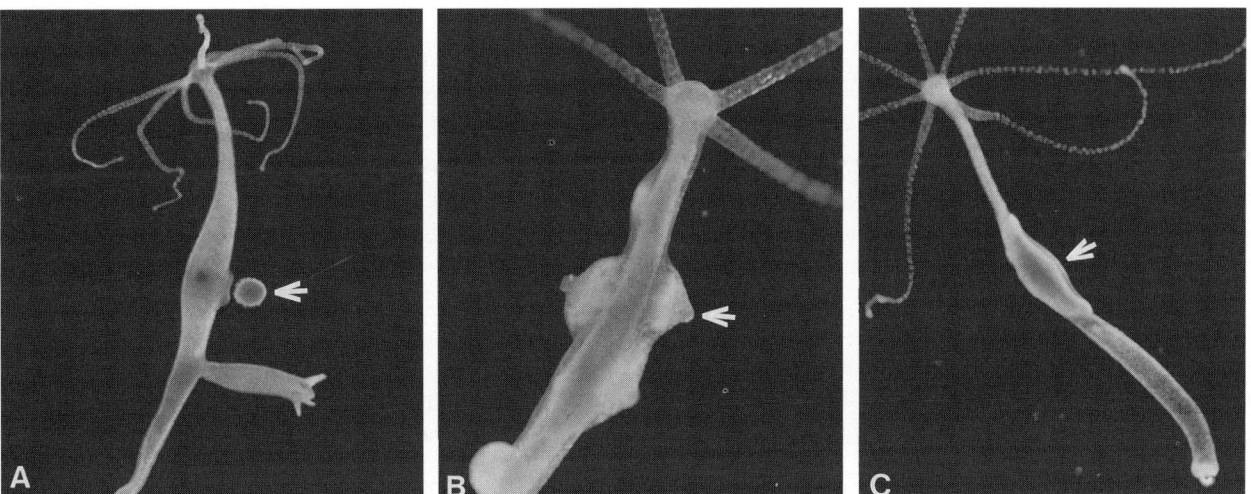

FIGURE 6 (A) Budding hydra with an egg (arrow) (×20); (B) male hydra with testes (arrow) (×30); (C) female hydra forming an egg (arrow) (×10).

These intermediates undergo meiosis to form spermatids and sperm. The time required to complete spermatogenesis is less than 3 days, and proliferation of interstitial cells proceeds for 2 weeks prior to the development of morphologically identifiable testes.

In females, gamete differentiation begins with the proliferation of interstitial cells. These interstitial cells produce multiple clusters containing 4–32 cells that are located in the interstitial spaces of the ectoderm between the mesoglea and the epitheliomuscular cells. During egg formation one of these interstitial cells becomes the oocyte, whereas the others start to synthesize yolk. These yolk-producing cells are called nurse cells. The oocyte increases in mass by extending finger-like processes in between the epitheliomuscular cells (Fig. 6C). These processes fuse

with and/or engulf large numbers of nurse cells. As the oocyte grows in mass it becomes spherical, distending the overlying ectoderm (Fig. 8A). Egg formation, beginning with the division of interstitial cells, requires 3 or 4 days for completion.

Once the egg becomes spherical, polar body formation occurs. The egg nucleus migrates to the apical end of the cell, where it undergoes meiosis, resulting in the extrusion of two polar bodies. In the region of polar body discharge, the overlying ectoderm ruptures exposing the egg (Fig. 7). This ruptured ectoderm recedes around the edge of the egg, forming a raised ring of tissue at the base of the egg called the egg cup. Each egg measures 100–500 μm in diameter and is loosely attached to the egg cup by thin strands of tissue. Eggs are now fertilizable for 2 hr. If they

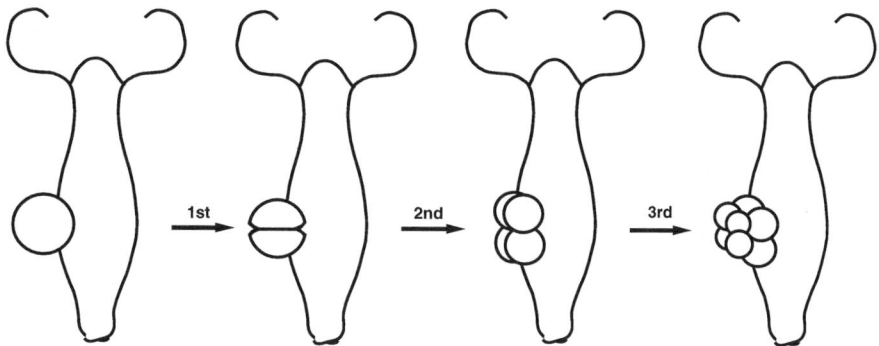

FIGURE 7 Early development of hydra showing an uncleaved egg and the first three cleavages.

are not fertilized during this time they swell and disintegrate within the day.

2. Fertilization

Fertilization occurs at the site of emission of the polar bodies. In this area of the egg the surface is indented to form a cuplike depression, the fertilization pit. Below this pit is the egg pronucleus. A sperm enters the fertilization pit and fuses with the plasma membrane of the egg. During fusion both the sperm head and the midpiece are incorporated into the egg. Fusion of male and female pronuclei occurs 8–10 min after insemination, and the fertilization pit disappears 15 min after insemination.

3. Events of Embryogenesis

i. Cleavage and Blastulation First cleavage begins 2 hr after the egg breaks through the mother's body column (Figs. 7 and 8B). First cleavage and all subsequent divisions are unipolar; that is, the cleavage furrow progresses inward from one side of each cell. First cleavage is equal holoblastic; the cleavage furrow starts at the apical end of the egg (the side away from the egg cup and the site of polar body discharge) and moves inward to the basal pole of the egg (perpendicular to the mother's body column), producing two blastomeres (Fig. 8C). Second cleavage begins shortly before the first division is complete and also occurs in an apical–basal direction, perpendicular to the first cleavage (Fig. 7). It is equal holoblastic producing four equal-sized blastomeres (Fig. 8D). The third division is equatorial (parallel to the mother's body column) and begins at the end of the embryo closest to the head of the mother hydra (Figs. 7 and 8E). This division occurs much closer to the apical end of the embryo, resulting

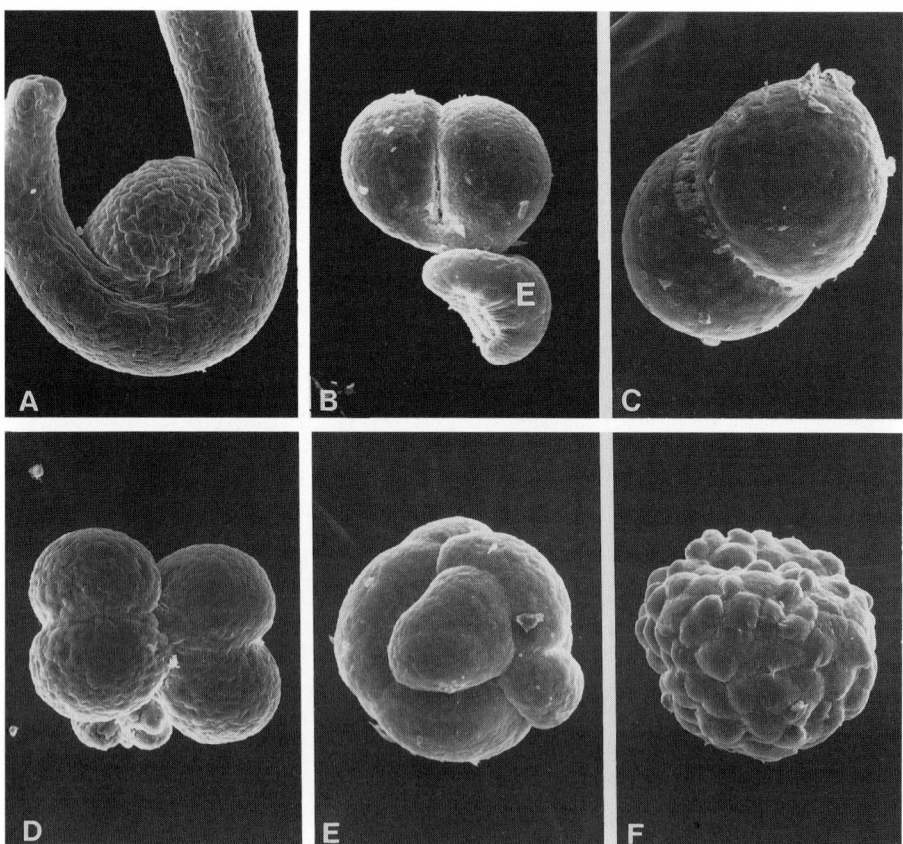

FIGURE 8 (A) Hydra forming an egg. As the egg grows, it distends the overlying ectoderm of the parent (×75). (B) Hydra egg undergoing first cleavage. E, egg cup (×100). (C) Two-cell stage (×190). (D) Four-cell embryo (×190). (E) Forming eight-cell embryo showing two tiers of unequal-sized cells (×110). (F) Late cleavage stage that resembles a morula (×130).

in two tiers of unequal-sized blastomeres. At the end of the third cleavage, the embryo consists of four large cells (closest to the egg cup) and four small cells (farthest from the egg cup) (Fig. 7). Thereafter, cleavage planes become oblique and cell divisions asynchronous, with those near the apical end smaller than those at the basal end. Cleavage divisions occur every 60 min up to the 32-cell stage. Thereafter, until the formation of the blastula, the cell division rate slows down, and the cells become more uniform in size (Fig. 8F).

A hollow spherical blastula is formed by 6–8 hr after fertilization (Fig. 9A). The wall of the blastula is a single layer of cells surrounding a fluid-filled blastocoel. The outer surface of the blastula resembles a patchwork quilt because the outlines of the cells are clearly delineated by rows of short microvilli.

ii. Gastrulation
Gastrulation occurs via multipolar ingression. Individual cells of the wall of the blastula detach from their neighbors and move into the blastocoel in a manner similar to that of primary mesenchyme in sea urchin embryos. Ingression begins at the apical end of the embryo, opposite the egg cup, and progresses outward from that point all around the embryo. The process of ingression occurs in about 4 hr, and when complete, the embryo consists of an outer layer of columnar cells and a central mass of unorganized spherical cells (Fig. 9B). No cavity remains. Gastrulation is completed by 10–12 hr postfertilization. There is little visible activity for the next 10–12 hr.

iii. Cuticle Deposition
The next stage involves the formation of a cuticle, a thick protective layer that is also referred to as the embryotheca (Figs. 9C

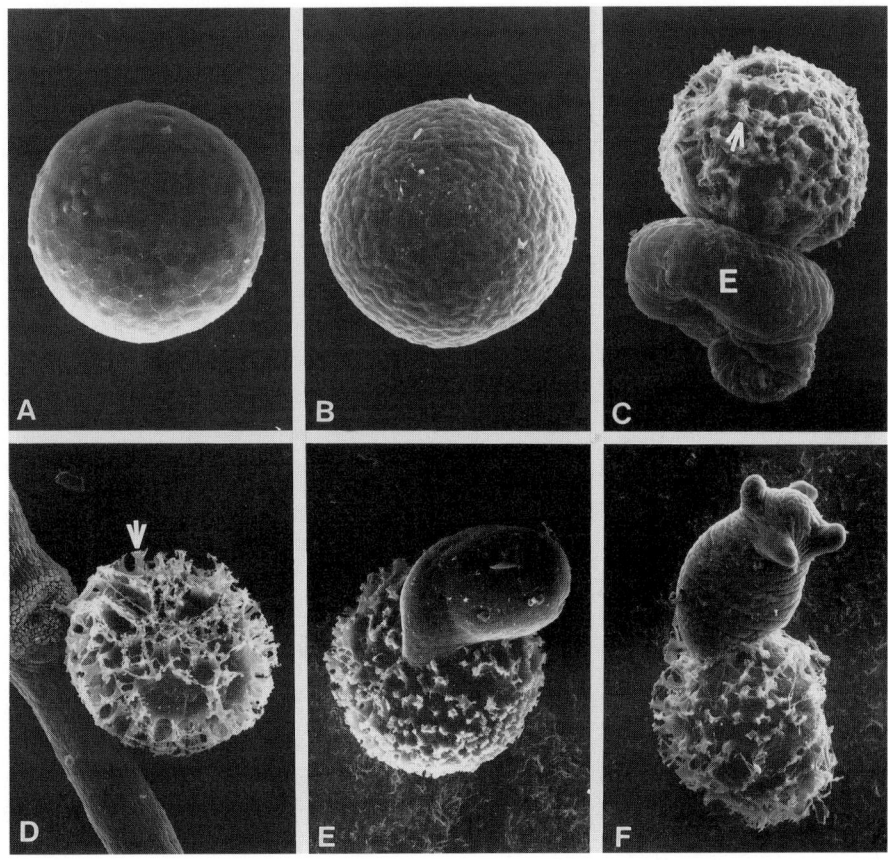

FIGURE 9 (A) Blastula stage (×200); (B) gastrula stage (×150); (C) Early cuticle stage showing deposition of the cuticle (arrow). E, egg cup (×130); (D) hydra embryo encased in the cuticle. Thick spines (arrow) adorn the cuticle (×75); (E) hydra hatchling emerging from the cuticle (×80); (F) hydra hatchling with four tentacle primordia (×95).

and 9D). At 20–24 hr postfertilization, the cells of the outer layer extend filopodia into the surrounding medium. Cuticular material is deposited in layers around the filopodia and on the apical surfaces of the cells. The material builds up producing a multi-layered structure over the surface of the embryo with ornate spines 25–50 μm long forming where filopodia are located (Fig. 9D). Toward the end of cuticle deposition, the filopodia are retracted and the resulting channels filled with cuticular material. The process is complete 40–48 hr after fertilization. By this time the outer layer of cells has flattened against the cuticle.

Cuticle deposition starts on the basal side of the embryo in the egg cup region and slowly progresses outward around the embryo in an apical direction. The final distribution of material is asymmetric, with the cuticle being thickest (60 μm) at the basal end and thinnest (45 μm) at the apical end. After deposition is complete a second very thin layer is deposited by the outer cells beneath the cuticle. Once both cuticle layers are formed the embryo detaches from the parent, although detachment may occur at any point from the two-cell stage onward. When detachment occurs has no bearing on the progress of embryogenesis.

iv. Cuticle Stage The length of time the embryo spends in the cuticle is variable, ranging in the laboratory from 2 to 24 weeks. This corresponds to a dormant overwintering period of the embryo in nature. Two events occur during this stage: The cells of the outer layer acquire characteristics of the epithelial cells of the adult ectoderm, and the cells of the interstitial cell lineage first appear.

v. Bilayer Formation During the last stage of embryogenesis, two epithelial layers that closely resemble the layers of the adult animal and a mesoglea are formed. Because the cuticle stage is of variable length, the timing of the onset of bilayer formation is not easy to predict. However, an embryo in the final stage can be recognized by a clear morphological change 2 days before hatching. Before this time the embryo is opaque due to the refractility of the several thousand nurse cell nuclei; however, 2 days before hatching, the outer layer of the embryo becomes translucent due to the expansion of the outer layer of cells that are devoid of nurse cell nuclei. This outer layer now consists of low columnar epithelial cells, large numbers of interstitial cells, a few nematoblasts (differentiating nematocytes), and some neurons. This layer closely resembles the ectoderm of the adult.

Once the ectoderm has formed, cells of the unorganized interior mass line up along the ectoderm and change their shape from spherical to columnar. This alignment occurs in different parts of the embryo simultaneously, thus producing a complete endodermal layer. Cells not incorporated into the forming endodermal layer degenerate, producing an internal cavity, the future gastrovascular cavity of the adult. Once the two epithelial layers are formed, the mesoglea is synthesized, and the cuticle of the embryo begins to loosen.

vi. Hatching Once the bilayer has formed, the embryo begins to pulsate rhythmically, which continues until hatching is complete. As the embryo pulsates, perforations and channels appear in the cuticle, indicating that it is beginning to break down. Within 24 hr of the formation of the two epithelial layers, the cuticle cracks open on its thinnest side, the original apical side of the embryo, where the head forms. The spherical embryo emerges from the cuticle through this crack (Fig. 9E). For the next 2.5 hr the embryo elongates and contracts, and the shape of the animal changes from a compact sphere to an elongate cylinder. During this time the embryo is enveloped in the innermost thin layer of the cuticle, which eventually ruptures as the hatchling increases in length. Upon rupture of the layer, the hatchling increases its activity, gyrating and wiggling until it completely frees itself of the cuticle.

During the final stages of hatching the apical end of the hatchling undergoes dramatic morphological changes. Before the rupture of the inner thin layer of the cuticle, the apical end has the shape of a smooth dome (Fig. 9E). Within 15 min after rupture, the apical end narrows into a more conical shape, and one to five tentacle primordia evaginate in a ring below the apical tip (Fig. 9F). At this point the head of the hatchling is morphologically complete. The foot is also completely functional because, upon removing itself from the cuticle, the hatchling is able to stick to the surface of a petri dish. Young hydra bud within 10 days of hatching and become sexual 4–8 weeks after hatching.

IV. HYDRA AS A MODEL DEVELOPMENTAL SYSTEM

Hydra has long played a central role in our understanding of basic developmental processes within the metazoa. Many significant scientific "firsts" were discovered using hydra. In 1702 Anton van Leeuwenhoek published the first description of hydra and described the first example of asexual reproduction (budding) ever observed in animals. Abraham Trembley in 1744 published *Memoires*, a notable paper in which he described experiments using hydra that demonstrated asexual reproduction by budding, the first controlled experiments on animal regeneration, the first successful animal grafts, the first study of phototaxis in animals lacking eyes, and the first vital staining of tissues. In modern times, using electron microscopy, septate desmosomes (junctional complexes between cells) and microtubules (major organelles of most cells) were first described in hydra. Furthermore, some of the first signaling molecules demonstrated to function as morphogens were isolated from hydra during the 1970s. Today hydra continues to serve as a model research animal. Because of its simplicity it is a favorite animal of developmental biologists for studying regeneration, patterning, differentiation of multipotent stem cells, embryogenesis, neural development, and gene function during development.

In recent years studies on the molecular aspects of patterning and cell fate specification in the adult polyp and on the embryogenesis of hydra have confirmed that hydra uses many of the same developmental molecules and mechanisms for these processes as do animals which arose later in evolution. Homeobox genes have been identified in a variety of metazoans, from sponges to arthropods and vertebrates, and have been shown to play roles in forma-

tion of body axes and cell and tissue specifications. Several homeobox genes have been isolated from hydra, and the expression patterns of some of these genes indicate they play roles in patterning along the oral–aboral axis. Members of the Achaete-scute family of basic helix–loop–helix transcription factors are involved in cell fate specification in vertebrates and invertebrates. A cnidarian *achaete-scute* homolog, *CnASH*, has been isolated from hydra that shares a high degree of amino acid sequence similarity with the Achaete-scute proteins of vertebrates and *Drosophila*. Finally, embryogenesis in hydra displays a number of features that are characteristic of deuterostomes. The cleavage pattern of hydra is radial for the first three divisions and irregular thereafter. Both are characteristic of some vertebrates, such as amphibians. The hydra blastula consists of a single layer of cells surrounding a blastocoel, as is found in sea urchins. Also, as with many deuterostomes, hydra embryos are regulative during early cleavage stages. The two cells and four cells resulting from the first two cleavages sometime separate, with each blastomere developing individually into a normal animal. Hydra embryos use ingression for gastrulation. The endoderm and mesoderm of the chick embryo are also formed by ingression, as are the primary mesenchyme cells of the sea urchin embryo.

Hydra, like vertebrates, has several stem cell systems that arise during embryogenesis. Both the ectoderm and the endoderm of hydra are epithelial stem cell systems that are similar to the epidermis and intestinal epithelium of vertebrates. All consist of stem cells and differentiation products. Furthermore, the hydra interstitial cell system has many similarities to the hemopoietic system of the mouse. In both cases the cells of the tissue are migratory, and a multipotent stem cell gives rise to a number of differentiation products.

Acknowledgments

I thank Hans Bode and Bill Archer for their support, advice, and fruitful discussions. This work was supported by National Science Foundation Grants DCB-8702212, DCB-8942149, DCB-9046094, and DUE-9552116 and Career Advancement Award DCB-8711245.

See Also the Following Articles

Bibliography

Bode, H. (1992). Neuron determination in the ever-changing nervous system of Hydra. In *Determinants of Neuronal Identity* (M. Shankland and E. Macagno, Eds.), pp. 323–357. Academic Press, San Diego.

Bode, H. (1996). The interstitial cell lineage of hydra: A stem cell system that arose early in evolution. *J. Cell Sci.* **109**, 1155–1164.

Grens, A., Mason, E., Marsh, J., and Bode, H. (1995). Evolutionary conservation of a cell fate specification gene: The Hydra *achaete-scute* homolog has proneural activity in *Drosophila. Development* **121**, 4027–4035.

Grens, A., Gee, L., Fisher, D., and Bode, H. (1996). CnNK-2, an NK-2 homeobox gene, has a role in patterning the basal end of the axis in hydra. *Dev. Biol.* **180**, 473–488.

Hyman, L. (1940). *The Invertebrates: Protozoa through Ctenophora*, pp. 365–661. McGraw-Hill, New York.

Javois, L. (1992). Biological features and morphogenesis of Hydra. In *Morphogenesis, An Analysis of the Development of Biological Form* (E. Rossomando and S. Alexander, Eds.), pp. 93–127. Dekker, New York.

Littlefield, L. (1994). Cell–cell interactions and the control of sex determination in hydra. *Sem. Dev. Biol.* **5**, 13–20.

Martin, V. (1997). Cnidarians, the jellyfish and hydras. In *Embryology, Constructing the Organism* (S. Gilbert and A. Raunio, Eds.), pp. 57–86. Sinauer, Sunderland, MA.

Martin, V., Littlefield, C., Archer, W., and Bode, H. (1997). Embryogenesis in Hydra. *Biol. Bull.* **192**, 345–363.

Hyenas

Christine M. Drea, Elizabeth M. Coscia, and Stephen E. Glickman

University of California, Berkeley

I. Reproduction in the Hyenidae: An Overview
II. Reproduction in the Spotted Hyena
III. Sexual Differentiation in the Spotted Hyena
IV. Summary: Mechanisms and Implications of Sexual Differentiation in the Spotted Hyena

GLOSSARY

androstenedione An androgenic steroid secreted by the ovaries and adrenals that is involved in the synthesis of testosterone.

clitoris A small, sensitive erectile organ in the female corresponding to the male penis and located at the ventral end of the vulva.

masculinize To cause an individual to develop the secondary sexual characteristics of a mature male, as by the administration of androgens.

placenta A structure formed by fusion of the chorion with the wall of the uterus that serves to attach the embryo to the uterine wall and to exchange nutrients, wastes, and gases between the maternal blood and the embryonic blood.

prohormone A precursor of a hormone.

sexual differentiation The process by which the two sexes become different; whereby a bipotential embryo develops along either male or female lines, as determined initially by the chromosomes and subsequently by hormones.

The family Hyenidae, a lineage that separated from viverrid ancestors 25–30 million years ago, comprises four extant species: striped hyenas (*Hyaena hyaena*), brown hyenas (*Hyaena brunnea*), spotted hyenas (*Crocuta crocuta*), and the aardwolf (*Pro-*

teles cristatus). Although there is substantial variation in the details of mating systems, gestation, parental care, and diet, three of these species present traditional, sexually dimorphic, external genitalia. In the fourth species, the spotted hyena, females exhibit remarkably "masculinized" external genitalia. Because of the extraordinary similarity between male and female genital morphology, the common belief that these animals were hermaphrodites persisted well into the nineteenth century. Recently, it was discovered that female spotted hyenas are also behaviorally male-like, being very aggressive and totally dominating adult males in social situations. This suite of "masculine" characters has directed attention to the process of sexual differentiation in the female spotted hyena: Either there are androgens circulating during fetal life that create the unusual external genitalia of these animals (and might enhance their aggressiveness) or there is a hitherto unrecognized mechanism for the production of male-like genital morphology in a female mammal. This article is, accordingly, focused on reproduction and sexual differentiation in the spotted hyena, although first a brief outline is provided of the major features of reproductive behavior in the four extant hyenid species.

I. REPRODUCTION IN THE HYENIDAE: AN OVERVIEW

In all four hyenid species (Table 1)—aardwolf (*Proteles cristatus*), striped hyenas (*Hyaena hyaena*), brown hyenas (*Hyaena brunnea*), and spotted hyenas (*Crocuta crocuta*)— males have internal testes, there is no baculum, and both sexes have anal scent glands contained within an anal pouch. Males and females of all four species engage in scent-marking behavior, which presumably plays a role in sexual advertisement. Regarding generalizations on reproduction, the similarities across species end here. Teat number varies individually and by species. Notably, in spotted hyenas, only the caudal pair of teats is functional, a factor contributing to sibling rivalry in triplet litters. Also, only spotted hyenas show sexual size dimorphism favoring the female and only female crocuta display highly specialized sexual organs. In all other hyenids, adults are sexually size monomorphic and the sexual organs are not specialized.

A. The Aardwolf

For such a small family, hyenids show fairly diverse reproductive behavior and mating strategies, including both monogamy and promiscuity. The least typical member of Hyenidae, the aardwolf, is socially monogamous, forming pair bonds that can last from 2 to 5 years. Aardwolves are the smallest, the shortest lived, and the only seasonal breeders among Hyenidae. Females have a single yearly estrus in July and give birth to between one and four cubs in October. Because of this species' exclusive termite diet, cubs are entirely dependent on maternal milk for the first 4 months of life. Though fathers guard and defend their young, they do not provide food.

B. Striped Hyenas and Brown Hyenas

Striped and brown hyenas are the most closely related and, consequently, the most similar in their reproductive biology. They are of equal size and longevity and have similar lifestyles (both are solitary foragers and scavengers). In contrast to the aardwolf, these two species are aseasonal breeders, with females being polyestrous. Neonatal mass in striped and brown hyenas is at least threefold that in the aardwolf, but cubs of all three species are born with their eyes closed. The most prominent differences in reproduction between striped and brown hyenas are evidenced by adult behavior. Though generally loners, striped hyenas form short-term pair bonds for breeding and the resulting family unit may endure for several years. There is biparental care as both parents provision their young. The "single family" approach of striped hyenas contrasts with the "cooperative venture" of brown hyenas. The latter occur in small social groups in which members are related to each other. Females mate with nongroup-living, nomadic males; nevertheless, resident males provide care for the young they did not sire. Mothers are also cooperative, suckling the cubs of other females along with their own. Although striped and brown

TABLE 1
Comparative Reproductive Biology and Life Histories of Hyenids

Feature	Hyenid species			
	Aardwolf (Proteles cristatus)	Striped hyena (Hyaena hyaena)	Brown hyena (Hyaena brunnea)	Spotted hyena (Crocuta crocuta)
Age at maturity (years)	M = 1.8 F = 1.8	M = 2–3 F = 2–3	M = ? F = 2	M = 2 F = 3
Life span (years)	Wild: 7 Captive: 15	Wild: >8 Captive: 24	Wild: 12 Captive: 20	Wild: 20–25 Captive: >40
Adult mass (kg)	Size monomorphism M = 7.8–10 F = 7.7–10	Size monomorphism M = 28.0–43.2 F = 27.5–36.0	Size monomorphism M = 35–49.5 F = 28–47.5	Size dimorphism M = 40–62.5 F = 44–75.0
Sexual organs and secondary sexual characteristics	Unspecialized sexual dimorphism Internal testes No baculum Anal glands and pouch	Unspecialized sexual dimorphism Internal testes No baculum Anal glands and pouch Teats: 2–3 pairs	Unspecialized sexual dimorphism Internal testes No baculum Anal glands and pouch Teats: 2–6 pairs	Specialized sexual monomorphism Internal testes No baculum Anal glands and pouch Teats: 1–2 pairs
Mating system	Monogamy	Monogamy	Promiscuity (F mates with nongroup-living M)	Promiscuity (F mates with group-living M)
Breeding	Seasonal	Aseasonal	Aseasonal	Aseasonal
Estrus cycle	Monoestrous Estrus: 1–3 days	Polyestrous Estrus: 1 day	Polyestrous Estrus: 1 week	Polyestrous Estrus: unknown
Mating behavior	Multiple mounting	Multiple mounting No thrusting	Multiple mounting No thrusting	Multiple mounting Thrusting
Gestation (days)	90	91	96	110
Litter size	1–4 (x = 3)	1–6 (x = 2.4)	1–5 (x = 2.3)	1–3 (x = 2)
Neonatal mass	223 g	600–700 g	630–812 g	1.5 kg
Parental care	F suckles own cubs Biparental care No provisioning of cubs	F suckles own cubs Biparental care Provisioning by both parents	F allosuckles Cooperative care Provisioning by mother and adults	F suckles own cubs No paternal care Limited or no provisioning
Age at weaning (months)	4	4–5	10–12	12–18
Interlitter interval (months)	12	14	12–41	16–19

hyenas have multiple pairs of teats, only the two caudal pairs are functional.

C. The Spotted Hyena

Unlike their relatives, spotted hyenas are pack hunters and live in the largest social groups of any carnivore. Within this complex social network, there is no extended pair bonding of the sort seen in other hyenas: Mating is promiscuous and there is no paternal care. Spotted hyenas are aseasonal breeders, but there is little information on their estrus cycle. They have an exceptionally long period of gestation (97–132 days) and give birth to young that are unusually precocial. Despite this "head start," spotted hyenas have longer lactation periods than do most other carnivores, which places a substantial energetic burden on mothers and, by extension, puts lower-ranking females at a disadvantage. Not surprisingly, therefore, mothers suckle their own young exclusively. Details of spotted hyena reproductive biology are described in the following sections.

II. REPRODUCTION IN THE SPOTTED HYENA

A. Morphology

In the female, the clitoris has hypertrophied, approximating the size and shape of the male penis. It is traversed by a central urogenital canal, through which the female spotted hyena urinates, copulates, and gives birth. There is no external vagina: The external labia have fused to form a pseudoscrotum, marked by two pads of fibrous/fatty tissue which are visible externally and might be mistaken for testes. Moreover, this elongated clitoris is fully erectile and females display erections similar to those of the male. The internal reproductive system follows the standard mammalian pattern. Careful examination of the external genitalia reveals sex differences in "phallic" structure. The clitoris is shorter and thicker than the penis, and the central urogenital canal ends in an opening (the urogenital meatus) that is larger and more elastic in the female than in the male. This elasticity is prerequisite for mating and birth. The shape of the glans is also sexually dimorphic: The glans clitordis has a rounded contour, whereas the glans penis has a more angular profile. These differences in the shape of the glans are used by field researchers to distinguish male from female spotted hyenas.

B. Mating Behavior

Mating occurs in the social context of the hyena clan. Unlike the social canids, in which breeding is commonly restricted to the dominant female, all female spotted hyenas mate and give birth. Females achieve sexual maturity at approximately 30–36 months of age and typically reproduce in the clan of their birth. Males generally disperse at the time of puberty (18–24 months of age) and mate only after joining a new clan. Currently, little is known

of the spotted hyena estrous or ovulatory cycle (if, in fact, there is a cycle). The presence of spines on the glans penis of spotted hyenas is characteristic of other species of induced ovulators, suggesting the possibility of induced ovulation in spotted hyenas. There is no sharply defined breeding season. Although there are seasonal peaks in frequency of births, associated with rainy seasons in Sub-Saharan Africa, mating occurs throughout the year.

Because of their unusual external anatomy, the act of mating requires unusual cooperation on the part of the female and some agility on the part of the male (Fig. 1). During the sexual act, the female retracts her

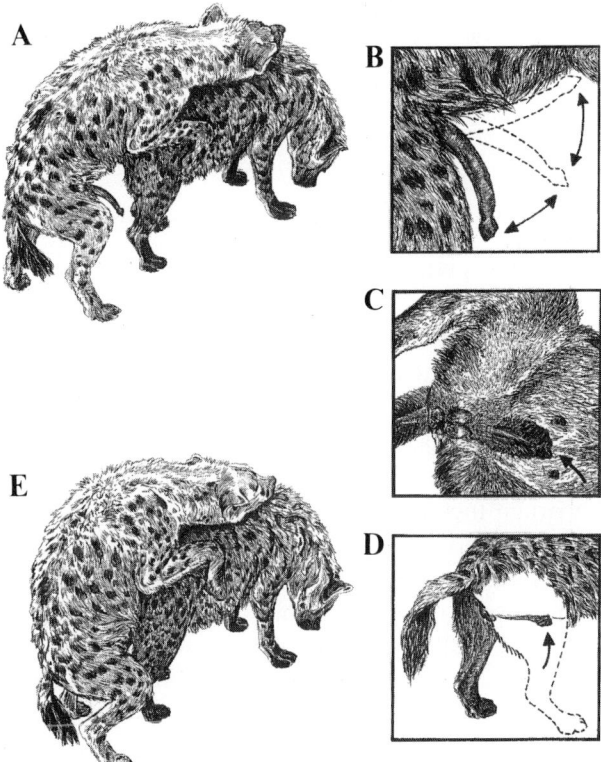

FIGURE 1 The mating sequence in spotted hyenas. (A) The male mounts the female, who stands still with her head lowered and her clitoris retracted into her abdomen. (B) The male shows penile "flips," as indicated by arrows. (C) Underside view and (D) lateral view of a female spotted hyena, with arrows showing the site of the clitoral opening, which is located anterior to the point of entry in other female mammals. (E) Once the male locates the opening, he squats down and scoots forward to enable intromission, which is followed by pelvic thrusting (illustration by Christine M. Drea).

clitoris into the abdomen and, standing exceptionally still, assumes a head-down posture. The male mounts the female and stands on his hindlegs with his forepaws on her flanks (Fig. 1A). He then "flips" his erect penis against her lower abdomen (Fig. 1B), searching for the clitoral opening (Fig. 1C), which is well anterior to the position normally occupied by the external vagina (Fig. 1D). To achieve entry, the male must squat down on his haunches, move forward, and reposition his pelvis (Fig. 1E). After achieving entry there is an interval of pelvic thrusting followed by ejaculation. During the minutes following ejaculation, the male maintains insertion and often assumes a posture in which he leans forward resting his head on the back of the female. A number of additional ejaculatory sequences may follow, with the female taking an active role in soliciting the male's attention.

C. Gestation and Parturition

As noted previously, gestation is prolonged in the spotted hyena, relative to the other hyenids. Elevated concentrations of plasma progesterone can be detected within the first 2 weeks of gestation. As pregnancy proceeds, plasma levels of estrogen and testosterone, as well as the peptide hormone relaxin, also exhibit a substantial rise from prepregnancy values. A marked increase in the size of the nipples is often the first external indicator of pregnancy, particularly in primiparous animals. In nature, as the female proceeds toward term, she locates a site removed from the communal den of the clan and gives birth in this "natal den." The young will be nursed and maintained in this den for a variable period (2–4 weeks) before being carried by the mother to the communal den and introduced to the clan.

D. Costs of Clitoral Delivery

Giving birth to a large (1–2 kg) fetus through the clitoris is a mechanically challenging feat. The urogenital meatus, although highly elastic, is not adequately large to permit passage of the fetus and must tear before an initial birth can occur. The process is further complicated by the presence of a short umbilical cord and a rather tortuous route from the point of placental attachment in a uterine horn. The fetus must travel through the uterus and upper vagina before finally reaching the clitoris. If delivery of the fetus does not proceed in a timely fashion, the infant will be stillborn (presumably due to anoxia, consequent to detachment of the placenta from the uterine wall). In captivity, approximately 60% of first births result in stillborn offspring due to the time required for passage of the fetus. Moreover, in some primiparous females, the fetus was lodged in the birth canal and would have taken the life of the mother without veterinary intervention. Such intervention is obviously not available for female hyenas in nature. In subsequent pregnancies, after a fetus has successfully passed through the clitoris, infant mortality (and indeed maternal mortality) is much lower. Because there are substantial costs to clitoral delivery in this species, it is probably not surprising that this process has evolved only once.

E. Behavior of Infants

The twins that generally constitute a hyena litter are quite precocial. At birth, their eyes are open, they are highly mobile, and they possess an impressive array of needle-like teeth. In addition, within minutes of the birth of the second cub, the twins begin to fight. The fighting is intense and, if not interrupted, wounds result from the vigorous biting. At a minimum, fighting during the first days of life establishes a dominance relationship between the siblings that lasts, at least, for several years. In nature, if the dominant sibling can keep its young twin sequestered in a burrow and prevent it from emerging to nurse, death will result from a combination of starvation and wounding. If both infants survive, however, the intense aggression is largely replaced by playful, prosocial interactions. The development of these prosocial behavior patterns is also precocial, emerging in the second week of life and increasing over subsequent months. This schedule ensures that both aggressive and prosocial behavior patterns will be in the infant's repertoire when it joins the clan. As outlined later, the exposure of all fetuses to high levels of androgens during fetal life might be expected to facilitate the aggression observed at birth in this species.

III. SEXUAL DIFFERENTIATION IN THE SPOTTED HYENA

A. The Search for Androgens in Adult Female Hyenas

Contemporary understanding of sexual differentiation requires the secretion of androgens, by the fetal testes, for the production of masculine external genitalia from embryological precursors. Such androgens are also known to affect the developing nervous system, modifying hypothalamic mechanisms responsible for sex-typical profiles of hormone secretion as well as structures modulating sexually dimorphic behavior patterns. Given the exceptionally masculine external genitalia of the spotted hyena observed at birth, it is not surprising that early research was directed toward a search for elevated testosterone in female hyenas. The size, aggressiveness, and dominance of these animals also suggested that substantial concentrations of this androgen might be found in adult females. It soon became clear, however, that nonpregnant, adult female hyenas have relatively low plasma concentrations of testosterone. By contrast, androstenedione, primarily of ovarian origin, is the primary androgen found in high concentrations in female spotted hyenas. Androstenedione is an interesting steroid. It is not known to bind to conventional estrogen or androgen receptors, but it is readily converted to an active estrogen or testosterone in the presence of appropriate enzymes. Following conversion, androstenedione could exert powerful hormonal effects on a developing fetus or on the postnatal nervous system. In accord with this possibility, removal of the ovaries eliminates mating behavior and drastically reduces female aggressiveness.

B. The Search for Androgens in Fetal Hyenas

If the fetal ovary of the spotted hyena secreted androstenedione in the same manner as the adult ovary, this could explain the exceptional masculinization of the external genitalia, provided the appropriate enzymes (17β-hydroxysteroid dehydrogenase and 5α-reductase) and androgen receptors were present in target tissues. However, anatomical studies revealed that the penile clitoris of the female spotted hyena is fully formed at an early stage (before 36 days) of gestation, prior to maturation of the fetal ovary.

A clue to potential sources of fetal testosterone was provided by the high concentrations of this steroid observed in the circulation of pregnant hyenas. It seemed possible that the gradual increment in maternal testosterone concentration during pregnancy reflected two processes: the metabolic conversion of maternal ovarian androstenedione to testosterone by the placenta and the absolute growth in size of this structure during the course of gestation. Subsequent studies of placental activity supported this hypothesis and verified that maternal androstenedione, converted to testosterone in the placenta, was transferred to the developing fetus via the umbilical circulation (and the maternal circulation via the uterine venous drainage of the placenta).

It is clear that a mechanism is in place that could account for masculinization of the external genitalia in this unique species. This possibility is reinforced by the appearance of several human cases in which genetic female infants were born with masculinized genitalia due to a defect in aromatase activity, i.e., a defect in which an abnormal human placenta acted, metabolically, like a normal hyena placenta.

C. Evidence for Significant Nonandrogenic Processes during Prenatal and Postnatal Life

Testing the preceding line of thought required interfering with the putative action of placental testosterone on the developing female fetus. Antiandrogens administered to pregnant female dogs during appropriate stages of pregnancy result in male offspring with a very small penis, a urethral opening near the base rather than the tip of the penis (i.e., hypospadia), and a blind vaginal pouch in the midline of the scrotal area. By contrast, antiandrogens (flutamide and finasteride) administered to pregnant female spotted hyenas had a rather different phenotypic effect. Male offspring exposed to these compounds as early as Day 12 of gestation were indeed "feminized," but their genitalia now approximated those of a normal female spotted hyena rather than a generic female mammal. Such males had a short, thick penis with a large urogenital meatus, and the glans was rounded like the glans clitordis rather than angular like a normal glans penis. Female offspring exposed to antiandrogens were still more feminized, with an exceptionally short, thick clitoris and a still larger urogenital opening at the tip. In summary, the data from antiandrogen experiments suggest that androgens normally modulate fetal development in both male and female spotted hyenas. However, the robust development of a penile clitoris from the genital tubercle and a fully formed scrotum from the genital swellings, in the absence of androgenic activity, suggests that a nonandrogenic mechanism may be involved in the initial differentiation of these structures.

Further evidence for the existence of androgen-independent mechanisms regulating the growth of the clitoris and the penis is provided by examining the effects of prepubertal gonadectomy. In other mammalian species, there is typically a period of accelerated growth of the genitalia accompanying puberty. Moreover, castration or ovariectomy prior to puberty inhibits such growth. In both male and female spotted hyenas, the normal growth curve actually decelerates during puberty, with more rapid development of both genitalia and overall body size occurring prior to the pubertal period. Prepubertal gonadectomy removes more than 95% of testosterone and 90% of androstenedione from the circulation; however, there are minimal effects of such surgery on growth of the clitoris or penis (aside from some changes in the elasticity and thickness of the clitoris, which are normally induced by the presence of estrogens in females).

IV. SUMMARY: MECHANISMS AND IMPLICATIONS OF SEXUAL DIFFERENTIATION IN THE SPOTTED HYENA

A. Sexual Differentiation and Development: Integration of Androgenic and Androgen-Independent Mechanisms

It appears that both androgenic and nonandrogenic mechanisms are involved in the development of the penile clitoris of the female spotted hyena. The basic formation of the clitoris, including enclosure of the urethra within the genital tubercle, can apparently proceed without the intervention of androgens; however, placental testosterone does play a role in the early development of the clitoris and penis *in utero*. Paradoxically, the changes produced in the normal hyena clitoris by naturally circulating testosterone (i.e., a smaller urogenital opening and a thinner clitoris) would serve to make a very difficult initial parturition still more costly.

B. General Implications from an Unusual Animal

Despite the very unusual events that underlie formation of the clitoris and scrotum in the spotted hyena, the contemporary theory of sexual differentiation survives intact. This theory provides an explanation of how one arrives at a male phenotype instead of a female phenotype for each species. Presumably, the sex differences in genital morphology observed in spotted hyenas at birth are the result of both placental and testicular testosterone acting on the male embryo, whereas only placental testosterone is circulating in the female embryo. Further separation of male and female genital morphology is produced by the different hormonal environments that are present postnatally, particularly the activity of estrogens in the female but not in the male. Consideration of the spotted hyena calls attention to the absence of a substantive theory of feminine development. There is wide variation in clitoral morphology within mammalia, including such species as lemurs and European moles that also display a penile clitoris. To date, development of female genitalia often has been treated as a passive process. This situation has to change in the future.

Studies of the spotted hyena also stress the potential significance of androstenedione as a significant prohormone in all female (and male) mammals. Often dismissed in the contemporary literature as a "weak androgen," studies of the spotted hyena and other species (e.g., the redwing blackbird) demonstrate the capacity of androstenedione to act as a very powerful steroid following conversion to estrogen or testosterone.

Finally, the role of placental metabolism is typically ignored in accounts of sexual differentiation. After all, within a species, such metabolic effects are assumed to be the same for male and female fetuses. This oversight may be rectified in the future, as the recent accounts of human pseudohermaphroditism, produced by a defect in placental aromatase activity, direct attention to the impact of placental metabolism on development. Studies of the hyena also emphasize the role of species differences in placental metabolism in modulating structural development in embryos of both sexes. Examination of "experiments of nature," such as the spotted hyena, reveal common processes in other, less exotic, species.

Bibliography

Conte, A. F., Grumbach, R. M., Ito, Y., Fisher, C. R., and Simpson, E. R. (1994). A syndrome of female pseudohermaphrodism, hypergonadotropic hypogonadism, and multicystic ovaries associated with missense mutations in the gene encoding aromatase (P450arom). *J. Clin. Endocrinol. Metabol.* **78**, 1287–1292.

Drea, C. M., Weldele, M. L., Forger, N. G., Coscia, E. M., Frank, L. G., Licht, P., and Glickman, S. E. (1998). Androgens and masculinization of genitalia in the spotted hyaena (*Crocuta crocuta*). 2. Effects of prenatal anti-androgens. *J. Reprod. Fertil.*, in press.

Frank, L. G. (1997). Evolution of genital masculinization: Why do female hyaenas have such a large "penis"? *Trends Ecol. Evol.* **12**, 58–62.

Frank, L. G., Holekamp, K. E., and Smale, L. (1995). Dominance, demography, and reproductive success of female spotted hyenas. In *Serengeti II: Dynamics, Management, and Conservation of an Ecosystem* (A. R. E. Sinclair and P. Arcese, Eds.), pp. 364–384. Univ. of Chicago Press, Chicago.

Glickman, S. E., Coscia, E. M., Frank, L. G., Licht, P., Weldele, M. L., and Drea, C. M. (1998). Androgens and masculinization of genitalia in the spotted hyaena (*Crocuta crocuta*). 3. Effects of juvenile gonadectomy. *J. Reprod. Fertil.*, in press.

Kruuk, H. (1972). *The Spotted Hyena: A Study of Predation and Social Behavior.* Univ. of Chicago Press, Chicago.

Licht, P., Hayes, T., Tsai, P.-S., Cuhna, G., Kim, H.-S., Golbus, M., Hayward, S., Martin, M. C., Jaffe, R. B., and Glickman, S. E. (1998). Androgens and masculinization of genitalia in the spotted hyaena (*Crocuta crocuta*). 1. Urogenital morphology and placental androgen production during fetal life. *J. Reprod. Fertil.*, in press.

Mills, M. G. L. (1990). *Kalahari Hyaenas: Comparative Behavioural Ecology of Two Species.* Unwin-Hyman, London.

Yalcinkaya, T. M., Siiteri, P. K., Vigne, J.-L., Licht, P., Pavgi, S., Frank, L. G., and Glickman, S. E. (1993). A mechanism for virilization of female spotted hyenas in utero. *Science* **260**, 1929–1931.

Hyperprolactinemia

Howard A. Zacur

Johns Hopkins Medical Institutions

I. Introduction
II. Prolactin Secretion
III. Biological Actions of Prolactin
IV. Measurement of Prolactin
V. Diagnosis of Hyperprolactinemia
VI. Management of the Hyperprolactinemic Patient
VII. Summary

GLOSSARY

decidual cells Stromal cells of the human endometrium which become eosinophilic and form a tile-like pattern after exposure to progesterone during pregnancy.

lactotropes Cells within the anterior pituitary gland which synthesize and secrete prolactin.

menotropins Luteinizing hormone and follicle-stimulating hormone obtained from the urine of menopausal women.

The ability of pituitary extracts to cause lactation in pseudopregnant rabbits led to the discovery of prolactin in the rabbit by Stricker and Grueter in 1928. It then required more than 40 years to prove that a human prolactin molecule existed. This proof resulted from the pioneering work within the laboratory of Friesen in Canada that led to the development of a radioimmunoassay for human prolactin. Prior to this discovery it had been assumed that the biological activity of prolactin observed in other species was provided by growth hormone in the human.

I. INTRODUCTION

Prolactin is a protein hormone consisting of 198 amino acids. It is a product of the prolactin gene located on chromosome 6. Lactotropes within the pituitary gland, decidual cells of the endometrium, and lymphocytes within the bloodstream are all capable of expressing the prolactin gene. Six exons comprise the prolactin gene. Exon 1a, located 5.8 kb upstream of exon 1b, is the start of transcription at extrapituitary sites. Exon 1b is the start of transcription in pituitary cells (Fig. 1). In cell types other than the pituitary the mRNA transcript of exon 1a joins the mRNA transcript of exon 1b to create a 5′ untranslated region of mRNA which is longer than that made by pituitary cells. Use of an alternative promotor at extrapituitary sites is believed to be responsible for differences in the regulation of prolactin secretion observed between extrapituitary and pituitary sites.

Expression of the human pituitary prolactin gene is dependent on the activities of proximal and distal promotors located within 1800 bp of the start site. The proximal promotor of the human pituitary prolactin gene contains three Pit-1 binding sites, whereas the distal promotor contains eight Pit-1 binding sites. Pit-1 is a pituitary-specific transcription factor containing a sequence of amino acids similar to that found in two other transcription factors known as Oct-1 and Unc-1. As a consequence, this particular amino acid sequence is known as the POU homeodomain.

An estrogen response element (ERE) is located within the distal promotor region of the pituitary prolactin gene and is situated near the first Pit-1 binding site of the distal promotor. Estrogen has been shown to stimulate transcription of the pituitary prolactin gene. A cyclic AMP response site has also been identified.

Decidual cells do not secrete prolactin in response to thyrotropin-releasing hormone (TRH) as do pituitary cells. Dopamine does not inhibit decidual cell prolactin secretion, although it does inhibit pituitary cell prolactin secretion. As previously mentioned, differences in the promoters of pituitary and decidual cells are believed to be responsible for these effects.

A single gene located on chromosome 5 codes for the human prolactin receptor. The prolactin receptor belongs to the hematopoietic/cytokine receptor family. In rats, "long" and "short" forms of the prolactin receptor have been identified, but only a long form has been demonstrated in the human.

From previous work performed in rodents it is believed that activation of the prolactin receptor requires the binding of one prolactin molecule to two prolactin receptors. A single molecule of prolactin will "dimerize" two prolactin receptor molecules. After dimerization occurs a tyrosine kinase within the intracellular portion of the prolactin receptor known as the Janus kinase becomes phosphorylated. Phosphorylation of other intracellular proteins then follows, with STAT (single transducer and activator of transcription) proteins being prominent. These proteins stimulate the synthesis of other cellular proteins.

It is believed that at high concentrations prolactin could saturate all of its cell surface receptors. This could prevent receptor dimerization and lead to reduced biological activity. This may explain why the biological responses observed after prolactin stimulation are often "bell shaped," i.e., too little or too much hormone results in reduced biological activity.

II. PROLACTIN SECRETION

Prolactin secretion from lactotrophs located within the anterior pituitary gland is chronically inhibited

Prolactin Gene Expression in Pituitary Cells

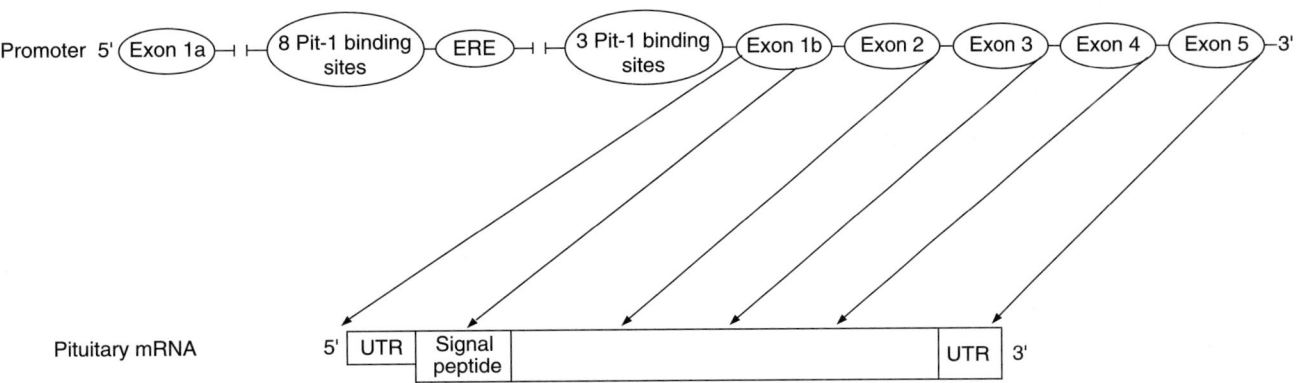

Prolactin Gene Expression in Decidual Cells

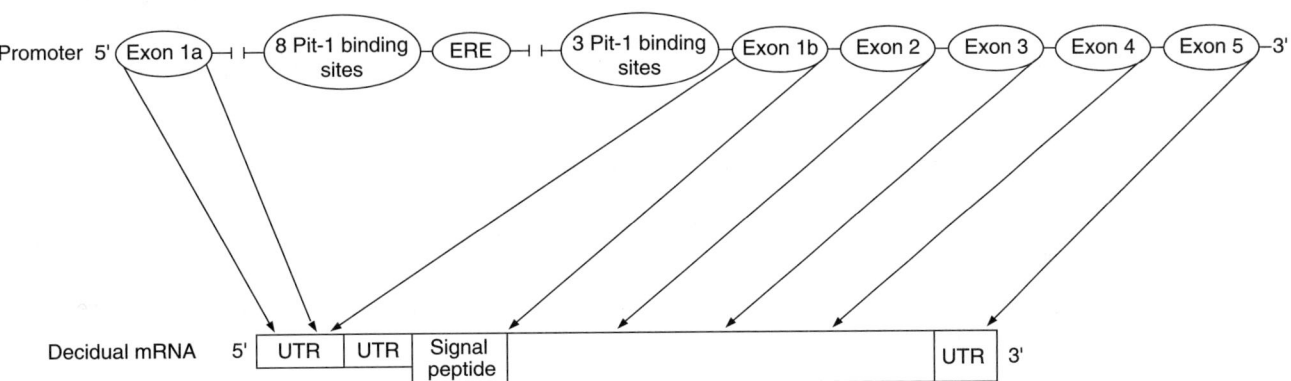

FIGURE 1 Transcripts of the prolactin gene from pituitary and decidual cells. UTR, untranslated region; ERE, estrogen response element.

since transection of the pituitary stalk or infundibulum results in an elevation in the periheral level of this hormone. Dopamine released from axons originating in the hypothalamus and terminating on pituitary portal blood vessels is believed to account for most of the inhibitory influence of the hypothalamus upon pituitary prolactin secretion. Lactotrope secretion of prolactin is dependent in part on intracellular cyclic AMP. Low concentrations are associated with diminished hormone release, whereas higher levels are associated with increased secretion.

Several types of dopamine receptors have been identified and are classified as D1–D4. Stimulation of the D2 receptor on the membrane of lactotrope cells is associated with lowering of the intracellular cyclic AMP level and inhibition of prolactin secretion.

Regulation of pituitary prolactin secretion is complex because of multifactorial control. In addition to the inhibitory influence exerted by dopamine, γ-aminobutyric acid and prolactin itself may inhibit prolactin secretion. Estradiol, TRH, vasoactive intestinal peptide, and gonadotropin-releasing hormone (GnRH) are examples of steroid and protein hormones which have been shown to stimulate pituitary prolactin secretion.

Agents which affect the synthesis or block the receptors of substances responsible for stimulating or inhibiting pituitary prolactin secretion alter circulating blood levels of this hormone. Examples include antihypertensive drugs which block the synthesis of dopamine or tranquilizers which block the dopamine receptor. Iatrogenic hypothyroidism created after use of lithium provided for treatment of manic-depressive illness will cause hyperprolactinemia presumably by increasing TRH.

As previously mentioned, decidual cell prolactin secretion is neither stimulated by TRH nor inhibited by dopamine. Progesterone appears to be largely responsible for stimulating decidual prolactin secretion, whereas lipocortin-1 and endothelin-1 inhibit decidual prolactin release. Secretion of prolactin by decidual cells into the systemic circulation does not appear to occur. During pregnancy uterine decidual cells secrete prolactin into the amniotic cavity. Regulation of prolactin secretion from cells of the immune system appears to occur by other mechanisms.

III. BIOLOGICAL ACTIONS OF PROLACTIN

In humans prolactin has been shown to have lactogenic, steroidogenic, and immunoregulatory functions. Prolactin receptors have been identified within membranes of mammary cells responsible for secretion of breast milk. Following binding of prolactin to these receptors cell signaling mechanisms utilizing tyrosine kinase and phosphatase are triggered and c-fos, c-jun, c-myb, and ornithine decarboxylase are activated. This ultimately results in the initiation of synthesis of the milk proteins casein, whey, lactoglobulin, and lactalbumin.

Using luteinized human granulosa cells in culture, prolactin has been shown to affect progesterone secretion. Conditions of prolactin excess (>20 ng/ml) or deficiency (0 ng/ml) within the culture media of these cells are associated with a marked decrease in the progesterone concentration found in the culture media. In rats, the importance of prolactin in maintaining corpus luteum function during early pregnancy is well-known.

The ability of prolactin to affect immunological function has recently been reported. Prolactin's ability to stimulate the proliferation of lymphoma cells taken from Nb2 rats has supplanted the pigeon crop sac assay as the biological assay system of choice for prolactin. In humans prolactin receptors on T and B cells have been identified as well as secretion of prolactin from these cells. Maintaining immunocompetence is the apparent role of prolactin in this circumstance. Cyclosporine is a well-known immunosuppressive drug given to humans receiving organ transplants. This drug has been shown to displace prolactin from its receptors. Hypophysectomized rats demonstrate diminished immunocompetence and die from infections following administration of antiserum to prolactin. Immunocompetence is restored in these animals when prolactin is administered.

In teleost fishes, prolactin has been shown to affect the permeability of water across the gills as these fish swim between salt and freshwater environments. Amniotic fluid prolactin concentrations increase during gestation and it has been postulated that prolactin may regulate water transport across fetal lungs or across fetal dermal surfaces not yet covered by squamous epithelium. An affect of prolactin on decreasing diffusion of water across the amnion from the fetal to maternal side has been shown *in vitro* and a role of prolactin in causing polyhydramnios postulated.

In humans, circulating prolactin levels steadily rise throughout pregnancy but begin declining prior to parturition. A relationship between decidual prolactin secretion, calcium, and prostaglandin levels affecting the onset of labor has been postulated to exist. An effect on behavior by prolactin has been demonstrated in birds. Nesting behavior in birds is induced following administration of prolactin. An association between prolactin levels and bone density exists. Humans with hyperprolactinemia have been shown to have reduced bone density. This relationship is believed to result from hypoestrogenism which coexists during states of hyperprolactinemia. The possibility of a direct effect of prolactin on bone turnover has been suggested from one previous study which was not confirmed in a subsequent study.

In humans, states of hyperprolactinemia are commonly associated with altered pituitary gonadotropin release. Reduced pulsatility of luteinizing hormone (LH) and follicle-stimulating hormone (FSH) is observed with rising serum/plasma prolactin concentrations. Inhibition of hypothalamic GnRH release as a result of increased hypothalamic release of dopamine and/or endorphin in response to prolactin is suspected. Retrograde portal blood flow from the pituitary to the hypothalamus has been demonstrated, providing a means whereby pituitary prolactin secretion may directly affect hypothalamic function.

During a physiological state of hyperprolactinemia, such as that which occurs during active breast-feeding immediately postpartum, a delay in the resumption of ovulation may occur. This appears to be mediated by prolactin in part at the level of the hypothalamic–pituitary axis as discussed previously and in part at the level of the ovary. Stimulation as well as inhibition of progesterone secretion from human granulosa cells by prolactin *in vitro* has been demonstrated. Amenorrhea occurring simultaneously with hyperprolactinemia at times unassociated with pregnancy or breast-feeding is believed to be caused by similar mechanisms.

In men, an effect of prolactin on testicular and prostate function has been suggested and the presence of prolactin in prostate fluid reported. Hyperprolactinemia is commonly diagnosed later in men as contrasted with an early diagnosis in women. In women, disturbances of the menstrual cycle will frequently lead to a visit to a physician followed by measurement of a prolactin level. In men, sustained elevated prolactin levels will in most instances go unnoticed until hypogonadism and impotence result. Because these are late clinical manifestations of hyperprolactinemia, large pituitary prolactin lesions are more frequently encountered in men than women.

IV. MEASUREMENT OF PROLACTIN

In humans immunoassays are commonly used to measure the serum/plasma prolactin concentration. Each laboratory performing this assay is responsible for determining a "normal" range of prolactin values using a standard reference. Unfortunately, when aliquots of the same serum sample are sent to different laboratories for prolactin measurement, markedly different values have been obtained. This variation in assay values must be remembered when assessing results from patients who have had blood samples taken for prolactin measurement at different laboratories.

The normal range for serum/plasma prolactin concentrations is frequently reported as 0–20 ng/ml in women during their reproductive years and 0–15 ng/ml in men. Prolactin concentrations in women during menopause fall to values approximating those of men. These changes in prolactin concentration are believed to result from corresponding changes in estrogen levels.

Differences between immunological and biological activity of prolactin measurements in the same blood sample have been reported. Normally the prolactin molecule circulates as a protein of 25 kDa. Existence of a large prolactin molecule (48–56 kDa) and a very large prolactin molecule (100 kDa) have been reported. These prolactin molecules may be dimers or tetramers of the native molecule or they may be native molecules complexed with a immunoglobulin molecule. While immunologic activity of large and very large prolactin is maintained, biologic activity of these molecules is not. Consequently, some women may exhibit hyperprolactinemia as determined by immunoassay but have no history or evidence of amenorrhea or galactorrhea due to the presence of these large prolactin molecules. Use of column chromatography followed by immunoassay and bioassay for prolactin would be required to detect these individuals.

V. DIAGNOSIS OF HYPERPROLACTINEMIA

Measurement of the serum/plasma prolactin concentration is indicated in women who experience galactorrhea, menstrual cycle irregularity including oligomenorrhea, luteal phase defects, and amenorrhea. In men serum/plasma prolactin measurements may be requested when galactorrhea, impotence, or a decline in libido are detected or complained of.

Diagnosis of "pathologic" hyperprolactinemia

should be made only after all expected causes of hyperprolactinemia are excluded. More than one blood sample obtained under rigorously defined conditions should be drawn to verify a diagnosis of hyperprolactinemia.

Expected causes of transient hyperprolactinemia are numerous, including drug use, chest wall irritation, sleep, food ingestion, menstrual cycle changes, pregnancy, stress, and renal failure. Sleep is a well-known cause of hyperprolactinemia. The biological significance of this change in the serum/plasma prolactin concentration remains unclear, but an association with immune function has been suggested. Food ingestion at lunch or dinner, but not at breakfast, will result in a transient rise in the serum/plasma prolactin concentration. Particular amino acids acting as neurotransmitters may be responsible for this effect. The biological significance of this event remains unknown. Irritation of the anterior chest wall will result in secretion of prolactin from the pituitary gland. This appears to result from afferent nerve impulses causing changes in hypothalamic catecholamine turnover. Suckling, chest wall trauma, or surgery, e.g., herpes zoster or breast augmentation, have all been reported in individuals with hyperprolactinemia. Some authors have asserted that a routine breast exam in women may result in transient hyperprolactinemia, whereas others have denied that this occurs.

A rise in prolactin during pregnancy is well-known with hormone concentrations reaching 400 ng/ml or higher at term. Stimulation of pituitary lactotrope cells by estrogen and other hormones released during pregnancy is believed to be responsible. A 70% increase in the volume of the pituitary gland has been observed during pregnancy, presumably reflecting lactotrope hyperplasia.

Postpartum, plasma estrogen levels decline rapidly following expulsion of the placenta. If breast-feeding does not occur prolactin concentrations are normalized by the end of the first postpartum week. If breast-feeding occurs, prolactin levels will remain elevated for varying time periods depending on the frequency, duration, and intensity of nipple stimulation. Sensitization of the prolactin receptor within milk-producing cells is believed to occur since lactating women postpartum may produce up to 1 liter of breast milk per day even though their circulating prolactin concentration is within the normal range. As a result of the previously mentioned interaction between prolactin and regulation of GnRH secretion, individuals who are fully breast-feeding may delay resumption of ovulation for up to 27 weeks postpartum.

Hypothyroidism is a well-known cause of hyperprolactinemia. A decrease in thyroxine is believed to result in an increase in hypothalamic TRH release stimulating pituitary TSH release. Hyperprolactinemia in this instance is believed to be "sympathetic" because it results from TRH stimulation of lactotropes. Normalization of the hypothyroid state results in the lowering of the prolactin concentration to normal. Measurement of the serum TSH is usually all that is required to diagnose primary hypothyroidism.

Drug use is a common cause of transient hyperprolactinemia. Use of pharmacological agents known to affect prolactin secretion may be elicited by taking a careful medical history. Drugs frequently causing hyperprolactinemia include tranquilizers, narcotics, antihypertensives, and antiemetics.

Patients with renal disease and in particular those undergoing renal dialysis may have hyperprolactinemia. This appears to result from the presence of a circulating pituitary prolactin stimulation factor rather than from a decline in the metabolic clearance rate of the hormone.

Before entertaining a clinical diagnosis of hyperprolactinemia all of the expected causes of hyperprolactinemia should be excluded before ordering additional diagnostic tests or providing treatment.

VI. MANAGEMENT OF THE HYPERPROLACTINEMIC PATIENT

Once a diagnosis of pathologic hyperprolactinemia has been made, radiographic imaging of the pituitary gland should be performed. Plain films of the skull are of limited value since only large or destructive pituitary lesions will be identified. Use of computerized tomography (CT) or CT scanning of the head will provide details about the bony architecture of

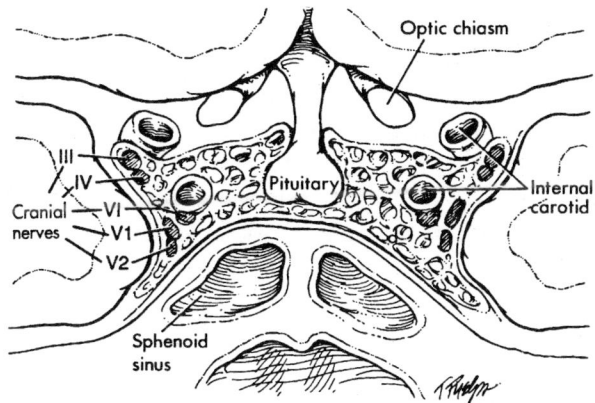

FIGURE 2 The pituitary gland and its relationship to adjacent important brain structures (reproduced with permission from Zacur and Hutchison, 1995).

the pituitary sella. Unfortunately, this method has mistaken normal bony variations of the sella for pituitary pathology. Small noncalcified lesions existing within the center of the pituitary gland or in the hypothalamus will not be identified by CT scanning. Repeated CT scanning without eye shielding has also been shown to increase the risk of cataract formation.

Magnetic resonance imaging (MRI) exquisitely displays the normal anatomy of the pituitary gland and its relationship to other important structures within the brain. (Fig. 2). Empty sella, suprasellar extension of pituitary lesions, and noncalcified brain lesions, e.g., germinomas, may be detected by this method. Use of contrast dyes will increase the sensitivity of the scan in detecting pituitary and hypothalamic lesions by MRI.

Over 40% of patients with an otherwise unexplained prolactin elevation above the normal range have been shown to have a detectable pituitary abnormality after MRI imaging. Hypothalamic lesions, lymphocytic hypophysitis, and non-prolactin-secreting tumors of the pituitary gland have all been identified in individuals with very mild elevations of the plasma/serum prolactin level. Disturbances of the normally inhibitory hypothalamic influence on pituitary prolactin secretion are suspected when hypothalamic or pituitary lesions are present. Reliance on a particular concentration of prolactin, e.g., 100 ng/ml, before ordering an imaging study of the hypothalamus and pituitary should be avoided.

Diagnosis of pituitary abnormalities following CT or MRI should be made carefully. Radiologists have relied in the past on the shape of the superior aspect of the pituitary gland and on the vertical height of the gland as reference measurements on which to make clinical diagnoses. Convex shape and gland height exceeding 7 mm were once thought to be reliable criteria "suggestive" of a pituitary lesion. It is now known that normal women are more likely to have a vertical gland height exceeding 7 mm and to also have a convex pituitary shape compared to normal men. The usefulness of these pituitary parameters as criteria for diagnosing pituitary disease is therefore questionable.

Lesions identified within the pituitary by imaging studies may be misunderstood by patients and physicians. Density changes within the pituitary gland identified by MRI are reported as "microadenomas" if their diameter is <10 mm or "macroadenomas" if the maximal diameter is >10 mm. Almost half of all pituitary lesions are prolactin secreting. The remaining pituitary lesions secrete other hormones or no hormone at all, e.g., chromophobe adenoma. Use of the terms pituitary adenoma or tumor are frightening to patients. Almost all prolactin-secreting pituitary lesions are benign. These lesions are neither "tumors" nor "adenomas." A tumor implies neoplastic cytoplasmic or nuclear changes which are detected histologically and not radiographically. Pituitary prolactin-secreting lesions exhibit increased secretory activity, i.e., enlarged cells. Rarely, if ever, do these cells demonstrate cytoplasmic or nuclear atypia when examined histologically. Therefore, they are not tumors. In rare cases the gland may expand into the cavernous or sphenoid sinus or compress the optic chiasm. Pituitary lesions rarely metastasize and are rarely encapsulated. Consequently, it is incorrect to even refer to these lesions as adenomas. Since pituitary prolactin lesions represent increased cellular secretory activity which is localized, it is probably best to refer to these lesions as "nodular hyperplasia of lactotrophs" or "prolactinomas."

Since pituitary lesions other than prolactinomas may be responsible for causing hyperprolactinemia, screening for these entities is required before recommending therapy. Pituitary lesions which do not secrete prolactin may cause hyperprolactinemia by blocking the inhibitory influence of the hypothalamus on lactotrope secretion of prolactin. Once a pituitary lesion has been detected by imaging, screening for growth hormone (GH), adrenocorticotropic hormone (ACTH), FSH, LH, and TSH lesions should be performed. Measurement of a normal serum/plasma somatomedin C or insulin growth factor-1 will exclude a GH lesion. A 24-hr urine collected to measure the free cortisol concentration may be ordered to exclude an ACTH secreting lesion because the cortisol level will be elevated in this instance. If a gonadotropin-secreting pituitary lesion is suspected, particularly in a menopausal woman with hyperprolactinemia, an α subunit may be ordered because this will frequently be found to be greatly elevated. Lastly, an elevated TSH measurement occurring with an elevated thyroxine level will be interpreted as suspicious for a pituitary lesion secreting TSH. If the pituitary lesion is suspected of being only prolactin secreting, further management of the hyperprolactinemic female patient will depend on the desire to conceive, concern about hypoestrogenism, and the size of the pituitary lesion. It is useful to classify the management of the hyperprolactinemic female patient using the parameters presented in the following sections.

A. The Absence of a Pituitary Lesion: Pregnancy Not Desired

Expectant management or dopamine agonist therapy may be offered to the hyperprolactinemic patient who does not have a pituitary lesion and does not wish to conceive. These patients should be informed that they have a one-third chance of having their elevated prolactin concentration return to the normal range within 5 years without therapy. They should also be advised that hyperprolactinemia is associated with a reduction in bone mass. A theoretical increased risk of heart disease resulting from hypoestrogenism is also possible.

Treatment of hyperprolactinemia is provided by prescribing medications which lower the prolactin concentration. These medications are dopamine agonists. In the United States bromocriptine (Parlodel)

and cabergoline (Dostinex) are approved by the United States Food and Drug Administration (FDA) for treatment of hyperprolactinemia. These drugs are derived from ergot precursors and stimulate both D1 and D2 dopamine receptors. Over 50% of patients taking these medications will experience minor side effects, such as orthostatic hypertension, lethargy, nausea, and nasal stuffiness. Bromocriptine is metabolized more quickly than cabergoline and must be taken on a daily basis with divided doses to maintain effectiveness. Carbergoline has a longer half-life and administration only twice per week may be required. Clinical studies performed in Europe that compared cabergoline to bromocriptine therapy reported that patients experienced fewer side effects when taking cabergoline.

In Europe a pure D2 dopamine agonist which is derived from apomorphine is available to treat hyperprolactinemia. This drug is quinagolide and is dispensed in Europe as Norprolac. Fewer side effects from this dopamine agonist compared to bromocriptine have been reported. Clinical trials using this medication have been performed in the United States, but the process of obtaining formal FDA approval has not been completed by the manufacturer. Pergolide is an ergot alkaloid which is a dopamine agonist and has been used in Europe to treat hyperprolactinemia. This drug may be dispensed in the United States as Permax, which is approved by the FDA only for the treatment of Parkinsonism.

Estrogen replacement or use of the combination oral contraceptive pill in hyperprolactinemic patients has been recommended by some authorities when patients are unable to tolerate dopamine agonist therapy. Protection against bone loss and heart disease are reasons given for this therapeutic approach. A stimulatory effect on pituitary prolactin secretion from estrogen exposure is well documented. Development of pituitary microadenomas or macroadenomas in hyperprolactinemic women taking oral contraceptive pills or hormone replacement therapy have been published in case reports. Hyperprolactinemic patients should therefore be informed of this possible risk before beginning estrogen therapy and also monitored during treatment.

B. The Absence of a Pituitary Lesion: Pregnancy Desired

Dopamine agonist therapy is available for women with hyperprolactinemia who do not have a pituitary lesion and who desire to conceive. Bromocriptine is the only FDA-approved dopamine agonist for this purpose. Children conceived when this drug is taken are not at increased risk of congenital malformations. In a recent report from Europe of over 100 women who conceived while taking cabergoline, no increased risk of congenital malformation was reported.

Once the prolactin concentration has been normalized by drug therapy, resumption of ovulation followed by menses should occur within 8 weeks. To minimize drug exposure to the fetus during pregnancy contraception should be used until menses resumes after beginning dopamine agonist treatment. When pregnancy is diagnosed drug therapy should be discontinued. There is no increased risk of pregnancy loss when dopamine agonist therapy is discontinued during early pregnancy.

Hyperprolactinemic women desiring to conceive who cannot tolerate dopamine agonist therapy may be induced to ovulate using ovarian-stimulating drugs. These include clomiphene citrate, which infrequently works, or menotropin therapy, which almost always results in successful ovulation induction. It is not necessary to use dopamine agonist treatment with menotropin ovulation induction.

Measurement of the serum prolactin concentration during pregnancy does not provide useful information for monitoring purposes due to the wide variation in prolactin values normally occurring during pregnancy. Visual field testing and pituitary imaging are not necessary since development of significant pituitary lesions during pregnancy in patients who did not have a pituitary lesion prior to conceiving has not been reported.

C. The Presence of a Pituitary Lesion: Pregnancy Not Desired

Exclusion of other hormone-producing lesions within the pituitary gland should be performed after a pituitary lesion has been detected in a patient with hyperprolactinemia. If the pituitary lesion is a microadenoma, enlargement over time rarely occurs. Even if a macroadenoma is detected, enlargement over time may be nonexistent or slow.

Expectant management or dopamine agonist therapy may be offered to patients with hyperprolactinemia and a pituitary lesion. The dopamine agonist drugs previously described may be prescribed. Following onset of drug therapy pituitary gland imaging at periodic intervals may be requested. For pituitary microadenomas the time interval between imaging studies may be expanded once stability of the lesion has been demonstrated. Imaging at more frequent intervals coupled with visual field testing is necessary if optic chiasm compression is suspected or if a macroadenoma has been identified. Imaging studies may be performed at similar intervals for patients with pituitary lesions who wish to be treated expectantly. Changes in bone density and adverse effects of the hypoestrogenic state on the heart should also be discussed.

Hyperprolactinemic women with pituitary lesions choosing expectant management should receive counseling regarding undesired pregnancy. Spontaneous ovulation is possible and due to preexisting amenorrhea the diagnosis of pregnancy may be delayed. Unmonitored pituitary enlargement for a protracted time period may therefore occur under these conditions.

Counseling patients about the risk of pituitary lesion enlargement following estrogen replacement therapy or oral contraceptive pill use should be provided if either of these therapeutic options is considered. Following the onset of dopamine agonist therapy in hyperprolactinemic women with pituitary lesions, it is not uncommon to find focal areas of hemorrhage within the pituitary gland. These small areas of infarction are believed to reflect the inhibitory effect of dopamine on lactotrope growth.

D. The Presence of a Pituitary Lesion: Pregnancy Desired

Exclusion of other pituitary lesions before making a diagnosis of a prolactinoma is required. Dopamine agonist therapy is commonly prescribed to women with micro- and macroadenomas of the pituitary who desire to conceive. Once pregnancy is diagnosed it is not necessary to continue dopamine agonist therapy throughout pregnancy as prophylaxis against pituitary enlargement.

Dopamine agonist therapy is stopped once pregnancy is diagnosed and patients may be monitored during their pregnancy by visual field testing after each trimester. Measurement of the serum/plasma prolactin concentration is not helpful due to normal fluctuation in the prolactin concentration during pregnancy. Administration of a dopamine agonist during the pregnancy of a hyperprolactinemic patient with a pituitary macroadenoma is indicated if optic nerve damage is suspected. If dopamine agonist therapy cannot be tolerated, other ovulation-inducing drugs may be prescribed if fertility is desired. Once pregnancy has been achieved, delivery should be determined by obstetrical indications and breast-feeding is not contraindicated in these patients.

Postpartum, it is not uncommon to see spontaneous infarction of the macroadenoma within the pituitary gland. This occurs as a result of the fall in the plasma concentration of estrogen present during pregnancy and its stimulating effect on pituitary lactotropes. Resumption of spontaneous menses with normalization of prolactin levels has even been reported in some of these patients who have experienced autoinfarction of their pituitary gland postpartum.

VII. SUMMARY

Prolactin is a critically important pituitary hormone because of its numerous biological actions, including lactogenic, steroidogenic, and immunologic functions. Its importance is made manifest as a consequence of its synthesis within important target tissues. As a result of the synthesis of prolactin in important organs, prolactin replacement following hypophysectomy is not required to sustain life.

Because of the variability of the concentration of this hormone in the serum or plasma of humans,

laboratory identification of pathologic hyperprolactinemia can be difficult. Once a truly hyperprolactinemic patient is identified, radiologic imaging of the pituitary of that patient is required to exclude the presence of other pituitary lesions which may cause sympathetic hyperprolactinemia.

Management of the hyperprolactinemic female patient will depend on her desire to conceive and the presence or absence of a pituitary lesion. Dopamine agonists are usually given to those patients wishing to conceive or who are concerned about hypoestrogenism. Ovulation-inducing agents may be provided to patients wishing to conceive who cannot tolerate dopamine agonist therapy.

Pituitary lesions which are prolactinomas are in most cases benign. Long-term follow-up of hyperprolactinemic patients with and without pituitary lesions has revealed that these patients do well clinically. This is reassuring information for both patients and physicians.

See Also the Following Articles

Lactogenesis; Pituitary Gland, Overview; Prolactin, Overview; Steroidogenesis, Overview

Bibliography

Ben-Jonathan, N., Mershon, J. L., Allen, D. L., and Steinmetz, R. W. (1996). Extrapituitary prolactin: Distribution, regulation, functions and clinical aspects. *Endocrinol. Rev.* 17, 639–669.

Cooke, N. E., and Liebhaber, S. A. (1995). Molecular biology of the growth hormone-prolactin gene system. *Vitamins Horm.* 50, 385–459.

Hwang, P., Guyda, H., and Friesen, H. (1971). A radioimmunoassay for human prolactin. *Proc. Natl. Acad. Sci. USA* 68, 1902–1906.

Jackson, J. C., Wortsman, J., and Malarkey, W. B. (1985). Characterization of a large molecular weight prolactin in women with idiopathic hyperprolactinemia and normal menses. *J. Clin. Endocrinol. Metab.* **61**, 258–264.

Macleod, R. M., Fontham, E. H., and Lehmeyer, J. E. (1970). Prolactin and growth hormone production as influenced by catecholamines and agents that affect brain catecholamines. *Neuroendocrinology* 6, 283–294.

McKnatty, K. P., Sawers, R. S., and McNeilly, A. S. (1974). A possible role for prolactin in control of steroid secretion by the human graafian follicle. *Nature* 250, 653–655.

Molitch, M. E. (1995). Prolactin. In *The Pituitary* (S. Melmed, Ed.), pp. 136–186. Blackwell, Cambridge, MA.

Owerbach, D., Rutter, W. J., Cooke, N. E., *et al.* (1981). The prolactin gene is located on chromosome 6 in humans. *Science* 212, 815–816.

Rui, H., Lebrun, J. J., Kirken, R. A., Kelly, P. A., and Farrar, W. L. (1994). JAK 2 activation and cell proliferation in-duced by antibody mediated prolactin receptor dimeriza-tion. *Endocrinology* 135, 1299–1306.

Stricker, P., and Grueter, F. (1928). Action du lobe anterieure de l'hypohyse sur la montee laiteuse. *C.R. Soc Biol* 99, 1978–1980.

Tyson, J. E. (1982). The evolutionary role of prolactin in mammalian osmoregulation. Effects on fetoplacental hy-dromineral transport. *Sem. Perinatal.* 6, 216–228.

Zacur, H. A., and Hutchison, S. (1995). Evaluation and ther-apy of hyperprolactinemia. In *Reproductive Medicine and Surgery* (E. E. Wallach and H. A. Zacur, Eds.), pp. 196–208. Mosby, St Louis, MO.

Hypogonadism

Andrew J. Friedman

Boston Regional Center for Reproductive Medicine

I. Embryology of the Male and Female Reproductive Systems
II. The Hypothalamic–Pituitary–Gonadal Axis
III. Hypogonadotropic Hypogonadism
IV. Hypergonadotropic Hypogonadism

GLOSSARY

anti-Müllerian hormone A product of fetal Sertoli cells that induces regression of the Müllerian or paramesonephric ducts.

granulosa cells Specialized cells within the ovarian follicle, controlled by gonadotropins and estradiol, that synthesize all three classes of sex steroids (estrogens, androgens, and progesterone), principally estrogen.

Leydig cells Specialized cells within the testes, controlled by luteinizing hormone and human chorionic gonadotropin, that secrete testosterone.

Müllerian ducts The paired structures of the paramesonephros that differentiate into the fallopian tubes, uterus, cervix, and upper one-third of the vagina in the absence of anti-Müllerian hormone.

Sertoli cells Specialized cells within the testes, controlled by follicle-stimulating hormone and testosterone, that secrete anti-Müllerian hormone and androgen-binding protein.

theca cells Specialized cells within the ovarian follicle, controlled by luteinizing hormone, that synthesize androgens (androstenedione and testosterone).

Wolffian ducts The paired structures of the mesonephros that differentiate into the epididymis, vas deferens, and seminal vesicles under the influence of testosterone.

Hypogonadism defines a state of absent or sub-normal gonadal function. Gonads have dual func-tions—gamete production and synthesis of sex ste-roids. These two functions are usually closely linked. The development of gonadal function depends on genetic, embryologic, endocrinologic, enzymatic, and environmental factors. Gonadal dysfunction may occur any time between embryologic development and advanced age. Signs and symptoms of hypogo-nadism are determined by the age of the individual at its onset and the degree of gonadal dysfunction.

I. EMBRYOLOGY OF THE MALE AND FEMALE REPRODUCTIVE SYSTEMS

The development of competent gonads is, in part, determined by the presence of a normal complement of sex chromosomes. Deletions of critical regions of the sex chromosomes, or abnormalities in the number of sex chromosomes, often lead to varying degrees of gonadal dysfunction. In normal embryonic development, the undifferentiated gonadal ridge develops at 5 weeks of gestation. The gonadal ridge contains primitive germ cells along with coelomic surface epithelium and an inner core of medullary mesenchymal tissue and is the only site where germ cells can survive. By the sixth week of gestation, the germ cells undergo mitotic cell multiplication to a total of 10,000 cells destined to become either sperm or ova. In this undifferentiated stage, the fetus contains both Wolffian and Müllerian ducts. One of these systems eventually differentiates, whereas the other regresses, giving rise to a gonadal male (Wolffian differentiation) or female (Müllerian differentiation). Differentiation of the testis occurs between 6 and 9 weeks of gestation and requires testis-determining factor, a gene product of the Y chromosome. In the absence of testicular development, the gonads will differentiate into ovaries.

The male phenotype depends on the products of the fetal testes, namely testosterone and anti-Müllerian hormone (AMH). Testosterone synthesis begins during the eighth week of gestation, peaks between Weeks 15 and 18, and then declines. Testosterone is produced by the fetal Leydig cells which are stimulated by human chorionic gonadotropin (hCG). Testosterone acts locally causing the differentiation of the Wolffian system (epididymis, vas deferens, and the seminal vesicles); a reduced form of testosterone, dihydrotestosterone (DHT), is required for development of the male external genitalia, urethra, and prostate. AMH is secreted by the Sertoli cells beginning at 7 weeks of gestation and results in the involution of the Müllerian system. Following differentiation along male phenotype lines, there is a relatively long period of quiescence that lasts until puberty. At that time, the maturation of the hypothalamic–

pituitary–testicular axis results in a secondary increase in testosterone production which, in turn, leads to the development of male secondary sexual characteristics—enlargement of the testes and phallus, acceleration of long bone growth and subsequent epiphysial closure, an increase in muscle mass, and pigmented hair growth in the male distribution.

In the absence of testosterone and AMH, female phenotypic development occurs. Rapid mitotic multiplication of germ cells begins at 6–8 weeks, reaching 6 or 7 million oogonia by 16–20 weeks. Oogonia are transformed into oocytes as they enter the first meiotic division and arrest in prophase under the influence of granulosa cell products. Thereafter, no further development of oocytes occurs and attrition of these germ cells begins and continues until they are exhausted causing ovarian failure. By birth, most females have 1 or 2 million oocytes. The number decreases to 300,000 or 400,000 by puberty, when reproductive competence is achieved.

Loss of germ cells may result from several mechanisms, including errors of mitosis and meiosis, failure of oocytes to become enveloped by granulosa cells, and failure of germ cell migration to the surface of the gonad. Chromosomal abnormalities can accelerate germ cell loss or atresia. The most common chromosomal abnormality leading to rapid germ cell attrition is 45,X, or Turner's syndrome, in which oogonia fail to undergo meiosis leading to the loss of all germ cells by birth.

In the absence of high local concentrations of testosterone, the Wolffian system regresses and the absence of AMH causes the Müllerian system to develop. The Müllerian system develops into the fallopian tubes, uterus, cervix, and upper one-third of the vagina. In the absence of DHT, the undifferentiated urogenital sinus, labioscrotal folds, and genital tubercle form the labia minora, labia majora, and clitoris, respectively. In addition, the urogenital sinus differentiates into the lower two-thirds of the vagina and the urethra.

As in the male, there is a relatively long period of quiescence of gonadal function in females. The maturation of the hypothalamic–pituitary–ovarian axis occurs during puberty, leading to cyclic follicular development (and, often, ovulation) which, in turn, leads to estrogen production and the development of secondary sexual characteristics including breast development, enlargement of the uterus and cervix, cornification of vaginal epithelium, and accelerated long bone growth and subsequent epiphyseal closure. An increase in adrenal androgen production around this time causes the development of coarse pigmented pubic and axillary hair.

II. THE HYPOTHALAMIC–PITUITARY–GONADAL AXIS

Hormonal stimulation of the gonads is necessary for gamete development and sex steroid synthesis and secretion. The highly integrated system responsible for gonadal function is composed of three major anatomic compartments: the arcuate nucleus in the hypothalamus, the anterior pituitary gonadotrophs, and the gonads.

The hypothalamus is the origin of the neurons which synthesize and secrete gonadotropin-releasing hormone (GnRH). During embryogenesis, GnRH-containing neurons migrate from the medial olfactory placode into the brain. In primates, the primary network of GnRH cell bodies is located within the medial basal hypothalamus. Most of these cell bodies are located in the arcuate nucleus. These neurons terminate in the median eminence where neuronal secretory products feed into a dense capillary network which, in turn, drains into the portal circulation. The portal vessels descend into the anterior pituitary where the GnRH decapeptide can act on the gonadotrophs.

GnRH induces the synthesis and secretion of the gonadotropins, luteinizing hormone (LH), and follicle-stimulating hormone (FSH). GnRH pulse frequency determines the relative amounts of LH and FSH that are released. A slow GnRH pulse generator favors FSH secretion, whereas a fast rate favors LH; extreme (i.e., very rapid or extremely slow) rates of GnRH release induce a hypogonadotropic state with low levels of LH and FSH secretion.

LH and FSH are released into the peripheral circulation and act on their target cells in the gonads. In general, LH acts on Leydig cells in the testes to secrete testosterone; LH acts on the theca and stroma in the ovary to synthesize the androgens androstenedione and testosterone. FSH stimulates the testicular Sertoli cells and germinal epithelium. A high testicular concentration of testosterone, together with FSH, act on the seminiferous tubules to induce spermatogenesis. In women, FSH causes proliferation of granulosa cells which synthesize estradiol from androgen precursors. FSH also induces oocyte selection and maturation. A sufficiently high and sustained circulating concentration of estradiol is necessary for the initiation of an LH surge, the hormonal trigger for resumption of meiosis in the oocyte, and its eventual release (ovulation) from the ovarian cortex. Following ovulation, GnRH pulsation slows and LH stimulates the corpus luteum to synthesize and secrete progesterone. In the absence of conception and implantation, the corpus luteum undergoes demise, usually within 12–16 days. This leads to a precipitous fall in circulating levels of estradiol and progesterone which, in turn, stimulates the sloughing of the functionalis layer of the endometrium—a process known as menstruation.

Levels of gonadotropin stimulation required to induce mature gonadal function in males and females are not achieved until puberty. Aberrations of GnRH, gonadotropin, and sex steroid synthesis can lead to gonadal dysfunction.

III. HYPOGONADOTROPIC HYPOGONADISM

Hypogonadotropic hypogonadism is a state of subnormal or absent gonadal function coexisting with low circulating levels of gonadotropins. Because a negative long-loop feedback mechanism is normally operative, low circulating concentrations of sex steroids will usually signal high levels of gonadotropin secretion. Thus, in hypogonadotropic hypogonadism, normal feedback mechanisms are not operative, usually due to a defect at the level of the hypothalamus or pituitary.

If aberrations of this hormone feedback mechanism occur before puberty, the child presents with delayed puberty. Constitutional delay of puberty, de-

TABLE 1
Causes of Hypogonadotropic Hypogonadism

Constitutional delay
Anatomic
    Tumors (e.g., prolactinoma)
    Craniopharyngiomas
    Infiltration disorders (e.g., hemochromatosis, sarcoidosis,
        tuberculosis, and histiocytosis X)
    Empty sella syndrome
    Pituitary apoplexy
    Sheehan's syndrome
Idiopathic
    Kallmann's syndrome
    Prader–Willi syndrome
    Laurence–Moon–Biedl syndrome
Functional
    Drugs (e.g., anabolic steroids and glucocorticoids)
    Exercise induced
    Ethanol
    Severe systemic illness (e.g., uremia, liver failure, and
        acquired immune deficiency syndrome)
    Aging
    Eating disorders (e.g., anorexia nervosa and morbid
        obesity)
    Psychogenic
Environmental
    Cranial radiation
    Cranial trauma
Pharmacologic
    GnRH agonists

fined as the absence of secondary sexual development at age 14 or the absence of menarche in girls by age 16, is the most common etiology of delayed puberty in boys and girls. Although many of these children are treated with exogenous steroids to initiate sexual maturation, pubertal milestones will eventually be reached spontaneously.

There are many individuals with hypogonadotropic hypogonadism who have other causes of hypogonadism. The clinician must rule out other etiologies of absent or inappropriate hormone feedback mechanisms including anatomic, idiopathic, functional, environmental, or pharmacologic causes (Table 1).

Hypoprolactinemia may cause symptoms of hypogonadism, amenorrhea, galactorrhea (in women), and visual impairment (due to compression of the optic chiasm by a pituitary tumor). Elevations in serum prolactin, with or without the presence of a tumor, may cause hypogonadism through a variety of mechanisms. Prolactin acts directly on the hypothalamus altering aminergic function, thereby inhibiting GnRH pulsations. In rare situations, very large tumors may displace or destroy gonadotrophs or compress the pituitary venous system, thereby inducing a functional hypophysectomy. Treatment of these individuals involves medication (e.g., dopamine agonists such as bromocriptine), ablative surgery, or both. Finally, significant elevations in serum prolactin may be secondary to drug use such as phenothiazines. Identification of the offending drug and its discontinuation will treat the problem.

Craniopharyngiomas, or tumors of Rathke's pouch, occur most commonly between ages 6 and 14 and may be associated with either hypogonadotropic hypogonadism or precocious puberty. This, and other infiltrative disorders, is thought to cause hypogonadotropic hypogonadism by interruption of hypothalamic–pituitary signaling.

Empty sella syndrome occurs when the sella turcica forms an extension of the subarachnoid space and is partially or completely filled with cerebrospinal fluid. The syndrome is most commonly seen in middle-aged women who usually present with nonspecific symptoms such as headache. The cause of empty sella syndrome is unknown and most individuals have no discernible endocrine abnormalities. However, some people may present with either deficiency or hypersecretion of a single or multiple pituitary hormones.

Pituitary apoplexy is a rare, but life-threatening, acute massive infarction of the pituitary gland. Pituitary apoplexy requires prompt recognition and medical intervention or it may prove fatal. It is most often associated with infarction of a chromophobe adenoma, most of which are endocrinologically inactive.

Sheehan's syndrome, first described in 1939, is a syndrome of hypopituitarism resulting from acute necrosis of the anterior (and, sometimes, posterior) pituitary following postpartum hemorrhage and shock. Clinical signs and symptoms develop in immediate postpartum period and include failure of lactation, fatigue, hypotension, and loss of pubic and axillary hair. As in pituitary apoplexy, prompt diagnosis and medical intervention is essential.

Of the idiopathic causes, Kallmann's syndrome, or isolated gonadotropin deficiency, has been the most widely studied. It is five to seven times more common in men and its incidence is estimated to be 1 in 10,000 male births. The syndrome is most often due to an isolated mutation, but there also exist inherited forms of the disease. The inherited forms have most often been autosomal dominant, although some families demonstrate autosomal recessive or X-linked modes of inheritance. A gene has been isolated from the critical region on Xp22.3 which has been implicated in this syndrome. Deletion or translocation of this gene locus can lead to the absence of the protein(s) responsible for GnRH cell migration and may be associated with other neurologic defects such as anosmia. Clinical features of Kallmann's syndrome vary widely as the degree of hypogonadism may be partial or complete. Individuals with complete forms of the syndrome are characterized by sexual infantilism, the absence of gametogenic function, and classic eunuchoidal features.

Treatment of the disorder may be accomplished by replacing hormones in any of the anatomic compartments in the hypothalamic–pituitary–gonadal axis. GnRH can be given in a pulsatile fashion either by intravenous or by subcutaneous infusions. GnRH delivered in this way will stimulate pituitary gonadotropin secretion which, in turn, induces gonadal sex steroid release and gametogenesis. Gonadotropin therapy with LH, FSH, or both can be delivered intramuscularly, avoiding the need for an indwelling catheter. Sex steroid supplementation is the least expensive therapy and may be delivered orally or parenterally. Sex steroid treatment results only in the development of secondary sexual characteristics and fails to achieve gametogenesis.

Prader–Willi and Laurence–Moon–Biedl syndromes are congenital disorders characterized by obesity, mental retardation, and delayed puberty or sexual infantilism, along with other traits unique to each syndrome. In both of these syndromes, the mechanism of hypogonadotropic hypogonadism is unclear.

Functional abnormalities of gonadotropin secretion are a heterogeneous group of disorders with the common end result of hypogonadotropic hypogonadism. Anabolic steroid use in men can lead to profoundly impaired spermatogenesis and decreased concentrations of testosterone and DHT. The mechanism for the hypogonadism appears to be due to a direct effect of the anabolic steroids on the hypothalamic–pituitary axis. Cessation of anabolic steroid use will usually reverse the hypogonadotropic hypogonadism, but it may take at least 16 weeks for gonadal function to return.

The effect of exercise on gonadal function is complex and appears to depend on multiple variables including the sex of the individual; the type, intensity, and duration of exercise; body composition; psychologic background; and the intensity of stress. In general, the effects of exercise on the hypothalamic–pituitary–gonadal axis are less dramatic in males than in females. In highly physically conditioned males, serum testosterone concentration may be suppressed, but frank hypogonadism is rarely seen.

In females, the type and intensity of physical activity can have a profound impact on gonadal function. For example, high-intensity runners and ballet dancers have a higher incidence of amenorrhea (40–50%) compared to elite swimmers (approximately 12%). The delay in menarche in elite gymnasts and ballet dancers averages 3 years and the incidence of chronic anovulation later in life is significantly higher than that in sedentary females.

Body composition, especially the percentage of body fat, appears to play a critical role in the onset and maintenance of menses. It has been hypothesized that the proportion of body fat required for normal menstrual function must be at least 22%. The percentage of body fat among competitive runners and ballet dancers averages 15% compared with 20% in competitive swimmers.

Acute exercise also results in alterations in the hypothalamic neuroendocrine environment. Exercise increases corticotropin-releasing hormone

(CRH) which, in turn, increases central opioidergic tone. An increase in opioid secretion causes a decrease in GnRH pulse amplitude and frequency, leading to a significant decrease in LH secretion without altering FSH release. The end result, with repetitive, chronic episodes of intense exercise, is decreased testosterone production in men and suppressed estrogen production and amenorrhea in women. Psychologic stress, which is often present in elite athletes, can also result in chronic elevations in CRH, leading to hypogonadism. Interruptions in training schedules and/or weight gain favoring an increase in the percentage of body fat may reverse some of the signs and symptoms of hypogonadism.

The effects of ethanol ingestion as a direct gonadal toxin are well documented. Although less certain and more controversial, there are data to support an effect of alcohol on the hypothalamus. Alcohol-induced hypogonadism may be completely or partially reversible if alcohol intake is stopped early in the course of gonadal dysfunction.

Severe chronic systemic illness is often associated with malnutrition and a profound decrease in GnRH pulse frequency and amplitude. In some illnesses, such as acquired immune deficiency syndrome, the pituitary can be secondarily involved by infectious processes leading to hypogonadism.

The effects of aging on gonadal function are markedly different in males and females. In women, gamete development and sex steroid secretion are closely linked and depend on the available pool of follicles. Oocytes and follicles are nonregenerative and their supply is eventually exhausted leading to secondary gonadal failure known as menopause. In men, spermatogenesis and Leydig cell function (e.g., testosterone synthesis) are not directly linked and new sperm are always being generated. Both spermatogenesis and Leydig cell function do decline with age (e.g., there is a 35% decline in total serum testosterone concentration between the ages of 20 and 80 years). The decrease in Leydig cell function appears to be due to decrease in the number of Leydig cells and a decrease in LH pulse amplitude but not pulse frequency. Although mean sperm concentrations may also decline with age, there are numerous examples of men well into their eighties and nineties who are able to father children.

Eating disorders, such as anorexia nervosa, have profound effects on virtually all hypothalamic and pituitary hormones. The mechanisms involved for these effects are numerous and involve psychogenic factors, body composition, and fuel intake. In general, girls with this disorder have high central opioidergic tone, possibly due to chronic hypersecretion of CRH, leading to a reduction of gonadotropin secretion comparable with that seen in prepubertal girls. Treatment of anorexia nervosa is complex, lengthy, and often involves a multidisciplinary approach with dietary, psychologic, and hormonal interventions.

Cranial irradiation is sometimes performed to treat pituitary tumors. Whereas pituitary cells are relatively radioresistant, hypothalamic neurons are more radiosensitive. Changes in pituitary function following cranial irradiation may be attributable to hypothalamic neuronal damage and may progress to panhypopituitarism. The effects of irradiation are often slow to develop and may take as long as 10 years to be manifested.

GnRH agonists are synthetic analogs of GnRH with enhanced potency. In the synthesis of GnRH agonists, the native GnRH decapeptide is preserved except for two substitutions: (i) glycine in position 10 is replaced by either N-ethylamide ($NH-CH_2CH_3$) or aza-Gly ($-NHNHCO-$) moieties, and (2) glycine in position 6 is replaced by a D-amino acid. The resultant synthetic GnRH analog has greater affinity for the GnRH receptor and is resistant to proteolytic degradation. Repetitive or constant GnRH agonist administration will therefore mimic a continuous, high-dose GnRH infusion. GnRH agonist treatment will cause an initial surge of gonadotropin release, followed by a sustained state of hypogonadotropism due to pituitary desensitation and downregulation of GnRH receptors. The end result is a pharmacologically induced hypogonadotropic hypogonadism that may be used to treat endocrine-sensitive malignancies (e.g., prostatic carcinoma) or benign disease states (e.g., endometriosis, leiomyomas, and precocious puberty). Hypogonadotropism is reversible, with restoration of normal gonadotropin secretion within 1–6 months after cessation of treatment.

IV. HYPERGONADOTROPIC HYPOGONADISM

Elevations of serum gonadotropins in the face of hypogonadism (i.e., hypergonadotropic hypogonadism) suggest gonadal failure. Gonadal failure may be congenital (primary) or acquired (secondary). The age of onset of gonadal failure will determine the clinical presentation of the affected individual. Gonadal failure before puberty will result in sexual infantilism; gonadal failure following puberty will have more profound effects on reproductive function than on secondary sexual characteristics. A list of causes of hypergonadotropic hypogonadism is shown in Table 2.

Klinefelter's syndrome is the most common cause of hypergonadotropic hypogonadism in males. It is a type of testicular dysgenesis characterized by one or more supernumerary X chromosomes. The classic form, 47,XXY, occurs in approximately 1 in 500 males. The most common cause of Klinefelter's syndrome is nondisjunction during the first or second meiotic division of parental oogenesis or spermatogenesis. Other less common causes include nondisjunction during the first postzygotic mitosis or anaphase lag during mitosis or meiosis.

The phenotypic characteristics of classic Klinefelter's syndrome include markedly decreased testicular volume (<1.5 ml), gynecomastia, and an exaggerated growth of the lower extremities resulting in a decreased crown to pubis/pubis to floor ratio. These abnormal skeletal proportions differ from those seen in individuals with other forms of prepubertal hypogonadism in which the absence of sex steroids leads to delayed long bone epiphysial closure and eunuchoidal stature (i.e., arm span at least 6 cm > height). The prostate is small, sexual hair is sparse, and intelligence, especially regarding verbal skills, is often reduced. Other psychiatric manifestations may also be present, including timidity, introspection, and withdrawn social behavior. Men with classic Klinefelter's syndrome are irreversibly azoospermic. These men are infertile, but androgen therapy may have some masculinizing effects and improve some of the behavioral problems.

TABLE 2
Causes of Hypogonadotropic Hypogonadism

Genetic
  Klinefelter's syndrome (e.g., 47,XXY)
  Turner's syndrome (e.g., 45,X)
  Gonadal dysgenesis (e.g., 46,XX; 46,XY; 46,XX/45,X; and 46,XY/45,X)
  Swyer's syndrome (e.g., 46,XY female)
Idiopathic
  Sertoli cell only syndrome
Enzymatic defects
  20,22-desmolase
  3β-ol-dehydrogenase
  17α-hydroxylase
  17,20-desmolase
  17-ketosteroid reductase
  5α-reductase
Anatomic
  Anorchia
  Cryptorchidism
  Surgical extirpation of gonads
Infectious
  Postpubertal orchitis/oophoritis (e.g., mumps and leprosy)
Environmental
  Gonadal irradiation
  Chemotherapy (e.g., alkylating agents)
  Industrial hydrocarbons
  Gonadal trauma
  Ethanol
  Marijuana
Immunologic
  Autoimmune oophoritis

Other less common genetic causes of hypergonadotropic hypogonadism include 46,XX males and 46,XX/45,X gonadal dysgenesis. Myotonic dystrophy, characterized by an inability to relax striated muscles after contraction, is an autosomal dominant disorder characterized by frontal balding, cataracts, and testicular failure. Since testicular failure may not occur until the man is 30–40 years old, the disorder may be passed from father to son. The cause of testicular failure is unknown in this disorder.

Sertoli cell only syndrome is characterized by small testes, which, on biopsy, demonstrate no seminiferous tubules. These men are all sterile, although androgen levels are often in the normal range.

Enzymatic defects in testosterone synthesis occur *in utero*, often leading to ambiguous (or female appearing) external genitalia at birth. Depending on the type and severity of the enzymatic defect, these individuals may be raised as males or females. If the individual is to be raised as a male, testosterone therapy and reconstructive surgery are usually required. Some enzymatic defects in testosterone production involve 20,22-desmolase, 3β-ol-dehydrogenase, 17α-hydroxylase, 17,20-desmolase, and 17-ketosteroid reductase. Males with 5α-reductive deficiency are unable to convert testosterone to DHT, the hormone responsible for development of the scrotum, prostate, and testes. These males appear as girls until puberty, when a significant increase in testicular testosterone production induces masculinization of the individual.

Anorchia and cryptorchidism occur in phenotypic males with a 46,XY karyotype and no palpable testes. In anorchia, or vanishing testes syndrome, the testes involute *in utero* at some point after they initially secrete testosterone and AMH. A male phenotype results with regression of the Müllerian system. Males with anorchia are sterile and have no spontaneous development of secondary sexual characteristics and must be treated with testosterone. In cryptorchidism, one or both testes are intraabdominal in location. Up to 10% of males are cryptorchid at birth. However, 70–90% of these individuals will have spontaneous testicular descent and the incidence falls to 0.3–0.4% following puberty. Treatment of cryptorchid testes may be hormonal (e.g., hCG treatment) or surgical (i.e., orchiopexy). Up to 70% of males with cryptorchidism will have infertility because their testes may never have normal spermatogenic function. Treatment before the age of 5 improves the chances of fertility. Untreated individuals have an 8% chance of malignant change of the intraabdominal testes; early treatment will lower the chance of malignancy.

Postpubertal orchitis, seen in 25% of cases of mumps, will cause oligo- or azoospermia in 60% of males. In addition to impaired spermatogenesis, these individuals will have significantly elevated serum FSH concentrations. In some cases, testosterone levels are decreased and LH is increased, suggesting involvement of the Leydig cells. Other infectious causes of testicular failure, such as leprosy, are quite rare.

Traumatic loss of both testes will result in sterility and hypogonadism requiring testosterone treatment. Testicular irradiation, often done for certain malignancies (e.g., Hodgkin's disease and prostatic carcinoma), can lead to impaired spermatogenesis and Leydig cell dysfunction. A dose of 15 rads will induce oligospermia and 50 rads will cause transient azoospermia. Sperm production will return in these individuals unless repeated treatments are performed. A cumulative dose of 400–600 rads will result in irreversible azoospermia; a dose of 2000 or 3000 rads will result in azoospermia and Leydig cell failure.

Gonadal toxins, such as cytotoxic drugs used for treating malignancies or nephrotic syndrome, may result in testicular damage. Alkylating agents are the most common offenders in this group of drugs. Marijuana and alcohol, when used chronically and in large quantities, may lead to a reduction in sperm count and motility as well as a decreased serum testosterone concentration. Industrial hydrocarbon exposure may also lead to impaired spermatogenesis.

In females, gonadal dysgenesis is the most common abnormality leading to hypergonadotropic hypogonadism. Of females who present with gonadal dysgenesis, half will have Turner's syndrome (i.e., 45,X), 25% will be mosaics (e.g., 46,XX/45,X), and the remainder will have a normal female karyotype. It is important to perform a karyotype to evaluate females with primary amenorrhea and elevated gonadotropins to rule out the presence of a Y chromosome which would predispose the individual to gonadal tumors.

In classic Turner's syndrome, the phenotypic characteristics include short stature, webbed neck, shield chest, sexual infantilism, increased carrying angle at the elbow, and short fourth metacarpals. Many Turner's individuals have coarctation of the aorta and a horseshoe kidney. Although the number of oocytes reaches 6 or 7 million in these females by midgestation, the entire oocyte pool undergoes atresia prior to birth. Essentially all these females are sterile and require exogenous sex steroids to induce secondary sexual characteristics. Pregnancy is possible only with donated oocytes. Of note is that 98 or 99% of 45,X females are aborted, making this karyotype the single largest genetic abnormality in abortuses.

Individuals with mosaicism (the presence of multiple cell lines of varying sex chromosome composition) may present with varying degrees of hypogonadism and the stigmata of Turner's syndrome. Some of these females may have normal pubertal development, menstruate regularly, and achieve fertility only to present with premature ovarian failure in their twenties or thirties. Ovarian failure occurs because of accelerated follicular atresia. These women are usually shorter than females with a 46,XX karyotype and rarely achieve final height of >160 cm (63 in.).

A rare condition in females resulting in a lack of sexual development and elevated gonadotropins along with a 46,XY karyotype is Swyer's syndrome. The presence of a Y chromosome greatly increases the likelihood of malignant transformation of the gonads, necessitating surgical extirpation of the gonadal streaks as soon as the diagnosis is made.

Premature ovarian failure (POF), defined a s secondary amenorrhea and elevated gonadotropins before age 40, occurs in 1% of women in the United States. Although the etiology of POF is usually unknown, a thorough evaluation must be done to rule out genetic (e.g., mosaicism), autoimmune, infectious (e.g., mumps oophoritis), environmental (e.g., gonadal irradiation), and pharmacologic (e.g., chemotherapy) causes. Gonadal surgery, with removal of all or a portion of one or both ovaries, will significantly reduce the follicle pool and may lead to premature menopause. Autoimmune POF may be associated with autoimmune polyglandular failure and the clinician must rule out hypothyroidism, hypoadrenalism (e.g., Addison's disease), hypoparathyroidism, diabetes mellitus, and pernicious anemia.

The effect of gonadal irradiation is dependent on age and X-ray dose. In general, ovaries in older women are more sensitive to radiation and require a lower X-ray dose to induce sterility. Ovarian doses of 150 rads will sterilize some women over the age of 40; doses over 800 rads will permanently sterilize all females over age 15. As exposure increases between 150 and 800 rads, an increasing proportion of women between ages 15 and 40 will become sterilized.

Chemotherapy drugs, especially alkylating agents, are toxic to gonads. As with radiation, there is an inverse relationship between the dose required to induce ovarian failure and age at the start of therapy. Occasionally, as with radiation, the development of POF may be delayed, suggesting that a significant proportion of the viable follicle pool was damaged during treatment.

In summary, hypogonadism may be caused by many different factors any time between conception and old age. The type and severity of symptoms are determined by the severity of gonadal dysfunction and the age of the individual when hypogonadism occurs. In general, treatment of individuals with hypogonadotropic hypogonadism can restore fertility and develop and/or maintain secondary sexual characteristics. Treatment of individuals with hypergonadotropic hypogonadism may induce and maintain secondary sexual characteristics but cannot restore fertility. The successful diagnosis and treatment of hypogonadism is essential because some causes may have other profound health implications. In addition to medical and/or surgical intervention, psychological support is an integral part of management of these individuals.

See Also the Following Articles

Granulosa Cells; Kallmann's Syndrome; Leydig Cells; Wolffian Ducts

Bibliography

Elias, A. N., and Wilson, A. F. (1993). Exercise and gonadal function. *Hum. Reprod.* 8, 1747–1761.

Frisch, R. E., Gotz-Welbergen, A. V., McArthur, J. W., Albright, T., Witschi, J., Bullen, B., Birnholz, J., Reed, R. B., and Herman, H. (1981). Delayed menarche and amenorrhea of college athletes in relation to age of onset of training. *J. Am. Med. Assoc.* 246, 1559.

Matsumoto, A. M. (1994). Hormonal therapy of male hypogonadism. *Clin. Androl.* 23, 857–875.

Plymate, S. (1994). Hypogonadism. *Clin. Androl.* 23, 749–773.

Vermeulen, A., and Kaufman, J. M. (1995). Ageing of the hypothalamo–pituitary–testicular axis. *Menop. Horm. Res.* **43**, 25–28.

Whitcomb, R. W., and Crowley, W. F. (1993). Male hypogonadotropic hypogonadism. *Neuroendocrinology II* **22**, 125–143.

White, B. J., Rogol, A. D., Brown, K. S., Lieblich, J. M., and Rosen, C. W. (1983). The syndrome of anosmia with hypo-gonadotropic hypogonadism: A genetic study of 18 new families and a review. *Am. J. Med. Genet.* **15**, 417.

Williams, C. L., Nishihara, M., Thalabard, J.-C., Grosser, P. M., Hotchkiss, J., and Knobil, E. (1990). Corticotropin-releasing factor and gonadotropin-releasing hormone pulse generator activity in the rhesus monkey. *Neuroendocrinology* **52**, 133.

Yen, S. S. C. (1993). Female hypogonadotropic hypogonadism. *Neuroendocrinology II* **22**, 29–59.

Hypophysectomy

Donald C. Johnson
University of Kansas Medical Center

GLOSSARY

adenohypophysis The anterior lobe of the pituitary gland, made up of the pars distalis (pars glandularis); the pars tuberalis, which surrounds the infundibular stalk of the neurohypophysis; and the pars intermedia, which in several species separates the glandularis from the neurohypophysis. The adenohypophysis consists of five kinds of cells: the somatotroph for growth hormone, the lactotroph for prolactin, gonadotrophs for follicle-stimulating hormone and luteinizing hormone, thyrotrophs for thyroid-stimulating hormone, and corticotrophs for adrenocroticotropin.

diaphragma sellae The dura mater covering the pituitary. It contains an opening of variable size through which the infundibular stalk passes.

hypophysis cerebri The anatomical name for what is commonly called the pituitary gland. It consists of two lobes: the anterior or *adenohypophysis,* an epithelial derivative of the embryonic Rathke's pouch, and the posterior lobe, or *neurohypophysis,* which is derived from the floor of the diencephalon of the brain.

sella turcica A shallow depression of the sphenoid bone in the base of the skull that contains the pituitary gland.

Hypophysectomy refers to the removal—by surgical, chemical, or pharmacological means—of the function of the adenohypophysis, the glandular portion of the vertebrate hypophysis, which is responsible for the production and secretion of several important hormones including somatotropin (growth hormone), prolactin, follicle-stimulating hormone (FSH), luteinizing hormone (LH), thyrotropin- or thyroid-stimulating hormone (TSH), and corticotropin (ACTH). Hypophysectomy also removes the pars intermedia, which constitutes a separate lobe in some species, as well as a large part of the neurohypophysis, or posterior pituitary, which is a neural structure that is a reservoir for the hormones vasopressin and oxytocin. Partial hypophysectomy was originally defined as incomplete surgical removal or chemical destruction of the adenohypophysis but has come to include reduction or removal of specific hypophyseal hormones by pharmacologic or genetic methods.

I. INTRODUCTION

Interest in hypophysectomy began in the later part of the nineteenth century after the French neurologist Pierre Marie associated excessive body growth (gigantism and acromegaly) with pituitary tumors. Although his idea that growth was inhibited by the pituitary, and that this inhibition was lost after tumor development, proved to be wrong it stimulated interest in the effects of removal of the gland. However, the relative inaccessibility of the pituitary proved a considerable surgical challenge and the initial high mortality rates for humans and animals discouraged use of the operation. Although there were only a few successes with the early operations they established that the pituitary was not essential for life. On the other hand, these hypophysectomies were of little value for clarifying the physiological function of the pituitary because of the many disparities in their consequences. We can now understand some of the postoperative effects in terms of damage to neural tissue, particularly the hypothalamus, associated with manipulation of the brain to gain access to the pituitary. Additional complicating factors included the lack of a means for controlling bacterial infections and the inability to control the effects of loss of adrenocortical steroids following the severe stress of the operation. Because of the latter, hypophysectomy in the human for treatment of nonpituitary diseases, such as breast or prostatic cancer, was not attempted until after the introduction of synthetic adrenal corticosteroids in the 1950s.

During the first quarter of the twentieth century investigations into the physiologic functions of hypophyseal hormones, particularly those involved in growth and reproduction, intensified and removal of the source of the endogenous hormones involved in these functions became essential for further progress. Investigators quickly realized that there were basically only two approaches to the gland, either from above, the so-called intracranial route, or from below through the base of the skull. That is, the sella turcica could be reached through a hole in the roof of the oral cavity (the transbuccal approach) and the simplicity of this procedure in amphibians stimulated interest in using it for other animals. However, the major advance in the study of hypophyseal physiol-ogy accompanied the introduction, by P. E. Smith in the late 1920s, of hypophysectomy of rats and mice by a parapharyngeal approach. Subsequently hypophysectomy of these animals became a routine procedure for many laboratories and now is offered by commercial suppliers of experimental animals.

II. COMPLETE HYPOPHYSECTOMY

A. Surgical

Although careful histologic examination of the sella turcica after the operation reveals that small fragments of the pituitary may remain, complete hypophysectomy means removal of the anterior, and intermediate if present, lobes of the hypophysis as well as most of the posterior lobe. A portion of the pituitary stalk and associated pars tuberalis remains. Because of the disruption of the blood supply to the tissue and loss of its hypothalamic control, the functional significance of these remnants is considered minor, particularly in acute studies. However, several investigators have suggested washing the empty sella turcica with a fixative to ensure complete destruction of adenohypophyseal cells. Lack of growth of young animals is often used as an indication of complete removal of the pituitary but this can be an unreliable index. Even though sensitive immunoassays of hypophyseal hormones, or even their subunits, are now available for verifying the effectiveness of the operation, definitive proof of completeness depends on histologic examination of serial sections through the entire sella turcica.

All the surgical procedures for hypophysectomy require special attention to the care of the animal after the operation. Included in these are minimizing environmental stresses and provision for suitable nourishment and recognizing that trauma to the oral or pharyngeal area may affect the animal's ability to eat. For example, the hypoglycemia that can accompany loss of the pituitary accounts for the usual use of a 5–10% solution of dextrose for drinking water following the operation. Abundant consumption of this solution, at least for a few days, is assured by the polydipsia that occurs following removal of the neurohypophyseal antidiuretic hormone vasopressin. Associated with the polydipsia is polyuria, which necessitates attention to maintaining a clean and dry environment.

The intracranial approach, also referred to as the transcranial approach, was the first attempted for hypophysectomy in humans and experimental animals. Although mortality originally was as high as 80%, it was reduced to <5% after considerable experience. This approach, which requires a craniotomy and extensive retraction of cortical tissues, has been extensively used for humans but it is not recommended for most experimental animals. However, a procedure developed for the human that used a transfrontal approach with excision of an anterior medial section of a frontal lobe to better visualize the optic chiasm and pituitary was successfully applied to nonhuman primates. Another method that has had a great deal of use for human hypophysectomy involves a sublabial transseptal transsphenoidal approach. This involves accessing the sella turcica from underneath using an initial approach through the subnasal region and as such is more related to the parapharyngeal than the intracranial approach.

Although described by some as a troublesome and time-consuming procedure, the parapharyngeal has been, without doubt, the most frequently used approach for hypophysectomy of a large variety of experimental animals, including rats, mice, hamsters, ferrets, guinea pigs, dogs, sheep, goats, monkeys, birds, and fish. The procedure evolved as a modification of the transbuccal approach where the pituitary is reached through the roof of the oral cavity. Use of the latter approach is limited by the size and shape of the mouth and has the disadvantage of crossing a septic field. Several minor variations of P. E. Smith's original description of parapharyngeal hypophysectomy have been proposed by investigators and these should be consulted by anyone contemplating the surgery. The objective of the procedure is to expose the bone under the pituitary, drill a hole directly below the center of the gland, avoiding the extensive vascularity of the region, and completely remove the gland with negative pressure.

Because of the short time needed for the surgery of rats and mice, inhalation anesthesia is preferred, but longer acting injectable anesthetics, such as tribromoethanol or barbiturates, can also be used. The anesthetized animal on its back is fastened by its legs to a restraining board and its head firmly stabilized. The pituitary is approached using a midventral skin incision in the neck beginning just anterior to the posterior angle of the jaw. Using blunt dissection, the skin and submaxillary salivary glands are retracted. An invaluable surgical instrument for the operation is a cross-action forceps with jaws bent 90° to the angle of the handle. The forceps is held in the left hand and kept parallel to the long axis of the animal but with the end of the handle slightly raised. The jaws of the forceps are inserted between the fibers of the sternohyoideus muscle just lateral of the trachea and at the point where the muscle goes beneath the hyoid bone. As the jaws of the forceps are pushed ventrally through the muscle, they also move slightly anteriorly because the handle is slightly elevated. When the forceps encounters the resistance of bone, the jaws are gently separated while a microspatula, in the right hand, exposes the junction between the basisphenoid and occipital bones, which is the landmark for the position of the pituitary. The two bones in young animals are separated by a blue band of cartilage that is easily seen using a small amount of magnification. Because the open jaws of the cross-action forceps will place pressure on the trachea, they must be relaxed, but not removed, every few seconds to allow the animal to breathe. Using a drill with a flexible shaft and a straight handpiece, which allows continuous visibility, a hole is drilled exactly in the midline of the skull. In young animals the blue cartilage suture line should be the posterior border of the drilled hole, i.e., most of the hole should be anterior to the suture line. In older animals the pituitary is slightly more posterior so the suture line should be in the center of the hole. The size of the dental burr needed for the drill depends on the size of the animal, but a No. 8 burr is the most common size. After puncturing the dura mater that covers the pituitary, the anterior lobe will bulge into the drill hole and can be removed with negative pressure supplied to a suction tube. With practice the entire procedure requires approximately 2 min and with inhaled anesthesia the animal will be up and about within another 2 or 3 min. The procedure for rats and mice is also used for hypophysectomy of hamsters. However, the skull bones in hamsters are generously supplied with blood sinuses and bleeding is usually encountered when drilling the hole for removing the pituitary. Cotton dental pellets held by a watchmakers forceps are helpful for controlling bleeding. In goats, sheep, and other ruminants the pituitary is nearly encapsulated by a complex vascular network consisting of venous sinuses and the carotid rete mirabile. For these animals extreme care is needed when drilling in the sphenoid bone in order to avoid serious hemorrhage.

Shortly after the introduction of the parapharyngeal approach, use of the auditory canal for approaching the hypophysis was described and is the method generally used not only by commercial suppliers of animals but also by a large number of investigators. Unlike the parapharyngeal approach, in which an empty sella turcica can be seen, the transauricular is a blind approach and thus one needs to visualize the pituitary tissue removed when using this procedure. The critical tool needed for transauricular hypophysectomy of rats or mice is a hypodermic needle of appropriate size attached to a source of negative pressure. The needle must be fitted with a sheath to control the depth of penetration into the skull. The amount of needle that extends beyond the sheath depends on the size of the animal, i.e., the width of the skull. The needle with its sheath is attached to a source of negative pressure, which can be supplied by a glass syringe containing a small amount of water. The advantage of the syringe is that the pituitary tissue that was removed can be easily seen. For the procedure the anesthetized animal is placed flat on the table with its head toward the operator. The head is firmly held by the left hand while the needle, bevel side down, is inserted into the animal's left auditory canal. After puncturing the eardrum the needle is pointed slightly down and continually advanced to the extent of its exposed length. Vacuum is applied to the needle and the pituitary is sucked out. With practice the entire procedure takes less than 1 min after the animal is completely anesthetized.

In order to study the role of pituitary hormones in development, removal of the gland in embryos or neonates by a variety of methods has been attempted. The adenohypophysis develops from an invagination of oral ectoderm, referred to as Rathke's pouch, that comes in contact with the developing midbrain. Removal of this primordial tissue from the embryo prevents differentiation of the pituitary. Early studies used larval amphibians because the presumptive pituitary tissue was so easily accessible, but subsequent experiments used birds and mammals. Pituitary development in the chicken was prevented by removing the anterior part of the head fold after 33–38 hr of incubation. The chick embryos did not develop eyes nor upper bill and did not hatch from the egg. Hypophysectomy of rodent or rabbit embryos has been accomplished only by decapitation *in utero* with autopsy done near the time of delivery. However, sheep fetuses have been successfully hypophysectomized by a transbuccal approach when gestation was about half complete. In these embryos the head was exteriorized through a maternal hysterotomy and a hole drilled in the basisphenoid bone. The pituitary was removed and the fetus returned to the uterus to allow development until near term: None of the animals were allowed to deliver. Thus, hypophysectomy of embryos has been useful only for studies on the effects of the fetal pituitary on development. A problem with these studies is that maternal hormones may complicate interpretations.

In contrast to embryos, newborn rats and mice have been successfully hypophysectomized for chronic studies on the effects of loss of pituitary hormones. Although rats as young as 6 days old have been hypophysectomized by the parapharyngeal approach, it is, in general, an unsatisfactory procedure with few animals surviving. The trauma of the surgery causes inflammation and swelling of the pharyngeal region which seriously interferes with the animal's ability to suckle and subsequent malnourishment contributes to difficulties for survival as well as for interpretation of results. On the other hand, a modified transauricular approach has proved quite successful for hypophysectomy of rats 4–6 days old. For this operation, a 20-gauge hypodermic needle is fitted with a cylindrical plastic or Plexiglas spacer disc so that 9 mm of unsheathed needle is exposed. Because the size of the animal, and therefore the dimensions of the skull, can be quite variable depending on growth rate, litter size needs to be adjusted to nine pups on the day of birth. The neonatal rats are anesthetized by hypothermia and the needle, attached to a glass syringe containing water, is used to probe the presumptive ear canal, which is slightly ventral to the developing pinna, and then gently pushed through the skull. After entry the needle is raised so that its beveled end rests on the floor of the skull. The angle of the needle is slowly lowered as it progresses inward to the extent of the spacer disc. Slight negative pressure via the syringe plunger removes the pituitary. Some mortality is expected with this procedure due to hypothalamic damage and intracranial bleeding, but with practice the procedure is effective in more than 80% of the animals.

B. Nonsurgical Hypophysectomy

Several hypophysectomy techniques have been described as nonsurgical but most of them require some dissection under surgical conditions. M. G. Marienesco was the first to attempt destruction of the pituitary of the cat by use of chromic acid and the early experiments of P. E. Smith involved injection of 3–5% chromic acid into the rat pituitary gland that had been exposed by an intracranial approach. Although mortality was very high, results obtained with those animals that survived were qualitatively similar to those obtained by surgical hypophysectomy except for an unusual deposition of body fat. Smith concluded that it was not possible to limit exposure of the chromic acid only to the pituitary and that hypothalamic damage was inevitable, which accounted for the obesity. This technique has not been taken up again as a method for hypophysectomy of experimental animals.

Removal of the pituitary function in humans for therapeutic purposes has found the most use in so-called nonsurgical hypophysectomy. Radioactive destruction of the pituitary with the β-emitter yttrium 90 or the γ-emitter gold-198 has been accomplished several times, but because of serious complications of destruction of adjacent tissue use of this procedure has been curtailed. Focus of high-energy X-irradiation to pituitary tumors of humans has proved to be of significant value, but is not a procedure for hypophysectomy in experimental animals.

The consequences of complete hypophysectomy depend to a large degree on the stage of development when the operation is performed. For example, in young prepubertal animals loss of the pituitary inhibits further somatic growth as well as gonadal functions necessary for puberty. In both young and adult animals the adrenal cortex undergoes atrophy after pituitary removal, except for the zona glomerulosa, which is under the control of hormones produced by the kidneys. The metabolic effects of hypophysectomy are complex because growth hormone from the pituitary and the glucocorticoids from the adrenal are important counterregulatory hormones to insulin for the control of carbohydrate and lipid metabolism. Much of our understanding of the physiology of pituitary hormones depended on the effects of hormonal replacement following hypophysectomy. However, as we have learned more about the patterns of pituitary hormone secretion, and their control by the central nervous system, interpretations of these earlier studies have had to be revisited.

III. PARTIAL HYPOPHYSECTOMY

In the older literature partial hypophysectomy was synonymous with incomplete hypophysectomy. Clinicians found that total removal of the gland was not required for improvement in humans suffering from the effects of pituitary tumors. However, because so little anterior lobe is required for nearly complete physiologic function in most animals an incomplete hypophysectomy is of little, if any, use for experimental purposes. Typically, incompletely hypophysectomized rats or mice cannot be distinguished physiologically from sham-operated control animals. On the other hand, selective partial hypophysectomy with removal of specific hypophyseal hormones provides valuable probes not only for physiologic effects of hypophyseal hormones but also for nonsurgical therapeutic regimens. Progress in understanding control mechanisms for synthesis and release of hypophyseal hormones has provided rationales for removing the activity of specific hormones. This is an area of very active research and as more details of the signal transduction systems for hormone synthesis and release become available we can expect more methods of control.

A. Pharmacologic Methods

The aim of selective hypophysectomy is to remove one or more hypophyseal hormones without significantly interfering with production or secretion of the other hormones. The pharmacologic approach for accomplishing this end has largely relied on the use of agonists or antagonists of hypothalamic hypophyseal-releasing hormones that have specific receptors on the various pituitary cells. For example, selective removal of pituitary gonadotropins has been accomplished by use of highly potent analogs of gonadotropin-releasing hormone (GnRH). GnRH is an hypothalamic-derived decapeptide that binds to specific receptors on adenohypophyseal gonadotroph cells to stimulate production and release of both FSH and LH. The initial response to GnRH agonists is a large increase in serum gonadotropins, but with repeated dosing both FSH and LH levels are greatly reduced, presumably because of desensitization of the GnRH receptor. This method of producing a "medical hypophysectomy" has become routine in assisted reproductive protocols and is also used in some therapeutic regimens.

A "gonadotropin hypophysectomy" has also been accomplished by use of GnRH antagonists. The mechanism of action of antagonists does not depend on desensitization but rather on competitive inhibition of the GnRH receptor. Thus, there is no initial outpouring of gonadotropin with the antagonists, which is a distinct advantage for some studies. One such antagonist (MI-1544) has been shown to lower serum LH to undetectable levels in ovariectomized rats, whereas FSH levels were essentially unaffected. In intact animals treated with the antagonist, the ovaries became enlarged and produced estrogen, indicating an adequate FSH level, but follicles did not ovulate or form corpora lutea, which accounted for the low serum progesterone levels. The ability to selectively remove LH, even though it is a product of the same cells that produce FSH, supports the contention that secretion of FSH is under the control of systems other than GnRH and that these may also become amenable to pharmacologic control.

Selective removal of hypophyseal adrenocorticotropin by use of antagonistic analogs of corticotropin-releasing hormone (CRH) would seem a formidable task because the 41-amino acid peptide amide is produced in several organs other than the brain. However, the pituitary corticotrophs have a CRH receptor (CRH-R1) that is distinct from those found in peripheral tissues and provides for hypophyseal selectivity of antagonist actions. Several peptide analogs of CRH, both agonists and antagonists, have proved valuable for probing the physiologic effects of this hormone within the pituitary and central nervous system. Particularly promising in early studies is a nonpeptidic antagonist of CRH called Antalarmin (butyl-ethyl-[2,5-dimethyl-7-(2,4,6-trimethylphenyl)-7H-pyrrolo[2,3-d]pyrimidin-4-yl]amine). The compound binds the CRH-R1 receptors and inhibits CRH-stimulated ACTH release, thus proving valuable for studying the control of the latter as well as evaluating other central nervous effects of the releasing hormone.

One of the first pharmacologic agents used to selectively remove a single hypophyseal hormone was bromocriptine, an ergot alkaloid derivative. Bromocriptine is a dopamine agonist and as such inhibits the production and release of prolactin from hypophyseal lactotroph cells. Bromocriptine has proved a valuable tool for research into pituitary physiology as well as for control of the infertility associated with hyperprolactinemia and prolactin-secreting pituitary adenomas. Although bromocriptine has also been used for suppression of growth hormone from pituitary adenomas, in cases of acromegaly doses greatly in excess of those used for control of prolactin are required and its effectiveness for growth hormone is less than that of more specific agents.

Because of the serious physiological consequences of excess growth hormone, i.e., gigantism or acromegaly, selective removal of this hormone has been the objective of many studies. Production and secretion of growth hormone (GH; somatotropin) is regulated by hypothalamic stimulatory [growth hormone-releasing hormone (GHRH)] and inhibitory (somatostatin) hormones, suggesting that control could be accomplished by at least two pharmacologic approaches. However, difficulties arise because both GHRH and somatostatin are produced in tissues other than the hypothalamus and therefore unexpected effects can be associated with interference with these agents. The first method used to selectively reduce GH levels in the blood was neutralization of GHRH action using a polyclonal antiserum against recombinant rat GHRH. A second method used antagonists of GHRH, but their relatively short biological half-life restricts their effectiveness. This is an area of intensive investigation and new analogs appear frequently. Growth hormone release can also be reduced using somatostatin or its analogs. Somatostatin is a 14-amino acid cyclic peptide that has inhibitory action on the pituitary, gastrointestinal tract, exocrine and endocrine pancreas, and the immune system. This multitude of action sites has created major problems for designing somatostatin agonists. The biological half-life of somatostatin is less than 3 min, making it of no value as an exogenous inhibitor of GH. However, several smaller peptides have been synthesized that are potent agonists of somatostatin action and have been shown effective at reducing basal as well as pulsatile levels of GH in the serum. Octreotide (Sandostatin) is a octapeptide with somatostatin activity. It can be administered subcutaneously, intravenously, or by continuous infusion but still has a half-life of only 113 min. To be really effective as a selective hypophysectomy agent should be active over a longer period and thus Sandostatin-LAR, a long-acting form given by intramuscular injection on a monthly basis, was developed. As would be expected, there are several side effects, including those on the gastrointestinal system. Currently, five distinct somatostatin receptor types are recognized and future research must focus on development of somatostatin agonists that recognize receptors unique to pituitary somatotrophs.

B. Genetic Methods

Recent advances in molecular genetics have provided additional opportunities for partial hypophysectomy as a physiological probe in the study of pituitary function and the mechanisms for pituitary cell differentiation. The rationale for use of genetic methods was suggested by the many examples of selective loss of hypophyseal hormones that occur in nature, such as types of dwarfism or hypogonadotropic hypogonadism. Failure of animals or humans to grow normally has been recognized for centuries but understanding the bases for their dwarfism is

relatively recent. In many cases, there is a loss of ability of peripheral tissues to respond to pituitary hormones, but in others the problem is with the hypothalamic–hypophyseal axis. For example, a recessively inherited missense mutation in the extracellular domain of the GHRH receptor gene has been found to be associated with the mouse dwarfism trait called little. A similar mutation has been found in some dwarf humans. Because GHRH plays a role in the growth and differentiation of somatotrophs, the anterior pituitary of animals lacking GHRH receptors also shows a decrease in growth hormone. However, there are GH secretagogues that do not function via the GHRH receptor and these may be of use for providing the hormone in these dwarfs.

The selective loss of CRH action by synthetic antagonists was discussed in the preceding section, but loss of this hypothalamic hormone has also been addressed using gene targeting technology. Mice heterozygous for a null CRH allele were mated to produce homozygous CRH-deficient animals (CRH knockout mice). These animals had very low serum levels of corticosterone, the principal mouse glucocorticoid, but showed little effect of the loss, i.e., they grew normally and were fertile. Although the adrenal cortex, particularly the zona fasciculata, was hypoplastic, the pituitary content of corticotrophs appeared normal, i.e., CRH is not needed for corticotroph development. Increases in corticosterone production following exposure to restraint, fasting, or ether stresses were greatly reduced in females and almost nonexistent in males. Mating of homozygous null CRH males and females produced normal-sized litters but all animals died within the first 12 hr because of lack of lung maturation. This loss could be completely prevented by providing the pregnant homozygous female with corticosterone via the drinking water after Day 12 of pregnancy. This is an example of being able to differentiate the effects of loss of a specific pituitary hormone in the fetus from those with loss in both mother and fetus. That is, loss of CRH in the fetus was of no consequence if the mother had CRH, but it was lethal if both mother and fetus lacked the hormone. The technology of producing selected hormone deficiencies by use of knockout animals is rapidly progressing and can be expected to include each of the pituitary hormones.

Nature has produced a multiple knockout mouse. The pituitaries of Snell, Jackson and Ames dwarf mutant mice contain few, if any, somatotrophs, lactotrophs, or thyrotrophs. DNA analyses of these animals have shown that although they all have the same phenotype, they have different genotypes. These animals have provided valuable probes for investigating the genetic mechanisms involved in the ontogeny of the five different kinds of hypophyseal cells. On the basis of these findings, transgenic animals carrying specific mutations can be produced that should mimic nature and provide selective partial hypophysectomies.

Another experimental approach for production of partial hypophysectomies relies on the ability to target specific pituitary cells with cytotoxic agents. The first studies used cell-specific expression of the A chain of diphtheria toxin to ablate somatotrophs in transgenic mice. The animals developed normally but their fertility was much reduced, which is a serious drawback for producing a pure animal line for experimentation. All the pituitary hormones in these transgenics were not studied but histologic examination of the pituitary suggested loss of only somatotrophs and lactotrophs, consistent with the view that these two kinds of cells are derived from the same stem cell. This was further supported by the finding that somatotrophs and lactotrophs, but not gonadotrophs, thyrotrophs, or corticotrophs, were ablated in transgenic mice expressing a thymidine kinase obliteration system controlled by a growth hormone promoter. More selective destruction of somatotrophs was accomplished using transgenic mice expressing a transcriptionally inactive mutant of the cyclic adenosine monophosphate response element-binding protein fused to a fragment of the rat growth hormone promoter. In these animals, lactotroph populations were essentially normal.

Using similar techniques, pituitary gonadotrophs have been ablated in transgenic mice. The gonadotropins, LH and FSH, along with thyroid-stimulating hormone, are glycoproteins made up of two subunits, a common α subunit and a cell-specific β subunit. A fragment of an α-subunit promoter fused to the A chain of diphtheria toxin gene was used to specifi-cally remove gonadotrophs in transgenic mice. As expected, the animals were sterile due to lack of gonadal development but other hypophyseal hormones appeared quite normal; prolactin was reduced, but this could be explained by a lack of ovarian estrogen, which is an essential ingredient for lactotroph proliferation. The use of genetics for producing selective partial hypophysectomies will expand as we gain more information about the mechanisms involved in pituitary cell differentiation and function. Particularly valuable will be techniques to provide qualitative, quantitative, and temporally selective removal of hypophyseal hormones.

See Also the Following Articles

GnRH (Gonadotropin-Releasing Hormone); Hypothalamic-Hypophysial Complex; Pituitary Gland, Overview

Bibliography

Glasscock, G. F., Gelber, S. E., Lamson, G., McGee-Tekula, R., and Rosenfeld, R. G. (1990). Pituitary control of growth in the neonatal rat: Effects of neonatal hypophysectomy on somatic and organ growth, serum insulin-like growth factors (IGF)-I and II levels, and expression of IGF binding proteins. *Endocrinology* **127**, 1792–1803.

Jacobsohn, D. (1966). The techniques and effects of hypophysectomy, pituitary stalk section and pituitary transplantation in experimental animals. In *The Pituitary Gland* (G. W. Harris and B. T. Donovan, Eds.), Vol. 2. Univ. of California Press, Berkeley.

Kendall, S. K., Saunders, T. L., Jin, L., Lloyd, R. V., Glode, L. M., Nett, T. M., Keri, R. A., Nilson, J. H., and Camper, S. A. (1991). Targeted ablation of pituitary gonadotrophs in transgenic mice. *Mol. Endocrinol.* **5**, 2025–2036.

Loy, R. A. (1994). The pharmacology and the potential applications of GnRH antagonists. *Curr. Opin. Obstet. Gynecol.* **6**, 262–268.

Mugila, L., Jacobson, L., Dikkes, P., and Majzoub, A. (1995). Corticotropin-releasing hormone deficiency reveals major fetal but not adult glucocorticoid need. *Nature* **373**, 427–432.

Treier, M., and Rosenfeld, M. G. (1996). The hypothalamic–pituitary axis: Codevelopment of two organs. *Curr. Opin. Cell Biol.* **8**, 833–843.

Hypopituitarism

William W. Hurd

Indiana University School of Medicine

GLOSSARY

diabetes insipidus An antidiuretic hormone insufficiency resulting in polyuria and polydipsia. This condition may be either a result of hypopituitarism or secondary to a kidney defect and can be mimicked by excessive fluid intake related to psychiatric causes.

panhypopituitarism The inadequate or absent secretion of all anterior pituitary hormones.

partial hypopituitarism The inadequate or absent secretion of one or more (but not all) anterior pituitary hormones. Used sometimes to refer to inadequate secretion of antidiuretic hormone from the posterior pituitary (diabetes insipidus).

pituitary adenoma A benign neoplasm of the anterior pituitary, usually made up of functioning secretary cell of one of the types normally found in the pituitary.

pituitary apoplexy Hemorrhagic infarction of the anterior pituitary, usually related to infarction of a preexisting pituitary adenoma. This condition usually presents with severe headaches and visual disturbances and may culminate in coma.

pituitary crisis Acute systemic decompensation occurring in a patient with untreated panhypopituitarism that may be fatal if not rapidly diagnosed and treated. Presenting signs and symptoms may be those of both adrenal insufficiency (Addisonian crisis) and thyroid insufficiency (Myxedema coma) and may include diarrhea, vomiting, hypotension, hypoglycemia, coma, and, in some cases, fever.

Sheehan's syndrome Panhypopituitarism secondary to pituitary necrosis as a result of hypovolemic shock, usually related to postpartum hemorrhage. Classic symptoms include amenorrhea, failure of puerperal lactation, hypothyroidism, and adrenal insufficiency.

Hypopituitarism is the inadequate or absent secretion of one or more of the hormones of the pituitary gland. This term is most often applied to inadequate secretion of hormones of the anterior pituitary gland, including the gonadotropins, follicle-stimulating hormone, and luteinizing hormone; thyroid-stimulating hormone; corticotropin; and growth hormone. This term is sometimes used to refer to inadequate secretion of antidiuretic hormone from the posterior pituitary.

I. ETIOLOGY

A. General Considerations

The etiology of hypopituitarism can be related to either a pituitary or a hypothalamic abnormality since all anterior pituitary hormones are under hormonal regulation of the hypothalamus and the posterior pituitary hormones originate in the hypothalamus.

In some cases, inadequate or absent secretion of all anterior pituitary hormones (panhypopituitarism) may occur. However, it is more common for one or more (but not all) hormones to be affected (partial hypopituitarism), and thus the clinical presentation may be quite variable.

The etiologic mechanism of hypopituitarism may also obscure the clinical presentation because hyposecretion of one pituitary hormone is often associated with hypersecretion of a different pituitary hormone. There are two possible reasons for this. First, one of the most common causes of hypopituitarism is a pituitary adenoma, and the majority of these actively secrete a pituitary hormone. Second, although the majority of pituitary hormones are under positive regulatory control by the hypothalamus [follicle-stimulating hormone (FSH), luteinizing hormone (LH), thyroid-stimulating hormone (TSH), and corticotropin (ACTH)], PRL is under negative regulatory control. Thus, any process that interferes with this hypothalamic control may result in decreases in most pituitary hormones while simultaneously resulting in increased PRL secretion.

A final etiologic characteristic that often makes diagnosis difficult is the variation in the rate of onset of symptoms. The majority of processes result in the insidious onset of symptoms of slowly progressive hypopituitarism. In contrast, pituitary apoplexy may present dramatically with symptoms of profound hypopituitarism.

Taken together, these factors result in an extremely variable presentation of symptoms. Knowledge of the various causes of hypopituitarism, the diverse symptoms associated with hypopituitarism, and understanding of standard screening tests for pituitary function may help avoid unnecessary delays in diagnosing these conditions.

B. Pituitary Causes

Although there are multiple central nervous system (CNS) lesions and conditions that can cause hypopituitarism (Table 1), the single most common cause is a pituitary adenoma. These benign tumors result in hypopituitarism by one of two mechanisms. Direct pressure resulting from growth of the adenoma in the limited space available in the sella turcica can destroy other hormone-secreting pituitary cells. Alternatively, interference of blood flow from the hypothalamus can block stimulating hormones from reaching the pituitary via the portal circulation from the hypothalamus.

As noted previously, pituitary adenomas are usually functional and secrete prolactin approximately 70% of the time. Less commonly, these may secrete growth hormone (GH) or ACTH. In rare cases, adenomas secreting TSH, FSH, or LH have been reported. Approximately 25% are nonsecreting and commonly referred to null cell or chromophobe adenomas.

TABLE 1
Etiology of Selective or Panhypopituitarism

Pituitary abnormalities	Hypothalamic abnormalities
Pituitary tumors	Hypothalamic tumors
Adenomas	Benign
Prolactinoma	Craniopharyngioma
Nonfunctioning	Teratoma
Other tumors	Malignant
Pituitary infarction	Astrocytoma
Sheehan's syndrome	Glioma
Pituitary apoplexy	Congenital GnRH deficiency
Sickle cell anemia	
Empty sella syndrome	Kallmann's syndrome
Pituitary destruction	Systemic disorders
Transsphenoidal surgery	Weight loss
Radiation therapy	Excess exercise
Head trauma	Renal failure
Infection	Liver disease
Tuberculosis	Psychiatric disorders
Symphilis	Anorexia nervosa
Inflammatory disease	Anxiety
Lymphocytic hypophysitis	Infection
Adenoanterior pituitary	Tuberculosis
Infundibuloposterior pituitary	Syphilis
	Encephalitis
	Infiltrative diseases
	Sarcoidosis
	Histiocytosis X

The rare pituitary malignancies may first appear to be simple adenomas. However, the ominous nature of these lesions is usually indicated by rapid tumor growth and progression of symptoms. The second most common pituitary lesion to cause hypopituitarism is pituitary infarction. Probably one of the best described causes of pituitary infarction in women is Sheehan's syndrome. In these cases, pituitary infarction and necrosis are related to obstetric hemorrhage and hypotension. During the puerperium, the pituitary is much more susceptible to hypoperfusion because of pregnancy-related pituitary growth and a resultant increase in intrasellar pressure. Symptoms include postpartum amenorrhea, failure to lactate, hypothyroidism, and adrenal insufficiency.

A more common cause of pituitary infarction that can occur in both men and women is pituitary apoplexy, or pituitary hemorrhage related to infarction of a pituitary adenoma. The clinical presentation is usually dramatic, with the sudden onset of severe headaches, often with visual symptoms such ophthalmoplegia of peripheral visual loss. In severe cases, resulting increases in intracranial pressure can result in coma, and emergency surgical decompression may be required.

A less dramatic cause of pituitary infarction can be chronic systemic disease. Sickle cell anemia can result in infarction by vascular sludging. Diabetes mellitus can result in infarction secondary to vascular disease.

A unique pituitary etiology is the empty sella syndrome. In this condition, the sella turcica appears enlarged but is empty on CT or MRI. In this situation, the pituitary is compressed up against the base of the brain due to a congenital defect in the sella membrane or a suprasellar arachnoid cyst that allows extension of the arachnoid space into the sella turcica. In either case, CNS fluid displaces the pituitary from the sella turcica.

Destruction of the pituitary body or stalk can result in hypopituitarism. Surgical destruction of the pituitary may occur during transsphenoidal resection of a pituitary adenoma. Radiation therapy, a common treatment for aggressive brain tumors, may also result in hypopituitarism. Sever head trauma of any sort may injure the pituitary body or result in transection of the pituitary stalk.

Finally, any pituitary inflammation can result in hypopituitarism. This is may be infectious as a result in tuberculosis or syphilis. Noninfectious inflammatory conditions such as lymphocytic hypophysitis (autoimmune) or granulomatous hypophysitis (usually of unknown etiology) may also result in hypopituitarism.

C. Hypothalamic Causes

Hypopituitarism can also be due to interruption of normal hormonal signaling from the hypothalamus.

Most commonly, a tumor is responsible, causing pituitary stalk compression or damage to the hypothalamus. Common congenital tumors of the hypothalamus include craniopharyngiomas and teratomas. Symptoms may occur during childhood or be delayed to adulthood.

Malignant tumors of the hypothalamus may also present as hypopituitarism. These may be metastatic in origin or may be primary CNS tumors such as astrocytomas, meningiomas, or gliomas. Because of the potentially life-threatening nature of lesions in this area, neurosurgical consultation is always required when a hypothalamic lesion is discovered.

An interesting hypothalamic cause of hypopituitarism is Kallman's syndrome. In these patients, delayed puberty is accompanied by a decreased sense of smell, which can be either hyposmia or anosmia. This congenital, and sometimes familial condition results from failed migration to the hypothalamus of both olfactory neurons and neurons that secrete gonadotropin-releasing hormone (GnRH).

Any condition that causes severe metabolic or psychiatric stress or both can cause hypopituitarism via a hypothalamic pathway. Systemic diseases such as kidney failure or liver failure will often result in hypopituitarism. Extreme weight loss or excessive exercise can also result in decreased secretion of pituitary hormones. Anoxia nervosa, or psychogenic weight loss related to a distorted body image, is a classic cause of hypogonadotropic hypogonadism.

Inflammatory diseases may lead to hypothalamic dysfunction. Noninfectious causes include sarcoidosis and histiocytosis X. Infectious causes include encephalitis and meningitis of any etiology. Infectious agents are multiple but can include tuberculosis and syphilis.

II. SPECIFIC HORMONAL DEFICIENCIES

As discussed previously, hypopituitarism often presents as deficiency of more than one hormone or a combination of pituitary hormone deficiencies and excesses. However, for clarity, individual hormone deficiencies are considered separately.

A. Gonadotropins (FSH and LH)

1. Women

Inadequate secretion of FSH and LH, resulting in hypogonadotropic hypogonadism, is a relatively common condition in women of reproductive age. In the adolescent, this may manifest as primary amenorrhea, usually accompanied by lack of secondary sexual development. The patient's sense of smell should be evaluated to rule out Kallman's syndrome. In some cases, a behavioral cause may be found, such as excessive exercise or the malnutrition associated with anorexia nervosa. However, most cases are idiopathic or familial and may present as either primary or secondary amenorrhea. With long-standing hypoestrogenemia, vaginal atrophy and decreased breast size will also be apparent.

Hormonal evaluation of patients with primary or secondary amenorrhea should include both PRL and TSH since hyperprolactinemia and hypothyroidism can both present with amenorrhea. Estradiol, FSH, and LH should be evaluated in all patients with primary amenorrhea and in any patient with secondary amenorrhea associated with signs of hypoestrogenemia. These signs can include either vaginal atrophy or lack of withdrawal bleeding after progesterone administration. If hypoestrogenemia is documented, low or normal FSH and LH levels are diagnostic for hypogonadotropic hypogonadism. Evaluation of the pituitary and hypothalamus using either CT or MRI is recommended in these patients.

If a pituitary adenoma is discovered, other pituitary hormones should be measured both to ruled out hypersecretion and to determine if other hormones are affected by the lesion. Useful tests include ACTH, cortisol (drawn in the morning to screen for adrenal insufficiency), and GH.

As mentioned previously, it is imperative to rule out both hypothyroidism and hyperprolactinemia since both of these conditions can result in hypogonadotropic hypogonadism. It is estimated that one-third of women with secondary amenorrhea will have elevated PRL levels. However, only one-third of these patients will actually have galactorrhea, and thus this symptom is a relatively insensitive marker for hyperprolactinemia.

An interesting situation occurs when both PRL and TSH are elevated in a patient with primary hypothyroidism. In this condition, elevated hypothalamic TRH acts as a potent stimulator of both PRL-secreting and TSH-secreting pituitary cells. Cases have been reported in which pituitary prolactinomas have regressed with appropriate treatment of primary hypothyroidism.

Young women with delayed puberty are treated with low doses of estrogen and cyclic progestins to protect the endometrium. Older women interested in fertility are treated with either injectable menotropins or pulsatile GnRH administered by pump. Excellent pregnancy rates can be expected for these patients with this therapy. Women not interested in pregnancy are treated with either continuous estrogens with cyclic progestins or oral contraceptives if birth control is desired. Spontaneous remission of hypogonadotropic hypogonadism is uncommon.

2. Men

In boys, hypogonadotropic hypogonadism will present as delayed puberty. In men, hypogonadotropic hypogonadism will result in decreased virilism, with decrease libido and decrease growth of hair that is androgen dependent (e.g., face, chest, axilla, and genitals). In some men, the first signs of hypogonadotropic hypogonadism may be an abnormal semen analysis.

In men in whom there is a suspicion of this condition, a total serum testosterone should be checked as well as FSH and LH. Because prolactin-secreting tumors are common, a PRL and TSH are also helpful in these cases. Low testosterone in the presence of a low or normal FSH and LH is diagnostic of hypogonadotropic hypogonadism. Imaging of the pituitary and hypothalamus is indicated to rule out anatomical CNS causes of this condition.

Infertility treatment for hypogonadotropic men consists of testicular stimulation with gonadotropins. A combination of FSH and LH is most often given by injection. In men not interested in fertility, androgen replacement in the form of testosterone injections is the most common therapy.

B. TSH

Hypothyroid is one of the most common endocrinologic conditions and is much more common in women than in men. However, the majority of these cases are due to thyroid gland dysfunction (primary hyperthyroid). In a minority of cases, inadequate secretion of pituitary TSH will be found, either as a result if intrinsic pituitary dysfunction (secondary hypothyroid) or as a result of hypothalamic dysfunction (tertiary hypothyroidism). Although the initial presentation may be identical for all three types of hypothyroidism, the treatment and prognosis vary considerably.

In children of either sex, hypothyroidism can be associated with delayed puberty. However, precocious puberty may also occur, especially with primary hypothyroidism. One clue of hypothyroid as the initiating event in precocious puberty is bone age. With most causes of precocious puberty, bone age is advanced. However, when precocious puberty is related to hypothyroidism, bone age will be found to be delayed.

In adult women, a common presenting symptom is metrorrhagia, which occurs in almost half of the women found to have hypothyroidism. In other women, amenorrhea may occur. Patients also complain of weight gain, lethargy, and neurologic aberrations including paresthesias. Increased spontaneous abortions have also been reported. A common secondary finding is galactorrhea since increased TRH can stimulate PRL-secreting cells as well as TSH-secreting cells. Other physical findings may include thick, dry skin, edema of hands and face, enlarged tongue, and "thickened" speech.

In men with hypothyroidism, systemic signs and symptoms are common, as noted previously. In addition, men often complain of decreased libido. However, the semen analyses in these patients are usually normal. Untreated cases in either gender may exhibit signs of chronic hypothyroidism, including pretibial edema, puffy face, cardiomegaly, and eventually, coma.

Initial laboratory evaluation in patients suspected of having hypothyroidism is a TSH since an elevated TSH is diagnostic of primary hypothyroidism. If TSH is found to be low or low-normal in a patient who is clinically hypothyroid, a direct measure of thyroid function is important to rule out secondary or tertiary hypothyroidism. A traditional measure of thyroid function is the free thyroxin index (FTI), which is calculated as the product of total T_4 and T_3 resin uptake. However, with the development of relatively inexpensive methods for measuring unbound thyroid hormones, free T_4 (FT_4) may be determined as a direct measure of thyroid function. PRL should be measured as well since this may be dramatically elevated in hypothyroid patients as a result of TRH from the hypothalamus on PRL-secreting cells. Cases have been reported in which PRL-secreting macroadenomas have regressed when TRH is decreased by thyroid hormone therapy.

Once a diagnosis of secondary or tertiary hypothyroid is made, CNS imaging is required to search for CNS pathology. Either a CT or an MRI will effectively evaluate the pituitary and hypothalamic areas. The presence of a pituitary abnormality requires thorough evaluation of pituitary function to determine if hypo- or hypersecretion of any of the other pituitary hormones is present. A hypothalamic lesion is an indication for referral to a neurosurgeon for further evaluation.

As in primary hypothyroidism, treatment consists of oral thyroid hormone replacement. However, in contradistinction to primary hypothyroid, TSH cannot be followed as a measure of appropriate replacement. Rather, an FTI or a FT_4 must be measured to document appropriate replacement therapy.

C. ACTH

Adrenal insufficiency is probably the most dangerous sequela of hypopituitarism. Analogous to hypothyroidism, primary adrenal insufficiency (Addison's disease) is the most common cause of adrenal insufficiency and is related to intrinsic adrenal gland dysfunction. However, insufficient ACTH secretion resulting from either pituitary dysfunction (secondary adrenal insufficiency) or hypothalamic dysfunction (tertiary adrenal insufficiency) is also possible. Although isolated ACTH deficiency may be relatively uncommon, symptoms of adrenal insufficiency can be the first sign of panhypopituitarism.

Any of the previously discussed causes of panhypopituitarism can result in this condition. In addition, a sudden drop of a previously elevated circulating corticosteroids can result in acute secondary adrenal insufficiency. Most commonly, this is due to discontinuation of long-term corticosteroid therapy. However, it can also be seen after the removal of either a corticosteroid-producing adrenal tumor or an ACTH-producing pituitary adenoma.

The symptoms of secondary or tertiary adrenal insufficiency are identical to those of Addison's disease and include weakness, weight loss, hypotension, and gastrointestinal symptoms including diarrhea and vomiting. Menstrual abnormalities are also common, including amenorrhea, oligomenorrhea, and menorrhagia. Pubic and axillary hair loss sometimes occurs as a result of decreased adrenal androgen secretion. Hyperpigmentation may also be present, as in Addison's disease.

Acute adrenal insufficiency is much more dramatic, with nausea and vomiting accompanied by hypotension and, in extreme cases, shock and coma. This may be associated with either hypothermia or hyperthermia. The classic laboratory findings in a hypotensive patient are hypoglycemia, hyponatremia, and hyperkalemia.

Chronic adrenal insufficiency symptoms may be subtle. A screening test for patients suspected of possibly having adrenal insufficiency is a serum cortisol obtained in the morning since cortisol is highest at this time. A low morning value is an indication to measure both ACTH and DHEAS.

In acute adrenal insufficiency, rapid treatment may be life-saving. This treatment consists of intravenous administration of a glucocorticoid (e.g., hydrocortisone, 100–300 mg).

Chronic adrenal insufficiency is usually treated with the combination of glucocorticoid and mineralocorticoids. Supplemental glucocorticoids must be given during times of physiologic stress (e.g., surgery) to avoid adrenal crisis.

D. GH

Growth hormone deficiency in children will result in growth deficiency and delayed puberty. This is usually detected by evaluation of serum GH and insulin-like growth factor-I levels. With the availability of growth hormone therapy, relatively normal growth patterns can often be achieved in these children.

In adults, there is no specific syndrome associated with GH deficiency. However, GH secretion decreases in the elderly, and there is some question whether inadequate GH secretion may actually have health effects in some adults. A pituitary adenoma that secretes GH can result in acromegaly and cause inadequate secretion of other pituitary hormones, as described previously.

E. PRL

Prolactin deficiency is an uncommon clinical entity. A classic example is Sheehan's syndrome, in which postpartum hypotension has resulted in pituitary infarction. This condition is characterized by puerperal amenorrhea and failure to lactate in the immediate postpartem period. Any condition that leads to panhypopituitarism as a result of pituitary destruction can likewise cause inadequate prolactin secretion. Except for failure to lactate, no other symptoms are associated with the lack of PRL secretion.

F. Antidiuretic Hormone

Inadequate secretion of antidiuretic hormone (ADH) by the posterior pituitary is not always included in discussions of hypopituitarism. This is probably because diabetes insipidus is not usually associated with inappropriate secretion of anterior pituitary hormones. Diabetes insipidus (DI) can result from either inadequate ADH secretion (neurogenic DI) or renal insensitivity to ADH (nephrogenic DI). As with other disease resulting from inadequate secretion of pituitary hormones, the causes of neurogenic DI are multiple and may originate in either the pituitary or the hypothalamus.

Clinically, diabetes insipidus presents as a combination of polydipsia and polyuria. Nocturia is usually present as well. Urinalysis will demonstrate decreases in urine specific gravity and osmolality. A 24-hr urine collection will show a volume of >2500 ml in patients with DI. Untreated patients will eventually develop significant dehydration and neurologic symptoms, including encephalopathy. Further evaluation for DI routinely involves water deprivation under careful observation. Neurogenic diabetes insipidus is usually effectively treated by intranasal administration of a long-acting ADH analog.

G. Panhypopituitarism

Inadequate secretion of all hormones of the anterior pituitary can be a life-threatening condition. In women with puerperal amenorrhea and failure to lactate, Sheehan's syndrome should be suspected. Other patients may present with symptoms of both hypothyroidism and adrenal insufficiency.

One of the most common causes of panhypopituitarism is infarction of a pituitary adenoma, termed pituitary apoplexy. Many of these patients will have symptoms suggestive of a preexisting adenoma, such as galactorrhea, acromegaly, or myxedema. More acute symptoms of pituitary apoplexy may include the recent onset of severe headaches or visual symptoms. Some patients may present with unexplained coma as the first evidence of panhypopituitarism. Systemic collapse associated with pituitary infarction is referred to as pituitary crisis and may be indistinguishable from adrenal crisis from any etiology. As discussed under ACTH, the most notable symptoms may be hypotension, hypoglycemia, and coma. After ruling out other more common causes of profound hypotension, such as myocardial infarction and septic shock, treatment with intravenous glucocorticoid may be life-saving.

See Also the Following Articles

Gonadotropins; Kallman's Syndrome; Sheehan's Syndrome

Bibliography

Dexter, R. N. (1995). Hypoituitarism. In *Principles and Practices of Endocrinology and Metabolism* (K. L. Becker, Ed.), 2nd ed. Lippincott-Raven, Philadelphia.

Houlden, R. L., and Reid, R. L. (1995). Disorders of the pituitary thyrotroph. In *Reproductive Endocrinology, Surgery, and Technology* (E. Y. Adashi, J. A. Rock, and Z. Rozenwaks, Eds.). Lippincott-Raven, Philadelphia.

Itskovitz-Eldor, J., and Kol, S. (1995). Disorders of the pituitary gonadotroph. In *Reproductive Endocrinology, Surgery, and Technology* (E. Y. Adashi, J. A. Rock, and Z. Rozenwaks, Eds.). Lippincott-Raven, Philadelphia.

Melmed, S. (1994). *The Pituitary*. Blackwell, Oxford, UK.

Militch, M. E. (1995). Disorders of the pituitary lactotroph. In *Reproductive Endocrinology, Surgery, and Technology* (E. Y. Adashi, J. A. Rock, and Z. Rozenwaks, Eds.). Lippincott-Raven, Philadelphia.

Scherzer, W. J. (1995). Disorders of the pituitary somatotroph. In *Reproductive Endocrinology, Surgery, and Technology* (E. Y. Adashi, J. A. Rock, and Z. Rozenwaks, Eds.). Lippincott-Raven, Philadelphia.

Sheehan, H. L. (1982). *Post-Partum Hypopituitarism*. Charles C Thomas, Springfield, IL.

Yen, S. S. C. (1991). Chronic anovulation due to CNS–hypothalamic–pituitary dysfuntion. In *Reproductive Endocrinology* (S. S. C. Yen and R. B. Jaffe, Eds.), 3rd ed. Saunders, Philadelphia.

Hypospadias

Ranjiv Mathews and Steven Docimo

Johns Hopkins School of Medicine

GLOSSARY

anesthetic blocks Direct injection of anesthetic agents at the location of known nerves, providing anesthesia to the tissues supplied by the nerve.

chordee Bending of the penis, most noticeable during erection. Significant bends may affect the ability to have normal intercourse.

fistula A connection between a closed organ or space in the body and the outside. It is usually kept open by a leakage of fluid through its track.

genital tubercle The embryological primordium of the phallus; seen in both males and females, it enlarges in the male to form the penis. In the female significant growth does not occur and it forms the clitoris.

hernia Bulging of the peritoneum through a defect in the abdominal wall. It can occur in the groin (inguinal hernia) or abdominal wall (ventral hernia).

intersexuality Gender ambiguity noted in some congenital enzyme deficiencies (e.g., congenital adrenal hyperplasia) or with sex chromosomal abnormalities.

prepuce The fold of loose skin that covers the glans of the penis. It is excised during circumcision.

raphe of the penis A ventral ridge on the penis denoting the edge of fusion of the two urethral folds.

urethral meatus The tip of the urethra.

urogenital sinus The precursor of the bladder. Division of the cloaca into the posterior rectum and anterior urogenital sinus occurs during the fourth week of development of the fetus.

vesicoureteral reflux Reflux of urine from the bladder toward the kidney; due to abnormalities in the ureteral insertion in the bladder.

Hypospadias is a congenital anomaly characterized by failure of the urethra to form completely. The urethral meatus can be located anywhere on the ventral aspect of the shaft of the penis, from the glans to the perineum. The lack of the ventral aspect of the prepuce causes a hooded appearance of the dorsal prepuce (Fig 1b). Hypospadias is usually associated with some degree of penile bending (chordee).

I. INCIDENCE

Hypospadias (Fig. 1) has an incidence of 3.2 per 1000 live male births, making it one of the more common congenital anomalies in boys. Eight percent of patients have affected fathers and 14% have affected male siblings. It is more common in whites and has a higher incidence in monozygotic twins.

II. EMBRYOLOGY

Masculinization of the male fetus is due to the action of androgen in the male fetus. Testosterone is the major androgen secreted by the testis; however, it has relatively few actions of its own. It is converted to dihydrotestosterone in target tissues of the external genitalia by an 5α reductase. Dihydrotestosterone is required for normal masculinization of the male external genitalia.

The primordia of the external genitalia are the genital tubercle, urethral folds, genital swellings, and

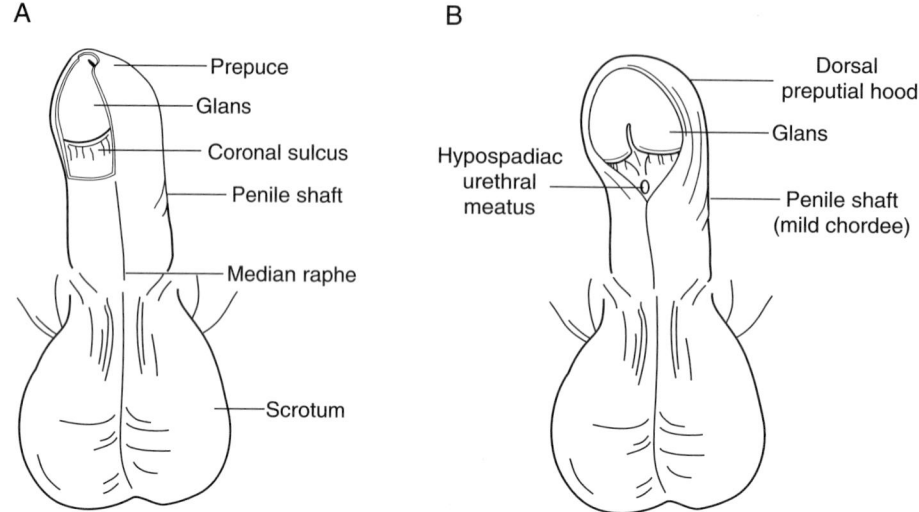

FIGURE 1   (a) Normal uncircumcised penile anatomy. The foreskin is cut away anteriorly to demonstrate the glans and coronal sulcus. (b) Anatomy of the penis in hypospadias.

the urogenital sinus. Under the action of dihydrotestosterone the genital tubercle enlarges to form the glans penis and the urethral folds fuse to form the urethra. Incomplete fusion of the urethral folds leads to hypospadias. The embryology of the urethra in the glans is debated. It has been suggested that the urethra within the glans forms as a solid core that later canalizes. Glanular hypospadias may therefore have a different embryological basis.

III. CLASSIFICATION

Hypospadias is classified by the location of the urethral meatus as follows:

Anterior hypospadias: The urethral meatus is located on the glans penis or on the corona of the penis. This is the most frequently noted type of hypospadias and comprises 50% of all cases. This can be further subdivided into glanular (meatus on the glans) or coronal (meatus on the corona) forms.

Middle hypospadias: This is further subdivided into distal penile, middle penile, and proximal penile types. This constitutes 30% of hypospadias cases and subdivision is dependent on the location of the meatus after release of chordee.

Proximal hypospadias: Constitutes 20% of cases; includes the penoscrotal, scrotal, and perineal forms of hypospadias.

Complex hypospadias: Patients failing prior reconstruction constitute a special group. Reconstruction is difficult due to the lack of available skin and may require skin transfer or other maneuvers for cosmetic closure.

IV. ASSOCIATED MALFORMATIONS

Penile torsion: Twisting of the penis is noted in 15% of patients and is due to the deviation of the raphe toward the right, where it joins the preputial insertion. It results in a counterclockwise twist.

Undescended testes: Nine percent of children with hypospadias have undescended testes, i.e., the testes do not descend into the scrotal sac and may be located within the abdomen, in the canal leading to the scrotum, or in an ectopic location. In proximal hypospadias this incidence may be as high as 31%. The combination of undescended testes and hypospadias should raise concerns of intersexuality (gender ambiguity).

Inguinal hernia: The incidence of inguinal hernia is also related to the severity of hypospadias, noted in 9% of distal and 17% of proximal hypospadias.

Enlarged utricle: An enlarged utricle was noted in 57% of children with proximal and 10% of patients with distal hypospadias. The etiology for this malformation is debated but has been postulated to be due to inadequate production of Müllerian-inhibiting substance.

Vesicoureteral reflux: Ten to 17% of patients with hypospadias are noted to have vesicoureteral reflux (urine refluxes from the bladder into the ureter or kidney due to an abnormality of the opening of the ureter), most of low-grade and probably minimal significance.

V. TREATMENT CONSIDERATIONS

Outpatient one-stage repair: Early efforts at hypospadias reconstruction were multistage operative procedures with significant morbidity. Improvements in surgical techniques and instrumentation have made outpatient one-stage repair a reality for most forms of hypospadias. Proximal and complex forms of hypospadias may still require more than one operation for complete reconstruction.

Timing of operation: Surgical repair is usually performed between 6 and 18 months of age. The push toward early correction has been based on the emotional effects that operations on the genitalia could potentially have on young male patients. Penile size, however, makes early operation more technically demanding. With current improvements in techniques, earlier repair has not been associated with decreased success and the only limiting factor is penile size.

Chordee: Ventral curvature of the penis is noted in 35% of children with hypospadias. Severity of chordee is proportional to the degree of hypospadias. Chordee may be present independent of hypospadias due to disproportionate length of the corporal bodies.

Artificial erection: Introduced in 1974, this was a major advancement in the repair of hypospadias. The ability to determine complete correction of chordee

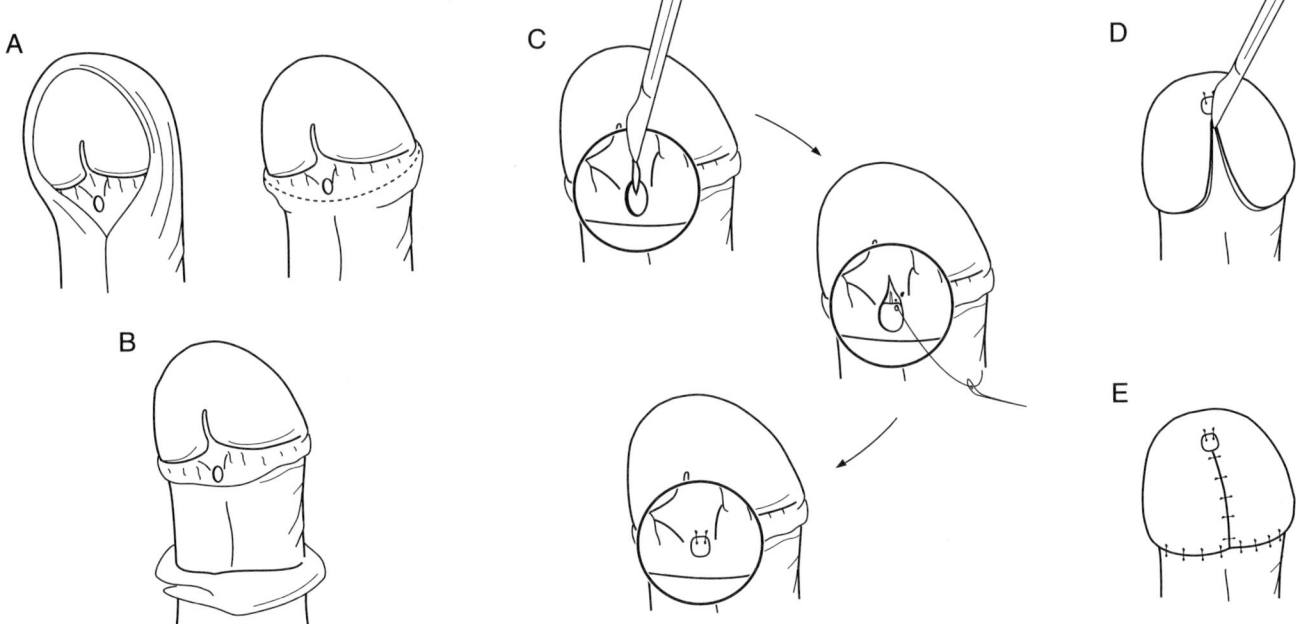

FIGURE 2 MAGPI procedure. (a, b) Circumferential incision of the foreskin and degloving of the penis. (c) Incision of the urethral plate extending into the dorsal lip of the urethra. This incision is closed horizontally to fix the urethra to the urethral plate and widen the urethral meatus. (d) Glans edge is elevated with skin hook bringing the urethral meatus to the tip of the penis. Glans wings are closed to each other, maintaining the urethral meatus at its final location on the tip of the penis. (e) Completed repair.

A

C

B

D

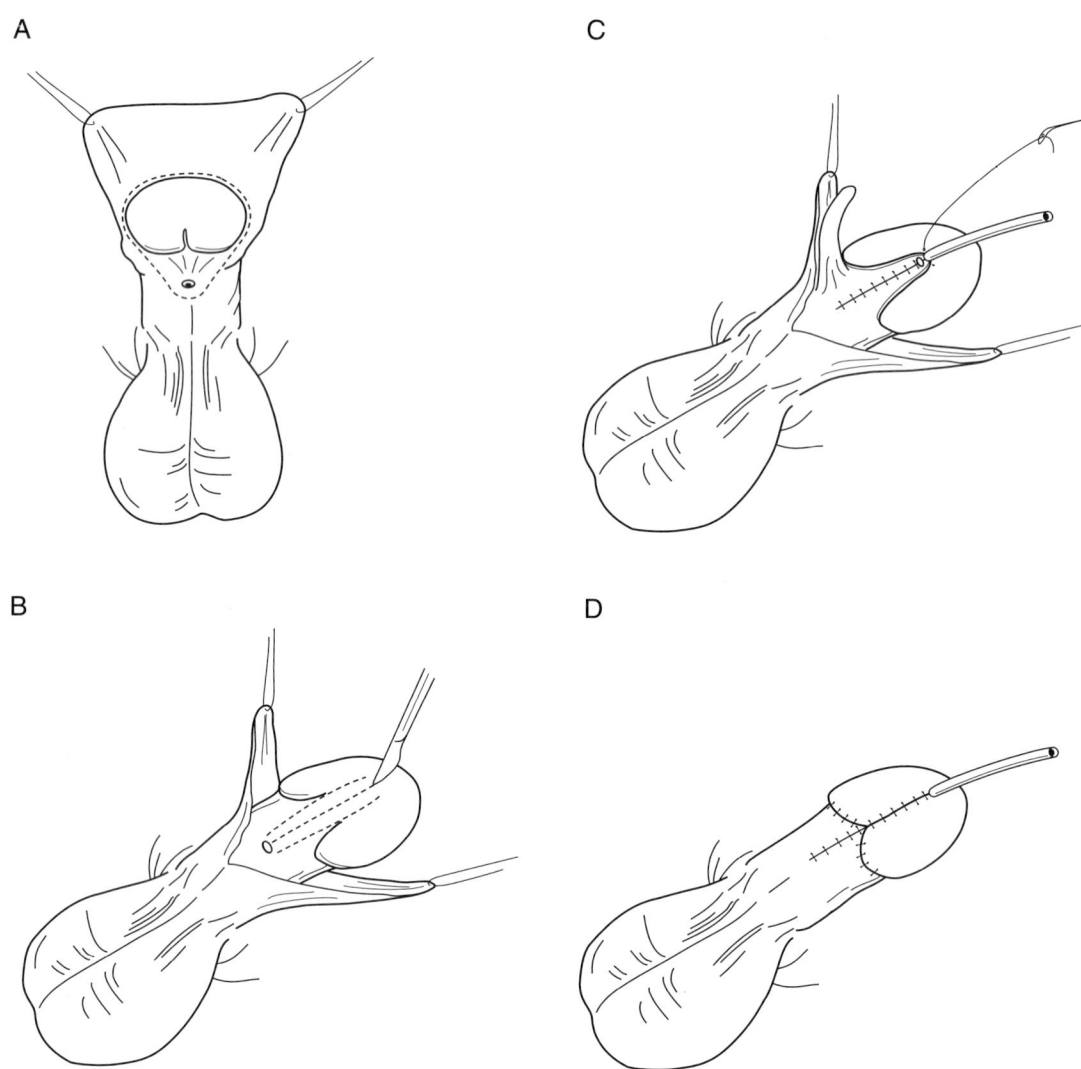

FIGURE 3 Tubularized incised plate urethroplasty (Snodgrass repair). (a) Initial incision extending around the urethral meatus and paralleling the coronal sulcus. (b) Urethral plate incised in the midline dorsally and on the edges. (c) Incised urethral plate tubularized around a stent. A flap of inner preputial tissue is laid over the completed urethral closure. (d) Completed repair following closure of the glans wings over the urethra and rearrangement of penile skin.

is essential to one-stage hypospadias repair. A tourniquet is placed at the base of the penis and erection is induced with the injection of normal saline.

Testosterone stimulation: Use of testosterone cream or injection has been tried to increase penile size prior to repair. Benefits of testosterone have been questioned.

Prior circumcision and/or prior surgery: An intact prepuce is important for hypospadias reconstruction

because it can be used for construction of the urethra or for penile skin coverage. Prior failed hypospadias repair and paucity of skin require harvesting of other tissue for construction of the urethra. Other tissue that has been used successfully include skin grafts from non-hair-bearing locations (forearm and thigh), buccal mucosa, bladder mucosa, scrotal skin, or the use of tissue expansion to generate skin for reconstruction.

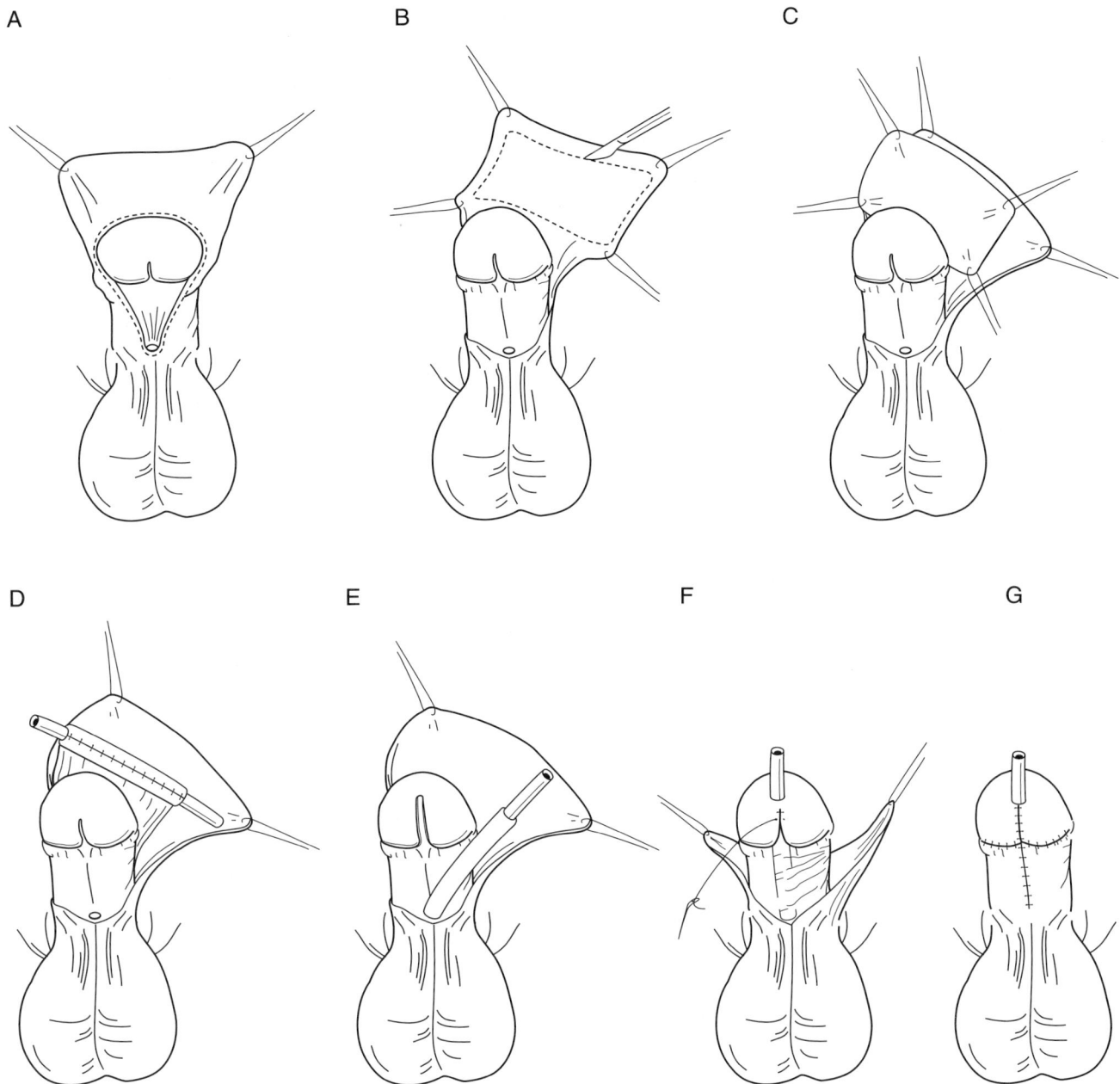

FIGURE 4 Transverse preputial tubularized flap technique. (a) Initial circumcising incision extending around the urethral plate. (b) Preputial flap outlined on inner surface of the prepuce. (c) Separation of inner and outer preputial layers with their vasculature intact. (d) Inner preputial flap tubularized around a stent. (e) Proximal anastomosis of the tubularized neourethra. (f) Glans wings sutured over the tubularized neourethra. (g) Completed repair following penile skin rearrangement.

Analgesia: Discomfort postoperatively is managed with oral pain medications. The use of local anesthetics in the form of penile blocks or caudal blocks may help to reduce discomfort in the immediate postoperative phase.

VI. SURGICAL TECHNIQUES

Multiple techniques, each with specific merits and indications, have evolved with surgical experience and improvements in instrumentation. Discussion of all these techniques is beyond the scope of this text; however, commonly used techniques for each of the forms of hypospadias are described with accompanying illustrations.

Meatal advancement and glanuloplasty (MAGPI) technique: MAGPI is best performed only in the most distal forms, i.e., glanular and coronal forms. Additionally, mobility of the urethra is critical to the success of this procedure (Fig. 2).

Tubular incised plate hypospadias repair (Snodgrass technique): Initially presented in 1994, this technique utilizes tubularization of the urethral plate following dorsal incision. This procedure has been utilized successfully in the correction of subcoronal and midpenile hypospadias and allows the urethral meatus to be positioned in the center of the glans. The multiinstitutional experience recently reported attests to the versatility of this technique (Fig. 3).

Transverse preputial tubularized island flap technique: Used for the correction of more proximal forms of hypospadias, this technique uses the inner face of the prepuce with an intact blood supply to construct a tubularized urethra. This is rotated into place to form the distal urethra. Large urethral defects can be bridged using this technique (Fig. 4).

Two-stage repairs: While single-stage repairs continue to be the goal of the pediatric surgeon, very proximal forms of hypospadias and complex hypospadias may be best managed with a staged approach of chordee correction with harvest and transposition of tissue that will be used to complete urethral reconstruction at a second procedure.

VII. RESULTS

With the advent of modern microsurgical techniques, results have been exceptional, with fistulas and meatal stenosis being the two most frequently reported complications.

A. Fistulae

Fistulae usually result from technical error or ischemic necrosis of the flaps at the suture lines. An incidence of 10–15% has been reported in most series. Avoidance of suture lines that overlie one another, adherence to microsurgical techniques, and tension-free closures will prevent most fistulae.

B. Meatal Stenosis

The majority of cases of meatal stenosis appear to involve the glanular urethra. It has been proposed that urethral compression within the glans leads to meatal stenosis and removal of some of the glanular tissue should offset this complication. Making large glans flaps that can be reconstructed without undue urethral compression may also prevent this complication.

VIII. SUMMARY

Hypospadias is a common congenital anomaly of the male penis. Advances in surgical techniques allow earlier and more cosmetic repair of this deformity. In patients diagnosed with hypospadias, circumcision should be withheld until genital reconstruction is performed.

See Also the Following Article

Penis

Bibliography

Altemus, A. R., and Hutchins, G. M. (1991, October). Development of the human anterior urethra. *J Urol.* **146**, 1085–1093.

Belman, A. B. (1992). Hypospadias and other urethral abnormalities. In *Clinical Pediatric Urology* (P. P. Kelalis, L. R. King, and A. B. Belman, Eds.), 3rd ed., pp. 619–664. Saunders, Philadelphia.

Duckett, J. W. (1990). Hypospadias repair. In *Operative Paediatric Urology* (J. D. Frank and J. H. Johnston, Eds.), pp. 197–208. Churchill Livingstone, Edinburgh, UK.

Duckett, J. W. (1992). Hypospadias. In *Campbell's Urology* (P. C. Walsh, A. B. Retik, T. A. Stamey, and E. Darracott Vaughan, Eds.), 6th ed., pp. 1893–1916. Saunders, Philadelphia.

Hinman, F., Jr. (1994). *Atlas of Pediatric Urologic Surgery,* Chaps. 114–123. Saunders, Philadelphia.

Hypothalamic–Hypophysial Complex (Pituitary Portal System)

Robert B. Page

M. S. Hershey Medical Center of the Pennsylvania State University

I. Arterial Supply
II. Primary Capillary Plexus
III. Portal Vessels
IV. Secondary Capillary Plexus
V. Venous Drainage

GLOSSARY

adenohypophysis The glandular organ made up of secretory epithelial cells that is applied to the neurohypophysis and divided into a pars tuberalis, a pars intermedia, and a pars distalis.

infundibular process The caudal region of the neurohypophysis which lies within the sella turcica; also called the *neural lobe*.

infundibular stem The neurohypophysial contribution to the pituitary stalk.

infundibulum The most rostral region of the neurohypophysis; also called the *median eminence*.

median eminence The most rostral region of the neurohypophysis; also called the *infundibulum*. It is divided into an ependymal layer, an internal zone, and an external zone. The supraopticohypophysial tract passes through the internal zone in its fiber layer to pass on to the neural lobe. The tuberohypophysial tract terminates in the external zone.

neural lobe The caudal region of the neurohypophysis which lies within the sella turcica also called the *infundibular process*.

neurohypophysis The diverticulum from the hypothalamus which contains neurosecretory axon terminals, modified astrocytes, microglia, capillaries, and pericytes but lacks a blood–brain barrier and neuronal cell bodies and which is subdivided into the infundibulum (median eminence), the infundibular stem, and infundibular process.

pars distalis The region of the adenohypophysis which lies within the sella turcica and below the diaphragm sella and is separated from the infundibular process by the hypophysial cleft.

pars intermedia The region of the adenohypophysis which lies within the sella turcica applied to the lower infundibular stem and the infundibular process and is separated from the pars distalis by the infundibular cleft.

pars tuberalis The region of the adenohypophysis which lies above the sella turcica and is applied to the median eminence.

primary capillary plexus The capillary bed of the neurohypophysis which extends throughout the infundibulum (median eminence), the infundibular stem, and infundibular process (neural lobe).

The pituitary portal system carries blood from the neurohypophysis to the adenohypophysis. It is supplied by hypophysial arteries and drained by hypophysial veins. It functions to transport hypothalamic-releasing and -inhibiting hormones from the median eminence, infundibular stem, and infundibular process to secretory epithelial cells in the pars tuberalis, pars intermedia, and pars distalis and to transport the secretions of adenohypophysial epithelial cells to the general circulation.

I. ARTERIAL SUPPLY

The primary capillary plexus receives arterial supply from three sources. The superior hypophysial arteries arise from the intracranial internal carotid arteries and from the vessels of the circle of Willis to supply the primary capillary plexus in the infundibulum (median eminence). The major contribution, in the human, is from a dominant single vessel which arises from each internal carotid artery. In all species studied the paired superior hypophysial arteries form an anastomotic ring around the upper infundibular stem and the infundibulum (median eminence). The paired middle hypophysial arteries, sometimes called the trabecular artery, supply the infundibular stem. The paired inferior hypophysial arteries arise from the intracavernous portion of the internal carotic arteries and supply the infundibular process (neural lobe).

II. PRIMARY CAPILLARY PLEXUS

The primary capillary plexus extends throughout the neurohypophysis. Observation of the pattern of blood flow within it reveals that a bolus of dye injected into the cervical internal carotid artery of a pig first enters the infundibular process (neural lobe) from the inferior hypophysial arteries and then enters the infundibulum (median eminence) from the superior and middle hypophysial arteries. Within the primary plexus blood flows rostrally from the infundibular process and caudally from the infundibulum to meet in a watershed zone in the infundibular stem.

Blood flow into the neurohypophysial capillary bed (the primary capillary plexus) is high as determined by the microsphere technique in the sheep and by the iodoantipyrine technique in the rat. In the sheep mean regional blood flow $\pm$ standard error in the infundibulum (median eminence) was 436 ± 39 and 461 ± 41 cc/100 g/ min in the neural lobe. The values were not significantly different and were 10 times the values in the hypothalamus. As in the cerebrum, blood flow autoregulated with changing blood pressure, i.e., it remained constant between mean arterial pressures of 60–110 Torr, and it decreased with hypocarbia. In the neural lobe blood flow increased with increased (neurosecretory) function. It is of considerable interest that meta-arterioles and precapillary sphincters lie on the surface of the median eminence in several species. These sites of resistance, and hence of regulation of blood flow, lie in the extracellular space of neurosecretory axon terminals which secrete vasoactive peptides. This arrangement

raises the possibility that regional secretory patterns within the median eminence regulate local blood flow.

The primary capillary plexus lacks a blood–brain barrier. Its capillaries are fenestrated. Although there is no capillary barrier to the passage of neurohormones, pituicytes are frequently interposed between neurosecretory axon terminals and the (porus) walls of primary plexus capillaries in a resting state. Upon increased neurosecretory demand, such as increased lactation with suckling, these glial processes are withdrawn, leaving axon terminals in the perivascular space facing the fenestrated capillary walls.

The primary plexus within the median eminence has been subdivided into an external (superficial) and an internal (deep) plexus (Fig. 1). The external plexus is continuous with the capillaries of the infundibular stem and process. Its morphology differs little from species to species. It receives the arterial supply to the median eminence. It drains to the internal plexus and to the portal vessels. It lies partially embedded within the surface of the median eminence and is oriented horizontal to the plane of the its oral surface, and it is made up of repeating hexagonal arrays of capillary tubes which enclose posts of neurosecretory tissue. The capillary arrangement resembles that found in the stomach and lung. These organs also rapidly exchange materials between their capillaries and their parenchyma. The pulmonary vascular space has been described as resembling an underground garage, consisting of a floor, ceiling, and supporting pillars which are all covered with endothelium, and it has been argued that blood flows as a sheet through a continuous space lined by endothelium. In the median eminence a similar pattern of flow may be present because its angioarchitecture resembles that of the lung.

The internal plexus is oriented perpendicular to the external plexus. In the floor of the median eminence it consists of simple loops that penetrate for varying distances into the median eminence and then retreat back to its surface. In some species arborization occurs at the apices of the loops to form a horizontally oriented plexus of vessels beneath the ependymal surface. In the wall of the median eminence the loops are more complex. In the rabbit, for example, the capillary loops in the walls consist of a central vessel surrounded by a helical coil. In the rabbit and

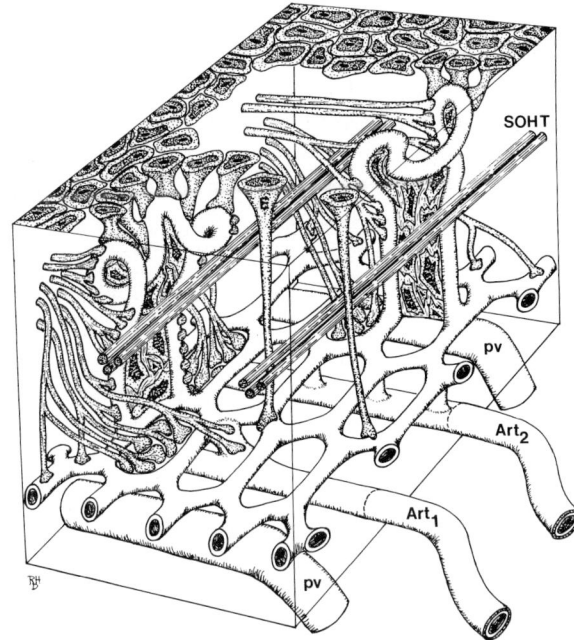

FIGURE 1 Neurovascular organization in the medial third of the (rabbit) median eminence. The external plexus is embedded in the oral surface of the median eminence and is composed of hexagonal arrays of capillaries and receives arterial blood from arteriolar branches of the superior hypophysial arteries (Art₁ and Art₂). The regions enclosed by the hexagons are filled with axon terminals of the tuberoinfundibular tract. These enclosed microdomains of neurosecretory terminals resemble posts of tissue surrounded by porus channels that contain flowing blood. Internal plexus capillary loops rise from the external plexus and arborize in the subependymal zone. Connective tissue fills the spaces between the ascending and descending limbs of the internal plexus loops and the spaces between the surface of the external zone and the pars tuberalis at the ventral aspect of the median eminence. Nonciliated ependymal line the infundibular zone recess (E). The basal processes of some ependymal cells end in the perivascular space around the apex of capillary loops and form an ependymal cuff around them in the hypendymal layer. Some ependymal processes extend to the external plexus. Parvicellular axons terminate on external plexus capillaries and around internal plexus capillaries to form an axonal cuff around them deep in the fiber layer and in the palisade zone. These axons are peptidergic and aminergic. It appears that capillary loops arise from the external plexus by invaginating into the substance of the median eminence and carrying a "mesentery" of connective tissue between the ascending and descending limbs into the neuropil. Magnocellular systems (SOHT) pass through the fiber layer and do not send collaterals to the capillary loops. Arterioles originate from the superior hypophysial arteries by tapering downs from terminals to

gerbil there is a cuff of axon terminals that surrounds the capillary loops of the internal plexus. The stroma lying between the ascending and descending limbs is composed of connective tissue. It appears that in development these loops invaginate from the surface into the substance of the median eminence and carry lining of axon terminals with them (similar to pushing one's finger into a softly inflated balloon).

III. Portal Vessels

The rostral region of the neurohypophysis (the infundibulum or median eminence) drains by capillary routes to the pars tuberalis and by capillary and long portal routes to the pars distalis. In species such as the human and the rat, in which the pars distals is some distance away from the median eminence, long portal routes predominate. In species in which the pituitary stalk is very short and the median eminence sits atop the pars distalis (e.g., the cat or dog) capillary routes are plentiful. Portal vessels are arrayed on the surface of the pituitary stalk. They are fed by coalescing external plexus capillaries. They drain into the secondary plexus of capillaries in the pars distalis. These vessels are fenestrated and have smooth muscle cells in their walls. Direct observation has demonstrated that blood flows down the portal vessels from the median eminence into the pars distalis. Whether or not humoral mechanisms exits that can act on the smooth muscle cells to regulate blood flow in the portal vessels has not been established.

The infundibular stem and process are drained by capillary and short portal routes. In many mammalian species a pars intermedia is present. In humans it is present in the fetus and in the pregnant female. The rat pars intermedia is less well vascularized than

precapillary arterioles (Art₁) and/or by sending small (precapillary arteriolar) branches at right angles from the terminal arterioles (Art₂). Arterioles lie in close proximity to the external plexus and to posts (microdomains) of neurosecretory terminals. Portal vessels (pv) are fenestrated and do not contain a continuous layer of smooth muscle in their walls. (Modified from Page and Dovey-Hartman, 1984b, © Wiley-Liss, a subsidiary of John Wiley and Sons, Inc.)

the pars distalis. Murakami *et al.* described a deep plexus of vessels in the regions of the pars intermedia which face the infundibular process. A superficial plexus of vessels lies in the regions facing the hypophysial cleft and the pars distalis. Murakami *et al.* proposed that blood flows from the neural lobe to the pars intermedia to the pars distalis.

Blood flows from the infundibular process (neural lobe) to the adjacent pars distalis on the posterior surface of the pituitary via short portal routes in the pig and presumably in other species in which short portal vessels have been described.

IV. SECONDARY CAPILLARY PLEXUS

The secondary capillary plexus lies in the pars distalis. It is supplied by long and short portal vessels and by capillaries. There is no convincing evidence that a direct arterial supply to the pars distalis is present in any of the species studied. Capillaries in the pars distalis are fenestrated and the organ lacks a blood–brain barrier. Blood flow in this region is difficult to measure but appears to be in the range of 60 cc/100 g/min.

V. VENOUS DRAINAGE

Lateral hypophysial vein drain the pars distalis and the. Confluent Y-shaped veins which send one limb to the pars distalis and one to the infundibular process (neural lobe) course to the adjacent cavernous sinus in the monkey and presumably drain both these structures.

See Also the Following Articles

Anterior Pituitary; Neurohypophysial Hormones

Bibliography

Bergland, R. M., and Page, R. B. (1978). Can the pituitary secrete directly to the brain? (Affirmative anatomical evidence). *Endocrinology* **102**, 1325–1338.

Bryan, R. M., Myers, C. L., and Page, R. B. (1988). Regional neurohypophyseal and hypothalamic blood flow in rats during hypercapnea. *Am. J. Physiol.* **255**, R295–R302.

Daniel, P. M., and Pritchard, M. M. L. (1975). Studies of the hypothalamus and the pituitary gland. With special reference to the effects of transection of the pituitary stalk. *Acta Endocr.* **80**(Suppl. 201), 1–216.

Duvernoy, H. (1972). The vascular architecture of the median eminence. In *Brain–Endocrine Interaction. Median Eminence; Structure and Function*, International Symposium, Munich 1971, pp. 79–108. Karger, Basel.

Green, J. D. (1951). The comparative anatomy of the hypophysis, with special reference to its blood supply and innervation. *Am. J. Anat.* **88**, 225–311.

Green, J. D., and Harris, G. W. (1947). The neurovascular link between the neurohypophysis and adenohypophysis. *Endocrinology* **5**, 136–146.

Green, J. D., and Harris, G. W. (1949). Observation of the hypophysio-portal vessels of the living rat. *J. Physiol.* **108**, 359–361.

Hatton, G. I. (1990). Emerging concepts of structure–function dynamics in adult brain: The hypothalamo-neurohypophysial system progress. *Neurobiology* **34**, 437–504.

Monroe, B. G. A. (1967). Comparative study of the ultrastructure of the median eminence, infundibular stem and neural lobe of the hypophysis of the rat. *Z. Zellforsch.* **76**, 405–432.

Murakami, T. (1976). Pliable methacrylate cast of blood vessels: Use of a scanning electron microscope study of the microcirculation in the rat hypophysis. *Arch. Histol. Jpn.* **38**, 151–168.

Murakami, T., Ohtsuka, A., Taguchi, T., and Ohtani, O. (1985). Blood vascular bed of the rat pituitary intermediate lobe, with special reference to its development and portal drainage into the anterior lobe. A scanning electron microscopic study of vascular casts. *Arch. Histol. Jpn.* **48**, 69–87.

Page, R. B. (1983). Directional pituitary blood flow: A microcinephotographic study. *Endocrinology* **112**, 157–165.

Page, R. B., and Bergland, R. M. (1977). The neurohypophyseal capillary bed. Part I. Anatomy and arterial supply. *Am. J. Anat.* **148**, 345–358.

Page, R. B., and Dovey-Hartman, B. J. (1984a). Resistance vessels in the tuber cinereum of the rabbit, rat and cat. *Anat. Rec.* **210**, 647–655.

Page, R. B., and Dovey-Hartman, B. J. (1984b). Neurohemal contact in the internal zone of the rabbit median eminence. *J. Comp. Neurol.* **226**, 274–288.

Page, R. B., Munger, B. L., and Bergland, R. M. (1976). Scanning microscopy of pituitary vascular casts: The rabbit pituitary portal system revisited. *Am. J. Anat.* **146**, 273–301.

Page, R. B., Leure-Dupree, A. E., and Bergland, R. M. (1978). The neurohypophyseal capillary bed. Part II. Specializations within median eminence. *Am. J. Anat.* **153**, 33–66.

Page, R. B., Funsch, D. J., Brennan, R. W., and Hernandez, M. J. (1981). Regional neurohypophyseal blood flow and its control in adult sheep. *Am. J. Physiol.* **241R**, 36–43.

Porter, J. C., Hines, M. F. M., Smith, K. R., Repass, R. L., and Smith, A. J. K. (1967). Quantitative evaluation of local blood flow of the adenohypophysis in rats. *Endocrinology* **80**, 583–598.

Redecker, P. (1991). Ultrastructural demonstration of neurohemal contacts in the internal zone of the median eminence of the Mongolian gerbil (*Meriones unguiculatus*): Correlation with synaptophysin immunohistochemistry. *Histochemistry* **95**, 503–511.

Rinne, U. K. (1966). Ultrastructure of the median eminence of the rat. *Z. Zellforsch.* **74**, 98–122.

Sobin, S. S., and Tremer, H. M. (1977). Three-dimensional organization of microvascular beds as related to function. In *Microcirculation* (G. Kalsz and B. Altura, Eds.), Vol. 1, pp. 43–57. University Park Press, Baltimore.

Wislocki, G. B. (1938). The vascular supply of the hypophysis cerebri of the rhesus monkey and man. *Res. Publ. Assoc. Res. Nerv. Ment. Dis.* **17**, 48–68.

Wislocki, G. B., and King, L. S. (1936). The permeability of the hypophysis and hypothalamus to vital dyes, with a study of the hypophyseal vascular supply. *Am. J. Anat.* **58**, 421–472.

Ziedonis, D. M., Severs, W. B., Brennan, R. W., and Page, R. B. (1986). Blood flow and functional response correlate in the ovine neural lobe. *Brain Res.* **373**, 27–34.

Hypoxia, Effect on Reproduction

Charles A. Ducsay

Loma Linda University School of Medicine

I. Introduction
II. Puberty
III. Fertility
IV. Pregnancy
V. Summary and Conclusions

GLOSSARY

acclimatization/adaptation The physiological adjustment to a new environment.

altitude The increase in terrestrial elevation that leads to a decrease in atmospheric oxygen.

fertility The ability to conceive or induce conception.

gametogenesis The development of female and male sex cells, or gametes.

hypothalamic–pituitary–gonadal axis The interaction of the hypothalamus and pituitary gland with the gonads to affect reproductive changes.

hypoxia The reduced availability of oxygen to the body tissues.

pregnancy The condition of having a developing fetus or embryo in the body.

reproduction The ability of a species to propagate.

Altitude has profound effects on the process of reproduction at various stages, including puberty, fertility, and the ovarian cycle, and it affects the mother and fetus during pregnancy. The principal effect is mediated through hypoxia, the reduced availability of oxygen that occurs with increased altitude. The stress of hypoxia can alter these parameters of reproductive capacity, whereas adaptation helps to balance these effects and is imperative for successful continuation of the species. Although it is necessary to exert caution in extrapolating research findings from animal studies to the human condition, the results are largely complimentary.

I. INTRODUCTION

Worldwide, over 40 million people reside at altitudes higher than 2400 m and a similar number visit such altitudes each year while traveling. This height is considered the standard "high altitude" since this is the point where most individual's arterial oxygen saturation falls below 90%. In other words, at this altitude, the decreased availability of oxygen becomes apparent. Residing at altitude even temporarily can have profound effects on an individual. Therefore, adaptation and acclimatization are very important. Natives that are indigenous to high-altitude environments demonstrate a number of adaptive changes that allow them to function under apparently adverse conditions. These include hyperventilation, increased erythropoesis and hematocrit, elevated cardiac output, as well as greater vascularization of critical tissues and redistribution of blood flow away from regions with relatively low oxygen demands.

According to Hochachka, there are three principal types of adaptation with respect to time available for mounting a response to high altitude:

1. *Acute*: Literally instantaneous and depends on physiological systems already in place. It is the output of the system that is altered, not the actual nature of the system. All these responses are designed to protect the system from the limited availability of oxygen.

2. *Acclimatory:* These changes take place over a period of time ranging from days to weeks and maybe even longer. Since there is time for these responses to occur, there is time for reorganization or restructuring. There may be either up- or downregulation of already available macromolecular components, generation of new receptors, enzymes, or membrane proteins permitting an integration of physiological responses that are not possible in acute circumstances. In this case, both functional organization and output are regulated by the hypoxic stress.

3. *Long-term or phylogenetic:* These include biochemical reorganization and restructuring that involve generations of time to become a fixture of the species genome.

Although a number of physiological changes occur following exposure to altitude, the mechanisms of these adaptations are not well understood. Even less is known regarding the effects of altitude on reproduction. Many descriptive and correlative studies have been conducted, but there is still little information available on the actual mechanisms responsible for the observed changes in reproductive parameters at altitude.

II. PUBERTY

Puberty is the stage of sexual differentiation in which the individual attains full reproductive capability. This includes the maturation of sexual organs and formation of oocytes and mature spermatozoa as well as development of secondary sexual characteristics, all under the influence of steroid hormones. The initiation of puberty is mediated through the central nervous system, specifically through the frequency and amplitude of gonadotropin hormone release which, in turn, activates gonadal growth and function. Before the central nervous system activates puberty, a specific level of neural maturity must be reached. It has also been clearly shown that a critical body mass is an important component. Underweight girls and boys, as well as adolescents with poor nutrition, experience delayed puberty. Chronic stress also can lead to inhibition of the onset of puberty. Since the effects of residing at altitude can be considered chronic stress, it is not surprising that puberty is affected.

There are ample data in the literature to suggest that puberty is delayed following exposure to high altitude. Early studies of the 1940's clearly showed that rats maintained at 7600 m for only 4 hr per day beginning on Day 14 after birth exhibited delayed puberty, a decrease in spermatogenesis (males), and a decrease in gonad/body weight (both males and females). In humans, male adolescents at altitude achieve puberty at a significantly later age. In addition, they also exhibit a more prolonged process of sexual maturation than control subjects living at sea level. In females, the onset of breast development and menarche occur at significantly older age at high altitude than at sea level. Additionally, at puberty, serum follicle-stimulating hormone (FSH) concentrations increased later in life when compared to women at sea level.

Although the mechanisms involved in the delay of puberty at high altitude are unknown, it has been suggested that the chronic stress of life at altitude may be a predisposing factor. Since restraint of the onset of puberty resides in the central nervous system, hypoxia may have a direct effect on the hypothalamic–pituitary–gonadal axis. It is possible that there is a decreased pubertal gonadal response to gonadotropin stimulation since abnormal responses of the gonads and adrenals have been reported in high-altitude natives.

III. FERTILITY

The hypothalamic–pituitary–gonadal axis also plays a key role in reproductive function. Since it is regulated by a complex interaction of neural, metabolic, and hormonal signals, it is not surprising that environmental factors can have a profound influence on other aspects of this elaborate system.

A. Gametogenesis

Studies in men at 4300 m for 4–14 days found no changes in concentrations of FSH, luteinizing hormone (LH), or testosterone in plasma or urinary gonadotropins. However, other studies of slightly longer duration did detect a decrease in plasma testosterone levels. In addition, urinary testosterone responses to human chorionic gonadotropin administration were decreased in men at 4200 m and both LH and FSH responses to gonadotropin-releasing hormone were decreased. Taken together, these results suggest that long-term hypoxic exposure may be responsible for the hypogonadotropism/hypoganadism observed in men living at altitude and perhaps even patients with chronic hypoxemia resulting from lung disease.

Additional studies have shown that there is decreased testicular size, Leydig cell number, and sperm viability in men at high altitude. Although the

mechanisms responsible for these alterations are not well defined, steroid hormones regulate these systems. Since steroidogenesis involves pathways that require oxygen (cytochrome P450 enzymes), one may infer that the reduced oxygen availability at high altitude may decrease steroid production. However, there is little experimental evidence that this is indeed the case. It appears that only nonphysiologically low levels of oxygen will consistently decrease the production of androgen precursors. Cortisol synthesis also seems to be relatively resistant to the effects of low-oxygen tension.

Studies in rats have shown that in the early stages of exposure to high altitude, there is hypertrophy of the interstitial tissue of the testes and epithelial portion of the seminal vesicles. In longer term studies, such alterations were not apparent. It seems that the duration of the hypoxic effects of high altitude exposure is a key determinant of the effect. Likewise, spermatogenesis appears to be affected only on a transient basis in rats and domestic farm animals. In men, as well, there is just a temporary reduction in sperm count and increased proportion of abnormal sperm upon ascent to altitude. These temporary alterations are accompanied by a transient decline in urinary testosterone excretion. In comparison, no such changes are observed in men that are native to altitude. Such observations imply that the effect of altitude on the reproductive capacity of males is a short-lived phenomenon and only occurs during the transition to altitude.

B. Ovarian Cycle

Because of all the effects of altitude on reproductive parameters described previously, it seems likely that there are also effects on fertility and the ovarian cycle in females. Female rats exposed to hypobaric hypoxia (simulated altitude of 5500 m) demonstrated an *in vitro* conversion of labeled pregnenelone to estradiol that was 140% higher than that for the normoxic, sea level control rats. Furthermore, ovarian weight was 63% higher at high altitude. Although the incidence of estrous cycles was also higher in the high-altitude rats, fertility was actually lower. There was an increased incidence of polycystic ovarian follicles in this group; however, it is not know if this was a direct effect of hypoxia on the ovaries or an effect on the hypophyso–hypothalamic axis. Additionally, rodents have a reduced rate of copulation when exposed to high-altitude stress which suggests a lower incidence of ovulation. Induced ovulation in mice at altitude also revealed a reduction in the number of ova shed during estrus.

Studies of women in the Andes indicated a number of differences in the menstrual cycle. In women at high altitude, the follicular phase of the cycle is longer than that in women at sea level, whereas the diameter of preovulatory follicles is reduced. Furthermore, serum levels of estradiol, progesterone, and prolactin were all lower in women at altitude. Although the mechanism of these alterations remains unknown, these dramatic alterations in hormone levels appear to have a profound effect. Interestingly, women that are transported from low to high altitude also experience some irregularities of the menstrual cycle.

Despite the delay in puberty and the changes in the menstrual cycle induced by life at high altitude, there does not appear to be adverse effects on fertility in women. The global fecundity rate of women residing at high altitude is actually higher than that for women at lower altitudes or at sea level. These results indicate that the adaptation that has taken place over generations for women at altitude, which is in marked contrast to the more acute exposure response of women to high altitude. The lack of adaptation was reported by the early Spanish settlers in Peru, who found that their fertility rates were severely suppressed for almost a generation. This was also noted for their domestic animals, which also experienced a high mortality rate among newborns. This may have actually been one of the principal reasons that the capital city of Peru was moved from Juaja (3300 m) to Lima (150 m).

IV. PREGNANCY

Pregnancy places a tremendous physiological strain on the mother. This is compounded at altitude by reduced maternal (and therefore fetal) oxygen availability. Despite this apparent disadvantage, pregnancy at altitude can proceed to a successful completion. A number of adaptive mechanisms facilitate this process, including an elevated level of maternal and fetal hemoglobin and increased size of the placenta. The obvious purpose of these alterations is to increase the efficiency of oxygen delivery to the developing conceptus. There are, however, additional effects of altitude on pregnancy that may be more subtle.

A. Placenta

Many gas-exchange systems experience structural adaptations as the result of hypoxia. The goal of these changes is to increase the diffusion capacity of the system and this may be accomplished by increases in size, a reduction in the thickness of the diffusion barrier, or an increase in capillary density. The placenta, being the primary gas-exchange system for the fetus, is quite susceptible to the effects of hypoxia. In the guinea pig, for example, chronic hypoxia is responsible for increased branching and coiling of capillaries in the placenta. In guinea pig, it was also noted that at high compared to low altitude, systemic vascular resistance, involved in the regulation of blood pressure, was increased during pregnancy. Although the mechanism for this change has not been elucidated, it does not appear to be the result of alterations in prostaglandin production.

A relatively large body of data exists on development of the human placenta at altitude. Placentae collected from high-altitude pregnancies have thinner villous membranes, which increases the diffusing capacity of the placenta, and there are reports that they are "hypervascularized" or "hypercapillarized," with increased branching of the capillaries in the terminal villi of the placenta.

Although the mechanisms responsible for these changes remain undefined, it appears that there may be some effects of growth factors. Placental cells produce angiogenic cytokines that stimulate vascular growth. Secretion of vascular endothelial growth factor (VEGF) is increased under hypoxic conditions, which stimulates development of the placental vasculature to allow it to meet its metabolic needs. This effect of VEGF may also be responsible for the increased incidence of chorioangioma in high-altitude pregnancies. This disorder is a benign tumor of the placenta that is characterized by excessive proliferation of the villous capillaries.

Another factor that may contribute to the altitude-induced changes in the placenta is a change in placental circulatory dynamics. Under hypoxic stress conditions, redistribution of fetal blood flow from the peripheral to umbilical circulation has been reported. The increased wall stretch on the villous capillaries may then lead to endothelial cell remodeling.

B. Fetus

The classic work of Barcroft suggested that the supply of oxygen to the fetus is similar to the ascent of Mt. Everest. That is, in normal circumstances, fetal PO_2 is greatly reduced and effects of high altitude reduce oxygen availability even further. As described previously, chronic exposure to altitude results in maternal hyperventilation or rapid breathing which causes an excess loss of carbon dioxide. This leads to a mild respiratory alkalosis with a compensatory decrease in serum bicarbonate levels. Since studies from animal models have shown a direct correlation between maternal and fetal acid-base status, the effects of altitude on the maternal system are generally mirrored in the fetal system. The early studies of placental gas exchange in sheep at high altitude revealed an increase in fetal umbilical cord blood pH; however, in the unstressed fetus, such alterations would not be of physiologic significance. Blood gas measurements made in the umbilical cord blood of human neonates at altitude suggested similar results. The differences in blood gas values for the high-altitude neonates differed very little from those considered normal for healthy newborns at sea level.

In addition to physiologic compensation of acid-base status, there are other mechanisms that help the fetus deal with the reduced oxygen availability at high altitude. Following exposure to altitude, maternal hemoglobin concentrations as well as plasma pH increase. The higher hemoglobin concentrations result in a greater oxygen-carrying capacity of red blood cells, whereas the elevated pH leads to greater affinity of hemoglobin for oxygen. A number of studies have clearly shown an increase in fetal hemoglobin concentrations.

Although many physiological mechanisms are aimed at compensating for the reduction in inspired maternal oxygen at high altitude, the compensation is apparently incomplete. Long-term maternal exposure to altitude has a profound and well-documented effect on birth weight. Studies with high-altitude populations in South America and Tibet have shown that between 0 and 5000 m, there is an average reduction in birth weight of 115 g/1000 m of elevation gain. In addition, the incidence of low-birth-weight infants (birth weight <2.5 kg) increases significantly and this is accompanied by an increase in neonatal and infant mortality. These results are consistent with those from studies of a variety of animal species that also experience decreased fetal weights associated with a reduction in inspired maternal oxygen. However, despite a generalized effect of altitude on birth weight, the magnitude of the reduction in birth weight varies among and even within species.

Taken together, available results indicate a chronic impairment of oxygen/nutrient delivery to the fetus at high altitude. Is the decrease in birth weight solely the result of decreased oxygen availability or are there additional alterations in the perfusion of the uterus/placenta? Studies of oxygen delivery to the uterus and placenta have shown that the arterial oxygen content is similar in pregnant women at high altitude compared to that of women who reside at low altitudes. It does not appear, therefore, that oxygen availability is the key problem. Rather, available information suggests that there is a decrease in utero-placental blood flow at high altitude.

C. Mother

During the normal course of human pregnancy, there is an elevation in uterine arterial blood flow. In early pregnancy, the increase is the result of both increased uterine artery diameter and the velocity of blood flow. However, in the latter stages of gestation, increased blood flow velocity is the principal factor responsible for increased uterine blood flow. Studies comparing uterine arterial blood flow in women at high altitude (3100 m) with low-altitude counterparts (1600 m) found that the effects of pregnancy on blood flow were quantitatively similar. However, high altitude significantly affected the magnitude of the observed changes. There was less of an increase in volumetric flow and uterine artery diameter as well as a smaller percentage of common iliac flow directed to the uterine artery in women at high altitude compared to those at low altitude. The mechanisms responsible for these alterations in uterine blood flow may involve changes in blood volume during pregnancy.

Low blood volume has clearly been associated with intrauterine fetal growth retardation as well as pre-eclampsia, or pregnancy-induced hypertension. Both of these abnormalities are much more prevalent at high than at low altitude. It seems likely that blood volume expansion may also be affected by low oxygen availability at high altitude. Studies in Colorado found that pregnant women at high altitude indeed had lower blood volumes that women at moderate altitude; however, this was the result of lower blood volumes in the nonpregnant state.

D. Labor and Delivery

As previously described, hypoxic exposure as the result of high altitude can be considered a chronic stress. It is common knowledge that stress can have profound effects on the initiation of labor. One would expect, therefore, to observe an increased incidence of preterm labor following exposure to altitude. However, studies of pregnant women at altitude have detected a definite increase in the incidence of intrauterine growth retardation but no evidence for higher numbers of preterm births. Studies of chronic high-altitude hypoxia in ewes also failed to detect an increased incidence of premature births.

The initiation of parturition in sheep occurs through activation of the fetal hypothalamic–pituitary–adrenal axis. Following long-term exposure to high altitude during pregnancy, there is suppression of fetal adrenal responsiveness to ACTH which may prevent a premature rise in cortisol as well as labor and delivery. Since there is no apparent change in adrenal weight or uterine blood flow be-

tween animals at high and low altitude, substrate availability does not appear to be an explanation. Together, these results suggest that adrenals of chronically hypoxemic fetuses fail to mature in terms of endocrine responsiveness. Since the fetal adrenal in lambs plays a key role in the initiation of labor, this "delayed adrenal maturation" may prevent preterm delivery in ewes exposed to chronic hypoxia at altitude.

Anecdotal reports in other species, such as the rat, suggest that chronic hypoxia may even delay parturition. In rats, there appears to be a direct effect of hypoxia on the myometrium to decrease contractile sensitivity to oxytocin that is mediated by a significant reduction in myometrial oxytocin receptors. Studies have also shown that blunting of myometrial responsiveness also occurs in chronically hypoxic sheep. For survival of the species at high altitude, it is imperative that the mechanisms involved in triggering parturition be attenuated so that the birth process is not stimulated prematurely in response to this stressor. It appears that there is an adaptive response by both mother and fetus to prevent preterm delivery in response to chronic stress at high altitude.

V. SUMMARY AND CONCLUSIONS

Altitude, with its accompanying hypoxia, can have profound effects on all aspects of the reproductive process. In response, the individual, as well as the species itself, can mount a series of adaptations to achieve successful reproduction at high altitude. Knowledge of the effects of altitude on reproduction is based largely on observational data collected by studying populations exposed to the chronic stress of high altitude compared to individuals at sea level. Additionally, animal studies have added actual experimental data to this knowledge base. However, the mechanisms responsible for the observed changes in reproductive strategies of high altitude remain relatively undefined. The application of more sophisticated biochemical and molecular techniques to the study of high-altitude reproductive biology will undoubtedly further understanding of this very complex and important area of scientific research.

Bibliography

Cogswell, M. E., and Yip, R. (1995). The influence of fetal and maternal factors on the distribution of birthweight. *Sem. Perinatol.* **19**, 222–240.

Hochachka, P. W. (1996). Metabolic defense adaptations to hypobaric hypoxia in man. In *Handbook of Physiology, Section 4: Environmental Physiology* (M. J. Fregly and C. M. Blatteis, Eds.), Vol. 2, pp. 1115–1123. Oxford Univ. Press, Oxford, UK.

Hoff, C. J., and Abelson, A. E. (1976). Fertility. In *Man in the Andes, a Multidiciplinary Study of High Altitude Quecheua* (P. T. Baker and M. A. Little, Eds.), US/IBP Synthesis Series 1, pp. 128–146. Dowden, Hutchinson, & Ross, Stroudsburg, PA.

Monge, C. (1948). *Acclimatization in the Andes*, pp. 26–46. Johns Hopkins Univ. Press, Baltimore.

Raff, H. (1996). Endocrine adaptations to hypoxia. In *Handbook of Physiology, Section 4: Environmental Physiology* (M. J. Fregly and C. M. Blatteis, Eds.), Vol. 2, pp. 1259–1275. Oxford Univ. Press, Oxford, UK.

Hysterectomy

Howard T. Sharp

University of Utah School of Medicine

I. History
II. Indications for Hysterectomy
III. Surgical Approach
IV. Complications
V. Surgical Alternatives to Hysterectomy

GLOSSARY

cervical conization The removal of a cone-shaped wedge of cervix which involves both ectocervix and endocervix.

fibroid A benign smooth muscle tumor of the uterus.

hysterectomy The surgical removal of the uterus.

laparoscopically assisted vaginal hysterectomy The use of a laparoscope and endoscopic instrumentation inserted through small ports in the abdominal wall to assist in the vaginal removal of the uterus.

myomectomy The surgical removal of a fibroid.

parametrium The tissues adjacent to the uterus and cervix which include the broad ligament, bladder pillars, uterosacral ligaments, and cardinal ligaments.

placenta accreta The erosion of placental tissue into the endometrial lining of the uterus.

salpingo-oophorectomy The surgical removal of the fallopian tube and ovary.

uterine atony The inability of the uterus to maintain its muscular tone in the postpartum state usually resulting in hemorrhage.

In 1975, hysterectomy was the most commonly performed major surgical procedure in the United States. It is now the second most common major surgery and is performed in 6.1–8.6/1000 women. Hysterectomy refers to the surgical removal of the uterus. There are generally three types of hysterectomy: total, subtotal, and radical. Total hysterectomy refers to removal of the uterine corpus and cervix, whereas subtotal hysterectomy refers to removal of the uterine corpus only. A radical hysterectomy refers to removal of the parametrium along with the uterine corpus and cervix. If the fallopian tubes and ovaries are to be removed, this distinction is designated by bilateral or unilateral salpingo-oophorectomy. The surgical approach to hysterectomy may be vaginal, abdominal, laparoscopic, or laparoscopically assisted.

I. HISTORY

The first vaginal hysterectomy is credited by many to Jacopo Berengario de Capri of Bologna in 1517. This operation was first performed in the United States by John Collins Warren in 1829. The first abdominal hysterectomy was performed by Ephraim McDowell in 1809. In the early days of American surgery, morbidity from hysterectomy was extremely high due to hemorrhage and shock. In 1864, Koeberle of France contributed significantly to hemostasis by introducing a method of securing vascular pedicles. Individual pedicle clamping and ligation was further developed by Stimson and Kelly in 1889 and 1891, respectively. With modern techniques and antibiotics, mortality from hysterectomy is now 1 or 2 per 1000.

II. INDICATIONS FOR HYSTERECTOMY

Indications to perform hysterectomy should be aimed to save life, relieve suffering, and correct deformity (Table 1). In general, in the absence of gynecologic malignancy, medical or conservative management should be performed prior to hysterectomy. Conditions whereby a hysterectomy is necessary as a life-saving measure usually include intractable hemorrhage and malignancy.

A. Intractable Hemorrhage

Intractable uterine hemorrhage as an indication for hysterectomy rarely occurs outside of pregnancy

TABLE 1
Indications for Hysterectomy

Intractable uterine hemorrhage
Malignancy
Refractory dysfunctional uterine bleeding
Symptomatic leiomyomata
Refractory uterine infection
Uterine prolapse
Chronic pelvic pain of presumed uterine origin

---

and malignancy. During pregnancy, severe hemorrhage can occur as a result of abnormal placentation such as placenta accreta, refractory uterine atony, and uterine rupture. In cases of uterine atony, uterotonic agents such as oxytocin, ergot alkaloids, and 15-methyl-F-2-α prostaglandin should be tried prior to hysterectomy. Uterine artery and hypogastric artery ligation may also be tried prior to hysterectomy.

B. Malignancy

Treatment of premalignant and malignant tumors of the cervix, uterus, and often ovaries may be indications for hysterectomy. In the case of cervical cancer, the clinical stage of disease should be assessed prior to hysterectomy. In general, stage IA1 disease (<3 mm invasion) without lymph or vascular space involvement may be treated by total hysterectomy or if future fertility is desired by conization of the cervix. A radical hysterectomy would be appropriate for stage IA2 (3–5 mm invasion) or with lymph or vascular space involvement, IB (greater than stage IA2 but confined to the cervix), or IIA disease (extension to the upper vagina with no parametrial involvement). Tumors assessed to be stage IIB or greater should be treated by radiation therapy. Endometrial adenocarcinomas and uterine sarcomas are also treated by hysterectomy. In the case of ovarian cancer with local spread, hysterectomy is performed as part of optimal debulking.

C. Dysfunctional Uterine Bleeding

Dysfunctional uterine bleeding is defined as bleeding which occurs in the absence of demonstrable pathology. It typically occurs in the perimenarchal and perimenopausal years. Excessive uterine bleeding is usually defined as lasting more than 8 days during a single cycle or profuse bleeding with large clots or gushes requiring additional protection. If this indication for hysterectomy is to be used, other causes such as endometrial polyps, uterine fibroids, coagulopathy, cervical and endometrial neoplasia, and pregnancy should be excluded. Nonsteroidal antiinflammatory drugs may be effective in treating dysfunctional uterine bleeding. Hormonal therapy using estrogen–progestogen combinations should be tried prior to hysterectomy in appropriate patients. An alternative to hysterectomy is to perform an endometrial ablative procedure which is described in greater detail later.

D. Leiomyoma

Leiomyomata or uterine fibroids are the most common indication for hysterectomy. They are the most common uterine tumor and are found in approximately 50% of women in autopsy series. Indications for hysterectomy due to fibroids include excessive bleeding or anemia, symptoms due to mass effect, and rapid uterine growth to a size $\geqq 12$ weeks size. The contemporary management of uterine fibroids allows for several alternatives to hysterectomy (see below). Leiomyomata arise from a single smooth muscle cell and are virtually always benign. The incidence of sarcomatous degeneration in surgically removed specimens ranges from 0.07 to 0.5%. They may occur in several uterine locations, including beneath the mucosal and serosal lining, or within the myometrium.

E. Infection

In the case of tuboovarian abscess refractory to broad-spectrum intravenous antibiotic therapy or percutaneous drainage, hysterectomy is warranted. Severe postpartum endomyometritis or postabortal infection may result in life-threatening septic shock. Hysterectomy is usually only necessary in the case of failure to respond to uterine curettage and appropriate medical therapy, poor response to surgical drainage and debridement, and in the presence of gas gangrene (clostridial necrotizing myometritis).

F. Pelvic Relaxation

Uterine prolapse may cause symptoms of pressure, fullness, and bearing down. Conservative management may include lifestyle changes by avoiding heavy lifting or excessive bearing down by correction of constipation or chronic cough. Estrogen therapy may also by beneficial to restore vaginal thickness, elasticity, and blood flow. Alternatives to hysterectomy include pessary use and uterine suspension. If hysterec-

tomy is to be performed, it is usually performed vaginally with the concomitant repair of existing vaginal support defects.

G. Chronic Pelvic Pain

Prior to performing hysterectomy for chronic pelvic pain, musculoskeletal, gastrointestinal, and urological causes of pain should be excluded. Pelvic pain should be documented for at least 6 months with a negative impact on the patient's quality of life. A psychiatric or psychosexual evaluation is also recommended. Laparoscopy is recommended prior to hysterectomy to exclude remediable causes for pain. Endometriosis refractory to conservative surgery and/or medical management in a patient who has completed her child bearing is also considered an indication for hysterectomy due to pain.

III. SURGICAL APPROACH

A. Abdominal Hysterectomy

An abdominal hysterectomy is performed by making a vertical or transverse incision in the abdomen. Traditionally, this approach has been the choice for patients with gynecologic cancer, large uterine fibroids, endometriosis, adnexal masses, contracted bony pelvis, and in some cases of inadequate uterine descent or mobility. The decision to perform an abdominal hysterectomy for benign indications opposed to a vaginal hysterectomy is often a function of one's practice style and training.

B. Vaginal Hysterectomy

In the absence of contraindication, a vaginal hysterectomy offers many advantages over abdominal hysterectomy because it is mostly an extraperitoneal surgery. Advantages include quicker recovery time, shorter hospital stay, less need for postoperative analgesia, decreased cost and charge, fewer postoperative adhesions, and less incisional morbidity. Despite the many advantages of vaginal hysterectomy, abdominal hysterectomy is still performed three to one over vaginal hysterectomy in North America.

C. Laparoscopic and Laparoscopically Assisted Vaginal Hysterectomy

With the advent of advanced laparoscopic technology, an attempt has been made to convert the abdominal hysterectomy into an assisted vaginal hysterectomy. There is controversy to date whether a significant benefit has been fully realized. Most studies show that this type of hysterectomy is more expensive than either the abdominal or the vaginal hysterectomy. Specific procedures that can be performed laparoscopically include ligation of the infundibulopelvic ligament for ovary removal, lysis of adhesions, and treatment of endometriosis. In the case of total laparoscopic hysterectomy, the entire uterus and cervix is removed laparoscopically. With the laparoscopically assisted vaginal hysterectomy, a portion of the hysterectomy is performed laparoscopically, then the uterus is subsequently removed vaginally.

D. Radical Hysterectomy

Radical hysterectomy is usually reserved for the treatment of early invasive cervical malignancies (stages IA2, IB, and IIA) and some endometrial cancers. It involves removal of the uterus with adjacent portions of the vagina, bladder pillars, cardinal ligaments, and uterosacral ligaments (parametria). It involves extensive dissection of the ureters and is usually done in conjunction with pelvic lymphadenectomy. An alternative to hysterectomy is radiation therapy, which confers a similar survival rate but may be associated with complications from radiation injury.

E. Subtotal Hysterectomy

Prior to the 1940s, subtotal hysterectomy was common. The total hysterectomy gained favor in the mid-1940s due to the development of antibiotics and safe blood transfusions and an effort to prevent cervical cancer. Some are now advocating subtotal hysterectomy because of improved screening methods for cervical cancer, claiming that the subtotal hysterectomy results in fewer complications. To date, there is no definitive evidence that performing subtotal

hysterectomy results in less mortality, sexual dysfunction, or surgical complications.

IV. COMPLICATIONS

Complications associated with hysterectomy may include intraoperative complications such as hemorrhage, injury to abdominal organs (bowel, bladder, and ureter), and death. Postoperative complications include infection and chronic bladder, bowel, or neurologic leg dysfunction.

V. SURGICAL ALTERNATIVES TO HYSTERECTOMY

Assuming medical treatments have failed, several surgical alternatives to hysterectomy are now available. In the case of dysfunctional uterine bleeding, endometrial ablative procedures may be performed; in the case of leiomyomata, myomectomy, myolysis, cryomyolysis, and embolization may be performed.

A. Ablative Procedures of the Endometrium

Endometrial ablative procedures include using operative hysteroscopic techniques to cauterize the endometrium using a Nd:YAG laser or electrocautery (rollerball) or to resect endometrium (loop electrode). Using these methods, approximately 70–90% of women will experience decreased blood loss; however, only 30–50% of patients with achieve amenorrhea. Nonhysteroscopic methods include cryoablation and thermoablation, which involve the insertion of ablative devises directly adjacent to the endometrium.

B. Conservative Surgery for Uterine Fibroids

Conservative surgery usually refers to uterine sparring surgery. Myomectomy may be performed by laparotomy, hysteroscopy, and in some cases by laparoscopy. Indications for myomectomy include uterine size $\geq$12 weeks, abnormal uterine bleeding, pain or pressure symptoms, rapid growth, infertility, and ureteral obstruction. The goal of surgery is to remove the entire fibroid, restore anatomy (adequate myometrial and serosal closure), and minimize blood loss. The decision as to which surgical approach to take depends on the myoma size and location and the operator's experience and training. In general, for submucous myomas 5 cm or less, the operation of choice is hysteroscopic resection, whereas larger myomas and intramural and subserosal myomas are usually removed by an abdominal approach. Laparoscopic myomectomy has not gained widespread acceptance due to the technical difficulty associated with this procedure.

Destructive procedures are also available for the treatment of fibroid tumors. Myolysis involves the use of bipolar coagulation needles inserted directly into the myoma by a laparoscopic approach. Similarly, cryomyolysis uses a cryo probe inserted laparoscopically into the fibroid to cause destruction. Recently, embolization has been used to restrict the vascular supply to fibroids. This is performed by some interventional radiologists by inserting a transcutaneous catheter through the femoral artery into the uterine artery for embolization. The long-term effects of these procedures are not yet known.

See Also the Following Articles

CERVIX; LEIOMYOMA; MENSTRUAL DISORDERS; UTERUS, HUMAN

Bibliography

Bachmann, G. A. (1990). Hysterectomy: A critical review. *J. Reprod. Med.* **35**, 839.

Benrubi, G. I. (1988). History of hysterectomy. *J. Fla. Med. Assoc.* **75**, 533.

Goldrath, M. P., Fuller, T., and Segal, S. (1988). Laser photovaporization of endometrium for treatment of menorrhagia. *Am. J. Obstet. Gynecol.* **140**, 14.

Hasson, H. H. (1993). Cervical removal at hysterectomy for benign disease: Risks and benefits. *J. Reprod. Med.* **38**, 781.

Kovac, S. R. (1995). Guidelines to determine the route of hysterectomy. *Obstet. Gynecol.* **85**, 18.

Lomano, J. M., Feste, J. R., Loffler, F. D., and Goldrath, M. H. (1986). Ablation of the endometrium with the Nd:YAG laser: A multicenter study. *Colposc. Laser Surg.* **2**, 203.

Schwartz, L. B., Diamond, M. P., and Schwartz, P. E. (1993). Leiomyosarcoma: Clinical presentation. *Am. J. Obstet. Gynecol.* **168**, 180.

Scott, J. R., Sharp, H. T., Dodson, M. K., Norton, P. A., and Warner, H. (1997). Supracervical hysterectomy: A decision analysis. *Am. J. Obstet. Gynecol.* **176**, 1186.

Thompson, J. D. (1992). Hysterectomy. In *TeLinde's Operative Gynecology, Seventh Edition* (J. D. Thompson and J. A. Rock, Eds.). Lippincott, Philadelphia.

Vollenhoven, B. J., Lawrence, A. S., and Healy, D. L. (1990). Uterine fibroids: A clinical review. *Br. J. Obstet. Gynaecol.* **97**, 285.

Wilcox, L. S., Koonin, L. M., Pokras, R., *et al.* (1994). Hysterectomy in the United States, 1988–1990. *Obstet. Gynecol.* **83**, 549.

IGF (Insulin-like Growth Factor)

James M. Hammond

Pennsylvania State University College of Medicine

GLOSSARY

IGF binding proteins A family of six or more soluble proteins secreted by IGF target tissues which bind the IGFs with high affinity and regulate their bioavailability.

IGF receptors A family of cell membrane proteins with high-affinity binding sites for the IGFs and signal-transducing capability, including the type I (IGF-I) receptor and the type II (IGF-II, mannose-6-phosphate) receptor.

insulin-like growth factors (IGFs) The peptide growth factors IGF-I and -II with homology to insulin and widespread expression and action in reproductive tissues.

I. INTRODUCTION

The insulin-like growth factors (IGFs) were initially postulated as circulating mediators of the actions of growth hormone and as insulin-like factors which were not neutralized with antibodies to insulin. Since that time, the conceptual system framework has evolved to include a multicomponent network of proteins which locally regulate the growth and differentiation of every tissue examined to date. Several mechanisms interact to determine the set point of this autocrine/paracrine system: expression and levels of IGF-I and/or -II, modulation of their bioavailability by IGF binding proteins (IGFBPs), and levels of their cell surface receptors and of their postreceptor signaling pathways. In addition, there is an endocrine component with circulating IGFs.

II. OVERVIEW OF THE IGF SYSTEM: COMPONENTS AND THEIR INTERACTION

A. Ligands

IGF-I and -II are the only defined members of this growth factor family. As depicted in Fig. 1, they bear striking homology to proinsulin (55%) as well as approximately 70% homology to each other. Despite the relative simplicity of the mature peptides, their biosynthetic pathway is complex, utilizing two functionally discrete promoters, alternative transcription start and termination sites, and variable RNA processing steps which can yield multiple RNA and peptide precursors. The latter have variable N-terminal leader peptides and/or alternative C-terminal extension, "trailer" domains. Despite considerable effort and speculation, the control and physiological importance of these alternative biosynthetic pathways have not been settled.

B. IGFBPs

During the past decade, study of the IGF system has been dominated by the discovery and characterization of a family of more than six binding proteins.

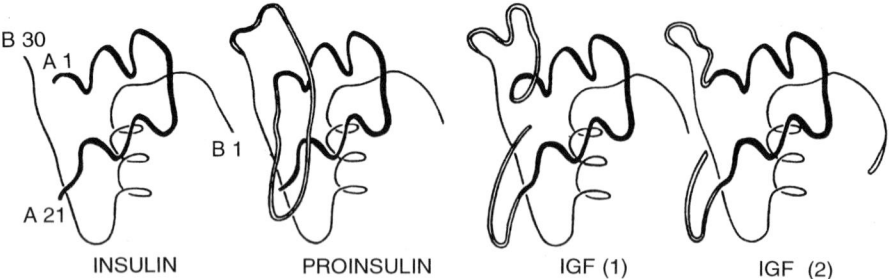

FIGURE 1   Projected tertiary structure of insulin, proinsulin, IGF-I, and IGF-II.

These proteins appear to be ubiquitous and are present in molar excess of the IGFs. They extend the half-life of circulating IGFs from minutes to hours. In tissues they regulate the bioavailability and presentation of the IGFs to their target cells. Structural homology among the six characterized proteins is greatest in the N- and C-terminal regions, both of which are required for IGF binding and in the 15–18 intrachain sulfide bridges (Fig. 2). The remainder of the linear sequence, in the midportion of the molecule, is more variable, but several recurring features have been described which appear to represent cell or matrix attachment sites.

The mechanism by which the IGFBPs govern the tissue availability of IGFs is a subject of active investigation. A working model for these complex interactions is presented in Fig. 3. Activity at each of the enumerated sites has been clearly established. However, the precise role of these activities in tissue trafficking and bioactivity of the IGFs remains incompletely defined. In addition to binding IGFs with high affinity, most of the IGFBPs can attach to target tissues. Two types of binding interactions have been identified. The most widespread involves binding to heparin-like moieties in cell or matrix glycosoaminoglycans. IGFBP-2, -3, and -5 exhibit this type of binding. In addition, binding to cell surface sites which may have signal-transducing capabilities (in-

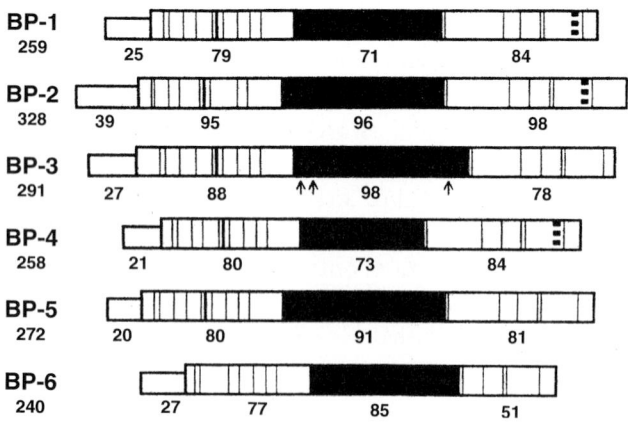

FIGURE 2   Linear structures of IGFBPs. Structures are aligned with regard to signal peptides, N-terminal and C-terminal domains (where multiple cystines are shown), and central region (From Clemnious, 1993).

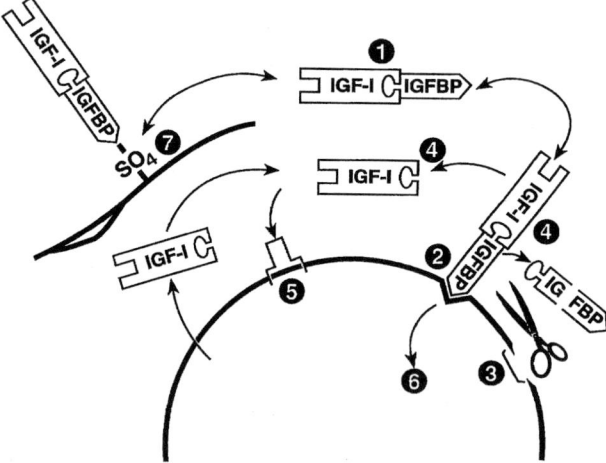

FIGURE 3   Projected interactions of IGFs, IGFBPs, and their receptors in the pericellular space. Enumerated sites depict established functions thought to modulate IGF effects. 1, IGF–IGFBP-soluble complex in extracellular space; 2, cell surface IGFBP binding site (receptor); 3, cell-associated IGFBP proteolytic site; 4, free IGF and modified, low-affinity IGFBP resulting from proteolysis; 5, type I IGF receptor; 6, putative signaling through IGFBP receptor; 7, heparin-like IGFBP binding site on target cells or matrix.

dependent of the IGFs) has been described. Thus, IGFBP-1 can bind to integrins and influence cell growth and migration. IGFBP-3 interacts with incompletely characterized cellular sites which may mediate IGF-independent inhibitory actions of this protein. Apparently, IGFBP-3 can be internalized with its ligand. In addition, both IGFBP-3 and IGFBP-5 are released into the extracellular space when they are bound by the IGFs. In contrast, the attachment of IGFBP-2 to glycoproteins appears to be enhanced by occupation of the IGF binding site. Formation of IGF–IGFBP complexes results in a labile pool of potentially active IGFs in the pericellular space. Such pools could target IGFs to their signal-transducing receptors in a "delayed paracrine" mechanism.

Posttranslational modifications of the IGFBPs can change their affinity for the IGFs and for their cell attachment sites. The two processes that have been most clearly defined include phosphorylation, which occurs on IGFBP-1 and IGFBP-3, and proteolysis, which impacts on all the IGFBPs. In some tissues, proteolysis appears to be the predominant mechanism whereby levels of these proteins are regulated. Serine proteases and or metallothionine proteases are the classes of enzymes most often involved. These mechanisms have obvious potential for influencing IGF and IGFBP trafficking.

With the exception of IGFBP-4, all the identified members of this gene family can both enhance and inhibit the actions of the IGFs. Studies demonstrating inhibition of IGF actions predominate, and this is particularly true in reproductive tissues. Almost certainly, this effect derives from the ability of these high-affinity binders to sequester the IGFs and block their access to signal-transducing receptors. The stimulatory interactions between the IGFs and their BPs are more challenging to explain. The most popular theory assumes that these proteins deliver IGFs to target cells as described previously. For IGFBP-3, attachment of the ligand and consequent dissociation of the complex from the cell surface could negate the inhibitory action of the IGFBP and result in cell stimulation. This novel mechanism has been invoked to explain IGFBP-3 and IGF actions in cells which lack IGF receptors.

C. Regulation and Importance of Circulating IGFs

In contrast to most tissue growth factors, IGF-I and -II circulate in serum in high concentrations. Hepatic production can count for most of the circulating IGFs. At least in rodents, there is a major shift from IGF-II to IGF-I early in neonatal life. In other species, the two forms are closer to equal in abundances. Most of the IGF-I in the circulation is bound in a high-molecular-weight ternary complex comprising IGF-I or -II, IGFBP-3, and an acid labile subunit. A small proportion circulates bound to other IGFBPs, and only a very small fraction is free. Most likely, the ternary complex cannot exit the vascular space. However, it can be processed at the endothelial surface or in serum to allow access of circulating IGFs and/or IGFBPs to tissues. The main hormonal regulator of circulating IGF-I levels is growth hormone (GH). However, these levels are also influenced by the reproductive system. This is most obvious at the time of puberty, when there is a substantial increase in circulating IGF-I levels, because gonadal steroids enhance the secretion of growth hormone. In addition, these steroids amplify the effects of GH on hepatic IGF-I production. In the ovary, the evidence suggests that serum IGFs and/or their BPs reach tissues in significant concentrations. However, the data in this and other reproductive tissues are insufficient to allow detailed analysis.

D. IGF Receptors and Signal Transduction

Two cell surface binding sites must be considered in dealing with the actions of the IGFs on their target cells (Fig. 4). The type I IGF receptor is composed of a heterodimer of 2α and 2β subunits which contribute mainly to the extracellular ligand-binding domain and the intracellular tyrosine kinase signaling domain, respectively. This receptor has striking homology to the insulin receptor and the two receptors share signal transduction pathways. In addition, the type II (IGF-II; mannose-6-phosphate) receptor binds IGF-II and lysosomal enzymes with high affinity. It lacks a clear intracellular signaling domain but can be associated with intracellular kinases. In most

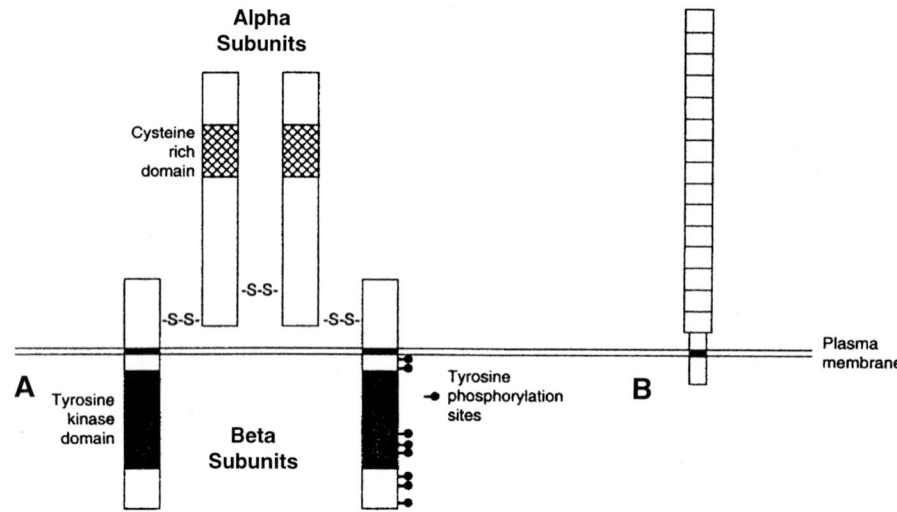

FIGURE 4   Structure of IGF receptors. (A) Type I IGF-I receptor; (B) type II IGF-II, mannose-6-phosphate receptor. See text for details.

tissues, signaling for both IGF-I and- II can be accounted for by binding to the IGF-I (type I) receptor. Mutations of the type II receptor have not had profound reproductive consequences. Postreceptor signaling pathways coupled to the type I receptor have not been completely delineated. However, activation of the receptor results in tyrosine autophosphorylation on the β chain of the receptor as well as on, presumably adjacent, cytosolic proteins. Subsequent to these initial events, there is a complex and variable cascade of tyrosine and serine/threonine phosphorylation steps which ultimately lead to activation of preexisting proteins and transcription of IGF-dependent genes. Proteins prominently and frequently involved in this pathway include insulin receptor substrate (IRS)-I and -II, phosphatitdylinoisotol-3-kinase, shc, Ras, Raf, and the MAP kinase system.

III. IMPORTANCE OF IGF SYSTEM IN REPRODUCTION

Studies in reproductive tissues have often been critical in developing the conceptual framework presented previously. Detailed information about the expression of these components and their control by hormones has been presented for several such tissues. Stimulatory actions of IGFs on cells from several reproductive organs have also been established. In the aggregate, these data have led to the strong presumption that the IGF system is crucial for normal reproductive function. Recently, disruption of the IGF-I gene in mice has been shown to have profound reproductive consequences. These data provide additional compelling evidence for the importance of this system.

A. Ovarian IGF System

The ovary is probably the reproductive tissue about which the most information is available on the IGF system. The data in the rat, pig, and human are the most complete, but relatively comprehensive studies are also available in mouse, cow, and sheep. There have been detailed descriptive studies *in vivo* utilizing *in situ* hybridization, immunohistochemistry, and measurement of system proteins in ovarian follicular fluid. Numerous studies of the expression and action of these proteins in cultured ovarian cells lead to the overarching concept that the IGFs serve as a local amplification mechanism for gonadotropin action facilitating follicle development. In contrast, the IGFBPs appear to inhibit this process, leading to atresia. The experiments which have supported these concepts most directly have been those in which gonadotropin actions in cell or organ culture have been blocked by coincubation with antibodies to the IGFs and/or the naturally occurring IGFBPs. *In vivo* disruption of the IGF-I gene in mice blocks follicle development, follicle-stimulating hormone (FSH) action, and the development of FSH receptors. In subsequent paragraphs, the data on each of the IGF system components in the ovary are briefly summarized.

1. Ligands

Both IGF-I and IGF-II are expressed in the ovary of each of the species which has been adequately examined. However, the localization of these peptides, their abundance, and their regulation appear to differ among species. In the human, IGF-II is the dominant ligand produced, the principal IGF in the granulosa cell, and the one most clearly linked to follicular development. In domestic animals and rodents, these roles are subsumed by IGF-I. *In vivo*, the expression of the IGFs has been colocalized with markers of follicular health and development such as gonadotropin receptors, steroidogenic enzymes, and cell replication markers. These data, coupled with the well-described actions of the IGFs on cultured ovarian cells (see below), support a critical role for these peptides in follicle growth and selection. Gonadotropin and growth hormone treatment *in vivo* increases ovarian IGF-I (in some species). Studies with cultured granulosa cells and/or follicles of several species have shown that gonadotropins, cAMP, and GH amplify the expression of IGFs (on both the message and peptide level). These data suggest that locally produced IGFs could mediate or amplify gonadotropin actions in growing or selected follicles.

2. IGFBPs

All the IGFBPs described to date have been found in the ovary of at least some species. Many of them were initially purified from porcine ovarian follicular fluid. The most abundant IGFBPs in the ovary are IGFBP-2, -3, and -4. Important differences between mRNA and peptide levels suggest regulatory mechanisms in addition to biosynthetic rates. For example, IGFBP-3 is the most abundant IGFBP in ovarian

ovarian follicular fluid, but its mRNA is largely confined to corpora lutea and endothelial cells. This protein appears to gain access to the ovary from serum, raising the possibility of a significant role for circulating IGFs and their BPs. In contrast, mRNA for IGFBP-4 and -5 is abundant in some follicles in which the peptides are absent, suggesting major postsynthetic regulation, probably by proteolysis. Levels of the proteins and/or the mRNA for several of the binding proteins undergo important changes during follicular growth *in vivo*. Levels of IGFBP-2 and -4 correlate with follicular degeneration or atresia in several species. In contrast, IGFBP-3 in ovarian follicular fluid increases during the preovulatory period. Cultured granulosa cells produce each of the binding proteins described to date, at least under some experimental conditions. However, the dominant secretory proteins are IGFBP-2, -4, and -5. Luteinizing porcine granulosa cells also produce IGFBP-3, and luteinizing human granulosa cells secrete IGFBP-1. In general, levels of the IGFBPs in culture are reduced by gonadotropins and cAMP, whereas they are stimulated by IGF-I. The effects of FSH and cAMP are dose dependent and sometimes biphasic, particularly with IGFBP-4. The stimulatory effects of IGF-I appear to be through coordinated actions on expression and postsynthetic stabilization.

The physiological effects of all of the IGFBPs on cultured ovarian cells have been inhibitory. To date, it has been impossible to show the stimulatory interactions between the IGFs and their BPs which have been demonstrated with other cells. Nonetheless, several of the putative regulatory mechanisms depicted in Fig. 3 have been demonstrated in ovarian cells. The IGFBPs can become associated with ovarian cells; in fact, approximately half of IGF binding to such cells is to IGFBPs rather than IGF receptors. Cultured granulosa cells have also been shown to process IGFBPs by proteolysis; IGFBP-3 and -4 have been studied in detail. Cell binding and protease activity could enhance IGF delivery. The increase in IGFBP-3 in preovulatory follicles would fit with this concept. However, such effects have been difficult to demonstrate directly. In contrast, the potent inhibitory actions of the IGFBPs in culture, coupled with their localization in atretic follicles (especially

IGFBP-2 and -4), argue for significant inhibitory actions.

3. IGF Action, Receptors, and Signal Transduction

On cultured ovarian cells IGFs are global stimulators, enhancing more than a dozen tissue-specific functions. In general, these peptides amplify the effects of other agonists which often have more potent intrinsic stimulatory activity. Thus, IGFs stimulate steroidogenesis and luteinizing hormone (LH)-receptor expression, but the effect is most impressive in the presence of FSH. IGFs are also important ovarian cell mitogens, but their effects are more dramatic in the presence of EGF and related peptides. When cultured follicles or granulosa cells are deprived of IGFs by antibodies or IGFBPs, they undergo apoptosis or atresia. Almost certainly, the effects of the IGFs in cultured ovarian cells are mediated by the (type I) IGF-I receptor. The dose–response characteristics of the several ligands fit with this model. *In vitro* studies suggest that this may be an important control locus since these structures can be induced by gonadotropins and gonadal steroids. *In vivo*, changes in binding have also correlated with follicle development. However, studies with *in situ* hybridization have suggested that the expression of the receptor is widespread if not ubiquitous (including atretic follicles). In the aggregate, these suggest that expression and bioavailability of the ligand is more likely to be rate limiting in the determination of follicular life or death. The tyrosine kinase inhibitors block the actions of IGF in cultured ovarian cells. However, the details of the more distal phosphorylation cascade which leads to the mitogenic and/or differentiative action of these growth factors have yet to be established.

B. Testicular IGF System

Studies of the expression and role of the IGF system in testes have generally paralleled those in the ovary. The results are similar, as expected. However, there are some significant differences. As proved to be the case in the female, knockout of the IGF-I, but not IGF-II, gene in male mice resulted in profound reproductive abnormalities and infertility. Morphologically, the testes in these animals were small but proportionate to body weight. Secondary sex organs, which depend on androgens for differentiation, were vestigial. Nonetheless, when sperm were harvested mechanically and used for *in vitro* fertilization, they were functional. In the aggregate, these results suggest a block in steroidogenesis and a consequent reduction in the ability of male animals to inseminate their mates. In addition to these recent observations, the testicular IGF system has been documented for more than a decade with cultured testicular cells and with histochemical techniques. These data are reviewed below for the several components comprising this system.

1. Ligands

As is true in the ovary, the dominant IGF in the testes of humans appears to be IGF-II. In most of the other species examined, IGF-I seems more important. However, both peptides are expressed in a regulated fashion in some animals, e.g., the pig. Several sites of IGF production have been detected. Early studies with cultured Sertoli cells detected IGF-like activity in the medium and suggested FSH stimulation. With immunohistochemistry, IGF-I peptide was also found in several cell types in the seminiferous tubules including Sertoli cells and spermatogonia. Recent studies using sensitive molecular techniques have clearly demonstrated more IGF mRNA in interstitial and Leydig cells. Gonadotropins increase IGF content in the testes *in vivo* and this effect was significantly reduced by pretreatment with toxins which selectively destroyed Leydig cells. In contrast, IGF-I expression by cultured Leydig cells was not stimulated by hCG and, surprisingly, one thorough study showed an inhibitory effect.

2. IGFBPs

Most of the IGFBPs are expressed in testes, with IGFBP-2, -3, and -4 being the most abundant. By hybridization histochemistry, IGFBP-3 is expressed predominantly in endothelial cells, but it binds to Sertoli cells. The others are expressed both in the interstitial compartment and in the seminiferous tubules. In microdissected testes, the three IGFBPs were enriched in the interstitial compartment, but also found in the tubules. FSH decreases IGFBP-3 abundance both *in vivo* and *in vitro* (in cultured Sertoli cells). hCG decreased the secretion of IGFBP-2 by Leydig cells. As for the ovary, the actions of the IGFBPs appear to be inhibitory on both gonadotropin and IGF effects. The relative potency of these factors as inhibitors was IGFBP-3 > IGFBP-4 > IGFBP-2.

3. IGF Action, Physiology, and Signaling

IGFs have demonstrated effects on cells of all the major testicular compartments. The most frequently studied is the interaction between LH/hCG and IGF-I on Leydig cell steroidogenesis. This interaction, which parallels similar events in the ovary, may be mediated in part by mutual receptor induction. IGF-I enhances hCG binding to these cells and reciprocal effects have also been demonstrated. Effects on cells from the seminiferous tubules include stimulation of metabolism (lactate production in Sertoli cells) and of DNA synthesis and morphological differentiation in spermatogonia. Somewhat surprisingly, IGF-I inhibits aromatase activity in Sertoli cells. However, since loss of this enzyme is part of the age-dependent maturation of these cells *in vivo*, this action may be viewed as stimulation of the programmed differentiation of these cells. The major receptor mediating these effects seems to be the type I IGF receptor. Effects of insulin on testicular systems show a diminished potency characteristic of this response.

C. Uterine IGF System

The uterus is another well-studied reproductive tissue in which the IGF system is abundant and important. It appears that the abundance of mRNA for IGF-I in the uterus is second only to liver in the rat and it is more abundant in uterus than liver in the pig. A considerable number of studies have implicated the IGF system in two interrelated but somewhat discrete processes: (i) mediation of estrogen/progesterone-dependent growth and differentiation during the reproductive cycle and (ii) maternal/fetal communication during implantation and fetal growth. Knockout studies of the mouse, one of the best studied species, indicate uterine hypoplasia, particularly of the myo-

metrium, in the absence of IGF-I. The defect is in excess of that which could be accounted for by the concomitant estrogen deficiency arising in the ovary. Embryos which carry homozygous gene disruptions for IGF-I, IGF-II, or the type I receptor can implant, but they have significant intrauterine growth retardation. The results suggest that each of these peptides has a unique role to play, but probably not in implantation. As in other tissues, the distribution of and control of multiple proteins contributing to the system is complex. In subsequent paragraphs, we have emphasized data in the mouse, which has been particularly well studied, with some comments regarding important species differences.

1. Ligands

Both IGF-I and -II are expressed in the reproductive cycle of mature animals. IGF-I expression appears to be enriched in myometrium, but it is also expressed in endometrium. IGF-II is only enriched in endometrium. The expression of IGF-I in the uterus is highly responsive to estrogens; thus, these growth factors have been put forth as major mediators of estrogen-dependent growth and differentiation during the reproductive cycle. During the peri-implantation period and maternal recognition of pregnancy, there is coordinate expression of IGF-I and -II in both endometrium and the embryo. The specific pattern varies significantly from species to species. In the mouse, IGF-II seems particularly enriched in the placenta; in addition, IGF-II has been detected in preimplantation embryos as early as the two-cell stage. In this species, IGF-I transcripts are in the embryo after implantation.

2. IGFBPs

All the IGFBPs are expressed in the uterus of most of the species examined. The human is relatively unique in that it expresses very high concentrations of IGFBP-1. This protein was purified as a decidual marker protein from that species before its IGF-binding characteristics were recognized. Physiological changes in these several proteins have been described in several species. In the mouse, they are expressed very early in the developing embryo. Differences in distribution of mRNA and peptides as well as colocalization of the IGF-I or -II with their binding proteins suggest targeting functions.

3. IGF Actions and Signal Transduction

Based on the knockout studies mentioned previously, it has been inferred that the effects of IGF-I and -II during pregnancy and embryonic growth are mediated through the type I receptor. In addition, studies of animals with double knockouts suggest additional signal transduction pathways. Expression of the type I receptor has been well studied and appears to be present in myometrium, several cell types in the endometrium, and in periimplantation embryos. Estrogen administration upregulates the IGF-I receptor in myometrium. However, it appears to be virtually ubiquitous in most species and under most conditions studied. Histochemical studies have shown coexpression of IGF-I with other growth-related genes in endometrium and conceptus. These results plus demonstrated IGF effects on cultured uterine cells and related cell lines have implicated IGF action in a number of important uterine functions: aromatase, c-*fos*, c-*jun*, progestin receptors, and IGFBPs.

D. Function of IGF System Components in Other Reproductive Tissues

Since the IGF system is present in all tissues, the expression of this family of proteins in other reproductive tissues is predictable. In addition, data from gene disruption experiments suggest that it is important. The accessory male sex organs in mice with IGF-I knockouts are vestigial. In these animals, the relative importance of steroid versus growth factor deprivation has not been entirely resolved. However, the available data suggest that the peptides are important. Prostate deserves special mention because it has all of the IGF proteins, including both ligands, all the binding proteins, and the receptors. In addition, IGFBP proteases have been studied in detail. Interestingly, the prostate-specific antigen, used as a prostate tumor marker in humans, functions as a potent IGFBP protease. This enzyme is secreted into seminal plasma where it reduces the affinity of IGFBP-3 for the IGFs; in related systems it can block the inhibitory effects of IGFBP-3 on IGF-induced prostate cell growth. IGFs are potent growth promoters on cultured prostate epithelial cells and this action is inhibited by the IGFBPs. Hormonal regulation of IGF and binding protein expression is not completely understood, but available data are consistent with the IGFs acting as mediators of androgen-dependent growth.

In the female reproductive tract, the oviduct has also been studied in several species. The expression of most of the IGFs and IGFBPs appears to be lower than that in uterus and ovary, but it is still easily measurable. These factors have been postulated to impact the preimplantation embryo.

IV. SUMMARY AND CONCLUSIONS

The IGF system, comprising IGF-I, -II, the IGF-I receptor, and a large family of IGF binding proteins, is highly expressed in reproductive tissues. Each of these components is subject to physiological regulation which is dependent mainly on the hormonal signals of the reproductive cycle emanating from the pituitary or gonads. The general model is that tropic hormones from the pituitary or gonad impact on this system to amplify the actions of IGF-I or -II. This overall effect is achieved through several mutually facilitory events: induction of the ligands; enhanced IGF sensitivity by increased function of the cell surface receptor or through amplification of postreceptor signaling mechanisms; and by reduced expression of the IGF-binding proteins. The latter proteins exert a dominant negative effect in the reproductive tissues examined to date. However, histochemical and culture studies also suggest a shuttle or targeting function which could enhance the availability of IGFs at their target cells. It is clear that most of these activities are restricted to growing and differentiating compartments within each of these tissues. This localization is undoubtedly crucial to orderly function of these organs. However, the biochemical mechanisms which ultimately lead to this cellular specificity are incompletely understood.

See Also the Following Articles

CYTOKINES; GROWTH FACTORS

Bibliography

Baker, J., Hardy, M. P., Zhou, J., Bondy, C., Lupu, F., Bellvé, A. R., and Efstratiadis, A. (1996). Effects of an *Igf1* gene null mutation on mouse reproduction. *Mol. Endocrinol.* **10**, 903–917.

Clemmons, D. R. (1993). IGF binding proteins and their functions. *Mol. Reprod. Dev.* **35**, 368–375.

Cohen, P., Peehl, D. M., Bhala, A., Dong, G., Hintz, R. L., and Rosenfeld, R. G. (1994). The IGF axis in prostatic disease. In *The Insulin-like Growth Factors and Their Regulatory Proteins* (R. C. Baxter, P. D. Gluckman, and R. G. Rosenfeld, Eds.), pp. 369–377. Elsevier, Amsterdam.

Collett-Solberg, P. F., and Cohen, P. (1996). The role of the insulin-like growth factor binding proteins and the IGFBP proteases in modulating IGF action. *Endocrinol. Metab. Clin. North Am.* **25**, 591–614.

D'Ercole, A. J. (1996). Insulin-like growth factors and their receptors in growth. *Endocrinol. Metab. Clin. North Am.* **25**, 573–590.

Giudice, L. C. (1994). Growth factors and growth modulators in human uterine endometrium: Their potential relevance to reproductive medicine. *Fertil. Steril.* **61**, 1–17.

LeRoith, D. (Ed.) (1996). *The Role of Insulin-like Growth Factors in Ovarian Physiology,* Frontiers in Endocrinology, Vol. 19. Ares-Serono Symposia.

Peehl, D. M., Cohen, P., and Rosenfeld, R. G. (1995). The insulin-like growth factor system in the prostate. *World J. Urol.* **13**, 306–311.

Simmen, R. C. M., Green, M. L., and Simmen, F. A. (1995). IGF system in periimplantation uterus and embryonic development. In *Molecular and Cellular Aspects of Periimplantation Processes* (S. K. Dey, Ed.), pp. 185–204. Springer-Verlag, New York.

Smith, E. P., and Conti, M. (1996). Growth factors and testicular function: Relevance to disorders of spermatogenesis in humans. *Sem. Reprod. Endocrinol.* **14**, 209–217.

Spiteri-Grech, J., and Nieschlag, E. (1992). The role of growth hormone and insulin-like growth factor I in the regulation of male reproductive function. *Horm. Res.* **38**(Suppl. 1), 22–27.

Tazuke, S. I., and Guidice, L. C. (1996). Growth factors and cytokines in endometrium, embryonic development, and maternal: Embryonic interactions. *Sem. Reprod. Endocrinol.* **14**, 231–245.

Immunocytochemistry

William C. Okulicz

University of Massachusetts Medical School

I. Tissue Preparation and Fixation
II. Antibodies to Antigens of Interest
III. Direct and Indirect Immunocytochemical Methods
IV. Detection Systems
V. Immunocytochemistry Controls/Troubleshooting
VI. Other Applications: *In Situ* Hybridization
VII. Sample Protocol for Immunocytochemistry
VIII. Summary and Conclusions

GLOSSARY

antibody A specific immunoglobulin made in the host animal in response to foreign (antigenic) material, predominantly circulating γ immunoglobulins of the G-class, IgG, which recognizes and binds with high specificity to the antigen.

antigen Foreign material (most often proteinaceous) that enters an animal's body and elicits production of an antibody.

epitope The portion or structural element of an antigen that an antibody recognizes (binds).

The ability to produce highly specific antibodies to protein substrate targets of interest provides the basis for the immunological detection of their presence. Application of immunological recognition techniques has led to powerful quantitative and semi-quantitative assay systems, including radioimmunoassay, enzyme-linked immunoassay, and immunocytochemistry. These techniques have important implications for basic biological research and clinical diagnostic pathology. Important applications of immunocytochemical analysis include determination of tissue and cell type-specific expression of proteins, the presence of microorganism-specific antigens, and the presence of embryonic origin of cells. Immunocytochemistry is the visual detection of protein substrates of interest in cells and tissues by light microscopy.

I. TISSUE PREPARATION AND FIXATION

The initial steps in immunocytochemistry involve appropriate preparation of the specimen for presentation to immunological reagents. Paraffin or frozen tissue sections (4–8 μm), cytologic smears, or cell cultures grown on coverslips are the most frequently studied specimens for immunocytochemistry. Paraffin sections must be carefully dewaxed in xylene and rehydrated through descending ethanol washes (100, 95, 70, and 50%) and finally in phosphate-buffered saline (PBS). Frozen sections are thaw-mounted on glass slides. To ensure good section adhesion, specimens should be mounted on surfaces that promote attachment and adhesion, e.g., polylysine-coated slides. Proper fixation is necessary to retain the antigen(s) of interest within tissues/cells without blocking epitope sites or preventing accessibility of the antibody to antigen-specific sites. Good fixation must be balanced with the degree of morphology required for a particular application. Paraffin tissue sections provide better morphology than frozen tissue sections but suffer from inaccessibility or blocking of antigen epitopes, most often due to excessive fixation. Recently, a variety of antigen retrieval procedures have greatly improved antigen detection in paraffin and frozen sections. These approaches most commonly involve brief heat treatment of the tissue sections, most commonly a microwave oven. Although these techniques have greatly improved immunocytochemical analyses of archival and routinely processed formalin-fixed paraffin sections, frozen sections provide the best latitude for fixation-sensitive antigens. This is particularly true whenever specimens may be used in the future for immunocytochemical detection of antigens not currently known. Cytologic smears or cell cultures require a brief treatment with a detergent, e.g., 0.1% Triton X-100, during or after fixation to allow antibody penetration for detection of cytoplasmic or nuclear antigens.

Although physical means of fixation have been used, e.g., heat and freezing, chemical means of fixation have gained wide popularity because of their consistency and reproducibility for immunocytochemical analysis. A list of commonly used chemical fixatives, their composition, conditions of fixation, and applications are presented in Table 1. Because cell and tissue architecture is primarily determined by protein constituents, preservation of these components is essential. These protein constituents include lipoproteins (membranes), glycoproteins (collagen and basement membranes), cytoplasmic globular proteins, and nucleoproteins. Nucleic acid preservation is also very important for many applications. All fixatives listed in Table 1 will preserve protein constituents and nucleic acids. Preservation of lipids is more problematic since agents that solubilize them are often used.

The appropriate fixative and length of fixation must be determined for the antigen of interest. Specimens embedded in paraffin are often fixed in formalin (37–40% by weight concentration of formaldehyde gas), but the most commonly used fixative for

TABLE 1
Commonly Used Fixatives for ICC

Fixative	Concentration (%)	Temperature	Time	Fixation	Action
Formaldehyde	2–4	4°C to RT[a]	15 min to 24 hr	Somewhat rapid	Cross-linking
Alcohols	70–100	−5°C to −20°C	10 min to 30′	Very rapid	Water displacement
Ethanol					
Methanol					
Acetone					
Picric acid	0.5–5	4°C to RT	15 min to 24 hr	Slow	Protein salts (precipitates)

[a] RT, room temperature.

frozen sections or cells is freshly prepared paraformaldehyde (solid form of formaldehyde), which deteriorates during storage to result in poor fixation. Formaldehyde cross-links proteins, primarily via methylene bridges between ε-amino groups of lysine and amide groups in peptide linkages. Although most lipids are preserved by formaldehyde fixation, carbohydrates and glycogen are often lost. Alcohols displace water and denature most proteins. Nucleic acids and lipids remain soluble and most will not be retained. Ethanol and methanol preserve glycogen, whereas acetone does not. The dramatic displacement of water by these agents also leads to considerable shrinkage of the specimen. Despite these drawbacks, the rapid fixation and simplicity of alcohols is useful in many applications. Fixatives are often combined with other agents that offset their deleterious effects on cell/tissue architecture.

Picric acid forms protein salts (picrates) which precipitate. Because of the low pH of picric acid, it hydrolyzes DNA and RNA to make the specimen unsuitable for studies of nucleic acids. It is also a dangerous explosive and must be stored and maintained under water and handled with great care. Picric acid is rarely used as the sole fixative, similar to alcohols, and is most often combined with other fixatives. The most common combination is Bouin's solution, a combination of picric acid, formaldehyde, and alcohol.

The ultimate fixation allows preservation of specimens in their natural state. Unfortunately, intrinsic chemical characteristics of fixatives often alter some physical property of the specimen, e.g., shrinkage, swelling, or excessive hardening. Mixtures of the fixatives shown in Table 1 and/or sequential fixation steps are commonly used to provide the best preservation of the specimen and the antigen of interest. The tissue source of the specimen, the cellular localization of the antigen of interest, and the degree of morphology required will determine the fixative of choice, temperature of fixation, and concentration of the fixative. Unless information is available in the literature on the antigen and/or tissue of interest, the appropriate parameters must be determined empirically. If no information is available, a good starting point for frozen tissue sections would be 4% paraformaldehyde for 15 min to 1 hr at room temperature. For cell smears or cultures, 2% paraformaldehyde containing 0.1% Triton X-100 at a similar time and temperature of fixation will generally provide adequate fixation and morphology.

II. ANTIBODIES TO ANTIGENS OF INTEREST

Antibodies (immunoglobulins) are the essential ingredients in immunological reactions for immunocytochemical detection methods. Antibodies elicited in an animal in response to foreign material are predominately of the G-class of γ immunoglobulins (IgGs). The IgG molecule is composed of two long chains [heavy (H)] joined by a disulfide bond (Fig. 1). The distal end of each of these two H chains contains two light chains that are also joined by disulfide bonds. These structural characteristics result in the formation of a Y junction at the distal end of the molecule. These circulating antibodies are also

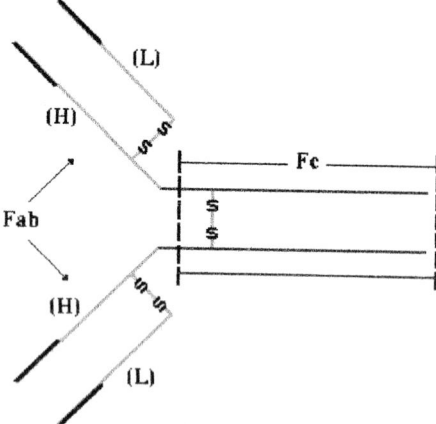

FIGURE 1 Diagrammatic representation of an IgG antibody. The antibody is composed of two heavy (H) chains and two light (L) chains joined by a disulfide link (–S–S–). The two heavy chains are joined by a disulfide link in the Fc region (constant protein sequence for a given species). The Fab regions (outside the Fc region) contain variable regions (thick lines) at their ends that provide specificity to an antigenic epitope (prepared by Eric Merithew).

cations from many commercial vendors. Although antibody specificity for the species under study may be demonstrated, the antibody may not necessarily be appropriate for immunocytochemistry. For example, cross-reactivity tested only in neutralization or immunoblotting procedures may not be applicable to immunocytochemistry. Monoclonal antibodies which often perform well in the previous applications will not necessarily be useful for immunocytochemical analyses. This may occur despite the exquisite specificity of monoclonal antibodies for a particular antigen. In this case, polyclonal antibodies which recognize multiple epitopes on the antigen are an important alternative. The use of polyclonal antibodies may, however, result in increased background staining due to the wide spectrum of epitopes recognized. The choice of monoclonal or polyclonal antibodies must usually be determined empirically unless there is previously published work in the same species for use of a specific antibody.

functionally composed of a constant (Fc) region and two arms (Fab regions). The Fc region is an invariant protein for a given species. Antibodies (secondary) to species-specific IgGs recognize the Fc region of these immunoglobulins. The Fabs are variable regions that specifically recognize a cognate epitope in a specific antigen. The two Fab arms can interact with a specific epitope, and therefore the IgG antibody can recognize two antigenic domains (the antibody is bivalent). These structural and functional properties of the IgG molecule provide the basis for the application and development of various immunocytochemical methods.

There are available literally thousands of antibodies for a myriad of antigens from individual laboratories and commercial sources. There are several resources for identification of an antibody for an antigen of interest. These include *Linscott's Directory of Immunological and Biological Reagents* (Mill Valley, CA), which is updated annually, and The Antibody Resource Page website (*http://www.antibodyresource.com*). The latter source allows the user to search a number of antibody directories for product specifi-

III. DIRECT AND INDIRECT IMMUNOCYTOCHEMICAL METHODS

A. Direct Method

Although the availability of specific antibodies to protein targets is essential for immunocytochemistry, it is necessary to visually detect their location. The first detection system used successfully was fluorescence. This direct method (Fig. 2A) involves labeling (conjugation) the primary antibody with a fluorescent label that emits at a specific wavelength (color) when excited at a different but specific wavelength. Subsequently, in the late 1960s, enzyme-linked primary antibodies were introduced that permitted deposition of a chromogenic substrate at the site of enzyme activity. Horseradish peroxidase (HRP) was the first immunoenzyme used to label primary antibodies and it remains an important component in the arsenal of detection systems currently available. These initial approaches validated the usefulness of immunological techniques for localization of specific proteins in tissues and cells. The direct method suf-

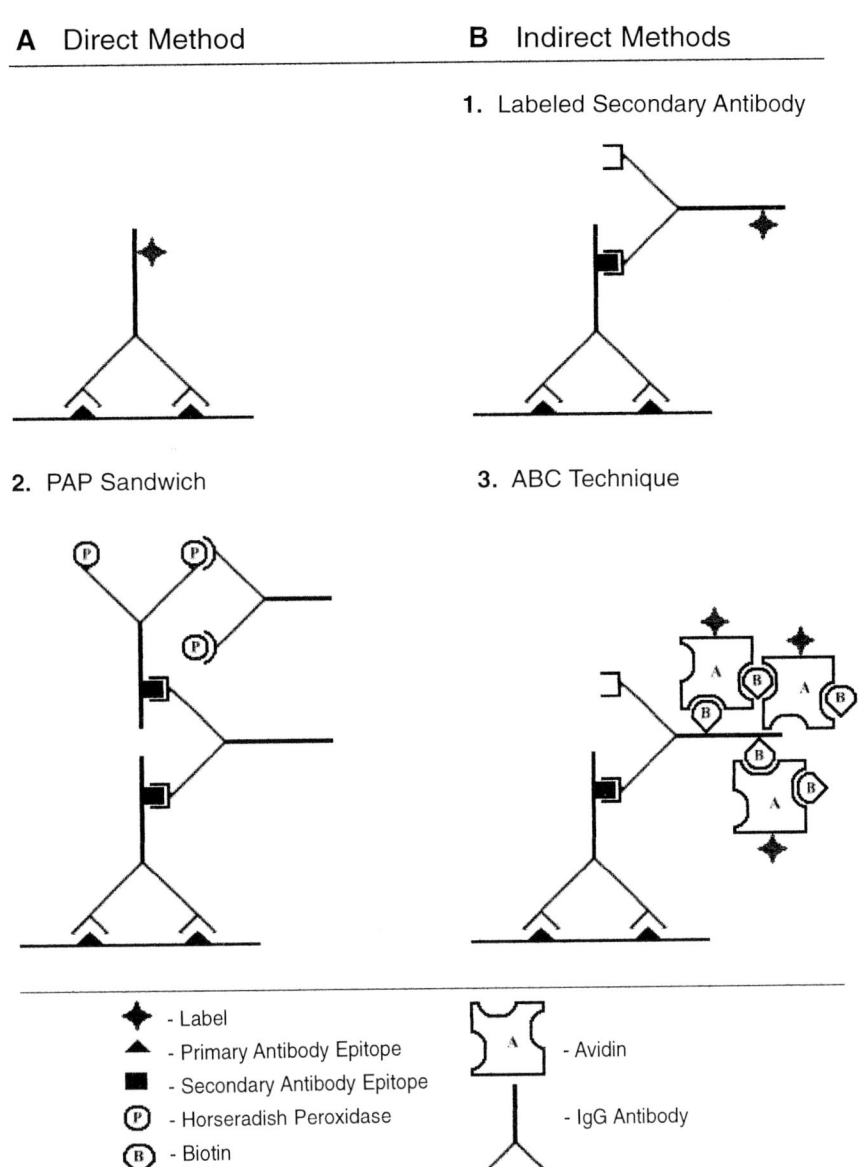

A Direct Method   **B** Indirect Methods

1. Labeled Secondary Antibody

2. PAP Sandwich   **3.** ABC Technique

◆ - Label
▲ - Primary Antibody Epitope
■ - Secondary Antibody Epitope
Ⓟ - Horseradish Peroxidase
Ⓑ - Biotin

Ⓐ - Avidin

- IgG Antibody

FIGURE 2 Pictorial representation of basic immunocytochemical techniques. Although the direct method is rarely used, there are numerous variations of the sandwich and ABC technique. See text for further details on all the techniques depicted here (prepared by Eric Merithew).

fered because of the necessity to label every primary antibody and the method's limited sensitivity. Indeed, labeling of primary antibodies, as a means of detection, has been replaced by approaches that permit an increase in label concentration at the site of the primary antibody with appropriate secondary antibodies and other reagents.

B. Indirect Methods

1. Labeled Secondary Antibodies

The direct detection method requires that the primary antibody directed against the antigen of interest carries the label or means for chromogenic detection (Fig. 2A). Sensitivity was improved, however, using an indirect technique by the application of a species-specific secondary antibody directed against the primary antibody. The secondary antibody is conjugated with a fluorescent or enzymatic label and avoids the difficulty of obtaining labeled and purified primary antibodies (Fig. 2B, 1). This approach improved sensitivity and eliminated the difficult task of labeling and purifying precious primary antibodies. In addition, HRP- or fluorescent-conjugated species-specific antibodies are now readily available from commercial sources.

2. Peroxidase-Antiperoxidase Technique

The indirect detection method was further developed by the introduction of a horseradish peroxidase antiperoxidase (PAP) procedure by Sternberger. This approach also uses a species-specific secondary antibody but this antibody is unlabeled and serves as a "bridge" for a tertiary antibody (Fig. 2B, 2). The tertiary antibody is an antibody made to HRP in the same species as the primary antibody. Thus, the unlabeled secondary antibody recognizes and binds to both the primary and tertiary antibodies with one of the two arms of its recognition domains to form a "sandwich." Additional bridges can form to increase the ratio of HRP moieties to primary antibody, resulting in greater sensitivity. In addition, background is generally improved since secondary antibodies not specific for the primary antibody and which bind to other tissue antigens will not bind PAP.

3. Avidin-Biotin Complex Technique

The avidin-biotin complex (ABC) technique has become one of the most commonly used methods for immunocytochemical detection of antigens in tissues or cells (Fig. 2B, 3). The technique uses a bridging antibody similar to the PAP procedure to amplify the concentration of chromogenic reporters at the site of antibody-primary antigen interaction. The secondary, or bridging, antibody is also species specific for the animal in which the primary antibody was made. Instead of a label or a binding site for a tertiary antibody (PAP), this secondary antibody is biotinylated. Avidin (streptavidin and avidin D) possesses up to four very high-affinity binding sites for biotin ($K_D = 10^{-15}$ M) that allow very tight binding

of the biotin moieties conjugated to the secondary antibody.

Avidin is readily available and can be conjugated to numerous labels, e.g., FITC, HRP, Texas Red, and alkaline phosphatase. In addition, labeled avidin conjugates are readily available commercially. The addition of labeled avidin–biotin complexes following binding of a biotinylated species-specific antibody results in an amplification of label at the initial site of antigen–antibody interaction. Not only can the labeled ABCs bind to several biotin moieties on the biotinylated secondary antibody but also, because of the number of biotin binding sites, they can bind to other ABCs that possess available biotin interaction sites (Fig. 2B, 3). The resultant amplification provides considerable improvement in sensitivity over the PAP procedure.

Although the sensitivity of this procedure is excellent, there is a caveat as a result of this sensitivity. Some tissues or cells contain endogenous biotin-binding sites that may lead to increased background staining. This can in part be overcome by preincubation of the specimen with biotin blocking reagents.

IV. DETECTION SYSTEMS

There are two basic labels that permit detection of an antigen–antibody interaction: fluorescence and enzymatic (Table 2). The method of choice depends on the intended application of the investigator. A few guidelines can be used to help identify the better approach. Immunofluorescent detection is highly specific, relatively easy to perform, and can be visually dramatic, particularly when applied in double-staining procedures. Morphological details of the specimen are, however, often lost due to the incompatibility of counterstains that may autofluoresce. In addition, fluorescence fades during exposure and storage, necessitating photographic documentation of the initial staining pattern. There are special mounting media, purported to reduce fading during storage or exposure, available from several commercial sources. Enzymatic methods of detection provide fewer of these drawbacks. Counterstaining of specimens is straightforward and staining is most often permanent and permits reexamina-

TABLE 2
Commonly Used Fluorescent and Enzymatic Labels for ICC

Label	Excitation (nm)	Emission (nm)	Color
Fluorescent labels			
Fluorescein	495	515	Green
Texas Red	595	615	Red
Rhodamine red/orange	550	575	
Phycoerythrin red/orange	450–570	574	
Coumarin (AMCA)	350	450	Blue
Enzymatic labels			
	Substrate/chromogen		
Horseradish peroxidase	Hydrogen peroxide\DAB		Brown
	Hydrogen peroxide\(+Ni)		Gray/black
	Hydrogen peroxide\AEC		Red
Alkaline phosphatase	BCIP\NBT		Blue/violet
Glucose oxidase	Glucose\NBT		Purple/blue

Note. Abbreviations used: AMCA, 7-amino-4-methylcoumarin-3-acetic acid; DAB, 3,3'-diaminobenzidine; AEC, 3-amino-9-ethylcarbazole; BCIP, 5-bromo-4-chloro-3-indolyl phosphate; NBT, nitoblue tetrazolium.

tion and analysis of immunostained specimens. Double-immunostaining is not, however, straightforward unless the two antigens of interest are in different cells or in different regions of a cell. Semiquantitive analysis of staining is often problematic because of the difficulty of controlling the multiple steps required in these procedures. Fluorescent labels require an appropriately equipped microscope and specific filter sets that permit excitation (absorption) and emission in an indirect manner (specimen emits epifluorescence). With the appropriate equipment and fluorescent conjugates, the procedures are fairly straightforward.

Fluorescein is the most common fluorescent label and various conjugates with secondary antibodies or avidin are available commercially. Minor autofluorescence of specimens at the excitation wavelength of fluorescein allows easy localization of cells/sections on slides. Texas Red is an increasingly popular fluorescent label due to its intensity, low background staining, and compatibility with the fluorescein excitation/emission parameters for double-immunofluorescent staining. Rodamine and phycoerythrin conjugates are similar to those of

Texas Red and can also be used for dual staining with fluorescein. AMCA (coumarin) provides an intense blue fluorescence with little autofluorescence and can be used in combination with other fluors in dual-staining procedures. Although the previously discussed fluors constitute the majority of fluorescent labels in use, there are many other fluors for specialty applications (see Haugland, 1996).

Horseradish peroxidase is the benchmark for immunocytochemical detection by enzymatic chromogenic deposition. Its enzymatic action has been studied in detail and this information has been used to identify new chromogens for enhanced sensitivity and contrast. The substrate for HRP is hydrogen peroxide, and in the presence of electron donors such as 3,3'-diaminobenzidine (DAB), it forms a complex which subsequently dissociates, leaving the enzyme and a brown precipitate. In the presence of other agents, such as nickel or AEC, different colored precipitates will be formed (gray/black and red, respectively). There are additional modifications from commercial sources for enhancement of staining intensity and color deposition. The use of biotinylated tyramine, for example, increases the sensitivity of

HRP detection up to 1000-fold (New England Nuclear, Boston). A drawback to the use of HRP is the potential for background staining from endogenous peroxidase, particularly tissue samples containing red blood cells, with high peroxidase activity. Endogenous peroxidase activity can be quenched by pretreatment of the specimen with 3% hydrogen peroxide in methanol.

Alkaline phosphatase, glucose oxidase, and β-galactosidase make use of their enzymatic properties to reduce tetrazolium salts to insoluble compounds at the site of their localization. Because phosphatases are present in various proportions in tissues and cells, they may present a potential background problem. To eliminate background phosphatase activity, specimens are usually pretreated with levamisole, an inhibitor of phosphatase activity. Glucose oxidase, on the other hand, is not a mammalian enzyme and presents no background difficulties. The action of glucose oxidase is dependent on the oxidation of glucose and the coreduction of tetrazolium salts.

V. IMMUNOCYTOCHEMISTRY CONTROLS/TROUBLESHOOTING

Because of the number of procedural steps that are inherent to immuncytochemical techniques, controls are essential to confirm the specificity of immunostaining. A first control step to avoid problems should be careful dilution of the primary antibody to determine the highest dilution of antibody that provides adequate staining. This is not only cost-saving but also will detect and limit potential background staining due to the primary antibody. In a dilution series large changes in antibody dilution may result in negligible differences in staining intensity. Other common controls and approaches include the following:

1. Incubation of the specimen with PBS alone (no primary antibody) to control for all other steps in the procedure, i.e., background staining due to secondary antibodies, autofluoresence, endogenous enzyme activity, or endogenous biotin binding sites. If background staining is present with omission of the primary antibody, selective omission of each step should identify the culprit. Appropriate dilution or substitution, e.g., another source of secondary antibody, or a more purified form of the reagent may overcome background problems while maintaining antigenic staining. Blocking of endogenous peroxidase (3% hydrogen peroxide in methanol) or phosphatase (levamisole) may be required. Endogenous biotin binding sites can be blocked by biotin-blocking reagents (Vector Laboratories, Burlingame, CA). The use of a blocking serum (from the species in which the secondary antibody was elicited) may paradoxically contribute to nonspecific staining. Deletion of this conventional step may eliminate background staining.

2. Establish specificity by substitution of the primary antibody with a comparable dilution of antisera or appropriate immunoglobulin concentration (μg/ml protein) for the primary antibody. Negative staining will confirm specificity of the antibody.

3. Replace the primary antibody with an antibody not found in the cells/tissue, e.g., an antibody to a plant antigen not present in mammalian tissues when studying mammalian tissues. This approach further confirms antibody specificity.

4. If available, use a known positive cell or tissue for the antigen of interest. This will provide additional support for the usefulness and applicability of the chosen primary antibody. Preimmunoadsorption of the primary antibody with purified antigen (if available) and subsequent incubation with the tissue should significantly decrease the staining signal.

VI. OTHER APPLICATIONS: *IN SITU* HYBRIDIZATION

Traditional *in situ* hybridization to identify and localize specific nucleic acid sequences, most commonly mRNAs, has relied on radiolabeled nucleic acid probes for detection. This approach provides the highest sensitivity for the detection of single-copy gene products and rare RNAs. Radioactive waste disposal, contamination, and safety issues are drawbacks to these techniques. Nonradioactive labeling and detection of nucleic acid probes have gained acceptance because of new methods and reagents for incorporation of several haptens, such as biotin, digoxigenin, and fluorescein, in relatively high numbers into nucleic acid probes. Highly specific antibodies to these agents are readily available commercially and the presence of nucleic acid probes containing these molecules can be determined by indirect immunocytochemical techniques. Incorporation of these haptens transforms the nucleic acid probe into a molecule possessing specific antigenic domains. A description of the many labeling procedures available, is provided in some of the references listed in the bibliography.

Radiolabeled probes are analogous to labeled primary antibodies in the direct method of immunocytochemical detection. After appropriate preparation of the tissue/cells for *in situ* hybridization, a nonradioactive-labeled (antigenic) probe serves as a "primary antibody" and indirect immunocytochemical techniques are used to identify its presence and location. For example, labeled "secondary" antibodies to the incorporated hapten are used to visualize their presence by fluorescent or enzymatic techniques. In addition, amplification techniques such as the labeled avidin–biotin complex give comparable sensitivity to that of radiolabeled probes for several mRNAs.

VII. SAMPLE PROTOCOL FOR IMMUNOCYTOCHEMISTRY

Our laboratory has published several reports on the immuncytochemical analysis of the estrogen receptor (ER), progesterone receptor (PR), and the Ki-67 antigen (a marker of proliferation) by the HRP ABC technique. A typical protocol is shown below.

Basic reagents
  PBS: pH 7.2
    9 g NaCl (Sigma No. S-9888)
    1.3 g Sodium phosphate dibasic anhydrous (Sigma No. S-0876)
    4.0 g Sodium phosphate monobasic anhydrous (Sigma No. S-0751)

q.s. (quantum sufficient) to 1 liter final volume with glass distilled water (GDW)
4% Paraformaldehyde: pH 7.2
4 g Paraformaldehyde/100 ml PBS (7.2) (Fisher No. T353)
  Mix and heat to 60°C
  Add NaOH (5 *M*) drop by drop until clear
  Filter
  Cool, pH (7.2)
DAB (Sigma Chemical–ISO PAC No. 9015): light sensitive
  Stock: Reconstitute ISO PAC with 20 ml 0.1 *M* Tris (Fisher No. BP152) (pH 7.2) for final con   centration of 5 mg/ml
  Mix well and vacuum filter
  Aliquot 200 μl/microfuge tube
  Store at −20°C
  For applications, dilute as follows:
    800 μl Tris HCL (0.1 *M*, pH 7.2)
    2 μl 10% hydrogen peroxide (0.25% final) (Sigma No. H-1009)
    200 μl DAB stock

DAB is a suspected carcinogen; therefore, handle with gloves and inactivate solutions (working solution and GDW wash) with bleach before disposal.

VIII. SUMMARY AND CONCLUSIONS

Immunocytochemical techniques are used widely in the scientific and medical community for basic and diagnostic applications, e.g., identification and localization of proteins and detection of specific nucleic acids. The development of these techniques over the past four decades has focused on improvements in sensitivity and the enhancement of detection systems to localize smaller amounts of antigen. Further improvements and refinements will lead to even more sensitive, faster, and more reliable approaches.

See Also the Following Article

Immunology of Reproduction

Bibliography

Gosselin, E. J., Cate, C. C., Pettengill, O. S., and Sorenson, G. D. (1986). Immunocytochemistry: Its evolution and criteria for its application in the study of epon-embedded cells and tissue. *Am. J. Anat.* **175**, 134–160.

Haugland, R. P. (Ed.) (1996). *Handbook of Fluorescent Probes and Research Chemicals*, 6th ed. Molecular Probes, Eugene, OR.

Kiernan, J. A. (1981). *Histological and Histochemical Methods.* Pergamon, New York.

Login, G. R., and Dvorak, A. M. (1994). Methods of microwave fixation for microscopy. A review of research and clinical applications: 1970–1992. *Prog. Histochem. Cytochem.* **27**, 1–127.

Okulicz, W. C., and Balsamo, M. (1993). A double immunofluorescent method for the simultaneous analysis of progesterone-dependent changes in proliferation (Ki-67) and the estrogen receptor in the rhesus endometrium. *J. Reprod. Fertil.* **99**, 545–549.

Okulicz, W. C., Balsamo, M., and Tast, J. (1993). Progesterone regulation of endometrial estrogen receptor and proliferation during the late proliferative and secretory phase in artificial menstrual cycles in the rhesus monkey. *Biol. Reprod.* **49**, 24–32.

Polar, J. M., and Van Noordan, S. (1983). Immunocytochemistry. In *Practical Applications in Pathology and Biology* (P. S. G. Wright, Ed.). Littleton, MA.

Shi, S.-R., Gu, J., Kalra, K. L., Chen, T., Cote, R. J., and Taylor, C. R. (1995). Antigen retrieval technique: A novel approach to immunohistochemistry on routinely processed tissue sections. *Cell Vision* **2**, 6–22.

Sternberger, L. A. (1986). *Immunocytochemistry*, 3rd ed. Wiley, New York.

Zeller, R., Rogers, M., and Watkins, S. (1992). In situ hybridization and immunohistochemistry. In *Current Protocols in Molecular Biology* (F. M. Ausubel, R. Brent, R. E. Kingston, D. D. Moore, J. G. Seidman, J. Smith, and K. Struhl, Eds.). Wiley, New York.

Immunology of Reproduction

Joan S. Hunt
University of Kansas School of Medicine

Peter M. Johnson
University of Liverpool Faculty of Medicine

I. Introduction
II. Basic Features of the Immune System
III. Gamete Immunology
IV. Mucosal Immunity
V. Immunology of Implantation and Pregnancy
VI. Immune Defense in the Fetus
VII. Summary and Future Initiatives

GLOSSARY

antibodies Antigen-specific products of B lymphocytes that bind to and facilitate destruction of antigens.

antigens Substances that are foreign to the host and induce immune responses.

B cells Lymphocytes that produce antibodies following encounter with foreign substances (antigens).

cytokines Soluble or cell-bound polypeptides that mediate cell–cell communication and modulate gene expression via binding to membrane receptors and activation of specific intracellular pathways.

immune privileged sites Tissues and organs that forbid the entrance of immune cells. These include the ovary, testes, uterus, and placenta.

mammalian major histocompatibility complexes (MHC) These contain genes coding for the transplantation antigens. In humans these are called *human leukocyte antigens*.

T cell receptors Multiunit structures on T lymphocytes that recognize antigenic peptides within the context of MHC antigens.

T cells Multifunctional lymphocytes that can be subdivided into helper (Th) cells identified by the presence of CD4 on their surfaces and cytotoxic T lymphocytes (CTLs) identified by CD8 on their surfaces.

Th cells These can be functionally subdivided into Th1, which produce proinflammatory cytokines and help CTLs, and Th2, which produce antiinflammatory cytokines and help B cells.

I. INTRODUCTION

The reproductive and immune systems interact extensively by cell to cell contact and exchange of soluble mediators throughout embryonic and adult life. Early reproductive immunologists focused primarily on understanding and alleviating the interactions between these two systems that had a detrimental impact on fertility and pregnancy. For example, the underlying causes of hemolytic disease of the newborn caused by Rh factor incompatibility between the mother and the baby (*erythroblastosis fetalis*) were identified and an effective treatment was devised. This pathological condition highlights a central conundrum: How might the embryo, which expresses paternal chromosomes coding for substances foreign to the mother, be protected from destruction by the maternal immune system? An early transplantation immunologist, Sir Peter Medawar, was the first to articulate the immunological paradox posed by the success of the fetal semiallograft and to suggest potential mechanisms. These included anatomical separation between the mother and the fetus, antigenic immaturity of the fetus, and immunological tolerance in the mother. It now appears that all the features suggested by Medawar, as well as a number of others, combine effectively to prevent immune rejection, and pregnancy is most often successful despite genetic differences between the mother and fetus. Furthermore, evidence is mounting for the idea that some genetic disparity has beneficial effects, particularly regarding prevention of maternal microchimerism.

Although research in reproductive immunology has been driven by a need to understand how pregnancy goes forward in a potentially hostile environment, immunological aspects of gamete development, fertilization, implantation, delivery of passive immunity to the fetus, and development of the fetal immune system also represent important aspects of reproductive immunology. In this article, comments on these subjects will focus on humans and will include a limited discussion of the potential consequences of breakdown of normal mechanisms.

II. BASIC FEATURES OF THE IMMUNE SYSTEM

The following are intended to provide a framework for understanding specific and nonspecific host defenses. For further details the reader should consult a general immunology textbook.

A. Specific Immunity

Specific immunity in mammals rests on the effective functioning of its two distinct components, the humoral immune system and the cell-mediated immune system (Fig. 1). Antibodies produced by activated B lymphocytes comprise the effector mechanism in the first system and killer T lymphocytes (CTLs) comprise the effectors in the second. Helper T (Th) cells participate in the generation of both arms. The immune response is initiated when antigen-specific receptors (TcRs) on helper T cells encounter matching peptides displayed by antigen presenting cells (APCs), such as dendritic cells, macrophages, and B cells, within the context of major histocompatibility complex (MHC)-derived antigens. Cytokines are produced that assist lymphocytes to expand clonally into antibody-producing plasma cells and effector T cells. Some powerful substances such as lipopolysaccharide (LPS) from gram-negative bacteria bypass the requirement for TcRs and directly activate B and T cells by stimulating production of cytokines.

1. MHC Antigens and Recognition by Lymphocyte Subsets, CTLs, and Ths

Human MHC antigens, termed human leukocyte antigens (HLAs), are divided into two major classes, HLA class I and HLA class II (HLA-D). HLA class I

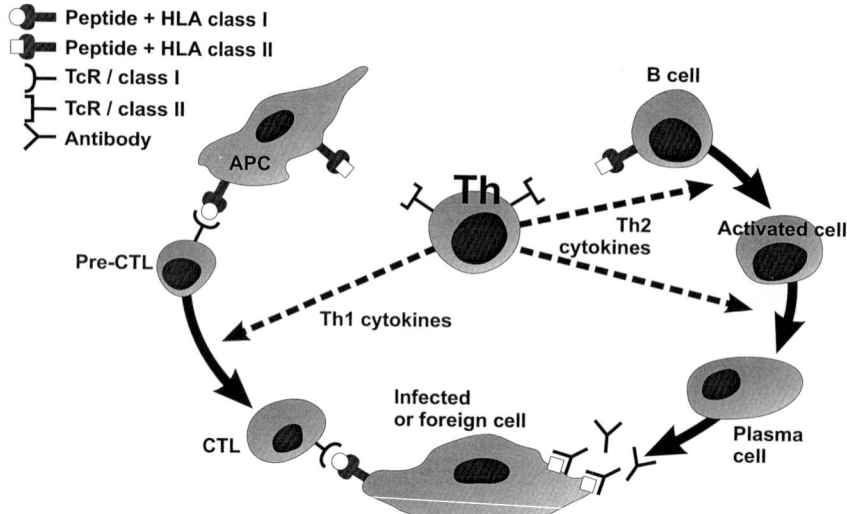

FIGURE 1 The generation of specific immunity includes development of cytotoxic T lymphocytes (CTL) from precursor T cells and maturation of B lymphocytes into plasma cells, which produce antibodies. Development of both cellular and humoral immunity is facilitated by helper T lymphocytes (Th), which recognize antigenic peptides within the context of HLA class II molecules. Th1-type and Th2-type cytokines promote CTL and B lymphocyte maturation, respectively.

is subdivided into the highly polymorphic class Ia antigens, which include HLA-A, -B, and -C, and the strikingly less polymorphic class Ib antigens, such as HLA-E and -G. HLA class I and class II antigens are recognized by different subpopulations of T cells. CTLs, which exhibit membrane CD8 and antigen receptors (TcRs) containing α/β chains, recognize antigenic peptides presented within the cleft of assembled HLA class I antigens. These cells are responsible for destroying altered, infected, and foreign cells. Th cells, which display CD4 and α/β TcRs, recognize antigenic peptides within the context of HLA class II antigens. Their major function is to assist in the maturation and generation of other lymphocytes. APC/TcR recognition is facilitated by secondary binding and signaling systems that include CD4 and CD8 as well as the antigen B7 on APCs and CD28 on T lymphocytes.

2. Cytokines and Th Subsets

The recognition phase is followed by synthesis of a host of cytokines, cellular binding via specific receptors, and transduction of signals in receptive cells. This stimulates clonal expansion of the lymphocytes. APCs secrete interleukin-1 (IL-1), which

activates Th cells. Th cells respond by synthesizing IL-2, an important autocrine cytokine which promotes proliferation and leads to production of other cytokines and growth factors. In mice, one subdivision of Th cells, the Th1 cells, produces primarily proinflammatory cytokines, such as IL-2 and interferon-γ (IFN-γ), that promote CTL development, whereas a second, the Th2 cells, produces antiinflammatory cytokines, such as IL-4 and IL-10, that promote antibody production. Th1 cytokines may inhibit Th2 cytokines and vice versa. Certain of these and other cytokines promote maturation of APCs and natural killer (NK) cells.

3. Tolerance and Immune Privilege

In two instances, antigen may not stimulate immune responses. The first instance is when tolerance has been established as occurs with self-antigens. During fetal and early postnatal life, thymocytes reacting to autologous antigens are deleted or immobilized. Tolerance may be prevented or overcome as a consequence of inflammation, by genetic defects, and during the aging process. The result is autoreactivity culminating in autoimmune diseases such as rheumatoid arthritis and system lupus erythematosis.

In the second case, T cells are killed before reaching the antigen. This is called immune privilege and is a remarkable property of certain organs and tissues in which lymphocyte reactivity could be detrimental. Immune privileged tissues have several common features, such as lack of vasculature and the presence of epithelial cell tight junctions. Recent studies in testis, eye, and the uterus and placenta suggest that in these sites, immune privilege is established at least in part by the expression of a cytotoxic molecule, Fas ligand (FasL, CD95L), on cells bordering blood and lymphatic spaces. FasL is a member of the tumor necrosis factor (TNF) gene family that delivers a death signal through interaction with its specific receptor, FasR (CD95). FasR is expressed at high levels on activated hematopoietic cells, including T and B lymphocytes, neutrophils, and monocyte/macrophages.

B. Nonspecific Host Defense

Non-antigen-specific host defense is provided by hematopoietic cells, including polymorphonuclear leukocytes, macrophages, and NK cells, and by a group of soluble proteins collectively referred to as the complement system.

1. Cellular Defenses

Hematopoietic cells mature in the bone marrow, circulate in the blood, and migrate into tissues in response to chemotactic stimuli. While polymorphonuclear leukocytes comprise the initial vigorous host response to injury and inflammation, these are short-lived cells and do not reside for long in normal tissues. By contrast, macrophages and NK cells are seeded into certain organs and tissues early in fetal life and their numbers are supplemented during injury. Both lineages are programmed by local environmental signals to differentiate *in situ* and perform highly individualized functions. Macrophages serve as APCs and secrete a wide range of cytokines, growth factors, and bioactive lipids. NK cells, which are related to lymphocytes, guard against aberrant cells lacking a normal complement of HLA antigens and exhibiting other abnormalities. As with T cells, NK cells display several types of receptors for HLA

class I, but in contrast to T cells, recognition of class I antigens may prevent rather than provoke killing through cytotoxic signals.

2. Complement

A second type of nonspecific host defense is provided by a series of interacting serum proteases termed the complement cascade. This cascade is usually initiated by antigen–antibody complexes and results in disruption of membrane integrity, release of the intracellular contents, and cell death. Complement also facilitates binding and phagocytosis. Host cells are protected by cell membrane expression of complement regulatory proteins, such as CD46, CD55, and CD59, which interrupt specific steps of the cascade.

III. GAMETE IMMUNOLOGY

Gametes exhibit novel antigens toward which the host has not developed tolerance. In consequence, germ cells must be continuously sequestered from hematopoietic cells in blood and tissues. Human oocytes are encased in avascular follicles composed of epitheloid granulosa cells surrounded by a thick basement membrane, and maturing sperm and nurse-like Sertoli cells are sequestered together in tubules that are also avascular and encased by a basement membrane. The follicles and tubules comprise immune privileged sites where residency of immune cells is forbidden. Recent experiments indicate that Sertoli cells can destroy any activated lymphocytes that might penetrate the tubule by killing through the FasL/FasR cytotoxicity pathway. It is not known whether granulosa cells have the same capability.

Female and male germ cells as well as fertilized eggs are protected from complement-mediated lysis by the display of complement regulatory proteins on their cell membranes. Although sperm also exhibit proteins foreign to the female and might therefore be susceptible to killing by immune cells in the female reproductive tract, such cells are scarce in the lumen of the oviduct and uterus. Furthermore, sperm do not exhibit HLA antigens, which are primary recognition structures for the cytotoxic antibodies and CTLs that mediate graft rejection.

Immunocontraception research has focused on germ cells and fertilized preimplantation embryos in the hope of developing effective but reversible methods for preventing pregnancy. These have not yet been successful although one approach, immunization against a trophoblast cell hormone, chorionic gonadotropin, is undergoing clinical trials.

IV. MUCOSAL IMMUNITY

Mucosal immunity in the female and male reproductive tracts is provided primarily by antibodies. In females, IgG- and IgA-producing plasma cells are present to a greater or lesser degree in all segments of the genital tract, with fewer in the endometrium than in the cervix and low numbers in the vagina. Seminal fluid also contains both classes of antibody, which may originate in the prostate or in the seminal vesicles. In both female and male reproductive tracts antibodies found in secretions are frequently serum transadates.

Local immunity is difficult to stimulate in the female reproductive tract; high concentrations of antigens and repeated immunizations are required. This reduced sensitivity to antigenic stimuli may protect against the development of immunity to sperm. Antisperm antibodies are well documented causes of infertility. Because sexually transmitted diseases are extremely common, major efforts are under way to improve local immunity.

V. IMMUNOLOGY OF IMPLANTATION AND PREGNANCY

Implantation requires that the fertilized egg, now developed to the blastocyst stage, breach the luminal epithelial cell layer and burrow into the endometrial stroma. This process is dependent on production of proteases by trophoblast cells derived from the trophectoderm layer of the blastocyst, resembles tumor invasion, and has a mild inflammatory effect. Implantation and the development of hemochorial placentas in humans and rodents potentially places the reproductive and immune systems into direct conflict.

Throughout gestation, layers of trophoblast cells intervene between the embryo and maternal cells, thus shielding the embryo from the maternal immune system. The barrier role of these unique cells is readily appreciated, but trophoblast also contributes importantly to its own protection in other ways that include modulation of HLA antigen expression as well as production of immunomodulatory cytokines and hormones (Table 1). The mother cooperates in the development of a nonthreatening environment by developing a new hematopoietic cell profile in the decidualized endometrium and by synthesizing protective substances. Furthermore, the Fas/FasL system is in position to prevent extensive trafficking of activated leukocytes between the mother and the fetus.

A. Trophoblast Cell Subpopulations

Cytotrophoblast cells serve as progenitor cells for all other human trophoblast cell subpopulations (Fig. 2). In early gestation, one subpopulation of these cells establishes direct cell to cell communication with maternal decidual cells, which have differ-

TABLE 1
Immune Protection at the Maternal–Fetal Interface[a]

	Compartment	
	Maternal	Fetal
Nonantigen-specific hematopoietic cells		
preferred	+	+
Cell membranes		
Regulation of HLA genes	0	+
Complement regulatory proteins	0	+
Fas ligand	+	+
Soluble molecules		
HLA-G	?	+
Interleukin-10	+?	+
Interleukin-4	0	+
Prostaglandin E$_2$	+	+
Progesterone	+	+
Transforming growth factor-β	+	+

[a] Restriction on residency of leukocyte subpopulations, unusual patterns of expression of membrane proteins, and production of various soluble substances are believed to protect against maternal immunological rejection of the fetal semiallograft.

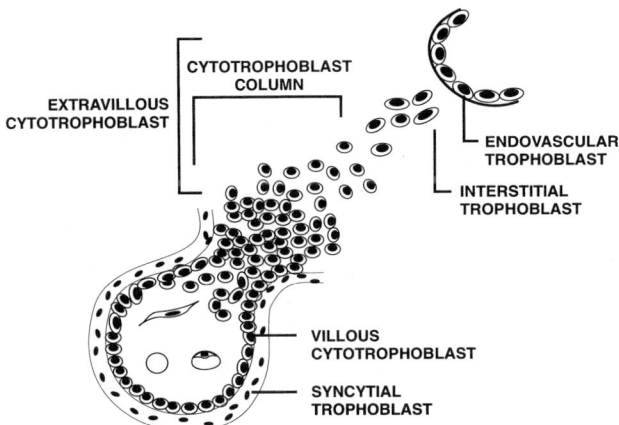

FIGURE 2 Schematic illustration of trophoblast subpopulations in first-trimester placentas. A villus composed of an outer layer of syncytialized trophoblast and an inner layer of cytotrophoblast cells is shown in cross section. Cytotrophoblast cells form a column that attaches to and invades maternal decidua. Individual cells (interstitial trophoblast) intermingle with decidual cells, and some cells (endovascular trophoblast) replace the endothelial cells of the spiral arteries (reprinted with permission from R. G. Landes, Austin, TX).

entiated from endometrial stromal cells under the influence of progesterone. A second subpopulation forms an uninterrupted cell layer that is continuously exposed to maternal blood. These two subpopulations are termed the extravillous cytotrophoblast cells and the syncytiotrophoblast layer, respectively. Figure 2 illustrates some additional, anatomically defined subpopulations. As gestation progresses, the supply of progenitor cytotrophoblast cells is exhausted and syncytiotrophoblast comprises the main population in the placenta proper, whereas extravillous trophoblast cells regress to form the chorion membrane. A basement membrane and mesenchyme containing fibroblasts and macrophages underlie the trophoblast cells. In the placental mesenchyme, fetal blood vessels feeding into the umbilical cord and embryo are also prominent.

B. Immunoevasive Strategies of Trophoblast Cells

Extravillous trophoblast cells, which are in intimate juxtaposition with decidual cells, and syncytio-

trophoblast, which is continuously washed in maternal blood, use similar but not identical means to protect against maternal immune mechanisms. For example, both subpopulations of trophoblast cells express high levels of the complement regulatory proteins on their cell membranes, which is believed to prevent complement-mediated lysis, but expression of individual proteins (CD46, CD55, and CD59) is not the same. All trophoblast cells also tightly regulate cell membrane expression of HLA antigens, but different pathways are used in extravillous and syncytiotrophoblast cells as required by their environments.

In addition to regulating membrane proteins, trophoblast cells produce soluble substances such as progesterone, prostaglandins (particularly prostaglandin E$_2$), and immunomodulatory cytokines that establish an immunosuppressive environment within which lymphocyte proliferation is discouraged.

1. Regulated Expression of HLA Genes

Products of the HLA genes were first named "transplantation antigens" because they serve as the major targets for graft-rejecting CTLs and cytotoxic antibodies. Failure of trophoblast cells to express HLA antigens that the mother could recognize as foreign is believed to protect against killing by maternal immune cells.

Trophoblast cells do not express HLA class II antigens and cannot be induced to do so by either demethylating agents or activating agents such as IFN-γ. Therefore, a major system for amplifying immune responses is absent. Trophoblast cells may or may not express HLA class I antigens. Neither syncytiotrophoblast nor the progenitor villous cytotrophoblast cell subpopulation exhibit cell membrane class I antigens. Repression has been traced to failure of synthesis of certain transcription factors and reduced transcription of the genes coding for processing components such as the transporter for antigen processing-1 protein. By contrast, extravillous cytotrophoblast cells express high levels of cell membrane antigen encoded by the HLA-G gene, which has few polymorphic variants. Recent experiments suggest that these cells may also exhibit an HLA-C-like molecule. The class I antigens protect against lysis by decidual large granular endometrial lymphocytes (LGLs) and probably also facilitate binding of growth factors such as insulin.

HLA-G mRNA has at least five alternative splice sites and is synthesized in both membrane and soluble forms. The functions of soluble HLA-G remain to be discovered but could include blocking of receptors as well as induction of apoptosis in CD8$^+$ CTL.

2. Uteroplacental Cytokines

A wide range of cytokines and growth factors are produced in the pregnant uterus. Synthesis is not limited to Th cells; both immune cells and nonimmune cells contribute cytokines, and placental trophoblast cells are particularly prominent sites of production. Both Th1- and Th2-type cytokines are produced at the maternal–fetal interface but the cellular sites and kinetics of production relative to stage of gestation differ. Proinflammatory, Th1-type cytokines, such as TNF and IL-1β, produced by immune cells are most prominent in early postimplantation tissues and near parturition. These cytokines may assist in maternal recognition of pregnancy and timing of parturition. The same cytokines are produced by nonimmune cells at low levels throughout pregnancy and may have growth- and differentiation-regulating functions.

Experiments in mice first suggested that Th2-type cytokines dominate over Th1-type cytokines at the maternal–fetal interface. There is also evidence for this phenomenon in human pregnancy. Antiinflammatory, Th2-type cytokines, such as IL-4, IL-10, and transforming growth factor-β (TGF-β), appear to be synthesized throughout gestation in both mice and women. The local effect and, to a limited extent, the systemic effect is to diminish proinflammatory conditions and actions of CTLs while promoting humoral immunity.

C. Protection in the Decidua

Human cycling endometrium contains a variety of hematopoietic cells that include classical T and B cells assembled into lymphoid aggregates, macrophages, and LGLs. The LGLs are similar in many respects to rodent uterine granulated metrial gland cells, which comprise a subset of NK cells and are now referred to as uterine NK (uNK) cells. T and B cells become scarcer as the menstrual cycle progresses through the secretory phase, whereas macrophages are slightly increased and LGLs are dramatically increased.

Following implantation, the cellular milieux of the decidualized endometrium is dramatically altered; classical lymphoid cells are difficult to locate, whereas cells mediating non-antigen-specific host defense, macrophages, and LGLs are abundant. This may reflect changes in vascular adhesion molecules, chemotactic substances, or endometrial extracellular matrices or may be due to increased production of substances such as prostaglandin E$_2$ that have inhibitory effects on lymphocytes. Macrophages remain a stable subpopulation throughout pregnancy, comprising approximately 10% of all cells in the decidua. Their functions are known to include phagocytosis and protection against infection as well as production of cytokines such as TNF and IL-1β, prostaglandins, and proteases. Presumably, these contribute to the integrity and well-being of the maternal–fetal interface. However, macrophages are activated by LPS and other mediators of inflammation to produce exceedingly high levels of cytokines, and this is believed to contribute to infection-associated preterm labor.

In contrast to macrophages, LGLs are exceedingly numerous in early gestation, comprising more than 40% of decidual cells, but their numbers decline thereafter. Although there appears to be a lineage relationship between LGLs and NK cells, the LGLs express an array of cell surface markers (CD56bright, CD3-, and CD16-) that are different from classical NK cells in blood. Also, the cells do not kill normal NK targets unless deliberately activated by IL-2. It is therefore of importance that IL-2 is entirely absent from the cytokine milieux of the pregnant uterus, nor do the LGLs kill migrating trophoblast cells; this appears to be prohibited by the presence of HLA-G and possibly also HLA-C on the extravillous trophoblast cell membranes. However, the presence of killer substances, such as perforin and TNF, in LGL granules suggests that the cells may have a role in defending against infected and/or abnormal cells, and recent studies on their counterparts in transgenic mice indicate that they may supply factors essential to placental development such as nitric oxide to relax smooth muscle of the arterial bed.

D. Immune Privilege

Migratory cells, macrophages, and LGLs are the dominant hematopoietic cell subpopulations in decidua during pregnancy. However, trafficking of cells between the mother and the fetus is limited. Extensive chimerism in pregnant women may be prevented by maternal recognition and destruction of HLA-expressing fetal cells. Bidirectional migration of activated leukocytes may be restricted by FasL, which has been identified on mouse and human trophoblast cells and, in mice, on decidual cells. Whether fetal or maternal cells that escape killing mechanisms and establish microchimerism have long-term effects on immunity remains to be determined.

E. Systemic Modulation of the Maternal Immune System

Although the immune alterations that occur during pregnancy are most profound at the maternal–fetal interface, systemic changes have also been noted. Maternal antipaternal T lymphocytes are difficult to identify and there are reports of modest changes in the proportions of T cell subsets in maternal blood. Autoimmune diseases dominated by cell-mediated immunity such as rheumatoid arthritis are usually alleviated during pregnancy. By contrast, antibodies to fetopaternal HLA antigens are abundantly produced and no overall diminution in the vigor of maternal humoral immune responses is seen. Autoimmune diseases dominated by antibody formation such as systemic lupus erythematosis are exacerbated during pregnancy. Although this suggests a bias against systemic cell-mediated immunity during pregnancy (which could be due to the predominance of Th2-type cytokines), it does not cripple maternal host defenses; women and pregnant rodents are fully capable of rejecting skin grafts and killing intracellular pathogens.

VI. IMMUNE DEFENSE IN THE FETUS

The fetus is vulnerable to infection because of immaturity of its immune system. As pregnancy progresses, the fetal immune system undergoes many developmental and maturational steps that occur in a predictable pattern. For example, weak expression of HLA class I antigens is typical of cells in first-trimester fetal tissue, whereas class II antigens are undetectable. During the second trimester, HLA class I antigens increase and class II antigens appear. Cord blood lymphocytes taken at different stages of pregnancy show progressive immune competence in terms of their ability to produce an adult-type repertoire of cytokines.

The fetal immune system is supplemented by maternal antibody transferred across the placenta and supplied in colostrum and breast milk. The syncytiotrophoblast IgG receptor has recently been identified as a novel, HLA class I-like protein.

VII. Summary and Future Initiatives

Investigative work over the past two decades has provided a basic framework for understanding the maternal–fetal immunological relationship. It is now abundantly clear that the mother and growing embryo cooperate to develop a safe harbor, and that multiple mechanisms are involved. However, critical features remain unresolved, including elucidation of the role(s) of immune cells and proinflammatory cytokines in the promotion of fertility and parturition, the functions of uNK cells, uses for soluble and cell membrane-bound class I antigens, and the effects of microchimerism on maternal and fetal auto- and alloimmunity. Development of the fetal immune system is poorly understood.

Clinical aspects also require considerable attention. While some research findings have been effectively translated into improvements in reproductive performance, the practice of fertility medicine often relies on inadequate factual bases, methods for improving mucosal immunity for protection against sexually transmitted disease are not well developed, and infection-associated preterm labor remains a major problem.

See Also the Following Articles

Cytokines; Erythroblastosis Fetalis; Implantation; Trophoblast to Human Placenta

Bibliography

Bronson, R. A., Alexander, N. J., Anderson, D. J., Branch, D. W. and Kutteh, W. H. (Eds). (1996). *Reproductive Immunology*. Blackwell, Cambridge, MA.

Herr, J. C. (1996). Immunocontraceptive approaches. In *Contraceptive Research and Development* (P. F. Harrison and A. Rosenfield, Eds.). Natl. Acad. Press, Washington, DC.

Hunt, J. S. (1989). Cytokine networks in the uteroplacental unit: Macrophages as pivotal regulatory cells. *J. Reprod. Immunol.* **16**, 1–17.

Hunt, J. S. (Ed.) (1996). *HLA and the Maternal–Fetal Relationship*. Landes, Austin, TX.

Hunt, J. S., and Robertson, S. A. (1996). Uterine macrophages and environmental programming for pregnancy success. *J. Reprod. Immunol.* **32**, 1–25.

Medawar, P. B. (1953). Some immunological and endocrinological problems raised by the evolution of viviparity in vertebrates. *Symp. Soc. Exp. Biol.* **44**, 320–338.

Mitchell, M. D., Trautman, M. S., and Dudley, D. J. (1993). Cytokine networking in the placenta. *Placenta* **14**, 249–275.

Parr, M. B., and Parr, E. L. (1994). Mucosal immunity in the female and male reproductive tracts. In *Handbook of Mucosal Immunology* (P. L. Ogra, J. Mestecky, M. E. Lamm, W. Strober, J. R. McGhee, and J. Bienenstock, Eds.), pp. 677–689. Academic Press, San Diego.

Pudney, J., and Anderson, D. J. (1994). Immunity in the human male reproductive tract. In *Immunobiology of Reproduction* (J. S. Hunt, Ed.), pp. 14–22. Springer-Verlag, New York.

Roitt, I. M., and Delves, P. J. (Eds.) (1992). *Encyclopedia of Immunology*. Academic Press, London.

Solana, R., and Pena, J. (Eds.) (1994). *MHC Antigens and NK Cells*. Landes, Austin, TX.

Vince, G. S., and Johnson, P. M. (1996). Reproductive immunology—Conception, contraception and the consequences. *Immunologist* **4/5**, 172–178.

Whitelaw, P. F., and Croy, B. A. (1996). Granulated lymphocytes of pregnancy. *Placenta* **17**, 533–543.

Implantation

Daniel D. Carson

M. D. Anderson Cancer Center

I. Introduction
II. Cell Biology
III. Uterine Epithelia
IV. Trophectoderm
V. Stroma/Decidua

GLOSSARY

blastocoel The fluid-filled cavity of the blastula that develops from the cleavage of a fertilized egg.

blastocyst stage The modified blastula stage of mammalian embryos made up of the inner cell mass and a thin trophoblast layer that encloses the blastocoel.

blastula The stage of an embryo following cleavage and preceding gastrulation.

invasive Describing a species in which the trophoblast invades the uterine wall to establish direct contact with maternal circulation and initiate placentation, e.g., humans and rodents.

noninvasive Describing a species in which the trophoblast does not invade the uterine wall to establish direct contact with maternal circulation but remains essentially in the uterine lumen and placentation involves only superficial contact with the maternal tissue, e.g., ruminants.

trophectoderm The outermost layer of cells surrounding the blastocoel during the blastocyst stage of embryonic development.

trophoblast A layer of extraembryonic ectodermal tissue in mammals that attaches the embryo to the uterine wall and supplies nourishment to the embryo.

Implantation is the process by which the mammalian embryo initiates intimate contact with the mother. In most circumstances, this event is highly coordinated with respect to both development of the embryo and maturation of the uterine endometrium.

I. INTRODUCTION

In the process of implantation, the embryo develops from a one-cell zygote to the blastocyst stage within the glycoprotein coat of the zona pellucida (Fig. 1). This process entails a series of cell divisions which result in the generation of two tissues in the blastocyst stage embryo, the totipotential cells of the inner cell mass and the first differentiated cell type of the embryo, the trophectoderm. The trophectoderm surrounds the blastocoelic cavity and the inner cell mass and it is these cells which establish physical and biochemical contact with maternal uterine endometrium. Shortly after development to the blastocyst stage, the embryo hatches from the zona pellucida and becomes attachment competent. Until this time, the blastocyst is not capable of attaching to various substrates, even when the nonadhesive coat of the zona pellucida is removed. Other observations also indicate that development to an attachment-competent state is both distinct from and subsequent to

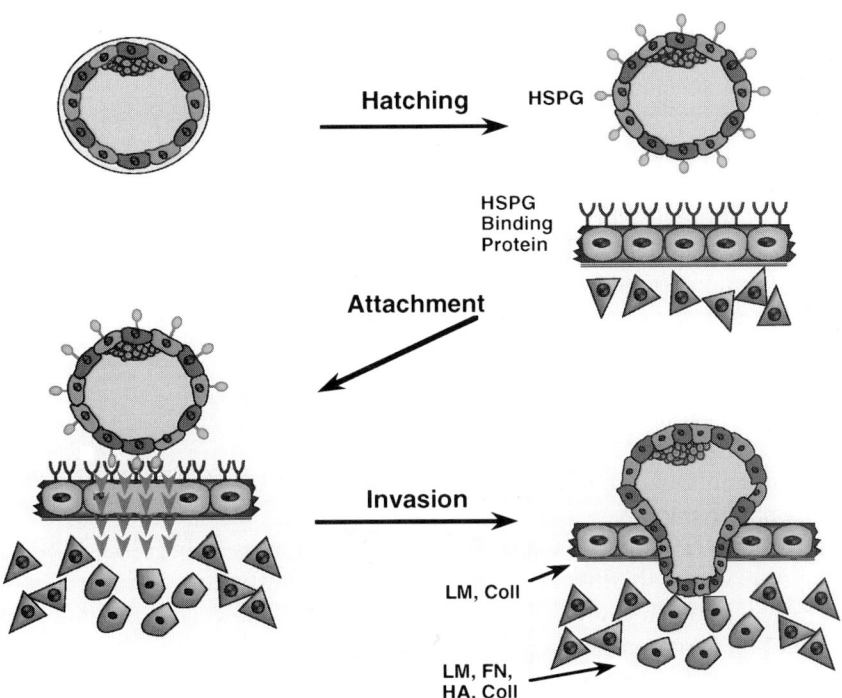

Binding → Signal Transduction → Decidual Response

FIGURE 1 Blastocyst implantation. The diagram depicts a blastocyst hatching from its zona pellucida. Shortly thereafter, the blastocyst becomes attachment competent and begins to express adhesion molecules at the external trophectodermal surface such as heparan sulfate proteoglycans (HSPG). Coordinately, the uterine epithelium loses antiadhesive mucins (not shown), creating access to adhesion-promoting molecules, e.g., HSPG-binding proteins, at the apical surface. Following attachment, signals are transmitting from the blastocyst and the uterine epithelium at the attachment site that may locally or systemically trigger maternal responses. One example of such a response is the decidual response in which uterine stromal cells surrounding the implantation site are stimulated to undergo a series of physical and biochemical changes. In invasive species, the blastocyst embryo penetrates the uterine epithelium, the underlying basal lamina containing laminin (LM) and collagen type IV (Coll), and finally into the underlying decidua containing a variety of extracellular matrix components that support blastocyst adhesion and migration, such as laminin, collagens, fibronectin (FN), and hyaluronate (HA).

hatching from the zona pellucida. In some species, blastocysts can be maintained in a state of implantation delay during which the embryo hatches from the zona pellucida but fails to implant. Implantation can be triggered by appropriate endogenous or exogenous hormonal stimuli. As discussed later, these hormones have important actions on the endometrium; however, the blastocyst also responds to these stimuli by initiating production of receptors for growth factors and cell adhesion molecules that are expressed during the normal progression of blastocyst development.

II. CELL BIOLOGY

Following attachment, trophectodermal cells continue a program of development into trophoblast cells. Trophoblast of various species, including rodents and humans, are highly invasive cells which secrete high levels of extracellular matrix degradative enzymes which facilitate penetration into the uterine endometrium. However, trophoblast of many species, e.g., ruminants, do not penetrate even the lumenal epithelial layer of the endometrium. Thus, such degradative enzymes must play other roles, such as nutrition of the blastocyst or remodeling of embryonic structures. In invasive species, the trophoblast invades the uterine wall to establish direct contact with maternal circulation and to initiate placentation. In noninvasive species, the trophoblast remain essentially in the uterine lumen and placentation involves only superficial physical contact with the maternal tissue. In general, trophoblast are highly active in producing lipid and polypeptide hormones. These hormones strongly affect maternal physiology, although they may have actions on the conceptus as well.

In concert with development of the blastocyst, the uterus must transition to a state in which it will permit attachment and support development of the blastocyst. This state is referred to as the "receptive" state. A vast array of cyclic changes in the pattern of expression of growth factors, cytokines, and corresponding receptor systems occur in uterine tissue in response to ovarian steroid hormones (estrogen and progesterone). After ovulation and before implantation, the uterus exists in a prereceptive state. In this

state, uterine cells can respond to the actions of polypeptide and steroid hormones by controlled proliferation of particular cell types and expression of differentiated functions through the activation, or repression, of specific genes and gene products. This program of responses leads to the generation of a transient receptive phase during which attachment-competent embryos can attach to the uterine lumenal surface. Once initiated, the receptive phase persists for a limited time which varies from species to species. In most cases, the maturation of the uterus to a receptive state is coordinated with the embryonic development by the actions of ovarian steroids; however, blastocysts of some species are capable of undergoing implantation delay during which the blastocysts may pause at the hatched blastocyst stage while the uterus maintains a prereceptive state. Following the receptive period, the uterus becomes refractory and even hostile to the blastocyst. The uterus cannot reenter a prereceptive state until ovarian steroid hormone levels decline. At this point, in menstruating animals, a large portion of the endometrium is sloughed off; however, many, if not most, mammals do not menstruate. In nonmenstruating animals, reentry into the prereceptive state is characterized largely by tissue remodeling, limited cellular loss through apoptosis, and cell replenishment through proliferation.

III. UTERINE EPITHELIA

Under most conditions, the apical surface of the uterine epithelium is covered with a thick glycoprotein coat; however, in many species, this coat is greatly diminished during the receptive phase. Much of this coat appears to consist of high-molecular-weight mucin glycoproteins similar to those coating other mucosal surfaces. Such mucins are thought to provide a protective barrier to microbial infection and the actions of degradative enzymes. Thus, loss of mucins would be expected to compromise these uterine functions. In this regard, it has been shown that the uterus is more susceptible to microbial infection and invasion by tumor cells during the receptive phase. In addition to loss of nonadhesive mucin glycoproteins, the abundant microvilli that characterize the apical surface of uterine epithelia retract and the apical actin filament network becomes disorganized. Increased endocytosis/pinocytosis appears to occur during the receptive phase and, in some species, pinopod-like structures are observed at the apical surface of luminal epithelia during the receptive phase. Collectively, these alterations create a uterine surface that is considerably more flattened and accessible than that observed under most other physiological states. These characteristics may facilitate interactions between embryonic and uterine adhesion molecules.

IV. TROPHECTODERM

Embryonic adhesion molecules such as heparan sulfate proteoglycans can bind various ligands on uterine epithelial cell surfaces via binding motifs for heparan sulfate polysaccharides found in a variety of cell surface and extracellular matrix proteins. Other polysaccharide motifs potentially expressed by multiple glycoproteins and glycolipids may be involved in the initial phases of blastocyst attachment. Growth factor receptors, e.g., epidermal growth factor receptor, expressed on trophectodermal surfaces also may support blastocyst adhesion by binding ligands with membrane-spanning motifs on the surfaces of uterine cells. Integrins appear to participate in multiple steps of this process, particularly in interactions of trophoblast with extracellular matrix components encountered during penetration of extracellular matrices. In humans, integrins expressed by uterine epithelia may support blastocyst attachment, although the blastocyst ligands for these integrins are unknown.

V. STROMA/DECIDUA

In many species, the uterus responds to the implanting blastocyst through local morphological and biochemical alterations in a process called the decidual cell response. In this process, the implantation site undergoes increased vascular permeability and alters the pattern of expression of a variety of genes encoding growth factors, extracellular matrix components, and intracellular proteins. The nature of the biochemical signals that trigger the decidual cell response is not clear. In fact, isolated uterine stromal cells of many species will spontaneously undergo this response *in vitro* in the absence of blastocyst signals. Furthermore, under the appropriate hormonal conditions, many of these responses can be replicated *in vivo* in the absence of blastocysts by physical stimuli. Thus, it is possible that uterine stroma or epithelia respond to mechanical signals in the form of mechanoreceptors or alterations in cytoskeletal structure. Nonetheless, it is possible that blastocysts trigger this response through biochemical interactions with receptor systems on uterine epithelia. In any event, the uterus is remodeled at many levels at the implantation site. In addition to promoting the further development of the conceptus by facilitating contact with maternal growth regulatory factors and nutrients, uterine remodeling may be necessary to protect maternal tissue from excessive assault by the highly invasive blastocyst. Thus, during the implantation process a balance is created between the need to foster embryonic and placental development and the need to protect maternal tissue integrity.

See Also the Following Articles

Blastocyst; Decidua; Trophoblast to Human Placenta

Bibliography

Bazer, F. W., and Roberts, R. M. (1983). Biochemical aspects of conceptus–uterine interactions. *J. Exp. Zool.* **228**, 373–383.

Bell, S. C. (1983). Decidualization: Regional differentiation and associated functions. *Oxford Rev. Reprod. Biol.* **5**, 220–271.

Cross, J. C., Werb, Z., and Fisher, S. J. (1994). Implantation and the placenta: Key pieces of the development puzzle. *Science* **266**, 1508–1518.

Das, S. K., Wang, X.-N., Paria, B. C., Damm, D., Abraham, J. A., Klagsbrun, M. A., Andrews, G. K., and Dey, S. K. (1994). Heparin-binding EGF-like growth factor gene is induced in the mouse uterus temporally by the blastocyst solely at the site of its apposition: A possible ligand for interaction with blastocyst EGF-receptor at implantation. *Development* **120**, 1071–1083.

Finn, C. A. (1986). Implantation, menstruation and inflammation. *Biol. Rev.* **61**, 313–328.

Hey, N. A., Graham, R. A., Seif, M. W., and Aplin, J. D.

(1994). The polymorphic epithelial mucin, MUC1, in human endometrium is regulated with maximal expression in the implantation phase. *J. Clin. Endocrinol. Metab.* **78**, 337–342.

Lessey, B. A., Castelbaum, A. J., Buck, C. A., Lei, Y., Yowell, C. W., and Sun, J. (1994). Further characterization of endometrial integrins during the menstrual cycle and in pregnancy. *Fertil. Steril.* **62**, 497–506.

Mead, R. A. (1993). Embryonic diapause in vertebrates. *J. Exp. Zool.* **266**, 629–641.

Tabibzadeh, S. S. (1991). Human endometrium: An active site of cytokine production and action. *Endocrinol. Rev.* **12**, 272–289.

Wegner, C. C., and Carson, D. D. (1994). Cell adhesion processes in embryo implantation. *Oxford Rev. Reprod. Biol.* **16**, 87–135.

Impotence

Lawrence S. Hakim
University of Miami School of Medicine

Ajay Nehra
Mayo Clinic

Irwin Goldstein
Boston University School of Medicine

GLOSSARY

corporal venoocclusive mechanism The mechanism whereby blood is stored in the phallus during erection due to lacunar smooth muscle relaxation and passive compression of the draining subtunical venules.

impotence The consistent inability to attain and maintain a penile erection sufficient to permit satisfactory sexual intercourse. Also, *erectile dysfunction*.

inflatable penile prosthesis An artificial device placed into the corpora to achieve girth and rigidity for the treatment of erectile dysfunction.

low-flow priapism A painful, unremitting venoocclusive state of erection, typically seen following intracorporal therapy or sickle cell disease.

Peyronie's disease Abnormal penile curvature due to fibrotic plaque in the tunical lining of the corpora, typically as a result of injury.

self-injection therapy Use of intracavernosal vasoactive agents for the treatment of impotence.

Male sexual dysfunction may be classified into erectile dysfunction; problems of libido; dysfunction of emission, ejaculation, or orgasm; Peyronie's disease; and priapism. One of the more common types of sexual dysfunction is erectile dysfunction or impotence. Erectile dysfunction has been defined by the NIH Consensus Panel as the consistent inability to attain and maintain a penile erection sufficient to permit satisfactory sexual intercourse. Numerous advances have been made in understanding the physiologic and biochemical mechanisms controlling penile erection. In addition, improved clinical techniques for the diagnosis and treatment of erectile dysfunction have been developed. This article reviews the contemporary state of knowledge of erectile dysfunction, focusing on the pathophysiology of impotence and the current diagnostic and treatment options for this disorder.

I. DEMOGRAPHICS OF ERECTILE DYSFUNCTION

In 1948, Kinsey and associates estimated by epidemiological data that between 5 and 10 million American males, or approximately 1 in 10 men, suffer from erectile dysfunction. In 1994, Goldstein and associates reported on the Massachusetts Male Aging Study, an epidemiological investigation that examined the medical and psychosocial correlates of erectile dysfunction. This large, random, community-based study suggested that the prevalence of impotence from ages 40 to 70 years is over 50%. Contemporary studies indicate, therefore, that impotence afflicts over 30 million American men.

Certain groups of patients have been found to have a particularly high prevalence of erectile dysfunction. In the Massachusetts Male Aging Study, aging, hypertension, heart disease, and treated diabetes were among several physiologic variables found to strongly predict impotence. With regard to diabetic men, impotence is an age-dependent disorder which occurs earlier and is accelerated in this population. In addition, men with treated diabetes are more than three times more likely to have impotence than men without diabetes.

II. ANATOMY OF THE PENIS

The penis consists of two paired corpora cavernosa and a corpus spongiosum, which surrounds the urethra and, distally, forms the glans penis. Each corpus cavernosum is surrounded by a thick fibrous sheath, the tunica albuginea, which encases the sponge-like cavernosal tissue. The erectile tissue consists of lacunae which are multiple, interconnected, and lined by vascular endothelium. The walls of the lacunae, the trabeculae, are composed of thick bundles of smooth muscle and a fibroelastic frame, consisting of fibroblasts, collagen, and elastin.

The right and left cavernosal arteries, which are terminal branches of the hypogastric arteries, supply blood to the paired corpora cavernosa, respectively. Numerous muscular helicine arteries branch off each cavernosal artery and open directly into the lacunar spaces. Venous drainage from the corporal bodies is accomplished via multiple subtunical venules located between the periphery of the erectile tissue and the tunica albuginea. The numerous subtunical venules coalesce to form the emissary veins, which pierce the tunica albuginea and drain out of the penis.

The peripheral innervation of the penis consists of sympathetic nerves arising from the 11th thoracic to the second lumbar spinal cord segments and parasympathetic and somatic nerves arising from the second, third, and fourth sacral spinal cord segments. Somatic innervation is via the pudendal nerve, which is composed of efferent fibers innervating the striated musculature of the perineum and of afferent fibers from the penile and perineal skin.

III. PHYSIOLOGY OF ERECTION

A. Mechanism of Erection

Penile erections are initiated by local sensory stimulation of the genital organs (reflexogenic erections) and by central psychogenic stimuli received by or generated within the brain (psychogenic erections). A variety of stimuli processed in several regions of the brain, including the thalamic nuclei, the rhinencephalon, and the limbic structures, can elicit supraspinal erectile responses. The medial preoptic–anterior hypothalamic area appears to integrate input from these diverse regions and send projections to the thoracolumbar sympathetic and sacral parasympathetic centers of the spinal cord. The afferent limb for reflexogenic erections, the pudendal nerve, collects somatic sensation from the genital skin. The autonomic nerve fibers that arise from the sacral parasympathetic center (S2–S4) make up the efferent limb for this reflex, giving innervation to the penile smooth muscle. Reflexogenic and psychogenic erectile mechanisms probably act synergistically in the control of penile erection.

Erection follows arterial and trabecular smooth muscle relaxation. Dilation of the cavernosal and helicine arteries increases blood flow into the lacunar spaces. Relaxation of the trabecular smooth muscle dilates the lacunar spaces, accommodating a larger volume of blood, with engorgement of the penis. The expansion of the relaxed trabecular walls against the tunica albuginea compresses the plexus of subtunical venules. This results in increased outflow resistance and increased intracorporal pressure, making the penis rigid. This reduction of venous outflow by the mechanical compression of subtunical venules is known as the corporal venoocclusive mechanism.

Detumescence results following penile smooth muscle contraction. Activation of sympathetic constrictor nerves causes an increase in the tone of the smooth muscle of the helicine arteries and the trabeculae. This results in a reduction of arterial inflow and collapse of the lacunar spaces with decompression of subtunical venules and increased corporal venous outflow returning the penis to the flaccid state.

B. Neurogenic and Endothelium-Mediated Control of Penile Smooth Muscle

Many investigations have been performed to understand the mechanism of neurogenic and endothelial control of the tone of penile blood vessels. Histochemical studies of the cavernosal and helicine arteries have demonstrated the presence of adrenergic nerves, acetylcholinesterase-containing (probably cholinergic) nerves, as well as vasoactive intestinal polypeptide and neuropeptide Y immunoreactive nerves. The local control of the trabecular smooth muscle has also been studied.

Adrenergic nerves constrict corporal smooth muscle via norepinephrine acting on α adrenoreceptors. Relaxation is controlled by cholinergic and nonadrenergic, noncholinergic neurotransmitters (NANC). Cholinergic nerves appear to have a modulatory role over the other two neuroeffector systems and not a direct effect on the smooth muscle. Vaso-

active intestinal polypeptide (VIP), a 28-amino acid peptide, had been proposed as the NANC in penile smooth muscle and was supported by the observation that VIP-immunoreactive fibers densely innervate the trabecular smooth muscle and VIP elicits relaxation of the trabecular smooth muscle.

However, NANC-mediated relaxation of corpus cavernosal smooth muscle was shown to release nitric oxide and simultaneously cause the accumulation of cyclic guanosine monophosphate (cGMP) in this tissue. Nitric oxide is a neurotransmitter in the peripheral nervous system in various organs. Nitric oxide synthase has been localized in the peripheral autonomic nerves of several organs which contain smooth muscle. In preparations of human corpus cavernosum in which the endothelium has been removed, the NANC neurogenic relaxation is inhibited by substances that interfere with the synthesis or effects of nitric oxide. Such evidence suggests that nitric oxide mediates NANC-mediated relaxation in trabecular smooth muscle.

Smooth muscle relaxation in the corpus cavernosum, which is induced by various vasodilators and physical stimuli, requires the presence of functional endothelium. In human corpus cavernosum, the endothelium-derived relaxing factor has the chemical properties of nitric oxide. Nitric oxide stimulates guanylate cyclase with accumulation of (cGMP), which then leads to smooth muscle relaxation. This second messenger pathway appears to be of key importance for trabecular smooth muscle relaxation since nitric oxide mediates both endothelium- and nerve-mediated relaxation.

IV. PATHOPHYSIOLOGY OF IMPOTENCE

The underlying etiology of erectile dysfunction is primarily one of organic origin rather than psychogenic. In fact, studies on the natural history of impotence revealed that erectile dysfunction is rarely reversible, supporting a progressive, organic, nonreversible cause. Organic erectile dysfunction can often be attributed to vascular factors, neurogenic factors, various drugs or medications, endocrinologic or hormonal factors, or trauma. Vascular insufficiency by far accounts for the majority of patients with organic impotence.

Diabetes mellitus is a major primary organic cause for erectile dysfunction. Commonly, erectile dysfunction in diabetes mellitus develops insidiously over a period of months to years. Patients frequently describe diminished rigidity and reduced ability to sustain an erection. Erectile dysfunction, however, is not always a late complication of the disease but can occur early in the natural history of the disease. Some investigators have reported that impotence is the first clinical finding which occasionally leads to the discovery of unrecognized diabetes mellitus. Since neuropathy and vasculopathy are common complications associated with the natural history of diabetes mellitus, it has been hypothesized that cavernosal artery insufficiency, corporal venoocclusive dysfunction, and/or autonomic neuropathy are the major organic pathophysiological mechanisms leading to persistent erectile impairment in these patients.

A. Vasculogenic Impotence

Vascular disease, including both large- and small-vessel vasculopathy, has long been recognized as a major contributor to erectile dysfunction. Small-vessel changes, such as those seen in diabetic patients, can result in alteration of local microvascular blood flow. Progressive venule dilation, arteriolar vasoconstriction, and sclerosis of the walls of the arterioles, capillaries, and venules are seen, contributing to poor vascular flow and impotence. Alterations in endothelial cell metabolism and function, decreased oxygen transport, as well as thickened vessel wall basement membranes may also contribute to the underlying dysfunction.

Atherosclerotic vascular disease and erectile dysfunction are strongly related. This results from the intimal, medial, and luminal changes of large-vessel disease observed in obliterative atherosclerosis. In fact, arteriosclerosis is the most common organic disorder leading to impotence. Up to 50% of men with clinically significant peripheral arterial disease complain of some degree of erectile dysfunction and

in 80% of these cases, the cause of the impotence is primarily organic. Erectile dysfunction develops when more than 50% of the major arterial supply to penis is involved by atherosclerotic occlusive disease. Autopsy studies in impotent diabetic men revealed numerous atherosclerotic lesions in the penile arterial bed, including intimal fibrosis and calcification with narrowing and obliteration of the lumen. Vascular alterations in the penile arteries decrease blood flow to the corporal bodies and contribute to erectile dysfunction.

Men with organic erectile dysfunction may also have other associated vascular risk factors, such as cigarette smoking, hypertension, and hyperlipidemia. Cigarette smoking is a statistically significant independent risk factor in the development of atherosclerotic arterial occlusive disease to the hypogastric cavernous arterial bed. Five-, 10-, and 20-pack-year histories of cigarette smoking exposure are associated with 15, 30, and 70% incidence of arterial occlusive disease within the common penile artery. The presence of hypertension has been demonstrated in up to 45% of impotent men, whereas hyperlipidemia, hypercholesterolemia, or other disturbances of lipid metabolism have been found in 40–50% of men with erectile dysfunction.

The mechanism of vasculogenic erectile dysfunction involves decreased arterial perfusion pressures and lower arterial inflow to the lacunar spaces of the corpora cavernosa. The clinical consequences of these hemodynamic changes are decreased rigidity of the erect penis as well as prolongation of the time to maximum erection. Vasculogenic erectile dysfunction may also occur via disruption of the corporal venoocclusive mechanism. Corporal venoocclusive dysfunction is due to a structural alteration in the fibroelastic components of the trabeculae. Loss of compliance of the fibroelastic frame of the penis may be the result of altered synthesis of collagen, elastin, or tissue cellularity. Another explanation may be as a result of alterations in the reactivity of the corpus cavernosum smooth muscle and lacunar space endothelial cells. Decreased compliance of the penile fibroelastic frame and/or an inability to achieve complete smooth muscle relaxation may result in an inability to expand the trabeculae against the tunica albuginea and compress the subtunical venules. The clinical consequence of such hemodynamic alterations in the "trapping" mechanism is excessive outflow of lacunar blood through the subtunical venules. This culminates in decreased penile rigidity and a diminished ability to sustain an erection.

B. Neurogenic Impotence

A role for penile autonomic neuropathy in the pathophysiology of erectile dysfunction is supported by several indirect clinical observations as well as laboratory investigations. First and foremost, patients with peripheral neuropathy (such the diabetic male) are frequently impotent. The incidence of peripheral and autonomic neuropathy is significantly higher in males reporting erectile dysfunction as opposed to potent males. Erectile failure is, in fact, the most common feature of diabetic autonomic neuropathy and studies have shown that the development of impotence was significantly associated with the appearance of neuropathy.

Clinical tests of the integrity of the autonomic nerves to the corpora were often performed by indirect testing. This involves recording impaired nocturnal penile tumescence activity, in conjunction with evidence of intact hemodynamic integrity, as assessed by the erectile response to the intracavernosal administration of vasoactive agents. The presence of bladder areflexia and concomitant bladder or bowel dysfunction may provide further indirect support for the impairment of the motor efferent autonomic cavernosal nerves. Since the bladder and the penis receive autonomic innervation from common origins, the hypogastric sympathetic and the pelvic parasympathetic nerves, it is assumed that neuropathy of the bladder will coexist with neuropathy of the penis.

Vascular reflexes (e.g., single-breath beat-to-beat variation) provide an indirect measurement of autonomic parasympathetic neuropathy. A normal variation of heart rate during shallow breathing is 10 or more beats per minute. Therefore, autonomic parasympathetic neuropathy is considered when a heart rate variation during breathing of <10 beats per minute is noted. Studies have documented abnormal vascular reflexes in patients with neurogenic impotence.

Direct testing of the autonomic innervation of the corpora may be achieved by recording the electrical activity of the corporal smooth muscle utilizing intracavernosal electromyographic needles or from surface electrodes on the penile shaft. Such direct testing offers the ability to identify neurologic abnormalities at the level of the autonomic neuron–corporal smooth muscle interface. Using single potential analysis of this cavernous electrical activity, waveforms of a defined duration and amplitude have been recorded in a reproducible pattern from normal subjects. Following visual sexual stimulation, these potentials decrease in amplitude and duration and increase in frequency. Following smooth muscle relaxation induced by intracavernosal vasoactive drug administration, electrical silence of these potentials has been noted. In patients with peripheral neuropathy, the single potential analysis of cavernous electrical activity may reveal potentials that are either of short duration and high amplitude or of low amplitude.

Several tests are used in the evaluation of the presence of neuropathy in the sensory afferent nerves from the penile skin and the motor efferent nerves to the perineal skeletal musculature. These tests include perineal electromyography, sacral latency testing, dorsal nerve somatosensory-evoked potential evaluation, and vibration perception sensitivity testing. Abnormalities in these tests in patients with neurogenic erectile dysfunction have been commonly reported. These tests do not record the integrity of the efferent autonomic nerves responsible for penile erection. An abnormal result in somatic testing may suggest but does not prove the coexistence of autonomic neuropathy in the corpora cavernosa.

C. Endocrinologic Impotence

Abnormal endocrinologic factors, primarily hypogonadotropic hypogonadism, may be an associated finding in erectile dysfunction, although the etiologic significance of the hypothalamic–pituitary–testicular axis in erectile dysfunction is unclear. Androgens influence the growth and development of the male reproductive tract and secondary sexual characteristics. Their effect on libido and sexual behavior is well established, but the effect of androgens on normal erectile physiology remains poorly understood. Patients with castrate levels of testosterone can achieve erections in response to visual sexual stimulation which are comparable in quality to erections in men with normal levels of testosterone. In addition, total serum testosterone values are not consistently diminished in men with organic erectile dysfunction. This suggests that the neurovascular mechanisms which control erection are functional in the presence of low levels of androgens. Further studies are required to understand how abnormal androgen function results in erectile impotence.

Although hypogonadism is the most common endocrine cause of erectile dysfunction, thyroid disease, other pituitary disorders, adrenal disease, and hyperprolactinemia may all be associated with erectile dysfunction. Varying levels of serum prolactin have been observed in impotent men, and in patients with hyperprolactinemia, declining libido and erectile function can be early symptoms. Hyperprolactinemia is associated with low circulating testosterone levels, secondary to the inhibition of normal gonadotropin-releasing hormone pulsation by elevated prolactin. In impotent patients with hyperprolactinemia and low testosterone levels, potency is not restored in approximately half of the patients, despite the normalization of serum testosterone levels, thus possibly implying some degree of antagonism by prolactin to the peripheral action of testosterone. Erectile dysfunction may also be associated with abnormal thyroid function. Hyperthyroidism is commonly associated with diminished libido and less often with impotence. In hypothyroid states, erectile dysfunction may be secondary to concomitant low testosterone secretion and elevated prolactin levels.

D. Medications

A number of medications are associated with sexual dysfunction. Various antihypertensive agents, most notably the β-blockers, have been shown to cause impotence, possibly via a decrease in corporal perfusion pressure. Digoxin, a frequently utilized cardiac inotropic medication, can cause impotence via its effect on calcium transport in the corporal smooth muscle cell. In fact, recent studies have supported its use for the prevention of recurrent episodes of low-flow priapism.

Various antidepressant medications have been associated with erectile dysfunction. The class of medications known as serotonin-reuptake inhibitors (SRIs) may cause decreased erectile function. In addition, the SRI drugs can lead to an increased latency to ejaculation, making the culmination of the sexual act more difficult. An application of this "side effect" of the SRI family of medications is to treat the patient suffering from premature ejaculation.

Various illicit drugs, including cocaine, marijuana, and alcohol, are also associated with increased erectile dysfunction, possibly via a central neurologic effect. Recent studies have shown that cocaine is also associated with a higher incidence of priapism via its effect centrally and peripherally on calcium transport and smooth muscle relaxation. In these patients, impotence may be the culmination of repeated incidents of venoocclusive priapism, resulting in corporal fibrosis and abnormal vascular changes.

E. Trauma

Although more commonly reported in men over age 40, erectile dysfunction can effect men of all ages. While systemic vascular diseases, such as that due to hypertension, atherosclerosis, diabetes, or hypercholesterolemia, are often risk factors in older men, vascular insufficiency due to penile or perineal trauma is often associated with erectile dysfunction in young men.

Any type of blunt traumatic injury which affects the penile vessels, corpora, or neurologic supply can lead to abnormal corporal blood flow and diminished nerve supply. Bicycle riding, and the blunt trauma which may be associated with the constant perineal pressure from a narrow, unpadded bicycle seat or cross-bar, has been implicated as the underlying cause of erectile dysfunction in many men in this population. Sports injuries to the genitalia and scrotum, such as kicks or hits seen in football, hockey, lacrosse, and other contact sports, are often found to be the culprit in these typically young, otherwise healthy men after complete vascular testing has been performed.

Once this type of dysfunction is suspected by history, a number of specific vascular diagnostic tests can be utilized to make an accurate diagnosis. In selected cases of traumatic arterial obstruction, microscopic reconstructive vascular surgical techniques may be utilized to enhance normal erectile function.

F. Pelvic Surgery

Radical pelvic surgery, such as that seen with patients undergoing radical retropubic prostatectomy or cystoprostatectomy for prostate or bladder cancer, is often the cause of erectile dysfunction in previously potent men. Despite a "nerve-sparing" approach, many men, especially those over age 60, will experience some decrease in their potency. This loss of function is most likely a result of a combination of neurogenic and vasculogenic factors. Fortunately, there are excellent treatment alternatives available to these men to help restore them to a level of good sexual function.

V. DIAGNOSTIC EVALUATION OF ORGANIC ERECTILE DYSFUNCTION

The initial evaluation for erectile dysfunction begins with a sexual, psychosocial, and medical history, a complete physical examination, and specific laboratory tests. The results of the initial evaluation may be further corroborated by a variety of diagnostic examinations such as neurologic, psychological, hormonal, and vascular testing. The algorithm is illustrated in Fig. 1.

With more sophisticated diagnostic evaluations, a primary organic cause of erectile dysfunction is often found. However, it is important to remember that either a primary or secondary psychogenic problem may contribute to the sexual dysfunction. In order to maximize patient success, both the psychogenic and organic factors should be addressed.

It must be remembered that impotence is a "couple's disease." Although we often deal exclusively with the "patient," with regard to sexual dysfunction, the partner is a vital part of the equation. Efforts to include the partner in all aspects of the diagnostic evaluation and treatment decision process are extremely paramount to obtaining a successful outcome.

Patient history
  Medical, sexual, and psychosocial history
  Determine level of libido
  Rule out: Ejaculatory or orgasmic dysfunction
Physical examination
  Complete exam
  Abnormal penile curvature
  Palpable corporal fibrosis
Endocrinologic and laboratory evaluation
  Routine hematology and chemistry profile, fasting blood glucose, HgA1C
  Lipid/cholesterol profile
  Hormonal profile (testosterone, luteinizing hormone, follicle-stimulating hormone)
  Rule out: Hyperprolactinemia (serum prolactin level)
  Rule out thyroid disease, other pituitary disorders, adrenal disease
Vascular evaluation (*in selected cases)
  Penile blood flow study
  Visual sexual stimulation*
  Penile duplex Doppler ultrasonography*
  Dynamic infusion pharmacocavernosography and cavernosography*
  Selective internal pudendal arteriography*
  Cavernosal tissue biopsy*
Neurologic evaluation (performed in selected patients)
  Biothesiometry (vibration perception sensitivity testing)
  Nocturnal penile tumescence testing
  Single-breath beat-to-beat variation (autonomic parasympathetic neuropathy)
  Single potential analysis: cavernous electrical activity (peripheral neuropathy)
  Dorsal nerve somatosensory evoked potentials
Sacral latency testing

FIGURE 1   Erectile dysfunction: diagnostic algorithm.

VI. THERAPEUTIC MANAGEMENT OF ORGANIC ERECTILE DYSFUNCTION

A variety of therapeutic alternatives are utilized for the man with organic erectile dysfunction. New oral medications allow for improved erectile function simply by taking a pill prior to sexual intercourse. These medications may be effective in 40–70% of patients tested. For patients in whom this is unsatisfactory, penile self-injection therapy or transurethral prostaglandin E-1 may be effective. External vacuum erection devices, inflatable penile implants, and pe-

Nonhormonal therapy
  Type-5 specific phosphodiesterase inhibitors (sildenafil, Viagra)—oral
  α2-Adrenergic blocking agents (yohimbine hydrochloride)—oral
  Centrally acting medications (i.e., Apomorphine)—sublingual
  Phentolamine—oral
Hormonal therapy
  Hypogonadotropic hypogonadism: Testosterone replacement
  Parenteral/intramuscular
  Transdermal patch
  Hyperprolactinemia/pituitary tumor
  Cessation of causative medications (e.g., estrogen and α-methyldopa); Bromocryptine
  Extirpative surgery
Noninvasive therapy:
  Vacuum erection devices
  Constricting ring
Minimally invasive therapy
  Intracavernosal injection of vasoactive agents (Caverject, Edex, Trimix: Papaverine, phentolamine, prostaglandin E1)
  Intraurethral delivery of vasoactive agents (MUSE)
Invasive therapy
  Penile prosthesis (malleable versus inflatable device)
  Microvascular arterial bypass surgery

FIGURE 2 Erectile dysfunction: therapeutic options.

nile revascularization are also effective treatment alternatives for selected patients. The treatment options are summarized in Fig. 2.

Hormone replacement therapy should only be used in the treatment of erectile dysfunction in the presence of concomitant hypogonadal disorders (hypogonadotropic hypogonadism or hypergonadotropic hypogonadism) or hyperprolactinemia. In these cases, testosterone replacement is employed to maintain normal serum levels of testosterone in an attempt to restore potency and libido. The indiscriminate use in men without hypogonadism should be avoided due in part to the risk of hepatotoxicity and adenocarcinoma of the prostate. Before beginning testosterone replacement therapy in any man over the age of 50, serum prostate-specific antigen levels, a digital rectal examination, and possibly transrectal ultrasound studies should be performed.

VII. SUMMARY

Numerous advances have been made in understanding the physiologic and biochemical mechanisms controlling penile erection. This knowledge has led to improved diagnostic evaluations and better therapeutic alternatives for the treatment of erectile dysfunction. Advances in dynamic vascular testing, effective pharmacotherapy, prosthetic devices, and microsurgical revascularization have allowed us to enter a new and exciting era in the quest for a more complete understanding of erectile dysfunction.

See Also the Following Articles

Erection; Penis

Bibliography

Feldman, H. A., Goldstein, I., Hatzichristou, D. G., *et al.* (1994). Impotence and its medical and psychosocial correlates: Results of the Massachusetts Male Aging Study. *J. Urol.* **151,** 54.

Goldstein, I., and Krane, R. J. (1993). Diagnosis and therapy of erectile dysfunction. In *Campbell's Urology* (P. C. Walsh, R. F. Gittes, A. D. Perlmutter, and T. A. Stamey, Eds.), p. 2591. Saunders, Philadelphia.

Hakim, L. S., Hashmat, A. I., and Macchia, R. J. (1994). Priapism. In *Sickle Cell Anemia: Basic Principles to Clinical Practice* (S. H. Embury, Ed.), p. 633. Raven Press, New York.

Hakim, L. S., Nehra, A., Kulaksizoglu, H., *et al.* (1995). Penile microvascular arterial bypass surgery. *Microsurgery* **16,** 1.

Hellstrom, W. J. G. (Ed.) (1997). *Male Infertility and Sexual Dysfunction.* Springer-Verlag, New York.

Krane, R. J., Goldstein, I., and Saenz de Tejada, I. (1989). Medical progress: Impotence. *N. Engl. J. Med.* **321,** 1648.

Munarriz, R. M., Qingwei, R. Y., Nehra, A., *et al.* (1995). Blunt trauma: The pathophysiology of hemodynamic injury leading to erectile dysfunction. *J. Urol.* **153,** 1831.

NIH Consensus Development Panel on Impotence (1993). Impotence. *J. Am. Med. Assoc.* **270,** 83.

Rosen, M. P., Greenfield, A. J., Walker, T. G., *et al.* (1991). Cigarette smoking: An independent risk factor for atherosclerosis in the hypogastric–cavernous arterial bed of men with arteriogenic impotence. *J. Urol.* **146,** 453.

Infections in Pregnancy

Jack Ludmir

Harvard Medical School

GLOSSARY

asymptomatic infection The presence of bacteria, virus, or protozoa in the mother and/or fetus without symptoms or manifestations of clinical disease.

bacteriuria The presence of bacteria in the urine without clinical symptoms of infection.

congenital infection Infection on the baby occurring *in utero* and present at birth; may be asymptomatic or symptomatic.

congenital infection syndrome A specific pattern of clinical manifestations on the baby secondary to infection *in utero*.

fetal hydrops Accumulation of fluid in the entire body of the fetus.

serologic testing Laboratory determination of antibody titers against specific infectious agents. Antibodies can be of the IgG and IgM type, and their presence correlates with timing of infection.

teratogen An agent or factor that causes physical defects in the developing embryo.

vertical transmission Transmission of an infectious agent from mother to fetus either through the placenta or blood or through the birth canal at time of birth.

Infections in pregnancy can cause significant maternal and perinatal morbidity and mortality. The normal physiologic changes in pregnancy that allow for the maintenance of the placenta and the fetus without maternal signs of rejection include an altered immune state. However, this state of immune tolerance may predispose mother and baby to severe illnesses caused by various infectious agents. Some of these infections may be life-threatening for the mother, whereas others may cause minimal disease in the pregnant woman but can cause tremendous tragic sequalae in the fetus by altering organogenesis, fetal sensory and motor development, and central nervous system function. The pregnant woman and her fetus are prone to infections caused by a variety of different bacteria, viruses, and protozoa. Sexually transmitted diseases continue to be a prevalent problem in society and place the pregnant woman and her baby at great risk for serious compromise. In this article, the most common infections affecting pregnancy are discussed.

I. BACTERIAL INFECTIONS

A. Urinary Tract Infections

Urinary tract infection (UTI) is the most common bacterial infection in pregnancy. It may be asymptomatic or symptomatic [acute cystitis (bladder infection) and pyelonephritis (renal infection)]. UTI is 14 times more common in the female than the male. The physiologic and anatomic changes that occur in the urinary tract in pregnancy include dilatation of the renal pelvis and urinary stasis. These changes facilitate bacterial migration and bladder infection. Asymptomatic bacteriuria affects 6% of pregnant women, and if untreated it can result in pyelonephritis in 30–50% of pregnant women. Furthermore, several studies support the role of asymptomatic bacteriuria in the development of preterm labor and delivery of low-birth-weight infant. Pyelonephritis is characterized by fever, chills, nausea, vomiting, and flank pain. It can cause generalized sepsis and respiratory compromise, and it may place the patient at great risk for delivering prematurely. For these reasons, every pregnant women should have a urinalysis as a screening test at the first prenatal visit. If leukocytes or nitrates are present, a urine culture should be performed. The most common organisms causing UTIs are bacteria present in the normal fecal flora. *Escherichia coli* accounts for approximately 80% of all cases. The remaining organisms include gram-negative bacteria such as *Klebsiella, Proteus, Enterobacter,* and *Pseudomonas* species. The remaining UTIs are caused by group B streptococci, *Enterococcus* species, and *Staphylococcus saprophyticus.* The most common agents for treatment of asymptomatic bacteriuria include nitrofurantoin, ampicillin, and sulfonamides. Single-dose therapy will cure the large majority of UTIs in pregnancy.

B. Group B Streptococcus

The proportion of pregnant women colonized with group B streptococcus (GBS) ranges from approximately 10 to 30%. Although genital colonization is usually asymptomatic, GBS is responsible for significant peripartum infection. This organism has become recognized over the past two decades as one of the most important causes of neonatal infection. An estimated 7600 cases of GBS sepsis in newborns occur annually in the United States. This infection is responsible for 300 deaths annually in infants <90 days of age. GBS infection in the newborn is characterized by sepsis, pneumonia, and meningitis. Vertical transmission of this bacteria during labor and delivery may result in invasive infection in the newborn during the first week of life (known as early onset GBS infection). Risk factors associated with early onset disease include preterm birth (<37 weeks gestation), prolonged interval between rupture of membranes and delivery, maternal fever in labor, GBS bacteriuria during pregnancy, and previous delivery of an infant with GBS disease. Two strategies to prevent early onset GBS disease in the newborn have been developed by the American College of Obstetricians and Gynecologists, the American Academy of Pediatrics, and the Centers for Disease Control. The first is based on the risk factors mentioned previously. If the patient has any of these risk factors, intrapartum antibiotics preferably with penicillin should be given. The second strategy is based on performance of cultures for GBS in all pregnant patients at 35–37 weeks and intrapartum treatment with antibiotics for all patients with positive cultures. Regardless of the strategy, intrapartum antibiotic prophylaxis can reduce the rate of vertical transmission of infection to the neonate but cannot prevent all cases of neonatal sepsis. Currently, no single screening protocol has been demonstrated to be superior to another.

II. VIRAL INFECTIONS

A. Cytomegalovirus

Cytomegalovirus (CMV) is a ubiquitous double-stranded DNA virus in man. Approximately 50–85% of pregnant women will show evidence of prior antigenic exposure. This virus represents the most common cause of congenital infection in humans, with 1% of all neonates demonstrating its presence. Of the 40,000 infants born infected in the United States, 10% will show evidence of syptomatic disease characterized by low birth weight, microcephaly, intracranial calcifications, chorioretinitis, and mental and motor retardation. An additional 3% may develop hearing loss in the first few years of life. Unfortunately, maternal immunity to CMV does not prevent recurrence (reactivation), nor does it prevent congenital infection. In fact, most infections during pregnancy are recurrent and are responsible for the majority of babies affected with congenital infection. Fortunately, most of the infections resulting from reactivation of the virus result in lesser clinical sequalae for the infant. Establishing seroconversion, and demonstrating the presence of IgM CMV antibody, is the most accurate method for documenting primary maternal infection. Isolation of the virus from either amniotic fluid or fetal blood can be helpful to detect fetal CMV infection. The lack of vaccines to prevent infection, and the inability to provide effective therapy against this virus, precludes the use of routine serologic testing of either mothers or neonates.

B. Rubella

Rubella, also called German measles, is an exanthematous disease caused by a single-stranded RNA virus. This virus can cause a mild illness in the mother, but it is responsible for major morbidity and mortality in the fetus. In 1942, the Australian opthalmologist Gregg was the first to describe the congenital rubella syndrome, which is characterized by eye lesions, including cataracts, glaucoma, and microphthalmia; heart anomalies; sensorineural deafness; central nervous system defects; fetal growth retardation; and hepatitis and pneumonitis. Surveillance for susceptibility to rubella infection is essential in prenatal care. Serologic testing for immunity is the primary mode of diagnosis. Immunity to this virus will prevent infection, and seropositive mothers do not need further testing regardless of their subsequent history of exposure. If rubella is diagnosed during pregnancy, the patient should be advised of the risks of fetal infection: in particular, if the infection occurs during the period of embryogenesis. Widespread vaccination of children and susceptible adults will prevent outbreaks of rubella and reduce the risk for congenital infection. In rubella-susceptible women of reproductive age, vaccination is highly effective and has few side effects. Following vaccination, women should be advised to avoid conception for 3 months. If a woman is inadvertently vaccinated early in gestation, she should be counseled of the theoretical risk to the fetus caused by the live attenuated virus. However, a registry of women vaccinated during pregnancy has failed to reveal a single case of congenital rubella syndrome secondary to maternal vaccination.

C. Parvovirus

Human parvovirus B_{19} causes erythema infectiosum, or fifth disease. Clinical features include a bright red macular rash that affects the face, giving a slapped cheek appearance. This small single-stranded DNA virus replicates in rapidly proliferating cells such as erythroblast precursors; if it infects the fetus, it can cause anemia, which can result in fetal hydrops and death. Although there is no evidence that parvovirus is teratogenic, it can cause abortion. Because of widespread asymptomatic infection in both children and adults, pregnant women who learn that they have been in contact with children with active infection should be offered the option of serologic testing. Diagnosis is confirmed by specific IgM parvovirus antibodies. For women with positive serology, ultrasonographic evaluation of the fetus is indicated. If hydrops fetalis is present, fetal transfusion should be considered.

D. Varicella

Varicella zoster virus is a DNA virus of the herpes family. It is responsible for chicken pox, a common childhood viral disease; however, this virus has the potential to remain dormant for years in the dorsal root ganglia and may be reactivated years later causing herpes zoster or shingles. Most adults acquired chicken pox during childhood and are immune. The virus is highly contagious for nonimmune persons. Varicella infection tends to be more severe in adults than in children. Pregnant women acquiring varicella are at greater risk to develop severe varicella pneumonia that could be life-threatening. Administration of varicella zoster immunoglobulin (VZIG) will either prevent or attenuate varicella infection in susceptible individuals if given within 96 hr from exposure. Chicken pox during early pregnancy is associated rarely with congenital infection. A congenital varicella syndrome, characterized by fetal cutaneous scarring and limb hypoplasia, has been described. However, if clinical maternal infection occurs within 5 days before or 2 days after delivery, the baby is at risk for developing severe varicella. For this reason, VZIG should be given to the neonate as soon as possible. Due to the highly contagious nature of this virus, hospitalized patients must be kept under airbone and contact precautions. Recently, a live-attenuated vaccine has been licensed and is routinely recommended for susceptible children, adolescents, and adults in high-risk categories.

E. Hepatitis A

Hepatitis A virus has no major effect in pregnancy, and it is not transmitted vertically to the fetus. There is no evidence that this virus is a teratogen.

F. Hepatitis B

Hepatitis B virus (HBV) is a DNA virus accounting for 35% of reported cases of viral hepatitis. In the normal adult 90% of acute HBV infections resolved completely within 6 months. The remaining 10%

will develop chronic infection and continue to manifest HBV surface antigen (HBsAg). HBV is not considered a teratogen, and cases of a congenital HBV syndrome have not been described. Vertical transmission of the virus to the baby occurs primarily during delivery. Babies born to mothers that are antigen positive are at great risk for infection, and if infected they have a significant chance of becoming chronic carriers of HBV. For this reason all pregnant women should be screened for HBsAg at the same time that other prenatal screening is accomplished. Universal HBV immunization is recommended for all babies. For babies born to HBsAg-negative mothers, three intramuscular doses are required to provide effective protection. For babies born to HBsAg-positive mothers, vaccination should be done shortly after birth. In addition they should receive one dose of HBV immunoglobulin, preferably within 12 hr of birth. This regimen can prevent perinatal HBV infection in 95% of neonates.

G. Hepatitis C

Hepatitis C virus (HCV) is the principal cause of non-A, non-B hepatitis. Before the implementation of universal screening for HCV in donors of blood, this virus accounted for 95% of cases of posttransfusion hepatitis. A significant number of individuals affected by this virus can develop chronic liver disease, and cirrhosis ultimately develops in 20–25% of cases. This virus has not be proven to be a teratogen, but maternal to fetal transmission can occur in 5% of patients. The risk of transmission is correlated with maternal HCV RNA titer levels and appears to be increased in women infected with human immunodeficiency virus. To date, the significance of perinatal transmission to the neonate has not been established.

III. SEXUALLY TRANSMITTED DISEASES

A. Syphilis

Syphilis is an infectious disease cause by the spirochete *Treponema pallidum*. This spirochete can cross the placenta and cause congenital disease and stillbirth. Thus, it is important to screen and treat all pregnant women demonstrating infection. The incidence of syphilis in the United States continues to increase and, unfortunately, the number of babies with congenital syphilis has also risen. Syphilis is divided into three stages: primary, secondary, and tertiary. The primary lesion follows an incubation period of 10–90 days, and it is characterized by a perineal and/or cervical chancre. If untreated, this lesion will heal spontaneously, followed by a generalized macular papular rash in the palms and soles, generalized lymphadenopathy, and genital condyloma lata. These findings resolve spontaneously followed by a latent period in which manifestations of the disease are absent. Untreated maternal primary and secondary syphilis can result in congenital syphilis in up to 50% of babies and can result in a significant number of premature babies and stillbirths. The spectrum of congenital disease includes stillbirth, fetal hydrops, early congenital syphilis characterized by maculopapular rash, rhinitis or "snuffles," hepatosplenomegaly chorioretinitis, and osteochondritis. Late congenital syphilis is significant for the presence of Hutchinson's teeth, interstitial keratitis, eight nerve deafness, saddle nose, saber shins, and cardiovascular lesions. The diagnosis of syphilis in pregnancy usually depends on serologic testing. The most common screening tests are the rapid plasma reagin test and the venereal disease research laboratory test and detect nontreponemal antibodies. The diagnosis should be confirmed by specific antitreponemal antibody tests such as the fluorescent treponemal antibody-absorption test. Pregnant women with syphilis should be treated with a penicillin regimen appropriate to the stage of infection. The most critical feature of the management of the pregnant patient with syphilis is follow-up care. Serial serologic titer determinations are crucial to confirm cure of the disease.

B. Gonorrhea

Gonorrhea is a common sexually transmitted disease caused by *Neisseria gonorrhoeae*, a gram-negative diplococci. The prevalence of gonorrhea in pregnancy varies from 0.5 to 7% and reflects the risk status of the population. Risk factors for gonococcal infection include young age, multiple sexual partners, low socioeconomic status, drug abuse, and being single. Gonococcal infection is associated with concomitant chlamydial infection in 40% of pregnant women. In most pregnant women, gonococcal infection is limited to the cervix, urethra, periurethral, and vestibular glands. Gonococcal cervicitis has been associated with postabortion and postpartum endometritis, premature rupture of membranes, chorioamnionitis, and preterm delivery. The neonate may develop gonococcal ophthalmia if exposed to maternal infection and if ocular prophylaxis is not given. Pregnant women with risk factors for gonococcal infection should have cervical cultures for *N. gonorrhoeae* performed at an early prenatal visit. The treatment for gonorrhea includes a regimen of ceftriaxone plus erythromycin. All infants should be given prophylaxis against eye infection.

C. Chlamydia

Chlamydia trachomatis is one of the most prevalent sexually transmitted bacterial infections in the world. Chlamydia has been detected in 2–13% of pregnant women, with a prevalence as high as 37% in sexually active adolescent girls. Unrecognized infection is common. This bacteria may cause urethritis and mucopurulent cervicitis in the nonpregnant state. The role of maternal chlamydial infection in pregnancy is controversial. Several studies have found an association between cervical infection and premature rupture of the membranes, premature delivery, and low birth weight infants, but the strength of this association needs to be determined. However, 50–60% of infants delivered vaginally to women infected with chlamydia will be colonized with this bacteria. The most common manifestations of neonatal infection are inclusion conjuctivitis and pneumonia. Treatment should be administered to women who have chlamydia infection with a regimen of erythromycin. Simultaneous treatment of male partners is an important component of the therapeutic regimen.

D. Herpes Simplex

The herpes simplex virus (HSV) is a DNA virus which can cause significant disease in man. HSV-1 causes orolabial herpes, traditionally known as fever blisters. HSV-2 causes genital herpes, a sexually transmitted disease. All pregnant women should be questioned about a history of genital herpes. Patients with an active primary genital herpetic lesion who deliver vaginally have a high risk of transmitting the infection to the baby (33–50%), whereas patients with recurrent herpetic lesions have a very low risk of transmitting the virus to their progeny (2%). Congenital (intrauterine) infection is extremely rare, and a congenital HSV syndrome has not been described. In contrast, vertical transmission to the fetus at time of delivery can result in generalized infection in the neonate, involving the central nervous system, skin, eyes, and mouth. The benefits of prophylactic administration of antiviral drugs such as acyclovir to pregnant women with genital HSV have not been established. This drug may be administered to treat maternal life-threatening HSV infection such as encephalitis and pneumonitis. To date, there is no evidence that acyclovir causes adverse effects on the fetus. Pregnant women with a history of HSV infection should undergo a careful perineal examination when they present for delivery; in the presence of an active HSV genital lesion, cesarean delivery is recommended to avoid contact between the baby and the virus. Serial cervical and vaginal HSV cultures for screening asymptomatic pregnant women have not been shown to predict the risk of newborn infection.

E. Human Papillomavirus

Several types of human papillomavirus (HPV) cause condylomata acuminata (genital warts), many cases of cervical intraepithelial neoplasia, and juvenile laryngeal papillomatosis. Most genital HPV infections are sexually transmitted and usually asymptomatic. In pregnancy genital warts may proliferate and become increasingly friable. Treatments with cryotherapy and trichloroacetic acid are not contraindicated in pregnancy. The risk for a neonate whose mother has genital HPV to develop laryngeal papillomatosis is very small, and cesarean section for prevention is not indicated.

F. Human Immunodeficiency Virus

The virus responsible for acquired immunodeficiency syndrome (AIDS) is discussed elsewhere in this encyclopedia.

IV. OTHER INFECTIONS

A. Toxoplasmosis

Toxoplasmosis infection is caused by the protozoan *Toxoplasma gondii*. Infection is acquired by eating infected raw or undercooked meat and through contact with infected cat feces. As many as one third of women in the United States have antibodies to this organism and are immune to infection during pregnancy. In the adult, infection can result in fatigue, muscle pains, and lymphadenopathy, but most often it is subclinical. Primary infection in pregnancy may result in abortion or significant fetal disease. Although congenital infection is more common after maternal infection in the third trimester, the sequelae from first-trimester infection are more severe. The classical tetrad of congenitally acquired toxoplasmosis consists of chorioretinitis, microcephaly, cerebral calcifications, and clinically evident cerebral damage. Overall, 75% of infected infants are asymptomatic at birth. Currently, there is no reliable way to predict the outcome for these babies. The diagnosis of maternal infection is based on serologic antibody testing. Seroconversion is the best method to diagnose acute infection. A significant rise in IgG titer or the presence of very high titers most often indicate recent or current infection. The presence of IgM antibody is suggestive of acute infection, but IgM antibodies may persist for years. Confirmation of fetal infection can be obtained by culturing amniotic fluid or fetal blood for toxoplasma or detecting the protozoan's DNA by polymerase chain reaction. Treatment of infected mothers with the macrolide antibiotic spiramycin may reduce the incidence of fetal infection, but it may not modify its severity. Combination therapy with pyrimethamine and sulfadiazine has been used; however, the efficacy of this treatment has not been proven.

B. Vaginitis

Vaginal discharge is one of the most common complaints of pregnant patients. This discharge may be the result of normal physiologic changes of pregnancy or may result from infectious vaginitis. The most common causes of vaginitis in pregnancy are bacterial vaginosis, candidiasis, and trichomoniasis. These infections are usually asymptomatic, but recently bacterial vaginosis has been associated with premature rupture of the membranes, amniotic fluid infection, preterm labor, and preterm delivery. In bacterial vaginosis infection, there is an increased prevalence of *Gardnerella vaginalis* and selected anaerobes in the vagina. The mechanisms by which bacterial vaginosis may cause prematurity are currently unknown. However, the strong association between this vaginal infection and prematurity risk warrants treatment of symptomatic patients and patients at risk for preterm delivery and asymptomatic vaginal infection with either metronidazole or clindamycin cream.

See Also the Following Articles

Fetal Anomalies; Fetal Monitoring and Testing; HIV Infection and AIDS; Immunology of Reproduction; Puerperal Infections; Sexually-Transmitted Diseases

Bibliography

American Academy of Pediatrics (AAP) (1997). *Red Book: Report of the Committee on Infectious Diseases* (G. Peter, Ed.), 24th ed. AAP, Elk Grove Village, IL.

American College of Obstetricians and Gynecologists (ACOG) (1988). Antimicrobial therapy for obstetric patients, ACOG Technical Bulletin No. 117. ACOG, Washington, DC.

American College of Obstetricians and Gynecologists (ACOG) (1993). Perinatal viral and parasitic infections, ACOG Technical Bulletin No. 177. ACOG, Washington, DC.

American College of Obstetricians and Gynecologists (ACOG) (1997). *Guidelines for Perinatal Care*. ACOG, Washington, DC.

Centers for Disease Control and Prevention (1993). 1993 sexually transmitted diseases treatment guidelines. *Morb. Mortal. Weekly Rep.* 42(RR-14), 27–47.

Centers for Disease Control and Prevention (1996). Prevention of perinatal group B streptococcal disease: A public health perspective. *Morb. Mortal. Weekly Rep.* 45(RR-7), 1–24.

Charles, D. (1993). *Obstetric and Perinatal Infections*. Mosby/ Year Book, St. Louis, MO.

Gilstrap, L. C., and Faro, S. (1997). *Infections in Pregnancy*, 2nd ed. Wiley, New York.

Gonik, B. (1994). *Viral Diseases in Pregnancy*. Springer-Verlag, New York.

Sweet, R. L., and Gibbs, R. S. (1995). *Infectious Diseases of the Female Genital Tract*, 3rd ed. Williams & Williams, Baltimore.

Infertility

Laurence C. Udoff and Eli Y. Adashi

University of Utah

primary infertility No history of a conception.
secondary infertility At least one previous documented conception.

GLOSSARY

fecundability The probability of achieving a pregnancy within one menstrual cycle.

fecundity The ability to achieve a live birth within one menstrual cycle.

infertility The inability to conceive following 12 months of regular coitus without contraception.

intrauterine insemination A process that utilizes special laboratory techniques (such as washing, swimup, or gradients) to separate from a semen sample a population of sperm that may have a higher fertilization potential. These sperm are resuspended in a small volume of media and are then introduced into the uterus by the placement of a small catheter through the cervix. *Artificial insemination* is a more general term that describes this process wherein the sample may be inserted into the vagina, cervix, or uterus. The sperm may be from the husband [artificial insemination husband (AIH)] or a donor [artificial insemination donor (AID)].

in vitro fertilization An assisted reproductive technology wherein oocytes are retrieved from the ovaries, usually after pharmacological stimulation to produce multiple mature oocytes. The oocytes are fertilized in the laboratory and allowed to grow and develop in culture for 2–4 days. Usually, two to four of the highest quality embryos are then chosen for placement into the uterus with the goal that one will successfully implant.

laparoscopy An outpatient surgical procedure in which a fiber-optic scope is inserted into the intraabdominal cavity through a 1-cm umbilical incision. Accessory incisions of similar or less size may also be required if additional instrumentation is needed. This allows for a visual inspection of the pelvis as well as the ability to treat many of the common abnormalities encountered.

Infertility is defined as the inability to conceive after 12 months of unprotected intercourse. Primary infertility is diagnosed if the couple has never had a conception. Secondary infertility is the term applied to couples that have a history of at least one documented conception but who currently have been unable to conceive despite 12 months of regular intercourse and no contraception.

I. INTRODUCTION

The probability of a normal couple conceiving in a given menstrual cycle is called fecundability and is estimated to be 25%. The probability of a live birth resulting from the conception is designated as the fecundity, which takes into account an estimated 25% spontaneous abortion rate. At the end of a year, it is expected that over 90% of normal couples will have conceived. For those who do not conceive the diagnosis of infertility does not mean that conception cannot occur. Several epidemiologic studies have demonstrated that there is an incidence of spontaneous conception in infertile couples, especially in the first 3 years of infertility. It has been estimated that in patients with a history of infertility of 1 year's duration, up to a 50% spontaneous pregnancy rate can be expected in the following year. These statistics illustrate that in addition to identifying those couples that have a serious pathologic condition, the role of the physician in many instances is to speed up the period of time it takes to conceive.

II. CAUSES OF INFERTILITY

Infertility is thought to affect 10–15% of all reproductive-aged married couples in the United States. The etiology of infertility can be divided into three major categories: (i) male factor, (ii) female factor, and (iii) undetermined etiology. In approximately 40% of couples in which a cause for infertility is uncovered, the etiology is primarily male factor. Female factor is the major underlying disorder in another 40%. The remaining 20% of couples have infertility that is due to a combination of male and female factors. In 10–20% of couples presenting for evaluation, despite a thorough investigation, no specific etiology is found.

A. Male Factor

The major causes of male factor infertility can be subdivided into pretesticular, testicular, and posttesticular. Table 1 outlines the common diagnosis in each of these categories. It is important to note that approximately 50% of cases have no specific identifiable etiology. Additionally, in 75% of patients, there is no treatment to directly improve the abnormality.

Fortunately, technology is available that can overcome the deficit in the majority of cases.

B. Female Factor

In approximately 40% of infertile women, the main etiological factor is the failure to ovulate. In another 40%, the cause is due to tubal damage or other pelvic pathology such as pelvic adhesions (i.e., scar tissue) or endometriosis. Unusual causes of infertility, such as uterine or cervical abnormalities, occur in 10% of patients. Unexplained infertility is found in approximately 10% of women with infertility.

III. EVALUATION OF INFERTILITY

A. History and Physical Examination

The investigation of an infertile couple begins with obtaining a complete medical history and performing a physical exam. Ideally, both partners should be present. The goal of the interview is to try to uncover facts that may guide the evaluation. Questions should be directed toward evaluating the integrity of the

TABLE 1
Etiologic Factors in Male Infertility[a]

Pretesticular	Testicular	Posttesticular
Chromosomal (Klinefelter's syndrome)	Congenital	Obstructive
	Cryptorchidism	Epididymal
	Immotile cilia syndrome	Congenital
		Infective
Hormonal		
Hypogonadotropic hypogonadism	Infective (orchitis)	Vasal
Hyperprolactinemia		Congenital
	Vascular	Acquired
Coital disorders	Torsion	
Coital frequency	Varicocele	Epididymal hostility
Erectile dysfunction		
Ejaculatory failure	Gonadotoxic agents	Accessory gland infection
	Immunologic	Immunologic
	Idiopathic	

[a] Modified from D. M. de Krester, and H. W. G. Baker, Human infertility: The male factor, In *Reproductive Endocrinology, Surgery and Technology* (E. Y. Adashi, J. A. Rock, and Z. Rosenwaks, Eds.). Lippencott-Raven, New York, 1996.

entire reproductive axis (i.e., hypothalamus, pituitary, ovary, testes, Fallopian tubes, uterus, cervix, and vagina) since this must be intact for normal fertility. For example, if the history includes irregular, unpredictable menses, anovulation is suspected and the initial evaluation should focus on the causes of anovulation. Likewise, the physical exam should evaluate each individual component of the reproductive system. Particular attention is paid to examination of the thyroid for abnormalities, the skin for evidence of hirsutism or acne (signs of androgen excess), and the breast for evidence of galactorrhea (which suggests hyperprolactinemia). Pelvic examination can yield clues regarding ovarian function (e.g., cervical mucous characteristics and vaginal mucosa appearance may correlate with the level of estrogen production) and the presence of pelvic pathology (e.g., a nonmobile retroverted uterus suggests the presence of pelvic adhesions from past infections, surgeries, or endometriosis).

B. Tests

As noted, laboratory testing should be guided by clinical judgment based on the history and physical exam. However, in many situations, the history and exam are unremarkable. In these cases a series of tests are routinely performed to assess the basics of the reproductive axis.

1. Ovulation

The documentation of ovulation can be accomplished in several different ways. Commonly, women are asked to record their basal body temperatures (BBT). This involves taking an oral temperature immediately upon awakening (i.e., before any activity) on a daily basis and graphing the results. Under the influence of circulating progesterone produced by the corpus luteum, the BBT rises after ovulation, creating a biphasic curve. This elevation (a mean increase of at least 0.4° Fahrenheit) is noted approximately 1 day after ovulation. The advantages of the BBT are the low cost, the ability to confirm ovulation in most women, as well as the assessment of the length of the luteal phase. Drawbacks include that it does not accurately prospectively predict ovulation, it can be bothersome to record, and it may be

difficult to interpret or fail to document ovulation in a normal patient.

Home urinary luteinizing hormone (LH) surge testing is another common diagnostic tool to assess ovulation. As the follicle and egg matures, circulating estradiol levels increase, exerting feedback to the hypothalamic–pituitary axis, eventually leading to the release of a surge of LH from the pituitary that stimulates the final maturation process of the egg and eventual ovulation. Various commercial kits are readily available that can detect the occurrence of the LH surge by utilizing an enzyme-linked immunosorbent assay technique to measure LH levels in the urine. Most kits record a color change in a test strip when the LH surge is detected, signaling that ovulation is expected to occur in approximately 24 hr.

Serum progesterone levels obtained in the midluteal phase may also be used to assess ovulation. Values above 4 ng/ml obtained between Days 19 and 23 of the menstrual cycle are considered to be consistent with ovulation. Pitfalls of a single measurement are that progesterone is secreted in a pulsatile fashion, creating the potential for sampling error, and that a single progesterone level is a poor predictor of luteal phase adequacy.

2. Cervical Mucous

The postcoital test provides information regarding the quality of the cervical mucous as well as the sperm/cervical mucous interaction. Just prior to ovulation, circulating estrogen levels peak, providing stimulation to the endocervical glands to produce copious amounts of clear, watery mucous. Before or 2 or 3 days after ovulation, the mucous is thick, viscid, and opaque. The postcoital test is performed by instructing the patient to have intercourse around the time of the LH surge and ovulation. An exam is done within 2–8 hr to check the quality of the mucous [e.g., appearance, pH, stretchability (spinnbarkeit), and sperm–mucous interaction]. A normal postcoital test should find several progressively motile spermatozoa with microscopic examination. Abnormal findings include thick viscid mucous which suggests inadequate estrogen effect or an incorrectly timed test. Additionally, sperm may be rare or not seen, which could suggest oligospermia or poor coital technique. Sperm may also appear not to be motile or clumped together with no progressive movement, suggesting anti-sperm antibodies or a "hostile" cervical mucous environment. A postcoital test should not be used in place of a semen analysis and an abnormal test should be repeated since the incidence of incorrectly timed tests is high. Treatment for abnormal postcoital tests includes the use of preovulation estrogen therapy as well as mucolytic agents such as guaifenesin. Ultimately, intrauterine inseminations may be performed to bypass the cervical mucous effect. In common practice, due to the rarity of true cervical factor infertility, the poor predictive value of the test, the frequent use of intrauterine insemination, and the inconveniences of the test, the postcoital test is often omitted.

3. *Luteal Phase Adequacy*

Implantation of the fertilized egg occurs approximately 1 week after ovulation. The success of implantation is due in part to the readiness of the endometrium. Correct maturation of the endometrium requires sufficient hormonal stimulation, primarily by progesterone, as well as normal end organ responsiveness. In some cases of infertility (3–5%) it is thought that the endometrium is improperly prepared and a luteal phase inadequacy or luteal phase defect is present. This is diagnosed by an endometrial biopsy, a procedure usually performed in the office by passing a small catheter into the uterus to obtain a tissue sample of the uterine lining. This tissue is then analyzed by a pathologist who "dates" the endometrium, based on the development of the endometrial glands and stroma. If there is more than a 2-day discrepancy between the pathologist's impression and the actual day of cycle the tissue was obtained, a luteal phase defect may exist. Since this finding has occurred commonly in normal women (up to 30%), a second biopsy should be performed to confirm the suspicion. Treatment of this defect involves progesterone supplementation in the luteal phase and/or the use of clomiphene citrate. Due to the rarity of this diagnosis, and the discomfort and expense involved in attempting to make the diagnosis, in common practice this test is usually only offered to patients who have significant risk factors such as short luteal phases and unexplained infertility.

4. *Hysterosalpingogram*

Tubal abnormalities (e.g., occlusion) are a relatively common finding in infertile women. They are particularly likely if there is a history of pelvic inflammatory disease, ruptured appendix, tubal surgery, ectopic pregnancy, or septic abortion. Tubal disease may be diagnosed either by a hysterosalpingogram or by laparoscopy. Since the former is a nonsurgical technique, it is usually performed first, in the initial stages of evaluation. Additionally, the hysterosalpingogram allows for evaluation of the contour and adequacy of the uterine cavity. It is performed by instilling contrast dye into the uterus, through a cannula placed into the cervix, under image intensification fluoroscopy. X-ray films are taken to document the outline of the uterine cavity as well as the caliber and patency of the Fallopian tubes. Though hysterosalpingograms have been a standard part of the infertility evaluation for many years, there are limitations to their interpretations. For example, tubal patency does not guarantee the absence of pelvic adhesions or endometriosis. Also, the inability to demonstrate free flow of dye from a tube may represent a technical artifact. Tubal abnormalities found on hysterosalpingograms are most often evaluated by laparoscopy. Uterine cavity abnormalities may be further evaluated and treated by hysteroscopy (a fiber-optic scope inserted into the uterine cavity to visualize or allow for the surgical treatment of endometrial abnormalities).

5. *Male Factor*

As previously noted, abnormalities in sperm number, quality, and/or function may be the primary cause of infertility in 40% of couples. For this reason, a semen analysis should be one of the first tests obtained in an evaluation. Specimens are usually obtained by masturbation after an interval of abstinence of 2–5 days. At least two samples should be examined several weeks apart since there can be considerable variability in quality. A semen analysis should at a minimum examine the number of sperm [total number and count (number of sperm/ml)], sperm motility, and sperm morphology. Though normal values vary significantly from lab to lab depending on the criteria used, in general a sperm count should be over 20 million, with 40% or more of sperm having good motility and 60% or more demonstrating normal forms. As in the evaluation of the female partner, an abnormality in the semen analysis should be evaluated by a specialist in the field (often a urologist) who inspects the entire male reproductive axis. Other specialized sperm function tests are available, such as a test for anti-sperm antibodies (often seen in patients with a prior vasectomy reversal procedure) or a sperm penetration assay (especially useful when the semen analysis suggests fertilization potential may be compromised).

6. *Laparoscopy*

Laparoscopy is an outpatient surgical procedure in which a fiber-optic scope is inserted in to the intraabdominal cavity through a 1-cm incision in the umbilicus. The pelvic anatomy is inspected, with particular attention paid to the appearance of the ovaries, tubes, and uterus. Accessory incisions are often required in the lower abdomen to allow for the manipulation of pelvic structures with probes and grasper to ensure an adequate inspection. Dye can be passed through a cannula in the cervix to confirm tubal patency. Common abnormal findings at laparoscopy include the presence of adhesions or endometriosis (ectopic endometrial tissue often found on the tubes, ovaries, or peritoneal surfaces that are associated with infertility). Due to the expense, risk (including general anesthesia), and invasive nature of laparoscopy, it is usually reserved as the last step in the infertility evaluation unless the history and exam dictate otherwise.

IV. TREATMENT OF INFERTILITY

Treatment strategies depend on the specific etiology of infertility. For example, anatomical defects, such as damaged tubes or Müllerian anomalies, may be corrected surgically. Endocrinopathies such as a luteal phase deficiency are treated by progesterone supplementation. Other examples include thyroid hormone replacement for hypothyroid patients and the use of bromocriptine in hyperprolactinemic patients. The following sections review in greater detail some of the more commonly employed treatments for other etiologies of infertility.

A. Ovulation Induction

Approximately 10–15% of infertile females are anovulatory. If potential causes of anovulation have been addressed and the women remains anovulatory (e.g., failed to ovulate despite adequate treatment of hypothyroidism or hyperprolactinemia), attempts at medical induction of ovulation are reasonable. Several different medications may be used to induce ovulation, including clomiphene citrate, human menopausal gonadotropins [or purified follicle-stimulating hormone (FSH)], and gonadotropin-releasing hormone (GnRH).

The most commonly used medication for ovulation induction is clomiphene citrate. It is a relatively inexpensive oral preparation with few serious side effects (e.g., patients may experience hot flashes and/or, less commonly, headaches). Despite years of research and a wealth of literature on the subject, the specific mechanism of action of clomiphene is not fully understood. Classically, clomiphene is thought to act via its capacity as an antiestrogen, displacing estrogen from estrogen receptor sites. This may cause the hypothalamus to sense a low estrogen environment and to respond by signaling the pituitary via pulsatile GnRH to increase gonadotropin (FSH and LH) release to produce follicular development and subsequent estrogen synthesis. With an intact hypothalamic–pituitary axis, clomiphene has a 90% success rate in inducing ovulation. If there are no other infertility-related factors present, pregnancy rates are approximately 65%. Eighty percent of patients who conceive on clomiphene do so within three cycles of therapy, suggesting that protracted courses of therapy are not warranted. Additionally, recent data indicate that patients receiving more than a year of therapy may have an increased risk for ovarian tumors. Other risks of therapy include an increased risk for multiple pregnancy (approximately 5–7%), with the overwhelming majority being twin gestations, and a small risk for ovarian hyperstimulation syndrome (1% or less).

Human menopausal gonadotropins (hMG) or the newly available pure FSH compounds made through recombinant DNA technology may also be used to induce ovulation. These medications must be given by injection and are very expensive. The mechanism of action is direct stimulation of the ovarian follicles by FSH. The response can be very vigorous, especially in anovulatory patients, and requires very close monitoring with serum estradiol levels and transvaginal ultrasound to observe the number of follicles that are developing. When a follicle or follicles appear mature by ultrasound (i.e., diameter between 16 and 18 mm) an injection of human chorionic gonadotropin (hCG) is given to stimulate ovulation. This is effective due to the similarities between the hCG and LH molecules, with this injection essentially simulating the LH surge that would occur naturally prior to ovulation. The risks of treatment include an elevated risk for multiple pregnancy and the risk of ovarian hyperstimulation syndrome. Since the patient is being monitored with ultrasounds, the risks can be tempered by canceling cycles in which too many follicles mature and/or the serum estradiol level is unacceptably high. This same treatment strategy may be employed in patients that ovulate normally with the hope of stimulating more than one follicle to mature, thereby improving the chances of conception. This "superovulation" tactic is often used in couples with unexplained infertility and other situations of low estimated fecundity prior to resorting to *in vitro* fertilization (IVF), which is more costly and invasive.

GnRH is another option for ovulation induction. Its utility has been limited due to the route of administration (intravenous infusion pump) and its effectiveness mainly in patients with hypogonadotropic hypogonadal anovulation. The medication is given in pulsatile boluses, not unlike the pulsatile secretion of GnRH that occurs naturally with a normally functioning hypothalamus. An intact pituitary ovarian axis is required. Patients are monitored by ultrasound for follicular development. Treatment is more likely to result in the development of only one mature follicle, lowering the risk of multiple gestation and ovarian hyperstimulation syndrome.

B. Treatment of Endometriosis

Endometriosis is the ectopic growth of both endometrial glands and stroma that can only be diagnosed by direct visualization and/or biopsy (e.g., through laparoscopy or laparotomy). It most commonly af-

fects the ovaries, the pelvic peritoneal surfaces, the Fallopian tubes, and the bowel, though it can be found in almost any location. The ectopic tissue is hormonally sensitive, especially to stimulation by estrogen, just like the endometrial tissue that lines the uterus. For this reason when endometriosis is associated with pain or other symptoms, the symptoms are usually cyclical in nature, timed with the menstrual cycle. It is not clearly understood how endometriosis may cause infertility, but it is known to occur with greater frequency in women with infertility (approximately 30–40%) and when treated fecundity improves.

The treatment of endometriosis in infertile patients may involve expectant management, medical therapies, surgical procedures, combined medical/surgical approaches, or assisted reproductive technologies. Expectant management may be appropriate in those couples with very minimal disease and no other etiologies for infertility since spontaneous pregnancy rates have been reported to be approximately 7% per cycle. However, a recent randomized clinical trial has suggested that treatment by ablation, coagulation, or resection at the time of the diagnostic surgery may significantly improve fecundity. Medical options, once the diagnosis has been made, include oral contraceptive pills, danazol, progestins, and GnRH agonists. These compounds act directly or indirectly to decrease the stimulation of the ectopic endometrial tissue and have been shown to decrease the amount of endometriotic disease and associated pelvic pain. However, insufficient data are available to unequivocally prove effectiveness in treating infertility. Each of these medical options is associated with specific side effects. One that is common to all the choices is the inability to attempt conception since all these strategies usually prevent ovulation. Surgical options include laparoscopic ablation or coagulation of tissue as well as the resection of lesions. Laparotomy, previously the mainstay of treatment for pelvic endometriosis, is now usually reserved for patients with severe or extensive disease in which adnexal masses, extensive adhesions, or intestinal involvement precludes the use of laparoscopy. The combination of medical and surgical approaches has been addressed in numerous studies. Unfortunately, no consensus has been reached as to the utility of pre- or postoperative medical therapy in the treatment of endometriosis in patients with infertility. Assisted reproductive technologies are another treatment option for infertile patients with endometriosis. These include the use of superovulation with or without intrauterine insemination, IVF, or gamete intrafallopian transfer (GIFT). These procedures, which will be discussed in detail later, have in some instances been shown to improve fecundity rates in patients with endometriosis, especially in cases of tubal damage and distortion of pelvic anatomy that are treated by IVF. However, unequivocal data are lacking to suggest the appropriate treatment regimen in less obvious cases and the role of medical pretreatment is incompletely understood.

C. Treatment of Male Factor Infertility

As noted, in approximately 75% of cases, no specific treatment is available to reliably improve male factor abnormalities. Table 2 outlines those occa-

TABLE 2
Treatment of Male Factor Infertility According
to Diagnosis[a]

Sperm autoimmunity
 Glucocorticoids
 AIH
 IVF
 Surgery for associated genital tract obstruction or autoimmune orchitis
Gonadotropin deficiency
 Gonadotropins or pulsatile GnRH therapy
Coital disorders
 Sexual behavior therapy, intrapenile injections, AIH with sperm collected by other means
Genital tract obstruction
 Bypass surgery
 Sperm aspiration for IVF
Low sperm count, motility, and abnormal morphology
 Remove potential torsion, improve poor health, varicocelectomy
 IVF

[a] Modified from D. M. de Krester, and H. W. G. Baker, Human infertility: The male factor, In *Reproductive Endocrinology, Surgery and Technology* (E. Y. Adashi, J. A. Rock, and Z. Rosenwaks, Eds.). Lippencott-Raven, New York, 1996.

sional conditions in which treatment regimens have been recommended. One example is a patient with an abnormal semen analysis and a varicocele on examination. Varicoceles are abnormal dilations (varices) of scrotal veins. They are usually caused by deficient valves in the left internal spermatic vein and are associated with abnormalities in sperm motility and morphology. A varicocelectomy (ligation of the internal spermatic vein) may be performed. However, this usually results in only marginal increases in pregnancy rates, rendering this treatment controversial. Likewise, genital tract obstruction may be bypassed surgically or patients with gonadotropin deficiencies may be treated hormonally. In many cases, especially when the semen parameters are severely compromised, male factor infertility is treated by advanced reproductive technologies such as IVF and intracytoplasmic sperm injection (ICSI).

D. IVF/GIFT/ZIFT

Indications for IVF include tubal factor infertility, endometriosis, male factor infertility, immunologic infertility, and unexplained infertility. Treatment most commonly starts with the downregulation of endogenous gonadotropin release by a GnRH agonist to prevent interference with the ovulation induction process. Then the ovaries are stimulated to produce maximal numbers of oocyte-containing follicles using hMG and/or a pure FSH compound. All these medications must be given by injection (either subcutaneous or intramuscular). Patients are monitored by ultrasound and through the evaluation of serum estradiol levels. When the follicles are thought to be mature an injection of hCG is given as in superovulation cycles. The distinct difference is that in IVF, 34–36 hr after this injection the eggs are retrieved, usually by a transvaginal ultrasound guided approach. The oocytes are than fertilized with sperm *in vitro*. Depending on the degree of male factor, this may involve simply coincubation of sperm and egg or the use of micromanipulation techniques. One technique that has revolutionized the treatment of male factor infertility is ICSI. This procedure introduces a single sperm directly into the cytoplasm of the egg and can produce fertilization even in cases with very few poor-quality sperm. Fertilized eggs are than allowed to grow and divide in culture. Currently, most IVF centers will transfer embryos into the uterus on the third day after the egg retrieval, when they are often at the eight-cell stage. This is accomplished by simply placing a catheter through the cervix and into the uterus and expelling the embryos into the uterine cavity. The number of embryos transferred to the uterus is determined by many factors, including the embryo quality, age of the patient, and concern for multiple pregnancy. In general, pregnancy rates rise as the number of embryos replaced is increased; however, beyond three or four embryos, the percentage of patients with high-order multiple pregnancies increases sharply. Due to the poor outcomes of these pregnancies, most centers limit the number of embryos transferred to two or three. Excess embryos may be cryopreserved for future attempts at conception, though conception rates in these cycles are low. After transfer, the patient is given luteal phase support with progesterone. A serum pregnancy test is obtained 2 weeks after the date of oocyte retrieval. The Society for Assisted Reproductive Technologies registry in 1995 showed 281 programs performing 41,087 treatment cycles of IVF with an average delivery rate per oocyte retrieval of 22.5%. Success rates vary significantly by diagnosis and age. For example, some centers report delivery rates of 50% or more in couples under the age of 40 with tubal factor infertility, whereas patients over the age of 40 may experience success rates of <15%.

ZIFT is the same procedure as IVF except that embryos are transferred into the Fallopian tubes rather than the uterus. The main advantage of this technique is the exposure of the embryos to the possible beneficial effects of the tubal environment which *in vivo* is thought to be an important component of the conception process. Disadvantages include the need for normal tubes and the need for laparoscopy to perform the transfer. ZIFT is a much less commonly employed procedure compared to IVF.

GIFT utilizes the same oocyte stimulation protocol as IVF; however, once the eggs are retrieved, they are mixed with sperm and immediately transferred into the tubes. This does not allow for documentation of fertilization and also requires laparoscopy.

E. Gamete Donation

In couples with inadequate gametogenesis, the use of donor sperm or donor eggs is possible. Donor sperm is easily obtained through the many commercial sperm banks. An insemination of the sperm sample via a catheter introduced into the uterus is performed on the day of ovulation as predicted by LH surge detection or other means. Donor egg therapy requires IVF. In this case, a donor undergoes oocyte stimulation and egg retrieval with the resultant embryos being placed in the recipients uterus. Since the donors are usually fertile and young, this form of therapy is often more successful than standard IVF.

F. Future Therapies

Several significant developments appear likely to occur in the near future to improve the success rates of infertility treatment as well as to broaden the choices of therapy. These advances include the availability of recombinant DNA-produced gonadotropins and the GnRH antagonist, both of which are likely to improve ovulation induction and superovulation treatments. With respect to IVF, preliminary data suggest that the ability to mature embryos to the blastocyst stage prior to uterine transfer will significantly improve pregnancy rates. Since the implantation rate per embryo may also be higher, fewer embryos can be replaced, which it is hoped will reduce the multiple pregnancy rate associated with IVF. In addition, the ability to select embryos for transfer based on genetic information obtained through preimplantation genetic diagnostic technology may decrease the spontaneous pregnancy loss rate seen with IVF. For patients who need to delay childbearing or who are going to undergo gonadotoxic therapy (e.g., radiation or chemotherapy), oocyte cryopreservation is likely to be available in the future. This may also simplify and expand the availability of donor egg therapy.

G. Adoption/Childless Living

In many instances, the major responsibility of the physician is to suggest to a couple when to discontinue attempts at conception. Appropriate referrals to adoption agencies or to counselors should be a consideration in most cases.

See Also the Following Articles

ENDOMETRIOSIS; FERTILITY AND FECUNDITY; *IN VITRO* FERTILIZATION; STERILITY

Bibliography

Adashi, E. Y., Rock, J. A., and Rosenwaks, Z. (1996). Human infertility: Diagnosis and management. In *Reproductive Endocrinology, Surgery and Technology* (E. Y. Adashi, J. A. Rock, and Z. Rosenwaks, Eds.). Lippencott-Raven, New York.

Assisted reproductive technology in the United States and Canada: 1995 results generated from the American Society for Reproductive Medicine/Society for Assisted Reproductive Technology Registry (1998, March). *Fertil. Steril.* 69, 389–398.

Chandra, A., and Stephen, E. H. (1988, January). Impaired fecundity in the United States; 1982–1995. *Family Planning Perspect.* 30, 34–42.

Speroff, L., Glass, R. K., and Kase, N. G. (1994). Female infertility. In *Clinical Gynecological Endocrinology and Infertility*. Williams & Wilkins, Baltimore.

Inhibin

Ralph H. Schwall

Genentech, Incorporated

I. Introduction
II. Structure and Biosynthesis
III. Measurement
IV. Gene Expression and Biological Activities
V. Mechanism of Action
VI. Inhibin-α Knockout Mice

GLOSSARY

inhibin "A glycoprotein hormone consisting of two dissimilar, disulfide-linked subunits, which inhibits pituitary gonadotropin production and/or secretion, preferentially that of FSH" (H. G. Burger, M. Igarashi, *et al.*, *Endocrinology* 122, 1701, 1988).

knockout mice A family of mice created from embryonic stem cells in which a specific gene has been deleted by homologous recombination.

Müllerian-inhibiting substance A glycoprotein hormone of the inhibin/activin, transforming growth factor-β superfamily that is produced by the fetal testis and causes involution of the Müllerian ducts in the male fetus.

It has been known for centuries that castration leads to regression of the secondary sex characteristics, but in the early 1900s, with the advent of improved histological methods, several groups described another, more subtle, change—hypertrophy of cells within the pituitary gland. In 1932, McCullagh reported that an organic extract of the testis restored the secondary sex characteristics but had no effect on the pituitary hypertrophy. More important for this discussion, he also found that the pituitary hypertrophy could be reversed by an aqueous testicular extract, which had no effect on the secondary sex characteristics. He concluded that "the testicle secretes a hitherto unrecognized water soluble hormone, one function of which is a control of the pituitary gland," and he named this water-soluble hormone "inhibin."

I. INTRODUCTION

With the isolation of follicle-stimulating hormone (FSH) and luteinizing hormone (LH) and the development of gonadotropin-specific immunoassays, inhibin preparations were found to selectively inhibit the secretion of FSH. Purification of inhibin, however, proved extremely difficult. It was more than 50 years from the time McCullagh coined the term inhibin (1932) before the chemical nature of gonadal inhibin was finally resolved. A major breakthrough was the development of pituitary cell culture systems in which the effects of inhibin preparations on FSH secretion could be reproducibly quantitated, providing a reliable bioassay. That led, in 1985, to the purification of inhibin to homogeneity and the determination of its subunit structure by several groups, most notably the teams of Henry Burger, David de Kretser, and David Robertson in Australia, Roger Guillemin and Nicholas Ling at the Salk Institute, and Wylie Vale and Jean Rivier at the Salk Institute. Cloning of the inhibin subunits rapidly followed in 1986. Cornelia Channing and Neena Schwartz also are acknowledged for their tireless efforts, often in the face of severe criticism from peers, in trying to understand the biology of inhibin before its chemical nature was elucidated. Their work established the principles that led to the development of the pituitary bioassays that enabled purification.

The objective of this article is to provide a broad overview of the inhibin field. The focus is on gonadal inhibin. The reader should be aware that there is a large literature on inhibin, much of which involves the use of biological fluids and gonadal extracts. Because of the complexity of these fluids, caution needs to be used when reviewing data published before 1985 or any study that uses anything other than purified inhibin. In addition, there is another literature on seminal inhibins, which are small peptides that are biochemically and functionally distinct from the gonadal inhibins described here. Their physiological significance is a matter of controversy.

II. STRUCTURE AND BIOSYNTHESIS

Isolation of inhibin was only achieved by the use of chaotropic reagents, such as urea, that prevent intermolecular aggregation as well as adhesion to chromatography resins. Not only do the molecules behave poorly in standard purification methods but

also there are a number of biological aspects of this system that further complicated the isolation. First, inhibin is a complex group of molecules consisting of inhibin-A and inhibin-B, each of which exists in several forms. Thus, inhibin activity can be found in several regions of a chromatographic eluate. Second, starting materials such as follicular fluid also contain activin, which functionally antagonizes the effects of inhibin on FSH secretion in the pituitary cell bioassay. Third, the starting material contains yet another molecule, follistatin, that has inhibin-like activity in pituitary cells. With this hindsight, it is no wonder that the purification of inhibin proved intractable for so long.

Although the term inhibin is used here and elsewhere in the singular, it describes a complex family of proteins (Fig. 1). The inhibin molecules that have activity on FSH secretion are heterodimers that contain an α subunit that is covalently coupled by disulfide bond to either a βA or βB subunit to yield inhibin-A or inhibin-B, respectively, much like FSH and LH are composed of a common α subunit combined with similar but distinct β subunits. The inhibin α subunit is glycosylated to varying degrees, generating two isoforms, but glycosylation is not required for bioactivity.

The α, βA, and βB subunits are encoded by distinct but related genes, which are members of a much larger gene family that includes transforming growth factor-β (TGF-β), Müllerian-inhibiting substance (MIS), and many more. The primary transcript of each gene is a large precursor molecule. Dimerization between α and β subunits takes place early during biosynthesis, and the heterodimers are then processed by sequential removal of portions from the N terminus. A number of intermediate inhibin species are formed, all of which have been identified in biological fluids and many of which have activity on pituitary cells. The smallest inhibin that has FSH suppressing activity is the 32-kDa species, which consists of the C-terminal 18-kDa portion of the α subunit and the C-terminal 14-kDa portion of the β subunit. There was conjecture that activity of larger molecular weight forms might be due to local processing to the 32-kDa form; however, that concept was convincingly negated by Tony Mason and colleagues, who showed that biological activity is retained by a 56-kDa inhibin containing a mutation in

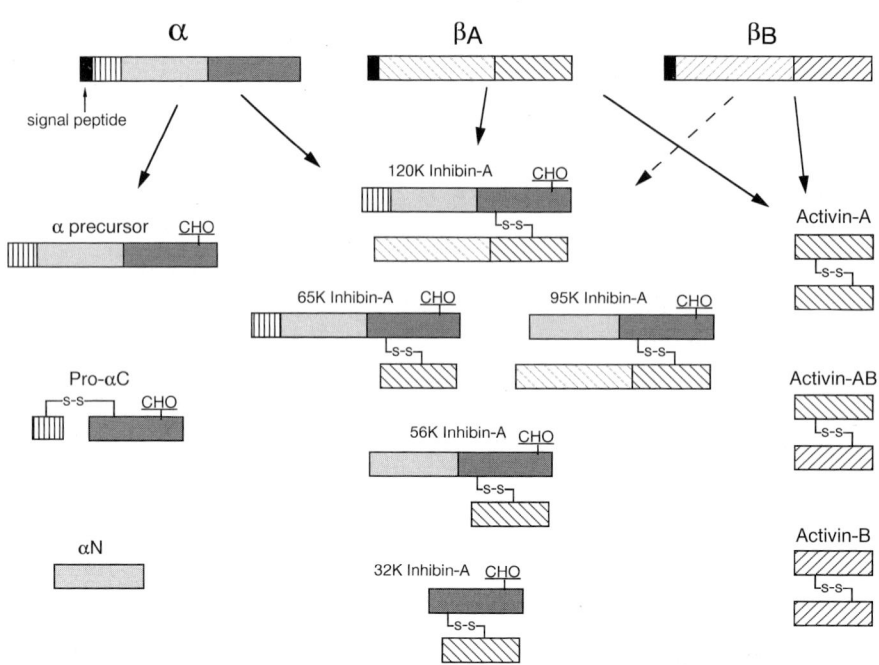

FIGURE 1 Inhibin structure and biosynthesis. The α, βA, and βB subunits are each encoded by mRNAs that encode for much larger precursors. The initial step in processing is cleavage of the signal peptide, which is represented by the black region at the N terminus. The α precursor can then be secreted intact or it can be further processed to Pro-αC and αN. The β subunits can homodimerize to form three forms of activin or they can combine with α to form inhibin. Formation and processing of the α–βA heterodimer is shown, but α–βB heterodimers can also form in a similar manner. CHO represents glycosylation.

the α subunit that prevents processing to the 32-kDa form.

The α and both β subunits contain a large number of cysteine residues. The placement of the cysteine residues is conserved among members of this super-family. The crystal structure of one member, TGF-β2, has been solved. It shows that the dimer is held by a single disulfide bond, with the rest of the cysteines making intrachain cross-links to form a tight "disulfide knot" in the center of each subunit. This explains the stability of inhibin against denaturation by the chaotropic agents that were used in the purification.

In addition to forming inhibin, the β subunits can also homodimerize to form activin-A, activin-B, or activin-AB (Fig. 1). The formation of inhibin vs activin within a cell is regulated by the relative expression level of the α and β subunits; expression of β in the absence of α leads to secretion of activin, whereas expression of increasing amounts of α with a fixed amount of β leads a progressive increase is the inhibin:activin ratio. All of the β subunit becomes incorporated into inhibin when the α:β ratio is 10:1. Two novel β subunit genes, βC and βE, have recently been cloned, but whether they can associate with the α subunit to form novel inhibins has yet to be determined.

Unlike the β subunits, the α subunit does not form homodimers when expressed in the absence of β. However, the α subunit precursor can be secreted intact, or it can undergo processing to yield two other inhibin-related proteins (Fig. 1). One is pro-αC, which contains a 6-kDa portion of the extreme N terminus of the pro-α molecule linked by disulfide bond to the mature 18-kDa C-terminal portion. The other is the intermediate peptide released by this processing, referred to as αN.

III. MEASUREMENT

The plethora of inhibins and inhibin-related molecules has made the development of specific immunoassays very difficult. The first antisera were produced against peptides corresponding to the N terminus of the 18-kDa form of the α subunit. Although these antisera recognize dimeric inhibin and could detect immunoactivity in biological fluids, they also cross-react with other α subunit derivatives such as full-length α precursor and pro-αC. In the past few years dimer-specific monoclonal antibodies to inhibin-A and to inhibin-B have been developed. These antibodies have been incorporated into ELISAs that are now beginning to permit the measurement of dimeric inhibin-A and inhibin-B in biological fluids in several circumstances.

A. Menstrual Cycle

Inhibin-A is the major form of circulating inhibin in the female, and the ovary is the major source in women, as serum levels become undetectable after menopause. Before that point is reached, however, serum inhibin-A fluctuates during each normal menstrual cycle, as illustrated in Fig. 2. Inhibin-A rises toward the end of the follicular phase, in parallel with estradiol, as FSH concomitantly declines. This is consistent the long-held hypothesis that the dominant follicle is a major source of circulating inhibin and that inhibin functions in suppressing FSH secretion. Inhibin levels in serum are elevated even further in cycles stimulated with gonadotropins, and inhibin levels are a useful marker for follicular maturation for *in vitro* fertilization.

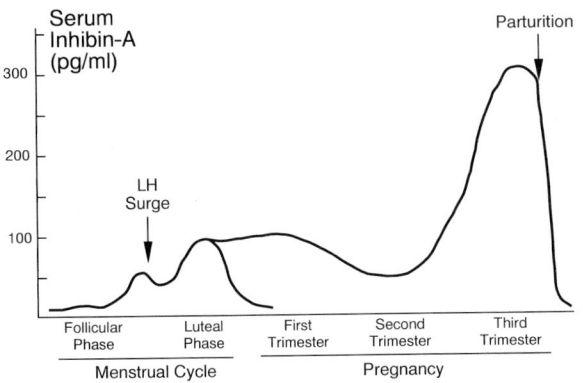

FIGURE 2 Circulating inhibin levels during the menstrual cycle and pregnancy. Compilation based on the data of Woodruff and Mather (1995) and others.

Previous results had indicated, quite unexpectedly, that inhibin levels were very high in the luteal phase, but conclusions were tenuous because of cross-reactivity of the early assays with free α subunit. The new, specific immunoassays have confirmed that inhibin levels are high in the luteal phase, exceeding those in the follicular phase.

B. Pregnancy

As illustrated in Fig. 2, serum inhibin-A levels in nonfertile cycles fall to baseline at the end of the luteal phase. In fertile cycles, however, inhibin-A is maintained and tends to increase in the first trimester. Similar patterns are observed when the corpus luteum is pharmacologically rescued by daily injections of human chorionic gonadotropin (hCG). Inhibin-A falls in the second trimester but then rises strikingly in the third trimester, especially near term, when levels can be 50-fold greater than those in the normal cycle. Inhibin-A quickly drops to follicular phase levels after delivery, suggesting that the placenta is the major source. For unknown reasons, maternal inhibin levels during the second trimester are significantly higher than normal when the fetus has Down's syndrome. In addition, inhibin levels are elevated in preeclampsia.

C. Males

Inhibin-B is the major inhibin in male serum, and in normal men, levels remain fairly constant throughout life. However, inhibin shows a reciprocal relation to circulating FSH in a number of circumstances, such as in gonadotropin-releasing hormone (GnRH)-deficient males as they undergo GnRH therapy.

D. Puberty

In females, inhibin levels are low at birth, rise throughout puberty, and then vary conversely with FSH as feedback mechanisms develop. Inhibin levels also rise in males as puberty progresses but then decline to steady-state levels in adults.

E. Pathology

Many granulosa cell tumors secrete large amounts of dimeric inhibin-A, whereas mucinous ovarian tumors make other α subunit forms. Thus, the measurement of inhibins has been touted as a marker for tumor burden in ovarian cancer. This is especially helpful in postmenopausal women, in which inhibin levels are normally undetectable. Maternal inhibin-A levels are also gaining favor as a marker for Down's syndrome fetuses, in which, in the second trimester, they are two- or threefold higher than normal. Although a two- or threefold difference does not seem great, 75% of Down's syndrome fetuses were identified based only on maternal inhibin-A levels in a recent study. Also, when combined with other conventional markers, the false-positive rate was only 5%. The mechanism for the increased inhibin is unclear, but the fact that it is not manifested until the second trimester suggests that it is the result of some aspect of the trisomy.

IV. GENE EXPRESSION AND BIOLOGICAL ACTIVITIES

Although inhibin was isolated from gonadal fluids based on its activity on pituitary gonadotrophs, the α and β subunits are expressed in a wide variety of tissues, and the inhibins have been found to have a number of other biological activities. Because of the assay difficulties outlined previously, many studies have been limited to analyzing expression of mRNAs for the individual subunits. However, this type of approach tells nothing about what protein products are produced. That situation should change in the near future as the newly developed specific immunoassays become more widely available.

The existence of two inhibins suggests that they may have unique functions, but in the biological systems that have been examined to date, inhibin-A and inhibin-B have similar biological activities. They may have differential effects on gametogenesis or gonadal function—based on the fact that inhibin-A is the predominant form in the ovary, whereas inhibin-B is the major form in the testis—but this is purely speculation.

A. The Pituitary

Inhibin suppresses basal (unstimulated) FSH secretion from pituitary gonadotropes and also opposes the stimulatory effects of activin. FSH-β mRNA levels are reduced, and there is evidence for enhanced mRNA degradation and for reduced transcription of the FSH-β gene. The importance of inhibin in the physiological regulation of FSH secretion is demonstrated by the fact that replacement therapy with inhibin immediately after castration prevents the rise in FSH levels. In addition, immunization with a peptide corresponding to the N terminus of the 18-kDa form of the α subunit generates antibodies that cross-react with and neutralize inhibin, the appearance of which is correlated with increased serum FSH in the host. These animals also superovulate and recurrently deliver larger litters. Attempts are being made to exploit this mechanism to enhance production in commercially valuable domestic species.

Inhibin was initially thought to be produced by the gonad and to act on the pituitary in a classical endocrine manner, but it is now known that inhibin/activin subunits are also expressed within the pituitary, suggesting that there may also be paracrine functions. Local changes in inhibin secretion within the pituitary may therefore provide a mechanism to fine-tune FSH secretion.

There are a number of pieces of evidence that implicate intrapituitary activin-B as a major positive regulator of FSH secretion. First, βB subunit mRNA is expressed within the pituitary gland. Second, activin-B is secreted by pituitary cells *in vitro*. Third, pituitary βB mRNA levels increase markedly after castration, accompanying the increase in serum FSH. Fourth, the activin receptor is expressed in the pituitary. Fifth, a neutralizing antibody to activin-B selectively suppresses FSH secretion but not LH *in vitro*. Moreover, the effects of the antibody are of the same magnitude as those of inhibin, and maximal concentrations of the antibody and inhibin provide no greater suppression than either agent alone. The observations with this antibody strongly suggest that the inhibitory effects of inhibin on basal FSH secretion are due to blockade of endogenous pituitary activin-B.

B. The Ovary

Inhibin levels are high in follicular fluid, correlate with estradiol, and are a good predictor of follicle maturity. The ovary expresses the α subunit at higher levels than any other tissue. However, expression of all three subunits is tightly regulated during follicular development. In primates and women, small antral follicles express high levels of βB but very little α or βA, This pattern reverses as the follicle matures to the preovulatory stage. Expression of α and βA mRNAs is stimulated by FSH, through a cAMP-dependent pathway, in granulosa cells in small follicles and becomes responsive to LH as the follicle matures. There is also a gonadotropin-independent component, as α, βA, and βB subunits are all expressed, though at low levels, in the ovaries in the hypogonadal (hpg) mutant mouse. Interestingly, the Wilm's tumor protein WT-1 is expressed at high levels in primordial, primary, and secondary follicles but decreases during subsequent follicular development. This protein can suppress α mRNA expression but its role in the regulation of α subunit expression is unclear. Expression of the α subunit is rapidly downregulated after the LH surge, and in the rat, the transient loss of inhibin may play a role in the secondary FSH surge. Subunit expression then rises again in the corpus luteum, and serum inhibin levels parallel progesterone.

As in the pituitary, inhibin also has paracrine actions within the ovary. It stimulates oocyte maturation in primates and enhances LH-induced androgen synthesis by theca cells, providing a mechanism for granulosa–theca communication and contributing to the rapid growth of the preovulatory follicle.

In addition to dimeric inhibins, the ovary contains other inhibin-related molecules (Fig. 1). Follicular fluid contains significant amounts ($\sim$500 ng/ml) of pro-αC, and serum levels of this peptide vary from $\sim$150 pg/ml in the early follicular phase to $\sim$700 pg/ml in the midluteal phase and as high as $\sim$2000 pg/ml in IVF cycles. However, its function is unclear. The 56-kDa α subunit precursor, which can inhibit FSH binding to granulosa cells, is also present in follicular fluid. Lastly, the αN fragment of the α subunit precursor seems to have a role in ovulation because immunization against it leads to reduced fertility associated with impaired oocyte release and the development of cystic corpora lutea.

C. The Testis

McCullagh proposed that inhibin was probably produced by the germinal epithelium and that prediction has been borne out. The testis produces both inhibins, but inhibin-B is the predominant form. Sertoli cells secrete a significant amount of inhibin *in vitro*, and expression of α subunit is highly responsive to FSH, via a cAMP-dependent mechanisms, and to TGF-β and activin. In contrast, TNF suppresses α subunit expression but this effect can be overcome by FSH. Subunit expression in Sertoli cells varies during the cycle of the seminiferous tubule, with maximal bioactive inhibin secreted at stages XIV–XVI. Germ cells seem to be involved in a complex paracrine interaction. Addition of spermatids to Sertoli cell cultures increases inhibin secretion, and α expression increases *in vivo* following removal of spermatocytes by radiation. Conversely, local infusion of inhibin suppresses sperm maturation in the absence of any changes in circulating FSH.

Inhibin secretion is likely to be an index of Sertoli function. However, spermatogenesis is a complex, multifaceted process. In addition, Leydig cells also express α subunit, and inhibin secretion in men can be stimulated by FSH and hCG *in vivo*. Thus, the correlation between serum inhibin levels and fertility in males is poor.

D. The Placenta

Inhibin-A is present in high concentration in maternal blood during pregnancy, and the placenta is likely to be the major source because levels fall very rapidly at delivery. Secretion of inhibin by trophoblast cells is stimulated by prostaglandin (PGE and PGF2α) and by epidermal growth factor and TGF-α. The stimulatory effects of prostaglandins are mimicked by cyclic AMP analogs. The high inhibin levels during pregnancy are probably important in suppressing follicular development, but inhibin also enhances secretion of placental lactogen II in the mouse.

E. The Adrenal

In his seminal manuscript, McCullagh also noted that the adrenal glands were hypertrophied after castration, suggesting that inhibin may also be important in regulation of the adrenal cortex. Consistent with that concept, expression of the α subunit was detected in the adrenals shortly after the cDNA was cloned. A 2.5-kb fragment of 5′ flanking region of the α subunit gene can direct expression of a marker gene to the adult adrenal, whereas expression in immature adrenal requires a larger, 6-kb fragment. The level of α expression is stimulated by adenocorticotropin hormone, again through a cAMP-dependent process. The βA subunit is also expressed in the adrenal but its regulation is not understood. Inhibin levels in the adrenal are normally much lower than in the ovary but secretion rates and serum levels are markedly elevated in adenomas that cause Cushing syndrome. Mice in which the α subunit has been genetically deleted develop gonadal tumors at puberty, but if gonadectomized prior to puberty, they subsequently develop adrenal tumors.

F. Cancer

Inhibin has been identified as a tumor suppressor based on the observation that genetic deletion of the α subunit leads to gonadal tumors at very high incidence in mice. In women, many ovarian tumors secrete large amounts of inhibin and free α subunit, which has led to the use of serum inhibin as a marker for tumor burden in these patients. It is unclear whether the α subunit in these tumors has any causative function or simply reflects hyperplasia of a tissue that normally secretes this gene product.

V. MECHANISM OF ACTION

The first activity ascribed to inhibin was its ability to selectively inhibit the secretion of FSH but not LH from pituitary cells. This can be explained as follows: Proteins can be secreted by a regulated pathway, which typically involves secretory granules that can be rapidly induced to exocytose their product by a secretagogue or by a constitutive pathway. Although both FSH and LH can be found in secretory granules, GnRH has a much larger effect on LH secretion than FSH. In contrast, basal FSH secretion is much greater than LH secretion. Thus, the primary mode of LH secretion is the regulated pathway, whereas FSH is primarily secreted through the constitutive pathway. This makes sense physiologically because sustained FSH levels are needed to stimulate follicular development, whereas LH needs to be released primarily as a bolus to induce ovulation. Based on this difference in secretion mechanism, it is predicted that agents that affect gonadotropin synthesis would have much larger effects on FSH and LH. Along these lines, nonspecific protein synthesis inhibitors such as cycloheximide selectively suppress FSH secretion and are indistinguishable from inhibin in terms of their effects of gonadotropin secretion. These data indicate that inhibin acts primarily by inhibiting gonadotropin synthesis. Thus, FSH selectivity of inhibin action is related more to differences in the mechanisms of FSH and LH secretion than in the ability of inhibin to selectively target FSH.

At a molecular level, the biological effects of inhibin are primarily due to its antagonism of activin, which is often produced locally in tissues such as the pituitary. This antagonism occurs at the cell surface. Inhibin can bind to the activin RII receptor subunit but cannot form a complex with the RI subunit, so no intracellular signal is transduced. By occupying the RII subunit, however, signaling by activin is prevented. Recently, an inhibin-binding protein has been identified, but characterization of this component is still in its infancy.

VI. INHIBIN-α KNOCKOUT MICE

Mice in which the α subunit gene has been genetically deleted develop gonadal tumors shortly after puberty. The mice develop cachexia and die soon thereafter. The cachexia is associated with lesions in the liver that are similar to lesions observed in animals infused with exogenous activin. As noted previously, expression of β subunits in the absence of α leads to secretion of activin, and these mice have extremely high levels of activin in their blood. Moreover, cachexia does not develop in mice in which both the α subunit and the activin type II receptor genes have been inactivated. Thus, the cachexia is caused by excess activin secreted by the tumors.

If the gonads are surgically removed from these mice before puberty, nearly all develop adrenal tumors, which then also lead to cachexia. When crossed with mice that lack the gene for MIS, tumors develop earlier and grow more rapidly, suggesting that MIS may also have a tumor-suppressor function.

The incidence of tumors in the α subunit KO mice is nearly 100%, indicating a major effect on tumorigenesis, but the nature of this effect is probably indirect. Tumors fail to develop when the α subunit knockouts are bred onto the hpg background, suggesting that the tumors may arise secondarily to sustained gonadotropin secretion. Indeed, gonadal tumors also arise in rats when the ovary is transplanted to the spleen. In this setting, ovarian steroids are rapidly metabolized by the liver and gonadotropin levels are markedly elevated. By analogy, it is likely that the adrenal tumors that arise in gonadectomized α knockout mice are also secondary to sustained trophic stimulation.

See Also the Following Articles

ACTIVIN AND ACTIVIN RECEPTORS; GONADOTROPIN BIOSYNTHESIS; TRANSGENIC ANIMALS

Bibliography

Aitken, D. A., Wallace, E. M., Crossley, J. A., Swanston, I. A., van Pareren, Y., van Maarle, M., Groome, N. P., Macri, J. N., and Connor, J. M. (1996). Dimeric inhibin A as a marker for Down's syndrome in early pregnancy. *N. Engl. J. Med.* **334**, 1231–1236.

Burger, H. G. (1994). Inhibin as a tumour marker. *Clin. Endocrinol.* **41**, 151–153. [Comment]

Hillier, S. G., and Miro, F. (1993). Inhibin, activin, and follistatin. Potential roles in ovarian physiology. *Ann. N. Y. Acad. Sci.* **687**, 29–38.

Kumar, T. R., Wang, Y., and Matzuk, M. M. (1996). Gonadotropins are essential modifier factors for gonadal tumor development in inhibin-deficient mice. *Endocrinology* **137**, 4210–4216.

Lebrun, J. J., and Vale, W. W. (1997). Activin and inhibin have antagonistic effects on ligand-dependent heteromerization of the type I and type II activin receptors and human erythroid differentiation. *Mol. Cell. Biol.* **17**, 1682–1691.

Matzuk, M. M., Finegold, M. J., Su, J. G., Hsueh, A. J., and Bradley, A. (1992). Alpha-inhibin is a tumour-suppressor gene with gonadal specificity in mice. *Nature* **360**, 313–319.

McCullagh, D. R. (1932). Dual endocrine activity of the testis. *Science* **76**, 19–20.

Moore, A., Krummen, L. A., and Mather, J. P. (1994). Inhibins, activins, their binding proteins and receptors: Interactions underlying paracrine activity in the testis. *Mol. Cell. Endocrinol.* **100**, 81–86.

Nishi, Y., Haji, M., Takayanagi, R., Yanase, T., Ikuyama, S., and Nawata, H. (1995). In vivo and in vitro evidence for the production of inhibin-like immunoreactivity in human adrenocortical adenomas and normal adrenal glands: Relatively high secretion from adenomas manifesting Cushing's syndrome. *Eur. J. Endocrinol.* **132**, 292–299.

Plant, T. M., Winters, S. J., Attardi, B. J., and Majumdar, S. S. (1993). The follicle stimulating hormone–inhibin feedback loop in male primates. *Hum. Reprod.* **8**(Suppl. 2), 41–44.

Qu, J., and Thomas, K. (1995). Inhibin and activin production in human placenta. *Endocr. Rev.* **16**, 485–507.

Woodruff, T. K., and Mather, J. P. (1995). Inhibin, activin and the female reproductive axis. *Annu. Rev. Physiol.* **57**, 219–244.

Xu, J., McKeehan, K., Matsuzaki, K., and McKeehan, W. L. (1995). Inhibin antagonizes inhibition of liver cell growth by activin by a dominant-negative mechanism. *J. Biol. Chem.* **270**, 6308–6313.

Insect Accessory Glands

Cedric Gillott

University of Saskatchewan

I. Introduction
II. Glands of Female Insects
III. Glands of Male Insects

GLOSSARY

ectadenia Accessory glands of male insects derived from ectoderm.

mesadenia Accessory glands of male insects derived from mesoderm.

The accessory (= collateral) glands of insects include a variety of discrete, almost always paired, tubular secretory structures that, in most species, release their product into the common genital tract at its anterior end or at some point along its length. Generally, in female insects they arise from ectodermal elements, whereas in males they are mesodermal. Their products are many and varied and may be used directly in some facet of reproductive biology or may have taken on functions only indirectly or not at all related to reproduction.

I. INTRODUCTION

With rare exceptions, the functions of accessory glands reflect the terrestrial evolution of the Insecta and the specific reproductive problems that were thereby created, namely, the need to evolve a mechanism for transfer of sperm from male to female in the absence of a watery medium (internal fertilization) and the need to protect the eggs after laying from desiccation (as well as from predators, parasites, and pathogens). However, given the evolutionary diversity of this huge class of animals, it is not surprising to discover that, in some species, the glands have taken on new functions which may be either only indirectly related or quite unrelated to reproductive activity.

II. GLANDS OF FEMALE INSECTS

Accessory glands occur in females of most orders of insects, though they are absent (sometimes secondarily, as indicated by their brief appearance in embryogenesis) in Ephemeroptera (mayflies), Plecoptera (stoneflies), Dermaptera (earwigs), Psocoptera (booklice), Heteroptera (true bugs), and most Coleoptera (beetles).

A. Structure

The glands are typically paired (sometimes bi- or multipaired) tubules that join the common genital tract just posterior to the opening of the spermatheca. However, in Acrididae (locusts and short-horned grasshoppers) the glands are anterior extensions of the lateral oviducts.

Except for those of Acrididae, which are mesodermal, the accessory glands of female insects are ectodermal (see Section II,B) and therefore are lined with a thin cuticle (intima). Outside the intima is an epithelium, typically one-cell thick, set on a basal lamina. A thin layer of muscle may occur around the lamina. When the epithelium is a single layer of cells, these have an apical brush border and produce both the intima and the accessory gland secretion. When two or more layers of cells make up the epithelium, the cells are differentiated into columnar secretory cells having a large central cavity and cuticulogenic cells that secrete the intima and the thin lining of the central cavity and its duct. These two basic patterns are widespread, though in multipaired accessory glands subtle changes in ultrastructure may be seen among the various tubule types, reflecting presumably the varied nature of the secretions that they produce.

B. Development, Differentiation, and Maturation

The female accessory glands normally arise in embryogenesis as an invagination of the body wall behind the ninth abdominal sternite. However, throughout embryonic and most postembryonic development, the anlagen remain small and undifferentiated. Growth and differentiation occur in the final juvenile instar and are dependent on the changes in hormone balance that occur at this time. Specifically, the major decrease in the juvenile hormone titer, along with one or more surges in the level of ecdysteroids in the blood, permits the growth and differentiation of the glands. Normally, in the newly emerged adult female, the glands contain no secretion; rather, during sexual maturation, the glands are stimulated to synthesize secretion under the influence of juvenile hormone, following renewed activity of the corpora allata.

In muscomorph Diptera the accessory glands arise from single or paired imaginal discs on the eighth abdominal sternite. Like the accessory gland anlagen of other species, the discs grow and differentiate only at metamorphosis.

For most adult insects, a period of sexual maturation is required after emergence when the components of the reproductive system become competent to carry out their various roles. For the accessory glands, this means the production and accumulation of secretory material. The regulation of secretory activity in female accessory glands has been examined in only a few species. Removal of the corpora allata (and hence JH) prevents the accumulation of the protein and calcium oxalate in the left accessory gland of cockroaches but appears to be unimportant in the production of β-glucosidase by the right accessory gland (see Section II,C). In tsetse flies (*Glossina* spp.), which are viviparous, changes in the volume

of the corpora allata can be correlated with the pregnancy cycle, and removal of the corpora allata leads to reduced larval growth, an effect that can be reversed by topical application of JH. Similarly, removal of the corpora allata of butterflies prevents normal secretory activity of the accessory glands. Subsequent injection of JH into these insects restores the development of the accessory glands.

C. Functions

Reproduction-related functions of the female accessory glands include production of a protective coat around the eggs [e.g., the ootheca in Dictyoptera (cockroaches and mantids) and the gelatinous mass of some aquatic species], secretion of a cement for sticking the eggs to a substrate (often the food plant of the juvenile stage of the insect's life), assisting in the movement of sperm into the spermatheca, involvement in the egg-fertilization process, lubrication of the egg-laying canal, and the production of "milk" for nourishment of the larva in tsetse flies.

In oviparous cockroaches the eggs are enclosed in a hardened case, the ootheca, that is carried about by the female until shortly before hatching. The case is produced when the secretions of the paired, but dissimilar, accessory glands are mixed in the female genital tract. The left accessory gland is much larger than the right and produces the protein that forms the bulk of the ootheca as well as the precursor of a tanning agent, laccase, and often calcium oxalate. In contrast, the right gland secretes only β-glucosidase. When mixing of the secretions occurs, the β-glucosidase hydrolyzes the precursor to release a phenolic compound. The latter is then oxidized by the laccase to an orthoquinone that can cross-link proteins by binding to their amino residues. The function of the calcium oxalate remains unclear; it may serve to create the optimum pH for the tanning process, strengthen the ootheca, provide protection against microorganisms, or simply be an excretory product. As the biochemical changes are taking place, the secretions, which form a thick paste, are moulded into shape by the ovipositor valves.

In mantids, a pair of long tubular glands and a pair of short tubular glands occur, corresponding to the left and right glands, respectively, of cockroaches.

What work has been done suggests that the biochemistry of the process of ootheca formation is similar between cockroaches and mantids.

In stoneflies, mayflies, caddisflies, some dragonflies, and Chironomidae (Diptera), the eggs when laid become coated with secretions of the accessory glands. The secretions swell on contact with water to form a gelatinous material which presumably protects the eggs.

The accessory glands of female Hemiptera (true bugs), Neuroptera (lacewings), and Lepidoptera (butterflies and moths) are widely presumed to secrete a cement that sticks the eggs to the substrate. However, it should be emphasized that there is little convincing evidence on which to base this presumption.

For a few species, the accessory gland secretion may have a role in sperm transfer from the spermatophore into the spermatheca. For example, in the blister beetle *Lytta nuttalli* the secretion may soften the hard jellylike material that initially surrounds the sperm, which is then moved into the spermatheca by peristalsis of the spermathecal duct wall. In other species, the secretion may serve as a lubricant that facilitates passage of the sperm into the spermatheca.

In the housefly *Musca domestica* the accessory gland secretion contains a variety of hydrolytic enzymes important in the digestion of the micropylar cap, a glycoproteinaceous material that covers the anterior end of the egg. As the egg moves through the fertilization chamber, spines within it initially break up the cap which is then dissolved by the accessory gland secretion, allowing sperm to penetrate the micropyles and fertilize the egg.

A unique function for the accessory glands ("milk glands") of tsetse flies has evolved, namely to provide nourishment ("uterine milk") for the developing larva in these viviparous species. In the glands of the newly emerged female, the secretory cells are small, flattened structures, largely occupied by the nucleus. However, after mating the cells undergo an enormous increase in size, eventually reaching 30–40 times their height at emergence, and become filled with the elements characteristic of cells engaged in secretory activity (Golgi complexes, numerous free ribosomes, and packets of flattened rough endoplasmic reticulum). After parturition, the cells autolyze, then redif-

ferentiate as a new pregnancy begins. The uterine milk is composed largely of protein, peptides, and amino acids (49% of the dry weight) and lipid (48% of the dry weight), principally phospholipids, cholesterol, and triglycerides. The lipid fraction appears to be synthesized in the insect's fat body and then transferred to the milk glands.

In Hymenoptera (bees, wasps, and ants), the female accessory glands have taken on functions only indirectly related or unrelated to reproduction. These include production of venoms, used in conjunction with the sting (modified ovipositor) to paralyze or kill prey or in social species to defend the colony, and production of lubricants for the ovipositor valves. This is particularly important in species in which the ovipositor is a long and very narrow structure and is drilled into plant tissues to locate the prey. In ants, the accessory glands produce trail-marking pheromones used by the workers (sterile females) to recruit nestmates to sources of food.

III. GLANDS OF MALE INSECTS

A. Structure

Accessory glands occur in most male Insecta; however, they are primitively absent in Thysanura (silverfish and firebrats), Ephemeroptera, Plecoptera, Dermaptera, and Odonata and have been secondarily lost in some Diptera (true flies). They are typically mesodermal in origin (mesadenia), though in the few insect groups in which secondary substitution of ectodermal for mesodermal tissues has occurred in the reproductive system (e.g., many Coleoptera), the accessory glands are ectodermal (ectadenia) or may include both mesodermal and ectodermal components. The accessory glands usually occur as a single pair of tubules whose cytology may show regional or intercellular differences. However, in Coleoptera there are two or three pairs of tubules, and in Thysanoptera (thrips) and Acrididae multiple pairs of tubules occur. In this arrangement, each pair of tubules has its characteristic cytology, though elements characteristic of cells producing protein for export (Golgi complexes, prominent rough endoplasmic reticulum, and secretory vesicles) are common features.

B. Development, Differentiation, and Maturation

In exopterygotes and some primitive endopterygotes, mesadenia develop from the terminal ampullae of the vasa deferentia, which themselves arise from the coelomic cavities of the ninth or tenth abdominal segment. The mesadenial anlagen remain embryonic through most of the juvenile period, and it is only in the final juvenile instar that organogenesis and differentiation begin. Differentiation is usually complete by the time the adult emerges. Ectadenia form by invagination of the ventral ectoderm, typically behind the ninth segment, though usually the ectadenial rudiments are not seen until late in the juvenile period. Like mesadenia, ectadenia develop and differentiate during the final juvenile instar. In some higher endopterygotes, notably Diptera and Hymenoptera, the accessory glands, along with all other ductal parts of the reproductive system, develop from genital imaginal discs that arise late in embryonic development. These groups of cells remain undifferentiated throughout larval life, and the reproductive elements develop from them only at metamorphosis. As was noted for the accessory glands of female insects, the growth and differentiation of mesadenia and ectadenia are regulated hormonally, being promoted by the major decline in JH and by the surges in ecdysteroid titer in the blood that take place in the final juvenile instar. Typically, differentiation is complete by the time the adult male emerges; however, in a few species studied (e.g., the mealworm beetle *Tenebrio molitor*), the process continues for the first few days of adulthood.

Maturation (the synthesis and accumulation of secretory material) in the male accessory glands generally occurs in the first few days after emergence and, for most species studied, is regulated hormonally, especially by JH. Numerous studies have shown that removal of the corpora allata prevents the normal accumulation of secretion in the accessory glands, an effect that is readily reversed by implantation of the endocrine glands or application of JH. The action of JH appears to be direct because *in vitro* cultures of accessory glands treated with hormone show enhanced synthetic activity. Though there are, as yet, only a few studies on the mode of action of JH,

the hormone appears to stimulate the synthesis of specific proteins in the accessory glands. Recently, a few reports have indicated that the male accessory glands contain or even synthesize ecdysteroids. Furthermore, it has been demonstrated that β-ecdysone can significantly stimulate or depress the synthesis of particular accessory gland proteins. For a given protein, the effect of β-ecdysone is usually opposite that of JH, and the effect of the hormone is concentration dependent.

C. Functions

The secretion of the male accessory glands, even of individual tubules, is a complex mixture, and, not surprisingly, a variety of functions have been ascribed to these structures, including spermatophore, mating plug, and seminal fluid formation; fecundity enhancement and receptivity inhibition; sperm activation; and supply of nutrients to the female.

The use of a spermatophore as a vehicle for effecting transfer of sperm from male to female appears to be the primitive condition in Insecta, and a series of steps can be traced from a primitive but structurally complex sac, formed in the male reproductive tract, that fully encloses the sperm through a simpler structure formed in the female tract and which also encloses the sperm, to a stage in which, in effect, no spermatophore is formed. Rather, the accessory gland secretions follow the sperm into the female reproductive tract where they may harden to form a temporary plug to prevent loss of semen (and further mating).

Though there are many early studies, typically using histology or occasionally surgical removal of individual tubules, of the mechanics of spermatophore formation, investigations of the biochemical events surrounding this process are still rare. Work on the spermatophores of the mealworm beetle and migratory grasshopper show that there are precursor structural proteins produced in specific tubules. On release from the tubules, these proteins may be modified (cleaved) by enzymes from other tubules, a process which then allows them to cross-link with other proteins or form a strong and inert material in a distinct layer within the spermatophore wall. Amino acid analysis of one of the spermatophore proteins from the mealworm beetle showed a high proportion (>25%) of proline residues, which is in accord with the large amounts of this amino acid reported for other insect structural proteins (in cuticle, egg shell, and ootheca) and for collagen.

Relatively little is known about the control of spermatophore formation. In almost all insects, production of a spermatophore begins only when the male has taken up the correct mating position. Sensory input, especially tactile but sometimes including chemical and visual, triggers the process, and initially at least the central nervous system must be intact. For example, in *Locusta migratoria* (the migratory locust) the head ganglia are essential for the first 15 min, presumably to interpret the sensory stimuli received. Beyond this point, only the most posterior abdominal ganglion remains essential; indeed, even the female can be dispensed with at this stage! In crickets a spermatophore is produced prior to mating, and its formation appears to be under circadian control; that is, a new spermatophore is produced during each 24-hr period. The "clock" appears to reside in the insect's optic lobes because their destruction leads to random generation of spermatophores.

Because spermatophore production can be an energetically expensive process, especially when there are several spermatophores produced in a single copulation or when the spermatophore is very large, it is perhaps not surprising that in some insect species, the female utilizes the energy-rich contribution that the empty spermatophore represents. Either the female removes and eats the spermatophore or it is digested *in situ* by enzymatic secretions.

For some insects, the accessory gland secretion has an important role in the transfer of sperm from the spermatophore into the spermatheca. This transfer may be achieved in one of two ways. In the bug *Rhodnius*, for example, the secretion from the opaque accessory gland induces peristalsis of the female genital tract to move sperm into the spermatheca. In contrast, in silk moths and some beetles, sperm are activated by accessory gland secretion and migrate actively into the spermatheca.

In addition to its obvious function of sperm transfer, in many insects mating induces major changes in the female's behavior and physiology, specifically her fecundity and receptivity. In some species of Orthoptera, Lepidoptera, and Diptera, it is components of the male accessory gland secretion that bring about these changes. The active molecules, termed fecundity-enhancing substances (FES) and receptivity-inhibiting substances (RIS), may be peptides (molecular weight 750–3000 Da) or proteins (molecular weight 13–60 kDa). In some Diptera the same molecule appears to serve both as a FES and as a RIS.

It is not yet clear how either FES or RIS work. FES typically induce egg laying but rarely ovulation. However, these processes may not occur immediately after mating, suggesting that the FES is not a myotropin. Peptidic myotropic factors are known to occur, however, in the insect brain, and the latter seems a likely site of action for the FES. A second role for FES in mosquitoes and fruitflies appears to be the enhancement of egg development. Again, this is probably an indirect action—the FES acting on the brain to trigger release of the hormones that regulate vitellogenesis. Even less is known about the site and mode of action of RIS, though the evidence available indicates that the brain or terminal abdominal ganglion is where the RIS acts. Though their mode of action remains unknown, ultimately both FES and RIS serve to inform a female that she has been inseminated. For FES, the significance rests with the fact that in insects, entry of sperm into the eggs occurs when the egg is being laid. Because unfertilized eggs are generally inviable, it is critical that oviposition occurs only after mating. The RIS, by rendering a female refractory, may promote new behaviors, including food-seeking and oviposition site selection, that may take place in locations quite distinct from where mating occurred. From the male perspective, both FES and RIS ensure that the "fittest" (i.e., successful) male's sperm will be used to fertilize the eggs and, therefore, that his genes will be passed on to the next generation.

See Also the Following Articles

Bibliography

Chen, P. S. (1984). The functional morphology and biochemistry of insect male accessory glands. *Annu. Rev. Entomol.* **29**, 233–255.

Gillott, C. (1988). Arthropoda-Insecta. in *Reproductive Biology of Invertebrates* (K. G. Adiyodi and R. G. Adiyodi, Eds.), Vol. III (Accessory Sex Glands). Wiley, New York.

Gillott, C. (1995). Insect male mating systems. In *Insect Reproduction* (S. R. Leather and J. Hardie, Eds.). CRC Press, Boca Raton, FL.

Gillott, C. (1996). Male insect accessory glands: Functions and control of secretory activity. *Invertebr. Reprod. Dev.* **30**, 199–205.

Gillott, C., and Gaines, S. B. (1992). Endocrine regulation of male accessory gland development and activity. *Can. Entomol.* **124**, 871–886.

Happ, G. M. (1984). Structure and development of male accessory glands in insects. In *Insect Ultrastructure* (R. C. King and H. Akai, Eds.), Vol. 2. Plenum, New York.

Happ, G. M. (1992). Maturation of the male reproductive system and its endocrine regulation. *Annu. Rev. Entomol.* **37**, 303–320.

Kaulenas, M. S. (1992). *Insect Accessory Reproductive Structures. Function, Structure and Development*. Springer, Berlin.

Insect Reproduction, Overview

K. G. Davey

York University

I. Introduction
II. The Production of the Egg
III. The Production of Spermatozoa
IV. The Development of Internal Fertilization
V. Bringing the Sexes Together
VI. Ovulation and Oviposition
VII. Hormonal Control of Reproduction
VIII. Unusual Reproductive Strategies
IX. Sex Determination

GLOSSARY

aedeagus The extensible portion of the intromittent organ or penis of most insects.

allatostatin A neuropeptide which acts to suppress the synthesis of juvenile hormone by the corpus allatum.

allatotropin A neuropeptide which stimulates the synthesis of juvenile hormone by the corpus allatum.

bursa copulatrix The portion of the female tract of the insect which receives the male genitalia, analogous to the vagina of mammals.

chorion The eggshell surrounding the insect egg.

corpus allatum A nonnervous endocrine gland of insects which secretes juvenile hormone.

corpus cardiacum An endocrine gland of insects consisting of the secretory terminals of neurosecretory cells in the brain together with intrinsic neurosecretory cells.

corpus luteum The remnant of the follicular epithelium, left behind at the base of the ovariole when the egg is ovulated; not known to have an endocrine function.

ecdysteroid A family of steroids related to ecdysone, the hormone controlling molting in insects.

follicle A chamber consisting of an oocyte surrounded by a single layer of epithelium, the follicle cells.

hemolymph The blood of insects, contained in an open system, the hemocoel.

juvenile hormone The hormone secreted by the corpus allatum of insects controlling metamorphosis and many aspects of reproduction.

ovariole One of the tubular structures of the ovary within which eggs are developed in a linear series. Panoistic ovarioles lack trophocytes. Meroistic ovarioles, which possess trophocytes, may be telotrophic, with the trophocytes at the apex of the ovariole, or polytrophic, with trophocytes included in each follicle.

spermatheca The organ in female insects which stores spermatozoa transferred by the male.

spermatophore A structure produced by the male accessory gland, usually during copulation, which encloses the semen.

trophocytes Sister cells of an oocyte which are specialized for nourishing the growing oocyte and which are connected to it by intercellular bridges.

vitellogenin A protein produced primarily in the fat body which is released into the hemocoel and taken up by the developing oocyte as vitellin, the principal yolk protein.

I. INTRODUCTION

Reproduction in the insects (the class Insecta of the phylum Arthropoda) is dominated by two major considerations. First, insects are essentially terrestrial organisms. Appearing first in the Palaeozoic, presumably from some aquatic or semiaquatic ancestor, they have exploited a wide variety of terrestrial environments, have returned to fresh water, and, in a few cases, have invaded the marine environment. Second, insects display an enormous diversity. Of the slightly more than 1.4 million known species of organisms, including plants, protists, fungi, bacteria, and viruses, a little over half are insects. Of the known species of metazoans, over 75% are insects; there are more species of beetles than all of the known noninsect metazoans combined. In terms of species, they are clearly the dominant terrestrial form.

Their spectacular success as terrestrial organisms is at least partly due to their enormous reproductive potential. For most insects, the reproductive strategy is based on the rapid production of enormous numbers of young, tolerating very high mortalities. Under favorable conditions, single species can occur in unimaginably large numbers. For example, a single swarm of locusts has been described as over a mile wide, 100 ft deep, and requiring 9 hr to pass a single point at about 6 miles per hour. The swarm was conservatively estimated to contain 10 billion insects.

The life cycle is characterized by metamorphosis so that larval and adult stages may exploit different aspects of the environment. In many cases the adult form does not feed, using the energy stored from the feeding larval stage to undertake the related tasks of distribution and reproduction. The great diversity of insects presents a considerable challenge when attempting to produce an overview of reproductive mechanisms. This overview emphasizes the adaptations which have permitted insects to exploit the land by producing very large numbers of offspring. These adaptations involve the development of some form of internal fertilization, mechanisms for accelerating egg production, the development of protective coatings for the egg, and mechanisms for timing reproduction. There are many specific entries related to insects elsewhere in this encyclopedia, and readers are referred to these for more detailed information.

II. THE PRODUCTION OF THE EGG

Insects produce yolky eggs already encased in a more or less waterproof chorion, or eggshell. As in many other organisms, the yolk in most, but not all, insects is produced outside the ovary as a yolk protein, or vitellogenin (Vg). The site of Vg synthesis is the fat body, the principal organ of storage and intermediary metabolism in insects. The Vg is released into the hemolymph from which it must gain entry to the oocyte. In a few insects, such as the tsetse fly, some or all of the yolk proteins are produced within the ovary by the follicle cells. In other insects, such as *Drosophila*, both the fat body and the ovary produce Vg.

The unit of egg production in the ovary is the ovariole, a tubular structure within which the oocytes develop in linear fashion from oogonia at the apex of the ovariole. As the maturing oocytes pass down the ovariole, they pass through a mass of follicular primordium and become surrounded by a single layer of follicle cells which have a number of functions.

Many insects increase the rate of egg production by assigning some of the functions involved in oogenesis to sister cells of the oocytes, termed trophocytes. Panoistic ovarioles in which there are no trophocytes are found in the more primitive orders. In those with trophocytes (meroisitic ovarioles), the trophocytes form or retain direct intercellular cytoplasmic communication with the oocytes. In telotrophic ovarioles the trophocytes remain near the apex of the ovariole around a common central cytoplasm, the trophic core, which communicates with the developing oocytes via cytoplasmic extensions, the trophic cords. In polytrophic ovarioles, the trophocytes are included in the follicular chamber together with their sister cell, the oocyte. The trophocytes ultimately contribute most of their contents to the developing oocyte, but it is clear that they have the particular function of contributing the ribosomal RNA, and in many meroistic ovarioles the oocytes do not produce new ribosomes.

The number of eggs produced at any time is increased by increasing the number of ovarioles in each ovary. The number is characteristic of species and varies from 1 or 2 in each ovary to as many as 1000 in some termites. Typically, the number in an ovary is about 10, with some Diptera possessing up to 100 or more.

The follicle cells surrounding each oocyte pass through a number of functional stages. As noted previously, they may produce some of the Vg. In most species they also produce other proteins which are incorporated into the oocyte. They regulate the entry of Vg into the follicle. The Vg passes between the cells of the follicular epithelium via large spaces which open between the cells, a process which is hormonally controlled. When vitellogenesis is complete, the spaces disappear, and the follicle cells secrete the eggshell or chorion and then disintegrate, producing a remnant known as the corpus luteum. Because the eggs are fertilized after the chorion is formed, one or more micropyles (tiny channels through the eggshell) are produced through which the spermatozoa gain entry to the oocyte. The mature eggs may be stored in the ducts for some time before oviposition, or they may be laid as soon as they are complete.

III. THE PRODUCTION OF SPERMATOZOA

The insect testis consists of a small number of seminiferous tubules contained within a connective tissue sheath. The basic pattern of spermatogenesis is not different from that in other animals: Indeed, many of the descriptions of spermatogenesis were first developed in insect models. In many insects, total sperm production is completed in the last larva or in the pupa, and testes may not be present in the adult. In others, mature sperm may be present in the seminal vesicles by the time of adult emergence. However, there are some important general differences in the morphology of insect spermatozoa. There is no midpiece because the mitochondrial apparatus is included in the long tail, frequently wrapped around the flagellum. The head of the sperm is not bulbous but rather it is slender and, hence, rather longer than in that other animals. In general, insect spermatozoa are very much longer and more slender than those in other animals. In *Rhodnius prolixus*, for example, the spermatozoa are about 600 μm in length, and in some insects, such as leafhoppers, the spermatozoa may exceed in length that of the insect which produced them!

This great elongation of the sperm into a very slender structure is no doubt an adaptation to the terrestrial environment. As noted previously, in almost all insects, the egg is fertilized after the chorion is laid down and the sperm gains entry via one or more micropyles in the chorion. Since the micropyle represents a source of potential water loss, there would be advantage in reducing its diameter—hence the diameter of the spermatozoa. Support for this view can be found in the spermatozoa of cockroaches. Cockroaches produce a secondary egg covering, the ootheca, in which groups of eggs are encased in a tanned and waterproof purse-like object. The individual eggs thus have a much thinner chorion and several micropyles of greater diameter. The sperms of cockroaches are much less slender and elongate.

The spermatozoa leave the testis via the vasa efferentia and are stored for a greater or lesser time in an expansion of the vas deferens, the seminal vesicle. While they acquire the capacity to be motile as they leave the testis, the spermatozoa in the seminal vesicle are typically completely or relatively nonmotile. It is not until they are transferred to the female that they exhibit great motility. They may reside in the spermatheca of the female for some time before they are used to fertilize the eggs: In the case of honeybees, the product of a single mating may be used by the queen for many months. The activation of the spermatozoa in Lepidoptera is known to be the result of the action of a peptide produced by the male ducts. The activation of the spermatozoa in cockroaches is accompanied by changes in the acrosomal region. In the spermatheca of the female the sperms are sustained by secretions of that organ: If the supply of secretions is reduced or halted, the spermatozoa die.

IV. THE DEVELOPMENT OF INTERNAL FERTILIZATION

Since gametes require an aqueous medium for their functioning; the development of some form of copulation is a sine qua non for life on land. Insects have approached this problem via the spermatophore. The males of many marine arthropods transfer sperm to the female encased in a gelatinous and membranous structure which often functions as a sperm storage organ within or upon the female until the sperms are used in fertilization. In the most primitive orders of insects, the spermatozoa are delivered to the female in a spermatophore, and the mode of delivery is often bizarre. In some Collembola, for example, the male produces a stalked spermatophore which the female inserts into her own reproductive tract. In some Thysanura a droplet of spermatozoa often encased in a gelatinous material is transferred from the long cerci of the male to that of the female and then to the reproductive tract. In the more primitive winged orders, a spermatophore produced by accessory glands of the male, usually during copulation, is deposited in the female reproductive tract. In the most primitive orders, such as the orthopteroid orders, the spermatophore protrudes from the female genital opening. It is frequently consumed by the female, and in stick insects, ants have been observed to remove the structure. It is therefore no surprise that in some slightly more advanced orders the spermatophore becomes protected by an invagination of the genital chamber forming a more or less capacious bursa copulatrix or vagina.

Once the spermatophore is internalized, it no longer functions to protect the spermatozoa. Two tendencies can be discerned. On the one hand, the spermatophore may be retained as a copulatory plug to hold the semen in place while contractions of the female ducts transport it to the spermathecae, as in *R. prolixus*. On the other hand, the spermatophore may be lost, and the intromittent organ of the male may be elongated so as to transport the semen directly to the spermathecae.

The external genitalia of male insects originate as a pair of outgrowths on the ventral surface of the ninth segment. These primary phallic lobes form the paired penes in the Ephemeroptera, but in Thysanoptera they unite to form a single intromittent organ. In all the remaining insects, each of the two primary phallic lobes divides to form a medial mesomere and a lateral paramere. The parameres eventually become clasping organs, but the two mesomeres fuse to form the aedeagus into which the ejaculatory duct opens via the gonopore. In those insects which form a spermatophore, the aedeagus may become an elaborate structure forming a mold for the spermatophore. In the cimicoid Hemiptera, in which the spermatozoa are deposited in the hemocoel, the male genitalia are highly modified.

V. BRINGING THE SEXES TOGETHER

Most insects fly as adults, and this great motility has resulted in development of a number of strategies for ensuring that the sexes meet at the appropriate time. Many insects have developed chemical signals of great sensitivity. In many Lepidoptera, gravid females "call" by emitting a volatile pheromone which can act over impressive distances. Males detecting the pheromone by chemosensory hairs on their antennae fly into the wind and thereby move closer to the source. Other insects signal by sound: Crickets and katydids and their relatives produce sound by rubbing the specially ribbed forewings together or against the hindfemur or by rubbing hindwing against forewing, and in cicadas sound is produced by the rapid oscillations of a membranous structure on the abdomen. In lampyrid beetles ("lightning bugs"), males and females communicate by flashes of light from abdominal organs, and the patterns of light pulses are often species specific. Many small flies produce swarms (large aggregations of a single species) often above a prominent landmark, such as a tree, post, or bush. The swarms may consist of one sex, with the other sex joining the swarm briefly until mating ensues, or both sexes may be included in the swarm.

Typically, females will not call until they are gravid, and once mated they will cease calling. In some lampyrid beetles, the female will cease her specific flashing pattern signaling that she is ready to mate and will substitute a second pattern that is attractive to the males of another species of Lampyridae. If males of the second species respond, they are captured and eaten by the female; this behavior has been termed "femme fatale behavior."

VI. OVULATION AND OVIPOSITION

Once the female is equipped with a supply of spermatozoa in her spermatheca and has developed her eggs, she can ovulate her eggs from her ovaries. In the insects that have been studied carefully, this appears to be a hormonally controlled event involving a myotropic peptide from the brain. Oviposition, however, involves both myotropic hormones and the nervous system because site selection for oviposition is frequently crucial.

The ovipositor in insects is remarkably varied: Indeed, not all structures designated as ovipositors are homologous. In the Ephemeroptera, there are paired gonopores opening on the ventral surface between the seventh and eighth abdominal segments. In the remaining orders, the single gonopore opens on either the eighth or the ninth segment. In those in which the opening is on the eighth segment, an ovipositor, composed of highly modified appendages of the eighth and ninth segments is present. The form of this ovipositor varies greatly from species to species. In many insects, however, the appendicular ovipositor is reduced or absent. In such insects, the opening of the gonopore tends to be on the ninth segment. The terminal abdominal segments may be elongated so as to aid in the placement of the eggs; such modifications of the abdomen are common in Lepidoptera, Coleoptera, and Diptera.

In the Hymenoptera, the appendicular ovipositor is well developed and in some cases it is associated with the injection of venoms. In some social insects,

such as bees and ants, the sting is an organ of defense or aggression, but in many wasps, particularly the solitary wasps, the venom is an important part of the oviposition process. Venoms may be injected at the time that an egg is laid in the host of some parasitoid wasps. The host is usually paralyzed but remains alive until the young hatch. In other cases the egg is laid on or near the paralyzed prey. Such venoms are frequently impressive: They immobilize the host, but paralyzed hosts remain alive. In many species oviposition or even egg development may be delayed until a migration has occurred.

VII. HORMONAL CONTROL OF REPRODUCTION

The timing of the development of the reproductive system in insects is determined by the action of the well-known developmental hormones, ecdysone and juvenile hormone (JH), which jointly control larval development and metamorphosis. In brief, ecdysone, secreted by the prothoracic glands in response to a neuropeptide from the brain, initiates the developmental program associated with molting. Juvenile hormone, from the corpus allatum, also controlled by cerebral neuropeptides, encourages the expression of larval characteristics and inhibits adult development. Thus, the gonads, their associated ducts, the accessory glands, and the external genitalia develop slowly or not at all until the final molt to the adult when JH is not secreted. Thus, most insects emerge as adults with a fully formed reproductive system, although additional events may take place in the adult, particularly females, which govern the functioning of the reproductive system. Most male insects emerge with a supply of mature spermatozoa, although their accessory glands, while fully formed, may not be fully functional. Since the secondary sexual characteristics of insects are fully developed at the time of adult emergence, it is not surprising that gonadal hormones have not been implicated in their appearance. In females, the development of the eggs may have occurred largely within the pupa, as in some Lepidoptera, but for a large number of species egg development, particularly in terms of vitellogenesis, may be delayed until the adult stage.

These further developmental events in the adult are under the control of a variety of hormones, and the details of the controls vary with the species. The corpus allatum becomes active again in the adult, and JH plays a central role in reproduction in most species. In general, JH has been identified as essential for the full expression of protein synthesis in the accessory glands of male insects, the synthesis of vitellogenin by the fat body, the entry of vitellogenin into the follicle, increasing the rate of receptor-mediated uptake of vitellogenin by the developing oocyte, the control of reproductive behavior, the control of reproductive diapause, and the control of wing muscle development.

The role of JH in directing Vg synthesis by the fat body has been controversial. Some clarity can be achieved by recognizing that JH has two broad actions. On the one hand, it has a priming or activating action, whereby it prepares a tissue for hormonal response by directing the synthesis and assembly of the cellular machinery necessary to carrying out particular functions in response to JH or some other hormone. On the other hand, it performs a regulatory function by controlling the rate of functioning of the primed tissue. Thus, the fat body of virtually all insects thus far examined is not capable of synthesizing Vg until it has been primed by exposure to JH. After this activating exposure to JH, Vg synthesis by the fat body may be directed by JH, as in *R. prolixus*, or by ecdysone produced by the follicle cells of the ovaries, as in mosquitoes. In other insects, such as *Drosophila*, both these hormones appear to be involved in the regulation of Vg synthesis. While a role for ovarian ecdysteroids in regulating Vg synthesis has been considered to be restricted to Diptera, recent evidence suggests that ovarian ecdysteroids may play a role in Vg synthesis in *Locusta migratoria*, formerly thought to rely exclusively on JH for regulation of Vg synthesis.

In virtually all insects that have been examined, the ovary produces ecdysteroids in response to a hormone from the brain. These ecdysteroids are sequestered by the developing oocytes, in which they may play a role in governing the later development of the embryo, but some of the ecdysteroids are also employed in regulating reproduction by means other than the control of Vg synthesis. Ovarian edysteroids frequently are used as signals to the brain, resulting,

for example, in the release of the myotropic neuropeptide governing oviposition and ovulation in *R. prolixus* or the inhibition of the corpus allatum via allatostatins in the viviparous cockroach *Diploptera punctata*.

In addition to these developmental hormones, a wide variety of neuropeptides released from the corpus cardiacum impinge on the reproductive function. Some neuropeptides from other parts of the nervous system, for example, the antigonadotropin in *R. prolixus*, have specialized functions in various insects.

VIII. UNUSUAL REPRODUCTIVE STRATEGIES

While the vast majority of insects reproduce by mating and laying eggs, there are a number of specialized reproductive modes. Some of these, such as hemocoelic impregnation, are dealt with under separate headings in this encyclopedia. Others warrant a brief description in this overview.

A. Viviparity

Several species deposit fully developed first-stage larvae rather than fertilized eggs. In most cases, this results from the incubation of the egg within the female, but no nutrition other than the vitellogenin in the completed egg is made available to the developing embryo. Such cases are examples of ovoviviparous development, which has arisen independently in various groups of insects.

In other insects, however, the developing embryo is nourished by the mother. In aphids, for example, the developing parthenogenetic eggs are retained in the ovarioles, and the chorion is absent. Nourishment for the very early embryo passes down the trophic cords, whereas later development is sustained by the passage of material across the follicular epithelium. In members of the remarkable ectoparasitic genus of earwigs, *Hemimerus*, the eggs develop at the lower end of the ovariole and the follicular epithelium undergoes hypertrophy to form a placenta-like structure. In a few insects, such as the viviparous cockroach *Diploptera* and *Hesperoctenes*, a cimicoid heteropteran parasitic on bats, the pleuropodia, which are normally evanescent appendages of the first abdominal segment of the embryo, persist and grow to form a placenta-like structure.

In tsetse flies (members of the genus *Glossina*) and in the Pupipara (a group of flies ectoparasitic on birds and mammals), single eggs are ovulated into a "uterus" in which they hatch, and the larvae grow and mature by feeding on "milk," a secretion of the accessory glands. The female gives birth to (larviposits) a fully mature larva which in tsetse flies weighs appreciably more than the mother. The larva does not feed further but burrows into soil to pupate.

In a few species, the eggs develop in the general body cavity, the hemocoel of the female. Many of these are pedogenetic forms, but in the order Strepsiptera, all the members of which are endoparasitic on other insects, the eggs develop within the female, which has developed within a host which it will never leave. The eggs develop into larvae which bear legs, and these larvae emerge from the female via up to five special genital canals. The abdomen of the female is already protruding from the host, and the active larvae crawl onto the surface of the host to await an opportunity to penetrate another host.

B. Parthenogenesis

Parthenogenesis, the development of unfertilized eggs, occurs in a wide variety of insects. It may be facultative, occurring only occasionally as in some grasshoppers, it may occur at specified times of the year, as in aphids and their relatives, or it may be obligatory such that males are exceedingly rare or unknown, as in some stick insects. Parthenogenesis may be thelytokous (female producing), arrhenotokous (male producing), or amphitokous (producing either sex).

The mechanism by which parthenogenesis occurs varies. In some species, such as most aphids, some Orthoptera, and some Diptera, there is no reduction in the number of chromosomes during oogenesis, and the resultant egg is diploid. This is known as apomictic parthenogenesis. By contrast, in automictic parthenogenesis there are normal meiotic divisions resulting in a haploid egg. The diploid number is restored by the fusion of the oocyte nucleus with one of the polar bodies. This arrangement occurs in some stick insects, some Homoptera, the Psychidae among the Lepidoptera, and some Diptera and Hyme-

noptera. Haploid parthenogenesis is a special case in which the oocytes undergo the normal meiotic divisions. If the egg is fertilized, the offspring is a female. If the egg remains unfertilized, then parthenogenetic development results in a male offspring, which is haploid in its somatic tissues. This arrangement occurs in many Hymenoptera and some Homoptera, Thysanoptera, and Coleoptera.

In a very few species, gynogenesis occurs in which penetration of the egg by the spermatozoa is required to initiate embryonic development, but the male nucleus does not fuse with the female nucleus and there is no genetic contribution from the male.

C. Pedogenesis

In some insects, the ovaries of larvae become active and develop parthenogenetically. This is often associated with parasitoids which live in the hemocoel of a host insect, and it represents a means of increasing the numbers of offspring resulting from the deposition of a single egg. However, free-living species also exhibit paedogenesis, and an extreme form is seen in many Homoptera, whereby the parthenogenetically developing embryos within the ovarioles of the mother contain parthenogenetically developing embryos such that a female aphid will contain her own granddaughters! In cecidomyid midges, the ovaries of larvae may become active and oocytes may develop by parthenogenesis. The resulting daughter larvae invade the hemocoel of the mother, feed on her tissues and emerge through holes in the body wall.

D. Polyembryony

In many parasitic Hymenoptera, a different strategy is used to increase the number of offspring. The female oviposits an egg into the host insect. This egg gives rise to more than one embryo. In many cases, each of the cells formed from the early cleavages gives rise to a separate embryo, and in other cases the division of the egg into several embryos is delayed until after a large mass of cells is formed. Polyembryony may result in a very large number of larvae—as many as 2000—produced from a single egg. It is associated with the development of a special nutritive tissue, the trophamnion, derived from extraembryonic sources. In some species the trophamnion originates from the two polar bodies of the egg which fuse and divide by amitosis to form a large tissue mass.

E. Alternation of Generations

In many insects which engage in one of the forms of parthenogenesis, there is an alternation of parthenogenetic generations with one or more generations which reproduce by normal fertilization. This alternation of generations may be influenced by season or other environmental factors. The seasonal control of the appearance of sexual reproduction is best understood in aphids. Such an arrangement represents the best of two worlds. On the one hand, favorable environmental conditions can be exploited by the rapid increase in numbers associated with parthenogenesis. On the other hand, the genetic advantages of sexual reproduction, such as increased heterosis, and the reassortment of genetic characters can be maintained by a periodic appearance of one or more sexual generations.

IX. SEX DETERMINATION

Sex determination in insects, as in most other animals, is largely a consequence of the distribution of the sex chromosomes (hetrochromosomes). The sex chromosome may be unpaired (XO) or it may be paired with a Y chromosome (XY). If the former, there will be an unpaired chromosome (monosome) in diploid cells of one sex and no X chromosome in the other sex. In many insects, the female has two X chromosomes, and the male has only the monosome or is XY. However, in many, perhaps most, Lepidoptera, the opposite is true: It is the female which has either the monosome or is XY and the male which is XX.

There are some special arrangements influencing sex determination in parthenogenetic insects. In aphids, for example, the overwintering egg produced as a result of fertilization always develops into a parthenogenetic female because all the spermatocytes which lack the X chromosome degenerate. The oocytes of parthenogenetic females of aphids generally do not undertake a second meiotic division, and

the eggs thus remain XX. In those cases in which a sexual generation is produced, males develop when one of the X chromosomes does not divide and is expelled in the polar body, thereby producing an XO individual. For those parthenogenetic species in which the male is heterogametic, no matter the method by which parthenogenetic development occurs, the offspring will normally be female. On the other hand, in Lepidoptera, it is the females which are heterogametic, and in those few cases in which parthenogenesis occurs normally, either males or females may be produced. In haploid parthenogenesis, as in honeybees and other Hymenoptera, unfertilized eggs remain haploid and are male producing, whereas fertilized eggs are female. This suggests that all the spermatozoa must bear an X chromosome since females are homogametic.

Sex determination is sometimes influenced by environmental conditions. High rearing temperatures may increase the frequency with which the X chromosome is assigned to the polar body in some female Lepidoptera, thereby increasing the frequency of XX (male) offspring. Delayed fertilization of *Drosophila* eggs increases the proportion of male offspring.

Bibliography

Büning, J. (1994). *The Insect Ovary*. Chapman & Hall, London.
Davey, K. G. (1965). *Reproduction in the Insects*. Oliver & Boyd, London
Davey, K. G. (1985a). The male reproductive tract. In *Comprehensive Insect Physiology, Biochemistry and Pharmacology* (G. A. Kerkut and L. I. Gilbert, Eds.), pp. 1–13. Pergamon, Oxford, UK.
Davey, K. G. (1985b). The female reproductive tract. In *Comprehensive Insect Physiology, Biochemistry and Pharmacology* (G. A. Kerkut and L. I. Gilbert, Eds.), pp. 15–36. Pergamon, Oxford, UK.
Davey, K. G. (1996). Hormonal control of the follicular epithelium during vitellogenin uptake. *Invertebr. Reprod. Dev.* 30, 249–254.
Engelmann, F. (1970). *The Physiology of Insect Reproduction.* Pergamon, Oxford, UK.
Wigglesworth, V. B. (1972). *The Principles of Insect Physiology,* 7th ed. Chapman & Hall, London.
Wyatt, G. R., and Davey, K. G. (1996). Cellular and molecular actions of juvenile hormone. II. Roles of juvenile hormone in adult insects. *Adv. Insect Physiol.* 26, 1–155.

Interferons

Troy L. Ott

University of Idaho

I. Classification
II. Interferons in Pregnancy Recognition

GLOSSARY

allograft Genetically unrelated tissue.

caruncle Structures found in regular rows on the uterine endometrium in ruminants which form the maternal half of placentomes during pregnancy.

chorioallantois The placental tissue layer formed following fusion of the chorionic epithelium and the allantois. Allantois supplies the vascular network.

conceptus The products of conception including embryo and associated membranes (placenta).

corpus luteum A structure that forms from the ovulated follicle on the ovary which secretes progesterone.

cytotrophoblast The outermost layer of the primate placenta in contact with maternal blood. It can be a part of placental villi or can be an extravillous migratory cell.

decidua The maternal component of the primate placenta with structural and endocrine roles in pregnancy. It is derived from uterine stromal cells during implantation of a hormonally primed uterine endometrium.

intron Gene sequences found between exons; noncoding DNA sequences found interspersed among coding sequences (exons) which are spliced out of pre-mRNA to yield mature mRNA.

maternal recognition of pregnancy Maternal physiologic changes, induced by signals emanating from the conceptus, which result in maintenance of corpus luteum function and facilitate establishment and maintenance of pregnancy.

transcription The process of copying DNA sequences either into DNA or RNA sequences occurring in the nucleus of eukaryotes.

translation The process of converting RNA sequences into proteins occurring in the cytoplasm of eukaryotes.

trophoblast The first differentiated tissue layer of the of the placenta which forms the outermost layer (chorion) of the placenta.

Interferons (IFNs) were first described by Isaacs and Lindenmann in 1957 as substances secreted by chicken chorioallantoic membrane in culture in response to treatment with inactivated influenza virus. This soluble factor rendered other membrane cultures resistant to virus challenge. This effect was termed virus "interference" and was the derivation of the name interferon. Interferons have subsequently be shown to be pleitrophic modulators of cellular functions with their defining characteristics being antiviral, antiproliferative, and immunomodulatory activities. In addition, in domestic ruminants, IFNs block the uterine mechanism responsible for corpus luteum regression (antiluteolytic effect) and alter the pattern of proteins secreted into the uterine lumen during the time of maternal recognition of pregnancy.

I. CLASSIFICATION

In general, interferons (IFNs) are classified as either type I (α, β, ω, and τ) or type II (γ). Type I IFNs all share moderate sequence homology, can all compete for the type I IFN receptor to varying degrees, and, except for IFN-β, are products of multiple genes (e.g., there are more than 15 different genes for human IFN-α). IFN-τ and IFN-ω are subfamilies in the IFN-α family. Type I IFNs are generally nonglycosylated, except for IFN-β, and in certain species (i.e., cattle and goats) both glycosylated and nonglycosylated forms of IFN-τ are present. All type I IFNs possess a characteristic stability at pH 2.0, are relatively heat stable, and their genes lack introns—properties which are not shared by the type II IFN, IFN-γ. IFN-γ shares little homology with type I IFNs, binds to a distinct type II IFN receptor, and, like IFN-β, is glycosylated and the product of a single gene. Alpha IFNs (includes α, τ, and ω) were initially referred to as "leukocyte" IFNs because they were first described as products of lymphocytes and monocytes infected with virus; however, this designation is increasingly less descriptive because other members of the IFN-α family are characterized as products of nonimmune cells (e.g., IFN-τ production by trophectoderm). Similarly, IFN-β was designated as "fibroblast" IFN and IFN-γ termed "immune" IFN. There is little doubt that IFNs serve a number of functions throughout pregnancy; however, these have yet to be clearly demonstrated in all but a few cases. Perhaps the best characterized function for IFNs in reproduction is the role that IFN-τ plays in maternal recognition of pregnancy in domestic ruminants.

II. INTERFERONS IN PREGNANCY RECOGNITION

A. Domestic Ruminants

Ruminant conceptuses (sheep, cattle, and goats) secrete IFN-τ, which serves as the signal for maternal recognition of pregnancy. First called trophoblastin or ovine trophoblast protein-1, and later ovine trophoblast interferon, IFN-τs were only recently officially designated as a distinct subclass of the IFN-α family and given the designation tau, reflecting their trophoblastic origin. IFN-τs are most closely related to the 172-amino acid IFN-ω and their expression is apparently limited to ruminant ungulates in which they serve as the conceptus (embryo and associated membranes)-produced signal for maternal recognition of pregnancy. IFN-τ is produced during a defined period of early pregnancy (Days 10–20 and 13–35 for sheep and cattle, respectively) by the mononuclear trophoblast cells (chorionic epithelium). IFN-τs bind receptors on the uterine endometrium and suppress transcription of the estrogen and oxytocin receptors genes to block pulsatile release of prostaglandin F2α (PGF). This allows for maintenance of corpus luteum (CL) function and continued production of progesterone. An immunologically related IFN was also detected in allantoic fluid and in culture medium conditioned by explants of chorioallantoic membrane between Days 25 and 40 of gestation in sheep. The nature and function of this IFN has not been determined.

IFN-τs are generally thought to be developmentally regulated, but both GM-CSF and synthetic double-stranded RNA can only modestly increase IFN-τ production *in vitro*, and viruses are poor inducers of IFN-τ. Onset of IFN-τ production coincides with outgrowth of the embryonic mesoderm (Days 11–14 in sheep and cattle) and IFN-τ production is greatly reduced or ceases as the trophectoderm attaches and the definitive cotyledonary placenta forms.

A specific endometrial receptor for IFN-τ has not been characterized. IFN-τ and IFN-α compete for the same binding sites on the uterine endometrium. This receptor is thought to be the same receptor which binds all type I IFNs (α, β, ω, and τ). The second messenger system activated by IFN-τ is unknown but is thought to be similar to the janus kinase (Jak)/signal transducers and activators of transcription (STAT) pathway activated by other type I IFNs. IFN-τ induces uterine expression of a number of proteins, including interferon regulatory factor-1 (IRF-1) and IRF-2, Mx protein, β2 microglobulin, ubiquitin-like protein, 2'5'-oligoadenylate synthase, and granulocyte chemotactic protein-2.

B. Pigs

Although pigs do not appear to utilize an IFN as the signal for pregnancy recognition, porcine conceptuses do secrete a number of proteins into the uterine lumen around the time of maternal recognition of pregnancy (Days 11–15 postmating). Included among these are both α (~25%) and γ (~75%) IFNs. In addition, the IFN-α produced by porcine conceptuses is unique in lacking the carboxyl terminal 16 or 23 amino acids characteristic of α and ω IFNs, respectively. The functions of these IFNs are not known.

C. Horses

Horse conceptuses between Days 13 and 25 postmating apparently do not express α or ω IFNs. It remains to be determined if other IFNs are expressed during early pregnancy or whether IFN expression commences later in pregnancy.

D. Primate

Although not thought to be required for pregnancy recognition, IFNs could play a number of important roles in primate pregnancy, including (i) protecting the fetus from transplacental transmission of maternal viruses, (ii) modulating the maternal immune system to protect the conceptus which is a semiallograft, and (iii) regulating proliferation and differentiation of the invading fetal trophoblast.

Human amniotic fluid collected between Week 16 of gestation and term contains IFN-α. In addition, IFNs have been demonstrated in fetal organs, blood, placenta, and decidua throughout gestation in primates. In humans, IFN-α, IFN-β, and IFN-γ were detected in normal pregnancy in villous mesenchyme and villous and extravillous cytotrophoblast and maternal decidua. Expression of IFNs, primarily IFN-β, is induced in response to challenge with virus or double-stranded RNA in amniotic membranes or cells derived from fetal trophoblast.

Whether the human placenta produces a τ IFN remains a point of controversy. Southern analysis of DNA from a wide variety of species suggests that IFN-τs diverged rather recently from IFN-ω and are only expressed by members of the suborder Ruminantia (i.e., cattle, sheep, and goats) within the order Artiodactyla. However, a human placental IFN (hpIFN) was cloned recently which was 73% identical to ovine IFN-τ. This is a greater level of similarity than between hpIFN and human IFN-ω (62%), suggesting that hpIFN belongs to the IFN-τ subfamily. Precise classification of the hpIFN awaits more detailed structural and functional analyses.

See Also the Following Articles

Allantochorion; Decidua; Pregnancy, Maternal Recognition of; Ruminants

Bibliography

Bazer, F. W., Spencer, T. E., and Ott, T. L. (1996). Placental interferons. *Am. J. Reprod. Immunol.* **35**, 297–308.

Briscoe, J., Guschin, D., Rogers, N. C., Watling, D., Muller, M., Horn, F., Heinrich, P., Stark, G. R., and Kerr, I. M. (1996). JAKs, STATs and signal transduction in response to the interferons and other cytokines. *Philos. Trans. R. Soc. London Ser. B* **351**(1336), 167–171.

Chaouat, G., Menu, E., Delage, G., Moreau, J. F., Khrishnan, L., Hui, L., Meliani, A. A., Martal, J., Raghupathy, R., and Lelaidier, C. (1995). Immuno-endocrine interactions in early pregnancy. *Hum. Reprod.* **10**(Suppl. 2), 55–59.

Dereuddre-Bosquet, N., Clayette, P., Martin, M., Mabondzo, A., Fretier, P., Gras, G., Martal, J., and Dormont, D. (1996). Anti-HIV potential of a new interferon, interferon-tau (trophoblastin). *J. Acquired Immune Deficiency Syndromes Hum. Retrovirol.* **11**(3), 241–246.

Formby, B. (1995). Immunologic response in pregnancy. Its role in endocrine disorders of pregnancy and influence on the course of maternal autoimmune diseases. *Endocrinol. Metab. Clin. North Am.* **24**(1), 187–205.

Hansen, P. J. (1995). Interactions between the immune system and the ruminant conceptus. *J. Reprod. Fertil. Suppl.* **49**, 69–82.

Kalvakolanu, D. V., and Borden, E. C. (1996). An overview of the interferon system: Signal transduction and mechanisms of action. *Cancer Invest.* **14**(1), 25–53.

Langer, J., Garotta, G., and Pestka, S. (1996). Interferon receptors. *Biotherapy* **8**(3/4), 163–174.

Larner, A., and Reich, N. C. (1996). Interferon signal transduction. *Biotherapy* **8**(3/4), 175–181.

Internal Fertilization in Birds and Mammals[1]

Sally D. Perreault

U.S. Environmental Protection Agency

John D. Kirby

University of Arkansas

GLOSSARY

cortical granules Organelles located near the surface of the egg. The contents of the cortical granules are released during fertilization and alter the zona pellucida making it impenetrable by other sperm and hence preventing polyspermy.

germinal disc A 2- or 3-mm region of the surface of the avian oocyte that contains the female pronucleus, cellular organelles, and the only continuous oolemma with microvilli.

magnum A region of the avian oviduct in which the preponderance of egg white proteins (albumen) are deposited on the nascent oocyte.

oviposition The process of expulsion of the fully formed egg from the oviduct and cloaca.

pronucleus The haploid nucleus formed from both the egg and sperm chromatin during zygote formation. The pronuclei undergo DNA synthesis and fuse to establish a diploid zygote in preparation for the first mitotic cleavage division.

protamines Highly basic proteins that replace somatic histones in the chromatin of maturing spermatids and facilitate packaging of the sperm nucleus in a compact and transcriptionally inert state. In mammals, protamines are stabilized by disulfide bonds.

sperm capacitation Changes undergone by mammalian sperm within the female tract that make the sperm capable of fertilization. Capacitation is a prerequisite for the acrosome reaction, a phenomenon (not unique to mammals) whereby the acrosomal cap and its lytic enzymes are shed and the sperm becomes able to fuse with the egg membrane (oolemma).

zona pellucida A noncellular glycoprotein layer that surrounds mammalian oocytes and plays roles in sperm recognition and binding, induction of the acrosome reaction, and the prevention of polyspermy.

Fertilization, or the union of gametes that gives rise to a zygote, evolved in simple aquatic organisms that shed their gametes into the water. This form of fertilization is said to be "external," or taking place outside the body. Movement of animals onto the land required that fertilization occur internally, or inside the body. In general, internal fertilization led to adaptations in both sexes to ensure the coordinated maturation of the gametes and to provide for deposition of sperm within the female tract, as well as transportation of the sperm to the site of fertilization. In addition, sperm became adapted to survive and remain fertile in the female reproductive tract for an adequate period of time. Thus, the requirement for internal fertilization is believed to have influenced the evolution of the reproductive organs, mating be-

[1] This document has been reviewed in accordance with the U.S. Environmental Protection Agency policy and approved for publication. Mention of trade names or commercial products does not constitute endorsement or recommendation for use.

havior, and the gametes themselves. Our understanding of fertilization in humans can be enhanced by considering how this process evolved in the various animal phyla, particularly in animals most like ourselves, the birds and mammals.

I. THE ORIGINS OF INTERNAL FERTILIZATION

A. Fertilization Always Takes Place in an Aqueous Microenvironment

Fertilization (and sexual reproduction) evolved in simple, aquatic organisms, in which gametes (sperm and eggs) were shed into the water at the appropriate time. Many simple organisms still reproduce in this manner and depend on environmental cues for gamete release. During the course of evolution to more complex organisms, the male gamete remained small and mobile, capable of wriggling or swimming to, recognizing, and penetrating the female gamete. In contrast, the egg became sessile but increased in size, thereby providing nutrition to the zygote and cleavage-stage embryo, until the embryo (larva) became capable of self-feeding. When animals moved from the oceans to the land and became adapted for terrestrial life, the requirement for sperm motility and fertilization in an aqueous solution remained. Land animals evolved means (structural and behavioral) of introducing sperm from the male into the female's body and of releasing the eggs internally at the appropriate time. This allowed internal fertilization to continue to occur in an aqueous local environment, with the liquid medium now within the female reproductive tract.

B. Fresh Eggs Are the Rule

A general feature found in all animals that reproduce by internal fertilization is the release of sperm ahead of the release of eggs. This ensures that fertilization occurs without undue postovulatory aging of the eggs. As a consequence of this general rule, highly varied and sometimes complex systems for sperm storage, both within the male track and particularly within the female tract, have also evolved across animal phyla.

C. Internal Fertilization Has Evolved Many Times

Internal fertilization appears to have evolved separately in most multicellular animal phyla as body plans became more complex and/or as animals moved to the land. External fertilization has been retained in the simple aquatic invertebrates, including the sponges (Porifera, with a few exceptions), the jellyfish, hydra, and corals (Coelenterata), and one more advanced phylum, Echinodermata (starfish and sea urchins), in which all representatives are marine species. Reproductive patterns are varied in worm phyla and may include oviparity with either external fertilization (particularly common in aquatic segmented worms) or internal fertilization (common in parasitic worms and the terrestrial earthworm) as well as viviparity, which requires internal fertilization. Among the molluscs, the more primitive gastropods (e.g., bivalves such as clams) release gametes into the ocean to be fertilized externally, but the more advanced gastropods (such as snails) and the cephalopods (octapuses and squids) have evolved unique structures and behaviors through which the male introduces a sperm packet into the body of the female. Indeed, internal fertilization in snails made land colonization possible for them. Internal fertilization is the rule in arthropods (both terrestrial and aquatic species), in which a wide array of specialized appendages and strategies have evolved for sperm storage and transfer during copulation, and the capability for long-term sperm storage in the female has reached an extreme (years in some queen bees). Even barnacles, which are sessile as adults, are able to transfer sperm through a specialized penis that is capable of considerable elongation.

Among the vertebrate classes, external fertilization has been retained in some of the fishes and in amphibians. Most bony fish release male and female gametes into the water during spawning (which may be accompanied by elaborate courtship behavior). In this case, the eggs are adapted to limit polyspermy; the micropyle, a small hole in the egg coat, provides admission of sperm and closes after fertilization. In turn, sperm in these species lack an acrosome, an organelle that is important for penetrating egg vestments in some animals. Other bony fish such as the guppy and many cartilaginous fish (sharks and rays) are viviparous so fertilization is internal, and males of these species have evolved specialized external genitalia and structures for grasping (positioning) the female during mating. In the cartilaginous fish, oocytes are fertilized in the female oviduct where they may be fed from the mother's bloodstream or become encased in an egg capsule for further development.

Amphibians (frogs, toads, and salamanders) spend part of their life cycle in the water as larvae (tadpoles) and their adulthood on the land. They breed in the water and fertilization is external. Males may have specialized grasping thumbs used to hold the females and induce egg laying as the sperm are shed.

On the other hand, the other vertebrate classes live their entire life cycle on land (or, in the case of the whales, have become secondarily adapted to an aquatic life). Internal fertilization is obligatory in reptiles, birds, and egg-laying mammals since it must precede shell deposition and in mammals, in which embryonic development is also internal. As in the invertebrate phyla, the requirement for internal fertilization in birds and mammals has provided selective pressure for the evolution of many specializations to the reproductive tract and to the gametes themselves.

II. REPRODUCTIVE ANATOMY, DUCT SYSTEMS, AND GAMETE TRANSPORT SYSTEMS FOR INTERNAL FERTILIZATION IN BIRDS AND MAMMALS

A. Sperm Deposition, Transport, and Storage in Birds

Unlike male reptiles, which have evolved penises for sperm transfer to the female, male birds lack an external phallus. Instead, complex behavioral patterns, some with elaborate courtship rituals, result in positioning of the cloacae such that sperm can be transferred directly into the female tract. Endocrine coordination of breeding provides for mating prior to the onset of ovulation.

Once in the female tract, sperm may be stored for long periods and are available to fertilize eggs laid sequentially over the course of several days. Sperm storage strategies are critical since the eggs of birds (and reptiles) must be fertilized immediately after ovulation, before the proteinaceous albumen is added in the oviduct and shell is deposited in the shell gland. Therefore, sperm must be transported to the infundibulum of the oviduct to be available for fertilization when the egg is ovulated.

In the hen, the most widely studied bird, sperm must be transported about 60 cm from the vagina to the site of fertilization. The hen breeds once but lays an egg each day over a number of days.

Since each successive egg is ovulated within only 30–45 min of oviposition (expulsion of the egg from the cloaca) and must be fertilized immediately, coordination of sperm transport and storage is essential. Rooster sperm are motile upon deposition in the vagina, and they traverse the vagina to reach the uterovaginal junction within 15–60 min. The uterovaginal junction is analogous to the cervix in mammals, helping to regulate both sperm transport prior to fertilization and expulsion of the egg at oviposition. Formed by a sphincter that holds the egg in the shell gland during shell deposition, it also helps expel the egg at oviposition. Its role in sperm transport involves sequestration of sperm in numerous sperm-storage tubules in which they may reside for a period of hours to weeks. These are complex tubules formed by invagination of the glandular oviductal epithelium that lines the inner face of the uterovaginal sphincter. Following oviposition, sperm are released from the sperm storage tubules into the lumen of the oviduct and are transported by ciliary action and contractions (reverse peristalsis) of the muscular oviduct. A portion of these sperm will arrive at the infundibulum prior to the next ovulation and are therefore available to fertilize the oocyte. Alternatively, they may be deposited in a mucosal fold of the oviduct for 1 or more days to be available to fertilize oocytes released on subsequent days. In this way, a single mating, occurring 1 or more days prior to a first ovulation, can result in numerous offspring. This strategy also lessens the requirement for precise determination of the time of ovulation and reduces the likelihood of the costly generation of unfertilized eggs. Particularly in wild species with smaller clutches of eggs and higher predation, unfertilized eggs could greatly reduce the likelihood of successful reproduction in a given year. As will be discussed later, this sperm transport and storage system may be successful in part because, unlike mammalian sperm, avian sperm do not have to undergo capacitation.

B. Sperm Deposition, Transport, and Storage in Mammals

The hormones that control the female mammal's reproductive cycle also control her readiness to mate. Consequently, mating nearly always occurs shortly before ovulation (except in bats, in which mating may precede ovulation by months, thus requiring extremely long sperm storage in the female tract). During mating, sperm deposition is facilitated by the male phallus, which becomes erect through vascular means during sexual arousal. In some groups of mammals (e.g., seals), the vascular erection of the phallus is augmented by the presence of a penile bone.

During ejaculation, sperm are mixed with seminal fluid and deposited into either the vagina or the uterus. Movement of sperm into the uterus and then into the oviducts is controlled by the female tract so that the number of sperm reaching the site of fertilization (the ampulla, or distal end of the oviduct) is quite limited. Primates, rabbits, and many ungulates deposit semen into the vagina, and the cervix limits the number of sperm entering the uterus. Cervical mucus, secreted during the preovulatory period (also under hormonal control), regulates passage of the sperm into the uterus. In humans, mating can occur at any time of the reproductive cycle; however, sperm transport through the cervix is reduced or absent except during the periovulatory interval when secretion of cervical mucus increases. In these species, crypts in the wall of the cervix provide a means of sperm storage. Fertility studies

in humans have indicated that sperm may be stored and remain fertile for up to a week.

In contrast, the cervix does not act as a barrier limiting sperm entry in most mammals. For example, male rodents deposit a concentrated suspension of sperm in the anterior vagina against the cervix and then release a seminal coagulum or "plug" to hold the semen in place. As a result, the uterus is distended with sperm within a few minutes. In dogs, horses, and pigs, the male penis penetrates partly into the cervix so that sperm enter the uterus more or less directly. Where the cervix does not act as a barrier to sperm transport, the uterotubal junction plays the major role in limiting the number of sperm that ascend into the oviducts. Motile sperm are more likely to cross this junction than immotile sperm, and their chances of passing the junction are greatly increased near the time of ovulation. Thus, the hormones that control mating and ovulation also help to control sperm transport. Regardless of the site of sperm deposition, the number of sperm that reach the upper end of the oviduct (the ampulla), where fertilization occurs, is limited. Of the millions of sperm typically deposited in the female tract, only tens to a few hundred typically reach the ampulla. In eutherian mammals, further controls at the oocyte level prevent all but a single sperm from entering the egg.

Recent attention has also been drawn to the uterine end of the oviduct (the isthmus) for its role in sperm storage and capacitation. By viewing directly through the transparent oviduct walls in mice, or by culturing sperm with oviduct epithelium, biologists have shown that mammalian sperm associate by their heads with the oviductal epithelium in this region. When they disassociate and swim free of the epithelium, they exhibit the hyperactivated motility that is characteristic of capacitated sperm. Only capacitated sperm bind and penetrate the oocyte. Thus, the oviduct is thought to be very important for the development of a functional state in sperm. This appears to be true in marsupials as well as in eutherian mammals. In some marsupial species sperm can be stored in subluminal crypts in the lower oviduct for days to weeks awaiting ovulation, and a period of oviductal residence appears to be required for sperm capacitation.

III. SPECIALIZATIONS OF GAMETES FOR INTERNAL FERTILIZATION

A. Avian Eggs Are Large and Polar

Avian eggs (like those of reptiles) must provide for all of the nutritional and gas exchange needs of the developing embryo and therefore must be large and yolky. Due to tremendous differences in hatchling size and developmental state in various species, bird oocytes vary greatly in size, weighing as little as 300 mg in hummingbirds and as much as 1.6 kg in ostriches. In addition, because birds are endothermic and bird embryos are incapable of sufficient endothermy to maintain an appropriate body temperature, birds must incubate their developing eggs. The tropical megapodes are an exception in this regard. The requirement for incubation of the embryo by the parent (or surrogate) coupled with

the restrictions that flight places on the maximum weight of the female bird's reproductive tract have been suggested as possible causes for the evolution of the bird's uniquely hard-shelled egg.

In contrast to variability in the size of the avian egg, the germinal disc region (the portion of the surface of the oocyte that contains the female's genetic material and the requisite cellular organelles) is relatively constant in size across species and occupies only a small portion (<1%) of the oocyte (Fig. 1). At fertilization, spermatozoa bind to receptors immediately on or proximal to the germinal disc, and the avian oocyte thereby becomes both anatomically and functionally a polar cell.

The avian oocyte is ovulated as a naked cell; that is, with no attached follicular cells. The transparent inner perivitelline membrane may, however, contain proteins important for sperm binding and therefore play a role similar to that of the mammalian zona pellucida. Sperm binding sites appear to be concentrated near the germinal disc. Howarth and coworkers have advanced the hypothesis that a specific

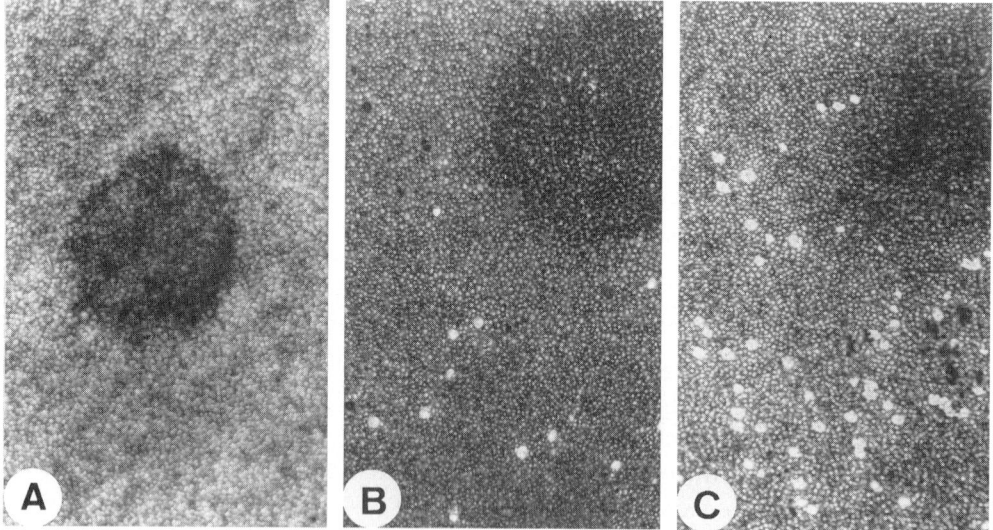

FIGURE 1 Fertilization of a turkey egg. (A) The germinal disc region of a freshly laid, unfertilized egg is a compact structure with a clearly defined edge. (B and C) Penetrating sperm appear as white holes through the perivitelline membrane. Spermatozoa bind to receptors on the inner perivitelline membrane which are primarily localized to the area of the germinal disc, undergo the acrosome reaction, and penetrate the layer in an attempt to fuse with the oolemma. In C, the fertilized oocyte clearly shows that physiological polyspermy occurs in birds, and that development can nevertheless progress. Note the loss of a clear blastodisc boundary relative to that seen in the unfertilized oocyte in A. Perivitelline membranes were isolated from freshly laid turkey eggs, fixed in buffered formalin, and stained with Schiff's reagent (photographs courtesy of Dr. Ann Donoghue, USDA/ARS, Beltsville, MD).

sperm-binding receptor is found at a higher density in the germinal disc region of the domestic fowl's egg. This hypothesis is supported by observations that the largest number of sperm penetrate the inner perivitelline membrane within a few millimeters of the germinal disc (Fig. 1). Rapid and focally targeted sperm binding is important since fertilization must take place within 10–15 min of entry of the oocytes into the infundibulum. If sperm–egg binding has not occurred within this limited window of time, fertilization will be blocked due to the deposition of the outer perivitelline layer to the oocyte. This thin, proteinaceous layer adheres to the inner perivitelline membrane and functions to mask the sperm binding sites over the germinal disc, thus acting as a mechanical block to excessive polyspermy.

When a freshly laid egg is evaluated for sperm penetration, the germinal disc appears as a darkly stained region on the surface of the oocyte (Fig. 1A). Sperm bind to the surface of the inner perivitelline membrane, undergo an acrosome reaction, and penetrate the fibrous keratin-like protein layer and then fuse with the oolemma (Figs, 1B and 1C). Once fusion has occurred, the fertilizing sperm cell's nucleus enters the oocyte and over time is transformed into the male pronucleus. As is evident in Figs. 1B and 1C, polyspermy is common in birds. However, it does not create an adverse situation since only one spermatozoon, probably the one closest to the newly formed mitotic spindle, becomes the paternal pronucleus. As in other vertebrates, pronucleus formation requires the removal of sperm nuclear protamines, which are highly basic, sperm cell-specific proteins that replace the somatic histones during spermiogenesis. The oocyte, in turn, is activated by the fertilizing sperm, the second meiotic division occurs, and the resultant polar body is released.

Immediately following ovulation the oocyte is collected by the fimbriae of the infundibulum. The subsequent events in egg formation (with or without fertilization) are similar in all avian species; however, the time involved may vary considerably. In the domestic fowl, the egg spends about 3 hr in the magnum, a tortuously folded glandular region of the oviduct in which the albumen is deposited, and then 1–1.5 hr in the isthmus, in which the inner shell membrane is produced. The egg then proceeds to the shell gland ("uterus") where it spends from 18 to 22 hr during eggshell deposition. Since sperm are physically blocked from ascending the oviduct during this entire period, storage of sperm in the oviduct is critical to ensure fertilization in birds with sequential ovulation.

B. Mammalian Eggs Are Small and Enclosed in the Zona Pellucida

The general scheme of fertilization is similar in mammals, except that the oocyte is surrounded by a zona pellucida and frequently by cumulus cells, and these vestments must be penetrated by the fertilizing spermatozoon (Fig. 2). The preovulatory surge of luteinizing hormone (LH) not only triggers ovulation but also induces the germinal vesicle oocyte (arrested in late prophase of meiosis) to mature and arrest at meiotic metaphase II. If mating has occurred, capacitated sperm await the oocytes as they enter the ampulla so that sperm bind to the zona, undergo the acrosome reaction (induced by one of the zona proteins), penetrate the zona, and then fuse with the oolemma. Gamete fusion triggers oocyte activation (completion of meiosis II), and the sperm and egg chromatin is remodeled into male and female pronuclei, giving rise to the zygote. Finally, the zygotic pronuclei undergo DNA synthesis in preparation for the first mitotic cleavage.

The features of the gametes that are adapted for internal fertilization in placental mammals can best be understood by comparing them with those seen in birds and egg-laying mammals (Table 1). There are three groups of mammals. The most primitive group is the prototheria or egg-laying mammals (the platypus and spiny anteater), whose gametes are reptilian in form. Prototheria are oviparous but otherwise meet the criteria for the class (hairy with mammary glands). The metatheria (opossums, kangaroos, etc.) are viviparous but have a truncated gestation with limited placental function and complete fetal development in the mother's pouch. Metatherian gametes are quite different from those of reptiles, birds, and prototheria and mark a transition to those of the eutheria or placental mammals.

In general, the transition from egg laying to viviparity in mammals was accompanied by a dramatic

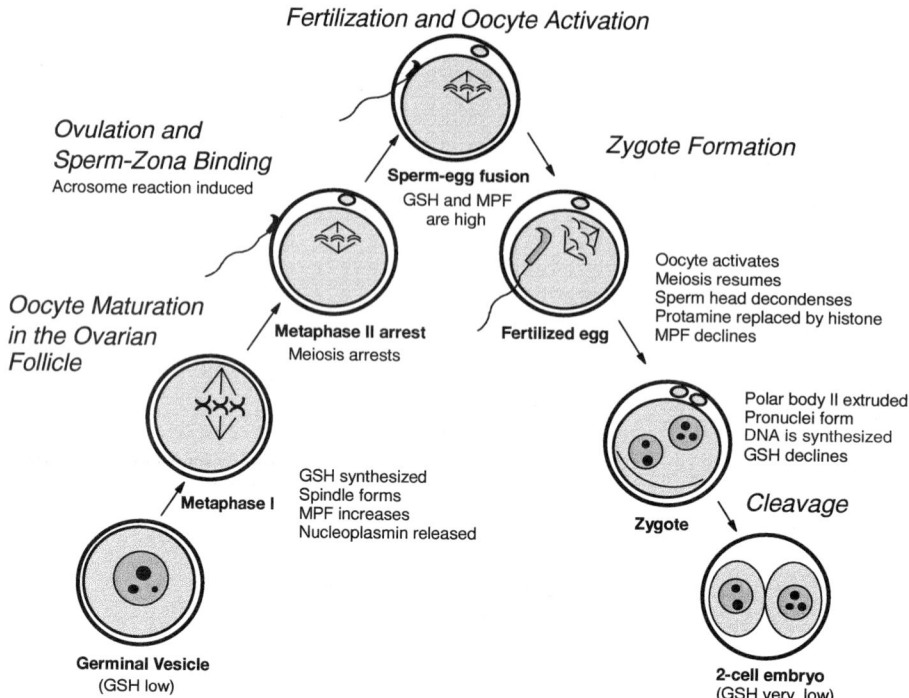

FIGURE 2 Key events in fertilization in mammals involve glutathione (GSH) and maturation promotion factor (MPF).

decrease in oocyte size and an increase in the rigidity and complexity of the sperm tail and its ability to propel the sperm (Table 1). Prototherian eggs resemble those of birds in that they are large and yolky, but they possess a functional, if very thin, zona pellucida. In contrast, the eggs of metatheria and eutheria are very small in comparison and are surrounded by a more robust zona pellucida. The eutherian zona pellucida is made up of several major glycoproteins and plays a number of critical roles in fertilization, acting as the highly species-specific sperm receptor, an inducer of the sperm acrosome reaction, and a physical block to polyspermy. In addition, eutherian oocytes are typically ovulated surrounded by cumulus cells suspended in a sticky matrix of hyaluronic acid. The cumulus has been suggested to aid in fimbrial pick up of the oocytes. There is also evidence that the cumulus may stimulate sperm capacitation and, in eutheria, may help to fill the expanded ampulla such that sperm moving through this region are more likely to encounter an oocyte. Metatherian oocytes do not have a cumulus, but the ampulla is as narrow as the oocyte in these

species. Some intriguing hypotheses have been proposed, particularly by J. Michael Bedford, concerning the coevolution of changes in sperm that parallel these developments in the oocytes and thus ensure continued reproductive success.

C. Spermatozoa Are Adapted for Fertilizing the Respective Oocyte

Birds and protherian mammals produce sperm that resemble those of reptiles (Table 1). Their vermiform nucleus, attached to a short flagellum, gives them a distinctly serpentine appearance and they exhibit limited locomotive power. While these sperm undergo some maturational membrane modifications after leaving the testes, they do not have to undergo capacitation in the female tract. However, as mentioned earlier, avian sperm can be stored and remain viable in the female tract for many days. Avian and protherian sperm swim in a progressive manner and do not appear to rely on specialized motion in order to penetrate their respective oocytes. Since the oviduct can transport avian sperm via cilia and reverse

TABLE 1
Comparison of Gamete Specializations in Birds and Mammals

Common name	*Formal name*	*Egg size and type*	*Zona pellucida*	*Sperm nucleus*	*Comments*
Birds	Avia	Large and yolky	Very thin, vitelline envelope	Filamentous (vermiform) shape; fragile[a]	Protein and shell added to egg after ovulation; sperm do not require capacitation
Platypus and spiny anteater	Mammalia, protheria	Large and yolky	Very thin (<0.5 μm)	Filamentous shape; fragile	Rudimentary outer dense fibers in sperm tail; sperm may not require capacitation
Opossums, kangaroos, etc.	Mammalia, metatheria	Small (120–240 μm diameter)	Thin (1–6 μm)	Somewhat oval in shape; fragile	Sperm tail inserts on midventral surface opposite a dorsal acrosome; tail stiffened by outer dense fibers; capacitation probably required
"True" mammals (rodents, ungulates, carnivores, insectivores, primates, etc.)	Mammalia, eutheria	Smallest (~75–150 μm diameter); cumulus enclosed	Thick, (~8–16 μm)	Oval or hooked and flattened; stabilized by protamine S–S bonds	Sperm tail stiffened by outer dense fibers; sperm require capacitation which results in hyperactivated motion; egg is capable of protamine S–S bond reduction

[a] Fragile: Nuclear protamines lack cysteine and are not disulfide cross-linked.

peristalsis, sperm may not even require progressive forward motility within the oviduct once they have entered the sperm storage tubules. Indeed, dead sperm placed within the oviduct above the uterovaginal junction can be found trapped between the inner and outer perivitelline membranes of subsequently laid eggs. Furthermore, when testicular sperm are surgically placed within the magnum they successfully fertilize eggs for a number of days postsurgery. Thus, the inner perivitelline membrane is readily breached, requiring little if any thrust by the sperm. The requirement for sperm motility, of any kind, for fertilization in birds is still under investigation. Both avian and protherian sperm must, nevertheless, undergo an acrosome reaction before fusing with the egg membrane via their inner acrosomal membranes.

In contrast, metatherian and eutherian sperm are more highly specialized, possibly due to coevolution with the oocyte vestments (Table 1). Metatherian sperm heads are somewhat oval in shape and flattened in a dorsal–ventral manner (opposite that of mammalian sperm). They differ from both protherian and eutherian sperm in that the flagellum inserts on the midventral surface rather than the trailing end of the nucleus, with the acrosome occupying

the dorsal surface on the opposite side. Sperm of American opossums pair up acrosome to acrosome and swim in tandem, a behavior which would tend to provide protection for the acrosome and appears to facilitate linear motion in the high viscosity fluids of the female tract. As in eutherian sperm, metatherian sperm fertilize better after having been in the oviducts for some time, a sign that they too undergo capacitation. They bind to the zona across the entire surface of the acrosome, which optimizes the utilization of their lytic acrosomal enzymes, including hyaluronidase and acrosin. Also as in eutherian sperm, metatherian sperm flagella are strengthened by nine outer dense fibers which provide sufficient rigidity for more prolonged progressive motility. By applying a combination of thrust and lytic enzymes released during the acrosome reaction, metatherian sperm make a relatively large hole in the relatively thick zona pellucida.

Eutherian sperm, on the other hand, may be adapted in multiple ways for zona binding and penetration. First, they look different, having more dense nuclei that are flattened like a spatula, and are shaped specifically by species (e.g., oval, tear shaped, scythelike, and hooked). Eutherian sperm undergo a complex series of modifications during their requisite maturation in the epididymis, including changes in their membrane proteins which are important for the acquisition of fertilizing ability. Extensive research, especially in the mouse, indicates that one or more sperm surface proteins serve as the primary sperm receptor for zona protein ZP3. For many years a simplistic model held that the acrosomal enzyme hyaluronidase helped the sperm penetrate the hyaluronic acid matrix of the cumulus oophorus and disperse the cumulus cells, whereas the trypsin-like enzyme, acrosin, acted as a lysin to cut a path though the zona. This model was based on observations that synthetic and natural inhibitors of these enzymes block sperm–egg binding and/or zona penetration *in vitro*, and that purified hyaluronidase can disperse the cumulus, whereas purified acrosin can dissolve the zona. However, several arguments can be made against this model. For example, although bovine sperm contain much hyaluronidase, bovine cumulus disperses shortly after ovulation (apparently without the assistance of sperm), and the eggs are fertilized in the absence of a cumulus layer. In addition, eutherian sperm have lower levels of both of these acrosomal enzymes in comparison with metatherian sperm, and yet they have to traverse a generally thicker zona. Also, the penetration slit made in the zona by eutherian sperm appears too narrow to have been made by lysis.

On the other hand, the force applied by the highly motile eutherian spermatozoon may play a more important role in zona penetration. In general, the structure of the mammalian sperm is more complex than the reptilian model, and eutherian sperm exhibit a unique whiplash motion called "hyperactivation" once they have undergone the process of capacitation. Disulfide binding in the sperm tail (outer dense fibers) is thought to modulate the rigidity of the flagellum for hyperactivated motion. A specialized structure at the tip or apex of the eutherian sperm nucleus, the perforatorium, is also stabilized by disulfide bonds, as is the eutherian sperm nucleus. Thus, the rigid structure of the eutherian sperm head, coupled with the increased thrust provided by hyperactivated motion, may allow these sperm to penetrate the zona without reliance on the lytic properties of acrosin. Alternative (or additional and perhaps more important) roles for acrosin in eutherian sperm include a role in the acrosome reaction (dispersion of the acrosomal matrix, at least in some species) and secondary sperm–zona binding, after the initial receptor-mediated binding and the acrosome reaction have occurred.

For many years, acrosin was thought to be required for fertilization in eutheria based largely on the enzyme inhibitor studies. Recently, however, the acrosin gene was genetically "knocked out" in mice. Mice homozygous for the mutated gene produce sperm that lack acrosin activity. Surprisingly, these mice are nevertheless fertile! Thus, acrosin appears not to be essential for fertilization. However, in competitive fertility trials its expression does appear to facilitate fertilization.

The zona pellucida also plays a critical role in preventing polyspermy in most eutherian mammals. When the sperm fuses with the oolemma, it triggers oocyte activation. One of the first consequences of sperm–egg fusion is the release of the contents of the oocyte's cortical granules. In the mouse, these contents diffuse across the perivitelline space and modify ZP2 (the zona protein thought to help sperm maintain their association with the zona during zona penetration). As a consequence, secondary binding can no longer occur. This so-called "zona hardening" is not universal in mammals since in rabbits the zona does not limit the number of sperm that cross the zona, and in pigs it is common to see multiple sperm in the perivitelline space. In these species, mechanisms associated with the oolemma must limit polyspermy.

Fusion of the sperm with the oolemma is also slightly, but uniquely, different in eutherian mammals. In birds and earlier mammals the inner acrosomal membrane, exposed as a consequence of the acrosome reaction, fuses with the egg membrane. In eutherian mammals, however, it is the plasma membrane overlying the equatorial segment of the acrosome that fuses with the oolemma. It has been proposed that this more complicated system coevolved with changes in the zona pellucida. As the zona became thicker, the sperm head became more rigid and acquired the perforatorium. Perhaps as the anterior sperm head became more rigid and better equipped for zona penetration, the inner acrosomal membrane lost its ability to fuse with the zona.

Changes in the sperm may, in turn, have put selective pressures on the oocyte to change. Vertebrate sperm DNA is packaged in a different way than that of invertebrates. In most vertebrates, with the exception of some fish, somatic histones are replaced by more basic proteins, the protamines. They are thought to package sperm DNA in a more stable manner that helps protect it from damage during the sperm's sojourn in the female tract. Eutherian protamine differs from that of other vertebrates in that it contains cysteine and is uniquely stabilized by protamine disulfide bonds. (The disulfide bonds form during epididymal maturation and may also help stabilize the sperm nucleus during storage in the male tract.) After the sperm enters the oocyte, these bonds must be reduced (broken) before the protamine can be replaced by somatic histones. Eutherian oocytes have thus evolved the capability to reduce these bonds as a prerequisite for sperm chromatin decondensation and subsequent remodeling of the sperm nucleus into the male pronucleus (Fig. 2). The uniqueness of this capability is demonstrated by the fact that amphibian egg cytoplasmic extracts can remodel a mammalian sperm nucleus into a male pronucleus, but only if a disulfide-reducing agent is added. Glutathione (GSH), the major free thiol in cells, has been implicated in providing this reducing power in the oocyte. The importance of GSH for this purpose has been demonstrated in experiments showing that GSH depletion in mature eutherian oocytes results in failure of the sperm chromatin to decondense normally. GSH is synthesized by eutherian oocytes shortly after the LH surge induces oocyte maturation (Fig. 2) and is twice as high in mature, ovulated oocytes as in immature, germinal vesicle-intact oocytes. Its levels decline to more typical levels in cleavage-stage embryos. Of course, other cell cycle modulators, such as maturation promotion factor (involved in chromosome condensation and spindle formation) and nucleoplasmin (involved in protamine–histone exchange), also increase and decrease as the oocyte progresses through the cell cycle (Fig. 2). However, these factors are common to all oocytes and are not a unique specialization in those of eutherian mammals.

In conclusion, internal fertilization appears to have evolved separately in vertebrates and was a necessary adaptive component in the evolution of the terrestrial classes: reptiles, birds, and mammals. It requires specializations in the external reproductive organs, the internal reproductive tract, and the gametes themselves. Reproductive patterns in mammals, particularly the eutherians, are generally more complex and coevolution of the gametes appears to have occurred. Some of these changes may also be related to adaptations for epididymal maturation and storage of eutherian sperm. Finally, our earlier perceptions of the requirement for sperm enzymes in penetrating the oocyte vestments have been modified in response to recent evidence which favors more varied and facilitative functions of these enzymes in eutherian fertilization. For example, the role of acrosin may have evolved from a primarily lytic function to include several roles in sperm–zona binding, acrosome reaction, and induction of a block to polyspermy.

See Also the Following Articles

Acrosome Reaction; Egg, Avian; Epididymis; Fertilization; Gametes, Overview; Ovarian Cycles and Follicle Development in Birds; Polyspermy; Sperm Capacitation; Sperm Transport

Bibliography

Bakst, M. R., Wishart, G., and Brilland, J. P. (1994). Oviductal sperm selection, transport and storage in poultry. *Poultry Sci. Rev.* **5**, 117–143.

Bedford, J. M. (1991). The coevolution of mammalian gametes. In *A Comparative Overview of Mammalian Fertilization* (B. S. Dunbar and M. G. O'Rand, Eds.), pp. 3–35. Plenum, New York.

Bedford, J. M. (1996). What marsupial gametes disclose about gamete function in eutherian mammals. *Reprod. Fertil. Dev.* **8**, 569–580.

Bellairs, R. (1993). Fertilization and embryonic development in poultry. *Poultry Sci.* **72**, 874–881.

Birkhead, T. R., Shelton, B. C., and Fletcher, F. (1994). A comparative study of sperm egg interactions in birds. *J. Reprod. Fertil.* **101**, 353–361.

Bramwell, R. K., and Howarth, B., Jr. (1992). Preferential binding of cock spermatozoa to the perivitelline layer directly over the germinal disc of the hen's ovum. *Biol. Reprod.* **47**, 1113–1117.

Etches, R. J. (1996). *Reproduction in Poultry.* CAB International, Oxon, UK.

Howarth, B., Jr. (1983). Fertilizing ability of cock spermatozoa from the testis, epididymis and vas deferens following intramagnal insemination. *Biol. Reprod.* **28**, 586–590.

Kirby, J. D., and Froman, D. P. (1998). Avian male reproduction. In *Sturkies Avian Physiology, 5th Edition* (G. C. Whittow, Ed.). Academic Press, New York.

Perreault, S. D. (1992). Chromatin remodeling in mammalian zygotes. *Mutat. Res.* **296**, 43–55.

Rodger, J. C. (1991). Fertilization in marsupials. In *A Comparative Overview of Mammalian Fertilization* (B. S. Dunbar and M. G. O'Rand, Eds.), pp. 117–135. Plenum, New York.

Snell, W. J., and White, J. M. (1996). The molecules of mammalian fertilization. *Cell* **85**, 629–637.

Yanagimachi, R. (1994). Mammalian fertilization. In *The Physiology of Reproduction* (E. Knobil and J. D. Neill, Eds.), 2nd ed., pp. 189–317. Raven Press, New York.

Interrenal Gland, Stress Response and Reproduction

W. Hanke

University of Karlsruhe

I. External Factors Influencing Reproduction Which May Activate the Adrenal Gland and Stress Responses
II. Correlations between Stress, Elevated Corticosteroid Levels, and Reduction of Gonadal Activity
III. Reciprocal Interactions between Adrenal and Gonadal Hormone Release and Actions
IV. Tissue and Cellular Sites of Interaction between Gonadal and Adrenal Hormone Systems

GLOSSARY

gonadal activity The growth and development of external and internal sexual characters as well as levels of hormones and events of reproduction.

interrenal gland Another term for adrenal cortex; so-called in nonmammalian vertebrates because of the distribution of steroid- and catecholamine-producing cells in these groups of vertebrates. Both cell types are more or less randomly mixed and not associated as a cortex and a medulla.

stress A phenomenon in which external factors called *stressors* influence an organism in an unusual way or with an abnormal strength regarding the adaptational conditions.

I. EXTERNAL FACTORS INFLUENCING REPRODUCTION WHICH MAY ACTIVATE THE ADRENAL GLAND AND STRESS RESPONSES

Despite the fact that a correlation between corticosteroid activity combined with stress phenomena and reproduction is well-known in mammals, clear experimental data in lower vertebrates on this topic are few. It is evident from observations in aquaculture that stressed fish living under suboptimal conditions reduce the production of fertile sperm or eggs and therefore the success of reproduction decreases.

In a general review of the influences of external factors on gonadal activity in fish, the effects of the factors influencing spawning are discussed. Water current, temperature, photoperiod, oxygen levels, pH, salinity, or overcrowding and social interaction are the most important inhibitors when they are abnormal and are associated with stress-like hormonal phenomena such as increased concentrations of catecholamines or corticosteroids in the blood. Particular attention has been paid to social factors and crowding that inhibit or retard spawning in fish. The adaptive ability of the species will determine whether one or more of these parameters is effective and not all species are responsive to all conditions.

Temperature may be used to demonstrate how changes outside the normal range for the species are detrimental to their reproduction. There is an optimal temperature for gametogenesis and ovulation, but temperatures outside the optimum strongly inhibit reproduction. Temperature optima are species and reproduction stage specific. In the same way, food availability, aquatic vegetation, water flow rate, and oxygen supply can act as both limiting and necessary factors for the development of germ cells and reproduction in aquatic animals.

The appropriate environmental conditions are detected by sense organs for the particular parameter and can modify internal reproductive responses via the hypothalamo–hypophysial–interrenal (HPI) axis. Modification may occur at different levels (the hypothalamus, the pituitary, or the gonad), and neuroendocrine activation of the brain–pituitary–interrenal or brain–sympathetic nervous systems leads to corticosteroid and catecholamine release. It should be noted, however, that activation of the interrenal gland is necessary for final oocyte maturation in normal circumstances.

II. CORRELATIONS BETWEEN STRESS, ELEVATED CORTICOSTEROID LEVELS, AND REDUCTION OF GONADAL ACTIVITY

Many reports of seasonal or annual changes in gonadal hormone and corticosteroid levels in the same animal exist and in all nonmammalian vertebrate groups plasma corticosteroids generally increase during the breeding season. Such an increase is normal for the demands during reproduction. While this is true, it is well established that when plasma corticosteroid levels increase in response to stress in all groups of vertebrates, reproductive activity decreases. Several environmental stressors also decrease the level of plasma sex steroids while at the same time elevating catecholamine levels.

The effects of environmental stress on catecholamine and cortisol release have been extensively described. In this context, the metabolic clearance rate of corticosteroids was higher in *Oncorhynchus kisutch* acclimated to seawater than in those acclimated to fresh water. This suggests that the influence of corticosteroids is stronger after entrance to fresh water for spawning during the last steps of oocyte maturation. Chronic and acute stress affect aspects of reproduction in fish. A reduction of the level of circulating estradiol and vitellogenesis is obvious.

The plasma steroid levels have been determined in plaice during the seasonal cycle. Generally, levels of cortisol and gonadal hormones are synchronous. The hormones peak before spawning and a gradual increase beginning in late autumn precedes this peak; low levels are found after spawning. There are some exceptions to this synchrony. Thus, cortisol titer peaks at the height of spawning when energy demand is high and gonadal hormone titers are already low. The duration of high cortisol values in winter depends on the amount of available food. This coincidence of both glucocorticoids and sex hormones is very well documented in the explosive breeding desert spadefoot toad, in which corticosterone levels are elevated in male toads in conjunction with elevated testosterone or dihydrotestosterone levels during different breeding events. However, in female toads, congruence between plasma estradiol and corticosterone changes was not found. Comparison of amphibian species with seasonal or aseasonal breeding patterns shows that those species with the highest levels of androgens also had high corticosterone concentrations in the blood. As with fish, it has been suggested that the energy demand of vocalization in amphibians might account for the corticosterone level. However, the intensity of calling is better correlated with the androgen levels. Thus, androgen levels determine the intensity of calling, but this activity requires metabolic resources and corticosterone. In breeding in male bullfrogs, the peak in corticosterone coincides with a transitory decline in androgen level at the start of chorusing, suggesting that corticosterone reduces the secretion and function of androgens.

Studies of the effect of captivity stress on plasma steroid levels in the green frog, *Rana esculenta*, showed that the increase of corticosterone due to the stress is high in the postreproductive period in both males and females. The response is seen about 24 hr after capture, and it is lower preceding reproduction and lowest during reproduction. Androgen and estrogen levels are reduced in males and females, respectively. The strongest suppression is found for androgens in the postreproductive period and for estrogens in the reproductive and postreproductive periods. This fits in the general picture that the extent of the corticosterone elevation after stress depends on the reproductive period and then secondarily inhibits androgen and estrogen secretion; males are more strongly influenced than females.

Studies of acute captivity stress in wild loggerhead sea turtles (*Caretta caretta*) support the hypothesis that breeding under severe conditions is associated with resistance to stress and inhibition of plasma corticosteroids. This is also suggested in Sonoran desert birds, in which suppression of the adrenocortical response occurs when exposed to stress during the breeding season. Wild birds frequently fail to reproduce in captivity, with females being more sensitive than males, and the most likely explanation for this is stress-induced activation of the adrenal through the central nervous system and pituitary.

Reproduction, stress, and behavior are also strongly correlated. Dominant males within a social group have higher gonadal activity than subordinates, but adrenal activation and stress hormone levels are higher in subordinates than in the dominants. Thus, elevated plasma androgens are mostly found in species of fish which defend a territory. The relationship between male social status and levels of androgens has been studied in the rainbow and brown trout during spawning. Typically, higher levels of androgens are found in dominant males, and in the subordinate fish the androgen level increases after removal of the dominants from the social groups.

In summary, the reports on the coincidence of changes in interrenal and gonadal activity point to two seemingly contrary interactions: (i) Stress and increasing corticosteroid titers suppress the gonadal activity and (ii) increases or decreases of corticosteroid titers and gonadal activity may occur synchronously because of the energy demands of germ cell production and events during reproduction, such as calling, amplexus, and increased locomotion activity.

It is clear that levels of glucocorticoids must be appropriately modulated to allow the full complex of reproductive activity.

III. RECIPROCAL INTERACTIONS BETWEEN ADRENAL AND GONADAL HORMONE RELEASE AND ACTIONS

The cellular sites of action of corticosteroids in modulating/inhibiting reproductive activity are many. Thus, glucocorticoids mediate the gonadotropin-induced oocyte maturation in fish. Furthermore, when glucocorticoid levels are high gonadotropins are inhibited, resulting in slower/reduced gonadal growth. Glucocorticoids may also act to inhibit liver responsiveness to estradiol so that yolk production is inhibited. This probably operates through hepatic estradiol receptions since a significant reduction of hepatic estradiol receptors occurs in rainbow trout with cortisol implants. Cytosolic as well as nuclear estradiol binding sites were diminished and plasma cortisol was elevated to chronically stressed levels by the implant. Plasma estradiol levels are not changed by the implantation. Since estradiol regulates liver synthesis of yolk protein (vitellogenesis), reduction in liver estrogen receptors may explain these results. Although most evidence points to an inhibition of gonadal hormones and reproductive capacity by glucocorticoids, gonadal steroids can suppress elevated interrenal activity in rainbow trout in response to confinement stress. A special link between the environment, corticosteroid levels, and gonadal activity exists in salmonid fish because of the migration from fresh to seawater and back in connection with spawning. The pacific salmon, which die after a single spawning episode, exhibit interrenal hyperactivity and increased cortisol secretion during sexual maturation and upstream migration. In contrast, such hypothalamic–pituitary–adrenal hyperactivity after stress is lacking in mature males in multiple-spawning rainbow and brown trout. Mature male rainbow trout have significantly lower basal levels of plasma adrenocorticotropic hormone (ACTH) compared to immature males. Depression of ACTH levels by circulating 11-keto-testosterone and a direct inhibitory action of 11-

ketotestosterone on corticosteroidogenic tissue are evident. It appears that the cortisol/ACTH feedback "set point" has been modified to a lower set point in mature rainbow trout. For salmonids, steroid treatment reduces stress response in castrate salmon and the effects are species and stage specific. Chinook salmon may have a reduced sensitivity to ACTH during spawning, but this was not observed in sockeye salmon. The reduced sensitivity to ACTH may be an adjustment to the elevated levels of ACTH and cortisol after estradiol treatment. In contrast to estradiol, 11-ketotestosterone significantly suppressed resting ACTH levels without affecting baseline cortisol concentration. Confinement stress increases plasma ACTH and cortisol levels, 11-ketotestosterone or testosterone attenuates this effect by 50%.

IV. TISSUE AND CELLULAR SITES OF INTERACTION BETWEEN GONADAL AND ADRENAL HORMONE SYSTEMS

The coincidence of changes of interrenal and gonadal activity in wild-living animals suggests that the hypothalamus–pituitary unit may be the common site of action and integration of the two systems. Corticosteroids are required to support gonadal function at an appropriate energetic level. Possibly, the basal set points in the hypothalamus for the release of both gonadotropin- and corticotropin-releasing hormone (GnRH and CRH) and the subsequent response of the pituitary are jointly regulated, and the level of this set point is changed appropriately according to reproductive activity. This linkage is interrupted by a stress response or a suddenly changed environment. While it is most likely that elevated corticosteroids suppress gonadal activity, the extent of the interrenal response depends reciprocally on the gonadal activity. These changes are likely mediated in the hypothalamus by inhibitory interactions between GnRH and CRH production and/or at the level of the pituitary by an inhibition of the response of the ACTH- or gonadotropin (GTH)-producing cells in the pituitary.

In addition to interactions of the hypothalamo–pituitary, adrenal, and gonad systems during stress, growth is also retarded. However, while growth hormone (GH) levels are suppressed in rainbow trout by acute stress, chronic stress (overcrowding) increases blood GH concentrations. The interactions are complex, however, because both types of stress result in an activation of the HPI axis yet the growth rate does not increase during chronic stress. These effects are species specific and differ from those found in goldfish. As for gonadotropin–corticotropin interactions, it is likely that activation/suppression of GH during stress is coordinated in the hypothalamus or pituitary. In addition, GnRH may be involved because stimulation of GH secretion by GnRH has been observed in some fish. Although the mechanism(s) responsible for these reciprocal interactions between GnRH, CRH, GH, GTH, and ACTH is unclear, both intrapituitary paracrine effects and hypothalamic neural inputs are likely involved.

See Also the Following Article

STRESS AND REPRODUCTION

Bibliography

Billard, R., Bry, C., and Gillet, C. (1981). Stress, environment and reproduction in teleost fish. In *Stress and Fish* (A. D. Pickering, Ed.), pp. 185–208. Academic Press, London.

Donaldson, E. M. (1981). The pituitary–interrenal axis as an indicator of stress in fish. In *Stress and Fish* (A. D. Pickering, Ed.), pp. 11–47. Academic Press, London.

Greenberg, N., and Wingfield, J. C. (1987). Stress and reproduction: Reciprocal relationships. In *Hormones and Reproduction in Fishes, Amphibians, and Reptiles* (D. O. Norris and R. E. Jones, Eds.), pp. 461–503. Plenum, New York.

Lam, T. J. (1983). Environmental influences on gonadal activity in fish. In *Fish Physiology, Vol. IX* (W. S. Hoar, D. J. Randall, and E. M. Donaldson, Eds.), pp. 65–116. Academic Press, London.

Mazeaud, M. M., and Mazeaud, F. (1981). Adrenergic responses to stress in fish. In *Stress and Fish* (A. D. Pickering, Ed.), pp. 49–75. Academic Press, London.

Schreck, C. B. (1981). Stress and compensation in teleostean fishes: Response to social and physical factors. In *Stress and Fish* (A. D. Pickering, Ed.), pp. 295–321. Academic Press, London.

Sumpter, J. P., Carragher, J., Pottinger, T. G., and Pickering, A. D. (1987). The interaction of stress and reproduction in trout. In *Reproductive Physiology of Fish* (D. R. Idler, L. W. Crim, and J. M. Walsh, Eds.), pp. 299–302. Memorial Univ. Press, St. John's, Newfoundland, Canada.

Intersexuality in Mammals

R. H. F. Hunter

Royal Veterinary and Agricultural University, Copenhagen

GLOSSARY

chimera An embryo or animal containing two or more genetically distinct cell lines. The distribution of cell lines within and between tissues is variable.

inguinal canal The bilateral conduits in the groin region of the body wall enabling the testes to descend from their initial abdominal location into the scrotal sac.

karyotype A microscopic method of displaying the chromosome constitution of a cell, usually involving a squash and stained preparation of a metaphase plate in diploid cells.

pampiniform plexus The highly convoluted arrangement of testicular artery and corresponding vein in immediate proximity to the gonad enabling a countercurrent transfer of heat and hormones across the walls of the two blood vessels.

tunica albuginea The connective tissue sheath encompassing each testis, usually in a scrotal location.

Intersexuality in this article will refer to animals that possess gonads of mixed composition with respect to phenotypic sex; that is, they possess ovotestes. One or both gonads may be ovotestes, and the proportion of testis-like tissue can vary from <10% in one gonad to both gonads appearing morphologically as testis-like structures. Although such tissue contains extensive interstitial cells of Leydig that secrete physiological concentrations of androgens, the seminiferous tubules never contain a germ cell line but only Sertoli-like cells. Most intersex animals described are genetic females possessing two X chromosomes, and most such animals have a largely female-type genital tract, although there may be modifications to the proximal portions of the Müllerian and Wolffian ducts.

I. INTRODUCTION

Intersex animals should not be thought of as hermaphrodites since there is no possibility of sperm production from an ovotestis, nor would self-fertilization be possible in a physical sense, even if spermatozoa were available. There is abundant evidence for a genetic component to the condition, seemingly autosomal in most instances rather than directly involving the sex chromosomes.

Intersexes have been reported widely in rodents, cats, dogs, and large farm species both in the domesticated state and in wild populations. They also exist in primates, including corresponding conditions in man. There is little doubt that they will be found throughout placental mammals, although usually at a low incidence.

While it is important to devote attention to the anatomy and physiology of intersex animals, and also to behavioral aspects and potential fertility, the more demanding question concerns the etiology of ovotestis formation. How have these tissues arisen within the same animal, and what might be the putative instructional anomalies that have led to derangement of gonadal formation? As will be pointed out, anomalies in the genital tract are thought to be a consequence of anomalies in the gonad(s), with the extent of genital tract perturbation seemingly being related to the time of imposition of gonadal derangement and also to its extent. Nonetheless, this article will commence with some descriptive detail before presenting histological and physiological aspects and then focusing on current molecular interpretations. It is worth mentioning at the outset that intersexuality in placental mammals has been the subject of a substantial monograph by Hunter. This book provides a wide range of references, including those containing relevant molecular information.

Finally, it should be noted that, in one sense, there is nothing novel in a discussion of intersexuality. Intersex animals and "men" were certainly recognized in ancient cultures, were widely depicted in ancient art, and both animals and "man" presenting unusual genital morphology found a privileged role in diverse societies. Indeed, intersex animals, especially pigs, were specifically bred in the New Hebrides and used symbolically in tribal fertility rites.

II. ANATOMY OF REPRODUCTIVE SYSTEM

At a scientific level, inquiry into the nature of intersexes probably commenced long ago with the studies of John Hunter. His report concentrated on the so-called "freemartin" condition in cattle. Reasonably extensive studies followed in the early part of the twentieth century, and these have continued to the present day. They have focused on the reproductive system rather than on the brain or other organs, and this emphasis will be maintained in this article.

Although both gonads may show morphological derangement, a classical finding in XX intersex animals is that one gonad is an ovotestis and the other an ovary. The gonad presenting as an ovotestis is almost always on the right-hand side; this is considered a key observation in trying to understand such wayward development. Similarly, the amount of testicular tissue within an individual gonad is variable and may extend to the contralateral gonad in the sense that both have become ovotestes (Fig. 1). One interpretation of this range of gonadal types might be the imposition or cessation of a masculinization program at different stages of development. Testis-like gonads in intersex animals may or may not possess a tunica albuginea and frequently have an uneven, "bumpy" surface morphology.

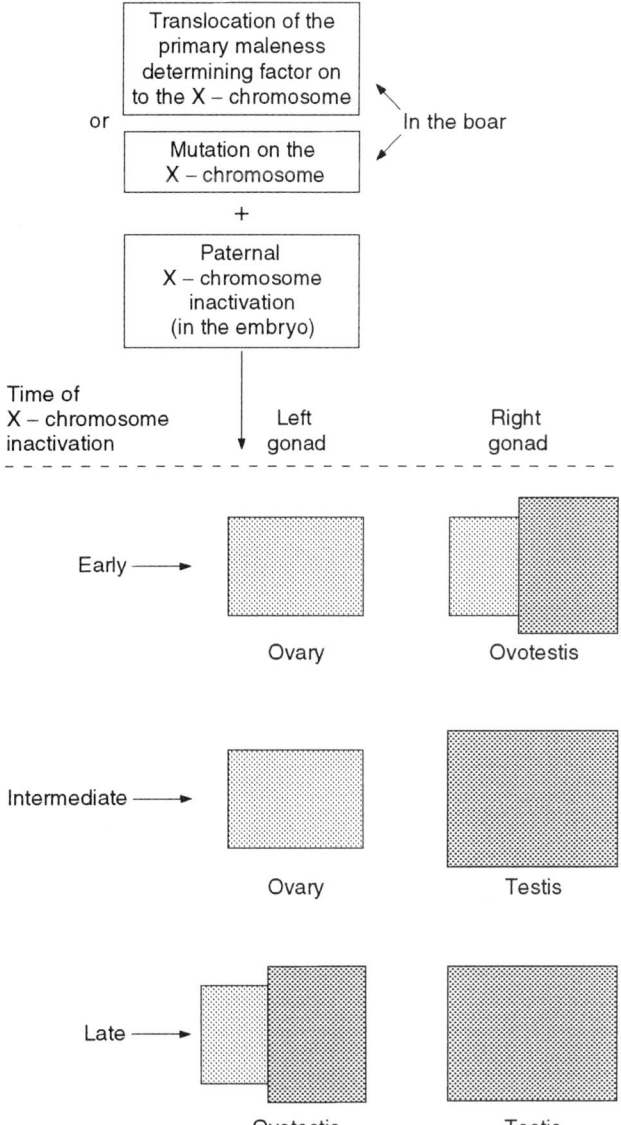

FIGURE 1 The model suggests the manner whereby varying gonadal constitutions could arise from translocation of male-determining sequences onto an X chromosome, or by an appropriate mutation on an X chromosome, in the boar followed by inactivation of the paternal X chromosome at different times in the embryo. A presumption in this model is that the right-hand gonad develops at a different rate than the left (adapted from Hunter *et al.*, 1988).

sess a tunica albuginea and frequently have an uneven, "bumpy" surface morphology.

Ovarian tissue may be functional in the presence of limited amounts of testicular tissue. Not only may an ovary show evidence of cyclic ovulation and corpus luteum formation when the contralateral gonad is an ovotestis but also ovarian tissue within an ovotestis may show such functional characteristics when the proportion of testicular tissue does not exceed 50%. However, a frequent observation, especially in domestic animals, is that the greater the extent of testicular tissue, the greater the suppression of ovarian follicular development. Not only are Graafian follicles restricted to prepuberal dimensions but also they do not respond detectably to injected gonadotropic hormones. Thus, in the presence of substantial amounts of testicular tissue, ovarian follicles (within an ovotestis) appear to have lost their receptors for gonadotropic hormones or these have become masked. Paradoxically, although not responding to injected gonadotropins such as pregnant mare serum gonadotropin in terms of follicular growth, ovotestis tissue may respond to challenges of adrenocorticotropin and secrete significantly elevated levels of diverse sex steroid hormones.

The gonads of intersex animals usually remain abdominal but may assume an inguinal location or even become scrotal. The latter is frequently associated with bilateral ovotestes containing predominantly or exclusively testicular tissue. Once again, however, the range of conditions is so diverse that, within a genetic female, one testis-like gonad may be abdominal and the other scrotal. The vascular supply of an ovotestis is extremely prominent and is reminiscent of the pampiniform plexus associated with a mature testis.

The reproductive tract of perturbed XX animals tends to be predominantly of female type but with specific regions of masculinization adjoining an ovotestis or testis-like structure. The most typical and interesting finding is that a portion of the proximal Wolffian duct develops as an epididymis, and a corresponding portion of the adjacent Müllerian duct, the Fallopian tube, is usually strongly suppressed, especially in its ampullary portion.

The epididymis develops morphologically just as in an authentic male and is frequently closely applied to the ovotestis or testis-like gonad, whether abdominal, inguinal, or scrotal. Its surface is well vascularized, and the actual epididymal duct is highly coiled but, upon histological examination, is found to be completely devoid of spermatozoa.

Regarding the Müllerian duct, suppression of Fallopian tube development appears approximately proportional to the amount of adjoining testicular-like tissue. It is difficult to refute the conclusion that anti-Müllerian hormone (AMH) (= Müllerian inhibiting substance) from the Sertoli-like cells must be acting locally in a limited manner to achieve the degree of suppression described. It is a consistent observation across species that the uterus invariably escapes the inhibition, so a local diffusion from the modified gonad appears reasonable. The Sertoli cells within testicular tissue of ovotestes or testis-like structures have been termed Sertoli-like because of their pale staining reaction. On histological grounds alone, one might question their molecular competence, but in conjunction with the very limited extent of Müllerian suppression, this conclusion seems inescapable, despite apparently normal quantitative populations of Sertoli cells. It is possible that such Sertoli-like cells secrete very limited amounts of AMH, secrete AMH on an inappropriate time scale, or both.

In instances of an inguinal or scrotal testis-like gonad, the descending gonad may have drawn with it a portion of the uterus out of the abdominal cavity, together with a regressed Fallopian tube. Indeed, a loop of uterine tissue may occasionally be located in the scrotal sac.

The vagina of intersex animals may occasionally be "blind," i.e., occluded by membranous tissue and/or fusion of the walls. However, the principal observation at the caudal extremity of the genital system is conspicuous masculinization of the clitoris, frequently into a very prominent organ. The vulva is also slightly modified, usually into an "upturned" arrangement in quadrupeds.

III. SEXUAL BEHAVIOR AND ENDOCRINE RESPONSES

As a reflection of their gonadal type and especially of gonadal steroid hormone secretion, intersex animals may exhibit typically female, cyclic sexual behavior, with periods of receptivity to a mature male, or lack of overt cyclic behavior or even typical male sexual behavior, with mounting of estrous females and classical pelvic thrusting. Similarly, hypothalamic and pituitary endocrine responses of intersex animals may range from typical female, with a dramatic preovulatory luteinizing hormone (LH) surge, to male type in which even a large estradiol challenge cannot provoke a female-type LH surge. Females with only a small amount of testicular tissue in one gonad may exhibit estrus, be mated, ovulate, and become pregnant. The spectrum of sexual and ovulatory behavior would appear to be well correlated with the spectrum of gonadal tissue types.

IV. CHROMOSOME CONSTITUTION

As already mentioned, a majority of intersex animals karyotype XX with respect to their sex chromosomes. Only a small proportion (<5%) will be revealed as mosaics or chimeras, and there is only very limited evidence from chromosome banding studies for translocation of Y chromosome material onto an X chromosome. Accordingly, how does one account for the intersex condition as revealed in the intersex nature of the gonad(s)? This remains the most topical and exacting question in this subject area. It would be prudent to assume that there is more than one answer.

V. MOLECULAR INTERPRETATIONS

A male-determining gene or testis-determining gene located on the Y chromosome has long been sought and was thought to have been identified in 1991. It has been termed *SRY* in man and *Sry* in mouse, and it refers to the sex-determining region on the Y chromosome. It is essential, however, to regard this as a male-determining gene rather than the only male-determining gene as initially proposed because it is only part of a complex sequence of instructional information. As shown in elegant transgenic experiments in mice, microinjection of an *Sry*-bearing gene construct into a pronuclear mouse egg enabled sex reversal of a female mouse embryo, karyotyped retrospectively, into a phenotypic male.

Penis, testes, and scrotum developed but the testicular tissue was devoid of germ cells. Such sex-reversed mice were therefore not functional males but rather irreversibly sterile.

In light of the condition of the seminiferous tubules of these transgenic animals, there may well be some relevance to the intersex condition. It is therefore of interest to establish whether the *Sry* gene is present in intersex animals, i.e., in genetic females with varying quantities of testicular tissue that has arisen naturally and not as a result of transgenic manipulations. Some of the most extensive studies have been undertaken in the domestic pig, which shows a remarkable susceptibility to the intersex condition. There are also appropriate studies in domestic goats, in which there is an association between the hornless (polled) condition and intersexuality. Irrespective of the amount of testicular tissue present in XX animals, the *Sry* gene is seldom ever detected. Therefore, how has the testicular tissue been programmed?

Various lines of evidence can be brought to bear on the problem. The first one comes from freemartins, female calves that bear virilized gonads and usually have inhibited development of the (female) genital tract. An explanation for the freemartin condition has been sought ever since the writings of John Hunter. A seemingly satisfactory answer now exists that is open to scientific validation. Many modern texts unfortunately still present incorrect and out-of-date interpretations.

Freemartins are genetic females that develop co-twinned to a male. There are three requirements: (i) There must be twin ovulations and normal fertilization of the resulting eggs, (ii) the developing embryos must be of different sex, and (iii) there must be chorioallantoic fusion with vascular anastomosis, thereby permitting conjoined circulations. Freemartins are formed most extensively in cattle (92% incidence under the previously mentioned conditions) but also exist in sheep and goats. They are extremely rare in pigs, seemingly due to fusion of placental sacs only late in gestation—too late to permit perturbed development of the embryonic reproductive system. Freemartins are claimed not to occur at all in equids and primates, despite instances of placental fusion in dizygotic heterosexual pregnancies.

Regarding the derivation of freemartins, the long-standing explanation has been that the conjoined circulations of the male and female cotwins permit androgens from the precociously developing male gonads to cross the placenta, damage development of the female gonads, and inhibit portions of the female genital duct system. However, there is no evidence that androgen can sex reverse an ovary, despite experiments with a variety of such sex steroids. By contrast, there is good evidence that testes develop sooner than ovaries in placental mammals, and that the activity of the Sertoli cells—a target of *Sry* gene action—may underlie the freemartin condition. AMH is thought to be the earliest, or one of the earliest, proteins synthesized and secreted by Sertoli cells and, as long suspected by Jost, appears to be at the root of the freemartin condition.

AMH from the fetal male cotwin crosses the circulation of the fused placentae and has a virilizing and stunting influence on the female gonads between 60–100 days of gestation. AMH production from the virilized ovary itself (at around 100 days of gestation) has a secondary masculinizing influence. However, this occurs after the period of loss of germ cells, development of a tunica albuginea, and formation of cords of Sertoli-like cells. XX germ cells do not survive in a testicular environment, so any remaining oogonia would degenerate. Similarly, spermatozoa cannot be generated in the presence of two X chromosomes. AMH also has a regressive influence on the embryonic female genital tract, as occurs in a normal male.

AMH may well play a key role in perturbing gonadal formation and development in other species such as dogs and moles. In domestic pigs, one of the current proposals is that the gene that regulates inhibin production in the ovary mutates to modify protein synthesis from the inhibin pathway to that for AMH. Inhibin and AMH are closely related members of the biologically active TGF-β family of glycoproteins. The different times of the mutation influence might explain the formation of different proportions of testicular tissue in the virilized ovaries. It is also worth bearing in mind that Graafian follicles themselves develop the ability to synthesize AMH although too late and in quantities too small to influence development of the female reproductive tract. Even so, with or without the inhibin gene mutation as proposed previously, a wayward time scale and quantity of gonadal AMH production in a genetic female might offer one explanation for ovotestis formation.

VI. GENETIC MODEL FOR OVOTESTIS FORMATION

Regarding karyotypes and chromosome banding, it is perhaps still valid to consider the model proposed by Hunter *et al*, even though evidence for relevant translocations remains slim. If a male-determining DNA sequence were to be translocated onto a sperm X chromosome during spermatogenesis, or if there were an appropriate mutation on such an X chromosome, coupled with a variable time of paternal X inactivation in female (XX) embryos (Fig. 1), then this might explain a variable influence on the nature of the gonads. The extent of ovotestis formation would depend on the influence of sequences on the paternal X chromosome if and before it underwent inactivation. Asynchronous development of the two gonads would be anticipated, and this would facilitate the formation of a range of gonadal types.

Although left–right morphological differences have long been a source of puzzlement, interpretation of these gonadal anomalies should be less perplexing now that a gene for left–right (lateral) asymmetry has been identified. As already noted, testicular tissue in an ovotestis or testis-like tissue in an XX intersex arises predominantly in the right-hand gonad; its precocious development may make it particularly vulnerable to derangement.

VII. GENE SEQUENCE MODEL

A recent model for testis tissue formation in a putative XX female (i.e., no Y-related DNA sequences detected) concerns gene sequences of variable length that influence the composition of the gonad. An analogy has been drawn with aircraft runway usage and minimum takeoff speeds (Fig. 2). An aircraft normally has available the full length of a runway to

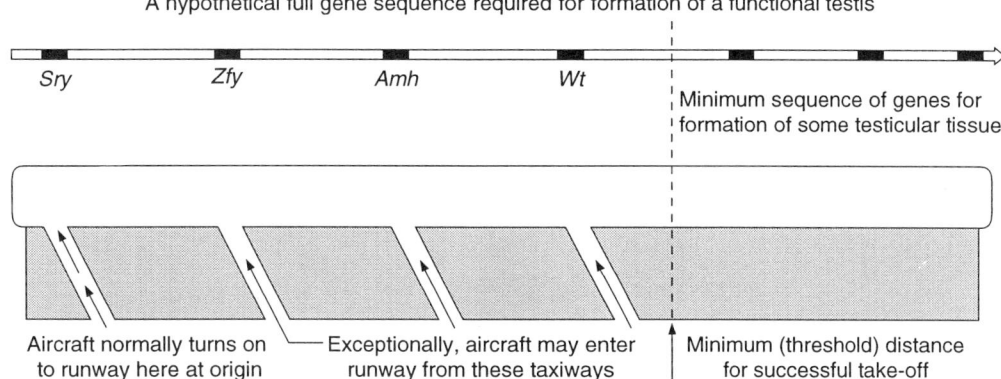

A hypothetical full gene sequence required for formation of a functional testis

Sry   *Zfy*   *Amh*   *Wt*

Minimum sequence of genes for formation of some testicular tissue

Aircraft normally turns on to runway here at origin — Exceptionally, aircraft may enter runway from these taxiways — Minimum (threshold) distance for successful take-off

FIGURE 2   Potential parallels between a minimum gene sequence required for formation of testicular tissue and runway usage by aircraft accelerating at different speeds over shorter distances to successful takeoff. Although a minimum sex-determining gene sequence is clearly crucial in this model, genes influencing rate of development in the genital ridges are also considered vital (from Hunter, 1996).

gather speed and become safely airborne, but there are circumstances in which a shorter length of runway may be appropriate. The shorter the stretch of runway to be used, the greater the acceleration required to reach takeoff speed. In this analogy, the complete runway length is equated with the normal sequence of genes required to prescribe formation of a functional testis. However, a shorter gene sequence may be able to generate testicular tissue and not involve, for example, *Sry, Zfy, Amh*, if the rate of development (takeoff speed) is sufficiently rapid in at least one of the presumptive gonads. Testicular tissue formed under the influence of such an abridged gene sequence will not be fully functional. Not only may there be no spermatogenesis but also Sertoli cell secretion of proteins such as AMH would be suboptimal. In this model, special emphasis needs to be given to rates of gonadal development and also to left–right asymmetries between the paired gonads.

Whether or not this model survives rigorous analysis, there is persuasive evidence in pigs, for example, that a boar generating intersex offspring carries an autosomal recessive gene specifically implicated in the anomaly.

VIII. CONCLUDING THOUGHTS

This article has made no attempt to be comprehensive in the sense of listing all ovotestis-like conditions and other sexual anomalies across a range of mammalian species. Nor does it make reference to spontaneously arising mutant genes, except in a hypothetical context. Much detail can be found in the book by Hunter. However, it should be clear from the previous discussion that abnormal development of the gonads seems to be more frequent in individuals with an XX sex chromosome constitution. This prompts the question as to whether the slightly less rapid female program of differentiation renders XX animals more vulnerable to developmental perturbations compared with males. This could in part be because the bipotential state of the gonad would exist for slightly longer in females. Moreover, although the gene *Sry* has received some emphasis in the context of testicular differentiation, autosomal genes interacting with or, more probably, downstream of *Sry* are thought to make a crucial contribution to both normal and abnormal gonadal development. Their identification is the topic of much current research, as is a potential involvement of the Wilms' tumor gene: deranged kidneys and deranged gonads might have common molecular pathways.

Acknowledgments

The writing of this article was supported by a Carlsberg Guest Professorship in Veterinary Physiology, for which grateful acknowledgment is made. Frances Anderson kindly prepared the manuscript.

See Also the Following Articles

Equine Chorionic Gonadotropin (ECG); Freemartin; Gonadogenesis, Female; Gonadogenesis, Male; Hermaphroditism; Pampiniform Plexus

Bibliography

Baker, J. R. (1925). On sex-intergrade pigs: Their anatomy, genetics, and developmental physiology. *J. Exp. Biol.* **2**, 247–263.

Chalmers, C., Cook, B., Foxcroft, G. R., and Hunter, R. H. F. (1989). Luteinizing hormone response to an oestradiol challenge in 5 intersex pigs possessing ovotestes. *J. Reprod. Fertil.* **87**, 455–461.

Hunter, J. (1779). An account of the free martin. *Philos. Trans. R. Soc. London* **69**, 279–293.

Hunter, R. H. F. (1995). *Sex Determination, Differentiation and Intersexuality in Placental Mammals*. Cambridge Univ. Press, Cambridge, UK.

Hunter, R. H. F. (1996). Aetiology of intersexuality in female (XX) pigs, with novel molecular interpretations. *Mol. Reprod. Dev.* **45**, 392–402.

Hunter, R. H. F., Chalmers, C., and Cavazos, F. (1988). Intersexuality in domestic pigs: A guide to mechanisms of gonadal differentiation. *Anim. Breed. Abstr.* **56**, 785–791.

Hyatt, B. A., Lohr, J. L., and Yost, H. J. (1996). Initiation of vertebrate left–right axis formation by maternal Vg1. *Nature (London)* **384**, 62–65.

Jiménez, R., Burgos, M., Sanchez, A., Sinclair, A. H., Alarcon, F. J., Marin, J. J., Ortega, E., and de la Guardia, R. F. (1993). Fertile females of the mole *Talpa occidentalis* are phenotypic intersexes with ovotestes. *Development* **118**, 1303–1311.

Jost, A. (1970). Hormonal factors in the sex differentiation of the mammalian foetus. *Philos. Trans. R. Soc. London B* **259**, 119–130.

Koopman, R., Gubbay, J., Vivian, N., Goodfellow, P., and Lovell-Badge, R. (1991). Male development of chromosomally female mice transgenic for *Sry*. *Nature (London)* **351**, 117–121.

Meyers-Wallen, V. N., MacLaughlin, D., Palmer, V., and Donahoe, P. K. (1994). Müllerian-inhibiting substance secretion is delayed in XX sex-reversed dog embryos. *Mol. Reprod. Dev.* **39**, 1–7.

Mittwoch, U. (1988). The race to be male. *New Scientist* **1635**, 38–42.

Pailhoux, E., Popescu, P. C., Parma, P., Boscher, L., Legault, C., Molteni, L., Fellous, M., and Cotinot, C. (1994a). Genetic analysis of 38,XX males with genital ambiguities and true hermaphrodites in pigs. *Anim. Genet.* **25**, 299–305.

Pailhoux, E., Cribiu, E. R., Chaffaux, S., Darre, R., Fellous, M., and Cotinot, C. (1994b). Molecular analysis of 60, XX pseudohermaphrodite polled goats for the presence of *Sry* and *Zfy* genes. *J. Reprod. Fertil.* **100**, 491–496.

Thomsen, P. D., and Poulsen, P. H. (1993). Analysis of the gonadal sex of five intersex pigs using Y chromosomal markers. *Hereditas* **119**, 205–207.

Vigier, B., Tran, D., Legeai, L., Bézard, J., and Josso, N. (1984). Origin of anti-Müllerian hormone in bovine freemartin foetuses. *J. Reprod. Fertil.* **70**, 473–479.

Vigier, B., Watrin, F., Magre, S., Tran, D., and Josso, N. (1987). Purified bovine AMH induces a characteristic freemartin effect in fetal rat prospective ovaries exposed to it *in vitro*. *Development* **100**, 43–55.

Intrauterine Growth Restriction and Mechanisms of Fetal Growth

Victor K. M. Han

The University of Western Ontario

GLOSSARY

apoptosis A naturally occurring phenomenon in developing or diseased tissues, in which scattered cells undergo a defined process of spontaneous death.

asymmetric IUGR Referring to an infant whose weight is below normal standard for gestational age but is of normal length and some body organs are more spared than others.

confined placental mosaicism A dichotomy between the chromosomal constitution of the placental tissues, both cytotrophoblast and chorionic connective tissue, and the embryonic/fetal tissues.

epigenetic factors Environmental factors nongenetic in origin.

genomic imprinting A phenomenon whereby a specific DNA segment is differentially modified during the gametogenesis and therefore functions differently depending on the parental origin of the chromosome.

intrauterine growth restriction/retardation (IUGR) A pregnancy in which the fetal size is less than the 10th percentile of expected estimated size for gestational age; a severe IUGR is defined when the pregnancy is associated with fetal size of <2 standard deviations below the expected mean (<3rd percentile). Also, fetal growth restriction/retardation.

small for gestational age Referring to infants whose birth weights are below the 10th percentile for their gestational age.

symmetric IUGR Referring to an infant who is proportionately small with both weight and length below the normal standards for gestational age, and whose organs are proportionately reduced in size.

Intrauterine growth restriction/retardation (IUGR), or fetal growth restriction/retardation, is defined as a pregnancy in which the fetal size is less than the 10th percentile of expected estimated size for gestational age. A severe IUGR is defined when the pregnancy is associated with fetal size of <2 standard deviations below the expected mean (<3rd percentile).

This is obviously an obstetric term used to define a pregnancy condition and the fetus is hence known to be growth retarded. Many now prefer the term "restricted" to "retarded" because of the connotation to mental retardation of the latter term. A newborn resulting from this pregnancy is termed a growth-restricted (retarded) infant, but most pediatricians would prefer to use the term small for gestational age (SGA) for those infants whose birth weights are below the 10th percentile for their gestational age. Those above the 90th percentile are called large for gestational age and those between the 10th and 90th percentiles are called appropriate for gestational age. Although IUGR as defined previously includes all fetuses <10th percentile, some have argued that, by strict statistical criteria, normal limits should be defined by +2 standard deviations below the mean. This definition would therefore restrict the terminology to infants below the 3rd percentile. However, it was argued that such a strict definition would leave out those infants between the 3rd and 10th percentiles who are at a similar risk of perinatal and neonatal mortality to those below the 3rd percentile. In view of such a risk, it is now accepted that the diagnosis of IUGR or SGA should be considered for all fetuses below the 10th percentile, even though its has been estimated that approximately 25% of these infants are within normal limits when other variables, such as maternal ethnic group, parity, and parental constitution, are considered. Irrespective of the birth weight percentile recommended for the diagnosis, those infants who are SGA are also considered to manifest intrauterine growth restriction. Even though "growth restricted/retarded" and "small for gestational age" are used interchangeably, it should be noted that SGA represents a mathematical or statistical description of a small infant, whereas IUGR is reserved for those fetuses or infants with clinical evidence of abnormal growth. Thus, IUGR is more of a pathological term. In addition, a fetus can develop IUGR even though it is not severe enough to result in fetal size of <10th percentile. For example, a term infant may weigh over 2500 g at birth, but it may be considered growth restricted if there is evidence that intrauterine growth rate has been falling gradually for a few weeks before birth, especially if the mother is known to produce good size infants in previous pregnancies or if a predisposing maternal, fetal, or placental pathology is present (e.g., preeclampsia). With this ongoing controversy, some have proposed a more precise definition of IUGR as one in which the growth of the fetus is less than its genetic growth potential. This would be ideal because the infant will be diagnosed as being growth restricted based on its own standard. However, genetic growth potential is impossible to define for an individual fetus because objective measurements do not exist. The neonatal growth assessment score is an attempt at this measurement, but it is too complex to be widely used. An accurate determination of gestational age is crucial for the diagnosis of IUGR. This requirement has been the argument in support of a routine ultrasound examination in late first trimester before biological variations begin to have a significant impact on its accuracy. The last menstrual period is a reliable index of gestational age if the mother with regular menstrual periods is seen early in pregnancy. After birth, neonatal assessment scores based on physical features and neurologic maturity can be used. However, these scores tend to overestimate the gestational age, and their usefulness in SGA infants has not been verified. Another criteria essential for the diagnosis of IUGR is the availability of relevant standards for fetal growth throughout gestation appropriate for the population. The latter consideration is important because race, altitude, and recent trends in societal behavior contribute to the small but definite differences between different standards. In the United States, the standards most widely used up until the 1970s were those derived in Denver, Colorado, which do not represent the birth weights of infants born at sea level nor reflect the increase in median birth weights that has occurred in the past two decades. Recent standards derived from Canadian data, consisting of over 1 million singletons and over 10,000 twin gestations between 1986 and 1988, are more likely to be relevant. Even this database is not entirely applicable to most U.S. centers because it does not include the different racial mix that is an appropriate representation of the population. The standards derived from a Californian database between 1970 and 1976 or Alabama database between 1984 and 1986 may be more applicable for a U.S. population.

I. INTRODUCTION

Intrauterine growth restriction (IUGR) and prematurity are the two predominant causes of fetal and neonatal mortality and morbidity in perinatal–neonatal medicine. The use of antenatal steroids, high-resolution ultrasound, and modern perinatal care, together with neonatal surfactant replacement therapy, alternative modes of ventilation in addition to conventional ventilation, and modern neonatal nursing and supportive care, have reduced neonatal mortality and morbidity significantly in premature newborns in the past two decades. With modern perinatal–neonatal care, some of the acute neonatal morbidities associated with IUGR pregnancies, such as birth asphyxia and biochemical disturbances, can be recognized and, to an extent, are preventable. The large improvement in perinatal and neonatal outcomes in term pregnancies have led some to question the validity of the diagnosis of "term IUGR." However, despite rapid development in technologies in the assessment and monitoring of fetal health and growth, perinatal mortality associated with IUGR is still associated with significantly higher perinatal mortality, particularly in preterm infants. The majority of the mortality occurred *in utero*, highlighting deficiencies in the diagnosis of fetal health in intrauterine growth-restricted fetuses and in the knowledge of fetal growth and its adaptation to adverse maternal uteroplacental environment.

A. Classification of IUGR

It is important to differentiate between IUGR and low-birth-weight infants: All IUGR infants are low birth weight but not all low-birth-weight infants are IUGR. The World Health Organization classifies all newborns weighing <2500 g as having low birth weight, irrespective of the gestational age. Many of the infants are preterm (<37 weeks' gestation) and some of them are both preterm and growth restricted. The latter are infants who are doubly compromised and they have significantly greater perinatal and neonatal mortality than age-matched appropriately grown infants.

From the clinical point of view, IUGR infants are classified as either being symmetric or asymmetric, even though the pathological basis from which these definitions are derived is questionable. Asymmetric IUGR refers to an infant whose weight is below normal standard for gestational age but is of normal length and some body organs are more spared than others (fetal liver tends to be disproportionately small compared to the brain, which is spared). It is due mainly to placental insufficiency resulting from hypertensive diseases of pregnancy, including preeclampsia, collagen vascular diseases, abnormal placenta, and advanced diabetes. Symmetric IUGR refers to an infant who is proportionately small, with both weight and length below the normal standards for gestational age, and whose organs are proportionately reduced in size. It is due to chromosomal abnormalities of the fetus and placenta, genetic disease of the musculoskeletal system, congenital infections, severe chronic maternal malnutrition, and smoking. Placental insufficiency caused by severe preeclamp-

sia, which may manifest early in the latter part of second trimester, can lead to symmetric growth restriction. Therefore, the differentiation between symmetric and asymmetric IUGR based on potential pathology is not entirely correct. The phenotype of the fetus will depend on the severity of the cause and the gestational age the growth restriction manifests, i.e., the earlier the pathology is initiated, the more likely the phenotype of the infant to be symmetric.

One measurement that can be used to differentiate between the two types of IUGR is the ponderal index (PI), which can be calculated by the formula:

$$PI = birth\ weight\ (g) \times 100\ crown\text{–}heel\ length\ (cm)^3$$

This measurement assesses whether the infant is relatively fat or thin and is independent of race, sex, birth rank, and gestational age. Infants with a PI <3rd percentile for gestational age are relatively long and appear wasted. The perinatal mortality rate of such infants is greater and the neonatal morbidity and mortality are higher than those of proportionately small infants. It has also been suggested that these infants are more likely to develop cardiovascular diseases in adult life.

II. NORMAL EMBRYONIC AND FETAL GROWTH

Fetal growth is not a uniform progression of cell replication but a series of fundamentally different, precisely integrated anabolic processes. Development prior to placental implantation involves rapid cell hyperplasia and pattern formation, which becomes apparent when primitive germ layers separate and presumptive organs can be first identified. Metabolism at this time is anaerobic, and cells communicate by cell–cell contacts and by the release of autocrine and paracrine factors, including growth factors. The implantation of the placenta allows aerobic respiration and the development of energy-expensive enzyme systems necessary for differentiative cell function. The appearance of fetoplacental circulation improves the delivery of nutrients to the tissues, and consequently the rate of cell multiplication rises. The placenta acts as an endocrine organ influencing both fetus and maternal physiology and also as an immunological barrier between the fetus and the mother to allow the "foreign" fetus to develop within the maternal immunologic environment. During early embryonic development, the pattern of growth is largely dictated by the fetal genome, but as the body size increases, the fetus becomes constrained by maternal and environmental influences such as uterine blood flow, maternal size, and maternal disease.

Embryogenesis involves three major processes: morphogenesis (the mechanics of generating shape), pattern formation (how the fate of each cell is specified into different tissues and organs of the body), and differentiation (the specialization of cells into different phenotypes). Although most mammals have a similar number of cell types, the process that differentiates between species in terms of their functional characteristics is the pattern formation that spatially organizes the cells. Therefore, the genetic makeup of the cell (lineage) together with the epigenetic mechanisms involving cell–cell, cell–matrix, and cell-soluble factor interactions are important in the ultimate development of the tissues and organs.

A. Morphogenesis and Pattern Formation

Morphogenesis and pattern formation are regulated by homeobox genes. They contain a common sequence element of 180 bp, the homeobox, which was first discovered in *Drosophila*. The homeobox encodes a 60-amino acid homeodomain that is responsible for sequence-specific DNA binding of much larger homeodomain proteins. These proteins act as transcriptional factors which activate or repress target genes. In vertebrates, over 170 different homeobox genes have been cloned, which can be divided into (i) the clustered homeobox genes called *Hox* genes or class I homeobox genes and (ii) the nonclustered or divergent homeobox genes, e.g., *Pax, Msx, Emx,* and *Reig,* named after their homologs in the fly. There are at least 39 mammalian *Hox* genes that are organized in four clusters called HoxA, HoxB, HoxC and HoxD, each localized to a different chromosome (7, 17, 12, and 2, respectively) and

comprising 9–11 genes. The individual *Hox* genes within the different cluster can be aligned with each other and with genes of the *Drosophila*. These genes are expressed as early as in the gastrula and are localized to all three germ layers with the overlapping domains that extend from the caudal end of the embryo to the sharp anterior limit which is specific for each *Hox* gene. Based on transgenic and gene targeting studies in mice involving the *Hox* genes, the function of these genes has been identified. As a result of these gain- and loss-of-function mutations, it is now clear that, as in *Drosophila,* the mammalian *Hox* genes of the posterior are more dominant with respect to function over the more anterior acting ones. In general, the *Hox* genes are important in the patterning of the axial skeleton, the appendicular skeleton, the reproductive and digestive tracts, and the hindbrain, and they are a component of the development of the craniofacial region. Recent studies suggest that the vertebral or axial specification is more likely achieved by the functional cooperation of a "combination" of *Hox* genes expressed at a given axial level than by the function of a single gene. The second type of homeobox genes, the noncluster type, are involved in the craniofacial and forebrain development (*Otx, Pax,* and *Emx*), hindbrain development (*En-1* and *En-2*), eye development (*Pax6*), and the development of many other organs such as kidney and heart. Mutation of human homologs of both clustered (*Hox*) and nonclustered homeobox genes has been associated with different congenital syndromes, such as synpolydactly and hand–foot–genital syndrome (*HoxD13* and *HoxA13*), schizencephaly (*Emx2*), Reiger's syndrome (*Reig*), and Boston-type craniosynostosis (*Msx2*), and may predispose the individual to cancer, such as acute myeloid leukemia (*HoxA9*), pre-B cell leukemia (*Pbx1*), T cell leukemia (*Hox11*), and rhabdomyosarcoma (*Pax3* and *Pax7*).

B. Cellular Growth and Differentiation

Once the general body plan has been specified, the development of the embryonic tissues and organs is regulated by local interactions between the cells and their extracellular matrix, which determine further morphogenesis and differentiation. Cellular prolifer-

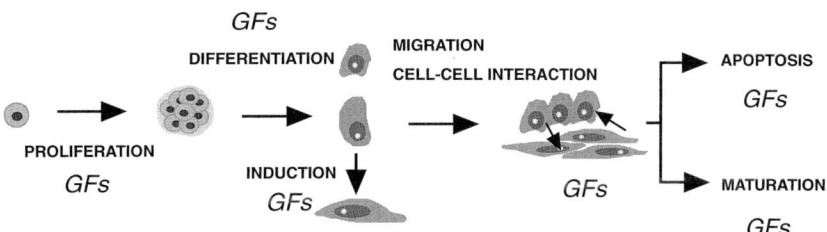

FIGURE 1 Schematic of cellular processes involved in ontogenesis (embryonic and fetal growth and development) and the role of growth factors (GFs) in the regulation of these processes.

ation and differentiation are just two aspects of a complex process by which the conceptus (organs and tissues inclusive) increases in size and function (Fig. 1). In addition, considerable cell migration occurs in the early embryogenesis to bring different cell types into intimate contact, which may at a certain phase of development lead to tissue induction (e.g., induction of mesoderm from ectoderm by endoderm). Following blastulation, epithelial–mesenchymal interactions become particularly important in shaping tissue morphology. With further proliferation, differentiation, migration, aggregation, and programmed cell death (apoptosis), a mature organ or tissue is formed. Even after the formation of a mature functioning organ, cells continue to die and require replacement with new functioning cells (maintenance). Due to certain pathological conditions, parts of the developed organ may be damaged and may need to be repaired (regeneration). In the majority of organs, the regeneration capacity of an organ is greatest during embryonic and fetal life compared to any other time during the life span of an animal or a human.

The epithelial–mesenchymal interaction represents a form of signaling between the germ layers, and the interaction can occur in either direction depending on the stage of embryonic development. For example, in the palate, the mesenchyme initially signals to the overlying epithelium to differentiate at an early stage, and later a reciprocal signaling occurs from the epithelium to the mesenchyme to cause osteogenic differentiation. These interactions can occur via direct cell–cell contact or via specialized molecules, such as cell adhesion molecules (both calcium dependent and independent), integrins, extracellular matrix molecules (both nonsolu-

ble and soluble), and growth factors. In many cases, these interactions occur via a combination of methods involving more than one molecule, such as extracellular matrix–growth factor–integrin interaction. For example, during palatal development, tumor growth factor-β1 (TGF-β1) stimulates the synthesis of an extracellular matrix molecule tenascin in the mesenchyme, and then the overlying epithelial cells interact with the matrix via integrins.

III. REGULATION OF FETAL GROWTH

A. Genetic Determinants

The effect of the embryonic/fetal genome and its relative contribution compared to the maternal uterine environment on the ultimate size of the newborn have been the subject of research for the past half a century or more. Classic studies in farm animals demonstrated that hybrids of shire horses and Shetland ponies have approximately a fourfold difference in their birth weights depending on the dams from which they were conceived. However, birth weights of the hybrids in shire dams were still significantly lower than the normal shire birth weight, illustrating that there is an intrinsic limit to growth set by the embryonic/fetal genome. Similar results were shown by comparable studies in cattle and sheep. However, it is simplistic to assume that the embryonic/fetal genome is mainly responsible for the limitation of fetal size. It has been estimated that fetal size is determined in 30–60% by embryonic/fetal genes. Chromosomal mosaicism of the fetus, which arises from a postzygotic error, may involve all or some of the tissues of the conceptus, including the placenta. Recent studies on the genotype of the placentae have identified the significance of placental genome on the growth and development of the organ and this may be an important cause of idiopathic intrauterine growth restriction.

1. Confined Placental Mosaicism

Confined placenta mosaicism (CPM) is defined as a dichotomy between the chromosomal constitution of the placental tissues, both cytotrophoblast and chorionic connective tissue, and the embryonic/fetal tissues. It is estimated to occur in about 1 or 2% of all pregnancies that are viable after the first trimester. Three types, termed types I–III, have been classified depending on the karyotype of the cytotrophoblasts and the chorionic mesodermal cells. A pregnancy with type II CPM (trisomy of the chorionic mesoderm) is usually uneventful, but sometimes it may be associated with unexplained fetal death or IUGR. A pregnancy with type III CPM (aneuploidy of both cytotrophoblasts and chorionic mesoderm) is often complicated by IUGR. Placental trisomy usually results from postzygotic loss of one of the trisomic chromosomes in the progenitor cells of the embryo leading to a uniparental disomy. This affects the fetal growth if it involves the chromosome that carries the imprinted genes. Genomic imprinting is a phenomenon whereby a specific DNA segment is differentially modified during the gametogenesis and therefore functions differently depending on the parental origin of the chromosome. Differential DNA methylation of the regulatory regions of the genes has been implicated as the potential mechanism of imprinting. In addition to CPM, generalized mosaicism involving both the placenta and the fetus may be the cause of either general or limited growth restriction.

2. Chromosomal Abnormalities of the Fetus

The most obvious and severe forms of IUGR of genetic origin in the fetus are chromosomal defects such as trisomies of chromosomes 13 and 18 and, less frequently, of 21. Trisomy 18 usually presents with severe symmetric IUGR and hydramnios and is usually detectable by ultrasound and then confirmed by fetal karyotyping. Trisomy 13 and chromosomal

abnormalities involving the X chromosome (Turner's syndrome and extra X) can present with IUGR. To date, no cellular or growth regulatory mechanism has been associated with the chromosomal abnormalities to explain the growth failure.

B. Epigenetic Determinants

The epigenetic or environmental factors include the general maternal environment, immediate maternal environment, maternal age and parity, and other unknown factors. In general, it is believed that the size of the fetus at birth is genetically predetermined based on the genetic factors described previously. However, the epigenetic or environmental factors influence the final birth size via the biochemical mediators that are elaborated into the circulation (endocrine factors) or synthesized locally (paracrine factors).

1. Endocrine Regulation

i. *Maternal Hormones*   Maternal hormones such as growth hormone, thyroid hormones, insulin, prolactin, and steroid hormones do not cross the placenta in physiologically important quantities. The placenta is permeable to cortisol, but this is predominantly converted to a biologically inactive form, cortisone, by the placental 11β-hydroxysteroid dehydrogenase. Factors regulating fetal growth are therefore largely self-contained within the developing fetus. However, growth and development of the fetus may be indirectly associated with maternal hormonal status since it is intimately related to the normal development of the maternal reproductive and uterine tissues as well as the placenta.

ii. *Fetal Hormones*   The data indicating that the fetal endocrine organs are important in the regulation of fetal growth are based mainly on animal experimentation on sheep fetuses in which the specific endocrine gland has been removed (Table 1). Whether the same regulatory mechanisms exist in the humans has been difficult to prove because the congenital defects of the absence of endocrine glands may not necessarily lead to similar phenotypic changes in the affected fetuses due to the presence

TABLE 1
The Effect of Removal of Fetal Ovine Endocrine Glands on Somatic and Organ Growth

Procedure	Gestational age (days)	Body weight (%)	CRL (%)	Body weight: CRL (%)	Organ
Px	115–120	<30	<10	<15	Spleen, thymus
Tx	80–95	<35	<10	<25	Skin, skeleton
Hx	100–110	<5–20	NC	NC	Skeleton, lungs, liver, GIT
Hx	70–79	<26	NC	NC–<20	Adrenal, heart, lungs, thyroid
Ax	110–115	NC	NC	NC	Lungs, liver, GIT

Note. Abbreviations used: Px, pancreatectomy; Tx, thyroidectomy; Hx, hypophysectomy; Ax, adrenalectomy; NC, no change; GIT, gastrointestinal tract; CRL, crown–rump length.

of small amounts of hormones from ectopic glands or the potential influence of placentally transferred maternal hormones. Of the major fetal hormones, insulin is the most relevant hormone associated with fetal growth both clinically and experimentally.

Insulin   Macrosomia in infants of diabetic mothers is due mainly to increased amounts of body fat, which are in part is due to excessive insulin secretion. Conversely, fetal insulin deficiency can arise from pancreatic agenesis, nutrient deprivation, transient diabetes, or reduced tissue sensitivity to insulin (leprechaunism). These conditions are all associated with fetal growth retardation, reduced body fat, and varying degrees of muscle wastage. In experimental animals, this condition is induced by surgical ablation of the fetal pancreas and by the fetal administration of diabetogenic drugs. Both procedures cause chronic insulin deficiency *in utero* and lead to fetal hyperglycemia and other metabolic disturbances. They also produce fetal growth retardation with reductions in body weight at term of 10–40% depending on species and gestational age at the onset of fetal hypoinsulinemia. In the sheep fetus, the decrease in body weight was accompanied by reductions in limb and crown–rump length. With insulin replacement treatment, growth rate and body size at term were restored to normal. The mechanisms by which insulin deficiency causes growth retardation include (i) alteration of fetal metabolic environment in favor of catabolism and (ii) reduction in insulin growth factor-I (IGF-I) concentrations and an increase in insulin-like growth factor binding protein-1 (IGFBP-1).

2. Autocrine/Paracrine Regulation

The relative lack of endocrine regulation of fetal growth suggests that local regulatory mechanisms must exist to control and coordinate the multifaceted processes of embryogenesis and fetal development, which involve not only cellular proliferation (hyperplasia) and differentiation (hypertrophy) but also other complex cellular events (Fig. 1). This integrated process requires precise intercellular communication. The local cell–cell and cell–matrix interactions occur through several effector molecules, including extracellular matrix, intercellular recognition molecules, and peptide growth factors.

i. Mechanisms of Regulation of Cellular Growth and Differentiation   At least three different control mechanisms govern early tissue interactions and responses: (i) the deposition and subsequent modification of extracellular matrix molecules, (ii) the temporal expression of cellular recognition molecules, and (iii) the appropriate expression of intercellular messengers such as peptide growth factors. These three basic control mechanisms are by no means independent but occur in an extremely interactive and interdependent manner. For example, growth factors regulate the expression of genes encoding many, if not all, extracellular matrix proteins, and the latter serves as a storage depot for many growth factors such as TGF-β; the expression of many cellular recognition molecule genes is regulated by growth factors (IGF-I and E-cadherin gene expression during differentiation from cytotrophoblast to syncytiotrophoblast).

Extracellular Matrix Extracellular matrix consists of two major classes of molecules: proteins such as collagens, laminin, fibronectin, and tenascin and mucopolysaccharides such as hyaluronic acid, heparin sulfate, and chondroitin and keratin sulfate. Most epithelia are tightly attached to basal lamina or basement membrane, which is rich in type IV collagen and laminin. Fibronectin and tenascin govern cell shape and proliferation rate *in vitro*; evidence is accumulating that they can determine cell migration and the onset of differentiation *in vivo*. Tenascin, in particular, has been shown to appear immediately along the migration tracts before the onset of migration of neural crest cells from around the neural tube and is removed immediately afterwards. Hyaluronic acid, a hydrophilic mucopolysaccharide, also forms an important component of the migratory tracts and forms a coating of the migrating cells.

Intercellular Recognition Molecules The intercellular recognition molecules are utilized by the cells to recognize their anatomical positions within developing structures and to sort out the major cell types (e.g., epithelial and mesenchymal cells). Two categories of recognition molecules exists: (i) extracellular matrix (ECM)-cell recognition molecules or integrins, which are utilized for interaction between ECM and cells, and (ii) the intercellular recognition molecules, which are utilized for interaction and recognition between cells. There are two forms of intercellular recognition molecules: (i) the cell adhesion molecules, which do not require calcium for their biological action, and (ii) the cadherins, which are calcium dependent. Several subclasses exist within each group according to their principal or first-observed anatomical sites of expression (e.g., epithelial cadherin, neural cadherin, and placental cadherin). Identical recognition molecules are expressed on the membranes of homotypic cells during tissue condensation or between heterotypic cells during mesenchymal–epithelial or other tissue interactions. Communication of cadherins with actin filament bundles within the cytoskeleton suggests that cells can respond directly to these molecular interactions, or lack of them, with movement. In addition, the cytoskeletal network may be utilized to initiate intracellular biochemical events following cell–cell interaction via cadherins on the cell surfaces.

Peptide Growth Factors

Chemistry: Peptide growth factors are proteins of less than 30 kDa which are widely synthesized and act within local tissue environment as paracrine or autocrine factors and serve to regulate various aspects of cellular growth and development (Fig. 1). Growth factors that are expressed in developing tissues and have been shown to have biological actions on developing cells are listed in Table 2. Growth factors that have been identified to act on the different aspects of cellular growth and development are shown in Table 3.

Synthesis: Unlike classical hormones, which are secreted by specific endocrine organs, each growth factor may be synthesized by many anatomically diverse cells. They are generally not stored intracellularly and their release is largely dependent on *de novo* synthesis. The capacity of many different developing cells to synthesize growth factors is an essential requisite of their autocrine/paracrine mode of action. Many are also released into the circulation, but it is doubtful if any of them has an endocrine mode of action like classical hormones.

TABLE 2
Growth Factors Expressed during Development

Insulin-like growth factors (IGFs)
Epidermal growth factor (EGF)
Transforming growth factor-α (TGF-α)
Transforming growth factor-β family (TGF-β)
Platelet-derived growth factor (PDGF)
Fibroblast growth factors (FGF)
Nerve growth factor family (NGF)
Other growth factors
   Hemopoietic growth factors
   Interferons
   Bombesin- and gastrin-releasing peptide
   Müllerian-inhibitory substance
   Inhibin/activin family

TABLE 3
Biologic Actions of Growth Factors on Various Developmental Processes

Developmental process	Growth factor
Proliferation	IGF-I and IGF-II
	EGF
	TGF-α
	TGF-β
	FGF
	PDGF
	Hemopoietic growth factors
Differentiation	IGF-I and IFG-II
	EGF
	FGF
	NGF
Induction	FGF
	TGF-β
Migration and aggregation	Cell adhesion molecules and chemotactic factors
	GFs
Apoptosis	NGF
	IGF-I, IGF-II
	TGF-β
	Inhibins
	Müllerian-inhibitory substance
Maintenance	IGF-I and IGF-II
	TGF-α
	TGF-β
Regeneration	IGF-I and IGF-II
	TGF-α
	TGF-β
	NGF
	PDGF

Receptors: Since growth factors are polypeptides or glycoproteins in nature, they are hydrophilic or lipid insoluble and therefore have to interact with specific protein receptors located on the cell membrane of target cells and communicate with second messenger systems by conformational changes. This often involves the autophosphorylation of tyrosine residues located on the intracellular domain of the receptor.

Signal transduction: The second messenger sys-

tems utilized are diverse and include changes to intracellular calcium levels, phosphorylation or dephosphorylation of a cascade of intracellular proteins by activation of kinases and phosphatases, and alterations in inositol phosphates. For some growth factors, the system is becoming defined. It is interesting that many of the diverse actions of growth factors, in terms of their mitogenic, differentiating, or anti-apoptotic effects, that follow the growth factor–receptor interaction are now being delineated by the differences in signal transduction pathways.

Actions: The net biological events are remarkably consistent and include a rapid stimulation of amino acid transport, glucose uptake and utilization, RNA and protein synthesis, and, most important, alterations in gene expression. For many but not all growth factors, this is followed by DNA synthesis and cell replication. Some growth factors, such as TGF-β and tumor necrosis factor, act predominantly as growth inhibitors.

Stimulation of DNA synthesis and mitogenesis are the prominent basic biologic functions of many growth factors. The basic mechanism by which this action is mediated is best described in studies on the cell cycle of 3T3 fibroblasts which showed that growth factors act in the G1 phase of the cell cycle. This is the phase in which the cell acquires the necessary nutrients, proteins, and enzymes to begin its preparation for DNA synthesis (S phase). Growth factors may act in the first or second half of this phase as competence factors [fibroblast growth factor (FGF) and platelet-derived growth factor (PDGF)] or progression factors [epidermal growth factor (EGF) and IGF-I]. However, in many other cell systems, the roles of different growth factors are not as well delineated. Various growth factors have been shown to be able to act as both competence and progression factors.

Tissue induction is an important action of growth factors. TGF-β and FGF have been shown to promote the induction of mesoderm from primitive ectoderm in the early embryo. Such induction in the amphibian embryo depends on diffusable morphogens from the vegetable pole ectoderm causing mesodermal development from the animal pole ectoderm. FGF has been shown to be the inducer for ventral mesoderm and TGF-β for the dorsal mesoderm.

Another important function of growth factors is to augment or antagonize the onset of differentiation. A potentiation of differentiation need not exclude a mitogenic role since an immediate action on proliferation may be followed by a later tendency to promote terminal differentiation. Separate growth factors can cooperate or compete during the onset of differentiation, whereas some show synergism with endocrine hormones. In some growth factors (e.g., IGFs), the induction of mitogenesis or differentiation may be a matter of concentration or dose. Physiological concentrations of IGF-I promote proliferation of newly isolated myoblasts, whereas higher concentrations promote differentiation into myotubes. Many of the cells continue to express growth factors even after they have differentiated, suggesting that growth factors continue to be involved in maintenance of certain differentiative functions of the cells. It has also been suggested that these cells may exhibit a regenerative capacity from pathological processes.

Recently, attention of many research groups has been focused on the cell survival effects of growth factors. Apoptosis, or programmed cell death, is a naturally occurring phenomenon in developing or diseased tissues, in which scattered cells undergo a defined process of spontaneous death. It plays an important role in regulating cell populations. Morphologically, the death process usually occurs in two stages. First, the cell undergoes nuclear and cytoplasmic condensation, eventually breaking up into a number of membrane-bound fragments containing structurally intact organelles. Second, these cell fragments, termed apoptotic bodies, are phagocytosed by neighboring cells and rapidly degraded. One characteristic feature of apoptosis that is distinct from cellular necrosis is the fact that no inflammatory or immune response is induced at the site of this occurrence. Biochemically, the characteristic hallmark feature of apoptosis is DNA fragmentation that occurs in multiples of 180–200 bp. This is because of the production of oligonucleosomal fragments as a result of endonuclease activity that cleaves chromatin at the linker DNA between nucleosomes. It occurs very early in the process—within a couple of hours of the initiating event even before cell viability starts

to decrease. Mediators of apoptosis in either a positive or a negative way (usually the latter) are mainly growth factors or hormones. Following binding to their respective receptors, these mediators initiate specific signal transduction pathways involving intracellular kinases such as protein kinase C, Akt kinase, PI-3 kinase, MAP kinases, and lysosomal protease cathepsin D. Recently, an oncogene, *bcl-2*, has been implicated in apoptosis. *bcl-2* is an inner membrane mitochondrial protein that may counteract apoptosis. Binding of another intracellular protein, *bad*, to *bcl-2* inhibits the action of *bcl-2* and promotes apoptosis. Accelerated apoptosis may be an important mechanism by which fetal growth retardation can occur in circumstances in which conditions such as chronic hypoxia may not be severe enough to cause cellular necrosis.

Since more than one growth factor may be synthesized locally within a developing organ or tissue, it is likely that significant interactions exist among the GFs. Such interactions have been demonstrated in many cell systems. The most prominent of which are muscle (IGFs and TGF-β), cartilage and skeletal system (IGFs, FGFs, and TGF-β), brain (IGFs and EGF/TGF-α), integumentary system (IGFs and TGF-α), and respiratory system (IGFs, FGFs, and PDGF). The final result of such interactions is the programmed development of a mature functioning organ.

IV. PATHOGENESIS OF IUGR

Except for a rare occurrence of pancreatic hypoplasia in the fetus as the cause for IUGR, fetal endocrine factors are not of significance in the pathogenesis of IUGR. However, many of the paracrine growth factors are involved in the pathogenesis of IUGR as local mediators of impaired fetal growth by inhibiting cellular proliferation and differentiation and promoting apoptosis (Fig. 2). Such cellular events can occur either within the fetal tissues or in the placenta. The latter may explain in part the poorly developed placenta due to impaired placentation or genetic reasons.

The most supportive evidence for the role of growth factors in fetal growth is provided by evidence

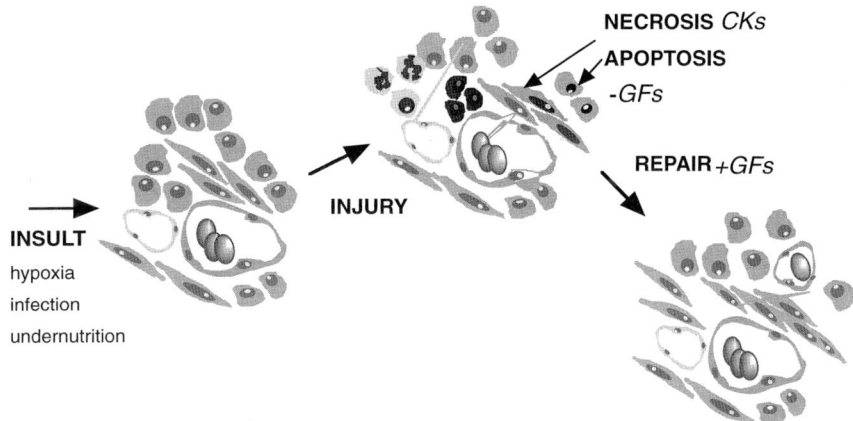

FIGURE 2 Schematic of cellular processes involved in the pathogenesis of IUGR and the role of cytokines (CKs) and growth factors (GFs) in the process.

of fetal and newborn growth retardation in mice with null mutations of IGF-I, IGF-II, and IGF-I receptor genes by gene targeting. Since this discovery, many other growth factors and their receptors have been targeted and they have demonstrated a growth retarded phenotype or none or unexpected observations. The absence of an expected phenotype is explained by genetic or biologic redundancy, compensation by other similar growth factors, or maternal rescue.

V. ETIOLOGY OF IUGR

A. Genetic Causes

Familial studies of low-birth-weight infants suggest that maternal family history of IUGR may predispose the woman to have small babies despite seemingly normal pregnancy. Paternal sibling history does not seem to increase the incidence of IUGR. Maternal genetic diseases such as phenylketonuria can also cause IUGR.

Fetal genetic makeup, particularly related to chromosomal disorders such as trisomy 18 and 13, XO, or Turner's syndrome, various dysmorphic syndromes associated with dwarfism and chondrodyplasia, or abnormal fetal brain development cause IUGR. It is estimated that approximately 25% of fetuses with early onset IUGR may have chromosomal abnormalities.

Newborns with congenital cardiac anomalies are frequently small for gestational age (SGA). The reason for this association is still unknown, and impaired systemic perfusion may not be an adequate explanation because anomalies such as atrial septal defect, which has normal fetal circulation, are frequently associated with IUGR.

B. Multiple Gestation

There is an indirect relationship between the birth weight and the number of offsprings in a pregnancy. In twin pregnancies, the growth rate peaks at 160–170 g/week at 28–32 weeks' gestation compared to singleton pregnancies in which the growth rate peaks at 220–240 g/week at 34 weeks' gestation. The decrease in weight of twin fetuses is due to cellular hypotrophy, similar to IUGR of poor maternal undernutrition or placental hypoperfusion, suggesting that it is due to the inability of the maternal intrauterine environment to meet the needs of the fetus. In monozygotic twins, the discordance in size is often attributable to twin–twin transfusion, leading to further growth restriction of the nondominant twin.

C. Maternal Undernutrition

Maternal undernutrition of protein or calories or both causes IUGR in both animals and humans. The type of IUGR observed will be dependent on the duration and the gestational age of onset of the nutritional insult. The famine in Holland during World War II led to approximately 10% reduction in birth weight when undernutrition occurred in the third trimester in previously well-nourished mothers. Recent studies indicate that the poor prepregnancy nutritional status and the inadequacy in weight gain in pregnancy are associated with an increased risk of IUGR. Limitation of specific substrates such as minerals (e.g., zinc) or vitamins (e.g., folate) can also be associated with IUGR. One important substrate for adequate fetal growth is oxygen, and its limitation due to high altitude, maternal cyanotic congenital heart disease, chronic pulmonary disease (e.g., poorly controlled asthma and cystic fibrosis), and maternal anemia may increase the risk of IUGR.

Nutritional supplementation, if targeted can be of benefit to improve the birth weight. In poorly nourished populations, caloric supplementation can lead to an average increase in birth weight of 50–225 m. However, high protein supplementation does not always lead to a desired improvement in birth weight, and in some studies it was demonstrated that it may have an adverse effect.

D. Placental Insufficiency and Disease

This is perhaps the most important group of IUGR pregnancies because of its clinical significance in terms of both diagnosis and management. The underlying pathophysiology in these conditions is the reduced placental perfusion which leads to impaired transfer of nutrients, substrates, and metabolites between the mother and the fetus. Not only does the placenta transport basic nutrients, such as glucose, amino acids, and lipids, from the mother to the fetus but also it is a metabolically active organ, accounting for more than 50% of the uptake of oxygen and glucose by the conceptus. The uptake of glucose by the placenta and its transfer to the fetus occur by facilitated diffusion mediated by specific transporters. Only Glut-1 and Glut-3 transporters are expressed in the placenta, neither of which is insulin sensitive. The energy produced by placental oxidative metabolism is used for the concentration of amino acids to very high levels within the placental cells, which then diffuse facilitatively into the fetal

plasma. Thus, placental metabolism plays a major role in providing nutritional requirement for fetal growth.

Clinically, uteroplacental blood flow is decreased in pregnancies complicated by maternal hypertensive diseases (preeclampsia and chronic hypertension), maternal vascular disease associated with antiphospholipid antibodies such as lupus anticoagulant factor and anticardiolipin antibodies, and atherosclerosis associated with advancing maternal age. In preeclampsia, the characteristic obstructive arterionecrosis of the maternal decidual vessels is often observed in those pregnancies associated with IUGR. Impairment of trophoblastic invasion and failure of trophoblasts to take up the maternal decidual spiral arteries are thought to be the primary pathological processes of preeclampsia. This leads to local placental hypoxia and ischemia and the release of an endothelial toxic factor into the maternal circulation. On the fetal side of the circulation, placental vessels undergo vasoconstriction and increase resistance to flow, which cause the characteristic abnormal flow wave patterns on umbilical Doppler waveforms. The increase in placental vascular resistance is positively correlated with fetal hypoxemia and IUGR.

Decreased placental perfusion by a high level intense exercise during pregnancy can reduce neonatal fat mass and therefore birth weight. Abnormal placentae (e.g., circumvallate placenta), abnormal uterus (e.g., unicornuate uterus), and recurrent antepartum hemorrhage from either abruption or placenta previa cause IUGR from impaired placentation and/or perfusion.

Twin–twin transfusion in monozygotic (monochorionic) twins with abnormal vascular anastomosis in the placental vessels is often associated with discordant twins due to a "steal" phenomenon. Abnormal insertion of the umbilical vessels and single umbilical vessel may also be associated with IUGR.

Several models of IUGR have been established in experimental animals by decreasing placental perfusion by ligating the maternal uterine artery, repetitive embolization of the maternal or fetal placental arteries, and reducing the maternal ambient oxygen. Each model has its own advantages as well as disadvantages in representing human IUGR resulting from uteroplacental insufficiency.

E. Cigarette Smoking and Toxins

Maternal cigarette smoking decreases birth weight by approximately 135–300 g, and the resulting newborn is usually symmetrically growth restricted, especially if smoking is continued throughout pregnancy. If smoking is stopped prior to the third trimester, its adverse effect on birth weight is reduced. The mechanism is unknown, but it may be related to chronic fetal hypoxemia as a result of increased carboxyhemoglobin or its direct effects on placenta function. Placentae from pregnancies with maternal smoking are large relative to fetal size.

Chronic maternal alcohol ingestion of more that two drinks per day, cocaine and drug (e.g., heroin) abuse, and prolonged use of drugs such as corticosteroids, dilantin, and coumadin are associated with IUGR. In such cases, Growth restriction appears to affect head size more than birth weight.

VI. DIAGNOSIS OF IUGR

A high index of suspicion in those women who are at risk of IUGR is an important key to diagnose IUGR pregnancies. A regular, and if necessary frequent, antenatal check-up is warranted for these pregnancies. However, the clinical tools currently available, such as the symphysis–fundal height measurement and clinical estimation of fetal size, have a high negative predictive value. It is therefore imperative that such clinical monitoring be accompanied by imaging and laboratory tests to not only monitor fetal growth but also predict fetal health.

A. Ultrasonography and Doppler Studies

Ultrasonography is the most reliable and preferred method currently available for the diagnosis and monitoring of fetal growth and health. If done at late first trimester, ultrasound measurements give the most reliable estimation of gestational age. The establishment of normograms has allowed relatively accurate assessment of fetal weight. Most estimates of fetal weight are derived from measurements of the biparietal diameter, abdominal circumference, and femur length, with the ratio of femur length to abdominal circumference having an improved diagnostic accuracy in asymmetric IUGR fetuses. In addition, the ability to measure several fetal dimensions allows not only the pattern of growth abnormality but also a potential cause of the growth failure.

Using head and abdominal circumference, the ratio of the two measurements, and the femur length, the clinician can determine whether the fetus is symmetrically or asymmetrically growth restricted. Insults occurring early in pregnancy, such as infection, chromosomal abnormalities, and maternal exposure to toxins and drugs, usually lead to symmetric growth restriction. In this type of IUGR, femur length and head circumference are low for gestational age, as are the abdominal circumference and estimated fetal weight. An examination done by a careful and experienced ultrasonographer on a good machine can detect subtle abnormalities, such as clinodactyly and abnormal facies, and therefore suggest potential causes of IUGR. An insult occurring later in pregnancy usually results in asymmetric growth restriction, and uteroplacental insufficiency is likely the cause. In this type, femur length and head circumference are within normal limits, but abdominal circumference is decreased due to subnormal liver growth. In some cases, differentiation between the two types of IUGR is not clear, and the patterns often merge. This is seen in diseases that cause maternal nutritional deprivation and placental underperfusion, starting early in pregnancy. Ultrasound measurement of amniotic fluid volume is a useful indicator of fetal health, and if associated with IUGR, it is a good indicator of either a significant fetal compromise or a congenital anomaly of the kidneys (e.g., agenesis).

Doppler studies of the fetal vasculature have enhanced the study of fetal cardiovascular adaptive response to growth-restrictive conditions and to date Doppler analysis is a useful adjunct tool to diagnose fetal health and/or distress. Its usefulness as a predictive measurement for fetal compromise, both short and long term, should be evaluated with caution. The flow velocity waveform in the umbilical artery is perhaps the most widely used parameter. However, both the sensitivity and specificity of the test are too low to be of clinical value. In addition, the signal from the umbilical artery should be analyzed only if it could be displayed simultaneously with the low-frequency signal from the umbilical vein. In the majority of IUGR fetuses, due to uteroplacental insufficiency there is an increase in pulsatility as a result of a decrease in diastolic flow velocity. The increase in pulsatility index, resistance index, or S/D ratio and a decrease or even reversal of end-diastolic flow indicate a rise in peripheral resistance fetal placental vessels and may indicate the severity of fetal compromise. Doppler analysis of the fetal cerebral vessels can add valuable information on cerebral resistances (and potentially flow) and "brain sparing effect" seen in asymmetric IUGR. Its usefulness as a predictive assessment of potential neurologic sequelae is not yet proven.

B. New Imaging Studies

New imaging techniques such as three-dimensional ultrasound and magnetic resonance imaging and spectroscopy have been described as potential tools of the future for the diagnosis and management of IUGR. These investigative tools are still in their infancy of development, but with improvement in the computation power and hardware (reduction in acquisition time), they will add to our diagnostic armamentarium.

C. Cordocentesis

Measurement of fetal blood gases and biochemical parameters on umbilical blood samples obtained by cordocentesis can be used to manage IUGR pregnancies. Normograms have been established, and it is confirmed that IUGR fetuses due to uteroplacental insufficiency are more frequently hypoxemic, hypercapneic, and acidotic than appropriately grown fetuses. Cordocentesis, if feasible, can also be used for fetal karyotyping.

D. Biochemical Measurements

The low to low-normal levels of maternal blood or urine estriol and the low maternal circulating human placental lactogen (hPL) have low predictive value in the diagnosis or management because of significant overlap between normal and IUGR pregnancies. Elevated maternal serum β-human chorionic gonadotropin concentration, maternal and fetal (IGFBP-1), and low maternal serum unconjugated estriol and fetal serum leptin and IGF-I concentrations are associated with IUGR. The usefulness of these tests in the diagnosis or management of IUGR awaits further large-scale studies.

VII. MANAGEMENT OF IUGR

The management of IUGR pregnancy is based on the principle of a timely delivery of the fetus once fetal pulmonary maturity is achieved to avoid intrauterine demise of a compromised fetus. Preterm delivery is not contemplated unless the IUGR fetus demonstrates fetal compromise (fetal biophysical profile is a useful monitoring tool) or evidence of lack of fetal growth. In recurrent IUGR pregnancies due to preeclampsia or maternal collagen vascular disease, low-dose aspirin therapy and dipyridamole may be used to decrease thromboxane A2 (thus increasing the prostacyclin to thromboxane ratio) to improve placental circulation. The prophylactic treatment can increase the fetal birth weight by approximately 500 g, but it also increases the risk of abruption.

Maneuvers such as maternal hyperoxia and nutritional supplementation have been used to improve the outcome in IUGR pregnancies, but the results are too variable to be used routinely. The most important component of management of IUGR pregnancy is the utilization of a group of tests that have been shown to have predictive value in fetal compromise to decide on the timely delivery of the fetus. These include (i) ultrasound measurement at early gestation to accurately determine the gestational age, (ii) serial ultrasound measurements of fetal parameters to monitor body and organ growth and amniotic fluid volume to prognosticate fetal health, (iii) serial fetal biophysical profile measurements to predict fetal compromise, and (iv) selective clinical and biochemical tests on the mother (e.g., blood pressure, platelet count, and 24-hr urinary protein for progression of preeclampsia) and fetus (e.g., chromosome test). Doppler ultrasound measurement, computerized heart rate tracing, and other biochemical tests are still considered research tools. Cordocentesis in a growth-restricted fetus is an operative challenge, with associated morbidity, and therefore it is not used routinely in many centers.

VIII. OUTCOME OF IUGR

A. Mortality

IUGR is described as having survival advantages because it is associated with advanced pulmonary maturity for gestational age. This is attributed to the fetal response to a stressful environment in which there is increased adrenal glucocorticoid production that leads to increased production of surface-active phospholipids and surfactant-associated proteins. Although it may result in decreased neonatal morbidity, it has not been shown to be associated with an improvement in neonatal mortality and morbidity since the other major organ systems are still compromised. In addition, this so-called "advantage" is of no benefit to the fetus, which is still at great risk for an intrauterine death (fetuses with estimated sizes of <3rd percentile for gestational age may have up to an 8-fold higher mortality rate than those within the 10th and 90th percentiles). With advancements in perinatal/neonatal care, the neonatal mortality of IUGR/SGA infants near or at term is not different from that of appropriate for gestational age (AGA) infants; however, the perinatal mortality rate is still 3–10 times higher. For example, in infants weighing <1500 g near term, the perinatal mortality rate may be increased up to 100-fold higher than that of appropriately grown infants. The IUGR fetus is also at risk for obstetric complications, such as oligohydramnios and fetal compression, intrapartum and neonatal asphyxia, meconium aspiration, and operative delivery.

B. Morbidity

In the neonatal period, IUGR infants, particularly those that are premature, are at risk for asphyxia (hypoxic ischemic encephalopathy, persistent pulmonary hypertension, and transient myocardial ischemia), metabolic complications (e.g., hypoglycemia, hypocalcemia, persistent metabolic acidosis and hy-

perbilirubinemia, hypothermia, and associated complications), polycythemia, and poor immune function. Most of these complications can be avoided by anticipatory management with good antenatal care and monitoring.

In infancy and early childhood, IUGR infants are at risk for growth failure and neurodevelopmental sequelae. Up to 50% of SGA infants, especially those who are premature, may continue to be small (below or lower range of normal). The most rapid catch-up growth occurs during the first year, and if they fail to do so, it is rare for them to catch up later. Neurologically, most SGA infants have normal IQs, but as high as 30% of the infants may exhibit a variety of minimal brain dysfunction, such as language delay, speech difficulties, and behavioral and school problems. Recent epidemiologic data indicate that SGA infants, especially ones that have low ponderal index, are at risk for developing cardiovascular diseases (hypertension, coronary heart disease, and stroke) and diabetes in adult life, suggesting that the fetal adaptation to growth-restrictive conditions may lead to "programming" of the cardiovascular and endocrine systems, with untoward long-term outcomes.

See Also the Following Articles

Apoptosis; Fetal Anomalies; Fetal Growth and Development; Preterm Labor and Delivery; Ultrasound

Bibliography

Baker, J., Liu, J.-P., Robertson, E. J., and Efstratiadis, A. (1993). Role of insulin-like growth factors in embryonic and postnatal growth. *Cell* 75, 83–95.

Barker, D. J. P., (1990). Fetal and infant origins of adult disease. *Br. Med. J.* 301, 1111–1113.

Creasy R. K., and Resnik, R. (1994). *Intrauterine Growth Restriction*, pp. 558–574.

Creasy, R. K., and Resnik, R. (Eds.) (1994). *Maternal Fetal Medicine*, 3rd ed. Saunders, Philadelphia.

Divon, M. Y. (1991). *Abnormal Fetal Growth*. Elsevier, New York.

Han, V. K. M. (1993). Growth factors in placental growth and development. In *Molecular Aspects of Placental and Fetal Membrane Autacoids* (G. E. Rice and S. P. Brennecke, Eds.), pp. 395–446. CRC Press, Boca Raton, FL.

Han, V. K. M., and Hill, D. J. (1994). Growth factors in fetal growth. In *Textbook of Fetal Physiology* (G. D. Thorburn and R. Harding, Eds.), pp. 48–69. Oxford Med. Pub., Oxford, UK.

Han, V. K. M., Lund, P. K., and D'Ercole, A. J. (1987). Cellular localization of somatomedin (insulin-like growth factor) messenger RNA in the human fetus. *Science* **236**, 193–197.

Hill, D. J., and Han, V. K. M. (1997). Growth factors in health and disease: Fetus and placenta (D. LeRoith and C. L. Bondy, Eds.), pp. 1–53. JAI, London.

Hynes, R. O. (1992). Integrins: Versatility, modulation, and signaling in cell adhesion. *Cell* **69**, 11–25.

Kurjak, A., and Beazley, J. M. (1989). *Fetal Growth Retardation: Diagnosis and Treatment*. CRC Press, Boca Raton, FL.

Liu, J.-P., Baker, J., Perkins, A. S., Robertson, E. J., and Efstratiadis, A. (1993). Mice carrying null mutations of the genes encoding insulin-like growth factor I (Igf-1) and type 1 IGF receptor (Igfr). *Cell* **75**, 73–82.

Mark, M., Rijli, F. M., and Chambon, P. (1997). Homeobox genes in embryogenesis and pathogenesis. *Pediatr. Res.* **42**, 421–429.

Osmond, C., Barker, D. J. P., Winter, P. D., Fall, C. H. D., and Simmons, S. J. (1993). Early growth and death from cardiovascular disease. *Br. Med. J.* **307**, 1519–1524.

Rappolee, D. A., Brenner, C. A., Schultz, R., Mark, D., and Werb, Z. (1988). Developmental expression of PDGF, TGF-α, and TGF-β genes in preimplantation mouse embryos. *Science* **241**, 1823–1825.

Robinson, J. S., Owens, J. A., and Owens, P. C. (1994). Fetal growth and fetal growth retardation. In *Textbook of Fetal Physiology* (G. D. Thorburn and R. Harding, Eds.), pp. 83–94. Oxford Med. Pub., Oxford, UK.

Schwartzman, R. A., and Cidlowski, J. A. (1993). Apoptosis: The biochemistry and molecular biology of programmed cell death. *Endocr. Rev.* **14**, 133–151.

Ward, R. H. T., Smith, S. K., and Donnai, D. (1994). *Early Fetal Growth and Development*. Royal College of Obstetricians and Gynaecologists Press, London.

Intrauterine Infections

see Pelvic Inflammatory Disease

Intrauterine Position Phenomenon

Frederick S. vom Saal
University of Missouri–Columbia

Lee C. Drickamer
Northern Arizona University

Mertice M. Clark and Bennett G. Galef, Jr.
McMaster University

John G. Vandenbergh
North Carolina State University

GLOSSARY

anogenital distance (AGD) The length of the perineal tissue, which becomes the scrotum in males, separating the anus and the genital tubercle (penis in males and clitoris in females). AGD is longer in males than in females.

estradiol The steroid hormone produced at high levels by cells in the ovary during the first half of the menstrual cycle (after puberty). During pregnancy in humans the placenta also produces estradiol. Estradiol is one of a family of hormones called estrogens.

intrauterine position The position of a fetus, when there is more than one fetus, within the uterus relative to adjacent male and females fetuses, with 2M animals being positioned between two male fetuses, 1MF animals being positioned between a male and a female fetus, and 2F animals being positioned between two female fetuses.

testosterone The steroid hormone produced at high levels by the testes both during fetal life (when it causes masculinization of the fetal brain and reproductive system) and after puberty (when it causes additional changes associated with puberty in boys). Much lower amounts of testosterone are also produced by the ovaries in females. Testosterone is one of a family of hormones called androgens.

I. HORMONAL TRANSPORT BETWEEN FETUSES

The intrauterine position (IUP) phenomenon is a result of hormonal transport between adjacent fetuses before birth and has been described in a range of litter-bearing mammals (including mice, rats, gerbils, and pigs) as well as in humans carrying twins (Fig. 1). To identify the IUP of animals for postnatal study, the timing of conception is determined and fetuses are removed from the uterus just before normal parturition, are marked for individual identification, and then reared by foster mothers. IUP effects have been extensively studied in house mice, a species in which male fetuses have higher serum levels of testosterone than do females and female fetuses have higher serum levels of estradiol than do males. These steroid hormones diffuse between adjacent fetuses causing both male and female fetuses occupying an IUP between two males (2M fetuses) to have higher blood concentrations of testosterone (by approximately 30%) than do their same-sex siblings occupying an IUP between two females (2F fetuses). 2F fetuses, on the other hand, have higher blood concentrations of estradiol than do 2M fetuses (also approximately a 30% difference). Those fetuses situated between a male and a female fetus (1MF fetuses) have intermediate serum concentrations of both testosterone and estradiol, and in adulthood are intermediate between 2M and 2F animals of the same sex in many morphological, physiological, and behavioral characteristics.

An important consequence of IUP research has been to increase awareness of the exquisite sensitivity of fetuses during critical periods in the development

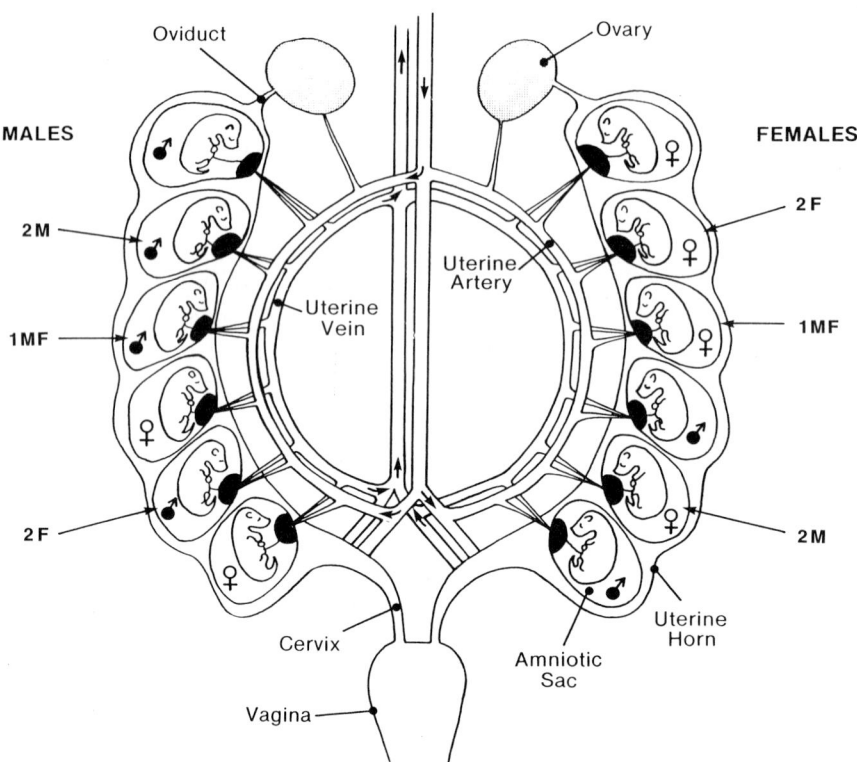

FIGURE 1 Schematic diagram of the two independent uterine horns of a mouse near the end of pregnancy. The labels 2M, 1MF, and 2F refer to the sex of adjacent fetuses, with 2M positioned between two male fetuses, 1MF between a male and a female fetus, and 2F between female fetuses.

of the brain and reproductive organs to very small differences in hormone concentrations. For example, a difference in serum estradiol between 2M and 2F male mouse fetuses of 23 pg (23 trillionths of a gram) per milliliter of serum is sufficient to induce a cascade of developmental events which, as indicated below, have profound consequences on subsequent reproductive life history.

The mechanism of transport of testosterone between fetuses has been examined directly in rats. Testosterone in a male fetus diffuses into the amniotic fluid, which both circulates through a fetus and surrounds it, thus picking up hormones from fetal blood. The testosterone in amniotic fluid diffuses across fetal membranes (amnion and chorion) surrounding each fetus and into the amniotic fluid and blood of adjacent fetuses. Consequently, at least in those litter-bearing mammals studied to date, being positioned *in utero* between siblings of the same or opposite sex leads to predictable differences in serum

concentrations of testosterone and/or estradiol at a time in fetal life when these hormones influence the differentiation of virtually all morphological, physiological, and behavioral traits relating to reproduction. Furthermore, because IUP is a stochastic aspect of development, and frequency of occurrence of stochastic events is predictable in large populations, phenotypic variability induced by IUP is guaranteed in litter-bearing species.

Even before birth, the length of the tissue which becomes the scrotum in males and is located between the embryonic external genital and anus (anogenital distance; AGD) is longer in male than in female rodents, and AGD also varies within sex; 2M females have a longer AGD than do 2F females. This finding, together with the observation that in cages containing 6 female mice, one female is typically more aggressive than the others, led to experiments examining the postnatal consequences for reproductive performance and behavior of prior IUP. In litters containing 12 mice, as is typical in the stock of mice used in the experiments described in this article, 1 in 6 females is a 2M female. It thus seemed reasonable to determine whether IUP might be the source of observed variability in aggression and possibly other traits relating to reproduction and, if so, what the adaptive significance of this variability might be. It is sometimes assumed, erroneously, that only variability due to genetic differences can play a role in evolutionary processes. However, an interest in evolution and the adaptive significance of IUP effects has led to research in natural environments testing evolutionary hypotheses based on laboratory findings concerning effects of IUP on reproduction and behavior.

We first briefly review the consequences, in both gerbils and mice, of developing as a 2M or 2F fetus. We then describe recent studies in which hypotheses concerning the potential impact of IUP on individual reproductive success and population dynamics have been tested in wild mice both in seminatural environments and by using mathematical models.

II. LABORATORY STUDIES ON THE PHENOTYPE OF MONGOLIAN GERBILS AND HOUSE MICE FROM DIFFERENT IUPS

One of the more interesting findings to emerge from the study of IUP effects on members of two different species of Old World rodent is that effects of IUP on adult phenotype vary from one species to the next. Such variability in IUP effects on development should, perhaps, have been expected because, although testosterone and estradiol influence development in all vertebrates, their specific effects on development are not likely to be exactly the same in any two species. Thus, it is not surprising that as measurements of IUP effects are extended across species, what might appear as inconsistencies emerge. We hypothesize that such inconsistencies in IUP effects reflect differences in the way that testosterone and estradiol regulate development in different animals, and that as our understanding of comparative endocrinology grows, apparent inconsistencies

across species in IUP effects on adult phenotypes will be explained.

A. IUP Effects in Males

1. Reproductive Organs

Adult 2M male gerbils are more responsive to testosterone than are 2F males of their species, suggesting that IUP influences the sensitivity of target tissues to gonadal hormones. In mice, an increased sensitivity to testosterone in adult 2M males, relative to 2F males, has also been demonstrated in terms of responsivity of some accessory reproductive organs, such as the seminal vesicles. However, results of some comparisons of organs of males from different IUPs initially appeared counter to predictions concerning the role of specific steroids in sexual differentiation. For example, as adults, 2F male mice have larger prostates and smaller seminal vesicles than do 2M male mice. The enlarged adult prostates of 2F males are associated with an increase in androgen receptors, whereas the smaller seminal vesicles of 2F males are associated with a decrease in 5α-reductase activity. These two organs develop from different embryonic tissues (prostate from embryonic urogenital sinus and seminal vesicles from embryonic Wolffian ducts), and these findings suggest that testosterone and estradiol have different effects in these two embryonic tissues. Thus, in addition to species differences due to IUP effects, within a species not all embryonic tissues respond the same to an increase or a decrease in a specific hormone.

2. Behavior

Effects of IUP on a number of behaviors that influence reproductive success have been examined. Elevation in estradiol in 2F male mouse fetuses correlates with an increase in some indices of adult sexual performance (frequency of mounting and intromitting) relative to 2M males, whereas in gerbils, 2M males, with elevated testosterone during fetal life, are more likely to impregnate females with which they are paired than are 2F males. While less sexually proficient than 2M male gerbils, 2F male gerbils make a greater parental effort and are more effective in reducing the burden that reproduction imposes on their mates than are 2M male gerbils. In this regard, 2M male gerbils are markedly different from 2M male mice, which are more parental toward young than are 2F males of their species. In fact, 2F male mice are more likely to be infanticidal than are 2M male mice. On the other hand, 2M male mice are more aggressive toward other adult males than are 2F males. In mice, therefore, elevated levels of testosterone during fetal life lead to an increase in intermale aggression (competition for territories and mates), whereas fetal exposure to elevated estradiol and decreased levels of testosterone are related to an increase in aggression toward newborn pups sired by other males.

Study of the effects of IUP in gerbils is complicated by the fact that 2M males, in addition to having elevated testosterone levels during fetal life, have higher plasma testosterone levels in adulthood than do 2F males. Clearly, functioning of the brain–pituitary–gonadal axis, or of enzyme systems and components of plasma involved in regulating clearance of steroids (such as plasma-binding proteins), can be permanently affected by an animal's IUP. Perhaps as a result of their higher circulating levels of testosterone in adulthood, 2M adult male gerbils are more attractive to females in estrus than are 2F males. In addition, possibly because of differences in their copulatory patterns and genital musculature, 2M males are also more successful than are 2F males in impregnating females they encounter for the first time. Further experiments will be required to separate effects on reproductive behavior of gerbils due to differences in gonadal steroid levels during fetal and adult life.

B. IUP Effects in Females

IUP influences numerous reproductive traits in both female mice and female gerbils. In relation to 2F female mice, 2M females enter puberty later. In mice, 2M females show longer estrous cycles and cease producing pups at a younger age than do 2F females. Thus, 2M female mice have a shorter period in life during which they are fertile relative to 2F females. Males choose to be near and prefer to mate with 2F female mice, and when mounted by males,

2M females are less likely to lordose (exhibit sexual receptivity) than are 2F females.

In contrast, 2M female mice are more aggressive toward other females before mating, and after parturition they are more aggressive toward conspecific intruders than are 2F females. This finding suggests that, as competition for resources increases, 2M female mice would have increasing advantage over 2F females. An important related observations is that, as population density increases, only the most aggressive female mice reproduce.

In gerbils, as in mice, 2M females reach puberty at a significantly later age than do 2F females, and late-maturing female gerbils have substantially fewer litters, smaller litters, and half the lifetime fecundity of their early maturing sisters. Late-maturing female gerbils are more attentive mothers and less willing to accept strange males as sexual partners than are early maturing females (perhaps related to the enhanced nest defense shown by nursing 2M female relative to 2F female mice).

Evidence for direct IUP effects on brain function has been provided by a study in which cytochrome oxidase activity in brain areas involved in reproduction in female gerbils was examined. There was increased cytochrome oxidase activity in the medial and posterior anterior hypothalamus of 2M compared with 2F females.

1. IUP Effects on Litter Sex Ratio

There is direct evidence in both gerbils and mice that a female's prior IUP affects the sex ratio of litters she produces: 2M females give birth to male-biased litters, whereas 2F females deliver female-biased litters. Consequently, daughters of both 2F and 2M females have a greater probability than would be expected by chance alone of maturing in IUPs concordant with those of their respective mothers. Although the mechanism mediating differences in the sex ratio of litters delivered by 2F and 2M female gerbils and mice is not known, it is unlikely to be the result of differential mortality either during or shortly following vaginal delivery in gerbils.

The right ovary of female gerbils is known to produce more male than female destined eggs, and the left ovary the reverse, even though the genotype of the sperm determines an embryo's sex. However, the relationship, if any, of this phenomenon to observed effects of IUP on the sex ratio of litters born to 2M and 2F gerbil dams is not clear. Possibly, the difference in sex ratio of litters gestated by 2M and 2F females results from a difference in their timing of copulation with respect to ovulation, though no studies have examined effects of copulatory timing on litter sex ratio of either Mongolian gerbils or mice.

Regardless of the causes of differences in the sex ratio of litters produced by female gerbils and mice from different IUPs, concordance in the IUPs of mothers and daughters results in a form of hormonally mediated transmission of acquired characteristics, producing concordance between mothers and daughters in those characteristics affected by prenatal level of testosterone and estradiol. A daughter's phenotype will tend to resemble that of her mother not only because mother and daughter share a relatively large proportion of genes but also because of a stochastic tendency toward congruence in the IUPs of mothers and daughters that produces similarities in their patterns of exposure to steroid hormones during fetal development.

III. IUP EFFECTS IN SWINE

Rhode-Parfet et al. report minimal effects of IUP on various aspects of reproductive physiology of domestic swine (*Sus scrofa domesticus*). However, even though Rhode-Parfet et al.'s data analysis failed to reveal significant effects of IUP on morphology, examination of their data describing the mean AGD of 2F and 2M female swine in their study reveals what appears to be a significant difference. Rhode-Parfet et al. placed in the same analysis data from animals from many different IUPs (ends of uterine horns, 1MF females, etc.) and this could have obscured a significant difference between 2F and 2M females.

Recently, a significant relationship was established between the proportion of males in litters of swine and the mean AGD of females in those litters. The greater the proportion of males in a litter (and consequently the greater the probability that females in that litter were from 2M IUPs), the larger the mean

AGD of females in that litter, suggesting a relationship between IUP and AGD in swine. Drickamer et al. examined possible reproductive consequences of gestation in a large litter with a high proportion of males. Females from litters of nine or more piglets containing 66.7% or more males were more likely to experience conception failure than were gilts and sows from litters with a lower percentage of males. This finding regarding sex ratio of swine litters and subsequent reproductive success has potentially important implications for management practices by swine producers.

IV. IUP EFFECTS IN HUMANS

In humans, females born with a male twin differ from females born with a female twin in the frequency of spontaneous acoustic emissions from their inner ears. Females that develop with a male twin show male-like levels of acoustic emissions. The interesting aspect of this particular finding is that levels of acoustic emissions are set by prenatal factors and do not change between birth and adulthood. Consequently, a measurement made in adulthood can provide accurate information concerning a prenatal event. Whether other traits differ in people who share their prenatal life with a twin of the same or opposite sex remains to be determined.

V. IUP EFFECTS IN HOUSE MICE UNDER FIELD CONDITIONS

Studies have been conducted with house mice to determine whether IUP effects would also occur under free-living conditions . Of particular interest were potential consequences of variation due to IUP with respect to life history traits. Zielinski et al. determined whether the IUP of wild female house mice influenced their home range size, survival, and reproduction. Wild mice were captured, bred in the laboratory for one generation, and then used as stock to produce animals of known IUP. As young adults, 2M and 2F females, along with 1MF males, were released into segments of a highway cloverleaf approximately 1 ha in size. The only effect of IUP on behavior found was that the home range size occupied by 2M females was approximately 45% larger than that of 2F females. The failure to detect a difference in survival rate or reproductive characteristics of females from different IUPs may have been the result of the relatively short duration of the 8-week experiment.

Under field conditions it is not possible to determine the actual IUP of female and male fetuses. However, because there is a relationship between the anogenital distance of female house mice and their prior IUP, it is possible to identify probable 2M and 2F females in wild populations. Using the AGD measure as an indicator of prior IUP, Drickamer studied the characteristics of wild house mice living in 0.1-ha outdoor enclosures. Male mice with larger AGDs were more aggressive and had larger home ranges than did males with shorter AGDs. Survival was not related to AGD for either sex. However, female mice with smaller AGDs were more likely to attain reproductive status, had a higher rate of pregnancy, and had more pregnancies, on average, than did females with larger AGDs. Thus, in the enclosures, female mice with larger AGDs were at a selective disadvantage with respect to reproductive success. These findings are consistent with results obtained in the laboratory relating IUP and aggression in male mice and IUP and fecundity in female mice.

VI. ELIMINATION OF IUP EFFECTS IN MICE BY MATERNAL STRESS

A number of studies have shown that in female mice, maternal stress can override variability in reproductive characteristics due to IUP. It is interesting that maternal stress results in the elevation of serum testosterone levels by a similar amount in 2M, 1MF, and 2F female mouse fetuses. However, only 2F female fetuses, with the lowest background serum levels of testosterone as a result of their IUP, show a marked change in traits relative to unstressed 2F females, such that stressed 2F females resemble 2M females in all traits that have been examined. In contrast, maternal stress appears to have no effect on the traits of 2M females, such that stressed and unstressed 2M females do not differ in their reproductive characteristics, even though maternal stress does significantly increase testosterone in 2M female fetuses. One possible explanation for this finding is that the high background levels of testosterone in unstressed 2M females renders them less sensitive to an increase in testosterone relative to unstressed 2F females.

Zielinski et al. examined effects of crowding of mothers on female offspring by placing pregnant mice into social groups with strangers at one of three densities: low (stud only), moderate (4 animals per cage), or high (10 mice per cage). There was a great deal of social turmoil in the high-density populations and essentially none in the low-density environments. All female pups, regardless of intrauterine position, born to mothers experiencing high social stress had anogenital distances similar to those of 2M females, whereas females born to mothers subjected to either moderate or low social stress during pregnancy showed the typical result (i.e., 2F females had short AGDs, 1MF females had intermediate AGDs, and 2M females had long AGDs). The findings of vom Saal et al. (1990) and Zielinski et al. (1991) that maternal stress renders all female offspring 2M-like in their reproductive traits are thus consistent, but the mechanisms by which this occurs remain unknown.

The finding that environmental stress appears to create hormonal signals within a pregnant female that can eliminate differences due to IUP suggests that the social stress that occurs at high population densities will reduce the degree of variability in reproductive traits in the next generation. Since natural selection operates on variability in phenotype, factors that alter phenotypic variability within a population have the potential to alter the course of evolution in the population. Also, effects on population dynamics would be expected. Because all female offspring of stressed mothers are prenatally androgenized and thus 2M-like females, they may be more likely to emigrate (or at least to have relatively large home ranges) and more likely to attain puberty at a later age than female offspring of nonstressed mothers. The suite of characteristics seen in the offspring of socially stressed dams could cause members of a population stressed by crowding to disperse.

VII. IMPLICATIONS OF THE IUP PHENOMENON FOR POPULATION DYNAMICS

As has been documented in both laboratory and field populations of rodents, a significant component of phenotypic variation can be attributed to effects of IUP. Taken together, the results of over two decades of work on IUP in rodents suggest that IUP can significantly affect the dynamics of rodent populations by affecting the reproductive behavior of population members. All traits relating to reproduction of small mammals can be affected by IUP. If, as seems to be the case, animals from different IUPs respond differentially to environmental changes, such as the stress produced by overcrowding, this will have consequences on their reproductive physiology and behavior. Therefore, changes in population size (cyclic changes, irruptions, or crashes) for rodents could be mediated, in part, via IUP effects interacting with environmental stressors. Such reasoning also leads to the notion that population cycles, such as those exhibited by some microtine rodents, could involve IUP effects. A scenario by which IUP effects could play a role in the population dynamics of rodents has been proposed, and this prediction has recently received support using mathematical modeling.

Aggression can be a key factor in population dynamics. Thus, IUP influences on levels of aggression could directly impact reproduction and thus affect population dynamics. Dennis Chitty and Charles Krebs had also proposed that shifts in the proportion of aggressive animals in a population could contribute to changes in population dynamics, but their assumption was that variation in traits, such as aggression and other sociosexual behaviors, was mediated solely by genes and that such traits would thus be heritable. In contrast, IUP effects appear to be mediated by levels of gonadal steroids during fetal life due to where fetuses happen to be positioned in the uterus, which is a random event. During fetal life hormones (and other intercellular signaling molecules) determine which genes in differentiating cells are turned on or off, and for genes that are turned on, hormones determine the rate of activity of the genes. These genetic imprinting events, once set, are irreversible, but they do not involve changing the genetic code. This would lead to the prediction that IUP effects are not directly heritable, and that aggressive animals in a population (due to developing in an IUP between male fetuses) would not produce aggressive offspring. In this regard, the intriguing finding that in both gerbils and mice, a mother's prior intrauterine position (and thus her fetal hormone levels) biases the sex ratio of her litters, leading to what amounts to heritable differences in the traits of her offspring, raises the possibility of as yet unexplored factors governing the heritability of traits.

See Also the Following Articles

Testosterone Biosynthesis; Twinning

Bibliography

Clark, M. M., and Galef, B. G., Jr. (1998). Effects of intrauterine position on the genital morphology and behavior of litter-bearing rodents. *Dev. Neuropsychol.*, in press.

Clark, M. M., Bishop, A. M., vom Saal, F. S., and Galef, B. G., Jr. (1993a). Responsiveness to testosterone of adult male gerbils from known intrauterine positions. *Physiol. Behav.* **53**, 1183–1187.

Clark, M. M., Karpiuk, P., and Galef, B. G., Jr. (1993b). Hormonally mediated inheritance of acquired characteristics. *Nature* **364**, 712.

Cowell, L. G., Crowder, L. B., and Kepler, T. B. (1998). Density-dependent prenatal androgen exposure as an endogenous mechanism for the generation of cycles in small mammal populations. *J. Theor. Biol.*, in press.

Drickamer, L. C. (1996). Intra-uterine position and anogenital distance in house mice: Consequences under field conditions. *Anim. Behav.* **51**, 925–934.

Drickamer, L. C., vom Saal, F. S., Marriner, L. M., and Mossman, C. A. (1995). Anogenital distance and dominance status in male house mice (*Mus domesticus*). *Aggr. Behav.* **21**, 301–309.

Drickamer, L. C., Arthur, R. D., and Rosenthal, T. L. (1997). Conception failure in swine: Importance of sex ratio of female's birth litter and tests of other factors. *J. Anim. Sci.* **75**, 2192–2196.

Jones, D., Gonzales-Lima, F., Crews, D., Galef, B. G., Jr., and Clark, M. M. (1996). Effects of intrauterine position on the metabolic capacity of the hypothalamus of female gerbils. *Physiol. Behav.* **61**, 513–519.

McFadden, D. (1993). A masculinizing effect on the auditory systems of human females having male co-twins. *Proc. Natl. Acad. Sci. USA* **90**, 11900–11904.

Rohde-Parfet, K. A., Lamberson, W. R., Rieke, A. R., Cantley, T., Ganjam, V. K., vom Saal, F. S., and Day, B. N. (1990). Intrauterine position effects in male and female swine: Subsequent survivability, growth rate, morphology and serum characteristics. *J. Anim. Sci.* **68**, 179–185.

Vandenbergh, J. G., and Huggett, C. L. (1995a). The anogenital distance index, a predictor of the intrauterine position effects on reproduction in female house mice. *Lab. Anim. Sci.* **45**, 567–573.

Vandenbergh, J. G., and Huggett, C. L. (1995b). Mother's prior intrauterine position affects the sex ratio of her offspring in house mice. *Proc. Natl. Acad. Sci. USA* **91**, 1155–1159.

vom Saal, F. S. (1984). The intrauterine position phenomenon: Effects on physiology, aggressive behavior and population dynamics in house mice. *Prog. Clin. Biol. Res.* **169**, 135–179.

vom Saal, F. S. (1989). Sexual differentiation in litter bearing mammals: Influence of sex of adjacent fetuses in utero. *J. Anim. Sci.* **67**, 1824–1840.

vom Saal, F. S., and Bronson, F. H. (1978). In utero proximity of female mouse fetuses to males: Effect on reproductive performance during later life. *Biol. Reprod.* **19**, 842–853.

vom Saal, F. S., Quadagno, D., Even, M., Keisler, L., Keisler, D., and Khan, S. (1990). Paradoxical effects of maternal stress on fetal steroids and postnatal reproductive traits in female mice from different intrauterine positions. *Biol. Reprod.* **43**, 751–761.

vom Saal, F. S., Timms, B. G., Montano, M. M., Palanza, P., Thayer, K. A., Nagel, S. C., Dhar, M. D., Ganjam, V. K., Parmigiani, S., and Welshons, W. V. (1997). Prostate enlargement in mice due to fetal exposure to low doses of estradiol or diethylstilbestrol and opposite effects at high doses. *Proc. Natl. Acad. Sci. USA* **94**, 2056–2061.

Zielinski, W. J., Vandenbergh, J. G., and Montano, M. M. (1991). Effects of social stress and intrauterine position on sexual phenotype in wild-type house mice (*Mus musculus*). *Physiol. Behav.* **49**, 117–123.

Zielinski, W. J., vom Saal, F. S., and Vandenbergh, J. G. (1992). The effect of intrauterine position on the survival, reproduction and home range size of female house mice (*Mus musculus*). *Behav. Ecol. Sociobiol.* **30**, 185–191.

In Vitro Fertilization

Lewis C. Krey and Alan S. Berkeley
New York University School of Medicine

I. Introduction
II. The IVF Cycle
III. Future Directions

GLOSSARY

blastocyst A stage of embryo development at 96–120 hr postfertilization in humans. The blastocyst consists of the inner cell mass, destined to become the fetus, and the trophectoderm, which will develop into the placenta.

blastomere A cell in the early preimplantation embryo.

cryopreservation The freezing of oocytes or embryos and their storage at low temperature.

GIFT Gamete intra-Fallopian transfer, a surgical procedure in which egg(s) and sperm are laparoscopically placed in the Fallopian tube.

gonadotropin A hormone secreted by the anterior pituitary to stimulate gonadal growth and development. Follicle-stimulating hormone (FSH) and luteinizing hormone (LH) stimulate oocyte growth and maturation through ovulation.

gonadotropin-releasing hormone (GnRH) A peptide neurohormone that regulates the synthesis and secretion of LH and FSH by the anterior pituitary gland. Synthetic GnRH analogs are administered to shut down endogenous gonadotropin secretion.

in vitro fertilization (IVF) Fertilization of oocyte(s) in the laboratory.

Encyclopedia of Reproduction VOLUME 2

oocyte The female gamete. In the germinal vesicle stage, the oocyte contains a 4n complement of chromosomes. During meiosis the chromosome complement is reduced to 2n and then 1n with the excess chromosomes being discharged in the polar body.

ovarian stimulation The induction of single or multiple follicular development by the administration of exogenous drugs.

preimplantation embryo The fertilized egg and developing embryo to the blastocyst stage.

ZIFT Zygote intra-Fallopian transfer, the laparoscopic placement of a fertilized egg into the Fallopian tube.

zona pellucida The glycoprotein shell that surrounds the oocyte which spontaneously splits during hatching of the blastocyst.

zygote A single-celled fertilized egg with male and female pronuclei.

Many findings generated by scientific research on the physiologic, biochemical, and molecular mechanisms underlying reproduction have been applied in one of two primary clinical directions: the development of safe and effective contraceptives to block unwanted pregnancy and the generation of diagnostic and treatment paradigms to identify and correct the endocrine, gynecologic, obstetrical, and urologic problems causing infertility. This latter aim has been addressed by the development of many different procedures, several of which have been discussed in other articles of this encyclopedia. However, the best publicized of these is the subject of this article—*in vitro* fertilization.

I. INTRODUCTION

When Drs. Robert Edwards and Patrick Steptoe announced in 1978 the birth of Louise Brown, the first "test-tube baby" born from an egg fertilized *in vitro*, *in vitro* fertilization (IVF) grabbed the interest of the lay public. It jogged the imagination of some who viewed the procedure as the ultimate application of cutting-edge science to clinical care; unfortunately, it also offended the ethics and sensibilities of others. In the subsequent 20 years, IVF has accounted for the births of tens of thousands of children

to previously infertile couples throughout the world. Registries in the United States and abroad are now available to report the number of IVF procedures conducted each year and their resulting pregnancy rates. These registries variously provide data about the rates of miscarriage and genetic anomalies associated with IVF cycles, rates which are not statistically different from those of naturally occurring pregnancies and which further document the usefulness of IVF to infertile patients. These registries also serve as a reminder that IVF and its associated assisted reproductive technologies are currently subject to regulatory oversight by various agencies, some self-imposed by practicing programs such as the Society for Assisted Reproductive Technologies (SART) in the United States, whereas others, such as the Human Fertilization and Embryo Authority in England, are established by law. Governmental regulation of IVF practice has continued to be discussed in the United States and several states currently have laws that directly impact on the day-to-day protocols of their IVF clinics.

Significantly, the rate of clinical pregnancy per IVF procedure has increased consistently during the past 20 years. The overall clinical pregnancy rate for IVF programs reporting to SART in 1995 was 32% per retrieval in young women (<35 years of age with a resultant live birth rate of 27.5%); however, several IVF programs in the United States are now experiencing a clinical pregnancy rate per retrieval of >50% for these young patients. Such an increase can be attributed to several procedural advances to be discussed later. These advances have also provided another important contribution—the development of human reproductive science into a active area of research. In fact, much of the basic information that is now being collected on human reproductive biology is a direct or indirect result of observations and research carried out by IVF programs. Important concepts, many unique to humans, have been proposed to explain hormonal regulation of ovarian follicle development and the influences of maternal age and other factors on oocyte maturation and development. In addition, our understanding of human embryology has progressed dramatically during these past few years as the biologically significant events and problems encountered during early embryonic development are noted and reported by IVF laboratories.

During the early 1980s, IVF was thought to be an appropriate treatment only for infertile patients afflicted with tubal disease. The fact that Louise Brown was the result of a naturally occurring ovulatory cycle led many physicians to limit IVF further to those women who normally ovulated. However, physicians soon appreciated the fact that IVF could also be successful when performed on infertile couples suffering from other gynecologic, endocrine, or male factor disorders. The development of stimulation protocols in which ovarian follicle development was stimulated by exogenous gonadotropins further expanded the range of treatable patients to include those suffering from ovulatory dysfunction including polycystic ovarian disease. Oocyte donation protocols further expanded the patient base to include those suffering from premature or perimenopausal ovarian failure. Finally, in the early 1990s micromanipulation procedures were developed that extended IVF to couples in whom the male partner had a profound infertility problem. These procedures improved the embryologist's ability to inseminate human eggs and included partial zonal dissection (PZD), subzonal sperm insertion (SUZI), and, recently, intracytoplasmic sperm injection (ICSI) in which eggs are inseminated individually by the microinjection of a single sperm. According to the SART registry, ICSI was performed in 12.3% of all IVF cycles in 1995, the first year this procedure was used extensively by IVF practices.

II. THE IVF CYCLE

The IVF cycle begins with stimulation of Graafian follicle development in the patient's ovaries and ends with the transfer of her zygotes or embryos into the uterus. It requires a collaboration between physicians, nursing staff, and technician scientists in the endocrine and embryology laboratories, each playing a crucial role to ensure a successful outcome. Significant improvements in treatment protocols and patient care in the treatment room, surgery, and laboratory have evolved during the past 20 years, and each of these improvements has contributed signifi-

cantly to the rise in pregnancy rates seen during this time. Although all IVF cycles consist of the following phases, there is no typical procedure; IVF programs incorporate procedural variations in one or more phases that make their IVF protocol unique.

A. Ovarian Follicle Stimulation

The first successful IVF cycle was a natural ovulatory cycle. As a result of Louise Brown's birth, most of the original IVF cycles did not utilize exogenous gonadotropin treatments; rather, an egg was retrieved prior to ovulation and fertilized *in vitro*. However, the Drs. H. and G. S. Jones in the United States and Dr. C. Wood in Australia quickly realized that the chance for a pregnancy was greater when there was a greater number of embryos available for transfer. As a result, urinary gonadotropin preparations and clomiphene citrate, drugs originally used to treat ovulatory dysfunction, were administered to women with normal ovulatory cycles to stimulate egg production. The success of these treatments for IVF had further ramifications as the same treatments were extended to women undergoing intrauterine insemination cycles. Unfortunately, these treatments not only increased the pregnancy rate associated with these procedures but also occasionally resulted in quadruplets and higher order multiple pregnancies.

Perhaps the most significant advance in ovarian stimulation procedure was the use of synthetic analogs of gonadotropinreleasing hormone (GnRH) to suppress the patient's ability to secrete gonadotropins. The observations that GnRH agonists actually display a biphasic action on the pituitary first to stimulate and then to suppress gonadotropin secretion led in 1987 to a report of the first baby born from an IVF cycle in which chronic GnRH agonist treatment was used. In current protocols GnRH agonist treatment usually begins during the luteal phase prior to the IVF cycle; as a result, gonadotropin secretion is suppressed at cycle onset and exogenous gonadotropin treatment can begin. The incorporation of GnRH agonists into follicle stimulation protocols had far-reaching consequences that impacted the way IVF was practiced. Since endogenous gonadotropin secretion was suppressed during chronic GnRH agonist treatment, there was a significant reduction in the number of patients whose IVF cycle was canceled due to an inappropriate periovulatory LH surge. Moreover, physicians were now able to schedule oocyte retrievals throughout the day, thereby increasing the number of patients that could be treated. More importantly, physicians gained the ability to control the timing of ovarian stimulation. Because follicle development is markedly suppressed in patients treated chronically with a GnRH agonist, their reproductive cycle is essentially "on hold" until the treatment is stopped or until exogenous gonadotropins are administered. As a result, an IVF physician could now reliably schedule the onset of his patients' IVF cycles in advance and effectively bypass any problems associated with the timing of their menstrual cycles. Together, these benefits led to development of large IVF programs with the capacity to treat as many as a 1000+ patient cycles a year. Because of the scope of their practice and patient base, these large programs often took the lead in IVF-related research, including the development and testing of new protocols. Today, daily injections of human follicle-stimulating hormone (hFSH) and human luteinizing hormone (hLH) beginning on Days 2 or 3 are used to stimulate the development of a cohort of follicles for oocyte retrieval. The hormone preparations are obtained from urinary extracts from postmenopausal women. These extracts contain both hFSH and hLH in ratios that range from 1:1 (e.g., Pergonal, Humegon, and Repronex) to >75:1 (e.g., Metrodin) depending on extraction and isolation procedures. With the recent decoding and reconstruction of the human genes for the α and β subunits of hFSH and hLH, recombinant gonadotropins are now being introduced into daily IVF practice.

Other follicle-stimulation protocols incorporate treatments which trigger endogenous gonadotropin release. Clomiphene citrate is often used either by itself or in combination with gonadotropin therapy. This antiestrogen elevates endogenous FSH secretion by disrupting the negative feedback actions of ovarian estrogens on FSH secretion. "Flare" paradigms of GnRH agonists have also been tested. These treatments take advantage of the initial response of the patient's pituitary gland to secrete FSH and LH. Both of these treatments are begun during the first 3 days of the patient's menstrual cycle.

Prior to the onset of follicle stimulation, circulating FSH levels are routinely monitored during the first 3 days of the menstrual cycle, especially in patients of increasing maternal age. An abnormally elevated FSH level is an indication of an anomaly in the feedback regulation of FSH secretion and suggests a poor prognosis for follicle stimulation. Analyses of the relationship between FSH, follicular responsiveness, and IVF success have indicated that often there is a disparity between patient age and "ovarian age," which is the physiologically relevant marker of ovarian follicle reserve. Ultrasound examinations prior to treatment onset are useful to detect an ovarian cyst(s), which must be reduced by extended GnRH agonist therapy before the cycle can begin.

Once gonadotropin treatment commences, follicle growth is monitored by serial serum estradiol assays and ovarian ultrasound exams. Together, these parameters provide a description of the number and size of the developing follicles as well as their secretory capacity. Ideally, each follicle should account for each 100–125 pg/ml elevation in circulating estradiol. Low numbers of follicles and poor estrogen secretion are cause for cycle cancellation as is the presence of a dominant follicle among the developing cohort. Extremely high numbers of follicles accompanied by high estradiol levels are also a cause for concern since they suggest a predisposition for ovarian hyperstimulation syndrome. This syndrome develops following luteinization of the follicles; the resultant tissue secretes high levels of progesterone, which interferes with salt and fluid balance to create a potentially life-threatening situation that often requires hospitalization.

When a sufficient number of follicles reach an appropriate size, from 17–19 mm in diameter on ultrasound examination for patients who are on GnRH analog and 15–17 mm for patients who are not, human chorionic hormone (hCG; 5000–10,000 IU) is injected to stimulate the final events of oocyte maturation. Should this injection increase circulating estradiol further the next day, the eggs will be aspirated approximately 36 hr after hCG administration; a decrease of >25% in estradiol levels is cause to cancel the retrieval. Because of the similarity of the α and β subunits of hCG and LH, the hCG injection serves as a replacement for the periovulatory LH surge. It activates LH receptors within the ovary to set in motion the intercellular events that initiate the first stage of meiosis in the oocytes. During this stage of meiosis chromosome number is reduced from 4n to 2n, with the supernumerary chromosomes being discharged with a small amount of cytoplasm as a polar body. These events take approximately 36 hr to develop and are necessary for fertilization to occur.

Over the years there has been a marked improvement of the efficacy of ovarian stimulation protocols. This was due in large part to the fact that physicians were quickly able to see the results of new protocols, i.e., the oocytes. This immediate feedback of information encouraged experimentation and the development of several variations to individualize stimulation on a diagnosis by diagnosis or even a patient by patient basis. These variations included "step down" protocols which reduce gonadotropin stimulation as the follicles mature, hCG injection when follicles are small, lower dose hCG injections to potentially hyperstimulated patients, and microdose GnRH agonist treatment for older patients with a small follicle reserve.

B. Oocyte Retrieval

The retrieval of cumulus–oocyte complexes takes place during a surgical procedure conducted under general or local anesthesia approximately 36 hr following the hCG injection. Initially, laparoscopic and ultimately transbladder and transvaginal routes have been used to gain access to the ovarian follicles. Because visual access to the ovaries can be blocked during laparoscopy, assessment laparoscopies were necessary, especially since most patients at this time had tubal disease. Often, occlusion of the cornua of the Fallopian tubes was also performed during the assessment laparoscopy. This precautionary step was taken to minimize the chances for an ectopic pregnancy and was adopted because the first IVF pregnancy was, in fact, an ectopic pregnancy.

Ultrasound-directed retrievals were first described in 1984. Initially, the aspiration needle was passed through the bladder because of sterility concerns about a transvaginal approach. Experience has shown that such concerns were groundless, and oocyte retrieval is most often performed transvaginally

with the needle attached to an ultrasound probe inserted into the vagina. The aspiration needle is connected to a suction device, either a syringe or a vacuum pump, with an intervening collection vessel in which to harvest the aspirate; follicles can be subsequently flushed with media to ensure oocyte capture. The aspirates are then immediately transferred into the embryology laboratory. Following oocyte retrieval and continuing through embryo transfer, the patient is placed on a brief regimen of antibiotics and antiinflammatory steroids to decrease the chance that the patient may experience an inflammatory response to her embryos. In order to prepare the uterine lining for embryo implantation, daily progesterone supplementation is also begun at this time. The steroid is administered either orally, by intramuscular injection, or by vaginal suppositories or creme until the presence of a fetal heart is confirmed approximately 5 weeks after oocyte retrieval; it is discontinued immediately upon a negative pregnancy test for hCG.

C. Embryology

There are two basic approaches to the work conducted in the embryology laboratory during an IVF cycle. The first approach is simply observational— eggs and embryos are examined microscopically at key intervals for cell markers (Figs. 1–3). At retrieval, each oocyte may be examined for a polar body which indicates that it has completed the first meiotic division and is ready for fertilization either by exposure to a solution of motile sperm or by ICSI (Fig. 1). Approximately 15–18 hr later, on Day 1, the eggs are rechecked for evidences of fertilization; the presence of two pronuclei indicates that the egg has been fertilized normally. Polyspermic fertilized eggs contain three or more pronuclei and are discarded because such a genetic makeup is incompatible with life. During the next few days the embryos are reexamined at varying intervals to verify that early embryonic development has proceeded normally (Fig. 2). Embryos should consist of two to four cells, or blastomeres, on Day 2 and approximately eight cells on Day 3. They should reach the blastocyst stage by Days 5 or 6. In some eggs and embryos, however, the embryologist may also see anomalies such as

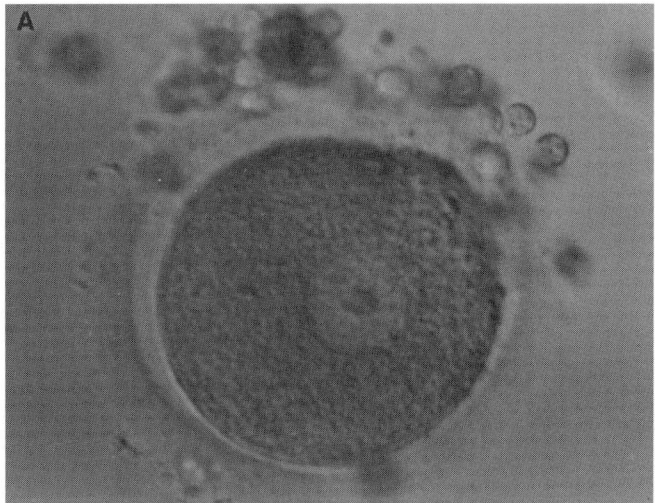

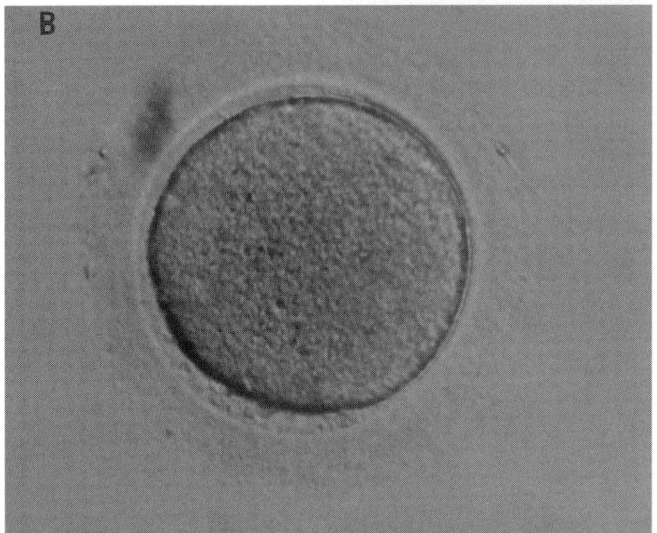

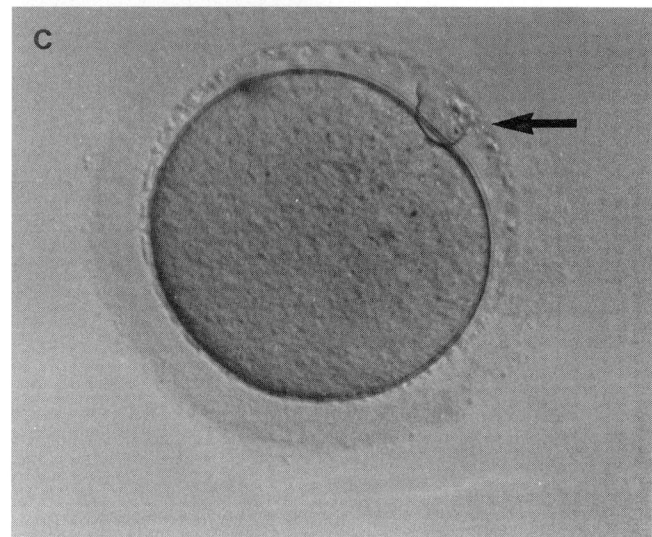

FIGURE 1 Maturational stages of human oocytes at retrieval. These oocytes have been pretreated with hyaluronidase to strip off cumulus cells prior to ICSI. (A) Immature germinal vesicle stage. (B) Metaphase of the first meiotic division; the germinal vesicle has broken down and the chromosomes have migrated peripherally. (C) Metaphase of the second meiotic division as indicated by the presence of a polar body (arrow). The oocytes in A and B must mature further *in vitro* before they can be fertilized successfully. Magnification, ×400.

vacuoles, multinucleation, and anucleate fragmentation (Fig. 3). After taking into consideration the presence or absence of all these markers the embryologist assigns each egg or embryo an evaluation or rating that will eventually play an important role in determining which eggs or embryos are selected for transfer or cryopreservation.

Recently, there have been two major advances in the manner in which embryos are cultured. The length of time the eggs and resultant embryos spend in the laboratory can vary dramatically and can influence the pregnancy rate. The eggs may be mixed with sperm immediately following retrieval and injected laparascopically into the Fallopian tubes [gamete intra-Fallopian transfer (GIFT)]. Alternatively, the fertilized egg, or zygote, may be similarly placed on the following day [zygote intra-Fallopian transfer (ZIFT)]. Originally, embryos were transferred to the

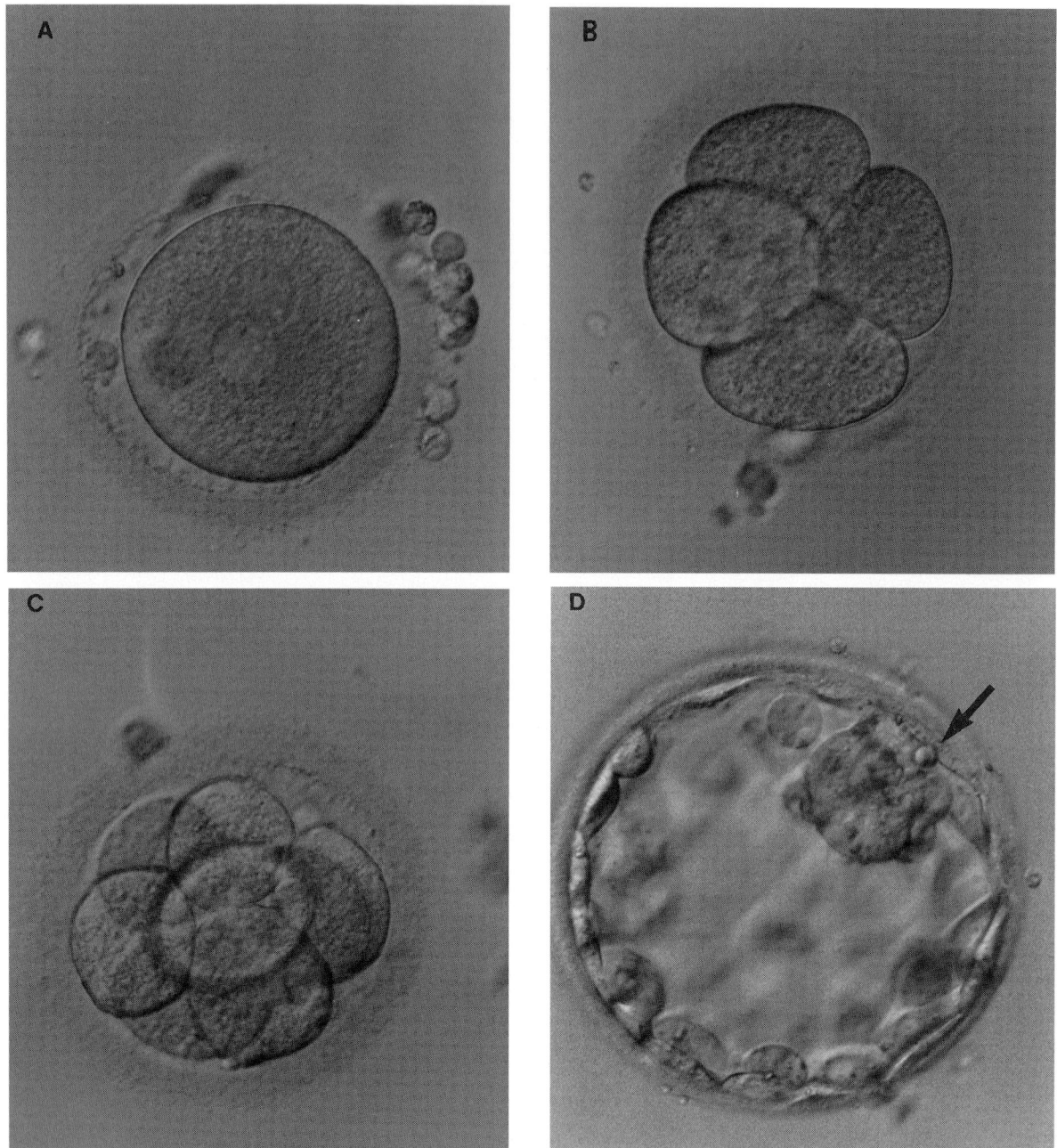

FIGURE 2 Maturational stages of early embryonic development observed during an IVF cycle. (A) A zygote with two pronuclei viewed approximately 18 hr after fertilization on Day 1. (B) A four-cell embryo viewed on Day 2. (C) An eight-cell embryo viewed on Day 3. (D) A blastocyst viewed on Day 5; note the inner cell mass (arrow). Magnification, ×400.

uterus on Day 2, approximately 48 hr after retrieval. However, in 1993 Cohen and associates noted that transfer on Day 3 resulted in significantly higher pregnancy rates. This observation was made during a study testing the effects of assisted hatching, a procedure that could only be performed on Day 3

when intercellular junctions formed. Recent studies suggest that even higher pregnancy rates may be possible when embryos are cultured to blastocyst stage prior to transfer.

Throughout their stay in the laboratory, eggs and embryos are cultured in media in tubes or dishes

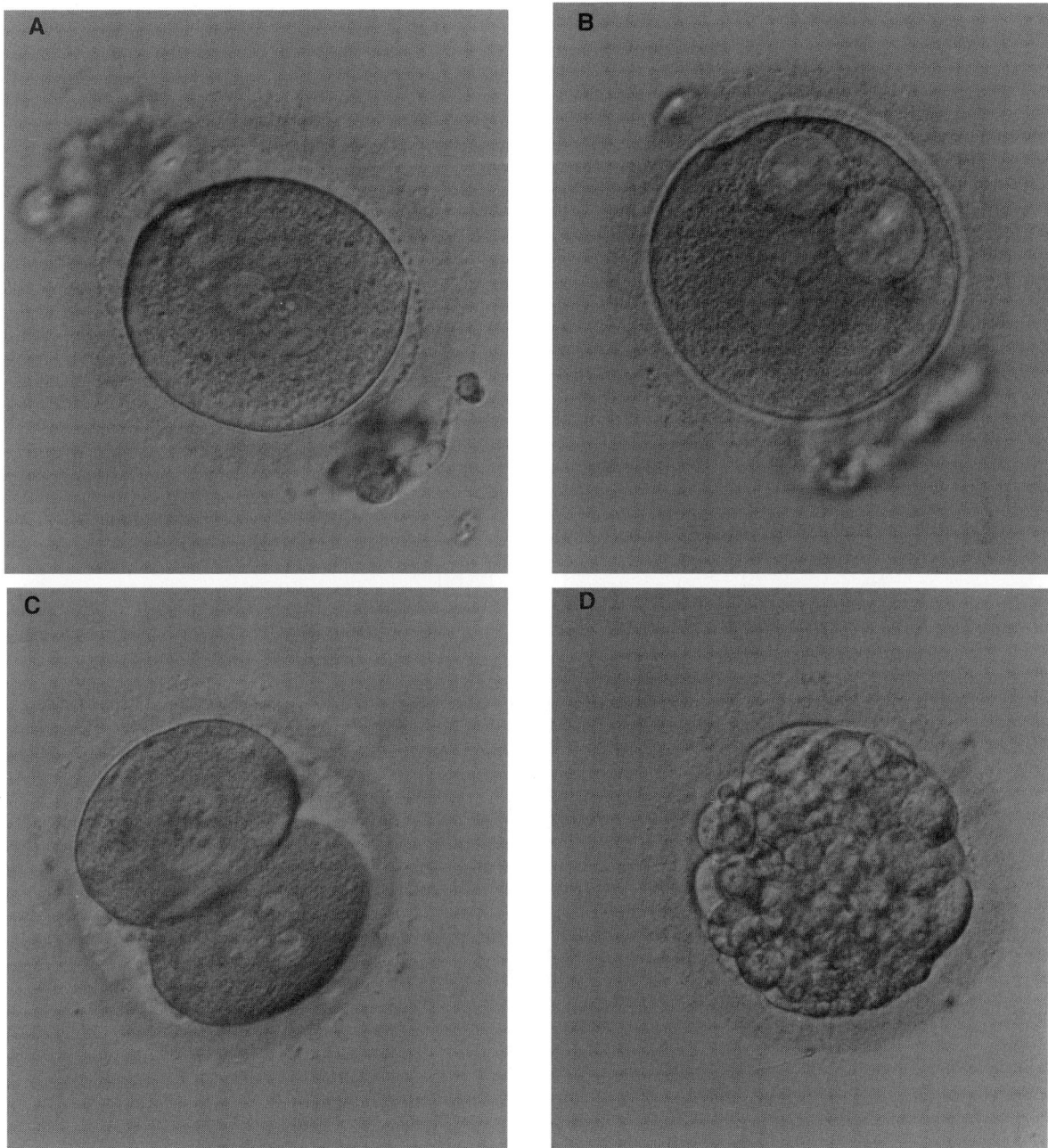

FIGURE 3 Anomalies in human embryos observed during an IVF cycle. (A) Abnormally fertilized zygote; note the three pronuclei. (B) A vacuolated normally fertilized zygote; the two pronuclei are centrally located, whereas the large vacuole is more peripheral. (C) Multinucleated blastomere in a two-cell embryo. (D) Anucleate fragmentation in a Day 3 embryo. None of these embryos are suitable for transfer or cryopreservation. Magnification, ×400.

stored in an incubator (37°C; 5% CO$_2$). In many laboratories microdrops of media are kept in dishes under light paraffin or mineral oil to minimize the environmental changes that occur when the dishes are removed from the incubators for microscopic examination or micromanipulation procedures. New defined media have been developed specifically for egg and embryo culture. These media are based on chemical analyses of the Fallopian tube fluids that normally bathe the egg and embryo *in vivo* as well

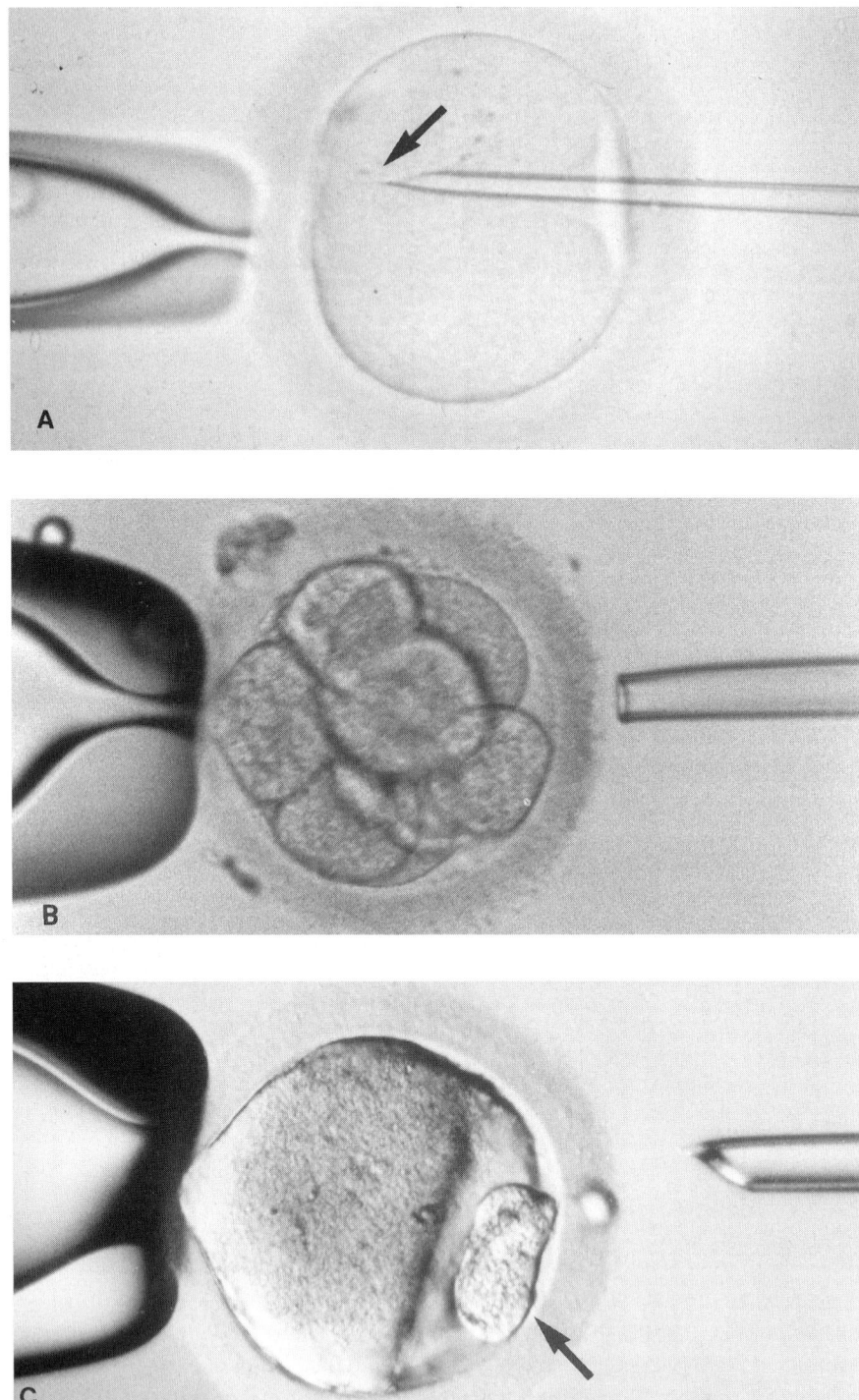

FIGURE 4. Micromanipulation procedures often performed during an IVF cycle. (A) Intracytoplasmic sperm injection (ICSI) in which a single sperm (arrow) is injected into an oocyte; and (B) Assisted hatching in which acidified media is used to drill a hole through the zona pelliucida. (C) In germinal vesicle transfer electrofusion is used to incorporate a germinal vesicle karyoplast (arrow) into an enucleated immature oocyte; this and other micromanipulation procedures are under development. Magnification ×400.

as the metabolic changes that occur during early embryonic life. These new media provide better support for early embryonic development and appear to have contributed to recent increases in IVF pregnancy rates.

The second laboratory approach is more "hands-on." The eggs and embryos are actually manipulated to ensure that particular developmental events will occur. Such procedures are conducted under a microscope using semirobotic micromanipulation devices and specially designed microtools. ICSI is performed to ensure fertilization when there are few motile sperm in the male partner's semen or frozen specimen, or if sperm has to be obtained either by aspiration of the seminal vesicle or by testis biopsy, microsurgical procedures are often performed on the same day as oocyte retrieval. Although other micromanipulation procedures were initially designed to achieve a similar goal, i.e., PZD and SUZI, ICSI now appears to be the method of choice (Fig. 4). Pregnancy rates for ICSI cycles do not appear to be significantly different than those when egg and sperm are mixed together. A second micromanipulation procedure performed by many embryology laboratories is assisted hatching. In this procedure, the zona pellucida is pierced by lance or laser or drilled by a localized microinjection of acidified medium. This procedure is routinely performed the day of embryo transfer and is thought to provide an initiation site to facilitate blastocyst hatching 2 or 3 days later and/or improve the transport of growth factors important for embryo development (Fig. 4). In many laboratories assisted hatching is only performed on selected embryos, i.e., those of poor quality or those from older patients. A final micromanipulation procedure is preimplantation genetic diagnosis. The goal of this procedure is to biopsy the egg or embryo for polar bodies or blastomeres, respectively, for diagnosis of chromosomal translocations or genetically transmitted diseases.

D. Embryo Transfer

The transfer of embryos to the uterus is perhaps the least appreciated aspect of IVF. However, it is the last step before the embryos are returned to *in vivo* conditions, and a poor embryo transfer can negate all prior efforts in the treatment room and laboratory to achieve a successful pregnancy.

Embryos are loaded into a thin catheter in a small volume of media, and the catheter is threaded through the vagina and cervix into the uterine cavity. In some IVF programs the positioning of the catheter is confirmed by ultrasound. After the embryos have been expelled into the uterine cavity the catheter is removed and examined microscopically to verify that the embryos have been transferred. If embryos are retained, the process is repeated.

There have been several improvements in catheter design and construction that have also led to higher pregnancy rates. More pliable plastics have been used to minimize damage and irritation to the uterine lining and to improve catheter placement. Finer milling and polishing of the catheter tip and lumen have reduced significantly the number of surface imperfections and sharp edges that might damage embryos during loading and discharge.

E. Cryopreservation

When there are more embryos than are recommended for uterine transfer, the excess embryos are often cryopreserved for a subsequent frozen embryo transfer cycle(s). Many IVF practices select for cryopreservation only good quality embryos which historically have been shown to be capable of surviving a freeze and thaw. Other IVF programs do not have this option and, as dictated by local legislation, must freeze all excess embryos. Successful pregnancies have been reported following the transfer of embryos cryopreserved at virtually all stages of embryonic development, from zygote to blastocyst.

In most IVF practices embryos are cryopreserved in specially designed freezing units which generate a gradual and steady decline in temperature from room temperature to -40 to $-100°C$. To minimize ice crystal damage at freezing, the embryos are routinely pretreated with increasing concentrations of a cryoprotectant such as dimethylsulfoxide, 1,2-propanediol, or glycerol; this process is reversed when the cryoprotectants are cleared from the embryo following thawing. Cryopreserved embryos are routinely stored in liquid nitrogen at $-196°C$. In many IVF practices there is a limit on the length of time the embryos can be stored; in some cases this policy is determined by local legislation.

III. FUTURE DIRECTIONS

A. New Ovarian Stimulation Protocols

There are several new directions reproductive endocrinologists are taking to stimulate follicular development more effectively. Recombinant gonadotropins, such as Follistim and Gonal F, have already been mentioned. Unlike the urinary gonadotropin preparations, these genetically engineered hFSH preparations are free of hLH contamination. As a result, they can be administered in combination with recombinant hLH in different ratios and at different intervals to identify the quantitative and temporal relationships necessary to stimulate large numbers of optimal quality oocytes.

A second direction is the development and use of GnRH antagonists. Since these analogs suppress endogenous gonadotropin secretion immediately, they can be administered late in the treatment cycle to selectively block the periovulatory surge. This results in several potential advantages. First, the amount of exogenous gonadotropin can be reduced since endogenous gonadotropin secretion continues unabated during the early days of the cycle. Second, several side effects of chronic GnRH suppression are avoided. Finally, the early days of the cycle are now open for testing of endogenous factors, e.g., FSH, estradiol, currently unidentified growth factors, that may indicate the quality of the oocytes that will be collected. Such protocols do have a downside: Patient flow may be difficult to regulate, but this may be overcome by the administration of oral contraceptives prior to cycle onset.

A final direction is *in vitro* maturation of oocytes, an approach that offers the ability to collect multiple oocytes without the benefit of massive gonadotropic stimulation. Several programs in the United States and abroad are performing IVF using immature eggs collected from unstimulated patients and matured *in vitro*; live births have already been reported with this approach. *In vitro* maturation may be the ideal protocol to obtain eggs from patients suffering from polycystic ovarian disease or who are at risk for ovarian hyperstimulation syndrome.

B. Advanced Reproductive Technologies

There are two new micromanipulation procedures currently under investigation that have the potential to expand the patient base for IVF: preimplantation genetic diagnosis and oocyte "reconstruction" by germinal vesicle or cytoplasm transfer.

The biopsy of six- to eight-cell embryos to detect genetically transmitted diseases prior to uterine transfer is currently being conducted in several IVF programs throughout the world. To date, approximately 200 children have been born from embryos tested in this manner. In this procedure one or two nucleated cells are removed from a six- to eight-cell embryo and tested for genetic mutations or aneuploidy using molecular techniques. Fluorescence *in situ* hybridization is used to count the numbers of specific chromosomes or to identify chromosomal translocations. The polymerase chain reaction, alone or followed by endonuclease digestion of its products, is used to identify single point mutations in specific genes. Both techniques can be applied to determine the sex of the embryo. Only the normal, i.e., nonaffected, embryos are transferred.

Oocyte reconstruction offers the potential for improved treatment in many groups of patients. Patients of advanced maternal age, those who suffer multiple miscarriages due to aneuploidy, or those who generate poor quality oocytes (e.g., fragmentation and slow growth rate) may all benefit. Cytoplasmic anomalies are thought to contribute to these developmental problems. Thus, cytoplasmic and nuclear components from two oocytes are recombined to construct a presumably better quality egg. This may be accomplished by injecting the cytoplasm of a donor egg into a patient's eggs. Alternatively, the nucleus of the patient's egg may be removed during the germinal vesicle stage prior to the onset of meiosis and transferred to an enucleated oocyte from a young donor in the same developmental stage (Fig. 4). Current studies indicate that, in both instances, the developmental capacity of the "reconstructed" oocyte is not compromised.

C. Addressing Current Problems in IVF

The quality of IVF practice will continue to improve, especially if two major problems can be addressed: the rapidly growing number of cryopreserved embryos in storage and the high multiple pregnancy rate associated with IVF procedures. Legislative action has been taken in some countries to address these problems and will only increase unless clinical and scientific solutions are found. Oocyte freezing and *in vitro* maturation as well as blastocyst transfer are experimental procedures that may offer solutions.

1. Minimizing the Number of Cryopreserved Embryos

The number of cryopreserved embryos has increased dramatically as IVF has grown. This is particularly so in areas in which local regulations require that all excess embryos be cryopreserved. Many of these embryos now appear to be "abandoned," as patients move and lose contact with their IVF program. Legislation has been passed to limit the length of cryostorage and to provide for the disposition of abandoned embryos.

An option of oocyte cryopreservation and banking may ultimately reduce the demand for embryo cryopreservation. However, the freezing of mature eggs has not been particularly successful to date because of damage to the meiotic spindle. However, this outcome may be avoided if immature eggs are cryopreserved prior to spindle organization. Such cryopreserved eggs can then be matured *in vitro* prior to fertilization and embryo culture. Protocols have been developed to mature eggs *in vitro* following retrieval early in unstimulated or stimulated IVF cycles. More developmental work is needed to understand the physiologic events underlying these final stages of oocyte maturation as well as their metabolic requirements. Blastocyst culture may minimize the numbers of embryos for cryopreservation since few embryos in a cohort will progress to the blastocyst stage.

2. Maximizing Pregnancy Rate and Minimizing Multiple Birth Rate

In an IVF procedure the pregnancy rate will increase when one increases the number of embryos transferred to the uterus. However, such an increase at transfer should also dramatically increase the multiple pregnancy rate. The SART registry reports that the multiple gestation for IVF cycles conducted in 1995 ranged from 20 to 40+% depending on maternal age and that 6.2% of all deliveries following IVF were triplets or quadruplets. Since multiple pregnancies are accompanied by a high rate of obstetrical and neonatal complications, such a high multiple rate is unacceptable.

Recent studies suggest that culturing embryos to the blastocyst stage prior to transfer may increase the frequency of singleton pregnancies. Because most embryos fail to reach this stage of embryonic development because of biologic deficiencies, blastocyst culture is an effective way of identifying those embryos with the best potential for implantation. A recent study by Gardner and coworkers suggests that an implantation rate of approximately 50% is possible. Thus, it should be possible to limit the number of blastocysts to one or two at transfer without compromising the pregnancy rate. Excess blastocysts, should they even develop, could be cryopreserved for subsequent IVF cycles. Louise Brown's birth followed an IVF cycle in which only a single embryo was transferred. It would be most appropriate that the practice of IVF comes full circle to improve future patient care by developing protocols that ensure the success of a single embryo transfer.

See Also the Following Articles

ARTIFICIAL INSEMINATION, IN HUMANS; CHORIONIC GONADOTROPIN, HUMAN; CRYOPRESERVATION OF EMBRYOS; EMBRYO TRANSFER; FOLLICULAR DEVELOPMENT; GRAAFIAN FOLLICLE; INFERTILITY; REPRODUCTIVE TECHNOLOGIES, OVERVIEW; ULTRASOUND

Bibliography

Assisted reproductive technology in the United States and Canada: 1995 results generated from the American Society for Reproductive Medicine/ Society for Assisted Reproductive Technology Registry. *Fertil. Steril.* **69**, 389–398.

Cohen, J., Malter, H. E., Talansky, B. E., and Grifo, J. (1992). *Micromanipulation of Human Gametes and Embryos.* Raven Press, New York.

Edwards, R. G., and Brody, S. A. (1995). *Principles and Practice of Assisted Human Reproduction.* Saunders, Philadelphia.

Gardner, D. K., Vella, P., Lane, M., Wagley, L., Schlenker, T., and Schoolcraft, W. B. (1998). Culture and transfer of human blastocysts increases implantation rates and reduces the need for multiple embryo transfers. *Fertil. Steril.* **69**, 84.

Handyside, A. H., Kontogianni, H., Hardy, K., and Winston, R. M. L. (1990). Pregnancies from biopsied human preimplantation embryos sexed by Y-specific DNA amplification. *Nature* **344**, 368.

Palermo, G., Joris, H., Devroey, P., and Van Steirteghem, A. C. (1992). Pregnancies after intracytoplasmic injection of single spermatozoon into an oocyte. *Lancet* **340**, 17.

Steptoe, P. C., and Edwards, R. G. (1978). Birth after the reimplantation of a human embryo. *Lancet* **2**, 366.

Veeck, L. L. (1998). *An Atlas of Human Gametes and Conceptuses: An Illustrated Reference for Assisted Reproductive Technology.* Parthenon, New York.

Juvenile Hormone

Gerard R. Wyatt

Queen's University

GLOSSARY

corpus allatum An endocrine gland, usually paired but sometimes fused, located behind the insect brain and linked to it by nerve tracts; the primary source of juvenile hormone.

ecdysteroid The molting hormone of insects and other arthropods, such as Crustacea, comprising several closely related hydroxylated C_{27} steroids, produced in the prothoracic gland; the major active form is 20-hydroxyecdysone.

fat body A mesodermal tissue in the form of lobes dispersed in the insect body cavity, to some degree analogous with the liver of vertebrates; both a storage organ for nutrient reserves and the major source for synthesis of hemolymph proteins.

hemolymph The circulating fluid or blood of insects which is pumped by a dorsal tubular heart and circulates freely in the body cavity or hemocoel; it serves to transport nutrients, immunity factors, and hormones; several types of hemocytes, but no erythrocytes, are present.

vitellogenin The major yolk precursor protein which is produced in the fat body, transported in the hemolymph, and taken up into the growing oocytes within the ovary.

Juvenile hormone (JH) is one of the two lipoidal developmental hormones of insects, the other being the steroidal molting hormone, ecdysone. The JHs comprise a small group of closely related terpenoids, sharing common activity, which are produced by the corpus allatum under control of the insect brain. JH plays key roles in the regulation of both metamorphosis and reproduction. During the larval stage, JH serves to maintain larval character and prevent the expression of adult features at each molt, and it is for this activity that the hormone is named. At the completion of larval development, JH production ceases and the hormone disappears from the hemolymph, permitting metamorphosis to the adult form to take place. In adult insects, JH reappears to perform a new role, the stimulation and regulation of a wide variety of processes that are essential for reproduction, including the synthesis of yolk proteins. Certain of the actions of JH are intermediated by its binding to cell membrane receptors, but most of the JH activities that have been studied involve modulation of gene transcription, probably by direct action in the nucleus. Some synthetic chemical analogs of JH, termed "insect growth regulators," can override normal regulatory processes, resulting in sterility or death, and are effective in the control of certain pest species.

I. DISCOVERY AND CHARACTERIZATION OF JH

The basic framework of endocrine regulation of insect molting, metamorphosis, and reproduction was elucidated in the 1930s by the British insect physiologist V. B. Wigglesworth using the blood-sucking bug, *Rhodnius prolixus* (a vector of the trypanosome causing Chagas' disease). By simple but ingenious experiments involving microsurgery and grafting of insects taken at different times after a blood meal (which initiates molting in the larva and reproduction in the adult), Wigglesworth first showed that molting and reproductive maturation depend on endocrine secretion from the brain. It was later demonstrated that the role of the brain is to transduce the input of stimuli such as food intake to the release of a neurohormone which activates a gland in the thorax (prothoracic gland) to produce the steroidal molting hormone, 20-OH-ecdysone, and this steroid in turn induces molt-related activities in the various body tissues (Fig. 1). Other experiments showed that the character of the molt induced by ecdysone was determined by another hormone secreted by the corpus allatum, a gland located be-

hind the brain. When corpus allatum secretion was present, larval characters were retained so that the molt produced a larger larva, but when this gland was removed the ensuing molt was accompanied by the appearance of adult characters which signaled metamorphosis. Since the corpus allatum hormone supported the expression of larval characters and prevented adult differentiation, it was named the juvenile hormone.

Other experiments, however, showed that in adult *Rhodnius* the corpus allatum hormone played a distinct role related to reproduction. When the corpus allatum was removed in females, the eggs failed to mature because yolk was not deposited and the surrounding follicular epithelium degenerated. In males, the accessory reproductive gland failed to fill with secretion. Subsequent research has found that in the great majority of insects, a wide variety of processes directly or indirectly related to reproduction are dependent on stimulation by JH (see Section III). JH is a major coordinator of reproductive physiology in insects, and were it not for the history of its discovery it might well be called the reproductive hormone. It may be compared to the sex hormones of vertebrates, with the important distinction that in insects the hormone is identical in the two sexes and it is the cellular responses to its presence that differ between males and females.

The purification of JH, leading to its chemical characterization, followed the discovery by C. M. Williams, working in the laboratory of Wigglesworth in 1956, that adult males of the giant silk moth *Hyalophora cecropia* contained a large reserve of active JH that was extractable with lipid solvents. After major efforts in several laboratories, the first pure JH, designated JH I, was obtained by H. Röller and colleagues in 1967, and its structure was determined as a sesquiterpenoid epoxide methyl ester. This was followed by the identification from insect sources of several homologs and closely related compounds, constituting a small family of natural JHs (Fig. 2). The usual form present in most insect orders is JH III. JH I is found, often together with some JH II, chiefly in the Lepidoptera (butterflies and moths), in which certain stages may also contain minor amounts of the 3,7,11-triethyl homolog (JH 0) and

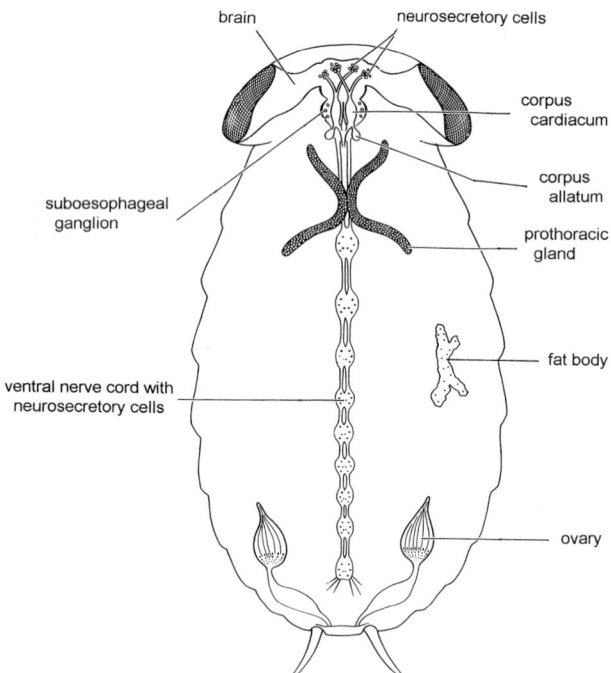

FIGURE 1 A diagrammatic insect, similar to a cockroach, showing the chief endocrine centers and some associated organs. The suboesophageal ganglion and the corpus cardiacum are neuroendocrine organs responsible for the synthesis and release to the hemolymph of peptide hormones not mentioned in this chapter (after P. J. Gullan and P. S. Cranston, *The Insects, an Outline of Entomology*, Chapman & Hall, London, 1994).

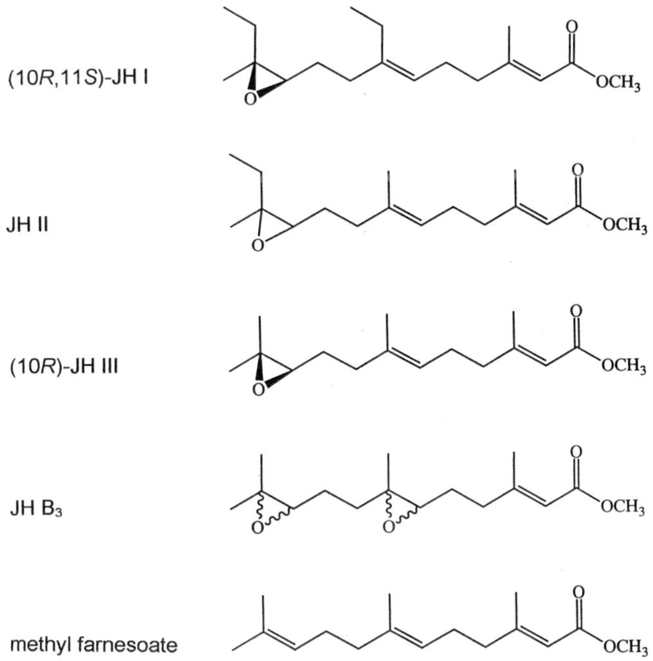

FIGURE 2 Structures of the principal juvenile hormones of insects and Crustacea.

its isomer, 4-methyl JH I. The 6-7,10-11-bisepoxide, JHB$_3$, has been discovered recently as the principal form of JH in certain Diptera (flies). Additional variants may remain to be identified. In Crustacea, methyl farnesoate, corresponding to JH III without the epoxide function, has been identified and shown to have some activities similar to those of JH, and it may represent the JH analog regulating reproductive processes in this taxonomic group.

Chemically, the JHs are unique among animal hormones. JH III (C$_{16}$) is the methyl ester of a sesquiterpene acid, and the C$_{17}$ and C$_{18}$ homologs, JH II and JH I, are homosesquiterpenoids. The JHs thus fall in the same metabolic family as the steroids, but with half the carbon chain length and lacking carbon ring structures. Like steroids, they are lipoidal in character, with low (but appreciable) water solubility. The natural configurations at the centers of asymmetry are essential for biological activity: *trans, trans* at the two double bonds, 10*R,* and 11*S* in JH I. In this chapter, the term JH is used to comprise the several naturally occurring forms.

II. PRODUCTION, METABOLISM, AND TRANSPORT OF JH

JH is produced principally in the corpora allata, although the final steps of its synthesis can occur in certain other organs. JH III is synthesized from acetyl CoA by the isoprenoid pathway via mevalonate, isopentenyl pyrophosphate, and farnesyl pyrophosphate. In the vertebrates, this pathway leads to cholesterol, but insects do not synthesize the steroid nucleus, and the C$_{15}$ farnesyl moieties, instead of condensing to form C$_{30}$ squalene, are oxidized to farnesoic acid. This is converted in two steps to the 10,11-epoxide methyl ester, JH III. In JH II and JH I, the ethyl side chains are produced by an unusual variant of the pathway, commencing with the condensation of propionyl CoA with acetyl CoA, leading to homomevalonate. The adult male cecropia silkmoth is exceptional in that the corpora allata produce JH I acid, which is converted to JH I and stored in the accessory reproductive gland, creating the source from which this hormone was initially purified. The

JHs are metabolized and inactivated by specific enzymes: JH esterase, which hydrolyzes the ester to an acid, and epoxide hydrolase, which cleaves the epoxide group to the diol.

JH levels in the hemolymph vary widely between insect species and during their life cycles. In adult female locusts in a gonadotrophic cycle, for example, JH titers of the order of 200 nM are observed, whereas the cockroach, *Diploptera punctata*, attains the exceptional level of 6 μM. In the regulation of circulating titers, the primary factor is the rate of synthesis in the corpora allata, and a secondary control is exercised by the rate of inactivation of circulating hormone. In adult insects of several species, the corpus allatum activity, assayed *in vitro*, has been found to be correlated with the vitellogenic phase of ovarian cycles. The brain is the site of endocrine coordination transducing external and internal stimuli, such as day length, temperature, mating, nutritional state, and circulating titers of the ecdysteroid molting hormone and JH itself, to modulation of JH production. There is evidence for a variety of mechanisms which differ in different species and between the larval and adult stages of development.

JH production in the corpus allatum can be influenced both positively and negatively by neurohormones that may reach the gland either by direct axonal transport from the brain or via the circulation. In certain species, such as locusts and the tobacco hornworm moth (*Manduca sexta*), positively acting factors termed allatotropins, which stimulate JH output from corpora allata *in vitro*, have been identified. The allatotropin purified from *M. sexta* brains is a 13-amino acid, terminally amidated peptide of which the N-terminal 5 amino acids could be removed without loss of activity (Taylor *et al.*, 1996). Responsiveness to this neuropeptide appears to be confined to the corpora allata of adult Lepidoptera. Sequence analysis of the gene shows that the allatotropin is encoded within a prohormone together with other peptides of unknown function. The gene is expressed, as revealed by *in situ* hybridization and immunohistochemistry, in cells in the brain, frontal ganglion, and terminal ganglion.

In other instances, allatostatins, which inhibit JH synthesis in the corpora allata, appear to be important. Allatostatins have been characterized from cockroaches and some other insects both by isolation and analysis and by sequencing of the DNAs that encode them. They comprise a family of peptides ranging from 6 to 18 amino acids with a conserved C-terminal sequence. Allatostatins are produced in specific neuroendocrine cells in the brain but also in some other ganglia and in cells of the midgut, and they appear to exercise additional endocrine functions as well as the regulation of JH synthesis. The mechanisms by which JH synthesis is regulated by the integrated actions of allatotropins, allatostatins, and other factors remain to be elucidated.

Hemolymph JH titers can also be influenced by the rate of metabolic inactivation. Of special importance is hydrolysis by a specific JH esterase which is produced in the fat body and released into the hemolymph, where it may show great developmental changes in activity. In last-stage larval Lepidoptera, a period of reduced JH synthesis is followed by a JH esterase activity peak which appears to clear the hormone from the hemolymph, allowing ecdysone to bring about commitment of the tissues to metamorphosis. JH esterase may also be involved in JH titer regulation in the adult stage, and in cockroaches during the reproductive cycle there is a negative correlation between hormone level and esterase activity. Synthesis of JH esterase in the fat body is regulated by several mechanisms which include induction by JH itself, thus providing a feedback on its titer.

In the hemolymph, JH is principally bound to carrier proteins that possess specific, high-affinity binding sites. Surprisingly, at least three distinct, unrelated types of JH-binding proteins are found in different orders of insects. The Lepidoptera possess low-molecular-weight (~30 kDa) proteins that bind JH I preferentially; certain Orthoptera such as locusts have high-molecular-weight, hexameric (subunits about 75 kDa) proteins; and in several other orders a type of lipophorin (a lipid-transport protein of about 500 kDa) with specific binding sites for JH has been found. In both of the high-molecular-weight types the binding preference is for JH III. All these carrier proteins show enantiomeric selectivity for the ligand. Their hemolymph concentrations and binding affinities are high enough so that the circulating JH exists almost totally in the bound state and must be presented to the target cells in this form. The functions of the JH-binding proteins may include protection from degradation by nonspecific esterases and prevention of loss to nonspecific sites such as lipid droplets and lipoprotein membranes. The same JH-binding proteins have been found within target cells and they may also function in intracellular transport.

III. ROLES OF JH IN INSECT DEVELOPMENT AND REPRODUCTION

In the early insect embryo, JH is absent, and application of JH analogs can produce toxic effects. During the latter part of embryonic development, however, after the corpora allata differentiate, JH is produced and is required for normal organogenesis.

The biological role for which JH was named is the maintenance of the larval (juvenile) body form and physiology during each molt prior to the final metamorphic molt in hemimetabolous insects (those having incomplete metamorphosis, such as cockroaches and grasshoppers) or the two metamorphic molts that produce successively the pupa and the adult in holometabolous species (those with complete metamorphosis, such as moths and flies). Each molt is induced by release into the hemolymph of a peak of the steroid hormone, 20-OH-ecdysone. In all stages before metamorphosis, there is also an effective titer of JH, and the presence of JH when the tissues are exposed to ecdysteroid is the signal for retention of larval characters. Cell division and differentiation are then directed toward repeating the expression of larval features. Near the beginning of the final larval stage, the JH titer falls sharply so that JH is absent at the next appearance of ecdysteroid, and this exposure to ecdysteroid without JH is the signal that initiates metamorphosis, with expression of pupal or adult features. In the Hemimetabola, the adult differs from the larva chiefly in the sculpturing and pigmentation of the cuticle as well as in the development of wings, reproductive organs, and genitalia, and this change is achieved in one molt. In the Holometabola, however, metamorphosis involves the adoption of a totally new body form, with reconstruction of most of the organ systems, which requires two molting steps. In these insects, a peak of JH toward the end of the last larval instar, after commitment to metamorphosis, prevents the premature expression of adult characters and ensures the formation of a pupa. The adult is then produced by subsequent molting of the pupa in the absence of JH. JH, interacting with ecdysone, is thus the major determinant of morphogenesis.

In the adult stage, after the absence of JH due to inactivity of the corpora allata during metamorphosis, these glands resume activity to produce JH which, in most kinds of insects, is essential for stimulation and coordination of processes essential for reproduction. The timing of renewed JH output is subject to a variety of signals appropriate for the onset of reproduction that may include age, nutrition, mating, day length, and temperature. Through JH, reproduction may be coordinated with phases of migration, diapause, or as a succession of gonadotrophic cycles. In certain insects, however, such as the silk moths, in which oogenesis is completed by the time of adult emergence and which are ready to reproduce without need for feeding or other environmental signal, mating and oviposition proceed without any hormonal stimulus.

Virtually all the organs of adult insects can be targets for JH action, reflecting altered cellular responsiveness that is a product of metamorphosis. A prominent consequence of JH action in adult females of most kinds of insects is the induction of synthesis in the fat body of massive amounts of yolk precursor proteins (vitellogenins), which are released into the hemolymph and taken up into the developing oocytes. The process of vitellogenin uptake through the ovarian follicular epithelium is also dependent on activation by JH. JH also stimulates growth and selective protein synthesis in the accessory reproductive glands which, in females, may produce proteins needed for protection of the deposited eggs and, in males, produce protein components of the semen. Other examples of JH-dependent processes related to insect reproduction are the production and release of pheromones (sex attractants), the histolysis or regeneration (in different species, according to their lifestyles) of flight muscles, and some aspects of behavior. Behavioral effects include female calling and sexual responsiveness, oviposition behavior, male sexual activity (in certain species), and migratory flight. In the adult stage, in contrast to the effects observed before metamorphosis, JH acts independently of ecdysteroids to produce these results. Spermatogenesis and the previtellogenic phases of oogenesis generally do not require JH.

IV. CELLULAR AND MOLECULAR MODES OF ACTION OF JH

In its premetamorphic actions, which have been most intensively studied in the last larval stage of the tobacco hornworm, *M. sexta,* by L. M. Riddiford and co-workers, the role of JH is to modify the tissues' responses to the ecdysteroid molting hormone so that the cells are committed to larval-type synthesis if JH is present together with ecdysteroid and to pupal- or adult-type if JH is absent. Some genes for cuticular proteins (produced in the epidermis) have been identified which are expressed in the presence of both hormones but repressed by exposure to ecdysteroid alone. Certain storage proteins have genes that are repressed by JH, and these proteins are produced in the fat body at the onset of metamorphosis when JH disappears from the hemolymph. These effects are observed as altered mRNA levels and are chiefly due to modulation of transcription. Both negative and positive effects of JH on specific gene transcription have been experimentally demonstrated. Some of the JH effects on these target genes may be indirect, intermediated by modulation of key transcription factors, as indicated by a time lag and a requirement for protein synthesis. In the *Manduca* larval epidermis, the early induction of specific transcription factors by ecdysteroid has been shown to be modulated by the simultaneous presence of JH.

In studies on adult insects related to reproduction, two phases of JH action on target tissues can be recognized: (i) priming or preparatory action and (ii) the evocation of specific, overt responses. For example, *Rhodnius* ovarian follicle cells will shrink within minutes in response to JH *in vitro* only if previously exposed to JH *in vivo* during adult maturation; similarly, the grasshopper male accessory gland shows rapid elevation of protein synthesis in response to JH *in vitro* only after prolonged exposure to JH *in vivo*. Priming undoubtedly includes a variety of processes. One component is the proliferation of ribosomes and endoplasmic reticulum in preparation for enhanced protein synthesis, whereas prolonged deprivation of JH leads to depletion of ribosomes and massive accumulation of stored lipid in the fat body. Other aspects of priming are more specific and may include up-regulation of JH receptors and the production of transcription factors or signal transduction components required for particular cellular responses.

Although JH is a small lipoidal substance which can readily penetrate cells, some of its effects are clearly linked to primary action at the cell membrane. In the ovarian follicular epithelium of the bug *Rhodnius,* studied by K. G. Davey, the cells shrink (thus permitting intercellular passage of yolk proteins to the oocyte) within minutes of application of JH I due to activation of membrane Na^+/K^+-ATPase, not dependent on RNA or protein synthesis. Isolated follicle cell membranes possess specific, high-affinity binding sites for JH I, and the effects of enzyme activators and inhibitors indicate the participation of a signal transduction pathway using protein kinase C. The follicular cells of the locust exhibit a similar response with JH III, the form of the hormone used by this species, instead of JH I. The stimulation of protein synthesis in the male accessory gland of the fruit fly, *Drosophila,* by JH also appears to depend on primary hormonal action at the cell membrane.

Most of the effects of JH that have been studied, on the other hand, involve modulation of nuclear activity and probably result from hormone action at the gene level, although to date the evidence for primary JH action within the nucleus is circumstantial rather than conclusive. Several genes have been identified whose expression in adult insects of different species is either induced or repressed by JH. The most detailed studies have been done in the laboratory of G. R. Wyatt with the genes for two locust yolk proteins, vitellogenin (the major, high-molecular-weight component of yolk) and Jhp21 (a minor, low-molecular-weight yolk component), which are expressed in the female fat body. Transcription is dependent on the presence of JH or an active analog but is initiated, in JH-deprived adults, only after a lag of about 12 hr; this time delay can be extended by inhibition of protein synthesis or shortened by prior application of a low dose of JH which is insufficient by itself to activate the target genes. These results suggest a requirement for a JH-induced transcription factor(s), and a JH-induced protein has been detected that binds to a DNA enhancer element located upstream of the *Jhp21* target gene (Zhang *et al.,* 1996). These findings suggest transcriptional activation through binding of a receptor with a hormone-response element, similar to the known mechanisms of nuclear steroid/thyroid/retinoid hormone action in vertebrates and *Drosophila* but the key components of the JH-regulated system remain to be identified. It should be noted that the requirement for a JH-induced transcription factor does not preclude JH action at the target gene itself, and the hormone could act at both genetic sites. It may be relevant that the vitellogenins of oviparous vertebrates (birds and amphibia) are regulated by estrogen using a mechanism that involves direct transcriptional activation of the vitellogenin genes by the estradiol–receptor complex.

JH, like other hormones, must exert its primary effect through specific binding to protein receptors, and many studies have sought to identify and characterize JH receptors. Proteins possessing high-affinity, specific JH binding sites have been purified and monitored by radioactively labeled JH or analogs, including photoaffinity analogs that link covalently to the binding site. This approach has achieved the purification of the JH carrier proteins that are ubiquitous in hemolymph and tissues, but no convincing evidence for the characterization of a nuclear JH receptor has yet been published. Nuclear hormone receptors generally exist at very low copy number per cell, making their purification from insect tissues a formidable task. A 35-kDa locust ovarian follicle cell membrane protein with specific JH-binding properties, however, is proposed as the receptor for JH action at the membrane. The analysis of primary mechanisms of JH action and the identification of nuclear JH receptors through gene cloning approaches are objectives of current research.

V. USE OF JH ANALOGS IN INSECT PEST MANAGEMENT

Since JH is a lipophilic molecule capable of penetrating the insect cuticle and interfering with normal development when applied in high doses or at abnormal stages, active synthetic JH analogs can be used as insecticides. Many such compounds, having greater chemical and biological stability than natural JH, have been produced and tested, and some have found practical application. They can interfere with metamorphosis and reproduction and are commercially designated by the euphemism, insect growth regulators (Fig. 3). The use of methoprene in preparations for domestic flea control is a familiar example. Whereas methoprene shows obvious structural relationship to the natural hormones, certain compounds in which this is not apparent, such as the phenyl ethers fenoxycarb and pyriproxyfen, are extremely active, and the relationship between their chemical structures and biological activity is not understood. While they appear to act as true JH mimics, some nonspecific effects may also be produced. In contrast to the situation with mammalian steroid hormones, effective direct antagonists of JH action have not been found. Characterization of JH receptors and their binding sites would place this field of research on a more rational footing. Precocene, a chromene derivative from plants which blocks JH synthesis in the corpora allata by interfering with a specific enzymic step, is a useful tool in insect endocrine research but has not proved effective in pest management.

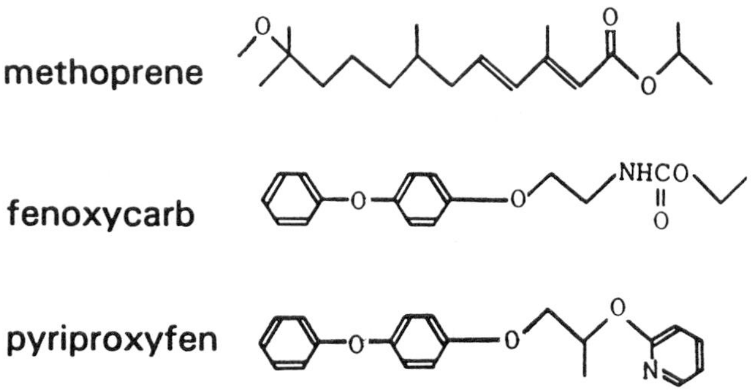

FIGURE 3 Some juvenile hormone agonists, or mimics, known as insect growth regulators, that are used in insect pest management.

See Also the Following Articles

Bibliography

Chang, E. S. (1993). Comparative endocrinology of molting and reproduction: Insects and crustaceans. *Annu. Rev. Entomol.* **38**, 161–180.

Goodman, W. G. (1990). Biosynthesis, titer regulation and transport of juvenile hormones. In *Morphogenetic Hormones of Arthropods* (A. P. Gupta, Ed.). Rutgers Univ. Press, New Brunswick, NJ.

Kerkut, G. A., and Gilbert, L. I. (Eds.) (1985). *Comprehensive Insect Physiology, Biochemistry and Pharmacology*, 13 Vols. Pergamon, Oxford, UK. (See especially Vol. 7, Chaps. 10–12 and 14 and Vol. 8, Chaps. 1 and 6)

Jones, G. (1995). Molecular mechanisms of action of juvenile hormone. *Annu. Rev. Entomol.* **40**, 147–169.

Nijhout, H. F. (1994). *Insect Hormones*. Princeton Univ. Press, Princeton, NJ.

Riddiford, L. M. (1994). Cellular and molecular actions of juvenile hormone. I. General considerations and premetamorphic actions. *Adv. Insect Physiol.* **24**, 213–274.

Riddiford, L. M. (1996). Juvenile hormone: The status of its "status quo" action. *Arch. Insect Biochem. Physiol.* **32**, 271–286.

Stay, B., Tobe, S. S., and Bendena, W. G. (1994). Allatostatins: Identification, primary structure, functions and distribution. *Adv. Insect Physiol.* **25**, 267–337.

Taylor, P. A., Bhatt, T. R., and Hordyski, F. M. (1996). Molecular characterization and expression analysis of *Manduca sexta* allatotropin. *Eur. J. Biochem.* **239**, 588–596.

Trowell, S. C. (1992). High affinity juvenile hormone carrier proteins in the haemolymph of insects. *Comp. Biochem. Physiol.* **103**, 795–807.

Wyatt, G. R. (1997). Juvenile hormone in insect reproduction—A paradox? *Eur. J. Entomol.*, **94**, 323–333.

Wyatt, G. R., and Davey, K. G. (1996). Cellular and molecular actions of juvenile hormone. II. Roles of juvenile hormone in adult insects. *Adv. Insect Physiol.* **26**, 1–155.

Zhang, J., Saleh, D. S., and Wyatt, G. R. (1996). Juvenile hormone regulation of an insect gene: a specific transcription factor and a DNA response element. *Mol. Cell. Endocrinol.* **122**, 15–20.

Kallmann's Syndrome

Lisa M. Halvorson

Tufts University School of Medicine

I. Introduction
II. Clinical Presentation
III. Genetics
IV. Diagnosis
V. Treatment

GLOSSARY

anosmia An inability ability to perceive odors.

autosomal dominant inheritance A disorder expressed in the presence of a single abnormal autosomal allele.

autosomal recessive A disorder apparent only when both alleles at a particular autosomal genetic locus are mutant.

autosome One of 22 paired chromosomes in the human as distinguished from the X and Y sex chromosomes.

gametogenesis The production of germ cells by the gonads (oocytes in the ovary and spermatozoa in the testes).

genotype The genetic makeup of an individual.

gonadotropin-releasing hormone (GnRH) A 10-amino acid peptide (decapeptide) which is secreted in pulses by cells within the arcuate nucleus of the medial basal hypothalamus and stimulates synthesis and secretion of the gonadotropins.

gonadotropin Any hormone having a stimulatory effect on the gonads; the anterior pituitary secretes two gonadotropins, luteinizing hormone and follicle-stimulating hormone, which differentially affect the various cell types in the ovary and testes.

hypogonadotropic hypogonadism A condition of decreased gonadal function due to a failure to secrete gonadotropins, often secondary to lack of appropriate pulsatile GnRH stimulation.

phenotype The observable physical or biochemical characteristics of an individual as determined by both genetic makeup and environment.

variable penetrance The variable expression of a heritable trait between individuals despite identical genotype.

X-linked inheritance A disorder due to an abnormal gene on the X-chromosome and therefore invariably expressed in males.

Kallmann's syndrome is a heritable disorder characterized by gonadotropin-releasing hormone deficiency and anosmia which usually presents as delayed or incomplete puberty. This syndrome has provided invaluable insight into normal hypothalamic control of the reproductive system in the human and has encouraged the development of highly successful treatment modalities for these patients and others with delayed puberty. This syndrome has also provided important clues into common pathways which govern the development of multiple organ systems.

I. INTRODUCTION

Kallmann's syndrome was originally reported by Maestre de San Juan in 1856, with the first familial cases described by Kallmann in 1944.

Patients with Kallmann's syndrome represent a subset of the disorder termed idiopathic hypogonadotropic hypogonadism (IHH). IHH patients lack hypothalamic gonadotropin-releasing hormone and therefore fail to stimulate the expression and secretion of gonadotropins from the pituitary (i.e., they are hypogonadotropic). This lack of gonadotropins, in turn, results in failure to initiate gonadal gameto-genesis and steroidogenesis (i.e., patients are hypo-gonadal) and thereby produces pubertal delay. By definition, in addition to hypogonadotropic hypogo-nadism, Kallmann's patients lack the sense of smell.

Kallmann's syndrome is more prevalent in males than in females. It is estimated that the incidence in males is approximately 1 in 10,000 with a male to female ratio of between 5:1 and 7:1 (Jones and Kem-man, 1976).

II. CLINICAL PRESENTATION

As just stated, patients with Kallmann's syndrome, by definition, are anosmic with hypogonadotropic hypogonadism resulting in delayed puberty. How-ever, in clinical practice, these patients present as a phenotypic continuum. Severely affected individuals may present in infancy with cryptorchidism (failure of testicular descent) and micropenis, whereas others may undergo at least the early stages of pubertal development and present with infertility. While anosmia or hyposmia may be demonstrable by test-ing, patients may be unaware of this deficit.

Interestingly, this disorder is also associated with abnormalities outside the olfactory and reproductive systems. As many as 13% of Kallmann's patients suf-fer from cleft lip and/or palate (with an estimated incidence of 0.1% in the general population). Renal agenesis, either unilateral or bilateral, and cardiac anomalies are also believed to occur more frequently in Kallmann's patients In addition, Kallmann's syn-drome has been associated with several neurologic disorders, including oculomotor abnormalities, sen-sorineural hearing loss, and cerebellar dysfunction. This wide array of defects suggests disruption of a common pathway required for normal development of multiple organ systems.

III. GENETICS

Kallmann's syndrome is genetically heteroge-neous, occurring either as a sporadic or, less com-monly, as an inherited defect. Of the inherited forms,

it has been estimated that the mode of transmission is X-linked in approximately one of five families, autosomal dominant in one of three families, and autosomal recessive in approximately one-half of families. The fact that inheritance can occur via either sex chromosome or autosomal mutations implies that at least two separate genes must be able to pro-duce the same phenotype. It has also been observed that the gene(s) which causes Kallmann's syndrome demonstrates variable penetrance. In other words, family members with identical genetic abnormalities may have differing phenotypes, ranging from hy-posmia alone to complete gonadotropin-releasing hormone (GnRH) deficiency and sexual infantilism.

The X-linked pattern of inheritance has been most fully characterized. In this subset of Kallmann's pa-tients, the observed abnormalities have been attrib-uted to mutations within a gene called *KALIG-1* or *ADMLX,* which is located within the distal region of the short arm of the X-chromosome (Xp22.3). The so-called KAL protein encoded by this gene is 680 amino acids in length with sequence homology to proteins known to be involved in cell adhesion and protease inhibition, functions necessary for neuronal migration. Biochemical characterization of this pro-tein has been hampered by its low level of expression; however, recent studies suggest that KAL is a glyco-sylated membrane protein which is cleaved to yield a 45-kDa diffusible component which may act as a chemoattractant. In embryologic studies, the mRNA encoding the KAL protein has been identified in the developing olfactory bulb, kidney, ocular muscles, and cerebellum. This developmental pattern of ex-pression suggests a unifying mechanism for the dis-parate organ involvement observed in Kallmann's pa-tients.

Recent research has demonstrated that the olfac-tory placode gives rise to neurons which migrate toward the developing forebrain, creating the acces-sory olfactory nerves. Contact between these neu-ronal projections and the overlying brain is required for induction of the olfactory bulb. GnRH-secreting neurons originate in the medial olfactory placode and use the accessory olfactory nerves as a guide toward their final destination in the hypothalamus. Therefore, it has been hypothesized that loss or muta-tion of the KAL protein could adversely affect olfactory neuronal migration and thus indirectly prevent appropriate GnRH-neuronal migration, resulting in a combined olfactory/reproductive abnormality. Consistent with this model, a Kallmann fetus was determined to have GnRH-containing cells in the nose but not in the brain, suggesting failure of these cells to migrate appropriately. As an extension of this theory, it has also been proposed that the degree of the neuronal migration failure may differ between individuals, producing the observed gradation in disease severity.

IV. DIAGNOSIS

While the primary reproductive defect in Kallmann's syndrome is GnRH deficiency, GnRH levels cannot be measured directly in these patients. GnRH is normally secreted as pulses into the hypothalamic–pituitary portal blood system, a location which cannot be reached in humans for obvious technical reasons. Due to a short half-life and the large dilutional effect, circulating GnRH levels are too low to be detected reliably. Fortunately, peripheral gonadotropin levels correlate closely with the degree of GnRH stimulation. Fortuitously, Kallman's patients, lacking GnRH, demonstrate very low levels of the gonadotropins, luteinizing hormone and follicle-stimulating hormone, with otherwise normal anterior pituitary function. Serum levels of the gonadal steroids, estrogen (in females) and testosterone (in males), will also be prepubertal (low). The inability to smell should be readily detectable by olfactory testing, such as with coffee grounds and perfume.

Radiologic evaluation has been shown to confirm the diagnosis and may be indicated to eliminate less common causes of pubertal delay, including congenital central nervous system defects and intracranial tumors. On magnetic resonance imaging, the majority of patients with Kallmann's syndrome demonstrate hypoplastic or aplastic olfactory bulbs and tracts, consistent with the postulated defect in olfactory neuronal migration. In contrast, the olfactory bulbs are normal in patients with other causes of delayed puberty, including IHH.

A detailed family history should also be obtained in these patients in an attempt to ascertain the pattern of inheritance. This information may be useful in predicting the presence of this disorder in other family members and their offspring, allowing for earlier treatment. While the *KALIG-1* gene may be screened for mutations in cases of X-linked inheritance, this testing is currently performed primarily for research purposes.

V. TREATMENT

Initial treatment requires the initiation of puberty followed by the maintenance of secondary sexual characteristics. This is most easily achieved by replacement of the gonadal steroids through exogenous estrogen treatment in females or testosterone treatment in males.

If fertility is desired, gonadotropin stimulation of the gonads is required in order to activate gametogenesis as well as gonadal steroid production. Most elegantly, pulsatile GnRH can be provided using a subcutaneous pump system (continuous GnRH desensitizes gonadotropes and therefore decreases gonadotropin production). Alternatively, the gonadotropins themselves can be replaced by daily injection with recombinant or purified urinary preparations. Fertility has been achieved in a high percentage of both males and females using either of these approaches with successful pregnancies reported.

See Also the Following Articles

FSH (Follicle-Stimulating Hormone); GnRH (Gonadotropin-Releasing Hormone); Hypogonadism; LH (Luteinizing Hormone)

Bibliography

Crowley, W. F., Jr., and McArthur, J. W. (1980). Simulation of the normal menstrual cycle in Kallman's syndrome by pulsatile administration of luteinizing hormone-releasing hormone (LHRH). *J. Clin. Endocrinol. Metab.* **51**, 173–175.

Delemarre-Van de Waal, H. A. (1993). Induction of testicular growth and spermatogenesis by pulsatile, intravenous administration of gonadotrophin-releasing hormone in patients with hypogonadotrophic hypogonadism. *Clin. Endocrinol.* **38**, 473–480.

Duke, V. M., Winyard, P. J. D., Thorogood, P., Soothill, P., Bouloux, P.-M. G., and Woolf, A. S. (1995). *KAL*, a gene mutated in Kallmann's syndrome, is expressed in the first trimester of human development. *Mol. Cell. Endocrinol.* **110**, 73–79.

Franco, B., Guioli, S., Pragliola, A., Incerti, B., Bardoni, B., Tonlorenzi, R., Carrozzo, R., Maestrini, E., Pieretti, M., Taillon-Miller, P., Brown, C. J., Willard, H. F., Lawrence, C., Persico, M. G., Camerino, G., and Ballabio, A. (1991). A gene deleted in Kallman's syndrome shares homology with neural cell adhesion and axonal path-finding molecules. *Nature* **353**, 529–536.

Georgopoulos, N. A., Pralong, F. P., Seidman, C. E., Seidman, J. G., Crowley, W. F., Jr., and Vallejo, M. (1997). Genetic heterogeneity evidenced by low incidence of KAL-1 gene mutations in sporadic cases of gonadotropin-releasing hormone deficiency. *J. Clin. Endocrinol. Metab.* **82**, 213–217.

Hill, J., Elliott, C., and Colquhoun, I. (1992). Audiological, vestibular and radiological abnormalities in Kallman's syndrome. *J. Laryngol. Otol.* **106**, 530–534.

Jones, J., and Kemman, E. (1976). Olfacto-genital dysplasia in the female. *Obstet. Gynecol. Ann.* **5**, 443.

Kallmann, F. J., Schoenfeld, W. A., and Barrera, S. E. (1944). The genetic aspects of primary eunochoidism. *Am. J. Mental Deficiency* **48**, 203–236.

Kirk, J. M. W., Grant, D. B., Besser, G. M., Shalet, S., Quinton, R., Smith, C. S., White, M., Edwards, O., and Bouloux, P.-M. G. (1994). Unilateral renal aplasia in X-linked Kallmann's syndrome. *Clin. Genet.* **46**, 260–262.

Kirk, J. M. W., Savage, M. O., Grant, D. B., Bouloux, P.-M. G., and Besser, G. M. (1994). Gonadal function and response to human chorionic and menopausal gonadotrophin therapy in male patients with idiopathic hypogonadotrophic hypogonadism. *Clin. Endocrinol.* **41**, 57–63.

Kousta, E., White, D. M., Piazza, A., Loumaye, E., and Franks, S. (1996). Successful induction ovulation and completed pregnancy using recombinant human luteinizing hormone and follicle stimulating hormone in a woman with Kallmann's syndrome. *Hum. Reprod.* **11**, 70–71.

Legouis, R., Hardelin, J.-P., Levilliers, J., Claverie, J.-M., Compain, S., Wunderle, V., Millasseau, P., Le Paslier, D., Cohen, D., Caterina, D., Bougueleret, L., Delemarre-Van de Waal, H., Lutfalla, G., Weissenbach, J., and Petit, C. (1991). The candidate gene for the X-linked Kallmann syndrome encodes a protein related to adhesion molecules. *Cell* **67**, 423–435.

Quinton, R., Duke, V. M., De Zoysa, P. A., Platts, A. D., Valentine, A., Kendall, B., Pickman, S., Kirk, J. M. W., Gesser, G. M., Jacobs, H. S., and Bouloux, P.-M. G. (1996). The neuroradiology of Kallman's syndrome: A genotypic and phenotypic analysis. *J. Clin. Endocrinol. Metab.* **81**, 3010–3017.

Rugarli, E. I., Ghezzi, C., Valsecchi, V., and Ballabio, A. (1996). The Kallmann syndrome gene product expressed in COS cells is cleaved on the cell surface to yield a diffusible component. *Hum. Mol. Genet.* **5**, 1109–1115.

Saal, W., Happ, J., Cordes, U., Baum, R. P., and Schmidt, M. (1991). Subcutaneous gonadotropin therapy in male patients with hypogonadotropic hypogonadism. *Fertil. Steril.* **56**, 319–324.

Schopohl, J. (1993). Pulsatile gonadotrophin releasing hormone versus gonadotrophin treatment of hypothalamic hypogonadism in males. *Hum. Reprod.* **8**(Suppl. 2), 175–179.

Schwanzel-Fukuda, M., and Pfaff, D. W. (1989). Origin of luteinizing hormone-releasing hormone neurons. *Nature* **338**, 161–164.

Schwanzel-Fukuda, M., Bick, D., and Pfaff, D. W. (1989). Luteinizing hormone-releasing hormone (LHRH)-expressing cells do not migrate normally in an inherited hypogonadal (Kallmann) syndrome. *Brain Res. Mol. Brain Res.* **6**, 311–326.

Tompach, P. C., and Zeitler, D. L. (1995). Kallmann syndrome with associated cleft lip and palate: Case report and review of the literature. *J. Oral Maxillofac. Surg.* **53**, 85–87.

Truwit, C. L., Barkovich, A. J., Grumbac, M. M., and Martini, J. J. (1993). MR imaging of Kallmann syndrome, a genetic disorder of neuronal migration affecting the olfactory and genital systems. *Am. J. Neuroradiol.* **14**, 827–838.

Waldstreicher, J., Seminara, S. B., Jameson, J. L., Geyer, A., Nachtigall, L. B., Boepple, P. A., Holmes, L. B., and Crowley, W. F., Jr. (1996). The genetic and clinical heterogeneity of gonadotropin-releasing hormone deficiency in the human. *J. Clin. Endocrinol. Metab.* **81**, 4388–4395.

Kamptozoa (Entoprocta)

Kerstin Wasson

Humboldt State University

I. Kamptozoa (Entoprocta)
II. Sexual Reproduction
III. Asexual Reproduction

GLOSSARY

atrium A deep indentation of the calyx in the ventral region above the stomach between the mouth and anus; this enclosure is not technically an internal body space because it is outside the body wall.

calyx The bowl-shaped apical portion of kamptozoan zooid containing the gut and other internal organs.

tholophore A kamptozoan larva, typically shaped like a hat.

zooid A term usually reserved for the modular unit of a colony, but in the kamptozoan literature, it is applied to the clonal as well as colonial unit consisting of the calyx, stalk, and basal attachment structure.

Kamptozoans (entoprocts) comprise a small clade of suspension feeding invertebrates. While superficially resembling hydroids or bryozoans, kamptozoans have a body plan that is unique among metazoans. All kamptozoans reproduce asexually via budding, forming clonal aggregations or colonies. All kamptozoans also reproduce sexually, with internal fertilization and brooding. For a relatively small group of animals, there is a surprising amount of diversity in larval form and in life history patterns within the Kamptozoa.

I. KAMPTOZOA (ENTOPROCTA)

Kamptozoans are small tentaculate suspension feeders. About 150 species have been described worldwide (1 from freshwater and the rest marine), but kamptozoan diversity probably exceeds 500 species. While they are widespread and are quite abundant in some microhabitats, most of the world's kamptozoans are poorly characterized because most species are tiny and easily overlooked.

All kamptozoans reproduce asexually via budding, forming genetically identical zooids. Each zooid has the shape of a wine glass: A bowl-shaped calyx is supported by a slender, flexible stalk that attaches basally to the substratum (Fig. 1). The calyx is ringed by ciliated feeding tentacles and contains a U-shaped gut, a small ganglion, a pair of protonephridia, and one or two pairs of gonads. By convention, the region above the stomach is considered to be ventral (this region was below the stomach in the larva) and the bottom of the calyx and stalk as dorsal. Both mouth and anus are located ventrally, at opposite sides of the calyx, regarded as anterior and posterior, respectively. Between the mouth and anus, there is a cave-like indentation forming the so-called atrium. The calyx is bilaterally symmetrical; a vertical plane through mouth and anus divides the calyx into right and left mirror images. Kamptozoans lack a coelom. The cavity surrounding the calycal organs and extending into the tentacles and stalk is filled by a loose fluid matrix of mesenchyme cells, which acts as a hydrostatic skeleton.

There are two major groups of kamptozoans, the Solitaria and the Stolonata. The Solitaria (family Loxosomatidae), which forms clonal aggregations by calycal budding (Fig. 1a), is considered the most plesiomorphic. The highly contractile zooids are often very small (<1 mm high). Calyx and stalk are not sharply separated, and longitudinal musculature is continuous between them. The calyx and tentacles are generally oriented obliquely to the stalk. Loxosomatids have a specialized basal organ—either a suction disc or a differentiated "foot" with an associated gland—

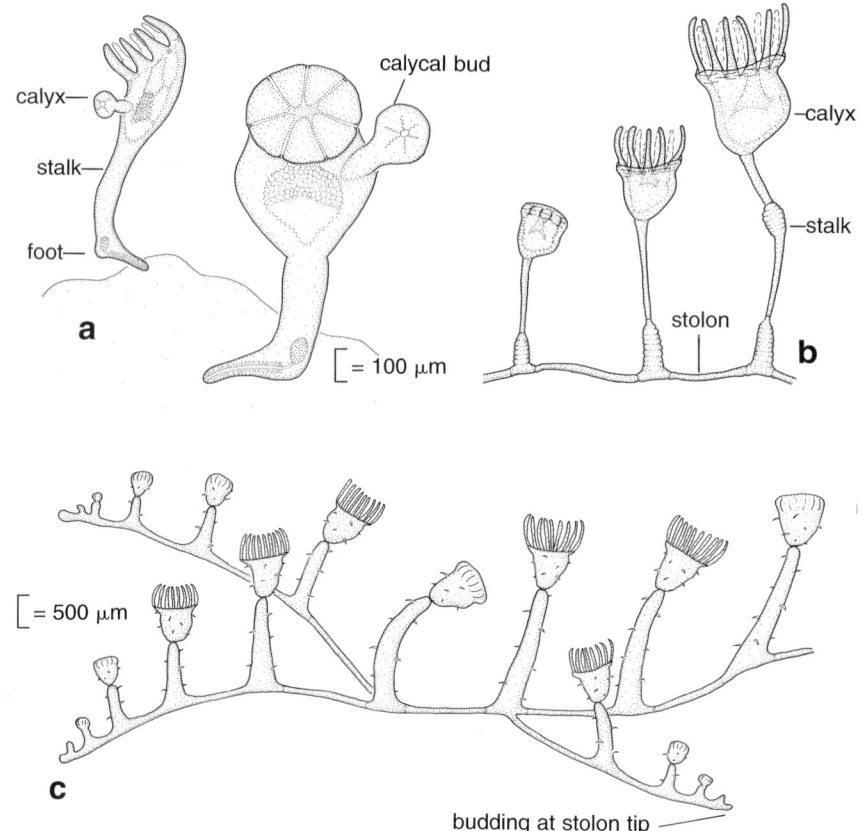

FIGURE 1 Structure of kamptozoan zooids from the three major families: (a) Loxosomatidae, (b) Barentsiidae, and (c) Pedicellinidae.

with which they attach to invertebrate hosts. Loxosomatids occur on polychaetes, sponges, sipunculans, and a variety of other invertebrates; many species appear to be host specific.

The Stolonata (families Pedicellinidae and Barentsiidae) comprises colonial forms, in which zooids remain permanently interconnected by basal stolons. Colonies adhere to various living and nonliving substrata with glandular epithelia of the stolons; they are not host specific. Calyx and stalk are separated by a cuticular indentation, and the calyx–stalk junction is spanned by the circulatory star-cell organ; the longitudinal musculature of the stalk is not continuous with that of the calyx. Stolonate zooids are generally larger than solitary ones and usually have the tentacles extended parallel rather than oblique to the stalk. The family Pedicellinidae (Fig. 1c) is considered more plesiomorphic than the Barentsiidae;

pedicellinid zooids retain a fairly simple zooidal structure, with undifferentiated stalks that have continuous musculature. The family Barentsiidae (Fig. 1b) is characterized by the division of the stalk into wide, flexible, muscular nodes and narrow, rigid, nonmuscular rods. There is a minimum of one basal node and one rod apical to it, but many species have multiple alternating nodes and rods, lending a segmented appearance to the stalk.

Kamptozoan zooids actively bend and twist; their characteristic motion is reflected in the phylum's scientific name (Greek: *kamptestai* = to bend) and its common name, "nodding heads." Another name for the phylum, Entoprocta, is less appropriate because it suggests an affiliation with the Bryozoa (Ectoprocta) and it implies erroneously that the anus is completely enclosed by the tentacular ciliation. Kamptozoans bear only a superficial resemblance to bryozoans, with which they were once grouped. Developmentally, kamptozoans are spiralians, but their phylogenetic relationships to other metazoans remain enigmatic. A recent 18S rRNA analysis suggests kamptozoans may be closely affiliated with polychaete annelids. The oldest known kamptozoan is a barentsiid fossilized by bioimmuration in upper Jurassic rocks. This Mesozoic fossil sets a minimum time for the divergence of what is probably the most derived family, suggesting that ancestral members of the phylum may date back much farther.

II. SEXUAL REPRODUCTION

A. Gonochorism, Hermaphroditism, and Germ Cells

The sexual mode of most kamptozoans is not known. Individual zooids have only very rarely been followed over time, in order to distinguish gonochorism from sequential hermaphroditism; consequently, reports of gonochorism in the literature are mostly unreliable.

It appears that sequential hermaphroditism at the calyx level is the most common sexual mode for loxosomatids; their calyces are often protandric. In these sequentially hermaphroditic species, timing of sex change is not synchronized between clonemates, and so the genetic individual composed of them is simultaneously hermaphroditic. Calyx gonochorism has also been reported, but not substantiated, for loxosomatids.

Pedicellinids and barentsiids display a diversity of sexual modes. Many species have gonochoric calyces, either in gonochoric colonies (containing calyces of only one sex) or in simultaneously hermaphroditic colonies (containing calyces of both sexes). In the latter situation, the sex of calyces is environmentally determined. A few species have simultaneously hermaphroditic calyces.

Male maturation typically precedes female maturation in the calyx, colony, or population. Testes mature before ovaries in hermaphroditic calyces, male calyces mature before female ones in hermaphroditic colonies, and male colonies become sexually mature before female ones in populations of gonochoric col-

onies. This suggests that the threshold (temperature, food levels, calyx size, etc.) for male maturity is lower than that for female maturity.

The reproductive system is rather simple in both sexes. The germ cells are mesenchymally derived; gonad anlagen first appear above the stomach as a pair of tiny oval translucent vesicles around the time at which the developing calyx begins to feed. The gonad anlagen eventually grow into a pair of ovoid sacs lying ventral to the stomach, consisting of a one-layered epithelium which is the germinal layer from which the gametes arise. In simultaneously hermaphroditic calyces, a pair of testes lies posterior to the pair of ovaries. In sequentially hermaphroditic calyces, the same pattern holds, except that the testes degenerate as the ovaries are developing. Each gonad feeds into a gonoduct, and the right and left gonoducts merge at the ventral midline to open to the atrium through a common gonopore posterior to the ganglion.

The reproductive period of most kamptozoans is not known. Some species reproduce year-round, whereas others reproduce only seasonally.

B. Spermatogenesis

The pair of sacciform testes grow rapidly and may fill much of the calyx (Fig. 2a). Atrophied (nonfunctional) glandular cells surround the spermiduct; these are apparently homologous to the shell gland of females. In a few species, there is a seminal vesicle for the storage of ripe sperm located at the distal end of the spermiduct.

Spermatozoa at various stages of differentiation are typically found in the testes, with no particular organization. Mature spermatozoa are filiform, about 75 μm in length, of the modified type usually associated with internal fertilization. The lance-shaped head accounts for about a third of the length; no acrosome has been observed in this nuclear region in mature spermatozoa. The middle third of the spermatozoon contains a single elongated mitochondrion surrounding a rod of electron-dense material and is accompanied by the axoneme. The final third of the spermatozoon consists of the tail flagellum.

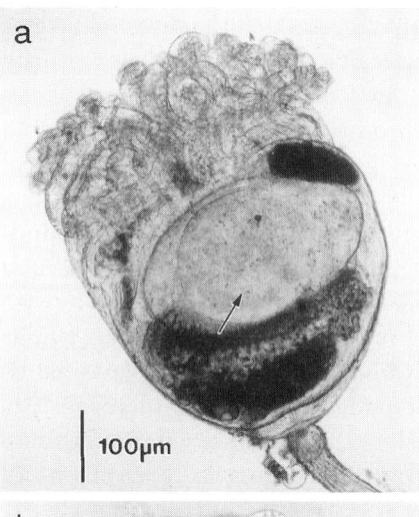

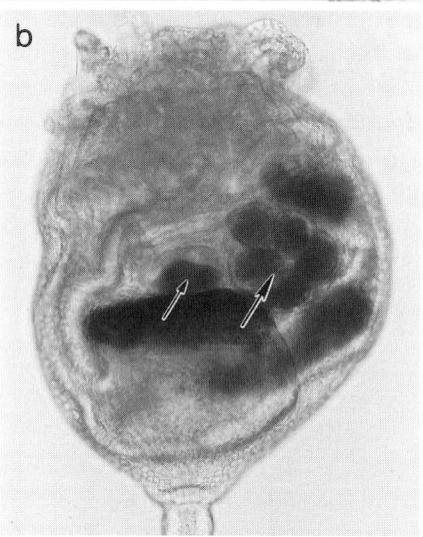

100μm

FIGURE 2   Sexually mature *Barentsia* calyces. (a) Male calyx in right-side view; right testis marked by arrow. (b) Female calyx in right-side view; right ovary marked by small arrow and brood chamber containing embryos marked by large arrow. Scale bar applies to both a and b (reprinted from Wasson, 1997. © Marine Biological Laboratory).

C. Oogenesis

The ovaries generally remain much smaller than the testes (Fig. 2b). Each ovary generally contains several ooyctes at various stages of oogenesis, with the larger, older ooyctes near the blind end and the smaller, younger ones near the oviduct. Follicular cells surround the ooyctes; their contribution to oogenesis is not known. As they grow, the oocytes accumulate large amounts of yolk spheres and lipid droplets. The mature eggs are yolky but fairly small, about 30–100 μm depending on species. Eosinophilous shell glands surrounds the oviduct.

D. Fertilization and Brooding

Spawning by males has rarely been observed; apparently a cloud of spermatozoa is released following a sudden contraction of the calyx. Germinal vesicle breakdown and fertilization occur in the ovary, but it is not known how the spermatozoa get there from the water column surrounding a male. Spermatozoa are found even in small, immature ovaries, suggesting that there is a capacity for sperm storage (there is no indication that these spermatozoa are due to spermatogenesis in the ovary). Nothing is known about the occurrence of self-fertilization (within hermaphroditic calyces or between gonochoric calyces within hermaphroditic colonies).

All kamptozoans brood their embryos and release fully formed larvae. In most species, fertilized eggs are ovulated at metaphase of the first meiotic division. As the zygote passes through the oviduct, the shell glands secrete a pliant envelope, which encloses the embryo and extends into a cord that tethers the embryo to the deepest part of the atrium after it emerges from the gonopore. One or a few zygotes per day are ovulated in alternation from the two ovaries, with the youngest embryos pushing the older ones farther from the gonopore. The atrium contains many embryos in a regular succession of stages from cleaving eggs to contractile larvae. The tethered embryos, like a bouquet of successively enlarged balloons, can occupy a substantial portion of their mother's calyx (Fig. 2b). While the position of the larvae inside the atrium appears superficially to represent internal brooding, one must recall that the larvae are merely inside an indentation, not enclosed by the maternal body wall.

When larvae hatch out of their envelopes, they remain attached to the atrial floor by the cord, with their mouth and ciliary band upward, allowing them to feed on particles in their mother's current. The wall of the atrium enclosing the larvae often becomes conspicuously thickened. There are unsubstantiated reports of nutrient-laden cells breaking free from the atrial walls and being ingested by larvae. In general, however, there appears to be little or no nutritive transfer from the mother to the developing embryos; brooding provides only protection and oxygenation by the mother's ciliary currents. The fully formed larva is typically two or three times as large as the fertilized egg; increase in size can be attributed either to the yolk in the egg or to feeding by the larva itself while it is still in the atrium. About a week after fertilization, the oldest larva in the atrium breaks its tether, by its own contractions or contractions of the maternal calyx, and swims away.

In a few species, the pattern of brooding and development deviates from the previous description. In these species, the eggs are very small and not yolky. A single developing embryo is held in an ectodermal pocket formed by the atrial wall. This pocket serves as a placenta, transporting nutrients to the embryo, which grows considerably as a result.

E. Embryology

Most kamptozoans show typical spiralian, determinate development. Cleavage is holoblastic, spiral, and equal or almost so for the first three divisions. At the 16-cell stage, the four blastomeres at the animal pole and those at the vegetal pole are slightly larger than the eight equatorial ones, of which some give rise to the trochoblast cells. At about the 64-cell stage, a small blastocoel appears. The mesentoblasts (4d cells) are still undivided at this point. The arrangement of cells at the animal pole resembles a annelidan rather than a molluscan cross.

Gastrulation begins by invagination at about the 90-cell stage. The blastopore eventually becomes occluded and a stomodeum forms nearby from an anterior group of cells. A proctodeum develops as a small posterior invagination. Both the stomodeum and proctodeum become linked to the previously invaginated archenteron, forming a complete, endodermally derived, U-shaped gut.

At this stage, the mesentoblasts divide to form a pair of loose lateral rows of mesodermal cells, but these soon become scattered. The resulting mesenchyme eventually gives rise to the muscles, connective tissue, and germ cells. There is never any hint of coelom formation. The embryo extends in the apical–anterior axis. An indentation between the mouth and anus becomes the future atrium. An apical sensory organ develops at the apical plate and is connected by neuromuscular strands to a frontal organ. A girdle of locomotory cilia develops from the trochoblast cells.

Deviations from this pattern of spiralian development are observed in those few known species that have placental nourishment. In *Loxosomella vivipara*, the spiral pattern can be recognized only up to the eight-cell stage. No gastrulation has been observed, and there is never any trace of a gut. A large internal bud develops inside the larva. However, typical larval sensory organs are formed.

F. Larval Form and Types

Kamptozoan larvae are hat shaped, and so the name tholophora (Greek: *tholos* = dome; *tholia* = straw hat) has been proposed. The hyposphere of the larva is deeply indented into the prominent, hat-like episphere. The curve of the U-shaped gut is in the upper part of the hat; mouth and anus open on the ventral surface.

At the top of the hat, there is an apical organ (Fig. 3a) consisting of a ciliary bundle and an underlying ganglion. This is connected by nerve fibers to a frontal organ (Fig. 3a) of similar structure located on the front of the hat; the frontal organ in turn is connected to a nerve ring approximately at the brim of the hat. In loxosomatids, the frontal organ is paired and often carries a pair of photoreceptors, each consisting of a cup-shaped pigment cell, a lens cell, and a sensory cell. The structure of the eye is unusual in that light enters perpendicular to rather than parallel to the long axes of the sensory cilia. In stolonate kamptozoans, the paired frontal ganglia have become fused, and there are no photoreceptors.

Around the brim of the hat-like body of the larva is a girdle of long compound cilia (Fig. 3a). Inside (ventral to) this ring, there is a second band of shorter cilia in the shape of a horseshoe, with the opening of the horseshoe at the anus. These two ciliary bands beat in opposition and capture particles that are then transported to the mouth (which lies between the two ciliary bands) by short cilia in the atrial grooves, which run between the two bands of longer cilia

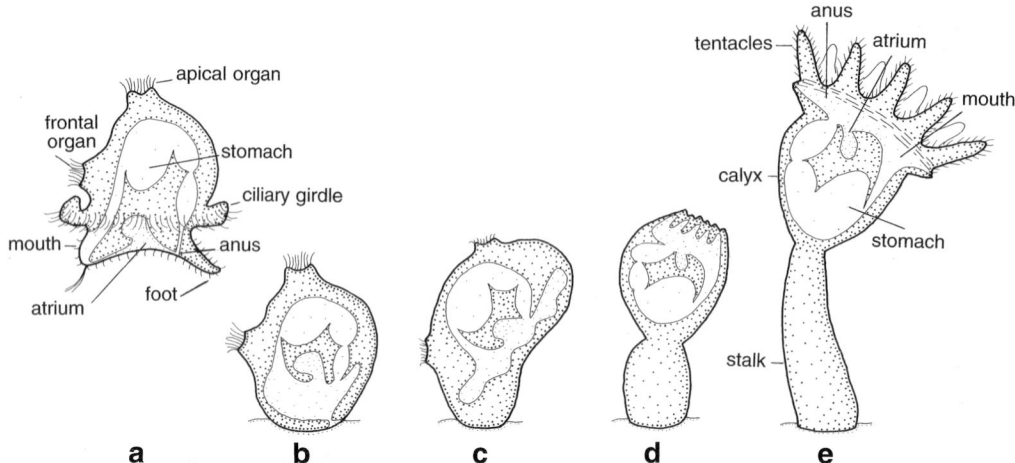

FIGURE 3 Schematic representation of metamorphosis in *Pedicellina cernua*. (a) Swimming larva; (b) newly settled larva; (c) period of vigorous anterior growth; (d) zooid with separation between stalk and calyx, tentacles forming; and (e) feeding zooid.

from anus to mouth on both sides, as in the adults. The larva also has an atrium (Fig. 3a). Often there is a ciliated creeping foot (Fig. 3a) in the ventral area between mouth and anus. A pair of protonephridia are also found in this ventral region.

Tholophores are very contractile. As they move about in the brood chamber or in the water column, they retract and extrude their sensory organs and foot and deform their hat-like shape. Tholophores are often covered with an adhering layer of detritus, perhaps as a result of glandular secretions coating their external surfaces. The function, if any, of this adhering layer is not known. Some tholophores show other unusual features—stalked vesicles, a spider-web pattern of raised ridges, eversible glandular pouches, etc.—that are not yet understood.

Tholophores resemble the trochophores of some spiralians. The downstream-collecting ciliary bands of tholophores are similar to those of trochophores in cell lineage, structure, and function. The apical organs of tholophores also resemble those of trochophores. However, unlike trochophores, most tholophores have a frontal organ and a ciliated foot, and their hyposphere is deeply indented into the episphere. Regardless of whether tholophores are homologous to or convergent with trochophores, it is clear that the strongest resemblance of tholophores is to adult kamptozoan calyces. Larva and adult share the same shape, structure of the digestive system, atrium with atrial grooves, and a very similar ventral ciliary feeding mechanism. This similarity has led to speculation about the origin of the adult kamptozoan body plan from a neotenous trochophore-type larva.

G. Settlement and Metamorphosis

After breaking the tethers holding them in the maternal atrium, tholophores may swim and feed in the plankton for a while before beginning the phase of benthic substrate exploration that eventually results in permanent settlement. In some loxosomatids, the planktonic phase appears to be very long; some loxosomatid larvae have been maintained in the laboratory for over a month and have been obtained in plankton tows far from any shallow benthic habitats. In many species, however, the planktonic phase is almost certainly very short, perhaps only a matter of hours or days. Tholophores of stolonate species in particular have been observed settling within hours of release in the immediate vicinity of their mothers. Some stolonate larvae apparently do not have strong enough ciliary bands for swimming and can only creep on the benthos. In these species with extremely short-lived larvae, the larva's feeding while still in the maternal atrium may actually be more important for obtaining the necessary reserves to undergo metamorphosis than subsequent feeding after the larva leaves the atrium.

Tholophores apparently employ strategies to locate and choose an appropriate settlement site. During their benthic phase, tholophores creep actively on the substrate, probing it with their frontal organs. Most loxosomatids live on invertebrate hosts and are fairly host specific. Many stolonate species are gregarious, settling close to conspecifics or even to any other colonial kamptozoans. The nature of the settlement cues, and the mechanism of their effect on the tholophores, is not known.

Metamorphosis has been carefully described in a few kamptozoan species. The larva attaches to the substrate with the anterior part of the ephisphere, adhering with a ring of cells around the frontal organ (loxosomatids) or with three pairs of cement glands in the foot (stolonates). The apical and frontal organs, the foot, the hyposphere, and the ciliary girdle are all retracted, and the atrium becomes sealed off by the contracted margins of the episphere (Fig. 3b). The larval sensory organs and nervous system disintegrate, and a new ganglion forms. It is not known whether the protonephridia are retained or whether they form anew. Meanwhile, the anterior region of the ephisphere grows more rapidly than the posterior region. As a consequence, the atrium and digestive tract are rotated (Figs. 3c and 3d) nearly 180° in stolonates but only about 90° in loxosomatids. Besides changing in orientation, the gut undergoes few changes between larval and adult forms.

Next, a separation forms between the calyx and the nascent stalk, and the latter elongates (Fig. 3d). Ciliated tentacles form as ectodermal protuberances at the periphery of the atrium (Fig. 3d) roughly in the location of the degenerating larval ciliary bands. It is not known whether the same cells that formed the larval ciliary girdle contribute to the ciliary bands on the adult tentacles. Finally, the atrium breaks open, releasing the tentacles, and feeding begins (Fig. 3e). This whole process takes about 1–8 days, depending on the species.

While in all stolonates and many loxosomatids the larva does metamorphose directly into the adult, some loxosomatids (all members of the genus *Loxosoma* and some members of *Loxosomella*) have precocious budding in which the larva does not metamor-

phose but instead dies as the buds it bears grow and are released. In effect, the larval bud, rather than the larva itself, is the route to adulthood in these species. In the most extreme cases, the larval gut is absent, and the larva is completely consumed by an internal bud that forms while the larva is still within its parent. Some remarkable species display further heterochrony: The buds themselves already have buds in turn or even are sexually mature while still contained in the larva.

III. ASEXUAL REPRODUCTION

All kamptozoans reproduce asexually by budding. In loxosomatids, buds form in two anterior or anterolateral regions of the calyx (Fig. 1a), often approximately level with the top of the stomach. Buds may be produced alternately or simultaneously at the two budding sites. The basal part of the bud's stalk develops an attachment organ. The bud may remain attached to its "parent" for some time, feeding and even becoming sexually mature, but it eventually breaks away, often attaching to a nearby spot on the invertebrate host.

Stolonate kamptozoans also bud at the anterior face of zooids, but budding occurs earlier in the life of zooids than in loxosomatids. The zooids producing buds are often themselves still tiny buds; each stolon tip is a bud primordium forming anterior to the next youngest bud (Fig. 1c). As the buds grow and differentiate into fully formed zooids, they are separated by intercalating growth of the stolon. Eventually this growth ceases and a septum forms on each side of the zooid, partitioning the stolon into fertile (zooid bearing) and sterile (without zooids) segments. The anterior side of every zooid along a stolon faces the growing stolon tip. Complex, upright, and branching colonies are formed by some barentsiids, which bud from specialized stalk regions as well as at stolon tips. Additionally, in some stolonate species, resting buds (hibernacula) are formed at stolon tips. These undifferentiated buds are enclosed in single or multiple chambers and are covered by a thick cuticle. They germinate only after the stolonic connection to the rest of the colony is severed and following exposure to cold temperatures. No other type of zooidal polymorphism occurs in kamptozoans, perhaps because colonies are not sufficiently integrated to sustain nonfeeding zooidal types.

Pedicellinids and barentsiids, unlike most loxosomatids, have deciduous calyces. Calyces degenerate and are shed and are replaced by a budding process at the apical stalk tip comparable to that at stolon tips. Injured barentsiid zooids can regenerate new calyces and stalks even from basal stalk and stolon remnants.

While the main type of asexual reproduction for stolonate kamptozoans is colonial growth, clonal propagation may also occur. In the freshwater kamptozoan (*Urnatella*), creeping propagation stolons released from older colonies establish new ones. Marine colonies may become passively fragmented by damage, which separates colonies into clonal replicates. Hibernacula also result in the formation of separate, genetically identical colonies.

Patterns of bud formation at the histological level are very similar in all kamptozoans. An epidermal proliferation of the anterior body wall of a zooid results in an evagination that forms the bud primordium. Budding is essentially an ectodermal process; whereas some mesenchyme cells migrate from the "parent" into the bud, no endoderm is contributed. At the apex of the bud primordium, an invagination forms then constricts into upper and lower vesicles, which become the atrium and the digestive tract, respectively. A narrow passage connecting the vesicles becomes the mouth, whereas the anus breaks through at a later stage. A constriction soon separates calyx and stalk, and the latter elongates. Eventually the atrial cavity breaks through, freeing the tentacles, and the bud begins to feed.

See Also the Following Articles

Marine Invertebrate Larvae; Marine Invertebrates, Modes of Reproduction in

Bibliography

Emschermann, P. (1982). Les kamptozoaires. État actuel de nos connaissances sur leur anatomie, leur développement, leur biologie et leur position phylogénétique. *Bull. Soc. Zool. France* **107**(2), 317–344.

Franzen, Å. (1983a). Bryozoa Entoprocta. In *Reproductive Biology of Invertebrates* (K. G. Adiyodi and R. G. Adiyodi, Eds.), Vol. 1, pp. 561–569. Wiley, New York.

Franzen, Å. (1983b). Bryozoa Entoprocta. In *Reproductive Biology of Invertebrates* (K. G. Adiyodi and R. G. Adiyodi, Eds.), Vol. 2, pp. 505–517. Wiley, New York .

Marcus, E. (1939). Briozoários marinhos brasileiros. III. *Boletins Faculdade Filosofia Ciências Letras* (Universidade São Paulo) *Zool.* **3**, 111–295, plates 1–30.

Mariscal, R. N. (1975). Entoprocta. In *Reproduction of Marine Invertebrates* (A. C. Giese and J. S. Pearse, Eds.), Vol. 2, pp. 1–41. Academic Press, New York.

Mukai, H., and Makioka, T. (1980). Some observations on the sex differentiation of an entoproct, *Barentsia discreta* (Busk). *J. Exp. Zool.* **213**, 45–59.

Nielsen, C. (1971). Entoproct life-cycles and the entoproct/ectoproct relationship. *Ophelia* **9**(2), 209–341.

Nielsen, C. (1990). Bryozoa Entoprocta. In *Reproductive Biology of Invertebrates* (K. G. Adiyodi and R. G. Adiyodi, Eds.), Vol. 4b, pp. 201–209. Wiley, New York.

Wasson, K. (1997). Sexual modes in the colonial kamptozoan genus *Barentsia. Biol. Bull.* **193**, 163–170.

Kangaroos

see Marsupials

Kinorhyncha

Birger Neuhaus

Humboldt-Universität of Berlin

GLOSSARY

penile spines One to three pairs of spines at the posterior end of males only; supposed to be involved in copulation as a sensory organ or to assist in transmission of sperm.

scalids Regularly arranged spines at the head of juvenile and adult Kinorhyncha; protrusion and retraction of the head allows forward locomotion.

The minute Kinorhyncha live in the interstices of marine sediments throughout the world. They are gonochoric, reproduce sexually, and often show sexual dimorphism. Sperm from internal fertilization is stored in the paired receptacula seminis of the female. Here, the sperm undergoes significant morphological differentiation. Direct development includes six juvenile stages which are separated by molts.

I. DESCRIPTION OF THE KINORHYNCHA

The Kinorhyncha are a taxon of exclusively marine, microscopic animals (up to 1 mm in length). They inhabit muddy or sandy sediments all over the world from the wadden sea to the deep sea. The animals feed on bacteria, diatoms, microscopic algae, or detritus. Kinorhyncha possess 11 trunk segments, a neck segment, and a head segment (= introvert) (Figs. 1A–1C). Locomotion is facilitated by protrusion and withdrawal of the spiny head region. Contraction of the segmental dorsoventral muscles increases the internal body pressure and, therefore, pushes the head forward. Action of the retractor muscles moves the head inside the trunk. So far, about 120 species have been described and traditionally grouped into the Cyclorhagida (with 10 genera) and the Homalorhagida (with 4 genera). Most certainly, the Priapulida or the Loricifera represent the closest relatives of the Kinorhyncha. These 3 taxa are often included in the Aschelminthes (= Nemathelminthes) together with the Nematomorpha, Nematoda, Gastrotricha, Rotifera, and Acanthocephala. The phylogenetic relationships between these groups are unresolved.

II. MODE OF REPRODUCTION

The Kinorhyncha are gonochoric and reproduce sexually. Signs of asexual reproduction are missing.

Primary external sexual dimorphic characters include penile spines in males. In *Echinoderes* (Cyclorhagida), *Paracentrophyes*, *Neocentrophyes*, and *Kinorhynchus* (Homalorhagida), one or occasionally two pairs of regular spines of the last trunk segment are modified to penile spines. However, two pairs of penile spines have evolved *de novo* ventrally in *Pycnophyes* (Fig. 2A) and *Kinorhynchus* and laterally in *Echinoderes*. Such penile spines have probably been integrated into sexual reproduction several times independently because the spines vary both in position and in morphology.

Secondary external dimorphic characters occur mainly in the last two trunk segments. Middorsal,

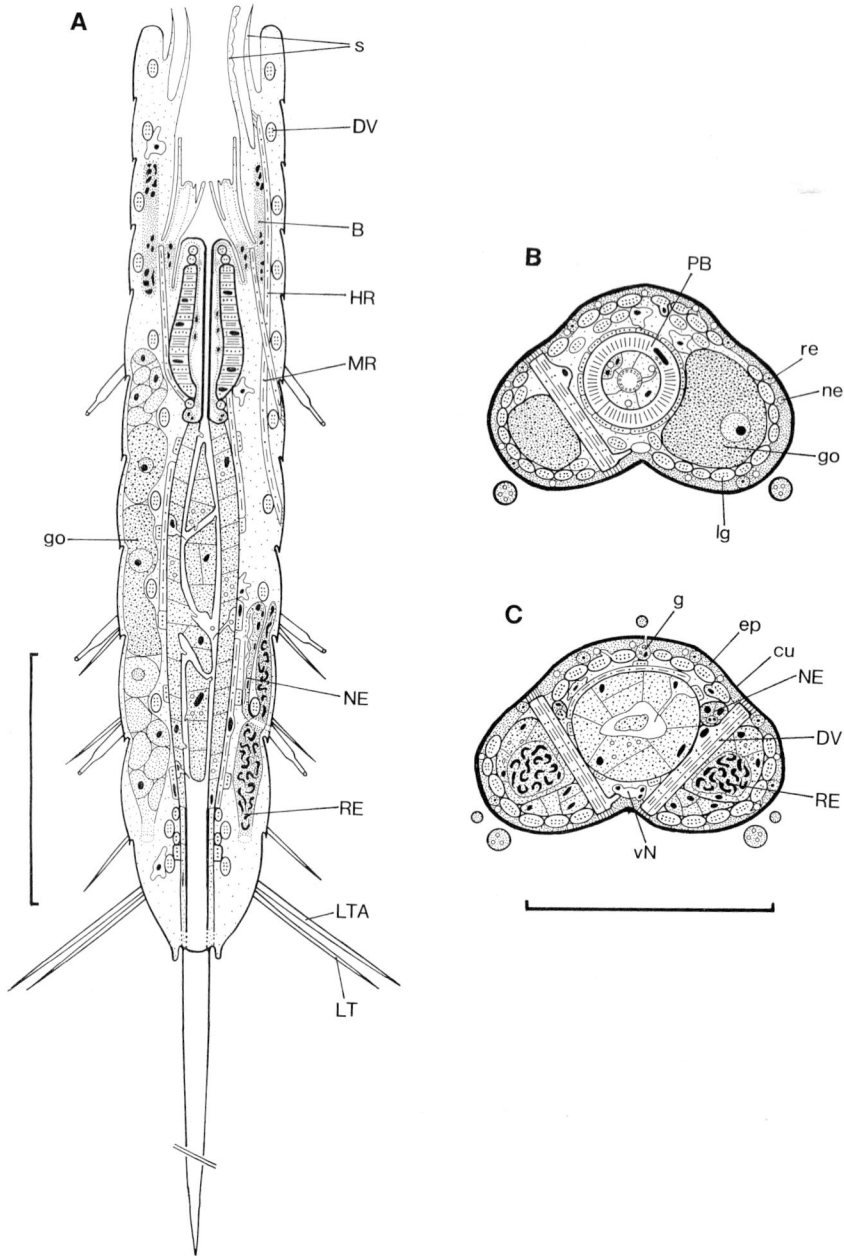

FIGURE 1   Anatomy of female *Zelinkaderes floridensis* (Cyclorhagida), head retracted. (A) Longitudinal section. Ovary, receptaculum seminis, and protonephridium drawn on one side only. (B, C) Cross sections through segment 7 (B) and segments 10 (C, left) and 11 (C, right), respectively. Scale bars = 100 µm in A and 50 µm in B and C. B, brain; cu, cuticle; DV, dorsoventral muscle; ep, epidermis; g, gland; go, gonad (ovary); HR, head retractor; lg, longitudinal muscle; LTA, lateroterminal accessory spine; LT, lateroterminal spine; MR, mouth cone retractor; ne, nerve cells; NE, protonephridium; PB, pharyngeal bulb; re, receptor; RE, receptaculum seminis; s, scalid; vN, ventral nerve cord (modified from B. Neuhaus, *Microfauna Mar.* **9**, 61–156, 1994, with permission by Fischer-Verlag).

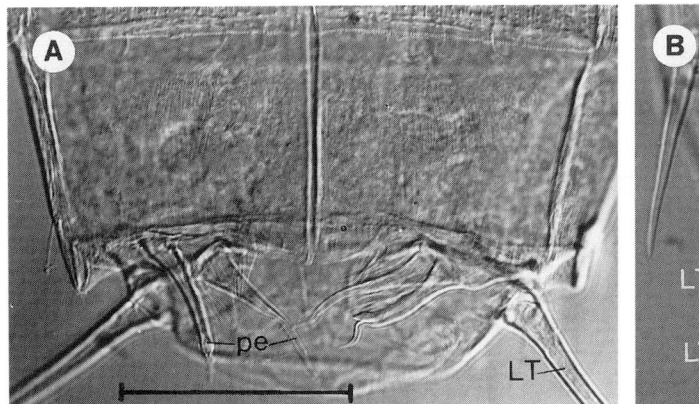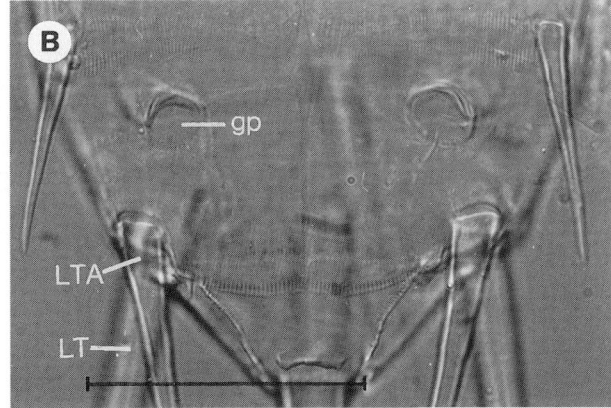

FIGURE 2 Ventral side of segments 12 and 13 of young male *Pycnophyes kielensis* (Homalorhagida) (A) and female *Antygomonas* sp. (Cyclorhagida) (B). Notice rigid penile spines in A and distinct female gonopores in B. (A, B) Differential interference contrast. Scale bar =50 μm. gp, gonopore; LTA, lateroterminal accessory spine; LT, lateroterminal spine; pe, penile spine.

lateral, or laterodorsal spines are flexible in males of many Cyclorhagida, *Paracentrophyes,* and *Neocentrophyes* but are rigid or missing in females (Fig. 2B) of these taxa. One pair of prominent, possibly adhesive, tubules exists ventrally in the second trunk segment of male *Pycnophyes* and *Kinorhynchus* only. Arrangement and number of sensory organs or tubules on the trunk may differ between males and females. Female gonopores are stronger sclerotized in the Cyclorhagida than in the Homalorhagida but quite indistinctive in males of both taxa.

III. ANATOMY OF THE REPRODUCTIVE SYSTEM

Kinorhyncha possess one pair of saccate gonads extending between the dorsoventral muscles and the epidermis. One pair of receptacula seminis joins the gonoducts from dorsally close to the genital aperture at the border of the 12th and 13th segment (Figs. 1A–1C). Duct cells show numerous microvilli anteriorly but a thin cuticular lining posteriorly. An epithelium envelops the receptacula and seems to surround the gonads in some species; extracellular matrix alone covers the gonads in other species. Ultrastructural observations are limited to *Echinoderes aquilonius, Kinorhynchus phyllotropis,* and three species of *Pycnophyes.*

A single, huge oocyte with numerous yolk vesicles and an enormous nucleus and nucleolus develops in the center of each ovary (Fig. 1A). Smaller cells anterior and posterior of the large oocyte appear to be immature germ cells which seem to be resorbed during maturation of the gonad. Sperm cells in the receptaculum seminis differ considerably from sperm in the spermatophore and testis. In the receptaculum, 20–30 irregularly shaped sperm cells exhibit condensed or uncondensed polymorphic nuclei, vesicles are reduced in size and number, and the tubes occur in the periphery of the cells. These dramatic changes in the anatomy of sperm cells led to the conclusion that *Pycnophyes flaveolatus* and *P. communis* show "morphological hermaphroditism." However, wall cells of the receptaculum may have been misinterpreted as spermatogonia which were assumed to differentiate to the sperm cells in the lumen of the receptaculum.

Mature spermatozoa are sausage shaped and measure up to one-fourth of the body length. In three species of the Homalorhagida, a central elongated nucleus is surrounded by three types of vesicles. Opposite the vesicles, tubes with central electron-dense filaments run along the entire length of each spermatozoon. A short cilium with a 9×2 + 2 axoneme originates at the distal end. An acrosomal structure is missing. One type of vesicle derives from mitochondria and may serve as a storage organelle for the female, the zygote, or for movements of the spermatozoon. The latter suggestion is supported by the observations that a middle piece is lacking and that the locomotory activity is limited to the anterior third of the cell. Spermatogonia at the anterior end of the male gonad differentiate gradually to mature sperm cells in the posterior part via spermatocytes and spermatids.

IV. REPRODUCTIVE PHYSIOLOGY AND ENDOCRINAL CONTROL

Virtually nothing is known about the physiology and endocrinal control of reproduction or about how Kinorhyncha find partners for copulation.

V. MODE OF FERTILIZATION

Most certainly, fertilization is internal. Penile spines of *Echinoderes, Pycnophyes,* and *Kinorhynchus* appear to be rigid (Fig. 2A). These spines are filled with epidermal cells and ciliary receptor cells, leaving no space for a duct; muscles do not attach to them. Possibly, penile spines are inserted into the female gonopores, if the dorsoventral muscles of the trunk contract and increase the internal body pressure. In this way, the spines may serve to keep the genital apertures open and to anchor males and females together. The flexible spines of *Paracentrophyes* and *Neocentrophyes* may function as exclusively sensory during copulation.

Females of *Pycnophyes* and *Kinorhynchus* occasion-

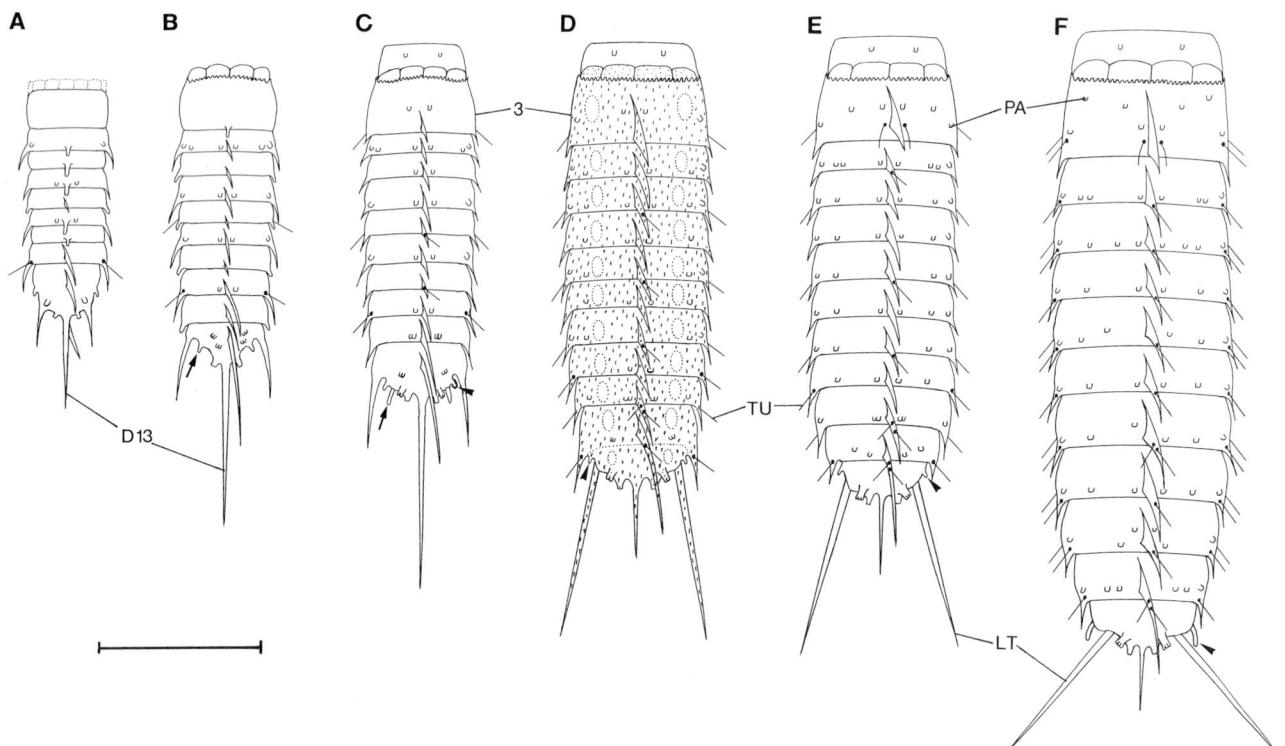

FIGURE 3 Juvenile stages 1 (A) through 6 (F) of *Paracentrophyes praedictus* (Homalorhagida) from dorsal side. Distribution of cuticular hairs and insertion area of dorsoventral muscles indicated in juvenile stage 4 (D). Arrows mark anlagen of lateroterminal spines; arrowheads point to anlagen of accessory lateroterminal spines. Scale bar =100 μm. D13, dorsal spinose process of segment 13; LT, lateroterminal spine; PA, sensory papilla; 3, segment 3 (= first trunk segment); TU, tubule (modified from Neuhaus, 1995, with permission).

ally bear a brownish, mucous mass covered by detritus at their posterior end which seems to contain one or two spherical bodies filled with about 140 intertwined spermatozoa and approximately the same number of spermatids. Consequently, this mass has been interpreted as a spermatophore.

Copulation has been perceived once in *P. kielensis* by the author. Here, the ventral posterior ends of male and female are directed toward each other, with the heads of specimens facing opposite directions. A brownish mucous mass surrounds the posterior ends. From this observation, it is concluded that the spermatophore indeed originates from the substance secreted during copulation of the kinorhynchs.

VI. MODE OF DEVELOPMENT

Eggs are probably deposited into the sediment. No information is available on the embryological development. Postembryonic ontogeny is direct. From the egg, the first of a series of six juvenile stages hatches, exhibiting 11 segments (Figs. 3A–3F). Two new trunk segments are introduced in a subcaudal growing zone during the ongoing development. The number of head spines (= scalids) as well as of trunk sensory organs and tubules increases from stage to stage. New scalids occur as spinose anlagen (= protoscalids) in a subfrontal growing zone and differentiate fully in the following juvenile stage.

Gonads start development during the last juvenile stages, but gonopores do not become visible and are not functional before the final molt to the adult stage. Growing of the gonads starts in the terminal segments and continues anteriorly. Molting of adult specimens has been reported only once in *Antygomonas oreas* and *Zelinkaderes floridensis*, respectively.

Neotenic development has been assumed for species such as *Paracentrophyes praedictus*, which possess as an adult a translucent, weakly sclerotized body cuticle without clearly defined articulations between segmental dorsal and ventral cuticular plates.

A cuticle with these features is also found in juvenile stages of species with a strongly sclerotized cuticle, such as *Echinoderes*, *Pycnophyes*, and *Kinorhynchus*. However, it has been demonstrated recently that *Paracentrophyes* does not become mature at an earlier stage of its ontogeny but rather develops via the same six juvenile stages (Figs. 3A–3F) as do other kinorhynchs. In connection with ultrastructural data, it is highly improbable that neotenic development exists in Kinorhyncha.

See Also the Following Articles

Acanthocephala; Marine Invertebrates, Modes of Reproduction in; Nematodes and Related Phyla; Nematomorpha; Priapulida; Rotifera

Bibliography

Brown, R. (1983). Spermatophore transfer and subsequent sperm development in a homalorhagid kinorhynch. *Zool. Scripta* 12, 257–266.

Higgins, R. P. (1974). Kinorhyncha. In *Reproduction of Marine Invertebrates, Vol. 1: Acoelomate and Pseudocoelomate Metazoans* (A. C. Giese and J. S. Pearce, Eds.), pp. 507–518. Academic Press, New York.

Higgins, R. P. (1990). Zelinkaderidae, a new family of cyclorhagid Kinorhyncha. *Smithsonian Contrib. Zool.* 500, 1–26.

Kristensen, R. M., and Higgins, R. P. (1991). Kinorhyncha. In *Microscopic Anatomy of Invertebrates, Vol. 4: Aschelminthes* (F. W. Harrison and E. E. Ruppert, Eds.), pp. 377–404. Wiley-Liss, New York.

Needham, A. E. (1989). Kinorhyncha. In *Reproductive Biology of Invertebrates, Vol. IV, Part A: Fertilization, Development, and Parental Care* (K. G. Adiyodi and R. G. Adiyodi, Eds), pp. 207–217. Wiley, New York.

Neuhaus, B. (1995). Postembryonic development of *Paracentrophyes praedictus* (Homalorhagida): Neoteny questionable among the Kinorhyncha. *Zool. Scripta* 24, 179–192.

Nyholm, K.-G., and Nyholm, P.-G. (1983). Kinorhyncha. In *Reproductive Biology of Invertebrates, Vol. II: Spermatogenesis and Sperm Function* (K. G. Adiyodi and R. G. Adiyodi, Eds.), pp. 207–220. Wiley, New York.

Klinefelter's Syndrome

Fady I. Sharara

University of Maryland School of Medicine

GLOSSARY

fluorescent in situ hybridization A molecular technique that uses chromosome-specific DNA probes for the detection of abnormal chromosomal numbers.

germ cells Precursor cells located in the seminiferous tubules that give rise to spermatozoa.

intracytoplasmic sperm injection A laboratory technique used in men with severe male factor infertility to aid fertilization. It involves the use of a single spermatozoon that is microinjected into the oocyte's cytoplasm using a micromanipulator.

karyotype The chromosomal composition of the analyzed cells. It is usually performed on T lymphocytes and skin fibroblasts.

Klinefelter's syndrome is a clinical entity characterized by gynecomastia, tall extremities, small atrophic testicles, elevated gonadotropins, and usually azoospermia. The underlying defect is the presence of an extra X chromosome.

I. INTRODUCTION

This syndrome was first described by Klinefelter *et al.* in 1942. The etiology of this syndrome was unknown until 1956 when several investigators found that a high proportion of men with this syndrome were X-chromosome positive, but it was not until 1959 that the first 47,XXY chromosomal complement was first reported. Since the initial publication by Klinefelter, it has become clear that this is a heterogeneous syndrome with variable clinical and histologic manifestations.

II. CLINICAL MANIFESTATIONS

A. Frequency and Genetic Findings:

Klinefelter's syndrome is found in 1:400–1:500 newborn males. Unlike Turner's syndrome, its incidence in spontaneous miscarriages is only 0.1%. The classic variety has the 47,XXY karyotype and the most common mosaic complement is 46,XY/47,XXY. The extra X chromosome can arise through nondisjunction of the sex chromosomes during the first or second meiotic division in either parent, or, less frequently, from mitotic nondisjunction in the zygote at or after fertilization. Both X chromosomes are of maternal origin in two-thirds of the cases and one X is of paternal origin in one-third of the cases. This suggests that nondisjunction during oogenesis is the most frequent etiology. This is supported by the higher incidence of Klinefelter cases with advanced maternal age. Paternal nondisjunction is not age dependent.

The diagnosis of mosaicism is established when at least 5% of cells in blood, skin, or gonads are 46,XY in which the second cell line is 47,XXY. 46,XY/47,XXY arises from nondisjunction or anaphase lag in a 47,XXY zygote. In mosaic 46,XY/47,XXY, some cell lines are 46,XY and others are 47,XXY.

B. Clinical Features

Characteristic findings are shown in Table 1. Children with Klinefelter's syndrome have lower birth weights, smaller head circumferences, and a slightly increased rate of major and minor congenital anomalies, especially clindactyly. As these children get

TABLE 1
Features of Classic Kleinfelter's Syndrome

Karyotype	47,XXY
Inheritence	Sporadic, associated with advanced maternal age
Incidence	1:400–1:500
Genitalia	Male, small phallus
Gonads	Small, firm testes; extensive hyalinization of seminiferous tubules; azoospermia; relative Leydig cell hyperplasia
Habitus	Poor to normal virilization, gynecomastia, disproportionate long legs
Hormones	Normal to low testosterone, elevated gonadotropins, normal to elevated estradiol, elevated SHBG
Associated findings	Diabetes, breast cancer, varicose veins

older, their height percentiles increase, mainly because of disproportionate increase in growth of long bones, resulting in a span : height ratio <1. This increase in long bone growth is not related to hypoandrogenism and delayed epiphyseal closure. There is a slight impairment in verbal IQ, leading to a higher incidence of problems with speech development and causing aberrant social behavior and personality disorders. Mental retardation is not common except in those with the poly-X syndrome.

Bilateral gynecomastia is present in up to 90% of patients secondary to an increase in free estradiol. However, the gynecomastia in men with Klinefelter's syndrome is unusual in that it is caused by hyperplasia of the interductal tissue as opposed to hyperplastic ducts in gynecomastia caused by estrogen excess. The androgen concentrations are normal to low, resulting in decreased facial hair growth, decreased muscle mass, and a small phallus. With decreasing testosterone production with aging, impotence becomes common.

Clinical manifestations are variable in mosaic variants. In the 46,XY/47,XXY variant, testicular size is larger, testosterone production is higher, and spermatogenesis and fertility can occur. Patients with the poly-X variants have more severe abnormalities than the classic variant, with a high incidence of mental retardation and somatic anomalies.

C. Testicular Findings

In the classic variant, testicular failure occurs in 80% of affected individuals usually between 30 and 40 years of age. The testicles are small and firm with a volume <3 cc. Histologic changes include extensive seminiferous tubule hyalinization with azoospermia and relative Leydig cell hyperplasia. However, there is some variation among patients and even among testes in the same patient. These findings become more pronounced with aging. It appears that a normal or near-normal complement of germ cells is present in early fetal life, but accelerated atresia of germ cells occurs secondary to the extra X chromosome. Rarely, spermatogenesis is noted in few isolated tubules. In these cases, germ cell mosaicism could be present.

Even before the onset of puberty, progressive hyalinization is noted resulting in small, firm testes. With the degeneration of the seminiferous tubules and the decreased to absent elastic tissue in the peritubular compartment, there is relative Leydig cell hypertrophy. These histologic changes progress with age, and tubules are rarely identified in older patients.

D. Endocrinologic Changes

The plasma concentrations of follicle-stimulating hormone and luteininizing hormone (LH) and the LH response to gonadotropin-releasing hormone are normal before puberty. With the onset of puberty, there is a progressive decrease in testosterone production and an accelerated destruction of germ cells resulting in steadily increasing gonadotropin levels. Testosterone production in response to hCG diminishes with aging. Serum estradiol concentrations are normal to high. The role that elevated gonadotropins play in accelerating the testicular changes is unknown, and it is interesting to note that 47,XXY patients with gonadotropin deficiency do not exhibit the characteristic changes in testicular histology. In addition, central precocious puberty is associated with a more rapid destruction of seminiferous tubules compared to Klinefelter boys without precocious puberty.

E. Associated Anomalies

There is a higher incidence of diabetes mellitus in men with Klinefelter's syndrome, which is usually mild and diagnosed before the age of 50. Clinically significant thyroid disease is uncommon, despite a decreased thyroid response to provocative stimuli. There is also a higher incidence of breast cancer with infiltrating ductal carcinoma being the most common neoplasm. A higher incidence of varicose veins with stasis ulcers is seen, which is increased in patients with the poly-X variety. A higher incidence of germ cell neoplasms is also reported.

III. KLINEFELTER VARIANTS

Variants of the classic syndrome have been reported. These include 46,XY/47,XXY, 48,XXYY, the poly-X syndromes (48,XXXY and 49, XXXYY), and 46,XX males. As noted previously, the severity of the clinical and histologic changes are exaggerated with increasing X-chromosome numbers, especially mental retardation and somatic anomalies.

IV. FERTILITY POTENTIAL

A. Natural Fertility

Men with Klinefelter's syndrome are usually azoospermic and therefore sterile. However, as discussed previously, azoospermia is not a constant feature even in men with the classic 47,XXY karyotype, and pregnancies with proven paternity have been reported. Spermatogenesis is noted in up to 7% in men with the classic variant and in 50% of those with mosaic Klinefelter. Pregnancies in the 46,XY/47,XXY mosaic variant are more common compared to the classic variety.

B. Results With Intracytoplasmic Sperm Injection

Intracytoplasmic sperm injection (ICSI) has been successfully used in men with severe male factor infertility. This technique has also been applied using testicular sperm. Harari and co-workers (1995) showed a high fertilization rate in a 35-year-old man

with "mosaic" Klinefelter (however, 96% of the cells evaluated showed a 47,XXY karyotype with two cells showing 46,XY and two showing a 48,XXXY karyotype). The authors evaluated two cycles in the 28-year-old spouse. In the first cycle, seven oocytes were retrieved, and four were injected. All four were fertilized; however, three preembryos failed to cleave by 60 hr. One preembryo was replaced but no pregnancy ensued. In the second cycle, eight oocytes were retrieved and four were mature and injected. All four fertilized, and one preembryo failed to cleave. Two preembryos were replaced and one was frozen. Again, the patient failed to conceive.

Staessen *et al.* (1996) performed ICSI on three men with the classic variant and subjected all resulting preembryos to preimplantation genetic diagnosis using fluorescent *in situ* hybridization prior to replacement. Fertilization rate of the 22 injected oocytes was 42%, and five preembryos (62.5%) reached the 6-cell stage and were biopsied. All were normal. One patient had a biochemical pregnancy, and the others failed to conceive.

First Deliveries Using ICSI

ICSI has revolutionized the treatment of severe male infertility, and it was only a matter of time before the first successful birth to men with classic Klinefelter's syndrome was reported. Two separate groups, one in Australia and the other in the United States, recently reported successful births using ICSI. In the first reported case, Bourne *et al.* used frozen sperm from multiple ejaculated samples from a 29-year-old man with the classic variant to achieve the first birth. In addition, the twin pregnancy was obtained by the transfer of two frozen pre-embryos. The couple had refused prenatal genetic testing, but both male and female infants had normal karyotypes. The same group reported that another woman, whose male partner had classic Klinefelter's syndrome conceived but had an ectopic pregnancy.

The second case was reported by Palermo *et al.* in which testicular biopsies were obtained in two men with classic Klinefelter's syndrome. The first couple conceived on their second attempt and had a singelton gestation. Amniocentesis at 20 weeks showed a normal 46,XY karyotype, and a healthy male was delivered at 38.5 weeks. The second couple had a twin gestation. Amniocentesis at 20 weeks showed normal 46,XY and 46,XX karyotypes, and a healthy male and female were delivered at 35.5 weeks. The authors also reported an ongoing pregnancy in another couple with classic Klinefelter's syndrome. None of the couples agreed to preimplantation genetic diagnosis. Of interest, the men were treated with testolactone (an aromatase inhibitor) in an attempt to optimize sperm production.

V. FUTURE DIRECTIONS

Men with Klinefelter's syndrome have a heterogeneous spectrum, depending on their karyotypes and age at presentation. This explains the various seminal and testicular parameters seen in the literature. The introduction of ICSI could present a powerful tool to help those men with evidence of spermatogenesis father a child. However, we must ensure that the conceptus does not suffer from the introduction of abnormal genetic material. ICSI in patients with this syndrome should only be performed in centers with proven success in preimplantation genetic diagnosis.

See Also the Following Articles

Gynecomastia; Reproductive Technologies; Sterility

Bibliography

Bourne, H., Stern, K., Clarke, G., Pertile, M., Speirs, A., Gordon-Baker, H. W. (1997). Delivery of normal twins following the intracytoplasmic injection of spermatozoa from a patient with 47,XXY Klinefelter's syndrome. *Hum. Reprod.* **12,** 2447–450.

Cozzi, J., Cheveret, E., Rousseaux, S., *et al.* (1994). Achievement of meiosis in XXY germ cells: A study of 543 sperm karyotypes from an XY/XXY mosaic patient. *Hum. Reprod.* **9,** 533–537.

Foss, G. L., and Lewis, F. J. (1971). A study of four cases with Klinefelter's syndrome showing motile spermatozoa in their ejaculates. *J. Reprod. Fertil.* **25,** 401–408.

Grumbach, M. M., and Conte, F. A. (1992). Disorders of sexual differentiation. In *Williams Textbook of Endocrinology,* 8th ed., pp. 879–884. Saunders, Philadelphia.

Harari, O., Bourne, H., Baker, G., Gronow, M., and Johnston, I. (1995). High fertilization rate with intracytoplasmic sperm injection in mosaic Klinefelter's syndrome. *Fertil. Steril.* **63,** 182–184.

Laron, Z., Dickerman, Z., Zamir, R., and Galatzer, A. (1963). Paternity in Klinefelter's syndrome. *Lancet* **1,** 506.

Palermo, G. D., Schlegel, P. N., Sills, E. S., Veeck, L. L., Zaninovic, N., Menendez, S., Rosenwaks, Z. (1998). Births after intracytoplasmic injection of sperm obtained by testicular extraction from men with nonmosaic Klinefelter's syndrome. *N. Engl. J. Med.* **338,** 588–590.

Paulsen, C. A., Gordon, D. L., Carpenter, R. W., *et al.* (1968). Klinfelter's syndrome and its variants: A hormonal and chromosomal study. *Recent Prog. Horm. Res.* **24,** 321–363.

Staessen, C., Coonen, E., Van Assche, E., Tournaye, H., *et al.* (1996). Preimplantation diagnosis for X and Y normality in embryos from three Klinefelter patients. *Hum. Reprod.* **11,** 1650–1653.

Terzol, G., Lalatta, F., Lobbiani, A., Simoni, G., and Colucci, G. (1992). Fertility in 47,XXY patient: Assessment of biological paternity by deoxyribonucleic acid fingerprinting. *Fertil. Steril.* **58,** 821–822.

Tournaye, H., Staessen, C., Liebaers, I., Van Assche, E., *et al.* (1996). Testicular sperm recovery in nine 47,XXY Klinefelter patients. *Hum. Reprod.* **11,** 1644–1649.

Knockout Organisms

see Transgenic Animals

Labor and Delivery, Human

P. W. Nathanielsz and Gordon C. S. Smith

Cornell University

GLOSSARY

amniotomy Artificial rupture of the fetal membranes.

anencephaly Congenital absence of the brain.

Braxton–Hicks contractions Low-amplitude uterine contractions in human pregnancy that occur throughout pregnancy and give way to labor contractions at delivery; probably correspond to contractures recorded directly from the myometrium of pregnant nonhuman primates.

cervical effacement The process by which the uterine cervix changes from a 3-or 4-cm fibrous cylinder to a lumen in continuity with the lower pole of the uterus.

contractions Efficiently coordinated epochs of myometrial activity that occur at labor, generally lasting <1 min.

contractures Long-lasting, low-amplitude epochs of myometrial activity that occur throughout pregnancy and can be recorded in experimental animals using electromyogram electrodes or intrauterine pressure.

hypothalamo–pituitary–adrenal axis A collective term for the hypothalamus, pituitary gland, and adrenal gland whose endpoint is secretion of steroid hormones from the adrenal gland.

prostaglandins Derivatives of arachidonic acid formed by the enzyme, cyclooxygenase.

I. THE PROCESS OF LABOR

A. Definition

Labor is a state characterized by regular painful uterine contractions associated with dilatation of the cervix leading to expulsion of the conceptus. The term "labor" is usually only used at gestational ages associated with fetal viability: In the first half of pregnancy expulsion of the conceptus is usually referred to as spontaneous abortion.

B. Stages of Labor

Human labor is conventionally divided into three stages: The first stage occurs from its onset to achieving full dilatation of the cervix, the second stage occurs from achieving full dilatation of the cervix to delivery of the fetus, and the third stage occurs from delivery of the fetus to delivery of the placenta, fetal membranes, and umbilical cord.

1. The First Stage of Labor

The onset of labor is sometimes very difficult to establish with certainty. Many women present both preterm and at term with symptoms of the onset of labor, including cyclical pain, passing of cervical mucus vaginally, or suspected loss of liquor. In the presence of such ambiguity, the diagnosis may become apparent or might be confirmed by serial vaginal examination. Progressive dilatation and effacement of the cervix establishes the diagnosis. However, cervical dilatation initially may progress slowly. This has been referred to as the latent phase of the first stage of labor and is highly variable in its duration.

Once established, however insidious the onset, labor is characterized by painful contractions typically lasting <1 min with a frequency of three or four contractions in a 10-min period. The rate of progress is highly variable between women, but progress is

TABLE 1
Progression of Spontaneous Labor[a]

Parameter	Mean or median	5th Centile
Nulliparas		
Total duration	10.1 hr	25.8 hr
Stages		
First	9.7 hr	24.7 hr
Second	33.0 min	117.5 min
Third	5.0 min	30.5 min
Latent phase (duration)	6.4 hr	20.6 hr
Maximal dilatation (rate)	3.0 cm/hr	1.2 cm/hr
Descent (rate)	3.3 cm/hr	1.0 cm/hr
Multiparas		
Total duration	8.2 hr	19.5 hr
Stages		
First	8.0 hr	18.8 hr
Second	8.5 min	46.5 min
Third	5.0 min	30.0 min
Latent phase (duration)	4.8 hr	13.6 hr
Maximal dilatation (rate)	5.7 cm/hr	1.5 cm/hr
Descent (rate)	6.6 cm/hr	2.1 cm/hr

[a] Reproduced with permission from Gabbe et al., 1991.

generally more rapid in those who have labored previously (Table 1). Uterine contractions push the presenting part of the fetus against the cervix, promoting its dilatation. As labor progresses, the cervix is incorporated into the lower segment of the uterus and the presenting part, usually the fetal head, provides a stable point for the rest of the uterus to exert tension on the cervix and lower uterine segment. The fetal head descends into the pelvis and this is usually associated with flexion of the fetal neck. The fetal occiput normally turns to face the anterior abdominal wall of the mother as it descends into the pelvis. If the fetal neck is partially extended, the fetal occiput will tend to rotate to face the maternal sacrum (occipitoposterior position), which often delays the progress of the first and second stages of labor. If the neck is very extended, the fetus may present by the face.

2. The Second Stage of Labor

Once the cervix is fully dilated, uterine contractions continue to push the fetus into the pelvis. As the head delivers, the maternal perineum is stretched,

the fetal neck extends, and finally the head is delivered. After delivery of the head, the fetus rotates and the shoulders are delivered with the head facing either maternal leg. The head and shoulders represent the main physical obstruction to delivery of a normal fetus.

3. The Third Stage of Labor

Following delivery of the fetus, the uterus contracts and reduces its volume considerably. Contraction favors separation of the placenta and membranes, which are usually delivered soon after the fetus (in terms of minutes) but which can be delayed (hours or even days in the absence of medical intervention). It is during the third stage that considerable bleeding is most likely to be encountered, and this is usually due to failure of the uterus to contract completely following delivery. This can be prevented and treated by a range of drugs promoting uterine contraction.

II. THE PHYSIOLOGICAL INITIATION OF LABOR

The control systems which initiate labor in the human remain obscure. It is not even established with certainty whether labor is normally initiated by the fetus, the other products of conception, or the mother. Given the difficulties involved in invasive studies on pregnant women, this level of uncertainty is perhaps not surprising. However, animal models have shed considerable light on the potential mechanisms regulating parturition in women.

A. Changing Patterns of Uterine Activity

Studies in several species of nonhuman primate have demonstrated that the uterus is contracting throughout pregnancy. The pattern of activity is referred to as contractures, which are low-amplitude (i.e., a small rise in intrauterine pressure), relatively prolonged periods (up to 15 min) of uterine activation. The human correlate of these is probably the Braxton–Hicks contraction. Labor is associated with a switch to high-amplitude, short-duration (<1 min) activity, referred to as contractions. In pregnant non-human primates, labor is preceded by nocturnal switches from contractures to contractions in the days leading up to labor. The clinical correlate of this is probably women who present with "false alarms" prior to presenting in established labor. It is also noted in the human prior to induction that the cervix of some women has already undergone some softening and dilatation. It has been suggested that there is a prelabor state characterized by increasing uterine activity secondary to increasing sensitivity to oxytocin. The 24-hr rhythms of uterine activity in pregnant women warrant further study.

B. The Physiological Basis of Contractions

One of the key features of uterine activity in labor is that contractions are coordinated as well as forcible. It has been suggested that there is a pacemaker in the human uterus and this is located in the cornua. If this is the case, it is much less well-defined than in the heart. Continuing the parallel with the heart, the activating signal, wherever initiated, needs to be propagated rapidly to allow synchronous contraction of the whole uterus. Propagation of activation between myocytes is effected by gap junctions, which are complex protein structures which allow the passage of electrical signals between adjacent uterine smooth muscle cells. The number of gap junctions connecting uterine muscle cells increases with labor and there has been no state of labor described in which gap junction expression did not increase. However, gap junctions merely allow the propagation of a contractile signal, and other factors must initiate contraction. There are many hormones which promote uterine contraction, but the main factors identified thus far are prostaglandins and oxytocin.

Prostaglandins are oxygenated derivatives of the 20-carbon chain fatty acid arachidonate and their main precursor is formed from arachidonate by the enzyme cyclooxygenase. Most prostaglandins act to stimulate uterine activity and cervical dilatation, but the human and primate uterus, in contrast to that of the sheep, also expresses prostaglandin receptors coupled to inhibitory pathways. The main sources of prostaglandins within the uterus are the fetal membranes (especially the amnion), the decidua, and the placenta. Toward term the fetus has high circulating concentrations of prostaglandin E_2 (PGE_2) which may have an endocrine role in promoting parturition. Administration of exogenous PGE_2 and $PGF_{2\alpha}$ can stimulate labor throughout gestation and inhibitors of prostaglandin synthesis prolong gestation and are used therapeutically in the treatment of preterm labor.

Oxytocin is a polypeptide hormone secreted by the maternal posterior pituitary. It contracts the uterus and the sensitivity of the uterus to this effect increases with advancing gestation. Nocturnal increases in maternal plasma oxytocin concentrations have been demonstrated in late pregnancy in women and rhesus monkeys. These surges in oxytocin are responsible for the nocturnal myometrial switch from contractures to contractions. In monkeys and baboons this switch can be inhibited by oxytocin antagonists. Oxytocin is also produced by the decidua in human pregnancy. Oxytocin stimulates prostaglandin production. Prostaglandins cause further myometrial contraction. Furthermore, release of oxytocin is stimulated by the pressure of the fetal head on the maternal pelvic floor (Fergusson reflex). This is an example of one of the many positive feed-forward loops that rapidly increase the rate of progress of normal term labor.

C. The Endocrinology of Parturition

In contrast to the human, the initiation of labor in the sheep is clearly understood. It is clear that in the sheep the trigger for the initiation of labor comes from the fetus. The fetal paraventricular nucleus stimulates the fetal pituitary, and the pituitary stimulates the fetal adrenal to produce cortisol. Cortisol induces the production of placental enzymes (principally 17α hydroxylase) which result in an increased ratio of estrogen to progesterone in maternal and fetal plasma. Increasing estrogen and decreasing progesterone promote uterine contraction both directly (e.g., by depolarizing uterine smooth muscle cells) and indirectly by increasing sensitivity to contractile agonists, such as prostaglandins and oxytocin. It is possible that in certain circumstances (e.g., intrauterine growth retardation) physiological stress of the fetus activates the fetal hypothalamo–pituitary–adrenal (HPA) axis and promotes preterm delivery. Administration of exogenous corticosteroid stimulates labor in the sheep, and interruption of the fetal HPA axis delays the onset of labor at term. Corticosteroid also stimulates maturation of numerous fetal systems for the critical adaptations that occur at birth.

In pregnant women, there is no fall in circulating concentrations of progesterone prior to labor. Furthermore, although estrogen rises toward term, there is no abrupt rise in estrogen as occurs in the sheep. The human placenta lacks the enzyme 17α hydroxylase. Neither cortisol nor estrogen induce labor in the human, and progesterone does not inhibit labor.

A number of "natural experiments" are cited which may shed light on the control systems involved in initiating labor in women. However, as one might expect, these do not give definitive information due to confounding variables. For instance, an anencephalic human fetus necessarily has an interrupted HPA. The duration of gestation is not consistently affected in pregnancies with such complications, although very prolonged pregnancy is a clearly recognized complication of anencephaly. However, this condition can be complicated by excessive liquor (polyhydramnios). Excessive stretch stimulates the uterus and may precipitate labor. Pregnancy may only be prolonged in anencephalics in the absence of polyhydramnios.

In order to understand term labor in women, it is worth searching for parallels between primates and sheep. For example, it is clear in both species that corticosteroids play an essential role in maturing vital fetal organ systems. As in the sheep, progesterone is critical in the human for maintaining uterine quiescence throughout gestation because administration of a progesterone receptor antagonist (such as mifepristone, RU 486) profoundly increases uterine sensitivity to contractile agonists in all three trimesters. Various cellular mechanisms of progesterone withdrawal have been proposed to explain the absence of any fall in plasma progesterone preceding labor in the human. These potential functional progesterone withdrawal mechanisms include altered local synthesis of $17\beta,\alpha$-hydroxysteroid dehydrogenase, proges-

terone receptor, heat shock proteins, or modulation of progesterone-responsive genes by other procontractile agonists (such as transforming growth factor-β) The possible role of these mechanisms in the initiation of labor in the human and nonhuman primate remain to be resolved.

While there is no evidence for withdrawal of the maintenance function of progesterone, there are very clear parallels between the sheep and human in the case of estrogen. A clear, marked increase in late-gestation maternal estrogen concentrations has been demonstrated in pregnant women, monkeys, and baboons. In primates, estrogen is synthesized from precursors formed in the fetal adrenal. The primate adrenal is very large relative to its body weight compared with that of the adult and has a unique "fetal zone." It synthesizes large quantities of androgens which are converted in the placenta to estrogens. There is striking similarity between the increase in fetal plasma androgen in the rhesus monkey and fetal cortisol in the sheep. There is a rise in the circulating concentrations of estrogen approaching term and the ratio of unconjugated estriol to progesterone (measured in maternal saliva) increases before both term and preterm labor. While administration of exogenous estrogen does not induce labor in the human or nonhuman primate, maternal administration of androstenedione, an androgenic precursor of estrogen, induces preterm labor in the rhesus monkey. It is proposed that the production of estrogen at a site local to its synthesis is critical. It is noteworthy that topical estradiol is as effective as PGE_2 in priming the cervix prior to induction of labor. The view that estrogen is the factor that converts the androgen signal from the fetus to permit stimulation of the essential mechanisms of labor is supported by firm evidence that estrogen stimulates many of the factors necessary for successful completion of labor—oxytocin, oxytocin receptors, connexins, and prostaglandin production.

While the precise events initiating labor in the human remain to be resolved, it seems likely that increasing maternal estrogen from androgenic precursors from the fetal adrenal plays a central role. Given the strong stimulatory mechanisms put in motion by estrogen, it is possible to conceive of a situation in which progesterone withdrawal is not necessary. The maintenance function is simply overcome.

III. THE THERAPEUTIC INDUCTION OF LABOR

A. Surgical Induction

Surgical induction of labor is effected by artificial rupture of the fetal membranes (amniotomy). This can be difficult in women who have not previously labored because the cervix is often completely closed prior to labor. Although rupture of the fetal membranes can be used to induce labor, this is not a necessary condition for the initiation of labor or even for delivery of the fetus because delivery of the infant in an intact sac is well recognized.

B. Pharmacological Induction

The characterization and synthesis of oxytocin allowed its use to stimulate uterine activity in the induction of labor. Pharmacological induction was advanced further with the availability of synthetic prostaglandins. Both PGE_2 and $PGF_{2\alpha}$ can be used to induce labor, although the former is by far the more potent and widely employed. PGE_2 is generally used topically on the cervix. It is particularly useful in women with an unfavorable cervix (i.e., long, firm, and closed) because it alters the connective tissue of the cervix to promote dilatation and effacement. PGE_2 can also be used vaginally in the presence of a favorable cervix as an adjunct to amniotomy and oxytocin. Its use in this context is associated with a reduced incidence of assisted vaginal delivery and cesarean section. PGE_2 can also be administered intravenously or intraamniotically. However, these routes are usually only used in the first and second trimester and even then have been superseded by synthetic prostaglandin derivatives administered vaginally.

A number of other agents are being investigated for the induction of labor, in particular progesterone receptor blockers such as mifepristone. These drugs induce labor in a proportion of women at term and

increase uterine sensitivity to oxytocin and PGE_2. Such drugs are particularly useful when inducing labor in the second trimester, usually in the context of fetal death or termination of pregnancy.

Drugs acting on the uterus (principally oxytocin, ergometrine, and/or synthetic prostanoids) are central to the management of the third stage of labor in the prevention of primary postpartum hemorrhage (PPH). The use of an oxytocic decreases the incidence by about 40%. In the developing world, about 125,000 women die from PPH each year. This is due to lack of access to both blood transfusion and oxytocics.

IV. NORMAL AND ABNORMAL DELIVERY

The basic mechanisms of an occipitoanterior (OA) vertex delivery were described in Section I,B,2. The fetus may also be delivered presenting by the vertex (i.e., the area bounded by the parietal eminences and anterior and posterior fontanelles) with the head in an occipitoposterior (OP) orientation. The fetal head is not a perfect sphere and, therefore, its dimensions differ according to its presentation and position. An OP delivery results in a wider diameter of the fetal head passing through the maternal pelvis than an OA delivery, and the former is associated with prolonged labor and difficult or even impossible delivery.

The fetal head may also be arrested in a transverse position (i.e., occipitolateral). In cases in which malposition of the fetal head obstructs delivery, rotation may be effected manually, by forceps, or by using vacuum extraction. Frequently, malposition is associated with ineffective uterine activity and perhaps the most effective treatment is prevention by use of an oxytocin infusion as an early response to slow progress in the first stage of labor. Manual rotation would usually be performed immediately prior to a forceps delivery. The operator's fingers use the sutures of the fetal skull gently to exert rotational force and place the fetal head in OA or near OA position. Conventional obstetric forceps (metal instruments consisting of a blade, shank, and handle) are then applied to either side of the fetal head over its parietal bones exerting traction on the occiput and delivery is effected by traction with uterine contractions. Specialized forceps have been developed for rotation; probably the most widely utilized are Kielland's forceps. The shank and handles of these forceps are in a straight line to prevent trauma to maternal tissues during rotation. They are probably the most effective means of delivering a fetus vaginally which is truly stuck due to malposition. The vacuum extractor consists of a cup (formerly metal but now more commonly silastic) which is placed on the fetal vertex and connected by a flexible tube to an adjustable vacuum. A maximum vacuum of -0.8 kg/cm^2 is applied and often rotation occurs with appropriate degree and direction of traction. Forceps and vacuum are commonly employed in nonrotational-assisted deliveries. Indications include maternal exhaustion and fetal heart rate patterns indicative of fetal distress.

In 3 to 4% of labors at term, the fetus presents by the breech. The frequency of breech presentation varies inversely with gestational age (e.g., 25% are breech at 28 weeks or less). There is no consensus on the management of breech presentation. Many obstetricians favor elective cesarean section. There is an increased perinatal mortality associated with vaginal breech delivery, but this is at least in part secondary to associations with prematurity and fetal abnormality. The principal hazards to the otherwise healthy fetus are the potential for delivery through an incompletely dilated cervix and the fact that the head is a major physical obstruction to delivery and a narrow pelvis or large baby may only be apparent after delivery of the legs and body. When diagnosing a breech at term, it is important to exclude important causes such as fetal abnormality or a low-lying placenta. The art of delivering a breech is beyond the scope of this chapter. However, as with vertex deliveries, flexion of the fetal head ensures the more favorable diameters of the skull. Extension of the neck can result from misguided efforts to exert traction following delivery of the legs or lower body. Many operators favor the use of forceps for the aftercoming head and there is evidence that this improves the oxygenation of the baby.

Cesarean section, delivery of the fetus by a surgical procedure through the maternal abdomen, is commonly performed in the course of labor, frequently for slow or arrested progress. Some of these procedures may be prevented by judicious use of oxytocin to augment uterine activity. Cesarean section may be performed for more immediate causes such as asphyxia of the baby. In some cases, such as prolapse of the umbilical cord, the risk is unambiguous. One of the most common indications is fetal distress as diagnosed by fetal heart rate monitoring. In most cases there is little evident distress in the newborn and the widespread use of the procedure for this indication is often criticized. The rate of cesarean section for fetal distress can be reduced by sampling of fetal scalp blood for hydrogen ion concentration.

See Also the Following Articles

Apgar Score; Cervix; Cesarean Delivery; Oxytocin; Postdate (Postterm) Pregnancy; Preterm Labor and Delivery; Uterine Contraction

Bibliography

Beazley, J. M. (1995). Normal labour and its active management. In *Dewhurst's Textbook of Obstetrics and Gynaecology for Postgraduates* (C. R. Whitfield, Ed.). Blackwell, Oxford, UK.

Buster, J. E., Ghang, R. H., Preston, D. L., Elashoff, R. M., Cousins, L. M., Abraham, G. E., Hotel, C. J., and Marshall, J. R. (1979). Interrelationships of circulating maternal steroid concentrations in third trimester pregnancies II. C18 and C19 steroids: Estradiol, estriol, dehydroepiandrosterone, dehydroepiandrosterone sulfate, 5-androstenediol, 4-androstenedione, testosterone, and dihydrotestosterone. *J. Clin. Endocrinol. Metab.* 48, 139–142.

Challis, J. R. G., and Lye, S. L. (1994). Parturition. In *The Physiology of Reproduction* (E. Knobil and J. D. Neill, Eds.), 2nd ed., pp. 985–1031. Raven Press, New York.

Chard, T., and Grudzinkas, J. G. (Eds.) (1994). *The Uterus.* Cambridge Univ. Press, Cambridge, UK.

Drife, J. O., and Calder, A. A. (Eds.) (1992). *Prostaglandins and the Uterus.* Springer-Verlag, London.

Enkin, M., Keirse, M. J. N. C., Renfrew, M., and Neilson, J. (1995). *A Guide to Effective Care in Pregnancy and Childbirth,* 2nd ed. Oxford Univ. Press, Oxford, UK.

Gabbe, S. G., Niebyl, J. R., and Simpson, J. L. (Eds.). (1991). *Obstetrics: Normal and Program Pregnancies,* 2nd ed., p. 428. Churchill Livingston, London.

Lopez-Bernal, A., Europe-Finner, N., Phaneuf, S., and Watson, S. P. (1995). Pre-term labor: A pharmacological challenge. *Trends Pharmacol. Sci.* **16**, 129–133.

McGregor, J. A., Jackson, G. M., Lachelin, G. C. L., Goodwin, T. M., Artal, R., Hastings, C., and Dullien, V. (1995). Salivary estriol as risk assessment for preterm labor: A prospective trial. *Am. J. Obstet. Gynecol.* **173**, 1337–1342.

Mecenas, C. A., Giussani, D. A., Owiny, J. R., Jenkins, S. L., Wu, W. X., Honnebier, M. B. O. M., Lockwood, C. J., Kong, L., Guller, S., and Nathanielsz, P. W. (1996). Production of premature delivery in pregnant rhesus monkeys by androstenedione infusion. *Nat. Med.* **2**(4), 443–448.

Mitchell, M. D., Romero, R. J., Edwin, S. S., and Trautman, M. S. (1995). Prostaglandins and parturition. *Reprod. Fertil. Dev.* **7**, 623–632.

Nathanielsz, P. W. (1996). *Life before Birth: The Challenges of Fetal Development.* Freeman, New York.

Olson, D. M., Mijovic, J. E., and Sadowsky, D. W. (1995). Control of human parturition. *Sem. Perinatol.* **19**, 52–63.

Romero, R., Mazor, M., Wu, Y. K., Sirtori, M., Oyarzun, E., Mitchell, M. D., and Hobbins, J. C. (1988). Infection in the pathogenesis of preterm labor. *Sem. Perinatol.* **12**, 262–279.

Lactational Amenorrhea

Amy Banulis and William Schlaff

University of Colorado Health Sciences Center

I. Introduction
II. Intrapartum and Postpartum Breast Physiology
III. Suckling and the Hypothalamic–Pituitary–Ovarian Axis
IV. Lactation and Postpartum Infertility
V. The Lactational Amenorrhea Method of Contraception
VI. Summary

GLOSSARY

amenorrhea The absence or cessation of menstrual periods.

estrogen The female sex steroid hormone produced by the ovaries that is responsible for secondary sexual development and is necessary for ovulation.

fecundity The probability of pregnancy occurring in a single ovarian cycle.

gonadotropins The anterior pituitary hormones (luteinizing hormone and follicle-stimulating hormone) necessary for ovarian follicular development and steroidogenesis.

lactation The production and secretion of milk by the mammary glands of the breast.

ovulation The release of an ovum by a mature ovarian follicle.

progesterone The female sex steroid hormone produced by the ovary that is necessary for preparing the uterus for pregnancy.

Lactational amenorrhea refers to the delay in return of menses in postpartum women who are breast-feeding their infants. The length of amenorrhea is variable and depends on a number of factors such as breast-feeding frequency and duration, interval between feeds, and whether or not supplemental feeding is provided. Lactational amenorrhea is associated with, but does not guarantee, infertility.

I. INTRODUCTION

It has long been recognized that breast-feeding contributes significantly to reduction in infant morbidity and mortality as well as to birth spacing and regulation of fertility worldwide. Demographic data show that in developing countries, breast-feeding is employed as a contraceptive method more commonly than all other reversible forms of family planning combined. The contraceptive effect of nursing after parturition is primarily due to the hormonal impact of suckling on the hypothalamic–pituitary–ovarian feedback system. However, the mechanisms of lactational infertility and the precise determinants of its efficacy as a contraceptive method have not been well studied until recently. Nevertheless, there are now data to support the use of lactational amenorrhea as a contraceptive method if certain parameters regarding length of amenorrhea and breast-feeding patterns are met.

II. INTRAPARTUM AND POSTPARTUM BREAST PHYSIOLOGY

During pregnancy, increasing levels of prolactin (the principal hormone required for milk synthesis), cortisol, human placental lactogen, estrogen, and progesterone combine to stimulate the secretory apparatus of the breast. Increased levels of estradiol, the most potent human estrogen, cause hypertrophy and hyperplasia of prolactin-secreting cells of the pituitary (lactotrophs) and also promote prolactin release by inhibiting hypothalamic production of dopamine [also called prolactin inhibiting factor (PIF)]. These factors result in an increase in prolactin levels from 5–20 ng/ml in early pregnancy to 300 ng/ml

at term. Paradoxically, despite the increasing levels of prolactin during gestation, there is no engorgement or milk letdown during pregnancy. This is likely due to the inhibitory effect of high estradiol and progesterone concentrations both of which are produced in huge quantities by the placenta. Estradiol is thought to exert its inhibitory effect at the level of the prolactin receptor and progesterone inhibits one of the enzymes necessary for production of lactose, or milk sugar.

With delivery there is a rapid decline in the concentration of estradiol and progesterone, and therefore the inhibitory effect on milk production and letdown is reversed. When suckling occurs, sensory signals are sent from the nipple via afferent pathways in the spinal cord to the hypothalamus resulting in an acute release of prolactin and oxytocin, the latter an important stimulator of milk letdown. This neuroendocrine reflex has three consequences, namely, induction of lactogenesis, milk flow, and inhibition of pulsatile gonadotropin-releasing hormone (GnRH) secretion with resultant amenorrhea (Fig. 1).

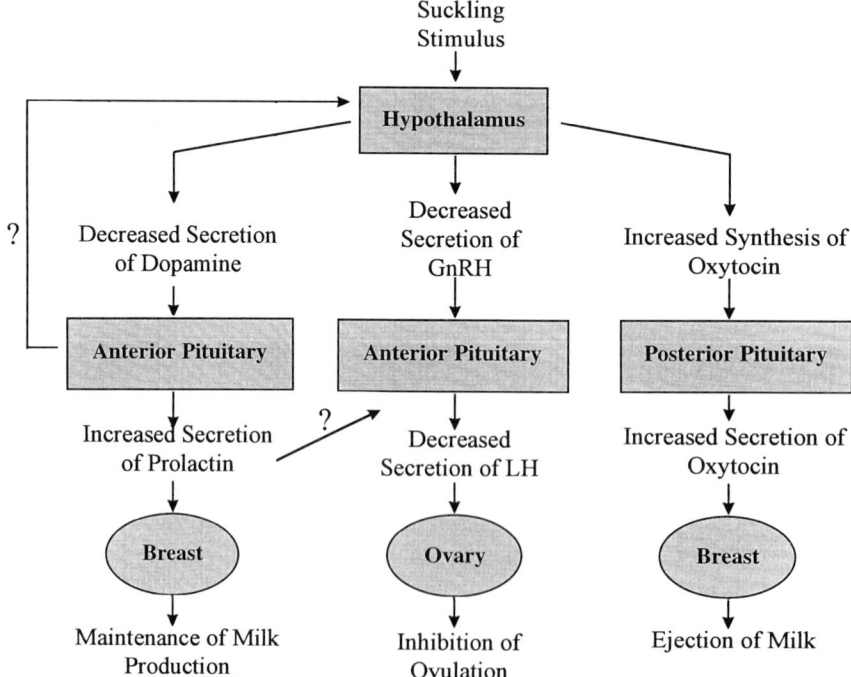

FIGURE 1 Physiological mechanisms involved in lactational infertility (adapted with permission from R. V. Short, Breast feeding, *Sci. Am.* **250**(4), 38, 1994).

Lactogenesis, or milk production, occurs acutely in response to suckling. This is associated with a 10- to 20-fold increase in prolactin compared to basal postpartum levels of 40–50 ng/ml. Suckling also causes a decrease in hypothalamic dopamine (PIF) which is normally released into the portal system to the anterior pituitary, where it causes suppression of lactotroph cells. The amount of prolactin released with each feeding is sufficient to maintain lactogenesis and promote adequate milk supply for the next feeding. Milk flow itself is mediated through oxytocin release from the posterior pituitary. Oxytocin causes contraction of myoepithelial cells in the breast which in turn causes release of previously secreted and stored milk from the alveolar lumina.

III. SUCKLING AND THE HYPOTHALAMIC–PITUITARY–OVARIAN AXIS

Follicle growth and ovulation are dependent on organized, cyclic stimulation of the ovary by the pituitary gonadotropins follicle-stimulating hormone (FSH) and luteinizing hormone (LH). FSH initiates and sustains follicle growth by causing a proliferation of follicle cells and increased enzymatic efficiency within the follicle. Production and release of both FSH and LH occur in response to pulsatile secretion of GnRH from the hypothalamus. Estrogen secretion increases as the ovarian follicle grows over the first part of the ovulatory cycle under the influence of FSH. Once sustained elevations of estrogen are maintained for several days near midcycle, LH is released from the pituitary gland and causes the physical changes in the ovarian follicle wall which result in ovulation (release of the mature egg). Also under the influence of LH, the follicle cells remaining in the ovary after ovulation begin to secrete progesterone, the hormone required to maintain mature endometrium and allow embryo implantation to occur. The postovulatory follicular cells become yellow in color and therefore this is called a corpus luteum (yellow body).

Amenorrhea in lactating women is caused by suppression of gonadotropin secretion, which produces a state in which follicle development, ovulation, and corpus luteum function do not occur. Suckling is the most important mechanism of gonadotropin suppression and lactational amenorrhea. Suckling patterns vary considerably and individually. Variability in duration of nursing, frequency, interval, and use of supplemental feedings may significantly influence hormonal patterns and profoundly affect the duration of lactational amenorrhea.

The mechanism of gonadotropin suppression by suckling is not as well understood as the physiology of milk production and flow. It was previously hypothesized that elevated prolactin levels during lactation caused amenorrhea by decreasing the responsiveness of the anterior pituitary to GnRH or by attenuating normal ovarian response to gonadotropin stimulation. However, several observations undermine the universal application of this hypothesis. For example, prolactin levels are elevated in the immediate postpartum period but remain high for only 3 or 4 months. During this time FSH and LH levels are somewhat suppressed. However, normal gonadotropin patterns and normal menstrual cyclicity do not necessarily return as prolactin levels decrease over time. Conversely, some lactating women may have normal hormonal patterns and ovulatory cycles despite persistent hyperprolactinemia. Thus, other factors, such as endorphins or serotonin, must contribute to prolactin-mediated gonadotropin suppression, at least in some instances. It is likely that reduced GnRH pulsatile secretion is a common denominator of low gonadotropin levels. This hypothesis is supported by studies showing that normal ovulation can be induced in breast-feeding women with high prolactin levels by infusing them with pulsatile GnRH.

In addition to suppressing pulsatile GnRH and LH secretion, suckling also alters the normal positive feedback of estrogen on LH. Additionally, there is an enhanced negative feedback of estrogen on both LH and FSH. Therefore, while suckling continues, follicles which start to develop may be inhibited by the estrogen produced by early folliculogenesis. This will result in low estrogen levels which are inadequate to produce the LH surge. This combination of factors helps to explain observations regarding menstrual patterns in lactating women. While many women have complete lactational amenorrhea due to profound suppression of gonadotropins, others may have some follicular activity and a moderate estrogen level. However, even when follicular development is seen, ovulation and normal fertility are unlikely, presumably due to the elimination of the LH surge as described.

IV. LACTATION AND POSTPARTUM INFERTILITY

It is widely recognized that breast-feeding suppresses normal ovarian cyclicity via alteration in pulsatile GnRH and gonadotropin secretion, thus conferring a period of natural infertility. The duration of infertility varies with suckling patterns and the sensitivity of the mother to the inhibitory effects of suckling on the pituitary. It is also becoming clear that even after resumption of normal menstrual cycles and ovulation, there can be an extended period of reduced fecundity possibly secondary to inadequate luteal function. Poor luteal function also represents part of the continuum of suppressed follicular development. In those women who have transient follicular activity there may be sufficiently prolonged intervals of reasonably high estrogen to induce an LH surge and ovulation. However, the estrogen levels are often lower than those in nonlactating women, and the peaks may be short-lived. This results in blunted LH secretion, inadequate corpus luteum development, and poor endometrial proliferation. These events are often manifest by relatively short cycles and a reduced amount of flow compared to normal ovulatory cycles in the same woman.

International studies have shown that 3–10% of women conceive during lactational amenorrhea if not using any other contraceptive method. These percentages are derived from studies that were not necessarily controlled for interval since delivery, frequency of breast-feeding, use of supplemental feedings, etc. One study of 236 women in Chile found a probability of pregnancy at 6 months of 0.9% among women who exclusively breast-fed and remained amenorrheic (Perez *et al.*, 1992). There is currently no reliable way to predict when ovulation will resume in women who are breast-feeding. Nonetheless, certain parameters associated with return of ovarian cyclicity have been established which appear to hold

true cross-culturally. Some women will experience their first postpartum ovulation and conceive during the period of lactational amenorrhea. One study of 60 breast-feeding mothers in Baltimore and 41 in Manila demonstrated that the risk of ovulation was reduced by higher frequency and longer duration of feeds and by using less supplementary feeds. During the first 6 months postpartum, amenorrheic women had an ovulatory frequency of <10% with partial breast-feeding but only 1–5% with exclusive breast-feeding (Gray *et al.*, 1990). The risk of pregnancy increases significantly once menstruation has occurred or after 6 months postpartum. In studies in the United States, United Kingdom, Australia, and Denmark, where average suckling frequency is five or six times/day, it has been demonstrated that ovarian activity and ovulation occur as a result of decreased frequency or duration of nursing, often at the time when supplemental food is introduced. Abrupt weaning was also clearly associated with resumption of ovarian function.

V. THE LACTATIONAL AMENORRHEA METHOD OF CONTRACEPTION

In 1988, a number of researchers convened to discuss the lactational amenorrhea method of contraception which produced the Bellagio Consensus. The goal of this group was to determine the impact of various patterns and observations of breast-feeding women worldwide in order to synthesize a statement on the reliability of lactational amenorrhea as a contraceptive method. Data were compiled from 13 prospective studies in seven countries to determine the birth-spacing effect of breast-feeding. The most important conclusion was that the pregnancy rate in the first 6 months postpartum was 2% or less if women were fully breast-feeding and remained amenorrheic. Other important conclusions from this consensus were that (i) breast-feeding should be regarded as a potential family planning method in all maternal and child health programs in developed and developing countries, (ii) women should be offered a choice of using breast-feeding as a means of family planning and counseled on how to maximize its contraceptive efficacy, and (iii) guidelines specific to a particular culture or population can be developed based on breast-feeding practices.

From the Bellagio Consensus the lactational amenorrhea method (LAM) of contraception was developed to serve as a practical guide for family planning organizations. Three criteria must be met for a woman to know she has maximal contraceptive efficacy from the LAM: (i) The woman must be amenorrheic, (ii) she must be fully or nearly fully breast-feeding, and (iii) she must be within 6 months postpartum. The specific criteria for "fully breast-feeding" stipulate that there should be no intervals >4–6 hr between feeds and that only small, infrequent supplemental feeds with water or juice may be provided if at all. If all criteria are met, the woman has less than a 2% risk of pregnancy. When any of the three criteria is no longer met, an additional contraceptive method should be initiated in order to provide the same level of contraceptive certainty. The method emphasizes that less than full breast-feeding is associated with decreased duration of lactational amenorrhea and an increased occurrence of ovulation prior to the first menses. In many cultures, women breast-feed for significantly longer than 6 months and therefore investigators have begun to look at an extended LAM. In a study in Rwanda, 419 women who self-selected to continue LAM for 9 months had no pregnancies. The criteria were the same with the specific stipulation that if any supplemental feeding was provided, it was done so after nursing.

VI. SUMMARY

Women who breast-feed their infants have a substantial delay in the resumption of menses and return of normal fertility. The worldwide impact on birth spacing and improved maternal and child health is undisputed. The variability in duration of amenorrhea and infertility is related to patterns of suckling and use of supplemental feeding. It is widely accepted that suckling suppresses the normal pattern of pulsatile GnRH release and therefore prevents normal follicular growth, ovulation, and/or luteal function. Additionally, during lactation estradiol exerts an enhanced negative feedback on LH and FSH in the follicular phase, and the normal positive feedback of estradiol on LH release is blunted. The exact mechanisms whereby suckling exerts these effects on the hypothalamus remain unclear. However, the use of lactational amenorrhea as a method of family planning has been well studied and if certain criteria are met has been shown consistently to confer at least a 98% protection from pregnancy in the first 6 months postpartum. This contraceptive efficacy is comparable to currently available methods of contraception. The return of menses usually heralds resumption of normal ovulatory function and therefore an alternative contraceptive method should be initiated.

See Also the Following Articles

Breastfeeding; Contraceptive Methods and Devices, Female; Gonadotropin Secretion, Control of; Lactation, Human; Lactogenesis; Mammary Gland, Overview

Bibliography

Birkenfield, A., and Kase, N. (1994). Functional anatomy and physiology of the human breast. *Obstet. Gynecol. Clin. North Am.* **21**, 433–443.

Gray, R., *et al.* (1990). The risk of ovulation during lactation. *Lancet* **335**, 25–29.

Kennedy, K. (1993). Fertility, sexuality, and contraception during lactation. In *Breastfeeding and Human Lactation* (J. Riordan and K. Auerbach, Eds.). Jones & Bartlett, Boston.

Kennedy, K., Rivera, R., and McNeilly, A. (1989). Consensus statement on the use of breastfeeding as a family planning method. *Contraception* **39**, 477–491.

Labbock, M., Perez, A., Valdes, V., *et al.* (1994). The lactational amenorrhea method (LAM): A postpartum introductory family planning method with policy and program implications. *Adv. Contraception* **10**, 93–109.

McNeilly, A. S. (1993). Lactational amenorrhea. *Endocrinol. Metabol. Clin. North Am.* **22**, 59–71.

Perrez, A., Labbock, M., and Queenan, J. (1992). Clinical study of the lactational amenorrhea method for family planning. *Lancet* **339**, 968–970.

Speroff, L., Glass, R., and Kase, N. (1994). *Clinical Gynecologic Endocrinology and Infertility*, pp. 547–582. Williams & Wilkins, Baltimore.

Yen, S., and Jaffe, R. (1991). *Reproductive Endocrinology*, pp. 371–375. Saunders, Philadelphia.

Zinaman, M., *et al.* (1995). Pulsatile GnRH stimulates normal cyclic ovarian function in amenorrheic lactating postpartum women. *J. Clin. Endocrinol. Metab.* **80**, 2088–2093.

Lactational Anestrus

Jeffrey S. Stevenson
Kansas State University

GLOSSARY

anestrus or anovulation The absence of ovulatory estrous cycles.

energy balance The difference in total nutrients ingested as feed inputs minus the amount of nutrients excreted as the result of various physiological functions (i.e., urination, digestion, milk synthesis, and respiration).

follicular waves Cohorts of ovarian follicles that emerge every 8 or 9 days from which one follicle becomes dominant and eventually is capable of ovulation.

lactational or postpartum anestrus The absence of ovulatory estrous cycles during lactation.

Lactation is a natural extension of the culminating event of successful reproduction: birth of live offspring. In domestic farm animals, onset and establishment of lactation and estrous cycles are concomitant, energy-competing processes. Because metabolic events essential to milk secretion compete for available nutrients that support processes leading to the first postpartum estrus and subsequent fertility, understanding the interdependency of lactation and reproduction is essential to optimize reproductive efficiency and profitability of livestock enterprises. Most of the information in this article focuses on cattle as the model in the study of lactational anestrus.

I. INTRODUCTION

Nearly all farm animals experience a period of postpartum anestrus or anovulation of varying duration that is dependent on a number of factors, including whether or not parturition occurs during the normal breeding season (Table 1). For example, most sheep give birth near the end of their normal seasonal breeding season. Some spring-lambing ewes in northern latitudes may experience lactational anestrus before eventually beginning estrous cycles before the end of the breeding season or give birth during their seasonal anestrus. Mares rarely have lactational anestrus because most show a fertile ovulatory estrus (foal heat) early after parturition. Most sows are anestrous during the 2–4 weeks they nurse their litters, but a fertile, postweaning estrus is observed between 2 and 10 days after weaning of their litters. If the period of lactation exceeds 4 weeks or litter size is reduced abruptly, sows will overcome suckling-induced anovulation and express estrus.

Intervals to first postpartum ovulation are less than but proportionally similar to intervals to estrus in milked (by machine) and suckled (by calf) cows. Cows that are neither milked nor suckled after parturition have shorter periods of anovulation than those that are milked or suckled. When calves are removed from their mothers (about 30 days after calving) and milk removal is either terminated temporarily or permanently, cows generally return to estrus in a few days (Table 1). In all livestock, lactational anestrus eventually gives way to the onset of normal estrous cycles when milk removal is terminated, energy balance becomes more positive, or other suckling-mediated inhibitors wane.

The terms anovulation and anestrus often are used interchangeably to describe lactational anestrus. Al-

TABLE 1

Effect of Lactation and Weaning on the Onset of Postpartum Estrous Cycles in Domestic Livestock

Farm animal	Begin estrous cycles in lactation	Effect of weaning
Mare	Foal heat: 8–15 days	No effect
Sow	Only if lactation exceeds 4 or 5 weeks	Estrus in 2–10 days
Ewe	3 or 4 weeks only if lambing occurs during the breeding season	Estrus in 2–10 days only if lambing occurs during the breeding season
Suckled cow	5 or 6 weeks minimum	Estrus in 2–10 days
Milked cow	3 or 4 weeks minimum in cows milked twice daily	Does not apply

though most pre-1970 studies relied solely on behavioral estrus to define the end of lactational anestrus, the widespread use of specific immunoassays to measure progesterone has since been used to determine when first luteal function becomes established based on the first increase in concentrations of progesterone in blood or milk. In addition, use of intrarectal ultrasonography in cattle allows visualization of ovarian follicles and corpora lutea during the reestablishment of estrous cycles. Once detection of luteal function was possible by either method, periods of anovulation could be measured and monitored independently of expression of estrus. The interval to first behavioral estrus is most important to livestock producers because detected estrus is the cue for artificial insemination or natural mating. Duration of postpartum anestrus largely determines the likelihood of pregnancy establishment during the breeding season or postpartum breeding period.

II. FREQUENCY OF MILK REMOVAL

Increased frequency of milking or suckling increases the duration of the anovulatory period. Average intervals to first ovulation and estrus are presented (Table 2) from two recent studies in which dairy cows were milked daily three times, six times (during 3 or 6 weeks postpartum and then switched to three times), or three times milked plus three times suckled (during 6 weeks postpartum then switched to three times milked). Increased milking frequency further prolonged the interval to first ovulation, with suckling plus milking having the greatest effect. In contrast, intervals to first estrus and conception are prolonged more by increasing milking frequencies than by short-term suckling plus milking at similar milk removal frequencies.

Suckling restricted to 1 time daily, but not 2 times daily, generally reduces the interval to first estrus or first ovulation in cattle, suggesting that 2 times daily suckling bouts are sufficient to prolong postpartum anovulation. Increasing suckling intensity to two calves per cow prolonged anestrus in older cows nursing twins. Cows may be nursed by calves from 2 to 23 times daily during the first 6 weeks postpartum. Complete weaning, short-term weaning (48 hr), or partial weaning (suckling restricted to a few times daily) alter onset of first estrus after parturition. Compared to weaning, other factors such as age, nutrition, and genotype of cow and age of calf affect duration of anestrus to a lesser extent.

Combining suckling and milking during early lactation increases milk yield and prolongs postpartum anovulation. Cows that were machine milked in the morning and suckled by calves in the evening during the first 8 weeks of lactation produced 16% more milk (total of milking plus suckling) during a 300-day lactation than cows milked twice daily. Suckling plus milking apparently increases milk yield by reducing the amount of residual milk in the udder after each suckling event during the first 4 days after calving but not at later periods. Cows nursing their own calf plus one foster calf during the first 8 or 9 weeks postpartum produced 18–55% more milk than cows with single calves, even after the foster calf was weaned. Holstein cows milked three times daily plus suckled three times daily by two calves during the first 6 weeks of lactation produced more milk than

TABLE 2

Intervals to First Ovulation, First Estrus, and Conception in Milked Dairy Cows
and Milked Plus Suckled Dairy Cows[a]

Trait	Milked 3×	Milked 6× for 3 weeks	Milked 6× for 6 weeks	Milked 3× + suckled 3×
Experiment 1[b]				
Days to first ovulation	26[x]	—	37[y]	48[z]
Days to first estrus	46[x]	—	76[y]	48[x]
Days to conception	112[x]	—	184[y]	100[x]
6-Week milk yield,[c] (kg/day)	35.3[x]	—	42.6[y]	50.0[z]
Nadir of energy balance (Mcal/day)	−9.1	—	−13.6	−19.4[z]
Week at zero energy balance	6	—	10	8
Experiment 2[d]				
Days to first ovulation	22[x]	30[y]	39[z]	—
Days to first estrus	45[x]	61[y]	75[z]	—
Days to conception	105[x]	148[y]	189[z]	—
6-Week milk yield (kg/day)	35[x]	40[y]	42[z]	—
6-Week loss in weight (kg)	24[x]	29[y]	32[z]	—
Week at zero energy balance	6	—	10	—

[a] Data from Bar-Peled *et al.* (1995) plus other unpublished results.

[b] Cows were milked 3×, 6×, or 3× milked plus 3× suckled by two calves during the first 6 weeks postpartum and then milked 3× daily thereafter.

[c] Milk yield from 3× daily suckling periods was included (estimated by weekly weigh–suckle–weight).

[d] Cows were milked 3×, 6× for 3 weeks, 3× for 6 weeks, and then milked 3× daily thereafter.

[x,y,z] Means within row with uncommon superscript letters differ ($P < 0.05$).

contemporary Holsteins that were milked six times daily (Table 2).

III. MAGNITUDE OF MILK YIELD

Prolonged anovulation generally is associated with increasing milk yield in milked cows, but greater milk yield does not predispose cows to prolonged anovulation. Because a negative relationship ($r = -0.80$) exists between milk yield and energy balance, and most variation in energy balance during the first weeks after calving is the result of nutrient intake ($r = 0.73–0.83$) or body condition (visual appraisal of fatness) rather than milk yield ($r = -0.25$), milk yield per se is not a good indicator of the duration of anovulation. In fact, cows milked twice daily with the longest anovulatory periods are those that fail to consume adequate nutrients to support lactation and therefore produce less milk and are less fertile. Models that best accounted for variation in days to first

ovulation in Holsteins and Jerseys milked twice daily provide little support that milk yield increased interval to first ovulation. Age at calving and periparturient clinical abnormalities had a greater influence on interval to first ovulation than any criterion of milk yield.

In contrast to cows milked only twice daily, milk production alone in more frequently milked cows (three and six times daily) during the first 6 weeks of lactation accounted for 58% of the variation in days to first ovulation when changes in body weight and body condition (reflections of adequate or inadequate nutrient intake) also were included in regression models. Furthermore, changes in body condition alone accounted for 55% of the variation in intervals to first observed estrus. Variation in reproductive performance (days to onset of estrous cycles and subsequent fertility) is explained by milk yield and its consequences (negative energy balance and losses in body weight and body condition).

When comparing twice-milked cows that were

sired by either average or superior sires from a 20-year genetic selection study (selection based on milk production), mean days to first ovulation between the two genetic groups differed by only 2 days, whereas days to first visually detected estrus differed by 23 days. Thus, greater milk production, regardless of milking frequency, is antagonistic to behavioral expression of estrus but not to the reestablishment of estrous cycles after calving unless milking frequency exceeds twice daily. Similar conclusions were reached for twice milked cows given various galactopoietic doses of somatotropin in which intervals to first ovulation were unaffected by increasing somatotropin dose, but days to first service were prolonged and rate of detected estrus was reduced. Thus, approximately 60% of the variation in days to conception in milked cows is explained by events that occurred during the first 60 days after parturition.

IV. REMOVAL OR DENERVATION OF THE MAMMARY GLANDS

Much of the early literature about suckling-induced anestrus in cattle was based on the assumption that the presence of the mammary gland was an essential component of the suckling stimulus. The mammary gland obviously is essential for milk synthesis, but its contribution to postpartum suppression of ovulation could only be assessed in models of mammary ablation. When the mammary gland was removed and cows were maintained without their calf, the period of anestrus was shorter (25 versus 12 days) than that in udder-intact, nonsuckled cows. In the same experiment, mastectomized cows (no calf present) and nonsuckled udder-intact cows had shorter periods of anestrus than udder-intact suckled cows (65 days). Therefore, the mere presence of functional mammary tissue slightly delayed onset of estrous cycles.

Removal of the mammary glands resulted in first postpartum ovulations occurring approximately 2 weeks after calving when calves were removed from their mothers at birth. In contrast, when mastectomized mothers were maintained continuously with their natural-born calf, the period of anovulation is prolonged as in suckled, udder-intact cows. Summa-

TABLE 3
Days to First Postpartum Ovulation Based on Daily Samples of Progesterone in Serum of Beef Cattle

Treatment[a]	Experiment			
	1	2	3	4
Mastectomized mother + calf weaned (MCW)	13.9	16.0[c]	15.2[d]	22.6[d]
Mastectomized mother + calf restricted (MCR)	—	—	—	29.1[d]
Mastectomized mother + calf present (MCP)	—	52.2[b]	37.0	36.1
Udder-intact mother + calf present (UICP)	—	—	28.2	41.6
± Pooled standard error	1.6	1.5	4.2	2.6

[a] Treatments were initiated within 24 hr after calving. MCW, mastectomized cows with calves weaned within 24 hr of birth; MCR, mastectomized cows with restricted contact with calves (calf was maintained in smaller pen within the individual cow pen with no inguinal contact with the cow); MCP, mastectomized cows with unrestricted presence of their calf; UICP, udder-intact cows with unrestricted presence of their calf.

[b] Calves were "weaned" from mastectomized cows from 46 to 53 days postpartum.

[c] Different ($P < 0.05$) from MCP treatment within experiment.

[d] Different ($P < 0.05$) from UICP treatment within experiment.

rized in Table 3 are intervals to first ovulation from four studies utilizing mastectomized and/or udder-intact cows. In total, these studies indicate that the continuous presence of the calf with its mother delays onset of ovulatory cycles as long as the calf was not restricted from the inguinal region of its mastectomized mother.

How does the presence of a nonsuckling calf prolong anovulation in its mastectomized mother? This question may be answered in part by the behavior of calves and cows following twice-daily bottle-feeding of calves maintained continuously with their mastectomized mothers. Following overnight separation of cow and calf, and immediately after feeding calves, calves attempt to suckle their mastectomized mothers for as long as other calves successfully suckled their udder-intact mothers. The behavior of udder-intact and mastectomized cows during these suckling or pseudosuckling bouts is remarkably similar. Calves exhibiting this chronic pseudosuckling possibly stimulated sensory nerves innervating the skin of the inguinal area where the mammary gland normally is attached. It is known that the ventral branch of the first lumbar nerve and a secondary branch of the second lumbar nerve innervate the gland and skin of the forequarters; in addition, the perineal nerve (branch of the pudic nerve) innervates the skin of the posterior part of the bovine udder. This pseudosuckling involved the calf positioning itself in a normal suckling position and bunting and orally manipulating the skin of the inguinal region of the cow. Concentrations of cortisol, oxytocin, and prolactin measured in cows during these suckling bouts were increased for up to 12 min after reunion of cow and calf (in mastectomized and udder-intact cows). Pseudosuckling of mastectomized cows by their calf triggered secretion of hormones normally associated with suckling events in udder-intact, suckled cows.

When mammary glands were denervated (ventral branches of lumbar nerves 1–4) in beef cows that were maintained with their calf, postpartum intervals to first estrus were similar to those in udder-intact suckled cows but longer than those of nonsuckled cows. These findings indicate that udder-intact cows or cows with denervated udders, maintained in the presence of their calf, had similar postpartum intervals to estrus, regardless of whether or not the mother received a normal tactile suckling stimulus via teat manipulation. Mechanical, electrical, and thermal hyperstimulation of sensory neurons in the teat fail to simulate the effects of suckling. Therefore, suckling-mediated anovulation is not dependent on mammary somatosensory cues from the calf, although prolactin, but not oxytocin release was reduced in suckled, mammary-denervated cows.

V. THE PRESENCE OF OFFSPRING

When calves were removed early (early weaning at 4 weeks), estrus and/or ovulation occurred within a few days and concentrations and pulses of luteinizing hormone (LH) were increased. The continuous presence of a nonsuckling (muzzled) calf further prolonged (by 24 days) the interval to first postpartum estrus compared to cows whose calf was removed 72 hr after birth. In addition, in cows whose calf was fitted with nose plates after 30 days postpar-

tum to prevent suckling, interval to first postpartum estrus was similar to that in suckled controls but 4 weeks longer than that in cows whose calf was weaned at 30 days. Spatially restricting the calf to prohibit suckling by housing it within a pen and maintaining the continuous presence with its mother prolonged anovulation as well. Therefore, continuously maintaining a nonsuckling calf with its udder-intact mother prolongs the period of anovulation.

Cow–calf bonding is a critical component in the process by which anovulation is prolonged. Because under natural conditions the majority of cows allow only their own calf to suckle, recognition by the cow of her own calf involves olfactory, visual, and auditory cues from the calf. An acute temporal increase in blood concentrations of LH occurs in cows whose calf is weaned and in cows suckled by unrelated calves but not in cows suckled every 6 hr for 144 hr by their own calf. Olfaction plays a key role in maternal bonding in sheep because after ablation or functional impairment of the olfactory bulbs, ewes allow foster lambs to suckle, whereas normally they reject all but their own. These results in sheep indicate that olfactory cues are critical components in mother–offspring "self" recognition. Suppressed recognition of the cow's own calf was confirmed by eliminating both visual and olfactory cues and exposing cows to either their own or an unrelated calf. Olfaction and vision were equally effective in permitting calf identification, but elimination of both senses prevented calf recognition by its mother and the negative effects of suckling on acute LH release; thus allowing ovulation to occur.

If unrelated calves are transferred to foster cows within 2 days of birth, the percentage of cows exhibiting estrus by 60 or 90 days postpartum was not different from that of cows each nursing their own single calf. The importance of cow–calf bonding and its relationship to postpartum anovulation was demonstrated when cows were suckled by their own calf until 15 days postpartum, then calves were (i) weaned, (ii) restricted from the udder of their mother, (iii) maintained continuously with their mother, (iv) removed from their mother and placed continuously with a foster mother, or (v) restricted from the udder of their own mother while their mother was suckled by an unrelated calf. Based on a defined cow–calf interaction test, cows suckled by

TABLE 4
Days to First Postpartum Ovulation in Beef Cows Milked or Suckled Twice (2×) Daily in the Presence or Absence of Their Own Nonsuckling (Restricted) Calf

Treatment[a]	Average days to first ovulation
Calf weaned (CW)	14.1[x]
CW + 2× milked (CW + M)	13.0[x]
Calf restricted from udder contact (CR)	14.2[x]
CR + 2× milked (CR + M)	17.2[x]
CR + 2× suckled (CR + S)	33.9[x]
Calf present (CP)	34.7[x]
± Pooled standard error	3.2

[a] Cows were suckled by their own calf until treatments were initiated at 13–18 days after calving.

[x,y] Means with uncommon superscript letters differ ($P < 0.05$).

an unrelated calf but maintained in the presence of their own nonsuckling calf remained bonded to their own calf after 4 weeks of treatment. In contrast, cows nursing only an unrelated calf formed a new bond with that calf and no longer recognized their own natural-born calf from which they were separated for 4 weeks. Cows that readily fostered an unrelated calf, or remained bonded to their own restricted calf while suckled by an unrelated foster calf, had average intervals to first postpartum ovulation that were the same as those of control cows suckled by their own calf (Table 4). Therefore, cow–calf bonding (mother to her own calf) plus milk removal (by suckling) is sufficient to delay first ovulation in the udder-intact cow.

In contrast, when beef cows are maintained with their own udder-restricted calf but milked twice daily rather than being suckled by an unrelated calf, first ovulation is not delayed (Table 4). Therefore, when a bond is established between a cow and its calf or reformed with an unrelated calf, milk must be removed by twice suckling but not twice milking to prolong anovulation.

VI. NUTRIENT AVAILABILITY

Combined effects of various metabolic changes and nutrient repartitioning, essential for lactogenesis and galactopoiesis in cattle, compete for nutrients that are essential to reinitiate estrous cycles after parturition. Nutrient intake, adequate availability of nutrients, diet composition, acute changes in body weight or body condition before or after calving, variously related metabolites, metabolic hormones, and overall energy balance are factors that correlate with the reestablishment of estrous cycles and may be indicators of intervals to first postpartum ovulation.

A. Nutrient Partitioning

Metabolic functions are regulated in mammals to achieve two major purposes: self-survival and perpetuation of their species. Homeostatic mechanisms achieve the first purpose because they regulate the internal biochemical environment. Homeorhetic regulations achieve the second purpose by altering metabolic support to sustain growth, pregnancy, or lactation. The concept of homeorhesis explains how various hormonal mechanisms orchestrate coordinated changes in metabolic pathways in specific body tissues necessary to support an existing physiological function such as lactation or pregnancy at the expense of other less vital bodily functions. For example, when lactation begins, various priorities for nutrient utilization to support lactation are established. These priorities involve redirecting homeorhetic mechanisms that may be mediated by prolactin, somatotropin, insulin, insulin-like growth factor-I (IGF-I), estradiol-17β, glucocorticoids, and other hormones.

The approximate ranking priority for use of available energy in ruminants is (i) basal metabolism, (ii) activity, (iii) growth, (iv) energy reserves, (v) pregnancy, (vi) lactation, (vii) additional energy reserves, (viii) estrous cycles and initiation of pregnancy, and (ix) excess reserves. Based on this prioritized utilization of available dietary nutrients, reinitiation of the estrous cycle and all its associated components (e.g., gonadotropin secretion, ovarian follicular development, and ovulation) occurs only after needs for lactation and minimal energy reserves are met. Nutrient repartitioning for milk production and reproduction is the result of complex interactions between diet quantity and quality; nutrient reserves (body condition); and the demand for growth and metabolism.

B. Energy Balance

Differences in management of milked and suckled cattle of various genotypes account for some of the differences in postpartum reproductive recrudescence, but, as previously discussed, their lactational management is the major determinant. In general, if adequately fed, cows nursing a calf rarely experience a significant negative energy balance or extensive loss in body weight that occurs in full-fed, milked cows producing up to five times more milk. The effect of nutrition in suckled cows depends somewhat on whether nutrition is adequate before or after parturition. In general, prepartum nutrient status, as estimated by body condition at parturition, is more important than that after calving. The relationship between body condition at calving and the interval to first postpartum estrus in suckled beef cows is nonlinear. That is, the effects of poor body condition are greatest at very low body condition scores (<4 on a 10-point body condition scoring system) and

become less significant as body condition increases to a value ≧7. When nutrition is adequate in suckled cows, milk yield has little effect on postpartum anestrus, but when nutrition is limiting, particularly for cows having genotypes for high milk production, increasing level of milk delays first ovulation and estrus. In milked cows, such a relationship of interval to first estrus and body condition has not been established. However, the magnitude of nutrient deficit (negative energy balance) may be an important factor that inhibits first ovulation and has some relationship to the number of days to first ovulation.

Postpartum changes in energy balance generally are predictive of when first ovulation will occur in milked cows. In milked cows, the nadir of energy balance occurs during the first or second week after calving and recovers at a variable rate, with first ovulation occurring approximately 10–15 days after the nadir (Fig. 1). Days to first ovulation are correlated positively ($r = 0.76$) with days to nadir of

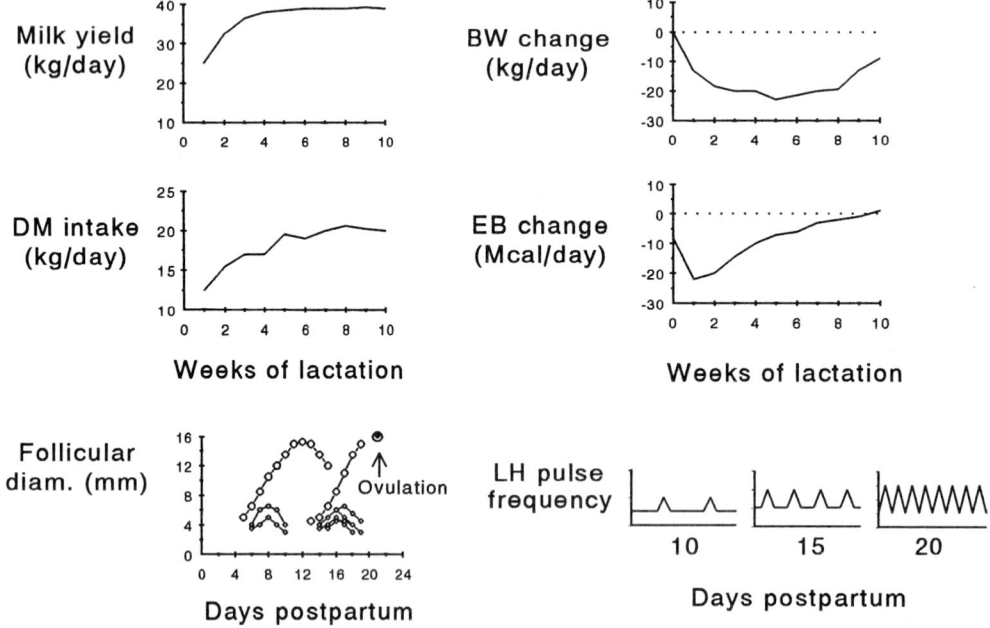

FIGURE 1 Postpartum physiological changes in milk yield, nutrient (dry matter; DM) intake, body weight (BW), and energy balance (EB) during the first 10 weeks of lactation in milked cows. Onset of ovarian follicular waves with their emerging dominant follicles begins in the first or second week after parturition. Ovulation usually occurs 10–15 days after the nadir in energy balance and in response to the onset of more frequently released pulses of LH from the anterior pituitary. Note that milk yield reaches a maximum earlier than maximum DM intake (*ad libitum* feed offerings) accounting for the loss in body weight and overall negative energy balance.

energy balance. Furthermore, because milk fat percentage is high near the nadir of energy balance, it may be a good indicator of the severity of nutrient deficit. During the enormous metabolic demands of high milk production in early lactation, major amounts of nutrients are required for mammary synthesis of lactose, proteins, and triglycerides in the mammary gland that cannot be met by dietary intake. Nevertheless, milked cows with greater dry matter intake, despite having a negative energy balance, produce more milk, lose less body weight, and ovulate earlier postpartum than those with lower intakes. Cows with greater intakes also reach the nadir of energy balance earlier and experience a more severe, but shorter, period of negative energy balance, suggesting that when cows are more efficient at partitioning dietary and stored nutrients toward milk synthesis, they are also better able to recover ovarian cyclicity.

Milked plus suckled dairy cows have greater milk yield than those milked at similar milk removal frequencies but their voluntary nutrient intake was less. Energy balance was most negative and loss of body weight also was greatest in the milked plus suckled cows (Table 2, Experiment 1). Reasons for the failure of milked plus suckled cows to increase their dietary intake to match output of milk are not known. Particularly perplexing was their superior reproductive performance (subsequent estrus detection and fertility) after suckling was terminated (three times milking continued) in comparison to the six times milked cows (switched to three times milking), despite their greater losses in body weight and body condition and more negative energy balance (Table 2). Better expression of estrus and earlier conception indicate that acute positive changes in energy balance (resulting from termination of suckling) are beneficial to subsequent reproductive performance.

VII. POSTPARTUM PHYSIOLOGICAL EVENTS

Studies in primates and other mammals indicate that an intermittent or discontinuous secretion of gonadotropins is required to establish and maintain ovarian cycles. In cattle, pulsatile release of LH is requisite for normal cyclicity and for reestablishment of estrous cycles in postpartum cattle. Interaction between the hypothalamus, pituitary, and ovary occurs, beginning with intermittent delivery of hypothalamic-secreted gonadotropin-releasing hormone (GnRH) to the pituitary gland via the hypophyseal portal vessels. Delivery of pituitary-derived follicle-stimulating hormone (FSH) and LH to the ovary stimulates production of androgen and estrogen in theca and granulosa cells of follicles, respectively, and leads to a surge release of LH and ovulation. Pregnancy interrupts this cyclic interaction, and because of large quantities of estrogen and progesterone produced by the placenta, pituitary stores of LH become depleted. Although restoration of pituitary stores of LH occurs within 1 or 2 weeks after parturition and is paralleled by increases in releasable LH (Fig. 1), which appear to reach a maximum between 2 and 4 weeks postpartum, periods of anovulation often continue. In short, suckling, and to a lesser extent milking, inhibits or delays the return of the requisite pulsatile secretion of gonadotropins, particularly LH.

A. Ovarian Follicular Dynamics and LH Pulse Frequency

Early postpartum resumption of follicular waves is usually not limiting in either suckled or milked cattle (Fig. 1). However, dominant follicles that are present simply fail to ovulate. With a more exacerbated negative energy balance or inadequate nutrition, fewer cows ovulated a dominant follicle from the first or subsequent postpartum follicular wave.

In milked cows, follicular development before first ovulation resulted in ovulation of the first or second dominant follicle, multiple waves of nonovulatory dominant follicles, or formation of an ovarian cyst. Development of this first follicular wave seems to occur in milked cows, regardless of differences in mean daily energy balance (Fig. 1). Atresia of the first-wave dominant follicle is associated with a delayed nadir of negative energy balance, reduced LH pulse frequency, and low circulating concentrations of IGF-I and plasma estradiol. Therefore, the number of follicular waves that occurs before a dominant follicle ovulates is probably a function of energy bal-

TABLE 5

Relative Changes in Various Metabolites or Metabolic Hormones in Blood and Their Changing Status as Postpartum Energy Balance Becomes More Positive

Metabolic signal	Change relative to parturition	Correlation with restored LH pulse frequency
Nonesterified fatty acids	Decreasing	−
β-Hydroxybutyrate	Decreasing	?[a]
Ketones	Decreasing	?
Glucose	Increasing	+
Propionate : acetate	Increasing	?
Insulin	Increasing	?
IGF-1	Increasing	+
Somatotropin	Increasing	?
Tyrosine	Increasing	+
Large neutral amino acids	Increasing	−
Aspartate transaminase	Decreasing	−
Urea nitrogen	Increasing	?
3-Methylhistidine	Decreasing	?

[a] Unknown relationship.

ance or the nutritional status in suckled and milked cows. The dominant follicle that ovulates follows and responds to the reestablishment of increased pulsatile LH secretion. Basal and pulse frequencies of LH secretion increase around the nadir in negative energy balance in milked cows (Fig. 1). Some effects of undernutrition or negative energy balance may involve reduced ovarian responsiveness to LH because milked cows with a more negative energy balance had similar LH patterns but ovulated later than nonmilked postpartum cows with a positive energy balance.

B. Hormonal and Metabolic Signals

Because nutritional status and energy balance influence intermediary metabolism, various metabolites or metabolic hormones may serve as signals to stimulate or inhibit reestablishment of pulsatile LH secretion. Summarized in Table 5 are several of these candidates, their relative changes as energy balance becomes more positive (increasing days after parturi-

tion), and reported positive or negative correlations with the pulse frequency of LH. Although blood concentrations of nonesterified fatty acids (NEFA), large neutral amino acids, and aspartate transaminase are correlated negatively with LH pulse frequency, concentrations of glucose, IGF-I, and tyrosine are correlated positively and may be potential metabolic signals to the hypothalamic–pituitary axis. Plasma aspartate transaminase, a measure of increased gluconeogenesis during various catabolic conditions, decreases significantly and the proportion of tyrosine to total large amino acids increases during 12 days before first ovulation in milked cows. These findings also indicate that decreased gluconeogenesis from amino acids is associated with approaching ovulation. Tyrosine may participate in metabolic signaling to control ovarian function because it is a potential stimulator of LH secretion. Evidence for tyrosine as a metabolic signal to increase LH pulses is supported by studies in which intake of rumen-protected tyrosine by anovulatory milked (dairy) cows accelerated follicular growth and induced estrus and ovulation.

See Also the Following Articles

Cattle (Bovidae); Epitheliochorial Placentation; Pregnancy, Maternal Recognition of; Seasonal Reproduction, Mammals

Bibliography

Bar-Peled, U., Maltz, E., Bruckental, I., Folman, Y., Kali, Y., Gacitua, H., Lehrer, A. R., Knight, C. H., Robinzon, B., Voet, H., and Tagari, H. (1995). Relationship between frequent milking or suckling in early lactation and milk production of high producing dairy cows. *J. Dairy Sci.* **78**, 2726–2736.

Butler, W. R., and Smith, R. D. (1989). Interrelationships between energy balance and postpartum reproductive function in dairy cattle. *J. Dairy Sci.* **72**, 767–783.

Lucy, M. C., Savio, J. D., Badinga, L., De La Sota, R. L., and Thatcher, W. W. (1992). Factors that affect ovarian follicular dynamics in cattle. *J. Anim. Sci.* **70**, 3615–3626.

Randel, R. D. (1990). Nutrition and postpartum rebreeding in cattle. *J. Anim. Sci.* **68**, 853–862.

Schillo, K. K. (1992). Effects of dietary energy on control of luteinizing hormone secretion in cattle and sheep. *J. Anim. Sci.* **70**, 1271–1282.

Short, R. E., Bellows, R. A., Staigmiller, R. B., Berardinelli, J. G., and Custer, E. E. (1990). Physiological mechanisms controlling anestrus and fertility in postpartum beef cattle. *J. Anim. Sci.* **68**, 799–816.

Stevenson, J. S., Lamb, G. C., Hoffman, D. P., and Minton, J. E. (1997). Interrelationships of lactation and postpartum anovulation in suckled and milked cows. *Livestock Prod. Sci.* **50**, 57–74.

Williams, G. L. (1990). Suckling as a regulator of postpartum rebreeding in cattle: A review. *J. Anim. Sci.* **68**, 831–852.

Zurek, E., Foxcroft, G. R., and Kennelly, J. J. (1995). Metabolic status and interval to first ovulation in postpartum dairy cows. *J. Dairy Sci.* **78**, 1909–1920.

Lactation, Human

Margaret C. Neville

University of Colorado Health Sciences Center

I. Anatomy of the Breast
II. Mammary Development
III. The Volume and Composition of Human Milk
IV. Cellular Mechanisms of Milk Secretion
V. Lactogenesis
VI. Lactation
VII. Maternal Adaptation to Lactation

GLOSSARY

involution The process by which the alveolar structure of the lactating breast reverts to the primarily ductule structure of the breast found in the nonpregnant, nonlactating woman.

lactogenesis The onset of milk secretion.

let-down reflex A neuroendocrine reflex whose afferent arm is provided by suckling and whose efferent arm is mediated by oxytocin release from the posterior pituitary. The result of this reflex is ejection of milk from the breast.

mammogenesis The development of the mammary gland under the influence of the hormones of puberty and pregnancy.

milk ejection Contraction of myoepithelial cells surrounding the alveoli and ducts that brings about expulsion of milk.

The basic mechanisms by which milk is secreted are fundamentally the same in all placental mammals. However, humans differ sufficiently in detail that a separate discussion of human lactation is justified. Lactation has been best studied in dairy cattle and rodents. In these species the mammary ducts terminate in a cistern or single teat canal; in humans, the major ducts terminate directly on the nipple. Other major differences are the timing of lactogenesis, the minimal increase in maternal metabolism necessary to support exclusive breast-feeding of a single infant, and, of course, milk composition, which is always tailored to meet the demands of each species. In this discussion we proceed from a description of the anatomy of the breast to a discussion of mammary development. The hormonal controls important through the four developmental stages of mammogenesis, lactogenesis, lactation, and involution are emphasized. A brief outline of the volume and composition of human milk is followed by a more detailed description of the mechanisms by which the major milk components are secreted. Lactogenesis, lactation, and involution are then considered in more detail with particular attention to the role of the let-down reflex in milk ejection and the maintenance of milk secretion. Finally, the interactions of lactation with maternal metabolism, reproductive status, and risk of breast cancer are briefly described.

I. ANATOMY OF THE BREAST

The epithelial compartment of the mammary gland of the adult woman consists of 10–15 lobules, each of which is drained by a single duct that opens directly on the nipple (Fig. 1). These lobuloalveolar complexes are embedded in a fatty stroma that is in turn supported by heavy ligamentous bands that transverse and subdivide the adipose tissue and glandular elements. The nipple is surrounded by a specialized pigmented dermis, the areola. The melanophores of the areola usually darken during pregnancy and retain their dark coloration thereafter. The areola also contains sebaceous glands that hypertrophy and form papillae during pregnancy (Montgomery's follicles), sweat glands, and accessory mammary glands with miniature duct openings through the stratified squamous areolar epithelium. The area contains a dense intradermal nerve plexus supply with numerous sensory end organs including Meissner's corpuscles and Merkel's discs as well as free nerve endings. This sensory innervation is mediated by the fourth intercostal nerve and is critical for prolactin secretion during lactation and usually mediates the afferent arm of the let-down reflex.

Milk is secreted by a lobulotubular epithelium intimately associated with the stromal fat pad, a copious blood supply, and elements of the immune system. For reasons that are not yet understood, mammary epithelium cells require the mammary fat pad, initially laid down during fetal life, for development. Myoepithelial cells surround the ducts and alveoli. These contract during milk let-down to expel stored milk. The blood supply provides the multitude of nutrients necessary for milk synthesis and secretion as well as the hormones that coordinate mammary development, milk secretion, and milk ejection. During lactation plasma cells migrate to the interstitial spaces of the mammary gland where they reside and produce immunoglobulins. Although the gland receives sympathetic innervation, transplantation experiments in goats showed that the only neurons necessary for milk secretion are those that carry signals from tactile stimulation of the nipple and areola to effect the let-down reflex and prolactin secretion.

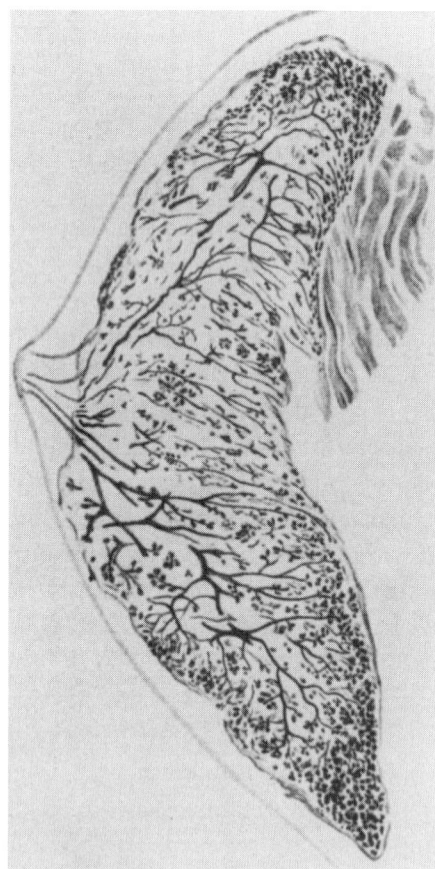

FIGURE 1 Anatomy of the breast of a 19-year-old nulliparous woman. Several ducts emanate from the nipple and course through the adipose and connective tissue stroma to terminate in alveolar structures near the borders of the mammary fat pad (from A. Dabelow, *Morphol. J.* **85**, 361–416, 1941).

II. MAMMARY DEVELOPMENT

Postembryonic development of the mammary gland can be divided into four major stages: mammogenesis, lactogenesis, lactation, and involution. Mammogenesis takes place in two phases as the gland responds sequentially to the hormones of puberty and pregnancy. At puberty estrogen and a pituitary factor that may well be growth hormone stimulate the growth of the mammary ducts into the preexisting mammary fat pad. The branching ducts, the product of complex interactions of stromal and epithelial growth factors, form a tree-like pattern that extends to the edges of the fat pad (Fig. 1). Limited alveolar development is stimulated by the progesterone secreted in the luteal phase of the menstrual cycle. Although there is a small amount of mammary development during each luteal phase, the gland tends to regress with the onset of the menses and the loss of hormonal support. However, there is a gradual accretion of epithelial tissue with each successive cycle.

Full alveolar development and maturation of the epithelium must await the higher concentrations of hormones in pregnancy, particularly progesterone, to complete the process of alveolar development . It is likely that a lactogenic hormone from either the pituitary (e.g., prolactin) or placenta (e.g., human placental lactogen) aids in this process although the growth effects of these lactogenic hormones in the normal mammary gland are still controversial. By midpregnancy the gland is competent to secrete milk and, indeed, small amounts of secretion product are formed. However, the mammary gland continues to develop until parturition with the secretory process held in check by the high circulating concentrations of progesterone.Lactogenesis, the onset of milk secretion, occurs in two stages. About midpregnancy, lactogenesis stage 1 occurs. The gland becomes competent to secrete milk and small amounts of secretion product are often formed. Lactogenesis stage 2 is the onset of copious milk secretion and is a multistep process that proceeds during the first 4 days after parturition as plasma progesterone and estrogen fall from the peak levels attained at the end of pregnancy. In humans progesterone does not decline prepartum as in most animal species but, along with estrogens, falls relatively slowly after birth thereby prolonging the colostral phase. Lactogenesis takes place only if the mammary gland is fully developed and the high plasma prolactin levels present during pregnancy are maintained. It does not occur if the pregnancy is terminated prior to midgestation and can be prevented by dopaminergic drugs that inhibit prolactin secretion. The orderly developmental program followed by the mammary gland during lactogenesis is discussed in detail in Section V.

Lactation is the process of milk secretion and is prolonged as long as milk is removed from the gland on a regular basis. In women both prolactin and oxytocin are required to maintain milk secretion.

Prolactin, produced by the anterior pituitary in response to the suckling stimulus, acts on the mammary epithelial cells to stimulate milk synthesis. Unlike dairy cows, in women inhibitors of prolactin secretion such as dopaminergic drugs inhibit milk synthesis at all stages of lactation. Oxytocin, produced by the posterior pituitary, acts on the myoepithelial cells, causing them to contract and bring about milk ejection. The effects of oxytocin on milk secretion are indirect; in the absence of the let-down reflex milk stasis inhibits the secretory activity of the mammary epithelium. It is currently thought that a feedback inhibitor of lactation (FIL) secreted into milk is responsible for this inhibition. As discussed in greater detail in Sections III and VI, FIL is also postulated to act as a local regulator of milk volume secretion.

Involution takes place when regular removal of milk from the gland ceases. Like lactogenesis, this stage involves an orderly sequence of events progressing from cessation of milk secretion to (i) an increase in milk sodium and chloride signaling the opening of the tight junctions between mammary epithelial cells, (ii) increased secretion of lactoferrin, (iii) secretion of proteases, and (iv) apoptosis, or programmed cell death, of the mammary epithelium. The gland then returns to its prepregnant state. After menopause, with the loss of support of the ovarian hormones the gland involutes further, mainly losing alveolar structures.

III. THE VOLUME AND COMPOSITION OF HUMAN MILK

The major macronutrients in human as well as animal milk can be classified as carbohydrates, proteins, lipids, minerals, and organic anions. The concentrations of the major human milk components are compared to those of bovine milk in Table 1. The carbohydrates represented by lactose and oligosaccharides provide about 8% of human milk by weight and are responsible for two-thirds of the osmolarity. The major proteins are casein, lactoferrin, secretory immunoglobulin A (sIgA), and α-lactalbumin, each present at a concentration of about 2 g per liter. Many other proteins, including plasma pro-

TABLE 1

Comparison of the Macronutrient Contents of Human and Bovine Milk

Component	Human milk	Bovine milk
Carbohydrates		
Lactose	73 g/liter (203 mM)	40 g/liter (111 mM)
Oligosaccharides	12 g/liter	1 g/liter
Proteins		
Caseins	2 g/liter	26 g/liter
α-Lactalbumin	2 g/liter	2 g/liter
Lactoferrin	2 g/liter	Trace
Secretory IgA	2 g/liter	Trace
β-Lactoglobulin	0	5 g/liter
Lipids		
Triglycerides	40 g/liter	40 g/liter
Phospholipids	0.4 g/liter	0.4 g/liter
Cholesterol	180 g/liter	140 g/liter
Minerals and other ionic constituents		
Sodium	5.0 mM	15 mM
Potassium	15.0 mM	43 mM
Chloride	15.0 mM	24 mM
Calcium	7.5 mM	30 mM
Magnesium	1.4 mM	5 mM
Phosphate	1.8 mM	11 mM
Bicarbonate	6.0 mM	5 mM
Citrate	2.5 mM	9 mM

teins, enzymes, hormones, and growth factors, are present at lower concentrations. Lipids, mainly in the form of triacylglycerol, provide about 40% of the caloric content of human milk and also include phospholipids, cholesterol, and glycosphingolipids. Minerals include sodium, potassium, chloride, calcium, magnesium, and phosphate. Citrate and bicarbonate are the major organic anions. The pH of milk within the breast is 6.8. Because bicarbonate is a major buffer in human milk (in contrast to cow's milk, in which citrate is much higher), when CO_2 is lost to the air as when milk is expressed, the pH rises rapidly to 7.2 or higher. Human milk contains many minor components of importance or potential importance to the neonate, including enzymes such as bile salt-stimulated lipase, vitamins, trace elements, hormones, and growth factors. The function and secretion mechanisms of these milk components are not well understood. A number of agents in human milk are thought to protect against gastrointesti-

nal and respiratory infections including the oligosaccharides, which were recently shown to interact specifically with a number of pathogens or their receptors, lactoferrin and secretory IgA. All these compounds are present at very low concentrations in bovine milk.

A meta-analysis of the volume of milk secreted by exclusively breast-feeding women showed that milk volume is remarkable constant at about 800 ml/day in populations throughout the world (Fig. 2). Milk volume is increased 5–15% in women with very low body fat who secrete milk with a lower lipid content, which decreases caloric density up to 15%. This observation illustrates the principle that the volume of milk secreted in lactating women is regulated by infant demand. Thus, when the milk has a lower caloric density, increased suckling by the infant is thought to result in increased emptying of the breast, in turn bringing about an increase in milk secretion. Similarly, mothers of twins, and occasionally even

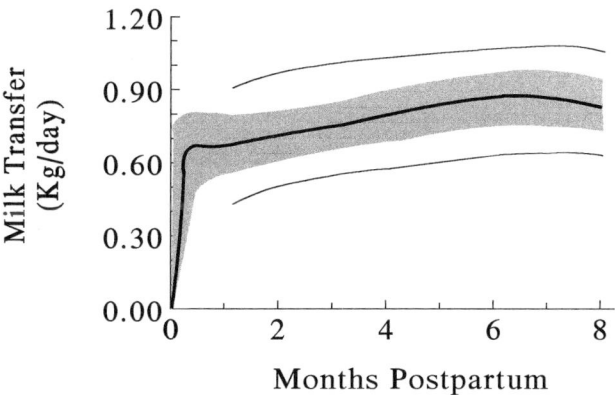

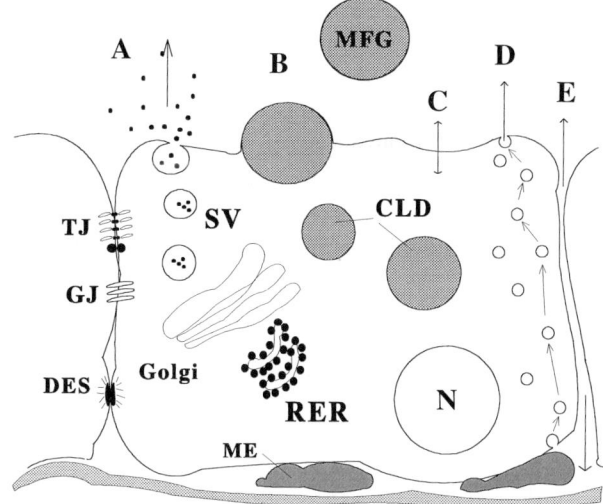

FIGURE 2 Volume of milk transferred to exclusively breast-fed infants as a function of time postpartum [data are from a meta-analysis in Chapter 3 of *Handbook of Milk Composition* (R. G. Jensen, Ed.), Academic Press, San Diego, 1995].

FIGURE 3 The pathways for milk synthesis and secretion by the mammary epithelial cell. (A) Exocytosis of milk protein, lactose, and other components of the aqueous phase in Golgi-derived secretory vesicles. (B) Milk fat secretion via the milk fat globule. (C) Direct movement of monovalent ions, water, and glucose across the apical membrane of the cell. (D) Transcytosis of components of the interstitial space. (E) The paracellular pathway for plasma components and leukocytes; open only during pregnancy, involution, and in inflammatory states such as mastitis. SV, secretory vesicle; RER, rough endoplasmic reticulum; MFG, milk fat globule; N, nucleus; TJ, tight junction; GJ, gap junction; DES, desmosome; CLD, cytoplasmic lipid droplet; ME, myoepithelial cell [modified from Fig. 1 in M. C. Neville *et al.*, The mechanisms of milk secretion, In *Lactation: Physiology, Nutrition and Breastfeeding* (M. C. Neville and M. R. Neifert, Eds.), Plenum Press, New York, 1983].

triplets, are able to produce volumes of milk sufficient for complete nutrition of their multiple infants. Studies of wet nurses, completed in the 1930s, show that at least some women are capable of producing up to 3.5 liters of milk per day. On the other hand, if infants are supplemented with foods other than breast milk, milk secretion is proportionately reduced. For example, in countries such as Peru and the Gambia where infants are customarily supplemented with small amounts of food at mealtimes but given several breast-feeds a day, the daily milk production remains at about 600 ml/day for 12 months or longer.

IV. CELLULAR MECHANISMS OF MILK SECRETION

Five distinct pathways contribute to the production of the diverse components of milk. These pathways are shown diagrammatically in Fig. 3, in which a model mammary epithelial cell is positioned on a basement membrane with processes of myoepithelial cells shown in cross section at the basal surface. Tight junctions join this cell to its neighbors, providing a tightly sealed gasket that, during lactation, prevents direct exchange between the interstitial fluid compartment and the milk space through the paracellular pathway. During pregnancy and under certain other

conditions the paracellular pathway is open, allowing movement of interstitial fluid components into the milk space and the flux of milk components in the other direction. Four major transcellular pathways are responsible for the secretion of milk components during lactation.

A. Exocytosis

The exocytotic pathway secretes milk components by a mechanism common to all endocrine and exocrine glands (Fig. 3A). Casein, lactoferrin, and α-lactalbumin are synthesized by ribosomes and inserted into the lumen of the endoplasmic reticulum. After initial processing they are transported to the Golgi system where they are further processed and packaged into secretory vesicles. These vesicles travel to the plasma membrane, where they fuse and release their components into the interstitial space. This pathway, which appears to be largely constitutive, is also responsible for secretion of lactose, calcium, phosphate, citrate, and most other components of the aqueous phase of milk. The enzyme galactosyl transferase uses α-lactalbumin as an essential coenzyme and glucose and UDP-galactose as substrates to synthesize lactose in the terminal portion of the Golgi vesicles. The osmotic effects of the high concentration of lactose draw water into the secretory vesicles. Galactosyl transferase and a variety of other glycosyl transferases present in the Golgi vesicles are responsible for the formation of the wide variety of complex oligosaccharides present in human milk.

B. Lipid Secretion

The mechanism for the secretion of milk lipid is unique to the mammary epithelium (Fig. 3B). Triacylglycerols are synthesized in the mammary alveolar cell from glycerol and free fatty acids. The free fatty acids may be derived from three sources: (i) Plasma triacylglycerol in chylomicra or very low-density lipoprotein is hydrolyzed by the enzyme lipoprotein lipase within the lumen of mammary capillaries, and the released fatty acids travel into the mammary alveolar cell; (ii) plasma nonesterified fatty acids carried on serum albumin are directly utilized by the alveolar cell during postabsorptive and fasting states; and (iii) fatty acids can be synthesized within the alveolar cell from glucose as saturated fatty acids usually containing 12 or 14 carbons rather than the long-chain fatty acids derived from the plasma. After formation, the triacylglycerols form droplets that merge and gradually move to the apical membrane that contains a specialized complement of proteins including butyrophilin and xanthine oxidase. These proteins are thought to be important in the interaction of the cytoplasmic lipid droplet with the membrane so that when the droplet reaches the apical surface of the cell it complexes with the apical membrane forming a protrusion that is gradually enveloped in membrane.

Eventually, the milk fat globule pinches off as a membrane-bound package of lipid. The membrane prevents coalescence of the fat droplets and serves to deliver phospholipids to the infant.

The composition of milk triacylglycerols depends on the diet. On a typical Western diet with 30–40% of calories consumed as fat, long-chain fatty acids predominate (Table 2). The relatively high proportion of unsaturated fatty acids reflects the consumption of vegetable oils in the current American diet. Under these conditions only about 10% of the milk fatty acids are synthesized in the mammary gland itself as medium-chain fatty acids (C_8–C_{14}). On high-carbohydrate, low-fat diets, such as those consumed by a group of Nigerian women whose milk lipid composition is also given in Table 2, the proportion of fatty acids synthesized in the mammary gland increases, leading to an increase in medium-chain fatty acids. Because these Nigerian women consume relatively large quantities of fish, their milk also contains a higher proportion of the n-3 long-chain polyunsaturated acids present in fish oils. On reducing or low-calorie diets, much of the milk fat is derived from adipose tissue and the fatty acid composition of milk lipid resembles the depot fat. Human milk contains rather large quantities of very long-chain polyunsaturated fatty acids such as arachidonic acid and docosahexaenoic acid. Currently, it is not clear whether these apparently important milk components originate in the liver or are synthesized in the mammary gland.

C. Transmembrane Transport

There are transport mechanisms for only a few milk components in the apical membrane of the mammary alveolar cell (Fig. 3C). Studies in goats showed this membrane to be permeable to monovalent ions such as sodium, potassium, and chloride as well as to glucose. The membrane is impermeable to disaccharides and divalent cations. The mechanisms that control the substantial concentration gradients of monovalent cations across the apical membrane are not understood. It is also important to remember that plasma membranes are permeable to many types of pharmacological agents. These compounds equilibrate across the membranes of the

TABLE 2
Major Fatty Acids of Human Milks (wt%)

		Human milk	
Fatty acid		Western diet[a]	Low-fat diet[a,b]
Saturated fatty acids			
Intermediate and medium chain (formed in mammary gland)			
8:0	Octanoic acid	0.46	—
10:0	Decanoic acid	1.03	0.54
12:0	Lauric acid	4.40	8.34
14:0	Myristic acid	6.27	9.57
Long chain			
16:0	Palmitic acid	22.00	23.35
18:0	Stearic acid	8.06	10.15
Monounsaturated fatty acids			
16:1 n-7 (*cis*)	Palmitoleic acid	3.29	0.91
18:1 n-9 (*cis*)	Oleic acid	31.30	18.52
18:1 n-9 (*trans*)		2.67	0.86
Polyunsaturated fatty acids (PUFA) (essential fatty acids)			
18:2 n-6	Linoleic acid	10.76	11.06
18:3 n-3	Linolenic acid	0.81	1.41
Long-chain PUFA (n-6)			
18:3 n-6	α-Linolenic acid	0.16	0.12
20:2 n-6		0.34	0.26
20:3 n-6	Dihomolinolenic acid	0.26	0.49
20:4 n-6	Arachidonic acid	0.36	0.82
Long-chain PUFA (n-3)			
20:5 n-3	Eicosapentaenoic acid	0.04	0.48
22:5 n-3		0.17	0.39
22:6 n-3	Docosahexaenoic acid	0.22	0.93

[a] Data from Jensen (1995).
[b] Data from Nigerian women who had a diet high in fish.

mammary alveolar cell producing, under some conditions, relevant concentrations in breast milk.

D. The Transcytotic Pathway

Proteinaceous substances from the interstitial space find their way into milk via a transcytotic pathway (Fig. 3D). One of these pathways, that for the secretion of IgA, has been relatively well characterized. An IgA receptor on the basolateral membrane of the cell interacts with dimeric IgA made by the plasma cells that reside in the interstitial spaces of the mammary gland. The IgA is endocytosed and transported in a vesicle to the apical membrane, where it is secreted with the extracellular domain of its transporter ("secretory component") into the milk. Secretory component renders the protein resistant to proteases, and the IgA thus has a reasonable survival time in the infant's intestine, where it may contribute to protection from infectious agents. Many other proteins that are not synthesized in the mammary alveolar cell enter milk via transcytosis, including hormones such as insulin and prolactin, growth factors such as IGF-1, and serum albumin.

E. Paracellular Transport

During lactation, tight junctions (Fig. 3E) form a seal between alveolar cells that closes the paracellular pathway and completely isolates the alveolar lumina from the interstitial spaces. These junctions are highly impermeable as shown by the high transepithelial potential (up to 35 mV) recorded between the plasma and milk space in a number of species and the low concentrations of sodium and chloride in normal human milk. Small molecules such as sucrose, injected up the teat of goats or mice into the milk space, are not able to reach the plasma in the fully lactating animal. In pregnancy, after involution and during mastitis, the tight junctions open by mechanisms that are not well understood and even large proteins can be transferred between the milk and interstitial spaces. Under these conditions plasma components such as sodium, chloride, and albumin enter the milk space directly from the interstitial space and the concentrations of sodium and chloride in the milk are high. Milk components such as lactose and possibly milk proteins also enter the bloodstream. Immune cells pass into milk via the paracellular pathway. They appear to be able to cross the paracellular pathway in the lactating gland without opening the tight junctions to the passage of other interstitial components.

V. LACTOGENESIS

Lactogenesis stage 2 represents a profound series of changes in the activity of the mammary epithelial cells from a quiescent state during pregnancy to the fully active secretory state characteristic of lactation. During lactogenesis the secretion product of the breasts is called colostrum. The composition of colostrum changes rapidly throughout the period of lactogenesis, representing a highly programmed series of

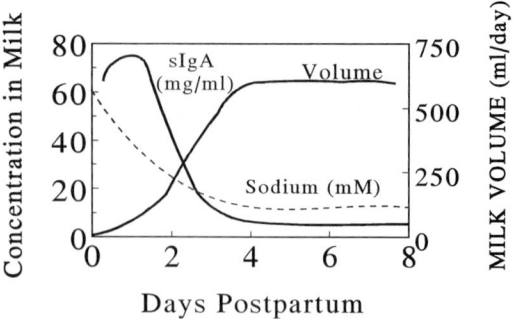

FIGURE 4 Lactogenesis stage 2. The time course of changes in sodium and secretory IgA concentrations and volume of milk during the first 8 days postpartum takes place with a 36-hr delay (graph based on data from 12 women from M. C. Neville *et al., Am. J. Clin. Nutr.* 54, 81–93, 1991).

events that lead to the secretion of mature milk. Slower changes in composition take place between Days 5 and 10 postpartum, leading many authorities to refer to the milk of this period as transitional milk. However, slow changes in milk composition occur throughout lactation and it is not clear that the changes that occur during the transitional period should be singled out for particular attention.

The first change to occur is a fall in the sodium (Fig. 4) and chloride concentrations in the milk and an increase in the lactose concentration signaling closure of the tight junctions, largely complete by 72 hr postpartum. With closure of the tight junctions, lactose, which is made by the epithelial cells, can no longer pass into the plasma and sodium and chloride can no longer pass from the interstitial space into the lumen of the mammary alveolus. The next change to occur is an increase in the rates of secretion of sIgA (Fig. 4) and lactoferrin. The concentrations of these two important protective proteins remain high for the first 48 hr after birth, together comprising as much as 10% by weight of the milk. They fall rapidly after Day 2, a consequence both of dilution as milk volume increases and of an actual decrease in the rate of their secretion. By 8 days postpartum, together they make up <1% of the milk; however, their secretion rate is still substantial, amounting to nearly 2 g per day for each protein.

Finally, about 40 hr postpartum, milk secretion begins in earnest with a 10-fold increase in volume, from about 50 to 500 ml/day, occurring over the subsequent 48 hr. This volume increase is perceived by the parturient woman as the "coming in" of the milk and is brought about by a massive increase in the rates of synthesis and/or secretion of almost all the components of mature milk.

The early changes in activity of the mammary gland are entirely the result of hormonal changes that take place with the birth of the infant. However, maintenance of milk secretion requires that milk be removed from the breast from approximately Day 3 postpartum onward. If milk removal does not occur irreversible involution begins to take place.

VI. LACTATION

With the completion of lactogenesis the breast enters into the stage of lactation, sometimes referred to as galactopoiesis. This stage persists as long as the infant continues to suckle at least twice a day. The daily milk volume transferred to the fully breast-fed infant increases on average from 500 ml on Day 5 to about 650 ml at 1 month and about 800 ml at 6 months. Milk volume is regulated mainly by infant demand as discussed previously. The mechanisms appear to involve an interaction between prolactin and local control that is only now beginning to be understood.

A. Regulation of Milk Secretion

Prolactin is necessary for milk secretion in women and suckling promotes its secretion. However, the volume of milk secreted is not directly related to the concentration of prolactin in the blood. Rather, local mechanisms within the mammary gland related to the amount of milk removed by the infant are thought to be responsible for the day-to-day regulation of milk volume. A protein factor called FIL is secreted with other milk components into the alveolar lumen. If milk is not completely removed from the gland, this factor, whose identity and mechanism of action are not yet entirely clear, interacts with the mammary alveolar cell and inhibits milk secretion possibly by altering the sensitivity of the cells to prolactin. Thus, increased emptying of the mammary gland brings about an increase in the rate of milk synthesis over

a period of days. Conversely, decreased emptying brings about a reduction in milk synthesis.

B. Regulation of Milk Ejection

Suckling initiates a neuroendocrine reflex essential for removal of milk from the breast. Afferent impulses travel via sensory neurons from the areola to the hypothalamus where they stimulate magnocellular neurons to fire, sending an impulse down their axons into the posterior pituitary where the hormone oxytocin is released. This hormone travels through the bloodstream to the breast where it causes contraction of the myoepithelial cells, forcing the milk from the alveoli into the mammary ducts and sinuses from whence it can be removed by the suckling infant. The activity of the magnocellular neurons can be profoundly influenced by higher brain centers. For this reason, emotional distress can inhibit let-down. Conversely, let-down is subject to conditioning so that women often release oxytocin at the sound of their (or someone else's) infant's cry or in response to a picture of their infant. Without an adequate let-down reflex, complete removal of milk from the breast is not possible. As stated previously, decreased emptying leads to milk stasis and inhibition of milk secretion. Thus, in the absence of an adequate let-down reflex, involution of the mammary epithelium can result.

VII. MATERNAL ADAPTATION TO LACTATION

The lactating breast is largely an autonomous organ dependent only on permissive levels of prolactin and other hormones, such as insulin and hydrocortisone, for continued milk secretion. The amount of milk produced is under local control and depends on the rate of milk removal from the breast as discussed previously. The mammary gland removes nutrients from blood according to its needs for substrate to synthesize milk components and is usually only marginally affected by changes in maternal dietary intake and other maternal factors such as fluid intake, body composition, illness, and exercise. The limits to which lactation is maintained in starvation and severe illness are unknown since experimental observations under such conditions are difficult. A single case report suggests that 7 days of total fasting does not significantly impact human milk secretion. The composition of the diet and/or maternal depletion does, however, affect some milk components, most notably water-soluble vitamins and some trace minerals such as selenium. The effects of diet on lipid-soluble vitamins are not yet entirely clear. The effects of the lipid composition of the diet have been discussed previously. It does not appear that fluid intake is directly linked to milk production because Gambian women who took no water during the day during the religious period of Ramadan produced normal volumes of slightly hypertonic milk. The total fat content of milk does not appear to be affected by dietary lipid intake but is inversely proportional to body fat. Diabetics may have difficulty initiating lactation but, if lactation can be established, have no problem nursing an infant. Mastitis alters milk composition by opening the tight junctions so that the sodium and chloride content of the milk is increased and the protein content decreased. Pregnancy anecdotally reduces milk volume, although there are no good data on this point. Exercise has not been observed to alter either milk volume or composition.

Compared to species such as rodents or dairy animals, whose body lipid reserves are depleted within a short time after lactogenesis by their very copious milk production, adaptations to lactation in the woman are relatively small except in the area of reproductive function. Exclusive breast-feeding inhibits the release of gonadotropin-releasing hormone from the hypothalamus, which tends to prevent the return of reproductive function after birth of the infant. However, even exclusive breast-feeding does not necessarily offer protection against a subsequent pregnancy. Bone calcium has been shown to decline during lactation in several studies, probably due to lack of estrogen during the period of postpartum amenorrhea, but appears to recover to normal values once menses resume. Women with marginal protein nutrition may suffer some deterioration of their protein status. The same is true for folate and the folate stores of women with marginal folate status may be depleted, a situation that may negatively impact a subsequent pregnancy. There is little evidence that lactation in women has any effect on insulin resistance, as it does in ruminants. This finding may be related to the fact that growth hormone levels are not elevated in lactating women as they are in dairy species.

Finally, one factor that has been shown in epidemiological studies to decrease the risk of breast cancer, particularly of the premenopausal variety, is pregnancy prior to the age of 20. If lactation follows that pregnancy, current evidence suggests that the risk is further reduced. This important observation signals a linkage between the differentiated function of the mammary gland and its susceptibility to abnormal proliferative states. Further research on the mechanisms underlying this linkage is likely to enhance our understanding of the origin of breast cancer as well as that of potential strategies for its prevention.

See Also the Following Articles

Breastfeeding; GnRH; Lactogenesis; Mammary Gland, Overview; Milk, Composition and Synthesis; Milk Ejection; Oxytocin

Bibliography

Dils, R. R. (1986). Comparative aspects of milk fat synthesis. *J. Dairy Sci.* **69**, 904–910.

Jensen, R. G. (1995). *Handbook of Milk Composition.* Academic Press, San Diego.

Lawrence, R. A. (1989). Breastfeeding and medical disease. *Med. Clin. North Am.* **73**, 583–603.

McNeilly, A. S., Tay, C. C. K., and Glasier, A. (1994). Physiological mechanisms underlying lactational amenorrhea. *Ann. N. Y. Acad. Sci.* **709**, 145–155.

Medina, D. (1996). The mammary gland: A unique organ for the study of development and tumorigenesis. *J. Mammary Gland Biol. Neoplasia* **1**, 5–20.

Neville, M. C. (1990). The physiological basis of milk secretion. *Ann. N. Y. Acad. Sci.* **586**, 1–11.

Neville, M. C., and Picciano, M. F. (1997). Regulation of milk lipid synthesis and secretion. *Annu. Rev. Nutr.* **17**, 159–183.

Neville, M. C., and Daniel, C. W. (1987). *The Mammary Gland; Development, Regulation and Function.* Plenum Press, New York.

Neville, M. C., and Neifert, M. R. (1983). *Lactation: Physiology, Nutrition and Breastfeeding.* Plenum Press, New York.

Newburg, D. S. (1996). Oligosaccharides and glycoconjugates in human milk: Their role in host defense. *J. Mammary Gland Biol. Neoplasia* **1**, 271–285.

Peaker, M., and Wilde, C. J. (1996). Feedback control of milk secretion from milk. *J. Mammary Gland Biol. Neoplasia*

Lactation, Nonhuman

Robert J. Collier

Monsanto Company

GLOSSARY

adaptation The physical or metabolic alteration of a species to permit the occupation of a specific ecological niche.

alveolus The basic secretory unit of the mammary gland. Each alveolus is spherical and composed of a single layer of epithelial cells that surround a cavity or lumen.

colostrum The first secretion of the mammary gland after parturition. It contains high concentrations of immunoglobulins and growth factors.

homeorhesis Orchestrated changes in the metabolism of a species to support a specific biological function.

involution The coordinated regression of the secretory component of the mammary gland at the cessation of milk secretion and removal. When complete, only the ductal structure of the mammary gland is apparent.

lactogenesis The initiation of milk secretion at parturition.

lactose The carbohydrate found in milk which is composed of a molecule of glucose and galactose. It is the primary osmotic determinant of milk.

Lactation is the process by which maternal members of the class Mammalia supply nutrients, immunity (to varying degrees), and growth regulatory compounds to the growing neonate. Milk is the collective term for this form of nourishment and milk is essential to the survival of every newborn mammal. The composition of milk is quite variable depending on the species, stage of development of the neonate, and environmental factors. Mammary development and onset and regulation of milk secretion are closely tied to the reproductive process. The extensive output of nutrients required to meet maintenance and growth requirements of the neonate makes lactation the most metabolically expensive phase of the reproductive process in mammals. Although the basic structure of the mammary gland has changed little across the class Mammalia, several species have evolved specific metabolic adaptations to meet the metabolic demands of lactation in their specific ecological niche.

I. INTRODUCTION

The mammary gland is a defining characteristic of the class Mammalia. Since egg-laying mammals produce milk, evolution of the mammary gland preceded placental gestation. Both the placenta and mammary gland confer advantages in reproductive efficiency and allow extended immaturity of offspring leading to increased neural development in young. In many mammals the mammary gland also facilitates passive disease resistance by way of colostrum, the first lacteal secretion following parturition. Many growth regulatory compounds are also found in milk which accelerate maturation of the digestive tract and internal organs. By way of immunity transfer, continued nutritional support, and growth regulation, the mammary gland is essential to the survival of newborn mammals. Interestingly, despite considerable dispersion of the class Mammalia across habitats and remarkable variation in the process of reproduction in this class, the basic structure of the mammary gland has changed very little. However, some species have evolved metabolic adaptations that allows them to meet the metabolic demands of lactation in their specific ecological niche.

II. MILK COMPOSITION

Milk is the only food source of a newborn mammal and therefore must contain all the key nutrients for normal growth and development. However, since requirements vary considerably across species depending on size, growth rate, and metabolic rate, the composition of milk from various mammals reflects these differences (Table 1). Milk composition is usually defined in terms of macronutrients and micronutrients. The macronutrients of milk are water, lactose, fat, protein, and minerals. The micronutrients in milk are trace elements, vitamins, hormones, nucleotides, immunoglobulins, free amino acids, and phospholipids.

Milk composition varies by species, stage of lactation, environment, diet, and age of the mother. Major differences among species can be attributed to differing nutrient requirements of the offspring. Typically, the milk produced by a dam during lactation is relatively consistent across all the mammary glands. In the case of marsupials, each mammary gland produces a milk of different composition due to differing ages of neonates undergoing development in the brood pouch.

The discovery that colostrum and milk contain several hormones and growth factors was the first evidence implicating these peptides in maternal involvement in postnatal growth regulation. The initial evidence indicating growth factors were present in mammary secretions led to the discovery of mitogens that were capable of causing cell division in several cell types. Since these reports, several laboratories have described the biochemical and functional characteristics of growth factors isolated from mammary secretions. Among the cell types used to identify these mitogens were fibroblasts, chondrocytes, muscle cells, epithelial cells, and mammary carcinoma cells. Clearly, a wide variety of mammalian cells respond to growth factors found in milk with increased cell division. Currently, more than 30 hormones and growth factors have been identified in milk from domestic animals and humans. However,

TABLE 1
Phylogenetic Variation in Mammalian Milk Composition[a]

Order	Species	Dry matter	Fat	Protein	Sugar	Ash
Marsupialia	Red kangaroo	22.8	4.9	6.7	9.8	1.4
Insectivora	White-toothed shrew	51.2	31.9	9.7	ND	2.6
Primates	Baboon	14.0	4.6	1.5	7.7	0.3
	Human	12.4	4.1	0.8	6.8	0.2
Lagomorpha	Rabbit	31.2	15.2	10.3	1.8	1.8
Rodentia	Brown rat	22.1	8.8	8.1	3.8	1.2
Carnivora	Dog	22.7	9.5	7.5	3.8	1.1
	Brown bear	33.6	18.5	8.5	2.3	1.5
	Weddell seal	57.2	42.1	15.8	1.0	—
Proboscidea	African elephant	17.3	5.0	4.0	5.3	0.7
Perissodactyla	Horse	10.5	1.3	1.9	6.9	0.4
Artiodayctyla	Pig	20.1	8.3	5.6	5.0	0.9
	Cow	12.4	3.7	3.2	4.6	0.7
	Water buffalo	16.8	6.5	4.3	4.9	0.8

[a] Adapted from Oftedal, 1984, pp. 50–56. ND, not determined.

little is known concerning growth factor concentrations in other species.

III. MAMMOGENESIS

The mammary gland is a modified sweat gland, and therefore a skin gland, and of ectodermal origin. The basic structure and location of mammary glands is established during embryonic development. The glands develop from bilateral mammary lines or ridges on the ventrolateral surface of the early embryo. These lines extend from the thoracic to the inguinal region. Mammary lines then differentiate into mammary buds, and the number and location of these buds along the mammary line varies between species. Mammary buds develop in the inguinal region in ungulates (deer, cattle, goats, sheep, and camels) and equidae (horses). In primates, they develop in the pectoral region and in litter-bearing species, such as the carnivores (cats and dogs), lagomorpha (rabbits), and suidae (pigs), they develop in the thoracic, abdominal and inguinal regions. The cetacea or sea mammals (dolphins, whales, seals, and manatees) develop mammary buds in either the pectoral or the inguinal region. The location of mammary buds then determines the ultimate location of the mammary glands and associated structures, such as the nipples or teats, blood vessels, and connective tissue support elements. Although there is a positive relationship between the number of mammary glands and the number of offspring, there is no apparent relationship between placental type and numbers of mammary glands (Table 2).

Further development of the mammary gland involves outgrowth of primary sprouts from the mammary bud into the surrounding mesenchyme. Secondary sprouts then arise from the primary sprouts. Both primary and secondary sprouts canalize to form true ducts. Supporting structures from the mammary gland also develop during this period, including connective tissue, mammary fat pads, and blood and lymph vessels. The size of the mammary fat pad in some species influences the ultimate extent of the mammary gland because the ductule structure grows out into this fat pad. The ductule structures open to the exterior at the nipple or teat. The number of teat openings in each mammary gland also varies between species from 1 to more than 10. In species with a single opening there is a collecting cistern to allow pooling of milk from several ducts before exiting the gland. The gland cistern, teat, and primary ducts are present at birth. However, extensive ductule development and formation of the secretory units (alveoli) continues into later life. At birth, male and female mammals differ very little in extent of mammary development; however, exposure to testosterone in males during fetal life markedly reduces subsequent mammary development after birth. Thus, mammalian males uniformly demonstrate very little growth

TABLE 2
Variation in Placental Type and Number of Mammary Glands in the Class Mammalia

Order	Species	Placental type	Total glands
Carnivora	Dog	Endotheliochorial	10
Rodentia	Rat	Hemochorial	12
Lagomorpha	Rabbit	Hemochorial	10
Cetacea	Whale	Epitheliochorial	2
Perissodactyla	Horse	Epitheliochorial	2
Artiodactyla	Cow	Epitheliochorial	4
Artiodactyla	Sheep	Epitheliochorial	2
Artiodactyla	Goat	Epitheliochorial	2
Artiodactyla	Pig	Epitheliochorial	12

of the mammary gland at subsequent puberty and maturity.

The mammary gland undergoes periods of mammary growth and regression throughout the lifetime of the female mammal associated with reproductive cycles. The mammary gland is hormone dependent and continued mammary development in female mammals is regulated primarily via circulating and local hormonal signals, transmitted into individual cells by binding to specific hormone receptors. The three general classes of hormones involved in regulating mammary development are peptide hormones, steroid hormones, and prostaglandins.

IV. LACTOGENESIS

Long recognized as a sign of impending parturition, onset of milk synthesis and secretion (lactogenesis) has proved to be a fascinating field of biology. Until recently, the majority of research in this area focused on the endocrine changes associated with lactogenesis and changes in mammary cell metabolism associated with the onset of milk synthesis and secretion.

In the 1970s it was recognized that parturition and lactogenesis were associated with massive shifts in nutrient partitioning in the female mammal to support nourishment of the growing neonate. This process was identified as a specific physiological state known as homeorhesis. Homeorhesis is defined as the coordinated regulation of metabolism to support a specific biological function. This is contrasted with homeostasis, which is the regulation of a specific body function around a set point. Two examples of homeorhesis are pregnancy and lactation. In each case, the metabolism of the mother is altered to support the new metabolic demands imposed by the growing fetus (pregnancy) or growing neonate (lactation). A key result of the homeorhetic process is the proper coordination of metabolism to avoid metabolic disease. This is most critical at lactogenesis because the shift in nutrient demand is rapid and extensive. This is even more impressive when considering that several species neither eat nor drink during lactation. Hibernating animals such as bears and the aquatic mammals that go ashore avoid eating and drinking during lactation to secure a safe environment for their pups.

In many species the process of lactogenesis occurs in two stages. In stage I, beginning several days prior to parturition, protein-rich secretion (colostrum) builds up in alveolar lumina as maternal immunoglobulins are transported from blood across the mammary cell into the aveolar lumen. In stage II, occurring just hours prior to parturition, mammary secretory cells rapidly increase their synthetic rate and milk fat, protein, and lactose are secreted into the aveolar lumen. This causes the characteristic engorgement of the mammary gland at parturition. In humans, stage II lactogenesis occurs after parturition. Both stages of lactogenesis take place in response to specific hormonal signals.

V. LACTATION

Following lactogenesis, the mammary gland is anatomically and biochemically competent to synthesize and secrete milk. Synthesis and secretion of milk products are also controlled by hormonal signals in all species studied to date. Physiologically, three sets of stimuli must be maintained to ensure lactation success: (i) those stimuli that maintain cell numbers, (ii) those stimuli that maintain secretory activity, and (iii) those stimuli associated with milk removal.

The final stage of lactation is milk ejection or removal. Stimuli to initiate milk ejection involve both neural and endocrine signals, making this a neuroendocrine response. The efferent and afferent pathways controlling the milk ejection reflex are presented in Fig. 1. The basic pathway in all placental mammals involves the following: an external signal such as touch, sight, or smell; nervous stimulation of the neurohypophysis; oxytocin release into blood and transport to the mammary gland; and subsequent contraction of alveolar myoepithelial cells, which contract and force milk from the alveolus into the collecting duct. In prototherian mammals, such as the platypus and echidna, there is no external nipple or teat and the milk oozes from gland openings at the base of hair follicles. However, the young do exhibit suckling behavior that may influence the rate of milk flow.

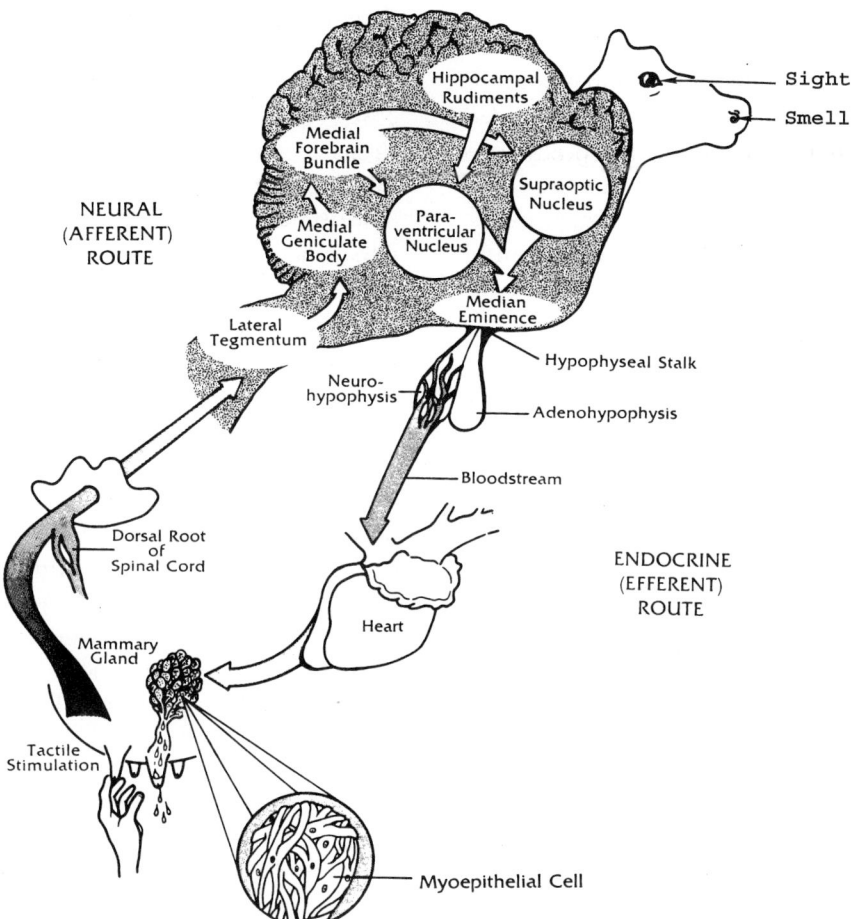

FIGURE 1 Neuroendocrine pathway from tactile stimulation of the mammary gland to oxytocin release at the posterior pituitary into the bloodstream and subsequent milk ejection.

VI. INVOLUTION

Lactation is terminated with weaning of the offspring. Since there is no further need to synthesize milk, the remaining alveolar structures are lost through apoptosis or cell death unless a concurrent pregnancy and impending parturition prevents their regression. As involution proceeds, the mammary ductule tree regresses to a rudimentary structure awaiting the next round of growth in response to a new pregnancy. In most species the normal state of a mature female is pregnancy and many species initiate a new pregnancy prior to or immediately following the weaning of the neonate. Thus, in most female mammals the mammary gland never fully regresses before the next round of mammary growth is initiated.

VII. ADAPTATIONS

Despite the fact that mammary gland anatomy and development has been conserved across species, adaptations have taken place that have permitted variation in the pattern and metabolic control of lactation and alteration in the length of lactation.

Lactation length is an indicator of parental investment in the young as well as the dependency of the offspring on milk for survival. The range of lactation length is approximately three orders of magnitude, from 2 to 5 days in guinea pigs, hooded seals, elephant shrews, and spiny rats to over 900 days in some of the apes, such as chimpanzees and orangutan. For mammals, length of lactation is positively correlated with adult female mass, although several exceptions to this pattern exist. Availability of first solid food, which is highly influenced by habitat and environment, would also limit lactation length since it would provide an alternate nutrient source. Finally, lactation length is influenced by gestation length. Animals with longer pregnancies have longer intervals to provide nourishment to offspring prior to birth.

Examples of metabolic adaptations include the following:

1. The individual regulation of mammary function in marsupials: Marsupial neonates are only a fraction of maternal weight at birth and suckle for extended periods before leaving their mother's pouch. During this interval other pregnancies can occur since the gestation period in marsupials is quite short. Thus, a marsupial having four mammary glands can have four suckling young at different stages of development. Since the nutrient requirements change quite dramatically as body size increases in these neonates, the milk composition in each gland can differ as well. Factors controlling asynchronous lactation in marsupials are not well defined but clearly involve an interaction with the neonate.

2. Lactations carried out during estivation and hibernation in sea lions and bears: These species seek changes in their habitat to undergo parturition and carry out lactation. Bears hibernate during winter and during this hibernation deliver the cub and carry out a lactation without eating or drinking. Likewise, sea lions, gray seals, and harbor seals leave the ocean to calve ashore and do not return to the water to eat or drink during the entire lactation. The ability to coordinate their body metabolism to allow rapid increases in body weight of the suckling neonate (up to 238 kg in elephant seals) while they fast is an excellent example of homeorhesis in action.

3. Alteration in metabolism of mammary tissue in ruminants: In ruminants glucose consumed in the diet is quickly utilized by rumen bacteria as an energy source, preventing any glucose from reaching the lower digestive tract. Mammary glands produce all their glucose from the volatile fatty acid proprionate, a by-product of bacterial metabolism in the rumen. During lactation there is a dramatic increase in glucose requirement to meet demands for lactose synthesis in mammary tissue. Alterations in ruminant mammary tissue metabolism prevent glucose from contributing carbons or reducing equivalents for fatty acid synthesis, which spares glucose for lactose synthesis.

4. Alteration in lactose synthesis to decrease water content of milk: In many marine species the nutrient requirements of the neonate are quite high due to the environment (whales and dolphins), to high growth rate requirements (elephant seals, gray seals, and harbor seals), or because the lactation is of short duration. In these species the water content of milk is greatly reduced, and the protein and fat content is quite high, whereas the synthesis of lactose in mammary tissue is not measurable or very low. Since lactose is the major osmotic determinant of milk production, the reduction in lactose synthesis results in low water and high solids content of milk. These rich milks allow rapid weight gain in the neonate.

See Also the Following Articles

Mammary Gland Development; Marsupials; Milk, Composition and Synthesis; Ruminants

Bibliography

Akers, R. M. (1985). Lactogenic hormones: Binding sites, mammary growth, secretory cell differentiation and milk biosynthesis in ruminants. *J. Dairy Sci.* **68**, 501–519.

Bauman, D. E., and Currie, W. B. (1980). Control of nutrient partitioning in lactating ruminants. In *Biochemistry of Lactation* (T. B. Mepham, Ed.), p. 437. Elsevier, Amsterdam.

Bonner, N. W. (1984). Lactation strategies in pinnipeds: Problems for a marine mammalian group. *Symp. Zool. Soc. London* **51**, 253–272.

Collier, R. J., McGrath, M. F., Byatt, J. C., and Zurfluh, L. L. (1992). Regulation of mammary growth by peptide hormones: Involvement of receptors, growth factors and binding proteins. *Livestock Prod. Sci.* **35**, 21–33.

Forsyth, I. A. (1991). The mammary gland. *Baillieres Clin. Endocr.-Metab.* **5**, 809–831.

Griffiths, M. (1978). *The Biology of Monotremes.* Academic Press, New York.

Hayssen, V. (1993). Empirical and theoretical constraints on the evolution of lactation. *J. Dairy Sci.* **76**, 3213–3233.

Jenson, R. G. (Ed.) (1989). *Handbook of Milk Composition.* Academic Press, New York.

Klagsbrun, M. (1978). Human milk stimulates DNA synthesis and cellular proliferation in cultured fibroblasts. *Proc. Natl. Acad. Sci. USA* **75**, 5057–5061.

Knight, C. H. (1984). Mammary growth and development: Strategies of animals and investigators. In *Physiological Strategies in Lactation* (M. Peaker, R. G. Vernon, and C. H. Knight, Eds.), pp. 147–166. Academic Press, London.

Koldovsky, O. (1989). Search for role of milk-borne biologically active peptides for the suckling. *J. Nutr.* **119**, 1543–1551.

Oftedal, O. T. (1984). Milk composition, milk yield and energy output at peak lactation: A comparative review. In *Physiological Strategies in Lactation* (M. Peaker, R. G. Vernon, and C. H. Knight, Eds.), pp. 33–74. Academic Press, London.

Renner, E. (Ed.) (1989). *Micronutrients in Milk and Milk-Based Food Products.* Elsevier, New York.

Vernon, R. G., and Flint, D. J. (1983). Control of fatty acid synthesis in lactation. *Proc. Nutr. Soc.* **42**, 315–331.

Lactogenesis

R. Michael Akers

Virginia Polytechnic Institute and State University

GLOSSARY

alveolus A milk-producing unit of the mammary gland of a lactating animal; composed of a hollow sphere of secretory epithelial cells; surrounded by a network of myoepithelial cells and blood vessels; and drained by a terminal duct.

caseins A major class of specific milk proteins made by the secretory cells of the alveolus. They account for 82% of the specific milk proteins and are packaged into colloidal micelles for secretion.

α-lactalbumin A specific milk whey protein that is induced by prolactin; part of the enzyme lactose synthetase necessary for the production of lactose (the primary milk carbohydrate).

mammary ducts Hollow tubes composed of a double layer of nonsecreting epithelial cells, which direct mammary secretions from the alveolar lumen to the teat or nipple.

mammary explant A small portion of intact mammary tissue (~3 mg) removed from the mammary gland and often used in tissue culture to determine the effects of hormones and other substances on milk component biosynthesis or secretory cell differentiation.

myoepithelial cells Elongated, stellate cells which form a mesh-like network surrounding the alveolus. In response to oxytocin the myoepithelial cells contract to initiate milk letdown.

I. INTRODUCTION

With the possible exceptions of bottle-fed humans and milk replacer-fed dairy calves, the success of reproduction cannot be declared with the birth of healthy offspring, but rather only after initiation of milk synthesis and secretion and suckling of the neonate. Development of the mammary gland to allow growth of potential secretory cells during gestation and subsequent differentiation of these cells to allow onset of milk synthesis and secretion in conjunction with parturition is indeed a biological marvel—a hallmark of mammals. Milk of all mammals contains variable amounts of proteins, carbohydrates, and fats in an aqueous medium. Although there are species differences with regard to details of milk composition, having the birth of the neonate and functionality of the mammary gland coincide is obviously critical. The purpose of this article is to provide an overview of final stages of mammary development during gestation and in particular the dramatic, acute changes in secretory cell structure and function as the gland prepares for onset of copious milk secretion referred to as lactogenesis.

II. OVERVIEW OF MAMMARY STRUCTURE

As perhaps best illustrated in a mammary whole mount prepared from the mammary gland of a virgin mouse (Fig. 1), mammary ducts composed of hollow tubes of epithelial cells elongate throughout the mammary fat pad in the period before and after puberty. These ducts provide a framework for the subsequent appearance of side branches and alveolar structures largely during gestation. The situation is grossly

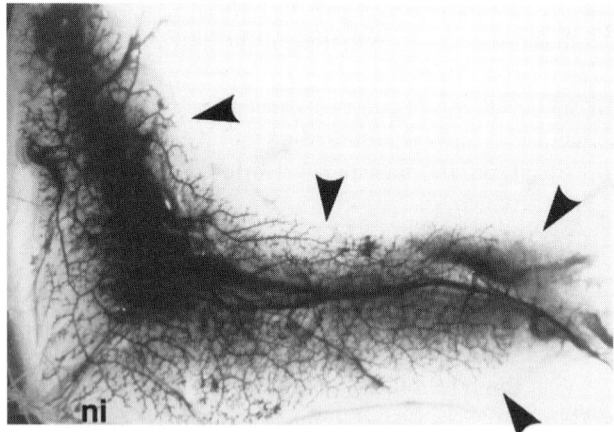

FIGURE 1   Fixed, defatted, stained whole mount of fourth inguinal mammary gland of a virgin mouse. Note that the gland has been completely defatted to allow observation of the epithelial elements of the gland. Note network of ducts which radiate throughout the entire fat pad. Magnification approximately 3×. Small arrows indicate the margin of the mammary fat pad. ni, nipple.

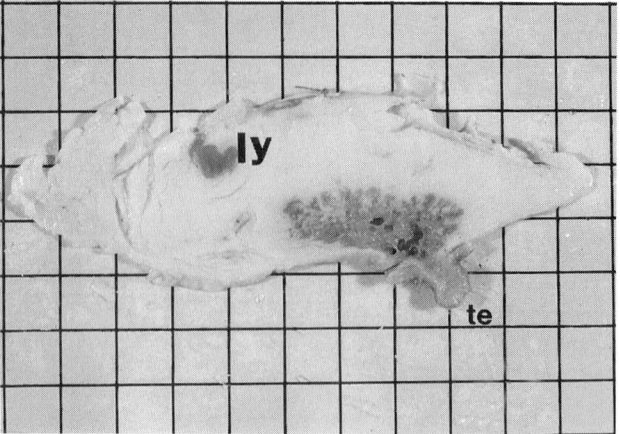

FIGURE 2   Saggital slice through the mammary gland of a 12-week-old ewe. Note the dark compact mass of mammary parenchyma in the region above and to the left of the teat. Slices were photographed against a background of 1.27 cm² grid squares. Despite the appearance of a large area of stromal tissue surrounding in the developing parenchyma, there is in reality much less fatty stromal directly associated with the developing epithelial tissue than in the mouse. ly, lymph node; te, teat (unpublished photomicrograph by Ellis and Akers, 1995).

similar in the ruminant except that the mammary stromal tissue contains many fewer adipocytes than in rodents and proliferation of ducts into the mammary stroma follows a much more dense, compact pattern such that the stroma is not filled during the peripubertal period (Fig. 2). Regardless, especially during the second half of gestation, alveolar formation predominates mammogenesis as new alveoli are formed and existing alveoli increase in size. Connected via a terminal duct to progressively larger ducts and ultimately the teat or nipple, it is the epithelial cells of the alveoli that synthesize and secrete milk. Once lactation is established, milk secreted by these epithelial cells is stored in the lumenal spaces of the hollow alveoli and ducts between milking and suckling episodes (Fig. 3). There is also storage of milk in the gland and teat cisterns of those animals with mammary glands arranged into an udder (ruminants). Thus, it is the single layer of epithelial cells of the internal surface of the alveoli that is responsible for milk synthesis and secretion and it is the onset of these functions that is the focus of lactogenesis. Specifically, lactogenesis is most of-

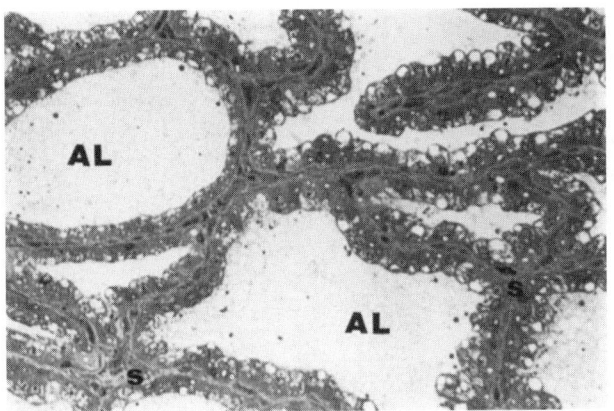

FIGURE 3 Light microscope view of a section of mammary parenchymal tissue from a lactating cow. Several alveoli cut in cross section are illustrated. Lighter stained areas are alveolar lumena. Secretory cells form a single layer around the periphery of each alveolus. Because of accumulation of secretions within the lumenal spaces, alveoli are closely packed together with little apparent stomal or vascular tissue. The following abbreviations are used in various figures: AL, alveolar lumen; s, stroma.

ten described as a two-stage process. Stage I consists of limited structural and functional differentiation of the secretory epithelium during the last third of pregnancy. Stage II involves completion of differentiation during the immediate periparturient period coinciding with onset of copious milk synthesis and secretion.

III. SECRETORY CELL DIFFERENTIATION

The successful lactation requires at least three distinct events: (i) the prepartum proliferation of alveolar epithelial cells, (ii) biochemical and structural differentiation of these cells, and (iii) synthesis and secretion of milk constituents. During lactation these secretory cells synthesize and secrete copious amounts of carbohydrate, protein, and lipid. Production of this complex mixture of nutrients requires close coordination between biochemical pathways to supply synthesis intermediates and secretory pathways for secretion. To illustrate, the disaccharide lactose is the predominate sugar in milk. Membrane-bound galactosyltransferase and the whey protein α-lactalbumin, which form the enzyme complex necessary for lactose synthesis, combine in the Golgi apparatus to form lactose synthetase, which serves to combine glucose and galactose and thereby form lactose. Full activation of the α-lactalbumin gene and subsequent synthesis of α-lactalbumin is most closely associated with stage II of lactogenesis. Indeed, continuing synthesis of lactose is essential to maintain milk volume and composition. This is because lactose becomes trapped in secretory vesicles and water is osmotically drawn into the vesicles. These lactose- and protein-containing vesicles fuse both singly and in chains with the apical plasma membrane and release their contents into the alveolar lumen via exocytosis. Since the plasma membranes of the secretory and ductular cells are also impermeable to lactose, the osmolarity of the secretory vesicles is maintained in secreted milk and water remains within the lumens of the alveoli and ducts. It is generally accepted that some minimal level of lactose production is likely essential to maintain the relative fluidity of milk for efficient milk removal either by the sucking young or by the milking machine. This is perhaps best illustrated by recent data for transgenic mice in which prevention of α-lactalbumin synthesis essentially prevented lactation because sucking mice failed to survive despite the presence of milk proteins and fat in the alveolar lumina of mammary tissue of "lactating" mothers.

Although it is clear that biochemical differentiation of the secretory cells is required for onset of milk secretion, it should also be appreciated that the cells must also acquire the structural machinery needed to synthesize, package, and secrete milk constituents. When alveolar cells first appear during midgestation, they exhibit few of the organelles needed for copious milk biosynthesis or secretion (Fig. 4A). The cells are characterized by a sparse cytoplasm with few polyribosomes, few clusters of free ribosomes, limited rough endoplasmic reticulum, rudimentary Golgi usually in close apposition to the nucleus, some isolated mitochondria, and occasional widely dispersed vesicles. Individual cells often contain large lipid droplets (especially during later stages of gestation) that along with irregularly shaped nuclei account for much of the cellular area. Soon after the

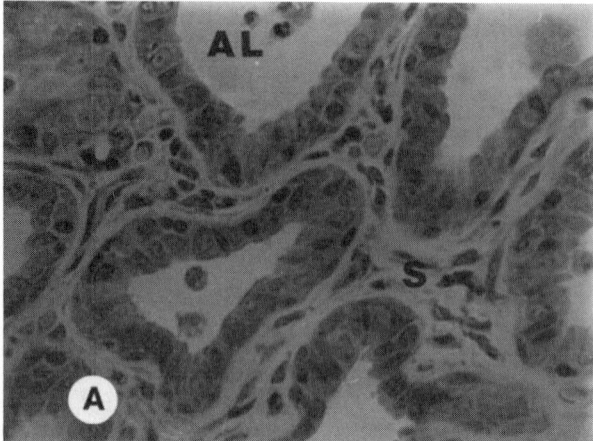
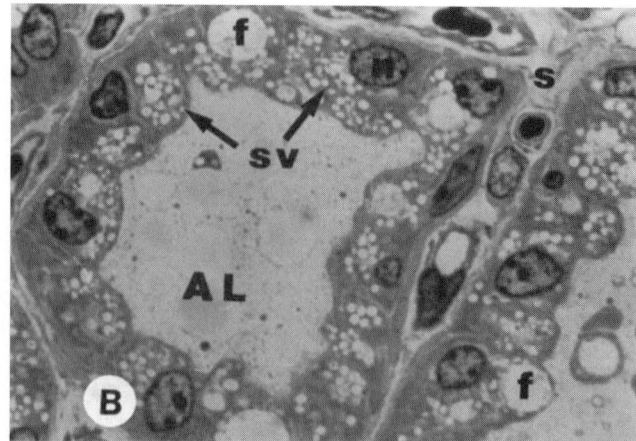

FIGURE 4 (A) Mammary tissue taken from a first-gestation heifer approximately 4 months prepartum. Several alveoli are shown. Secretory cells are poorly differentiated with no evidence of secretory activity. (B) Mammary tissue for the same heifer taken 1 week postpartum. Illustrated are parts of two alveoli. The secretory cells are well differentiated, basal areas of the cells are darkly stained, and apical regions have a distinct lacy appearance because of the presence of numerous secretory vesicles and lipid droplets. Approximate magnification, 350 and 900×, respectively, for A and B. The following abbreviations are used in various figures: AL, alveolar lumen; s, stroma; sv, secretory vesicles; f, fat droplets.

alveolar structures appear, lumenal spaces accumulate increasing volumes of fluid and progressively increasing concentrations of serum-derived proteins. Accumulated secretions result in formation of immunoglobulin-rich colostrum, which—depending on the species—may be essential for the survival of the offspring. As parturition approaches the cells undergo a dramatic structural transformation as undifferentiated cells characteristic of the early gestation appear less frequently. Instead, cell nuclei become more rounded and positioned in the basal area of the cells. Lateral and basal regions of the cell become filled with abundant arrays of rough endoplasmic reticulum and small lipid droplets. The apical area becomes populated with swollen arrays of Golgi membranes, developing secretory vesicles, and small lipid droplets. Even in the light microscope these changes are evident (Fig 4B). A lacy appearance highlights the apical region of the cell because of the abundance of secretory vesicles in contrast with the darkly stained basal–lateral cytoplasm (ergastoplasm). The fully differentiated cell becomes decidedly polarized with the basolateral area devoted to the uptake of precursors and synthesis of proteins and lipids and the apical cytoplasm, with now abundant Golgi, devoted to posttranslational modification of proteins and packaging of proteins and lactose for secretion from the cell (Fig. 5).

IV. CONTROL OF LACTOGENESIS

As confirmed in numerous species, now classic mammary explant culture studies demonstrated that the major positive regulators of structural differentiation of the secretory cells are glucocorticoids and prolactin. Mammary tissue explants from pregnant donors of essentially all tested animals were shown to exhibit both biochemical and structural differentiation when incubated in a combination of insulin, glucocorticoids, and prolactin. It has been generally believed that insulin is necessary for mammary tissue maintenance in culture, but species variation in insulin sensitivity of the mammary gland in intact animals cast some doubt on this belief. Recent data support the idea that insulin-mediated effects on mammary cells in culture may actually represent effects more appropriately ascribed to the insulin-like growth factors (IGF-I and IGF-II). This is because mammary epithelial cells have specific IGF-I receptors and insulin (especially at higher concentrations typical of culture experiments) is likely to bind to the IGF-I

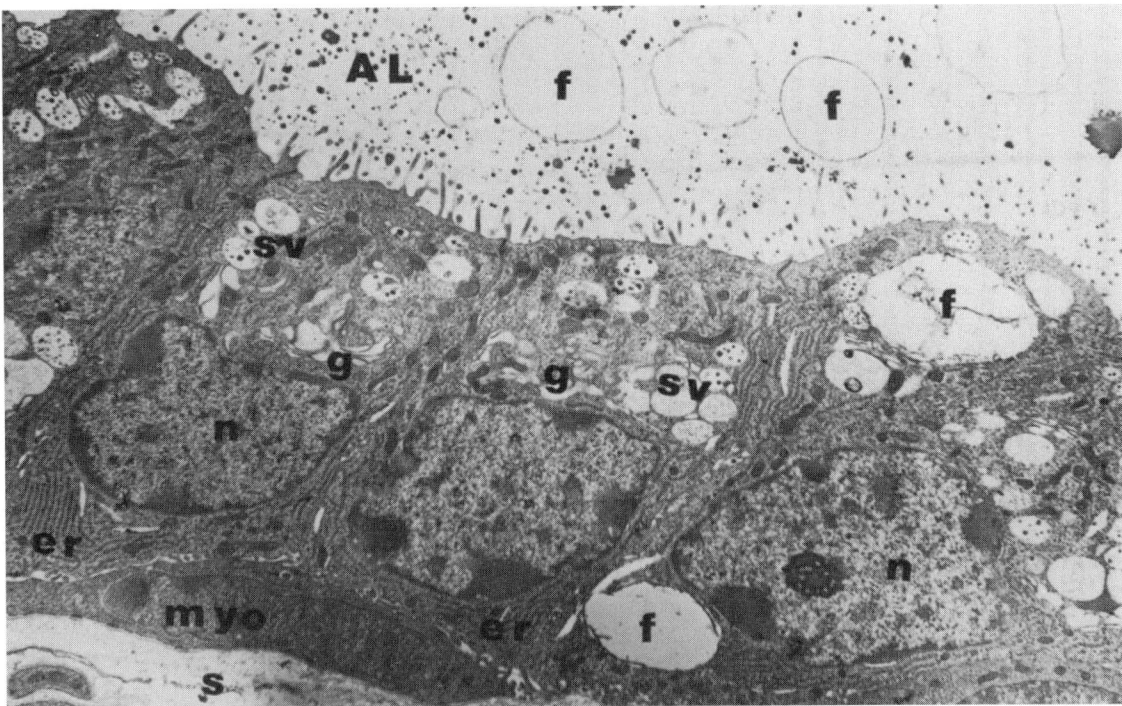

FIGURE 5 Transmission electron microscope view of mammary alveolar cells taken from a lactating cow 2 days postpartum. The profiles of three epithelial cells illustrate a stage of intermediate structural differentiation of the alveolar epithelium as stage II lactogenesis is initiated. Compared with cells prior to parturition, cell nuclei are more rounded and basally displaced, secretory vesicles appear juxtapositioned to the Golgi, and the basal area of the cells contains many strands of rough endoplasmic reticulum. Numerous casein micelles (black particles) and fat globules are present in the alveolar lumen. Approximate magnification, 6000×. The following abbreviations are used in various figures: AL, alveolar lumen; f, fat droplets; sv, secretory vesicles; g, Golgi apparatus; n, nucleus; er, endoplasmic reticulum; s, stroma; myo, myo-epithelial cell.

receptor. In general terms, glucocorticoids are most closely associated with development of rough endoplasmic reticulum and prolactin with maturation of the Golgi apparatus and appearance of secretory vesicles.

Widespread development of molecular techniques applied to mammary gland biology has served to solidify the idea that prolactin and glucocorticoids are primary stimulators of mammary cell differentiation. For example, with these techniques both prolactin and glucocorticoid response elements were found within the promoter regions of the genes for several mammary-specific milk proteins. Similarly, induction of both mRNA and specific milk proteins in response to addition of prolactin or glucocorticoids in isolated mammary epithelial cells indicates the importance of these hormones in lactogenesis.

On the other hand, despite the continuous pres-

ence of both prolactin and glucocorticoids in blood circulation during gestation, stage II of lactogenesis is held in check until the period just prior to parturition. Given that removal of mammary secretion prepartum can induce premature lactogenesis, assay of serum concentrations of the milk protein α-lactalbumin provides an excellent noninvasive illustration of each of the two phases of lactogenesis. As shown in Fig. 6, for dairy heifers in their first pregnancy, α-lactalbumin concentrations are undetectable before Day 200 prepartum but reach levels of about 20 ng/ml by 120 days, then increase exponentially during the last 30 days prepartum (40 to ~300 ng/ml) within a day of calving. Similar changes are also evident among dairy animals induced into lactation with exogenous hormones.

Clearly the endocrine system is a major regulator of stage II lactogenesis. Essentially, the combined

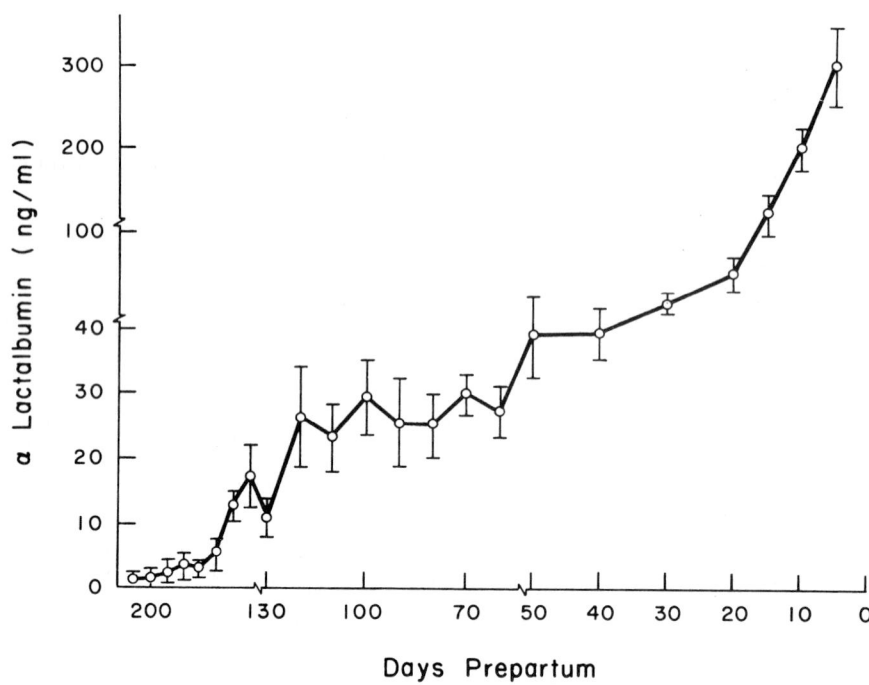

FIGURE 6 Mean concentrations of α-lactalbumin in sera of pregnant heifers prior to parturition. Each point is the mean ± SE from ≧7 heifers. Serum α-lactalbumin concentrations (ng/ml) averaged 7.3 ± 1.3 (n = 146) at >120 days, 29.9 ± 1.8 (n = 305) at Days 120–130, and 133.2 ± 11.8 (n = 115) at <30 days prepartum, respectively. Note breaks in the scales for both the x and y axes. Clearly, there is an increase in α-lactalbumin at about 120 days prepartum (stage I lactogenesis) and a further marked increase about 15 days prepartum (stage II lactogenesis) (data from McFadden, Akers, and Kazmer, *J. Dairy Sci.* **70**, 259–264, 1987).

effects of positive stimulators (prolactin, glucocorticoids, growth hormone, and estradiol) and the overriding negative influence of progesterone interact to determine the timing of the onset of copious milk secretion. A wealth of studies have reported changes in blood concentrations of these hormones in correspondence with parturition. Fundamentally, in cattle, for example, there are relatively consistent large increases in concentrations of serum prolactin in the period of several days around parturition and more acute increases in concentrations of glucocorticoids in closer association with the actual birth process. Concentrations of estradiol progressively increase during late gestation to a maximum within a few days of parturition. In contrast, in association with luteolysis of the corpus luteum, concentrations of progesterone abruptly decline 3 or 4 days prior to parturition. Changes in other hormones (prostaglandin F, which also increases at parturition) as well as changes in hormone and growth factor concentrations in mammary secretions may also serve to regulate stage II lactogenesis. In particular, there are marked changes in concentrations of IGF-binding proteins during the periparturient period. However, it remains to be seen if IGFs are acutely involved in lactogenesis. It is also possible that such changes correspond to a decrease in mammary cell proliferation with the onset of lactation, given that IGFs are potent stimulators of mammary cell proliferation. In rodents at least, there is also evidence that epidermal growth factor may act to modify the ability of prolactin to stimulate milk protein synthesis and secretion. It should not be forgotten that many of the growth factors and other biologically active substances in milk, especially during the period early postpartum, may have evolved for a role in gastrointestinal tract physiology quite independent of possible roles in maternal mammary physiology.

On the other hand, even though progesterone certainly can inhibit lactogenesis, simple removal of progesterone does not necessarily induce onset of lactation. It is also clear that for many species, lactogenesis is well under way before the marked decline in progesterone in blood occurs immediately before parturition. Also, prepartum milk removal is associated with premature differentiation of the mammary epithelium and subsequent appearance of mammary secretions of essentially normal composition sometimes several weeks before parturition.

Some of the best evidence for the importance of increased periparturient secretion of prolactin in stage II lactogenesis has come from experiments in which the administration of a dopamine agonist has been used to inhibit prolactin secretion and correspondingly prevent lactation. However, for many nondairy species these data are confounded by the lack of suckling, which as data for prepartum milking suggest, is also critical in the early phases of stage II lactogenesis. In ruminants in which postpartum milking continues, administration of a dopamine agonist reduced basal prolactin concentration about 80% and prevented the usual periparturient rise as well as milking-induced prolactin during the first week postpartum. Milk production was reduced 45% during the first 10 days postpartum. Lost milk production was associated with reduced synthesis of α-lactalbumin, lactose, and fatty acids, as well as impaired structural differentiation of the mammary secretory cells. Cows treated with exogenous prolactin in addition to the agonist (to replace the periparturient surge in prolactin) showed no loss of milk production or effects on milk component biosynthesis or alveolar cell differentiation. Clearly, prolactin plays a primary role in mammary cell differentiation and lactogenesis. That the periparturient period is critical can also be gleaned from experiments in which intramammary infusion of the microtubule inhibitor colchicine during the week prior to parturition also markedly inhibited subsequent structural and biochemical differentiation of the secretory epithelium. This suggests that *in vivo* mammary cell differentiation normally occurs in a relatively short period of time during the periparturient period and that disruption of the process during this period may have long-term consequences on milk production.

In addition to circumstantial evidence related to changes in circulating hormone concentrations, there is a marked increase in numbers of mammary cell receptors for prolactin, IGF-I, and cortisol during late gestation. It is also relevant that progesterone receptor concentration is correspondingly reduced with the onset of lactation. Thus, simultaneous, coordinated changes in circulating hormones and receptors appear to regulate the timing of lactogenesis.

Data from a variety of tissue culture experiments also strongly support a role of these hormones in lactogenesis. For example, additions of estradiol or cortisol markedly enhance prolactin-induced secretion of α-lactalbumin by mammary explants taken from pregnant cows. Mammary explants from estrogen-primed or pregnant mice also require insulin, cortisol, and prolactin for the accumulation of casein as well as α-lactalbumin. However, some caution is advised with wholesale extrapolation of results from culture to the intact animal as well as uncritical extrapolation between species. For example, induction of the various milk proteins in culture does not necessarily reflect the timing of events *in vivo*. Nor do existing culture methods allow consistent synthesis and secretion of milk lipids. Finally, hormone concentrations employed may not accurately reflect the situation at the level of the mammary cell in the animal. Indeed, culture systems by their very nature represent relatively uncomplicated regulation compared with the intact animal.

See Also the Following Articles

Lactation; Mammary Gland; Milk, Composition and Synthesis

Bibliography

Akers, R. M. (1990). Lactation physiology: A ruminant animal perspective. *Protoplasma* **159**, 96–111.

Akers, R. M. (1994). Lactation. In *Encyclopedia of Agricultural Science*, Vol. 2, pp. 635–643. Academic Press, San Diego.

Tucker, H. A. (1994). Lactation and its hormonal control. In *The Physiology of Reproduction* (E. Knobil and J. D. Neill, Eds.), 2nd ed., pp. 1065–1098. Raven Press, New York.

Lactotrophs

Tom E. Porter

University of Maryland

GLOSSARY

adenohypophysis Another name for the anterior pituitary gland.

corticotroph An anterior pituitary cell type that produces adrenocorticotropic hormone.

gonadotroph An anterior pituitary cell type that produces luteinizing hormone and/or follicle-stimulating hormone.

lactotroph An anterior pituitary cell type that secretes prolactin.

prolactin One of six anterior pituitary hormones responsible for maintaining milk production in mammals and nesting behavior in birds.

Rathke's pouch The embryonic fold in oral ectoderm that grows to form the anterior pituitary gland.

somatotroph An anterior pituitary cell type, closely related to lactotrophs, which secretes growth hormone.

thyrotroph An anterior pituitary cell type that produces thyroid-stimulating hormone.

Lactotrophs comprise the most predominant cell type in the anterior pituitary gland or adenohypophysis. These cells are also referred to as lactotropes, mammotropes, or mammotrophs, although the more commonly used term lactotroph will be used throughout this article. Lactotrophs secrete the anterior pituitary hormone prolactin, which functions to regulate a number of physiological processes, most notably lactation in mammals and incubation behavior in birds. This article will focus on several aspects of pituitary lactotrophs not related directly to the secretion or function of prolactin. Specifically, this article will describe the differentiation of lactotrophs during embryonic development, the alterations in lactotroph abundance associated with changes in physiologic condition, and the existence of functional subpopulations of lactotrophs in adults.

I. DIFFERENTIATION OF LACTOTROPHS

A. Rathke's Pouch and Organogenesis

The anterior pituitary forms during embryonic development from an evagination of oral ectoderm known as Rathke's pouch. Formation of Rathke's pouch occurs by Day 8 and Week 5 of gestation in rats and humans, respectively, and by Day 4 of embryonic development in chickens. The cells forming the anterior surface of Rathke's pouch proliferate to form the anterior pituitary. The posterior surface of Rathke's pouch grows to form the intermediate lobe of the pituitary in some mammals including rats. Although the regulation of these events is not well understood, two homeobox gene products have been strongly implicated as controlling the initial formation of the anterior pituitary. These proteins, Rpx1 and Lhx3, are similar in structure to proteins known to control the formation of various body structures in *Drosophila* (fruit flies). Although Rpx1 is present only during the initial formation and proliferation of Rathke's pouch, Lhx3 persists through differentiation of the various anterior pituitary cell types into adulthood. Repression of Rpx1 expression appears necessary for normal pituitary development and cellular differentiation, and this process is controlled by the df gene product, which is absent in the Ames dwarf mouse. Thus, formation of the anterior pituitary involves the timed and orderly expression of a series of developmental gene products.

B. Pituitary Cell Differentiation

The anterior pituitary secretes six major hormones: adrenocorticotropin (ACTH), follicle-stimulating hormone, luteinizing hormone, thyroid-stimulating hormone, growth hormone (GH), and prolactin. In general, each hormone is produced by a separate cell type. The first recognizable cells to functionally differentiate are the corticotrophs, which regulate adrenal function through secretion of ACTH. Corticotrophs appear shortly after formation of Rathke's pouch. Next to differentiate are the gonadotrophs, thyrotrophs, somatotrophs, and lactotrophs, in roughly that order. Thus, lactotrophs, the most abundant cell type in adults, are the last to differentiate. Lactotroph differentiation begins by birth in rats, by Embryonic Day 17 in chickens, and by Week 16 of gestation in humans. The precise signal resulting in lactotroph differentiation is unknown. Evidence that anencephalic human fetuses possess lactotrophs clearly indicates that the hypothalamus or higher portions of the brain are not necessary for lactotroph differentiation. Similarly, lactotrophs do not differentiate spontaneously in cultures of mammalian pituitary cells. Thus, the lactotroph-differentiating signal must arise from specific cell–cell contacts within the anterior pituitary or from factors present in the systemic circulation. In support of this latter possibility are findings that differentiation of lactotrophs in embryonic pituitaries transplanted beneath the renal capsule of host rats was dependent on the estrogen status of the host animal. In addition, lactotroph differentiation in neonatal rats appears to be regulated in part by a factor transferred to the circulation of nursing pups from their mothers' milk. Thus, the precise signal that initiates lactotroph differentiation from its precursor cell type remains unknown. It may emanate from the pituitary itself, from another endocrine organ, or even from mothers' milk.

C. Lactotroph Precursor

Unlike each of the other anterior pituitary cell types, a definitive precursor for lactotrophs has been identified. The initial lactotrophs to differentiate during development also release GH, providing evidence for a functional lineage between somatotrophs and lactotrophs. These two cell types also express a cell type-specific transcription factor necessary for transcription of the prolactin and growth hormone genes. This transcription factor, *Pit-1*, is found exclusively in thyrotrophs, somatotrophs, and lactotrophs. Natural or induced mutation of the *Pit-1* gene has resulted in deficiencies in both somatotrophs and lactotrophs during development and in adults. Through the use of transgenic mice, somatotrophs have been confirmed as the precursors for lactotrophs. In these experiments, destruction of somatotrophs during embryonic development led to a nearly complete elimination of lactotrophs as well. Thus, lactotrophs arise from somatotrophs during pituitary development.

II. ABUNDANCE OF LACTOTROPHS

A. Progression to Adulthood

Although lactotroph differentiation begins during embryonic development, the size of the lactotroph population remains low until sexual maturation, when it continues to increase into adulthood. Generally, the lactotroph population accounts for fewer than 10% of all pituitary cells at birth. No differences in lactotroph abundance are found between genders through the first month of postnatal life in rats. With the onset of gonadal growth and sexual maturation, the lactotroph population increases to reach approximately 30 and 50% of all pituitary cells in male and female adults, respectively. This increase concomitant with increased gonadal activity suggests that gonadal steroid hormones may regulate the size of the lactotroph population. Because more lactotrophs are found in females than in males, ovarian estrogens are strongly implicated in this process.

B. Fluctuations during Reproduction

The size of the lactotroph population is not static in adult animals. The number and size of lactotrophs increase during times of increased prolactin secretion. Indeed, for years it was thought that lactotrophs only existed in lactating animals. We now know that lactotrophs are quite abundant in all adult animals, but that the lactotrophs become relatively easy to detect during lactation due to the large amounts of prolactin that they produce to support milk production. However, in addition to an increased production of prolactin by each existing lactotroph, the number of lactotrophs present during lactation also increases dramatically. In lactating rats the size of the lactotroph population can increase to nearly 70% of all anterior pituitary cells. Similar changes occur in incubating chickens, and in nesting turkey hens the portion of the pituitary committed to prolactin production increases to more than half of the anterior lobe. In addition, the abundance of lactotrophs in the bovine pituitary gland increases temporarily during the early luteal phase of the estrous cycle. Thus, during periods of heightened need for pituitary prolactin secretion, lactotrophs become the predominant pituitary cell type. After weaning of the young in mammals or hatching of the eggs in birds, prolactin secretion rapidly decreases and so does the size of the lactotroph population. This decrease in the number of pituitary lactotrophs occurs within 3 days of the cessation of lactation or nesting and involves autolytic processes within the lactotrophs and phagocytosis by pituitary folliculo-stellate cells. Thus, the size of the pituitary lactotroph population is quite plastic in adult animals, fluctuating up or down depending on the physiologic needs of the animal.

With each increase or decrease in the size of the lactotroph population, commensurate but reciprocal decreases or increases in the size of the somatotroph population occur. These changes in lactotroph and somatotroph populations likely involve mitotic as well as postmitotic mechanisms. During lactation in mammals and nesting in birds, changes also occur in the abundance of a subpopulation of lactotrophs that secretes both GH and prolactin. These mammosomatotrophs may serve as functional intermediates for the conversion of somatotrophs to lactotrophs during lactation or nesting and back again to somatotrophs after weaning of the young or hatching of the eggs.

C. Potential Mitogens or Regulatory Factors

The dramatic growth of the lactotroph population between birth and adulthood suggests that the size of the lactotroph population is controlled by one or more mitogens. Although any of the numerous hormones and neuropeptides known to regulate prolactin secretion are candidates as lactotroph mitogens, the strongest evidence exists for ovarian estrogens. As noted earlier, lactotrophs are more abundant in females than in males. Neonatal ovariectomy of female rats results in adult lactotroph populations similar in size to those found in males. Furthermore, treatment of ovariectomized rats or pituitary cells from neonatal rats with estrogen increases the abundance of lactotrophs. This estrogen-induced increase in lactotrophs likely results from mitosis of existing lactotrophs because lactotroph DNA synthesis is stimulated by estrogen. Evidence also exists that the α subunit of luteinizing hormone can stimulate proliferation of lactotrophs present during embryonic and neonatal development. In addition, another minor product of gonadotrophs, released in response to luteinizing hormone-releasing hormone, the pro-opiomelanocortin (POMC) gene product POMC-(1-74), can stimulate proliferation of lactotrophs in culture. In contrast to estrogen and POMC-(1-74), fibroblast growth factor, epidermal growth factor, vasoactive intestinal peptide, and several other smaller POMC gene products, including α-melanocyte-stimulating hormone (α-MSH) and β-endorphin (β-END), increase the abundance of lactotrophs in culture through a mechanism not involving mitosis. These factors may function to stimulate prolactin gene expression and protein production in a population of existing nonlactotroph cells. These lactotroph precursors are most likely somatotrophs, which decrease commensurate with increases in lactotroph abundance.

In contrast to the proliferative effects of these factors on the lactotroph population, other factors may suppress lactotroph proliferation. Dopaminergic stimulation reduces lactotroph mitosis, and transforming growth factor-β (TGF-β) also suppresses prolactin secretion and lactotroph proliferation in culture. Interestingly, the lactotroph mitogen estrogen decreases lactotroph expression of TGF-β receptors.

III. SUBPOPULATIONS OF LACTOTROPHS

A. Growth Hormone Release by Mammosomatotrophs

Secretion of prolactin and growth hormone by the same cells has been documented in many physiological and pathological conditions. These cells that release both hormones are called mammosomatotrophs, and this subpopulation of lactotrophs can be quite rare to relatively abundant. Precisely how a single cell regulates secretion of both hormones independently is not known. Mammosomatotrophs have been found in birds and mammals, including rats and humans. During development of rats, mammosomatotrophs are known to serve as a transitional cell type for the differentiation of lactotrophs. Essentially, release of growth hormone seems to persist after initiation of prolactin secretion. Thus, mammosomatotrophs are a relatively large subpopulation of lactotrophs at the time of lactotroph differentiation. In adults, mammosomatotrophs are most abundant during the onset of periods of increased prolactin secretion, such as lactation in mammals and incubation in birds. In these states of physiologic transition, mammosomatotrophs constitute about 1 in 10 lactotrophs. Pathologically, mammosomatotrophs can be very abundant. Most prolactin-secreting pituitary tumors also produce growth hormone. Similarly, most prolactin and growth hormone-producing tumor cell lines were originally derived from cells that secreted both hormones. Thus, the abundance of mammosomatotrophs, pathologically and physiologically, has made this lactotroph subpopulation one of the more interesting and well studied.

B. Regional Distribution of Lactotrophs

Lactotrophs are not distributed homogeneously throughout the anterior pituitary gland. They are found in clusters in most mammals, often surrounding a single or pair of gonadotrophs. In rats lactotrophs are found predominantly near the junction of the anterior and intermediate lobes, at the ventral surface of the gland, and in the lateral tips or wings of the anterior pituitary. In birds lactotrophs are restricted to the cephalic portion of the gland and are generally absent from the caudal region of the anterior pituitary. In rats, most of the lactotrophs located in the inner zone toward the intermediate lobe contain small secretory granules, whereas the majority of the lactotrophs in the outer zones contain larger granules, indicating that the relative amounts of prolactin stored by these lactotroph subpopulations differ. In agreement with this contention, lactotrophs cultured from the inner zone secrete more prolactin under basal conditions than those from the outer regions. Moreover, lactotrophs from the inner region are more sensitive to the inhibitory effects of dopamine, whereas outer zone lactotrophs are more responsive to thyrotropin-releasing hormone (TRH) stimulation. Regional differences in lactotroph location within the pituitary have led to speculation that these subpopulations may be regulated by different hypothalamic influences. For example, those lactotrophs located adjacent to the intermediate lobe in rats would receive much of their blood supply from the short portal vessels running from the neurointermediate lobe to the anterior lobe. Thus, these lactotrophs would be regulated predominantly by the tuberohypophyseal dopaminergic (THDA) neurons terminating in the posterior pituitary. The lactotrophs located in other regions of the anterior lobe would receive most of their blood supply from long portal vessels extending from the hypothalamus down the pituitary stalk and into the anterior lobe. These lactotrophs would be regulated primarily by the tuberoinfundibular dopaminergic (TIDA) neurons terminating in the median eminence. Because the THDA and TIDA neurons originate in different hypothalamic nuclei, it follows that their control over the different pituitary lactotroph subpopulations would also differ. Although intriguing, this hypothesis requires further investigation before it should be openly accepted.

C. Variations in Prolactin Secretion and Responsiveness

Not all lactotrophs respond equally to stimulatory or inhibitory secretagogues, release equivalent amounts of prolactin, or secrete prolactin with the same bioactive potency. This heterogeneity and microheterogeneity in the lactotroph population is well documented. While dopamine release from the hypothalamus and high prolactin levels in the circulation are known to suppress pituitary prolactin secretion in mammals, these responses appear to be mediated by subpopulations of lactotrophs that decrease their release of prolactin in response to these agents. Other lactotrophs do not respond to inhibitory regulation. Similarly, while TRH is a potent stimulator of prolactin release from the pituitary, not all lactotrophs increase their release of prolactin in response to TRH. β-END and α-MSH, products of the POMC gene, stimulate prolactin secretion in rats. However, these responses are limited to those lactotrophs proximal to the intermediate lobe of the pituitary, the source of these POMC gene products. Responsiveness of prolactin production to estrogen stimulation is greater in the highly granulated cells of the peripheral region than in the lightly granulated cells of the inner region of the anterior pituitary, near the intermediate lobe in rats. Thus, dramatic variations exist among subpopulations of lactotrophs in their responsiveness to various secretagogues.

Variations also exist in the amount of prolactin released by each lactotroph. Using reverse-hemolytic plaque assays, which detect hormone release from individual cells, >100-fold differences in the amount of prolactin released by individual lactotrophs have been detected. Thus, not all lactotrophs contribute equally to total pituitary prolactin secretion. Furthermore, the amount of prolactin secreted by lactotrophs varies over time. While some lactotrophs remain relatively constant in their rate of prolactin secretion, others vary dramatically, releasing small amounts on one day and large amounts on another day. The occurrence of lactotrophs with variable prolactin release over time is more common in males than in females. In addition, the ability of the prolactin released from individual lactotrophs to stimulate mammary epithelial cell casein production has been shown to vary among lactotrophs. This microheterogeneity in lactotroph function may be due to differences in the posttranslational modification of the prolactin protein, such as phosphorylation and glycosylation, which are known to affect its potency and stability.

IV. CONCLUSION

Lactotrophs demonstrate the complexity of the physiological mechanisms regulating pituitary hormone secretion and action. In addition to positive and negative trophic regulation of prolactin secretion by hypothalamic secretagogues, lactotroph abundance also varies with the physiological needs of the animal during development and changes in reproductive state. This form of regulation is complex, involving several factors of paracrine and endocrine origin and in a manner specific to anterior pituitary regions and lactotroph subpopulations.

See Also the Following Articles

LACTATION, HUMAN; PARENTAL BEHAVIOR, BIRDS; PROLACTIN SECRETION, REGULATION OF

Bibliography

Begeot, M., Dubois, M. P., and Dubois, P. M. (1984). Evolution of lactotropes in normal and anencephalic human fetuses. *J. Clin. Endocrinol. Metab.* 58, 726–730.

Borrelli, E., Heyman, R. A., Arias, C., Sawchenko, P. E., and Evans, R. M. (1989). Transgenic mice with inducible dwarfism. *Nature* 339, 538–541.

Castano, J. P. (1994). Real-time measurement of prolactin secretion from individual lactotropes. *Neuroprotocols* 5, 216–220.

Frawley, L. S., and Boockfor, F. R. (1991). Mammosomatotrophs: Presence and functions in normal and neoplastic pituitary tissue. *Endocr. Rev.* 12, 337–355.

Goluboff, L. G., and Ezrin, C. (1969). Effect of pregnancy on the somatotroph and the prolactin cell of the human adenohypophysis. *J. Clin. Endocrinol.* 29, 1533–1538.

Hoeffler, J. P., Boockfor, F. R., and Frawley, L. S. (1985). Ontogeny of prolactin cells in neonatal rats: Initial prolactin secretors also release growth hormone. *Endocrinology* 117, 187–195.

Porter, T. E., and Frawley, L. S. (1993). Regulation of lactotrope differentiation in neonatal rats by a milk-borne peptide: A review. *Endocr. Regul.* 27, 125–131.

Porter, T. E., Hill, J. B., Wiles, C. D., and Frawley, L. S. (1990). Is the mammosomatotrope a transitional cell for the functional interconversion of growth hormone- and prolactin-secreting cells? Suggestive evidence from virgin, gestating, and lactating rats. *Endocrinology* 127, 2789–2794.

Takahashi, S. (1995). Development and heterogeneity of prolactin cells. *Int. Rev. Cytol.* 157, 33–98.

Voss, W., and Rosenfeld, M. G. (1992). Anterior pituitary development: Short tales from dwarf mice. *Cell* 70, 527–530.

Lagomorpha

see Rabbits

Leiomyoma

Salli Tazuke and Linda Giudice

Stanford University

I. Introduction
II. Epidemiology
III. Role of Estrogen and Progesterone
IV. Cytogenetic Abnormalities
V. Growth Factors
VI. Clinical Presentation
VII. Diagnosis
VIII. Treatment
IX. Summary

GLOSSARY

aneuploidy Abnormality in the total number of chromosomes in a cell (normal is 46).

endometrium Tissue lining the uterine cavity where implantation of the conceptus occurs; in the absence of pregnancy, it is shed and then regenerated during each cycle.

endoscopy An operative procedure in which a fiberoptic camera is used to visualize any potential body cavity.

estromedin A factor that mediates the effect of estrogen.

follicular phase The first half of the menstrual cycle in which the oocyte develops in the ovarian follicle prior to ovulation. Estrogen synthesis by the follicle raises the level in the serum.

GnRH gonadotropin-releasing hormone, a hypothalamic hormone that regulates the synthesis and release of gonadotropins which in turn stimulate ovarian steroidogenesis and regulate the menstrual cycle.

GnRH agonists Synthetic molecules that bind to GnRH receptors with a longer half-life than endogenous GnRH. Initially, administration of GnRH agonists stimulates synthesis and release of gonadotropins. After about 1 week, suppression of the pituitary gonadotropes occurs resulting in low estrogenic milieu.

hysteroscopy Endoscopy of the uterine cavity.

leiomyoma Benign smooth muscle neoplasm (tumor) of the uterine origin. Also called *myoma* or *fibroid*.

luteal phase The second half of the menstrual cycle after ovulation. The follicle, from which the oocyte ovulated, converts to the corpus luteum and produces both estrogen and progesterone.

mitogen Any factor that induces mitosis or cell replication.

myomectomy Surgical removal of leiomyoma without the removal of the entire uterus.

nulliparity No childbirth.

progestin A synthetic progesterone analog.

progestomedin A factor that mediates the effects of progesterone.

sonohysterography A procedure in which the uterine cavity is filled with saline and evaluated by tranvaginal ultrasound.

Uterine leiomyomas are benign smooth muscle neoplasms and comprise the most common solid pelvic tumors occurring in women. Leiomyomas are synonymous with "fibroids," "myomas," or "fibromyomas." The enlargement of this neoplasm during the reproductive years of women can lead to significant morbidity, necessitating medical and surgical intervention. The condition is highly prevalent and is one of the most common indications for hysterectomy in the United States.

I. INTRODUCTION

In recent years, increasing attention has been turned to the biology of myomas with the hope of developing better treatment or preventive modalities. Although the exact etiology of uterine leiomyoma remains unknown, our understanding surrounding the role of ovarian steroid hormones, growth factors, and cytogenetic abnormalities in fibroids has greatly increased in the recent years. Current treatment regimens attempt to regulate growth of this neoplasm by altering the female hormonal milieu or eliminate the mass by surgical means.

II. EPIDEMIOLOGY

Textbooks of gynecologic pathology cite that uterine myomas occur in 20–30% of women over 30 years old, and the need for treatment of fibroids peaks at around age 45. Leiomyomas are rarely detected before menarche. Whereas the previously noted prevalence has been estimated from autopsy specimens at the turn of the century, in a recent study that examined all uteri from hysterectomies with detailed 2-mm sections, fibroids were identified in 77% of all specimens. The study noted that the myomas were detected as frequently in women whose uteri were removed for indications other than leiomyoma. Precise prevalence is difficult to determine since <50% of women with fibroids are estimated to be asymptomatic and the tumors remain clinically occult. Nevertheless, fibroids are significant from a public health standpoint since 10–15% of all women undergo hysterectomy between the ages of 25 and 64 years for myomas.

Approximately one-third of women with myomas develop significant symptoms such as menorrhagia, metrorrhagia, pelvic pain, pelvic pressure, dyspareunia, and urinary frequency or constipation. Other conditions associated with myomas include reproductive dysfunctions such as infertility, recurrent miscarriages, and complications during pregnancy. How myomas actually give rise to these reproductive dysfunctions is, however, difficult to define precisely. The findings of myoma may be purely coincidental or coexists with other factors contributing to the reproductive dysfunction. Because treatment itself can cause significant morbidity, gynecologists are frequently challenged regarding management of myomas. For example, while 27% of women undergoing surgical removal of myomas (myomectomy) reported a history of infertility in one study, only 2% had myomas as the sole abnormality detected during their workup.

Factors associated with increased risk for leiomyoma include nulliparity or low parity, increasing years since last child birth, obesity, and African racial origin. Decreased risk ratio was observed with exercise, smoking, and use of oral contraceptives (Table 1). Such a pattern is reminiscent of the positive and negative risk factors associated with estrogen-dependent neoplasias, such as endometrial cancer, and suggests that steroid hormones may be essential for the growth of this neoplasm.

TABLE 1
Risk Factors for Leiomyoma

Factors associated with increased risk
   Nulliparity
   Low parity
   Increasing years since last child birth
   Obesity
   African racial origin
Factors associated with decreased risk
   Exercise
   Smoking
   Oral contraceptive use

III. ROLE OF ESTROGEN AND PROGESTERONE

Steroid hormones act via intracellular nuclear receptors which are themselves transcription factors that induce expression of genes. For estrogen, the genes induced are frequently those that direct cellular growth and proliferation, and thus numerous investigations have been conducted to test the hypothesis that estrogen causes uterine fibroid growth. Leiomyoma cells do contain increased concentrations of estrogen receptors compared to normal myometrium of the same uterus. The differences were greatest during the follicular phase when the ovary predominantly produces estrogen. When these cells were tested *in vitro*, leiomyoma cells exhibited a greater transcription response to estrogen compared to that of the myometrium. These differences likely account for the 20% more estradiol per milligram of cytoplasmic protein found in leiomyoma compared to normal myometrium. Thus, fibroids maintain a relative hyperestrogenic milieu which may further induce proliferation and growth of its own cells. Clinically, when women with fibroids encounter a hypoestrogenic state, such as menopause or treatment with a gonadotropin-releasing hormone (GnRH) agonist, a reduction in the size of leiomyoma up to 50% can be observed.

While estrogen is a mitogen, and the previous evidence implicates the role of estrogen in fibroids, the growth of leiomyomas may be more affected by progesterone. The mitotic rate of leiomyoma cells varies during the menstrual cycle. It is highest during the progesterone-dominant luteal phase and does not differ significantly from that of the myometrium in the estrogen-dominant follicular phase. Similarly, leiomyoma from women receiving high-dose progestin treatment also contain higher mitotic rates than the myometrium, whereas leiomyoma and myometrium from recipients of combined estrogen and progestin treatments had similar mitotic rates. A clinical observation that women who were treated with a combination of GnRH agonists and progestin did not show any significant regression of the leiomyoma also suggests that progesterone may be implicated as the stimulus for leiomyoma growth. This is further supported by the finding that progesterone receptor concentrations in leiomyoma were increased, compared to matched myometrium, throughout the menstrual cycle. Furthermore, RU 486 (mifepristone), a progesterone receptor antagonist, induced shrinkage of the leiomyoma, suggesting that progesterone may be more important than estrogen physiologically in regulating the growth of leiomyoma. The net effect is likely to be a combined one since both steroids affect the expression of steroid receptors in the uterus. For instance, estrogen induces the synthesis of the progesterone receptor, making the cells more sensitive to progesterone and, conversely, progesterone suppresses estrogen receptor expression, making the cells more resistant to estrogen.

IV. CYTOGENETIC ABNORMALITIES

Although both ovarian steroids appear to be involved with the biology of leiomyoma, the fact that all women are exposed to both estrogen and progesterone throughout their reproductive years suggests that other factors must be involved with the initial neoplastic transformation of the uterine smooth muscle. The pathogenesis of leiomyomas has been difficult to determine, particularly since leiomyomas are rare in other mammalian species, and currently there are no working animal models that mimic this physiological condition seen in humans.

Somatic mutations have been implicated in neoplastic transformations. In the case of leiomyoma, nonrandom chromosomal changes, such as translocation, deletion, and aneuploidy, have been identified. Two-thirds of women with a fibroid uterus have multiple myomas of various size and location, and these myomas are cytogenetically distinct from one another. Growth factor genes, protooncogenes, and chromosomal fragile sites that may be present at the sites of mutations may lead to dysregulated proliferation, contributing to the initial tumorigenesis and subsequent growth. Of interest, some of the cytogenetic abnormalities observed in myomas have also been seen in other tumors. For instance, translocation of 12q14–15 exists in leiomyomas, lipomas, and pleiomorphic salivary gland tumors, whereas trisomy 12 is associated with B cell chronic lymphocytic leukemia, ovarian fibroma, ovarian fibrothecoma, and ovarian adenocarcinoma. However, only 21–63% of leiomyomas contain cytogenetic abnormalities, and no single finding is characteristic. Even different myomas from the same uteri can contain different cytogenetic abnormalities. While some argue that somatic mutations may result in the loss of growth regulation and subsequent clonal expansion leading to neoplastic proliferation, others believe that somatic mutations may be secondary events subsequent to the primary events that induced the neoplastic transformation leading to the clonality of each leiomyoma.

V. GROWTH FACTORS

Growth factors and cytokines can mediate the effects of estrogen and progesterone (Table 2). They can also act independently to regulate cellular growth, proliferation, angiogenesis, and apoptosis. Research regarding the role of various growth factors and cytokines in leiomyoma is extensive. While many of them are implicated in supporting the growth and proliferation of leiomyoma smooth muscle cells, none singularly explains adequately the mechanism of pathogenesis of leiomyoma.

A. Epidermal Growth Factor

Epidermal growth factor (EGF) is a 53-amino acid, single-chain polypeptide with well-documented mitogenic and differentiating effects in various organs. In the rat and mouse, EGF has been identified to act as an estromedin in the endometrium and, thus, its

TABLE 2
Possible Mediators of Growth Stimulus

Hormones
    Estrogen
    Progesterone
    Prolactin
    Growth hormone
Growth factors
    Epidermal growth factor
    Insulin
    Insulin-like growth factor-I and -II
    Platelet-derived growth factor

potential role in the growth of leiomyoma has been studied extensively. In the human uterus, EGF mRNA is detected throughout the menstrual cycle in the myometrium and leiomyoma, but the expression is somewhat greater in the leiomyoma during the luteal phase. EGF acts by binding to a type I tyrosine kinase receptor on the cell membrane, and in the uterus, EGF receptor is equally present in both the myometrium and the leiomyoma throughout the cycle. Leiomyoma from women receiving GnRH agonist treatment and who are hypoestrogenic contained decreased levels of EGF receptors compared to myometrium. *In vitro,* EGF induced DNA synthesis in both the myometrium and leiomyoma cells, but the induction was greater in the myometrium. These data suggest that EGF may indeed be an estromedin but do not explain why there are more mitotic cells in the leiomyoma during the progesterone-dominant luteal phase. EGF may prime the leiomyoma for the progesterone effect during the luteal phase.

B. Insulin-Like Growth Factors

The insulin-like growth factor (IGF) family consists of IGF-I, IGF-II, type I and type II IGF receptors, and insulin-like growth factor-binding proteins (IGFBP-1–6). IGF-I and IGF-II have growth-promoting activities and thus have been investigated for their possible roles in leiomyoma pathophysiology. In human endometrium, IGF-I mRNA is detected in the follicular phase, whereas IGF-II transcripts are more abundant in the luteal, compared to the proliferative, phase. In the myometrium and leiomyoma, IGF-I mRNA has been detected in greater amounts in leiomyoma, particularly during the follicular phase. The uterine muscle cells themselves secrete IGF-I, as demonstrated by an explant culture system, in which leiomyoma explants synthesize greater amounts than the myometrium. The phasic pattern of IGF-I transcripts suggests that IGF-I may mediate estrogen's effects. IGF-II, on the other hand, displayed no menstrual cyclicity, and transcripts for IGF-II are present at higher levels in leiomyoma than myometrium. The same explant culture system as described previously also displayed increased secretion of IGF-II by the leiomyoma tissue compared to the myometrium. Treatment of women with a GnRH agonist, resulting in hypoestrogenism, abolished the differences in the amount of IGF-I and IGF-II secreted by leiomyoma and myometrium explant cultures. In terms of the receptors, both type I and type II IGF receptors are present in the uterine muscle, but leiomyoma cells contained greater concentrations of the type I receptor than normal myometrial cells which may enhance the growth-promoting action of IGFs present at higher concentrations in leiomyoma.

IGFBPs modify IGF action by binding to them and can be either inhibitory or enhancing. In the uterus, IGFBPs 3–6 have been detected in both the myometrium and the leiomyoma with no menstrual cyclicity noted. Only IGFBP-3 levels are decreased in leiomyoma compared to myometrium. IGFBP-3 is thought to inhibit IGF action, and thus decreased levels could increase the bioavailability of IGFs locally in the tissue. The various patterns of expression of the IGF family members in the uterus suggest a likely role of IGFs in mediating the effects of estrogen and supporting the growth of leiomyoma cells in these neoplasms.

Insulin is also a mitogen and, *in vitro,* has been shown to stimulate DNA synthesis by both leiomyoma and myometrial cells. Insulin displays synergistic action with EGF, but the effect is most prominent in myometrium compared to leiomyoma cells.

C. Platelet-Derived Growth Factor

Platelet-derived growth factor (PDGF) is another potent growth factor, particularly for cells of mesenchymal lineage such as the uterine smooth muscle. PDGF has been shown to correlate with the physiological hypertrophy of myometrium during preg-

nancy, and thus great interest in its possible role in the pathogenesis of leiomyoma exists. In a limited study, however, the levels of PDGF expression do not differ between leiomyoma and myometrium, whereas more PDGF receptor sites were identified in primary cultures of leiomyoma compared to that of myometrium. *In vitro,* PDGF induces DNA synthesis and mitosis independently and synergistically with EGF in the myometrium.

D. Prolactin

Prolactin is typically known as a pituitary hormone but is synthesized by other organs, including uterine myometrium and leiomyoma. Prolactin can act as a mitogen, and its secretion by uterine muscle cells increases in the presence of estrogen. In the explant culture system, greater amounts of prolactin are secreted by leiomyoma than myometrium. Human chorionic gonadotropin, produced by the syntiotrophoblast cells of the placenta, also induces prolactin secretion by the uterine myometrium and threefold more by leiomyoma. The prolactin effect is not prominent in a hypoestrogenic environment such as with GnRH agonist treatment, suggesting that estrogen may be required for maximum prolactin effect. While prolactin is known to be a mitogen, the receptor for prolactin has not been detected in the myometrium or leiomyoma, and a precise functional role of prolactin in the pathophysiology of uterine myoma has not been clearly identified.

E. Growth Hormone

Another pituitary hormone, growth hormone (GH), has also been implicated in promoting the growth of myoma. Peripheral levels of growth hormone are increased in women with leiomyoma, and peak GH levels are greater in patients with leiomyoma during a hypoglycemia challenge test. Women undergoing GnRH agonist treatment, on the other hand, have decreased fasting levels of GH compared to controls. In rats and guinea pigs, exogenous growth hormone treatment increases uterine weight, independent of changes in ovarian steroid levels and synergistically when given with estrogen. Growth hormone injections lead to increased estradiol receptors in leiomyoma and myometrium, and acromegalies have a higher incidence of fibroids compared to women with normal GH levels, suggesting that GH either directly or through its mediator, IGF-I, potentiates estrogen action.

VI. CLINICAL PRESENTATION

Approximately one-third of women with fibroids develop significant symptoms necessitating treatment (Table 3). The most common chronic symptoms associated with uterine fibroids include abnormal uterine bleeding with or without significant anemia (approximately one-third), abdominal pain in another one-third, and symptoms secondary to mass effects, such as abdominal distortion, genitourinary compression, gastrointestinal compression, and sensation of pelvic pressure. Other presenting conditions are recurrent pregnancy loss, infertility, or complications during pregnancy, labor, and delivery. In general, the likelihood of developing symptoms and reproductive dysfunction increases with greater size and number of myomas involved.

The role of myomas in reproductive dysfunction such as infertility, recurrent pregnancy loss, and obstetrical complications is difficult to define. While 40% of women undergoing myomectomy reported a history of infertility in one study, only 2 or 3% had myoma as the sole abnormality identified during their workup for infertility. Several mechanisms can be invoked in which leiomyoma can affect fertility. For example, submucous myomas may be associated

with local changes in vasculature, endocrine milieu, and endometrial lining, making nidation unfavorable. If myomas are located in the cornual area near the tubal ostia, the conduit may be obstructed for either the sperm or the fertilized embryo. Cervical or low uterine segment myoma may also interfere with sperm reservoir or transport. General uterine cavity enlargement secondary to leiomyoma may increase the distance for sperm to travel, making it prohibitively long. Also, other myomas, such as intraligamentous ones, may distort the tuboovarian relationship, impeding with ovum capture by the tubal fimbriae.

With regard to association of myomas with recurrent pregnancy loss, while a significant reduction in spontaneous abortion rates following myomectomy has been reported in one study, whether leiomyomas per se were responsible for the difference is difficult to determine. Because surgical treatment of leiomyomas can be detrimental to reproductive function, if leiomyomata are identified during the workup for infertility and recurrent pregnancy loss, other possible causes are first ruled out prior to myomectomy unless there is clear distortion of the uterine cavity or a definite intracavitary location that could compromise fertility and successful pregnancy outcome.

The presence of myomas can be associated with various obstetrical complications, such as myoma infarction and severe pain necessitating hospitalization, preterm labor, premature rupture of membranes, malpresentation necessitating cesarean delivery, and placental abruption. Approximately one-third of the leiomyomas grow during pregnancy, which can increase the risk for these complications. However, myomas are identified only in about 2% of obstetrical patients by routine second-trimester ultrasounds, and only 10% of those with myomas developed complications during gestation.

VII. DIAGNOSIS

The diagnosis of leiomyoma is made by pelvic examination, ultrasound, magnetic resonance imaging (MRI), hysterosalpingogram, sonohysterography, or endoscopy. MRI is particularly accurate and informative with its ability to clearly delineate the

TABLE 3
Associated Clinical Conditions

Abnormal uterine bleeding
Anemia
Abdominal pain
Pelvic pressure
Infertility
Pregnancy loss
Myoma infarction and pain during pregnancy
Preterm labor
Premature rupture of membranes
Placental abruption
Malpresentation

location and distinguish leiomyoma from adenomyosis, which may otherwise be difficult to discern. MRI is, however, not utilized routinely for the management of fibroids due to the higher cost compared to pelvic ultrasound. Its use has been most advocated preoperatively prior to myomectomy to aid the surgeon in identifying the location of all the myomas and to identify the course of the ureter which could be significantly displaced secondary to an enlarged uterus. If genitourinary compression is suspected, insidious compression of the ureters may result in hydronephrosis and renal damage, and ultrasound of the kidneys, serum renal function chemistry tests, and/or intravenous pyelogram may be valuable. For submucous myomas, hysterosonography and hysteroscopy are excellent diagnostic tools. The latter has the advantage of allowing the clinician to simultaneously resect and remove the submucous myoma endoscopically.

VIII. TREATMENT

Treatment of leiomyomas traditionally has consisted of surgical removal, namely, myomectomy or hysterectomy. Prior to the era of pelvic ultrasound and MRI, myomas were difficult to discern from other pelvic masses, and thus size alone was an indication for surgical intervention. With greater efficacy of ultrasound in differentiating leiomyoma from other pathology, size alone is no longer sufficient as an indication for myomectomy or hysterectomy. Today, women with myomatous uterii have other options, including expectant management for those who are asymptomatic or minimally symptomatic with monitoring of the growth of the uterine myoma by pelvic examination or ultrasounds. While such an option is least morbid, patients should also be counseled regarding possible risks associated with reproductive dysfunction and complications during pregnancy.

For those experiencing symptoms such as abnormal uterine bleeding, even if fibroids are present, the clinician should rule out other causes, including endocrine perturbations such as hyperprolactinemia, thyroid dysfunction, and endometrial pathology such as hyperplasia or carcinoma. Abnormal bleeding, not resulting in significant anemia, may initially be regulated with nonsteroidal antiinflammatory drugs, low-dose estrogen–progestin combination oral contraceptive pills, or progestins to temporize the condition. However, given the evidence regarding the mitogenic stimulus that steroids have, particularly progestins, concern remains regarding induction of further growth. While progestins may induce amenorrhea by decidualizing the endometrium, and subsequently render it atrophic, significant uterine enlargement may occur which could increase the need for surgical intervention. The antiprogestin, RU 486 (mifepristone), has been shown, in a limited study, not only to not only induce amenorrhea by inhibiting ovulation but also to reduce uterine volume to 87% of original after 3 months of therapy. This finding needs to be further studied for long-term effects, and lack of availability of RU 486 and other progestins in the United States currently limits regular use.

Other hormonal manipulations that effectively result in the reduction of leiomyoma size consist of androgens and GnRH agonists. Gestrinone, a synthetic androgen, inhibits estrogen and progesterone effects, and after 8 weeks of treatment leads to amenorrhea in approximately half of patients. Total uterine volume is decreased in 70% of women treated with Gestrinone, and in 89% of these patients, the reduction is sustained 18 months after discontinuing the treatment. Gestrinone, however, is associated with significant androgenic side effects, such as acne, irreversible hirsutism, deepening voice, seborrhea, and weight gain. Another androgen, Danazol, has also been reported to succeed in a limited 3-month trial but shares similar side effect profiles with Gestrinone. These androgens can masculinize female fetuses, and thus patients must also utilize appropriate birth control methods while receiving androgen therapy.

GnRH agonists act by downregulating GnRH receptors in the pituitary gonadotrophs after an initial brief stimulation of GnRH release from these cells. The subsequent decrease in follicle-stimulating hormone (FSH) and luteinizing hormone (LH) synthesis and secretion induces a hypogonadotropic, hypoestrogenic state. Estrogen and progesterone production by the ovary is abolished, and uterine volume sharply decreases to 25–80% by 12 weeks of therapy. Unlike androgens, however, regrowth after cessation of ther-

TABLE 4
Treatment Options

Expectant management
  Medical
    Nonsteroidal antiinflammatory drugs
    Oral contraceptives
    Progestins
    GnRH agonist
    RU 486 (mifepristone)[a]
    Gestrinone[a]
  Surgical:
    Hysterectomy
    Myomectomy
      Laparotomy
      Laparoscopy
      Laparoscopic assisted
      Hysteroscopy
    Myloysis[a]
    Cryolysis[a]

[a] Experimental.

apy is rapid. Within 3 months, the volume returns to 88% of its original size. Side effects of GnRH agonists are identical to menopause and include hot flashes, night sweats, insomnia, vaginal dryness, urinary incontinence, irritability, and mood swings. Furthermore, the lipid profile becomes atherogenic and bone mineral density decreases with the hypoestrogenic state, and concern for long-term health effects limits the treatment with GnRH agonists alone to a total of 6 months. The main reduction in volume, however, is achieved within 3 months of treatment. To minimize the side effects and decrease the risk of osteoporosis and atherogenesis, estrogen add-back therapy has been explored but data are lacking regarding the risk of regrowth of leiomyoma.

When surgical intervention is indicated, myomectomy is an option to hysterectomy, allowing a woman to preserve fertility. Myomectomy may be approached endoscopically or via laparotomy. Hysteroscopic resection of submucosal myomas offers significant advantages over laparotomy by avoiding an abdominal incision, uterine incision, and lengthy postoperative recovery period. Both resection by a unipolar wire loop and various laser instruments and vaporization by electrocautery have been demonstrated to be feasible and should be chosen depending on the surgeon's training and preference. Concomitant laparoscopy is not always mandated. Successful pregnancy rates of 47–77% have been reported in patients with a history of infertility.

When the myoma size or location does not lend itself to the hysteroscopic approach, myomectomy can be performed laparoscopically or by laparotomy. Laparoscopic myomectomy, if feasible, offers the advantages of minimal access surgery to the patients. However, laparoscopic myomectomy can be technically very challenging and lengthy as a procedure, and concerns remain regarding the integrity of the uterine wall after laparoscopic suturing at the site of myomectomy. During second-look laparoscopy, uterine indentation at the site of prior laparoscopic myomectomy of intramural or deep subserosal myomas has been reported, and in one study that followed the course of 154 women, 6 developed fistulas. Of more concern is that 2 women reported cases of uterine dehiscence during pregnancy prior to labor following laparoscopic myomectomy.

A combined approach, called laparoscopic-assisted myomectomy, is an alternative whereby the myoma is first identified laparoscopically and isolated, then retracted to the abdominal wall with a corkscrew. A minilaparotomy incision over the retracted myoma leads to the removal of the tumor and allows a thorough multilayer repair as in conventional abdominal myomectomy. This approach also has the advantage of substantially reducing the operative time by eliminating the need for laparoscopic morcellation of the myoma. Because the identification and isolation of the myoma are done laparoscopically, the abdominal incision can be limited, and the postoperative hospitalization and recovery period between laparoscopic myomectomy and laparoscopic-assisted myomectomy are comparable. Overall, the laparoscopic approach to myomectomy is feasible with the advantages of minimal access surgery. However, more studies are needed to compare the efficacy, safety, and long-term outcome with those of the traditional abdominal myomectomy.

Postmyomectomy adhesion formation may result in chronic pelvic pain or reproductive dysfunction regardless of the surgical approach. Even with laparoscopic myomectomy, dense adhesions can form, particularly if suturing was necessary at the initial surgery. Barrier materials, such as oxidized regenerated cellulose [Interceed (TC7); Ethicon, Inc., Sommerville, NJ] and nonabsorbable expanded polytetrafluoroethylene (Gore-Tex Surgical Membrane; W. L. Gore and Associates, Inc., Flagstaff, AZ) have been reported to effectively reduce adhesion formation at the incision sites when used properly.

Pregnancy rates after myomectomy have been reported in the range of 16–82% for abdominal myomectomy and 25–44% for laparoscopic myomectomy. The interval from myomectomy to conception is short, with approximately 70% of the pregnancies occurring within the first year. If the patient had a history of infertility, the conception rate declines to 16–38% after myomectomy. Spontaneous miscarriage rate has also been reported to decline from 41% preoperatively to 19% postoperatively.

Besides resection of leiomyomata from the uterus, other alternative modalities such as destruction and necrosis of myoma by an electric coagulator (myolysis) or freeze–thaw process (cryolysis) have been tried with limited success. Both have been advocated as a minimal access outpatient surgical procedure but long-term outcome data are lacking for both. These methods currently are not in general use and do not represent "standard of care."

IX. SUMMARY

The fibroid uterus is a very common condition encountered in gynecology and is associated with significant morbidity and health care costs. The pathogenesis of leiomyomata remains uncertain but a number of factors, such as estrogen, progesterone, growth hormone, and growth factors, are likely contributors. Our increasing understanding of the biology of leiomyoma has led to development of newer treatment modalities in addition to the traditional surgical therapy. Surgical approaches have also enjoyed incorporation of endoscopic techniques, offering patients the advantage of minimal access surgery. Postoperative adhesion formation and adverse effects of uterine wall integrity remain concerns for myomectomy by all surgical modalities. It is hoped that further elucidation of the mechanism of leiomyoma pathogenesis and growth will generate more effective and less invasive medical treatment of fibroids as well as preventive measures for their development.

See Also the Following Articles

Hysterectomy; IGF (Insulin-Like Growth Factors); Prolactin, Overview; Tumors of the Female Reproductive System; Uterine Anomalies

Bibliography

Adashi, E., Rock, J., and Rosenwaks, Z. (1995). *Reproductive Endocrinology, Surgery, and Technology*, 1st ed., pp. 2160–2163, 2266–2270. Lippincott-Raven, Philadelphia.

Barbieri, R. (Ed.) (1992). Leiomyoma uteri. *Sem. Reprod. Endocrinol.* **10**(4).

Giudice, L. (Ed.) (1996). Growth factors and reproduction. *Sem. Reprod. Endocrinol.* **14**(3).

Hutchins, F., and Greenberg, M. (Eds.) (1995). Uterine fibroids. *Obstet. Gynecol. Clin. North Am.* **22**(4).

Kurman, R. J. (1987). *Blaustein's Pathology of the Female Genital Tract*, 3rd ed., pp. 374–384. Springer-Verlag, New York.

Nezhat, C., Nezhat, F., Luciano, A., Siegler, A., Metzger, D., Nezhat, C. (1995). *Operative Gynecologic Laparoscopy. Principles and Techniques*, 1st ed., pp. 205–238. McGraw-Hill, New York.

Thompson, J., and Rock, J. *Te Linde's Operative Gynecology*, 7th ed., pp. 647–662, 664–738.

Lemmings
see Cricetidae

Leukemia Inhibitory Factor

Colin L. Stewart

National Cancer Institute, Frederick Cancer Research and Development Center

GLOSSARY

decidua A mass of cells produced by the uterine stroma and also containing many blood cells derived from the bone marrow. The decidua is formed after implantation and surrounds and supports the early growth and development of the embryo. In mammals in which the trophoblast invades the uterus the decidua restricts trophoblast growth.

delayed implantation A physiological state that can be induced in about 40 mammalian species, such as mice, rats, and Roe deer. Blastocyst implantation is inhibited and embryo development arrested. By delaying implantation mammals regulate the duration of pregnancy, thus ensuring the young are born at the most optimal time and conditions for their survival.

embryonic stem (ES) cells Developmentally totipotent cells derived from the mouse embryo's inner cell mass (ICM). These cells are cultured *in vitro* and can differentiate into many other cell types or produce entire mice when reintroduced into blastocysts and transferred back into the uterus to develop to term. They are primarily used to derive lines of mice ("knockout" mice) carrying mutations in any gene of interest.

gene targeting A powerful technique used to introduce mutations into specific genes in ES cells grown *in vitro*. The ES cells are used to derive mice carrying the mutated gene so that its function can be studied in the whole animal.

inner cell mass A group of cells in the blastocyst that will form the embryo proper.

luminal epithelium The single layer of cells lining the cavity or lumen of the uterus. This tissue is central to regulating when blastocyst implantation occurs and in stimulating the underlying stromal cells to decidualize.

signal transduction The biochemical pathway by which cell surface or intracellular receptors send signals to the nucleus to regulate gene expression.

trophoblast An epithelial hollow sphere of cells enclosing the inner cell mass. The trophoblast initially mediates interactions between the developing embryo and the maternal uterus. After implantation, the trophoblast, together with cells from the embryo, forms the placenta.

It is long established that the ovarian steroid hormones, estrogen and progesterone, have a prime role in regulating the preparation of the uterus for implantation by stimulating both cell proliferation and differentiation. However, recent evidence has revealed that effects of these hormones on the uterus are mediated through locally produced peptide growth factors. One such peptide is leukemia inhibitory factor (LIF), which is essential for embryo implantation. The molecular biology of LIF, its receptor, and how it regulates implantation is the subject of this article.

I. INTRODUCTION

Implantation is a unique and essential stage in the development of the mammalian embryo. After fertilization, the embryo moves down the reproductive tract, undergoing a series of cleavage divisions resulting in the formation of the blastocyst shortly after entering the uterus. In most mammalian embryos, as typified by the mouse and human, the blastocyst consists of two cell types: the trophoblast, which will contribute to the placenta, and the inner cell mass (ICM), which will form the embryo proper. For the blastocyst to continue development, it establishes physical contact with the maternal uterine tissues; this process is referred to as implantation. At implantation the extent to which the trophoblast attaches to and then invades the uterine tissues varies between mammals. In humans and mice, the trophoblast is highly invasive, whereas in the pig it comes to lie close to the uterine epithelium but without penetration of the maternal tissues. For all mammals the net effect of implantation is the establishment of a more intimate association between the maternal tissues and the embryo, facilitating transfer of oxygen and nutrients to the embryo for its continued growth and removal of its metabolic waste products. An essential requirement for successful implantation is that the cellular and physiological changes occurring in the uterus during each reproductive or estrous cycle are coordinated with the development of the preimplantation embryo. Any asynchrony between the preparation of the uterus and development of the embryo results in failure of implantation and loss of the embryo.

II. MOLECULAR BIOLOGY OF LEUKEMIA INHIBITORY FACTOR AND ITS RECEPTOR

A. Leukemia Inhibitory Factor

Leukemia inhibitory factor (LIF) is a secreted monomeric, glycosylated protein with a molecular weight (MW) of 32–62 kDa. It consists of 180 amino acids that are folded into a 4α helix bundle topology. The molecular weight of the unglycosylated protein is ~20 kDa, with glycosylation being nonessential to LIF's biological activity. LIF is encoded by a single gene, 6 kb in length and consisting of three exons. The gene is transcribed as a 4.8-kb mRNA transcript, of which ~3.2 kb at the 3' end is untranslated. In the human the chromosomal locus of the LIF gene is 22q12.1–12.2 and in the mouse it maps to chromosome 11, band A1.

B. LIF Receptor α and gp130

LIF binds to cells via a heterodimeric high-affinity receptor consisting of two transmembrane proteins, the LIF receptor α (LIFrα) and gp130. The human LIFrα precursor consists of 1097 amino acids, of which the first 44 form the leader sequence that is subsequently cleaved to produce the mature form of the receptor. The extracellular domain consists of 789 amino acids, a transmemebrane domain of 26 amino acids, and an intracellular domain of 238 amino acids, resulting in the molecular weight of the unglycosylated form being 110 kDa with glycosylation increasing weight to ~200 kDa.

The LIF receptor gene is approximately 70 kb long and contains 20 exons. In humans the LIFrα locus maps to chromosome 5p12–13 and in the mouse to the proximal region of chromosome 15. At least three transcripts are transcribed from the gene in all adult murine and human tissues analyzed. The largest transcript is ~10 or 11 kb, and both it and the 5-kb transcript code for the full-length transmembrane protein. The other shorter form of 3 kb is generated by alternate splicing with the introduction of a novel stop codon truncating the mRNA, resulting in the synthesis of a soluble form of the receptor lacking both the transmembrane and intracellular domains. The soluble form is predominantly produced in the liver and uterus. During pregnancy synthesis of all three forms increases with the rise in liver expression resulting in elevated levels of the soluble form of the receptor appearing in the circulatory system that peak at midgestation. The function of the soluble form is unclear, although it may act as an antagonist by binding to and sequestering LIF from the transmembrane form of the receptor.

The other component of the high-affinity LIF receptor, gp130, is a 918-amino acid-long glycoprotein, with a 22-amino acid transmembrane domain and a 277-amino acid intracellular domain. This results in a predicted MW of 101 kDa, with glycosylation increasing the size to 130 kDa. In humans the locus encoding gp130 maps to chromosome 5q11, although a pseudogene is also located on chromosome 17p11. In humans the gene is transcribed as a single 8.5-kb transcript in all tissues analyzed, whereas in the mouse two transcripts of 7 and 10 kb are detected in all tissues.

The binding of LIF to the LIFrα results in the noncovalent association of the complex with gp130 to create the high-affinity receptor (Fig. 1). Although in most cell types the formation of the high-affinity complex between LIF, LIFrα, and gp130 is required

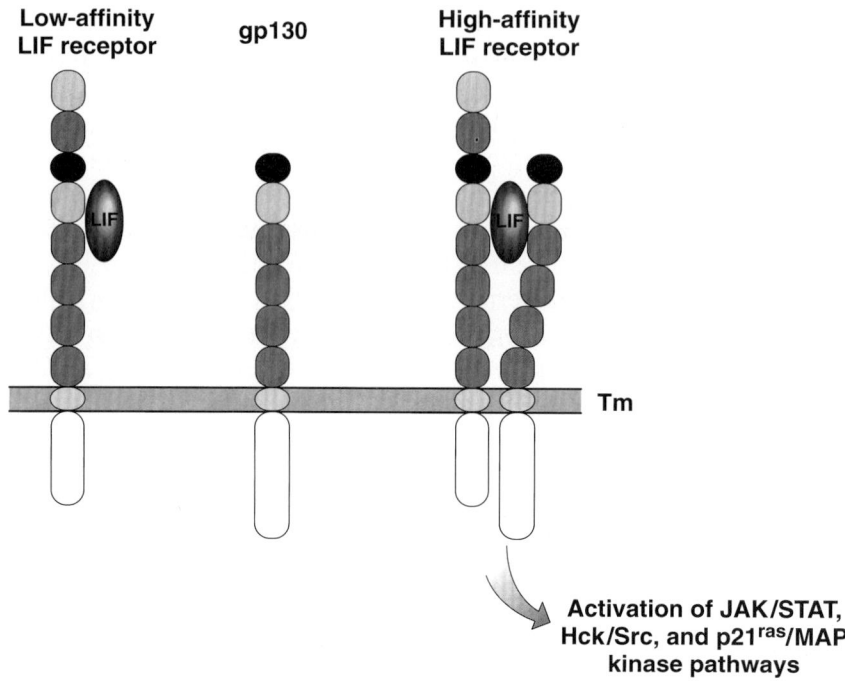

FIGURE 1 Structure of the LIF receptor. LIF binds to the low-affinity LIF receptor (LIFrα), which then forms a noncovalent association with gp130 to form the high-affinity receptor complex. This activates the signal-transducing pathways via the intracellular domains of the LIFrα and gp130.

for signal transduction, there is evidence that the LIFrα can function as a homodimer, without gp130, in inhibiting embryonic stem (ES) cell differentiation. The heterodimeric receptor complex then activates at least three known signal transduction pathways: the cytoplasmic tyrosine kinase pathways of the Janus (Jak) kinase family, in particular, Jak1, Jak2, and tyk2; the Src kinase family, in particular, Hck; and the p21ras/MAP kinase pathway. The activated Jak kinases phosphorylate the cytoplasmic STAT transcription factors, which then enter the nucleus and modulate gene expression.

III. REQUIREMENT FOR LIF IN REPRODUCTION

LIF was originally purified, cloned, and characterized because it induced the differentiation and cessation of proliferation of the myeloid leukemia cell line M1. However, an involvement of LIF in embryonic development was revealed when it was shown that LIF maintained the undifferentiated state of ES cells, the totipotent derivatives established from cultured

ICM cells of the blastocyst. In preimplantation embryos LIF transcripts are found principally in the trophoblast of the blastocyst. Following implantation, LIF is difficult to detect in the developing embryo, but following birth it reappears in many tissues, particularly the skin and small intestine. LIFrα transcripts are localized to the ICM as is gp130. In post-implantation embryos both the LIFrα and gp130 are detected in the developing embryo, particularly in the central nervous system, and in the adult they are found in all tissues analyzed.

In adults, the tissue in which LIF is most highly expressed is the uterus. In the mouse, uterine expression of LIF is biphasic, with the first peak of expression occurring at ovulation. Following mating the levels fall but then rise again, starting on the morning of the fourth day of pregnancy and peaking late on Day 4 before falling again to basal levels that persist throughout the rest of pregnancy. In the human uterus, LIF expression is first detected around Day 18 in the postovulatory phase of the menstrual cycle and persists until the end of the cycle (Fig. 2). However, unlike the mouse, there is no evidence of LIF expression occurring in the uterus at ovulation. In

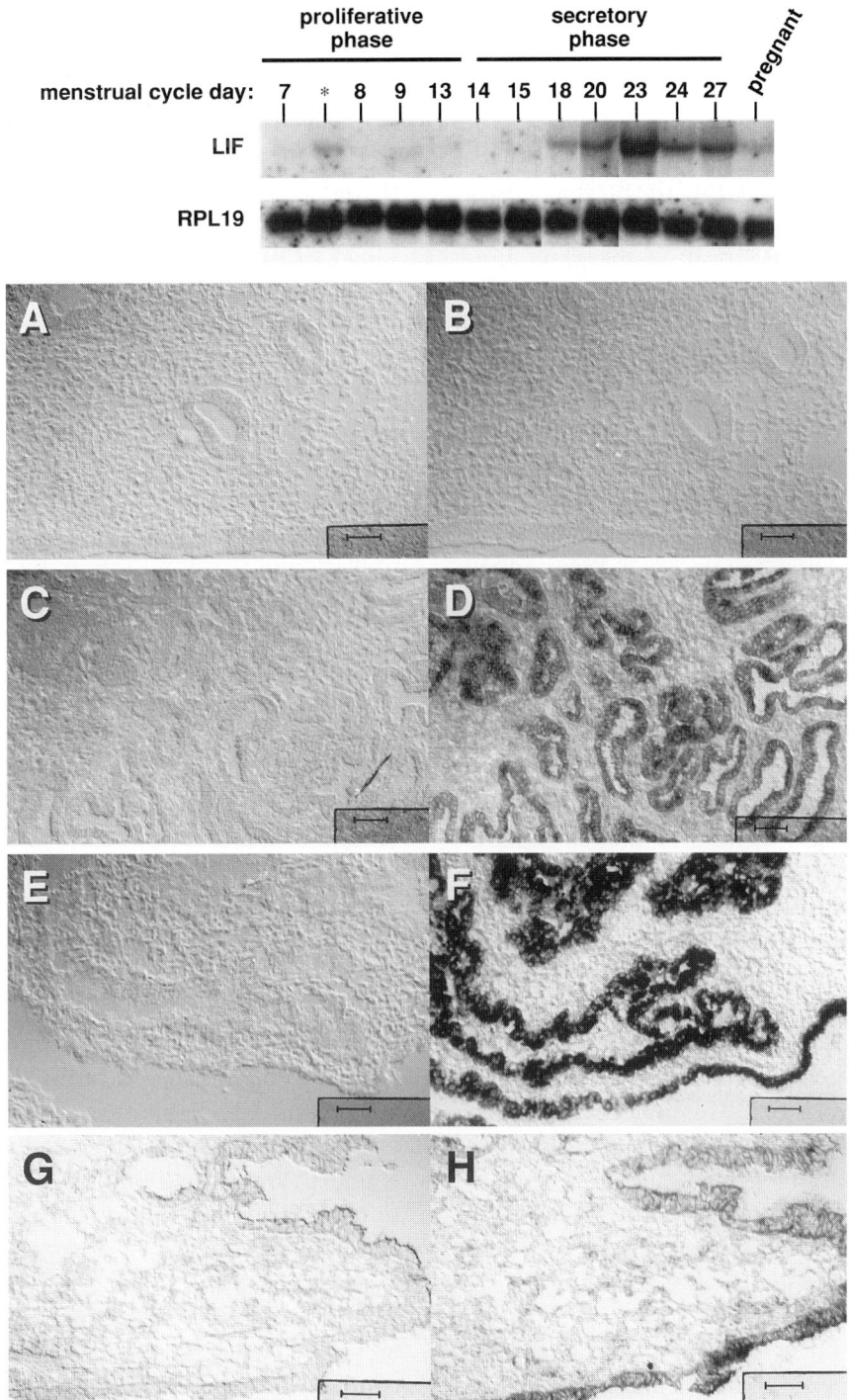

FIGURE 2 (Top) A Northern analysis showing the onset of LIF expression in the secretory or postovulatory phase of the human menstrual cycle. Expression commences around Day 18 of the cycle, the time at which embryo implantation starts. The lower RPL19 band is a control for the amount of RNA loaded in each lane. (A–H) *In situ* analyses of LIF expression in a sample at Day 14 (at about the time ovulation occurs) showing (A) a sense control, (B) no LIF expression in the endometrial glands, (C) a sense control for glandular epithelium from Day 20, (D) LIF expression localized to the glandular epithelium from a Day 20 sample, (E) a sense control for F, (F) gp130 expression in glandular and luminal epithelium, (G) a sense control for LIFrα, and (H) LIFrα in the luminal epithelium.

both the mouse and human uterus, LIF transcripts and protein are localized to the epithelium of the endometrial glands (Fig. 2D), although in the human LIF is also found in the luminal epithelium. In the human, LIF protein is detected in uterine washings, indicating that LIF is secreted into the uterine lumen.

It is of significance that in both humans and rodents uterine levels of LIF peak at the exact time that the uterus becomes receptive to the embryo to allow implantation. Similar increases in LIF expression around the time of embryo implantation have been reported in the rabbit, rat, pig, cow, rhesus monkey, and western spotted skunk, indicating that this pattern of expression is widespread among mammals. LIF expression is regulated by the maternal environment and does not depend on the presence of embryos. Thus, expression occurs in pseudopregnant female mice, in which no embryos are present, despite the female being physiologically primed for pregnancy. Furthermore, LIF is not detected in mice during lactationally induced implantational delay, in which implantation is arrested, with the blastocysts entering a state of physiological dormancy. Once delay is broken, by weaning or removal of the suckling pups, LIF is expressed and the embryos implant. Similarly, in ovariectomized progesterone-treated mice (experimentally induced delay), LIF expression is induced following a single injection of estrogen.

In ovariectomized mice, LIF expression is induced by estrogen or estrogen and progesterone combined. Progesterone alone has no apparent effect. In contrast, LIF expression in rabbits is induced by progesterone but not by estrogen. This reflects the differences in hormonal regulation in the rabbit, in which estrogen is not required for the induction of implantation. LIF may be directly regulated by the action of the steroid hormone receptors because its promoter contains steroid response elements located upstream of the transcription start site. In DNA transfection studies, the human LIF promoter linked to a reporter gene is activated when cells are treated with progesterone.

A. LIF in Embryo Implantation

The coincidence between LIF expression in the uterus with the onset of embryo implantation suggested that LIF might be essential to this process.

Direct evidence for this was established by the derivation of mice in which the LIF gene was mutated so that it was no longer functional. This was achieved by the technique of targeted mutagenesis by homologous recombination in embryonic stem cells, with the subsequent introduction of the mutated gene into the mouse's germline (so-called "knockout" mice). Mice lacking a functional LIF gene develop to adulthood. Homozygous LIF-deficient males are fertile; however, homozygous females never produce any offspring despite repeated matings to males of proven fertility. The females ovulate and the eggs are fertilized and develop to the blastocyst stage, but the blastocysts do not implant. Surgical transfer of the LIF-deficient blastocysts to the uteri of wild-type pseudopregnant recipients demonstrated they can implant and develop to term. Furthermore, reciprocal transfer of wild-type blastocysts to pseudopregnant LIF-deficient females revealed an inability of the wild-type blastocysts to implant in the uterus of LIF-deficient females. This showed that the failure of implantation was due to some maternal defect rather than a defect in the blastocysts and that uterine expression of LIF was essential for blastocyst implantation.

IV. THE ACTION OF LIF ON UTERINE TISSUES

In normal mice, implantation proceeds through a series of stages. By the fourth day of pregnancy the blastocysts are distributed along the length of the uterine lumen. The mechanism by which this occurs is poorly understood but probably involves contractions of the uterine musculature as there is no evidence for the existence of specific attachment sites in the uterus. The shape of the lumen changes to a slit, with the top of the lumen orientated to the mesometrial side of the uterus, the side attached to connective tissue, fat, and the blood vessels supplying the uterus. The blastocysts are always located at the opposite (antimesometrial) side of the lumen and orientated such that the ICM faces the mesometrial side (Fig. 3A). The luminal epithelium is in close contact with the blastocyst and often surrounds the embryo. However, no firm attachment is established between the embryo and uterus at this stage because

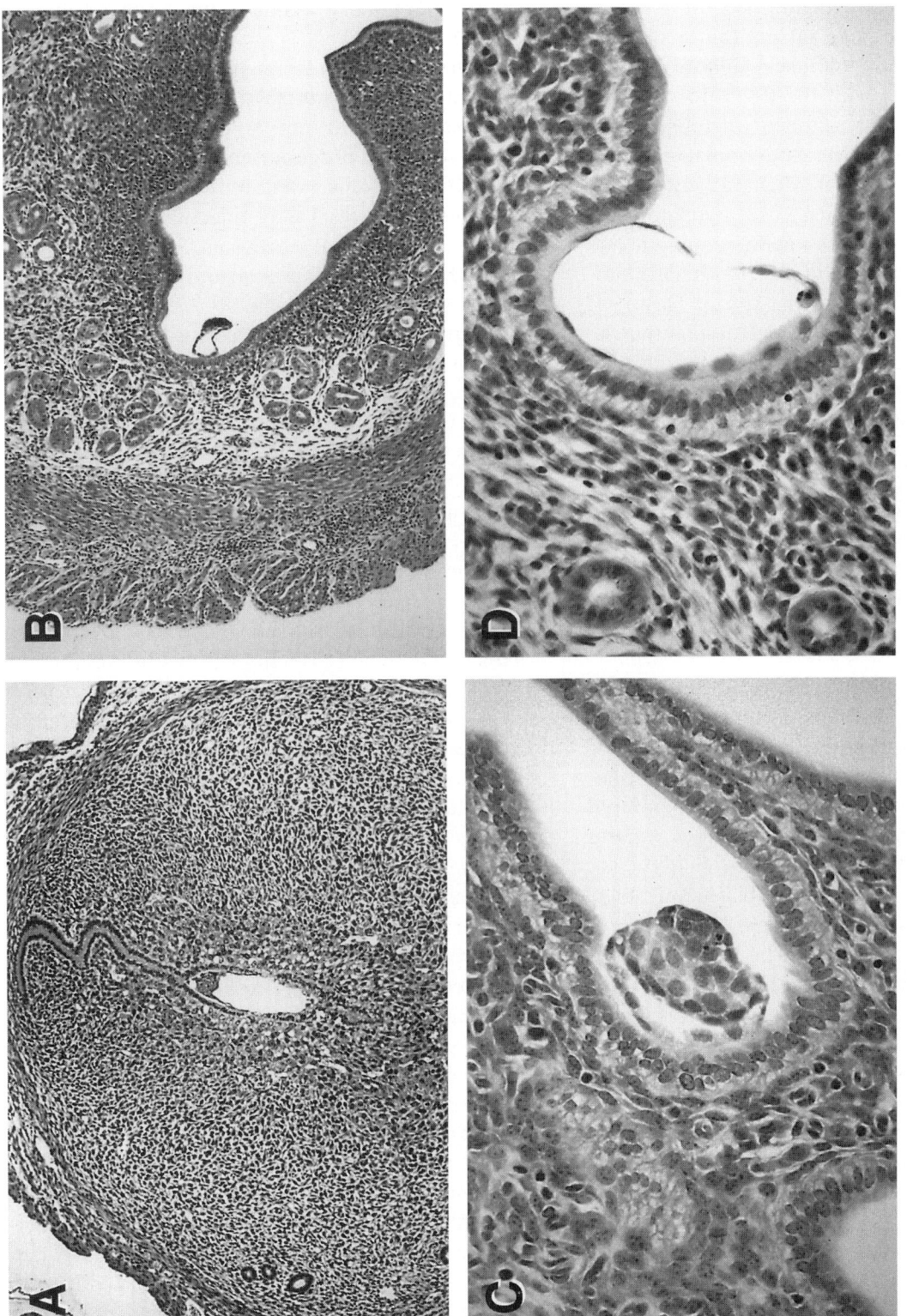

FIGURE 3 (A) A cross section through a normal mouse uterus showing a blastocyst implanting on the fifth day of pregnancy. The blastocyst is orientated toward the mesometrial side of the uterus. The luminal epithelium adjacent to the trophoblast is undergoing apoptosis and the stromal cells have started to decidualize. Note the primary decidual zone next to the blastocyst and luminal epithelium. (B) A cross section through the uterus from a LIF-deficient mouse on the sixth day of pregnancy. The blastocyst is located in the antimesometrial side of the uterus and correctly orientated to the mesometrial side. The luminal epithelium is still intact and the stroma has not decidualized. (C and D) Two different blastocysts in LIF-deficient mice on Day 6 of pregnancy. They are in close appositional contact with the luminal epithelium, which is still intact, and the underlying stromal cells show a nondecidualized fibroblastic morphology.

the embryos can be readily flushed from the uterus; hence, this phase of implantation is referred to as the appositional stage. Late on the fourth day full attachment or adhesion commences with the luminal epithelium, in contact with the trophoblast, lifting up from the underlying basement membrane and separating from the fibroblastic or stromal cells. The trophoblast cells start to interdigitate between the epithelial cells which then die apopototicaly, allowing full access to the basement membrane and stroma by the invading trophoblast. By the fifth day, the stromal cells adjacent to the site of blastocyst attachment start to proliferate and differentiate to form the primary decidual zone, which is accompanied by a localized increase in capillary permeability and edema. The differentiation of the stroma into a decidua is characterized by a change in morphology, with the stromal cells increasing in size, undergoing polyploidy, becoming more closely packed, and expressing alkaline phosphatase. The antimesometrial stromal cells in proximity to the embryo form the primary decidual zone, with the more mesometrial stromal cells subsequently forming the secondary deciduum (Fig. 3A).

Histological examination of uteri in LIF-deficient mice over the same time period revealed that, as with normal mice, the blastocysts are located at the antimesometrial side of the uterine lumen and orientated correctly with the ICM facing the mesometrial side of the lumen (Fig. 3B). The blastocysts were in appositional contact with the uterine lumen, which in some instances enveloped the blastocyst (Figs. 3C and 3D), and as with wild-type blastocysts in appositional contact, they could be readily flushed from the uterus, even as late as Day 7 of pregnancy. There was, however, no overt indication of the uterus responding to the blastocyst by the epithelium undergoing apoptosis, the stromal cells starting to decidualize, or a localized increase in edema. Attempts to induce decidualization in the LIF-deficient females using a variety of artificial procedures in either naturally mated or ovariectomized hormonally primed mice failed, indicating that the uteri are unresponsive to conditions that would normally stimulate decidualization.

The failure of the uterus to respond to the blastocysts and to other stimuli that induce stromal decidu-alization could be due to the uterus failing to proliferate or differentiate properly in the period leading up to implantation. This is unlikely because the LIF-deficient uteri, when treated with estrogen and progesterone, undergo normal patterns of cell proliferation and changes in morphology. Furthermore, following ovulation progesterone levels rise at the same rate and to the same extent in the LIF-deficient mice as in normal females. Lastly, injection of recombinant LIF into LIF-deficient females on Day 4 of pregnancy induced embryo implantation and uterine decidualization. Thus, sterility in the LIF-deficient females is due to the otherwise normal uterus being unresponsive to signals inducing implantation and decidualization.

A. Cellular Targets for LIF's Action

In situ and immunohistochemical analysis of normal uteri from both wild-type and LIF-deficient mice, as well as from uterine biopsy samples from women of proven fertility, localized the LIFRα and gp130 to both the glandular and luminal epithelium in humans (Figs. 2F and 2H). In mice gp130 is expressed at low levels in the luminal epithelium. In neither species is there any indication that the LIF receptor or gp130 are expressed in stromal cells.

The presence of the LIF receptor in the luminal epithelium strongly suggests that this is the tissue responsive to LIF secreted from the endometrial glands. Thus, the stroma responds indirectly to LIF due to some secondary signal(s) originating from the luminal epithelium. A model for the action of LIF is proposed in Fig. 4 where LIF, secreted from the endometrial glands, binds to receptors on the luminal epithelium. This in turn stimulates the epithelium to respond to the already closely apposed blastocyst by allowing trophoblast invasion of the uterus and for a second signal(s) to induce the underlying stromal cells to start decidualization. This model is consistent with previous data indicating that stromal decidualization depends on the presence of an intact luminal epithelium. The identity of the second signal(s) required for decidualization is not yet established. Recent evidence from S. K. Dey and colleagues revealed that the enzyme cyclooxygenase-2 (*Ptgs2* or *Cox-2*), the major inducible enzyme regulating

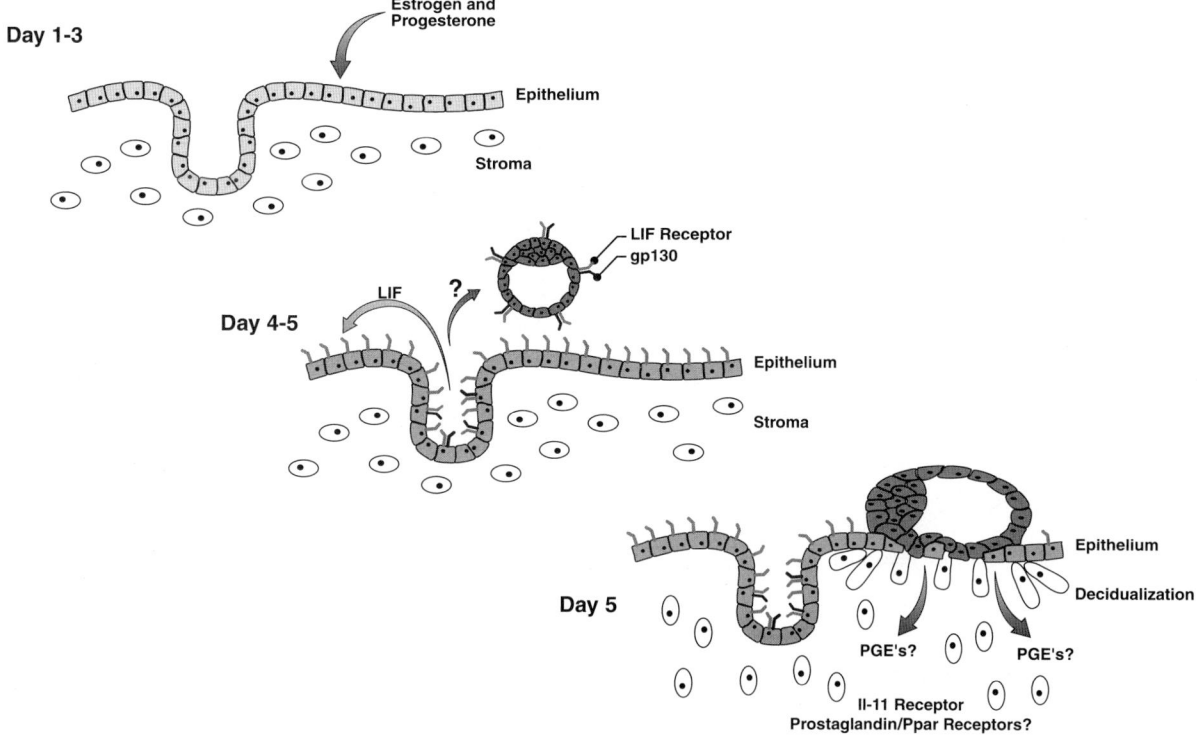

FIGURE 4 A model for the proposed molecular events occurring during implantation in mice. From Days 1 to 3 the stromal and epithelial cells of the uterus are induced to proliferate and differentiate by the combined actions of estrogen and progesterone. On Days 4 or 5 LIF is secreted from the glandular epithelium into the uterine lumen. Whether the secreted LIF has an effect on the blastocyst, which also expresses the LIFrα and gp130, is unknown. LIF, however, may bind to and activate the LIFrα in the luminal epithelium. In turn, this induces the luminal epithelium to become fully responsive to the blastocyst, which early on Day 5 starts to invade the uterus. Invasion is accompanied by the production of other signals (possibly prostaglandins) originating from the epithelium, stimulating the underlying stroma to decidualize. Once the decidua is formed its continued growth is dependent on the cytokine IL-11.

prostaglandin synthesis, is required for decidualization. This is consistent with previous observations, largely made by Tom Kennedy, that prostaglandins can enhance decidualization in rodents and that inhibitors of prostaglandin synthesis, such as indomethacin, suppress decidualization. *Cox-2* is transiently expressed in the luminal epithelium at the site of embryo implantation and therefore could be regulated by LIF. The prostaglandins may be required for the initiation of decidualization; however, female mice that carry a mutated receptor for the cytokine interleukin-11 (IL-11) are also infertile as recently shown by Glen Begley, Achim Gossler, and colleagues. In these mice, embryo implantation and development continues to Day 7 of gestation, before the embryos die. In these mice, the decidua initially forms but fails to continue to grow and proliferate, indicating that its later proliferation or survival depends on IL-11, another locally produced growth factor.

V. CONCLUSIONS AND PROSPECTS

It has been established that cellular proliferation and differentiation in the uterus, as well as embryo implantation, are regulated by the steroid hormones estrogen and progesterone, which in turn are under the control of the hormones of the hypothalamic–pituitary axis. However, it is becoming increasingly apparent that the complex cellular events at implantation are not only regulated by a few hormones secreted into the circulatory system but also by a cascade of locally produced peptide growth factors and other small molecules, such as prostaglandins. This multitude of factors act to "amplify" the cellular and tissue responses to estrogen and progesterone by regulating the expression of a large number of genes through different signal transduction pathways. An increasing understanding of just which factors are crucial, such as LIF, how they interact with each other, and the genes involved will provide new opportunities to both regulate the fertility of many mammals and gain a deeper understanding to some of the causes of infertility in humans.

Acknowledgment

Research sponsored by the National Cancer Institute, DHHS, under contract with ABL.

See Also the Following Articles

BLASTOCYST; DECIDUA; IMPLANTATION; TROPHOBLAST TO HUMAN PLACENTA

Bibliography

Cross, J. C., Werb, Z., and Fisher, S. J. (1994). Implantation and the placenta: Key pieces of the development puzzle. *Science* **266**, 1508–1518.

Cullinan, E. B., Abbondanzo, S. J., Anderson, P. S., Pollard, J. W., Lessey, B. A., and Stewart, C. L. (1996). Leukemia inhibitory factor (LIF) and LIF receptor expression in human endometrium suggests a potential autocrine/paracrine function in regulating embryo implantation. *Proc. Natl. Acad. Sci. USA* **93**, 3115–3120.

Finn, C. A. (1977). The implantation reaction. In *Biology of the Uterus* (R. M. Wynn, Ed.), pp. 245–308. Plenum, New York.

Lim, H., Paria, B. C., Das, S. K., Dinchuk, J. E., Langenbach, R., Trzaskos, J. M., and Dey, S. K. (1997). Multiple female reproductive failures in cyclooxygenase 2-deficient mice. *Cell* **91**, 197–208.

Schultz, G. A., and Edwards, D. R. (1997). Biology and genetics of implantation. *Dev. Genet.* **21**, 1–5.

Stewart, C. L., Kaspar, P., Brunet, L. J., Bhatt, H., Gadi, I., Kontgen, F., and Abbondanzo, S. J. (1992). Blastocyst implantation depends on maternal expression of leukaemia inhibitory factor. *Nature* **359**, 76–79.

Leydig and Sertoli Cells, Nonmammalian

Jeffrey Pudney
Harvard Medical School

I. Introduction
II. Organization of the Vertebrate Testis
III. The Leydig cell
IV. The Sertoli cell
V. Conclusion

GLOSSARY

androgens Steroid hormones involved in regulating and maintaining spermatogenesis, male secondary sex characteristics, and mating behavior.

Leydig cells Cells present in the interstitial tissue that synthesize androgens.

Sertoli cell A somatic cell present in the germinal compartment; responsible for regulating spermatogenesis.

spermatocysts The unit of spermatogenesis present in anamniote testis.

Leydig and Sertoli cells both play pivotal roles in the maintenance and regulation of spermatogenesis in the vertebrate testis.

I. INTRODUCTION

The organization of the testis is highly conserved throughout the vertebrate series and consists of two separate compartments. One is the germinal (also referred to as seminiferous) epithelium, which contains developing and maturing germ cells that undergo a complex series of morphological changes to ultimately form the male gametes or spermatozoa. Also present in the germinal compartment is a somatic cell called the Sertoli cell. The association of germinal cells with a somatic element of the germinal epithelium is a consistent feature of the testis occurring in all classes of vertebrates. The Sertoli cell controls germ cell development and maturation in an unknown manner. The second compartment in the testis lies outside the germinal epithelium. This is called the interstitial tissue, which apart from connective tissue and blood vessels, contains Leydig cells which produce the male hormones or androgens. Androgens are, in part, responsible for initiating and maintaining spermatogenesis as well as secondary sex characteristics, sexual behavior, and libido. Leydig cells have been reported to be present in the testes of representative species of every extant vertebrate class. Although spermatogenesis proceeds in a similar fashion in all vertebrates, there is variation in how this is accomplished. In order to understand and appreciate the role Sertoli and Leydig cells play in regulating spermatogenesis, it is necessary to be cognizant of how the organization of the testis varies between nonmammalian vertebrates.

II. ORGANIZATION OF THE VERTEBRATE TESTIS

In the phylum Chordata, nonmammalian vertebrates comprise the jawless fishes, lampreys, and hagfish (Agnatha); cartilaginous fishes, sharks, rays, and chimeras (Chondrichthyes); bony fishes (Teleostei); newts, salamanders, frogs, and toads, (Amphibia); turtles, lizards, snakes, crocodiles, and alligators (Reptilia); and birds (Aves). The organization of the testis in these different classes of vertebrates can be divided into two distinct patterns, one found in anamniotes (fish and amphibia) and the other in amniotes (reptiles, birds, and mammals).

A. Anamniotes

In these vertebrates the unit of spermatogenesis is a spermatocyst, usually referred to as a cyst. Nascent cysts are formed when a primary spermatogonium becomes associated with a Sertoli cell that envelops the germ cell with cytoplasmic processes (Fig. 1). Spermatogenesis begins and the cyst develops and grows to appear as a mass of germ cells surrounded by Sertoli cytoplasm which forms the wall of the cyst. Finally, in the mature cyst a clone of isogeneic spermatozoa appear that are released by rupture of the cyst wall. Following spermiation the cyst, i.e., the Sertoli cell, degenerates.

In Agnatha and Chondrichthyes, the testis is composed of a mass of isolated cysts and is termed polyspermatocystic (Fig. 1). For Teleosts and Amphibia, however, cysts develop within a germinal compartment that has been variously referred to as lobules, tubules, ampullae, or follicles (Fig. 1). A consistent and unifying terminology has now been adopted concerning this germinal compartment, which is now called a lobule. Interspersed between the cysts or lobules is a connective tissue containing Leydig cells.

B. Amniotes

In amniotes there is no cystic form of spermatogenesis; instead, the germinal epithelium is contained within structures called seminiferous tubules. During spermatogenesis amniote Sertoli cells are always associated with several layers of germ cells at different stages of differentiation (Fig. 4). Immature germ cells (spermatogonia) are located at the base of the epithelium, whereas the more mature germ cells (spermatocytes and spermatids) occur at successively higher levels with spermatozoa bordering the lumen of the seminiferous tubules into which they will be released. This pattern of spermatogenesis results in a distinct stratification of germ cells which are associated with different regions of the Sertoli cell as they develop. The amniote Sertoli cell does not degenerate after spermiation, and it represents a permanent feature of the germinal epithelium. Leydig cells are pres-

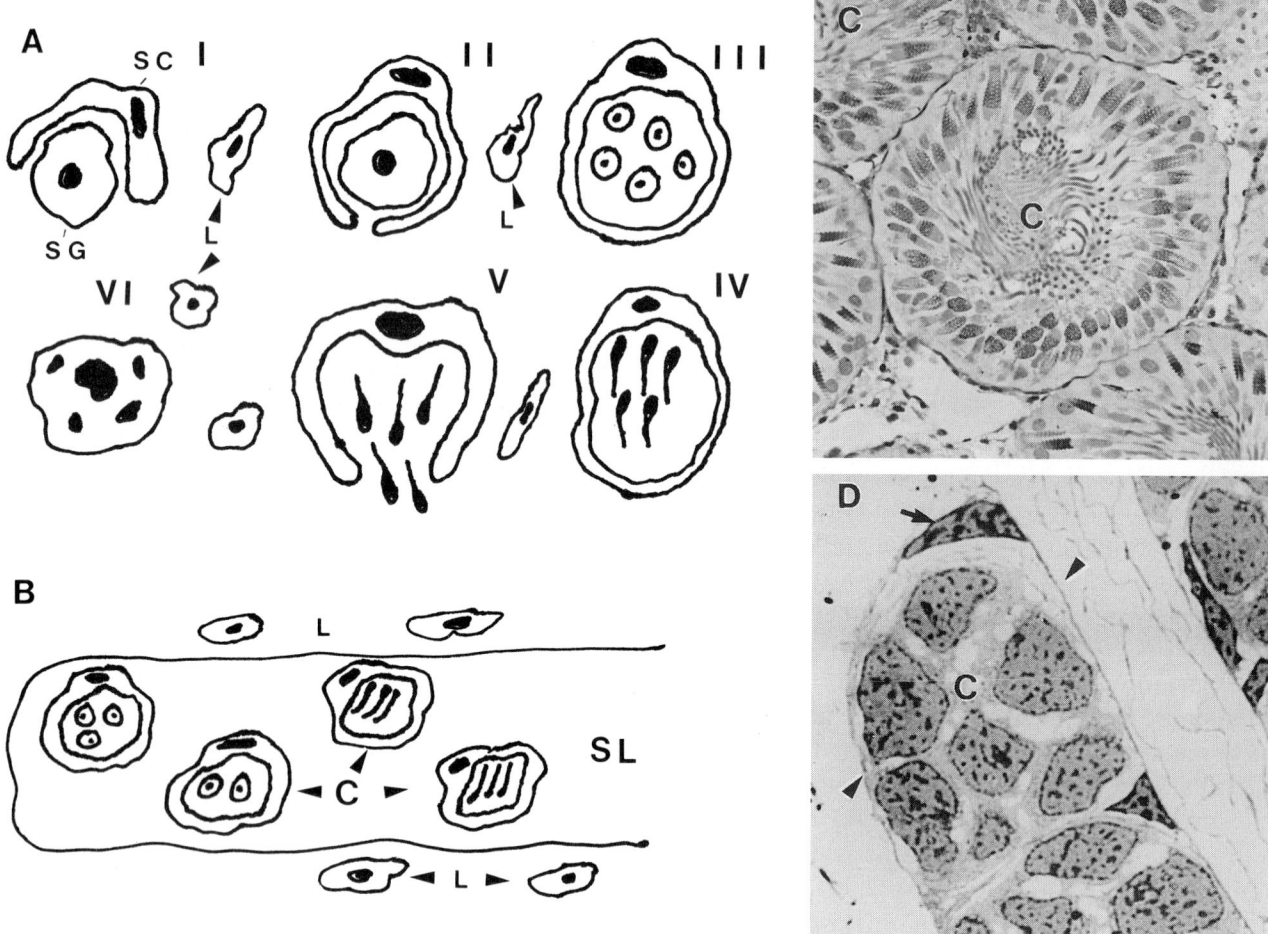

FIGURE 1 Schematic representation of the phylogenetic organization of the testis. (A) Development of cysts in polyspermatocystic mode of spermatogenesis, i.e., agnatha and elasmobranchs. (I) Initially, a Sertoli cell precursor (SC) becomes associated with a spermatogonium (SG). (II) The Sertoli cell envelops the germ cell to form the nascent cyst. (III) The spermatogonium then enters the spermatogenic cycle, resulting in growth of the cyst with attenuated Sertoli cytoplasm forming the walls of the cyst. (IV) Spermatogenesis progresses, with eventual production of an isogenetic clone of spermatozoa with their heads (usually) orientated toward the nucleus of the Sertoli cell. (V) Spermiation occurs when the thin walls of the cyst rupture. (VI) The Sertoli cell eventually collapses and undergoes fatty degeneration. Leydig cells (L) are found in the connective tissue surrounding the cysts. (B) In teleosts and amphibia the testes are composed of seminiferous lobules (SL) that contain cysts (C) formed by Sertoli cells. Leydig cells (L) are located in the interlobular tissue. (C) Light micrograph of polyspermatocystic testis of the elasmobranch *Squalas acanthias*. Cysts (C) contain bundles of maturing spermatids. (D) Light micrograph of seminiferous lobule of the urodele *Necturus maculosus* containing cysts (C) filled with developing germ cells. The Sertoli cell cytoplasm (arrowheads) forming the walls of the cyst is attenuated. Arrow points to the Sertoli cell nucleus. Notice that it is difficult to detect Leydig cells in the testes of either *S. acanthias* or *N. maculosus* at this level of resolution.

ent in the interstitial tissue surrounding the seminiferous tubules.

III. THE LEYDIG CELL

Franz Leydig was born on May 21, 1821 in a small town in Germany. After receiving his medical degree in 1847 he became a lecturer in anatomy at Wurzburg University in 1849. It was about this time that Leydig published a study on the comparative histology of the mammalian male reproductive tract. This article contained the first description of testicular cells, found between the seminiferous tubules, that later came to bear his name.

Initially there was controversy as to the exact function of Leydig cells. Leydig, in fact, thought they represented a type of connective tissue acting to structurally support the seminiferous tubules. Other theories ranged from the fanciful, i.e., Leydig cells were juvenile Sertoli cells, to the more logical, i.e., Leydig cells somehow maintained spermatogenesis, male secondary sex characteristics, and mating behaviors.

Eventually, comparative studies demonstrated Leydig cells had an endocrine function and secreted a hormone involved in establishing and maintaining the reproductive activity of male animals. Following the concept of the testis as an endocrine gland, a great deal of interest was centered on isolating and characterizing the secreted hormone. Classical experiments of many notable endocrinologists during the early twentieth century resulted in the identification of the hormone secreted by the testis which was found to be a steroid that was termed testosterone. In amniotes testosterone is the major androgen produced by the testis, whereas in fish it is 11-keto testosterone and for amphibia it is dihydrotestosterone. Early studies on the testis using the light microscope and routine histological stains often failed to identify Leydig cells in the interstitial tissue surrounding the germinal compartment (Fig. 1). This is because at this level of analysis no distinguishing cytological feature could be used to specifically detect Leydig cells. It was recognized that the testis secreted a steroid hormone long before the morphological characteristics of steroid-producing cells were de-

scribed. During this period, the only means available for identifying putative Leydig cells were initially by histological techniques for staining lipid or cholesterol and later by tests developed for detecting enzymes [mainly 3β-hydroxysteroid dehydrogenase (3β-HSD)] involved in steroid synthesis.

A number of problems are associated with the early histological investigation on Leydig cells in the nonmammalian testis. First, due to the poor resolving power of the light microscope (0.2 μm) it was difficult to determine the exact location of the lipid/cholesterol or 3β-HSD-positive cells in the testis. Second, frozen sections are required to histochemically stain for lipid/cholesterol or 3β-HSD and this often results in poor structural preservation of tissue. This further compounds the difficulty of precisely determining in which testicular compartment, i.e., germinal or interstitial, the lipid/cholesterol/3β-HSD-positive cells reside.

These problems were resolved with the development of the electron microscope. This allowed analysis of the fine structure of steroid-producing cells, such as Leydig cells (Fig. 2). It was soon realized that cells specializing in the synthesis of steroid hormones possessed common cytological features. An organelle found to be characteristic of steroid-producing cells was the agranular or smooth endoplasmic reticulum (SER). The SER represents a discrete compartment in the cytoplasm delineated by single membranes. It is termed SER because it lacks ribosomes attached to another membranous system involved in protein synthesis called the granular or rough endoplasmic reticulum. The SER is morphologically versatile and can exist in a variety of configurations. Normally, the SER occurs as a system of tubules which often branch and anastomose to form a complex compartment within the cell. Other forms of the SER have been described in steroid-producing cells, including Leydig cells, such as cisternal whorls of membranes, fenestrated cisternae, and parallel arrays of tubules.

Another structural hallmark of steroid-producing cells is the internal organization of the mitochondria, in which the cristae are tubular in shape rather than straight or lamellar as in other cells. Ultrastructurally, these tubules have a vesicular profile in cross section and are, therefore, referred to as tubulovesicular (Fig. 2). Lipid droplets are also a common feature of

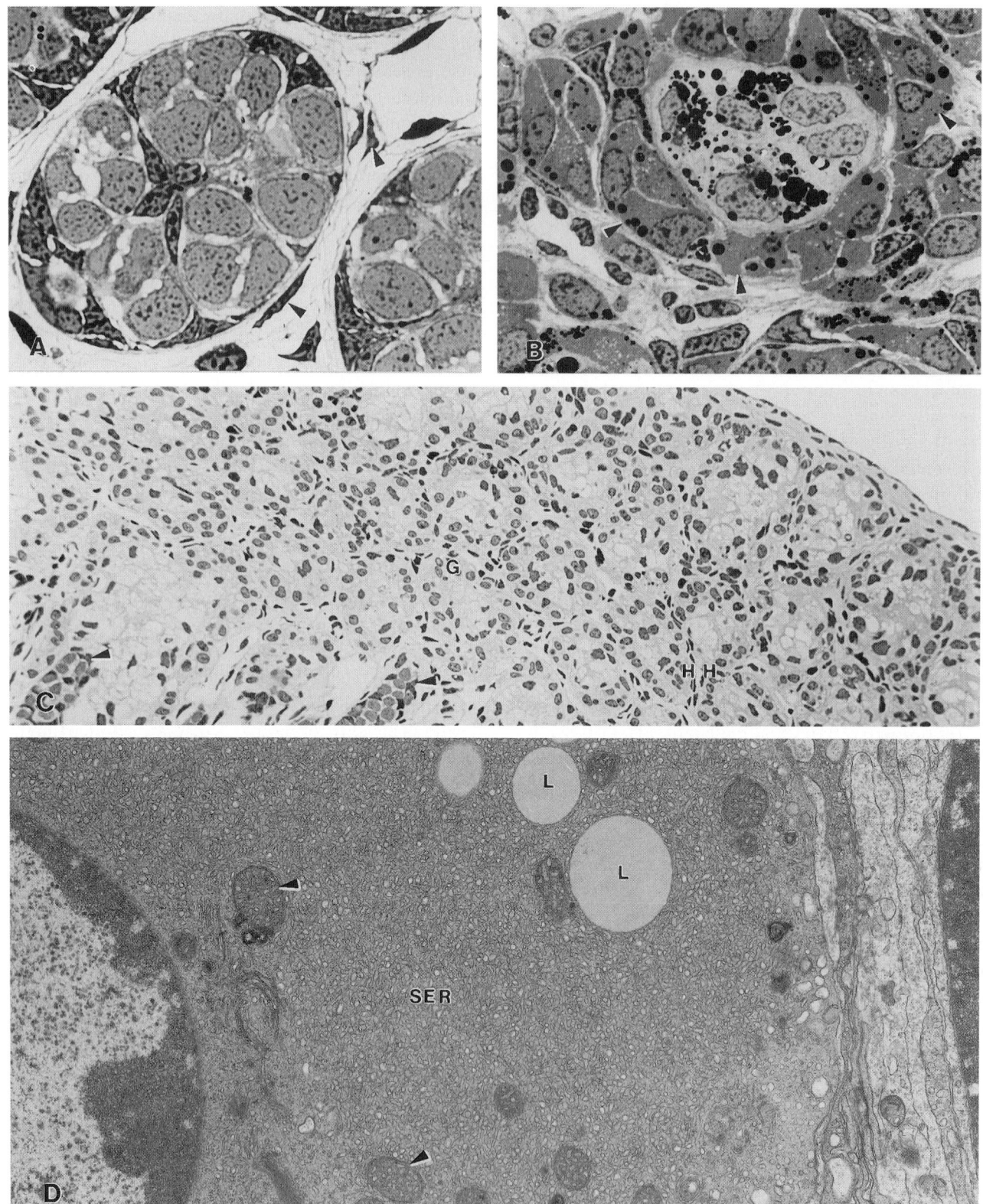

steroid-producing cells. In Leydig cells, as shall be discussed later, lipid droplets vary in both number and size during the breeding season.

The fine structure of steroid-producing cells, e.g., Leydig cells, is related to the synthesis of steroid compounds, e.g., androgens. Lipid droplets actually represent stores of cholesterol, a precursor of steroid hormones. Cholesterol can also be synthesized by enzymes located on the SER. This cholesterol then enters the mitochondria, where steroidogenic enzymes present on the tubulovesicular cristae covert it to a steroid called pregnenolone. This steroid is then transported to the SER, where a series of enzymes residing on these membranes are responsible for the synthesis of steroid hormones such as androgens. Changes in abundance and morphological appearance of these organelles involved in the synthesis of steroid hormones have been used as an index of the functional activity of Leydig cells.

A. Anamniotes

As mentioned previously, the use of histochemical tests for lipid cholesterol/3β-HSD in frozen sections of the testis at the light microscope level to identify Leydig cells makes it difficult to resolve the exact location of these cells. In the past, this has led to confusion and controversy as to whether or not Leydig cells occur in the testes of different anamniote species. This situation arose from studies reporting that Leydig cells could not be detected in the testes of several species of teleost fishes. Instead, a cell was described, present in the boundary tissue of the seminiferous lobules, that was considered homologous with Leydig cells. These cells were called lobule boundary cells. It was suggested, therefore, that in teleosts two different structural types of testes existed. For one group the interlobular tissue contained

Leydig cells homologous to those found in other vertebrate species. In the other group Leydig cells could not be recognized and instead lobule boundary cells were considered to be Leydig cell homologs. Lobule boundary cells were suggested to arise from fibroblasts in the boundary wall that became seasonally modified to produce steroid hormones. Evidence of homology between lobule boundary cells and Leydig cells was empirical, being based on the degree of sudanophilia (lipid/cholesterol content) exhibited by these cells. Subsequently, critical analysis of the teleost testis using the electron microscope demonstrated that this sudanophilia, purported to reside in the Leydig cell homolog the lobule boundary cell, was actually present in the intralobular Sertoli cell.

The identity of Leydig cells in the anamniote testis has been further complicated by the description of lobule boundary cells in the testes of urodele amphibians. These reports, again based on light microscopic studies, have been refuted by more critical examination of the urodele testis using the electron microscope. It was shown that the so-called lobule boundary cells are structurally homologous with Leydig cells of other vertebrates species and do not at any stage of development form part of the boundary wall but rather appear as a permanent feature of the interlobular tissue. Based on the preceding observations it has been proposed that the concept of the lobule boundary cell is no longer appropriate to describe Leydig cells in either the teleost or urodelean testis. This term, however, continues to be perpetuated in the literature.

There has also been debate concerning the presence of Leydig cells in the chondrichthyan testis. Thus, it has been reported, even within the same species, that Leydig cells are absent, rare, or present. All these studies, carried out at the light microscope level, mostly relied on histochemical tests for de-

FIGURE 2 Light micrographs showing development of glandular tissue in the testis of the urodele *N. maculosus*. (A). Leydig cells (arrowheads) surrounding spermatogenically active seminiferous lobules are poorly developed. (B) Following spermiation when Sertoli cells degenerate the Leydig cells (arrowheads) hypertrophy. (C) At the end of the breeding season, fully developed Leydig cells form a glandular tissue (G) situated between the periphery of the testis and the seminiferous lobules (arrowheads). (D) Electron micrograph illustrating the fine structure of a typical nonmammalian vertebrate Leydig cell in the glandular tissue of *N. maculosus*. As for all steroid-producing cells, Leydig cells contain abundant smooth endoplasmic reticulum (SER) organized as a mass of anastomosing tubules and mitochondria with tubulovesicular cristae (arrowheads) and lipid droplets (L).

tecting lipid/cholesterol/3β-HSD as criteria for identifying Leydig cells. As mentioned previously, the caveats associated with these methods can preclude exact detection of testicular Leydig cells. Furthermore, the problem is confounded by the fact that (i) 3β-HSD activity has been reported to occur in Sertoli cells of several species of elasmobranchs and (ii) variation in abundance of Leydig cells has been reported in the testes of a variety of elasmobranchs not only between species but also between individual specimens of the same species. All these differences could explain the discrepancies between various studies concerning the presence or absence of Leydig cells in the elasmobranch testis. Once more, however, analysis of the elasmobranch testis at the electron microscope level resolved these problems. It was observed that cells present in the interstitial tissue, between cysts, contained cells that possessed cytological features of steroid-producing cells, thus identifying them as Leydig cells.

Leydig cells have been reported to occur in the testes of representative species of each class of anamniotes. All anamniotes undergo a reproductive seasonal cycle that is reflected in the functional activity and morphological appearance of Leydig cells. During the breeding season, for most species of anamniotes, Leydig cells are large and often visually conspicuous in the interstitial tissue surrounding the germinal compartment. At this time Leydig cells secrete high titers of androgens needed for spermatogenesis as well as development of secondary sex characteristics. Structurally, the Leydig cells are hypertrophied with an abundance of organelles associated with steroid production (Fig. 2). Typically, the SER is extremely well developed and usually present as a mass of anastomising tubules, a configuration usually related with active steroid synthesis (Fig. 2). Mitochondria posses tubulovesicular cristae and large numbers of lipid droplets often occur in the cytoplasm of the Leydig cell. Species variation occurs in the abundance of these organelles present in Leydig cells as well as in the organization of the SER, which can appear as a loose system of tubules, as whorls of tubules, or as isolated tubules. Following the breeding season the Leydig cells of most anamniotes undergo structural involution. This involves a dramatic reduction in the volume of SER, which

appears as a small number of tubules. For many species the tubulovesicular cristae of the mitochondria also revert to a lamellar or straight configuration. There are also distinct changes in the lipid content of Leydig cells for many anamniote species following the breeding season that involve an accumulation of cholesterol. This build up of steroid substrate is presumably due to the decreased androgen production of Leydig cells after the breeding season is over. During the nonbreeding season this cholesterol gradually disappears.

Most anamniote Leydig cells occur between the cysts or seminiferous lobules. For some species, however, Leydig cells can also occur in testicular regions separate from the germinal compartment. In a number of teleosts, Leydig cells are present concentrated around the periphery of the testis and/or surrounding testicular excurrent ducts, e.g., *Molliensia latipinna* and *Lebistes reticulatus*. This disassociation of Leydig cells from the germinal tissue reaches a climax in gobies. For these species Leydig cells occur either as a central glandular mass surrounded by seminiferous lobules, e.g., *Glossogobius olivaceus*, or as a large accumulation lying along the entire length of the mesorchial side of the testis, e.g., *Gobius jozo*.

An interesting development of Leydig cells occur in the testes of urodelean amphibia. Unlike most anamniotes, the Leydig cells are poorly developed during the period of active spermatogenesis (Fig. 2). Only following spermiation, when Sertoli cells degenerate, is there development of the Leydig cells (Fig. 2). In urodeles spermatogenesis starts at the posterior end of the testis and progresses in an anterior direction. Thus, seminiferous lobules at sequential stages of germ cell development occur along the length of the testis. This caudocephalic differentiation or spermatogenic wave is restricted to the mature, terminal portions of the seminiferous lobules. Urodelean Leydig cells only hypertrophy after spermiation; therefore, at the end of the breeding season there is a wave of Leydig cell development following in the wake of spermatogenesis. This eventually forms a mass of highly differentiated Leydig cells possessing abundant organelles associated with steroid synthesis, called the glandular tissue, located at the periphery of the testis (Fig. 2). This glandular tissue then gradually regresses prior to the next breeding season. It is not known how the spermiated, degenerating portions of the seminiferous lobules stimulate the differentiation of Leydig cells associated with these regions.

In several species of teleosts and urodele amphibia, an association between Leydig cells and neurons has been described. This suggests that, in these anamniotes, the endocrine function of Leydig cells may be modulated by the nervous system. A more direct connection between external stimuli associated with mating behaviors via the nervous system could influence the release of sex hormones by Leydig cells to affect reproductive events such as spermiation.

B. Amniotes

Leydig cells tend to be histologically conspicuous in the interstitial tissue occupying the interstices of the amniote seminiferous tubules. Essentially, these Leydig cells are structurally similar to those described for anamniotes since they are specialized for the production of steroid hormones. Like anamniotes, amniotes also undergo seasonal cycles of reproductive activity resulting, for most species, in fully developed Leydig cells containing abundant organelles associated with steroid synthesis during the period of active spermatogenesis. The Leydig cells of amniotes also undergo a seasonal accumulation and depletion of cholesterol-rich lipids. Species differences in the structure of the Leydig cell involve the abundance of organelles related to steroid synthesis and the organization of the SER. Also, as was found in anamniotes, some amniotes have unusual arrangements of Leydig cells. This occurs in several species of teiid lizards of the genus *Cnemidophorous*. For these reptiles a circumtesticular layer of Leydig cells is formed just beneath the tunica albuginea. Few Leydig cells occur in the more normal location between the seminiferous tubules in these species. For some reptilian species spermatogenesis and Leydig cell activity are temporally separated. Thus, structurally and functionally Leydig cells are at their peak when spermatogenesis has ceased. In these species mating is accomplished by discharge of spermatozoa produced from the previous spermatogenic cycle and stored in the epididymides. Thus, accessory sexual structures have to be maintained by androgens long

after spermatogenesis has ended since spermatozoa are stored for several months prior to copulation.

IV. THE SERTOLI CELL

Enrico Sertoli was born June 6th, 1842 in Italy. Following his graduation from the University of Pavia in 1865, Sertoli published a report describing, for the first time, a columnar cell with branching processes which enveloped germ cells in the seminiferous tubules of the human testis. This cellular element of the germinal epithelium was subsequently variously described as a branched, supporting, or accessory cell until eventually it became the eponym for Sertoli himself. Sertoli believed these cells did not produce spermatozoa, but that their function was linked somehow to germ cell development. These observations by Sertoli were remarkable given the crude histological methods used at the time. However, his basic description of these cells was accurate and his interpretation of their function not appreciated until much later.

Despite Sertoli's early contention concerning the function of the cells he discovered, there was, during the nineteenth century, debate concerning whether one cell type (germinal or somatic) or two cell types (germinal and somatic) existed in the testis. The manufacture of better lenses improved the resolution of the light microscope and revealed the true structural relationship between germ cells and Sertoli cells. It was eventually realized that Sertoli cells served a supportive and nutritive role with respect to germ cells. Eventually, with the development of the electron microscope, a long-lasting argument was resolved concerning whether Sertoli cells were syncytial or single entities. Fine structural studies not only showed Sertoli cells were single cells but also revealed the morphological complexity and diversity exhibited by Sertoli cells of different vertebrate species.

A greater understanding of spermatogenesis and the organization of the germinal epithelium promoted analyses of cytological changes in Sertoli cells which could be related to germ cell development. In the case of nonmammalian vertebrates many studies were carried out to identify somatic cells in the germinal epithelium in order to detect cytological features suggesting analogy or homology with mammalian Sertoli cells. Contemporary investigations, once purely descriptive, now tend to focus on the structural composition of Sertoli cells in relation to their function, i.e., supporting and regulating spermatogenesis.

Two important functions common to vertebrate Sertoli cells are the phagocytosis of residual bodies and formation of the blood–testis barrier. Residual bodies represent remnants of cytoplasm sloughed by mature spermatids during the process of spermiation. These residual bodies are engulfed by Sertoli cells presumably for recycling of the cytoplasmic contents. The blood–testis barrier results from the formation of extremely tight junctions developed between adjacent Sertoli cells. These Sertoli–Sertoli junctional specializations exclude the entry and exit of many large molecules into and out of the germinal compartment to provide a specialized milieu for germ cell development. The blood–testis barrier is also formed to sequester autoantigens, present on maturing germ cells, from the rest of the body and thus prevent an autoimmune response. The cytological appearance of the Sertoli–Sertoli tight junctions differs between vertebrate groups. Structurally, they are simpler in nonmammalian vertebrates compared to mammalian species. Also, the developmental phase at which germ cells become isolated from the systemic milieu varies between vertebrates, i.e., postmeotic germ cells for anamniotes and premeotic germ cells for amniotes.

A. Anamniotes

As mentioned previously, the anamniote Sertoli cell forms the wall of the cyst that contains the developing germ cells (Fig. 3). For most anamniote species the morphological changes Sertoli cells undergo during spermatogenesis are similar. Variations occur in the presence, abundance, and temporal appearance of cytoplasmic organelles and inclusions. During the early stages of germ cell development Sertoli cells appear to be poorly developed (Fig. 3). As germ cells divide and develop, the cyst increases in size. The Sertoli cell, forming the walls of the cyst, therefore becomes attenuated. Despite this, Sertoli cells un-

dergo differentiation and hypertrophy. For some anamniotes, particularly species of elasmobranchs, this involves the appearance and development of organelles associated with steroidogenesis (Fig. 3). This suggests that for many nonmammalian vertebrates, the Sertoli cell plays a role in the hormonal regulation of spermatogenesis. The involvement of Sertoli cells in steroid synthesis has been partially confirmed by the histochemical detection of key steroidogenic enzymes (particularly 3β-HSD) in these cells for a number of anamniote species. This is in contrast to the mammalian testes in which the consensus is that Leydig cells are responsible for the major production of androgens. Following spermiation Sertoli cells of most anamniote species undergo fatty degeneration. For elasmobranchs, however, the apical region of the Sertoli cell is sloughed along with mature bundles of spermatids. These Sertoli cell remnants present in the semen of elasmobranchs have been called cytoplasts. They are thought to be responsible for the steroidogenic activity demonstrated in the semen of different elasmobranch species since they contain abundant SER, lipid droplets, and mitochondria with tubulovesicular cristae. A population of precursor Sertoli cells persists in the anamniote testis that develop to form the next generation of cysts.

In some anamniote species the relationship between Sertoli cells and germ cells is quite complex. For instance, in the Holocephali (a small group of chondrichthyes referred to as rat fish) the Sertoli cell develops an extensive system of actomyosin filaments thought to act as contractile structures involved in the organization and alignment of spermatids into bundles within the cyst. Dramatic variations can occur in the cytology of Sertoli cells (presumably reflecting different functions) between anamniote species. A group of teleosts called Atheriniformes are an excellent example of this diversity in Sertoli cell structure and function because some species practice external fertilization, whereas others engage in internal fertilization. An adaptation associated with internal fertilization is the production of spermatozoa packaged either in naked (spermatozeugmata) or encapsulated (spermatophores) bundles, which are introduced into the female reproductive tract at insemination. This results in the development of structural relationships between spermatids and Sertoli cells

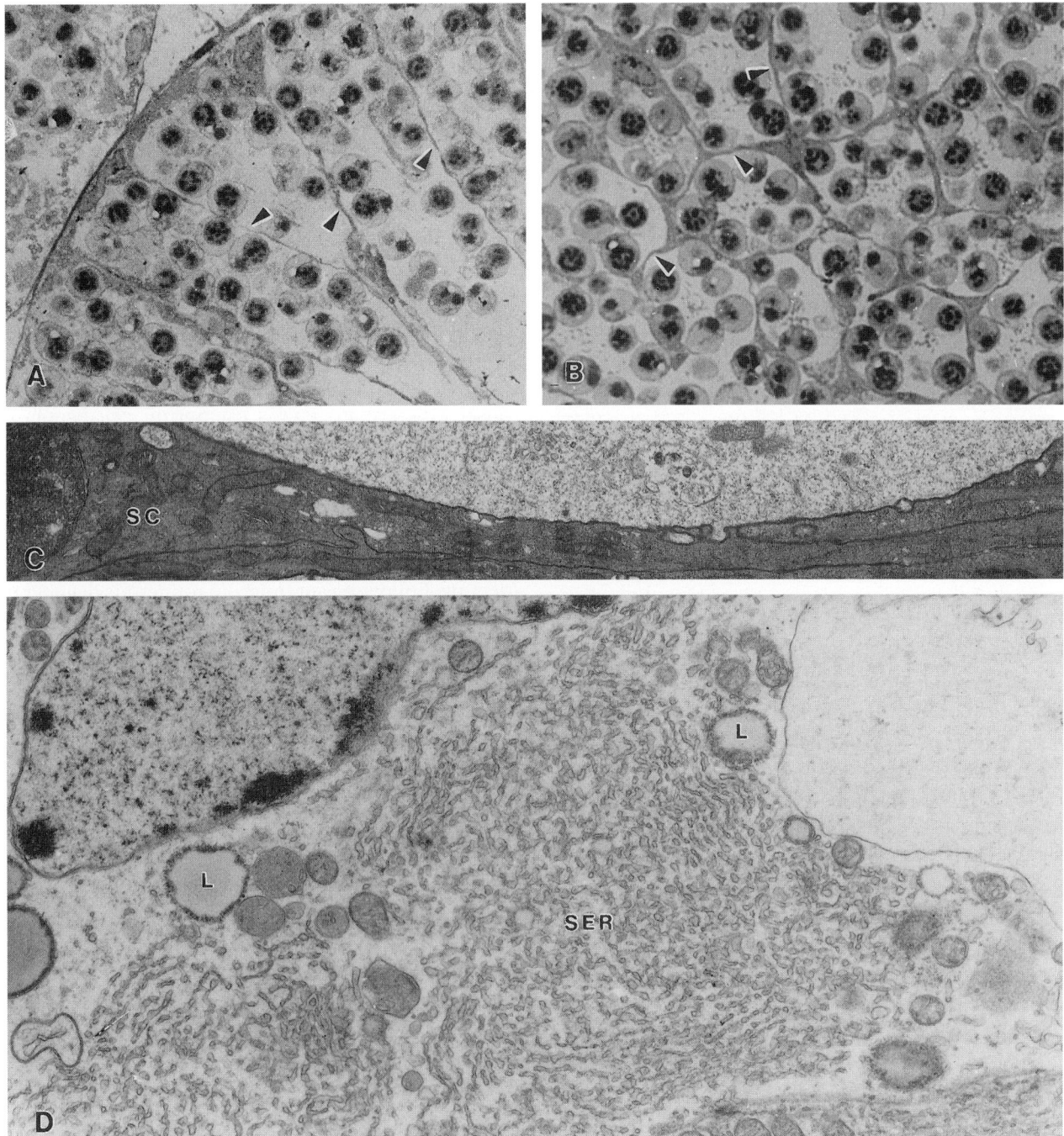

FIGURE 3 Light micrographs of anamniote Sertoli cells. (A) Longitudinal and (B) cross section of cysts of *S. acanthias* containing developing germ cells. Attenuated Sertoli cytoplasm (arrowheads) form the walls of these cysts. (C) Electron micrograph of Sertoli cell (Sc) of *N. maculosus* showing these cells are poorly differentiated during the early stages of spermatogenesis. (D) Seroli cell of *S. acanthias* demonstrating that at later stages of spermatogenesis cells become highly differentiated and develop abundant organelles associated with steroid synthesis such as smooth endoplasm reticulum (SER) mitochondria with tubulovesicular cristae and lipid droplets (L).

not seen in species which undergo external fertilization.

For species that use external fertilization the Sertoli cell undergoes the typical anamniote cycle of development, i.e., poorly differentiated during early spermatogenesis but hypertrophied at later stages. Spermatophores are formed within cysts and the walls of these structures, which surround the spermatozoa, are derived from a Sertoli cell secretion. At the height of their development, Sertoli cells appear highly differentiated for the synthesis of proteins and actively release a secretion into the lumen of the cyst. Rupture of the cyst results in the transfer of spermatophores into the efferent ducts.

Spermatozeugmata develop within cysts when spermatozoa align themselves with either their heads, e.g., peociliids, or tails, e.g., goodeids, next to the apical Sertoli cytoplasm. Junctional specializations do not develop between Sertoli cells and spermatozoa. Sertoli cells do react, however, by producing many finger-like apical processes which insinuate themselves between the spermatozoa and act to anchor the germ cells at the lumenal margin of the cyst. As spermatogenesis proceeds the cysts migrate down the seminiferous lobule until they reach the efferent ducts, at which time spermiation occurs and the spermatozeugmata are released.

The amphibian Sertoli cell essentially resembles that previously described for other species of anamniotes. The cytological development of the anuran Sertoli cell during maturation of cysts, however, differs when compared to Sertoli cells of urodeles. Urodeles are more fish-like since spermiation results by rupture of the cyst, whereas in anura the cyst wall breaks down much earlier, i.e., as soon as the spermatid flagellum begins to develop. These open cysts form the wall of the seminiferous lobule, which is lined by a layer of highly differentiated Sertoli cells with a bundle of spermatids deeply embedded in their apical cytoplasma. It has been shown that spermiation in anura occurs when Sertoli cells take up fluid, resulting in extreme swelling of the apical cytoplasm containing mature spermatids. This expansion of the apical Sertoli cytoplasm causes a displacement of spermatids toward the lumen of the seminiferous lobule into which they are eventually shed.

B. Amniotes

Few comprehensive studies have been carried out on the structure or function of amniote Sertoli cells. In reptiles and birds reproduction is cyclical. During the period of testicular regression the seminiferous tubules contain a layer of spermatogonia and Sertoli cells. At this time, the typical amniote Sertoli cell is morphologically poorly developed, containing an SER composed of loosely organized tubules and, usually, residual remnants of degenerating germ cells from the previous spermatogenic cycle (Fig. 4). For many species Sertoli cells possess a number of lipid droplets and glycogen granules (Fig. 4). With the onset of spermatogenesis the Sertoli cell begins to differentiate and there is a marked increase in the abundance of SER, and other organelles, such as lysosomes and mitochondria, contain distinctly tubulovesicular cristae (Fig. 4). At this time, the amounts of lipid and glycogen within the Sertoli cell decline. Following spermiation numerous degenerating germ cells and lipid droplets occur in the Sertoli cells. With continued testicular regression Sertoli cells revert back to a structurally involuted state. The amniote Sertoli cell undergoes a pronounced seasonal lipid cycle. Depending on the species, the Sertoli cell can possess few or many lipid droplets during the breeding season. With postnuptial testicular regression, however, the Sertoli cell becomes filled with lipid droplets that are positive for cholesterol. This lipid disappears from the Sertoli cell during the nonbreeding season. As for other nonmammalian vertebrates, the amniote Sertoli cell develops a blood–testis barrier. These junctional specializations are formed between adjacent Sertoli cells at the base of the seminiferous tubule. This results in a permeable basal compartment containing spermatogonia and a "tight" adlumenal compartment containing germ cells undergoing meosis as well as postmeotic germ cells.

Unlike nonmammalian vertebrates with cystic spermatogenesis, the stratification of germ cells in the amniote seminiferous tubules results in specific germ cell/Sertoli cell associations. This can produce structural features developed by Sertoli cells in relationship with discrete germ cell populations. An ex-

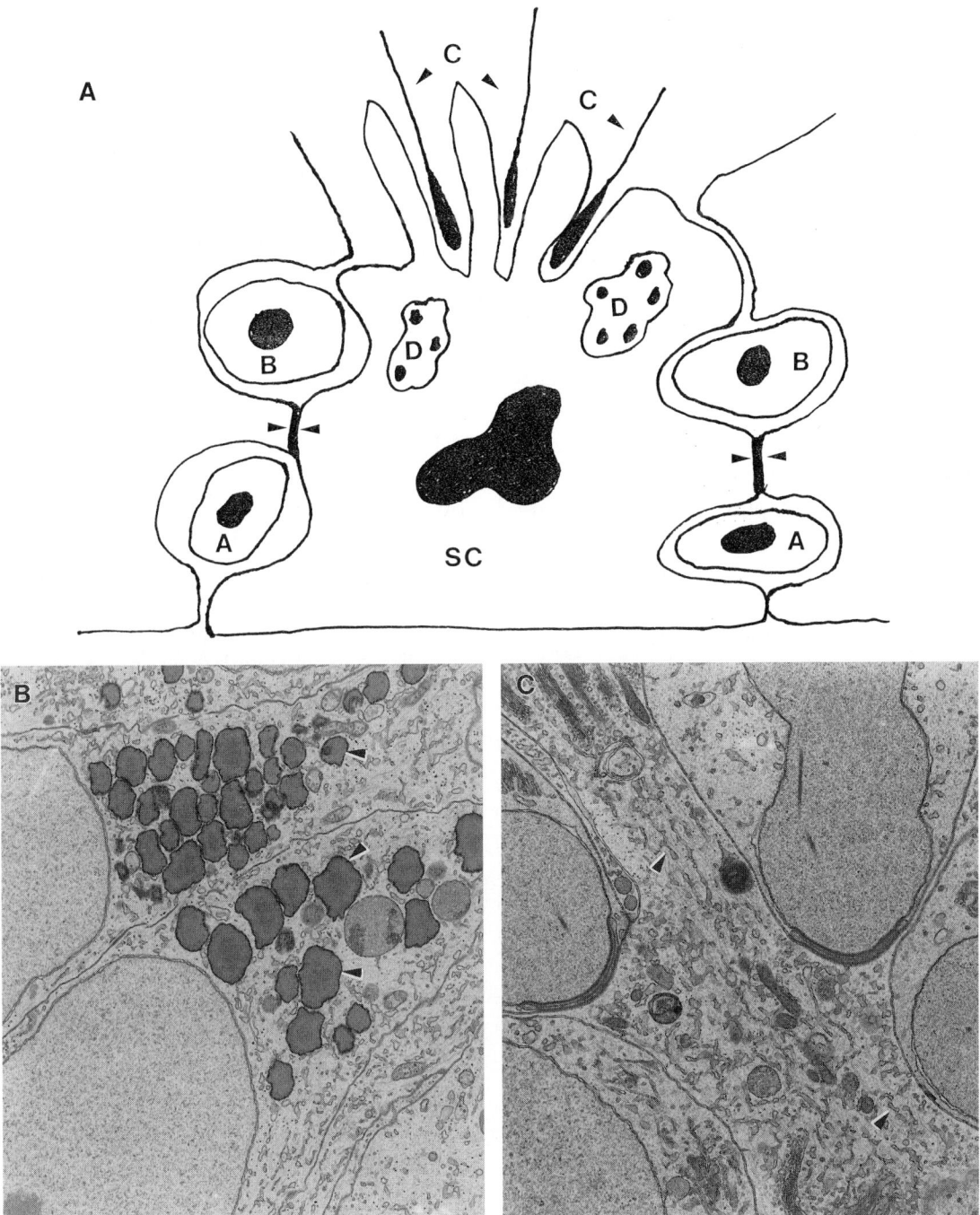

FIGURE 4 (A) Schematic diagram of amniote Sertoli cell. The Sertoli cell (SC) is columnar in shape and extends cytoplasmic processes to surround germ cells. Stratification of germ cells occurs with spermatogonia (A) present at the base, with the more mature germ cells, spermatocytes (B) and spermatids (C), located toward the lumen. Residual bodies (D) released by spermatids at spermiation are phagocytosed by Sertoli cells. Also, adjacent Sertoli cells develop tight junctions (arrowheads) to form the blood–testis barrier, resulting in a permeable basal compartment containing premeotic germ cells and a "tight" adlumenal compartment in which postmeotic germ cells are sequestered from the systemic circulation. (B) Electron micrograph showing typical appearance of the amniote Sertoli cell during the nonbreeding season in the testis of the turtle, *Chrysemys picta*. At this time, Sertoli cells are poorly developed and contain many lipid droplets (arrowheads). (C) During the breeding season the amniote Sertoli cell differentiates and often contains organelles related with steroid synthesis such as smooth endoplasmic reticulum (arrowheads). This Sertoli cell is associated with elongating spermatids in the testis.

ample of this is the formation of ectoplasmic specializations by Sertoli cytoplasm adjacent maturing spermatids. These ectoplasmic specializations are characterized by a prominent layer of filaments plus deeper elements of SER present in Sertoli cytoplasm adjacent to the head of maturing spermatids. Another structural specialization peculiar to reptilian and avian Sertoli cells is a complex relationship with elongating spermatids called the accessory cap or mantle. This is formed when a projection of Sertoli cytoplasm forms a cap or mantle which envelops the spermatid acrosome and is in turn then covered by a layer of spermatid cytoplasm. The functional significance of these morphological interactions between Sertoli cells and developing germ cells is essentially unknown.

V. CONCLUSION

More information is available concerning the structure and function of mammalian Sertoli and Leydig cells compared to nonmammalian Sertoli and Leydig cells. Most investigations of the nonmammalian testis are descriptive in nature and were carried out to identify Sertoli and Leydig cells and determine how analogous or homologous these cells are with their mammalian counterparts. Given the large number of species that make up the different classes of nonmammalian vertebrates, relatively few have been adequately studied in terms of the structure and function of Sertoli and Leydig cells. It is not known, therefore, how representative these description are for the various groups of vertebrates. Thus, it is difficult to make generalized statements concerning the structure and function of Sertoli and Leydig cells in many classes or even orders of nonmammalian vertebrates.

Due to the unusual organization of the testis and/or pattern of spermatogenesis, many nonmammalian vertebrates provide excellent models for studying the role Sertoli and Leydig cells play in regulating germ cell development. In mammalian species the mode of spermatogenesis results in a complex germinal epithelium in which the Sertoli cell is associated with several successive generations of germ cells at any one particular point in time. Thus, it is very difficult to dissect out the role the Sertoli cell plays in controlling the process of spermatogenesis. For nonmammalian vertebrates with cystic spermatogenesis, however, the Sertoli cell, during spermatogenesis, is associated with individual clones of germ cells. This makes it easier to determine how the structure and function of the Sertoli cell changes as each generation of germ cells develop and mature. Also in the mammalian testis, it is difficult and tedious to obtain a pure population of Leydig cells for *in vitro* studies. By contrast, in the testes of many nonmammalian vertebrates there is a distinct segregation of Leydig cells from the germinal tissue and examples were described in this article. This obviously allows the opportunity to obtain pure populations of Leydig cells for experimental analysis.

Evidence is also presented in this article that nonmammalian vertebrate Sertoli cells possess, during the period of spermatogenic activity, abundant organelles associated with the synthesis of steroid compounds. This suggests that the Sertoli cell in nonmammalian vertebrates plays an active role in the hormonal control of spermatogenesis, whereas Leydig cells possibly function to maintain systemic levels of androgen needed for development of accessory reproductive structures and mating behaviors. This is in contrast to mammals, in which it is assumed that Leydig cells are totally responsible for androgen production needed to initiate and maintain spermatogenesis as well as any accessory reproductive structures and mating.

These examples illustrate the importance of studying unconventional animal models, such as nonmammalian vertebrates, to further understand the role Sertoli and Leydig cells play in the regulation of spermatogenesis.

See Also the Following Articles

Leydig Cells; Sertoli Cells, Overview; Spermatogenesis, in Nonmammals

Bibliography

Payne, A. H., Hardy, M. P., and Russell, L. D. (1996). *The Leydig Cell.* Cache River Press, FL.
Russell, L. D., and Griswold, M. D. (1993). *The Sertoli Cell.* Cache River Press, FL.

Leydig Cells

Benson T. Akingbemi, Ren-Shan Ge, and Matthew P. Hardy

Population Council

GLOSSARY

adult Leydig cell A steroidogenic cell in the adult testis that contains a large volume of smooth endoplasmic reticulum, is highly responsive to luteinizing hormone, and produces testosterone for the maintenance of secondary sex characteristics and spermatogenesis.

fetal Leydig cell A Leydig cell that differentiates from Leydig stem cells during embryogenesis and contains numerous lipid droplets and low aromatase activity; it produces testosterone needed for masculinization during fetal and neonatal life.

immature Leydig cell A cell that differentiates directly from the progenitor Leydig cell and contains high levels of testosterone-metabolizing enzyme activities; it produces more 5α-androstane-$3\alpha,17\beta$-diol than testosterone.

interstitial tissue Loose connective tissue that lies between the seminiferous tubules and contains mesenchymal cells; fibroblasts; Leydig cells; blood-derived cells such as macrophages, lymphocytes, monocytes, and mast cells; and blood and lymphatic vessels.

Leydig stem cell A mesenchymal cell from which the Leydig cell lineage originates and that does not yet express specific proteins that are characteristic of adult Leydig cells.

mesenchymal cell A spindle-shaped cell and principal constituent of mesenchyme. It is the first tissue to appear in the embryo; it has the ability to differentiate into all connective tissue cell types, including Leydig cells in the testis.

progenitor Leydig cell The first recognizable cell in the adult Leydig cell lineage, possessing luteinizing hormone receptors and 3α- and 3β-hydroxysteroid dehydrogenase activi-

ties but little smooth endoplasmic reticulum; nevertheless, it produces androgens, primarily androsterone, rather than testosterone.

steroidogenesis The biochemical synthesis of steroids from cholesterol, starting with the initial conversion of cholesterol to pregnenolone by cytochrome P450 cholesterol side-chain cleavage enzyme, which is present in all steroidogenic cells, including Leydig cells.

testosterone The major steroid produced by fetal and adult Leydig cells in a series of enzymatic conversions catalyzed by four steroidogenic enzymes that are situated in the mitochondria and smooth endoplasmic reticulum. It is the primary hormone supporting the male phenotype.

Leydig cells constitute 4–6% of testicular volume and secrete the primary male sex hormone, testosterone. Testosterone is required in prenatal life to stimulate male sexual differentiation. Postnatally, testosterone promotes development of male secondary sex characteristics, causes hormonal imprinting of extratesticular tissues such as the liver, prostate, and hypothalamus, and maintains spermatogenesis. Most of the basic information on Leydig cells provided in this article was collected from studies in rodents. An extensive literature of human studies allows additional coverage of clinical and pathological aspects of the Leydig cell.

I. INTRODUCTION

The testis consists of two compartments: seminiferous tubules and interstitium. A seminiferous tubule is an elongated tube, measuring several meters in most species and lined by layers of germ cells and supporting Sertoli cells. The tubule is enveloped with a sheath of connective tissue and contractile peritubular myoid cells. The blood–testis barrier, formed by specialized junctions between adjacent Sertoli cells, partitions the seminiferous tubule into basal and adluminal zones and prevents immune recognition of antigens present on adluminal germ cells by white blood cells in the basal zone. The seminiferous tubule functions in the development and exocrine release of male germ cells.

The second compartment consists of the interstitium, which fills the space between the seminiferous tubules. The interstitium contains blood and lymphatic vessels and various cell types, including Leydig cells, fibroblasts, macrophages, lymphocytes, monocytes, and mast cells. The cellular composition of the interstitium varies between species, and in the rat, mouse, and guinea pig, much of the interstitial space is given up to lymphatic drainage. This arrangement facilitates the diffusion of steroids and other cell products in the interstitial space. In pigs, Leydig cells fill up most of the space available. Testosterone secreted by the Leydig cells passively diffuses across short distances (tens of micrometers) until it becomes associated with carrier proteins for transport to more remotely located tissues in the body.

The Leydig cell was definitively identified in 1850 by the German anatomist, Franz Leydig. Between 1897 and 1904, Bouin and Ancel established that Leydig cells were the source of a masculinizing factor. The masculinizing factor was chemically identified as testosterone by David and coworkers in 1935. In 1958, Wattenberg demonstrated the steroidogenic capacity of Leydig cells by histochemical localization of a key steroidogenic enzyme, 3β-hydroxysteroid dehydrogenase (3β-HSD).

II. MORPHOLOGY

The Leydig cell is round or flattened in shape, depending on its location and relationship to neighboring cells. Adjacent Leydig cells are connected by communicating (gap) junctions, which contain a group of proteins named connexins. Since these junctions are fewer between steroidogenically inactive cells, it is thought that secretory activity in individual Leydig cells is coordinated through gap junctions.

The size of the Leydig cell varies with species, its diameter ranging from 12 μm in the rat to as large as 30 μm in the pig.

The smooth endoplasmic reticulum (SER) and mitochondria, organelles that are involved in steroidogenesis, are especially prominent in Leydig cells (Fig. 1). The cytochrome P450 cholesterol side-chain cleavage enzyme (P450$_{scc}$) and its accessory proteins, adrenodoxin and adrenodoxin reductase, which serve as electron carriers, are present in the inner mitochondrial membrane. Three testosterone biosynthetic enzymes, 3β-HSD, cytochrome P450 17α-hydroxylase/17,20 lyase (P450$_{17\alpha}$), and 17β-hydroxysteroid dehydrogenase (17β-HSD), are localized in the SER. The abundance of SER in the Leydig cell is its most significant ultrastructural characteristic. Studies by Ewing and coworkers established that the total surface area of SER in Leydig cells is highly correlated with their capacity for testosterone production. In contrast, the rough endoplasmic reticulum comprises only one-sixth the surface area of SER in Leydig cells. Mitochondria are present as spherical, oval, or rod-shaped bodies. The majority of these have tubular membranous infoldings (cristae) in contrast to the lamellar cristae found in nonsteroidogenic cells.

The Golgi complex, which is typically located near the nucleus, glycolysates proteins, including oxytocin, corticotropin-releasing factor, enkephalins, dynorphins, and proopiomelanocortin-derived peptides, that are secreted by Leydig cells. Lysosomes, endosomes, and multivesicular bodies are present in the Leydig cell and take part in the autodigestion of exogenous and endogenous material. Nonmetabolized material accumulates in the cytoplasm in the form of lipofuscin granules as Leydig cells age. Peroxisomes are found in close association with the SER and are sites of cholesterol synthesis from lanosterol. Sterol carrier protein-2, which is highly concentrated in peroxisomes, associates with cholesterol and provides an intracellular shuttle to the outer mitochondrial membrane, in conjunction with microtubules, microfilaments, and intermediate filaments. Lipid droplets are scarce in rat and human Leydig cells but are abundant in the mouse and monkey and represent sites of cholesterol storage and/or synthesis. They are also thought to be an indicator of the

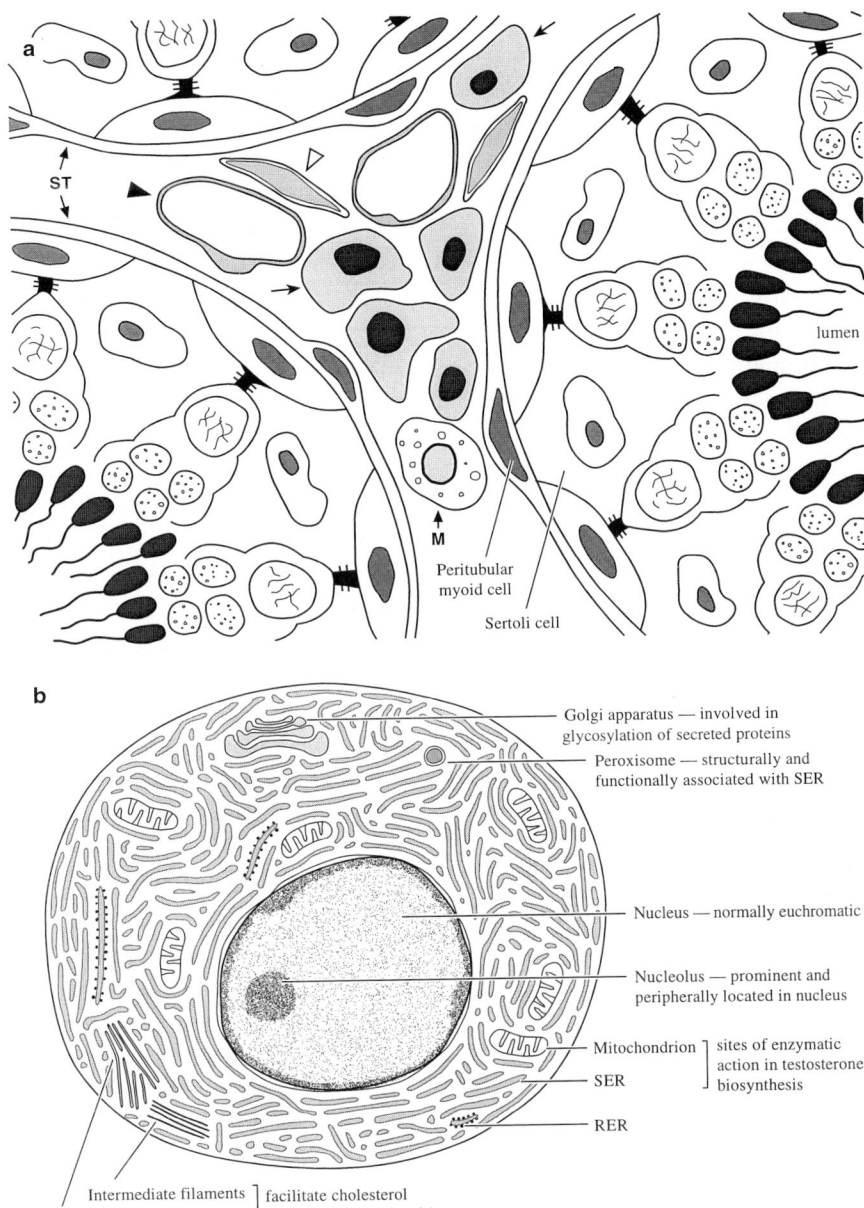

FIGURE 1 Diagram of the testis (a) showing the location of the interstitium between the seminiferous tubules (ST). Prominent within the interstitium are Leydig cells (arrows), macrophages (M), fibroblasts (open arrowhead), and blood vessels (closed arrowhead). The organelles involved in Leydig cell function, including smooth endoplasmic reticulum (SER), mitochondria, peroxisomes, Golgi complex, microtubules, and intermediate filaments, are as annotated in b. One steroidogenic enzyme, $P450_{scc}$, is located in mitochondria, and three others, 3β-HSD, $P450_{17\alpha}$, 17β-HSD, are situated in the SER. Sterol carrier protein-2, which, together with microtubules and intermediate filaments, aids the movement of cholesterol to mitochondria, is present in peroxisomes. Intracellular membranes of the rough endoplasmic reticulum (RER) are less voluminous in Leydig cells compared to the SER (one-sixth the surface area) and are the sites of protein translation.

level of steroidogenic activity in the Leydig cells of seasonally breeding species, becoming more numerous during the nonbreeding season when testosterone biosynthesis occurs at a slower rate. Human Leydig cells contain crystalloid bodies named Reinke's crystals which increase with advancing age but are not associated with any specific activity.

The Leydig cell contains a single nucleus with a prominent nucleolus. Multinucleate Leydig cells have been reported in aging humans. The nucleus is predominantly euchromatic, with a low affinity for basic stains and with patches of heterochromatin, which occupy about 13% of the nuclear volume, distributed peripherally next to the nuclear membrane.

III. DEVELOPMENT

Two populations of Leydig cells are sequentially generated during testicular development. The first, fetal Leydig cells, are formed during embryogenesis; the second, adult Leydig cells, differentiate postnatally. Fetal Leydig cells produce testosterone required for male sexual differentiation and neuroendocrine functions. Later, testosterone secreted by adult Leydig cells stimulates male sexual characteristics and spermatogenesis.

A. Fetal Leydig Cells

Fetal Leydig cells differentiate after the testis segregates into interstitium and epithelial testicular cords which are the primordia of the seminiferous tubules. At the onset of testicular development, 13 days of gestation in the rat and 6 weeks in the human, cells from the neighboring mesonephros and associated coelomic epithelium invade the testis and occupy the presumptive interstitium. The ultimate origin of the stem cells is the subject of continuing debate, but the preponderance of evidence indicates that Leydig cells are derived from mesenchymal cells of embryonic mesoderm. It has also been suggested that Leydig stem cells are derived from migratory neuroectodermal cells because Leydig cells express proteins that are found in the nervous system.

In the rat, fetal Leydig cells differentiate between 15 and 17 days of gestation, reaching a maximum number of 500,000 by 19 days. Androgen secretion begins by 15 days and is independent of luteinizing hormone (LH). Fetal Leydig cells contain numerous lipid droplets, are highly steroidogenic, and secrete more testosterone on a per cell basis than adult Leydig cells.

Fetal Leydig cells also show greater resistance to desensitization due to LH or its analog, human chorionic gonadotropin (hCG), *in vitro*. This phenomenon results from internalization and proteolytic cleavage of LH receptors present on the cell surface and their dissociation from adenylate cyclase as well as reduction in LH/hCG receptor mRNA. Estradiol could be involved in this process because estrogen antagonists block LH-induced desensitization. Under physiological conditions, Leydig cell unresponsiveness to LH is prevented by the pulsatile or episodic release of LH from the pituitary.

In humans, precursors of fetal Leydig cells proliferate between 8 and 14 weeks and gradually differentiate during Weeks 14–18. The maximum number of fetal Leydig cells, 48 million per testis, is attained between 13 and 16 weeks of gestation. Between 18 weeks and birth at 38 weeks, precursor cells cease to proliferate and the number and volume of fetal Leydig cells is reduced by half. In the human, fetal Leydig cells persist into the neonatal period, but from 3 months after birth until puberty at 10 years of age, few Leydig cells are present in the interstitium.

Fetal Leydig cells lack 11β-hydroxysteroid dehydrogenase (11β-HSD) activity and, therefore, are unable to inactivate glucocorticoids oxidatively. It is possible that glucocorticoids play a stimulatory role in Leydig cell development, although metabolism of this hormone by placental 11β-HSD has not been excluded.

B. Adult Leydig Cells

In the rat, the lineage that gives rise to adult Leydig cells is undetectable until Day 14, when stem cells convert to progenitor Leydig cells which express 3β-HSD activity. Progenitor Leydig cells are spindle shaped, resembling the stem cells from which they are derived. Progenitor cells have a high capacity for proliferation, contain little SER, possess androgen receptors but few LH receptors, and differentiate into a second intermediate by Day 28, designated the immature Leydig cell. Immature Leydig cells are round in shape, contain SER and lipid droplets, and have a lower capacity to proliferate. Immature Leydig cells transform into adult Leydig cells by Day 56. Adult Leydig cells are larger than progenitor and immature Leydig cells, spherical in shape, contain a large volume of SER and numerous LH receptors, and do not divide; occasional Leydig cell mitoses in adult testes are attributed to a residual number of immature Leydig cells. The turnover rate of adult Leydig cells is estimated to exceed the average 2-year life span of the rat but is accelerated experimentally by the exogenous administration of LH/hCG or a cytotoxin, ethane dimethanesulfonate, which selectively kills Leydig cells.

IV. TESTOSTERONE BIOSYNTHESIS

A. Sources of Cholesterol

Testosterone biosynthesis is the primary function of the Leydig cell, and as with all steroids, testosterone is synthesized from cholesterol. Cholesterol is obtained by receptor-mediated endocytosis of serum lipoproteins, *de novo* synthesis from acetate in the SER, or hydrolysis of intracellular cholesterol esters. Under physiological conditions, Leydig cells synthesize cholesterol from acetate, whereas adrenal and ovarian tissues are more dependent on serum lipoproteins. The first enzyme of testosterone biosynthesis, $P450_{scc}$, is localized in the inner mitochondrial membrane, and cholesterol must be transported from the cytosol into mitochondria. The LH-stimulated movement of cholesterol to the inner mitochondrial membrane is now known to be performed primarily by steroidogenic acute regulatory protein (StAR). Studies by Stocco and colleagues have identified StAR as a family of cycloheximide-sensitive mitochondrial proteins, 30 kDa in size. A precursor of StAR interacts with a protein on the outer mitochondrial membrane, causing the precursor to convert into an active inter-mediate that is threaded through contact sites formed by apposition of the outer and inner mitochondrial membranes. In one congenital disease, lipoid adrenal hyperplasia, StAR is not synthesized and steroidogenesis is consequently blocked. Another polypeptide, diazepam-binding inhibitor, is thought to facilitate cholesterol transfer under basal conditions, binding to peripheral-type benzodiazepine receptor (PBR) and opening a channel for cholesterol in the outer mitochondrial membrane.

B. LH Signaling Pathways

Luteinizing hormone is the stimulus for testosterone biosynthesis and secretion. Leydig cells are the principal site of LH binding in the testis; in addition, recently discovered LH receptors on testicular endothelial cells may facilitate transport of the hormone into the interstitium. Human Leydig cells contain fewer LH receptors compared to rats, but the maximal rate of testosterone biosynthesis after LH stimulation is similar for both species, probably because a maximal rate of steroidogenesis is achieved when only 10% of available LH receptors are occupied. The LH receptor belongs to the family of glycosylated-protein-coupled receptors with seven transmembrane domains, and a serine/threonine-rich cytoplasmic region possessing a phosphorylation site. The major ligand-binding domain resides in the large N-terminal extracellular region. The binding of LH to the LH receptor activates both guanyl nucleotide stimulatory protein and adenylate cyclase. Adenylate cyclase catalyzes the synthesis of cyclic adenosine 3′,5′-monophosphate (cAMP), which in turn triggers a cascade of intracellular events leading to cholesterol transport into the mitochondria and testosterone biosynthetic enzyme activities. In addition to cAMP, other signaling systems modulate LH action in the Leydig cell. These include calcium and its binding protein calmodulin, arachidonic acid, leukotrienes, guanylate cyclase, and free radicals generated from hydrogen peroxide and nitric oxide (Fig. 2). Some testosterone is produced by Leydig cells in the absence of LH stimulation due to growth factors that also increase intracellular cAMP.

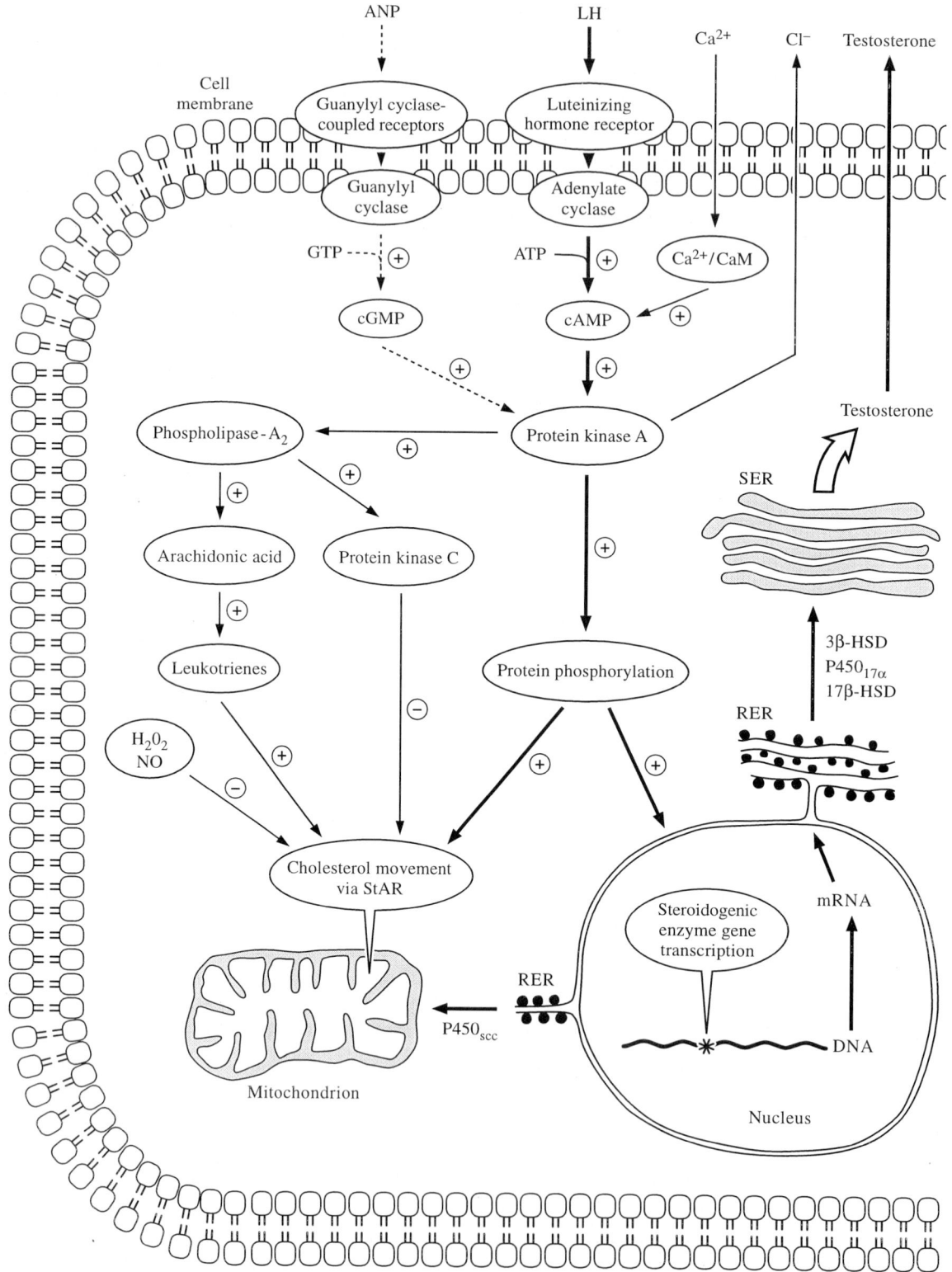

C. Biosynthesis of Testosterone from Cholesterol

Conversion of cholesterol to testosterone occurs in the mitochondria and SER and is catalyzed by four enzymes: $P450_{scc}$, 3β-HSD, $P450_{17\alpha}$, and 17β-HSD. In the inner mitochondrial membrane, adrenodoxin and adrenodoxin reductase transfer electrons from reduced pyridine nucleotides to $P450_{scc}$, which then catalyzes the conversion of cholesterol to pregnenolone. Pregnenolone is thought to move from the mitochondrion to SER by passive diffusion. The remaining steps of testosterone biosynthesis occur in the SER, and the three steroidogenic enzymes, 3β-HSD, $P450_{17\alpha}$, and, 17β-HSD, are adjacent, with each steroid intermediate passing directly to the next reaction. There are two alternative biosynthetic pathways, depending on the species. With the Δ^4 pathway followed in rodents, 3β-HSD first catalyzes dehydrogenation of pregnenolone at the 3β-hydroxy group and then isomerizes the double bond between carbons 5 and 6 to carbons 4 and 5, forming progesterone. Subsequently, $P450_{17\alpha}$ catalyzes the two-step conversion of progesterone into androstenedione by 17α-hydroxylation followed by cleavage between carbons 17 and 20. Androstenedione is then reduced by 17β-HSD to form testosterone. This last reaction is reversible; the oxidative and reductive directions are catalyzed by different isoforms of 17β-HSD and regulated by both substrate and product concentra-

tions. In rabbit and human Leydig cells, testosterone biosynthesis follows the Δ^5 pathway, and in the first reaction, $P450_{17\alpha}$ catalyzes the two-step conversion of pregnenolone into 17α-hydroxypregnenolone, and further into dehydroepiandrosterone. Dehydroepiandrosterone is converted by 3β-HSD into androstenedione, which is then converted to testosterone by 17β-HSD (pathways are summarized in Fig. 3).

Newly synthesized testosterone immediately diffuses out of the Leydig cell, and there is no evidence of intracellular storage. With a few exceptions, testosterone biosynthetic enzymes are exclusively localized in Leydig cells. The Sertoli cell expresses 17β-HSD and the metabolizing enzyme 5α-reductase.

D. Testosterone Metabolism

Testosterone is metabolized by two groups of enzymes: (i) 5α-reductase, 3α-HSD, and cytochrome P450 aromatase ($P450_{arom}$) are expressed in the testis and various tissues in the body and convert testosterone into other steroids; and (ii) 7α-hydroxylase and conjugating enzymes remove biological activity and clear testosterone from the blood through conversion into water-soluble derivatives that can be excreted in the urine.

In the rat, progenitor Leydig cells, at 21 days postpartum, produce more androsterone than testosterone because 17β-HSD activity is low while 5α-reductase and 3α-HSD activities are high. Compared to

---

FIGURE 2 Signal transduction pathways in the Leydig cell. Coupling of LH to its receptor on the cell membrane causes the formation of cAMP from ATP and activation of protein kinase A. Protein kinase A promotes protein phosphorylation and opening of chloride ion (Cl^-) channels and may activate phospholipase A_2. Efflux of Cl^- ions has a permissive effect on LH stimulation. Under physiological conditions, Leydig cell function is maintained by cAMP-mediated LH stimulation (bold pathway in figure). Acute LH stimulation causes the rapid movement of cholesterol into mitochondria, facilitated by steroidogenic acute regulatory protein (StAR), and increased testosterone biosynthesis. Additional responses in Leydig cells are induced by chronic LH stimulation, including an increase in the synthesis of protein regulators such as sterol carrier protein-2 and increased steroidogenic enzyme gene expression. Calcium (Ca^{2+}) and its binding protein, calmodulin, enhance cAMP-mediated LH stimulation of Leydig cells. Additionally, phospholipase A_2 is able to activate protein kinase C and cause the formation of leukotrienes from arachidonic acid. The main action of protein kinase C is inhibitory, acting primarily at the level of cholesterol movement. Leukotrienes generally enhance LH-stimulated testosterone production. Increased generation of intracellular free radicals from hydrogen peroxide (H_2O_2) and nitric oxide (NO) in the Leydig cell interfere with electron transfer and cause loss of activity of the cytochrome P450 family of enzymes, thus downregulating steroidogenesis. Other signaling systems modulate LH regulation of Leydig cells. For example, atrial natriuretic peptide (ANP), a vasoactive compound produced by cardiac atrial muscle cells, can bind guanylyl cyclase-coupled receptors on the cell surface, activating guanylyl cyclase and the formation of cyclic guanidine 3′,5′-monophosphate (cGMP) from guanidine triphosphate (GTP). Cyclic GMP acts in concert with cAMP to increase protein phosphorylation. The physiological relevance of other regulators such as ANP on overall Leydig cell function requires further analysis.

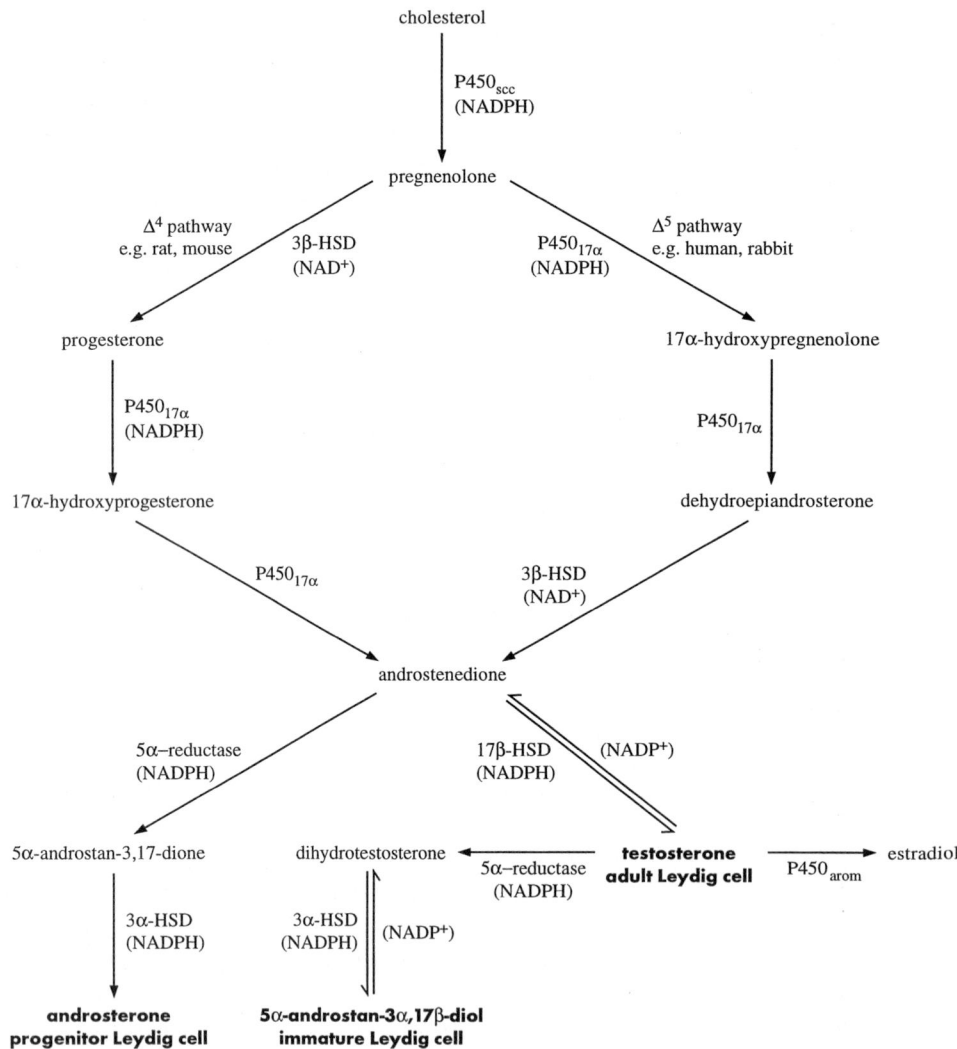

FIGURE 3 Testosterone biosynthetic and metabolic pathways in the Leydig cell. (Top) The major steps involved in testosterone biosynthesis: conversion of cholesterol to pregnenolone in mitochondria and synthesis of testosterone from pregnenolone in the smooth endoplasmic reticulum (SER). Conversion of cholesterol into pregnenolone occurs by side-chain cleavage, in a reaction that is catalyzed by P450$_{scc}$ in the inner mitochondrial membrane. Subsequent reactions, which occur in the SER, follow the Δ4 (rat and mouse) or Δ5 (rabbit and human) pathway. The enzymes involved and direction of reactions are as shown, with cofactors in parentheses (bottom). After synthesis, there is no intracellular storage, and testosterone diffuses out of the cell, bound to androgen-binding protein in seminiferous tubules and to sex hormone-binding globulin for transport in the blood. Unlike adult Leydig cells, which primarily produce testosterone due to high 17β-HSD and low levels of androgen metabolism by 5α-reductase and 3α-HSD, androsterone is the major steroid produced by rat progenitor Leydig cells. Immature Leydig cells produce 5α-androstane-3α,17β-diol because they express high 5α-reductase and 3α-HSD activities.

progenitor Leydig cells, immature Leydig cells produce more 5α-androstane-3α,17β-diol (3α-DIOL) than testosterone due to increased 17β-HSD. Testosterone production prevails in adult Leydig cells as a result of sharp declines in the levels of 5α-reductase and 3α-HSD. The $P450_{arom}$ aromatizes testosterone to 17β-estradiol. Leydig and Sertoli cells both express $P450_{arom}$ depending on the age. In rats, aromatase activity is localized in Sertoli cells prepubertally and in Leydig cells during adulthood. Leydig cells appear to be the only source of testicular estrogens in humans. Among the domestic species, the stallion and boar have unusually high levels of aromatization in adult Leydig cells and, consequently, circulating estrogen levels are twofold higher than testosterone in these species.

Circulating testosterone is bound to plasma proteins, primarily albumin and the sex hormone-binding globulin. The amount of free testosterone available for androgen receptor binding is small, about 2 or 3% of the total concentration in humans. In rodents, androgen-binding protein secreted by Sertoli cells binds testosterone in the seminiferous tubular fluid. In this way, high levels of androgen accumulate in the testis; it is estimated that testosterone concentrations in seminiferous tubular fluid are 20-fold higher than those in general circulation.

In the rat, the enzyme 7α-hydroxylase, a member of the cytochrome P450 family of enzymes that is present in the liver and testis, catalyzes degradation of testosterone to 7α-hydroxylated metabolites, which are weak androgens. There are two groups of conjugating enzymes: sulfotransferases are found mainly in the liver, testes, and adrenals, whereas glucuronyltransferases are present only in the liver. These enzymes conjugate testosterone with sulfates or glucuronic acid prior to excretion in the kidney.

V. HORMONAL CONTROL OF LEYDIG CELLS

A. Regulation of Leydig Cell Proliferation during Development

1. Embryogenesis

Embryonic differentiation of steroidogenic tissues is mediated by steroidogenic factor 1 (SF-1). The ligand has not been identified, but SF-1 is known to be a nuclear transcription factor that recognizes a sequence in the promoter regions of all genes encoding P450 enzymes and the α and β subunits of follicle-stimulating hormone (FSH) and LH. It is likely that SF-1 is also needed for differentiation of adult Leydig cells postnatally. Recent reports suggest that two polypeptides, vasoactive intestinal peptide and pituitary adenylate cyclase-activating polypeptide, produced primarily by intestinal epithelial cells and hypothalamic neurons but also in small amounts by testicular cells, stimulate differentiation of fetal Leydig cells during embryogenesis.

2. FSH

The initial stimulus for Leydig cell differentiation is not LH, because progenitor Leydig cells have few LH receptors. Even though FSH receptors are present in Sertoli cells, and absent in Leydig cells, a role for FSH in the postnatal development of Leydig cells has been proposed. Rising FSH levels in the prepubertal period coincide with a progressive increase in the number of Leydig cells, and it is possible that FSH stimulates Sertoli cells to secrete unidentified growth factors that cause progenitor Leydig cells to proliferate. There is evidence that estradiol inhibits division of progenitor Leydig cells, and immature Sertoli cells secrete estradiol. Moreover, estradiol administration prevents FSH-induced increases in the numbers of Leydig cell precursors in immature hypophysectomized rats. However, estradiol secretion by rat Sertoli cells declines precipitously between 15 and 21 days of age, when progenitor Leydig cells actively divide. These observations led to the hypothesis that FSH action on Sertoli cells could determine the final number of Leydig cells in the testis.

3. Androgen

Acting as an autocrine regulator, androgen promotes Leydig cell proliferation. The presence of a large number of androgen receptors in progenitor Leydig cells, which contain few LH receptors, indicates that developing Leydig cells are initially more responsive to androgen stimulation and later acquire sensitivity to LH.

4. Thyroid Hormone

While hypothyroidism has long been associated with delayed onset of puberty in most species, the role of thyroid hormones in Leydig cell development became apparent only recently. In the rat, transient neonatal hypothyroidism increases adult testis weight and Leydig, Sertoli, and germ cell numbers but reduces LH receptor numbers. Since thyroid hormone receptors are present in Leydig cell precursors, further studies are warranted to analyze the role of thyroid hormone in Leydig cell development.

B. Regulation of Testosterone Biosynthesis in Adult Leydig Cells

1. Hypothalamic–Pituitary–Testicular Axis

In all mammalian species, testicular function is dependent on gonadotropins secreted by the pituitary gland. Neurons in the hypothalamus synthesize and episodically secrete a gonadotropin-releasing hormone (GnRH) into the network of capillary vessels that form the hypothalamo–hypophysial portal system. GnRH is transported through the portal system to reach the pituitary, where it stimulates gonadotropes to release LH and FSH into the general circulation. The two gonadotropins are transported in the blood to the testis, where LH directly stimulates Leydig cells to produce testosterone. Testosterone secreted into the general circulation inhibits LH secretion, acting both at the level of gonadotropes in the pituitary and in the central nervous system to slow the frequency and amplitude of the hypothalamic pulse generator.

2. LH

It is generally agreed that factors other than LH are important for Leydig cell proliferation, and that Leydig cells require LH to maintain their fully differentiated structure and function. Acute LH stimulation increases cholesterol transport and, therefore, testosterone biosynthesis. Prolonged LH stimulation causes increased steroidogenic enzyme gene transcription. For example, in studies with isolated purified rat adult Leydig cells, a twofold increase in 3β-HSD and $P450_{17\alpha}$ enzyme activities occurs following LH stimulation. Similarly, cAMP, the chief intracellular mediator of LH action, has been shown to enhance the expression of genes that encode for $P450_{scc}$, 3β-HSD, and $P450_{17\alpha}$ in *vitro*. Under physiological conditions, lack of LH stimulation causes Leydig cells to atrophy and to lose cellular volume, SER, and steroidogenic enzyme activities.

3. FSH

There is a positive correlation between serum FSH levels and the steroidogenic response of Leydig cells to LH stimulation during sexual maturation in both the human and rat. In the mouse and hamster, FSH enhancement of Leydig cell responsiveness to LH stimulation in *vivo* and in *vitro* is associated with increased activities of $P450_{scc}$, $P450_{17\alpha}$, and 17β-HSD. When Sertoli and Leydig cells are cultured together in the presence of FSH, basal and LH-stimulated testosterone production by Leydig cells is increased in all species, indicating that FSH plays a role in regulating Leydig cell steroidogenesis.

4. Androgen

Androgen is able to self-regulate its production because it inhibits adult Leydig cell testosterone production by an androgen receptor-mediated reduction in 3β-HSD and $P450_{17\alpha}$ gene transcription. In contrast, testosterone stimulates expression of 17β-HSD genes in the pubertal rat. This disparity in expression of 17β-HSD could result from differences in substrate specificity between the adult and immature rat Leydig cell, with testosterone metabolism being more active in the latter. Further evidence for the role of androgens in Leydig cell steroidogenesis is provided by testicular-feminized male (*Tfm*) mice in which a loss-of-function mutation of the androgen receptor results in reduced capacity for testosterone production. This deficiency in *Tfm* Leydig cells is caused by lower $P450_{17\alpha}$ and 17β-HSD activities compared to Leydig cells with normal androgen receptors.

5. Glucocorticoid

Increased levels of glucocorticoids in pathologic conditions, such as Cushing's syndrome, are associated with reduced circulating testosterone levels and reproductive dysfunction. Leydig cells contain glucocorticoid receptors, and glucocorticoids are known to repress both basal and cAMP-induced synthesis of $P450_{scc}$, resulting in reduced testosterone production. However, the inhibitory effects of increased glucocorticoids, due to either stress or disease, are alleviated by 11β-HSD-mediated oxidation. Rat Leydig cells synthesize 11β-HSD by Day 25 postpartum, and expression of this enzyme is highest in adult Leydig cells. Therefore, the pubertal rise in testosterone production is related, in part, to a decline in glucocorticoid-induced inhibition.

6. Growth Factors

In addition to hormones, several growth factors moderate Leydig cell steroidogenic activity. Insulin-like growth factor-1 stimulates immature and adult Leydig cell testosterone production. In contrast, basic fibroblast growth factor suppresses both basal and LH-stimulated 3β-HSD activity in immature rat Leydig cells, and epidermal growth factor and transforming growth factor-β both cause a decrease in testosterone production in pig Leydig cells. Numerous factors present in plasma, including the prostaglandin $PGF_{2\alpha}$, fatty acids, and peptides such as endothelins and angiotensin II, moderate the steroidogenic response of Leydig cells to LH stimulation but have no effect on basal testosterone production. The significance of their actions *in vivo* requires further elucidation.

VI. CLINICAL ASPECTS OF LEYDIG CELL FUNCTION

A. Leydig Cell Tumors

Leydig cell tumors have been reported for both rodents and humans. Spontaneously occurring Leydig cell tumors are most common in the Fischer strain of rat, with lower incidences in other strains and in mice. In humans, 1–3% of testicular neoplasms are of Leydig cells, occurring typically during the ages of 30–40 years. In adults, 10% of the cases are malignant but the clinical course in children is benign. Clinically, children show precocious pseudopuberty, whereas adults present with testicular swelling and gynecomastia. Leydig cell tumors are generally associated with lower than normal LH levels, and they secrete both androgenically and estrogenically active steroids. An excess of secreted estrogens is thought to be the cause of gynecomastia in affected adults. Compared to normal Leydig cells, tumor Ley-

dig cells are larger and show obvious variation in nuclear size and configuration and increased heterochromatin, which suggests reduced transcriptional activity. In humans, the cytoplasm of tumor Leydig cells contain large amounts of Reinke's crystals. In the interstitium, tumor Leydig cells provoke an inflammatory reaction, manifested by an increase in the number of macrophages. Studies in mice show that there is a genetic predisposition to tumor Leydig cell growth, but the precipitating factors are unknown. However, tumors can be experimentally induced by the administration of high levels of gonadotropins, prolonged treatment with estrogens or cadmium, and exposure to X-irradiation. Recently, many environmental toxicants, including kepone, methoxychlor, and bisphenol A, have been credited with the ability to bind the estrogen receptor, causing cellular proliferation and possibly promoting Leydig cell tumor growth.

B. The Effect of Aging on Leydig Cell Function

Reduced reproductive capacity has been reported in aging rats. Studies with the Brown Norway rat showed reduced testosterone production by Leydig cells, but serum LH levels were normal, indicating that in this strain of rats, age-related changes occur at the testicular level. In other rat strains, however, serum LH levels decline with age, suggesting that changes at the hypothalamic–pituitary axis possibly contribute to reduced fertility seen in old rats. Age-related decreases in Leydig cell testosterone production in the rat are now associated with degenerative changes in Leydig cells, reduced steroidogenic enzyme activities, and chronic understimulation of Leydig cells by LH.

In humans, the process of aging is accompanied by reduced serum testosterone levels with a decline of 35–50% in free testosterone levels between ages 20 and 80. The metabolic clearance of testosterone from the blood does not increase with age, and the lower serum levels of testosterone in elderly men is caused by impaired Leydig cell steroidogenesis, resulting in diminished sperm production. Reduced steroidogenesis during aging is related to structural alterations and increased cytoplasmic content of lipofuscin granules and Reinke's crystals in the Leydig cell, decreased oxygen supply to the interstitial tissue, and generation of free oxygen radicals as a result of reduced testicular blood flow.

C. Clinical Evaluation of Leydig Cell Function

Measurement of blood levels of testosterone is routine for clinical management of male infertility. Many diseases, including altered thyroid function and obesity, perturb the ratio of bound to unbound steroid. Since measurements of serum testosterone concentration by radioimmunoassay ordinarily include both bound and unbound fractions, it is useful to determine free testosterone concentration and/or the sex hormone binding capacity in order to detect abnormalities of testosterone concentrations that are due to binding rather than biosynthesis. The levels of LH and FSH are measured to differentiate primary Leydig cell dysfunction from lesions arising at the level of the hypothalamus and pituitary. Testosterone secretion exhibits pulsatility and a diurnal rhythm. Therefore, in order to derive a correct assessment, several replicate blood samples are collected from the same subject at the same time of day.

D. Male Contraception

While deficient testosterone production is a consideration for patients with male infertility, uncontrolled population growth in many parts of the world highlights the need for fertility regulation. Other than the use of condoms and vasectomy, there is a dearth of male methods. Since the Leydig cell synthesizes testosterone required for spermatogenesis, it is an obvious target for contraceptive development. Attempts have been made to devise male contraceptives that are safe, easy to use, convenient, and reversible. In order to make these methods acceptable to men, it is necessary to provide androgen levels that are insufficient to support germ cell development but adequate to maintain sex drive and skeletal muscle mass. One approach has been to inhibit testicular function by suppressing LH and FSH secretion from the pituitary, combined with simultaneous androgen supplementation to prevent androgen deficiency in

other tissues. Substantial progress had been made with a number of products such as GnRH antagonists in combination with testosterone preparations. A synthetic androgen, 7α-methyl-19-nortestosterone, has generated interest because it is incapable of being 5α-reduced to DHT and, therefore, provides lower levels of androgen action in tissues that abundantly express 5α-reductase, notably the prostate. Long-acting testosterone preparations have also proved promising because exogenous androgen exerts a negative feedback on the pituitary and brain.

VII. SUMMARY

Leydig cells form 4–6% of the testicular volume and represent the endocrine portion of the testis. Developmentally, Leydig cells appear as two distinct populations; fetal Leydig cells prenatally and adult Leydig cells after birth. Leydig cells secrete testosterone for sexual differentiation during fetal development and stimulate secondary sex characteristics and sperm production in adulthood. The Leydig cell has ultrastructural characteristics associated with steroidogenesis, including abundant smooth endoplasmic reticulum and mitochondria with tubular cristae.

Fetal Leydig cells produce testosterone at a higher rate than adult Leydig cells and are resistant to LH-induced desensitization. Initially, steroidogenesis in the rat and human is independent of LH. During postnatal development, adult Leydig cells in all species differentiate from intermediates that are morphologically and functionally distinct. Proliferation of Leydig cell intermediates is regulated by several factors, including FSH and the Sertoli cell, whereas LH is the main stimulator of adult Leydig cell differentiated function.

Testosterone is synthesized from cholesterol in a series of reactions catalyzed by four enzymes. The first enzyme, $P450_{scc}$, converts cholesterol to pregnenolone and is located in the inner mitochondrial membrane. Therefore, movement of cholesterol from the Leydig cell cytosol to this site is a critical rate-determining step and is performed by StAR. The three other biosynthetic enzymes, 3β-HSD, $P450_{17\alpha}$, and 17β-HSD, are situated in the SER. Testosterone is metabolized into other steroids primarily by three enzymes: 5α-reductase, 3α-HSD, and P450$_{arom}$. In the rat, due to differential regulation of the biosynthetic and metabolizing enzymes, progenitor Leydig cells produce mainly androsterone, immature Leydig cells show higher activity of the metabolizing enzymes and produce primarily 3α-DIOL, and the major steroid produced by adult Leydig cells is testosterone.

Leydig cell tumors have a low incidence in humans, constituting only 1–3% of testicular neoplasm, with the highest incidence in men between the ages of 30 and 40. Of more general concern is the age-related decline in testosterone biosynthesis in Leydig cells. As the source of testosterone necessary for spermatogenesis, Leydig cells are a target for treatment of male infertility and contraception. Control of male fertility through suppression of Leydig cells will require identifying a level of androgen that is insufficient for spermatogenesis while maintaining libido and skeletal muscle mass.

Acknowledgments

The authors are grateful for the support of Grants TW 05350, HD 32588, and HD 33000 from the National Institutes of Health during the preparation of the manuscript. Dr. Dianne Hardy provided helpful comments on an early draft, and the illustrations were drawn by Evan Read.

See Also the Following Articles

BLOOD–TESTIS BARRIER; STEROIDOGENESIS; TESTICULAR DEVELOPMENT; TESTOSTERONE BIOSYNTHESIS

Bibliography

Dufau, M. L. (1988). Endocrine regulation and communicating functions of the Leydig cell. *Annu. Rev. Physiol.* **50**, 483–508.

Ewing, L. L., and Keeney, D. S. (1993). Leydig cells: Structure and function. In *Cell and Molecular Biology of the Testis* (C. Desjardins and L. L. Ewing, Eds.), pp. 137–165. Oxford Univ. Press, New York.

Ewing, L. L., and Zirkin, B. R. (1983). Leydig cell structure and steroidogenic function. *Recent Prog. Horm. Res.* **39**, 599–635.

Gnessi, L., Fabri, A., and Spera, G. (1997). Gonadal peptides as mediators of development and functional control of the testis: An integrated system with hormones and local environment. *Endocr. Rev.* **18**, 541–609.

Hardy, M. P., and Zirkin, B. R. (1997). Leydig cell function. In *Infertility in the Male* (L. I. Lipschultz and S. S. Howards, Eds.), 3rd ed., pp. 59–70. Mosby, New York.

Hardy, M. P., Nonneman, D., Ganjam, V. K., and Zirkin, B. R. (1993). Hormonal control of Leydig cell differentiation and mature function. In *Understanding Male Infertility: Basic and Clinical Approaches* (R. W. Whitcomb and B. R. Zirkin, Eds.), Serono Symposia Publ. No. 98, pp. 125–142. Raven Press, New York.

Huhtaniemi, I., Warren, D. W., and Catt, K. J. (1984). Functional maturation of rat testis Leydig cells. *Ann. N. Y. Acad. Sci.* **438**, 283–303.

Jegou, B. S., and Sharpe, R. M. (1993). Paracrine mechanisms in testicular control. In *Molecular Biology of the Male Reproductive System* (D. de Kretser, Ed.), pp. 271–310. Academic Press, New York.

Payne, A. H., Quinn, P. G., and Rani, C. S. J. (1985). Regulation of microsomal cytochrome P450 enzymes and testosterone production in Leydig cells. *Recent Prog. Horm. Res.* **41**, 153–197.

Russell, L. D. (1996). Mammalian Leydig cell structure. In *The Leydig Cell* (A. H. Payne, M. P. Hardy, and L. D. Russell, Eds), pp. 43–96. Cache River Press, Vienna, IL.

Saez, J. M. (1994). Leydig cells: Endocrine, paracrine, and autocrine regulation. *Endocr. Rev.* **15**, 574–626.

Sharpe, R. M. (1993). Experimental evidence for Sertoli–germ cell and Sertoli–Leydig cell interactions. In *The Sertoli Cell* (L. D. Russell and M. D. Griswold, Eds.), pp. 392–418. Cache River Press, Clearwater, FL.

Skinner, M. K. (1991). Cell–cell interactions in the testis. *Endocr. Rev.* **12**, 45–77.

LH (Luteinizing Hormone)

George R. Bousfield

Wichita State University

I. Introduction
II. LH Structural Features
III. LH Function
IV. LH Assays

GLOSSARY

α and β subunits The dissimilar subunits that comprise the glycoprotein hormones. The α subunit is common to all four mammalian hormones because it is the product of a single gene. The luteinizing hormone (LH) β subunit determines the hormone specificity and most other properties of the LH heterodimer.

cystine knot A protein structural motif consisting of a small ring created by two disulfide bonds connecting two antiparallel β strands through which a third disulfide bond connecting two other strands passes. Three loops result from this motif: two hairpin loops, L1 and L3, at one side of the knot and a single long loop, L2, at the other end.

determinant loop The small loop created between Cys residues 93 and 100 that is essential for subunit dimerization and which is an important LH specificity determinant.

oligosaccharide Branched carbohydrate side chains consisting of 8–20 monosaccharides attached to Asn residues cotranslationally or smaller chains consisting of 2 or more monosaccharides attached to Ser and Thr residues posttranslationally.

seat belt loop The fourth loop in the β subunit consisting of residues 90–110 that stabilizes the noncovalent association of α and β subunits by wrapping around the α subunit long loop L2.

species designation The species of origin for a specific LH is usually designated by a lowercase prefix. When few sequences were known, a single letter was sufficient to unambiguously identify each one. The abbreviations employed in this article will be given in the legends to Figs. 2 and 4.

Luteinizing hormone (LH) is a member of the classical glycoprotein hormone family, which also includes two other anterior pituitary hormones, follicle-stimulating hormone and thyroid-stimulating hormone, and a placental hormone, chorionic gonadotropin. All three pituitary hormones are found in most vertebrates, whereas CG is restricted to primates and equids. The glycoprotein hormone family has recently been added to the cystine knot growth factor superfamily by virtue of the fact that both glycoprotein hormone α and β subunits that comprise the functional hormone possess the cystine knot structural motif. LH is produced in the anterior pituitary gland of all vertebrates. Its functions include stimulation of steroidogenesis in the gonads of both males and females and induction of ovulation and luteinization in females. Its name originated from the latter function because ovulation is accompanied by the conversion of the relatively avascular, fluid-filled ovarian follicle to the solid, highly vascularized corpus luteum. Alternative names for LH include interstitial cell-stimulating hormone and, in fish, gonadotropic hormone-II. LH will be employed throughout this article.

I. INTRODUCTION

Luteinizing hormone (LH) plays a central role in the regulation of the reproductive system. In many species the female reproductive cycle is divided into two parts by a preovulatory surge of LH; a phase of follicular development leading up to ovulation followed by a postovulatory phase dominated by the corpus luteum. Because of its abundance and relative ease of purification, many of the early structure–

function relationships of the glycoprotein hormones were first established for LH and subsequently observed in the other glycoprotein hormones. In recent years many important structural studies have been performed with the functionally related glycoprotein hormone, human chorionic gonadotropin (hCG). Because of the high degree of sequence homology between LH and CG and because they share the same LH receptor, structural features observed in hCG are likely to be very similar, if not identical, in LH. Significant progress has been made in characterizing the glycoprotein hormones in fish which has overturned earlier ideas about the evolution of this hormone family. This article focuses on the structural aspects of LH.

II. LH STRUCTURAL FEATURES

A. Gene Structure

The organization of the known mammalian and teleost fish glycoprotein hormone α and β subunit genes is illustrated in Fig. 1. A quite striking difference between the mammalian α and β subunit genes is the greater size of the α subunit genes compared to the β subunit genes (Figs. 1A and 1B). This is largely because of larger introns in the former, particularly intron 1, which ranges in size from 5.4 kilobase pairs (kbp) in the rat to 14.7 kbp in the rhesus monkey. Organization of the fish α subunit genes is very different in that several species, including chum and masu salmon, carp, and goldfish, have been demonstrated to possess two α subunit genes. Two α subunits have also been obtained from purified bonito LH preparations; however, it is not known if these are also separate gene products. The two carp α subunit genes possess much smaller introns resulting in a smaller gene similar in size to the β subunit gene (a potential third α subunit gene or pseudogene lacks intron 3). The same number and position of α subunit introns has been conserved between fish and mammalian α subunit genes. Because of an extended 3′ untranslated region, the mature carp α subunit mRNA is larger than the mammalian α subunit mRNA (Fig. 1C). The salmon LHβ gene is similar in size and organization to the mammalian LHβ subunit genes. Its two introns are slightly smaller than their mammalian counterparts and salmon LHβ intron 1 is located in the 5′ untranslated region, as are the corresponding introns in the follicle-stimulating hormone (FSH) β and thyroid-stimulating hormone (TSH) β genes. In mammalian LHβ genes, the first intron interrupts the coding sequence for the signal peptide and the same pattern is observed in the cpLHβ gene, which is in this regard more similar in structure to the mammalian LHβ genes.

The mature mRNAs for the α subunits are larger than those of the β subunits (Fig. 1C). The α subunit mRNAs possess a larger 5′ untranslated region, about 100 nucleotides compared with 7 nucleotides for the β subunit. The leader sequence regions for the α subunits are roughly the same size for all species, ranging from 69 to 72 nucleotides, encoding a 23- or 24-residue signal peptide. Greater variability is observed for the β subunit leader sequences, which range from 60 nucleotides for the mammalian α subunits to 117 nucleotides for avian β subunit mRNAs. The β subunit coding region is larger because LHβ consists of 105–149 amino acid residues, whereas the α subunit is composed of 88–97 residues. The 3′ untranslated region is smaller for LHβ mRNA, ranging from 62 to 90 nucleotides, whereas those of the mammalian α subunit mRNAs are 255–290 nucleotides and that of the carp is 404 nucleotides.

Based on conserved protein structure, Hunt and Dayhoff suggested that the α and β subunits arose by duplication of an ancestral gene. This argument was strengthened recently by the discovery that hCG is a cystine knot protein. Although primary sequence homology between hCGα and hCGβ is only 10%, both proteins possess this structural motif, thereby illustrating the concept that three-dimensional structure is more highly conserved than primary structure. However, an early review of gonadotropin molecular biology pointed out that the different positions of the introns in the hCGα and hCGβ genes provided evidence against a common origin for these genes. The same review has been cited in support of a common origin for the α and β genes in subsequent reviews. The structures of the fish α subunit genes

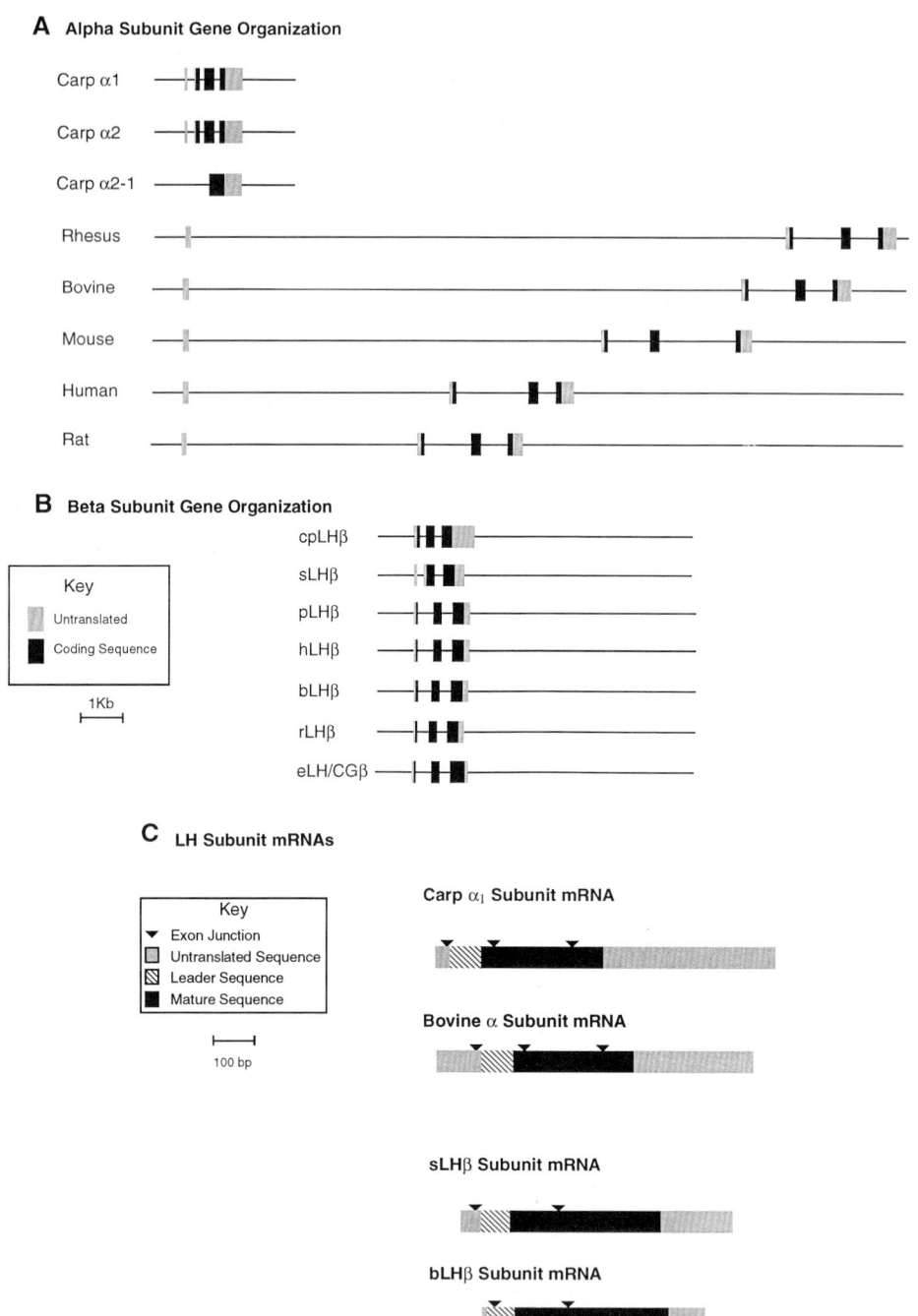

FIGURE 1 Comparison of the gene structure and mature mRNAs for the mammalian and teleost fish LH α and β subunits. (A) Gene organization for the α subunits illustrating the increase in size of the α subunit genes which is primarily due increased size of the α subunit introns. (B) Similar-sized carp, salmon, and mammalian LHβ genes. The different locations of the introns can be seen more clearly in the mature mRNAs. (C) The sizes of the mature mRNAs are much more similar than the genes. The β subunit coding region is larger because LHβ subunits consist of 105–149 residues, whereas the α subunits consist of 89–97 residues.

suggest that the ancestral α subunit gene was similar in size to the ancestral β subunit gene and the potential for loss of the third intron has been observed for one of the carp α subunit genes. A certain degree of structural flexibility exists in cystine knot proteins because the lengths of the three loops created by the cystine knot can vary. However, the α and β subunit introns differ in their locations relative to the cystine knot. In the β subunit, three of the six cystine knot cysteine residues are encoded in exon 2, whereas the other three are encoded in exon 3. In the α subunit gene, four of the cystine knot cysteine residues are encoded in exon 3, whereas the remaining two are found in exon 4.

B. Protein Structure

1. α Subunit

The primary structures of the α subunits from those species that possess a characterized LH are shown in Fig. 2. In most species, a single α subunit has been identified. However, in several species of teleost fish a second α subunit has been found. Interestingly, in salmon it is only associated with the FSH homolog, gonadotropic hormone-I (GTH-I) and not with GTH-II, the LH homolog. In the bonito, both α subunits are associated with bonLH and only one of them with bonFSH. Only the sequence of the bonito α subunit that combined with both LH and FSH β subunits was determined. No sequence data were reported for the other α subunit and the only difference described between the completely sequenced α subunit and the second α subunit was that the latter lacked the oligosaccharide normally found in loop L2. Therefore, it is not known if two genes exist in this species or if the second α subunit is simply a partially glycosylated α subunit variant. In carp, only one of the two potential α subunits was isolated from cpLH and cpFSH. Moreover, a recombinant version of the second α subunit combined with carp LHβ produced an inactive heterodimer. Because of these variations in the relationships of α and LHβ subunits in various fish, both α subunits are shown in Fig. 2 because it is not known for all species which α subunit is a component of LH. It is interesting that there is sometimes less sequence identity between the two α subunits found in a single

species than is found within α subunit sequences for other vertebrate groups. For example, the mammalian α subunit sequences average 84% identity, whereas the avian α subunit sequences are 98% identical. The homology between the two chum salmon α subunits is relatively low (72%). Amino acid sequence identity between α subunits obtained from the same species varies from 68% for the two carp α subunits to 91% for both goldfish α subunits.

The glycoprotein hormone α subunits consist of 88–97 residues. There are five highly conserved disulfide bonds indicated by connecting lines in Fig. 3. Three of these comprise the cystine knot and are indicated by heavy lines. The two other disulfides appear not to be essential because no loss of function accompanied selective reduction of the 11–35 disulfide or elimination of it or the 63–91 disulfide by site-directed mutagenesis. Nevertheless, all five of these disulfide bonds have remained absolutely conserved over the course of vertebrate evolution. There are two highly conserved N-glycosylation sites and each oligosaccharide serves different functions. The Asn56 oligosaccharide is the most important oligosaccharide for coupling receptor binding to signal transduction, whereas the Asn82 oligosaccharide is important for a subunit folding. In some of the fish α subunits, currently goldfish $\alpha2$ and Pike eel α, an Asp residue has been proposed for position 56. In addition, the bonito α subunit associated exclusively with bonLHβ appeared to lack this oligosaccharide as well. However, this may revert to Asn when additional data are obtained, as has already happened with the carp α subunit. In view of the important role for this oligosaccharide in LH function and because of the high sequence conservation of this glycosylation site across all species, it is likely that these other α subunits will be found to be glycosylated at this site. Alternatively, since Asp has universally been suggested as the substitution at this position, its presence during protein sequencing studies might result from the activity of a pituitary peptide–N-glycanase (PNGase) rather than indicating no glycosylation. PNGase has been found in the endoplasmic reticulum of some cells and it has been demonstrated to selectively remove oligosaccharide from intact α subunit Asn56, converting Asn to Asp in the process.

Comparison of all the α subunit sequences reveals

three variable regions and two conserved regions. The first variable region is the N-terminal 8 residues, which appears functionally unimportant because up to 7 residues are frequently missing. In the case of oLHα, it is not known if the truncated forms represent major or minor forms because the yields are generally low (<10% of theoretical) for all N-terminal amino acids during protein sequencing experiments. Because the N-terminal Phe residue is often blocked at its amino group, it is likely that these truncated forms represent a minority of the LH molecules contained in a given preparation. In the equine α subunit the N-terminal Phe accounts for 83% of the amino acids detected during the first cycle of Edman degradation; however, some of this is contributed by Phe6. Truncated residues account for 11% or less of the total amino acid derivatives recovered. A full-length α subunit amino terminus was found to be necessary only for the renotropic activity associated with oLH preparations. This proved to be due to an oLH fraction in which the α subunit N terminus was intact. Isoforms of oLH exhibiting N-terminal α subunit heterogeneity were devoid of renotropic activity. Whether this activity is mediated through the classical LH receptor is not known. The second variable region consists of residues 15–33, whereas the third variable region consists of residues 68–85. Comparison with the crystal structure for hCGα reveals that these two variable regions correspond to the hairpin loops L1 and L3. These represent the sites of many α subunit epitopes for monoclonal antibodies and the variability in these regions probably explains the species specificity of α subunit antibodies. The conserved regions consist of long-loop L2 and the C terminus. Both possess putative receptor binding regions defined by peptide-walking experiments for hCG and hFSH and both are involved in subunit association. It is believed that because the α subunit must combine with as many as four different β subunits, subunit association regions are highly conserved. Indeed, α subunit from one species can be associated with any glycoprotein hormone β subunit from any other species. A functional hormone does not always result from such combinations, but the subunits do combine. Because individual subunits are inactive and the functional receptor binding site is assembled from two discontinuous segments in

A

Species	-25		-20		-15		-10		-5		-1
Mammalia											
Ovine	M D Y Y R	K Y A A A	I L A I L	S L F L Q	I L H S		
Bovine	M D Y Y R	K Y A A V	I L A I L	S L F L Q	I L H S		
Porcine	M D Y Y R	K Y A A V	I L A I L	S V F L Q	I L H S		
Rabbit							
Equine							
Donkey	M D Y Y R	K H A A V	I L A T L	S V F L H	I L H S		
Mouse	M D Y Y R	K Y A A V	I L V M L	S M F L H	I L H S		
Rat	M D C Y R	R Y A A V	I L V M L	S M V L H	I L H S		
Whale							
Rhesus	M D Y Y R	K Y A A V	I L V T L	S V F L H	I L H S		
Human	M D Y Y R	K Y A A I	F L V T L	S V F L H	V L H S		
Aves							
Chicken	M D C Y R	K Y A A V	T L T I L	S V F L Q	L L H T		
Quail	M D C Y R	K Y A A V	T L T I L	S V F L H	L L H T		
Ostrich							
Turkey	M D C Y R	K Y A A V	T L T I L	S V F L H	L L H T		
Amphibia							
Bullfrog							
Actinopterigii							
Bonito							
Common carp 1	M F W T R	Y R G A S	I L L F F	M L I R L	G Q L		
Common carp 2	M F W T R	Y R G A S	V L L F L	M L I H L	G Q L		
Grass carp	M F W T R	Y A G A S	I L L F L	M L I H L	G Q V		
Silver carp	M F W T R	Y A G A S	I L L F L	M L I H L	G Q V		
Afr Catfish							
Goldfish1	M F W T R	Y A G A S	I L L L L	M L I E L	G Q L		
Goldfish2	M F W T R	Y A G A S	I L L F L	M L I E L	G Q L		
csGTH Ia1							
csGTH Ia2	M C L L K	S T G - -	L S L I L	S A L L V	I - A D S		
Masu sal 1	M C L L K	S I G V G	V S L I L	S I L L Y	M - A D S		
Masu sal 2	M C L L K	S T G - -	L S L I L	S A L L V	I - A D S		
Striped bass	M G S V K	S A G - -	L S L L L	L S F I L	Y V V D S		
Tuna							
Yellowfin Porgy	M G S V K	S A G - -	L S L L L	L S F L L	Y V A D S		
Yellowtail							
Jap. Eel	M M V C P	G K P G A	S L L M L	S M L F H	I I D S		
Pike Eel							
Eur. Eel	M M V C P	G K P G A	S L L M L	S M L F H	I I D S		
Consensus	M • • • r	• • • a •	i l l i l	s • f l •	• • • • •		

FIGURE 2 Glycoprotein hormone pre-α subunit amino acid sequences. With one exception, only those species for which a LHβ subunit has been characterized are included. The exception is the rhesus α subunit because it illustrates that the four-residue deletion in the human α gene is not typical of all primates. Both α subunit sequences are provided for teleost fish species that possess more than one because it has not yet been determined which α subunit is found in the LH in all these species. Single-letter code for the amino acids is employed as follows: A, Ala; B, either Asp or Asn; C, Cys; D, Asp; E, Glu; F, Phe; G, Gly; H, His; I, Ile; K, Lys; L, Leu; M, Met; N, Asn; P, Pro; Q, Gln; R, Arg; S, Ser; T, Thr; V, Val; W, Trp; Y, Tyr; Z, either Glu or Gln. Residue numbering follows that for oLHα, a 96-amino acid residue α subunit. The consensus sequence indicates residues that are 100% conserved by use of an uppercase letter, the majority residue is indicated by a lowercase letter, and those positions for which no residue is present in half the species are indicated by a dot.

each subunit that combine to form a contiguous surface, subunit association and receptor binding regions are intimately associated.

Circular dichroism studies on oLH indicated that a conformational change occurred when oLHα was dissociated from oLHβ, whereas the conformation

B

```
                      5        10       15       20       25       30       35       40       45       50
                      |        |        |        |        |        |        |        |        |        |
      Mammalia
Ovine           F P D G E F T M Q G C P E C K L K E N - K Y F S K P D A P I Y Q C M G C C F S R A Y - P T P A R S K K T
Bovine          F P D G E F T M Q G C P E C K L K E N - K Y F S K P D A P I Y Q C M G C C F S R A Y - P T P A R S K K T
Porcine         F P D G E F T M Q G C P E C K L K E N - K Y F S K L G A P I Y Q C M G C C F S R A Y - P T P A R S K K T
Rabbit          F P D G E F A M Q G C P E C K L K E N - K Y F S K L G A P I Y Q C M G C C F S R A Y - P T P A R S K K T
Equine          F P D G E F T T Q D C P E C K L R E N - K Y F F K L G V P I Y Q C K G C C F S R A Y - P T P A R S R K T
Donkey          F P D G E F T T Q D C P E C K L K K N - K Y F S K L G V P I Y Q C M G C C F S R A Y - P T P A R S K K T
Mouse           L P D G D F I I Q G C P E C K L K E N - K Y F S K L G A P I Y Q C M G C C F S R A Y - P T P A R S K K T
Rat             L P D G D L I I Q G C P E C K L K E N - K Y F S K L G A P I Y Q C M G C C F S R A Y - P T P A R S K K T
Whale           F P N G E F T M Q G C P E C K L K Q N - K Y F S K L G A P I Y Q C M G C C F S R A Y - P T P A R S K K T
Rhesus          F P D G E F T M Q D C P E C K P R E N - K F F S K P G A P I Y Q C M G C C F S R A Y - P T P V R S K K T
Human           A P D - - - - V Q D C P E C T L Q E N - P F F S Q P G A P I L Q C M G C C F S R A Y - P T P L R S K K T
      Aves
Chicken         F P D G E F L M Q G C P E C K L G E N - R F F S K P G A P I Y Q C T G C C F S R A Y - P T P M R S K K T
Quail           F P D G E F L M Q G C P E C K L G E N - R F F S K P G A P I Y Q C T G C C F S R A Y - P T P M R S K K T
Ostrich         F P D G E F L M Q G C P E C K L G E N - R F F S K P G A P V Y Q C T G C C F S R A Y - P T P L R S K K T
Turkey          F P D G E F L M Q G C P E C K L G E N - R F F S K P G A P I Y Q C T G C C F S R A Y - P T P M R S K K T
      Amphibia
Bullfrog        F P D D N F L T P G C P E C R L K E N L R F S N M G I G R I Y Q C S G C C Y S R A Y - P T P M R S K K T
      Actinopterigii
Bonito          Y P N V D L S N M G C E E C T L K K N - N V F S N R D P P I Y Q C M G C C F S R A Y F P T P L K A M K T
Common carp 1   Y P R N D M N N F G C E E C K L K E N - N I F S K P G A P V Y Q C M G C C F S R A Y - P T P L R S K K T
Common carp 2   Y P R N Y M N N F G C E E C K L K E N - N I F S K P G A P V Y Q C M G C C F S R A Y - P T P L R S K K T
Grass carp      W P R N D M T N F G C E E C K L K E N - N I F S K P G A P V Y Q C M G C C F S R A Y - P T P L R S K K T
Silver carp     Y P R N D I T N F G C E E C K L K E N - N I F S K P G A P V Y Q C M G C C F S R A Y - P T P L R S K K T
Afr Catfish     Y P N N D - - - F G C E E C K L K E N - N I F S K P G A P V Y Q C M G C C F S R A Y - P T P L R S K K T
Goldfish1       Y P R N D - N H F G C E E C K L K E N - N I F S R P G A P V Y Q C M G C C F S R A Y - P T P L R S K K T
Goldfish2       Y P R N Y M N N F G C E E C K L K E N - N I F S K P G A P V Y Q C M G C C F S R A Y - P T P L R S K K T
csGTH Ia1       Y Q N S D M T N V G C E E C K L K E N - K V F S N P G A P V Y Q C T G C C F S R A Y - P T P L Q S K K A
csGTH Ia2       Y P N S D K T N M G C E E C T L K P N - T I F P N - - - - I M Q C T G C C F S R A Y - P T P L R S K Q T
Masu sal 1      Y P N S D M T N V G C E E C K L K E N - K L F S N P G A P V Y Q C T G C C F S R A Y - P T P L Q S K K A
Masu sal 2      Y P N S D K T N M G C E E C T L K P N - T I F P N - - - - I I Q C T G C C F S R A Y - P T P L R S K Q T
Striped bass    Y P S M D L S N M G C E E C T L R K N - S V F S R D R P - V Y Q C M G C C F S R A Y - P T P L K A M K T
Tuna            Y P N V D L S N M G C E E C T L K K N - N V F S N R D P - I Y Q C M G C C F S R A F - P T P L K A M K T
Yellowfin Porgy Y P N T D L S N M G C E A C T L K K N - T V F S R D R P - I Y Q C M G C C F S R A Y - P T P L K A M K T
Yellowtail      Y P N I D S S N M G C E E C T L R K N - S L F P N - - - - V Y Q C M G C C F S R A Y - P T P L K A M K T
Jap. Eel        Y P N N E M A R G G C D E C R L Q E N - N I F S K P S A P I F Q C V G C C F S R A Y - P T P L R S K K T
Pike Eel        Y P N N E I S R G G C D E C R L K D N - K F F S K P S A P I F Q C V G C C F S R A Y - P T P L R S K K T
Eur. Eel        Y P N N E M A R G G C D E C R L Q E N - K I F S K P S A P I F Q C V G C C F S R A Y - P T P L R S K K T

Consensus       • p • • • • • • • g C • e C r l k e N - • • f s k p g a p • y Q C m G C C f S R A Y   P T P l r s k k t
```

C

```
                     55       60       65       70       75       80       85       90       95
                     |        |        |        |        |        |        |        |        |
      Mammalia
Ovine           M L V P K N I T S E A T C C V A K A F T K A T V M G N V R V E N H T E C H C S T C Y Y H K S
Bovine          M L V P K N I T S E A T C C V A K A F T K A T V M G N A R V E N H T E C H C S T C Y Y H K S
Porcine         M L V P K N I T S E A T C C V A K A F T K A T V M G N A R V E N H T E C H C S T C Y Y H K S
Rabbit          M L V P K N I T S E A T C C V A K A F T K A T V M G N A K V E N H T E C H C S T C Y Y H H K S
Equine          M L V P K N I T S E S T C C V A K A F I R V T V M G N I K L E N H T Q C Y C S T C Y H H K I
Donkey          M L V P K N I T S E A T C C V A K A F I R V T L M G N I R L E N H T Q C Y C S T C Y H H K I
Mouse           M L V P K N I T S E A T C C V A K A F T K A T V M G N A R V E N H T E C H C S T C Y Y H K S
Rat             M L V P K N I T S E A T C C V A K S F T K A T V M G N A R V E N H T D C H C S T C Y Y H K S
Whale           M L V P K N I T S E A T C C V A K A F T K A T V M G N A R V Q N H T Z C H C S T C Y Y H K S
Rhesus          M L V Q K N V T S E S T C C V A K S L T R V M V M G S V R V E N H T E C H C S T C Y Y H K F
Human           M L V Q K N V T S E S T C C V A K S Y N R V T V M G G F K V E N H T A C H C S T C Y Y H K S
      Aves
Chicken         M L V P K N I T S E A T C C V A K A F T K I T L K D N V K I E N H T D C H C S T C Y Y H K S
Quail           M L V P K N I T S E A T C C V A K A F T K I T L K D N V K I E N H T D C H C S T C Y Y H K S
Ostrich         M L V P K N I T S E A T C C V A K A F T K I T L K D N V K I E N H T D C H C S T C Y Y H K S
Turkey          M L V P K N I T S E A T C C V A K A F T K I T L K D N V K I E N H T D C H C S T C Y Y H K S
      Amphibia
Bullfrog        M L V P K N I T S E A K C C V A K T Q Y R V T V M D N V K I E N H T A C H C S T C L Y H K S
      Actinopterigii
Bonito          M T I P K N I T S E A T C C V A K H S Y E T E V - A G I R V R N H T D C H C S T C Y F H K S
Common carp 1   M L V P K N I T S E A T C C V A K - E V K R V L V N D V K L V N H T D C H C S T C Y Y H K S
Common carp 2   M L V P K N I T S E A T C C V A K - E F K Q V L V N D I K L V N H T D C H C S T C Y Y H K S
Grass carp      M L V P K N I T S E A T C C V A K - E V K R V L V N D V K L V N H T D C H C S T C Y Y H K S
Silver carp     M L V P K N I T S E A T C C V A K - E V K R V L V N D V K L V N H T D C H C S T C Y Y H K S
Afr Catfish     M L V P K N I T S E A T C C V A K - E V K R V I V N D V K L V N H T D C H C S T C Y Y H K F
Goldfish1       M L V F R N I T S E A T C C V A K - E V K R V L V N D V R L V N H T D C H C S T C Y Y H K S
Goldfish2       M L V P K D I T S E A T C C V A K - E V K R V L V D D V K L V N H T D C H C S T C Y Y H K S
csGTH Ia1       M L V P K N I T S E A T C C V A K E G E R V V V - D N I K L T N H T E C W C N T C Y Y H H K S
csGTH Ia2       M L V P K N I T S E A T C C V A K E G E R V T T K D G F P V T N H T E C H C S T C Y Y H K S
Masu sal 1      M L V P K N I T S E A T C C V A K E G E R V V V - D N I K - L T N H T E C W C N T C Y Y H H K S
Masu sal 2      M L V P K N I T S E A T C C V A K E G E R V V T T K D G F P V T N H T E C W C N T C Y Y H H K S
Striped bass    M T I P K N I T S E A T C C V A K H S Y E T E V A G I K - V R N H T D C H C S T C Y F H K I
Tuna            M T I P K N I T S E A T C C V A K H S Y E T E V A G I R - V R N H T D C H C S T C Y F H K S
Yellowfin Porgy M T I P K N I T S E A T C C V A K H V Y E T E V A G I R - V R N H T D C H C S T C Y Y H K I
Yellowtail      M T I P K N I T S E A T C C V A K H S Y E T E V A G I I - V K N H T D C H C S T C Y F H K S
Jap. Eel        M L V P K N I T S E A T C C V A R E V T R L - - - D N M K L E N H T D C H C S T C Y Y H K F
Pike Eel        M L V P K D I T S E A T C C V A R E V T K L - - - D N M K L E N H T D C H C S T C Y Y H K S
Eur. Eel        M L V P K N I T S E A T C C V A R E V T R L - - - D N M K L E N H T D C H C S T C Y Y H K F

Consensus       M l v p k N i T S E a t C C V A k • • • k • • • • • • n • k • • N H T • C h C s T C y y H K s
```

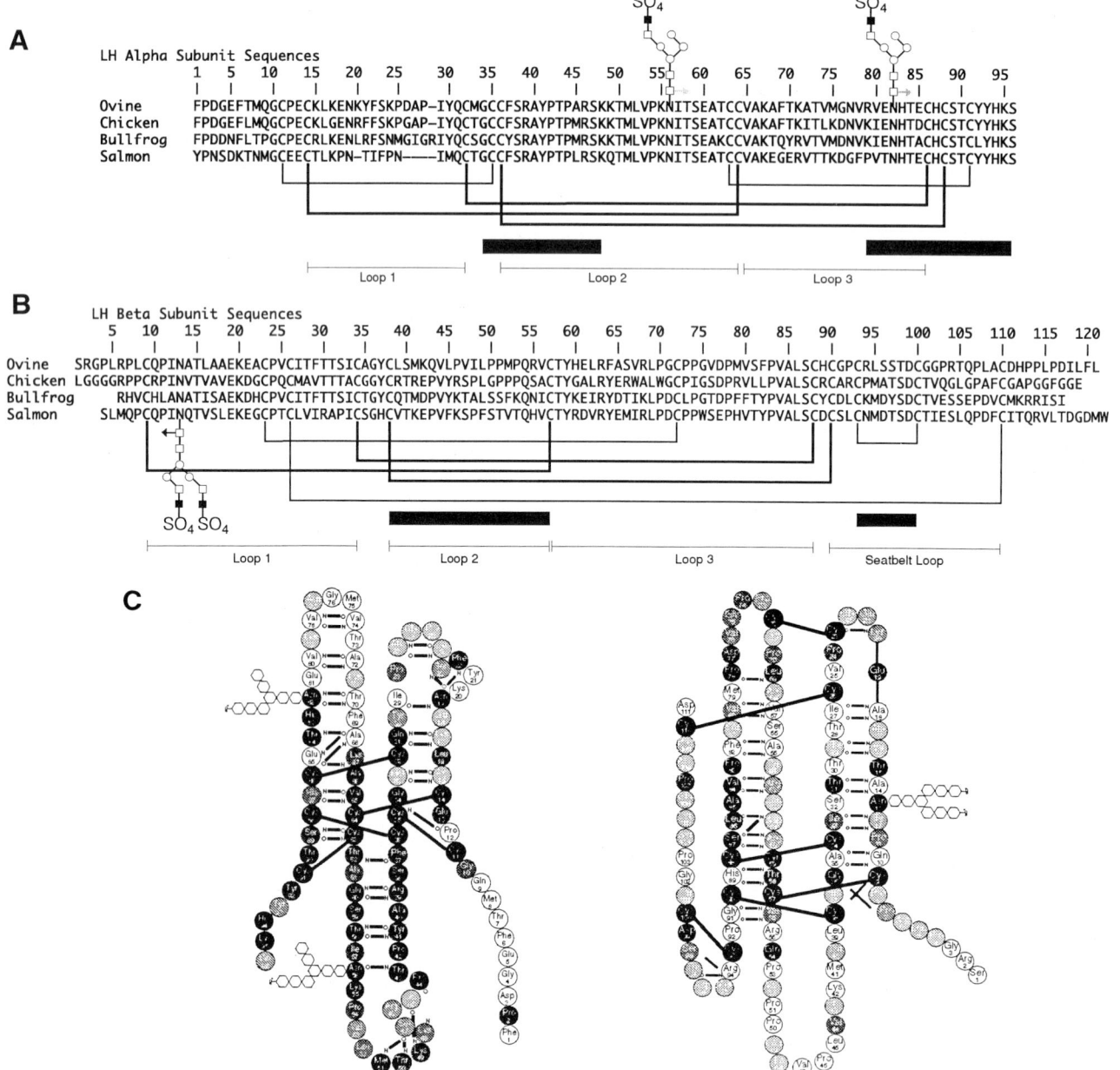

FIGURE 3 Comparison of α and β subunit sequences for four of five of the major vertebrate groups. (A) Amino acid sequences of representative mammalian, avian, amphibian, and teleost α subunits. (B) Amino acid sequences of representative LHβ subunits from the same species as in A. The single-letter amino acid code is employed as described in the legend for Fig. 2. Glycosylation sites are indicated diagrammatically as described in the legend for Fig. 6. Lines connecting Cys residues indicate disulfide bonds according to those determined in the crystal structure of hCG. The high conservation of Cys residues and interchangeability of subunits suggests that the pattern of disulfide bonds is the same for all glycoprotein hormones. Solid bars indicate regions implicated in LH receptor binding by peptide-walking experiments. The three cystine knot loops are indicated as loops 1–3. For the LHβ subunit, the seat belt loop is also indicated. (C) Bead diagram showing the disulfide bonds, glycosylation sites, and sequence conservation indicated by the consensus sequences in Figs. 2 and 4. The absolutely conserved residues are indicated by solid circles. Open circles indicate positions in which no single amino acid residue is found in at least 50% of the sequences. Gray circles indicate residues that are present in more than 50% of the sequences as indicated by the consensus sequence in Fig. 2, and the intensity of the shading indicates the percentage identity. Disulfide bonds are indicated by the heavy lines. Hydrogen bonds indicate those found in hCG. Hydrogen bonds in α subunit loop 2 appear to be disrupted when α subunit is dissociated from the heterodimer.

of oLHβ was affected to a lesser degree. A recent nuclear magnetic resonance study revealed that all secondary structure in the long-loop L2 was lost when hCGα was dissociated from hCGβ. The secondary structure of this region consists of two β strands and a two-turn α helix. These data are consistent with the protease sensitivity of a portion of the α subunit long-loop L2 as contrasted with the protease resistance exhibited by loops L1 and L3. In addition, monoclonal antibodies to loop L2 tend to recognize linear epitopes and are α-subunit specific because the epitope is buried in the heterodimer. Conformational epitopes are unlikely to survive the change in conformation associated with dimerization. However, endoproteinase Arg-C cleavage of oLHα produces a change in the α-helical region of its CD spectrum, suggesting that the α helix is present in isolated oLHα but absent following nicking. Furthermore, the α helix does not reappear in the nicked heterodimer, which exhibits only 3% receptor-binding activity.

2. β Subunit

The LHβ subunit consists of a 20- to 39-amino acid residue leader sequence and a mature sequence of 105–149 amino acid residues (Fig. 4). There are six disulfide bonds in the mature β subunit (Fig. 3) and the functions of five of them are recognized. Three disulfide bonds, 9–57, 34–87, and 38–90, create the β subunit cystine knot that dictates the basic fold of the protein. A fourth, the 93–100 disulfide bond, is essential for stabilizing the structure of the determinant loop, which plays an essential role in subunit dimerization and is important for LH and FSH specificity. The 93–100 disulfide bond is the fifth to form during hCGβ biosynthesis and confers the ability to dimerize with the α subunit. It appears to play the same role in LH because a proteolytic nick in the eLHβ determinant loop completely eliminates its ability to associate with eLHα. The sixth

disulfide bond to form, 26–110, stabilizes a β subunit-specific region, called the seat belt loop, which embraces the α subunit long-loop L2. *In vivo*, this disulfide bond is formed following subunit dimerization. *In vitro*, it appears to remain intact because intermolecular disulfide bonds would form at the high protein concentrations required for subunit association creating high-molecular-weight aggregates. The seat belt loop possesses the LH and FSH specificity determinants, although the relatively high intrinsic FSH activity in eLH and eCG may also involve an α subunit component.

As is true for all other glycoprotein hormones, the relative potency, species specificity, and hormone specificity of LH preparations are primarily determined by the β subunit. These structure–function relationships were largely derived from studies involving hybrid hormones prepared by combining complementary LH or CG subunits obtained from different species. The rat has been used extensively for gonadotropin bioassays because it exhibits little species specificity. Nevertheless, considerable differences in potency have been reported for mammalian LH in this species. Human CG and eLH are the most potent preparations commonly employed in mammalian LH studies. Intact hLH preparations exhibit similar high activities. Typical hLH preparations possessing β subunit nicks exhibit reduced potencies that vary with the extent of nicking. Ovine and bovine LH are significantly less active than these other hormone preparations and pLH is even less active. Significant species specificity has been reported for the recombinant hLH receptor with only hLH and hCG exhibiting high-affinity binding. Even eLH was found to have low receptor-binding activity when recombinant hLH receptors were employed. An extensive series of studies involving hybrids composed of subunits of hCG and most common mammalian LH preparations revealed that the potency of the β subunit determined the potency of the hybrid hor-

FIGURE 4 (overleaf) Amino acid sequences for all LHβ subunits. The single-letter amino acid code was employed as described in the legend for Fig. 2. Residues are numbered according to oLHβ. The species are indicated by the lowercase letter prefixes as follows: e, equine; dk, donkey; rhn, rhinoceros; h, human; o, ovine; b, bovine; p, porcine; l, lagomorph (rabbit); r, rat; m, mouse; w, whale; cam, camel; ha, hamster; c, chicken; q, quail; t, turkey; os, ostrich; bf, bullfrog; cp, common carp; gc, grass carp; chnok s, Chinook salmon; chum s, chum salmon; ms, masu salmon; bo, *Baikal omul*; k, killifish; stbass, striped bass; bon, bonito; cf, African catfish; gf, goldfish; tu, tuna; y, yellowfin; el, eel; eel, European eel; je, Japanese eel; d, dog; sc, silver carp; spst, spotted seatrout.

A

```
                 -40              -30              -20              -10              -1
                  |                |                |                |                |
eLH  β           M E T L Q G L L L W M L L S V G G V W A
dkLH β           M E M L Q G L L L W M L L S V G G V W A
rhnLH β          M E M L Q G L L L W L L L S V G G V W A
hLH  β           M E M L Q G L L L L L L L S M G G A W A
oLH  β           M E M L Q G L L L W L L L G V A G V W A
bLH  β           M E M F Q G L L L W L L L G V A G V W A
pLH  β           M E M L Q G L L L W L L L S V A G V W A
lLH  β
rLH  β           M E R L Q G L L L W L L L S P S V V W A
mLH  β           M E R L Q L L S P S V V W A
wLH  β
dLH  β           * * * L Q G L L L W L L L S V G G V W A
(NT) camLH β
(NT) haLH  β

cLH  β           M G G A Q V L V L M T L L G T P P A T T G N P P V A V D P P L A V V G P P M G
qLH  β
tLH  β           M G G A Q V L V L M T L L G T P P V T T G T P P V V V D P S V A V V G P P L G
osLH β
bfLH β

cpLH β           M G T P V K I L V V R N H I L F S V V V L L A V A Q S
gcLH β           T G T P V K I L V V R N I L L L L F C L V V L L V F A Q S
scLH β           M L A V R N N I L L L L F C L V V L L V F A Q S
chnok sLH β      M L G L H V G T L I S L F L C I L L E P I E G
chum sLH β       M L G L H V G T L I S L F L C I L L E P V E G
msLH β           M L G L H V G T L I S L S L C I L L E P V E G
boLH β           M L G L H V G T L M I S L F L C I L L E P V E G
kLH  β           M V C L F L G A S S F I W S L A P A A A A
stbassLH β       M A V Q A S R V M F P L V L S L F L G A T S D I W P L
bonLH β
cfLH β
gfLH β           M G T P V K I L V     V L F S V I V L L A V A Q S
tuLH β
yLH  β
spstLH β

elLH β
eeLH β           M S V Y P E C T W L L F V C L C H L L V S A G G
jeLH β           M S V Y P E C T W L L F V C L C H L L V S A G G

Consensus LH     m • • l • • l l l • • l l • • • • • • • • •
```

B

```
                              5      10     15     20     25     30     35     40     45     50
                              |      |      |      |      |      |      |      |      |      |
eLH  β              S R G P L R P L C R P I N A T L A A E K E A C P I C I T F T T S I C A G Y C P S M V R V M P A A L P
dkLH β              S R G P L R P L C R P I N A T L A A E K E A C P I C I T F T T S I C A G Y C R S M V R V M P A A L P
hLH  β              S R E P L R P W C H P I N A I L A V E K E G C P V C I T V N T T I C A G Y C P T M M R V L Q A V L P
rhnLH β             S K G P L R P L C R P I N A T L A A E N E A C P V C I T F T T S I C A G Y C P S M V R V L P A A L P
oLH  β              S R G P L R P L C Q P I N A T L A A E K E A C P V C I T F T T S I C A G Y C L S M K Q V L P V I L P
bLH  β              S R G P L R P L C Q P I N A T L A A E N E A C P V C I T F T T S I C A G Y C P S M K R V L P V I L P
pLH  β              S R G P L R P L C R Z P I N A T L A A E N E A C P V C I T F T T S I C A G Y C P S M V R V L P A A L P
lLH  β        pE P A R G P L R P L C R P V N A T L A A E N E A C P V C I T F T T S I C A G Y C P S M V R V L P A A L P
rLH  β              S R G P L R P L C R P V N A T L A A E N E F C P V C I T F T T S I C A G Y C P S M V R V L P A A L P
mLH  β              S R G P L R P L C R P V N A T L A A E N E F C P V C I T F T T S I C A G Y C P S M V R V L P A A L P
wLH  β              P R G P L R P L C R P I N A T L A A Q N Z A C P V C I T F T T S I C A G Y C P S M V R V L P A A L P
dLH  β              S R G P L R P L C R P I N A T L A A E N E A C P V C I T F T T T I C A G Y C P S M V R V L P A A L P
(NT) camLH β        S R G P L R P L C R P I N A T L A A E N E A C P V C I T F T T S I C A G Y C P S M R R V L P A A L P
(NT) haLH β         S R G P L R P L C R P I N A

cLH  β              L G G G G R P P C R P I N V T V A V E K D G C P Q C M A V T T T A C G G Y C R T R E P V Y R S P L G
qLH  β              M G G S G R P P C R P I N V T V A V E K E E C P Q C M A V T T T A C G G Y C R T R E P V Y R S P L G
tLH  β              L G G G G R P P C R P I N V T V A V E K D E C P Q C M A V T T T A C G G Y C R T R E P V Y R S P L G
osLH β  V P L G V P V A L G V P P S R P P C R P V N V T V A A E K D E C P Q C L A V T T T A C G G Y C R T R E P V Y R S P L G
bfLH β                        R H V C H L A N A T I S A E K D H C P V C I T F T T S I C T G Y C Q T M D P V Y K T A L S

cpLH β              S Y L P P C E P V N E T V A V E K E G C P K C L V L Q T T I C S G H C L T K E P V Y K S P F S
gcLH β              S F L P P C E P V N E T V A V E K E G C P K C L V F Q T T I C S G H C L T K E P V Y K S P F S
scLH β              S F L P P C E P V N E T V A V E K E G C P K C L V F Q T T I C S G H C L T K E P V Y K S P F S
chnok sLH β         S L M Q P C Q P I N Q T V S L E K E G C P T C L V I Q T P I C S G H C V T K E P V F K S P F S
chum sLH β          S L M Q P C Q P I N Q T V S L E K E G C P T C L V I Q T P I C S G H C V T K E P V F K S P F S
msLH β              S L M Q P C Q P I N Q T V S L E K E G C P T C L V I Q T P I C S G H C I T K E P V F R S P F S
boLH β              S L M Q P C Q P I N Q T V S L E K E G C P T C L V I Q T P I C S G H C F T K E L V F K S P F S
kLH  β              F Q L P R C Q L L N Q T I S L E K R G C S G C H R V E T T I C S G Y C A T K D P N Y K T S Y N
stbassLH β  A P A E A F Q L P P C Q L I N Q T V S L E K E G C P K C H P V E T T I C S G H C I T K D P V I K I P F S
bonLH β             F Q L P P C Q L I N Q T V S V E K E G C A T C I P E V T T I C S G H C I T K D P V I K I P F S
cfLH β              Y L L T H C E P - N E T V S V E K D G C P K C L A F Q T S I C S G H C F T K E P V Y K S P F S
gfLH β              S Y L P P C E P V N E T V A V E K E G C P K C L V L Q T T I C S G H C L T K E P V Y K S P F S
tuLH β              F Q L P P C Q L I N Q T V S V E K E G C A S C H P V E T T I C S G H C I T K D P V I K I P F S
yLH  β              F Q L P P C Q L I N Q T V S L E K E G C P K C H A V E T T I C S G H C I T K D P V I K I P F S
spstLH β            S Q L P P C Q L I I Q T V S L

elLH β              S V L Q P C Q P I N E T I S V E K D G C P K C L V F Q T S I C S G H C I T K D P S Y K S P L S
eeLH β              S L L L P C E P I N E T I S V E K D G C P K C L V F Q T S I C S G H C I T K D P S Y K S P L S
jeLH β              S L L L P C E P I N E T I S V E K D G C P K C L V F Q T S I C S G H C I T K D P S Y K G P L S

Consensus LH        • • • • l • p p C • p i N • T v a • e k e g C p • C • • f • T • i C • G h C • t • • • p v • • • p l •
```

C

```
                 55        60        65        70        75        80        85        90        95        100
                 |         |         |         |         |         |         |         |         |         |
eLH β            A I P Q P V C T Y R E L R F A S I R L P G C P P G V D P M V S F P V A L S C H C G P C Q I K T T D C
dkLH β           P I P Q P V C T Y R E L R F G S I R L P G C P P G V D P M V S F P V A L S C R C G P C R L K T T D C
rhnLH β          P A P Q P V C T Y H E L R F A S I R L P G C P P G V D P M V S F P V A L S C R C G P C R L S S S D C
hLH β            P L P Q V V C T Y R D V R F E S I R L P G C P P G V D P V V S F P V A L S C R C G P C R R S T S D C
oLH β            P M P Q R V C T Y H E L R F A S V R L P G C P P G V D P M V S F P V A L S C H C G P C R L S S T D C
bLH β            P M P Q R V C T Y H E L R F A S V R L P G C P P G V D P M V S F P V A L S C H C G P C R L S S T D C
pLH β            P V P Q P V C T Y R E L S F A S I R L P G C P P G V D P T V S F P V A L S C H C G P C R L S S S D C
lLH β            P V P Q P V C T Y R E L R F A S I R L P G C P P G V D P E V S F P V A L S C R C G P C R L S S S D C
rLH β            P V P Q P V C T Y R E L R F A S V R L P G C P P G V D P I V S F P V A L S C R C G P C R L S S S D C
mLH β            P V P Q P V C T Y R E L A F A S V R L P G C P P G V D P I V S F P V A L S C R C G P C R L S S S D C
wLH β            P V P Z P V C T Y R Q L R F A S I R L P G C P P G V N P M V S F P V A L S C H C G P C R L S S S D C
dLH β            P V P Q P V C T Y H E L H F A S I R L P G C P P G V D P M V S F P V A L S C R C G P C R L S N S D C
(NT)camLH β      P V P Q R V C T Y R E L R F A S S
(NT)haLH β

cLH β            P P P Q S A C T Y G A L R Y E R W A L W G C P I G S D P R V L L P V A L S C R C A R C P M A T S D C
qLH β            P P P Q S S C T Y G A L R Y E R W D L W G C P I G S D P K V I L P V A L S C R C A R C P I A T S D C
tLH β            R P P Q S S C T Y G A L R Y E R W A L W G C P I G S D P R V L L P V A L S C R C A R C P I A T S D C
osLH β           G P A Q Q A C G Y G A L R Y E R L A L P G C A P G A D P T V A V P V A L S C R C A R C P M A T A D C
bfLH β           S F K Q N I C T Y K E I R Y D T I K L P D C L P G T D P F F T Y P V A L S C Y C D L C K M D Y S D C

cpGTH β          T V Y Q H V C T Y R D V R Y E T V R L P D C P P G V D P H I T Y P V A L S C D C S L C T M D T S D C
gcGTH β          T V Y Q H V C T Y R D V R Y E T V R L P D C P P G V D P H I T Y P V A L S C D C S L C T M D T S D C
scGTH β          T V Y Q H V C T Y R D V R Y E T V R L P D C P P G V D P H I T Y P V A L S C D C S L C T M D T S D C
chnok sLH β      T V T Q H V C T Y R D V R Y E M I R L P D C P W W V D P H V T Y P V A L S C D C S L C N M D T S D C
chum sLH β       T V I Q H V C T Y R D V R Y E T I R L P D C P P W V D P H V T Y P V A L S C D C S L C N M D T S D C
msLH β           T V Y Q H V C T Y R D V R Y E M I R L P D C P W W V D P H V T Y P V A L S C D C S L C N M D T S D C
boLH β           T V Y Q H V C T Y R D V R Y E T I C L P D C S P W V D P H V T Y P V A L S C D C S L C N M D T S D C
kLH β            K A I Q H V C T Y G D L Y Y K T F E F P E C V P G V D P V V T Y P V A L S C R C G G C A M A T S D C
stbassLH β       N V Y Q H V C T Y R D L H Y K T F E L P D C P P G V D P T V T Y P V A Q S C H C G R C A M D T S D C
bonLH β          K V Y Q H V C T Y R D F Y Y K T F E L P D C P P G V D P T V T Y P V A L S C H C G R C A M D T S D C
cfLH β           S I Y Q H V C T Y R D V R Y E T I R L P D C R P G V D P H V T Y P V A L S C E C S L C T M D T S D C
gfLH β           T V Y Q K V C T Y R D V R Y E T V R L P D C P P G V D P H I T Y P V A L S C D C S L C T M D T S D C
tuLH β           K V Y Q H V C T Y R D F Y Y K T F E L P D C P P G V D P T V T Y P V A L S C H C G R C A M D T S D C
yLH β            N V Y Q H V C T Y R D L Y Y R T F D L P D C P P D V D P T V T Y P V A L S C H C G R C A M D T S D C
spstLH β

elLH β           T V Y Q R V C T Y R D V R Y E T V R L P D C R P G V D P H V T F P V A L S C D C N L C T M D T S D C
eeLH β           T V Y Q R V C T Y R D V R Y E T V R L P D C R P G V D P H V T F P V A L S C D C N L C T M D T S D C
jeLH β           T V Y Q R V C T Y R D V R Y E T V R L P D C R P G V D P H V T F P V A L S C D C N L C T M D T S D C

Consensus LH     • v • Q • v C t Y r d l r y • • • r l p d C p p • v d P • v t • P V A l S C • C • • • C • m d t s D C
```

D

```
                 105       110       115       120       125       130       135       140       145       150
                 |         |         |         |         |         |         |         |         |         |
eLH β            G V F R D Q P L A C A P Q A S S S S K D P P S Q P L T S T S T P T G A S R R S S H P L P I K T S
dkLH β           G G P R D H P L A C A P Q T S S S C K D P P S Q P L T S T S T P T G A S R R S S H P L P I N T S
rhnLH β          G G P R A Q P L A C D R P P L P G L L F L
hLH β            G G P K D H P L T C D H P Q L S G L L F L
oLH β            G G P R T Q P L A C D H P P L P D I L F L
bLH β            G G P R T Q P L A C D H P P L P D I L F L
pLH β            G G P R A Q P L A C D R P L L P G L L F L
lLH β            G G P R A E P L A C D L P H L P G
rLH β            G G P R T Q P M T C D L P H L P G L L L F
mLH β            G G P R T Q P M A C D L P H L P G L L L L
wLH β            G P G R A Q P L A C N R S P R P G L
dLH β            G G P R A Q S L A C D R P L L P G L L F L
(NT) camLH β
(NT) haLH β

cLH β            T V Q G L G P A F C G A P G G F G G E
qLH β            T V Q G L G P A F C G A P G G F G G Q
tLH β            T V Q G L G P A F C G A P G G F G I G E
osLH β           T V A G L G P A F C G A P A G F G P Q
bfLH β           T V E S S E P D V C M K R R I S I

cpLH β           T I E S L Q P D F C M S Q R E D F L V Y
gcLH β           T I E S L Q P D F C M S Q R E D F P V Y
scLH β           T I E S L Q P D Y C M S Q R E D F P V Y
chnok sLH β      T I E S L Q P D F C I T Q R V L T D G D M W
chum sLH β       T I E S L Q P D F C I T Q R V L T D G D M W
msLH β           T I E S L Q P D F C I T Q R V L T D G D M W
boLH β           T I E S L Q P D L C M T Q R V L A D G M W
kLH β            T F E S L Q P D F C M N D I P F Y H
stbassLH β       T F E S L Q P N F C M N D I P F Y Y
bonLH β          T F E S L Q P D F C M N D I P F Y Y
cfLH β           T I E S L N P D F C M T Q K E F I L D Y
gfLH β           T I E S L Q P D F C K S Q R E D F L V Y
tuLH β           T F E S L Q P D F C M N D I P F Y Y
yLH β            T F E S L Q P D F C M N D I L F Y Y
SpStLH β

elLH β           A I Q S L R P D F C M S Q R A P
eeLH β           A I Q S L R P D F C M S Q R A S L P A
jeLH β           A I Q S L R P D F C M S Q R A S L P A

Consensus LH     t • • S l q p d f C • • • • • • t d g d l
```

mone preparation. Thus, those hybrids possessing hCGβ exhibited the highest activity, whereas those hybrids possessing pLHβ were the least active. A secondary modulatory effect of the α subunit was observed only with the most active hybrids possessing hCGβ. Studies involving eLH subunits combined with hCG, oLH, and pLH subunits revealed a significant stimulatory effect of the eLH α subunit. In this study, eLHα hybrids, including either oLHβ or pLHβ, were significantly more active than either oLH or pLH.

The predominant role of the β subunit in determining the species specificity of LH has been demonstrated in studies involving nonmammalian LH preparations. As a general rule, nonmammalian LH preparations exhibit very low activities in conventional mammalian LH bioassays. For example, Licht and colleagues measured the LH potencies of a variety of purified nonmammalian LH preparations— turkey LH, snapping turtle LH, sea turtle LH, alligator LH, and bullfrog LH—in the ovarian ascorbic acid depletion assay. The potencies of these nonmammalian LH preparations ranged from 0.6 to 3% that of NIH-LH-S18. In an *in vitro* steroidogenesis assay employing rabbit testis, the LH potencies of these nonmammalian preparations were even lower, ranging from 0.03 to 0.8% that of NIH-LH-S18. In contrast, mammalian LH preparations are active in some nonmammalian bioassays but not in others. Thus, hCG and oLH have been used to stimulate ovulation and spermiation in frogs. Ovine LH is more often active in nonmammalian bioassays than hLH and hCG. An hLH preparation reported to have a potency of 4.1× NIH-LH-S1 in the ovarian ascorbic acid depletion assay was only 0.05× NIH-LH-S1 in the *Xenopus laevis* ovulation assay. In avian bioassays, oLH has been reported to be highly active in stimulating the testicular uptake of ^{32}P in day-old chicks, stimulating steroid production in cultured chick embryonic ovaries, and stimulating testosterone production by minced rooster testes. The human gonadotropins hCG and hLH exhibited little biological activity in the chicken and studies involving hybrid preparations of oLH and hCG subunits revealed that the β subunit determined whether the hybrid would be active in the rooster testis steroidogenesis assay; hCGα–oLHβ was active, whereas oLHα–hCGβ was

inactive. When turkey–ovine LH hybrids were tested in the rat testis Leydig cell assay, turkey LH was only about 0.1% as active as oLH, and the hybrid hormone preparation oLHα–turkey LHβ was even less active. The activity of the turkey LHα–oLHβ hybrid was 100-fold greater than that of turkey LH and about 10% that of oLH. Experiments performed in reptiles exhibited greater species specificity. In an assay measuring androgen production by turtle testis, oLH was only 1.4% as active as turtle LH. The oLH α subunit fully substituted for turtle LHα; oLHα–turtle LHβ exhibited low activity in rat ovary LH radioligand assay but displayed full LH activity in the turtle testis androgen assay. The oLHβ hybrid, turtle (t) LHα–oLHβ, was only 2.5% as active as tLH in the turtle testis assay; however, full activity was not recovered in the mammalian LH bioassays.

FSH preparations display very low levels of activity in LH bioassays; however, the converse is not true. The intrinsic FSH activity of eCG (formerly, pregnant mare serum gonadotropin) has been exploited by physiologists for decades. Intrinsic FSH activity has subsequently been demonstrated for eLH, which differs only in its glycosylation from eCG. Avian LH also expresses intrinsic FSH activity. Three groups have reported that chicken LH preparations exhibited greater activity in mammalian FSH bioassays than in mammalian LH bioassays and a similar pattern of activity has been reported for turkey LH.

Progress has been made toward defining the specific regions of LHβ responsible for these functional variations between LH. The first receptor binding region was identified in the eCGβ amino acid sequence as the small octapeptide loop between residues 93 and 100, designated the determinant loop. It was established that eCG was not an LH/FSH chimera as anticipated; rather, it possessed the primary structure of a conventional LHβ subunit except in the determinant loop, where it possessed a unique amino acid sequence not observed in either LH or FSH. Results of peptide-walking experiments using synthetic peptides based on the amino acid sequence of hCGβ identified two regions of the β subunit as potential LH receptor binding sites: the determinant loop and the residues between Cys residues 38 and 57, the long-loop L2. Studies with hCG/hFSH chimeras indicated that the determinant loop was important for LH receptor binding but could coexist with the adjacent FSH determinant (residues 101–110), resulting in hormones that could bind both LH and FSH receptors. Chicken LH is a naturally occurring LH/FSH chimera which exhibits greater FSH activity than LH activity in mammalian bioassay systems. It was found to be 28–44× NIH-FSH-S1 in rat testis radioligand assays employing ^{125}I-oFSH or ^{125}I-eFSH tracers but only 2× NIH-FSH-S1 when chicken testis receptors were employed. This apparently results because its β subunit determinant loop has a net negative charge of −1 rather than the neutral or +1 positive charge required for binding to the LH receptor and because it shares homology with mammalian FSHβ between Cys residues 100 and 110 (Fig. 5). Turkey LH also exhibits significant intrinsic FSH activity, consistent with it having a primary structure similar to that of cLH. However, experiments employing iodinated avian gonadotropin preparations indicated a lack of specificity on the part of the avian FSH receptor. The lack of FSH specificity has not been observed in studies with eLH and eCG because both hormones exhibit significantly reduced FSH receptor-binding activities in chicken testis radioligand assays that employ ^{125}I-eFSH tracer. Studies with recombinant chimeric hCG–hFSH β subunits which had confirmed the role of the determinant loop in LH receptor binding failed to confirm the role of long-loop L2 in receptor specificity because substitution of an FSHβ sequence for the corresponding hCGβ sequence had no effect on receptor-binding specificity. The importance of the long-loop L2 region for receptor binding was supported by the observation that hLH preparations partially nicked in this

loop were less active than intact hLH preparations. Experimental nicking of oLHβ in this region followed by recombination with oLHα produced an oLH derivative exhibiting only 1% the receptor-binding activity of intact oLH. The integrity of long-loop L2 was also demonstrated to be important for subunit association because nicked hLHβ dissociated from hLH by 6 M guanidine–HCl was incapable of reassociating with hLHβ. Although nicked oLHβ could dimerize with intact oLHα (nicked oLHβ cannot combine with nicked oLHα), following overnight incubation with 6 M guanidine–HCl its ability to associate with oLHα was greatly reduced. As with other regions playing an important role in receptor binding, it has not been possible to distinguish between their contribution to heterodimer formation and receptor binding.

Conserved LHβ residues are primarily Cys residues responsible for its three-dimensional structure. Even in the putative LH determinant loop, sequence conservation is limited to the two Cys residues forming the 93–100 disulfide bond and Asp99, which are conserved in all glycoprotein hormones. Mammalian LH determinant loops are characterized by the presence of one or two basic residues and three hydroxyl residues, usually Ser. Avian and fish LH determinant loops entirely lack the basic residue. In addition, fish LHβ subunits have an Asp residue at position 96 which makes them more similar to mammalian FSHβ subunits that have an Asp residue at the same position. However, an LH preparation obtained from a teleost fish, *Tilapia mossambica*, exhibited a relatively high potency for a nonmammalian LH in the rat testis Leydig cell steroidogenesis assay, 0.002× NIH-LH, that was 17-fold higher than the activity of a bullfrog LH preparation. Unfortunately, the amino acid sequence of LH from this species is not known and not all fish LH determinant loops have the same amino acid sequence. Thus, further work will be required to identify the sequence variations in the receptor binding regions that contribute to the variations in potency observed between LH preparations.

Glycosylation of the β subunit is more variable than that of the α subunit. Almost all LH β subunits are exclusively N-glycosylated at Asn13 with two notable exceptions. The first exception is hLHβ, which is glycosylated at Asn30 rather than at Asn13.

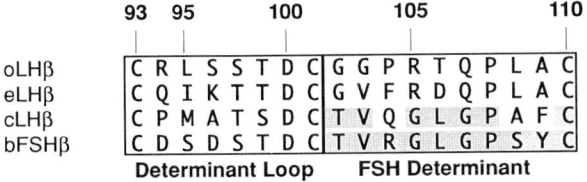

FIGURE 5 Comparison of the seat belt loop amino acid sequences of cLHβ with those of oLHβ, eLHβ, and bFSHβ. Two adjacent determinants, the LH-specific determinant loop and the FSH-specificity determinant, are boxed. Shaded residues highlight the similarities between cLHβ and bFSHβ in the FSH-specific region.

The second is eLH, which is N-glycosylated at Asn13 but is additionally O-glycosylated at its C-terminal extension. The amino acid sequence predicted from the cDNA sequence of donkey LHβ predicts O-glycosylation of its C-terminal extension as well. The functions of the β subunit oligosaccharides will be discussed in the following section.

3. Quaternary Structure

LH exists as a heterodimer held together by noncovalent interactions. Its subunits can be readily dissociated and recombined to reconstitute active LH with 100% recovery of biological activity provided proteolytic damage to the subunits is avoided during subunit isolation and reassociation. The nature of the noncovalent interaction between the α and β subunits was revealed in the crystal structure of hCG. Because both subunits are cystine knot proteins they possess the three loops dictated by that structural motif. The β subunit possesses a fourth loop, the so-called seat belt loop, which embraces the long-loop L2 of the α subunit. Experimental elimination of this loop destabilizes the heterodimer, and modification of the determinant loop, which is contained within the seat belt loop, prevents dimerization altogether. It is interesting that *in vitro* dimerization of mammalian LH subunits does not proceed at 4°C, whereas fish LH subunits can combine at this temperature. The αAsn56 oligosaccharide provides a potential steric barrier to dimerization *in vitro* because the disulfide bonds remain intact. This is apparent in the increased rate of subunit dimerization following removal of the Asn56 oligosaccharide from oLHα and the equine gonadotropin α subunits. Even intact LH subunits combine readily because they possess hybrid high mannose/complex oligosaccharides consisting of one Man(α1,3)Man branch, whereas the Man(α1,6)Man branch is limited to a single Man residue or one or two Man residues branching from it. Elimination of the branching Man residues and their replacement by addition of a second antenna to Man(α1,6)Man inhibits association with LHβ, whereas the presence of a third oligosaccharide antenna appears to block dimerization. Bullfrog LH oligosaccharides have been demonstrated to be truncated relative to their mammalian counterparts. Such truncated oligosaccharides are potentially permissive

for *in vitro* dimerization because increasing length of oligosaccharide branches has been accompanied by increased inhibition of subunit dimerization. It is interesting to speculate that this oligosaccharide difference may account for the differential properties of mammalian and nonmammalian subunits. The limited information that exists regarding fish glycoprotein hormones indicates that the most abundant oligosaccharides present are neutral, consistent with truncated carbohydrate structures.

C. Carbohydrate Structure

1. Location of Glycosylation Sites

Glycosylation sites are highly conserved in the α subunit and somewhat variable in the β subunit (Fig. 6). This is probably related to their different functions. The highly conserved α subunit N-linked oligosaccharides appear to provide essential functions for hormone action. The β subunit oligosaccharide roles have not been characterized for LH. Experiments with hCG have indicated that the βAsn13 oligosaccharide contributes to the coupling of receptor binding to signal transduction and the O-linked oligosaccharides contribute to the longer circulatory survival observed for this hormone. A study of recombinant FSH glycosylation mutants indicated that FSHβ oligosaccharides determined the metabolic clearance rate for the hormone, whereas FSHα oligosaccharides had no influence on the clearance rate. Until fish gonadotropins began to be characterized, 100% glycosylation was the rule for all three LH N-glycosylation sites. In several fish species, Asn56 appeared to be partially deglylcosylated, perhaps by means of PNGase. Variable eLH O glycosylation permitted its amino acid sequence to be determined because none of its potential O-glycosylation sites were 100% glycosylated. This variability produces the very broad band associated with eLHβ following sodium dodecyl sulfate– (SDS)– polyacrylamide gel electrophoresis.

2. Carbohydrate Structures

In contrast to the classical, complex, biantennary, sialylated oligosaccharides found in hCG, oligosaccharides attached to LH are typically terminated with sulfate residues. The enzymes responsible for substi-

Alpha Subunit Glycosylation

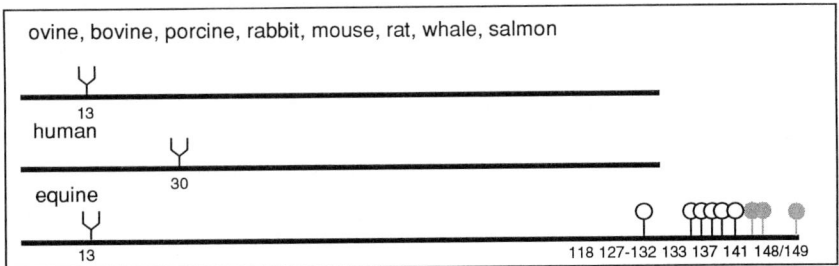

Patterns of LH Beta Subunit Glycosylation

FIGURE 6 Location of glycosylation sites in LH α and β subunits. N-glycosylation sites are highly conserved in the α subunits of all species. The numbering of the glycosylation sites is illustrated for oLHα, a typical mammalian α subunit, and for the human α subunit, which, due to an N-terminal deletion, is four amino acid residues shorter than all other mammalian α subunits. The latter is shown because of the extensive use of its residue numbering due to its clinical relevance. Three patterns of glycosylation are observed for LHβ. Of the two positions N-glycosylated in the glycoprotein hormone β subunits, most LHβ subunits are glycosylated at Asn13. The single exception to this is hLHβ, which is glycosylated at Asn30. The horse and donkey LHβ subunits are both N-glycosylated at Asn13 but also possess an O-glycosylated C-terminal extension previously believed to be restricted to the placental CGβ subunits.

tution of SO4–GalNAc for the NANA–Gal typically found at the nonreducing termini in complex oligosaccharides are present in pituitary glands of all five groups of vertebrates. This modification significantly delayed characterization of LH oligosaccharides because sulfate inhibited exoglycosidases routinely employed in oligosaccharide structure determination. Oligosaccharide structure is dependent on the subunit and the specific glycosylation site if there are two sites as exist in the α subunit. In general, the presence of fucose in a majority of oligosaccharides and a higher abundance of biantennary oligosaccharides are associated with LHβ oligosaccharides, whereas those obtained from the α subunit are predominantly monoantennary and lack fucose. The examples of carbohydrate heterogeneity in the following sections illustrate the potential for generating many isoforms of LH.

3. Carbohydrate Heterogeneity

i. oLH Glycosylation patterns at each individual glycosylation site have been determined for both oLH and hLH. The pattern for oLH (Fig. 7A) is relatively simple. Most of its oligosaccharides are terminated with sulfate because there is little sialylation. The subunit and site-specific patterns of glycosylation

A   Distribution and Relative Abundance of oLH Oligosaccharides

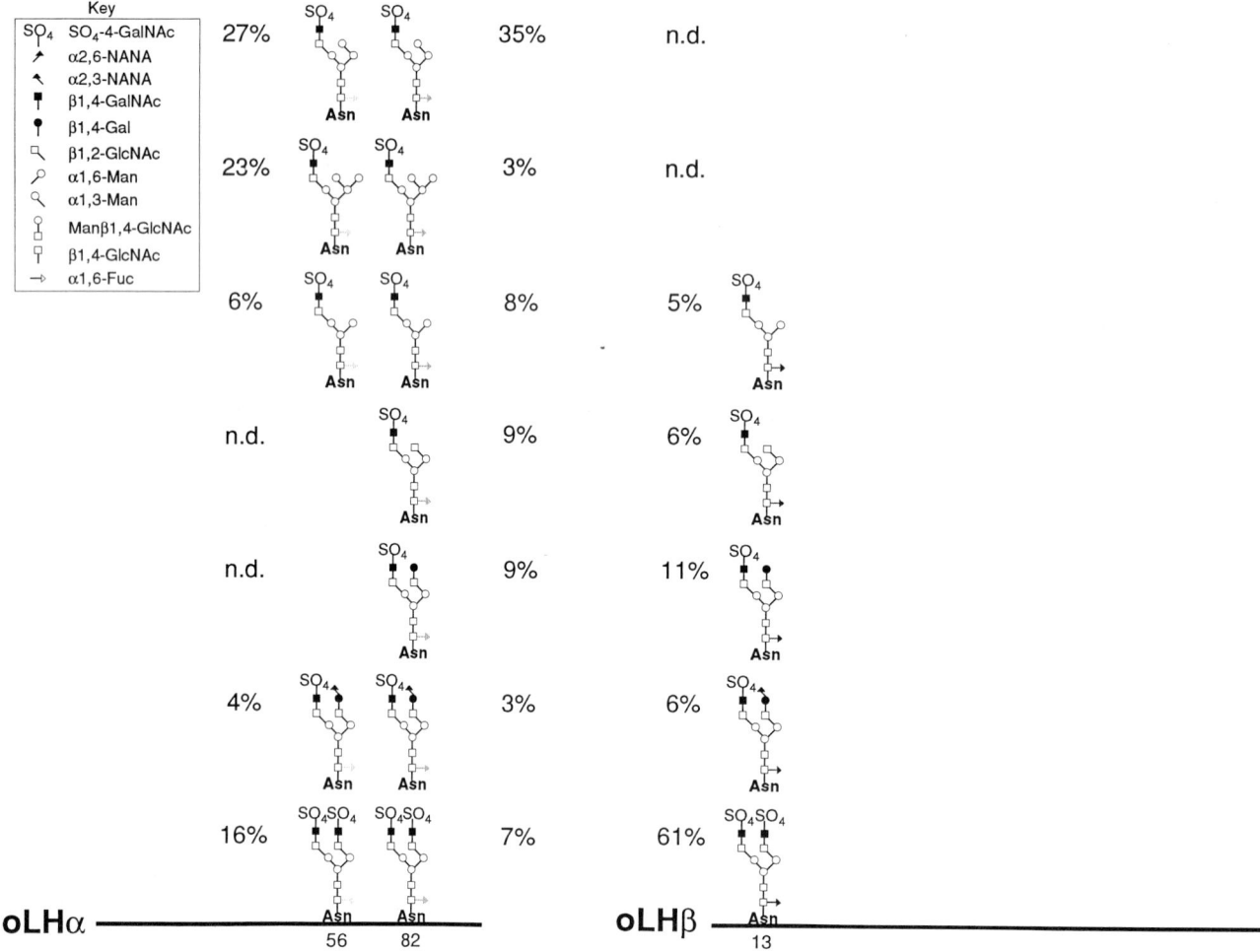

FIGURE 7   Patterns of oligosaccharide heterogeneity observed for oLH and hLH. (A) Oligosaccharides determined to exist at each glycosylation site of oLH are shown diagrammatically and the relative abundance at each site is indicated. (B) Human LH oligosaccharides are more heterogeneous. The most abundant oligosaccharide at each glycosylation site is indicated on the lines representing the α and β subunits. Because all the oligosaccharide structures are found on each site, they are indicated diagrammatically on the left. The relative abundance of each oligosaccharide in percentage is indicated above each glycosylation site.

include the absence of fucose in most α subunit oligosaccharides and an abundance of monoantennary oligosaccharides attached to αAsn56. The β subunit oligosaccharides are typically biantennary; only 11% are monoantennary. The most abundant biantennary oligosaccharides possess terminal sulfate residues on both branches. A small percentage of hybrid biantennary oligosaccharides exist in which one branch is terminated with sialic acid while the other is terminated with sulfate. The most heteroge-

neous collection of oligosaccharides is found attached to αAsn82. These represent all the structures found on the two other sites. The latter only have a subset of the total oligosaccharide population. The most abundant oLHβ oligosaccharides possess biantennary disulfated structures.

ii. hLH   For hLH, glycosylation is very heterogeneous at all three sites (Fig. 7B). Each site appears to possess all the structures characterized for the

B Distribution and Relative Abundance of hLH Oligosaccharides

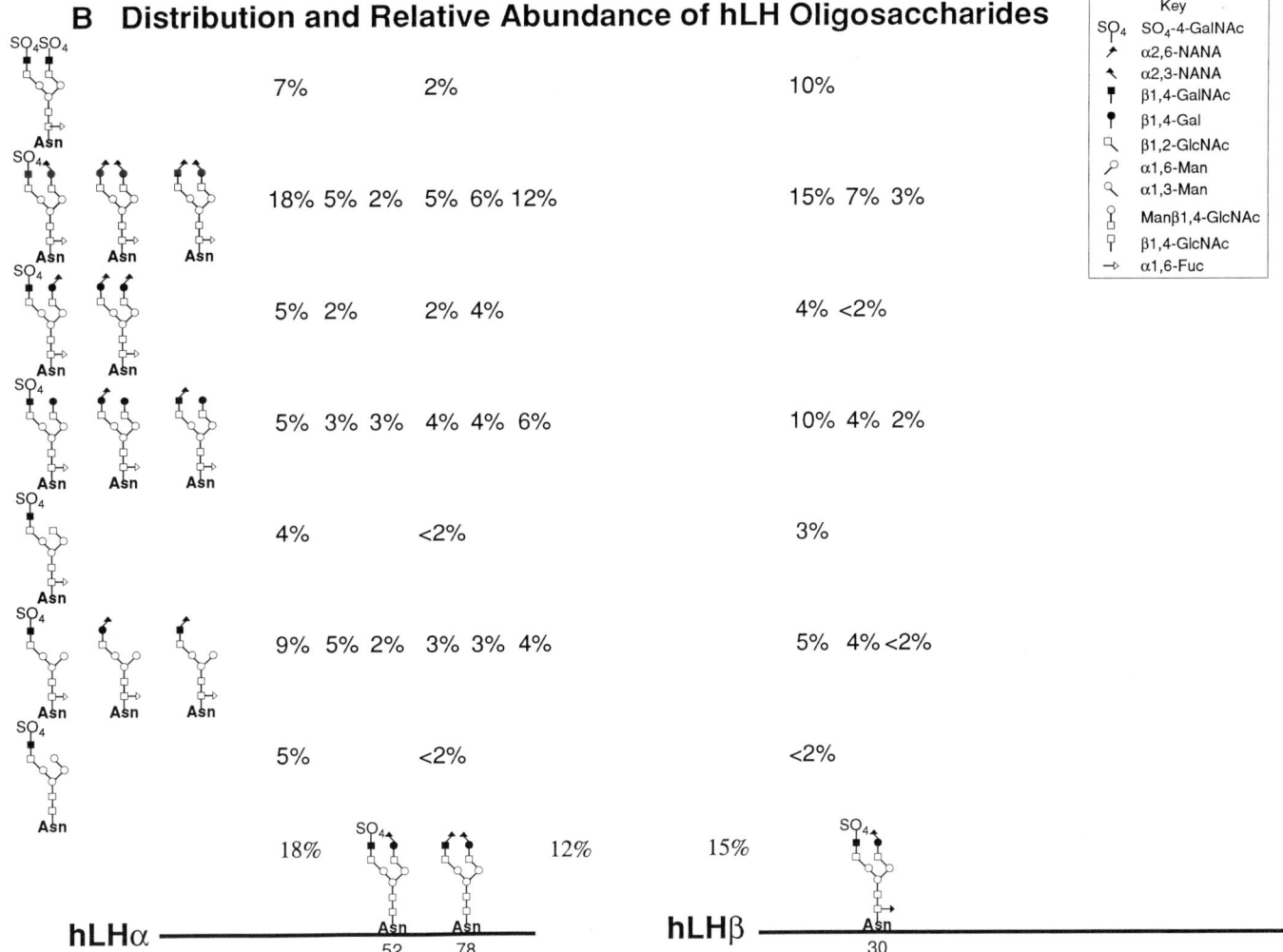

hormone; however, the relative abundance of each structure varies somewhat between glycosylation sites. The oligosaccharides are categorized into five structural classes. Three classes of oligosaccharide structures were classified on the basis of the structure of the Man(α1–3)Man antenna in biantennary oligosaccharides. Two classes of uncharacterized oligosaccharides include the neutral oligosaccharides and those possessing three negative charges indicating triantennary structures. The three biantennary classes include those possessing a sulfated Man(α1–3)Man branch, those possessing a classical, complex sialylated branch, and those in which sialic acid is attached to GalNAc instead of sulfate. The last struc-

tural class presents the potential for regulating the metabolic clearance rate for hLH because sulfated oligosaccharide-bearing glycoproteins are cleared rapidly by a specific liver lectin, whereas sialylated glycoproteins must be desialylated before the asialoglycoprotein receptor can eliminate them from circulation.

4. Carbohydrate Function

Each glycosylation site appears to play a different role. The αAsn82 oligosaccharide promotes folding of the α subunit. This appears to involve only one monosaccharide residue, the GlcNAc(β1–N)Asn82 residue which is located in the cleft between hairpin loops αL1 and αL2 and forms hydrogen bonds with residues in both loops. Because it may be crucial that the α subunit folds more rapidly than the β subunit, this oligosaccharide is potentially very important for LH biosynthesis. The remainder of the oligosaccharide appears to exert a stabilizing effect on the α subunit because its removal results in a progressive aggregation of the deglycosylated subunit. The oligosaccharide attached to αAsn56 is critical for coupling LH receptor binding to the signal transduction machinery. In the absence of this oligosaccharide, significantly more hormone is required to stimulate signal transduction and submaximal stimulation of steroidogenesis occurs. This probably results from submaximal activation of early stages of the signal transduction pathway because cAMP increase is also submaximal. The Asn56 oligosaccharide also plays an inhibitory role on receptor binding. It is located closest to the putative LH receptor binding site in hCG and increased size of the Man(α1–6)Man antenna causes a progressive inhibition of receptor binding. It is interesting that the inhibitory effect is reduced for FSH, suggesting that in addition to having a slightly different receptor binding site, LH and FSH may bind their receptors with a slightly different orientation that is responsible for the differing influence of their Asn56 oligosaccharides on receptor binding. The role for LH βAsn13 oligosaccharide from most species and βAsn30 for hLH is currently undefined. A potential role for Asn13 oligosaccharide in coupling LH receptor binding to signal transduction was indicated in studies on hCG; however, this could only be demonstrated after the hαAsn52 glycosylation site had been eliminated. Studies on recombinant hFSH indicated that the β subunit oligosaccharides, but not the α subunit oligosaccharides, dictated the metabolic clearance rates for this hormone. Whether this is true for LH is not known. The O-glycosylated C-terminal extension of eLH is known to increase circulatory half-life of this hormone. The half-life is 24 hr in the horse, although this pales in comparison with the 6-day half-life for eCG in this species. Glycosylation of eCGβ differs from that of eLHβ primarily in the extent of O glycosylation. Transfer of the O-glycosylated region of hCGβ to recombinant hFSHβ also increased the circulatory survival for this FSH derivative.

Little information is available concerning nonmammalian LH glycosylation. A single study involving bullfrog LH indicated that its oligosaccharides lacked terminal residues associated with mammalian oligosaccharides (Fig. 8). Evidence for sulfated oligosaccharides has been reported for salmon LH, although neutral oligosaccharides were the most abundant and the monoantennary oligosaccharides which appeared to be the next most abundant were equally divided between those terminating in sulfate and those terminated with sialic acid. The same study reported that the potential for sulfated oligosaccharides existed in frogs as indicated by the presence of detectable GalNAc transferase and sulfotransferase; however, no analysis of frog LH was undertaken.

D. Circulating Forms

Most information on LH structure is derived from material isolated from pituitary glands. This source contains hormones at various stages of the biosynthetic pathway; however, most LH is contained in secretory vesicles and, therefore, represents com-

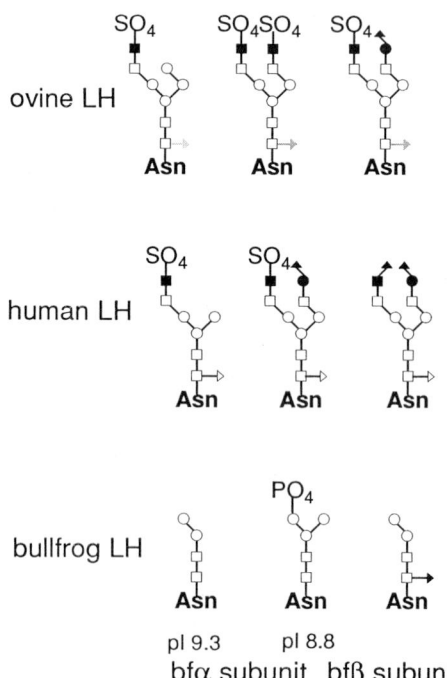

FIGURE 8 Comparison of the three characterized bullfrog LH oligosaccharides with typical hLH and oLH oligosaccharides.

pletely processed hormone. O-glycosylated LH glycoforms, such as donkey and horse LH, have an additional source of variability because O-glycosylation is likely to be completed just prior to packaging in the secretory vesicles.

1. Nicked Forms

The potency of LH preparations often varies significantly and two sources of variation result from the action of proteases. N-terminal heterogeneity of the α subunit and C-terminal heterogeneity of the β subunit are well-known in mammalian LH preparations but appear to have little effect on their biological activity in the gonad. The presence of proteolytic nicks in long-loop L2 of both LHα and LHβ has more drastic consequences. Loop αL2 nicks result in a 97% reduction in receptor-binding activity and a corresponding reduction in steroidogenic potency. Because this region of the α subunit is partially shielded by the clasped embrace of the β subunit seat belt loop, αL2 nicks are relatively uncommon on the heterodimer. Nicking here primarily occurs during subunit isolation, particularly during gel filtration. The ability of αL2-nicked α subunit preparations to combine with β subunit is impaired but not prevented. The stability of the heterodimer in the presence of 0.1% SDS is reduced, although the physiological relevance of this observation is not known. Loop βL2 nicks result in 99% loss of receptor-binding activity. Nicks in this region are common in hLH and account for the relatively low biological activities of many hLH preparations. The nicking is assumed to be the result of postmortem degradation. Isoforms of hCG have also been found having nicks in this region. It is not known if they exist in circulation or are the result of passing through the kidneys. Nicked hLHβ preparations do not reassociate with intact hLHα. This appears to be the result of irreversible denaturation during subunit dissociation. Nicking of oLHβ with Arg-C protease resulted in reduced efficiency of dimerization but did not prevent it. However, treatment of the nicked oLHβ with 6 M guanidine–HCl produced a preparation with reduced dimerization ability and a greater tendency for aggregation, indicating denaturation. Association of both nicked α and nicked β subunits produces no heterodimer at all because loop L2 in at least one subunit must remain intact.

2. Altered Glycosylation

Separation of serum samples by techniques such as chromatofocusing reveals the heterogeneity of circulating LH isoforms. Peter Stanton and colleagues have isolated 41 hLH isoforms and Fig. 7B shows the potential for an even greater number of isoforms. These can vary under altered physiological conditions, and it has been proposed that such glycosylation changes are physiologically relevant. Clearance of glycoproteins from serum involves two receptors localized to the liver. The classical asialoglycoprotein receptor rapidly clears sialylated glycoproteins from which the sialic acid has been removed. Although neuraminidase is widespread in its distribution, the site and mechanism of serum glycoprotein desialylation remains undetermined. A recently discovered liver lectin binds sulfated oligosaccharides. Because no modification of the oligosaccharide moiety is required for recognition by this lectin, it can rapidly clear newly synthesized glycoproteins and it has been argued that this accounts for the more rapid clearance of LH from the circulation than hormones that possess a greater abundance of oligosaccharides terminated with sialic acid, such as hCG, eCG, and hFSH. The sialylated GalNAc residues found in hLH are interesting in this regard because the replacement of sulfate with sialic acid can potentially block binding to the sulfated oligosaccharide lectin, thereby prolonging hLH survival in circulation. There is evidence that during the midluteal phase of the menstrual cycle the half-life of hLH decreases compared with its half-life during the follicular phase. A switch from alternative terminal glycosylation with sialic acid to sulfate may provide a mechanism to explain this phenomenon. At the cellular level, the structures of the αAsn56 oligosaccharides are very important for coupling receptor binding to signal transduction, and they are also important because they are located closest to the putative LH receptor binding site. Most of these oligosaccharides are monoantennary, whereas some are biantennary. Recently, it has been demonstrated for eLH that the presence and size of the second oligosaccharide branch can have a significant inhibitory influence on its receptor-binding activity. In static cultures of rat pituitaries continuous stimulation with gonadotropin-releasing hormone resulted in increased incorporation of labeled galactose, but not mannose, into LH. Because the former is a marker of distal glycosylation, the significantly increased incorporation of this monosaccharide indicates either replacement of terminal sulfate–GalNAc with NANA–Gal or increased branching of oligosaccharides. The former is likely to mean increased survival in the circulation. On the other hand, if increased branching occurs at αAsn56, a reduction in LH receptor-binding activity is likely. The uncharacterized nature of the rLH oligosaccharides makes interpretation of these results difficult.

III. LH FUNCTION

A. Receptor Binding

LH acts primarily in the gonads of both sexes. Specific target cells within the gonads can be identified because they possess a specific LH receptor in their plasma membranes. The LH receptor is an unusual member of the G-protein-coupled receptor family because it possesses a large extracellular domain that provides the high-affinity binding site for LH. A low-affinity binding site has been identified in the transmembrane domain that appears to be necessary for signal transduction. The extracellular domain shares sequence homology with leucine-rich repeat proteins. The crystal structure of an exclusively leucine-rich repeat protein, ribonuclease inhibitor, has revealed that the leucine-rich repeats form an alternating helix-β strand structure. Several leucine-rich repeats combine to form a horseshoe-shaped molecule consisting of an outer surface of α helices and an inner surface of β strands. Several attempts to create models of the LH receptor extracellular domain based on the ribonuculease inhibitor structure have been reported. These models differ in the number of leucine-rich repeats they incorporate and in how hCG interacts with the receptor model. Low doses of LH stimulate adenylyl cyclase through the Gαs subunit, whereas high doses of LH, such as those occurring during the preovulatory surge, can stimulate phospholipase C via Gαq. There is some

evidence provided by studies on FSH and TSH that glycosylation can influence which G protein isoform is activated by glycoprotein hormone binding. The presence of fucose in TSH oligosaccharides appears to be a requirement for activation of Gαq and it is interesting that this monosaccharide is generally in low abundance in LHα Asn56 oligosaccharides. The mechanism by which oligosaccharides modulate LH action is not known. In the past, oligosaccharides were envisioned to lie alongside the polypeptide chain. Alterations in oligosaccharide structure were believed to induce alterations in protein conformation resulting in altered activity. However, it was recently demonstrated that the hCGα Asn52 sticks straight out from the isolated hCGα and intact hCG with no detectable interaction with the protein moieties of either subunit. Another possibility is that the oligosaccharide causes microaggregation of LH–LH receptor complexes. The reversal of the inactivation of chemically deglycosylated hCG by wheat germ agglutinin supports this model.

Recently, LH receptors have been identified in a variety of other tissues, including placenta, uterus, Fallopian tubes, umbilical cord, fetal membranes, prostate, brain, and various cell lines derived from tumors. The physiological relevance of the presence of the LH receptor in these extragonadal tissues has not been demonstrated. Renotropic activity in oLH preparations was attributed to a population of oLH that possessed an intact α subunit amino terminus. It is not known if the renotropic activity is mediated through the classical LH receptor because an intact amino terminus is not required for high-affinity binding of LH. Direct extragonadal effects of LH have long been known to occur in birds. The weaver finch LH bioassay is based on the ability of LH to influence the pigmentation of feathers replacing those plucked from the breast of these birds. A colored bar on the new feather results from exposure to LH.

B. Steroidogenesis

LH stimulates steroidogenesis in the ovary and testis. In the ovary, it stimulates increased blood flow to the gland as well as activating steroidogenesis by target cells. Target cells vary with the ovarian cycle. During follicular development, LH works synergistically with FSH to stimulate estrogen production by the preovulatory follicle. LH stimulates theca cells, the vascularized tissue lying outside the basement membrane separating these cells from the relatively avascular granulosa cell layer. According to the two-cell model for estrogen synthesis, LH stimulation of theca cells produces androgens which diffuse to the granulosa cells, where they are converted to estrogens by the enzyme aromatase. In species in which this mechanism is operative, both LH and FSH are necessary for estrogen production. In mammalian follicles, granulosa cells are initially devoid of LH receptors. FSH stimulation subsequently induces LH receptor expression in these cells, making them LH responsive. In the cow, expression of the LH receptor is one of the earliest indications of which follicle will become the dominant follicle. In the mammalian corpus luteum (CL), LH stimulates progesterone secretion by this highly vascularized derivative of the postovulatory follicle. In species such as rats, LH appears to work in conjunction with prolactin to maintain luteal function. In other species, such as pigs, LH appears to operate alone. In humans, the LH analog, hCG, preserves the critical function of the CL until placental progesterone synthesis becomes adequate to maintain pregnancy.

In the male, the primary role of LH is to stimulate testosterone production by testicular interstitial Leydig cells. LH recruits interstitial Leydig cells during pubertal development of the testis, and in the mature male LH stimulates testosterone synthesis and release by these cells.

C. Effects on Germ Cells

LH induces luteinization and ovulation in the mammalian ovary. Theca interna and granulosa cells are luteinized to form the steroidogenic cells of the CL that replaces the ruptured follicle. The mechanism by which LH stimulates ovulation remains unclear. It may involve the combined actions of matrix metalloproteases and their inhibitors which weaken a restricted site on the ovary, eventually leading to follicular rupture. The oocyte is carried off in the follicular fluid that escapes through the hole in the follicle wall.

The effects of LH on spermatogenesis are mediated by testosterone, the primary steroid hormone produced by the Leydig cells in response to LH stimulation. LH stimulates spermiation in lower vertebrates, such as amphibians.

IV. LH ASSAYS

LH levels in serum are routinely determined by radioimmunoassay. This technique measures the displacement of radiolabeled hormone from an LH antiserum by unlabeled LH in serum samples. Recently developed sandwich assays capture LH from samples using one antibody adsorbed to an insoluble surface and detect the captured LH with a radiolabeled monoclonal antibody. This has resulted in increased sensitivity of LH detection.

In vitro bioassays consist of receptor-binding assays or the stimulation of testosterone synthesis by freshly isolated Leydig cells. A mouse Leydig cell tumor line, MA-10, is also widely employed for *in vitro* bioassays. Receptor-binding assays involve the competitive displacement of radiolabeled hormone from its receptor by unlabeled hormone, much the same as is done in radioimmunoassays. Although possessing adequate sensitivity, receptor-binding assays are not used to measure LH in serum due to interference by serum factors.

The classical *in vivo* bioassays are still employed in the characterization of LH reference preparations and LH preparations employed for use in animals. The two generally accepted LH bioassays measure the stimulation of the ventral prostate in male rats or the depletion of ovarian ascorbic acid in female rats. The former is an indirect effect mediated by testosterone, whereas the latter measures a direct effect of LH on the ovary. If the LH oligosaccharide moieties are intact, there is generally good correspondence between the results of *in vivo* and *in vitro* LH assays.

The development of monoclonal antibodies and cloning of the LH receptor has led to the development of additional assays. The so-called BioIRMA assay involves the capture of the hormone by the LH receptor, followed by detection of the receptor-bound hormone by a radiolabeled monoclonal antibody. Expression of the LH receptor in a cell line in principle permits the creation of homologous bioassays for any LH of interest, regardless of the availability of testicular tissue. For example, expression of the human LH receptor in human fetal kidney 293 cells permitted development of a homologous hLH receptor binding assay. A hLH bioassay employing this cell line used cAMP as the endpoint.

See Also the Following Articles

Chorionic Gonadotropin, Human; Equine Chorionic Gonadotropin (eCG); FSH (Follicle-Stimulating Hormone Gonadotropins, Overview; Luteinization; Steroidogenesis, Overview

Bibliography

Bousfield, G. R., Perry, W. M., and Ward, D. N. (1994). Gonadotropins: Chemistry and biosynthesis. In *The Physiology of Reproduction* (E. Knobil and J. D. Neill, Eds.), pp. 1749–1792. Raven Press, New York.

Chin, W. W. (1986). Glycoprotein hormone genes. In *Molecular Cloning of Hormone Genes* (J. F. Habener, Ed.), pp. 137–172. Humana Press, Clifton, NJ.

Combarnous, Y. (1992). Molecular basis of the specificity of binding of glycoprotein hormones to their receptors. *Endocr. Rev.* 13, 670–691.

Fiddes, J. C., and Talmadge, K. (1984). Structure, expression and evolution of the genes for the human glycoprotein hormones. *Recent Prog. Horm. Res.* 40, 43–74.

Ishii, S. (1993). The molecular biology of avian gonadotropin. *Poultry Sci.* 72, 856–866.

Licht, P., Papkoff, H., Farmer, S. W., Muller, C. H., Tsui, H. W., and Crews, D. (1977). Evolution of gonadotropin structure and function. *Recent Prog. Horm. Res.* 33, 168–248.

Moyle, W. R., and Campbell, R. K. (1996). Gonadotropins. In *Reproductive Endocrinology, Surgery, and Technology* (E. Y. Adashi, J. A. Rock, and Z. Rosenwaks, Eds.), pp. 683–724. Lippincott-Raven, Hagerstown, PA.

Murphy, B. D., and Martinuk, S. D. (1991). Equine chorionic gonadotropin. *Endocr. Rev.* 12, 27–44.

Pierce, J. G., and Parsons, T. F. (1981). Glycoprotein hormones: Structure and function. *Annu. Rev. Biochem.* 50, 465–490.

Quérat, B. (1993). Molecular evolution of the glycoprotein hormones in vertebrates. In *Perspectives in Comparative Endocrinology* (K. G. Davey, R. E. Peter, and S. S. Tobe, Eds.), pp. 27–35. National Research Council of Canada, Toronto.

Ryan, R. J., Keutmann, H. T., Charlesworth, M. C., McCormick, D. J., Milius, R. P., Calvo, F. O., and Vutyavananich, T. (1987). Structure–function relationships of gonadotropins. *Recent Progress in Hormone Research,* 43, 383–429.

Ryan, R. J., Charlesworth, M. C., McCormick, D. J., Milius, R. P., and Keutmann, H. T. (1988). The glycoprotein hormones: Recent studies of structure–function relationships. *FASEB J.* 2, 2661–2669.

Ward, D. N., Bousfield, G. R., and Moore, K. H. (1991). Gonadotropins. In *Reproduction in Domestic Animals* (P. T. Cupps, Ed.), pp. 25–80. Academic Press, San Diego.

LIF

see Leukemia Inhibitory Factor

Lizards

see Reptilian Reproduction

Local Control Systems in Reproduction

John A. McCracken

Worcester Foundation for Biomedical Research

GLOSSARY

apocrine A type of secretion in which the apical portion of a cell is shed, e.g., the secretion of fat into milk by cells of the mammary gland.

autocrine A system in which a substance secreted by a cell acts on cell surface receptors of the same cell which secreted it, e.g., growth factors.

countercurrent transfer A close anatomical apposition of arteries and veins which allows substances secreted into the venous return of an organ to pass directly into the adjacent arterial supply and thus feed back to the same or adjacent organ at a higher concentration than would reach it via the systemic circulation, e.g., the countercurrent transfer of the luteolytic hormone, prostaglandin $F_{2\alpha}$, from a uterine horn to the adjacent ovary.

endocrine The secretion of a substance, usually a hormone, directly into the bloodstream which is transported to a distant target organ or tissue, e.g., secretion of luteinizing hormone from the anterior pituitary into the bloodstream which acts on the ovaries or testes.

exocrine The secretion of a substance into a glandular duct which opens onto an epithelial surface, i.e., not into the bloodstream; for example, secretion of proteins by uterine glands into the uterine lumen.

holocrine A type of secretion in which the whole cell is shed filled with its component substances, e.g., the secretion of sebaceous glands.

intracrine A system in which a substance produced by a cell acts on a receptor or membrane within the cell without being secreted by the cell, e.g., second messengers.

juxtacrine A system in which a substance produced by a cell is bound to the plasma membrane or extracellular matrix of the cell and acts on an adjacent cell by direct contact, e.g., growth hormones.

paracrine A system in which a substance secreted by a cell acts locally on an adjacent cell, sometimes of a different type, e.g., neurotransmitters.

portal system A vascular arrangement whereby a substance secreted into the venous return of one organ or tissue is transported directly in the venous blood to an adjacent organ or tissue on which it then acts, e.g., the transport of gonadotropin-releasing hormone from the hypothalamus to the anterior pituitary.

transcytosis A process by which a substance in the bloodstream is bound and transported across endothelial cells to target cells or tissues, e.g., endothelial transport of gonadotropins in the gonads.

$\mathbf{L}$ocal control systems in reproduction are mechanisms which have evolved to permit hormones, growth factors, and other substances to act locally at, or close to, their sites of production, thus allowing reproductive systems to function in a more integrated and efficient manner.

I. INTRODUCTION

Although local control systems are found throughout the body, these mechanisms are particularly well developed in reproductive processes. Local control systems are advantageous since they permit biological events to operate with greater economy in terms of quantities of mediating substances produced. Several control systems in reproductive processes function in a nonlocal manner via the systemic circulation (endocrine mode) such as the gonadotropins, which regulate the secretion of steroid hormones by the gonads. Also, an exocrine mode of secretion is important in reproduction, e.g., in the glandular secretions of the male and female reproductive tracts. It has also been proposed that some substances can be secreted in an endocrine mode and then change to an exocrine mode. Such a mechanism is thought to occur during early pregnancy in the pig when there is a reduction of the endocrine secretion of prostaglandin $F_{2\alpha}$ ($PGF_{2\alpha}$) by the endometrium, accompanied by an increase in the exocrine secretion of prostaglandin $F_{2\alpha}$ into the uterine lumen. At the tissue level, locally secreted peptide growth factors and other factors are also involved in reproductive processes. These local tissue factors resemble traditional hormones in that they act through specific receptors, use similar signal transduction systems, and are subject to control by other circulating hormones. There are also several specialized local control systems involving the vasculature which serve to enhance reproductive efficiency. These specialized systems include transcytosis of gonadotropins across endothelial cells in the gonads, the portal transport of releasing factors in the hypothalamic–pituitary system, and the countercurrent transfer of prostaglandins, steroids, and possibly other substances in the vasculature of the male and female reproductive systems.

II. DESCRIPTION OF LOCAL CONTROL SYSTEMS

A. Growth and Related Local Factors

Most growth factors and local factors do not act in an endocrine manner. Rather, they act in a local manner on the tissue or cells of origin. As shown in Fig. 1, these local factors can act on neighboring cells (paracrine), on the cell of origin (autocrine), within the cell of origin (intracrine), or by membrane contact with an adjacent cell (juxtacrine). The term "paracrine" was originally proposed for the action of hormone-like substances secreted by certain cells of the gastrointestinal tract which act on adjacent cells of the gut. Later, the concept of autocrine control was introduced to explain the production of autostimulatory growth factors by transformed cells. Although originally coined for the autoregulatory control of malignant cells, it soon became apparent that autocrine control was also important in the function of normal cells, including those of reproductive tissues.

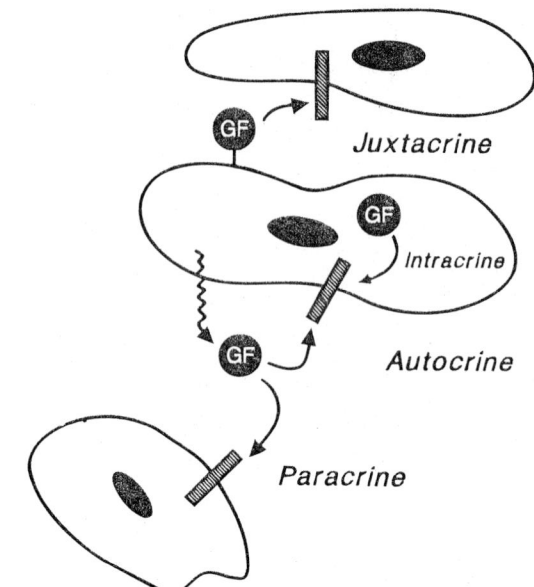

FIGURE 1   Cellular actions of growth factors: autocrine, paracrine, juxtacrine, and intracrine pathways (from Rotwein, 1997).

Some growth factors also appear in the bloodstream, e.g., insulin-like growth factor-1 after the administration of growth hormone. However, growth factor levels in blood do not accurately reflect, either quantitatively or temporally, the local tissue levels of growth factors. Some local tissue mediators such as platelet-activating factor (PAF) are produced by a variety of different cells including those in reproductive tissues. However, PAF may not be completely secreted by some cell types so that, like growth factors, PAF may act by several different local modes, i.e., intracrine, autocrine, paracrine, and, possibly, juxtacrine. Examples of autocrine and paracrine control are also found in the process of follicular steroidogenesis. Under the influence of luteinizing hormone (LH), thecal cells, which surround the outside of the follicle wall, synthesize androgens which pass across the basement membrane of the follicle and act on the avascular granulosa cell layer in a paracrine manner. There, under the influence of follicle-stimulating hormone (FSH), the androgens, testosterone and androstenedione, are converted by the aromatase enzyme in the granulosa cells into estradiol-17β and estrone, respectively. Estrogens produced by the granulosa cells may act in an autocrine manner to regulate granulosa cell division and growth. In the endometrium (the lining of the uterus), there is evidence that steroid hormones can stimulate stromal cells to produce growth factors including epidermal growth factor. Such growth factors may act on the stromal cells themselves (autocrine effect) or pass into the adjacent glandular epithelial cells where they may stimulate secretion and cell division (paracrine effect). Similar autocrine and paracrine mechanisms may operate in the oviduct. Endothelin-1 (ET-1) is a 21-amino acid peptide produced by endothelial cells and other cell types. There is evidence that ET-1 is produced by the endometrium under the influence of oxytocin and vasopressin. Endometrial ET-1 is thought to act in a paracrine fashion to cause contraction of the adjacent myometrium via receptors for ET-1 which appear to be controlled by estrogen and progesterone. ET-1 produced by amnion cells may also act on ET-1 receptors in the myometrium during labor and may play a role in closure of the ductus arteriosus at birth. Mesenchymal/epithelial interaction is thought to occur in the male during organogenesis of the prostate and other accessory glands. During organogenesis of the reproductive system, there is evidence that local factors are involved in the structural differentiation of the male and female reproductive tracts. The lack of androgens in the female fetus prevents the development of the Wolffian duct, whereas a specific factor secreted in the male fetus causes regression of the Müllerian duct. This substance, termed Müllerian duct-inhibiting substance (MIS), is responsible for the regression of the Müllerian ducts in the male during embryogenesis. MIS produced by fetal sertoli cells in the male is thought to act via a specific MIS receptor in the Müllerian ducts. In the adult male, testosterone from the Leydig cells is considered to promote spermatogenesis in the adjacent seminiferous tubules in a paracrine manner.

B. Transcytosis

Binding of LH to ovarian and testicular blood vessels has been described in the rat. In addition, endothelial cells in the ovarian and the testicular microvasculature have been reported to bind and transport human chorionic gonadotropin (hCG) to adjacent endocrine cells by a process termed "transcytosis." Endothelial cells of other tissues, such as heart, lung,

and diaphragm, did not exhibit a similar capacity for transcytosis. It is proposed that hCG binds to its receptor on the luminal side of the endothelial cell and is internalized by coated pits and vesicles. The hormone–receptor complex is then localized in the endosomal compartment and subsequently it is delivered by smooth vesicles into the subendothelial space. Antibodies against the receptor were similarly transported by the same mechanism. Moreover, the rate of transport of the receptor was increased by the administration of hCG. It was also found that the transendothelial transport system was saturable as evidenced by the reduction of gold-labeled hCG in the presence of excess unlabeled hCG. When bovine serum albumin–gold conjugate was perfused through the testes at a concentration similar to that of hCG–gold conjugate, there was essentially no uptake of particles on the luminal side of the endothelium. Currently, it is not known whether transcytosis plays an essential role in ovarian or testicular physiology. However, transcytosis is potentially an attractive regulatory mechanism in gonadal function. For example, it is known that the luteolytic hormone $PGF_{2\alpha}$ causes a rapid block of gonadotropin uptake *in vivo* in the rat ovary. If a transcytotic mechanism exists in endothelial cells of the corpus luteum, then an effect of $PGF_{2\alpha}$ on the transcytosis of gonadotropins could explain such a rapid blockade of gonadotropin uptake *in vivo* by rat luteal cells.

C. Portal Transport of Gonadotropin-Releasing Hormone

The transport of nutrients from the gastrointestinal tract via the hepatic portal veins to the liver has long been recognized. The pioneering work of Geoffrey Harris and others established that a portal system also exists between the hypothalamus and the anterior pituitary. This portal system conveys releasing factors, including gonadotropin-releasing hormone (GnRH), from the hypothalamus to the anterior pituitary, where they stimulate the release of anterior pituitary hormones such as LH and FSH into the bloodstream (Fig. 2). In contrast, the posterior pituitary hormones, oxytocin and vasopressin, are released directly into the systemic circulation from nerve terminals located in the posterior pituitary. The release of GnRH can occur by a neural pathway

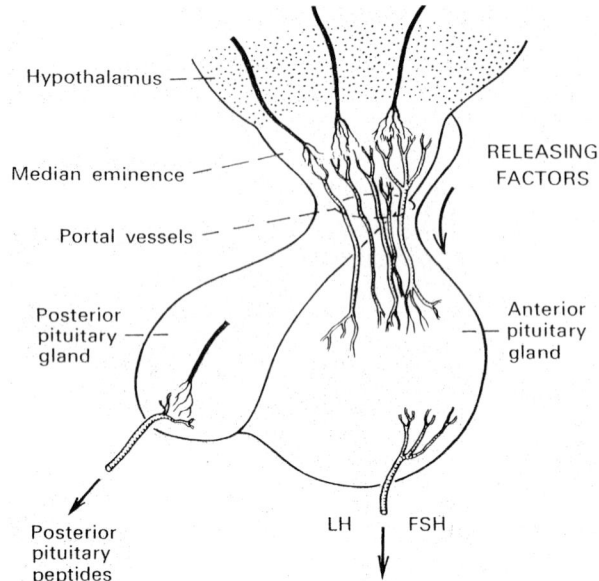

Hypothalamus

Median eminence

Portal vessels

Posterior
pituitary
gland

RELEASING
FACTORS

Anterior
pituitary
gland

Posterior
pituitary
peptides

LH   FSH

FIGURE 2 Releasing factors secreted by nerve terminals in the hypothalamus are conveyed by portal veins to the anterior pituitary where they stimulate the secretion of trophic hormones such as LH and FSH. The posterior pituitary hormones, oxytocin and vasopressin, are secreted directly into the bloodstream from nerve terminals in the posterior pituitary (from Fawcett *et al.*, 1968).

via stimulation of the reproductive tract in reflex ovulators, such as the rabbit and the cat, or can be mediated by the preovulatory rise of estrogens in spontaneously ovulating mammals, such as primates and domestic animals. GnRH is a decapeptide whose pulsatile secretion by the hypothalamus is considered to be controlled in part by negative and positive feedback effects of gonadal steroids. GnRH is synthesized as part of a larger peptide, the prepro-GnRH precursor, which gives rise to GnRH and a GnRH-associated peptide called GAP. However, no specific function has been attributed to GAP. Following portal transport, GnRH acts on specific receptors in the anterior pituitary. The GnRH receptor has a seven-transmembrane structure and is subject to downregulation by both GnRH and its synthetic agonists and antagonists, the latter being used therapeutically to regulate fertility. In addition to its action on the gonadotrophes in the anterior pituitary, GnRH can elicit lordosis in the female and mounting behavior in the male when injected into the ventromedial nucleus of the hypothalamus. Also, GnRH can modulate steroidogenesis by the ovary *in vitro*, but the physio-

logical significance of this action remains to be determined.

D. Countercurrent Transfer Mechanism

In the vascular supply of both the testes and ovaries in most mammals, there is an intimate apposition and intertwining of the arteries and veins. Such an arrangement is often termed a pampiniform plexus (intertwining network). This complicated arrangement of blood vessels is not peculiar to the gonads since similar arrangements are seen in the kidney, the limbs, and the brain of some species. The first function assigned to such a vascular arrangement was the countercurrent transfer of heat in the limbs of various birds and marine mammals for the purpose of conserving heat in the extremities in cold conditions. Subsequently, it was established that there is a similar countercurrent mechanism for heat transfer in the testes. The cooler venous blood returning from the testis cools the incoming arterial blood so that the testes are maintained at a temperature 2–5°C below body core temperature. Such a cooling mechanism is thought to provide optimal conditions for spermatogenesis and the storage of viable sperm. A similar vascular arrangement between arteries and veins occurs in the uteroovarian vascular pedicle of a number of female mammals (Fig. 3). It was suggested that such a close apposition of arteries and veins might be of functional significance at least in the bovine species, particularly since it was noted that a marked thinning of the vessel walls occurred at their areas of contact. Subsequently, as shown in Fig. 4, it was demonstrated that, during luteolysis in the sheep, about 1% of the $PGF_{2\alpha}$ secreted into the uteroovarian vein is transferred to the closely adherent and highly convoluted ovarian artery. Such a mechanism permits a luteolytic level of $PGF_{2\alpha}$ to reach the ovary directly without passing through the systemic circulation where otherwise $PGF_{2\alpha}$, secreted by the uterus, would be rapidly metabolized, especially in the pulmonary vascular bed. In some species, such as the horse and the rabbit, there is no evidence for a vascular arrangement in the uteroovarian pedicle which would be consistent with a countercurrent transfer mechanism. In these species, the luteolytic hormone $PGF_{2\alpha}$, or one of its metabolites, appears to act via

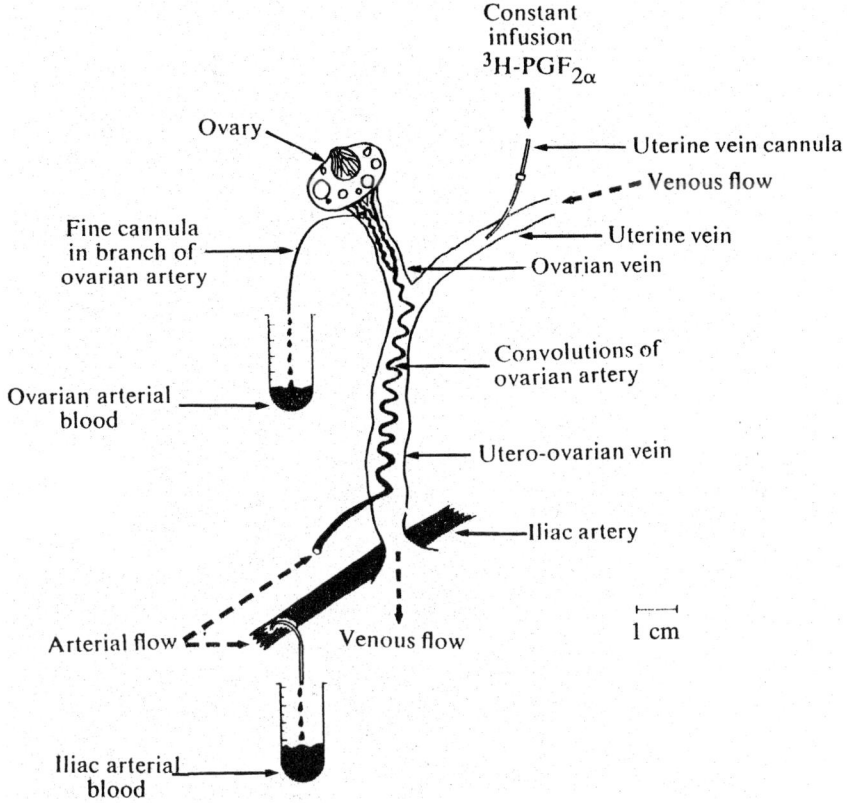

FIGURE 3 The experimental design for the counter-current study is shown diagrammatically. A constant amount of $[^3H]PGF_{2\alpha}$ was infused into the uterine vein and the radioactivity was sampled in ovarian arterial blood and in an equal volume of iliac arterial blood (from McCracken *et al.*, 1972)

the systemic circulation. In other species, such as the cow and the pig, there is evidence that $PGF_{2\alpha}$ acts both locally and via the systemic circulation.

Other substances, such as xenon, krypton, and several ovarian steroids, were later shown to be transferred in the ovarian vascular pedicle, whereas larger molecules such as bovine serum albumin were not transferred. In the case of steroids, it was shown that the ketonic forms were transferred more efficiently than their hydroxyl counterparts (Fig. 5). This finding is consistent with the observation that hydroxyl groups retard the passage of molecules through biological membranes and is consistent with Overton's theory of lipid solubility which states that the more penetrating substances are, in general, the more lipid soluble. There is also evidence that a countercurrent transfer mechanism exists in the reproductive tract in women, although the evidence suggests that the transfer occurs locally at the ovarian level. The exact mechanism for the transfer of substances in the vasculature of the reproductive system has not been

established. However, the term countercurrent transfer has been applied because of (i) the observed close apposition of the arteries and veins, (ii) extensive convolutions of the arteries on the veins which increase the potential surface area for transfer, (iii) the opposite direction of blood flow in the two vessels, and (iv) a concentration gradient of transferred substances between the two vessels. There is evidence that the lymphatics may also contribute to the countercurrent transfer of various substances including steroids and prostaglandins. However, although the concentration of steroids is higher in ovarian lymph than in ovarian venous blood, the proportion of steroids secreted in ovarian lymph is <10% of total ovarian secretion, suggesting that lymphatic transport may play a relatively minor role. The physiological importance of the countercurrent transfer of steroids in the ovarian vascular pedicle is not yet established, but it may serve as a pathway connecting different compartments of the ovary and/or provide a local connection between the ovary and the Fallopian

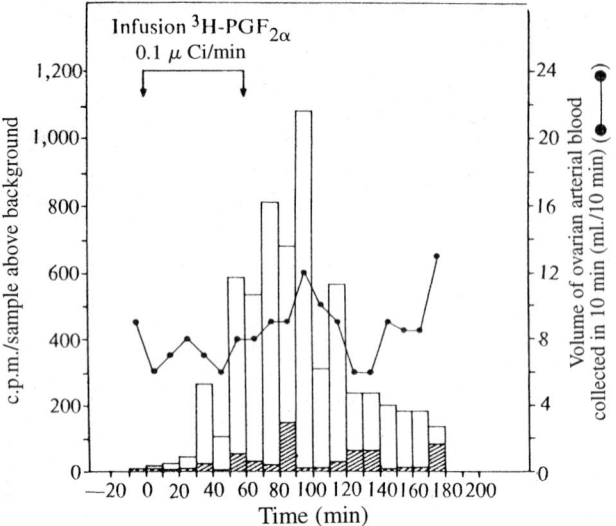

FIGURE 4 The time course of the counter-current transfer of [³H]PGF$_{2\alpha}$ from the utero-ovarian vein to the adherent ovarian artery. White areas, cpm ovarian arterial plasma; hatched areas, cpm iliac arterial plasma (from McCracken *et al.*, 1972).

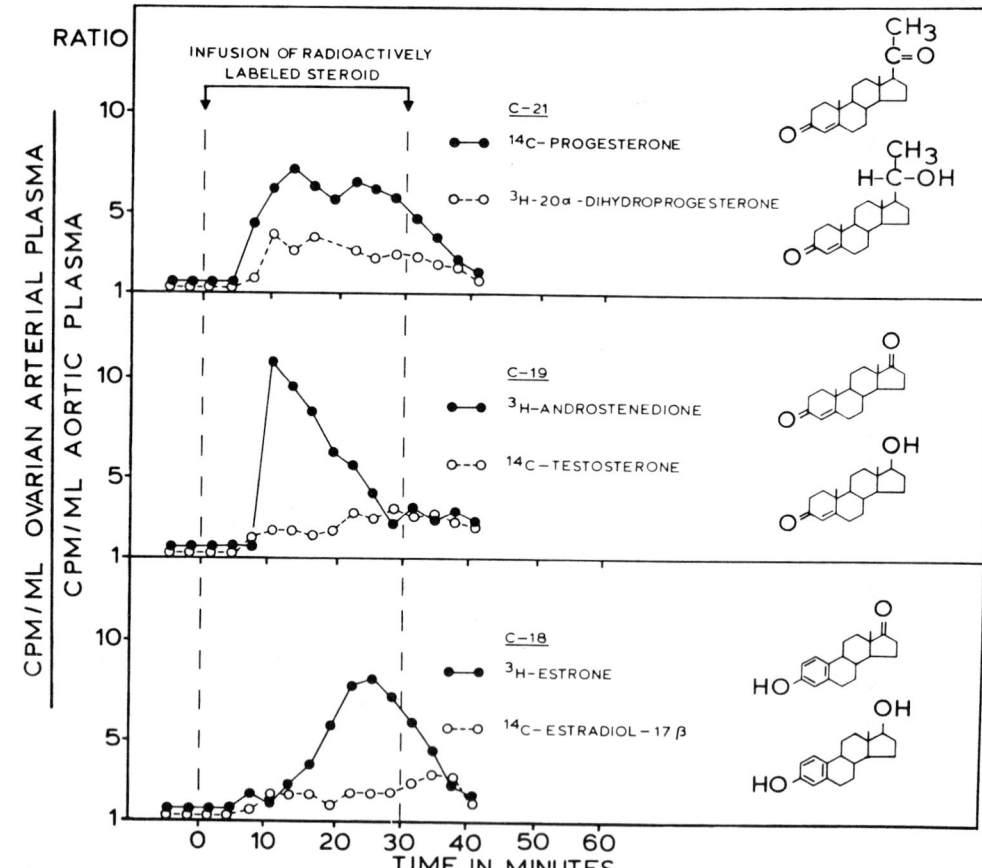

FIGURE 5 Ratio of counts between samples of ovarian arterial and aortic plasma before, during, and after a 30-min infusion of tracer amounts of three pairs of related labeled steroids into an ovarian vein in three sheep. In each case, the less polar (ketonic) pair member was transferred to the ovarian artery more efficiently than its hydroxyl counterpart (from McCracken *et al.*, 1984).

tube and uterus. In the male, in addition to the transfer of heat, a countercurrent transfer in the testicular vasculature has been reported for xenon, krypton, tritiated water, and testosterone. In the latter case, a countercurrent mechanism may ensure a relatively high concentration of testosterone within the testis to promote spermatogenesis and to ensure a relatively high concentration of testosterone in the epididymis.

III. CONCLUSIONS

This brief survey of local control systems in reproductive processes illustrates the complexities of regulatory mechanisms in both female and male reproduction. In some species, certain local control systems become critical for the reproductive system to function efficiently. For example, the countercurrent transfer of $PGF_{2\alpha}$ from the uterus to the ovary is an absolute requirement for luteolysis in the sheep and the guinea pig which is in keeping with the extremely rapid clearance rate of $PGF_{2\alpha}$ in the peripheral circulation in both these species. In other species, such as the cow and pig, a relatively large proportion of $PGF_{2\alpha}$ passes unchanged across the lungs in keeping with the finding that $PGF_{2\alpha}$ may act both locally and systemically in these species. In the alpaca, a member of the camel family indigenous to South America, there is evidence that the luteolytic effect of the uterus is mediated locally in the right uterine horn but is mediated, both systemically and locally, in the left uterine horn. Such species variation in local mechanisms illustrates how different animals can adapt a system to suit their special requirements for reproductive processes.

See Also the Following Articles

GnRH (GONADOTROPIN-RELEASING HORMONE); IGF (INSULIN-LIKE GROWTH FACTORS); LUTEOLYSIS; NEUROSECRETION

Bibliography

Bazer, F. W., and Thatcher, W. W. (1977). Theory of maternal recognition of pregnancy in swine based on estrogen controlled endocrine versus exocrine secretion of prostaglandin $F_{2\alpha}$ by the uterine endometrium. *Prostaglandins* **14**, 397–401.

Behrman, H. R., and Hichens, M. (1976). Rapid block of gonadotropin uptake by corpora lutea *in vivo* induced by prostaglandin $F_{2\alpha}$. *Prostaglandins* **12**, 83–95.

Bendz, A. (1977). The anatomical basis for a possible counter current exchange mechanism in the human adnex. *Prostaglandins* **13**, 355–362.

Bukovsky, A., Chen, T. T., Wimalasena, J., and Caudle, M. R. (1993). Cellular localization of luteinizing hormone receptor immunoreactivity in the ovaries of immature, gonadotropin-primed and normal cycling rats. *Biol. Reprod.* **48**, 1367–1382.

Carlsson, B., Hillensjo, T., Nilsson, A., Tormell, J., and Billig, B. (1993). Expression of insulin-like growth factor (IGF-1) in the rat fallopian tube: Possible autocrine and paracrine action on the fallopian tube and on the preimplantation embryo. *Endocrinology* **133**, 2031–2039.

Cunha, G. R., *et al.* (1987). *Endocr. Rev.* **8**, 338–362.

Donahoe, P. K., Cate, R. L., MacLaughlin, D. T., *et al.* (1987). Mullerian inhibiting substance: Gene expression and mechanism of action of a fetal regressor. *Rec. Prog. Horm. Res.* **43**, 431–467.

Everett, J. W. (1994). Pituitary and hypothalamus: Perspectives and overview. In *Physiology of Reproduction* (E. Knobil and J. D. Neill, Eds.), 2nd Ed., Vol. 1, pp. 1509–1526. Raven Press, New York.

Fawcett, C. P., Reed, M., Charlton, H. M., and Harris, G. W. (1968). The purification of luteinizing-hormone-releasing factor with some observations on its properties. *Biochem. J.* **106**, 229–236.

Fernendez-Baca, S., Hansel, W., Saatman, R., Sumar, J., and Novoa, C. (1979). Differential luteolytic effects of right and left uterine horns in the alpaca. *Biol. Reprod.* **20**, 586–595.

Ferreira, S. H., and Vane, J. R. (1967). Prostaglandins: Their disappearance from and release into the circulation. *Nature* **216**, 868–873.

Ghinea, N., Hai, M. T. V., Groyer-Picard, M., and Milgrom, E. (1994). How protein hormones reach their target cells: Receptor-mediated transcytosis of hCG through endothelial cells. *J. Cell Biol.* **125**, 87–97.

Granström, E., and Kindahl, H. (1985). Species differences in circulating prostaglandin metabolites relevance for the assay of prostaglandin release. *Biochem. Biophys. Acta* **713**, 555–569.

Harris, G. W., and Naftolin, F. (1970). The hypothalamus and control of ovulation. *Br. Med. Bull.* **26**, 3–9.

Heap, R. B., Fleet, I. R., and Hamon, M. (1985). Prostaglandin $F_{2\alpha}$ is transferred from the uterus to the ovary in the sheep by lymphatic and blood vascular pathways. *J. Reprod. Fertil.* **74**, 645–656.

Kotwica, J. (1980). Mechanism of prostaglandin $F_{2\alpha}$ penetration from the horn of the uterus to the ovaries in pigs. *J. Reprod. Fertil.* **59**, 237–241.

Krey, L. C., Gulyas, B. J., and McCracken, J. A. (Eds.) (1989). *Autocrine and Paracrine Mechanisms in Reproductive Endocrinology.* Plenum, New York.

Lindner, H. R., Sass, M. B., and Morris, B. (1964). Steroids in the ovarian lymph and blood of conscious ewes. *J. Endocrinol.* **30**, 361–376.

Masaki, T. (1993). Endothelins: Homeostatic and compensatory actions in the circulatory and endocrine systems. *Endocr. Rev.* **14**, 256–268.

McCracken, J. A. (1997). Prostaglandins and leukotrienes—Locally acting agents. In *Endocrinology: Basic and Clinical Principles* (P. M. Conn and S. Melmed, Eds.), pp. 101–114. Humana Press, Totowa, NJ.

McCracken, J. A., Custer, E. E., and Lamsa, J. C. (1998). Luteolysis—A neuroendocrine-mediated event. *Physiol. Rev.,* in press.

McCracken, J. A., Carlson, J. C., Glew, M. E., *et al.* (1972). Prostaglandin $F_{2\alpha}$ identified as a luteolytic hormone in the sheep. *Nature (London)* **238**, 129–134.

McCracken, J. A., Schramm, W., and Einer-Jensen, N. (1984). The structure of steroids and their diffusion through blood vessel walls in a counter-current system. *Steroids* **143**, 293–303.

Peplow, P. V., and Mikahilidis, D. P. (1990). Platelet-activating factor (PAF) and its relation to prostaglandins, leukotrienes and other aspects of arachidonic metabolism. *Prostaglandin Leukocyte EFA* **41**, 71–82.

Rotwein, M. D. (1997). Peptide growth factors. In *Endocrinology: Basic and Clinical Principles* (P. M. Conn and S. Melmed, Eds.), pp. 79–99. Humana Press, Totowa, NJ.

Setchell, B. P., Maddocks, S., and Brooks, D. E. (1994). Anatomy, vasculature, innervation and fluids of the male reproductive tract. In *Physiology of Reproduction* (E. Knobil and J. D. Neill, Eds.), 2nd Ed., Vol. 1, pp. 1063–1176. Raven Press, New York.

Tsafriri, A. and Adashi, E. Y. (1994). Local non-steroidal regulators of ovarian function. In *Physiology of Reproduction* (J. Knobil and J. D. Neill, Eds.), 2nd Ed., Vol. 1, pp. 817–860. Raven Press, New York.

Underwood, L. E., and Van Wyk, J. J. (1990). Normal and aberrant growth. In *Williams Textbook of Endocrinology, 8th Edition* (J. D. Wilson and D. W. Foster, Eds.), pp. 1079–1138. Saunders, Philadelphia.

Vollmerhaus, B. (1964). Investigation of the vascular architecture of the bovine female genital tract. *Zentralbl. Veterinarmed.* **11**, 597–646.

Locusts

Cedric Gillott

University of Saskatchewan

GLOSSARY

reproductive diapause Arrested development of the reproductive system induced by environmental factors.

Locusts are perhaps the best known and oldest recorded insect pests. Their voracious appetites, phenomenal migrations, and wide geographic distribution have given them legendary status, and despite huge efforts to control them, they remain capable of developing into massive plagues that cause billions of dollars' worth of damage to crops. In large part because of their importance as pests, but also because they are excellent model insects and easy to rear in culture, they rank among the most-researched of insects. Though all aspects of their biology have been studied, their reproductive biology perhaps is the most understood component. Studies of reproduction in locusts (and their North American counterparts, the pestiferous grasshoppers) have played a major role in establishing the basic pattern by which this process occurs in the Insecta.

I. INTRODUCTION

The term "locust" is loosely applied to about 20, mostly Old World, species (representing several subfamilies) in the family Acrididae of the order Orthoptera. In common, these species have the ability, under appropriate conditions, to form dense populations that may migrate for considerable distances, feeding and reproducing en route, thereby causing immense damage to crops and other vegetation. Under such conditions, the locusts are said to be in the gregarious phase. However, under conditions that are less suitable for growth and reproduction, populations become less dense and are nonmigratory, and the so-called solitarious phase locust develops. These two phases differ significantly in their development, behavior, color, morphology, and physiology (see Box).

Among the locusts, two species stand out because of their historic and economic importance: namely, the migratory locust, *Locusta migratoria*, and the desert locust, *Schistocerca gregaria* (most probably the locust of biblical fame). This importance has generated an enormous volume of research, both in the field and in the laboratory, and of the applied and basic kind. The migratory locust especially, with its large size, fecund nature, and relatively simple dietary, photoperiodic, and temperature requirements, has been adopted as the research insect of choice by many laboratories in Western Europe and North America. This is particularly true where biochemical extraction and characterization of microquantities of key molecules such as neuropeptides are being undertaken.

Though there are no true locusts in North America, there are several species of grasshopper, very closely related to locusts taxonomically, capable of forming dense populations, migrating, and becoming major cropland or rangeland pests under suitable climatic conditions. Indeed, because of these habits, a number of North American species were commonly referred to as locusts in the early literature. However, for a complex of reasons—not the least of which is the presence in the life history of most species of a dormant, overwintering egg stage that requires a lengthy period of chilling before development can continue—relatively few North American grasshoppers have been the subject of extensive laboratory research. The principal exception is the lesser migratory grasshopper, *Melanoplus sanguinipes*, in which a nondormant strain was developed by Agriculture Canada scientists in the early 1960s. The reproductive biology of this species, particularly that of the male, has been well studied and, where appropriate, will be included in this review.

Given the economic importance of locusts and grasshoppers, it is not surprising that most of the early studies on their reproductive biology focused on the extrinsic factors that affect sexual maturation, mating behavior, and egg laying, as these are most relevant to control. Recently, the emphasis has shifted toward the use of these insects as models for understanding the intrinsic factors, especially the endocrine axes, that regulate reproduction.

II. STRUCTURE OF THE REPRODUCTIVE SYSTEM

A. Female

The reproductive system of female locusts and grasshoppers includes a pair of ovaries, paired lateral oviducts whose anterior ends are extended forward to form accessory glands, a median common oviduct, and a spermatheca with its very elongate duct (Fig. 1). Each ovary includes a species- and phase-specific number of panoistic ovarioles (e.g., in the desert locust, about 110 in the gregarious phase and about 150 in the solitary phase, and in the migratory grasshopper, from 26 to 31) that open linearly along the lateral oviduct. Within each ovariole, the oocytes are arranged linearly with, at any one time, only the terminal oocyte (the one adjacent to the lateral oviduct) accumulating yolk (see Section III,A,1). Each oocyte, as it matures, becomes enclosed within a one-cell thick layer of follicular epithelium. The accessory glands tend to be pinkish in color, especially in mature females, due to the accumulation of secretion. The latter is released as a foamy material during egg laying; it surrounds the eggs and hardens to form a

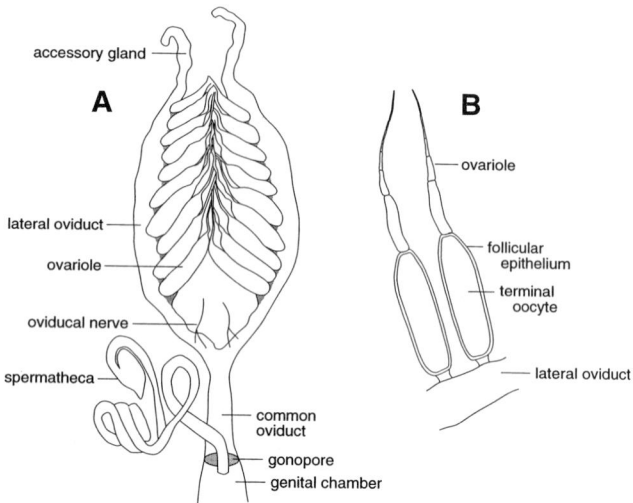

FIGURE 1 Female reproductive system of locust. (A) Dorsal view of system. (B) Entry of ovarioles into lateral oviduct.

protective pod within the soil. The common oviduct, which is lined with cuticle and has a thick muscular wall, opens to the exterior at the gonopore within the genital chamber. The spermathecal duct also opens into the genital chamber, just dorsal to the gonopore. The spermathecal duct is long (>3 cm in the migratory locust) and terminates at the spermatheca where sperm is stored. In many Acrididae a pair of Comstock–Kellogg glands can be found at the anterior end of the genital chamber. The function of these sac-like structures, which are sometimes everted before and during copulation, remains unclear. It has been suggested that they may produce a sexual pheromone, though a more likely role for their secretion appears to be to promote tanning of newly laid eggs and the froth that forms the egg pod.

B. Male

The male system includes a pair of testes, a pair of vasa deferentia that open into the ejaculatory duct, and the paired accessory gland complex, comprising 15 (in *Locusta*) or 16 (in *Melanoplus*) tubules on each side (Fig. 1A under Male Reproductive System, Insects). These tubules can be readily separated into four groups: (i) 9 (*Locusta*) or 10 (*Melanoplus*) short to medium-length hyaline tubules, (ii) 4 white tubules, (iii) a centrally placed seminal vesicle, and (iv) a very long and coiled hyaline tubule. Each tubule within a group has its individual ultrastructural

and staining characteristics and opens separately into the anterior part of the ejaculatory duct. Analysis of the secretions of these tubules reveals that each includes a complex of proteinaceous materials ranging in size from small peptides to large conjugated proteins. Many of these larger molecules are structural proteins and enzymes used to form the spermatophore, in which sperm are transferred to the female, though it should be noted that there is very limited information on either the specific molecules involved or the biochemistry of the process. Other components are physiologically active; for example, in *Melanoplus* and *Locusta* a protein with a molecular weight of about 60 and 13 kDa, respectively, has been reported to stimulate egg laying when injected into gravid females. The precise mode of action of this oviposition-stimulating protein is far from clear; however, it may be speculated that its role is to prime the female neuroendocrine system so that, at the appropriate time, it can release oviposition-stimulating neurohormone to initiate egg laying (see Section V). *Locusta* accessory glands also contain a number of peptides that invoke contractions of the oviduct muscle when tested *in vitro*. However, when injected into whole insects, they do not induce egg laying and their true function remains unclear.

The testes are bound closely together within a connective tissue sheath, and within each organ there is a large number of follicles (300–400 in *Locusta*; 80–95 in *Melanoplus*). In each follicle is a conspicuous apical cell thought to provide nutrients to the spermatogonia that surround it. Along the length of each follicle can be distinguished groups of germ cells in their various stages of maturation (Fig. 1B, under Male Reproductive System, Insects). Mature sperm, both in the basal region of each follicle and in the seminal vesicle, remain in bundles of several hundred, their heads invested in a gelatinous cap that disintegrates only after insemination. The ejaculatory duct is a thick-walled, muscular tube lined with cuticle, in which spermatophore formation occurs.

III. SEXUAL MATURATION

Except for spermatogenesis, which largely occurs during the final juvenile stage several processes occur in newly emerged female and male Acrididae that collectively constitute sexual maturation. These processes are structural, physiological, and behavioral and, while typically extending over a period of a few days (e.g., about 5–7 days in *M. sanguinipes* and about 12–14 days in *L. migratoria*), sexual maturation may not occur for several months in species that undergo reproductive diapause. A variety of environmental factors can affect sexual maturation; these factors typically exert their influence via the endocrine system.

A. Female

In females sexual maturation encompasses the production of eggs (including vitellogenesis and vitelline membrane and chorion formation), development of characteristic body coloration, maturation of pheromone-producing glands, growth of the reproductive tract (including the accessory glands), and an increase in receptivity (willingness to mate).

1. Egg Production

Only the terminal oocyte within each ovariole typically accumulates yolk during each ovarian cycle (the production of a batch of eggs). Insect egg yolk is made up almost entirely of spherical droplets of material. Most droplets are membrane bound and contain protein; the rest are without a membrane and contain either lipid or glycogen. The fat body (the insect analog of the liver) is the source of the yolk proteins (vitellogenins). These large, conjugated proteins are transported by the blood to the ovary, reaching the oocyte via channels that run between the follicular cells, and are then taken into the oocyte by pinocytosis. Where the yolk lipid and glycogen are produced remains unclear, though the oocyte itself or the follicular epithelium would seem the most likely source. In addition to these macromolecules, the yolk also contains relatively huge amounts of the hormone ecdysone. This is produced by the follicle cells and stored in an inactive form in the yolk, to be used periodically during embryonic development when it triggers bouts of cuticle production associated with the several molts that embryonic grasshoppers and locusts undergo.

In many insect species, under stress conditions (e.g., lack of or poor-quality food and the absence of males), resorption of some developing eggs can occur. However, in locusts, even under favorable conditions, a proportion of oocytes are always resorbed (in the desert locust, about 22%), mostly in the later stages of vitellogenesis. Experiments have shown that in the desert locust resorption is the result of competition among oocytes for both protein and juvenile hormone (JH) (see Section III,C). The resorbed materials are presumably made available to other developing oocytes so that at least some eggs develop completely even under less than optimal conditions.

When vitellogenesis is complete, but before ovulation, formation of the vitelline membrane and chorion takes place. Both structures are produced as secretions of the follicle cells. In addition, during egg laying, the common oviduct secretes material that forms a third layer around the egg, the extra-chorion.

2. Other Components of Female Sexual Maturation

In locusts there are typically marked changes in body coloration during sexual maturity; for example, in gregarious *L. migratoria* the immature females are pale, creamy brown whereas mature specimens are brown, and in gregarious *S. gregaria* fledgling adults of both sexes are a beautifully patterned grayish-brown-pink color which changes to yellow as the insects mature.

Another important aspect of sexual maturity in females of some locusts and grasshoppers is the maturation of pheromone-producing glands, typically located on the abdomen. Several distinct pheromones may be produced—namely, maturation-inhibiting pheromone, group-oviposition pheromone, and sex-attractant pheromone—though it must be noted that for none of these is the evidence strong. For example, it seems that the presence of newly fledged adults of both sexes can inhibit maturation of older (but sexually immature) males. The long-standing observation that an egg-laying female *S. gregaria* attracts other, ready-to-lay females is now known to depend in part on the release of a contact pheromone (though visual attraction also occurs in daylight). Physical contact with the ovipositing female is necessary. In contrast, in *L. migratoria*, females are strongly attracted to sand in which other females have oviposited already, suggesting that the pheromone is associated with the egg pod rather than the surface of the abdomen as in *S. gregaria*. Much anecdotal evidence suggests that some female locusts and grasshoppers produce sex-attractants, but for only a handful of species is their presence confirmed. Nothing appears to be known about the chemical nature of these pheromones.

During sexual maturation, growth and development of secretory activity occur in the female reproductive tract. Growth serves not only to accommodate the developing eggs but also to facilitate egg laying, sperm transfer, and sperm storage. The spermathecal epithelium shows massive development of secretory activity at this time, presumably in preparation for receiving sperm, and the accessory glands grow and accumulate secretory material for forming the egg pod.

B. Male

The major elements of sexual maturation in male locusts and grasshoppers are growth and development of the reproductive tract, especially the accessory glands, development of pheromone-secreting glands in some species, and the evolution of mating behavior. In contrast to the situation in mammals, spermatogenesis is not a major event in sexual maturation because it has taken place mostly in the final juvenile stage.

The development of secretory activity in the accessory glands has been followed in several species of locusts and grasshoppers, from both ultrastructural and biochemical perspectives. Though, as noted in Section II,B, the accessory glands of acridids are multitubular, the general changes in cell ultrastructure are common to all tubules except the seminal vesicle and typical of cells producing protein for export; that is, there are massive increases in the amounts of rough endoplasmic reticulum, free ribosomes, Golgi complexes, and secretory vesicles. Though the cells of the seminal vesicle develop some rough endoplasmic reticulum and limited numbers of Golgi complexes, their predominant features are the massive accumulation of glycogen and numerous mitochondria in their basal cytoplasm. Each tubule

type produces an array of distinct proteins during sexual maturation; however, both the onset of production and the amount produced are specific for each protein.

In male locusts there are typically marked changes in body coloration during sexual maturity; for example, in *L. migratoria* immature males are pale, creamy brown whereas mature insects are strongly yellow, and in *S. gregaria* fledgling males are grayish-brown-pink color but become yellow as they mature.

Another important aspect of sexual maturity in male locusts is the development of pheromone-producing glands located on the abdomen. It seems that two distinct pheromones may be produced, namely, maturation-accelerating pheromone and maturation-inhibiting pheromone. Good evidence exists that, in both *L. migratoria* and *S. gregaria*, mature males produce a pheromone that accelerates maturation in younger adults of both sexes. Conversely, though direct evidence is lacking, it seems that newly fledged adults of both sexes can inhibit maturation of older (but sexually immature) males. Almost nothing is known about the chemical nature of these pheromones. However, a possible candidate for the maturation-accelerating pheromone produced by male *Schistocerca* is Δ^1-pyrroline, which is present in the air around these males and in their accessory glands. It must be emphasized, however, that the critical experiments to demonstrate a pheromonal function for Δ^1-pyrroline have not been performed.

In parallel with the changes in their body coloration and the onset of pheromone secretion, male acridids begin to develop elements of sexual behavior during sexual maturation. For example, in *S. gregaria* signs of courtship and attempts to mount begin approximately 6–12 days after adult emergence.

C. Control of Sexual Maturation

The rate at which sexual maturation occurs in insects is dependent on a variety of environmental factors. In most instances, these environmental factors exert their effect via the insects' neuroendocrine system. Because of their ability to undergo, under suitable conditions, rapid increases in population density and swarming, the conditions that govern these outbreaks in locusts have been examined in detail under both field and laboratory conditions. Furthermore, the endocrine system that "translates" these environmental effects into reproductive potential has proved to be highly amenable to experimentation in these large, generalized insects.

1. Environmental Factors Affecting Sexual Maturation

Parameters known to modulate sexual maturation in locusts include quality and quantity of food available, temperature, population density and structure, and photoperiod. Though availability of food and temperature directly affect sexual maturation by controlling the supply of nutrients and metabolic rate, respectively, these and the other factors also affect the activity of the locusts' endocrine system, as is outlined below.

Feeding, by stretching the wall of the foregut, triggers the synthesis and release of neurosecretory material from the locust brain–corpus cardiacum complex. The neurosecretory material, either directly or indirectly (by activating other endocrine centers), stimulates protein synthesis in a variety of tissues (e.g., digestive enzymes in the midgut, vitellogenins in the fat body, and proteins in the accessory glands).

Locusts exposed to temperatures that fluctuate about a mean value have been shown to become sexually mature more rapidly than insects kept constantly at the mean temperature. Apparently, the fluctuating conditions promote release of neurosecretion, hence the increased rate of sexual development.

Both population density and population structure can influence sexual maturation rate in locusts, and these effects are species specific. Thus, gregarious phase *S. gregaria* females mature eggs more rapidly than those in the solitary phase. However, in *L. migratoria* and *Nomadacris septemfasciata* (the red locust) the greater the population density, the slower the rate of egg maturation. The effect of population structure (i.e., age profile) is exerted through the release of the maturation-accelerating and maturation-inhibiting pheromones described in Sections II,A,2 and II,B. Though direct evidence is not available, it is assumed that these population influences

are achieved through modification of the insects' endocrine activity.

For a number of locust species, photoperiod exerts a major influence on sexual maturation, and by doing so serves to synchronize the species' reproductive activity with seasonal availability of resources (e.g., food for egg development and/or larval growth and suitable egg-laying sites). When conditions for reproduction are unsuitable, the insect remains in reproductive diapause. The influence of photoperiod is species specific and will vary according to the species' geographic distribution and climate. For example, in the Egyptian tree locust (*Anacridium aegyptium*) reproductive diapause is induced by decreasing day lengths experienced in the fall and is maintained for about 4 months (hibernation). Termination of diapause, brought about by increasing day lengths in spring, is correlated with renewed availability of oviposition sites and food for the juvenile stages. In contrast, *S. gregaria* estivates; that is, long-day conditions cause females to enter reproductive diapause, whereas short-day lengths terminate diapause and promote reproductive activity. This strategy is correlated with hot, dry weather (hence lack of food and egg laying sites) in the summer and cooler conditions with intermittent rainfall in winter in the arid areas of Africa and Asia where the species lives. Like other environmental factors, photoperiod influences locust reproductive activity via the brain neurosecretory system and, in turn, the corpora allata.

2. Endocrinology of Reproduction

Though the pioneering work of Weed-Pfeiffer in the United States, carried out in the 1930s on *Melanoplus differentialis* (the differential grasshopper), was one of the first demonstrations of the importance of the corpora allata in controlling egg production and male accessory gland development, it was not until the late 1950s that locusts were discovered to be ideal models for insect endocrinology. By that time, cultures of *L. migratoria* and *S. gregaria* had been established in laboratories in the United Kingdom, France, Germany, and The Netherlands, and in the early 1960s the nondiapausing strain of *M. sanguinipes* was developed in Canada. Since then, the use of locusts, specifically for studies of the endocrine regulation of reproduction, has spread, notably to laboratories in Canada and Belgium.

Much of our basic understanding of the regulation of egg development came from the work of Highnam and his students on *Schistocerca* during the 1960s. However, the majority of researchers since then have used the more easily cultured *Locusta*—for example, the Girardies, Wyatt's group, Hoffmann and associates, and De Loof and colleagues. Major contributions to knowledge of the endocrinology of males have been made in the laboratories of Pener (*Schistocerca* and *Locusta*) and Gillott (*Melanoplus*).

Highnam and co-workers determined that the neurosecretory cells of the female locust brain appeared to produce both tropic factors (notably allatotropic hormone that activates the corpora allata) and hormones that directly affected egg development (specifically by modifying the protein-synthesizing capacity of the fat body). However, until relatively recently, the nature and role of these neurosecretory materials in locusts had proven difficult to determine. By contrast, the corpora allata have been thoroughly investigated, and the nature, role, and mode of action of their product, JH, are well understood in both males and females.

The corpora allata of locusts and grasshoppers produce only one form of JH—JH III ($C_{16}JH$). This sesquiterpenoid molecule is transported to its site of action bound to specific blood proteins. In both sexes, JH has a multifunctional role. In females it stimulates the fat body to synthesize vitellogenins. These large glycoproteins are transported in the blood to the ovary where JH exerts a second major effect—namely, to promote the formation of intercellular channels between the follicular epithelial cells so that the maturing oocyte, now bathed by the blood, can take up the vitellogenins by pinocytosis for incorporation into yolk. Currently, there is much interest in the JH control of vitellogenin production in *Locusta*, using this as a model system for determining how JH acts at the molecular level. A third responsibility for JH in females is to stimulate growth and secretory activity of the lateral oviducts and accessory glands. Not surprisingly, in view of these major effects of JH on the reproductive physiology of the female, the activity of the corpora allata regulates the onset and termination of reproductive diapause in species that exhibit this phenomenon. In addition, in some but not all acridids, JH is required for female sexual behavior.

A major breakthrough in our knowledge of locust neurohormones came at the beginning of the 1990s when the Girardies discovered and characterized a gonadotropic neurohormone produced by the B-type median neurosecretory cells. This 7-kDa peptide, named ovary maturing parsin (Lom OMP), can induce the appearance of vitellogenin in the blood as does JH, though its mode of action is clearly different to and independent of that of JH. Apparently, Lom OMP does not act directly on the fat body but via the ovary where it may trigger the synthesis of ecdysone, which in turn promotes vitellogenesis. Though Hoffmann and colleagues had discovered, in the late 1970s, that the ovarian follicle cells of *Locusta* were a major site of ecdysone synthesis, a role for this hormone in locust vitellogenesis, comparable to that in Diptera had remained in doubt. (In developing eggs of locusts ecdysone is accumulated in large quantity in the yolk and is used gradually to trigger the several molts undertaken by the developing embryo.)

In males JH exerts some parallel effects to those noted in females. In terms of male reproductive physiology, the primary site of action of JH is the accessory glands where it stimulates the synthesis of some major specific secretory proteins. The most notable of these is LHPI, a 72-kDa glycoprotein that makes up about 50% of the total soluble protein in the long hyaline tubule of *Melanoplus*. Paralleling the use of the *Locusta* female fat body–vitellogenin system as a model, the *Melanoplus* male accessory gland system has been used to determine the site and mode of action of JH. A second, though less important site of action for JH, at least in male *M. sanguinipes*, is the fat body, in which the hormone regulates the synthesis of some proteins that subsequently are accumulated by the accessory glands. As in females, JH regulates reproductive diapause in males. The hormone also controls production of the maturation-accelerating pheromone in male *S. gregaria*. The importance of the corpora allata in regulating male sexual behavior varies widely. For example, in *S.*

gregaria and *N. septemfasciata*, the glands are essential for development of mating behavior. An intermediate condition exists in gregarious phase *L. migratoria*, subspecies *migratorioides* (African migratory locust), and in *M. sanguinipes*, in which sexual behavior is considerably delayed and reduced in intensity after removal of the corpora allata. At the opposite extreme, the glands are unimportant in the control of male sexual behavior in the grasshoppers *Gomphocerippus rufus*, *Euthystira brachyptera*, *Syrbula fuscovittata*, and *Chorthippus curtipennis*.

Recently, it has become apparent that ecdysone also has a role in male reproductive biology. In male *Melanoplus* ecdysone is produced in the testes and the accessory glands, where it likely regulates spermatogenesis and the synthesis of specific proteins, respectively.

IV. MATING BEHAVIOR

Mating behavior includes the events that encompass and relate to the act of mating. For convenience, it may be divided into four sequential components—location and recognition of a mate, courtship, copulation, and postcopulatory behavior—though it should be noted that signals used in one phase of mating behavior may be continued as the next phase is begun.

Obviously, in gregarious phase locusts, mate location and recognition presents no difficulty. However, solitarious locusts and grasshoppers utilize a range of mechanisms to ensure that males and females establish contact. In a few species males appear simply to wait for a female to come close enough to be seen and then "ambush" her by leaping onto or close to her. More typically, however, males attract receptive females by auditory or visual cues. Males usually stridulate (produce the "attraction song") by rubbing their hind femora against a specialized region of the wings. In some species the female merely moves toward the male; in others, the female stridulates in response to the male's song to facilitate the approach. Visual attraction includes stationary wing fluttering and display flights. The former is carried out by females, typically with brightly colored wings,

while they rest on the ground, whereas display flights are generally a male behavior. Again, the wings often are brightly colored or have reflecting properties, and in some species sounds are also produced by the wings as they beat or by stridulation after the male has returned to his original perch.

It is difficult to generalize about courtship in acridids because a wide range of strategies are evident both among and within the various subfamilies. Some species appear to have no preliminaries to copulation, with the male simply attempting to jump onto the female, who either accepts or repulses him, the latter typically by vigorous kicking movements of the hindlegs. Males of other species may engage in a "without contact" courtship display that may include specific body language (e.g., movements of the antennae, mouthparts, and legs and wing fluttering) and sound production by stridulation. Typically, the courtship song is highly species specific and may serve to prevent hybridization between two closely related species that occupy the same habitat.

Females are generally passive, though in many species receptive females evert their Comstock–Kellogg glands; however, the significance of this action in terms of mating behavior remains unclear. Conversely, in some species females signal their unwillingness to mate by stridulation or by "drumming" their hindlegs on the ground. The male's courtship activities are usually terminated by his attempting to jump onto the back of the female. However, even if he successfully achieves the mounted position, mating may not immediately follow. Often, further courtship behavior, especially palpation of the female's head and pronotum by the male's mouthparts and/or antennae, is necessary to render a female receptive.

During copulation, sperm is transferred to the female in one or more spermatophores. The length of time spent in copulation, the frequency of copulation, the number of spermatophores produced per copulation, and the complexity and fate of the spermatophore vary widely among species, and it is difficult to see clear trends. For example, in *L. migratoria* copulation takes 6–18 hr, yet only a single spermatophore is produced that extends the full length of the spermathecal duct. The spermatophore remains in the female tract for some time and appears to be slowly digested. The plugging of the female tract with the spermatophore may ensure the successful male's paternity of the next eggs laid. In contrast, mature *M. sanguinipes* mate almost daily and the males transfer, on average, seven spermatophores during each copulation, which lasts from a few minutes to more than 3 hr. However, the spermatophore has only a short anterior tube that reaches to the opening of the spermathecal duct; thus, it is easily expelled from the female prior to formation of the next one.

Little attention has been paid to postcopulatory behavior. Males of some species (e.g., *S. gregaria*) may remain on the female for some time after copulation has terminated, even until egg laying has occurred, and this may be another mechanism for ensuring paternity. In other species the partners separate after copulation (the female may vigorously dislodge the male by kicking with the hindlegs!). However, the female may remain unreceptive to mating for some time, and for one grasshopper species it has been documented that this is due to the mechanical stimulation of the spermatheca by the spermatophore.

V. OVULATION, FERTILIZATION, AND OVIPOSITION

Mature (i.e., chorionated) eggs are released into the lateral oviduct and may be held there for several days if the female is unmated or if suitable egg-laying sites are not available (see below).

The great majority of locusts and grasshoppers lay their eggs in soil, and a good deal of attention has been paid to this aspect of their reproductive biology. The factors that determine the selection of an egg-laying site are many and varied and may be interrelated; they include soil temperature, density and height of vegetation, soil moisture, the physical and chemical properties of the soil, and the occurrence of other laying females (the group-laying effect).

Many field studies have reported that locusts oviposit in soil that is within a "preferred" temperature range; for example, it was observed that in very hot weather, *S. gregaria* females laid eggs in the shade provided by shrubs, and *L. migratoria migratorioides* laid eggs at three different sites throughout the day— dense, tall reeds when the temperature was highest, dense grass at intermediate temperatures, and short, sparse grass at the lowest temperatures. It should be cautioned, however, that these observations may simply be the result of the insects congregating at a preferred temperature, and not specifically a preferred temperature for egg laying.

Prior to egg laying, a female investigates a potential site by tapping the ground with the tip of the abdomen and touching it with her antennae and palps. It is assumed that this behavior provides sensory information on the physical and chemical qualities of the soil. Subsequently, a female carries out probing behavior—that is, digs a series of test holes without actually laying in them, probably to ascertain the qualities of the soil, including its moisture content, at the depth where eggs will be left. Both field and laboratory studies have shown that, generally speaking, acridids prefer to lay in soil that is of medium compactness (i.e., neither too fine nor too coarse), nonsaline, and with a pH in the range 4–10. The preferred moisture content seems to vary widely among species, with some preferring very dry sand, others (including *Schistocerca* and *Locusta*) laying best in soil with a 15% (ml/100 g) moisture content, and yet others selecting soils with a very high water content (e.g., *Chorthippus montanus*, which lives in damp habitats, preferring soil with 43–71% water). Water is required during embryonic development and, therefore, the strategy of ovipositing in dry soil may be an adaptation in species that require a seasonal delay in development.

Though, for the reasons explained previously, females might be expected to concentrate in a suitable area, it has long been known that dense groups of ovipositing locusts form even on quite uniform substrate. Egg pod densities of several thousand per square meter have been recorded, with such numbers being possible (because of the area covered by an individual insect) only as a result of laying in virtually the same spot by successive females. Experimental studies of this group-laying effect indicate that females about to lay are initially visually attracted to ovipositing females; however, a nonvolatile pheromone released by the epidermal glands is also involved, being perceived by tactile contact among females. It is now appreciated that the group-laying effect is not restricted to gregarious phase locusts but occurs also in solitarious forms and in many grasshoppers. Recent studies on *Schistocerca* have shown that high egg pod densities promote the development of both the characteristic behavioral attributes of gregarious hoppers and their typical dark color patterns. Analysis has revealed that the foam plugs of the egg pods of gregarious (but not solitarious) females contain a "gregarizing factor" which predisposes the hatchlings to take on gregarious characteristics. Apparently, the converse is not true; that is, no "solitarization factor" is present in the foam produced by laying solitarious females. The gregarizing factor appears to be a small (<3 kDa), water-soluble molecule that is active for only about 1 day after oviposition.

For solitary locusts and grasshoppers, the significance of the group-laying effect may be equally subtle—namely, that the earliest-laying females may make the site more suitable. The disturbance brought about by their digging (even of test holes), and the possibility that the egg pods may be hygroscopic, causes water in the soil to rise toward the surface, thereby improving egg survivability and the chances of successful embryonic development.

To dig the hole in which the eggs are to be deposited, a female raises its body, then arches its abdomen downwards until it is more or less vertical. Repeated opening and closing movements of the ovipositor valves push the soil outwards and pull the abdomen into the soil. As the hole deepens, the abdomen gradually elongates as a result of the unfolding of the intersegmental membranes linking segments 4–7. These membranes (but not those joining other segments) are extensively folded in normal circumstances but can be stretched considerably during digging (e.g., *Schistocerca* can dig to a depth of 14 cm). To maintain the pressure required for abdominal extension and to compensate for the increase in body volume, the insect swallows air into its anterior gut and increases the volume of its air sacs, respectively.

Periodically while digging, a female rotates her abdomen from side to side to smooth and compact the sides of the hole.

When egg laying per se begins, the eggs leave the lateral oviducts singly and from alternating sides, entering the common oviduct with their micropylar end first. As the micropyles pass the opening of the spermathecal duct, sperm are released (though there is almost no information on the control and coordination of this process in acridids). The oviducal muscles, which are responsible for moving the eggs to the exterior, have been studied extensively with respect to both physiology and pharmacology. These muscles have a myogenic rhythm of contraction and relaxation that can be modified by both hormonal and neural inputs. During egg laying, a neurosecretory factor (oviposition-stimulating hormone) is released from the brain that enhances both the frequency and the amplitude of the contractions. However, the posterior region of the lateral oviducts and the common oviduct also receive motor neurons from the terminal abdominal ganglion (Fig. 1). Stimulation of these neurons produces sustained, strong contractions of the muscles of these regions, thus preventing exit of the eggs. Furthermore, these neurons show bursts of action potentials during digging and if the female is interrupted in the process of oviposition. Thus, it has been proposed that this neural control will prevent egg laying at inappropriate times.

As more eggs fill the hole, the abdomen slowly returns to its original length and is withdrawn. Depending on the species, secretions of the accessory glands and lateral oviducts may or may not be released synchronously with the eggs so that considerable variation in the final structure of the egg pod is seen across the group. After withdrawal of the abdomen, the female typically scrapes soil over the top of the pod using her hindlegs; however, *Melanoplus* spp. use the tip of the abdomen for scraping and *Schistocerca* makes no attempt to cover the egg pod. The entire process of oviposition may take up to 2 hr, of which perhaps only 10% of the time is spent in actual laying of the eggs.

PHASE TRANSFORMATION AND LOCUST PLAGUES

A characteristic of locusts is that they can exist in two extreme forms (phases), the gregarious and solitarious. The gregarious phase is seen under conditions of high population density, whereas the solitarious phase is found in low-density populations. So great are the differences between the two extremes that they were considered originally to be different species [e.g., the solitarious phase of the Asiatic migratory locust (*Locusta migratoria migratoria*) was first named *L. danica*]. The solitarious and gregarious phases differ both in their juvenile and in their adult stages with respect to many aspects of their morphology, behavior, ecology, and physiology (including reproductive biology) (Table 1). Locusts can change from one phase to the other over a number of generations, with locusts of the intervening generations having features that are intermediate between those of gregarious and solitarious forms; such intermediate forms are described as "transient phase" locusts. Thus, the gregarious and solitarious phases may be considered as the extremes of polygenic differences, and all phase states must be regarded as adaptive in nature.

Given the importance of JH in the regulation of insect metamorphosis it is perhaps not surprising that many early studies involving artificial manipulation of the JH titer (e.g., by removing or implanting additional corpora allata) led to claims that JH played a central role in the expression of phase characters. Specifically, due to the presence of greater titers of this hormone, solitarious phase locusts were neotenic compared to the gregarious phase. Recently, critical analysis of these earlier data combined with the use of modern techniques for assaying JH titers and JH biosynthesis by the corpora allata have led to the conclusion that, though there may be differences in the endocrine milieu between the two phases, such differences are not likely to be at the start of the chain of events that leads to phase differentiation.

The "phase theory" for the polymorphic nature of locust species was proposed originally (in 1921) by the late Sir Boris Uvarov in an attempt to explain the origin and disappearance of locust plagues. However, it has become clear over time that changes in phase may occur independently of plagues; that is, plagues represent an extreme and long-term expression of the effects of gregarization and have been suggested to be a type of "ecological opportunism."

The transformation from the solitarious to the gregarious phase occurs as population densities increase, the primary factor for the increase being a lengthy period of suitable conditions (notably adequate rainfall, temperatures, and vegetation) for breeding and juvenile development. Moreover, gregarization is accelerated because the developing juveniles (hoppers) are thought to release a gregarization pheromone which induces the

TABLE 1
Comparison of Some Characters of Gregarious and Solitary Phase Desert Locusts (*Schistocerca gregaria*)

Character	Gregarious Phase	Solitarious Phase
Hopper color	Dark	Pale green
Hopper behavior	Strong tendency to aggregate	Little tendency to aggregate
	Greater tendency to march	Little tendency to march
Adult color	Strong yellow	Pale creamy yellow
Sexual maturation	Faster	Slower
Reproduction	Lower fecundity (fewer eggs/pod, fewer pods/female, greater proportion of nonfunctional ovarioles, smaller overall number of eggs in lifetime)	Higher fecundity (more eggs/pod, more pods/female, lower proportion of nonfunctional ovarioles, greater overall number of eggs in lifetime)

darkening of hoppers, stimulates the aggregation and marching behaviors of hopper bands, and brings about the morphological changes and swarming tendencies seen in adults. Though both hoppers and adults migrate in enormous numbers (estimates of 10^{10} or more individuals per swarm are common) in search of food and breeding areas, covering perhaps thousands of square kilometers, this in itself does not constitute a plague. Plagues may be defined as the occurrence during the same year, and in each of several successive years, of numerous swarms and bands. For example, during the period 1950–1970, there were five plagues of the desert locust, each one lasting from 1 to 5 or 6 years, and the last plague of the African migratory locust extended from 1928 to 1941. That there has been no plague of migratory locusts for more than 50 years appears to be due to increased agricultural development and the presence of locust control organizations in its major outbreak area, the Middle Niger River flood plains in Mali. By contrast, the threat of desert locust plagues remains because this species occupies a much larger range than the African migratory locust. The desert locust does not have permanent outbreak areas; rather, there are many potential sites within its range from which an outbreak may erupt, and much of the desert locust's range has not become agriculturalized and, hence, is less easily monitored.

See Also the Following Articles

ACCESSORY GLANDS, INSECTS; DIAPAUSE; JUVENILE HORMONE; METAMORPHOSIS; OVULATION AND OVIPOSITION, INSECTS; PHEROMONES, INSECT

Bibliography

Chapman, R. F., and Joern, A. (Eds.) (1990). *Biology of Grasshoppers.* Wiley, New York.

Hemming, C. F., and Taylor, T. H. C. (Eds.) (1972). *Proceedings of the International Study Conference on the Current and Future Problems of Acridology, London, UK, 6–16 July, 1970.*

Pener, M. P. (1983). Endocrine research in orthopteran insects, Occasional Papers of the Pan American Acridological Society, No. 1, pp. 1–44.

Pener, M. P. (1986). Endocrine effects on mating behavior in acridids, Proceedings of the 4th Triennial Meeting, Pan American Acridological Society, Saskatoon, Canada, 29 July–2 August, 1985, pp. 9–26.

Uvarov, B. (1966/1977). *Grasshoppers and Locusts: A Handbook of General Acridology,* Vols. 1 and 2. Centre for Overseas Pest Research, London.

Lordosis

Donald W. Pfaff
The Rockefeller University

GLOSSARY

estradiol The most important of a class of hormones called estrogens, secreted in large amounts by the ovaries and in small amounts by the adrenals.

hypothalamus A small region comprising a collection of nerve cell groups at the very bottom of the forebrain, just above the pituitary gland. These cell groups coordinate endocrine controls over the pituitary with behavior and with autonomic physiology.

progesterone A steroid hormone secreted by both the ovaries and the adrenals. Following estrogen priming, this hormone potentiates the effects of estrogen. Given before estrogen priming, it defeats the actions of estrogen.

I. ROLE OF LORDOSIS BEHAVIOR

The broadest definition of lordosis behavior is that it is a standing response coupled with vertebral dorsiflexion. In human terms it would be called "a swayback posture." It would thus be distinguished from scoliosis, which is bending to one side. However, in all four-footed animals, lordosis behavior is the primary mating behavior of the female. Holding still in a rigid standing posture coupled with vertebral dorsiflexion allows penetration by the male, which is essential for fertilization. In a variety of species, male mounting behavior and female lordosis behavior follow a long series of hormone-dependent communications, termed "a hormone-dependent behavioral funnel" (Pfaff, 1998). That is, a long series of communications influenced by sex hormone treatment ensures that only reproductively competent conspecifics will get together.

In a reproductive context, lordosis behavior is performed by females but not by normal males. In terms of mechanism, the main reason for the insensitivity of the adult male animal to the hormones driving lordosis behavior is the failure by the male to respond to progesterone.

II. CONTROL OF LORDOSIS BY STEROID HORMONES

Lordosis behavior by the female in response to adequate cutaneous stimuli by the male depends on a long priming period with estrogens, followed by progesterone. There is an impressive collection of estrogen-binding neurons in the hypothalamus in the females of all vertebrate species including humans. Especially important for lordosis behavior is the binding of estradiol by neurons in a restricted hypothalamic cell group, termed the ventromedial nucleus of the hypothalamus. Turning on genes in these neurons and subsequently turning on electrical activity is the way in which estrogens facilitate lordosis behavior.

III. CONTROL OF LORDOSIS BY GONADOTROPIN-RELEASING HORMONE

Gonadotropin-releasing hormone (GnRH) is a small fragment of a protein manufactured at the bottom of the forebrain, just in front of the hypothalamus. It is a sequence of only 10 amino acids, but, as shown by Knobil and collaborators, pulsatile release of GnRH is essential for the proper operation of the pituitary, which in turn controls the testes and the ovaries.

In still another indication of the "unity of the body," GnRH, which controls reproduction through the pituitary, also fosters reproductive behavior. Following migration from the nasal epithelium during development, GnRH neurons take up residence in the preoptic area and the anterior hypothalamus. While some of their axons go to the small protrusion from the hypothalamus leading to the pituitary stalk, other axons lead to other hypothalamic regions and even back to the midbrain. These other projections, to other neurons, underlie the ability of GnRH to foster lordosis behavior. Importantly, the fact that one compound controls both endocrine and behavioral events helps to ensure that reproductive behavior in such animals will occur only when the peripheral preparations for reproduction are indeed complete.

IV. NEURAL CIRCUITRY CONTROLLING LORDOSIS

The neural circuit for lordosis behavior was the first discovered for any mammalian behavior. Following cutaneous stimuli by the male on the female, electrical activity in the female is carried by nerves over several dorsal roots into the lumbar and sacral regions of the spinal cord. However, spinal cord tissue by itself is not sufficient for the behavior. The ascending fibers of the obligatory supraspinal loop go up specific columns of the spinal cord to the hindbrain and the midbrain. There, the ascending signals join with hormone-dependent signals coming back from the hypothalamus. That is, for the circuit to work properly, a facilitation from hypothalamic neurons as a result of estrogen and progesterone action on those neurons is necessary. Together, the ascending sensory signals and the hormonal signals synergize to send the "enabling" signal for lordosis behavior back to the hindbrain. There, reticulospinal and vestibulospinal axons go back down the anterolateral columns of the spinal cord to lumbar and sacral levels. At that point the facilitatory descending signal synergizes with the sensory input coming in from the male in order to enable motor neurons in the ventromedial corner of the ventral horn to activate deep back muscles. Contraction of these deep back muscles leads to the "swayback posture", the vertebral dorsiflexion of lordosis behavior. As a result of the lordosis posture, the male can penetrate and fertilize.

V. GENETIC INFLUENCES

It is very clear that normal expression of specific genes in hypothalamic neurons is necessary for the normal performance of lordosis behavior. First, if the gene for the classical estrogen receptor is knocked out, lordosis behavior will not occur. Second, if the gene for the classical progesterone receptor is knocked out, progesterone facilitation will not occur. Third, if the gene for GnRH is damaged, estrous cycles will not occur and thus the hormonal preparations for lordosis behavior are absent. These can be restored by implantation of normal GnRH neurons into the hypothalamus. Finally, even those genetic abnormalities leading to a blocked migration of developing GnRH neurons into the brain can influence adult reproductive behavior.

Thus, genetic influences on this complicated social behavior have been proven beyond doubt. On the other hand, in every case, a single gene product does not work by itself. It cooperates with other gene products and with stimuli coming in from the environment to determine lordosis behavior.

See Also the Following Articles

Copulation, Mammals; Estrogen Action, Behavior; GnRH (Gonadotropin-Releasing Hormone); Mating Behaviors, Mammals; Progesterone Actions, Behavior

Bibliography

Ogawa, S., Taylor, J., Lubahn, D. B., Korach, K. S., and Pfaff, D. W. (1996). Reversal of sex roles in genetic female mice by disruption of estrogen receptor gene. *Neuroendocrinology* 64, 467–470.

Pfaff, D. W. (1998). *Drive: Neurobiological and Molecular Un-* derpinnings of Sexual Motivation in Humans and Other Animals. MIT Press, Cambridge.

Pfaff, D. W., Schwartz-Giblin, S., McCarthy, M. M., and Kow, L.-M. (1994). Cellular and molecular mechanisms of female reproductive behaviors. In *The Physiology of Reproduction* (E. Knobil and J. D. Neill, Eds.), 2nd ed., pp. 107–220. Raven Press, New York.

Low Birth Weight

see Intrauterine Growth Restriction

Luteinization

Anthony J. Zeleznik

University of Pittsburgh School of Medicine

I. Structural Changes Associated with Luteinization
II. Hormonal Control of Luteinization
III. Functional Changes Associated with Luteinization
IV. Molecular Aspects of Luteinization
V. Summary

GLOSSARY

angiogenesis The growth of new blood vessels.

gonadotropins Protein hormones of pituitary and placental origin which stimulate the gonads.

granulosa cells Steroid-producing cells of the ovarian follicle which lie inside the basement membrane of the follicle.

luteinization The process by which the granulosa and theca cells undergo structural and functional remodeling to form the corpus luteum.

theca cells Steroid-producing cells of the ovarian follicle which lie immediately outside the basement membrane of the follicle.

L uteinization involves the structural and functional transition of the ovarian follicle into the corpus luteum that occurs following ovulation. The process of luteinization is associated with a dramatic remodeling of the ovary that is evident at the cellular level and in the architecture of the tissue itself. The principal endocrine change that accompanies luteinization is a switch from the ovary's production of estrogen during the follicular (proliferative) phase of the menstrual cycle to the production of progesterone during the luteal (secretory) phase, a transition that is obligatory for successful implantation of the fertilized oocyte in the uterus and its continued postimplantation development. The goal of this article is to summarize extant information regarding the endocrine and cellular control mechanisms that govern the structural and functional changes that occur during the process of luteinization.

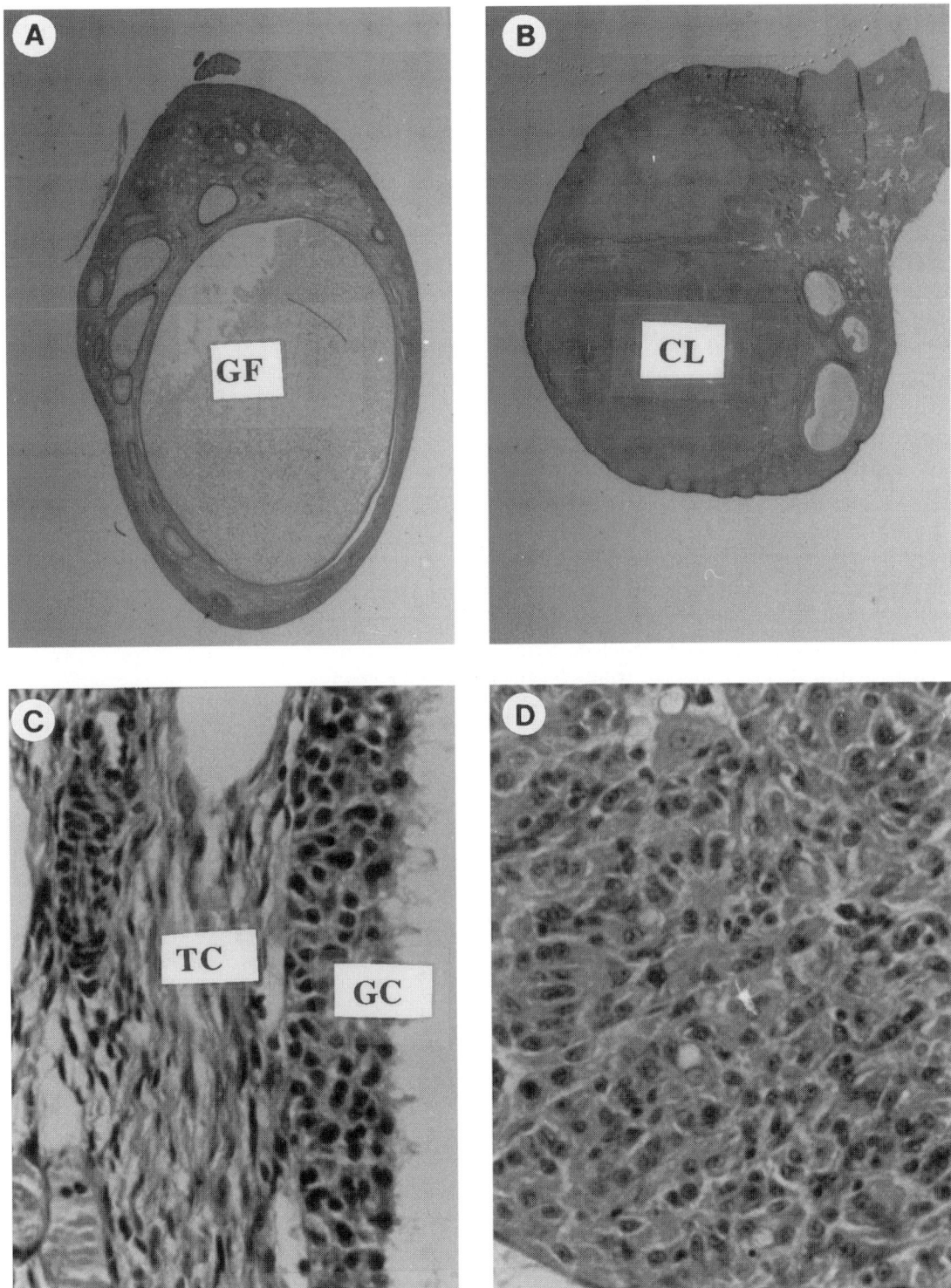

FIGURE 1 Macroscopic and microscopic anatomy of a monkey ovary. (A and C) Low-power (20×) and high-power (400×) views of a monkey ovary obtained prior to ovulation and luteinization. (B and D) low-power (20×) and high-power (400×) views of a monkey ovary collected after ovulation and luteinization. Abbreviations: GF, Graffian follicle; CL, corpus luteum; GC, granulosa cell; TC, theca cell.

I. STRUCTURAL CHANGES ASSOCIATED WITH LUTEINIZATION

Figure 1 illustrates macroscopic and microscopic changes that occur during luteinization. Figure 1A shows a low-power photomicrograph of an ovary obtained from a macaque monkey during the late follicular phase of the menstrual cycle, whereas Fig. 1B shows a low-power view of an ovary removed during the luteal phase. On a macroscopic level, the most notable changes associated with luteinization are the disappearance of the follicular antrum (Fig. 1A) and the physical expansion of the thin layer of follicular granulosa cells that borders the antral cavity into the compact mass of cells that comprises the corpus luteum (Fig. 1B). On a microscopic level, the increase in the volume of the corpus luteum is due principally to a hypertrophy of the granulosa cell layer. In the rhesus monkey, granulosa cells collected from preovulatory follicles are approximately 10 μm in diameter with a low cytoplasmic to nuclear ratio (Fig. 1C), whereas luteal cells collected from newly formed corpora lutea are approximately 25 μm in diameter with a greater nuclear to cytoplasmic ratio than that of granulosa cells (Fig. 1D). At the ultrastructural level, luteinization is accompanied by cellular remodeling and is associated with increased amounts of smooth endoplasmic reticulum, well-formed Golgi apparatus, and numerous mitochondria with tubular cristae. The structural changes associated with luteinization are similar among many mammalian species. To date, little is known regarding the mechanisms by which this dramatic cellular remodeling occurs during the process of luteinization.

In addition to the remodeling of granulosa cells, a dramatic change in the vasculature of the ovary occurs during luteinization. Granulosa cells are completely avascular because the capillary network that supplies the maturing follicle abruptly ends at the basement membrane that separates the theca interna cells from the granulosa cells (Fig. 1C). In contrast, after ovulation blood vessels sprout from the thecal layer and penetrate the luteinizing granulosa cells such that an extensive capillary network is formed within the corpus luteum (Fig. 1D) and by the completion of the process of luteinization, the corpus luteum becomes one of the most highly vascularized tissues in the body. The notion that the luteinizing follicle secretes substances capable of stimulating capillary growth (angiogenic factors) was convincingly demonstrated by experiments which demonstrated that luteinizing follicles or extracts of luteinizing follicles stimulated the proliferation of blood vessels in a variety of angiogenesis assays including rabbit cornea and the chicken chorioallantoic membrane. A likely candidate for the angiogenic factor(s) responsible for the neovascularization of the corpus luteum is a family of heparin-binding endothelial cell growth factors named vascular endothelial growth factor (VEGF) and vascular permeability factor (VPF). Members of this protein family affect both capillary proliferation and vascular permeability. mRNA for VEGF was shown the be present in the rat corpus luteum and thereafter studies in cynomolgus monkeys revealed selective expression of mRNA for VEGF in both the preovulatory follicle and the corpus luteum.

II. HORMONAL CONTROL OF LUTEINIZATION

The process of luteinization is temporally coordinated with the resumption of meiosis of the developing oocyte as well as its release from the follicle during ovulation. The number of follicles which ovulate is species specific and depends on the endocrine and cellular mechanisms that govern follicle selection. Those follicles which are "selected" to ovulate during the follicular phase are those which have been stimulated appropriately by follicle-stimulating hormone (FSH). A major difference between FSH-stimulated follicles which are destined to ovulate and form corpora lutea and lesser mature follicles which will not ovulate is the presence of cell surface receptors for luteinizing hormone (LH) on their granulosa cells. The presence of LH receptors on the granulosa cells of mature follicles endow these follicles with the ability to respond to the midcycle elevation of LH that coordinately triggers oocyte maturation, ovulation, and luteinization. Follicles which have not undergone FSH-dependent maturation and whose granulosa cells do not posess LH receptors are incapable of ovulating and forming corpora lutea in response to the midcycle elevation in LH.

A primary intracellular signaling system activated by LH in the preovulatory follicle is the stimulation of adenylyl cyclase and the resultant generation of increased intracellular concentrations cAMP. A role for LH and cAMP in luteinization was convincingly demonstrated by studies which involved dissecting preovulatory follicles from ovaries, incubating the isolated follicles *in vitro* in various test agents, and transferring the follicles beneath the kidney capsules of recipient animals. With this model system, the number of follicles which formed ectopic corpora lutea could be directly counted and these ectopic corpora lutea could be dissected from under the kidney capsules and assessed functionally and histologically to verify luteinization. Exposure of isolated follicles *in vitro* to either LH or cAMP analogs resulted in a full manifestation of luteinization including cellular hypertrophy, neovascularization, and enhanced steroidogenesis.

While it is certain that LH and cAMP are primary stimuli associated with luteinization, the secondary pathway(s) involved in the cascade of events that are involved in the synchronization of oocyte maturation, ovulation, and luteinization appears to be more complex. Extant information indicates that other signaling molecules, including prostaglandins and ovarian steroids, are involved in these processes. Moreover, it now appears that the control mechanisms that are responsible for oocyte maturation, ovulation, and luteinization can be functionally separated. For example, the prostaglandin synthase inhibitor indomethacin inhibits ovulation but does not inhibit luteinization of granulosa cells.

In a number of species, including subhuman primates, inhibition of progesterone synthesis by blocking the conversion of pregnenolone to progesterone results in an inhibition of both ovulation and luteinization and that administration of exogenous progestins overcomes the ovulatory defect. Identical conclusions regarding the role of progesterone in ovulation and luteinization were obtained from studies using mice in which the progesterone receptor was rendered nonfunctional by genetic mutations, i.e., the progesterone receptor knockout mouse. Similarly, female mice bearing a null mutation of the estrogen receptor also fail to ovulate and form corpora lutea. The precise roles of prostaglandins and ovarian steroids in the process of luteinization are currently unknown.

Granulosa cells taken from mature preovulatory follicles and placed in tissue culture undergo spontaneous luteinization as reflected by their production of progesterone. In contrast, granulosa cells taken from less mature follicles failed to undergo spontaneous luteinization. These findings suggest that the local environment within the antral fluids of mature follicles may suppress luteinization and that removal of the granulosa cells from this suppressive environment permits them to undergo luteinization. Extending this notion to the luteinization process *in vivo* suggests that the dispersal of follicular fluid and its contents that occurs following the rupture of the follicle at ovulation may remove a local inhibitor and permit the granulosa cells to undergo luteinization. However, the finding that luteinization can occur without ovulation and release of follicular fluid and the oocyte (i.e., in indomethacin-treated animals) would appear to argue against the hypothesis of local inhibition of luteinization unless atresia of the follicle-enclosed oocyte precedes luteinization of the unruptured follicle, which currently is not known.

III. FUNCTIONAL CHANGES ASSOCIATED WITH LUTEINIZATION

A. Steroidogenesis

The pattern of plasma concentrations of FSH, LH, estrogen, and progesterone in humans during transition between the follicular phase and the luteal phase of the menstrual cycle is illustrated in Fig. 2. The principal ovarian steroid produced during the follicular phase of the menstrual cycle is estradiol, the production of which requires the participation of both theca cells and granulosa cells of the follicle. Theca cells, under the control of LH, produce the aromatizable androgens testosterone and androstenedione but lack the aromatase enzyme complex necessary to convert aromatizable androgens to estrogen. Gran-

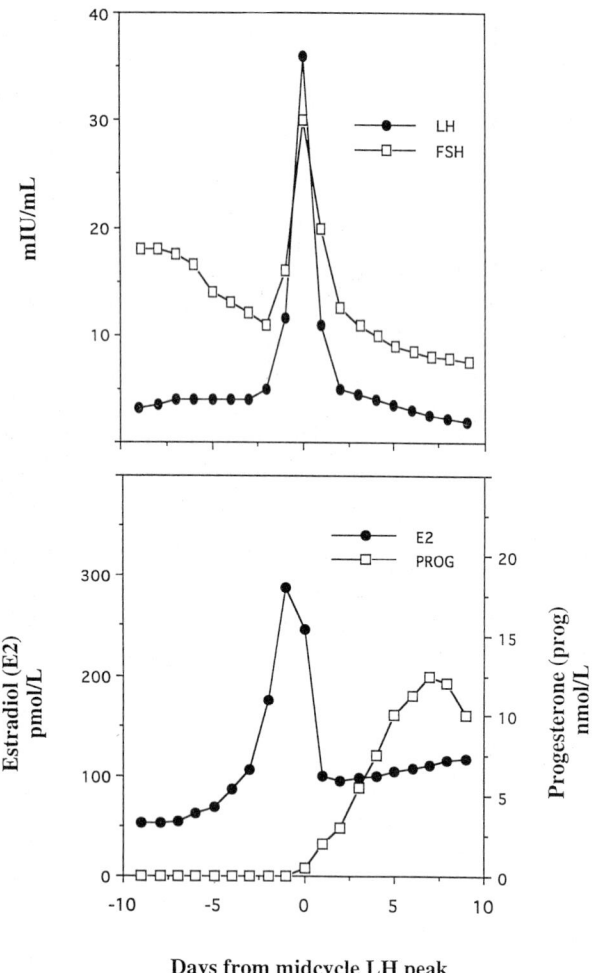

FIGURE 2 Serum gonadotropin (FSH and LH) and ovarian steroid (estradiol and progesterone) concentrations during the human menstrual cycle. Day 0 represents the time at which luteinization begins.

ulosa cells, like theca cells, are able to metabolize cholesterol to progesterone but, unlike theca cells, lack the enzyme necessary to metabolize progesterone to aromatizable androgens (17α-hydroxylase, 17-20 lyase; P45017α). Following stimulation by FSH, granulosa cells acquire the aromatase enzyme required to convert aromatizable androgens to estrogens. Hence, production of estrogen by the maturing follicle requires the participation of both theca cells and granulosa cells as well as their respective regulation by LH and FSH—the two cell–two gonadotropin model for estrogen production.

The abrupt rise in plasma LH concentrations at midcycle is necessary for ovulation of the follicle and its luteinization. Shortly after the midcycle gonadotropin surge, estrogen secretion declines precipitously. The decline in estrogen secretion during this period is seen in most, if not all, mammalian species and is thought to be due principally to a decline in the activity of P45017α and the resultant loss in available testosterone and androstenedione as substrate for aromatase. Following ovulation and luteinization, plasma concentrations of progesterone begin to increase and becomes the predominant ovarian steroid produced during the luteal phase. At its zenith during the midluteal phase of the menstrual cycle, the human corpus luteum produces approximately 25 mg progesterone per day.

Although serum progesterone concentrations do not rise appreciably before ovulation and luteinization, the acquisition of the ability of the luteal cell to produce progesterone actually occurs during the latter stages of follicular development by the stimulatory actions of FSH on mRNA synthesis for the enzymes involved in the conversion of cholesterol to progesterone; 3β-hydroxysteroid dehydrogenase $\Delta^{D5/E4}$ isomerase (3β HSD) and cytochrome P450 cholesterol side-chain cleavage enzyme (P450scc). The midcycle gonadotropin surge results in a further increase in 3β HSD and P450scc in the luteinizing granulosa cells. In addition, the process of luteinization is associated with an increase in receptors for low-density lipoproteins which facilitate the intracellular transport of cholesterol that serves as substrate for progesterone production. The production of estrogen by the corpus luteum differs between species. The corpus luteum of humans and subhuman primates produces estradiol during the luteal phase, whereas the corpus luteum of other species such as the rat and sheep does not produce appreciable amounts of estrogen. In humans and subhuman primates, it appears that the cellular compartmentalization of P45017α and cytochrome P450 aromatase (P450arom) as immunocytochemical localization of P45017α is restricted to the outer boundaries of the corpus luteum occupied by the theca lutein cells, whereas aromatase-expressing cells are localized within the area of the corpus luteum occupied by granulosa lutein cells.

An interesting paradox is that *in vitro* studies of luteal cells removed from human corpora lutea throughout the luteal phase have shown that progesterone secretion by luteal cells collected during the early luteal phase produce equivalent or greater amounts of progesterone than do cells isolated during the midluteal phase of the menstrual cycle. However, as shown in Fig. 2, progesterone concentrations in plasma during the early luteal phase are substantially lower than those seen during the midluteal phase. The best explanation for the discrepancy between *in vitro* biosynthetic capacity of luteal cells for progesterone and actual progesterone production *in vivo* is that the luteal cells present in early corpora lutea are substrate limited due to incomplete vascularization and hence there is insufficient delivery of low-density lipoproteins which serve as the major source of substrate (cholesterol) for progesterone biosynthesis. After appropriate vascularization of the corpus luteum occurs, cholesterol delivery no longer is limiting and maximal progesterone production occurs.

B. Gonadotropic Control

While both FSH and LH are required for the maturation of the preovulatory follicle and its production of estrogen, the corpus luteum loses its requirement for FSH (as a result of the FSH-mediated induction of LH receptors on granulosa cells) and is maintained by LH. The recent identification of the steroidogenic acute regulatory protein (StAR) and the elucidation of its role in facilitating the entry of cholesterol to the inner mitochondrial membrane provides a likely intracellular target of the LH, cAMP, and protein kinase A (PKA) pathway for the acute regulation of progesterone production by the corpus luteum.

IV. MOLECULAR ASPECTS OF LUTEINIZATION

As noted previously, the transition of the ovarian follicle into a corpus luteum involves changes in its macroscopic, microscopic, functional (steroidogenic), and regulatory properties which endow the corpus luteum with the capacity to produce the large quantities of progesterone which are essential for successful implantation and maintenance of pregnancy. In addition to these changes which impact on the functional activity of the corpus luteum, the process of luteinization is associated with the terminal differentiation of luteal cells and the exit of the granulosa–lutein cells from the cell cycle. In the developing follicle, FSH results in both proliferation and differentiation of granulosa cells. However, following luteinization, luteal cells produce steroid hormones under the influence of LH but their ability to proliferate in response to gonadotropic stimulation is lost. An enigma in understanding the regulation of granulosa cell and luteal cell function is why granulosa cells proliferate in response to FSH but luteal cells fail to proliferate in response to LH despite the fact that both FSH and LH, at least in part, act by activating the cAMP–PKA intracellular signaling system.

The contribution of the cAMP–PKA signaling system to ovarian cellular function is illustrated in Fig. 3. Intracellular accumulation of cAMP in response to FSH or LH binding to their receptors results in the activation of PKA by causing the dissociation of the regulatory subunit of PKA from its catalytic subunit. The active catalytic subunit then is able to phosphorylate cytoplasmic proteins which cause a rapid stimulation of steroidogenesis in ovarian cells by facilitating the entry of cholesterol to the inner mitochondrial membrane via the recently identified StAR protein. In addition to this rapid action on steroidogenesis, which occurs within minutes, the catalytic subunit of PKA may also regulate long-term nuclear events following its translocation to the nucleus and its phosphorylation of PKA-dependent nuclear transcription factors such as CREB (cAMP *re*sponse *e*lement *b*inding protein), CREM (cAMP *r*esponse *e*lement *m*odulating protein), and ATF-1 (activating transcription factor-1). The acute actions of cAMP on the control of steroidogenesis may be separated from the long-term actions on nuclear transcription, the latter of which are mediated by activated PKA which results in the phosphorylation of nuclear regulatory proteins. Studies in macaque monkeys have revealed that the expression of the cAMP and PKA-dependent nuclear transcription factor CREB is lost following luteinization. Studies in

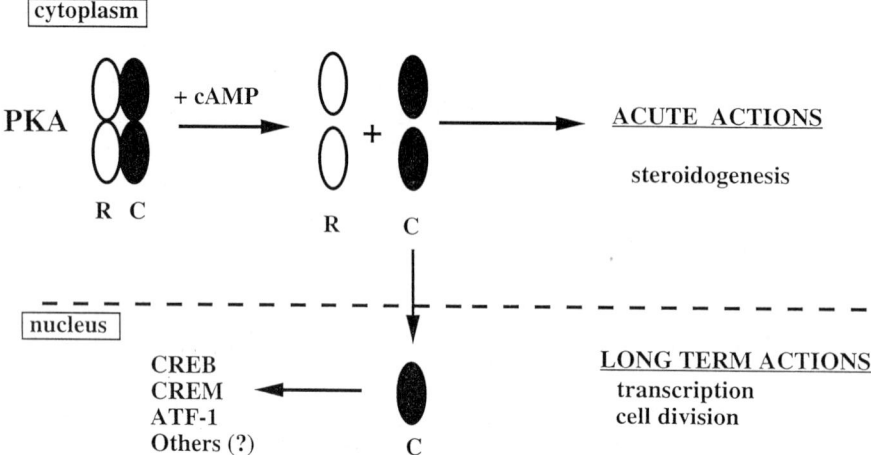

FIGURE 3 Model of a cAMP signaling system in the primate ovary. Upon elevation of intracellular cAMP levels following stimulation of cells by FSH or LH, the regulatory (R) and catalytic (C) subunits of protein kinase A (PKA) dissociate from one another, which results in the activation of the catalytic subunit. Within the cytoplasm of the cell, the activated C subunits phosphorylate cytoplasmic proteins which lead to the rapid mobilization of cholesterol to the inner mitochondria of cells thereby initiating the production of progesterone. Activated C subunits may also be translocated to the nucleus and phosphorylate nuclear transcription factors CREB, CREM, ATF-1, and others, which in turn may activate gene transcription leading to changes in cellular protein levels and cell proliferation.

other cell systems have shown that experimental manipulations interfering with the ability of CREB to function as an activated transcription factor leads to defects in cellular proliferation. Thus, the loss of CREB or other cAMP-dependent nuclear signaling pathways could constitute a mechanism which results in the cessation of gonadotropin (and cAMP)-stimulated proliferation following luteinization. The still operative cAMP–PKA signaling pathway in the cytoplasm would continue to function to allow rapid production of progesterone in response to LH stimulation by facilitating cholesterol delivery to the inner mitochondrial membrane.

Potential candidate molecules involved in cell cycle regulation whose activity (in some cell types) appears to be controlled by cAMP and/or CREB include proliferating cell nuclear antigen (PCNA) and cyclin D2. PCNA is an accessory protein to DNA polymerase-Δ; and is essential for both replicative and repair DNA synthesis. Recent studies have shown that the promoter region of the human PCNA gene contains a CREB-dependent regulatory site. Cyclin D2, which is essential for the progression of cells through the G_1 phase of the cell cycle, is regu-

lated in granulosa cells by the cAMP–PKA signaling pathway. However, it is not known if the expression of cyclin D2 in the ovary is regulated directly by CREB or other cAMP-responsive transcription factors.

V. SUMMARY

The process of luteinization is accompanied by changes in the morphology, the function, the tissue architecture, as well as the cell cycle of the growing preovulatory follicle as it is transformed into the terminally differentiated corpus luteum. As such, this process provides an excellent model to investigate the endocrine and cellular signaling systems that control cellular proliferation and differentiation.

See Also the Following Articles

Corpus Luteum; Granulosa Cells; Steroidogenesis; Theca Cells

Bibliography

Benyo, D. F., and Zeleznik , A. J. (1997). cAMP signaling in the primate corpus luteum: Maintenance of protein kinase A activity throughout the luteal phase of the menstrual cycle. *Endocrinology*, in press.

Corner, G. W. (1974). The early history of progesterone. *Gynecol. Invest.* **5**, 106–112.

Espey L. L., and Lipner, H. (1994). Ovulation. In *The Physiology of Reproduction* (E. Knobil and J. D. Neill, Eds.), 2nd ed., Vol. 1, pp. 725–780. Raven Press, New York.

Koos R. D. (1993). Ovarian angiogenesis. In *The Ovary* (E. Y. Adashi and P. C. K. Leung, Eds.), pp. 433–453. Raven Press, New York.

Meyer, T. E., and Habener, J. F. (1993). Cyclic adenosine 3′,5′-monophosphate response element binding protein (CREB) and related transcription-activating deoxyribonucleic acid-binding proteins. *Endocr. Rev.* **14**, 269–290.

Richards J. S. (1994). Hormonal control of gene expression in the ovary. *Endocr. Rev.* **15**, 725–751.

Stocco, D. M., and Clark, B. J. (1996). Regulation of the acute production of steroids in steroidogenic cells. *Endocr. Rev.* **17**, 221–244.

Tsafriri, A., and Adashi, E. Y. (1994). Local nonsteroidogenic regulators of ovarian function. In *The Physiology of Reproduction* (E. Knobil and J. D. Neill, Eds.), 2nd ed., Vol. 1, pp. 817–860. Raven Press, New York.

Zeleznik, A. J., and Benyo, D. F. (1994). Control of follicular development, corpus luteum function, and the recognition of pregnancy in higher primates. In *The Physiology of Reproduction* (E. Knobil and J. D. Neill, Eds.), 2nd ed., Vol. 1, pp. 751–782. Raven Press, New York.

Luteolysis

John A. McCracken

Worcester Foundation for Biomedical Research

GLOSSARY

corpus luteum The yellow-colored tissue which fills the cavity of the ruptured ovulatory follicle(s). It was named by early anatomists from the Latin *corpus* (body) and *luteum* (yellow).

estrous cycle The reproductive cycle of nonprimates initiated by estrogen produced by the ovary, culminating in estrus, a time of acceptance of the male (from Greek *oistros* = to excite).

follicular phase The initial stage of the reproductive cycle arising from the growth of preovulatory follicles and the secretion of estrogen.

functional luteolysis The decline of progesterone secretion by the corpus luteum which is the first sign of impending luteolysis.

luteal phase The postovulatory stage of the reproductive cycle due to the development of corpora lutea and the secretion of progesterone.

luteolysis The involution of the corpus luteum which defines the end of the luteal phase of the cycle as well as, in some mammals, the end of pregnancy or pseudopregnancy. Also, *corpus luteum regression*, *corpus luteum demise*, and *corpus luteum involution*.

menstrual cycle The reproductive cycle of Old World primates characterized by monthly shedding of the endometrium (lining of the uterus) resulting in a period of vaginal bleeding (from Latin *mensa* = a month).

progesterone The major steroid hormone secreted by the corpus luteum which is essential for the maintenance of pregnancy (from Latin *pro* = for and *gestio* = pregnancy).

structural luteolysis Later involution of the corpus luteum in the ovary to form a small white connective tissue scar called the corpus albicans (from Latin *corpus* = body and *alba* = white).

Luteolysis is the process by which the corpus luteum (in monovulatory species) or corpora lutea (in polyovulatory species) undergo regression. The corpus luteum forms in the ovary from the cells lining the ovulatory follicle(s) after ovulation has occurred under the influence of luteinizing hormone. The corpus luteum is an unusual structure because of its transient existence as an endocrine gland. The only other structures which function in such a temporary manner are the preovulatory follicle and the placenta, although the latter structure has a much longer functional life span. The main secretory product of the corpus luteum is the steroid hormone progesterone, which maintains pregnancy in mammals by inducing uterine quiescence and by suppressing maternal immune responses to fetal antigens. In addition, progesterone influences tubal transport of fertilized ova and creates a uterine milieu which supports blastocyst development and implantation. Progesterone also reduces cyclic ovarian activity in most mammals during pregnancy and contributes to mammary development. If conception does not occur in species with a functional luteal phase (which includes primates, domestic animals, and the guinea pig), the corpus luteum undergoes luteolysis (involution or regression) so that a new ovarian cycle can be initiated and, hence, another opportunity for conception is achieved.

I. INTRODUCTION

The term "luteolysis" is something of a misnomer since lysis of the cells of the corpus luteum does not occur, at least not during the early stages of luteolysis when progesterone secretion from the corpus luteum diminishes. Such a decline in progesterone production is termed "functional luteolysis" as differentiated from "structural or morphological luteolysis" which, as the name suggests, signifies the more gradual change in the cellular composition of the gland and its involution to form the corpus albicans, a small residual structure in the ovary composed of connective tissue and collagen. The relative occurrence of luteolysis in different types of mammalian reproductive cycles is shown in Fig. 1. Each type of cycle is described in more detail in the following sections.

A. Primates

In primates, ovulation and formation of the corpus luteum occur under the influence of gonadotropic hormones. Toward the end of the luteal phase of the cycle the corpus luteum, which produces both progesterone and estrogen, undergoes luteolysis, thus terminating the cycle. Luteolysis in primates is not dependent on the presence of the uterus and is currently considered to involve an as yet unidentified intraovarian mechanism. However, in nonprimates, luteolysis is dependent in the presence of the uterus. In Old World primates, the decline in progesterone due to luteolysis causes shedding of the endometrium which results in menstruation, an external sign of completion of the cycle. Menstruation occurs only in Old World primates including humans since, unlike New World primates, a spiral arteriolar complex exists in the endometrium, the contraction of which induces menstruation. In nonhuman primates, the midcycle preovulatory increase in estrogen induces an increase in sexual receptivity (estrus), whereas in the human female, there is no manifestation of any increase in sexual receptivity at this time. Thus, nonhuman primates exhibit a transitional form of both estrous and menstrual cycles. When a fertile mating occurs in primates, luteolysis is subverted and the life span of the corpus luteum is extended by chorionic gonadotropin secreted by the implanting blastocyst. Pregnancy is maintained initially by progesterone from the corpus luteum until the placenta assumes this function (the luteoplacental shift).

B. Domestic Animals

As in primates, ovulation and corpus luteum formation are initiated by pituitary gonadotropins. During a relatively short follicular phase, the preovulatory rise in estrogen induces a brief period of sexual receptivity (estrus) and induces the preovulatory surge of luteinizing hormone (LH). Thus, the time of mating is synchronized with the time of ovulation. If the cycle is infertile, the corpus luteum undergoes luteolysis which terminates the luteal phase and a new ovarian cycle is initiated. If a fertile mating occurs, the luteolytic signal emanating from the uterus is abrogated and the corpus luteum continues to

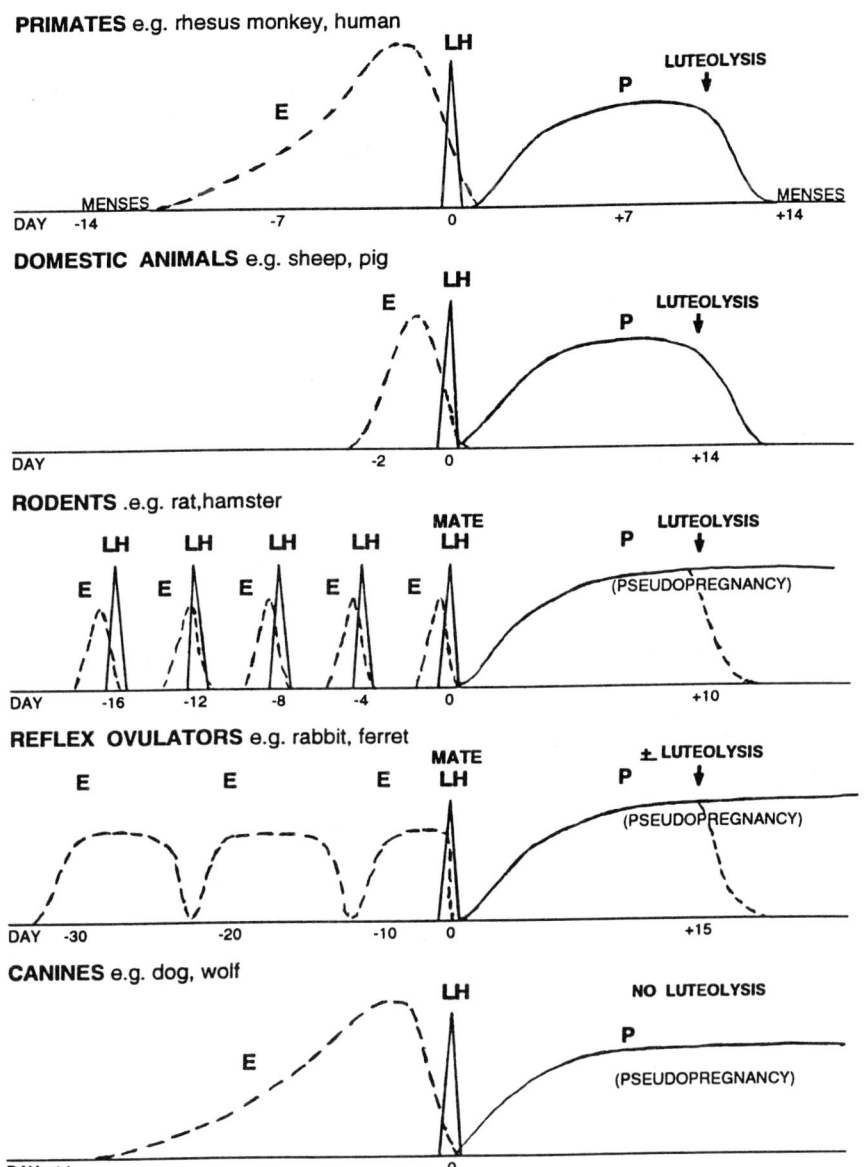

FIGURE 1 The role of luteolysis in controlling the life span of the corpus luteum during the cycle or during pseudopregnancy in different species (see text for details) (from McCracken et al., 1998).

secrete progesterone for pregnancy maintenance. In some species, e.g., goat and pig, the corpora lutea remain the sole source of progesterone throughout pregnancy. The initiation of labor (parturition) in some species, such as the goat, may involve a luteolytic mechanism. In other species, such as the horse and sheep, the placenta later becomes the principal source of progesterone for pregnancy maintenance. The guinea pig is included in this category because its reproductive cycle is very similar to that of domestic animals.

C. Rodents

In rodents (e.g., rat, mouse, and hamster), corpora lutea form after ovulation but require a mating stimulus to become functional under the influence of pituitary prolactin. If the mating is fertile, corpora lutea persist throughout pregnancy (22 days), but if mating is infertile the corpora lutea become temporarily functional, i.e., pseudopregnancy. However, most rodents develop a luteolytic mechanism which curtails the length of pseudopregnancy. By not spontaneously developing a functional luteal phase after ovulation unless they are mated, rodents are reproductively efficient since they have an opportunity to conceive every 4 days. In unmated rodents, corpora lutea from several 4- or 5-day estrous cycles may accumulate in the ovary and may contribute progesterone for the maintenance of pregnancy or pseudopregnancy after a mating stimulus.

D. Induced Ovulators

The induced, or reflex, ovulatory species, such as the rabbit, the cat, the mink, the ferret, and the camel, have waves of follicular development during which there is a relatively long period of sexual receptivity. A mating stimulus is required for the neurogenic release of LH and, thus, the induction of ovulation and corpus luteum development. The corpora lutea are maintained in both fertile (pregnancy) and infertile matings (pseudopregnancy). The latter may last as long as a normal pregnancy (e.g., ferret) or be terminated by a luteolytic mechanism (e.g., rabbit). The induced ovulators are reproductively quite efficient since ovulation and conception are achieved only by a mating stimulus.

E. Canine Cycle

Canines (e.g., dog and wolf) exhibit only about two ovulatory cycles per year and ovulation and corpora lutea are formed spontaneously under gonadotropic control. The corpora lutea secrete progesterone in both fertile and infertile matings, and in the latter case (pseudopregnancy), progesterone production persists for a period approximating that of pregnancy. It appears that a luteolytic mechanism has not developed in these animals to curtail the life span of corpora lutea when pregnancy is not established and no signal is required from the embryo to extend the life span of the corpora lutea. These species are thus not very efficient reproductively since they do not have repetitive cycles and only have a chance to conceive about twice per year.

II. IDENTIFICATION OF PROSTAGLANDIN F$_{2\alpha}$ AS THE MAMMALIAN LUTEOLYSIN

The first indication that the uterus played a role in corpus luteum regression came from the work of Leo Loeb, who showed that extirpation of the uterus in guinea pigs prevented regression of corpora lutea at the end of the cycle. A similar persistence of corpora lutea after hysterectomy was reported in most of the domestic animals and in a variety of pseudopregnant rodents and induced ovulators. These species, unlike primates, all have a bicornate uterus. In most of them, the absence of one uterine horn (either congenitally or surgically) results in the persistence of the corpora lutea only in the ovary adjacent to the absent uterine horn. However, in primates, congenital absence of the uterus or its surgical removal has no effect on the life span of corpora lutea. The failure of corpora lutea to regress normally at the end of the luteal phase following hysterectomy in nonprimates indicated that the uterus produced a factor which caused luteolysis at the end of the cycle.

Selective ligation of ovarian and uterine blood vessels of the ovary and uterus indicated that, in many of these species, vascular connections were essential for the local transport of the putative luteolysin from the uterus to the ovary. Moreover, in the sheep, transplantation of one ovary with vascular anastomoses to a jugulocarotid skin loop, with the uterus remaining *in situ* in the abdomen, prevented luteolysis. Transplantation of one ovary and its contiguous uterine horn to a jugulocarotid skin loop resulted in regular cyclical regression of the corpus luteum. These findings proved unequivocally that the luteolytic factor from the uterus in the sheep acted locally on the ovary and could not be transmitted via the systemic circulation. Possible candidates for such a luteolytic factor were the class of substances called prostaglandins, acidic lipids originally isolated from seminal plasma of sheep and several other species. Two of these new substances, prostaglandin (PG) F$_{2\alpha}$

and PGE$_2$, were soon identified in human menstrual fluid and were shown to be produced in large quantities by the endometrium. Based on their abundance in the uterus and their known vasoactive properties, prostaglandins were suggested as potential candidates for the uterine luteolytic factor. It was found that high levels of PGF$_{2\alpha}$ administered subcutaneously to female rats caused a shortening of pseudopregnancy and a reduction in the progesterone content of the corpora lutea. In the rat study, it was unclear whether the administered PGF$_{2\alpha}$ acted directly or indirectly on the ovary, e.g., via the pituitary. However, the infusion of small amounts of PGF$_{2\alpha}$ directly into the arterial supply of the transplanted ovary of the sheep caused an immediate and sustained decline in ovarian progesterone secretion followed by the onset of estrus and an LH peak, whereas the same amount of PGF$_{2\alpha}$ infused systemically had no effect on ovarian progesterone secretion. Because PGF$_{2\alpha}$ was known to be rapidly metabolized in the peripheral circulation particularly by the lungs, these infusion experiments in the sheep strengthened the proposal that the locally acting uterine luteolytic factor in the sheep might be PGF$_{2\alpha}$. Because of its known rapid clearance rate in the systemic circulation, PGF$_{2\alpha}$ would only be effective if delivered locally to the ovary from the uterus. Subsequent experiments indicated that a small fraction (about 1%) of radioactively labeled PGF$_{2\alpha}$ infused into a uterine vein passed directly via countercurrent transfer into the closely adherent ovarian artery and thus bypassed the systemic route. Such a mechanism provided an explanation for the local effect on the ovary of the putative uterine luteolysin PGF$_{2\alpha}$. Using a combination of gas chromatography and mass spectrometry, it was found that PGF$_{2\alpha}$ levels increased approximately 20-fold in uterine vein blood of the sheep coincident with the onset of luteolysis. Moreover, when the calculated amount of PGF$_{2\alpha}$ produced by the uterus during luteolysis was infused into the uterine vein adjacent to the ovary containing the corpus luteum, premature luteolysis was consistently observed but not when the same amount was infused into the peripheral circulation. These uterine vein infusions of physiological levels of PGF$_{2\alpha}$ not only confirmed the local countercurrent transfer of PGF$_{2\alpha}$

in the uteroovarian vasculature, but also proved the identity of PGF$_{2\alpha}$ as the local uterine luteolytic hormone in the sheep. Further studies in the sheep, with high-frequency sampling of uterine vein blood, revealed that PGF$_{2\alpha}$ was secreted by the uterus as a series of five to eight pulses, each lasting about 1 hr and occurring at intervals of 6–9 hr over the course of luteolysis. The infusion of low levels of PGF$_{2\alpha}$ in the form of 1-hr pulses into the arterial supply of the sheep ovary indicated that the pulsatile infusion of PGF$_{2\alpha}$ caused luteolysis with only 1/40th of the minimal amount of PGF$_{2\alpha}$ which caused luteolysis by a continuous infusion. Thus, a pulsatile pattern of PGF$_{2\alpha}$ appears to be advantageous for luteolysis. Pulses of the primary metabolite of PGF$_{2\alpha}$, 15-keto-13,14-dihydro-PGF$_{2\alpha}$ (PGFM) were observed in the peripheral blood of sheep over the course of luteolysis with the same frequency as pulses of PGF$_{2\alpha}$ in uterine vein blood. The role of PGF$_{2\alpha}$ as a uterine luteolytic hormone was further supported by the finding that administration of indomethacin to several different species, either systemically or via the intrauterine route, delayed or prevented luteolysis. Also, immunization against PGF$_{2\alpha}$, either passively or actively, delayed luteolysis in the sheep. The role of PGF$_{2\alpha}$ as the uterine luteolytic hormone has been confirmed in several other species, including the cow and other bovidae, the goat, the pig, the mare, and the guinea pig. Also, a luteolytic role for PGF$_{2\alpha}$ as a means of curtailing the length of pseudopregnancy has been established in the rat, rabbit, hamster, and mouse. The identification of PGF$_{2\alpha}$ as the luteolytic hormone in these species was based on one or more of the following: (i) a luteolytic effect of administered PGF$_{2\alpha}$, (ii) an elevation of PGF$_{2\alpha}$ in uterine vein blood or PGFM in peripheral blood during luteolysis, (iii) prolonged cycles in animals treated with indomethacin or immunized against PGF$_{2\alpha}$, and (iv) increased uterine secretion of PGF$_{2\alpha}$ following treatment of ovariectomized animals with estrogen or progesterone. There is, however, a species difference as to the degree of local control of the corpus luteum by the adjacent uterine horn. In the guinea pig, as in sheep, there appears to be an absolute requirement for a local transfer of uterine PGF$_{2\alpha}$ to the adjacent ovary. Although direct measurement of a local transfer of

$PGF_{2\alpha}$ has not been demonstrated in the guinea pig, other indirect evidence supports this conclusion, e.g., the unilateral effect of hysterectomy, the correct vascular anatomy in the uteroovarian pedicle, and, importantly, the extremely rapid clearance of $PGF_{2\alpha}$ in the peripheral circulation of the guinea pig.

However, in some species, such as the horse and the rabbit, it appears that $PGF_{2\alpha}$, or one of its metabolites, reaches the ovary via the systemic circulation because the same amount of $PGF_{2\alpha}$ is equally effective when given by the systemic or intrauterine routes. Moreover, in these two species, there is no evidence of a vascular arrangement in the uteroovarian vessels which would be consistent with a countercurrent transfer mechanism. In other species, such as the pig and the cow, there is evidence for both a local and a systemic route of transfer. Such a conclusion is supported by the finding that in these two species, a relatively large proportion of radioactively labeled $PGF_{2\alpha}$ (about 40%) passes through the pulmonary circulation unchanged, whereas in the sheep, <1% passes unchanged. In the alpaca, a member of the camel family indigenous to South America, it appears that the luteolytic effect of the uterus is mediated locally in the right uterine horn, but is mediated both systemically and locally in the left uterine horn. In the sheep, progesterone and several other ovarian steroids are transferred from the ovarian vein to the adjacent ovarian artery, indicating the presence of an ovarian–ovarian countercurrent system in addition to a uteroovarian transfer system in this species. An ovarian–ovarian countercurrent transfer system has also been reported to exist in the human ovarian vascular pedicle with no evidence for a uteroovarian transfer mechanism, which is in keeping with the observed lack of an effect of the uterus on primate ovarian function.

III. HORMONAL REGULATION OF UTERINE $PGF_{2\alpha}$ SYNTHESIS

Early studies established that the endometrium was the site of $PGF_{2\alpha}$ synthesis which appeared to be controlled by both estrogen and progesterone. Estrogen alone had a modest stimulatory effect but a prior period of progesterone treatment enhanced estrogen-stimulated production of $PGF_{2\alpha}$ in several species. However, in the sheep, it subsequently became apparent that estrogen and progesterone indirectly controlled $PGF_{2\alpha}$ synthesis by regulating receptors for oxytocin in the endometrium. This was based on the finding that mechanical stimulation of the uterus evoked the secretion of $PGF_{2\alpha}$ only early and especially late in the luteal phase of the cycle, whereas no effect was seen during the midluteal phase. Since mechanical stimulation of the female reproductive tract causes the release of neurohypophyseal oxytocin via the Ferguson reflex, and since exogenous oxytocin had been shown to cause a shortening of the cycle in the cow, it seemed likely that oxytocin released by mechanical stimulation of the uterus mediated the observed stimulatory effect on $PGF_{2\alpha}$ secretion. Indeed, it was found that infusions of physiological levels of oxytocin into a uterine artery mimicked the cyclical pattern of uterine $PGF_{2\alpha}$ secretion caused by mechanical stimulation of the uterus in the sheep. Therefore, it appeared that the cyclical variation in the ability of oxytocin to evoke the uterine secretion of $PGF_{2\alpha}$ was due to a cyclical variation in the concentration of oxytocin receptors in the endometrium. Such a proposal was strengthened by reports that other target sites for oxytocin action, such as the mammary gland and the oviduct, contained high-affinity oxytocin-binding sites which appeared to be increased by estrogen. Subsequently, it was shown that the stimulation of $PGF_{2\alpha}$ by oxytocin from ovine endometrium *in vitro* was positively correlated with the relative abundance of oxytocin receptors in this tissue. A model for hormonal regulation of $PGF_{2\alpha}$ in the ovine uterus is depicted in Fig. 2. It is proposed that estrogen enhances the formation of endometrial oxytocin receptors while, during the luteal phase, progesterone reduces the concentration of oxytocin receptors by blocking the action of estrogen. However, since progesterone catalyzes the destruction of its own receptor, progesterone action will decline toward the end of the luteal phase, thus allowing the return of estrogen action and the synthesis of oxytocin receptors. The greatly enhanced production of $PGF_{2\alpha}$ by oxytocin at the end of the luteal phase most likely results from the priming action of progesterone on lipid precursors in the endometrium during the luteal phase. There is now evidence that

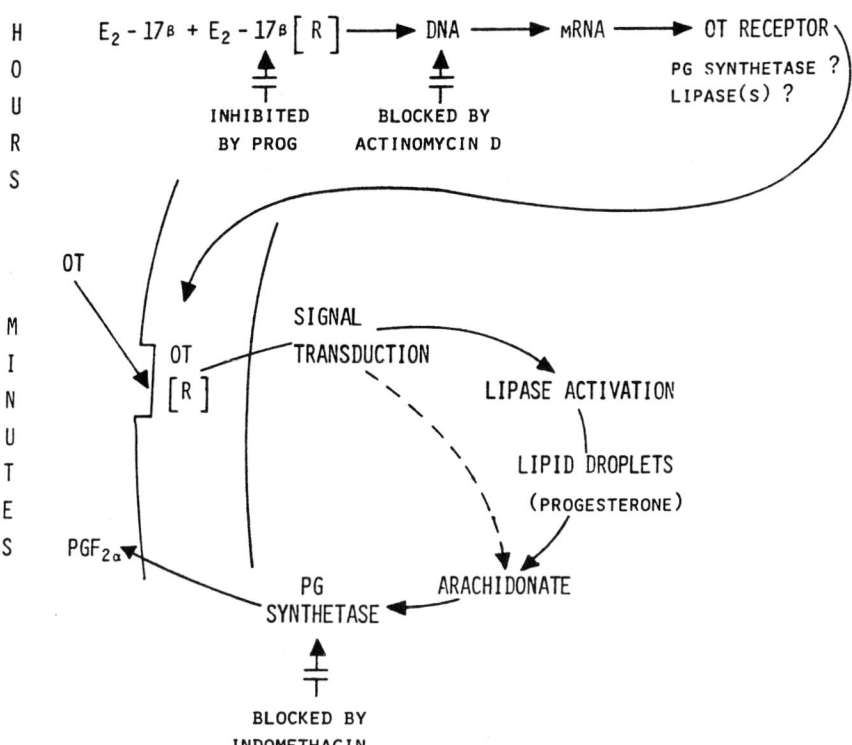

FIGURE 2 A model for the endocrine control of PGF$_{2\alpha}$ synthesis in the endometrial cell of the sheep during luteolysis (see text for explanation) (from McCracken *et al.*, 1995).

similar hormonal mechanisms regulate endometrial synthesis of PGF$_{2\alpha}$ in other species, such as the cow, the sow, and the mare. However, in primates and guinea pigs, the evidence points to a direct control of endometrial PGF$_{2\alpha}$ by estrogen and progesterone.

IV. REGULATION OF PULSATILE UTERINE PGF$_{2\alpha}$ SECRETION

In most species, PGF$_{2\alpha}$ is released from the uterus in a pulsatile manner which appears to be advantageous for luteolysis. Therefore, it was important to determine how such pulsatility was controlled. The model shown in Fig. 2 explains the cellular regulation of PGF$_{2\alpha}$ synthesis via the hormonal induction of endometrial oxytocin receptors. However, it does not explain the role of circulating levels of oxytocin. In the sheep and other ruminants, the regulation of blood levels of oxytocin is more complex because,

in addition to the neurohypophysis, the corpus luteum is an ectopic site of oxytocin synthesis and secretion. Recent work indicates that a finite store of oxytocin in the corpus luteum acts to supplement periodic discharges of oxytocin from the neurohypophysis and thus amplifies the magnitude of each luteolytic pulse of PGF$_{2\alpha}$ from the uterus. A model for the endocrine control of the pulsatile secretion of PGF$_{2\alpha}$ from the uterus during luteolysis is shown in Fig. 3 (see numbers in model):

1. At the end of the luteal phase, loss of progesterone action both in the hypothalamus and in the uterus, due to catalysis of its own receptor, results in the return of estrogen action in these tissues.

2. Estrogen increases endometrial oxytocin receptors and also causes changes in the frequency of the central oxytocin pulse generator.

3. Low levels of PGF$_{2\alpha}$ will be generated in the uterus.

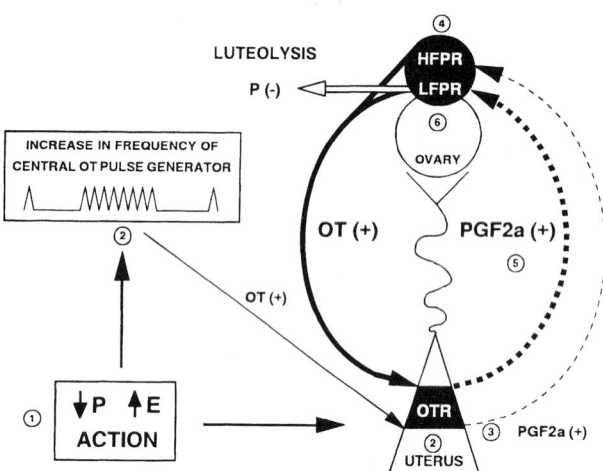

LUTEOLYSIS

P (-)

INCREASE IN FREQUENCY OF
CENTRAL OT PULSE GENERATOR

↓P ↑E
ACTION

HFPR
LFPR

OVARY

OT (+)   PGF2a (+)

OT (+)

OTR

UTERUS

PGF2a (+)

FIGURE 3 A model for the neuroendocrine control of pulsatile uterine $PGF_{2\alpha}$ secretion during luteolysis in the sheep (see text for explanation of numbers) (from McCracken *et al.*, 1995).

4. These initial low levels of $PGF_{2\alpha}$, acting on a high-sensitivity state of the luteal $PGF_{2\alpha}$ receptor, appear to initiate the supplemental release of luteal oxytocin.

5. Luteal oxytocin will serve to enhance uterine $PGF_{2\alpha}$ secretion, which will now inhibit progesterone synthesis via a low-sensitivity state of the $PGF_{2\alpha}$ receptor and, at the same time, release more oxytocin in a closed-loop fashion.

The response of the corpus luteum to $PGF_{2\alpha}$ appears to become refractory after about 1 hr, which may help to curtail each pulse of $PGF_{2\alpha}$. Additional luteolytic pulses of $PGF_{2\alpha}$ will follow when $PGF_{2\alpha}$ receptors and/or second messengers systems have recovered and/or when endometrial oxytocin receptors are further upregulated in time to respond to the next discharge of oxytocin via the central oxytocin pulse generator. It remains to be determined whether the amplification of uterine $PGF_{2\alpha}$ secretion by luteal oxytocin is an absolute requirement for luteolysis in ruminants. In species such as the sow and the mare that do not synthesize oxytocin in the ovary, it is likely that oxytocin secreted by the neurohypophysis alone may be sufficient to generate the observed pulses of uterine $PGF_{2\alpha}$. Thus, in the sheep and other species, the uterus can be regarded as a transducer which, under appropriate conditions,

converts neural signals (oxytocin) into local hormonal signals ($PGF_{2\alpha}$) to cause luteolysis. In addition to the well-established interplay between ovarian steroids and the anterior pituitary/hypothalamic axis for the initiation of the ovarian cycle (via the gonadotropins), there is now good evidence for an interplay between ovarian steroids and the posterior pituitary/hypothalamic system for the termination of the reproductive cycle (via oxytocin).

Although the primate uterus is an abundant source of prostaglandins, the presence of the uterus is not required for luteolysis since ovarian cyclicity is normal in hysterectomized subjects. However, many of the components in the sheep uteroovarian system are present within the primate ovary itself, namely, small amounts of oxytocin, oxytocin receptors, $PGF_{2\alpha}$ and receptors for $PGF_{2\alpha}$, as well as different luteal cell types. Thus, the local production of arachidonic acid and/or one of its metabolites could account for an intraovarian process of luteolysis in the absence of the uterus. Potential differences in the control of luteolysis in different classes of regularly cycling mammals are depicted in Fig. 4 and are described in the following sections.

A. Ruminants

The uterus can be regarded as a transducer which converts neurohypophyseal oxytocin signals, generated by the central oxytocin pulse generator, into pulses of $PGF_{2\alpha}$, which then causes luteolysis. However, in ruminants, a finite store of oxytocin in the corpus luteum may act to supplement oxytocin signals from the central oxytocin pulse generator and hence amplify $PGF_{2\alpha}$ pulses from the uterus. It remains to be determined whether such an amplification of $PGF_{2\alpha}$ by luteal oxytocin is a necessary requirement for luteolysis in ruminants.

B. Nonruminants

Since the corpora lutea of most nonruminants contain very little oxytocin, it is likely that these species rely mainly on the neurohypophysis as a source of oxytocin. Thus, the central pulse generator signals of oxytocin are transduced by the uterus into luteolytic $PGF_{2\alpha}$ pulses without amplification by supplemental releases of luteal oxytocin. However, oxytocin has

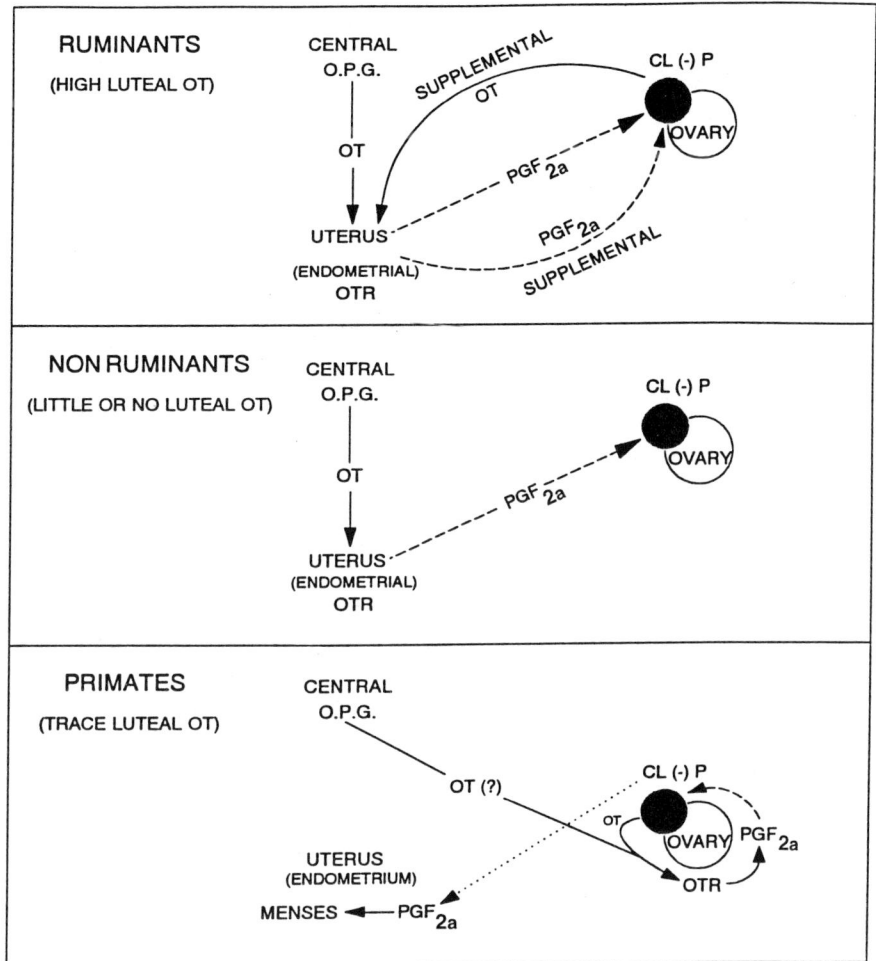

FIGURE 4 Comparative models for the control of luteolysis in different species (see text for explanation) (from McCracken et al. 1998).

recently been reported to be produced by the endometrium in the sow and the mare. Therefore, it is possible that the endometrium could act as a supplemental source of oxytocin during luteolysis in these species.

C. Primates

Since hysterectomy is without effect on ovarian cyclicity, and since oxytocin and oxytocin receptors have been identified in the primate corpus luteum and/or ovary, oxytocin from the ovary and/or a central oxytocin pulse generator could act on ovarian oxytocin receptors to generate low levels of intraovarian $PGF_{2\alpha}$ which may then cause luteolysis. The transducing function of the uterus would thus be bypassed in this paradigm. However, when the uterus is present, as in the intact subject, the withdrawal of progesterone action in the uterus due to intraovarian luteolysis acts as a major stimulus for endometrial $PGF_{2\alpha}$ synthesis and hence the initiation of the vascular and contractile events associated with menstruation.

V. MECHANISM OF LUTEOLYTIC ACTION OF $PGF_{2\alpha}$

An extensive number of mechanisms have been proposed to explain the luteolytic action of $PGF_{2\alpha}$ in the corpus luteum. Broadly, these can be divided into factors mediating functional luteolysis (the reduction of progesterone secretion or antisteroidogenic action) and structural luteolysis (the physical disappearance of the corpus luteum). However, it is currently unclear whether functional and structural luteolysis are truly separate events. Functional luteolysis is considered to be due to a blockade of LH action (the major luteotropic hormone in most species) potentially mediated by a variety of substances, such as adenosine, calcium, hydrogen peroxide, leukotrienes, heat shock proteins, microtubules, reactive oxygen, and, recently, endothelin-1, sterol carrier protein-2, and steroidogenic acute regulatory protein (StAR). Agents mediating structural luteolysis include apoptosis (from the Greek *apo* = leaf and *ptosis* = to drop), immune cells, tumor necrosis factor-α, and gap junctions.

The luteolytic action of $PGF_{2\alpha}$ is considered to be mediated via specific receptors for $PGF_{2\alpha}$ on the plasma membrane of luteal cells. Specific binding has been demonstrated in luteal cell membrane preparations and the $PGF_{2\alpha}$ receptor has been cloned from luteal tissue in a variety of species. Some studies report the presence of both high- and low-affinity sites for $PGF_{2\alpha}$ in luteal tissue, whereas others report only high-affinity sites. This may be due to the existence of high- or low-affinity states of the receptor or possibly to different isoforms of the receptor. There is also evidence in the sheep of a differential sensitivity of the corpus luteum to $PGF_{2\alpha}$ *in vivo*. Subluteolytic levels of $PGF_{2\alpha}$ can selectively evoke the release of luteal oxytocin without a significant effect on progesterone secretion, whereas much higher levels of $PGF_{2\alpha}$ are required to depress progesterone secretion. The differential sensitivity of the corpus luteum to $PGF_{2\alpha}$ could be due to different affinity states of the $PGF_{2\alpha}$ receptor, to isoforms of the $PGF_{2\alpha}$ receptor, or to a dose-dependent effect on second messenger systems. A key step in the process of steroidogenesis is the cleavage of cholesterol to pregnenolone by the P450 side chain cleavage enzyme (P450scc) located on the inner mitochondrial membrane of steroidogenic cells. Several studies have indicated an effect of $PGF_{2\alpha}$ on cholesterol transport to and into the mitochondria of luteal cells. In the rat, $PGF_{2\alpha}$ treatment decreases the luteal cell level of mRNA and the level of sterol carrier protein-2, which is considered to mediate intracellular transport of cholesterol to mitochondria. Also of particular interest is the poten-

tial action of $PGF_{2\alpha}$ on StAR, which is considered to transport cholesterol across the mitochondrial membrane. A reduction in StAR message has been demonstrated after high doses of $PGF_{2\alpha}$ in several species, an effect which may be mediated by heat shock proteins. It is also possible that $PGF_{2\alpha}$ could either reduce translation of StAR message to the StAR protein or block the cholesterol transporting function of the protein. Moreover, the message for StAR is reduced via the removal of LH by hypophysectomy and restored by the administration of exogenous LH. Thus, the abrogation of LH action at the cellular level, mediated by a reduction in StAR, could provide an attractive explanation for the antisteroidogenic action of $PGF_{2\alpha}$. Other potential mediators of luteolysis, such as endothelin-1 produced by endothelial cells in the corpus luteum, should also be considered. For example, the luteolytic action of $PGF_{2\alpha}$ in bovine steroidogenic luteal cells could only be demonstrated when they were cocultured with endothelial cells, an effect which was blocked by an endothelin-1 receptor antagonist. Moreover, endothelins could also play a role in the reduction in blood flow which follows functional regression. Overall, it seems likely that, during the process of luteolysis, $PGF_{2\alpha}$ exerts its effects at several loci in the cell which may be controlled by multiple mediators.

VI. ABROGATION OF LUTEOLYSIS IN EARLY PREGNANCY

In animals which normally undergo luteolysis toward the end of the reproductive cycle, mechanisms have evolved to prevent luteolysis when the animal becomes pregnant. Such an abrogation of luteolysis allows the continued secretion of luteal progesterone, which is necessary for pregnancy maintenance either throughout gestation (e.g., goat and pig) or until the placenta assumes this function (e.g., sheep and primate). In the event of pregnancy, several strategies have emerged to prevent luteolysis in different species. Early studies in sheep demonstrated that the transfer of an embryo, or its secreted proteins, to a synchronized donor prevented luteolysis even late in the cycle. Since $PGF_{2\alpha}$ was identified as the proximate cause of luteolysis in sheep, it seemed that the ability of an embryo to prevent luteolysis depended on either its ability to inhibit the production of $PGF_{2\alpha}$ (antiprostaglandin-secreting effect) or the production of a substance which protected the corpus luteum against the luteolytic action of $PGF_{2\alpha}$ (luteoprotective effect). There is, indeed, evidence that one or both mechanisms may be operative in a number of species.

A. Antiprostaglandin-Secreting Effect

In species such as the domestic animals and the guinea pig, in which pulses of uterine $PGF_{2\alpha}$ have been identified as the proximate cause of luteolysis, there is a marked diminution or absence of peaks of $PGF_{2\alpha}$ in the early pregnant animal. However, in some species, the basal levels of $PGF_{2\alpha}$ are higher in early pregnancy in the presence of diminished pulsatile $PGF_{2\alpha}$ secretion. The suppression of $PGF_{2\alpha}$ pulses in ruminants was found to be due to a reduction in the concentration of oxytocin receptors in the endometrium. Later studies indicated that a protein, secreted by the embryo, identified as IFN-τ, blocked the formation of endometrial oxytocin receptors by suppressing the transcription of the estrogen receptor gene. In the pig, estrogen secreted by the implanting blastocyst is thought to change endometrial $PGF_{2\alpha}$ secretion from an endocrine mode to an exocrine mode, thus preventing the secretion of $PGF_{2\alpha}$ into the uterine venous system. The equine blastocyst does not elongate rapidly as it does in ruminants and pigs, but rather maintains contact with the endometrium by transversing the uterus every 2 hr during the establishment of pregnancy. Intrauterine migration by the blastocyst is thought to inhibit $PGF_{2\alpha}$ secretion but the mechanism involved is not known. In the guinea pig, $PGF_{2\alpha}$ production by the endometrium is suppressed during early pregnancy but the mechanism of suppression is not known. However, unlike domestic animals in which the suppression of $PGF_{2\alpha}$ secretion is a local phenomenon, there is evidence that the unknown factor can act systemically in the guinea pig. In the rat, it appears that a decidual luteotropin overcomes the potential luteolytic effect of $PGF_{2\alpha}$ rather that acting as a suppressor of $PGF_{2\alpha}$ production. Likewise in the rabbit, a placental luteotropin in combination with follicular estrogen is thought to maintain the corpus luteum during pregnancy. However, the increase in uterine $PGF_{2\alpha}$

production seen during pseudopregnancy is also suppressed by an unknown mechanism.

B. Luteoprotective Effects of Pregnancy

In addition to the suppression of endometrial $PGF_{2\alpha}$ synthesis during early pregnancy, there is also evidence that in some species the corpus luteum is resistant to the luteolytic action of $PGF_{2\alpha}$ at this time. For example, in early pregnant sheep, $PGF_{2\alpha}$ is less efficient in causing luteolysis when injected into the ovarian artery or into ovarian follicles adjacent to the corpus luteum compared to similar injections in nonpregnant sheep. Also, the luteolytic effect of exogenous estrogen is only partially effective in early pregnant sheep compared to nonpregnant sheep. The ovine corpus luteum was found to be more resistant to a $PGF_{2\alpha}$ analog when multiple blastocysts were present compared to the presence of a single blastocyst. Similarly, early pregnant sheep showed a resistance to exogenous $PGF_{2\alpha}$ compared to the nonpregnant animals on Days 13–15 of the cycle. It has been suggested that PGE_2, which is produced by the blastocyst or endometrium in the sheep and pig, may counteract the luteolytic effect of $PGF_{2\alpha}$ since PGE_2 delays luteolysis in nonpregnant sheep when infused into the uterus or into an ovarian artery. Based on its ability to stimulate cyclic AMP production and its known vasodilator effect, PGE_2 may be one of the factors which protects the corpus luteum against the potential luteolytic effect of $PGF_{2\alpha}$ in early pregnancy. In primates, progesterone from the corpus luteum is necessary to maintain pregnancy in its early stages until placental production of progesterone begins (termed the luteoplacental shift, the timing of which varies among different primate species). The extension of the life span of the corpus luteum during early pregnancy in primates is caused by the secretion of chorionic gonadotropin by the blastocyst around the time of implantation. Thus, during early pregnancy in primates, the rescue of the corpus luteum is primarily a luteotropic event which overrules luteolysis. In domestic animals, however, the rescue of the corpus luteum is primarily due to the prevention of uterine $PGF_{2\alpha}$ secretion and possibly, in part, to a luteoprotective effect. The corpus luteum of early pregnant monkeys is relatively resistant to the

luteolytic effect of administered prostaglandin analogs compared to the nonpregnant monkey, suggesting that one function of chorionic gonadotropin
in primates is to override the potential luteolytic
effect of endogenous prostaglandins. Moreover, the
administration of human chorionic gonadotropin to
rhesus monkeys during the midluteal phase of the
cycle prevents the induction of corpus luteum regression by the intraluteal infusion of $PGF_{2\alpha}$. These findings suggest that, in primates, chorionic gonadotropin may rescue the corpus luteum not by suppressing
the synthesis of potential luteolysins such as prostaglandins but rather by overriding their action by a
luteotropic process.

See Also the Following Articles

Corpora Lutea of Nonmammalian Species; Corpus Luteum;
Estrous Cycle; Local Control Systems in Reproduction;
Pregnancy, Maternal Recognition of; Progesterone Ac-
tions on Reproductive Tract

Bibliography

Anderson, L. L., Bland, K. P., and Melampy, R. M. (1969).
Comparative aspects of uterine–luteal relationships. *Recent
Prog. Horm. Res.* **25**, 57–104.

Auletta, F. J., and Flint, A. P. F. (1988). Mechanisms controlling corpus luteum function in sheep, cows, non-human
primates, and women especially in relation to the time of
luteolysis. *Endocr. Rev.* **9**, 88–105.

Behrman, H. R., Endo, T., Aten, R., and Musiki, B. (1993).
Corpus luteum function and regression. *Reprod. Med. Rev.*
2, 153–160.

Clark, B. J., Wells, J., King, S. R., and Stocco, D. M. (1994). The
purification, cloning, and expression of a novel luteinizing
hormone-induced mitochondrial protein in MA-10 mouse
Leydig tumor cells. Characterization of the steroidogenic
acute regulatory protein (StAR). *J. Biol. Chem.* **269**, 28314–
28322.

Eglinton, G., Raphael, R. A., Smith, G. N., *et al.* (1963).
Isolation and identification of two smooth muscle stimu-
lants from menstrual fluid. *Nature* **200**, 963–965.

Ferreira, S. H., and Vane, J. R. (1967). Prostaglandins: Their
disappearance from and release into the circulation. *Nature*
216, 868–873.

Girsh, E., Wang, W., Mamluk, R., Arditi, F., Friedman, F.,
Milvae, R. A., and Meidan, E. (1996). Regulation of endo-
thelin-1 expression in the bovine corpus luteum: Elevation
by prostaglandin $F_{2\alpha}$. *Endocrinology* **137**, 5191–5196.

Liu, Z., and Stocco, D. M. (1997). Heat shock-induced inhibition of acute steroidogenesis in MA-10 cells is associated
with inhibition of the synthesis of the steroidogenic acute
regulatory protein. *Endocrinology* **138**, 2722–2728.

Loeb, L. (1923). The effect of extirpation of the uterus on
the life and function of the corpus luteum in the guinea
pig. *Proc. Soc. Exp. Biol. Med.* **20**, 441–443.

McCracken, J. A. (1997). Prostaglandins and leukotrienes—
Locally acting agents. In *Endocrinology: Basic and Clinical
Principles* (P. M. Conn and S. Melmed, Eds.), pp. 101–114.
Humana Press, Clifton, NJ..

McCracken, J. A., Custer, E. E., and Lamsa, J. C. (1998).
Luteolysis—A neuroendocrine-mediated event. *Physiol.
Rev.*, in press.

McCracken, J. A., Carlson, J. C., Glew, M. E., Goding, J.
R., Baird, D. T., Green, K., and Samuelsson, B. (1972).
Prostaglandin $F_{2\alpha}$ identified as a luteolytic hormone in
sheep. *Nature* **238**, 129–134.

McCracken, J. A., Custer, E. E., Lamsa, J. C., and Robinson,
A. G. (1995). The central oxytocin pulse generator: A pacemaker for luteolysis. *Adv. Exp. Med. Biol.* **395**, 133–154.

McLean, M. P., Billheimer, J. T., Warden, K. J., and Irby, R.
B. (1995). Prostaglandin $F_{2\alpha}$ mediates ovarian sterol carrier
protein-2 expression during luteolysis. *Endocrinology*
136, 4963–4972.

Michael, A. E., Abayasekara, D. R. E., and Webley, G. E.
(1994). Cellular mechanisms of luteolysis. *Mol. Cell. Endocrinol.* **99**, R1–R9.

Niswender, G. D., and Nett, T. M. (1994). Corpus luteum
and its control in infraprimates. In *The Physiology of Reproduction* (E. Knobil and J. D. Neill, Eds.), 2nd ed., vol. 2,
pp. 781–816. Raven Press, New York.

Pharriss, B. B., and Wyngarden, L. J. (1969). The effect of
prostaglandin $F_{2\alpha}$ on the progestogen content of ovaries
from pseudopregnant rats. *Proc. Soc. Exp. Biol. Med.*
130, 92–94.

Poyser, N. L. (1995). The control of prostaglandin production
by the endometrium in relation to luteolysis and menstruation. *Prosta. Leuk. Ess. Fatty Acids* **53**, 147–195.

Roberts, R. M., Xie, S., and Mathialgan, N. (1996). Maternal
recognition of pregnancy. *Biol. Reprod.* **54**, 294–302.

Stouffer, R. L. and Duffy, D. M. (1995). Receptors for sex
steroids in the primate corpus luteum: New insight into
gonadotropin and steroid action. *Trends Endocrinol. Metab.*
6, 83–89.

Zeleznik, A. J., and Fairchild-Benyo, D. (1994). Control of
follicular development, corpus luteum function, and the
recognition of pregnancy in higher primates. In *The Physiology of Reproduction* (E. Knobil and J. D. Neill, Eds.), 2nd
ed., vol. 2, pp. 751–782. Raven Press, New York.

Luteotropic Hormones

Gilbert S. Greenwald

University of Kansas Medical Center

GLOSSARY

corpus luteum The postovulatory endocrine organ with progesterone as its principal steroid product; derived from ovarian follicle(s).

luteolysis Regression of the corpus luteum; occurs in two phases: functional luteolysis involves drastic reduction in progesterone synthesis followed by structural luteolysis, which entails morphological destruction of the corpus luteum. In many, but not all, mammalian species uterine production of prostaglandin $F_{2\alpha}$ triggers luteal regression.

luteotropic factors Factors intrinsic to the corpus luteum, acting as local luteotropins by either paracrine or autocrine pathways.

luteotropic hormones Blood-borne substances acting directly on the corpus luteum to sustain the secretion of progesterone, thereby maintaining it in a functional state.

The corpus luteum (CL) represents the postovulatory fate of the ovarian follicle, distinguished from its precursor by its extreme vascularity, conversion of granulosa and thecal cells into luteal cells, which are characterized by hypertrophy and postmitotic differentiation. The most important product of the CL is progesterone, which is essential for the maintenance of pregnancy. Unlike the rather stereotyped hormonal controls involved in follicular development and function, there are myriad species variations in the factors regulating the life span of the CL. This article focuses on comparative aspects of the luteotropic hormones and factors acting directly on the CL to sustain the secretion of progesterone. The CL varies considerably during its life history in its ability to produce progesterone and other hormones. Moreover, there are temporal shifts in the luteotropic hormones required for luteal maintenance. Depending on the species, the source of progesterone may vary as pregnancy progresses. In recent years, considerable attention has been devoted to establishing the functions of small and large luteal cells which show species variations. There is no common denominator for luteotropic hormone: Although luteinizing hormone has been considered a universal luteotropin, numerous species differences exist.

I. INTRODUCTION

The corpus luteum is an ephemeral structure, the end product of follicular growth and differentiation. Unlike the follicle, it is heavily vascularized, an essential feature to supply its enlarged parenchymal mass with substrates and to release numerous products. Hormonal regulation of follicular development in the mammalian ovary is fairly consistent from species to species. In contrast, luteal control shows considerable species variations which must be taken into account. Therefore, this article will focus on comparative aspects of luteal function.

The luteotropins are a very diversified group of compounds running the gamut from catecholamines, eicosanoids, peptides, proteins, glycoproteins, steroids, growth factors. They are equally varied in their source including among others, pituitary, ovarian follicles, intrinsic factors in the corpus luteum, uterus, conceptus, placenta.

Various methods are used to establish whether a compound is a luteotropin (Lth): hypophysectomy, pituitary stalk sectioning, chemical hypophysectomy by gonadotropin-releasing hormone (GnRH) agonists and antagonists, the presence of appropriate receptors in the corpus luteum (CL), isolation and chemical identity of the Lth from luteal tissue, binding studies and Scatchard analysis of the suspected Lth from CL, administration of Lth antibodies and consequent effects on luteal function, and extirpation and replacement with the suspected Lth. In addition, a great deal has been learned by *in vitro* incubation of luteal cells with appropriate Lths and by measuring progesterone and other products in the medium.

II. LIFE HISTORY OF THE CORPUS LUTEUM

The corpus luteum varies considerably during its life history in its morphological and physiological development, ability to secrete progesterone and other steroids, and presumably the nature of luteotropic requirements. For example, the corpus luteum of the estrous cycle of rat, mouse, and hamster is an evanescent structure with a functional life span of 2 or 3 days. Their autonomous secretion of progesterone is triggered by ovulation and the preovulatory surge of follicle-stimulating hormone (FSH) and luteinizing hormone (LH). Cervical stimulation by a vasectomized male or mechanical stimulation converts the CL of the estrous cycle to the CL of pseudopregnancy, with the duration of progesterone synthesis extended for 8 days in the hamster and 10–12 days in the mouse and rat. If successful mating occurs, the CL of the cycle is converted into the CL of pregnancy, with an approximate doubling in half-life due to the superimposition, at the halfway point of pregnancy, of the placenta as a source of luteotropins. In the mouse and rat, the pituitary is dispensable at the midpoint of gestation. Finally, parturition in mouse and rat, but not hamster, leads to postpartum ovulation and formation of the CL of lactation.

These variations in luteal development in laboratory rodents are paralleled by significant changes in species with longer spontaneous cycles. The estrous cycle of large domestic species and primate menstrual cycle correspond to the rodent CL of pseudopregnancy with functional CL lasting about 2 weeks. In the menstrual cycle a relatively long follicular phase is followed by a luteal phase of about 2 weeks in which the CL secretes progesterone, estradiol, and androgens. In contrast, in sheep the luteal phase is the dominant one, with a relatively short follicular stage at the end of the cycle. In ruminants, but not primates, the CL of the cycle is converted into the CL of pregnancy by "maternal recognition of pregnancy," i.e., production of conceptus-derived proteins that are antiluteolytic and that act by dampening $PGF_{2\alpha}$ synthesis. In the primates, the development of the placenta as a steroidogenic organ secreting chorionic gonadotropin eliminates dependency on pituitary LH. In fact, the primate placenta as a source of progesterone and estrogen and other hormones reaches its peak and is capable of maintaining gestation after hypophysectomy.

Postpartum estrus in sow and ewe is anovulatory in contrast to that in the mare. The first postpartum ovulation in the cow invariably leads to a CL with a shorter life span than the CL of the cycle, contrary to the pig in which weaning leads to ovulation and formation of a normal CL. It is intriguing that the CL of pregnancy in human and macaque is rejuvenated at term and renews, secreting significant levels of progesterone.

III. TEMPORAL SHIFT IN SOURCES OF PROGESTERONE AND LUTEOTROPINS

For all these species, there are significant temporal shifts in the nature and source of luteotropins as the CL evolves. For example, in the early pregnant rat, pituitary prolactin initially is the predominant Lth. In the second week of gestation, a decidual Lth—a prolactin-like hormone—and pituitary LH become the dominant luteotropins. Finally, from Days 12 to 21, placental prolactin-like hormones predominate along with placental production of androgens converted to estrogens by luteal aromatase. The CLs of lactation in the postpartum rat are controlled by the intensity of the suckling stimulus. Dams nursing two pups resume ovulating by Day 12, whereas animals nursing eight pups do not ovulate for at least 20 days and have functional CLs over that time span. This prolonged pseudopregnancy is associated with elevated levels of prolactin and consistently reduced amounts of LH; the CLs therefore resemble the luteotropic profile of the first week of pregnancy.

Changing source and nature of luteotropic hormones is typical of most pregnant mammals. For example, a luteal–placental shift occurs in the human with pituitary LH superseded by placental human chorionic gonadotropin (hCG). Similarly, in the ewe the CL is only required during the first 90 days; thereafter, the placenta is the major source of progesterone. In contrast, for the maintenance of pregnancy in the sow the CL is indispensable and there is a significant increase in luteal prolactin and LH receptors in the second half of gestation.

IV. CELL TYPES IN THE CORPUS LUTEUM AND THEIR INTERACTIONS

The CL is a heterogenous organ composed of several cell types in addition to luteal cells. The most accurate method to determine percentage and number of cells per CL is by ultrastructural morphometric analysis. In all three species summarized in Table 1, nonluteal cells, especially endothelial cells, far exceed the number of luteal cells. This correlates with the extremely high luteal blood flow which makes the ovary one of the most vascular organs in the body.

The close proximity of luteal and nonluteal cells may influence the production of progesterone. The newly formed murine CL is infiltrated with macrophages. Luteal cells from dispersed CL—depleted of macrophages—produce progesterone for several days; when recombined with macrophages, there is a five- to eightfold increase in progesterone. The demise of the CL is frequently associated with an influx of macrophages, indicative of the onset of structural luteolysis but in the early stages of luteal development macrophages may act in a paracrine fashion as luteotropic factors.

Similarly, dispersed large luteal cells from Day 9 pregnant rats synthesize progesterone over an extended period when provided with lipoproteins. Co-culturing with nonluteal cells, presumably mainly endothelial cells, results in a several-fold increase in progesterone production without the necessity of exogenous cholesterol. Endothelial cells produce prostacyclin and TGF-β_1 factors, among others, which may interact with luteal cells.

As shown in Table 1, there are about sevenfold more small luteal cells than large luteal cells in the ovine and bovine CL and there is usually intimate contact between a large luteal cell and several small luteal cells. When luteal cells of pregnant sows are separated into the two populations and then recombined, progesterone accumulation in the medium is about twice as high as the amount released by each cell type alone. Moreover, in a superfusion system

---

TABLE 1
Cell Types in the Corpus Luteum

Species (day of cycle [c] or pregnancy [p])	Total number of cells	Endothelial cells	Luteal cells		Fibrocytes	Other cell types
			Large	Small		
Sheep (D13c)[a]	212.5×10^{-6}	113.4×10^{-6}	7.7×10^{-6}	47.5×10^{-6}	15.6×10^{-6}	28.3×10^{-6}
Cow (D12c)[a]	1477.6×10^{-6}	779.8×10^{-6}	51.5×10^{-6}	392.4×10^{-6}	147.5×10^{-6}	106.4×10^{-6}
Rat (D16p)[b]	—	925×10^{-3}	337×10^{-3}		—	—

[a] From O'Shea (1987).
[b] From Dharmarajan *et al.* (1993).

the small luteal cells upstream are responsible for increasing progesterone production by the large cells.

On the other hand, incubation of small and large ovine luteal cells—separately or combined—does not show any synergism in progesterone production. A similar lack of response of bovine large and small luteal cells has been reported anecdotally but differences in experimental design make it difficult to determine whether there are any interactions between the two luteal cell types. An interesting species difference is that the primate CL shows some semblance of separation of large and small luteal cells, with the paraluteal cells representing mainly the thecal derivatives.

V. LARGE VERSUS SMALL LUTEAL CELLS

The mammalian CL contains two distinct populations of luteal cells which differ morphologically and functionally: large luteal cells usually $\geqq 20$–25 μm in diameter and small luteal cells (12–20 μm). It is believed that they are initially derived from granulosa and thecal cells, respectively, from the newly ovulated follicle. For several species, however, it is now known that even the granulosa compartment is heterogeneous, composed of large and small cells that conceivably might be capable of contributing to both large and small luteal cells. It is equally feasible that stem or stromal cells may ultimately add to the luteal pool.

Immunological evidence, based on monoclonal antibodies to bovine thecal and granulosa cells, favors thecal derivation of small luteal cells and granulosal origin of large luteal cells. By midcycle, it is believed that large luteal cells represent transformed small luteal cells, but after enzymatic dissociation and dispersal, cells are undoubtedly lost. Ultrastructural analysis probably gives a much more accurate estimate of cell numbers.

The CL is enzymatically dissociated followed by flow cytometry, elutriation, unit gravity sedimentation, or ficoll gradients to separate large and small luteal cells. The consensus is that the enzymatic dispersal and the various separation procedures are harsh enough to destroy significant numbers of luteal cells. There are enough variations in techniques to

obscure whether species differences or artifacts account for the diverse findings.

In sheep and cow, the large luteal cells secrete the bulk of progesterone, oxytocin, and relaxin consistent with structural features indicative of steroidogenic and protein synthesis: smooth endoplasmic reticulum and secretory granules. The large luteal cells possess LH receptors but do not further increase progesterone and cAMP production in response to LH. In contrast, the small luteal cells also contain LH receptors and do increase progesterone secretion when LH is added. However, in the rhesus monkey, at midluteal phase the large luteal cells react to LH or cAMP by secreting progesterone and are the principal source of P_{450} aromatase. The CL of the early pregnant rat shows still another pattern: The lipid-filled large cells secrete the most progesterone, androgen, and estradiol and both large and small luteal cells contain prolactin and LH receptors and respond in long-term culture to LH with parallel increases in progesterone and androgens. The pseudopregnant superovulated rabbit has twice as many large luteal cells as small ones and unstimulated large cells secrete 28-fold more progesterone. In response to LH or isoproterenol, progesterone production by large luteal cells is increased ~3-fold but the increase is not significant. An acute steroidogenic response is elicited by both secretogogues in the small cell population but the levels are still only about one-sixth of those of the larger cells.

VI. LUTEOTROPIC HORMONES

There are a bewildering variety of luteotropic hormones regulating the mammalian CL with considerable species differences.

A. Luteinizing Hormone

LH has been termed the universal luteotropin based on its widespread distribution. LH receptors are present in the CL of the laboratory rodents, rabbits, primates, ewe, cow, sow, and mare. The related placental product, equine chorionic gonadotropin, serves a similar role in the pregnant mare as does human chorionic gonadotropin in the human. Briefly, the major effects of LH as a luteotropic hormone are mediated by binding to its membrane receptor and activation of adenylate cyclase to form cAMP, which then activates various protein kinase A pathways.

B. Prolactin

Prolactin (PRL) has long been recognized as a major Lth in hamster, rat, and mouse, but its importance in other species has been slighted. Added to large porcine luteal cells, prolactin stimulates progesterone secretion. Receptors for PRL are present in the CL of primates, ewe, and sow, but luteotropic activity has only been established for the sow. In the pregnant rat, prolactin prevents the conversion of progesterone to the inactive metabolite 20α-hydroxysteroid progesterone by downregulating 20α-hydroxysteroid dehydrogenase. In early corpora lutea of the pregnant hamster, mouse, and rat, *in vitro* addition of PRL stimulates progesterone secretion to the same extent as LH. Prolactin also increases the number of LH receptors in rat CL and maintains luteal estrogen receptors.

C. Follicle-Stimulating Hormone

Until recently, it has been difficult to establish a luteotropic role for FSH. Recent *in vitro* studies from this laboratory have shown that ovine FSH or recombinant FSH—devoid of LH activity—are capable of increasing progesterone synthesis in hamster, rat, and mouse luteal cells in 24-hr incubations. At the early luteal phase of the sow, *in vitro* FSH stimulates progesterone synthesis. Moreover, FSH receptors are present in the human and bovine CL. Progesterone production is stimulated in bovine luteal cells by FSH.

D. Growth Hormone

Growth hormone (GH) and IGF_1 receptors are found in porcine CL and GH receptors are also present in the bovine CL, localized in large luteal cells. Growth hormone, prolactin, and IGF_1 also increase progesterone in the media of porcine large luteal cells. In sheep, GH increases gene expression of P450 side chain cleavage enzyme and 3β-hydroxysteroid dehydrogenase; consequently, luteal weight and pro-

gesterone concentration are enhanced. Human GH also increases progesterone production in cultured small luteal cells of nonpregnant women.

E. Estradiol

Estradiol receptors are found in the CL of numerous species: rabbit, rat, ewe, and cow, but not human or macaque. The luteotropic effect of estrogen is especially interesting in the rabbit. The principal source of the hormone is from large ovarian follicles, present throughout pregnancy. It has been known for over 50 years that luteal function is maintained by estrogens in the hypophysectomized rabbit; receptors for both LH and estrogen are present in the CL but progesterone synthesis is insensitive to hCG or cAMP. *In vitro* rabbit luteal cells are autonomous and maintain progesterone production without any hormonal support. However, *in vivo* withdrawal of estrogen leads to luteal regression. The paradox has been resolved by recent findings which reveal that the rate-limiting mitochondrial protein—steroidogenic acute regulating protein (StAR)—is reduced after *in vivo* estrogen deprivation and restored by exogenous estrogen. This stimulates the reappearance of StAR and restores progesterone synthesis by the CL.

A variant on the luteotropic role of estrogen is provided by the pregnant rat. In the second half of gestation, administration of massive doses of estrogen maintains luteal progesterone synthesis in hypophysectomized–hysterectomized animals. Estradiol acts in multiple ways on the rat CL by increasing the number of binding sites for high-density lipoproteins (HDL) and thus enhancing the uptake of circulating HDL and its incorporation for progesterone synthesis. Estradiol evidently also acts by increasing mitochondrial content of StAR and thus transferring cholesterol from the outer to inner mitochondrial membranes. Thus, in the rat intraluteal estrogen plays multiple roles rather than reliance on a follicular source as in the rabbit.

F. Progesterone

Rothchild theorized in 1981 that progesterone facilitates its own secretion and that luteotropins and luteolysins exert their effects through this self-stimulating system. The theory was based on research of the rat CL but it is now known that the rat CL does not express the progesterone receptor. However, this does not rule out the possibility that progesterone may act via other pathways.

On the other hand, progesterone receptors have been identified in the primate CL in both human and rhesus monkey. The receptors are detected by messenger RNA and immunohistochemistry and it is especially intriguing that the progesterone receptor colocalizes with the steroidogenic enzyme 3β-hydroxysteroid dehydrogenase in macaque luteal cells, suggesting a possible autocrine regulating role for progesterone.

Two recent studies with very different designs point to the possible role of progesterone as a luteotropic hormone. Intraluteal injection of RU 486—an inhibitor of progesterone synthesis—at different stages of rat pregnancy enhances serum progesterone levels at the beginning and end of pregnancy but significantly lowers levels from Days 7 to 14. More direct evidence is provided by incubating macaque luteinized granulosa cells in the presence of hCG or trilostane—a 3β-hydroxy inhibitor. Progesterone receptor expression is enhanced by hCG, whereas trilostane reduces the percentage of positive cells and drastically reduces progesterone production. The concept that progesterone can reinforce its secretion is intriguing but obviously requires further study.

G. Luteal Peptides

Since the late 1970s, a variety of peptides synthesized by the CL have been identified: oxytocin, vasopressin, relaxin, and inhibin. For example, oxytocin is found in the CL of cow, ewe, pig, goat, monkey, human, and baboon. Oxytocin is usually localized in large luteal cells and although it is usually involved in luteolysis, in young porcine and bovine CL intraluteal infusion of oxytocin increases progesterone release. Short-term incubation of dispersed luteal cells from cow and human respond to low doses of oxytocin by increasing progesterone accumulation, whereas larger doses are inhibitory.

There is no evidence to substantiate a luteotropic role for any other luteal peptide, including relaxin (pig, cow, rat, mouse, and human) and inhibin (human, cynomolgus monkey, and ewe).

H. Catecholamines

Epinephrine-sensitive adenylate cyclase activity exists in the CL of several species including cow, rabbit, rat, and human. *In vitro* addition of epinephrine to luteal slices or cell suspensions stimulates progesterone synthesis and the effect is blocked by β-adrenergic antagonists. Levels of noradrenaline increase in rat and human corpora lutea toward the end of pseudopregnancy and menstrual cycle, respectively, and $β_2$-adrenergic receptors are present on rat luteal membranes. The luteotropic effects of catecholamine have only been demonstrated *in vitro*. The fact that denervated rabbit ovaries still maintain functional CLs casts doubts on the physiological significance of catecholamines as luteotropins.

I. Prostaglandins

Although prostaglandin $F_{2α}$ acts as a luteolysin in numerous species, it is now apparent that other derivatives of the cyclooxygenase pathway act as luteotropins. Prostacyclin (PGI_2) injected directly into the bovine CL causes a rapid and prolonged increase in peripheral progesterone but no change in circulating LH levels. The bovine CL has greater levels of PGI_2 synthetase than any other tissue. PGI_2 regulates cAMP levels. In women, prostaglandin E_2 (PGE_2) is one of the major luteotropins during the menstrual cycle and early pregnancy. It is synthesized by the human CL to the same extent as $PGF_{2α}$ and receptors for PGE_2 are present in the CL of the menstrual cycle. In isolated early CL, PGE_2 significantly increases cAMP production and synergizes with hCG to produce maximal levels of progesterone.

J. Growth Factors

Insulin-like growth factor-I (IGF-I) is a mitogenic polypeptide which increases progesterone production by porcine large luteal cells but not small cells. It interacts with prolactin or growth hormone to further increase progesterone accumulation by porcine large luteal cells. In the pregnant rat, IGF-I and IGF-I receptors are abundantly expressed in the CL. Similar to the pig, *in vitro* addition of IGF-I enhances progesterone synthesis by large rat luteal cells but not small cells and IGF-I interacts with estradiol to produce maximal levels of progesterone. IGF-I also stimulates *in vitro* progesterone production by bovine and rabbit luteal cells.

Receptors for epidermal growth factor (EGF) are present in bovine and human CL and in the cow they are eightfold more abundant than in granulosa cells. Human granulosa–luteal cells exposed for 2 days to EGF show substantial increases in basal progesterone synthesis and even greater increases when incubated with either hCG or PGE_2. Evidently, both growth factors (and others) act as local mediators of luteotropic hormones.

VII. LUTEOTROPIC COMPLEX

The luteotropic factors considered previously have been considered as acting alone, without significant interactions, but this is most likely an oversimplification for many species. In the pregnant hamster, rat, and hamster, a luteotropic complex involves PRL, FSH, and LH *in vivo* and *in vitro*. *In vitro*, the three Lths have additive, but not synergistic, effects on progesterone synthesis. In the pregnant hamster, the presence of luteal receptors for all three pituitary hormones supports the *in vitro* results. Whether similar Lth interactions exist in other species has rarely been studied but deserves consideration.

See Also the Following Articles

CORPUS LUTEUM OF PREGNANCY; FSH (FOLLICLE-STIMULATING HORMONE); LH (LUTEINIZING HORMONE); LUTEOLYSIS; PROGESTERONE ACTIONS ON REPRODUCTIVE TRACT; PROLACTIN

Bibliography

Auletta, F. J., and Flint, A. P. F. (1988). Mechanisms controlling corpus luteum function in sheep, cows, non-human primates and women especially in relation to the time of luteolysis. *Endocr. Rev.* **9**, 88–105.

Brannian, J. D., and Stouffer, R. L. (1991). Cellular approaches to understanding the function and regulation of the primate corpus luteum. *Sem. Reprod. Endocrinol.* **9**, 341–351.

Dharmarajan, A. M., Meyer, G. T., and Bruce, N. W. (1993). Morphometric analysis of the corpus luteum of 16-day pregnant rats: The effect of preparative procedures on volume of luteal cell, interstitial, and vascular compartments. *Am. J. Anat.* **168**, 51–65.

Gibori, G. (1993). The corpus luteum of pregnancy. In *The Ovary* (E. A. Adashi and P. Leung, Eds.), pp. 261–317. Raven Press, New York.

Hansel, W., and Dowd, J. P. (1986). New concepts of the control of corpus luteum function. *J. Reprod. Fertil.* **78**, 755–768.

Holt, J. A. (1989). Regulation of progesterone production in the rabbit corpus luteum. *Biol. Reprod.* **40**, 201–208.

Niswender, G. D., and Nett, T. M. (1994). Corpus luteum and its control in infraprimate species. In *Physiology of Reproduction* (E. Knobil and J. D. Neill, Eds.), 2nd ed., pp. 781–816. Raven Press, New York.

O'Shea, J. D. (1987). Heterogenous cell types in the corpus luteum of sheep, goats and cattle. *J. Reprod. Fertil. Suppl.* **34**, 71–85.

Richardson, M. (1986). Hormonal control of ovarian luteal cells. In *Oxford Review of Reproductive Biology*, Vol. 8, pp. 321–378. Oxford Univ. Press, London

Rothchild, I. (1996). The corpus luteum revisited: Are the paradoxical effects of RU486 a clue to how progesterone stimulates its own secretion. *Biol. Reprod.* **55**, 1–4.

Smith, M. F., McIntosh, E. W., and Smith, G. W. (1994). Mechanisms associated with corpus luteum development. *J. Anim. Sci.* **72**, 1857–1872.

Stormshak, F., Zelinski-Wooten, M. B., and Abdulgadir, S. E. (1987). Comparative aspects of the regulation of corpus luteum function in various species. In *Regulation of Ovarian and Testicular Function* (V. B. Mahesh, D. S. Dhindsa, E. Anderson, and S. P. Kalra, Eds.), Experimental Medicine & Biology, Vol. 219, pp. 327–360. Plenum, New York.

Stouffer, R. L., and Brannian, J. D. (1993). The function and regulation of cell populations composing the corpus luteum of the ovarian cycle. In *The Ovary* (E. A. Adashi and P. Leung, Eds.), pp. 245–260. Raven Press, New York.

Towson, D. H., Wang, X. J., Keyes, P. L., Kostyo, J. L., and Stocco, D. M. (1996). Expression of the steroidogenic acute regulatory protein in the corpus luteum of the rabbit: Dependence upon the luteotropic hormone, estradiol-17β. *Biol. Reprod.* **55**, 868–874.

Wiltbank, M. C. (1994). Cell types and hormonal mechanisms associated with mid-cycle corpus luteum function. *J. Anim. Sci.* **72**, 1873–1883.

Yuan, W., and Greenwald, G. S. (1997). Progesterone production *in vitro* by mouse luteal cells: Response to follicle stimulating hormone, luteinizing hormone and prolactin. *Proc. Soc. Exp. Biol. Med.* **21**, 265–270.

Lymphokines

Wenbin Tuo

Washington State University

Fuller W. Bazer

Texas A&M University

Wendy C. Brown

Washington State University

I. Lymphokine Receptors
II. Lymphokines in Reproduction

GLOSSARY

antigen-presenting cells Cells that take up, process, and present antigens in a form that can stimulate lymphocytes.

chemokines Chemotactic cytokines including three families, α (CXC), β (CC), and γ (C), based on the presence and position of the conserved cysteine residues. The first two cysteines of members of the α family are separated by another amino acid and in the β family they are located next to each other. Members of the γ family contain one cysteine at the N terminus.

chemotaxis Directional migration of cells especially in response to concentration gradients of chemotactic factors.

cytokines Secretory proteins which control the survival, growth, differentiation, and effector functions of cells, including colony-stimulating factors, growth factors, lymphokines, interleukins, monokines, chemokines, and interferons.

interleukins Polypeptides involved in signaling between cells of the immune system.

monokines Polypeptides produced by the monocyte/macrophage lineages that mediate signaling in a number of cellular systems; however, some traditional monokines are from other cell lineages.

Th1, Th2, and Th0 concepts CD4 helper T cells were originally defined as Th1 and Th2 based on their functional expression of cytokines upon stimulation in the mouse. Murine Th1 cells produce exclusively Th1-type cytokines (IL-2, IFN-γ, and TNF-β) and Th2 cells produce Th2-type cytokines (IL-4, IL-5, and IL-10). The cross-regulating nature of Th1 and Th2 cytokine networks and their expression of either type of these specific cytokines tends to make them mutually exclusive (at least in the mouse). T cells from other species, such as human and cattle, make both Th1 and Th2 cytokines, with Th1 cells making relatively more Th1 cytokines and Th2 cells more Th2 cytokines. T cells that produce both Th1 and Th2 cytokines simultaneously are termed Th0 cells. Other T cells, including CD8 and $\gamma\delta$ T cells, can also produce the same patterns of cytokines, leading to a broader definition of type 1 or type 2 immune responses.

The term lymphokine has been used to describe soluble mediators derived primarily from T and B lymphocytes during immunological reactions. However, lymphokines can also be derived from a variety of nonimmune cells. Lymphokines play a key role in regulating T and B lymphocyte differentiation, growth, and effector functions. Lymphokines include interleukins (IL)-2, -4 through -7, -10, and -12 through -18; interferon-γ; and tumor necrosis factor-α and -β. The term lymphokine is often used interchangeably with cytokines, monokines, chemokines, interferons, and hematopoietic growth factors since many cytokines are derived from multiple sources and their functions are pleiotropic.

I. LYMPHOKINE RECEPTORS

The hemopoietin/interferon receptor (HIR) and the tumor necrosis factor (TNF)/nerve growth factor (NGF) receptor families belong to the cytokine

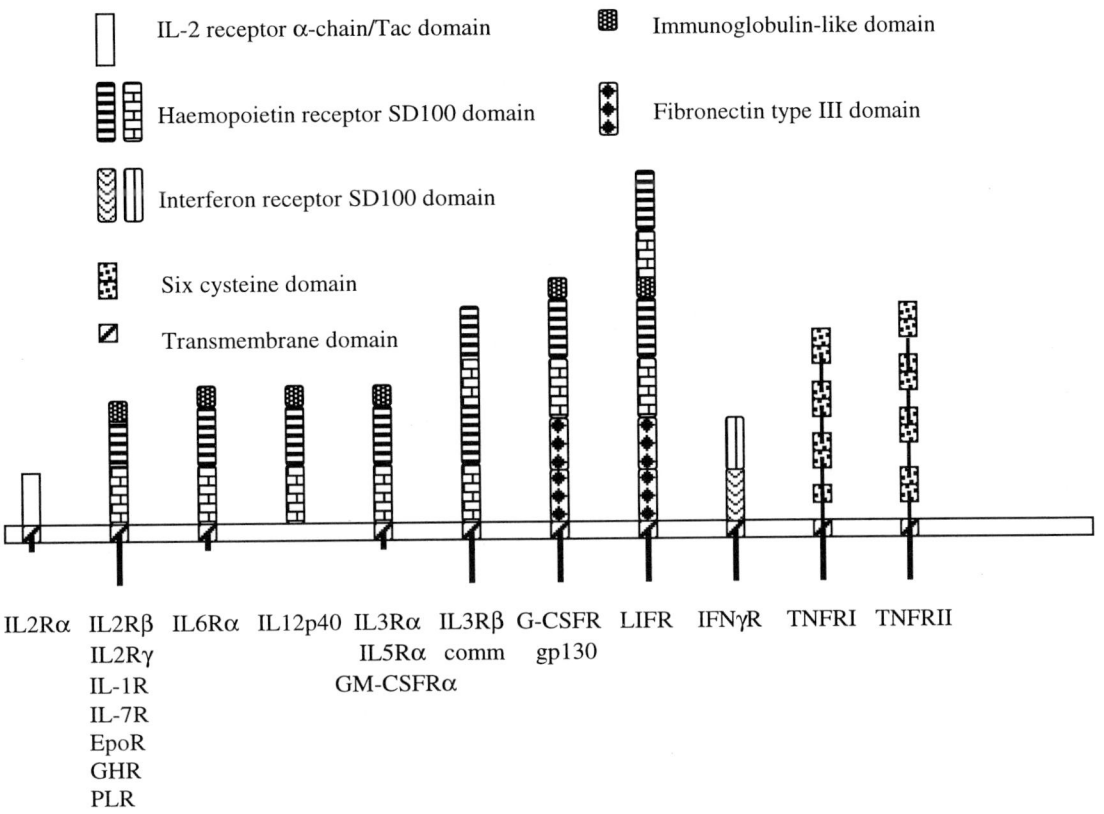

FIGURE 1 Schematic diagram of the hematopoietic/interferon receptor and NGF-R/TNF-R families. IL, interleukin; R, receptor; Epo, erythropoietin; GH, growth hormone; PL, prolactin; G-CSF, granulocyte-colony stimulating factor; GM-CSF, granulocyte macrophage-CSF; comm, common chain to IL-3, IL-5, and GM-CSF; LIF, leukemia inhibitory factor; IFN, interferon; TNF, tumor necrosis factor (adapted from G. P. Gearing and S. F. Ziegler, *Current Opinion in Hemotology*, 138–148, 1993. Lippincott Williams and Wilkins).

receptor superfamily based on sequence similarity in their extracellular domains. Cytokines/lymphokines that interact with HIRs include interferon (IFN)-α, -β, and -γ; interleukin (IL)-2 through -7 and -9 through -11, and -13; leukemia inhibitory factor (LIF), oncostatin-M; erythropoietin; ciliary neurotrophic factor; growth hormone; and prolactin. IL-2, IL-4, IL-7, IL-9, and IL-13 receptors share the same γ chain (or IL-2Rγ) to form their functional receptor complex. IL-3, IL-5, and GM-CSF receptor complexes use a common β chain, whereas IL-6, LIF, and IL-11 receptors use the gp130 as part of their receptor complex. A schematic diagram depicting the relationship between these receptors is shown in Fig. 1).

II. LYMPHOKINES IN REPRODUCTION

A. Current Hypotheses on Allopregnancy in Relationship to Lymphokines

The following are current hypotheses on allopregnancy in relationship to lymphokines:

1. The uterus is an immunologically privileged site, but extrauterine pregnancies do exist. Recent studies with eyes and testis (immunologically privileged sites) indicate that Fas/Fas ligand (FasL) may play a critical role in keeping these sites free of in-

flammatory cells and damaging immune responses. Since both Fas and FasL are present at the fetal–maternal interface, the same mechanism may exist to protect the fetus from potential attacks by maternal immune cells.

2. Immunosuppression is hypothesized to be either a systemic or local suppression of maternal immune function by soluble factors and/or suppressor T cells. However, this hypothesis alone does not explain the observations that placental growth and function are improved by autoimmunization and T cell cytokines produced by activated T cells and other nonimmune cells.

3. The placental immunotrophism hypothesis was proposed by Thomas G. Wegmann and states that maternal T cell recognition of foreign alloantigens expressed on fetal trophoblast cells at the maternal–fetal interface leads to lymphokine (e.g., GM-CSF, CSF-1, and IL-3) release from those T cells to improve placental growth and function. However, current results indicate that cytokines that have "immunotrophic" effects on placental cells are derived primarily from the conceptus and/or the uterus (nonimmune sources).

4. The hypothesis that successful pregnancy is a Th2 phenomenon states that the conceptus protects itself by secreting Th2 cytokines which downregulate the harmful inflammatory cytokines (or Th1 cytokines). Consequently, the maternal immune system during pregnancy preferentially mounts Th2-biased responses, resulting in increased susceptibility to certain autoimmune diseases and intracellular infections. Available results from mouse models demonstrated that IL-4, IL-10, and TGF-β are present at high levels and may be critical for development of a Th2-biased environment in the uterus during normal pregnancy.

B. Lymphokines in the Reproductive System

Increasing numbers of cytokines and lymphokines have been discovered and characterized in the past few decades. However, only those lymphokines (IL-2, IL-4, IL-5, IL-7, IL-10, IL-12, IL-15, IFN-γ, TNF-α, and TNF-β), selected chemokines (IL-8), inflammatory cytokines (IL-1, IL-6, and TNF-α), and hematopoietic growth factors (IL-11) that may play a role in regulating reproductive processes in different species are discussed.

1. Interleukin-1

i. Alternative Names   Alternative names include catabolin, endogenous pyrogen, osteoclast-activating factor, lymphocyte-activating factor, proteolysis-inducing factor, hemopoietin (H-1), epidermal cell-derived thymocyte-activating factor, serum amyloid A inducer, hepatocyte-stimulating factor (HSF), leukocyte endogenous mediator, and fibroblast-activating factor.

ii. Interleukin-1 and Its Receptor   The IL-1 family includes three polypeptide members: IL-1α, IL-1β, and IL-1 receptor antagonist (IL-1ra). The genes encoding for IL-1 have been cloned and sequenced for humans, mice, rats, cows, sheep, goats, rabbits, pigs, red deer, horses, cynomolgus monkeys, red-crowned mangabeys, rhesus monkeys, and pig-tailed macaques. The amino acid sequence homology between IL-1α, IL-1β, and IL-1ra is only 19–26%. All three members have molecular weights of 17–20 kDa and bind to both IL-1 receptor I and II (IL-RI and IL-1RII), with IL-1ra binding with highest affinity to IL-1R1 and IL-1β binding with highest affinity to IL-1RII. Binding of IL-1ra to its receptor does not trigger signal transduction. Recombinant IL-1RI is also an effective antagonist of IL-1 action. IL-1 can be produced by a variety of cell types including monocytes/macrophages, Kupffer cells, synovial cells, peritoneal macrophages, astrocytes, microglia, glioma cells, smooth muscle cells, endothelial cells, keratinocytes, neutrophils, fibroblasts, chondrocytes, epithelial cells, Langerhans cells, renal mesangial cells, T cells, B cells, and natural killer (NK) cells. IL-1 has a wide range of target cells including pre- and mature B cells, mature T cells, monocytes and neutrophils, hematopoeitic precursors, fibroblasts, smooth muscle cells, keratinocytes, osteoclasts, chondrocytes, and hepatocytes. IL-1 can induce fever, bone resorption, muscle proteolysis, and cachexia and increase acute phase proteins *in vivo*.

iii. Expression and Function of IL-1 in the Reproductive System   IL-1 Function in Male Reproduc-

TABLE 1
IL-1 Expression in the Testes

Tissue/cell	Regulation	IL-1α/β	Species
Epididymis, interstitial area of the testes, interstitial cells, testicular interstitial fluid and cytosolic preparations, cytoplasm of the epithelium of epididymal ducts		IL-1	Rat
Primary spermatocytes, early spermatids		IL-1	Rat
Sertoli cells	Increased by LPS, T4, TNF-α, NGF, phagocytosis of latex beads, residual bodies, and cytoplasts from elongating and elongated spermatids	IL-1α	Rat
Testicular macrophages	Low levels	IL-1	Rat
Leydig cells	Increased by LHs and CG Inhibits T4 release by blocking hCG binding	IL-1α, IL-1β	Rat, human

Note. Abbreviations used: LPS, lipopolysaccharide; T4, testosterone; NGF, nerve growth factor; CG, choriogonodotropin; hCG, human CG; LH, luteinizing hormone.

tion   IL-1 is produced by the testes and IL-1α stimulates DNA synthesis and cell proliferation. The current hypothesis derived from studies in rats states that phagocytosis of residual bodies by Sertoli cells triggers Sertoli cells to produce IL-1α which, in turn, stimulates DNA synthesis in preleptotene spermatocytes. These results support the theory that IL-1 may play an important role in regulating spermatogenesis.

The effect of IL-1 on Leydig cells in steroidogenesis is not conclusive (Tables 1 and 3).

IL-1 Function in Ovarian Physiology   The ovary contains the complete IL-1 system, including IL-1, IL-1R, and IL-1ra (Tables 2 and 3). The ovulation process is considered an inflammatory-like reaction and IL-1 is a mediator of the cascade of events leading

TABLE 2
IL-1 Expression in the Ovary

Tissue/cell	Regulation	Species
Theca–interstitial cells, granulosa–luteal cells, cultured surface epithelium	LPS	Human, rat, mouse
Preovulatory follicular fluid, follicular fluid, granulosa and cumulus cells	GnRH	Human, pig
Ovarian cells at luteal phase	Menstrual cycle	Human
Theca–interstitial cells	Increased by PMSG and hCG *in vivo*	Mouse
Ovarian cells, theca–interstitial cells	IL-1β	Rat
Ovarian monocytes/macrophages		Human
Whole ovarian dispersates	Stimulates PGF$_{2\alpha}$ and PGE$_2$ production Increased PGE$_2$ production by granulosa–theca cell interactions	

Note. Abbreviations used: LPS, lipopolysaccharide; GnRH, gonadotropin-releasing hormone; hCG, human choriogonadotropin; PMSG, pregnant mare serum gonadotropin; PGF$_{2\alpha}$, prostaglandin F$_{2\alpha}$; PGE$_2$, prostaglandin E$_2$.

TABLE 3

TABLE 3
IL-1R and IL-1ra Expression in Reproductive Organs

Tissue/cell	Regulation	Receptor type	Species
Whole ovary, follicular aspirates, theca–interstitial layer of growing follicles, oocyte, granulosa cells, and granulosa–luteal cells of the corpus luteum	Increased by forskolin	IL-1RI	Mouse
Whole ovarian material, macrophage-free preovulatory fluid aspirates		IL-1ra	Human
Sertoli cells		IL-RI and -II	Human, rat
Interstitial tissue		IL-RI and -II	Mouse
Epithelium	Menstrual cycle, maximal at luteal phase	IL-1RI	Human

to ovulation. In conjunction with human choriogonadotropin (hCG), IL-1β stimulates ovulation *in vitro* in the rat, indicating that IL-1 participates directly in the ovulation process.

IL-1 present in the brain may be inhibitory to gonadotropin-releasing hormone (GnRH) and lutenizing hormone (LH) secretion but not follicular-stimulating hormone (FSH) release and is therefore antigonadotrophic. Local IL-1 stimulates the proliferation of granulosa cells but blocks their differentiation and secretion of progesterone (P$_4$). Therefore, the local or immune cell-derived IL-1 modulates ovarian functions through autocrine and paracrine pathways in conjunction with ovarian steroid hormones.

IL-1 Function in the Female Reproductive Tract: Implantation and Fetal–Maternal Communication Successful pregnancy in the mouse, human, and pig is associated with high levels of IL-1 at the fetal–maternal interface (Table 4) and exogenous human IL-1ra blocks murine embryo attachment and implantation by blocking the IL-1RI expressed by the endoemetrial epithelium. Therefore, the IL-1 system is in part responsible for fetal–maternal communications during normal pregnancy in several species.

2. Interleukin-2

i. Alternative Name T cell growth factor (TCGF) is the alternative name for IL-2.

ii. IL-2 and IL-2 Receptor IL-2 is a 15-kDa glycoprotein produced by antigen- or mitogen-activated T lymphocytes and exerts regulatory effects on essentially all cell types participating in immunological reactions. The IL-2 genes have been cloned and sequenced for humans, mice, rats, Mongolian gerbils, cattle, sheep, horses, goats, pigs, dogs, rabbits, red deer, pig-tailed macaques, red-crowned mangabeys,

TABLE 4
IL-1 Expression in the Female Reproductive Tract

Tissue/cell	Regulation	Species
Epithelium, stroma	Estrous/menstrual cycle	Mouse, human
Trophoblasts, macrophages	P$_4$, E$_2$	Mouse, human
Preimplantation conceptuses		Human, mouse, pig
Epithelium, macrophages, endothelium	Gestation, P$_4$, E$_2$	Human, mouse
Amnion	Gestational age	Human

Note. Abbreviations used: P$_4$, progesterone; E$_2$, estradiol.

cynomolgus monkeys, rhesus monkeys, gibbon apes, and chickens. After binding to its receptor through paracrine and autocrine mechanisms, IL-2 promotes clonal expansion of antigen-specific effector T cells, stimulates growth and differentiation of T and B cells, activates and induces proliferation of cytotoxic effector cells including NK cells and lymphokine-activated killer (LAK) cells, and activates monocytes/macrophages. Lymphokine production by lymphocytes, NK cells, and macrophages can also be regulated by IL-2. The IL-2 receptor (IL-2R) complex consists of three functional subunits: the α, β, and γ chains. The α and β chains are specific for IL-2 and the γ chain is shared by other lymphokine receptors including IL-4, IL-7, IL-9, and IL-13. IL-15 also utilizes the IL-2R α and β chains as part of its receptor complex to mediate its biological effects. The cytoplasmic domains of β and γ chains play a critical role in coupling ligand binding to signaling pathways through tyrosine kinases and other intracellular components. The α chain has a shorter cytoplasmic domain and may not play a functional role in signal transduction; however, it combines with β and γ chains to form high-affinity IL-2R.

iii. Expression and Function of IL-2 and IL-2R in the Reproductive System

IL-2 is produced by human fetal membranes, syncytiotrophoblast cells, decidual cells of patients with preeclampsia, and murine placenta. IL-2R is expressed by human fetal membranes. The IL-2Rγ, but not IL-2Rα or IL-2Rβ, is expressed by human trophoblast. Horse trophoblast only produces TNF-α, not IL-2, IL-4, or IFN-γ. Infertile men with poor sperm counts, mobility, and morphology have high levels of IL-2 and soluble IL-2R in the semen, but the source of IL-2 in the semen is unknown.

IL-2 may regulate uterine leukocyte-like cells including murine granulated metrial gland (GMG) cells. IL-2 promotes migration of GMG cells out of murine decidual explants and stimulates peripheral blood or decidual NK cell cytotoxicity toward choriocarcinoma target cells. This effect is receptor mediated since IL-2-induced NK cell activity is suppressed by inhibiting IL-2R expression on target cells. However, decidual NK cell proliferation is not IL-2 dependent and lysis of porcine trophoblast cells by endo-

metrial NK cells does not require IL-2. Recent studies have shown that an upregulated type 2 cytokine profile (IL-4, IL-5, and IL-10) is associated with a downregulated type 1 cytokine profile (IL-2 and IFN-γ, TNF) during normal pregnancy. Administration of exogenous TNF-α, IFN-γ, and IL-2 increases fetal resorption and implantation failure in abortion-prone and non-abortion-prone murine models, indicating that Th1 cytokines, such as IL-2, are deleterious to pregnancy. Pregnancy-associated proteins also regulate IL-2R expression by lymphocytes including NK and LAK cells and may modulate adverse effects of IL-2 during pregnancy. TGF-β derived from mouse uterus suppresses IL-2-stimulated lymphocyte proliferation and cytotoxicity.

Mice transgenic for the murine MT-1 promoter–human IL-2 fusion protein gene have atrophic testes and motor ataxia due to infiltration of lymphocytes. In these mice, the testes are depleted of spermatogenic cells without evidence of an immune response; therefore, the relationship between IL-2 function and these symptoms is unknown. It is evident, however, that mice lacking either IL-2 or IL-2R are healthy and fertile, indicating functional cytokine redundancy.

Human placental IL-2 is unique in that the 5' untranslated region of the human placental IL-2 gene is about 247 bp longer than that of lymphocyte-derived IL-2, although the coding sequences and 3' untranslated regions for lymphocyte IL-2 are identical to those for placental IL-2. The extended 5' untranslated sequence of placental IL-2 may provide alternate promoter utilization. To summarize, IL-2 may play a role in reproduction, but the precise functions of IL-2 in reproduction remain to be determined.

3. Interleukin-4

i. Alternative Names

Alternative names for IL-4 include mast cell growth factor-II (MCGF-II), B cell growth factor-I, B cell stimulatory factor-I, and T cell growth factor-II.

ii. IL-4 and IL-4 Receptor

IL-4 is a glycoprotein of 18–20 kDa derived primarily from T cell (CD4$^+$, CD8$^+$, and $\gamma\delta$ TcR$^+$) and basophil/mast cell lineages. The IL-4 genes have been cloned from humans, red-crowned mangabeys, mice, rats, Syrian hamsters, gerbils, cattle, sheep, goats, cats, and pigs. The induction of IL-4 in IL-4-producing cells requires cross-linkage of either T cell receptors on T cells or IgE Fc receptors on basophil/mast cells. IL-4 is a growth costimulator for T cells, B cells, myeloid progenitors, and erythroid progenitors, and it induces B cells to switch to IgE synthesis and enhances secretion of human IgG4 and mouse IgG antibodies. One of the most important functions of IL-4 is to regulate induction and differentiation of Th2 cells that primarily modulate humoral immunity, eosinophilia, mast cell development, and deactivation of inflammatory macrophages. The IL-4R complex consists of IL-4Rα and IL-2Rγ chains and is a member of the cytokine receptor superfamily. IL-4R is expressed by T and B lymphocytes, mast cells, cells from myeloid and monocytic origin, fibroblasts, and endothelial cells.

iii. Expression and Function of IL-4 in the Reproductive System

IL-4 and other Th2 cytokines, including IL-5 and IL-10, are present in cell supernatants from murine fetal–placental units. During normal pregnancy, higher levels of IL-4 and IL-10 are produced by maternal T cells when compared with those for IL-2 and IFN-γ. The presence of Th2 cytokines at the fetal–maternal interface may be required for normal pregnancy in humans and mice. The IL-2Rγ chain is expressed by human trophoblast cells, but IL-2Rα and IL-2Rβ chains are not detected in these cells. IL-4 and IL-4Rα are localized to human trophoblast, decidua, and amniochorionic membranes as well as to human choriocarcinoma (BeWo, JEG-3, and Jar) and amnion (AV3) cell lines. The expression of IL-4Rα and IL-2Rγ chains is functional since IL-4 can stimulate production of hCG by human trophoblast cells. IL-4 identified in human amnion epithelial cells has heparin-binding ability and binds components of the extracellular matrix. However, horse trophoblast does not express either IL-4 or IFN-γ. Progesterone may be partially responsible for the induction of Th2-type response at the fetal–maternal interface by inducing IL-4 and other Th2 cytokines during normal pregnancy. IL-4 blocks IL-2-induced NK cell activity of decidual NK cells by inhibiting expression of IL-2R. IL-4 also plays a critical role in the induction of neonatal tolerance by

biasing development and maturation of CD4 cells toward the Th2 phenotype and by preventing maturation of alloreactive CD8 cytotoxic T lymphocytes (CTLs). Overall, IL-4 is considered a protective lymphokine both locally and systemically. However, IL-4-deficient mice are fertile and healthy, suggesting that the effect of IL-4 is either redundant or subtle.

IL-4 may also be involved in pathology of the reproductive system, such as preterm labor, particularly in association with uterine infection. Amniotic fluid IL-4 is elevated in women with preterm labor and is often associated with chorioamnionitis. Amniotic fluid IL-4 may act on amnion to induce production of cyclooxygenase-2 and secretion of PGE$_2$ to cause uterine contractions. Macrophage inhibitory protein (MIP)-1 production can be induced by IL-1, IL-4, and TNF-α in human decidual cells *in vitro*. MIP-1 induction in the uterus may be the early event in the pathology of infection-associated preterm labor.

4. Interleukin-5

i. Alternative Names

Alternative names include B cell growth factor-II, T cell replacing factor (TRF), eosinophil differentiation factor (EDF), and eosinophil colony-stimulating factor (Eo-CSF).

ii. IL-5 and IL-5 Receptor

IL-5 is a 40- to 45-kDa homodimer primarily produced by Th2 CD4$^+$ T cells and other T cells. IL-5 genes have been cloned and sequenced for humans, rhesus monkeys, mice, rats, gerbils, domestic guinea pigs, horses, sheep, and cattle. Structural analysis demonstrated that IL-5 is similar to other cytokines and most closely resembles IL-4 and GM-CSF. IL-5 induces production of eosinophils from both mouse and human bone marrow and controls eosinophilia. IL-5R is a heterodimer consisting of two subunits, both belonging to the cytokine receptor superfamily.

iii. Expression and Function of IL-5 in the Reproductive System

IL-5, and other Th2 cytokines such as IL-4 and IL-10, is present in lysates of freshly isolated decidual and placental cells and cell supernatants from fetal–placental units throughout gestation in mice. Administration of estradiol (E$_2$) to ovariectomized mice results in a dramatic influx of eosinophils into the uterine tissue, which can be blocked by antibody to IL-5. The local effects of IL-5 and eosinophils in the uterus are known. Pregnancy impairs the systemic resistance of C57BL/6 mice to *Leishmania major* infection by causing increased levels of Th2 cytokines (IL-4, IL-5, and IL-10) by spleen and lymph node CD4 T cells. These results indicate that IL-5, along with other Th2 cytokines, is beneficial to successful pregnancy. Further studies are required to define the role(s) of IL-5 at the fetal–maternal interface.

5. Interleukin-6

i. Alternative Names
Alternative names include IFN-β_2, B cell differentiation factor, B cell stimulatory factor-2, 26-kDa protein, hybridoma growth factor, and hepatocyte-stimulating factor (HSF).

ii. IL-6 and IL-6 Receptor
IL-6 is a 21- to 26-kDa glycoprotein derived from many cellular sources. Genes coding for IL-6 have been cloned and sequenced for humans, rhesus monkeys, cynomolgus monkeys, mice, rats, woodchucks, cattle, sheep, horses, dogs, minks, and Kaposi's sarcoma-associated herpes-like virus. IL-6 can stimulate immunoglobulin secretion by B cells, production of acute protein in liver cells, proliferation of B cells, maturation and differentiation of Th2 cells from naive CD4$^+$ T cells, neuronal differentiation, and osteoclast activation. The IL-6R system consists of a ligand-binding protein (IL-6Rα) and a ligand nonbinding signal transduction protein, gp130, designated IL-6Rβ, both of which belong to the cytokine receptor superfamily. Upon binding of IL-6 to its receptor, association of IL-6R to gp130 forms a high-affinity IL-6-binding complex. LIF, oncostatin, ciliary neurotrophic factor, IL-11, and cardiotrophin-1 also use the common IL-6β chain, gp130, as part of their receptor signal transduction mechanism.

iii. Expression and Function of IL-6 in the Reproductive System
Cells of both male and female reproductive systems produce IL-6. The diverse functions of IL-6 in reproduction can be similar to or different from those inflammatory cytokines (IL-1 and TNF-α) and IL-6-type cytokines (LIF). IL-6-deficient mice are healthy and fertile.

Function of IL-6 in Male Reproduction IL-6 is a pleiotropic lymphokine that modulates growth and differentiation of many cell types in the immune and nonimmune systems. IL-6 is present in the testes (Table 5) and its production can be upregulated by testicular IL-1. IL-6 increases basal levels of and FSH-induced transferrin in seminiferous tubule segments and inhibits onset of DNA synthesis in preleptotene spermatocytes and DNA synthesis in A3 type B spermatogomia. These results indicate that IL-6 is involved in the complex regulation of spermatogenesis in conjunction with FSH and other hormones and can be considered an inhibitor of germ cell growth and differentiation at specific stages of the spermatogenic cycle. The role of IL-6 as a mediator between Sertoli cells, Leydig cells, and spermatogenic cells at all stages of spermatogenesis warrants further investigation.

Function of IL-6 in Female Reproduction Both IL-6 and IL-6R are localized to the female reproductive tract and the conceptus (Table 6). However, the function of this lymphokine in female reproduction is not clear. It is known that IL-6 increases production of hCG by human placental cells and may influence the signaling for establishment and mainte-

---

TABLE 5
IL-6 Expression and Regulation in the Testes

Tissue/cell type	Regulation	Species
Peritubule cells, Sertoli cells, Leydig cells, testicular macrophages	Increased with aging and by phagocytosis of latex beads and residual bodies, LPS, IL-1, IFN-γ, FSH	Rat, human

Note. Abbreviations used: LPS, lipopolysaccharide; FSH, follicle-stimulating hormone.

TABLE 6
IL-6 Expression in the Female Reproductive Tract

IL-6R	Species	Tissue/cell type	Regulation
IL-6R	Human, sheep, pig, cattle	Embryos	
IL-6	Human	Trophoblasts and placental macrophages	
IL-6	Rat, pig, human	Follicular fluid prior to or during ovulation	Inhibits aromatase activity, LH receptor expression, P_4 production of FSH-induced granulosa cells
IL-6	Human	Decidual cells	Stimulated by P_4, RU486
Both	Human	Endometrial explants and stromal cells	Increased by IL-1, TNF-α, IFN-γ, placental proteins Inhibited by E_2 Suppresses endometrial stromal cell proliferation
IL-6	Human	Fetal membrane cells, amniotic fluid, chorion	Stimulated by IL-1, LPS, labor, preterm labor, intrauterine infection
IL-6	Human	Vaginal secretions	Increases as pregnancy proceeds and during labor
IL-6	Human	Peritoneal fluid	Endometriosis
IL-6		Ovarian fluid	Reduces FSH binding capacity Inhibits aromatase activity, LH-R expression, and P_4 production Increases PG synthesis Stimulates angiogenesis
IL-6	Human	Leukocytes	Increases as pregnancy proceeds

Note. Abbreviations used: LH, lutenizing hormone; FSH, follicle-stimulating hormone; LPS, lipopolysaccharide; P_4, progesterone; RU486, a progesterone receptor antagonist; PG, prostaglandin; E_2, estradiol.

nance of pregnancy in primates. However, transgenic mice lacking IL-6 have normal embryogenesis and fetal development. This phenomenon may be a function of cytokine redundancy and compensation for IL-6 by other IL-6-type lymphokines. More detailed studies are warranted to investigate the role of IL-6 during gestation.

Inflammatory cytokines, including IL-6, IL-1, and TNF-α, can also be involved in parturition during normal or abnormal situations. IL-6 derived from fetal membranes and the uterus the can be upregulated by intrauterine infections and lipopolysaccharide (LPS), IL-1, or TNF-α. Elevated IL-6 levels in maternal serum and amniotic fluid are associated with spontaneous labor and onset of labor. IL-6 may increase the production of oxytotics, including prostanoids, endothelins, and leukotrienes, that can cause uterine contraction and the cascade of events that lead to labor or preterm labor.

6. Interleukin-7

i. Alternative Names   Alternative names include pre-B-cell growth factor and lymphopoietin.

ii. IL-7 and IL-7 Receptor   IL-7 is a 25-kDa glycoprotein whose primary sources are adherent stromal cells of bone marrow, spleen, thymus, and kidney. Genes encoding IL-7 have been cloned and sequenced for humans, mice, rats, and sheep. IL-7 stimulates proliferation, differentiation, and development of B cells and T cells and induces cytotoxicity of CTLs, LAK cells, and monocytes. It also induces monocytes to release cytokines such as IL-1, IL-6, and TNF-α. IL-7R is a member of the cytokine receptor family and is expressed by B cells, thymocytes, mature T cells, and monocytes.

iii. Expression and Function of IL-7 in the Reproductive System   IL-7 mRNA is detected in murine unfertilized eggs and preimplantation embryos. IL-7Rα and IL-2Rγ mRNA are also present in human trophoblast cells and IL-7 enhances production of hCG by these cells. The functions of IL-7 in reproduction remain to be determined.

7. Interleukin-8

i. Alternative Names
Alternative names include neutrophil activating protein-1, alveolar macrophage-derived chemotactic factor, lung giant cell carcinoma-derived chemotactic protein, TPA repressed gene 1, monocyte-derived neutrophil chemotactic factor, and tumor necrosis factor-stimulated gene 1.

ii. IL-8 and IL-8 Receptor
IL-8 is a 8- to 10-kDa chemokine produced by many cell types, including monocytes, macrophages, T cells, fibroblasts, granulocytes, keratinocytes, endothelial cells, hepatocytes, NK cells, astrocytoma cells, glioblastoma cells, and chondrocytes. IL-8 genes have been cloned and sequenced for humans, mice, rats, cattle, sheep, pigs, dogs, cats, red-crowned mangabeys, pig-tailed macaques, and rhesus monkeys. IL-8 is a chemotactic factor for T cells, neutrophils, monocytes/macrophages, basophils, fibroblasts, and smooth muscle cells. Its actions on neutrophils are to cause degranulation, increase adherence to endothelium and expression of surface markers, and stimulate lysosomal enzyme release. Two IL-8Rs have been identified, type A and type B. Both type A and type B bind IL-8 with high affinity and can also bind other chemokines, including MGSA, GRO, and NAP-2.

iii. Expression and Function of IL-8 in the Reproductive System
IL-8 is constitutively produced by human placenta, fetal membranes, endometrium, and uterine cervix, often in response to intrauterine infection and stimulation by inflammatory cytokines. IL-8 has been localized to the trophoblasts and macrophage-like cells in the placenta and smooth muscle of arterioles and stromal cells of the endometrium. IL-8Rs (types A and B) are expressed by fetal membranes, the placenta, and myometrium and their expression increases after initiation of parturition. The human placenta is not permeable to circulating IL-8. The production of IL-8 by human fetal membranes, the placenta, or endometrium is induced by bacterial endotoxic LPS and reduced by both P_4 and dexamethasone, indicating that P_4 limits the influx of leukocytes into the pregnant uterus. IL-8 can also be induced by IL-1 and TNF-α in human endometrial stromal cells. Withdrawal of P_4 initiates a cascade of events involving local production of inflammatory mediators in the pregnant uterus. IL-8 and IL-1 may be involved in parturition by inducing cervical ripening during late pregnancy. The mechanism by which IL-8 and other inflammatory cytokines control menstruation and parturition is to attract an influx of monocytes and neutrophils into the uterus, which provides proteases and collagenases for cervical ripening. IL-8 induces production of uterine hyaluronic acid involved in cervical ripening. The production of IL-8 is also believed to be initiated and stimulated by stretching of the amnion, amniochorion, and muscles of the lower uterine segment in humans. Figure 2 illustrates the mechanisms of action mediated by IL-8 during parturition.

8. Interleukin-10

i. Alternative Names
Alternative names include cytokine synthesis inhibitory factor, mast cell growth factor (MCGF), and thymocyte growth-promoting factor.

ii. IL-10 and IL-10 Receptor
IL-10 is a secreted 17- to 21-kDa protein which is not glycosylated in humans and exists as a noncovalently linked homodimer. The IL-10 genes have been cloned and sequenced for humans, cynomolgus monkeys, mice, rats, gerbils, red deer, cats, dogs, rabbits, horses, cattle, sheep, pigs, and poxvirus (orf virus). IL-10 is a pleiotropic immunomodulatory cytokine controlling the functions of myeloid and lymphoid cells. It is primarily a product of T cells (mainly Th2 in the mouse), B cells, and macrophages after activation by antigen or bacterial products. IL-10 plays a key role in the establishment and maintenance of Th2 immune response by suppressing Th1-dependent cell-mediated immune responses. As an antiinflammatory cytokine, IL-10 inhibits synthesis of proinflammatory cytokines, including IL-1, IL-6, IL-12, and TNF-α. IL-10 can directly and indirectly suppress the effector

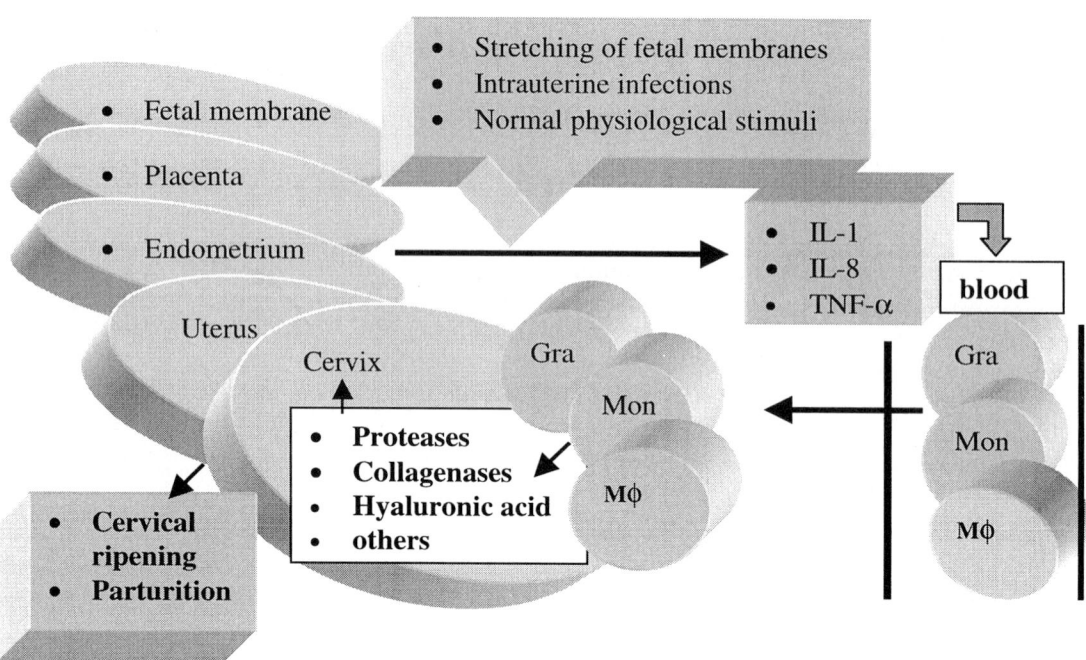

FIGURE 2   A mechanism by which IL-8 and other inflammatory cytokines regulate parturition. Gra, granulocyte; mon, monocyte; Mθ, macrophage.

functions of macrophages and NK and T cells. IL-10 also regulates proliferation and differentiation of B cells, mast cells, and thymocytes. IL-10R is composed of two chains, IL-10Rα or IL-10R1 and IL-10Rβ and IL-10R2 (CRFB4). IL-10Rα alone can bind IL-10, but it does not induce signal transduction. IL-10Rβ has been considered an accessory molecule for IL-10 signal transduction. The IL-10 homodimer binds the IL-10Rα/IL-10Rβ complex and initiates signal transduction events through the Jak–Stat pathway. IL-10Rβ (CRFB4) is genetically linked to the IFN-α receptor region on human chromosome 21; therefore, a function of IL-10Rβ in the type I IFN receptor complex has been speculated.

iii. Expression and Function of IL-10 in the Reproductive System   Th2 cytokines, including IL-10, are produced by human endometrium throughout the menstrual cycle and early pregnancy and by cytotrophoblasts at all stages of pregnancy. Levels of IL-4 and IL-10 are increased in the placenta and peripheral blood lymphocytes of pregnant mice. Elevated concentrations of IL-10 are also detected in human amniotic fluid. IL-10 is detected in human

oocytes and embryos and administration of IL-10 or stimulation of IL-10 production benefits normal pregnancy in mice. Using an abortion-prone mouse model deficient in IL-4 and IL-10, autoimmunization increases production of both IL-4 and IL-10 and administration of recombinant IL-10 prevents fetal resorption, which can be reversed by giving anti-IL-10 antibody. Human seminal plasma is a potent stimulator of IL-10 by peripheral blood lymphocytes, causing a marked increase in IL-10/IL-12 ratio *in vitro*. IL-10 production by T cells and placenta can be stimulated by the trophoblast IFN-τ. Implantation failure is increased in pregnant mice with Th1 responses against *L. major*, and this phenomena is correlated with increased TNF and IFN-γ and reduced IL-10.

9. Interleukin-11

i. Alternative Name   An alternative name for IL-11 is adipogenesis inhibitory factor.

ii. IL-11 and IL-11 Receptor   IL-11 is a 19-kDa stromal cell-derived protein that belongs to the cytokine family. The genes encoding IL-11 and IL-

11R have been cloned for humans and mice. IL-11 mediates signal transduction through gp130, a component of the IL-6R signal transduction system, but IL-11 has very little direct effect on T or B lymphocytes. IL-11 has functions in the hematopoietic systems (primitive, erythroid, and megakaryocytic progenitors), hepatic system (inducer of acute phase protein response), stromal cell system (inhibition of adipocyte differentiation), and intestinal epithelial system.

iii. Expression and Function of IL-11 in the Reproductive System IL-11 is present in conditioned medium from granulosa cells and human follicular fluid, but amounts are not correlated with concentrations of either ovarian steroids and IGF-I or follicular maturity. IL-11 mRNA is also detected in endometrial samples of women suffering from endometriosis. Cytokines, including IL-11, that use gp130 for signal transduction suppress secretion of placental lactogen-II and stimulate production of placental lactogen-II by cultured mouse placental cells in a stage-specific fashion.

10. Interleukin-12

i. Alternative Names Alternative names include T cell stimulatory factor, natural killer cell stimulatory factor, and cytotoxic lymphocyte maturation factor.

ii. IL-12 and IL-12 Receptor IL-12 is a heterodimeric glycoprotein composed of 35 kDa (p35) and 40 kDa (p40) subunits. The production of IL-12p35 is constitutive, whereas IL-12p40 expression is regulated. If overexpressed, IL-12p40 can compete with heterodimeric IL-12 for IL-12-specific receptors on the cell surface and thus suppress IL-12 bioactivity. The genes encoding IL-12 have been cloned and sequenced for humans, mice, rats, red-crowned mangabeys, pigs, rhesus monkeys, dogs, red deer, cattle, and woodchucks. Coexpression of both chains is necessary for biological activity of IL-12. IL-12 is derived primarily from monocytes stimulated by bacteria and intracellular parasites or their products. IL-12 stimulates production of IFN-γ by T cells and NK cells and activates NK cell activity. IL-12 is one of the key Th1 cytokines to regulate differentiation of Th1 cells from progenitors and to suppress IgE production.

iii. Expression and Function of IL-12 in the Reproductive System IL-12 stimulates release of IFN-γ by human decidual large granulated lymphocytes cells and its presence at the fetal–maternal interface may result in a Th1 immune response and therefore compromise the outcome of pregnancy. IL-12 levels are elevated in sera of women with severe preeclampsia and HELLP syndrome (a special form of preeclampsia with hemolysis, elevated liver enzymes, and low platelets). Human preeclampsia may represent a Th1 immunologic state; however, this is highly controversial. The roles of IL-12 in physiology and pathology of reproduction must be investigated further.

11. Interleukin-15

i. Alternative Name An alternative name for IL-15 is IL-T.

ii. IL-15 and IL-15 Receptor IL-15 is a 14- or 15-kDa polypeptide lymphokine produced by placenta, skeletal muscle, kidney, lung, heart, fibroblast cells, epithelial cells, and activated macrophages, but not by T cells. The genes coding for IL-15 have been cloned and sequenced for humans, rhesus monkeys, cynomolgus monkeys, mice, rats, cattle, goats, and pigs. IL-15 shares many properties with IL-2 in that both use the IL-2Rβ and IL-2Rγ chains and they mediate similar biological functions on T, B, and NK cells. The IL-15R complex consists of IL-15Rα, IL-2Rβ, and IL-2Rγ chains. Mast cells express an IL-15R signal transduction pathway distinct from that utilized by IL-2-and IL-15-activated T cells. Therefore, mast cells are regulated by IL-15, but not IL-2, because IL-15 uses a novel receptor/signal transduction system in mast cells which does not include IL-2R chains. The primary function of IL-15 is to stimulate proliferation of T cells and NK cells to increase IFN-γ production by these cells in synergy with IL-12. IL-15 promotes cytotoxicity of T and LAK cells and regulates immunoglobulin synthesis by B cells.

iii. Expression and Function of IL-15 in the Reproductive System   IL-15 and its receptor are expressed by the placenta and pregnant uterus. The expression of IL-15 and IL-15R by the pregnant mouse uterus coincides with expression of cytolytic mediators in differentiated GMG cells. IL-15 induces expression of perforin and granzymes in the pregnant uterus and may regulate differentiation of GMG cells in the murine pregnant uterus. A direct role of IL-15 in the maintenance of pregnancy remains to be determined.

12. Tumor Necrosis Factor-α

i. Alternative Names   Alternative names include cachectin, necrocin, macrophage cytotoxin, differentiation-inducing factor, cytotoxin, hemorrhagic factor, and macrophage cytotoxic factor.

ii. TNF-α and TNF-α Receptor   TNF-α is a 17-kDa polypeptide produced by a variety of cell types, including macrophages, NK cells, B cells, granulocytes, fibroblast cells, mast cells, smooth muscle cells, breast and glial tumor cells, ovarian cells, astrocytes, Kupffer's cells, epidermal cells, and adipocytes. The genes coding for TNF-α have been cloned and sequenced for humans, mice, rats, cattle, cats, red deer, horses, sheep, pigs, rabbits, dogs, cynomolgus monkeys, *Colobus guereza*, *Gorilla gorilla*, rhesus monkeys, *Pongo pygmaeu*, baboons, chimpanzees, domestic guinea pigs, *Peromyscus leucopus*, and cowpox virus. The production of TNF-α is stimulated by gram-negative and gram-positive bacteria and their products, viruses, mycoplasma, immune complexes, IL-1, IL-2, GM-CSF, IFN-γ, tumor cells, complement, protein kinase activators, reactive oxygen species, cyclooxygenase inhibitors, platelet-activating factor (PAF), and neropeptides. The production of TNF-α is suppressed by dexamethasone, PGE_2, cyclosporin, TGF-β, retinoid acid, IL-4, IL-6, IL-10, and PAF receptor antagonist. TNF-α is a multifunctional cytokine which exhibits antitumor activities, induces septic shock and fever, and is involved in a number of pathological processes. Two TNF-Rs (type I or type B or p55 and type II or type A or p75) with molecular weights of 55–60 and 75–80 kDa have been described. The p55 TNF-R is primarily expressed on epithelial cells, whereas p75 is preferentially expressed on myeloid cells. Both TNF-Rs are members of the TNF receptor family including NGF-R, CD27, CD30, CD40, and Fas antigen. Both TNF-α and -β bind to both receptors with equal affinity.

iii. Expression and Function of TNF-α in the Reproductive System   TNF-α is one of the multifunctional inflammatory cytokines produced in tissues/cells of the reproductive organs. Tables 7–9 summarize the distribution and regulation of TNF-α and TNF-R in the ovary, embryos, and reproductive tract. Exogenous TNF-α has detrimental effects on fetal survival, but administration of TNF-α antibody induces abnormalities of fetal development. Further studies are required to elucidate the role of TNF-α during pregnancy. Transgenic mice lacking TNF-R are normal and reproduce, which may indicate that TNF-α is functional in reproduction but not absolutely necessary since cytokines are redundant and can be compensated for by related molecules.

Potential Functions of TNF-α in Ovarian Physiology   *Growth Regulation.*   TNF-α stimulates cell proliferation in several cell types. TNF-α alone has no effect, but in combination with IFN-γ it is toxic to bovine luteal cells. TNF-α alone stimulates proliferation of human granulosa–luteal cells, whereas established luteal cells are sensitive to the cytolytic activity of TNF-α in combination with IFN-γ. During corpus luteum formation and development, TNF-α may stimulate proliferation of luteal cells. TNF-α has diverse effects on ovarian epithelial cell lines *in vitro*. TNF-α or IFN-γ alone promote proliferation of some ovarian cancer cell lines, but IFN-γ and TNF-α together suppress proliferation of these same ovarian cancer cell lines. Overall, data suggest that TNF-α may enhance growth of ovarian cancers.

Follicular Development.   The production of TNF-α by oocytes and its ability to induce clustering of rat theca–interstitial cells *in vitro* suggests a role for TNF-α in organizing pregranulosa and theca cells around oocytes. The identification of TNF-α receptors on some ovarian cells supports this hypothesis. Oocytic TNF-α modulates follicular development, in a gradient radiating from the oocyte, by inhibiting steroidogenesis in granulosa and theca cells. Thus,

TABLE 7
TNF-α Expression in the Ovary and Ovarian Functions Regulated by TNF-α

Cell/compartment	Species	Regulation by TNF-α
Cumulus cells	Human	
Oocytes	Mouse, rat, cow, human	
Granulosa cells	Mouse, rat, human	Inhibits FSH-stimulated aromatase, adenyl cyclase, LH receptor expression, P_4 accumulation, cAMP accumulation
Thecal cells	Sheep, rat, human	Inhibits LH-induced LH receptor expression, cAMP accumulation, P_4 accumulation, hydroxylase activity
		Suppresses cAMP-stimulated protein kinase A, androstenedione
		Increases clustering of cells *in vitro*, protein kinase C
Luteal cells	Rabbit, sheep	
Macrophage-like cells	Rabbit, sheep	
Follicular fluid	Cow, human,	
Preovulatory follicles		Stimulates P_4, E_2, $PGF_{2\alpha}$, and PGE_2
		Inhibits FSH-induced aromatase

Note. Abbreviations used: FSH, follicle-stimulating hormone; LH, lutenizing hormone; P_4, progesterone; E_2, estradiol; PG, prostaglandin.

cells closest to the oocyte, such as cumulus cells, have low activities compared to granulosa and theca cells. In the rat, this mechanism may protect oocytes from estrogen which could cause adverse developmental changes in embryos during prolonged estrous cycles.

Luteal Development. TNF-α derived from different sources in the ovary may stimulate development of the corpus luteum. However, TNF-α released from infiltrating macrophages may also stimulate the local release of $PGF_{2\alpha}$ that contributes to luteal regression in species including rats, cows, and humans (Tables 7 and 9).

Potential Functions of TNF-α in the Reproductive Tract *Uterus.* In the nonpregnant uterus, low concentrations of E_2 stimulate TNF-α expression during early (proliferation phase) stages of the menstrual cycle, which in turn stimulates DNA synthesis in endometrial epithelial and stromal cells. Higher levels of TNF-α derived from the endometrium in response to both E_2 and P_4 may mediate cytolysis and menstruation. TNF-α may also affect uterine cell apoptosis and leukocyte migration, as well as growth, differentiation, angiogenesis, and tissue remodeling of the endometrium and trophoblasts during pregnancy.

Placenta. The role of TNF-α in placental function and differentiation is unclear. TNF-α may facilitate migration and differentiation of trophoblast cells and stimulate trophoblast cells to proliferate in an autocrine fashion. *In vitro*, TNF-α does not interfere with blastocyst attachment, trophoblast outgrowth, and hCG production, but it suppresses proliferation of trophoblast cell lines. TNF-α mediates apoptosis in cultured cytotrophoblast cells through TNF-Rp55 exclusively, indicating that TNF-α may damage placental cells or facilitate the rapid turnover of these cells. TNF-α levels are increased in women with spontaneous term labor or preterm labor and may be involved in parturition by stimulating placental/fetal membrane production of PGF, IL-8, and IL-6 that can initiate uterine contractions and cervical dilation.

Embryo. TNF-α has not been detected in preimplantation embryos; however, it is expressed by a number of organs of the fetus after midpregnancy. TNF-α derived from oocytes may be important for the initial stages of embryonic growth and survival and fetal growth and lymphoid organ development. A deficiency in this cytokine may cause retardation of these developmental processes. However, TNF-α and IL-1 are considered inducers of stress-related abortion since administration of soluble receptors

TABLE 8
TNF-α Expression and Regulation in Embryos and Reproductive Tract

Organ	Tissues/cells	Regulation	Species
Oviduct	Luminal epithelium, stroma, smooth muscle, macrophage-like cells	Stimulated by P_4 and infection, but not by CSF-1	Mouse
Embryo	Fetal organs	Developmental	Mouse
Uterus	Luminal and glandular epithelium, stroma, smooth muscle, macrophages	Regulated by estrous/menstrual cycle Stimulated by E_2 and P_4, infection, or LPS Stimulates $PGF_{2\alpha}$, PGE_2 production	Rat, human, mouse
Placenta	Syncytiotrophoblasts, extravillous cytotrophoblasts, placental stromal cells (or macrophages) in the human Human term placenta, placental core mesenchymal cells, placental trophoblasts in the mouse and rat	Regulated by IL-1, P_4, pregnancy-specific proteins Stimulates G-CSF, IL-8 release Induces apoptosis through TNFR (p55)	Rat, human, mouse
Endometrium	Epithelial cells	IL-1, P_4, pregnancy-associated protein Increases PGE_2 and $PGF_{2\alpha}$ at luteal phase	Human
Tumor cell lines	Endometrial adenocarcinoma cells, choriocarcinoma cells (JAR, JEG-3, PL/B6)	Enhances HCG release Increases HLA-G expression	Human, mouse
Decidua	Granulated leukocyte clones	Produce more TNF-α and other cytokines than peripheral NK cells	Human

for both cytokines partially prevents stress-triggered fetal loss.

Systemic Administration of TNF-α. Administration of exogenous TNF-α increases fetal resorption in the mouse abortion model; however, immunoneutralization of systemic TNF-α causes fetal growth retardation and lymphoid organ abnormality. There is also evidence that transgenic mice lacking TNF-RI can carry out pregnancy successfully.

13. Tumor Necrosis Factor-β

i. Alternative Names Alternative names include lymphotoxin, lymphotoxin-α, and cytotoxic factor.

ii. TNF-β and TNF-β Receptor TNF-β is a 25-kDa protein secreted primarily by lymphocytes. It is related to TNF-α, lymphotoxin-β (LT-β), CD40 ligand, Fas ligand, CD27 ligand, and CD30 ligand. TNF-β genes have been cloned and sequenced for human, mouse, rat, cattle, rabbit, and chicken. TNF-β can be linked to a transmembrane-binding protein, LT-β, to form a membrane heteromeric complex. TNF-β binds to both TNF-Rs (p55 and p75) with high affinity and mediates a broad array of biological functions including cytotoxicity and cell proliferation and differentiation.

iii. Expression and Function of TNF-β in the Reproductive System TNF-β mRNA is present in human term placental macrophages and trophoblast cells. Plasma levels of TNF-β are decreased in pregnant women at term in labor when compared to women at term not in labor. Serum and amniotic fluid TNF-β levels are not affected in women with preterm labor or during pregnancy, although TNF-β levels in amniotic fluid are much lower than those in maternal plasma. Soluble TNF-β and LT-β are not detectable in the human amnion, choriodecidua, or placental explants, although mRNAs for both TNF-β and LT-β are present in these tissues. These results suggest that TNF-β expression is suppressed during labor and may not be involved in parturition. The function of TNF-β during pregnancy warrants further study.

TABLE 9
TNF-R Distribution in the Ovary, Embryos, and the Reproductive Tract

Organ	TNF-RI/TNF-RII	Species	Regulation
Oocytes	Both	Human	
Embryonic tissue	Both	Mouse, human	
Uterus	Both	Human, mouse	Estrous/menstrual cycle, E_2 and P_4
Choriocarcinoma cells	TNF-RI	Human	
Placenta	Both	Human	TNF-RI is expressed by all cell types and increases as pregnancy advances to term
			TNF-RII is restricted to trophoblasts during early pregnancy and shifted to stromal cells later during gestation
Placenta	Both	Mouse	Constitutive expression of TNF-RI and regulated expression of TNF-RII during pregnancy
Ovary	Both	Mouse, human	

14. Interferon-γ

i. Alternative Names

Alternative names include type II interferon; immune interferon; and macrophage-activating factor.

ii. IFN-γ and IFN-γ Receptor

IFN-γ is a 20- to 25-kDa monomeric glycoprotein produced primarily by T cells and NK cells. The active form of IFN-γ is a homodimer, formed by noncovalant interactions, that binds two IFN-γR molecules. The clustering of its receptor after binding may be critical for signal transduction. The IFN-γ genes have been cloned for humans, pig-tailed macaques, mice, rats, gerbils, cattle, goats, pigs, and sheep. The activity of IFN-γ is species specific since the level of homology among IFN-γs from different species is low. The functions of IFN-γ include antiviral, antiproliferative, MHC class I and II antigen-inducing, and macrophage-activating effects. IFN-γ is also an important Th1 cytokine which is critical for Th1-type immune responses.

iii. Expression and Function of IFN-γ in the Reproductive System

IFN-γ is expressed by the first-trimester human placenta and porcine trophoblast during early pregnancy. Metrial glands do not produce IFN-γ when examined at the mRNA level. IFN-γR is also detected in human and mouse trophoblast cells. The function of IFN-γ at the maternal–fetal interface is still unknown. In general, IFN-γ is considered to be harmful to fetuses and perhaps to the mother. IFN-γ and TNF-α are cytotoxic to human term cytotrophoblast cells, implying that these lymphokines may induce apoptosis during trophoblast development. In an abortion-prone mouse model, supernatants from mixed lymphocyte–placental reactions contain elevated levels of IFN-γ. This is consistent with the observation that administration of IFN-γ to an abortion-prone model increases fetal resorptions. These results are also consistent with those from *in vitro* studies indicating that IFN-γ suppresses trophoblast outgrowth and proliferation of trophoblast cell lines. Mice infected with *L. major* exhibit an antiparasite Th1 response (high levels of IL-2 and IFN-γ production by T cells) that increases implantation failure and fetal resorption. In a separate study with mice, IFN-γ induced expression of MHC class I and II antigens in the pregnant uterus but not in the conceptus. Increased MHC class I antigen expression in the placenta is generally associated with increased implantation failure and fetal resorption. IFN-γ can also increase the release of inflammatory cytokines such as IL-6 from human endometrial explant cultures, release of IL-1, IL-6, and TNF-α from macrophages of rat testis, and secretion of GM-CSF by murine uterus. Some of these

IFN-γ-inducible cytokines are deleterious (TNF-α) to embryos and others are beneficial (IL-1 and GM-CSF). IFN-γ increases both steady-state levels of HLA-G mRNA and protein expression in macrophage cell lines and blood macrophages. This selective induction of HLA-G by IFN-γ in placental macrophages may influence the repertoire of peptide antigens presented during pregnancy to prevent recognition of paternal antigens by maternal T cells, and therefore protect the fetal allograft. The role(s) of IFN-γ in reproduction is complex and deserving of further research.

See Also the Following Articles

CYTOKINES; ENDOMETRIOSIS; IMMUNOLOGY OF REPRODUCTION; INFECTIONS IN PREGNANCY; INTERFERONS

Bibliography

Adashi, E. Y. (1996). Immune modulators in the context of the ovulatory process: A role for interleukin-1. *Am. J. Reprod. Immunol.* **35**, 190–194.

Chaouat, G., Assal-Meliani, A., Martal, J., Raghupathy, R., Elliot, J., Mosmann, T., and Wegmann, T. G. (1995). IL-10 prevents naturally occurring fetal loss in the CBA × DBA/2 mating combination, and local defect in IL-10 production in this abortion-prone combination is corrected by in vivo injection of IFN-τ. *J. Immunol.* **154**, 4261–4268.

Guilbert, L. (1996). There is a bias against type I (inflammatory) cytokine expression and function in pregnancy. *J. Reprod. Immunol.* **32**, 105–110.

Hunt, J. S., Chen, H. L., and Miller, L. (1996). Tumor necrosis factors: Pivotal components of pregnancy. *Biol. Reprod.*, **54**, 554–562.

Mackiewicz, A., Koj, A., and Sehgal, P. B. (1995). Interleukin-6-type cytokines. *Ann. N. Y. Acad. Sci.* **762**.

Mosmann T. R., and Sad, S. (1996). The expanding universe of T cell subsets—Th1, Th2 and more. *Immunol. Today* **17**, 138–146.

Nicola, N. A. (1994). *Guidebook to Cytokines and Their Receptors.* Sambrook & Tooze, New York.

Tagaya, Y., Bamford, R. N., DeFilippis, A. P., and Waldmann, T. A. (1996). IL-15: A pleiotropic cytokine with diverse receptor/signaling pathways whose expression is controlled at multiple levels. *Immunity* **4**, 329–336.

Terranova, P. F., Hunter, V. J., Roby, K. F., and Hunt, J. S. (1995). Tumor necrosis factor-α in the female reproductive tract. *Proc. Soc. Exp. Biol. Med.* **209**, 325–342.

Vaddi, K., Keller, M. and Newton, R. C. (1997). *The Chemokine Fact Book.* Academic Press, San Diego.

Wegmann, T. G. (1987). Placental immunotrophism: Maternal T cells enhance placental growth and function. *Am. J. Reprod. Immunol.* **15**, 67–70.

Wegmann, T. G., Lin, H., Guilbert, L., and Mosmann, T. R. (1993). Bidirectional cytokine interactions in the maternal–fetal relationship: Is successful pregnancy a Th2 phenomenon? *Immunol. Today* **14**, 353–356.

Ye, W., Zheng, L. M., Young, J. D., and Liu, C. C. (1996). The involvement of interleikin-15 in regulating the differentiation of granulated metrial gland cells in mouse pregnant uterus. *J. Exp. Med.* **184**, 2405–2410.